Exponents and Radicals

$x^m x^n = x^{m+n}$

$(x^m)^n = x^{mn}$

$(xy)^n = x^n y^n$

$x^{1/n} = \sqrt[n]{x}$

$\sqrt[n]{xy} = \sqrt[n]{x}\,\sqrt[n]{y}$

$\sqrt[m]{\sqrt[n]{x}} = \sqrt[n]{\sqrt[m]{x}} = \sqrt[mn]{x}$

$\dfrac{x^m}{x^n} = x^{m-n}$

$x^{-n} = \dfrac{1}{x^n}$

$\left(\dfrac{x}{y}\right)^n = \dfrac{x^n}{y^n}$

$x^{m/n} = \sqrt[n]{x^m} = (\sqrt[n]{x})^m$

$\sqrt[n]{\dfrac{x}{y}} = \dfrac{\sqrt[n]{x}}{\sqrt[n]{y}}$

Special Products

$(x + y)^2 = x^2 + 2xy + y^2$

$(x - y)^2 = x^2 - 2xy + y^2$

$(x + y)^3 = x^3 + 3x^2y + 3xy^2 + y^3$

$(x - y)^3 = x^3 - 3x^2y + 3xy^2 - y^3$

Factoring Formulas

$x^2 - y^2 = (x + y)(x - y)$

$x^2 + 2xy + y^2 = (x + y)^2$

$x^2 - 2xy + y^2 = (x - y)^2$

$x^3 + y^3 = (x + y)(x^2 - xy + y^2)$

$x^3 - y^3 = (x - y)(x^2 + xy + y^2)$

Quadratic Formula

If $ax^2 + bx + c = 0$, then

$$x = \frac{-b \pm \sqrt{b^2 - 4ac}}{2a}$$

Inequalities and Absolute Value

If $a < b$ and $b < c$, then $a < c$.

If $a < b$, then $a + c < b + c$.

If $a < b$ and $c > 0$, then $ca < cb$.

If $a < b$ and $c < 0$, then $ca > cb$.

If $a > 0$, then

$|x| = a$ means $x = a$ or $x = -a$.

$|x| < a$ means $-a < x < a$.

$|x| > a$ means $x > a$ or $x < -a$.

Geometric Formulas

Formulas for area A, perimeter P, circumference C, and volume V:

Rectangle

$A = lw$

$P = 2l + 2w$

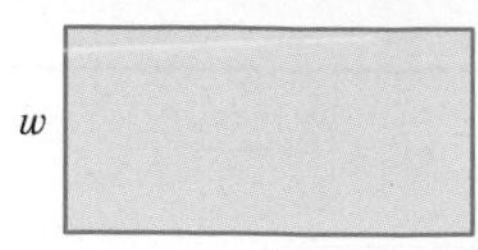

Box

$V = lwh$

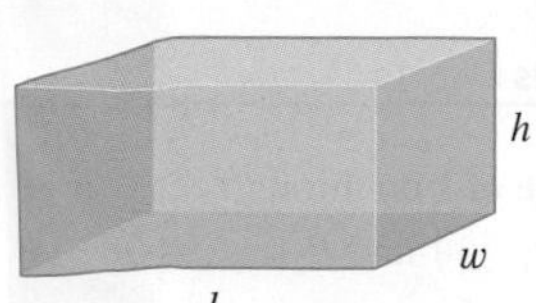

Triangle

$A = \frac{1}{2}bh$

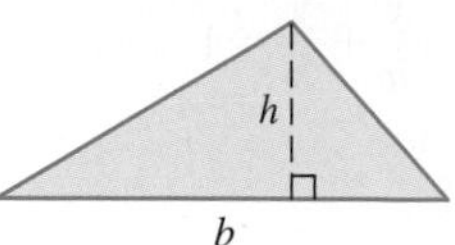

Pyramid

$V = \frac{1}{3}ha^2$

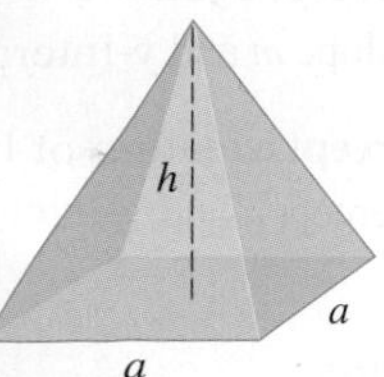

Circle

$A = \pi r^2$

$C = 2\pi r$

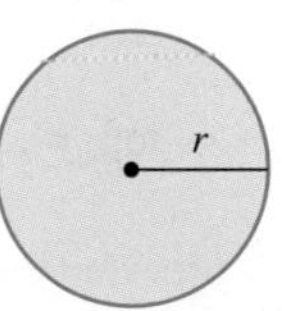

Sphere

$V = \frac{4}{3}\pi r^3$

$A = 4\pi r^2$

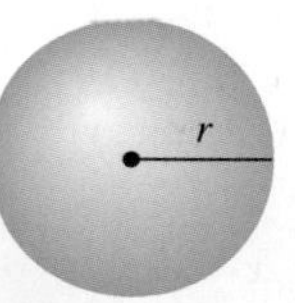

Cylinder

$V = \pi r^2 h$

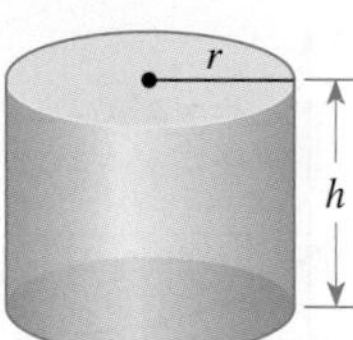

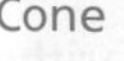

Cone

$V = \frac{1}{3}\pi r^2 h$

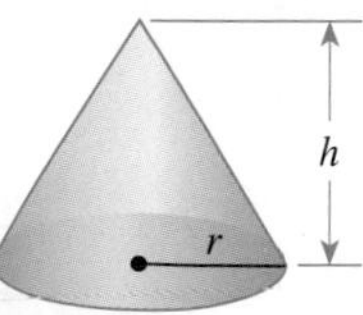

Heron's Formula

Area $= \sqrt{s(s - a)(s - b)(s - c)}$

where $s = \dfrac{a + b + c}{2}$

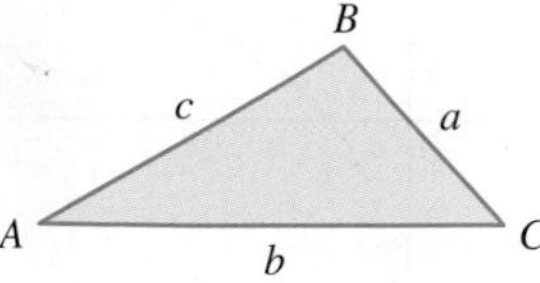

Distance and Midpoint Formulas

Distance between $P_1(x_1, y_1)$ and $P_2(x_2, y_2)$:

$$d = \sqrt{(x_2 - x_1)^2 + (y_2 - y_1)^2}$$

Midpoint of $\overline{P_1P_2}$: $\left(\dfrac{x_1 + x_2}{2}, \dfrac{y_1 + y_2}{2}\right)$

Lines

Slope of line through $P_1(x_1, y_1)$ and $P_2(x_2, y_2)$ $\qquad m = \dfrac{y_2 - y_1}{x_2 - x_1}$

Point-slope equation of line through $P_1(x_1, y_1)$ with slope m $\qquad y - y_1 = m(x - x_1)$

Slope-intercept equation of line with slope m and y-intercept b $\qquad y = mx + b$

Two-intercept equation of line with x-intercept a and y-intercept b $\qquad \dfrac{x}{a} + \dfrac{y}{b} = 1$

Logarithms

$y = \log_a x$ means $a^y = x$

$\log_a a^x = x \qquad a^{\log_a x} = x$

$\log_a 1 = 0 \qquad \log_a a = 1$

$\log x = \log_{10} x \qquad \ln x = \log_e x$

$\log_a xy = \log_a x + \log_a y \qquad \log_a\left(\dfrac{x}{y}\right) = \log_a x - \log_a y$

$\log_a x^b = b \log_a x \qquad \log_b x = \dfrac{\log_a x}{\log_a b}$

Exponential and Logarithmic Functions

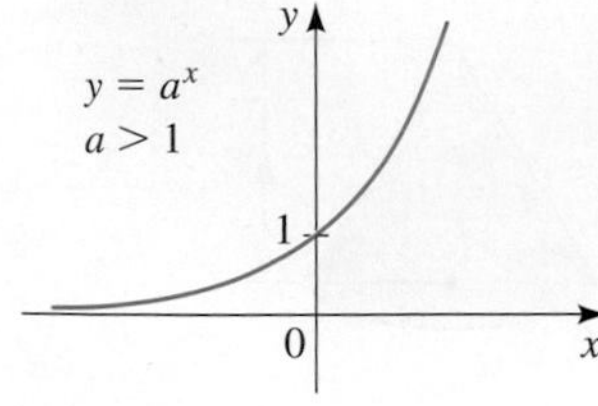

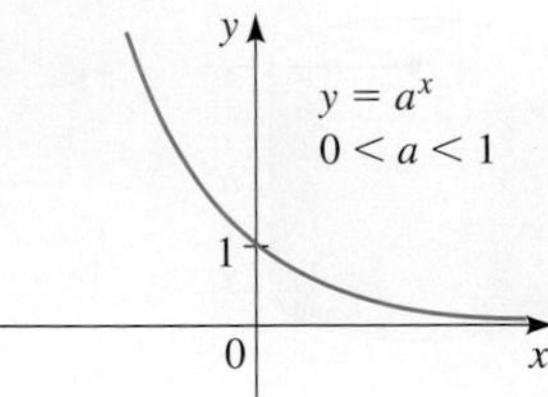

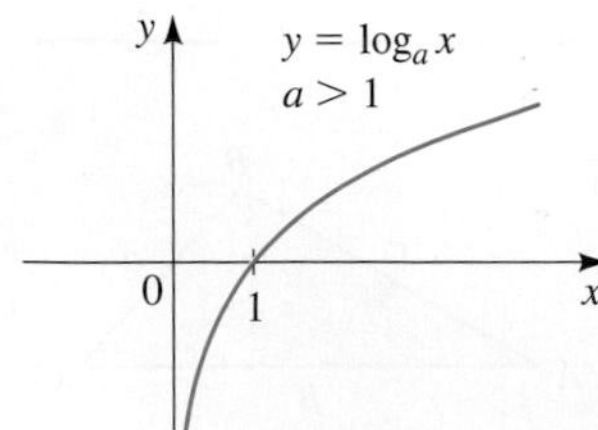

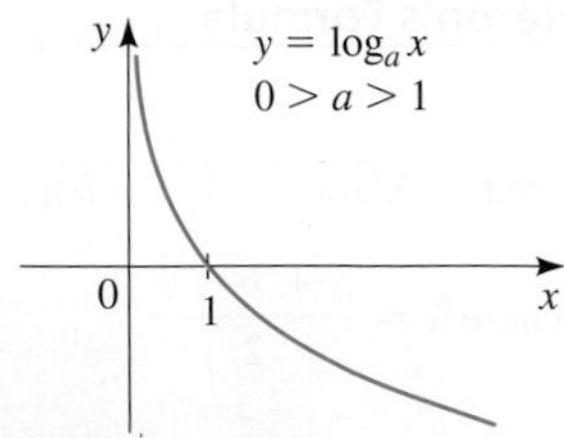

Graphs of Functions

Linear functions: $f(x) = mx + b$

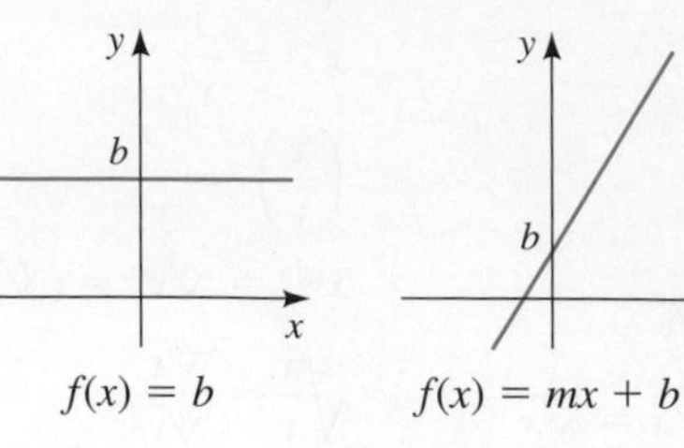

$f(x) = b \qquad f(x) = mx + b$

Power functions: $f(x) = x^n$

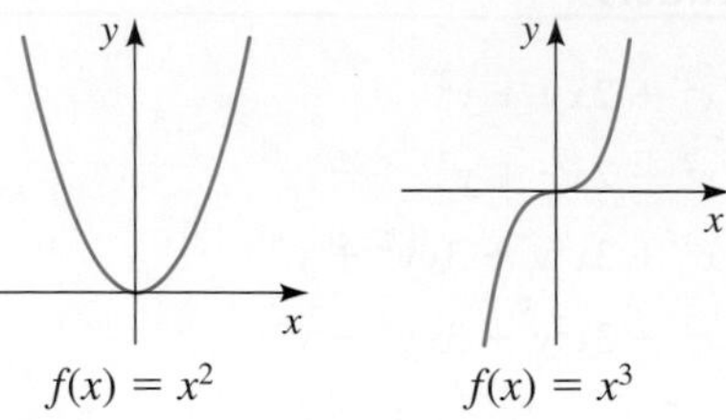

$f(x) = x^2 \qquad f(x) = x^3$

Root functions: $f(x) = \sqrt[n]{x}$

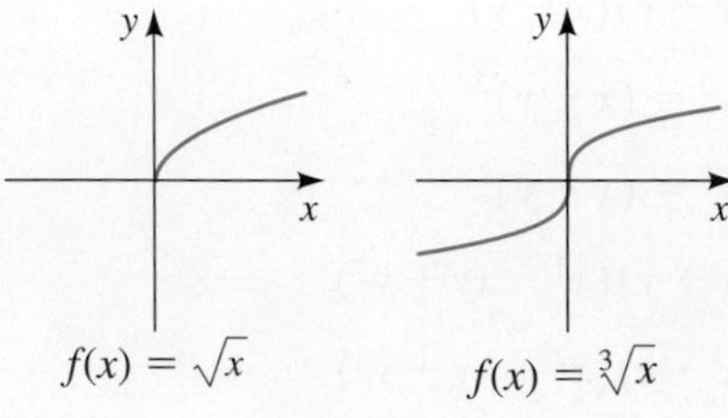

$f(x) = \sqrt{x} \qquad f(x) = \sqrt[3]{x}$

Reciprocal functions: $f(x) = 1/x^n$

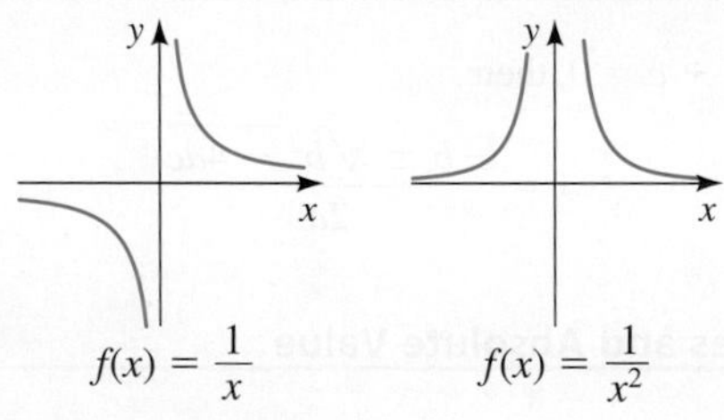

$f(x) = \dfrac{1}{x} \qquad f(x) = \dfrac{1}{x^2}$

Absolute value function $\qquad$ Greatest integer function

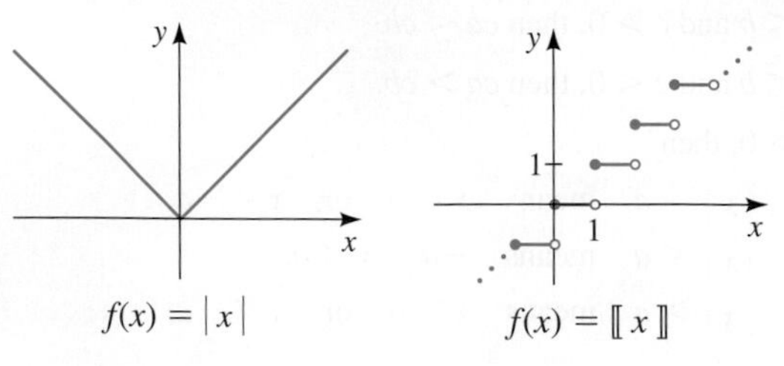

$f(x) = |x| \qquad f(x) = [\![x]\!]$

Complex Numbers

For the complex number $z = a + bi$

the **conjugate** is $\bar{z} = a - bi$

the **modulus** is $|z| = \sqrt{a^2 + b^2}$

the **argument** is θ, where $\tan\theta = b/a$

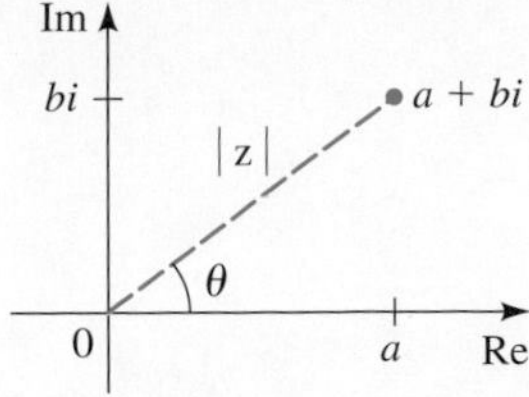

Polar form of a complex number

For $z = a + bi$, the **polar form** is

$$z = r(\cos\theta + i\sin\theta)$$

where $r = |z|$ is the modulus of z and θ is the argument of z

De Moivre's Theorem

$$z^n = [r(\cos\theta + i\sin\theta)]^n = r^n(\cos n\theta + i\sin n\theta)$$

$$\sqrt[n]{z} = [r(\cos\theta + i\sin\theta)]^{1/n}$$

$$= r^{1/n}\left(\cos\frac{\theta + 2k\pi}{n} + i\sin\frac{\theta + 2k\pi}{n}\right)$$

where $k = 0, 1, 2, \ldots, n - 1$

Rotation of Axes

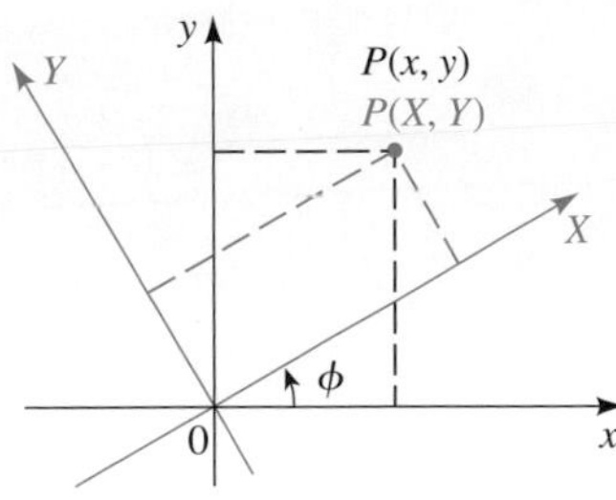

Rotation of axes formulas

$$x = X\cos\phi - Y\sin\phi$$

$$y = X\sin\phi + Y\cos\phi$$

Angle-of-rotation formula for conic sections

To eliminate the xy-term in the equation

$$Ax^2 + Bxy + Cy^2 + Dx + Ey + F = 0$$

rotate the axis by the angle ϕ that satisfies

$$\cot 2\phi = \frac{A - C}{B}$$

Polar Coordinates

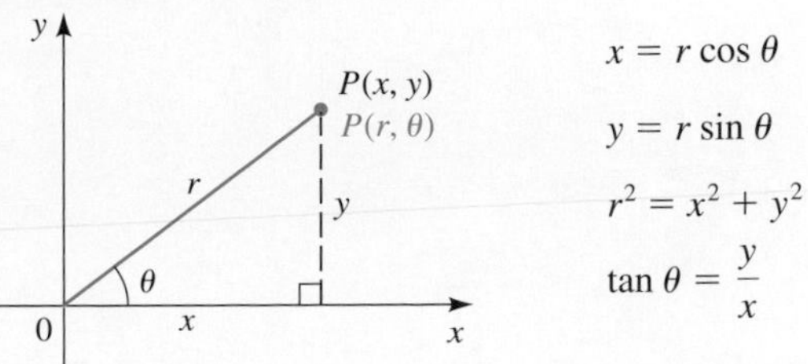

$$x = r\cos\theta$$

$$y = r\sin\theta$$

$$r^2 = x^2 + y^2$$

$$\tan\theta = \frac{y}{x}$$

Sums of Powers of Integers

$$\sum_{k=1}^{n} 1 = n \qquad \sum_{k=1}^{n} k = \frac{n(n+1)}{2}$$

$$\sum_{k=1}^{n} k^2 = \frac{n(n+1)(2n+1)}{6} \qquad \sum_{k=1}^{n} k^3 = \frac{n^2(n+1)^2}{4}$$

The Derivative

The **average rate of change** of f between a and b is

$$\frac{f(b) - f(a)}{b - a}$$

The **derivative** of f at a is

$$f'(a) = \lim_{x\to a}\frac{f(x) - f(a)}{x - a}$$

$$f'(a) = \lim_{h\to 0}\frac{f(a+h) - f(a)}{h}$$

Area Under the Graph of f

The **area under the graph of f** on the interval $[a, b]$ is the limit of the sum of the areas of approximating rectangles

$$A = \lim_{n\to\infty}\sum_{k=1}^{n} f(x_k)\,\Delta x$$

where

$$\Delta x = \frac{b - a}{n}$$

$$x_k = a + k\,\Delta x$$

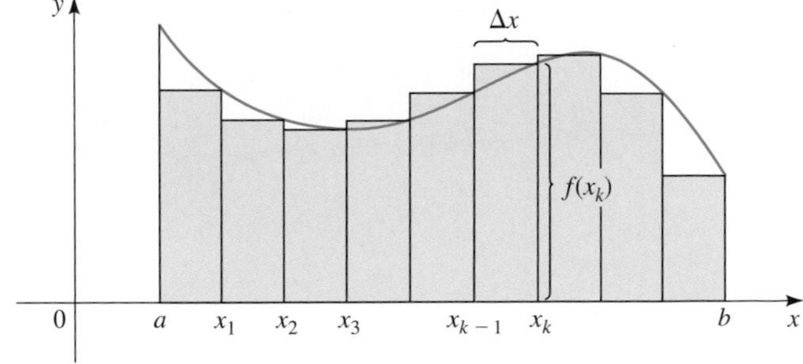

Precalculus

Mathematics for Calculus

METRIC VERSION

Eighth Edition

About the Authors

James Stewart received a Master of Science degree from Stanford University and a PhD in mathematics from the University of Toronto. He subsequently did research at the University of London. His research areas are harmonic analysis and the connections between mathematics and music. At Stanford, he was influenced by the mathematician George Polya, who is famous for his work on the heuristics of problem-solving. James Stewart was professor of mathematics at McMaster University and the University of Toronto for many years. He is the author of the bestselling calculus textbook series *Calculus, Calculus: Early Transcendentals*, and *Calculus: Concepts and Contexts*, published by Cengage Learning.

Lothar Redlin received a Bachelor of Science degree from the University of Victoria and a PhD in mathematics from McMaster University. He subsequently did research and taught at the University of Washington, California State University Long Beach, and the University of Waterloo, where he worked with Janos Aczel, one of the leading researchers in information theory. His research areas are functional equations and topology. Lothar Redlin was professor of mathematics at The Pennsylvania State University, Abington Campus, for many years.

Saleem Watson received a Bachelor of Science degree from Andrews University in Michigan. He did graduate studies at Dalhousie University and at McMaster University where he received a PhD in mathematics. He subsequently did research at the Mathematics Institute at the University of Warsaw. His research areas are functional analysis and topology. He also taught and did research at The Pennsylvania State University for several years. Saleem Watson is now professor emeritus of mathematics at the California State University Long Beach.

Stewart, Redlin, and Watson have also published *College Algebra, Trigonometry, Algebra and Trigonometry,* and (with Phyllis Panman) *College Algebra: Concepts and Contexts.*

About the Cover

The cover photograph shows a portion of the Trade Buildings, an architectural complex for offices with private garden in Barcelona, Spain. The complex was designed by architect Jose Antonio Coderech and Sentmenat. The building consists of four identical towers with a curved curtain wall on all its facades. The design and construction of such curves requires precise geometric descriptions of the curves; mathematical formulas are used for computing the structural stability and hence the feasibility of constructing such large curves in steel and concrete. In this book we explore geometric descriptions of certain curves that can help in constructing them at any scale. (See *Focus on Modeling*, Conics in Architecture, following Chapter 10.)

Precalculus

Mathematics for Calculus

METRIC VERSION
Eighth Edition

James Stewart
McMaster University and University of Toronto

Lothar Redlin
The Pennsylvania State University

Saleem Watson
California State University, Long Beach

With the assistance of Phyllis Panman

Cengage

Australia • Brazil • Canada • Mexico • South Africa • Singapore • United Kingdom • United States

***Precalculus: Mathematics for Calculus*, Metric Version, Eighth Edition**
James Stewart, Lothar Redlin, Saleem Watson

Publisher, Global Editions: Marinda Louw

Editorial Assistant: Veronique Botha

Associate Content Project Manager: Narmada Kaushal

Production Service/Compositor: MPS Limited

Cover Designer: Varpu Lauchlan

Intellectual Property Analyst: Brittani Morgan

Manager, Global IP Integration: Eleanor Rummer

Manufacturing Buyer: Eyvett Davis

Cover Image Source: Cavan Images/ Alamy Stock Photo

Previous Editions: © 2017, © 2012

Library of Congress Control Number: 2022922046

ISBN: 979-8-214-03181-1

Cengage Learning International Offices

Asia
www.cengageasia.com
tel: (65) 6410 1200

Australia/New Zealand
www.cengage.com.au
tel: (61) 3 9685 4111

Brazil
www.cengage.com.br
tel: (55) 11 3665 9900

India
www.cengage.co.in
tel: (91) 11 4364 1111

Latin America
www.cengage.com.mx
tel: (52) 55 1500 6000

UK/Europe/MiddleEast/Africa
www.cengage.co.uk
tel: (44) 0 1264 332 424

Canada
www.cengage.ca
tel: (416) 752 9100

Cengage Learning is a leading provider of customized learning solutions with office locations around the globe, including Singapore, the United Kingdom, Australia, Mexico, Brazil, and Japan. Locate your local office at: **www.cengage.com/global**.

For product information: **www.cengage.com/international**
Visit your local office: **www.cengage.com/global**
Visit our corporate website: **www.cengage.com**

Printed in China by RR Donnelley
Print Number: 01 Print Year: 2023

Contents

6 Trigonometric Functions: Right Triangle Approach 485

7 Analytic Trigonometry 553

8 Polar Coordinates, Parametric Equations, and Vectors 603

9 Systems of Equations and Inequalities 665

10 Conic Sections 765

11 Sequences and Series 827

12 Limits: A Preview of Calculus 877

Preface

The art of teaching is the art of assisting discovery.

Mark Van Doren

This International Metric Version differs from the US Version of *Precalculus: Mathematics for Calculus, Eighth Edition* as follows: The units used in most of the examples and exercises have been converted from the US Customary System (USCS) of units (also referred to as English or Imperial Units) to Metric Units.

What do students really need to know to be prepared for calculus? What tools do instructors really need to assist their students in preparing for calculus? These two questions have motivated the writing of this book.

To be prepared for calculus a student needs not only technical skill but also a clear understanding of concepts. Indeed, *conceptual understanding* and *technical skill* go hand in hand, each reinforcing the other. A student also needs to gain experience in *problem-solving*, as well as an appreciation for the power and utility of mathematics in *modeling* the real world. Every feature of this textbook is devoted to fostering these goals.

In this Eighth Edition our objective is to further enhance the effectiveness of the book as a teaching tool for instructors and as a learning tool for students. Many of the changes in this new edition are a result of suggestions we have received from instructors and students who are using the current edition, recommendations from reviewers, insights we have gained from our own experience teaching from the text, and conversations we have had with colleagues and students. In all the changes, both small and large, we have retained the features that have contributed to the success of this book.

What's New in the Eighth Edition?

The overall structure of the book remains largely the same, but we have made many improvements. Some of the changes are as follows.

- **Exercises** More than 20% of the exercises are new, including new types of conceptual and skill exercises.
- **Emphasis on Problem-Solving** The Prologue *Principles of Problem Solving* contains several principles that can be used to tackle any problem. This edition includes new exercises designed to give students the opportunity to experience the process of problem-solving for themselves; each such exercise is identified by the icon PS and includes a suggestion for which problem-solving principle to try. (See, for instance, Exercises 1.4.105, 1.9.116 and 118, 2.3.76, 2.6.103, 2.7.93, 6.1.71 and 74, 6.2.75, and 6.3.75–76.)
- **Special Application Exercises** In many sections there are new real-word application exercises that include some background information about the formulas or data used; we trust these exercises will capture students's interest. (See, for instance, Exercises 1.2.99, 1.8.127, 1.9.114, 1.12.53, 2.1.86, 2.5.48, 2.6.96, 3.7.62, 4.2.41, 4.5.96, and 4.6.29–31.)
- **Preparing for Calculus** Exercises that highlight skills which are needed in a calculus course are now identified by the icon $\int$. (See, for instance, Exercises 1.3.135–138, 2.4.19–24, 2.7.65–78, 4.2.25–28, 4.3.87–90, and 5.2.75–82.) The presentations, notation, features, and tone of this book are designed to be consistent with the Stewart Calculus textbooks.
- **Equations and Their Graphs** In this edition, we increase our emphasis on the essential relationship between an equation and its graph. (See, for instance,

Example 1.11.1 and Exercise 2.R.83.) Where appropriate, the end-of-chapter review exercises now include a new exercise in which students are asked to match equations with their graphs and give reasons for their choices. (See, for instance, the Review Exercises 1.R.160, 2.R.111, 3.R.109, 4.R.111, and 5.R.75.)

- **Discovery Projects** Several new *Discovery Projects* have been added. Each project is briefly described in the appropriate section. [See, for instance, the projects on Weighing the Whole World (Section 1.5), Hours of Daylight (Section 5.6), Collision (Section 8.4), and Symmetry (Section 10.4).] The projects are available at the book companion website **www.stewartmath.com**.
- **Expanded Problems in WebAssign** This edition's new online Expanded Problems further reinforce understanding by asking students to show the specific steps of their work or to explain the reasoning behind the answers they've provided.
- **Geometry Review** Appendix A *Geometry Review* has been expanded to contain all the geometry background referred to in the textbook. The new topics include theorems on parallel lines and theorems on circles.
- **Chapter 1 Fundamentals** In this chapter there are new groups of exercises with the title *Putting it All Together* that are intended to encourage students to first recognize the type of problem and then decide which method to use to solve it. (See, for instance, Exercises 1.2.77–80 and 1.5.103–118.)
- **Chapter 2 Functions** This chapter now includes a subsection on relations, which is introduced in the context of answering the following question: Which tables of values (or relations) represent functions? This topic complements the preceding subsections Which Graphs Represent Functions? and Which Equations Represent Functions?
- **Chapter 4 Exponential and Logarithmic Functions** This chapter now introduces the concept of logistic growth as a real-world application of exponential functions.
- **Chapter 5 Trigonometric Functions: Unit Circle Approach** The material on graphing trigonometric functions has been restructured. The chapter now includes step-by-step guidelines for graphing transformations of trigonometric functions.
- **Chapter 7 Analytic Trigonometry** The exercises on proving trigonometric identities have been restructured; these exercises now include a greater variety of trigonometric identities for students to try to prove.
- **Chapter 8 Polar Equation, Parametric Equations, Vectors** This is a new chapter that combines the topics of polar coordinates and parametric equations, together with the topic of vectors in the plane (previously in Chapter 9).

Teaching with the Help of This Book

We are keenly aware that good teaching comes in many forms and that there are many effective approaches to teaching and learning the concepts and skills of precalculus. The organization and exposition of the topics in this book are designed to accommodate different teaching and learning styles. In particular, each topic is presented algebraically, graphically, numerically, and verbally, with emphasis on the relationships between these different representations.

The features in this book—*Focus on Modeling* sections, *Discovery Projects*, *Discuss/Discover/Prove/Write* exercises, historical insights, and *Mathematics in the Modern World* vignettes—provide a variety of enhancements to a central core of fundamental concepts and skills. Our aim is to provide instructors and their students with the tools they need to navigate their own course toward discovering precalculus mathematics.

The following are some special features that can be used to complement different teaching and learning approaches.

Exercise Sets The most important way to foster conceptual understanding and hone technical skill is through the problems that the instructor assigns. To that end we have provided a wide selection of exercises.

- **Concept Exercises** These exercises are at the beginning of each exercise set and ask students to state and use basic facts about the topics of each section.
- **Skills Exercises** These exercises reinforce and provide practice with all the learning objectives of each section. They comprise the core of each exercise set.
- **Skills Plus Exercises** The *Skills Plus* exercises contain challenging problems that often require the synthesis of previously learned material with new concepts.
- **Applications Exercises** We have included substantial applied problems from many different real-world contexts. We believe that these exercises will capture students' interest.
- **Discovery, Writing, and Group Learning** Each exercise set ends with a block of exercises labeled *Discuss* ■ *Discover* ■ *Prove* ■ *Write*. These exercises are designed to encourage students to experiment, preferably in groups, with the concepts developed in the section and then to write about what they have learned rather than simply looking for the answer. The *Prove* exercises highlight the importance of deriving a formula. (See for instance Exercises 2.6.97–103 and 2.8.108–112.)
- **Problem-Solving Exercises** In each chapter there are challenging exercises which may be solved by using one of the principles described in the Prologue: *Principles of Problem Solving*. Each such exercise is identified by the icon PS and is accompanied by a hint that recommends a particular problem-solving principle to try. These exercises are intended to give students the opportunity to experience the process of problem-solving. (See for instance Exercises 2.7.94, 3.2.93, and 5.2.95.)
- **Now Try Exercise . . .** At the end of each example in the text the student is directed to one or more similar exercises in the section that help to reinforce the concepts and skills developed in that example.
- **Check Your Answer** Students are encouraged to check whether an answer they obtained is reasonable. This is emphasized throughout the text in numerous *Check Your Answer* sidebars that accompany the examples. (See for instance Examples 1.5.1 and 10, 1.7.1, and 9.1.7.)

A Complete Review Chapter We have included an extensive review chapter primarily as a handy reference for the basic concepts that are preliminary to this course.

- **Chapter 1 Fundamentals** This is the review chapter; it contains the fundamental concepts from algebra and analytic geometry that a student needs in order to begin a precalculus course. As much or as little of this chapter can be covered in class as needed, depending on the background of the students.
- **Diagnostic Test** The test at the end of Chapter 1 is designed as a diagnostic test for determining what parts of this review chapter need to be taught. It also serves to help students gauge exactly what topics they need to review.

Flexible Approach to Trigonometry The trigonometry chapters of this text have been written so that either the right triangle approach or the unit circle approach may be taught first. Putting these two approaches in different chapters, each with its relevant applications, helps to clarify the purpose of each approach. The chapters introducing trigonometry are as follows.

- **Chapter 5 Trigonometric Functions: Unit Circle Approach** This chapter introduces trigonometry through the unit circle. This approach emphasizes that the trigonometric functions are functions of real numbers, just like the polynomial and exponential functions with which students are already familiar.
- **Chapter 6 Trigonometric Functions: Right Triangle Approach** This chapter introduces trigonometry through right triangles. This approach builds on the foundation of a conventional high-school course in trigonometry.

Another way to teach trigonometry is to intertwine the two approaches. Some instructors teach this material in the following order: Sections 5.1, 5.2, 6.1, 6.2, 6.3, 5.3, 5.4, 5.5, 5.6, 6.4, 6.5, and 6.6. Our organization makes it easy to do this without obscuring the fact that the two approaches involve distinct representations of the same functions.

Graphing Devices We make use of graphing devices (graphing calculators, computers, and math apps) throughout the book. The examples in which technology is used are always preceded by examples in which students must graph or calculate by hand so that they can understand precisely what the device is doing when they later use it to simplify the routine part of their work. The exercises that require technology are marked with the special symbol ▦; these are optional and may be omitted without loss of continuity. (See *Technology in the Eighth Edition* following this preface, p. xix.)

Focus on Modeling The theme of modeling has been used throughout to unify and explain the many applications of precalculus. We have made a special effort to clarify the essential process of translating problems from words into the language of mathematics. (See, for instance, Examples 1.7.1 and 9.1.7.) Each chapter concludes with a *Focus on Modeling* section that highlights how the topics learned in the chapter are used in modeling real-world situations.

Review Sections and Chapter Tests Each chapter ends with an extensive review section that includes the following.

- **Properties and Formulas** The *Properties and Formulas* section at the end of each chapter contains a summary of the main formulas and procedures of the chapter.
- **Concept Check and Concept Check Answers** The *Concept Check* at the end of each chapter is designed to get the students to think about and explain each concept presented in the chapter and then to use the concept in a given problem. This provides a step-by-step review of all the main concepts in the chapter. Answers to the *Concept Check* questions are available at the book companion website.
- **Review Exercises** The *Review Exercises* at the end of each chapter recapitulate the basic concepts and skills of the chapter and include exercises that combine the different ideas learned in the chapter. Where appropriate, the last exercise is about matching equations with their graphs. The exercise requires students to relate properties of an equation to the corresponding properties of the graph.
- **Chapter Test** Each review section concludes with a *Chapter Test* designed to help students gauge their progress.
- **Cumulative Review Tests** *Cumulative Review Tests* following selected chapters are available at the book companion website. These tests contain problems that combine skills and concepts from the preceding chapters. The problems are intended to highlight the connections between the topics in these related chapters.
- **Answers** Brief answers to odd-numbered exercises in each section (including the *Review Exercises*) and to all questions in the *Concepts* exercises and *Chapter Tests,* are given in the back of the book.

Mathematical Vignettes Throughout the book we make use of the margins to provide historical notes, key insights, or applications of mathematics in the modern world. These serve to enliven the material and show that mathematics is an important, vital activity and that it is fundamental to everyday life.

- **Biographical Vignettes** These vignettes include biographies of interesting mathematicians and often include a key insight that the mathematician discovered. [See, for instance, the vignettes on Viète (Section 1.5), Turing (Section 2.6), and Thales (Section 6.6).]
- **Mathematics in the Modern World** This series of vignettes emphasizes the central role of mathematics in current advances in technology, the sciences, and the

social sciences. [See, for instance Unbreakable Codes (Section 3.6), Global Positioning System (Section 9.8), Looking Inside Your Head (Section 10.1), and Fair Division of Assets (Section 11.2).]

Book Companion Website A website that accompanies this book can be found at **www.stewartmath.com**. The site includes many resources for teaching precalculus, including the following.

- **Discovery Projects** *Discovery Projects* for each chapter are available at the book companion website. The projects are referenced in the text in the appropriate sections. Each project provides a challenging yet accessible set of activities that enable students (perhaps working in groups) to explore in greater depth an interesting aspect of the topic they have just learned (see, for instance, the Discovery Projects *Visualizing a Formula, Every Graph Tells a Story, Toricelli's Law, Mapping the World, and Will the Species Survive?,* referenced in Sections 1.3, 2.3, 3.1, 8.1, and 9.4).
- **Focus on Problem Solving** Several *Focus on Problem Solving* sections are available on the website. Each such section highlights one of the problem-solving. principles introduced in the Prologue and includes several challenging problems. (See, for instance, *Recognizing Patterns, Using Analogy, Introducing Something Extra, Taking Cases,* and *Working Backward*.)
- **Cumulative Review Tests** *Cumulative Review Tests* following Chapters 4, 7, 10, and 12 are available at the book companion website.
- **Appendix B: Calculations and Significant Figures** This appendix, available at the book companion website, contains guidelines for rounding when working with approximate values.
- **Appendix C: Graphing with a Graphing Calculator** This appendix, available at the book companion website, includes general guidelines on graphing with a graphing calculator as well as guidelines on how to avoid common graphing pitfalls.
- **Appendix D: Using the TI-83/84 Graphing Calculator** In this appendix, available at the book companion website, we provide simple, easy-to-follow, step-by-step instructions for using the TI-83/84 graphing calculators.
- **Additional Topics** Several additional topics are available at the book companion website, including material on three-dimensional coordinate geometry, mathematics of finance, and probability and statistics.

Ancillaries

Instructor Resources

Additional instructor resources for this product are available online. Instructor assets include an Instructor's Manual, Educator's Guide, PowerPoint® slides, and a test bank powered by Cognero®. Sign up or sign in at **https://faculty.cengage.com/** to search for and access this product and its online resources.

Solutions and Answers Guide

The Solutions and Answers Guide provides worked-out solutions to all of the problems in the text. Located at the book companion website.

Instructor's Manual

The Instructor's Manual contains points to stress, suggested time to allot, text discussion topics, core materials for lecture, workshop/discussion suggestions, group work exercises in a form suitable for handout, and suggested homework problems. Located at the book companionn website.

Cengage Learning Testing Powered by Cognero

Cengage Learning Testing (CLT) is a flexible online system that allows the instructor to author, edit, and manage test bank content; create multiple test versions in an instant; and deliver tests from your Learning Management System (LMS), your classroom, or wherever you want. This is available online via **https://faculty.cengage.com/**.

WebAssign

With *WebAssign*, deliver your course all in one place with online activities and secure testing, plus eTextbook and study resources for students. Whether building their prerequisite knowledge or honing their problem-solving skills, you have flexible settings and rich content to customize your course to fuel deeper student learning—and confidence—in any course format.

Student Resources

Student Solutions Manual

The Student Solutions Manual contains fully worked-out solutions to all odd-numbered exercises in the text, giving students a way to check their answers and ensure that they took the correct steps to arrive at an answer. Located at the book companion website.

Note-Taking Guide

The Note-Taking Guide is an innovative study aid that helps students develop a section-by-section summary of key concepts. Located at the book companion website.

WebAssign

Ever wonder if you have studied enough? *WebAssign* from Cengage can help.

WebAssign is an online learning platform for your Math, Statistics, Physical Sciences, and Engineering courses. It helps you practice, focus your study time, and absorb what you learn. When class comes—you're way more confident.

With *WebAssign*, you will:

- Get instant feedback and grading on your assignments
- Know how well you understand concepts
- Watch videos and tutorials when you're stuck
- Perform better on in-class activities

Acknowledgments

We feel fortunate that all those involved in the production of this book have worked with exceptional energy, intense dedication, and passionate interest. It is surprising how many people are essential in the production of a mathematics textbook, including content editors, reviewers, faculty colleagues, production editors, copy editors, permissions editors, solutions and accuracy checkers, artists, photo researchers, text designers, typesetters, compositors, proofreaders, printers, and the technology team that creates the electronic version of the book and implements the online homework. We thank them all. We particularly mention the following.

Reviewers for the Seventh Edition Mary Ann Teel, University of North Texas; Natalia Kravtsova, The Ohio State University; Belle Sigal, Wake Technical Community College; Charity S. Turner, The Ohio State University; Yu-ing Hargett, Jefferson State Community College–Alabama; Alicia Serfaty de Markus, Miami Dade College;

Cathleen Zucco-Teveloff, Rider University; Minal Vora, East Georgia State College; Sutandra Sarkar, Georgia State University; Jennifer Denson, Hillsborough Community College; Candice L. Ridlon, University of Maryland Eastern Shore; Alin Stancu, Columbus State University; Frances Tishkevich, Massachusetts Maritime Academy; Phil Veer, Johnson County Community College; Phillip Miller, Indiana University–Southeast; Mildred Vernia, Indiana University–Southeast; Thurai Kugan, John Jay College–CUNY.

Reviewers for the Eighth Edition Mary Ann Barber, University of North Texas; Stephanie Garofalo, Georgia State University, Perimeter College; Kirk Mehtlan, Pima Community College—East Campus; Lora Merchant, Auburn University; Ed Migliore, University of California, Santa Cruz; Debra Prescott, Central Texas College; Karin Pringle, University of Tennessee; Hasan Z. Rahim, San Jose City College; Candice Ridlon, University of Maryland Eastern Shore; Steven Safran, Rutgers University

We thank all those who have contributed to this edition—and there are many—as well as those whose input in previous editions continues to enhance this book. We extend special thanks to our colleagues Joseph Bennish, Linda Byun, Bruce Chaderjian, David Gau, Daniel Hernandez, Robert Mena, Kent Merryfield, Florence Newberger, Joshua Sack, and Alan Safer, all at California State University, Long Beach; to Karen Gold, Betsy Huttenlock, Cecilia McVoy, and Samir Ouzomgi, at Penn State, Abington; to Gloria Dion, of Educational Testing Service, Princeton, New Jersey. We thank Aaron Watson at the University of California, Santa Barbara, for suggesting several application exercises; Jessica Shi for solving all the *Focus on Problem Solving* problems and making valuable suggestions; Nakhle Asmar and Mark Ashbaugh at the University of Missouri, and Tariq Rizvi at the University of Ohio for many conversations about problem-solving; and Zidslav Kovarik at McMaster University for examples illustrating the limitations of calculators. We are grateful to Frances Gulick at the University of Maryland for critically reviewing the entire book. We thank Jonathan Watson at Andrews University in Michigan for working out many preliminary trials of the discovery projects and suggesting improvements. We have received many comments from students; we particularly thank Weihua "Benny" Zeng for his insightful observations on the exercise about beehives.

We are particularly grateful to Daniel Clegg at Palomar College for reading through the manuscript and checking all the exercise answers for accuracy. His suggestions for improving the manuscript are greatly appreciated. We thank Andrew Bulman-Fleming for his accuracy and engaging style in authoring the solutions manual. We thank Doug Shaw at the University of Northern Iowa for creating the Instructor Manual.

We especially thank Kathi Townes, our copyeditor, for her expert editing of the manuscript. Her exceptional skill and her immense experience in editing all the Stewart textbooks, have contributed greatly to the seamless transition between the Precalculus and Calculus books in the Stewart textbook series.

At Cengage Learning, we thank Lynh Pham for his skillful coordination of all aspects of the creation and production of the original US edition of the book; Varpu Lauchlan for her elegant cover design for the Metric Version and Timothy Biddick for the design of the interior of the book; Laura Gallus for vetting the entire original manuscript and making valuable comments. At WebAssign we thank Megan Gore for her capable leadership of the team that creatively transforms each exercise into an appropriate format for online homework.

Finally, this textbook has benefited greatly since its first edition from the advice and guidance of some of the best mathematics editors anywhere: Jeremy Hayhurst, Gary Ostedt, Bob Pirtle, and Gary Whalen, our current editor. Gary Whalen's broad knowledge of current issues in the teaching of mathematics, his focused interactions with mathematics faculty across the country, his tireless work as editor for all the Stewart textbooks, and his continual search for better ways of using technology as a teaching and learning tool have been invaluable resources in the creation of this book.

A Tribute to Lothar Redlin

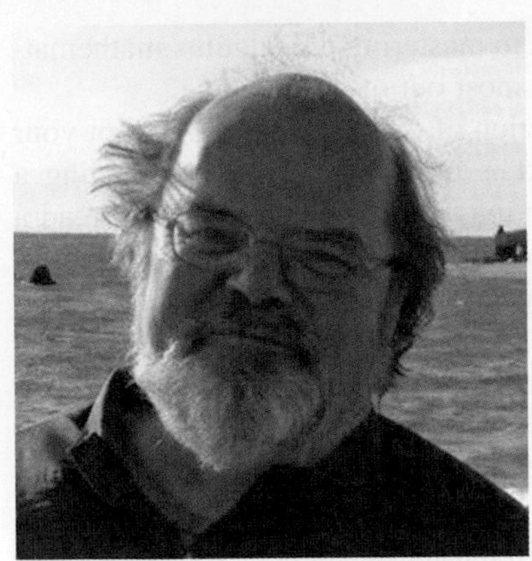

Lothar Redlin had an extraordinary talent for teaching mathematics. The classes that he taught were always full to capacity with students eager to learn from him. Lothar's presence in the classroom put everyone at ease, thus generating a wonderful learning atmosphere. Lothar held everyone's attention with his unassuming manner and his crystal clear explanations, which were drawn from his deep understanding of the meaning and purpose of the mathematics he was teaching.

Lothar attributed his vibrant mathematical intuition to the way his professors taught him; consequently, he always strived to transmit his personal insights into mathematics to his own students—in his teaching and in his textbooks—and his insights about precalculus continue to enhance this book. Lothar would complete teaching all the topics in a course well before the end of the school term, and his students would understand the subject exceptionally well—a tremendous witness to his unique way of understanding and explaining mathematics. Over the years, several of Lothar's former students, by then scientists, engineers, or financial analysts themselves, would contact him to discuss a mathematical problem that they came across in their work. Some of these discussions were the inspiration for applied exercises or discovery projects in this book.

Lothar grew up on Vancouver Island where he spent the summers helping his father, a professional fisherman, on their fishing boat (a 52-foot troller which Lothar had helped his father build). In the many hours they spent out at sea, Lothar's mind was not idle; he later recalled that the things he thought about were essentially mathematical (for instance, he figured out the mathematical basis for the Long-Range Navigation system they were using on the boat). Later, in high school, Lothar's talent for mathematics was recognized by his teachers and they encouraged him to attend the University of Victoria. He became the first member of his extended family to go to college.

I met Lothar when we were both graduate students in mathematics at McMaster University, and where James Stewart was one of our professors. Subsequently, we all worked together on various projects—first on research and later on writing textbooks. It has been my great fortune to have known and worked with Lothar for all the intervening years—until his untimely passing in 2018. Lothar's unique insights into mathematics, his humble and unassuming manner, his boundless generosity, and his earnest concern for everyone he met have enriched my life and the lives of his many students in inimitable ways; he is profoundly missed.

Saleem Watson

To the Student

This textbook was written for you to use as a guide to mastering precalculus mathematics. Here are some suggestions to help you get the most out of your course.

First of all, you should read the appropriate section of text *before* you attempt your homework problems. Reading a mathematics text is quite different from reading a novel, a newspaper, or even another textbook. You may find that you have to reread a passage several times before you understand it. Pay special attention to the examples, and work them out yourself with pencil and paper as you read. Then do the linked exercises referred to in "*Now Try Exercise . . .*" at the end of each example. With this kind of preparation you will be able to do your homework much more quickly and with more understanding.

Don't make the mistake of trying to memorize every single rule or fact you may come across. Mathematics doesn't consist simply of memorization. Mathematics is a *problem-solving art*, not just a collection of facts. To master the subject you must solve problems—lots of problems. Do as many of the exercises as you can. Be sure to write your solutions in a logical, step-by-step fashion. Don't give up on a problem if you can't solve it right away. Try to understand the problem more clearly—reread it thoughtfully and relate it to what you have learned from your teacher and from the examples in the text. Struggle with it until you solve it. Once you have done this a few times you will begin to understand what mathematics is really all about.

Answers to the odd-numbered exercises, as well as all the answers (even and odd) to the concept exercises and chapter tests, appear at the back of the book. If your answer differs from the one given, don't immediately assume that you are wrong. There may be a calculation that connects the two answers and makes both correct. For example, if you get $1/(\sqrt{2} - 1)$ but the answer given is $1 + \sqrt{2}$, your answer *is* correct, because you can multiply both numerator and denominator of your answer by $\sqrt{2} + 1$ to change it to the given answer. In rounding approximate answers, follow the guidelines in Appendix B: *Calculations and Significant Figures.*

The symbol ⊘ is used to warn against committing an error. We have placed this symbol in the margin to point out situations where we have found that many of our students make the same mistake.

Abbreviations

The following abbreviations are used in the text.

cm	centimeter	**kPa**	kilopascal	**N**	Newton
dB	decibel	**L**	liter	**psi**	pounds per in^2
dL	deci-liter	**lb**	pound	**qt**	quart
F	farad	**lm**	lumen	**oz**	ounce
ft	foot	**mol**	mole	**s**	second
g	gram	**m**	meter	**Ω**	ohm
gal	gallon	**mg**	milligram	**V**	volt
h	hour	**MHz**	megahertz	**W**	watt
H	henry	**mi**	mile	**yd**	yard
Hz	Hertz	**min**	minute	**yr**	year
in.	inch	**mL**	milliliter	**°C**	degree Celsius
J	Joule	**mm**	millimeter	**°F**	degree Fahrenheit
kcal	kilocalorie	**MW**	megawatts	**K**	Kelvin
kg	kilogram	**nm**	nanometers	$\Rightarrow$	implies
km	kilometer	**Å**	angstrom	$\Leftrightarrow$	is equivalent to

Technology in the Eighth Edition

Several different technology resources are available for computing and graphing. These include graphing and computing tools on the Internet, math apps for smartphones and tablets, and graphing calculators. In many situations, using technology eliminates the routine work of computing and graphing, thus allowing us to more sharply focus our attention on understanding the concepts of precalculus. For instance, we can use technology to quickly graph a function, modify the function and graph it again, in order to gain insight into the behavior of the function.

There are graphing and computing programs available free of charge on the Internet or as apps for smartphones or tablets. These include Desmos, GeoGebra, WolframAlpha, and others. Of course, graphing calculators can also be used to perform all the needed computing and graphing in the exercises, although many of the web-based math apps are more intuitive to use.

In this edition, rather than refer to a specific math app or graphing device, we use the icon ⊞ to indicate that technology is needed to complete the exercise. For such exercises you are expected to use technology to draw a graph or perform a calculation. You are encouraged to use technology on other exercises as well, to check your work or to explore related problems.

Prologue

Principles of Problem Solving

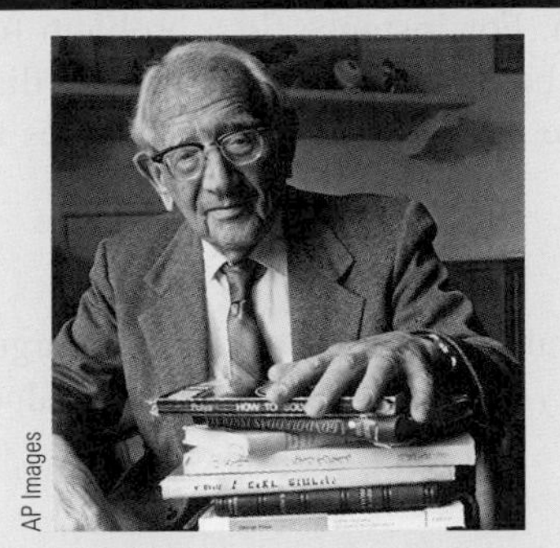

AP Images

GEORGE POLYA (1887–1985) is famous among mathematicians for his ideas on problem-solving. His lectures on problem-solving at Stanford University attracted overflow crowds, and he held his audience on the edges of their seats, leading them to discover solutions for themselves. He was able to do this because of his deep insight into the psychology of problem-solving. His well-known book *How To Solve It* has been translated into 15 languages. He said that Euler (see Section 1.6) was unique among great mathematicians because he explained how he found his results. Polya often said to his students and colleagues, "Yes, I see that your proof is correct, but how did you discover it?" In the preface to *How To Solve It,* Polya writes, "A great discovery solves a great problem but there is a grain of discovery in the solution of any problem. Your problem may be modest; but if it challenges your curiosity and brings into play your inventive faculties, and if you solve it by your own means, you may experience the tension and enjoy the triumph of discovery."

The ability to solve problems is a highly prized skill in many aspects of our lives; it is certainly an important part of any mathematics course. There are no hard and fast rules that will ensure success in solving problems. However, in this prologue we outline some general steps in the problem-solving process and we give principles that are useful in solving certain types of problems. These steps and principles are just common sense made explicit. They have been adapted from George Polya's insightful book *How To Solve It.*

1. Understand the Problem

The first step is to read the problem and make sure that you understand it. Ask yourself the following questions:

What is the unknown?

What are the given quantities?

What are the given conditions?

For many problems it is useful to

draw a diagram

and identify the given and required quantities on the diagram. Usually, it is necessary to

introduce suitable notation

In choosing symbols for the unknown quantities, we often use letters such as a, b, c, m, n, x, and y, but in some cases it helps to use initials as suggestive symbols, for instance, V for volume or t for time.

2. Think of a Plan

Find a connection between the given information and the unknown that enables you to calculate the unknown. It often helps to ask yourself explicitly: "How can I relate the given to the unknown?" If you don't see a connection immediately, the following ideas may be helpful in devising a plan.

■ Try to Recognize Something Familiar

Relate the given situation to previous knowledge. Look at the unknown and try to recall a more familiar problem that has a similar unknown.

■ Try to Recognize Patterns

Certain problems are solved by recognizing that some kind of pattern is occurring. The pattern could be geometric, numerical, or algebraic. If you can see regularity or repetition in a problem, then you might be able to guess what the pattern is and then prove it.

■ Use Analogy

Try to think of an analogous problem, that is, a similar or related problem but one that is easier than the original. If you can solve the similar, simpler problem, then it might give you the clues you need to solve the original, more difficult one. For instance, if a problem involves very large numbers, you could first try a similar problem with smaller numbers. Or if the problem is in three-dimensional geometry, you could look for something similar in two-dimensional geometry. Or if the problem you start with is a general one, you could first try a special case.

■ Introduce Something Extra

You may sometimes need to introduce something new—an auxiliary aid—to make the connection between the given and the unknown. For instance, in a problem for which a diagram is useful, the auxiliary aid could be a new line drawn in the diagram. In a more algebraic problem the aid could be a new unknown that relates to the original unknown.

■ Take Cases

You may sometimes have to split a problem into several cases and give a different argument for each case. We often have to use this strategy in dealing with absolute value.

■ Work Backward

Sometimes it is useful to imagine that your problem is solved and work backward, step by step, until you arrive at the given data. Then you might be able to reverse your steps and thereby construct a solution to the original problem. This procedure is commonly used in solving equations. For instance, in solving the equation $3x - 5 = 7$, we suppose that x is a number that satisfies $3x - 5 = 7$ and work backward. We add 5 to each side of the equation and then divide each side by 3 to get $x = 4$. Since each of these steps can be reversed, we have solved the problem.

■ Establish Subgoals

In a complex problem it is often useful to set subgoals (in which the desired situation is only partially fulfilled). If you can attain or accomplish these subgoals, then you may be able to build on them to reach your final goal.

■ Indirect Reasoning

Sometimes it is appropriate to attack a problem indirectly. In using **proof by contradiction** to prove that P implies Q, we assume that P is true and Q is false and try to see why this cannot happen. Somehow we have to use this information and arrive at a contradiction to what we absolutely know is true.

■ Mathematical Induction

In proving statements that involve a positive integer n, it is frequently helpful to use the Principle of Mathematical Induction, which is discussed in Section 11.4.

3. Carry Out the Plan

In Step 2, a plan was devised. In carrying out that plan, you must check each stage of the plan and write the details that prove that each stage is correct.

4. Look Back

Having completed your solution, it is wise to look back over it, partly to see whether any errors have been made and partly to see whether you can discover an easier way to solve the problem. Looking back also familiarizes you with the method of solution, which may be useful for solving a future problem. Descartes said, "Every problem that I solved became a rule which served afterwards to solve other problems."

We illustrate some of these principles of problem-solving with an example.

Problem ■ Average Speed

A driver sets out on a journey. For the first half of the distance, the driver's speed is 30 km/h; during the second half the speed is 60 km/h. What is the average speed on this trip?

Thinking About the Problem

It is tempting to take the average of the speeds and say that the average speed for the entire trip is $\dfrac{30 + 60}{2} = 45$ km/h. But is this approach really correct?

Try a special case. ➤ Let's look at an easily calculated special case. Suppose that the total distance traveled is 120 km. Since the first 60 km is traveled at 30 km/h, it takes 2 h. The second 60 km is traveled at 60 km/h, so it takes one hour. Thus, the total time is $2 + 1 = 3$ hours and the average speed is $\dfrac{120}{3} = 40$ km/h. So our guess of 45 km/h was wrong.

Solution

Understand the problem. ➤ We need to look more carefully at the meaning of average speed. It is defined as

$$\text{average speed} = \frac{\text{distance traveled}}{\text{time elapsed}}$$

Introduce notation. ➤ Let d be the distance traveled on each half of the trip. Let t_1 and t_2 be the times taken for the first and second halves of the trip. Now we can write down the information we

State what is given. ➤ have been given. For the first and second halves of the trip we have

$$30 = \frac{d}{t_1} \qquad 60 = \frac{d}{t_2}$$

Identify the unknown. ➤ Now we identify the quantity that we are asked to find:

$$\text{average speed for entire trip} = \frac{\text{total distance}}{\text{total time}} = \frac{2d}{t_1 + t_2}$$

Connect the given with the unknown. ➤ To calculate this quantity, we need to know t_1 and t_2, so we solve the above equations for these times:

$$t_1 = \frac{d}{30} \qquad t_2 = \frac{d}{60}$$

Now we have the ingredients needed to calculate the desired quantity:

$$\text{average speed} = \frac{2d}{t_1 + t_2} = \frac{2d}{\dfrac{d}{30} + \dfrac{d}{60}}$$

$$= \frac{60(2d)}{60\left(\dfrac{d}{30} + \dfrac{d}{60}\right)} \qquad \text{Multiply numerator and denominator by 60}$$

$$= \frac{120d}{2d + d} = \frac{120d}{3d} = 40$$

So the average speed for the entire trip is 40 km/h.

Look back ➤ This answer agrees with the answer we obtained for the special cases we tried above in *Thinking About the Problem.* ■

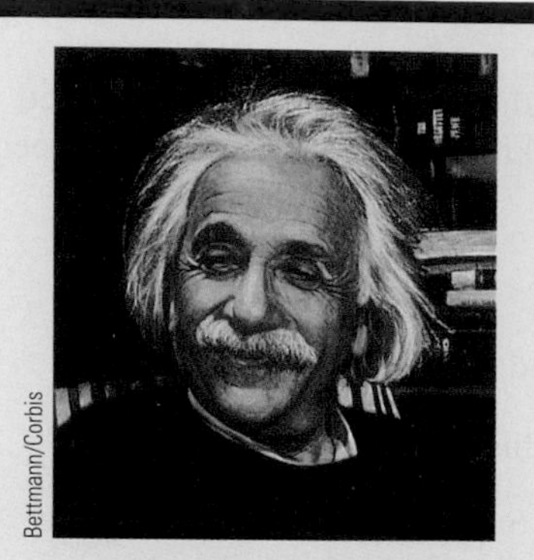

Don't feel bad if you can't solve these problems right away. Problems 1 and 4 were sent to Albert Einstein by his friend Wertheimer. Einstein (and his friend Bucky) enjoyed the problems and wrote back to Wertheimer. Here is part of his reply:

> Your letter gave us a lot of amusement. The first intelligence test fooled both of us (Bucky and me). Only on working it out did I notice that no time is available for the downhill run! Mr. Bucky was also taken in by the second example, but I was not. Such drolleries show us how stupid we are!

(See *Mathematical Intelligencer*, Spring 1990, page 41.)

■ Problems

1. **Distance, Time, and Speed** An old car has to travel a 2-kilometer route, uphill and down. Because it is so old, the car can climb the first kilometer—the ascent—no faster than an average speed of 15 km/h. How fast does the car have to travel the second kilometer—on the descent it can go faster, of course—to achieve an average speed of 30 km/h for the trip?

2. **Comparing Discounts** Which price is better for the buyer, a 40% discount or two successive discounts of 20%?

3. **Cutting up a Wire** A piece of wire is bent as shown in the figure. You can see that one cut through the wire (represented by the red lines) produces four pieces and two parallel cuts produce seven pieces. How many pieces will be produced by 142 parallel cuts? Write a formula for the number of pieces produced by n parallel cuts.

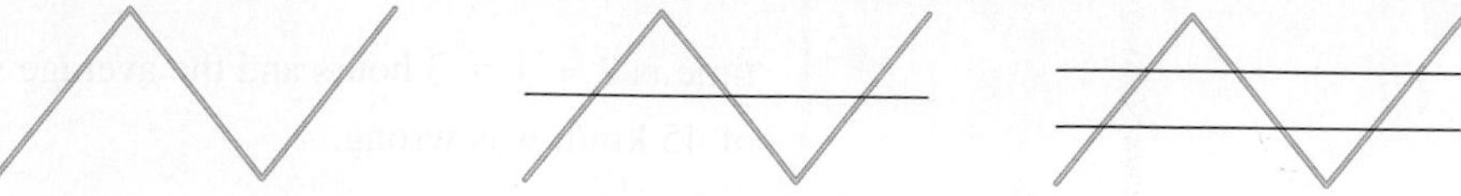

4. **Amoeba Propagation** An amoeba propagates by simple division; each split takes 3 minutes to complete. When such an amoeba is put into a glass container with a nutrient fluid, the container is full of amoebas in one hour. How long would it take for the container to be filled if we start with not one amoeba, but two?

5. **Batting Averages** Player A has a higher batting average than player B for the first half of the baseball season. Player A also has a higher batting average than player B for the second half of the season. Is it necessarily true that player A has a higher batting average than player B for the entire season?

6. **Coffee and Cream** A spoonful of cream is taken from a pitcher of cream and put into a cup of coffee. The coffee is stirred. Then a spoonful of this mixture is put into the pitcher of cream. Is there now more cream in the coffee cup or more coffee in the pitcher of cream?

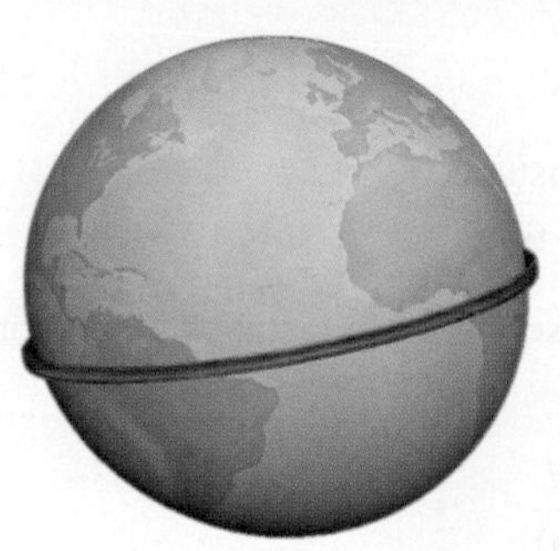

7. **Wrapping the World** A ribbon is tied tightly around the earth at the equator. How much more ribbon would you need if you raised the ribbon 1 m above the equator everywhere? (You don't need to know the radius of the earth to solve this problem.)

8. **Ending Up Where You Started** A woman starts at a point P on the earth's surface and walks 1 km south, then 1 km east, then 1 km north, and finds herself back at P, the starting point. Describe all points P for which this is possible. [*Hint:* There are infinitely many such points, all but one of which lie in Antarctica.]

You can apply these problem-solving principles in solving any problem or exercise. The icon PS next to an exercise indicates that one or more of the strategies discussed here is particularly useful for solving the exercise. You can find more examples and problems that highlight these principles at the book companion website **www.stewartmath.com**.

Rawpixel.com/Shutterstock.com

1 Fundamentals

In this first chapter we review the real numbers, equations, and the coordinate plane. You are probably already familiar with these concepts, but it is useful to get a fresh look at how these ideas work together to help you solve problems and model (or describe) real-world situations.

In the *Focus on Modeling* at the end of the chapter we learn how to find linear trends in data and how to use these trends to make predictions about the object or process being modeled.

1.1 Real Numbers

■ Real Numbers ■ Properties of Real Numbers ■ Addition and Subtraction ■ Multiplication and Division ■ The Real Line ■ Sets and Intervals ■ Absolute Value and Distance

In the real world we use numbers to measure and compare quantities. For example, we measure temperature, length, height, weight, distance, speed, acceleration, energy, force, angles, pressure, cost, and so on. Figure 1 illustrates some situations in which numbers are used. Numbers also allow us to express relationships between different quantities—for example, relationships between the radius and volume of a ball, between kilometers driven and gas used, or between education level and starting salary.

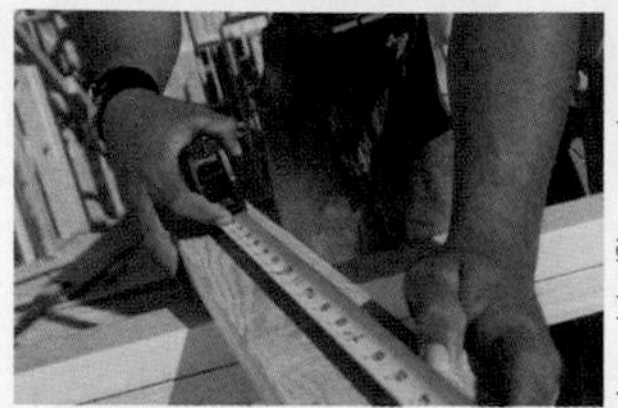
sirtravelalot/Shutterstock.com

Length

Andrey Burstein/Shutterstock.com

Weight

Oleksiy Mark/Shutterstock.com

Speed

KhanunHaHa/Shutterstock.com

Pressure

Figure 1 | Measuring with real numbers

■ Real Numbers

Let's review the types of numbers that make up the real number system. We start with the **natural numbers**:

$$1, 2, 3, 4, \ldots$$

The different types of real numbers were invented to meet specific needs. For example, natural numbers are needed for counting, negative numbers for describing debt or below-zero temperatures, rational numbers for concepts like "half a gallon of milk," and irrational numbers for measuring certain distances, like the diagonal of a square.

The **integers** consist of the natural numbers together with their negatives and 0:

$$\ldots, -3, -2, -1, 0, 1, 2, 3, 4, \ldots$$

We construct the **rational numbers** by taking ratios of integers. Thus any rational number r can be expressed as

$$r = \frac{m}{n}$$

where m and n are integers and $n \neq 0$. Examples are

$$\tfrac{1}{2} \qquad -\tfrac{3}{7} \qquad 46 = \tfrac{46}{1} \qquad 0.17 = \tfrac{17}{100}$$

(Recall that division by 0 is always ruled out, so expressions like $\frac{3}{0}$ and $\frac{0}{0}$ are undefined.) There are also real numbers, such as $\sqrt{2}$, that cannot be expressed as a ratio of integers and are therefore called **irrational numbers**. It can be shown, with varying degrees of difficulty, that these numbers are also irrational:

$$\sqrt{3} \qquad \sqrt{5} \qquad \sqrt[3]{2} \qquad \pi \qquad \frac{3}{\pi^2}$$

The set of all real numbers is usually denoted by the symbol $\mathbb{R}$. When we use the word *number* without qualification, we will mean "real number." Figure 2 is a diagram of the types of real numbers that we work with in this book.

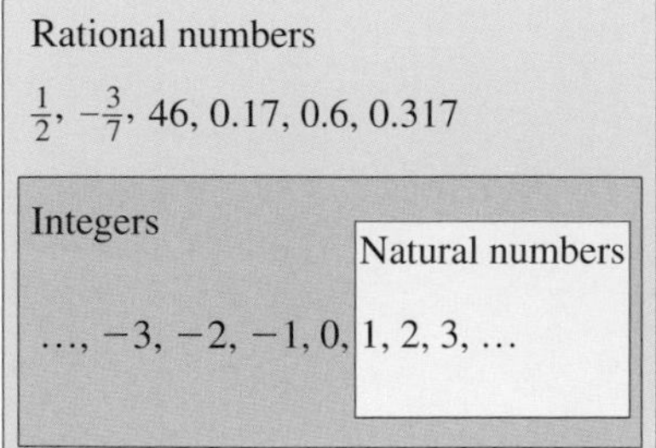

Figure 2 | The real number system

Every real number has a decimal representation. If the number is rational, then its corresponding decimal is repeating. For example,

$$\tfrac{1}{2} = 0.5000\ldots = 0.5\overline{0} \qquad \tfrac{2}{3} = 0.66666\ldots = 0.\overline{6}$$

$$\tfrac{157}{495} = 0.3171717\ldots = 0.3\overline{17} \qquad \tfrac{9}{7} = 1.285714285714\ldots = 1.\overline{285714}$$

A repeating decimal such as

$$x = 3.5474747...$$

is a rational number. To convert it to a ratio of two integers, we write

$$\begin{aligned} 1000x &= 3547.47474747... \\ 10x &= \underline{35.47474747...} \\ 990x &= 3512.0 \end{aligned}$$

Thus $x = \frac{3512}{990}$. (The idea is to multiply x by appropriate powers of 10 and then subtract to eliminate the repeating part.) See also Example 11.3.7 and Exercise 11.3.77.

(The bar indicates that the sequence of digits repeats indefinitely.) If a given number is irrational, its decimal representation is nonrepeating:

$$\sqrt{2} = 1.414213562373095... \qquad \pi = 3.141592653589793...$$

If we stop the decimal expansion of any number at a certain place, we get an approximation to the number. For instance, we can write

$$\pi \approx 3.14159265$$

where the symbol $\approx$ is read "is approximately equal to." The more decimal places we retain, the better our approximation.

■ Properties of Real Numbers

We all know that $2 + 3 = 3 + 2$, and $5 + 7 = 7 + 5$, and $513 + 87 = 87 + 513$, and so on. In algebra we express all these (infinitely many) facts by writing

$$a + b = b + a$$

where a and b stand for any two numbers. In other words, "$a + b = b + a$" is a concise way of saying that "when we add two numbers, the order of addition doesn't matter." This fact is called the *Commutative Property* of addition. From our experience with numbers we know that the properties in the following box are also valid.

Properties of Real Numbers

Property	Example	Description
Commutative Properties		
$a + b = b + a$	$7 + 3 = 3 + 7$	When we add two numbers, order doesn't matter.
$ab = ba$	$3 \cdot 5 = 5 \cdot 3$	When we multiply two numbers, order doesn't matter.
Associative Properties		
$(a + b) + c = a + (b + c)$	$(2 + 4) + 7 = 2 + (4 + 7)$	When we add three numbers, it doesn't matter which two we add first.
$(ab)c = a(bc)$	$(3 \cdot 7) \cdot 5 = 3 \cdot (7 \cdot 5)$	When we multiply three numbers, it doesn't matter which two we multiply first.
Distributive Property		
$a(b + c) = ab + ac$	$2 \cdot (3 + 5) = 2 \cdot 3 + 2 \cdot 5$	When we multiply a number by a sum of two numbers, we get the same result as we get if we multiply the number by each of the terms and then add the results.
$(b + c)a = ab + ac$	$(3 + 5) \cdot 2 = 2 \cdot 3 + 2 \cdot 5$	

The Distributive Property applies whenever we multiply a number by a sum. Figure 3 explains why this property works for the case in which all the numbers are positive integers, but the property is true for any real numbers a, b, and c.

The Distributive Property is crucial because it describes the way addition and multiplication interact with each other.

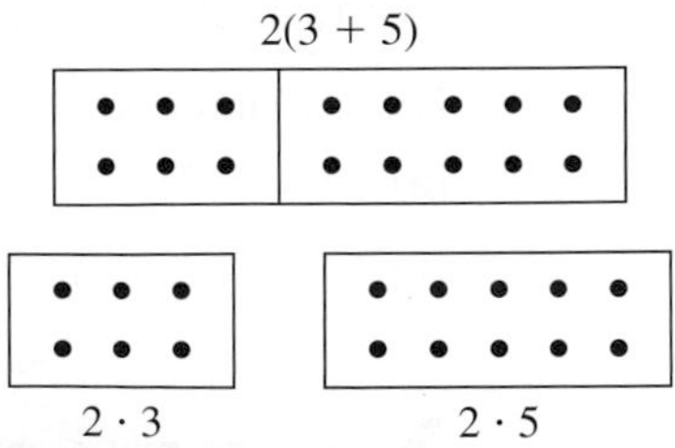

Figure 3 | The Distributive Property

Example 1 ■ Using the Distributive Property

(a) $2(x + 3) = 2 \cdot x + 2 \cdot 3$	Distributive Property
$= 2x + 6$	Simplify
(b) $(a + b)(x + y) = (a + b)x + (a + b)y$	Distributive Property
$= (ax + bx) + (ay + by)$	Distributive Property
$= ax + bx + ay + by$	Associative Property of Addition

In the last step we removed all the parentheses because, according to the Associative Property, the order of addition doesn't matter.

 Now Try Exercise 15 ■

■ Addition and Subtraction

The number 0 is special for addition; it is called the **additive identity** because $a + 0 = a$ for any real number a. Every real number a has a **negative**, $-a$, that satisfies $a + (-a) = 0$. **Subtraction** is the operation that undoes addition; to subtract a number from another, we simply add the negative of that number. By definition

$$a - b = a + (-b)$$

To combine real numbers involving negatives, we use the following properties.

Don't assume that $-a$ is a negative number. Whether $-a$ is negative or positive depends on the value of a. For example, if $a = 5$, then $-a = -5$, a negative number, but if $a = -5$, then $-a = -(-5) = 5$ (Property 2), a positive number.

Properties of Negatives

Property	**Example**
1. $(-1)a = -a$	$(-1)5 = -5$
2. $-(-a) = a$	$-(-5) = 5$
3. $(-a)b = a(-b) = -(ab)$	$(-5)7 = 5(-7) = -(5 \cdot 7)$
4. $(-a)(-b) = ab$	$(-4)(-3) = 4 \cdot 3$
5. $-(a + b) = -a - b$	$-(3 + 5) = -3 - 5$
6. $-(a - b) = b - a$	$-(5 - 8) = 8 - 5$

Properties 5 and 6 follow from the Distributive Property.

Property 6 states the intuitive fact that $a - b$ and $b - a$ are negatives of each other. Property 5 is often used with more than two terms:

$$-(a + b + c) = -a - b - c$$

Example 2 ■ Using Properties of Negatives

Let x, y, and z be real numbers.

(a) $-(x + 2) = -x - 2$	Property 5: $-(a + b) = -a - b$
(b) $-(x + y - z) = -x - y - (-z)$	Property 5: $-(a + b) = -a - b$
$= -x - y + z$	Property 2: $-(-a) = a$

Now Try Exercise 23 ■

■ Multiplication and Division

The number 1 is special for multiplication; it is called the **multiplicative identity** because $a \cdot 1 = a$ for any real number a. Every nonzero real number a has an **inverse**, $1/a$, that satisfies $a \cdot (1/a) = 1$. **Division** is the operation that undoes multiplication; to divide by a number, we multiply by the inverse of that number. If $b \neq 0$, then, by definition,

$$a \div b = a \cdot \frac{1}{b}$$

We write $a \cdot (1/b)$ as simply a/b. We refer to a/b as the **quotient** of a and b or as the **fraction** a over b; a is the **numerator** and b is the **denominator** (or **divisor**). To combine real numbers using the operation of division, we use the following properties.

Properties of Fractions

Property	Example	Description
1. $\dfrac{a}{b} \cdot \dfrac{c}{d} = \dfrac{ac}{bd}$	$\dfrac{2}{3} \cdot \dfrac{5}{7} = \dfrac{2 \cdot 5}{3 \cdot 7} = \dfrac{10}{21}$	When **multiplying fractions**, multiply numerators and denominators.
2. $\dfrac{a}{b} \div \dfrac{c}{d} = \dfrac{a}{b} \cdot \dfrac{d}{c}$	$\dfrac{2}{3} \div \dfrac{5}{7} = \dfrac{2}{3} \cdot \dfrac{7}{5} = \dfrac{14}{15}$	When **dividing fractions**, invert the divisor and multiply the numerators and denominators.
3. $\dfrac{a}{c} + \dfrac{b}{c} = \dfrac{a + b}{c}$	$\dfrac{2}{5} + \dfrac{7}{5} = \dfrac{2 + 7}{5} = \dfrac{9}{5}$	When **adding fractions** with the **same denominator**, add the numerators.
4. $\dfrac{a}{b} + \dfrac{c}{d} = \dfrac{ad + bc}{bd}$	$\dfrac{2}{5} + \dfrac{3}{7} = \dfrac{2 \cdot 7 + 3 \cdot 5}{35} = \dfrac{29}{35}$	When **adding fractions** with **different denominators**, find a common denominator. Then add the numerators.
5. $\dfrac{ac}{bc} = \dfrac{a}{b}$	$\dfrac{2 \cdot 5}{3 \cdot 5} = \dfrac{2}{3}$	**Cancel** numbers that are **common factors** in numerator and denominator.
6. If $\dfrac{a}{b} = \dfrac{c}{d}$, then $ad = bc$	$\dfrac{2}{3} = \dfrac{6}{9}$, so $2 \cdot 9 = 3 \cdot 6$	**Cross-multiply**.

When adding fractions with different denominators, we don't usually use Property 4. Instead we rewrite the fractions so that they have the smallest possible common denominator (often smaller than the product of the denominators), and then we use Property 3. This denominator is the Least Common Denominator (LCD) described in the next example.

Example 3 ■ Using the LCD to Add Fractions

Evaluate: $\dfrac{5}{36} + \dfrac{7}{120}$

Solution Factoring each denominator into prime factors gives

$$36 = 2^2 \cdot 3^2 \qquad \text{and} \qquad 120 = 2^3 \cdot 3 \cdot 5$$

We find the least common denominator (LCD) by forming the product of all the prime factors that occur in these factorizations, using the highest power of each prime factor. Thus the LCD is $2^3 \cdot 3^2 \cdot 5 = 360$. So

$$\frac{5}{36} + \frac{7}{120} = \frac{5 \cdot 10}{36 \cdot 10} + \frac{7 \cdot 3}{120 \cdot 3} \qquad \text{Use common denominator}$$

$$= \frac{50}{360} + \frac{21}{360} = \frac{71}{360} \qquad \text{Property 3: Adding fractions with the same denominator}$$

Now Try Exercise 29 ■

■ The Real Line

The real numbers can be represented by points on a line, as shown in Figure 4. The positive direction (toward the right) is indicated by an arrow. We choose an arbitrary reference point O, called the **origin**, which corresponds to the real number 0. Given any convenient unit of measurement, each positive number x is represented by the point on the line a distance of x units to the right of the origin, and the corresponding negative number $-x$ is represented by the point x units to the left of the origin. The number associated with the point P is called the coordinate of P, and the line is then called a **coordinate line**, or a **real number line**, or simply a **real line**. Often we identify the point with its coordinate and think of a number as being a point on the real line.

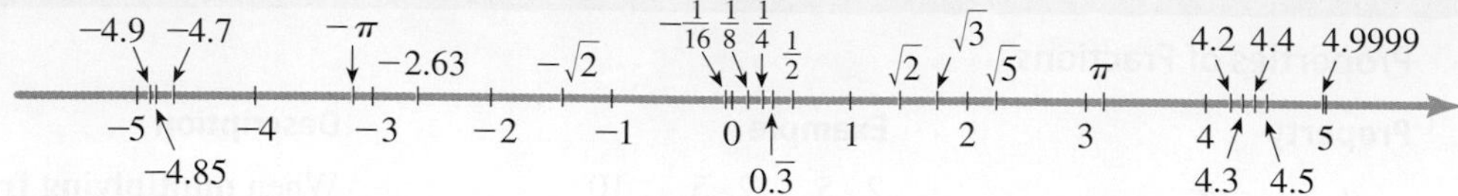

Figure 4 | The real line

The real numbers are *ordered*. We say that ***a* is less than *b*** and write $a < b$ if $b - a$ is a positive number. Geometrically, this means that a lies to the left of b on the number line. Equivalently, we can say that ***b* is greater than *a*** and write $b > a$. The symbol $a \le b$ (or $b \ge a$) means that either $a < b$ or $a = b$ and is read "a is less than or equal to b." For instance, the following are true inequalities (see Figure 4):

$$-5 < -4.9 \qquad -\pi < -3 \qquad \sqrt{2} < 2 \qquad 4 < 4.4 < 4.9999$$

■ Sets and Intervals

A **set** is a collection of objects, and these objects are called the **elements** of the set. If S is a set, the notation $a \in S$ means that a is an element of S, and $b \notin S$ means that b is not an element of S. For example, if Z represents the set of integers, then $-3 \in Z$ but $\pi \notin Z$.

Some sets can be described by listing their elements within braces. For instance, the set A that consists of all positive integers less than 7 can be written as

$$A = \{1, 2, 3, 4, 5, 6\}$$

We could also write A in **set-builder notation** as

$$A = \{x \mid x \text{ is an integer and } 0 < x < 7\}$$

which is read "A is the set of all x such that x is an integer and $0 < x < 7$."

If S and T are sets, then their **union** $S \cup T$ is the set that consists of all elements that are in S *or* T (or in both). The **intersection** of S and T is the set $S \cap T$ consisting of all

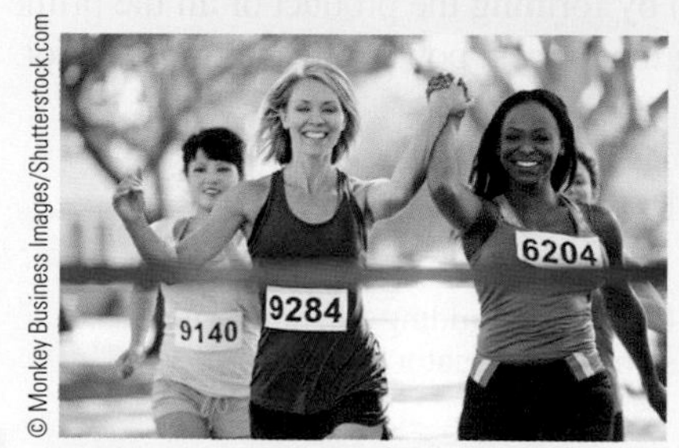

© Monkey Business Images/Shutterstock.com

Discovery Project ■ Real Numbers in the Real World

Real-world measurements often involve units. For example, we usually measure distance in feet, miles, centimeters, or kilometers. Some measurements involve different types of units. For example, speed is measured in miles per hour or meters per second. We often need to convert a measurement from one type of unit to another. In this project we explore types of units used for different purposes and how to convert from one type of unit to another. You can find the project at **www.stewartmath.com**.

elements that are in both S *and* T. In other words, $S \cap T$ is the common part of S and T. The **empty set**, denoted by $\varnothing$, is the set that contains no element.

Example 4 ■ Union and Intersection of Sets

If $S = \{1, 2, 3, 4, 5\}$, $T = \{4, 5, 6, 7\}$, and $V = \{6, 7, 8\}$, find the sets $S \cup T$, $S \cap T$, and $S \cap V$.

Solution

$$S \cup T = \{1, 2, 3, 4, 5, 6, 7\} \qquad \text{All elements in } S \text{ or } T$$

$$S \cap T = \{4, 5\} \qquad \text{Elements common to both } S \text{ and } T$$

$$S \cap V = \varnothing \qquad S \text{ and } V \text{ have no element in common}$$

T: 1, 2, 3, **4, 5, 6, 7**, 8 — S = 1, 2, 3, 4, 5; V = 6, 7, 8

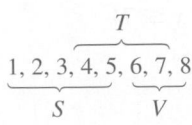

Now Try Exercise 41

Certain sets of real numbers, called **intervals**, occur frequently in calculus and correspond geometrically to line segments. If $a < b$, then the **open interval** from a to b consists of all numbers between a and b and is denoted (a, b). The **closed interval** from a to b includes the endpoints and is denoted $[a, b]$. Using set-builder notation, we can write

$$(a, b) = \{x \mid a < x < b\} \qquad [a, b] = \{x \mid a \le x \le b\}$$

Figure 5 | The open interval (a, b)

Note that parentheses ( ) in the interval notation and open circles on the graph in Figure 5 indicate that endpoints are *excluded* from the interval, whereas square brackets [ ] and solid circles in Figure 6 indicate that the endpoints are *included*. Intervals may also include one endpoint but not the other, or they may extend infinitely far in one direction or both directions. The following table lists the possible types of intervals.

Figure 6 | The closed interval $[a, b]$

Notation	Set Description	Graph
(a, b)	$\{x \mid a < x < b\}$	a b
$[a, b]$	$\{x \mid a \le x \le b\}$	a b
$[a, b)$	$\{x \mid a \le x < b\}$	a b
$(a, b]$	$\{x \mid a < x \le b\}$	a b
(a, ∞)	$\{x \mid a < x\}$	a
$[a, \infty)$	$\{x \mid a \le x\}$	a
$(-\infty, b)$	$\{x \mid x < b\}$	b
$(-\infty, b]$	$\{x \mid x \le b\}$	b
$(-\infty, \infty)$	$\mathbb{R}$ (set of all real numbers)	

The symbol ∞ ("infinity") does not stand for a number. The notation (a, ∞), for instance, simply indicates that the interval has no endpoint on the right but extends infinitely far in the positive direction.

Example 5 ■ Graphing Intervals

Express each interval in terms of inequalities, and then graph the interval.

(a) $[-1, 2) = \{x \mid -1 \le x < 2\}$

(b) $[1.5, 4] = \{x \mid 1.5 \le x \le 4\}$

(c) $(-3, \infty) = \{x \mid -3 < x\}$

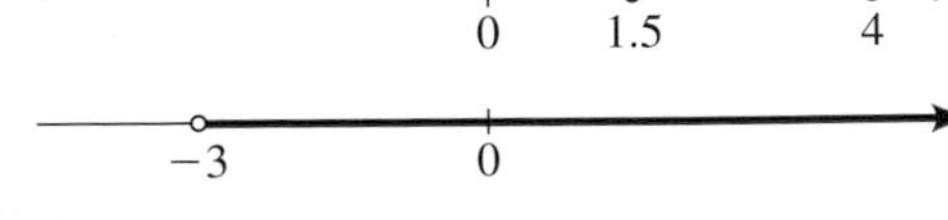

Now Try Exercise 47

No Smallest or Largest Number in an Open Interval
Any interval contains infinitely many numbers—every point on the graph of an interval corresponds to a real number. In the closed interval $[0, 1]$, the smallest number is 0 and the largest is 1, but the open interval $(0, 1)$ contains no smallest or largest number. To see this, note that 0.01 is close to zero, but 0.001 is closer, 0.0001 is closer yet, and so on. We can always find a number in the interval $(0, 1)$ closer to zero than any given number. Since 0 itself is not in the interval, the interval contains no smallest number. Similarly, 0.99 is close to 1, but 0.999 is closer, 0.9999 closer yet, and so on. Since 1 itself is not in the interval, the interval has no largest number.

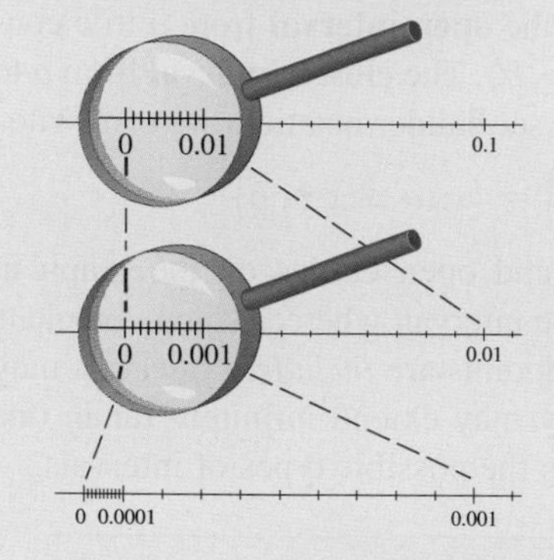

Example 6 ■ Finding Unions and Intersections of Intervals

Graph each set.

(a) $(1, 3) \cap [2, 7]$ **(b)** $(1, 3) \cup [2, 7]$

Solution

(a) The intersection of two intervals consists of the numbers that are in both intervals; geometrically, this is where the intervals overlap. Therefore

$$\begin{aligned}(1, 3) \cap [2, 7] &= \{x \mid 1 < x < 3 \text{ and } 2 \le x \le 7\} \\ &= \{x \mid 2 \le x < 3\} = [2, 3)\end{aligned}$$

This set is illustrated in Figure 7.

(b) The union of two intervals consists of the numbers that are in either one interval or the other (or both). Therefore

$$\begin{aligned}(1, 3) \cup [2, 7] &= \{x \mid 1 < x < 3 \text{ or } 2 \le x \le 7\} \\ &= \{x \mid 1 < x \le 7\} = (1, 7]\end{aligned}$$

This set is illustrated in Figure 8.

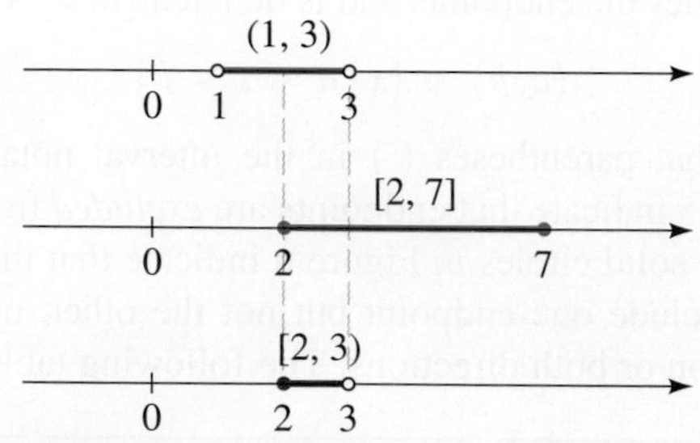

Figure 7 | $(1, 3) \cap [2, 7] = [2, 3)$

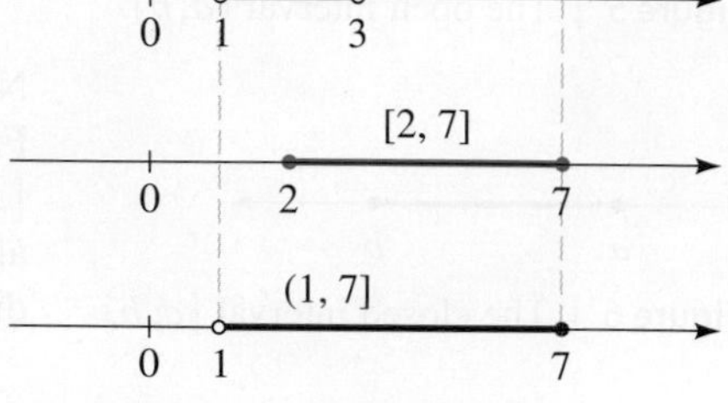

Figure 8 | $(1, 3) \cup [2, 7] = (1, 7]$

Now Try Exercise 61 ■

■ Absolute Value and Distance

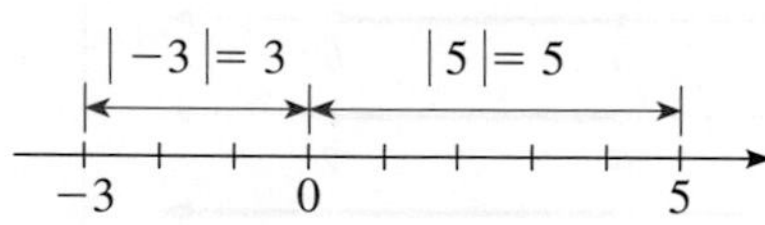

Figure 9

The **absolute value** of a number a, denoted by $|a|$, is the distance from a to 0 on the real number line (see Figure 9). Distance is always positive or zero, so we have $|a| \ge 0$ for every number a. Remembering that $-a$ is positive when a is negative, we have the following definition.

Definition of Absolute Value

If a is a real number, then the **absolute value** of a is

$$|a| = \begin{cases} a & \text{if } a \ge 0 \\ -a & \text{if } a < 0 \end{cases}$$

Example 7 ■ Evaluating Absolute Values of Numbers

(a) $|3| = 3$

(b) $|-3| = -(-3) = 3$

(c) $|0| = 0$

(d) $|3 - \pi| = -(3 - \pi) = \pi - 3 \quad (\text{since } 3 < \pi \quad \Rightarrow \quad 3 - \pi < 0)$

Now Try Exercise 67 ■

When working with absolute values, we use the following properties.

Properties of Absolute Value

Property	Example	Description
1. $\lvert a \rvert \geq 0$	$\lvert -3 \rvert = 3 \geq 0$	The absolute value of a number is always positive or zero.
2. $\lvert a \rvert = \lvert -a \rvert$	$\lvert 5 \rvert = \lvert -5 \rvert$	A number and its negative have the same absolute value.
3. $\lvert ab \rvert = \lvert a \rvert \lvert b \rvert$	$\lvert -2 \cdot 5 \rvert = \lvert -2 \rvert \lvert 5 \rvert$	The absolute value of a product is the product of the absolute values.
4. $\left\lvert \frac{a}{b} \right\rvert = \frac{\lvert a \rvert}{\lvert b \rvert}$	$\left\lvert \frac{12}{-3} \right\rvert = \frac{\lvert 12 \rvert}{\lvert -3 \rvert}$	The absolute value of a quotient is the quotient of the absolute values.
5. $\lvert a + b \rvert \leq \lvert a \rvert + \lvert b \rvert$	$\lvert -3 + 5 \rvert \leq \lvert -3 \rvert + \lvert 5 \rvert$	Triangle Inequality

What is the distance on the real line between the numbers -2 and 11? From Figure 10 we see that the distance is 13. We arrive at this by finding either $\lvert 11 - (-2) \rvert = 13$ or $\lvert (-2) - 11 \rvert = 13$. From this observation we make the following definition.

13

−2 0 11

Figure 10

Distance between Points on the Real Line

If a and b are real numbers, then the **distance** between the points a and b on the real line is

$$d(a, b) = \lvert b - a \rvert$$

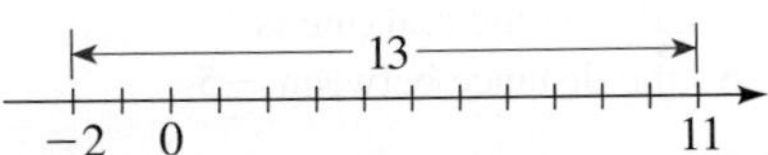

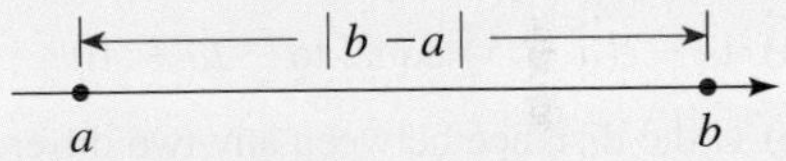

From Property 6 of negatives it follows that

$$\lvert b - a \rvert = \lvert a - b \rvert$$

This confirms that, as we would expect, the distance from a to b is the same as the distance from b to a.

Example 8 ■ Distance Between Points on the Real Line

The distance between the numbers 2 and -8 is

$$d(a, b) = \lvert -8 - 2 \rvert = \lvert -10 \rvert = 10$$

We can check this calculation geometrically, as shown in Figure 11.

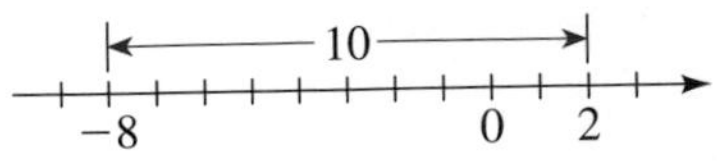

Figure 11

Now Try Exercise 75 ■

1.1 Exercises

Concepts

1. Give an example of each of the following:

(a) A natural number

(b) An integer that is not a natural number

(c) A rational number that is not an integer

(d) An irrational number

2. Complete each statement by using a property of real numbers and name the property you have used.

(a) $ab =$ ________; ________ Property

(b) $a + (b + c) =$ ________; ________ Property

(c) $a(b + c) =$ ________; ________ Property

3. Express the set of real numbers between but not including -3 and 5 as follows.

(a) In set-builder notation: ________________

(b) In interval notation: ________________

(c) As a graph: ________________

4. The symbol $|x|$ stands for the ________ of the number x. If x is not 0, then the sign of $|x|$ is always ________.

5. The distance between a and b on the real line is $d(a, b) =$ ________. So the distance between -5 and 2 is ________.

6–8 ■ *Yes or No?* If *No*, give a reason. Assume that a and b are nonzero real numbers.

6. (a) Does an interval always contain infinitely many numbers?

(b) Does the interval $(5, 6)$ contain a smallest element?

7. (a) Is $a - b$ equal to $b - a$?

(b) Is $-2(a - 5)$ equal to $-2a - 10$?

8. (a) Is the distance between any two different real numbers always positive?

(b) Is the distance between a and b the same as the distance between b and a?

Skills

9–10 ■ Real Numbers List the elements of the given set that are

(a) natural numbers

(b) integers

(c) rational numbers

(d) irrational numbers

9. $\{-1.5, 0, \frac{5}{2}, \sqrt{7}, 2.71, -\pi, 3.1\overline{4}, 100, -8\}$

10. $\{4.5, \frac{1}{3}, 1.6666\ldots, \sqrt{2}, 2, -\frac{100}{2}, \sqrt{9}, \sqrt{3.14}, 10\}$

11–18 ■ Properties of Real Numbers State the property of real numbers being used.

11. $3 + 7 = 7 + 3$

12. $4(2 + 3) = (2 + 3)4$

13. $(x + 2y) + 3z = x + (2y + 3z)$

14. $2(A + B) = 2A + 2B$

15. $(5x + 1)3 = 15x + 3$

16. $(x + a)(x + b) = (x + a)x + (x + a)b$

17. $2x(3 + y) = (3 + y)2x$

18. $7(a + b + c) = 7(a + b) + 7c$

19–22 ■ Properties of Real Numbers Rewrite the expression using the given property of real numbers.

19. Commutative Property of Addition, $x + 3 =$

20. Associative Property of Multiplication, $7(3x) =$

21. Distributive Property, $4(A + B) =$

22. Distributive Property, $5x + 5y =$

23–28 ■ Properties of Real Numbers Use properties of real numbers to write the expression without parentheses.

23. $-2(x + y)$

24. $(a - b)(-5)$

25. $5(2xy)$

26. $\frac{4}{3}(-6y)$

27. $-\frac{5}{2}(2x - 4y)$

28. $(3a)(b + c - 2d)$

29–32 ■ Arithmetic Operations Perform the indicated operations and simplify. Express your answer as a single fraction.

29. (a) $\frac{2}{3} + \frac{5}{7}$ **(b)** $\frac{5}{12} - \frac{3}{8}$

30. (a) $\frac{2}{5} - \frac{3}{8}$ **(b)** $\frac{3}{2} - \frac{5}{8} + \frac{1}{6}$

31. (a) $\frac{2}{3}(6 - \frac{3}{2})$ **(b)** $(3 + \frac{1}{4})(1 - \frac{4}{5})$

32. (a) $\dfrac{2}{\frac{2}{3}} - \dfrac{\frac{2}{3}}{2}$ **(b)** $\dfrac{\frac{2}{5} + \frac{1}{2}}{\frac{1}{10} + \frac{3}{15}}$

33–34 ■ Inequalities Place the correct symbol ($<$, $>$, or $=$) in the blue box.

33. (a) $3 \;\square\; \frac{7}{2}$ **(b)** $-3 \;\square\; -\frac{7}{2}$ **(c)** $3.5 \;\square\; \frac{7}{2}$

34. (a) $\frac{2}{3} \;\square\; 0.67$ **(b)** $\frac{2}{3} \;\square\; -0.67$ **(c)** $|0.6| \;\square\; |-0.6|$

35–38 ■ Inequalities State whether each inequality is true or false.

35. (a) $-3 < -4$ **(b)** $3 < 4$

36. (a) $\sqrt{3} > 1.7325$ **(b)** $1.732 \geq \sqrt{3}$

37. (a) $\frac{10}{2} \geq 5$ **(b)** $\frac{6}{10} \geq \frac{5}{6}$

38. (a) $\frac{7}{11} \geq \frac{8}{13}$ **(b)** $-\frac{3}{5} > -\frac{3}{4}$

39–40 ■ Inequalities Write each statement in terms of inequalities.

39. (a) x is positive.

(b) t is less than 4.

(c) a is greater than or equal to π.

(d) x is less than $\frac{1}{3}$ and greater than -5.

(e) The distance from p to 3 is at most 5.

40. **(a)** y is negative.
(b) z is at least 3.
(c) b is at most 8.
(d) w is positive and less than or equal to 17.
(e) y is at least 2 units from π.

41–44 ■ Sets Find the indicated set if

$$A = \{1, 2, 3, 4, 5, 6, 7\} \qquad B = \{2, 4, 6, 8\}$$
$$C = \{7, 8, 9, 10\}$$

41. **(a)** $A \cup B$ **(b)** $A \cap B$
42. **(a)** $B \cup C$ **(b)** $B \cap C$
43. **(a)** $A \cup C$ **(b)** $A \cap C$
44. **(a)** $A \cup B \cup C$ **(b)** $A \cap B \cap C$

45–46 ■ Sets Find the indicated set if

$$A = \{x \mid x \geq -2\} \qquad B = \{x \mid x < 4\}$$
$$C = \{x \mid -1 < x \leq 5\}$$

45. **(a)** $B \cup C$ **(b)** $B \cap C$
46. **(a)** $A \cap C$ **(b)** $A \cap B$

47–52 ■ Intervals Express the interval in terms of inequalities, and then graph the interval.

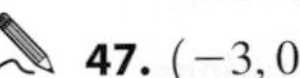

47. $(-3, 0)$ **48.** $(2, 8]$
49. $[2, 8)$ **50.** $[-6, -\frac{1}{2}]$
51. $[2, \infty)$ **52.** $(-\infty, 1)$

53–58 ■ Intervals Express the inequality in interval notation, and then graph the corresponding interval.

53. $x \leq 1$ **54.** $1 \leq x \leq 2$
55. $-2 < x \leq 1$ **56.** $x \geq -5$
57. $x > -1$ **58.** $-5 < x < 2$

59–60 ■ Intervals Express each set in interval notation.

59. **(a)** **(b)** **(c)**

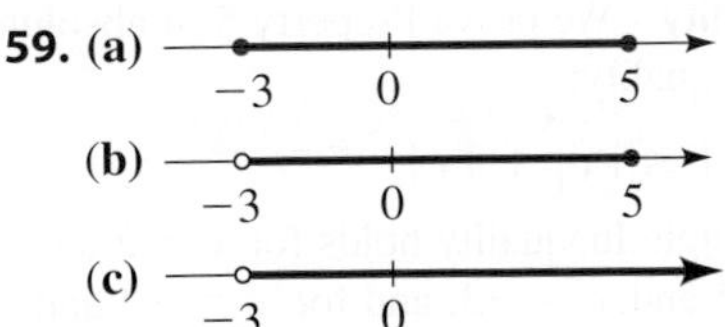

60. **(a)** **(b)** **(c)**

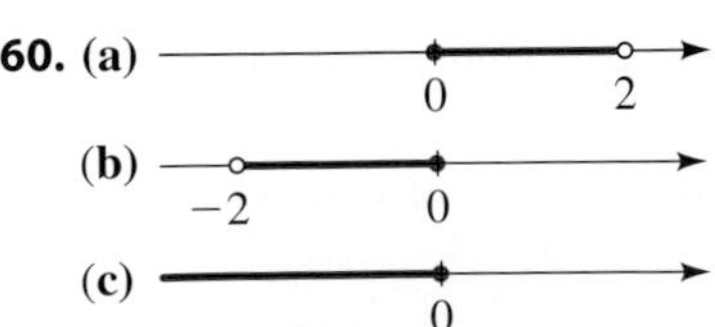

61–66 ■ Intervals Graph the set.

61. $(-2, 0) \cup (-1, 1)$ **62.** $(-2, 0) \cap (-1, 1)$
63. $[-4, 6] \cap [0, 8)$ **64.** $[-4, 6) \cup [0, 8)$
65. $(-\infty, -4) \cup (4, \infty)$ **66.** $(-\infty, 6] \cap (2, 10)$

67–72 ■ Absolute Value Evaluate each expression.

67. **(a)** $|50|$ **(b)** $|-13|$
68. **(a)** $|2 - 8|$ **(b)** $|8 - |-2||$
69. **(a)** $||-6| - |-4||$ **(b)** $\dfrac{-1}{|-1|}$
70. **(a)** $|2 - |-12||$ **(b)** $-1 - |1 - |-1||$
71. **(a)** $|(-2) \cdot 6|$ **(b)** $|(-\frac{1}{3})(-15)|$
72. **(a)** $\left|\dfrac{-6}{24}\right|$ **(b)** $\left|\dfrac{7-12}{12-7}\right|$

73–76 ■ Distance Find the distance between the given numbers.

73.
74.

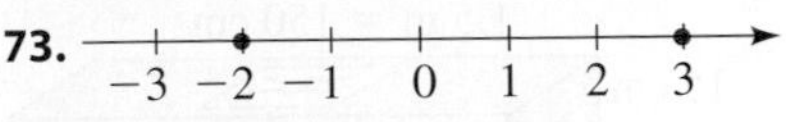

75. **(a)** 2 and 17 **(b)** -3 and 21 **(c)** $\frac{11}{8}$ and $-\frac{3}{10}$
76. **(a)** $\frac{7}{15}$ and $-\frac{1}{21}$ **(b)** -38 and -57 **(c)** -2.6 and -1.8

Skills Plus

77–78 ■ Repeating Decimal Express each repeating decimal as a fraction. (See the margin note in the subsection "Real Numbers.")

77. **(a)** $0.\overline{7}$ **(b)** $0.2\overline{8}$ **(c)** $0.\overline{57}$
78. **(a)** $5.\overline{23}$ **(b)** $1.3\overline{7}$ **(c)** $2.1\overline{35}$

79–82 ■ Simplifying Absolute Value Express the quantity without using absolute value.

79. $|\pi - 3|$ **80.** $|1 - \sqrt{2}|$
81. $|a - b|$, where $a < b$
82. $a + b + |a - b|$, where $a < b$

83–84 ■ Signs of Numbers Let a, b, and c be real numbers such that $a > 0$, $b < 0$, and $c < 0$. Find the sign of each expression.

83. **(a)** $-a$ **(b)** bc **(c)** $a - b$ **(d)** $ab + ac$
84. **(a)** $-b$ **(b)** $a + bc$ **(c)** $c - a$ **(d)** ab^2

Applications

85. Area of a Garden A backyard vegetable garden measures 6 m by 9 m, so its area is $6 \times 9 = 54\ \text{m}^2$. The garden needs to be made longer, as shown in the figure, so that the area increases to $A = 6(9 + x)$. Which property of real numbers tells us that the new area can also be written $A = 54 + 6x$?

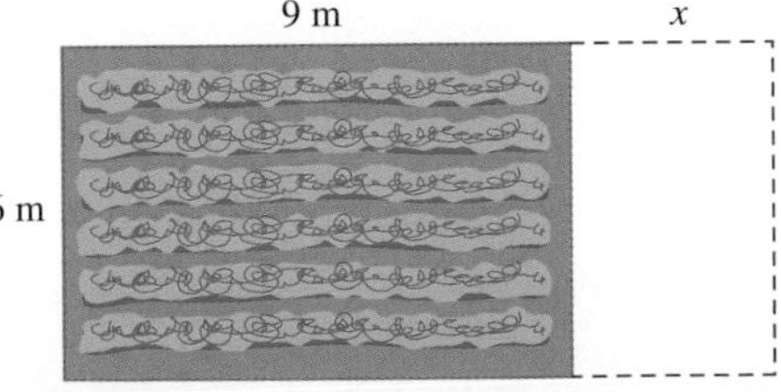

86. Mailing a Package The post office will accept only packages for which the length plus the "girth" (distance around) is no more than 274 cm. Thus for the package in the figure, we must have

$$L + 2(x + y) \leq 108$$

(a) Will the post office accept a package that is 15 cm wide, 20 cm deep, and 1.5 m long? What about a package that measures 0.6 m by 0.6 m by 1.2 m?

(b) What is the greatest acceptable length for a package that has a square base measuring 22.5 cm by 22.5 cm?

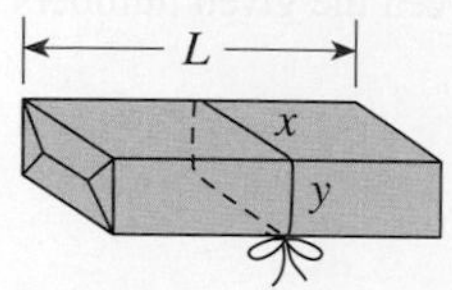

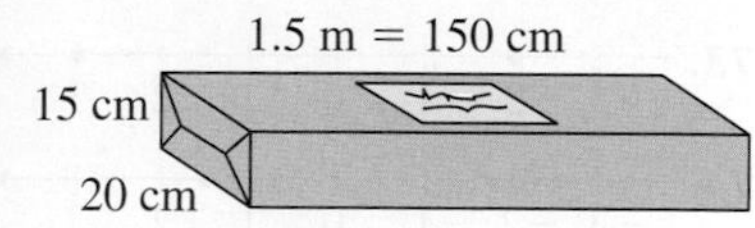

■ Discuss ■ Discover ■ Prove ■ Write

87. Discuss: Sums and Products of Rational and Irrational Numbers Explain why the sum, the difference, and the product of two rational numbers are rational numbers. Is the product of two irrational numbers necessarily irrational? What about the sum?

88. Discover ■ Prove: Combining Rational and Irrational Numbers Is $\frac{1}{2} + \sqrt{2}$ rational or irrational? Is $\frac{1}{2} \cdot \sqrt{2}$ rational or irrational? Experiment with sums and products of other rational and irrational numbers. Prove the following.

(a) The sum of a rational number r and an irrational number t is irrational.

(b) The product of a non-zero rational number r and an irrational number t is irrational.

PS *Indirect reasoning.* For part (a), suppose that $r + t$ is a rational number q, that is, $r + t = q$. Show that this leads to a contradiction. Use similar reasoning for part (b).

89. Discover: Limiting Behavior of Reciprocals Complete the tables. What happens to the size of the fraction $1/x$ as x gets large? As x gets small?

x	$1/x$
1	
2	
10	
100	
1000	

x	$1/x$
1.0	
0.5	
0.1	
0.01	
0.001	

90. Discover: Locating Irrational Numbers on the Real Line Using the figures below, explain how to locate the point $\sqrt{2}$ on a number line. Can you locate $\sqrt{5}$ by a similar method? How can the circle shown in the figure help us locate π on a number line? List some other irrational numbers that you can locate on a number line.

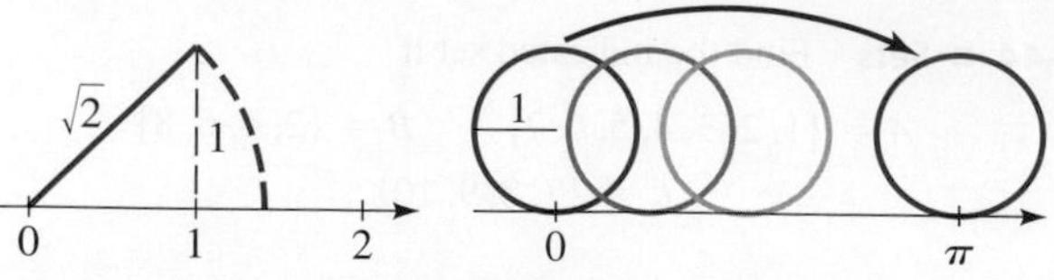

91. Prove: Maximum and Minimum Formulas Let max(a, b) denote the maximum and min(a, b) denote the minimum of the real numbers a and b. For example, $\max(2, 5) = 5$ and $\min(-1, -2) = -2$.

(a) Prove that $\max(a, b) = \dfrac{a + b + |a - b|}{2}$.

(b) Prove that $\min(a, b) = \dfrac{a + b - |a - b|}{2}$.

PS *Take cases.* Write these expressions without absolute values. See Exercises 81 and 82.

92. Write: Real Numbers in the Real World Write a paragraph describing several real-world situations in which you would use natural numbers, integers, rational numbers, and irrational numbers. Give examples for each type of situation.

93. Discuss: Commutative and Noncommutative Operations We have learned that addition and multiplication are both commutative operations.

(a) Is subtraction commutative?

(b) Is division of nonzero real numbers commutative?

(c) Are the actions of putting on your socks and putting on your shoes commutative?

(d) Are the actions of putting on your hat and putting on your coat commutative?

(e) Are the actions of washing laundry and drying it commutative?

94. Prove: Triangle Inequality We prove Property 5 of absolute values, the Triangle Inequality:

$$|x + y| \leq |x| + |y|$$

(a) Verify that the Triangle Inequality holds for $x = 2$ and $y = 3$, for $x = -2$ and $y = -3$, and for $x = -2$ and $y = 3$.

(b) Prove that the Triangle Inequality is true for all real numbers x and y.

PS *Take cases.* Consider the signs of x and y.

1.2 Exponents and Radicals

■ Integer Exponents ■ Rules for Working with Exponents ■ Scientific Notation
■ Radicals ■ Rational Exponents ■ Rationalizing the Denominator; Standard Form

In this section we give meaning to expressions such as $a^{m/n}$ in which the exponent m/n is a rational number. To do this, we need to recall some facts about integer exponents, radicals, and nth roots.

■ Integer Exponents

A product of identical numbers is usually written in exponential notation. For example, $5 \cdot 5 \cdot 5$ is written as 5^3. In general, we have the following definition.

Exponential Notation

If a is any real number and n is a positive integer, then the ***n*th power** of a is

$$a^n = \underbrace{a \cdot a \cdot \cdots \cdot a}_{n \text{ factors}}$$

The number a is called the **base**, and n is called the **exponent**.

Example 1 ■ Exponential Notation

Note the distinction between $(-3)^4$ and -3^4. In $(-3)^4$ the exponent applies to -3, but in -3^4 the exponent applies only to 3.

(a) $\left(\frac{1}{2}\right)^5 = \left(\frac{1}{2}\right)\left(\frac{1}{2}\right)\left(\frac{1}{2}\right)\left(\frac{1}{2}\right)\left(\frac{1}{2}\right) = \frac{1}{32}$

(b) $(-3)^4 = (-3)\cdot(-3)\cdot(-3)\cdot(-3) = 81$

(c) $-3^4 = -(3\cdot 3\cdot 3\cdot 3) = -81$

Now Try Exercise 9 ■

We can state several useful rules for working with exponential notation. To discover the rule for multiplication, we multiply 5^4 by 5^2:

$$5^4 \cdot 5^2 = \underbrace{(5 \cdot 5 \cdot 5 \cdot 5)}_{4 \text{ factors}}\underbrace{(5 \cdot 5)}_{2 \text{ factors}} = \underbrace{5 \cdot 5 \cdot 5 \cdot 5 \cdot 5 \cdot 5}_{6 \text{ factors}} = 5^6 = 5^{4+2}$$

It appears that *to multiply two powers of the same base, we add their exponents*. In general, for any real number a and any positive integers m and n, we have

$$a^m a^n = \underbrace{(a \cdot a \cdot \cdots \cdot a)}_{m \text{ factors}}\underbrace{(a \cdot a \cdot \cdots \cdot a)}_{n \text{ factors}} = \underbrace{a \cdot a \cdot a \cdot \cdots \cdot a}_{m+n \text{ factors}} = a^{m+n}$$

Thus $a^m a^n = a^{m+n}$.

We would like this rule to be true even when m and n are 0 or negative integers. For instance, we must have

$$2^0 \cdot 2^3 = 2^{0+3} = 2^3$$

But this can happen only if $2^0 = 1$. Likewise, we want to have

$$5^4 \cdot 5^{-4} = 5^{4+(-4)} = 5^{4-4} = 5^0 = 1$$

and this will be true if $5^{-4} = 1/5^4$. These observations lead to the following definition.

Zero and Negative Exponents

If $a \neq 0$ is a real number and n is a positive integer, then

$$a^0 = 1 \qquad \text{and} \qquad a^{-n} = \frac{1}{a^n}$$

Example 2 ■ Zero and Negative Exponents

(a) $\left(\frac{4}{7}\right)^0 = 1$

(b) $x^{-1} = \dfrac{1}{x^1} = \dfrac{1}{x}$

(c) $(-2)^{-3} = \dfrac{1}{(-2)^3} = \dfrac{1}{-8} = -\dfrac{1}{8}$

Now Try Exercise 11 ■

■ Rules for Working with Exponents

Familiarity with the following rules is essential for our work with exponents and bases. In In these laws the bases a and b are real numbers, and the exponents m and n are integers.

Laws of Exponents

Law	Example	Description
1. $a^m a^n = a^{m+n}$	$3^2 \cdot 3^5 = 3^{2+5} = 3^7$	To multiply two powers of the same number, add the exponents.
2. $\dfrac{a^m}{a^n} = a^{m-n}$	$\dfrac{3^5}{3^2} = 3^{5-2} = 3^3$	To divide two powers of the same number, subtract the exponents.
3. $(a^m)^n = a^{mn}$	$(3^2)^5 = 3^{2 \cdot 5} = 3^{10}$	To raise a power to a new power, multiply the exponents.
4. $(ab)^n = a^n b^n$	$(3 \cdot 4)^2 = 3^2 \cdot 4^2$	To raise a product to a power, raise each factor to the power.
5. $\left(\dfrac{a}{b}\right)^n = \dfrac{a^n}{b^n}$	$\left(\dfrac{3}{4}\right)^2 = \dfrac{3^2}{4^2}$	To raise a quotient to a power, raise both numerator and denominator to the power.
6. $\left(\dfrac{a}{b}\right)^{-n} = \left(\dfrac{b}{a}\right)^n$	$\left(\dfrac{3}{4}\right)^{-2} = \left(\dfrac{4}{3}\right)^2$	To raise a fraction to a negative power, invert the fraction and change the sign of the exponent.
7. $\dfrac{a^{-n}}{b^{-m}} = \dfrac{b^m}{a^n}$	$\dfrac{3^{-2}}{4^{-5}} = \dfrac{4^5}{3^2}$	To move a number raised to a power from numerator to denominator or from denominator to numerator, change the sign of the exponent.

Proof of Law 3 If m and n are positive integers, we have

$$(a^m)^n = (\underbrace{a \cdot a \cdot \cdots \cdot a}_{m \text{ factors}})^n$$

$$= \underbrace{(\underbrace{a \cdot a \cdot \cdots \cdot a}_{m \text{ factors}})(\underbrace{a \cdot a \cdot \cdots \cdot a}_{m \text{ factors}}) \cdots (\underbrace{a \cdot a \cdot \cdots \cdot a}_{m \text{ factors}})}_{n \text{ groups of factors}}$$

$$= \underbrace{a \cdot a \cdot \cdots \cdot a}_{mn \text{ factors}} = a^{mn}$$

The cases for which $m \leq 0$ or $n \leq 0$ can be proved by using the definition of negative exponents. ■

Mathematics and Mathematicians

Mathematicians have been sought after in every civilization to use their craft in the service of society. Several brief biographies of notable mathematicians are highlighted in margin notes in this book, but keep in mind that their discoveries always depended on the insights of thousands of others who came before them. Isaac Newton put it this way: "If I have seen further than others it is by standing on the shoulders of giants." And David Hilbert noted that, "Mathematics knows no races or geographic boundaries; for mathematics, the cultural world is one country."

You will see in the biographies that some great mathematicians did not at first find mathematics interesting, but were later attracted to it for one reason or another. Pascal liked mathematics because of "the clarity of its reasoning," Hermann Hankel, because of the permanence of its theorems. Hankel said, "In most sciences one generation tears down what another has built. . . . In mathematics alone each generation adds a new story to the old structure." For example, the elements of Aristotle (ca. 300 BC)— earth, wind, fire, water—are now replaced by the Periodic Table, whereas the theorem of Pythagoras (ca. 500 BC) continues to be valid. Joseph Fourier's reason for being drawn to mathematics was that "Mathematics compares the most diverse phenomena and discovers the secret analogies that unite them." For example, we'll see that quadratic functions model both the path of a projectile and the relationship between rainfall and crop yield. Perhaps you too will find reasons for exploring mathematics.

Proof of Law 4 If n is a positive integer, we have

$$(ab)^n = \underbrace{(ab)(ab)\cdots(ab)}_{n\text{ factors}} = \underbrace{(a \cdot a \cdot \cdots \cdot a)}_{n\text{ factors}} \cdot \underbrace{(b \cdot b \cdot \cdots \cdot b)}_{n\text{ factors}} = a^n b^n$$

Here we have used the Commutative and Associative Properties repeatedly. If $n \le 0$, Law 4 can be proved by using the definition of negative exponents. ■

You are asked to prove Laws 2, 5, 6, and 7 in Exercises 106 and 107.

Example 3 ■ Using Laws of Exponents

(a) $x^4x^7 = x^{4+7} = x^{11}$ Law 1: $a^ma^n = a^{m+n}$

(b) $y^4y^{-7} = y^{4-7} = y^{-3} = \dfrac{1}{y^3}$ Law 1: $a^ma^n = a^{m+n}$

(c) $\dfrac{c^9}{c^5} = c^{9-5} = c^4$ Law 2: $\dfrac{a^m}{a^n} = a^{m-n}$

(d) $(b^4)^5 = b^{4\cdot5} = b^{20}$ Law 3: $(a^m)^n = a^{mn}$

(e) $(3x)^3 = 3^3x^3 = 27x^3$ Law 4: $(ab)^n = a^nb^n$

(f) $\left(\dfrac{x}{2}\right)^5 = \dfrac{x^5}{2^5} = \dfrac{x^5}{32}$ Law 5: $\left(\dfrac{a}{b}\right)^n = \dfrac{a^n}{b^n}$

 **Now Try Exercises 19, 21, and 23** ■

Example 4 ■ Simplifying Expressions with Exponents

Simplify:

(a) $(2a^3b^2)(3ab^4)^3$ **(b)** $\left(\dfrac{x}{y}\right)^3\left(\dfrac{y^2x}{z}\right)^4$

Solution

(a)
$$\begin{aligned}(2a^3b^2)(3ab^4)^3 &= (2a^3b^2)[3^3a^3(b^4)^3] && \text{Law 4: } (ab)^n = a^nb^n\\ &= (2a^3b^2)(27a^3b^{12}) && \text{Law 3: } (a^m)^n = a^{mn}\\ &= (2)(27)a^3a^3b^2b^{12} && \text{Group factors with the same base}\\ &= 54a^6b^{14} && \text{Law 1: } a^ma^n = a^{m+n}\end{aligned}$$

(b)
$$\begin{aligned}\left(\frac{x}{y}\right)^3\left(\frac{y^2x}{z}\right)^4 &= \frac{x^3}{y^3}\frac{(y^2)^4x^4}{z^4} && \text{Laws 5 and 4}\\ &= \frac{x^3}{y^3}\frac{y^8x^4}{z^4} && \text{Law 3}\\ &= (x^3x^4)\left(\frac{y^8}{y^3}\right)\frac{1}{z^4} && \text{Group factors with the same base}\\ &= \frac{x^7y^5}{z^4} && \text{Laws 1 and 2}\end{aligned}$$

 **Now Try Exercises 25 and 29** ■

When simplifying an expression, you will find that many different methods will lead to the same result; feel free to use any of the rules of exponents to arrive at your own method. In the next example we see how to simplify expressions with negative exponents.

Mathematics in the Modern World

Although we are often unaware of its presence, mathematics permeates nearly every aspect of our lives. With the advent of technology, mathematics plays an ever greater role. Today you were probably awakened by a digital alarm, sent a text, searched the Internet, watched a streaming video, listened to music on your cell phone, then slept in a room whose temperature is controlled by a digital thermostat. In each of these activities mathematics is crucially involved. In general, a property such as the intensity or frequency of sound, the colors in an image, or the temperature in your home is transformed into sequences of numbers by complex mathematical algorithms. These numerical data, which usually consist of many millions of bits (the digits 0 and 1), are then transmitted and reinterpreted. Dealing with such huge amounts of data was not feasible until the invention of computers, machines whose logical processes were invented by mathematicians.

The contributions of mathematics in the modern world are not limited to technological advances. The logical processes of mathematics are now used to analyze complex problems in the social, political, and life sciences in new and surprising ways.

In other *Mathematics in the Modern World*, we will describe in more detail how mathematics affects all of us in our everyday activities.

Example 5 ■ Simplifying Expressions with Negative Exponents

Eliminate negative exponents and simplify each expression.

(a) $\dfrac{6st^{-4}}{2s^{-2}t^2}$ **(b)** $\left(\dfrac{y}{3z^3}\right)^{-2}$

Solution

(a) We use Law 7, which allows us to move a number raised to a power from the numerator to the denominator (or vice versa) by changing the sign of the exponent.

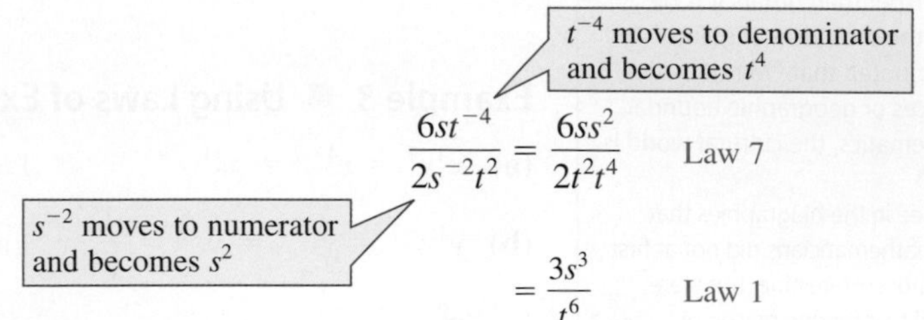

$$\frac{6st^{-4}}{2s^{-2}t^2} = \frac{6ss^2}{2t^2t^4} \qquad \text{Law 7}$$

$$= \frac{3s^3}{t^6} \qquad \text{Law 1}$$

(b) We use Law 6, which allows us to change the sign of the exponent of a fraction by inverting the fraction.

$$\left(\frac{y}{3z^3}\right)^{-2} = \left(\frac{3z^3}{y}\right)^2 \qquad \text{Law 6}$$

$$= \frac{9z^6}{y^2} \qquad \text{Laws 5 and 4}$$

 **Now Try Exercise 31** ■

■ Scientific Notation

Scientists use exponential notation as a compact way of writing very large numbers and very small numbers. For example, the nearest star beyond the sun, Proxima Centauri, is approximately 40,000,000,000,000 km away. The mass of a hydrogen atom is about 0.000 000 000 000 000 000 000 001 66 g. Such numbers are difficult to read and to write, so scientists usually express them in *scientific notation*.

Scientific Notation

A positive number x is said to be written in **scientific notation** if it is expressed as follows:

$$x = a \times 10^n \qquad \text{where } 1 \le a < 10 \text{ and } n \text{ is an integer}$$

For instance, when we state that the distance to the star Proxima Centauri is 4×10^{13} km, the positive exponent 13 indicates that the decimal point should be moved 13 places to the *right*:

$$4 \times 10^{13} = 40{,}000{,}000{,}000{,}000$$

Move decimal point 13 places to the right

When we state that the mass of a hydrogen atom is 1.66×10^{-24} g, the exponent -24 indicates that the decimal point should be moved 24 places to the *left*:

$$1.66 \times 10^{-24} = 0.000\,000\,000\,000\,000\,000\,000\,001\,66$$

Move decimal point 24 places to the left

Example 6 ■ Changing from Decimal to Scientific Notation

Write each number in scientific notation.

(a) 56,920 **(b)** 0.000093

Solution

(a) $\underbrace{56{,}920}_{\text{4 places}} = 5.692 \times 10^4$ **(b)** $0.\underbrace{00009}_{\text{5 places}}3 = 9.3 \times 10^{-5}$

 Now Try Exercise 81 ■

Example 7 ■ Changing from Scientific Notation to Decimal Notation

Write each number in decimal notation.

(a) 6.97×10^9 **(b)** 4.6271×10^{-6}

Solution

(a) $6.97 \times 10^9 = 6\underbrace{{,}970{,}000{,}000}_{\text{9 places}}$ Move decimal 9 places to the right

(b) $4.6271 \times 10^{-6} = 0.\underbrace{000004}_{\text{6 places}}6271$ Move decimal 6 places to the left

 Now Try Exercise 83 ■

Note Scientific notation is often used on a graphing device or calculator to display a very large or very small number. For instance, if we use a calculator to square the number 1,111,111, the display panel may show the approximation

`1.234568 12` or `1.234568 E12`

Here the final digits indicate the power of 10, and we interpret the result as

$$1.234568 \times 10^{12}$$

Example 8 ■ Calculating with Scientific Notation

If $a \approx 0.00046$, $b \approx 1.697 \times 10^{22}$, and $c \approx 2.91 \times 10^{-18}$, use a calculator to approximate the quotient ab/c.

Solution We could enter the data using scientific notation, or we could use laws of exponents as follows:

$$\begin{aligned}\frac{ab}{c} &\approx \frac{(4.6 \times 10^{-4})(1.697 \times 10^{22})}{2.91 \times 10^{-18}} \\ &= \frac{(4.6)(1.697)}{2.91} \times 10^{-4+22+18} \\ &\approx 2.7 \times 10^{36}\end{aligned}$$

For guidelines on working with significant figures, see Appendix B, *Calculations and Significant Figures*. Go to **www.stewartmath.com**.

We state the answer rounded to two significant figures because the least accurate of the given numbers is stated to two significant figures.

 Now Try Exercises 87 and 89 ■

■ Radicals

We know what 2^n means whenever n is an integer. To give meaning to a power, such as $2^{4/5}$, whose exponent is a rational number, we need to discuss radicals.

The symbol $\sqrt{}$ means "the positive square root of." Thus

$$\sqrt{a} = b \quad \text{means} \quad b^2 = a \quad \text{and} \quad b \geq 0$$

It is true that the number 9 has two square roots, 3 and -3, but the notation $\sqrt{9}$ is reserved for the *positive* square root of 9 (sometimes called the *principal square root* of 9). If we want the negative root, we must write $-\sqrt{9}$, which is -3.

Since $a = b^2 \geq 0$, the symbol $\sqrt{a}$ makes sense only when $a \geq 0$. For instance,

$$\sqrt{9} = 3 \quad \text{because} \quad 3^2 = 9 \quad \text{and} \quad 3 \geq 0$$

Square roots are special cases of *n*th roots. The *n*th root of x is the number that, when raised to the *n*th power, gives x.

Definition of *n*th Root

If n is any positive integer, then the **principal *n*th root** of a is defined as follows:

$$\sqrt[n]{a} = b \quad \text{means} \quad b^n = a$$

If n is even, then we must have $a \geq 0$ and $b \geq 0$.

For example,

$$\sqrt[4]{81} = 3 \quad \text{because} \quad 3^4 = 81 \quad \text{and} \quad 3 \geq 0$$

$$\sqrt[3]{-8} = -2 \quad \text{because} \quad (-2)^3 = -8$$

But $\sqrt{-8}$, $\sqrt[4]{-8}$, and $\sqrt[6]{-8}$ are not defined. (For instance, $\sqrt{-8}$ is not defined because the square of every real number is nonnegative.)

Notice that

$$\sqrt{4^2} = \sqrt{16} = 4 \quad \text{but} \quad \sqrt{(-4)^2} = \sqrt{16} = 4 = |-4|$$

So the equation $\sqrt{a^2} = a$ is not always true; it is true only when $a \geq 0$. However, we can always write $\sqrt{a^2} = |a|$. This last equation is true not only for square roots, but for any even root. This and other rules used in working with *n*th roots are listed in the following box. In each property we assume that all the given roots exist.

Properties of *n*th Roots

Property	Example
1. $\sqrt[n]{ab} = \sqrt[n]{a}\sqrt[n]{b}$	$\sqrt[3]{-8 \cdot 27} = \sqrt[3]{-8}\sqrt[3]{27} = (-2)(3) = -6$
2. $\sqrt[n]{\dfrac{a}{b}} = \dfrac{\sqrt[n]{a}}{\sqrt[n]{b}}$	$\sqrt[4]{\dfrac{16}{81}} = \dfrac{\sqrt[4]{16}}{\sqrt[4]{81}} = \dfrac{2}{3}$
3. $\sqrt[m]{\sqrt[n]{a}} = \sqrt[mn]{a}$	$\sqrt{\sqrt[3]{729}} = \sqrt[6]{729} = 3$
4. $\sqrt[n]{a^n} = a$ if n is odd	$\sqrt[3]{(-5)^3} = -5$, $\sqrt[5]{2^5} = 2$
5. $\sqrt[n]{a^n} = \|a\|$ if n is even	$\sqrt[4]{(-3)^4} = \|-3\| = 3$

Example 9 ■ Simplifying Expressions Involving *n*th Roots

(a)

$$\begin{aligned} \sqrt[3]{x^4} &= \sqrt[3]{x^3 x} && \text{Factor out the largest cube} \\ &= \sqrt[3]{x^3}\sqrt[3]{x} && \text{Property 1: } \sqrt[3]{ab} = \sqrt[3]{a}\sqrt[3]{b} \\ &= x\sqrt[3]{x} && \text{Property 4: } \sqrt[3]{a^3} = a \end{aligned}$$

(b) $\sqrt[4]{81x^8y^4} = \sqrt[4]{81}\sqrt[4]{x^8}\sqrt[4]{y^4}$ Property 1: $\sqrt[4]{abc} = \sqrt[4]{a}\sqrt[4]{b}\sqrt[4]{c}$

$= 3\sqrt[4]{(x^2)^4}\,|y|$ Property 5: $\sqrt[4]{a^4} = |a|$

$= 3x^2|y|$ Property 5: $\sqrt[4]{a^4} = |a|,\ |x^2| = x^2$

 **Now Try Exercises 33 and 35** ■

It is frequently useful to combine like radicals in an expression such as $2\sqrt{3} + 5\sqrt{3}$. This can be done by using the Distributive Property. For example,

$$2\sqrt{3} + 5\sqrt{3} = (2 + 5)\sqrt{3} = 7\sqrt{3}$$

The next example further illustrates this process.

Avoid making the following error:

$$\sqrt{a+b} \not= \sqrt{a} + \sqrt{b}$$

For instance, if we let $a = 9$ and $b = 16$, then we see the error:

$$\sqrt{9+16} \stackrel{?}{=} \sqrt{9} + \sqrt{16}$$

$$\sqrt{25} \stackrel{?}{=} 3 + 4$$

$$5 \stackrel{?}{=} 7 \quad \text{Wrong!}$$

Example 10 ■ Combining Radicals

(a) $\sqrt{32} + \sqrt{200} = \sqrt{16 \cdot 2} + \sqrt{100 \cdot 2}$ Factor out the largest squares

$= \sqrt{16}\sqrt{2} + \sqrt{100}\sqrt{2}$ Property 1: $\sqrt{ab} = \sqrt{a}\sqrt{b}$

$= 4\sqrt{2} + 10\sqrt{2} = 14\sqrt{2}$ Distributive Property

(b) If $b > 0$, then

$\sqrt{25b} - \sqrt{b^3} = \sqrt{25}\sqrt{b} - \sqrt{b^2}\sqrt{b}$ Property 1: $\sqrt{ab} = \sqrt{a}\sqrt{b}$

$= 5\sqrt{b} - b\sqrt{b}$ Property 5, $b > 0$

$= (5 - b)\sqrt{b}$ Distributive Property

(c) $\sqrt{49x^2 + 49} = \sqrt{49(x^2 + 1)}$ Factor out the perfect square

$= 7\sqrt{x^2 + 1}$ Property 1: $\sqrt{ab} = \sqrt{a}\sqrt{b}$

 **Now Try Exercises 37, 39, and 41** ■

■ Rational Exponents

To define what is meant by a *rational exponent* or, equivalently, a *fractional exponent* such as $a^{1/3}$, we need to use radicals. To give meaning to the symbol $a^{1/n}$ in a way that is consistent with the Laws of Exponents, we would have to have

$$(a^{1/n})^n = a^{(1/n)n} = a^1 = a$$

So by the definition of nth root,

$$a^{1/n} = \sqrt[n]{a}$$

In general, we define rational exponents as follows.

Definition of Rational Exponents

For any rational exponent m/n in lowest terms, where m and n are integers and $n > 0$, we define

$$a^{m/n} = (\sqrt[n]{a})^m \qquad \text{or equivalently} \qquad a^{m/n} = \sqrt[n]{a^m}$$

If n is even, then we require that $a \geq 0$.

With this definition it can be proved that *the Laws of Exponents also hold for rational exponents.*

DIOPHANTUS lived in Alexandria about 250 A.D. His book *Arithmetica* is considered the first book on algebra. In it he gives methods for finding integer solutions of algebraic equations. *Arithmetica* was read and studied for more than a thousand years. Fermat (see Section 1.11) made some of his most important discoveries while studying this book. Diophantus's major contribution is the use of symbols to stand for the unknowns in a problem. Although his symbolism is not as simple as the symbols we use today, it was a major advance over writing everything in words. In Diophantus's notation the equation

$$x^5 - 7x^2 + 8x - 5 = 24$$

is written

$$\Delta K^{\gamma}\alpha\varsigma\eta\pitchfork\Delta^{\gamma}\zeta\overset{\circ}{M}\varepsilon\iota^{\sigma}\kappa\delta$$

Our modern algebraic notation did not come into common use until the 17th century.

Example 11 ■ Using the Definition of Rational Exponents

(a) $4^{1/2} = \sqrt{4} = 2$

(b) $8^{2/3} = (\sqrt[3]{8})^2 = 2^2 = 4$ Alternative solution: $8^{2/3} = \sqrt[3]{8^2} = \sqrt[3]{64} = 4$

(c) $125^{-1/3} = \dfrac{1}{125^{1/3}} = \dfrac{1}{\sqrt[3]{125}} = \dfrac{1}{5}$

 **Now Try Exercises 51 and 53** ■

Example 12 ■ Using the Laws of Exponents with Rational Exponents

(a) $a^{1/3}a^{7/3} = a^{8/3}$ Law 1: $a^m a^n = a^{m+n}$

(b) $\dfrac{a^{2/5}a^{7/5}}{a^{3/5}} = a^{2/5+7/5-3/5} = a^{6/5}$ Law 1, Law 2: $\dfrac{a^m}{a^n} = a^{m-n}$

(c) $(2a^3b^4)^{3/2} = 2^{3/2}(a^3)^{3/2}(b^4)^{3/2}$ Law 4: $(abc)^n = a^n b^n c^n$

$= (\sqrt{2})^3 a^{3(3/2)}b^{4(3/2)}$ Law 3: $(a^m)^n = a^{mn}$

$= 2\sqrt{2}a^{9/2}b^6$

(d) $\left(\dfrac{2x^{3/4}}{y^{1/3}}\right)^3\left(\dfrac{y^4}{x^{-1/2}}\right) = \dfrac{2^3(x^{3/4})^3}{(y^{1/3})^3}\cdot(y^4x^{1/2})$ Laws 5, 4, and 7

$= \dfrac{8x^{9/4}}{y}\cdot y^4x^{1/2}$ Law 3

$= 8x^{11/4}y^3$ Laws 1 and 2

 **Now Try Exercises 57, 59, 61, and 63** ■

Example 13 ■ Simplifying by Writing Radicals as Rational Exponents

(a) $\dfrac{1}{\sqrt[3]{x^4}} = \dfrac{1}{x^{4/3}}$ Definition of rational exponents

(b) $(2\sqrt{x})(3\sqrt[3]{x}) = (2x^{1/2})(3x^{1/3})$ Definition of rational exponents

$= 6x^{1/2+1/3} = 6x^{5/6}$ Law 1

(c) $\sqrt{x\sqrt{x}} = (xx^{1/2})^{1/2}$ Definition of rational exponents

$= (x^{3/2})^{1/2}$ Law 1

$= x^{3/4}$ Law 3

 **Now Try Exercises 67 and 71** ■

■ Rationalizing the Denominator; Standard Form

It is often useful to eliminate the radical in a denominator by multiplying both numerator and denominator by an appropriate expression. This procedure is called **rationalizing the denominator**. If the denominator is of the form $\sqrt{a}$, then we multiply numerator and denominator by $\sqrt{a}$. In doing this we multiply the given quantity by 1, so we do not change its value. For instance,

$$\frac{1}{\sqrt{a}} = \frac{1}{\sqrt{a}}\cdot 1 = \frac{1}{\sqrt{a}}\cdot\frac{\sqrt{a}}{\sqrt{a}} = \frac{\sqrt{a}}{a}$$

Note that the denominator in the last fraction contains no radical. In general, if the denominator is of the form $\sqrt[n]{a^m}$ with $m < n$, then multiplying the numerator and denominator by $\sqrt[n]{a^{n-m}}$ will rationalize the denominator, because (for $a > 0$)

$$\sqrt[n]{a^m}\sqrt[n]{a^{n-m}} = \sqrt[n]{a^{m+n-m}} = \sqrt[n]{a^n} = a$$

A fractional expression whose denominator contains no radicals is said to be in **standard form**.

Example 14 ■ Rationalizing Denominators

Put each fractional expression into standard form by rationalizing the denominator.

(a) $\dfrac{2}{\sqrt{3}}$ **(b)** $\dfrac{1}{\sqrt[3]{5}}$ **(c)** $\sqrt[7]{\dfrac{1}{a^2}}$

Solution

(a) $\dfrac{2}{\sqrt{3}} = \dfrac{2}{\sqrt{3}} \cdot \dfrac{\sqrt{3}}{\sqrt{3}}$ This equals 1 — Multiply by $\dfrac{\sqrt{3}}{\sqrt{3}}$

$= \dfrac{2\sqrt{3}}{3}$ $\sqrt{3} \cdot \sqrt{3} = 3$

(b) $\dfrac{1}{\sqrt[3]{5}} = \dfrac{1}{\sqrt[3]{5}} \cdot \dfrac{\sqrt[3]{5^2}}{\sqrt[3]{5^2}}$ Multiply by $\dfrac{\sqrt[3]{5^2}}{\sqrt[3]{5^2}}$

$= \dfrac{\sqrt[3]{25}}{5}$ $\sqrt[3]{5} \cdot \sqrt[3]{5^2} = \sqrt[3]{5^3} = 5$

(c) $\sqrt[7]{\dfrac{1}{a^2}} = \dfrac{1}{\sqrt[7]{a^2}}$ Property 2: $\sqrt[n]{\dfrac{a}{b}} = \dfrac{\sqrt[n]{a}}{\sqrt[n]{b}}$

$= \dfrac{1}{\sqrt[7]{a^2}} \cdot \dfrac{\sqrt[7]{a^5}}{\sqrt[7]{a^5}}$ Multiply by $\dfrac{\sqrt[7]{a^5}}{\sqrt[7]{a^5}}$

$= \dfrac{\sqrt[7]{a^5}}{a}$ $\sqrt[7]{a^2} \cdot \sqrt[7]{a^5} = \sqrt[7]{a^7} = a$

Now Try Exercises 73 and 75 ■

1.2 | Exercises

Concepts

1. (a) Using exponential notation, we can write the product $5 \cdot 5 \cdot 5 \cdot 5 \cdot 5 \cdot 5$ as ______.

(b) In the expression 3^4 the number 3 is called the ______ and the number 4 is called the ______.

2. (a) When we multiply two powers with the same base, we ______ the exponents. So $3^4 \cdot 3^5 =$ ______.

(b) When we divide two powers with the same base, we ______ the exponents. So $\dfrac{3^5}{3^2} =$ ______.

3. To move a number raised to a power from numerator to denominator or from denominator to numerator, we change the sign of the ______. So $a^{-2} =$ ______, $\dfrac{1}{b^{-2}} =$ ______, $\dfrac{a^{-3}}{b^2} =$ ______, and $\dfrac{6a^2}{b^{-3}} =$ ______.

4. (a) Using exponential notation, we can write $\sqrt[3]{5}$ as ______.

(b) Using radicals, we can write $5^{1/2}$ as ______.

(c) Is there a difference between $\sqrt{5^2}$ and $(\sqrt{5})^2$? Explain.

5. Explain what $4^{3/2}$ means, then calculate $4^{3/2}$ in two different ways:

$(4^{1/2})^{\square} =$ ______ or $(4^3)^{\square} =$ ______

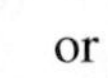

6. Explain how we rationalize a denominator, then complete the following steps to rationalize $\dfrac{1}{\sqrt{3}}$:

$$\frac{1}{\sqrt{3}} = \frac{1}{\sqrt{3}} \cdot \frac{\square}{\square} = \frac{\square}{\square}$$

7. Find the missing power in the following calculation: $5^{1/3} \cdot 5^{\square} = 5$.

8. *Yes or No*? If *No*, give a reason.

(a) Is there a difference between $(-5)^4$ and -5^4?

(b) Is the expression $(x^2)^3$ equal to x^5?

(c) Is the expression $(2x^4)^3$ equal to $2x^{12}$?

(d) Is the expression $\sqrt{4a^2}$ equal to $2a$?

Skills

9–18 ■ Radicals and Exponents Evaluate each expression.

9. (a) -2^6 **(b)** $(-2)^6$ **(c)** $\left(\frac{1}{5}\right)^2 \cdot (-3)^3$

10. (a) $(-5)^3$ **(b)** -5^3 **(c)** $(-5)^2 \cdot \left(\frac{2}{5}\right)^2$

11. (a) $\left(\frac{5}{3}\right)^0 \cdot 2^{-1}$ **(b)** $\dfrac{2^{-3}}{3^0}$ **(c)** $\left(\frac{2}{3}\right)^{-2}$

12. (a) $-2^3\cdot(-2)^0$ (b) $-2^{-3}\cdot(-2)^0$ (c) $\left(\dfrac{-3}{5}\right)^{-3}$

13. (a) $5^3\cdot 5$ (b) $5^4\cdot 5^{-2}$ (c) $(2^2)^3$

14. (a) $3^8\cdot 3^5$ (b) $\dfrac{10^7}{10^4}$ (c) $(3^5)^4$

15. (a) $3\sqrt[3]{16}$ (b) $\dfrac{\sqrt{18}}{\sqrt{81}}$ (c) $\sqrt{\frac{27}{4}}$

16. (a) $2\sqrt[3]{81}$ (b) $\dfrac{\sqrt{18}}{\sqrt{25}}$ (c) $\sqrt{\frac{12}{49}}$

17. (a) $\sqrt{3}\sqrt{15}$ (b) $\dfrac{\sqrt{48}}{\sqrt{3}}$ (c) $\sqrt[3]{24}\sqrt[3]{18}$

18. (a) $\sqrt{10}\sqrt{32}$ (b) $\dfrac{\sqrt{54}}{\sqrt{6}}$ (c) $\sqrt[3]{15}\sqrt[3]{75}$

19–24 ■ Exponents Simplify each expression and eliminate any negative exponents.

 **19.** (a) t^5t^2 (b) $(4z^3)^2$ (c) $x^{-3}x^5$

20. (a) a^4a^6 (b) $(-2b^{-3})^3$ (c) $-2y^{10}y^{-11}$

21. (a) $x^{-5}\cdot x^3$ (b) $w^{-2}w^{-4}w^5$ (c) $\dfrac{x^{16}}{x^{10}}$

22. (a) $y^2\cdot y^{-5}$ (b) $z^5z^{-3}z^{-4}$ (c) $\dfrac{y^7y^0}{y^{10}}$

23. (a) $\dfrac{a^9a^{-2}}{a}$ (b) $(a^2a^4)^3$ (c) $\left(\dfrac{x}{2}\right)^3(5x^6)$

24. (a) $\dfrac{z^2z^4}{z^3z^{-1}}$ (b) $(2a^3a^2)^4$ (c) $(-3z^2)^3(2z^3)$

25–32 ■ Exponents Simplify each expression and eliminate any negative exponents.

 **25.** (a) $(3x^3y^2)(2y^3)$ (b) $(5w^2z^{-2})^2(z^3)$

26. (a) $(8m^{-2}n^4)(\frac{1}{2}n^{-2})$ (b) $(3a^4b^{-2})^3(a^2b^{-1})$

27. (a) $\dfrac{x^2y^{-1}}{x^{-5}}$ (b) $\left(\dfrac{a^3}{2b^2}\right)^3$

28. (a) $\dfrac{y^{-2}z^{-3}}{y^{-1}}$ (b) $\left(\dfrac{x^3y^{-2}}{x^{-3}y^2}\right)^{-2}$

 **29.** (a) $\left(\dfrac{a^2}{b}\right)^5\left(\dfrac{a^3b^2}{c^3}\right)^3$ (b) $\dfrac{(u^{-1}v^2)^2}{(u^3v^{-2})^3}$

30. (a) $\left(\dfrac{x^4z^2}{4y^5}\right)\left(\dfrac{2x^3y^2}{z^3}\right)^2$ (b) $\dfrac{(rs^2)^3}{(r^{-3}s^2)^2}$

31. (a) $\dfrac{8a^3b^{-4}}{2a^{-5}b^5}$ (b) $\left(\dfrac{y}{5x^{-2}}\right)^{-3}$

32. (a) $\dfrac{5xy^{-2}}{x^{-1}y^{-3}}$ (b) $\left(\dfrac{2a^{-1}b}{a^2b^{-3}}\right)^{-3}$

33–36 ■ Radicals Simplify each expression. Remember to use Property 5 of nth roots where appropriate.

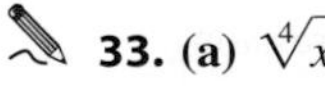 **33.** (a) $\sqrt[4]{x^4}$ (b) $\sqrt[4]{16x^8}$

34. (a) $\sqrt[5]{x^{10}}$ (b) $\sqrt[3]{x^3y^6}$

35. (a) $\sqrt[3]{8x^9y^3}$ (b) $\sqrt[4]{8x^6y^2}\sqrt[4]{2x^2y^2}$

36. (a) $\sqrt{16x^4y^6z^2}$ (b) $\sqrt{\sqrt[3]{512x^9}}$

37–42 ■ Radical Expressions Simplify each expression. Assume that all letters denote positive real numbers.

37. (a) $\sqrt{32}+\sqrt{18}$ (b) $\sqrt{75}+\sqrt{48}$

38. (a) $\sqrt{125}+\sqrt{45}$ (b) $\sqrt[3]{54}-\sqrt[3]{16}$

39. (a) $\sqrt{9a^3}+\sqrt{a}$ (b) $\sqrt{16x}+\sqrt{x^5}$

40. (a) $\sqrt[3]{x^4}+\sqrt[3]{8x}$ (b) $4\sqrt{18rt^3}+5\sqrt{32r^3t^5}$

41. (a) $\sqrt{36x^2+36x^4}$ (b) $\sqrt{81x^2+81y^2}$

42. (a) $\sqrt{27a^3+63a^2}$ (b) $\sqrt{25t^2+100t^2}$

43–50 ■ Radicals and Exponents Write each radical expression using exponents, and each exponential expression using radicals.

	Radical expression	Exponential expression
43.	$\sqrt{10}$	
44.	$\sqrt[5]{6}$	
45.		$7^{3/5}$
46.		$6^{-5/2}$
47.	$\dfrac{1}{\sqrt{5}}$	
48.		$5^{-3/4}$
49.		$y^{-1.5}$
50.	$\dfrac{1}{\sqrt[3]{x^2}}$	

51–56 ■ Rational Exponents Evaluate each expression, without using a calculator.

51. (a) $16^{1/4}$ (b) $-8^{1/3}$ (c) $9^{-1/2}$

52. (a) $27^{1/3}$ (b) $(-8)^{1/3}$ (c) $-(\frac{1}{8})^{1/3}$

53. (a) $32^{2/5}$ (b) $(\frac{4}{9})^{-1/2}$ (c) $(\frac{16}{81})^{3/4}$

54. (a) $125^{2/3}$ (b) $(\frac{25}{64})^{3/2}$ (c) $27^{-4/3}$

55. (a) $5^{2/3}\cdot 5^{1/3}$ (b) $\dfrac{3^{3/5}}{3^{2/5}}$ (c) $(\sqrt[3]{4})^3$

56. (a) $3^{2/7}\cdot 3^{12/7}$ (b) $\dfrac{7^{2/3}}{7^{5/3}}$ (c) $(\sqrt[5]{6})^{-10}$

57–64 ■ Rational Exponents Simplify each expression and eliminate any negative exponents. Assume that all letters denote positive numbers.

57. (a) $x^{3/4}x^{5/4}$ (b) $y^{2/3}y^{4/3}$

58. (a) $(4b)^{1/2}(8b^{1/4})$ (b) $(3a^{3/4})^2(5a^{1/2})$

59. (a) $\dfrac{w^{4/3}w^{2/3}}{w^{1/3}}$ (b) $(3x^{1/2}y^{1/3})^6$

60. (a) $(8y^3)^{-2/3}$ (b) $(u^4v^6)^{-1/3}$

61. (a) $(8a^6b^{3/2})^{2/3}$ (b) $(4a^{-4}b^6)^{3/2}$

62. (a) $(x^{-5}y^{1/3})^{-3/5}$ (b) $(u^3v^2)^{1/3}(16u^6v^{2/3})^{1/2}$

63. (a) $\left(\dfrac{3x^{1/4}}{y}\right)^2\left(\dfrac{x^{-1}}{y^4}\right)^{1/2}$ (b) $\left(\dfrac{3w^{-1/3}}{z^{-1}}\right)^{-4}\left(\dfrac{2w^{1/3}}{z^{1/2}}\right)^2$

64. (a) $\left(\dfrac{y^{2/3}}{x^{-1}}\right)^3\left(\dfrac{x}{y^{-2}}\right)^{-1}$ (b) $\left(\dfrac{4y^3z^{2/3}}{x^{1/2}}\right)^2\left(\dfrac{x^{-3}y^6}{8z^4}\right)^{1/3}$

65–72 ■ Radicals Write each expression using rational exponents and simplify. Eliminate any negative exponents. Assume that all letters denote positive numbers.

65. (a) $\sqrt{x^3}$ (b) $\sqrt[5]{x^6}$

66. (a) $\sqrt{x^5}$ (b) $\sqrt[4]{x^6}$

67. (a) $\sqrt[6]{y^5}\sqrt[3]{y^2}$ (b) $(5\sqrt[3]{x})(2\sqrt[4]{x})$

68. (a) $\sqrt[4]{b^3}\sqrt{b}$ (b) $(2\sqrt{a})(\sqrt[3]{a^2})$

69. (a) $\sqrt{4st^3}\sqrt[6]{s^3t^2}$ (b) $\dfrac{\sqrt[4]{x^7}}{\sqrt[4]{x^3}}$

70. (a) $\sqrt[5]{x^3y^2}\sqrt[10]{x^4y^{16}}$ (b) $\dfrac{\sqrt[3]{8x^2}}{\sqrt{x}}$

71. (a) $\sqrt[3]{y\sqrt{y}}$ (b) $\sqrt{\dfrac{18u^5v}{2u^3v^3}}$

72. (a) $\sqrt{s\sqrt{s^3}}$ (b) $\sqrt[3]{\dfrac{54x^2y^4}{2x^5y}}$

73–76 ■ Rationalize Put each fractional expression into standard form by rationalizing the denominator. Assume that all letters denote positive real numbers.

73. (a) $\dfrac{1}{\sqrt{6}}$ (b) $\sqrt{\frac{3}{2}}$ (c) $\dfrac{9}{\sqrt[4]{2}}$

74. (a) $\dfrac{12}{\sqrt{3}}$ (b) $\sqrt{\frac{12}{5}}$ (c) $\dfrac{8}{\sqrt[3]{5^2}}$

75. (a) $\dfrac{1}{\sqrt{5x}}$ (b) $\sqrt{\dfrac{x}{5}}$ (c) $\sqrt[5]{\dfrac{1}{x^3}}$

76. (a) $\sqrt{\dfrac{s}{3t}}$ (b) $\dfrac{a}{\sqrt[6]{b^2}}$ (c) $\dfrac{1}{c^{3/5}}$

77–80 ■ Putting It All Together Simplify the expression and eliminate any negative exponents. Assume that all letters denote positive numbers unless otherwise stated. Convert to rational exponents where it is helpful. If necessary, rationalize the denominator.

77. (a) $\sqrt[4]{\frac{1}{4}}\sqrt[4]{\frac{1}{64}}$ (b) $\dfrac{\sqrt{5}}{\sqrt{40}}$

78. (a) $\left(\dfrac{x^{3/2}}{y^{-1/2}}\right)^4$ (b) $(25u^2v^{-4})^{1/2}$, $(u < 0,\ \ v > 0)$

79. (a) $\sqrt{y}\sqrt[4]{y^2}$ (b) $(81\sqrt[4]{w^8z^8})^{1/2}$, $(w > 0,\ \ z < 0)$

80. (a) $\left(\dfrac{x^{3/2}}{y^{-1/2}}\right)^4\left(\dfrac{x^{-2}}{y^3}\right)$ (b) $\left(\dfrac{w^4}{\sqrt[3]{w^3z^6}}\right)^{1/2}$

81–82 ■ Scientific Notation Write each number in scientific notation.

81. (a) 69,300,000 (b) 7,200,000,000,000
(c) 0.000028536 (d) 0.0001213

82. (a) 129,540,000 (b) 7,259,000,000
(c) 0.0000000014 (d) 0.0007029

83–84 ■ Decimal Notation Write each number in decimal notation.

83. (a) 3.19×10^5 (b) 2.721×10^8
(c) 2.670×10^{-8} (d) 9.999×10^{-9}

84. (a) 7.1×10^{14} (b) 6×10^{12}
(c) 8.55×10^{-3} (d) 6.257×10^{-10}

85–86 ■ Scientific Notation Write the number indicated in each statement in scientific notation.

85. (a) A light-year, the distance that light travels in one year, is about 9,460,000,000,000 km.
(b) The diameter of an electron is about 0.0000000000004 cm.
(c) A drop of water contains more than 33 billion billion molecules.

86. (a) The distance from the earth to the sun is about 150 million kilometers.
(b) The mass of an oxygen molecule is about 0.000000000000000000000053 g.
(c) The mass of the earth is about 5,970,000,000,000,000,000,000,000 kg.

87–92 ■ Scientific Notation Use scientific notation, the Laws of Exponents, and a calculator to perform the indicated operations. State your answer rounded to the number of significant digits indicated by the given data.

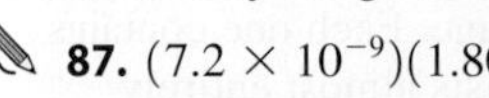

87. $(7.2 \times 10^{-9})(1.806 \times 10^{-12})$

88. $(1.062 \times 10^{24})(8.61 \times 10^{19})$

89. $\dfrac{1.295643 \times 10^9}{(3.610 \times 10^{-17})(2.511 \times 10^6)}$

90. $\dfrac{(73.1)(1.6341 \times 10^{28})}{0.0000000019}$

91. $\dfrac{(0.0000162)(0.01582)}{(594{,}621{,}000)(0.0058)}$ 92. $\dfrac{(3.542 \times 10^{-6})^9}{(5.05 \times 10^4)^{12}}$

Skills Plus

93. **Sign of an Expression** Let a, b, and c be real numbers with $a > 0$, $b < 0$, and $c < 0$. Determine the sign of each expression.
(a) b^5 (b) b^{10} (c) ab^2c^3
(d) $(b - a)^3$ (e) $(b - a)^4$ (f) $\dfrac{a^3c^3}{b^6c^6}$

94. **Comparing Roots** Without using a calculator, determine which number is larger in each pair.
(a) $2^{1/2}$ or $2^{1/3}$ (b) $\left(\frac{1}{2}\right)^{1/2}$ or $\left(\frac{1}{2}\right)^{1/3}$
(c) $7^{1/4}$ or $4^{1/3}$ (d) $\sqrt[3]{5}$ or $\sqrt{3}$

Applications

95. **Distance to the Nearest Star** Proxima Centauri, the star nearest to our solar system, is 4.3 light-years away. Use the information in Exercise 85(a) to express this distance in kilometers.

96. Speed of Light The speed of light is about 300,000 km/s. Use the information in Exercise 86(a) to find how long it takes for a light ray from the sun to reach the earth.

97. Volume of the Oceans The average ocean depth is 3.7×10^3 m, and the surface area of the oceans is 3.6×10^{14} m^2. What is the total volume of the ocean in liters? (One cubic meter contains 1000 liters.)

98. National Debt In 2020, the population of the United States was 3.3145×10^8, and the national debt was 2.670×10^{13} dollars. How much was each person's share of the debt? [*Source:* U.S. Census Bureau and U.S. Department of the Treasury]

99. Number of Atoms in the Observable Universe The *Hubble Deep Field* is a long exposure image of a tiny region of the sky (about 2.6 minutes of arc). Each dot or smudge in the image is an entire galaxy; there are over ten thousand galaxies in this one image. Each one contains billions of stars, each of which consists almost entirely of hydrogen atoms. Use the following information to give an estimate of the number of atoms in the observable universe.

Mass of typical star: 1.77×10^{30} kg
Mass of hydrogen atom: 1.67×10^{-27} kg
Number of stars in typical galaxy: 2×10^{11}
Number of galaxies in observable universe: 1.5×10^{12}

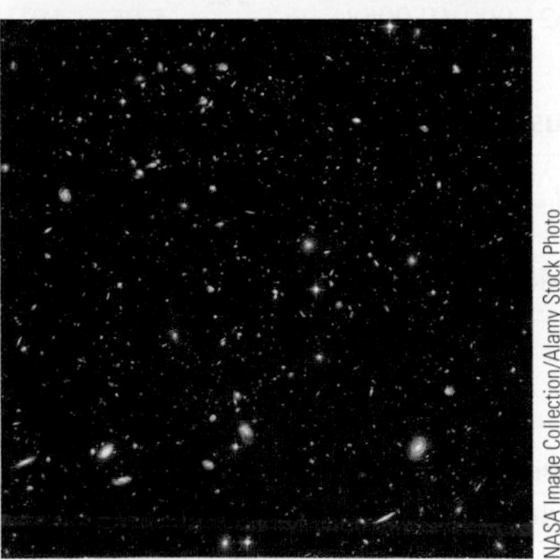
NASA Image Collection/Alamy Stock Photo

100. How Far Can You See? Because of the curvature of the earth, the maximum distance D that you can see from the top of a tall building of height h is estimated by the formula

$$D = \sqrt{2rh + h^2}$$

where $r = 6370$ km is the radius of the earth and D and h are also measured in kilometers. On a clear day, how far can you see from the observation deck of the Toronto CN Tower, 346 m above the ground?

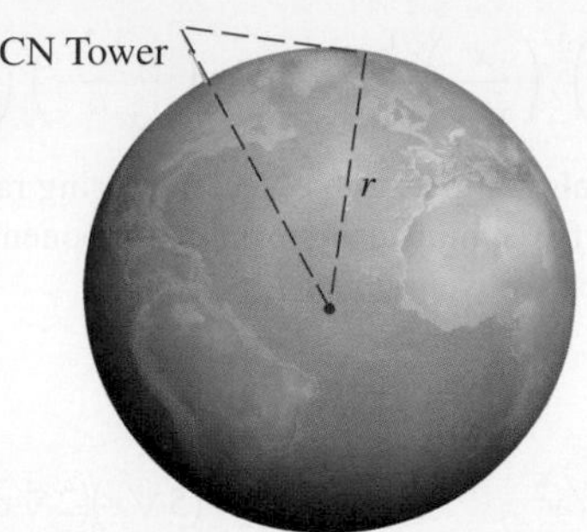

101. Speed of a Skidding Car Police use the formula $s = \sqrt{30fd}$ to estimate the speed s (in km/h) at which a car is traveling if it skids d meters after the brakes are applied suddenly. The number f is the coefficient of friction of the road, a measure of the "slipperiness" of the road. The table below gives some typical estimates for f.

	Tar	Concrete	Gravel
Dry	1.0	0.8	0.2
Wet	0.5	0.4	0.1

(a) If a car skids 65 m on wet concrete, how fast was it moving when the brakes were applied?

(b) If a car is traveling at 50 km/h, how far will it skid on wet tar?

102. Distance from the Earth to the Sun It follows from **Kepler's Third Law** of planetary motion that the average distance from a planet to the sun (in meters) is

$$d = \left(\frac{GM}{4\pi^2}\right)^{1/3} T^{2/3}$$

where $M = 1.99 \times 10^{30}$ kg is the mass of the sun, $G = 6.67 \times 10^{-11}$ N·m^2/kg^2 is the gravitational constant, and T is the period of the planet's orbit (in seconds). Use the fact that the period of the earth's orbit is about 365.25 days to find the distance from the earth to the sun.

■ Discuss ■ Discover ■ Prove ■ Write

103. Discuss: How Big is a Billion? If you had a million (10^6) dollars in a suitcase, and you spent a thousand (10^3) dollars each day, how many years would it take you to use all the money? Spending at the same rate, how many years would it take you to empty a suitcase filled with a *billion* (10^9) dollars?

104. Discover: Limiting Behavior of Powers Complete the following tables. What happens to the nth root of 2 as n gets large? What about the nth root of $\frac{1}{2}$?

n	$2^{1/n}$
1	
2	
5	
10	
100	

n	$\left(\frac{1}{2}\right)^{1/n}$
1	
2	
5	
10	
100	

Construct a similar table for $n^{1/n}$. What happens to the nth root of n as n gets large?

105. Discuss: Easy Powers that Look Hard Calculate these expressions in your head. Use the Laws of Exponents to help you.

(a) $\dfrac{18^5}{9^5}$ **(b)** $20^6 \cdot (0.5)^6$

106. Prove: Laws of Exponents Prove the following Laws of Exponents for the case in which m and n are positive integers and $m > n$.

(a) Law 2: $\dfrac{a^m}{a^n} = a^{m-n}$ **(b)** Law 5: $\left(\dfrac{a}{b}\right)^n = \dfrac{a^n}{b^n}$

107. Prove: Laws of Exponents Prove the following Laws of Exponents.

(a) Law 6: $\left(\dfrac{a}{b}\right)^{-n} = \dfrac{b^n}{a^n}$ **(b)** Law 7: $\dfrac{a^{-n}}{b^{-m}} = \dfrac{b^m}{a^n}$

1.3 Algebraic Expressions

■ Adding and Subtracting Polynomials ■ Multiplying Algebraic Expressions ■ Special Product Formulas ■ Factoring Common Factors ■ Factoring Trinomials ■ Special Factoring Formulas ■ Factoring by Grouping Terms

A **variable** is a letter that can represent any number from a given set of numbers. If we start with variables, such as x, y, and z, and some real numbers and combine them using addition, subtraction, multiplication, division, powers, and roots, we obtain an **algebraic expression.** Here are some examples:

$$2x^2 - 3x + 4 \qquad \sqrt{x} + 10 \qquad \frac{y - 2z}{y^2 + 4}$$

In the first expression we say that 2 is the coefficient of x^2 and -3 is the coefficient of x. In general, a *coefficient* is any number that is multiplied with a variable.

A **monomial** is an expression of the form ax^k, where a is a real number and k is a nonnegative integer. A **binomial** is a sum of two monomials and a **trinomial** is a sum of three monomials. In general, a sum of monomials is called a *polynomial.* For example, the first expression listed above is a polynomial, but the other two are not.

Polynomials

A **polynomial** in the variable x is an expression of the form

$$a_n x^n + a_{n-1}x^{n-1} + \cdots + a_1 x + a_0$$

where $a_0, a_1, \ldots, a_n$ are real numbers, and n is a nonnegative integer. If $a_n \neq 0$, then the polynomial has **degree** $\boldsymbol{n}$. The monomials $a_k x^k$ that make up the polynomial are called the **terms** of the polynomial.

Note that the degree of a polynomial is the highest power of the variable that appears in the polynomial.

Polynomial	Type	Terms	Degree
$2x^2 - 3x + 4$	trinomial	$2x^2, -3x, 4$	2
$x^8 + 5x$	binomial	$x^8, 5x$	8
$8 - x + x^2 - \frac{1}{2}x^3$	four terms	$-\frac{1}{2}x^3, x^2, -x, 8$	3
$5x + 1$	binomial	$5x, 1$	1
$9x^5$	monomial	$9x^5$	5
6	monomial	6	0

■ Adding and Subtracting Polynomials

Distributive Property

$ac + bc = (a + b)c$

We **add** and **subtract** polynomials using the properties of real numbers discussed in Section 1.1. The idea is to combine **like terms** (that is, terms with the same variables raised to the same powers) using the Distributive Property. For instance,

$$5x^7 + 3x^7 = (5 + 3)x^7 = 8x^7$$

In subtracting polynomials, we have to remember that if a minus sign precedes an expression in parentheses, then the sign of every term within the parentheses is changed when we remove the parentheses:

$$-(b + c) = -b - c$$

[This is simply a case of the Distributive Property, $a(b + c) = ab + ac$, with $a = -1$.]

Example 1 ■ Adding and Subtracting Polynomials

(a) Find the sum $(x^3 - 6x^2 + 2x + 4) + (x^3 + 5x^2 - 7x)$.

(b) Find the difference $(x^3 - 6x^2 + 2x + 4) - (x^3 + 5x^2 - 7x)$.

Solution

(a) $(x^3 - 6x^2 + 2x + 4) + (x^3 + 5x^2 - 7x)$

$= (x^3 + x^3) + (-6x^2 + 5x^2) + (2x - 7x) + 4$ Group like terms

$= 2x^3 - x^2 - 5x + 4$ Combine like terms

(b) $(x^3 - 6x^2 + 2x + 4) - (x^3 + 5x^2 - 7x)$

$= x^3 - 6x^2 + 2x + 4 - x^3 - 5x^2 + 7x$ Distributive Property

$= (x^3 - x^3) + (-6x^2 - 5x^2) + (2x + 7x) + 4$ Group like terms

$= -11x^2 + 9x + 4$ Combine like terms

Now Try Exercises 17 and 19 ■

■ Multiplying Algebraic Expressions

To find the **product** of polynomials or other algebraic expressions, we need to use the Distributive Property repeatedly. In particular, using it three times on the product of two binomials, we get

$$(a + b)(c + d) = a(c + d) + b(c + d) = ac + ad + bc + bd$$

This says that we multiply the two factors by multiplying each term in one factor by each term in the other factor and adding these products. Schematically, we have

$$(a + b)(c + d) = \underset{\mathbf{F}}{ac} + \underset{\mathbf{O}}{ad} + \underset{\mathbf{I}}{bc} + \underset{\mathbf{L}}{bd}$$

The acronym **FOIL** helps us remember that the product of two binomials is the sum of the products of the **F**irst terms, the **O**uter terms, the **I**nner terms, and the **L**ast terms.

In general, we can multiply two algebraic expressions by using the Distributive Property and the Laws of Exponents.

Example 2 ■ Multiplying Binomials Using FOIL

$$(2x + 1)(3x - 5) = \underset{\mathbf{F}}{6x^2} - \underset{\mathbf{O}}{10x} + \underset{\mathbf{I}}{3x} - \underset{\mathbf{L}}{5}$$ Distributive Property

$$= 6x^2 - 7x - 5$$ Combine like terms

Now Try Exercise 27 ■

When we multiply trinomials or polynomials with more terms, we use the Distributive Property. It is also helpful to arrange our work in table form. The following example illustrates both methods.

Example 3 ■ Multiplying Polynomials

Find the product: $(2x + 3)(x^2 - 5x + 4)$

Solution 1: Using the Distributive Property

$$
\begin{aligned}
(2x + 3)(x^2 - 5x + 4) &= 2x(x^2 - 5x + 4) + 3(x^2 - 5x + 4) && \text{Distributive Property}\\
&= (2x \cdot x^2 - 2x \cdot 5x + 2x \cdot 4) + (3 \cdot x^2 - 3 \cdot 5x + 3 \cdot 4) && \text{Distributive Property}\\
&= (2x^3 - 10x^2 + 8x) + (3x^2 - 15x + 12) && \text{Laws of Exponents}\\
&= 2x^3 - 7x^2 - 7x + 12 && \text{Combine like terms}
\end{aligned}
$$

Solution 2: Using Table Form

$$
\begin{array}{rrrrl}
 & x^2 & -\ 5x & +\ 4 & \\
 & & 2x & +\ 3 & \\
\hline
 & 3x^2 & -\ 15x & +\ 12 & \text{Multiply } x^2 - 5x + 4 \text{ by } 3\\
2x^3 & -\ 10x^2 & +\ 8x & & \text{Multiply } x^2 - 5x + 4 \text{ by } 2x\\
\hline
2x^3 & -\ 7x^2 & -\ 7x & +\ 12 & \text{Add like terms}
\end{array}
$$

 Now Try Exercise 47 ■

■ Special Product Formulas

Certain types of products occur so frequently that you should memorize them. You can verify the following formulas by performing the multiplications.

Special Product Formulas

If A and B are any real numbers or algebraic expressions, then

1. $(A + B)(A - B) = A^2 - B^2$	Sum and difference of same terms
2. $(A + B)^2 = A^2 + 2AB + B^2$	Square of a sum
3. $(A - B)^2 = A^2 - 2AB + B^2$	Square of a difference
4. $(A + B)^3 = A^3 + 3A^2B + 3AB^2 + B^3$	Cube of a sum
5. $(A - B)^3 = A^3 - 3A^2B + 3AB^2 - B^3$	Cube of a difference

The key idea in using these formulas (or any other formula in algebra) is the **Principle of Substitution**: We may substitute any algebraic expression for any letter in a formula. For example, to find $(x^2 + y^3)^2$ we use Product Formula 2, substituting x^2 for A and y^3 for B, to get

$$(x^2 + y^3)^2 = (x^2)^2 + 2(x^2)(y^3) + (y^3)^2$$

$$(A + B)^2 = A^2 + 2AB + B^2$$

Mathematics in the Modern World

Changing Words, Sound, and Pictures into Numbers

Pictures, sound, and text are routinely transmitted from one place to another via the Internet, fax machines, or modems. How can such things be transmitted electronically? The key to doing this is to change them into numbers or bits (the digits 0 or 1). It's easy to see how to change text to numbers. For example, we could use the correspondence A = 00000001, B = 00000010, C = 00000011, D = 00000100, E = 00000101, and so on. The word "BED" then becomes 000000100000010100000100. By reading the digits in groups of eight, it is possible to translate this number back to the word "BED."

Changing sound to bits is more complicated. A sound wave can be graphed on an oscilloscope or a computer. The graph is then broken down mathematically into simpler components corresponding to the different frequencies of the original sound. (A branch of mathematics called Fourier analysis is used here.) The intensity of each component is a number, and the original sound can be reconstructed from these numbers. For example, music is stored in a digital file as a sequence of bits; it may look like 101010001010010100101010100000101111101010001010111.... (One second of music requires 1.5 million bits!) The computer reconstructs the music from the numbers in the digital file.

Changing pictures into numbers involves expressing the color and brightness of each dot (or pixel) into a number. This is done very efficiently using a branch of mathematics called wavelet theory. The FBI uses wavelets as a compact way to store the millions of fingerprints they have on file.

Example 4 ■ Using the Special Product Formulas

Use the Special Product Formulas to find each product.

(a) $(3x + 5)^2$ **(b)** $(x^2 - 2)^3$

Solution

(a) Substituting $A = 3x$ and $B = 5$ in Product Formula 2, we get

$$(3x + 5)^2 = (3x)^2 + 2(3x)(5) + 5^2 = 9x^2 + 30x + 25$$

(b) Substituting $A = x^2$ and $B = 2$ in Product Formula 5, we get

$$\begin{aligned}(x^2 - 2)^3 &= (x^2)^3 - 3(x^2)^2(2) + 3(x^2)(2)^2 - 2^3 \\ &= x^6 - 6x^4 + 12x^2 - 8\end{aligned}$$

 **Now Try Exercises 33 and 45** ■

Example 5 ■ Recognizing a Special Product Formula

Find each product.

(a) $(2x + \sqrt{y})(2x - \sqrt{y})$ **(b)** $(x + y - 1)(x + y + 1)$

Solution

(a) Substituting $A = 2x$ and $B = \sqrt{y}$ in Product Formula 1, we get

$$(2x + \sqrt{y})(2x - \sqrt{y}) = (2x)^2 - (\sqrt{y})^2 = 4x^2 - y$$

(b) If we group $x + y$ together and think of this as one algebraic expression, we can use Product Formula 1 with $A = x + y$ and $B = 1$.

$$\begin{aligned}(x + y - 1)(x + y + 1) &= [(x + y) - 1][(x + y) + 1] \\ &= (x + y)^2 - 1^2 && \text{Product Formula 1} \\ &= x^2 + 2xy + y^2 - 1 && \text{Product Formula 2}\end{aligned}$$

 **Now Try Exercises 59 and 63** ■

■ Factoring Common Factors

We use the Distributive Property to expand algebraic expressions. We sometimes need to reverse this process (again using the Distributive Property) by **factoring** an expression as a product of simpler ones. For example, we can write

EXPANDING →

$$(x - 2)(x + 3) = x^2 + x - 6$$

← FACTORING

We say that $x - 2$ and $x + 3$ are **factors** of $x^2 + x - 6$.

The easiest type of factoring occurs when the terms have a common factor.

Example 6 ■ Factoring Out Common Factors

Factor each expression.

(a) $3x^2 - 6x$ **(b)** $8x^4y^2 + 6x^3y^3 - 2xy^4$ **(c)** $(2x + 4)(x - 3) - 5(x - 3)$

Solution

(a) The greatest common factor of the terms $3x^2$ and $-6x$ is $3x$, so we have

$$3x^2 - 6x = 3x(x - 2)$$

Check Your Answer

Multiplying gives

$3x(x - 2) = 3x^2 - 6x$ ✓

(b) We note that

$$8, 6, \text{ and } -2 \text{ have the greatest common factor } 2$$

$$x^4, x^3, \text{ and } x \text{ have the greatest common factor } x$$

$$y^2, y^3, \text{ and } y^4 \text{ have the greatest common factor } y^2$$

So the greatest common factor of the three terms in the polynomial is $2xy^2$, and we have

$$8x^4y^2 + 6x^3y^3 - 2xy^4 = (2xy^2)(4x^3) + (2xy^2)(3x^2y) + (2xy^2)(-y^2)$$
$$= 2xy^2(4x^3 + 3x^2y - y^2)$$

Check Your Answer

Multiplying gives

$$2xy^2(4x^3 + 3x^2y - y^2)$$
$$= 8x^4y^2 + 6x^3y^3 - 2xy^4 \checkmark$$

(c) The two terms have the common factor $x - 3$.

$$(2x + 4)(x - 3) - 5(x - 3) = [(2x + 4) - 5](x - 3) \quad \text{Distributive Property}$$
$$= (2x - 1)(x - 3) \quad \text{Simplify}$$

 **Now Try Exercises 65, 67, and 69** ■

■ Factoring Trinomials

To factor a trinomial of the form $x^2 + bx + c$, we note that

$$(x + r)(x + s) = x^2 + (r + s)x + rs$$

so we need to choose numbers r and s so that $r + s = b$ and $rs = c$.

Example 7 ■ Factoring $x^2 + bx + c$ by Trial and Error

Factor: $x^2 + 7x + 12$

Solution We need to find two integers whose product is 12 and whose sum is 7. By trial and error we find that the two integers are 3 and 4. Thus the factorization is

$$x^2 + 7x + 12 = \underbrace{(x + 3)(x + 4)}_{\text{factors of 12}}$$

Check Your Answer

Multiplying gives

$$(x + 3)(x + 4) = x^2 + 7x + 12 \checkmark$$

 **Now Try Exercise 73** ■

To factor a trinomial of the form $ax^2 + bx + c$ with $a \neq 1$, we look for factors of the form $px + r$ and $qx + s$:

$$ax^2 + bx + c = (px + r)(qx + s) = pqx^2 + (ps + qr)x + rs$$

Therefore we try to find numbers p, q, r, and s such that $pq = a$, $rs = c$, $ps + qr = b$. If these numbers are all integers, then we will have a limited number of possibilities to try for p, q, r, and s.

factors of a (arrows pointing to p and q); factors of c (arrows pointing to r and s):

$$ax^2 + bx + c = (px + r)(qx + s)$$

Discovery Project ■ Visualizing a Formula

b | ab | b^2
a | a^2 | ab
a | b

Many of the Special Product Formulas in this section can be "seen" as geometrical facts about length, area, and volume. For example, the formula about the square of a sum can be interpreted to be about areas of squares and rectangles. The ancient Greeks always interpreted algebraic formulas in terms of geometric figures. Such figures give us special insight into how these formulas work. You can find the project at **www.stewartmath.com**.

Example 8 ■ Factoring $ax^2 + bx + c$ by Trial and Error

Factor: $6x^2 + 7x - 5$

Solution We can factor 6 as $6 \cdot 1$ or $3 \cdot 2$, and -5 as $-5 \cdot 1$ or $5 \cdot (-1)$. By trying these possibilities, we arrive at the factorization

Check Your Answer

Multiplying gives

$(3x + 5)(2x - 1) = 6x^2 + 7x - 5$ ✓

$$6x^2 + 7x - 5 = (\overset{\text{factors of } 6}{\overset{\downarrow}{3x}} + \underset{\underset{\text{factors of } -5}{\uparrow}}{5})(\overset{\downarrow}{2x} - \underset{\uparrow}{1})$$

Now Try Exercise 75 ■

Example 9 ■ Recognizing the Form of an Expression

Factor each expression.

(a) $x^2 - 2x - 3$ **(b)** $(5a + 1)^2 - 2(5a + 1) - 3$

Solution

(a) $x^2 - 2x - 3 = (x - 3)(x + 1)$ Trial and error

(b) This expression is of the form

$$\square^2 - 2\square - 3$$

where $\square$ represents $5a + 1$. This is the same form as the expression in part (a), so it will factor as $(\square - 3)(\square + 1)$.

$$\begin{aligned}(5a + 1)^2 - 2(5a + 1) - 3 &= [(5a + 1) - 3][(5a + 1) + 1] \\ &= (5a - 2)(5a + 2)\end{aligned}$$

Now Try Exercises 79 and 131 ■

■ Special Factoring Formulas

Some special algebraic expressions can be factored by using the following formulas. The first three are simply Special Product Formulas written in reverse.

Special Factoring Formulas

Formula	Name
1. $A^2 - B^2 = (A + B)(A - B)$	Difference of squares
2. $A^2 + 2AB + B^2 = (A + B)^2$	Perfect square
3. $A^2 - 2AB + B^2 = (A - B)^2$	Perfect square
4. $A^3 - B^3 = (A - B)(A^2 + AB + B^2)$	Difference of cubes
5. $A^3 + B^3 = (A + B)(A^2 - AB + B^2)$	Sum of cubes

Example 10 ■ Recognizing a Difference of Squares

Factor each expression.

(a) $4x^2 - 25$ **(b)** $(x + y)^2 - z^2$

Terms and Factors

When we multiply two numbers together, each of the numbers is called a **factor** of the product. When we add two numbers together, each number is called a **term** of the sum.

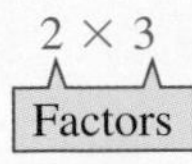

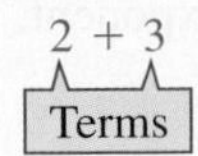

If a factor is common to each term of an expression, we can factor it out. The following expression has two terms.

$$ax + 2ay$$

a is a factor of each term

Each term contains the factor a, so we can factor out a and write the expression as

$$ax + 2ay = a(x + 2y)$$

Solution

(a) Using the Difference of Squares Formula with $A = 2x$ and $B = 5$, we have

$$4x^2 - 25 = (2x)^2 - 5^2 = (2x + 5)(2x - 5)$$

$$A^2 - B^2 = (A + B)(A - B)$$

(b) We use the Difference of Squares Formula with $A = x + y$ and $B = z$.

$$(x + y)^2 - z^2 = (x + y - z)(x + y + z)$$

Now Try Exercises 81 and 115 ■

A trinomial is a perfect square if it is of the form

$$A^2 + 2AB + B^2 \qquad \text{or} \qquad A^2 - 2AB + B^2$$

So we **recognize a perfect square** if the middle term ($2AB$ or $-2AB$) is plus or minus twice the product of the square roots of the two outer terms.

Example 11 ■ Recognizing Perfect Squares

Factor each trinomial.

(a) $x^2 + 6x + 9$ **(b)** $4x^2 - 4xy + y^2$

Solution

(a) Here $A = x$ and $B = 3$, so $2AB = 2 \cdot x \cdot 3 = 6x$. Since the middle term is $6x$, the trinomial is a perfect square. By the Perfect Square Formula we have

$$x^2 + 6x + 9 = (x + 3)^2$$

(b) Here $A = 2x$ and $B = y$, so $2AB = 2 \cdot 2x \cdot y = 4xy$. Since the middle term is $-4xy$, the trinomial is a perfect square. By the Perfect Square Formula we have

$$4x^2 - 4xy + y^2 = (2x - y)^2$$

Now Try Exercises 111 and 113 ■

Example 12 ■ Factoring Differences and Sums of Cubes

Factor each polynomial.

(a) $27x^3 - 1$ **(b)** $x^6 + 8$

Solution

(a) Using the Difference of Cubes Formula with $A = 3x$ and $B = 1$, we get

$$\begin{aligned} 27x^3 - 1 &= (3x)^3 - 1^3 = (3x - 1)[(3x)^2 + (3x)(1) + 1^2] \\ &= (3x - 1)(9x^2 + 3x + 1) \end{aligned}$$

(b) Using the Sum of Cubes Formula with $A = x^2$ and $B = 2$, we have

$$x^6 + 8 = (x^2)^3 + 2^3 = (x^2 + 2)(x^4 - 2x^2 + 4)$$

Now Try Exercises 83 and 85 ■

When we factor an expression, the result can sometimes be factored further. In general, *we first factor out common factors*, then inspect the result to see whether it can be factored by any of the other methods of this section. We repeat this process until we have factored the expression completely.

Example 13 ■ Factoring an Expression Completely

Factor each expression completely.

(a) $2x^4 - 8x^2$ **(b)** $x^5y^2 - xy^6$

Solution

(a) We first factor out the power of x with the smallest exponent.

$$2x^4 - 8x^2 = 2x^2(x^2 - 4) \quad \text{Common factor is } 2x^2$$
$$= 2x^2(x + 2)(x - 2) \quad \text{Factor } x^2 - 4 \text{ as a difference of squares}$$

(b) We first factor out the powers of x and y with the smallest exponents.

$$x^5y^2 - xy^6 = xy^2(x^4 - y^4) \quad \text{Common factor is } xy^2$$
$$= xy^2(x^2 + y^2)(x^2 - y^2) \quad \text{Factor } x^4 - y^4 \text{ as a difference of squares}$$
$$= xy^2(x^2 + y^2)(x + y)(x - y) \quad \text{Factor } x^2 - y^2 \text{ as a difference of squares}$$

 Now Try Exercises 121 and 123 ■

In the next example we factor out variables with fractional exponents. This type of factoring occurs in calculus.

Example 14 ■ Factoring Expressions with Fractional Exponents

Factor each expression.

(a) $3x^{3/2} - 9x^{1/2} + 6x^{-1/2}$ **(b)** $(2 + x)^{-2/3}x + (2 + x)^{1/3}$

Solution

(a) Factor out the power of x with the *smallest exponent*, that is, $x^{-1/2}$.

$$3x^{3/2} - 9x^{1/2} + 6x^{-1/2} = 3x^{-1/2}(x^2 - 3x + 2) \quad \text{Factor out } 3x^{-1/2}$$
$$= 3x^{-1/2}(x - 1)(x - 2) \quad \text{Factor the quadratic } x^2 - 3x + 2$$

To factor out $x^{-1/2}$ from $x^{3/2}$, we *subtract* exponents:

$$x^{3/2} = x^{-1/2}(x^{3/2-(-1/2)})$$
$$= x^{-1/2}(x^{3/2+1/2})$$
$$= x^{-1/2}(x^2)$$

(b) Factor out the power of $2 + x$ with the *smallest exponent*, that is, $(2 + x)^{-2/3}$.

$$(2 + x)^{-2/3}x + (2 + x)^{1/3} = (2 + x)^{-2/3}[x + (2 + x)] \quad \text{Factor out } (2 + x)^{-2/3}$$
$$= (2 + x)^{-2/3}(2 + 2x) \quad \text{Simplify}$$
$$= 2(2 + x)^{-2/3}(1 + x) \quad \text{Factor out 2}$$

Check Your Answers

To see that you have factored correctly, multiply using the Laws of Exponents.

(a) $3x^{-1/2}(x^2 - 3x + 2)$
$= 3x^{3/2} - 9x^{1/2} + 6x^{-1/2}$ ✓

(b) $(2 + x)^{-2/3}[x + (2 + x)]$
$= (2 + x)^{-2/3}x + (2 + x)^{1/3}$ ✓

 Now Try Exercises 97 and 99 ■

■ Factoring by Grouping Terms

Polynomials with at least four terms (quadrinomials) can sometimes be factored by grouping terms. The following example illustrates the idea.

Example 15 ■ Factoring by Grouping

Factor each polynomial.

(a) $x^3 + x^2 + 4x + 4$ **(b)** $x^3 - 2x^2 - 9x + 18$

Solution

(a) $x^3 + x^2 + 4x + 4 = (x^3 + x^2) + (4x + 4)$ Group terms

$= x^2(x + 1) + 4(x + 1)$ Factor out common factors

$= (x^2 + 4)(x + 1)$ Factor $x + 1$ from each term

(b) $x^3 - 2x^2 - 9x + 18 = (x^3 - 2x^2) - (9x - 18)$ Group terms

$= x^2(x - 2) - 9(x - 2)$ Factor out common factors

$= (x^2 - 9)(x - 2)$ Factor $x - 2$from each term

$= (x + 3)(x - 3)(x - 2)$ Factor completely

Now Try Exercises 89 and 125 ■

1.3 Exercises

Concepts

1. The greatest common factor in the expression $18x^3 + 30x$ is ________, and the expression factors as ___ (___ + ___).

2. Consider the polynomial $2x^3 + 3x^2 + 10x$.

(a) How many terms does this polynomial have? ________

List the terms: ________________.

(b) What factor is common to all the terms? ________

Factor the polynomial: ________________.

3. To factor the trinomial $x^2 + 8x + 12$, we look for two integers whose product is ______ and whose sum is ______. These integers are ______ and ______, so the trinomial factors as ________________.

4. The Special Product Formula for the "square of a sum" is $(A + B)^2 =$ ________________.
So $(2x + 3)^2 =$ ________________.

5. The Special Product Formula for the "product of the sum and difference of terms" is $(A + B)(A - B) =$ ________________.
So $(6 + x)(6 - x) =$ ________________.

6. The Special Factoring Formula for the "difference of squares" is $A^2 - B^2 =$ ________________. So $49x^2 - 9$ factors as ________________.

7. The Special Factoring Formula for a "perfect square" is $A^2 + 2AB + B^2 =$ ________________. So $x^2 + 10x + 25$ factors as ________________.

8. *Yes or No?* If *No*, give a reason.

(a) Is the expression $(x + 5)^2$ equal to $x^2 + 25$?

(b) When you expand $(x + a)^2$, where $a \neq 0$, do you get three terms?

(c) Is the expression $(x + 5)(x - 5)$ equal to $x^2 - 25$?

(d) When you expand $(x + a)(x - a)$, where $a \neq 0$, do you get three terms?

Skills

9–14 ■ Polynomials Complete the following table by stating whether the polynomial is a monomial, binomial, or trinomial; then list its terms and state its degree.

Polynomial	Type	Terms	Degree
9. $5x^3 + 6$			
10. $-2x^2 + 5x - 3$			
11. -8			
12. $\frac{1}{2}x^7$			
13. $x - x^2 + x^3 - x^4$			
14. $\sqrt{2}x - \sqrt{3}$			

15–26 ■ Polynomials Find the sum, difference, or product.

15. $(12x - 7) - (5x - 12)$ **16.** $(5 - 3x) + (2x - 8)$

17. $(-2x^2 - 3x + 1) + (3x^2 + 5x - 4)$

18. $(3x^2 + x + 1) - (2x^2 - 3x - 5)$

19. $(5x^3 + 4x^2 - 3x) - (x^2 + 7x + 2)$

20. $3(x - 1) + 4(x + 2)$

21. $8(2x + 5) - 7(x - 9)$

22. $4(x^2 - 3x + 5) - 3(x^2 - 2x + 1)$

23. $x^3(x^2 + 3x) - 2x(x^4 - 3x^2)$

24. $4x(1 - x^3) + 3x^3(x^3 - x)$

25. $4x(x - 2) - 2(x^2 - 4x) + 2x^2(x - 1)$

26. $6x^3(x^2 - 1) - 2x(2 + 3x^2) + 2(2x - 4)$

27–32 ■ Using FOIL Multiply the algebraic expressions using the FOIL method and simplify.

27. $(3t - 2)(7t - 4)$ **28.** $(4s - 1)(2s + 5)$

29. $(3x + 5)(2x - 1)$ **30.** $(7y - 3)(2y - 1)$

31. $(x + 3y)(2x - y)$ **32.** $(4x - 5y)(3x - y)$

33–46 ■ Using Special Product Formulas Multiply the algebraic expressions using a Special Product Formula and simplify.

33. $(4x + 3)^2$ **34.** $(2 - 7y)^2$

35. $(y - 3x)^2$ **36.** $(5x - y)^2$

37. $(2x + 3y)^2$ **38.** $(r - 2s)^2$

39. $(w + 7)(w - 7)$ **40.** $(5 - y)(5 + y)$

41. $(3x - 4)(3x + 4)$ **42.** $(2y + 5)(2y - 5)$

43. $(\sqrt{x} + 2)(\sqrt{x} - 2)$ **44.** $(\sqrt{y} + \sqrt{2})(\sqrt{y} - \sqrt{2})$

45. $(y + 2)^3$ **46.** $(x - 3)^3$

47–64 ■ Multiplying Algebraic Expressions Perform the indicated operations and simplify.

47. $(x + 2)(x^2 + 2x + 3)$ **48.** $(x + 1)(2x^2 - x + 1)$

49. $(2x - 5)(x^2 - x + 1)$ **50.** $(1 + 2x)(x^2 - 3x + 1)$

51. $\sqrt{x}(x - \sqrt{x})$ **52.** $x^{3/2}\left(\sqrt{x} - \frac{1}{\sqrt{x}}\right)$

53. $y^{1/3}(y^{2/3} + y^{5/3})$ **54.** $x^{1/4}(2x^{3/4} - x^{1/4})$

55. $(x^{1/2} - y^{1/2})^2$ **56.** $\left(\sqrt{u} + \frac{1}{\sqrt{u}}\right)^2$

57. $(x^2 - a^2)(x^2 + a^2)$

58. $(x^{1/2} + y^{1/2})(x^{1/2} - y^{1/2})$

59. $(\sqrt{a} - b)(\sqrt{a} + b)$

60. $(\sqrt{h^2 + 1} + 1)(\sqrt{h^2 + 1} - 1)$

61. $[(x - 1) + x^2][(x - 1) - x^2]$

62. $[x + (2 + x^2)][x - (2 + x^2)]$

63. $(2x + y - 3)(2x + y + 3)$

64. $(x + y + z)(x - y - z)$

65–72 ■ Factoring a Common Factor Factor out the common factor.

65. $4x^4 - x^2$ **66.** $3x^4 - 6x^3 - x^2$

67. $y(y - 6) + 9(y - 6)$ **68.** $(z + 2)^2 - 5(z + 2)$

69. $4x^3y^2 - 6xy^3 + 8x^2y^4$ **70.** $-7x^4y^2 + 14xy^3 + 21xy^4$

71. $(x + 3)^5(x + 2)^2 - (x + 3)^4(x + 2)^3$

72. $3(2x - 1)^3(x^2 + 1)^4 - (2x - 1)^4(x^2 + 1)^3$

73–80 ■ Factoring Trinomials Factor the trinomial.

73. $z^2 - 11z + 18$ **74.** $x^2 + 4x - 5$

75. $10x^2 - 19x + 6$ **76.** $6y^2 + 11y - 21$

77. $3x^2 - 16x + 5$ **78.** $5x^2 - 7x - 6$

79. $(3x + 2)^2 + 8(3x + 2) + 12$

80. $2(a + b)^2 + 5(a + b) - 3$

81–88 ■ Using Special Factoring Formulas Use a Special Factoring Formula to factor the expression.

81. $36a^2 - 49$ **82.** $(x + 3)^2 - 4$

83. $27x^3 + y^3$ **84.** $a^3 - b^6$

85. $8s^3 - 125t^3$ **86.** $1 + 1000y^3$

87. $x^2 + 12x + 36$ **88.** $16z^2 - 24z + 9$

89–94 ■ Factoring by Grouping Factor the expression by grouping terms.

89. $x^3 + 4x^2 + x + 4$ **90.** $3x^3 - x^2 + 6x - 2$

91. $5x^3 + x^2 + 5x + 1$ **92.** $18x^3 + 9x^2 + 2x + 1$

93. $x^3 + x^2 + x + 1$ **94.** $x^5 + x^4 + x + 1$

95–100 ■ Fractional Exponents Factor the expression by factoring out the lowest power of each common factor.

95. $x^{2/3} + 3x^{5/3}$ **96.** $x^{3/4} - 5x^{-1/4}$

97. $x^{-3/2} - x^{-1/2} + x^{1/2}$ **98.** $x^{5/3} + x^{2/3} + 2x^{-1/3}$

99. $(x^2 + 1)^{1/2} + 2(x^2 + 1)^{-1/2}$

100. $x^{-1/2}(x + 1)^{1/2} + x^{1/2}(x + 1)^{-1/2}$

101–134 ■ Factoring Completely Recognize the type of expression and use an appropriate method to factor it completely.

101. $2x + 12x^3$ **102.** $12x^2 + 3x^3$

103. $x^2 - 2x - 8$ **104.** $x^2 - 14x + 48$

105. $2x^2 + 5x + 3$ **106.** $2x^2 + 7x - 4$

107. $9x^2 - 36x - 45$ **108.** $8x^2 + 10x + 3$

109. $49 - 4y^2$ **110.** $4t^2 - 9s^2$

111. $t^2 - 6t + 9$ **112.** $x^2 + 10x + 25$

113. $y^2 - 10yz + 25z^2$ **114.** $r^2 - 6rs + 9s^2$

115. $(a + b)^2 - (a - b)^2$ **116.** $\left(1 + \frac{1}{x}\right)^2 - \left(1 - \frac{1}{x}\right)^2$

117. $x^2(x^2 - 1) - 9(x^2 - 1)$

118. $(a^2 - 1)(b - 2)^2 - 4(a^2 - 1)$

119. $8x^3 - 125$ **120.** $x^6 + 64$

121. $x^3 + 2x^2 + x$ **122.** $3x^3 - 27x$

123. $x^4y^3 - x^2y^5$ **124.** $18y^3x^2 - 2xy^4$

125. $3x^3 - x^2 - 12x + 4$ **126.** $9x^3 + 18x^2 - x - 2$

127. $x^{-3/2} + 2x^{-1/2} + x^{1/2}$

128. $(x - 1)^{7/2} - (x - 1)^{3/2}$

129. $(x - 1)(x + 2)^2 - (x - 1)^2(x + 2)$

130. $y^4(y + 2)^3 + y^5(y + 2)^4$

131. $(a^2 + 1)^2 - 7(a^2 + 1) + 10$

132. $(a^2 + 2a)^2 - 2(a^2 + 2a) - 3$

133. $(x^2 + 3)^2(x - 1)^3 - (4x + 1)^2(x - 1)^3$

134. $(x + 1)^2(x + 2) - 6(x + 1)(x + 2) + 9(x + 2)$

135–138 ■ Factoring Completely Factor the expression completely. (This type of expression arises in calculus when using the "Product Rule.")

135. $5(x^2 + 4)^4(2x)(x - 2)^4 + (x^2 + 4)^5(4)(x - 2)^3$

136. $3(2x - 1)^2(2)(x + 3)^{1/2} + (2x - 1)^3(\tfrac{1}{2})(x + 3)^{-1/2}$

137. $(x^2 + 3)^{-1/3} - \frac{2}{3}x^2(x^2 + 3)^{-4/3}$

138. $\frac{1}{2}x^{-1/2}(3x + 4)^{1/2} - \frac{3}{2}x^{1/2}(3x + 4)^{-1/2}$

Skills Plus

139–140 ■ Verifying Formulas Show that the following formulas hold.

139. (a) $ab = \frac{1}{2}[(a + b)^2 - (a^2 + b^2)]$
(b) $(a^2 + b^2)^2 - (a^2 - b^2)^2 = 4a^2b^2$

140. $(a^2 + b^2)(c^2 + d^2) = (ac + bd)^2 + (ad - bc)^2$

141. Factoring Completely Factor completely:

$$4a^2c^2 - (a^2 - b^2 + c^2)^2$$

142. Factoring $x^4 + ax^2 + b$ A trinomial of the form $x^4 + ax^2 + b$ can sometimes be factored easily. For example,

$$x^4 + 3x^2 - 4 = (x^2 + 4)(x^2 - 1)$$

But $x^4 + 3x^2 + 4$ cannot be factored in this way. Instead, we can use the following method.

$x^4 + 3x^2 + 4 = (x^4 + 4x^2 + 4) - x^2$	Add and subtract x^2
$= (x^2 + 2)^2 - x^2$	Factor perfect square
$= [(x^2 + 2) - x][(x^2 + 2) + x]$	Difference of squares
$= (x^2 - x + 2)(x^2 + x + 2)$	

Factor each trinomial, using whichever method is appropriate.

(a) $x^4 + x^2 - 2$ **(b)** $x^4 + 2x^2 + 9$
(c) $x^4 + 4x^2 + 16$ **(d)** $x^4 + 2x^2 + 1$

Applications

143. Volume of Concrete A culvert is constructed out of large cylindrical shells cast in concrete, as shown in the figure. Using the formula for the volume of a cylinder given on the inside front cover of this book, explain why the volume of the cylindrical shell is

$$V = \pi R^2h - \pi r^2h$$

Factor to show that

$$V = 2\pi \cdot \text{average radius} \cdot \text{height} \cdot \text{thickness}$$

Use the "unrolled" diagram to explain why this makes sense geometrically.

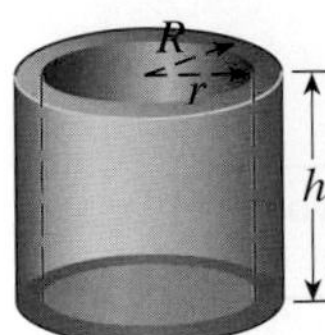

144. Mowing a Field A square field in a certain state park is mowed around the edges every week. The rest of the field is kept unmowed to serve as habitat for birds and small animals (see the figure). The field measures b meters by b meters, and the mowed strip is x meters wide.

(a) Explain why the area of the mowed portion is $b^2 - (b - 2x)^2$.

(b) Factor the expression in part (a) to show that the area of the mowed portion is also $4x(b - x)$. Do you see that the region with mowed grass consists of four rectangles, each with area $x(b - x)$?

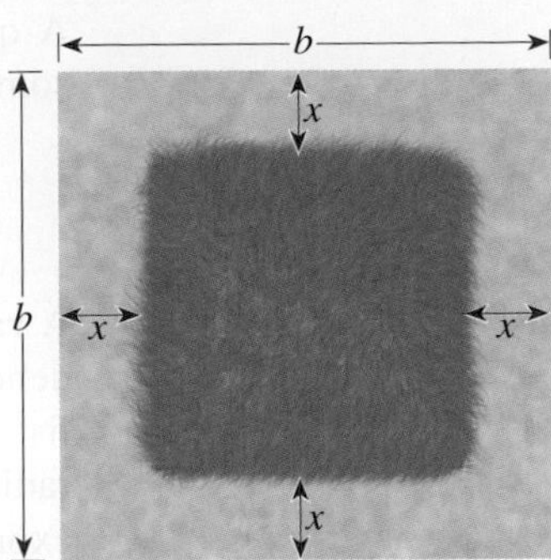

Discuss ■ Discover ■ Prove ■ Write

145. Discover: Degree of a Sum or Product of Polynomials Make up several pairs of polynomials, then calculate the sum and product of each pair. On the basis of your experiments and observations, answer the following questions.

(a) How is the degree of the product related to the degrees of the original polynomials?

(b) How is the degree of the sum related to the degrees of the original polynomials?

146. Discuss: The Power of Algebraic Formulas Use the formula $A^2 - B^2 = (A + B)(A - B)$ to evaluate the following without using a calculator.

(a) $528^2 - 527^2$
(b) $1020^2 - 1010^2$
(c) $501 \cdot 499$
(d) $1002 \cdot 998$

147. Discover: Factoring $A^n - 1$

(a) Verify the following formulas by expanding and simplifying the right-hand side.

$$A^2 - 1 = (A - 1)(A + 1)$$
$$A^3 - 1 = (A - 1)(A^2 + A + 1)$$
$$A^4 - 1 = (A - 1)(A^3 + A^2 + A + 1)$$

(b) Try to recognize a pattern for the formulas listed in part (a). On the basis of your pattern, how do you think $A^5 - 1$ factors? Verify your conjecture. Now generalize the pattern you have observed to obtain a factoring formula for $A^n - 1$, where n is a positive integer.

148. Prove: Special Factoring Formulas Prove each of the following formulas by expanding the right-hand side.

(a) Difference of Cubes:
$A^3 - B^3 = (A - B)(A^2 + AB + B^2)$

(b) Sum of Cubes:
$A^3 + B^3 = (A + B)(A^2 - AB + B^2)$

1.4 Rational Expressions

■ The Domain of an Algebraic Expression ■ Simplifying Rational Expressions ■ Multiplying and Dividing Rational Expressions ■ Adding and Subtracting Rational Expressions ■ Compound Fractions ■ Rationalizing the Denominator or the Numerator ■ Avoiding Common Errors

A quotient of two algebraic expressions is called a **fractional expression**. Here are some examples:

$$\frac{2x}{x-1} \qquad \frac{y-2}{y^2+4} \qquad \frac{x^3-x}{x^2-5x+6} \qquad \frac{x}{\sqrt{x^2+1}}$$

A **rational expression** is a fractional expression in which both the numerator and the denominator are polynomials. For example, the first three expressions in the above list are rational expressions, but the fourth is not because its denominator contains a radical. In this section we learn how to perform algebraic operations on rational expressions.

■ The Domain of an Algebraic Expression

Expression	Domain
$\frac{1}{x}$	$\{x \mid x \neq 0\}$
$\sqrt{x}$	$\{x \mid x \geq 0\}$
$\frac{1}{\sqrt{x}}$	$\{x \mid x > 0\}$

In general, an algebraic expression may not be defined for all values of the variable. The **domain** of an algebraic expression is the set of real numbers that the variable is permitted to have. The table in the margin gives some basic expressions and their domains.

Example 1 ■ Finding the Domain of an Expression

Find the domain of each of the following expressions.

(a) $2x^2 + 3x - 1$ **(b)** $\frac{x}{x^2 - 5x + 6}$ **(c)** $\frac{\sqrt{x}}{x-5}$

Solution

(a) This polynomial is defined for every x. Thus the domain is the set $\mathbb{R}$ of real numbers.

(b) We first factor the denominator.

$$\frac{x}{x^2-5x+6} = \frac{x}{(x-2)(x-3)}$$

Denominator would be 0 if $x = 2$ or $x = 3$

Since the denominator is zero when $x = 2$ or 3, the expression is not defined for these numbers. The domain is $\{x \mid x \neq 2 \text{ and } x \neq 3\}$.

(c) For the numerator to be defined, we must have $x \geq 0$. Also, we cannot divide by zero, so $x \neq 5$.

Must have $x \geq 0$ to take square root

$$\frac{\sqrt{x}}{x-5}$$

Denominator would be 0 if $x = 5$

Thus the domain is $\{x \mid x \geq 0 \text{ and } x \neq 5\}$.

Now Try Exercise 13 ■

■ Simplifying Rational Expressions

To **simplify a rational expression**, we factor both numerator and denominator and use the following property of fractions:

$$\frac{AC}{BC} = \frac{A}{B}$$

This allows us to **cancel** common factors from the numerator and denominator.

Example 2 ■ Simplifying Rational Expressions by Cancellation

Simplify: $\dfrac{x^2 - 1}{x^2 + x - 2}$

Solution

 We can't cancel the x^2 terms in $\dfrac{x^2 - 1}{x^2 + x - 2}$ because x^2 is not a factor.

$$\frac{x^2 - 1}{x^2 + x - 2} = \frac{(x - 1)(x + 1)}{(x - 1)(x + 2)} \qquad \text{Factor}$$

$$= \frac{x + 1}{x + 2} \qquad \text{Cancel common factors}$$

Now Try Exercise 21 ■

■ Multiplying and Dividing Rational Expressions

To **multiply rational expressions**, we use the following property of fractions:

$$\frac{A}{B} \cdot \frac{C}{D} = \frac{AC}{BD}$$

This says that to multiply two fractions, we multiply their numerators and multiply their denominators.

Example 3 ■ Multiplying Rational Expressions

Perform the indicated multiplication and simplify: $\dfrac{x^2 + 2x - 3}{x^2 + 8x + 16} \cdot \dfrac{3x + 12}{x - 1}$

Solution We first factor.

$$\frac{x^2 + 2x - 3}{x^2 + 8x + 16} \cdot \frac{3x + 12}{x - 1} = \frac{(x - 1)(x + 3)}{(x + 4)^2} \cdot \frac{3(x + 4)}{x - 1} \qquad \text{Factor}$$

$$= \frac{3(x - 1)(x + 3)(x + 4)}{(x - 1)(x + 4)^2} \qquad \text{Property of fractions}$$

$$= \frac{3(x + 3)}{x + 4} \qquad \text{Cancel common factors}$$

Now Try Exercise 29 ■

To **divide rational expressions**, we use the following property of fractions:

$$\frac{A}{B} \div \frac{C}{D} = \frac{A}{B} \cdot \frac{D}{C}$$

This says that to divide one fraction by another fraction, we invert the divisor and multiply.

Example 4 ■ Dividing Rational Expressions

Perform the indicated division and simplify: $\dfrac{x-4}{x^2-4} \div \dfrac{x^2-3x-4}{x^2+5x+6}$

Solution

$$\frac{x-4}{x^2-4} \div \frac{x^2-3x-4}{x^2+5x+6} = \frac{x-4}{x^2-4} \cdot \frac{x^2+5x+6}{x^2-3x-4} \qquad \text{Invert and multiply}$$

$$= \frac{(x-4)(x+2)(x+3)}{(x-2)(x+2)(x-4)(x+1)} \qquad \text{Factor}$$

$$= \frac{x+3}{(x-2)(x+1)} \qquad \text{Cancel common factors}$$

 Now Try Exercise 35 ■

■ Adding and Subtracting Rational Expressions

To **add or subtract rational expressions**, we first find a common denominator and then use the following property of fractions:

$$\frac{A}{C} + \frac{B}{C} = \frac{A+B}{C}$$

Avoid making the following error:

$$\frac{A}{B+C} \;\text{✗}\; \frac{A}{B} + \frac{A}{C}$$

For instance, if we let $A = 2$, $B = 1$, and $C = 1$, then we see the error:

$$\frac{2}{1+1} \stackrel{?}{=} \frac{2}{1} + \frac{2}{1}$$

$$\frac{2}{2} \stackrel{?}{=} 2 + 2$$

$$1 \stackrel{?}{=} 4 \quad \text{Wrong!}$$

Although any common denominator will work, it is most efficient to use the **least common denominator** (LCD), as explained in Section 1.1. The LCD is found by factoring each denominator and taking the product of the distinct factors, using the highest power that appears in any of the factors.

Example 5 ■ Adding and Subtracting Rational Expressions

Perform the indicated operations and simplify.

(a) $\dfrac{3}{x-1} + \dfrac{x}{x+2}$ **(b)** $\dfrac{1}{x^2-1} - \dfrac{2}{(x+1)^2}$

Solution

(a) Here the LCD is simply the product $(x-1)(x+2)$.

$$\frac{3}{x-1} + \frac{x}{x+2} = \frac{3(x+2)}{(x-1)(x+2)} + \frac{x(x-1)}{(x-1)(x+2)} \qquad \text{Write fractions using LCD}$$

$$= \frac{3x+6+x^2-x}{(x-1)(x+2)} \qquad \text{Add fractions}$$

$$= \frac{x^2+2x+6}{(x-1)(x+2)} \qquad \text{Combine terms in numerator}$$

(b) The LCD of $x^2 - 1 = (x - 1)(x + 1)$ and $(x + 1)^2$ is $(x - 1)(x + 1)^2$.

$$\frac{1}{x^2 - 1} - \frac{2}{(x + 1)^2} = \frac{1}{(x - 1)(x + 1)} - \frac{2}{(x + 1)^2} \qquad \text{Factor}$$

$$= \frac{(x + 1) - 2(x - 1)}{(x - 1)(x + 1)^2} \qquad \text{Combine fractions using LCD}$$

$$= \frac{x + 1 - 2x + 2}{(x - 1)(x + 1)^2} \qquad \text{Distributive Property}$$

$$= \frac{3 - x}{(x - 1)(x + 1)^2} \qquad \text{Combine terms in numerator}$$

 **Now Try Exercises 45 and 47** ■

■ Compound Fractions

A **compound fraction** is a fraction in which the numerator, the denominator, or both, are themselves fractional expressions.

Example 6 ■ Simplifying a Compound Fraction

Simplify: $\dfrac{\dfrac{x}{y} + 1}{1 - \dfrac{y}{x}}$

Solution 1 We combine the terms in the numerator into a single fraction. We do the same in the denominator. Then we invert and multiply.

$$\frac{\dfrac{x}{y} + 1}{1 - \dfrac{y}{x}} = \frac{\dfrac{x + y}{y}}{\dfrac{x - y}{x}} = \frac{x + y}{y} \cdot \frac{x}{x - y}$$

$$= \frac{x(x + y)}{y(x - y)}$$

Mathematics in the Modern World

NASA

Error-Correcting Codes

The pictures sent back by the spacecraft rover *Perseverance* from the surface of Mars are astoundingly clear. But few viewing these pictures are aware of the complex mathematics used to accomplish this feat. The distance to Mars is enormous, and the background noise (or static) is many times stronger than the original signal emitted by the spacecraft. So when scientists receive the signal, it is full of errors. To get a clear picture, the errors must be found and corrected. This same problem of errors is routinely encountered in transmitting data over the Internet.

To understand how errors are found and corrected, we must first understand that to transmit pictures, sound, or text, we transform them into bits (the digits 0 or 1; see Mathematics in the Modern World, Section 1.3). To help the receiver recognize errors, the message is "coded" by inserting additional bits. For example, suppose you want to transmit the message "10100." A very simple-minded code is as follows: Send each digit a million times. The person receiving the message reads it in blocks of a million digits. If the first block is mostly 1's, he concludes that you are probably trying to transmit a 1, and so on. To say that this code is not efficient is a bit of an understatement; it requires sending a million times more data than the original message. Another method inserts "check digits." For example, for each block of eight digits insert a ninth digit; the inserted digit is 0 if there is an even number of 1's in the block and 1 if there is an odd number. So if a single digit is wrong (a 0 changed to a 1 or vice versa), the check digits allow us to recognize that an error has occurred. This method does not tell us where the error is, so we can't correct it. Modern error-correcting codes use mathematical algorithms that require inserting relatively few digits but that allow the receiver to not only recognize, but also correct, errors. The first error-correcting code was developed in the 1940s by Richard Hamming at MIT. The English language has a built-in error-correcting mechanism; to test it, try reading this error-filled sentence: Gve mo libty ox giv ne deth.

Solution 2 We find the LCD of all the fractions in the numerator and denominator of the expression, then multiply numerator and denominator by it. In this example the LCD of all the fractions is xy. Thus

$$\frac{\frac{x}{y}+1}{1-\frac{y}{x}} = \frac{\frac{x}{y}+1}{1-\frac{y}{x}}\cdot\frac{xy}{xy} \quad \text{Multiply numerator and denominator by } xy$$

$$= \frac{x^2+xy}{xy-y^2} \quad \text{Distribute and simplify}$$

$$= \frac{x(x+y)}{y(x-y)} \quad \text{Factor}$$

Now Try Exercises 61 and 67 ■

The next two examples show situations in calculus that require the ability to work with fractional expressions.

Example 7 ■ Simplifying a Compound Fraction

Simplify: $\dfrac{\frac{1}{a+h}-\frac{1}{a}}{h}$

Solution We begin by combining the fractions in the numerator using a common denominator.

We can also simplify by multiplying the numerator and the denominator by $a(a+h)$.

$$\frac{\frac{1}{a+h}-\frac{1}{a}}{h} = \frac{\frac{a-(a+h)}{a(a+h)}}{h} \quad \text{Combine fractions in the numerator}$$

$$= \frac{a-(a+h)}{a(a+h)}\cdot\frac{1}{h} \quad \text{Property of fractions (invert divisor and multiply)}$$

$$= \frac{a-a-h}{a(a+h)}\cdot\frac{1}{h} \quad \text{Distributive Property}$$

$$= \frac{-h}{a(a+h)}\cdot\frac{1}{h} \quad \text{Simplify}$$

$$= \frac{-1}{a(a+h)} \quad \text{Property of fractions (cancel common factors)}$$

Now Try Exercise 75 ■

Example 8 ■ Simplifying a Compound Fraction

Simplify: $\dfrac{(1+x^2)^{1/2}-x^2(1+x^2)^{-1/2}}{1+x^2}$

Solution 1 Factor $(1+x^2)^{-1/2}$ from the numerator.

Factor out the power of $1+x^2$ with the *smallest* exponent, in this case $(1+x^2)^{-1/2}$.

$$\frac{(1+x^2)^{1/2}-x^2(1+x^2)^{-1/2}}{1+x^2} = \frac{(1+x^2)^{-1/2}[(1+x^2)-x^2]}{1+x^2}$$

$$= \frac{(1+x^2)^{-1/2}}{1+x^2} = \frac{1}{(1+x^2)^{3/2}}$$

Solution 2 Since $(1 + x^2)^{-1/2} = 1/(1 + x^2)^{1/2}$ is a fraction, we can clear all fractions by multiplying numerator and denominator by $(1 + x^2)^{1/2}$.

$$\frac{(1 + x^2)^{1/2} - x^2(1 + x^2)^{-1/2}}{1 + x^2} = \frac{(1 + x^2)^{1/2} - x^2(1 + x^2)^{-1/2}}{1 + x^2} \cdot \frac{(1 + x^2)^{1/2}}{(1 + x^2)^{1/2}}$$

$$= \frac{(1 + x^2) - x^2}{(1 + x^2)^{3/2}} = \frac{1}{(1 + x^2)^{3/2}}$$

 Now Try Exercise 83 ■

■ Rationalizing the Denominator or the Numerator

In Section 1.2 we learned how to rationalize a denominator of the form $\sqrt{C}$.

If a fraction has a denominator of the form $A + B\sqrt{C}$, we can rationalize the denominator by multiplying numerator and denominator by the **conjugate radical** $A - B\sqrt{C}$. This works because, by Special Product Formula 1 in Section 1.3, the product of the denominator and its conjugate radical does not contain a radical:

$$(A + B\sqrt{C})(A - B\sqrt{C}) = A^2 - B^2C$$

Example 9 ■ Rationalizing the Denominator

Rationalize the denominator: $\dfrac{1}{1 + \sqrt{2}}$

Solution We multiply both the numerator and the denominator by the conjugate radical of $1 + \sqrt{2}$, which is $1 - \sqrt{2}$.

$$\frac{1}{1 + \sqrt{2}} = \frac{1}{1 + \sqrt{2}} \cdot \frac{1 - \sqrt{2}}{1 - \sqrt{2}}$$ Multiply numerator and denominator by the conjugate radical

$$= \frac{1 - \sqrt{2}}{1^2 - (\sqrt{2})^2}$$ Special Product Formula 1

Special Product Formula 1
$(A + B)(A - B) = A^2 - B^2$

$$= \frac{1 - \sqrt{2}}{1 - 2} = \frac{1 - \sqrt{2}}{-1} = \sqrt{2} - 1$$

Now Try Exercise 87 ■

Example 10 ■ Rationalizing the Numerator

Rationalize the numerator: $\dfrac{\sqrt{4 + h} - 2}{h}$

Solution We multiply both numerator and denominator by the conjugate radical $\sqrt{4 + h} + 2$.

$$\frac{\sqrt{4 + h} - 2}{h} = \frac{\sqrt{4 + h} - 2}{h} \cdot \frac{\sqrt{4 + h} + 2}{\sqrt{4 + h} + 2}$$ Multiply numerator and denominator by the conjugate radical

$$= \frac{(\sqrt{4 + h})^2 - 2^2}{h(\sqrt{4 + h} + 2)}$$ Special Product Formula 1

Special Product Formula 1
$(A + B)(A - B) = A^2 - B^2$

$$= \frac{4 + h - 4}{h(\sqrt{4 + h} + 2)}$$

$$= \frac{h}{h(\sqrt{4 + h} + 2)} = \frac{1}{\sqrt{4 + h} + 2}$$ Property of fractions (cancel common factors)

Now Try Exercise 93 ■

■ Avoiding Common Errors

 Don't make the mistake of applying properties of multiplication to the operation of addition. Many of the common errors in algebra involve doing just that. The following table states several properties of multiplication and illustrates the error in applying them to addition.

Multiplication Property	Common Error with Addition
$(a \cdot b)^2 = a^2 \cdot b^2$	$(a+b)^2 \neq a^2 + b^2$
$\sqrt{a \cdot b} = \sqrt{a}\sqrt{b} \quad (a, b \geq 0)$	$\sqrt{a+b} \neq \sqrt{a} + \sqrt{b}$
$\sqrt{a^2 \cdot b^2} = a \cdot b \quad (a, b \geq 0)$	$\sqrt{a^2+b^2} \neq a + b$
$\dfrac{1}{a} \cdot \dfrac{1}{b} = \dfrac{1}{a \cdot b}$	$\dfrac{1}{a} + \dfrac{1}{b} \neq \dfrac{1}{a+b}$
$\dfrac{ab}{a} = b$	$\dfrac{a+b}{a} \neq b$
$(a \cdot b)^{-1} = a^{-1} \cdot b^{-1}$	$(a+b)^{-1} \neq a^{-1} + b^{-1}$

To verify that the equations in the right-hand column are wrong, simply substitute numbers for a and b and calculate each side. For example, if we take $a = 2$ and $b = 2$ in the fourth error, we get different values for the left- and right-hand sides:

$$\frac{1}{a} + \frac{1}{b} = \frac{1}{2} + \frac{1}{2} = 1 \qquad \frac{1}{a+b} = \frac{1}{2+2} = \frac{1}{4}$$

Left-hand side Right-hand side

Since $1 \neq \frac{1}{4}$, the stated equation is wrong. You should similarly convince yourself of the error in each of the other equations. (See Exercises 103 and 104.)

1.4 Exercises

Concepts

1. What is a rational expression? Which of the following are rational expressions?

(a) $\dfrac{3x}{x^2 - 1}$ (b) $\dfrac{\sqrt{x+1}}{2x+3}$ (c) $\dfrac{x(x^2-1)}{x+3}$

2. To simplify a rational expression, we cancel *factors* that are common to the ________ and ________. So the expression

$$\frac{(x+1)(x+2)}{(x+3)(x+2)}$$

simplifies to ________.

3. To multiply two rational expressions, we multiply their ________ together and multiply their ________ together.

So $\dfrac{2}{x+1} \cdot \dfrac{x}{x+3}$ is the same as ________.

4. Consider the expression $\dfrac{1}{x} - \dfrac{2}{x+1} - \dfrac{x}{(x+1)^2}$.

(a) How many terms does this expression have?

(b) Find the least common denominator of all the terms.

(c) Perform the addition and simplify.

5–6 ■ *Yes or No?* If *No*, give a reason. (Disregard any value that makes a denominator zero.)

5. (a) Is the expression $\dfrac{x(x+1)}{(x+1)^2}$ equal to $\dfrac{x}{x+1}$?

(b) Is the expression $\sqrt{x^2+25}$ equal to $x + 5$?

6. (a) Is the expression $\dfrac{3+a}{3}$ equal to $1 + \dfrac{a}{3}$?

(b) Is the expression $\dfrac{2}{4+x}$ equal to $\dfrac{1}{2} + \dfrac{2}{x}$?

Skills

7–16 ■ Domain Find the domain of the expression.

7. $4x^2 - 10x + 3$

8. $-x^4 + x^3 + 9x$

9. $\dfrac{x^2 - 1}{x - 3}$

10. $\dfrac{2t^2 - 5}{3t + 6}$

11. $\sqrt{x+3}$

12. $\dfrac{1}{\sqrt{x-1}}$

13. $\dfrac{x^2+1}{x^2 - x - 2}$

14. $\dfrac{x}{x^2 - 4}$

15. $\dfrac{\sqrt{x-2}}{x+3}$

16. $\dfrac{\sqrt{x-2}}{x^2 - 9}$

17–26 ■ Simplify Simplify the rational expression.

17. $\dfrac{(x-5)(x+5)}{2x-10}$ **18.** $\dfrac{x^3-2x}{x^2+x}$

19. $\dfrac{x-2}{x^2-4}$ **20.** $\dfrac{x^2-x-2}{x^2-1}$

21. $\dfrac{x^2-7x-8}{x^2-10x+16}$ **22.** $\dfrac{x^2-x-12}{x^2+5x+6}$

23. $\dfrac{y^2+y}{y^2-1}$ **24.** $\dfrac{y^3-5y^2-24y}{y^3-9y}$

25. $\dfrac{2x^3-x^2-6x}{2x^2-7x+6}$ **26.** $\dfrac{1-x^2}{x^3-1}$

27–40 ■ Multiply or Divide Perform the multiplication or division and simplify.

27. $\dfrac{4x}{x^2-4}\cdot\dfrac{x+2}{16x}$ **28.** $\dfrac{x^2-25}{x^2-16}\cdot\dfrac{x+4}{x+5}$

29. $\dfrac{x^2+2x-15}{x^2-25}\cdot\dfrac{x-5}{x+2}$ **30.** $\dfrac{x^2+2x-3}{x^2-2x-3}\cdot\dfrac{3-x}{3+x}$

31. $\dfrac{2t+3}{t^2+9}\cdot\dfrac{2t-3}{4t^2-9}$ **32.** $\dfrac{3y^2+9y}{y^3-9y}\cdot\dfrac{y^2-9}{2y^2+3y-9}$

33. $\dfrac{x^3-2x^2-8x}{x^2+8x+12}\cdot\dfrac{x^2+2x-24}{x^3-16x}$

34. $\dfrac{2x^2+xy-y^2}{x^2+xy-2y^2}\cdot\dfrac{x^2-2xy+y^2}{2x^2-3xy+y^2}$

35. $\dfrac{x+3}{4x^2-9}\div\dfrac{x^2+7x+12}{2x^2+7x-15}$

36. $\dfrac{2x+1}{2x^2+x-15}\div\dfrac{6x^2-x-2}{x+3}$

37. $\dfrac{\dfrac{x^3}{x+1}}{\dfrac{x}{x^2+2x+1}}$ **38.** $\dfrac{\dfrac{2x^2-3x-2}{x^2-1}}{\dfrac{2x^2+5x+2}{x^2+x-2}}$

39. $\dfrac{x/y}{z}$ **40.** $\dfrac{x}{y/z}$

41–60 ■ Add or Subtract Perform the addition or subtraction and simplify.

41. $1+\dfrac{1}{x+3}$ **42.** $\dfrac{3x-2}{x+1}-2$

43. $\dfrac{1}{x+5}+\dfrac{2}{x-3}$ **44.** $\dfrac{1}{x+1}+\dfrac{1}{x-1}$

45. $\dfrac{3}{x+1}-\dfrac{1}{x+2}$ **46.** $\dfrac{x}{x-4}-\dfrac{3}{x+6}$

47. $\dfrac{5}{2x-3}-\dfrac{3}{(2x-3)^2}$ **48.** $\dfrac{x}{(x+1)^2}+\dfrac{2}{x+1}$

49. $u+1+\dfrac{u}{u+1}$ **50.** $\dfrac{2}{a^2}-\dfrac{3}{ab}+\dfrac{4}{b^2}$

51. $\dfrac{1}{x^2}+\dfrac{1}{x^2+x}$ **52.** $\dfrac{1}{x}+\dfrac{1}{x^2}+\dfrac{1}{x^3}$

53. $\dfrac{2}{x+3}-\dfrac{1}{x^2+7x+12}$ **54.** $\dfrac{x}{x^2-4}+\dfrac{1}{x-2}$

55. $\dfrac{1}{x+3}+\dfrac{1}{x^2-9}$ **56.** $\dfrac{x}{x^2+x-2}-\dfrac{2}{x^2-5x+4}$

57. $\dfrac{2}{x}+\dfrac{3}{x-1}-\dfrac{4}{x^2-x}$ **58.** $\dfrac{1}{x^3y^3z}+\dfrac{1}{xy^3z^3}+\dfrac{1}{x^3yz^3}$

59. $\dfrac{1}{x^2(x+1)}+\dfrac{1}{x^2(x+1)^2}+\dfrac{1}{x^3(x+1)^2}$

60. $\dfrac{1}{x+1}-\dfrac{2}{(x+1)^2}+\dfrac{3}{x^2-1}$

61–74 ■ Compound Fractions Simplify the compound fractional expression.

61. $\dfrac{1+\dfrac{1}{x}}{\dfrac{1}{x}-2}$ **62.** $\dfrac{1-\dfrac{2}{y}}{\dfrac{3}{y}-1}$

63. $\dfrac{1+\dfrac{1}{x+2}}{1-\dfrac{1}{x+2}}$ **64.** $\dfrac{1+\dfrac{1}{c-1}}{1-\dfrac{1}{c-1}}$

65. $\dfrac{\dfrac{1}{x-1}+\dfrac{1}{x+3}}{x+1}$ **66.** $\dfrac{\dfrac{x-3}{x-4}-\dfrac{x+2}{x+1}}{x+3}$

67. $\dfrac{x-\dfrac{x}{y}}{y-\dfrac{y}{x}}$ **68.** $\dfrac{x+\dfrac{y}{x}}{y+\dfrac{x}{y}}$

69. $\dfrac{\dfrac{x}{y}-\dfrac{y}{x}}{\dfrac{1}{x^2}-\dfrac{1}{y^2}}$ **70.** $x-\dfrac{y}{\dfrac{x}{y}+\dfrac{y}{x}}$

71. $\dfrac{x^{-2}-y^{-2}}{x^{-1}+y^{-1}}$ **72.** $\dfrac{1}{1+u}+\dfrac{1}{1+\dfrac{1}{u}}$

73. $1-\dfrac{1}{1-\dfrac{1}{x}}$ **74.** $1+\dfrac{1}{1+\dfrac{1}{1+x}}$

75–80 ■ Expressions Found in Calculus Simplify the fractional expression.

75. $\dfrac{\dfrac{1}{1+x+h}-\dfrac{1}{1+x}}{h}$ **76.** $\dfrac{\dfrac{1}{\sqrt{x+h}}-\dfrac{1}{\sqrt{x}}}{h}$

77. $\dfrac{\dfrac{1}{(x+h)^2}-\dfrac{1}{x^2}}{h}$

78. $\dfrac{(x+h)^2+3(x+h)-x^2-3x}{h}$

79. $\sqrt{1+\left(\dfrac{x}{\sqrt{1-x^2}}\right)^2}$ **80.** $\sqrt{1+\left(x^3-\dfrac{1}{4x^3}\right)^2}$

81–86 ■ Expressions Found in Calculus Simplify the expression. (This type of expression arises in calculus when using the "quotient rule.")

81. $\dfrac{2(x-3)(x+5)^3 - 3(x+5)^2(x-3)^2}{(x+5)^6}$

82. $\dfrac{4x^3(1-x)^3 - 3(1-x)^2(-1)(x^4)}{(1-x)^6}$

83. $\dfrac{2(1+x)^{1/2} - x(1+x)^{-1/2}}{x+1}$ **84.** $\dfrac{(1-x^2)^{1/2} + x^2(1-x^2)^{-1/2}}{1-x^2}$

85. $\dfrac{3(1+x)^{1/3} - x(1+x)^{-2/3}}{(1+x)^{2/3}}$ **86.** $\dfrac{(7-3x)^{1/2} + \frac{3}{2}x(7-3x)^{-1/2}}{7-3x}$

87–92 ■ Rationalize Denominator Rationalize the denominator.

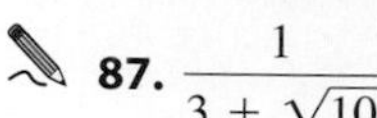

87. $\dfrac{1}{3+\sqrt{10}}$ **88.** $\dfrac{3}{2-\sqrt{5}}$

89. $\dfrac{2}{\sqrt{5}-\sqrt{3}}$ **90.** $\dfrac{1}{\sqrt{x}+1}$

91. $\dfrac{y}{\sqrt{3}+\sqrt{y}}$ **92.** $\dfrac{2(x-y)}{\sqrt{x}-\sqrt{y}}$

93–98 ■ Rationalize Numerator Rationalize the numerator.

93. $\dfrac{2-\sqrt{5}}{5}$ **94.** $\dfrac{\sqrt{3}+\sqrt{5}}{2}$

95. $\dfrac{\sqrt{r}+\sqrt{2}}{5}$ **96.** $\dfrac{\sqrt{x}-\sqrt{x+h}}{h\sqrt{x}\sqrt{x+h}}$

97. $\sqrt{x^2+1} - x$ **98.** $\sqrt{x+1} - \sqrt{x}$

Applications

99. Electrical Resistance If two electrical resistors with resistances R_1 and R_2 are connected in parallel (see the figure), then the total resistance R is given by

$$R = \frac{1}{\dfrac{1}{R_1} + \dfrac{1}{R_2}}$$

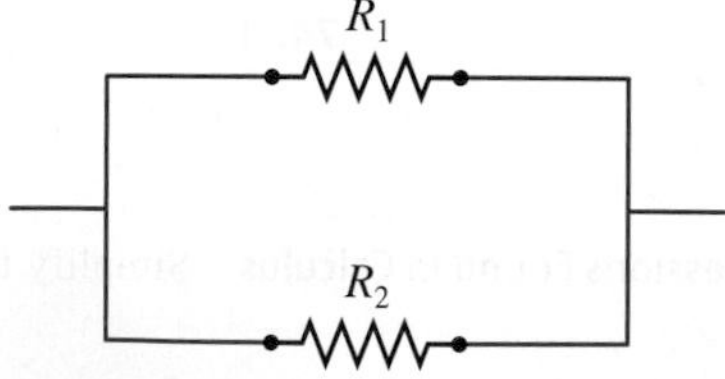

(a) Simplify the expression for R.

(b) If $R_1 = 10$ ohms and $R_2 = 20$ ohms, what is the total resistance R?

100. Average Cost A clothing manufacturer finds that the cost of producing x shirts is $500 + 6x + 0.01x^2$ dollars.

(a) Explain why the average cost per shirt is given by the rational expression

$$A = \frac{500 + 6x + 0.01x^2}{x}$$

(b) Complete the table by calculating the average cost per shirt for the given values of x.

x	Average Cost
10	
20	
50	
100	
200	
500	
1000	

Discuss ■ Discover ■ Prove ■ Write

101. Discover: Limiting Behavior of a Rational Expression The rational expression

$$\frac{x^2-9}{x-3}$$

is not defined for $x = 3$. Complete the tables, and determine what value the expression approaches as x gets closer and closer to 3. Why is this reasonable? Factor the numerator of the expression and simplify to see why.

x	$\dfrac{x^2-9}{x-3}$
2.80	
2.90	
2.95	
2.99	
2.999	

x	$\dfrac{x^2-9}{x-3}$
3.20	
3.10	
3.05	
3.01	
3.001	

102. Discuss ■ Write: Is This Rationalization? In the expression $2/\sqrt{x}$ we would eliminate the radical if we were to square both numerator and denominator. Is this the same thing as rationalizing the denominator? Explain.

103. Discuss: Algebraic Errors The left-hand column of the table lists some common algebraic errors. In each case, give an example using numbers that shows that the formula is not valid. An example of this type, which shows that a statement is false, is called a *counterexample*.

Algebraic Errors	Counterexample
$\dfrac{1}{a} + \dfrac{1}{b} \times \dfrac{1}{a+b}$	$\dfrac{1}{2} + \dfrac{1}{2} \neq \dfrac{1}{2+2}$
$(a+b)^2 \times a^2 + b^2$	
$\sqrt{a^2+b^2} \times a + b$	
$\dfrac{a+b}{a} \times b$	
$\dfrac{a}{a+b} \times \dfrac{1}{b}$	
$\dfrac{a^m}{a^n} \times a^{m/n}$	

104. Discuss: Algebraic Errors Determine whether the given equation is true for all values of the variables. If not, give a counterexample. (Disregard any value that makes a denominator zero.)

(a) $\frac{5+a}{5} = 1 + \frac{a}{5}$ **(b)** $\frac{x+1}{y+1} = \frac{x}{y}$

(c) $\frac{x}{x+y} = \frac{1}{1+y}$ **(d)** $2\left(\frac{a}{b}\right) = \frac{2a}{2b}$

(e) $\frac{-a}{b} = -\frac{a}{b}$ **(f)** $\frac{1+x+x^2}{x} = \frac{1}{x} + 1 + x$

105. Discover ■ Prove: Values of a Rational Expression For $x > 0$, consider the expression

$$x + \frac{1}{x}$$

(a) Fill in the table, and try other values for x. What do you think is the smallest possible value for this expression?

x	1	3	$\frac{1}{2}$	$\frac{9}{10}$	$\frac{99}{100}$	
$x + \frac{1}{x}$						

(b) Prove that for $x > 0$,

$$x + \frac{1}{x} \geq 2$$

PS *Work backward.* Assume the inequality is valid; multiply by x, move terms to one side, and then factor to arrive at a true statement. Note that each step you made is reversible.

1.5 Equations

■ Solving Linear Equations ■ Formulas: Solving for One Variable in Terms of Others ■ Solving Quadratic Equations ■ Other Types of Equations

An equation is a statement that two mathematical expressions are equal. For example,

$$3 + 5 = 8$$

is an equation. Most equations that we study in algebra contain variables, which are symbols (usually letters) that stand for numbers. In the equation

$$4x + 7 = 19$$

the letter x is the variable. We think of x as the "unknown" in the equation, and our goal is to find the value of x that makes the equation true. The values of the unknown that make the equation true are called the **solutions** or **roots** of the equation, and the process of finding the solutions is called **solving the equation**.

$x = 3$ is a solution of the equation $4x + 7 = 19$ because substituting $x = 3$ makes the equation true:

$x = 3$

$4(3) + 7 = 19$ ✓

Two equations with exactly the same solutions are called **equivalent equations**. To solve an equation, we try to find a simpler, equivalent equation in which the variable stands alone on one side of the equal sign. Here are the properties that we use to solve an equation. (In these properties, A, B, and C stand for any algebraic expressions, and the symbol $\Leftrightarrow$ means "is equivalent to.")

Properties of Equality

Property	Description
1. $A = B \quad \Leftrightarrow \quad A + C = B + C$	Adding the same quantity to both sides of an equation gives an equivalent equation.
2. $A = B \quad \Leftrightarrow \quad CA = CB \quad (C \neq 0)$	Multiplying both sides of an equation by the same nonzero quantity gives an equivalent equation.

These properties require that you *perform the same operation on both sides of an equation* when solving it. Thus if we say "*add* -7" when solving an equation, that is just a short way of saying "*add* -7 to each side of the equation."

■ Solving Linear Equations

The simplest type of equation is a *linear equation*, or first-degree equation, which is an equation in which each term is either a constant or a nonzero multiple of the variable.

Linear Equations

A **linear equation** in one variable is an equation equivalent to one of the form

$$ax + b = 0$$

where a and b are real numbers and x is the variable.

Here are some examples that illustrate the difference between linear and nonlinear equations.

Linear equations	**Nonlinear equations**	
$4x - 5 = 3$	$x^2 + 2x = 8$	Not linear; contains the square of the variable
$2x = \frac{1}{2}x - 7$	$\sqrt{x} - 6x = 0$	Not linear; contains the square root of the variable
$x - 6 = \dfrac{x}{3}$	$\dfrac{3}{x} - 2x = 1$	Not linear; contains the reciprocal of the variable

Example 1 ■ Solving a Linear Equation

Solve the equation $7x - 4 = 3x + 8$.

Solution We solve this equation by changing it to an equivalent equation with all terms that have the variable x on one side and all constant terms on the other.

$7x - 4 = 3x + 8$	Given equation
$(7x - 4) + 4 = (3x + 8) + 4$	Add 4
$7x = 3x + 12$	Simplify
$7x - 3x = (3x + 12) - 3x$	Subtract $3x$
$4x = 12$	Simplify
$\frac{1}{4} \cdot 4x = \frac{1}{4} \cdot 12$	Multiply by $\frac{1}{4}$
$x = 3$	Simplify

Check Your Answer

$x = 3$:

$x = 3$ → LHS $= 7(3) - 4 = 17$

$x = 3$ → RHS $= 3(3) + 8 = 17$

LHS = RHS ✓

Because it is important to CHECK YOUR ANSWER, we do this in many of our examples. In these checks, LHS stands for "left-hand side" and RHS stands for "right-hand side" of the original equation.

 **Now Try Exercise 17** ■

■ Formulas: Solving for One Variable in Terms of Others

In mathematics and the sciences, the term *formula* commonly refers to an equation that relates different variables. Examples include the formula $A = \pi r^2$, which relates the area of a circle to its radius, and the formula $PV = nRT$, which relates the pressure, volume, and temperature of an ideal gas. We can solve for any variable in a formula to find out how that variable relates to the other variables. For example, solving for r in the formula for the area of a circle gives $r = \sqrt{A/\pi}$ and solving for P in the ideal gas formula gives $P = nRT/V$. In the next example we solve for a variable in Newton's Law of Gravity.

Example 2 ■ Solving for One Variable in Terms of Others

This formula is Newton's Law of Gravity. It gives the gravitational force F between two masses m and M that are a distance r apart. The constant G is the universal gravitational constant.

Solve for the variable M in the equation

$$F = G\frac{mM}{r^2}$$

Solution Although this equation involves more than one variable, we solve it as usual by isolating M on one side and treating the other variables as we would numbers.

$$F = \left(\frac{Gm}{r^2}\right)M \qquad \text{Factor } M \text{ from RHS}$$

$$\left(\frac{r^2}{Gm}\right)F = \left(\frac{r^2}{Gm}\right)\left(\frac{Gm}{r^2}\right)M \qquad \text{Multiply by reciprocal of } \frac{Gm}{r^2}$$

$$\frac{r^2F}{Gm} = M \qquad \text{Simplify}$$

The solution is $M = \dfrac{r^2F}{Gm}$.

 Now Try Exercise 29 ■

Example 3 ■ Solving for One Variable in Terms of Others

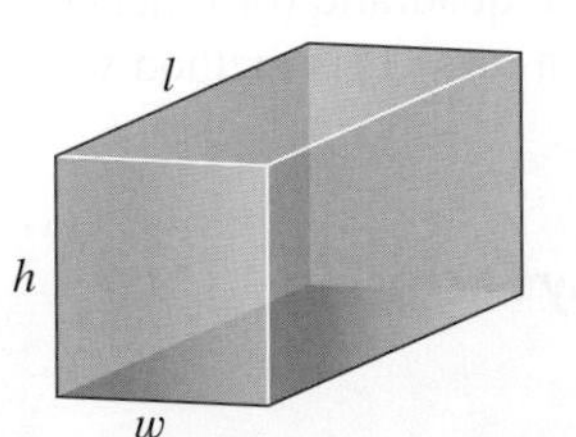

Figure 1 | A closed rectangular box

The surface area A of the closed rectangular box shown in Figure 1 can be calculated from the length l, the width w, and the height h according to the formula

$$A = 2lw + 2wh + 2lh$$

Solve for w in terms of the other variables in this equation.

Solution Although this equation involves more than one variable, we solve it as usual by isolating w on one side, treating the other variables as we would numbers.

$$A = (2lw + 2wh) + 2lh \qquad \text{Collect terms involving } w$$

$$A - 2lh = 2lw + 2wh \qquad \text{Subtract } 2lh$$

$$A - 2lh = (2l + 2h)w \qquad \text{Factor } w \text{ from RHS}$$

$$\frac{A - 2lh}{2l + 2h} = w \qquad \text{Divide by } 2l + 2h$$

The solution is $w = \dfrac{A - 2lh}{2l + 2h}$.

 Now Try Exercise 31 ■

Discovery Project ■ Weighing the Whole World

Have you ever wondered how much the world weighs? In this project you will answer this question by using Newton's formula for gravitational force. That such questions can be answered by simply using formulas shows the remarkable power of formulas. Edsger Dijkstra, one of the founders of computer science, once said, "A picture may be worth a thousand words, a formula is worth a thousand pictures." You can find the project at **www.stewartmath.com**.

■ Solving Quadratic Equations

Linear equations are first-degree equations like $2x + 1 = 5$ or $4 - 3x = 2$. Quadratic equations are second-degree equations like $x^2 + 2x - 3 = 0$ or $2x^2 + 3 = 5x$.

Quadratic Equations

$x^2 - 2x - 8 = 0$

$3x + 10 = 4x^2$

$\frac{1}{2}x^2 + \frac{1}{3}x - \frac{1}{6} = 0$

Quadratic Equations

A **quadratic equation** is an equation of the form

$$ax^2 + bx + c = 0$$

where a, b, and c are real numbers with $a \neq 0$.

Some quadratic equations can be solved by factoring and using the following basic property of real numbers.

Zero-Product Property

$$AB = 0 \quad \text{if and only if} \quad A = 0 \quad \text{or} \quad B = 0$$

This means that if we can factor the left-hand side of a quadratic (or other) equation, then we can solve it by setting each factor equal to 0 in turn. This method works only when the right-hand side of the equation is 0.

Example 4 ■ Solving a Quadratic Equation by Factoring

Find all real solutions of the equation $x^2 + 5x = 24$.

Solution We must first rewrite the equation so that the right-hand side is 0.

$$x^2 + 5x = 24$$

$$x^2 + 5x - 24 = 0 \qquad \text{Subtract 24}$$

$$(x - 3)(x + 8) = 0 \qquad \text{Factor}$$

$$x - 3 = 0 \quad \text{or} \quad x + 8 = 0 \qquad \text{Zero-Product Property}$$

$$x = 3 \qquad\qquad x = -8 \qquad \text{Solve}$$

The solutions are $x = 3$ and $x = -8$.

Check Your Answers

$x = 3$:

$(3)^2 + 5(3) = 9 + 15 = 24$ ✓

$x = -8$:

$(-8)^2 + 5(-8) = 64 - 40 = 24$ ✓

Now Try Exercise 41 ■

Do you see why one side of the equation must be 0 in Example 4? Factoring the equation as $x(x + 5) = 24$ does not help us find the solutions because 24 can be factored in infinitely many ways, such as $6 \cdot 4$, $\frac{1}{2} \cdot 48$, $\left(-\frac{2}{5}\right) \cdot (-60)$, and so on.

A quadratic equation of the form $x^2 - c = 0$, where c is a positive constant, factors as $(x - \sqrt{c})(x + \sqrt{c}) = 0$, so the solutions are $x = \sqrt{c}$ and $x = -\sqrt{c}$. We often abbreviate this as $x = \pm\sqrt{c}$.

Solving a Simple Quadratic Equation

The solutions of the equation $x^2 = c$ are $x = \sqrt{c}$ and $x = -\sqrt{c}$.

Example 5 ■ Solving Simple Quadratic Equations

Find all real solutions of each equation.

(a) $x^2 = 5$ **(b)** $(x - 4)^2 = 5$

Solution

(a) From the principle in the preceding box we get $x = \pm\sqrt{5}$.

(b) We can take the square root of each side of this equation as well.

$$(x - 4)^2 = 5$$

$$x - 4 = \pm\sqrt{5} \qquad \text{Take the square root}$$

$$x = 4 \pm \sqrt{5} \qquad \text{Add 4}$$

The solutions are $x = 4 + \sqrt{5}$ and $x = 4 - \sqrt{5}$.

Now Try Exercises 47 and 49 ■

See Section 1.3 for how to recognize when a quadratic expression is a perfect square.

As we saw in Example 5, part (b), if a quadratic equation is of the form $(x \pm a)^2 = c$, then we can solve it by taking the square root of each side. In an equation of this form, the left-hand side is a *perfect square*: that is, the square of a linear expression in x. So if a quadratic equation does not factor readily, then we can solve it by using the technique of **completing the square**. This means that we add a constant to an expression to make it a perfect square. For example, to make $x^2 - 6x$ a perfect square, we must add 9 because $x^2 - 6x + 9 = (x - 3)^2$.

Completing the Square

The area of the blue region is

$$x^2 + 2\left(\frac{b}{2}\right)x = x^2 + bx$$

Add a small square of area $(b/2)^2$ to "complete" the square.

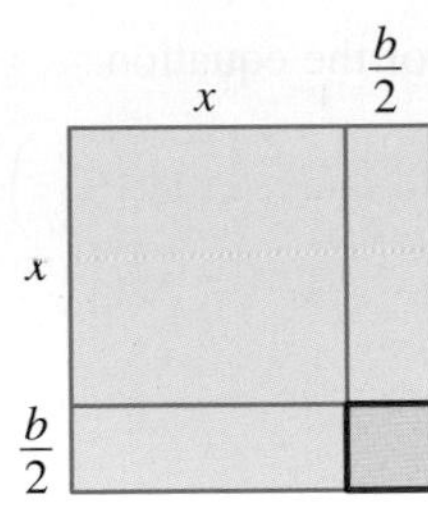

Completing the Square

To make $x^2 + bx$ a perfect square, add $\left(\frac{b}{2}\right)^2$, the square of half the coefficient of x. This gives the perfect square

$$x^2 + bx + \left(\frac{b}{2}\right)^2 = \left(x + \frac{b}{2}\right)^2$$

Example 6 ■ Solving Quadratic Equations by Completing the Square

Find all real solutions of each equation.

(a) $x^2 - 8x + 13 = 0$ **(b)** $3x^2 - 12x + 6 = 0$

Solution

(a)

$$x^2 - 8x + 13 = 0 \qquad \text{Given equation}$$

$$x^2 - 8x = -13 \qquad \text{Subtract 13}$$

$$x^2 - 8x + 16 = -13 + 16 \qquad \text{Complete the square: add } \left(\frac{-8}{2}\right)^2 = 16$$

$$(x - 4)^2 = 3 \qquad \text{Perfect square}$$

$$x - 4 = \pm\sqrt{3} \qquad \text{Take square root}$$

$$x = 4 \pm \sqrt{3} \qquad \text{Add 4}$$

⊘ When you complete the square, make sure the coefficient of x^2 is 1. If it isn't, then you must factor this coefficient from both terms that contain x:

$$ax^2 + bx = a\left(x^2 + \frac{b}{a}x\right)$$

Then complete the square inside the parentheses. Remember that the term added inside the parentheses is multiplied by a.

(b) After subtracting 6 from each side of the equation, we must factor the coefficient of x^2 (the 3) from the left side to put the equation in the correct form for completing the square.

$$3x^2 - 12x + 6 = 0 \qquad \text{Given equation}$$

$$3x^2 - 12x = -6 \qquad \text{Subtract 6}$$

$$3(x^2 - 4x) = -6 \qquad \text{Factor 3 from LHS}$$

Now we complete the square by adding $(-2)^2 = 4$ *inside* the parentheses. Since everything inside the parentheses is multiplied by 3, this means that we are actually adding $3 \cdot 4 = 12$ to the left side of the equation. Thus we must add 12 to the right side as well.

$$3(x^2 - 4x + 4) = -6 + 3 \cdot 4 \qquad \text{Complete the square: add 4}$$
$$3(x - 2)^2 = 6 \qquad \text{Perfect square}$$
$$(x - 2)^2 = 2 \qquad \text{Divide by 3}$$
$$x - 2 = \pm\sqrt{2} \qquad \text{Take square root}$$
$$x = 2 \pm \sqrt{2} \qquad \text{Add 2}$$

Now Try Exercises 53 and 57 ■

We can use the technique of completing the square to derive a formula for the solutions of the general quadratic equation $ax^2 + bx + c = 0$.

The Quadratic Formula

The solutions of the general quadratic equation $ax^2 + bx + c = 0$, where $a \neq 0$, are

$$x = \frac{-b \pm \sqrt{b^2 - 4ac}}{2a}$$

Proof First, we divide each side of the equation by a and move the constant to the right side, giving

$$x^2 + \frac{b}{a}x = -\frac{c}{a} \qquad \text{Divide by } a$$

We now complete the square by adding $(b/2a)^2$ to each side of the equation:

$$x^2 + \frac{b}{a}x + \left(\frac{b}{2a}\right)^2 = -\frac{c}{a} + \left(\frac{b}{2a}\right)^2 \qquad \text{Complete the square: Add } \left(\frac{b}{2a}\right)^2$$
$$\left(x + \frac{b}{2a}\right)^2 = \frac{-4ac + b^2}{4a^2} \qquad \text{Perfect square}$$
$$x + \frac{b}{2a} = \pm\frac{\sqrt{b^2 - 4ac}}{2a} \qquad \text{Take square root}$$
$$x = \frac{-b \pm \sqrt{b^2 - 4ac}}{2a} \qquad \text{Subtract } \frac{b}{2a}$$
■

The Quadratic Formula can be used to solve any quadratic equation. You should confirm that the Quadratic Formula gives the same solutions when applied to the equations in Examples 4 and 6.

Library of Congress Prints and Photographs Division [LC-USZ62-62123]

FRANÇOIS VIÈTE (1540–1603) had a successful political career before taking up mathematics late in life. He became one of the most famous French mathematicians of the 16th century. Viète introduced a new level of abstraction in algebra by using letters to stand for *known* quantities in an equation. Before Viète's time, each equation had to be solved on its own. For instance, the quadratic equations

$$3x^2 + 2x + 8 = 0$$
$$5x^2 - 6x + 4 = 0$$

had to be solved separately by completing the square. Viète's idea was to consider all quadratic equations at once by writing

$$ax^2 + bx + c = 0$$

where a, b, and c are known quantities. Thus he made it possible to write a *formula* (in this case the Quadratic Formula) involving a, b, and c that can be used to solve all such equations in one fell swoop.

Viète's mathematical genius proved valuable during a war between France and Spain. To communicate with their troops, the Spaniards used a complicated code that Viète managed to decipher. Unaware of Viète's accomplishment, the Spanish king, Philip II, protested to the Pope, claiming that the French were using witchcraft to read his messages.

Example 7 ■ Using the Quadratic Formula

Find all real solutions of each equation.

(a) $3x^2 - 5x - 1 = 0$ **(b)** $4x^2 + 12x + 9 = 0$ **(c)** $x^2 + 2x = -2$

Solution

(a) In this quadratic equation $a = 3$, $b = -5$, and $c = -1$.

$b = -5$

$$3x^2 - 5x - 1 = 0$$

$a = 3$ $c = -1$

By the Quadratic Formula,

$$x = \frac{-(-5) \pm \sqrt{(-5)^2 - 4(3)(-1)}}{2(3)} = \frac{5 \pm \sqrt{37}}{6}$$

If approximations are desired, we can use a calculator to obtain

$$x = \frac{5 + \sqrt{37}}{6} \approx 1.8471 \quad \text{and} \quad x = \frac{5 - \sqrt{37}}{6} \approx -0.1805$$

Another Method

$4x^2 + 12x + 9 = 0$

$(2x + 3)^2 = 0$

$2x + 3 = 0$

$x = -\frac{3}{2}$

(b) Using the Quadratic Formula with $a = 4$, $b = 12$, and $c = 9$ gives

$$x = \frac{-12 \pm \sqrt{(12)^2 - 4 \cdot 4 \cdot 9}}{2 \cdot 4} = \frac{-12 \pm 0}{8} = -\frac{3}{2}$$

This equation has only one solution, $x = -\frac{3}{2}$.

(c) We first write the equation in the form $x^2 + 2x + 2 = 0$. Using the Quadratic Formula with $a = 1$, $b = 2$, and $c = 2$ gives

$$x = \frac{-2 \pm \sqrt{2^2 - 4 \cdot 2}}{2} = \frac{-2 \pm \sqrt{-4}}{2} = \frac{-2 \pm 2\sqrt{-1}}{2} = -1 \pm \sqrt{-1}$$

In Section 1.6 we study the complex number system, in which the square roots of negative numbers are defined.

Since the square of any real number is nonnegative, $\sqrt{-1}$ is undefined in the real number system. So this equation has no real solution.

Now Try Exercises 61, 67, and 71 ■

The quantity $b^2 - 4ac$ that appears under the square root sign in the Quadratic Formula is called the *discriminant* of the equation $ax^2 + bx + c = 0$ and is given the symbol D. If $D < 0$, then $\sqrt{b^2 - 4ac}$ is undefined, and the quadratic equation has no real solution, as in Example 7(c). If $D = 0$, then the equation has only one real solution, as in Example 7(b). Finally, if $D > 0$, then the equation has two distinct real solutions, as in Example 7(a). The following box summarizes these observations.

The Discriminant

The **discriminant** of the general quadratic equation $ax^2 + bx + c = 0$ $(a \neq 0)$ is $D = b^2 - 4ac$.

1. If $D > 0$, then the equation has two distinct real solutions.
2. If $D = 0$, then the equation has exactly one real solution.
3. If $D < 0$, then the equation has no real solution.

Example 8 ■ Using the Discriminant

Use the discriminant to determine how many real solutions each equation has.

(a) $x^2 + 4x - 1 = 0$ **(b)** $4x^2 + 12x + 9 = 0$ **(c)** $\frac{1}{3}x^2 - 2x + 4 = 0$

Solution

(a) The discriminant is $D = 4^2 - 4(1)(-1) = 20 > 0$, so the equation has two distinct real solutions.

Compare part (b) with Example 7(b).

(b) The discriminant is $D = (12)^2 - 4 \cdot 4 \cdot 9 = 0$, so the equation has exactly one real solution.

(c) The discriminant is $D = (-2)^2 - 4(\frac{1}{3})4 = -\frac{4}{3} < 0$, so the equation has no real solution.

Now Try Exercises 73, 75, and 77 ■

This formula depends on the fact that acceleration due to gravity is constant near the earth's surface. Here we neglect the effect of air resistance.

Now let's consider a real-life situation that can be modeled by a quadratic equation.

Example 9 ■ The Path of a Projectile

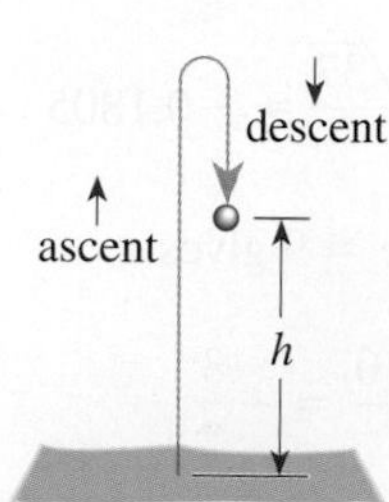
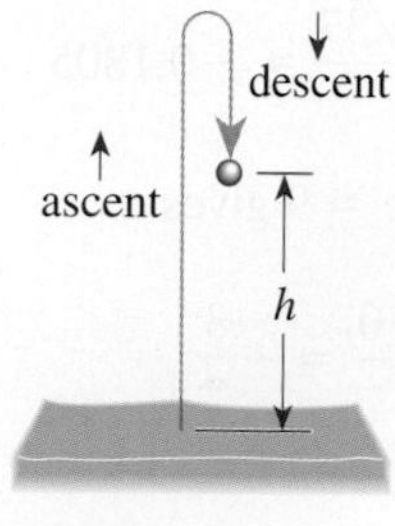

Figure 2

An object thrown or fired straight upward at an initial speed of v_0 m/s will reach a height of h meters after t seconds, where h and t are related by the formula

$$h = -4.9t^2 + v_0 t$$

Suppose that a bullet is shot straight upward with an initial speed of 245 m/s. Its path is shown in Figure 2.

(a) When does the bullet fall back to ground level?

(b) When does the bullet reach a height of 1960 m?

(c) When does the bullet reach a height of 3.2 km?

(d) How high is the highest point the bullet reaches?

Solution Since the initial speed in this case is $v_0 = 245$ m/s, the formula is

$$h = -4.9t^2 + 245t$$

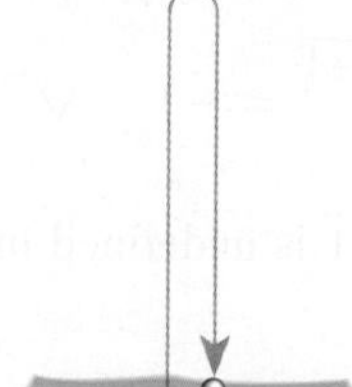

(a) Ground level corresponds to $h = 0$, so we must solve the equation

$$0 = -4.9t^2 + 245t \qquad \text{Set } h = 0$$

$$0 = -4.9t(t - 50) \qquad \text{Factor}$$

Thus $t = 0$ or $t = 50$. This means the bullet starts ($t = 0$) at ground level and returns to ground level after 50 s.

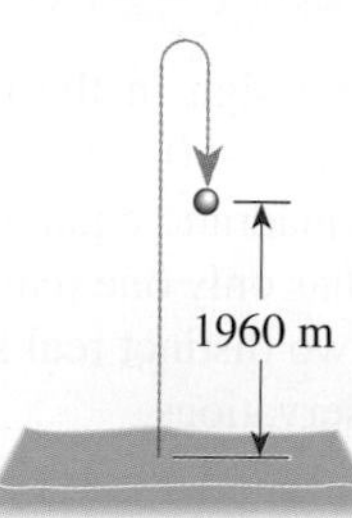

(b) Setting $h = 1960$ gives the equation

$$1960 = -4.9t^2 + 245t \qquad \text{Set } h = 1960$$

$$4.9t^2 - 245t + 1960 = 0 \qquad \text{All terms to LHS}$$

$$t^2 - 50t + 400 = 0 \qquad \text{Divide by 4.9}$$

$$(t - 10)(t - 40) = 0 \qquad \text{Factor}$$

$$t = 10 \quad \text{or} \quad t = 40 \qquad \text{Solve}$$

The bullet reaches 1960 m after 10 s (on its ascent) and again after 40 s (on its descent).

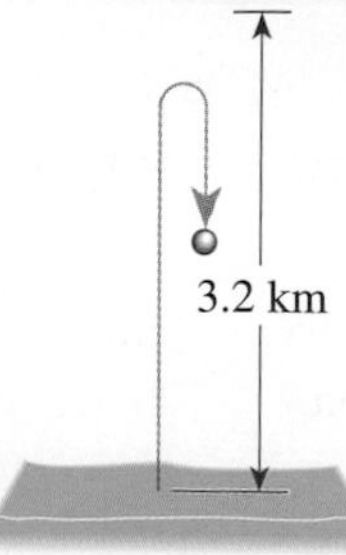

(c) 3.2 kilometers is $3.2 \times 1000 = 3{,}200$ m.

$$3{,}200 = -4.9t^2 + 245t \qquad \text{Set } h = 3{,}200$$

$$4.9t^2 - 245t + 3200 = 0 \qquad \text{All terms to LHS}$$

$$t^2 - 50t + 653 = 0 \qquad \text{Divide by 4.9}$$

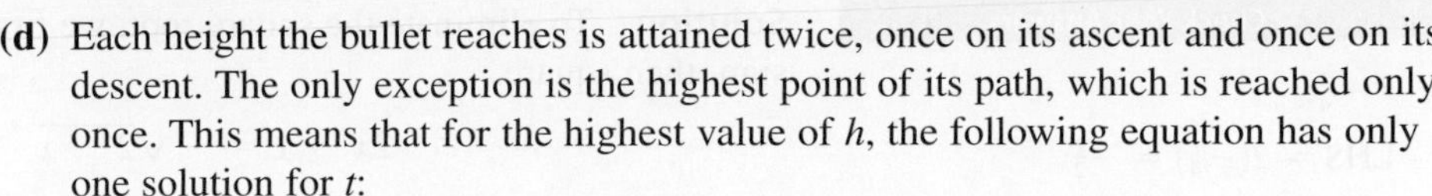

The discriminant of this equation is $D = (-50)^2 - 4(653) = -112$, which is negative. Thus the equation has no real solution. The bullet never reaches a height of 3.2 km.

2,500 m

(d) Each height the bullet reaches is attained twice, once on its ascent and once on its descent. The only exception is the highest point of its path, which is reached only once. This means that for the highest value of h, the following equation has only one solution for t:

$$h = -4.9t^2 + 245t$$

$$4.9t^2 - 245t + h = 0 \qquad \text{All terms to LHS}$$

This in turn means that the discriminant D of the equation is 0, so

$$D = (-245)^2 - 4(4.9)h = 0$$

$$60{,}025 - 24h = 0$$

$$h = 2{,}500$$

The maximum height reached is 2,500 m.

 **Now Try Exercise 131** ■

■ Other Types of Equations

So far we have learned how to solve linear and quadratic equations. Now we study other types of equations, including those that involve higher powers, fractional expressions, and radicals.

When we solve an equation that involves fractional expressions or radicals, we must be especially careful to check our answers. The next two examples demonstrate why.

Example 10 ■ An Equation Involving Fractional Expressions

Solve the equation $\dfrac{3}{x} - \dfrac{2}{x-3} = \dfrac{-12}{x^2-9}$.

Solution We eliminate the denominators by multiplying each side by the lowest common denominator.

$$\left(\frac{3}{x} - \frac{2}{x-3}\right)x(x^2-9) = \frac{-12}{x^2-9}x(x^2-9) \qquad \text{Multiply by LCD, } x(x^2-9)$$

$$3(x^2-9) - 2x(x+3) = -12x \qquad \text{Simplify}$$

$$3x^2 - 27 - 2x^2 - 6x = -12x \qquad \text{Expand LHS}$$

$$x^2 - 6x - 27 = -12x \qquad \text{Add like terms on LHS}$$

$$x^2 + 6x - 27 = 0 \qquad \text{Add } 12x$$

$$(x-3)(x+9) = 0 \qquad \text{Factor}$$

$$x - 3 = 0 \quad \text{or} \quad x + 9 = 0 \qquad \text{Zero-Product Property}$$

$$x = 3 \qquad\qquad x = -9 \qquad \text{Solve}$$

Check Your Answers

$x = 3$:

$$\text{LHS} = \frac{3}{3} - \frac{2}{3-3} \quad \text{undefined}$$

$$\text{RHS} = \frac{-12}{3^2-9} \quad \text{undefined} \quad ✖$$

$x = -9$:

$$\text{LHS} = \frac{3}{-9} - \frac{2}{-9-3} = -\frac{1}{6}$$

$$\text{RHS} = \frac{-12}{(-9)^2-9} = -\frac{1}{6}$$

LHS = RHS ✔

We must check our answer because multiplying by an expression that contains the variable can introduce extraneous solutions. From *Check Your Answers* we see that the only solution is $x = -9$.

 **Now Try Exercise 83** ■

Example 11 ■ An Equation Involving a Radical

Solve the equation $2x = 1 - \sqrt{2 - x}$.

Solution To eliminate the square root, we first isolate it on one side of the equal sign, then square.

$2x - 1 = -\sqrt{2 - x}$	Subtract 1
$(2x - 1)^2 = 2 - x$	Square each side
$4x^2 - 4x + 1 = 2 - x$	Expand LHS
$4x^2 - 3x - 1 = 0$	Add $-2 + x$
$(4x + 1)(x - 1) = 0$	Factor
$4x + 1 = 0$ or $x - 1 = 0$	Zero-Product Property
$x = -\frac{1}{4}$ $\quad$ $x = 1$	Solve

Check Your Answers

$x = -\frac{1}{4}$:

$\text{LHS} = 2(-\frac{1}{4}) = -\frac{1}{2}$

$\text{RHS} = 1 - \sqrt{2 - (-\frac{1}{4})}$

$= 1 - \sqrt{\frac{9}{4}}$

$= 1 - \frac{3}{2} = -\frac{1}{2}$

LHS = RHS 

$x = 1$:

$\text{LHS} = 2(1) = 2$

$\text{RHS} = 1 - \sqrt{2 - 1}$

$= 1 - 1 = 0$

LHS ≠ RHS ⊗

The values $x = -\frac{1}{4}$ and $x = 1$ are only potential solutions. We must check them to see whether they satisfy the original equation. From *Check Your Answers* we see that $x = -\frac{1}{4}$ is a solution but $x = 1$ is not. The only solution is $x = -\frac{1}{4}$.

Now Try Exercise 89 ■

Note When we solve an equation, we may end up with one or more **extraneous solutions**, that is, potential solutions that do not satisfy the original equation. In Example 10 the value $x = 3$ is an extraneous solution, and in Example 11 the value $x = 1$ is an extraneous solution.

In the case of equations involving fractional expressions, potential solutions may be undefined in the original equation and hence are extraeous solutions. In the case of equations involving radicals, extraneous solutions may be introduced when we square each side of an equation because the operation of squaring can turn a false equation into a true one. For example, $-1 \neq 1$, but $(-1)^2 = 1^2$. Thus the squared equation  may be true for more values of the variable than the original equation. That is why you must always check your answers to make sure that each satisfies the original equation.

An equation of the form $aW^2 + bW + c = 0$, where W is an algebraic expression, is an equation of **quadratic type**. We solve equations of quadratic type by substituting for the algebraic expression, as we see in the next two examples.

Example 12 ■ An Equation of Quadratic Type

Find all solutions of the equation $x^4 - 8x^2 + 8 = 0$.

Solution If we set $W = x^2$, then we get a quadratic equation in the new variable W.

$(x^2)^2 - 8x^2 + 8 = 0$	Write x^4 as $(x^2)^2$
$W^2 - 8W + 8 = 0$	Let $W = x^2$
$W = \dfrac{-(-8) \pm \sqrt{(-8)^2 - 4 \cdot 8}}{2} = 4 \pm 2\sqrt{2}$	Quadratic Formula
$x^2 = 4 \pm 2\sqrt{2}$	$W = x^2$
$x = \pm\sqrt{4 \pm 2\sqrt{2}}$	Take square roots

So there are four solutions:

$$\sqrt{4 + 2\sqrt{2}} \qquad \sqrt{4 - 2\sqrt{2}} \qquad -\sqrt{4 + 2\sqrt{2}} \qquad -\sqrt{4 - 2\sqrt{2}}$$

Using a calculator, we obtain the approximations $x \approx 2.61, 1.08, -2.61, -1.08$.

 **Now Try Exercise 91** ■

CPA Media Pte Ltd/Alamy Stock Photo

AL-KHWARIZMI (780–850 CE) was a Persian mathematician, astronomer, and geographer; he was a scholar at the House of Wisdom in Baghdad. He is known today as the "father of algebra," because his book *Hisāb al-Jabr w'al-muqābala* was the first to deal with the rules of algebra. The title roughly translates to "Calculation by Completion and Balancing," which are the operations he used to solve algebraic equations. He described his book as containing "what is easiest and most useful in arithmetic." Among other things, the book contains the method of solving quadratic equations by completing the square. In Latin translations the title of the book was shortened to *Al-Jabr* from which we get the word *algebra*. Al Khwarizmi's name itself made its way into the English language in the word *algorithm*. On the 1200th anniversary of his birth, a stamp (shown here) was issued by the former USSR to celebrate his birth.

Example 13 ■ An Equation Involving Fractional Powers

Find all solutions of the equation $x^{1/3} + x^{1/6} - 2 = 0$.

Solution This equation is of quadratic type because if we let $W = x^{1/6}$, then $W^2 = (x^{1/6})^2 = x^{1/3}$.

$$x^{1/3} + x^{1/6} - 2 = 0$$

$$(x^{1/6})^2 + x^{1/6} - 2 = 0 \qquad \text{Write } x^{1/3} \text{ as } (x^{1/6})^2$$

$$W^2 + W - 2 = 0 \qquad \text{Let } W = x^{1/6}$$

$$(W - 1)(W + 2) = 0 \qquad \text{Factor}$$

$$W - 1 = 0 \quad \text{or} \quad W + 2 = 0 \qquad \text{Zero-Product Property}$$

$$W = 1 \qquad W = -2 \qquad \text{Solve}$$

$$x^{1/6} = 1 \qquad x^{1/6} = -2 \qquad W = x^{1/6}$$

$$x = 1^6 = 1 \qquad x = (-2)^6 = 64 \qquad \text{Take the 6th power}$$

From *Check Your Answers* we see that $x = 1$ is a solution but $x = 64$ is not. The only solution is $x = 1$.

Check Your Answers

$x = 1$:

$\text{LHS} = 1^{1/3} + 1^{1/6} - 2 = 0$

$\text{RHS} = 0$

$\text{LHS} = \text{RHS}$ ✓

$x = 64$:

$\text{LHS} = 64^{1/3} + 64^{1/6} - 2$
$= 4 + 2 - 2 = 4$

$\text{RHS} = 0$

$\text{LHS} \neq \text{RHS}$ ✗

Now Try Exercise 95 ■

When solving an absolute-value equation, we use the following property

$$|X| = C \quad \text{is equivalent to} \quad X = C \quad \text{or} \quad X = -C$$

where X is any algebraic expression. This property says that to solve an absolute-value equation, we must solve two separate equations.

Example 14 ■ An Absolute-Value Equation

Solve the equation $|2x - 5| = 3$.

Solution By the definition of absolute value, $|2x - 5| = 3$ is equivalent to

$$2x - 5 = 3 \quad \text{or} \quad 2x - 5 = -3$$

$$2x = 8 \qquad 2x = 2$$

$$x = 4 \qquad x = 1$$

The solutions are $x = 1$, $x = 4$.

Now Try Exercise 99 ■

1.5 | Exercises

■ Concepts

1. *Yes or No?* If *No*, give a reason.
   (a) When you add the same number to each side of an equation, do you always get an equivalent equation?
   (b) When you multiply each side of an equation by the same nonzero number, do you always get an equivalent equation?
   (c) When you square each side of an equation, do you always get an equivalent equation?

2. What is a logical first step in solving the equation?
(a) $(x + 5)^2 = 64$ **(b)** $(x + 5)^2 + 5 = 64$
(c) $x^2 + x = 2$

3. Explain how you would use each method to solve the equation $x^2 - 6x - 16 = 0$.
(a) By factoring: ______
(b) By completing the square: ______
(c) By using the Quadratic Formula: ______

4. (a) The Zero-Product Property says that if $a \cdot b = 0$, then either a or b must be ______.
(b) The solutions of the equation $x^2(x - 4) = 0$ are ______.
(c) To solve the equation $x^3 - 4x^2 = 0$, we ______ the left side.

5. Solve the equation $\sqrt{2x} + x = 0$ by doing the following steps.
(a) Isolate the radical: ______.
(b) Square both sides: ______.
(c) The solutions of the resulting quadratic equation are ______.
(d) The solution(s) that satisfy the original equation are ______.

6. The equation $(x + 1)^2 - 5(x + 1) + 6 = 0$ is of ______ type. To solve the equation, we set $W =$ ______. The resulting quadratic equation is ______.

7. To eliminate the denominators in the equation $\frac{3}{x} + \frac{5}{x+2} = 2$, we multiply each side by the lowest common denominator ______ to get the equivalent equation ______.

8. To eliminate the square root in the equation $2x + 1 = \sqrt{x + 1}$, we ______ each side to get the equation ______.

Skills

9–12 ■ Solution? Check whether the given value is a solution of the equation.

9. $4x + 7 = 9x - 3$
(a) $x = -2$ **(b)** $x = 2$

10. $1 - [2 - (3 - x)] = 4x - (6 + x)$
(a) $x = 2$ **(b)** $x = 4$

11. $\frac{1}{x} - \frac{1}{x - 4} = 1$
(a) $x = 2$ **(b)** $x = 4$

12. $\frac{x^{3/2}}{x - 6} = x - 8$
(a) $x = 4$ **(b)** $x = 8$

13–28 ■ Linear Equations The given equation is either linear or equivalent to a linear equation. Solve the equation.

13. $8x + 13 = 5$ **14.** $1 - x = 10$
15. $\frac{1}{2}x - 8 = 1$ **16.** $3 + \frac{1}{3}x = 5$
17. $-x + 3 = 4x$ **18.** $2x + 3 = 7 - 3x$
19. $\frac{x}{3} - 2 = \frac{5}{3}x + 7$ **20.** $\frac{2}{5}x - 1 = \frac{3}{10}x + 3$
21. $2(1 - x) = 3(1 + 2x) + 5$
22. $(3y + 2) - 5(2y - 1) = 2(3 - y) + 1$
23. $x - \frac{1}{3}x - \frac{1}{2}x - 5 = 0$ **24.** $2x - \frac{x}{2} + \frac{x + 1}{4} = 6x$
25. $\frac{2x - 1}{x + 2} = \frac{4}{5}$ **26.** $\frac{3}{x - 1} = \frac{5}{x + 2} - \frac{1}{x - 1}$
27. $\frac{1}{x} = \frac{4}{3x} + 1$ **28.** $\frac{3}{x + 1} - \frac{1}{2} = \frac{1}{3x + 3}$

29–40 ■ Solving for a Variable in a Formula Solve the equation for the indicated variable.

29. $E = \frac{1}{2}mv^2$; for m **30.** $F = G\frac{mM}{r^2}$; for m
31. $P = 2l + 2w$; for w **32.** $\frac{1}{R} = \frac{1}{R_1} + \frac{1}{R_2}$; for R_1
33. $\frac{ax + b}{cx + d} = 2$; for x
34. $a - 2[b - 3(c - x)] = 6$; for x
35. $a^2x + (a - 1) = (a + 1)x$; for x
36. $\frac{a + 1}{b} = \frac{a - 1}{b} + \frac{b + 1}{a}$; for a
37. $V = \frac{1}{3}\pi r^2 h$; for r **38.** $F = G\frac{mM}{r^2}$; for r
39. $a^2 + b^2 = c^2$; for b **40.** $A = P\left(1 + \frac{i}{100}\right)^2$; for i

41–46 ■ Solving by Factoring Find all real solutions of the equation by factoring.

41. $x^2 + x = 12$ **42.** $x^2 + 9x = -20$
43. $x^2 + 13x - 30 = 0$ **44.** $x^2 - 13x + 30 = 0$
45. $4x^2 - 4x - 15 = 0$ **46.** $2y^2 + 7y + 3 = 0$

47–52 ■ Solving Simple Quadratics Find all real solutions of the equation.

47. $2x^2 = 8$ **48.** $3x^2 - 27 = 0$
49. $(2x - 5)^2 = 81$ **50.** $(5x + 1)^2 + 3 = 10$
51. $8(x + 5)^2 = 16$ **52.** $2(3x - 1)^2 - 3 = 11$

53–60 ■ Completing the Square Find all real solutions of the equation by completing the square.

53. $x^2 + 10x + 2 = 0$ **54.** $x^2 - 4x + 2 = 0$
55. $x^2 - 6x - 11 = 0$ **56.** $x^2 + 3x - \frac{7}{4} = 0$
57. $5x^2 + 10x = 1$ **58.** $3x^2 - 6x - 1 = 0$
59. $4x^2 - x = 0$ **60.** $x^2 = \frac{3}{4}x - \frac{1}{8}$

61–72 ■ Using the Quadratic Equation Find all real solutions of the quadratic equation.

61. $x^2 - 2x - 15 = 0$ **62.** $x^2 - 13x + 42 = 0$
63. $2x^2 + x - 3 = 0$ **64.** $3x^2 + 7x + 4 = 0$
65. $3x^2 + 6x - 5 = 0$ **66.** $x^2 - 6x + 1 = 0$
67. $9x^2 + 12x + 4 = 0$ **68.** $4x^2 - 4x + 1 = 0$
69. $4x^2 + 16x = 9$ **70.** $4x = x^2 + 1$
71. $7x^2 = 2x - 4$ **72.** $z(z - 3) = 5$

73–78 ■ Discriminant Use the discriminant to determine the number of real solutions of the equation. Do not solve the equation.

73. $x^2 - 6x + 1 = 0$ **74.** $3x^2 = 6x - 9$

75. $x^2 + 2.20x + 1.21 = 0$ **76.** $x^2 + 2.21x + 1.21 = 0$

77. $4x^2 + 5x + \frac{13}{8} = 0$ **78.** $x^2 + rx - s = 0 \quad (s > 0)$

79–102 ■ Other Equations Find all real solutions of the equation.

79. $\dfrac{x^2}{x + 100} = 50$ **80.** $\dfrac{1}{x-1} - \dfrac{2}{x^2} = 0$

81. $\dfrac{y+1}{y^2+1} = \dfrac{2}{y+2}$ **82.** $\dfrac{3w+1}{w} = \dfrac{w-3}{w-1}$

83. $\dfrac{1}{x-1} + \dfrac{1}{x+2} = \dfrac{5}{4}$ **84.** $\dfrac{x+5}{x-2} = \dfrac{5}{x+2} + \dfrac{28}{x^2-4}$

85. $5 = \sqrt{4x - 3}$ **86.** $\sqrt{8x - 1} = 3$

87. $\sqrt{2x - 1} = \sqrt{3x - 5}$ **88.** $\sqrt{3 + x} = \sqrt{x^2 + 1}$

89. $\sqrt{2x + 1} + 1 = x$ **90.** $2x + \sqrt{x + 1} = 8$

91. $x^4 - 13x^2 + 40 = 0$ **92.** $x^6 - 2x^3 - 3 = 0$

93. $(x + 2)^4 - 5(x + 2)^2 + 4 = 0$

94. $2x^4 + 4x^2 + 1 = 0$

95. $x^{4/3} - 5x^{2/3} + 6 = 0$

96. $4(x + 1)^{1/2} - 5(x + 1)^{3/2} + (x + 1)^{5/2} = 0$

97. $x^{1/2} + 3x^{-1/2} = 10x^{-3/2}$ **98.** $x - 5\sqrt{x} + 6 = 0$

99. $|3x + 5| = 1$ **100.** $|2x| = 3$

101. $|x - 4| = 0.01$ **102.** $|x - 6| = -1$

103–118 ■ Putting It All Together Recognize the type of the equation and find all real solutions of the equation using an appropriate method.

103. $2x + \frac{1}{2} = \frac{1}{4}x - 3$ **104.** $x^2 - 7x + 12 = 0$

105. $\dfrac{4 - a}{3a - 2} = a$ **106.** $\dfrac{3x - 5}{x + 5} = 8$

107. $\dfrac{2}{3}y + \dfrac{1}{2}(y - 3) = \dfrac{y + 1}{4}$ **108.** $6x(x - 1) = 21 - x$

109. $x - \sqrt{9 - 3x} = 0$ **110.** $\sqrt{3}x + \sqrt{12} = \dfrac{x + 5}{\sqrt{3}}$

111. $\sqrt{x} - 3\sqrt[4]{x} - 4 = 0$ **112.** $x^6 - x^3 - 6 = 0$

113. $|3x - 10| = 29$ **114.** $x^{1/2} - 3x^{1/3} = 3x^{1/6} - 9$

115. $(z - 1)^2 - 8(z - 1) + 15 = 0$

116. $(t - 4)^2 = (t + 4)^2 + 32$

117. $|3x^2 - 1| - 4 = 7$

118. $\left(\dfrac{x}{x+1}\right)^2 - 6\left(\dfrac{x}{x+1}\right) + 8 = 0$

Skills Plus

119–124 ■ More on Solving Equations Find all real solutions of the equation.

119. $\dfrac{1}{x^3} + \dfrac{4}{x^2} + \dfrac{4}{x} = 0$ **120.** $4x^{-4} - 16x^{-2} + 4 = 0$

121. $\sqrt{\sqrt{x + 5} + x} = 5$ **122.** $\sqrt[3]{4x^2 - 4x} = x$

123. $x^2\sqrt{x + 3} = (x + 3)^{3/2}$

124. $\sqrt{11 - x^2} - \dfrac{2}{\sqrt{11 - x^2}} = 1$

125–128 ■ More on Solving Equations Solve the equation for the variable x. The constants a and b represent positive real numbers.

125. $x^4 - 5ax^2 + 4a^2 = 0$ **126.** $a^3x^3 + b^3 = 0$

127. $\sqrt{x + a} + \sqrt{x - a} = \sqrt{2}\sqrt{x + 6}$

128. $\sqrt{x} - a\sqrt[3]{x} + b\sqrt[6]{x} - ab = 0$

Applications

129–130 ■ Falling-Body Problems Suppose an object is dropped from a height h_0 above the ground. Then its height after t seconds is given by $h = -4.8t^2 + h_0$, where h is measured in meters. Use this information to solve the problem.

129. If a ball is dropped from 88 m above the ground, how long does it take the ball to reach ground level?

130. A ball is dropped from the top of a building 29 m tall.
(a) How long will it take the ball to fall half the distance to ground level?
(b) How long will it take the ball to fall to ground level?

131–132 ■ Falling-Body Problems Use the formula $h = -4.8t^2 + v_0t$ discussed in Example 9.

131. A ball is thrown straight upward at an initial speed of $v_0 = 12$ m/s.
(a) When does the ball reach a height of 7.2 m?
(b) When does the ball reach a height of 14.4 m?
(c) What is the greatest height reached by the ball?
(d) When does the ball reach the highest point of its path?
(e) When does the ball hit the ground?

132. How fast would a ball have to be thrown upward to reach a maximum height of 30 m? [*Hint:* Use the discriminant of the equation $4.8t^2 - v_0t + h = 0$.]

133. Shrinkage in Concrete Beams As concrete dries, it shrinks: the higher the water content, the greater the shrinkage. If a concrete beam has a water content of w kg/m³, then it will shrink by a factor

$$S = \frac{0.032w - 2.5}{10{,}000}$$

where S is the fraction of the original beam length that disappears due to shrinkage.

(a) A beam 12.025 m long is cast in concrete that contains 250 kg/m³ water. What is the shrinkage factor S? What length will the beam be when it has dried?

(b) A beam is 10.014 m long when wet. We want it to shrink to 10.009 m, so the shrinkage factor should be $S = 0.00050$. What water content will provide this amount of shrinkage?

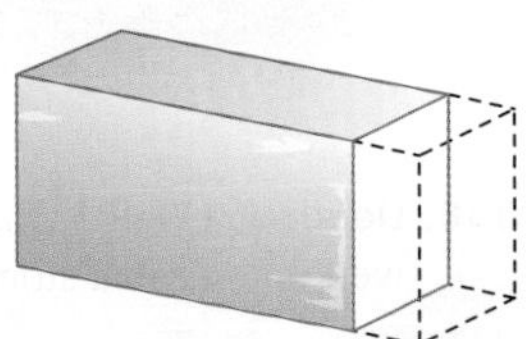

134. The Lens Equation If F is the focal length of a convex lens and an object is placed at a distance x from the lens, then its image will be located a distance y from the lens, where F, x, and y are related by the *lens equation*

$$\frac{1}{F} = \frac{1}{x} + \frac{1}{y}$$

Suppose that a lens has a focal length of 4.8 cm and that the image of an object is 4 cm closer to the lens than the object itself. How far from the lens is the object?

135. The Bernoulli Equation If a fluid with density ρ (kg/m^3) is pumped with velocity v (m/s) through a given pipe, then the pressure P(Pa) in the pipe satisfies Bernoulli's equation

$$\tfrac{1}{2}\rho v^2 + \rho g h + P = C$$

where h is the height (m) above some fixed reference point, g is acceleration due to gravity (9.81 m/s^2), and C is a constant specific to the given pipe. Suppose water ($\rho = 1000$ kg/m^3) is flowing through a pipe at height 1 m, with velocity 2 m/s and pressure 200,000 Pa.

(a) Find the constant C for the pipe.

(b) Solve for the variable P in Bernoulli's equation. If the height increases to 5 m, what is the new pressure? If the height returns to 1 m and the velocity increases to 4 m/s, what is the new pressure?

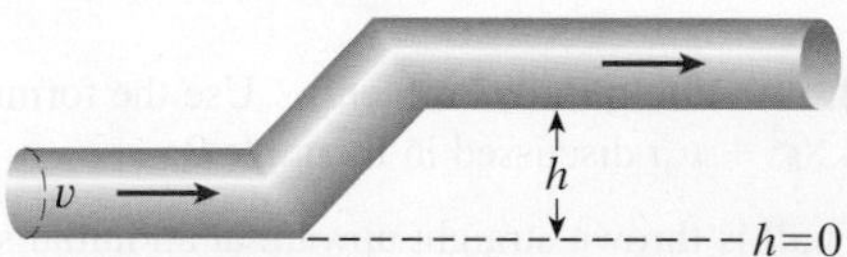

136. Profit A small-appliance manufacturer finds that the profit P (in dollars) generated by producing x microwave ovens per week is given by the formula $P = \frac{1}{10}x(300 - x)$, provided that $0 \le x \le 200$. How many ovens must be manufactured in a given week to generate a profit of \$1250?

137. Gravity If an imaginary line segment is drawn between the centers of the earth and the moon, then the net gravitational force F acting on an object situated on this line segment is

$$F = \frac{-K}{x^2} + \frac{0.012K}{(239 - x)^2} \qquad (0 < x < 239)$$

where $K > 0$ is a constant and x is the distance of the object from the center of the earth, measured in thousands of kilometers. How far from the center of the earth is the "dead spot" where no net gravitational force acts upon the object? (Express your answer to the nearest thousand kilometers.)

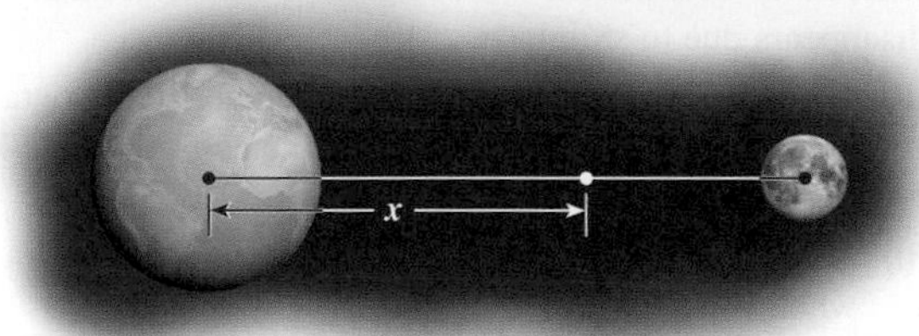

138. Depth of a Well One method for determining the depth of a well is to drop a stone into it and then measure the time it takes until the splash is heard. If d is the depth of the well (in meters) and t_1 the time (in seconds) it takes the stone to fall, then $d = 4.9t_1^2$, so $t_1 = \sqrt{d}/2.21$. Now if t_2 is the time it takes the sound to travel back up, then $d = 332t_2$ because the speed of sound is 332 m/s. So $t_2 = d/332$. Thus the total time elapsed between dropping the stone and hearing the splash is

$$t_1 + t_2 = \frac{\sqrt{d}}{2.21} + \frac{d}{332}$$

How deep is the well if this total time is 3 seconds?

■ **Discuss** ■ **Discover** ■ **Prove** ■ **Write**

139. Discuss: A Family of Equations The equation

$$3x + k - 5 = kx - k + 1$$

is really a **family of equations**, because for each value of k, we get a different equation with the unknown x. The letter k is called a **parameter** for this family. What value should we pick for k to make the given value of x a solution of the resulting equation?

(a) $x = 0$ (b) $x = 1$ (c) $x = 2$

140. Discuss: Proof that 0 = 1? The following steps appear to give equivalent equations, which seem to prove that $1 = 0$. Find the error.

$x = 1$	Given
$x^2 = x$	Multiply by x
$x^2 - x = 0$	Subtract x
$x(x - 1) = 0$	Factor
$\dfrac{x(x-1)}{x-1} = \dfrac{0}{x-1}$	Divide by $x - 1$
$x = 0$	Simplify
$1 = 0$	Given $x = 1$

141. Discover ■ Prove: Relationship Between Solutions and Coefficients The Quadratic Formula gives us the solutions of a quadratic equation from its coefficients. We can also obtain the coefficients from the solutions.

(a) Find the solutions of the equation $x^2 - 9x + 20 = 0$, and show that the product of the solutions is the constant term 20 and the sum of the solutions is 9, the negative of the coefficient of x.

(b) Use the Quadratic Formula to prove that, in general, if the equation $x^2 + bx + c = 0$ has solutions r_1 and r_2, then $c = r_1 r_2$ and $b = -(r_1 + r_2)$.

142. Discover ■ Prove: Depressed Quadratics A quadratic equation is called *depressed* if it is missing the x-term. For example, $x^2 - 5 = 0$ is a depressed quadratic.

(a) For the quadratic equation $x^2 + bx + c = 0$, show that the substitution $x = u - b/2$ transforms it into a depressed quadratic in the variable u:

$$u^2 - \left(\frac{b^2}{4} - c\right) = 0$$

(b) Solve the equation $x^2 + 5x - 6 = 0$ by first transforming it into a depressed quadratic, as described in part (a).

143. Discuss: Solving an Equation in Different Ways We have learned several ways to solve an equation in this section. Some equations can be tackled by more than one method. Solve the following equations using both methods indicated, and show that you get the same final answers.

(a) $x - \sqrt{x} - 2 = 0$ quadratic type; solve for the radical, and square

(b) $\dfrac{12}{(x-3)^2} + \dfrac{10}{x-3} + 1 = 0$ quadratic type; multiply by LCD

1.6 Complex Numbers

■ Arithmetic Operations on Complex Numbers ■ Square Roots of Negative Numbers
■ Complex Solutions of Quadratic Equations

In Section 1.5 we saw that if the discriminant of a quadratic equation is negative, the equation has no real solution. For example, the equation

$$x^2 + 4 = 0$$

has no real solution. If we try to solve this equation, we get $x^2 = -4$, so

$$x = \pm\sqrt{-4}$$

But this is impossible, since the square of any real number is positive. [For example, $(-2)^2 = 4$, a positive number.] Thus negative numbers don't have real square roots.

See the note on Cardano in Section 3.5 for an example of how complex numbers are used to find real solutions of polynomial equations.

To make it possible to solve *all* quadratic equations, mathematicians invented an expanded number system, called the *complex number system*. First they defined the new number

$$i = \sqrt{-1}$$

This means that $i^2 = -1$. A complex number is then a number of the form $a + bi$, where a and b are real numbers.

Definition of Complex Numbers

A **complex number** is an expression of the form

$$a + bi$$

where a and b are real numbers and $i^2 = -1$. The **real part** of this complex number is a, and the **imaginary part** is b. Two complex numbers are **equal** if and only if their real parts are equal and their imaginary parts are equal.

Note that both the real and imaginary parts of a complex number are real numbers.

Example 1 ■ Complex Numbers

The following are examples of complex numbers.

$3 + 4i$	Real part 3, imaginary part 4
$\frac{1}{2} - \frac{2}{3}i$	Real part $\frac{1}{2}$, imaginary part $-\frac{2}{3}$
$6i$	Real part 0, imaginary part 6
-7	Real part -7, imaginary part 0

Now Try Exercises 7 and 11 ■

A number such as $6i$, which has real part 0, is called a **pure imaginary number**. A real number such as -7 can be thought of as a complex number with imaginary part 0.

In the complex number system every quadratic equation has solutions. The numbers $2i$ and $-2i$ are solutions of $x^2 = -4$ because

$$(2i)^2 = 2^2i^2 = 4(-1) = -4 \quad \text{and} \quad (-2i)^2 = (-2)^2i^2 = 4(-1) = -4$$

Note Although we use the term *imaginary* in this context, imaginary numbers should not be thought of as any less "real" (in the ordinary rather than the mathematical sense of that word) than negative numbers or irrational numbers. All numbers (except possibly the positive integers) are creations of the human mind—the numbers -1 and $\sqrt{2}$ as well as the number i. We study complex numbers because they complete, in a useful and elegant fashion, our study of the solutions of equations. In fact, imaginary numbers are useful not only in algebra and mathematics, but in the other sciences as well. For example, in electrical theory complex numbers are used in modeling electrical *impedance*.

■ Arithmetic Operations on Complex Numbers

Complex numbers are added, subtracted, multiplied, and divided just as we would any number of the form $a + b\sqrt{c}$. The only difference that we need to keep in mind is that $i^2 = -1$. Thus the following calculations are valid.

$$\begin{aligned}(a + bi)(c + di) &= ac + (ad + bc)i + bdi^2 && \text{Multiply and collect like terms}\\ &= ac + (ad + bc)i + bd(-1) && i^2 = -1\\ &= (ac - bd) + (ad + bc)i && \text{Combine real and imaginary parts}\end{aligned}$$

We therefore define the sum, difference, and product of complex numbers as follows.

Adding, Subtracting, and Multiplying Complex Numbers

Definition	Description
Addition $(a + bi) + (c + di) = (a + c) + (b + d)i$	To add complex numbers, add the real parts and the imaginary parts.
Subtraction $(a + bi) - (c + di) = (a - c) + (b - d)i$	To subtract complex numbers, subtract the real parts and the imaginary parts.
Multiplication $(a + bi) \cdot (c + di) = (ac - bd) + (ad + bc)i$	Multiply complex numbers like binomials, using $i^2 = -1$.

Example 2 ■ Adding, Subtracting, and Multiplying Complex Numbers

When using FOIL to multiply complex numbers, the terms are real or imaginary as follows:

imaginary
FOIL
real

Express each of the following in the form $a + bi$.

(a) $(3 + 5i) + (4 - 2i)$ **(b)** $(3 + 5i) - (4 - 2i)$

(c) $(3 + 5i)(4 - 2i)$ **(d)** i^{23}

Solution

(a) According to the definition, we add the real parts and we add the imaginary parts:

$$(3 + 5i) + (4 - 2i) = (3 + 4) + (5 - 2)i = 7 + 3i$$

(b) $(3 + 5i) - (4 - 2i) = (3 - 4) + [5 - (-2)]i = -1 + 7i$

(c) $(3 + 5i)(4 - 2i) = [3 \cdot 4 - 5(-2)] + [3(-2) + 5 \cdot 4]i = 22 + 14i$

(d) $i^{23} = i^{22+1} = (i^2)^{11}i = (-1)^{11}i = (-1)i = -i$

Now Try Exercises 19, 23, 29, and 47 ■

Complex Conjugates

Number	Conjugate
$3 + 2i$	$3 - 2i$
$1 - i$	$1 + i$
$4i$	$-4i$
5	5

Division of complex numbers is much like rationalizing the denominator of a radical expression, which we considered in Section 1.2. For the complex number $z = a + bi$ we define its **complex conjugate** to be $\overline{z} = \overline{a + bi} = a - bi$. Note that

$$z \cdot \overline{z} = (a + bi)(a - bi) = a^2 + b^2$$

So the product of a complex number and its conjugate is always a nonnegative real number. We use this property to divide complex numbers.

Dividing Complex Numbers

To simplify the quotient $\dfrac{a + bi}{c + di}$, multiply the numerator and the denominator by the complex conjugate of the denominator:

$$\frac{a + bi}{c + di} = \left(\frac{a + bi}{c + di}\right)\left(\frac{c - di}{c - di}\right) = \frac{(ac + bd) + (bc - ad)i}{c^2 + d^2}$$

Rather than memorizing this entire formula, it is easier to just remember the first step and then multiply out the numerator and the denominator as usual.

Example 3 ■ Dividing Complex Numbers

Express each of the following in the form $a + bi$.

(a) $\dfrac{3 + 5i}{1 - 2i}$ **(b)** $\dfrac{7 + 3i}{4i}$

Solution We multiply both the numerator and denominator by the complex conjugate of the denominator to make the new denominator a real number.

(a) The complex conjugate of $1 - 2i$ is $\overline{1 - 2i} = 1 + 2i$. Therefore

$$\frac{3 + 5i}{1 - 2i} = \left(\frac{3 + 5i}{1 - 2i}\right)\left(\frac{1 + 2i}{1 + 2i}\right) = \frac{-7 + 11i}{5} = -\frac{7}{5} + \frac{11}{5}i$$

(b) The complex conjugate of $4i$ is $-4i$. Therefore

$$\frac{7 + 3i}{4i} = \left(\frac{7 + 3i}{4i}\right)\left(\frac{-4i}{-4i}\right) = \frac{12 - 28i}{16} = \frac{3}{4} - \frac{7}{4}i$$

Now Try Exercises 39 and 43 ■

■ Square Roots of Negative Numbers

Just as every positive real number r has two square roots ($\sqrt{r}$ and $-\sqrt{r}$), every negative number has two square roots. If $-r$ is a negative number, then its square roots are $\pm\sqrt{r}i$, because $(\sqrt{r}i)^2 = ri^2 = -r$ and $(-\sqrt{r}i)^2 = r(-1)^2i^2 = -r$.

Square Roots of Negative Numbers

If $-r$ is negative, then the **principal square root** of $-r$ is

$$\sqrt{-r} = i\sqrt{r}$$

The two square roots of $-r$ are $\sqrt{r}i$ and $-\sqrt{r}i$.

Example 4 ■ Square Roots of Negative Numbers

(a) $\sqrt{-1} = \sqrt{1}i = i$ **(b)** $\sqrt{-16} = \sqrt{16}i = 4i$ **(c)** $\sqrt{-3} = \sqrt{3}i$

 Now Try Exercises 53 and 55 ■

Special care must be taken in performing calculations that involve square roots of negative numbers. Although $\sqrt{a}\cdot\sqrt{b} = \sqrt{ab}$ when a and b are positive, this is *not* true when both are negative. For example,

$$\sqrt{-2}\cdot\sqrt{-3} = \sqrt{2}i\cdot\sqrt{3}i = \sqrt{6}i^2 = -\sqrt{6}$$

but

$$\sqrt{(-2)(-3)} = \sqrt{6}$$

so

$$\sqrt{-2}\cdot\sqrt{-3} \neq \sqrt{(-2)(-3)}$$

⊘ When multiplying radicals of negative numbers, express them first in the form $\sqrt{r}i$ (where $r > 0$) to avoid errors of this type.

Example 5 ■ Using Square Roots of Negative Numbers

Evaluate $(\sqrt{12} - \sqrt{-3})(3 + \sqrt{-4})$ and express the result in the form $a + bi$.

Solution

$$\begin{aligned}(\sqrt{12} - \sqrt{-3})(3 + \sqrt{-4}) &= (\sqrt{12} - \sqrt{3}i)(3 + \sqrt{4}i)\\ &= (2\sqrt{3} - \sqrt{3}i)(3 + 2i)\\ &= (6\sqrt{3} + 2\sqrt{3}) + (2\cdot 2\sqrt{3} - 3\sqrt{3})i\\ &= 8\sqrt{3} + \sqrt{3}i\end{aligned}$$

 Now Try Exercise 57 ■

■ Complex Solutions of Quadratic Equations

We have already seen that if $a \neq 0$, then the solutions of the quadratic equation $ax^2 + bx + c = 0$ are

The Quadratic Formula

$$x = \frac{-b \pm \sqrt{b^2 - 4ac}}{2a}$$

If $b^2 - 4ac < 0$, then the equation has no real solution. But in the complex number system this equation will always have solutions, because negative numbers have square roots in this expanded setting.

Library of Congress Prints and Photographs Division

LEONHARD EULER (1707–1783) was born in Basel, Switzerland, the son of a pastor. When Euler was 13, his father sent him to the University at Basel to study theology, but Euler soon devoted himself to the sciences. Besides theology he studied mathematics, medicine, astronomy, physics, and Asian languages. It is said that Euler could calculate as effortlessly as "men breathe or as eagles fly." One hundred years before Euler, Fermat (see Section 1.11) had conjectured that $2^{2^n} + 1$ is a prime number for all n. The first five of these numbers are

5 17 257 65537 4,294,967,297

It is easy to show that the first four are prime. The fifth was also thought to be prime until Euler, with his phenomenal calculating ability, showed that it is the product 641 × 6,700,417 and so is not prime. Euler published more than any other mathematician in history. His collected works comprise 75 large volumes. Although he was blind for the last 17 years of his life, he continued to work and publish. In his writings he popularized the use of the symbols π, e, and i, which you will find in this textbook. One of Euler's enduring contributions is his development of complex numbers.

Example 6 ■ Quadratic Equations with Complex Solutions

Solve each equation.

(a) $x^2 + 9 = 0$ **(b)** $x^2 + 4x + 5 = 0$

Solution

(a) The equation $x^2 + 9 = 0$ means $x^2 = -9$, so

$$x = \pm\sqrt{-9} = \pm\sqrt{9}\,i = \pm 3i$$

The solutions are therefore $3i$ and $-3i$.

(b) By the Quadratic Formula, we have

$$x = \frac{-4 \pm \sqrt{4^2 - 4 \cdot 5}}{2} = \frac{-4 \pm \sqrt{-4}}{2} = \frac{-4 \pm 2i}{2} = -2 \pm i$$

So the solutions are $-2 + i$ and $-2 - i$.

 **Now Try Exercises 61 and 63** ■

We see from Example 6 that if a quadratic equation with real coefficients has complex solutions, then these solutions are complex conjugates of each other. So if $a + bi$ is a solution, then $a - bi$ is also a solution.

Example 7 ■ Complex Conjugates as Solutions of a Quadratic

Show that the solutions of the equation

$$4x^2 - 24x + 41 = 0$$

are complex conjugates of each other.

Solution We use the Quadratic Formula to get

$$x = \frac{-(-24) \pm \sqrt{(-24)^2 - 4(4)(41)}}{2(4)} = \frac{24 \pm \sqrt{-80}}{8} = \frac{24 \pm 4\sqrt{5}\,i}{8} = 3 \pm \frac{\sqrt{5}}{2}i$$

So the solutions are $3 + \frac{\sqrt{5}}{2}i$ and $3 - \frac{\sqrt{5}}{2}i$, and these are complex conjugates.

 **Now Try Exercise 69** ■

1.6 Exercises

■ Concepts

1. The imaginary number i has the property that $i^2 =$ ______.

2. For the complex number $3 + 4i$ the real part is ______ and the imaginary part is ______.

3. (a) The complex conjugate of $3 + 4i$ is $\overline{3 + 4i} =$ ______.

(b) $(3 + 4i)(\overline{3 + 4i}) =$ ______.

4. If $3 + 4i$ is a solution of a quadratic equation with real coefficients, then ______ is also a solution of the equation.

5–6 ■ *Yes or No?* If *No*, give a reason.

5. Is every real number also a complex number?

6. Is the sum of a complex number and its complex conjugate a real number?

■ Skills

7–16 ■ Real and Imaginary Parts Find the real and imaginary parts of the complex number.

7. $3 - 8i$ **8.** $-(5 - i)$

9. $\dfrac{-2 - 5i}{3}$ **10.** $\dfrac{4 + 7i}{2}$

11. 3 **12.** $-\frac{1}{2}$

13. $-\frac{2}{3}i$ **14.** $\sqrt{3}i$

15. $\sqrt{3} + \sqrt{-4}$ **16.** $2 - \sqrt{-5}$

17–26 ■ Sums and Differences Evaluate the sum or difference and write the result in the form $a + bi$.

17. $(3 + 2i) + 5i$ **18.** $3i - (2 - 3i)$

19. $(5 - 3i) + (-4 - 7i)$ **20.** $(-3 + 4i) - (2 - 5i)$

21. $(-6 + 6i) + (9 - i)$ **22.** $(3 - 2i) + (-5 - \frac{1}{3}i)$

23. $(7 - \frac{1}{2}i) - (5 + \frac{3}{2}i)$ **24.** $(-4 + i) - (2 - 5i)$

25. $(-12 + 8i) - (7 + 4i)$ **26.** $6i - (4 - i)$

27–36 ■ Products Evaluate the product and write the result in the form $a + bi$.

27. $4(-1 + 2i)$ **28.** $-2(3 - 4i)$

29. $(3 - 4i)(2 + 5i)$ **30.** $(-5 + i)(6 - 2i)$

31. $(6 + 5i)(2 - 3i)$ **32.** $(-2 + i)(3 - 7i)$

33. $(3 - 2i)(3 + 2i)$ **34.** $(10 + i)(10 - i)$

35. $(3 - 2i)^2$ **36.** $(10 + i)^2$

37–46 ■ Quotients Evaluate the quotient and write the result in the form $a + bi$.

37. $\dfrac{1}{i}$ **38.** $\dfrac{1}{1 + i}$

39. $\dfrac{1 - 3i}{1 + 2i}$ **40.** $\dfrac{2 - i}{1 - 3i}$

41. $\dfrac{10i}{1 - 2i}$ **42.** $(2 - 3i)^{-1}$

43. $\dfrac{4 + 6i}{3i}$ **44.** $\dfrac{-3 + 5i}{15i}$

45. $\dfrac{1}{1 + i} - \dfrac{1}{1 - i}$ **46.** $\dfrac{(1 + 2i)(3 - i)}{2 + i}$

47–52 ■ Powers Evaluate the power, and write the result in the form $a + bi$.

47. i^3 **48.** i^{10}

49. $(3i)^5$ **50.** $(2i)^4$

51. i^{1000} **52.** i^{1002}

53–60 ■ Radical Expressions Evaluate the radical expression and express the result in the form $a + bi$.

53. $\sqrt{-25}$ **54.** $\dfrac{\sqrt{-8}}{\sqrt{2}}$

55. $\sqrt{-4}\sqrt{-9}$ **56.** $\sqrt{\frac{1}{2}}\sqrt{-32}$

57. $(2 + \sqrt{-1})(3 - \sqrt{-3})$

58. $(\sqrt{3} - \sqrt{-4})(\sqrt{6} - \sqrt{-8})$

59. $\dfrac{2 + \sqrt{-8}}{1 + \sqrt{-2}}$ **60.** $\dfrac{\sqrt{-36}}{\sqrt{-2}\sqrt{-9}}$

61–72 ■ Quadratic Equations Find all solutions of the equation and express them in the form $a + bi$.

61. $x^2 + 25 = 0$ **62.** $2x^2 + 5 = 0$

63. $x^2 - 6x + 13 = 0$ **64.** $x^2 + 2x + 2 = 0$

65. $2x^2 - 2x + 5 = 0$ **66.** $8x^2 - 12x + 5 = 0$

67. $x^2 + x + 1 = 0$ **68.** $x^2 - 3x + 3 = 0$

69. $9x^2 - 4x + 4 = 0$ **70.** $t + 2 + \dfrac{6}{t} = 0$

71. $6x^2 + 12x + 7 = 0$ **72.** $x^2 + \frac{1}{2}x + 1 = 0$

■ Skills Plus

73–76 ■ Conjugates Evaluate the given expression for $z = 3 - 4i$ and $w = 5 + 2i$.

73. $\overline{z} + \overline{w}$ **74.** $\overline{z + w}$

75. $z \cdot \overline{z}$ **76.** $\overline{z} \cdot \overline{w}$

77–84 ■ Conjugates If $z = a + bi$ and $w = c + di$, show that each statement is true.

77. $\overline{z + w} = \overline{z} + \overline{w}$ **78.** $\overline{zw} = \overline{z} \cdot \overline{w}$

79. $(\overline{z})^2 = \overline{z^2}$ **80.** $\overline{\overline{z}} = z$

81. $z + \overline{z}$ is a real number.

82. $z - \overline{z}$ is a pure imaginary number.

83. $z \cdot \overline{z}$ is a real number.

84. $z = \overline{z}$ if and only if z is real.

■ Discuss ■ Discover ■ Prove ■ Write

85. Prove: Complex Conjugate Solutions Suppose that the equation $ax^2 + bx + c = 0$ has real coefficients and complex solutions. Why must the solutions be complex conjugates of each other? [*Hint:* Think about how you would find the solutions using the Quadratic Formula.]

86. Discuss: Powers of *i* Explain how you would calculate any whole number power of i, and then calculate i^{4446}.

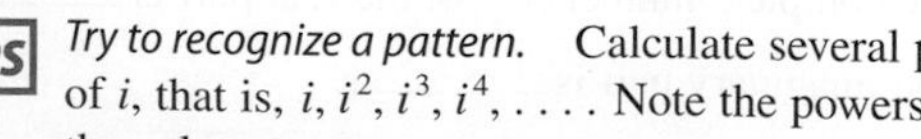

PS *Try to recognize a pattern.* Calculate several powers of i, that is, $i, i^2, i^3, i^4, \ldots$. Note the powers at which the values repeat.

1.7 Modeling with Equations

■ Making and Using Models ■ Problems About Interest ■ Problems About Area or Length ■ Problems About Mixtures ■ Problems About the Time Needed to Do a Job ■ Problems About Distance, Rate, and Time

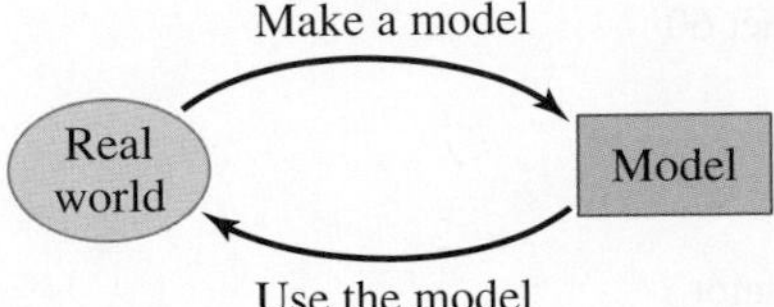

In this section a **mathematical model** is an equation that describes a real-world object or process. Modeling is the process of finding such equations. Once the model or equation has been found, it is then used to obtain information about the thing being modeled. The process is described in the diagram in the margin. In this section we learn how to make and use models to solve real-world problems.

■ Making and Using Models

We will use the following guidelines to help us set up equations that model situations described in words. To show how the guidelines can help you set up equations, we refer to them as we work each example in this section.

Guidelines for Modeling with Equations

1. **Identify the Variable.** Identify the quantity that the problem asks you to find. This quantity can usually be determined by a careful reading of the question that is posed at the end of the problem. Then **introduce notation** for the variable (call it x or some other letter).
2. **Translate from Words to Algebra.** Read each sentence in the problem again, and express all the quantities mentioned in the problem in terms of the variable you defined in Step 1. To organize this information, it is sometimes helpful to **draw a diagram** or **make a table.**
3. **Set Up the Model.** Find the crucial fact in the problem that gives a relationship between the expressions you listed in Step 2. **Set up an equation** (or **model**) that expresses this relationship.
4. **Solve the Equation and Check Your Answer.** Solve the equation, check your answer, and **state your answer** as a sentence.

The following example illustrates how these guidelines are used to translate a "word problem" into the language of algebra.

Example 1 ■ Renting a Car

A car rental company charges \$30 a day and 15¢ a kilometer for renting a car. A tourist rents a car for two days, and the bill comes to \$108. How many kilometers was the car driven?

Solution **Identify the variable.** We are asked to find the number of kilometers driven. So we let

$$x = \text{number of kilometers driven}$$

Translate from words to algebra. Now we translate all the information given in the problem into the language of algebra.

In Words	In Algebra
Number of kilometers driven	x
Mileage cost (at \$0.15 per km)	$0.15x$
Daily cost (at \$30 per day)	$2(30)$

Set up the model. Now we set up the model.

mileage cost	+	daily cost	=	total cost

$$0.15x + 2(30) = 108$$

Solve. Now we solve for x.

$$0.15x = 48 \quad \text{Subtract 60}$$

$$x = \frac{48}{0.15} \quad \text{Divide by 0.15}$$

$$x = 320 \quad \text{Calculator}$$

The rental car was driven 320 kilometers.

Check Your Answer

total cost = mileage cost + daily cost

$= 0.15(320) + 2(30)$

$= 108$ 

Now Try Exercise 21 ■

In the examples and exercises that follow, we construct equations that model problems in many different real-life situations.

■ Problems About Interest

Compound interest is studied in Section 4.1.

When you borrow money from a bank or when a bank "borrows" your money by keeping it for you in a savings account, the borrower in each case must pay for the privilege of using the money. The fee that is paid is called **interest**. The most basic type of interest is **simple interest**, which is just an annual percentage of the total amount borrowed or deposited. The amount of a loan or deposit is called the **principal** P. The annual percentage paid for the use of this money is the **interest rate** r. We will use the variable t to stand for the number of years that the money is deposited (or borrowed) and the variable I to stand for the total interest earned (or paid). The following **simple interest formula** gives the amount of interest I when a principal P is deposited (or borrowed) for t years at an interest rate r.

$$I = Prt$$

 When using this formula, remember to convert r from a percentage to a decimal. For example, in decimal form, 5% is 0.05. So at an interest rate of 5%, the interest paid on a \$1000 deposit over a 3-year period is $I = Prt = 1000(0.05)(3) = \$150$.

Example 2 ■ Interest on an Investment

An amount of \$100,000 is invested in two certificates of deposit. One certificate pays 6% and the other pays $4\frac{1}{2}\%$ simple interest annually. If the total interest is \$5025 per year, how much money is invested at each rate?

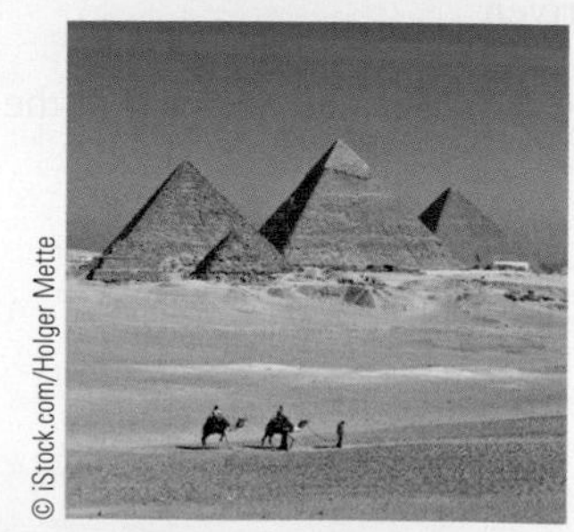

Discovery Project ■ Equations Through the Ages

Equations have always been important in solving real-world problems. Very old manuscripts from Babylon, Egypt, India, and China show that ancient peoples used equations to solve real-world problems that they encountered. In this project we discover that they also solved equations just for fun or for practice. You can find the project at **www.stewartmath.com**.

Solution **Identify the variable.** The problem asks for the amount invested at each rate. So we can let

$$x = \text{the amount invested at } 6\%$$

Translate from words to algebra. Since the total amount is \$100,000, it follows that $100{,}000 - x$ is invested at $4\frac{1}{2}\%$. We translate all the information given into the language of algebra.

In Words	In Algebra
Amount invested at 6%	x
Amount invested at $4\frac{1}{2}\%$	$100{,}000 - x$
Interest earned at 6%	$0.06x$
Interest earned at $4\frac{1}{2}\%$	$0.045(100{,}000 - x)$

Set up the model. We use the fact that the total interest earned is \$5025 to set up the model.

$$\boxed{\text{interest at } 6\%} + \boxed{\text{interest at } 4\tfrac{1}{2}\%} = \boxed{\text{total interest}}$$

$$0.06x + 0.045(100{,}000 - x) = 5025$$

Solve. Now we solve for x.

$$0.06x + 4500 - 0.045x = 5025 \qquad \text{Distributive Property}$$

$$0.015x + 4500 = 5025 \qquad \text{Combine the } x\text{-terms}$$

$$0.015x = 525 \qquad \text{Subtract 4500}$$

$$x = \frac{525}{0.015} = 35{,}000 \qquad \text{Divide by 0.015}$$

So \$35,000 is invested at 6% and the remaining \$65,000 at $4\frac{1}{2}\%$.

Check Your Answer

$$\text{total interest} = 6\% \text{ of } \$35{,}000 + 4\tfrac{1}{2}\% \text{ of } \$65{,}000$$

$$= \$2100 + \$2925 = \$5025 \quad ✓$$

Now Try Exercise 25 ■

■ Problems About Area or Length

You can find formulas for area and perimeter in the front endpapers of this book.

When we use algebra to model a physical situation, we must sometimes use basic formulas from geometry. For example, we may need a formula for an area or a perimeter, or the formula that relates the sides of similar triangles, or the Pythagorean Theorem. The next two examples use these geometric formulas to solve some real-world problems.

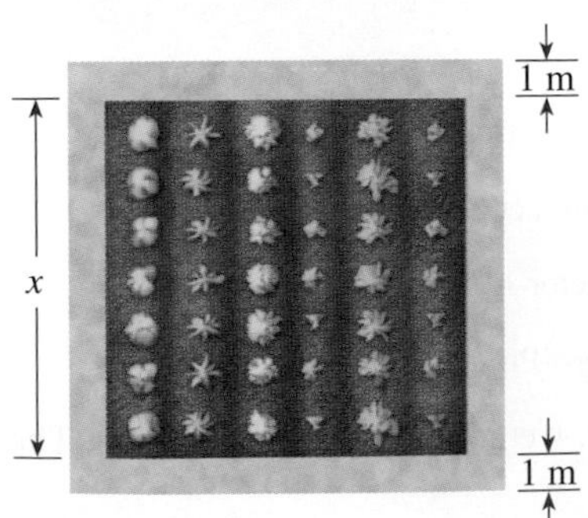

Figure 1

Example 3 ■ Dimensions of a Garden

A square garden has a walkway 1 m wide around its outer edge, as shown in Figure 1. If the area of the entire garden, including the walkway, is 1672 m^2, what are the dimensions of the planted area?

Solution **Identify the variable.** We are asked to find the length and width of the planted area. So we let

$$x = \text{the length of the planted area}$$

Translate from words to algebra. Next, translate the information from Figure 1 into the language of algebra.

In Words	In Algebra
Length of planted area	x
Length of entire garden	$x + 2$
Area of entire garden	$(x + 2)^2$

Set up the model. We now set up the model.

$$\boxed{\text{area of entire garden}} = \boxed{1672 \text{ m}^2}$$

$$(x + 2)^2 = 1672$$

Solve. Now we solve for x.

$$x + 2 = \sqrt{1672} \qquad \text{Take square roots}$$

$$x = \sqrt{1672} - 2 \qquad \text{Subtract 2}$$

$$x \approx 39$$

The planted area of the garden is about 39 m by 39 m.

 Now Try Exercise 49 ■

Example 4 ■ Dimensions of a Building Lot

A rectangular building lot is 8 m longer than it is wide and has an area of 2900 m². Find the dimensions of the lot.

Solution **Identify the variable.** We are asked to find the width and length of the lot. So let

$$w = \text{width of lot}$$

Figure 2

Translate from words to algebra. Then we translate the information given in the problem into the language of algebra (see Figure 2).

In Words	In Algebra
Width of lot	w
Length of lot	$w + 8$

Set up the model. Now we set up the model.

$$\boxed{\begin{matrix}\text{width}\\ \text{of lot}\end{matrix}} \cdot \boxed{\begin{matrix}\text{length}\\ \text{of lot}\end{matrix}} = \boxed{\begin{matrix}\text{area}\\ \text{of lot}\end{matrix}}$$

$$w(w + 8) = 2900$$

Solve. Now we solve for w.

$$w^2 + 8w = 2900 \qquad \text{Expand}$$

$$w^2 + 8w - 2900 = 0 \qquad \text{Subtract 2900}$$

$$(w - 50)(w + 58) = 0 \qquad \text{Factor}$$

$$w = 50 \quad \text{or} \quad w = -58 \qquad \text{Zero-Product Property}$$

Since the width of the lot must be a positive number, we conclude that $w = 50$ m. The length of the lot is $w + 8 = 50 + 8 = 58$ m.

Now Try Exercise 41 ■

Example 5 ■ Determining the Height of a Building Using Similar Triangles

A person needs to find the height of a certain four-story building and observes that the shadow of the building is 8.5 m long. The person is 1.8 m tall and has a shadow 1 m long when standing next to the building. How tall is the building?

Solution **Identify the variable.** The problem asks for the height of the building, so let

$$h = \text{the height of the building}$$

Similar triangles are studied in Appendix A, *Geometry Review*.

Translate from words to algebra. We use the fact that the triangles in Figure 3 are similar. Recall that for any pair of similar triangles the ratios of corresponding sides are equal. Now we translate these observations into the language of algebra.

In Words	In Algebra
Height of building	h
Ratio of height to base in large triangle	$\frac{h}{8.5}$
Ratio of height to base in small triangle	$\frac{1.8}{1}$

Figure 3

Set up the model. Since the large and small triangles are similar, we get the equation

$$\boxed{\text{ratio of height to base in large triangle}} = \boxed{\text{ratio of height to base in small triangle}}$$

$$\frac{h}{8.5} = \frac{1.8}{1}$$

Solve. Now we solve for h.

$$h = \frac{1.8 \cdot 8.5}{1} = 15.3 \qquad \text{Multiply by 8.5}$$

The building is 15.3 m tall.

Now Try Exercise 53 ■

■ Problems About Mixtures

Many real-world problems involve mixing different types of substances. For example, construction workers may mix cement, gravel, and sand; fruit juice concentrate may involve a mixture of different types of juices. Problems involving mixtures and concentrations make use of the fact that if an amount x of a substance is dissolved in a solution with volume V, then the concentration C of the substance is given by

$$C = \frac{x}{V}$$

So if 10 g of sugar is dissolved in 5 L of water, then the resulting sugar concentration is $C = 10/5 = 2$ g/L. Solving a mixture problem usually requires us to analyze the amount x of the substance in the solution. When we solve for x in this equation, we see that $x = CV$. In many mixture problems the concentration C is expressed as a percentage, as in the next example.

Example 6 ■ Mixtures and Concentration

A manufacturer of soft drinks advertises their orange soda as "naturally flavored," although it contains only 5% orange juice. A new federal regulation stipulates that to be called "natural," a drink must contain at least 10% fruit juice. How much pure orange juice must this manufacturer add to 900 L of orange soda to conform to the new regulation?

Solution **Identify the variable.** The problem asks for the amount of pure orange juice to be added. So let

$$x = \text{the amount (in liters) of pure orange juice to be added}$$

Translate from words to algebra. In any problem of this type—in which two different substances are to be mixed—drawing a diagram helps us to organize the given information (see Figure 4). The information in the figure can be translated into the language of algebra, as follows.

In Words	In Algebra
Amount of orange juice to be added	x
Amount of the new mixture	$900 + x$
Amount of orange juice in the first vat	$0.05(900) = 45$
Amount of orange juice in the second vat	$1 \cdot x = x$
Amount of orange juice in the new mixture	$0.10(900 + x)$

Figure 4

Set up the model. To set up the model, we use the fact that the total amount of orange juice in the new mixture is equal to the orange juice in the first two vats.

amount of orange juice in first vat	+	amount of orange juice in second vat	=	amount of orange juice in the new mixture

$$45 + x = 0.1(900 + x) \qquad \text{From Figure 4}$$

Solve. Now we solve for x.

$$45 + x = 90 + 0.1x \qquad \text{Distributive Property}$$

$$0.9x = 45 \qquad \text{Subtract } 0.1x \text{ and } 45$$

$$x = \frac{45}{0.9} = 50 \qquad \text{Divide by } 0.9$$

The manufacturer should add 50 L of pure orange juice to the soda.

Check Your Answer

amount of juice before mixing = 5% of 900 L + 50 L pure juice

= 45 L + 50 L = 95 L

amount of juice after mixing = 10% of 950 L = 95 L

Amounts are equal. 

Now Try Exercise 55 ■

■ Problems About the Time Needed to Do a Job

When solving a problem that involves determining how long it takes several workers to complete a job, we use the fact that if a person or machine takes H time units to complete the task, then in one time unit the fraction of the task that has been completed is $1/H$. For example, if a worker takes 5 hours to mow a lawn, then in 1 hour the worker will mow $1/5$ of the lawn.

Example 7 ■ Time Needed to Do a Job

Because of an anticipated heavy rainstorm, the water level in a reservoir must be lowered by 1 m. Opening spillway A (see the diagram in the margin) lowers the level by this amount in 4 hours, whereas opening the smaller spillway B does the job in 6 hours. How long will it take to lower the water level by 1 m if both spillways are opened?

Solution **Identify the variable.** We are asked to find the time needed to lower the level by 1 m if both spillways are opened. So let

x = the time (in hours) it takes to lower the water level by 1 m if both spillways are opened

Translate from words to algebra. Finding an equation relating x to the other quantities in this problem is not easy. Certainly x is not simply $4 + 6$ because that would mean that together the two spillways require longer to lower the water level than either spillway alone. Instead, *we look at the fraction of the job that can be done in 1 hour by each spillway*.

In Words	**In Algebra**
Time it takes to lower level 1 m with A and B together	x h
Distance A lowers level in 1 h	$\frac{1}{4}$ m
Distance B lowers level in 1 h	$\frac{1}{6}$ m
Distance A and B together lower levels in 1 h	$\frac{1}{x}$ m

Set up the model. Now we set up the model.

fraction done by A + fraction done by B = fraction done by both

$$\frac{1}{4} + \frac{1}{6} = \frac{1}{x}$$

Solve. Now we solve for x.

$$3x + 2x = 12 \quad \text{Multiply by the LCD, } 12x$$

$$5x = 12 \quad \text{Add}$$

$$x = \frac{12}{5} \quad \text{Divide by 5}$$

It will take $2\frac{2}{5}$ hours, or 2 h 24 min, to lower the water level by 1 m if both spillways are opened.

 Now Try Exercise 63 ■

■ Problems About Distance, Rate, and Time

The next example deals with distance, rate (speed), and time. The formula to keep in mind here is

distance = rate × time

where the rate is either the constant speed or average speed of a moving object. For example, driving at 60 km/h for 4 hours takes you a distance of $60 \cdot 4 = 240$ km.

Example 8 ■ A Distance-Speed-Time Problem

A jet flew from New York to Los Angeles, a distance of 4200 km. The speed for the return trip was 100 km/h faster than the outbound speed. If the total trip took 13 hours of flying time, what was the jet's speed from New York to Los Angeles?

Solution **Identify the variable.** We are asked for the speed of the jet from New York to Los Angeles. So let

$$s = \text{speed from New York to Los Angeles}$$

Then $\quad s + 100 = \text{speed from Los Angeles to New York}$

Translate from words to algebra. Now we organize the information in a table. We fill in the "Distance" column first, since we know that the cities are 4200 km apart. Then we fill in the "Speed" column, since we have expressed both speeds (rates) in terms of the variable s. Finally, we calculate the entries for the "Time" column, using

$$\text{time} = \frac{\text{distance}}{\text{rate}}$$

	Distance (km)	Speed (km/h)	Time (h)
N.Y. to L.A.	4200	s	$\dfrac{4200}{s}$
L.A. to N.Y.	4200	$s + 100$	$\dfrac{4200}{s + 100}$

Set up the model. The total trip took 13 hours, so we have the model

$$\boxed{\text{time from N.Y. to L.A.}} + \boxed{\text{time from L.A. to N.Y.}} = \boxed{\text{total time}}$$

$$\frac{4200}{s} + \frac{4200}{s + 100} = 13$$

Solve. Multiplying by the common denominator, $s(s + 100)$, we get

$$4200(s + 100) + 4200s = 13s(s + 100)$$

$$8400s + 420{,}000 = 13s^2 + 1300s$$

$$0 = 13s^2 - 7100s - 420{,}000$$

Although this equation does factor, with numbers this large it is probably quicker to use the Quadratic Formula and a calculator.

$$s = \frac{7100 \pm \sqrt{(-7100)^2 - 4(13)(-420{,}000)}}{2(13)}$$

$$= \frac{7100 \pm 8500}{26}$$

$$s = 600 \qquad \text{or} \qquad s = \frac{-1400}{26} \approx -53.8$$

Since s represents speed, we reject the negative answer and conclude that the jet's speed from New York to Los Angeles was 600 km/h.

 **Now Try Exercise 69** ■

Example 9 ■ Energy Expended in Bird Flight

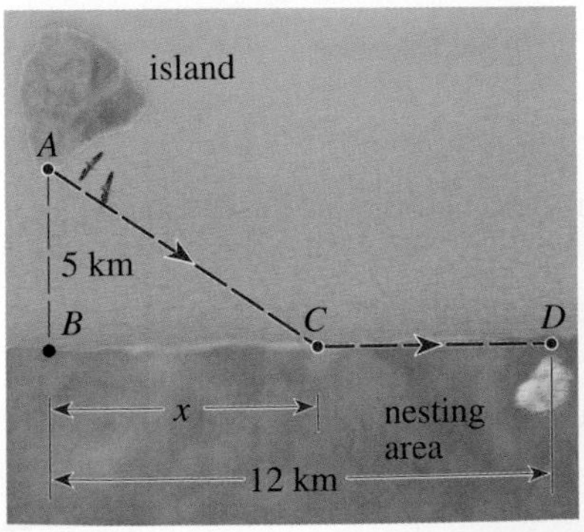

Ornithologists have determined that some species of birds tend to avoid flights over large bodies of water during daylight hours, because air generally rises over land and falls over water in the daytime, so flying over water requires more energy. A bird is released from point A on an island, 5 km from B, the nearest point on a straight shoreline. The bird flies to a point C on the shoreline and then flies along the shoreline to its nesting area D, as shown in Figure 5. Suppose the bird has 170 kJ of energy in reserve. It uses 10 kJ/km flying over land and 14 kJ/km flying over water.

(a) Where should the point C be located so that the bird uses exactly 170 kJ of energy during its flight?

(b) Does the bird have enough energy in reserve to fly directly from A to D?

Figure 5

HYPATIA OF ALEXANDRIA (circa 350–370, died 415) was a Greek mathematician, astronomer, and philosopher. She is best known for her mathematical work on conic sections. She also edited the sevenvolume work *On Conics* by the Greek mathematician Apollonius of Perga (c. 240–c. 290 BC). Hypatia was the daughter of the prominent philosopher Theon, so she received an education rarely afforded girls in those days; she soon excelled as a mathematician and philosopher. Socrates Scholasticus, a church historian and contemporary of Hypatia, wrote that Hypatia "made such achievements in literature and science, as to far exceed all the philosophers of her own time." Hypatia was also a talented teacher; letters from her student Synesius (who later became the Bishop of Ptolemais) show his admiration for Hypatia's mathematical knowledge. Hypatia wrote about learning: "To understand the things that are at our door is the best preparation for understanding those that lie beyond." The 2009 film *Agora* is a historical drama about the life of Hypatia.

Solution

(a) **Identify the variable.** We are asked to find the location of C. So let

$$x = \text{distance from } B \text{ to } C$$

Translate from words to algebra. From the figure, and from the fact that

$$\text{energy used} = \text{energy per kilometer} \times \text{kilometers flown}$$

we determine the following:

In Words	In Algebra	
Distance from B to C	x	
Distance flown over water (from A to C)	$\sqrt{x^2 + 25}$	Pythagorean Theorem
Distance flown over land (from C to D)	$12 - x$	
Energy used over water	$14\sqrt{x^2 + 25}$	
Energy used over land	$10(12 - x)$	

Set up the model. Now we set up the model.

$$\boxed{\text{total energy used}} = \boxed{\text{energy used over water}} + \boxed{\text{energy used over land}}$$

$$170 = 14\sqrt{x^2 + 25} + 10(12 - x)$$

Solve. To solve this equation, we eliminate the square root by first bringing all other terms to the left of the equal sign and then squaring each side.

$170 - 10(12 - x) = 14\sqrt{x^2 + 25}$	Isolate square-root term on RHS
$50 + 10x = 14\sqrt{x^2 + 25}$	Simplify LHS
$(50 + 10x)^2 = (14)^2(x^2 + 25)$	Square each side
$2500 + 1000x + 100x^2 = 196x^2 + 4900$	Expand
$0 = 96x^2 - 1000x + 2400$	Move all terms to RHS

This equation could be factored, but because the numbers are so large, it is easier to use the Quadratic Formula and a calculator.

$$x = \frac{1000 \pm \sqrt{(-1000)^2 - 4(96)(2400)}}{2(96)}$$

$$= \frac{1000 \pm 280}{192} = 6\tfrac{2}{3} \quad \text{or} \quad 3\tfrac{3}{4}$$

Point C should be either $6\frac{2}{3}$ km or $3\frac{3}{4}$ km from B so that the bird uses exactly 170 kJ of energy during its flight.

See Appendix A, *Geometry Review*, for the Pythagorean Theorem.

(b) By the Pythagorean Theorem the length of the route directly from A to D is $\sqrt{5^2 + 12^2} = 13$ km, so the energy the bird requires for that route is $14 \times 13 = 182$ kJ. This is more energy than the bird has available, so it can't use this route.

Now Try Exercise 85 ■

1.7 | Exercises

Concepts

1. Explain in your own words what it means for an equation to model a real-world situation, and give an example.

2. In the formula $I = Prt$ for simple interest, P stands for ________, r for ________, and t for ________.

3. Give a formula for the area of the geometric figure.
   (a) A square of side x: $A =$ ________.
   (b) A rectangle of length l and width w: $A =$ ________.
   (c) A circle of radius r: $A =$ ________.

4. Balsamic vinegar contains 5% acetic acid, so a 960 mL bottle of balsamic vinegar contains ________ milliliters of acetic acid.

5. A mason builds a wall in x hours, so the fraction of the wall that the mason builds in 1 hour is ________.

6. The formula $d = rt$ models the distance d traveled by an object moving at the constant rate r in time t. Find formulas for the following quantities.

   $r =$ ________ $t =$ ________

Skills

7–20 ■ Using Variables Express the given quantity in terms of the indicated variable.

7. The sum of three consecutive integers; n = first integer of the three

8. The sum of three consecutive integers; n = middle integer of the three

9. The sum of three consecutive even integers; n = first integer of the three

10. The sum of the squares of two consecutive integers; n = first integer of the two

11. The average of three test scores if the first two scores are 78 and 82; s = third test score

12. The average of four quiz scores if each of the first three scores is 8; q = fourth quiz score

13. The interest obtained after 1 year on an investment at $2\frac{1}{2}\%$ simple interest per year; x = number of dollars invested

14. The total rent paid for an apartment if the rent is \$945 a month; n = number of months

15. The area (in m^2) of a rectangle that is four times as long as it is wide; w = width of the rectangle (in m)

16. The perimeter (in cm) of a rectangle that is 6 cm longer than it is wide; w = width of the rectangle (in cm)

17. The time (in hours) that it takes to travel a given distance at 55 km/h; d = given distance (in km)

18. The distance (in km) that a car travels in 45 min; s = speed of the car (in km/h)

19. The concentration (in g/L) of salt in a mixture of 3 L of brine containing 750 g of salt to which some pure water has been added; x = volume of pure water added (in L)

20. The value (in cents) of the change in a purse that contains twice as many nickels as pennies, four more dimes than nickels, and as many quarters as dimes and nickels combined; p = number of pennies

Applications

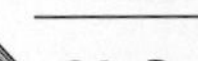

21. **Renting a Truck** A rental company charges \$65 a day and 40 cents a kilometer for the rental of one of their trucks. A truck was rented for 3 days, and the total rental charge was \$283. How many kilometers was the truck driven?

22. **Travel Cell Plan** A cell phone company offers a travel plan for cell phone usage in countries outside the United States. The travel plan has a monthly fee of \$100 for the first 250 minutes of talk outside the United States, and \$0.25 for each additional minute (or part thereof). A tourist using this plan receives a bill of \$120.50 for the month of June. How many minutes of talk did the tourist use that month?

23. **Average** A student received scores of 82, 75, and 71 on their midterm algebra exams. If the final exam counts twice as much as a midterm, what score must the student make on their final exam to earn an average score of 80? (Assume that the maximum possible score on each test is 100.)

24. **Average** In a class of 25 students, the average score is 84. Six students in the class each received a maximum score of 100, and three students each received a score of 60. What is the average score of the remaining students?

25. **Investments** An amount of \$12,000 was invested in two accounts, each earning simple interest—one earned $2\frac{1}{2}\%$ per year and the other earned 3% per year. After one year the total interest earned on these investments was \$318. How much money was invested at each rate?

26. **Investments** If an amount of \$8000 is invested at a simple interest rate of $3\frac{1}{2}\%$ per year, how much additional money must be invested at a simple interest rate of 5% per year to ensure that the interest each year is 4% of the total amount invested?

27. **Investments** What annual rate of interest would you have to earn on an investment of \$3500 to ensure receiving \$262.50 interest after one year?

28. **Investments** A financial advisor invests \$3000 at a certain annual interest rate, and another \$5000 at an annual rate that is one-half percent higher. If the total interest earned in one year is \$265, at what interest rate is the \$3000 invested?

29. **Salaries** An executive in an engineering firm earns a monthly salary plus a Christmas bonus of \$7300. If the executive earns a total of \$180,100 per year, what is the monthly salary?

30. **Salaries** A factory foreman earns 15% more than their assistant. Together they make \$113,305 per year. What is the assistant's annual salary?

31. **Overtime Pay** A lab technician earns \$18.50 an hour and works 35 hours per week. For any hours worked more than this, the pay rate increases to $1\frac{1}{2}$ times the regular hourly rate for the overtime hours worked. If the technician's pay was \$814 for one week, how many overtime hours were worked that week?

32. Labor Cost A plumber and a cabinetmaker work together to remodel a kitchen. The plumber charges \$150 an hour for labor and the cabinetmaker charges \$80 an hour for labor. The cabinetmaker works nine times as long as the plumber on the remodeling job, and the labor charge on the final invoice is \$2610. How long did the cabinetmaker work on this job?

33. A Riddle A movie star, unwilling to reveal their age, posed the following riddle to a gossip columnist: "Seven years ago, I was eleven times as old as my daughter. Now I am four times as old as she is." How old is the movie star?

34. Career Home Runs During his major league career, Hank Aaron hit 41 more home runs than Babe Ruth hit during his career. Together they hit 1469 home runs. How many home runs did Babe Ruth hit?

35. Value of Coins A change purse contains an equal number of nickels, dimes, and quarters. The total value of the coins is \$2.80. How many coins of each type does the purse contain?

36. Value of Coins You have \$3.00 in nickels, dimes, and quarters. If you have twice as many dimes as quarters and five more nickels than dimes, how many coins of each type do you have?

37. Length of a Garden A rectangular garden is 25 m wide. If its area is 1125 m^2, what is the length of the garden?

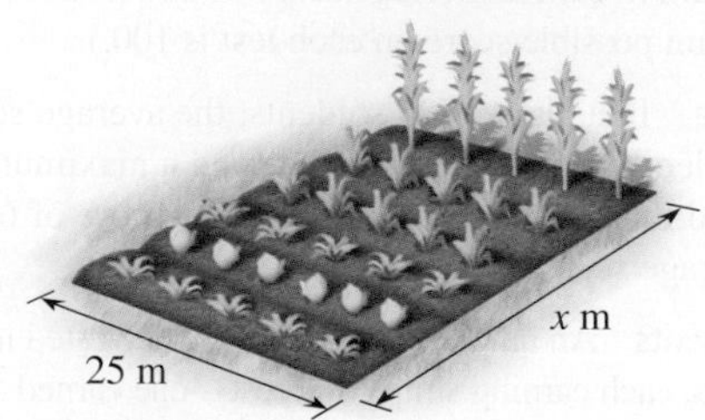

38. Width of a Pasture A pasture is three times as long as it is wide. Its area is 115,200 m^2. How wide is the pasture?

39. Dimensions of a Lot A half-acre building lot is five times as long as it is wide. What are its dimensions? [*Note:* 1 acre = 4046 m^2.]

40. Dimensions of a Lot A square plot of land has a building 60 m long and 40 m wide at one corner. The rest of the land outside the building forms a parking lot. If the parking lot has area 12,000 m^2, what are the dimensions of the entire plot of land?

41. Dimensions of a Garden A rectangular garden is 10 m longer than it is wide. Its area is 875 m^2. What are its dimensions?

42. Dimensions of a Room A rectangular bedroom is 4 m longer than it is wide. Its area is 21 m^2. What is the width of the room?

43. Dimensions of a Garden A farmer has a rectangular garden plot surrounded by 200 m of fence. Find the length and width of the garden if its area is 2400 m^2.

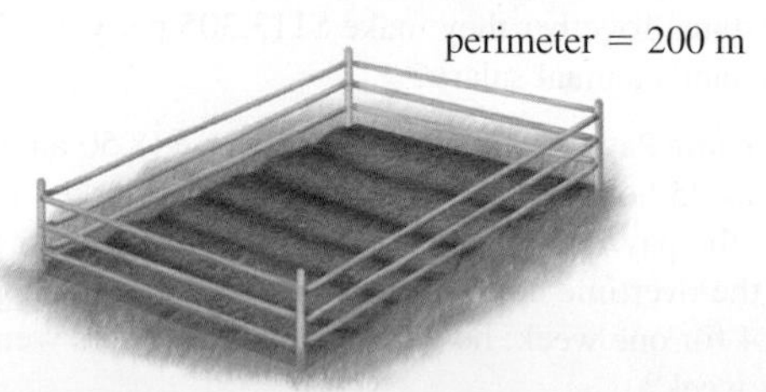

44. Dimensions of a Lot A parcel of land is 6 m longer than it is wide. Each diagonal from one corner to the opposite corner is 174 m long. What are the dimensions of the parcel?

45. Dimensions of a Lot A rectangular parcel of land is 50 m wide. The length of a diagonal between opposite corners is 10 m more than the length of the parcel. What is the length of the parcel?

46. Dimensions of a Track A running track has the shape shown in the figure, with straight sides and semicircular ends. If the length of the track is 440 m and the two straight parts are each 110 m long, what is the radius of the semicircular parts (to the nearest meter)?

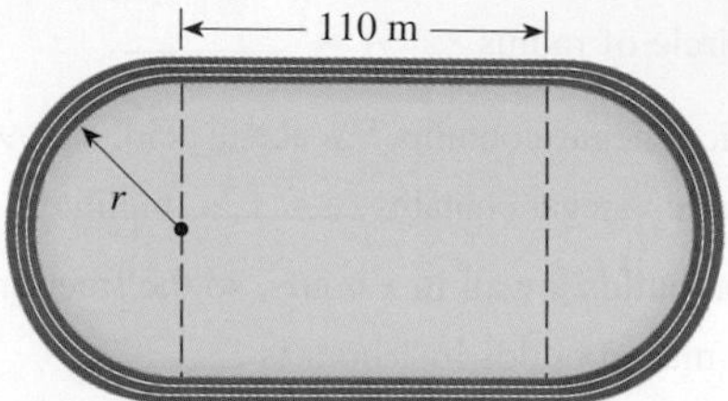

47. Length and Area Find the length x in each figure. The area of the shaded region is given.

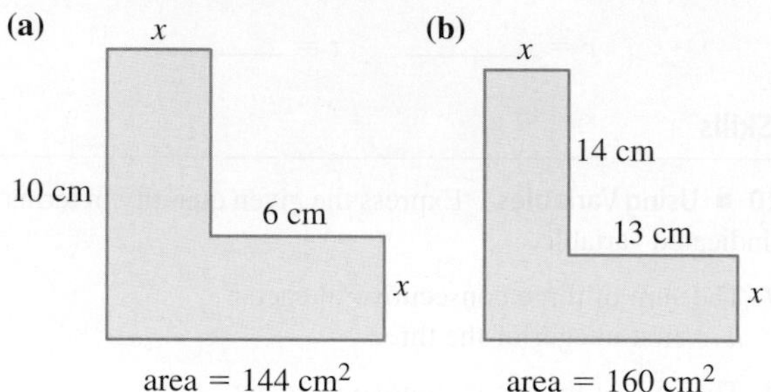

48. Length and Area Find the length y in each figure. The area of the shaded region is given.

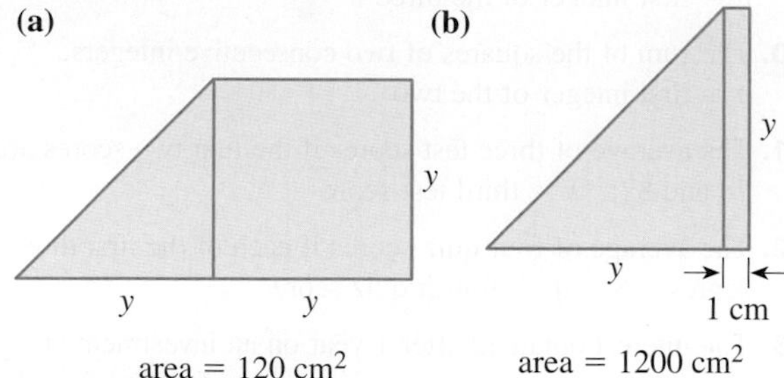

49. Framing a Painting An artist paints with watercolors on a sheet of paper 50 cm wide by 38 cm high. This sheet is then placed on a mat so that a uniformly wide strip of the mat shows all around the picture. The perimeter of the mat is 256 cm. How wide is the strip of the mat showing around the picture?

50. Dimensions of a Poster A poster has a rectangular printed area 100 cm by 140 cm and a blank strip of uniform width around the edges. The perimeter of the poster is $1\frac{1}{2}$ times the perimeter of the printed area. What is the width of the blank strip?

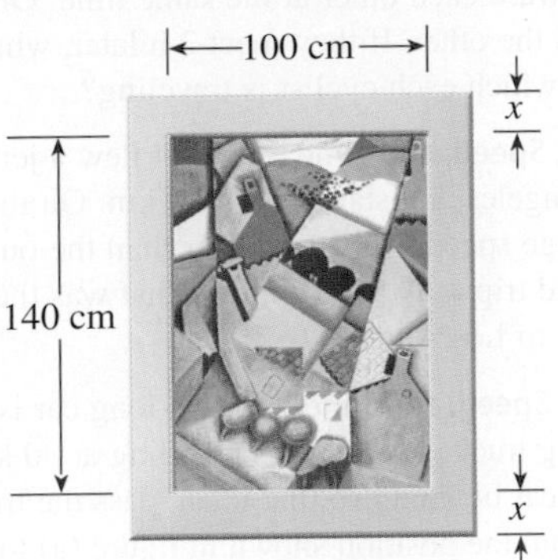

51. Reach of a Ladder A 6 m ladder leans against a building. The base of the ladder is 2.3 m from the building. How high up the building does the ladder reach?

52. Height of a Flagpole A flagpole is secured on opposite sides by two guy wires, each of which is 1.5 m longer than the pole. The distance between the points where the wires are fixed to the ground is equal to the length of one guy wire. How tall is the flagpole (to the nearest cm)?

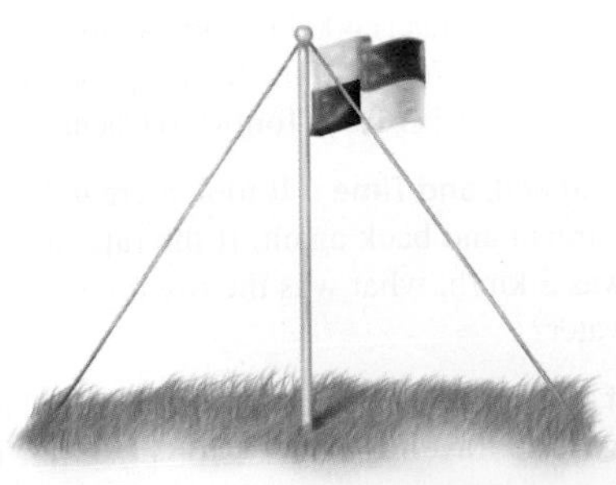

53. Length of a Shadow A person is walking away from a lamppost with a light source 6 m above the ground. The person is 2 m tall. How long is the person's shadow when the person is 10 m from the lamppost? [*Hint:* Use similar triangles.]

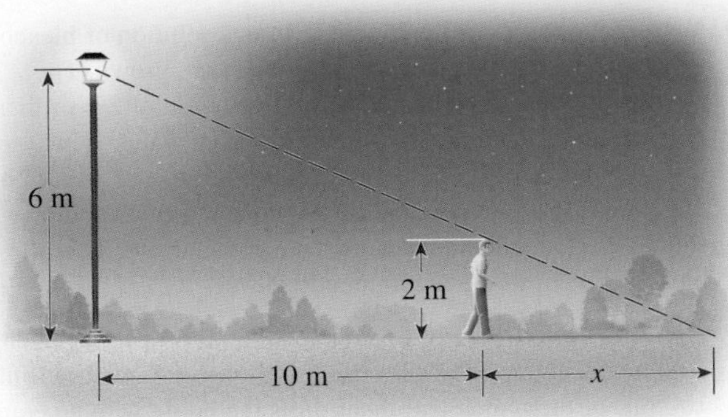

54. Height of a Tree A woodcutter determines the height of a tall tree by first measuring the height of a smaller one, 37.5 m away, then moving so that the tops of the two trees are in the same line of sight (see the figure). Suppose the small tree is 6 m tall, the woodcutter is 7.5 cm from the small tree, and the woodcutter's eye level is 1.5 m above the ground. How tall is the taller tree?

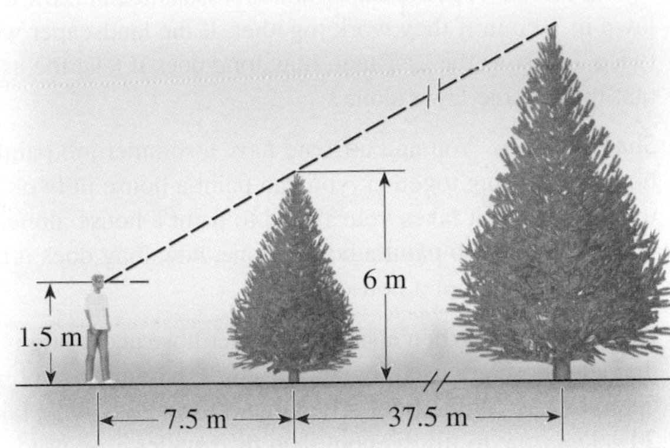

55. Mixture Problem What amount of a 60% acid solution must be mixed with a 30% solution to produce 300 mL of a 50% solution?

56. Mixture Problem What amount of pure acid must be added to 300 mL of a 50% acid solution to produce a 60% acid solution?

57. Mixture Problem A jeweler has five rings, each weighing 18 g, made of an alloy of 10% silver and 90% gold. The jeweler decides to melt down the rings and add enough silver to reduce the gold content to 75%. How much silver should be added?

58. Mixture Problem A pot contains 6 L of brine at a concentration of 120 g/L. How much of the water should be boiled off to increase the concentration to 200 g/L?

59. Mixture Problem The radiator in a car is filled with a solution of 60% antifreeze and 40% water. The manufacturer of the antifreeze suggests that for summer driving, optimal cooling of the engine is obtained with only 50% antifreeze. If the capacity of the radiator is 3.6 L, how much coolant should be drained and replaced with water to reduce the antifreeze concentration to the recommended level?

60. Mixture Problem A health clinic uses a solution of bleach to sterilize petri dishes in which cultures are grown. The sterilization tank contains 100 L of a solution of 2% ordinary household bleach mixed with pure distilled water. New research indicates that the concentration of bleach should be 5% for complete sterilization. How much of the solution should be drained and replaced with bleach to increase the bleach content to the recommended level?

61. Mixture Problem A bottle contains 750 mL of fruit punch with a concentration of 50% pure fruit juice. A person drinks 100 mL of the punch and then refills the bottle with an equal amount of a cheaper brand of punch. If the concentration of juice in the bottle is now reduced to 48%, what was the concentration in the cheaper punch that was added?

62. Mixture Problem A merchant blends tea that sells for $3.00 a packet with tea that sells for $2.75 a packet to produce 80 packets of a mixture that sells for $2.90 a packet. How many packets of each type of tea does the merchant use in the blend?

63. Sharing a Job Two friends work together to wash a car. If one takes 25 min to wash the car and the other takes 35 min, how long does it take the two friends when they work together?

64. Sharing a Job A landscaper and an assistant can mow a lawn in 10 min if they work together. If the landscaper works twice as fast as the assistant, how long does it take the assistant to mow the lawn alone?

65. Sharing a Job You and a friend have a summer job painting houses. Working together, you can paint a house in two-thirds the time it takes your friend to paint a house alone. If it takes you 7 h to paint a house alone, how long does it take your friend to paint a house alone?

66. Sharing a Job When a small-diameter hose and a large-diameter hose are used together to fill a swimming pool, it takes 16 h to fill the pool. The larger hose, used alone, takes 20% less time to fill the pool than the smaller hose used alone. How much time is required to fill the pool by each hose alone?

67. Sharing a Job You and your roommate can clean all the windows in your place in 1 h and 48 min. Working alone, it takes your roommate $1\frac{1}{2}$ h longer than it takes you to do the job. How long does it take each person working alone to wash all the windows?

68. Sharing a Job You, your manager, and an assistant deliver advertising flyers in a small town. If you each work alone, it takes you 4 h to deliver all the flyers, and it takes the assistant 1 h longer than it takes your manager. Working together, you can deliver all the flyers in 40% of the time it takes your manager working alone. How long does it take your manager to deliver all the flyers alone?

69. Distance, Speed, and Time A commuter travels from Davenport to Omaha, a distance of 480 km. The first part of the trip is traveled by bus, and the remainder is completed by train. The bus averages 64 km/h and the train averages 96 km/h. The entire trip takes $5\frac{1}{2}$ h. How long does the commuter spend on the train?

70. Distance, Speed, and Time Two cyclists, 90 km apart, start riding toward each other at the same time. One cycles twice as fast as the other. If they meet 2 h later, what is the average speed at which each cyclist is traveling?

71. Distance, Speed, and Time A pilot flew a jet from Montreal to Los Angeles, a distance of 4000 km. On the return trip, the average speed was 20% faster than the outbound speed. The round trip took 9 h 10 min. What was the speed from Montreal to Los Angeles?

72. Distance, Speed, and Time A 4-m-long car is passing a 10-m-long truck. The truck is traveling at 80 km/h. How fast must the car be going so that it can pass the truck completely in 6 s, from the position shown in figure (a) to the position shown in figure (b)? [*Hint:* Use meters and seconds instead of kilometers and hours.]

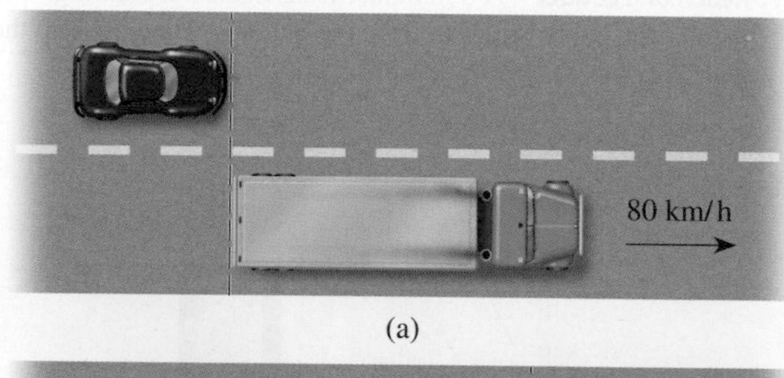

(a)

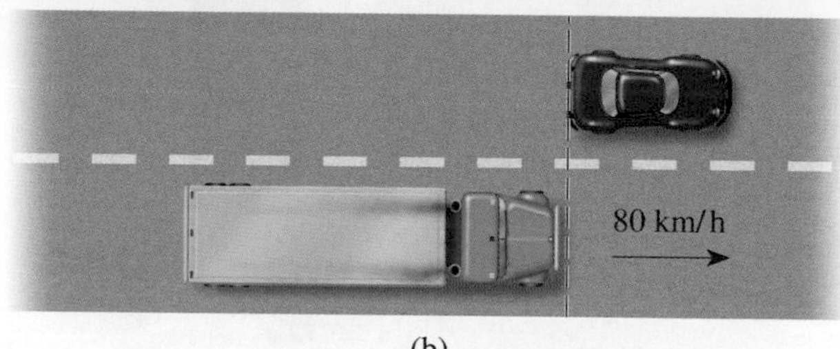

(b)

73. Distance, Speed, and Time A salesperson drives from Ajax to Berrington, a distance of 192 km, and then from Berrington to Collins, a distance of 240 km. The salesperson travels at a constant rate for the first leg of the trip and 16 km/h faster for the second leg of the trip. If the second leg of the trip took 6 min longer than the first leg, how fast was the salesperson driving on the first leg of the trip?

74. Distance, Speed, and Time A trucker drove from Tortula to Dry Junction via Cactus. On the second leg of the trip, the trucker drove 16 km/h faster than on the first leg. The distance from Tortula to Cactus is 400 km and the distance from Cactus to Dry Junction is 576 km. If the total trip took 11 h, what was the speed of the trucker from Tortula to Cactus?

75. Distance, Speed, and Time It took a crew 2 h 40 min to row 6 km upstream and back again. If the rate of flow of the stream was 3 km/h, what was the rowing speed of the crew in still water?

76. Speed of a Boat Two fishing boats depart a harbor at the same time, one traveling east, the other south. The eastbound boat travels at a speed 3 km/h faster than the southbound boat. After 2 h the boats are 30 km apart. Find the speed of the southbound boat.

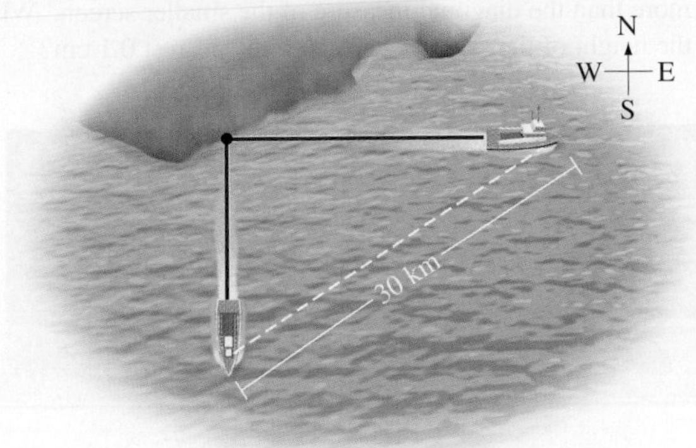

77. Law of the Lever The figure shows a lever system, similar to a seesaw that you might find in a children's playground. For the system to balance, the product of the weight and its distance from the fulcrum must be the same on each side; that is,

$$w_1x_1 = w_2x_2$$

This equation is called the **law of the lever** and was first discovered by Archimedes (see Section 10.1).

Two young friends are playing on a seesaw. One weighs 45 kg and the other 56 kg. The 45 kg-friend sits 2 m from the fulcrum. If the see saw is to be balanced, where should the other friend sit?

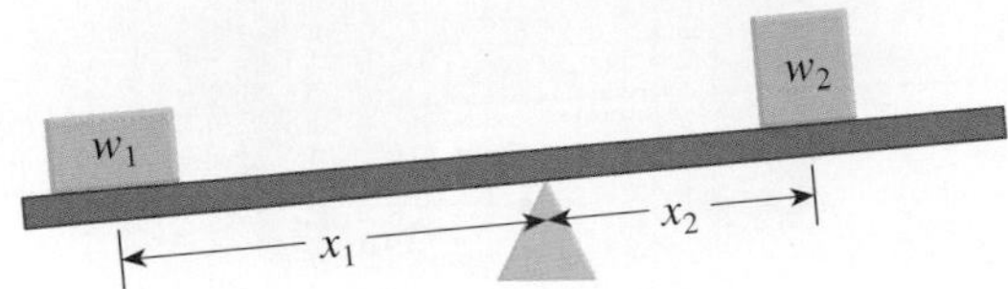

78. Law of the Lever A plank 9 m long rests on top of a flat-roofed building, with 1.5 m of the plank projecting over the edge, as shown in the figure. A worker weighing 110 kg sits on one end of the plank. What is the largest weight that can be hung on the projecting end of the plank if it is to remain in balance? (Use the law of the lever stated in Exercise 77.)

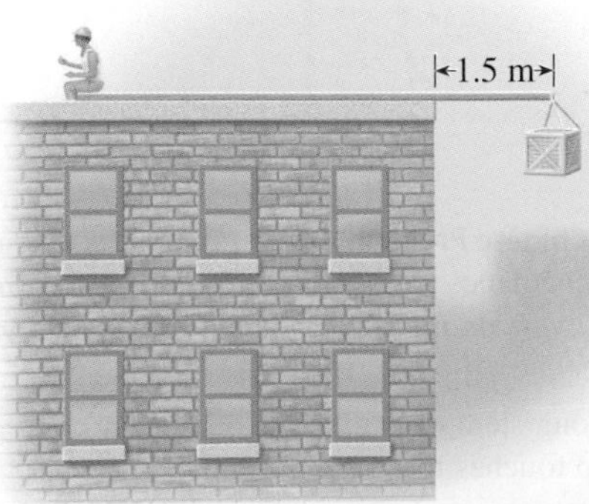

79. Dimensions of a Box A large plywood box has a volume of 180 m^3. Its length is 9 m greater than its height, and its width is 4 m less than its height. What are the dimensions of the box?

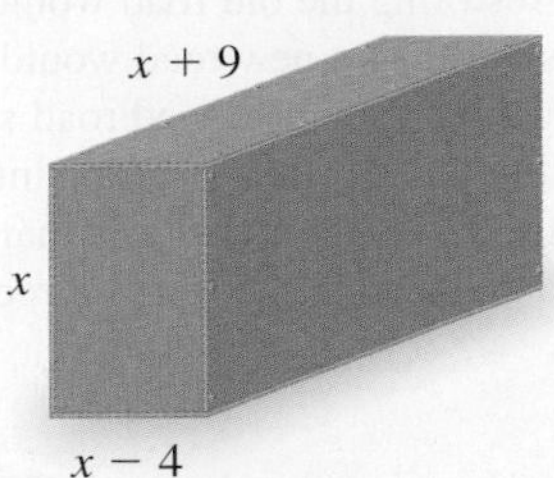

80. Radius of a Sphere A jeweler has three small solid spheres made of gold, of radius 2 mm, 3 mm, and 4 mm. The jeweler decides to melt these and make just one sphere out of them. What will the radius of this larger sphere be?

81. Dimensions of a Box A box with a square base and no top is to be made from a square piece of cardboard by cutting 4-cm squares from each corner and folding up the sides, as shown in the figure. The box is to hold 100 cm^3. How big a piece of cardboard is needed?

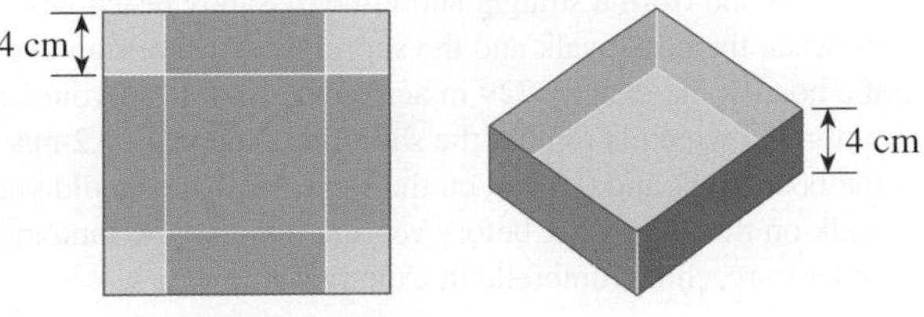

82. Dimensions of a Can A cylindrical can has a volume of 40π cm^3 and is 10 cm tall. What is its diameter? [*Hint:* Use the volume formula listed on the inside front cover of this book.]

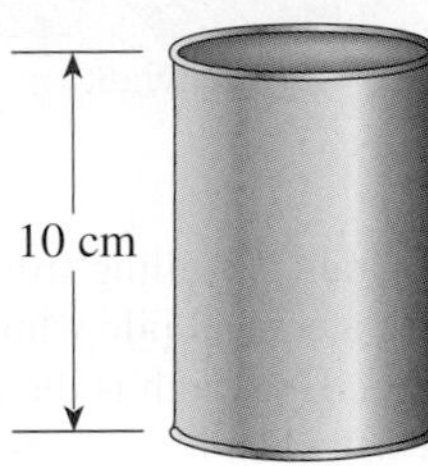

83. Radius of a Tank A spherical tank has a capacity of 2840 L. Using the fact that one liter is about 0.001 m^3, find the radius of the tank (to the nearest hundredth of a meter).

84. Dimensions of a Lot A city lot has the shape of a right triangle whose hypotenuse is 7 m longer than one of the other sides. The perimeter of the lot is 392 m. How long is each side of the lot?

85. Construction Costs The town of Foxton lies 10 km north of an abandoned east-west road that runs through Grimley, as shown in the figure. The point on the abandoned road closest to Foxton is 40 km from Grimley. County officials are about to build a new road connecting the two towns. They have determined that restoring the old road would cost \$100,000 per km, whereas building a new road would cost \$200,000 per km. How much of the abandoned road should be used (as indicated in the figure) if the officials intend to spend exactly \$6.8 million? Would it cost less than this amount to build a new road connecting the towns directly?

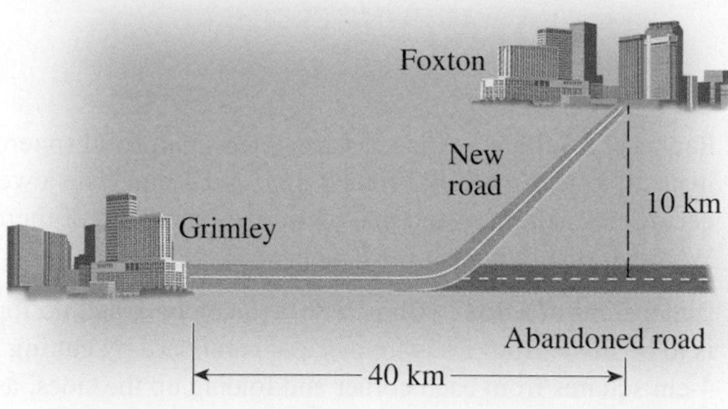

86. Distance, Speed, and Time A boardwalk is parallel to and 64 m inland from a straight shoreline. A sandy beach lies between the boardwalk and the shoreline. You are standing on the boardwalk, exactly 229 m across the sand from your beach umbrella, which is right at the shoreline. You walk 1.2 m/s on the boardwalk and 0.6 m/s on the sand. How far should you walk on the boardwalk before veering off onto the sand in order to reach the umbrella in exactly 4 min 45 s?

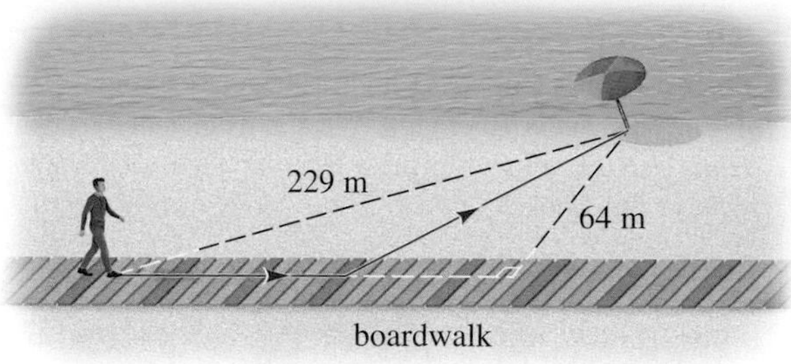

87. Volume of Grain Grain is spilling from a chute onto the ground, forming a conical pile whose diameter is always three times its height. How high is the pile (to the nearest hundredth of a meter) when it contains 28 m^3 of grain?

88. Computer Monitors Two computer monitors have the same screen height. One has a screen that is 17.5 cm wider than it is high. The other has a wider screen that is 1.8 times as wide as it is high. The diagonal measure of the wider screen is 7.5 cm more than the diagonal measure of the smaller screen. What is the height of the screens, correct to the nearest 0.1 cm?

89. Dimensions of a Structure A storage bin for corn consists of a cylindrical section made of wire mesh, surmounted by a conical tin roof, as shown in the figure. The height of the roof is one-third the height of the entire structure. If the total volume of the structure is 40π m^3 and its radius is 3 m, what is its height? [*Hint:* Use the volume formulas listed on the inside front cover of this book.]

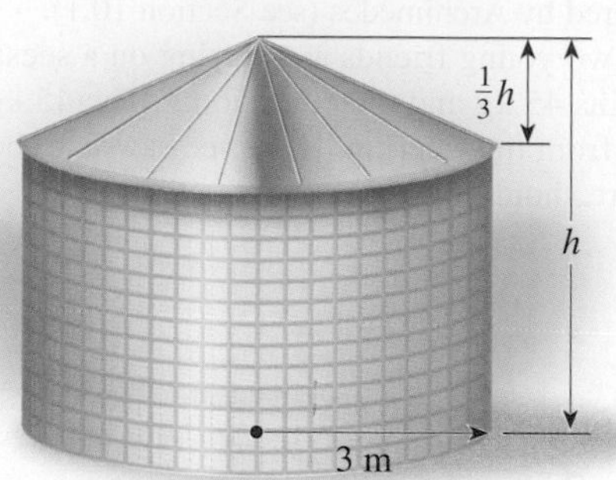

90. Comparing Areas A wire 360 cm long is cut into two pieces. One piece is formed into a square, and the other is formed into a circle. If the two figures have the same area, what are the lengths of the two pieces of wire (to the nearest tenth of an centimeter)?

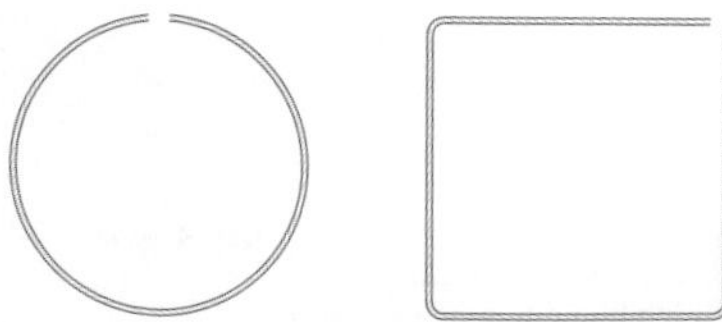

91. An Ancient Chinese Problem This problem is taken from a Chinese mathematics textbook called *Chui-chang suan-shu*, or *Nine Chapters on the Mathematical Art*, which was written about 250 BC.

A 10-m-long stem of bamboo is broken in such a way that its tip touches the ground 3 m from the base of the stem, as shown in the figure. What is the height of the break?

[*Hint:* Use the Pythagorean Theorem.]

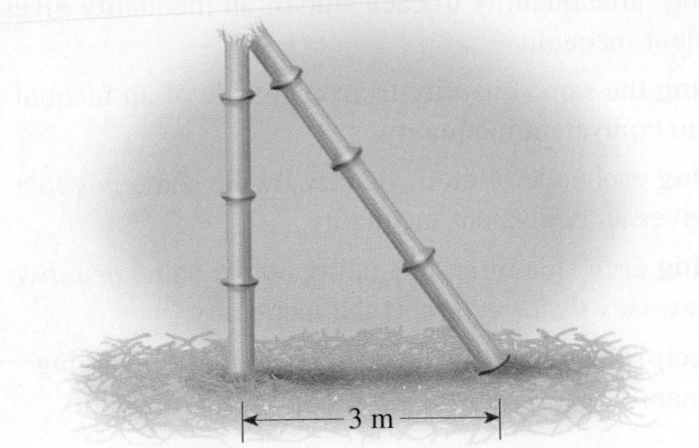

■ Discuss ■ Discover ■ Prove ■ Write

92. Write: Real-world Equations In this section we learned how to translate words into algebra. In this exercise we try to find real-world situations that could correspond to an algebraic equation. For instance, the equation $A = (x + y)/2$ could model the average amount of money in two bank accounts, where x represents the amount in one account and y the amount in the other. Write a story that could correspond to the given equation, stating what the variables represent.

(a) $C = 20{,}000 + 4.50x$

(b) $A = w(w + 10)$

(c) $C = 10.50x + 11.75y$

93. Discuss: A Babylonian Quadratic Equation The ancient Babylonians knew how to solve quadratic equations. Here is a problem from a cuneiform tablet found in a Babylonian school dating back to about 2000 BC.

> I have a reed, I know not its length. I broke from it one cubit, and it fit 60 times along the length of my field. I restored to the reed what I had broken off, and it fit 30 times along the width of my field. The area of my field is 375 square nindas. What was the original length of the reed?

Solve this problem. Use the fact that 1 ninda = 12 cubits.

1.8 Inequalities

■ Solving Linear Inequalities ■ Solving Nonlinear Inequalities ■ Absolute-Value Inequalities ■ Modeling with Inequalities

Some problems in algebra lead to **inequalities** instead of equations. An inequality looks just like an equation, except that in the place of the equal sign is one of the symbols $<$, $>$, $\le$, or $\ge$. Here is an example of an inequality:

$$4x + 7 \le 19$$

x	$4x + 7 \le 19$
1	$11 \le 19$ ✓
2	$15 \le 19$ ✓
3	$19 \le 19$ ✓
4	$23 \le 19$ ✗
5	$27 \le 19$ ✗

The table in the margin shows that some numbers satisfy the inequality and some numbers don't.

To **solve** an inequality that contains a variable means to find all values of the variable that make the inequality true. Unlike an equation, an inequality generally has infinitely many solutions, which form an interval or a union of intervals on the real line. The following illustration shows how an inequality differs from its corresponding equation:

		Solution	Graph
Equation:	$4x + 7 = 19$	$x = 3$	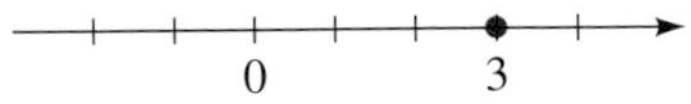
Inequality:	$4x + 7 \le 19$	$x \le 3$	

To solve inequalities, we use the following rules to isolate the variable on one side of the inequality sign. These rules tell us when two inequalities are *equivalent* (the symbol $\Leftrightarrow$ means "is equivalent to"). In these rules the symbols A, B, and C stand for real numbers or algebraic expressions. Here we state the rules for inequalities involving the symbol $\le$, but they apply to all four inequality symbols.

Rules for Inequalities

Rule	Description
1. $A \le B \iff A + C \le B + C$	**Adding** the same quantity to each side of an inequality gives an equivalent inequality.
2. $A \le B \iff A - C \le B - C$	**Subtracting** the same quantity from each side of an inequality gives an equivalent inequality.
3. If $C > 0$, then $A \le B \iff CA \le CB$	**Multiplying** each side of an inequality by the same *positive* quantity gives an equivalent inequality.
4. If $C < 0$, then $A \le B \iff CA \ge CB$	**Multiplying** each side of an inequality by the same *negative* quantity *reverses the direction* of the inequality.
5. If $A > 0$ and $B > 0$, then $A \le B \iff \dfrac{1}{A} \ge \dfrac{1}{B}$	**Taking reciprocals** of each side of an inequality involving *positive* quantities *reverses the direction* of the inequality.
6. If $A \le B$ and $C \le D$, then $A + C \le B + D$	Inequalities can be added.
7. If $A \le B$ and $B \le C$, then $A \le C$	Inequality is transitive.

Pay special attention to Rules 3 and 4. Rule 3 says that we can multiply (or divide) each side of an inequality by a *positive* number, but Rule 4 says that if we multiply each side of an inequality by a *negative* number, then we reverse the direction of the inequality. For example, if we start with the inequality

$$3 < 5$$

and multiply by 2, we get

$$6 < 10$$

but if we multiply by -2, we get

$$-6 > -10$$

■ Solving Linear Inequalities

An inequality is **linear** if each term is constant or a multiple of the variable. To solve a linear inequality, we isolate the variable on one side of the inequality sign.

Example 1 ■ Solving a Linear Inequality

Solve the inequality $3x < 9x + 4$, and sketch the solution set.

Solution We isolate x on one side of the inequality sign.

$3x < 9x + 4$	Given inequality
$3x - 9x < 9x + 4 - 9x$	Subtract $9x$
$-6x < 4$	Simplify
$(-\frac{1}{6})(-6x) > (-\frac{1}{6})(4)$	Multiply by $-\frac{1}{6}$ and reverse inequality
$x > -\frac{2}{3}$	Simplify

Multiplying by the negative number $-\frac{1}{6}$ *reverses* the direction of the inequality.

The solution set consists of all numbers greater than $-\frac{2}{3}$. In other words, the solution of the inequality is the interval $(-\frac{2}{3}, \infty)$. It is graphed in Figure 1.

Figure 1

Now Try Exercise 21 ■

Example 2 ■ Solving a Pair of Simultaneous Inequalities

Solve the inequalities $4 \le 3x - 2 < 13$.

Solution The solution set consists of all values of x that satisfy both of the inequalities $4 \le 3x - 2$ and $3x - 2 < 13$. Using Rules 1 and 3, we see that the following inequalities are equivalent:

$4 \le 3x - 2 < 13$	Given inequality
$6 \le 3x < 15$	Add 2
$2 \le x < 5$	Divide by 3

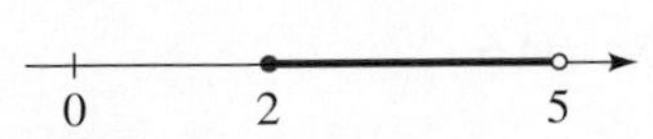

Figure 2

Therefore the solution set is $\{x \mid 2 \le x < 5\} = [2, 5)$, as shown in Figure 2.

Now Try Exercise 31 ■

■ Solving Nonlinear Inequalities

To solve inequalities involving squares and other powers of the variable, we use factoring, together with the following principle.

The Sign of a Product or Quotient

- If a product or a quotient has an *even* number of *negative* factors, then its value is *positive*.
- If a product or a quotient has an *odd* number of *negative* factors, then its value is *negative*.

For example, to solve the inequality $x^2 \le 5x - 6$, we first move all terms to the left-hand side and factor to get

$$(x - 2)(x - 3) \le 0$$

This form of the inequality says that the product $(x - 2)(x - 3)$ must be negative or zero, so to solve the inequality, we must determine where each factor is negative or positive (because the sign of a product depends on the sign of the factors). The details are explained in Example 3, in which we use the following guidelines.

Guidelines for Solving Nonlinear Inequalities

1. **Move All Terms to One Side.** If necessary, rewrite the inequality so that all nonzero terms appear on one side of the inequality sign. If the nonzero side of the inequality involves quotients, bring them to a common denominator.
2. **Factor.** Factor the nonzero side of the inequality.
3. **Find the Intervals.** Determine the values for which each factor is zero. These numbers divide the real line into intervals. List the intervals that are determined by these numbers.
4. **Make a Table or Diagram.** Use **test values** to make a table or diagram of the signs of each factor on each interval. In the last row of the table determine the sign of the product (or quotient) of these factors.
5. **Solve.** Use the sign table to find the intervals on which the inequality is satisfied. Check whether the **endpoints** of these intervals satisfy the inequality. (This may happen if the inequality involves $\le$ or $\ge$.)

The factoring technique that is described in these guidelines works only if all nonzero terms appear on one side of the inequality symbol. If the inequality is not written in this form, first rewrite it, as indicated in Step 1.

Example 3 ■ Solving a Quadratic Inequality

Solve the inequality $x^2 \leq 5x - 6$.

Solution We follow the preceding guidelines.

Move all terms to one side. We move all the terms to the left-hand side.

$$x^2 \leq 5x - 6 \qquad \text{Given inequality}$$

$$x^2 - 5x + 6 \leq 0 \qquad \text{Subtract } 5x\text{, add } 6$$

Factor. Factoring the left-hand side of the inequality, we get

$$(x - 2)(x - 3) \leq 0 \qquad \text{Factor}$$

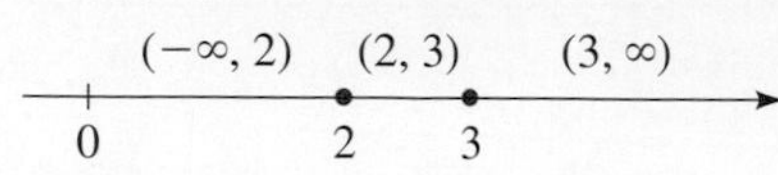

Figure 3

Find the intervals. The factors of the left-hand side are $x - 2$ and $x - 3$. These factors are zero when x is 2 and 3, respectively. As shown in Figure 3, the numbers 2 and 3 divide the real line into the three intervals

$$(-\infty, 2), \quad (2, 3), \quad (3, \infty)$$

The factors $x - 2$ and $x - 3$ change sign only at 2 and 3, respectively. So these factors maintain their sign on each of these three intervals.

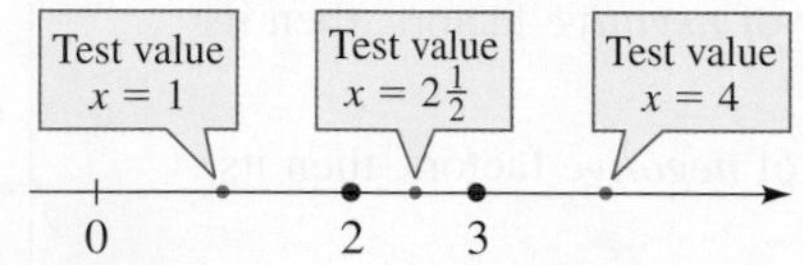

Figure 4

Make a table or diagram. To determine the sign of each factor on each of the intervals that we found, we use test values. We choose a number inside each interval and check the sign of the factors $x - 2$ and $x - 3$ at the number we have chosen. For the interval $(-\infty, 2)$, let's choose the test value 1 (see Figure 4). Substituting 1 for x in the factors $x - 2$ and $x - 3$, we get

$$x - 2 = 1 - 2 = -1 < 0$$
$$x - 3 = 1 - 3 = -2 < 0$$

So both factors are negative on this interval. Notice that we need to check only one test value for each interval because the factors $x - 2$ and $x - 3$ do not change sign on any of the three intervals we have found.

Using the test values $x = 2\frac{1}{2}$ and $x = 4$ for the intervals $(2, 3)$ and $(3, \infty)$ (see Figure 4), respectively, we construct the following sign table. The final row of the table is obtained from the fact that the expression in the last row is the product of the two factors.

Interval	$(-\infty, 2)$	$(2, 3)$	$(3, \infty)$
Sign of $x - 2$	−	+	+
Sign of $x - 3$	−	−	+
Sign of $(x - 2)(x - 3)$	+	−	+

If you prefer, you can represent this information on a real line, as in the following sign diagram. The vertical lines indicate the points at which the real line is divided into intervals:

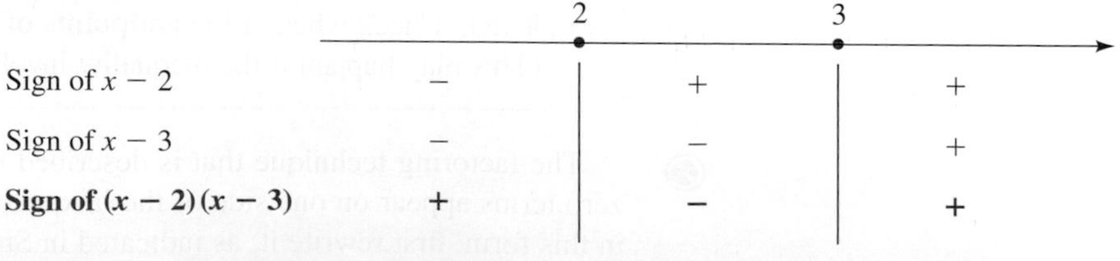

Solve. We read from the table or the diagram that $(x - 2)(x - 3)$ is negative on the interval $(2, 3)$. You can check that the endpoints 2 and 3 satisfy the inequality, so the solution is

$$\{x \mid 2 \le x \le 3\} = [2, 3]$$

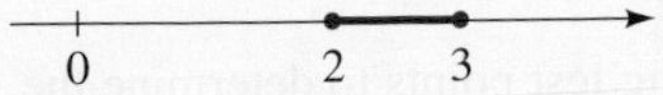

Figure 5

The solution is illustrated in Figure 5.

Now Try Exercise 43 ■

Example 4 ■ Solving an Inequality with Repeated Factors

Solve the inequality $x(x - 1)^2(x - 3) < 0$.

Solution All nonzero terms are already on one side of the inequality, and the nonzero side of the inequality is already factored. So we begin by finding the intervals for this inequality.

Find the intervals. The factors of the left-hand side are x, $(x - 1)^2$, and $x - 3$. These are zero when $x = 0, 1, 3$. These numbers divide the real line into the intervals

$$(-\infty, 0),\quad (0, 1),\quad (1, 3),\quad (3, \infty)$$

Make a diagram. We make the following diagram, using test points to determine the sign of each factor in each interval.

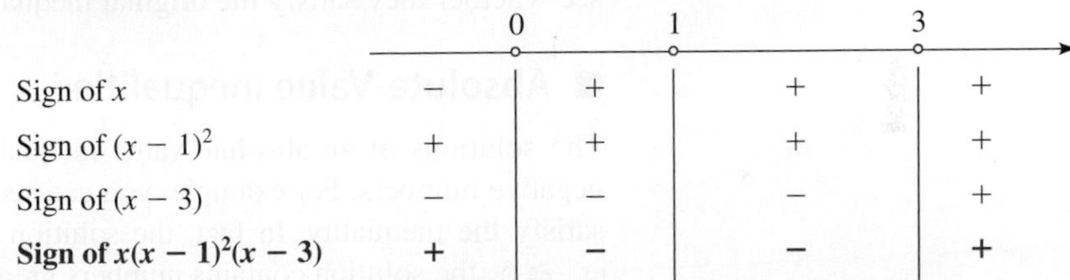

	$(-\infty, 0)$	$(0, 1)$	$(1, 3)$	$(3, \infty)$
Sign of x	−	+	+	+
Sign of $(x - 1)^2$	+	+	+	+
Sign of $(x - 3)$	−	−	−	+
Sign of $x(x - 1)^2(x - 3)$	+	−	−	+

Solve. From the diagram we see that the inequality is satisfied on the intervals $(0, 1)$ and $(1, 3)$. Since this inequality involves $<$, the endpoints of the intervals do not satisfy the inequality. So the solution set is the union of these two intervals:

$$\{x \mid 0 < x < 1 \quad \text{or} \quad 1 < x < 3\} = (0, 1) \cup (1, 3)$$

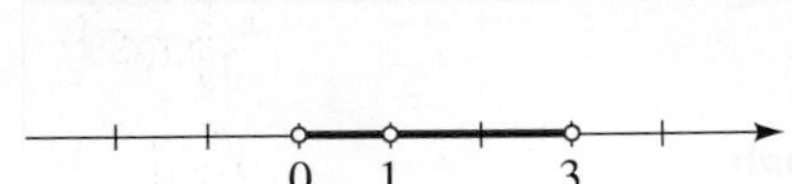

Figure 6

The solution set is graphed in Figure 6.

Now Try Exercise 55 ■

Example 5 ■ Solving an Inequality Involving a Quotient

Solve the inequality $\dfrac{1 + x}{1 - x} \ge 1$.

Solution **Move all terms to one side.** We move the terms to the left-hand side and simplify using a common denominator.

$$\frac{1 + x}{1 - x} \ge 1 \qquad \text{Given inequality}$$

$$\frac{1 + x}{1 - x} - 1 \ge 0 \qquad \text{Subtract 1}$$

$$\frac{1 + x}{1 - x} - \frac{1 - x}{1 - x} \ge 0 \qquad \text{Common denominator } 1 - x$$

$$\frac{1 + x - 1 + x}{1 - x} \ge 0 \qquad \text{Combine the fractions}$$

$$\frac{2x}{1 - x} \ge 0 \qquad \text{Simplify}$$

⊘ It is tempting to simply multiply both sides of the inequality by $1 - x$ (as you would if this were an *equation*). But this doesn't work because we don't know whether $1 - x$ is positive or negative, so we can't tell whether the inequality needs to be reversed. (See Exercise 135.)

Find the intervals. The factors of the left-hand side are $2x$ and $1 - x$. These are zero when x is 0 and 1. These numbers divide the real line into the intervals

$$(-\infty, 0), (0, 1), (1, \infty)$$

Make a diagram. We make the following diagram using test points to determine the sign of each factor in each interval.

	$(-\infty, 0)$	$(0, 1)$	$(1, \infty)$
Sign of $2x$	$-$	$+$	$+$
Sign of $1 - x$	$+$	$+$	$-$
Sign of $\dfrac{2x}{1-x}$	$-$	$+$	$-$

Solve. From the diagram we see that the inequality is satisfied on the interval $(0, 1)$. Checking the endpoints, we see that 0 satisfies the inequality but 1 does not (because the quotient in the inequality is not defined at 1). So the solution set is the interval $[0, 1)$. The solution set is graphed in Figure 7.

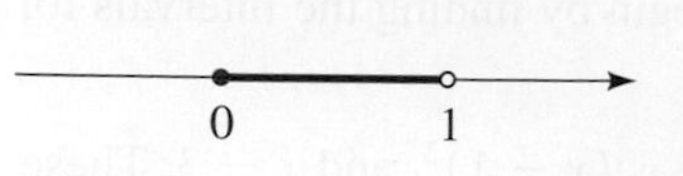

Figure 7

Now Try Exercise 63 ■

Example 5 shows that we should always check the endpoints of the solution set to see whether they satisfy the original inequality.

■ Absolute-Value Inequalities

The solutions of an absolute-value inequality like $|x| \le 5$ include both positive and negative numbers. For example, you can see that numbers like 0.5, -0.5, 1, -1, 3, -3 satisfy the inequality. In fact, the solution is the interval $[-5, 5]$. For the inequality $|x| \ge 5$, the solution contains numbers greater than 5 and their negatives (for example, 5.5, -5.5, 100, -100), so the solution is a union of intervals, $(-\infty, -5] \cup [5, \infty)$. In general we have the following properties.

These properties hold when x is replaced by any algebraic expression. (We assume that $c > 0$.)

Properties of Absolute-Value Inequalities

Inequality	Equivalent form	Graph
1. $\lvert x \rvert < c$	$-c < x < c$	$-c$ 0 c
2. $\lvert x \rvert \le c$	$-c \le x \le c$	$-c$ 0 c
3. $\lvert x \rvert > c$	$x < -c$ or $c < x$	$-c$ 0 c
4. $\lvert x \rvert \ge c$	$x \le -c$ or $c \le x$	$-c$ 0 c

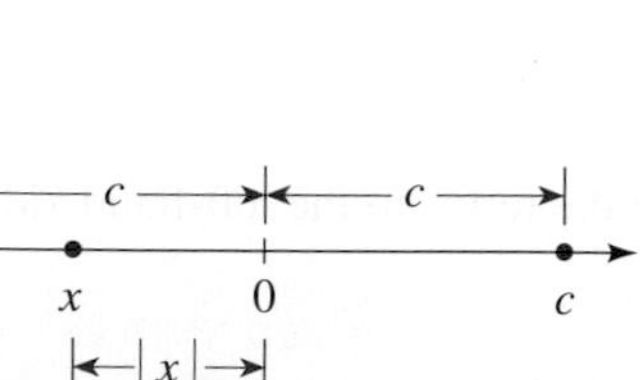

Figure 8

These properties can be proved using the definition of absolute-value. To prove Property 1, for example, note that the inequality $|x| < c$ says that the distance from x to 0 is less than c, and from Figure 8 you can see that this is true if and only if x is between $-c$ and c.

Example 6 ■ Solving an Absolute Value Inequality

Solve the inequality $|x - 5| < 2$.

Solution 1 The inequality $|x - 5| < 2$ is equivalent to

$$-2 < x - 5 < 2 \qquad \text{Property 1}$$

$$3 < x < 7 \qquad \text{Add 5}$$

The solution set is the open interval $(3, 7)$.

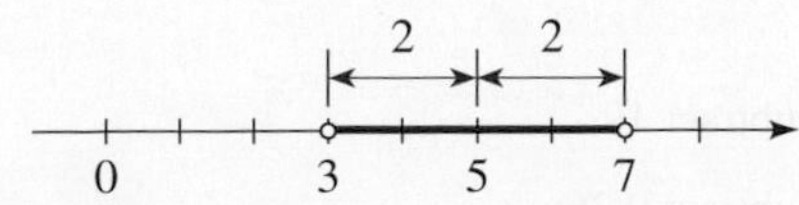

Figure 9

Solution 2 Geometrically, the solution set consists of all numbers x whose distance from 5 is less than 2. From Figure 9 we see that this is the interval $(3, 7)$.

Now Try Exercise 79 ■

Example 7 ■ Solving an Absolute Value Inequality

Solve the inequality $|3x + 2| \geq 4$.

Solution By Property 4 the inequality $|3x + 2| \geq 4$ is equivalent to

$$\begin{aligned} 3x + 2 &\geq 4 \quad \text{or} \quad & 3x + 2 &\leq -4 & & \\ 3x &\geq 2 & 3x &\leq -6 & & \text{Subtract 2} \\ x &\geq \tfrac{2}{3} & x &\leq -2 & & \text{Divide by 3} \end{aligned}$$

So the solution set is

$$\{x \mid x \leq -2 \quad \text{or} \quad x \geq \tfrac{2}{3}\} = (-\infty, -2] \cup [\tfrac{2}{3}, \infty)$$

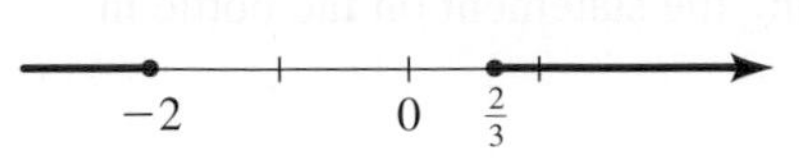

Figure 10

The set is graphed in Figure 10.

Now Try Exercise 81 ■

■ Modeling with Inequalities

Modeling real-life problems frequently leads to inequalities because we are often interested in determining when one quantity is more (or less) than another.

Example 8 ■ Truck Rental

Two truck rental companies offer the following pricing plans.

Company A: \$19 a day and \$0.40 per kilometer

Company B: \$68 a day and \$0.26 per kilometer

For a one-day rental, how many kilometers would you have to drive so that renting from Company B is less expensive than renting from Company A?

Solution **Identify the variable.** We are asked to find the number of kilometers you would have to drive so that renting from Company B is less expensive than renting from Company A. So let

$$x = \text{number of kilometers}$$

Translate from words to algebra. The information in the problem may be organized as follows:

In Words	In Algebra
Number of kilometers	x
Rental cost with Company A	$19 + 0.40x$
Rental cost with Company B	$68 + 0.26x$

Set up the model. Now we set up the model.

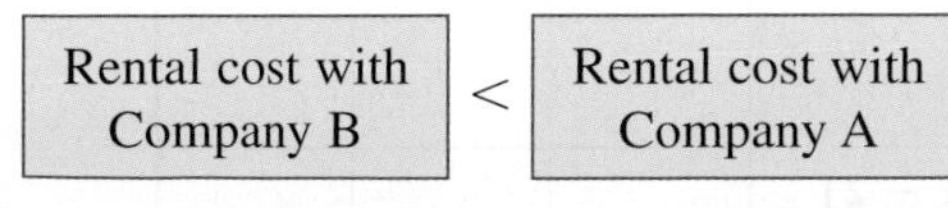

$$68 + 0.26x < 19 + 0.40x$$

Solve. We solve for x.

$$49 + 0.26x < 0.40x \quad \text{Subtract 19}$$
$$49 < 0.14x \quad \text{Subtract } 0.26x$$
$$350 < x \quad \text{Divide by } 0.14x$$

If you plan to drive *more than* 350 kilometers, renting from Company B would be less expensive.

Now Try Exercise 119 ■

Example 9 ■ Relationship Between Fahrenheit and Celsius Scales

The instructions on a bottle of medicine indicate that the bottle should be stored at a temperature between 5 °C and 30 °C. What range of temperatures does this correspond to on the Fahrenheit scale?

Solution The relationship between degrees Celsius (C) and degrees Fahrenheit (F) is given by the equation $C = \frac{5}{9}(F - 32)$. Expressing the statement on the bottle in terms of inequalities, we have

$$5 < C < 30$$

So the corresponding Fahrenheit temperatures satisfy the inequalities

$$5 < \tfrac{5}{9}(F - 32) < 30 \quad \text{Substitute } C = \tfrac{5}{9}(F - 32)$$
$$\tfrac{9}{5} \cdot 5 < F - 32 < \tfrac{9}{5} \cdot 30 \quad \text{Multiply by } \tfrac{9}{5}$$
$$9 < F - 32 < 54 \quad \text{Simplify}$$
$$9 + 32 < F < 54 + 32 \quad \text{Add 32}$$
$$41 < F < 86 \quad \text{Simplify}$$

The medicine should be stored at a temperature between 41°F and 86°F.

Now Try Exercise 117 ■

1.8 | Exercises

■ Concepts

1. Fill in each blank with an appropriate inequality sign.

(a) If $x < 5$, then $x - 3$ ______ 2.

(b) If $x \le 5$, then $3x$ ______ 15.

(c) If $x \ge 2$, then $-3x$ ______ -6.

(d) If $x < -2$, then $-x$ ______ 2.

2. To solve the nonlinear inequality $\dfrac{x+1}{x-2} \le 0$, we first observe that the numbers ____ and ____ are zeros of the numerator and denominator. These numbers divide the real line into three intervals. Complete the table.

Interval			
Sign of $x + 1$ **Sign of $x - 2$**			
Sign of $(x + 1)/(x - 2)$			

Do any of the endpoints fail to satisfy the inequality? If so, which one(s)? ______. The solution of the inequality is ______.

3. (a) Find three positive and three negative numbers that satisfy the inequality $|x| \le 3$, then express the solution as an interval: ______________. Graph the numbers you found and the solution:

0

(b) Find three positive and three negative numbers that satisfy the inequality $|x| \ge 3$, then express the solution as a union of intervals: ________ ∪ ________. Graph the numbers you found and the solution:

0

4. (a) The set of all points on the real line whose distance from zero is less than 3 can be described by the absolute value inequality $|x|$ ________.

(b) The set of all points on the real line whose distance from zero is greater than 3 can be described by the absolute value inequality $|x|$ ________.

5. *Yes or No?* If *No,* give an example.

(a) If $x(x+1) > 0$, does it follow that x is positive?

(b) If $x(x+1) > 5$, does it follow that $x > 5$?

6. What is a logical first step in solving the inequality?

(a) $3x \le 7$ (b) $5x - 2 \ge 1$ (c) $|3x + 2| \le 8$

Skills

7–12 ■ Solutions? Let $S = \{-5, -1, 0, \frac{2}{3}, \frac{5}{6}, 1, \sqrt{5}, 3, 5\}$. Determine which elements of S satisfy the inequality.

7. $-2 + 3x \ge \frac{1}{3}$ **8.** $1 - 2x \ge 5x$

9. $1 < 2x - 4 \le 7$ **10.** $-2 \le 3 - x < 2$

11. $\dfrac{1}{x} \le \dfrac{1}{2}$ **12.** $x^2 + 2 < 4$

13–36 ■ Linear Inequalities Solve the linear inequality. Express the solution using interval notation, and graph the solution set.

13. $2x \le 7$ **14.** $-4x \ge 10$

15. $2x - 5 > 3$ **16.** $3x + 11 < 5$

17. $7 - x \ge 5$ **18.** $5 - 3x \le -16$

19. $2x + 1 < 0$ **20.** $0 < 4x - 8$

21. $4x - 7 < 8 + 9x$ **22.** $5 - 3x \ge 8x - 7$

23. $x - \frac{3}{2} > \frac{1}{2}x$ **24.** $\frac{2}{5}x + 1 < \frac{1}{5} - 2x$

25. $\frac{1}{3}x + 2 < \frac{1}{6}x - 1$ **26.** $\frac{2}{3} - \frac{1}{2}x \ge \frac{1}{6} + x$

27. $4 - 3x \le -(1 + 8x)$ **28.** $2(7x - 3) \le 12x + 16$

29. $2 \le x + 5 < 4$ **30.** $-8 \le x - 3 \le 12$

31. $-1 < 2x - 5 < 7$ **32.** $1 < 3x + 4 \le 16$

33. $-2 < 8 - 2x \le -1$ **34.** $-3 \le 3x + 7 \le \frac{1}{2}$

35. $\dfrac{1}{6} < \dfrac{2x - 13}{12} \le \dfrac{2}{3}$ **36.** $-\dfrac{1}{2} \le \dfrac{4 - 3x}{5} \le \dfrac{1}{4}$

37–60 ■ Nonlinear Inequalities Solve the nonlinear inequality. Express the solution using interval notation, and graph the solution set.

37. $(x + 2)(x - 3) < 0$ **38.** $(x - 5)(x + 4) \ge 0$

39. $x(2x + 7) \ge 0$ **40.** $x(2 - 3x) \le 0$

41. $x^2 - 3x - 18 \le 0$ **42.** $x^2 - 8x + 7 > 0$

43. $3x^2 + 5x \ge 2$ **44.** $x^2 < x + 2$

45. $3x^2 - 3x < 2x^2 + 4$ **46.** $5x^2 + 3x \ge 3x^2 + 2$

47. $x^2 > 3(x + 6)$ **48.** $x^2 + 2x > 3$

49. $x^2 < 4$ **50.** $x^2 \ge 9$

51. $(x + 2)(x - 1)(x - 3) \le 0$

52. $(x - 5)(x - 2)(x + 1) > 0$

53. $(x - 4)(x + 2)^2 < 0$

54. $(x - 4)(x + 2)^2 > 0$

55. $(x + 3)^2(x - 2)(x + 5) \ge 0$

56. $4x^2(x^2 - 9) \le 0$

57. $x^3 - 4x > 0$ **58.** $9x \le x^3$

59. $x^4 > x^2$ **60.** $x^5 > x^2$

61–72 ■ Inequalities Involving Quotients Solve the nonlinear inequality. Express the solution using interval notation, and graph the solution set.

61. $\dfrac{x - 3}{x + 1} \ge 0$ **62.** $\dfrac{2x + 6}{x - 2} < 0$

63. $\dfrac{x}{x - 2} > 5$ **64.** $\dfrac{x - 4}{2x + 1} < 5$

65. $\dfrac{2x + 1}{x - 5} \le 3$ **66.** $\dfrac{3 + x}{3 - x} \ge 1$

67. $\dfrac{4}{x} < x$ **68.** $\dfrac{x}{x + 1} > 3x$

69. $1 + \dfrac{2}{x + 1} \le \dfrac{2}{x}$ **70.** $\dfrac{3}{x - 1} - \dfrac{4}{x} \ge 1$

71. $\dfrac{x + 2}{x + 3} < \dfrac{x - 1}{x - 2}$ **72.** $\dfrac{1}{x + 1} + \dfrac{1}{x + 2} \le 0$

73–88 ■ Absolute-Value Inequalities Solve the absolute-value inequality. Express the answer using interval notation, and graph the solution set.

73. $|5x| < 20$ **74.** $|16x| \le 8$

75. $|2x| > 7$ **76.** $\frac{1}{2}|x| \ge 1$

77. $|x - 3| \le 10$ **78.** $|x + 1| \ge 1$

79. $|3x + 2| < 4$ **80.** $|5x - 2| < 8$

81. $|3x - 2| \ge 5$ **82.** $|3x - 4| \ge 5$

83. $\left|\dfrac{x - 2}{3}\right| < 2$ **84.** $\left|\dfrac{x + 1}{2}\right| \ge 4$

85. $|x + 6| < 0.001$ **86.** $3 - |2x + 4| \le 1$

87. $8 - |2x - 1| \ge 6$ **88.** $7|x + 2| + 5 > 4$

89–98 ■ Putting It All Together Recognize the type of inequality and solve the inequality by an appropriate method. Express the answer using interval notation, and graph the solution set.

89. $0 < 5 - 2x$ **90.** $|x - 5| \le 3$

91. $16x \le x^3$ **92.** $\dfrac{6}{x - 1} - \dfrac{6}{x} \ge 1$

93. $x^2 + 5x + 6 > 0$ **94.** $5 \le 3x - 4 \le 14$

95. $(x + 3)^2(x + 1) > 0$ **96.** $\dfrac{x}{x + 1} > 3$

97. $1 \ge 3 - 2x \ge -5$ **98.** $|8x + 3| > 12$

99–102 ■ Absolute Value Inequalities A phrase describing a set of real numbers is given. Express the phrase as an inequality involving an absolute value.

99. All real numbers x less than 3 units from 0

100. All real numbers x more than 2 units from 0

101. All real numbers x at least 5 units from 7

102. All real numbers x at most 4 units from 2

103–108 ■ Absolute Value Inequalities A set of real numbers is graphed. Find an inequality involving an absolute value that describes the set.

103. 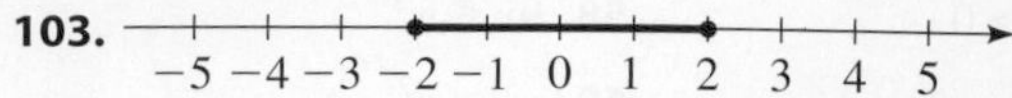

104. −5 −4 −3 −2 −1 0 1 2 3 4 5

105. −5 −4 −3 −2 −1 0 1 2 3 4 5

106. −5 −4 −3 −2 −1 0 1 2 3 4 5

107. −5 −4 −3 −2 −1 0 1 2 3 4 5

108. −5 −4 −3 −2 −1 0 1 2 3 4 5

109–112 ■ Domain Determine the values of the variable for which the expression is defined as a real number.

109. $\sqrt{x^2 - 9}$ **110.** $\sqrt{x^2 - 5x - 50}$

111. $\left(\dfrac{1}{x^2 - 3x - 10}\right)^{1/2}$ **112.** $\sqrt[4]{\dfrac{1 - x}{2 + x}}$

Skills Plus

113–116 ■ Inequalities Solve the inequality for x. Assume that a, b, and c are positive constants.

113. $a(bx - c) \geq bc$ **114.** $a \leq bx + c < 2a$

115. $a|bx - c| + d \geq 4a$ **116.** $\left|\dfrac{bx + c}{a}\right| > 5a$

Applications

117. Temperature Scales Use the relationship between C and F given in Example 9 to find the interval on the Fahrenheit scale corresponding to the temperature range $20 \leq C \leq 30$.

118. Temperature Scales What interval on the Celsius scale corresponds to the temperature range $50 \leq F \leq 95$?

119. RV Rental Cost An RV rental company offers two plans for renting an RV.

Plan A: $95 per day and $0.40 per kilometer
Plan B: $135 per day with free unlimited mileage

For what range of kilometers will Plan B save you money?

120. International Plans A phone service provider offers two international plans.

Plan A: $25 per month and 5¢ per minute
Plan B: $5 per month and 12¢ per minute

For what range of minutes of international calls would Plan B be financially advantageous?

121. Driving Cost It is estimated that the annual cost of driving a certain new car is given by the formula

$$C = 0.35m + 2200$$

where m represents the number of kilometers driven per year and C is the cost in dollars. A teacher has purchased such a car and decides to budget between $6400 and $7100 for next year's driving costs. What is the corresponding range of kilometers that can be driven within this budget?

122. Temperature and Elevation As dry air moves upward, it expands and, in so doing, cools at a rate of about 1°C for each 100-m rise, up to about 12 km.

(a) If the ground temperature is 20°C, write a formula for the temperature at elevation h.

(b) What range of temperatures can be expected if a plane takes off and reaches an elevation of 5 km?

123. Airline Ticket Price A charter airline finds that on its Saturday flights from Philadelphia to London all 120 seats will be sold if the ticket price is $200. However, for each $3 increase in ticket price, the number of seats sold decreases by one.

(a) Find a formula for the number of seats sold if the ticket price is P dollars.

(b) Over a certain period the number of seats sold for this flight ranged between 90 and 115. What was the corresponding range of ticket prices?

124. Accuracy of a Scale A coffee merchant sells a customer 3 kg of Hawaiian Kona at $6.50 per kilogram. The merchant's scale is accurate to within ±0.03 kg. By how much could the customer have been overcharged or undercharged because of possible inaccuracy in the scale?

125. Gravity The gravitational force F exerted by the earth on an object having a mass of 100 kg is given by the equation

$$F = \frac{4{,}000{,}000}{d^2}$$

where d is the distance (in km) of the object from the center of the earth and the force F is measured in newtons (N). For what distances will the gravitational force exerted by the earth on this object be between 0.0004 N and 0.01 N?

126. Bonfire Temperature In the vicinity of a bonfire the temperature T in °C at a distance of x meters from the center of the fire was given by

$$T = \frac{600{,}000}{x^2 + 300}$$

At what range of distances from the fire's center was the temperature less than 500°C?

127. Quarter-Mile Time Performance cars are often compared using their quarter-mile (0.4 km) time, that is, the time it takes to travel a quarter mile (or approximately 400 m) from a standing start. In a quarter-mile run the acceleration a is approximately constant, so the distance d the car travels in time t is

$$d = \tfrac{1}{2}at^2$$

where d is measured in feet, a in ft/s^2, and t in seconds.

(a) Find the values of acceleration that result in a quarter-mile time of less than 10 seconds. (One quarter mile equals 402 m.)

(b) Find the quarter-mile time for a car in free fall (acceleration $g = 10$ m/s^2).

128. Gas Mileage The gas mileage g (measured in km/L) for a particular vehicle, driven at v km/h, is given by the formula $g = 4.2 + 0.4v - 0.004v^2$, as long as v is between 15 km/h and 120 km/h. For what range of speeds is the vehicle's mileage 13 km/L or better?

129. Stopping Distance For a certain model of car the distance d required to stop the vehicle if it is traveling at v km/h is given by the formula

$$d = v + \frac{v^2}{20}$$

where d is measured in meters. You want your stopping distance not to exceed 70 m. At what range of speeds can you travel?

130. Manufacturer's Profit If a manufacturer sells x units of a certain product, revenue R and cost C (in dollars) are given by

$$R = 20x$$

$$C = 2000 + 8x + 0.0025x^2$$

Use the fact that

$$\text{profit} = \text{revenue} - \text{cost}$$

to determine how many units the manufacturer should sell to enjoy a profit of at least \$2400.

131. Fencing a Garden A determined gardener has 120 m of deer-resistant fence to enclose a rectangular vegetable garden with area at least 800 m^2. What range of values is possible for the length of the garden?

132. Thickness of a Laminate A company manufactures industrial laminates (thin nylon-based sheets) of thickness 0.05 cm with a tolerance of 0.0075 cm.

(a) Find an inequality involving absolute values that describes the range of possible thickness for the laminate.

(b) Solve the inequality you found in part (a).

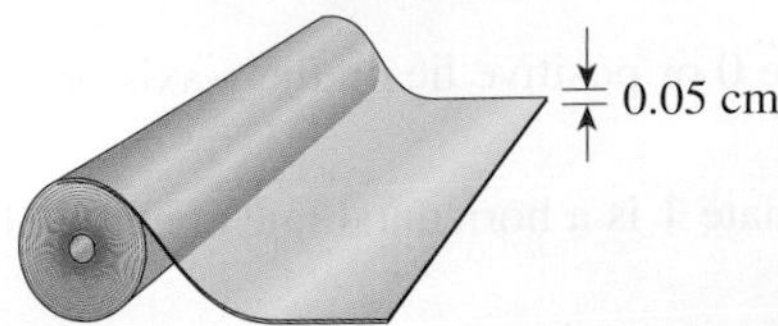

133. Range of Height The average height of adult males is 170.5 cm, and 95% of adult males have height h that satisfies the inequality

$$\left|\frac{h - 170.5}{7.25}\right| \le 2$$

Solve the inequality to find the range of heights.

■ Discuss ■ Discover ■ Prove ■ Write

134. Discuss ■ Discover: Do Powers Preserve Order? If $a < b$, is $a^2 < b^2$? (Check both positive and negative values for a and b.) If $a < b$, is $a^3 < b^3$? On the basis of your observations, state a general rule about the relationship between a^n and b^n when $a < b$ and n is a positive integer.

135. Discuss ■ Discover: What's Wrong Here? It is tempting to try to solve an inequality like an equation. For instance, we might try to solve $1 < 3/x$ by multiplying both sides by x to get $x < 3$, so the solution would be $(-\infty, 3)$. But that's wrong; for example, $x = -1$ lies in this interval but does not satisfy the original inequality. Explain why this method doesn't work (think about the *sign* of x). Then solve the inequality correctly.

136. Discuss ■ Discover: Using Distances to Solve Absolute Value Inequalities Recall that $|a - b|$ is the distance between a and b on the number line. For any number x, what do $|x - 1|$ and $|x - 3|$ represent? Use this interpretation to solve the inequality $|x - 1| < |x - 3|$ geometrically. In general, if $a < b$, what is the solution of the inequality $|x - a| < |x - b|$?

137–138 ■ Prove: Inequalities Use the rules for inequalities to prove the following.

137. Rule 6 for Inequalities: If a, b, c, and d are any real numbers such that $a < b$ and $c < d$, then $a + c < b + d$. [*Hint:* Use Rule 1 to show that $a + c < b + c$ and $b + c < b + d$. Use Rule 7.]

138. If a, b, c, and d are positive numbers such that $\frac{a}{b} < \frac{c}{d}$, then $\frac{a}{b} < \frac{a + c}{b + d} < \frac{c}{d}$.

PS *Establish subgoals.* First show that $\frac{ad}{b} + a < c + a$ and $a + c < \frac{cb}{d} + c$.

139. Prove: Arithmetic-Geometric Mean Inequality If $a_1, a_2, \ldots, a_n$ are nonnegative numbers, then their arithmetic mean is $\frac{a_1 + a_2 + \cdots + a_n}{n}$, and their geometric mean is $\sqrt[n]{a_1 a_2 \cdot \cdots \cdot a_n}$. The arithmetic-geometric mean inequality states that the geometric mean is always less than or equal to the arithmetic mean. In this problem we prove this in the case of two numbers x and y.

(a) Show that if x and y are nonnegative and $x \le y$, then $x^2 \le y^2$. [*Hint:* First use Rule 3 of Inequalities to show that $x^2 \le xy$ and $xy \le y^2$.]

(b) Prove the arithmetic-geometric mean inequality

$$\sqrt{xy} \le \frac{x + y}{2}$$

1.9 The Coordinate Plane; Graphs of Equations; Circles

■ The Coordinate Plane ■ The Distance and Midpoint Formulas ■ Graphs of Equations in Two Variables ■ Intercepts ■ Circles ■ Symmetry

The *coordinate plane* is the link between algebra and geometry. In the coordinate plane we can draw graphs of algebraic equations. The graphs, in turn, allow us to "see" the relationship between the variables in the equation.

■ The Coordinate Plane

The Cartesian plane is named in honor of the French mathematician René Descartes (1596–1650), although another Frenchman, Pierre Fermat (1607–1665), also invented the principles of coordinate geometry at the same time. (See their biographies in Section 1.11 and Section 2.6.)

Just as points on a line can be identified with real numbers to form the coordinate line, points in a plane can be identified with *ordered pairs* of numbers to form the **coordinate plane** or **Cartesian plane**. To do this, we draw two perpendicular real lines that intersect at 0 on each line. Usually, one line is horizontal with positive direction to the right and is called the ***x*-axis**; the other line is vertical with positive direction upward and is called the ***y*-axis**. The point of intersection of the *x*-axis and the *y*-axis is the **origin *O***, and the two axes divide the plane into four **quadrants**, labeled I, II, III, and IV in Figure 1. (The points that lie *on* the coordinate axes are not assigned to any quadrant.)

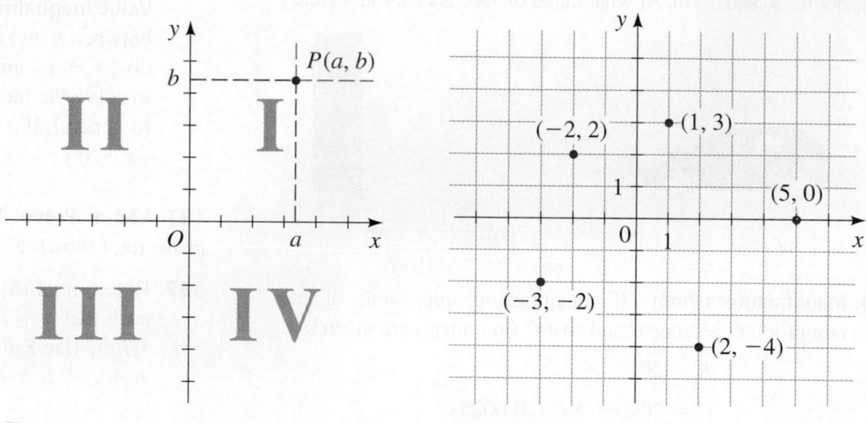

Figure 1

Figure 2

Although the notation for an ordered pair (a, b) is the same as the notation for an open interval (a, b), the context should make clear which meaning is intended.

Any point P in the coordinate plane can be located by a unique **ordered pair** of numbers (a, b), as shown in Figure 1. The first number a is called the ***x*-coordinate** of P; the second number b is called the ***y*-coordinate** of P. Several points are labeled with their coordinates in Figure 2. We can think of the ordered pair (a, b) as the "address" of the point P because (a, b) specifies the location of the point P in the plane. We often refer to the ordered pair (a, b) as the point (a, b).

Example 1 ■ Graphing Regions in the Coordinate Plane

Describe and sketch the regions given by each set.

(a) $\{(x, y) \mid x \geq 0\}$ **(b)** $\{(x, y) \mid y = 1\}$ **(c)** $\{(x, y) \mid |y| < 1\}$

Solution

(a) The points whose *x*-coordinates are 0 or positive lie on the *y*-axis or to the right of it, as shown in Figure 3(a).

(b) The set of all points with *y*-coordinate 1 is a horizontal line one unit above the *x*-axis, as shown in Figure 3(b).

Coordinates as Addresses

The coordinates of a point in the *xy*-plane uniquely determine its location. We can think of the coordinates as the "address" of the point. In Salt Lake City, Utah, the addresses of most buildings are in fact expressed as coordinates. The city is divided into quadrants with Main Street as the vertical (North-South) axis and S. Temple Street as the horizontal (East-West) axis. An address such as

1760 W 2100 S

indicates a location 17.6 blocks west of Main Street and 21 blocks south of S. Temple Street. (This is the address of the main post office in Salt Lake City.) With this logical system it is possible for someone unfamiliar with the city to locate any address as easily as one locates a point in the coordinate plane.

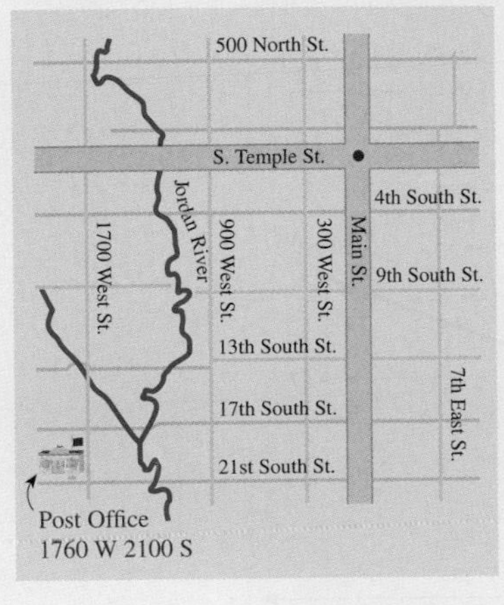

(c) Recall from Section 1.8 that

$$|y| < 1 \quad \text{if and only if} \quad -1 < y < 1$$

So the given region consists of those points in the plane whose y-coordinates lie between -1 and 1. Thus the region consists of all points that lie between (but not on) the horizontal lines $y = 1$ and $y = -1$. These lines are shown as broken lines in Figure 3(c) to indicate that the points on these lines are not included in the set.

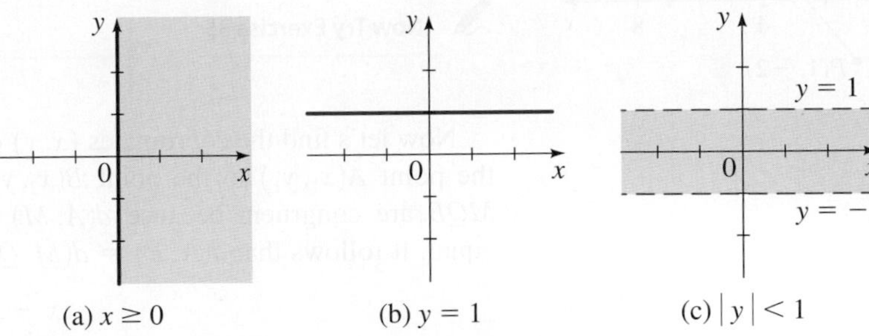

Figure 3

Now Try Exercises 15 and 17 ■

■ The Distance and Midpoint Formulas

We now find a formula for the distance $d(A, B)$ between two points $A(x_1, y_1)$ and $B(x_2, y_2)$ in the plane. Recall from Section 1.1 that the distance between points a and b on a number line is $d(a, b) = |b - a|$. So from Figure 4 we see that the distance between the points $A(x_1, y_1)$ and $C(x_2, y_1)$ on a horizontal line must be $|x_2 - x_1|$, and the distance between $B(x_2, y_2)$ and $C(x_2, y_1)$ on a vertical line must be $|y_2 - y_1|$.

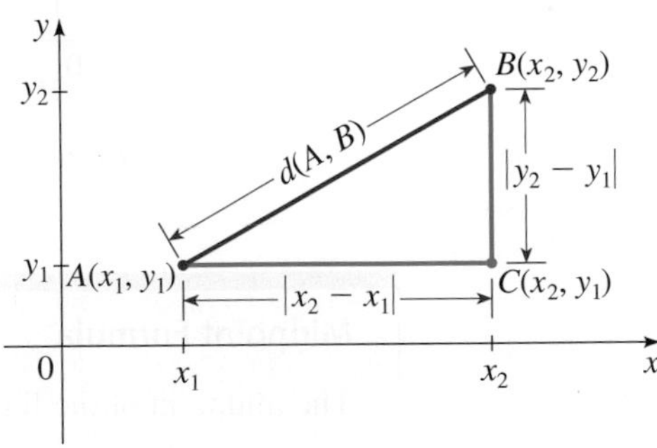

Figure 4

Since triangle ABC is a right triangle, the Pythagorean Theorem gives

$$d(A, B) = \sqrt{|x_2 - x_1|^2 + |y_2 - y_1|^2} = \sqrt{(x_2 - x_1)^2 + (y_2 - y_1)^2}$$

Distance Formula

The distance between the points $A(x_1, y_1)$ and $B(x_2, y_2)$ in the plane is

$$d(A, B) = \sqrt{(x_2 - x_1)^2 + (y_2 - y_1)^2}$$

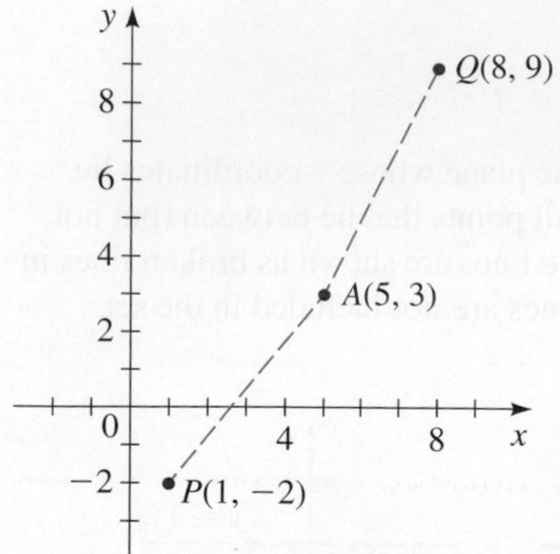

Figure 5

Example 2 ■ Applying the Distance Formula

Which of the points $P(1, -2)$ or $Q(8, 9)$ is closer to the point $A(5, 3)$?

Solution By the Distance Formula we have

$$d(P, A) = \sqrt{(5-1)^2 + [3-(-2)]^2} = \sqrt{4^2 + 5^2} = \sqrt{41}$$

$$d(Q, A) = \sqrt{(5-8)^2 + (3-9)^2} = \sqrt{(-3)^2 + (-6)^2} = \sqrt{45}$$

This shows that $d(P, A) < d(Q, A)$, so P is closer to A. (See Figure 5.)

Now Try Exercise 35 ■

Now let's find the coordinates (x, y) of the *midpoint M* of the line segment that joins the point $A(x_1, y_1)$ to the point $B(x_2, y_2)$. In Figure 6 notice that triangles APM and MQB are congruent because $d(A, M) = d(M, B)$ and the corresponding angles are equal. It follows that $d(A, P) = d(M, Q)$, so

$$x - x_1 = x_2 - x$$

Solving this equation for x, we get $2x = x_1 + x_2$, so

$$x = \frac{x_1 + x_2}{2} \qquad \text{Similarly,} \qquad y = \frac{y_1 + y_2}{2}$$

The x- and y-coordinates of the midpoint M are the averages of the x-coordinates of A and B and the y-coordinates of A and B, respectively.

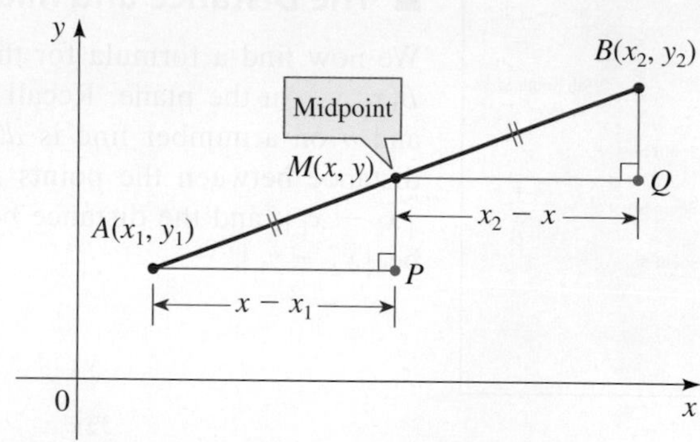

Figure 6

Midpoint Formula

The midpoint of the line segment from $A(x_1, y_1)$ to $B(x_2, y_2)$ is

$$\left(\frac{x_1 + x_2}{2}, \frac{y_1 + y_2}{2}\right)$$

Example 3 ■ Applying the Midpoint Formula

If the diagonals of a quadrilateral bisect each other, then the quadrilateral is a parallelogram. (See Appendix A, *Geometry Review*.)

Show that the quadrilateral with vertices $P(1, 2)$, $Q(4, 4)$, $R(5, 9)$, and $S(2, 7)$ is a parallelogram by proving that its two diagonals bisect each other.

Solution If the two diagonals have the same midpoint, then they must bisect each other. The midpoint of the diagonal PR is

$$\left(\frac{1+5}{2}, \frac{2+9}{2}\right) = \left(3, \frac{11}{2}\right)$$

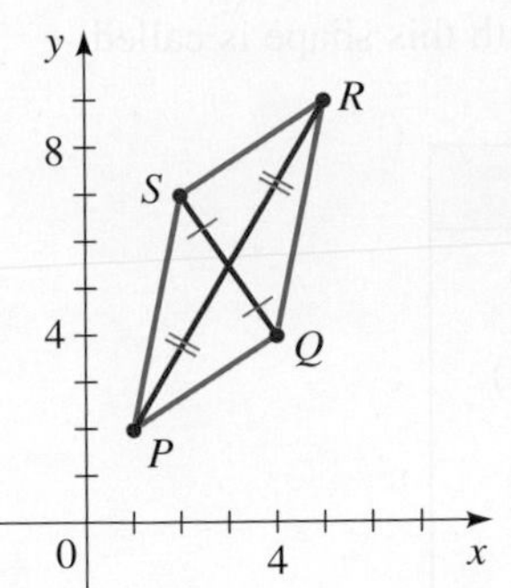

Figure 7

and the midpoint of the diagonal QS is

$$\left(\frac{4+2}{2}, \frac{4+7}{2}\right) = \left(3, \frac{11}{2}\right)$$

so each diagonal bisects the other, as shown in Figure 7.

Now Try Exercise 49 ■

■ Graphs of Equations in Two Variables

An **equation in two variables**, such as $y = x^2 + 1$, expresses a relationship between two quantities. A point (x, y) **satisfies** the equation if it makes the equation true when the values for x and y are substituted into the equation. For example, the point $(3, 10)$ satisfies the equation $y = x^2 + 1$ because $10 = 3^2 + 1$, but the point $(1, 3)$ does not, because $3 \neq 1^2 + 1$.

Fundamental Principle of Analytic Geometry
A point (x, y) lies on the graph of an equation if and only if its coordinates satisfy the equation.

The Graph of an Equation

The **graph** of an equation in x and y is the set of all points (x, y) in the coordinate plane that satisfy the equation.

The graph of an equation is a curve, so to graph an equation, we plot as many points as we can, then connect them by a smooth curve.

Example 4 ■ Sketching a Graph by Plotting Points

Sketch the graph of the equation $2x - y = 3$.

Solution We first solve the given equation for y to get

$$y = 2x - 3$$

This helps us calculate the y-coordinates in the following table.

x	$y = 2x - 3$	(x, y)
-1	-5	$(-1, -5)$
0	-3	$(0, -3)$
1	-1	$(1, -1)$
2	1	$(2, 1)$
3	3	$(3, 3)$
4	5	$(4, 5)$

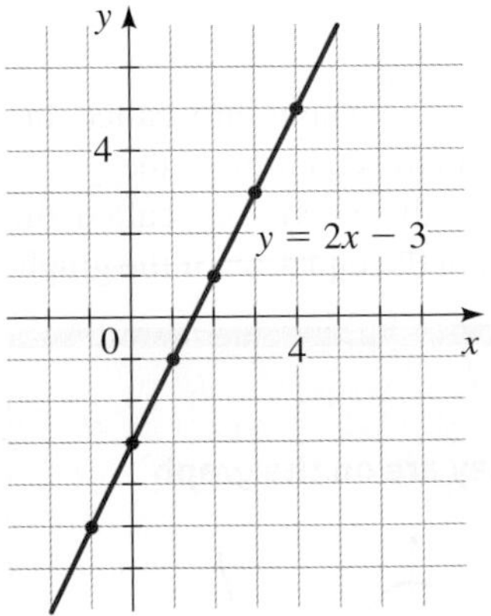

Figure 8

Of course, there are infinitely many points on the graph, and it is impossible to plot all of them. But the more points we plot, the better we can imagine what the graph represented by the equation looks like. In Figure 8 we plot the points corresponding to the ordered pairs we found in the table; they appear to lie on a line. So we complete the graph by joining the points with a line. (In Section 1.10 we verify that the graph of an equation of this type is indeed a line.)

 Now Try Exercise 55 ■

Detailed discussions of parabolas and their geometric properties are presented in Sections 3.1 and 10.1.

Example 5 ■ Sketching a Graph by Plotting Points

Sketch the graph of the equation $y = x^2 - 2$.

Solution We find some of the ordered pairs (x, y) that satisfy the equation in the table on the next page. In Figure 9 we plot the points corresponding to these ordered

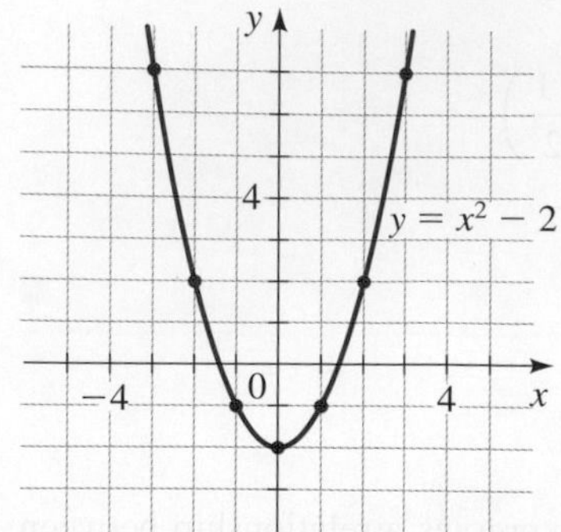

Figure 9

pairs and then connect them by a smooth curve. A curve with this shape is called a *parabola*.

x	$y = x^2 - 2$	(x, y)
−3	7	(−3, 7)
−2	2	(−2, 2)
−1	−1	(−1, −1)
0	−2	(0, −2)
1	−1	(1, −1)
2	2	(2, 2)
3	7	(3, 7)

Now Try Exercise 57 ■

Example 6 ■ Graphing an Absolute Value Equation

Sketch the graph of the equation $y = |x|$.

Solution We make a table of values:

x	$y = \|x\|$	(x, y)
−3	3	(−3, 3)
−2	2	(−2, 2)
−1	1	(−1, 1)
0	0	(0, 0)
1	1	(1, 1)
2	2	(2, 2)
3	3	(3, 3)

Figure 10

In Figure 10 we plot the points corresponding to the ordered pairs in the table and use them to sketch the graph of the equation.

Now Try Exercise 59 ■

■ Intercepts

The x-coordinates of the points where a graph intersects the x-axis are called the **x-intercepts** of the graph and are obtained by setting $y = 0$ in the equation of the graph. The y-coordinates of the points where a graph intersects the y-axis are called the **y-intercepts** of the graph and are obtained by setting $x = 0$ in the equation of the graph.

Definition of Intercepts

Intercepts	**How to find them**	**Where they are on the graph**
x-intercepts: The x-coordinates of points where the graph of an equation intersects the x-axis	Set $y = 0$ and solve for x	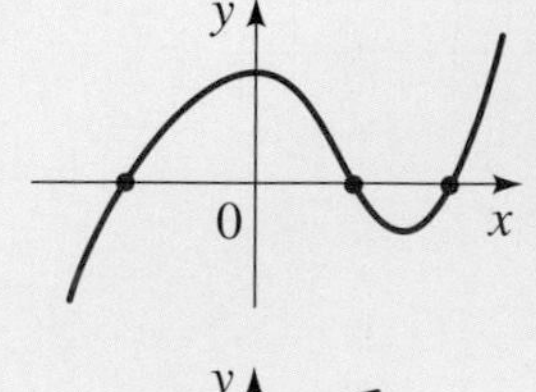
y-intercepts: The y-coordinates of points where the graph of an equation intersects the y-axis	Set $x = 0$ and solve for y	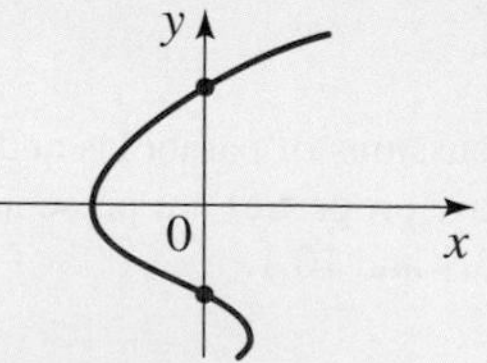

Example 7 ■ Finding Intercepts

Find the x- and y-intercepts of the graph of the equation $y = x^2 - 2$.

Solution To find the x-intercepts, we set $y = 0$ and solve for x. Thus

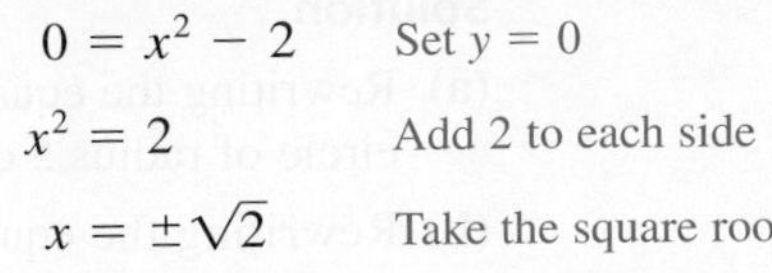

$$0 = x^2 - 2 \qquad \text{Set } y = 0$$

$$x^2 = 2 \qquad \text{Add 2 to each side}$$

$$x = \pm\sqrt{2} \qquad \text{Take the square root}$$

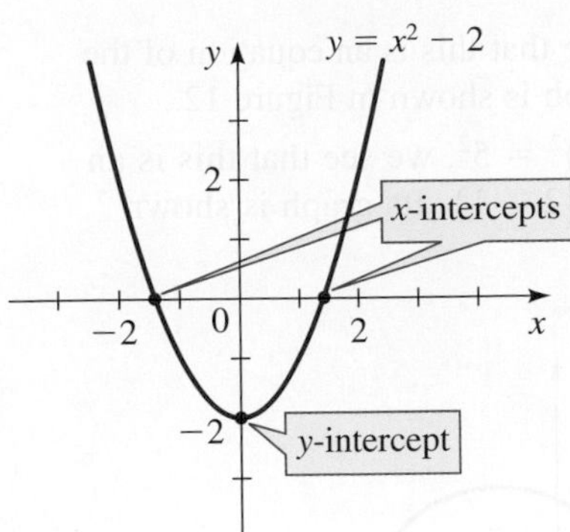

Figure 11

The x-intercepts are $\sqrt{2}$ and $-\sqrt{2}$.

To find the y-intercepts, we set $x = 0$ and solve for y. Thus

$$y = 0^2 - 2 \qquad \text{Set } x = 0$$

$$y = -2$$

The y-intercept is -2.

The graph of this equation was sketched in Example 5. It is repeated in Figure 11 with the x- and y-intercepts labeled.

Now Try Exercise 67 ■

■ Circles

So far, we have discussed how to find the graph of an equation in x and y. The converse problem is to find an equation of a graph, that is, an equation that represents a given curve in the xy-plane. Such an equation is satisfied by the coordinates of the points on the curve and by no other point. This is the other half of the fundamental principle of analytic geometry as formulated by Descartes and Fermat. The idea is that if a geometric curve can be represented by an algebraic equation, then the rules of algebra can be used to analyze the curve.

As an example of this type of problem, let's find the equation of a circle with radius r and center (h, k). By definition the circle is the set of all points $P(x, y)$ whose distance from the center $C(h, k)$ is r. Thus P is on the circle if and only if $d(P, C) = r$. From the distance formula we have

$$\sqrt{(x-h)^2 + (y-k)^2} = r$$

$$(x-h)^2 + (y-k)^2 = r^2 \qquad \text{Square each side}$$

This is the desired equation.

Equation of a Circle

An equation of the circle with center (h, k) and radius $r > 0$ is

$$(x-h)^2 + (y-k)^2 = r^2$$

This is called the **standard form** for the equation of the circle.

If the center of the circle is the origin $(0, 0)$, then the equation is

$$x^2 + y^2 = r^2$$

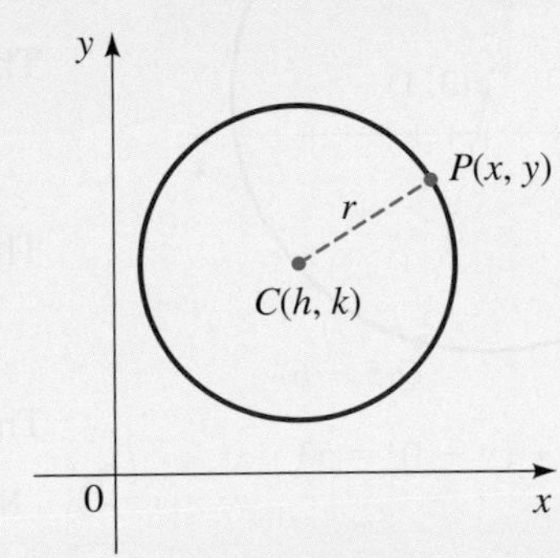

Example 8 ■ Graphing a Circle

Graph each equation.

(a) $x^2 + y^2 = 25$ **(b)** $(x - 2)^2 + (y + 1)^2 = 25$

Solution

(a) Rewriting the equation as $x^2 + y^2 = 5^2$, we see that this is an equation of the circle of radius 5 centered at the origin. Its graph is shown in Figure 12.

(b) Rewriting the equation as $(x - 2)^2 + (y + 1)^2 = 5^2$, we see that this is an equation of the circle of radius 5 centered at $(2, -1)$. Its graph is shown in Figure 13.

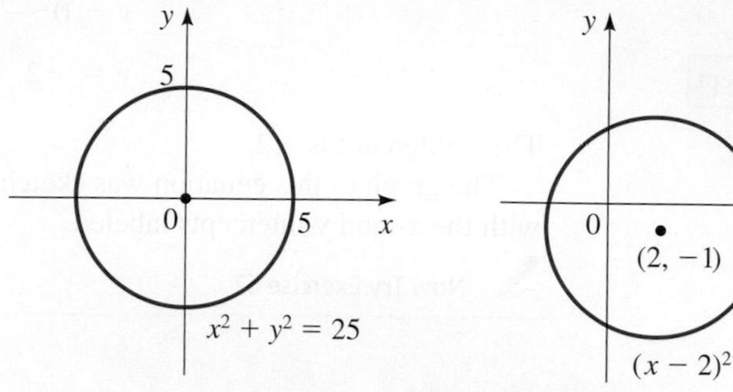

Figure 12 Figure 13

 **Now Try Exercises 75 and 77** ■

Example 9 ■ Finding an Equation of a Circle

(a) Find an equation of the circle with radius 3 and center $(2, -5)$.

(b) Find an equation of the circle that has the points $P(1, 8)$ and $Q(5, -6)$ as the endpoints of a diameter.

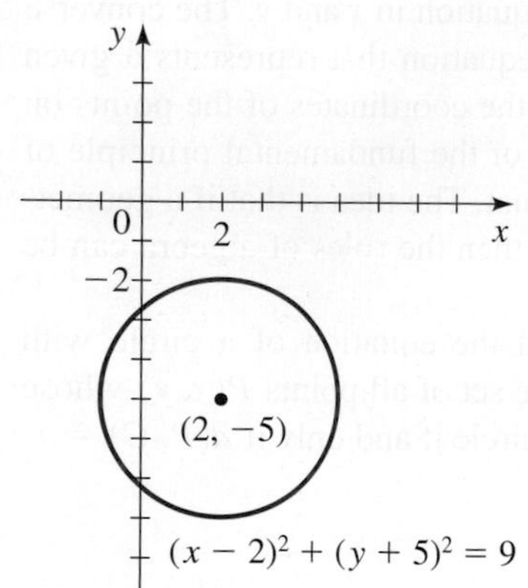

Figure 14

Solution

(a) Using the equation of a circle with $r = 3$, $h = 2$, and $k = -5$, we obtain

$$(x - 2)^2 + (y + 5)^2 = 9$$

The graph is shown in Figure 14.

(b) We first observe that the center is the midpoint of the diameter PQ, so by the Midpoint Formula the center is

$$\left(\frac{1 + 5}{2}, \frac{8 - 6}{2}\right) = (3, 1)$$

The radius r is the distance from P to the center, so by the Distance Formula

$$r^2 = (3 - 1)^2 + (1 - 8)^2 = 2^2 + (-7)^2 = 53$$

Therefore the equation of the circle is

$$(x - 3)^2 + (y - 1)^2 = 53$$

The graph is shown in Figure 15.

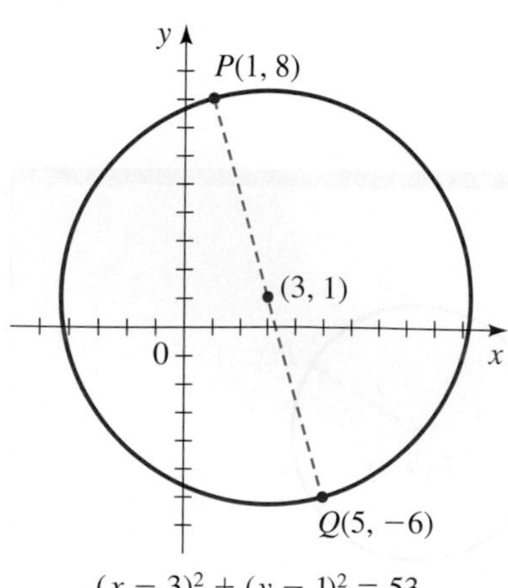

Figure 15

Now Try Exercises 81 and 85 ■

Let's expand the equation of the circle in the preceding example.

$(x-3)^2 + (y-1)^2 = 53$	Standard form
$x^2 - 6x + 9 + y^2 - 2y + 1 = 53$	Expand the squares
$x^2 - 6x + y^2 - 2y = 43$	Subtract 10 to get expanded form

Completing the square is used in many contexts in algebra. In Section 1.5 we used completing the square to solve quadratic equations.

Suppose we are given the equation of a circle in expanded form. Then to find its center and radius, we must put the equation back into standard form. That means that we must reverse the steps in the preceding calculation, and to do that, we need to know what to add to an expression like $x^2 - 6x$ to make it a perfect square—that is, we need to complete the square, as in the next example.

Example 10 ■ Identifying an Equation of a Circle

Show that the equation $x^2 + y^2 + 2x - 6y + 7 = 0$ represents a circle, and find the center and radius of the circle.

Solution We need to put the equation in standard form, so we first group the x-terms and y-terms. Then we complete the square within each grouping. That is, we complete the square for $x^2 + 2x$ by adding $\left(\frac{1}{2}\cdot 2\right)^2 = 1$, and we complete the square for $y^2 - 6y$ by adding $\left[\frac{1}{2}\cdot(-6)\right]^2 = 9$.

 We must add the same numbers to *each side* to maintain equality.

$(x^2 + 2x \quad) + (y^2 - 6y \quad) = -7$	Group terms
$(x^2 + 2x + 1) + (y^2 - 6y + 9) = -7 + 1 + 9$	Complete the square by adding 1 and 9 to each side
$(x+1)^2 + (y-3)^2 = 3$	Factor and simplify

Comparing this equation with the standard equation of a circle, we see that $h = -1$, $k = 3$, and $r = \sqrt{3}$, so the given equation represents a circle with center $(-1, 3)$ and radius $\sqrt{3}$.

Now Try Exercise 91 ■

Note An equation such as $(x-h)^2 + (y-k)^2 = c$, where $c \le 0$, may look like the equation of a circle but its graph is not a circle. For example, the graph of $x^2 + y^2 = 0$ is the single point $(0, 0)$ and the graph of $x^2 + y^2 = -4$ is empty (because the equation has no solution). (See Exercise 117.)

■ Symmetry

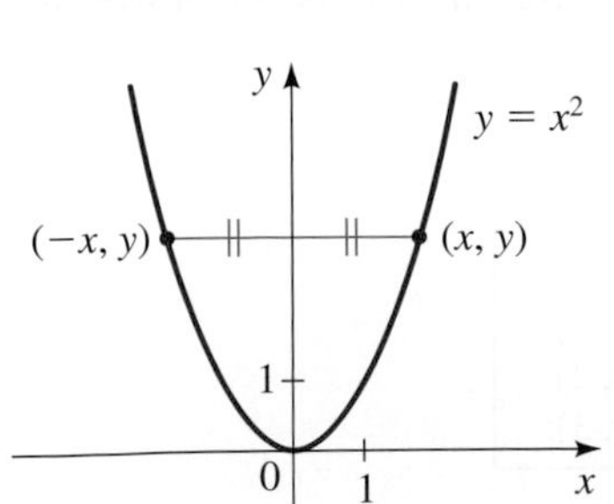

Figure 16

Figure 16 shows the graph of $y = x^2$. Notice that the part of the graph to the left of the y-axis is the mirror image of the part to the right of the y-axis. The reason is that if the point (x, y) is on the graph, then so is $(-x, y)$, and these points are reflections of each other about the y-axis. In this situation we say that the graph is **symmetric with respect to the y-axis**. Similarly, we say that a graph is **symmetric with respect to the x-axis** if whenever the point (x, y) is on the graph, then so is $(x, -y)$. A graph is **symmetric with respect to the origin** if whenever (x, y) is on the graph, so is $(-x, -y)$. (We often say symmetric "about" instead of "with respect to.")

Types of Symmetry

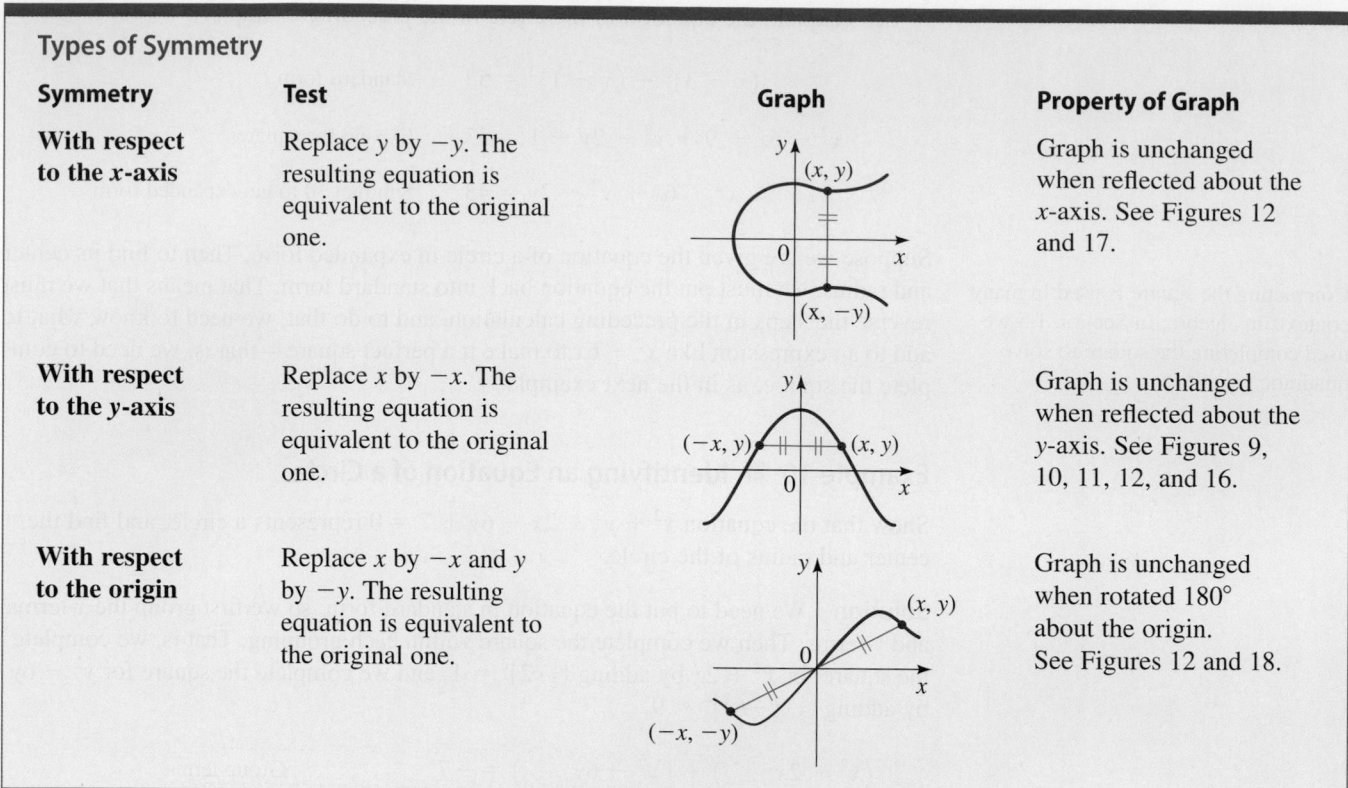

Symmetry	Test	Graph	Property of Graph
With respect to the x-axis	Replace y by $-y$. The resulting equation is equivalent to the original one.		Graph is unchanged when reflected about the x-axis. See Figures 12 and 17.
With respect to the y-axis	Replace x by $-x$. The resulting equation is equivalent to the original one.		Graph is unchanged when reflected about the y-axis. See Figures 9, 10, 11, 12, and 16.
With respect to the origin	Replace x by $-x$ and y by $-y$. The resulting equation is equivalent to the original one.		Graph is unchanged when rotated 180° about the origin. See Figures 12 and 18.

The remaining examples in this section show how symmetry helps us sketch the graphs of equations.

Example 11 ■ Using Symmetry to Sketch a Graph

Test the equation $x = y^2$ for symmetry and sketch the graph.

Solution If y is replaced by $-y$ in the equation $x = y^2$, we get

$$x = (-y)^2 \quad \text{Replace } y \text{ by } -y$$

$$x = y^2 \quad \text{Simplify}$$

and so the equation is equivalent to the original one. Therefore the graph is symmetric about the x-axis. But changing x to $-x$ gives the equation $-x = y^2$, which is not equivalent to the original equation, so the graph is not symmetric about the y-axis.

We use the symmetry about the x-axis to sketch the graph by first plotting points just for $y > 0$ and then reflecting the graph about the x-axis, as shown in Figure 17.

y	$x = y^2$	(x, y)
0	0	$(0, 0)$
1	1	$(1, 1)$
2	4	$(4, 2)$
3	9	$(9, 3)$

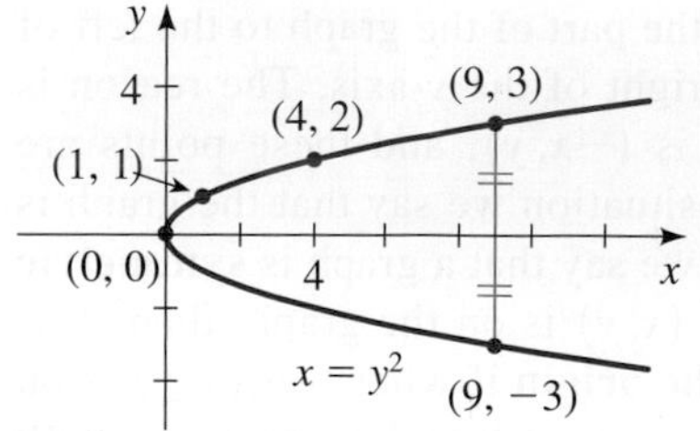

Figure 17

Now Try Exercises 97 and 103 ■

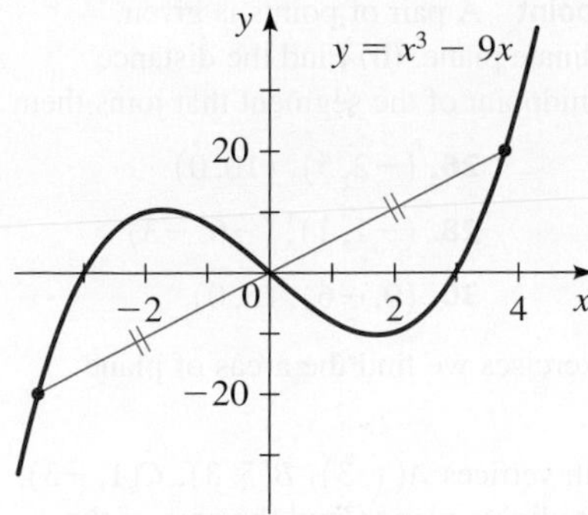

Figure 18

Example 12 ■ Testing an Equation for Symmetry

Test the equation $y = x^3 - 9x$ for symmetry.

Solution If we replace x by $-x$ and y by $-y$ in the equation, we get

$$-y = (-x)^3 - 9(-x) \qquad \text{Replace } x \text{ by } -x \text{ and } y \text{ by } -y$$

$$-y = -x^3 + 9x \qquad \text{Simplify}$$

$$y = x^3 - 9x \qquad \text{Multiply by } -1$$

and so the equation is equivalent to the original one. This means that the graph is symmetric with respect to the origin, as shown in Figure 18.

Now Try Exercise 99 ■

1.9 Exercises

Concepts

1. (a) The point that is 3 units to the right of the y-axis and 5 units below the x-axis has coordinates (____, ____).

(b) Is the point $(2, 7)$ closer to the x-axis or to the y-axis?

2. The distance between the points (a, b) and (c, d) is __________. So the distance between (1, 2) and (7, 10) is ______.

3. The point midway between (a, b) and (c, d) is ________. So the point midway between (1, 2) and (7, 10) is ________.

4. If the point $(2, 3)$ is on the graph of an equation in x and y, then the equation is satisfied when we replace x by ________ and y by ________. Is the point $(2, 3)$ on the graph of the equation $2y = x + 1$? Complete the table, and sketch a graph.

x	y	(x, y)
−2		
−1		
0		
1		
2		

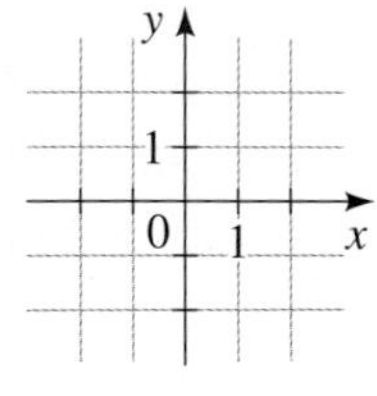

5. (a) To find the x-intercept(s) of the graph of an equation, we set ________ equal to 0 and solve for ________. So the x-intercept of $2y = x + 1$ is ________.

(b) To find the y-intercept(s) of the graph of an equation, we set ________ equal to 0 and solve for ________. So the y-intercept of $2y = x + 1$ is ________.

6. (a) The graph of the equation $(x - 1)^2 + (y - 2)^2 = 9$ is a circle with center (____, ____) and radius ________.

(b) Find the equation of the circle with center $(3, 4)$ that just touches (at one point) the y-axis.

7. (a) If a graph is symmetric with respect to the x-axis and (a, b) is on the graph, then (____, ____) is also on the graph.

(b) If a graph is symmetric with respect to the y-axis and (a, b) is on the graph, then (____, ____) is also on the graph.

(c) If a graph is symmetric about the origin and (a, b) is on the graph, then (____, ____) is also on the graph.

8. The graph of an equation is shown below.

(a) The x-intercept(s) are ________, and the y-intercept(s) are ________.

(b) The graph is symmetric about the ________ (x-axis/y-axis/origin).

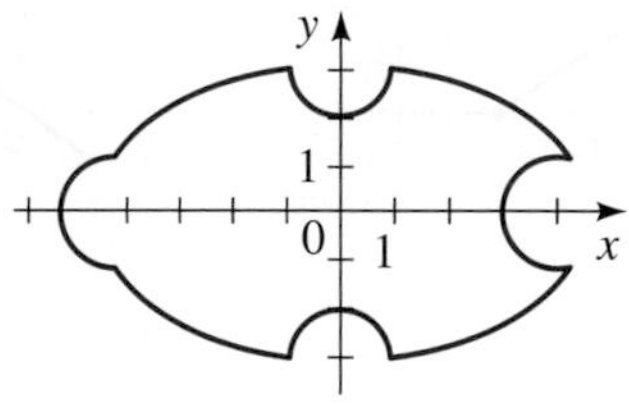

9–10 ■ *Yes or No*? If *No*, give a reason.

9. If the graph of an equation is symmetric with respect to both the x- and y-axes, is it necessarily symmetric with respect to the origin?

10. If the graph of an equation is symmetric with respect to the origin, is it necessarily symmetric with respect to the x- or y-axes?

Skills

11–12 ■ Points in a Coordinate Plane Refer to the figure below.

11. Find the coordinates of the points shown.

12. List the points that lie in either Quadrant I or Quadrant III.

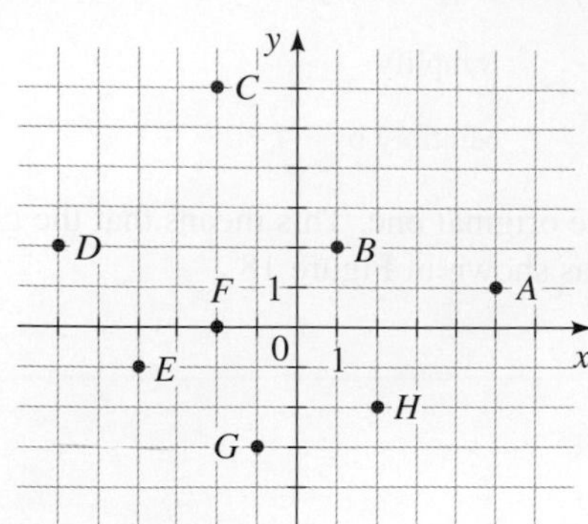

13–14 ■ Points in a Coordinate Plane Plot the given points in a coordinate plane.

13. $(0, 5), (-1, 0), (-1, -2), (\frac{1}{2}, \frac{2}{3})$

14. $(-5, 0), (2, 0), (2.6, -1.3), (-2.5, 3.5)$

15–20 ■ Sketching Regions Sketch the region given by each set.

15. (a) $\{(x, y) \mid x \geq 2\}$ **(b)** $\{(x, y) \mid y = 2\}$

16. (a) $\{(x, y) \mid y \leq -1\}$ **(b)** $\{(x, y) \mid x = 3\}$

17. (a) $\{(x, y) \mid -2 \leq x \leq 4\}$ **(b)** $\{(x, y) \mid |x| \leq 1\}$

18. (a) $\{(x, y) \mid 0 \leq y \leq 2\}$ **(b)** $\{(x, y) \mid |y| > 2\}$

19. (a) $\{(x, y) \mid -1 < x < 1 \text{ and } y \leq 4\}$
(b) $\{(x, y) \mid xy < 0\}$

20. (a) $\{(x, y) \mid |x| < 3 \text{ and } |y| > 2\}$
(b) $\{(x, y) \mid xy > 0\}$

21–24 ■ Distance and Midpoint A pair of points is graphed. **(a)** Find the distance between them. **(b)** Find the midpoint of the segment that joins them.

21.

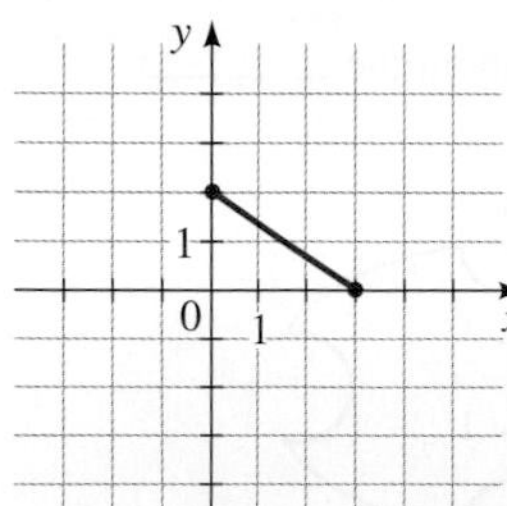

22.

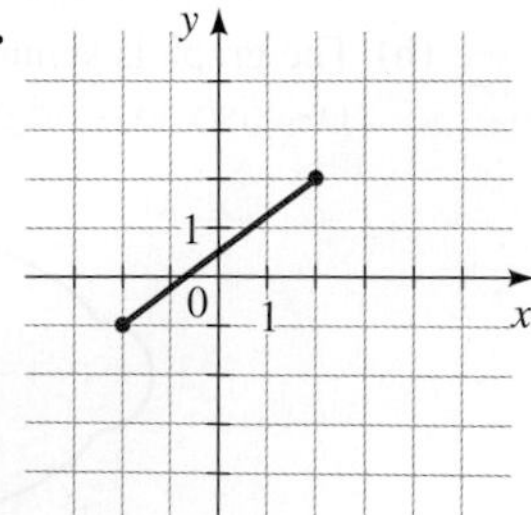

23.

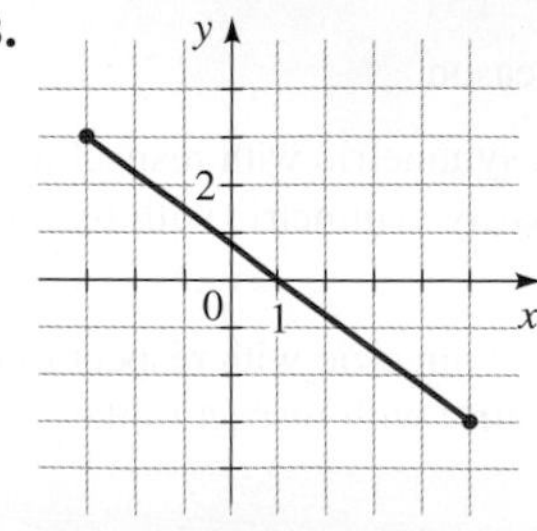

24.

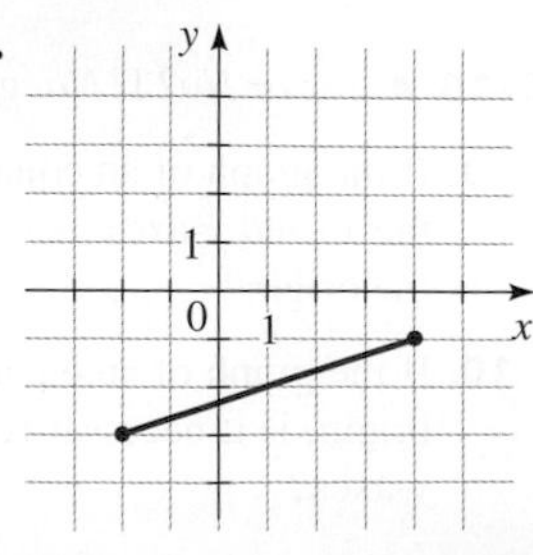

25–30 ■ Distance and Midpoint A pair of points is given. **(a)** Plot the points in a coordinate plane. **(b)** Find the distance between them. **(c)** Find the midpoint of the segment that joins them.

25. $(0, 8), (6, 16)$ **26.** $(-2, 5), (10, 0)$

27. $(3, -2), (-4, 5)$ **28.** $(-1, 1), (-6, -3)$

29. $(6, -2), (-6, 2)$ **30.** $(0, -6), (5, 0)$

31–34 ■ Area In these exercises we find the areas of plane figures.

31. Draw the rectangle with vertices $A(1, 3)$, $B(5, 3)$, $C(1, -3)$, and $D(5, -3)$ on a coordinate plane. Find the area of the rectangle.

32. Draw the parallelogram with vertices $A(1, 2)$, $B(5, 2)$, $C(3, 6)$, and $D(7, 6)$ on a coordinate plane. Find the area of the parallelogram.

33. Plot the points $A(1, 0)$, $B(5, 0)$, $C(4, 3)$, and $D(2, 3)$ on a coordinate plane. Draw the segments AB, BC, CD, and DA. What kind of quadrilateral is $ABCD$, and what is its area?

34. Plot the points $P(5, 1)$, $Q(0, 6)$, and $R(-5, 1)$ on a coordinate plane. Where must the point S be located so that the quadrilateral $PQRS$ is a square? Find the area of this square.

35–39 ■ Distance Formula In these exercises we use the Distance Formula.

35. Which of the points $A(6, 7)$ or $B(-5, 8)$ is closer to the origin?

36. Which of the points $C(-6, 3)$ or $D(3, 0)$ is closer to the point $E(-2, 1)$?

37. Which of the points $P(3, 1)$ or $Q(-1, 3)$ is closer to the point $R(-1, -1)$?

38. (a) Show that the points $(7, 3)$ and $(3, 7)$ are the same distance from the origin.
(b) Show that the points (a, b) and (b, a) are the same distance from the origin.

39. Show that the triangle with vertices $A(0, 2)$, $B(-3, -1)$, and $C(-4, 3)$ is isosceles.

40. Area of Triangle Find the area of the triangle shown in the figure.

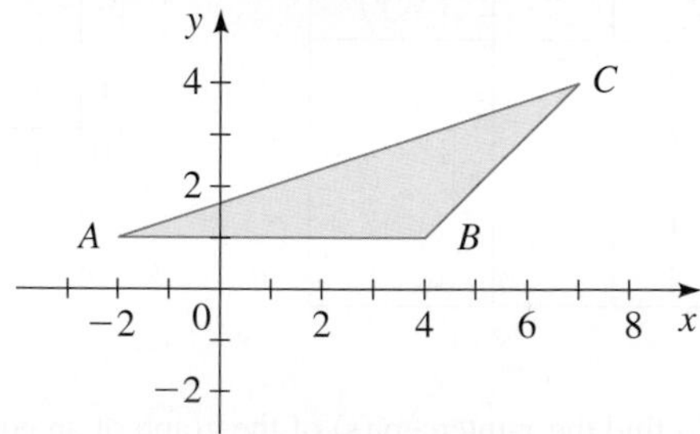

41–42 ■ Pythagorean Theorem In these exercises we use the converse of the Pythagorean Theorem (Appendix A) to show that the given triangle is a right triangle.

41. Refer to triangle ABC in the figure.
(a) Show that triangle ABC is a right triangle by using the converse of the Pythagorean Theorem.

(b) Find the area of triangle ABC.

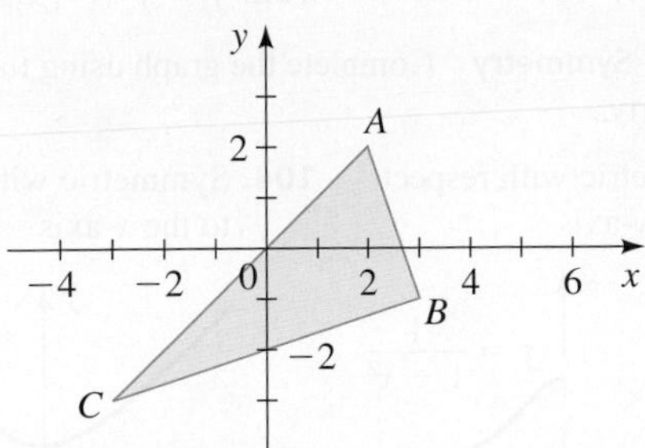

42. Show that the triangle with vertices $A(6, -7)$, $B(11, -3)$, and $C(2, -2)$ is a right triangle by using the converse of the Pythagorean Theorem. Find the area of the triangle.

43–45 ■ Distance Formula In these exercises we use the Distance Formula.

43. Show that the points $A(-2, 9)$, $B(4, 6)$, $C(1, 0)$, and $D(-5, 3)$ are the vertices of a square.

44. Show that the points $A(-1, 3)$, $B(3, 11)$, and $C(5, 15)$ are collinear by showing that $d(A, B) + d(B, C) = d(A, C)$.

45. Find a point on the y-axis that is equidistant from the points $(5, -5)$ and $(1, 1)$.

46–50 ■ Distance and Midpoint Formulas In these exercises we use the Distance Formula and the Midpoint Formula.

46. Find the lengths of the medians of the triangle with vertices $A(1, 0)$, $B(3, 6)$, and $C(8, 2)$. (A *median* is a line segment from a vertex to the midpoint of the opposite side.)

47. Plot the points $P(-1, -4)$, $Q(1, 1)$, and $R(4, 2)$ on a coordinate plane. Where should the point S be located so that the figure $PQRS$ is a parallelogram?

48. If $M(6, 8)$ is the midpoint of the line segment AB and if A has coordinates $(2, 3)$, find the coordinates of B.

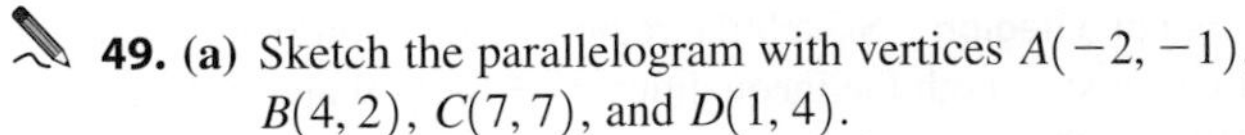

49. (a) Sketch the parallelogram with vertices $A(-2, -1)$, $B(4, 2)$, $C(7, 7)$, and $D(1, 4)$.

(b) Find the midpoints of the diagonals of this parallelogram.

(c) From part (b) show that the diagonals bisect each other.

50. The point M in the figure is the midpoint of the line segment AB. Show that M is equidistant from the vertices of triangle ABC.

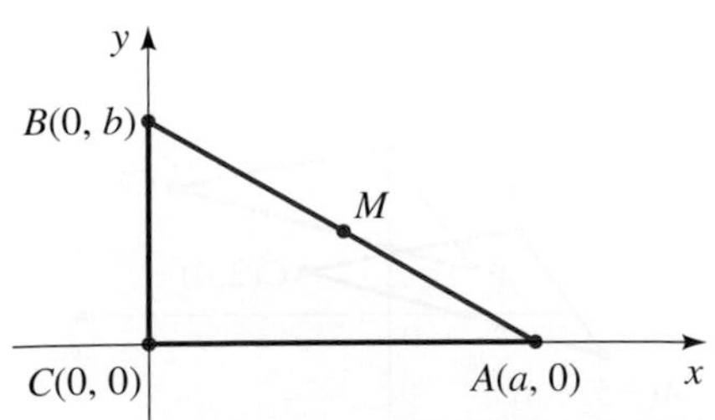

51–54 ■ Points on a Graph? Determine whether the given points are on the graph of the equation.

51. $3x - y + 5 = 0$; $(0, 5), (2, 1), (-2, -1)$

52. $y(x^2 + 1) = 1$; $(1, 1), (1, \frac{1}{2}), (-1, \frac{1}{2})$

53. $x^2 + xy + y^2 = 4$; $(0, -2), (1, -2), (2, -2)$

54. $x^2 + y^2 = 1$; $(0, 1), \left(\frac{1}{\sqrt{2}}, \frac{1}{\sqrt{2}}\right), \left(\frac{\sqrt{3}}{2}, \frac{1}{2}\right)$

55–60 ■ Graphing Equations Make a table of values, and sketch the graph of the equation.

55. $4x + 5y = 40$ **56.** $2x - 3y = 12$

57. $y = x^2 - 3$ **58.** $y = 3 - x^2$

59. $y = |x| - 1$ **60.** $y = |x + 1|$

61–66 ■ Graphing Equations Make a table of values, and sketch the graph of the equation. Find the x- and y-intercepts, and test for symmetry.

61. (a) $2x - y = 6$ **(b)** $y = 2(x - 1)^2$

62. (a) $x - 4y = 8$ **(b)** $y = -x^2 + 4$

63. (a) $y = \sqrt{x} + 2$ **(b)** $y = -|x|$

64. (a) $y = \sqrt{x - 4}$ **(b)** $x = |y|$

65. (a) $y = \sqrt{4 - x^2}$ **(b)** $y = x^3 - 4x$

66. (a) $y = -\sqrt{4 - x^2}$ **(b)** $x = y^3$

67–70 ■ Intercepts Find the x- and y-intercepts of the graph of the equation.

67. (a) $y = x + 6$ **(b)** $y = x^2 - 5$

68. (a) $4x^2 + 25y^2 = 100$ **(b)** $x^2 - xy + 3y = 1$

69. (a) $9x^2 - 4y^2 = 36$ **(b)** $y - 2xy + 4x = 1$

70. (a) $y = \sqrt{x^2 - 16}$ **(b)** $y = \sqrt{64 - x^3}$

71–74 ■ Intercepts An equation and its graph are given. Find the x- and y-intercepts from the graph. Check that your answers satisfy the equation.

71. $y = 4x - x^2$ **72.** $\frac{x^2}{9} + \frac{y^2}{4} = 1$

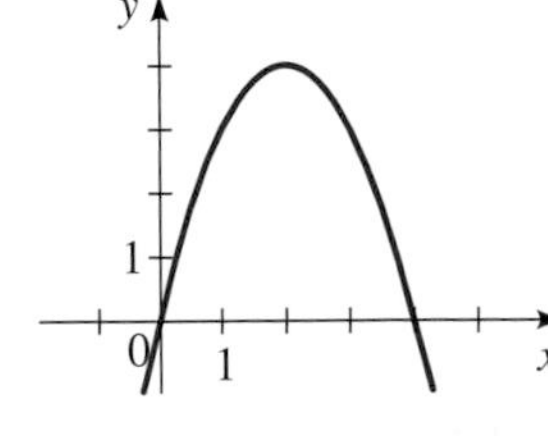

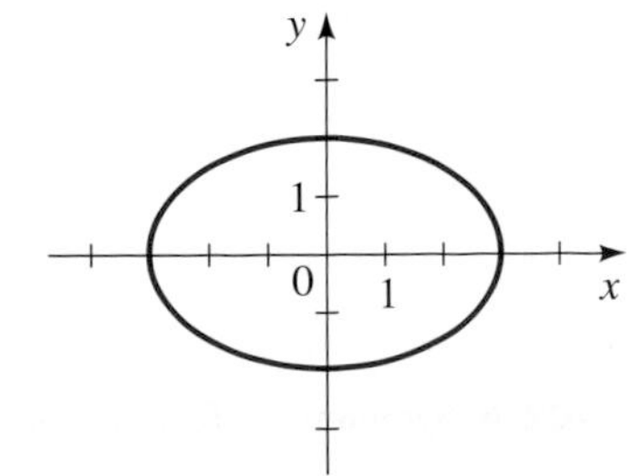

73. $x^4 + y^2 - xy = 16$

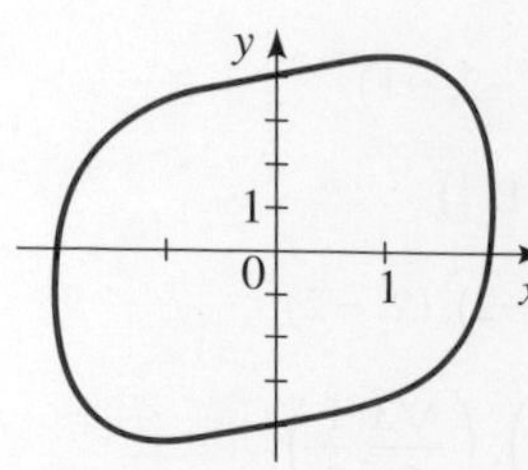

74. $x^2 + y^3 - x^2y^2 = 64$

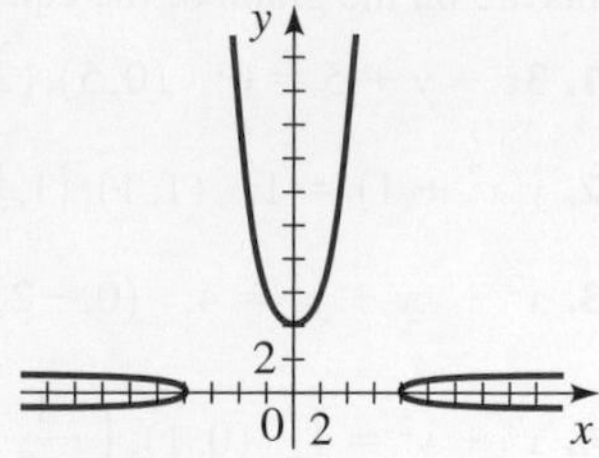

75–80 ■ Graphing Circles Find the center and radius of the circle, and sketch its graph.

75. $x^2 + y^2 = 9$

76. $x^2 + y^2 = 5$

77. $(x - 2)^2 + y^2 = 9$

78. $x^2 + (y + 1)^2 = 4$

79. $(x + 3)^2 + (y - 4)^2 = 25$

80. $(x + 1)^2 + (y + 2)^2 = 36$

81–88 ■ Equations of Circles Find an equation of the circle that satisfies the given conditions.

81. Center $(-3, 1)$; radius 2

82. Center $(2, -5)$; radius 3

83. Center at the origin; passes through $(4, 7)$

84. Center $(-1, 5)$; passes through $(-4, -6)$

85. Endpoints of a diameter are $P(-1, 1)$ and $Q(5, 9)$

86. Endpoints of a diameter are $P(-1, 3)$ and $Q(7, -5)$

87. Center $(7, -3)$; tangent to the x-axis

88. Circle lies in the first quadrant, tangent to both x- and y-axes; radius 5

89–90 ■ Equations of Circles Find the equation of the circle shown in the figure.

89.

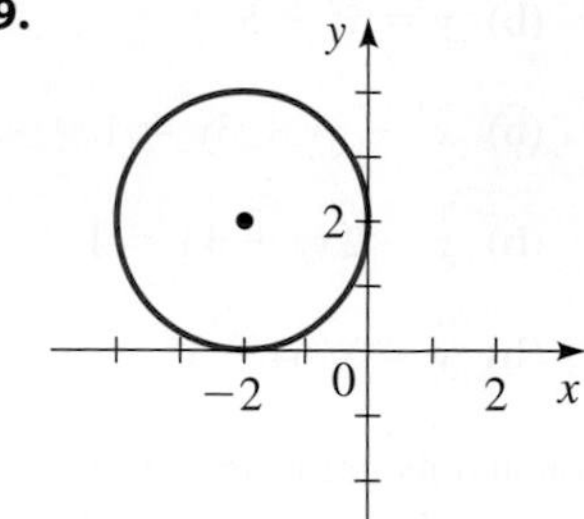

90.

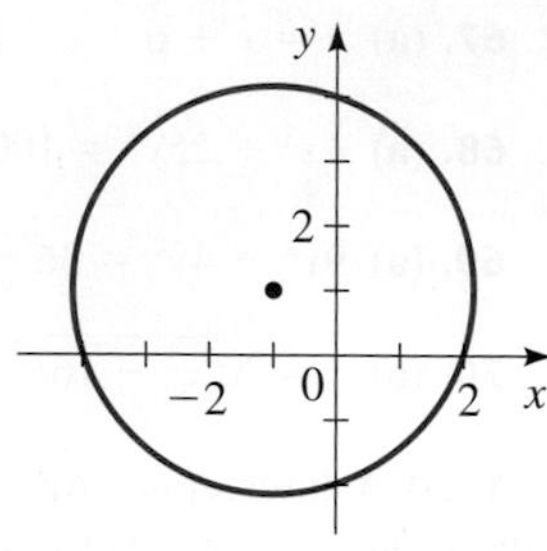

91–96 ■ Equations of Circles Show that the equation represents a circle, and find the center and radius of the circle.

91. $x^2 + y^2 + 4x - 6y + 12 = 0$

92. $x^2 + y^2 + 8x + 5 = 0$

93. $x^2 + y^2 - \frac{1}{2}x + \frac{1}{2}y = \frac{1}{8}$

94. $x^2 + y^2 + \frac{1}{2}x + 2y + \frac{1}{16} = 0$

95. $2x^2 + 2y^2 - 3x = 0$

96. $3x^2 + 3y^2 + 6x - y = 0$

97–102 ■ Symmetry Test the equation for symmetry.

97. $y = x^4 + x^2$

98. $x = y^4 - y^2$

99. $x^2y^4 + x^4y^2 = 2$

100. $x^3y + xy^3 = 1$

101. $y = x^3 + 10x$

102. $y = x^2 + |x|$

103–106 ■ Symmetry Complete the graph using the given symmetry property.

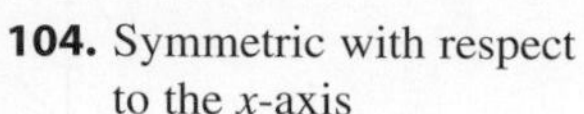

103. Symmetric with respect to the y-axis

104. Symmetric with respect to the x-axis

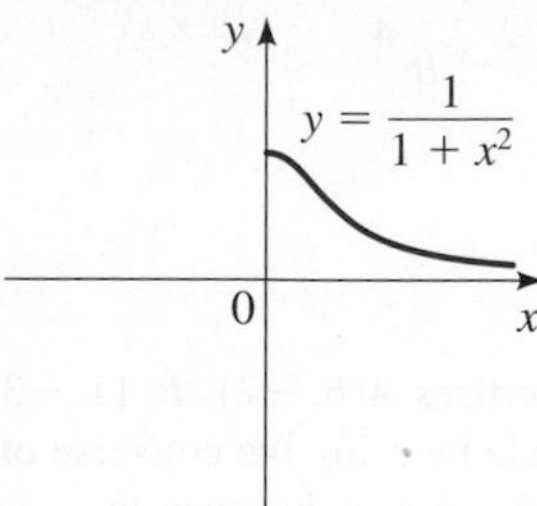

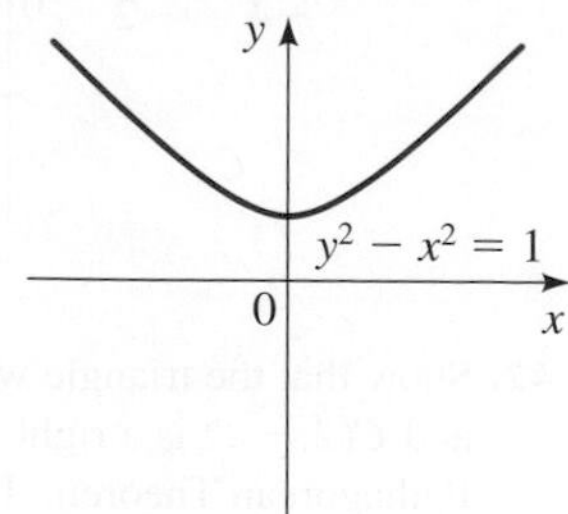

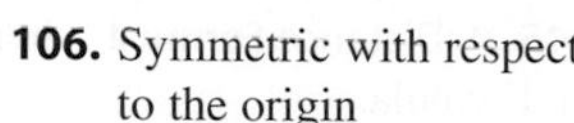

105. Symmetric with respect to the origin

106. Symmetric with respect to the origin

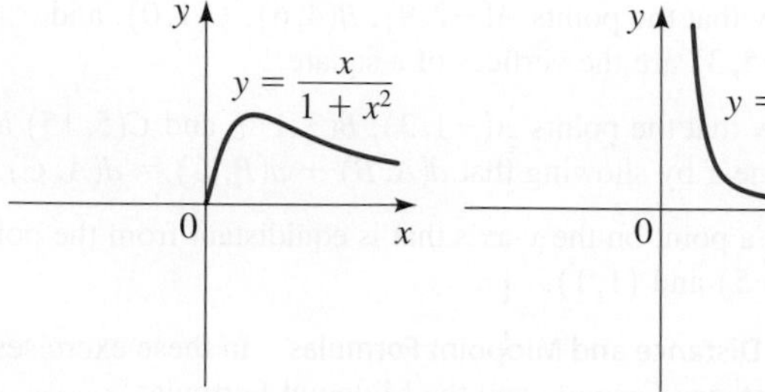

Skills Plus

107–108 ■ Graphing Regions Sketch the region given by the set.

107. $\{(x, y) \mid x^2 + y^2 \le 1\}$

108. $\{(x, y) \mid x^2 + y^2 > 4\}$

109. Area of a Region Find the area of the region that lies outside the circle $x^2 + y^2 = 4$ but inside the circle

$$x^2 + y^2 - 4y - 12 = 0$$

110. Area of a Region Sketch the region in the coordinate plane that satisfies both the inequalities $x^2 + y^2 \le 9$ and $y \ge |x|$. What is the area of this region?

111. Shifting the Coordinate Plane Suppose that each point in the coordinate plane is shifted 3 units to the right and 2 units upward.

(a) The point $(5, 3)$ is shifted to what new point?

(b) The point (a, b) is shifted to what new point?

(c) What point is shifted to $(3, 4)$?

(d) Triangle ABC in the figure has been shifted to triangle $A'B'C'$. Find the coordinates of the points A', B', and C'.

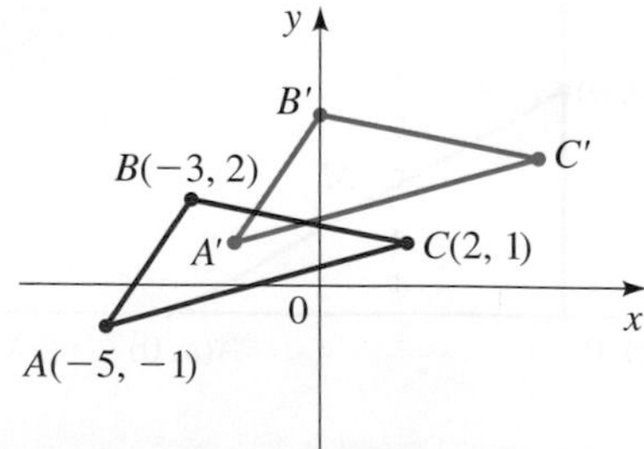

112. Reflecting in the Coordinate Plane Suppose that the y-axis acts as a mirror that reflects each point to the right of it into a point to the left of it.

(a) The point $(3, 7)$ is reflected to what point?

(b) The point (a, b) is reflected to what point?

(c) What point is reflected to $(-4, -1)$?

(d) Triangle ABC in the figure is reflected to triangle $A'B'C'$. Find the coordinates of the points A', B', and C'.

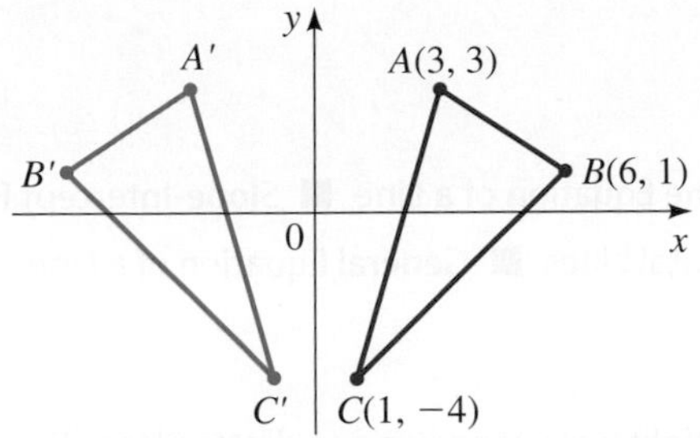

113. Making a Graph Symmetric The graph in the figure is not symmetric about the x-axis, the y-axis, or the origin. Add line segments to the graph so that it exhibits the indicated symmetry. In each case, add as little as possible.

(a) Symmetry about the x-axis

(b) Symmetry about the y-axis

(c) Symmetry about the origin

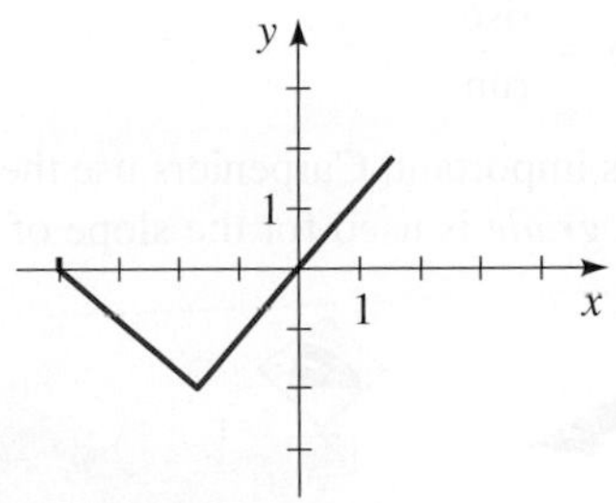

■ Applications

114. Lunar Gateway Orbit The *Lunar Gateway* (or simply *Gateway*) is a space station expected to be launched in 2024 to orbit the moon in a halo orbit. The station is intended to serve as a waypoint for missions to the moon and Mars. Its highly elliptical orbit forms a plane that is always visible from Earth. We can model the orbit by the equation.

$$\frac{(x - 32.5)^2}{1420} + \frac{y^2}{213} = 1$$

where x and y are measured in megameters (Mm).

(a) From the graph, estimate the closest and farthest distances of the satellite from the center of the moon. Use the equation to confirm your answers.

(b) There are two points in the orbit with y-coordinate 10. Find the x-coordinates of these points, and determine their distances from the center of the moon.

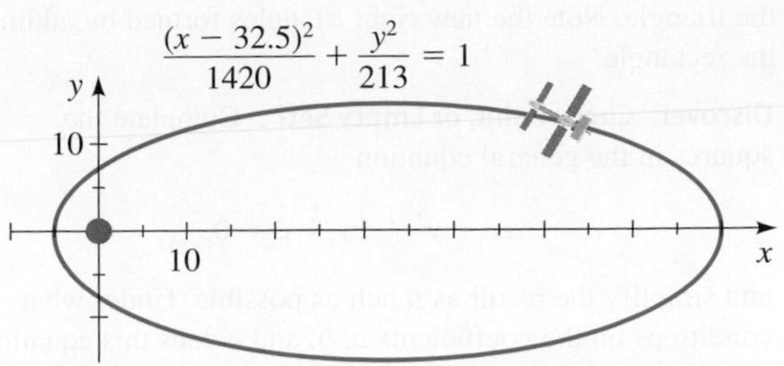

115. Distances in a City A city has streets that run north and south and avenues that run east and west, all equally spaced. Streets and avenues are numbered sequentially, as shown in the figure. The *walking* distance between points A and B is 7 blocks—that is, 3 blocks east and 4 blocks north. To find the *straight-line* distance d, we must use the Distance Formula.

(a) Find the straight-line distance (in blocks) between A and B.

(b) Find the walking distance and the straight-line distance between the corner of 4th St. and 2nd Ave. and the corner of 11th St. and 26th Ave.

(c) What must be true about the points P and Q if the walking distance between P and Q equals the straight-line distance between P and Q?

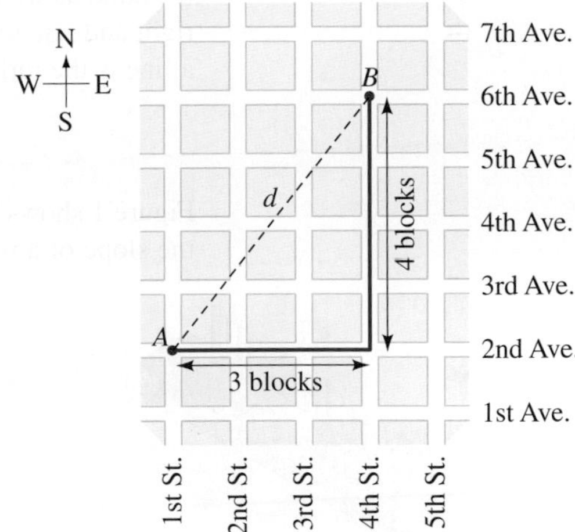

■ Discuss ■ Discover ■ Prove ■ Write

116. Discuss ■ Discover: Area of a Triangle in the Coordinate Plane Find the area of the triangle in the figure.

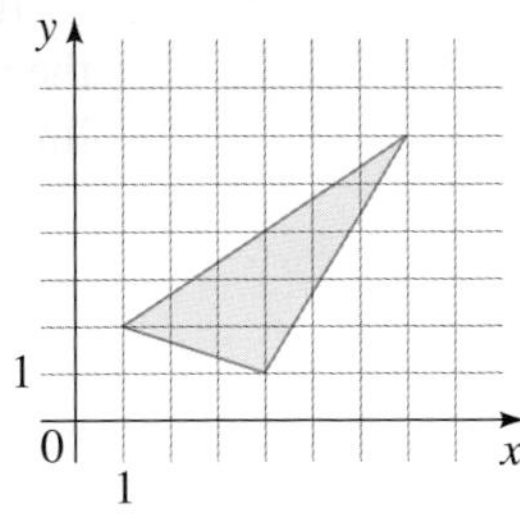

PS *Introduce something extra.* Draw the rectangle (with sides parallel to the coordinate axes) that circumscribes the triangle. Note the new right triangles formed by adding the rectangle.

117. Discover: Circle, Point, or Empty Set? Complete the squares in the general equation

$$x^2 + ax + y^2 + by + c = 0$$

and simplify the result as much as possible. Under what conditions on the coefficients a, b, and c does this equation represent a circle? A single point? The empty set? In the case in which the equation does represent a circle, find its center and radius.

118. Prove: Coloring the plane Suppose that each point in the plane is colored either red or blue. Show that there must exist two points of the same color that are exactly one unit apart.

PS *Introduce something extra.* Imagine an equilateral triangle with side length 1 in the plane.

1.10 Lines

■ The Slope of a Line ■ Point-Slope Form of the Equation of a Line ■ Slope-Intercept Form of the Equation of a Line ■ Vertical and Horizontal Lines ■ General Equation of a Line ■ Parallel and Perpendicular Lines

In this section we find equations for straight lines lying in a coordinate plane. The equations will depend on how the line is inclined, so we begin by discussing the concept of slope.

■ The Slope of a Line

We first need a way to measure the "steepness" of a line, or how quickly it rises (or falls) as we move from left to right. We define *run* to be the distance we move to the right and *rise* to be the corresponding distance that the line rises (or falls). The *slope* of a line is the ratio of rise to run:

$$\text{slope} = \frac{\text{rise}}{\text{run}}$$

Figure 1 shows situations in which slope is important. Carpenters use the term *pitch* for the slope of a roof or a staircase; the term *grade* is used for the slope of a road.

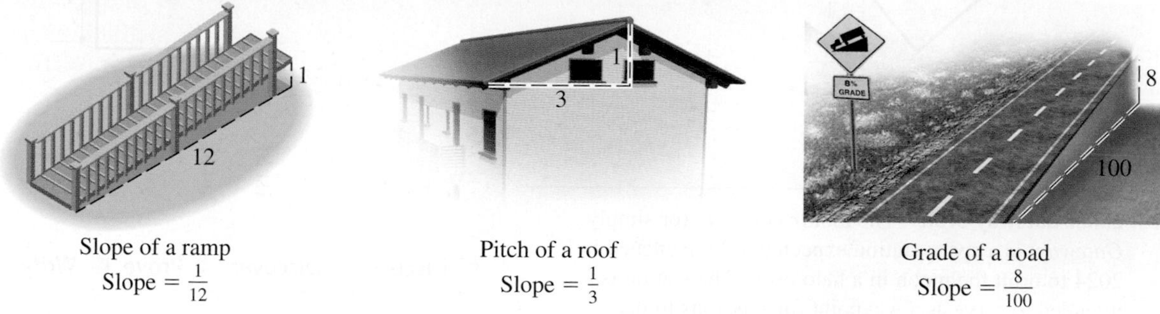

Slope of a ramp
Slope $= \frac{1}{12}$

Pitch of a roof
Slope $= \frac{1}{3}$

Grade of a road
Slope $= \frac{8}{100}$

Figure 1

If a line lies in a coordinate plane, then the **run** is the change in the x-coordinate and the **rise** is the corresponding change in the y-coordinate between any two points on the line (see Figure 2). This gives us the following definition of slope.

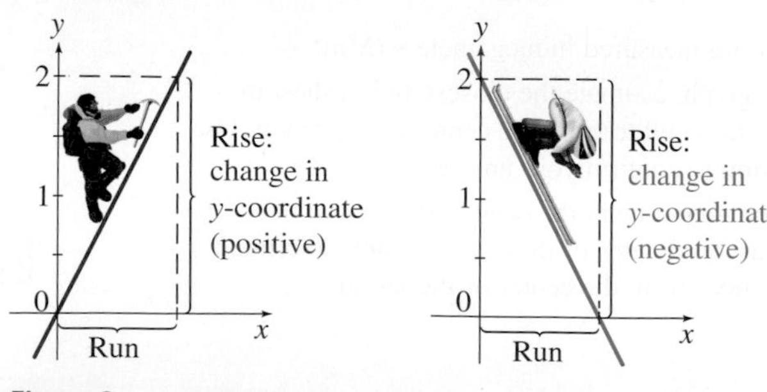

Figure 2

Slope of a Line

The **slope** m of a nonvertical line that passes through the points $A(x_1, y_1)$ and $B(x_2, y_2)$ is

$$m = \frac{\text{rise}}{\text{run}} = \frac{y_2 - y_1}{x_2 - x_1}$$

The slope of a vertical line is not defined.

The slope is independent of which two points are chosen on the line. We can see that this is true from the similar triangles in Figure 3.

$$\frac{y_2 - y_1}{x_2 - x_1} = \frac{y_2' - y_1'}{x_2' - x_1'}$$

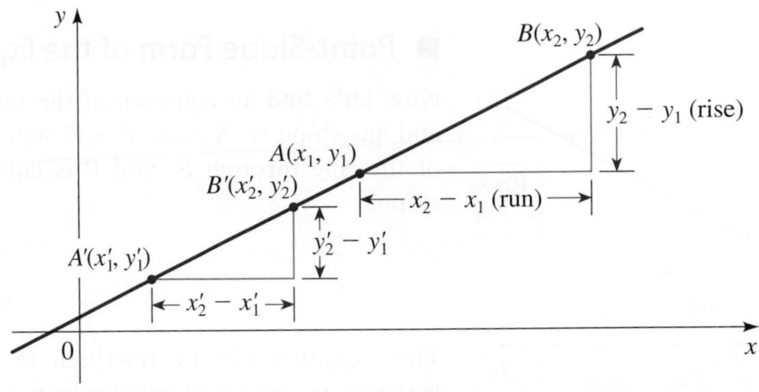

Figure 3

The figures in the box below show several lines labeled with their slopes. Noticc that lines with positive slope slant upward to the right, whereas lines with negative slope slant downward to the right. The steepest lines are those for which the absolute value of the slope is the largest; a horizontal line has slope 0. The slope of a vertical line is undefined (it has a 0 denominator), so we say that a vertical line has no slope.

Slope of a Line

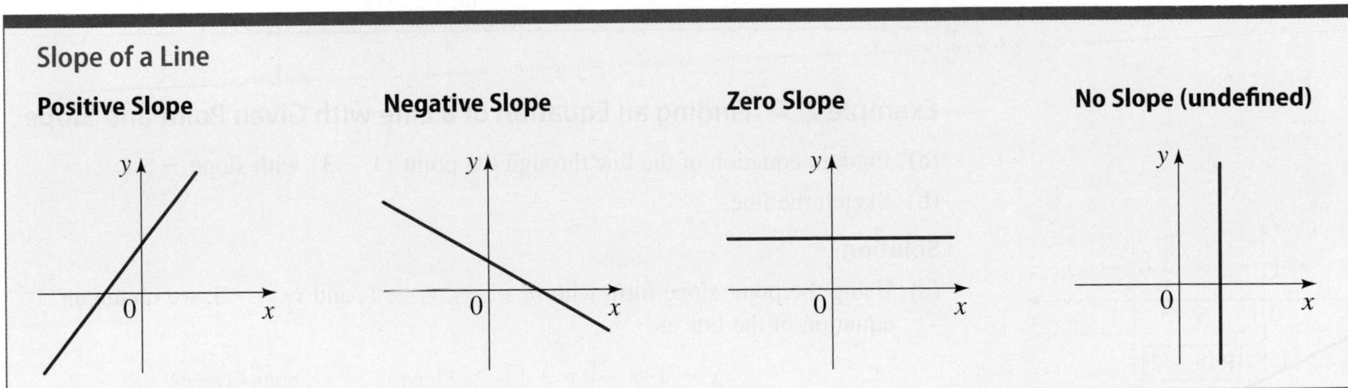

Example 1 ■ Finding the Slope of a Line Through Two Points

Find the slope of the line that passes through the points $P(2, 1)$ and $Q(8, 5)$.

Solution Since any two different points determine a line, only one line passes through these two points. From the definition the slope is

$$m = \frac{y_2 - y_1}{x_2 - x_1} = \frac{5 - 1}{8 - 2} = \frac{4}{6} = \frac{2}{3}$$

This says that for every 3 units we move to the right, the line rises 2 units. The line is drawn in Figure 4.

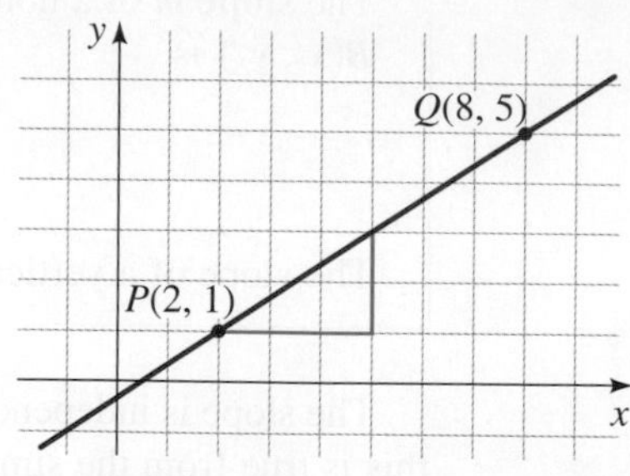

Figure 4

 Now Try Exercise 9 ■

■ Point-Slope Form of the Equation of a Line

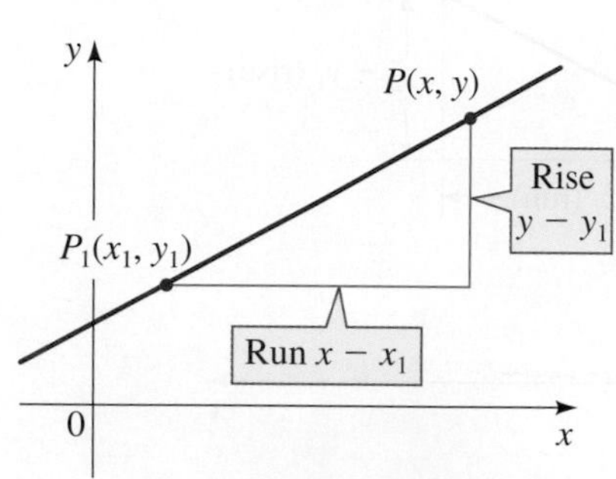

Figure 5

Now let's find an equation of the line that passes through a given point $P_1(x_1, y_1)$ and has slope m. A point $P(x, y)$ with $x \neq x_1$ lies on this line if and only if the slope of the line through P_1 and P is equal to m. (See Figure 5.) By the definition of slope,

$$\frac{y - y_1}{x - x_1} = m$$

This equation can be rewritten in the form $y - y_1 = m(x - x_1)$; note that the equation is also satisfied when $x = x_1$ and $y = y_1$. Therefore it is an equation of the given line.

Point-Slope Form of the Equation of a Line

An equation of the line that passes through the point (x_1, y_1) and has slope m is

$$y - y_1 = m(x - x_1)$$

Example 2 ■ Finding an Equation of a Line with Given Point and Slope

(a) Find an equation of the line through the point $(1, -3)$ with slope $-\frac{1}{2}$.

(b) Sketch the line.

Solution

(a) Using the point-slope form with $m = -\frac{1}{2}$, $x_1 = 1$, and $y_1 = -3$, we obtain an equation of the line as

$$y + 3 = -\tfrac{1}{2}(x - 1) \qquad \text{Slope } m = -\tfrac{1}{2}, \text{ point } (1, -3)$$

$$2y + 6 = -x + 1 \qquad \text{Multiply by 2}$$

$$x + 2y + 5 = 0 \qquad \text{Rearrange}$$

(b) The fact that the slope is $-\frac{1}{2}$ tells us that when we move to the right 2 units, the line drops 1 unit. This enables us to sketch the line in Figure 6.

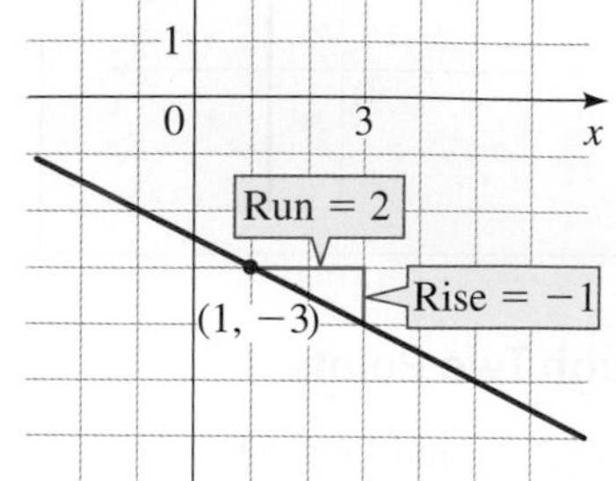

Figure 6

Now Try Exercise 25 ■

Example 3 ■ Finding an Equation of a Line Through Two Given Points

Find an equation of the line through the points $(-1, 2)$ and $(3, -4)$.

Solution The slope of the line is

$$m = \frac{-4 - 2}{3 - (-1)} = -\frac{6}{4} = -\frac{3}{2}$$

We can use *either* point, $(-1, 2)$ *or* $(3, -4)$, in the point-slope equation. We will end up with the same final answer.

Using the point-slope form with $x_1 = -1$ and $y_1 = 2$, we obtain

$$y - 2 = -\tfrac{3}{2}(x + 1) \qquad \text{Slope } m = -\tfrac{3}{2}, \text{ point } (-1, 2)$$

$$2y - 4 = -3x - 3 \qquad \text{Multiply by 2}$$

$$3x + 2y - 1 = 0 \qquad \text{Rearrange}$$

Now Try Exercise 29 ■

■ Slope-Intercept Form of the Equation of a Line

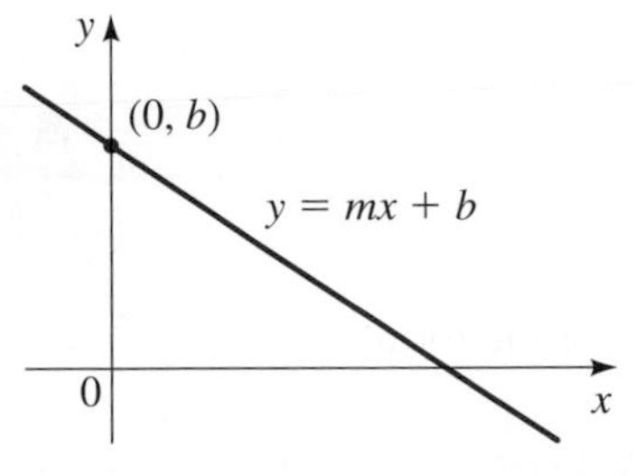

Figure 7

Suppose a nonvertical line has slope m and y-intercept b. (See Figure 7.) This means that the line intersects the y-axis at the point $(0, b)$, so the point-slope form of the equation of the line, with $x = 0$ and $y = b$, becomes

$$y - b = m(x - 0)$$

This simplifies to $y = mx + b$, which is called the **slope-intercept form** of the equation of a line.

> **Slope-Intercept Form of the Equation of a Line**
>
> An equation of the line that has slope m and y-intercept b is
>
> $$y = mx + b$$

Example 4 ■ Lines in Slope-Intercept Form

(a) Find an equation of the line with slope 3 and y-intercept -2.

(b) Find the slope and y-intercept of the line $3y - 2x = 1$.

Solution

(a) Since $m = 3$ and $b = -2$, from the slope-intercept form of the equation of a line we get

$$y = 3x - 2$$

(b) We first write the equation in the form $y = mx + b$.

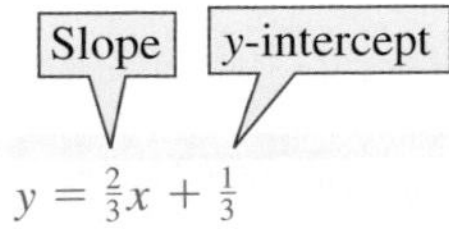

$$3y - 2x = 1$$

$$3y = 2x + 1 \qquad \text{Add } 2x$$

$$y = \tfrac{2}{3}x + \tfrac{1}{3} \qquad \text{Divide by 3}$$

From the slope-intercept form of the equation of a line, we see that the slope is $m = \frac{2}{3}$ and the y-intercept is $b = \frac{1}{3}$.

Now Try Exercises 23 and 61 ■

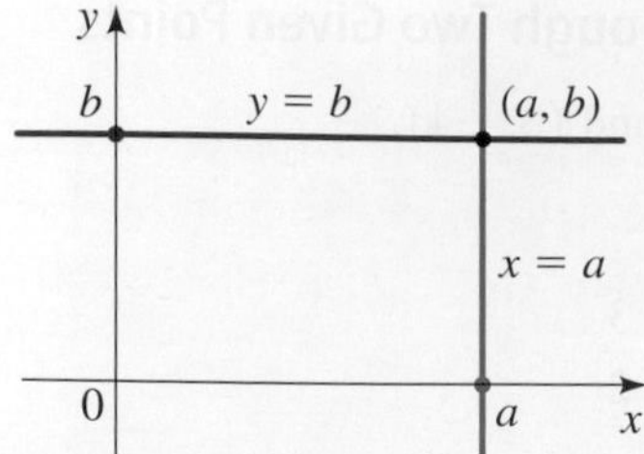

Figure 8

■ Vertical and Horizontal Lines

If a line is horizontal, its slope is $m = 0$, so its equation is $y = b$, where b is the y-intercept (see Figure 8). A vertical line does not have a slope, but we can write its equation as $x = a$, where a is the x-intercept, because the x-coordinate of every point on the line is a.

Vertical and Horizontal Lines

- An equation of the vertical line through (a, b) is $x = a$.
- An equation of the horizontal line through (a, b) is $y = b$.

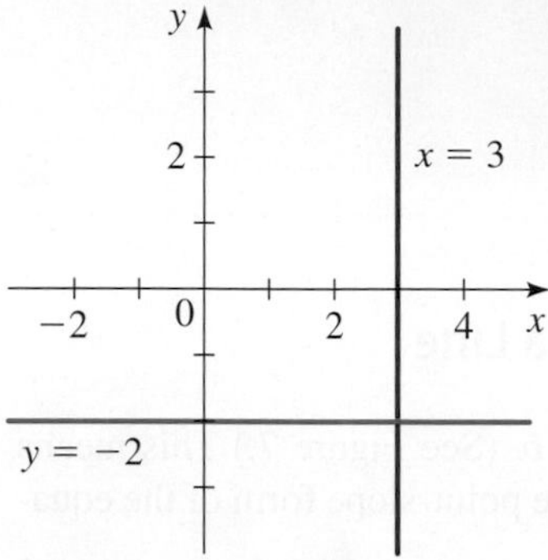

Figure 9

Example 5 ■ Vertical and Horizontal Lines

(a) An equation for the vertical line through $(3, 2)$ is $x = 3$.

(b) The graph of the equation $x = 3$ is a vertical line with x-intercept 3.

(c) An equation for the horizontal line through $(4, -2)$ is $y = -2$.

(d) The graph of the equation $y = -2$ is a horizontal line with y-intercept -2.

The lines are graphed in Figure 9.

Now Try Exercises 35, 37, 63, and 65 ■

■ General Equation of a Line

A **linear equation** in the variables x and y is an equation of the form

$$Ax + By + C = 0$$

where A, B, and C are constants and A and B are not both 0. An equation of a line is a linear equation:

- A nonvertical line has the equation $y = mx + b$ or $-mx + y - b = 0$, which is a linear equation with $A = -m$, $B = 1$, and $C = -b$.
- A vertical line has the equation $x = a$ or $x - a = 0$, which is a linear equation with $A = 1$, $B = 0$, and $C = -a$.

Conversely, the graph of a linear equation is a line:

- If $B \neq 0$, the equation becomes

  $$y = -\frac{A}{B}x - \frac{C}{B} \qquad \text{Divide by } B$$

  and this is the slope-intercept form of the equation of a line (with $m = -A/B$ and $b = -C/B$).
- If $B = 0$, the equation becomes

  $$Ax + C = 0 \qquad \text{Set } B = 0$$

  or $x = -C/A$, which represents a vertical line.

We have proved the following.

General Equation of a Line

The graph of every **linear equation**

$$Ax + By + C = 0 \qquad (A, B \text{ not both zero})$$

is a line. Conversely, every line is the graph of a linear equation.

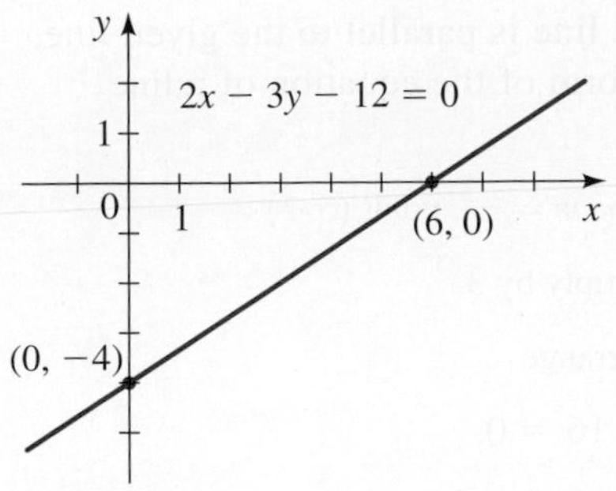

Figure 10

Example 6 ■ Graphing a Linear Equation

Sketch the graph of the equation $2x - 3y - 12 = 0$.

Solution 1 Since the equation is linear, its graph is a line. To draw the graph, it is enough for us to find any two points on the line. The intercepts are the easiest points to find.

x-intercept: Substitute $y = 0$ to get $2x - 12 = 0$, so $x = 6$

y-intercept: Substitute $x = 0$ to get $-3y - 12 = 0$, so $y = -4$

With these points we sketch the graph in Figure 10.

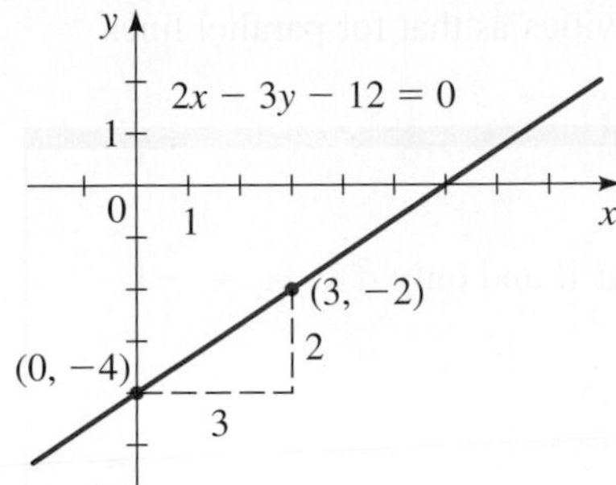

Figure 11

Solution 2 We write the equation in slope-intercept form.

$$2x - 3y - 12 = 0$$

$$2x - 3y = 12 \qquad \text{Add 12}$$

$$-3y = -2x + 12 \qquad \text{Subtract } 2x$$

$$y = \tfrac{2}{3}x - 4 \qquad \text{Divide by } -3$$

This equation is in the form $y = mx + b$, so the slope is $m = \frac{2}{3}$ and the y-intercept is $b = -4$. To sketch the graph, we plot the y-intercept and then move 3 units to the right and 2 units upward as shown in Figure 11.

Now Try Exercise 67 ■

■ Parallel and Perpendicular Lines

Since slope measures the steepness of a line, it seems reasonable that parallel lines should have the same slope. In fact, we can prove this.

Parallel Lines

Two nonvertical lines are parallel if and only if they have the same slope.

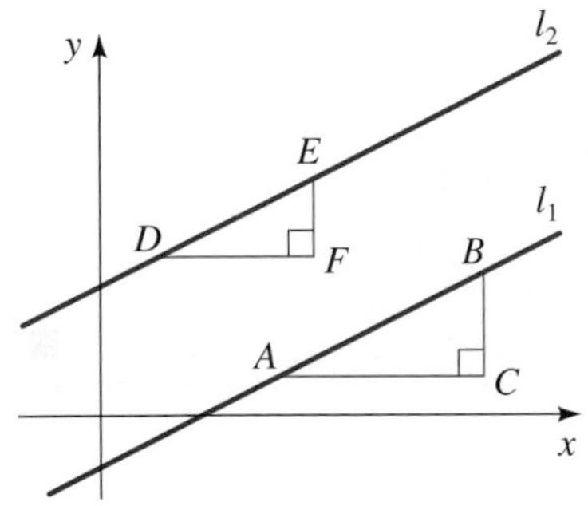

Figure 12

Proof Let the lines l_1 and l_2 in Figure 12 have slopes m_1 and m_2. If the lines are parallel, then the right triangles ABC and DEF are similar, so

$$m_1 = \frac{d(B, C)}{d(A, C)} = \frac{d(E, F)}{d(D, F)} = m_2$$

Conversely, if the slopes are equal, then the triangles will be similar, so $\angle BAC = \angle EDF$ and hence the lines are parallel. ■

Example 7 ■ Finding an Equation of a Line Parallel to a Given Line

Find an equation of the line through the point $(5, 2)$ that is parallel to the line $4x + 6y + 5 = 0$.

Solution First we write the equation of the given line in slope-intercept form.

$$4x + 6y + 5 = 0$$

$$6y = -4x - 5 \qquad \text{Subtract } 4x + 5$$

$$y = -\tfrac{2}{3}x - \tfrac{5}{6} \qquad \text{Divide by 6}$$

So the line has slope $m = -\frac{2}{3}$. Since the required line is parallel to the given line, it also has slope $m = -\frac{2}{3}$. From the point-slope form of the equation of a line we get

$$y - 2 = -\tfrac{2}{3}(x - 5) \qquad \text{Slope } m = -\tfrac{2}{3}, \text{ point } (5, 2)$$
$$3y - 6 = -2x + 10 \qquad \text{Multiply by 3}$$
$$2x + 3y - 16 = 0 \qquad \text{Rearrange}$$

Thus an equation of the required line is $2x + 3y - 16 = 0$.

Now Try Exercise 43 ■

The condition for perpendicular lines is not as obvious as that for parallel lines.

Perpendicular Lines

Two lines with slopes m_1 and m_2 are perpendicular if and only if $m_1 m_2 = -1$, that is, their slopes are negative reciprocals:

$$m_2 = -\frac{1}{m_1}$$

Also, a horizontal line (slope 0) is perpendicular to a vertical line (no slope).

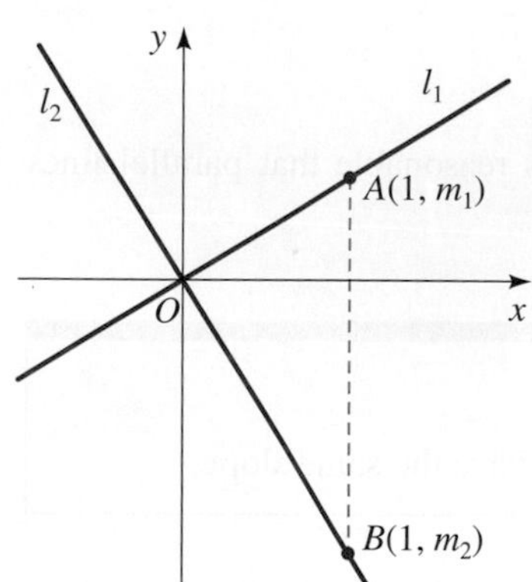

Figure 13

Proof In Figure 13 we show two lines intersecting at the origin. (If the lines intersect at some other point, we consider lines parallel to these that intersect at the origin. These lines have the same slopes as the original lines.)

If the lines l_1 and l_2 have slopes m_1 and m_2, then their equations are $y = m_1 x$ and $y = m_2 x$. Notice that $A(1, m_1)$ lies on l_1 and $B(1, m_2)$ lies on l_2. By the Pythagorean Theorem and its converse (see Appendix A) $OA \perp OB$ if and only if

$$[d(O, A)]^2 + [d(O, B)]^2 = [d(A, B)]^2$$

By the Distance Formula this becomes

$$(1^2 + m_1^2) + (1^2 + m_2^2) = (1 - 1)^2 + (m_2 - m_1)^2$$
$$2 + m_1^2 + m_2^2 = m_2^2 - 2m_1 m_2 + m_1^2$$
$$2 = -2m_1 m_2$$
$$m_1 m_2 = -1$$
■

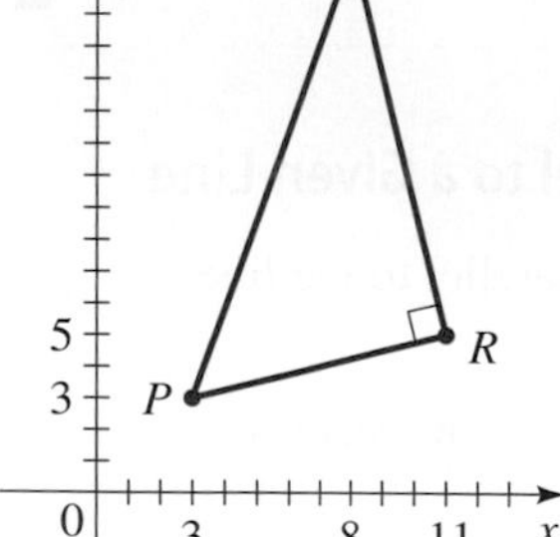

Figure 14

Example 8 ■ Perpendicular Lines

Show that the points $P(3, 3)$, $Q(8, 17)$, and $R(11, 5)$ are the vertices of a right triangle.

Solution The slopes of the lines containing PR and QR are, respectively,

$$m_1 = \frac{5 - 3}{11 - 3} = \frac{1}{4} \qquad \text{and} \qquad m_2 = \frac{5 - 17}{11 - 8} = -4$$

Since $m_1 m_2 = -1$, these lines are perpendicular, so PQR is a right triangle. It is sketched in Figure 14.

Now Try Exercise 81 ■

Example 9 ■ Finding an Equation of a Line Perpendicular to a Given Line

Find an equation of the line that is perpendicular to the line $4x + 6y + 5 = 0$ and passes through the origin.

Solution In Example 7 we found that the slope of the line $4x + 6y + 5 = 0$ is $-\frac{2}{3}$. Thus the slope of a perpendicular line is the negative reciprocal, that is, $\frac{3}{2}$. Since the required line passes through $(0, 0)$, the point-slope form gives

$$y - 0 = \tfrac{3}{2}(x - 0) \qquad \text{Slope } m = \tfrac{3}{2}, \text{ point } (0, 0)$$

$$y = \tfrac{3}{2}x \qquad \text{Simplify}$$

 Now Try Exercise 47 ■

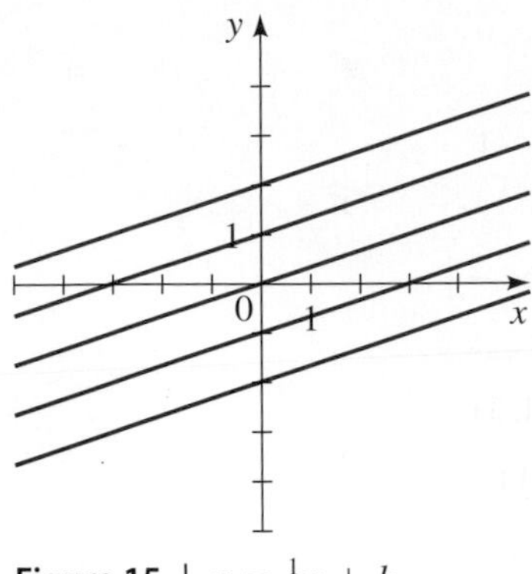

Figure 15 | $y = \frac{1}{2}x + b$

Example 10 ■ A Family of Lines

Graph the family of lines. What property do the lines share in common?

$$y = \tfrac{1}{2}x + b \qquad \text{for} \quad b = -2, -1, 0, 1, 2$$

Solution For $b = -2$ we get the line $y = \frac{1}{2}x - 2$, which has slope $\frac{1}{2}$ and y-intercept -2. In Figure 15, we graph this line along with the lines corresponding to $b = -1, 0, 1, 2$. From the equations we see that the lines all have slope $\frac{1}{2}$, so they are parallel. The graphs in Figure 15 confirm this observation.

 Now Try Exercise 53 ■

Example 11 ■ Application: Interpreting Slope

A hose is being used to fill a swimming pool. The water depth y (in m) in the pool t hours after the hose is turned on is given by

$$y = 1.5t + 2$$

(a) Find the slope and y-intercept of the graph of this equation.

(b) What do the slope and y-intercept represent?

Solution

(a) This is the equation of a line with slope 1.5 and y-intercept 2.

(b) The slope represents an increase of 1.5 m in water depth for every hour. The y-intercept indicates that the water depth was 2 m at the time the hose was turned on.

 Now Try Exercise 89 ■

1.10 | Exercises

■ Concepts

1. We find the "steepness," or slope, of a line passing through two points by dividing the difference in the ____-coordinates of these points by the difference in the ____-coordinates. So the line passing through the points (0, 1) and (2, 5) has slope ________.

2. A line has the equation $y = 3x + 2$.

(a) This line has slope ________.

(b) Any line parallel to this line has slope ________.

(c) Any line perpendicular to this line has slope ________.

3. The point-slope form of the equation of the line with slope 3 passing through the point (1, 2) is ________.

4. For the linear equation $2x + 3y - 12 = 0$, the x-intercept is ________ and the y-intercept is ________. The equation in slope-intercept form is $y =$ ________________. The slope of the graph of this equation is ________.

5. The slope of a horizontal line is ________. The equation of the horizontal line passing through (2, 3) is ________.

6. The slope of a vertical line is ________. The equation of the vertical line passing through (2, 3) is ________.

7. *Yes or No*? If *No*, give a reason.
   (a) Is the graph of $y = -3$ a horizontal line?
   (b) Is the graph of $x = -3$ a vertical line?
   (c) Does a line perpendicular to a horizontal line have slope 0?
   (d) Does a line perpendicular to a vertical line have slope 0?

8. Sketch a graph of the lines $y = -3$ and $x = -3$. Are the lines perpendicular?

Skills

9–16 ■ Slope Find the slope of the line through P and Q.

9. $P(-1, 2)$, $Q(0, 0)$
10. $P(0, 0)$, $Q(3, -1)$
11. $P(-3, 2)$, $Q(3, -3)$
12. $P(-5, 1)$, $Q(3, -2)$
13. $P(5, 4)$, $Q(0, 4)$
14. $P(4, -1)$, $Q(-2, -3)$
15. $P(8, 3)$, $Q(6, 5)$
16. $P(3, -2)$, $Q(6, -2)$

17. **Slope** Find the slopes of the lines l_1, l_2, l_3, and l_4 in the figure below.

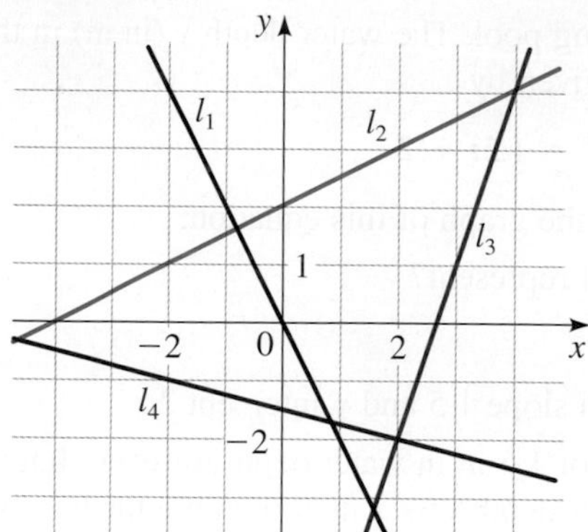

18. **Slope**
   (a) Sketch lines through (0, 0) with slopes 1, 0, $\frac{1}{2}$, 2, and -1.
   (b) Sketch lines through (0, 0) with slopes $\frac{1}{3}$, $\frac{1}{2}$, $-\frac{1}{3}$, and 3.

19–22 ■ Equations of Lines Find an equation for the line whose graph is sketched.

19.

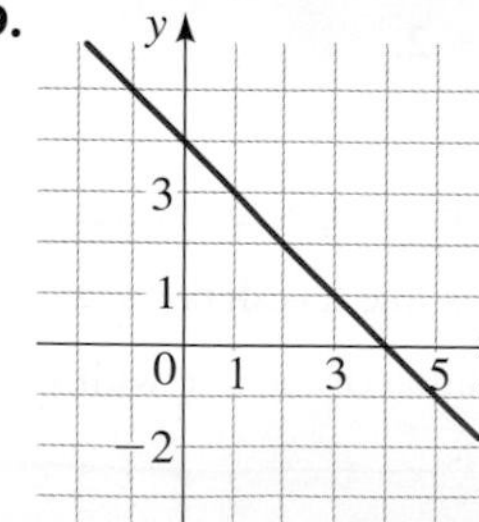

20.

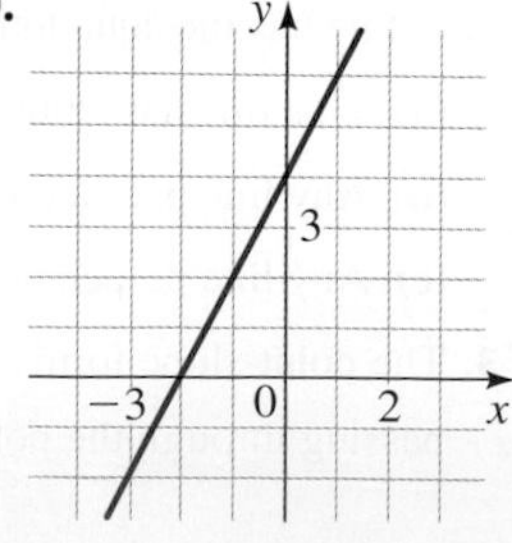

21.

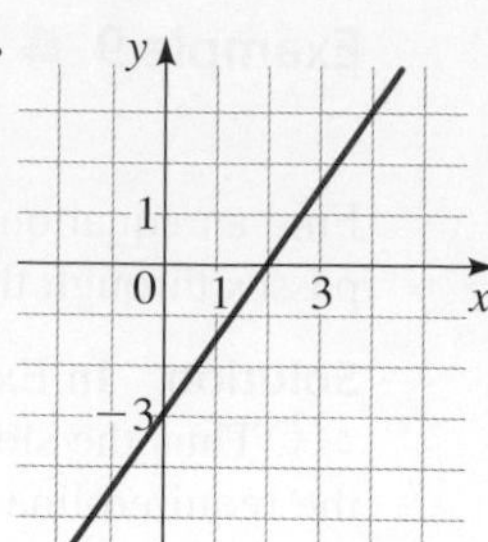

22.

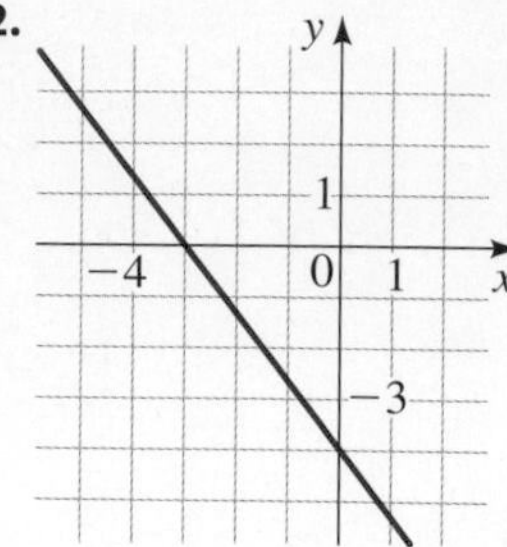

23–50 ■ Finding Equations of Lines Find an equation of the line that satisfies the given conditions.

23. Slope 3; y-intercept -2
24. Slope $\frac{2}{5}$; y-intercept 4
25. Through (4, 1); slope 3
26. Through $(-2, 4)$; slope -1
27. Through (1, 7); slope $\frac{2}{3}$
28. Through $(-3, -5)$; slope $-\frac{7}{2}$
29. Through (2, 1) and (1, 6)
30. Through $(-1, -2)$ and (4, 3)
31. Through $(2, -5)$ and (5, 1)
32. Through (1, 7) and (4, 7)
33. x-intercept 1; y-intercept -3
34. x-intercept -8; y-intercept 6
35. Through (1, 3); slope 0
36. Through $(3, -2)$; slope undefined
37. Through $(2, -1)$; slope undefined
38. Through (5, 1); slope 0
39. Through $(-1, 4)$; parallel to the line $y = 2x + 8$
40. Through $(-3, 2)$; perpendicular to the line $y = -\frac{1}{2}x + 7$
41. Through (4, 5); parallel to the x-axis
42. Through (4, 5); parallel to the y-axis
43. Through $(-3, -4)$; parallel to the line $3x + 2y = 4$
44. y-intercept 6; parallel to the line $2x + 3y + 4 = 0$
45. Through $(-1, 2)$; parallel to the line $x = 5$
46. Through (2, 6); perpendicular to the line $y = 1$
47. Through $(-2, 1)$; perpendicular to the line $3x + 4y + 7 = 0$
48. Through $(\frac{1}{2}, -\frac{2}{3})$; perpendicular to the line $4x - 8y = 1$
49. Through (1, 7); parallel to the line passing through (2, 5) and $(-2, 1)$
50. Through $(4, -10)$; perpendicular to the line passing through the points $(-3, 5)$ and (6, 2)

51. **Finding Equations of Lines and Graphing**
   (a) Sketch the line with slope $\frac{3}{2}$ that passes through the point $(-2, 1)$.
   (b) Find an equation for this line.

52. Finding Equations of Lines and Graphing

(a) Sketch the line with slope -2 that passes through the point $(4, -1)$.

(b) Find an equation for this line.

53–56 ■ Families of Lines Graph the given family of lines. What do the lines have in common?

53. $y = -2x + b$ for $b = 0, \pm 1, \pm 3, \pm 6$

54. $y = mx - 3$ for $m = 0, \pm\frac{1}{4}, \pm\frac{3}{4}, \pm\frac{3}{2}$

55. $y = m(x - 3)$ for $m = 0, \pm\frac{1}{4}, \pm\frac{3}{4}, \pm\frac{3}{2}$

56. $y = 2 + m(x + 3)$ for $m = 0, \pm\frac{1}{2}, \pm 1, \pm 2, \pm 6$

57–66 ■ Using Slopes and y-Intercepts to Graph Lines Find the slope and y-intercept of the line, and draw its graph.

57. $y = x - 4$

58. $y = -\frac{1}{2}x - 1$

59. $-2x + y = 7$

60. $2x - 5y = 0$

61. $4x + 5y = 10$

62. $3x - 4y = 12$

63. $y = 4$

64. $x = -5$

65. $x = 3$

66. $y = -2$

67–72 ■ Using x- and y-Intercepts to Graph Lines Find the x- and y-intercepts of the line, and draw its graph.

67. $3x - 2y - 6 = 0$

68. $6x - 7y - 42 = 0$

69. $\frac{1}{2}x - \frac{1}{3}y + 1 = 0$

70. $\frac{1}{3}x - \frac{1}{5}y - 2 = 0$

71. $y = 6x + 4$

72. $y = -4x - 10$

73–78 ■ Parallel and Perpendicular Lines The equations of two lines are given. Determine whether the lines are parallel, perpendicular, or neither.

73. $y = 2x + 3$; $2y - 4x - 5 = 0$

74. $y = \frac{1}{2}x + 4$; $2x + 4y = 1$

75. $2x - 5y = 8$; $10x + 4y = 1$

76. $15x - 9y = 2$; $3y - 5x = 5$

77. $7x - 3y = 2$; $9y + 21x = 1$

78. $6y - 2x = 5$; $2y + 6x = 1$

Skills Plus

79–82 ■ Using Slopes Verify the given geometric property.

79. Use slopes to show that $A(1, 1)$, $B(7, 4)$, $C(5, 10)$, and $D(-1, 7)$ are vertices of a parallelogram.

80. Use slopes to show that $A(-3, -1)$, $B(3, 3)$, and $C(-9, 8)$ are vertices of a right triangle.

81. Use slopes to show that $A(1, 1)$, $B(11, 3)$, $C(10, 8)$, and $D(0, 6)$ are vertices of a rectangle.

82. Use slopes to determine whether the given points are collinear (lie on a line).

(a) $(1, 1), (3, 9), (6, 21)$ (b) $(-1, 3), (1, 7), (4, 15)$

83. Perpendicular Bisector Find an equation of the perpendicular bisector of the line segment joining the points $A(1, 4)$ and $B(7, -2)$.

84. Area of a Triangle Find the area of the triangle formed by the coordinate axes and the line

$$2y + 3x - 6 = 0$$

85. Two-Intercept Form

(a) Show that if the x- and y-intercepts of a line are nonzero numbers a and b, then the equation of the line can be written in the form

$$\frac{x}{a} + \frac{y}{b} = 1$$

This is called the **two-intercept form** of the equation of a line.

(b) Use part (a) to find an equation of the line whose x-intercept is 6 and whose y-intercept is -8.

86. Tangent Line to a Circle

(a) Find an equation for the line tangent to the circle $x^2 + y^2 = 25$ at the point $(3, -4)$. (See the figure.)

(b) At what other point on the circle will a tangent line be parallel to the tangent line given in part (a)?

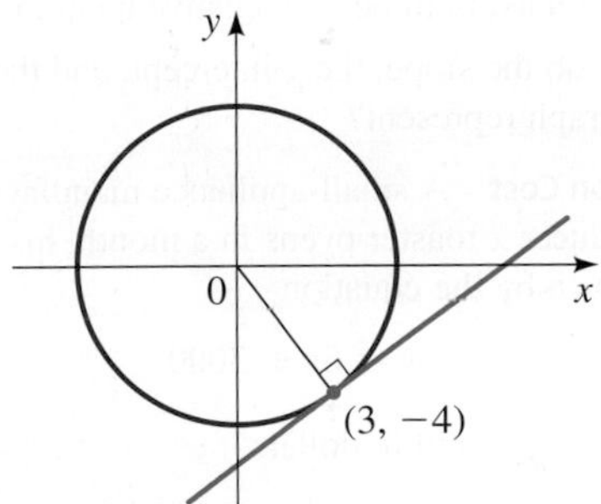

87. Triangle Midsegment Theorem Prove the following theorem from geometry: The line segment connecting the midpoints of two sides of a triangle is parallel to the third side and is half its length. [*Hint*: Place the triangle in a coordinate plane as shown in the figure, then use formulas from this section to verify the conclusions of the theorem.]

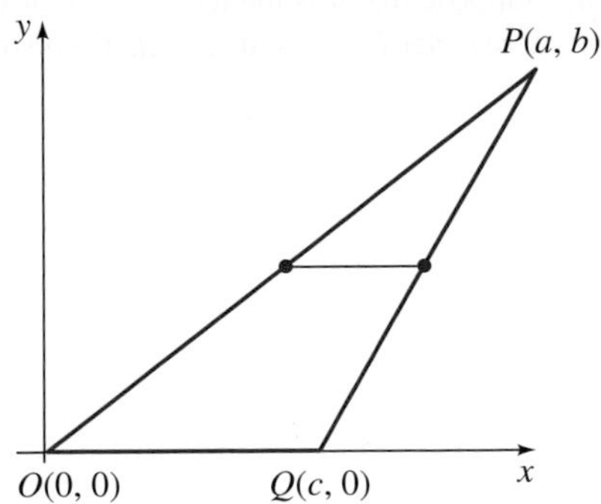

Applications

88. Global Warming Some scientists believe that the average surface temperature of the world has been rising steadily. The average surface temperature can be modeled by

$$T = 0.02t + 15.0$$

where T is temperature in °C and t is years since 1950.

(a) What do the slope and T-intercept represent?

(b) Use the equation to predict the average global surface temperature in 2050.

89. Drug Dosages If the recommended adult dosage for a drug is D (in mg), then to determine the appropriate dosage c for a child of age a, pharmacists use the equation

$$c = 0.0417D(a + 1)$$

Suppose the dosage for an adult is 200 mg.

(a) Find the slope. What does it represent?

(b) What is the dosage for a newborn?

90. Flea Market The manager of a flea market knows from past experience that if she charges x dollars for a rental space at the flea market, then the number y of spaces she can rent is given by the equation $y = 200 - 4x$.

(a) Sketch a graph of this linear equation. (Remember that the rental charge per space and the number of spaces rented must both be nonnegative quantities.)

(b) What do the slope, the y-intercept, and the x-intercept of the graph represent?

91. Production Cost A small-appliance manufacturer finds that if he produces x toaster ovens in a month, his production cost is given by the equation

$$y = 6x + 3000$$

where y is measured in dollars.

(a) Sketch a graph of this linear equation.

(b) What do the slope and y-intercept of the graph represent?

92. Temperature Scales The relationship between the Fahrenheit (F) and Celsius (C) temperature scales is given by the equation $F = \frac{9}{5}C + 32$.

(a) Complete the following table to compare the two scales at the given values.

(b) Find the temperature at which the scales agree. [*Hint:* Suppose that a is the temperature at which the scales agree. Set $F = a$ and $C = a$. Then solve for a.]

C	F
−30°	
−20°	
−10°	
0°	
	50°
	68°
	86°

93. Crickets and Temperature Biologists have observed that the chirping rate of crickets of a certain species is related to temperature, and the relationship appears to be very nearly linear. A cricket produces 120 chirps per minute at 21°C and 168 chirps per minute at 27°C.

(a) Find the linear equation that relates the temperature t and the number of chirps per minute n.

(b) If the crickets are chirping at 150 chirps per minute, estimate the temperature.

94. Depreciation A small business buys a computer for \$4000. After 4 years the value of the computer is expected to be \$200. For accounting purposes the business uses *linear depreciation* to assess the value of the computer at a given time. This means that if V is the value of the computer at time t, then a linear equation is used to relate V and t.

(a) Find a linear equation that relates V and t.

(b) Sketch a graph of this linear equation.

(c) What do the slope and V-intercept of the graph represent?

(d) Find the depreciated value of the computer 3 years from the date of purchase.

95. Pressure and Depth At the surface of the ocean the water pressure is the same as the air pressure above the water, about 103 kPa. Below the surface the water pressure increases by 29.9 kPa for every 3 m of descent.

(a) Find an equation for the relationship between pressure P and depth d below the ocean surface.

(b) Sketch a graph of this linear equation.

(c) What do the slope and d-intercept of the graph represent?

(d) At what depth is the pressure 690 kPa?

Discuss ■ Discover ■ Prove ■ Write

96. Discuss: What Does the Slope Mean? Suppose that the graph of the outdoor temperature over a certain period of time is a line. How is the weather changing if the slope of the line is positive? If it is negative? If it is zero?

97. Discuss: Collinear Points Suppose that you are given the coordinates of three points in the plane and you want to see whether they lie on the same line. How can you do this using slopes? Using the Distance Formula? Can you think of another method?

1.11 Solving Equations and Inequalities Graphically

■ Using Graphing Devices ■ Solving Equations Graphically ■ Solving Inequalities Graphically

If you are using a graphing device such as a math app on a computer or smartphone, be sure to familiarize yourself with how the app works. If you are using a graphing calculator, see Appendix C, *Graphing with a Graphing Calculator*, or Appendix D, *Using the TI-83/84 Graphing Calculator*. Go to **www.stewartmath.com**.

In the preceding two sections we studied the concept of the graph of an equation in two variables. We learned how to identify equations whose graphs are lines or circles and how to graph other equations by plotting points. **Graphing devices**, including math apps and graphing calculators, can perform the routine work of plotting the graph of an equation. In this section we use graphing devices to quickly draw graphs of equations and then use the graphs to obtain useful information about the equations. In particular, we'll solve equations and inequalities graphically.

■ Using Graphing Devices

Graphing devices draw the graph of an equation by plotting points, much as you would. They display a rectangular portion of the graph in a display window or viewing screen, which we call a **viewing rectangle**. If we choose the x-values to range over the interval $[a, b]$ and the y-values to range over the interval $[c, d]$ then the displayed portion of the graph lies in the rectangle

$$[a, b] \times [c, d] = \{(x, y) \mid a \le x \le b, c \le y \le d\}$$

as shown in Figure 1. We refer to this as the $[a, b]$ by $[c, d]$ viewing rectangle.

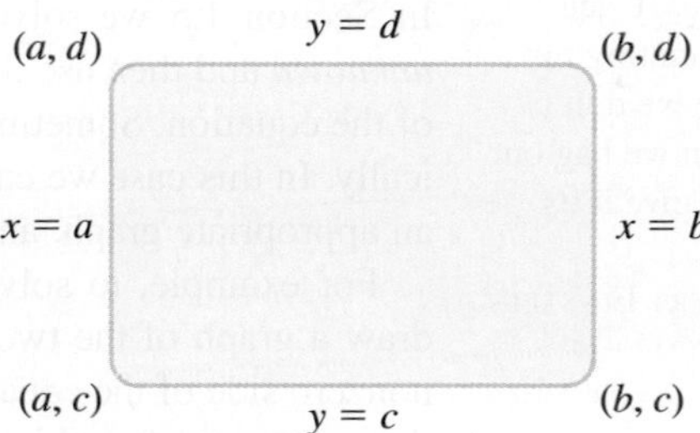

Figure 1 | The viewing rectangle $[a, b]$ by $[c, d]$

The device plots points of the form (x, y) for a certain number of values of x, equally spaced between a and b. If the equation is not defined for an x-value or if the corresponding y-value lies outside the viewing rectangle, the device ignores this value and moves on to the next x-value. The device connects each point to the preceding plotted point to form a representation of the graph of the equation.

Example 1 ■ Graphing an Equation with a Graphing Device

The graph in Figure 2 was obtained by using a graphing device to graph the equation

$$y = \frac{1}{1 + x^2}$$

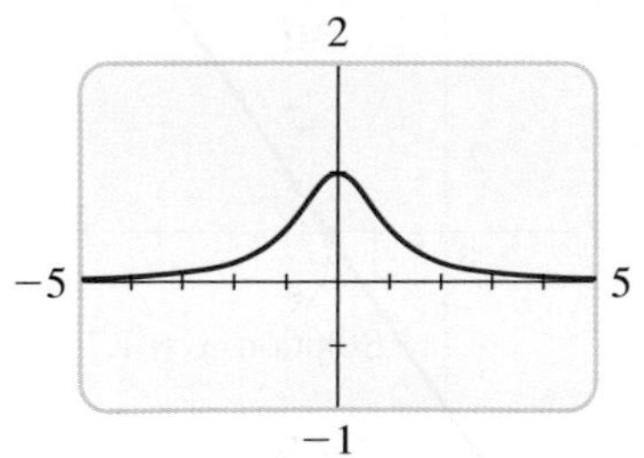

Figure 2 | Graph of $y = \dfrac{1}{1 + x^2}$

in the viewing rectangle $[-5, 5]$ by $[-1, 2]$. Let's compare the graph with the equation. The graph has y-intercept 1, which is confirmed by the equation (because setting $x = 0$ in the equation gives $y = 1$). The graph appears to be symmetric about the y-axis and this is also confirmed by the equation (because replacing x by $-x$ in the equation leaves the equation unchanged). For points (x, y) on the graph (with $x > 0$), the larger the x-coordinate, the smaller the y-coordinate; this observation can be seen from the equation because larger values of x correspond to larger values of $1 + x^2$ and hence to smaller values of $y = 1/(1 + x^2)$. So we see that the graph gives visual confirmation of the algebraic properties of the equation.

Now Try Exercise 5 ■

Example 2 ■ Two Graphs on the Same Screen

Graph the equations $y = 3x^2 - 6x + 1$ and $y = 0.23x - 2.25$ together in the viewing rectangle $[-1, 3]$ by $[-3, 2]$. Do the graphs intersect?

Solution Figure 3(a) shows the essential features of both graphs. One is a parabola and the other is a line. It looks as if the curves intersect near the point $(1, -2)$. However, if we zoom in on the area around this point as shown in Figure 3(b), we see that although the graphs almost touch, they do not actually intersect.

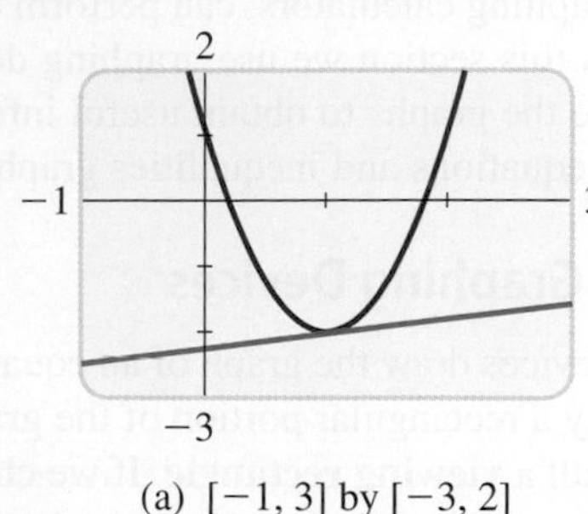

(a) $[-1, 3]$ by $[-3, 2]$

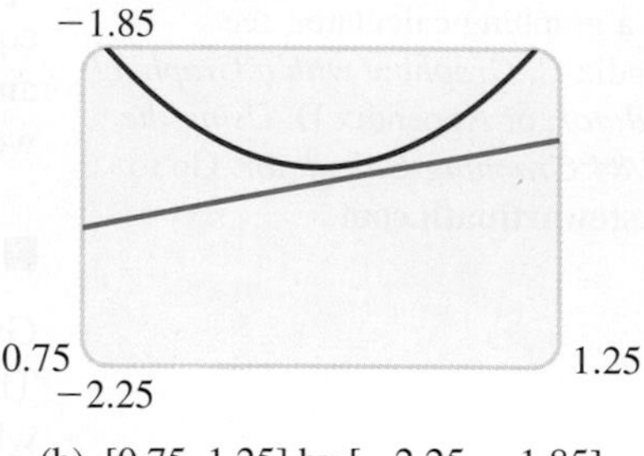

(b) $[0.75, 1.25]$ by $[-2.25, -1.85]$

Figure 3 | Graphs of the two equations in different viewing rectangles

Now Try Exercises 9 ■

■ Solving Equations Graphically

"Algebra is a merry science," Uncle Jakob would say. "We go hunting for a little animal whose name we don't know, so we call it *x*. When we bag our game we pounce on it and give it its right name."

ALBERT EINSTEIN

In Section 1.5 we solved equations *algebraically*. In this method, we view x as an *unknown* and then use the rules of algebra to "hunt it down" by isolating it on one side of the equation. Sometimes an equation may be difficult or impossible to solve algebraically. In this case we can solve it *graphically*. That is, we view x as a *variable*, sketch an appropriate graph, and get the solutions from the graph.

For example, to solve the one-variable equation $3x - 5 = 0$ graphically, we first draw a graph of the two-variable equation $y = 3x - 5$ that is obtained by setting the nonzero side of the equation equal to y. The solution(s) of the equation $y = 3x - 5$ are the values of x for which y is equal to zero. That is, the solutions are the x-intercepts of the graph.

Solving an Equation

Algebraic Method

Use the rules of algebra to isolate the unknown x on one side of the equation.

Example: $3x - 4 = 1$

$3x = 5$ Add 4

$x = \frac{5}{3}$ Divide by 3

The solution is $x = \frac{5}{3}$.

Graphical Method

Move all terms to one side, and set that side equal to y. Graph the resulting equation, and find the x-intercepts.

Example: $3x - 4 = 1$

$3x - 5 = 0$

Set $y = 3x - 5$ and graph. From the graph we see that the solution is $x \approx 1.7$

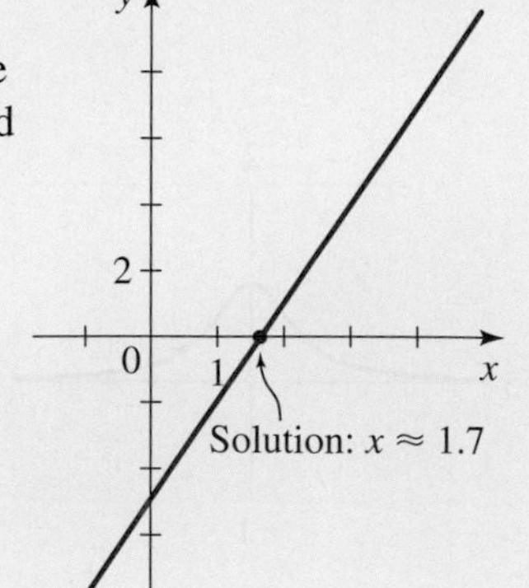

The advantage of the algebraic method is that it gives exact answers. Also, the process of unraveling the equation to arrive at the answer helps us understand the algebraic structure of the equation. On the other hand, for many equations it is difficult or impossible to isolate x.

The *Discovery Project* referenced in Section 3.4 describes a numerical method for solving equations.

The graphical method gives a numerical approximation to the answer—an advantage when a numerical answer is desired. (For example, an engineer might find an answer expressed as $x \approx 2.6$ more immediately useful than $x = \sqrt{7}$.) Also, graphing an equation helps us visualize how the solution is related to other values of the variable.

Example 3 ■ Solving a Quadratic Equation Algebraically and Graphically

Find all real solutions of each quadratic equation. Use both the algebraic method and the graphical method.

(a) $x^2 - 4x + 2 = 0$ **(b)** $x^2 - 4x + 4 = 0$ **(c)** $x^2 - 4x + 6 = 0$

Solution 1: Algebraic

The Quadratic Formula is discussed in Section 1.5.

You can check that the Quadratic Formula gives the following solutions.

(a) There are two real solutions, $x = 2 + \sqrt{2}$ and $x = 2 - \sqrt{2}$.

(b) There is one real solution, $x = 2$.

(c) There is no real solution. (The two complex solutions are $x = 2 + \sqrt{2}i$ and $x = 2 - \sqrt{2}i$.)

Solution 2: Graphical

The x-intercepts are the x-values for which the corresponding y-values of the equation are equal to 0.

We use a graphing device to graph the equations $y = x^2 - 4x + 2$, $y = x^2 - 4x + 4$, and $y = x^2 - 4x + 6$ in Figure 4. By determining the x-intercepts of the graphs, we find the following solutions.

(a) The two x-intercepts give the two solutions $x \approx 0.6$ and $x \approx 3.4$.

(b) The one x-intercept gives the one solution $x = 2$.

(c) There is no x-intercept, so the equation has no real solution.

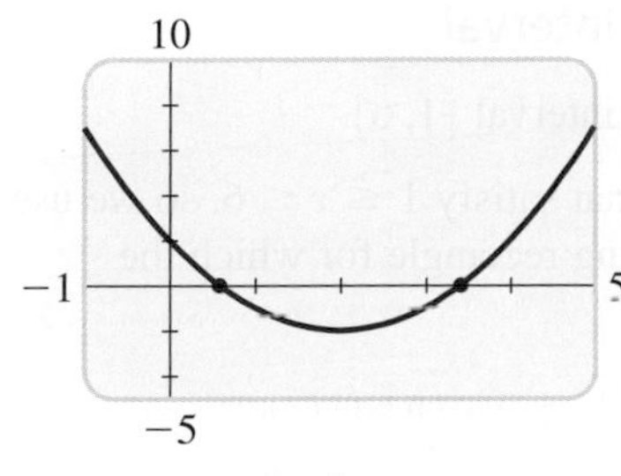

(a) $y = x^2 - 4x + 2$

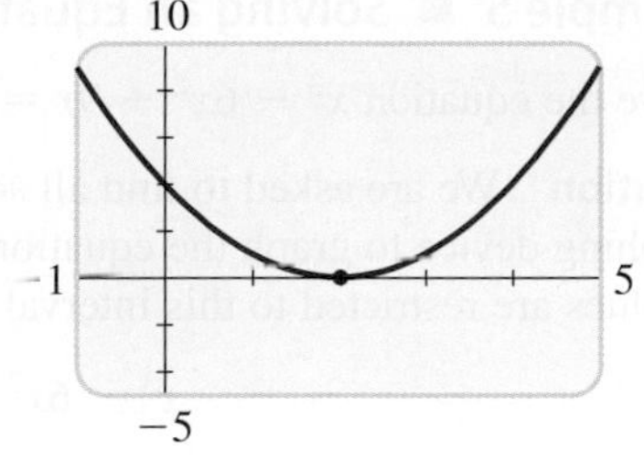

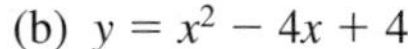
(b) $y = x^2 - 4x + 4$

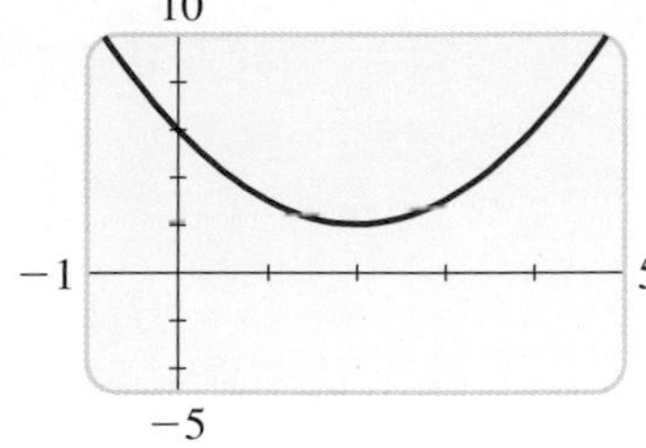

(c) $y = x^2 - 4x + 6$

Figure 4

Now Try Exercises 17, 19, and 23 ■

The graphs in Figure 4 show visually why a quadratic equation may have two solutions, one solution, or no real solution. We proved this fact algebraically in Section 1.5 when we studied the discriminant.

PIERRE DE FERMAT (1607–1665) was a French lawyer who became interested in mathematics at the age of 30. Because of his job as a magistrate, Fermat had little time to write complete proofs of his discoveries and often wrote them in the margin of whatever book he was reading at the time. After his death his copy of Diophantus's *Arithmetica* (see Section 1.2) was found to contain a particularly tantalizing comment. Where Diophantus discusses the solutions of $x^2 + y^2 = z^2$ (for example, $x = 3$, $y = 4$, and $z = 5$), Fermat states in the margin that for $n \geq 3$ there are no natural number solutions to the equation $x^n + y^n = z^n$. In other words, it's impossible for a cube to equal the sum of two cubes, a fourth power to equal the sum of two fourth powers, and so on. Fermat writes, "I have discovered a truly wonderful proof for this but the margin is too small to contain it." All the other margin comments in Fermat's copy of *Arithmetica* have been proved. This one, however, remained unproved, and it came to be known as "Fermat's Last Theorem."

In 1994, Andrew Wiles of Princeton University announced a proof of Fermat's Last Theorem, an astounding 350 years after it was conjectured. His proof is one of the most widely reported mathematical results in the popular press.

Example 4 ■ Another Graphical Method

Solve the equation algebraically and graphically: $5 - 3x = 8x - 20$

Solution 1: Algebraic

$5 - 3x = 8x - 20$	Given equation
$-3x = 8x - 25$	Subtract 5
$-11x = -25$	Subtract $8x$
$x = \dfrac{-25}{-11} = 2\frac{3}{11}$	Divide by –11 and simplify

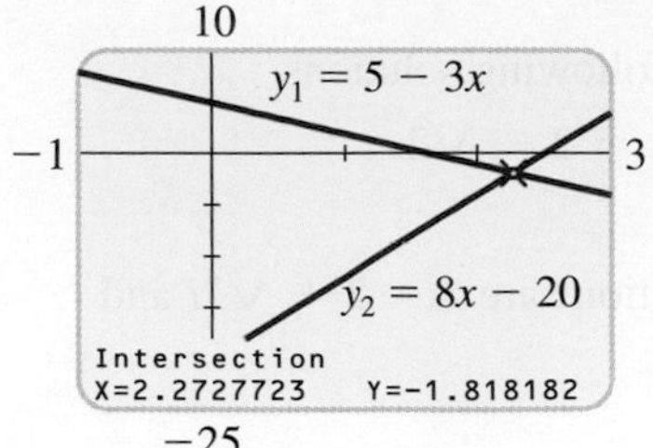

Figure 5

Solution 2: Graphical

We could move all terms to one side of the equal sign, set the result equal to y, and graph the resulting equation. Alternatively, we can graph the following two equations:

$$y_1 = 5 - 3x \quad \text{and} \quad y_2 = 8x - 20$$

The solution of the original equation will be the value of x that makes y_1 equal to y_2; that is, the solution is the x-coordinate of the intersection point of the two graphs. We see from the graph in Figure 5 that the solution is $x \approx 2.27$.

Now Try Exercise 13 ■

In the next example we use the graphical method to solve an equation that is very difficult to solve algebraically.

Example 5 ■ Solving an Equation in an Interval

Solve the equation $x^3 - 6x^2 + 9x = \sqrt{x}$ in the interval $[1, 6]$.

Solution We are asked to find all solutions x that satisfy $1 \le x \le 6$, so we use a graphing device to graph the equation in a viewing rectangle for which the x-values are restricted to this interval.

$x^3 - 6x^2 + 9x = \sqrt{x}$	Given equation
$x^3 - 6x^2 + 9x - \sqrt{x} = 0$	Subtract $\sqrt{x}$

Figure 6 shows the graph of the equation $y = x^3 - 6x^2 + 9x - \sqrt{x}$ in the viewing rectangle $[1, 6]$ by $[-5, 5]$. There are two x-intercepts in this viewing rectangle; zooming in, we see that the solutions are $x \approx 2.18$ and $x \approx 3.72$.

Graphing devices can accurately locate the intercepts of a graph and the points of intersection of two graphs.

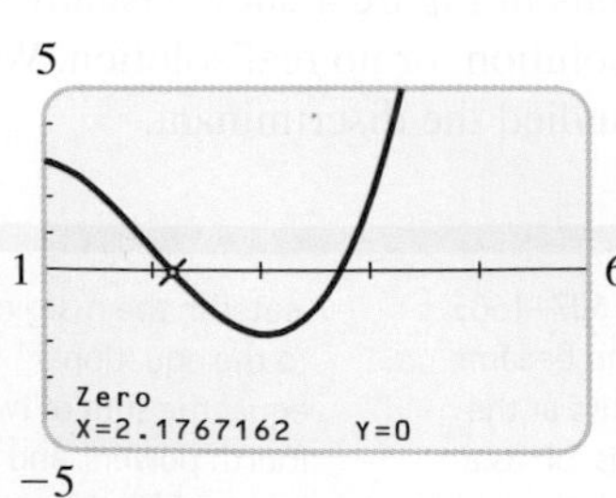

5
1
6
Zero
X=3.7200502 Y=0
−5

Figure 6

Now Try Exercise 25 ■

The equation in Example 5 actually has four solutions. You are asked to find the other two in Exercise 54.

■ Solving Inequalities Graphically

To solve a one-variable inequality such as $3x - 5 \geq 0$ graphically, we first draw a graph of the two-variable equation $y = 3x - 5$ that is obtained by setting the nonzero side of the inequality equal to a variable y. The solutions of the given inequality are the values of x for which y is greater than or equal to 0. That is, the solutions are the values of x for which the graph is above the x-axis.

Solving an Inequality

Algebraic Method

Use the rules of algebra to isolate the unknown x on one side of the inequality.

Example: $3x - 4 \geq 1$

$3x \geq 5$ Add 4

$x \geq \frac{5}{3}$ Divide by 3

The solution is $\left[\frac{5}{3}, \infty\right)$.

Graphical Method

Move all terms to one side, and set that side equal to y. Graph the resulting equation, and find the values of x where the graph is above or on the x-axis.

Example: $3x - 4 \geq 1$

$3x - 5 \geq 0$

Set $y = 3x - 5$ and graph. From the graph we see that the solution is approximately $[1.7, \infty)$.

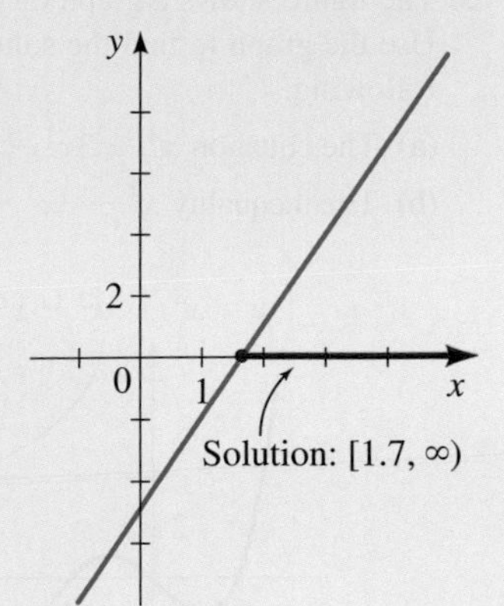

Example 6 ■ Solving an Inequality Graphically

Solve the inequality $x^3 - 5x^2 \geq -8$.

Solution We write the inequality as

$$x^3 - 5x^2 + 8 \geq 0$$

We then use a graphing device to graph the equation

$$y = x^3 - 5x^2 + 8$$

and find the x-intercepts, as shown in Figure 7. The solution of the inequality consists of those intervals on which the graph lies on or above the x-axis. From the graph in Figure 7 we see that, rounded to one decimal place, the solution is $[-1.1, 1.5] \cup [4.6. \infty)$.

Figure 7 | $x^3 - 5x^2 + 8 \geq 0$

 **Now Try Exercise 41** ■

Example 7 ■ Another Graphical Method

Solve the inequality $3.7x^2 + 1.3x - 1.9 \leq 2.0 - 1.4x$.

Solution We use a graphing device to graph the equations

$$y_1 = 3.7x^2 + 1.3x - 1.9 \qquad \text{and} \qquad y_2 = 2.0 - 1.4x$$

The graphs are shown in Figure 8. We are interested in those values of x for which $y_1 \leq y_2$; these are points for which the graph of y_2 lies on or above the graph of y_1. To determine the appropriate interval, we look for the x-coordinates of points where the graphs intersect. We conclude that the solution is (approximately) the interval $[-1.45, 0.72]$.

Figure 8 | $y_1 = 3.7x^2 + 1.3x - 1.9$, $y_2 = 2.0 - 1.4x$

 **Now Try Exercise 45** ■

1.11 Exercises

Concepts

1. The solutions of the equation $x^2 - 2x - 3 = 0$ are the ______-intercepts of the graph of $y = x^2 - 2x - 3$.

2. The solutions of the inequality $x^2 - 2x - 3 > 0$ are the x-coordinates of the points on the graph of $y = x^2 - 2x - 3$ that lie ________ the x-axis.

3. The figure shows a graph of $y = x^4 - 3x^3 - x^2 + 3x$. Use the graph to find the solution(s) of each of the following.

(a) The equation $x^4 - 3x^3 - x^2 + 3x = 0$

(b) The inequality $x^4 - 3x^3 - x^2 + 3x \le 0$

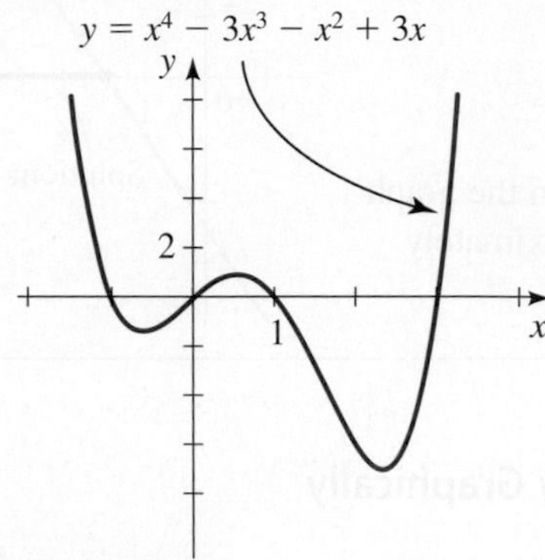

4. The figure shows the graphs of $y = 5x - x^2$ and $y = 4$. Use the graphs to find the solution(s) of each of the following.

(a) The equation $5x - x^2 = 4$

(b) The inequality $5x - x^2 > 4$

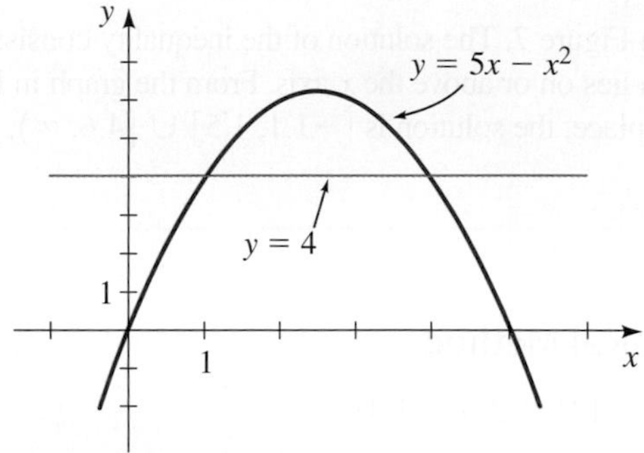

Skills

5–8 An equation is given.

(a) Use a graphing device to graph the equation in the given viewing rectangle.

(b) Find the x- and y-intercepts from the graph and confirm your answers algebraically (from the equation).

(c) If the graph appears to be symmetric, confirm that the equation satisfies the corresponding symmetry property.

5. $y = x^3 - x^2$; $[-2, 2]$ by $[-1, 1]$

6. $y = x^4 - 2x^3$; $[-2, 3]$ by $[-3, 3]$

7. $y = -\dfrac{2}{x^2 + 1}$; $[-5, 5]$ by $[-3, 1]$

8. $y = \sqrt[3]{1 - x^2}$; $[-5, 5]$ by $[-5, 3]$

9–12 Do the given graphs intersect in the indicated viewing rectangle? If so, how many points of intersection are there?

9. $y = -3x^2 + 6x - \frac{1}{2}$, $y = \sqrt{7 - \frac{7}{12}x^2}$; $[-4, 4]$ by $[-1, 3]$

10. $y = \sqrt{49 - x^2}$, $y = \frac{1}{5}(41 - 3x)$; $[-8, 8]$ by $[-1, 8]$

11. $y = 6 - 4x - x^2$, $y = 3x + 18$; $[-6, 2]$ by $[-5, 20]$

12. $y = x^3 - 4x$, $y = x + 5$; $[-4, 4]$ by $[-15, 15]$

13–24 ■ Equations Solve the equation both algebraically and graphically.

13. $3x + 2 = 5x - 4$

14. $\frac{2}{3}x + 11 = 1 - x$

15. $\dfrac{2}{x} + \dfrac{1}{2x} = 7$

16. $\dfrac{4}{x + 2} - \dfrac{6}{2x} = \dfrac{5}{2x + 4}$

17. $4x^2 - 8 = 0$

18. $x^3 + 10x^2 = 0$

19. $x^2 + 9 = 0$

20. $x^2 + 3 = 2x$

21. $81x^4 = 256$

22. $2x^5 - 243 = 0$

23. $(x - 5)^4 - 80 = 0$

24. $3(x + 5)^3 = 72$

25–32 ■ Equations Solve the equation graphically in the given interval. State each solution rounded to two decimals.

25. $x^2 - 11x + 30 = 0$; $[2, 8]$

26. $x^2 - 0.75x + 0.125 = 0$; $[-2, 2]$

27. $x^3 - 6x^2 + 11x - 6 = 0$; $[-1, 4]$

28. $16x^3 + 16x^2 = x + 1$; $[-2, 2]$

29. $x - \sqrt{x + 1} = 0$; $[-1, 5]$

30. $1 + \sqrt{x} = \sqrt{1 + x^2}$; $[-1, 5]$

31. $x^{1/3} - x = 0$; $[-3, 3]$

32. $x^{1/2} + x^{1/3} - x = 0$; $[-1, 5]$

33–36 ■ Equations Use the graphical method to solve the equation in the indicated exercise from Section 1.5.

33. Exercise 87

34. Exercise 88

35. Exercise 89

36. Exercise 90

 **37–40 ■ Equations** Find all real solutions of the equation, and state each rounded to two decimals.

37. $x^3 - 2x^2 - x - 1 = 0$

38. $x^4 - 8x^2 + 2 = 0$

39. $x(x - 1)(x + 2) = \frac{1}{6}x$

40. $x^4 = 16 - x^3$

41–48 ■ Inequalities Find the solutions of the inequality by drawing appropriate graphs. State each solution rounded to two decimals.

 **41.** $x^2 \le 3x + 10$

42. $0.5x^2 + 0.875x \le 0.25$

43. $x^3 + 11x \le 6x^2 + 6$

44. $16x^3 + 24x^2 > -9x - 1$

 **45.** $x^{1/3} < x$

46. $\sqrt{0.5x^2 + 1} \le 2|x|$

47. $(x + 1)^2 < (x - 1)^2$

48. $(x + 1)^2 \le x^3$

 **49–52 ■ Inequalities** Use the graphical method to solve the inequality in the indicated exercise from Section 1.8.

49. Exercise 45 **50.** Exercise 46

51. Exercise 55 **52.** Exercise 56

■ Skills Plus

53. Another Graphical Method In Example 4 we solved the equation $5 - 3x = 8x - 20$ by drawing graphs of two equations. Solve the equation by drawing a graph of only one equation. Compare your answer to the one obtained in Example 4.

 **54. Finding More Solutions** In Example 5 we found two solutions of the equation $x^3 - 6x^2 + 9x = \sqrt{x}$ in the interval $[1, 6]$. Find two more solutions, and state each rounded to two decimals.

■ Applications

 **55. Estimating Profit** An appliance manufacturer estimates that the profit y (in dollars) generated by producing x cooktops per month is given by the equation

$$y = 10x + 0.5x^2 - 0.001x^3 - 5000$$

where $0 \le x \le 450$.

(a) Graph the equation.

(b) How many cooktops must be produced to begin generating a profit?

(c) For what range of values of x is the company's profit greater than \$15,000?

 **56. How Far Can You See?** If you stand on a ship in a calm sea, then your height x (in m) above sea level is related to the farthest distance y (in km) that you can see by the equation

$$y = \sqrt{1.5x + \left(\frac{x}{5280}\right)^2}$$

(a) Graph the equation for $0 \le x \le 100$.

(b) How high above sea level do you have to be standing to be able to see 16 km?

 Discuss ■ Discover 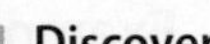 Prove ■ Write

57. Write: Algebraic and Graphical Solution Methods Write a short essay comparing the algebraic and graphical methods for solving equations. Make up your own examples to illustrate the advantages and disadvantages of each method.

58. Discuss: Enter Equations Carefully A student wishes to graph the equations

$$y = x^{1/3} \quad \text{and} \quad y = \frac{x}{x + 4}$$

on the same screen, so the student enters the following information into a graphing device:

```
Y1 = X^1/3    Y2 = X/X + 4
```

The device graphs two lines instead of the desired equations. What went wrong?

1.12 Modeling Variation

■ Direct Variation ■ Inverse Variation ■ Combining Different Types of Variation

When scientists talk about a *mathematical model* for a real-world phenomenon, they often mean an equation or formula that describes the relationship of one physical quantity to another. For instance, the model may be a formula that gives the relationship between the pressure and volume of a gas. In this section we study a kind of modeling—which occurs frequently in the sciences—called *variation.*

■ Direct Variation

One type of variation is called *direct variation*; it occurs when one quantity is a constant multiple of the other. We use a formula of the form $y = kx$ to model this relationship.

> **Direct Variation**
>
> If the quantities x and y are related by an equation
>
> $$y = kx$$
>
> for some constant $k \neq 0$, then we say that y **varies directly as** x, or y is **directly proportional to** x, or simply y **is proportional to** x. The constant k is called the **constant of proportionality**.

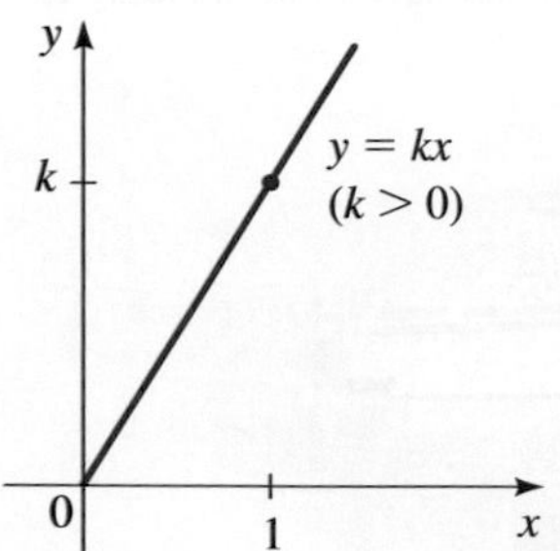

Figure 1

Recall that the graph of an equation of the form $y = mx + b$ is a line with slope m and y-intercept b. So the graph of an equation $y = kx$ that describes direct variation is a line with slope k and y-intercept 0. (See Figure 1.)

Example 1 ■ Direct Variation

During a thunderstorm you see the lightning before you hear the thunder because light travels much faster than sound. The distance between you and the storm varies directly as the time interval between the lightning and the thunder.

(a) Suppose that the thunder from a storm 1660 m away takes 5 s to reach you. Determine the constant of proportionality, and write the equation for the variation.

(b) Sketch the graph of this equation. What does the constant of proportionality represent?

(c) If the time interval between the lightning and thunder is now 8 s, how far away is the storm?

Solution

(a) Let d be the distance from you to the storm, and let t be the length of the time interval. We are given that d varies directly as t, so

$$d = kt$$

where k is a constant. To find k, we use the fact that $t = 5$ when $d = 1660$. Substituting these values in the equation, we get

$$1660 = k(5) \qquad \text{Substitute}$$

$$k = \frac{1660}{5} = 332 \qquad \text{Solve for } k$$

Substituting this value of k in the equation for d, we obtain

$$d = 332t$$

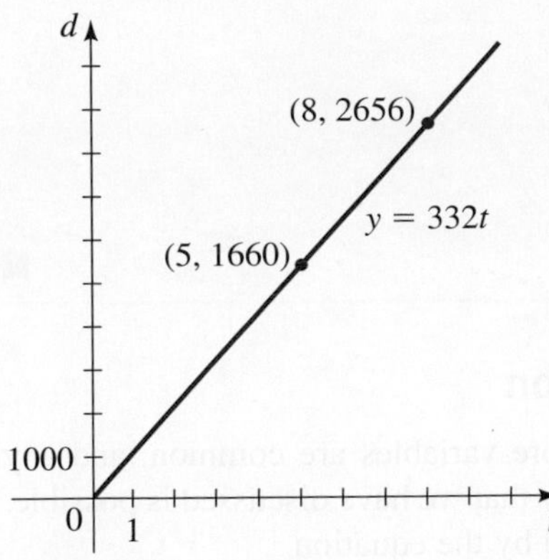

Figure 2

(b) The graph of the equation $d = 332t$ is a line through the origin with slope 332, as shown in Figure 2. The constant $k = 332$ is the approximate speed of sound (in m/s).

(c) When $t = 8$, we have

$$d = 332 \cdot 8 = 2656$$

So the storm is 2656 m $\approx$ 2.5 km away.

Now Try Exercises 19 and 35 ■

■ Inverse Variation

Another formula that is frequently used in mathematical modeling is $y = k/x$, where k is a constant.

Inverse Variation

If the quantities x and y are related by the equation

$$y = \frac{k}{x}$$

for some constant $k \neq 0$, then we say that y **varies inversely as** x or y **is inversely proportional to** x. The constant k is called the **constant of proportionality**.

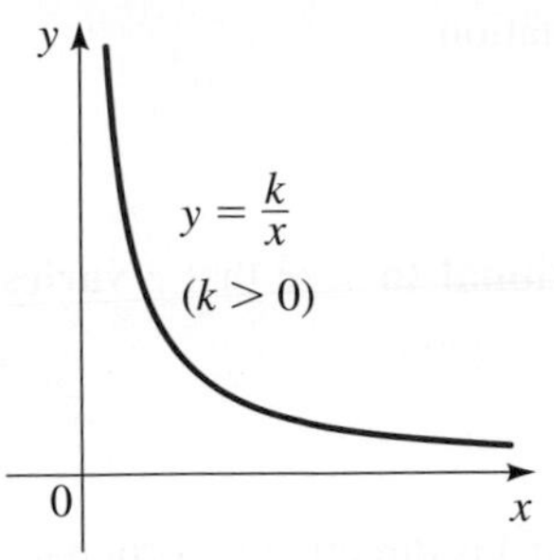

Figure 3 | Inverse variation

The graph of $y = k/x$ for $x > 0$ is shown in Figure 3 for the case $k > 0$. It gives a picture of what happens when y is inversely proportional to x.

Example 2 ■ Inverse Variation

Boyle's Law states that when a sample of gas is compressed at a constant temperature, the pressure of the gas is inversely proportional to the volume of the gas.

(a) Suppose the pressure of a sample of air that occupies 0.106 m^3 at 25°C is 50 kPa. Find the constant of proportionality, and write the equation that expresses the inverse proportionality. Sketch a graph of this equation.

(b) If the sample expands to a volume of 0.3 m^3, find the new pressure.

Solution

(a) Let P be the pressure of the sample of gas, and let V be its volume. Then, by the definition of inverse proportionality, we have

$$P = \frac{k}{V}$$

where k is a constant. To find k, we use the fact that $P = 50$ when $V = 0.106$. Substituting these values in the equation, we get

$$50 = \frac{k}{0.106} \qquad \text{Substitute}$$

$$k = (50)(0.106) = 5.3 \qquad \text{Solve for } k$$

Putting this value of k in the equation for P, we have

$$P = \frac{5.3}{V}$$

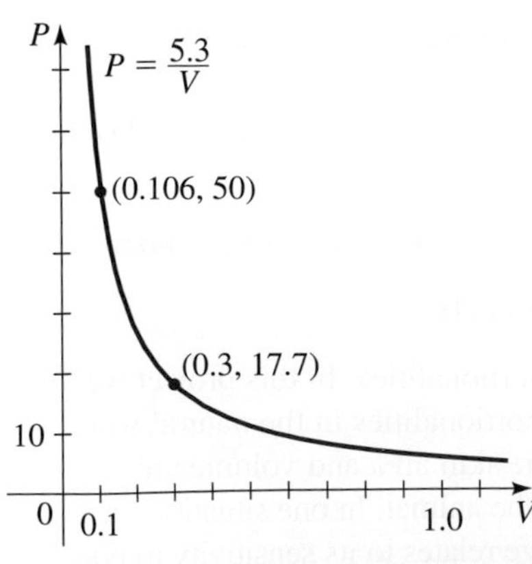

Figure 4

Since V represents volume (which is never negative), we sketch only the part of the graph for which $V > 0$. The graph is shown in Figure 4.

(b) When $V = 0.3$, we have

$$P = \frac{5.3}{0.3} \approx 17.7$$

so the new pressure is about 17.7 kPa.

Now Try Exercises 21 and 43 ■

■ Combining Different Types of Variation

In the sciences, relationships between three or more variables are common, and any combination of the different types of proportionality that we have discussed is possible. For example, if the quantities x, y, and z are related by the equation

$$z = kxy$$

then we say that z is **proportional to the product** of x and y. We can also express this relationship by saying that z **varies jointly** as x and y or that z **is jointly proportional to** x and y. If the quantities x, y, and z are related by the equation

$$z = k\frac{x}{y}$$

we say that z **is proportional to** x **and inversely proportional to** y or that z **varies directly as** x **and inversely as** y.

Example 3 ■ Combining Variations

The apparent brightness B of a light source (measured in W/m^2) is directly proportional to the luminosity L (measured in watts, W) of the light source and inversely proportional to the square of the distance d from the light source (measured in meters).

(a) Write an equation that expresses this variation.

(b) If the distance is doubled, by what factor will the brightness change?

(c) If the distance is cut in half and the luminosity is tripled, by what factor will the brightness change?

Solution

(a) Since B is directly proportional to L and inversely proportional to d^2, we have

$$B = k\frac{L}{d^2} \qquad \text{Brightness at distance } d \text{ and luminosity } L$$

where k is a constant.

(b) To obtain the brightness at double the distance, we replace d by $2d$ in the equation we obtained in part (a).

$$B = k\frac{L}{(2d)^2} = \frac{1}{4}\left(k\frac{L}{d^2}\right) \qquad \text{Brightness at distance } 2d$$

Comparing this expression with that obtained in part (a), we see that the brightness will be $\frac{1}{4}$ the original brightness.

Discovery Project ■ Proportionality: Shape and Size

Many real-world quantities are related by proportionalities. In this project we use the proportionality symbol $\propto$ to relate proportionalities in the natural world. For example, for animals of the same shape, the skin area and volume are proportional, in different ways, to the length of the animal. In one situation we use proportionality to determine how a frog's size relates to its sensitivity to pollutants in the environment. You can find the project at **www.stewartmath.com**.

(c) To obtain the brightness at half the distance d and triple the luminosity L, we replace d by $d/2$ and L by $3L$ in the equation we obtained in part (a).

$$B = k\frac{3L}{(\frac{1}{2}d)^2} = \frac{3}{\frac{1}{4}}\left(k\frac{L}{d^2}\right) = 12\left(k\frac{L}{d^2}\right)$$ Brightness at distance $\frac{1}{2}d$ and luminosity $3L$

Comparing this expression with that obtained in part (a), we see that the brightness will be 12 times the original brightness.

 **Now Try Exercises 23 and 45** ■

The relationship between apparent brightness, actual brightness (or luminosity), and distance is used in estimating distances to stars (see Exercise 56).

Example 4 ■ Newton's Law of Gravity

Newton's Law of Gravity says that two objects with masses m_1 and m_2 attract each other with a force F that is jointly proportional to their masses and inversely proportional to the square of the distance r between the objects. Express Newton's Law of Gravity as an equation.

Solution Using the definitions of joint and inverse variation and the traditional notation G for the gravitational constant of proportionality, we have

$$F = G\frac{m_1 m_2}{r^2}$$

 **Now Try Exercises 31 and 47** ■

If m_1 and m_2 are fixed masses, then the gravitational force between them is $F = C/r^2$ (where $C = Gm_1m_2$ is a constant). Figure 5 shows the graph of this equation for $r > 0$ with $C = 1$. Observe how the gravitational attraction decreases with increasing distance.

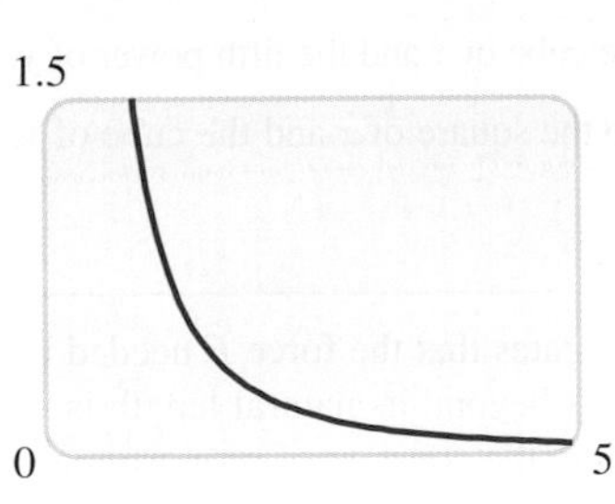

Figure 5 | Graph of $F = \frac{1}{r^2}$

Note Like the Law of Gravity, many laws of nature are *inverse square laws*. There is a geometric reason for this. Imagine a force or energy originating from a point source and spreading its influence equally in all directions, just like the light source in Example 3 or the gravitational force exerted by a planet in Example 4. The influence of the force or energy at a distance r from the source is spread out over the surface of a sphere of radius r, which has area $A = 4\pi r^2$. (See Figure 6.) So the intensity I at a distance r from the source is the source strength S divided by the area A of the sphere:

$$I = \frac{S}{4\pi r^2} = \frac{k}{r^2}$$

where k is the constant $S/(4\pi)$. Thus point sources of light, sound, gravity, electromagnetic fields, and radiation must all obey inverse square laws, simply because of the geometry of space.

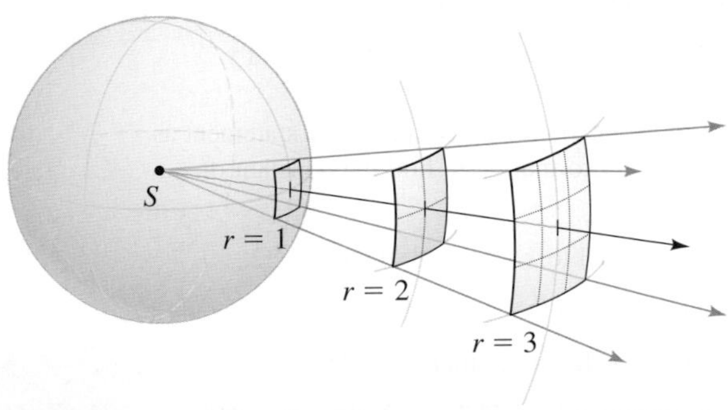

Figure 6 | Energy from a point source S

1.12 Exercises

Concepts

1. If the quantities x and y are related by the equation $y = 5x$, then we say that y is ________ ________ to x and the constant of ________ is 5.

2. If the quantities x and y are related by the equation $y = \frac{5}{x}$, then we say that y is ________ ________ to x and the constant of ________ is 5.

3. If the quantities x, y, and z are related by the equation $z = 5\frac{x}{y}$, then we say that z is ________ ________ to x and ________ ________ to y.

4. If z is directly proportional to the product of x and y and if z is 10 when x is 4 and y is 5, then x, y, and z are related by the equation $z =$ ________.

5–6 ■ In each equation, is y directly proportional, inversely proportional, or not proportional to x?

5. (a) $y = 3x$ **(b)** $y = 3x + 1$

6. (a) $y = \frac{3}{x+1}$ **(b)** $y = \frac{3}{x}$

Skills

7–18 ■ **Equations of Proportionality** Write an equation that expresses the statement.

7. T varies directly as x.

8. P is directly proportional to w.

9. v is inversely proportional to z.

10. w is proportional to the product of m and n.

11. y is proportional to s and inversely proportional to t.

12. P varies inversely as T.

13. z is proportional to the square root of y.

14. A is proportional to the square of x and inversely proportional to the cube of t.

15. V is proportional to the product of l, w, and h.

16. S is proportional to the product of the squares of r and θ.

17. R is proportional to the product of the squares of P and t and inversely proportional to the cube of b.

18. A is jointly proportional to the square roots of x and y.

19–30 ■ **Constants of Proportionality** Express the statement as an equation. Use the given information to find the constant of proportionality.

19. y is directly proportional to x. If $x = 8$, then $y = 32$.

20. w is inversely proportional to t. If $t = 8$, then $w = 3$.

21. A varies inversely as r. If $r = 5$, then $A = 15$.

22. P is directly proportional to T. If $T = 72$, then $P = 60$.

23. A is directly proportional to x and inversely proportional to t. If $x = 7$ and $t = 3$, then $A = 42$.

24. S is proportional to the product of p and q. If $p = 7$ and $q = 20$, then $S = 350$.

25. W is inversely proportional to the square of r. If $r = 3$, then $W = 24$.

26. t is proportional to the product of x and y and inversely proportional to r. If $x = 10$, $y = 15$, and $r = 12$, then $t = 125$.

27. C is jointly proportional to l, w, and h. If $l = w = h = 2$, then $C = 128$.

28. H is jointly proportional to the squares of l and w. If $l = 2$ and $w = \frac{1}{3}$, then $H = 36$.

29. R is inversely proportional to the square root of x. If $x = 121$, then $R = 2.5$.

30. M is jointly proportional to a, b, and c and inversely proportional to d. If a and d have the same value and if b and c are both 2, then $M = 128$.

31–34 ■ **Proportionality** A statement describing the relationship between the variables x, y, and z is given. **(a)** Express the statement as an equation of proportionality. **(b)** If x is tripled and y is doubled, by what factor does z change? (See Example 3.)

31. z varies directly as the cube of x and inversely as the square of y.

32. z is directly proportional to the square of x and inversely proportional to the fourth power of y.

33. z is jointly proportional to the cube of x and the fifth power of y.

34. z is inversely proportional to the square of x and the cube of y.

Applications

35. Hooke's Law Hooke's Law states that the force F needed to keep a spring stretched x units beyond its natural length is directly proportional to x. Here the constant of proportionality is called the **spring constant**.

(a) Write Hooke's Law as an equation.

(b) If a spring has a natural length of 5 cm and a force of 30 N is required to maintain the spring stretched to a length of 9 cm, find the spring constant.

(c) What force is needed to keep the spring stretched to a length of 11 cm?

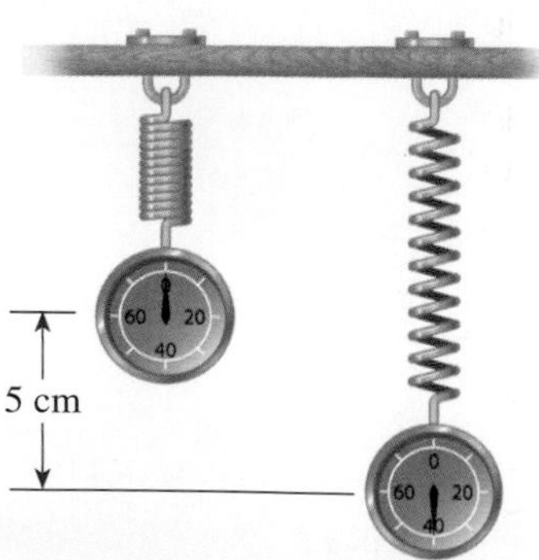

36. Printing Costs The cost C of printing a magazine is jointly proportional to the number of pages p in the magazine and the number of magazines printed m.

(a) Write an equation that expresses this joint variation.

(b) Find the constant of proportionality if the printing cost is $60,000 for 4000 copies of a 120-page magazine.

(c) How much would the printing cost be for 5000 copies of a 92-page magazine?

37. Power from a Windmill The power P that can be obtained from a windmill is directly proportional to the cube of the wind speed s.

(a) Write an equation that expresses this variation.

(b) Find the constant of proportionality for a windmill that produces 96 watts of power when the wind is blowing at 20 km/h.

(c) How much power will this windmill produce if the wind speed increases to 30 km/h?

38. Solubility of Carbon Dioxide The amount of carbon dioxide (CO_2) that can be dissolved in water is inversely proportional to the temperature of the water (in kelvins, K). An open glass of soda at a temperature of 273 K can dissolve about 3 g of CO_2. How much CO_2 can dissolve in the same bottle at the warmer temperature of 298 K?

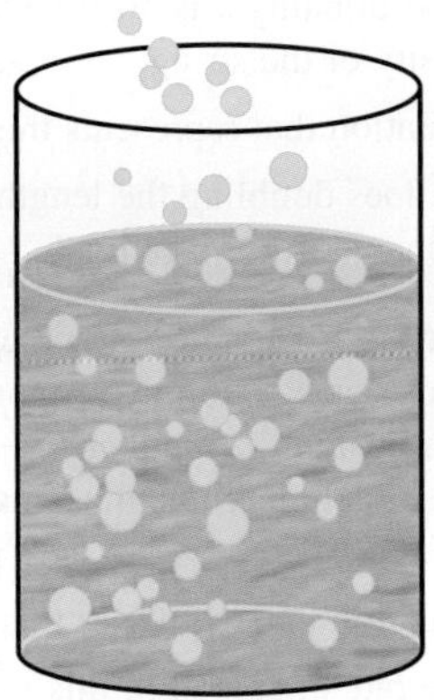

39. Stopping Distance The stopping distance D of a car after the brakes have been applied varies directly as the square of the speed s. A certain car traveling at 60 km/h can stop in 45 m. What is the maximum speed it can be traveling if it needs to stop in 60 m?

40. Aerodynamic Lift The lift L on an airplane wing at takeoff varies jointly as the square of the speed s of the plane and the area A of its wings. A plane with a wing area of 46 m^2 traveling at 80 km/h experiences a lift of 5950 N. How much lift would a plane with a wing area of 56 m^2 traveling at 64 km/h experience?

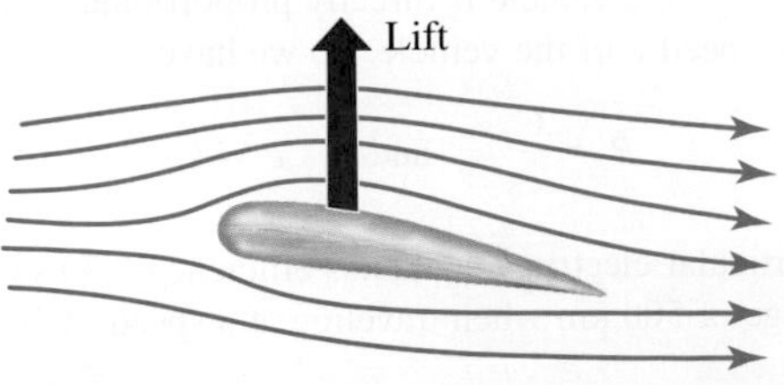

41. Drag Force on a Boat The drag force F on a boat is jointly proportional to the wetted surface area A on the hull and the square of the speed s of the boat. A boat experiences a drag force of 980 N when traveling at 8 km/h with a wetted surface area of 4 m^2. How fast must a boat be traveling if it has 2.5 m^2 of wetted surface area and is experiencing a drag force of 780 N?

42. Kepler's Third Law Kepler's Third Law of planetary motion states that the square of the period T of a planet (the time it takes for the planet to make a complete revolution about the Sun) is directly proportional to the cube of its average distance d from the Sun.

(a) Express Kepler's Third Law as an equation.

(b) Find the constant of proportionality by using the fact that for planet Earth the period is about 365 days and the average distance is about 149 million kilometers.

(c) The planet Neptune is about 4.46×10^9 km from the Sun. Find the period of Neptune.

43. Ideal Gas Law The pressure P of a sample of gas is directly proportional to the temperature T and inversely proportional to the volume V.

(a) Write an equation that expresses this variation.

(b) Find the constant of proportionality if 100 L of gas exerts a pressure of 33.2 kPa at a temperature of 400 K (absolute temperature measured on the Kelvin scale).

(c) If the temperature is increased to 500 K and the volume is decreased to 80 L, what is the pressure of the gas?

44. Skidding in a Curve A car is traveling on a curve that forms a circular arc. The force F needed to keep the car from skidding is jointly proportional to the weight w of the car and the square of its speed s and is inversely proportional to the radius r of the curve.

(a) Write an equation that expresses this variation.

(b) A car weighing 768 kg travels around a curve at 100 km/h. The next car to round this curve weighs 1200 kg and requires the same force as the first car to keep from skidding. How fast is the second car traveling?

45. Loudness of Sound The loudness L of a sound (measured in decibels, dB) is inversely proportional to the square of the distance d from the source of the sound.

(a) Write an equation that expresses this variation.

(b) Find the constant of proportionality if a person 10 ft from a lawn mower experiences a sound level of 70 dB.

(c) If the distance in part (b) is doubled, by what factor is the loudness changed?

(d) If the distance in part (b) is cut in half, by what factor is the loudness changed?

46. A Jet of Water The power P of a jet of water is jointly proportional to the cross-sectional area A of the jet and to the cube of the velocity v.

(a) Write an equation that expresses this variation.

(b) If the velocity is doubled and the cross-sectional area is halved, by what factor is the power changed?

(c) If the velocity is halved and the cross-sectional area is tripled, by what factor is the power changed?

47. Electrical Resistance The resistance R of a wire varies directly as its length L and inversely as the square of its diameter d.

(a) Write an equation that expresses this joint variation.

(b) Find the constant of proportionality if a wire 1.2 m long and 0.005 m in diameter has a resistance of 140 ohms.

(c) Find the resistance of a wire made of the same material that is 3 m long and has a diameter of 0.008 m.

(d) If the diameter is doubled and the length is tripled, by what factor is the resistance changed?

48. Growing Cabbages In the short growing season of the Canadian arctic territory of Nunavut, some gardeners find it possible to grow gigantic cabbages in the midnight sun. Assume that the final size of a cabbage is proportional to the amount of nutrients it receives and inversely proportional to the number of other cabbages surrounding it. A cabbage that received 600 g of nutrients and had 12 other cabbages around it grew to 15 kg. What size would it grow to if it received 300 g of nutrients and had only 5 cabbage "neighbors"?

49. Radiation Energy The total radiation energy E emitted by a heated surface per unit area varies as the fourth power of its absolute temperature T. The temperature is 6000 K at the surface of the sun and 300 K at the surface of the earth.

(a) How many times more radiation energy per unit area is produced by the sun than by the earth?

(b) The radius of the earth is 6340 km, and the radius of the sun is 696,000 km. How many times more total radiation does the sun emit than the earth?

50. Law of the Pendulum The period of a pendulum (the time elapsed during one complete swing of the pendulum) varies directly with the square root of the length of the pendulum.

(a) Express this relationship by writing an equation.

(b) To double the period, how would we have to change the length l?

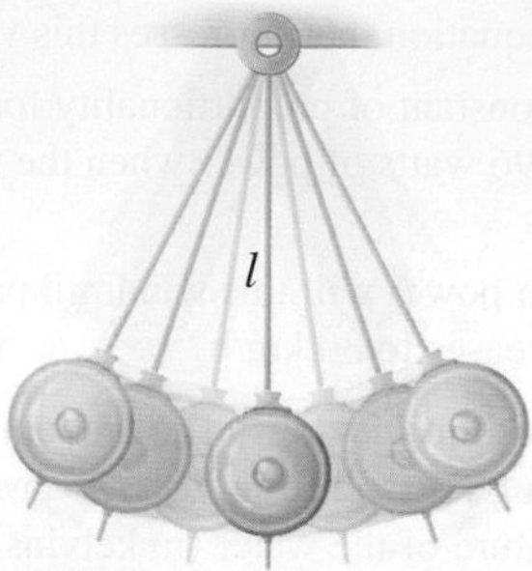

51. Frequency of Vibration The frequency f of vibration of a violin string is inversely proportional to its length L. The constant of proportionality k is positive and depends on the tension and density of the string.

(a) Write an equation that represents this variation.

(b) What effect does doubling the length of the string have on the frequency of its vibration?

52. Spread of a Disease The rate r at which a disease spreads in a population of size P is jointly proportional to the number x of infected people and the number $P - x$ who are not infected. An infection erupts in a small town that has population $P = 5000$.

(a) Write an equation that expresses this variation.

(b) Compare the rate of spread of this infection when 10 people are infected to the rate of spread when 1000 people are infected. Which rate is larger? By what factor?

(c) Calculate the rate of spread when the entire population is infected. Why does this answer make intuitive sense?

53. Electric Vehicle The range R of an electric vehicle is the distance (in km) the vehicle can travel on one complete charge C (in kWh). The efficiency rating e (measured in kilowatt-hours per km, kWh/km) of the vehicle is the amount of electric charge it uses per kilometer, so $e = C/R$. Due to increased aerodynamic drag at higher speeds, the efficiency of a vehicle is directly proportional to the square of the speed v of the vehicle. So we have

$$R = \frac{C}{e} \qquad \text{and} \qquad e = kv^2$$

A particular electric vehicle has efficiency 0.20 kWh/km and a range of 500 km when traveling at a speed of 100 km/h.

Estimate the range of the car when traveling at speeds of 130 km/h and 80 km/h.

Nerthuz/Shutterstock.com

54. Mass Flow Rate A fluid with density ρ is initially being pumped through a pipe with cross-sectional area A_0 at a constant velocity v_0. The *mass flow rate* of the fluid is $A_0 v_0 \rho$ and remains constant as the cross-sectional area of the pipe changes. (That is, if the cross-sectional area A of the pipe changes, the flow rate v also changes in such a way that the quantity $Av\rho$ remains unchanged.)

(a) Suppose that the cross-sectional area A_0 of a pipe is constricted to an area A, as shown in the figure. Use the fact that the mass flow rate is constant to show that the velocity v of the fluid becomes $v = (A_0 v_0)/A$ and conclude that the velocity of the fluid is inversely proportional to the cross-sectional area of the pipe.

(b) If a fluid is pumped with velocity 5 m/s through a pipe with radius 0.6 m, find the velocity of the fluid through a constricted part of the pipe with radius 0.2 m.

55–56 ■ Combining Variations Solve the problem using the relationship between brightness B, luminosity L, and distance d derived in Example 3. The proportionality constant is $k = 0.080$.

55. Brightness of a Star The luminosity of a star is given by $L = 2.5 \times 10^{26}$ W, and its distance from the earth is $d = 2.4 \times 10^{19}$ m. How bright does the star appear on the earth?

56. Distance to a Star The luminosity of a star is given by $L = 5.8 \times 10^{30}$ W, and its brightness as viewed from the earth is $B = 8.2 \times 10^{-16}$ W/m^2. Find the distance of the star from the earth.

■ Discuss ■ Discover ■ Prove ■ Write

57. Discuss: Is Proportionality Everything? A great many laws of physics and chemistry are expressible as proportionalities. Give at least one example of a relationship that occurs in the sciences that is *not* a proportionality.

Chapter 1 Review

Properties and Formulas

Properties of Real Numbers | Section 1.1

Commutative: $a + b = b + a$

$ab = ba$

Associative: $(a + b) + c = a + (b + c)$

$(ab)c = a(bc)$

Distributive: $a(b + c) = ab + ac$

Absolute Value | Section 1.1

$$|a| = \begin{cases} a & \text{if } a \geq 0 \\ -a & \text{if } a < 0 \end{cases}$$

$$|ab| = |a||b|$$

$$\left|\frac{a}{b}\right| = \frac{|a|}{|b|}$$

Distance between a and b:

$$d(a, b) = |b - a|$$

Exponents | Section 1.2

$$a^m a^n = a^{m+n}$$

$$\frac{a^m}{a^n} = a^{m-n}$$

$$(a^m)^n = a^{mn}$$

$$(ab)^n = a^n b^n$$

$$\left(\frac{a}{b}\right)^n = \frac{a^n}{b^n}$$

Radicals | Section 1.2

$\sqrt[n]{a} = b$ means $b^n = a$

$$\sqrt[n]{ab} = \sqrt[n]{a}\sqrt[n]{b}$$

$$\sqrt[n]{\frac{a}{b}} = \frac{\sqrt[n]{a}}{\sqrt[n]{b}}$$

$$\sqrt[m]{\sqrt[n]{a}} = \sqrt[mn]{a}$$

$$a^{m/n} = \sqrt[n]{a^m}$$

If n is odd, then $\sqrt[n]{a^n} = a$.

If n is even, then $\sqrt[n]{a^n} = |a|$.

Special Product Formulas | Section 1.3

Sum and difference of same terms:

$$(A + B)(A - B) = A^2 - B^2$$

Square of a sum or difference:

$$(A + B)^2 = A^2 + 2AB + B^2$$

$$(A - B)^2 = A^2 - 2AB + B^2$$

Cube of a sum or difference:

$$(A + B)^3 = A^3 + 3A^2B + 3AB^2 + B^3$$

$$(A - B)^3 = A^3 - 3A^2B + 3AB^2 - B^3$$

Special Factoring Formulas | Section 1.3

Difference of squares:

$$A^2 - B^2 = (A + B)(A - B)$$

Perfect squares:

$$A^2 + 2AB + B^2 = (A + B)^2$$

$$A^2 - 2AB + B^2 = (A - B)^2$$

Sum or difference of cubes:

$$A^3 - B^3 = (A - B)(A^2 + AB + B^2)$$

$$A^3 + B^3 = (A + B)(A^2 - AB + B^2)$$

Rational Expressions | Section 1.4

We can cancel common factors:

$$\frac{AC}{BC} = \frac{A}{B}$$

To multiply two fractions, we multiply their numerators together and their denominators together:

$$\frac{A}{B} \times \frac{C}{D} = \frac{AC}{BD}$$

To divide fractions, we invert the divisor and multiply:

$$\frac{A}{B} \div \frac{C}{D} = \frac{A}{B} \times \frac{D}{C}$$

To add fractions, we find a common denominator:

$$\frac{A}{C} + \frac{B}{C} = \frac{A + B}{C}$$

Properties of Equality | Section 1.5

$A = B \quad \Leftrightarrow \quad A + C = B + C$

$A = B \quad \Leftrightarrow \quad CA = CB \quad (C \neq 0)$

Linear Equations | Section 1.5

A **linear equation** in one variable is an equation of the form $ax + b = 0$.

Zero-Product Property | Section 1.5

If $AB = 0$, then $A = 0$ or $B = 0$.

Completing the Square | Section 1.5

To make $x^2 + bx$ a perfect square, add $\left(\frac{b}{2}\right)^2$. This gives the perfect square

$$x^2 + bx + \left(\frac{b}{2}\right)^2 = \left(x + \frac{b}{2}\right)^2$$

Quadratic Formula | Section 1.5

A **quadratic equation** is an equation of the form

$$ax^2 + bx + c = 0$$

Its solutions are given by the **Quadratic Formula**:

$$x = \frac{-b \pm \sqrt{b^2 - 4ac}}{2a}$$

The **discriminant** is $D = b^2 - 4ac$.

If $D > 0$, the equation has two real solutions.

If $D = 0$, the equation has one solution.

If $D < 0$, the equation has two complex solutions.

Complex Numbers | Section 1.6

A **complex number** is a number of the form $a + bi$, where $i = \sqrt{-1}$.

The **complex conjugate** of $a + bi$ is

$$\overline{a + bi} = a - bi$$

To **multiply** complex numbers, treat them as binomials and use $i^2 = -1$ to simplify the result.

To **divide** complex numbers, multiply numerator and denominator by the complex conjugate of the denominator:

$$\frac{a + bi}{c + di} = \left(\frac{a + bi}{c + di}\right) \cdot \left(\frac{c - di}{c - di}\right) = \frac{(a + bi)(c - di)}{c^2 + d^2}$$

Inequalities | Section 1.8

Adding the same quantity to each side of an inequality gives an equivalent inequality:

$$A < B \quad \Leftrightarrow \quad A + C < B + C$$

Multiplying each side of an inequality by the same *positive* quantity gives an equivalent inequality. Multiplying each side by the same *negative* quantity reverses the direction of the inequality:

$$\text{If } C > 0, \text{ then } A < B \quad \Leftrightarrow \quad CA < CB$$

$$\text{If } C < 0, \text{ then } A < B \quad \Leftrightarrow \quad CA > CB$$

Absolute-Value Inequalities | Section 1.8

To solve absolute-value inequalities, we use

$$|x| < C \quad \Leftrightarrow \quad -C < x < C$$

$$|x| > C \quad \Leftrightarrow \quad x < -C \text{ or } x > C$$

The Distance Formula | Section 1.9

The distance between the points $A(x_1, y_1)$ and $B(x_2, y_2)$ is

$$d(A, B) = \sqrt{(x_2 - x_1)^2 + (y_2 - y_1)^2}$$

The Midpoint Formula | Section 1.9

The midpoint of the line segment from $A(x_1, y_1)$ to $B(x_2, y_2)$ is

$$\left(\frac{x_1 + x_2}{2}, \frac{y_1 + y_2}{2}\right)$$

Intercepts | Section 1.9

To find the ***x*-intercepts** of the graph of an equation, set $y = 0$ and solve for x.

To find the ***y*-intercepts** of the graph of an equation, set $x = 0$ and solve for y.

Circles | Section 1.9

The circle with center (0, 0) and radius r has equation

$$x^2 + y^2 = r^2$$

The circle with center (h, k) and radius r has equation

$$(x - h)^2 + (y - k)^2 = r^2$$

Symmetry | Section 1.9

The graph of an equation is **symmetric with respect to the *x*-axis** if the equation remains unchanged when y is replaced by $-y$.

The graph of an equation is **symmetric with respect to the *y*-axis** if the equation remains unchanged when x is replaced by $-x$.

The graph of an equation is **symmetric with respect to the origin** if the equation remains unchanged when x is replaced by $-x$ and y by $-y$.

Slope of a Line | Section 1.10

The slope of the nonvertical line that contains the points $A(x_1, y_1)$ and $B(x_2, y_2)$ is

$$m = \frac{\text{rise}}{\text{run}} = \frac{y_2 - y_1}{x_2 - x_1}$$

Equations of Lines | Section 1.10

If a line has slope m, has y-intercept b, and contains the point (x_1, y_1), then:

the **point-slope form** of its equation is

$$y - y_1 = m(x - x_1)$$

the **slope-intercept form** of its equation is

$$y = mx + b$$

The equation of any line can be expressed in the **general form**

$$Ax + By + C = 0$$

where A and B can't both be 0.

Vertical and Horizontal Lines | Section 1.10

The **vertical** line containing the point (a, b) has the equation $x = a$.

The **horizontal** line containing the point (a, b) has the equation $y = b$.

Parallel and Perpendicular Lines | Section 1.10

Two lines with slopes m_1 and m_2 are

parallel if and only if $m_1 = m_2$

perpendicular if and only if $m_1 m_2 = -1$

Variation | Section 1.12

If y is **directly proportional** to x, then

$$y = kx$$

If y is **inversely proportional** to x, then

$$y = \frac{k}{x}$$

Concept Check

1. (a) What does the set of natural numbers consist of? What does the set of integers consist of? Give an example of an integer that is not a natural number.

(b) What does the set of rational numbers consist of? Give an example of a rational number that is not an integer.

(c) What does the set of irrational numbers consist of? Give an example of an irrational number.

(d) What does the set of real numbers consist of?

2. A property of real numbers is given. State the property and give an example in which the property is used.

(i) Commutative Property

(ii) Associative Property

(iii) Distributive Property

3. Explain the difference between the open interval (a, b) and the closed interval $[a, b]$. Give an example of an interval that is neither open nor closed.

4. Give the formula for finding the distance between two real numbers a and b. Use the formula to find the distance between 103 and -52.

5. Suppose $a \neq 0$ is any real number.

(a) In the expression a^n, which is the base and which is the exponent?

(b) What does a^n mean if n is a positive integer? What does 6^5 mean?

(c) What does a^{-n} mean if n is a positive integer? What does 3^{-2} mean?

(d) What does a^n mean if n is zero?

(e) If m and n are positive integers, what does $a^{m/n}$ mean? What does $4^{3/2}$ mean?

6. State the first five Laws of Exponents. Give examples in which you would use each law.

7. When you multiply two powers of the same number, what should you do with the exponents? When you raise a power to a new power, what should you do with the exponents?

8. (a) What does $\sqrt[n]{a} = b$ mean?

(b) Is it true that $\sqrt{a^2}$ is equal to $|a|$? Try values for a that are positive and negative.

(c) How many real nth roots does a positive real number have if n is even? If n is odd?

(d) Is $\sqrt[4]{-2}$ a real number? Is $\sqrt[3]{-2}$ a real number? Explain why or why not.

9. Explain the steps involved in rationalizing a denominator. What is the logical first step in rationalizing the denominator of the expression $\frac{5}{\sqrt{3}}$?

10. Explain the difference between expanding an expression and factoring an expression.

11. State the Special Product Formulas used for expanding each of the given expressions.

(a) $(a + b)^2$ **(b)** $(a - b)^2$ **(c)** $(a + b)^3$

(d) $(a - b)^3$ **(e)** $(a + b)(a - b)$

Use the appropriate formula to expand $(x + 5)^2$ and $(x + 5)(x - 5)$.

12. State the following Special Factoring Formulas.

(a) Difference of Squares

(b) Perfect Square

(c) Sum of Cubes

Use the appropriate formula to factor $x^2 - 9$.

13. If the numerator and the denominator of a rational expression have a common factor, how would you simplify the expression? Simplify the expression $\frac{x^2 + x}{x + 1}$.

14. Explain each of the following.

(a) How to multiply and divide rational expressions.

(b) How to add and subtract rational expressions.

(c) What LCD do we use to perform the addition in the expression $\frac{3}{x - 1} + \frac{5}{x + 2}$?

15. What is the logical first step in rationalizing the denominator of $\frac{3}{1 + \sqrt{x}}$?

16. What is the difference between an algebraic expression and an equation? Give examples.

17. Write the general form of each type of equation.

(a) Linear equation

(b) Quadratic equation

18. What are the three ways to solve a quadratic equation?

19. State the Zero-Product Property. Use the property to solve the equation $x(x - 1) = 0$.

20. What do you need to add to $ax^2 + bx$ to complete the square? Complete the square for the expression $x^2 + 6x$.

21. State the Quadratic Formula for the quadratic equation $ax^2 + bx + c = 0$, and use it to solve the equation $x^2 + 6x - 1 = 0$.

22. What is the discriminant of the quadratic equation $ax^2 + bx + c = 0$? Find the discriminant of the equation $2x^2 - 3x + 5 = 0$. How many real solutions does this equation have?

23. What is the logical first step in solving the equation $\sqrt{x - 1} = x - 3$? Why is it important to check your answers when solving equations of this type?

24. What is a complex number? Give an example of a complex number, and identify the real and imaginary parts.

25. What is the complex conjugate of a complex number $a + bi$?

26. (a) How do you add complex numbers?

(b) How do you multiply $(3 + 5i)(2 - i)$?

(c) Is $(3 - i)(3 + i)$ a real number?

(d) How do you simplify the quotient $\frac{3 + 5i}{3 - i}$?

27. State the guidelines for modeling with equations.

28. Explain how to solve the given type of problem.

(a) Linear inequality: $2x \geq 1$

(b) Nonlinear inequality: $(x - 1)(x - 4) < 0$

(c) Absolute-value equation: $|2x - 5| = 7$

(d) Absolute-value inequality: $|2x - 5| \leq 7$

29. (a) In the coordinate plane, what is the horizontal axis called and what is the vertical axis called?

(b) To graph an ordered pair of numbers (x, y), you need the coordinate plane. For the point $(2, 3)$, which is the x-coordinate and which is the y-coordinate?

(c) For an equation in the variables x and y, how do you determine whether a given point is on the graph? Is the point $(5, 3)$ on the graph of the equation $y = 2x - 1$?

30. (a) What is the formula for finding the distance between the points (x_1, y_1) and (x_2, y_2)?

(b) What is the formula for finding the midpoint between (x_1, y_1) and (x_2, y_2)?

31. How do you find x-intercepts and y-intercepts of a graph of an equation?

32. (a) Write an equation of the circle with center (h, k) and radius r.

(b) Find the equation of the circle with center $(2, -1)$ and radius 3.

33. (a) How do you test whether the graph an equation is symmetric with respect to the (i) x-axis, (ii) y-axis, and (iii) origin?

(b) What type of symmetry does the graph of the equation $xy^2 + y^2x^2 = 3x$ have?

34. (a) What is the slope of a line? How do you compute the slope of the line through the points $(-1, 4)$ and $(1, -2)$?

(b) How do you find the slope and y-intercept of the line $6x + 3y = 12$?

(c) How do you write the equation for a line that has slope 3 and passes through the point $(1, 2)$?

35. Give an equation of both a vertical line and a horizontal line that passes through the point $(2, 3)$.

36. State the general equation of a line.

37. Given lines with slopes m_1 and m_2, how you can tell whether the lines are (i) parallel? (ii) perpendicular?

38. How do you solve an equation (i) algebraically? (ii) graphically?

39. How do you solve an inequality (i) algebraically? (ii) graphically?

40. Write an equation that expresses each relationship.
(a) y is directly proportional to x.
(b) y is inversely proportional to x.
(c) z is jointly proportional to x and y.

Answers to the Concept Check can be found at the book companion website stewartmath.com.

Exercises

1–4 ■ Properties of Real Numbers State the property of real numbers being used.

1. $3x + 2y = 2y + 3x$
2. $(a + b)(a - b) = (a - b)(a + b)$
3. $4(a + b) = 4a + 4b$
4. $(A + 1)(x + y) = (A + 1)x + (A + 1)y$

5–6 ■ Intervals Express the interval in terms of inequalities, and then graph the interval.

5. $[-2, 6)$
6. $(-\infty, 4]$

7–8 ■ Intervals Express the inequality in interval notation, and then graph the corresponding interval.

7. $x \geq 5$
8. $-1 < x \leq 5$

9–16 ■ Evaluate Evaluate the expression.

9. $|1 - |-4||$
10. $5 - |10 - |-4||$
11. $2^{1/2}8^{1/2}$
12. $2^{-3} - 3^{-2}$
13. $216^{-1/3}$
14. $64^{2/3}$
15. $\dfrac{\sqrt{242}}{\sqrt{2}}$
16. $\sqrt{2}\sqrt{50}$

17–20 ■ Radicals and Exponents Simplify the expression.

17. (a) $(a^2)^{-3}(a^3b)^2(b^3)^4$ (b) $(3xy^2)^3(\frac{2}{3}x^{-1}y)^2$

18. (a) $\dfrac{x^4(3x)^2}{x^3}$ (b) $\left(\dfrac{r^2s^{4/3}}{r^{1/3}s}\right)^6$

19. (a) $\sqrt[3]{(x^3y)^2y^4}$ (b) $\sqrt{w^8z^{10}}$

20. (a) $\dfrac{8r^{1/2}s^{-3}}{2r^{-2}s^4}$ (b) $\left(\dfrac{ab^2c^{-3}}{2a^3b^{-4}}\right)^{-2}$

21–24 ■ Scientific Notation These exercises involve scientific notation.

21. Write the number 78,250,000,000 in scientific notation.

22. Write the number 2.08×10^{-8} in ordinary decimal notation.

23. If $a \approx 0.00000293$, $b \approx 1.582 \times 10^{-14}$, and $c \approx 2.8064 \times 10^{12}$, use a calculator to approximate the number ab/c.

24. If your heart beats 80 times per minute and you live to be 90 years old, estimate the number of times your heart beats during your lifetime. State your answer in scientific notation.

25–38 ■ Factoring Factor the expression completely.

25. $x^2 + 5x - 14$
26. $12x^2 + 10x - 8$
27. $x^4 - 2x^2 + 1$
28. $4y^2z^3 + 10y^3z - 12y^5z^2$
29. $16 - 4t^2$
30. $2y^6 - 32y^2$
31. $x^6 - 1$
32. $16a^4b^2 + 2ab^5$
33. $-3x^{-1/2} + 2x^{1/2} + 5x^{3/2}$
34. $7x^{-3/2} - 8x^{-1/2} + x^{1/2}$
35. $5x^3 + 15x^2 - x - 3$
36. $w^3 - 3w^2 - 4w + 12$
37. $(a + b)^2 - 3(a + b) - 10$
38. $(3x + 2)^2 - (3x + 2) - 6$

39–50 ■ Operations with Algebraic Expressions Perform the indicated operations and simplify.

39. $(2y - 7)(2y + 7)$
40. $(1 + x)(2 - x) - (3 - x)(3 + x)$
41. $x^2(x - 2) + x(x - 2)^2$
42. $\sqrt{x}(\sqrt{x} + 1)(2\sqrt{x} - 1)$
43. $\dfrac{x^2 - 4x - 5}{x^2 - 25} \cdot \dfrac{x^2 + 12x + 36}{x^2 + 7x + 6}$
44. $\dfrac{x^2 - 2x - 15}{x^2 - 6x + 5} \div \dfrac{x^2 - x - 12}{x^2 - 1}$
45. $\dfrac{2}{x} + \dfrac{1}{x - 2} + \dfrac{3}{(x - 2)^2}$
46. $\dfrac{1}{x + 2} + \dfrac{1}{x^2 - 4} - \dfrac{2}{x^2 - x - 2}$
47. $\dfrac{\frac{1}{x} - \frac{1}{2}}{x - 2}$
48. $\dfrac{\frac{1}{x} - \frac{1}{x + 1}}{\frac{1}{x} + \frac{1}{x + 1}}$
49. $\dfrac{\sqrt{x} - 2}{\sqrt{x} + 2}$ (rationalize the denominator)
50. $\dfrac{\sqrt{x + h} - \sqrt{x}}{h}$ (rationalize the numerator)

51–54 ■ Rationalizing Rationalize the denominator and simplify.

51. $\dfrac{1}{\sqrt{11}}$
52. $\dfrac{3}{\sqrt{6}}$
53. $\dfrac{5}{1 + \sqrt{2}}$
54. $\dfrac{\sqrt{6}}{\sqrt{3} + \sqrt{2}}$

55–72 ■ Solving Equations Find all real solutions of the equation.

55. $7x - 6 = 4x + 9$
56. $8 - 2x = 14 + x$
57. $\dfrac{3x + 6}{x + 3} = \dfrac{3x}{x + 1}$
58. $(x + 2)^2 = (x - 4)^2$
59. $x^2 - 9x + 14 = 0$
60. $x^2 + 24x + 144 = 0$

61. $2x^2 + x = 1$

62. $\sqrt{x - 1} = \sqrt{x^2 - 3}$

63. $4x^3 - 25x = 0$

64. $x^3 - 2x^2 - 5x + 10 = 0$

65. $3x^2 + 4x - 1 = 0$

66. $\dfrac{1}{x} + \dfrac{2}{x - 1} = 3$

67. $\dfrac{x}{x - 2} + \dfrac{1}{x + 2} = \dfrac{8}{x^2 - 4}$

68. $x^4 - 8x^2 - 9 = 0$

69. $(x - 3)^2 - 3(x - 3) - 4 = 0$

70. $9x^{-1/2} - 6x^{1/2} + x^{3/2} = 0$

71. $|x - 7| = 4$

72. $|2x - 5| = 9$

73–76 ■ Complex Numbers Evaluate the expression and write in the form $a + bi$.

73. **(a)** $(2 - 3i) + (1 + 4i)$ **(b)** $(2 + i)(3 - 2i)$

74. **(a)** $(3 - 6i) - (6 - 4i)$ **(b)** $4i(2 - \frac{1}{2}i)$

75. **(a)** $\dfrac{4 + 2i}{2 - i}$ **(b)** $(1 - \sqrt{-1})(1 + \sqrt{-1})$

76. **(a)** $\dfrac{2 - 3i}{1 + i}$ **(b)** $\sqrt{-10} \cdot \sqrt{-40}$

77–82 ■ Real and Complex Solutions Find all real and complex solutions of the equation.

77. $x^2 + 16 = 0$

78. $x^2 = -12$

79. $x^2 + 6x + 10 = 0$

80. $2x^2 - 3x + 2 = 0$

81. $x^4 - 256 = 0$

82. $x^3 - 2x^2 + 4x - 8 = 0$

83. Mixtures A bulk food store sells raisins for \$3.20 per kilogram and nuts for \$2.40 per kilogram. The store sells a 50 kg mix of raisins and nuts for \$2.72 per kilogram. What quantities of raisins and nuts are in the mixture?

84. Distance and Time A district supervisor leaves Kingstown at 2:00 P.M. and drives to Queensville, 256 km distant, at 72 km/h. At 2:15 P.M. a store manager leaves Queensville and drives to Kingstown at 64 km/h. At what time do they pass each other on the road?

85. Distance and Time An athlete has a daily exercise program of cycling and running. Their cycling speed is 8 km/h faster than their running speed. Every morning the athlete cycles 4 km and runs $2\frac{1}{2}$ km, for a total of one hour of exercise. How fast does the athlete run?

86. Geometry The hypotenuse of a right triangle has length 20 cm. The sum of the lengths of the other two sides is 28 cm. Find the lengths of the other two sides of the triangle.

87. Doing the Job An interior decorator paints twice as fast as the assistant and three times as fast as the apprentice. If it takes them 60 min to paint a room with all three working together, how long would it take the interior decorator to paint the room working alone?

88. Dimensions of a Garden A homeowner wishes to fence in three adjoining garden plots, as shown in the figure. If each plot is to be 80 m² in area and there is 88 m of fencing material at hand, what dimensions should each plot have?

89–96 ■ Inequalities Solve the inequality. Express the solution using interval notation, and graph the solution set on the real number line.

89. $3x - 2 > -11$

90. $-7 \le 3x - 1 \le -1$

91. $x^2 - 7x - 8 > 0$

92. $x^2 \le 1$

93. $\dfrac{x - 4}{x^2 - 4} \le 0$

94. $\dfrac{5}{x^3 - x^2 - 4x + 4} < 0$

95. $|x - 5| \le 3$

96. $|x - 4| < 0.02$

97–98 ■ Coordinate Plane Two points P and Q are given. **(a)** Plot P and Q on a coordinate plane. **(b)** Find the distance from P to Q. **(c)** Find the midpoint of the segment PQ. **(d)** Sketch the line determined by P and Q, and find its equation in slope-intercept form. **(e)** Sketch the circle that passes through Q and has center P, and find the equation of this circle.

97. $P(2, 0),\ Q(-5, 12)$

98. $P(7, -1),\ Q(2, -11)$

99–100 ■ Graphing Regions Sketch the region given by the set.

99. $\{(x, y) \mid x \ge 4 \text{ or } y \ge 2\}$

100. $\{(x, y) \mid |x| < 1 \text{ and } |y| < 4\}$

101. Distance Formula Which of the points $A(4, 4)$ or $B(5, 3)$ is closer to the point $C(-1, -3)$?

102–104 ■ Circles In these exercises we find equations of circles.

102. Find an equation of the circle that has center $(-3, 4)$ and radius $\sqrt{6}$.

103. Find an equation of the circle that has center $(-5, -1)$ and passes through the origin.

104. Find an equation of the circle that contains the points $P(2, 3)$ and $Q(-1, 8)$ and has the midpoint of the segment PQ as its center.

105–108 ■ Circles **(a)** Complete the square to determine whether the equation represents a circle or a point or has no graph. **(b)** If the equation is that of a circle, find its center and radius, and sketch its graph.

105. $x^2 + y^2 - 8x + 2y + 13 = 0$

106. $2x^2 + 2y^2 - 2x + 8y = \frac{1}{2}$

107. $x^2 + y^2 + 72 = 12x$

108. $x^2 + y^2 - 6x - 10y + 34 = 0$

109–114 ■ Graphing Equations Sketch the graph of the equation by making a table and plotting points.

109. $y = 2 - 3x$

110. $2x - y + 1 = 0$

111. $y = 16 - x^2$

112. $8x + y^2 = 0$

113. $x = \sqrt{y}$

114. $y = -\sqrt{1 - x^2}$

115–120 ■ Symmetry and Intercepts (a) Test the equation for symmetry with respect to the x-axis, the y-axis, and the origin. (b) Find the x- and y-intercepts of the graph of the equation.

115. $x = 16 - y^2$

116. $x^2 + 4y^2 = 9$

117. $x^2 - 9y = 9$

118. $(x + 1)^2 + y^2 = 4$

119. $x^2 + 4xy + y^2 = 1$

120. $x^3 + xy^2 = 5$

121–124 ■ Graphing Equations (a) Use a graphing device to graph the equation in an appropriate viewing rectangle. (b) Use the graph to find the x- and y-intercepts.

121. $y = x^2 - 6x$ **122.** $y = \sqrt{5 - x}$

123. $y = x^3 - 4x^2 - 5x$ **124.** $\dfrac{x^2}{4} + y^2 = 1$

125–132 ■ Lines A description of a line is given. (a) Find an equation for the line in slope-intercept form. (b) Find an equation for the line in general form. (c) Graph the line.

125. The line that has slope 2 and y-intercept 6

126. The line that has slope $-\frac{1}{2}$ and passes through the point $(6, -3)$

127. The line that passes through the points $(-3, -2)$ and $(1, 4)$

128. The line that has x-intercept 4 and y-intercept 12

129. The vertical line that passes through the point $(3, -2)$

130. The horizontal line with y-intercept 5

131. The line that passes through the origin and is parallel to the line containing $(2, 4)$ and $(4, -4)$

132. The line that passes through the point $(2, 3)$ and is perpendicular to the line $x - 4y + 7 = 0$

133. Stretching a Spring Hooke's Law states that if a weight w is attached to a hanging spring, then the stretched length s of the spring is linearly related to w. For a particular spring we have

$$s = 0.3w + 2.5$$

where s is measured in cm and w in kilograms.

(a) What do the slope and s-intercept in this equation represent?

(b) How long is the spring when a 5-kg weight is attached?

134. Annual Salary A tax accountant is hired by a firm at a salary of \$60,000 per year. Three years later the annual salary is increased to \$70,500. Assume that the accountant's salary increases linearly.

(a) Find an equation that relates the accountant's annual salary S and the number of years t that they have worked for the firm.

(b) What do the slope and S-intercept of the salary equation represent?

(c) What will be the accountant's annual salary after working 12 years with the firm?

135–140 ■ Equations and Inequalities Graphs of the equations $y = x^2 - 4x$ and $y = x + 6$ are given. Use the graphs to solve the equation or inequality.

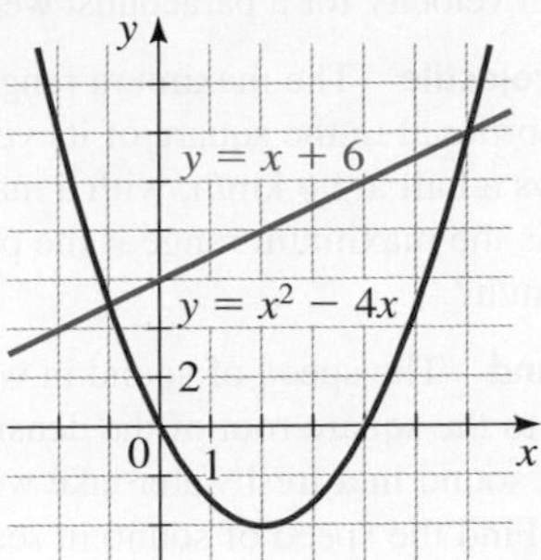

135. $x^2 - 4x = x + 6$ **136.** $x^2 - 4x = 0$

137. $x^2 - 4x \le x + 6$ **138.** $x^2 - 4x \ge x + 6$

139. $x^2 - 4x \ge 0$ **140.** $x^2 - 4x \le 0$

141–144 ■ Equations Solve the equation graphically. State each answer rounded to two decimals.

141. $x^2 - 4x = 2x + 7$ **142.** $\sqrt{x + 4} = x^2 - 5$

143. $x^4 - 9x^2 = x - 9$ **144.** $||x + 3| - 5| = 2$

145–148 ■ Inequalities Solve the inequality graphically. State each answer rounded to two decimals.

145. $x^2 > 12 - 4x$ **146.** $x^3 - 4x^2 - 5x > 2$

147. $x^4 - 4x^2 < \frac{1}{2}x - 1$ **148.** $|x^2 - 16| - 10 \ge 0$

149–150 ■ Circles and Lines Find an equation for both the circle and the line in the figure.

149.

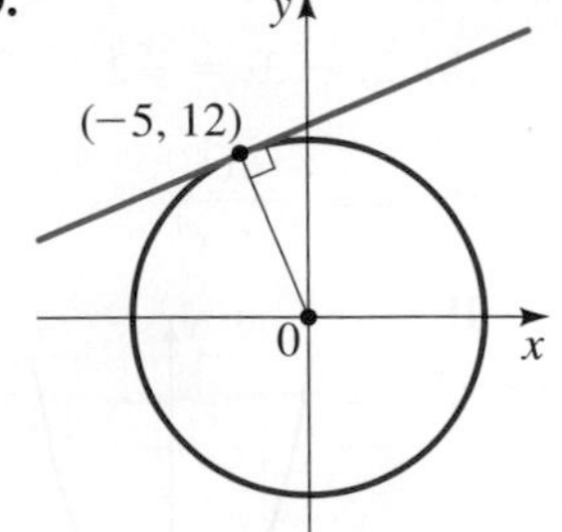

150.

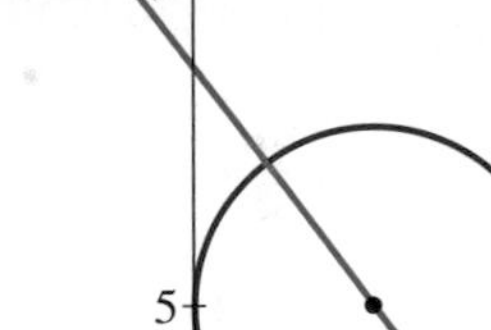

151. Variation Suppose that M varies directly as z, and $M = 120$ when $z = 15$. Write an equation that expresses this variation.

152. Variation Suppose that z is inversely proportional to y, and that $z = 12$ when $y = 16$. Write an equation that expresses z in terms of y.

153. Light Intensity The intensity of illumination I from a light varies inversely as the square of the distance d from the light.

(a) Write this statement as an equation.

(b) Determine the constant of proportionality if it is known that a lamp has an intensity of 1000 candles at a distance of 8 m.

(c) What is the intensity of this lamp at a distance of 20 m?

154. Vibrating String The frequency of a vibrating string under constant tension is inversely proportional to its length. If a violin string 12 centimeters long vibrates 440 times per second, to what length must it be shortened to vibrate 660 times per second?

155. Terminal Velocity The terminal velocity of a parachutist is directly proportional to the square root of their weight. A 70-kg parachutist attains a terminal velocity of 14 km/h. What is the terminal velocity for a parachutist weighing 105 kg?

156. Range of a Projectile The maximum range of a projectile is directly proportional to the square of its velocity. A baseball pitcher throws a ball at 96 km/h, with a maximum range of 74 m. What is the maximum range if the pitcher throws the ball at 112 km/h?

157. Speed of Sound The speed of sound in water is inversely proportional to the square root of the density of the water. The speed of sound in a freshwater lake with density 1 g/mL is 1480 m/s. Find the speed of sound in seawater with density 1.0273 g/mL.

158. Resolution of a Telescope The *resolving power* of a telescope is its ability to distinguish (or resolve) objects that are separated by a small angular distance θ (the angle between the two objects as viewed from the telescope). The angle θ that a telescope can resolve is given by the formula

$$\theta = 1.22\frac{\lambda}{D}$$

where λ is the wavelength of light and D is the diameter of the telescope lens.

(a) For a fixed diameter, is the angular distance θ that can be resolved by a telescope smaller for shorter or for longer wavelengths of light?

(b) For a fixed wavelength, if the diameter D of the mirror is doubled how does the angular distance θ of the telescope change?

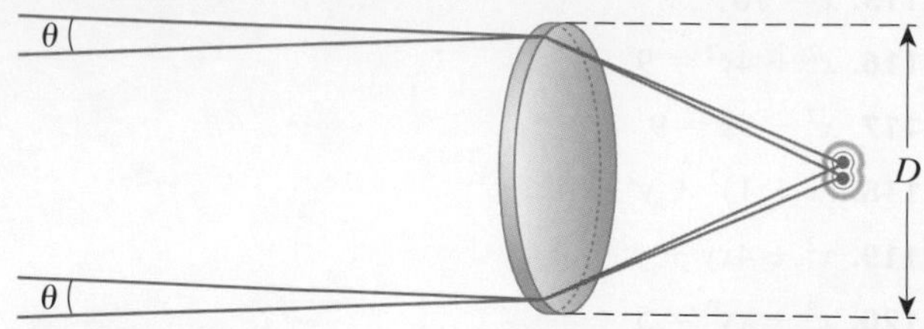

159. Hubble's Law The *redshift* z and *recessional* velocity v (km/s) of a galaxy are related by the formula

$$1 + z = \sqrt{\frac{c + v}{c - v}}$$

where $c = 3 \times 10^5$ km/s is the speed of light. Hubble's Law states that v varies directly with the distance D (in megalight-years, Mly) to the galaxy by the formula $v = H_0D$, where $H_0 \approx 20.8$ (km/s)/Mly. For a galaxy with $z = 2$, find v and D.

Matching

160. Equations and Their Graphs Match each equation with its graph. Give a reason for each answer. (Don't use a graphing device.)

(a) $y = 2|x| - 3$ **(b)** $2y - 3x = -2$ **(c)** $y = x^4$ **(d)** $(x + 1)^2 + (y + 1)^2 = 9$

(e) $6y + x = 3$ **(f)** $y = \dfrac{6x}{1 + x^4}$ **(g)** $x = y^3$ **(h)** $x^2 - 2x + y^2 - 4y + 1 = 0$

I

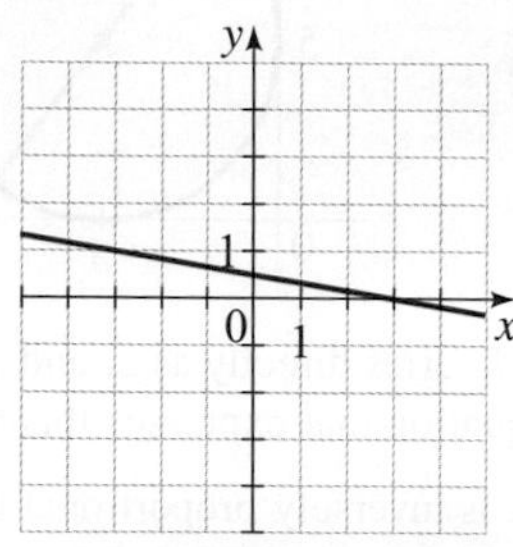

II

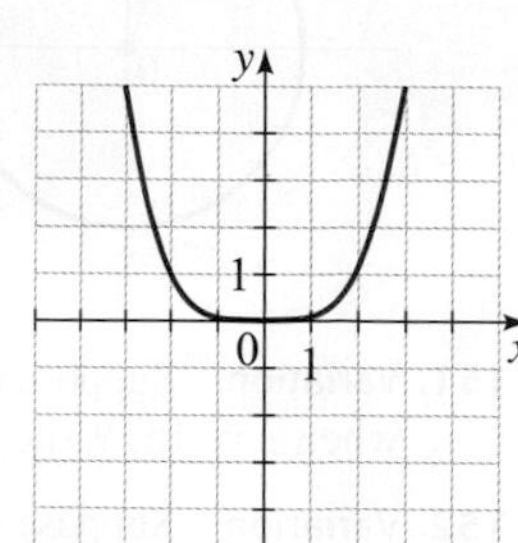

III

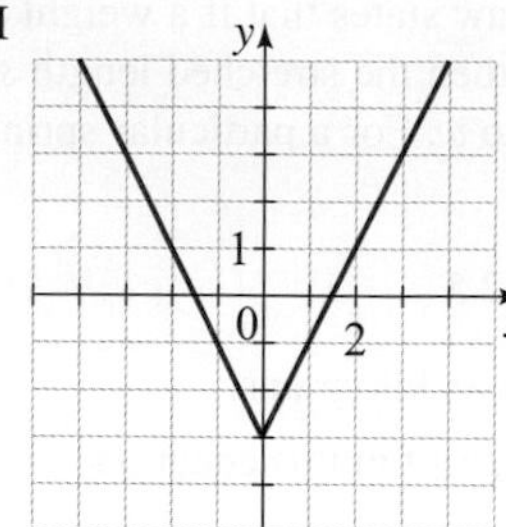

IV

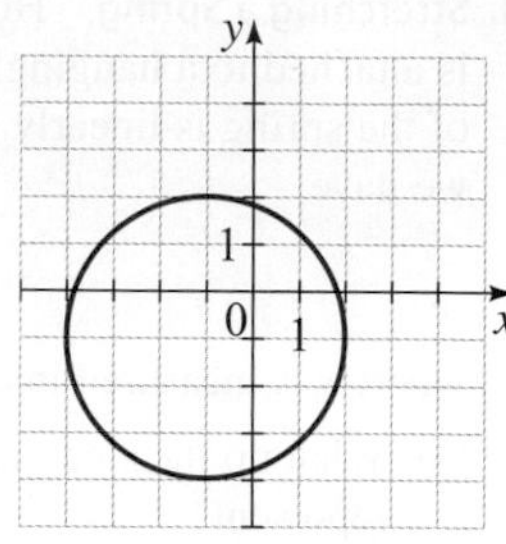

V

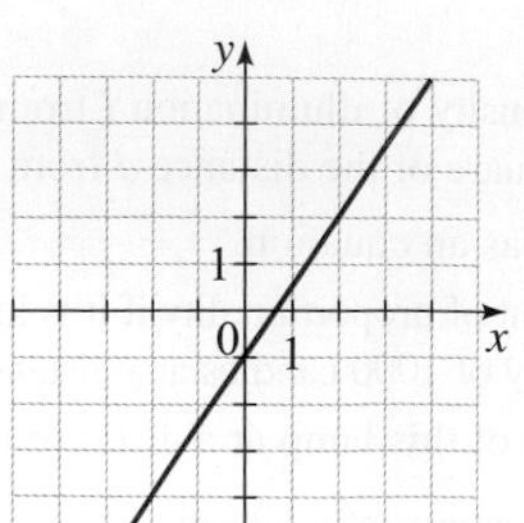

VI

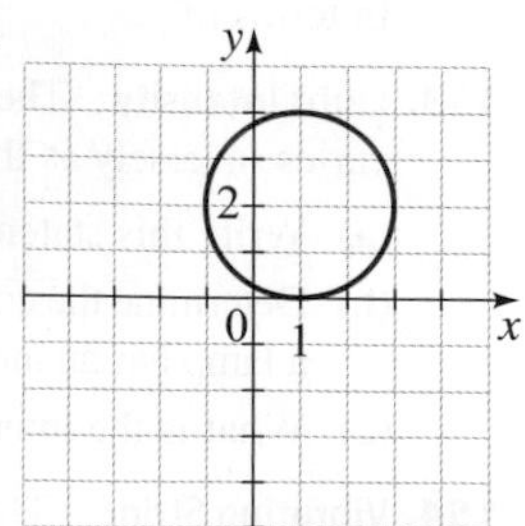

VII

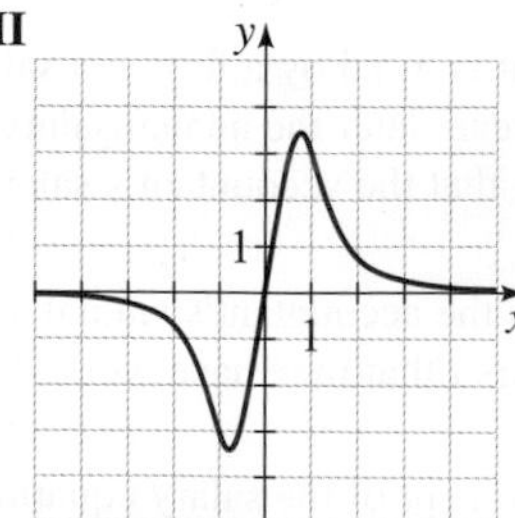

VIII

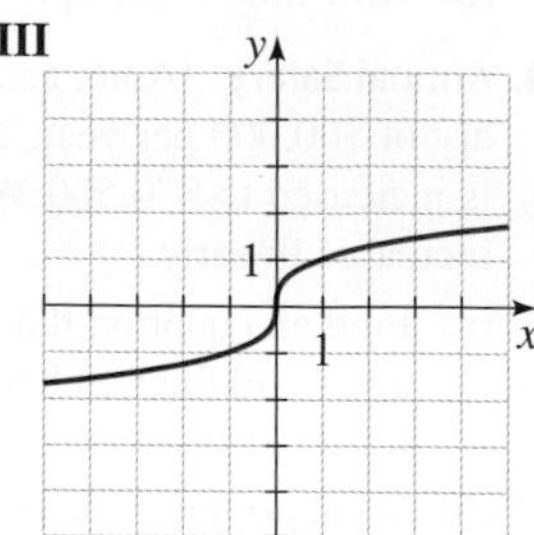

Chapter 1 | Test

1. (a) Graph the intervals $(-5, 3]$ and $(2, \infty)$ on the real number line.
   (b) Express the inequalities $x \le 3$ and $-1 \le x < 4$ in interval notation.
   (c) Find the distance between -7 and 9 on the real number line.

2. Evaluate each expression.
   (a) $(-3)^4$ (b) -3^4 (c) 3^{-4} (d) $\dfrac{3^{75}}{3^{72}}$ (e) $\left(\dfrac{2}{3}\right)^{-2}$ (f) $16^{-3/4}$

3. Write each number in scientific notation.
   (a) 186,000,000,000 (b) 0.0000003965

4. Simplify each expression. Write your final answer without negative exponents.
   (a) $\sqrt{200} - \sqrt{32}$ (b) $(3a^3b^3)(4ab^2)^2$ (c) $\left(\dfrac{4x^9y^3}{xy^7}\right)^{-1/2}$

5. Perform the indicated operations and simplify.
   (a) $z(4z - 3) + 2z(3 - 2z)$ (b) $(x + 3)(4x - 5)$ (c) $(\sqrt{a} + \sqrt{b})(\sqrt{a} - \sqrt{b})$
   (d) $(2x + 3)^2$ (e) $(x + 2)^3$

6. Factor each expression completely.
   (a) $4x^2 - 25$ (b) $2x^2 + 5x - 12$ (c) $x^3 - 3x^2 - 4x + 12$
   (d) $x^4 + 27x$ (e) $2x^{3/2} + 8x^{1/2} - 10x^{-1/2}$ (f) $x^4y^2 - 9x^2y^2$

7. Simplify each expression.
   (a) $\dfrac{w^2 + 4w + 3}{w^2 - 2w - 3}$ (b) $\dfrac{x^2}{x^2 - 4} - \dfrac{x + 1}{x + 2}$ (c) $\dfrac{\dfrac{y}{x} - \dfrac{x}{y}}{\dfrac{1}{y} - \dfrac{1}{x}}$

8. Rationalize the denominator and simplify: $\dfrac{\sqrt{2}}{4 - \sqrt{2}}$

9. Find all real solutions.
   (a) $x + 5 = 14 - \frac{1}{2}x$ (b) $\dfrac{2x}{x + 1} = \dfrac{2x - 1}{x}$ (c) $x^2 - x - 12 = 0$
   (d) $2x^2 + 4x + 1 = 0$ (e) $\sqrt{3 - \sqrt{x + 5}} = 2$ (f) $x^4 - 3x^2 + 2 = 0$
   (g) $3|x - 4| = 10$

10. Perform the indicated operations and write the result in the form $a + bi$.
    (a) $(3 - 2i) + (4 + 3i)$ (b) $(3 - 2i) - (4 + 3i)$
    (c) $(3 - 2i)(4 + 3i)$ (d) $\dfrac{3 - 2i}{4 + 3i}$
    (e) i^{48} (f) $(\sqrt{2} - \sqrt{-2})(\sqrt{8} + \sqrt{-2})$

11. Find all real and complex solutions of the equation $2x^2 + 4x + 3 = 0$.

12. A trucker drove from Amity to Belleville at a speed of 80 km/h. On the way back, the trucker drove at 96 km/h. The total trip took $4\frac{2}{5}$ h of driving time. Find the distance between these two cities.

13. A rectangular parcel of land is 70 m longer than it is wide. Each diagonal between opposite corners is 130 m. What are the dimensions of the parcel?

14. Solve each inequality. Write the answer using interval notation, and sketch the solution on the real number line.
    (a) $-4 < 5 - 3x \le 17$ (b) $x(x - 1)(x + 2) > 0$
    (c) $|x - 4| < 3$ (d) $\dfrac{3 - 2x}{2 - x} < x$

15. A bottle of medicine is to be stored at a temperature between 5°C and 10°C. What range does this correspond to on the Fahrenheit scale? [*Note:* Fahrenheit (F) and Celsius (C) temperatures satisfy the relation $C = \frac{5}{9}(F - 32)$.]

16. For what values of x is the expression $\sqrt{6x - x^2}$ defined as a real number?

17. (a) Plot the points $P(0, 3)$, $Q(3, 0)$, and $R(6, 3)$ in the coordinate plane. Where must the point S be located so that $PQRS$ is a square?

(b) Find the area of $PQRS$.

18. (a) Sketch the graph of $y = 4 - x^2$.

(b) Find the x- and y-intercepts of the graph.

(c) Is the graph symmetric about the x-axis, the y-axis, or the origin?

19. Let $P(-3, 1)$ and $Q(5, 6)$ be two points in the coordinate plane.

(a) Plot P and Q in the coordinate plane.

(b) Find the distance between P and Q.

(c) Find the midpoint of the segment PQ.

(d) Find an equation for the circle for which the segment PQ is a diameter.

20. Find the center and radius of each circle, and sketch its graph.

(a) $x^2 + y^2 = 5$ **(b)** $(x + 1)^2 + (y - 3)^2 = 4$ **(c)** $x^2 + y^2 - 10x + 16 = 0$

21. Write the linear equation $2x - 3y = 15$ in slope-intercept form, and sketch its graph. What are the slope and y-intercept?

22. Find an equation for the line with the given property.

(a) Passes through the points $(6, 7)$ and $(1, -3)$

(b) Passes through the point $(3, -6)$ and is parallel to the line $3x + y - 10 = 0$

(c) Has x-intercept 6 and y-intercept 4

23. A geologist measures the temperature T (in °C) of the soil at various depths below the surface and finds that at a depth of x centimeters, the temperature is given by $T = 0.08x - 4$.

(a) What is the temperature at a depth of 1 m (100 cm)?

(b) Sketch a graph of the linear equation.

(c) What do the slope, the x-intercept, and T-intercept of the graph represent?

24. Solve each equation or inequality graphically. State your answer rounded to two decimal places.

(a) $x^3 - 9x - 1 = 0$ **(b)** $x^2 - 1 \le |x + 1|$

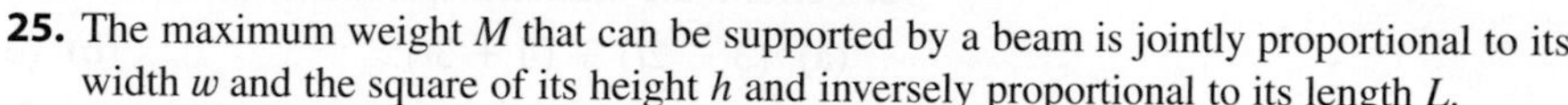

25. The maximum weight M that can be supported by a beam is jointly proportional to its width w and the square of its height h and inversely proportional to its length L.

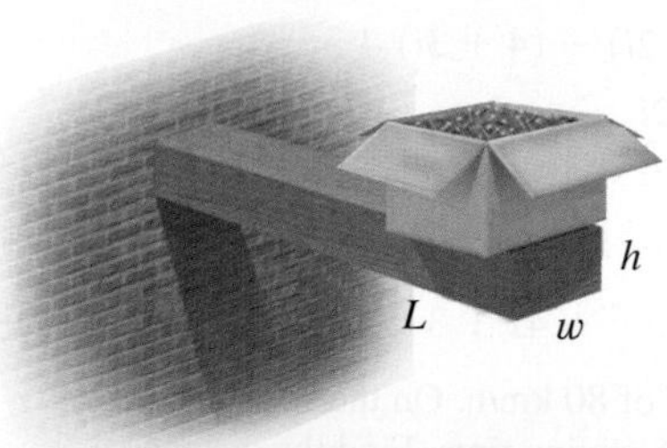

(a) Write an equation that expresses this proportionality.

(b) Determine the constant of proportionality if a beam 10 cm wide, 15 cm high, and 4 m long can support a maximum weight of 21,000 N.

(c) If a 3-m beam made of the same material is 7.5 cm wide and 25 cm high, what is the maximum weight it can support?

If you had difficulty with any of these problems, you may wish to review the section of this chapter indicated below.

Problem	Section	Problem	Section
1	Section 1.1	12, 13	Section 1.7
2, 3, 4	Section 1.2	14, 15, 16	Section 1.8
5, 6	Section 1.3	17, 18, 19, 20	Section 1.9
7, 8	Section 1.4	21, 22, 23	Section 1.10
9	Section 1.5	24	Section 1.11
10, 11	Section 1.6	25	Section 1.12

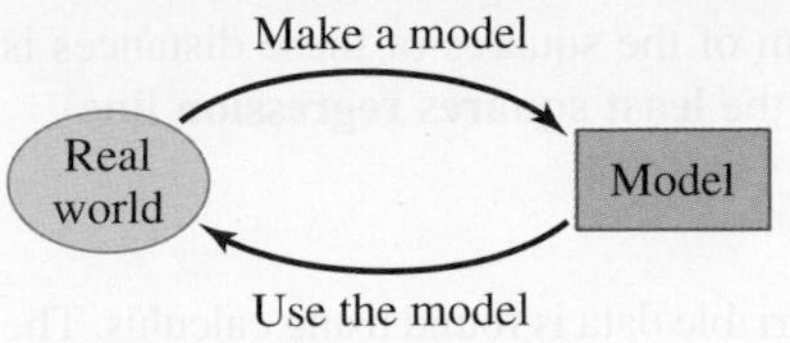

A model is a representation of an object or process. For example, a toy Ferrari is a model of the actual car; a road map is a model of the streets in a city. A **mathematical model** is a mathematical representation (usually an equation) of an object or process. Once a mathematical model has been made, it can be used to obtain useful information or make predictions about the thing being modeled. The process is described in the diagram in the margin. In these *Focus on Modeling* sections we explore different ways in which mathematics is used to model real-world phenomena.

■ The Line That Best Fits the Data

The body-mass index (BMI) is defined as the body mass (weight) divided by the square of the height; the units are kg/m^2.

In Section 1.10 we used linear equations to model relationships between varying quantities. In practice, such relationships are often discovered by collecting and analyzing data. Real-world data seldom fall into a precise line. For example, the two-variable data in Table 1 are from a study on childhood health; the table gives average screen time (hours/day) and body mass index (BMI) for each of several adolescent subjects. A **scatter plot** of the data is shown in Figure 1(b). Of course, we would not expect the data to lie exactly on a line, as shown in Figure 1(a). But the scatter plot in Figure 1(b) shows a linear trend: the more hours of screen time, the correspondingly higher the BMI tends to be. (You can further explore how scatter plots can reveal hidden relationships in data in the *Discovery Project*: *Visualizing Data* at **stewartmath.com**.)

Table 1
Screen time-BMI

x (Hours)	y BMI
0.25	23
0.5	20
0.5	24
1	18
1.5	26
2	22
2.5	24
3	19
3.5	25
4	32
4.5	28
4.5	33
5	29

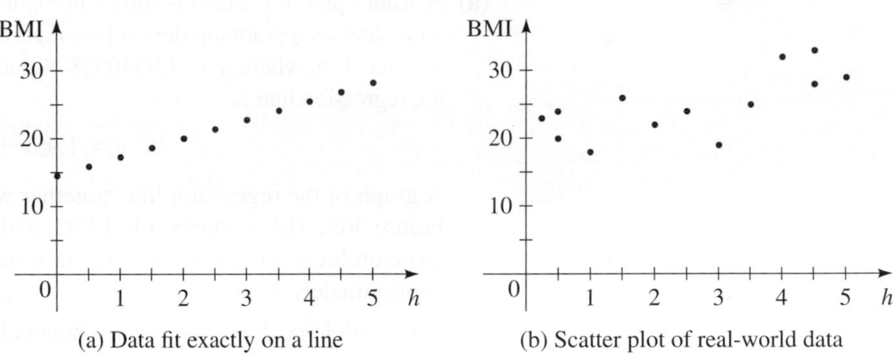

(a) Data fit exactly on a line

(b) Scatter plot of real-world data

Figure 1

The scatter plot in Figure 1(b) shows that the data lie roughly on a straight line. We can try to fit a line visually to approximate the data points, but since the data aren't exactly linear, there are many lines that may seem to fit the data. Figure 2(a) shows two such attempts at "eyeballing" a line to fit the data.

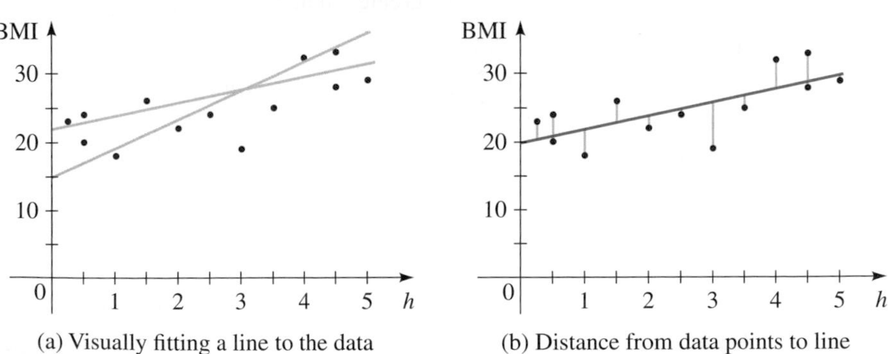

(a) Visually fitting a line to the data

(b) Distance from data points to line

Figure 2

Of all the lines that run through these data points, there is one that "best" fits the data. It seems reasonable that the line that best fits the data is the line that is as close as possible to all the data points. This is the line for which the sum of the vertical distances from the data points to the line is as small as possible, as shown in Figure 2(b). For technical reasons it is better to use the line for which the sum of the squares of these distances is smallest. This line is called the **regression line** (or the **least squares regression line**).

■ Examples of Linear Regression

The formula for the regression line is given in Exercise 9.1.77.

The formula for the regression line for a set of two-variable data is found using calculus. The formula is programmed into most graphing calculators; also, several Internet apps (such as Desmos or GeoGebra) can find the equation of the regression line for a given set of data.

Example 1 ■ Regression Line for the Screen Time-BMI Data

(a) Use a graphing device to draw a scatter plot of the data in Table 1 and find the regression line for the data. Graph the regression line and the scatter plot on the same screen. What does the y-intercept of the regression line represent?

(b) What does the model you found in part (a) predict about the BMI of an adolescent whose average screen time is 6 hours a day? 2.25 hours a day?

(c) Is a linear model reasonable for these data? What, if any, are the limitations of the model?

Solution

(a) A scatter plot of the data is shown in Figure 3(a). Using the `Linear Regression` command on a graphing device [see Figure 3(b)] we get the regression line $y = ax + b$, where $a = 1.954035874$ and $b = 19.92348655$. So, the equation of the regression line is

$$y = 1.95x + 19.92$$

A graph of the regression line, together with the scatter plot, is shown in Figure 3(c). The y-intercept is 19.92, which shows that according to this model, those children who spend no time in front of a screen would have a BMI of approximately 20.

(b) For an adolescent who averages 6 hours of screen time a day, $x = 6$. Substituting 6 for x in the equation of the regression line, we get $y = 1.95(6) + 19.92 \approx 31.6$. So, we would expect such an adolescent to have a BMI of about 32. For an average screen time of 2.25 the model predicts a BMI of $y = 1.95(2.25) + 19.92 \approx 24.3$.

(c) This model has some limitations. To begin, the model is based on too few data points; we need more data to increase the reliability of the model. Also, the model was based on data for screen times between 0 and 5 hours; the model should not be used to make predictions about screen times that are much greater than 5.

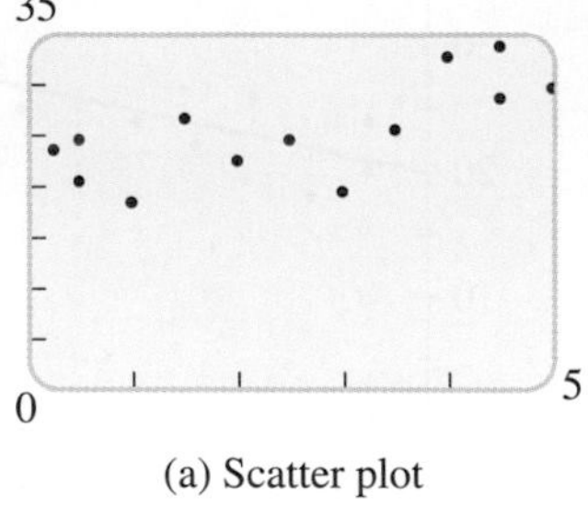

(a) Scatter plot

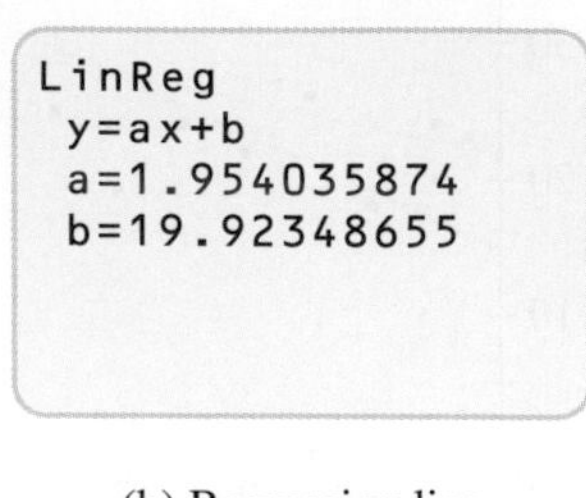

(b) Regression line

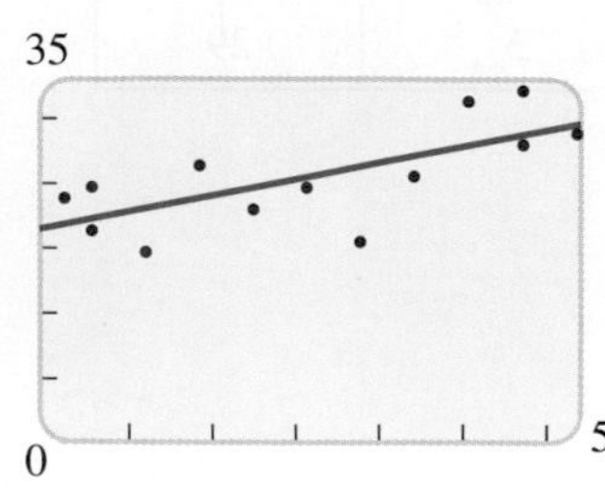

(c) Scatter plot and regression line

Figure 3

■

Extrapolation and **interpolation** refer to different ways we can use a model to predict values that are not included in the data. In Example 1, estimating the BMI associated with 6 hours of screen time is *extrapolation* (because 6 is "extra," or greater than 5, the largest x-value in the data); estimating the BMI associated with 2.25 hours of screen time is *interpolation* (because 2.25 is "inter," or between, data points).

Note that we can fit a line through any set of two-variable data, but a linear model is not always appropriate. In subsequent *Focus on Modeling* sections, we fit different types of curves to model data.

Table 2
Temperature-Energy Consumption

x (°C)	y (MWh/day)	x (°C)	y (MWh/day)
−5	24	8	19
−3	23	8	18
0	23	10	19
0	22	11	17
1	21	12	18
1	22	13	17
3	20	15	19
5	20	16	16
6	21	18	17

LStockStudio/Shutterstock.com

Example 2 ■ Regression Line for Temperature-Energy Consumption Data

The data in Table 2 give several daily high temperatures (in °C) for a city in Austria and the amount of electricity (in MWh/day) the city residents used on each day.

(a) Use a graphing device to draw a scatter plot of the data in Table 2 and find the regression line for the data. Graph the regression line and the scatter plot on the same screen. What does the slope of the regression line tell us about the relationship between temperature and electricity usage?

(b) If a weather forecaster predicts a high temperature of $-6°C$ for a particular day, how much demand for electricity (in MWh/day) should city officials prepare for?

Solution

(a) A scatter plot of the data is shown in Figure 4(a). Using the `Linear Regression` command on a graphing device [see Figure 4(b)], we get the regression line $y = ax + b$, where $a = -0.3262912457$ and $b = 21.93492546$. So, the regression line is approximately

$$y = -0.33x + 21.93$$

A graph of the scatter plot and regression line is shown in Figure 4(c). The negative slope indicates that the warmer the temperature, the less the demand for electricity (because there is less need for heating).

(b) Substituting -6 for x gives $y = -0.33(-6) + 21.93 = 23.91$. So, city officials should prepare for a demand of about 24 MWh for that day.

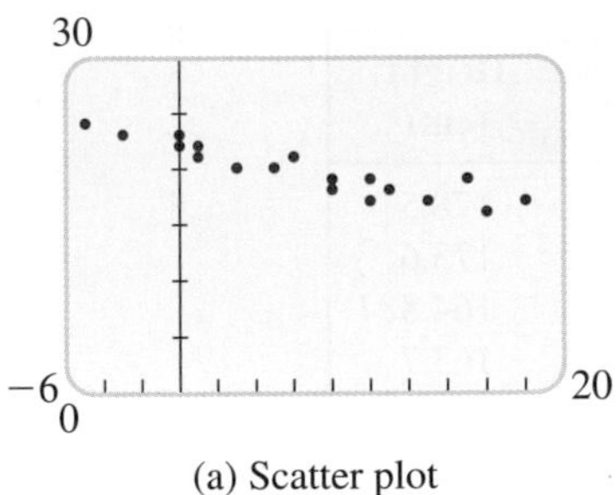

(a) Scatter plot

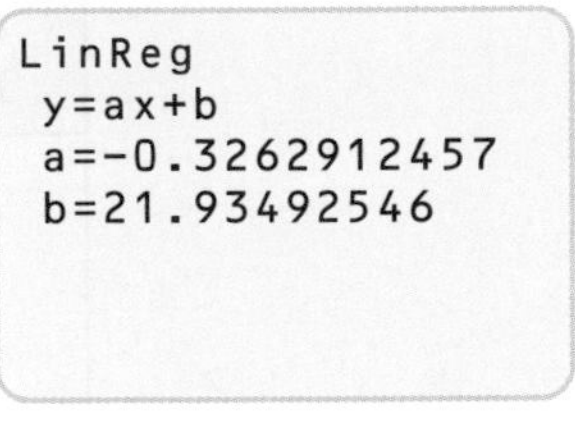

(b) Regression line

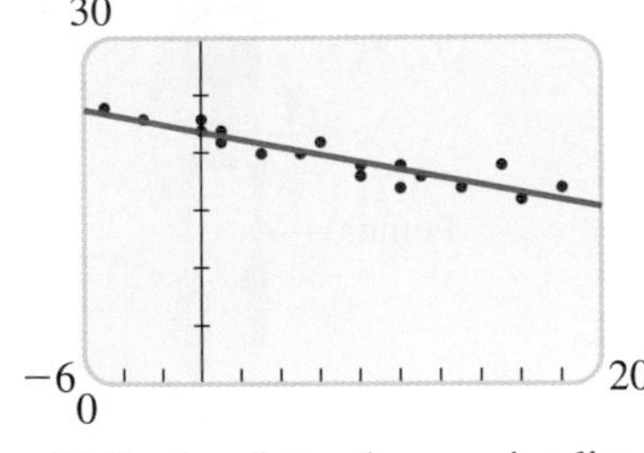

(c) Scatter plot and regression line

Figure 4

■

■ How Good Is the Fit? The Correlation Coefficient

For *any* given set of two-variable data, the regression formula produces a regression line, even if the trend is not linear. So how closely do the data fall along a line? To answer this question, statisticians have invented the **correlation coefficient**, a number

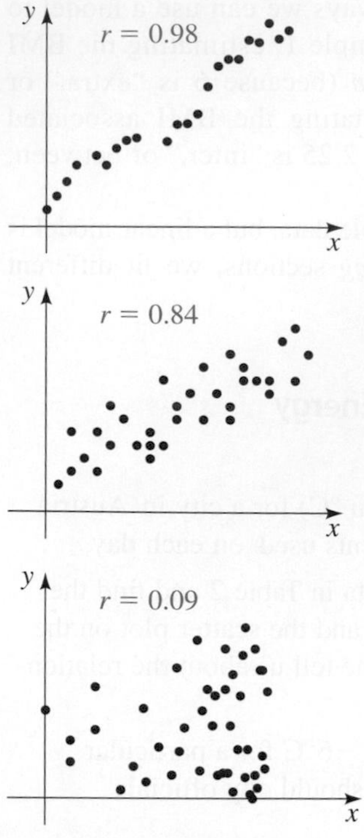

Figure 5

usually denoted by r that satisfies $-1 \le r \le 1$. The correlation coefficient measures how closely the data follow the regression line—in other words, how strongly the variables are *correlated*. Graphing devices usually give the value of r along with the equation of the regression line. If r is close to -1 or 1, then the variables are strongly correlated—that is, the scatter plot follows the regression line closely. If r is close to 0, then the variables are weakly correlated or not correlated at all. (The sign of r depends on the slope of the regression line.) Figure 5 shows the scatter plots of different data sets together with their correlation coefficients.

There are no hard and fast rules for deciding the values of r for which the correlation is "significant." The correlation coefficient as well as the number of data points serve as guides for deciding how well the regression line fits the data.

In Example 1 the correlation coefficient is 0.70, which shows that there is a good correlation between screen time and BMI. Does this mean that more screen time *causes* a higher BMI? In general, if two variables are correlated, it does not necessarily follow that a change in one variable causes a change in the other. The mathematician John Allen Paulos points out that shoe size is strongly correlated to mathematics scores among schoolchildren. Does this mean that big feet cause high math scores? Certainly not—both shoe size and math skills increase independently as children get older. So, it is important not to jump to conclusions: Correlation and causation are not the same thing. You can explore this topic further in the *Discovery Project: Correlation and Causation* at the book companion website **stewartmath.com**.

Correlation is a useful tool in bringing important cause-and-effect relationships to light; but to prove causation, we must explain the mechanism by which one variable affects the other. For example, the link between smoking and lung cancer was observed as a correlation long before science found the mechanism through which smoking can cause lung cancer.

Problems

1. **Femur Length and Height** Anthropologists use a linear model that relates femur length to height. The model allows an anthropologist to determine the height of an individual when only a partial skeleton (including the femur) is found. In this problem we find the model by analyzing the data on femur length and height for the eight males whose data is given in the table.
   (a) Find the regression line for the data in the table, and graph the regression line and scatter plot on the same screen.
   (b) An anthropologist finds a femur of length 58 cm. Use the model you found in part (a) to estimate how tall the person was.

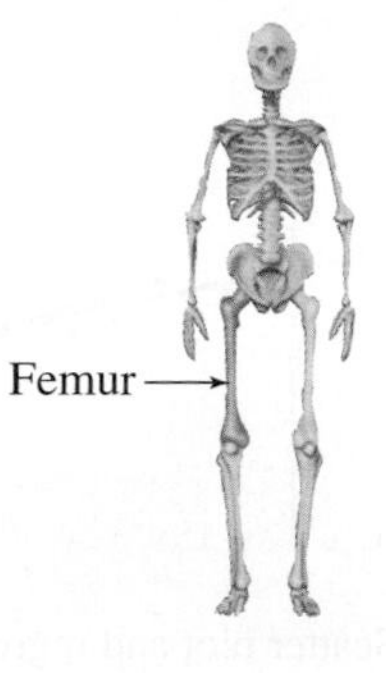

Femur Length (cm)	Height (cm)
50.1	178.5
48.3	173.6
45.2	164.8
44.7	163.7
44.5	168.3
42.7	165.0
39.5	155.4
38.0	155.8

2. **GDP and Carbon Dioxide Emissions** A 2016 study examined the relationship between per capita GDP (gross domestic product) and per capita emissions of greenhouse gases (CO_2) for a one-year period. The table shows data collected in the study for various countries.
   (a) Find the regression line for the data in the table, and graph the regression line and scatter plot on the same screen.

(b) What does the model that you found in part (a) predict about the per capita carbon emissions for a country with a per capita GDP of $80,000? A country with per capita GDP of $32,000?

(c) Does a linear model appear reasonable? What, if any, are the limitations of the model?

GDP per Capita (× $1000)	Carbon Dioxide Emissions (tonnes per capita)	GDP per Capita (× $1000)	Carbon Dioxide Emissions (tonnes per capita)
2	2	19	7
4	1	21	8
5	2	25	8
9	3	30	6
10	2	34	7
11	3	47	11
13	5	49	9
16	6	51	10

Source: The World Bank DataBank

Diameter (cm)	Age (yr)
6.3	15
10	24
15	32
20	56
22.5	49
23.8	76
31.3	90
38.8	89

3. Tree Diameter and Age To estimate ages of trees, forest rangers use a linear model that relates tree diameter to age. The model is useful because tree diameter is much easier to measure than tree age (which requires special tools for extracting a representative cross-section of the tree and counting the rings). To find the model, use the data in the table, which were collected for a certain variety of oaks.

(a) Find the regression line for the data in the table, and graph the regression line and scatter plot on the same screen.

(b) Use the model that you found in part (a) to estimate the age of an oak whose diameter is 45 cm.

Temperature (°C)	Chirping Rate (chirps/min)
9	20
12	46
15	79
18	91
21	113
24	140
27	173
30	198
32	211

4. Temperature and Chirping Crickets Biologists have observed that the chirping rate of crickets of a certain species appears to be related to temperature. The table in the margin shows the chirping rates for various temperatures.

(a) Find the regression line for the data in the table, and graph the regression line and scatter plot on the same screen.

(b) Use the linear model that you found in part (a) to estimate the chirping rate at 35°C.

5. Extent of Arctic Sea Ice The National Snow and Ice Data Center monitors the amount of ice in the Arctic year round. The table below gives approximate values for the sea ice extent in millions of square kilometers from 1994 to 2020, in two-year intervals.

(a) Find the regression line for the data in the table, and graph the regression line and scatter plot on the same screen.

(b) Use the linear model that you found in part (a) to estimate the sea ice extent in 2019. Compare your answer with the actual value of 4.4 that was measured for 2019.

(c) What limitations do you think this model has? Can this model be used to predict sea ice extent for many years in the future?

Year	Ice Extent (million km^2)	Year	Ice Extent (million km^2)
1994	7.2	2008	4.7
1996	7.9	2010	4.9
1998	6.6	2012	3.6
2000	6.3	2014	5.2
2002	6.0	2016	4.5
2004	6.0	2018	4.8
2006	5.9	2020	3.9

Source: National Snow and Ice Data Center

Flow Rate (%)	Mosquito Positive Rate (%)
0	22
10	16
40	12
60	11
90	6
100	2

6. Mosquito Prevalence The table in the margin lists the relative abundance of mosquitoes (as measured by the mosquito positive rate) versus the flow rate (measured as a percentage of maximum flow) of canal networks in Saga City, Japan.

(a) Find the regression line for the data in the table, and graph the regression line and scatter plot on the same screen.

(b) Use the linear model that you found in part (a) to estimate the mosquito positive rate if the canal flow is 70% of maximum.

7. Noise and Intelligibility Audiologists study the intelligibility of spoken sentences under different noise levels. Intelligibility, the MRT score, is measured as the percent of a spoken sentence that the listener can decipher at a certain noise level (in decibels, dB). The table shows the results of one such test.

(a) Find the regression line for the data in the table, and graph the regression line and scatter plot on the same screen.

(b) Find the correlation coefficient. Do you think a linear model is appropriate?

(c) Use the linear model that you found in part (a) to estimate the intelligibility of a sentence at a 94-dB noise level.

Noise Level (dB)	MRT Score (%)
80	99
84	91
88	84
92	70
96	47
100	23
104	11

Would you buy a candy bar from a vending machine if the price were as indicated?

Price	Yes or No
$1.00	
$1.25	
$1.50	
$1.75	
$2.00	
$2.50	
$3.00	

8. Shoe Size and Height Do you think that shoe size and height are correlated? Find out by surveying the shoe sizes and heights of at least ten students in your class. Find the correlation coefficient.

9. Demand for Candy Bars In this problem you will determine a linear demand equation that describes the demand for candy bars in your class. Survey your classmates to determine what price they would be willing to pay for a candy bar. Your survey form might look like the sample shown at the left.

(a) Make a table of the number of respondents who answered "yes" at each price level.

(b) Make a scatter plot of your data.

(c) Find and graph the regression line $y = mp + b$, which gives the number of respondents y who would buy a candy bar if the price were p cents. This is the *demand equation*. Why is the slope m negative?

(d) What is the p-intercept of the demand equation? What does this intercept tell you about pricing candy bars?

Graiki/Moment/Getty Images

2 Functions

A *function* is a rule that describes how one quantity depends on another. Many real-world situations follow precise rules, so they can be modeled by functions. For example, there is a rule that relates the distance skydivers fall to the time they have been falling. So the distance traveled by a skydiver is a *function* of time. Knowing this function allows skydivers to determine when to open their parachutes. In this chapter we study functions and their graphs, as well as many real-world applications of functions. In the *Focus on Modeling* at the end of the chapter we explore different real-world situations that can be modeled by functions.

2.1 Functions

■ Functions All Around Us ■ Definition of Function ■ Evaluating a Function
■ The Domain of a Function ■ Four Ways to Represent a Function

In this section we introduce the concept of a *function* and explore four different ways of describing a function—verbally, numerically, graphically, and algebraically.

■ Functions All Around Us

In nearly every physical phenomenon we observe that one quantity depends on another. For example, your height depends on your age, the temperature depends on the date, the cost of mailing a package depends on its weight (see Figure 1). We use the term *function* to describe this dependence of one quantity on another. That is, we say the following:

- Height is a function of age.
- Temperature is a function of date.
- Cost of mailing a package is a function of weight.

The US Post Office uses a simple rule to determine the cost of mailing a first-class parcel on the basis of its weight. But it's not so easy to describe the rule that relates height to age or the rule that relates temperature to date.

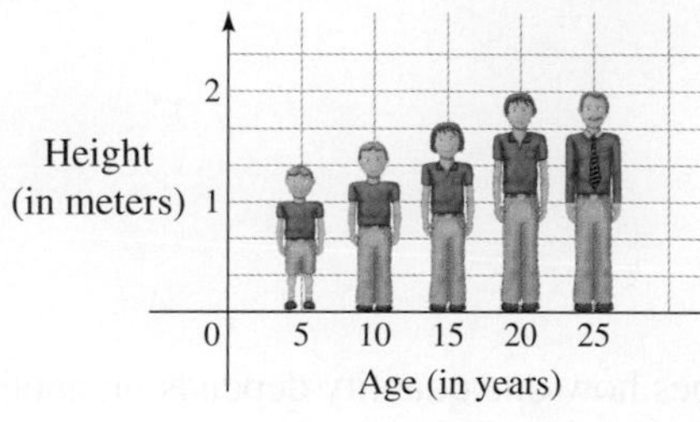

Height is a function of age.

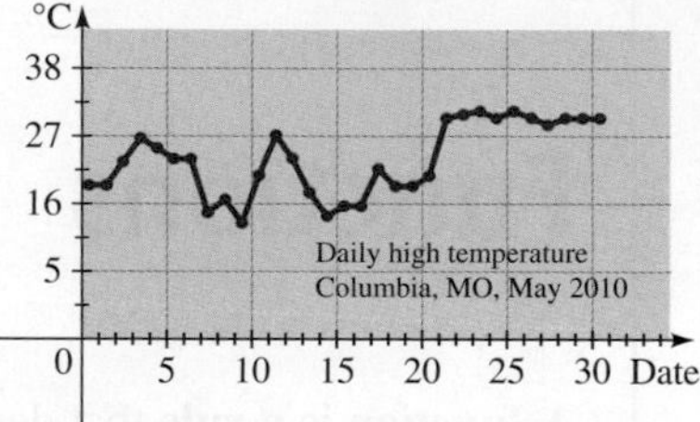

Temperature is a function of date.

w (grams)	2014 Postage (dollars)
$0 < w \le 30$	0.98
$30 < w \le 60$	1.19
$60 < w \le 90$	1.40
$90 < w \le 120$	1.61
$120 < w \le 150$	1.82
$150 < w \le 180$	2.03

Postage is a function of weight.

Figure 1

Can you think of other functions? Here are some more examples:

- The area of a circle is a function of its radius.
- The number of bacteria in a culture is a function of time.
- The weight of an astronaut is a function of elevation.
- The price of a commodity is a function of the demand for that commodity.

The rule that describes how the area A of a circle depends on its radius r is given by the formula $A = \pi r^2$. Even when a precise rule or formula describing a function is not available, we can still describe the function by a graph, as in the next example.

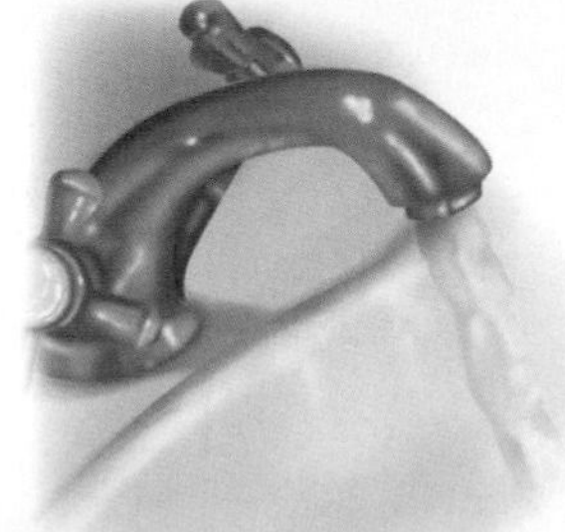

Example 1 ■ Describing a Function by a Graph

When you turn on a hot-water faucet that is connected to a hot-water tank, the temperature T of the water depends on how long the water has been running. So we can say:

- The temperature of water from the faucet is a function of time.

We can sketch a rough graph of the temperature T of the water as a function of the time t that has elapsed since the faucet was turned on, as shown in Figure 2. The graph shows that the initial temperature of the water is close to room temperature. When the water from the hot-water tank reaches the faucet, the water's temperature T increases quickly.

In the next phase, T is constant at the temperature of the water in the tank. When the tank is drained, T decreases to the temperature of the cold-water supply.

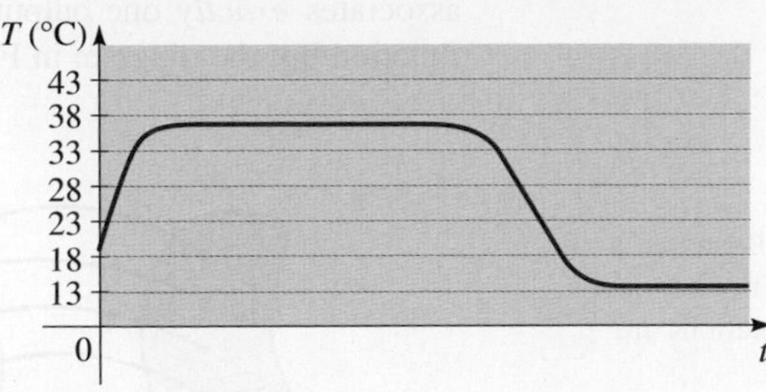

Figure 2 | Graph of water temperature T as a function of time t

Now Try Exercise 95 ■

■ Definition of Function

We have previously used letters to stand for numbers. Here we do something quite different: We use letters to represent *rules*.

A function is a rule. To talk about a function, we need to give it a name. We will use letters such as $f, g, h, \ldots$ to represent functions. For example, we can use the letter f to represent a rule as follows:

"f" is the rule "square the number"

When we write $f(2)$, we mean "apply the rule f to the number 2." Applying the rule gives $f(2) = 2^2 = 4$. Similarly, $f(3) = 3^2 = 9$, $f(4) = 4^2 = 16$, and in general $f(x) = x^2$.

> **Definition of a Function**
>
> A **function** f is a rule that assigns to each element x in a set A exactly one element, called $f(x)$, in a set B.

The square root key $\boxed{\sqrt{\ }}$ on your calculator is a good example of a function as a machine. First you input x into the display. Then you press the key labeled $\boxed{\sqrt{\ }}$. If $x < 0$, then x is not in the domain of this function; that is, x is not an acceptable input, and the calculator will indicate an error. If $x \geq 0$, then an *approximation* to $\sqrt{x}$ appears in the display, correct to a certain number of decimal places. (Thus the $\boxed{\sqrt{\ }}$ key on your calculator is not quite the same as the exact mathematical function f defined by $f(x) = \sqrt{x}$.)

We usually consider functions for which the sets A and B are sets of real numbers. The symbol $f(x)$ is read "f of x" or "f at x" and is called the **value of f at x**, or the **image of x under f**. The set A is called the **domain** of the function. The **range** of f is the subset of B that consists of all possible values of $f(x)$ as x varies throughout the domain, that is,

$$\text{range of } f = \{f(x) \mid x \in A\}$$

The symbol that represents an arbitrary number in the domain of a function f is called an **independent variable**. The symbol that represents a number in the range of f is called a **dependent variable**. So if we write $y = f(x)$, then x is the independent variable and y is the dependent variable.

It is helpful to think of a function as a **machine** (see Figure 3). If x is in the domain of the function f, then when x enters the machine, it is accepted as an **input** and the machine produces an **output** $f(x)$ according to the rule of the function. Thus we can think of the domain as the set of all possible inputs and the range as the set of all possible outputs.

Figure 3 | Machine diagram of f

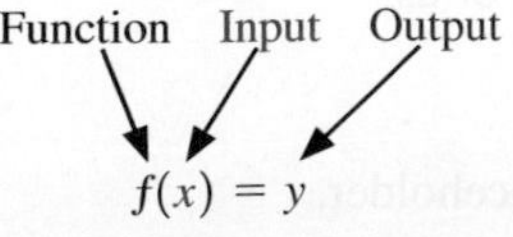

Note The notation $y = f(x)$ is called **function notation**. The letter f is the name of the function (or rule), x is the input, and y is the corresponding output. See the figure in the margin.

Another way to picture a function f is by an **arrow diagram** as in Figure 4(a). Each arrow associates an input from A to the corresponding output in B. Since a function associates *exactly* one output to each input, the diagram in Figure 4(a) represents a function but the diagram in Figure 4(b) does *not* represent a function.

The correspondence illustrated in the diagram in part (b) is not a function, but it is a *relation*. Relations are defined in Section 2.2.

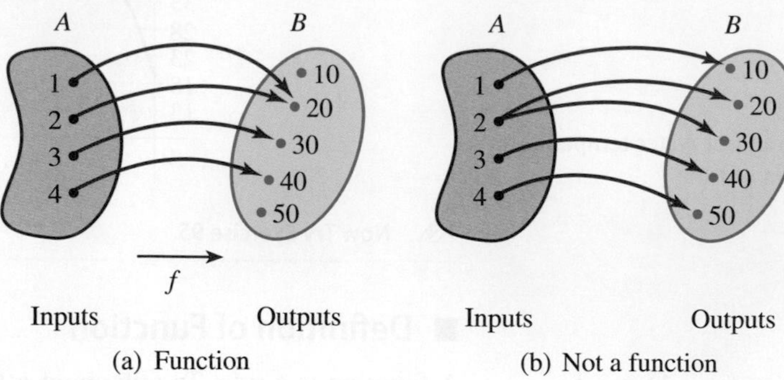

Figure 4 | Arrow diagrams

Example 2 ■ Analyzing a Function

A function f is defined by the formula

$$f(x) = x^2 + 4$$

(a) Express in words how f acts on the input x to produce the output $f(x)$.

(b) Evaluate $f(3)$, $f(-2)$, and $f(\sqrt{5})$.

(c) Find the domain and range of f.

(d) Draw a machine diagram for f.

Solution

(a) The formula tells us that f first squares the input x and then adds 4 to the result. So f is the function

"square, then add 4"

(b) The values of f are found by substituting for x in the formula $f(x) = x^2 + 4$.

$$f(3) = 3^2 + 4 = 13 \qquad \text{Replace } x \text{ by } 3$$

$$f(-2) = (-2)^2 + 4 = 8 \qquad \text{Replace } x \text{ by } -2$$

$$f(\sqrt{5}) = (\sqrt{5})^2 + 4 = 9 \qquad \text{Replace } x \text{ by } \sqrt{5}$$

(c) The domain of f consists of all possible inputs for f. Since we can evaluate the formula $f(x) = x^2 + 4$ for every real number x, the domain of f is the set $\mathbb{R}$ of all real numbers.

The range of f consists of all possible outputs of f. Because $x^2 \geq 0$ for all real numbers x, we have $x^2 + 4 \geq 4$, so for every output of f we have $f(x) \geq 4$. Thus the range of f is $\{y \mid y \geq 4\} = [4, \infty)$.

(d) A machine diagram for f is shown in Figure 5.

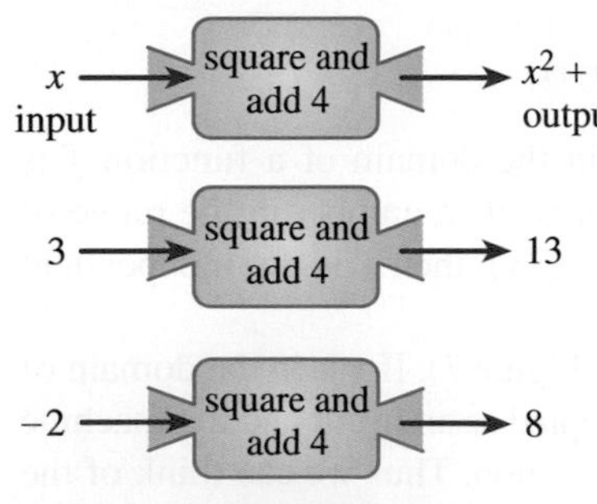

Figure 5 | Machine diagrams

Now Try Exercises 13, 17, 21, and 53 ■

■ Evaluating a Function

In the definition of a function the independent variable x plays the role of a placeholder. For example, the function $f(x) = 3x^2 + x - 5$ can be thought of as

$$f(\;\blacksquare\;) = 3 \cdot \blacksquare^2 + \blacksquare - 5$$

To evaluate f at a number, we substitute the number for the placeholder.

Example 3 ■ Evaluating a Function

Let $f(x) = 3x^2 + x - 5$. Evaluate each function value.

(a) $f(-2)$ **(b)** $f(0)$ **(c)** $f(4)$ **(d)** $f(\frac{1}{2})$

Solution To evaluate f at a number, we substitute the number for x in the definition of f.

(a) $f(-2) = 3 \cdot (-2)^2 + (-2) - 5 = 5$

(b) $f(0) = 3 \cdot 0^2 + 0 - 5 = -5$

(c) $f(4) = 3 \cdot (4)^2 + 4 - 5 = 47$

(d) $f(\frac{1}{2}) = 3 \cdot (\frac{1}{2})^2 + \frac{1}{2} - 5 = -\frac{15}{4}$

 **Now Try Exercise 23** ■

A **piecewise-defined function** (or a **piecewise function**) is a function that is defined by different formulas on different parts of its domain, as in the next example.

Example 4 ■ A Piecewise-Defined Function

A cell phone plan costs \$39 a month. The plan includes 5 gigabytes (GB) of free high-speed data and charges \$15 per gigabyte for any additional high-speed data used. The monthly charges are a function of the number of gigabytes of data used, given by

$$C(x) = \begin{cases} 39 & \text{if } 0 \le x \le 5 \\ 39 + 15(x - 5) & \text{if } x > 5 \end{cases}$$

Find $C(0.5)$, $C(5)$, and $C(8)$.

Solution Remember that a function is a rule. Here is how we apply the rule for this function. First we look at the value of the input, x. If $0 \le x \le 5$, then the value of $C(x)$ is 39. On the other hand, if $x > 5$, then the value of $C(x)$ is $39 + 15(x - 5)$.

Since $0.5 \le 5$, we have $C(0.5) = 39$.

Since $5 \le 5$, we have $C(5) = 39$.

Since $8 > 5$, we have $C(8) = 39 + 15(8 - 5) = 84$.

Thus the plan charges \$39 for 0.5 GB, \$39 for 5 GB, and \$84 for 8 GB.

 **Now Try Exercises 33 and 91** ■

From Examples 3 and 4 we see that the values of a function can change from one input to another. The **net change** in the value of a function f as the input changes from a to b (where $a \le b$) is given by

$$f(b) - f(a)$$

The next example illustrates this concept.

Example 5 ■ Finding Net Change

You can check that the values of the function in Example 5 decrease and then increase between -2 and 2, but the net change from -2 to 2 is 0 because $f(-2)$ and $f(2)$ have the same value.

Let $f(x) = x^2$. Find the net change in the value of f between the given inputs.

(a) From 1 to 3 **(b)** From -2 to 2

Solution

(a) The net change is $f(3) - f(1) = 9 - 1 = 8$.

(b) The net change is $f(2) - f(-2) = 4 - 4 = 0$.

 **Now Try Exercise 41** ■

Example 6 ■ Evaluating a Function

If $f(x) = 2x^2 + 3x - 1$, evaluate the following.

Expressions like the one in part (d) of Example 6 occur frequently in calculus; they are called *difference quotients*, and they represent the average change in the value of f between $x = a$ and $x = a + h$. (See Section 2.4.)

(a) $f(a)$ **(b)** $f(-a)$ **(c)** $f(a + h)$ **(d)** $\dfrac{f(a + h) - f(a)}{h}, \quad h \neq 0$

Solution

(a) $f(a) = 2a^2 + 3a - 1$

(b) $f(-a) = 2(-a)^2 + 3(-a) - 1 = 2a^2 - 3a - 1$

(c)
$$\begin{aligned} f(a + h) &= 2(a + h)^2 + 3(a + h) - 1 \\ &= 2(a^2 + 2ah + h^2) + 3(a + h) - 1 \\ &= 2a^2 + 4ah + 2h^2 + 3a + 3h - 1 \end{aligned}$$

(d) Using the results from parts (c) and (a), we have

$$\begin{aligned} \frac{f(a + h) - f(a)}{h} &= \frac{(2a^2 + 4ah + 2h^2 + 3a + 3h - 1) - (2a^2 + 3a - 1)}{h} \\ &= \frac{4ah + 2h^2 + 3h}{h} = 4a + 2h + 3 \end{aligned}$$

 **Now Try Exercise 45** ■

A **table of values** for a function is a table with two headings, one for inputs and one for the corresponding outputs. A table of values helps us to analyze a function numerically, as shown in the next example.

Example 7 ■ The Weight of an Astronaut

If an astronaut weighs 60 kg on the earth, then the astronaut's weight h kilometers above the surface of the earth is given by the function

$$w(h) = 60\left(\frac{6340}{6340 + h}\right)^2$$

(a) What is the astronaut's weight 160 km above the earth?

(b) Construct a table of values for the function w that gives the astronaut's weight at heights from 0 to 800 km. What do you conclude from the table?

(c) Find the net change in the astronaut's weight from ground level to a height of 800 km.

The weight of an object on or near the earth is the gravitational force that the earth exerts on it. When in orbit around the earth, an astronaut experiences the sensation of "weightlessness" because the centripetal force that keeps the astronaut in orbit is exactly the same as the gravitational pull of the earth.

Solution

(a) We want the value of the function w when $h = 100$; that is, we must calculate $w(100)$:

$$w(160) = 60\left(\frac{6340}{6340 + 160}\right)^2 \approx 57.08$$

So at a height of 160 km the astronaut weighs about 57 kg.

(b) The table gives the astronaut's weight, rounded to the nearest kilogram, at 160-km increments. The values in the table are calculated like the one in part (a).

h	$w(h)$
0	60
160	57
320	54
480	52
640	50
800	47

The table indicates that the astronaut's weight decreases as the height above the surface of the earth increases.

(c) The net change in the astronaut's weight from $h = 0$ to $h = 800$ is

$$w(800) - w(0) = 47 - 60 = -13$$

The negative sign indicates that the astronaut's weight *decreased* by about 13 kg.

 **Now Try Exercise 83** ■

■ The Domain of a Function

Recall that the *domain* of a function is the set of all inputs for the function. The domain of a function may be stated explicitly. For example, if we write

$$f(x) = x^2 \qquad 0 \le x \le 5$$

then the domain is the set of all real numbers x for which $0 \le x \le 5$. If the function is given by an algebraic expression and the domain is not stated explicitly, then by convention *the domain of the function is the domain of the algebraic expression—that is, the set of all real numbers for which the expression is defined as a real number*. For example, consider the functions

Domains of algebraic expressions are discussed in Section 1.4.

$$f(x) = \frac{1}{x - 4} \qquad g(x) = \sqrt{x}$$

The function f is not defined at $x = 4$, so its domain is $\{x \mid x \neq 4\}$. The function g is not defined for negative x, so its domain is $\{x \mid x \ge 0\}$.

Example 8 ■ Finding Domains of Functions

Find the domain of each function.

(a) $f(x) = \dfrac{1}{x^2 - x}$ **(b)** $g(x) = \sqrt{9 - x^2}$ **(c)** $h(t) = \dfrac{t}{\sqrt{t + 1}}$

Solution

(a) A rational expression is not defined when the denominator is 0. Since

$$f(x) = \frac{1}{x^2 - x} = \frac{1}{x(x - 1)}$$

we see that $f(x)$ is not defined when $x = 0$ or $x = 1$. Thus the domain of f is

$$\{x \mid x \neq 0, x \neq 1\}$$

The domain may also be written in interval notation as

$$(-\infty, 0) \cup (0, 1) \cup (1, \infty)$$

(b) We can't take the square root of a negative number, so we must have $9 - x^2 \ge 0$. Using the methods of Section 1.8, we can solve this inequality to find that $-3 \le x \le 3$. Thus the domain of g is

$$\{x \mid -3 \le x \le 3\} = [-3, 3]$$

(c) We can't take the square root of a negative number, and we can't divide by 0, so we must have $t + 1 > 0$, that is, $t > -1$. So the domain of h is

$$\{t \mid t > -1\} = (-1, \infty)$$

Now Try Exercises 59, 67, and 73 ■

■ Four Ways to Represent a Function

We have used machine and arrow diagrams to help us understand what a function is. We can describe a specific function in the following four ways:

- verbally (by a description in words)
- algebraically (by an explicit formula)
- visually (by a graph)
- numerically (by a table of values)

A single function may be represented in all four ways, and it is often useful to go from one representation to another to gain insight into the function. However, certain functions are described more naturally by one method than by the others. We summarize all four methods in the following box.

Four Ways to Represent a Function

Verbal Using words:

"To convert from Celsius to Fahrenheit, multiply the Celsius temperature by $\frac{9}{5}$, then add 32."

Relation between Celsius and Fahrenheit temperature scales

Algebraic Using a formula:

$$A(r) = \pi r^2$$

Area of a circle

Visual Using a graph:

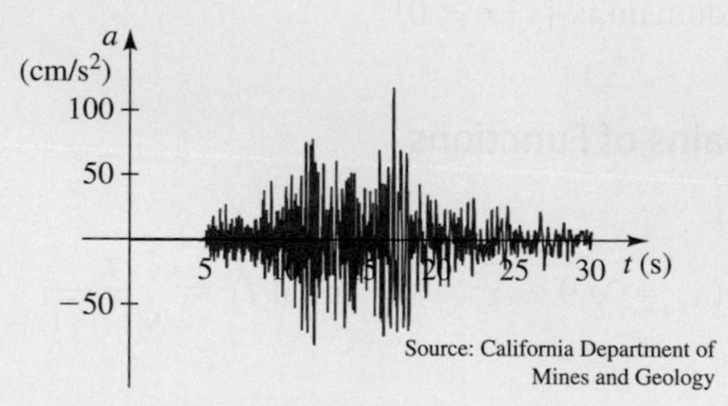

Vertical acceleration during an earthquake

Numerical Using a table of values:

w (grams)	$C(w)$ (dollars)
$0 < w \le 30$	\$0.98
$1 < w \le 60$	\$1.19
$2 < w \le 90$	\$1.40
$3 < w \le 120$	\$1.61
$4 < w \le 150$	\$1.82
⋮	⋮

Cost of mailing a large first-class envelope

Example 9 ■ Representing a Function Verbally, Algebraically, Numerically, and Graphically

Let $F(C)$ be the Fahrenheit temperature corresponding to the Celsius temperature C. (Thus F is the function that converts Celsius inputs to Fahrenheit outputs.) This function is described verbally by "multiply the Celsius temperature by $\frac{9}{5}$, then add 32". Find ways to represent this function in the other three ways:

(a) Algebraically (using a formula)

(b) Numerically (using a table of values)

(c) Visually (using a graph)

Solution

For the function F, the inputs are the temperatures in Celsius and the outputs are the corresponding temperatures in Fahrenheit.

(a) The verbal description tells us that we should first multiply the input C by $\frac{9}{5}$ and then add 32 to the result. So we get

$$F(C) = \tfrac{9}{5}C + 32$$

(b) We use the algebraic formula for F that we found in part (a) to construct a table of values:

C (Celsius)	F (Fahrenheit)	(C, F)
-10	14	$(-10, 14)$
0	32	$(0, 32)$
10	50	$(10, 50)$
20	68	$(20, 68)$
30	86	$(30, 86)$
40	104	$(40, 104)$

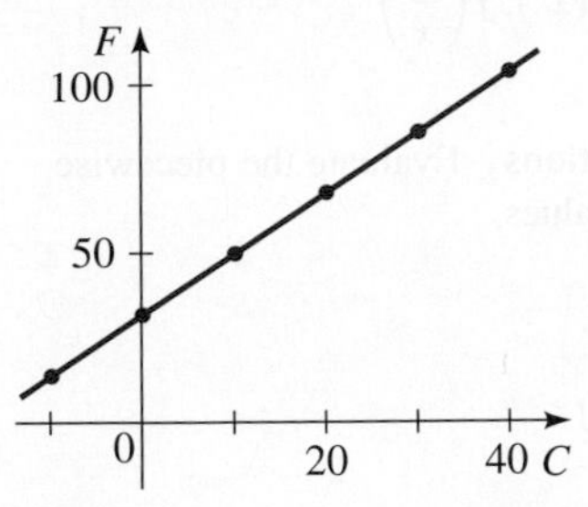

Figure 6 | Celsius and Fahrenheit

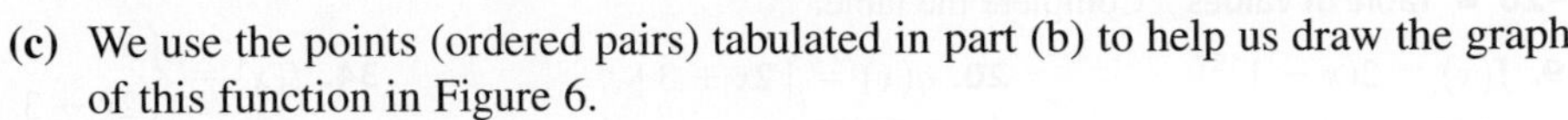

(c) We use the points (ordered pairs) tabulated in part (b) to help us draw the graph of this function in Figure 6.

Now Try Exercise 77 ■

2.1 | Exercises

Concepts

1. If $f(x) = x^3 + 1$, then

(a) the value of f at $x = -1$ is $f(\_\_\_) = \_\_\_\_$.

(b) the value of f at $x = 2$ is $f(\_\_\_) = \_\_\_\_$.

(c) the net change in the value of f between $x = -1$ and $x = 2$ is $f(\_\_\_) - f(\_\_\_) = \_\_\_\_$.

2. For a function f, the set of all possible inputs is called the ____ of f, and the set of all possible outputs is called the ____ of f.

3. (a) Which of the following functions have 5 in their domain?

$$f(x) = x^2 - 3x \qquad g(x) = \frac{x - 5}{x} \qquad h(x) = \sqrt{x - 10}$$

(b) For the functions from part (a) that *do* have 5 in their domain, find the value of the function at 5.

4. A function is given algebraically by the formula $f(x) = (x - 4)^2 + 3$. Complete these other ways to represent f:

(a) *Verbal:* "Subtract 4, then ____ and ____.

(b) *Numerical:*

x	$f(x)$
0	19
2	
4	
6	

5. A function f is a rule that assigns to each element x in a set A exactly ____ element(s) called $f(x)$ in a set B. Which of the following diagrams represents a function?

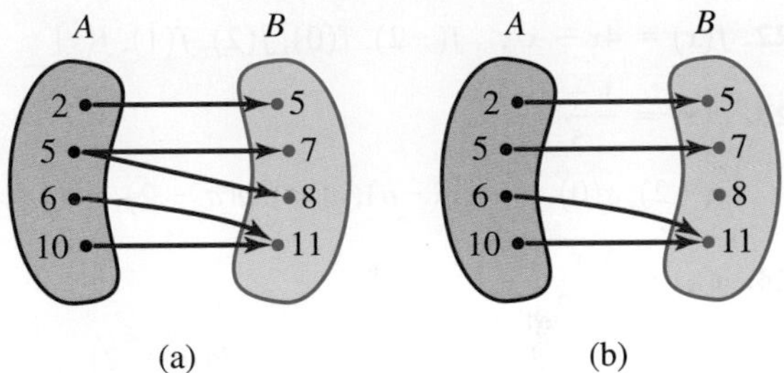

6. A function f is given by a table of values.

(a) From the table, $f(-1) = \_\_\_\_$ and $f(2) = \_\_\_\_$

(b) Can a function have the same output for two different inputs?

x	-2	-1	0	1	2	3
$f(x)$	5	4	-3	2	4	0

7–8 ■ *Yes or No*? If *No*, give a reason. Let f be a function.

7. Is it possible that $f(1) = 5$ and $f(2) = 5$?

8. Is it possible that $f(1) = 5$ and $f(1) = 6$?

Skills

9–12 ■ **Function Notation** Express the rule in function notation. (For example, the rule "square, then subtract 5" is expressed as the function $f(x) = x^2 - 5$.)

9. Multiply by 3, then subtract 5

10. Add 2, then multiply by 5

11. Square, add 1, then take the square root

12. Add 1, take the square root, then divide by 6

13–16 ■ Functions in Words Express the function (or rule) in words.

13. $f(x) = 5x + 1$ **14.** $g(x) = 4(x^2 - 2)$

15. $h(x) = \dfrac{\sqrt{x} - 4}{3}$ **16.** $k(x) = \dfrac{\sqrt{x^2 + 9}}{2}$

17–18 ■ Machine Diagram Draw a machine diagram for the function.

17. $f(x) = \sqrt{x - 1}$ **18.** $f(x) = \dfrac{3}{x - 2}$

19–20 ■ Table of Values Complete the table.

19. $f(x) = 2(x - 1)^2$

x	$f(x)$
−1	
0	
1	
2	
3	

20. $g(x) = |2x + 3|$

x	$g(x)$
−3	
−2	
0	
1	
3	

21–32 ■ Evaluating Functions Evaluate the function at the indicated values.

21. $f(x) = 3x^2 + 1$; $f(-2), f(2), f(0), f(\frac{1}{3}), f(\sqrt{5})$

22. $f(x) = 4x - x^3$; $f(-2), f(0), f(2), f(1), f(\frac{1}{2})$

23. $g(x) = \dfrac{1 - x}{5}$;

$g(-2), g(0), g(2), g(-a), g(x^2), g(a - 2)$

24. $h(x) = \dfrac{\sqrt{x + 3}}{2}$;

$h(-1), h(0), h(1), h(a), h(x - 2), h(a^2 - 2)$

25. $f(x) = x^2 + 2x$;

$f(0), f(3), f(-3), f(a), f(-x), f\left(\dfrac{1}{a}\right)$

26. $h(t) = t + \dfrac{1}{t}$;

$h(-1), h(2), h(\frac{1}{2}), h(x - 1), h\left(\dfrac{1}{x}\right)$

27. $g(x) = \dfrac{1 - x}{1 + x}$;

$g(2), g(-1), g(\frac{1}{2}), g(a), g(a - 1), g(x^2 - 1)$

28. $g(t) = \dfrac{t + 2}{t - 2}$;

$g(-2), g(2), g(0), g(a), g(a^2 - 2), g(a + 1)$

29. $k(x) = 3x^2 - x + 1$;

$k(-1), k(0), k(2), k(\sqrt{5}), k(a - 1), k(x^2)$

30. $k(x) = x^4 - x^3$;

$k(-2), k(-1), k(1), k\left(\dfrac{a}{3}\right), k(a^2), k\left(\dfrac{1}{t}\right)$

31. $f(x) = 2|x - 1|$;

$f(-2), f(0), f(\frac{1}{2}), f(2), f(x + 1), f(x^2 + 2)$

32. $f(x) = \dfrac{|x|}{x}$;

$f(-2), f(-1), f(0), f(5), f(x^2), f\left(\dfrac{1}{x}\right)$

33–36 ■ Piecewise-Defined Functions Evaluate the piecewise defined function at the indicated values.

33. $f(x) = \begin{cases} 3x + 1 & \text{if } x < 5 \\ x^2 - 1 & \text{if } x \geq 5 \end{cases}$

$f(-5), f(0), f(\frac{1}{3}), f(5), f(6)$

34. $f(x) = \begin{cases} 5 & \text{if } x \leq 2 \\ 2x - 3 & \text{if } x > 2 \end{cases}$

$f(-3), f(0), f(2), f(3), f(5)$

35. $f(x) = \begin{cases} x^2 + 2x & \text{if } x \leq -1 \\ x & \text{if } -1 < x \leq 1 \\ -1 & \text{if } x > 1 \end{cases}$

$f(-4), f(-\frac{3}{2}), f(-1), f(0), f(25)$

36. $f(x) = \begin{cases} 3x & \text{if } x < 0 \\ x + 1 & \text{if } 0 \leq x \leq 2 \\ (x - 2)^2 & \text{if } x > 2 \end{cases}$

$f(-5), f(0), f(1), f(2), f(5)$

37–40 ■ Evaluating Functions Use the function to evaluate the indicated expressions and simplify.

37. $f(x) = x^2 + 1$; $f(x + 2), f(x) + f(2)$

38. $f(x) = 3x - 1$; $f(2x), 2f(x)$

39. $f(x) = x + 4$; $f(x^2), (f(x))^2$

40. $f(x) = 6x - 18$; $f\left(\dfrac{x}{3}\right), \dfrac{f(x)}{3}$

41–44 ■ Net Change Find the net change in the value of the function between the given inputs.

41. $f(x) = 3x - 2$; from 1 to 5

42. $f(x) = 4 - 5x$; from 3 to 5

43. $g(t) = 1 - t^2$; from −2 to 5

44. $h(t) = t^2 + 5$; from −3 to 6

45–52 ■ Difference Quotient Find $f(a)$, $f(a + h)$, and the difference quotient $\dfrac{f(a + h) - f(a)}{h}$, where $h \neq 0$.

45. $f(x) = 3 - x$ **46.** $f(x) = x^2 - 4x$

47. $f(x) = 5$ **48.** $f(x) = \dfrac{1}{x + 1}$

49. $f(x) = \dfrac{x}{x + 1}$ **50.** $f(x) = \dfrac{x - 1}{x}$

51. $f(x) = 3 - 5x + 4x^2$ **52.** $f(x) = x^3$

53–58 ■ Domain and Range Find the domain and range of the function.

53. $f(x) = 3x$

54. $f(x) = 5x^2 + 4$

55. $f(x) = |x| + 3$

56. $f(x) = 2 + \sqrt{x - 1}$

57. $f(x) = 3x, \quad -2 \le x \le 6$

58. $f(x) = 5x^2 + 4, \quad 0 \le x \le 2$

59–76 ■ Domain Find the domain of the function.

59. $f(x) = \dfrac{2}{3 + x}$

60. $f(x) = \dfrac{x}{4 - x}$

61. $f(x) = \dfrac{x + 2}{x^2 - 1}$

62. $f(x) = \dfrac{x^4}{x^2 + x - 6}$

63. $f(t) = \sqrt{2 - t}$

64. $g(t) = \sqrt{t^2 + 9}$

65. $f(t) = \sqrt[3]{2t + 5}$

66. $g(x) = \sqrt{7 - 3x}$

67. $f(t) = \sqrt{t^2 - 25}$

68. $g(t) = \sqrt{36 - t^2}$

69. $g(x) = \dfrac{\sqrt{2 + x}}{3 - x}$

70. $g(x) = \dfrac{\sqrt{x}}{2x^2 + x - 1}$

71. $g(x) = \sqrt[4]{x^2 - 6x}$

72. $g(x) = \sqrt{x^2 - 2x - 8}$

73. $f(x) = \dfrac{4}{\sqrt{2 - x}}$

74. $g(x) = \dfrac{3x}{\sqrt{x + 2}}$

75. $f(x) = \dfrac{(x + 1)^2}{\sqrt{2x - 1}}$

76. $f(x) = \dfrac{x}{\sqrt[4]{9 - x^2}}$

77–80 ■ Four Ways to Represent a Function A verbal description of a function is given. Find **(a)** algebraic, **(b)** numerical, and **(c)** graphical representations for the function.

77. To evaluate $f(x)$, square the input and add 1 to the result.

78. To evaluate $f(x)$, add 2 to the input and square the result.

79. Let $T(x)$ be the amount of sales tax charged in Lemon County on a purchase of x dollars. To find the tax, take 8% of the purchase price.

80. Let $V(d)$ be the volume of a sphere of diameter d. To find the volume, take the cube of the diameter, then multiply by π and divide by 6.

Skills Plus

81–82 ■ Domain and Range Find the domain and range of f.

81. $f(x) = \begin{cases} 1 & \text{if } x \text{ is rational} \\ 5 & \text{if } x \text{ is irrational} \end{cases}$

82. $f(x) = \begin{cases} 1 & \text{if } x \text{ is rational} \\ 5x & \text{if } x \text{ is irrational} \end{cases}$

Applications

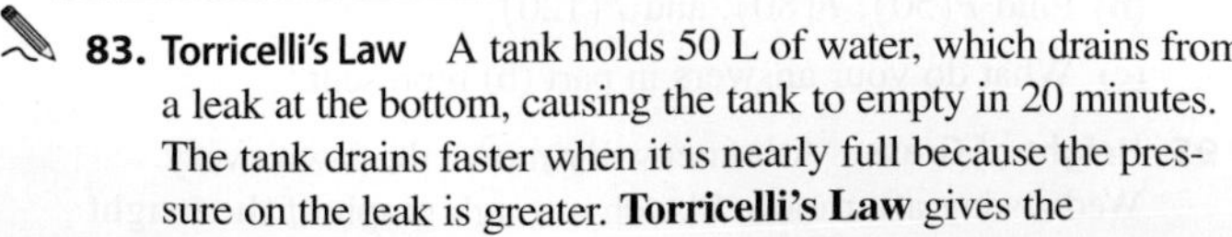

83. Torricelli's Law A tank holds 50 L of water, which drains from a leak at the bottom, causing the tank to empty in 20 minutes. The tank drains faster when it is nearly full because the pressure on the leak is greater. **Torricelli's Law** gives the volume of water remaining in the tank after t minutes as

$$V(t) = 50\left(1 - \frac{t}{20}\right)^2 \qquad 0 \le t \le 20$$

(a) Find $V(0)$ and $V(20)$.

(b) What do your answers to part (a) represent?

(c) Make a table of values of $V(t)$ for $t = 0, 5, 10, 15, 20$.

(d) Find the net change in the volume V as t changes from 0 min to 20 min.

84. Area of a Sphere The surface area S of a sphere is a function of its radius r given by

$$S(r) = 4\pi r^2$$

(a) Find $S(2)$ and $S(3)$.

(b) What do your answers in part (a) represent?

85. Relativity According to the Theory of Relativity, the length L of an object is a function of its velocity v with respect to an observer. For an object whose length at rest is 10 meters, the function is given by

$$L(v) = 10\sqrt{1 - \frac{v^2}{c^2}}$$

where c is the speed of light (300,000 km/s).

(a) Find $L(0.5c)$, $L(0.75c)$, and $L(0.9c)$.

(b) How does the length of an object change as its velocity increases?

86. Blackbody Radiation A *blackbody* is an ideal object that absorbs all electromagnetic radiation. A blackbody with temperature above 0 K radiates heat. The radiance L emitted by a blackbody is a function of the temperature T (in kelvins) given by

$$L(T) = \frac{a}{\pi}T^4$$

where $a = 5.67 \times 10^{-8}\,\mathrm{W \cdot m^{-2} \cdot K^{-4}}$.

(a) Find $L(300)$, $L(350)$, and $L(1000)$.

(b) How does the radiance change as temperature increases?

(c) The sun is well approximated by a blackbody. Find the radiance L of the sun given that its surface temperature is about 5778 K.

87. Pupil Size When the brightness x of a light source is increased, the eye reacts by decreasing the radius R of the pupil. The dependence of R on x is given by the function

$$R(x) = \sqrt{\frac{13 + 7x^{0.4}}{1 + 4x^{0.4}}}$$

where R is measured in millimeters and x is measured in appropriate units of brightness.

(a) Find $R(1)$, $R(10)$, and $R(100)$.

(b) Make a table of values of $R(x)$.

(c) Find the net change in the radius R as x changes from 10 to 100.

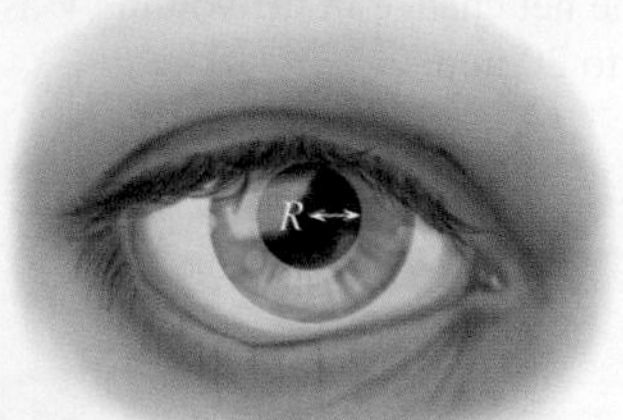

88. Blood Flow As blood moves through a vein or an artery, its velocity v is greatest along the central axis and decreases as the distance r from the central axis increases (see the figure). The formula that gives v as a function of r is called the **law of laminar flow**. For an artery with radius 0.5 cm, the relationship between v (in cm/s) and r (in cm) is given by the function

$$v(r) = 18{,}500(0.25 - r^2) \qquad 0 \le r \le 0.5$$

(a) Find $v(0.1)$ and $v(0.4)$.

(b) What do your answers to part (a) tell you about the flow of blood in this artery?

(c) Make a table of values of $v(r)$ for $r = 0, 0.1, 0.2, 0.3, 0.4, 0.5$.

(d) Find the net change in the velocity v as r changes from 0.1 cm to 0.5 cm.

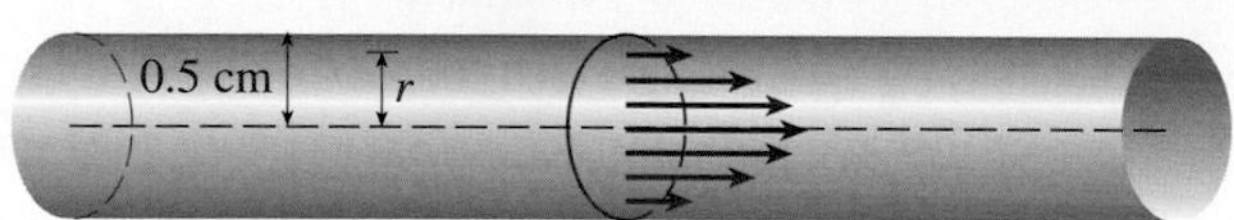

89. How Far Can You See? Because of the curvature of the earth, the maximum distance D that you can see from the top of a tall building or from an airplane at height h is given by the function

$$D(h) = \sqrt{2rh + h^2}$$

where $r = 6340$ km is the radius of the earth and D and h are measured in kilometers.

(a) Find $D(0.1)$ and $D(0.2)$.

(b) How far can you see from the observation deck of Toronto's CN Tower, 346 m above the ground?

(c) Commercial aircraft fly at an altitude of about 11 km. How far can the pilot see?

(d) Find the net change in the value of distance D as h changes from 346 m to 11 km.

90. Population Growth The population P (in millions) of the state of Arizona is a function of the year t. The table gives the population at ten-year intervals from 1960 to 2020.

(a) Find $P(1960)$, $P(1980)$, and $P(2020)$.

(b) Find the net change in the population from 1960 to 1980 and from 1980 to 2020.

t	1960	1970	1980	1990	2000	2010	2020
P	1.3	1.8	2.7	3.7	5.1	6.4	7.2

Source: US Census Bureau

91. Income Tax In a certain country, income tax T is assessed according to the following function of income x:

$$T(x) = \begin{cases} 0 & \text{if } 0 \le x \le 10{,}000 \\ 0.08(x - 10{,}000) & \text{if } 10{,}000 < x \le 20{,}000 \\ 800 + 0.15(x - 20{,}000) & \text{if } 20{,}000 < x \end{cases}$$

(a) Find $T(5{,}000)$, $T(12{,}000)$, and $T(25{,}000)$.

(b) What do your answers in part (a) represent?

92. Internet Purchases An online bookstore charges \$9 shipping for orders under \$50 but provides free shipping for orders of \$50 or more. The cost C of an order is a function of the total price x of the books purchased, given by

$$C(x) = \begin{cases} x + 9 & \text{if } x < 50 \\ x & \text{if } x \ge 50 \end{cases}$$

(a) Find $C(25)$, $C(45)$, $C(50)$, and $C(65)$.

(b) What do your answers in part (a) represent?

93. Cost of a Hotel Stay A hotel chain charges \$114 each night for the first two nights and \$99 for each additional night's stay. The total cost T is a function of the number of nights x that a guest stays.

(a) Complete the expressions in the following piecewise-defined function.

$$T(x) = \begin{cases} \rule{2em}{0.8em} & \text{if } 0 \le x \le 2 \\ \rule{2em}{0.8em} & \text{if } x > 2 \end{cases}$$

(b) Find $T(2)$, $T(3)$, and $T(5)$.

(c) What do your answers in part (b) represent?

94. Speeding Tickets In a certain state the maximum speed permitted on freeways is 105 km/h, and the minimum is 65 km/h. The fine F for violating these limits is \$15 for every kilometer above the maximum or below the minimum.

(a) Complete the expressions in the following piecewise-defined function, where x is the speed at which you are driving.

$$F(x) = \begin{cases} \rule{2em}{0.8em} & \text{if } 0 < x < 65 \\ \rule{2em}{0.8em} & \text{if } 65 \le x \le 105 \\ \rule{2em}{0.8em} & \text{if } x > 105 \end{cases}$$

(b) Find $F(50)$, $F(80)$, and $F(120)$.

(c) What do your answers in part (b) represent?

95. Height of Grass A homeowner mows the lawn every Wednesday afternoon. Sketch a rough graph of the height of the grass as a function of time over the course of a four-week period beginning on a Sunday.

96. Temperature Change You place a frozen pie in an oven and bake it for an hour. Then you take the pie out and let it cool before eating it. Sketch a rough graph of the temperature of the pie as a function of time.

97. Outdoor Temperature Sketch a rough graph of the outdoor temperature as a function of time during a typical spring day.

98. Price of a Car Sketch a rough graph of the market value of a car as a function of the number of years since the car was purchased, over a period of 40 years. Assume the car is well maintained. How would your graph change if the car becomes a collectible antique?

■ Discuss ■ Discover ■ Prove ■ Write

99. Discuss: Examples of Functions At the beginning of this section we discussed three examples of everyday, ordinary functions: Height is a function of age, temperature is a function of date, and postage cost is a function of weight. Give three other examples of functions from everyday life.

100. Discuss: Four Ways to Represent a Function Think of a function that can be represented in all four ways described in this section, and give the four representations.

101. Discuss: Piecewise Defined Functions In Exercises 91–94 we worked with real-world situations modeled by piecewise defined functions. Find other examples of real-world situations that can be modeled by piecewise defined functions, and express the models in function notation.

2.2 Graphs of Functions

■ Graphing Functions by Plotting Points ■ Graphing Functions with Graphing Devices ■ Graphing Piecewise-Defined Functions ■ Which Graphs Represent Functions? The Vertical Line Test ■ Which Equations Represent Functions? ■ Which Relations Represent Functions?

In Section 2.1 we explored how a function can be represented by a graph. In this section we investigate in more detail the concept of the graph of a function.

■ Graphing Functions by Plotting Points

To graph a function f, we plot the points $(x, f(x))$ in a coordinate plane. In other words, we plot the points (x, y) whose x-coordinate is an input and whose y-coordinate is the corresponding output of the function.

> **The Graph of a Function**
>
> If f is a function with domain A, then the **graph** of f is the set of ordered pairs
>
> $$\{(x, f(x)) \mid x \in A\}$$
>
> plotted in a coordinate plane. In other words, the graph of f is the set of all points (x, y) such that $y = f(x)$; that is, the graph of f is the graph of the equation $y = f(x)$.

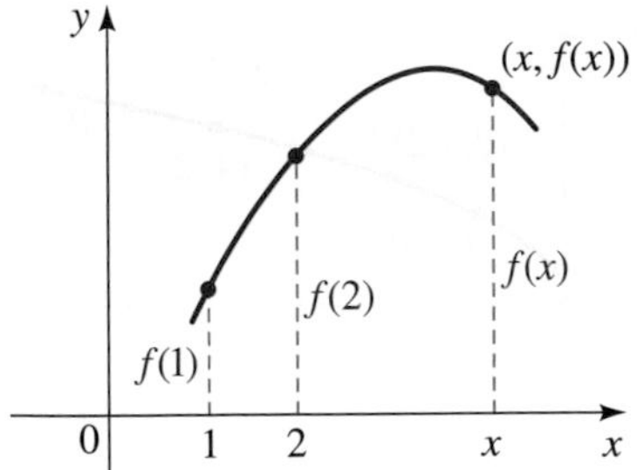

Figure 1 | The height of the graph above x is the value of $f(x)$.

The graph of a function f gives a picture of the behavior or "life history" of the function. We can read the value of $f(x)$ from the graph as being the height of the graph above x (see Figure 1).

Equations of lines are studied in Section 1.10.

A function f of the form $f(x) = mx + b$ is called a **linear function** because its graph is the graph of the equation $y = mx + b$, which represents a line with slope m and y-intercept b. A special case of a linear function occurs when the slope is $m = 0$. The function $f(x) = b$, where b is a given number, is called a **constant function** because all its values are the same number, namely, b. Its graph is the horizontal line $y = b$. Figure 2 shows the graphs of the constant function $f(x) = 3$ and the linear function $f(x) = 2x + 1$.

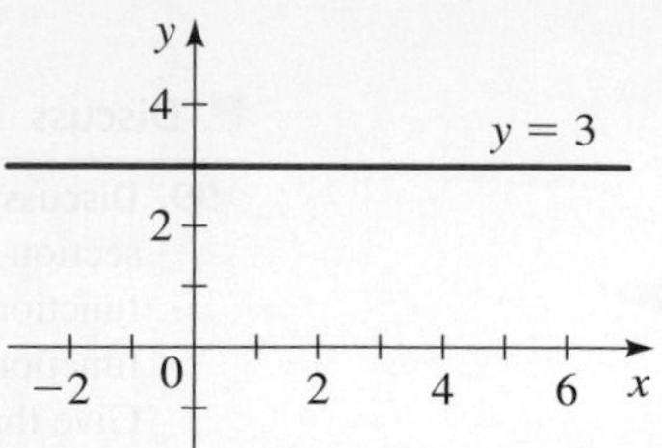

Figure 2 The constant function $f(x) = 3$ The linear function $f(x) = 2x + 1$

Functions of the form $f(x) = x^n$ are called **power functions**, and functions of the form $f(x) = x^{1/n}$ are called **root functions**. In the next example we graph two power functions and a root function.

Example 1 ■ Graphing Functions by Plotting Points

Sketch a graph of each of the following functions.

(a) $f(x) = x^2$ **(b)** $g(x) = x^3$ **(c)** $h(x) = \sqrt{x}$

Solution The graphs of these functions are the graphs of the equations $y = x^2$, $y = x^3$, and $y = \sqrt{x}$. To graph each of these equations, we make a table of values, plot the points corresponding to the ordered pairs in the table, and then join them by a smooth curve. The graphs are sketched in Figure 3.

x	$y = x^2$	(x, y)
-2	4	$(-2, 4)$
-1	1	$(-1, 1)$
$-\frac{1}{2}$	$\frac{1}{4}$	$(-\frac{1}{2}, \frac{1}{4})$
0	0	$(0, 0)$
$\frac{1}{2}$	$\frac{1}{4}$	$(\frac{1}{2}, \frac{1}{4})$
1	1	$(1, 1)$
2	4	$(2, 4)$

x	$y = x^3$	(x, y)
-2	-8	$(-2, -8)$
-1	-1	$(-1, -1)$
$-\frac{1}{2}$	$-\frac{1}{8}$	$(-\frac{1}{2}, -\frac{1}{8})$
0	0	$(0, 0)$
$\frac{1}{2}$	$\frac{1}{8}$	$(\frac{1}{2}, \frac{1}{8})$
1	1	$(1,1)$
2	8	$(2, 8)$

x	$y = \sqrt{x}$	(x, y)
0	0	$(0, 0)$
1	1	$(1, 1)$
2	$\sqrt{2}$	$(2, \sqrt{2})$
3	$\sqrt{3}$	$(3, \sqrt{3})$
4	2	$(4, 2)$
5	$\sqrt{5}$	$(5, \sqrt{5})$
6	$\sqrt{6}$	$(6, \sqrt{6})$

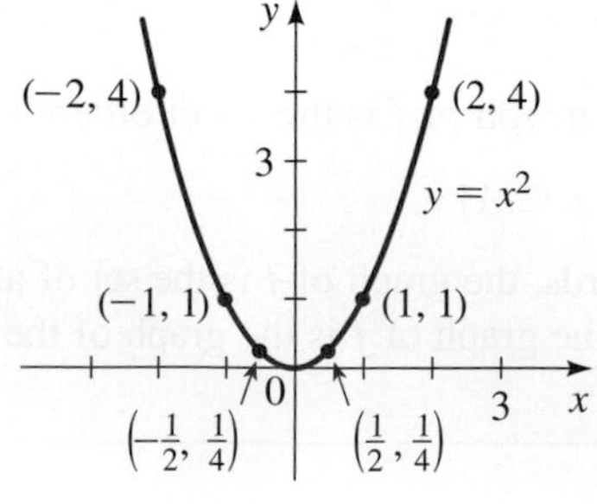

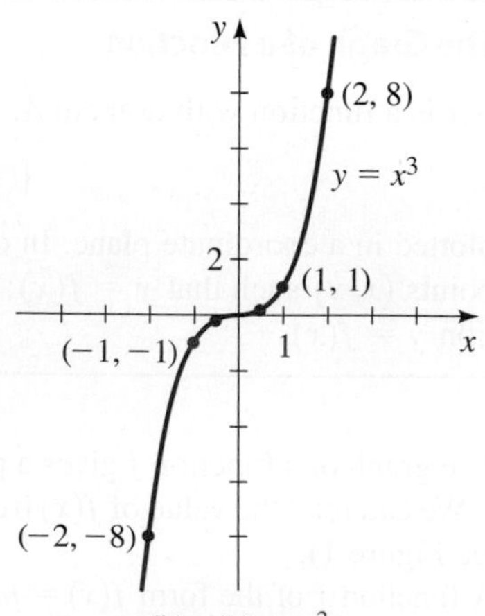

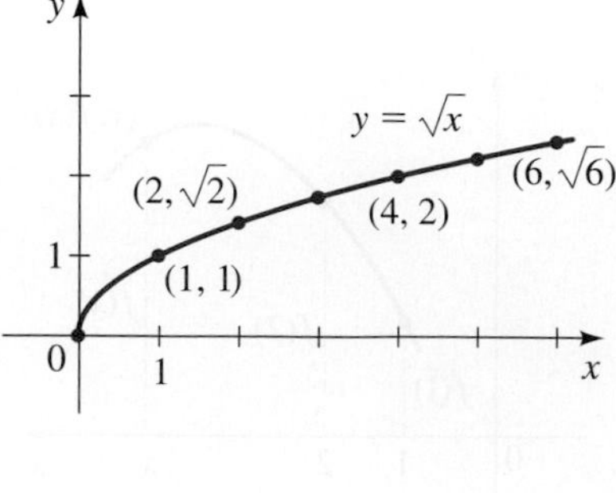

Figure 3 (a) $f(x) = x^2$ (b) $g(x) = x^3$ (c) $h(x) = \sqrt{x}$

Now Try Exercises 13, 19, and 23 ■

■ Graphing Functions with Graphing Devices

Graphing devices and viewing rectangles are introduced in Section 1.11. Graphing devices include graphing calculators as well as math apps for computers and smartphones. Familiarize yourself with the operation of the device you are using. If you are using a graphing calculator see Appendix C, *Graphing with a Graphing Calculator,* or Appendix D, *Using the TI-83/84 Graphing Calculator.* Go to **www.stewartmath.com.**

A convenient way to graph a function is to use a graphing device. To graph the function f, we use a device to graph the equation $y = f(x)$.

Example 2 ■ Graphing a Function with a Graphing Device

Use a graphing device to graph the function $f(x) = x^3 - 8x^2$ in an appropriate viewing rectangle.

Solution To graph the function $f(x) = x^3 - 8x^2$, we must graph the equation $y = x^3 - 8x^2$. A graphing device displays the graph in a default viewing rectangle, such as the one shown in Figure 4(a). But this graph appears to spill over the top and bottom of the screen. We need to expand the vertical axis to get a better representation of the graph. The viewing rectangle $[-4, 10]$ by $[-100, 100]$ gives a more complete picture of the graph, as shown in Figure 4(b).

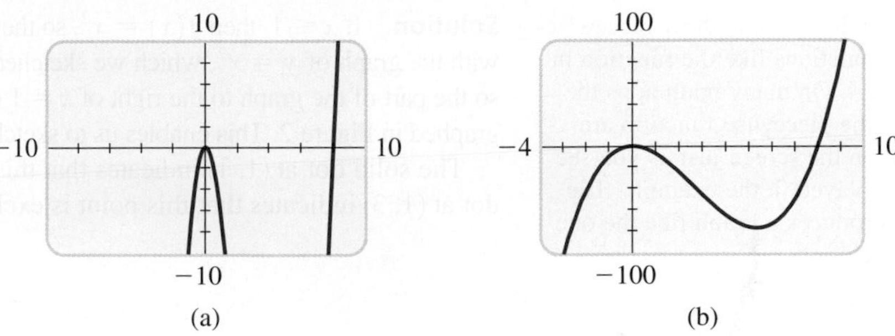

Figure 4 | Graphing the function $f(x) = x^3 - 8x^2$

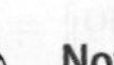 **Now Try Exercise 33** ■

Example 3 ■ A Family of Power Functions

(a) Graph the functions $f(x) = x^n$ for $n = 2, 4,$ and 6 in the viewing rectangle $[-2, 2]$ by $[-1, 3]$.

(b) Graph the functions $f(x) = x^n$ for $n = 1, 3,$ and 5 in the viewing rectangle $[-2, 2]$ by $[-2, 2]$.

(c) What conclusions can you make from these graphs?

Solution To graph the function $f(x) = x^n$, we graph the equation $y = x^n$. The graphs for parts (a) and (b) are shown in Figure 5.

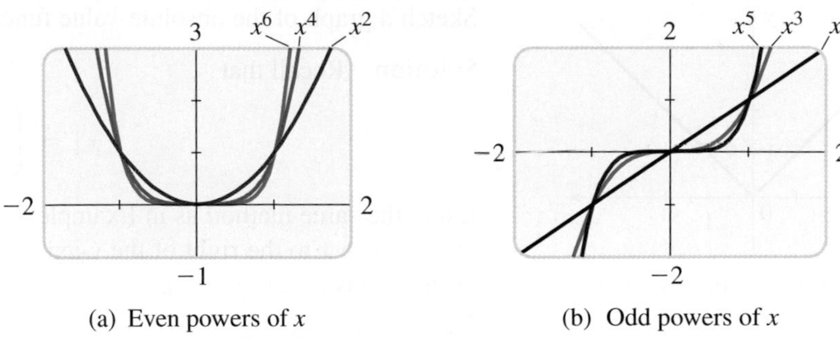

Figure 5 | A family of power functions: $f(x) = x^n$

(c) We see that the general shape of the graph of $f(x) = x^n$ depends on whether n is even or odd.

If n is even, the graph of $f(x) = x^n$ is similar to the parabola $y = x^2$.

If n is odd, the graph of $f(x) = x^n$ is similar to that of $y = x^3$.

Now Try Exercise 71 ■

Notice from Figure 5 that as n increases, the graph of $y = x^n$ becomes flatter near 0 and steeper when $x > 1$. When $0 < x < 1$, the lower powers of x are the "bigger" functions. But when $x > 1$, the higher powers of x are the dominant functions.

■ Graphing Piecewise-Defined Functions

A piecewise-defined function is defined by different formulas on different parts of its domain. As you might expect, the graph of such a function consists of separate pieces.

Example 4 ■ Graph of a Piecewise-Defined Function

Sketch the graph of the function

$$f(x) = \begin{cases} x^2 & \text{if } x \leq 1 \\ 2x + 1 & \text{if } x > 1 \end{cases}$$

Solution If $x \leq 1$, then $f(x) = x^2$, so the part of the graph to the left of $x = 1$ coincides with the graph of $y = x^2$, which we sketched in Figure 3. If $x > 1$, then $f(x) = 2x + 1$, so the part of the graph to the right of $x = 1$ coincides with the line $y = 2x + 1$, which we graphed in Figure 2. This enables us to sketch the graph in Figure 6.

The solid dot at $(1, 1)$ indicates that this point is included in the graph; the open dot at $(1, 3)$ indicates that this point is excluded from the graph.

Graphing devices can draw piecewise-defined functions like the function in Example 4. On many math apps the parts of the piecewise function are entered on the screen just as you see them displayed in the example. The device produces a graph like the one shown below.

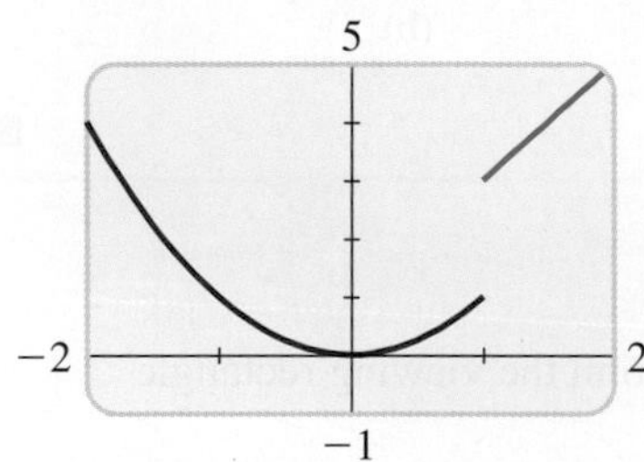

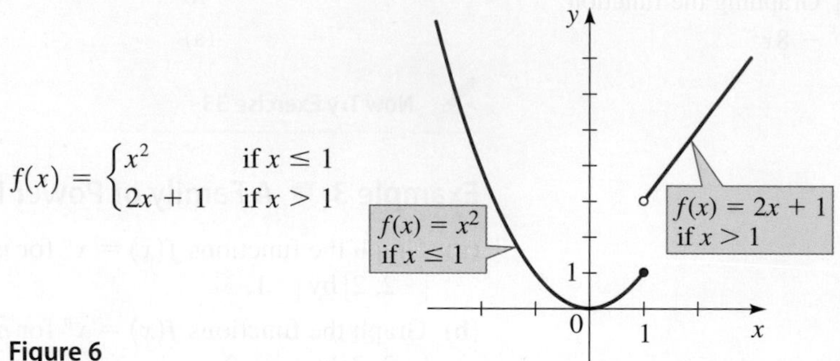

Figure 6

Now Try Exercise 37 ■

Example 5 ■ Graph of the Absolute-Value Function

Sketch a graph of the absolute-value function $f(x) = |x|$.

Solution Recall that

$$|x| = \begin{cases} x & \text{if } x \geq 0 \\ -x & \text{if } x < 0 \end{cases}$$

Using the same method as in Example 4, we note that the graph of f coincides with the line $y = x$ to the right of the y-axis and coincides with the line $y = -x$ to the left of the y-axis (see Figure 7).

Figure 7 | Graph of $f(x) = |x|$

 Now Try Exercise 27 ■

The greatest integer function $[\![x]\!]$ is also called the **floor function** and denoted by $\lfloor x \rfloor$.

The **greatest integer function** is defined by

$$[\![x]\!] = \text{greatest integer less than or equal to } x$$

For example, $[\![2]\!] = 2$, $[\![2.3]\!] = 2$, $[\![1.999]\!] = 1$, $[\![0.002]\!] = 0$, $[\![-3.5]\!] = -4$, and $[\![-0.5]\!] = -1$.

Example 6 ■ Graph of the Greatest Integer Function

Sketch a graph of $f(x) = [\![x]\!]$.

Solution The table shows the values of f for some values of x. Note that $f(x)$ is constant between consecutive integers, so the graph between integers is a horizontal line segment, as shown in Figure 8.

x	$[\![x]\!]$
$\vdots$	$\vdots$
$-2 \le x < -1$	-2
$-1 \le x < 0$	-1
$0 \le x < 1$	0
$1 \le x < 2$	1
$2 \le x < 3$	2
$\vdots$	$\vdots$

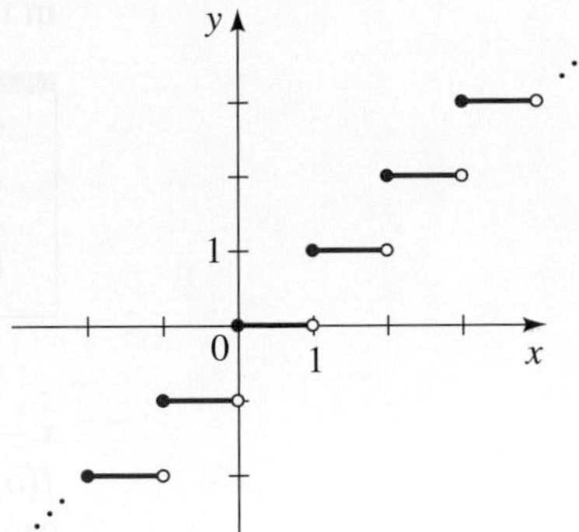

Figure 8 | The greatest integer function, $y = [\![x]\!]$ ■

The greatest integer function is an example of a **step function**. The next example gives a real-world application of a step function.

Example 7 ■ The Cost Function for a Cell Phone Travel Plan

A cell phone company offers a travel plan for cell phone data usage in countries outside the United States. The travel plan has a monthly fee of \$100 for the first 5 gigabytes of data used, and \$20 for each additional gigabyte (or portion thereof). Draw a graph of the cost C (in dollars) as a function of the number of gigabytes x used per month.

Solution Let $C(x)$ be the cost of using x gigabytes of data in a month. Since $x \ge 0$ the domain of the function is $[0, \infty)$. From the given information we have

$$\begin{aligned} C(x) &= 100 & &\text{if } 0 < x \le 5 \\ C(x) &= 100 + 20(1) = 120 & &\text{if } 5 < x \le 6 \\ C(x) &= 100 + 20(2) = 140 & &\text{if } 6 < x \le 7 \\ C(x) &= 100 + 20(3) = 160 & &\text{if } 7 < x \le 8 \\ &\vdots & &\vdots \end{aligned}$$

The graph is shown in Figure 9.

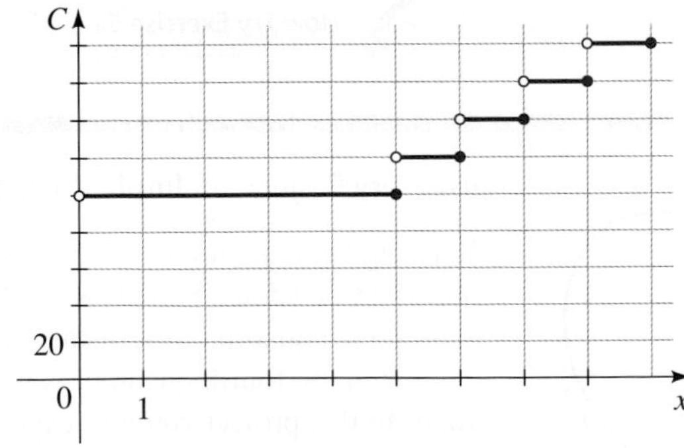

Figure 9 | Cost of data usage

Now Try Exercise 85 ■

A function is called **continuous** if its graph has no "break" or "hole." The functions in Examples 1, 2, 3, and 5 are continuous; the functions in Examples 4, 6, and 7 are not continuous.

■ Which Graphs Represent Functions? The Vertical Line Test

The graph of a function is a curve in the xy-plane. But the question arises: Which curves in the xy-plane are graphs of functions? This is answered by the following test.

> **The Vertical Line Test**
>
> A curve in the coordinate plane is the graph of a function if and only if no vertical line intersects the curve more than once.

We can see from Figure 10 why the Vertical Line Test is true. If each vertical line $x = a$ intersects a curve only once at (a, b), then exactly one function value is defined by $f(a) = b$. But if a line $x = a$ intersects the curve twice, at (a, b) and at (a, c), then the curve cannot represent a function because a function cannot assign two different values to a.

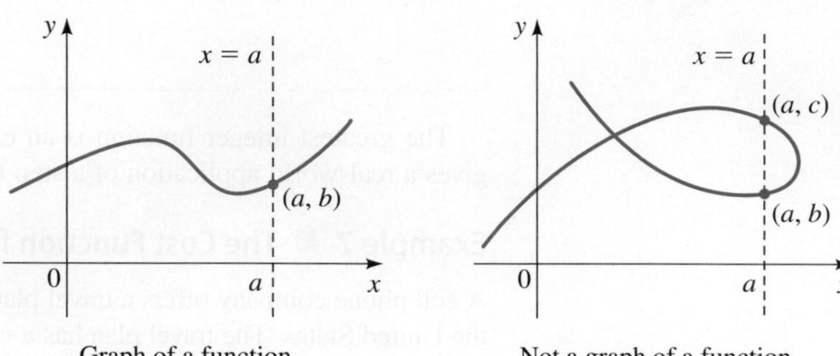

Figure 10 | Vertical Line Test

Example 8 ■ Using the Vertical Line Test

Using the Vertical Line Test, we see that the curves in parts (b) and (c) of Figure 11 represent functions, whereas those in parts (a) and (d) do not.

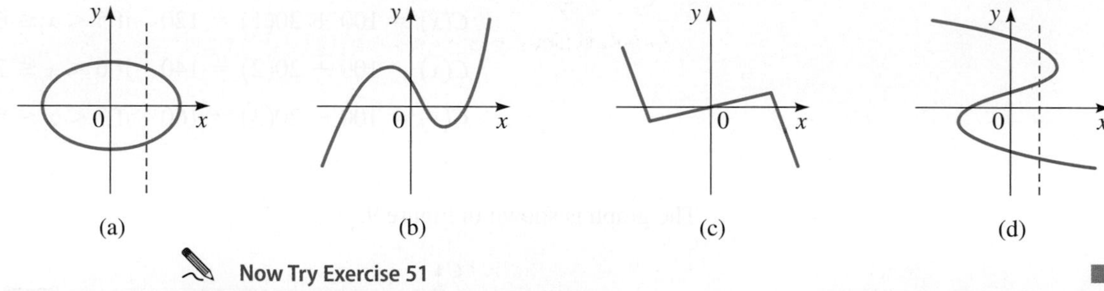

Figure 11

Now Try Exercise 51 ■

> **Discovery Project ■ Implicit Functions**
>
> Graphing the equation $(x^2 + y^2 - 1)^3 - x^2y^3 = 0$ with a graphing device gives the heart-shaped graph shown here. It is not the graph of a function, but parts of the graph (the part in the fourth quadrant, for instance) do represent functions. We say that the function whose graph is in the fourth quadrant is *implicitly* defined by the equation that produced the graph. In this project you will experiment with equations that produce interesting graphs and determine whether the equation can be solved to find *explicit* formulas for the functions that are implicit in the equation. You can find the project at **www.stewartmath.com**.

DONALD KNUTH was born in Milwaukee in 1938 and is Professor Emeritus of Computer Science at Stanford University. When Knuth was a high school student, he became fascinated with graphs of functions and laboriously drew many hundreds of them because he wanted to see the behavior of a great variety of functions. (Today, of course, it is far easier to use computers and graphing devices to do this.) While still a graduate student at Caltech, he started writing a monumental series of books entitled *The Art of Computer Programming*.

Knuth is famous for his invention of TEX, a system of computer-assisted typesetting. This system was used in the preparation of the manuscript for this textbook.

Knuth has received numerous honors, among them election as an associate of the French Academy of Sciences and as a Fellow of the Royal Society. Asteroid 21656 Knuth was named in his honor.

■ Which Equations Represent Functions?

Any equation in the variables x and y defines a relationship between these variables. For example, the equation

$$y - x^2 = 0$$

defines a relationship between y and x. Does this equation define y as a *function* of x? To find out, we solve for y and get

$$y = x^2 \qquad \text{Equation form}$$

We see that the equation defines a rule, or function, that gives one value of y for each value of x. We can express this rule in function notation as

$$f(x) = x^2 \qquad \text{Function form}$$

But not every equation defines y as a function of x, as the following example shows.

Example 9 ■ Equations That Define Functions

Does the equation define y as a function of x?

(a) $y - x^2 = 2$ **(b)** $x^2 + y^2 = 4$

Solution

(a) Solving for y in terms of x gives

$$y - x^2 = 2$$

$$y = x^2 + 2 \qquad \text{Add } x^2$$

The last equation is a rule that gives one value of y for each value of x, so it defines y as a function of x. We can write the function as $f(x) = x^2 + 2$.

(b) We try to solve for y in terms of x.

$$x^2 + y^2 = 4$$

$$y^2 = 4 - x^2 \qquad \text{Subtract } x^2$$

$$y = \pm\sqrt{4 - x^2} \qquad \text{Take square roots}$$

The last equation gives two values of y for a given value of x. Thus the equation does not define y as a function of x.

 **Now Try Exercises 53 and 59** ■

The graphs of the equations in Example 9 are shown in Figure 12. The Vertical Line Test shows graphically that the equation in Example 9(a) defines a function but the equation in Example 9(b) does not.

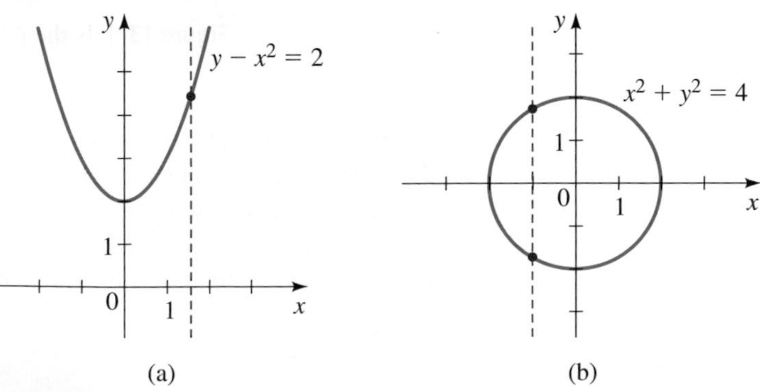

Figure 12

■ Which Relations Represent Functions?

A function f can be represented as a two-column table (as shown in Example 2.1.7) or equivalently as a set of ordered pairs (x, y) where x is the input and $y = f(x)$ is the output. Conversely, given a two-column table (or a set of ordered pairs), how can we tell whether it represents a function? Let's begin by considering collections of ordered pairs.

> **Relations**
>
> Any collection of ordered pairs (x, y) is called a **relation.**

If we denote the ordered pairs in a relation by (x, y), then the set of x-values (or inputs) is the **domain** of the relation and the set of y-values (or outputs) is the **range** of the relation. Every function is a relation consisting of the ordered pairs (x, y), where $y = f(x)$. However, not every relation is a function, because a relation can have more than one output for a given input.

> A relation is a function if each input corresponds to exactly one output.

Here are two examples of relations:

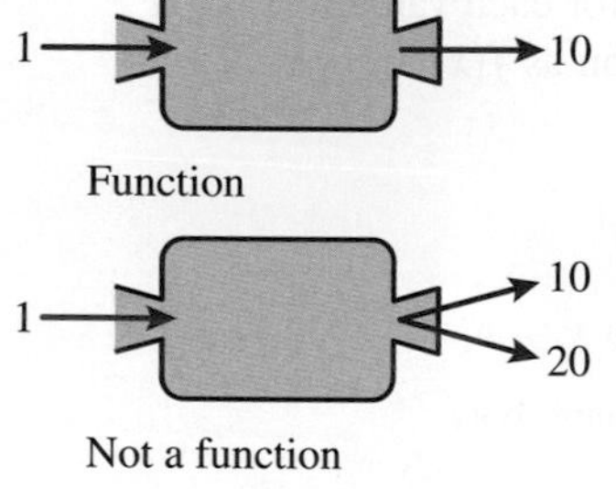

Relation 1: $\{(1, 10), (2, 20), (3, 40), (4, 40)\}$

Relation 2: $\{(1, 10), (1, 20), (2, 30), (3,30), (4, 40)\}$

The diagrams in Figure 13 are visual representations of these relations. The first relation is a function because each input corresponds to exactly one output; the second relation is not a function because the input 1 corresponds to two different outputs, 10 and 20 (see the machine diagrams in the margin).

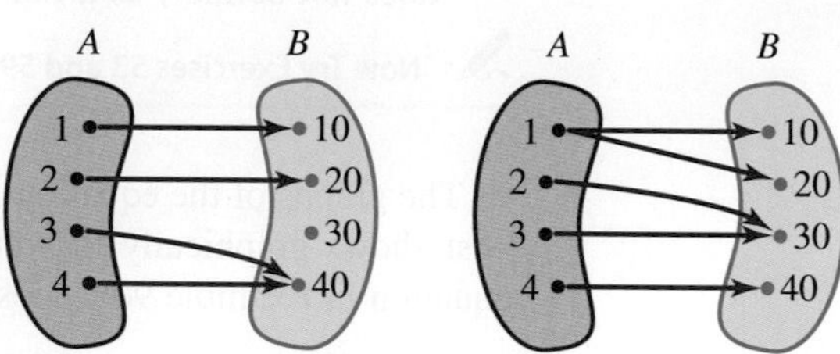

Figure 13 | Is the relation a function?

We can graph relations in the same way we graph functions. To graph a relation, plot the ordered pairs in the relation in a coordinate plane. Note that any graph in the coordinate plane defines a relation, namely the relation that consists of the set of coordinates (x, y) of the points on the graph. Also, any equation in the variables x and y defines a relation, namely the relation that consists of the ordered pairs (x, y) that satisfy the equation.

Example 10 ■ Is the Relation a Function?

Is the relation a function? Confirm your answer graphically. State the domain and range of the relation.

(a) $\{(-1, 2), (0, 3), (1, 1), (1, 2), (2, 3), (3, 1)\}$

(b) $\{(-1, 3), (0, 1), (1, 2), (2, 1), (3, 1), (4, 3)\}$

Solution

(a) Since the ordered pairs $(1, 1)$ and $(1, 2)$ are in the relation, the input 1 corresponds to the two different outputs, 1 and 2. It follows that the relation is not a function. We can also verify this conclusion graphically by making a table of values and graphing the relation, as in Figure 14(a). From the graph we see that the vertical line $x = 1$ intersects the graph at the two points $(1, 1)$ and $(1, 2)$, so the Vertical Line Test confirms that the relation is not a function. The domain is $\{-1, 0, 1, 2, 3\}$ and the range is $\{1, 2, 3\}$.

(b) The relation is a function because each input corresponds to exactly one output. That is, for each ordered pair the first component corresponds to exactly one second component. A graph of the relation is shown in Figure 14(b). Applying the Vertical Line Test to the graph shows that the relation is a function. The domain is $\{-1, 0, 1, 2, 3, 4\}$ and the range is $\{1, 2, 3\}$.

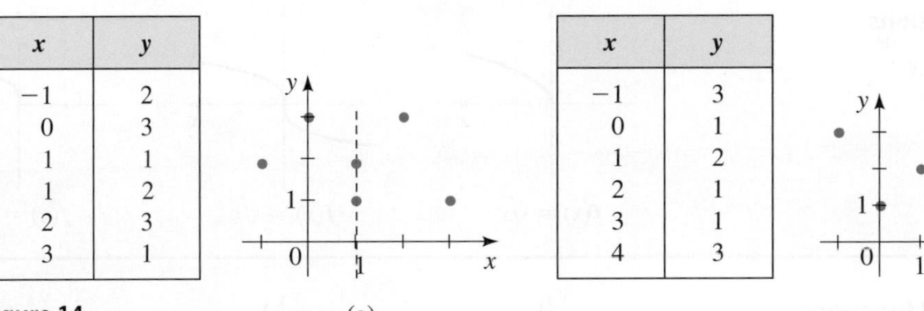

x	y
−1	2
0	3
1	1
1	2
2	3
3	1

x	y
−1	3
0	1
1	2
2	1
3	1
4	3

Figure 14

Now Try Exercise 67 ■

Example 11 ■ Is the Relation a Function?

A relation consists of the ordered pairs (x, y) that satisfy the equation $x^2 + y^2 = 4$. Show that the relation is not a function.

Solution The relation is not a function because there are ordered pairs that satisfy the equation and have different outputs for the same input; for instance, $(0, 2)$ and $(0, -2)$ or $(1, \sqrt{3})$ and $(1, -\sqrt{3})$.

Now Try Exercise 69 ■

Note There are different ways to show that an equation does not represent a function. For instance, consider the equation $x^2 + y^2 = 4$. In Example 11 we showed that the relation defined by this equation is not a function. In Example 9(b) we solved the equation for y and found that the equation does not define y as a function of x. And, in Figure 12(b) we used the Vertical Line Test to show that the graph of the equation is not the graph of a function.

The following box shows the graphs of some functions that you will see frequently in this book.

Some Functions and Their Graphs

Linear functions

$f(x) = mx + b$

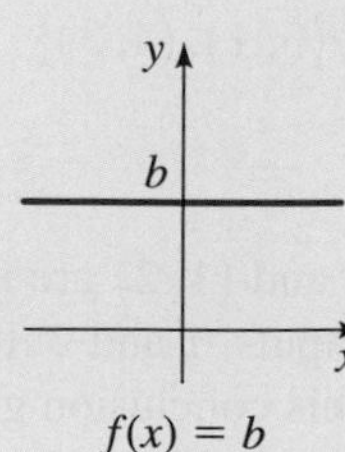

$f(x) = b$

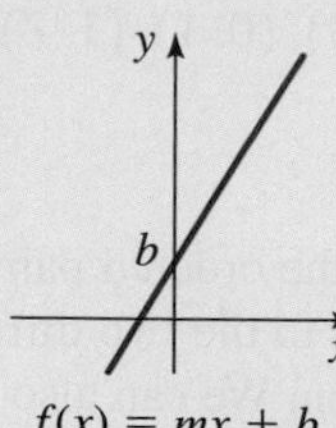

$f(x) = mx + b$

Power functions

$f(x) = x^n$

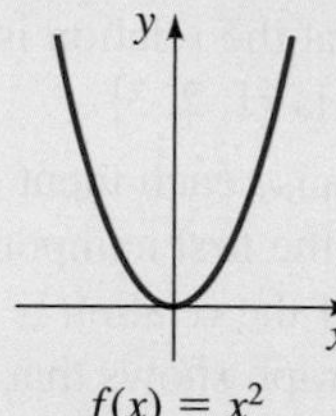

$f(x) = x^2$

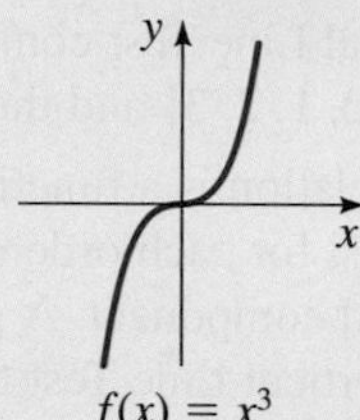

$f(x) = x^3$

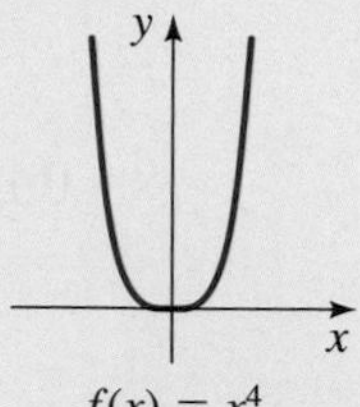

$f(x) = x^4$

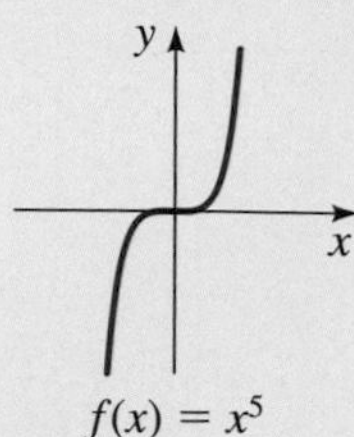

$f(x) = x^5$

Root functions

$f(x) = \sqrt[n]{x}$

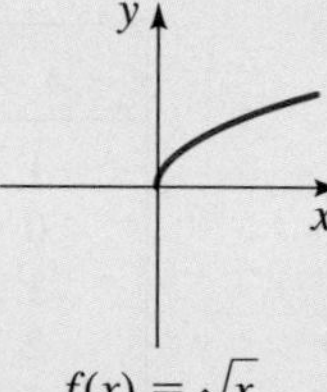

$f(x) = \sqrt{x}$

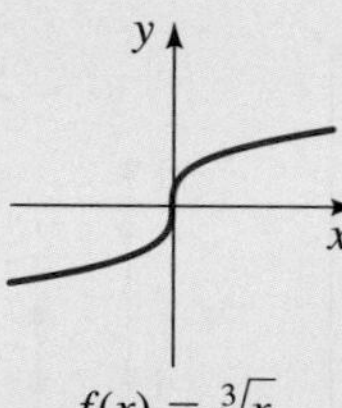

$f(x) = \sqrt[3]{x}$

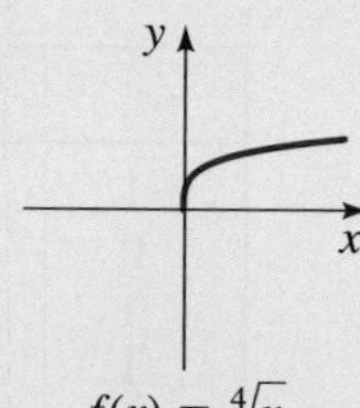

$f(x) = \sqrt[4]{x}$

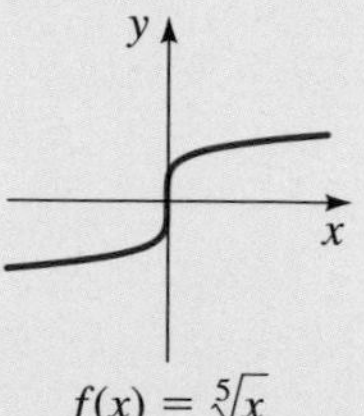

$f(x) = \sqrt[5]{x}$

Reciprocal functions

$f(x) = \dfrac{1}{x^n}$

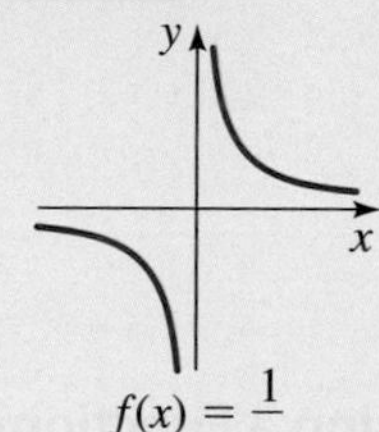

$f(x) = \dfrac{1}{x}$

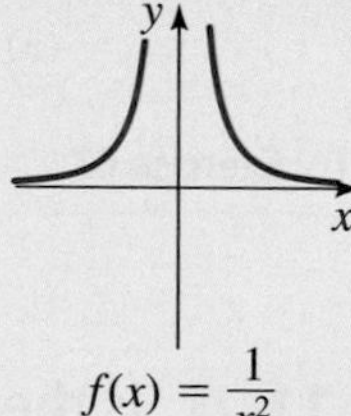

$f(x) = \dfrac{1}{x^2}$

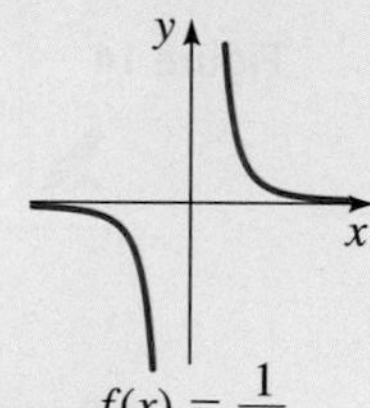

$f(x) = \dfrac{1}{x^3}$

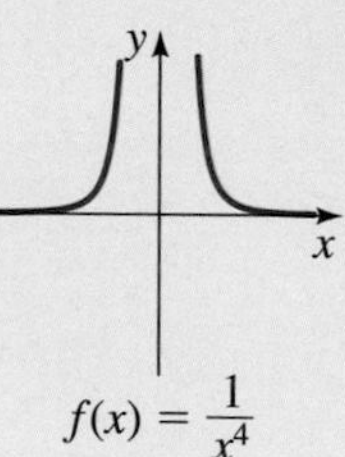

$f(x) = \dfrac{1}{x^4}$

Absolute-value function

$f(x) = |x|$

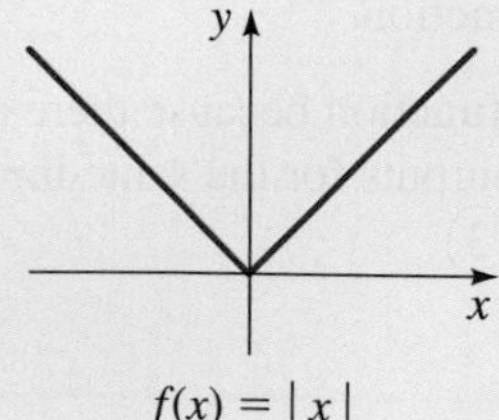

$f(x) = |x|$

Greatest integer function

$f(x) = [\![x]\!]$

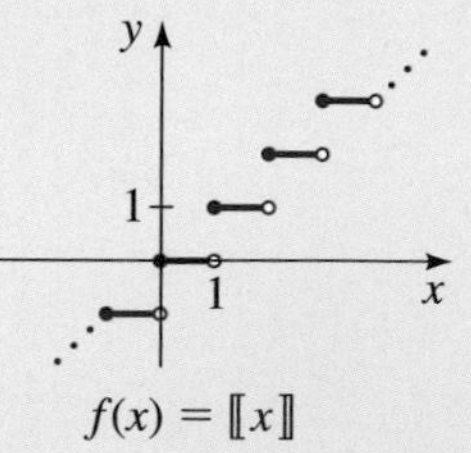

$f(x) = [\![x]\!]$

2.2 Exercises

Concepts

1. To graph the function f, we plot the points $(x, \underline{\qquad})$ in a coordinate plane. To graph $f(x) = x^2 - 2$, we plot the points $(x, \underline{\qquad})$. So the point $(3, \underline{\qquad})$ is on the graph of f. The height of the graph of f above the x-axis when $x = 3$ is ______. Complete the table, and sketch a graph of f.

x	$y = f(x)$	(x, y)
−2		
−1		
0		
1		
2		

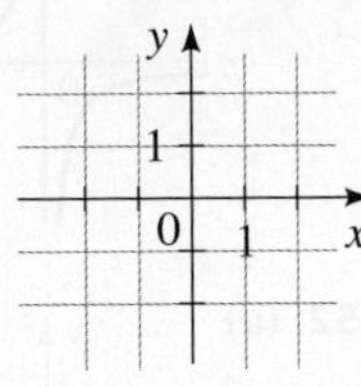

2. If $f(4) = 10$ then the point $(4, \_\_\_)$ is on the graph of f.

3. If the point $(3, 7)$ is on the graph of f, then $f(3) =$ ______.

4. Match the function with its graph.

(a) $f(x) = x^2$ **(b)** $f(x) = x^3$

(c) $f(x) = \sqrt{x}$ **(d)** $f(x) = |x|$

I

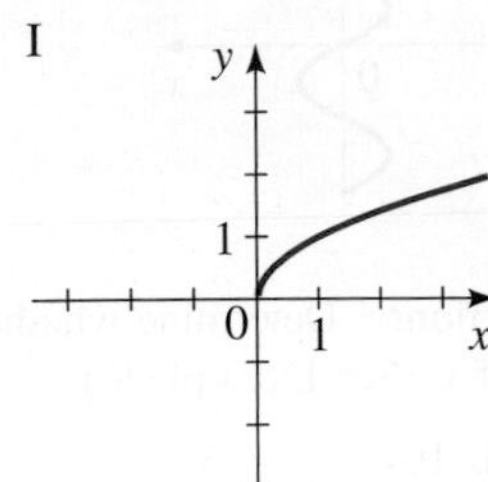

II

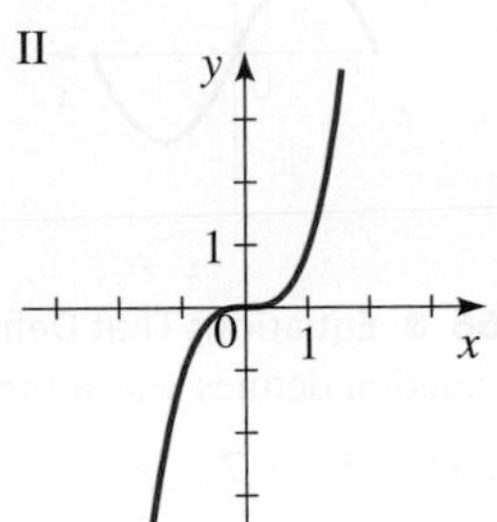

III

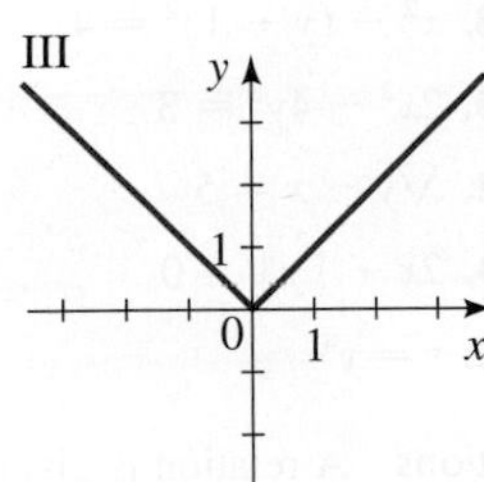

IV

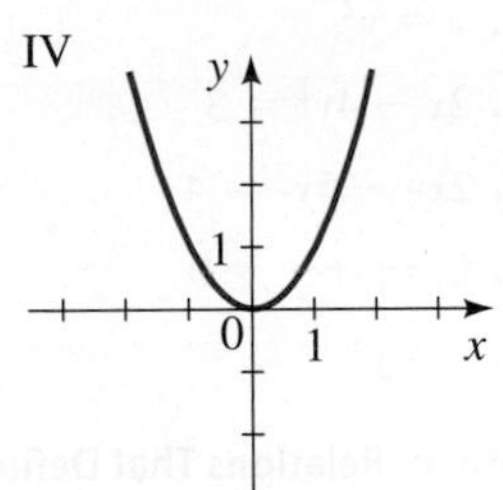

5–8 ■ Explain why the given graph, equation, set of ordered pairs (x, y), or table does *not* define y as a function of x.

5. $\{(1, 1), (1, 2), (2, 3), (3, 4)\}$

6. $x = 4y^2$

7.

x	y
8	11
10	15
2	11
10	10

8.

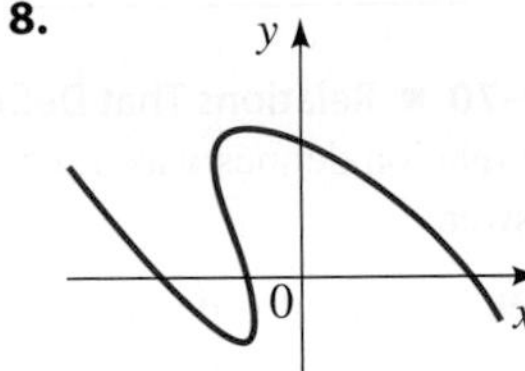

Skills

9–32 ■ Graphing Functions Sketch a graph of the function by first making a table of values.

9. $f(x) = x + 2$ **10.** $f(x) = 4 - 2x$

11. $f(x) = -x + 3, \quad -3 \le x \le 3$

12. $f(x) = \dfrac{x - 3}{2}, \quad 0 \le x \le 5$

13. $f(x) = -x^2$ **14.** $f(x) = x^2 - 2$

15. $g(x) = x^2 - 6x + 9$ **16.** $g(x) = -(x + 3)^2$

17. $r(x) = 3x^4$ **18.** $r(x) = 20 - x^4$

19. $g(x) = x^3 + 8$ **20.** $g(x) = (x + 2)^3$

21. $k(x) = \sqrt[3]{-x}$ **22.** $k(x) = -\sqrt[3]{x}$

23. $f(x) = 2 - \sqrt{x}$ **24.** $f(x) = \sqrt{x + 4}$

25. $C(t) = \dfrac{1}{t^2}$ **26.** $C(t) = -\dfrac{1}{t + 1}$

27. $H(x) = |2x|$ **28.** $H(x) = |x - 2|$

29. $G(x) = |x| + x$ **30.** $G(x) = |x| - x$

31. $f(x) = |2x - 2|$ **32.** $f(x) = \dfrac{x}{|x|}$

33–36 ■ Graphing Functions Graph the function in each of the given viewing rectangles, and select the one that produces the most appropriate graph of the function.

33. $f(x) = 8x - x^2$

(a) $[-5, 5]$ by $[-5, 5]$

(b) $[-10, 10]$ by $[-10, 10]$

(c) $[-2, 10]$ by $[-5, 20]$

(d) $[-10, 10]$ by $[-100, 100]$

34. $f(x) = x^2 - 4x - 32$

(a) $[-3, 3] \times [-5, 5]$

(b) $[-10, 10] \times [-10, 10]$

(c) $[-7, 7] \times [-30, 5]$

(d) $[-6, 10] \times [-40, 5]$

35. $f(x) = 3x^3 - 9x - 20$

(a) $[-2, 2] \times [-10, 10]$

(b) $[-5, 5] \times [-10, 20]$

(c) $[-5, 5] \times [-20, 20]$

(d) $[-3, 3] \times [-40, 20]$

36. $f(x) = x^4 - 10x^2 + 5x$

(a) $[-1, 1] \times [-10, 10]$

(b) $[-5, 5] \times [-10, 10]$

(c) $[-5, 5] \times [-40, 20]$

(d) $[-10, 10] \times [-20, 20]$

37–46 ■ Graphing Piecewise-Defined Functions Sketch a graph of the piecewise-defined function.

37. $f(x) = \begin{cases} 2x & \text{if } x \le 0 \\ 2 & \text{if } x > 0 \end{cases}$ **38.** $f(x) = \begin{cases} 1 & \text{if } x \le 1 \\ x + 1 & \text{if } x > 1 \end{cases}$

39. $f(x) = \begin{cases} x & \text{if } x \le 0 \\ x + 1 & \text{if } x > 0 \end{cases}$

40. $f(x) = \begin{cases} 2x + 3 & \text{if } x < -1 \\ 3 - x & \text{if } x \ge -1 \end{cases}$

41. $f(x) = \begin{cases} 3 - x & \text{if } x \le -1 \\ 2x^2 & \text{if } x > -1 \end{cases}$

42. $f(x) = \begin{cases} 4 - x^2 & \text{if } x \le 1 \\ x - 5 & \text{if } x > 1 \end{cases}$

43. $f(x) = \begin{cases} 0 & \text{if } |x| \le 2 \\ 3 & \text{if } |x| > 2 \end{cases}$

44. $f(x) = \begin{cases} x^2 & \text{if } |x| \le 1 \\ 1 & \text{if } |x| > 1 \end{cases}$

45. $f(x) = \begin{cases} 4 & \text{if } x < -2 \\ x^2 & \text{if } -2 \le x \le 2 \\ -x + 6 & \text{if } x > 2 \end{cases}$

46. $f(x) = \begin{cases} -1 & \text{if } x < -1 \\ x & \text{if } -1 \le x \le 1 \\ 1 & \text{if } x > 1 \end{cases}$

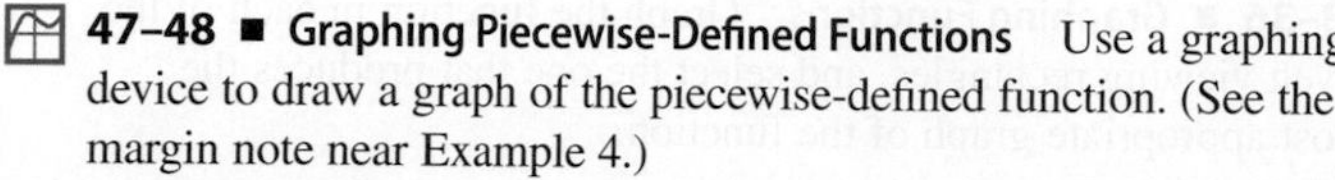

47–48 ■ Graphing Piecewise-Defined Functions Use a graphing device to draw a graph of the piecewise-defined function. (See the margin note near Example 4.)

47. $f(x) = \begin{cases} x^2 - 6x + 12 & \text{if } x \le 2 \\ 6 - x & \text{if } x > 2 \end{cases}$

48. $f(x) = \begin{cases} 2x - x^2 & \text{if } x > 1 \\ (x - 1)^3 & \text{if } x \le 1 \end{cases}$

49–50 ■ Finding Piecewise-Defined Functions A graph of a piecewise-defined function is given. Find a formula for the function in the indicated form.

49.

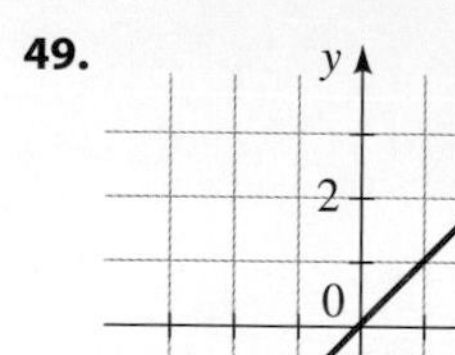

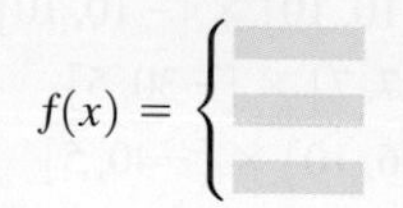

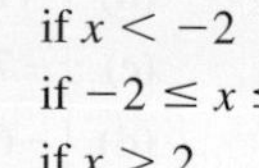

$f(x) = \begin{cases} \_\_\_ & \text{if } x < -2 \\ \_\_\_ & \text{if } -2 \le x \le 2 \\ \_\_\_ & \text{if } x > 2 \end{cases}$

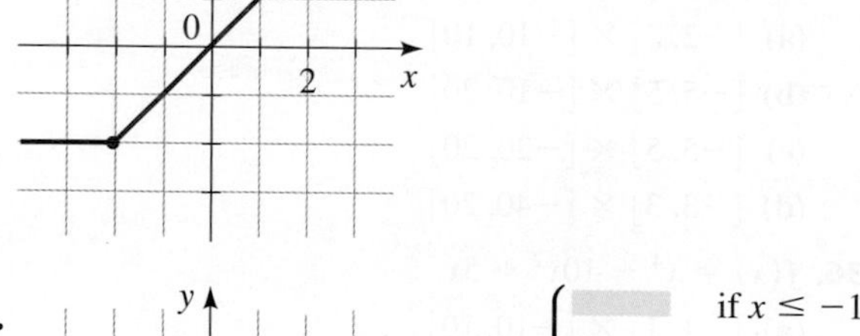

50.

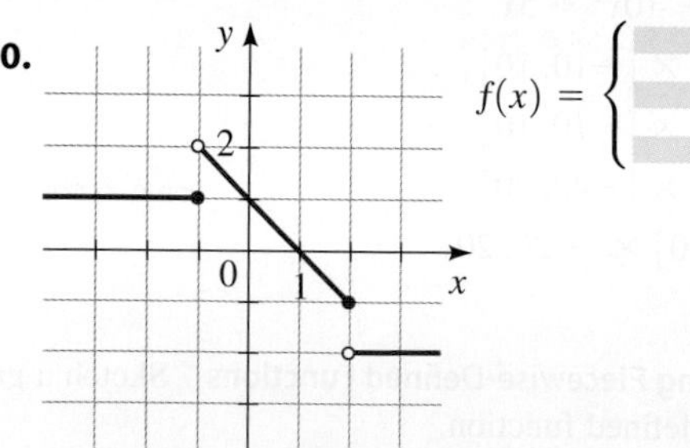

$f(x) = \begin{cases} \_\_\_ & \text{if } x \le -1 \\ \_\_\_ & \text{if } -1 < x \le 2 \\ \_\_\_ & \text{if } x > 2 \end{cases}$

51–52 ■ Graphs that Define Functions Use the Vertical Line Test to determine whether the curve is a graph of a function of x.

51. (a)

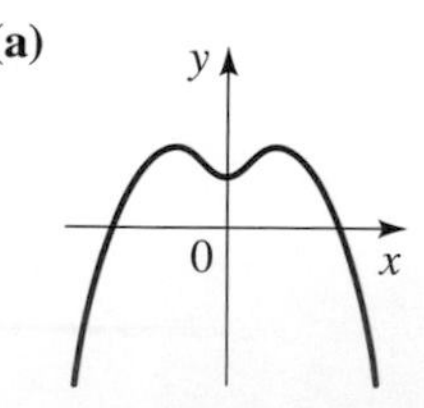

(b)

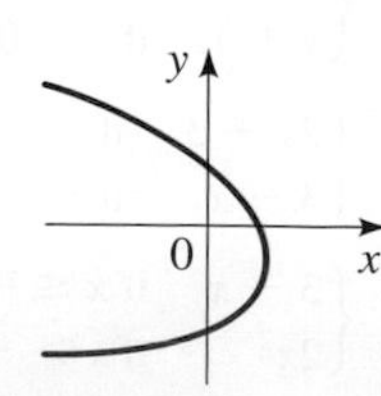

(c) (d)

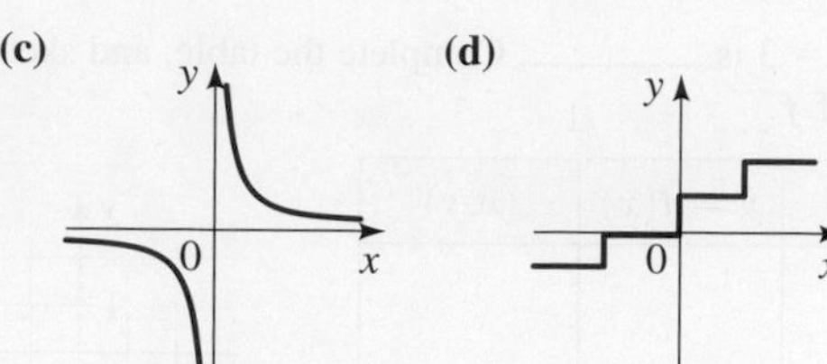

52. (a) (b)

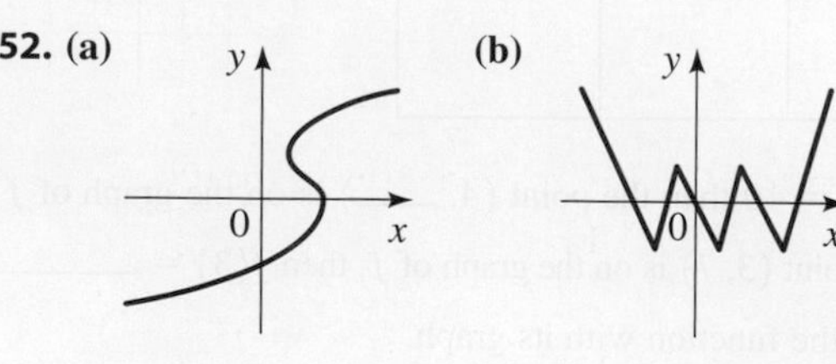

(c) (d)

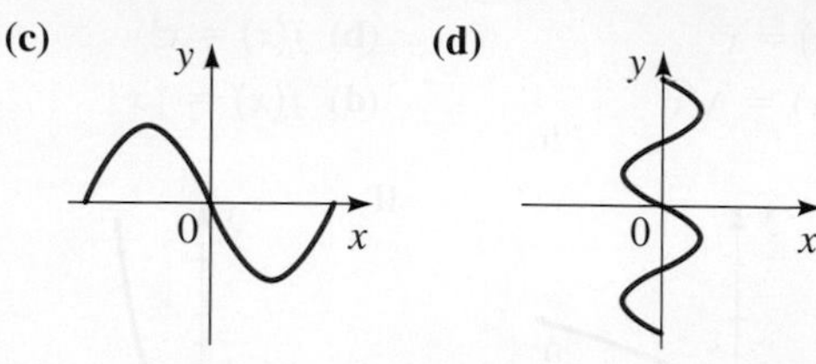

53–66 ■ Equations That Define Functions Determine whether the equation defines y as a function of x. (See Example 9.)

53. $x^2 - 3y = 7$

54. $10x - y = 5$

55. $y^3 - x = 5$

56. $x^2 - y^{1/3} = 1$

57. $x = y^2$

58. $x^2 + (y - 1)^2 = 4$

59. $2x - 4y^2 = 3$

60. $2x^2 - 4y^2 = 3$

61. $2xy - 5y^2 = 4$

62. $\sqrt{y} - x = 5$

63. $2|x| + y = 0$

64. $2x + |y| = 0$

65. $x = y^3$

66. $x = y^4$

67–68 ■ Relations That Define Functions A relation is given by a table or a set of ordered pairs. Graph the relation to determine whether it defines y as a function of x. State the domain and range of the relation.

67. $\{(0, 1), (1, 2), (1, 3), (4, 1), (5, 1), (6, 1)\}$

68.

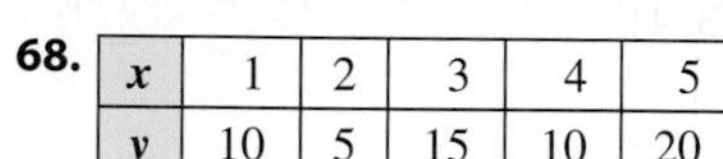

x	1	2	3	4	5
y	10	5	15	10	20

69–70 ■ Relations That Define Functions Determine whether the relation defines y as a function of x. Give reasons for your answer.

69. The set of ordered pairs (x, y) that satisfy the equation $(y + 1)^3 = x$.

70. The set of ordered pairs of natural numbers (x, y) for which y/x is a natural number.

71–74 ■ Families of Functions A family of functions is given. In parts (a) and (b) graph all the given members of the family in the viewing rectangle indicated. In part (c) state the conclusions that you can make from your graphs.

71. $f(x) = x^2 + c$

(a) $c = 0, 2, 4, 6;\quad [-5, 5]$ by $[-10, 10]$

(b) $c = 0, -2, -4, -6$; $[-5, 5]$ by $[-10, 10]$

(c) How does the value of c affect the graph?

72. $f(x) = (x - c)^2$

(a) $c = 0, 1, 2, 3$; $[-5, 5]$ by $[-10, 10]$

(b) $c = 0, -1, -2, -3$; $[-5, 5]$ by $[-10, 10]$

(c) How does the value of c affect the graph?

73. $f(x) = cx^2$

(a) $c = 1, \frac{1}{2}, 2, 4$; $[-5, 5]$ by $[-10, 10]$

(b) $c = 1, -1, -\frac{1}{2}, -2$; $[-5, 5]$ by $[-10, 10]$

(c) How does the value of c affect the graph?

74. $f(x) = x^{1/n}$

(a) $n = 2, 4, 6$, $[-1, 4]$ by $[-1, 3]$

(b) $n = 3, 5, 7$, $[-3, 3]$ by $[-2, 2]$

(c) How does the value of n affect the graph?

■ Skills Plus

75–78 ■ Finding Functions for Certain Curves Find a function whose graph is the given curve.

75. The line segment joining the points $(-2, 1)$ and $(4, -6)$

76. The line segment joining the points $(-3, -2)$ and $(6, 3)$

77. The top half of the circle $x^2 + y^2 = 9$

78. The bottom half of the circle $x^2 + y^2 = 9$

■ Applications

79. Weather Balloon As a weather balloon is inflated, the thickness T of its rubber skin is related to the radius of the balloon by

$$T(r) = \frac{0.5}{r^2}$$

where T and r are measured in centimeters. Graph the function T for values of r between 10 and 100.

80. Power from a Wind Turbine The power produced by a wind turbine depends on the speed of the wind. If a windmill has blades 3 meters long, then the power P produced by the turbine is modeled by

$$P(v) = 14.1v^3$$

where P is measured in watts (W) and v is measured in meters per second (m/s). Graph the function P for wind speeds between 1 m/s and 10 m/s.

81–84 ■ Graphing Applied Functions Graph the indicated function from Exercises 2.1 on the given interval.

81. Exercise 83; $0 \le t \le 20$ (Torricelli's law)

82. Exercise 86; $0 \le T \le 1000$ (Blackbody radiation)

83. Exercise 85; $0 < v \le 300{,}000$ (Relativity)

84. Exercise 88; $0 \le r \le 0.5$ (Blood flow)

85. Postage Rates The 2022 domestic postage rate for first-class letters weighing 105 g or less is 49 cents for the first 30 g (or less), plus 21 cents for each additional 30 g (or part). Express the postage P as a piecewise-defined function of the weight x of a letter, with $0 < x \le 105$, and sketch a graph of this function.

86. Utility Rates Westside Energy charges its electric customers a base rate of \$10.00 per month, plus 12¢ per kilowatt-hour (kWh) for the first 300 kWh used and 17¢ per kWh for all usage over 300 kWh. Suppose a customer uses x kWh of electricity in one month.

(a) Express the monthly cost E as a piecewise-defined function of x.

(b) Graph the function E for $0 \le x \le 600$.

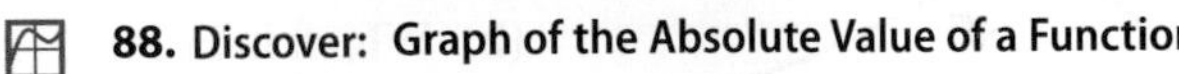

■ Discuss ■ Discover ■ Prove ■ Write

87. Discover: When Does a Graph Represent a Function? For every integer n, the graph of the equation $y = x^n$ is the graph of a function, namely $f(x) = x^n$. Explain why the graph of $x = y^2$ is *not* the graph of a function of x. Is the graph of $x = y^3$ the graph of a function of x? If so, what function of x is it the graph of? Determine for what integers n the graph of $x = y^n$ is a graph of a function of x.

88. Discover: Graph of the Absolute Value of a Function

(a) Draw graphs of the functions

$$f(x) = x^2 + x - 6$$

and $$g(x) = |x^2 + x - 6|$$

How are the graphs of f and g related?

(b) Draw graphs of the functions $f(x) = x^4 - 6x^2$ and $g(x) = |x^4 - 6x^2|$. How are the graphs of f and g related?

(c) In general, if $g(x) = |f(x)|$, how are the graphs of f and g related? Draw graphs to illustrate your answer.

89. Discuss: Everyday Relations In everyday life we encounter many relations. For example, we match people with their telephone number(s), baseball players with their batting averages, or married people with their partners. Discuss whether the following everyday relations define y as a functions of x.

(a) x is the sibling of y.

(b) x is the birth mother of y.

(c) x is a student in your school and y is their ID number.

(d) x is the age of a student and y is their shoe size.

90. Discuss ■ Discover: Is x a Function of y? The equation $x = y^2$ does not define y as a function of x. In other words, if we view x as the independent variable (the input) and y as the dependent variable (the output), then the equation does not define a function. On the other hand, if we view y as the independent variable and x as the dependent variable, then the equation does define x as a function of y (because in this case each value of y determines exactly one value of x, namely y^2). Determine whether the relations in Exercises 5–8 define x as a function of y.

2.3 Getting Information from the Graph of a Function

■ Values of a Function; Domain and Range ■ Comparing Function Values: Solving Equations and Inequalities Graphically ■ Increasing and Decreasing Functions ■ Local Maximum and Minimum Values of a Function

Many properties of a function are more easily obtained from a graph than from the rule that describes the function. We see in this section how a graph tells us whether the values of a function are increasing or decreasing and also where the maximum and minimum values of a function are located.

■ Values of a Function; Domain and Range

A complete graph of a function contains all the information about a function because the graph tells us which input values correspond to which output values. To analyze the graph of a function, we must keep in mind that *the height of the graph is the value of the function.* So we can read off the values of a function from its graph.

Example 1 ■ Finding the Values of a Function from a Graph

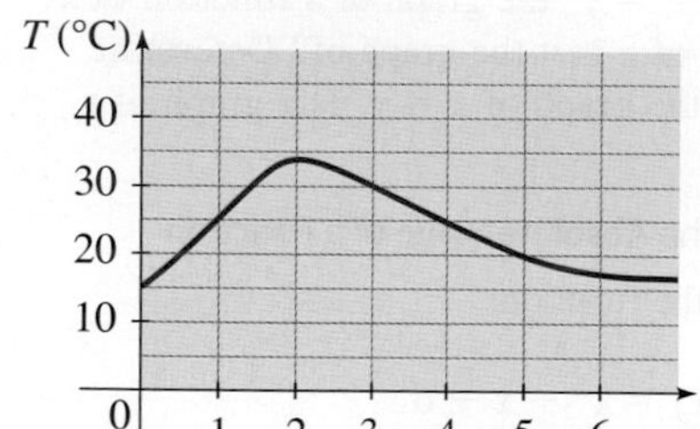

Figure 1 | Temperature function

The function T graphed in Figure 1 gives the temperature between noon and 6:00 P.M. at a certain weather station.

(a) Find $T(1)$, $T(3)$, and $T(5)$.

(b) Which is larger, $T(2)$ or $T(4)$?

(c) Find the value(s) of x for which $T(x) = 25$.

(d) Find the value(s) of x for which $T(x) \geq 25$.

(e) Find the net change in temperature from 1 P.M. to 3 P.M.

Solution

(a) $T(1)$ is the temperature at 1:00 P.M. It is represented by the height of the graph above the x-axis at $x = 1$. Thus $T(1) = 25$. Similarly, $T(3) = 30$ and $T(5) = 20$.

(b) Since the graph is higher at $x = 2$ than at $x = 4$, it follows that $T(2)$ is larger than $T(4)$.

(c) The height of the graph is 25 when x is 1 and when x is 4. In other words, the temperature is 25 at 1:00 P.M. and at 4:00 P.M.

(d) The graph is higher than 25 for x between 1 and 4. In other words, the temperature was 25 or greater between 1:00 P.M. and 4:00 P.M.

Net change is defined in Section 2.1.

(e) The net change in temperature is

$$T(3) - T(1) = 30 - 25 = 5$$

So there was a net increase of 5°C from 1 P.M. to 3 P.M.

Now Try Exercises 7 and 63 ■

The graph of a function helps us to picture the domain and range of the function on the x-axis and y-axis, as shown in the box below.

Domain and Range from a Graph

The **domain** and **range** of a function

$$y = f(x)$$

can be obtained from a graph of f, as shown in the figure. The domain is the set of all x-values for which f is defined, and the range is all the corresponding y-values.

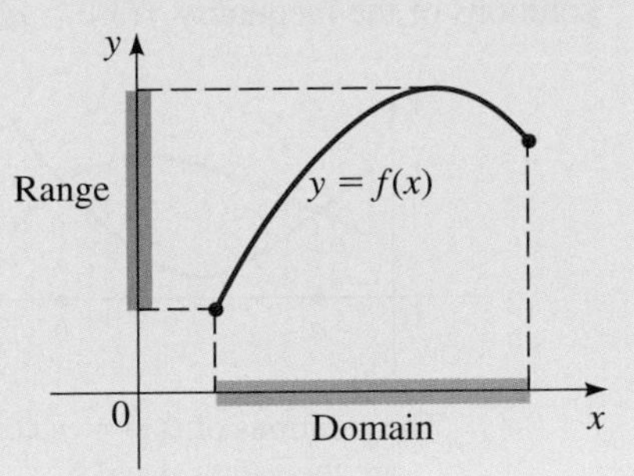

Example 2 ■ Finding the Domain and Range from a Graph

(a) Use a graphing device to draw the graph of $f(x) = \sqrt{4 - x^2}$.

(b) Find the domain and range of f.

Solution

(a) The graph is shown in Figure 2.

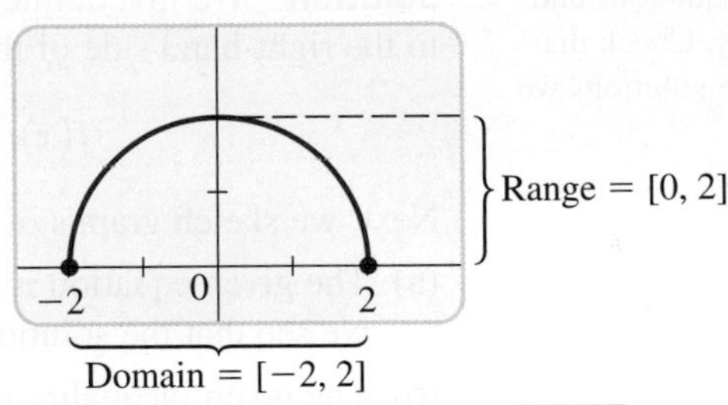

Figure 2 | Graph of $f(x) = \sqrt{4 - x^2}$

(b) From the graph in Figure 2 we see that the domain is $[-2, 2]$ and the range is $[0, 2]$.

 **Now Try Exercise 25** ■

■ Comparing Function Values: Solving Equations and Inequalities Graphically

We can compare the values of two functions f and g visually by drawing their graphs. The points at which the graphs intersect are the points where the values of the two functions are equal. So the solutions of the equation $f(x) = g(x)$ are the values of x at which the two graphs intersect. The points at which the graph of g is higher than the graph of f are the points where the values of g are greater than the values of f. So the solutions of the inequality $f(x) < g(x)$ are the values of x at which the graph of g is *higher than* the graph of f.

Solving Equations and Inequalities Graphically

The **solution(s) of the equation** $f(x) = g(x)$ are the values of x where the graphs of f and g intersect.

The **solution(s) of the inequality** $f(x) < g(x)$ are the values of x where the graph of g is higher than the graph of f. The solutions of the inequality $f(x) > g(x)$ are the values of x where the graph of f is higher than the graph of g.

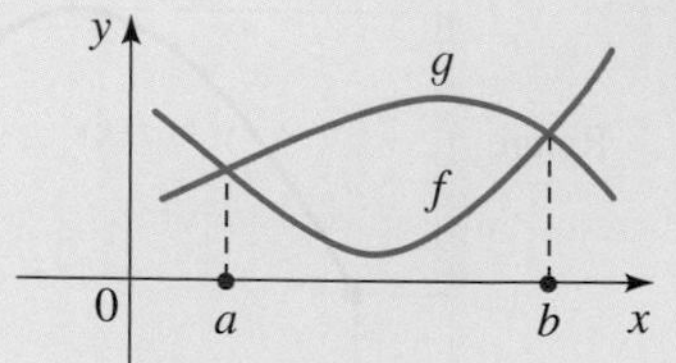

The solutions of $f(x) = g(x)$ are the values a and b.

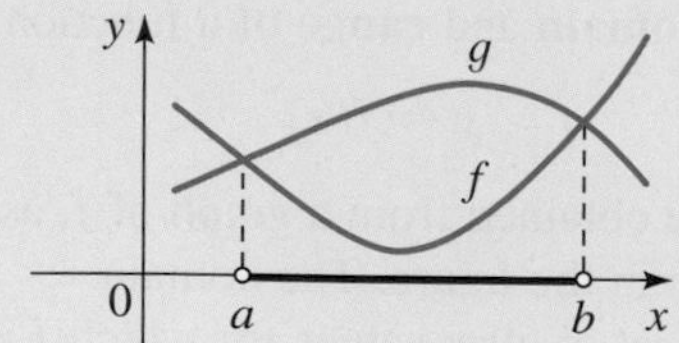

The solution of $f(x) < g(x)$ is the interval (a, b).

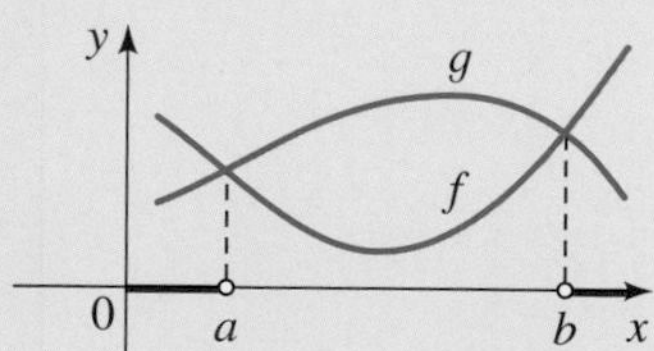

The solution of $f(x) > g(x)$ is a union of intervals $(-\infty, a) \cup (b, \infty)$.

We can use these observations to solve equations and inequalities graphically, as the next example illustrates.

Example 3 ■ Solving Graphically

Solve the given equation or inequality graphically.

(a) $2x^2 + 3 = 5x + 6$ **(b)** $2x^2 + 3 \leq 5x + 6$ **(c)** $2x^2 + 3 > 5x + 6$

You can also solve the equations and inequalities algebraically. Check that your solutions match the solutions we obtained graphically.

Solution We first define functions f and g that correspond to the left-hand side and to the right-hand side of the equation or inequality. So we define

$$f(x) = 2x^2 + 3 \qquad \text{and} \qquad g(x) = 5x + 6$$

Next, we sketch graphs of f and g on the same set of axes.

(a) The given equation is equivalent to $f(x) = g(x)$. From the graph in Figure 3(a) we see that the solutions of the equation are $x = -0.5$ and $x = 3$.

(b) The given inequality is equivalent to $f(x) \leq g(x)$. From the graph in Figure 3(b) we see that the solution is the interval $[-0.5, 3]$.

(c) The given inequality is equivalent to $f(x) > g(x)$. From the graph in Figure 3(c) we see that the solution is $(-\infty, -0.5) \cup (3, \infty)$.

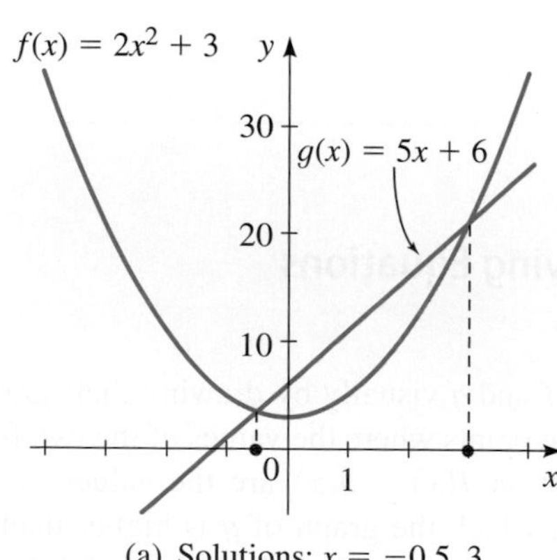

(a) Solutions: $x = -0.5, 3$

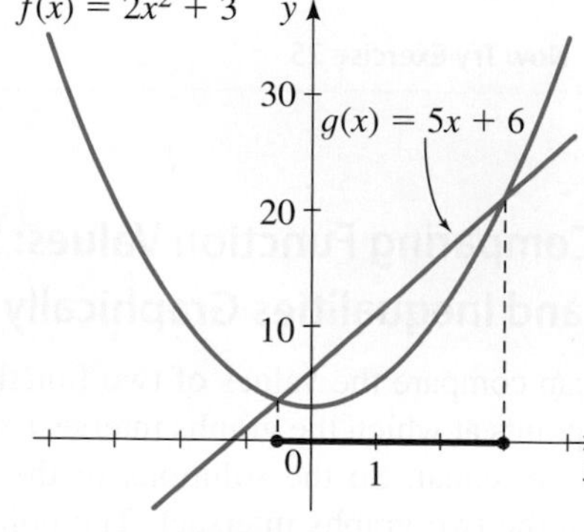

(b) Solution: $[-0.5, 3]$

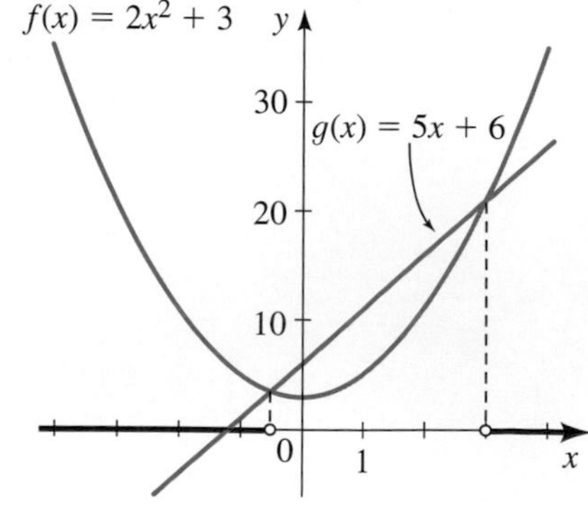

(c) Solution: $(-\infty, -0.5) \cup (3, \infty)$

Figure 3 | Graphs of $f(x) = 2x^2 + 3$ and $g(x) = 5x + 6$

Now Try Exercises 9 and 29 ■

Another way to solve an equation graphically is to first move all terms to one side of the equation and then graph the function that corresponds to the nonzero side of the equation. In this case the solutions of the equation are the x-intercepts of the graph. We can use this same method to solve inequalities graphically, as the following example shows.

Example 4 ■ Solving Graphically

Solve the given equation or inequality graphically.

(a) $x^3 + 6 = 2x^2 + 5x$ **(b)** $x^3 + 6 \geq 2x^2 + 5x$

Solution We first move all terms to one side to obtain an equivalent equation (or inequality). For the equation in part (a) we obtain

$$x^3 - 2x^2 - 5x + 6 = 0 \qquad \text{Move terms to LHS}$$

Then we define a function f by

$$f(x) = x^3 - 2x^2 - 5x + 6 \qquad \text{Define } f$$

Next, we use a graphing device to graph f, as shown in Figure 4.

(a) The given equation is the same as $f(x) = 0$, so the solutions are the x-intercepts of the graph. From Figure 4(a) we see that the solutions are $x = -2$, $x = 1$, and $x = 3$.

(b) The given inequality is the same as $f(x) \geq 0$, so the solutions are the x-values at which the graph of f is on or above the x-axis. From Figure 4(b) we see the solution is $[-2, 1] \cup [3, \infty)$.

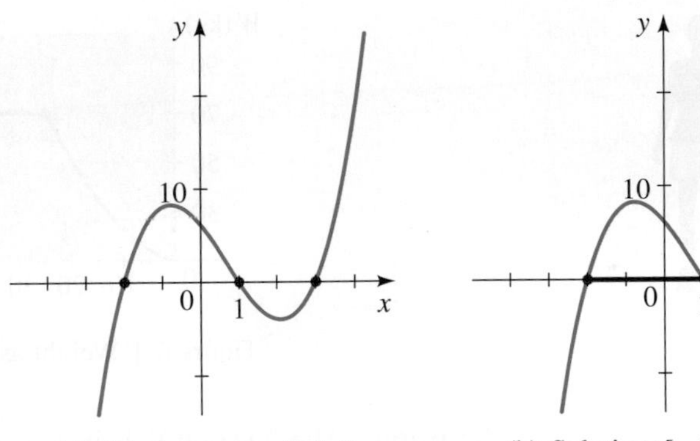

Figure 4 | Graphs of $f(x) = x^3 - 2x^2 - 5x + 6$

Now Try Exercise 33 ■

■ Increasing and Decreasing Functions

It is very useful to know where the graph of a function rises and where it falls. The graph shown in Figure 5 rises, falls, then rises again as we move from left to right: It rises from A to B, falls from B to C, and rises again from C to D. The function f is said to be *increasing* when its graph rises and *decreasing* when its graph falls.

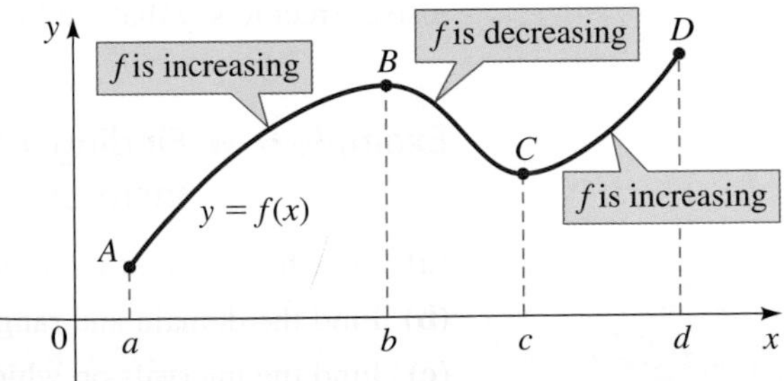

Figure 5 | f is increasing on (a, b) and (c, d); f is decreasing on (b, c)

We have the following definition.

From the definition we see that a function increases or decreases *on an interval*. It does not make sense to apply these definitions at a single point.

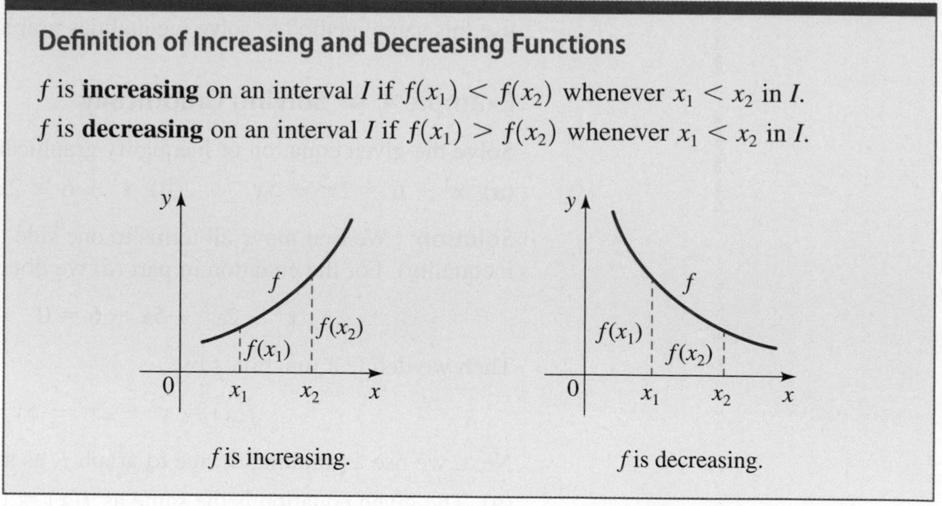

Definition of Increasing and Decreasing Functions

f is **increasing** on an interval I if $f(x_1) < f(x_2)$ whenever $x_1 < x_2$ in I.

f is **decreasing** on an interval I if $f(x_1) > f(x_2)$ whenever $x_1 < x_2$ in I.

Example 5 ■ Intervals on Which a Function Increases or Decreases

The graph in Figure 6 gives the weight W of a person at age x. Determine the intervals on which the function W is increasing and on which it is decreasing.

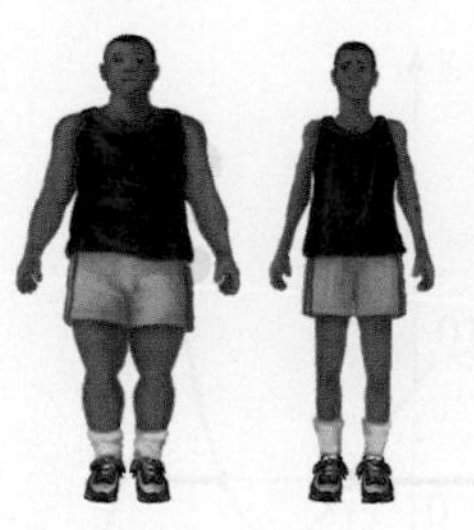

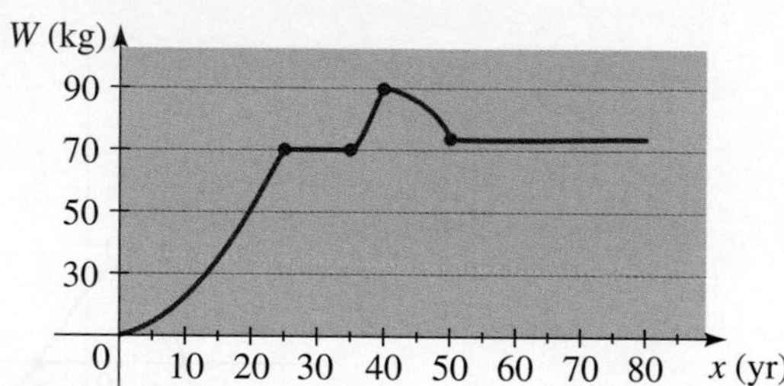

Figure 6 | Weight as a function of age

Solution The function W is increasing on the intervals $(0, 25)$ and $(35, 40)$. It is decreasing on $(40, 50)$. The function W is constant (neither increasing nor decreasing) on $(25, 35)$ and $(50, 80)$. This means that the person gained weight until age 25, then gained weight again between ages 35 and 40, and then lost weight between ages 40 and 50.

 **Now Try Exercise 65** ■

Note By convention we write the intervals on which a function is increasing or decreasing as open intervals. (It would also be true to say that the function is increasing or decreasing on the corresponding closed interval. So for instance, it is also correct to say that the function W in Example 5 is decreasing on $[40, 50]$.)

Example 6 ■ Finding Intervals on Which a Function Increases or Decreases

(a) Sketch a graph of the function $f(x) = 12x^2 + 4x^3 - 3x^4$.

(b) Find the domain and range of f.

(c) Find the intervals on which f is increasing and on which f is decreasing.

DOROTHY VAUGHAN (1910–2008) is famous for her work in leading a team of "human computers" at NASA. Vaughan was recognized as a talented mathematician and teacher when she was hired in 1943 (at the height of World War II) to work with a group of mathematicians at NASA. Vaughn believed that the job was a temporary war job, but her exceptional expertise was soon recognized, and she was chosen to lead the team, becoming the first African American woman to do so. Vaughn used complex numerical methods to calculate flight trajectories of rockets and flight characteristics of aircraft. Electronic computers had not yet been invented, so computations were done using pencil and paper, hence the term "human computers." When NASA introduced digital computers, Vaughn taught her team how to compute with the new technology. Vaughan said that working as a "computer" during the Space Race felt like being on "the cutting edge of something very exciting." In 2019 Vaughn was posthumously awarded the Congressional Gold Medal for her work at NASA.

The 2016 hit film *Hidden Figures* features the work of Dorothy and her colleagues Kathrine Johnson and Mary Jackson as "human computers" at NASA.

Solution

(a) We use a graphing device to sketch the graph in Figure 7 in an appropriate viewing rectangle.

(b) The domain of f is $\mathbb{R}$ because f is defined for all real numbers. From the graph, we find that the highest value is $f(2) = 32$. So the range of f is $(-\infty, 32]$.

(c) From the graph we see that f is increasing on the intervals $(-\infty, -1)$ and $(0, 2)$ and is decreasing on $(-1, 0)$ and $(2, \infty)$.

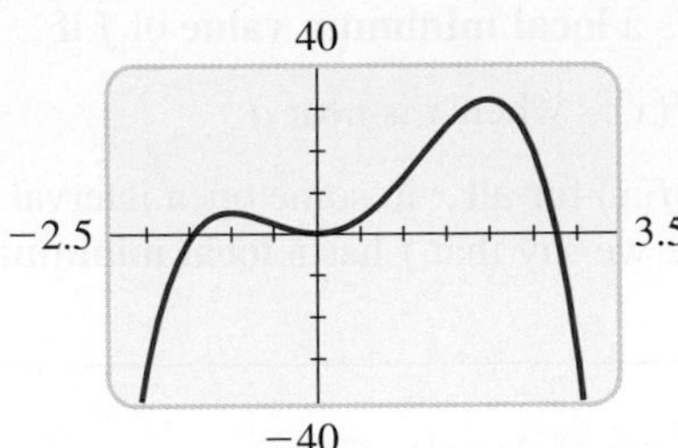

Figure 7 | Graph of $f(x) = 12x^2 + 4x^3 - 3x^4$

Now Try Exercise 41 ■

Example 7 ■ Finding Intervals Where a Function Increases and Decreases

(a) Sketch the graph of the function $f(x) = x^{2/3}$.

(b) Find the domain and range of the function.

(c) Find the intervals on which f is increasing and on which f is decreasing.

Solution

(a) We use a graphing device to sketch the graph in Figure 8.

(b) From the graph we observe that the domain of f is $\mathbb{R}$ and the range is $[0, \infty)$.

(c) From the graph we see that f is decreasing on $(-\infty, 0)$ and increasing on $(0, \infty)$.

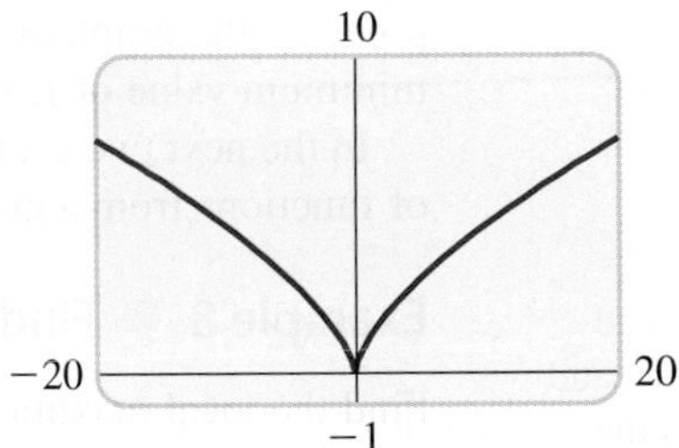

Figure 8 | Graph of $f(x) = x^{2/3}$

 Now Try Exercise 47 ■

■ Local Maximum and Minimum Values of a Function

Finding the largest or smallest values of a function is important in many applications. For example, if a function represents revenue or profit, then we are interested in its maximum value. For a function that represents cost, we would want to find its minimum value. (See *Focus on Modeling: Modeling with Functions* at the end of this chapter for many such examples.) We can easily find these values from the graph of a function. We first define what we mean by a local maximum or minimum.

Local Maxima and Minima of a Function

1. The function value $f(a)$ is a **local maximum value** of f if

$$f(a) \geq f(x) \quad \text{when } x \text{ is near } a$$

(This means that $f(a) \geq f(x)$ for all x in some open interval containing a.) In this case we say that f has a **local maximum** at $x = a$.

2. The function value $f(a)$ is a **local minimum value** of f if

$$f(a) \leq f(x) \quad \text{when } x \text{ is near } a$$

(This means that $f(a) \leq f(x)$ for all x in some open interval containing a.) In this case we say that f has a **local minimum** at $x = a$.

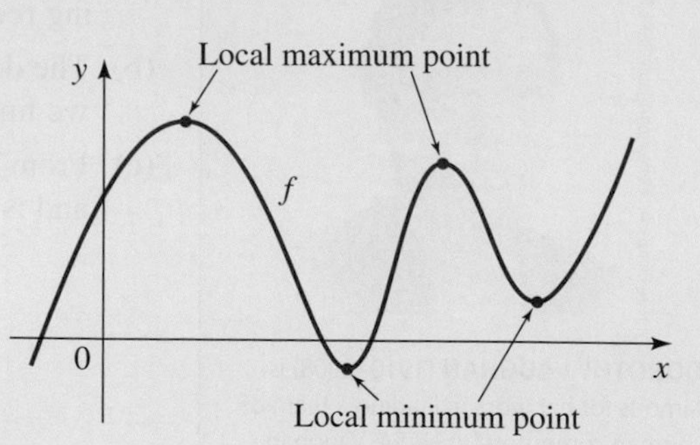

We can find the local maximum and minimum values of a function using a graphing device. If there is a viewing rectangle such that the point $(a, f(a))$ is the highest point on the graph of f *within* the viewing rectangle (not on the edge), then the number $f(a)$ is a local maximum value of f. (See Figure 9.) Notice that $f(a) \geq f(x)$ for all numbers x that are close to a.

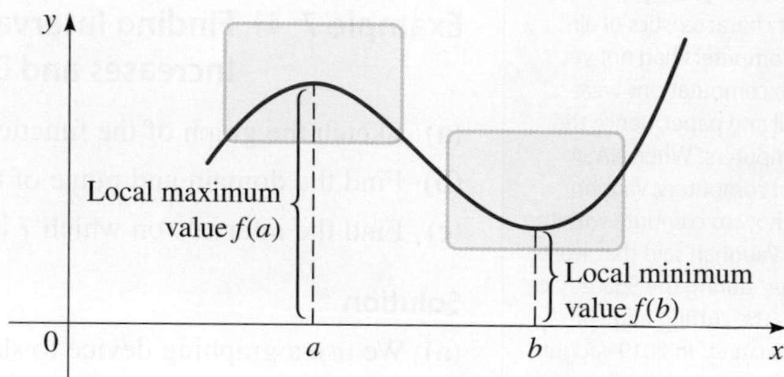

Figure 9

Similarly, if there is a viewing rectangle such that the point $(b, f(b))$ is the lowest point on the graph of f within the viewing rectangle, then the number $f(b)$ is a local minimum value of f. In this case $f(b) \leq f(x)$ for all numbers x that are close to b.

In the next two examples we use graphing devices to find local maxima and minima of functions from a graph.

On many math apps, you can find the coordinates of local maxima and minima by simply clicking on the appropriate point on the graph of the function.

Example 8 ■ Finding Local Maxima and Minima from a Graph

Find the local maximum and minimum values of the function $f(x) = x^3 - 8x + 1$, rounded to three decimal places.

Solution The graph of f is shown in Figure 10. There appears to be one local maximum between $x = -2$ and $x = -1$, and one local minimum between $x = 1$ and $x = 2$.

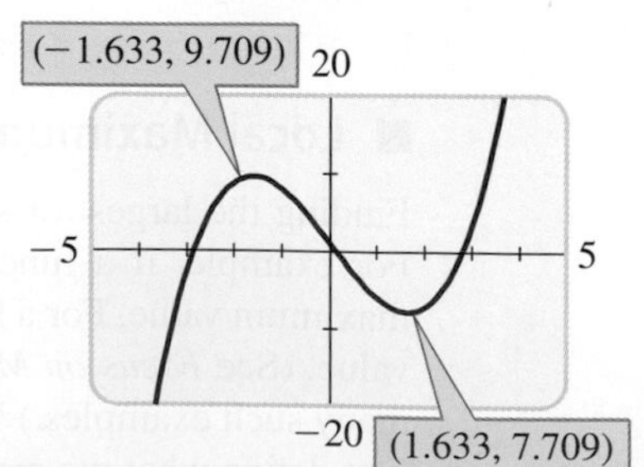

Figure 10 | Graph of $f(x) = x^3 - 8x + 1$

We see from the graph that the local maximum value of y is 9.709, and this value occurs when x is -1.633, rounded to three decimal places. The local minimum value is about -7.709, and this value occurs when $x \approx 1.633$.

Now Try Exercise 55 ■

Example 9 ■ A Model for Managing Traffic

See the *Discovery Project* referenced in Section 3.6, for how this model is obtained.

A highway engineer develops a formula to estimate the number of cars that can safely travel a particular highway at a given speed. The engineer assumes that each car is 5 m long, travels at a speed of x km/h, and follows the car in front of it at the safe following distance for that speed. The number N of cars that can pass a given point per minute can be modeled by the function

$$N(x) = \frac{25x}{5 + 5\left(\frac{x}{20}\right)^2}$$

Graph the function in the viewing rectangle $[0, 100]$ by $[0, 60]$.

(a) Find the intervals on which the function N is increasing and on which it is decreasing.

(b) Find the maximum value of N. What is the maximum carrying capacity of the road, and at what speed is it achieved?

Solution The graph is shown in Figure 11.

(a) From the graph we see that the function N is increasing on $(0, 20)$ and decreasing on $(20, \infty)$.

(b) There appears to be a maximum between $x = 19$ and $x = 21$. From the graph we see that the maximum value of N is about 50, and it occurs when x is 20. So the maximum carrying capacity is about 50 cars per minute at a speed of 20 km/h.

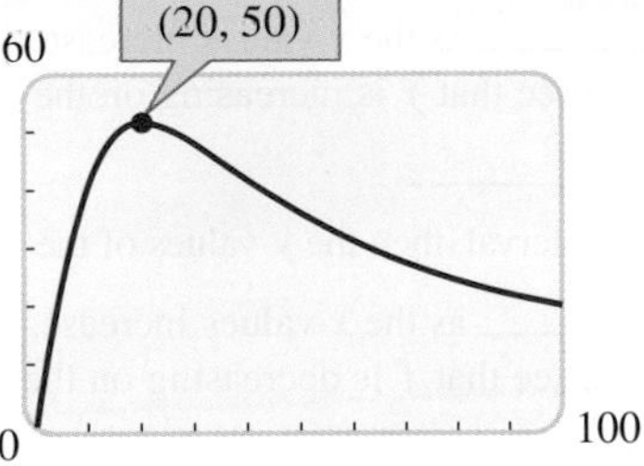

Figure 11 | Highway capacity at speed x

Now Try Exercise 73 ■

Discovery Project ■ Every Graph Tells a Story

Dave Carpenter/CartoonStock.com

A graph can often describe a real-world "story" much more quickly and effectively than many words. For example, the stock market crash of 1929 is effectively described by a graph of the Dow Jones Industrial Average. No words are needed to convey the message in the cartoon shown here. In this project we describe, or tell the story, that corresponds to a given graph as well as make graphs that correspond to a real-world story. You can find the project at **www.stewartmath.com**.

2.3 Exercises

Concepts

1–5 ■ Refer to the graph of f shown below.

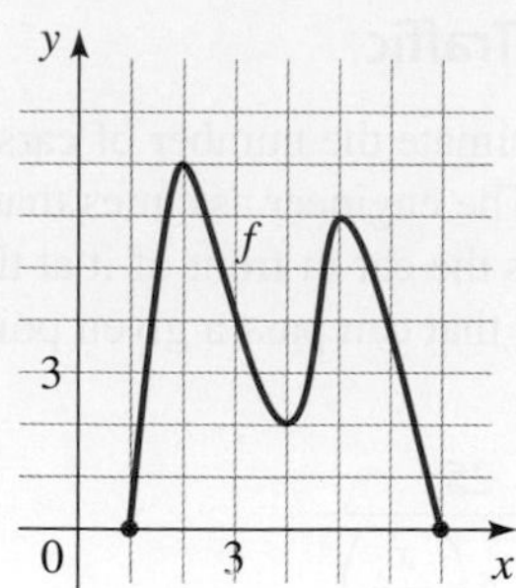

1. To find a function value $f(a)$ from the graph of f, we find the height of the graph above the x-axis at $x =$ ______.

From the graph of f we see that $f(3) =$ ______ and $f(5) =$ ____. The net change in f between $x = 3$ and $x = 5$ is $f($____$) - f($____$) =$ ____.

2. The domain of the function f is all the ____-values of the points on the graph, and the range is all the corresponding ____-values. From the graph of f we see that the domain of f is the interval ______ and the range of f is the interval ______.

3. (a) If f is increasing on an interval, then the y-values of the points on the graph ______ as the x-values increase. From the graph of f we see that f is increasing on the intervals ______ and ______.

(b) If f is decreasing on an interval, then the y-values of the points on the graph ______ as the x-values increase. From the graph of f we see that f is decreasing on the intervals ______ and ______.

4. (a) A function value $f(a)$ is a local maximum value of f if $f(a)$ is the ______ value of f on some open interval containing a. From the graph of f we see that there are two local maximum values of f: One local maximum is ______, and it occurs when $x = 2$; the other local maximum is ______, and it occurs when $x =$ ______.

(b) The function value $f(a)$ is a local minimum value of f if $f(a)$ is the ______ value of f on some open interval containing a. From the graph of f we see that there is one local minimum value of f. The local minimum value is ______, and it occurs when $x =$ ______.

5. The solutions of the equation $f(x) = 0$ are the ____-intercepts of the graph of f. The solution of the inequality $f(x) \geq 0$ is the set of x-values at which the graph of f is on or above the ____-axis. From the graph of f we find that the solutions of the equation $f(x) = 0$ are $x =$ ____ and $x =$ ____, and the solution of the inequality $f(x) \geq 0$ is ______.

6. (a) To solve the equation $2x + 1 = -x + 4$ graphically, we graph the functions $f(x) =$ ______ and $g(x) =$ ______ on the same set of axes and determine the values of x at which the graphs of f and g intersect. Graph f and g below, and use the graphs to solve the equation. The solution is $x =$ ______.

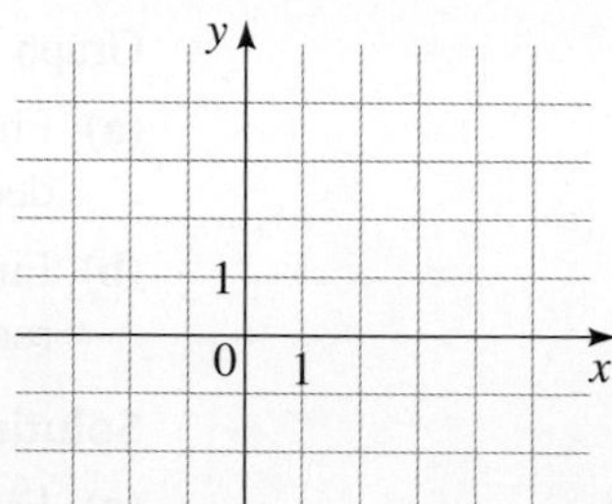

(b) To solve the inequality $2x + 1 < -x + 4$ graphically, we graph the functions $f(x) =$ ______ and $g(x) =$ ______ on the same set of axes and find the values of x at which the graph of g is ______ (higher/lower) than the graph of f. From the graphs in part (a) we see that the solution of the inequality is the interval (____, ____).

Skills

7. Values of a Function The graph of a function h is given.

(a) Find $h(-2)$, $h(0)$, $h(2)$, and $h(3)$.

(b) Find the domain and range of h.

(c) Find the values of x for which $h(x) = 3$.

(d) Find the values of x for which $h(x) \leq 3$.

(e) Find the net change in h between $x = -3$ and $x = 3$.

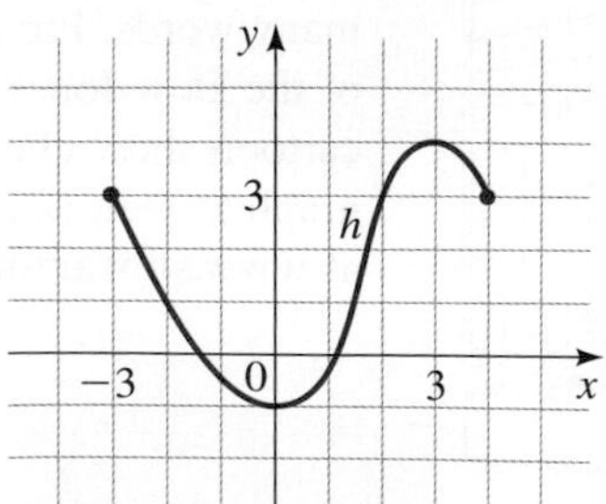

8. Values of a Function The graph of a function g is given.

(a) Find $g(-4)$, $g(-2)$, $g(0)$, $g(2)$, and $g(4)$.

(b) Find the domain and range of g.

(c) Find the values of x for which $g(x) = 3$.

(d) Estimate the values of x for which $g(x) \leq 0$.

(e) Find the net change in g between $x = -1$ and $x = 2$.

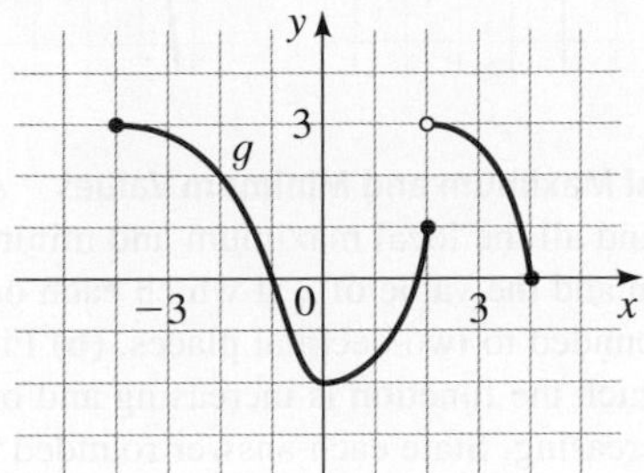

9. Solving Equations and Inequalities Graphically Graphs of the functions f and g are given.

(a) Which is larger, $f(0)$ or $g(0)$?

(b) Which is larger, $f(-1)$ or $g(-1)$?

(c) For which values of x is $f(x) = g(x)$?

(d) Find the values of x for which $f(x) \leq g(x)$.

(e) Find the values of x for which $f(x) > g(x)$.

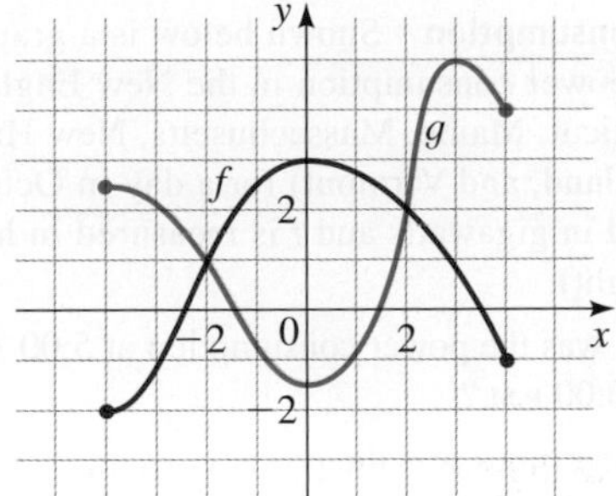

10. Solving Equations and Inequalities Graphically Graphs of the functions f and g are given.

(a) Which is larger, $f(6)$ or $g(6)$?

(b) Which is larger, $f(3)$ or $g(3)$?

(c) Find the values of x for which $f(x) = g(x)$.

(d) Find the values of x for which $f(x) \leq g(x)$.

(e) Find the values of x for which $f(x) > g(x)$.

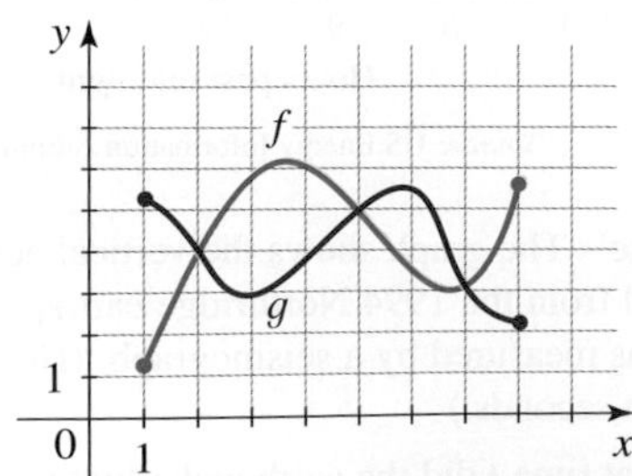

11–14 ■ Domain and Range from a Graph The graph of a function f is given. Use the graph to find the domain and range of f.

11.

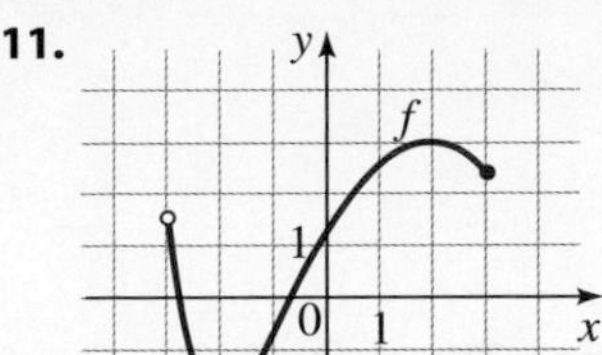

12.

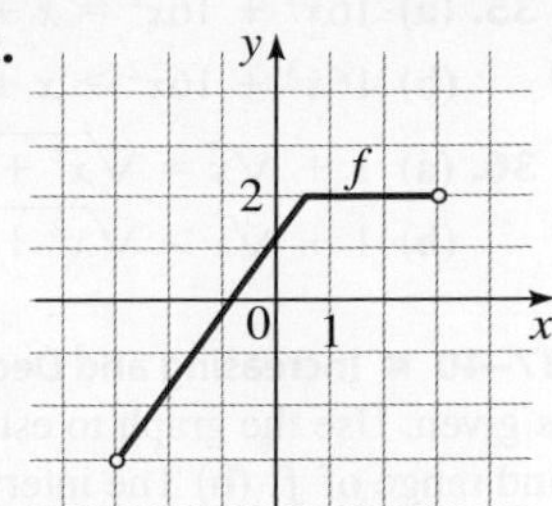

13.

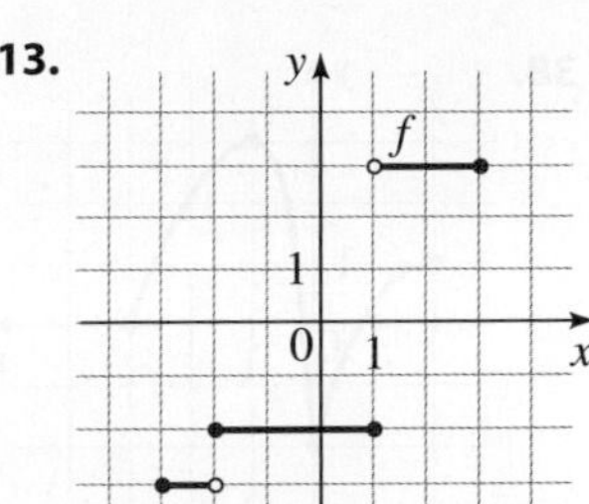

14.

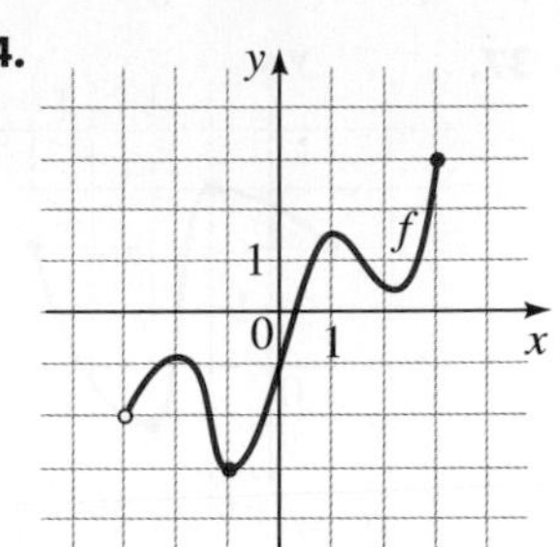

15–22 ■ Domain and Range from a Graph A function f is given. **(a)** Sketch a graph of f. **(b)** Use the graph to find the domain and range of f.

15. $f(x) = 2x + 3$ **16.** $f(x) = 3x - 2$

17. $f(x) = x^2 - 3$ **18.** $f(x) = x^2 + 2$

19. $f(x) = x - 2, \quad -2 \leq x \leq 5$

20. $f(x) = 4 - 2x, \quad 1 < x < 4$

21. $f(x) = x^2 - 1, \quad -3 \leq x \leq 3$

22. $f(x) = x^3 - 1; \quad -3 \leq x \leq 3$

23–28 ■ Finding Domain and Range Graphically A function f is given. **(a)** Use a graphing device to draw the graph of f. **(b)** Find the domain and range of f from the graph.

23. $f(x) = x^2 + 4x + 3$ **24.** $f(x) = \sqrt{x - 1}$

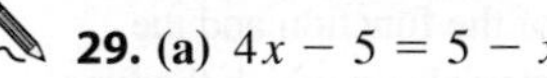

25. $f(x) = -\sqrt{36 - x^2}$ **26.** $f(x) = \dfrac{4}{x^2 + 2}$

27. $f(x) = \dfrac{2x}{x^2 + 1}$ **28.** $f(x) = x^4 - 6x^2$

29–32 ■ Solving Equations and Inequalities Graphically Solve the given equation or inequality graphically.

29. (a) $4x - 5 = 5 - x$ **(b)** $4x - 5 < 5 - x$

30. (a) $3 - 4x = 8x - 9$ **(b)** $3 - 4x \geq 8x - 9$

31. (a) $x^2 = 2 - x$ **(b)** $x^2 \leq 2 - x$

32. (a) $-x^2 = 3 - 4x$ **(b)** $-x^2 \geq 3 - 4x$

33–36 ■ Solving Equations and Inequalities Graphically Solve the given equation or inequality graphically. State your answers rounded to two decimals.

33. (a) $x^3 + 3x^2 = -x^2 + 3x + 7$

(b) $x^3 + 3x^2 \geq -x^2 + 3x + 7$

34. **(a)** $5x^2 - x^3 = -x^2 + 3x + 4$
(b) $5x^2 - x^3 \le -x^2 + 3x + 4$

35. **(a)** $16x^3 + 16x^2 = x + 1$
(b) $16x^3 + 16x^2 \ge x + 1$

36. **(a)** $1 + \sqrt{x} = \sqrt{x^2 + 1}$
(b) $1 + \sqrt{x} > \sqrt{x^2 + 1}$

37–40 ■ Increasing and Decreasing The graph of a function f is given. Use the graph to estimate the following. **(a)** The domain and range of f. **(b)** The intervals on which f is increasing and on which f is decreasing.

37.

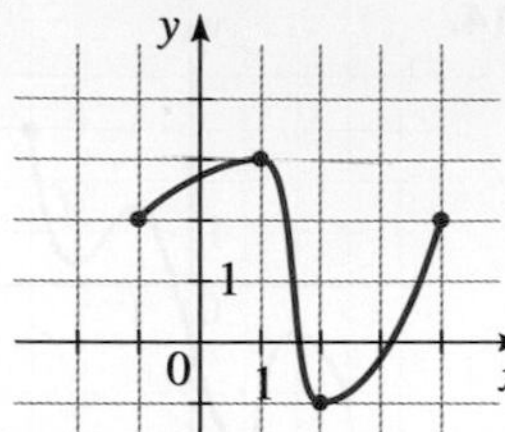

38.

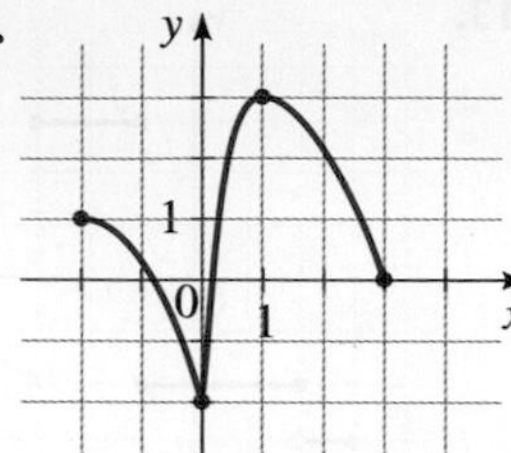

39.

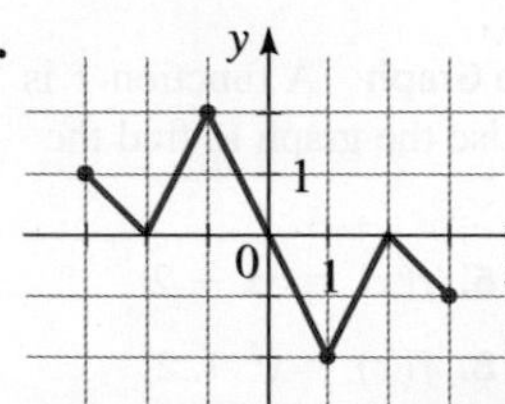

40.

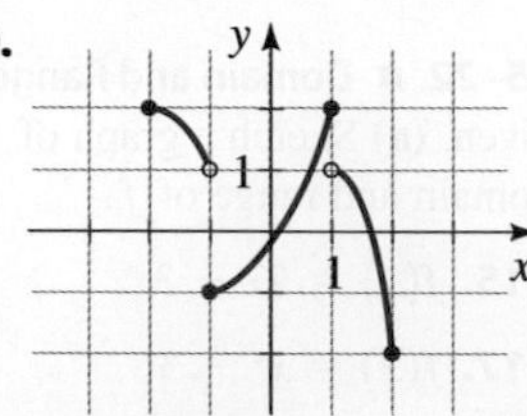

41–50 ■ Increasing and Decreasing A function f is given. **(a)** Use a graphing device to draw the graph of f. **(b)** Find the domain and range of f. **(c)** State approximately the intervals on which f is increasing and on which f is decreasing.

41. $f(x) = x^2 - 5x$ **42.** $f(x) = x^3 - 4x$

43. $f(x) = 2x^3 - 3x^2 - 12x$ **44.** $f(x) = x^4 - 16x^2$

45. $f(x) = x^3 + 2x^2 - x - 2$

46. $f(x) = x^4 - 4x^3 + 2x^2 + 4x - 3$

47. $f(x) = x^{2/5}$ **48.** $f(x) = 4 - x^{2/3}$

49. $f(x) = 2 + \sqrt{x + 3}$ **50.** $f(x) = \sqrt{25 - x^2}$

51–54 ■ Local Maximum and Minimum Values The graph of a function f is given. Use the graph to estimate the following: **(a)** All the local maximum and minimum values of the function and the value of x at which each occurs. **(b)** The intervals on which the function is increasing and on which the function is decreasing.

51.

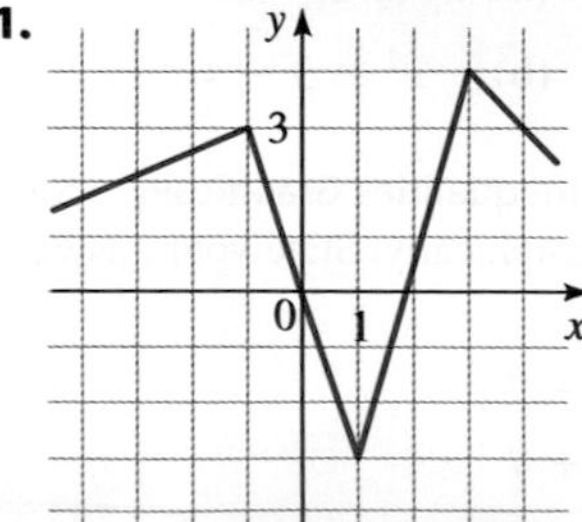

52.

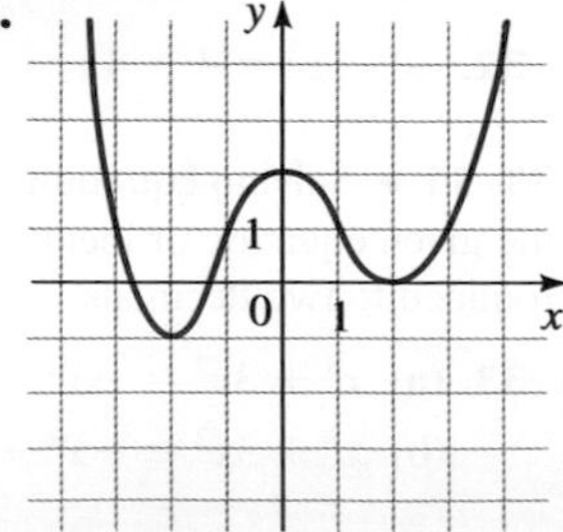

53. **54.**

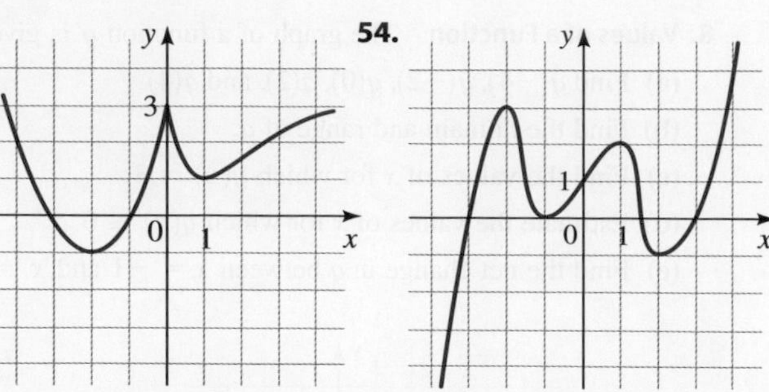

55–62 ■ Local Maximum and Minimum Values A function is given. **(a)** Find all the local maximum and minimum values of the function and the value of x at which each occurs. State each answer rounded to two decimal places. **(b)** Find the intervals on which the function is increasing and on which the function is decreasing. State each answer rounded to two decimal places.

55. $f(x) = x^3 - x$ **56.** $f(x) = 3 + x + x^2 - x^3$

57. $g(x) = x^4 - 2x^3 - 11x^2$ **58.** $g(x) = x^5 - 8x^3 + 20x$

59. $U(x) = x\sqrt{6 - x}$ **60.** $U(x) = x\sqrt{x - x^2}$

61. $V(x) = \dfrac{1 - x^2}{x^3}$ **62.** $V(x) = \dfrac{1}{x^2 + x + 1}$

Applications

63. Power Consumption Shown below is a graph of the electric power consumption in the New England states (Connecticut, Maine, Massachusetts, New Hampshire, Rhode Island, and Vermont) for a day in October (P is measured in gigawatts and t is measured in hours starting at midnight).

(a) What was the power consumption at 5:00 A.M.? At 10:00 P.M.?

(b) At what times was the power consumption the lowest? The highest?

(c) Find the net change in the power consumption from 5:00 A.M. to 10:00 P.M.

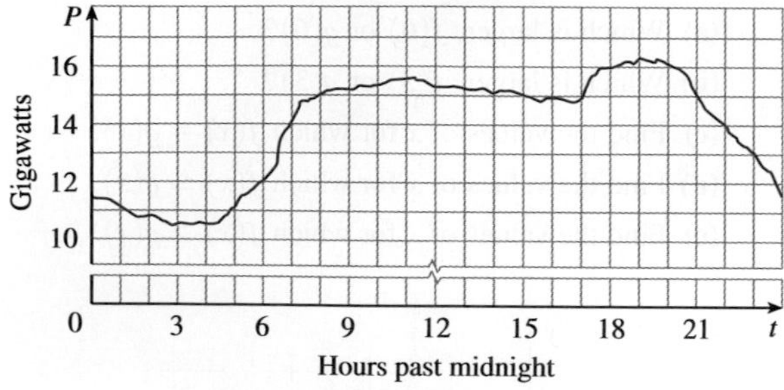

Source: US Energy Information Administration

64. Earthquake The graph shows the vertical acceleration of the ground from the 1994 Northridge earthquake in Los Angeles, as measured by a seismograph. (Here t represents the time in seconds.)

(a) At what time t did the earthquake first make noticeable movements of the earth?

(b) At what time t did the earthquake seem to end?

(c) At what time t was the maximum intensity of the earthquake reached?

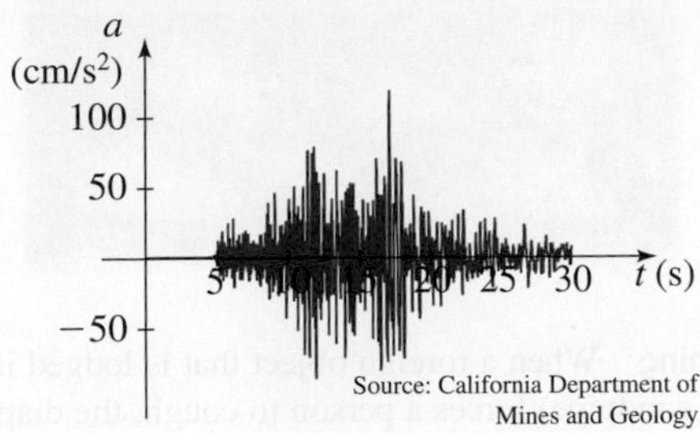

65. Weight Function The graph gives the weight W of a person at age x.

(a) Determine the intervals on which the function W is increasing and those on which it is decreasing.

(b) What do you think happened when this person was 30 years old?

(c) Find the net change in the person's weight W from age 10 to age 20.

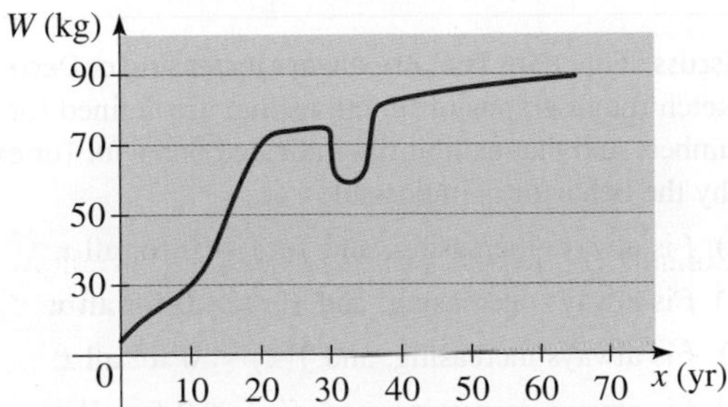

66. Distance Function The graph gives a sales representative's distance from home as a function of time on a certain day.

(a) Determine the time intervals on which the distance from home was increasing and those on which it was decreasing.

(b) Describe in words what the graph indicates about the representative's travels on this day.

(c) Find the net change in distance from home between noon and 1:00 P.M.

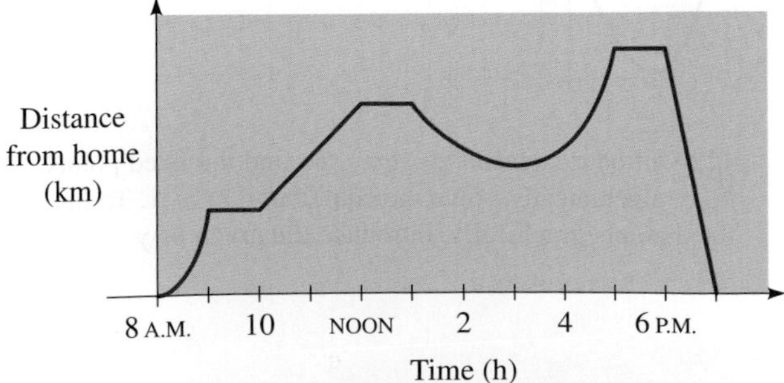

67. Changing Water Levels The graph shows the depth of water W in a reservoir over a one-year period as a function of the number of days x since the beginning of the year.

(a) Determine the intervals on which the function W is increasing and on which it is decreasing.

(b) At what value of x does W achieve a local maximum? A local minimum?

(c) Find the net change in the depth W from 100 days to 300 days.

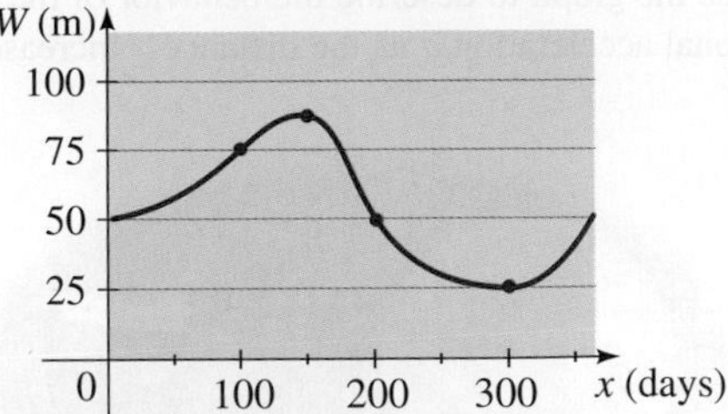

68. Population Growth and Decline The graph shows the population P in a small industrial city from 1970 to 2020. The variable x represents the number of years since 1970.

(a) Determine the intervals on which the function P is increasing and on which it is decreasing.

(b) What was the maximum population, and in what year was it attained?

(c) Find the net change in the population P from 1990 to 2010.

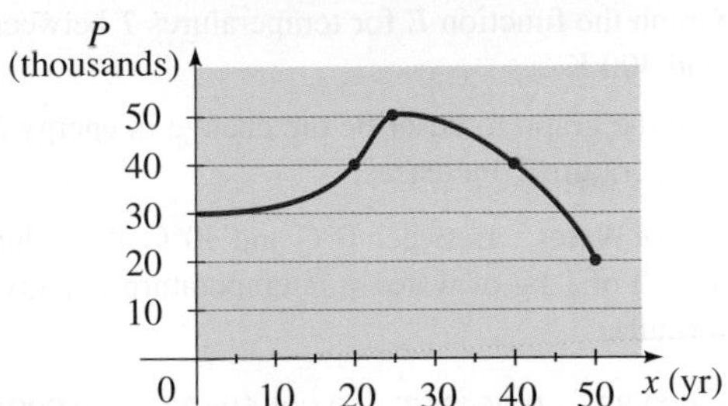

69. Hurdle Race Three runners compete in a 100-meter hurdle race. The graph depicts the distance run as a function of time for each runner. Describe in words what the graph tells you about this race. Who won the race? Did each runner finish the race? What do you think happened to Runner B?

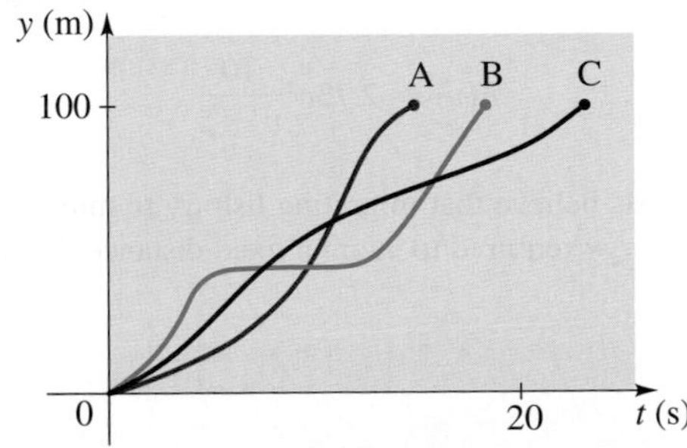

70. Gravity Near the Moon We can use Newton's Law to calculate the acceleration a due to gravity of a spacecraft at a distance x km from the center of the moon:

$$a(x) = \frac{5 \times 10^6}{x^2}$$

where a is measured in m/s^2.

(a) The radius r of the moon is 1737 km. Graph the function a for values of x between r and $2r$.

(b) Use the graph to describe the behavior of the gravitational acceleration a as the distance x increases.

71. Radii of Stars Astronomers infer the radii of stars using the Stefan Boltzmann Law:

$$E(T) = (5.67 \times 10^{-8})T^4$$

where E is the energy radiated per unit of surface area measured in Watts per square meter (W/m^2) and T is the absolute temperature measured in kelvins (K).

(a) Graph the function E for temperatures T between 100 K and 300 K.

(b) Use the graph to describe the change in energy E as the temperature T increases.

72. Volume of Water Between 0°C and 30°C, the volume V (in cm^3) of 1 kg of water at a temperature T is given by the formula

$$V = 999.87 - 0.06426T + 0.0085043T^2 - 0.0000679T^3$$

Find the temperature at which the volume of 1 kg of water is a minimum.

[*Source: Physics* by D. Halliday and R. Resnick]

73. Migrating Fish A fish swims at a speed v relative to the water, against a current of 5 km/h. Using a mathematical model of energy expenditure, it can be shown that the total energy E required to swim a distance of 10 km is given by

$$E(v) = 2.73v^3 \frac{10}{v-5}$$

Biologists believe that migrating fish try to minimize the total energy required to swim a fixed distance. Find the value of v that minimizes the energy required.

[*Note:* This result has been verified; migrating fish swim against a current at a speed 50% greater than the speed of the current.]

74. Coughing When a foreign object that is lodged in the trachea (windpipe) forces a person to cough, the diaphragm thrusts upward, causing an increase in pressure in the lungs. At the same time, the trachea contracts, causing the expelled air to move faster and increasing the pressure on the foreign object. According to a mathematical model of coughing, the velocity v (in cm/s) of the airstream through an average-sized person's trachea is related to the radius r of the trachea (in cm) by the function

$$v(r) = 3.2(1 - r)r^2 \qquad \tfrac{1}{2} \le r \le 1$$

Determine the value of r for which v is a maximum.

■ Discuss ■ Discover ■ Prove ■ Write

75. Discuss: Functions That Are Always Increasing or Decreasing Sketch rough graphs of functions that are defined for all real numbers and that exhibit the indicated behavior (or explain why the behavior is impossible).

(a) f is always increasing, and $f(x) > 0$ for all x.

(b) f is always decreasing, and $f(x) > 0$ for all x.

(c) f is always increasing, and $f(x) < 0$ for all x.

(d) f is always decreasing, and $f(x) < 0$ for all x.

76. Discuss ■ Discover: Fixed Points A *fixed point* of a function f is a number x for which $f(x) = x$. How would you go about finding the fixed points of a function algebraically? Graphically? Find the fixed points of the function f.

(a) $f(x) = 5x - x^2$

(b) $f(x) = x^3 + x + 1$

(c)

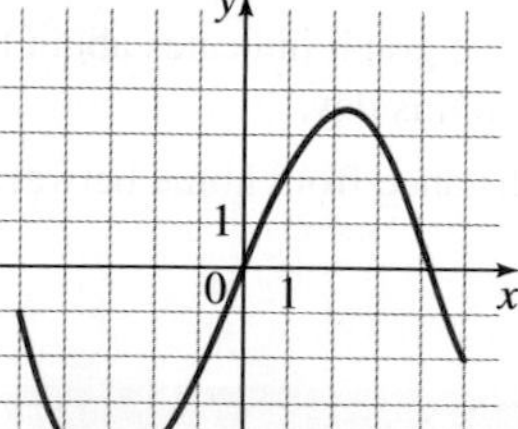

(d)

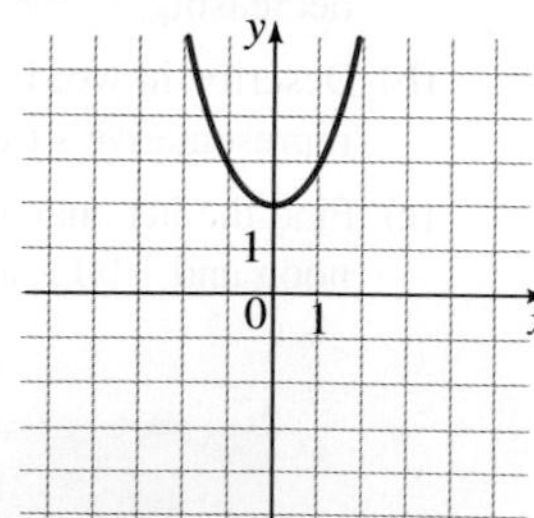

PS *Introduce something extra.* To find the fixed points algebraically, solve the equation $f(x) = x$. To find fixed points graphically, introduce the graph of $y = x$.

2.4 Average Rate of Change of a Function

■ Average Rate of Change ■ Linear Functions Have Constant Rate of Change

Functions are often used to model changing quantities. In this section we learn how to find the rate at which the values of a function change as the input variable changes.

■ Average Rate of Change

We are all familiar with the concept of speed: If you drive a distance of 120 kilometers in 2 hours, then your average speed, or rate of travel, is $\frac{120\text{ km}}{2\text{ h}} = 60$ km/h. Now suppose you take a car trip and record the distance that you travel every few minutes. The distance s you have traveled is a function of the time t:

$$s(t) = \text{total distance traveled at time } t$$

We graph the function s as shown in Figure 1. The graph shows that you have traveled a total of 50 kilometers after 1 hour, 75 kilometers after 2 hours, 140 kilometers after 3 hours, and so on. To find your *average* speed between any two points on the trip, we divide the distance traveled by the time elapsed.

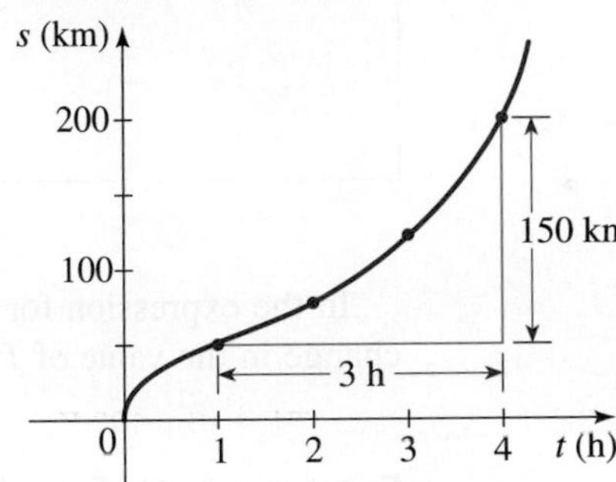

Figure 1 | Average speed

Let's calculate your average speed between 1:00 P.M. and 4:00 P.M. The time elapsed is $4 - 1 = 3$ hours. To find the distance you traveled, we subtract the distance at 1:00 P.M. from the distance at 4:00 P.M., that is, $200 - 50 = 150$ km. Thus your average speed is

$$\text{average speed} = \frac{\text{distance traveled}}{\text{time elapsed}} = \frac{150\text{ km}}{3\text{ h}} = 50\text{ km/h}$$

The average speed we have just calculated can be expressed by using function notation:

$$\text{average speed} = \frac{s(4) - s(1)}{4 - 1} = \frac{200 - 50}{3} = 50\text{ km/h}$$

Observe that the average speed is different over different time intervals. For example, between 2:00 P.M. and 3:00 P.M. we find that

$$\text{average speed} = \frac{s(3) - s(2)}{3 - 2} = \frac{140 - 75}{1} = 65\text{ km/h}$$

Finding average rates of change is important in many contexts. For instance, we might be interested in knowing how quickly the air temperature is dropping as a storm approaches or how fast revenues are increasing from the sale of a new product. So we need to know how to determine the average rate of change of the functions that model these quantities. In fact, the concept of average rate of change can be defined for any function.

Average Rate of Change

The **average rate of change** of the function $y = f(x)$ between $x = a$ and $x = b$ is

$$\text{average rate of change} = \frac{\text{change in } y}{\text{change in } x} = \frac{f(b) - f(a)}{b - a}$$

The expression $\dfrac{f(b) - f(a)}{b - a}$ is called a *difference quotient.*

The average rate of change is the slope of the **secant line** between $x = a$ and $x = b$ on the graph of f, that is, the line that passes through $(a, f(a))$ and $(b, f(b))$.

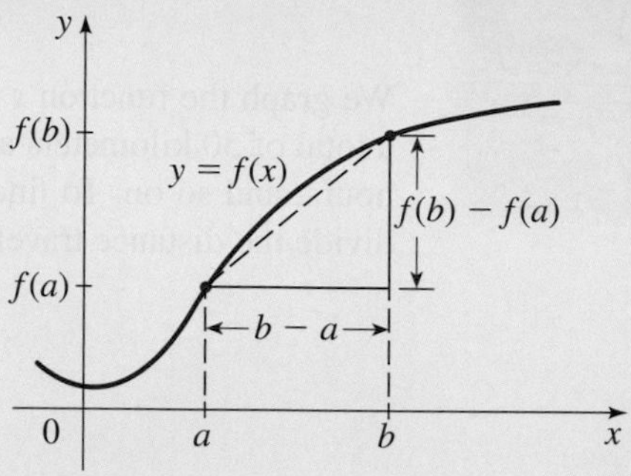

In the expression for average rate of change, the numerator $f(b) - f(a)$ is the net change in the value of f between $x = a$ and $x = b$ (see Section 2.1).

Example 1 ■ Calculating the Average Rate of Change

For the function $f(x) = (x - 3)^2$, whose graph is shown in Figure 2, find the net change and the average rate of change between the following values of x.

(a) $x = 1$ and $x = 3$

(b) $x = 4$ and $x = 7$

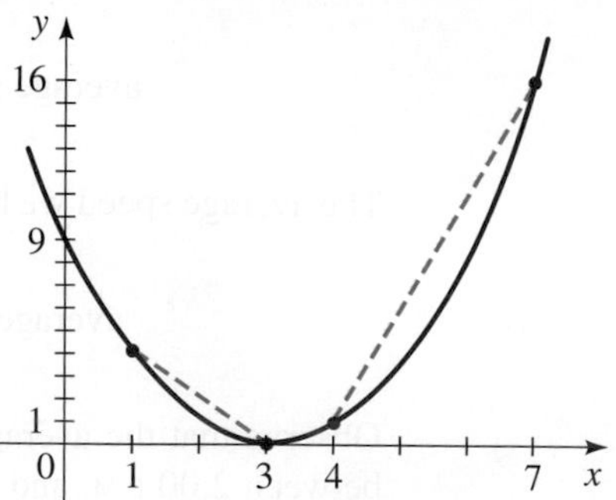

Figure 2 | $f(x) = (x - 3)^2$

Solution

(a)

Net change $= f(3) - f(1)$	Definition
$= (3 - 3)^2 - (1 - 3)^2$	Use $f(x) = (x - 3)^2$
$= -4$	Calculate

$$\text{Average rate of change} = \frac{f(3) - f(1)}{3 - 1} \qquad \text{Definition}$$

$$= \frac{-4}{2} = -2 \qquad \text{Calculate}$$

(b) $$\text{Net change} = f(7) - f(4) \qquad \text{Definition}$$

$$= (7 - 3)^2 - (4 - 3)^2 \qquad f(x) = (x - 3)^2$$

$$= 15 \qquad \text{Calculate}$$

$$\text{Average rate of change} = \frac{f(7) - f(4)}{7 - 4} \qquad \text{Definition}$$

$$= \frac{15}{3} = 5 \qquad \text{Calculate}$$

Now Try Exercise 13 ■

Example 2 ■ Average Speed of a Falling Object

If an object is dropped from a high cliff or a tall building, then the distance (in meters) it has fallen after t seconds is given by the function $d(t) = 5t^2$. Find its average speed (average rate of change) over the following intervals.

(a) Between 1 s and 5 s **(b)** Between $t = a$ and $t = a + h$

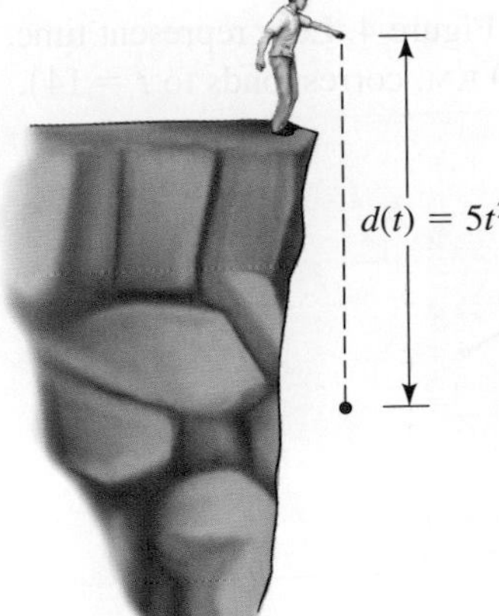

Function: In t seconds the stone falls $5t^2$ m.

Solution

(a) $$\text{Average rate of change} = \frac{d(5) - d(1)}{5 - 1} \qquad \text{Definition}$$

$$= \frac{5(5)^2 - 5(1)^2}{5 - 1} \qquad d(t) = 5t^2$$

$$= \frac{125 - 5}{4} \qquad \text{Calculate}$$

$$= 30 \text{ m/s} \qquad \text{Calculate}$$

(b) $$\text{Average rate of change} = \frac{d(a + h) - d(a)}{(a + h) - a} \qquad \text{Definition}$$

$$= \frac{5(a + h)^2 - 5(a)^2}{(a + h) - a} \qquad d(t) = 5t^2$$

$$= \frac{5(a^2 + 2ah + h^2 - a^2)}{h} \qquad \text{Expand and factor 5}$$

$$= \frac{5(2ah + h^2)}{h} \qquad \text{Simplify numerator}$$

$$= \frac{5h(2a + h)}{h} \qquad \text{Factor } h$$

$$= 5(2a + h) \qquad \text{Simplify}$$

Now Try Exercises 15 and 19 ■

Note The expression for the average rate of change (or the difference quotient) as in Example 2(b) is used in calculus to calculate an *instantaneous rate of change* (see Exercise 39 and Section 12.3).

The graphs in Figure 3 show that if a function is increasing on an interval, then the average rate of change between any two points in that interval is positive, whereas if a function is decreasing on an interval, then the average rate of change between any two points in that interval is negative.

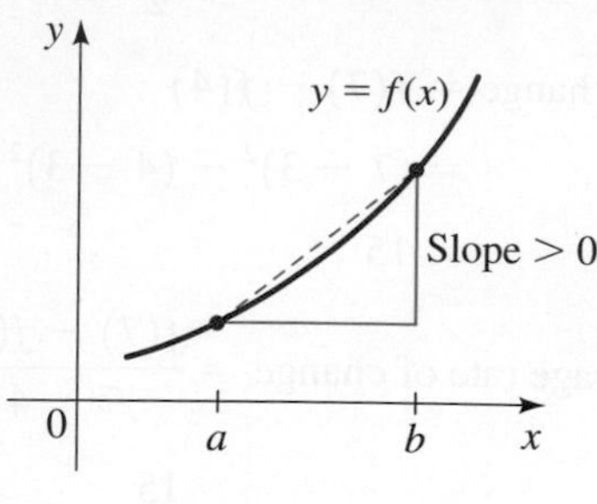

f increasing
Average rate of change positive

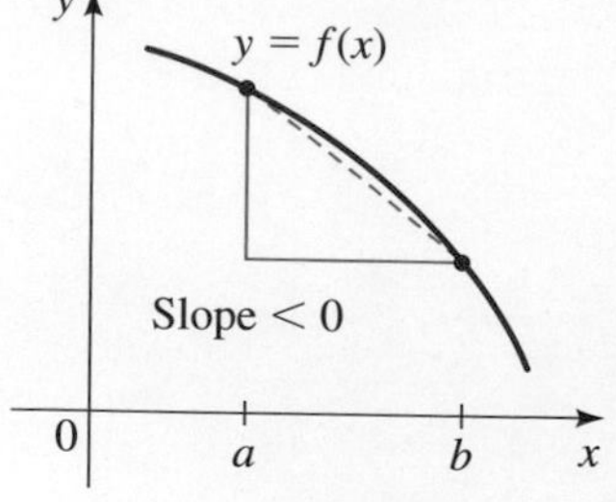

f decreasing
Average rate of change negative

Figure 3

Example 3 ■ Average Rate of Temperature Change

Time	Temperature (°C)
8:00 A.M.	3
9:00 A.M.	4
10:00 A.M.	6
11:00 A.M.	10
12:00 NOON	13
1:00 P.M.	16
2:00 P.M.	18
3:00 P.M.	19
4:00 P.M.	17
5:00 P.M.	14
6:00 P.M.	13
7:00 P.M.	11

The table in the margin gives the outdoor temperatures observed by a science student on a spring day. Draw a graph of the data, and find the average rate of change of temperature between the following times.

(a) 8:00 A.M. and 9:00 A.M.

(b) 1:00 P.M. and 3:00 P.M.

(c) 4:00 P.M. and 7:00 P.M.

Solution A graph of the temperature data is shown in Figure 4. Let t represent time, measured in hours since midnight (so, for example, 2:00 P.M. corresponds to $t = 14$). Define the function F by $F(t) =$ temperature at time t.

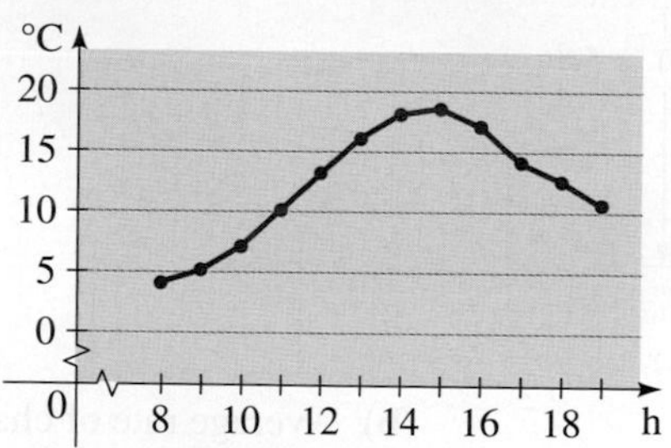

Figure 4

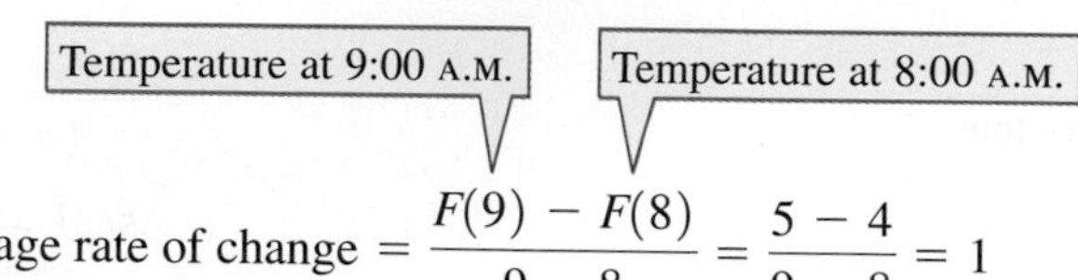

(a) $$\text{Average rate of change} = \frac{F(9) - F(8)}{9 - 8} = \frac{5 - 4}{9 - 8} = 1$$

The average rate of change was 1°C per hour.

(b) $$\text{Average rate of change} = \frac{F(15) - F(13)}{15 - 13} = \frac{19 - 16}{2} = 1.5$$

The average rate of change was 1.5°C per hour.

(c) $$\text{Average rate of change} = \frac{F(19) - F(16)}{19 - 16} = \frac{11 - 17}{3} \approx -2$$

The average rate of change was about −2°C per hour during this time interval. The negative sign indicates that the temperature was dropping.

Now Try Exercise 31 ■

■ Linear Functions Have Constant Rate of Change

Recall that a function of the form $f(x) = mx + b$ is a linear function (see Section 2.2). Its graph is a line with slope m. On the other hand, if a function f has constant rate of change, then it must be a linear function. (You are asked to prove these facts in Exercises 2.5.53 and 2.5.54) In general, the average rate of change of a linear function between any two x-values is the constant m. In the next example we find the average rate of change for a particular linear function.

Example 4 ■ Linear Functions Have Constant Rate of Change

Let $f(x) = 3x - 5$. Find the average rate of change of f between the following values of x.

(a) $x = 0$ and $x = 1$

(b) $x = 3$ and $x = 7$

(c) $x = a$ and $x = b$

What conclusion can you draw from your answers?

Solution

(a) $$\text{Average rate of change} = \frac{f(1) - f(0)}{1 - 0} = \frac{(3\cdot 1 - 5) - (3\cdot 0 - 5)}{1} = \frac{(-2) - (-5)}{1} = 3$$

(b) $$\text{Average rate of change} = \frac{f(7) - f(3)}{7 - 3} = \frac{(3\cdot 7 - 5) - (3\cdot 3 - 5)}{4} = \frac{16 - 4}{4} = 3$$

(c) $$\text{Average rate of change} = \frac{f(b) - f(a)}{b - a} = \frac{(3b - 5) - (3a - 5)}{b - a} = \frac{3b - 3a - 5 + 5}{b - a} = \frac{3(b - a)}{b - a} = 3$$

It appears that the average rate of change is always 3 for this function. In fact, part (c) proves that the rate of change between any two arbitrary points $x = a$ and $x = b$ is 3.

 **Now Try Exercise 25** ■

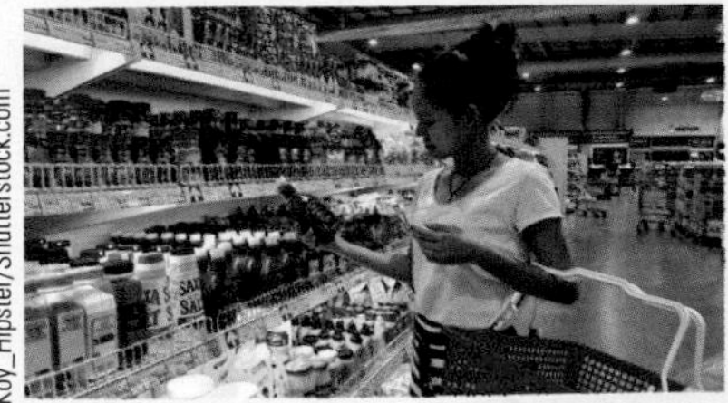

Koy_Hipster/Shutterstock.com

Discovery Project ■ Concavity: When Rates of Change Change

In the real world, rates of change often themselves change. A statement like "inflation is rising, but at a slower rate" involves a change of a rate of change. Also, the speed (rate of change of distance) of a car increases when a driver accelerates and decreases when a driver decelerates. From Example 4 we see that functions whose graph is a line (linear functions) have constant rate of change. In this project we explore how the shape of a graph corresponds to a changing rate of change. You can find the project at **www.stewartmath.com.**

2.4 Exercises

Concepts

1. If you travel 100 kilometers in two hours, then your average speed for the trip is

average speed = $\dfrac{\quad}{\quad}$ = ______

2. The average rate of change of a function f between $x = a$ and $x = b$ is

average rate of change = $\dfrac{\quad}{\quad}$

3. The average rate of change of the function $f(x) = x^2$ between $x = 1$ and $x = 5$ is

average rate of change = $\dfrac{\quad}{\quad}$ = ______

4. (a) The average rate of change of a function f between $x = a$ and $x = b$ is the slope of the ______ line between $(a, f(a))$ and $(b, f(b))$.

(b) The average rate of change of the linear function $f(x) = 3x + 5$ between any two x-values is ______.

5–6 ■ *Yes or No*? If *No*, give a reason.

5. (a) Is the average rate of change of a function between $x = a$ and $x = b$ the slope of the secant line through $(a, f(a))$ and $(b, f(b))$?

(b) Is the average rate of change of a linear function the same for all intervals?

6. (a) Can the average rate of change of an increasing function ever be negative?

(b) If the average rate of change of a function between $x = a$ and $x = b$ is negative, then is the function necessarily decreasing on the interval (a, b)?

Skills

7–10 ■ Net Change and Average Rate of Change The graph of a function is given. Determine **(a)** the net change and **(b)** the average rate of change between the indicated points on the graph.

7.

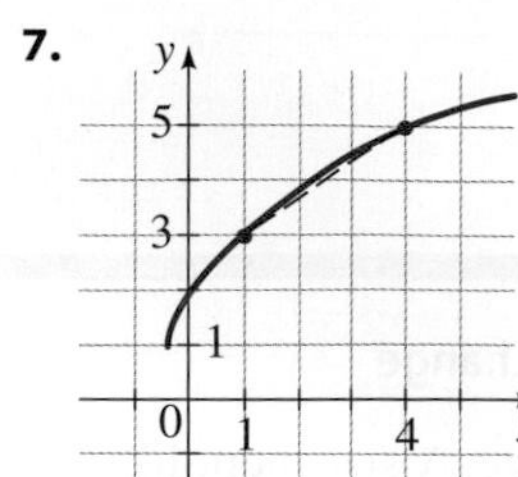

8.

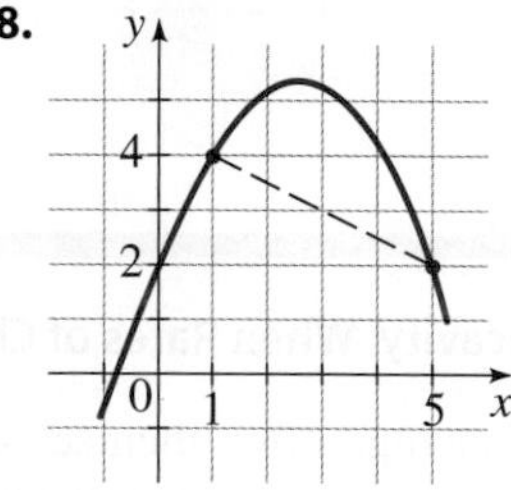

9.

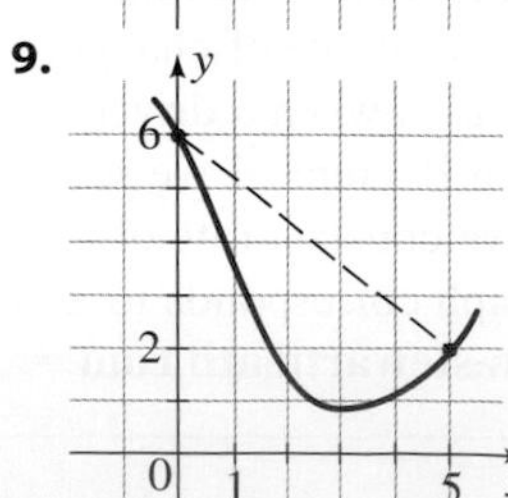

10.

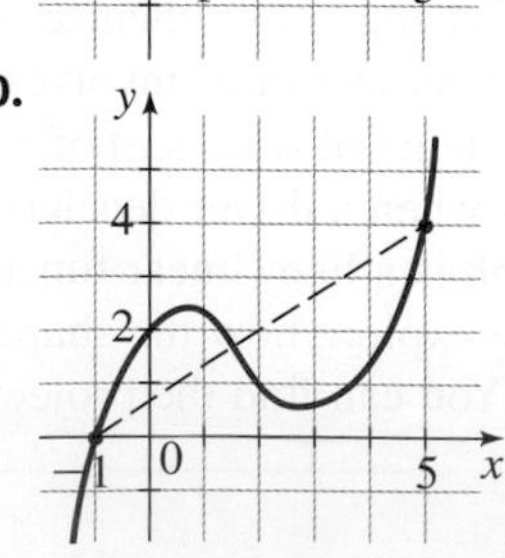

11–18 ■ Net Change and Average Rate of Change A function is given. Determine **(a)** the net change and **(b)** the average rate of change between the given values of the variable.

11. $f(t) = 5t + 3;\quad t = 4, t = 7$

12. $s(t) = 4 - 2t;\quad t = 1, t = 5$

13. $g(x) = 2 - \frac{1}{2}x;\quad x = -6, x = 10$

14. $h(x) = \dfrac{3x - 4}{5};\quad x = 2, x = 5$

15. $f(t) = 3t^2 + t;\quad t = 1, t = 3$

16. $f(z) = 3z - z^2;\quad z = -3, z = 0$

17. $f(x) = x^3 - 4x^2;\quad x = 0, x = 10$

18. $g(t) = t^4 - t^3 + t^2;\quad t = -2, t = 2$

19–24 ■ Difference Quotient A function is given. Find an expression for the difference quotient (or average rate of change) between $x = a$ and $x = a + h$. Simplify your answer.

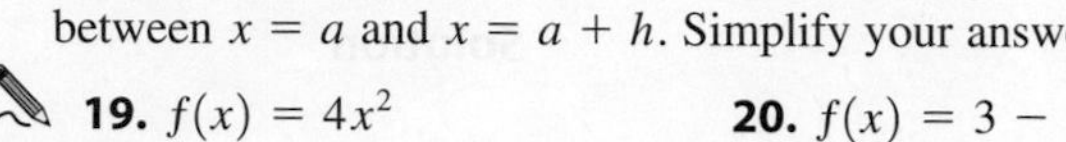

19. $f(x) = 4x^2$

20. $f(x) = 3 - 10x^2$

21. $f(x) = \dfrac{1}{x}$

22. $f(x) = \dfrac{2}{x + 1}$

23. $f(x) = \sqrt{x}$

24. $f(x) = \dfrac{2}{x^2}$

25–26 ■ Average Rate of Change of a Linear Function A linear function is given. **(a)** Find the average rate of change of the function between $x = a$ and $x = b$. **(b)** Show that the average rate of change is the same as the slope of the line.

25. $f(x) = \frac{1}{2}x + 3$

26. $g(x) = -4x + 2$

Skills Plus

27. Average Rate of Change The graphs of the functions f and g are shown. The function ___ (f or g) has a greater average rate of change between $x = 0$ and $x = 1$. The function ___ (f or g) has a greater average rate of change between $x = 1$ and $x = 2$. The functions f and g have the same average rate of change between $x =$ ___ and $x =$ ___.

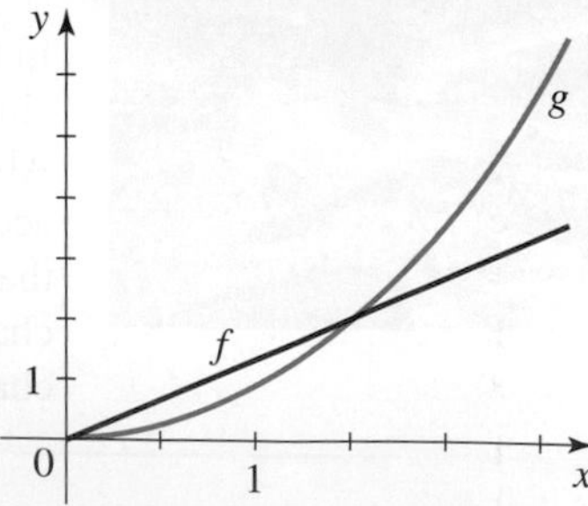

28. Average Rate of Change Graphs of the functions f, g, and h are shown below. What can you say about the average rate of change of each function on the successive intervals $[0, 1]$, $[1, 2]$, $[2, 3]$, . . .?

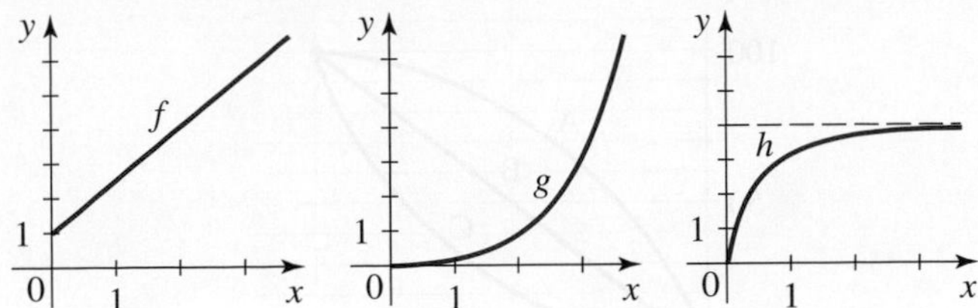

Applications

29. Changing Water Levels The graph shows the depth of water W in a reservoir over a one-year period as a function of the number of days x since the beginning of the year.

(a) What was the average rate of change of W between $x = 100$ and $x = 200$? What does the sign of your answer indicate?

(b) Identify an interval where the average rate of change is 0.

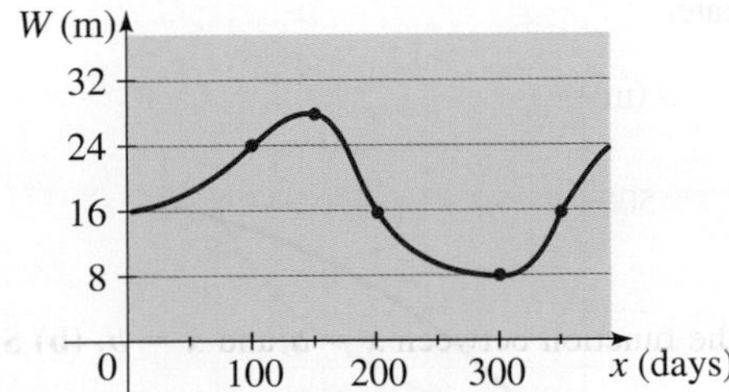

30. Population Growth and Decline The graph shows the population P in a small industrial city from 1970 to 2020. The variable x represents the number of years since 1970.

(a) What was the average rate of change of P between $x = 20$ and $x = 25$?

(b) Interpret the value of the average rate of change that you found in part (a).

(c) Identify a time period where the average rate of change is 0.

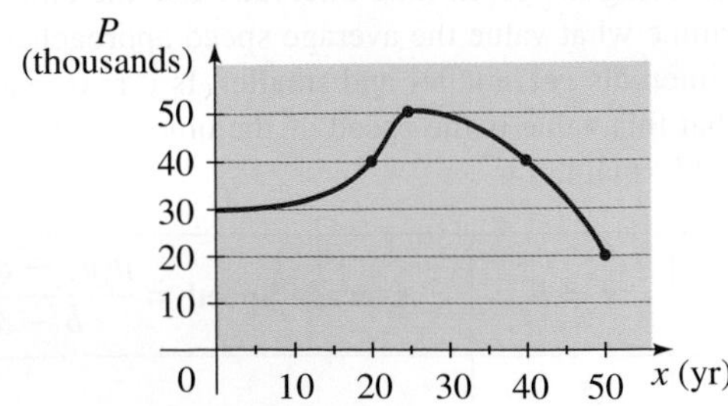

31. Population Growth and Decline The table gives the population in a small coastal community for the period 2002–2020. Figures shown are for January 1 in each year.

(a) What was the average rate of change of population between 2008 and 2012?

(b) What was the average rate of change of population between 2014 and 2018?

(c) For what period of time was the population increasing?

(d) For what period of time was the population decreasing?

Year	Population
2002	3220
2004	3645
2006	4357
2008	4869
2010	5871
2012	6375
2014	6288
2016	5318
2018	4921
2020	4636

32. Running Speed A runner is sprinting on a circular track that is 200 m in circumference. An observer uses a stopwatch to record the runner's time at the end of each lap, obtaining the data in the following table.

(a) What was the runner's average speed (rate) between 68 s and 152 s?

(b) What was the runner's average speed between 263 s and 412 s?

(c) Calculate the runner's speed for each lap. Is the runner slowing down, speeding up, or neither?

Time (s)	Distance (m)
32	200
68	400
108	600
152	800
203	1000
263	1200
335	1400
412	1600

33. Snack Cake Sales The table shows the number of creme-filled snack cakes sold in a small convenience store for the period 2010–2020.

(a) What was the average rate of change of sales between 2010 and 2020?

(b) What was the average rate of change of sales between 2011 and 2012?

(c) What was the average rate of change of sales between 2012 and 2013?

(d) Between which two successive years did snack cake sales *increase* most quickly? *Decrease* most quickly?

Year	Snack Cakes Sold
2010	1146
2011	1042
2012	638
2013	1145
2014	1738
2015	1804
2016	1121
2017	1987
2018	2533
2019	2983
2020	3629

34. Book Collection Between 2000 and 2020 a rare book collector purchased books at the rate of 60 books per year. Use this information to complete the following table. (Note that not every year is given in the table.)

Year	Number of Books	Year	Number of Books
2000	530	2015	
2001	590	2017	
2002		2018	
2006		2019	
2010		2020	1730
2012			

35. Cooling Soup When a bowl of hot soup is left in a room, the soup eventually cools to room temperature. The temperature T of the soup is a function of time t. The table below gives the temperature (in °C) of a bowl of soup t minutes after it was set on the table. Find the average rate of change of the temperature of the soup over the first 20 minutes and over the next 20 minutes. During which interval did the soup cool more quickly?

t (min)	T (°C)	t (min)	T (°C)
0	93	35	34
5	78	40	32
10	65	50	27
15	56	60	25
20	48	90	22
25	42	120	21
30	38	150	21

36. Farms in the United States The graph gives the number of farms in the United States from 1850 to 2020.

(a) Estimate the average rate of change in the number of farms between the years (i) 1860 and 1890 and (ii) 1950 and 1980.

(b) In which decade did the number of farms experience the greatest average rate of decline?

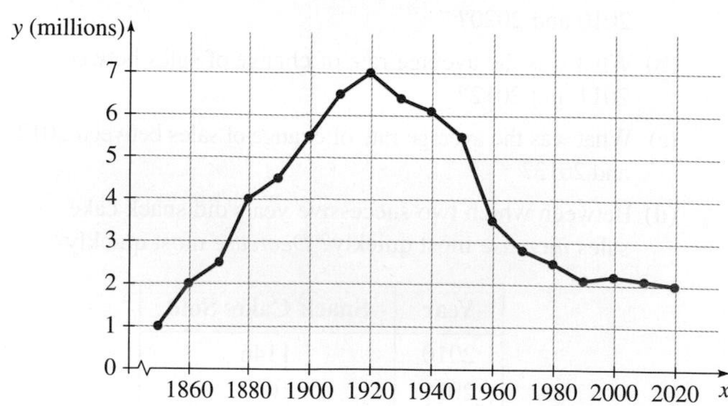

37. Three-Way Tie A downhill skiing race ends in a three-way tie for first place. The graph shows distance as a function of time for each of the three winners, A, B, and C.

(a) Find the average speed for each skier.

(b) Describe the differences between the ways in which the three participants skied the race.

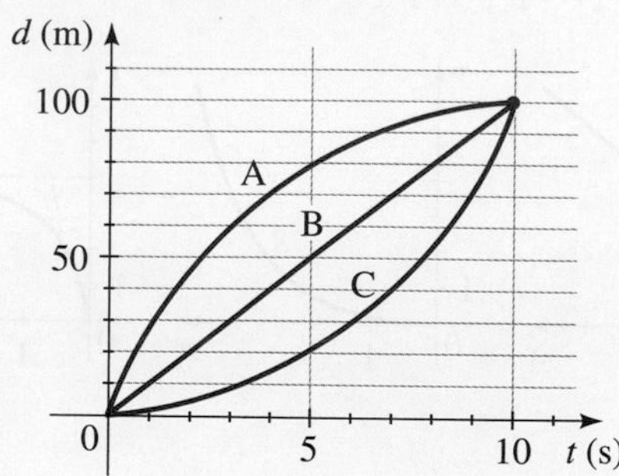

38. Speed Skating Two speed skaters, A and B, are racing in a 500-meter event. The graph shows the distance they have traveled as a function of the time from the start of the race.

(a) Who won the race?

(b) Find the average speed during the first 10 s for each skater.

(c) Find the average speed during the last 15 s for each skater.

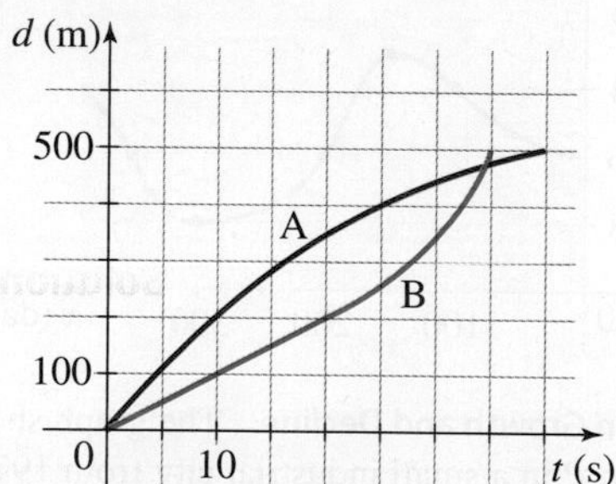

■ Discuss ■ Discover ■ Prove ■ Write

39. Discuss ■ Discover: Limiting Behavior of Average Speed An object is dropped from a high cliff, and the distance (in meters) it has fallen after t seconds is given by the function $d(t) = 5t^2$. Complete the table to find the average speed during the given time intervals. Use the table to determine what value the average speed approaches as the time intervals get smaller and smaller. Is it reasonable to say that this value is the speed of the object at the instant $t = 3$? Explain.

$t = a$	$t = b$	Average Speed $= \dfrac{d(b) - d(a)}{b - a}$
3	3.5	
3	3.1	
3	3.01	
3	3.001	
3	3.0001	

2.5 Linear Functions and Models

■ Linear Functions ■ Slope and Rate of Change ■ Making and Using Linear Models

In this section we study the simplest functions that can be expressed by an algebraic expression: linear functions.

■ Linear Functions

Recall that a *linear function* is a function of the form $f(x) = ax + b$. So in the expression defining a linear function the variable occurs to the first power only. We can also express a linear function in equation form as $y = ax + b$. From Section 1.10 we know that the graph of this equation is a line with slope a and y-intercept b.

> **Linear Functions**
>
> A **linear function** is a function of the form $f(x) = ax + b$.
>
> The graph of a linear function is a line with slope a and y-intercept b.

Example 1 ■ Identifying Linear Functions

Determine whether the given function is linear. If the function is linear, express the function in the form $f(x) = ax + b$.

(a) $f(x) = 2 + 3x$ **(b)** $g(x) = 3(1 - 2x)$

(c) $h(x) = x(4 + 3x)$ **(d)** $k(x) = \dfrac{1 - 5x}{4}$

Solution

(a) We have $f(x) = 2 + 3x = 3x + 2$. So f is a linear function in which a is 3 and b is 2.

(b) We have $g(x) = 3(1 - 2x) = -6x + 3$. So g is a linear function in which a is -6 and b is 3.

(c) We have $h(x) = x(4 + 3x) = 4x + 3x^2$, which is not a linear function because the variable x is squared in the second term of the expression for h.

(d) We have $k(x) = \dfrac{1 - 5x}{4} = -\dfrac{5}{4}x + \dfrac{1}{4}$. So k is a linear function in which a is $-\frac{5}{4}$ and b is $\frac{1}{4}$.

Now Try Exercise 7 ■

Example 2 ■ Graphing a Linear Function

Let f be the linear function defined by $f(x) = 3x + 2$.

(a) Make a table of values, and sketch a graph.

(b) What is the slope of the graph of f?

Solution

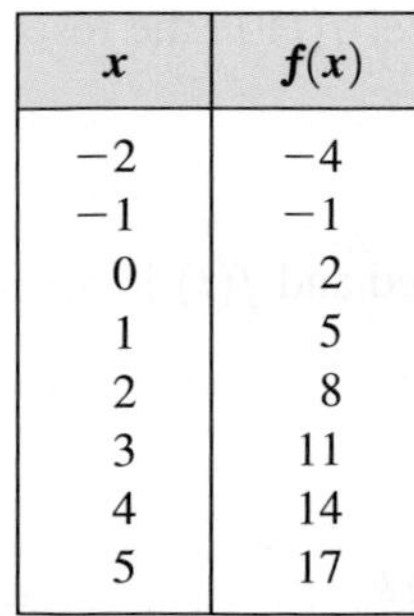

x	$f(x)$
−2	−4
−1	−1
0	2
1	5
2	8
3	11
4	14
5	17

(a) A table of values is shown in the margin. Since f is a linear function, its graph is a line. So to obtain the graph of f, we plot any two points from the table and draw the straight line that contains the points. We use the points $(1, 5)$ and $(4, 14)$. The graph is the line shown in Figure 1. You can check that the other points in the table of values also lie on the line.

From the box on the previous page, you can see that the slope of the graph of $f(x) = 3x + 2$ is 3.

(b) Using the points given in Figure 1, we see that the slope is

$$\text{slope} = \frac{14 - 5}{4 - 1} = 3$$

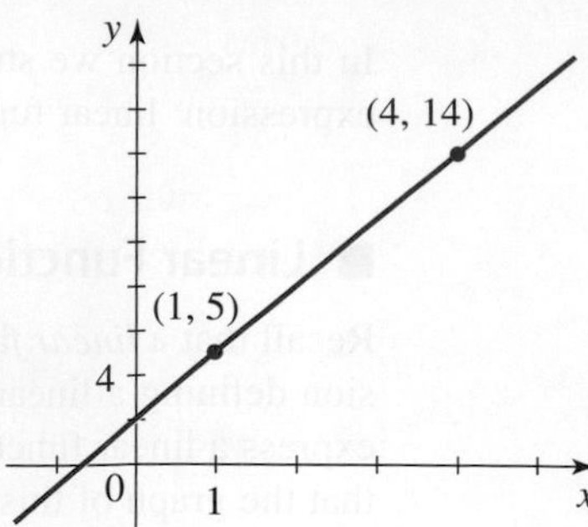

Figure 1 | Graph of the linear function $f(x) = 3x + 2$

Now Try Exercise 15 ■

■ Slope and Rate of Change

In Exercise 54 we prove that every function with constant rate of change is linear.

Let $f(x) = ax + b$ be a linear function. If x_1 and x_2 are two different values for x and if $y_1 = f(x_1)$ and $y_2 = f(x_2)$, then the points (x_1, y_1) and (x_2, y_2) lie on the graph of f. From the definitions of slope and average rate of change we have

$$\text{slope} = \frac{y_2 - y_1}{x_2 - x_1} = \frac{f(x_2) - f(x_1)}{x_2 - x_1} = \text{average rate of change}$$

From Section 1.10 we know that the *slope* of a linear function is the same between any two points. From the equation above we conclude that the *average rate of change* of a linear function is the same between any two points. Moreover, the average rate of change is equal to the slope (see Exercise 53). Since the average rate of change of a linear function is the same between any two points, it is called simply the **rate of change.**

Slope and Rate of Change

For the linear function $f(x) = ax + b$, the slope of the graph of f and the rate of change of f are both equal to a, the coefficient of x.

$$a = \text{slope of graph of } f = \text{rate of change of } f$$

Note The difference between "slope" and "rate of change" is simply a difference in point of view. For example, to describe how a reservoir fills up over time, it is natural to talk about the rate at which the water level is rising, but we can also think of the slope of the graph of the water level (see Example 3). To describe the steepness of a staircase, it is natural to talk about the slope of the trim board of the staircase, but we can also think of the rate at which the stairs rise (see Example 5).

Example 3 ■ Slope and Rate of Change

A dam is built on a river to create a reservoir. The water level $f(t)$ in the reservoir at time t is given by

$$f(t) = 4.5t + 28$$

where t is the number of years since the dam was constructed and $f(t)$ is measured in meters.

(a) Sketch a graph of f.

(b) What is the slope of the graph?

(c) At what rate is the water level in the reservoir changing?

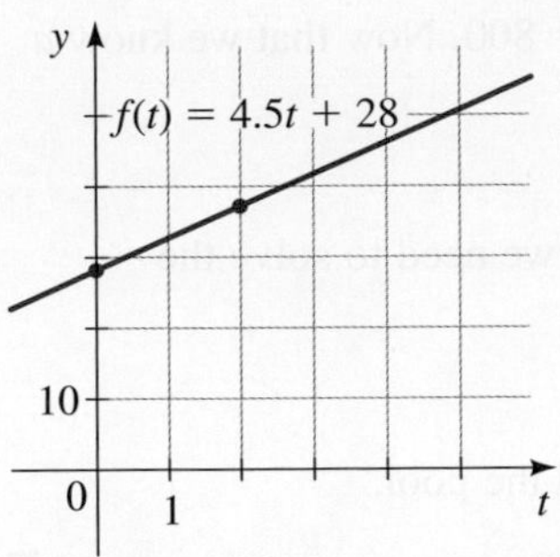

Figure 2 | Water level as a function of time

Solution

(a) A graph of f is shown in Figure 2.

(b) The graph is a line with slope 4.5, the coefficient of t.

(c) The rate of change of f is 4.5, the coefficient of t. Since time t is measured in years and the water level $f(t)$ is measured in meters, the water level in the reservoir is changing at the rate of 4.5 m per year. Since this rate of change is positive, the water level is rising.

Now Try Exercises 19 and 39 ■

■ Making and Using Linear Models

When a linear function is used to model the relationship between two quantities, the slope of the graph of the function is the rate of change of the one quantity with respect to the other. For example, the graph in Figure 3(a) gives the amount of gas in a tank that is being filled. The slope between the indicated points is

$$a = \frac{6 \text{ L}}{3 \text{ min}} = 2 \text{ L/min}$$

The slope is the rate at which the tank is being filled, 2 L per minute.

In Figure 3(b) the tank is being drained at the rate of 0.03 L per minute, and the slope is -0.03.

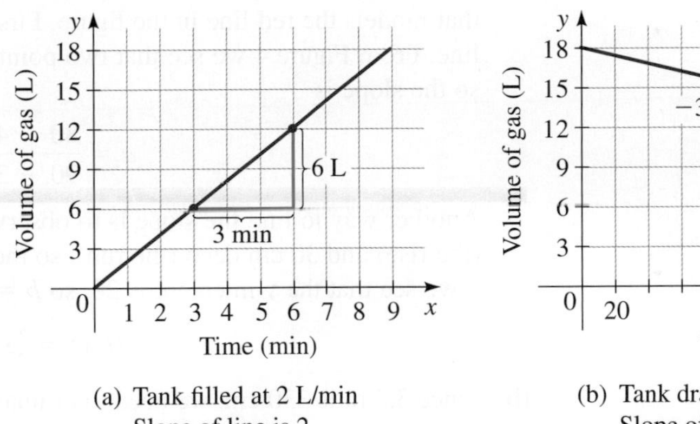

(a) Tank filled at 2 L/min
Slope of line is 2

(b) Tank drained at 0.03 L/min
Slope of line is -0.03

Figure 3 | Amount of gas as a function of time

In the following examples we model real-world situations using linear functions. In each of these examples the model involves a constant rate of change (or a constant slope).

Example 4 ■ Making a Linear Model from a Rate of Change

Water is being pumped into a swimming pool at the rate of 20 L per min. Initially, the pool contains 800 L of water.

(a) Find a linear function V that models the volume of water in the pool at any time t.

(b) If the pool has a capacity of 2400 L, how long does it take to fill the pool?

There are 800 L of water in the pool at time $t = 0$.

Solution

(a) We need to find a linear function

$$V(t) = at + b$$

that models the volume $V(t)$ of water in the pool after t minutes. The rate of change of volume is 20 L per min, so $a = 20$. Since the pool contains 800 L to begin with, we have $V(0) = a \cdot 0 + b = 800$, so $b = 800$. Now that we know a and b, we get the model

$$V(t) = 20t + 800$$

(b) We want to find the time t at which $V(t) = 2400$. So we need to solve the equation

$$2400 = 20t + 800$$

Solving for t, we get $t = 80$. So it takes 80 min to fill the pool.

 **Now Try Exercise 41** ■

Example 5 ■ Making a Linear Model from a Slope

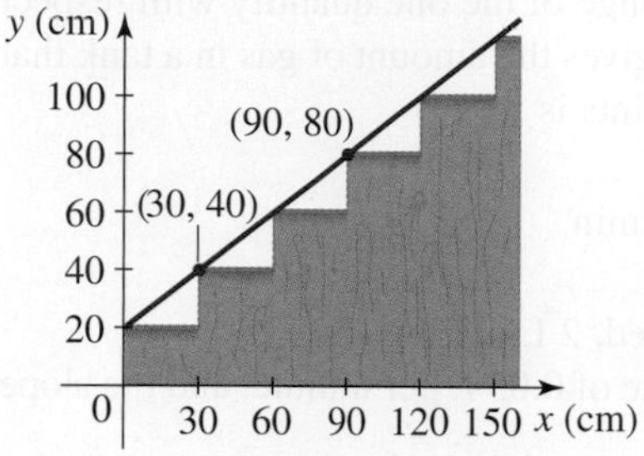

Figure 4 | Slope of a staircase

In Figure 4 we have placed a staircase in a coordinate plane, with the origin at the bottom left corner. The red line in the figure is the edge of the trim board of the staircase.

(a) Find a linear function H that models the height of the trim board above the floor.

(b) If the space available to build a staircase is 3.3 m wide, how high does the staircase reach?

Solution

(a) We need to find a function

$$H(x) = ax + b$$

that models the red line in the figure. First we find the value of a, the slope of the line. From Figure 4 we see that two points on the line are (30, 40) and (90, 80), so the slope is

$$a = \frac{80 - 40}{90 - 30} = \frac{2}{3}$$

Another way to find the slope is to observe that each of the steps is 20 cm high (the rise) and 30 cm deep (the run), so the slope of the line is $\frac{20}{30} = \frac{2}{3}$. From Figure 4 we see that the y-intercept is 20, so $b = 20$. So the model we want is

$$H(x) = \tfrac{2}{3}x + 20$$

(b) Since 3.3 m is 330 cm, we need to evaluate the function H when x is 330. We have

$$H(330) = \tfrac{2}{3}(330) + 20 = 240$$

So the staircase reaches a height of 240 cm, or 2.4 m.

 **Now Try Exercise 43** ■

Example 6 ■ Making Linear Models Involving Speed

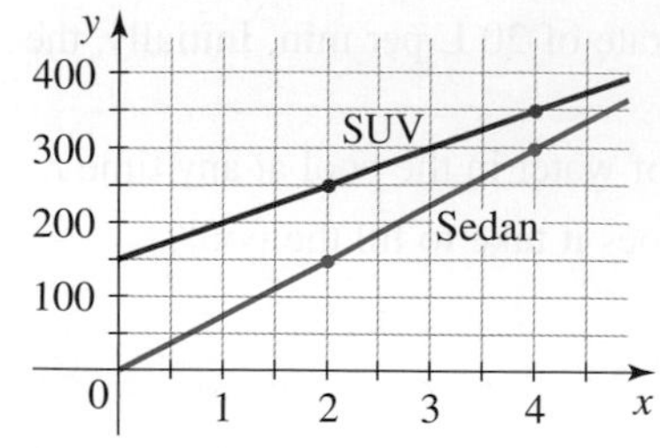

Figure 5

An SUV and a sedan are traveling westward along I-76 at constant speeds. The graphs in Figure 5 show the distance y (in kilometers) that they have traveled from Philadelphia at time x (in hours), where $x = 0$ corresponds to noon. (Note that at noon the SUV has a 150-km head start.)

(a) At what speed is each vehicle traveling? Which vehicle is traveling faster, and how does this show up in the graph?

(b) For each vehicle, find a function that models the distance traveled as a function of x.

(c) How far has each vehicle traveled at 5:00 P.M.?

(d) For what time period is the sedan behind the SUV? Will the sedan overtake the SUV? If so, at what time?

Solution

(a) From the graph we see that the SUV has traveled 250 km at 2:00 P.M. and 350 km at 4:00 P.M. The speed is the rate of change of distance with respect to time. So the speed is the slope of the graph. Therefore the speed of the SUV is

$$\frac{350\text{ km} - 250\text{ km}}{4\text{ h} - 2\text{ h}} = 50\text{ km/h} \qquad \text{Speed of SUV}$$

The sedan has traveled 150 km at 2:00 P.M. and 300 km at 4:00 P.M., so we calculate the speed of the sedan to be

$$\frac{300\text{ km} - 150\text{ km}}{4\text{ h} - 2\text{ h}} = 75\text{ km/h} \qquad \text{Speed of sedan}$$

The sedan is traveling faster than the SUV. We can see this from the graph because the line for the sedan is steeper (has greater slope) than the one for the SUV.

(b) Let $f(x)$ be the distance the SUV has traveled at time x. Since the speed (average rate of change) is constant, it follows that f is a linear function. Thus we can write f in the form $f(x) = ax + b$. From part (a) we know that the slope a is 50, and from the graph we see that the y-intercept b is 150. Thus the distance that the SUV has traveled at time x is modeled by the linear function

$$f(x) = 50x + 150 \qquad \text{Model for SUV distance}$$

Similarly, the sedan is traveling at 75 km/h, and the y-intercept is 0. Thus the distance the sedan has traveled at time x is modeled by the linear function

$$g(x) = 75x \qquad \text{Model for sedan distance}$$

(c) Replacing x by 5 in the models that we obtained in part (b), we find that at 5:00 P.M. the SUV has traveled $f(5) = 50(5) + 150 = 400$ km and the sedan has traveled $g(5) = 75(5) = 375$ km.

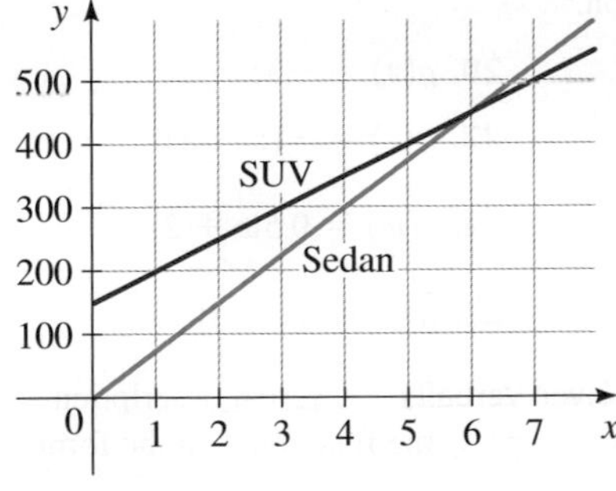

Figure 6

(d) The sedan overtakes the SUV at the time when each has traveled the same distance, that is, at the time x when $f(x) = g(x)$. So we must solve the equation

$$50x + 150 = 75x \qquad \text{SUV distance} = \text{Sedan distance}$$

Solving this equation, we get $x = 6$. So the sedan overtakes the SUV after 6 h, that is, at 6:00 P.M. We can confirm our solution graphically by drawing the graphs of f and g on a larger domain as shown in Figure 6. The graphs intersect when $x = 6$. From the graph we see that the sedan is behind the SUV (has traveled a shorter distance) from $x = 0$ to $x = 6$, that is, from noon until 6:00 P.M.

Now Try Exercise 45 ■

Deep sea divers know that the deeper they dive, the higher the water pressure becomes. But how deep can one dive before water pressure becomes dangerously high? How much pressure must a submarine be able to withstand if it is to dive to the deepest parts of the ocean? These questions can be answered using the model in the next example.

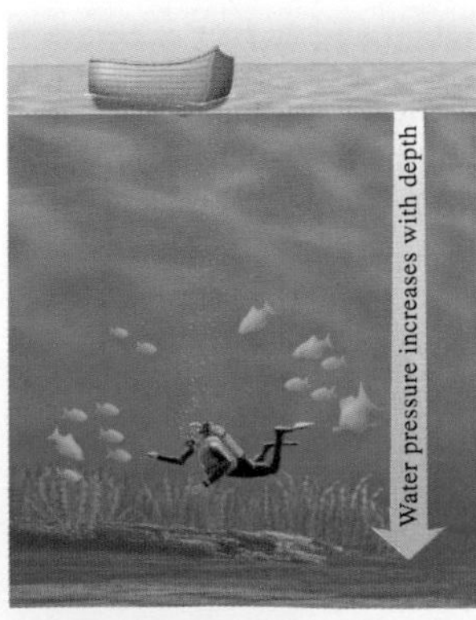

Example 7 ■ Modeling Pressure at Depth

The pressure at the surface of the ocean is 101.325 kPa and the pressure increases by 100.55 kPa for every 10-meters descent below the surface.

(a) Find a linear function P that models the water pressure at depth x meters below the surface of the ocean.

(b) What is the pressure one kilometer below the surface?

Solution

(a) We want to find a function $P(x) = ax + b$ that models the pressure $P(x)$ at depth x. The rate of change of pressure is $(100.55 \text{ kPa})/(10 \text{ m}) = 10.055 \text{ kPa/m}$, so $a = 10.055$. Because the pressure is 101.325 kPa when $x = 0$, it follows that $b = 101.325$ kPa. So the model is

$$P(x) = 10.055x + 101.325$$

(b) We can use the model to estimate water pressure at any ocean depth. Because 1 km = 1000 m, the pressure one kilometer below the surface is

$$P(1000) = 10.055(1000) + 101.325 \approx 10{,}156 \text{ kPa}$$

 **Now Try Exercise 49** ■

2.5 Exercises

Concepts

1. Let f be a function with constant rate of change. Then
(a) f is a ______ function and f is of the form $f(x) =$ ___ $x +$ ___.
(b) The graph of f is a ______.

2. Let f be the linear function $f(x) = -5x + 7$.
(a) The rate of change of f is ______.
(b) The graph of f is a ______ with slope ______ and y-intercept ______.

3–4 ■ A swimming pool is being filled. The graph shows the number of liters y in the pool after x minutes.

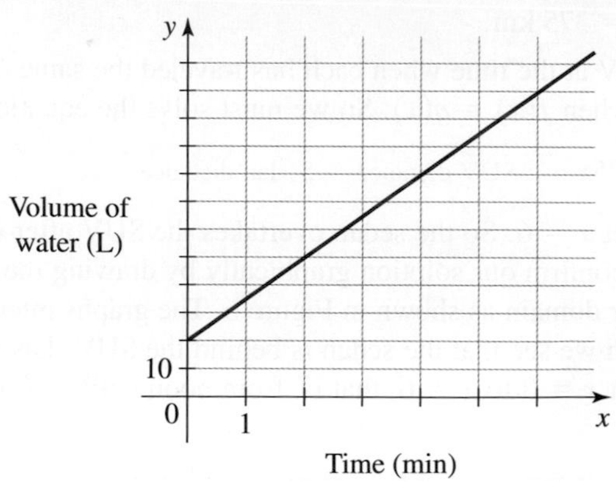

3. What is the slope of the graph?

4. At what rate is the pool being filled?

5. If a linear function has positive rate of change, does its graph slope upward or downward?

6. Is $f(x) = 3$ a linear function? If so, what are the slope and the rate of change?

Skills

7–14 ■ **Identifying Linear Functions** Determine whether the given function is linear. If the function is linear, express the function in the form $f(x) = ax + b$.

7. $f(x) = \sqrt{5} + 2x$ **8.** $f(x) = \frac{1}{2}(x + 4)$

9. $f(x) = \dfrac{20 - x}{5}$ **10.** $f(x) = \dfrac{2x - 4}{x}$

11. $f(x) = x(2 - 3x)$ **12.** $f(x) = -3(6 - 5x)$

13. $f(x) = \sqrt{x} + 1$ **14.** $f(x) = (2x - 5)^2$

15–18 ■ **Graphing Linear Functions** For the given linear function, make a table of values and use it to sketch the graph. What is the slope of the graph?

15. $f(x) = -2x + 3$ **16.** $g(x) = 3x - 1$

17. $r(t) = -\frac{2}{3}t + 2$ **18.** $h(t) = \frac{1}{2} - \frac{3}{4}t$

19–26 ■ **Slope and Rate of Change** A linear function is given. **(a)** Sketch the graph. **(b)** Find the slope of the graph. **(c)** Find the rate of change of the function.

19. $f(x) = 2x - 6$ **20.** $g(z) = -3z - 9$

21. $f(x) = 2 - 3x$ **22.** $g(z) = -(z - 3)$

23. $h(t) = \dfrac{5 - 2t}{10}$ **24.** $s(w) = 0.5w + 2$

25. $f(t) = -\frac{3}{2}t + 2$ **26.** $g(x) = \frac{5}{4}x - 10$

27–30 ■ **Linear Functions Given Verbally** A verbal description of a linear function f is given. Express the function f in the form $f(x) = ax + b$.

27. The linear function f has rate of change 5 and initial value 10.

28. The linear function f has rate of change -3 and initial value -1.

29. The graph of the linear function f has slope $\frac{1}{2}$ and y-intercept 3.

30. The graph of the linear function f has slope $-\frac{4}{5}$ and y-intercept -2.

31–32 ■ **Linear Functions Given Numerically** A table of values for a linear function f is given. **(a)** Find the rate of change of f. **(b)** Express f in the form $f(x) = ax + b$.

31.

x	$f(x)$
0	7
2	10
4	13
6	16
8	19

32.

x	$f(x)$
-3	11
0	2
2	-4
5	-13
7	-19

33–36 ■ Linear Functions Given Graphically The graph of a linear function f is given. **(a)** Find the rate of change of f. **(b)** Express f in the form $f(x) = ax + b$.

33.

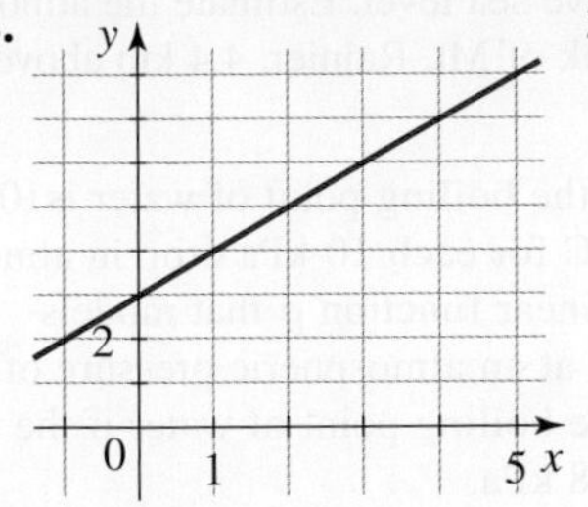

34.

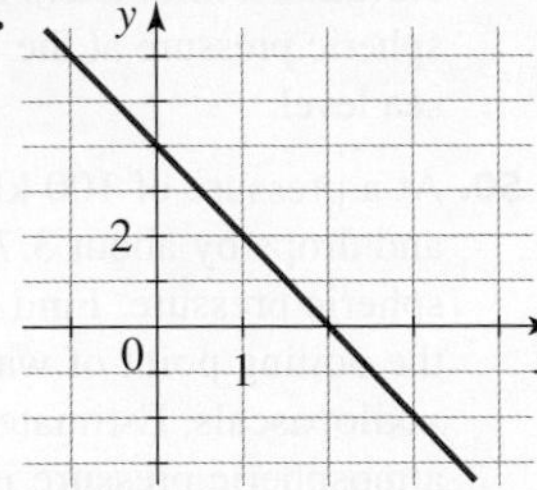

35.

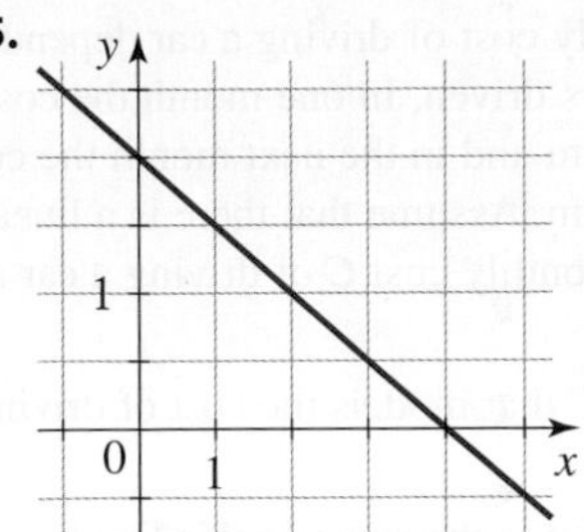

36.

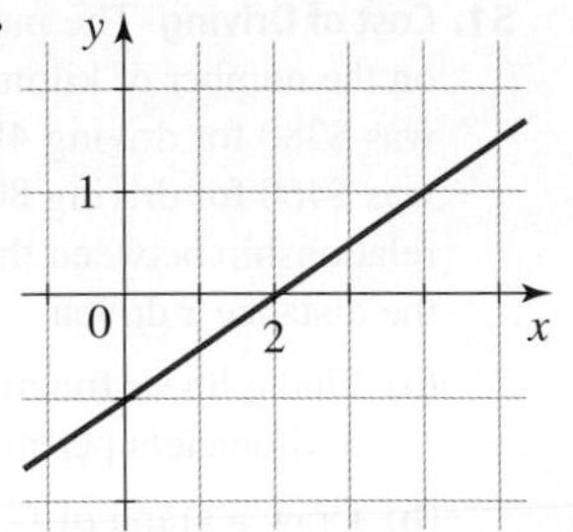

Skills Plus

37. Families of Linear Functions Graph $f(x) = ax$ for $a = \frac{1}{2}$, $a = 1$, and $a = 2$, all on the same set of axes. How does increasing the value of a affect the graph of f? What about the rate of change of f?

38. Families of Linear Functions Graph $f(x) = x + b$ for $b = \frac{1}{2}$, $b = 1$, and $b = 2$, all on the same set of axes. How does increasing the value of b affect the graph of f? What about the rate of change of f?

Applications

39. Landfill The amount of trash in a county landfill is modeled by the function

$$T(x) = 150x + 32$$

where x is the number of years since 2010 and $T(x)$ is measured in thousands of tons.

(a) Sketch a graph of T.

(b) What is the slope of the graph?

(c) At what rate is the amount of trash in the landfill increasing per year?

40. Copper Mining The amount of copper ore produced from a copper mine in Arizona is modeled by the function

$$f(x) = 200 + 32x$$

where x is the number of years since 2015 and $f(x)$ is measured in thousands of tons.

(a) Sketch a graph of f.

(b) What is the slope of the graph?

(c) At what rate is the amount of ore produced changing?

41. Weather Balloon Weather balloons are filled with hydrogen and released at various sites to measure and transmit data about conditions such as air pressure and temperature. A weather balloon is filled with hydrogen at the rate of 0.25 m/s. Initially, the balloon contains 2 m^3 of hydrogen.

(a) Find a linear function V that models the volume of hydrogen in the balloon at any time t.

(b) If the balloon has a capacity of 7.5 m^3, how long does it take to completely fill the balloon?

42. Filling a Pond A small koi pond is filled from a garden hose at the rate of 10 L/min. Initially, the pond contains 300 L of water.

(a) Find a linear function V that models the volume of water in the pond at any time t.

(b) If the pond has a capacity of 1300 L, how long does it take to completely fill the pond?

43. Wheelchair Ramp A local diner must build a wheelchair ramp to provide accessibility to the restaurant. Federal building codes require that a wheelchair ramp must have a maximum rise of 2.5 cm for every horizontal distance of 30 cm.

(a) What is the maximum allowable slope for a wheelchair ramp? Assuming that the ramp has maximum rise, find a linear function H that models the height of the ramp above the ground as a function of the horizontal distance x.

(b) If the space available to build a ramp is 375 cm wide, how high does the ramp reach?

44. Mountain Biking Two cyclists are traveling down straight roads with steep grades, one on a mountain bike and the other on a road bike. The graphs give a representation of the elevation of the road on which each of them cycles. Find the grade of each road. (The grade of a road is the slope expressed as a percentage.)

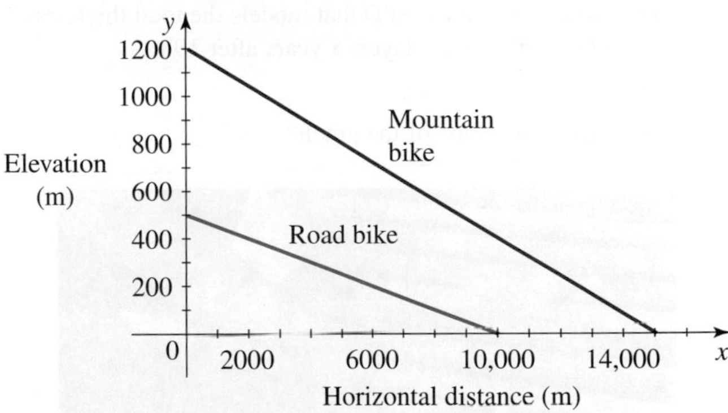

45. Commute to Work Two employees of a software company commute from the same housing complex each morning. One works as a software engineer and the other is a product manager. One morning the product manager left for work at 6:50 A.M., and the software engineer left 10 minutes later. Both drove at a constant speed. The following graphs show the distance (in kilometers) each of them had traveled on I-10 at time t (in minutes), where $t = 0$ is 7:00 A.M.

(a) Use the graph to decide which of them was traveling faster.

(b) Find the speed (in km/h) at which each of them was driving.

(c) Find linear functions f and g that model the distance each employee traveled as a function of time t (in minutes).

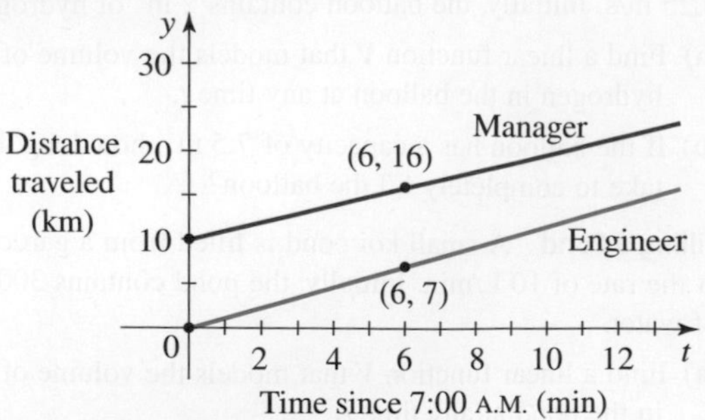

46. Distance, Speed, and Time A bus leaves Detroit at 2:00 P.M. and drives at a constant speed, traveling west on I-90. It passes Ann Arbor, 65 km from Detroit, at 2:50 P.M.

(a) Find a linear function d that models the distance (in km) the bus has traveled after t min.

(b) Draw a graph of d. What is the slope of this line?

(c) At what speed (in km/h) is the bus traveling?

47. Grade of Road West of Albuquerque, New Mexico, Route 40 eastbound is straight and makes a steep descent toward the city. The highway has a 6% grade, which means that its slope is $-\frac{6}{100}$. Driving on this road, you notice from elevation signs that you have descended a distance of 305 m. What is the change in your horizontal distance in kilometers?

48. Sedimentation Geologists study sedimentation by drilling tubular core samples. Studies show that the mean sedimentation rate at the bottom of Devil's Lake, North Dakota, is about 0.24 cm per year. In 1980, the total thickness of the sedimentary layers at a certain location was 20 cm.

(a) Find a linear function D that models the total thickness of the sedimentary layers x years after 1980.

(b) Sketch a graph of D.

(c) What is the slope of the graph?

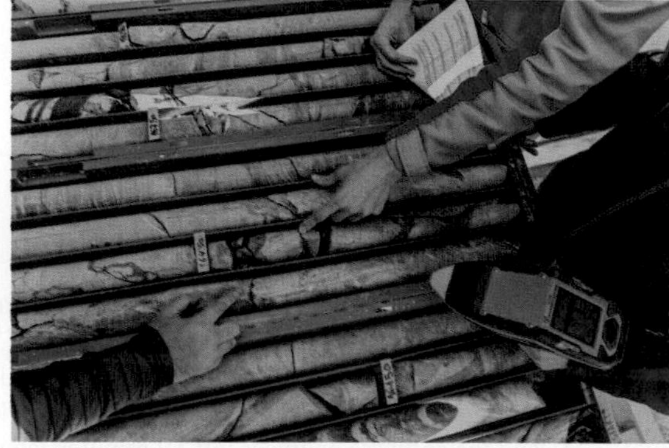

49–50 ■ Pressure, Boiling Point, and Elevation Mountain climbers know that the atmospheric pressure and the boiling point of water decrease as elevation increases. Although these properties depend on many factors, in these exercises we find approximate linear models that relate them. We measure the pressure in kilopascals (kPa), the boiling point in degrees Celsius (°C), and the elevation in kilometers (km).

49. If atmospheric pressure is 100 kPa at sea level and decreases by about 12 kPa for each kilometer increase in elevation, find a linear function f that models atmospheric pressure at elevation x kilometers above sea level. Estimate the atmospheric pressure at the peak of Mt. Rainier, 4.4 km above sea level.

50. At a pressure of 100 kPa the boiling point of water is100°C and drops by about 3.75°C for each 10-kPa drop in atmospheric pressure. Find a linear function g that models the boiling point of water at an atmospheric pressure of x kilopascals. Estimate the boiling point of water if the atmospheric pressure is 88 kPa.

51. Cost of Driving The monthly cost of driving a car depends on the number of kilometers driven. In one month the cost was \$380 for driving 480 km and in the next month the cost was \$460 for driving 800 km. Assume that there is a linear relationship between the monthly cost C of driving a car and the distance x driven.

(a) Find a linear function C that models the cost of driving x kilometers per month.

(b) Draw a graph of C. What is the slope of this line?

(c) At what rate does the cost increase for every additional kilometer driven?

52. Manufacturing Cost The manager of a furniture factory finds that it costs \$2200 to produce 100 chairs in one day and \$4800 to produce 300 chairs in one day.

(a) Assuming that the relationship between cost and the number of chairs produced is linear, find a linear function C that models the cost of producing x chairs in one day.

(b) Draw a graph of C. What is the slope of this line?

(c) At what rate does the factory's cost increase for every additional chair produced?

■ Discuss ■ Discover ■ Prove ■ Write

53. Prove: Linear Functions Have Constant Rate of Change Suppose that $f(x) = ax + b$ is a linear function.

(a) Use the definition of the average rate of change of a function to calculate the average rate of change of f between any two real numbers x_1 and x_2.

(b) Use your calculation in part (a) to show that the average rate of change of f is the same as the slope a.

54. Prove: Functions with Constant Rate of Change Are Linear Suppose that the function f has the same average rate of change c between any two values of the variable.

(a) Find the average rate of change of f between the values a and x to show that

$$c = \frac{f(x) - f(a)}{x - a}$$

(b) Rearrange the equation in part (a) to show that

$$f(x) = cx + [f(a) - ca]$$

How does this show that f is a linear function? What is the slope, and what is the y-intercept?

2.6 Transformations of Functions

■ Vertical Shifting ■ Horizontal Shifting ■ Reflecting Graphs ■ Vertical Stretching and Shrinking ■ Horizontal Stretching and Shrinking ■ Even and Odd Functions

In this section we study how certain transformations of a function affect its graph. This will give us a better understanding of how to graph functions. The transformations that we study are shifting, reflecting, and stretching.

■ Vertical Shifting

Adding a constant to a function shifts its graph vertically: upward if the constant is positive, and downward if it is negative.

In general, suppose we know the graph of $y = f(x)$. How do we obtain from it the graphs of

Recall that the graph of the function f is the same as the graph of the equation $y = f(x)$.

$$y = f(x) + c \quad \text{and} \quad y = f(x) - c \qquad (c > 0)$$

The y-coordinate of each point on the graph of $y = f(x) + c$ is c units above the y-coordinate of the corresponding point on the graph of $y = f(x)$. So we obtain the graph of $y = f(x) + c$ by shifting the graph of $y = f(x)$ upward c units. Similarly, we obtain the graph of $y = f(x) - c$ by shifting the graph of $y = f(x)$ downward c units.

Vertical Shifts of Graphs

Suppose $c > 0$.

To graph $y = f(x) + c$, shift the graph of $y = f(x)$ upward c units.

To graph $y = f(x) - c$, shift the graph of $y = f(x)$ downward c units.

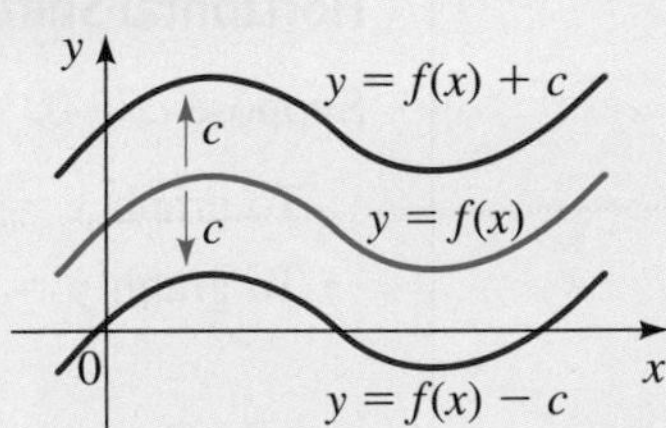

Example 1 ■ Vertical Shifts of Graphs

Use the graph of $f(x) = x^2$ to sketch the graph of each function.

(a) $g(x) = x^2 + 3$ **(b)** $h(x) = x^2 - 2$

Solution The function $f(x) = x^2$ was graphed in Example 2.2.1(a). It is sketched again in Figure 1.

(a) Observe that

$$g(x) = x^2 + 3 = f(x) + 3$$

So the y-coordinate of each point on the graph of g is 3 units above the corresponding point on the graph of f. This means that to graph g, we shift the graph of f upward 3 units, as in Figure 1.

(b) Similarly, to graph h we shift the graph of f downward 2 units, as shown in Figure 1.

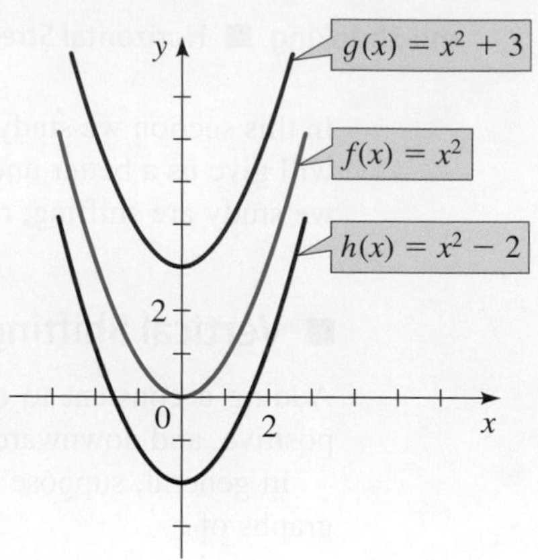

Figure 1

Now Try Exercises 27 and 29 ■

■ Horizontal Shifting

Suppose that we know the graph of $y = f(x)$. How do we use it to obtain the graphs of

$$y = f(x + c) \qquad \text{and} \qquad y = f(x - c) \qquad (c > 0)$$

The value of $f(x - c)$ at x is the same as the value of $f(x)$ at $x - c$. Since $x - c$ is c units to the left of x, it follows that the graph of $y = f(x - c)$ is just the graph of $y = f(x)$ shifted to the right c units. Similar reasoning shows that the graph of $y = f(x + c)$ is the graph of $y = f(x)$ shifted to the left c units.

Horizontal Shifts of Graphs

Suppose $c > 0$.

To graph $y = f(x - c)$, shift the graph of $y = f(x)$ to the right c units.

To graph $y = f(x + c)$, shift the graph of $y = f(x)$ to the left c units.

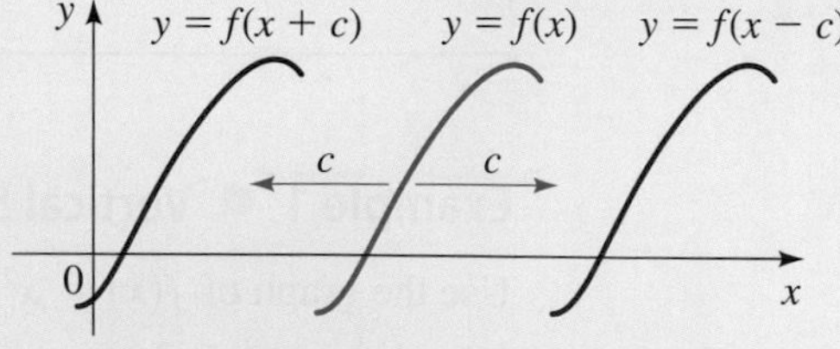

Example 2 ■ Horizontal Shifts of Graphs

Use the graph of $f(x) = x^2$ to sketch the graph of each function.

(a) $g(x) = (x + 4)^2$ **(b)** $h(x) = (x - 2)^2$

Solution

(a) To graph g, we shift the graph of f to the left 4 units.

(b) To graph h, we shift the graph of f to the right 2 units.

The graphs of g and h are sketched in Figure 2.

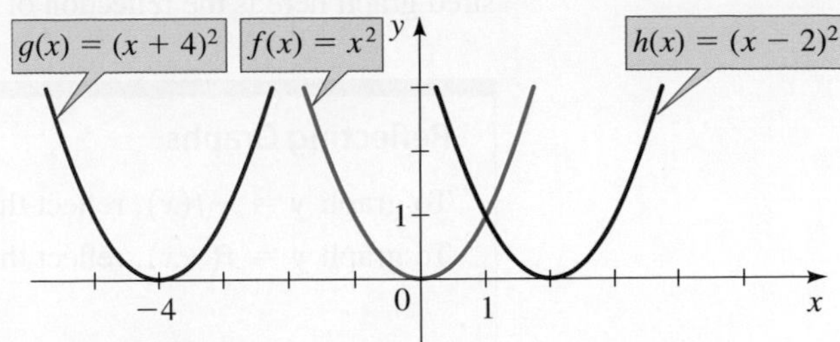

Figure 2

Now Try Exercises 31 and 33 ■

Example 3 ■ Combining Horizontal and Vertical Shifts

Sketch the graph of $f(x) = |x + 1| - 3$

Solution The equation involves two transformations of the graph of $y = |x|$ (Example 2.2.5). In Figure 3, we graph the equation using the following steps.

① Start with the graph of $y = |x|$.

② Shift the graph obtained in ① to the left 1 unit to obtain the graph of $y = |x + 1|$.

③ Shift the graph obtained in ② downward 3 units to obtain the graph of $y = |x + 1| - 3$.

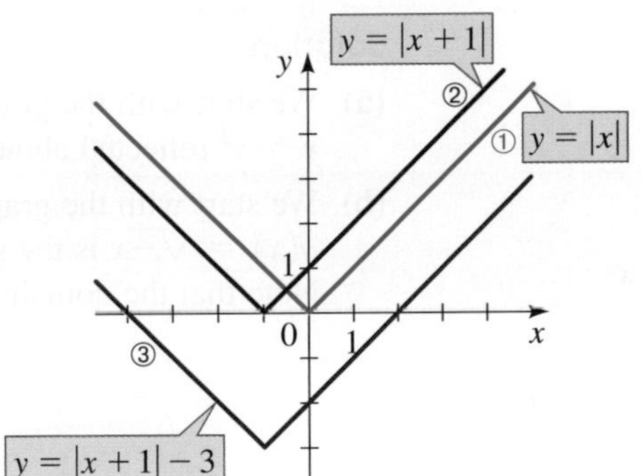

Figure 3

Now Try Exercise 43 ■

■ Reflecting Graphs

Suppose we know the graph of $y = f(x)$. How do we use it to obtain the graphs of $y = -f(x)$ and $y = f(-x)$? The y-coordinate of each point on the graph of $y = -f(x)$ is the negative of the y-coordinate of the corresponding point on the graph of $y = f(x)$. So the desired graph is the reflection of the graph of $y = f(x)$ about the x-axis. On the other

© TFoxFoto/Shutterstock.com

Discovery Project ■ Transformation Stories

If a real-world situation, or "story," is modeled by a function, how does transforming the function change the story? For example, if the distance traveled on a road trip is modeled by a function, then how does shifting or stretching the function change the story of the trip? How does changing the story of the trip transform the function that models the trip? In this project we explore some real-world stories and transformations of these stories. You can find the project at **www.stewartmath.com**.

hand, the value of $y = f(-x)$ at x is the same as the value of $y = f(x)$ at $-x$, so the desired graph here is the reflection of the graph of $y = f(x)$ about the y-axis.

Reflecting Graphs

To graph $y = -f(x)$, reflect the graph of $y = f(x)$ about the x-axis.
To graph $y = f(-x)$, reflect the graph of $y = f(x)$ about the y-axis.

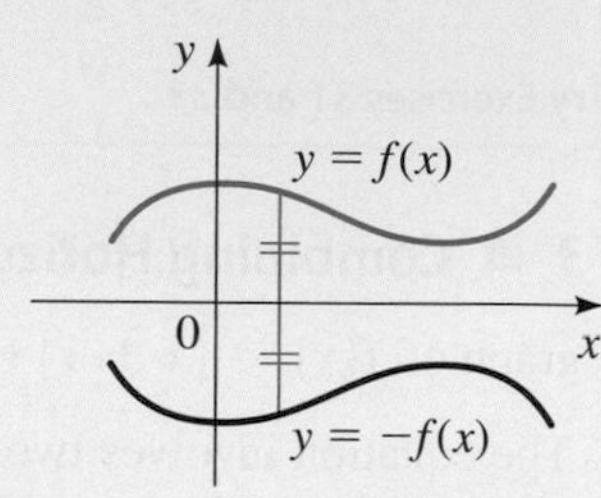

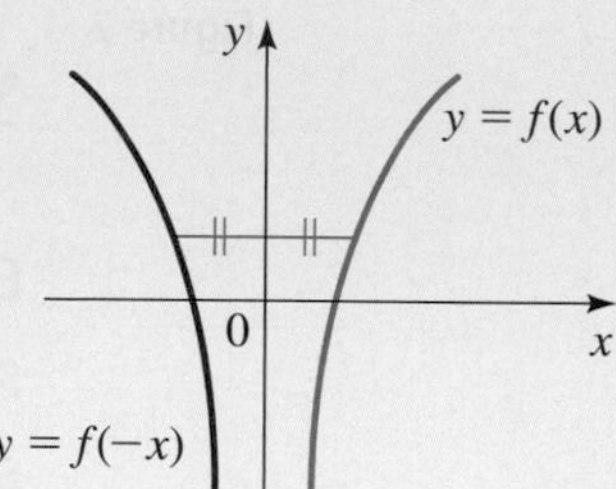

Example 4 ■ Reflecting Graphs

Sketch the graph of each function.

(a) $f(x) = -x^2$ **(b)** $g(x) = \sqrt{-x}$

Solution

(a) We start with the graph of $y = x^2$. The graph of $f(x) = -x^2$ is the graph of $y = x^2$ reflected about the x-axis (see Figure 4).

(b) We start with the graph of $y = \sqrt{x}$ [Example 2.2.1(c)]. The graph of $g(x) = \sqrt{-x}$ is the graph of $y = \sqrt{x}$ reflected about the y-axis (see Figure 5). Note that the domain of the function $g(x) = \sqrt{-x}$ is $\{x \mid x \le 0\}$.

Figure 4

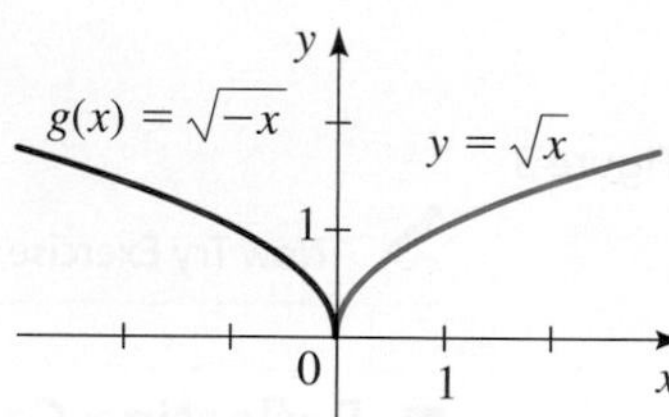

Figure 5

Now Try Exercises 35 and 37 ■

RENÉ DESCARTES (1596–1650) was born in the town of La Haye in France. From an early age Descartes liked mathematics because of "the certainty of its results and the clarity of its reasoning." He believed that to arrive at truth, one must begin by doubting everything, including one's own existence; this led him to formulate perhaps the best-known sentence in all of philosophy: "I think, therefore I am." In his book *Discourse on Method* he described what is now called the Cartesian plane. This idea of combining algebra and geometry enabled mathematicians for the first time to graph functions and thus "see" the equations they were studying. The philosopher John Stuart Mill called this invention "the greatest single step ever made in the progress of the exact sciences." Descartes liked to get up late and spend the morning in bed thinking and writing. He invented the coordinate plane while lying in bed watching a fly crawl on the ceiling, reasoning that he could describe the exact location of the fly by knowing its distance from two perpendicular walls. In 1649 Descartes became the tutor of Queen Christina of Sweden. She liked her lessons at 5 o'clock in the morning, when, she said, her mind was sharpest. However, the change from his usual habits and the ice-cold library where they studied proved too much for Descartes. In February 1650, after just two months of this regimen, he caught pneumonia and died. The 1974 television film *Cartesius* chronicles the life of Rene Descartes.

Library of Congress Prints and Photographs Division

■ Vertical Stretching and Shrinking

Suppose we know the graph of $y = f(x)$. How do we use it to obtain the graph of $y = cf(x)$? The y-coordinate of $y = cf(x)$ at x is the same as the corresponding y-coordinate of $y = f(x)$ multiplied by c. Multiplying the y-coordinates by c has the effect of vertically stretching or shrinking the graph by a factor of c (if $c > 0$).

Vertical Stretching and Shrinking of Graphs

To graph $y = cf(x)$:

If $c > 1$, stretch the graph of $y = f(x)$ vertically by a factor of c.

If $0 < c < 1$, shrink the graph of $y = f(x)$ vertically by a factor of c.

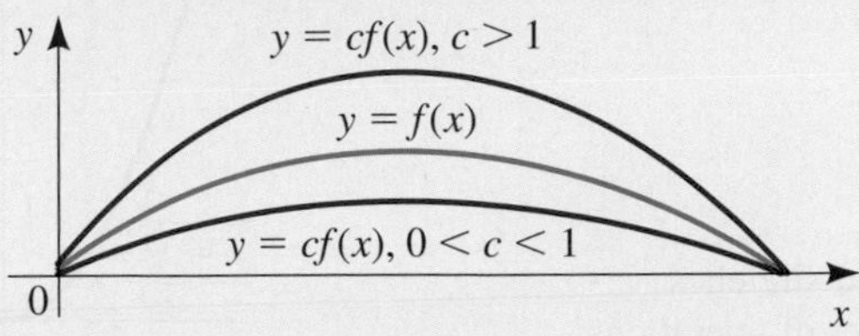

Example 5 ■ Vertical Stretching and Shrinking of Graphs

Use the graph of $f(x) = x^2$ to sketch the graph of each function.

(a) $g(x) = 3x^2$ **(b)** $h(x) = \frac{1}{3}x^2$

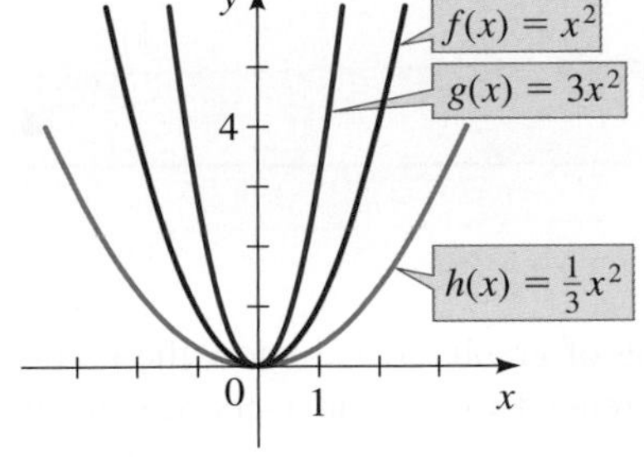

Figure 6

Solution

(a) The graph of g is obtained by multiplying the y-coordinate of each point on the graph of f by 3. That is, to obtain the graph of g, we stretch the graph of f vertically by a factor of 3. The result is the narrowest parabola in Figure 6.

(b) The graph of h is obtained by multiplying the y-coordinate of each point on the graph of f by $\frac{1}{3}$. That is, to obtain the graph of h, we shrink the graph of f vertically by a factor of $\frac{1}{3}$. The result is the widest parabola in Figure 6.

 **Now Try Exercises 39 and 41** ■

We illustrate the effect of combining shifts, reflections, and stretching in the following example.

Mathematics in the Modern World

Georgii Shipin/Shutterstock.com

Computers

For centuries machines have been designed to perform specific tasks. For example, a washing machine washes clothes, a weaving machine weaves cloth, an adding machine adds numbers, and so on. The computer has changed all that.

The computer is a machine that does nothing—until it is given instructions on what to do. So your computer can play games, draw pictures, or calculate π to a million decimal places; it all depends on what program (or instructions) you give the computer. The computer can do all this because it is able to accept instructions and logically change those instructions based on incoming data. This versatility makes computers useful in nearly every aspect of human endeavor.

The idea of a computer was described theoretically in the 1940s by the mathematician Alan Turing (see the biography in this section) in what he called a *universal machine*. In 1945 the mathematician John Von Neumann, extending Turing's ideas, built one of the first electronic computers.

Mathematicians continue to develop new theoretical bases for the design of computers. The heart of the computer is the "chip," (or CPU), which is capable of processing logical instructions. To get an idea of the chip's complexity, consider that modern computer chips contain more than ten billion transistors making up their logic circuits.

Example 6 ■ Combining Shifting, Stretching, and Reflecting

Sketch the graph of the function $f(x) = 1 - 2(x - 3)^2$.

Solution The equation involves several transformations of the graph of $y = x^2$. In Figure 7, we graph the equation using the following steps; in each step we transform the graph that we obtained in the preceding step.

① Start with the graph of $y = x^2$.

② Shift the graph in ① to the right 3 units to get the graph of $y = (x - 3)^2$.

③ Reflect the graph in ② about the x-axis and stretch vertically by a factor of 2 to obtain the graph of $y = -2(x - 3)^2$.

④ Finally, shift the graph in ③ upward 1 unit to get the graph of $y = 1 - 2(x - 3)^2$.

Note that the shifts and stretches follow the normal order of operations. In particular, the upward shift must be performed *last*.

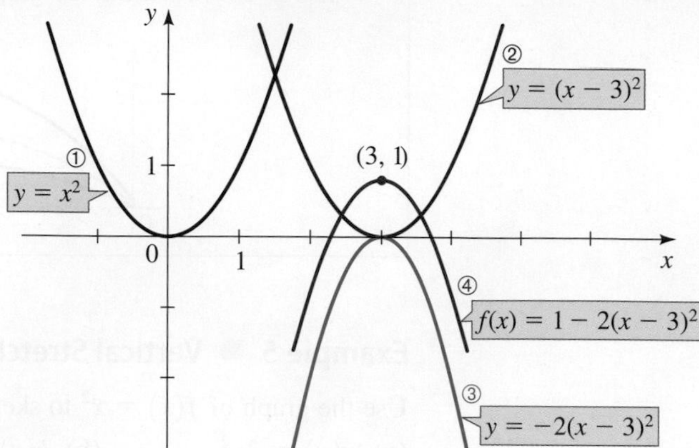

Figure 7

Now Try Exercise 45 ■

■ Horizontal Stretching and Shrinking

Now we consider horizontal shrinking and stretching of graphs. If we know the graph of $y = f(x)$, then how is the graph of $y = f(cx)$ related to it? The y-coordinate of $y = f(cx)$ at x is the same as the y-coordinate of $y = f(x)$ at cx. Thus the x-coordinates in the graph of $y = f(x)$ correspond to the x-coordinates in the graph of $y = f(cx)$ multiplied by c. Looking at this the other way around, we see that the x-coordinates in the graph of $y = f(cx)$ are the x-coordinates in the graph of $y = f(x)$ multiplied by $1/c$. In other words, to change the graph of $y = f(x)$ to the graph of $y = f(cx)$, we must shrink (or stretch) the graph horizontally by a factor of $1/c$ (if $c > 0$).

Horizontal Shrinking and Stretching of Graphs

To graph $y = f(cx)$:

If $c > 1$, shrink the graph of $y = f(x)$ horizontally by a factor of $1/c$.

If $0 < c < 1$, stretch the graph of $y = f(x)$ horizontally by a factor of $1/c$.

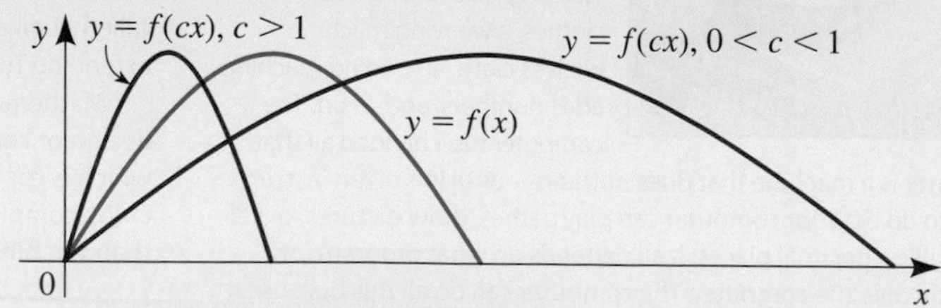

Science History Images/Alamy Stock Photo

ALAN TURING (1912–1954) was at the center of two pivotal events of the 20th century: World War II and the invention of computers. At the age of 23 Turing made his mark on mathematics by solving an important problem in the foundations of mathematics. In this research he invented a theoretical machine—now called a Turing machine,—which was the inspiration for modern digital computers. During World War II, Turing was in charge of the British effort to decipher secret German messages enciphered by the Enigma machine. His complete success in this endeavor played a decisive role in the Allied victory. To carry out the numerous logical steps that are required to break a coded message, Turing developed decision procedures similar to modern computer programs. After the war he helped to develop the first electronic computers in Britain. He also did pioneering work on artificial intelligence and computer models of biological processes. The 2014 film *The Imitation Game* is based on Turing's work in cracking the Enigma code.

Example 7 ■ Horizontal Stretching and Shrinking of Graphs

The graph of $y = f(x)$ is shown in Figure 8. Sketch the graph of each function.

(a) $y = f(2x)$ **(b)** $y = f(\frac{1}{2}x)$

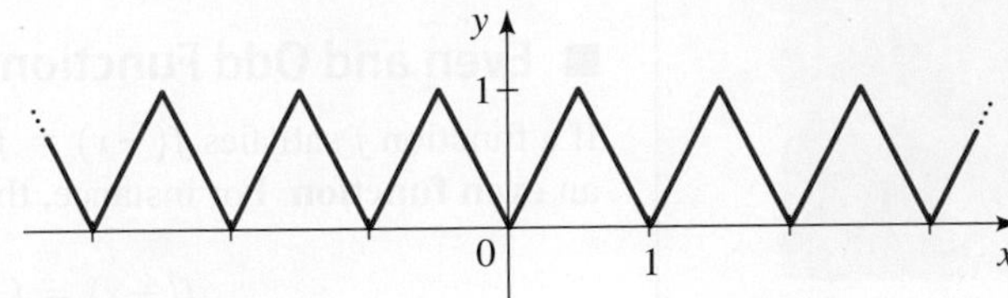

Figure 8 | $y = f(x)$

Solution

(a) We *shrink* the graph horizontally by the factor $\frac{1}{2}$ to obtain the graph in Figure 9.

(b) We *stretch* the graph horizontally by the factor 2 to obtain the graph in Figure 10.

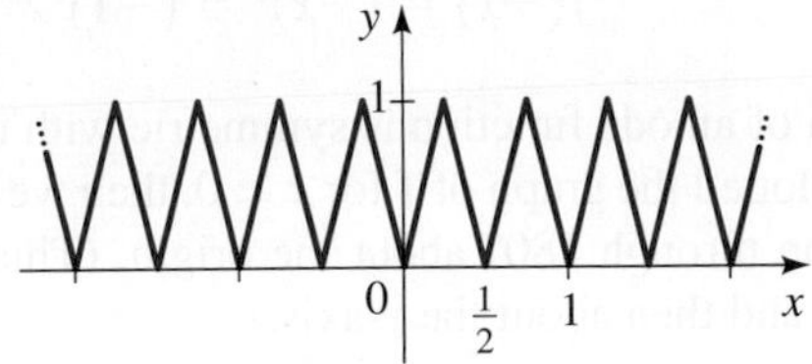

Figure 9 | $y = f(2x)$

y
1
−1
0
1
2
x

Figure 10 | $y = f(\frac{1}{2}x)$

Now Try Exercise 69 ■

Example 8 ■ Finding an Equation for a Combination of Transformations

Apply the following transformations (in order) to the graph of $f(x) = \sqrt{x}$: shift 1 unit to the left, shrink vertically by a factor of $\frac{1}{2}$, then reflect about the y-axis. Write an equation for the final transformed graph.

Solution We apply the given transformations in order, as shown in Figure 11. In each step we apply the required transformation to the graph obtained in the preceding step.

① Start with the graph of $y = \sqrt{x}$.

② Shift the graph 1 unit to the left, to obtain the graph of $y = \sqrt{x + 1}$.

③ Shrink the graph vertically by a factor of $\frac{1}{2}$, to obtain the graph of $y = \frac{1}{2}\sqrt{x + 1}$.

④ Finally, reflect the graph about the y-axis to obtain the graph of $y = \frac{1}{2}\sqrt{-x + 1}$, or $y = \frac{1}{2}\sqrt{1 - x}$.

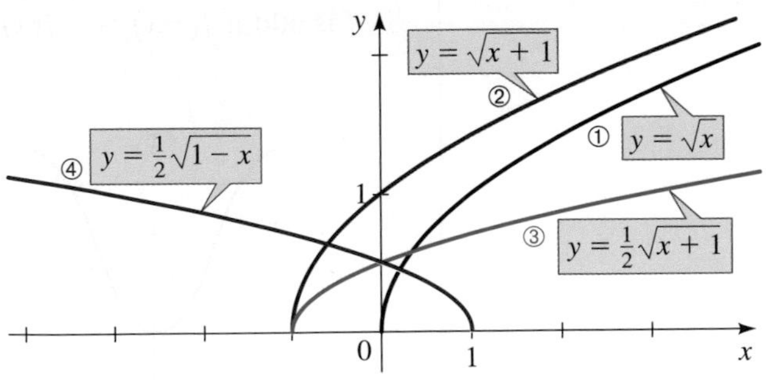

Figure 11

Now Try Exercises 51 and 59 ■

SOFYA KOVALEVSKAYA (1850–1891) is considered to be one of the most important mathematicians of the 19th century. She was born in Moscow to an aristocratic family. As a child, she was exposed to the principles of calculus in a very unusual way: Her bedroom was temporarily wallpapered with the pages of a calculus book. She later wrote that she "spent many hours in front of that wall, trying to understand it." Since Russian law forbade women from studying in universities, she entered a marriage of convenience, which allowed her to travel to Germany and obtain a doctorate in mathematics from the University of Göttingen. Eventually she was awarded a full professorship at the University of Stockholm, where she taught for eight years before dying in an influenza epidemic at the age of 41. Her research was instrumental in helping to put the ideas and applications of functions and calculus on a sound and logical foundation. She received many accolades and prizes for her research work. The 1983 film *A Hill on the Dark Side of the Moon* is about the life of Sofya Kovalevskaya.

Note When we apply more than one transformation to a graph, order can matter. For example, starting with the function $y = \sqrt{x}$, reflecting about the x-axis, then shifting upward 1 unit results in the equation $y = -\sqrt{x} + 1$, whereas if we first shift upward 1 unit, then reflect about the x-axis, we obtain the equation $y = -(\sqrt{x} + 1) = -\sqrt{x} - 1$.

■ Even and Odd Functions

If a function f satisfies $f(-x) = f(x)$ for every number x in its domain, then f is called an **even function**. For instance, the function $f(x) = x^2$ is even because

$$f(-x) = (-x)^2 = (-1)^2x^2 = x^2 = f(x)$$

The graph of an even function is symmetric with respect to the y-axis (see Figure 12). This means that if we have plotted the graph of f for $x \geq 0$, then we can obtain the entire graph simply by reflecting this portion about the y-axis.

If f satisfies $f(-x) = -f(x)$ for every number x in its domain, then f is called an **odd function**. For example, the function $f(x) = x^3$ is odd because

$$f(-x) = (-x)^3 = (-1)^3x^3 = -x^3 = -f(x)$$

The graph of an odd function is symmetric with respect to the origin (see Figure 13). If we have plotted the graph of f for $x \geq 0$, then we can obtain the entire graph by rotating this portion through 180° about the origin. (This is equivalent to reflecting first about the x-axis and then about the y-axis.)

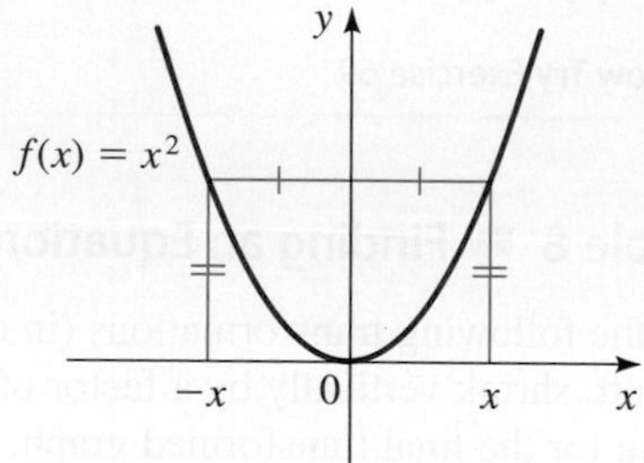

Figure 12 | $f(x) = x^2$ is an even function.

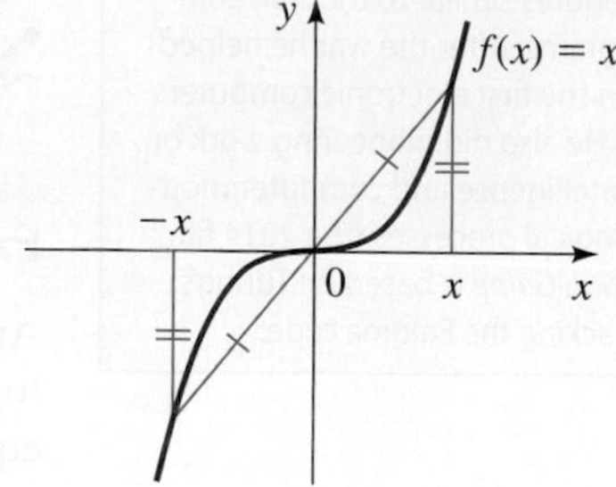

Figure 13 | $f(x) = x^3$ is an odd function.

Even and Odd Functions

Let f be a function.

f is **even** if $f(-x) = f(x)$ for all x in the domain of f.

f is **odd** if $f(-x) = -f(x)$ for all x in the domain of f.

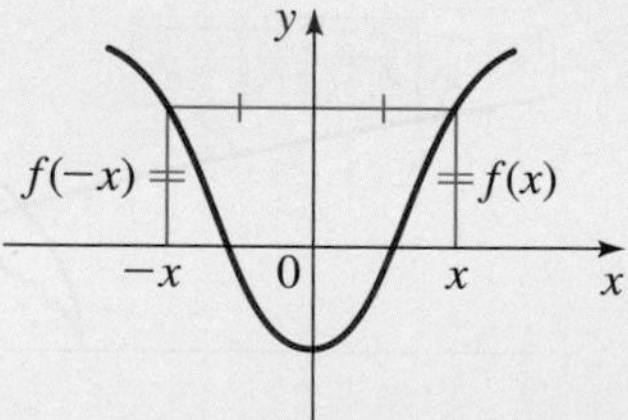

The graph of an even function is symmetric with respect to the y-axis.

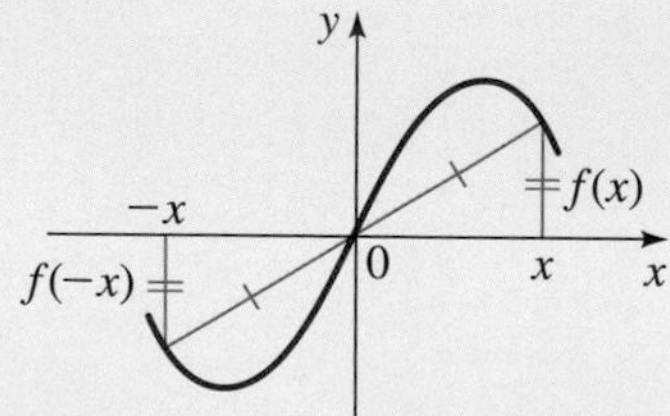

The graph of an odd function is symmetric with respect to the origin.

Example 9 ■ Even and Odd Functions

Determine whether the function is even, odd, or neither even nor odd.

(a) $f(x) = x^5 + x$ **(b)** $g(x) = 1 - x^4$ **(c)** $h(x) = 2x - x^2$

Solution

(a)
$$\begin{aligned} f(-x) &= (-x)^5 + (-x) \\ &= -x^5 - x = -(x^5 + x) \\ &= -f(x) \end{aligned}$$

Therefore f is an odd function.

(b) $g(-x) = 1 - (-x)^4 = 1 - x^4 = g(x)$

So g is even.

(c) $h(-x) = 2(-x) - (-x)^2 = -2x - x^2$

Since $h(-x) \neq h(x)$ and $h(-x) \neq -h(x)$, we conclude that h is neither even nor odd.

 **Now Try Exercises 81, 83, and 85** ■

The graphs of the functions in Example 9 are shown in Figure 14. The graph of f is symmetric with respect to the origin, and the graph of g is symmetric with respect to the y-axis. The graph of h is not symmetric with respect to either the y-axis or the origin.

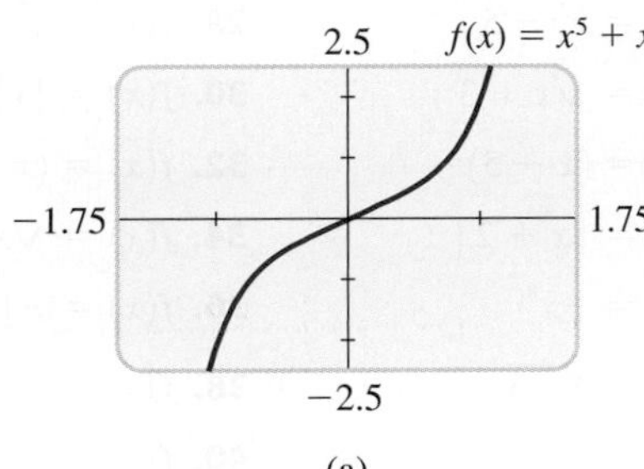

(a)

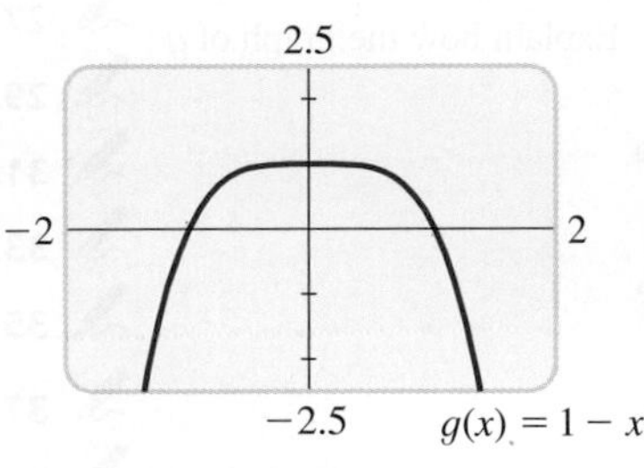

(b)

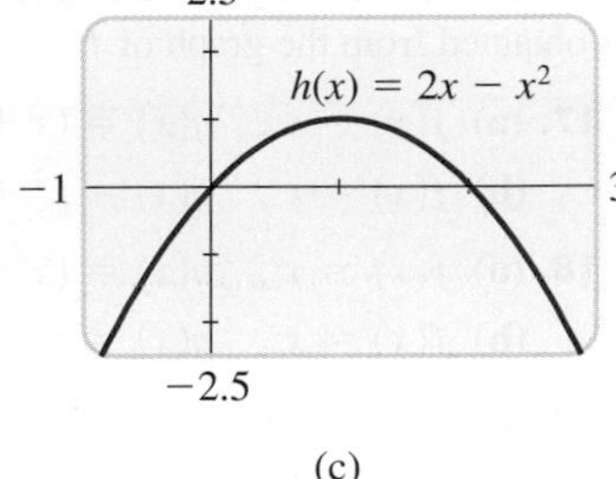

(c)

Figure 14

2.6 | Exercises

Concepts

1–2 ■ Fill in the blank with the appropriate direction (left, right, upward, or downward).

1. (a) The graph of $y = f(x) + 3$ is obtained from the graph of $y = f(x)$ by shifting ________ 3 units.

(b) The graph of $y = f(x + 3)$ is obtained from the graph of $y = f(x)$ by shifting ________ 3 units.

2. (a) The graph of $y = f(x) - 3$ is obtained from the graph of $y = f(x)$ by shifting ________ 3 units.

(b) The graph of $y = f(x - 3)$ is obtained from the graph of $y = f(x)$ by shifting ________ 3 units.

3. Fill in the blank with the appropriate axis (x-axis or y-axis).

(a) The graph of $y = -f(x)$ is obtained from the graph of $y = f(x)$ by reflecting about the ________.

(b) The graph of $y = f(-x)$ is obtained from the graph of $y = f(x)$ by reflecting about the ________.

4. A graph of a function f is given. Match each equation with one of the graphs labeled I–IV.

(a) $f(x) + 2$ **(b)** $f(x + 3)$

(c) $f(x - 2)$ **(d)** $f(x) - 4$

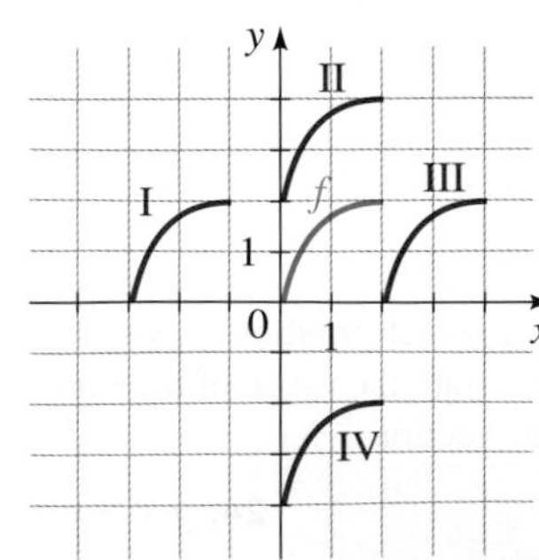

5. If a function f is an even function, then what type of symmetry does the graph of f have?

6. If a function f is an odd function, then what type of symmetry does the graph of f have?

Skills

7–16 ■ Describing Transformations Suppose the graph of f is given. Describe how the graph of each function can be obtained from the graph of f.

7. (a) $y = f(x) + 11$ (b) $y = f(x + 8)$

8. (a) $y = f(x - 7)$ (b) $y = f(x) - 10$

9. (a) $y = \frac{1}{4}f(-x)$ (b) $y = -5f(x)$

10. (a) $y = -6f(x)$ (b) $y = \frac{2}{3}f(-x)$

11. (a) $y = f(x - 1) - 5$ (b) $y = f(x + 2) - 4$

12. (a) $y = f(x - 4) + 6$ (b) $y = f(x + 2) + 9$

13. (a) $y = 5 + f(-x)$ (b) $y = 3 - \frac{1}{2}f(x + 2)$

14. (a) $y = 10 - f(x + 1)$ (b) $y = 4f(-x + 5) - 8$

15. (a) $y = 2 - f(5x)$ (b) $y = 1 + f(\frac{1}{2}(x + 1))$

16. (a) $y = f(\frac{1}{3}x) - 2$ (b) $y = f(2(x - 3)) - 1$

17–20 ■ Describing Transformations Explain how the graph of g is obtained from the graph of f.

17. (a) $f(x) = x^2$, $g(x) = (x + 2)^2$

(b) $f(x) = x^2$, $g(x) = x^2 + 2$

18. (a) $f(x) = x^3$, $g(x) = (x - 4)^3$

(b) $f(x) = x^3$, $g(x) = x^3 - 4$

19. (a) $f(x) = |x|$, $g(x) = |x + 2| - 2$

(b) $f(x) = |x|$, $g(x) = |x - 2| + 2$

20. (a) $f(x) = \sqrt{x}$, $g(x) = -\sqrt{x} + 1$

(b) $f(x) = \sqrt{x}$, $g(x) = \sqrt{-x} + 1$

21. Graphing Transformations Use the graph of $y = x^2$ in Figure 4 to graph the following equations.

(a) $g(x) = x^2 - 4$

(b) $g(x) = 2(x + 3)^2$

(c) $g(x) = 1 - x^2$

(d) $g(x) = (x + 1)^2 - 3$

22. Graphing Transformations Use the graph of $y = \sqrt{x}$ in Figure 5 to graph the given function.

(a) $g(x) = \sqrt{x - 2}$

(b) $g(x) = \sqrt{x} + 1$

(c) $g(x) = \sqrt{x + 2} + 2$

(d) $g(x) = -\sqrt{x} + 1$

23–26 ■ Identifying Transformations Match the function with its graph in I–IV, and state the range of the function. (See the graph of $y = |x|$ in Section 1.9.)

23. $y = |x + 1|$ **24.** $y = |x - 1|$

25. $y = |x| - 1$ **26.** $y = -|x|$

I

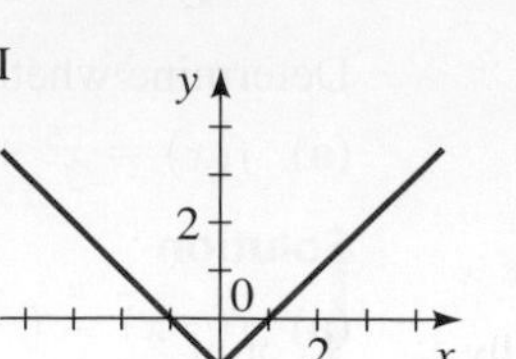

II

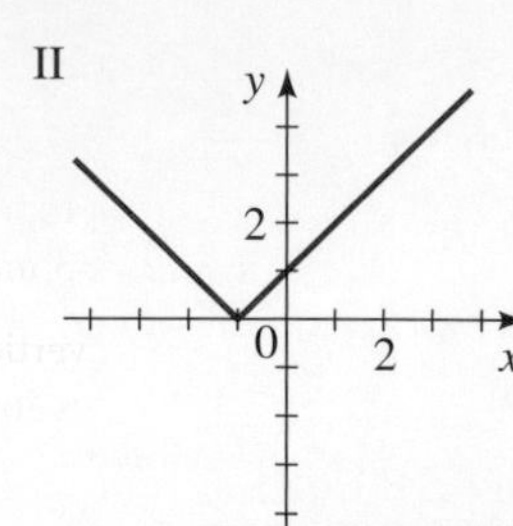

III

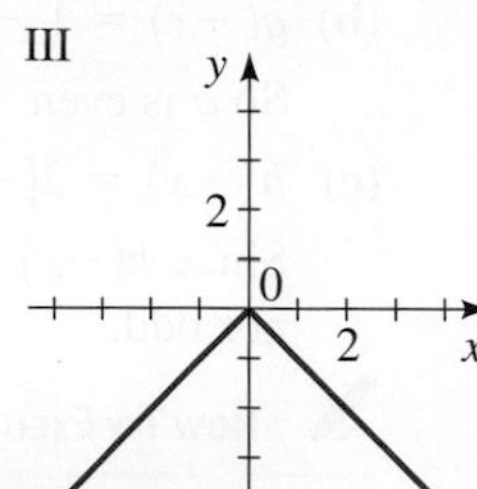

IV

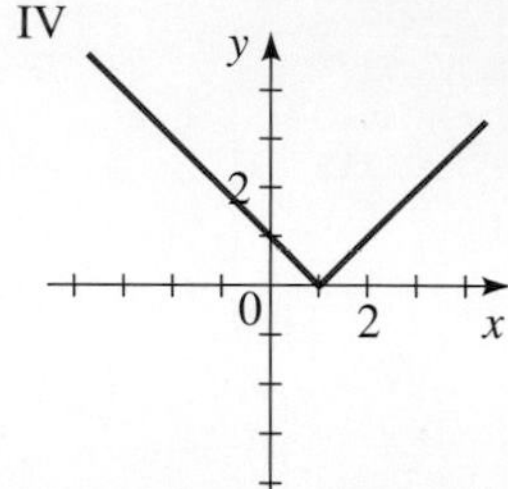

27–50 ■ Graphing Transformations Sketch the graph of the function, not by plotting points, but by starting with the graph of a standard function and applying transformations.

27. $f(x) = x^2 - 5$ **28.** $f(x) = x^2 + 2$

29. $f(x) = \sqrt{x} + 3$ **30.** $f(x) = |x| - 5$

31. $f(x) = (x - 5)^2$ **32.** $f(x) = (x + 1)^2$

33. $f(x) = |x + 2|$ **34.** $f(x) = \sqrt{x - 4}$

35. $f(x) = -x^3$ **36.** $f(x) = -|x|$

37. $f(x) = \sqrt[4]{-x}$ **38.** $f(x) = \sqrt[3]{-x}$

39. $f(x) = 5x^2$ **40.** $f(x) = \frac{1}{3}|x|$

41. $f(x) = -\frac{1}{5}\sqrt{x}$ **42.** $f(x) = 2\sqrt{-x}$

43. $f(x) = |x - 4| + 2$ **44.** $f(x) = (x + 1)^2 - 3$

45. $f(x) = -2\sqrt{x + 4} + 3$ **46.** $f(x) = 1 - \frac{1}{2}|x - 2|$

47. $f(x) = \frac{1}{2}(x + 2)^2 - 3$ **48.** $f(x) = 2\sqrt{x - 1} + 3$

49. $f(x) = \frac{1}{2}\sqrt{x + 4} - 3$ **50.** $f(x) = 3 - 2(x - 1)^2$

51–60 ■ Finding Equations for Transformations A function f is given, and the indicated transformations are applied to its graph (in the given order). Write an equation for the final transformed graph.

51. $f(x) = x^2$; shift upward 10 units

52. $f(x) = \sqrt{x}$; shift downward 4 units

53. $f(x) = x^4$; shift 3 units to the right

54. $f(x) = x^3$; shift 8 units to the left

55. $f(x) = |x|$; shift 2 units to the left and shift downward 5 units

56. $f(x) = |x|$; reflect about the x-axis, shift 4 units to the right, and shift upward 3 units

57. $f(x) = \sqrt[4]{x}$; reflect about the y-axis and shift upward 1 unit

58. $f(x) = x^2$; shift 2 units to the left and reflect about the x-axis

59. $f(x) = x^2$; stretch vertically by a factor of 2, shift downward 2 units, and shift 3 units to the right

60. $f(x) = |x|$; shrink vertically by a factor of $\frac{1}{2}$, shift 1 unit to the left, and shift upward 3 units

61–66 ■ Finding Formulas for Transformations The graphs of f and g are given. Find a formula for the function g.

61.

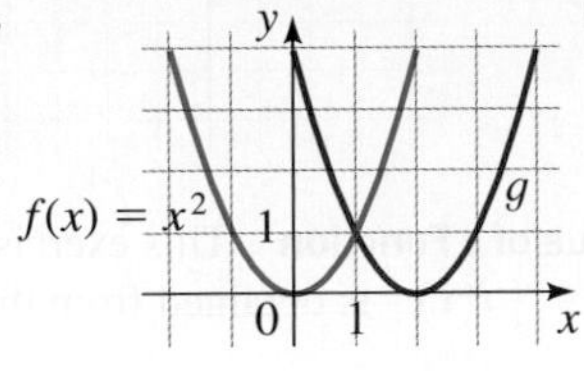

62.

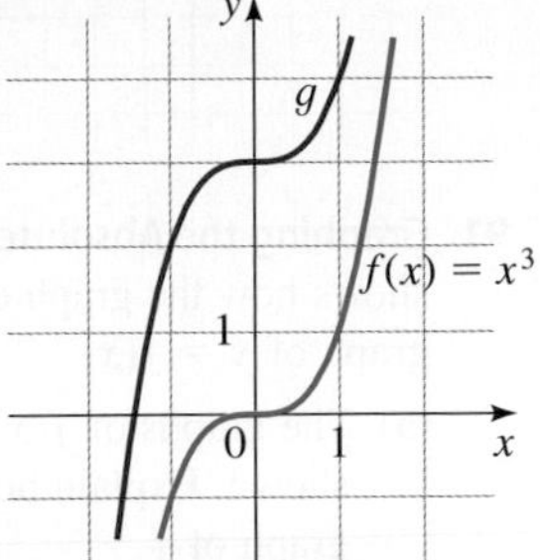

63.

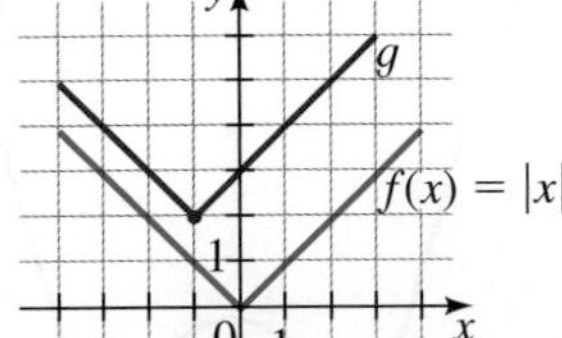

64.

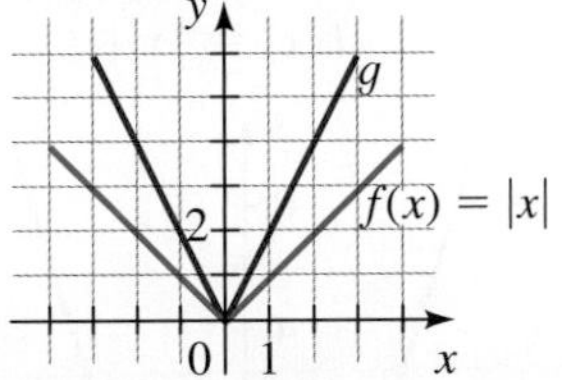

65.

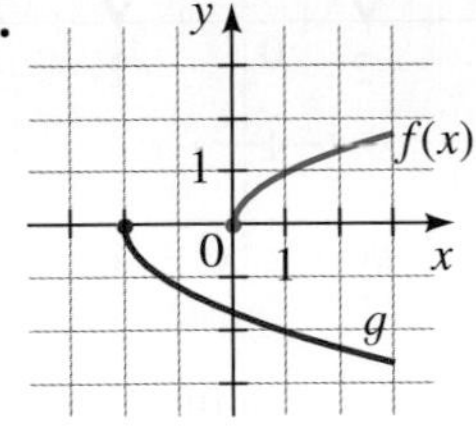

66.

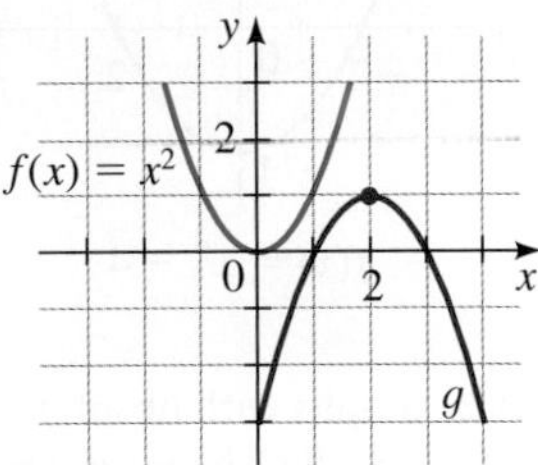

67–68 ■ Identifying Transformations The graph of $y = f(x)$ is given. Match each equation with its graph.

67. **(a)** $y = f(x - 4)$ **(b)** $y = f(x) + 3$
(c) $y = 2f(x + 6)$ **(d)** $y = -f(2x)$

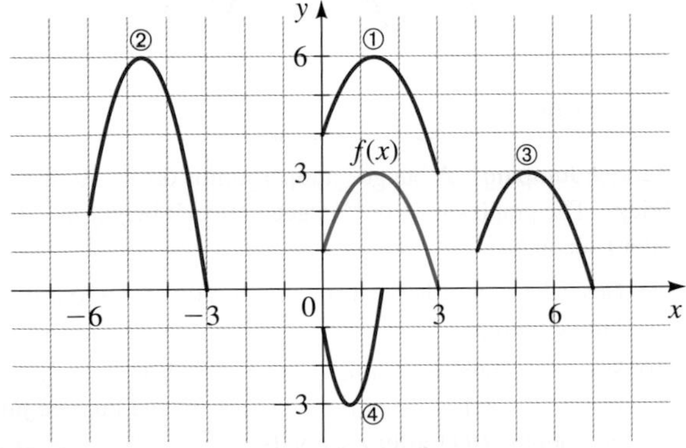

68. **(a)** $y = \frac{1}{3}f(x)$ **(b)** $y = -f(x + 4)$
(c) $y = f(x - 4) + 3$ **(d)** $y = f(-x)$

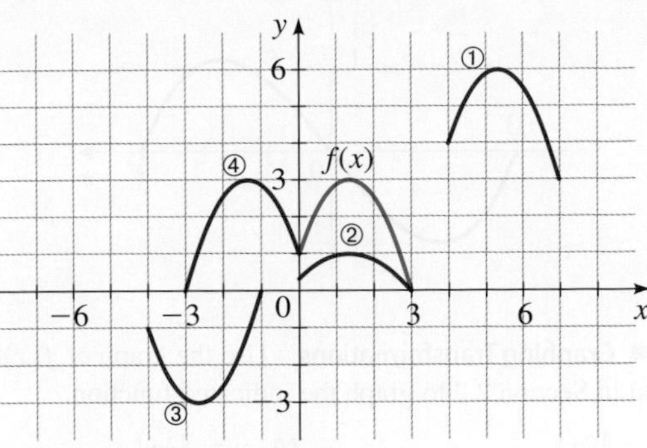

69–72 ■ Graphing Transformations The graph of a function f is given. Sketch the graphs of the following transformations of f.

69. **(a)** $y = f(x - 2)$ **(b)** $y = f(x) - 2$
(c) $y = 2f(x)$ **(d)** $y = -f(x) + 3$
(e) $y = f(-x)$ **(f)** $y = \frac{1}{2}f(x - 1)$

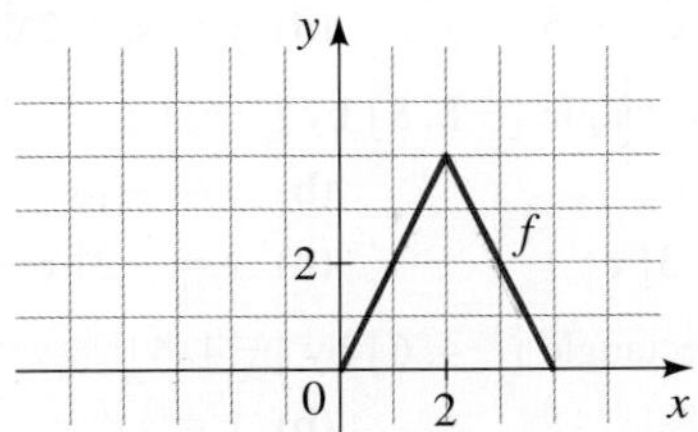

70. **(a)** $y = f(x + 1)$ **(b)** $y = f(-x)$
(c) $y = f(x - 2)$ **(d)** $y = f(x) - 2$
(e) $y = -f(x)$ **(f)** $y = 2f(x)$

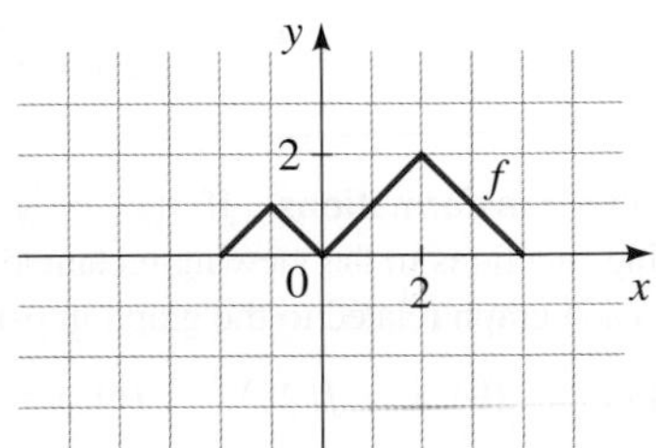

71. **(a)** $y = f(2x)$ **(b)** $y = f(\frac{1}{2}x)$
(c) $y = 2f(2x)$ **(d)** $y = -2f(\frac{1}{2}x)$

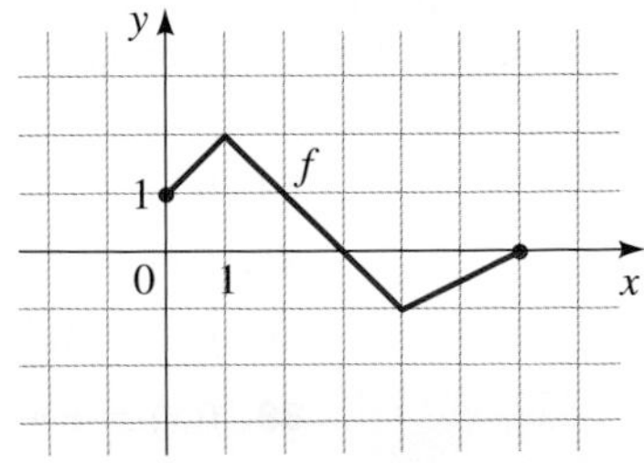

72. **(a)** $y = f(3x)$ **(b)** $y = f(\frac{1}{3}x)$
(c) $y = f(3(x + 1))$ **(d)** $y = 1 - f(\frac{1}{3}x)$

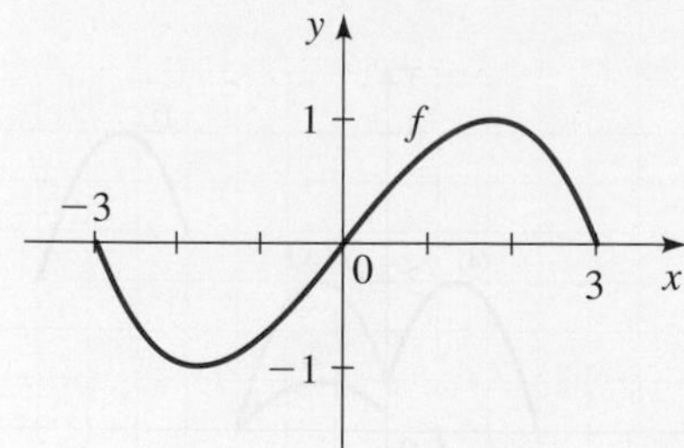

73–74 ■ Graphing Transformations Use the graph of $f(x) = [\![x]\!]$ described in Section 2.2 to graph the indicated function.

73. $y = [\![2x]\!]$ **74.** $y = [\![\frac{1}{4}x]\!]$

75–78 ■ Graphing Transformations Graph the functions on the same screen using the given viewing rectangle. How is each graph related to the graph in part (a)?

75. Viewing rectangle $[-8, 8]$ by $[-2, 8]$
(a) $y = \sqrt[4]{x}$ **(b)** $y = \sqrt[4]{x + 5}$
(c) $y = 2\sqrt[4]{x + 5}$ **(d)** $y = 4 + 2\sqrt[4]{x + 5}$

76. Viewing rectangle $[-8, 8]$ by $[-6, 6]$
(a) $y = |x|$ **(b)** $y = -|x|$
(c) $y = -3|x|$ **(d)** $y = -3|x - 5|$

77. Viewing rectangle $[-4, 6]$ by $[-4, 4]$
(a) $y = x^6$ **(b)** $y = \frac{1}{3}x^6$
(c) $y = -\frac{1}{3}x^6$ **(d)** $y = -\frac{1}{3}(x - 4)^6$

78. Viewing rectangle $[-6, 6]$ by $[-4, 4]$
(a) $y = \dfrac{1}{\sqrt{x}}$ **(b)** $y = \dfrac{1}{\sqrt{x + 3}}$
(c) $y = \dfrac{1}{2\sqrt{x + 3}}$ **(d)** $y = \dfrac{1}{2\sqrt{x + 3}} - 3$

79–80 ■ Graphing Transformations If $f(x) = \sqrt{2x - x^2}$, graph the following functions in the viewing rectangle $[-5, 5]$ by $[-4, 4]$. How is each graph related to the graph in part (a)?

79. **(a)** $y = f(x)$ **(b)** $y = f(2x)$ **(c)** $y = f(\frac{1}{2}x)$

80. **(a)** $y = f(x)$ **(b)** $y = f(-x)$
(c) $y = -f(-x)$ **(d)** $y = f(-2x)$
(e) $y = f(-\frac{1}{2}x)$

81–88 ■ Even and Odd Functions Determine whether the function f is even, odd, or neither. If f is even or odd, use symmetry to sketch its graph.

81. $f(x) = x^4$ **82.** $f(x) = x^3$
83. $f(x) = x^2 + x$ **84.** $f(x) = x^4 - 4x^2$
85. $f(x) = x^3 - x$ **86.** $f(x) = 3x^3 + 2x^2 + 1$
87. $f(x) = 1 - \sqrt[3]{x}$ **88.** $f(x) = x + \dfrac{1}{x}$

Skills Plus

89–90 ■ Graphing Even and Odd Functions The graph of a function defined for $x \geq 0$ is given. Complete the graph for $x < 0$ to make **(a)** an even function and **(b)** an odd function.

89.

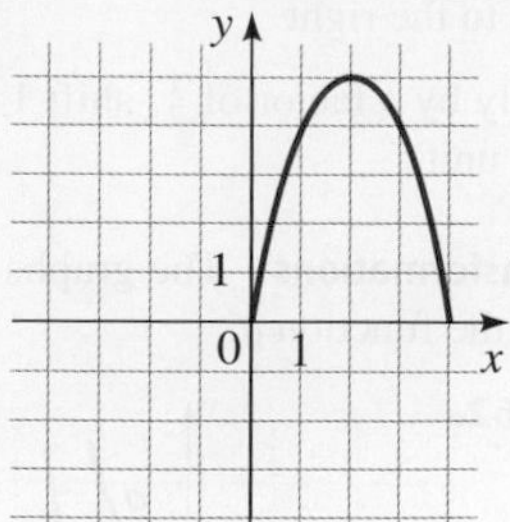

90.

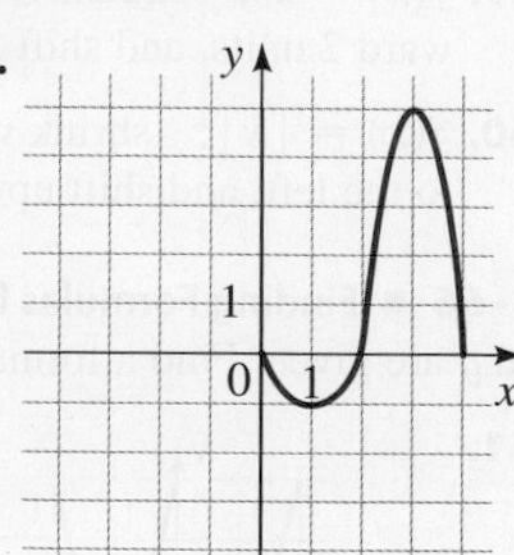

91. Graphing the Absolute Value of a Function This exercise shows how the graph of $y = |f(x)|$ is obtained from the graph of $y = f(x)$.

(a) The graphs of $f(x) = x^2 - 4$ and $g(x) = |x^2 - 4|$ are shown. Explain how the graph of g is obtained from the graph of f.

(b) Sketch the graph of the functions $g(x) = |4x - x^2|$ and $h(x) = |x^3|$.

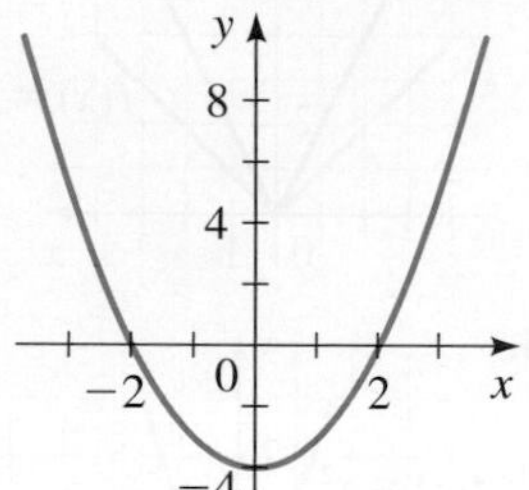

$f(x) = x^2 - 4$

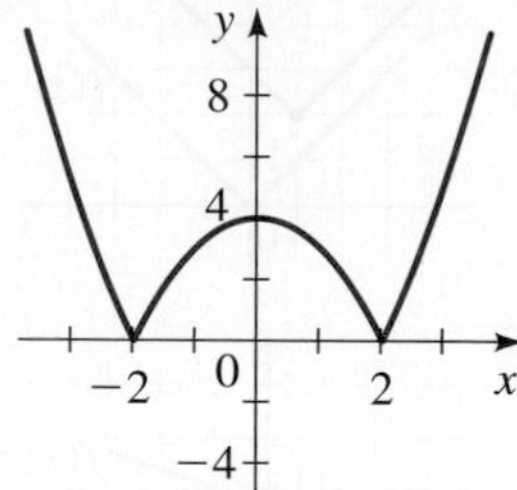

$g(x) = |x^2 - 4|$

92. Graphs with Absolute Value The graph of a function $y = f(x)$ is given. Draw graphs of **(a)** $y = |f(x)|$ and **(b)** $y = f(|x|)$.

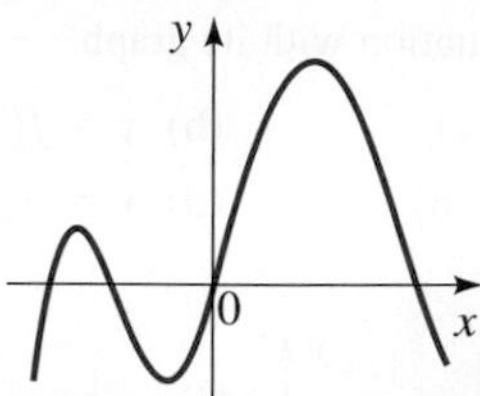

Applications

93. Bungee Jumping A bungee jumper jumps off a 150-m-high bridge. The graph shows the bungee jumper's height $h(t)$ (in m) after t seconds.

(a) Describe in words what the graph indicates about the bungee jump.

(b) Suppose a bungee jumper jumps off a 120-m-high bridge. Sketch a new graph that shows the bungee jumper's height $H(t)$ after t seconds.

(c) What transformation must be performed on the function h to obtain the function H? Express the function H in terms of h.

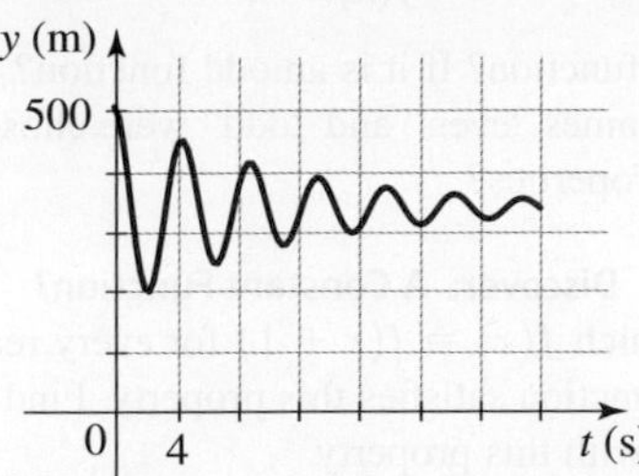

94. Swimming Laps The function $y = f(t)$ graphed below gives the distance (in meters) of a swimmer from the starting edge of a pool t seconds after the swimmer starts swimming laps.

(a) Describe in words the swimmer's lap practice. What is the average speed for the first 30 s?

(b) Graph the function $y = 1.2f(t)$. How is the graph of the new function related to the graph of the original function?

(c) What is the swimmer's new average speed for the first 30 s?

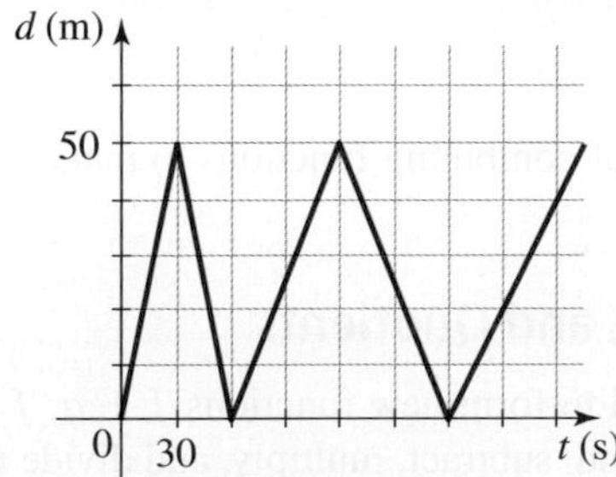

95. Field Trip A class of fourth graders walks to a park on a field trip. The function $y = f(t)$ graphed below gives their distance from school (in m) t minutes after they left the building.

(a) What is the average speed walking to the park? How long was the class at the park? How far away from the school is the park?

(b) Graph the function $y = 0.5f(t)$. How is the graph of the new function related to the graph of the original function? What is the average speed going to the new park? How far away from the school is the new park?

(c) Graph the function $y = f(t - 10)$. How is the graph of the new function related to the graph of the original function? How does the field trip descibed by this function differ from the original trip?

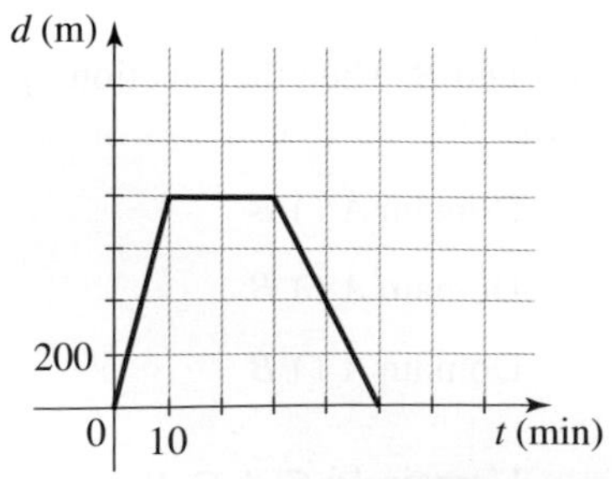

96. Redshift The colors in a light spectrum correspond to different wavelengths of light. If a galaxy is rapidly moving away from the earth, we see the spectrum of light from the galaxy "shifted" toward the red (long) end of the wavelength spectrum. A *redshift* $z > 0$ of a distant galaxy means that light emitted from the galaxy (the source) at wavelength x is measured from the earth to have wavelength $(1 + z)x$. (See also Exercise 1.R.159.) The spectrum shown below is of light from a distant galaxy. The function $y = f(x)$ graphed below gives the intensity (or brightness) of the light at wavelength x nanometers (nm) as measured from the earth. Let $y = g(x)$ be the intensity of the light at wavelength x as emitted from the source. Then $g(x) = f((1 + z)x)$. The largest measured redshift of any known galaxy is about $z = 11$. Assume $z = 11$ for the spectrum shown below.

(a) Describe how the graph of g is obtained from the graph of f.

(b) There is a dip in the graph at wavelength $x = 480$ nm. At what wavelength would this dip be at the source?

(c) Estimate the wavelengths at the source at several of the dips in the graph. Use your results to sketch a rough graph of the intensity function g.

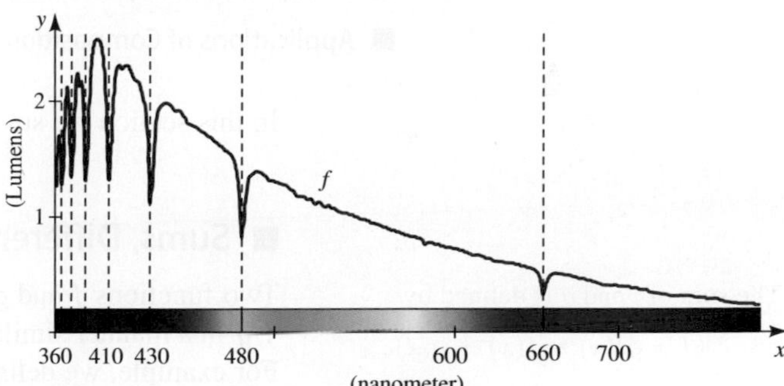

■ Discuss ■ Discover ■ Prove ■ Write

97–98 ■ Discuss: Obtaining Transformations Can the function g be obtained from f by transformations? If so, describe the transformations needed.

97. The functions f and g are described algebraically as follows:

$$f(x) = (x + 2)^2 \qquad g(x) = (x - 2)^2 + 5$$

98. The functions f and g are described graphically in the figure.

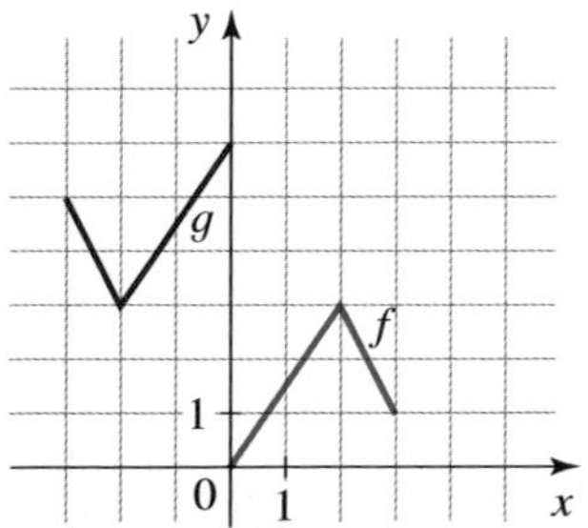

99. Discover: Stretched Step Functions Sketch graphs of the functions $f(x) = [\![x]\!]$, $g(x) = [\![2x]\!]$, and $h(x) = [\![3x]\!]$ on separate sets of axes. How are the graphs related? If n is a positive integer, what does a graph of $k(x) = [\![nx]\!]$ look like? What about the graphs of $k(x) = \left[\!\!\left[\frac{1}{n}x\right]\!\!\right]$?

100. Discuss: Sums of Even and Odd Functions If f and g are both even functions, is $f + g$ necessarily even? If both are odd, is their sum necessarily odd? What can you say about the sum if one is odd and one is even? In each case, prove your answer.

101. Discuss: Products of Even and Odd Functions Answer the same questions as in Exercise 100, except this time consider the product of f and g instead of the sum.

102. Discuss: Even and Odd Power Functions What must be true about the integer n if the function

$$f(x) = x^n$$

is an even function? If it is an odd function? Why do you think the names "even" and "odd" were chosen for these function properties?

103. Discuss ■ Discover: A Constant Function? Let f be a function for which $f(x) = f(x + 1)$ for every real number x. A constant function satisfies this property. Find a nonconstant function f with this property.

PS *Draw a diagram.* Think about defining a function graphically. That is, try to draw the graph of a function that satisfies the given condition.

2.7 Combining Functions

■ Sums, Differences, Products, and Quotients ■ Composition of Functions ■ Applications of Composition

In this section we study different ways of combining functions to make new functions.

■ Sums, Differences, Products, and Quotients

The sum of f and g is defined by

$$(f + g)(x) = f(x) + g(x)$$

The name of the new function is "$f + g$." So this $+$ sign stands for the operation of addition of *functions*. The $+$ sign on the right side, however, stands for addition of the *numbers* $f(x)$ and $g(x)$.

Two functions f and g can be combined to form new functions $f + g$, $f - g$, fg, and f/g in a manner similar to the way we add, subtract, multiply, and divide real numbers. For example, we define the function $f + g$ by

$$(f + g)(x) = f(x) + g(x)$$

The new function $f + g$ is called the **sum** of the functions f and g; its value at x is $f(x) + g(x)$. Of course, the sum on the right-hand side makes sense only if both $f(x)$ and $g(x)$ are defined, that is, if x belongs to the domain of f and also to the domain of g. So if the domain of f is A and the domain of g is B, then the domain of $f + g$ is the intersection of these domains, that is, $A \cap B$. Similarly, we can define the **difference** $f - g$, the **product** fg, and the **quotient** f/g of the functions f and g. Their domains are $A \cap B$, but in the case of the quotient we must remember not to divide by 0.

Algebra of Functions

Let f and g be functions with domains A and B. Then the functions $f + g$, $f - g$, fg, and f/g are defined as follows.

$(f + g)(x) = f(x) + g(x)$ Domain $A \cap B$

$(f - g)(x) = f(x) - g(x)$ Domain $A \cap B$

$(fg)(x) = f(x)g(x)$ Domain $A \cap B$

$\left(\dfrac{f}{g}\right)(x) = \dfrac{f(x)}{g(x)}$ Domain $\{x \in A \cap B \mid g(x) \neq 0\}$

Example 1 ■ Combinations of Functions and Their Domains

Let $f(x) = \dfrac{1}{x-2}$ and $g(x) = \sqrt{x}$.

(a) Find the functions $f + g$, $f - g$, fg, and f/g and their domains.

(b) Find $(f+g)(4)$, $(f-g)(4)$, $(fg)(4)$, and $(f/g)(4)$.

Solution

(a) The domain of f is $\{x \mid x \neq 2\}$, and the domain of g is $\{x \mid x \geq 0\}$. The intersection of the domains of f and g is

$$\{x \mid x \geq 0 \text{ and } x \neq 2\} = [0, 2) \cup (2, \infty)$$

Thus we have

$$(f+g)(x) = f(x) + g(x) = \frac{1}{x-2} + \sqrt{x} \qquad \text{Domain } \{x \mid x \geq 0 \text{ and } x \neq 2\}$$

$$(f-g)(x) = f(x) - g(x) = \frac{1}{x-2} - \sqrt{x} \qquad \text{Domain } \{x \mid x \geq 0 \text{ and } x \neq 2\}$$

$$(fg)(x) = f(x)g(x) = \frac{\sqrt{x}}{x-2} \qquad \text{Domain } \{x \mid x \geq 0 \text{ and } x \neq 2\}$$

$$\left(\frac{f}{g}\right)(x) = \frac{f(x)}{g(x)} = \frac{1}{(x-2)\sqrt{x}} \qquad \text{Domain } \{x \mid x > 0 \text{ and } x \neq 2\}$$

To divide fractions, invert the denominator and multiply:

$$\frac{1/(x-2)}{\sqrt{x}} = \frac{1/(x-2)}{\sqrt{x}/1} = \frac{1}{x-2} \cdot \frac{1}{\sqrt{x}} = \frac{1}{(x-2)\sqrt{x}}$$

Note that in the domain of f/g we exclude 0 because $g(0) = 0$.

(b) Each of these values exist because $x = 4$ is in the domain of each function:

$$(f+g)(4) = f(4) + g(4) = \frac{1}{4-2} + \sqrt{4} = \frac{5}{2}$$

$$(f-g)(4) = f(4) - g(4) = \frac{1}{4-2} - \sqrt{4} = -\frac{3}{2}$$

$$(fg)(4) = f(4)g(4) = \left(\frac{1}{4-2}\right)\sqrt{4} = 1$$

$$\left(\frac{f}{g}\right)(4) = \frac{f(4)}{g(4)} = \frac{1}{(4-2)\sqrt{4}} = \frac{1}{4}$$

Now Try Exercise 9 ■

The graph of the function $f + g$ can be obtained from the graphs of f and g by **graphical addition**. This means that we add corresponding y-coordinates, as illustrated in the next example.

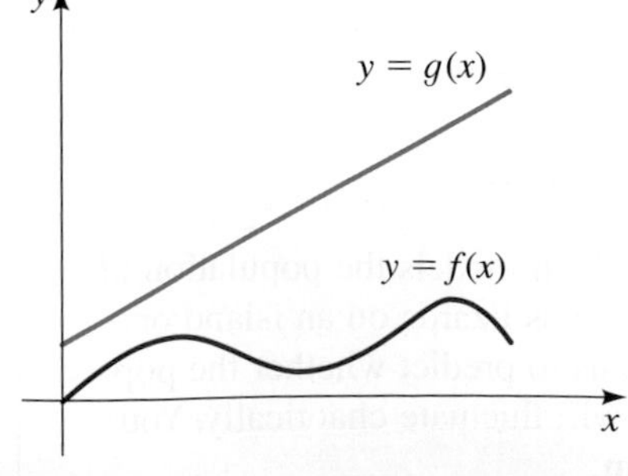

Figure 1

Example 2 ■ Using Graphical Addition

The graphs of f and g are shown in Figure 1. Use graphical addition to graph the function $f + g$.

Solution We obtain the graph of $f + g$ by "graphically adding" the value of $f(x)$ to $g(x)$ as shown in Figure 2. This is implemented by copying the line segment PQ on top of PR to obtain the point S on the graph of $f + g$.

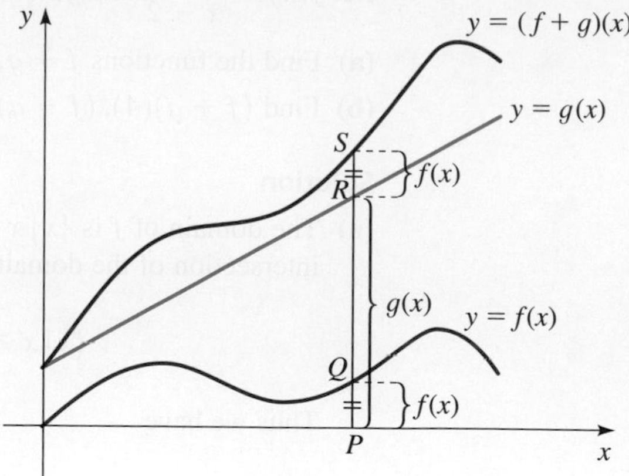

Figure 2 | Graphical addition

Now Try Exercise 21 ■

■ Composition of Functions

Now let's consider an important way of combining two functions to get a new function. Suppose $f(x) = \sqrt{x}$ and $g(x) = x^2 + 1$. We may define a new function h as

$$h(x) = f(g(x)) = f(x^2 + 1) = \sqrt{x^2 + 1}$$

The function h is made up of the functions f and g in an interesting way: Given a number x, we first apply the function g to it, then apply f to the result. In this case, f is the rule "take the square root," g is the rule "square, then add 1," and h is the rule "square, then add 1, then take the square root." In other words, we get the rule h by applying the rule g and then the rule f. Figure 3 shows a machine diagram for h.

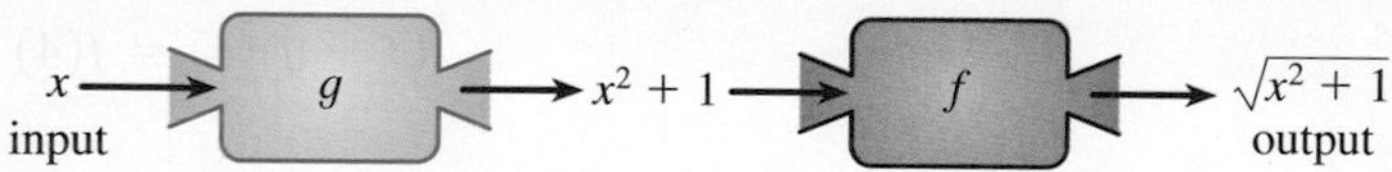

Figure 3 | The h machine is composed of the g machine (first) and then the f machine.

In general, given any two functions f and g, we start with a number x in the domain of g and find its image $g(x)$. If this number $g(x)$ is in the domain of f, we can then calculate the value of $f(g(x))$. The result is a new function $h(x) = f(g(x))$, which is obtained by substituting g into f. It is called the *composition* (or *composite*) of f and g and is denoted by $f \circ g$ ("f composed with g").

Discovery Project ■ Iteration and Chaos

The *iterates* of a function f at a point x are the numbers

$$f(x),\ f(f(x)),\ f(f(f(x))), \ldots$$

We examine iterates of the *logistic function,* which models the population of a species with limited potential for growth (such as lizards on an island or fish in a pond). Iterates of the model can help us to predict whether the population will eventually stabilize or whether it will fluctuate chaotically. You can find the project at **www.stewartmath.com**.

Composition of Functions

Given two functions f and g, the **composite function** $f \circ g$ (also called the **composition** of f and g) is defined by

$$(f \circ g)(x) = f(g(x))$$

The domain of $f \circ g$ is the set of all x in the domain of g such that $g(x)$ is in the domain of f. In other words, $(f \circ g)(x)$ is defined whenever both $g(x)$ and $f(g(x))$ are defined. We can picture $f \circ g$ using an arrow diagram (Figure 4).

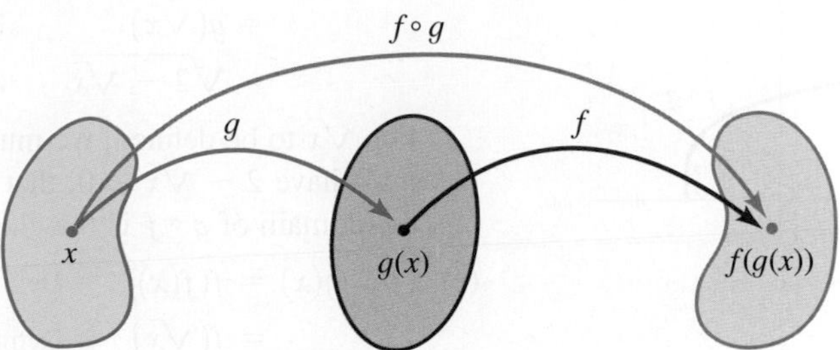

Figure 4 | Arrow diagram for $f \circ g$

Example 3 ■ Finding the Composition of Functions

Let $f(x) = x^2$ and $g(x) = x - 3$.

(a) Find the functions $f \circ g$ and $g \circ f$ and their domains.

(b) Find $(f \circ g)(5)$ and $(g \circ f)(7)$.

Solution

In Example 3, f is the rule "square," and g is the rule "subtract 3." The function $f \circ g$ *first* subtracts 3 and *then* squares; the function $g \circ f$ *first* squares and *then* subtracts 3.

(a) We have

$$\begin{aligned}(f \circ g)(x) &= f(g(x)) && \text{Definition of } f \circ g\\ &= f(x-3) && \text{Definition of } g\\ &= (x-3)^2 && \text{Definition of } f\end{aligned}$$

and

$$\begin{aligned}(g \circ f)(x) &= g(f(x)) && \text{Definition of } g \circ f\\ &= g(x^2) && \text{Definition of } f\\ &= x^2 - 3 && \text{Definition of } g\end{aligned}$$

The domains of both $f \circ g$ and $g \circ f$ are $\mathbb{R}$.

(b) We have

$$(f \circ g)(5) = f(g(5)) = f(2) = 2^2 = 4$$
$$(g \circ f)(7) = g(f(7)) = g(49) = 49 - 3 = 46$$

Now Try Exercises 27 and 49 ■

You can see from Example 3 that, in general, $f \circ g \neq g \circ f$. Remember that the notation $f \circ g$ means that the function g is applied first and then f is applied.

The graphs of f and g of Example 4, as well as those of $f \circ g$, $g \circ f$, $f \circ f$, and $g \circ g$, are shown below. These graphs indicate that the operation of composition can produce functions that are quite different from the original functions.

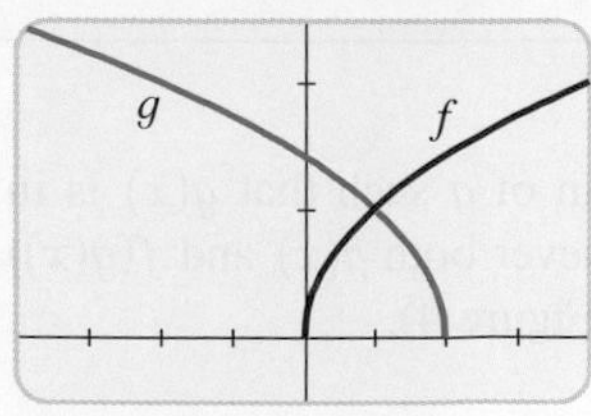

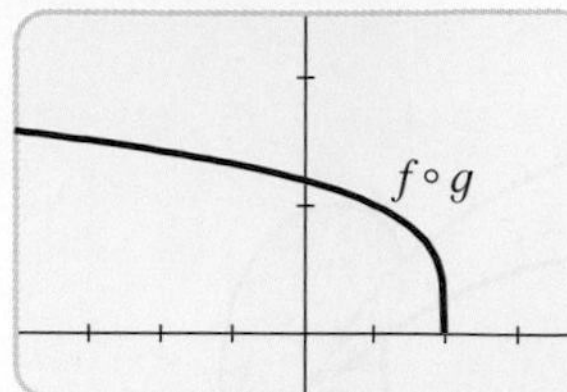

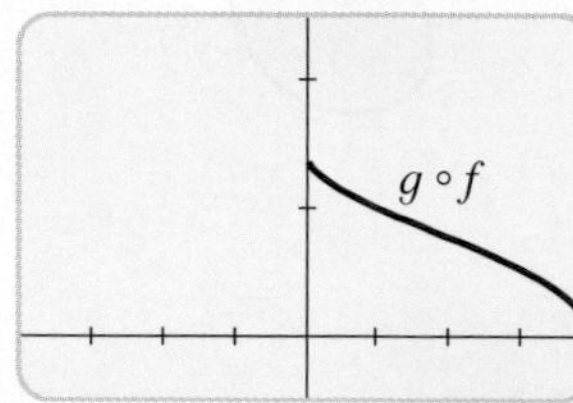

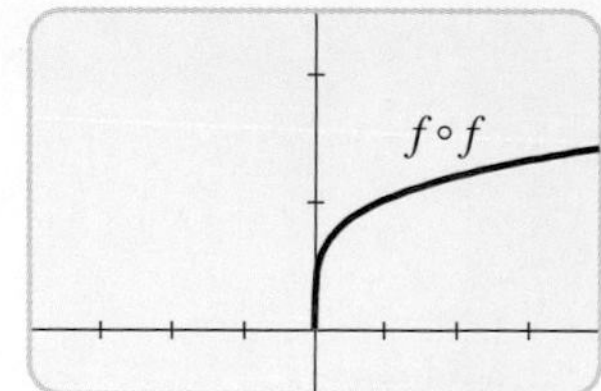

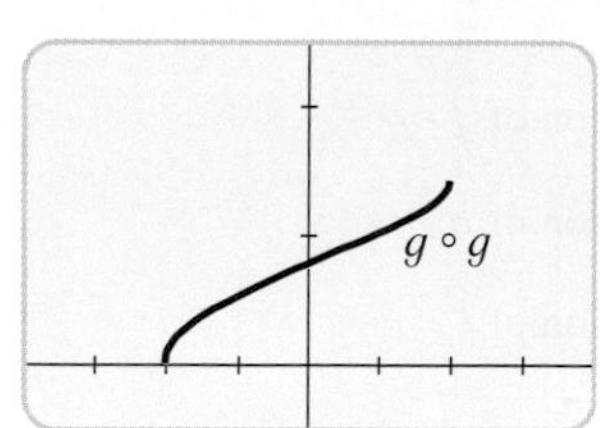

Example 4 ■ Finding the Composition of Functions

If $f(x) = \sqrt{x}$ and $g(x) = \sqrt{2 - x}$, find the following functions and their domains.

(a) $f \circ g$ **(b)** $g \circ f$ **(c)** $f \circ f$ **(d)** $g \circ g$

Solution

(a)
$$\begin{aligned}(f \circ g)(x) &= f(g(x)) && \text{Definition of } f \circ g\\ &= f(\sqrt{2 - x}) && \text{Definition of } g\\ &= \sqrt{\sqrt{2 - x}} && \text{Definition of } f\\ &= \sqrt[4]{2 - x} && \text{Property of } n\text{th roots}\end{aligned}$$

The domain of $f \circ g$ is $\{x \mid 2 - x \geq 0\} = \{x \mid x \leq 2\} = (-\infty, 2]$.

(b)
$$\begin{aligned}(g \circ f)(x) &= g(f(x)) && \text{Definition of } g \circ f\\ &= g(\sqrt{x}) && \text{Definition of } f\\ &= \sqrt{2 - \sqrt{x}} && \text{Definition of } g\end{aligned}$$

For $\sqrt{x}$ to be defined, we must have $x \geq 0$. For $\sqrt{2 - \sqrt{x}}$ to be defined, we must have $2 - \sqrt{x} \geq 0$, that is, $\sqrt{x} \leq 2$, or $x \leq 4$. Thus we have $0 \leq x \leq 4$, so the domain of $g \circ f$ is the closed interval $[0, 4]$.

(c)
$$\begin{aligned}(f \circ f)(x) &= f(f(x)) && \text{Definition of } f \circ f\\ &= f(\sqrt{x}) && \text{Definition of } f\\ &= \sqrt{\sqrt{x}} && \text{Definition of } f\\ &= \sqrt[4]{x} && \text{Property of } n\text{th roots}\end{aligned}$$

The domain of $f \circ f$ is $[0, \infty)$.

(d)
$$\begin{aligned}(g \circ g)(x) &= g(g(x)) && \text{Definition of } g \circ g\\ &= g(\sqrt{2 - x}) && \text{Definition of } g\\ &= \sqrt{2 - \sqrt{2 - x}} && \text{Definition of } g\end{aligned}$$

This expression is defined when both $2 - x \geq 0$ and $2 - \sqrt{2 - x} \geq 0$. The first inequality means $x \leq 2$, and the second is equivalent to $\sqrt{2 - x} \leq 2$, or $2 - x \leq 4$, or $x \geq -2$. Thus $-2 \leq x \leq 2$, so the domain of $g \circ g$ is $[-2, 2]$.

Now Try Exercise 53 ■

It is possible to take the composition of three or more functions. For instance, the composite function $f \circ g \circ h$ is found by first applying h, then g, and then f as follows:

$$(f \circ g \circ h)(x) = f(g(h(x)))$$

Example 5 ■ A Composition of Three Functions

Find $f \circ g \circ h$ if $f(x) = x/(x + 1)$, $g(x) = x^{10}$, and $h(x) = x + 3$.

Solution

$$\begin{aligned}(f \circ g \circ h)(x) &= f(g(h(x))) && \text{Definition of } f \circ g \circ h\\ &= f(g(x + 3)) && \text{Definition of } h\\ &= f((x + 3)^{10}) && \text{Definition of } g\\ &= \frac{(x + 3)^{10}}{(x + 3)^{10} + 1} && \text{Definition of } f\end{aligned}$$

Now Try Exercise 61 ■

So far, we have used composition to build complicated functions from simpler ones. In calculus it is also useful to be able to "decompose" a complicated function into simpler ones, as shown in the following example.

Example 6 ■ Recognizing a Composition of Functions

Given $F(x) = \sqrt[4]{x + 9}$, find functions f and g such that $F = f \circ g$.

Solution Since the formula for F says to first add 9 and then take the fourth root, we can let

$$g(x) = x + 9 \qquad \text{and} \qquad f(x) = \sqrt[4]{x}$$

Then

$$\begin{aligned} (f \circ g)(x) &= f(g(x)) && \text{Definition of } f \circ g \\ &= f(x + 9) && \text{Definition of } g \\ &= \sqrt[4]{x + 9} && \text{Definition of } f \\ &= F(x) \end{aligned}$$

 **Now Try Exercise 65** ■

■ Applications of Composition

When working with functions that model real-world situations, we can name the variables using letters that suggest the quantity being modeled. We may use t for time, d for distance, V for volume, and so on. For example, if air is being pumped into a balloon, then the radius R of the balloon is a function of the volume V of air pumped into the balloon, say, $R = f(V)$. Also the volume V is a function of the time t that the pump has been working, say, $V = g(t)$. It follows that the radius R is a function of the time t given by $R = f(g(t))$.

Example 7 ■ An Application of Composition of Functions

A ship is traveling at 20 km/h parallel to a straight shoreline. The ship is 5 km from shore. It passes a lighthouse at noon.

(a) Express the distance s between the lighthouse and the ship as a function of d, the distance the ship has traveled since noon; that is, find f so that $s = f(d)$.

(b) Express d as a function of t, the time elapsed since noon; that is, find g so that $d = g(t)$.

(c) Find $f \circ g$. What does this function represent?

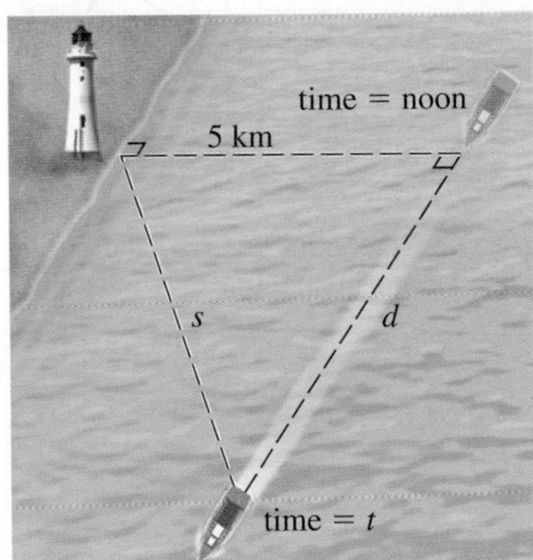

Figure 5

Solution We first draw a diagram as in Figure 5.

(a) We can relate the distances s and d by the Pythagorean Theorem. Thus s can be expressed as a function of d by

$$s = f(d) = \sqrt{25 + d^2}$$

(b) Since the ship is traveling at 20 km/h, the distance d that it has traveled is a function of t as follows:

distance = rate × time

$$d = g(t) = 20t$$

(c) We have

$$\begin{aligned} (f \circ g)(t) &= f(g(t)) && \text{Definition of } f \circ g \\ &= f(20t) && \text{Definition of } g \\ &= \sqrt{25 + (20t)^2} && \text{Definition of } f \end{aligned}$$

The function $f \circ g$ gives the distance of the ship from the lighthouse as a function of time.

Now Try Exercise 83 ■

2.7 Exercises

Concepts

1. From the graphs of f and g in the figure, we find

$(f+g)(2) =$ ________ $(f-g)(2) =$ ________

$(fg)(2) =$ ________ $\left(\dfrac{f}{g}\right)(2) =$ ________

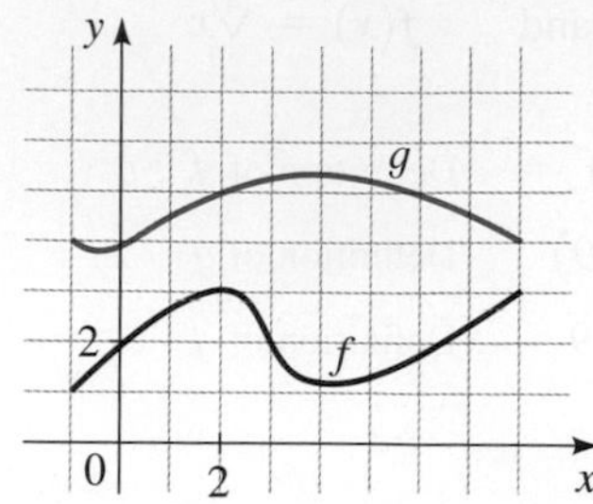

2. By definition, $(f \circ g)(x) =$ ________. So if $g(2) = 5$ and $f(5) = 12$, then $(f \circ g)(2) =$ ________.

3. If the rule of the function f is "add one" and the rule of the function g is "multiply by 2," then the rule of $f \circ g$ is

"________________________,"

and the rule of $g \circ f$ is

"________________________."

4. We can express the functions in Exercise 3 algebraically as

$f(x) =$ ________ $g(x) =$ ________

$(f \circ g)(x) =$ ________ $(g \circ f)(x) =$ ________

5–6 ■ Let f and g be functions.

5. (a) The function $(f+g)(x)$ is defined for all values of x that are in the domains of both ________ and ________.

(b) The function $(fg)(x)$ is defined for all values of x that are in the domains of both ________ and ________.

(c) The function $(f/g)(x)$ is defined for all values of x that are in the domains of both ________ and ________, and $g(x)$ is not equal to ________.

6. The composition $(f \circ g)(x)$ is defined for all values of x for which x is in the domain of ________ and $g(x)$ is in the domain of ________.

Skills

7–16 ■ Combining Functions Find $f+g$, $f-g$, fg, and f/g and their domains.

7. $f(x) = 3x, \quad g(x) = 1 - x$

8. $f(x) = 3 - 2x, \quad g(x) = 2x + 1$

9. $f(x) = x^3 + x^2, \quad g(x) = x^2$

10. $f(x) = x^2 + 1, \quad g(x) = x^2 + 2$

11. $f(x) = 5 - x, \quad g(x) = x^2 - 3x$

12. $f(x) = x^2 + 2x, \quad g(x) = 3x^2 - 1$

13. $f(x) = \sqrt{25 - x^2}, \quad g(x) = \sqrt{x + 3}$

14. $f(x) = \sqrt{16 - x^2}, \quad g(x) = \sqrt{x^2 - 1}$

15. $f(x) = \dfrac{1}{x+1}, \quad g(x) = \dfrac{3}{x-2}$

16. $f(x) = \dfrac{3}{x-3}, \quad g(x) = \dfrac{x}{x+3}$

17–20 ■ Domain Find the domain of the function.

17. $f(x) = \sqrt{x} + \sqrt{3 - x}$

18. $f(x) = \sqrt{x + 4} - \dfrac{\sqrt{1 - x}}{x}$

19. $h(x) = (x - 3)^{-1/4}$

20. $k(x) = \dfrac{\sqrt{x + 3}}{x - 1}$

21–22 ■ Graphical Addition Use graphical addition to sketch the graph of $f + g$.

21.

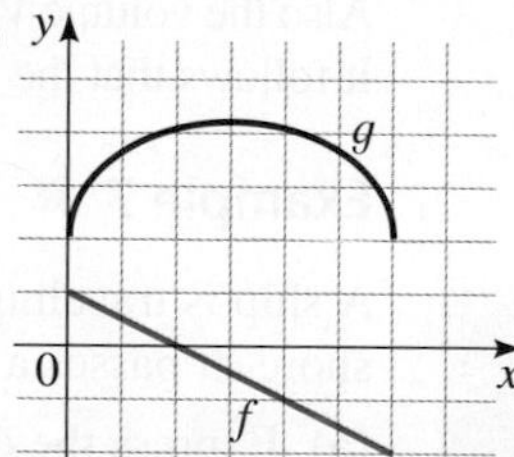

22.

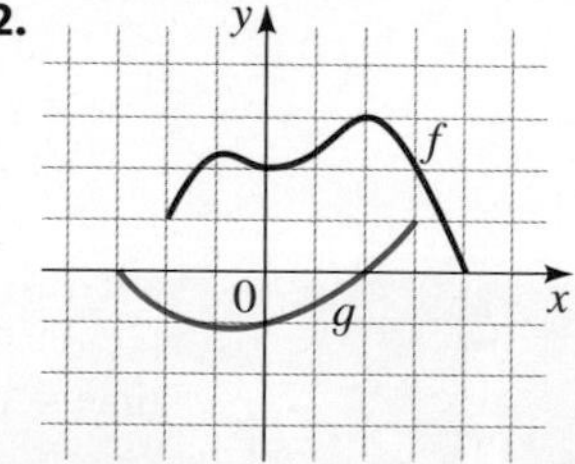

23–26 ■ Graphical Addition Draw the graphs of f, g, and $f + g$ on a common screen to illustrate graphical addition.

23. $f(x) = \sqrt{1 + x}, \quad g(x) = \sqrt{1 - x}$

24. $f(x) = x^2, \quad g(x) = \sqrt{x}$

25. $f(x) = x^2, \quad g(x) = \frac{1}{3}x^3$

26. $f(x) = \sqrt[4]{1 - x}, \quad g(x) = \sqrt{1 - \dfrac{x^2}{9}}$

27–32 ■ Evaluating Composition of Functions Use $f(x) = 4x + 5$ and $g(x) = x^2 + 2$ to evaluate the expression.

27. (a) $f(g(1))$ **(b)** $g(f(1))$

28. (a) $f(f(0))$ **(b)** $g(g(-1))$

29. (a) $(f \circ g)(-2)$ **(b)** $(g \circ f)(-1)$

30. (a) $(f \circ f)(1)$ **(b)** $(g \circ g)(0)$

31. (a) $(f \circ g)(x)$ **(b)** $(g \circ f)(x)$

32. (a) $(f \circ f)(x)$ **(b)** $(g \circ g)(x)$

33–38 ■ Composition Using a Graph Use the given graphs of f and g to evaluate the expression.

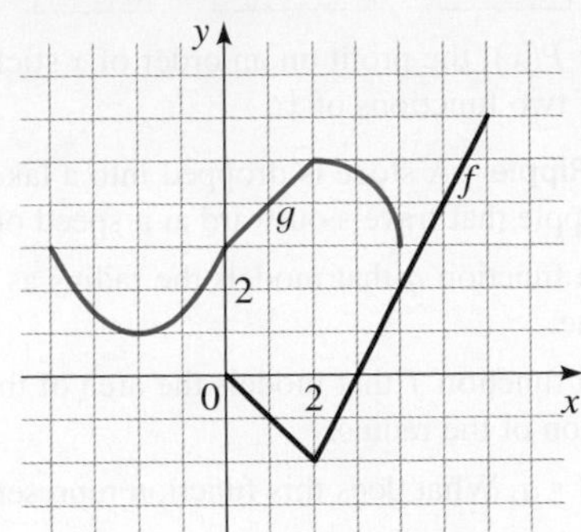

33. $f(g(2))$
34. $g(f(0))$
35. $(g \circ f)(4)$
36. $(f \circ g)(0)$
37. $(g \circ g)(-2)$
38. $(f \circ f)(4)$

39–46 ■ Composition Using a Table Use the table to evaluate the expression.

x	1	2	3	4	5	6
$f(x)$	2	3	5	1	6	3
$g(x)$	3	5	6	2	1	4

39. $f(g(2))$
40. $g(f(2))$
41. $f(f(1))$
42. $g(g(2))$
43. $(f \circ g)(6)$
44. $(g \circ f)(2)$
45. $(f \circ f)(5)$
46. $(g \circ g)(3)$

47–60 ■ Composition of Functions Find the functions $f \circ g$, $g \circ f$, $f \circ f$, and $g \circ g$ and their domains.

47. $f(x) = 2x + 3, \quad g(x) = 4x - 1$
48. $f(x) = 6x - 5, \quad g(x) = \dfrac{x}{2}$
49. $f(x) = x^2, \quad g(x) = x + 1$
50. $f(x) = \sqrt[3]{x}, \quad g(x) = \dfrac{1}{x^3}$
51. $f(x) = x^2 + 1, \quad g(x) = \dfrac{1}{\sqrt{x}}$
52. $f(x) = x - 4, \quad g(x) = |x + 4|$
53. $f(x) = \dfrac{x}{x+1}, \quad g(x) = 2x - 1$
54. $f(x) = \dfrac{x}{x+1}, \quad g(x) = \dfrac{1}{x}$
55. $f(x) = \dfrac{2}{x}, \quad g(x) = \dfrac{x}{x+2}$
56. $f(x) = \dfrac{1}{x-1}, \quad g(x) = \dfrac{1}{x^2+1}$
57. $f(x) = \dfrac{1}{\sqrt{x}}, \quad g(x) = x^2 - 4x$
58. $f(x) = x^2, \quad g(x) = \sqrt{x-3}$
59. $f(x) = 1 - \sqrt{x}, \quad g(x) = \sqrt[4]{x}$
60. $f(x) = \sqrt{x^2 - 1}, \quad g(x) = \sqrt{1 - x}$

61–64 ■ Composition of Three Functions Find $f \circ g \circ h$.

61. $f(x) = x - 1, \quad g(x) = \sqrt{x}, \quad h(x) = x - 1$
62. $f(x) = \dfrac{1}{x}, \quad g(x) = x^3, \quad h(x) = x^2 + 2$
63. $f(x) = x^4 + 1, \quad g(x) = x - 5, \quad h(x) = \sqrt{x}$
64. $f(x) = \sqrt{x}, \quad g(x) = \dfrac{x}{x-1}, \quad h(x) = \sqrt[3]{x}$

65–72 ■ Expressing a Function as a Composition Find functions f and g such that $F = f \circ g$.

65. $F(x) = (x - 9)^5$
66. $F(x) = \sqrt{x} + 1$
67. $F(x) = \dfrac{x^2}{x^2 + 4}$
68. $F(x) = \dfrac{1}{x + 3}$
69. $F(x) = |1 - x^3|$
70. $F(x) = \sqrt{1 + \sqrt{x}}$
71. $F(x) = 1 - \sqrt{x^3 + 1}$
72. $F(x) = \dfrac{1}{\sqrt[3]{x^2 + x + 1}}$

73–78 ■ Expressing a Function as a Composition Find functions f, g, and h such that $F = f \circ g \circ h$.

73. $F(x) = \dfrac{1}{x^2 + 1}$
74. $F(x) = \sqrt[3]{\sqrt{x} - 1}$
75. $F(x) = (4 + \sqrt[3]{x})^9$
76. $F(x) = \dfrac{2}{(3 + \sqrt{x})^2}$
77. $F(x) = \left(\dfrac{\sqrt{x}}{\sqrt{x} - 1}\right)^3$
78. $F(x) = \dfrac{1}{1 + \dfrac{1}{\sqrt{x^2 + 1}}}$

Skills Plus

79. **Composing Linear Functions** The graphs of the functions

$$f(x) = m_1 x + b_1$$
$$g(x) = m_2 x + b_2$$

are lines with slopes m_1 and m_2, respectively. Is the graph of $f \circ g$ a line? If so, what is its slope?

80. Combining Functions Graphically Graphs of the functions f and g are given. Match each combination of these functions given in parts (a)–(f) with its graph in I–VI. Give reasons for your answers.

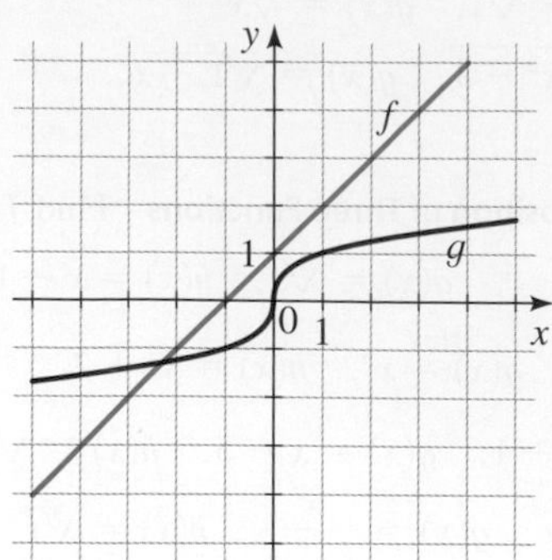

(a) $f(x) + g(x)$ (b) $f(x) - g(x)$ (c) $f(x)g(x)$

(d) $g^2(x)$ (e) $g(x)/f(x)$ (f) $(f \circ g)(x)$

I

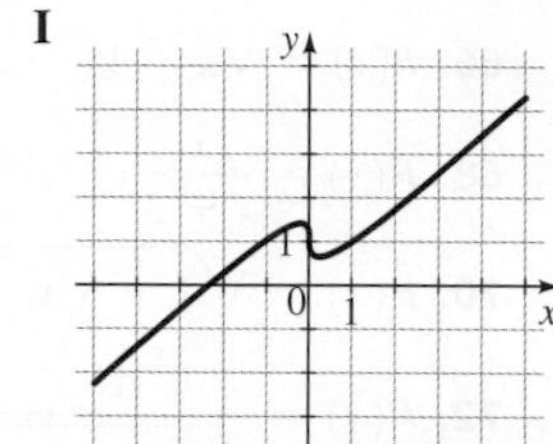

II

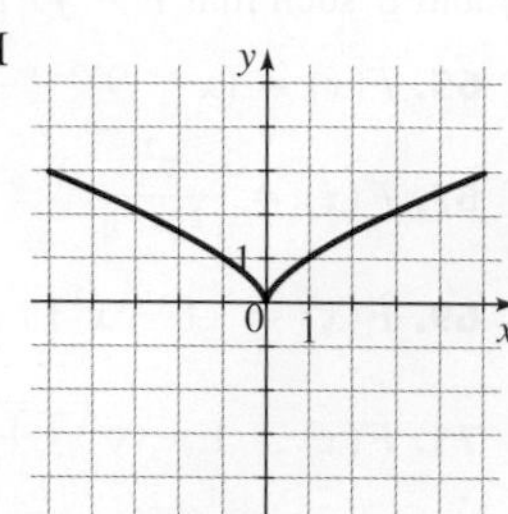

III

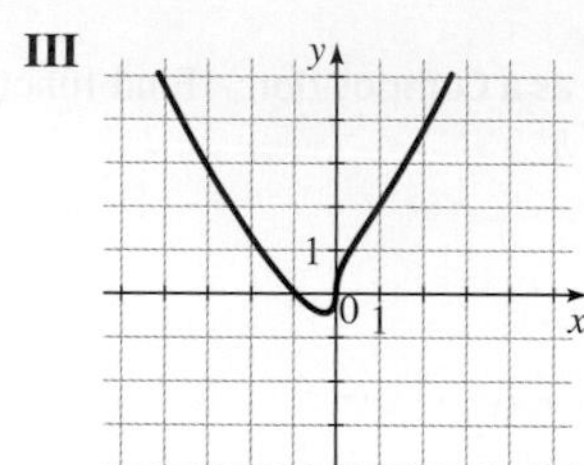

IV

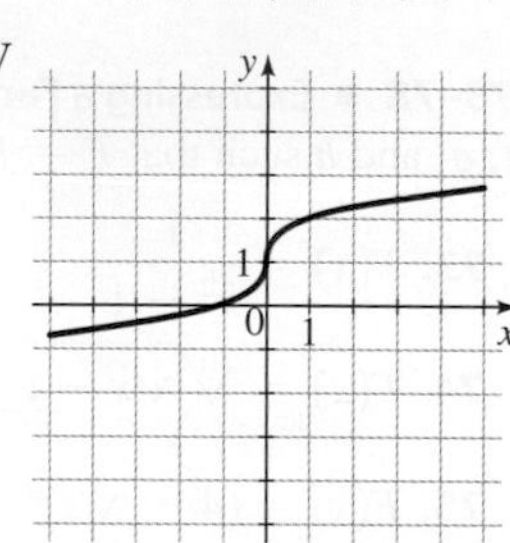

V

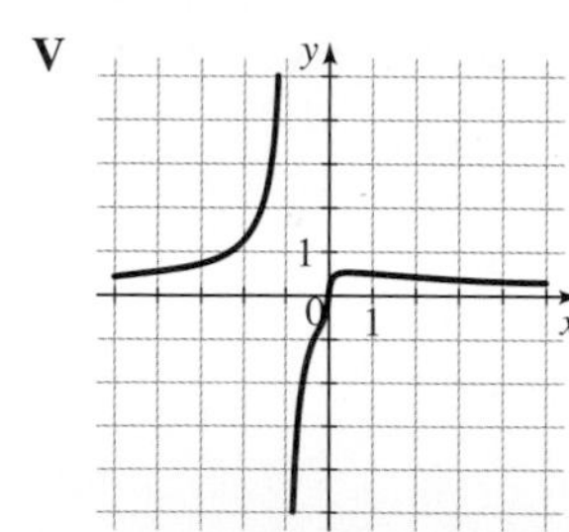

VI

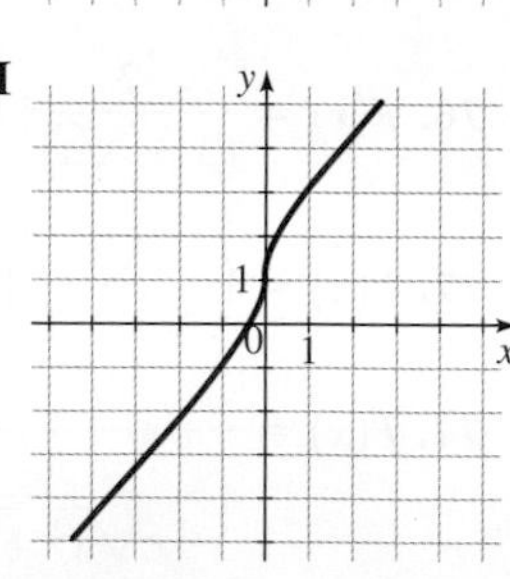

Applications

81–82 ■ Revenue, Cost, and Profit A print shop makes bumper stickers for election campaigns. If x stickers are ordered (where $x < 10{,}000$), then the price per bumper sticker is $0.15 - 0.000002x$ dollars, and the total cost of producing the order is $0.095x - 0.0000005x^2$ dollars.

81. Use the fact that

revenue = price per item × number of items sold

to express $R(x)$, the revenue from an order of x stickers, as a product of two functions of x.

82. Use the fact that

profit = revenue − cost

to express $P(x)$, the profit on an order of x stickers, as a difference of two functions of x.

83. Area of a Ripple A stone is dropped into a lake, creating a circular ripple that travels outward at a speed of 60 cm/s.

(a) Find a function g that models the radius as a function of time.

(b) Find a function f that models the area of the circle as a function of the radius.

(c) Find $f \circ g$. What does this function represent?

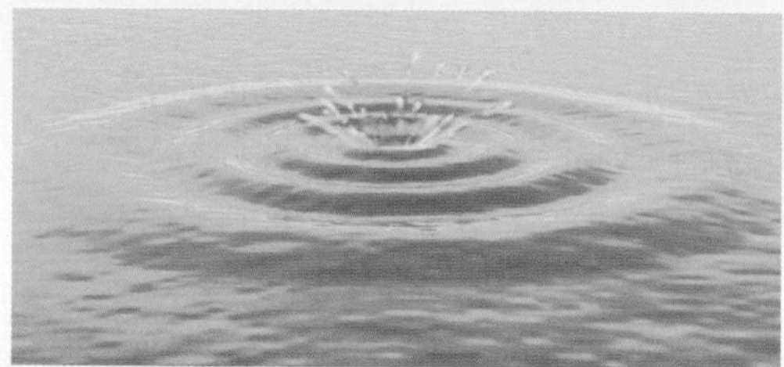

84. Inflating a Balloon A spherical balloon is being inflated. The radius of the balloon is increasing at the rate of 2 cm/s.

(a) Find a function f that models the radius as a function of time.

(b) Find a function g that models the volume as a function of the radius. (Formulas for volume are given on the inside front cover of this book.)

(c) Find $g \circ f$. What does this function represent?

85. Surface Area of a Balloon A spherical weather balloon is being inflated. The radius of the balloon is increasing at the rate of 2 cm/s. Express the surface area of the balloon as a function of time t (in seconds).

86. Elevation, Pressure, and Boiling Point At higher elevations, atmospheric pressure decreases, causing water to boil at lower temperatures than at sea level. Nineteenth century explorers used the boiling point of water to estimate elevation. In Exercises 2.5.49–50 you found a function f that models the atmospheric pressure at a given elevation and a function g that models the boiling point of water at a given atmospheric pressure.

(a) Find the function $h = g \circ f$. Describe the inputs and outputs of h. What does h model?

(b) Estimate the boiling point of water at the peak of Mt. Rainier, 4.4 km above sea level.

(c) If water boils at 91°C, estimate the elevation.

Thye-Wee Gn/Shutterstock.com

87. Multiple Discounts You have a \$50 coupon from the manufacturer that is good for the purchase of a cell phone. The store where you are purchasing your cell phone is offering a 20% discount on all phones. Let x represent the regular price of the phone.

(a) Suppose only the 20% discount applies. Find a function f that models the purchase price of the cell phone as a function of the regular price x.

(b) Suppose only the \$50 coupon applies. Find a function g that models the purchase price of the cell phone as a function of the full price x.

(c) If you can use both the coupon and the discount, then the purchase price is either $(f \circ g)(x)$ or $(g \circ f)(x)$, depending on the order in which they are applied to the price. Find both $(f \circ g)(x)$ and $(g \circ f)(x)$. Which composition gives the lower price?

88. Multiple Discounts An appliance dealer advertises a 10% discount on all washing machines. In addition, the manufacturer offers a \$100 rebate on the purchase of a washing machine. Let x represent the full price of the washing machine.

(a) Suppose only the 10% discount applies. Find a function f that models the purchase price of the washer as a function of the full price x.

(b) Suppose only the \$100 rebate applies. Find a function g that models the purchase price of the washer as a function of the full price x.

(c) Find $f \circ g$ and $g \circ f$. What do these functions represent? Which is the better deal?

89. Airplane Trajectory An airplane is flying at a speed of 560 km/h at an altitude of 1.6 kilometers. The plane passes directly above a radar station at time $t = 0$.

(a) Express the distance s (in kilometers) between the plane and the radar station as a function of the horizontal distance d (in km) that the plane has flown.

(b) Express d as a function of the time t (in hours) that the plane has flown.

(c) Use composition to express s as a function of t.

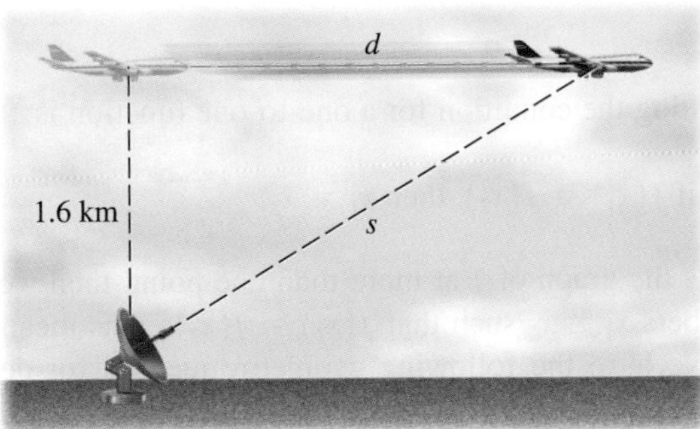

■ Discuss ■ Discover ■ Prove ■ Write

90. Discover: Compound Interest A savings account earns 5% interest compounded annually. If you invest x dollars in such an account, then the amount $A(x)$ of the investment after one year is the initial investment plus 5%; that is,

$$A(x) = x + 0.05x = 1.05x$$

Find

$$A \circ A$$

$$A \circ A \circ A$$

$$A \circ A \circ A \circ A$$

What do these compositions represent? Find a formula for what you get when you compose n copies of A.

91. Discover: Solving an Equation for an Unknown Function Suppose that

$$g(x) = 2x + 1$$
$$h(x) = 4x^2 + 4x + 7$$

Find a function f such that $f \circ g = h$. (Think about what operations you would have to perform on the formula for g to end up with the formula for h.) Now suppose that

$$f(x) = 3x + 5$$
$$h(x) = 3x^2 + 3x + 2$$

Use the same sort of reasoning to find a function g such that $f \circ g = h$.

92. Discuss: Compositions of Odd and Even Functions Suppose that

$$h = f \circ g$$

If g is an even function, is h necessarily even? If g is odd, is h odd? What if g is odd and f is odd? What if g is odd and f is even?

93. Discover: The Unit Step Function The *unit step function* u is defined by

$$u(x) = \begin{cases} 0 & \text{if } x < 0 \\ 1 & \text{if } x \geq 0 \end{cases}$$

Draw graphs of the following functions: $u(x)$, $2u(x)$, $u(x - 3)$, $u(x - 1) - u(x - 2)$, and $xu(x)$. Now find combinations and transformations involving the unit step function that yield the following graphs.

(a)

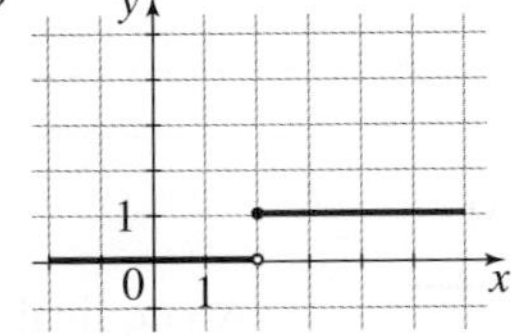

(b)

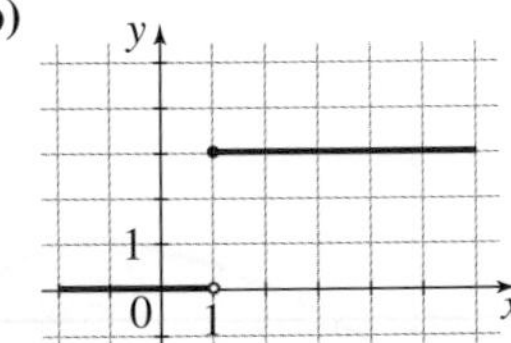

(c)

(d)

PS *Try to recognize something familiar.* We are familiar with what happens when we multiply a number by zero or one. Think about how subtracting two unit step functions or multiplying a function by a unit step function changes its graph.

94. Discover: Let $f_0(x) = 1/(1 - x)$ and $f_{n+1}(x) = (f_0 \circ f_n)(x)$ for $n = 0, 1, 2, \ldots$. Find $f_{1000}(x)$.

PS *Try to recognize a pattern.* First find $f_1(x)$, $f_2(x)$, $f_3(x)$, $f_4(x)$, . . . , simplify, and find a pattern.

2.8 One-to-One Functions and Their Inverses

■ One-to-One Functions : The Horizontal Line Test ■ The Inverse of a Function ■ Finding the Inverse of a Function ■ Graphing the Inverse of a Function ■ Applications of Inverse Functions

The *inverse* of a function is a rule that acts on the output of the function and produces the corresponding input. So the inverse "undoes" or reverses what the function has done. Not all functions have inverses; those that do are called *one-to-one*.

■ One-to-One Functions: The Horizontal Line Test

Let's compare the functions f and g whose arrow diagrams are shown in Figure 1. Note that f never takes on the same value twice (any two numbers in A have different images), whereas g does take on the same value twice (both 2 and 3 have the same image, 4). In symbols, $g(2) = g(3)$ but $f(x_1) \neq f(x_2)$ whenever $x_1 \neq x_2$. Functions that have this latter property are called *one-to-one*.

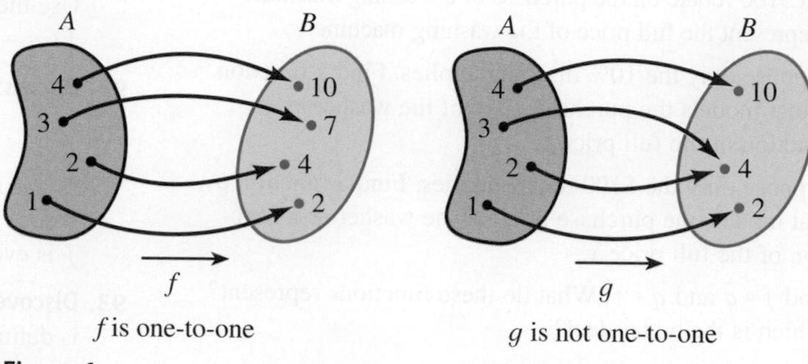

f is one-to-one   *g* is not one-to-one

Figure 1

Definition of a One-to-one Function

A function is called a **one-to-one function** if no two elements in its domain have the same image, that is,

$$f(x_1) \neq f(x_2) \quad \text{whenever } x_1 \neq x_2$$

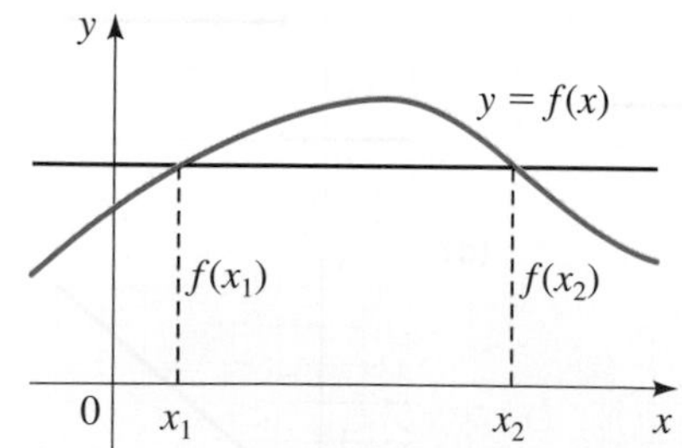

Figure 2 | This function is not one-to-one because $f(x_1) = f(x_2)$.

An equivalent way of writing the condition for a one-to-one function is:

$$\text{if } f(x_1) = f(x_2), \text{ then } x_1 = x_2.$$

If a horizontal line intersects the graph of f at more than one point, then we see from Figure 2 that there are numbers $x_1 \neq x_2$ such that $f(x_1) = f(x_2)$. This means that f is not one-to-one. Therefore we have the following geometric method for determining whether a function is one-to-one.

Horizontal Line Test

A function is one-to-one if and only if no horizontal line intersects its graph more than once.

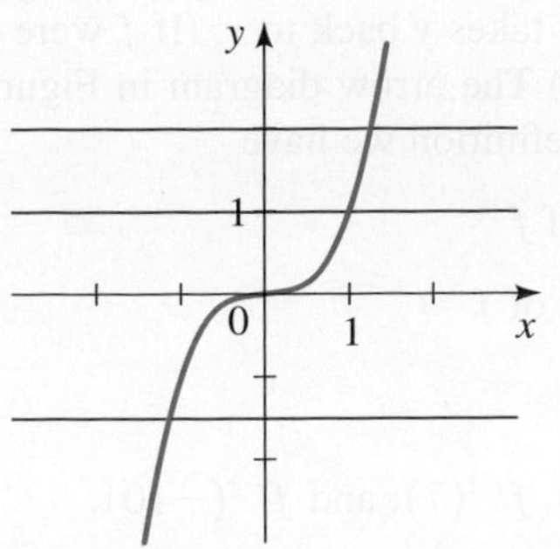

Figure 3 | $f(x) = x^3$ is one-to-one.

Example 1 ■ Deciding Whether a Function Is One-to-One

Is the function $f(x) = x^3$ one-to-one?

Solution 1 If $x_1 \neq x_2$, then $x_1^3 \neq x_2^3$ (two different numbers cannot have the same cube). Therefore $f(x) = x^3$ is one-to-one.

Solution 2 From Figure 3 we see that no horizontal line intersects the graph of $f(x) = x^3$ more than once. Therefore, by the Horizontal Line Test, f is one-to-one.

Now Try Exercise 15 ■

Notice that the function f of Example 1 is increasing and is also one-to-one. In fact, it can be proved that *every increasing function and every decreasing function is one-to-one.*

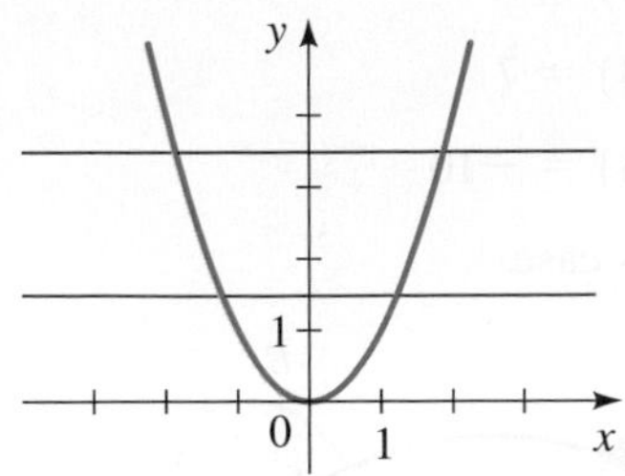

Figure 4 | $g(x) = x^2$ is not one-to-one.

Example 2 ■ Deciding Whether a Function Is One-to-One

Is the function $g(x) = x^2$ one-to-one?

Solution 1 This function is not one-to-one because, for instance,

$$g(1) = 1 \qquad \text{and} \qquad g(-1) = 1$$

so 1 and -1 have the same image.

Solution 2 From Figure 4 we see that there are horizontal lines that intersect the graph of g more than once. Therefore, by the Horizontal Line Test, g is not one-to-one.

Now Try Exercise 17 ■

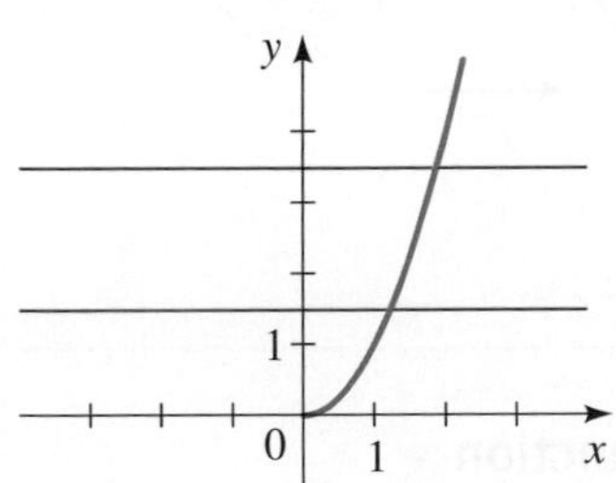

Figure 5 | $h(x) = x^2\ (x \geq 0)$ is one-to-one.

Note Although the function g in Example 2 is not one-to-one, it is possible to restrict its domain so that the resulting function *is* one-to-one. In fact, if we define

$$h(x) = x^2 \qquad (x \geq 0)$$

then h is one-to-one, as you can see from Figure 5 and the Horizontal Line Test.

Example 3 ■ Showing That a Function Is One-to-One

Show that the function $f(x) = 3x + 4$ is one-to-one.

Solution Suppose there are numbers x_1 and x_2 such that $f(x_1) = f(x_2)$. Then

$$\begin{aligned} 3x_1 + 4 &= 3x_2 + 4 && \text{Suppose } f(x_1) = f(x_2) \\ 3x_1 &= 3x_2 && \text{Subtract 4} \\ x_1 &= x_2 && \text{Divide by 3} \end{aligned}$$

Therefore f is one-to-one.

Now Try Exercise 13 ■

■ The Inverse of a Function

One-to-one functions are important because they are precisely the functions that possess inverse functions according to the following definition.

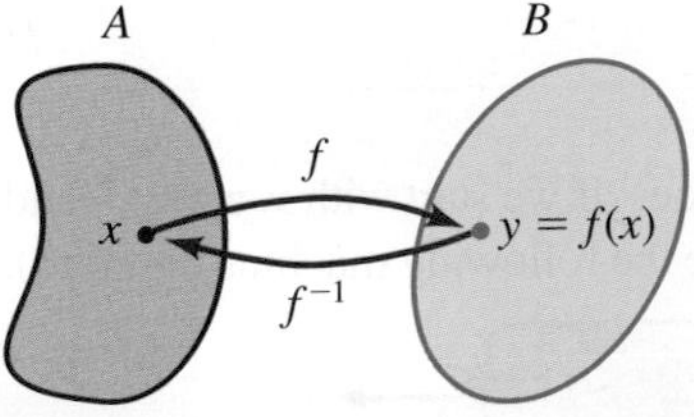

Figure 6

Definition of the Inverse of a Function

Let f be a one-to-one function with domain A and range B. Then its **inverse function** f^{-1} has domain B and range A and is defined by

$$f^{-1}(y) = x \quad \Leftrightarrow \quad f(x) = y$$

for any y in B.

This definition says that if f takes x to y, then f^{-1} takes y back to x. (If f were not one-to-one, then f^{-1} would not be defined uniquely.) The arrow diagram in Figure 6 indicates that f^{-1} reverses the effect of f. From the definition we have

$$\text{domain of } f^{-1} = \text{range of } f$$
$$\text{range of } f^{-1} = \text{domain of } f$$

Example 4 ■ Finding f^{-1} for Specific Values

 Don't mistake the -1 in f^{-1} for an exponent.

$$f^{-1}(x) \quad \textit{does not mean} \quad \frac{1}{f(x)}$$

The reciprocal $1/f(x)$ is written as $(f(x))^{-1}$.

If $f(1) = 5$, $f(3) = 7$, and $f(8) = -10$, find $f^{-1}(5)$, $f^{-1}(7)$, and $f^{-1}(-10)$.

Solution From the definition of f^{-1} we have

$$f^{-1}(5) = 1 \quad \text{because} \quad f(1) = 5$$
$$f^{-1}(7) = 3 \quad \text{because} \quad f(3) = 7$$
$$f^{-1}(-10) = 8 \quad \text{because} \quad f(8) = -10$$

Figure 7 shows how f^{-1} reverses the effect of f in this case.

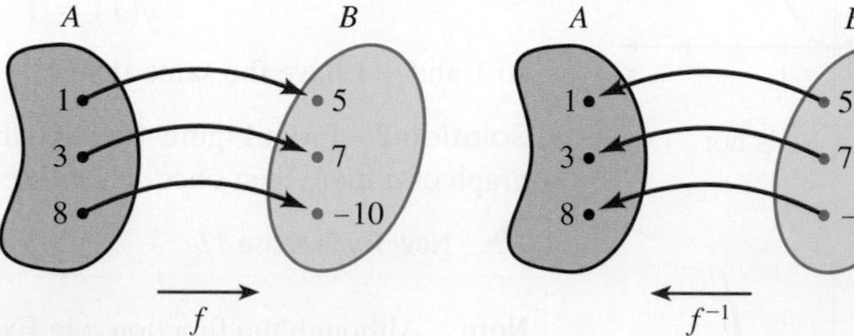

Figure 7

Now Try Exercise 25 ■

Example 5 ■ Finding Values of an Inverse Function

We can find specific values of an inverse function from a table or graph of the function itself. A graph of a function f is shown in Figure 8, as well as a table of values for f. From the table, we can see that $f^{-1}(3) = 4$ and $f^{-1}(5) = 7$. We can also "read" these values of f^{-1} from the graph.

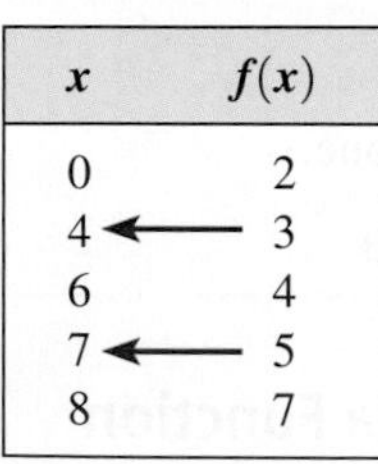

x	$f(x)$
0	2
4 ⟵	3
6	4
7 ⟵	5
8	7

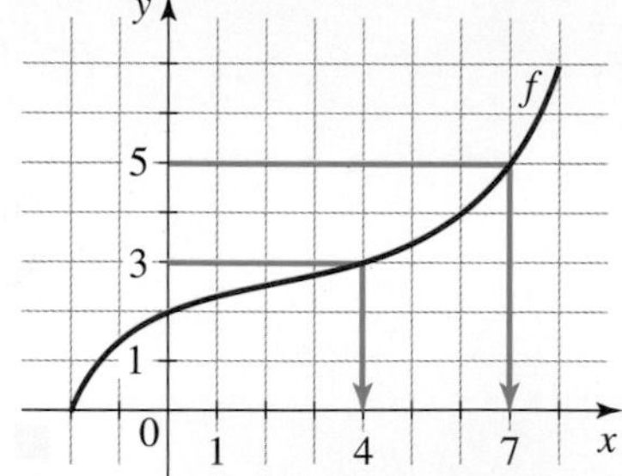

Figure 8 Finding values of f^{-1} from a table of f — Finding values of f^{-1} from a graph of f

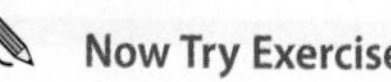 Now Try Exercises 29 and 31 ■

By definition the inverse function f^{-1} undoes what f does: If we start with x, apply f, and then apply f^{-1}, we arrive back at x, where we started (see the following machine diagram).

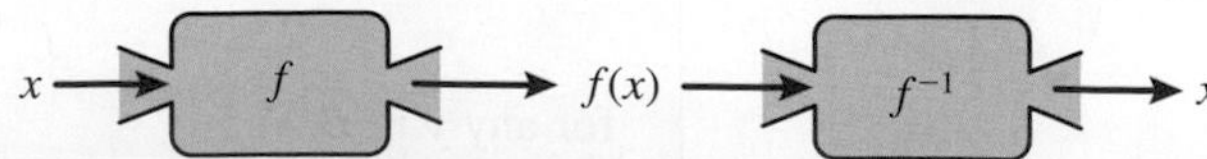

Similarly, f undoes what f^{-1} does. In general, any function that reverses the effect of f in this way must be the inverse of f. These observations are expressed precisely as follows.

Inverse Function Property

Let f be a one-to-one function with domain A and range B. The inverse function f^{-1} satisfies the following cancellation equations:

$$f^{-1}(f(x)) = x \quad \text{for every } x \text{ in } A$$

$$f(f^{-1}(x)) = x \quad \text{for every } x \text{ in } B$$

Conversely, any function f^{-1} satisfying these equations is the inverse of f.

These properties indicate that f is the inverse function of f^{-1}, so we say that f and f^{-1} are *inverses of each other*.

Example 6 ■ Verifying That Two Functions Are Inverses

Show that $f(x) = x^3$ and $g(x) = x^{1/3}$ are inverses of each other.

Solution Note that the domain and range of both f and g are $\mathbb{R}$. We show that f and g satisfy the cancellation equations of the Inverse Function Property. We have

$$g(f(x)) = g(x^3) = (x^3)^{1/3} = x$$

$$f(g(x)) = f(x^{1/3}) = (x^{1/3})^3 = x$$

So by the Inverse Function Property, f and g are inverses of each other. These equations simply say that the cube function and the cube root function, when composed, cancel each other.

 **Now Try Exercise 39** ■

■ Finding the Inverse of a Function

Now let's examine how we compute inverse functions. We first observe from the definition of f^{-1} that

$$y = f(x) \quad \Leftrightarrow \quad f^{-1}(y) = x$$

So if $y = f(x)$ and if we are able to solve this equation for x in terms of y, then we must have $x = f^{-1}(y)$. If we then interchange x and y, we have $y = f^{-1}(x)$, which is the desired equation.

How to Find the Inverse of a One-to-One Function

1. Write $y = f(x)$.
2. Solve this equation for x in terms of y (if possible).
3. Interchange x and y. The resulting equation is $y = f^{-1}(x)$.

Note that Steps 2 and 3 can be reversed. In other words, we can interchange x and y first and then solve for y in terms of x.

Example 7 ■ Finding the Inverse of a Function

Find the inverse of the function $f(x) = 3x - 2$.

Solution First we write $y = f(x)$.

$$y = 3x - 2$$

Then we solve this equation for x:

$$3x = y + 2 \qquad \text{Add 2}$$

$$x = \frac{y + 2}{3} \qquad \text{Divide by 3}$$

In Example 7 note how f^{-1} reverses the effect of f. The function f is the rule "Multiply by 3, then subtract 2," whereas f^{-1} is the rule "Add 2, then divide by 3."

Finally, we interchange x and y:

$$y = \frac{x + 2}{3}$$

Therefore the inverse function is $f^{-1}(x) = \dfrac{x + 2}{3}$.

Check Your Answer

We use the Inverse Function Property:

$$f^{-1}(f(x)) = f^{-1}(3x - 2) = \frac{(3x - 2) + 2}{3} = \frac{3x}{3} = x$$

$$f(f^{-1}(x)) = f\left(\frac{x + 2}{3}\right) = 3\left(\frac{x + 2}{3}\right) - 2 = x + 2 - 2 = x$$

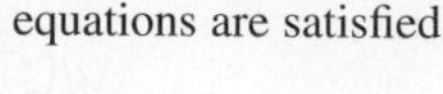

Both cancellation equations are satisfied. ✓

Now Try Exercise 49 ■

Example 8 ■ Finding the Inverse of a Function

Find the inverse of the function $f(x) = \dfrac{x^5 - 3}{2}$.

Solution We first write $y = (x^5 - 3)/2$ and solve for x.

$$y = \frac{x^5 - 3}{2} \qquad \text{Equation defining function}$$

$$2y = x^5 - 3 \qquad \text{Multiply by 2}$$

$$x^5 = 2y + 3 \qquad \text{Add 3 (and switch sides)}$$

$$x = (2y + 3)^{1/5} \qquad \text{Take fifth root of each side}$$

In Example 8 note how f^{-1} reverses the effect of f. The function f is the rule "Take the fifth power, subtract 3, then divide by 2," whereas f^{-1} is the rule "Multiply by 2, add 3, then take the fifth root."

Then we interchange x and y to get $y = (2x + 3)^{1/5}$. Therefore the inverse function is $f^{-1}(x) = (2x + 3)^{1/5}$.

Check Your Answer

We use the Inverse Function Property:

$$f^{-1}(f(x)) = f^{-1}\left(\frac{x^5 - 3}{2}\right) = \left[2\left(\frac{x^5 - 3}{2}\right) + 3\right]^{1/5} = (x^5 - 3 + 3)^{1/5} = (x^5)^{1/5} = x$$

$$f(f^{-1}(x)) = f((2x + 3)^{1/5}) = \frac{[(2x + 3)^{1/5}]^5 - 3}{2} = \frac{2x + 3 - 3}{2} = \frac{2x}{2} = x$$

Both cancellation equations are satisfied. ✓

Now Try Exercise 63 ■

A **rational function** is a function defined by a rational expression. (See Section 3.6.) In the next example we find the inverse of a rational function.

Example 9 ■ Finding the Inverse of a Rational Function

Find the inverse of the function $f(x) = \dfrac{2x + 3}{x - 1}$.

Solution We first write $y = (2x + 3)/(x - 1)$ and solve for x.

$$y = \frac{2x + 3}{x - 1} \qquad \text{Equation defining function}$$

$$y(x - 1) = 2x + 3 \qquad \text{Multiply by } x - 1$$

$$yx - y = 2x + 3 \qquad \text{Expand}$$

$$yx - 2x = y + 3 \qquad \text{Bring } x\text{-terms to LHS}$$

$$x(y - 2) = y + 3 \qquad \text{Factor } x$$

$$x = \frac{y + 3}{y - 2} \qquad \text{Divide by } y - 2$$

Therefore the inverse function is $f^{-1}(x) = \dfrac{x + 3}{x - 2}$.

 Now Try Exercise 57 ■

■ Graphing the Inverse of a Function

The principle of interchanging x and y to find the inverse function also gives us a method for obtaining the graph of f^{-1} from the graph of f. If $f(a) = b$, then $f^{-1}(b) = a$. Thus the point (a, b) is on the graph of f if and only if the point (b, a) is on the graph of f^{-1}. But we get the point (b, a) from the point (a, b) by reflecting about the line $y = x$ (see Figure 9). Therefore, as Figure 10 illustrates, the following is true.

> The graph of f^{-1} is obtained by reflecting the graph of f about the line $y = x$.

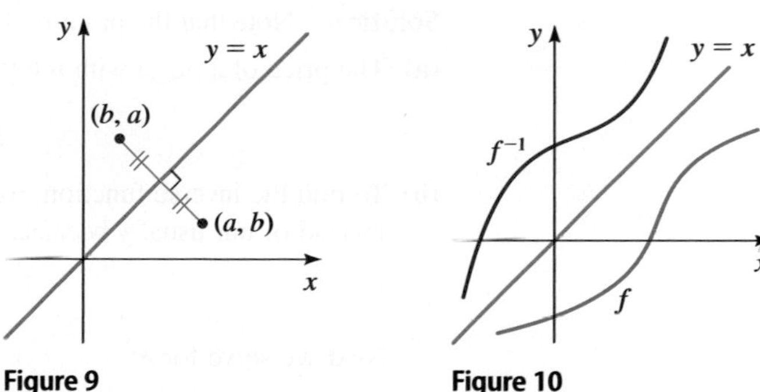

Figure 9 Figure 10

Example 10 ■ Graphing the Inverse of a Function

In Example 10 note how f^{-1} reverses the effect of f. The function f is the rule "Subtract 2, then take the square root," whereas f^{-1} is the rule "Square, then add 2."

(a) Sketch the graph of $f(x) = \sqrt{x - 2}$.

(b) Use the graph of f to sketch the graph of f^{-1}.

(c) Find an equation for f^{-1}.

Solution

(a) Using the transformations from Section 2.6, we sketch the graph of $y = \sqrt{x - 2}$ by plotting the graph of the function $y = \sqrt{x}$ [Example 2.2.1(c)] and shifting it 2 units to the right as shown in Figure 11.

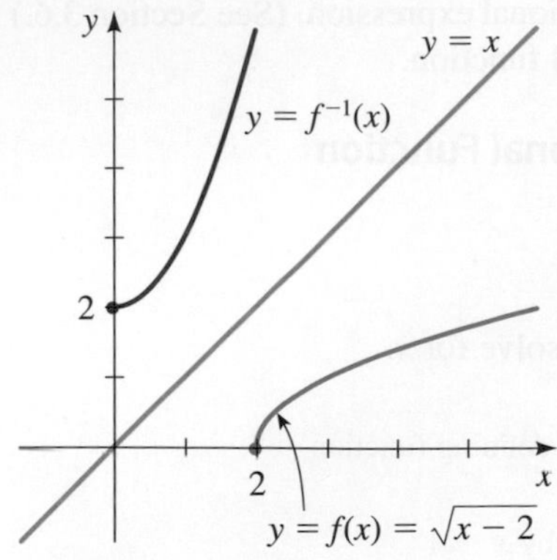

Figure 11

(b) The graph of f^{-1} is obtained from the graph of f in part (a) by reflecting it about the line $y = x$, as shown in Figure 11.

(c) Solve $y = \sqrt{x-2}$ for x, noting that $y \geq 0$.

$$\sqrt{x-2} = y$$

$$x - 2 = y^2 \qquad \text{Square each side}$$

$$x = y^2 + 2 \quad (y \geq 0) \qquad \text{Add 2}$$

Interchange x and y, as follows:

$$y = x^2 + 2 \quad (x \geq 0)$$

Thus $$f^{-1}(x) = x^2 + 2 \quad (x \geq 0)$$

This expression shows that the graph of f^{-1} is the right half of the parabola $y = x^2 + 2$, and from the graph shown in Figure 11 this seems reasonable.

Now Try Exercise 71 ■

■ Applications of Inverse Functions

When working with functions that model real-world situations, we name the variables using letters that suggest the quantity being modeled. For instance we may use t for time, d for distance, V for volume, and so on. When using inverse functions, we follow this convention. For example, suppose that the variable R is a function of the variable N, say, $R = f(N)$. Then $f^{-1}(R) = N$. So the function f^{-1} defines N as a function of R.

Example 11 ■ An Inverse Function

At a local pizza restaurant the daily special is \$12 for a plain cheese pizza plus \$2 for each additional topping.

(a) Find a function f that models the price of a pizza with n toppings.

(b) Find the inverse of the function f. What does f^{-1} represent?

(c) If a pizza costs \$22, how many toppings does it have?

Solution Note that the price p of a pizza is a function of the number n of toppings.

(a) The price of a pizza with n toppings is given by the function

$$f(n) = 12 + 2n$$

(b) To find the inverse function, we first write $p = f(n)$, where we use the letter p instead of our usual y because $f(n)$ is the price of the pizza. We have

$$p = 12 + 2n$$

Next we solve for n:

$$p = 12 + 2n$$

$$p - 12 = 2n$$

$$n = \frac{p - 12}{2}$$

So $n = f^{-1}(p) = \dfrac{p-12}{2}$. The function f^{-1} gives the number n of toppings for a pizza with price p.

(c) We have $n = f^{-1}(22) = (22 - 12)/2 = 5$. So the pizza has five toppings.

Now Try Exercise 99 ■

2.8 | Exercises

Concepts

1. A function f is one-to-one if different inputs produce ________ outputs. You can tell from the graph that a function is one-to-one by using the ______________ Test.

2. (a) For a function to have an inverse, it must be ________. Which one of the following functions has an inverse?

$$f(x) = x^2 \qquad g(x) = x^3$$

(b) What is the inverse of the function that you chose in part (a)?

3. A function f has the following verbal description: "Multiply by 3, add 5, and then take the third power of the result."

(a) Write a verbal description for f^{-1}.

(b) Find algebraic formulas that express f and f^{-1} in terms of the input x.

4. A graph of a function f is given. Does f have an inverse? If so, find $f^{-1}(1) =$ ________ and $f^{-1}(3) =$ ________.

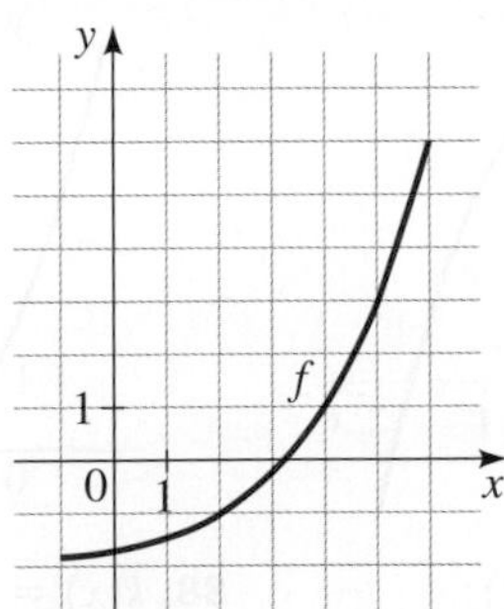

5. If the point $(3, 4)$ is on the graph of the function f, then the point (____, ____) is on the graph of f^{-1}.

6. *True or false?*

(a) If f has an inverse, then $f^{-1}(x)$ is always the same as $\dfrac{1}{f(x)}$.

(b) If f has an inverse, then $f^{-1}(f(x)) = x$.

Skills

7–12 ■ One-to-One Function? A graph of a function f is given. Determine whether f is one-to-one.

7.

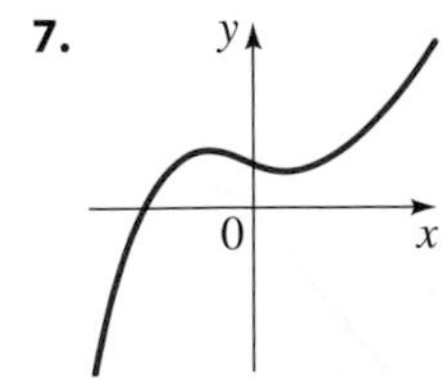

8.

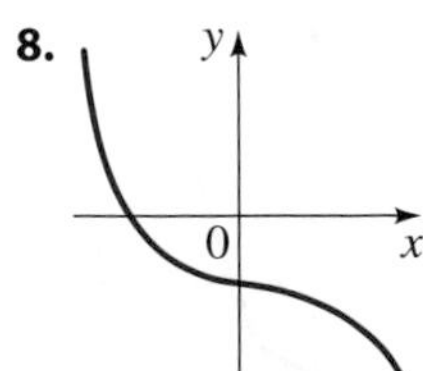

9.

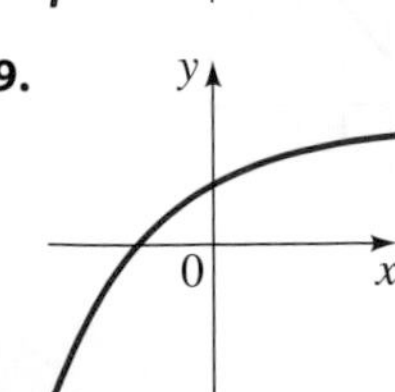

10.

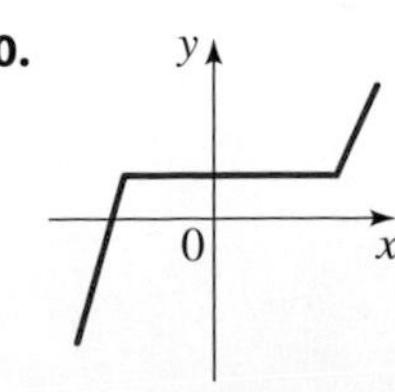

11.

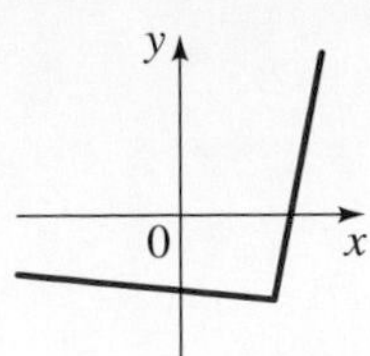

12.

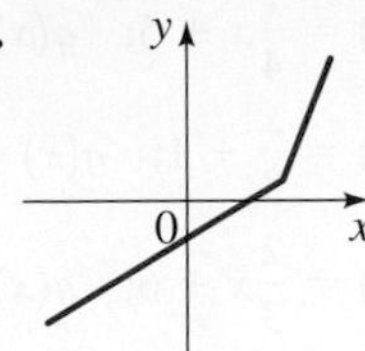

13–24 ■ One-to-One Function Determine whether the function is one-to-one.

13. $f(x) = -2x + 4$

14. $f(x) = 3x - 2$

15. $g(x) = \sqrt{x}$

16. $g(x) = |x|$

17. $h(x) = x^2 - 2x$

18. $h(x) = x^3 + 8$

19. $f(x) = x^4 + 5$

20. $f(x) = x^4 + 5, \quad 0 \le x \le 2$

21. $r(t) = t^6 - 3, \quad 0 \le t \le 5$

22. $r(t) = t^4 - 1$

23. $f(x) = \dfrac{1}{x^2}$

24. $f(x) = \dfrac{1}{x}$

25–28 ■ Finding Values of an Inverse Function Assume that f is a one-to-one function.

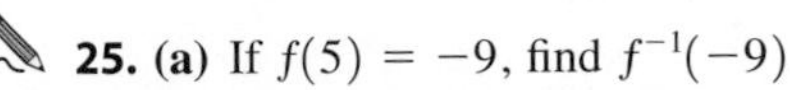

25. (a) If $f(5) = -9$, find $f^{-1}(-9)$.

(b) If $f^{-1}(10) = 0$, find $f(0)$.

26. (a) If $f(-3) = 6$, find $f^{-1}(6)$.

(b) If $f^{-1}(12) = 8$, find $f(8)$.

27. If $f(x) = 5 - 2x$, find $f^{-1}(3)$.

28. If $g(x) = x^2 + 4x$ with $x \ge -2$, find $g^{-1}(5)$.

29–30 ■ Finding Values of an Inverse from a Graph A graph of a function f is given. Use the graph to find the indicated values.

(a) $f^{-1}(2)$ **(b)** $f^{-1}(5)$ **(c)** $f^{-1}(6)$

29.

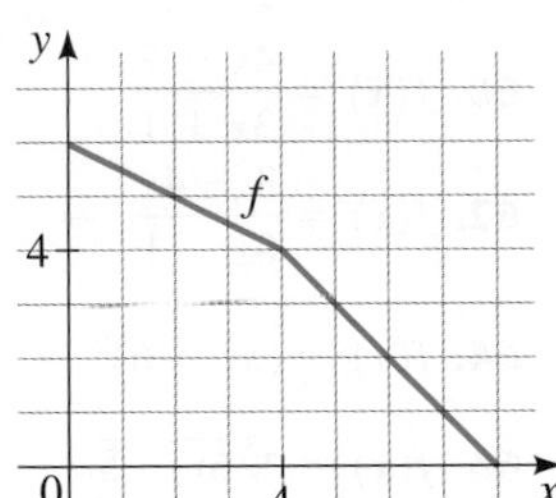

30.

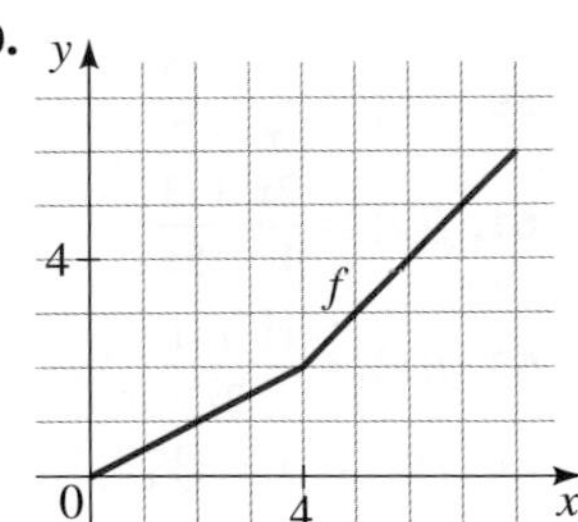

31–36 ■ Finding Values of an Inverse Using a Table A table of values for a one-to-one function is given. Find the indicated values.

31. $f^{-1}(5)$

32. $f^{-1}(0)$

33. $f^{-1}(f(1))$

34. $f(f^{-1}(6))$

35. $f^{-1}(f^{-1}(1))$

36. $f^{-1}(f^{-1}(0))$

x	1	2	3	4	5	6
$f(x)$	4	6	2	5	0	1

37–48 ■ Inverse Function Property Use the Inverse Function Property to show that f and g are inverses of each other.

37. $f(x) = \frac{1}{4}x + 5; \quad g(x) = 4(x - 5)$

38. $f(x) = \frac{x}{3} + 1; \quad g(x) = 3x - 3$

39. $f(x) = \frac{2}{3}x + 6; \quad g(x) = \frac{3}{2}x - 9$

40. $f(x) = 4x - 7; \quad g(x) = \frac{x + 7}{4}$

41. $f(x) = \frac{1}{x}; \quad g(x) = \frac{1}{x}$

42. $f(x) = x^5; \quad g(x) = \sqrt[5]{x}$

43. $f(x) = x^2 - 9, \quad x \geq 0; \quad g(x) = \sqrt{x + 9}, \quad x \geq -9$

44. $f(x) = x^3 + 1; \quad g(x) = (x - 1)^{1/3}$

45. $f(x) = \frac{1}{x - 1}; \quad g(x) = \frac{1}{x} + 1$

46. $f(x) = \sqrt{4 - x^2}, \quad 0 \leq x \leq 2;$
$g(x) = \sqrt{4 - x^2}, \quad 0 \leq x \leq 2$

47. $f(x) = \frac{x + 2}{x - 2}; \quad g(x) = \frac{2x + 2}{x - 1}$

48. $f(x) = \frac{x - 5}{3x + 4}; \quad g(x) = \frac{5 + 4x}{1 - 3x}$

49–68 ■ Finding Inverse Functions Find the inverse function of f. Check your answer by using the Inverse Function Property.

49. $f(x) = 3x + 15$

50. $f(x) = 8 - 3x$

51. $f(x) = \frac{3}{4}x - 12$

52. $f(x) = \frac{3 - x}{10}$

53. $f(x) = 5 - 4x^3$

54. $f(x) = 3x^3 + 8$

55. $f(x) = \frac{1}{x + 2}$

56. $f(x) = \frac{x - 2}{x + 2}$

57. $f(x) = \frac{x}{2 - x}$

58. $f(x) = \frac{4x}{x + 5}$

59. $f(x) = \frac{2x + 5}{x - 7}$

60. $f(x) = \frac{4x - 2}{3x + 1}$

61. $f(x) = \frac{2x + 3}{1 - 5x}$

62. $f(x) = \frac{3 - 4x}{8x - 1}$

63. $f(x) = \frac{x^3 + 1}{3}$

64. $f(x) = (x^5 - 6)^7$

65. $f(x) = 2 + \sqrt[3]{x}$

66. $f(x) = \sqrt[3]{6x - 5}$

67. $f(x) = x^{3/2} + 1$

68. $f(x) = (x - 2)^{3/5}$

69–74 ■ Graph of an Inverse Function A function f is given. **(a)** Sketch the graph of f. **(b)** Use the graph of f to sketch the graph of f^{-1}. **(c)** Find f^{-1}.

69. $f(x) = 3x - 6$

70. $f(x) = 16 - x^2, \quad x \geq 0$

71. $f(x) = x^3 - 1$

72. $f(x) = \sqrt{x + 1}$

73. $f(x) = 3 + \sqrt{x - 1}$

74. $f(x) = 2 + \sqrt{x + 1}$

75–80 ■ One-to-One Functions from a Graph Draw the graph of f, and use it to determine whether the function is one-to-one.

75. $f(x) = x^3 - x$

76. $f(x) = x^3 + x$

77. $f(x) = \frac{x + 12}{x - 6}$

78. $f(x) = \sqrt{x^3 - 4x + 1}$

79. $f(x) = |x| - |x - 6|$

80. $f(x) = x \cdot |x|$

81–84 ■ Finding Inverse Functions A one-to-one function is given. **(a)** Find the inverse of the function. **(b)** Graph both the function and its inverse on the same screen to verify that the graphs are reflections of each other about the line $y = x$.

81. $f(x) = 2 + x$

82. $f(x) = 2 - \frac{1}{2}x$

83. $g(x) = \sqrt{x + 3}$

84. $g(x) = x^2 + 1, \quad x \geq 0$

85–88 ■ Restricting the Domain The given function is not one-to-one. Restrict its domain so that the resulting function *is* one-to-one and has the same range as the given function. Find the inverse of the function with the restricted domain.

85. $f(x) = 4 - x^2$

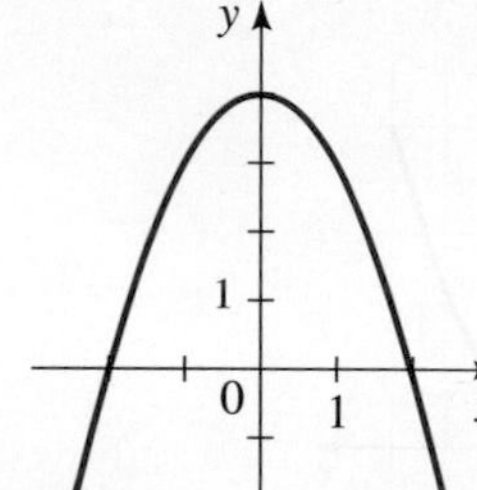

86. $g(x) = (x - 1)^2$

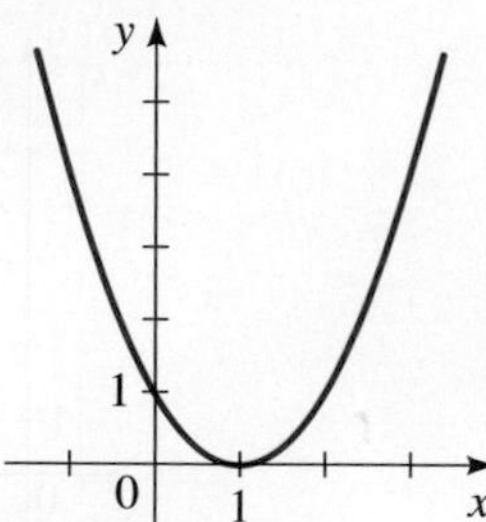

87. $h(x) = (x + 2)^2$

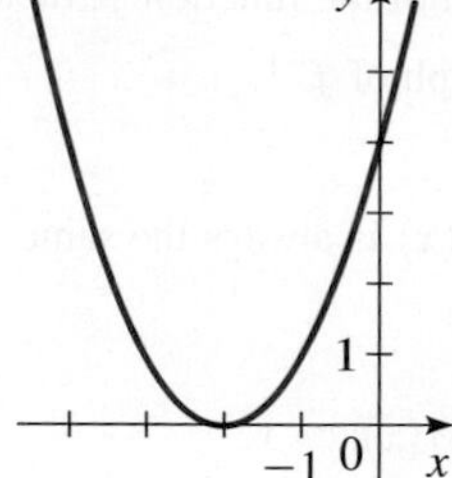

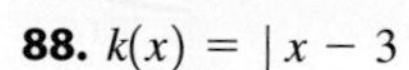

88. $k(x) = |x - 3|$

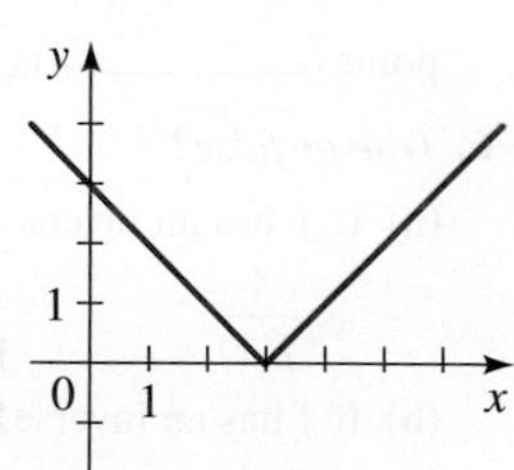

89–90 ■ Graph of an Inverse Function Use the graph of f to sketch the graph of f^{-1}.

89.

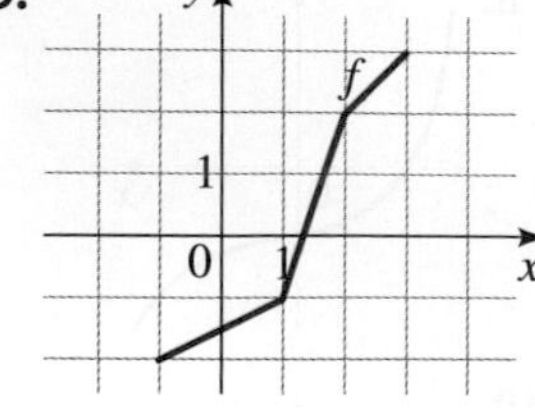

90.

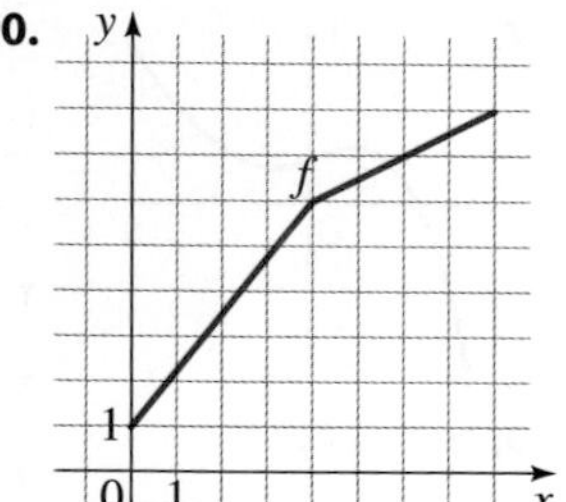

Skills Plus

91–96 ■ Inverse Functions Find the inverse function f^{-1} and state its domain. Check that the range of f is the same as the domain of the inverse function you found.

91. $f(x) = x^2 - 9;\ \ x \geq 0$

92. $f(x) = x^2 - 2x + 1;\ \ x \geq 1$

93. $f(x) = \dfrac{1}{x^4},\ \ x > 0$ **94.** $f(x) = \dfrac{1}{x^2 + 1};\ \ x \geq 0$

95. $f(x) = \sqrt{x},\ \ 0 \leq x \leq 9$ **96.** $f(x) = x^2 - 6x,\ \ x \geq 3$

97–98 ■ Functions That Are Their Own Inverse If a function f is its own inverse, then the graph of f is symmetric with respect to the line $y = x$. **(a)** Graph the given function. **(b)** Does the graph indicate that f and f^{-1} are the same function? **(c)** Find the function f^{-1}. Use your result to verify your answer to part (b).

97. $f(x) = \dfrac{1}{x}$ **98.** $f(x) = \dfrac{x + 3}{x - 1}$

Applications

99. Pizza Cost A popular pizza place charges a base price of \$16 for a large cheese pizza plus \$1.50 for each topping.

(a) Find a function f that models the price of a pizza with n toppings.

(b) Find the inverse of the function f. What does f^{-1} represent?

(c) If a pizza costs \$25, how many toppings does it have?

100. Fee for Service A private investigator requires a \$500 retainer fee plus \$80 per hour. Let x represent the number of hours the investigator spends working on a case.

(a) Find a function f that models the investigator's fee as a function of x.

(b) Find f^{-1}. What does f^{-1} represent?

(c) Find $f^{-1}(1220)$. What does your answer represent?

101. Torricelli's Law A tank holds 100 litres of water, which drains from a leak at the bottom, causing the tank to empty in 40 minutes. According to Torricelli's Law, the volume V of water remaining in the tank after t min is given by the function

$$V = f(t) = 100\left(1 - \frac{t}{40}\right)^2$$

(a) Find f^{-1}. What does f^{-1} represent?

(b) Find $f^{-1}(15)$. What does your answer represent?

102. Blood Flow As blood moves through a vein or artery, its velocity v is greatest along the central axis and decreases as the distance r from the central axis increases (see the figure below). For an artery with radius 0.5 cm, v (in cm/s) is given as a function of r (in cm) by

$$v = g(r) = 18{,}500(0.25 - r^2)$$

(a) Find g^{-1}. What does g^{-1} represent?

(b) Find $g^{-1}(30)$. What does your answer represent?

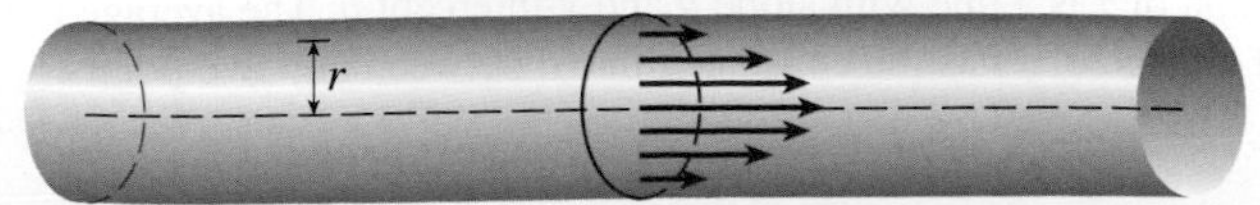

103. Demand Function The amount of a commodity that is sold is called the *demand* for the commodity. The demand D for a certain commodity is a function of the price given by

$$D = f(p) = -3p + 150$$

(a) Find f^{-1}. What does f^{-1} represent?

(b) Find $f^{-1}(30)$. What does your answer represent?

104. Temperature Scales The relationship between the Fahrenheit (F) and Celsius (C) scales is given by

$$F = g(C) = \tfrac{9}{5}C + 32$$

(a) Find g^{-1}. What does g^{-1} represent?

(b) Find $g^{-1}(86)$. What does your answer represent?

105. Exchange Rates The relative value of currencies fluctuates every day. When this problem was written, one Canadian dollar was worth 0.79 US dollars.

(a) Find a function f that gives the US dollar value $f(x)$ of x Canadian dollars.

(b) Find f^{-1}. What does f^{-1} represent?

(c) How much Canadian money would \$12,250 in US currency be worth?

106. Income Tax In a certain EU country the tax on incomes less than or equal to €20,000 is 10%. For incomes that are more than €20,000 the tax is €2000 plus 20% of the amount over €20,000. (Currency units are euros.)

(a) Find a function f that gives the income tax on an income x. Express f as a piecewise-defined function.

(b) Find f^{-1}. What does f^{-1} represent?

(c) How much income would require paying a tax of €10,000?

107. Multiple Discounts A car dealership advertises a 15% discount on all its new cars. In addition, the manufacturer offers a \$1000 rebate on the purchase of a new car. Let x represent the sticker price of the car.

(a) Suppose that only the 15% discount applies. Find a function f that models the purchase price of the car as a function of the sticker price x.

(b) Suppose that only the \$1000 rebate applies. Find a function g that models the purchase price of the car as a function of the sticker price x.

(c) Find a formula for $H = f \circ g$.

(d) Find H^{-1}. What does H^{-1} represent?

(e) Find $H^{-1}(13{,}000)$. What does your answer represent?

Discuss ■ Discover ■ Prove ■ Write

108. Discuss: Determining When a Linear Function Has an Inverse For the linear function $f(x) = mx + b$ to be one-to-one, what must be true about its slope? If it is one-to-one, find its inverse. Is the inverse linear? If so, what is its slope?

109. Discuss: Finding an Inverse "in Your Head" In the margin notes in this section we pointed out that the inverse of a function can be found by simply reversing the operations that make up the function. For instance, in Example 7 we saw that the inverse of

$$f(x) = 3x - 2 \qquad \text{is} \qquad f^{-1}(x) = \frac{x+2}{3}$$

because the "reverse" of "Multiply by 3 and subtract 2" is "Add 2 and divide by 3." Use the same procedure to find the inverse of the following functions.

(a) $f(x) = \dfrac{2x+1}{5}$ **(b)** $f(x) = 3 - \dfrac{1}{x}$

(c) $f(x) = \sqrt{x^3 + 2}$ **(d)** $f(x) = (2x - 5)^3$

Now consider another function:

$$f(x) = x^3 + 2x + 6$$

Is it possible to use the same sort of simple reversal of operations to find the inverse of this function? If so, do it. If not, explain what is different about this function that makes this task difficult.

110. Prove: The Identity Function The function $I(x) = x$ is called the *identity function*. Show that for any function f we have $f \circ I = f$, $I \circ f = f$, and $f \circ f^{-1} = f^{-1} \circ f = I$. (This means that the identity function I behaves for functions and composition just the way the number 1 behaves for real numbers and multiplication.)

111. Discuss: Solving an Equation for an Unknown Function In Exercise 2.7.91 you were asked to solve equations in which the unknowns are functions. Now that we know about inverses and the identity function (see Exercise 110), we can use algebra to solve such equations. For instance, to solve $f \circ g = h$ for the unknown function f, we perform the following steps:

$f \circ g = h$	Problem: Solve for f
$f \circ g \circ g^{-1} = h \circ g^{-1}$	Compose with g^{-1} on the right
$f \circ I = h \circ g^{-1}$	Because $g \circ g^{-1} = I$
$f = h \circ g^{-1}$	Because $f \circ I = f$

So the solution is $f = h \circ g^{-1}$. Use this technique to solve the equation $f \circ g = h$ for the indicated unknown function.

(a) Solve for f, where

$$g(x) = 2x + 1$$
$$h(x) = 4x^2 + 4x + 7$$

(b) Solve for g, where

$$f(x) = 3x + 5$$
$$h(x) = 3x^2 + 3x + 2$$

112. Prove: The Inverse of a Composition of Functions Show that the inverse function of $f \circ g$ is the function $g^{-1} \circ f^{-1}$.

PS *Try to recognize something familiar.* Show that these functions satisfy the cancellation properties of inverse functions.

Chapter 2 Review

Properties and Formulas

Function Notation | Section 2.1

If a function is given by the formula $y = f(x)$, then x is the independent variable and denotes the **input**; y is the dependent variable and denotes the **output**; the **domain** is the set of all possible inputs x; the **range** is the set of all possible outputs y.

Net Change | Section 2.1

The **net change** in the value of the function f between $x = a$ and $x = b$ is

$$\text{net change} = f(b) - f(a)$$

The Graph of a Function | Section 2.2

The graph of a function f is the graph of the equation $y = f(x)$ that defines f.

The Vertical Line Test | Section 2.2

A curve in the coordinate plane is the graph of a function if and only if no vertical line intersects the graph more than once.

Relations and Functions | Section 2.2

A **relation** is any collection of ordered pairs (x, y). The x-values are inputs and the corresponding y-values are outputs. A relation is a function if every input corresponds to exactly one output.

Increasing and Decreasing Functions | Section 2.3

A function f is **increasing** on an interval if $f(x_1) < f(x_2)$ whenever $x_1 < x_2$ in the interval.

A function f is **decreasing** on an interval if $f(x_1) > f(x_2)$ whenever $x_1 < x_2$ in the interval.

Local Maximum and Minimum Values | Section 2.3

The function value $f(a)$ is a **local maximum value** of the function f if $f(a) \geq f(x)$ for all x near a. In this case we also say that f has a **local maximum** at $x = a$.

The function value $f(b)$ is a **local minimum value** of the function f if $f(b) \leq f(x)$ for all x near b. In this case we also say that f has a **local minimum** at $x = b$.

Average Rate of Change | Section 2.4

The **average rate of change** of the function f between $x = a$ and $x = b$ is the slope of the **secant** line between $(a, f(a))$ and $(b, f(b))$:

$$\text{average rate of change} = \frac{f(b) - f(a)}{b - a}$$

Linear Functions | Section 2.5

A **linear function** is a function of the form $f(x) = ax + b$. The graph of f is a line with slope a and y-intercept b. The average rate of change of f has the constant value a between any two points.

$$a = \text{slope of graph of } f = \text{rate of change of } f$$

Vertical and Horizontal Shifts of Graphs | Section 2.6

Let c be a positive constant.

To graph $y = f(x) + c$, shift the graph of $y = f(x)$ **upward** by c units.

To graph $y = f(x) - c$, shift the graph of $y = f(x)$ **downward** by c units.

To graph $y = f(x - c)$, shift the graph of $y = f(x)$ **to the right** by c units.

To graph $y = f(x + c)$, shift the graph of $y = f(x)$ **to the left** by c units.

Reflecting Graphs | Section 2.6

To graph $y = -f(x)$, **reflect** the graph of $y = f(x)$ about the **x-axis**.

To graph $y = f(-x)$, **reflect** the graph of $y = f(x)$ about the **y-axis**.

Vertical and Horizontal Stretching and Shrinking of Graphs | Section 2.6

If $c > 1$, then to graph $y = cf(x)$, **stretch** the graph of $y = f(x)$ **vertically** by a factor of c.

If $0 < c < 1$, then to graph $y = cf(x)$, **shrink** the graph of $y = f(x)$ **vertically** by a factor of c.

If $c > 1$, then to graph $y = f(cx)$, **shrink** the graph of $y = f(x)$ **horizontally** by a factor of $1/c$.

If $0 < c < 1$, then to graph $y = f(cx)$, **stretch** the graph of $y = f(x)$ **horizontally** by a factor of $1/c$.

Even and Odd Functions | Section 2.6

A function f is

even if $f(-x) = f(x)$

odd if $f(-x) = -f(x)$

for every x in the domain of f.

Composition of Functions | Section 2.7

Given two functions f and g, the **composition** of f and g is the function $f \circ g$ defined by

$$(f \circ g)(x) = f(g(x))$$

The **domain** of $f \circ g$ is the set of all x for which both $g(x)$ and $f(g(x))$ are defined.

One-to-One Functions | Section 2.8

A function f is **one-to-one** if $f(x_1) \neq f(x_2)$ whenever x_1 and x_2 are *different* elements of the domain of f.

Horizontal Line Test | Section 2.8

A function is one-to-one if and only if no horizontal line intersects its graph more than once.

Inverse of a Function | Section 2.8

Let f be a one-to-one function with domain A and range B.

The **inverse** of f is the function f^{-1} defined by

$$f^{-1}(y) = x \quad \Leftrightarrow \quad f(x) = y$$

The inverse function f^{-1} has domain B and range A.

The functions f and f^{-1} satisfy the following **cancellation equations**:

$$f^{-1}(f(x)) = x \quad \text{for every } x \text{ in } A$$
$$f(f^{-1}(x)) = x \quad \text{for every } x \text{ in } B$$

Concept Check

1. Define each concept.

(a) Function

(b) Domain and range of a function

(c) Graph of a function

(d) Independent and dependent variables

2. Describe the four ways of representing a function.

3. Sketch graphs of the following functions by hand.

(a) $f(x) = x^2$ (b) $g(x) = x^3$

(c) $h(x) = |x|$ (d) $k(x) = \sqrt{x}$

4. What is a piecewise-defined function? Give an example.

5. What is a relation? How do you determine whether a relation is a function? Give an example of a relation that is not a function.

6. (a) What is the Vertical Line Test, and what is it used for?

(b) What is the Horizontal Line Test, and what is it used for?

7. Define each concept, and give an example of each.

(a) Increasing function

(b) Decreasing function

(c) Constant function

8. Suppose we know that the point $(3, 5)$ is a point on the graph of a function f. Explain how to find $f(3)$ and $f^{-1}(5)$.

9. What does it mean to say that $f(4)$ is a local maximum value of f?

10. Explain how to find the average rate of change of a function f between $x = a$ and $x = b$.

11. (a) What is the slope of a linear function? How do you find it? What is the rate of change of a linear function?

(b) Is the rate of change of a linear function constant? Explain.

(c) Give an example of a linear function, and sketch its graph.

12. Suppose the graph of a function f is given. Write an equation for each of the graphs that are obtained from the graph of f as follows.

(a) Shift upward 3 units

(b) Shift downward 3 units

(c) Shift 3 units to the right

(d) Shift 3 units to the left

(e) Reflect about the x-axis

(f) Reflect about the y-axis

(g) Stretch vertically by a factor of 3

(h) Shrink vertically by a factor of $\frac{1}{3}$

(i) Shrink horizontally by a factor of $\frac{1}{3}$

(j) Stretch horizontally by a factor of 3

13. (a) What is an even function? How can you tell that a function is even by looking at its graph? Give an example of an even function.

(b) What is an odd function? How can you tell that a function is odd by looking at its graph? Give an example of an odd function.

14. Suppose that f has domain A and g has domain B. What are the domains of the following functions?

(a) Domain of $f + g$

(b) Domain of fg

(c) Domain of f/g

15. (a) How is the composition function $f \circ g$ defined? What is its domain?

(b) If $g(a) = b$ and $f(b) = c$, then explain how to find $(f \circ g)(a)$.

16. (a) What is a one-to-one function?

(b) How can you tell from the graph of a function whether it is one-to-one?

(c) Suppose that f is a one-to-one function with domain A and range B. How is the inverse function f^{-1} defined? What are the domain and range of f^{-1}?

(d) If you are given a formula for f, how do you find a formula for f^{-1}? Find the inverse of the function $f(x) = 2x$.

(e) If you are given a graph of f, how do you find a graph of the inverse function f^{-1}?

Answers to the Concept Check can be found at the book companion website stewartmath.com.

Exercises

1–2 ■ Function Notation A verbal description of a function f is given. Find a formula that expresses f in function notation.

1. "Square, then subtract 5."

2. "Divide by 2, then add 9."

3–4 ■ Function in Words A formula for a function f is given. Give a verbal description of the function.

3. $f(x) = 3(x + 10)$ **4.** $f(x) = \sqrt{6x - 10}$

5–6 ■ Table of Values Complete the table of values for the given function.

5. $g(x) = x^2 - 4x$

x	$g(x)$
−1	
0	
1	
2	
3	

6. $h(x) = 3x^2 + 2x - 5$

x	$h(x)$
−2	
−1	
0	
1	
2	

7. Printing Cost A publisher estimates that the cost $C(x)$ of printing a run of x copies of a certain mathematics textbook is given by the function $C(x) = 5000 + 30x - 0.001x^2$.

(a) Find $C(1000)$ and $C(10{,}000)$.

(b) What do your answers in part (a) represent?

(c) Find $C(0)$. What does this number represent?

(d) Find the net change and the average rate of change of the cost C between $x = 1000$ and $x = 10{,}000$.

8. Earnings An electronics store pays each of their sales staff a weekly base salary plus a commission based on the retail price of the goods they have sold. If a salesperson sells x dollars of goods in a week, their earnings for that week are given by the function $E(x) = 400 + 0.03x$.

(a) Find $E(2000)$ and $E(15{,}000)$.

(b) What do your answers in part (a) represent?

(c) Find $E(0)$. What does this number represent?

(d) Find the net change and the average rate of change of the salesperson's earnings E between $x = 2000$ and $x = 15{,}000$.

(e) From the formula for E, determine what percentage the salesperson earns on the goods sold.

9–10 ■ Evaluating Functions Evaluate the function at the indicated values.

9. $f(x) = x^2 - 4x + 6$; $f(0), f(2), f(-2), f(a), f(-a), f(x + 1), f(2x)$

10. $f(x) = 4 - \sqrt{3x - 6}$; $f(5), f(9), f(a + 2), f(-x), f(x^2)$

11–12 ■ Difference Quotient Find $f(a)$, $f(a + h)$, and the difference quotient $\dfrac{f(a + h) - f(a)}{h}$.

11. $f(x) = x^2 + 8$

12. $f(x) = \dfrac{1}{x - 2}$

13. Functions Given by a Graph Which of the following figures are graphs of functions? Which of the functions are one-to-one?

(a)

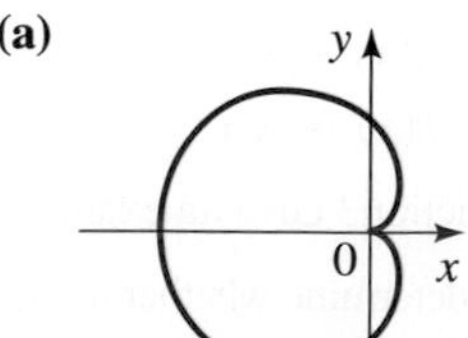

(b)

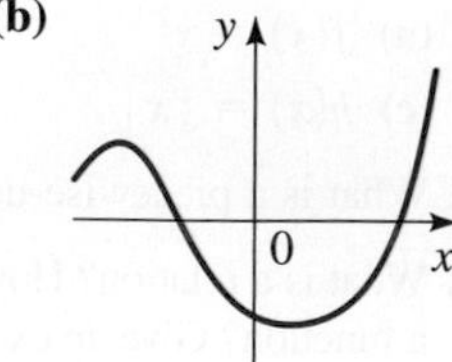

(c)

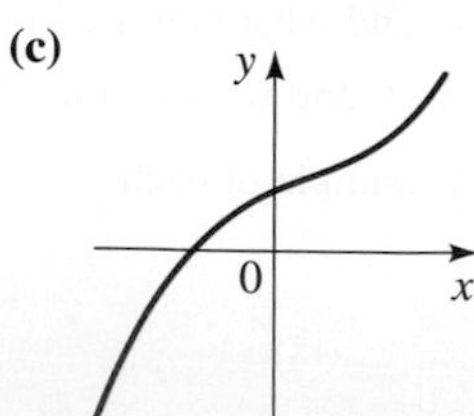

(d)

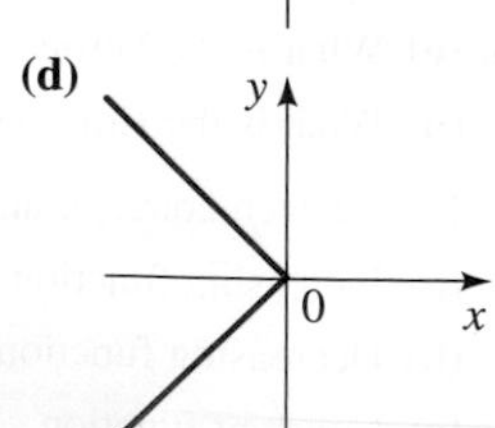

14. Getting Information from a Graph A graph of a function f is given.

(a) Find $f(-2)$ and $f(2)$.

(b) Find the net change and the average rate of change of f between $x = -2$ and $x = 2$.

(c) Find the domain and range of f.

(d) On what intervals is f increasing? On what intervals is f decreasing?

(e) What are the local maximum values of f?

(f) Is f one-to-one?

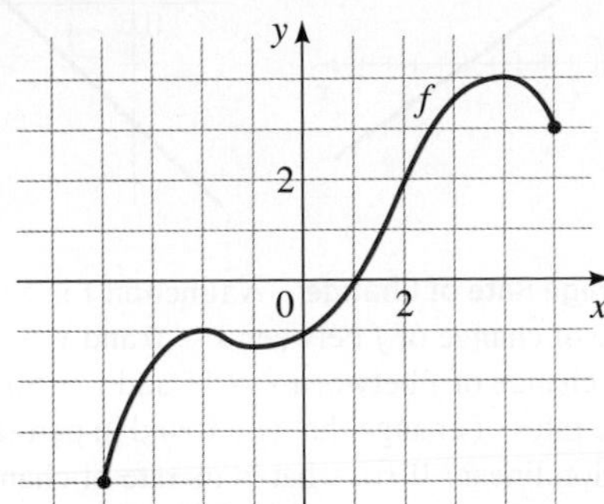

15–16 ■ Domain and Range Find the domain and range of the function.

15. $f(x) = \sqrt{x - 5}$ **16.** $f(x) = \dfrac{1}{x - 2}$

17–24 ■ Domain Find the domain of the function.

17. $f(x) = 7x + 15$ **18.** $f(x) = \dfrac{2x + 1}{2x - 1}$

19. $f(x) = \sqrt{x^2 + 4}$ **20.** $f(x) = 3x - \dfrac{2}{\sqrt{x + 1}}$

21. $f(x) = \dfrac{1}{x} + \dfrac{1}{x + 1} + \dfrac{1}{x + 2}$

22. $g(x) = \dfrac{2x^2 + 5x + 3}{2x^2 - 5x - 3}$

23. $h(x) = \sqrt{4 - x} + \sqrt{x^2 - 1}$

24. $f(x) = \dfrac{\sqrt[3]{2x + 1}}{\sqrt[3]{2x + 2}}$

25–42 ■ Graphing Functions Sketch a graph of the function. Use transformations of functions whenever possible.

25. $f(x) = 2 + \frac{3}{4}x$

26. $f(x) = 3(1 - 2x),\ -2 \le x \le 2$

27. $f(x) = -3x^2 + 4$ **28.** $f(x) = \frac{1}{2}x^2 - 8$

29. $f(x) = \sqrt{x - 5}$ **30.** $f(x) = \sqrt{3(x + 1)}$

31. $f(x) = \frac{1}{3}(x + 3)^2 + 2$ **32.** $f(x) = -2\sqrt{x + 4} - 3$

33. $f(x) = 4\sqrt{-x} - 2$ **34.** $f(x) = -\frac{1}{2}(x - 1)^2 + 2$

35. $f(x) = \frac{1}{2}x^3$ **36.** $f(x) = \sqrt[3]{-x}$

37. $f(x) = 5 - |x|$ **38.** $f(x) = 3 - |x + 2|$

39. $f(x) = -\dfrac{1}{x^2}$ **40.** $f(x) = \dfrac{1}{(x - 1)^3}$

41. $f(x) = \begin{cases} 1 - x & \text{if } x < 0 \\ 1 & \text{if } x \ge 0 \end{cases}$

42. $f(x) = \begin{cases} -x & \text{if } x < 0 \\ x^2 & \text{if } 0 \le x < 2 \\ 1 & \text{if } x \ge 2 \end{cases}$

43–46 ■ Equations That Represent Functions Determine whether the equation defines y as a function of x.

43. $x + y^2 = 14$ **44.** $3x - \sqrt{y} = 8$

45. $x^3 - y^3 = 27$ **46.** $2x = y^4 - 16$

47. Relations That Define Functions A relation is given by a table. List the ordered pairs in the relation, graph the relation, and determine whether the relation defines y as a function of x. State the domain and range of the relation.

(a)

x	y
−3	−3
−2	0
0	−1
2	3
3	−3

(b)

x	y
−3	3
2	1
0	−2
2	5
3	3

48. Relations That Define Functions The ordered pairs (x, y) in a relation are described. Determine whether the relation defines y as a function of x. Give reasons for your answer.

(a) $\{(-10, 20)(0, 70), (10, 50), (5, 30), (-5, 0)\}$

(b)

x Height (in)	64	68	72	71	68
y Student ID Number	23745	12933	10834	12772	91836

(c) The set of ordered pairs of real numbers (x, y) for which xy is an integer.

(d) The set of ordered pairs (x, y) that satisfy the equation $x = y^4$.

49–52 ■ Domain and Range from a Graph Draw a graph of the function f and use the graph to find the following.

(a) The domain and range of f

(b) The value(s) of x for which $f(x) = 0$

(c) The intervals on which $f(x) > 1$

49. $f(x) = \sqrt{9 - x^2}$

50. $f(x) = \sqrt{x^2 - 3}$

51. $f(x) = \sqrt{x^3 - 4x + 1}$

52. $f(x) = x^4 - x^3 + x^2 + 3x - 6$

53–58 ■ Getting Information From a Graph Draw a graph of the function f and use the graph to find the following.

(a) The local maximum and minimum values of f and the values of x at which they occur

(b) The intervals on which f is increasing and on which f is decreasing

53. $f(x) = 2x^2 - 4x + 5$

54. $f(x) = 1 - x - x^2$

55. $f(x) = 3.3 + 1.6x - 2.5x^3$

56. $f(x) = x^3 - 4x^2$

57. $f(x) = x^{2/3}(6 - x)^{1/3}$

58. $f(x) = |x^4 - 16|$

59–64 ■ Net Change and Average Rate of Change A function is given (either numerically, graphically, or algebraically). Find the net change and the average rate of change of the function between the indicated values.

59. Between $x = 4$ and $x = 8$

x	$f(x)$
2	14
4	12
6	12
8	8
10	6

60. Between $x = 10$ and $x = 30$

x	$g(x)$
0	25
10	−5
20	−2
30	30
40	0

61. Between $x = -1$ and $x = 2$

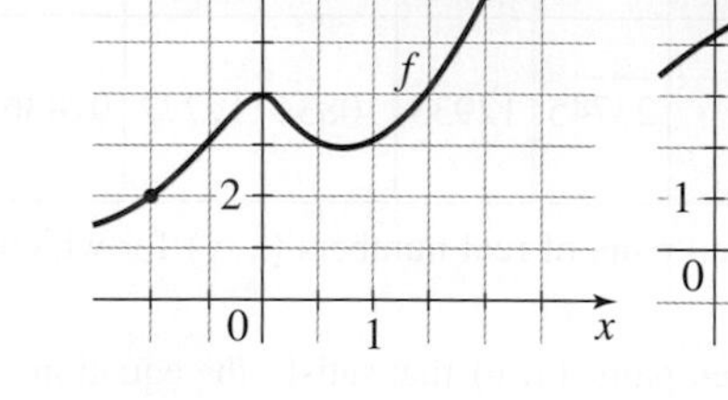

62. Between $x = 1$ and $x = 3$

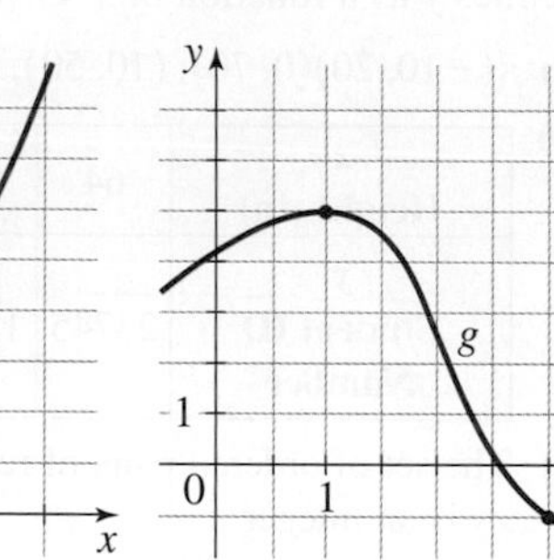

63. $f(x) = x^2 - 2x$; between $x = 1$ and $x = 4$

64. $g(x) = (x + 1)^2$; between $x = a$ and $x = a + h$

65–66 ■ Linear Functions? Determine whether the function is linear.

65. $f(x) = (2 + 3x)^2$

66. $f(x) = \dfrac{2x - 10}{\sqrt{5}}$

67–68 ■ Linear Functions A linear function is given. **(a)** Sketch a graph of the function. **(b)** What is the slope of the graph? **(c)** What is the rate of change of the function?

67. $f(x) = 3x + 2$

68. $g(x) = 3 - \frac{1}{2}x$

69–74 ■ Linear Functions A linear function is described either verbally, numerically, or graphically. Express f in the form $f(x) = ax + b$.

69. The function has rate of change -2 and initial value 3.

70. The graph of the function has slope $\frac{1}{2}$ and y-intercept -1.

71.

x	$f(x)$
0	3
1	5
2	7
3	9
4	11

72.

x	$f(x)$
0	6
2	5.5
4	5
6	4.5
8	4

73.

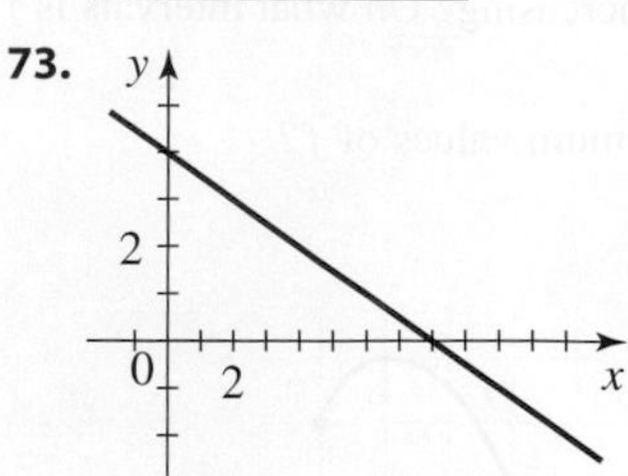

74.

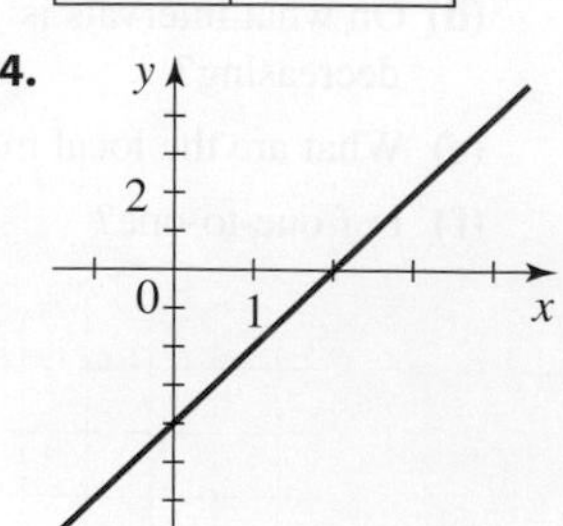

75–78 ■ Average Rate of Change A function f is given. **(a)** Find the average rate of change of f between $x = 0$ and $x = 2$, and the average rate of change of f between $x = 15$ and $x = 50$. **(b)** Were the two average rates of change that you found in part (a) the same? **(c)** Is the function linear? If so, what is its rate of change?

75. $f(x) = \frac{1}{2}x - 6$

76. $f(x) = 8 - 3x$

77. $f(x) = (x - 1)^2$

78. $f(x) = \dfrac{1}{x + 3}$

79. Transformations Suppose the graph of f is given. **(i)** Describe in words how the graph of each of the following functions can be obtained from the graph of f. **(ii)** Find a formula for the function you described in part (i) for the case $f(x) = x^3$.

(a) $y = f(x) + 8$
(b) $y = f(x + 8)$
(c) $y = 1 + 2f(x)$
(d) $y = f(x - 2) - 2$
(e) $y = f(-x)$
(f) $y = -f(-x)$
(g) $y = -f(x)$
(h) $y = f^{-1}(x)$

80. Transformations The graph of f is given. Draw the graph of each of the following functions.

(a) $y = f(x - 2)$
(b) $y = -f(x)$
(c) $y = 3 - f(x)$
(d) $y = \frac{1}{2}f(x) - 1$
(e) $y = f^{-1}(x)$
(f) $y = f(-x)$

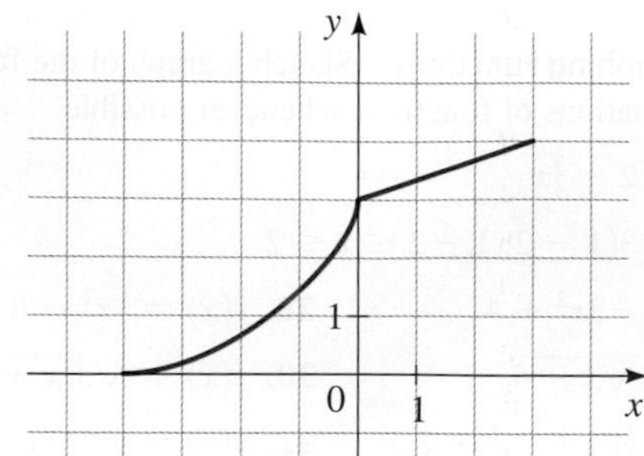

81. Even and Odd Functions Determine whether f is even, odd, or neither.

(a) $f(x) = 2x^5 - 3x^2 + 2$
(b) $f(x) = x^3 - x^7$
(c) $f(x) = \dfrac{1 - x^2}{1 + x^2}$
(d) $f(x) = \dfrac{1}{x + 2}$

82. Even and Odd Functions Determine whether the function in the figure is even, odd, or neither.

(a)

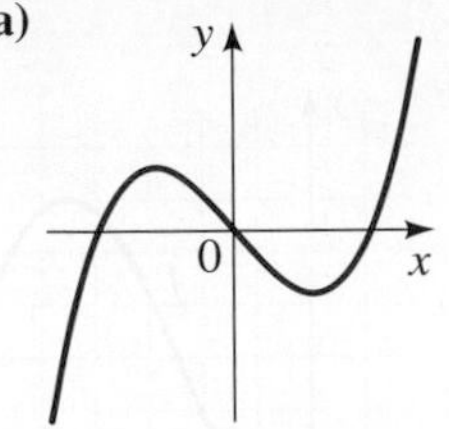

(b)

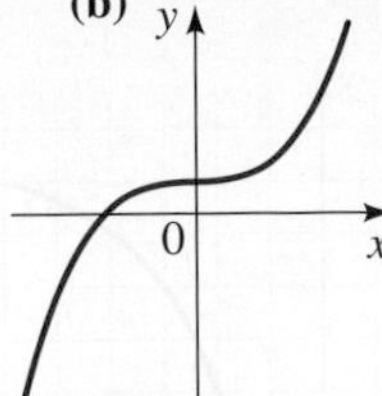

(c)

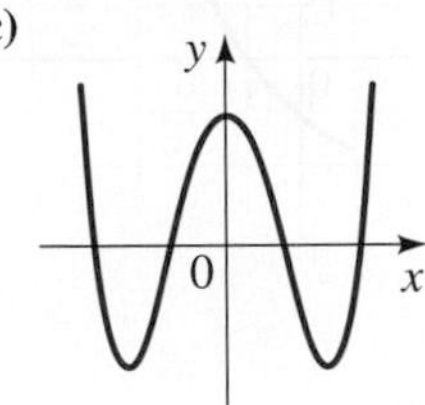

(d)

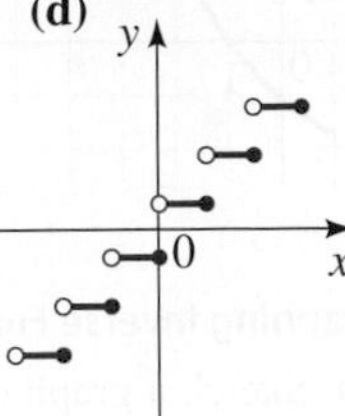

83. Getting Information From a Graph Match each description with the appropriate graph(s). Explain your choices. (A graph may satisfy more than one description.)

(a) Average rate of change is the same between any two points

(b) Increasing on $(-\infty, 2)$ and decreasing on $(2, \infty)$

(c) Domain $[1, \infty)$

(d) The function is even

(e) Decreasing on $(-\infty, \infty)$

(f) Has two local minima

(g) Has an inverse function

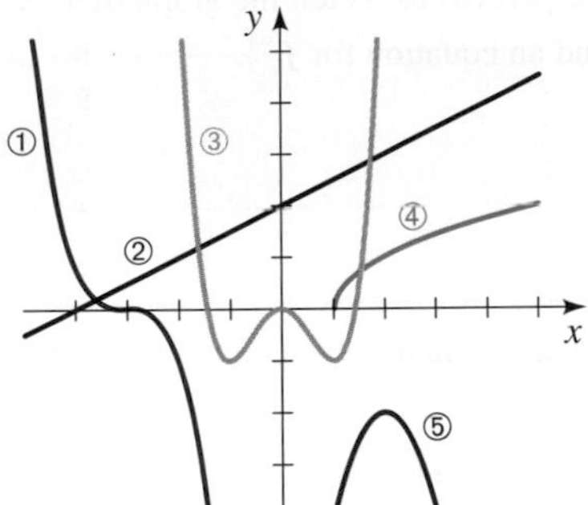

84. Maximum Height of Projectile A stone is thrown upward from the top of a building. Its height (in meters) above the ground after t seconds is given by

$$h(t) = -16t^2 + 48t + 32$$

What maximum height does it reach?

85–86 ■ Volume of Weather Balloon A pump is used to fill an approximately spherical weather balloon with helium at the rate of 0.25 m³/s. If the balloon is empty at time $t = 0$, then the function $V(t) = 0.25t$ models the volume of the balloon at time $t \geq 0$.

85. Find a transformation of V that models the given situation.

(a) The pump fills the balloon with helium at double the rate.

(b) The pump malfunctions for the first 4 s and then starts pumping.

(c) The balloon already contains 2 m³ of helium at $t = 0$.

86. As the balloon is being filled with helium, its radius increases.

(a) Find a function $r(V)$ that models the radius r of a sphere as a function of its volume V. [*Hint*: The volume of a sphere is $V = \frac{4}{3}\pi r^3$; solve for r.]

(b) Find $(r \circ V)(t)$. What does this composite function model?

(c) Find the radius of the balloon at time $t = 50$ s.

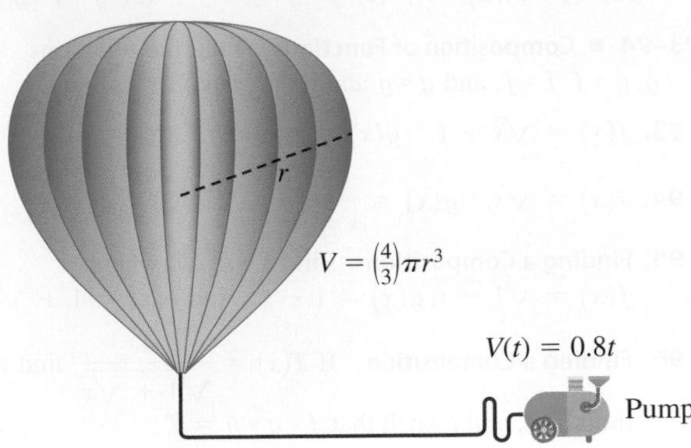

87. Weight of an Astronaut If an astronaut weighs 60 kg on the earth, then the astronaut's weight h kilometers above the surface of earth is given by the function

$$w(h) = 60\left(\frac{6340}{6340 + h}\right)^2$$

(a) Find w^{-1}. What does w^{-1} represent?

(b) Find $w^{-1}(64)$. Interpret your answer.

88. Crop Yield A model used for the yield Y (tons per acre) of an agricultural crop as a function of the nitrogen level x in the soil (measured in parts per million, ppm) is

$$Y(x) = \frac{kx}{1 + x^2} \qquad (x \geq 0)$$

where k is a constant that depends on the type of crop. A graph of this family of functions is shown for $k = 1, 2, 4, 6,$ and 7. Does the value of k affect the maximum yield? Does it affect the nitrogen level at which the maximum yield occurs? Find the maximum crop yield for a crop with $k = 5$.

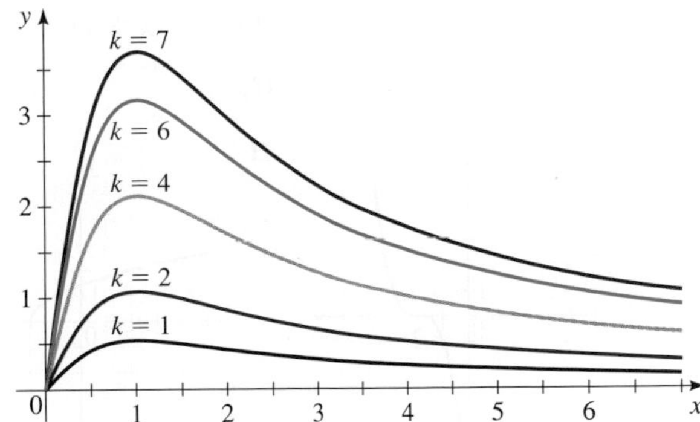

89–90 ■ Graphical Addition Functions, f and g are given. Draw graphs of f, g, and $f + g$ on the same coordinate axes to illustrate the concept of graphical addition.

89. $f(x) = x + 2, \quad g(x) = x^2$

90. $f(x) = x^2 + 1, \quad g(x) = 3 - x^2$

91. Combining Functions If $f(x) = x^2 - 3x + 2$ and $g(x) = 4 - 3x$, find the following functions.

(a) $f + g$ (b) $f - g$ (c) fg

(d) f/g (e) $f \circ g$ (f) $g \circ f$

92. If $f(x) = 1 + x^2$ and $g(x) = \sqrt{x - 1}$, find the following.

(a) $f \circ g$ **(b)** $g \circ f$ **(c)** $(f \circ g)(2)$

(d) $(f \circ f)(2)$ **(e)** $f \circ g \circ f$ **(f)** $g \circ f \circ g$

93–94 ■ Composition of Functions Find the functions $f \circ g, g \circ f, f \circ f$, and $g \circ g$ and their domains.

93. $f(x) = \sqrt{x} + 1, \quad g(x) = x - x^2$

94. $f(x) = \sqrt{x}, \quad g(x) = \dfrac{2}{x - 4}$

95. Finding a Composition Find $f \circ g \circ h$, where $f(x) = \sqrt{1 - x}, g(x) = 1 - x^2$, and $h(x) = 1 + \sqrt{x}$.

96. Finding a Composition If $T(x) = \dfrac{1}{\sqrt{1 + \sqrt{x}}}$, find functions f, g, and h such that $f \circ g \circ h = T$.

97–102 ■ One-to-One Functions Determine whether the function is one-to-one.

97. $f(x) = 3 + x^3$ **98.** $g(x) = 2 - 2x + x^2$

99. $h(x) = \dfrac{1}{x^4}$ **100.** $r(x) = 2 + \sqrt{x + 3}$

101. $p(x) = 3.3 + 1.6x - 2.5x^3$

102. $q(x) = 3.3 + 1.6x + 2.5x^3$

103–106 ■ Finding Inverse Functions Find the inverse of the function.

103. $f(x) = 3x - 2$ **104.** $f(x) = \dfrac{2x + 1}{3}$

105. $f(x) = (x + 1)^3$ **106.** $f(x) = 1 + \sqrt[5]{x - 2}$

107–108 ■ Inverse Functions From a Graph A graph of a function f is given. Does f have an inverse? If so, find $f^{-1}(0)$ and $f^{-1}(4)$.

107.

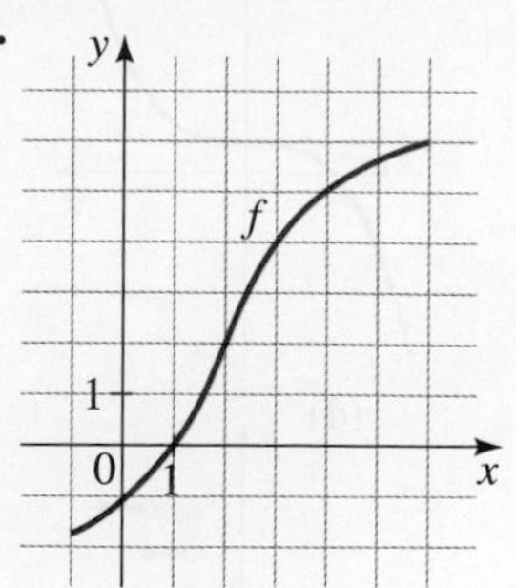

108.

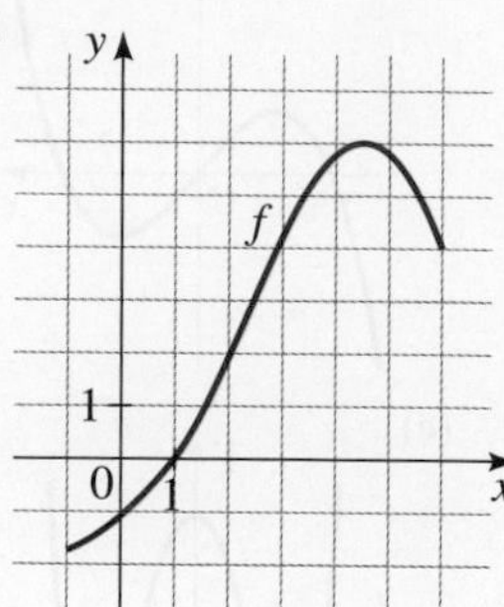

109. Graphing Inverse Functions

(a) Sketch a graph of the function

$$f(x) = x^2 - 4 \quad (x \geq 0)$$

(b) Use part (a) to sketch the graph of f^{-1}.

(c) Find an equation for f^{-1}.

110. Graphing Inverse Functions

(a) Show that the function $f(x) = 1 + \sqrt[4]{x}$ is one-to-one.

(b) Sketch the graph of f.

(c) Use part (b) to sketch the graph of f^{-1}.

(d) Find an equation for f^{-1}.

Matching

111. Equations and Their Graphs Match each equation with its graph in I–VIII, and state whether the equation defines y as a function of x. (Don't use a graphing device.)

(a) $y = |x| - 2$ **(b)** $y = (x + 1)^2 - 2$ **(c)** $(x - 2)^2 + (y + 1)^2 = 4$ **(d)** $y = (x - 2)^3$

(e) $x = y^2 - 3$ **(f)** $y = \dfrac{1}{x}$ **(g)** $y = -3(x + 3)^2 + 3$ **(h)** $y = \sqrt{2 - x}$

I

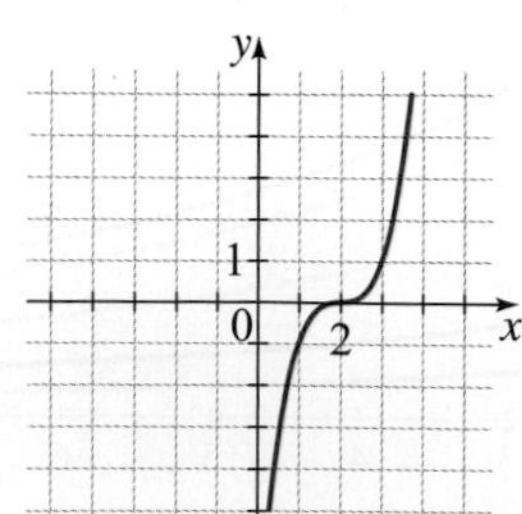

II

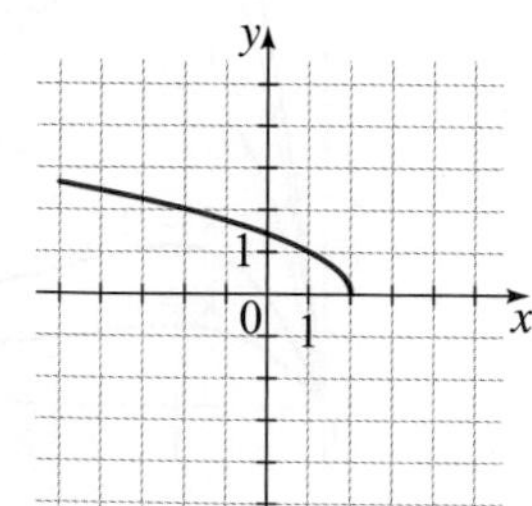

III

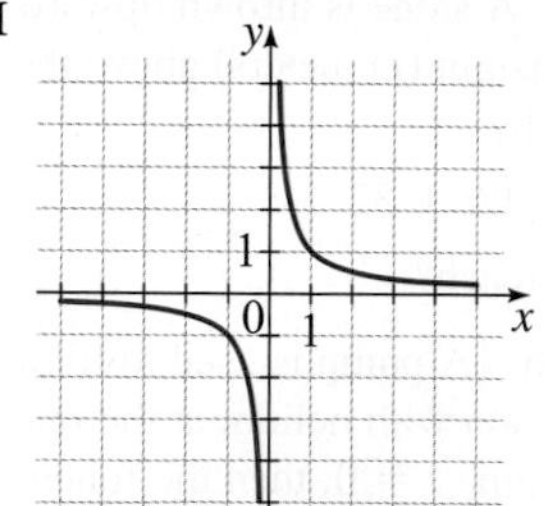

IV

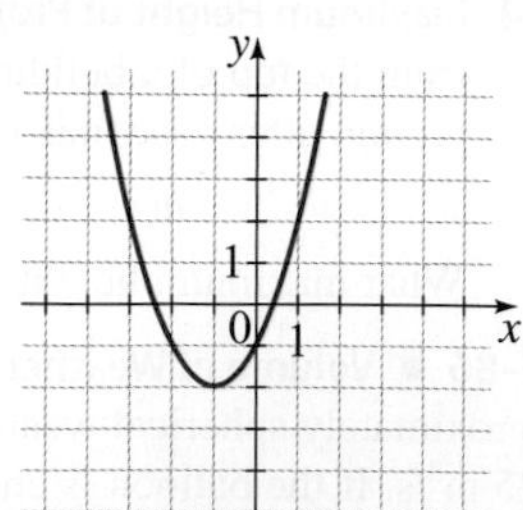

V

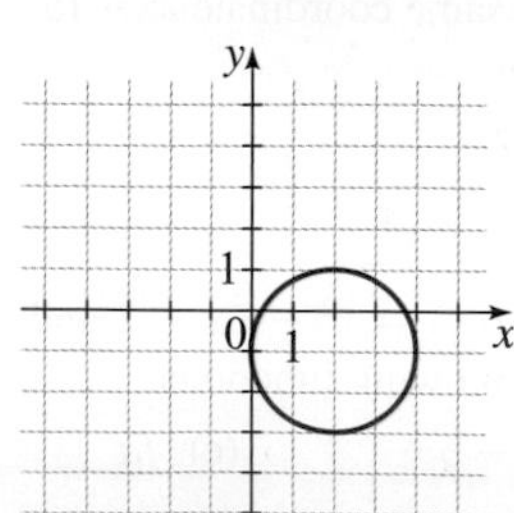

VI

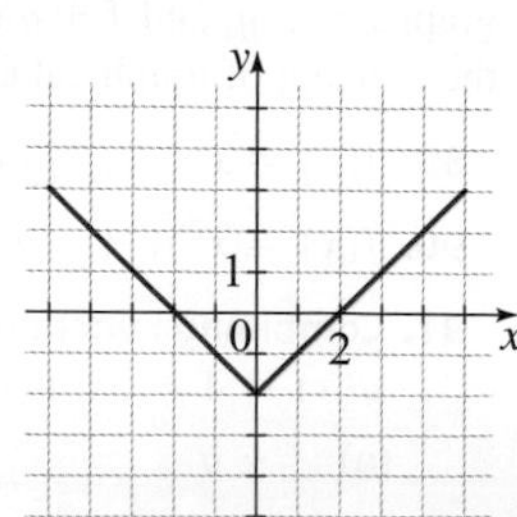

VII

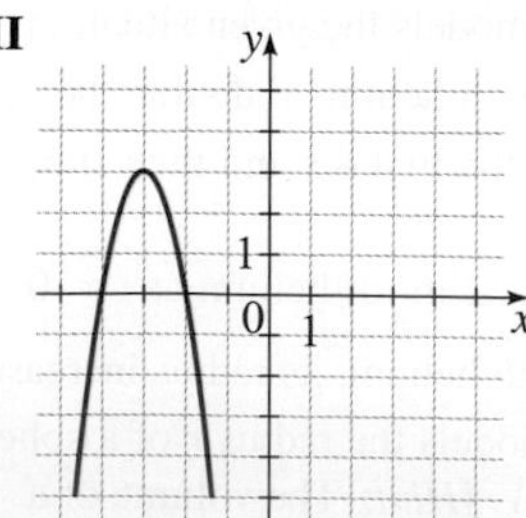

VIII

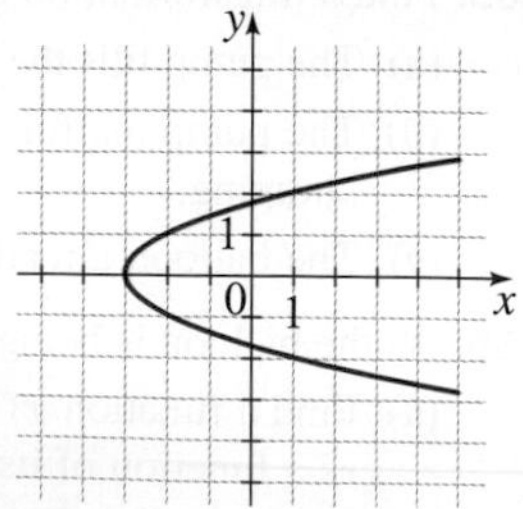

Chapter 2 | Test

1. Which of the following are graphs of functions? If the graph is that of a function, is it one-to-one?

(a)

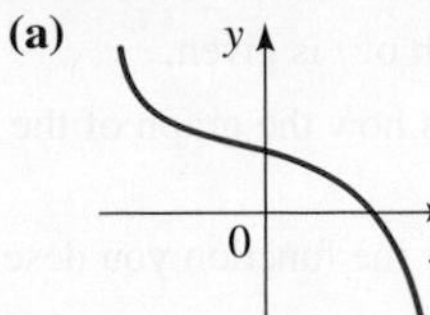

(b)

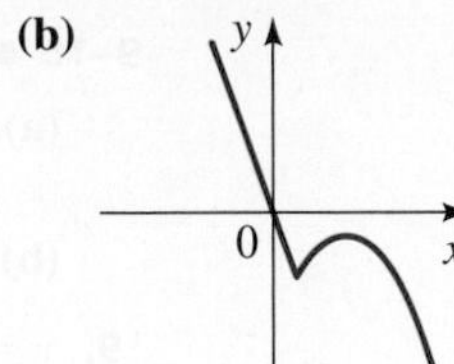

(c)

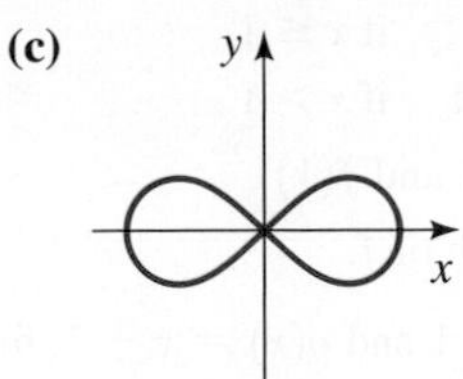

(d)

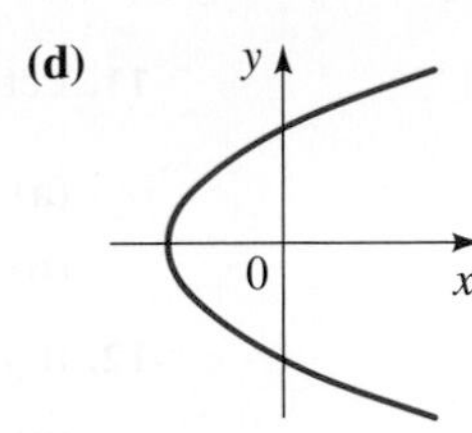

2. Let $f(x) = \dfrac{\sqrt{x}}{x+1}$.

(a) Evaluate $f(0)$, $f(2)$, and $f(a+2)$.

(b) Find the domain of f.

(c) What is the average rate of change of f between $x = 2$ and $x = 10$?

3. A function f has the following verbal description: "Subtract 2, then cube."

(a) Find a formula that expresses f algebraically.

(b) Make a table of values of f, for the inputs -1, 0, 1, 2, 3, and 4.

(c) Sketch a graph of f, using the table of values from part (b) to help.

(d) How do you know that f has an inverse? Give a verbal description for f^{-1}.

(e) Find a formula that expresses f^{-1} algebraically.

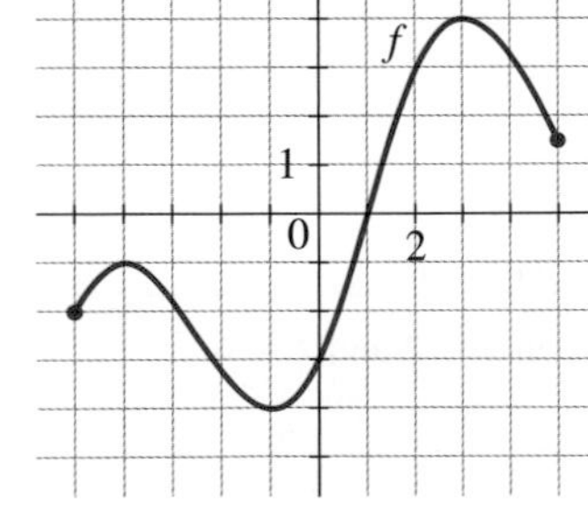

4. A graph of a function f is given in the margin.

(a) Find $f(-3)$ and $f(2)$.

(b) Find the net change and the average rate of change of f between $x = -3$ and $x = 2$.

(c) Find the domain and range of f.

(d) On what intervals is f increasing? On what intervals is f decreasing?

(e) What are the local maximum and local minimum values of f?

(f) Is f one-to-one? Give reasons for your answer.

5. A fund-raising group sells chocolate bars to finance a swimming pool for their school. The group finds that when they set the price at x dollars per bar (where $0 < x \le 5$), the total sales revenue (in dollars) is given by the function $R(x) = -500x^2 + 3000x$.

(a) Evaluate $R(2)$ and $R(4)$. What do these values represent?

(b) Use a graphing device to draw a graph of R. What does the graph tell you about what happens to revenue as the price increases from 0 to 5 dollars?

(c) What is the maximum revenue, and at what price is it achieved?

6. Determine the net change and the average rate of change for the function $f(t) = t^2 - 2t$ between $t = 2$ and $t = 2 + h$.

7. Let $f(x) = (x + 5)^2$ and $g(x) = 1 - 5x$.

(a) Only one of the two functions f and g is linear. Which one is linear, and why is the other one not linear?

(b) Sketch a graph of each function.

(c) What is the rate of change of the linear function?

8. (a) Sketch the graph of the function $f(x) = x^2$.

(b) Use part (a) to graph the function $g(x) = f(x) = (x + 4)^2 - 1$.

9–10 ■ Suppose the graph of f is given.

(a) Describe in words how the graph of the given function can be obtained from the graph of f.

(b) Find a formula for the function you described in part (a) for the case $f(x) = \sqrt{x}$.

9. $y = f(x - 3) + 2$ **10.** $y = f(-x)$

11. Let $f(x) = \begin{cases} 1 - x & \text{if } x \le 1 \\ 2x + 1 & \text{if } x > 1 \end{cases}$

(a) Evaluate $f(-2)$ and $f(1)$.

(b) Sketch the graph of f.

12. If $f(x) = x^2 + x + 1$ and $g(x) = x - 3$, find the following.

(a) $f + g$ (b) $f - g$ (c) $f \circ g$ (d) $g \circ f$

(e) $f(g(2))$ (f) $g(f(2))$ (g) $g \circ g \circ g$

13. Determine whether the function is one-to-one.

(a) $f(x) = x^3 + 1$ (b) $g(x) = |x + 1|$

14. Use the Inverse Function Property to show that $f(x) = \dfrac{1}{x - 2}$ is the inverse of $g(x) = \dfrac{1}{x} + 2$.

15. Find the inverse function of $f(x) = \dfrac{x - 3}{2x + 5}$.

16. (a) If $f(x) = \sqrt{3 - x}$, find the inverse function f^{-1}.

(b) Sketch the graphs of f and f^{-1} on the same coordinate axes.

17–22 ■ A graph of a function f is given below.

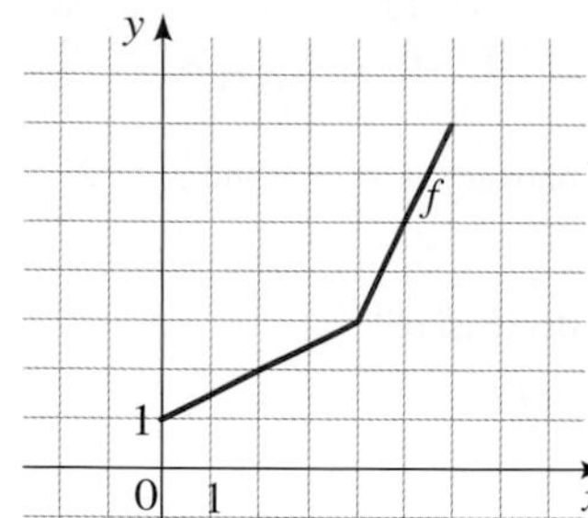

17. Find the domain and range of f.

18. Find $f(0)$ and $f(4)$.

19. Graph $f(x - 2)$ and $f(x) + 2$ on the same set of coordinate axes as f.

20. Find the net change and the average rate of change of f between $x = 2$ and $x = 6$.

21. Find $f^{-1}(1)$ and $f^{-1}(3)$.

22. Sketch the graph of f^{-1}.

23. Let $f(x) = 3x^4 - 14x^2 + 5x - 3$.

(a) Draw the graph of f in an appropriate viewing rectangle.

(b) Is f one-to-one?

(c) Find the local maximum and minimum values of f and the values of x at which they occur. State each answer correct to two decimal places.

(d) Use the graph to determine the range of f.

(e) Find the intervals on which f is increasing and on which f is decreasing.

Focus on Modeling | Modeling with Functions

Many objects or processes that are studied in the physical and social sciences involve understanding how one quantity varies with respect to another. Finding a function that describes the dependence of one quantity on another is called *modeling*. For example, a biologist who observes that the number of bacteria in a certain culture increases with time tries to model this phenomenon by finding the precise function (or rule) that relates the bacteria population to the elapsed time.

In this *Focus on Modeling* we learn how to find models that can be constructed using geometric or algebraic properties of the object or process under study. Once the model is found, we use it to analyze and predict properties of that object or process.

■ Modeling with Functions

We begin by giving some general guidelines for making a function model.

Guidelines for Modeling with Functions

1. **Express the Model in Words.** Identify the quantity you want to model, and express it, in words, as a function of the other quantities in the problem.
2. **Choose the Variable.** Identify all the variables that are used to express the function in Step 1. Assign a symbol, such as x, to one variable, and express the other variables in terms of this symbol.
3. **Set up the Model.** Express the function in the language of algebra by writing it as a function of the single variable chosen in Step 2.
4. **Use the Model.** Use the function to answer the questions posed in the problem. (To find a maximum or a minimum, use the methods described in Section 2.3.)

Example 1 ■ Fencing a Garden

A gardener has 140 meters of fencing to fence in a rectangular vegetable garden.

(a) Find a function that models the area of the garden that can be fenced.

(b) For what range of widths is the area greater than 825 m^2?

(c) Can the garden have a fenced area of 1250 m^2?

(d) Find the dimensions of the largest area that can be fenced.

Thinking About the Problem

If the gardener fences a plot with width 10 m, then the length must be 60 m, because $10 + 10 + 60 + 60 = 140$. So the area is

$$A = \text{width} \times \text{length} = 10 \cdot 60 = 600 \text{ m}^2$$

The table shows various choices for fencing the garden. We see that as the width increases, the fenced area increases, then decreases.

Width	Length	Area
10	60	600
20	50	1000
30	40	1200
40	30	1200
50	20	1000
60	10	600

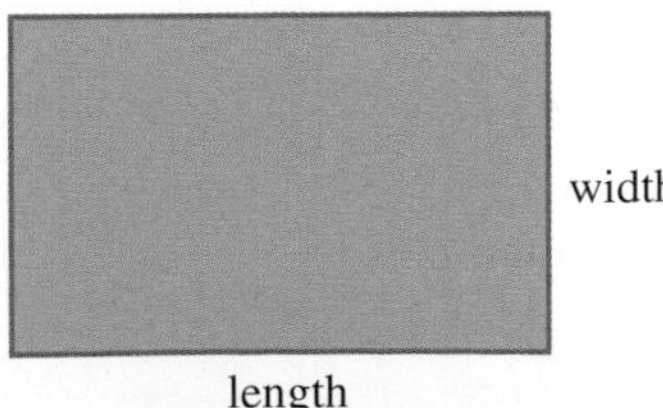

Solution

(a) The model that we want is a function that gives the area that can be fenced.

Express the model in words. We know that the area of a rectangular garden is

area = width × length

Choose the variable. There are two varying quantities: width and length. Because the function we want depends on only one variable, we let

$$x = \text{width of the garden}$$

Then we must express the length in terms of x. The perimeter is fixed at 140 m, so the length is determined once we choose the width. If we let the length be l, as in Figure 1, then $2x + 2l = 140$, so $l = 70 - x$. We summarize these facts:

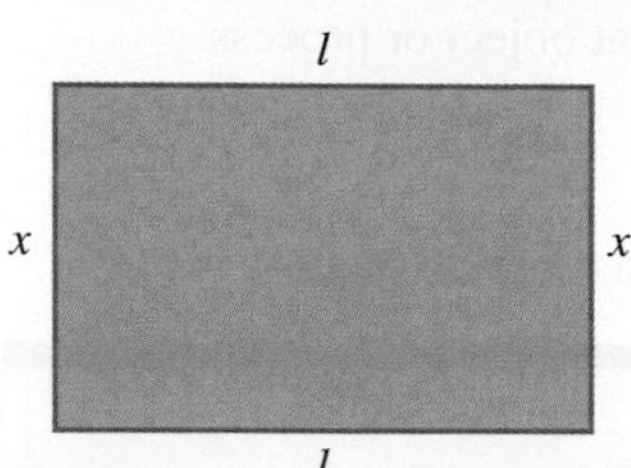

Figure 1

In Words	In Algebra
Width	x
Length	$70 - x$

Set up the model. The model is the function A that gives the area of the garden for any width x.

area = width × length

$$A(x) = x(70 - x)$$

$$A(x) = 70x - x^2$$

The area that can be fenced is modeled by the function $A(x) = 70x - x^2$.

Use the model. We use the model to answer the questions in parts (b)–(d).

(b) We need to solve the inequality $A(x) \geq 825$. To solve graphically, we graph $y = 70x - x^2$ and $y = 825$ in the same viewing rectangle (see Figure 2). We see that $15 \leq x \leq 55$.

(c) From Figure 3 we see that the graph of $A(x)$ always lies below the line $y = 1250$, so an area of 1250 m^2 is never attained.

Maximum values of functions are discussed in Section 2.3.

(d) We need to find where the maximum value of the function $A(x) = 70x - x^2$ occurs. Using a graphing device, we find that the function achieves its maximum value at $x = 35$ (see Figure 4). So the area that can be fenced is maximized when the garden's width is 35 m and its length is $70 - 35 = 35$ m. The maximum area then is $35 \times 35 = 1225$ m^2.

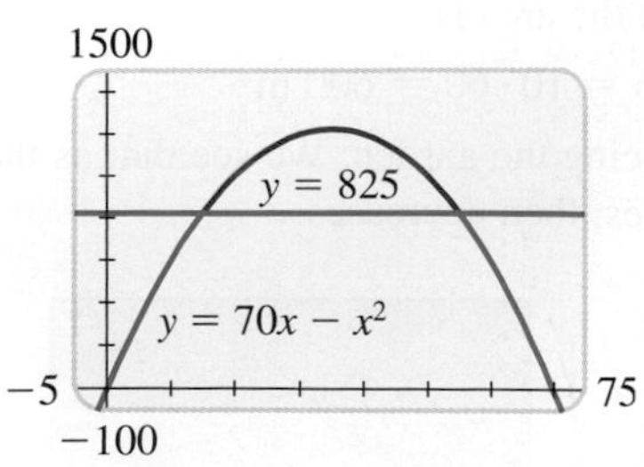

Figure 2

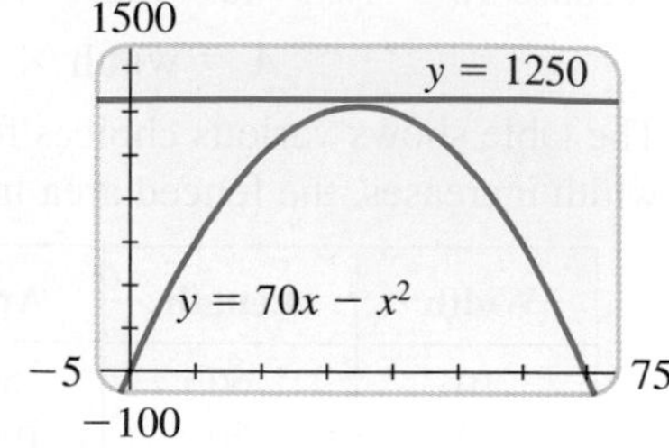

Figure 3

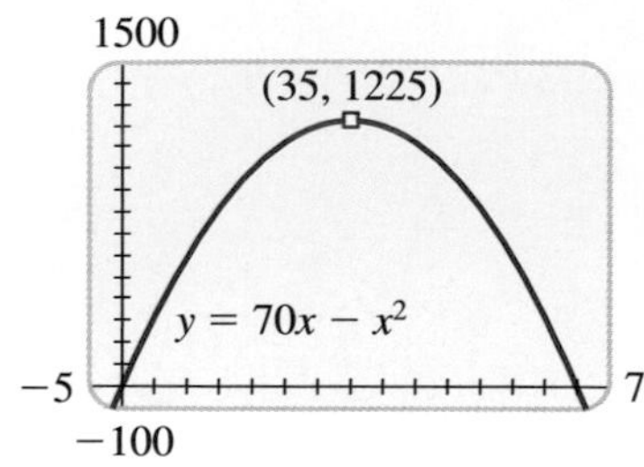

Figure 4

■

Example 2 ■ Minimizing the Metal in a Can

A manufacturer makes a cylindrical metal can that holds 1 L (liter) of oil. What radius minimizes the amount of metal in the can?

Thinking About the Problem

To use the least amount of metal, we must minimize the surface area of the can, that is, the area of the top, bottom, and the side. The combined area of the top and bottom is $2\pi r^2$ and the area of the side is $2\pi rh$ (see Figure 5), so the surface area of the can is

$$S = 2\pi r^2 + 2\pi rh$$

The radius and height of the can must be chosen so that the volume is exactly 1 L, or 1000 cm^3. If we want a small radius, say, $r = 3$, then the height must be just tall enough to make the total volume 1000 cm^3. In other words, we must have

$$\pi(3)^2 h = 1000 \qquad \text{Volume of the can is } \pi r^2 h$$

$$h = \frac{1000}{9\pi} \approx 35.37 \text{ cm} \qquad \text{Solve for } h$$

Now that we know the radius and height, we can find the surface area of the can:

$$\text{surface area} = 2\pi(3)^2 + 2\pi(3)(35.4) \approx 723.2 \text{ cm}^3$$

If we want a different radius, we can find the corresponding height and surface area in a similar fashion.

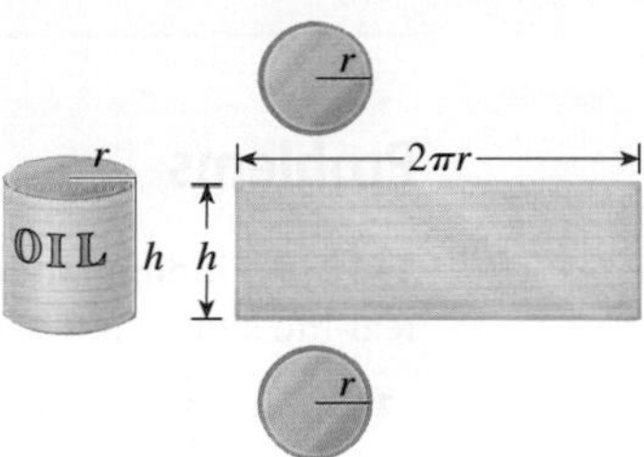

Figure 5

Solution The model that we want is a function that gives the surface area of the can.

Express the model in words. We know that for a cylindrical can

surface area	=	area of top and bottom	+	area of side

Choose the variable. There are two varying quantities: radius and height. Because the function we want depends on the radius, we let

$$r = \text{radius of can}$$

Next, we must express the height in terms of the radius r. Because the volume of a cylindrical can is $V = \pi r^2 h$ and the volume must be 1000 cm^3, we have

$$\pi r^2 h = 1000 \qquad \text{Volume of can is 1000 cm}^3$$

$$h = \frac{1000}{\pi r^2} \qquad \text{Solve for } h$$

We can now express the areas of the top, bottom, and side in terms of r only:

In Words	In Algebra
Radius of can	r
Height of can	$\dfrac{1000}{\pi r^2}$
Area of top and bottom	$2\pi r^2$
Area of side ($2\pi rh$)	$2\pi r\left(\dfrac{1000}{\pi r^2}\right)$

Set up the model. The model is the function S that gives the surface area of the can as a function of the radius r.

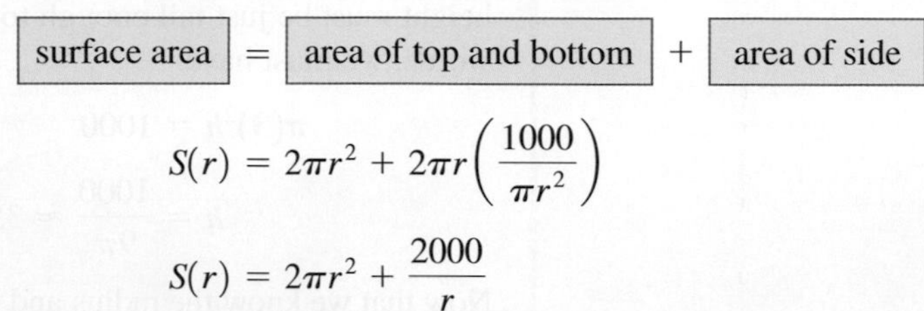

$$S(r) = 2\pi r^2 + 2\pi r\left(\frac{1000}{\pi r^2}\right)$$

$$S(r) = 2\pi r^2 + \frac{2000}{r}$$

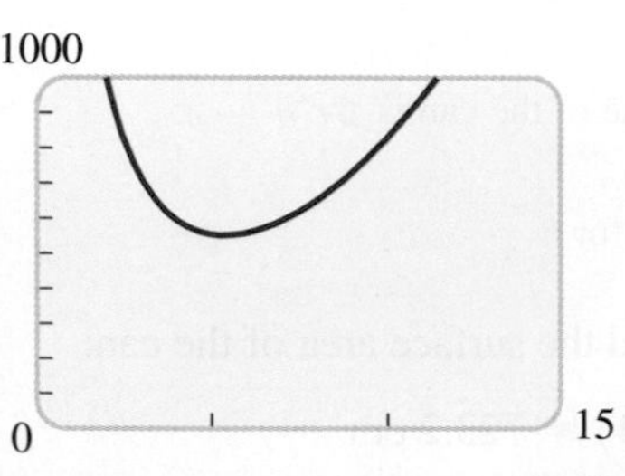

Figure 6 | $S(r) = 2\pi r^2 + \dfrac{2000}{r}$

Use the model. We use the model to find the minimum surface area of the can. We graph S in Figure 6 and find that the minimum value of S is about 554 cm^2 and this value occurs when the radius is about 5.4 cm. ■

Problems

1–18 ■ Finding Models In these problems you are asked to find a function that models a real-life situation. Use the principles of modeling described in this *Focus* to help you.

1. **Area** A rectangular building lot is three times as long as it is wide. Find a function that models its area A in terms of its width w.
2. **Area** A poster is 10 cm longer than it is wide. Find a function that models its area A in terms of its width w.
3. **Volume** A rectangular box has a square base. Its height is half the width of the base. Find a function that models its volume V in terms of its width w.
4. **Volume** The height of a cylinder is four times its radius. Find a function that models the volume V of the cylinder in terms of its radius r.
5. **Area** A rectangle has a perimeter of 20 m. Find a function that models its area A in terms of the length x of one of its sides.
6. **Perimeter** A rectangle has an area of 16 m^2. Find a function that models its perimeter P in terms of the length x of one of its sides.
7. **Area** Find a function that models the area A of an equilateral triangle in terms of the length x of one of its sides.
8. **Area** Find a function that models the surface area S of a cube in terms of its volume V.
9. **Radius** Find a function that models the radius r of a circle in terms of its area A.
10. **Area** Find a function that models the area A of a circle in terms of its circumference C.

PYTHAGORAS (circa 580–500 B.C.) founded a school in Croton in southern Italy, devoted to the study of arithmetic, geometry, music, and astronomy. The Pythagoreans, as they were called, were a secret society with peculiar rules and initiation rites. They wrote down nothing and were instructed not to reveal to anyone what they had learned from the Master. Although women were barred by law from attending public meetings, Pythagoras allowed women in his school, and his most famous student was Theano (whom he later married).

According to Aristotle, the Pythagoreans were convinced that "the principles of mathematics are the principles of all things." Their motto was "Everything is Number," by which they meant *whole* numbers. The outstanding contribution of Pythagoras is the theorem that bears his name: In a right triangle the area of the square on the hypotenuse is equal to the sum of the areas of the square on the other two sides.

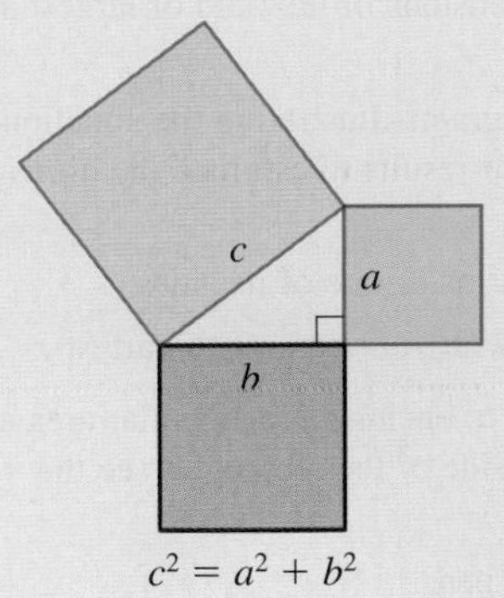

$$c^2 = a^2 + b^2$$

The converse of Pythagoras's Theorem is also true; that is, a triangle whose sides a, b, and c satisfy $a^2 + b^2 = c^2$ is a right triangle.

11. Area A rectangular box with a volume of 60 m^3 has a square base. Find a function that models its surface area S in terms of the length x of one side of its base.

12. Length A 1.5 m-tall person is standing near a street lamp that is 3.6 m tall, as shown in the figure. Find a function that models the length L of the person's shadow in terms of the distance d from the person to the base of the lamp.

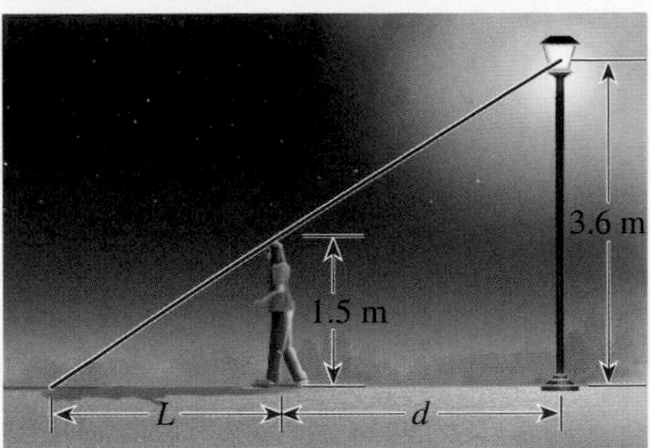

13. Distance Two ships leave port at the same time. One sails south at 15 km/h, and the other sails east at 20 km/h. Find a function that models the distance D between the ships in terms of the time t (in hours) elapsed since their departure from the port.

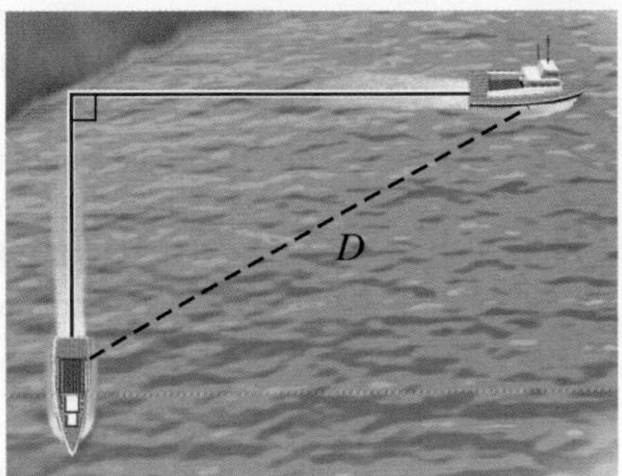

14. Product The sum of two positive numbers is 60. Find a function that models their product P in terms of x, one of the numbers.

15. Area An isosceles triangle has a perimeter of 8 cm. Find a function that models its area A in terms of the length of its base b.

16. Perimeter A right triangle has one leg twice as long as the other. Find a function that models its perimeter P in terms of the length x of the shorter leg.

17. Area A rectangle is inscribed in a semicircle of radius 10, as shown in the figure. Find a function that models the area A of the rectangle in terms of its height h.

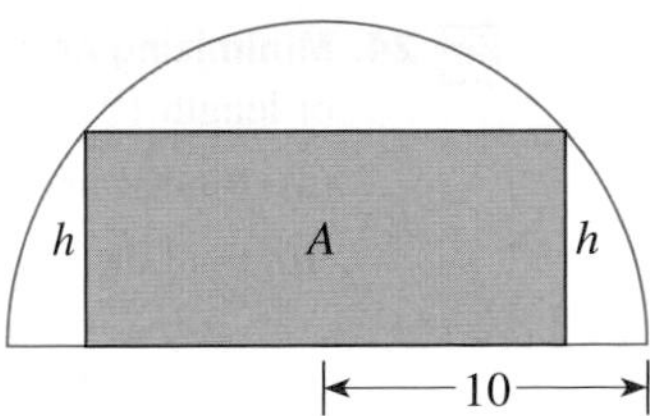

18. Height The volume of a cone is 100 in^3. Find a function that models the height h of the cone in terms of its radius r.

19–33 ■ Using Models to Find Maxima and Minima In these problems you are asked to find a function that models a real-life situation and then use the model to answer questions about the situation. Use the *Guidelines for Modeling with Functions* to help you answer these questions.

19. Maximizing a Product Consider the following problem: Find two numbers whose sum is 19 and whose product is as large as possible.

(a) Experiment with the problem by making a table like the one following, showing the product of different pairs of numbers that add up to 19. On the basis of the evidence in your table, estimate the answer to the problem.

First Number	Second Number	Product
1	18	18
2	17	34
3	16	48
⋮	⋮	⋮

(b) Find a function that models the product in terms of one of the two numbers.

(c) Use your model to solve the problem, and compare with your answer to part (a).

20. Minimizing a Sum Find two positive numbers whose sum is 100 and the sum of whose squares is a minimum.

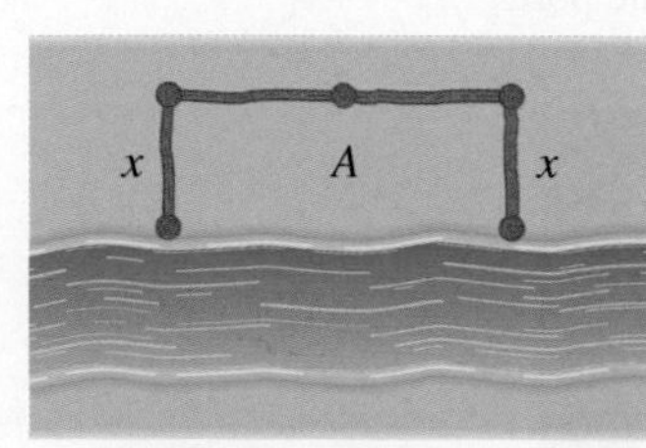

21. Fencing a Field Consider the following problem: A farmer has 2400 m of fencing and wants to fence off a rectangular field that borders a straight river. The farmer does not need a fence along the river (see the figure). What are the dimensions of the field of largest area that can be fenced?

(a) Experiment with the problem by drawing several diagrams illustrating the situation. Calculate the area of each configuration, and use your results to estimate the dimensions of the largest possible field.

(b) Find a function that models the area of the field in terms of one of its sides.

(c) Use your model to solve the problem, and compare with your answer to part (a).

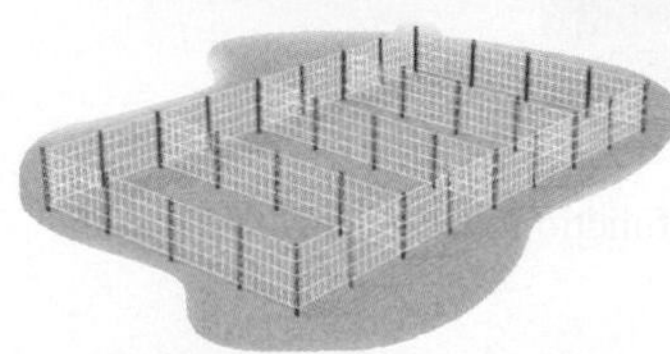

22. Dividing a Pen A rancher with 750 m of fencing wants to enclose a rectangular area and then divide it into four pens with fencing parallel to one side of the rectangle (see the figure).

(a) Find a function that models the total area of the four pens.

(b) Find the largest possible total area of the four pens.

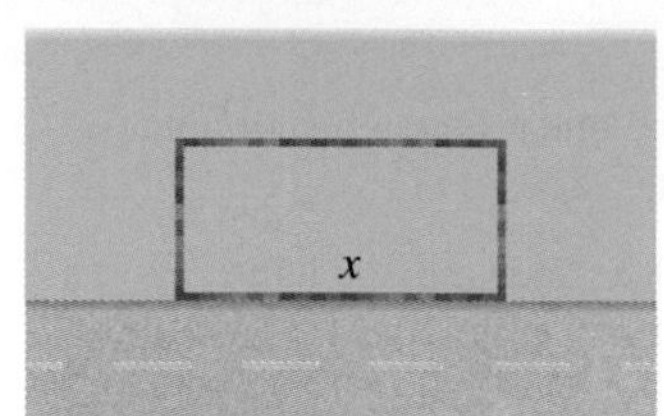

23. Fencing a Garden Plot A property owner wants to fence a garden plot adjacent to a road, as shown in the figure. The fencing next to the road must be sturdier and costs \$5 per meter, but the other fencing costs just \$3 per meter. The garden is to have an area of 1200 m^2.

(a) Find a function that models the cost of fencing the garden.

(b) Find the garden dimensions that minimize the cost of fencing.

(c) If the owner has at most \$600 to spend on fencing, find the range of lengths that can be fenced along the road.

24. Minimizing Area A wire 10 cm long is cut into two pieces, one of length x and the other of length $10 - x$, as shown in the figure. Each piece is bent into the shape of a square.

(a) Find a function that models the total area enclosed by the two squares.

(b) Find the value of x that minimizes the total area of the two squares.

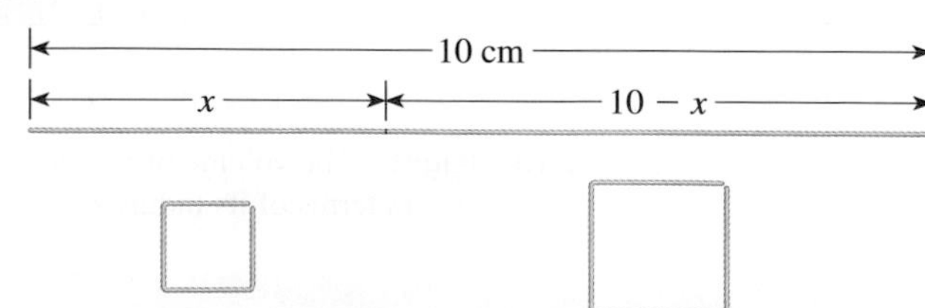

25. Light from a Window A Norman window has the shape of a rectangle surmounted by a semicircle, as shown in the figure to the left. A Norman window with perimeter 10 m is to be constructed.

(a) Find a function that models the area of the window.

(b) Find the dimensions of the window that admits the greatest amount of light.

26. Volume of a Box A box with an open top is to be constructed from a rectangular piece of cardboard with dimensions 12 cm by 20 cm by cutting out equal squares of side x at each corner and then folding up the sides (see the figure).

(a) Find a function that models the volume of the box.

(b) Find the values of x for which the volume is greater than 200 cm^3.

(c) Find the largest volume that such a box can have.

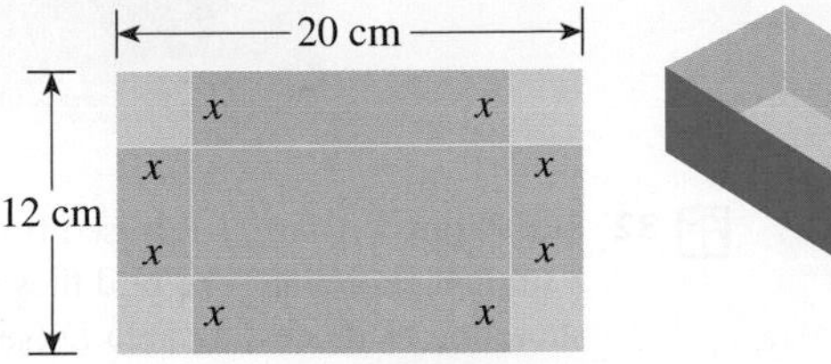

27. Area of a Box An open box with a square base of length x is to have a volume of 12 m^3.

(a) Find a function that models the surface area of the box.

(b) Find the box dimensions that minimize the amount of material used.

28. Inscribed Rectangle Find the dimensions that give the largest area for the rectangle shown in the figure. Its base is on the x-axis, and its other two vertices are above the x-axis, lying on the parabola $y = 8 - x^2$.

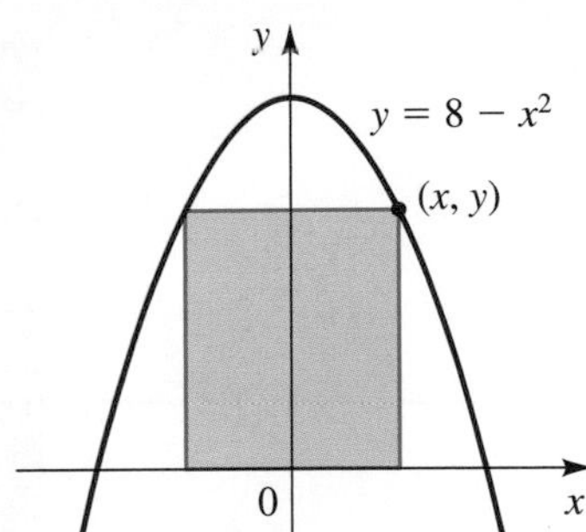

29. Minimizing Costs A rancher wants to build a rectangular pen with width x and an area of 100 m^2.

(a) Find a function that models the length of fencing required.

(b) Find the pen dimensions that require the minimum amount of fencing.

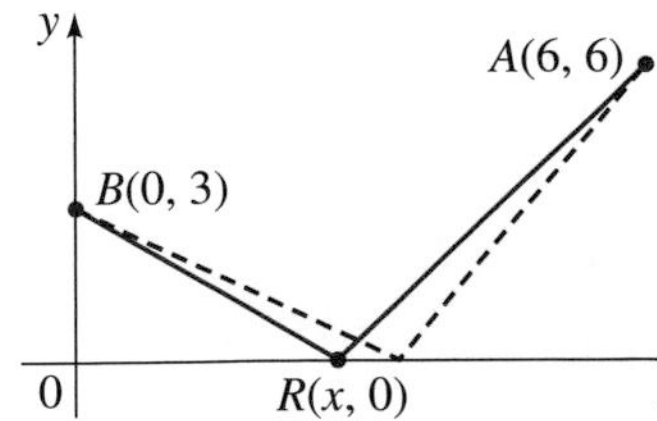

30. Minimizing a Distance Suppose that two villages are located at points $A(6, 6)$ and $B(0, 3)$, and the x-axis is a river. You are at point A and you need to walk to a point R on the river to get water and then walk to point B.

(a) Find a function of x that models the total distance from A to B via R.

(b) Find the value for x that minimizes the total distance you have to walk.

31. Minimizing Time You are standing at a point A on the bank of a straight river, that is 2 km wide. To reach point B, 7 km downstream on the opposite bank, you first row your boat to point P on the opposite bank and then walk the remaining distance x to B, as shown in the figure. You can row at a speed of 2 km/h and walk at a speed of 5 km/h.

(a) Find a function that models the time needed for your trip.

(b) Where should you land so that you reach B as soon as possible?

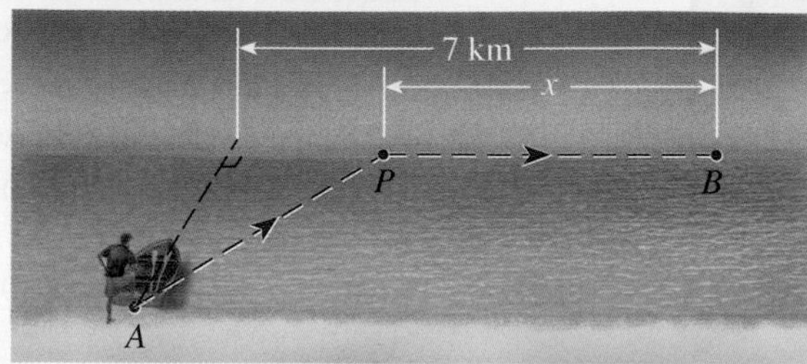

32. Bird Flight A bird is released from point A on an island, 5 km from the nearest point B on a straight shoreline. The bird flies to a point C on the shoreline and then flies along the shoreline to its nesting area D (see the figure). Suppose the bird requires 10 kJ/km of energy to fly over land and 14 kJ/km to fly over water.

(a) Use the fact that

$$\text{energy used} = \text{energy per kilometer} \times \text{kilometers flown}$$

to show that the total energy used by the bird is modeled by the function

$$E(x) = 14\sqrt{x^2 + 25} + 10(12 - x)$$

(b) If the bird instinctively chooses a path that minimizes its energy expenditure, what point does it fly to?

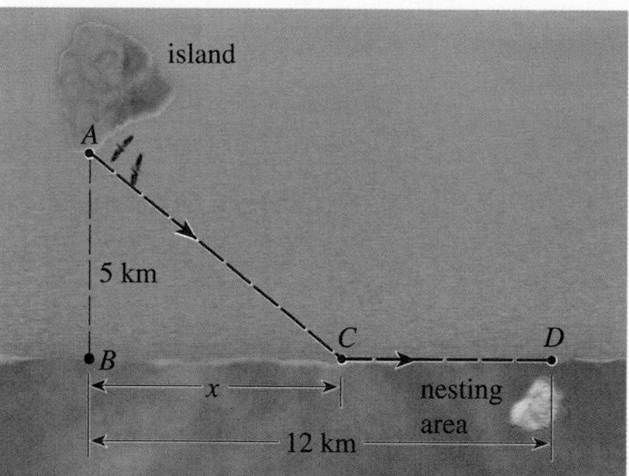

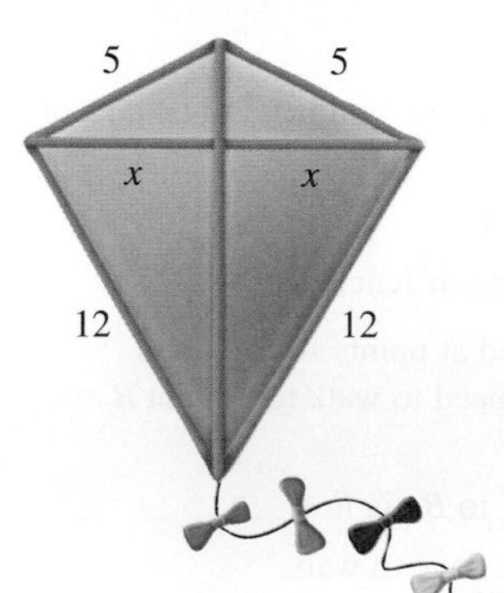

33. Area of a Kite A kite frame is to be made from six pieces of wood. The four pieces that form its border have been cut to the lengths indicated in the figure. Let x be as shown in the figure.

(a) Show that the area of the kite is given by the function

$$A(x) = x\left(\sqrt{25 - x^2} + \sqrt{144 - x^2}\right)$$

(b) How long should each of the two crosspieces be to maximize the area of the kite?

Aleksandr Ozerov/Shutterstock.com

3 Polynomial and Rational Functions

Functions defined by polynomial expressions are called *polynomial functions.* The graphs of polynomial functions can have many peaks and valleys. This property makes them suitable models for many real-world situations. For example, if a factory increases the number of workers, productivity increases, but if there are too many workers, productivity begins to decrease. This situation is modeled by a polynomial function of degree 2 (a quadratic function). The growth of many animal species follows a predictable pattern, beginning with a period of rapid growth, followed by a period of slow growth and then a final growth spurt. This variability in growth is modeled by a polynomial of degree 3.

In the *Focus on Modeling* at the end of this chapter we explore different ways in which polynomials are used to model real-world situations.

3.1 Quadratic Functions and Models

■ Graphing Quadratic Functions Using the Vertex Form ■ Maximum and Minimum Values of Quadratic Functions ■ Modeling with Quadratic Functions

A polynomial function is a function that is defined by a polynomial expression. So a **polynomial function of degree *n*** is a function of the form

$$P(x) = a_n x^n + a_{n-1}x^{n-1} + \cdots + a_1 x + a_0 \qquad (a_n \neq 0)$$

Polynomial expressions are defined in Section 1.3.

We have already studied polynomial functions of degree 0 and 1. These are functions of the form $P(x) = a_0$ and $P(x) = a_1 x + a_0$, respectively, whose graphs are lines. In this section we study polynomial functions of degree 2. These are called quadratic functions.

Quadratic Functions

A **quadratic function** is a polynomial function of degree 2. So a quadratic function is a function of the form

$$f(x) = ax^2 + bx + c \qquad (a \neq 0)$$

We see in this section how quadratic functions model many real-world phenomena. We begin by analyzing the graphs of quadratic functions.

■ Graphing Quadratic Functions Using the Vertex Form

For a geometric definition of parabolas, see Section 10.1.

If we take $a = 1$ and $b = c = 0$ in the quadratic function $f(x) = ax^2 + bx + c$, we get the quadratic function $f(x) = x^2$, whose graph is the parabola graphed in Example 2.2.1. In fact, the graph of any quadratic function is a **parabola**; it can be obtained from the graph of $f(x) = x^2$ by the transformations given in Section 2.6.

Vertex Form of a Quadratic Function

A quadratic function $f(x) = ax^2 + bx + c$ can be expressed in the **vertex form**

$$f(x) = a(x - h)^2 + k$$

by completing the square. The graph of f is a parabola with vertex (h, k).

If $a > 0$ the parabola **opens upward**

If $a < 0$ the parabola **opens downward**

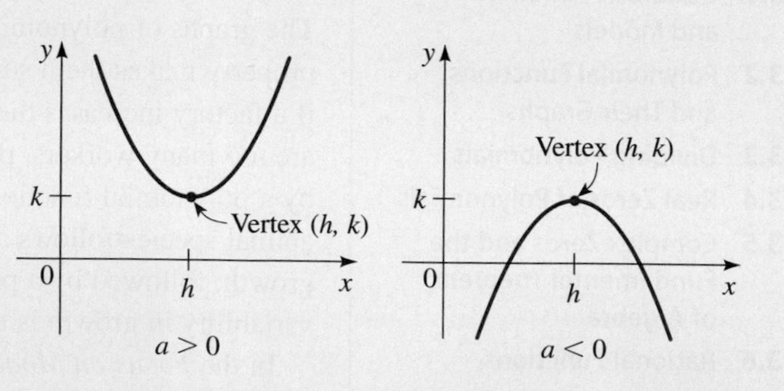

Example 1 ■ Vertex Form of a Quadratic Function

Let $f(x) = 2x^2 - 12x + 13$.

(a) Express f in vertex form.

(b) Find the vertex and x- and y-intercepts of f.

(c) Sketch a graph of f.

(d) Find the domain and range of f.

Solution

(a) Since the coefficient of x^2 is not 1, we must factor this coefficient from the terms involving x before we complete the square.

$$
\begin{aligned}
f(x) &= 2x^2 - 12x + 13 \\
&= 2(x^2 - 6x) + 13 && \text{Factor 2 from the } x\text{-terms} \\
&= 2(x^2 - 6x + 9) + 13 - 2\cdot 9 && \text{Complete the square: Add 9 inside parentheses, subtract } 2\cdot 9 \text{ outside} \\
&= 2(x - 3)^2 - 5 && \text{Factor and simplify}
\end{aligned}
$$

Completing the square is discussed in Section 1.5.

The vertex form is $f(x) = 2(x - 3)^2 - 5$.

(b) From the vertex form of f we can see that the vertex of f is $(3, -5)$. The y-intercept is $f(0) = 13$. To find the x-intercepts, we set $f(x) = 0$ and solve the resulting equation. We can solve a quadratic equation by any of the methods we studied in Section 1.5. In this case we solve the equation by using the Quadratic Formula.

$$
\begin{aligned}
0 &= 2x^2 - 12x + 13 && \text{Set } f(x) = 0 \\
x &= \frac{12 \pm \sqrt{144 - 4\cdot 2\cdot 13}}{4} && \text{Solve for } x \text{ using the Quadratic Formula} \\
x &= \frac{6 \pm \sqrt{10}}{2} && \text{Simplify}
\end{aligned}
$$

Thus the x-intercepts are $x = (6 \pm \sqrt{10})/2$, or approximately 1.42 and 4.58.

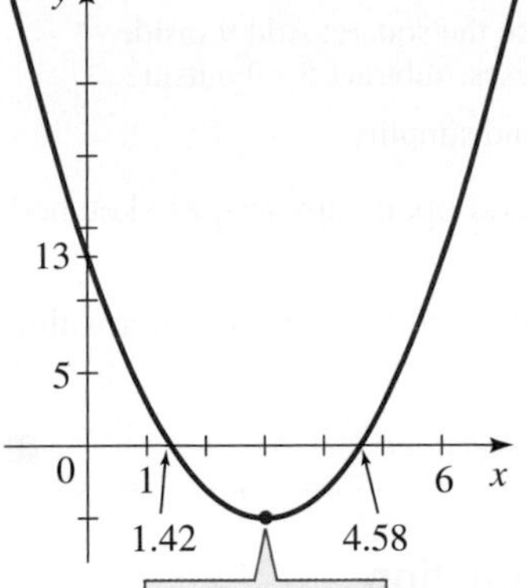

Figure 1 | $f(x) = 2x^2 - 12x + 13$

(c) The vertex form tells us that we get the graph of f by taking the parabola $y = x^2$, shifting it 3 units to the right, stretching it vertically by a factor of 2, and moving it downward 5 units. We sketch a graph of f in Figure 1, including the x- and y-intercepts found in part (b).

(d) The domain of f is the set of all real numbers $(-\infty, \infty)$. From the graph we see that the range of f is $[-5, \infty)$.

Obtaining the domain and range of a function from its graph is explained in Section 2.3.

Now Try Exercise 15 ■

■ Maximum and Minimum Values of Quadratic Functions

If a quadratic function has vertex (h, k), then the function has a minimum value at the vertex if its graph opens upward and a maximum value at the vertex if its graph opens downward. For example, the function graphed in Figure 1 has minimum value -5 when $x = 3$, since the vertex $(3, -5)$ is the lowest point on the graph.

Maximum or Minimum Value of a Quadratic Function

Let $f(x) = a(x - h)^2 + k$ be a quadratic function in vertex form. The maximum or minimum value of f occurs at $x = h$.

If $a > 0$ then the **minimum value** of f is $f(h) = k$

If $a < 0$ then the **maximum value** of f is $f(h) = k$

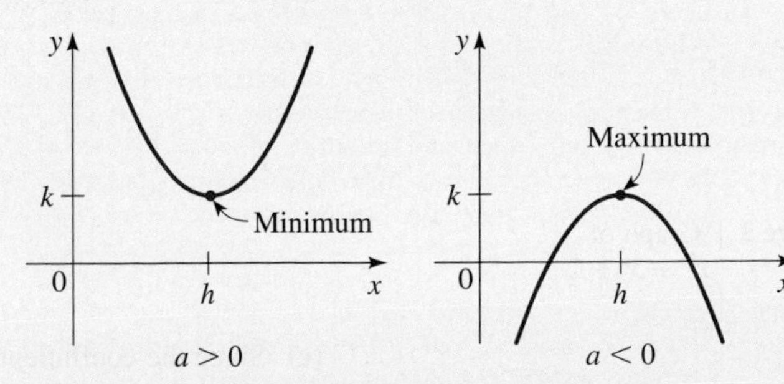

Example 2 ■ Minimum Value of a Quadratic Function

Consider the quadratic function $f(x) = 5x^2 - 30x + 49$.

(a) Express f in vertex form.

(b) Sketch a graph of f.

(c) Find the minimum value of f.

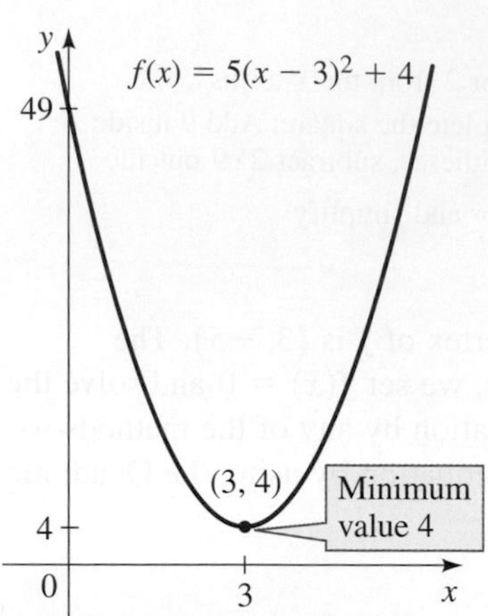

Figure 2

Solution

(a) To express this quadratic function in vertex form, we complete the square.

$$\begin{aligned} f(x) &= 5x^2 - 30x + 49 \\ &= 5(x^2 - 6x) + 49 && \text{Factor 5 from the } x\text{-terms} \\ &= 5(x^2 - 6x + 9) + 49 - 5 \cdot 9 && \text{Complete the square: Add 9 inside parentheses, subtract } 5 \cdot 9 \text{ outside} \\ &= 5(x - 3)^2 + 4 && \text{Factor and simplify} \end{aligned}$$

(b) The graph is a parabola that has its vertex at $(3, 4)$ and opens upward, as sketched in Figure 2.

(c) Since the coefficient of x^2 is positive, f has a minimum value. The minimum value is $f(3) = 4$.

Now Try Exercise 27 ■

Example 3 ■ Maximum Value of a Quadratic Function

Consider the quadratic function $f(x) = -x^2 + x + 2$.

(a) Express f in vertex form.

(b) Sketch a graph of f.

(c) Find the maximum value of f.

Solution

(a) To express this quadratic function in vertex form, we complete the square.

$$\begin{aligned} f(x) &= -x^2 + x + 2 \\ &= -(x^2 - x) + 2 && \text{Factor } -1 \text{ from the } x\text{-terms} \\ &= -(x^2 - x + \tfrac{1}{4}) + 2 - (-1)\tfrac{1}{4} && \text{Complete the square: Add } \tfrac{1}{4} \text{ inside parentheses, subtract } (-1)\tfrac{1}{4} \text{ outside} \\ &= -(x - \tfrac{1}{2})^2 + \tfrac{9}{4} && \text{Factor and simplify} \end{aligned}$$

In Example 3 you can check that the x-intercepts of the parabola are -1 and 2. These are obtained by solving the equation $f(x) = 0$.

(b) From the vertex form we see that the graph is a parabola that opens downward and has vertex $(\frac{1}{2}, \frac{9}{4})$. The graph of f is sketched in Figure 3.

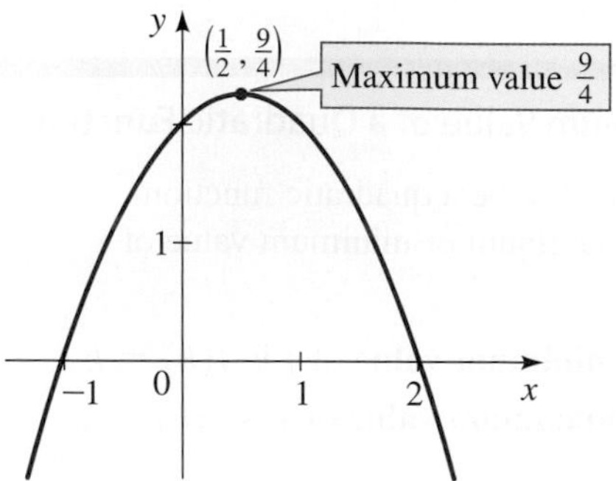

Figure 3 | Graph of $f(x) = -x^2 + x + 2$

(c) Since the coefficient of x^2 is negative, f has a maximum value, which is $f(\frac{1}{2}) = \frac{9}{4}$.

Now Try Exercise 29 ■

Expressing a quadratic function in vertex form helps us to sketch its graph as well as to find its maximum or minimum value. If we are interested only in finding the maximum or minimum value, then a formula is available for doing so. This formula is obtained by completing the square for the general quadratic function as follows.

$$f(x) = ax^2 + bx + c$$

$$= a\left(x^2 + \frac{b}{a}x\right) + c \qquad \text{Factor } a \text{ from the } x\text{-terms}$$

$$= a\left(x^2 + \frac{b}{a}x + \frac{b^2}{4a^2}\right) + c - a\left(\frac{b^2}{4a^2}\right) \qquad \text{Complete the square: Add } \frac{b^2}{4a^2} \text{ inside parentheses, subtract } a\left(\frac{b^2}{4a^2}\right) \text{ outside}$$

$$= a\left(x + \frac{b}{2a}\right)^2 + c - \frac{b^2}{4a} \qquad \text{Factor}$$

This equation is in vertex form with $h = -b/(2a)$ and $k = c - b^2/(4a)$. Since the maximum or minimum value occurs at $x = h$, we have the following result.

Maximum or Minimum Value of a Quadratic Function

The maximum or minimum value of a quadratic function $f(x) = ax^2 + bx + c$ occurs at

$$x = -\frac{b}{2a}$$

If $a > 0$, then the **minimum value** is $f\left(-\frac{b}{2a}\right)$.

If $a < 0$, then the **maximum value** is $f\left(-\frac{b}{2a}\right)$.

Example 4 ■ Finding Maximum and Minimum Values of Quadratic Functions

Find the maximum or minimum value of each quadratic function.

(a) $f(x) = x^2 + 4x$ **(b)** $g(x) = -2x^2 + 4x - 5$

Solution

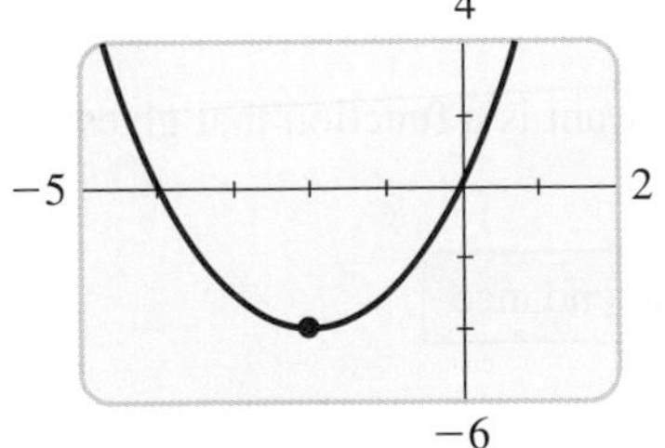

The minimum value occurs at $x = -2$.

(a) This is a quadratic function with $a = 1$ and $b = 4$. Thus the maximum or minimum value occurs at

$$x = -\frac{b}{2a} = -\frac{4}{2 \cdot 1} = -2$$

Since $a > 0$, the function has the *minimum* value

$$f(-2) = (-2)^2 + 4(-2) = -4$$

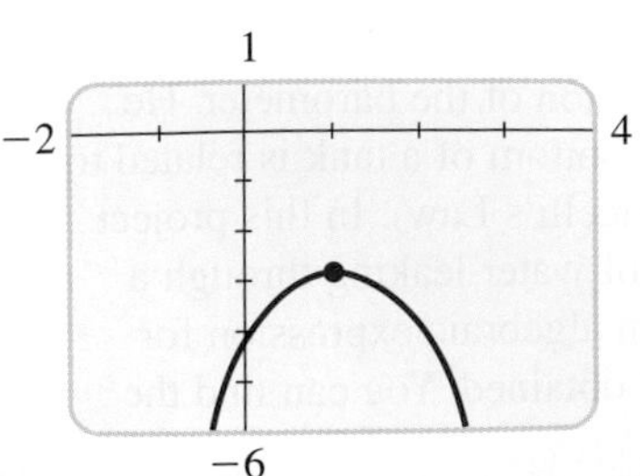

The maximum value occurs at $x = 1$.

(b) This is a quadratic function with $a = -2$ and $b = 4$. Thus the maximum or minimum value occurs at

$$x = -\frac{b}{2a} = -\frac{4}{2 \cdot (-2)} = 1$$

Since $a < 0$, the function has the *maximum* value

$$f(1) = -2(1)^2 + 4(1) - 5 = -3$$

Now Try Exercises 35 and 37 ■

■ Modeling with Quadratic Functions

We now study some examples of real-world phenomena that are modeled by quadratic functions. These examples and the *Applications* exercises for this section show some of the variety of situations that are naturally modeled by quadratic functions.

Example 5 ■ Maximum Gas Mileage for a Car

The function that models gas mileage is a quadratic function because air resistance is proportional to the square of the speed. You can feel air resistance by putting your hand out the window of a moving car.

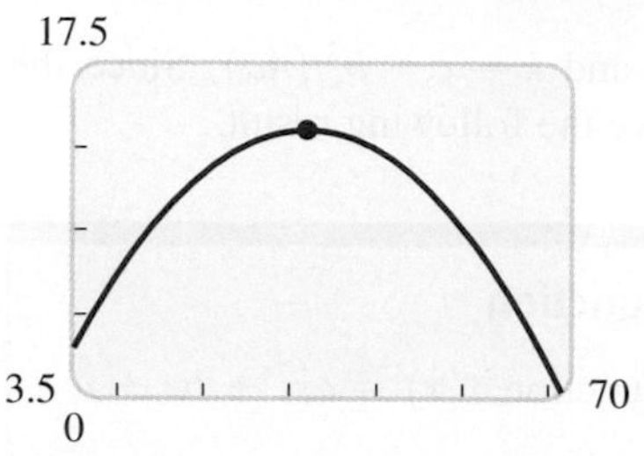

The maximum gas mileage occurs at 42 km/h

Most cars get their best gas mileage when traveling at relatively modest speeds. The gas mileage M for a certain new car is modeled by the function

$$M(s) = -0.0151s^2 + 1.2698s - 13.12 \qquad 6 \le s \le 30$$

where s is the speed in km/h and M is measured in km/L. What is the car's best gas mileage, and at what speed is it attained?

Solution The function M is a quadratic function with $a = -0.0151$ and $b = 1.2698$. Thus its maximum value occurs when

$$s = -\frac{b}{2a} = -\frac{1.2698}{2(-0.0151)} = 42$$

The maximum value is $M(42) = -0.0151(42)^2 + 1.2698(42) - 13.12 = 13.5$. So the car's best gas mileage is 14 km/L when it is traveling at 42 km/h.

 **Now Try Exercise 55** ■

Example 6 ■ Maximizing Revenue from Ticket Sales

A hockey team plays in an arena that has a seating capacity of 15,000 spectators. With the ticket price set at \$14, average attendance at recent games has been 9500. A market survey indicates that for each dollar the ticket price is lowered, the average attendance increases by 1000.

(a) Find a function that models the revenue in terms of ticket price.

(b) Find the price that maximizes revenue from ticket sales.

(c) What ticket price is so high that no one attends and so no revenue is generated?

Solution

(a) **Express the model in words.** The model that we want is a function that gives the revenue for any ticket price:

revenue = ticket price × attendance

Discovery Project ■ Torricelli's Law

Evangelista Torricelli (1608–1647) is best known for his invention of the barometer. He also discovered that the speed at which a fluid leaks from the bottom of a tank is related to the height of the fluid in the tank (a principle now called Torricelli's Law). In this project we conduct a simple experiment to collect data on the speed of water leaking through a hole in the bottom of a large soft-drink bottle. We then find an algebraic expression for Torricelli's Law by fitting a quadratic function to the data we obtained. You can find the project at **www.stewartmath.com**.

Choose the variable. There are two varying quantities: ticket price and attendance. Since the function we want depends on price, we let

$$x = \text{ticket price}$$

Next, we express attendance in terms of x.

In Words	In Algebra
Ticket price	x
Amount ticket price is lowered	$14 - x$
Increase in attendance	$1000(14 - x)$
Attendance	$9500 + 1000(14 - x)$

Set up the model. The model that we want is the function R that gives the revenue for a given ticket price x.

revenue	=	ticket price	×	attendance

$$R(x) = x \times [9500 + 1000(14 - x)]$$
$$R(x) = x(23{,}500 - 1000x)$$
$$R(x) = 23{,}500x - 1000x^2$$

(b) **Use the model.** Since R is a quadratic function with $a = -1000$ and $b = 23{,}500$, the maximum occurs at

$$x = -\frac{b}{2a} = -\frac{23{,}500}{2(-1000)} = 11.75$$

So a ticket price of \$11.75 gives the maximum revenue.

(c) **Use the model.** We want to find the ticket price for which $R(x) = 0$.

$$23{,}500x - 1000x^2 = 0 \qquad \text{Set } R(x) = 0$$
$$23.5x - x^2 = 0 \qquad \text{Divide by 1000}$$
$$x(23.5 - x) = 0 \qquad \text{Factor}$$
$$x = 0 \quad \text{or} \quad x = 23.5 \qquad \text{Solve for } x$$

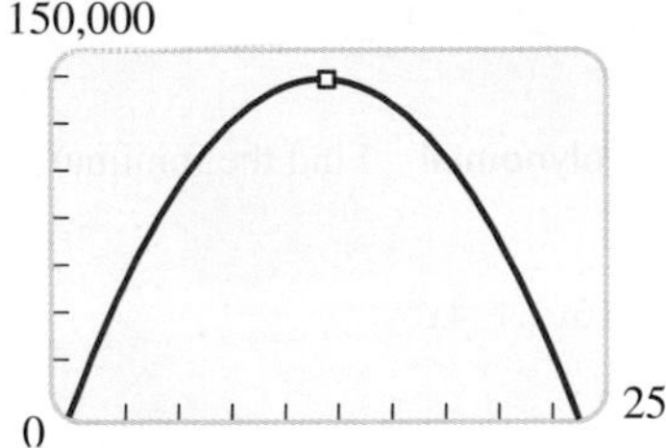

Maximum revenue occurs when ticket price is \$11.75.

So according to this model, a ticket price of \$23.50 is just too high; at that price no one attends to watch this team play. (Of course, revenue is also zero if the ticket price is zero.)

Now Try Exercise 65 ■

3.1 Exercises

Concepts

1. To put the quadratic function $f(x) = ax^2 + bx + c$ in vertex form, we complete the ______.

2. The quadratic function $f(x) = a(x - h)^2 + k$ is in vertex form.

(a) The graph of f is a parabola with vertex (____, ____).

(b) If $a > 0$, the graph of f opens ______. In this case $f(h) = k$ is the ______ value of f.

(c) If $a < 0$, the graph of f opens ______. In this case $f(h) = k$ is the ______ value of f.

3. The graph of $f(x) = 3(x - 2)^2 - 6$ is a parabola that opens ______, with its vertex at (____, ____), and $f(2) =$ ______ is the (minimum/maximum) ______ value of f.

4. The graph of $f(x) = -3(x - 2)^2 - 6$ is a parabola that opens ______, with its vertex at (____, ____), and $f(2) =$ ______ is the (minimum/maximum) ______ value of f.

Skills

5–8 ■ Graphs of Quadratic Functions The graph of a quadratic function f is given. **(a)** Find the coordinates of the vertex and the x- and y-intercepts. **(b)** Find the maximum or minimum value of f. **(c)** Find the domain and range of f.

5. $f(x) = -x^2 + 6x - 5$

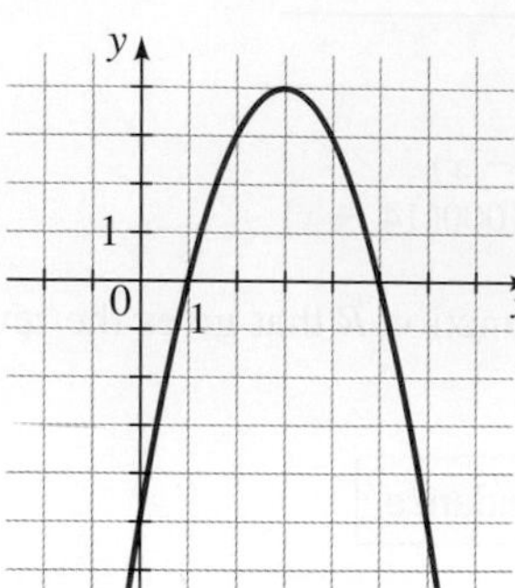

6. $f(x) = -\frac{1}{2}x^2 - 2x + 6$

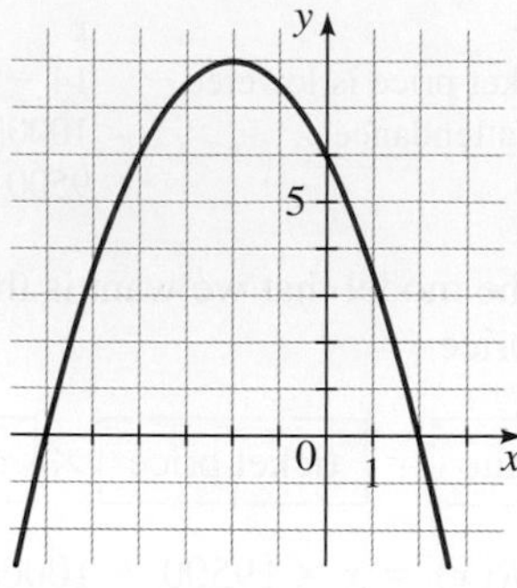

7. $f(x) = 2x^2 - 4x - 1$

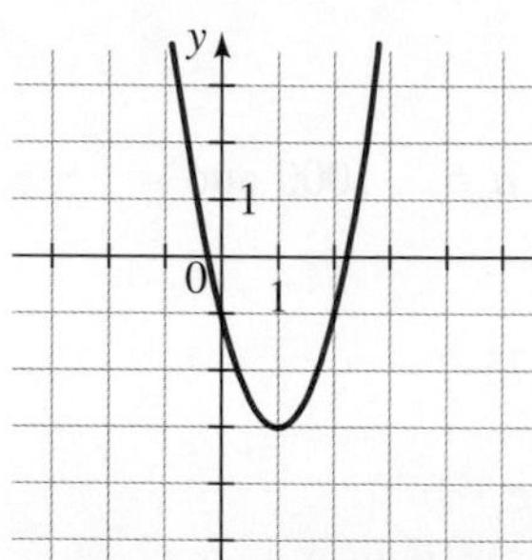

8. $f(x) = 3x^2 + 6x - 1$

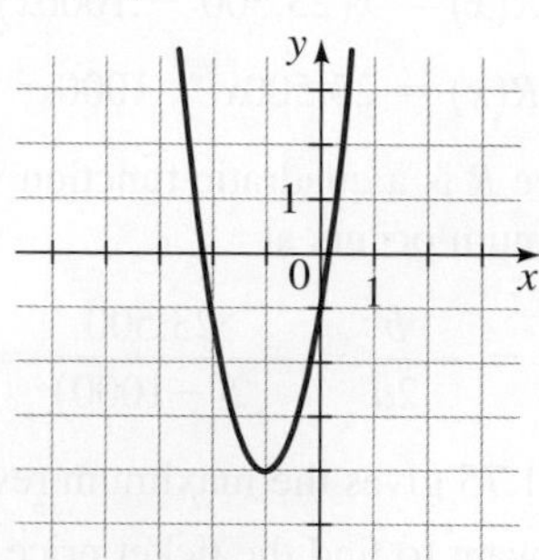

9–24 ■ Graphing Quadratic Functions A quadratic function f is given. **(a)** Express f in vertex form. **(b)** Find the vertex and x- and y-intercepts of f. **(c)** Sketch a graph of f. **(d)** Find the domain and range of f.

9. $f(x) = x^2 - 4x + 9$ **10.** $f(x) = x^2 + 6x + 8$

11. $f(x) = x^2 - 6x$ **12.** $f(x) = x^2 + 8x$

13. $f(x) = 3x^2 + 6x$ **14.** $f(x) = -x^2 + 10x$

15. $f(x) = x^2 + 4x + 3$ **16.** $f(x) = x^2 - 2x + 2$

17. $f(x) = -x^2 - 10x - 15$ **18.** $f(x) = -x^2 + 12x - 11$

19. $f(x) = 3x^2 - 6x + 7$ **20.** $f(x) = -3x^2 + 6x - 2$

21. $f(x) = 0.5x^2 + 6x + 16$ **22.** $f(x) = 2x^2 + 12x + 10$

23. $f(x) = -4x^2 - 12x + 1$ **24.** $f(x) = 3x^2 + 2x - 2$

25–34 ■ Maximum and Minimum Values A quadratic function f is given. **(a)** Express f in vertex form. **(b)** Sketch a graph of f. **(c)** Find the maximum or minimum value of f.

25. $f(x) = x^2 + 2x - 1$ **26.** $f(x) = x^2 - 8x + 8$

27. $f(x) = 4x^2 - 8x - 1$ **28.** $f(x) = 2x^2 - 12x + 14$

29. $f(x) = -x^2 - 3x + 3$ **30.** $f(x) = 1 - 6x - x^2$

31. $f(x) = 3x^2 - 12x + 13$ **32.** $f(x) = 2x^2 + 12x + 20$

33. $f(x) = 1 - x - x^2$ **34.** $f(x) = 3 - 4x - 4x^2$

35–44 ■ Formula for Maximum and Minimum Values Find the maximum or minimum value of the function.

35. $f(x) = -7x^2 + 14x - 5$ **36.** $f(x) = 6x^2 + 48x + 1$

37. $f(t) = 4t^2 - 40t + 110$ **38.** $g(x) = -5x^2 + 60x - 200$

39. $f(s) = s^2 - 1.2s + 16$ **40.** $g(x) = 100x^2 - 1500x$

41. $h(x) = \frac{1}{2}x^2 + 2x - 6$ **42.** $f(x) = -\frac{x^2}{3} + 2x + 7$

43. $f(x) = 3 - x - \frac{1}{2}x^2$ **44.** $g(x) = 2x(x - 4) + 7$

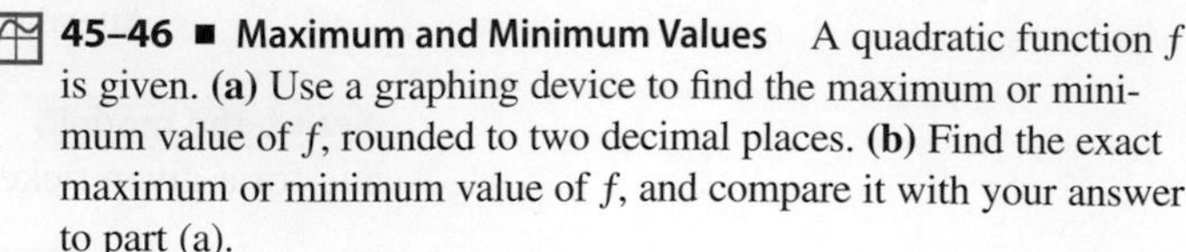

45–46 ■ Maximum and Minimum Values A quadratic function f is given. **(a)** Use a graphing device to find the maximum or minimum value of f, rounded to two decimal places. **(b)** Find the exact maximum or minimum value of f, and compare it with your answer to part (a).

45. $f(x) = x^2 + 1.79x - 3.21$

46. $f(x) = 1 + x - \sqrt{2}x^2$

Skills Plus

47–48 ■ Finding Quadratic Functions Find a function f whose graph is a parabola that has the given vertex and passes through the indicated point.

47. Vertex $(2, -3)$; point $(3, 1)$

48. Vertex $(-1, 5)$; point $(-3, -7)$

49. Maximum of a Fourth-Degree Polynomial Find the maximum value of the function

$$f(x) = 3 + 4x^2 - x^4$$

[*Hint:* Let $t = x^2$.]

50. Minimum of a Sixth-Degree Polynomial Find the minimum value of the function

$$f(x) = 2 + 16x^3 + 4x^6$$

[*Hint:* Let $t = x^3$.]

Applications

51. Height of a Ball If a ball is thrown directly upward with a velocity of 12.5 m/s, its height (in meters) after t seconds is given by $y = 12.5t - 5t^2$. What is the maximum height attained by the ball?

52. Path of a Ball A ball is thrown across a playing field from a height of 1.5 m above the ground at an angle of 45° to the horizontal at a speed of 6 m/s (See the figure.) It can be deduced from physical principles that the path of the ball is modeled by the function

$$y = -\frac{10}{(6)^2}x^2 + x + 1.5$$

where x is the distance (in meters) that the ball has traveled horizontally.

(a) Find the maximum height attained by the ball.

(b) Find the horizontal distance the ball has traveled when it hits the ground.

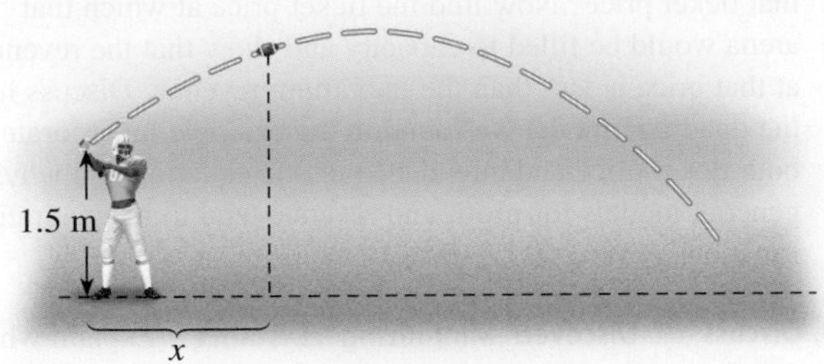

53. Revenue A manufacturer finds that the revenue generated by selling x units of a certain commodity is given by the function $R(x) = 80x - 0.4x^2$, where the revenue $R(x)$ is measured in dollars. What is the maximum revenue, and how many units should be manufactured to obtain this maximum?

54. Sales A soft-drink vendor at a popular beach analyzes sales records and finds that if x cans of soda are sold in one day, then the profit (in dollars) from soda sales is given by

$$P(x) = -0.001x^2 + 3x - 1800$$

What is the maximum profit per day, and how many cans must be sold to produce this maximum profit?

55. Advertising The effectiveness of a YouTube commercial depends on how many times a viewer watches it. After some experiments an advertising agency found that if the effectiveness E is measured on a scale of 0 to 10, then

$$E(n) = \tfrac{2}{3}n - \tfrac{1}{90}n^2$$

where n is the number of times a viewer watches a given commercial. For a commercial to have maximum effectiveness, how many times does a viewer need to watch it?

56. Pharmaceuticals When a certain drug is taken orally, the concentration of the drug in the patient's bloodstream after t minutes is given by $C(t) = 0.06t - 0.0002t^2$, where $0 \le t \le 240$ and the concentration is measured in mg/L. When is the maximum serum concentration reached, and what is that maximum concentration?

57. Agriculture The number of apples produced by each tree in an apple orchard depends on how densely the trees are planted. If n trees are planted on an acre of land, then each tree produces $900 - 9n$ apples. So the number of apples produced per acre is

$$A(n) = n(900 - 9n)$$

How many trees should be planted per acre to obtain the maximum yield of apples?

58. Agriculture At a certain vineyard each grape vine produces about 10 kg of grapes in a season when about 700 vines are planted per acre. For each additional vine that is planted, the production of each vine decreases by about 1 percent. So the number of kilograms of grapes produced per acre is modeled by

$$A(n) = (700 + n)(10 - 0.01n)$$

where n is the number of additional vines planted. Find the number of vines that should be planted to maximize grape production.

59–62 ■ Maximums and Minimums In the following problems from *Focus on Modeling: Modeling with Functions* at the end of Chapter 2, we found approximate maximum and minimum values graphically. Use the formulas of this section to find the exact maximum or minimum value for the indicated problem, and compare with the approximate value found graphically.

59. Problem 21 **60.** Problem 22

61. Problem 25 **62.** Problem 24

63. Fencing a Horse Corral A rancher has 2400 m of fencing to fence in a rectangular horse corral.

(a) Find a function that models the area of the corral in terms of the width x of the corral.

(b) Find the dimensions of the rectangle that maximize the area of the corral.

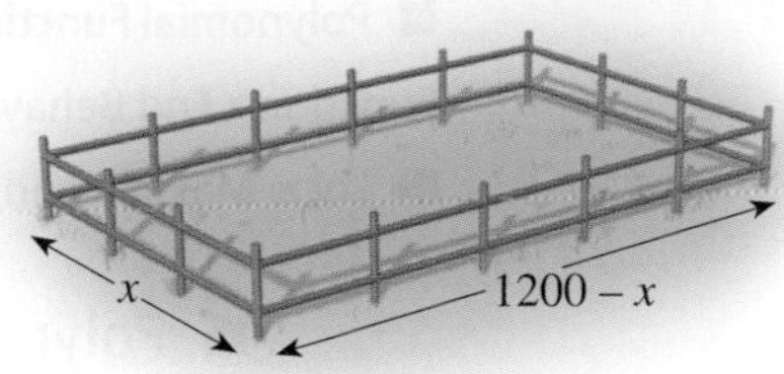

64. Making a Rain Gutter A rain gutter is formed by bending up the sides of a 30-cm-wide rectangular metal sheet as shown in the figure.

(a) Find a function that models the cross-sectional area of the gutter in terms of x.

(b) Find the value of x that maximizes the cross-sectional area of the gutter.

(c) What is the maximum cross-sectional area for the gutter?

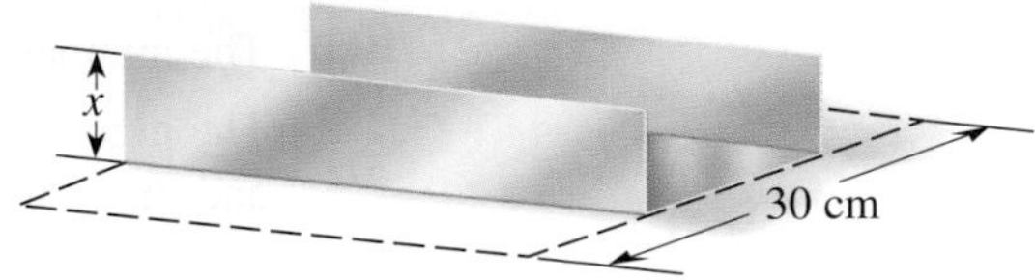

65. Stadium Revenue A baseball team plays in a stadium that holds 55,000 spectators. With the ticket price set at \$10, the average attendance at recent games has been 27,000. A market survey indicates that for every dollar the ticket price is lowered, attendance increases by 3000.

(a) Find a function that models the revenue in terms of ticket price.

(b) Find the price that maximizes revenue from ticket sales.

(c) What ticket price is so high that no revenue is generated?

66. Maximizing Profit A community bird-watching society makes and sells simple bird feeders to raise money for conservation activities. The materials for each feeder cost \$6, and the society sells an average of 20 feeders per week at a price of \$10 each. The society has been considering raising the price, so it conducts a survey and finds that for every dollar increase, it will lose 2 sales per week.

(a) Find a function that models weekly profit in terms of price per feeder.

(b) What price should the society charge for each feeder to maximize profit? What is the maximum weekly profit?

■ Discuss ■ Discover ■ Prove ■ Write

67. Discover: Vertex and x-Intercepts We know that the graph of the quadratic function $f(x) = (x - m)(x - n)$ is a parabola. Sketch a rough graph of what such a parabola looks like. What are the x-intercepts of the graph of f? Can you tell from your graph the x-coordinate of the vertex in terms of m and n? (Use the symmetry of the parabola.) Confirm your answer by expanding and using the formulas of this section.

68. Discuss ■ Discover: Maximizing Revenue In Example 6 we found that a ticket price of \$11.75 would maximize revenue for a sports arena. What is the attendance and revenue at that ticket price? Now find the ticket price at which that arena would be filled to capacity and show that the revenue at that price is less than the maximum revenue. Discuss how the quadratic model we found in the example incorporates both ticket price and attendance in one equation and why, in general, models for profit and revenue tend to be quadratic functions (Exercises 53–54).

69. Discuss ■ Discover: Minimizing a Distance Explain why the distance between the point $P(3, -2)$ and any point $Q(x, y)$ on the line $y = 2x - 3$ is given by the function

$$g(x) = \sqrt{5x^2 - 10x + 10}$$

Find the minimum value of the function g and the value of x at which the minimum is achieved. Find the point on the line that is closest to P. (Don't use a graphing device.)

PS *Try to recognize something familiar.* We know how to find the value of x that minimizes the quadratic expression under the square root sign. Argue that the minimum value of g occurs at this same value of x.

3.2 Polynomial Functions and Their Graphs

■ Polynomial Functions ■ Graphing Basic Polynomial Functions ■ Graphs of Polynomial Functions: End Behavior ■ Using Zeros to Graph Polynomials ■ Shape of the Graph Near a Zero ■ Local Maxima and Minima of Polynomials

■ Polynomial Functions

In this section we study polynomial functions of any degree. But before we work with polynomial functions, we must agree on some terminology.

Polynomial Functions

A **polynomial function of degree n** is a function of the form

$$P(x) = a_n x^n + a_{n-1}x^{n-1} + \cdots + a_1 x + a_0$$

where n is a nonnegative integer and $a_n \neq 0$.

The numbers $a_0, a_1, a_2, \ldots, a_n$ are called the **coefficients** of the polynomial.

The number a_0 is the **constant coefficient** or **constant term**.

The number a_n, the coefficient of the highest power, is the **leading coefficient**, and the term $a_n x^n$ is the **leading term**.

We often refer to polynomial functions simply as *polynomials*. The following polynomial has degree 5, leading coefficient 3, and constant term −6.

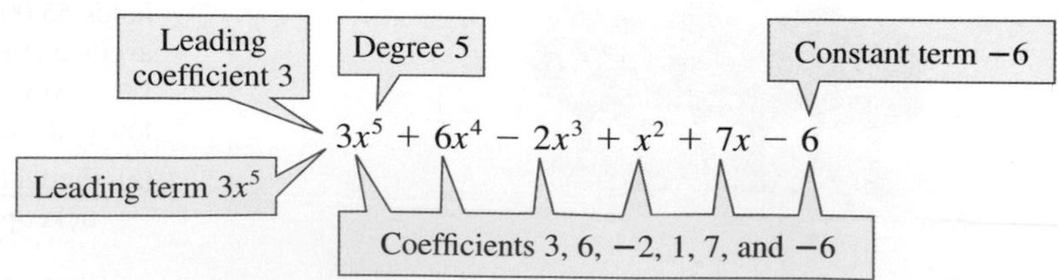

The table below lists some more examples of polynomials.

Polynomial	Degree	Leading Term	Constant Term
$P(x) = 4x - 7$	1	$4x$	-7
$P(x) = x^2 + x$	2	x^2	0
$P(x) = 2x^3 - 6x^2 + 10$	3	$2x^3$	10
$P(x) = -5x^4 + x - 2$	4	$-5x^4$	-2

If a polynomial consists of just a single term, then it is called a **monomial**. For example, $P(x) = x^3$ and $Q(x) = -6x^5$ are monomials.

■ Graphing Basic Polynomial Functions

The simplest polynomial functions are the monomials $P(x) = x^n$, whose graphs are shown in Figure 1. As the figure suggests, the graph of $P(x) = x^n$ has the same general shape as the graph of $y = x^2$ when n is even and the same general shape as the graph of $y = x^3$ when n is odd. However, as the degree n becomes larger, the graphs become flatter around the origin and steeper elsewhere.

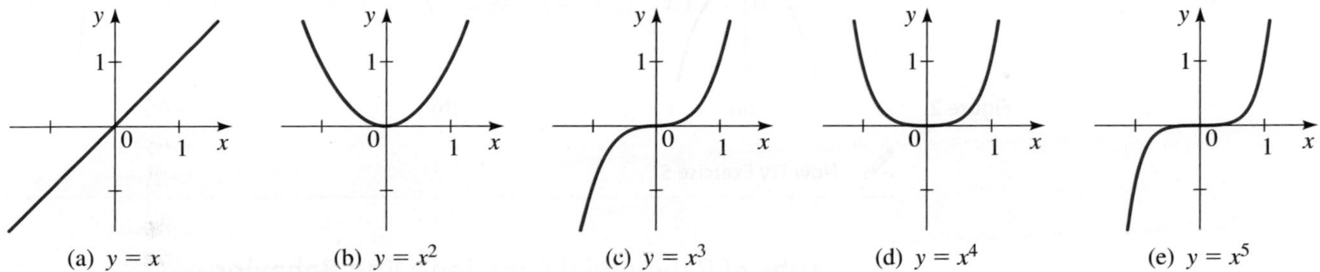

(a) $y = x$ (b) $y = x^2$ (c) $y = x^3$ (d) $y = x^4$ (e) $y = x^5$

Figure 1 | Graphs of monomials

Example 1 ■ Transformations of Monomials

Sketch the graph of each function, and state its domain and range.

(a) $P(x) = -x^3$ **(b)** $Q(x) = (x - 2)^4$

(c) $R(x) = -2x^5 + 4$

Mathematics in the Modern World

Splines

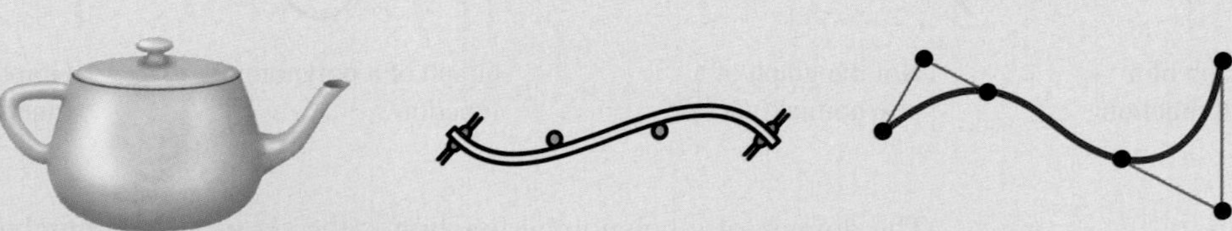

A spline is a long strip of wood or metal that is curved while held fixed at certain points. In the past shipbuilders used splines to create the curved shape of a boat's hull. Splines are also used to make the curves of a piano, a violin, or the spout of a teapot.

Mathematicians discovered that the shapes of splines can be obtained by piecing together parts of polynomials. For example, the graph of a cubic polynomial can be made to fit specified points by adjusting the coefficients of the polynomial (see Example 10). Curves obtained in this way are called cubic splines.

In computer design programs, such as Adobe Illustrator, a curve can be drawn between fixed points, called anchor points. Moving the anchor points amounts to adjusting the coefficients of a cubic polynomial.

Obtaining the domain and range of a function from its graph is explained in Section 2.3.

Solution We use the graphs in Figure 1 and transform them using the techniques of Section 2.6.

(a) The graph of $P(x) = -x^3$ is the reflection of the graph of $y = x^3$ about the x-axis, as shown in Figure 2(a). The domain is all real numbers and from the graph we see that the range is all real numbers.

(b) The graph of $Q(x) = (x - 2)^4$ is the graph of $y = x^4$ shifted 2 units to the right, as shown in Figure 2(b). The domain is all real numbers and from the graph we see that the range is $[0, \infty)$.

(c) We begin with the graph of $y = x^5$. The graph of $y = -2x^5$ is obtained by stretching the graph vertically by a factor of 2 and reflecting it about the x-axis (see the dashed blue graph in Figure 2(c)). Finally, the graph of $R(x) = -2x^5 + 4$ is obtained by shifting upward 4 units (see the red graph in Figure 2(c)). The domain is all real numbers and from the graph we see that the range is all real numbers.

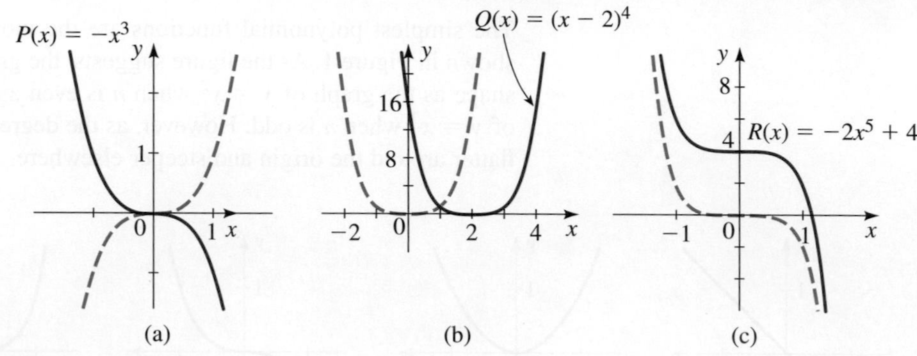

Figure 2

Now Try Exercise 5 ■

■ Graphs of Polynomial Functions: End Behavior

The graphs of polynomials of degree 0 or 1 are lines (Sections 1.10 and 2.5), and the graphs of polynomials of degree 2 are parabolas (Section 3.1). The greater the degree of a polynomial, the more complicated its graph can be. However, the graph of a polynomial function is **continuous**. This means that the graph has no break or hole (see Figure 3). Moreover, the graph of a polynomial function is a smooth curve; that is, it has no corner or sharp point (cusp) as shown in Figure 3.

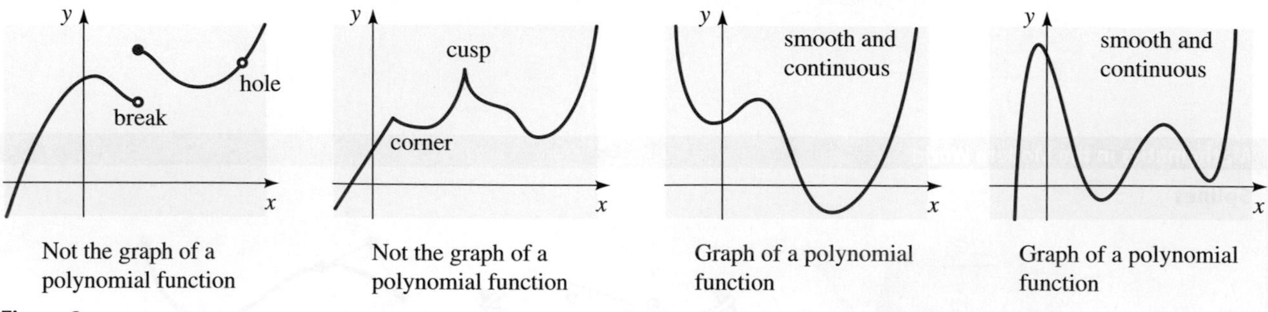

Figure 3

The domain of a polynomial function is the set of all real numbers, so we are able to sketch only a small portion of the graph. However, for values of x outside the portion of the graph we have drawn, we can describe the behavior of the graph.

The **end behavior** of a polynomial is a description of what happens as x becomes large in the positive or negative direction. To describe end behavior, we use the following **arrow notation**.

Symbol	Meaning
$x \to \infty$	x goes to infinity; that is, x increases without bound
$x \to -\infty$	x goes to negative infinity; that is, x decreases without bound

For example, the monomial $y = x^2$ in Figure 1(b) has the following end behavior.

$$y \to \infty \quad \text{as} \quad x \to \infty \qquad \text{and} \qquad y \to \infty \quad \text{as} \quad x \to -\infty$$

The monomial $y = x^3$ in Figure 1(c) has the following end behavior.

$$y \to \infty \quad \text{as} \quad x \to \infty \qquad \text{and} \qquad y \to -\infty \quad \text{as} \quad x \to -\infty$$

For any polynomial *the end behavior is determined by the term that contains the highest power of x*, because when x is large, the other terms are relatively insignificant in size. The following box shows the four possible types of end behavior, based on the highest power and the sign of its coefficient.

End Behavior of Polynomials

The end behavior of the polynomial $P(x) = a_n x^n + a_{n-1}x^{n-1} + \cdots + a_1 x + a_0$ is determined by the degree n and the sign of the leading coefficient a_n, as indicated in the following graphs.

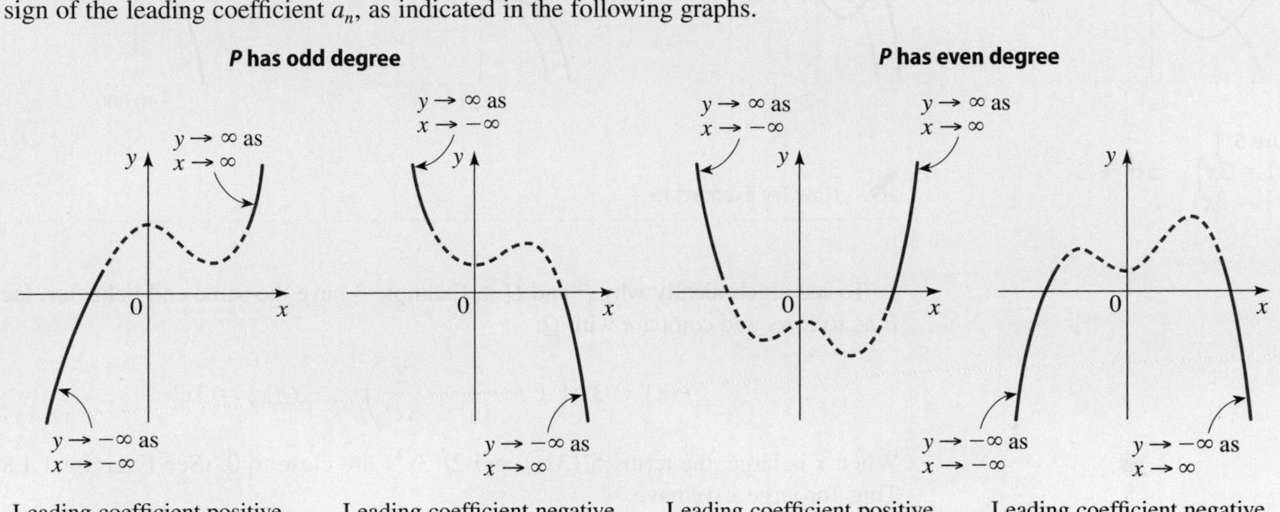

Example 2 ■ End Behavior of a Polynomial

Determine the end behavior of the polynomial

$$P(x) = -2x^4 + 5x^3 + 4x - 7$$

Solution The polynomial P has degree 4 and leading coefficient -2. Thus P has *even* degree and *negative* leading coefficient, so it has the following end behavior.

$$y \to -\infty \quad \text{as} \quad x \to -\infty \qquad \text{and} \qquad y \to -\infty \quad \text{as} \quad x \to \infty$$

The graph in Figure 4 illustrates the end behavior of P.

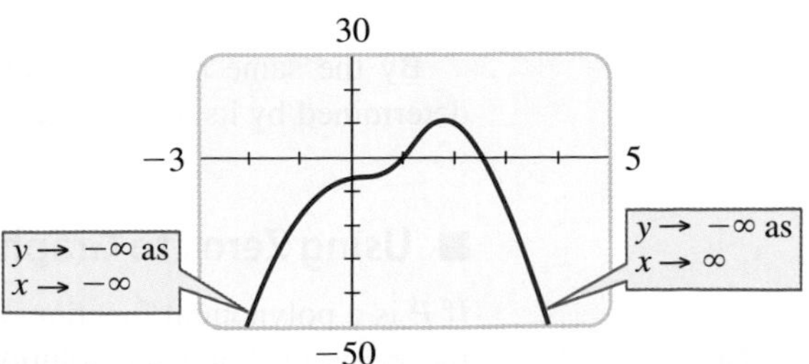

Figure 4 | $P(x) = -2x^4 + 5x^3 + 4x - 7$

Now Try Exercise 11 ■

Example 3 ■ End Behavior of a Polynomial

(a) Determine the end behavior of the polynomial $P(x) = 3x^5 - 5x^3 + 2x$.

(b) Confirm that P and its leading term $Q(x) = 3x^5$ have the same end behavior by graphing them together.

Solution

(a) Since P has odd degree and positive leading coefficient, it has the following end behavior.

$$y \to -\infty \quad \text{as} \quad x \to -\infty \qquad \text{and} \qquad y \to \infty \quad \text{as} \quad x \to \infty$$

(b) Figure 5 shows the graphs of P and Q in progressively larger viewing rectangles. The larger the viewing rectangle, the more the graphs look alike. This confirms that they have the same end behavior.

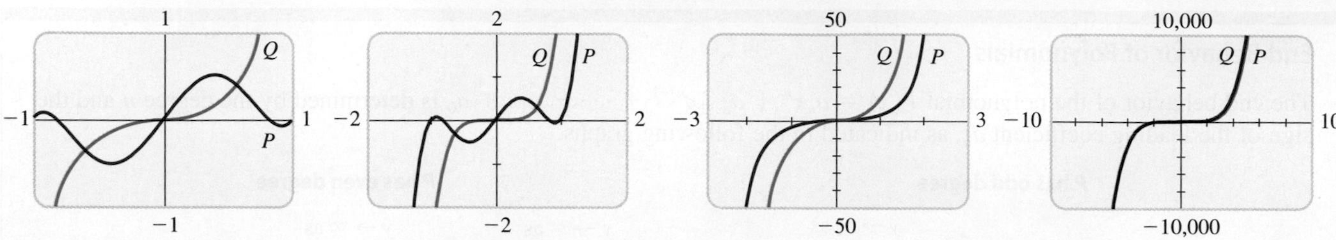

Figure 5 | $P(x) = 3x^5 - 5x^3 + 2x$
$Q(x) = 3x^5$

Now Try Exercise 45 ■

To see algebraically why P and Q in Example 3 have the same end behavior, factor P as follows and compare with Q.

$$P(x) = 3x^5\left(1 - \frac{5}{3x^2} + \frac{2}{3x^4}\right) \qquad Q(x) = 3x^5$$

When x is large, the terms $5/(3x^2)$ and $2/(3x^4)$ are close to 0. (See Exercise 1.1.89.) Thus for large x we have

$$P(x) \approx 3x^5(1 - 0 - 0) = 3x^5 = Q(x)$$

So when x is large, P and Q have approximately the same values. We can also see this numerically by making a table like the one shown below.

x	$P(x)$	$Q(x)$
15	2,261,280	2,278,125
30	72,765,060	72,900,000
50	936,875,100	937,500,000

By the same reasoning we can show that the end behavior of *any* polynomial is determined by its leading term.

■ Using Zeros to Graph Polynomials

If P is a polynomial function, then c is called a **zero** of P if $P(c) = 0$. In other words, the zeros of P are the solutions of the polynomial equation $P(x) = 0$. Note that if $P(c) = 0$, then the graph of P has an x-intercept at $x = c$, so the x-intercepts of the graph are the zeros of the function.

Real Zeros of Polynomials

If P is a polynomial and c is a real number, then the following are equivalent:

1. c is a zero of P.
2. $x = c$ is a solution of the equation $P(x) = 0$.
3. $x - c$ is a factor of $P(x)$.
4. c is an x-intercept of the graph of P.

To find the zeros of a polynomial P, we factor and then use the Zero-Product Property (see Section 1.5). For example, to find the zeros of $P(x) = x^2 + x - 6$, we factor P to get

$$P(x) = (x - 2)(x + 3)$$

From this factored form we see that

1. 2 is a zero of P.
2. $x = 2$ is a solution of the equation $x^2 + x - 6 = 0$.
3. $x - 2$ is a factor of $x^2 + x - 6$.
4. 2 is an x-intercept of the graph of P.

Similar facts are true for the other zero, -3.

The following theorem has many important consequences. (See, for instance, the *Discovery Project* referenced in Section 3.4.) Here we use it to help us graph polynomial functions.

Intermediate Value Theorem for Polynomials

If P is a polynomial function and $P(a)$ and $P(b)$ have opposite signs, then there exists at least one value c between a and b for which $P(c) = 0$.

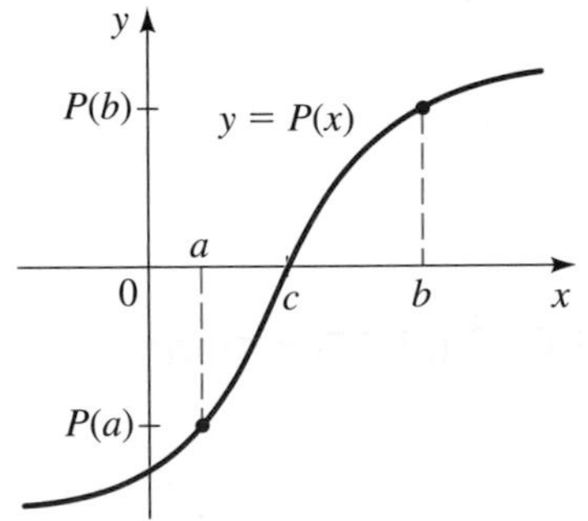

Figure 6

We will not prove this theorem, but Figure 6 shows why it is intuitively plausible.

One important consequence of this theorem is that between any two successive zeros the values of a polynomial are either all positive or all negative. That is, between two successive zeros the graph of a polynomial lies *entirely above* or *entirely below* the x-axis. To see why, suppose c_1 and c_2 are successive zeros of P. If P has both positive and negative values between c_1 and c_2, then by the Intermediate Value Theorem, P must have another zero between c_1 and c_2. But that's not possible because c_1 and c_2 are successive zeros. This observation allows us to use the following guidelines to graph polynomial functions.

Guidelines for Graphing Polynomial Functions

1. **Zeros.** Factor the polynomial to find all its real zeros: these are the x-intercepts of the graph.
2. **Test Points.** Make a table of values for the polynomial. Include test points to determine whether the graph of the polynomial lies above or below the x-axis on the intervals determined by the zeros. Include the y-intercept in the table.
3. **End Behavior.** Determine the end behavior of the polynomial.
4. **Graph.** Plot the intercepts and other points you found in the table. Sketch a smooth curve that passes through these points and exhibits the required end behavior.

Mathematics in the Modern World

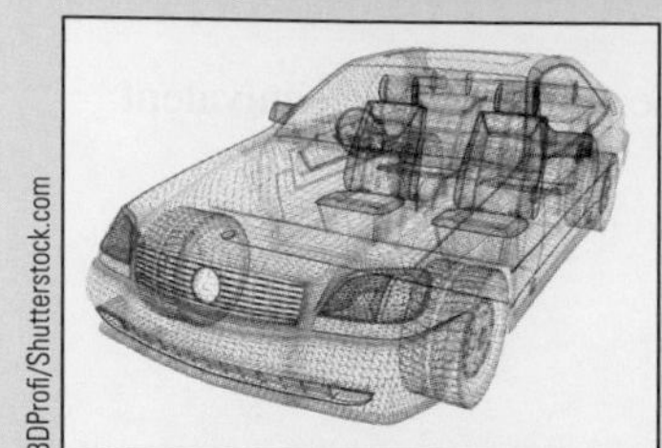

Automotive Design

Computer-aided design (CAD) has completely changed the way in which car companies design and manufacture cars. Before the 1980s automotive engineers would build a full-scale "nuts and bolts" model of a proposed new car; this was really the only way to tell whether the design was feasible. Today automotive engineers build a mathematical model, one that exists only in the memory of a computer. The model incorporates all the main design features of the car. Certain polynomial curves, called *splines* (see *Mathematics in the Modern World* earlier in this section), are used in shaping the body of the car. The resulting "mathematical car" can be tested for structural stability, handling, aerodynamics, suspension response, and more. All this testing is done before a prototype is built. As you can imagine, CAD saves car manufacturers millions of dollars each year. More importantly, CAD gives automotive engineers far more flexibility in design; desired changes can be created and tested within minutes. With the help of computer graphics, designers can see how good the "mathematical car" looks before they build the real one. Moreover, the mathematical car can be viewed from any perspective; it can be moved, rotated, or viewed from the inside. These manipulations of the car on the computer monitor translate mathematically into solving large systems of linear equations.

Example 4 ■ Using Zeros to Graph a Polynomial Function

Sketch the graph of the polynomial function $P(x) = (x + 2)(x - 1)(x - 3)$.

Solution The zeros are $x = -2$, 1, and 3. These determine the intervals $(-\infty, -2)$, $(-2, 1)$, $(1, 3)$, and $(3, \infty)$. Using test points in these intervals, we get the information in the following sign diagram (see Section 1.8).

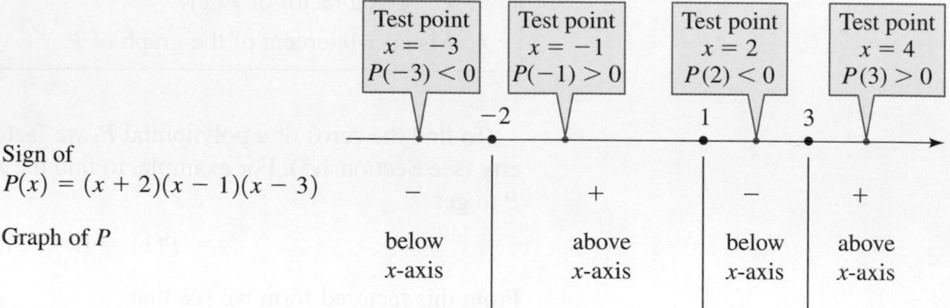

Plotting a few additional points and connecting them with a smooth curve helps us to complete the graph in Figure 7.

	x	$P(x)$
Test point →	-3	-24
	-2	0
Test point →	-1	8
	0	6
	1	0
Test point →	2	-4
	3	0
Test point →	4	18

Test point $P(-1) > 0$

Test point $P(4) > 0$

Test point $P(-3) < 0$

Test point $P(2) < 0$

Figure 7 | $P(x) = (x + 2)(x - 1)(x - 3)$

 **Now Try Exercise 17** ■

Example 5 ■ Finding Zeros and Graphing a Polynomial Function

Let $P(x) = x^3 - 2x^2 - 3x$.

(a) Find the zeros of P. **(b)** Sketch a graph of P.

Solution

(a) To find the zeros, we factor completely.

$$\begin{aligned} P(x) &= x^3 - 2x^2 - 3x \\ &= x(x^2 - 2x - 3) && \text{Factor } x \\ &= x(x - 3)(x + 1) && \text{Factor quadratic} \end{aligned}$$

Thus the zeros are $x = 0$, $x = 3$, and $x = -1$.

(b) The x-intercepts are $x = 0$, $x = 3$, and $x = -1$. The y-intercept is $P(0) = 0$. We make a table of values of $P(x)$, making sure that we choose test points between (and to the right and left of) successive zeros.

Since P is of odd degree and its leading coefficient is positive, it has the following end behavior:

$$y \to -\infty \quad \text{as} \quad x \to -\infty \qquad \text{and} \qquad y \to \infty \quad \text{as} \quad x \to \infty$$

We plot the points in the table and connect them by a smooth curve to complete the graph, as shown in Figure 8.

	x	$P(x)$
Test point →	-2	-10
	-1	0
Test point →	$-\frac{1}{2}$	$\frac{7}{8}$
	0	0
Test point →	1	-4
	2	-6
	3	0
Test point →	4	20

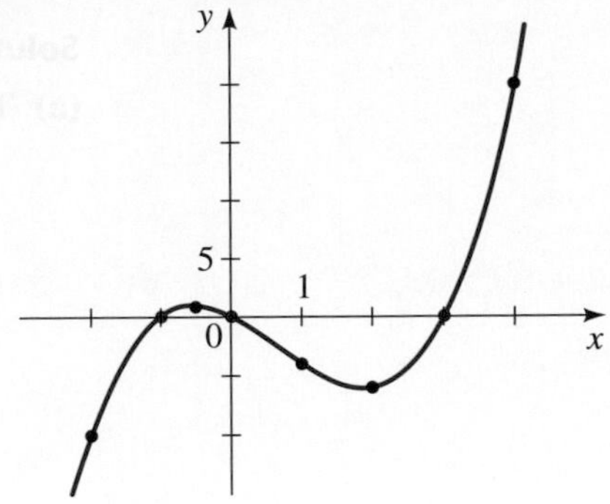

Figure 8 | $P(x) = x^3 - 2x^2 - 3x$

Now Try Exercise 31 ■

Example 6 ■ Finding Zeros and Graphing a Polynomial Function

Let $P(x) = -2x^4 - x^3 + 3x^2$.

(a) Find the zeros of P. **(b)** Sketch a graph of P.

Solution

(a) To find the zeros, we factor completely.

$$\begin{aligned} P(x) &= -2x^4 - x^3 + 3x^2 \\ &= -x^2(2x^2 + x - 3) && \text{Factor } -x^2 \\ &= -x^2(2x + 3)(x - 1) && \text{Factor quadratic} \end{aligned}$$

Thus the zeros are $x = 0$, $x = -\frac{3}{2}$, and $x = 1$.

(b) The x-intercepts are $x = 0$, $x = -\frac{3}{2}$, and $x = 1$. The y-intercept is $P(0) = 0$. We make a table of values of $P(x)$, making sure that we choose test points between (and to the right and left of) successive zeros.

Since P is of even degree and its leading coefficient is negative, it has the following end behavior.

$$y \to -\infty \quad \text{as} \quad x \to -\infty \qquad \text{and} \qquad y \to -\infty \quad \text{as} \quad x \to \infty$$

We plot the points from the table and connect the points by a smooth curve to complete the graph in Figure 9.

x	$P(x)$
-2	-12
-1.5	0
-1	2
-0.5	0.75
0	0
0.5	0.5
1	0
1.5	-6.75

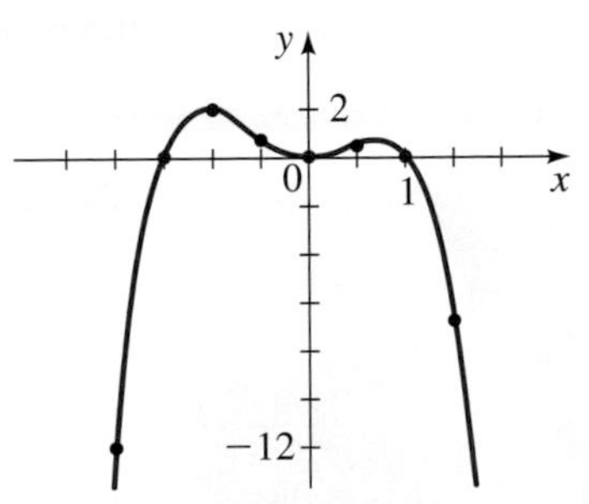

Figure 9 | $P(x) = -2x^4 - x^3 + 3x^2$

Now Try Exercise 35 ■

Example 7 ■ Finding Zeros and Graphing a Polynomial Function

Let $P(x) = x^3 - 2x^2 - 4x + 8$.

(a) Find the zeros of P. **(b)** Sketch a graph of P.

Solution

(a) To find the zeros, we factor completely.

$$\begin{aligned} P(x) &= x^3 - 2x^2 - 4x + 8 \\ &= x^2(x-2) - 4(x-2) && \text{Group and factor} \\ &= (x^2-4)(x-2) && \text{Factor } x-2 \\ &= (x+2)(x-2)(x-2) && \text{Difference of squares} \\ &= (x+2)(x-2)^2 && \text{Simplify} \end{aligned}$$

Thus the zeros are $x = -2$ and $x = 2$.

(b) The x-intercepts are $x = -2$ and $x = 2$. The y-intercept is $P(0) = 8$. The table gives additional values of $P(x)$.

Since P is of odd degree and its leading coefficient is positive, it has the following end behavior.

$$y \to -\infty \quad \text{as} \quad x \to -\infty \qquad \text{and} \qquad y \to \infty \quad \text{as} \quad x \to \infty$$

We connect the points by a smooth curve to complete the graph in Figure 10.

x	$P(x)$
−3	−25
−2	0
−1	9
0	8
1	3
2	0
3	5

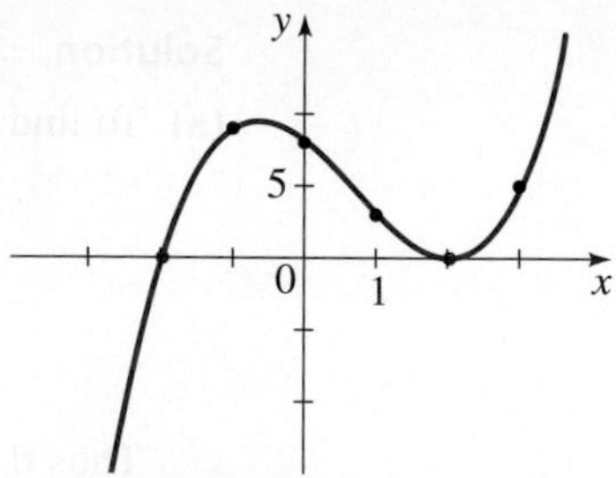

Figure 10 | $P(x) = x^3 - 2x^2 - 4x + 8$

 Now Try Exercise 37 ■

■ Shape of the Graph Near a Zero

Although $x = 2$ is a zero of the polynomial in Example 7, the graph does not cross the x-axis at the x-intercept 2. This is because the factor $(x - 2)^2$ corresponding to that zero is raised to an even power, so it doesn't change sign as we test points on either side of 2. In the same way the graph does not cross the x-axis at $x = 0$ in Example 6.

In general, if c is a zero of P and the corresponding factor $x - c$ occurs exactly m times in the factorization of P, then we say that c is a **zero of multiplicity** m. By

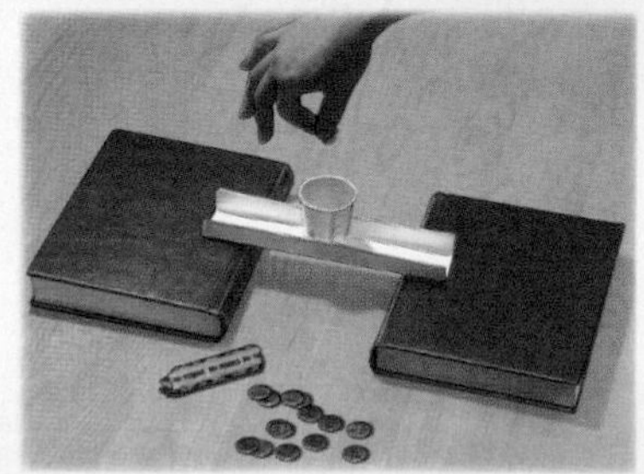

Discovery Project ■ Bridge Science

If you want to build a bridge, how can you be sure that your bridge design is strong enough to support the vehicles that will drive over it? In this project we perform a simple experiment using paper "bridges" to collect data on the weight our bridges can support. We model the data with linear and power functions to determine which model best fits the data. The model we obtain allows us to predict the strength of a large bridge *before* it is built. You can find the project at **www.stewartmath.com**.

considering test points on either side of the x-intercept c, we conclude that the graph crosses the x-axis at c if the multiplicity m is odd and does not cross the x-axis if m is even. Moreover, it can be shown by using calculus that near $x = c$ the graph has the same general shape as the graph of $y = A(x - c)^m$.

Shape of the Graph Near a Zero of Multiplicity m

If c is a zero of P of multiplicity m, then the shape of the graph of P near c is as follows.

Multiplicity of c | **Shape of the graph of P near the x-intercept c**

m odd, $m > 1$

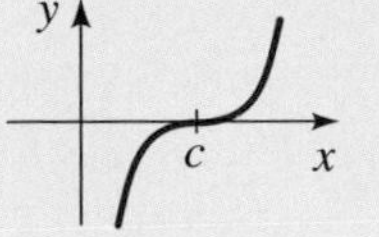

OR

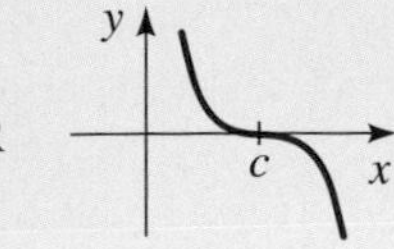

m even, $m > 1$

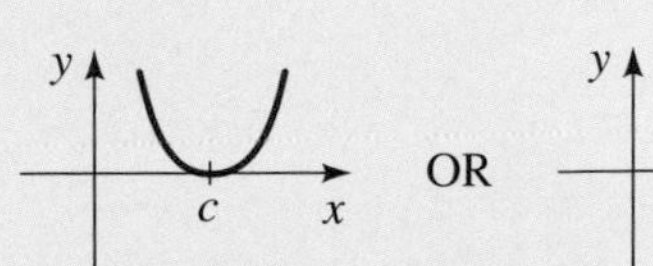

Example 8 ■ Graphing a Polynomial Function Using Its Zeros

Graph the polynomial $P(x) = x^4(x - 2)^3(x + 1)^2$.

Solution The zeros of P are -1, 0, and 2 with multiplicities 2, 4, and 3, respectively:

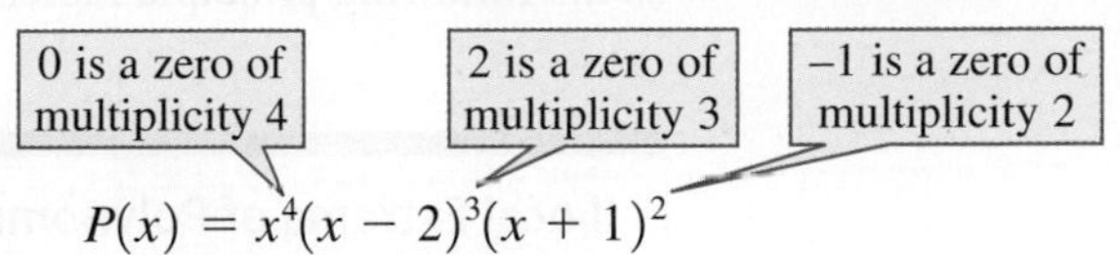

The zero 2 has *odd* multiplicity, so the graph crosses the x-axis at the x-intercept 2. But the zeros 0 and -1 have *even* multiplicities, so the graph does not cross the x-axis at the x-intercepts 0 and -1.

Since P is a polynomial of degree 9 and has positive leading coefficient, it has the following end behavior:

$$y \to -\infty \quad \text{as} \quad x \to -\infty \qquad \text{and} \qquad y \to \infty \quad \text{as} \quad x \to \infty$$

With this information and a table of values we sketch the graph in Figure 11.

x	$P(x)$
-1.3	-9.2
-1	0
-0.5	-0.2
0	0
1	-4
2	0
2.3	8.2

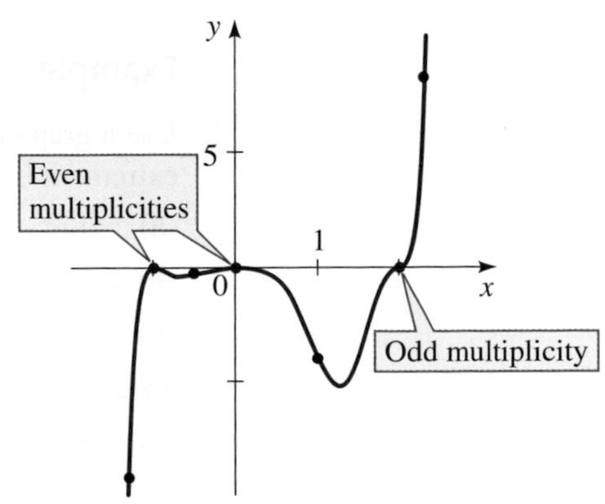

Figure 11 | $P(x) = x^4(x - 2)^3(x + 1)^2$

Now Try Exercise 29 ■

■ Local Maxima and Minima of Polynomials

Recall from Section 2.3 that if the point $(a, f(a))$ is the highest point on the graph of f within some viewing rectangle, then $f(a)$ is a local maximum value of f, and if $(b, f(b))$ is the lowest point on the graph of f within a viewing rectangle, then $f(b)$ is a local minimum value (see Figure 12). We say that such a point $(a, f(a))$ is a **local maximum point** on the graph and that $(b, f(b))$ is a **local minimum point**. The local maximum and minimum points on the graph of a function are called its **local extrema**.

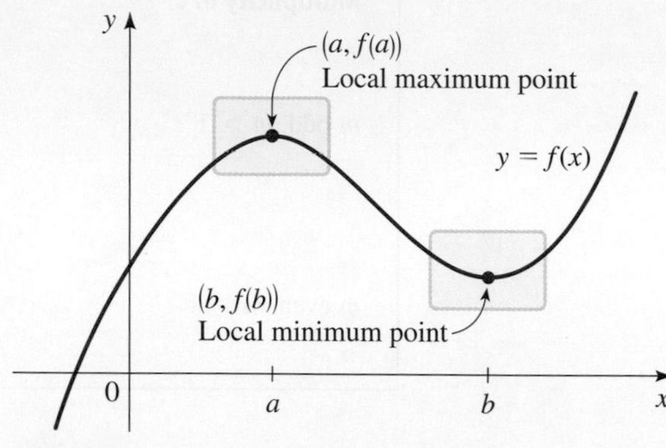

Figure 12

For a polynomial function the number of local extrema must be less than the degree, as the following principle indicates. (A proof of this principle requires calculus.)

> **Local Extrema of Polynomials**
>
> If $P(x) = a_n x^n + a_{n-1}x^{n-1} + \cdots + a_1 x + a_0$ is a polynomial of degree n, then the graph of P has at most $n - 1$ local extrema.

A polynomial of degree n may in fact have fewer than $n - 1$ local extrema. For example, $P(x) = x^5$ (graphed in Figure 1) has *no* local extrema, even though it is of degree 5. The preceding principle tells us only that a polynomial of degree n can have no more than $n - 1$ local extrema.

Example 9 ■ The Number of Local Extrema

Use a graphing device to graph the polynomial and determine how many local extrema it has.

(a) $P_1(x) = x^4 + x^3 - 16x^2 - 4x + 48$

(b) $P_2(x) = x^5 + 3x^4 - 5x^3 - 15x^2 + 4x - 15$

(c) $P_3(x) = 7x^4 + 3x^2 - 10x$

Solution The graphs are shown in Figure 13 on the next page.

(a) P_1 has two local minimum points and one local maximum point, for a total of three local extrema.

(b) P_2 has two local minimum points and two local maximum points, for a total of four local extrema.

(c) P_3 has just one local extremum, a local minimum.

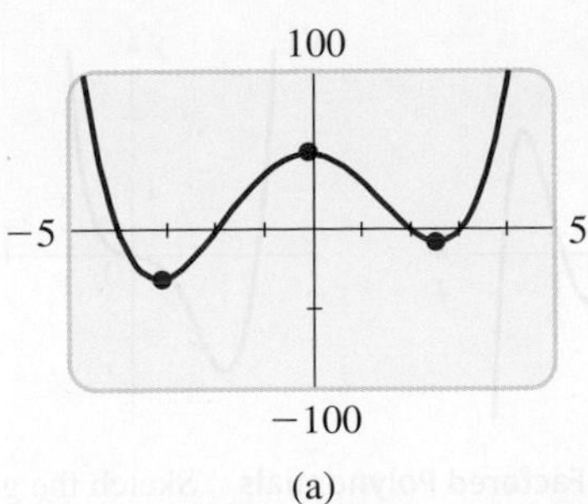

(a)
$P_1(x) = x^4 + x^3 - 16x^2 - 4x + 48$

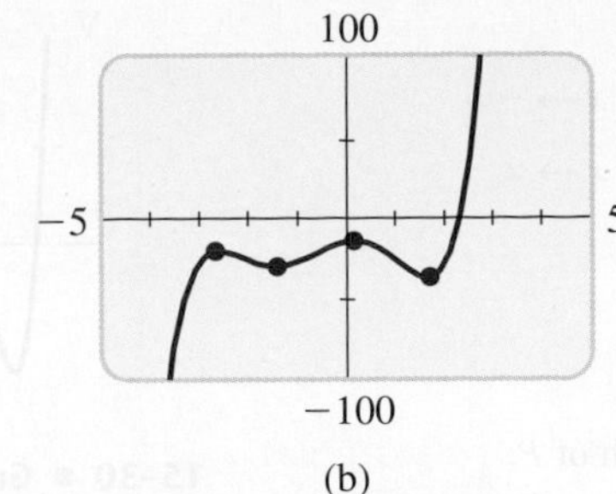

(b)
$P_2(x) = x^5 + 3x^4 - 5x^3 - 15x^2 + 4x - 15$

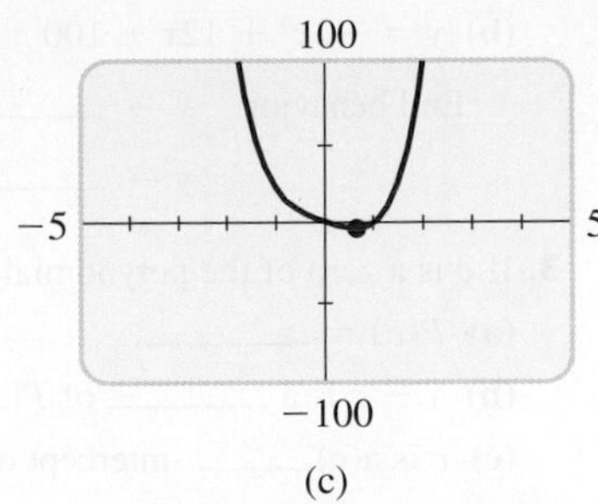

(c)
$P_3(x) = 7x^4 + 3x^2 - 10x$

Figure 13

Now Try Exercises 65 and 67 ■

With a graphing device we can quickly draw the graphs of many functions at once, on the same viewing screen. This allows us to see how changing a value in the definition of the functions affects the shape of its graph. In the next example we apply this principle to a family of third-degree polynomials.

Example 10 ■ A Family of Polynomials

Sketch the family of polynomials $P(x) = x^3 - cx^2$ for $c = 0$, 1, 2, and 3. How does changing the value of c affect the graph?

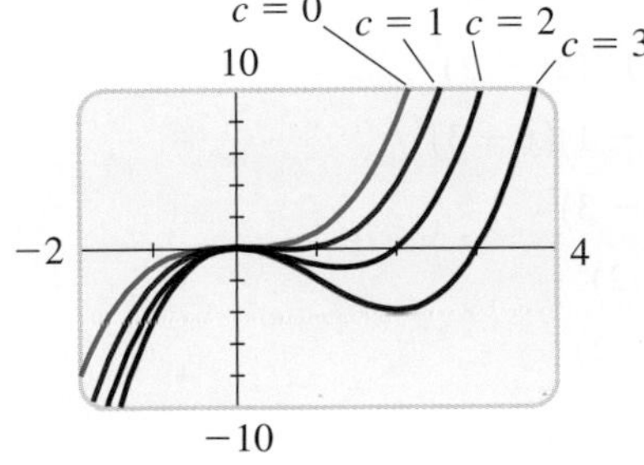

Figure 14 | A family of polynomials $P(x) = x^3 - cx^2$

Solution The polynomials

$$P_0(x) = x^3 \qquad P_1(x) = x^3 - x^2$$
$$P_2(x) = x^3 - 2x^2 \qquad P_3(x) = x^3 - 3x^2$$

are graphed in Figure 14. We see that increasing the value of c causes the graph to develop an increasingly deep "valley" to the right of the y-axis, creating a local maximum at the origin and a local minimum at a point in Quadrant IV. This local minimum moves lower and farther to the right as c increases. To see why this happens, factor $P(x) = x^2(x - c)$. The polynomial P has zeros at 0 and c, and the larger c gets, the farther to the right the minimum between 0 and c will be.

Now Try Exercise 75 ■

3.2 | Exercises

■ Concepts

1. Only one of the following four graphs could be the graph of a polynomial function. Which one? Why are the others not graphs of polynomials?

I
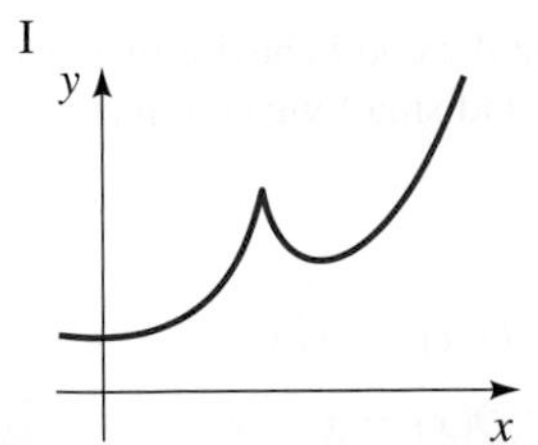

II
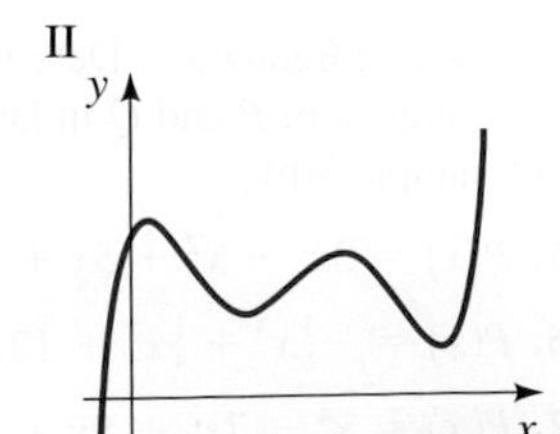

III
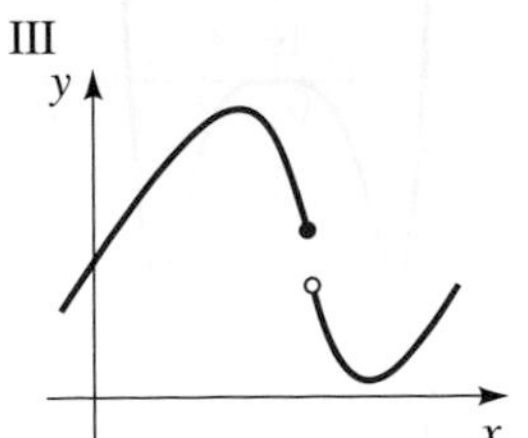

IV
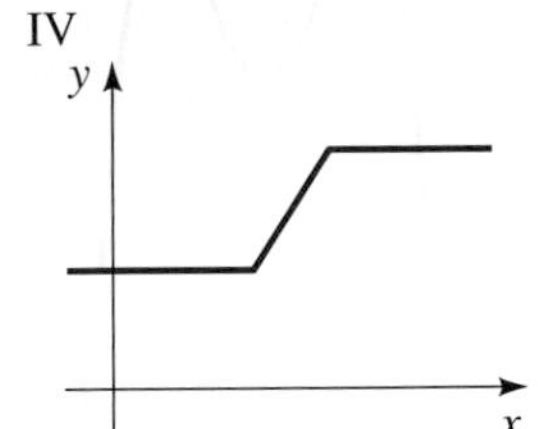

2. Describe the end behavior of each polynomial.

(a) $y = x^3 - 8x^2 + 2x - 15$

End behavior: $y \to$ ______ as $x \to -\infty$

$y \to$ ______ as $x \to \infty$

(b) $y = -2x^4 + 12x + 100$

End behavior: $y \to$ ______ as $x \to -\infty$

$y \to$ ______ as $x \to \infty$

3. If c is a zero of the polynomial P, then

(a) $P(c) =$ ______.

(b) $x - c$ is a ______ of $P(x)$.

(c) c is a(n) ______ -intercept of the graph of P.

4. Which of the following statements couldn't possibly be true about the polynomial function P?

(a) P has degree 3, two local maxima, and two local minima.

(b) P has degree 3 and no local maxima or minima.

(c) P has degree 4, one local maximum, and no local minima.

Skills

5–8 ■ Transformations of Monomials Sketch the graph of each function by transforming the graph of an appropriate function of the form $y = x^n$ from Figure 1. Indicate all x- and y-intercepts on each graph, and state the domain and range.

5. (a) $P(x) = \frac{1}{2}x^2 - 2$ **(b)** $Q(x) = 2(x - 3)^2$
(c) $R(x) = 4x^2 + 1$ **(d)** $S(x) = -(x + 2)^2$

6. (a) $P(x) = -x^4 + 1$ **(b)** $Q(x) = (x + 1)^4$
(c) $R(x) = 6x^4 - 6$ **(d)** $S(x) = \frac{1}{9}(x - 3)^4$

7. (a) $P(x) = x^3 - 8$ **(b)** $Q(x) = -x^3 + 27$
(c) $R(x) = -(x + 2)^3$ **(d)** $S(x) = \frac{1}{2}(x - 1)^3 + 4$

8. (a) $P(x) = (x + 3)^5$ **(b)** $Q(x) = 2(x + 3)^5 - 64$
(c) $R(x) = -\frac{1}{2}(x - 2)^5$ **(d)** $S(x) = -\frac{1}{2}(x - 2)^5 + 16$

9–14 ■ End Behavior A polynomial function is given.
(a) Describe the end behavior of the polynomial function.
(b) Match the polynomial function with one of the graphs I–VI.

9. $P(x) = x(x^2 - 4)$ **10.** $Q(x) = -x^2(x^2 - 4)$

11. $R(x) = -x^5 + 5x^3 - 4x$ **12.** $S(x) = \frac{1}{2}x^6 - 2x^4$

13. $T(x) = x^4 + 2x^3$ **14.** $U(x) = -x^3 + 2x^2$

I

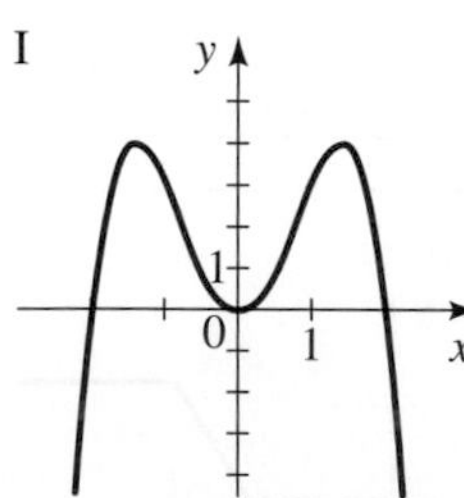

II

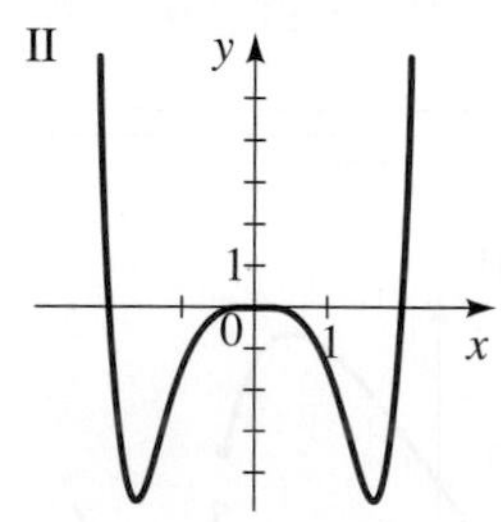

III

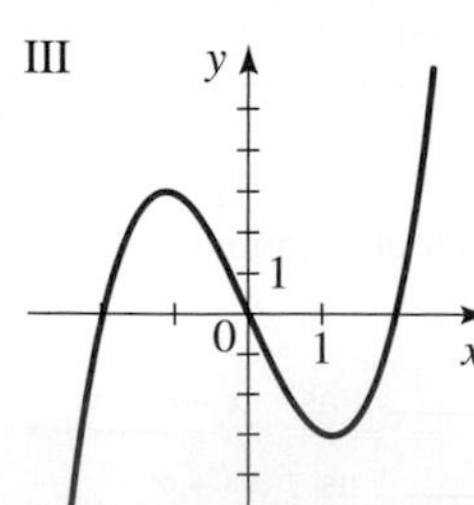

IV

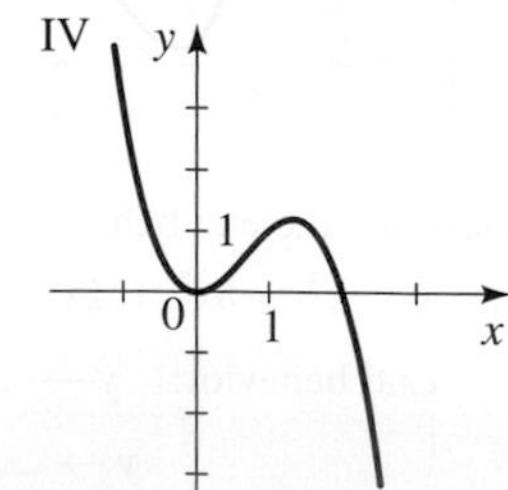

V

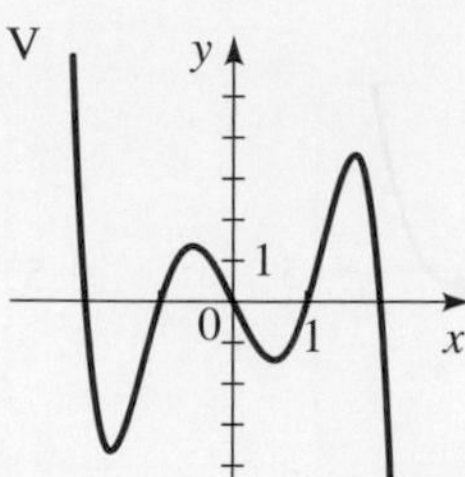

VI

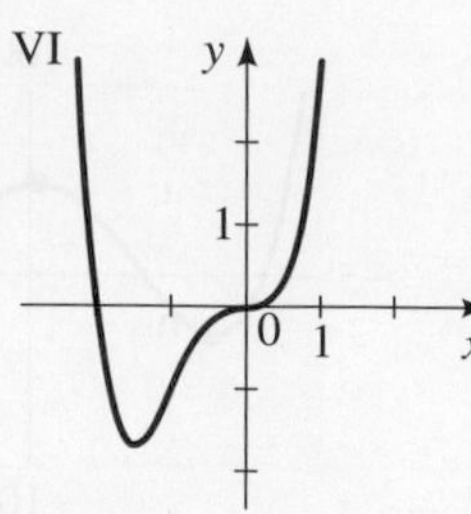

15–30 ■ Graphing Factored Polynomials Sketch the graph of the polynomial function. Make sure your graph shows all intercepts and exhibits the proper end behavior.

15. $P(x) = (x + 2)(x - 5)$

16. $P(x) = (4 - x)(x + 1)$

17. $P(x) = -x(x - 2)(x + 3)$

18. $P(x) = (x + 3)(x - 1)(x - 4)$

19. $P(x) = -(2x - 1)(x + 1)(x + 3)$

20. $P(x) = (x - 3)(x + 2)(3x - 2)$

21. $P(x) = (x + 2)(x + 1)(x - 2)(x - 3)$

22. $P(x) = x(x + 1)(x - 1)(2 - x)$

23. $P(x) = -2x(x - 2)^2$

24. $P(x) = \frac{1}{5}x(x - 5)^2$

25. $P(x) = (x + 2)(x + 1)^2(2x - 3)$

26. $P(x) = -(x + 2)^3(x - 1)(x - 3)^2$

27. $P(x) = \frac{1}{12}(x + 2)^2(x - 3)^2$

28. $P(x) = (x - 1)^2(x + 2)^3$

29. $P(x) = x(x - 2)^2(x + 2)^3$

30. $P(x) = \frac{1}{9}x^2(x - 3)^2(x + 3)^2$

31–44 ■ Graphing Polynomials Factor the polynomial and use the factored form to find the zeros. Then sketch the graph.

31. $P(x) = x^3 - x^2 - 6x$ **32.** $P(x) = x^3 + 2x^2 - 8x$

33. $P(x) = -x^3 + x^2 + 12x$ **34.** $P(x) = -2x^3 - x^2 + x$

35. $P(x) = x^4 - 3x^3 + 2x^2$ **36.** $P(x) = x^5 - 9x^3$

37. $P(x) = x^3 + x^2 - x - 1$

38. $P(x) = x^3 + 3x^2 - 4x - 12$

39. $P(x) = 2x^3 - x^2 - 18x + 9$

40. $P(x) = \frac{1}{8}(2x^4 + 3x^3 - 16x - 24)^2$

41. $P(x) = x^4 - 2x^3 - 8x + 16$

42. $P(x) = x^4 - 2x^3 + 8x - 16$

43. $P(x) = x^4 - 3x^2 - 4$ **44.** $P(x) = x^6 - 2x^3 + 1$

45–50 ■ End Behavior Determine the end behavior of P. Compare the graphs of P and Q in large and small viewing rectangles, as in Example 3(b).

45. $P(x) = 3x^3 - x^2 + 5x + 1$; $Q(x) = 3x^3$

46. $P(x) = -\frac{1}{8}x^3 + \frac{1}{4}x^2 + 12x$; $Q(x) = -\frac{1}{8}x^3$

47. $P(x) = x^4 - 7x^2 + 5x + 5$; $Q(x) = x^4$

48. $P(x) = -x^5 + 2x^2 + x;\quad Q(x) = -x^5$

49. $P(x) = x^{11} - 9x^9;\quad Q(x) = x^{11}$

50. $P(x) = 2x^2 - x^{12};\quad Q(x) = -x^{12}$

51–54 ■ Local Extrema The graph of a polynomial function is given. From the graph, find **(a)** the x- and y-intercepts, **(b)** the coordinates of all local extrema, and **(c)** the domain and range.

51. $P(x) = -x^2 + 4x$

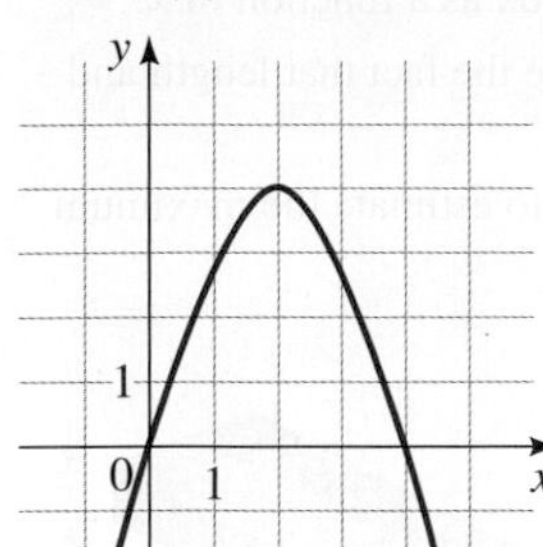

52. $P(x) = \frac{2}{9}x^3 - x^2$

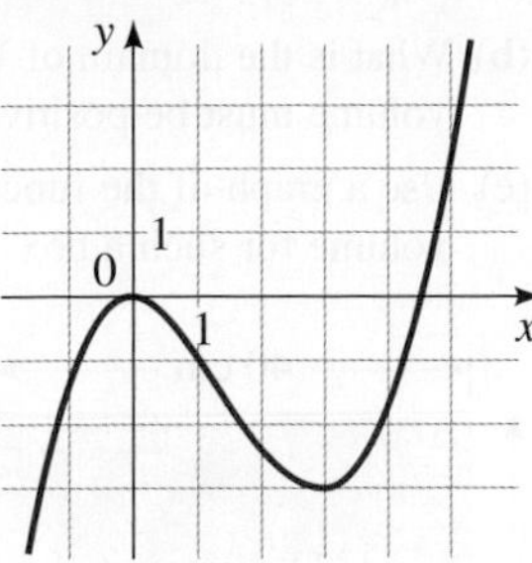

53. $P(x) = -\frac{1}{2}x^3 + \frac{3}{2}x - 1$

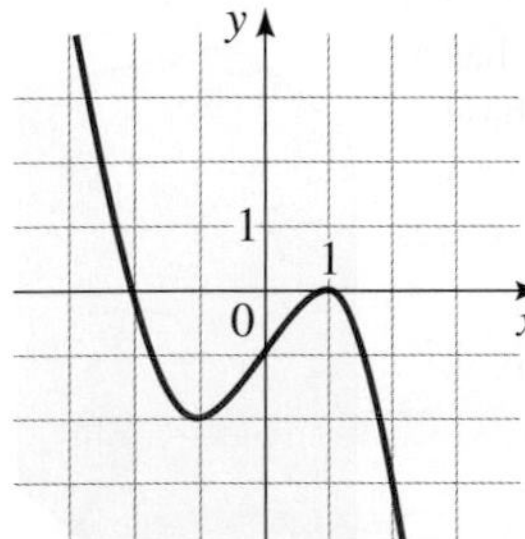

54. $P(x) = \frac{1}{9}x^4 - \frac{4}{9}x^3$

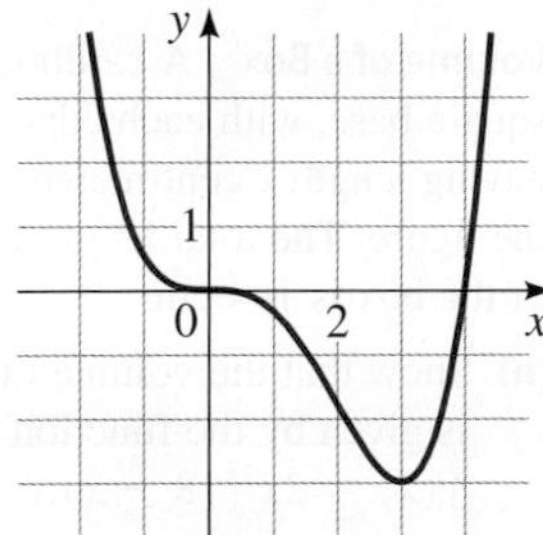

55–62 ■ Local Extrema Graph the polynomial in the given viewing rectangle. Find the coordinates of all local extrema, rounded to two decimal places. State the domain and range.

55. $y = 10x - x^2,\quad [-2, 12]$ by $[-15, 30]$

56. $y = x^3 - 3x^2,\quad [-2, 5]$ by $[-10, 10]$

57. $y = x^3 - 12x + 9,\quad [-5, 5]$ by $[-30, 30]$

58. $y = 2x^3 - 3x^2 - 12x - 32,\quad [-5, 5]$ by $[-60, 30]$

59. $y = x^5 - 9x^3,\quad [-4, 4]$ by $[-50, 50]$

60. $y = x^4 - 18x^2 + 32,\quad [-5, 5]$ by $[-100, 100]$

61. $y = 3x^5 - 5x^3 + 3,\quad [-3, 3]$ by $[-5, 10]$

62. $y = x^5 - 5x^2 + 6,\quad [-3, 3]$ by $[-5, 10]$

63–72 ■ Number of Local Extrema Use a graph to determine the number of local maximums and minimums the polynomial has.

63. $y = -2x^2 + 3x + 5$

64. $y = x^3 + 12x$

65. $y = x^3 - x^2 - x$

66. $y = 6x^3 + 3x + 1$

67. $y = x^4 - 5x^2 + 4$

68. $y = 1.2x^5 + 3.75x^4 - 7x^3 - 15x^2 + 18x$

69. $y = (x - 2)^5 + 32$

70. $y = (x^2 - 2)^3$

71. $y = x^8 - 3x^4 + x$

72. $y = \frac{1}{3}x^7 - 17x^2 + 7$

73–78 ■ Families of Polynomials Graph the family of polynomials in the same viewing rectangle, using the given values of c. Explain how changing the value of c affects the graph.

73. $P(x) = cx^3;\quad c = 1, 2, 5, \frac{1}{2}$

74. $P(x) = (x - c)^4;\quad c = -1, 0, 1, 2$

75. $P(x) = x^4 + c;\quad c = -1, 0, 1, 2$

76. $P(x) = x^3 + cx;\quad c = 2, 0, -2, -4$

77. $P(x) = x^4 - cx;\quad c = 0, 1, 8, 27$

78. $P(x) = x^c;\quad c = 1, 3, 5, 7$

Skills Plus

79. Intersection Points of Two Polynomials

(a) On the same coordinate axes, sketch graphs of the given functions as accurately as possible:

$$y = x^3 - 2x^2 - x + 2 \quad \text{and} \quad y = -x^2 + 5x + 2$$

(b) On the basis of your graphs in part (a), at how many points do the two graphs appear to intersect?

(c) Find the coordinates of all intersection points.

80. Power Functions Portions of the graphs of $y = x^2$, $y = x^3$, $y = x^4$, $y = x^5$, and $y = x^6$ are plotted in the figures. Determine which function belongs to each graph.

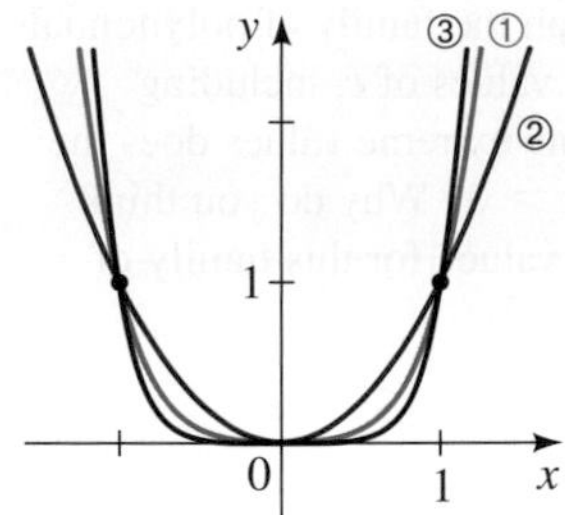

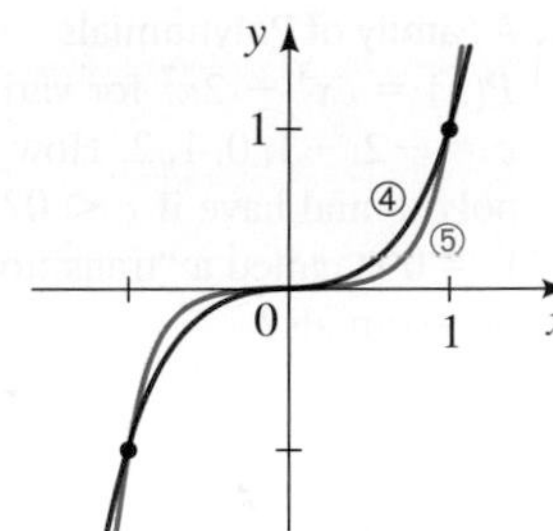

81. Odd and Even Functions Recall that a function f is *odd* if $f(-x) = -f(x)$ or *even* if $f(-x) = f(x)$ for all real x.

(a) Show that a polynomial $P(x)$ that contains only odd powers of x is an odd function.

(b) Show that a polynomial $P(x)$ that contains only even powers of x is an even function.

(c) Show that if a polynomial $P(x)$ contains both odd and even powers of x, then it is neither an odd nor an even function.

(d) Express the function

$$P(x) = x^5 + 6x^3 - x^2 - 2x + 5$$

as the sum of an odd function and an even function.

82. Number of Intercepts and Local Extrema

(a) How many x-intercepts and how many local extrema does the polynomial $P(x) = x^3 - 4x$ have?

(b) How many x-intercepts and how many local extrema does the polynomial $Q(x) = x^3 + 4x$ have?

(c) If $a > 0$, how many x-intercepts and how many local extrema does each of the polynomials $P(x) = x^3 - ax$ and $Q(x) = x^3 + ax$ have? Explain your answer.

83–84 ■ Local Extrema These exercises involve local maximums and minimums of polynomial functions.

83. **(a)** Graph the function $P(x) = (x - 1)(x - 3)(x - 4)$ and find all local extrema, correct to the nearest tenth.

(b) Graph the function

$$Q(x) = (x - 1)(x - 3)(x - 4) + 5$$

and use your answers to part (a) to find all local extrema, correct to the nearest tenth.

84. **(a)** Graph the function $P(x) = (x - 2)(x - 4)(x - 5)$ and determine how many local extrema it has.

(b) If $a < b < c$, explain why the function

$$P(x) = (x - a)(x - b)(x - c)$$

must have two local extrema.

85. Maximum Number of Local Extrema What is the smallest possible degree that the polynomial whose graph is shown can have? Explain.

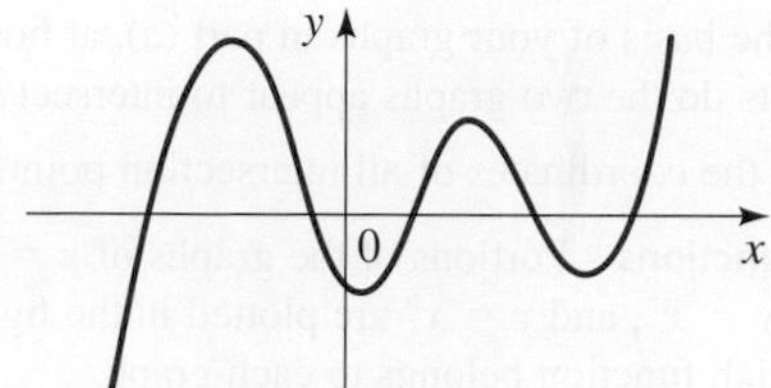

86. A Family of Polynomials Graph the family of polynomials $P(x) = cx^4 - 2x^2$ for various values of c, including $c = -2, -1, 0, 1, 2$. How many extreme values does the polynomial have if $c < 0$? If $c > 0$? Why do you think $c = 0$ is called a "transitional value" for this family of polynomials?

Applications

87. Market Research A market analyst working for a small-appliance manufacturer finds that if the firm produces and sells x blenders annually, the total profit (in dollars) is

$$P(x) = 8x + 0.3x^2 - 0.0013x^3 - 372$$

Graph the function P in an appropriate viewing rectangle and use the graph to answer the following questions.

(a) When just a few blenders are manufactured, the firm loses money (profit is negative). (For example, $P(10) = -263.3$, so the firm loses $263.30 if it produces and sells only 10 blenders.) How many blenders must the firm produce to break even?

(b) Does profit increase indefinitely as more blenders are produced and sold? If not, what is the largest possible profit the firm could have?

88. Population Change The rabbit population on a small island is observed to be given by the function

$$P(t) = 120t - 0.4t^4 + 1000$$

where t is the time (in months) since observations of the island began.

(a) What is the maximum population attained, and when is that maximum population?

(b) When does the rabbit population disappear from the island?

89. Volume of a Box An open box is to be constructed from a piece of cardboard 20 cm by 40 cm by cutting squares of side length x from each corner and folding up the sides, as shown in the figure.

(a) Express the volume V of the box as a function of x.

(b) What is the domain of V? (Use the fact that length and volume must be positive.)

(c) Use a graph of the function V to estimate the maximum volume for such a box.

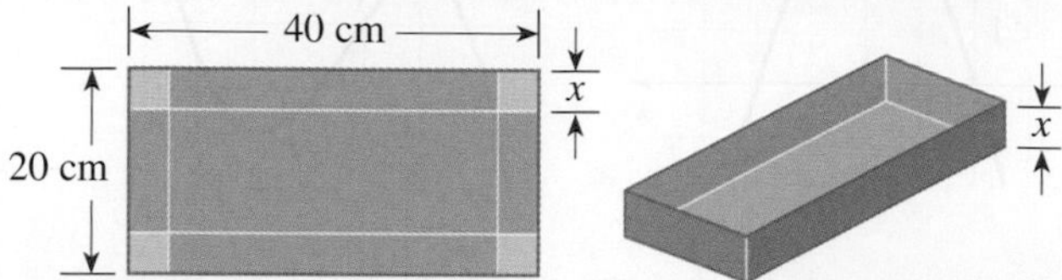

90. Volume of a Box A cardboard box has a square base, with each edge of the base having length x centimeters, as shown in the figure. The total length of all 12 edges of the box is 144 cm.

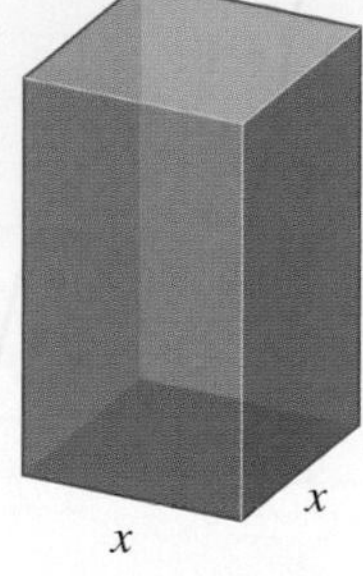

(a) Show that the volume of the box is given by the function $V(x) = 2x^2(18 - x)$.

(b) What is the domain of V? (Use the fact that length and volume must be positive.)

(c) Draw a graph of the function V and use it to estimate the maximum volume for such a box.

Discuss ■ Discover ■ Prove ■ Write

91. Discover: Graphs of Large Powers Graph the functions $y = x^2$, $y = x^3$, $y = x^4$, and $y = x^5$, for $-1 \leq x \leq 1$, on the same coordinate axes. What do you think the graph of $y = x^{100}$ would look like on this same interval? What about $y = x^{101}$? Make a table of values to confirm your answers.

92. Discuss ■ Discover: Possible Number of Local Extrema Is it possible for a third-degree polynomial to have exactly one local extremum? Is it possible for any polynomial to have two local maximums and no local minimum? Explain. Give an example of a polynomial that has six local extrema.

93. Discover ■ Prove: Fixed Points A *fixed point* of a function f is a number x for which $f(x) = x$. (See Exercise 2.3.76.) Show that if f is a continuous function with domain [0, 1] and range contained in [0, 1], then f has a fixed point in [0, 1].

PS *Draw a diagram.* Since f is continuous, its graph has no break or hole. Draw a graph of f and make a graphical argument.

3.3 Dividing Polynomials

■ Long Division of Polynomials ■ Synthetic Division ■ The Remainder and Factor Theorems

So far in this chapter we have been studying polynomial functions *graphically*. In this section we begin to study polynomials *algebraically*. Most of our work will be concerned with factoring polynomials, and to factor, we need to know how to divide polynomials.

■ Long Division of Polynomials

Dividing polynomials is much like the familiar process of dividing numbers. When we divide 38 by 7, the quotient is 5 and the remainder is 3. We write

$$\begin{array}{r} 5 \\ 7\overline{)38} \\ \underline{35} \\ 3 \end{array}$$

$$\frac{\overset{\text{Dividend}}{38}}{\underset{\text{Divisor}}{7}} = \underset{\text{Quotient}}{5} + \frac{\overset{\text{Remainder}}{3}}{7}$$

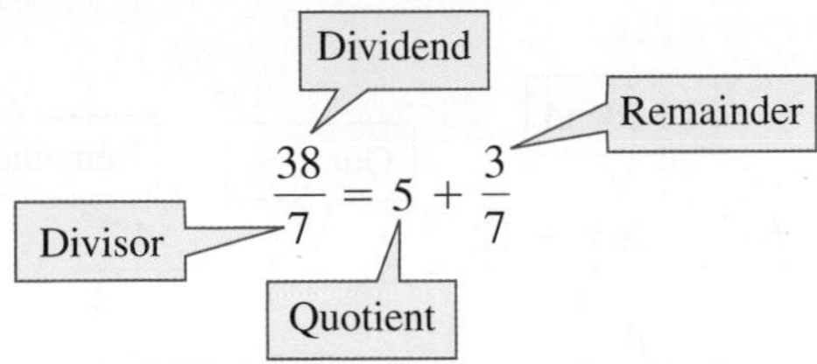

To divide polynomials, we use long division, as follows.

Division Algorithm

If $P(x)$ and $D(x)$ are polynomials, with $D(x) \neq 0$, then there exist unique polynomials $Q(x)$ and $R(x)$, where $R(x)$ is either 0 or of degree less than the degree of $D(x)$, such that

$$\frac{\overset{\text{Dividend}}{P(x)}}{\underset{\text{Divisor}}{D(x)}} = \overset{\text{Quotient}}{Q(x)} + \frac{\overset{\text{Remainder}}{R(x)}}{\underset{\text{Divisor}}{D(x)}} \qquad \text{or} \qquad \underset{\text{Dividend}}{P(x)} = \underset{\text{Divisor}}{D(x)} \cdot \underset{\text{Quotient}}{Q(x)} + \overset{\text{Remainder}}{R(x)}$$

The polynomials $P(x)$ and $D(x)$ are called the **dividend** and **divisor**, respectively; $Q(x)$ is the **quotient**, and $R(x)$ is the **remainder**.

Example 1 ■ Long Division of Polynomials

Divide $6x^2 - 26x + 12$ by $x - 4$. Express the result in each of the two forms shown in the preceding box.

Solution The *dividend* is $6x^2 - 26x + 12$, and the *divisor* is $x - 4$. We begin by arranging them as follows.

$$x - 4\overline{)6x^2 - 26x + 12}$$

Next we divide the leading term in the dividend by the leading term in the divisor to get the first term of the quotient: $6x^2/x = 6x$. Then we multiply the divisor by $6x$ and subtract the result from the dividend.

$$\begin{array}{rl} 6x & \text{Divide leading terms: } \dfrac{6x^2}{x} = 6x \\ x - 4\overline{)6x^2 - 26x + 12} & \\ \underline{6x^2 - 24x}\phantom{{} + 12} & \text{Multiply: } 6x(x - 4) = 6x^2 - 24x \\ -2x + 12 & \text{Subtract and “bring down” 12} \end{array}$$

We repeat the process using the last line, $-2x + 12$, as the dividend.

$$
\begin{array}{r@{}l}
 & \;\;6x - 2 \\
x - 4 & \overline{)\,6x^2 - 26x + 12} \\
 & \;\;\underline{6x^2 - 24x} \\
 & \qquad\;\; -2x + 12 \\
 & \qquad\;\; \underline{-2x + 8} \\
 & \qquad\qquad\quad 4
\end{array}
\qquad
\begin{array}{l}
\text{Divide leading terms: } \dfrac{-2x}{x} = -2 \\
\\
\\
\\
\text{Multiply: } -2(x - 4) = -2x + 8 \\
\text{Subtract}
\end{array}
$$

The division process ends when the last line is of lower degree than the divisor. The last line then contains the *remainder*, and the top line contains the *quotient*. The result of the division can be interpreted in either of two ways:

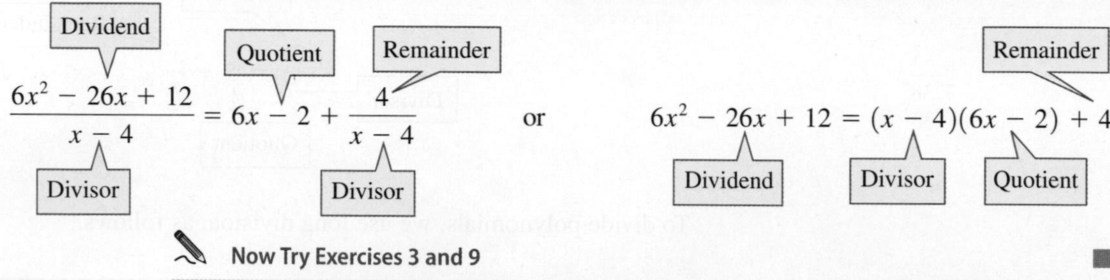

Now Try Exercises 3 and 9 ■

Example 2 ■ Long Division of Polynomials

Let $P(x) = 8x^4 + 6x^2 - 3x + 1$ and $D(x) = 2x^2 - x + 2$. Find polynomials $Q(x)$ and $R(x)$ such that $P(x) = D(x) \cdot Q(x) + R(x)$.

Solution We use long division after first inserting the term $0x^3$ into the dividend to ensure that the columns line up correctly.

$$
\begin{array}{r@{}l}
 & \;\;4x^2 + 2x \\
2x^2 - x + 2 & \overline{)\,8x^4 + 0x^3 + 6x^2 - 3x + 1} \\
 & \;\;\underline{8x^4 - 4x^3 + 8x^2} \\
 & \qquad\;\; 4x^3 - 2x^2 - 3x \\
 & \qquad\;\; \underline{4x^3 - 2x^2 + 4x} \\
 & \qquad\qquad\qquad\;\; -7x + 1
\end{array}
\qquad
\begin{array}{l}
\\
\\
\text{Multiply divisor by } 4x^2 \\
\text{Subtract} \\
\text{Multiply divisor by } 2x \\
\text{Subtract}
\end{array}
$$

The process is complete at this point because $-7x + 1$ is of lower degree than the divisor, $2x^2 - x + 2$. From the long division above we see that $Q(x) = 4x^2 + 2x$ and $R(x) = -7x + 1$, so

$$8x^4 + 6x^2 - 3x + 1 = (2x^2 - x + 2)(4x^2 + 2x) + (-7x + 1)$$

Now Try Exercise 19 ■

■ Synthetic Division

Synthetic division is a quick method of dividing polynomials; it can be used when the divisor is of the form $x - c$. In synthetic division we write only the essential parts of the long division. Compare the following long and synthetic divisions, in which we divide $2x^3 - 7x^2 + 5$ by $x - 3$. (We'll explain how to perform the synthetic division in Example 3.)

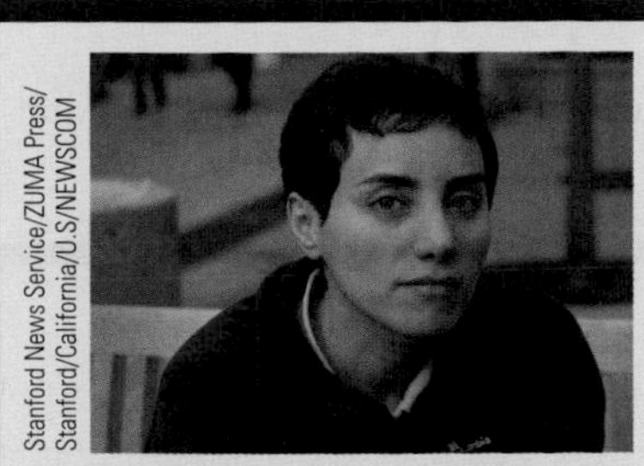

MARYAM MIRZAKHANI (1977–2017) was one of the top mathematicians in the world until her untimely death in 2017. In 2014 she was awarded the Fields Medal for her work on the geometry of surfaces. The Fields Medal is the top prize in mathematics (it's sometimes called the "Nobel Prize" of Mathematics).

Maryam was born in Iran, where she attended an all-girls school. She said, "I did poorly in math for a couple of years in middle school; I was just not interested in thinking about it." But she later became fascinated by problem solving; she said, "The beauty of mathematics only shows itself to patient followers." Maryam and her childhood friend Roya Beheshti were chosen to represent their country at the International Mathematics Olympiad, a worldwide competition, for high school students; they received the gold and silver medals. Roya is now also a professor of mathematics.

Maryam was the first woman to be awarded the Fields Medal. In accepting the award, she said, "I hope that this award will inspire lots more girls and young women ... to believe in their own abilities and aim to be the Fields Medalists of the future." The 2020 documentary film *Secrets of the Surface* features the life and mathematical vision of Maryam Mirzakhani.

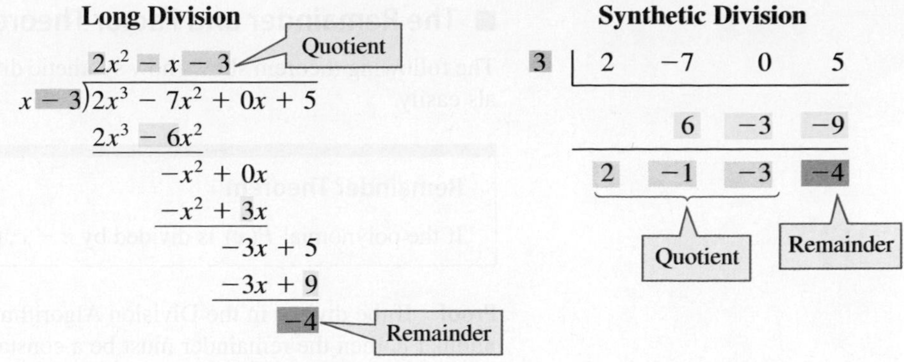

Long Division
$2x^2 - x - 3$
Quotient
$x - 3 \overline{)2x^3 - 7x^2 + 0x + 5}$
$2x^3 - 6x^2$
$-x^2 + 0x$
$-x^2 + 3x$
$-3x + 5$
$-3x + 9$
-4
Remainder
Synthetic Division
3 | 2 −7 0 5
6 −3 −9
2 −1 −3 −4
Quotient
Remainder

Note that in synthetic division we abbreviate $2x^3 - 7x^2 + 5$ by writing only the coefficients: 2 −7 0 5, and instead of $x - 3$, we simply write 3. (Writing 3 instead of −3 allows us to add instead of subtract, but this changes the sign of all the numbers that appear in the gold boxes.)

The next example shows how synthetic division is performed.

Example 3 ■ Synthetic Division

Use synthetic division to divide $2x^3 - 7x^2 + 5$ by $x - 3$.

Solution We begin by writing the appropriate coefficients to represent the divisor and the dividend:

Divisor $x - 3$ → 3 | 2 −7 0 5 ← Dividend $2x^3 - 7x^2 + 0x + 5$

We bring down the 2, multiply $3 \cdot 2 = 6$, and write the result in the middle row. Then we add.

```
3 | 2  -7   0   5
        6
  ---------------
    2  -1
```

Multiply: $3 \cdot 2 = 6$

Add: $-7 + 6 = -1$

We repeat this process of multiplying and then adding until the table is complete.

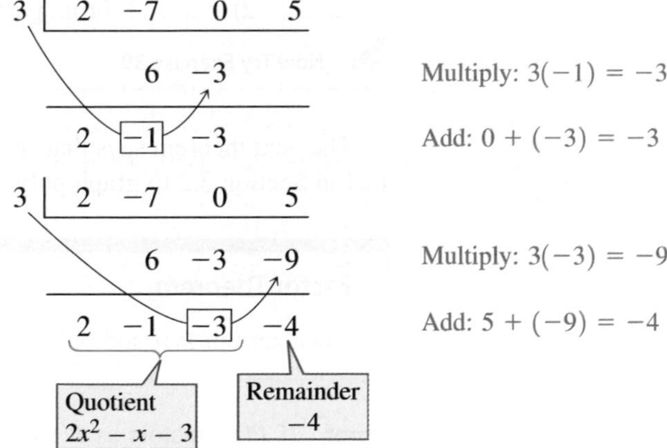

From the last line of the synthetic division we see that the quotient is $2x^2 - x - 3$ and the remainder is −4. Thus

$$2x^3 - 7x^2 + 5 = (x - 3)(2x^2 - x - 3) - 4$$

Now Try Exercise 31 ■

■ The Remainder and Factor Theorems

The following theorem shows how synthetic division can be used to evaluate polynomials easily.

> **Remainder Theorem**
>
> If the polynomial $P(x)$ is divided by $x - c$, then the remainder is the value $P(c)$.

Proof If the divisor in the Division Algorithm is of the form $x - c$ for some real number c, then the remainder must be a constant (since the degree of the remainder is less than the degree of the divisor). If we call this constant r, then

$$P(x) = (x - c) \cdot Q(x) + r$$

Replacing x by c in this equation, we get $P(c) = (c - c) \cdot Q(c) + r = 0 + r = r$, that is, $P(c)$ is the remainder r. ■

Example 4 ■ Using the Remainder Theorem to Find the Value of a Polynomial

Let $P(x) = 3x^5 + 5x^4 - 4x^3 + 7x + 3$.

(a) Find the quotient and remainder when $P(x)$ is divided by $x + 2$.

(b) Use the Remainder Theorem to find $P(-2)$.

Solution

(a) Since $x + 2 = x - (-2)$, the synthetic division for this problem takes the following form:

$$\begin{array}{r|rrrrrr} -2 & 3 & 5 & -4 & 0 & 7 & 3 \\ & & -6 & 2 & 4 & -8 & 2 \\ \hline & 3 & -1 & -2 & 4 & -1 & 5 \end{array}$$

Remainder is 5, so $P(-2) = 5$

The quotient is $3x^4 - x^3 - 2x^2 + 4x - 1$ and the remainder is 5.

(b) By the Remainder Theorem, $P(-2)$ is the remainder when $P(x)$ is divided by $x - (-2) = x + 2$. From part (a) the remainder is 5, so $P(-2) = 5$.

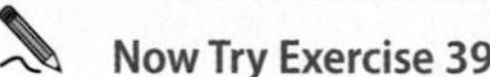
Now Try Exercise 39 ■

The next theorem says that *zeros* of polynomials correspond to *factors*. We used this fact in Section 3.2 to graph polynomials.

> **Factor Theorem**
>
> c is a zero of P if and only if $x - c$ is a factor of $P(x)$.

Proof If $P(x)$ factors as $P(x) = (x - c)Q(x)$, then

$$P(c) = (c - c)Q(c) = 0 \cdot Q(c) = 0$$

Conversely, if $P(c) = 0$, then by the Remainder Theorem

$$P(x) = (x - c)Q(x) + 0 = (x - c)Q(x)$$

so $x - c$ is a factor of $P(x)$. ■

Example 5 ■ Factoring a Polynomial Using the Factor Theorem

$$\begin{array}{r|rrrr} 1 & 1 & 0 & -7 & 6 \\ & & 1 & 1 & -6 \\ \hline & 1 & 1 & -6 & 0 \end{array}$$

$$\begin{array}{r} x^2 + x - 6 \\ x - 1 \overline{) x^3 + 0x^2 - 7x + 6} \\ \underline{x^3 - x^2} \qquad\qquad \\ x^2 - 7x \quad \\ \underline{x^2 - x} \quad \\ -6x + 6 \\ \underline{-6x + 6} \\ 0 \end{array}$$

Let $P(x) = x^3 - 7x + 6$. Show that $P(1) = 0$, and use this fact to factor $P(x)$ completely.

Solution Substituting, we see that $P(1) = 1^3 - 7 \cdot 1 + 6 = 0$. By the Factor Theorem this means that $x - 1$ is a factor of $P(x)$. Using synthetic or long division (shown in the margin), we see that

$$\begin{aligned} P(x) &= x^3 - 7x + 6 && \text{Given polynomial} \\ &= (x - 1)(x^2 + x - 6) && \text{See margin} \\ &= (x - 1)(x - 2)(x + 3) && \text{Factor quadratic } x^2 + x - 6 \end{aligned}$$

 Now Try Exercises 53 and 57 ■

Example 6 ■ Finding a Polynomial with Specified Zeros

Find a polynomial of degree 4 that has zeros -3, 0, 1, and 5, and the coefficient of x^3 is -6.

Solution By the Factor Theorem, $x - (-3)$, $x - 0$, $x - 1$, and $x - 5$ must all be factors of the desired polynomial. Let

$$\begin{aligned} P(x) &= (x + 3)(x - 0)(x - 1)(x - 5) \\ &= x^4 - 3x^3 - 13x^2 + 15x && \text{Expand} \end{aligned}$$

The polynomial $P(x)$ is of degree 4 with the desired zeros, but the coefficient of x^3 is -3, not -6. Multiplication by a nonzero constant does not change the degree, so the desired polynomial is a constant multiple of $P(x)$. If we multiply $P(x)$ by the constant 2, we get

$$Q(x) = 2x^4 - 6x^3 - 26x^2 + 30x$$

which is a polynomial with all the desired properties. The polynomial Q is graphed in Figure 1. Note that the zeros of Q correspond to the x-intercepts of the graph.

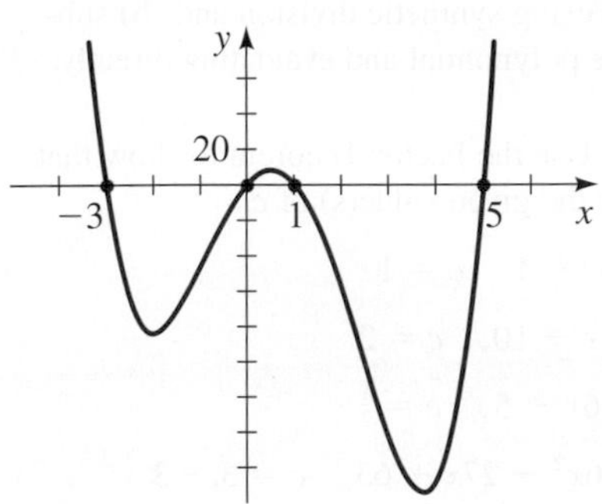

Figure 1 | $Q(x) = 2x(x + 3)(x - 1)(x - 5)$ has zeros -3, 0, 1, and 5, and the coefficient of x^3 is -6.

 Now Try Exercises 63 and 67 ■

3.3 | Exercises

■ Concepts

1. If we divide the polynomial P by the factor $x - c$ and we obtain the equation $P(x) = (x - c)Q(x) + R(x)$, then we say that $x - c$ is the divisor, $Q(x)$ is the ________, and $R(x)$ is the ________.

2. (a) If we divide the polynomial $P(x)$ by the factor $x - c$ and we obtain a remainder of 0, then we know that c is a ________ of P.

(b) If we divide the polynomial $P(x)$ by the factor $x - c$ and we obtain a remainder of k, then we know that $P(c) =$ ________.

■ Skills

3–8 ■ Division of Polynomials Two polynomials P and D are given. Use either synthetic division or long division to divide $P(x)$ by $D(x)$, and express the quotient $P(x)/D(x)$ in the form

$$\frac{P(x)}{D(x)} = Q(x) + \frac{R(x)}{D(x)}$$

3. $P(x) = 3x^2 - 6x + 4$, $D(x) = x - 3$

4. $P(x) = 2x^2 + 10x - 7$, $D(x) = x + 5$

5. $P(x) = 12x^3 - 16x^2 - x - 1$, $D(x) = 3x + 2$

6. $P(x) = 10x^3 + 2x^2 - 4x + 1$, $D(x) = 5x + 1$

7. $P(x) = 2x^4 - x^3 + 9x^2, \quad D(x) = x^2 + 4$

8. $P(x) = 2x^5 + x^3 - 2x^2 + 3x - 5, \quad D(x) = x^2 - 3x + 1$

9–14 ■ Division of Polynomials Two polynomials P and D are given. Use either synthetic division or long division to divide $P(x)$ by $D(x)$, and express P in the form

$$P(x) = D(x) \cdot Q(x) + R(x)$$

9. $P(x) = 3x^3 + 5x^2 + 5, \quad D(x) = x + 5$

10. $P(x) = 5x^4 - 10x^2 + 3x + 2, \quad D(x) = x - 2$

11. $P(x) = 2x^3 - 3x^2 - 2x, \quad D(x) = 2x - 3$

12. $P(x) = 4x^3 + 7x + 9, \quad D(x) = 2x + 1$

13. $P(x) = 8x^4 + 4x^3 + 6x^2, \quad D(x) = 2x^2 + 1$

14. $P(x) = 27x^5 - 9x^4 + 3x^2 - 3, \quad D(x) = 3x^2 - 3x + 1$

15–24 ■ Long Division of Polynomials Find the quotient and remainder using long division.

15. $\dfrac{x^2 - 3x + 7}{x - 2}$ **16.** $\dfrac{x^3 + 2x^2 - x + 1}{x + 3}$

17. $\dfrac{9x^3 - 6x^2 + x + 1}{3x - 1}$ **18.** $\dfrac{8x^3 - 2x^2 - 2x + 3}{4x + 3}$

19. $\dfrac{4x^3 + 2x^2 - 3}{x^2 + x - 1}$ **20.** $\dfrac{x^4 - 4x^3 - x + 3}{x^2 - 3x + 2}$

21. $\dfrac{6x^3 + 2x^2 + 22x}{2x^2 + 5}$ **22.** $\dfrac{9x^2 - x + 5}{3x^2 - 7x}$

23. $\dfrac{x^6 + x^4 + x^2 + 1}{x^2 + 1}$ **24.** $\dfrac{2x^5 - 7x^4 - 13}{4x^2 - 6x + 8}$

25–38 ■ Synthetic Division of Polynomials Find the quotient and remainder using synthetic division.

25. $\dfrac{2x^2 - 5x + 3}{x - 3}$ **26.** $\dfrac{-x^2 + x - 4}{x + 1}$

27. $\dfrac{3x^2 + x}{x + 1}$ **28.** $\dfrac{4x^2 - 3}{x - 2}$

29. $\dfrac{3x^3 - 2x^2 + x - 5}{x - 2}$ **30.** $\dfrac{5x^3 + 20x^2 - 30x + 10}{x + 5}$

31. $\dfrac{x^3 - 10x + 13}{x + 4}$ **32.** $\dfrac{x^4 - x^3 - 10x}{x - 3}$

33. $\dfrac{x^5 + 3x^3 - 6}{x - 1}$ **34.** $\dfrac{x^3 - 9x^2 + 27x - 27}{x - 3}$

35. $\dfrac{2x^3 + 3x^2 - 2x + 1}{x - \frac{1}{2}}$ **36.** $\dfrac{6x^4 + 10x^3 + 5x^2 + x + 1}{x + \frac{2}{3}}$

37. $\dfrac{x^3 - 27}{x - 3}$ **38.** $\dfrac{x^4 - 16}{x + 2}$

39–51 ■ Remainder Theorem Use synthetic division and the Remainder Theorem to evaluate $P(c)$.

39. $P(x) = x^4 - x^2 + 5, \quad c = -2$

40. $P(x) = 2x^2 + 9x + 1, \quad c = \frac{1}{2}$

41. $P(x) = x^3 + 3x^2 - 7x + 6, \quad c = 2$

42. $P(x) = x^3 - x^2 + x + 5, \quad c = -1$

43. $P(x) = x^3 + 2x^2 - 7, \quad c = -2$

44. $P(x) = 2x^3 - 21x^2 + 9x - 200, \quad c = 11$

45. $P(x) = 5x^4 + 30x^3 - 40x^2 + 36x + 14, \quad c = -7$

46. $P(x) = 6x^5 + 10x^3 + x + 1, \quad c = -2$

47. $P(x) = x^7 - 3x^2 - 1, \quad c = 3$

48. $P(x) = -2x^6 + 7x^5 + 40x^4 - 7x^2 + 10x + 112, \quad c = -3$

49. $P(x) = 3x^3 + 4x^2 - 2x + 1, \quad c = \frac{2}{3}$

50. $P(x) = x^3 - x + 1, \quad c = \frac{1}{4}$

51. $P(x) = x^3 + 2x^2 - 3x - 8, \quad c = 0.1$

52. Remainder Theorem Let

$$P(x) = 6x^7 - 40x^6 + 16x^5 - 200x^4 - 60x^3 - 69x^2 + 13x - 139$$

Calculate $P(7)$ by **(a)** using synthetic division and **(b)** substituting $x = 7$ into the polynomial and evaluating directly.

53–56 ■ Factor Theorem Use the Factor Theorem to show that $x - c$ is a factor of $P(x)$ for the given value(s) of c.

53. $P(x) = x^3 - 3x^2 + 3x - 1, \quad c = 1$

54. $P(x) = x^3 + 2x^2 - 3x - 10, \quad c = 2$

55. $P(x) = 2x^3 + 7x^2 + 6x - 5, \quad c = \frac{1}{2}$

56. $P(x) = x^4 + 3x^3 - 16x^2 - 27x + 63, \quad c = 3, -3$

57–62 ■ Factor Theorem Show that the given value(s) of c are zeros of $P(x)$, and find all other zeros of $P(x)$.

57. $P(x) = x^3 - 3x^2 - 18x + 40, \quad c = -4$

58. $P(x) = x^3 - 5x^2 - 2x + 10, \quad c = 5$

59. $P(x) = x^3 - x^2 - 11x + 15, \quad c = 3$

60. $P(x) = 3x^4 - x^3 - 21x^2 - 11x + 6, \quad c = -2, \frac{1}{3}$

61. $P(x) = 3x^4 - 8x^3 - 14x^2 + 31x + 6, \quad c = -2, 3$

62. $P(x) = 2x^4 - 13x^3 + 7x^2 + 37x + 15, \quad c = -1, 3$

63–66 ■ Finding a Polynomial with Specified Zeros Find a polynomial of the specified degree that has the given zeros.

63. Degree 3; zeros $-1, 1, 3$

64. Degree 4; zeros $-2, 0, 2, 4$

65. Degree 4; zeros $-1, 1, 3, 5$

66. Degree 5; zeros $-2, -1, 0, 1, 2$

67–70 ■ Polynomials with Specified Zeros Find a polynomial of the specified degree that satisfies the given conditions.

67. Degree 4; zeros $-2, 0, 1, 3$; coefficient of x^3 is 4

68. Degree 4; zeros $-1, 0, 2, \frac{1}{2}$; coefficient of x^3 is 3

69. Degree 4; zeros $-1, 1, \sqrt{2}$; integer coefficients and constant term 6

70. Degree 5; zeros $-2, -1, 2, \sqrt{5}$; integer coefficients and constant term 40

■ Skills Plus

71–74 ■ Finding a Polynomial from a Graph Find the polynomial of the specified degree whose graph is shown.

71. Degree 3

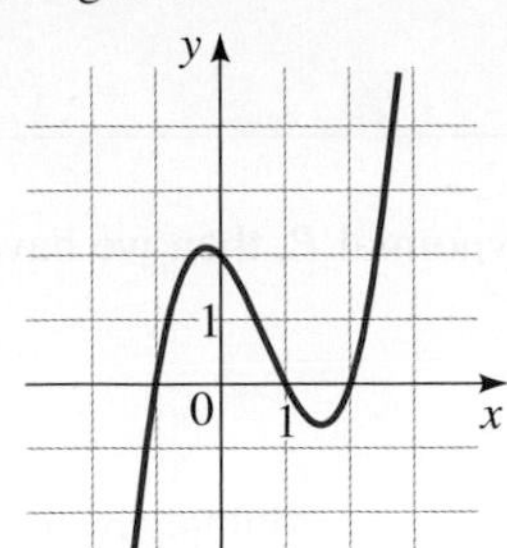

72. Degree 3

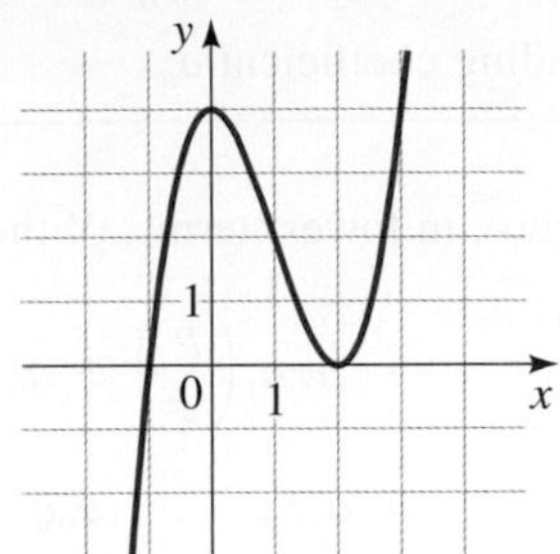

73. Degree 4

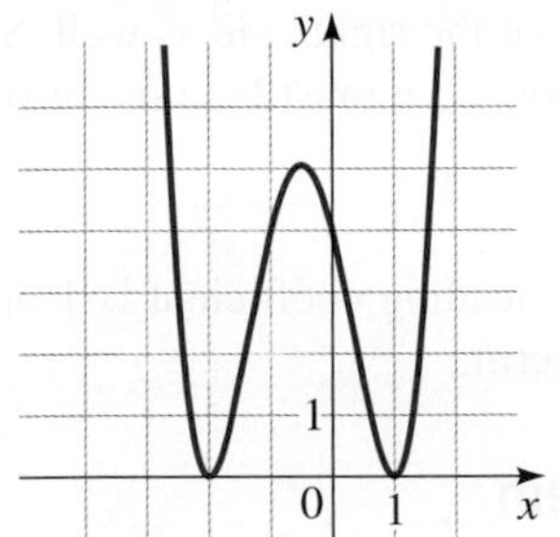

74. Degree 4

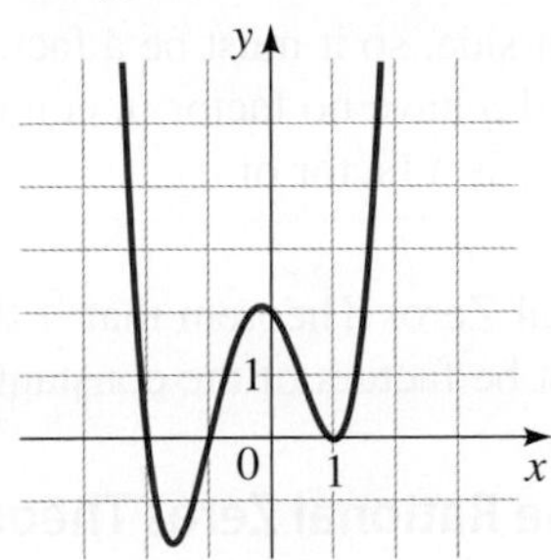

■ Discuss ■ Discover ■ Prove ■ Write

75. Discuss ■ Discover: Impossible Division? The following problems involve polynomials of very high degree. At first glance, it may seem impossible (or at least very time consuming) to solve them, but they can be solved quickly. Solve each problem.

(a) Find the remainder when $6x^{1000} - 17x^{562} + 12x + 26$ is divided by $x + 1$.

(b) Is $x - 1$ a factor of $x^{567} - 3x^{400} + x^9 + 1$?

PS *Try to recognize something familiar.* Use one or more of the theorems you have learned in this section to solve the problems without dividing.

76. Discover: Nested Form of a Polynomial Expand Q to prove that the polynomials P and Q are the same.

$$P(x) = 3x^4 - 5x^3 + x^2 - 3x + 5$$
$$Q(x) = (((3x - 5)x + 1)x - 3)x + 5$$

Try to evaluate $P(2)$ and $Q(2)$ in your head, using the forms given. Which is easier? Now write the polynomial

$$R(x) = x^5 - 2x^4 + 3x^3 - 2x^2 + 3x + 4$$

in "nested" form, like the polynomial Q. Use the nested form to find $R(3)$ in your head.

Do you see how calculating with the nested form follows the same arithmetic steps as calculating the value of a polynomial using synthetic division?

3.4 Real Zeros of Polynomials

■ Rational Zeros of Polynomials ■ Descartes's Rule of Signs ■ Upper and Lower Bounds Theorem ■ Using Algebra and Graphing Devices to Solve Polynomial Equations

The Factor Theorem tells us that finding the zeros of a polynomial is really the same thing as factoring it into linear factors. In this section we study some algebraic methods that help us find the real zeros of a polynomial and thereby factor the polynomial. We begin with the *rational* zeros of a polynomial.

■ Rational Zeros of Polynomials

To help us understand the next theorem, let's consider the polynomial

$$\begin{aligned} P(x) &= (x - 2)(x - 3)(x + 4) && \text{Factored form} \\ &= x^3 - x^2 - 14x + 24 && \text{Expanded form} \end{aligned}$$

From the factored form we see that the zeros of P are 2, 3, and -4. When the polynomial is expanded, the constant 24 is obtained by multiplying $(-2) \times (-3) \times 4$. This means that the zeros of the polynomial are all factors of the constant term. The following theorem generalizes this observation.

Rational Zeros Theorem

If the polynomial $P(x) = a_nx^n + a_{n-1}x^{n-1} + \cdots + a_1x + a_0$ has integer coefficients (where $a_n \neq 0$ and $a_0 \neq 0$), then every rational zero of P is of the form

$$\frac{p}{q}$$

where p and q are integers with p/q in lowest terms, and

p is a factor of the constant coefficient a_0

q is a factor of the leading coefficient a_n

We say that p/q is in *lowest terms* if p and q have no factor in common, other than 1.

Proof If p/q is a rational zero, in lowest terms, of the polynomial P, then we have

$$a_n\left(\frac{p}{q}\right)^n + a_{n-1}\left(\frac{p}{q}\right)^{n-1} + \cdots + a_1\left(\frac{p}{q}\right) + a_0 = 0$$

$$a_np^n + a_{n-1}p^{n-1}q + \cdots + a_1pq^{n-1} + a_0q^n = 0 \qquad \text{Multiply by } q^n$$

$$p(a_np^{n-1} + a_{n-1}p^{n-2}q + \cdots + a_1q^{n-1}) = -a_0q^n \qquad \text{Subtract } a_0q^n \text{ and factor LHS}$$

Now p is a factor of the left side, so it must be a factor of the right side as well. Since p/q is in lowest terms, p and q have no factor in common, so p must be a factor of a_0. A similar proof shows that q is a factor of a_n. ■

We see from the Rational Zeros Theorem that if the leading coefficient is 1 or -1, then the rational zeros must be factors of the constant term.

Example 1 ■ Using the Rational Zeros Theorem

Find the rational zeros of $P(x) = x^3 - 3x + 2$.

Solution Since the leading coefficient is 1, any rational zero must be a divisor of the constant term 2. So the possible rational zeros are ±1 and ±2. We test each of these possibilities.

$$P(1) = (1)^3 - 3(1) + 2 = 0$$

$$P(-1) = (-1)^3 - 3(-1) + 2 = 4$$

$$P(2) = (2)^3 - 3(2) + 2 = 4$$

$$P(-2) = (-2)^3 - 3(-2) + 2 = 0$$

The rational zeros of P are 1 and -2.

Now Try Exercise 15 ■

0 0.001 0.01

Discovery Project ■ Zeroing in on a Zero

We have learned how to find the zeros of a polynomial function algebraically and graphically. In this project we investigate a *numerical* method for finding the zeros of a polynomial. With this method we can approximate the zeros of a polynomial to any number of decimal places. The method involves finding smaller and smaller intervals that zoom in on a zero of a polynomial. You can find the project at **www.stewartmath.com**.

EVARISTE GALOIS (1811–1832) is one of the very few mathematicians to have an entire theory named in their honor. Not yet 21 when he died, he completely settled the central problem in the theory of equations by describing a criterion that reveals whether a polynomial equation can be solved by algebraic operations. Galois recognized that he was one of the greatest mathematicians in the world at that time, although no one knew it. He repeatedly sent his work to the eminent mathematicians Cauchy and Poisson, who either lost his letters or did not understand his ideas. Galois wrote in a terse style and included few details, which probably played a role in his failure to pass the entrance exams at the Ecole Polytechnique in Paris. A political radical, Galois spent several months in prison for his revolutionary activities. His brief life came to a tragic end when he was killed in a duel over a love affair. The night before his duel, fearing that he would die, Galois wrote down the essence of his ideas and entrusted them to his friend Auguste Chevalier. He concluded by writing "there will, I hope, be people who will find it to their advantage to decipher all this mess." The mathematician Camille Jordan did just that, 14 years later.

The following box explains how we use the Rational Zeros Theorem with synthetic division to factor a polynomial.

Finding the Rational Zeros of a Polynomial

1. **List Possible Zeros.** List all possible rational zeros, using the Rational Zeros Theorem.
2. **Divide.** Use synthetic division to evaluate the polynomial at candidates for the rational zeros that you found in Step 1, until you get a remainder 0. Note the quotient you have obtained when you get a remainder 0.
3. **Repeat.** Repeat Steps 1 and 2 for the quotient. Stop when you reach a quotient that is quadratic or factors easily, and use the quadratic formula or factor to find the remaining zeros.

Example 2 ■ Finding Rational Zeros

Write the polynomial $P(x) = 2x^3 + x^2 - 13x + 6$ in factored form, and find all its zeros.

Solution By the Rational Zeros Theorem the rational zeros of P are of the form

$$\text{possible rational zero of } P = \frac{\text{factor of constant term}}{\text{factor of leading coefficient}} = \frac{\text{factor of 6}}{\text{factor of 2}}$$

The factors of 6 are $\pm1, \pm2, \pm3, \pm6$, and the factors of 2 are $\pm1, \pm2$. Thus the possible rational zeros of P are

$$\pm\frac{1}{1}, \quad \pm\frac{2}{1}, \quad \pm\frac{3}{1}, \quad \pm\frac{6}{1}, \quad \pm\frac{1}{2}, \quad \pm\frac{2}{2}, \quad \pm\frac{3}{2}, \quad \pm\frac{6}{2}$$

Simplifying the fractions and eliminating duplicates, we get the following list of possible rational zeros:

$$\pm1, \quad \pm2, \quad \pm3, \quad \pm6, \quad \pm\frac{1}{2}, \quad \pm\frac{3}{2}$$

Now we use synthetic division to check these *possible* zeros until we find an *actual* zero (that is, when we get a remainder of 0).

Test whether 1 is a zero

$$\begin{array}{r|rrrr} 1 & 2 & 1 & -13 & 6 \\ & & 2 & 3 & -10 \\ \hline & 2 & 3 & -10 & -4 \end{array}$$

Remainder is *not* 0, so 1 is *not* a zero

Test whether 2 is a zero

$$\begin{array}{r|rrrr} 2 & 2 & 1 & -13 & 6 \\ & & 4 & 10 & -6 \\ \hline & 2 & 5 & -3 & 0 \end{array}$$

Remainder *is* 0, so 2 *is* a zero

From the last synthetic division we see that 2 is a zero of P and that P factors as

$$\begin{aligned} P(x) &= 2x^3 + x^2 - 13x + 6 && \text{Given polynomial} \\ &= (x - 2)(2x^2 + 5x - 3) && \text{From synthetic division} \\ &= (x - 2)(2x - 1)(x + 3) && \text{Factor } 2x^2 + 5x - 3 \end{aligned}$$

From the factored form we see that the zeros of P are 2, $\frac{1}{2}$, and –3.

 **Now Try Exercise 29** ■

Example 3 ■ Using the Rational Zeros Theorem and the Quadratic Formula

Let $P(x) = x^4 - 5x^3 - 5x^2 + 23x + 10$.

(a) Find the zeros of P. **(b)** Sketch a graph of P.

Solution

1	1	−5	−5	23	10
		1	−4	−9	14
	1	−4	−9	14	24

2	1	−5	−5	23	10
		2	−6	−22	2
	1	−3	−11	1	12

5	1	−5	5	23	10
		5	0	−25	−10
	1	0	−5	−2	0

(a) The leading coefficient of P is 1, so all the rational zeros are integers: They are divisors of the constant term 10. Thus the possible candidates are

$$\pm 1, \quad \pm 2, \quad \pm 5, \quad \pm 10$$

Using synthetic division (see the margin), we find that 1 and 2 are not zeros but that 5 is a zero and that P factors as

$$x^4 - 5x^3 - 5x^2 + 23x + 10 = (x - 5)(x^3 - 5x - 2)$$

We now try to factor the quotient $x^3 - 5x - 2$. Its possible zeros are the divisors of -2, namely,

$$\pm 1, \quad \pm 2$$

Since we already know that 1 and 2 are not zeros of the original polynomial P, we don't need to try them again. Checking the remaining candidates, -1 and -2, we see that -2 is a zero (see the margin), and P factors as

−2	1	0	−5	−2
		−2	4	2
	1	−2	−1	0

$$\begin{aligned} x^4 - 5x^3 - 5x^2 + 23x + 10 &= (x - 5)(x^3 - 5x - 2) \\ &= (x - 5)(x + 2)(x^2 - 2x - 1) \end{aligned}$$

Now we use the Quadratic Formula to solve $x^2 - 2x - 1 = 0$ and obtain the two remaining zeros of P:

$$x = \frac{2 \pm \sqrt{(-2)^2 - 4(1)(-1)}}{2} = 1 \pm \sqrt{2}$$

The zeros of P are $5, -2, 1 + \sqrt{2}$, and $1 - \sqrt{2}$.

(b) Now that we know the zeros of P, we can use the methods of Section 3.2 to sketch the graph. If we want to use a graphing device instead, then knowing the zeros allows us to choose an appropriate viewing rectangle—one that is wide enough to contain all the x-intercepts of P. The zeros of P are

$$5, \quad -2, \quad 1 + \sqrt{2} \approx 2.41, \quad 1 - \sqrt{2} \approx -0.41$$

So we choose the viewing rectangle $[-3, 6]$ by $[-50, 50]$ and draw the graph shown in Figure 1.

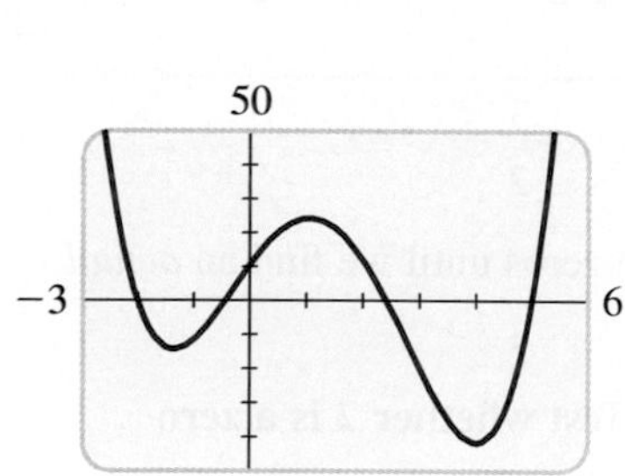

Figure 1 | $P(x) = x^4 - 5x^3 - 5x^2 + 23x + 10$

Now Try Exercises 45 and 55 ■

■ Descartes's Rule of Signs

In some cases, the following rule—discovered by the French philosopher and mathematician René Descartes around 1637 (see Section 2.6)—is helpful in eliminating candidates from lengthy lists of possible rational roots. To describe this rule, we need the concept of *variation in sign*. If $P(x)$ is a polynomial with real coefficients, written with descending powers of x (and omitting powers with coefficient 0), then a **variation in sign** occurs whenever adjacent coefficients have opposite signs. For example,

$$P(x) = 5x^7 - 3x^5 - x^4 + 2x^2 + x - 3$$

has three variations in sign.

Polynomial	Variations in Sign
$x^2 + 4x + 1$	0
$2x^3 + x - 6$	1
$x^4 - 3x^2 - x + 4$	2

Descartes's Rule of Signs

Let P be a polynomial with real coefficients.

1. The number of positive real zeros of $P(x)$ either is equal to the number of variations in sign in $P(x)$ or is less than that by an even whole number.
2. The number of negative real zeros of $P(x)$ either is equal to the number of variations in sign in $P(-x)$ or is less than that by an even whole number.

Multiplicity is discussed in Section 3.2.

In Descartes's Rule of Signs a zero with multiplicity m is counted m times. For example, the polynomial $P(x) = x^2 - 2x + 1$ has two sign changes and has the positive zero $x = 1$. But this zero is counted twice because it has multiplicity 2.

Example 4 ■ Using Descartes's Rule

Use Descartes's Rule of Signs to determine the possible number of positive and negative real zeros of the polynomial

$$P(x) = 3x^6 + 4x^5 + 3x^3 - x - 3$$

Solution The polynomial has one variation in sign, so it has one positive zero. Now

$$P(-x) = 3(-x)^6 + 4(-x)^5 + 3(-x)^3 - (-x) - 3$$
$$= 3x^6 - 4x^5 - 3x^3 + x - 3$$

So $P(-x)$ has three variations in sign. Thus $P(x)$ has either three or one negative zero(s), making a total of either two or four real zeros.

Now Try Exercise 63 ■

■ Upper and Lower Bounds Theorem

We say that a is a **lower bound** and b is an **upper bound** for the zeros of a polynomial if every real zero c of the polynomial satisfies $a \le c \le b$. The next theorem helps us find such bounds for the zeros of a polynomial.

The Upper and Lower Bounds Theorem

Let P be a polynomial with real coefficients.

1. If we divide $P(x)$ by $x - b$ (with $b > 0$) using synthetic division and if the row that contains the quotient and remainder has no negative entry, then b is an upper bound for the real zeros of P.
2. If we divide $P(x)$ by $x - a$ (with $a < 0$) using synthetic division and if the row that contains the quotient and remainder has entries that are alternately nonpositive and nonnegative, then a is a lower bound for the real zeros of P.

A proof of this theorem is suggested in Exercise 108. The phrase "alternately nonpositive and nonnegative" simply means that the signs of the numbers alternate, with 0 considered to be positive or negative as required.

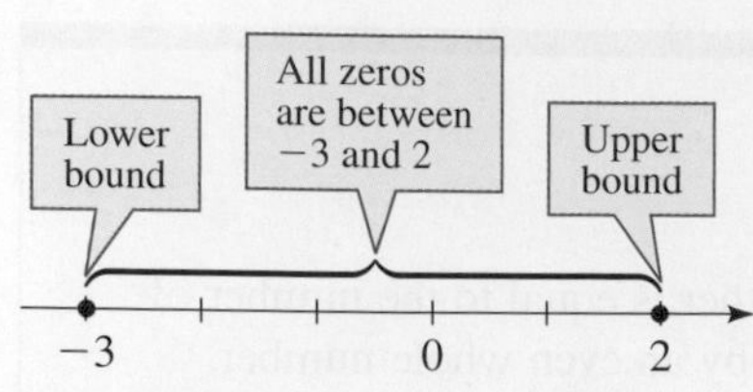

Example 5 ■ Upper and Lower Bounds for the Zeros of a Polynomial

Show that all the real zeros of the polynomial $P(x) = x^4 - 3x^2 + 2x - 5$ lie between -3 and 2.

Solution We divide $P(x)$ by $x - 2$ and $x + 3$ using synthetic division:

2	1	0	−3	2	−5
		2	4	2	8
	1	2	1	4	3

All entries nonnegative

−3	1	0	−3	2	−5
		−3	9	−18	48
	1	−3	6	−16	43

Entries alternate in sign

By the Upper and Lower Bounds Theorem -3 is a lower bound and 2 is an upper bound for the zeros. Since neither -3 nor 2 is a zero (the remainders are not 0 in the division table), all the real zeros lie between these numbers.

 **Now Try Exercise 69** ■

Example 6 ■ A Lower Bound for the Zeros of a Polynomial

Show that all the real zeros of the polynomial $P(x) = x^4 + 4x^3 + 3x^2 + 7x - 5$ are greater than or equal to -4.

Solution We divide $P(x)$ by $x + 4$ using synthetic division:

−4	1	4	3	7	−5
		−4	0	−12	20
	1	0	3	−5	15

Alternately nonnegative and nonpositive

Since 0 can be considered either nonnegative or nonpositive, the entries alternate in sign. So -4 is a lower bound for the real zeros of P.

 **Now Try Exercise 73** ■

Example 7 ■ Factoring a Fifth-Degree Polynomial

Factor completely the polynomial

$$P(x) = 2x^5 + 5x^4 - 8x^3 - 14x^2 + 6x + 9$$

Solution The possible rational zeros of P are $\pm\frac{1}{2}, \pm 1, \pm\frac{3}{2}, \pm 3, \pm\frac{9}{2}$, and ± 9. We check the positive candidates first, beginning with the smallest:

$\frac{1}{2}$	2	5	−8	−14	6	9
		1	3	$-\frac{5}{2}$	$-\frac{33}{4}$	$-\frac{9}{8}$
	2	6	−5	$-\frac{33}{2}$	$-\frac{9}{4}$	$\frac{63}{8}$

$\frac{1}{2}$ is not a zero

1	2	5	−8	−14	6	9
		2	7	−1	−15	−9
	2	7	−1	−15	−9	0

$P(1) = 0$

So 1 is a zero, and $P(x) = (x - 1)(2x^4 + 7x^3 - x^2 - 15x - 9)$. We continue by factoring the quotient. We still have the same list of possible zeros except that $\frac{1}{2}$ has been eliminated.

$$\begin{array}{r|rrrrr} 1 & 2 & 7 & -1 & -15 & -9 \\ & & 2 & 9 & 8 & -7 \\ \hline & 2 & 9 & 8 & -7 & -16 \end{array} \quad \text{1 is not a zero}$$

$$\begin{array}{r|rrrrr} \frac{3}{2} & 2 & 7 & -1 & -15 & -9 \\ & & 3 & 15 & 21 & 9 \\ \hline & 2 & 10 & 14 & 6 & 0 \end{array} \quad P(\tfrac{3}{2}) = 0, \text{ all entries nonnegative}$$

We see that $\frac{3}{2}$ is both a zero and an upper bound for the zeros of $P(x)$, so we do not need to check any further for positive zeros, because all the remaining candidates are greater than $\frac{3}{2}$.

$$\begin{aligned} P(x) &= (x - 1)(x - \tfrac{3}{2})(2x^3 + 10x^2 + 14x + 6) && \text{From synthetic division} \\ &= (x - 1)(2x - 3)(x^3 + 5x^2 + 7x + 3) && \text{Factor 2 from last factor, multiply into second factor} \end{aligned}$$

By Descartes's Rule of Signs, $x^3 + 5x^2 + 7x + 3$ has no positive zero, so its only possible rational zeros are -1 and -3:

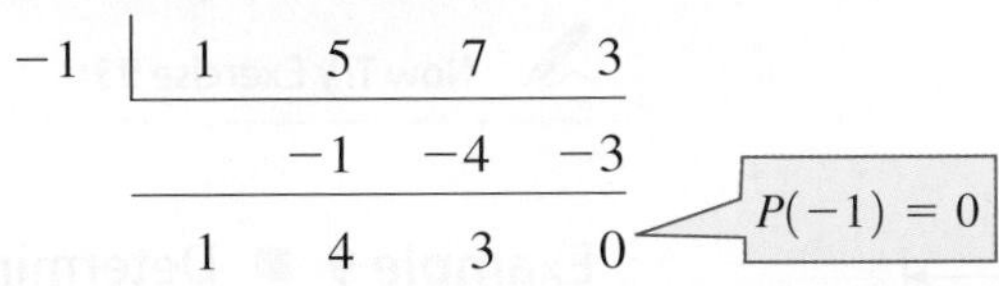

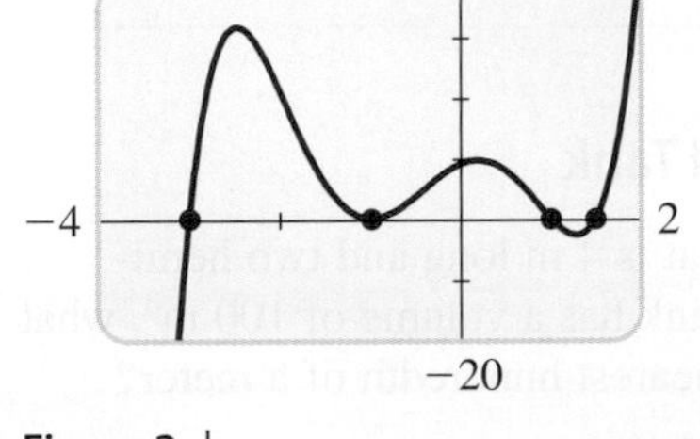

Figure 2 | $P(x) = 2x^5 + 5x^4 - 8x^3 - 14x^2 + 6x + 9 = (x - 1)(2x - 3)(x + 1)^2(x + 3)$

Therefore,

$$\begin{aligned} P(x) &= (x - 1)(2x - 3)(x + 1)(x^2 + 4x + 3) && \text{From synthetic division} \\ &= (x - 1)(2x - 3)(x + 1)^2(x + 3) && \text{Factor quadratic} \end{aligned}$$

This means that the zeros of P are 1, $\frac{3}{2}$, -1 (multiplicity 2), and -3. The graph of the polynomial is shown in Figure 2.

Now Try Exercise 81 ■

■ Using Algebra and Graphing Devices to Solve Polynomial Equations

In Section 1.11 we used graphing devices to solve equations graphically. When solving a polynomial equation graphically, we can use the algebraic techniques we've learned in this section to select an appropriate viewing rectangle (that is, a viewing rectangle that contains all the zeros of the polynomial).

Example 8 ■ Solving a Fourth-Degree Equation Graphically

Find all real solutions of the following equation, rounded to the nearest tenth:

$$3x^4 + 4x^3 - 7x^2 - 2x - 3 = 0$$

Solution To solve the equation graphically, we graph

$$P(x) = 3x^4 + 4x^3 - 7x^2 - 2x - 3$$

We use the Upper and Lower Bounds Theorem to see where the solutions can be found.

First we use the Upper and Lower Bounds Theorem to find two numbers between which all the solutions must lie. This allows us to choose a viewing rectangle that is certain to contain all the x-intercepts of P. We use synthetic division and proceed by trial and error.

To find an upper bound, we try the whole numbers, 1, 2, 3, . . . , as potential candidates. We see that 2 is an upper bound for the solutions:

2	3	4	−7	−2	−3	
		6	20	26	48	
	3	10	13	24	45	All positive

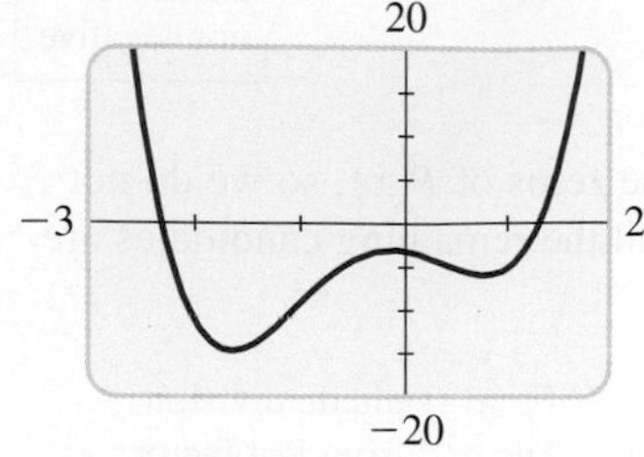

Figure 3 | $y = 3x^4 + 4x^3 - 7x^2 - 2x - 3$

Now we look for a lower bound, trying the numbers −1, −2, and −3,..., as potential candidates. We see that −3 is a lower bound for the solutions:

−3	3	4	−7	−2	−3	
		−9	15	−24	78	
	3	−5	8	−26	75	Entries alternate in sign

Thus all the solutions lie between −3 and 2. So the viewing rectangle $[-3, 2]$ by $[-20, 20]$ contains all the x-intercepts of P. The graph in Figure 3 has two x-intercepts, one between −3 and −2 and the other between 1 and 2. Zooming in, we find that the solutions of the equation, to the nearest tenth, are −2.3 and 1.3.

 **Now Try Exercise 95** ■

Example 9 ■ Determining the Size of a Fuel Tank

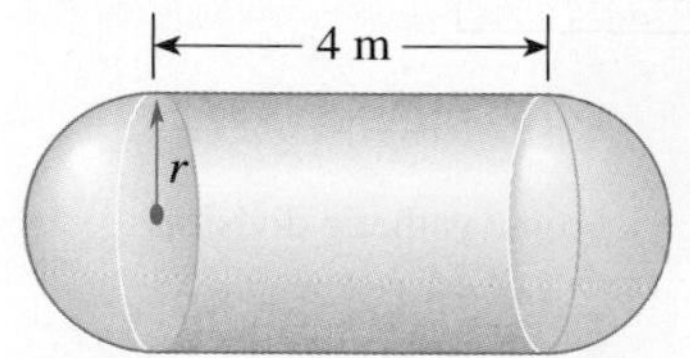
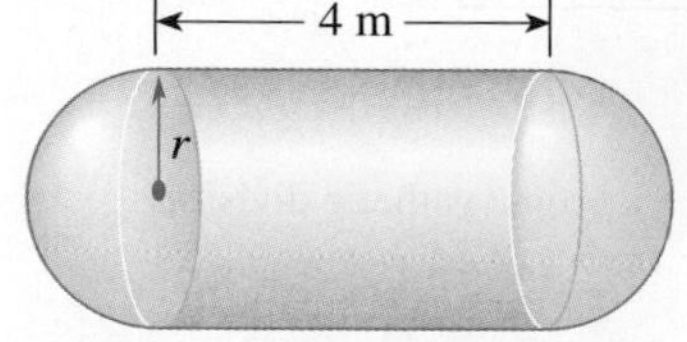

Figure 4

A fuel tank consists of a cylindrical center section that is 4 m long and two hemispherical end sections, as shown in Figure 4. If the tank has a volume of 100 m^3, what is the radius r (shown in the figure), rounded to the nearest hundredth of a meter?

Solution Using the volume formula listed on the inside front cover of this book, we see that the volume of the cylindrical section of the tank is

Volume of a cylinder: $V = \pi r^2 h$

$$\pi \cdot r^2 \cdot 4$$

The two hemispherical parts together form a complete sphere whose volume is

Volume of a sphere: $V = \frac{4}{3}\pi r^3$

$$\tfrac{4}{3}\pi r^3$$

Because the total volume of the tank is 100 m^3, we get the following equation:

$$\tfrac{4}{3}\pi r^3 + 4\pi r^2 = 100$$

A negative solution for r would be meaningless in this physical situation, and by substitution we can verify that $r = 3$ leads to a tank that is over 226 m^3 in volume, much larger than the required 100 m^3. Thus we know the correct radius lies somewhere between 0 and 3 m, so we use a viewing rectangle of $[0, 3]$ by $[50, 150]$ to graph the function $y = \frac{4}{3}\pi x^3 + 4\pi x^2$, as shown in Figure 5. Since we want the value of this function to be 100, we also graph the horizontal line $y = 100$ in the same viewing rectangle. The correct radius will be the x-coordinate of the point of intersection of the curve and the line. We see that at the point of intersection the x-coordinate is about $x \approx 2.15$. Thus the tank has a radius of about 2.15 m.

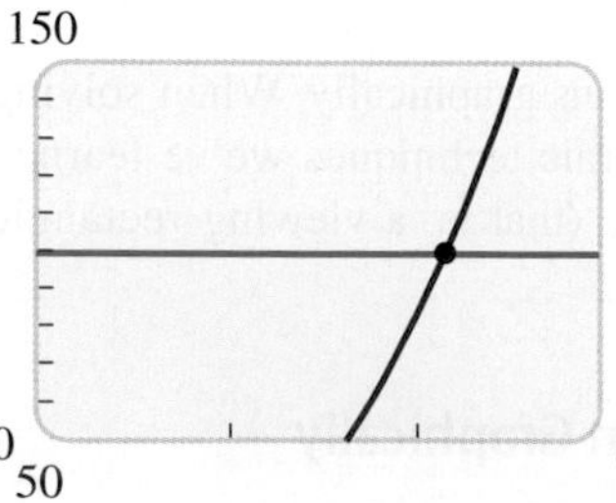

Figure 5 | $y = \frac{4}{3}\pi x^3 + 4\pi x^2$ and $y = 100$

Now Try Exercise 99 ■

Note that we also could have solved the equation in Example 9 by first writing it as

$$\tfrac{4}{3}\pi r^3 + 4\pi r^2 - 100 = 0$$

and then finding the x-intercept of the function $y = \frac{4}{3}\pi x^3 + 4\pi x^2 - 100$.

3.4 | Exercises

Concepts

1. If the polynomial function

$$P(x) = a_n x^n + a_{n-1}x^{n-1} + \cdots + a_1 x + a_0$$

has integer coefficients, then the only numbers that could possibly be rational zeros of P are all of the form $\frac{p}{q}$, where p is a factor of ________ and q is a factor of ________. The possible rational zeros of $P(x) = 6x^3 + 5x^2 - 19x - 10$ are ________________.

2. Using Descartes's Rule of Signs, we can tell that the polynomial $P(x) = x^5 - 3x^4 + 2x^3 - x^2 + 8x - 8$ has

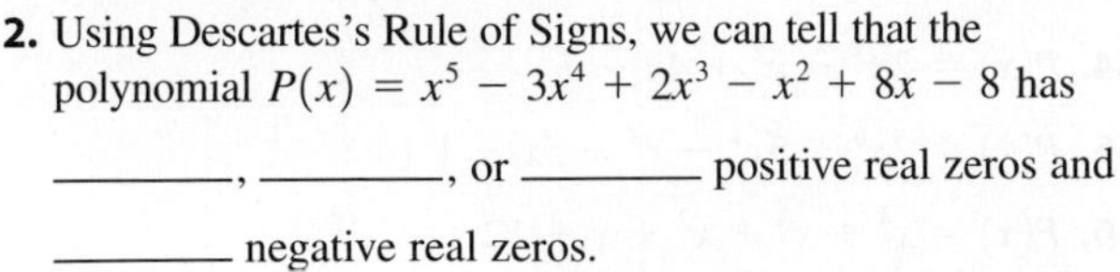

________, ________, or ________ positive real zeros and ________ negative real zeros.

3. *True or False?* If c is a real zero of the polynomial P, then all the other zeros of P are zeros of $P(x)/(x - c)$.

4. *True or False?* If a is an upper bound for the real zeros of the polynomial P, then $-a$ is necessarily a lower bound for the real zeros of P.

Skills

5–10 ■ Possible Rational Zeros List all possible rational zeros given by the Rational Zeros Theorem (but don't check to see which actually are zeros).

5. $P(x) = x^3 - 4x + 6$

6. $Q(x) = x^5 - 3x^2 + 5x + 10$

7. $R(x) = 3x^4 - 2x^3 + 8x^2 - 9$

8. $S(x) = 5x^6 - 3x^4 + 20x^2 - 15$

9. $T(x) = 6x^5 - 8x^3 + 5$

10. $U(x) = 12x^5 + 6x^3 - 2x - 8$

11–14 ■ Possible Rational Zeros A polynomial function P and its graph are given. **(a)** List all possible rational zeros of P given by the Rational Zeros Theorem. **(b)** From the graph, determine which of the possible rational zeros are actually zeros.

11. $P(x) = 5x^3 - x^2 - 5x + 1$

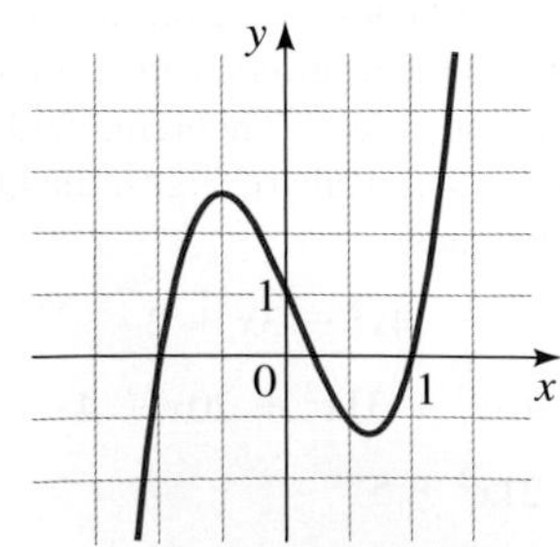

12. $P(x) = 3x^3 + 4x^2 - x - 2$

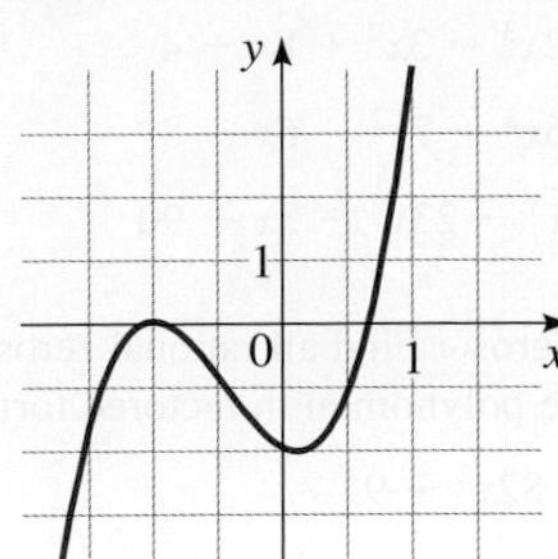

13. $P(x) = 2x^4 - 9x^3 + 9x^2 + x - 3$

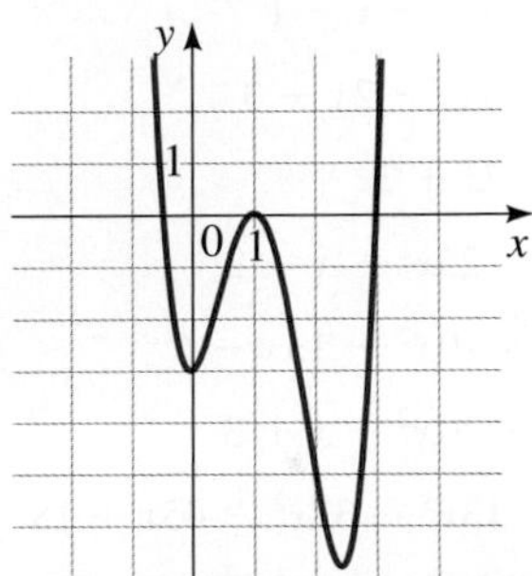

14. $P(x) = 4x^4 - x^3 - 4x + 1$

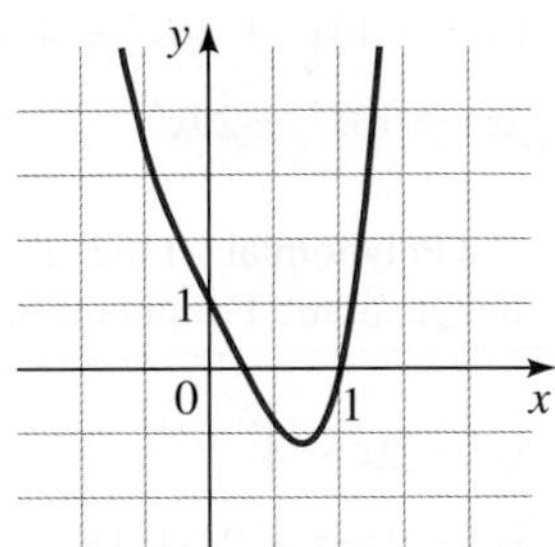

15–28 ■ Integer Zeros All the real zeros of the given polynomial are integers. Find the zeros, and write the polynomial in factored form.

15. $P(x) = x^3 - 5x^2 - 8x + 12$

16. $P(x) = x^3 - 4x^2 - 19x - 14$

17. $P(x) = x^3 - 5x^2 + 3x + 9$

18. $P(x) = x^3 - 3x - 2$

19. $P(x) = x^3 - 6x^2 + 12x - 8$

20. $P(x) = x^3 + 12x^2 + 48x + 64$

21. $P(x) = x^3 - 27x + 54$

22. $P(x) = x^3 + 5x^2 - 9x - 45$

23. $P(x) = x^3 + 3x^2 - x - 3$

24. $P(x) = x^3 - 4x^2 - 11x + 30$

25. $P(x) = x^4 - 5x^2 + 4$

26. $P(x) = x^4 - 2x^3 - 3x^2 + 8x - 4$

27. $P(x) = x^4 + 6x^3 + 7x^2 - 6x - 8$

28. $P(x) = x^4 - x^3 - 23x^2 - 3x + 90$

29–44 ■ Rational Zeros Find all rational zeros of the polynomial, and write the polynomial in factored form.

29. $P(x) = 9x^4 - 82x^2 + 9$

30. $P(x) = 6x^4 - 23x^3 - 13x^2 + 32x + 16$

31. $P(x) = 6x^4 + 7x^3 - 9x^2 - 7x + 3$

32. $P(x) = 6x^3 + 37x^2 + 5x - 6$

33. $P(x) = 4x^3 + 4x^2 - x - 1$

34. $P(x) = 2x^3 - 3x^2 - 2x + 3$

35. $P(x) = 4x^3 - 7x + 3$

36. $P(x) = 12x^3 - 25x^2 + x + 2$

37. $P(x) = 24x^3 + 10x^2 - 13x - 6$

38. $P(x) = 12x^3 - 20x^2 + x + 3$

39. $P(x) = 6x^4 + 13x^3 - 32x^2 - 45x + 18$

40. $P(x) = 2x^4 + 11x^3 + 11x^2 - 15x - 9$

41. $P(x) = x^5 + 3x^4 - 9x^3 - 31x^2 + 36$

42. $P(x) = x^5 - 4x^4 - 3x^3 + 22x^2 - 4x - 24$

43. $P(x) = 3x^5 - 14x^4 - 14x^3 + 36x^2 + 43x + 10$

44. $P(x) = 2x^6 - 3x^5 - 13x^4 + 29x^3 - 27x^2 + 32x - 12$

45–54 ■ Real Zeros of a Polynomial Find all the real zeros of the polynomial. Use the Quadratic Formula if necessary, as in Example 3(a).

45. $P(x) = 3x^3 + 5x^2 - 2x - 4$

46. $P(x) = 3x^4 - 5x^3 - 16x^2 + 7x + 15$

47. $P(x) = x^4 - 6x^3 + 4x^2 + 15x + 4$

48. $P(x) = x^4 + 2x^3 - 2x^2 - 3x + 2$

49. $P(x) = x^4 - 7x^3 + 14x^2 - 3x - 9$

50. $P(x) = x^5 - 4x^4 - x^3 + 10x^2 + 2x - 4$

51. $P(x) = 4x^3 - 6x^2 + 1$

52. $P(x) = 3x^3 - 5x^2 - 8x - 2$

53. $P(x) = 2x^4 + 15x^3 + 17x^2 + 3x - 1$

54. $P(x) = 4x^5 - 18x^4 - 6x^3 + 91x^2 - 60x + 9$

55–62 ■ Real Zeros of a Polynomial A polynomial P is given. **(a)** Find all the real zeros of P. **(b)** Sketch a graph of P.

55. $P(x) = -x^3 - 3x^2 + x + 3$

56. $P(x) = x^3 + 3x^2 - 6x - 8$

57. $P(x) = 2x^3 - 7x^2 + 4x + 4$

58. $P(x) = 3x^3 + 17x^2 + 21x - 9$

59. $P(x) = x^4 - 5x^3 + 6x^2 + 4x - 8$

60. $P(x) = -x^4 + 10x^2 + 8x - 8$

61. $P(x) = x^5 - x^4 - 5x^3 + x^2 + 8x + 4$

62. $P(x) = x^5 + 2x^4 - 8x^3 - 16x^2 + 16x + 32$

63–68 ■ Descartes's Rule of Signs Use Descartes's Rule of Signs to determine how many positive and how many negative real zeros the polynomial can have. Then determine the possible total number of real zeros.

63. $P(x) = x^3 - x^2 - x - 3$

64. $P(x) = 2x^3 - x^2 + 4x - 7$

65. $P(x) = 2x^6 + 5x^4 - x^3 - 5x - 1$

66. $P(x) = x^4 + x^3 + x^2 + x + 12$

67. $P(x) = x^5 + 4x^3 - x^2 + 6x$

68. $P(x) = x^8 - x^5 + x^4 - x^3 + x^2 - x + 1$

69–76 ■ Upper and Lower Bounds Show that the given values for a and b are lower and upper bounds for the real zeros of the polynomial.

69. $P(x) = 2x^3 + 5x^2 + x - 2;\quad a = -3, b = 1$

70. $P(x) = x^4 - 2x^3 - 9x^2 + 2x + 8;\quad a = -3, b = 5$

71. $P(x) = 8x^3 + 10x^2 - 39x + 9;\quad a = -3, b = 2$

72. $P(x) = 3x^4 - 17x^3 + 24x^2 - 9x + 1;\quad a = 0, b = 6$

73. $P(x) = x^4 + 2x^3 + 3x^2 + 5x - 1;\quad a = -2, b = 1$

74. $P(x) = x^4 + 3x^3 - 4x^2 - 2x - 7;\quad a = -4, b = 2$

75. $P(x) = 2x^4 - 6x^3 + x^2 - 2x + 3;\quad a = -1, b = 3$

76. $P(x) = 3x^4 - 5x^3 - 2x^2 + x - 1;\quad a = -1, b = 2$

77–80 ■ Upper and Lower Bounds Find integers that are upper and lower bounds for the real zeros of the polynomial.

77. $P(x) = x^3 - 3x^2 + 4$

78. $P(x) = 2x^3 - 3x^2 - 8x + 12$

79. $P(x) = x^4 - 2x^3 + x^2 - 9x + 2$

80. $P(x) = x^5 - x^4 + 1$

81–86 ■ Zeros of a Polynomial Find all rational zeros of the polynomial, and then find the irrational zeros, if any. Whenever appropriate, use the Rational Zeros Theorem, the Upper and Lower Bounds Theorem, Descartes's Rule of Signs, the Quadratic Formula, or other factoring techniques.

81. $P(x) = 2x^4 + 3x^3 - 4x^2 - 3x + 2$

82. $P(x) = 2x^4 + 15x^3 + 31x^2 + 20x + 4$

83. $P(x) = 4x^4 - 21x^2 + 5$

84. $P(x) = 6x^4 - 7x^3 - 8x^2 + 5x$

85. $P(x) = x^5 - 7x^4 + 9x^3 + 23x^2 - 50x + 24$

86. $P(x) = 8x^5 - 14x^4 - 22x^3 + 57x^2 - 35x + 6$

87–90 ■ Polynomials With No Rational Zeros Show that the polynomial does not have any rational zeros.

87. $P(x) = x^3 - x - 2$

88. $P(x) = 2x^4 - x^3 + x + 2$

89. $P(x) = 3x^3 - x^2 - 6x + 12$

90. $P(x) = x^{50} - 5x^{25} + x^2 - 1$

91–94 ■ Verifying Zeros Using a Graphing Device The real solutions of the given equation are rational. Use a graph of the polynomial in the given viewing rectangle and the list of possible rational zeros (given by the Rational Zeros Theorem) to determine which values in the list are actually solutions. (All solutions can be seen in the given viewing rectangle.)

91. $x^3 - 3x^2 - 4x + 12 = 0$; $[-4, 4]$ by $[-15, 15]$

92. $x^4 - 5x^2 + 4 = 0$; $[-4, 4]$ by $[-30, 30]$

93. $2x^4 - 5x^3 - 14x^2 + 5x + 12 = 0$; $[-2, 5]$ by $[-40, 40]$

94. $3x^3 + 8x^2 + 5x + 2 = 0$; $[-3, 3]$ by $[-10, 10]$

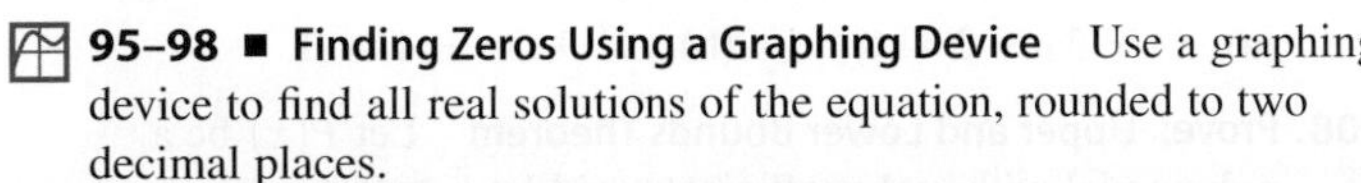

95–98 ■ Finding Zeros Using a Graphing Device Use a graphing device to find all real solutions of the equation, rounded to two decimal places.

95. $x^4 - x - 4 = 0$

96. $2x^3 - 8x^2 + 9x - 9 = 0$

97. $4.00x^4 + 4.00x^3 - 10.96x^2 - 5.88x + 9.09 = 0$

98. $x^5 + 2.00x^4 + 0.96x^3 + 5.00x^2 + 10.00x + 4.80 = 0$

Applications

99. Volume of a Silo A grain silo consists of a cylindrical main section surmounted by a hemispherical roof. If the total volume of the silo (including the part inside the roof section) is 432 m^3 and the cylindrical part is 10 m tall, what is the radius of the silo, rounded to the nearest tenth of a meter?

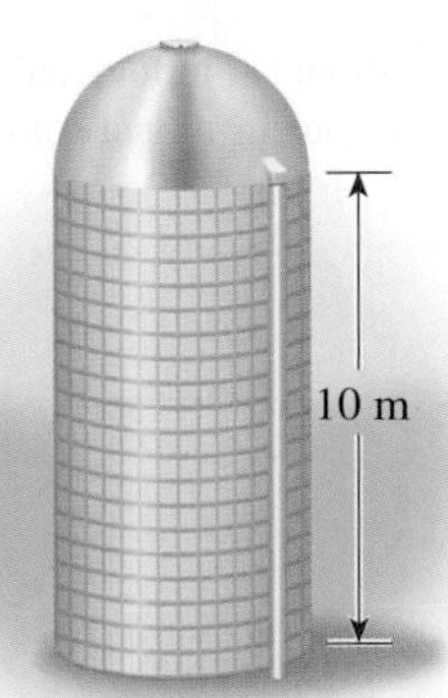

100. Volume of a Box An open box with a volume of 1500 cm^3 is to be constructed by taking a piece of cardboard 20 cm by 40 cm, cutting squares of side length x cm from each corner, and folding up the sides. Show that this can be done in two different ways, and find the exact dimensions of the box in each case.

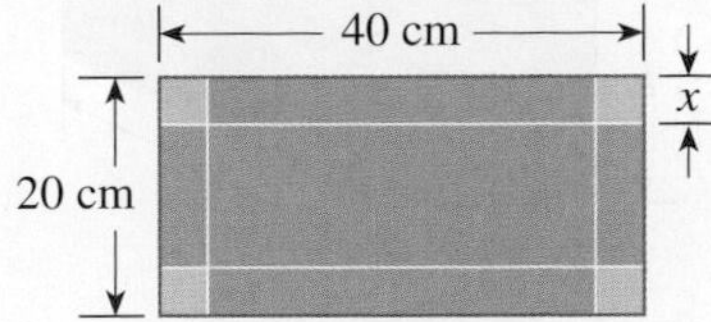

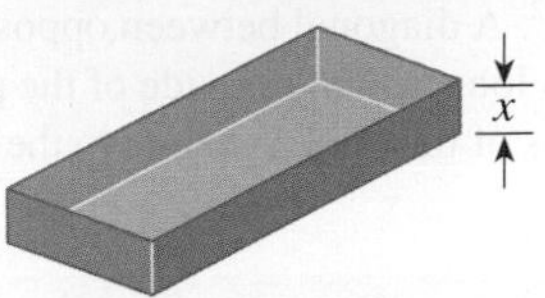

101. Volume of a Rocket A rocket consists of a right circular cylinder of height 20 m surmounted by a cone whose height and diameter are equal and whose radius is the same as that of the cylindrical section. What should this radius be (rounded to two decimal places) if the total volume is to be $500\pi/3$ m^3?

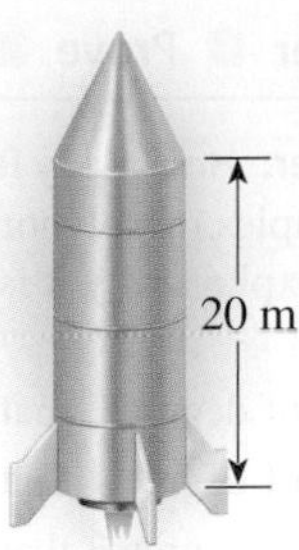

102. Volume of a Box A rectangular box with a volume of $2\sqrt{2}$ m^3 has a square base as shown below. The diagonal of the box (between a pair of opposite corners) is 1 m longer than each side of the base.

(a) If the base has sides of length x meters, show that

$$x^6 - 2x^5 - x^4 + 8 = 0$$

(b) Show that two different boxes satisfy the given conditions. Find the dimensions in each case, rounded to the nearest hundredth of a meter.

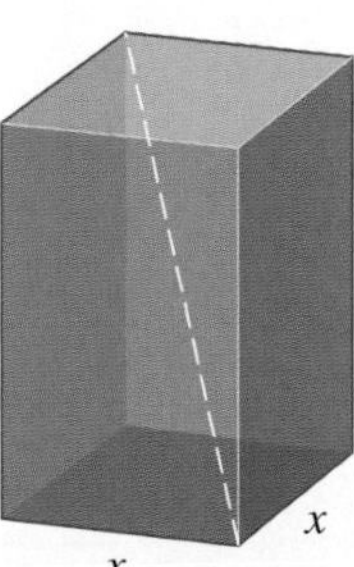

103. Girth of a Box A box with a square base has length plus girth of 108 cm. (Girth is the distance "around" the box.) What is the length of the box if its volume is 2200 cm^3?

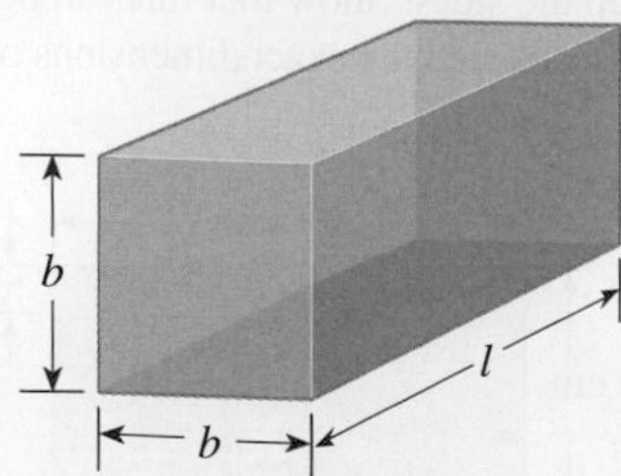

104. Dimensions of a Lot A rectangular parcel of land has an area of 5000 m^2. A diagonal between opposite corners is measured to be 10 m longer than one side of the parcel. What are the dimensions of the land, rounded to the nearest meter?

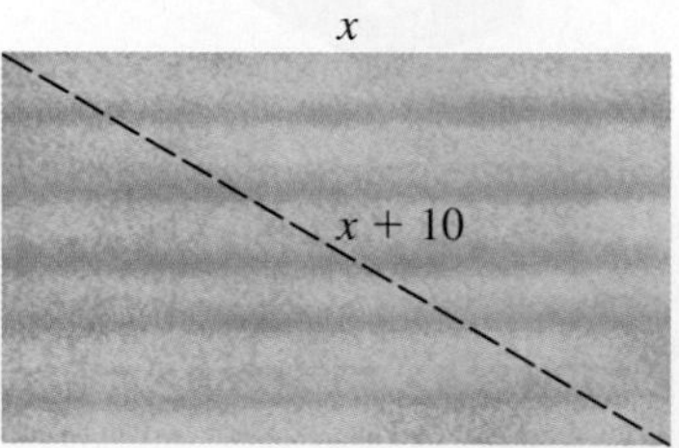

■ Discuss ■ Discover ■ Prove ■ Write

105. Discuss ■ Discover: How Many Real Zeros Can a Polynomial Have? Give examples of polynomials that have the following properties, or explain why it is impossible to find such a polynomial.

(a) A polynomial of degree 3 that has no real zero

(b) A polynomial of degree 4 that has no real zero

(c) A polynomial of degree 3 that has three real zeros, only one of which is rational

(d) A polynomial of degree 4 that has four real zeros, none of which is rational

What must be true about the degree of a polynomial with integer coefficients if it has no real zero?

106. Discuss ■ Prove: Depressed Cubics The most general cubic (third-degree) equation with rational coefficients can be written as

$$x^3 + ax^2 + bx + c = 0$$

(a) Show that the substitution $x = u - a/3$ transforms the general cubic into a *depressed cubic* (that is, a cubic in the variable u with no u^2 term):

$$u^3 + pu + q = 0$$

(See Exercise 1.5.142.)

(b) Use the procedure described in part (a) to depress the equation $x^3 + 6x^2 + 9x + 4 = 0$.

107. Discuss: The Cubic Formula The Quadratic Formula can be used to solve any quadratic (second-degree) equation. You might have wondered whether similar formulas exist for cubic (third-degree), quartic (fourth-degree), and higher-degree equations. For the depressed cubic $x^3 + px + q = 0$, Cardano (Section 3.5) found the following formula for one solution:

$$x = \sqrt[3]{\frac{-q}{2} + \sqrt{\frac{q^2}{4} + \frac{p^3}{27}}} + \sqrt[3]{\frac{-q}{2} - \sqrt{\frac{q^2}{4} + \frac{p^3}{27}}}$$

A formula for quartic equations was discovered by the Italian mathematician Ferrari in 1540. In 1824 the Norwegian mathematician Niels Henrik Abel proved that it is impossible to write a quintic formula, that is, a formula for fifth-degree equations. Finally, Galois (see the biography in this section) gave a criterion for determining which equations can be solved by a formula involving radicals.

Use the formula given above to find a solution for the following equations. Then solve the equations using the methods you learned in this section. Which method is easier?

(a) $x^3 - 3x + 2 = 0$ **(b)** $x^3 + 3x + 4 = 0$

108. Prove: Upper and Lower Bounds Theorem Let $P(x)$ be a polynomial with real coefficients, and let $b > 0$. Use the Division Algorithm (Section 3.3) to write

$$P(x) = (x - b) \cdot Q(x) + r$$

Suppose that $r \geq 0$ and that all the coefficients in $Q(x)$ are nonnegative. Let $z > b$.

(a) Show that $P(z) > 0$.

(b) Prove the first part of the Upper and Lower Bounds Theorem.

(c) Use the first part of the Upper and Lower Bounds Theorem to prove the second part. [*Hint:* Show that if $P(x)$ satisfies the second part of the theorem, then $P(-x)$ satisfies the first part.]

109. Prove: Number of Rational and Irrational Solutions Show that the equation

$$x^5 - x^4 - x^3 - 5x^2 - 12x - 6 = 0$$

has exactly one rational solution, and then prove that it must have either two or four irrational solutions.

3.5 Complex Zeros and the Fundamental Theorem of Algebra

■ The Fundamental Theorem of Algebra and Complete Factorization ■ Zeros and Their Multiplicities ■ Complex Zeros Occur in Conjugate Pairs ■ Linear and Quadratic Factors

We have already seen that an nth-degree polynomial can have at most n real zeros. In the complex number system an nth-degree polynomial has exactly n zeros (counting multiplicity) and so can be factored into exactly n linear factors. This fact is a consequence of the Fundamental Theorem of Algebra, which was proved by the German mathematician C. F. Gauss in 1799 (see the biography in this section).

■ The Fundamental Theorem of Algebra and Complete Factorization

The following theorem is the basis for much of our work in factoring polynomials and solving polynomial equations.

Fundamental Theorem of Algebra

Every polynomial

$$P(x) = a_n x^n + a_{n-1}x^{n-1} + \cdots + a_1 x + a_0 \qquad (n \geq 1, a_n \neq 0)$$

with complex coefficients has at least one complex zero.

Complex numbers are studied in Section 1.6.

Because any real number is also a complex number, the theorem applies to polynomials with real coefficients as well.

The Fundamental Theorem of Algebra and the Factor Theorem together show that a polynomial can be factored completely into linear factors, as we now prove.

Complete Factorization Theorem

If $P(x)$ is a polynomial of degree $n \geq 1$, then there exist complex numbers $a, c_1, c_2, \ldots, c_n$ (with $a \neq 0$) such that

$$P(x) = a(x - c_1)(x - c_2)\cdots(x - c_n)$$

Proof By the Fundamental Theorem of Algebra, P has at least one zero. Let's call it c_1. By the Factor Theorem (see Section 3.3), $P(x)$ can be factored as

$$P(x) = (x - c_1)Q_1(x)$$

where $Q_1(x)$ is of degree $n - 1$. Applying the Fundamental Theorem to the quotient $Q_1(x)$ gives us the factorization

$$P(x) = (x - c_1)(x - c_2)Q_2(x)$$

where $Q_2(x)$ is of degree $n - 2$ and c_2 is a zero of $Q_1(x)$. Continuing this process for n steps, we get a final quotient $Q_n(x)$ of degree 0, a nonzero constant that we will call a. This means that P has been factored as

$$P(x) = a(x - c_1)(x - c_2)\cdots(x - c_n)$$ ■

To find the complex zeros of an nth-degree polynomial, we usually first factor as much as possible, then use the Quadratic Formula on parts that we can't factor further.

Example 1 ■ Factoring a Polynomial Completely

Let $P(x) = x^3 - 3x^2 + x - 3$.

(a) Find all the zeros of P.

(b) Find the complete factorization of P.

Solution

(a) We first factor P as follows.

$$P(x) = x^3 - 3x^2 + x - 3 \qquad \text{Given}$$
$$= x^2(x - 3) + (x - 3) \qquad \text{Group terms}$$
$$= (x - 3)(x^2 + 1) \qquad \text{Factor } x - 3$$

We find the zeros of P by setting each factor equal to 0:

$$P(x) = (x - 3)(x^2 + 1)$$

This factor is 0 when $x = 3$

This factor is 0 when $x = i$ or $-i$

Setting $x - 3 = 0$, we see that $x = 3$ is a zero. Setting $x^2 + 1 = 0$, we get $x^2 = -1$, so $x = \pm i$. So the zeros of P are 3, i, and $-i$.

(b) Since the zeros are 3, i, and $-i$, the complete factorization of P is

$$P(x) = (x - 3)(x - i)[x - (-i)]$$
$$= (x - 3)(x - i)(x + i)$$

Now Try Exercise 7 ■

Example 2 ■ Factoring a Polynomial Completely

Let $P(x) = x^3 - 2x + 4$.

(a) Find all the zeros of P.

(b) Find the complete factorization of P.

Solution

−2	1	0	−2	4
		−2	4	−4
	1	−2	−2	0

(a) The possible rational zeros are the factors of 4, which are $\pm 1, \pm 2, \pm 4$. Using synthetic division (see the margin), we find that -2 is a zero, and the polynomial factors as

$$P(x) = (x + 2)(x^2 - 2x + 2)$$

This factor is 0 when $x = -2$

Use the Quadratic Formula to find when this factor is 0

To find the zeros, we set each factor equal to 0. Of course, $x + 2 = 0$ means that $x = -2$. We use the Quadratic Formula to find when the other factor is 0.

$$x^2 - 2x + 2 = 0 \qquad \text{Set factor equal to 0}$$

$$x = \frac{2 \pm \sqrt{4 - 8}}{2} \qquad \text{Quadratic Formula}$$

$$x = \frac{2 \pm 2i}{2} \qquad \text{Take square root}$$

$$x = 1 \pm i \qquad \text{Simplify}$$

So the zeros of P are -2, $1 + i$, and $1 - i$.

(b) Since the zeros are -2, $1 + i$, and $1 - i$, the complete factorization of P is

$$\begin{aligned} P(x) &= [x - (-2)][x - (1 + i)][x - (1 - i)] \\ &= (x + 2)(x - 1 - i)(x - 1 + i) \end{aligned}$$

Now Try Exercise 9 ■

■ Zeros and Their Multiplicities

In the Complete Factorization Theorem the numbers $c_1, c_2, \ldots, c_n$ are the zeros of P. These zeros need not all be different. If the factor $x - c$ appears k times in the complete factorization of $P(x)$, then we say that c is a zero of **multiplicity *k*** (see Section 3.2). For example, the polynomial

$$P(x) = (x - 1)^3(x + 2)^2(x + 3)^5$$

has the following zeros:

$$1 \text{ (multiplicity 3)} \qquad -2 \text{ (multiplicity 2)} \qquad -3 \text{ (multiplicity 5)}$$

The polynomial P has the same number of zeros as its degree: It has degree 10 and has 10 zeros, provided that we count multiplicities. This is true for all polynomials, as we prove in the following theorem.

Zeros Theorem

Every polynomial of degree $n \geq 1$ has exactly n zeros, provided that a zero of multiplicity k is counted k times.

Proof Let P be a polynomial of degree n. By the Complete Factorization Theorem

$$P(x) = a(x - c_1)(x - c_2) \cdots (x - c_n)$$

Now suppose that c is any given zero of P. Then

$$P(c) = a(c - c_1)(c - c_2) \cdots (c - c_n) = 0$$

Thus by the Zero-Product Property, one of the factors $c - c_i$ must be 0, so $c = c_i$ for some i. It follows that P has exactly the n zeros $c_1, c_2, \ldots, c_n$. ■

Example 3 ■ Factoring a Polynomial with Complex Zeros

Find the complete factorization and the five zeros of the polynomial

$$P(x) = 3x^5 + 24x^3 + 48x$$

Solution Since $3x$ is a common factor, we have

$$\begin{aligned} P(x) &= 3x(x^4 + 8x^2 + 16) \\ &= 3x(x^2 + 4)^2 \end{aligned}$$

This factor is 0 when $x = 0$

This factor is 0 when $x = 2i$ or $x = -2i$

Historical/Corbis Historical/Getty Images

CARL FRIEDRICH GAUSS (1777–1855) is considered the greatest mathematician of modern times. His contemporaries called him the "Prince of Mathematics." He was born into a poor family; his father made a living as a mason. As a very small child, Gauss found a calculation error in his father's accounts, the first of many incidents that gave evidence of his mathematical precocity. (See also Section 11.2.) At 19, Gauss demonstrated that the regular 17-sided polygon can be constructed with straight-edge and compass alone. This was remarkable because, since the time of Euclid, it had been thought that the only regular polygons constructible in this way were the triangle, the square, and the pentagon. Because of this discovery Gauss decided to pursue a career in mathematics instead of languages, his other passion. In his doctoral dissertation, written at the age of 22, Gauss proved the Fundamental Theorem of Algebra: A polynomial of degree n with complex coefficients has n zeros. His other accomplishments range over every branch of mathematics as well as physics and astronomy.

To factor $x^2 + 4$, note that $2i$ and $-2i$ are zeros of this polynomial. Thus $x^2 + 4 = (x - 2i)(x + 2i)$, so

$$\begin{aligned} P(x) &= 3x[(x - 2i)(x + 2i)]^2 \\ &= 3x(x - 2i)^2(x + 2i)^2 \end{aligned}$$

0 is a zero of multiplicity 1 | $2i$ is a zero of multiplicity 2 | $-2i$ is a zero of multiplicity 2

The zeros of P are 0, $2i$, and $-2i$. Since the factors $x - 2i$ and $x + 2i$ each occur twice in the complete factorization of P, the zeros $2i$ and $-2i$ are each of multiplicity 2 (or *double* zeros). Thus we have found all five zeros.

 **Now Try Exercise 31** ■

The following table gives further examples of polynomials with their complete factorizations and zeros.

Degree	Polynomial	Zero(s)	Number of Zeros
1	$P(x) = x - 4$	4	1
2	$P(x) = x^2 - 10x + 25$ $= (x - 5)(x - 5)$	5 (multiplicity 2)	2
3	$P(x) = x^3 + x$ $= x(x - i)(x + i)$	$0, i, -i$	3
4	$P(x) = x^4 + 18x^2 + 81$ $= (x - 3i)^2(x + 3i)^2$	$3i$ (multiplicity 2), $-3i$ (multiplicity 2)	4
5	$P(x) = x^5 - 2x^4 + x^3$ $= x^3(x - 1)^2$	0 (multiplicity 3), 1 (multiplicity 2)	5

Example 4 ■ Finding Polynomials with Specified Zeros

(a) Find a polynomial $P(x)$ of degree 4, with zeros i, $-i$, 2, and -2, and with $P(3) = 25$.

(b) Find a polynomial $Q(x)$ of degree 4, with zeros -2 and 0, where -2 is a zero of multiplicity 3.

Solution

(a) The required polynomial has the form

$$\begin{aligned} P(x) &= a(x - i)(x - (-i))(x - 2)(x - (-2)) \\ &= a(x^2 + 1)(x^2 - 4) && \text{Difference of squares} \\ &= a(x^4 - 3x^2 - 4) && \text{Multiply} \end{aligned}$$

We know that $P(3) = a(3^4 - 3 \cdot 3^2 - 4) = 50a = 25$, so $a = \frac{1}{2}$. Thus

$$P(x) = \tfrac{1}{2}x^4 - \tfrac{3}{2}x^2 - 2$$

(b) We require

$$\begin{aligned} Q(x) &= a[x - (-2)]^3(x - 0) \\ &= a(x + 2)^3x \\ &= a(x^3 + 6x^2 + 12x + 8)x && \text{Special Product Formula 4 (Section 1.3)} \\ &= a(x^4 + 6x^3 + 12x^2 + 8x) \end{aligned}$$

Since we are given no information about Q other than its zeros and their multiplicity, we can choose any number for a. If we use $a = 1$, we get

$$Q(x) = x^4 + 6x^3 + 12x^2 + 8x$$

Now Try Exercise 37 ■

Example 5 ■ Finding All the Zeros of a Polynomial

Find all four zeros of $P(x) = 3x^4 - 2x^3 - x^2 - 12x - 4$.

Solution Using the Rational Zeros Theorem from Section 3.4, we obtain the following list of possible rational zeros: $\pm 1, \pm 2, \pm 4, \pm\frac{1}{3}, \pm\frac{2}{3}, \pm\frac{4}{3}$. Checking these using synthetic division, we find that 2 and $-\frac{1}{3}$ are zeros, and we get the following factorization.

$$\begin{aligned} P(x) &= 3x^4 - 2x^3 - x^2 - 12x - 4 \\ &= (x-2)(3x^3 + 4x^2 + 7x + 2) && \text{Factor } x - 2 \\ &= (x-2)(x + \tfrac{1}{3})(3x^2 + 3x + 6) && \text{Factor } x + \tfrac{1}{3} \\ &= 3(x-2)(x + \tfrac{1}{3})(x^2 + x + 2) && \text{Factor 3} \end{aligned}$$

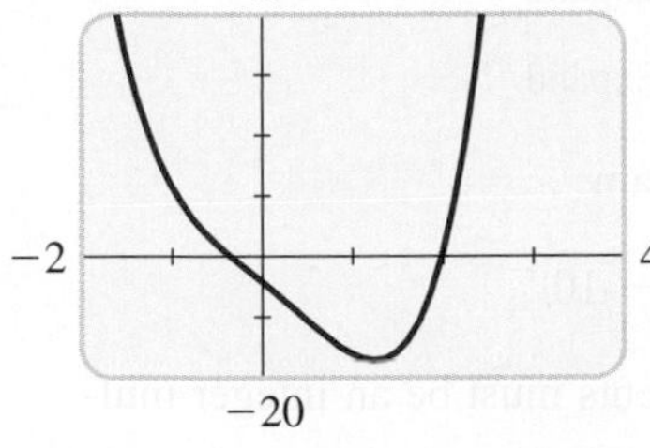

Figure 1 | $P(x) = 3x^4 - 2x^3 - x^2 - 12x - 4$

Figure 1 shows the graph of the polynomial P in Example 5. The x-intercepts correspond to the real zeros of P. The imaginary zeros cannot be determined from the graph.

The zeros of the quadratic factor are

$$x = \frac{-1 \pm \sqrt{1-8}}{2} = -\frac{1}{2} \pm \frac{\sqrt{7}}{2}i \qquad \text{Quadratic Formula}$$

so the zeros of $P(x)$ are

$$2, \quad -\frac{1}{3}, \quad -\frac{1}{2} + \frac{\sqrt{7}}{2}i, \quad \text{and} \quad -\frac{1}{2} - \frac{\sqrt{7}}{2}i$$

Now Try Exercise 47 ■

■ Complex Zeros Occur in Conjugate Pairs

As you may have noticed from the examples so far, the complex zeros of polynomials with real coefficients come in pairs. Whenever $a + bi$ is a zero, its complex conjugate $a - bi$ is also a zero.

Conjugate Zeros Theorem

If the polynomial P has real coefficients and if the complex number z is a zero of P, then its complex conjugate $\overline{z}$ is also a zero of P.

Proof Let

$$P(x) = a_n x^n + a_{n-1}x^{n-1} + \cdots + a_1 x + a_0$$

Properties of complex conjugates are stated in Exercises 1.6.77–84.

where each coefficient is real. Suppose that $P(z) = 0$. We must prove that $P(\overline{z}) = 0$. We use the facts that the complex conjugate of a sum of two complex numbers is the sum of the conjugates and that the conjugate of a product is the product of the conjugates.

$$\begin{aligned} P(\overline{z}) &= a_n(\overline{z})^n + a_{n-1}(\overline{z})^{n-1} + \cdots + a_1\overline{z} + a_0 \\ &= \overline{a_n}\,\overline{z^n} + \overline{a_{n-1}}\,\overline{z^{n-1}} + \cdots + \overline{a_1}\,\overline{z} + \overline{a_0} && \text{Because the coefficients are real} \\ &= \overline{a_n z^n} + \overline{a_{n-1}z^{n-1}} + \cdots + \overline{a_1 z} + \overline{a_0} \\ &= \overline{a_n z^n + a_{n-1}z^{n-1} + \cdots + a_1 z + a_0} \\ &= \overline{P(z)} = \overline{0} = 0 \end{aligned}$$

This shows that $\overline{z}$ is also a zero of $P(x)$, which proves the theorem. ■

North Wind Picture Archives/Alamy Stock Photo

GEROLAMO CARDANO (1501–1576) is certainly one of the most colorful figures in the history of mathematics. He was the best-known physician in Europe in his day, yet throughout his life he was plagued by numerous maladies, including ruptures, hemorrhoids, and an irrational fear of encountering rabid dogs. He was a doting father, but his beloved sons broke his heart—his favorite was eventually beheaded for murdering his own wife. Cardano was also a compulsive gambler; indeed, this vice might have driven him to write the *Book on Games of Chance*, the first study of probability from a mathematical point of view.

In Cardano's major mathematical work, the *Ars Magna*, he detailed the solution of the general third- and fourth-degree polynomial equations. At the time of its publication, mathematicians were uncomfortable even with negative numbers, but Cardano's formulas paved the way for the acceptance not just of negative numbers, but also of imaginary numbers, because they occurred naturally in solving polynomial equations. For example, for the cubic equation

$$x^3 - 15x - 4 = 0$$

one of his formulas gives the solution

$$x = \sqrt[3]{2 + \sqrt{-121}} + \sqrt[3]{2 - \sqrt{-121}}$$

(See Exercise 3.4.107.) This value for x actually turns out to be the *integer* 4, yet to find it, Cardano had to use the imaginary number $\sqrt{-121} = 11i$.

Example 6 ■ A Polynomial With a Specified Complex Zero

Find a polynomial $P(x)$ of degree 3 that has integer coefficients and zeros $\frac{1}{2}$ and $3 - i$.

Solution Since $3 - i$ is a zero, then so is $3 + i$ by the Conjugate Zeros Theorem. This means that $P(x)$ must have the following form.

$$\begin{aligned}
P(x) &= a\left(x - \tfrac{1}{2}\right)[x - (3 - i)][x - (3 + i)] \\
&= a\left(x - \tfrac{1}{2}\right)[(x - 3) + i][(x - 3) - i] && \text{Regroup} \\
&= a\left(x - \tfrac{1}{2}\right)[(x - 3)^2 - i^2] && \text{Difference of Squares Formula} \\
&= a\left(x - \tfrac{1}{2}\right)(x^2 - 6x + 10) && \text{Expand} \\
&= a\left(x^3 - \tfrac{13}{2}x^2 + 13x - 5\right) && \text{Expand}
\end{aligned}$$

To make all coefficients integers, we set $a = 2$ and obtain

$$P(x) = 2x^3 - 13x^2 + 26x - 10$$

Any other polynomial that satisfies the given requirements must be an integer multiple of this one.

 **Now Try Exercise 41** ■

■ Linear and Quadratic Factors

We have seen that a polynomial factors completely into linear factors if we use complex numbers. If we don't use complex numbers, then a polynomial with real coefficients can always be factored into linear and quadratic factors. (We use this property in Section 9.7 when we study partial fractions.) A quadratic polynomial with no real zero is called **irreducible** over the real numbers. Such a polynomial cannot be factored without using complex numbers.

Linear and Quadratic Factors Theorem

Every polynomial with real coefficients can be factored into a product of linear and irreducible quadratic factors with real coefficients.

Proof We first observe that if $c = a + bi$ is a complex number, then

$$\begin{aligned}
(x - c)(x - \bar{c}) &= [x - (a + bi)][x - (a - bi)] \\
&= [(x - a) - bi][(x - a) + bi] \\
&= (x - a)^2 - (bi)^2 \\
&= x^2 - 2ax + (a^2 + b^2)
\end{aligned}$$

The last expression is a quadratic with *real* coefficients.

Now, if P is a polynomial with real coefficients, then by the Complete Factorization Theorem

$$P(x) = a(x - c_1)(x - c_2) \cdots (x - c_n)$$

Since the complex roots occur in conjugate pairs, we can multiply the factors corresponding to each such pair to get a quadratic factor with real coefficients. This results in P being factored into linear and irreducible quadratic factors. ■

Example 7 ■ Factoring a Polynomial into Linear and Quadratic Factors

Let $P(x) = x^4 + 2x^2 - 8$.

(a) Factor P into linear and irreducible quadratic factors with real coefficients.

(b) Factor P completely into linear factors with complex coefficients.

Solution

(a)
$$\begin{aligned} P(x) &= x^4 + 2x^2 - 8 \\ &= (x^2 - 2)(x^2 + 4) \\ &= (x - \sqrt{2})(x + \sqrt{2})(x^2 + 4) \end{aligned}$$

The quadratic factor $x^2 + 4$ is irreducible because it has no real zero.

(b) To get the complete factorization, we factor the remaining quadratic factor:
$$\begin{aligned} P(x) &= (x - \sqrt{2})(x + \sqrt{2})(x^2 + 4) \\ &= (x - \sqrt{2})(x + \sqrt{2})(x - 2i)(x + 2i) \end{aligned}$$

 **Now Try Exercise 67** ■

3.5 Exercises

■ Concepts

1. The polynomial $P(x) = 5x^2(x - 4)^3(x + 7)$ has degree ________. It has zeros 0, 4, and ________. The zero 0 has multiplicity ________, and the zero 4 has multiplicity ________.

2. (a) If a is a zero of the polynomial P, then ________ must be a factor of $P(x)$.

(b) If a is a zero of multiplicity m of the polynomial P, then ________ must be a factor of $P(x)$ when we factor P completely.

3. A polynomial of degree $n \geq 1$ has exactly ________ zeros if a zero of multiplicity m is counted m times.

4. If the polynomial function P has real coefficients and if $a + bi$ is a zero of P, then ________ is also a zero of P. So if $3 + i$ is a zero of P, then ________ is also a zero of P.

5–6 ■ *True or False?* If *False*, give a reason.

5. Let $P(x) = x^4 + 1$.

(a) The polynomial P has four complex zeros.

(b) The polynomial P can be factored into linear factors with complex coefficients.

(c) Some of the zeros of P are real.

6. Let $P(x) = x^3 + x$.

(a) The polynomial P has three real zeros.

(b) The polynomial P has at least one real zero.

(c) The polynomial P can be factored into linear factors with real coefficients.

■ Skills

7–18 ■ **Complete Factorization** A polynomial P is given. **(a)** Find all zeros of P, real and complex. **(b)** Factor P completely.

7. $P(x) = x^4 + 4x^2$
8. $P(x) = x^5 + 9x^3$
9. $P(x) = x^3 - 2x^2 + 2x$
10. $P(x) = x^3 + x^2 + x$
11. $P(x) = x^4 + 2x^2 + 1$
12. $P(x) = x^4 - x^2 - 2$
13. $P(x) = x^4 - 16$
14. $P(x) = x^4 + 6x^2 + 9$
15. $P(x) = x^3 + 8$
16. $P(x) = x^3 - 8$
17. $P(x) = x^6 - 1$
18. $P(x) = x^6 - 7x^3 - 8$

19–36 ■ **Complete Factorization** Factor the polynomial completely, and find all its zeros. State the multiplicity of each zero.

19. $P(x) = x^4 + 16x^2$
20. $P(x) = 9x^6 + 16x^4$
21. $Q(x) = x^6 + 2x^5 + 2x^4$
22. $Q(x) = x^5 + x^4 + x^3$
23. $P(x) = x^3 + 4x$
24. $P(x) = x^3 - x^2 + x$
25. $Q(x) = x^4 - 1$
26. $Q(x) = x^4 - 625$
27. $P(x) = 16x^4 - 81$
28. $P(x) = x^3 - 64$
29. $P(x) = x^3 + x^2 + 9x + 9$
30. $P(x) = x^6 - 729$
31. $P(x) = x^6 + 10x^4 + 25x^2$
32. $P(x) = x^5 + 18x^3 + 81x$
33. $P(x) = x^4 + 3x^2 - 4$
34. $P(x) = x^5 + 7x^3$
35. $P(x) = x^5 + 6x^3 + 9x$
36. $P(x) = x^6 + 16x^3 + 64$

37–46 ■ Finding a Polynomial with Specified Zeros Find a polynomial with integer coefficients that satisfies the given conditions.

37. P has degree 2 and zeros $1 + i$ and $1 - i$.

38. P has degree 2 and zeros $1 + \sqrt{2}i$ and $1 - \sqrt{2}i$.

39. Q has degree 3 and zeros 3, $2i$, and $-2i$.

40. Q has degree 3 and zeros 0 and i.

41. P has degree 3 and zeros 2 and i.

42. Q has degree 3 and zeros -3 and $1 + i$.

43. R has degree 4 and zeros $1 - 2i$ and 1, with 1 a zero of multiplicity 2.

44. S has degree 4 and zeros $2i$ and $3i$.

45. T has degree 4, zeros i and $1 + i$, and constant term 12.

46. U has degree 5, zeros $\frac{1}{2}$, -1, and $-i$, and leading coefficient 4; the zero -1 has multiplicity 2.

47–64 ■ Finding Complex Zeros Find all zeros of the polynomial.

47. $P(x) = x^3 - 2x - 4$

48. $P(x) = x^3 - 7x^2 + 16x - 10$

49. $P(x) = x^3 - 2x^2 + 2x - 1$

50. $P(x) = x^3 + 7x^2 + 18x + 18$

51. $P(x) = x^3 - 3x^2 + 3x - 2$

52. $P(x) = x^3 - x - 6$

53. $P(x) = 2x^3 + 7x^2 + 12x + 9$

54. $P(x) = 2x^3 - 8x^2 + 9x - 9$

55. $P(x) = x^4 + x^3 + 7x^2 + 9x - 18$

56. $P(x) = x^4 - 2x^3 - 2x^2 - 2x - 3$

57. $P(x) = x^5 - x^4 + 7x^3 - 7x^2 + 12x - 12$
[*Hint:* Factor by grouping.]

58. $P(x) = x^5 + x^3 + 8x^2 + 8$ [*Hint:* Factor by grouping.]

59. $P(x) = x^4 - 6x^3 + 13x^2 - 24x + 36$

60. $P(x) = x^4 - x^2 + 2x + 2$

61. $P(x) = 4x^4 + 4x^3 + 5x^2 + 4x + 1$

62. $P(x) = 4x^4 + 2x^3 - 2x^2 - 3x - 1$

63. $P(x) = x^5 - 3x^4 + 12x^3 - 28x^2 + 27x - 9$

64. $P(x) = x^5 - 2x^4 + 2x^3 - 4x^2 + x - 2$

65–70 ■ Linear and Quadratic Factors A polynomial P is given. **(a)** Factor P into linear and irreducible quadratic factors with real coefficients. **(b)** Factor P completely into linear factors with complex coefficients.

65. $P(x) = x^3 - 5x^2 + 4x - 20$

66. $P(x) = x^3 - 2x - 4$

67. $P(x) = x^4 + 8x^2 - 9$

68. $P(x) = x^4 + 8x^2 + 16$

69. $P(x) = x^6 - 64$

70. $P(x) = x^5 - 16x$

■ Skills Plus

71. Number of Real and Nonreal Solutions By the Zeros Theorem, every nth-degree polynomial equation has exactly n solutions (including possibly some that are repeated). Some of these may be real, and some may be nonreal. Use a graphing device to determine how many real and nonreal solutions each equation has.

(a) $x^4 - 2x^3 - 11x^2 + 12x = 0$

(b) $x^4 - 2x^3 - 11x^2 + 12x - 5 = 0$

(c) $x^4 - 2x^3 - 11x^2 + 12x + 40 = 0$

72–74 ■ Real and Nonreal Coefficients So far, we have worked with polynomials that have only real coefficients. These exercises involve polynomials with real and imaginary coefficients.

72. Find all solutions of each equation.

(a) $2x + 4i = 1$ **(b)** $x^2 - ix = 0$

(c) $x^2 + 2ix - 1 = 0$ **(d)** $ix^2 - 2x + i = 0$

73. (a) Show that $2i$ and $1 - i$ are both solutions of the equation

$$x^2 - (1 + i)x + (2 + 2i) = 0$$

but that their complex conjugates $-2i$ and $1 + i$ are not.

(b) Explain why the result of part (a) does not violate the Conjugate Zeros Theorem.

74. (a) Find the polynomial with *real* coefficients of the smallest possible degree for which i and $1 + i$ are zeros and the coefficient of the highest power is 1.

(b) Find the polynomial with *complex* coefficients of the smallest possible degree for which i and $1 + i$ are zeros and the coefficient of the highest power is 1.

■ Discuss ■ Discover ■ Prove ■ Write

75. Discuss: Polynomials of Odd Degree The Conjugate Zeros Theorem says that the complex zeros of a polynomial with real coefficients occur in complex conjugate pairs. Explain how this fact proves that a polynomial with real coefficients and odd degree has at least one real zero.

76. Discuss ■ Discover: Factoring and Graphing Let's explore how a graph can help us factor a polynomial. Factor

$$P(x) = 2x^4 - 3x^3 + 6x^2 - 12x - 8$$

by first graphing the polynomial, then using the graph to identify the real zeroes of P. Use those zeros to factor the polynomial into linear and quadratic factors, and then find the complex zeroes. Finally, express the polynomial as a product of linear factors with complex coefficients.

3.6 Rational Functions

■ The Rational Function $f(x) = 1/x$ ■ Vertical and Horizontal Asymptotes ■ Finding Vertical and Horizontal Asymptotes of Rational Functions ■ Graphing Rational Functions ■ Common Factors in Numerator and Denominator ■ Slant Asymptotes and End Behavior ■ Applications

We studied rational expressions in Section 1.4. In this section we study functions that are defined by rational expressions.

Rational Functions

A **rational function** is a function of the form

$$r(x) = \frac{P(x)}{Q(x)}$$

where P and Q are polynomials. We assume that $P(x)$ and $Q(x)$ have no factor in common. The domain of r consists of all real numbers x except those values for which the denominator is zero. That is, the domain of r is $\{x \mid Q(x) \neq 0\}$.

Here are some examples of rational functions.

$$f(x) = \frac{1}{x} \qquad r(x) = \frac{2x}{x+3} \qquad s(x) = \frac{x+2}{x^2+5x+1} \qquad u(x) = \frac{x^3+3x^2-5}{4x^3+3x+4}$$

For instance, the function $f(x) = 1/x$ is a rational function because the numerator is the polynomial $P(x) = 1$ and the denominator is the polynomial $Q(x) = x$. Although rational functions are constructed from polynomials, their graphs can look quite different from the graphs of polynomials. We begin by graphing the rational function $f(x) = 1/x$ because its graph contains many of the main features of the graphs of rational functions.

■ The Rational Function $f(x) = 1/x$

When graphing a rational function, we must pay special attention to the behavior of the graph near those x-values for which the denominator is zero.

Example 1 ■ Graph of $f(x) = 1/x$

Graph the rational function $f(x) = 1/x$, and state the domain and range.

Solution The function f is not defined for $x = 0$. The following tables show that when x is close to zero, the value of $|f(x)|$ is large, and the closer x gets to zero, the larger $|f(x)|$ gets.

For positive real numbers,

$$\frac{1}{\text{BIG NUMBER}} = \text{small number}$$

$$\frac{1}{\text{small number}} = \text{BIG NUMBER}$$

x	$f(x)$
-0.1	-10
-0.01	-100
-0.00001	$-100,000$
Approaching 0^-	Approaching $-\infty$

x	$f(x)$
0.1	10
0.01	100
0.00001	100,000
Approaching 0^+	Approaching ∞

The first table shows that as x approaches 0 from the left, the values of $y = f(x)$ decrease without bound. We describe this behavior in symbols and in words as follows.

$$f(x) \to -\infty \quad \text{as} \quad x \to 0^-$$ "y approaches negative infinity as x approaches 0 from the left"

The second table on the previous page shows that as x approaches 0 from the right, the values of $f(x)$ increase without bound. In symbols,

$$f(x) \to \infty \quad \text{as} \quad x \to 0^+$$ "y approaches infinity as x approaches 0 from the right"

The next two tables show how $f(x)$ changes as $|x|$ becomes large.

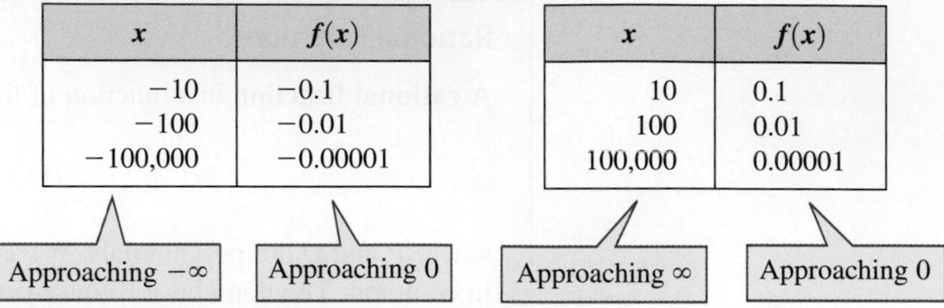

x	$f(x)$
-10	-0.1
-100	-0.01
$-100{,}000$	-0.00001

x	$f(x)$
10	0.1
100	0.01
100,000	0.00001

These tables show that as $|x|$ becomes large, the value of $f(x)$ gets closer and closer to zero. We describe this situation in symbols by writing

$$f(x) \to 0 \quad \text{as} \quad x \to -\infty \qquad \text{and} \qquad f(x) \to 0 \quad \text{as} \quad x \to \infty$$

Using the information in the preceding tables and plotting a few additional points, we obtain the graph shown in Figure 1.

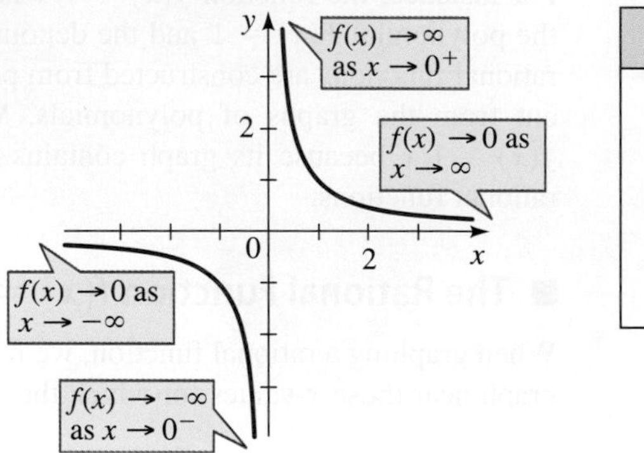

x	$f(x) = 1/x$
-2	$-\frac{1}{2}$
-1	-1
$-\frac{1}{2}$	-2
$\frac{1}{2}$	2
1	1
2	$\frac{1}{2}$

Figure 1 | $f(x) = 1/x$

Obtaining the domain and range of a function from its graph is explained in Section 2.3.

The function f is defined for all values of x other than 0, so the domain is $\{x \mid x \neq 0\}$. From the graph we see that the range is $\{y \mid y \neq 0\}$.

 **Now Try Exercise 11** ■

SJ Travel Photo and Video/Shutterstock.com

Discovery Project ■ Managing Traffic

A highway engineer wants to determine the optimal safe driving speed for a road. The higher the speed limit, the more cars the road can accommodate, but safety requires a greater following distance at higher speeds. In this project we find a rational function that models the carrying capacity of a road at a given traffic speed. The model can be used to determine the speed limit at which the road has its maximum carrying capacity. You can find the project at **www.stewartmath.com**.

In Example 1 we used the following **arrow notation**.

Symbol	Meaning
$x \to a^-$	x approaches a from the left
$x \to a^+$	x approaches a from the right
$x \to -\infty$	x goes to negative infinity; that is, x decreases without bound
$x \to \infty$	x goes to infinity; that is, x increases without bound

■ Vertical and Horizontal Asymptotes

Informally speaking, an *asymptote* of a function is a line that the graph of the function gets closer and closer to as one travels along that line. For instance, from Figure 1 we see that the line $x = 0$ is a *vertical asymptote* of the function $f(x) = 1/x$ and the line $y = 0$ is a *horizontal asymptote*. In general, we have the following definition.

Definition of Vertical and Horizontal Asymptotes

1. **Vertical Asymptotes.** The line $x = a$ is a **vertical asymptote** of the function $y = f(x)$ if y approaches $\pm\infty$ as x approaches a from the left or right.

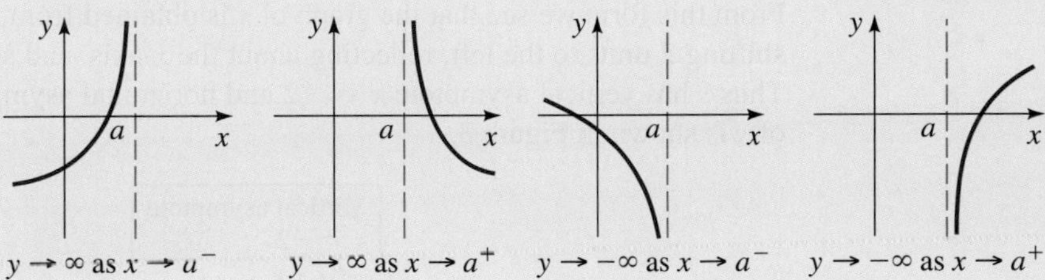

2. **Horizontal Asymptote.** The line $y = b$ is a **horizontal asymptote** of the function $y = f(x)$ if y approaches b as x approaches $\pm\infty$.

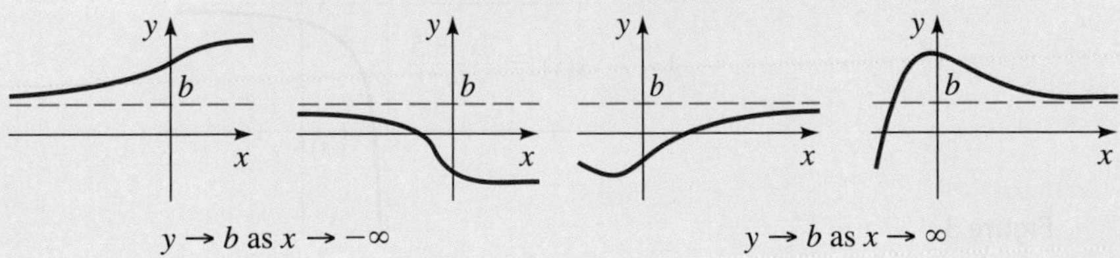

Transformations of functions are studied in Section 2.6.

In the following example we find the vertical and horizontal asymptotes of certain transformations of $f(x) = 1/x$. In general, a rational function of the form

$$r(x) = \frac{ax + b}{cx + d}$$

can be graphed by shifting, stretching, and/or reflecting the graph of $f(x) = 1/x$. (Functions of this form are called *linear fractional transformations.*)

Example 2 ■ Using Transformations to Graph Rational Functions

Graph each rational function, and state the domain and range.

(a) $r(x) = \dfrac{2}{x - 3}$ **(b)** $s(x) = \dfrac{3x + 5}{x + 2}$

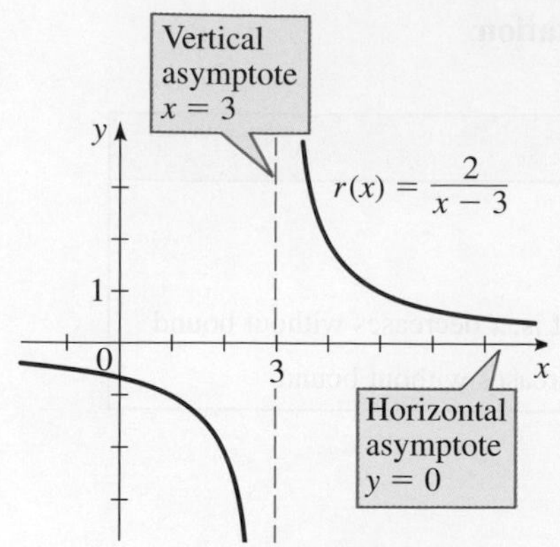

Figure 2

Solution

(a) Let $f(x) = 1/x$. Then we can express r in terms of f as follows:

$$\begin{aligned} r(x) &= \frac{2}{x-3} \\ &= 2\left(\frac{1}{x-3}\right) && \text{Factor 2} \\ &= 2(f(x-3)) && \text{Because } f(x) = 1/x \end{aligned}$$

From this form we see that the graph of r is obtained from the graph of f by shifting 3 units to the right and stretching vertically by a factor of 2. Thus r has vertical asymptote $x = 3$ and horizontal asymptote $y = 0$. The graph of r is shown in Figure 2.

The function r is defined for all x other than 3, so the domain is $\{x \mid x \neq 3\}$. From the graph we see that the range is $\{y \mid y \neq 0\}$.

$$x + 2\overline{)3x + 5} \quad 3 \qquad \frac{3x+6}{-1}$$

(b) Using long division we can express s in terms of f as follows.

$$\begin{aligned} s(x) &= 3 - \frac{1}{x+2} && \text{Long division (see margin)} \\ &= -\frac{1}{x+2} + 3 && \text{Rearrange terms} \\ &= -f(x+2) + 3 && \text{Since } f(x) = 1/x \end{aligned}$$

From this form we see that the graph of s is obtained from the graph of f by shifting 2 units to the left, reflecting about the x-axis, and shifting upward 3 units. Thus s has vertical asymptote $x = -2$ and horizontal asymptote $y = 3$. The graph of s is shown in Figure 3.

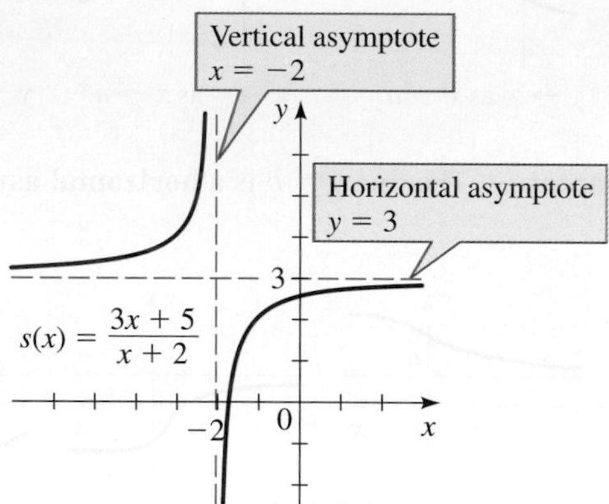

Figure 3

The function s is defined for all x other than -2, so the domain is $\{x \mid x \neq -2\}$. From the graph we see that the range is $\{y \mid y \neq 3\}$.

Now Try Exercises 17 and 19 ■

NOTE Example 2 shows that shifting the graph of a rational function vertically shifts the horizontal asymptotes; shifting the graph horizontally shifts the vertical asymptotes.

■ Finding Vertical and Horizontal Asymptotes of Rational Functions

In Example 2 we have found the asymptotes of a rational function by expressing the function in terms of transformations of the rational function $f(x) = 1/x$. But this method works only for rational functions for which both numerator and denominator are linear functions. For other rational functions we need to take a closer look at the behavior of the function near the values that make the denominator zero (for vertical asymptotes) as well as the behavior of the function as $x \to \pm\infty$ (for horizontal asymptotes).

Example 3 ■ Asymptotes of a Rational Function

Graph $r(x) = \dfrac{2x^2 - 4x + 5}{x^2 - 2x + 1}$, and state the domain and range.

Solution We first find the vertical and horizontal asymptotes of r.

Vertical asymptote. We first factor the denominator

$$r(x) = \frac{2x^2 - 4x + 5}{(x - 1)^2}$$

The line $x = 1$ is a vertical asymptote because the denominator of r is zero when $x = 1$.

To see what the graph of r looks like near the vertical asymptote, we make tables of values for x-values to the left and to the right of 1.

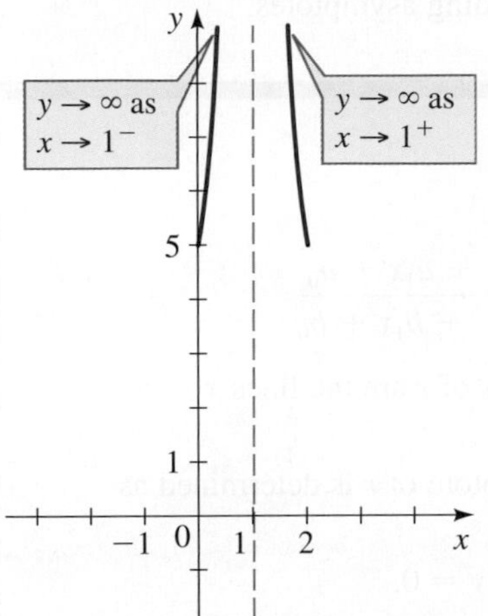

Figure 4

$x \to 1^-$

x	y
0	5
0.5	14
0.9	302
0.99	30,002

Approaching 1^- Approaching ∞

$x \to 1^+$

x	y
2	5
1.5	14
1.1	302
1.01	30,002

Approaching 1^+ Approaching ∞

From the tables we see that

$$y \to \infty \quad \text{as} \quad x \to 1^- \qquad \text{and} \qquad y \to \infty \quad \text{as} \quad x \to 1^+$$

Thus near the vertical asymptote $x = 1$, the graph of r has the shape shown in Figure 4.

Horizontal asymptote. The horizontal asymptote is the value that y approaches as $x \to \pm\infty$. To help us find this value, we divide both numerator and denominator by x^2, the highest power of x that appears in the denominator:

$$y = \frac{2x^2 - 4x + 5}{x^2 - 2x + 1} \cdot \frac{\frac{1}{x^2}}{\frac{1}{x^2}} = \frac{2 - \frac{4}{x} + \frac{5}{x^2}}{1 - \frac{2}{x} + \frac{1}{x^2}}$$

The fractional expressions $\frac{4}{x}, \frac{5}{x^2}, \frac{2}{x}$, and $\frac{1}{x^2}$ all approach 0 as $x \to \pm\infty$ (see Exercise 1.1.89). So as $x \to \pm\infty$, we have

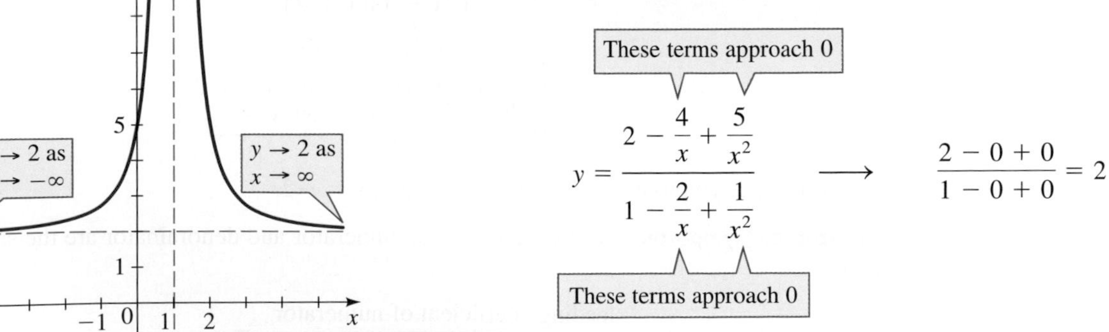

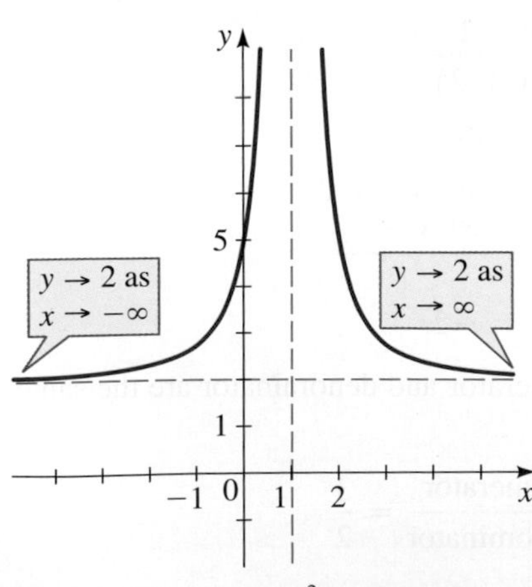

Figure 5 | $r(x) = \dfrac{2x^2 - 4x + 5}{x^2 - 2x + 1}$

Thus the horizontal asymptote is the line $y = 2$.

Since the graph must approach the horizontal asymptote, we can complete the graph as in Figure 5.

Domain and range. The function r is defined for all values of x other than 1, so the domain is $\{x \mid x \neq 1\}$. From the graph we see that the range is $\{y \mid y > 2\}$.

Now Try Exercise 47 ■

In Example 1 we saw that the vertical asymptotes are determined by the zeros of the denominator. In Example 3 we saw that the horizontal asymptote is determined by the leading coefficients of the numerator and denominator because, after dividing through by x^2 (the highest power of x), all other terms approach zero. In general, if $r(x) = P(x)/Q(x)$ and the degrees of P and Q are the same (both n, say), then dividing both numerator and denominator by x^n shows that the horizontal asymptote is

Recall that for a rational function $R(x) = P(x)/Q(x)$ we assume that $P(x)$ and $Q(x)$ have no factor in common.

$$y = \frac{\text{leading coefficient of } P}{\text{leading coefficient of } Q}$$

The following box summarizes the procedure for finding asymptotes.

Finding Asymptotes of Rational Functions

Let r be the rational function

$$r(x) = \frac{a_n x^n + a_{n-1}x^{n-1} + \cdots + a_1 x + a_0}{b_m x^m + b_{m-1}x^{m-1} + \cdots + b_1 x + b_0}$$

1. **Vertical Asymptotes** The vertical asymptotes of r are the lines $x = a$, where a is a zero of the denominator.
2. **Horizontal Asymptote** The horizontal asymptote of r is determined as follows:

   (a) If $n < m$, then r has horizontal asymptote $y = 0$.

   (b) If $n = m$, then r has horizontal asymptote $y = \dfrac{a_n}{b_m}$.

   (c) If $n > m$, then r has no horizontal asymptote.

Example 4 ■ Asymptotes of a Rational Function

Find the vertical and horizontal asymptotes of $r(x) = \dfrac{3x^2 - 2x - 1}{2x^2 + 3x - 2}$.

Solution

Vertical asymptotes. We first factor

$$r(x) = \frac{3x^2 - 2x - 1}{(2x - 1)(x + 2)}$$

This factor is 0 when $x = \frac{1}{2}$

This factor is 0 when $x = -2$

The vertical asymptotes are the lines $x = \frac{1}{2}$ and $x = -2$.

Horizontal asymptote. The degrees of the numerator and denominator are the same, and

$$\frac{\text{leading coefficient of numerator}}{\text{leading coefficient of denominator}} = \frac{3}{2}$$

Thus the horizontal asymptote is the line $y = \frac{3}{2}$.

To confirm our results, we graph r using a graphing device (see Figure 6).

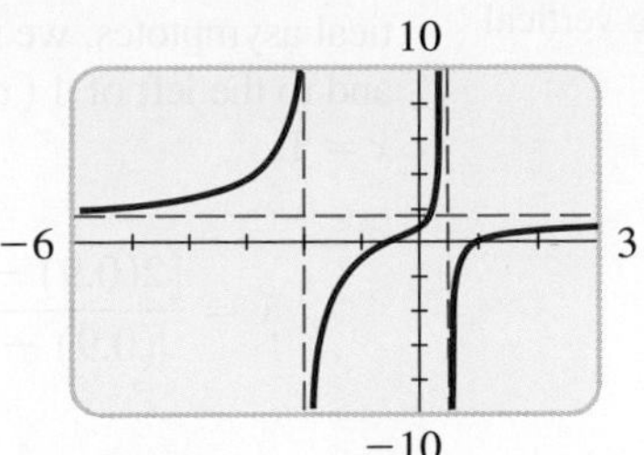

Figure 6 | $r(x) = \dfrac{3x^2 - 2x - 1}{2x^2 + 3x - 2}$

Now Try Exercises 35 and 37 ■

■ Graphing Rational Functions

In general, we use the following guidelines to graph rational functions.

A fraction is 0 only if its numerator is 0 (and the denominator is not 0).

Sketching Graphs of Rational Functions

1. **Factor.** Factor the numerator and denominator.
2. **Intercepts.** Find the x-intercepts by determining the zeros of the numerator and the y-intercept from the value of the function at $x = 0$.
3. **Vertical Asymptotes.** Find the vertical asymptotes by determining the zeros of the denominator, and then determine whether $y \to \infty$ or $y \to -\infty$ on each side of each vertical asymptote by using **test values.**
4. **Horizontal Asymptote.** Find the horizontal asymptote (if any), using the procedure described in the preceding box.
5. **Sketch the Graph.** Graph the information provided by steps 1–4. Then plot as many additional points as needed to fill in the rest of the graph of the function.

Example 5 ■ Graphing a Rational Function

Graph $r(x) = \dfrac{2x^2 + 7x - 4}{x^2 + x - 2}$, and state the domain and range.

Solution We factor the numerator and denominator, find the intercepts and asymptotes, and sketch the graph.

Factor. $y = \dfrac{(2x - 1)(x + 4)}{(x - 1)(x + 2)}$

x-Intercepts. The x-intercepts are the zeros of the numerator, $x = \frac{1}{2}$ and $x = -4$.

y-Intercept. To find the y-intercept, we substitute $x = 0$ into the original form of the function.

$$r(0) = \frac{2(0)^2 + 7(0) - 4}{(0)^2 + (0) - 2} = \frac{-4}{-2} = 2$$

The y-intercept is 2.

Vertical asymptotes. The vertical asymptotes occur where the denominator is 0, that is, where the function is undefined. From the factored form we see that the vertical asymptotes are the lines $x = 1$ and $x = -2$.

When choosing test values, we must make sure that there is no x-intercept between the test point and the vertical asymptote.

Behavior near vertical asymptotes. We need to know whether $y \to \infty$ or $y \to -\infty$ on each side of every vertical asymptote. To determine the sign of y for x-values near the vertical asymptotes, we use *test values*. For instance, as $x \to 1^-$, we use a test value close to and to the left of 1 ($x = 0.9$, say) to check whether y is positive or negative to the left of $x = 1$.

$$y = \frac{[2(0.9) - 1][(0.9) + 4]}{[(0.9) - 1][(0.9) + 2]} \quad \text{whose sign is} \quad \frac{(+)(+)}{(-)(+)} \quad \text{(negative)}$$

So $y \to -\infty$ as $x \to 1^-$. On the other hand, as $x \to 1^+$, we use a test value close to and to the right of 1 ($x = 1.1$, say), to get

$$y = \frac{[2(1.1) - 1][(1.1) + 4]}{[(1.1) - 1][(1.1) + 2]} \quad \text{whose sign is} \quad \frac{(+)(+)}{(+)(+)} \quad \text{(positive)}$$

So $y \to \infty$ as $x \to 1^+$. The other entries in the following table are calculated similarly.

As $x \to$	-2^-	-2^+	1^-	1^+
the sign of $y = \frac{(2x-1)(x+4)}{(x-1)(x+2)}$ is	$\frac{(-)(+)}{(-)(-)}$	$\frac{(-)(+)}{(-)(+)}$	$\frac{(+)(+)}{(-)(+)}$	$\frac{(+)(+)}{(+)(+)}$
so $y \to$	$-\infty$	∞	$-\infty$	∞

Horizontal asymptote. The degree of the numerator and the degree of the denominator are the same, and

$$\frac{\text{leading coefficient of numerator}}{\text{leading coefficient of denominator}} = \frac{2}{1} = 2$$

Thus the horizontal asymptote is the line $y = 2$.

Graph. We use the information we have found, together with some additional values, to sketch the graph in Figure 7.

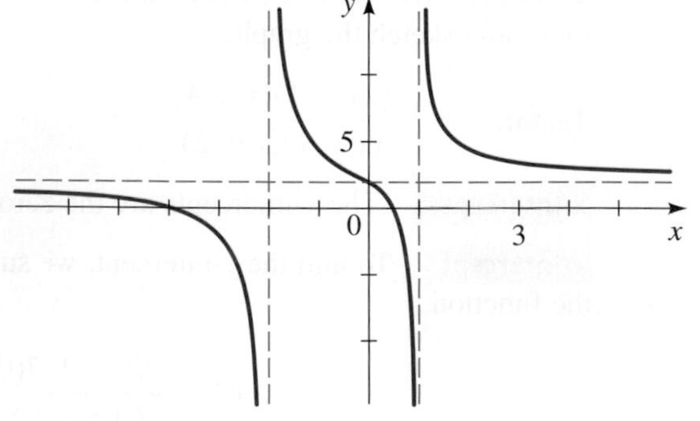

x	y
−6	0.93
−4	0
−3	−1.75
−1	4.50
0	2
0.5	0
1.5	6.29
2	4.50
6	2.75

Figure 7 | $r(x) = \dfrac{2x^2 + 7x - 4}{x^2 + x - 2}$

Domain and range. The domain is $\{x \mid x \neq 1, x \neq -2\}$. From the graph we see that the range is all real numbers.

Now Try Exercise 55 ■

Example 6 ■ Graphing a Rational Function

Graph the rational function $r(x) = \dfrac{x^2 - 4}{2x^2 + 2x}$. State the domain and estimate the range from the graph.

Solution

Factor. $y = \dfrac{(x + 2)(x - 2)}{2x(x + 1)}$

x-intercepts. -2 and 2, from $x + 2 = 0$ and $x - 2 = 0$

y-intercept. None, because $r(0)$ is undefined

Vertical asymptotes. $x = 0$ and $x = -1$, from the zeros of the denominator

Behavior near vertical asymptote.

As $x \to$	-1^-	-1^+	0^-	0^+
the sign of $y = \dfrac{(x+2)(x-2)}{2x(x+1)}$ is	$\dfrac{(+)(-)}{(-)(-)}$	$\dfrac{(+)(-)}{(-)(+)}$	$\dfrac{(+)(-)}{(-)(+)}$	$\dfrac{(+)(-)}{(+)(+)}$
so $y \to$	$-\infty$	∞	∞	$-\infty$

Horizontal asymptote. $y = \frac{1}{2}$ because the degree of the numerator and the degree of the denominator are the same and

$$\frac{\text{leading coefficient of numerator}}{\text{leading coefficient of denominator}} = \frac{1}{2}$$

Graph. We use the information we have found, together with some additional values, to sketch the graph in Figure 8.

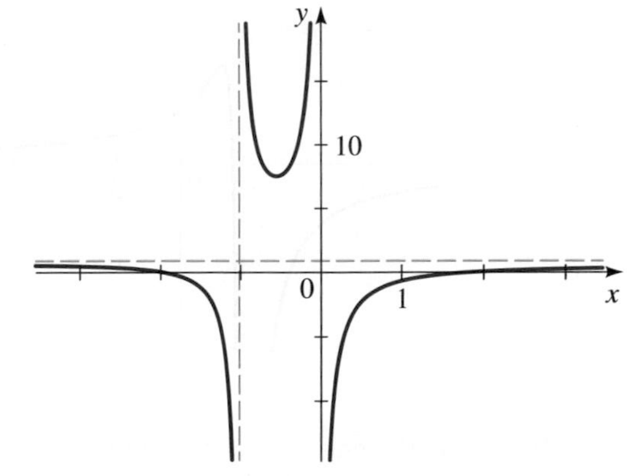

x	y
-2	0
-0.9	17.72
-0.5	7.50
-0.45	7.67
-0.4	8.00
-0.3	9.31
-0.1	22.17
2	0

Figure 8 | $r(x) = \dfrac{x^2 - 4}{2x^2 + 2x}$

Domain and range. The domain is $\{x \mid x \neq 0, x \neq -1\}$. From the graph we see that the range is approximately $\{y \mid y < \frac{1}{2} \text{ or } y > 7.5\}$.

 **Now Try Exercise 59** ■

Mathematics in the Modern World

Unbreakable Codes

If you read spy novels, you know about secret codes and how the hero "breaks" the code. Today secret codes have a much more common use. Most of the information that is stored on computers is coded to prevent unauthorized use. For example, your banking records, medical records, and school records are coded. Many cellular phones code the signal carrying your voice so that no one can listen in. Fortunately, because of recent advances in mathematics, today's codes are "unbreakable."

Modern codes are based on a simple principle: Factoring is much harder than multiplying. For example, try multiplying 78 and 93; now try factoring 9991. It takes a long time to factor 9991 because it is a product of two primes 97×103, so to factor it, we have to find one of these primes. Now imagine trying to factor a number N that is the product of two primes p and q, each about 200 digits long. Even the fastest computers would take many many years to factor such a number! But the same computer would take less than a second to multiply two such numbers. This fact was used by Ron Rivest, Adi Shamir, and Leonard Adleman in the 1970s to devise the RSA code. Their code uses an extremely large number to encode a message but requires us to know its factors to decode it; such a code is practically unbreakable.

The RSA code is an example of a "public key encryption" code. In such codes, anyone can code a message using a publicly known procedure based on N, but to decode the message, they must know p and q, the factors of N. When the RSA code was developed, it was thought that a carefully selected 80-digit number would provide an unbreakable code. But interestingly, recent advances in the study of factoring have made much larger numbers necessary.

Example 7 ■ Graphing a Rational Function

Graph $r(x) = \dfrac{5x + 21}{x^2 + 10x + 25}$. State the domain and estimate the range from the graph.

Solution

Factor. $y = \dfrac{5x + 21}{(x + 5)^2}$

x-Intercept. $-\dfrac{21}{5}$ from $5x + 21 = 0$

y-Intercept. $\dfrac{21}{25}$ because $r(0) = \dfrac{5 \cdot 0 + 21}{0^2 + 10 \cdot 0 + 25}$

$$= \frac{21}{25}$$

Vertical asymptote. $x = -5$, from the zeros of the denominator

Behavior near vertical asymptote.

As $x \to$	-5^-	-5^+
the sign of $y = \dfrac{5x + 21}{(x + 5)^2}$ is	$\dfrac{(-)}{(-)(-)}$	$\dfrac{(-)}{(+)(+)}$
so $y \to$	$-\infty$	$-\infty$

Horizontal asymptote. $y = 0$ because the degree of the numerator is less than the degree of the denominator

Graph. We use the information we have found, together with some additional values, to sketch the graph in Figure 9.

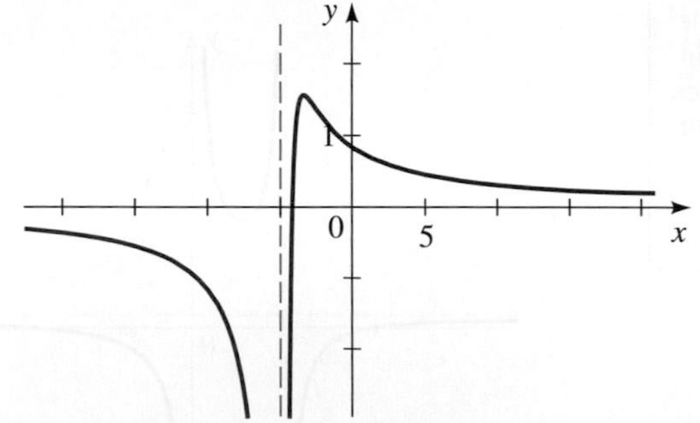

x	y
−15	−0.5
−4.2	0
−3	1.5
−1	1.0
0	0.84
3	0.6
10	0.3

Figure 9 | $r(x) = \dfrac{5x + 21}{x^2 + 10x + 25}$

Domain and range. The domain is $\{x \mid x \neq -5\}$. From the graph we see that the range is approximately the interval $(-\infty, 1.6]$.

Now Try Exercise 61 ■

From the graph in Figure 9 we see that, contrary to common misconception, a graph may cross a horizontal asymptote. The graph in Figure 9 crosses the x-axis (the horizontal asymptote) from below, reaches a maximum value near $x = -3$, and then approaches the x-axis from above as $x \to \infty$.

■ Common Factors in Numerator and Denominator

We have adopted the convention that the numerator and denominator of a rational function have no factor in common. If $s(x) = p(x)/q(x)$ and if p and q do have a factor in common, then we may cancel that factor, but only for those values of x for which that factor is *not zero* (because division by zero is not defined). Since s is not defined at those values of x, its graph has a "**hole**" at those points, as the following example illustrates.

Example 8 ■ Common Factor in Numerator and Denominator

Graph each of the following functions.

(a) $s(x) = \dfrac{x - 3}{x^2 - 3x}$ **(b)** $t(x) = \dfrac{x^3 - 2x^2}{x - 2}$

Solution

(a) We factor the numerator and denominator:

$$s(x) = \frac{x - 3}{x^2 - 3x} = \frac{\cancel{(x - 3)}}{x\cancel{(x - 3)}} = \frac{1}{x} \quad \text{for } x \neq 3$$

So s has the same graph as the rational function $r(x) = 1/x$ but with a "hole" when x is 3 because the function s is not defined when $x = 3$ [see Figure 10(a)].

(b) We factor the numerator and denominator:

$$t(x) = \frac{x^3 - 2x^2}{x - 2} = \frac{x^2\cancel{(x - 2)}}{\cancel{x - 2}} = x^2 \quad \text{for } x \neq 2$$

So the graph of t is the same as the graph of $f(x) = x^2$ but with a "hole" when x is 2 because the function t is not defined when $x = 2$ [see Figure 10(b)].

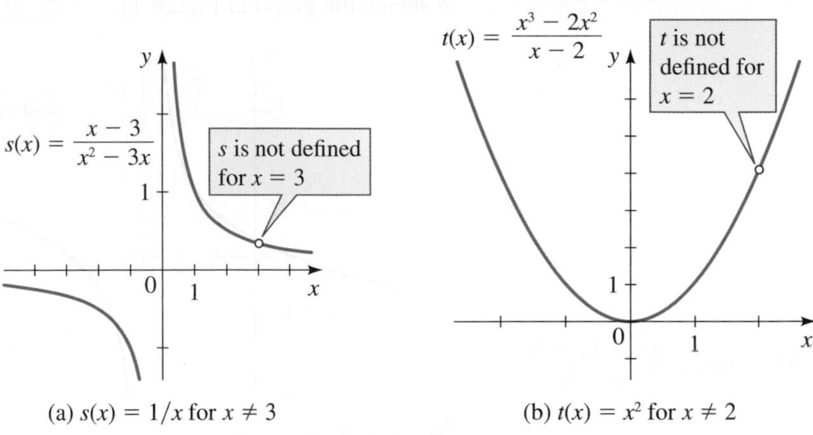

Figure 10 | Graphs with "holes"

(a) $s(x) = 1/x$ for $x \neq 3$

(b) $t(x) = x^2$ for $x \neq 2$

Now Try Exercise 65 ■

■ Slant Asymptotes and End Behavior

If $r(x) = P(x)/Q(x)$ is a rational function for which the degree of the numerator is one more than the degree of the denominator, we can use the Division Algorithm to express the function in the form

$$r(x) = ax + b + \frac{R(x)}{Q(x)}$$

where the degree of R is less than the degree of Q and $a \neq 0$. This means that as $x \to \pm\infty$, $R(x)/Q(x) \to 0$, so for large values of $|x|$ the graph of $y = r(x)$ approaches the graph of the line $y = ax + b$. In this situation we say that $y = ax + b$ is a **slant asymptote**, or an **oblique asymptote**.

Example 9 ■ A Rational Function with a Slant Asymptote

Graph the rational function $r(x) = \dfrac{x^2 - 4x - 5}{x - 3}$.

Solution

Factor. $y = \dfrac{(x + 1)(x - 5)}{x - 3}$

x-Intercepts. -1 and 5, from $x + 1 = 0$ and $x - 5 = 0$

y-Intercept. $\frac{5}{3}$ because $r(0) = \dfrac{0^2 - 4 \cdot 0 - 5}{0 - 3} = \dfrac{5}{3}$

Vertical asymptote. $x = 3$, from the zero of the denominator

Behavior near vertical asymptote. $y \to \infty$ as $x \to 3^-$ and $y \to -\infty$ as $x \to 3^+$

Horizontal asymptote. None, because the degree of the numerator is greater than the degree of the denominator

Slant asymptote. Because the degree of the numerator is one more than the degree of the denominator, the function has a slant asymptote. Dividing (see the margin), we obtain

$$\begin{array}{r} x - 1 \\ x - 3\,\overline{)\,x^2 - 4x - 5} \\ \underline{x^2 - 3x} \\ -x - 5 \\ \underline{-x + 3} \\ -8 \end{array}$$

$$r(x) = x - 1 - \frac{8}{x - 3}$$

Thus $y = x - 1$ is the slant asymptote.

Graph. We use the information we have found, together with some additional values, to sketch the graph in Figure 11.

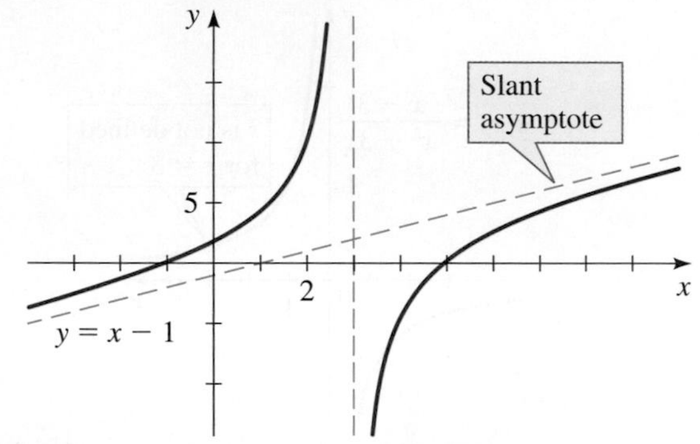

x	y
−2	−1.4
−1	0
0	1.67
1	4
2	9
4	−5
5	0
6	2.33

Figure 11 | $r(x) = \dfrac{x^2 - 4x - 5}{x - 3}$

Now Try Exercise 71 ■

So far, we have considered only horizontal and slant asymptotes as end behaviors for rational functions. In the next example we graph a function whose end behavior is like that of a parabola.

Example 10 ■ End Behavior of a Rational Function

Graph the rational function

$$r(x) = \frac{x^3 - 2x^2 + 3}{x - 2}$$

and describe its end behavior.

Solution

Factor. $y = \dfrac{(x + 1)(x^2 - 3x + 3)}{x - 2}$

x-Intercept. -1 from $x + 1 = 0$ (The other factor in the numerator has no real zeros.)

y-Intercept. $-\dfrac{3}{2}$ because $r(0) = \dfrac{0^3 - 2 \cdot 0^2 + 3}{0 - 2} = -\dfrac{3}{2}$

Vertical asymptote. $x = 2$, from the zero of the denominator

Behavior near vertical asymptote. $y \to -\infty$ as $x \to 2^-$ and $y \to \infty$ as $x \to 2^+$

Horizontal asymptote. None, because the degree of the numerator is greater than the degree of the denominator

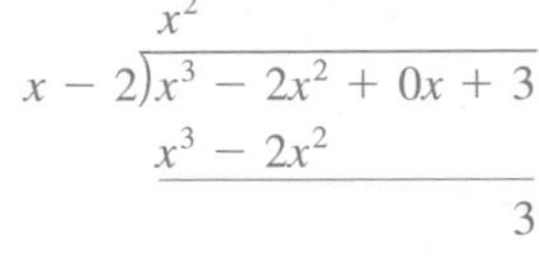

End behavior. Dividing (see the margin), we obtain

$$r(x) = x^2 + \frac{3}{x - 2}$$

This shows that the end behavior of r is like that of the parabola $y = x^2$ because $3/(x - 2)$ is small when $|x|$ is large. That is, $3/(x - 2) \to 0$ as $x \to \pm\infty$. This means that the graph of r will be close to the graph of $y = x^2$ for large $|x|$.

Graph. In Figure 12(a) we graph r in a small viewing rectangle; we can see the intercepts, the vertical asymptotes, and the local minimum. In Figure 12(b) we graph r in a larger viewing rectangle; here the graph looks almost like the graph of a parabola. In Figure 12(c) we graph both $y = r(x)$ and $y = x^2$; these graphs are very close to each other except near the vertical asymptote.

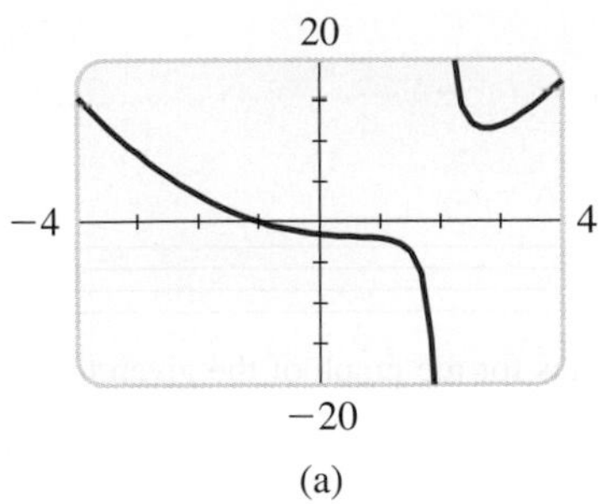

(a)

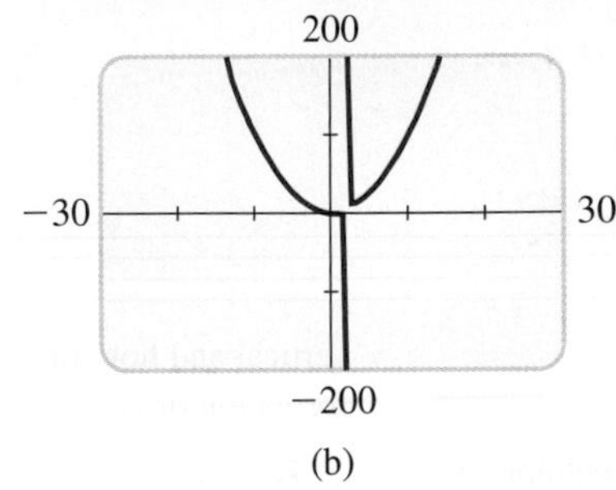

(b)

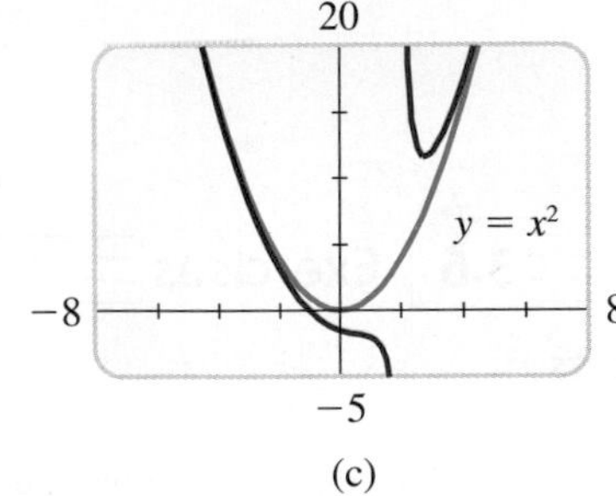

(c)

Figure 12 | $r(x) = \dfrac{x^3 - 2x^2 + 3}{x - 2}$

Now Try Exercise 79 ■

■ Applications

Rational functions occur frequently in scientific applications of algebra. In the next example we analyze the graph of a function from the theory of electricity.

Example 11 ■ Electrical Resistance

When two resistors with resistances R_1 and R_2 are connected in parallel, their combined resistance R is given by the formula

$$R = \frac{R_1 R_2}{R_1 + R_2}$$

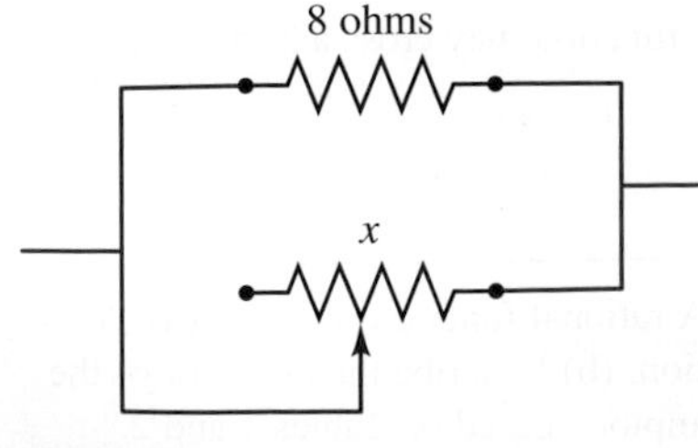

Figure 13

Suppose that a fixed 8-ohm resistor is connected in parallel with a variable resistor, as shown in Figure 13. If the resistance of the variable resistor is denoted by x, then the combined resistance R is a function of x. Graph R, and give a physical interpretation of the graph.

Solution Substituting $R_1 = 8$ and $R_2 = x$ into the formula gives the function

$$R(x) = \frac{8x}{8 + x}$$

Because resistance cannot be negative, this function has physical meaning only when $x > 0$. The function is graphed in Figure 14(a) using the viewing rectangle $[0, 20]$ by $[0, 10]$. The function has no vertical asymptote when x is restricted to positive values. The combined resistance R increases as the variable resistance x increases. If we widen the viewing rectangle to $[0, 100]$ by $[0, 10]$, we obtain the graph in Figure 14(b). For large x the combined resistance R levels off, getting closer and closer to the horizontal asymptote $R = 8$. No matter how large the variable resistance x, the combined resistance is never greater than 8 ohms.

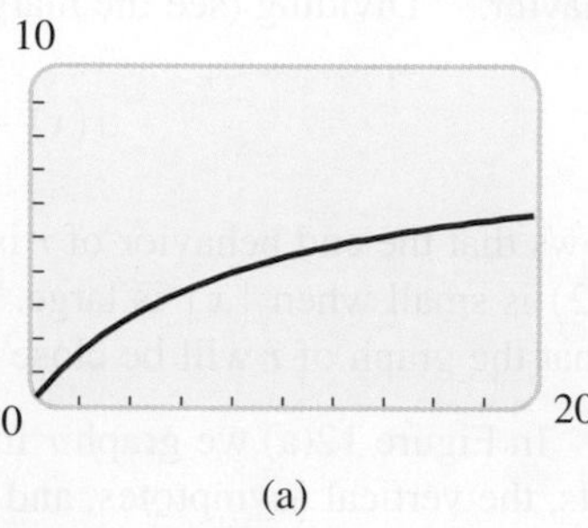

(a)

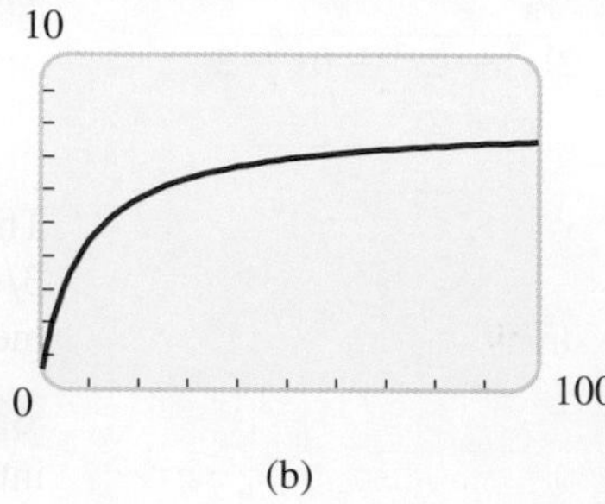

(b)

Figure 14 | $R(x) = \frac{8x}{8 + x}$

Now Try Exercise 93 ■

3.6 Exercises

Concepts

1. If the rational function $y = r(x)$ has the vertical asymptote $x = 2$, then as $x \to 2^+$, either $y \to$ ______ or $y \to$ ______.

2. If the rational function $y = r(x)$ has the horizontal asymptote $y = 2$, then $y \to$ ______ as $x \to \pm\infty$.

3–6 ■ The following questions are about the rational function

$$r(x) = \frac{(x + 1)(x - 2)}{(x + 2)(x - 3)}$$

3. The function r has x-intercepts ______ and ______.

4. The function r has y-intercept ______.

5. The function r has vertical asymptotes $x =$ ______ and $x =$ ______.

6. The function r has horizontal asymptote $y =$ ______.

7–8 ■ Suppose that the graph of a rational function r has vertical asymptote $x = 2$ and horizontal asymptote $y = 4$. Determine the vertical and horizontal asymptotes for the graph of the given transformation of r.

7. $s(x) = r(x - 1)$

8. $t(x) = r(x) - 5$

9–10 ■ *True or False?*

9. Let $r(x) = \dfrac{x^2 + x}{(x + 1)(2x - 4)}$. The graph of r has

(a) vertical asymptote $x = 2$.

(b) vertical asymptote $x = -1$.

(c) horizontal asymptote $y = 1$.

(d) horizontal asymptote $y = \frac{1}{2}$.

10. The graph of a rational function may cross a horizontal asymptote.

Skills

11–14 ■ **Table of Values** A rational function is given. **(a)** Complete each table for the function. **(b)** Describe the behavior of the function near its vertical asymptote, based on Tables 1 and 2. **(c)** Determine the horizontal asymptote, based on Tables 3 and 4.

Table 1

x	$r(x)$
1.5	
1.9	
1.99	
1.999	

Table 2

x	$r(x)$
2.5	
2.1	
2.01	
2.001	

Table 3

x	$r(x)$
10	
50	
100	
1000	

Table 4

x	$r(x)$
−10	
−50	
−100	
−1000	

11. $r(x) = \dfrac{x}{x - 2}$ **12.** $r(x) = \dfrac{4x + 1}{x - 2}$

13. $r(x) = \dfrac{3x - 10}{(x - 2)^2}$ **14.** $r(x) = \dfrac{3x^2 + 1}{(x - 2)^2}$

15–22 ■ Graphing Rational Functions Using Transformations Use transformations of the graph of $y = 1/x$ to graph the rational function, and state the domain and range, as in Example 2.

15. $r(x) = \dfrac{4}{x - 2}$ **16.** $r(x) = \dfrac{9}{x + 3}$

17. $s(x) = -\dfrac{2}{x + 1}$ **18.** $s(x) = -\dfrac{3}{x - 4}$

19. $t(x) = \dfrac{2x - 3}{x - 2}$ **20.** $t(x) = \dfrac{3x - 3}{x + 2}$

21. $r(x) = \dfrac{x + 2}{x + 3}$ **22.** $r(x) = \dfrac{2x - 9}{x - 4}$

23–28 ■ Intercepts of Rational Functions Find the x- and y-intercepts of the rational function.

23. $r(x) = \dfrac{x - 1}{x + 4}$ **24.** $s(x) = \dfrac{3x}{x - 5}$

25. $t(x) = \dfrac{x^2 - x - 2}{x - 6}$ **26.** $r(x) = \dfrac{2}{x^2 + 3x - 4}$

27. $r(x) = \dfrac{x^2 - 9}{x^2}$ **28.** $r(x) = \dfrac{x^3 + 8}{x^2 + 4}$

29–32 ■ Getting Information from a Graph From the graph, determine the x- and y-intercepts and the vertical and horizontal asymptotes.

29.

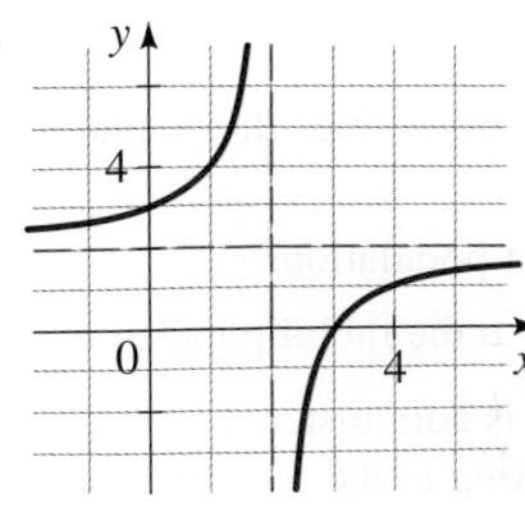

30.

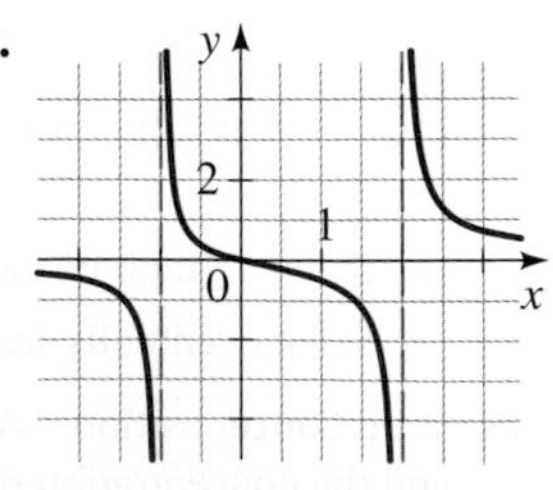

31.

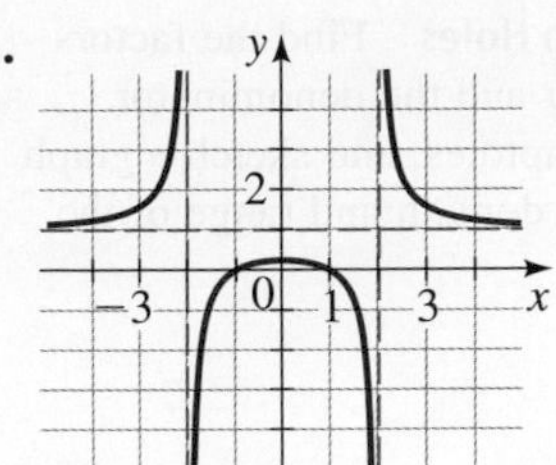

32.

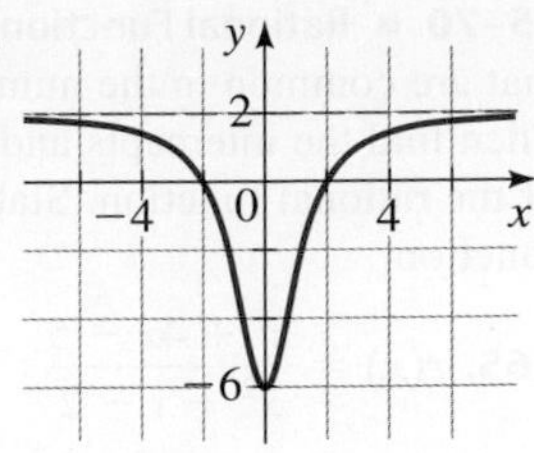

33–44 ■ Asymptotes Find all vertical and horizontal asymptotes (if any).

33. $r(x) = \dfrac{5}{x - 2}$ **34.** $r(x) = \dfrac{2x - 3}{x^2 - 1}$

35. $r(x) = \dfrac{3x + 10}{x^2 + 5}$ **36.** $r(x) = \dfrac{2x^3 + x^2}{x^4 - 16}$

37. $s(x) = \dfrac{10x^3 - 7}{x^3 - x}$ **38.** $s(x) = \dfrac{18x^2 + 9}{9x^2 + 1}$

39. $r(x) = \dfrac{(x + 1)(2x - 3)}{(x - 2)(4x + 7)}$ **40.** $r(x) = \dfrac{(x - 3)(x + 2)}{(5x + 1)(2x - 3)}$

41. $r(x) = \dfrac{6x^3 - 2}{2x^3 + 5x^2 + 6x}$ **42.** $r(x) = \dfrac{5x^3}{x^3 + 2x^2 + 5x}$

43. $t(x) = \dfrac{x^2 + 2}{x - 1}$ **44.** $r(x) = \dfrac{x^3 + 3x^2}{x^2 - 4}$

45–64 ■ Graphing Rational Functions Find the intercepts and asymptotes, and then sketch a graph of the rational function. State the domain and estimate the range from your graph.

45. $r(x) = \dfrac{2x + 2}{x - 1}$ **46.** $r(x) = \dfrac{1 - 3x}{2x + 4}$

47. $r(x) = \dfrac{3x^2 - 12x + 13}{x^2 - 4x + 4}$ **48.** $r(x) = \dfrac{-2x^2 - 8x - 9}{x^2 + 4x + 4}$

49. $r(x) = \dfrac{-x^2 + 8x - 18}{x^2 - 8x + 16}$ **50.** $r(x) = \dfrac{x^2 + 2x + 3}{2x^2 + 4x + 2}$

51. $s(x) = \dfrac{4x - 8}{(x - 4)(x + 1)}$ **52.** $s(x) = \dfrac{9}{x^2 - 5x + 4}$

53. $s(x) = \dfrac{9x - 18}{x^2 + x - 2}$ **54.** $s(x) = \dfrac{x + 2}{(x + 3)(x - 1)}$

55. $r(x) = \dfrac{(x - 1)(x + 2)}{(x + 1)(x - 3)}$ **56.** $r(x) = \dfrac{2x^2 + 10x - 12}{x^2 + x - 6}$

57. $r(x) = \dfrac{x^2 + 2x - 8}{x^2 + 2x}$ **58.** $r(x) = \dfrac{3x^2 + 6}{x^2 - 4x}$

59. $s(x) = \dfrac{x^2 - 2x + 1}{x^3 - 3x^2}$ **60.** $r(x) = \dfrac{x^2 - x - 6}{x^2 + 3x}$

61. $r(x) = \dfrac{x^2 - 2x + 1}{x^2 + 2x + 1}$ **62.** $r(x) = \dfrac{9x^2}{4x^2 + 4x - 8}$

63. $r(x) = \dfrac{5x^2 + 10x + 5}{x^2 + 6x + 9}$ **64.** $t(x) = \dfrac{x^3 - x^2}{x^3 - 3x - 2}$

65–70 ■ Rational Functions with Holes Find the factors that are common in the numerator and the denominator. Then find the intercepts and asymptotes, and sketch a graph of the rational function. State the domain and range of the function.

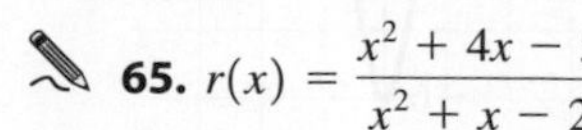

65. $r(x) = \dfrac{x^2 + 4x - 5}{x^2 + x - 2}$

66. $r(x) = \dfrac{x^2 + 3x - 10}{(x + 1)(x - 3)(x + 5)}$

67. $r(x) = \dfrac{x^2 - 2x - 3}{x + 1}$

68. $r(x) = \dfrac{x^3 - 2x^2 - 3x}{x - 3}$

69. $r(x) = \dfrac{x^3 - 5x^2 + 3x + 9}{x + 1}$

[*Hint:* Check that $x + 1$ is a factor of the numerator.]

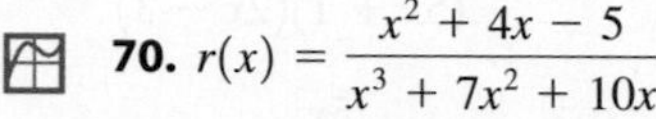

70. $r(x) = \dfrac{x^2 + 4x - 5}{x^3 + 7x^2 + 10x}$

71–78 ■ Slant Asymptotes Find the slant asymptote and the vertical asymptotes, and sketch a graph of the function.

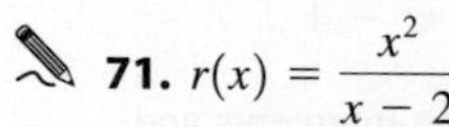

71. $r(x) = \dfrac{x^2}{x - 2}$

72. $r(x) = \dfrac{x^2 + 2x}{x - 1}$

73. $r(x) = \dfrac{x^2 - 2x - 8}{x}$

74. $r(x) = \dfrac{3x - x^2}{2x - 2}$

75. $r(x) = \dfrac{x^2 + 5x + 4}{x - 3}$

76. $r(x) = \dfrac{x^3 + 4}{2x^2 + x - 1}$

77. $r(x) = \dfrac{x^3 + x^2}{x^2 - 4}$

78. $r(x) = \dfrac{2x^3 + 2x}{x^2 - 1}$

Skills Plus

79–82 ■ End Behavior Graph the rational function f, and determine all vertical asymptotes from your graph. Then graph f and g in a sufficiently large viewing rectangle to show that they have the same end behavior.

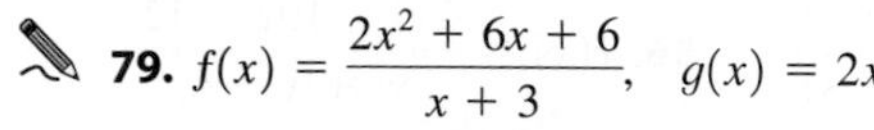

79. $f(x) = \dfrac{2x^2 + 6x + 6}{x + 3}$, $g(x) = 2x$

80. $f(x) = \dfrac{-x^3 + 6x^2 - 5}{x^2 - 2x}$, $g(x) = -x + 4$

81. $f(x) = \dfrac{x^3 - 2x^2 + 16}{x - 2}$, $g(x) = x^2$

82. $f(x) = \dfrac{-x^4 + 2x^3 - 2x}{(x - 1)^2}$, $g(x) = 1 - x^2$

83–88 ■ End Behavior Graph the rational function, and find all vertical asymptotes, x- and y-intercepts, and local extrema, correct to the nearest tenth. Then use long division to find a polynomial that has the same end behavior as the rational function, and graph both functions in a sufficiently large viewing rectangle to verify that the end behaviors of the polynomial and the rational function are the same.

83. $y = \dfrac{2x^2 - 5x}{2x + 3}$

84. $y = \dfrac{x^4 - 3x^3 + x^2 - 3x + 3}{x^2 - 3x}$

85. $y = \dfrac{x^5}{x^3 - 1}$

86. $y = \dfrac{x^4}{x^2 - 2}$

87. $r(x) = \dfrac{x^4 - 3x^3 + 6}{x - 3}$

88. $r(x) = \dfrac{4 + x^2 - x^4}{x^2 - 1}$

89–92 ■ Families of Rational Functions Draw the family of rational functions in the same viewing rectangle using the given values of c. What properties do the members of the family share? How do they differ?

89. $r(x) = \dfrac{cx}{x^2 + 1}$; $c = 1, 2, 3, 4$

90. $r(x) = \dfrac{x + 1}{x + c}$; $c = 2, 3, 4, 5$

91. $r(x) = \dfrac{cx^2}{x^2 - 1}$; $c = 1, 2, 3, 4$

92. $r(x) = \dfrac{x^2 - c}{x + 5}$; $c = 1, 4, 9, 16$

Applications

93. Average Cost A manufacturer of leather purses finds that the cost (in dollars) of producing x purses is given by the function $C(x) = 750 + 45x + 0.03x^2$.

(a) Explain why the average cost per purse is given by the rational function

$$A(x) = \frac{C(x)}{x}$$

(b) Graph the function A and interpret the graph.

94. Population Growth Suppose that the rabbit population on a farm follows the formula

$$p(t) = \frac{3000t}{t + 1}$$

where $t \geq 0$ is the time (in months) since the beginning of the year.

(a) Draw a graph of the rabbit population.

(b) What eventually happens to the rabbit population?

95. Drug Concentration A drug is administered to a patient, and the concentration of the drug in the bloodstream is monitored. At time $t \geq 0$ (in hours since giving the drug) the concentration (in mg/L) is given by

$$c(t) = \frac{5t}{t^2 + 1}$$

Graph the function c with a graphing device.

(a) What is the highest concentration of drug that is reached in the patient's bloodstream?

(b) What happens to the drug concentration after a long period of time?

(c) How long does it take for the concentration to drop below 0.3 mg/L?

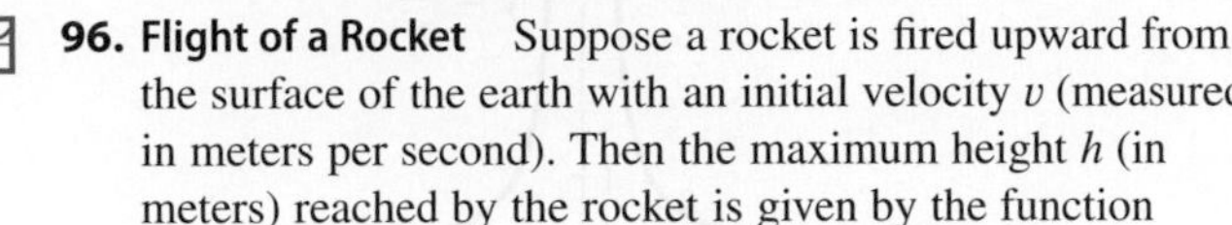

96. Flight of a Rocket Suppose a rocket is fired upward from the surface of the earth with an initial velocity v (measured in meters per second). Then the maximum height h (in meters) reached by the rocket is given by the function

$$h(v) = \frac{Rv^2}{2gR - v^2}$$

where $R = 6.4 \times 10^6$ m is the radius of the earth and $g = 9.8$ m/s^2 is the acceleration due to gravity. Use a graphing device to draw a graph of the function h. (Note that h and v must both be positive, so the viewing rectangle need not contain negative values.) What does the vertical asymptote represent physically?

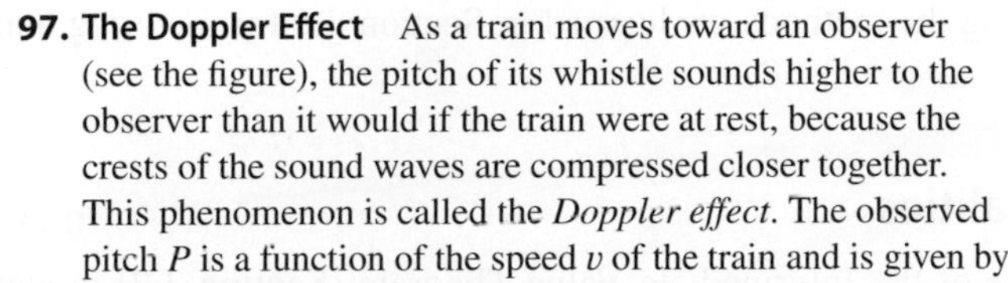

97. The Doppler Effect As a train moves toward an observer (see the figure), the pitch of its whistle sounds higher to the observer than it would if the train were at rest, because the crests of the sound waves are compressed closer together. This phenomenon is called the *Doppler effect*. The observed pitch P is a function of the speed v of the train and is given by

$$P(v) = P_0\left(\frac{s_0}{s_0 - v}\right)$$

where P_0 is the actual pitch of the whistle at the source and $s_0 = 332$ m/s is the speed of sound in air. Suppose that a train has a whistle pitched at $P_0 = 440$ Hz. Graph the function $y = P(v)$ using a graphing device. How can the vertical asymptote of this function be interpreted physically?

98. Focusing Distance For a camera with a lens of fixed focal length F to focus on an object located a distance x from the lens, the image sensor must be placed a distance y behind the lens, where F, x, and y are related by

$$\frac{1}{x} + \frac{1}{y} = \frac{1}{F}$$

(See the figure.) Suppose the camera has a 55-mm lens ($F = 55$).

(a) Express y as a function of x, and graph the function.

(b) What happens to the focusing distance y as the object moves far away from the lens?

(c) What happens to the focusing distance y as the object moves close to the lens?

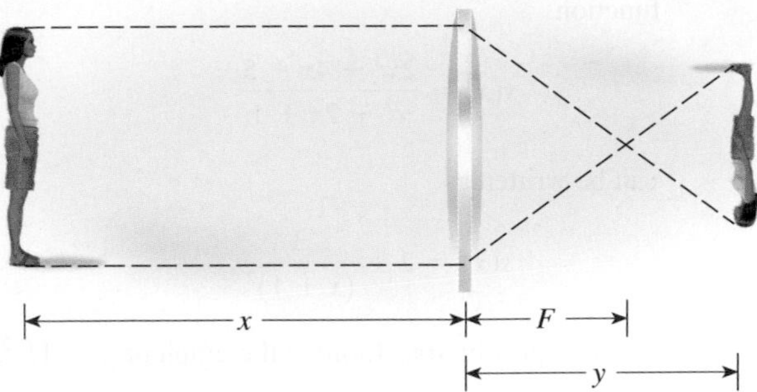

99. Salt Concentration A large tank is to be filled with brine. At time $t = 0$, the tank contains 100 liters of water and 4 kg of salt. Water is being pumped into the tank at the rate of 50 liters per minute and at the same time salt is poured into the tank at the rate of 5 kilograms per minute.

(a) Express the concentration C of salt (in kg/L) in the tank as a function of time t. Graph the function $C(t)$ for $t \geq 0$.

(b) What is the salt concentration in the tank after 10 minutes? What is the salt concentration when the tank has 1000 L of water?

(c) If the process of adding water and salt can continue indefinitely (t approaches infinity), what would the salt concentration in the tank approach?

■ Discuss ■ Discover ■ Prove ■ Write

100. Discuss: Constructing a Rational Function From Its Asymptotes Give an example of a rational function that has vertical asymptote $x = 3$. Now give an example of one that has vertical asymptote $x = 3$ *and* horizontal asymptote $y = 2$. Now give an example of a rational function with vertical asymptotes $x = 1$ and $x = -1$, horizontal asymptote $y = 0$, and x-intercept 4.

101. Discuss: A Rational Function With No Asymptote Explain how you can tell (without graphing it) that the function

$$r(x) = \frac{x^6 + 10}{x^4 + 8x^2 + 15}$$

has no x-intercept and no horizontal, vertical, or slant asymptote. What is its end behavior?'

102. Discover: Transformations of $y = 1/x^2$ In Example 2 we saw that some simple rational functions can be graphed by shifting, stretching, or reflecting the graph of $y = 1/x$. In this exercise we consider rational functions that can be graphed by transforming the graph of $y = 1/x^2$.

(a) Graph the function

$$r(x) = \frac{1}{(x-2)^2}$$

by transforming the graph of $y = 1/x^2$.

(b) Use long division and factoring to show that the function

$$s(x) = \frac{2x^2 + 4x + 5}{x^2 + 2x + 1}$$

can be written as

$$s(x) = 2 + \frac{3}{(x+1)^2}$$

Then graph s by transforming the graph of $y = 1/x^2$.

(c) One of the following functions can be graphed by transforming the graph of $y = 1/x^2$; the other cannot. Use transformations to graph the one that can be, and explain why this method doesn't work for the other one.

$$p(x) = \frac{2 - 3x^2}{x^2 - 4x + 4} \qquad q(x) = \frac{12x - 3x^2}{x^2 - 4x + 4}$$

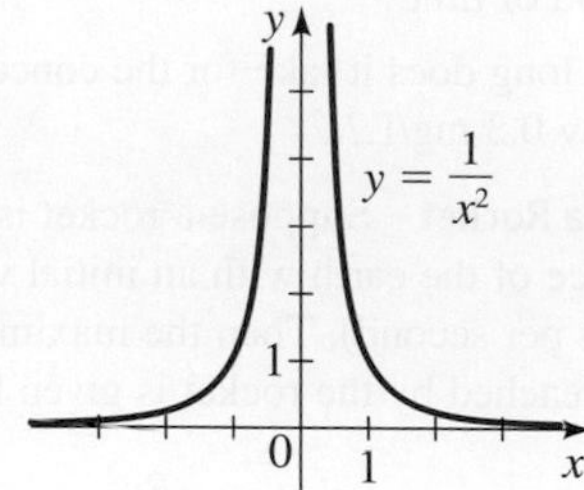

3.7 Polynomial and Rational Inequalities

■ Polynomial Inequalities ■ Rational Inequalities

In Section 1.8 we solved basic inequalities. In this section we solve polynomial and rational inequalities by using the methods we learned in Section 3.4 for factoring and graphing polynomials.

■ Polynomial Inequalities

An important consequence of the Intermediate Value Theorem (Section 3.2) is that the values of a polynomial function P do not change sign between successive zeros. In other words, the values of P between successive zeros are either all positive or all negative. Graphically, this means that between successive x-intercepts, the graph of P is entirely above or entirely below the x-axis. Figure 1 illustrates this property of polynomials. This property allows us to solve **polynomial inequalities** like $P(x) \geq 0$ by finding the zeros of the polynomial and using test values between successive zeros to determine the intervals that satisfy the inequality. We use the following guidelines.

$P(x) = 4(x-2)(x+1)(x-0.5)(x+2)$

Figure 1 | $P(x) > 0$ or $P(x) < 0$ for x between successive zeros of P

Solving Polynomial Inequalities

1. **Move All Terms to One Side.** Rewrite the inequality so that all nonzero terms appear on one side of the inequality symbol.
2. **Factor the Polynomial.** Factor the polynomial into irreducible factors, and find the **real zeros** of the polynomial.
3. **Find the Intervals.** List the intervals determined by the real zeros.
4. **Make a Table or Diagram.** Use **test values** to make a table or diagram of the signs of each factor in each interval. In the last row of the table determine the sign of the polynomial on that interval.
5. **Solve.** Determine the solutions of the inequality from the last row of the table. Check whether the **endpoints** of these intervals satisfy the inequality. (This may happen if the inequality involves $\leq$ or $\geq$.)

Example 1 ■ Solving a Polynomial Inequality

Solve the inequality $2x^3 + x^2 + 6 \geq 13x$.

Solution We follow the guidelines for solving polynomial inequalities.

Move all terms to one side. We move all terms to the left-hand side of the inequality to get

$$2x^3 + x^2 - 13x + 6 \geq 0$$

The left-hand side is a polynomial.

Factor the polynomial. This polynomial is factored in Example 3.4.2:

$$(x - 2)(2x - 1)(x + 3) \geq 0$$

The zeros of the polynomial are -3, $\frac{1}{2}$, and 2.

Find the intervals. The intervals determined by the zeros of the polynomial are

$$(-\infty, -3), \qquad \left(-3, \tfrac{1}{2}\right), \qquad \left(\tfrac{1}{2}, 2\right), \qquad (2, \infty)$$

Make a table or diagram. To determine the sign of each factor on each of the intervals we use test values. That is, we choose a number inside each interval and check the sign of each factor, as shown in the diagram.

See Section 1.8 for instructions on making a sign diagram.

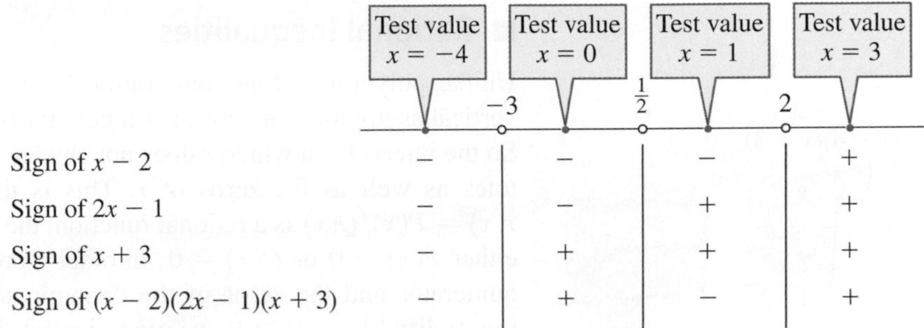

Test value $x = -4$
Test value $x = 0$
Test value $x = 1$
Test value $x = 3$
-3
$\frac{1}{2}$
2
Sign of $x - 2$: $-$ $-$ $-$ $+$
Sign of $2x - 1$: $-$ $-$ $+$ $+$
Sign of $x + 3$: $-$ $+$ $+$ $+$
Sign of $(x - 2)(2x - 1)(x + 3)$: $-$ $+$ $-$ $+$

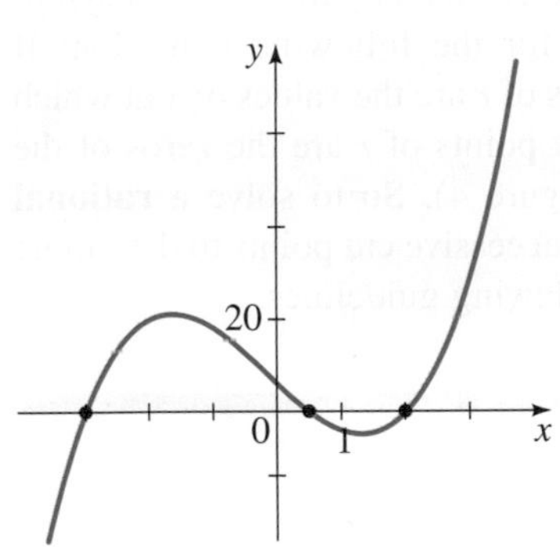

Figure 2

Solve. The last row of the diagram shows that the inequality

$$(x - 2)(2x - 1)(x + 3) \geq 0$$

is satisfied on the intervals $\left(-3, \frac{1}{2}\right)$ and $(2, \infty)$. Checking the endpoints, we see that -3, $\frac{1}{2}$, and 2 satisfy the inequality, so the solution is $\left[-3, \frac{1}{2}\right] \cup [2, \infty)$. The graph in Figure 2 confirms our solution.

 Now Try Exercise 7 ■

Example 2 ■ Solving a Polynomial Inequality

Solve the inequality $3x^4 - x^2 - 4 < 2x^3 + 12x$.

Solution We follow the guidelines for solving polynomial inequalities.

Move all terms to one side. We move all terms to the left-hand side of the inequality to get

$$3x^4 - 2x^3 - x^2 - 12x - 4 < 0$$

The left-hand side is a polynomial.

Factor the polynomial. This polynomial is factored into linear and irreducible quadratic factors in Example 3.5.5:

$$(x - 2)(3x + 1)(x^2 + x + 2) < 0$$

From the first two factors we obtain the zeros 2 and $-\frac{1}{3}$. The third factor has no real zeros.

Find the intervals. The intervals determined by the zeros of the polynomial are

$$\left(-\infty, -\tfrac{1}{3}\right), \quad \left(-\tfrac{1}{3}, 2\right), \quad (2, \infty)$$

Make a table or diagram. We make a sign diagram by using test values.

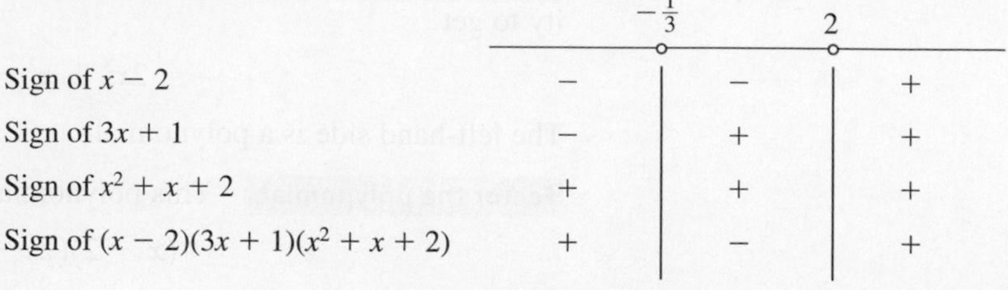

		$-\frac{1}{3}$		2	
Sign of $x - 2$	$-$		$-$		$+$
Sign of $3x + 1$	$-$		$+$		$+$
Sign of $x^2 + x + 2$	$+$		$+$		$+$
Sign of $(x - 2)(3x + 1)(x^2 + x + 2)$	$+$		$-$		$+$

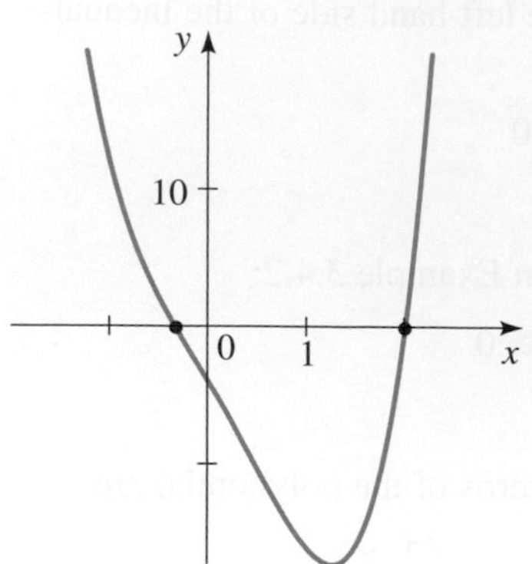

Figure 3

Solve. The inequality requires that the values of the polynomial be less than 0. From the last row of the sign diagram we see that the inequality is satisfied on the interval $\left(-\frac{1}{3}, 2\right)$. You can check that the two endpoints do not satisfy the inequality, so the solution is $\left(-\frac{1}{3}, 2\right)$. The graph in Figure 3 confirms our solution.

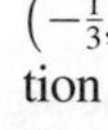 Now Try Exercise 11 ■

■ Rational Inequalities

Unlike polynomial functions, rational functions are not necessarily continuous. The vertical asymptotes of a rational function r break up the graph into separate "branches." So the intervals on which r does not change sign are determined by the vertical asymptotes as well as the zeros of r. This is the reason for the following definition: If $r(x) = P(x)/Q(x)$ is a rational function, the **cut points** of r are the values of x at which either $P(x) = 0$ or $Q(x) = 0$. In other words, the cut points of r are the zeros of the numerator and the zeros of the denominator (see Figure 4). So to solve a **rational inequality** like $r(x) \geq 0$, we use test points between successive cut points to determine the intervals that satisfy the inequality. We use the following guidelines.

$$r(x) = \frac{x - 2}{(x - 4)(x + 3)}$$

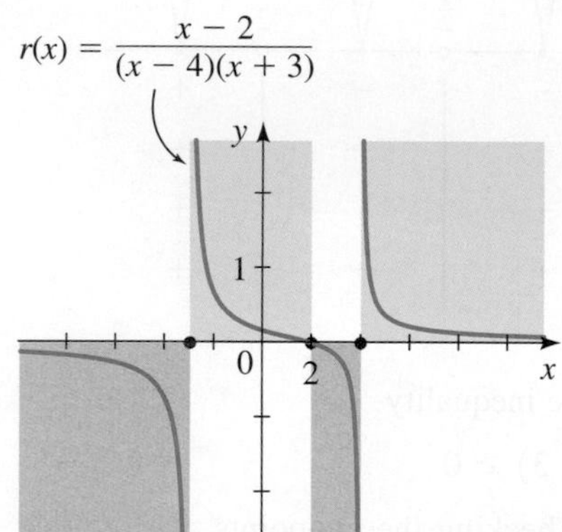

Figure 4 | $r(x) > 0$ or $r(x) < 0$ for x between successive cut points of r

Solving Rational Inequalities

1. **Move All Terms to One Side.** Rewrite the inequality so that all nonzero terms appear on one side of the inequality symbol. Bring all quotients to a common denominator.
2. **Factor Numerator and Denominator.** Factor the numerator and denominator into irreducible factors, and then find the **cut points**.
3. **Find the Intervals.** List the intervals determined by the cut points.
4. **Make a Table or Diagram.** Use **test values** to make a table or diagram of the sign of each factor in each interval. In the last row of the table determine the sign of the rational function on that interval.
5. **Solve.** Determine the solution of the inequality from the last row of the table. Check whether the **endpoints** of these intervals satisfy the inequality. (This may happen if the inequality involves $\leq$ or $\geq$.)

Example 3 ■ Solving a Rational Inequality

Solve the inequality

$$\frac{1 - 2x}{x^2 - 2x - 3} \geq 1$$

Solution We follow the guidelines for solving rational inequalities.

Move all terms to one side. We move all terms to the left-hand side of the inequality.

$$\frac{1-2x}{x^2-2x-3} - 1 \geq 0 \qquad \text{Move terms to LHS}$$

$$\frac{(1-2x)-(x^2-2x-3)}{x^2-2x-3} \geq 0 \qquad \text{Common denominator}$$

$$\frac{4-x^2}{x^2-2x-3} \geq 0 \qquad \text{Simplify}$$

The left-hand side of the inequality is a rational function.

Factor numerator and denominator. Factoring the numerator and denominator, we get

$$\frac{(2-x)(2+x)}{(x-3)(x+1)} \geq 0$$

The zeros of the numerator are 2 and -2, and the zeros of the denominator are -1 and 3, so the cut points are -2, -1, 2, and 3.

Find the intervals. The intervals determined by the cut points are

$$(-\infty, -2), \quad (-2, -1), \quad (-1, 2), \quad (2, 3), \quad (3, \infty)$$

Make a table or diagram. We make a sign diagram by using test values.

	$x<-2$	$-2<x<-1$	$-1<x<2$	$2<x<3$	$x>3$
Sign of $2-x$	+	+	+	−	−
Sign of $2+x$	−	+	+	+	+
Sign of $x-3$	−	−	−	−	+
Sign of $x+1$	−	−	+	+	+
Sign of $\dfrac{(2-x)(2+x)}{(x-3)(x+1)}$	−	+	−	+	−

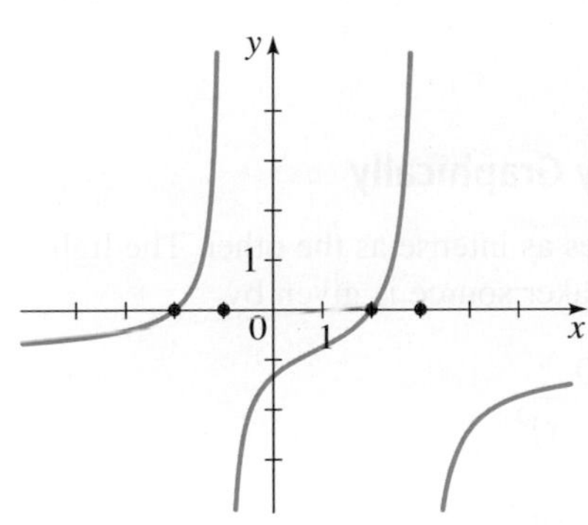

Figure 5

Solve. From the last row of the sign diagram we see that the inequality is satisfied on the intervals $(-2, -1)$ and $(2, 3)$. Checking the endpoints, we see that -2 and 2 satisfy the inequality, so the solution is $[-2, -1) \cup [2, 3)$. The graph in Figure 5 confirms our solution.

 Now Try Exercises 19 and 27 ■

Example 4 ■ Solving a Rational Inequality

Solve the inequality

$$\frac{x^2-4x+3}{x^2-4x-5} \geq 0$$

Solution Since all nonzero terms are already on one side of the inequality symbol, we begin by factoring.

Factor numerator and denominator. Factoring the numerator and denominator, we get

$$\frac{(x-3)(x-1)}{(x-5)(x+1)} \geq 0$$

The cut points are -1, 1, 3, and 5.

Find the intervals. The intervals determined by the cut points are

$$(-\infty, -1), \quad (-1, 1), \quad (1, 3), \quad (3, 5), \quad (5, \infty)$$

Make a table or diagram. We make a sign diagram.

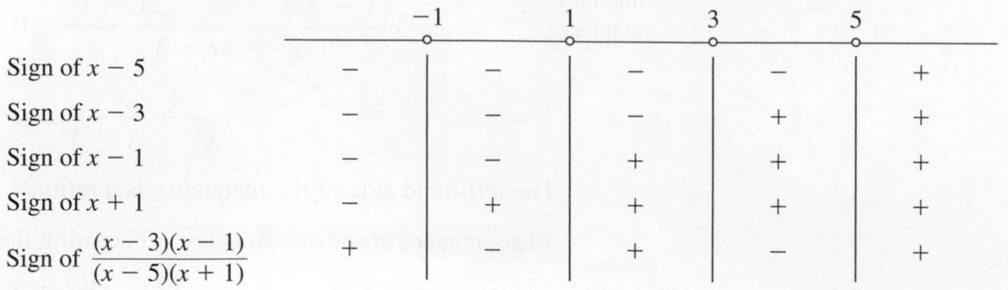

	$(-\infty,-1)$	$(-1,1)$	$(1,3)$	$(3,5)$	$(5,\infty)$
Sign of $x - 5$	−	−	−	−	+
Sign of $x - 3$	−	−	−	+	+
Sign of $x - 1$	−	−	+	+	+
Sign of $x + 1$	−	+	+	+	+
Sign of $\dfrac{(x-3)(x-1)}{(x-5)(x+1)}$	+	−	+	−	+

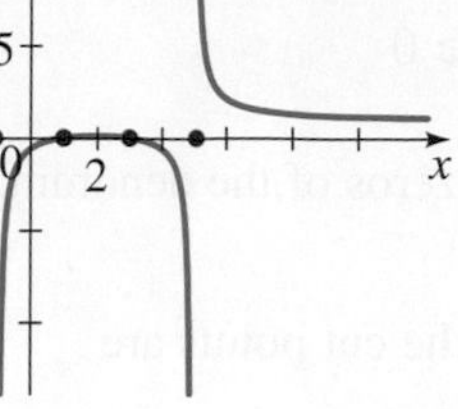

Figure 6

Solve. From the last row of the sign diagram we see that the inequality is satisfied on the intervals $(-\infty, -1)$, $(1, 3)$, and $(5, \infty)$. Checking the endpoints, we see that 1 and 3 satisfy the inequality, so the solution is $(-\infty, -1) \cup [1, 3] \cup (5, \infty)$. The graph in Figure 6 confirms our solution.

Now Try Exercises 23 and 29 ■

We can also solve polynomial and rational inequalities graphically (see Sections 1.11 and 2.3). In the next example we graph each side of the inequality and compare the values of left- and right-hand sides graphically.

Example 5 ■ Solving a Rational Inequality Graphically

Two light sources are 10 m apart. One is three times as intense as the other. The light intensity L (in lux) at a point x meters from the weaker source is given by

$$L(x) = \frac{10}{x^2} + \frac{30}{(10 - x)^2}$$

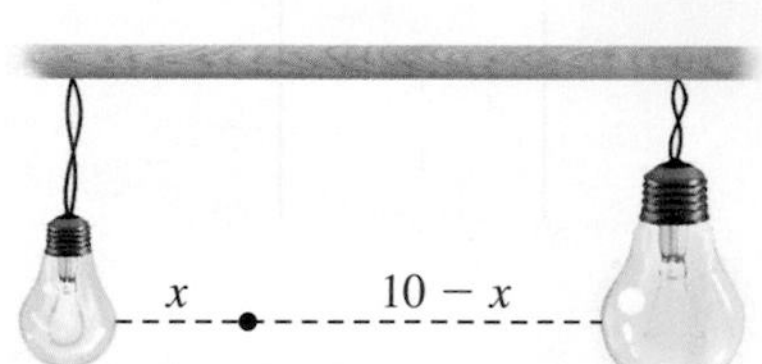

Figure 7

(See Figure 7.) Find the points at which the light intensity is 4 lux or less.

Solution We need to solve the inequality

$$\frac{10}{x^2} + \frac{30}{(10 - x)^2} \le 4$$

We solve the inequality graphically by graphing the two functions

$$y_1 = \frac{10}{x^2} + \frac{30}{(10 - x)^2} \qquad \text{and} \qquad y_2 = 4$$

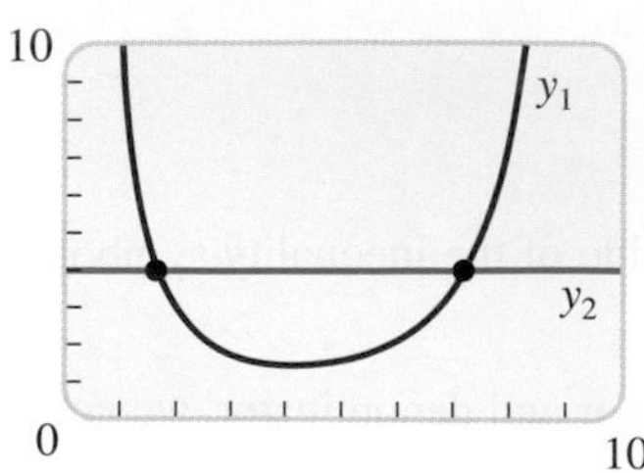

Figure 8

In this physical problem the possible values of x are between 0 and 10, so we graph the two functions in a viewing rectangle with x-values between 0 and 10, as shown in Figure 8. We want those values of x for which $y_1 \le y_2$. The graphs intersect at $x \approx 1.67$ and at $x \approx 7.19$, and between these x-values the graph of y_1 lies below the graph of y_2. So the solution of the inequality is the interval $(1.67, 7.19)$, rounded to two decimal places. Thus the light intensity is less than or equal to 4 lux when the distance from the weaker source is between 1.67 m and 7.19 m.

Now Try Exercises 49 and 59 ■

3.7 Exercises

Concepts

1. To solve a polynomial inequality, we factor the polynomial into irreducible factors and find all the real ________ of the polynomial. Then we find the intervals determined by the real ________ and use test points in each interval to find the sign of the polynomial on that interval. Let

$$P(x) = x(x + 2)(x - 1)$$

Fill in the diagram below to find the intervals on which $P(x) \geq 0$.

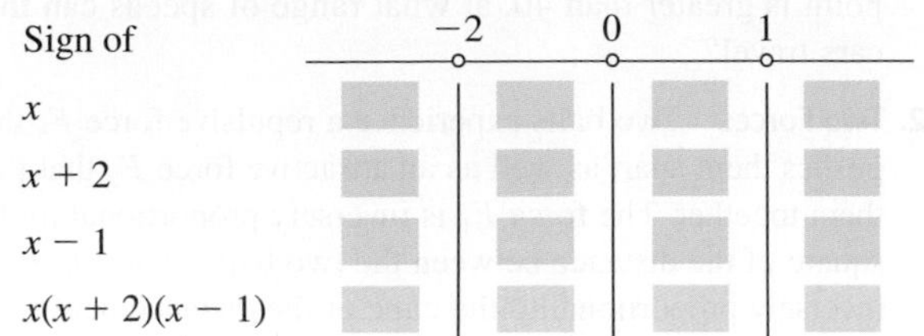

From the diagram above we see that $P(x) \geq 0$ on the intervals ________ and ________.

2. To solve a rational inequality, we factor the numerator and the denominator into irreducible factors. The cut points are the real ________ of the numerator and the real ________ denominator. Then we find the intervals determined by the ________ ________, and we use test points to find the sign of the rational function on each interval. Let

$$r(x) = \frac{(x + 2)(x - 1)}{(x - 3)(x + 4)}$$

Fill in the diagram below to find the intervals on which $r(x) \geq 0$.

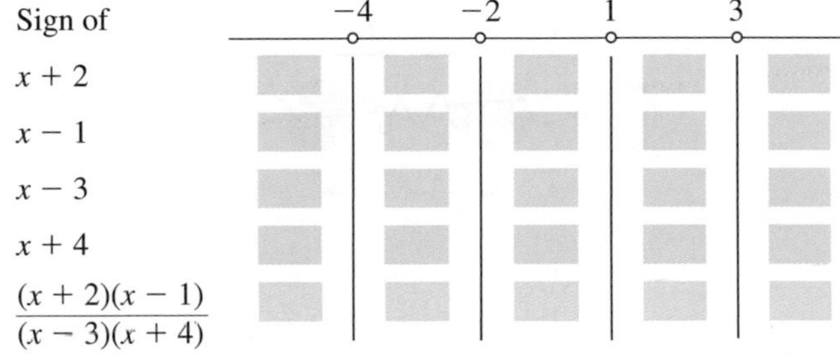

From the diagram we see that $r(x) \geq 0$ on the intervals ________, ________, and ________.

Skills

3–16 ■ Polynomial Inequalities Solve the inequality.

3. $(x - 3)(x + 5)(2x + 5) < 0$

4. $(x - 1)(x + 2)(x - 3)(x + 4) \geq 0$

5. $(x + 5)^2(x + 3)(x - 1) > 0$

6. $(2x - 7)^4(x - 1)^3(x + 1) \leq 0$

7. $x^3 + 4x^2 \geq 4x + 16$

8. $2x^3 - 18x < x^2 - 9$

9. $2x^3 - x^2 < 9 - 18x$

10. $x^4 + 3x^3 < x + 3$

11. $x^4 - 7x^2 - 18 < 0$

12. $4x^4 - 25x^2 + 36 \leq 0$

13. $x^3 + x^2 - 17x + 15 \geq 0$

14. $x^4 + 3x^3 - 3x^2 + 3x - 4 < 0$

15. $x(1 - x^2)^3 > 7(1 - x^2)^3$

16. $x^2(7 - 6x) \leq 1$

17–40 ■ Rational Inequalities Solve the inequality.

17. $\dfrac{x - 1}{x - 10} < 0$

18. $\dfrac{3x - 7}{x + 2} \leq 0$

19. $\dfrac{x - 3}{2x + 5} \geq 1$

20. $\dfrac{x + 4}{x - 5} \leq 4$

21. $\dfrac{5x + 7}{4x + 10} \leq 1$

22. $\dfrac{4x - 6}{x + 7} > 2$

23. $\dfrac{2x + 5}{x^2 + 2x - 35} \geq 0$

24. $\dfrac{4x^2 - 25}{x^2 - 9} \leq 0$

25. $\dfrac{x^2}{x^2 + 3x - 10} \leq 0$

26. $\dfrac{x - 3}{x^2 + 6x + 9} \leq 0$

27. $\dfrac{x^2 + 3}{x + 1} > 2$

28. $\dfrac{4x - 3}{x^2 + 1} \leq 1$

29. $\dfrac{x^2 + 2x - 3}{3x^2 - 7x - 6} > 0$

30. $\dfrac{x - 1}{x^3 + 1} \geq 0$

31. $\dfrac{x^3 + 3x^2 - 9x - 27}{x + 4} \leq 0$

32. $\dfrac{x^2 - 16}{x^4 - 16} < 0$

33. $\dfrac{(x - 1)^2}{(x + 1)(x + 2)} > 0$

34. $\dfrac{x^2 - 2x + 1}{x^3 + 3x^2 + 3x + 1} \leq 0$

35. $\dfrac{x}{2} \geq \dfrac{5}{x + 1} + 4$

36. $\dfrac{x + 2}{x + 3} < \dfrac{x - 1}{x - 2}$

37. $\dfrac{6}{x - 1} - \dfrac{6}{x} \geq 1$

38. $\dfrac{1}{x - 3} + \dfrac{1}{x + 2} \geq \dfrac{2x}{x^2 + x - 2}$

39. $\dfrac{1}{x + 1} - \dfrac{1}{x + 2} \leq \dfrac{1}{(x + 2)^2}$

40. $\dfrac{1}{x} + \dfrac{1}{x + 1} < \dfrac{2}{x + 2}$

41–44 ■ Graphs of Two Functions Find all values of x for which the graph of f lies above the graph of g.

41. $f(x) = x^2; \quad g(x) = 3x + 10$

42. $f(x) = \dfrac{1}{x}; \quad g(x) = \dfrac{1}{x - 1}$

43. $f(x) = 4x; \quad g(x) = \dfrac{1}{x}$

44. $f(x) = x^2 + x; \quad g(x) = \dfrac{2}{x}$

45–48 ■ Domain of a Function Find the domain of the given function.

45. $f(x) = \sqrt{6 + x - x^2}$

46. $g(x) = \sqrt{\dfrac{5 + x}{5 - x}}$

47. $h(x) = \sqrt[4]{x^4 - 1}$

48. $f(x) = \dfrac{1}{\sqrt{x^4 - 5x^2 + 4}}$

49–54 ■ Solving Inequalities Graphically Use a graphing device to solve the inequality, as in Example 5. Express your answer using interval notation, with the endpoints of the intervals rounded to two decimal places.

49. $x^3 - 2x^2 - 5x + 6 \geq 0$ **50.** $2x^3 + x^2 - 8x - 4 \leq 0$

51. $2x^3 - 3x + 1 < 0$ **52.** $x^4 - 4x^3 + 8x > 0$

53. $5x^4 < 8x^3$ **54.** $x^5 + x^3 \geq x^2 + 6x$

Skills Plus

55–56 ■ Rational Inequalities Solve the inequality. (These exercises involve expressions that arise in calculus.)

55. $\dfrac{(1-x)^2}{\sqrt{x}} \geq 4\sqrt{x}(x-1)$

56. $\frac{2}{3}x^{-1/3}(x+2)^{1/2} + \frac{1}{2}x^{2/3}(x+2)^{-1/2} < 0$

57. General Polynomial Inequality Solve the inequality

$$(x-a)(x-b)(x-c)(x-d) \geq 0$$

where $a < b < c < d$.

58. General Rational Inequality Solve the inequality

$$\frac{x^2 + (a-b)x - ab}{x+c} \leq 0$$

where $0 < a < b < c$.

Applications

59. Bonfire Temperature In the vicinity of a bonfire the temperature T (in °C) at a distance of x meters from the center of the fire is given by

$$T(x) = 25 + \frac{2500}{x^2 + 2}$$

At what range of distances from the fire's center is the temperature less than 300°C?

60. Stopping Distance For a certain model of car the distance d required to stop the vehicle if it is traveling at v km/h is given by the function

$$d(t) = v + \frac{v^2}{250}$$

where d is measured in meters. What range of speeds ensure that the stopping distance of the car does not exceed 55 m?

61. Managing Traffic A highway engineer develops a formula to estimate the number of cars that can safely travel a particular highway at a given speed x (in km/h). The engineer finds that the number N of cars that can pass a given point per minute is modeled by the function

$$N(x) = \frac{88x}{17 + 17\left(\frac{x}{20}\right)^2}$$

Graph the function in the viewing rectangle $[0, 100]$ by $[0, 60]$. If the number of cars that pass by the given point is greater than 40, at what range of speeds can the cars travel?

62. Two Forces Two balls experience a repulsive force F_1 that pushes them apart as well as an attractive force F_2 that pulls them together. The force F_1 is inversely proportional to the square of the distance between the two balls, whereas F_2 is inversely proportional to the cube of the distance between them. The constant of proportionality is -3 for force F_1 and 1 for F_2. The *net force* F between the balls is the sum of these two forces.

(a) Express F as a rational function of the distance x between the two balls and graph the function $F(x)$ for $x > 0$.

(b) From the graph, determine the distance x at which the net force is zero. For what distances is the net force attractive $[F(x) > 0]$? Repulsive $[F(x) < 0]$? At what distance does the net force achieve its greatest repulsive magnitude?

(c) Describe the net force F as the distance between the balls gets very small (approaches zero) or very large (approaches infinity). In each case state whether the net force is attractive or repulsive, and comment on the magnitude of the force.

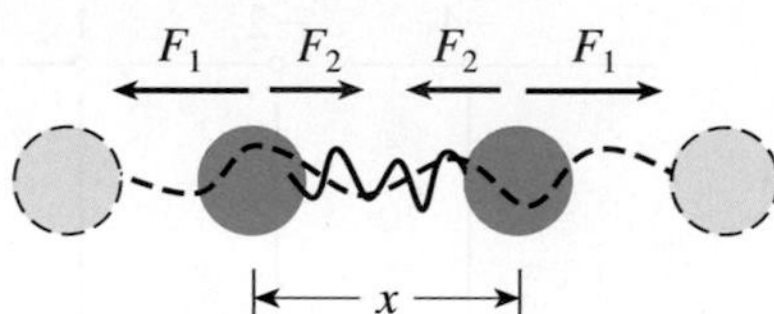

Chapter 3 Review

Properties and Formulas

Quadratic Functions | Section 3.1

A **quadratic function** is a function of the form

$$f(x) = ax^2 + bx + c$$

It can be expressed in the **vertex form**

$$f(x) = a(x-h)^2 + k$$

by completing the square.

The graph of a quadratic function in the vertex form is a **parabola** with **vertex** (h, k).

If $a > 0$, then the quadratic function f has the **minimum value** k at $x = h = -b/(2a)$.

If $a < 0$, then the quadratic function f has the **maximum value** k at $x = h = -b/(2a)$.

Polynomial Functions | Section 3.2

A **polynomial function** of **degree** n is a function P of the form

$$P(x) = a_n x^n + a_{n-1}x^{n-1} + \cdots + a_1 x + a_0$$

(where $a_n \neq 0$). The numbers a_i are the **coefficients** of the polynomial; a_n is the **leading coefficient**, and a_0 is the **constant coefficient** (or **constant term**).

The graph of a polynomial function is a smooth, continuous curve.

Real Zeros of Polynomials | Section 3.2

A **zero** of a polynomial P is a number c for which $P(c) = 0$. The following are equivalent ways of describing real zeros of polynomials:

1. c is a real zero of P.
2. $x = c$ is a solution of the equation $P(x) = 0$.
3. $x - c$ is a factor of $P(x)$.
4. c is an x-intercept of the graph of P.

Multiplicity of a Zero | Section 3.2

A zero c of a polynomial P has multiplicity m if m is the highest power for which $(x - c)^m$ is a factor of $P(x)$.

Local Maximums and Minimums | Section 3.2

A polynomial function P of degree n has $n - 1$ or fewer **local extrema** (i.e., local maximums and minimums).

Division of Polynomials | Section 3.3

If P and D are any polynomials [with $D(x) \neq 0$], then we can divide P by D using either **long division** or **synthetic division**. The result of the division can be expressed in either of the following equivalent forms:

$$P(x) = D(x) \cdot Q(x) + R(x)$$

$$\frac{P(x)}{D(x)} = Q(x) + \frac{R(x)}{D(x)}$$

In this division, P is the **dividend**, D is the **divisor**, Q is the **quotient**, and R is the **remainder**. When the division is continued to its completion, the degree of R is less than the degree of D [or $R(x) = 0$].

Remainder Theorem | Section 3.3

When $P(x)$ is divided by the linear divisor $D(x) = x - c$, the **remainder** is the constant $P(c)$. So one way to **evaluate** a polynomial function P at c is to use synthetic division to divide $P(x)$ by $x - c$ and observe the value of the remainder.

Rational Zeros of Polynomials | Section 3.4

If the polynomial P given by

$$P(x) = a_n x^n + a_{n-1}x^{n-1} + \cdots + a_1 x + a_0$$

has integer coefficients, then all the **rational zeros** of P have the form

$$x = \pm\frac{p}{q}$$

where p is a divisor of the constant term a_0 and q is a divisor of the leading coefficient a_n.

So to find all the rational zeros of a polynomial, we list all the *possible* rational zeros given by this principle and then check to see which are *actual* zeros by using synthetic division.

Descartes's Rule of Signs | Section 3.4

Let P be a polynomial with real coefficients. Then:

The number of positive real zeros of P either is the number of **changes of sign** in the coefficients of $P(x)$ or is less than that by an even number.

The number of negative real zeros of P either is the number of **changes of sign** in the coefficients of $P(-x)$ or is less than that by an even number.

Upper and Lower Bounds Theorem | Section 3.4

Suppose we divide the polynomial P by the linear expression $x - c$ and arrive at the result

$$P(x) = (x - c) \cdot Q(x) + r$$

If $c > 0$ and the coefficients of Q, followed by r, are all nonnegative, then c is an **upper bound** for the zeros of P.

If $c < 0$ and the coefficients of Q, followed by r (including zero coefficients), are alternately nonnegative and nonpositive, then c is a **lower bound** for the zeros of P.

The Fundamental Theorem of Algebra, Complete Factorization, and the Zeros Theorem | Section 3.5

Every polynomial P of degree n with complex coefficients has exactly n complex zeros, provided that each zero of multiplicity m is counted m times. P factors into n linear factors as follows:

$$P(x) = a(x - c_1)(x - c_2) \cdots (x - c_n)$$

where a is the leading coefficient of P and $c_1, c_1, \ldots, c_n$ are the zeros of P.

Conjugate Zeros Theorem | Section 3.5

If the polynomial P has real coefficients and if $a + bi$ is a zero of P, then its complex conjugate $a - bi$ is also a zero of P.

Linear and Quadratic Factors Theorem | Section 3.5

Every polynomial with real coefficients can be factored into linear and irreducible quadratic factors with real coefficients.

Rational Functions | Section 3.6

A **rational function** r is a quotient of polynomial functions:

$$r(x) = \frac{P(x)}{Q(x)}$$

We generally assume that the polynomials P and Q have no factors in common.

Asymptotes | Section 3.6

The line $x = a$ is a **vertical asymptote** of the function $y = f(x)$ if

$$y \to \infty \quad \text{or} \quad y \to -\infty \quad \text{as} \quad x \to a^+ \quad \text{or} \quad x \to a^-$$

The line $y = b$ is a **horizontal asymptote** of the function $y = f(x)$ if

$$y \to b \quad \text{as} \quad x \to \infty \quad \text{or} \quad x \to -\infty$$

Asymptotes of Rational Functions | Section 3.6

Let $r(x) = \dfrac{P(x)}{Q(x)}$ be a rational function.

(continued)

The vertical asymptotes of r are the lines $x = a$ where a is a zero of Q.

If the degree of P is less than the degree of Q, then the horizontal asymptote of r is the line $y = 0$.

If the degrees of P and Q are the same, then the horizontal asymptote of r is the line $y = b$, where

$$b = \frac{\text{leading coefficient of } P}{\text{leading coefficient of } Q}$$

If the degree of P is greater than the degree of Q, then r has no horizontal asymptote.

Polynomial and Rational Inequalities | Section 3.7

A **polynomial inequality** is an inequality of the form $P(x) \geq 0$, where P is a polynomial. We solve $P(x) \geq 0$ by finding the zeros of P and using test values between successive zeros to determine the intervals that satisfy the inequality.

A **rational inequality** is an inequality of the form $r(x) \geq 0$, where

$$r(x) = \frac{P(x)}{Q(x)}$$

is a rational function. The cut points of r are the values of x at which either $P(x) = 0$ or $Q(x) = 0$. We solve $r(x) \geq 0$ by using test points between successive cut points to determine the intervals that satisfy the inequality.

Concept Check

1. (a) What is the degree of a quadratic function f? What is the vertex form of a quadratic function? How do you put a quadratic function into vertex form?

(b) The quadratic function $f(x) = a(x - h)^2 + k$ is in vertex form. The graph of f is a parabola. What is the vertex of the graph of f? How do you determine whether $f(h) = k$ is a minimum value or a maximum value?

(c) Express $f(x) = x^2 + 4x + 1$ in vertex form. Find the vertex of the graph and the maximum value or minimum value of f.

2. (a) Give the general form of polynomial function P of degree n.

(b) What does it mean to say that c is a zero of P? Give two equivalent conditions that tell us that c is a zero of P.

3. Sketch graphs showing the possible end behaviors of polynomials of odd degree and polynomials of even degree.

4. What steps do you follow to graph a polynomial function P?

5. (a) What is a local maximum point or local minimum point of a polynomial P?

(b) How many local extrema can a polynomial P of degree n have?

6. When we divide a polynomial $P(x)$ by a divisor $D(x)$, the Division Algorithm tells us that we can always obtain a quotient $Q(x)$ and a remainder $R(x)$. State the two forms in which the result of this division can be written.

7. (a) State the Remainder Theorem.

(b) State the Factor Theorem.

(c) State the Rational Zeros Theorem.

8. What steps would you take to find the rational zeros of a polynomial P?

9. Let $P(x) = 2x^4 - 3x^3 + x - 15$.

(a) Explain how Descartes's Rule of Signs is used to determine the possible number of positive and negative real zeros of P.

(b) What does it mean to say that a is a lower bound and b is an upper bound for the zeros of a polynomial?

(c) Explain how the Upper and Lower Bounds Theorem is used to show that all the real zeros of P lie between -3 and 3.

10. (a) State the Fundamental Theorem of Algebra.

(b) State the Complete Factorization Theorem.

(c) State the Zeros Theorem.

(d) State the Conjugate Zeros Theorem.

11. (a) What is a rational function?

(b) What does it mean to say that $x = a$ is a vertical asymptote of $y = f(x)$?

(c) What does it mean to say that $y = b$ is a horizontal asymptote of $y = f(x)$?

(d) Find the vertical and horizontal asymptotes of

$$f(x) = \frac{5x + 3}{x^2 - 4}$$

12. Let s be the rational function

$$s(x) = \frac{a_n x^n + a_{n-1}x^{n-1} + \cdots + a_1 x + a_0}{b_m x^m + b_{m-1}x^{m-1} + \cdots + b_1 x + b_0}$$

(a) How do you find the vertical asymptotes of s?

(b) How do you find the horizontal asymptotes of s?

(c) Find the vertical and horizontal asymptotes of

$$r(x) = \frac{5x^3 + 3x^2 - x + 8}{(x - 3)(x - 1)(2x + 1)}$$

13. (a) Under what circumstances does a rational function have a slant asymptote?

(b) How do you determine the end behavior of a rational function?

14. (a) Explain how to solve a polynomial inequality.

(b) Solve the inequality $x^2 - 9 \leq 8x$.

15. (a) What are the cut points of a rational function? Explain how to solve a rational inequality.

(b) Solve the rational inequality $\dfrac{x}{x + 2} - \dfrac{1}{x} \leq 0$.

Answers to the Concept Check can be found at the book companion website stewartmath.com.

Exercises

1–4 ■ Graphs of Quadratic Functions A quadratic function is given. **(a)** Express the function in vertex form. **(b)** Graph the function.

1. $f(x) = x^2 + 6x + 2$ **2.** $f(x) = 2x^2 - 8x + 4$

3. $f(x) = 1 - 10x - x^2$ **4.** $g(x) = -2x^2 + 12x$

5–6 ■ Maximum and Minimum Values Find the maximum or minimum value of the quadratic function.

5. $f(x) = -x^2 + 3x - 1$ **6.** $f(x) = 3x^2 - 18x + 5$

7. Height of a Stone A stone is thrown upward from the top of a building. Its height (in meters) above the ground after t seconds is given by the function $h(t) = -16t^2 + 48t + 32$. What maximum height does the stone reach?

8. Profit The profit P (in dollars) generated by selling x units of a certain commodity is given by the function

$$P(x) = -1500 + 12x - 0.004x^2$$

What is the maximum profit, and how many units must be sold to generate it?

9–14 ■ Transformations of Monomials Graph the polynomial by transforming an appropriate graph of the form $y = x^n$. Show clearly all x- and y-intercepts, and state the domain and range.

9. $P(x) = -x^3 + 64$ **10.** $P(x) = 2x^3 - 16$

11. $P(x) = 2(x + 1)^4 - 32$ **12.** $P(x) = 81 - (x - 3)^4$

13. $P(x) = 32 + (x - 1)^5$ **14.** $P(x) = -3(x + 2)^5 + 96$

15–18 ■ Graphing Polynomials in Factored Form A polynomial function P is given. **(a)** Describe the end behavior. **(b)** Sketch a graph of P. Make sure your graph shows all intercepts.

15. $P(x) = (x - 3)(x + 1)(x - 5)$

16. $P(x) = -(x - 5)(x^2 - 9)(x + 2)$

17. $P(x) = -(x - 1)^2(x - 4)(x + 2)^2$

18. $P(x) = x^2(x^2 - 4)(x^2 - 9)$

19–20 ■ Graphing Polynomials A polynomial function P is given. **(a)** Find each zero of P and state its multiplicity. **(b)** Sketch a graph of P.

19. $P(x) = x^3(x - 2)^2$ **20.** $P(x) = x(x + 1)^3(x - 1)^2$

21–26 ■ Graphing Polynomials Use a graphing device to graph the polynomial. Find the x- and y-intercepts and the coordinates of all local extrema, rounded to one decimal place. Describe the end behavior of the polynomial.

21. $P(x) = -x^2 + 8x$

22. $P(x) = x^3 - 4x + 1$

23. $P(x) = -2x^3 + 6x^2 - 2$

24. $P(x) = x^4 + 4x^3$

25. $P(x) = 3x^4 - 4x^3 - 10x - 1$

26. $P(x) = x^5 + x^4 - 7x^3 - x^2 + 6x + 3$

27. Strength of a Beam The strength S of a wooden beam of width x and depth y is given by the formula $S = 13.8xy^2$. A beam is to be cut from a log of diameter 10 cm, as shown in the figure.

(a) Express the strength S of this beam as a function of x only.

(b) What is the domain of the function S?

(c) Draw a graph of S.

(d) What width will make the beam the strongest?

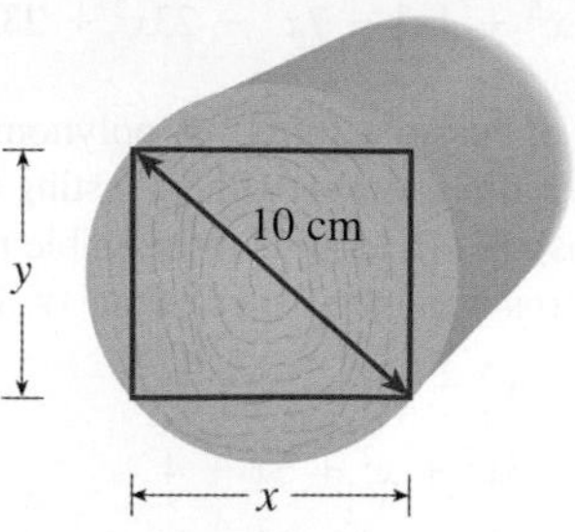

28. Volume A small shelter for delicate plants is to be constructed of thin plastic material. It will have square ends and a rectangular top and back, with an open bottom and front, as shown in the figure. The total area of the four plastic sides is to be 1200 cm^2.

(a) Express the volume V of the shelter as a function of the depth x.

(b) Draw a graph of V.

(c) What dimensions will maximize the volume of the shelter?

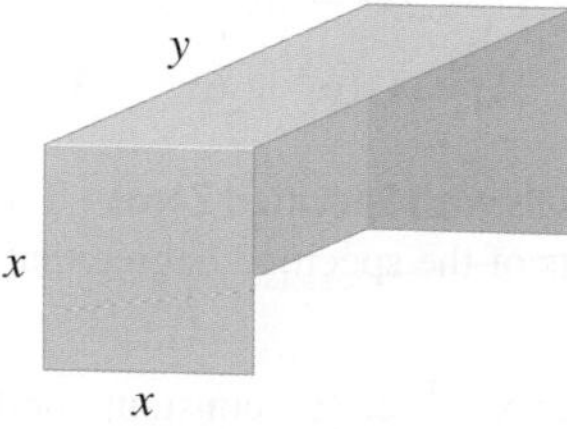

29–36 ■ Division of Polynomials Find the quotient and remainder.

29. $\dfrac{x^2 - 5x + 2}{x - 3}$ **30.** $\dfrac{3x^2 + x - 5}{x + 2}$

31. $\dfrac{2x^3 - x^2 + 3x - 4}{x + 5}$ **32.** $\dfrac{-x^3 + 2x + 4}{x - 7}$

33. $\dfrac{x^4 - 8x^2 + 2x + 7}{x + 5}$ **34.** $\dfrac{2x^4 + 3x^3 - 12}{x + 4}$

35. $\dfrac{2x^3 + x^2 - 8x + 15}{x^2 + 2x - 1}$ **36.** $\dfrac{x^4 - 2x^2 + 7x}{x^2 - x + 3}$

37–40 ■ Remainder Theorem These exercises involve the Remainder Theorem.

37. If $P(x) = 2x^3 - 9x^2 - 7x + 13$, find $P(5)$.

38. If $Q(x) = x^4 + 4x^3 + 7x^2 + 10x + 15$, find $Q(-3)$.

39. What is the remainder when the polynomial $P(x) = x^{500} + 6x^{101} - x^2 - 2x + 4$ is divided by $x - 1$?

40. What is the remainder when the polynomial $Q(x) = x^{101} - x^4 + 2$ is divided by $x + 1$?

41–42 ■ Factor Theorem Use the Factor Theorem to show that the given statement is true.

41. Show that $\frac{1}{2}$ is a zero of the polynomial

$$P(x) = 2x^4 + x^3 - 5x^2 + 10x - 4$$

42. Show that $x + 4$ is a factor of the polynomial

$$P(x) = x^5 + 4x^4 - 7x^3 - 23x^2 + 23x + 12$$

43–46 ■ Number of Possible Zeros A polynomial P is given. **(a)** List all possible rational zeros (without testing to see whether they are actual zeros). **(b)** Determine the possible number of positive and negative real zeros using Descartes's Rule of Signs.

43. $P(x) = x^5 - 6x^3 - x^2 + 2x + 18$

44. $P(x) = 6x^4 + 3x^3 + x^2 + 3x + 4$

45. $P(x) = 3x^7 - x^5 + 5x^4 + x^3 + 8$

46. $P(x) = 6x^{10} - 2x^8 - 5x^3 + 2x^2 + 12$

47–54 ■ Finding Real Zeros and Graphing Polynomials A polynomial P is given. **(a)** Find all real zeros of P, and state their multiplicities. **(b)** Sketch the graph of P.

47. $P(x) = x^3 - 16x$ **48.** $P(x) = x^3 - 3x^2 - 4x$

49. $P(x) = x^4 + x^3 - 2x^2$ **50.** $P(x) = x^4 - 5x^2 + 4$

51. $P(x) = x^4 - 2x^3 - 7x^2 + 8x + 12$

52. $P(x) = x^4 - 2x^3 - 2x^2 + 8x - 8$

53. $P(x) = 2x^4 + x^3 + 2x^2 - 3x - 2$

54. $P(x) = 9x^5 - 21x^4 + 10x^3 + 6x^2 - 3x - 1$

55–56 ■ Polynomials with Specified Zeros Find a polynomial with real coefficients of the specified degree that satisfies the given conditions.

55. Degree 3; zeros $-\frac{1}{2}, 2, 3$; constant coefficient 12

56. Degree 4; zeros 4 (multiplicity 2) and $3i$; integer coefficients; coefficient of x^2 is -25

57. Complex Zeros of Polynomials Does there exist a polynomial of degree 4 with integer coefficients that has zeros i, $2i$, $3i$, and $4i$? If so, find it. If not, explain why.

58. Polynomial with no Real Solutions Prove that the equation $3x^4 + 5x^2 + 2 = 0$ has no real solution.

59–70 ■ Finding Real and Complex Zeros of Polynomials Find all rational, irrational, and complex zeros (and state their multiplicities). Use Descartes's Rule of Signs, the Upper and Lower Bounds Theorem, the Quadratic Formula, or other factoring techniques to help you wherever appropriate.

59. $P(x) = x^3 - x^2 + x - 1$

60. $P(x) = x^3 - 8$

61. $P(x) = x^3 - 3x^2 - 13x + 15$

62. $P(x) = 2x^3 + 5x^2 - 6x - 9$

63. $P(x) = x^4 + 6x^3 + 17x^2 + 28x + 20$

64. $P(x) = x^4 + 7x^3 + 9x^2 - 17x - 20$

65. $P(x) = x^5 - 3x^4 - x^3 + 11x^2 - 12x + 4$

66. $P(x) = x^4 - 81$

67. $P(x) = x^6 - 64$

68. $P(x) = 18x^3 + 3x^2 - 4x - 1$

69. $P(x) = 6x^4 - 18x^3 + 6x^2 - 30x + 36$

70. $P(x) = x^4 + 15x^2 + 54$

71–74 ■ Finding Zeros Graphically Use a graphing device to find all real solutions of the equation, rounded to two decimal places.

71. $2x^2 = 5x + 3$

72. $x^3 + x^2 - 14x - 24 = 0$

73. $x^4 - 3x^3 - 3x^2 - 9x - 2 = 0$

74. $x^5 = x + 3$

75–76 ■ Complete Factorization A polynomial function P is given. Find all the real zeros of P, and factor P completely into linear and irreducible quadratic factors with real coefficients.

75. $P(x) = x^3 - 2x - 4$ **76.** $P(x) = x^4 + 3x^2 - 4$

77–80 ■ Transformations of $y = 1/x$ A rational function is given. **(a)** Find all vertical and horizontal asymptotes, all x- and y-intercepts, and state the domain and range. **(b)** Use transformations of the graph of $y = 1/x$ to sketch a graph of the function.

77. $r(x) = \dfrac{3}{x + 4}$ **78.** $r(x) = \dfrac{-1}{x - 5}$

79. $r(x) = \dfrac{3x - 4}{x - 1}$ **80.** $r(x) = \dfrac{2x + 5}{x + 2}$

81–86 ■ Graphing Rational Functions Graph the rational function. Show clearly all x- and y-intercepts and asymptotes, and state the domain and range of r.

81. $r(x) = \dfrac{3x - 12}{x + 1}$ **82.** $r(x) = \dfrac{1}{(x + 2)^2}$

83. $r(x) = \dfrac{x - 2}{x^2 - 2x - 8}$ **84.** $r(x) = \dfrac{x^3 + 27}{x + 4}$

85. $r(x) = \dfrac{x^2 - 9}{2x^2 + 1}$ **86.** $r(x) = \dfrac{2x^2 - 6x - 7}{x - 4}$

87–90 ■ Rational Functions with Holes Find the common factors of the numerator and denominator of the given rational function. Then find the intercepts and asymptotes, and sketch a graph. State the domain and range.

87. $r(x) = \dfrac{x^2 + 5x - 14}{x - 2}$

88. $r(x) = \dfrac{x^3 - 3x^2 - 10x}{x + 2}$

89. $r(x) = \dfrac{x^2 + 3x - 18}{x^2 - 8x + 15}$

90. $r(x) = \dfrac{x^2 + 2x - 15}{x^3 + 4x^2 - 7x - 10}$

91–94 ■ Graphing Rational Functions Use a graphing device to analyze the graph of the rational function. Find all x- and y-intercepts and all vertical, horizontal, and slant asymptotes. If the function has no horizontal or slant asymptote, find a polynomial that has the same end behavior as the given function.

91. $r(x) = \dfrac{x - 3}{2x + 6}$ **92.** $r(x) = \dfrac{2x - 7}{x^2 + 9}$

93. $r(x) = \dfrac{x^3 + 8}{x^2 - x - 2}$ **94.** $r(x) = \dfrac{2x^3 - x^2}{x + 1}$

95–98 ■ Polynomial Inequalities Solve the inequality.

95. $2x^2 \geq x + 3$ **96.** $x^3 - 3x^2 - 4x + 12 \leq 0$

97. $x^4 - 7x^2 - 18 < 0$ **98.** $x^8 - 17x^4 + 16 > 0$

99–102 ■ Rational Inequalities Solve the inequality.

99. $\dfrac{5}{x^3 - x^2 - 4x + 4} < 0$ **100.** $\dfrac{3x + 1}{x + 2} \leq \dfrac{2}{3}$

101. $\dfrac{1}{x - 2} + \dfrac{2}{x + 3} \geq \dfrac{3}{x}$

102. $\dfrac{1}{x + 2} + \dfrac{3}{x - 3} \leq \dfrac{4}{x}$

103–104 ■ Domain of a Function Find the domain of the given function.

103. $f(x) = \sqrt{24 - x - 3x^2}$

104. $g(x) = \dfrac{1}{\sqrt[4]{x - x^4}}$

105–106 ■ Solving Inequalities Graphically Use a graphing device to solve the inequality. Express your answer using interval notation, with the endpoints of the intervals rounded to two decimal places.

105. $x^4 + x^3 \leq 5x^2 + 4x - 5$

106. $x^5 - 4x^4 + 7x^3 - 12x + 2 > 0$

107. Application of Descartes's Rule of Signs We use Descartes's Rule of Signs to show that a polynomial $Q(x) = 2x^3 + 3x^2 - 3x + 4$ has no positive real zero.

(a) Show that -1 is a zero of the polynomial $P(x) = 2x^4 + 5x^3 + x + 4$.

(b) Use the information from part (a) and Descartes's Rule of Signs to show that the polynomial $Q(x) = 2x^3 + 3x^2 - 3x + 4$ has no positive real zero. [*Hint:* Compare the coefficients of the polynomial Q to your synthetic division table from part (a).]

108. Points of Intersection Find the coordinates of all points of intersection of the graphs of

$$y = x^4 + x^2 + 24x \quad \text{and} \quad y = 6x^3 + 20$$

Matching

109. Equations and Their Graphs Match each equation with its graph. Give reasons for your answers. (Don't use a graphing device.)

(a) $y = \dfrac{x^2 + 4}{x(x^2 - 4)}$ **(b)** $y = \frac{1}{16}(x - 2)^2(x + 2)^3$ **(c)** $y = (x + 1)(x - 2)^2$ **(d)** $y = 2x - x^2$

(e) $y = x^2(2x + 3)(x - 1)$ **(f)** $x = 2y - y^2$ **(g)** $y = \dfrac{x^2}{x^2 - 1}$ **(h)** $y = \sqrt{9 - x^2}$

I

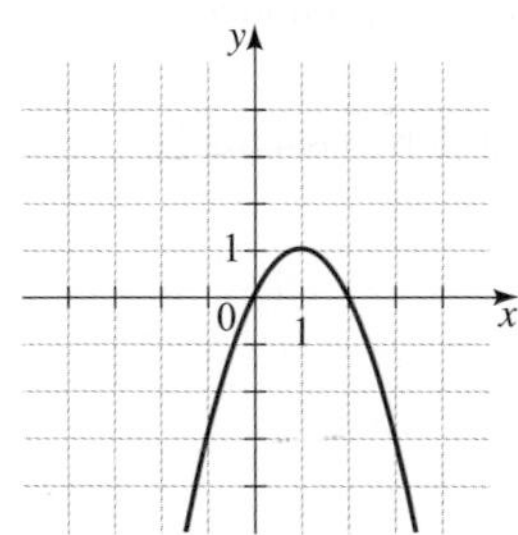

II

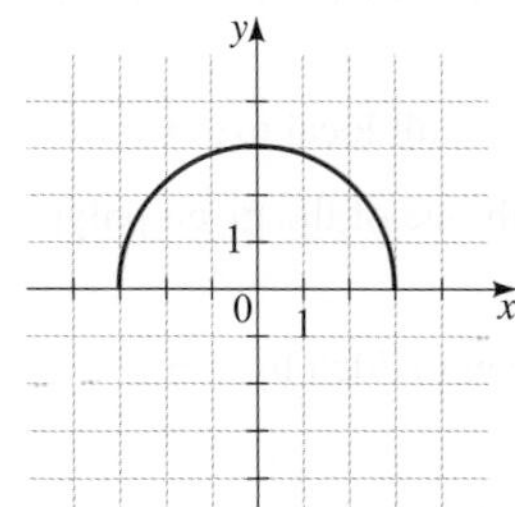

III

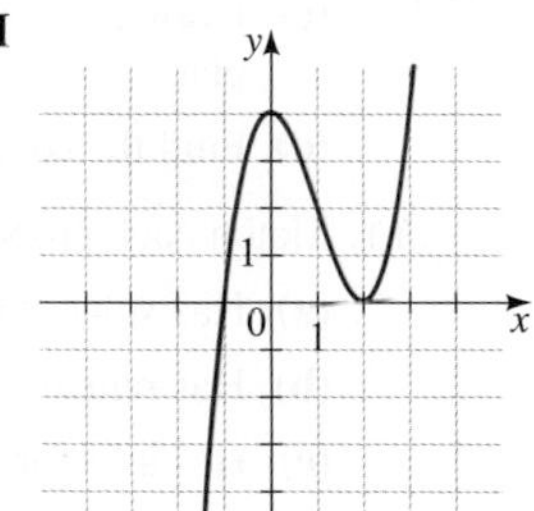

IV

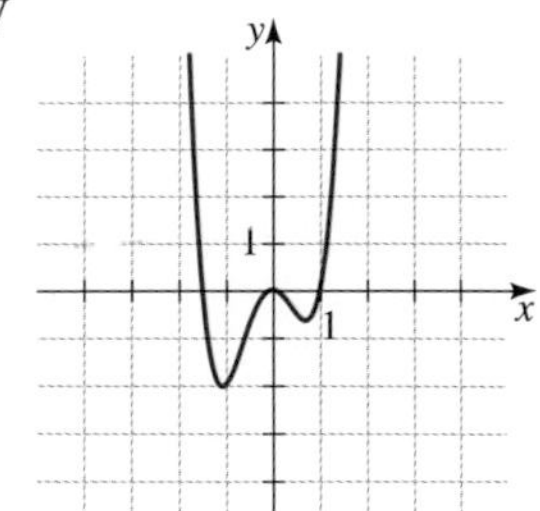

V

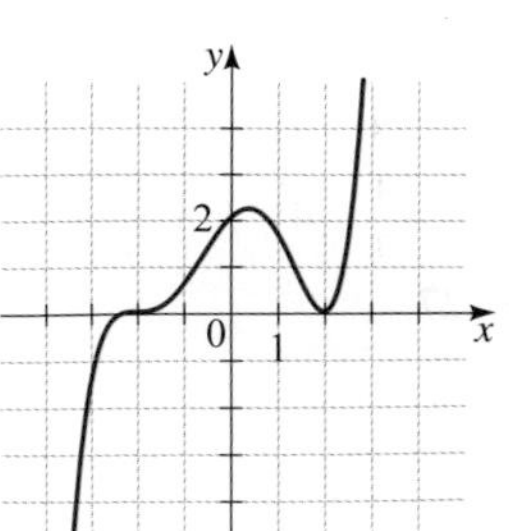

VI

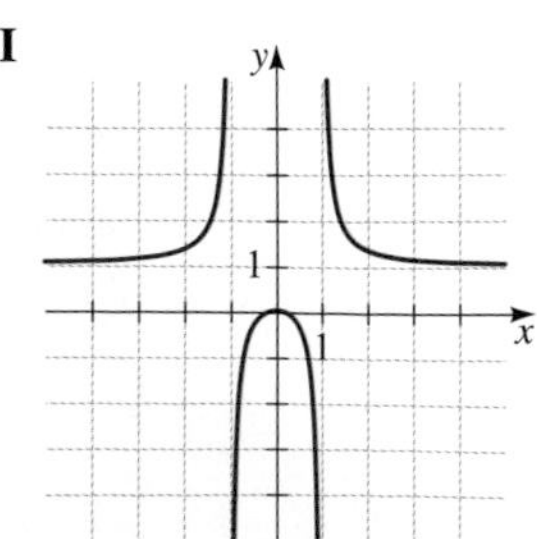

VII

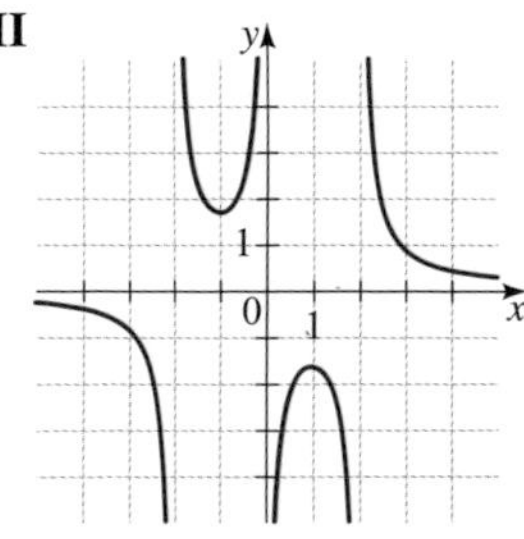

VIII

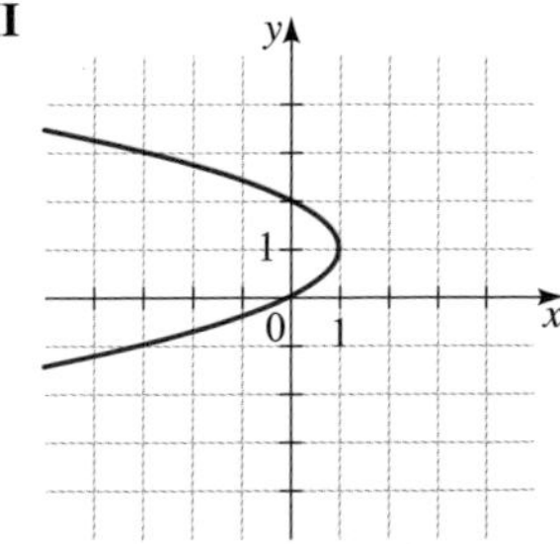

Chapter 3 | Test

1. Express the quadratic function $f(x) = x^2 - x - 6$ in vertex form, sketch its graph, and state the domain and range.

2. Find the maximum or minimum value of the quadratic function $g(x) = 2x^2 + 6x + 3$.

3. A cannonball fired out to sea from a shore battery follows a parabolic trajectory given by the graph of the function

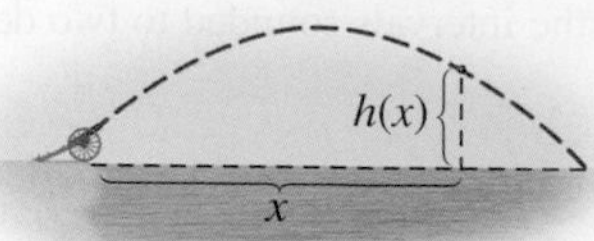

$$h(x) = 10x - 0.01x^2$$

where $h(x)$ is the height of the cannonball above the water when it has traveled a horizontal distance of x meters.

(a) What is the maximum height that the cannonball reaches?

(b) How far does the cannonball travel horizontally before splashing into the water?

4. Graph the polynomial $P(x) = -(x + 2)^3 + 27$, showing clearly all x- and y-intercepts.

5. (a) Use synthetic division to find the quotient and remainder of $x^4 - 4x^2 + 2x + 5$ divided by $x - 2$.

(b) Use long division to find the quotient and remainder of $2x^5 + 4x^4 - x^3 - x^2 + 7$ divided by $2x^2 - 1$.

6. Let $P(x) = 2x^3 - 5x^2 - 4x + 3$.

(a) List all possible rational zeros of P.

(b) Find the complete factorization of P.

(c) Find the zeros of P.

(d) Sketch the graph of P.

7. Find all real and complex zeros of $P(x) = x^3 - x^2 - 4x - 6$.

8. Find the complete factorization of $P(x) = x^4 - 2x^3 + 5x^2 - 8x + 4$.

9. Find a fourth-degree polynomial with integer coefficients that has zeros $3i$ and -1, with -1 a zero of multiplicity 2.

10. Let $P(x) = 2x^4 - 7x^3 + x^2 - 18x + 3$.

(a) Use Descartes's Rule of Signs to determine how many positive and how many negative real zeros P can have.

(b) Show that 4 is an upper bound and -1 is a lower bound for the real zeros of P.

(c) Draw a graph of P, and use it to estimate the real zeros of P, rounded to two decimal places.

(d) Find the coordinates of all local extrema of P, rounded to two decimal places.

11. Match each property with one of the given polynomials.

(a) Has value 0 at $x = 1$

(b) Has remainder 0 when divided by $x + 1$

(c) Has a factor $x + 2$

(d) Has a zero 3 of multiplicity 3

(e) Has zeros 2 and $2i$

$$P(x) = x^3 + 2x^2 + x \qquad Q(x) = x^3 - 2x^2 - 4x + 8$$

$$R(x) = x^3 + 4x - 5 \qquad S(x) = x^3 - 2x^2 + 4x - 8$$

$$T(x) = x^3 - 9x^2 + 27x - 27$$

12. Consider the following rational functions:

$$r(x) = \frac{2x - 1}{x^2 - x - 2} \qquad s(x) = \frac{x^3 + 27}{x^2 + 4} \qquad t(x) = \frac{x^3 - 9x}{x + 2}$$

$$u(x) = \frac{x^2 + x - 6}{x^2 - 25} \qquad w(x) = \frac{x^3 + 6x^2 + 9x}{x + 3}$$

(a) Which of these rational functions has a horizontal asymptote?

(b) Which of these functions has a slant asymptote?

(c) Which of these functions has no vertical asymptote?

(d) Which of these functions has a "hole"?

(e) What are the asymptotes of the function $r(x)$?

(f) Graph $y = u(x)$, showing clearly any asymptotes and x- and y-intercepts the function may have.

(g) Use long division to find a polynomial P that has the same end behavior as t. Graph P and t on the same screen to verify that they have the same end behavior.

13. Solve the rational inequality $x \le \dfrac{6 - x}{2x - 5}$.

14. Find the domain of the function $f(x) = \dfrac{1}{\sqrt{4 - 2x - x^2}}$.

15. (a) Graph the following function in an appropriate viewing rectangle and find all x-intercepts and local extrema, rounded to two decimal places.

$$P(x) = x^4 - 4x^3 + 8x$$

(b) Use your graph from part (a) to solve the inequality

$$x^4 - 4x^3 + 8x \ge 0$$

Express your answer in interval form, with the endpoints rounded to two decimal places.

Focus on Modeling | Fitting Polynomial Curves to Data

We have learned how to fit a line to data (see *Focus on Modeling* following Chapter 1). The line models the increasing or decreasing trend in the data. If the data exhibit more variability, such as an increase followed by a decrease, then to model the data, we need to use a curve rather than a line. Figure 1 shows a scatter plot with three possible models that appear to fit the data.

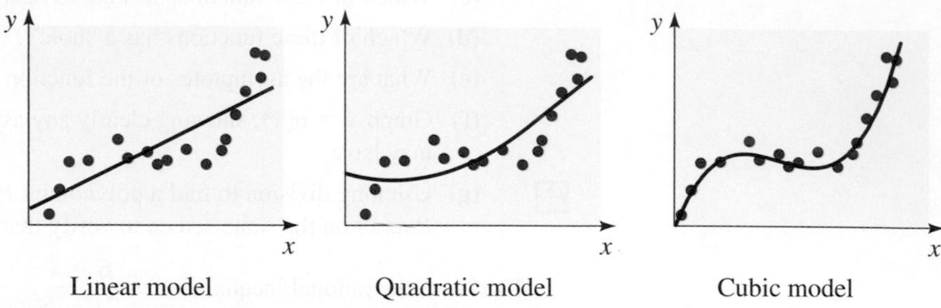

Figure 1

Although we can always find the linear model that best fits the data, it is clear from Figure 1 that sometimes a polynomial model of higher degree fits the data better. In general, it is important to choose an appropriate type of function for modeling real-world data; see the Discovery Project *The Art of Modeling* at **www.stewartmath.com**.

■ Polynomial Functions as Models

Polynomial functions are ideal for modeling data for which the scatter plot has peaks or valleys (that is, local maximums or minimums). Most graphing devices are programmed to find the **polynomial of best fit** of a specified degree. As is the case for lines, a polynomial of a given degree fits the data *best* if the sum of the squares of the vertical distances between the graph of the polynomial and the data points is minimized.

Dmac/Alamy Stock Photo

Example 1 ■ Rainfall and Crop Yield

Rain is essential for crops to grow, but too much rain can diminish crop yields. The data in the following table give rainfall and cotton yield per acre for 10 seasons in a certain county.

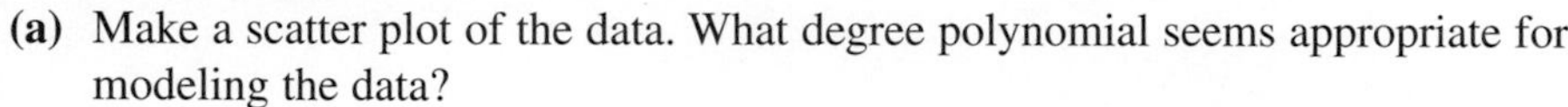

(a) Make a scatter plot of the data. What degree polynomial seems appropriate for modeling the data?

(b) Use a graphing device to find the polynomial of best fit. Graph the polynomial on the scatter plot.

(c) Use the model that you found to estimate the yield if there are 25 cm of rainfall.

Season	Rainfall (cm)	Yield (kg/acre)
1	23.3	5311
2	20.1	4382
3	18.1	3950
4	12.5	3137
5	30.9	5113
6	33.6	4814
7	35.8	3540
8	15.5	3850
9	27.6	5071
10	34.5	3881

Solution

(a) The scatter plot is shown in Figure 2(a). The data appear to have a peak, so it is appropriate to model the data by a quadratic polynomial (degree 2).

(b) Using a graphing device [see Figure 2(b)], we find that the quadratic polynomial of best fit is approximately

$$y = -12.6x^2 + 651.5x - 3283.2$$

This quadratic model together with the scatter plot are graphed in Figure 2(c).

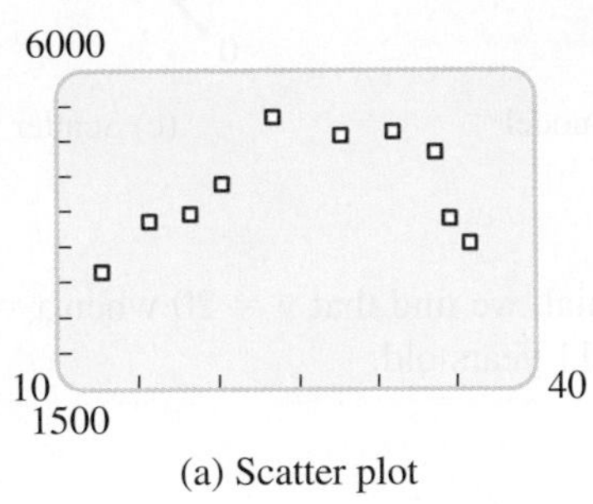

(a) Scatter plot

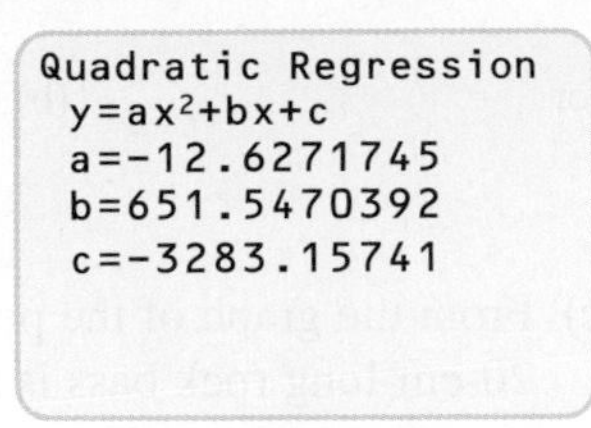

(b) Quadratic model

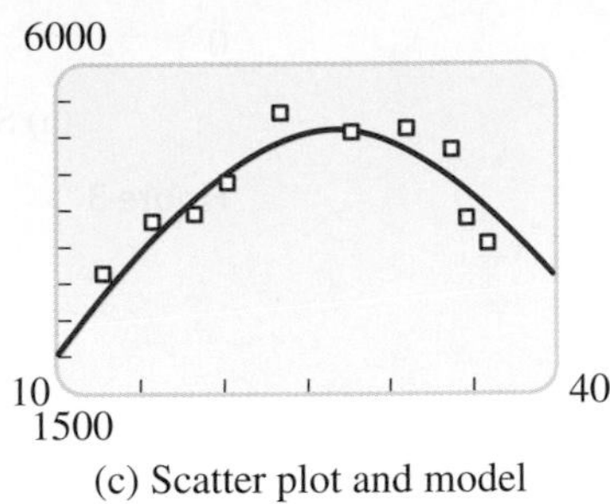

(c) Scatter plot and model

Figure 2

(c) Using the model with $x = 25$, we get

$$y = -12.6(25)^2 + 651.5(25) - 3283.2 \approx 5129.3$$

We estimate the yield to be about 5130 kg/acre. ■

Example 2 ■ Length-at-Age Data for Fish

Cod Redfish Hake

Otoliths for several fish species

Otoliths ("earstones") are tiny structures that are found in the heads of fish. Microscopic growth rings on the otoliths—not unlike growth rings on a tree—record the age of a fish. The following table gives the lengths of rock bass caught at different ages, as determined by the otoliths. Scientists have proposed a cubic polynomial to model such data.

(a) Use a graphing device to make a scatter plot of the data and to find the cubic polynomial of best fit.

(b) Graph the polynomial from part (a) on the same screen as the scatter plot.

(c) A fisherman catches a rock bass 20 cm long. Use the model to estimate its age.

Using a length-at-age model, marine biologists can estimate the age of a fish from its length (without having to examine its otoliths).

Age (yr)	Length (cm)	Age (yr)	Length (cm)
1	4.8	9	18.2
2	8.8	9	17.1
2	8.0	10	18.8
3	7.9	10	19.5
4	11.9	11	18.9
5	14.4	12	21.7
6	14.1	12	21.9
6	15.8	13	23.8
7	15.6	14	26.9
8	17.8	14	25.1

Solution

(a) Using a graphing device, we get the scatter plot in Figure 3(a), and we find the cubic polynomial of best fit [see Figure 3(b)]:

$$y = 0.0155x^3 - 0.372x^2 + 3.95x + 1.21$$

(b) The scatter plot of the data together with the cubic polynomial are graphed in Figure 3(c).

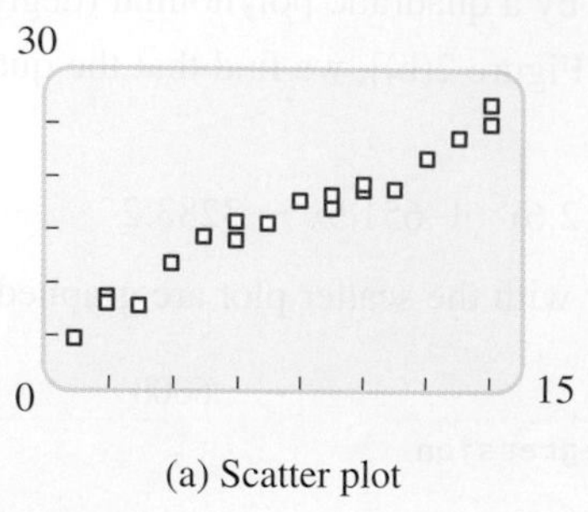

(a) Scatter plot

```
Cubic Regression
y=ax³+bx²+cx+d
a=.0154911306
b=-.372473323
c=3.94608636
d=1.21080418
```

(b) Cubic model

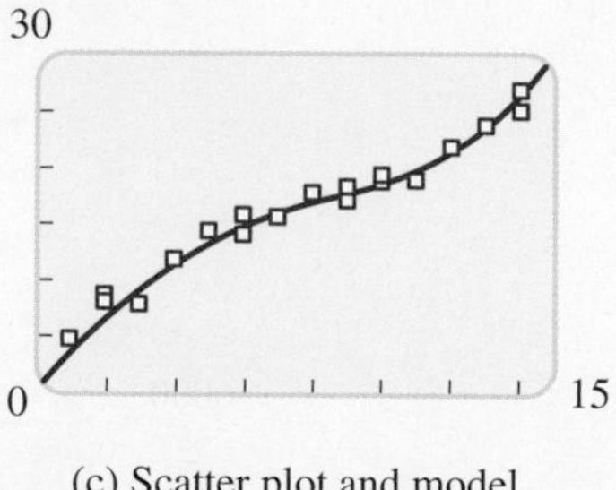

(c) Scatter plot and model

Figure 3

(c) From the graph of the polynomial, we find that $y = 20$ when $x \approx 10.8$. Thus a 20-cm-long rock bass is about 11 years old. ■

Problems

Pressure (kPa^2)	Tire Life (km)
180	50,000
193	66,000
214	78,000
241	81,000
262	74,000
290	70,000
310	59,000

1. Tire Inflation and Tread Wear Car tires need to be inflated properly. Overinflation or underinflation can cause premature tread wear. The data in the margin show tire life for different inflation values for a certain type of tire.

(a) Find the quadratic polynomial that best fits the data.

(b) Draw a graph of the polynomial from part (a) together with a scatter plot of the data.

(c) Use your result from part (b) to estimate the pressure that gives the longest tire life.

2. Too Many Corn Plants per Acre? The more corn a farmer plants per acre, the greater the yield, but only up to a point. Too many plants per acre can cause overcrowding and decrease yield. The data give crop yield per acre for various densities of corn planting, as found by researchers at a university test farm.

(a) Find the quadratic polynomial that best fits the data.

(b) Draw a graph of the polynomial from part (a) together with a scatter plot of the data.

(c) Use your result from part (b) to estimate the yield for 37,000 plants per acre.

Density (plants/acre)	15,000	20,000	25,000	30,000	35,000	40,000	45,000	50,000
Crop Yield (bushels/acre)	43	98	118	140	142	122	93	67

3. How Fast Can You List Your Favorite Things? If you are asked to make a list of objects in a certain category, how fast you can list them follows a predictable pattern. For example, if you try to name as many vegetables as you can, you'll probably think of several right away—for example, carrots, peas, beans, corn, and so on. Then after a pause you might think of some that you eat less frequently—perhaps zucchini, eggplant, or asparagus. Finally, a few more exotic vegetables might come to mind—artichokes, jicama, bok choy, and the like. A psychologist performs this experiment on a number of subjects. The table on the next page gives the average number of vegetables that the subjects named in a given number of seconds.

(a) Find the cubic polynomial that best fits the data.

(b) Draw a graph of the polynomial from part (a) together with a scatter plot of the data.

(c) Use your result from part (b) to estimate the number of vegetables that subjects would be able to name in 40 seconds.

(d) According to the model, how long (to the nearest 0.1 second) would it take a person to name five vegetables?

Seconds	Number of Vegetables
1	2
2	6
5	10
10	12
15	14
20	15
25	18
30	21

4. Height of a Baseball A baseball is thrown upward, and its height is measured at 0.5-second intervals using a strobe light. The resulting data are given in the table.

(a) Draw a scatter plot of the data. What degree polynomial is appropriate for modeling the data?

(b) Find a polynomial model that best fits the data, and graph it on the scatter plot.

(c) Find the times when the ball is 6 m above the ground.

(d) What is the maximum height attained by the ball?

Time (s)	Height (m)
0	1.3
0.5	8.0
1.0	12.2
1.5	14.0
2.0	13.4
2.5	10.3
3.0	4.8

5. Torricelli's Law Water in a tank will flow out of a small hole in the bottom faster when the tank is nearly full than when it is nearly empty. According to Torricelli's Law, the height $h(t)$ of water remaining at time t is a quadratic function of t.

A certain tank is filled with water and allowed to drain. The height of the water is measured at different times as shown in the table.

(a) Find the quadratic polynomial that best fits the data.

(b) Draw a graph of the polynomial from part (a) together with a scatter plot of the data.

(c) Use your graph from part (b) to estimate how long it takes for the tank to drain completely.

Time (min)	Height (m)
0	5.0
4	3.1
8	1.9
12	0.8
16	0.2

Andrzej Kubik/Shutterstock.com

4 Exponential and Logarithmic Functions

In this chapter we study *exponential functions*. These are functions like $f(x) = 2^x$, where the independent variable is in the exponent. Exponential functions are used in modeling many different real-world phenomena, such as the growth of a population, the growth of an investment that earns compound interest, the spread of an infectious disease, or the decay of a radioactive substance. We also see how exponential functions are used in modeling the growth of a population in an environment with limited resources (food, water, living space), such as the buffalo herd pictured here. Once an exponential model has been obtained, we can use the model to predict—for any future time—the size of a population, calculate the amount of an investment, estimate the number of infected individuals, or determine the remaining amount of a radioactive substance.

The inverse functions of exponential functions are called *logarithmic functions*. With exponential models and logarithmic functions, we can answer questions such as these: When will world population reach a given level? When will my bank account have a million dollars? When will the number of new infections level off? When will radiation from a radioactive spill decay to a safe level?

In the *Focus on Modeling* at the end of the chapter we learn how to fit exponential and power curves to data.

4.1 Exponential Functions

■ Exponential Functions ■ Graphs of Exponential Functions ■ Compound Interest

In this chapter we study a class of functions called *exponential functions*. For example,

$$f(x) = 2^x$$

is an exponential function (with base 2). Notice how quickly the values of this function increase.

$$f(3) = 2^3 = 8$$
$$f(10) = 2^{10} = 1024$$
$$f(30) = 2^{30} = 1{,}073{,}741{,}824$$

Exponential functions get their name from the fact that the variable is in the exponent.

Compare this with the function $g(x) = x^2$, where $g(30) = 30^2 = 900$. The point is that when the variable is in the exponent, even a small change in the variable can cause a dramatic change in the value of the function.

■ Exponential Functions

To study exponential functions, we must first define what we mean by the exponential expression a^x when x is any real number. In Section 1.2 we defined a^x for $a > 0$ and x a rational number, but we have not yet defined irrational powers. So what is meant by $5^{\sqrt{3}}$ or 2^{π}? To define a^x when x is irrational, we approximate x by rational numbers.

For example, since

$$\sqrt{3} \approx 1.73205\ldots$$

is an irrational number, we successively approximate $a^{\sqrt{3}}$ by the following rational powers:

$$a^{1.7},\ a^{1.73},\ a^{1.732},\ a^{1.7320},\ a^{1.73205},\ \ldots$$

Intuitively, we can see that these rational powers of a are getting closer and closer to $a^{\sqrt{3}}$. It can be shown by using advanced mathematics that there is exactly one number that these powers approach. We define $a^{\sqrt{3}}$ to be this number.

For example, using a calculator, we find

$$5^{\sqrt{3}} \approx 5^{1.732}$$
$$\approx 16.2411\ldots$$

The more decimal places of $\sqrt{3}$ we use in our calculation, the better our approximation of $5^{\sqrt{3}}$.

The Laws of Exponents are listed in Section 1.2.

It can be proved that the *Laws of Exponents are still true when the exponents are real numbers*.

Exponential Functions

The **exponential function with base** a is defined for all real numbers x by

$$f(x) = a^x$$

where $a > 0$ and $a \neq 1$.

We assume that $a \neq 1$ because the function $f(x) = 1^x = 1$ is just a constant function. Here are some examples of exponential functions:

$$f(x) = 2^x \qquad g(x) = 3^x \qquad h(x) = 10^x$$

Base 2 Base 3 Base 10

Example 1 ■ Evaluating Exponential Functions

We use a calculator to find values of the exponential function $f(x) = 3^x$. Check to make sure you get the following answers on your own calculator.

	Calculator keystrokes	Output
(a) $f(5) = 3^5 = 243$	3 ^ 5 ENTER	243
(b) $f(-\frac{2}{3}) = 3^{-2/3} \approx 0.4807$	3 ^ ((−) 2 ÷ 3) ENTER	0.4807498
(c) $f(\pi) = 3^\pi \approx 31.544$	3 ^ π ENTER	31.5442807
(d) $f(\sqrt{2}) = 3^{\sqrt{2}} \approx 4.7288$	3 ^ √ 2 ENTER	4.7288043

Now Try Exercise 7 ■

■ Graphs of Exponential Functions

We first graph exponential functions by plotting points. We will see that the graphs of such functions have an easily recognizable shape.

Example 2 ■ Graphing Exponential Functions by Plotting Points

Draw the graph of each function.

(a) $f(x) = 3^x$ **(b)** $g(x) = \left(\dfrac{1}{3}\right)^x$

Solution We calculate values of $f(x)$ and $g(x)$ and plot points to sketch the graphs in Figure 1.

x	$f(x) = 3^x$	$g(x) = (\frac{1}{3})^x$
−3	$\frac{1}{27}$	27
−2	$\frac{1}{9}$	9
−1	$\frac{1}{3}$	3
0	1	1
1	3	$\frac{1}{3}$
2	9	$\frac{1}{9}$
3	27	$\frac{1}{27}$

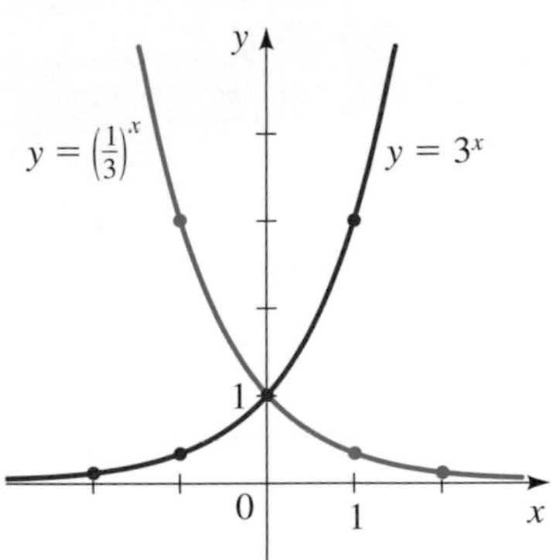

Figure 1

Now Try Exercise 17 ■

Reflecting graphs is explained in Section 2.6.

Note Figure 1 shows that the graph of g is the reflection of the graph of f about the y-axis. In general, the graph of $g(x) = (\frac{1}{a})^x$ is the reflection of the graph of $f(x) = a^x$ about the y-axis because

$$g(x) = \left(\frac{1}{a}\right)^x = \frac{1}{a^x} = a^{-x} = f(-x)$$

So, in Example 2 we could have obtained the graph of $g(x) = (\frac{1}{3})^x$ by reflecting the graph of $f(x) = 3^x$ about the y-axis.

The next figure shows graphs of the family of exponential functions $f(x) = a^x$ for various values of the base a. All these graphs have y-intercept 1 because $f(0) = a^0 = 1$

for $a \neq 0$. You can see from Figure 2 that there are two kinds of exponential functions: If $a > 1$, the exponential function increases rapidly, as in Figure 2(a). If $0 < a < 1$, the function decreases rapidly, as in Figure 2(b).

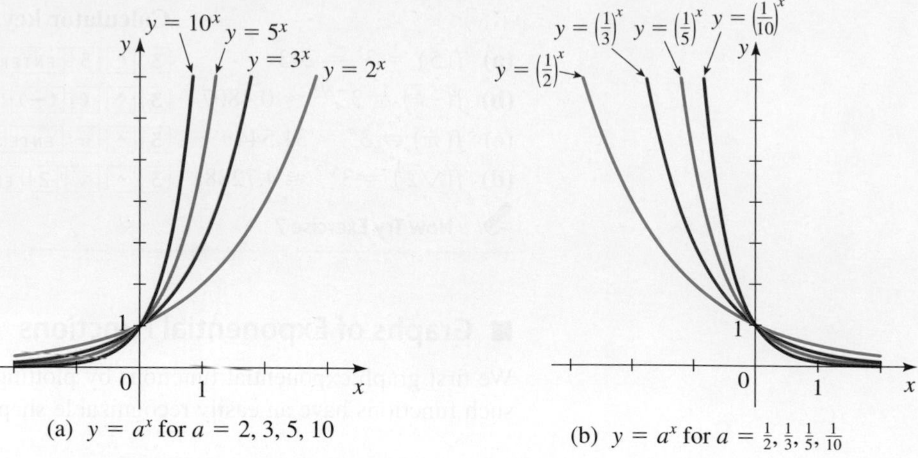

Figure 2 | Families of exponential functions

Arrow notation is explained in Section 3.6.

The x-axis is a horizontal asymptote for the exponential function $f(x) = a^x$. This is because when $a > 1$, we have $a^x \to 0$ as $x \to -\infty$, and when $0 < a < 1$, we have $a^x \to 0$ as $x \to \infty$. (See Figure 2.) Also, $a^x > 0$ for all $x \in \mathbb{R}$, so the function $f(x) = a^x$ has domain $\mathbb{R}$ and range $(0, \infty)$. These observations are summarized in the following box.

Graphs of Exponential Functions

The exponential function

$$f(x) = a^x \qquad (a > 0, a \neq 1)$$

has domain $\mathbb{R}$, range $(0, \infty)$, and y-intercept 1. The line $y = 0$ (the x-axis) is a horizontal asymptote of f. The graph of f has one of the following shapes.

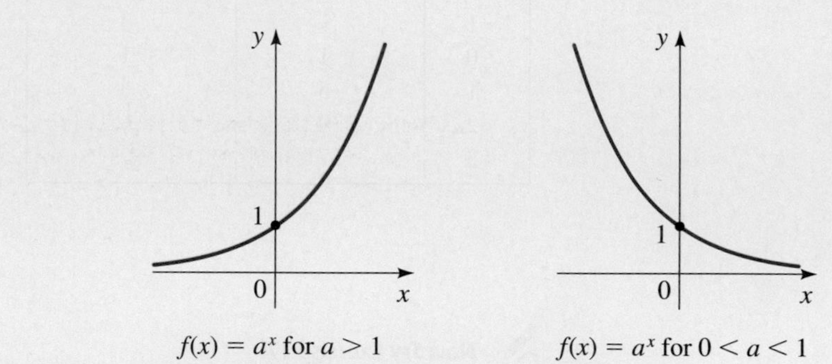

Example 3 ■ Identifying Graphs of Exponential Functions

Find the exponential function $f(x) = a^x$ whose graph is given.

(a) **(b)**

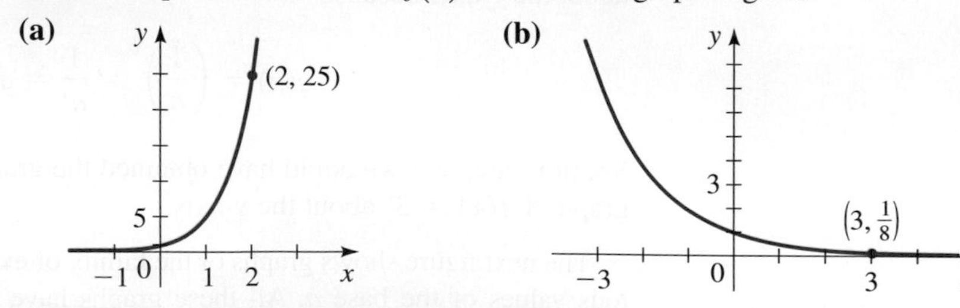

Solution

(a) Since $f(2) = a^2 = 25$, we see that the base is $a = 5$. So $f(x) = 5^x$.

(b) Since $f(3) = a^3 = \frac{1}{8}$, we see that the base is $a = \frac{1}{2}$. So $f(x) = \left(\frac{1}{2}\right)^x$.

Now Try Exercise 21 ■

In the next example we see how to graph certain functions, not by plotting points, but by taking the basic graphs of the exponential functions in Figure 2 and applying the shifting and reflecting transformations that we studied in Section 2.6.

Example 4 ■ Transformations of Exponential Functions

Use the graph of $f(x) = 2^x$ to sketch the graph of each function. State the y-intercept, domain, range, and horizontal asymptote.

(a) $g(x) = 1 + 2^x$ **(b)** $h(x) = -2^x$ **(c)** $k(x) = 2^{x-1}$

Solution

(a) To obtain the graph of $g(x) = 1 + 2^x$, we start with the graph of $f(x) = 2^x$ and shift it upward 1 unit to get the graph shown in Figure 3(a). The y-intercept is $y = g(0) = 1 + 2^0 = 2$. From the graph we see that the domain of g is the set $\mathbb{R}$ of real numbers, the range is the interval $(1, \infty)$, and the line $y = 1$ is a horizontal asymptote.

(b) Again we start with the graph of $f(x) = 2^x$, but here we reflect about the x-axis to get the graph of $h(x) = -2^x$ shown in Figure 3(b). The y-intercept is $y = h(0) = -2^0 = -1$. From the graph we see that the domain of h is the set $\mathbb{R}$ of all real numbers, the range is the interval $(-\infty, 0)$, and the line $y = 0$ is a horizontal asymptote.

(c) This time we start with the graph of $f(x) = 2^x$ and shift it 1 unit to the right to get the graph of $k(x) = 2^{x-1}$ shown in Figure 3(c). The y-intercept is $y = k(0) = 2^{0-1} = \frac{1}{2}$. From the graph we see that the domain of k is the set $\mathbb{R}$ of all real numbers, the range is the interval $(0, \infty)$, and the line $y = 0$ is a horizontal asymptote.

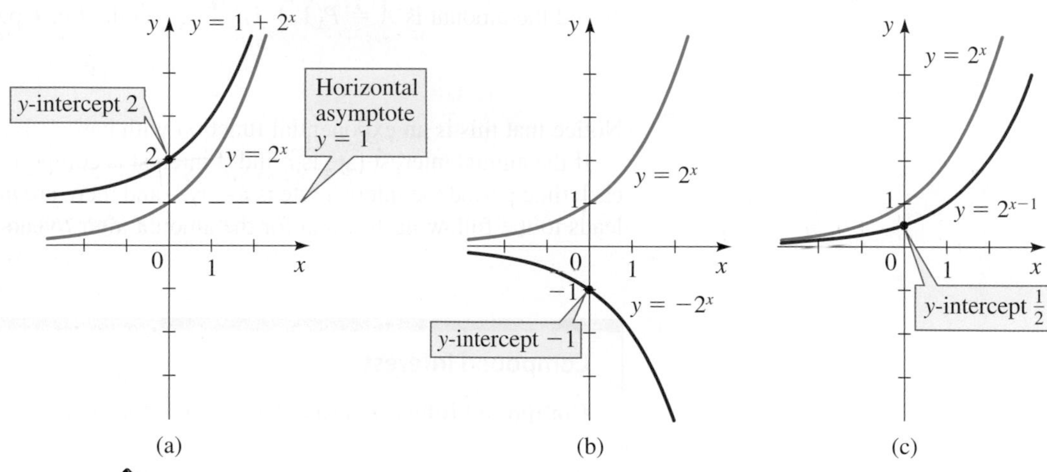

Figure 3

Now Try Exercises 27, 29, and 31 ■

Example 5 ■ Comparing Exponential and Power Functions

Compare the rates of growth of the exponential function $f(x) = 2^x$ and the power function $g(x) = x^2$ by drawing the graphs of both functions in the following viewing rectangles.

(a) $[0, 3]$ by $[0, 8]$ **(b)** $[0, 6]$ by $[0, 25]$ **(c)** $[0, 20]$ by $[0, 1000]$

Solution

(a) Figure 4(a) shows that the graph of $g(x) = x^2$ catches up with, and becomes higher than, the graph of $f(x) = 2^x$ at $x = 2$.

(b) The larger viewing rectangle in Figure 4(b) shows that the graph of $f(x) = 2^x$ overtakes that of $g(x) = x^2$ when $x = 4$.

(c) Figure 4(c) gives a more global view and shows that when x is large, $f(x) = 2^x$ is much larger than $g(x) = x^2$.

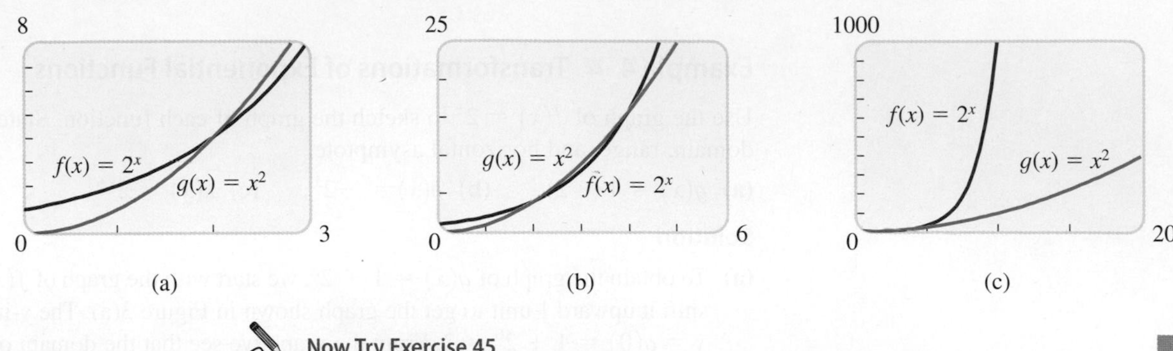

Figure 4 (a) (b) (c)

Now Try Exercise 45 ■

■ Compound Interest

Exponential functions occur in the calculation of compound interest. If an amount of money P, called the **principal**, is invested at an interest rate i per time period, then after one time period the interest is Pi, and the amount A of money is

$$A = P + Pi = P(1 + i)$$

If the interest is reinvested, then the new principal is $P(1 + i)$, and the amount after another time period is $A = P(1 + i)(1 + i) = P(1 + i)^2$. Similarly, after a third time period the amount is $A = P(1 + i)^3$. In general, after k periods the amount is

$$A = P(1 + i)^k$$

Notice that this is an exponential function with base $1 + i$.

If the annual interest rate is r and if interest is compounded n times per year, then in each time period the interest rate is $i = r/n$, and there are nt time periods in t years. This leads to the following formula for the amount after t years.

Compound Interest

Compound interest is calculated by the formula

$$A(t) = P\left(1 + \frac{r}{n}\right)^{nt}$$

where $A(t)$ = amount after t years

P = principal

r = interest rate per year

n = number of times interest is compounded per year

t = number of years

r is often referred to as the *nominal annual interest rate*.

Example 6 ■ Calculating Compound Interest

A sum of $1000 is invested at an interest rate of 12% per year. Find the amounts in the account after 3 years if interest is compounded annually, semiannually, quarterly, monthly, and daily.

Solution We use the compound interest formula with $P = \$1000$, $r = 0.12$, and $t = 3$.

Compounding	n	Amount After 3 Years
Annually	1	$1000\left(1 + \frac{0.12}{1}\right)^{1(3)} = \$1404.93$
Semiannually	2	$1000\left(1 + \frac{0.12}{2}\right)^{2(3)} = \$1418.52$
Quarterly	4	$1000\left(1 + \frac{0.12}{4}\right)^{4(3)} = \$1425.76$
Monthly	12	$1000\left(1 + \frac{0.12}{12}\right)^{12(3)} = \$1430.77$
Daily	365	$1000\left(1 + \frac{0.12}{365}\right)^{365(3)} = \$1433.24$

Now Try Exercise 57 ■

If an investment earns compound interest, then the **annual percentage yield** (APY) is the *simple* interest rate that yields the same amount at the end of one year.

Example 7 ■ Calculating the Annual Percentage Yield

Find the annual percentage yield for an investment that earns interest at a rate of 6% per year, compounded daily.

Solution After one year, a principal P will grow to the amount

$$A = P\left(1 + \frac{0.06}{365}\right)^{365} = P(1.06183)$$

Simple interest is studied in Section 1.7.

The formula for simple interest is

$$A = P(1 + r)$$

Comparing, we see that $1 + r = 1.06183$, so $r = 0.06183$. Thus the annual percentage yield is 6.183%.

Now Try Exercise 63 ■

Discovery Project ■ So You Want to Be a Millionaire?

In this project we explore how rapidly the values of an exponential function increase by examining some real-world situations. For example, if you save a penny today, two pennies tomorrow, four pennies the next day, and so on, how long do you have to continue saving in this way before you become a millionaire? You can find out the surprising answer to this and other questions by completing this discovery project. You can find the project at **www.stewartmath.com**.

4.1 | Exercises

Concepts

1. The function $f(x) = 5^x$ is an exponential function with base ______; $f(-2) =$ ______, $f(0) =$ ______, $f(2) =$ ______, and $f(6) =$ ______.

2. Match the exponential function with one of the graphs labeled I, II, III, or IV, shown below.

(a) $f(x) = 2^x$ **(b)** $f(x) = 2^{-x}$

(c) $f(x) = -2^x$ **(d)** $f(x) = -2^{-x}$

I

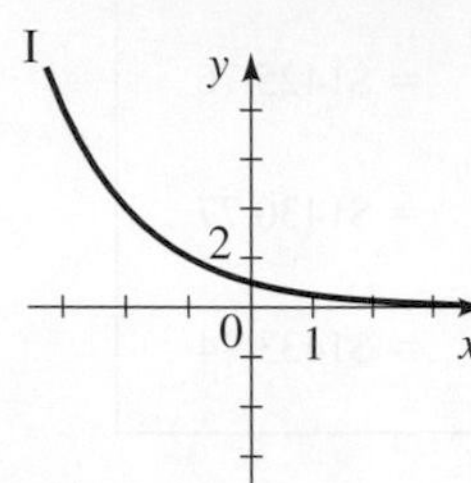

II

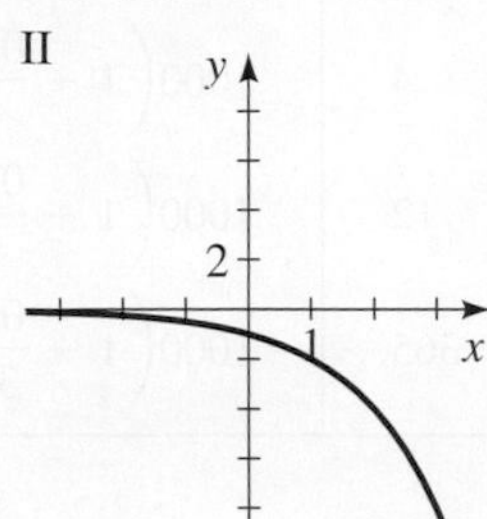

III

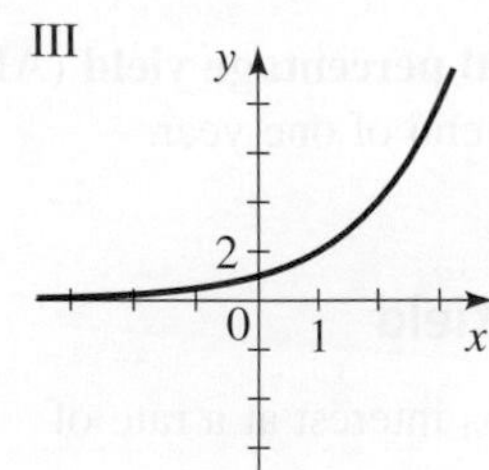

IV

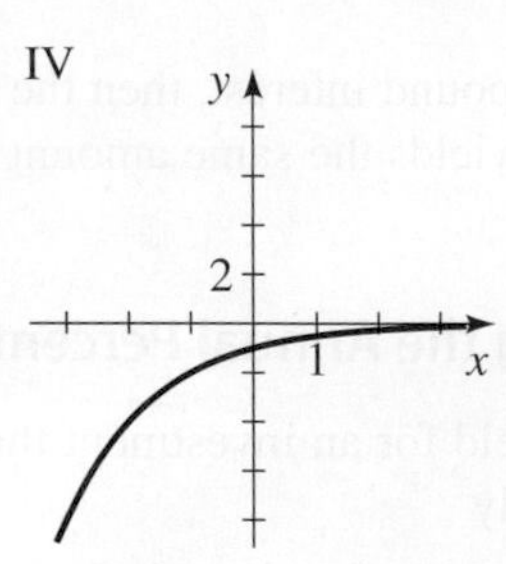

3. **(a)** To obtain the graph of $g(x) = 2^x - 1$, we start with the graph of $f(x) = 2^x$ and shift it ______ (upward/downward) 1 unit.

(b) To obtain the graph of $h(x) = 2^{x-1}$, we start with the graph of $f(x) = 2^x$ and shift it 1 unit to the ______ (left/right).

4. In the formula $A(t) = P(1 + \frac{r}{n})^{nt}$ for compound interest the letters P, r, n, and t stand for ______, ______, ______, and ______, respectively, and $A(t)$ stands for ______. So if \$100 is invested at an interest rate of 6% compounded quarterly, then the amount after 2 years is ______.

5. The exponential function $f(x) = (\frac{1}{2})^x$ has the ______ asymptote $y =$ ______. This means that as $x \to \infty$, we have $(\frac{1}{2})^x \to$ ______.

6. The exponential function $f(x) = (\frac{1}{2})^x + 3$ has the ______ asymptote $y =$ ______. This means that as $x \to \infty$, we have $(\frac{1}{2})^x + 3 \to$ ______.

Skills

7–10 ■ Evaluating Exponential Functions Use a calculator to evaluate the function at the indicated values. Round your answers to three decimal places.

7. $f(x) = 4^x$; $f(\frac{1}{2}), f(\sqrt{5}), f(-2), f(0.3)$

8. $f(x) = 3^{x-1}$; $f(\frac{1}{2}), f(2.5), f(-1), f(\frac{1}{4})$

9. $g(x) = (\frac{1}{3})^{x+1}$; $g(\frac{1}{2}), g(\sqrt{2}), g(-3.5), g(-1.4)$

10. $g(x) = (\frac{4}{3})^{3x}$; $g(-\frac{1}{2}), g(\sqrt{6}), g(-3), g(\frac{4}{3})$

11–16 ■ Graphing Exponential Functions Sketch the graph of the function by making a table of values. Use a calculator if necessary.

11. $f(x) = 6^x$ **12.** $f(x) = (0.1)^x$

13. $g(x) = (\frac{2}{3})^x$ **14.** $g(x) = 4^x$

15. $h(x) = 5(2.2)^x$ **16.** $h(x) = 4(\frac{5}{8})^x$

17–20 ■ Graphing Exponential Functions Graph both functions on one set of axes.

17. $f(x) = 4^x$ and $g(x) = 4^{-x}$

18. $f(x) = 8^{-x}$ and $g(x) = (\frac{1}{8})^x$

19. $f(x) = 4^x$ and $g(x) = 7^x$

20. $f(x) = (\frac{3}{4})^x$ and $g(x) = 1.5^x$

21–24 ■ Exponential Functions from a Graph Find the exponential function $f(x) = a^x$ whose graph is given.

21.

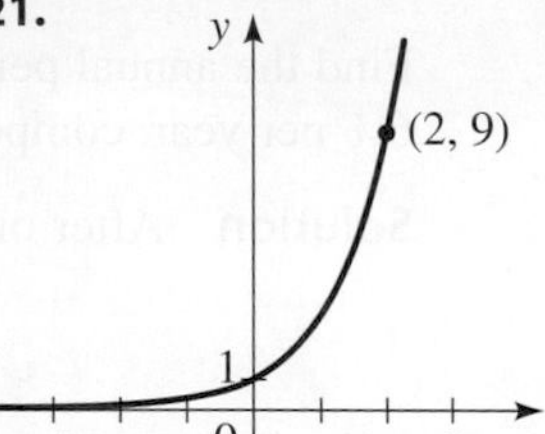

22.

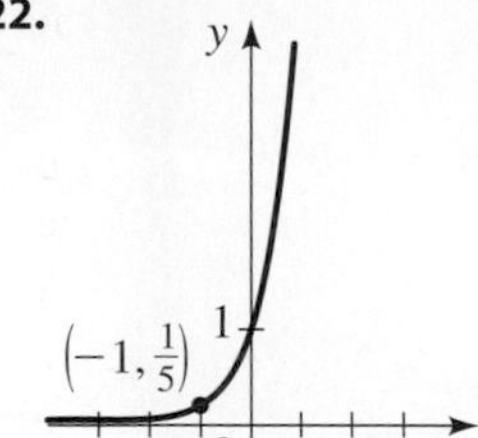

23.

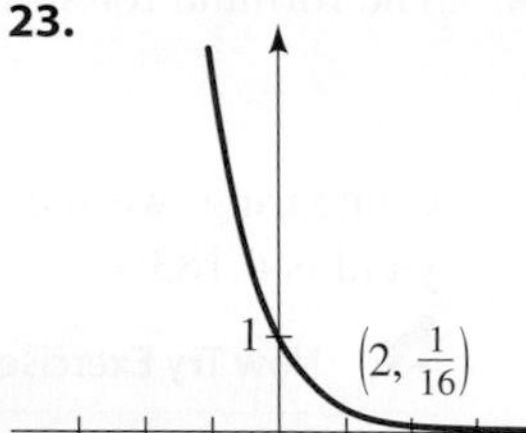

24.

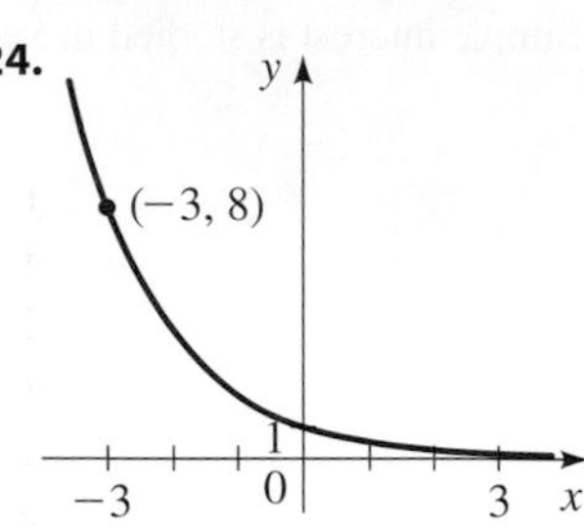

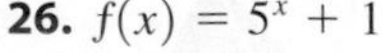

25–26 ■ Exponential Functions from a Graph Match the exponential function with one of the graphs labeled I or II.

25. $f(x) = 5^{x+1}$ **26.** $f(x) = 5^x + 1$

I

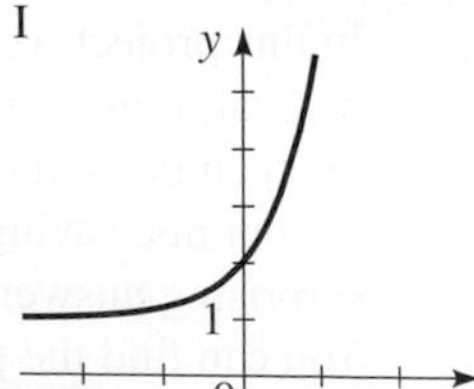

II

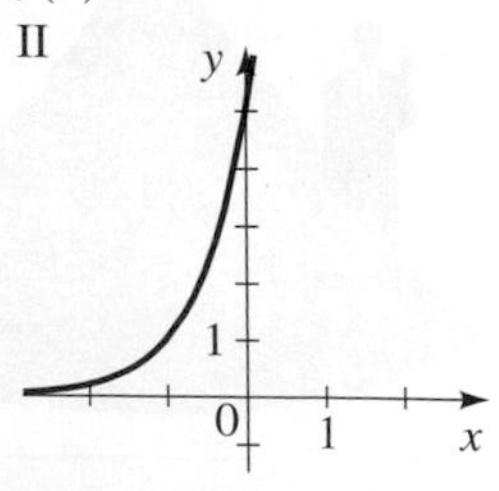

27–40 ■ Graphing Exponential Functions Graph the function, not by plotting points, but by starting from the graphs in Figure 2. State the y-intercept, domain, range, and horizontal asymptote.

27. $f(x) = 3^x + 1$ **28.** $f(x) = \left(\frac{1}{3}\right)^x - 2$

29. $g(x) = -5^x$ **30.** $g(x) = 5^{-x}$

31. $h(x) = 3^{x-2}$ **32.** $h(x) = 10^{x+1}$

33. $y = 2^{-x} + 3$ **34.** $y = -2^x + 3$

35. $y = -10^x - 1$ **36.** $y = \left(\frac{1}{2}\right)^x - 4$

37. $h(x) = 2^{x-4} + 1$ **38.** $y = 10^{x+1} - 5$

39. $g(x) = 1 - 3^{-x}$ **40.** $y = 3 - \left(\frac{1}{5}\right)^x$

41–42 ■ Comparing Exponential Functions In these exercises we compare the graphs of two exponential functions.

41. **(a)** Sketch the graphs of $f(x) = 2^x$ and $g(x) = 3(2^x)$.

(b) How are the graphs related?

42. **(a)** Sketch the graphs of $f(x) = 9^{x/2}$ and $g(x) = 3^x$.

(b) Use the Laws of Exponents to explain the relationship between these graphs.

43–44 ■ Comparing Exponential and Power Functions Compare the graphs of the power function f and exponential function g by evaluating both of them for $x = 0, 1, 2, 3, 4, 6, 8$, and 10. Then draw the graphs of f and g on the same set of axes.

43. $f(x) = x^3$; $g(x) = 3^x$ **44.** $f(x) = x^4$; $g(x) = 4^x$

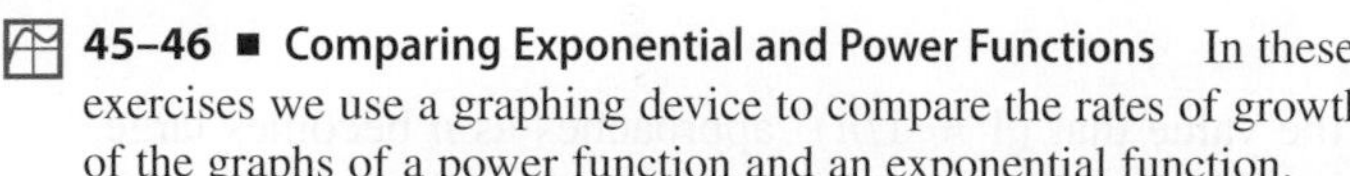

45–46 ■ Comparing Exponential and Power Functions In these exercises we use a graphing device to compare the rates of growth of the graphs of a power function and an exponential function.

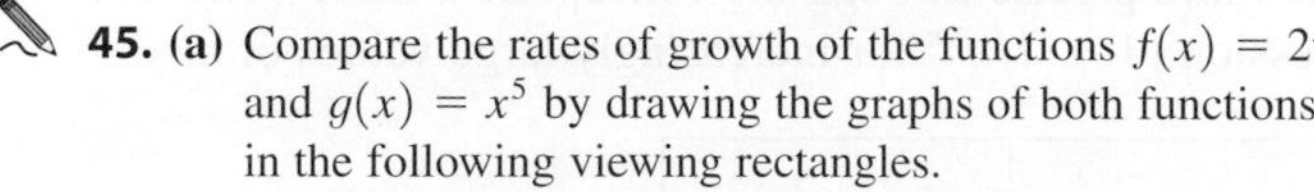

45. **(a)** Compare the rates of growth of the functions $f(x) = 2^x$ and $g(x) = x^5$ by drawing the graphs of both functions in the following viewing rectangles.

(i) $[0, 5]$ by $[0, 20]$

(ii) $[0, 25]$ by $[0, 10^7]$

(iii) $[0, 50]$ by $[0, 10^8]$

(b) Find the solutions of the equation $2^x = x^5$, rounded to one decimal place.

46. **(a)** Compare the rates of growth of the functions $f(x) = 3^x$ and $g(x) = x^4$ by drawing the graphs of both functions in the following viewing rectangles:

(i) $[-4, 4]$ by $[0, 20]$

(ii) $[0, 10]$ by $[0, 5000]$

(iii) $[0, 20]$ by $[0, 10^5]$

(b) Find the solutions of the equation $3^x = x^4$, rounded to two decimal places.

Skills Plus

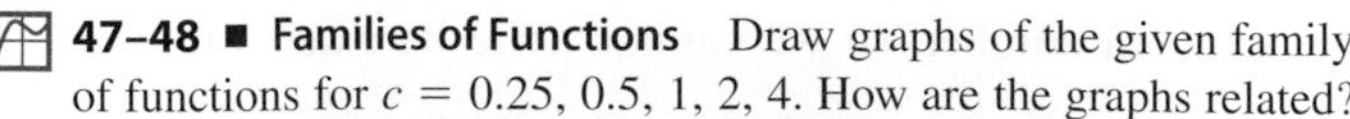

47–48 ■ Families of Functions Draw graphs of the given family of functions for $c = 0.25, 0.5, 1, 2, 4$. How are the graphs related?

47. $f(x) = c2^x$ **48.** $f(x) = 2^{cx}$

49–50 ■ Getting Information from a Graph Find, rounded to two decimal places, (a) the intervals on which the function is increasing or decreasing and (b) the range of the function.

49. $y = 10^{x-x^2}$ **50.** $y = x2^x$

51–52 ■ Difference Quotients These exercises involve a difference quotient for an exponential function.

51. If $f(x) = 10^x$, show that

$$\frac{f(x+h) - f(x)}{h} = 10^x\left(\frac{10^h - 1}{h}\right)$$

52. If $f(x) = 3^{x-1}$, show that

$$\frac{f(x+h) - f(x)}{h} = 3^{x-1}\left(\frac{3^h - 1}{h}\right)$$

Applications

53. Bacteria Growth A bacteria culture contains 1500 bacteria initially and doubles every hour.

(a) Find a function N that models the number of bacteria after t hours.

(b) Find the number of bacteria after 24 hours.

54. Mouse Population A certain breed of mouse was introduced onto a small island with an initial population of 320 mice, and scientists estimate that the mouse population is doubling every year.

(a) Find a function N that models the number of mice after t years.

(b) Estimate the mouse population after 8 years.

55–56 ■ Compound Interest An investment of \$5000 is deposited into an account in which interest is compounded monthly. Complete the table by filling in the amounts the investment grows to at the indicated times or interest rates.

55. $r = 4\%$

Time (years)	Amount
1	
2	
3	
4	
5	
6	

56. $t = 5$ years

Rate per Year	Amount
1%	
2%	
3%	
4%	
5%	
6%	

57. Compound Interest If \$8000 is invested at an interest rate of 6.25% per year, compounded semiannually, find the value of the investment after the given number of years.

(a) 5 years **(b)** 10 years **(c)** 15 years

58. Compound Interest If \$3500 is invested at an interest rate of 3.5% per year, compounded daily, find the value of the investment after the given number of years.

(a) 2 years **(b)** 3 years **(c)** 6 years

59. Compound Interest If \$1200 is invested at an interest rate of 2.75% per year, compounded quarterly, find the value of the investment after the given number of years.

(a) 1 year **(b)** 2 years **(c)** 10 years

60. Compound Interest If \$14,000 is borrowed at a rate of 5.25% interest per year, compounded quarterly, find the amount due at the end of the given number of years.

(a) 4 years **(b)** 6 years **(c)** 8 years

61–62 ■ Present Value The **present value** of a sum of money is the amount that must be invested now, at a given rate of interest, to produce the desired sum at a later date.

61. How much should be invested now (the present value) to have an amount of \$10,000, 3 years from now, if the amount is invested at an interest rate of 9% per year, compounded semiannually.

62. How much should be invested now (the present value) to have an amount of \$100,000, 5 years from now, if the amount is invested at an interest rate of 8% per year, compounded monthly.

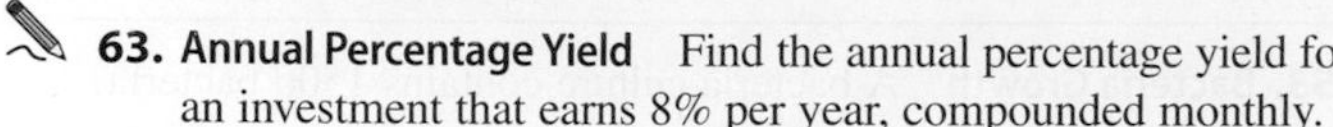

63. Annual Percentage Yield Find the annual percentage yield for an investment that earns 8% per year, compounded monthly.

64. Annual Percentage Yield Find the annual percentage yield for an investment that earns $5\frac{1}{2}\%$ per year, compounded quarterly.

■ Discuss ■ Discover ■ Prove ■ Write

65. Discuss ■ Discover: Exponential Functions Increase To see just how quickly the exponential function $f(x) = 2^x$ increases, perform the following thought experiment: Imagine that you have a sheet of paper that is 2.54×10^{-3} centimeters thick, then visualize repeatedly folding the paper in half (so the thickness doubles with each fold). How thick is the folded paper after 50 folds? (Express your answer in kilometers.)

66. Discuss ■ Discover: The Height of the Graph of an Exponential Function Your mathematics instructor asks you to sketch a graph of the exponential function

$$f(x) = 2^x$$

for x between 0 and 40, using a scale of 10 units to one centimeter. What are the dimensions of the sheet of paper you will need to sketch this graph?

4.2 The Natural Exponential Function

■ The Number e ■ The Natural Exponential Function ■ Continuously Compounded Interest

Any positive number can be used as a base for an exponential function. In this section we study the special base e, which is convenient for applications involving calculus.

■ The Number e

The number e is defined as the value that $(1 + 1/n)^n$ approaches as n becomes large. (In calculus this idea is made more precise through the concept of a limit.) The table shows the values of the expression $(1 + 1/n)^n$ for increasingly large values of n.

n	$\left(1 + \frac{1}{n}\right)^n$
1	2.00000
5	2.48832
10	2.59374
100	2.70481
1000	2.71692
10,000	2.71815
100,000	2.71827
1,000,000	2.71828

iStockphoto.com/Kubrak78

The **Gateway Arch** in St. Louis, Missouri, is shaped in the form of the graph of a combination of exponential functions (*not* a parabola, as it might first appear). Specifically, it is a **catenary**, which is the graph of an equation of the form

$$y = a(e^{bx} + e^{-bx})$$

(see Exercises 29 and 31). This shape was chosen because it is optimal for distributing the internal structural forces of the arch. Chains and cables suspended between two points (for example, the stretches of cable between pairs of telephone poles) hang in the shape of a catenary.

It appears that, rounded to five decimal places, $e \approx 2.71828$; in fact, the approximate value to 20 decimal places is

$$e \approx 2.71828182845904523536$$

It can be shown that e is an irrational number, so we cannot write its exact value in decimal form.

The notation e was chosen by Leonhard Euler (see Section 1.6), probably because it is the first letter of the word *exponential*.

■ The Natural Exponential Function

The exponential function with base e is called the *natural exponential function*. Why use such a strange base for an exponential function? It might seem at first that a base such as 10 is easier to work with. We will see, however, that in certain applications the

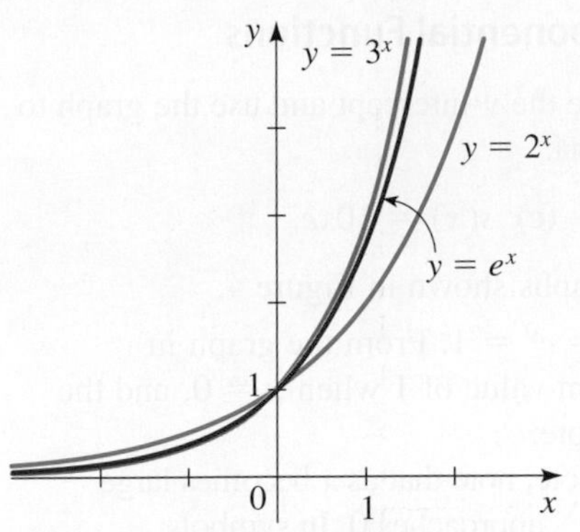

Figure 1 | Graph of the natural exponential function

number e is the best possible base. In this section we study how e occurs in the description of compound interest.

The Natural Exponential Function

The **natural exponential function** is the exponential function

$$f(x) = e^x$$

with base e. It is often referred to as *the* exponential function.

Since $2 < e < 3$, the graph of the natural exponential function lies between the graphs of $y = 2^x$ and $y = 3^x$, as shown in Figure 1.

Example 1 ■ Evaluating the Exponential Function

We use the $\boxed{e^x}$ key on a calculator to evaluate the natural exponential function. Check to make sure you get the following answers on your own calculator.

(a) $e^3 \approx 20.08554$ **(b)** $2e^{-0.53} \approx 1.17721$ **(c)** $e^{4.8} \approx 121.51042$

Now Try Exercise 3 ■

Example 2 ■ Graphing Exponential Functions

Sketch the graph of each function. State the y-intercept, domain, range, and horizontal asymptote.

(a) $f(x) = 3e^{0.5x}$ **(b)** $g(x) = e^{-x} - 2$

Solution

(a) We calculate several values, plot the resulting points, then connect the points with a smooth curve. The graph is shown in Figure 2. The y-intercept is $y = f(0) = 3e^0 = 3$. From the graph we see that the domain of f is the set $\mathbb{R}$ of all real numbers, the range is the interval $(0, \infty)$, and the line $y = 0$ is a horizontal asymptote.

x	$f(x) = 3e^{0.5x}$
-3	0.67
-2	1.10
-1	1.82
0	3.00
1	4.95
2	8.15
3	13.45

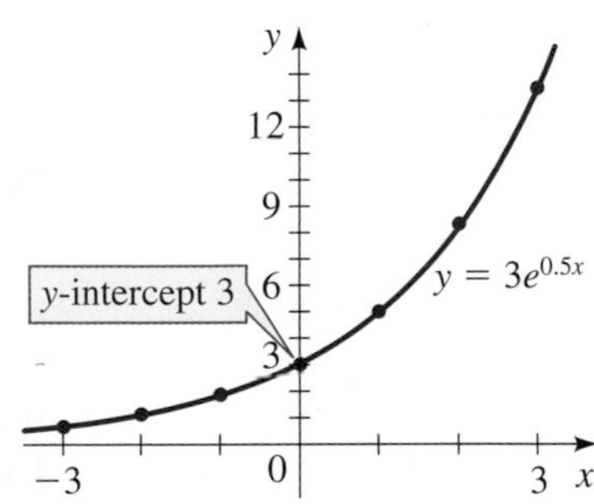

Figure 2

(b) We use transformations to sketch this graph. We start with the graph of $y = e^x$, reflect about the y-axis to obtain the graph of $y = e^{-x}$, and then shift downward 2 units to obtain the graph of $y = e^{-x} - 2$, as shown in Figure 3. The y-intercept is $y = g(0) = e^0 - 2 = -1$. From the graph we see that the domain of g is the set $\mathbb{R}$ of all real numbers, the range is the interval $(-2, \infty)$, and the line $y = -2$ is a horizontal asymptote.

Figure 3

Now Try Exercises 5 and 7 ■

When we combine the exponential function with other functions we get new functions with a variety of different graphs, as shown in the next example.

Example 3 ■ Combinations Involving Exponential Functions

Use a graphing device to graph each function. State the y-intercept and use the graph to find the horizontal asymptote and any local extrema.

(a) $g(x) = e^{-x^2}$ **(b)** $l(x) = \dfrac{5}{1 + e^{-x}}$ **(c)** $s(x) = 10xe^{-0.1x}$

Solution Using a graphing device we get the graphs shown in Figure 4.

(a) The y-intercept is 1: when $x = 0$, $y = g(0) = e^0 = 1$. From the graph in Figure 4(a) we see that g has a local maximum value of 1 when $x = 0$, and the line $y = 0$ (the x-axis) is a horizontal asymptote.

To confirm that $y = 0$ is a horizontal asymptote, note that as x becomes large in absolute value, e^{x^2} also becomes large, so $1/e^{x^2}$ approaches 0. In symbols: $y = g(x) = e^{-x^2} = 1/e^{x^2} \to 0$ as $x \to \pm\infty$.

See Section 3.6 for a discussion of arrow notation.

(b) The y-intercept is $\frac{5}{2}$: when $x = 0$, $y = l(0) = 5/(1 + e^0) = \frac{5}{2}$. From the graph in Figure 4(b) we see that the function l has no local extrema and that the lines $y = 0$ and $y = 5$ are horizontal asymptotes.

To confirm that $y = 5$ is a horizontal asymptote, note that as $x \to \infty$,

$$y = l(x) = \frac{5}{1 + e^{-x}} \to \frac{5}{1 + 0} = 5$$

Also $y = 0$ is a horizontal asymptote because as $x \to -\infty$, $e^{-x} \to \infty$, so $5/(1 + e^{-x}) \to 0$.

(c) The y-intercept is 0: when $x = 0$, $y = s(0) = 10 \cdot 0 \cdot e^0 = 0$. From the graph in Figure 4(c) we see that the function s has a local maximum value of approximately 36.8 at $x \approx 10.0$. The x-axis is a horizontal asymptote.

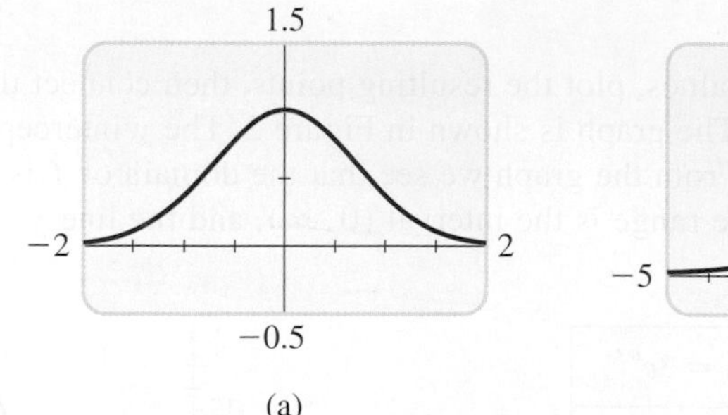

(a)

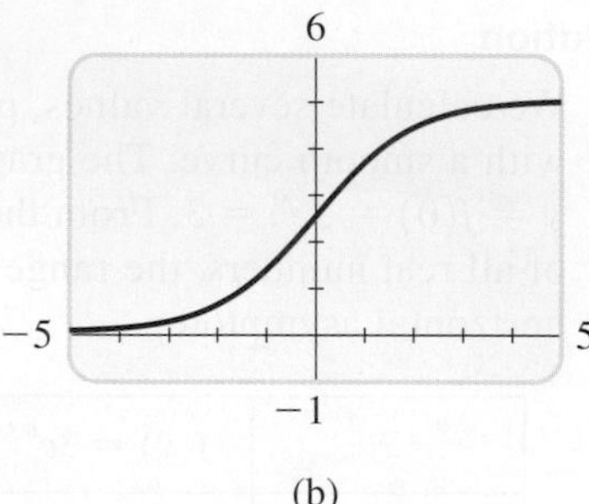

(b)

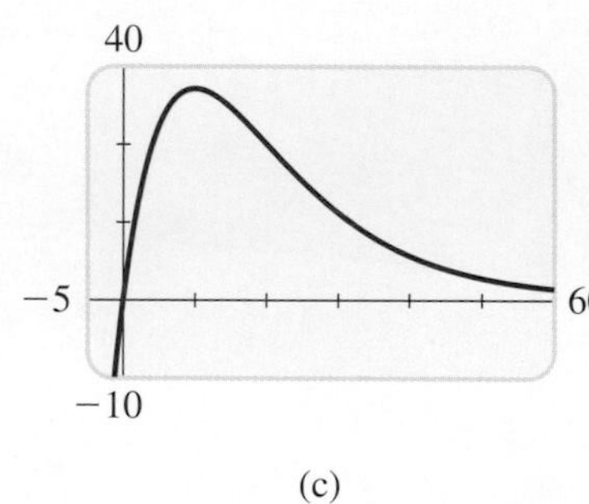

(c)

Figure 4

Now Try Exercises 19, 21, and 23 ■

The functions in Example 3 are used to model many real-world processes. The function g is a *Gaussian function* (see Exercise 38), the function l is a *logistic function* (see Exercise 41 and Section 4.6), and the function s is a *surge function* (see Exercise 37).

■ Continuously Compounded Interest

In Example 4.1.6 we saw that the interest paid on an investment increases as the number of compounding periods n increases. Let's see what happens as n increases indefinitely. If we let $m = n/r$, then

$$A(t) = P\left(1 + \frac{r}{n}\right)^{nt} = P\left[\left(1 + \frac{r}{n}\right)^{n/r}\right]^{rt} = P\left[\left(1 + \frac{1}{m}\right)^{m}\right]^{rt}$$

Recall that as m becomes large, the quantity $(1 + 1/m)^m$ approaches the number e. Thus the amount approaches $A = Pe^{rt}$, and this expression gives the amount when the interest is compounded at "every instant."

Continuously Compounded Interest

Continuously compounded interest is calculated by the formula

$$A(t) = Pe^{rt}$$

where $A(t)$ = amount after t years
P = principal
r = interest rate per year
t = number of years

Example 4 ■ Calculating Continuously Compounded Interest

Find the amount after 3 years if \$1000 is invested at an interest rate of 12% per year, compounded continuously.

Solution We use the formula for continuously compounded interest with $P = \$1000$, $r = 0.12$, and $t = 3$ to get

$$A(3) = 1000e^{(0.12)3} = 1000e^{0.36} = \$1433.33$$

Compare this amount with the amounts in Example 4.1.6.

 **Now Try Exercise 45** ■

4.2 Exercises

Concepts

1. The function $f(x) = e^x$ is called the ______ exponential function. The number e is approximately equal to ______.

2. In the formula $A(t) = Pe^{rt}$ for continuously compound interest, the letters P, r, and t stand for ______, ______, and ______, respectively, and $A(t)$ stands for ______. So if \$100 is invested at an interest rate of 6% compounded continuously, then the amount after 2 years is ______.

Skills

3–4 ■ Evaluating Exponential Functions Use a calculator to evaluate the function at the indicated values. Round your answers to three decimal places.

 **3.** $h(x) = e^x$; $h(1), h(\pi), h(-3), h(\sqrt{2})$

4. $h(x) = e^{-3x}$; $h(\frac{1}{3}), h(1.5), h(-1), h(-\pi)$

5–6 ■ Graphing Exponential Functions Complete the table of values, rounded to two decimal places, and sketch a graph of the function.

 **5.**

x	$f(x) = 1.5e^x$
−2	
−1	
−0.5	
0	
0.5	
1	
2	

6.

x	$f(x) = 4e^{-x/3}$
−3	
−2	
−1	
0	
1	
2	
3	

7–18 ■ Graphing Exponential Functions Graph the function, not by plotting points, but by starting from the graph of $y = e^x$ in Figure 1. State the y-intercept, domain, range, and horizontal asymptote.

 **7.** $f(x) = e^x + 3$

8. $f(x) = e^{-x} + 1$

9. $g(x) = e^{-x} - 3$

10. $h(x) = e^x - 4$

11. $f(x) = 3 - e^x$

12. $y = 2 - e^{-x}$

13. $y = 4 - 3e^{-x}$

14. $f(x) = 3e^x + 1$

15. $f(x) = e^{x-2}$

16. $y = e^{x-3} + 4$

17. $h(x) = e^{x+1} - 3$

18. $g(x) = -e^{x-1} - 2$

19–24 ■ Graphing Combinations of Exponential Functions Use a graphing device to graph the function. State the y-intercept, and use the graph to find the horizontal asymptote(s) and any local extrema.

19. $f(x) = 5e^{-x^2/2}$

20. $f(x) = 10e^{-(x-5)^2}$

21. $l(x) = \dfrac{20}{1 + 3e^{-1.5x}}$

22. $l(x) = \dfrac{100}{1 + 4e^{-0.5x}}$

23. $s(x) = 0.4xe^{-1.5x}$

24. $s(x) = 2.5xe^{-3.5x}$

25–28 ■ Expressing a Function as a Composition Find functions f and g such that $F = f \circ g$.

25. $F(x) = 2e^{(x-10)^2}$

26. $F(x) = e^x + e^{-x}$

27. $F(x) = \sqrt{1 + e^x}$

28. $F(x) = (3 + e^x)^3$

Skills Plus

29. Hyperbolic Cosine Function The *hyperbolic cosine function* is defined by

$$\cosh(x) = \frac{e^x + e^{-x}}{2}$$

(a) Sketch the graphs of the functions $y = \frac{1}{2}e^x$ and $y = \frac{1}{2}e^{-x}$ on the same axes, and use graphical addition (see Section 2.7) to sketch the graph of $y = \cosh(x)$.

(b) Use the definition to show that $\cosh(-x) = \cosh(x)$.

30. Hyperbolic Sine Function The *hyperbolic sine function* is defined by

$$\sinh(x) = \frac{e^x - e^{-x}}{2}$$

(a) Sketch the graph of this function using graphical addition as in Exercise 29.

(b) Use the definition to show that $\sinh(-x) = -\sinh(x)$.

31. Families of Functions

(a) Draw the graphs of the family of functions

$$f(x) = \frac{a}{2}(e^{x/a} + e^{-x/a})$$

for $a = 0.5$, 1, 1.5, and 2.

(b) How does a larger value of a affect the graph?

32. The Definition of e Illustrate the definition of the number e by graphing the curve $y = (1 + 1/x)^x$ and the line $y = e$ on the same screen, using the viewing rectangle [0, 40] by [0, 4].

33–34 ■ Local Extrema Find the local maximum and minimum values of the function and the value of x at which each occurs. State each answer rounded to two decimal places.

33. $g(x) = x^x \quad (x > 0)$

34. $g(x) = e^x + e^{-2x}$

Applications

35. Medical Drugs When a certain medical drug is administered to a patient, the number of milligrams remaining in the patient's bloodstream t hours after the drug is fully absorbed is modeled by

$$D(t) = 50e^{-0.2t}$$

How many milligrams of the drug remain in the patient's bloodstream after 3 hours?

36. Radioactive Decay A radioactive substance decays in such a way that the amount of mass remaining after t days is given by the function

$$m(t) = 13e^{-0.015t}$$

where $m(t)$ is measured in kilograms.

(a) Find the mass at time $t = 0$.

(b) How much of the mass remains after 45 days?

37. Drug Absorption When a certain drug is administered to a patient, the drug concentration in the bloodstream increases rapidly until it reaches a peak level, and then it decreases slowly. The concentration of the drug in the patient's bloodstream is modeled by the surge function

$$s(t) = 0.7te^{-1.2t}$$

where t is the time in hours since the drug was administered and $s(t)$ is the concentration of the drug in mg/mL.

(a) Draw a graph of the function s for $0 \le t \le 6$. Use the graph to describe how the drug's concentration in the bloodstream varies with time.

(b) After how many minutes does the amount of medication in the bloodstream reach its maximum level?

(c) Use the model to determine how long the patient must wait for the concentration of the drug to be only 0.01 mg/mL.

38. Distribution of Heights The distribution of the heights (in cm) of high-school players in a basketball league is modeled by the Gaussian function

$$g(x) = 0.005e^{-(x-185)^2/200}$$

(a) Draw a graph of the function g for $150 \le x \le 220$.

(b) Use the graph to estimate the average height of these players. (This is the value of x at which the graph g has a maximum.)

39. Skydiving A skydiver jumps from a reasonable height above the ground. The air resistance is proportional to the skydiver's velocity, and the constant of proportionality is 0.2. It can be shown that the downward velocity of the skydiver at time t is given by

$$v(t) = 54(1 - e^{-0.2t})$$

where t is measured in seconds (s) and $v(t)$ is measured in meters per second (m/s).

(a) Find the initial velocity of the skydiver.

(b) Find the velocity after 5 s and after 10 s.

(c) Draw a graph of the velocity function $v(t)$.

(d) The maximum velocity of a falling object with wind resistance is called its *terminal velocity*. From the graph in part (c), find the terminal velocity of this skydiver.

$v(t) = 54(1 - e^{-0.2t})$

40. Mixtures and Concentrations A 189-L barrel is filled completely with pure water. Salt water with a concentration of 36 g/L is then pumped into the barrel, and the resulting mixture overflows at the same rate. The amount of salt in the barrel at time t is given by

$$Q(t) = 6.8(1 - e^{-0.04t})$$

where t is measured in minutes and $Q(t)$ is measured in kilograms.

(a) How much salt is in the barrel after 5 minutes?

(b) How much salt is in the barrel after 10 minutes?

(c) Draw a graph of the function $Q(t)$.

(d) Use the graph in part (c) to determine the value that the amount of salt in the barrel approaches as t becomes large. Is this what you would expect?

41. Axolotl Salamander The axolotl is a paedomorphic salamander that is invaluable for medical research into the healing process: it has the unique ability to regenerate limbs, gills, and parts of its eyes and brain. The species is critically endangered in the wild, so scientists are planning to introduce the species into a lake where it can thrive. The expected number $n(t)$ of axolotls in the lake t years after they are introduced is modeled by the logistic function

$$n(t) = \frac{5000}{1 + 39e^{-0.04t}}$$

(a) Find the initial number of axolotls introduced into the lake.

(b) Draw a graph of the function n.

(c) What number does the axolotl population approach as time goes on?

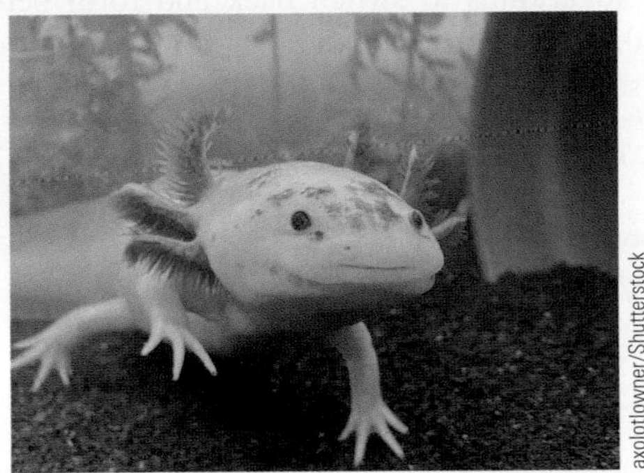

42. Tree Diameter For a certain type of tree the diameter D (in feet) depends on the tree's age t (in years), and is modeled by the logistic function

$$D(t) = \frac{5.4}{1 + 2.9e^{-0.01t}}$$

Find the diameter of a 20-year-old tree.

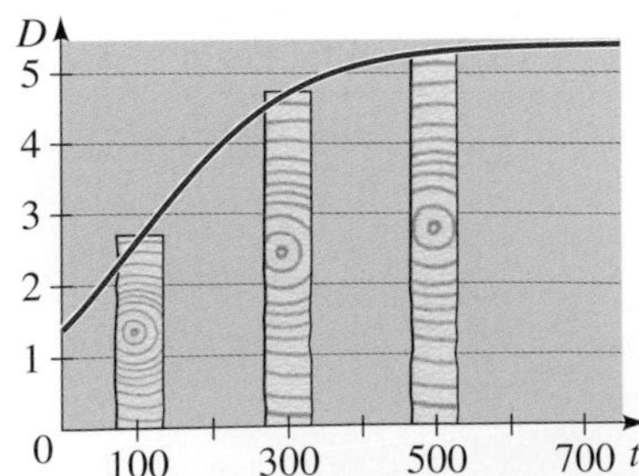

43–44 ■ Compound Interest An investment of \$7000 is deposited into an account for which interest is compounded continuously. Complete the table by filling in the amounts to which the investment grows at the indicated times or interest rates.

43. $r = 3\%$

Time (years)	Amount
1	
2	
3	
4	
5	
6	

44. $t = 10$ years

Rate per Year	Amount
1%	
2%	
3%	
4%	
5%	
6%	

45. Compound Interest If \$2000 is invested at an interest rate of 3.5% per year, compounded continuously, find the value of the investment after the given number of years.

(a) 2 years (b) 4 years (c) 12 years

46. Compound Interest If \$3500 is invested at an interest rate of 6.25% per year, compounded continuously, find the value of the investment after the given number of years.

(a) 3 years (b) 6 years (c) 9 years

47. Compound Interest If \$600 is invested at an interest rate of 2.5% per year, find the amount of the investment at the end of 10 years for each compounding method.

(a) Annual (b) Semiannual

(c) Quarterly (d) Continuous

48. Compound Interest If \$8000 is invested in an account for which interest is compounded continuously, find the amount of the investment at the end of 12 years for the given interest rates.

(a) 2% (b) 3% (c) 4.5% (d) 7%

49. Compound Interest Which of the given interest rates and compounding periods would provide the best investment?

(a) $2\frac{1}{2}\%$ per year, compounded semiannually

(b) $2\frac{1}{4}\%$ per year, compounded monthly

(c) 2% per year, compounded continuously

50. Compound Interest Which of the given interest rates and compounding periods would provide the better investment?

(a) $5\frac{1}{8}\%$ per year, compounded semiannually

(b) 5% per year, compounded continuously

51. Investment A sum of \$5000 is invested at an interest rate of 9% per year, compounded continuously.

(a) Find the value $A(t)$ of the investment after t years.

(b) Draw a graph of $A(t)$.

(c) Use the graph of $A(t)$ to determine when this investment will amount to \$25,000.

■ Discuss ■ Discover ■ Prove ■ Write

52. Discuss ■ Discover: Minimum Value Show that the minimum value of $f(x) = e^x + e^{-x}$ is 2. (Do not use a graphing device.)

PS *Establish subgoals.* First find the minimum value for $x \geq 0$.

4.3 Logarithmic Functions

■ Logarithmic Functions ■ Graphs of Logarithmic Functions ■ Common Logarithms ■ Natural Logarithms

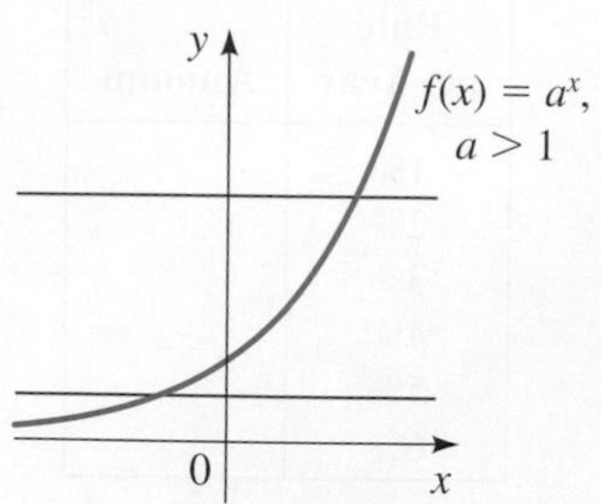

Figure 1 $f(x) = a^x$ is one-to-one.

In this section we study the inverse functions of exponential functions.

■ Logarithmic Functions

Every exponential function $f(x) = a^x$, with $a > 0$ and $a \neq 1$, is a one-to-one function by the Horizontal Line Test (see Figure 1 for the case $a > 1$) and therefore has an inverse function. The inverse function f^{-1} is called the *logarithmic function with base a* and is denoted by $\log_a$. Recall from Section 2.8 that f^{-1} is defined by

$$f^{-1}(x) = y \quad \Leftrightarrow \quad f(y) = x$$

This leads to the following definition of logarithmic functions.

We read $\log_a x = y$ as "log base a of x is y."

Definition of Logarithmic Functions

For $a > 0$ and $a \neq 1$, the **logarithmic function with base a**, denoted by $\log_a$, is defined by

$$\log_a x = y \quad \Leftrightarrow \quad a^y = x$$

So $\log_a x$ is the *exponent* to which the base a must be raised to give x.

By tradition the name of the logarithmic function is $\log_a$, not just a single letter. Also, we usually omit the parentheses in the function notation and write

$$\log_a(x) = \log_a x$$

When we use the definition of logarithms to switch back and forth between the **logarithmic form** $\log_a x = y$ and the **exponential form** $a^y = x$, it is helpful to notice that, in both forms, the base is the same.

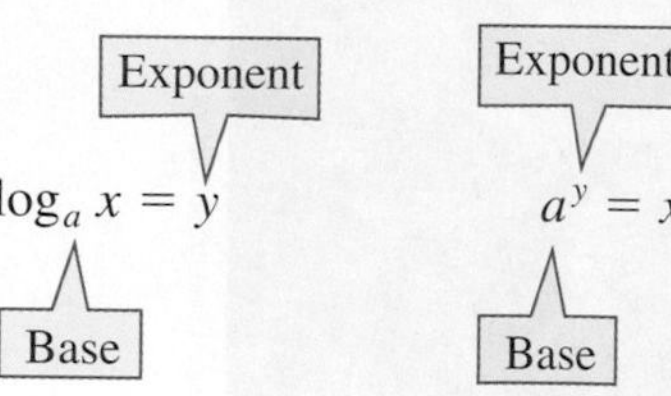

Example 1 ■ Logarithmic and Exponential Forms

The logarithmic and exponential forms are equivalent equations: If one is true, then so is the other. So we can switch from one form to the other as in the following illustrations.

In Example 1, the logarithmic form

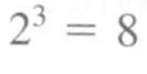

tells us that "2 raised to the power 3 is 8," so the exponential form is

$$2^3 = 8$$

Logarithmic Form	Exponential Form
$\log_{10} 100{,}000 = 5$	$10^5 = 100{,}000$
$\log_2 8 = 3$	$2^3 = 8$
$\log_2\left(\frac{1}{8}\right) = -3$	$2^{-3} = \frac{1}{8}$
$\log_{1/3} 9 = -2$	$\left(\frac{1}{3}\right)^{-2} = 9$

Now Try Exercise 7 ■

x	$\log_{10} x$
10^4	4
10^3	3
10^2	2
10	1
1	0
10^{-1}	-1
10^{-2}	-2
10^{-3}	-3
10^{-4}	-4

Note It is important to understand that $\log_a x$ is an *exponent*. For example, the numbers in the right-hand column of the table in the margin are the logarithms (base 10) of the numbers in the left-hand column. This is the case for all bases, as the following example illustrates.

Example 2 ■ Evaluating Logarithms

(a) $\log_{10} 1000 = 3$ because $10^3 = 1000$

(b) $\log_2 32 = 5$ because $2^5 = 32$

(c) $\log_{10} 0.1 = -1$ because $10^{-1} = 0.1$

(d) $\log_{16} 4 = \frac{1}{2}$ because $16^{1/2} = 4$

(e) $\log_{1/2} 16 = -4$ because $\left(\frac{1}{2}\right)^{-4} = 16$

Now Try Exercises 9 and 11 ■

Inverse Function Property:

$$f^{-1}(f(x)) = x$$

$$f(f^{-1}(x)) = x$$

When we apply the cancellation equations of the Inverse Function Property described in Section 2.8 to $f(x) = a^x$ and $f^{-1}(x) = \log_a x$, we get

$$\log_a(a^x) = x \qquad (x \in \mathbb{R})$$

$$a^{\log_a x} = x \qquad (x > 0)$$

We list these and other properties of logarithms discussed in this section.

Properties of Logarithms

Property	Reason
1. $\log_a 1 = 0$	We must raise a to the power 0 to get 1.
2. $\log_a a = 1$	We must raise a to the power 1 to get a.
3. $\log_a a^x = x$	We must raise a to the power x to get a^x.
4. $a^{\log_a x} = x$	$\log_a x$ is the power to which a must be raised to get x.

Example 3 ■ Applying Properties of Logarithms

We illustrate the properties of logarithms when the base is 5.

$\log_5 1 = 0$ Property 1 $\qquad \log_5 5 = 1$ Property 2

$\log_5 5^8 = 8$ Property 3 $\qquad 5^{\log_5 12} = 12$ Property 4

Now Try Exercises 25 and 31 ■

■ Graphs of Logarithmic Functions

We first graph logarithmic functions by plotting points.

Example 4 ■ Graphing a Logarithmic Function by Plotting Points

Sketch the graph of $f(x) = \log_2 x$.

Solution To make a table of values, we choose the x-values to be powers of 2 so that we can easily find their logarithms. We plot these points and connect them with a smooth curve as in Figure 2 (on the next page).

x	$\log_2 x$
2^3	3
2^2	2
2	1
1	0
2^{-1}	-1
2^{-2}	-2
2^{-3}	-3
2^{-4}	-4

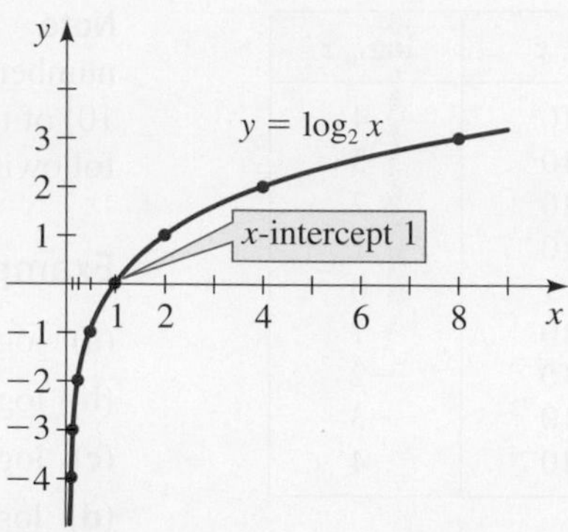

Figure 2

Now Try Exercise 49 ■

Figure 3 shows graphs of the family of logarithmic functions for various values of the base a.

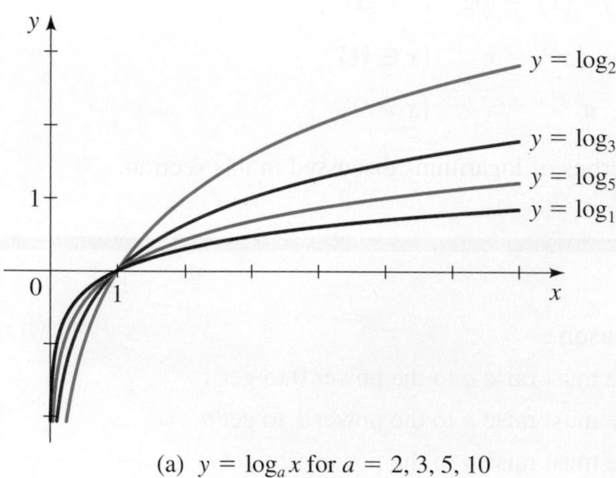

(a) $y = \log_a x$ for $a = 2, 3, 5, 10$

(b) $y = \log_a x$ for $a = \frac{1}{2}, \frac{1}{3}, \frac{1}{5}, \frac{1}{10}$

Figure 3 | Families of logarithmic functions

Note Each graph in Figure 3(b) is a reflection about the x-axis of the corresponding graph in Figure 3(a). This means that $\log_a x = -\log_{1/a} x$ (see Exercise 112). Thus, we mainly consider logarithmic functions with base $a > 1$.

Since $y = \log_a x$ is the inverse function of $y = a^x$, it follows that each logarithmic graph in Figure 3 is a reflection about the line $y = x$ of the corresponding exponential graph in Figure 4.1.2. This is illustrated in Figure 4 for the case $a > 1$. Also, since $f(x) = a^x$ has domain $\mathbb{R}$ and range $(0, \infty)$, it follows that its inverse function $y = \log_a x$ has domain $(0, \infty)$ and range $\mathbb{R}$.

Since $\log_a 1 = 0$, the x-intercept of the graph of $y = \log_a x$ is 1. The y-axis is a vertical asymptote because $\log_a x \to -\infty$ as $x \to 0^+$. We summarize these observations in the following box.

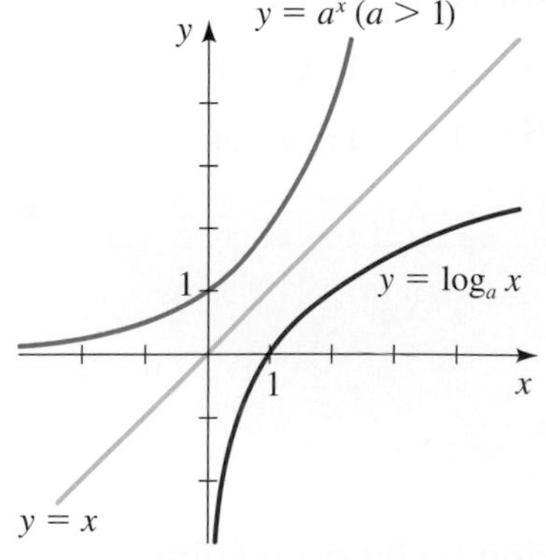

Figure 4 | Graph of the logarithmic function $f(x) = \log_a x$

Graphs of Logarithmic Functions ($a > 1$)

The logarithm function

$$f(x) = \log_a x \quad (a > 1)$$

has domain $(0, \infty)$, range $\mathbb{R}$, and x-intercept 1. The line $x = 0$ (the y-axis) is a vertical asymptote of f.

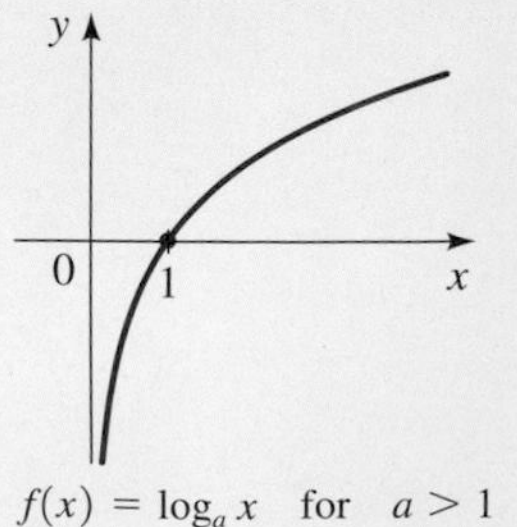

$f(x) = \log_a x$ for $a > 1$

Mathematics in the Modern World

Bettmann/Getty Images

Hulton Deutsch/Corbis Historical/Getty Images

Law Enforcement

Mathematics aids law enforcement in numerous and surprising ways, from the reconstruction of bullet trajectories to determining the time of death to calculating the probability that a DNA sample is from a particular person. One interesting use is in the search for missing persons. Someone who has been missing for several years might look quite different from their most recent available photograph. This is particularly true if the missing person is a child. Have you ever wondered what you will look like 5, 10, or 15 years from now?

Researchers have found that different parts of the body grow at different rates. For example, you have no doubt noticed that a baby's head is much larger relative to its body than an adult's. As another example, the ratio of arm length to height is $\frac{1}{3}$ in a child but about $\frac{2}{5}$ in an adult. By collecting data and analyzing the graphs, researchers are able to determine functions that model growth. As in all growth phenomena, exponential and logarithmic functions play a crucial role. For instance, the formula that relates arm length l to height h is $l = ae^{kh}$ where a and k are constants. By studying various physical characteristics of a person, mathematical biologists model each characteristic by a function that describes how it changes over time. Models of facial characteristics can be programmed into a computer to give a picture of how a person's appearance changes over time. These pictures aid law enforcement agencies in locating missing persons.

For $a > 1$, the fact that $y = a^x$ is a very rapidly increasing function for $x > 0$ implies that $y = \log_a x$ is a very slowly increasing function for $x > 1$ (see Exercise 108).

In the next two examples we graph logarithmic functions starting with the basic graph of $y = \log_a x$ for $a > 1$ and use the transformations of Section 2.6.

Example 5 ■ Reflecting Graphs of Logarithmic Functions

Sketch the graph of each function. State the x-intercept, domain, range, and vertical asymptote.

(a) $g(x) = -\log_2 x$ **(b)** $h(x) = \log_2(-x)$

Solution

(a) We start with the graph of $f(x) = \log_2 x$ and reflect about the x-axis to get the graph of $g(x) = -\log_2 x$ in Figure 5(a). In general, an x-intercept is unchanged when we reflect a graph about the x-axis, so in this case the x-intercept is 1. From the graph we see that the domain of g is $(0, \infty)$, the range is the set $\mathbb{R}$ of all real numbers, and the line $x = 0$ is a vertical asymptote.

(b) We start with the graph of $f(x) = \log_2 x$ and reflect about the y-axis to get the graph of $h(x) = \log_2(-x)$ in Figure 5(b). We observe from the graph that the x-intercept is -1 because the point $(-1, 0)$ is the reflection about the y-axis of the point $(1, 0)$. From the graph we see that the domain of h is $(-\infty, 0)$, the range is the set $\mathbb{R}$ of all real numbers, and the line $x = 0$ is a vertical asymptote.

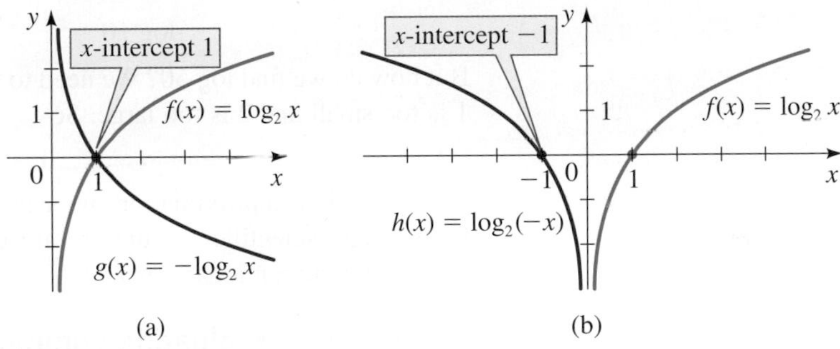

Figure 5

Now Try Exercise 63 ■

Example 6 ■ Shifting Graphs of Logarithmic Functions

Sketch the graph of each function. State the domain, range, and vertical asymptote.

(a) $g(x) = 2 + \log_5 x$ **(b)** $h(x) = \log_{10}(x - 3)$

Solution

(a) The graph of g is obtained from the graph of $f(x) = \log_5 x$ [Figure 3(a)] by shifting upward 2 units, as shown in Figure 6. Note, in particular, that the point $(1, 0)$ on the graph of f shifts to the point $(1, 2)$ on the graph of g. From the graph we see that the domain of g is $(0, \infty)$, the range is the set $\mathbb{R}$ of all real numbers, and the line $x = 0$ is a vertical asymptote.

(b) The graph of h is obtained from the graph of $f(x) = \log_{10} x$ [Figure 3(a)] by shifting to the right 3 units, as shown in Figure 7. Note, in particular, that the point $(1, 0)$ on the graph of f shifts to point $(4, 0)$ on the graph of h. From the

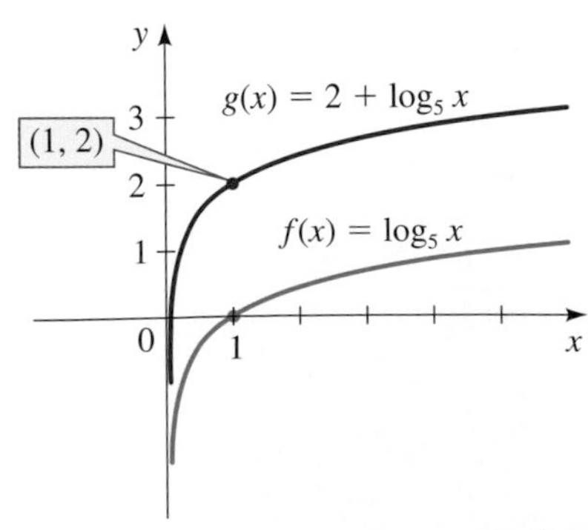

Figure 6

graph we see that the domain of h is $(3, \infty)$, the range is the set $\mathbb{R}$ of all real numbers, and the line $x = 3$ is a vertical asymptote.

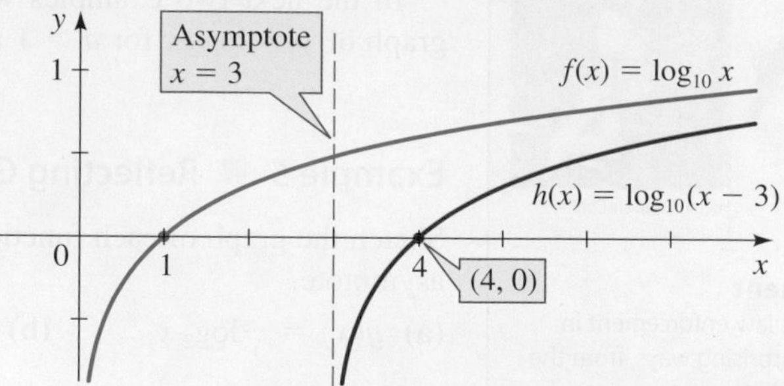

Figure 7

Now Try Exercises 65 and 69 ■

■ Common Logarithms

We now study logarithms with base 10.

Common Logarithm

The logarithm with base 10 is called the **common logarithm** and is denoted by omitting the base:

$$\log x = \log_{10} x$$

From the definition of logarithms we can find that

$$\log 10 = 1 \qquad \text{and} \qquad \log 100 = 2$$

But how do we find log 50? We need to find the exponent y such that $10^y = 50$. Clearly, 1 is too small and 2 is too large. So

$$1 < \log 50 < 2$$

To get a better approximation, we can experiment to find a power of 10 closer to 50. Fortunately, scientific calculators are equipped with a LOG key that directly gives values of common logarithms.

Example 7 ■ Evaluating Common Logarithms

Use a calculator to find appropriate values of $f(x) = \log x$, and use these values to sketch the graph.

Solution We make a table of values, using a calculator to evaluate the function at those values of x that are not powers of 10. We plot those points and connect them by a smooth curve as shown in Figure 8.

x	$\log x$
0.01	−2
0.1	−1
0.5	−0.301
1	0
4	0.602
5	0.699
10	1

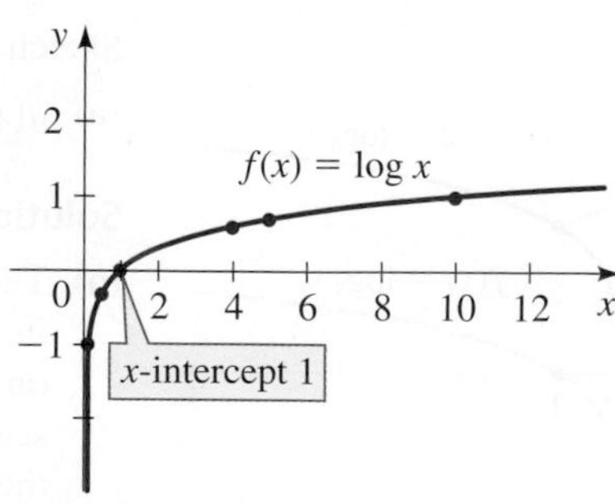

Figure 8

 Now Try Exercise 51 ■

Human response to sound and light intensity is logarithmic.

Scientists model human response to stimuli (such as sound, light, or pressure) using logarithmic functions. For example, the intensity of a sound must be increased manyfold before we "feel" that the loudness has simply doubled. The psychologist Gustav Fechner formulated the law as

$$S = k \log \frac{I}{I_0}$$

where S is the subjective intensity of the stimulus, I is the physical intensity of the stimulus, I_0 stands for the threshold physical intensity, and k is a constant that is different for each sensory stimulus.

Example 8 ■ Common Logarithms and Sound

We study the decibel scale in more detail in Section 4.7.

The perception of the loudness B (in decibels, dB) of a sound with physical intensity I (in W/m^2) is given by

$$B = 10 \log \frac{I}{I_0}$$

where I_0 is the physical intensity of a barely audible sound. Find the decibel level (loudness) of a sound whose physical intensity I is 100 times that of I_0.

Solution We find the decibel level B by using the fact that $I = 100I_0$.

$$\begin{aligned} B &= 10 \log \frac{I}{I_0} && \text{Definition of } B \\ &= 10 \log \frac{100I_0}{I_0} && I = 100I_0 \\ &= 10 \log 100 && \text{Cancel } I_0 \\ &= 10 \cdot 2 = 20 && \text{Definition of log} \end{aligned}$$

The loudness of the sound is 20 dB.

 **Now Try Exercise 103** ■

■ Natural Logarithms

Of all possible bases a for logarithms, it turns out that the most convenient choice for the purposes of calculus is the number e, which we defined in Section 4.2.

The notation ln is an abbreviation for the Latin name *logarithmus naturalis*.

Natural Logarithm

The logarithm with base e is called the **natural logarithm** and is denoted by **ln**:

$$\ln x = \log_e x$$

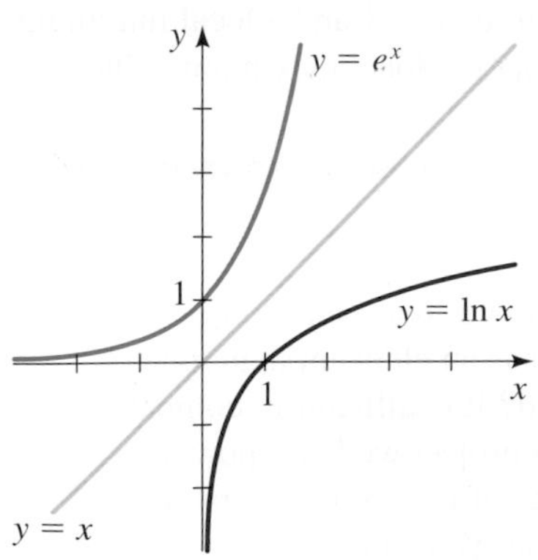

Figure 9 | Graph of the natural logarithmic function

The natural logarithmic function $y = \ln x$ is the inverse function of the natural exponential function $y = e^x$. Both functions are graphed in Figure 9. By the definition of inverse functions we have

$$\ln x = y \quad \Leftrightarrow \quad e^y = x$$

If we substitute $a = e$ and write "ln" for "$\log_e$" in the properties of logarithms mentioned earlier, we obtain the following properties of natural logarithms.

Properties of Natural Logarithms

Property	Reason
1. $\ln 1 = 0$	We must raise e to the power 0 to get 1.
2. $\ln e = 1$	We must raise e to the power 1 to get e.
3. $\ln e^x = x$	We must raise e to the power x to get e^x.
4. $e^{\ln x} = x$	$\ln x$ is the power to which e must be raised to get x.

Calculators are equipped with an LN key that directly gives the values of natural logarithms.

Example 9 ■ Evaluating the Natural Logarithm Function

(a) $\ln e^8 = 8$ — Definition of natural logarithm

(b) $\ln \dfrac{1}{e^2} = \ln e^{-2} = -2$ — Definition of natural logarithm

(c) $\ln 5 \approx 1.609$ — Calculator

Now Try Exercise 47 ■

Example 10 ■ Finding the Domain of a Logarithmic Function

Find the domain of the function $f(x) = \ln(4 - x^2)$.

Solution As with any logarithmic function, $\ln x$ is defined when $x > 0$. Thus, the domain of f is

$$\{x \mid 4 - x^2 > 0\} = \{x \mid x^2 < 4\} = \{x \mid |x| < 2\}$$
$$= \{x \mid -2 < x < 2\} = (-2, 2)$$

Now Try Exercise 75 ■

Example 11 ■ Drawing the Graph of a Logarithmic Function

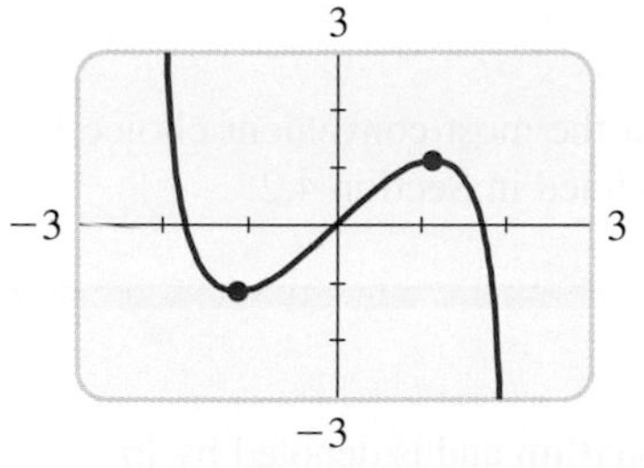

Figure 10 | $y = x\ln(4 - x^2)$

Draw the graph of the function $y = x\ln(4 - x^2)$, and use it to find the asymptotes and local maximum and minimum values.

Solution As in Example 10 the domain of this function is the interval $(-2, 2)$, so we choose the viewing rectangle $[-3, 3]$ by $[-3, 3]$. The graph is shown in Figure 10, and from it we see that the lines $x = -2$ and $x = 2$ are vertical asymptotes.

The function has a local maximum point to the right of $x = 1$ and a local minimum point to the left of $x = -1$. From the graph we find that the local maximum value is

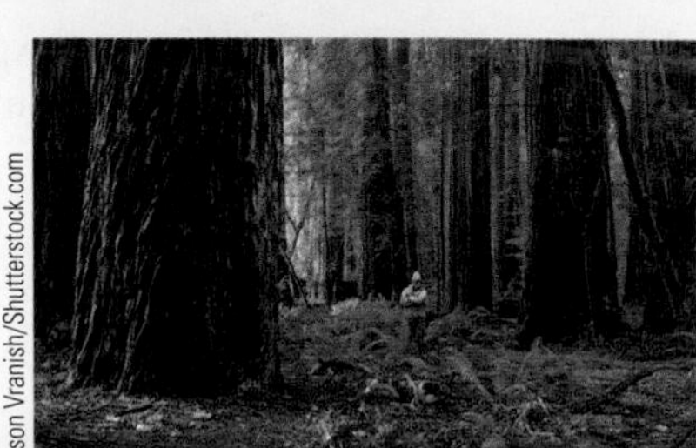

Discovery Project ■ Orders of Magnitude

In this project we explore how to compare the sizes of real-world objects using logarithms. For example, how much bigger is an elephant than a flea? How much smaller is a man than a giant redwood? It is difficult to compare objects of such enormously varying sizes. In this project we learn how logarithms can be used to define the concept of "order of magnitude," which provides a simple and meaningful way of comparison. You can find the project at **www.stewartmath.com**.

approximately 1.13 and this occurs when $x \approx 1.15$. Similarly (or by noticing that the function is odd), we find that the local minimum value is about -1.13, and it occurs when $x \approx -1.15$.

 Now Try Exercise 81 ■

4.3 Exercises

Concepts

1. $\log x$ is the exponent to which the base 10 must be raised to get ________. So we can complete the following table for $\log x$.

x	10^3	10^2	10^1	10^0	10^{-1}	10^{-2}	10^{-3}	$10^{1/2}$
$\log x$								

2. The function $f(x) = \log_9 x$ is the logarithm function with base ________. So $f(9) =$ ________, $f(1) =$ ________, $f(\frac{1}{9}) =$ ________, $f(81) =$ ________, and $f(3) =$ ________.

3. (a) $5^3 = 125$, so $\log$ ____ ____ $=$ ____

(b) $\log_5 25 = 2$, so ____ ____ $=$ ____

4. Match the logarithmic function with its graph.

(a) $f(x) = \log_2 x$ **(b)** $f(x) = \log_2(-x)$

(c) $f(x) = -\log_2 x$ **(d)** $f(x) = -\log_2(-x)$

I

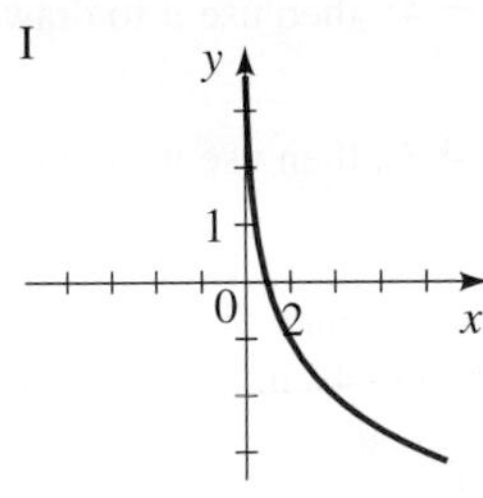

II

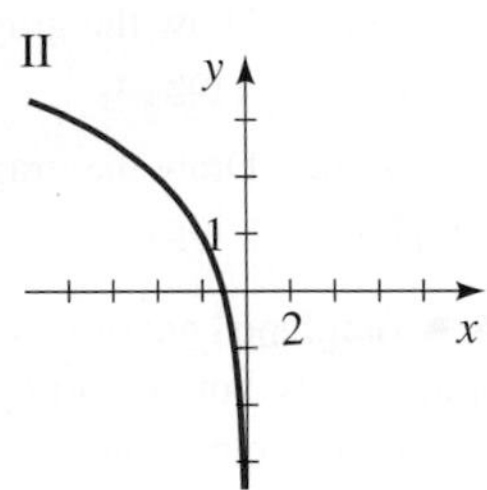

III

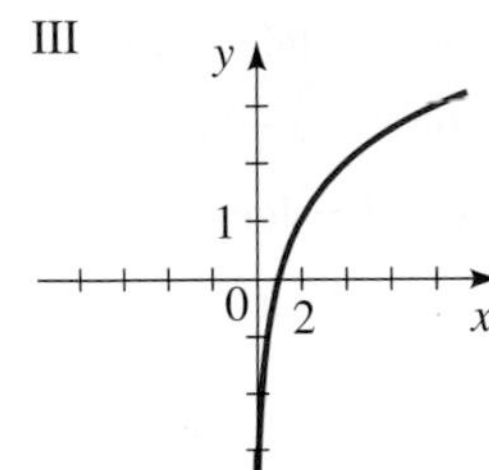

IV

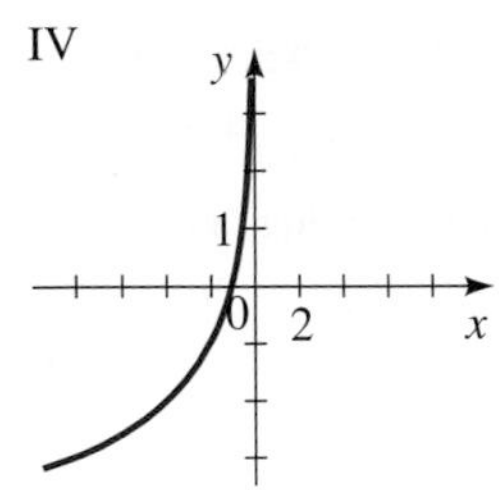

5. The natural logarithmic function $f(x) = \ln x$ has the ____________ asymptote $x =$ ____.

6. The logarithmic function $f(x) = \ln(x - 1)$ has the ____________ asymptote $x =$ ____.

Skills

7–8 ■ Logarithmic and Exponential Forms Complete the table by finding the appropriate logarithmic or exponential form of the equation, as in Example 1.

7.

Logarithmic Form	Exponential Form
$\log_8 8 = 1$	
$\log_8 64 = 2$	
	$8^{2/3} = 4$
	$8^3 = 512$
$\log_8(\frac{1}{8}) = -1$	
	$8^{-2} = \frac{1}{64}$

8.

Logarithmic Form	Exponential Form
	$4^3 = 64$
$\log_4 2 = \frac{1}{2}$	
	$4^{3/2} = 8$
$\log_4(\frac{1}{16}) = -2$	
$\log_4(\frac{1}{2}) = -\frac{1}{2}$	
	$4^{-5/2} = \frac{1}{32}$

9–16 ■ Exponential Form Express the equation in exponential form.

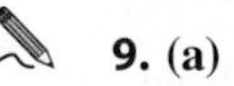

9. (a) $\log_3 81 = 4$ **(b)** $\log_{1/3} 1 = 0$

10. (a) $\log_5(\frac{1}{5}) = -1$ **(b)** $\log_{1/5} 5 = -1$

11. (a) $\log_8 2 = \frac{1}{3}$ **(b)** $\log_{10} 0.01 = -2$

12. (a) $\log_5(\frac{1}{125}) = -3$ **(b)** $\log_8 4 = \frac{2}{3}$

13. (a) $\log_3 5 = x$ **(b)** $\log_{1/6}(2y) = 3$

14. (a) $\log_{1/10} z = 2$ **(b)** $\log_{10} 3 = 2t$

15. (a) $\ln 10 = 2y$ **(b)** $\ln(3x + 1) = -2$

16. (a) $\ln(x - 2) = 3$ **(b)** $\ln(2x - 3) = 1$

17–24 ■ Logarithmic Form Express the equation in logarithmic form.

17. (a) $10^4 = 10{,}000$ **(b)** $5^{-2} = \frac{1}{25}$

18. (a) $6^2 = 36$ **(b)** $10^{-1} = \frac{1}{10}$

19. **(a)** $8^{-1} = \frac{1}{8}$ **(b)** $2^{-3} = \frac{1}{8}$

20. **(a)** $4^{-3/2} = 0.125$ **(b)** $\left(\frac{1}{2}\right)^{-3} = 8$

21. **(a)** $4^x = 70$ **(b)** $\left(\frac{1}{2}\right)^3 = w$

22. **(a)** $3^{2x} = 10$ **(b)** $10^{-4x} = 0.1$

23. **(a)** $e^x = 2$ **(b)** $e^3 = y$

24. **(a)** $e^{x+1} = 0.5$ **(b)** $e^{0.5x} = t$

25–34 ■ Evaluating Logarithms Evaluate the expression.

25. **(a)** $\log_2 2$ **(b)** $\log_5 1$ **(c)** $\log_{1/2} 2$

26. **(a)** $\log_3 3^7$ **(b)** $\log_4 64$ **(c)** $\log_{1/2} 0.25$

27. **(a)** $\log_6 36$ **(b)** $\log_9 81$ **(c)** $\log_7 7^{10}$

28. **(a)** $\log_2 32$ **(b)** $\log_5 5^{13}$ **(c)** $\log_6 1$

29. **(a)** $\log_3\left(\frac{1}{27}\right)$ **(b)** $\log_{1/3} 27$ **(c)** $\log_7 \sqrt{7}$

30. **(a)** $\log_5 125$ **(b)** $\log_{49} 7$ **(c)** $\log_9 \sqrt{3}$

31. **(a)** $3^{\log_3 5}$ **(b)** $5^{\log_5 27}$ **(c)** $e^{\ln 10}$

32. **(a)** $e^{\ln \sqrt{3}}$ **(b)** $e^{\ln(1/\pi)}$ **(c)** $10^{\log 13}$

33. **(a)** $\log_8 0.25$ **(b)** $\ln e^4$ **(c)** $\ln(1/e)$

34. **(a)** $\log_4 \sqrt{2}$ **(b)** $\log_4\left(\frac{1}{2}\right)$ **(c)** $\log_4 8$

35–44 ■ Logarithmic Equations Use the definition of the logarithmic function to find x.

35. **(a)** $\log_6 x = 2$ **(b)** $\log_{10} 0.001 = x$

36. **(a)** $\log_{1/3} x = 0$ **(b)** $\log_4 1 = x$

37. **(a)** $\ln x = 3$ **(b)** $\ln e^2 = x$

38. **(a)** $\ln x = -1$ **(b)** $\ln(1/e) = x$

39. **(a)** $\log_4\left(\frac{1}{64}\right) = x$ **(b)** $\log_{1/2} x = 3$

40. **(a)** $\log_9\left(\frac{1}{3}\right) = x$ **(b)** $\log_9 x = 0.5$

41. **(a)** $\log_2\left(\frac{1}{2}\right) = x$ **(b)** $\log_{10} x = -3$

42. **(a)** $\log_x 1000 = 3$ **(b)** $\log_x 25 = 2$

43. **(a)** $\log_x 16 = 4$ **(b)** $\log_x 8 = \frac{3}{2}$

44. **(a)** $\log_x 6 = \frac{1}{2}$ **(b)** $\log_x 3 = \frac{1}{3}$

45–48 ■ Evaluating Logarithms Use a calculator to evaluate the expression, correct to four decimal places.

45. **(a)** $\log 2$ **(b)** $\log 35.2$ **(c)** $\log\left(\frac{2}{3}\right)$

46. **(a)** $\log 50$ **(b)** $\log \sqrt{2}$ **(c)** $\log(3\sqrt{2})$

47. **(a)** $\ln 5$ **(b)** $\ln 25.3$ **(c)** $\ln(1 + \sqrt{3})$

48. **(a)** $\ln 27$ **(b)** $\ln 7.39$ **(c)** $\ln 54.6$

49–54 ■ Graphing Logarithmic Functions Sketch the graph of the function by plotting points.

49. $f(x) = \log_3 x$ **50.** $g(x) = \log_4 x$

51. $f(x) = 2 \log x$ **52.** $f(x) = \log_{1/2} x$

53. $g(x) = 1 + \log x$ **54.** $g(x) = -2 + \log_2 x$

55–58 ■ Finding Logarithmic Functions Find the function of the form $y = \log_a x$ whose graph is given.

55.

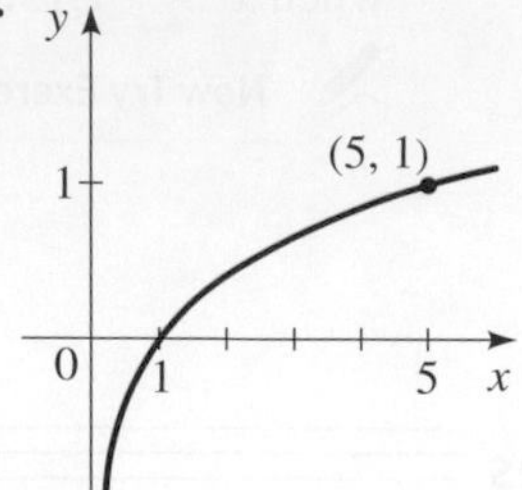

56.

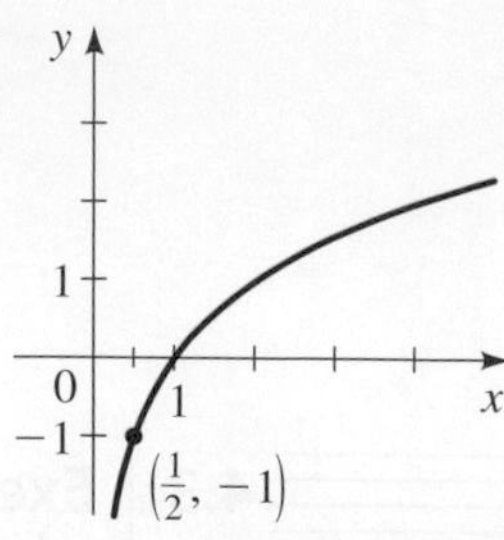

57.

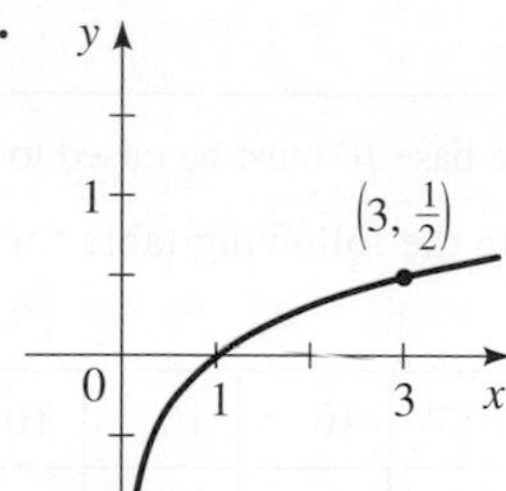

58.

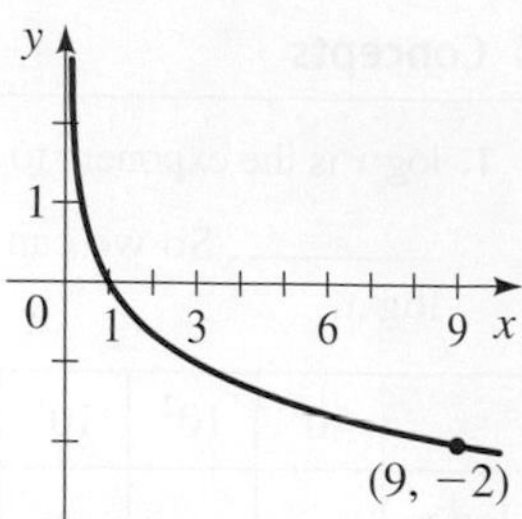

59–60 ■ Graphing Logarithmic Functions Match the logarithmic function with one of the graphs labeled I or II.

59. $f(x) = 2 + \ln x$ **60.** $f(x) = \ln(x - 2)$

I

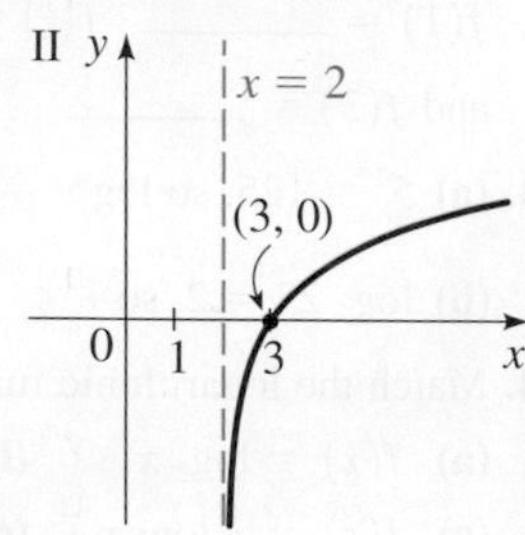

61. Graphing Draw the graph of $y = 4^x$, then use it to draw the graph of $y = \log_4 x$.

62. Graphing Draw the graph of $y = 3^x$, then use it to draw the graph of $y = \log_3 x$.

63–74 ■ Graphing Logarithmic Functions Graph the function, not by plotting points, but by starting from the graphs in Figures 3 and 9. State the domain, range, and vertical asymptote.

63. $g(x) = \log_5(-x)$ **64.** $f(x) = -\log_{10} x$

65. $f(x) = \log_2(x - 4)$ **66.** $g(x) = \ln(x + 2)$

67. $h(x) = \ln(x + 5)$ **68.** $f(x) = 2 - \log_{1/3} x$

69. $y = 2 + \log_3 x$ **70.** $y = 1 - \log_{10} x$

71. $y = \log_3(x - 1) - 2$ **72.** $y = 1 + \ln(-x)$

73. $y = |\ln x|$ **74.** $y = \ln|x|$

75–80 ■ Domain Find the domain of the function.

75. $f(x) = \log(x + 3)$ **76.** $f(x) = \log_5(8 - 2x)$

77. $g(x) = \log_3(x^2 - 1)$ **78.** $g(x) = \ln(x - x^2)$

79. $h(x) = \ln x + \ln(2 - x)$

80. $h(x) = \sqrt{x - 2} - \log_5(10 - x)$

 **81–86 ■ Graphing Logarithmic Functions** Draw the graph of the function in a suitable viewing rectangle, and use it to find the domain, the asymptotes, and the local maximum and minimum values.

81. $y = \log(1 - x^2)$ **82.** $y = \ln(x^2 - x)$

83. $y = x + \ln x$ **84.** $y = x(\ln x)^2$

85. $y = \dfrac{\ln x}{x}$ **86.** $y = x \log(x + 10)$

Skills Plus

87–90 ■ Expressing a Function as a Composition Find functions f and g such that $F = f \circ g$.

87. $F(x) = \ln(x^2 + 1)$ **88.** $F(x) = (\ln x)^3$

89. $F(x) = \sqrt{1 + |\ln x|}$ **90.** $F(x) = 5 - \log \sqrt{x}$

91–94 ■ Domain of a Composition Find the functions $f \circ g$ and $g \circ f$ and their domains.

91. $f(x) = 2^x, \quad g(x) = x + 1$

92. $f(x) = 3^x, \quad g(x) = x^2 + 1$

93. $f(x) = \log_2 x, \quad g(x) = x - 2$

94. $f(x) = \log x, \quad g(x) = x^2$

 **95. Rates of Growth** Compare the rates of growth of the functions $f(x) = \ln x$ and $g(x) = \sqrt{x}$ by drawing their graphs on a common screen using the viewing rectangle $[-1, 30]$ by $[-1, 6]$.

96. Rates of Growth

(a) By drawing the graphs of the functions

$$f(x) = 1 + \ln(1 + x) \quad \text{and} \quad g(x) = \sqrt{x}$$

in a suitable viewing rectangle, show that even when a logarithmic function starts out higher than a root function, it is ultimately overtaken by the root function.

(b) Find, rounded to two decimal places, the solutions of the equation $\sqrt{x} = 1 + \ln(1 + x)$.

97–98 ■ Family of Functions A family of functions is given. **(a)** Draw graphs of the family for $c = 1, 2, 3,$ and 4. **(b)** How are the graphs in part (a) related?

97. $f(x) = \log(cx)$ **98.** $f(x) = c \log x$

99–100 ■ Inverse Functions A function $f(x)$ is given. **(a)** Find the domain of the function f. **(b)** Find the inverse function of f.

99. $f(x) = \log_2(\log_{10} x)$ **100.** $f(x) = \ln(\ln(\ln x))$

101. Inverse Functions

(a) Find the inverse of the function $f(x) = \dfrac{2^x}{1 + 2^x}$.

(b) What is the domain of the inverse function?

Applications

102. Absorption of Light A spectrophotometer measures the concentration of a sample dissolved in water by shining a light through it and recording the amount of light that emerges. In other words, if we know the amount of light that is absorbed, we can calculate the concentration of the sample. For a certain substance the concentration (in moles per liter, mol/L) is found by using the formula

$$C = -2500 \ln \frac{I}{I_0}$$

where I_0 is the intensity of the incident light and I is the intensity of light that emerges. Find the concentration of the substance if the intensity I is 70% of I_0.

103. Carbon Dating The age of an ancient artifact can be determined by the amount of radioactive carbon-14 remaining in it. If D_0 is the original amount of carbon-14 and D is the amount remaining, then the artifact's age A (in years) is given by

$$A = -8267 \ln \frac{D}{D_0}$$

Find the age of an object if the amount D of carbon-14 that remains in the object is 73% of the original amount D_0.

104. Bacteria Colony A certain strain of bacteria divides every 3 hours. If a colony is started with 50 bacteria, then the time t (in hours) required for the colony to grow to N bacteria is

$$t = 3\frac{\log(N/50)}{\log 2}$$

Find the time required for the colony to grow to a million bacteria.

105. Investment The time required to double the amount of an investment at an interest rate r, compounded continuously, is

$$t = \frac{\ln 2}{r}$$

Find the time required to double an investment at 6%, 7%, and 8%.

106. Charging a Battery The rate at which a battery charges is slower the closer the battery is to its maximum charge C_0. The time (in hours) required to charge a fully discharged battery to a charge C is given by

$$t = -k \ln\left(1 - \frac{C}{C_0}\right)$$

where k is a positive constant that depends on the battery. For a certain battery, $k = 0.25$. If this battery is fully discharged, how long will it take to charge to 90% of its maximum charge C_0?

107. Difficulty of a Task The difficulty in "acquiring a target" (such as using a mouse to click on an icon on a computer screen) depends on the distance to the target and the size of the target. According to Fitts's Law, the index of difficulty (ID) is given by

$$\text{ID} = \frac{\log(2A/W)}{\log 2}$$

where W is the width of the target and A is the distance to the center of the target. Find the ID of clicking on an icon that is 5 mm wide and the ID for an icon that is 10 mm wide. In each case assume that the mouse pointer is 100 mm from the icon. Which task is more difficult?

■ Discuss ■ Discover ■ Prove ■ Write

108. Discuss: The Height of the Graph of a Logarithmic Function Suppose that the graph of $y = 2^x$ is drawn on a coordinate plane where the unit of measurement is a centimeter.

(a) Show that at a distance 0.24 meter to the right of the origin the height of the graph is about 168 km.

(b) If the graph of $y = \log_2 x$ is drawn on the same set of axes, how far to the right of the origin do we have to go before the height of the curve reaches 24 centimeters?

109. Discuss: The Googolplex A **googol** is 10^{100}, and a **googolplex** is 10^{googol}. Find

$$\log(\log(\text{googol})) \qquad \text{and} \qquad \log(\log(\log(\text{googolplex})))$$

110. Discuss: Comparing Logarithms Without using a calculator, determine which is larger, $\log_5 24$ or $\log_4 17$.

PS *Try to recognize something familiar.* Consider integers close to 24 or 17 for which you can easily calculate their respective logarithms.

111. Discuss ■ Discover: The Number of Digits in an Integer Compare log 1000 to the number of digits in 1000. Do the same for 10,000. How many digits does any number between 1000 and 10,000 have? Between what two values must the common logarithm of such a number lie? Use your observations to explain why the number of digits in any positive integer x is $[\![\log x]\!] + 1$. (The symbol $[\![n]\!]$ is the greatest integer function defined in Section 2.2.) How many digits does the number 2^{100} have?

112. Discover ■ Prove: A Logarithmic Identity For $a > 0$, prove that for all $x > 0$,

$$\log_{1/a} x = -\log_a x$$

How are the graphs of $f(x) = \log_a x$ and $g(x) = \log_{1/a} x$ related? Graph f and g for $a = 4$ to confirm your answer.

PS *Try to recognize something familiar.* Relate the familiar identity $a^{-1} = 1/a$ to the definition of the logarithm.

4.4 Laws of Logarithms

■ Laws of Logarithms ■ Expanding and Combining Logarithmic Expressions
■ Change of Base Formula

In this section we study properties of logarithms. These properties give logarithmic functions a wide range of applications, as we will see in Sections 4.6 and 4.7.

■ Laws of Logarithms

Since logarithms are exponents, the Laws of Exponents give rise to the following Laws of Logarithms.

Laws of Logarithms

Let a be a positive number, with $a \neq 1$. Let A, B, and C be any real numbers with $A > 0$ and $B > 0$.

Law	Description
1. $\log_a(AB) = \log_a A + \log_a B$	The logarithm of a product of numbers is the sum of the logarithms of the numbers.
2. $\log_a\left(\dfrac{A}{B}\right) = \log_a A - \log_a B$	The logarithm of a quotient of numbers is the difference of the logarithms of the numbers.
3. $\log_a(A^C) = C\log_a A$	The logarithm of a power of a number is the exponent times the logarithm of the number.

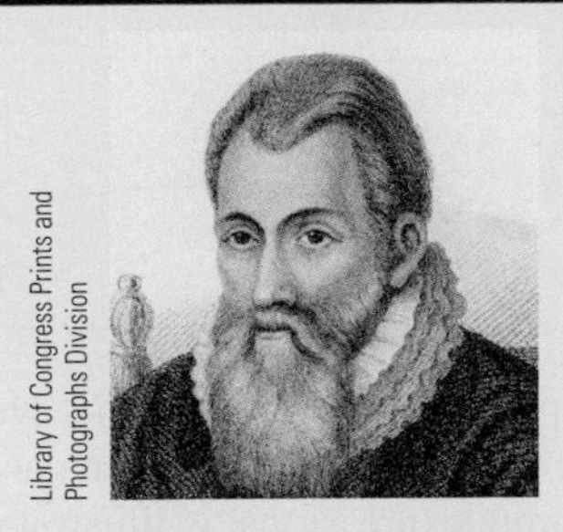

Library of Congress Prints and Photographs Division

JOHN NAPIER (1550–1617) was a Scottish landowner for whom mathematics was a hobby. We know him today because of his key invention: logarithms. He published his ideas in 1614 under the title *A Description of the Marvelous Rule of Logarithms*. In Napier's time, logarithms were used exclusively for simplifying complicated calculations. For example, to multiply two large numbers, we would write them as powers of 10. The exponents are simply the logarithms of the numbers. For instance,

$$\begin{aligned} &4532 \times 57783 \\ &\approx 10^{3.65629} \times 10^{4.76180} \\ &= 10^{8.41809} \\ &\approx 261{,}872{,}564 \end{aligned}$$

The idea is that multiplying powers of 10 is easy (we simply add their exponents). Napier produced extensive tables giving the logarithms (or exponents) of numbers. Since the advent of calculators and computers, logarithms are no longer used for this purpose. The logarithmic functions, however, have found many applications, some of which are described in this chapter.

Napier wrote on many topics. One of his more colorful works is a book entitled *A Plaine Discovery of the Whole Revelation of Saint John*, in which he predicted that the world would end in the year 1700.

Proof We make use of the property $\log_a a^x = x$ from Section 4.3.

Law 1 Let $\log_a A = u$ and $\log_a B = v$. When written in exponential form, these equations become

$$a^u = A \quad \text{and} \quad a^v = B$$

Thus
$$\begin{aligned} \log_a(AB) &= \log_a(a^u a^v) = \log_a(a^{u+v}) \\ &= u + v = \log_a A + \log_a B \end{aligned}$$

Law 2 Using Law 1, we have

$$\log_a A = \log_a\left[\left(\frac{A}{B}\right)B\right] = \log_a\left(\frac{A}{B}\right) + \log_a B$$

so
$$\log_a\left(\frac{A}{B}\right) = \log_a A - \log_a B$$

Law 3 Let $\log_a A = u$. Then $a^u = A$, so

$$\log_a(A^C) = \log_a(a^u)^C = \log_a(a^{uC}) = uC = C\log_a A$$

■

Example 1 ■ Using the Laws of Logarithms to Evaluate Expressions

Evaluate each expression (without using a calculator).

(a) $\log_4 2 + \log_4 32$ **(b)** $\log_2 80 - \log_2 5$ **(c)** $-\frac{1}{3}\log 1000$

Solution

(a) $\log_4 2 + \log_4 32 = \log_4(2 \cdot 32)$ Law 1

$= \log_4 64 = 3$ Because $64 = 4^3$

(b) $\log_2 80 - \log_2 5 = \log_2\left(\frac{80}{5}\right)$ Law 2

$= \log_2 16 = 4$ Because $16 = 2^4$

(c) $-\frac{1}{3}\log 1000 = \log 1000^{-1/3}$ Law 3

$= \log\frac{1}{10}$ Property of negative exponents

$= -1$ Because $\frac{1}{10} = 10^{-1}$

 Now Try Exercises 9, 11, and 13 ■

■ Expanding and Combining Logarithmic Expressions

The Laws of Logarithms allow us to write the logarithm of a product or a quotient as the sum or difference of logarithms. This process, called *expanding* a logarithmic expression, is illustrated in the next example.

Example 2 ■ Expanding Logarithmic Expressions

Use the Laws of Logarithms to expand each expression.

(a) $\log_2(6x)$ **(b)** $\log_5(x^3y^6)$ **(c)** $\ln\left(\frac{ab}{\sqrt[3]{c}}\right)$

Solution

(a) $\log_2(6x) = \log_2 6 + \log_2 x$ Law 1

(b) $\log_5(x^3y^6) = \log_5 x^3 + \log_5 y^6$ Law 1

$= 3\log_5 x + 6\log_5 y$ Law 3

(c) $\ln\left(\dfrac{ab}{\sqrt[3]{c}}\right) = \ln(ab) - \ln\sqrt[3]{c}$ Law 2

$= \ln a + \ln b - \ln c^{1/3}$ Law 1

$= \ln a + \ln b - \frac{1}{3}\ln c$ Law 3

 **Now Try Exercises 23, 31, and 37** ■

The Laws of Logarithms also allow us to reverse the process of expanding that was illustrated in Example 2. That is, we can write sums and differences of logarithms as a single logarithm. This process, called *combining* logarithmic expressions, is illustrated in the next example.

Example 3 ■ Combining Logarithmic Expressions

Use the Laws of Logarithms to combine each expression into a single logarithm.

(a) $3\log x + \frac{1}{2}\log(x+1)$ **(b)** $3\ln s + \frac{1}{2}\ln t - 4\ln(t^2+1)$

Solution

(a) $3\log x + \frac{1}{2}\log(x+1) = \log x^3 + \log(x+1)^{1/2}$ Law 3

$= \log(x^3(x+1)^{1/2})$ Law 1

(b) $3\ln s + \frac{1}{2}\ln t - 4\ln(t^2+1) = \ln s^3 + \ln t^{1/2} - \ln(t^2+1)^4$ Law 3

$= \ln(s^3t^{1/2}) - \ln(t^2+1)^4$ Law 1

$= \ln\dfrac{s^3\sqrt{t}}{(t^2+1)^4}$ Law 2

 **Now Try Exercises 51 and 53** ■

 **Warning** Although the Laws of Logarithms tell us how to compute the logarithm of a product or a quotient, *there is no corresponding rule for the logarithm of a sum or a difference.* For instance,

$$\log_a(x+y) \neq \log_a x + \log_a y$$

In fact, we know that the right side is equal to $\log_a(xy)$. Also, don't improperly simplify quotients or powers of logarithms. For instance,

$$\frac{\log 6}{\log 2} \neq \log\frac{6}{2} \quad \text{and} \quad (\log_2 x)^3 \neq 3\log_2 x$$

Logarithmic functions are used to model a variety of situations involving human behavior. One such behavior is how quickly we forget things we have learned. For example, if you learn algebra at a certain performance level (say, 90% on a test) and then don't use algebra for a while, how much will you retain after a week, a month, or a year? Hermann Ebbinghaus (1850–1909) studied this phenomenon and formulated the law described in the next example.

Forgetting what we've learned depends on how long ago we learned it.

Example 4 ■ The Law of Forgetting

If a task is learned at a performance level P_0, then after a time interval t the performance level P satisfies

$$\log P = \log P_0 - c\log(t+1)$$

where c is a constant that depends on the type of task and t is measured in months.

(a) Solve for P.

(b) If your score on a history test is 90, what score would you expect to get on a similar test after two months? After a year? (Assume that $c = 0.2$.)

Solution

(a) We first combine the right-hand side.

$$\log P = \log P_0 - c\log(t+1) \qquad \text{Given equation}$$

$$\log P = \log P_0 - \log(t+1)^c \qquad \text{Law 3}$$

$$\log P = \log \frac{P_0}{(t+1)^c} \qquad \text{Law 2}$$

$$P = \frac{P_0}{(t+1)^c} \qquad \text{Because log is one-to-one}$$

(b) Here $P_0 = 90$, $c = 0.2$, and t is measured in months.

In 2 months: $t = 2$ and $P = \dfrac{90}{(2+1)^{0.2}} \approx 72$

In 1 year: $t = 12$ and $P = \dfrac{90}{(12+1)^{0.2}} \approx 54$

Your expected scores after 2 months and after 1 year are 72 and 54, respectively.

 **Now Try Exercise 73** ■

■ Change of Base Formula

For some purposes we find it useful to change from logarithms in one base to logarithms in another base.

We may write the Change of Base Formula as

$$\log_b x = \frac{1}{\log_a b} \cdot \log_a x$$

So $\log_b x$ is just a constant multiple of $\log_a x$; the constant is $\dfrac{1}{\log_a b}$.

Change of Base Formula

$$\log_b x = \frac{\log_a x}{\log_a b} \qquad (a > 0, b > 0)$$

Proof Let $y = \log_b x$ and consider the following.

$$y = \log_b x \qquad \text{Given}$$

$$b^y = x \qquad \text{Exponential form}$$

$$\log_a b^y = \log_a x \qquad \text{Take } \log_a \text{ of each side}$$

$$y \log_a b = \log_a x \qquad \text{Law 3}$$

$$y = \frac{\log_a x}{\log_a b} \qquad \text{Divide by } \log_a b$$

So, if $y = \log_b x$, then $y = (\log_a x)/(\log_a b)$, and this proves the formula. ■

In particular, if we put $x = a$, then $\log_a a = 1$, and this formula becomes

$$\log_b a = \frac{1}{\log_a b}$$

We can now evaluate a logarithm to *any* base by using the Change of Base Formula to express the logarithm in terms of common logarithms or natural logarithms.

Example 5 ■ Evaluating Logarithms with the Change of Base Formula

Use the Change of Base Formula to express each logarithm in terms of common or natural logarithms, and then evaluate, rounded to five decimal places.

(a) $\log_8 5$ **(b)** $\log_9 20$

Solution

Check that you get the same answers in Example 5 if you use either log or ln with the Change of Base Formula.

(a) We use the Change of Base Formula with $b = 8$ and $a = 10$:

$$\log_8 5 = \frac{\log_{10} 5}{\log_{10} 8} \approx 0.77398$$

(b) We use the Change of Base Formula with $b = 9$ and $a = e$:

$$\log_9 20 = \frac{\ln 20}{\ln 9} \approx 1.36342$$

 Now Try Exercises 59 and 61 ■

Example 6 ■ Using the Change of Base Formula

Let $f(x) = \log_6 x$ and $g(x) = (\ln x)/(\ln 6)$. Explain why $f(x) = g(x)$, and confirm this graphically by graphing both functions on the same screen.

Solution By the Change of Base formula we have

$$f(x) = \log_6 x = \frac{\ln x}{\ln 6} = g(x)$$

The graphs of f and g in Figure 1 confirm that $f(x) = g(x)$ for every x in their domain.

 Now Try Exercise 67 ■

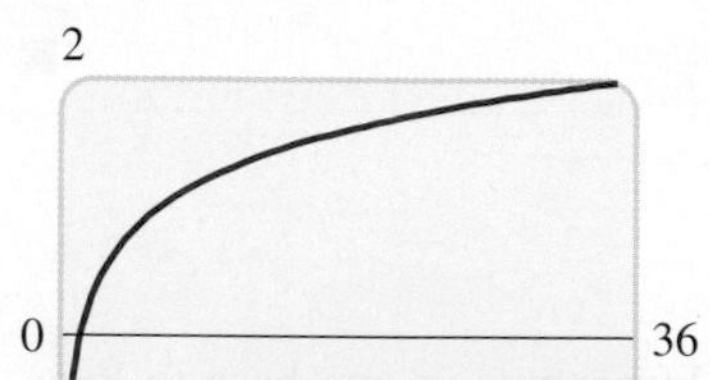

Figure 1 | $f(x) = \log_6 x = \dfrac{\ln x}{\ln 6} = g(x)$

4.4 | Exercises

■ Concepts

1. The logarithm of a product of two numbers is the same as the _______ of the logarithms of these numbers. So $\log_5(25 \cdot 125) =$ _______ + _______.

2. The logarithm of a quotient of two numbers is the same as the _______ of the logarithms of these numbers. So $\log_5\left(\frac{25}{125}\right) =$ _______ − _______.

3. The logarithm of a number raised to a power is the same as the _______ times the logarithm of the number. So $\log_5 25^{10} =$ _______ · _______.

4. We can expand $\log \dfrac{x^2 y}{z}$ to get _______________.

5. We can combine $2 \log x + \log y - \log z$ to get _______.

6. (a) To express $\log_7 12$ in terms of common logarithms, we use the Change of Base Formula to write

$$\log_7 12 = \frac{\log \square}{\log \square} \approx \_\_\_\_\_\_$$

(b) Do we get the same answer if we perform the calculation in part (a) using ln in place of log?

7–8 ■ *True or False?*

7. (a) $\log(A + B)$ is the same as $\log A + \log B$.

(b) $\log AB$ is the same as $\log A + \log B$.

8. (a) $\log \dfrac{A}{B}$ is the same as $\log A - \log B$.

(b) $\dfrac{\log A}{\log B}$ is the same as $\log A - \log B$.

Skills

9–22 ■ Evaluating Logarithms Use the Laws of Logarithms to evaluate the expression.

9. $\log 50 + \log 200$ **10.** $\log_6 9 + \log_6 24$

11. $\log_2 60 - \log_2 15$ **12.** $\log_3 135 - \log_3 45$

13. $\frac{1}{4}\log_3 81$ **14.** $-\frac{1}{3}\log_3 27$

15. $\log_5 \sqrt{5}$ **16.** $\log_5 \dfrac{1}{\sqrt{125}}$

17. $\log_2 6 - \log_2 15 + \log_2 20$

18. $\log_3 100 - \log_3 18 - \log_3 50$

19. $\log_4 16^{100}$ **20.** $\log_2 8^{33}$

21. $\log(\log 10^{10,000})$ **22.** $\ln(\ln e^{e^{200}})$

23–48 ■ Expanding Logarithmic Expressions Use the Laws of Logarithms to expand the expression.

23. $\log_3(8x)$ **24.** $\log_6(7r)$

25. $\log_3(2xy)$ **26.** $\log_5(4st)$

27. $\ln a^3$ **28.** $\log\sqrt{t^5}$

29. $\log_3\sqrt{xyz}$ **30.** $\log_5(xy)^6$

31. $\ln(a^3b^2)$ **32.** $\ln(y^2\sqrt{x})$

33. $\log_2\left(\dfrac{4a}{b}\right)$ **34.** $\log_6\left(\dfrac{y}{6z}\right)$

35. $\log_8\left(\dfrac{a^3b^2}{c}\right)$ **36.** $\log_7\left(\dfrac{3x^4y^2}{2z^3}\right)$

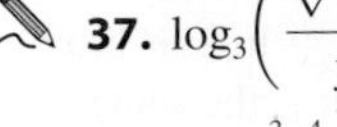

37. $\log_3\left(\dfrac{\sqrt{3x^5}}{y}\right)$ **38.** $\log\dfrac{y^3}{\sqrt{2x}}$

39. $\log\dfrac{x^3y^4}{z^6}$ **40.** $\log_a\left(\dfrac{x^2}{yz^3}\right)$

41. $\ln\sqrt{x^4+2}$ **42.** $\log\sqrt[3]{x^2+4}$

43. $\log\sqrt{\dfrac{x+z}{y}}$ **44.** $\ln\dfrac{4x^2}{x^2+3}$

45. $\ln\sqrt[3]{\dfrac{x^2+y^2}{x+y}}$ **46.** $\ln\dfrac{x^2}{\sqrt{x+1}}$

47. $\log\sqrt{\dfrac{x^2+4}{(x^2+1)(x^3-7)^2}}$ **48.** $\log\sqrt{x\sqrt{y\sqrt{z}}}$

49–58 ■ Combining Logarithmic Expressions Use the Laws of Logarithms to combine the expression.

49. $\log_4 6 + 2\log_4 7$

50. $\frac{1}{2}\log_2 5 - 2\log_2 7$

51. $2\log x - 3\log(x+1)$

52. $3\ln 2 + 2\ln x - \frac{1}{2}\ln(x+4)$

53. $\log(x+1) + \log(x-1) - 3\log x$

54. $\ln(x+3) - \ln(x^2-9)$

55. $\frac{1}{2}[\log_5(x+2) - \log_5(x^2+4) - \log_5 x]$

56. $4(\log_3 a - 3\log_3 b + 2\log_3 c)$

57. $\frac{1}{3}\log(x+2)^3 + \frac{1}{2}[\log x^4 - \log(x^2-x-6)^2]$

58. $\log_a b + c\log_a d - r\log_a s$

59–66 ■ Change of Base Formula Use the Change of Base Formula to express the given logarithm in terms of common or natural logarithms, and then evaluate. State your answer rounded to six decimal places.

59. $\log_5 10$ **60.** $\log_{14} 7$

61. $\log_9 4$ **62.** $\log_5 30$

63. $\log_7 2.61$ **64.** $\log_6 532$

65. $\log_4 125$ **66.** $\log_{12} 2.5$

67. Change of Base Formula Use the Change of Base Formula to show that

$$\log_3 x = \frac{\ln x}{\ln 3}$$

Then use this fact to draw the graph of the function $f(x) = \log_3 x$.

Skills Plus

68. Families of Functions Draw graphs of the family of functions $y = \log_a x$ for $a = 2, e, 5$, and 10 on the same screen, using the viewing rectangle $[0, 5]$ by $[-3, 3]$. How are these graphs related?

69. Change of Base Formula Use the Change of Base Formula to show that

$$\log e = \frac{1}{\ln 10}$$

70–72 ■ Logarithmic Identities Let $a, b, c, d > 0$ and n a positive integer. Prove the identity.

70. $(\log_a b)(\log_b c)(\log_c d) = \dfrac{\log d}{\log a}$

71. $\dfrac{1}{\log_a x} + \dfrac{1}{\log_b x} + \dfrac{1}{\log_c x} = \dfrac{1}{\log_{abc} x}$

72. $\log_{a^n} x = \dfrac{1}{n}\log_a x$

Applications

73. Wealth Distribution Vilfredo Pareto (1848–1923) observed that most of the wealth of a country is owned by a few members of the population. **Pareto's principle** is

$$\log P = \log c - k\log W$$

where W is the wealth level (how much money a person has) and P is the number of people in the population having that much money.

(a) Solve the equation for P.

(b) Assume that $k = 2.1$ and $c = 8000$ and that W is measured in millions of dollars. Use part (a) to find the number of people who have \$2 million or more. How many people have \$10 million or more?

74. Forgetting Use the Law of Forgetting (Example 4) to estimate a student's score on a biology test two years after the student got a score of 80 on a test covering the same material. Assume that $c = 0.3$ and t is measured in months.

75. Magnitude of Stars The magnitude M of a star is a measure of how bright a star appears to the human eye. It is defined by

$$M = -2.5 \log \frac{B}{B_0}$$

where B is the actual brightness of the star and B_0 is a constant.

(a) Expand the right-hand side of the equation.

(b) Use part (a) to show that the brighter a star, the less its magnitude.

(c) Betelgeuse is about 100 times brighter than Albiero. Use part (a) to show that Betelgeuse is 5 magnitudes less bright than Albiero.

76. Biodiversity Some biologists model the number of species S in a fixed area A (such as an island) by the species-area relationship

$$\log S = \log c + k \log A$$

where c and k are positive constants that depend on the type of species and habitat.

(a) Solve the equation for S.

(b) Use part (a) to show that if $k = 3$, then doubling the area increases the number of species eightfold.

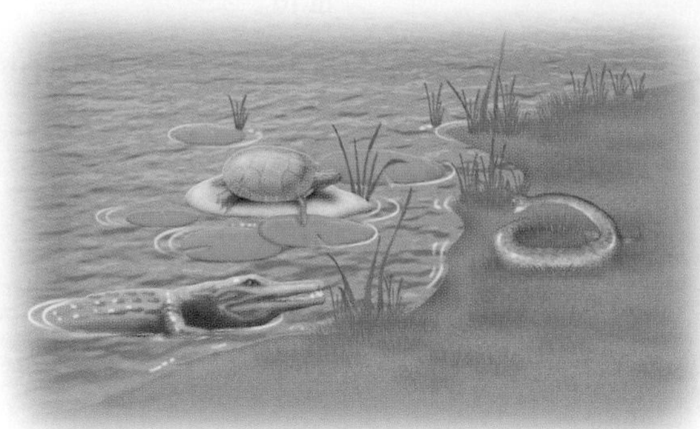

■ Discuss ■ Discover ■ Prove ■ Write

77. Discuss: True or False? Discuss each equation, and determine whether it is true for all possible values of the variables. (Ignore values of the variables for which any term is undefined.)

(a) $\log \dfrac{x}{y} = \dfrac{\log x}{\log y}$

(b) $\log_2(x - y) = \log_2 x - \log_2 y$

(c) $\log_5\left(\dfrac{a}{b^2}\right) = \log_5 a - 2\log_5 b$

(d) $\log 2^z = z \log 2$

(e) $(\log P)(\log Q) = \log P + \log Q$

(f) $\dfrac{\log a}{\log b} = \log a - \log b$

(g) $(\log_2 7)^x = x \log_2 7$

(h) $\log_a a^a = a$

(i) $\log(x - y) = \dfrac{\log x}{\log y}$

(j) $-\ln \dfrac{1}{A} = \ln A$

78. Discuss: Find the Error What is wrong with the following argument?

$$\begin{aligned} \log 0.1 &< 2 \log 0.1 \\ &= \log(0.1)^2 \\ &= \log 0.01 \\ \log 0.1 &< \log 0.01 \\ 0.1 &< 0.01 \end{aligned}$$

79. Prove: Shifting, Shrinking, and Stretching Graphs of Functions Let $f(x) = x^2$. Show that

$$f(2x) = 4f(x)$$

and explain how this shows that shrinking the graph of f horizontally has the same effect as stretching it vertically. Then use the identities $e^{2+x} = e^2 e^x$ and $\ln(2x) = \ln 2 + \ln x$ to show that for $g(x) = e^x$ a horizontal shifting is the same as a vertical stretching and for $h(x) = \ln x$ a horizontal shrinking is the same as a vertical shifting.

80. Prove: A Logarithmic Identity Show that

$$-\ln\left(x - \sqrt{x^2 - 1}\right) = \ln\left(x + \sqrt{x^2 - 1}\right)$$

PS *Work backward.* Assume the equation holds; then use the properties of logarithms and the rules of algebra to arrive at an equivalent true equation.

81–82 ■ Discuss ■ Prove: Linearizing Exponential and Power Curves These exercises are about "straightening" an exponential or power curve by applying a logarithmic function to the appropriate variable(s). Prove the statement.

81. If the points (x, y) are on the exponential curve $y = Ce^{kx}$, then the points $(x, \ln y)$ are on the line

$$Y = kX + \ln C$$

82. If the points (x, y) are on the power curve $y = ax^n$, then the points $(\ln x, \ln y)$ are on the line

$$Y = nX + \ln a.$$

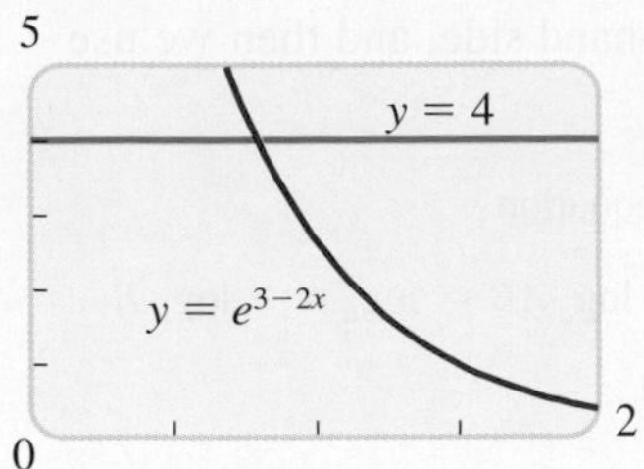

Figure 1

Solution 2: Graphical

We graph the equations $y = e^{3-2x}$ and $y = 4$ in the same viewing rectangle as in Figure 1. The solutions occur where the graphs intersect. In this case the only solution is $x \approx 0.81$.

 **Now Try Exercise 21** ■

Example 5 ■ An Exponential Equation of Quadratic Type

Solve the equation $e^{2x} - e^x - 6 = 0$.

Solution To isolate the exponential term, we factor.

If we let $w = e^x$, we get the quadratic equation

$$w^2 - w - 6 = 0$$

which factors as

$$(w - 3)(w + 2) = 0$$

$$e^{2x} - e^x - 6 = 0 \quad \text{Given equation}$$

$$(e^x)^2 - e^x - 6 = 0 \quad \text{Law of Exponents}$$

$$(e^x - 3)(e^x + 2) = 0 \quad \text{Factor (a quadratic in } e^x\text{)}$$

$$e^x - 3 = 0 \quad \text{or} \quad e^x + 2 = 0 \quad \text{Zero-Product Property}$$

$$e^x = 3 \qquad\qquad e^x = -2$$

The equation $e^x = 3$ leads to $x = \ln 3$. But the equation $e^x = -2$ has no solution because $e^x > 0$ for all x. Thus $x = \ln 3 \approx 1.0986$ is the only solution. You should check that this answer satisfies the original equation.

 **Now Try Exercise 37** ■

Example 6 ■ An Equation Involving Exponential Functions

Solve the equation $3xe^x + x^2e^x = 0$.

Solution First we factor the left side of the equation.

$$3xe^x + x^2e^x = 0 \quad \text{Given equation}$$

$$x(3 + x)e^x = 0 \quad \text{Factor out common factors}$$

$$x(3 + x) = 0 \quad \text{Divide by } e^x \text{ (because } e^x \neq 0\text{)}$$

$$x = 0 \quad \text{or} \quad 3 + x = 0 \quad \text{Zero-Product Property}$$

Thus the solutions are $x = 0$ and $x = -3$.

Check Your Answer

$x = 0$:

$3(0)e^0 + 0^2e^0 = 0$ ✓

$x = -3$:

$3(-3)e^{-3} + (-3)^2e^{-3}$

$= -9e^{-3} + 9e^{-3} = 0$ ✓

Now Try Exercise 43 ■

■ Logarithmic Equations

A *logarithmic equation* is an equation in which a logarithm of the variable occurs. Some logarithmic equations can be solved by using the fact that logarithmic functions are one-to-one. This means that

$$\log_a x = \log_a y \quad \Rightarrow \quad x = y$$

We use this property in the next example.

Example 7 ■ Solving a Logarithmic Equation

Solve the equation $\log_5(x^2 + 1) = \log_5(x - 2) + \log_5(x + 3)$.

Solution First we combine the logarithms on the right-hand side, and then we use the one-to-one property of logarithms.

$$\log_5(x^2+1) = \log_5(x-2) + \log_5(x+3) \qquad \text{Given equation}$$
$$\log_5(x^2+1) = \log_5[(x-2)(x+3)] \qquad \text{Law 1: } \log_a AB = \log_a A + \log_a B$$
$$\log_5(x^2+1) = \log_5(x^2+x-6) \qquad \text{Expand}$$
$$x^2+1 = x^2+x-6 \qquad \text{log is one-to-one (or raise 5 to each side)}$$
$$x = 7 \qquad \text{Solve for } x$$

The solution is $x = 7$. (You can check that $x = 7$ satisfies the original equation.)

 **Now Try Exercise 47** ■

The method of Example 7 is not suitable for solving an equation like $\log_5 x = 13$ because the right-hand side is not expressed as a logarithm (base 5). To solve such equations, we use the following guidelines.

Guidelines for Solving Logarithmic Equations

1. Isolate the logarithmic term on one side of the equation; you might first need to combine the logarithmic terms.
2. Write the equation in exponential form (or raise the base to each side of the equation).
3. Solve for the variable.

Example 8 ■ Solving Logarithmic Equations

Solve each equation for x.

(a) $\ln x = 8$ **(b)** $\log_2(25-x) = 3$

Solution

(a)

$$\ln x = 8 \qquad \text{Given equation}$$
$$x = e^8 \qquad \text{Exponential form}$$

Therefore $x = e^8 \approx 2981$.

We can also solve this problem another way.

$$\ln x = 8 \qquad \text{Given equation}$$
$$e^{\ln x} = e^8 \qquad \text{Raise } e \text{ to each side}$$
$$x = e^8 \qquad \text{Property of ln}$$

(b) The first step is to rewrite the equation in exponential form.

$$\log_2(25-x) = 3 \qquad \text{Given equation}$$
$$25 - x = 2^3 \qquad \text{Exponential form (or raise 2 to each side)}$$
$$25 - x = 8$$
$$x = 25 - 8 = 17$$

Check Your Answer

If $x = 17$, we get

$\log_2(25-17) = \log_2 8 = 3$ ✓

 **Now Try Exercises 53 and 57** ■

Example 9 ■ Solving a Logarithmic Equation

Solve the equation $4 + 3\log(2x) = 16$.

Solution We first isolate the logarithmic term. This allows us to write the equation in exponential form.

$$
\begin{aligned}
4 + 3\log(2x) &= 16 && \text{Given equation} \\
3\log(2x) &= 12 && \text{Subtract 4} \\
\log(2x) &= 4 && \text{Divide by 3} \\
2x &= 10^4 && \text{Exponential form (or raise 10 to each side)} \\
x &= 5000 && \text{Divide by 2}
\end{aligned}
$$

Now Try Exercise 59 ■

Check Your Answer

If $x = 5000$, we get

$$
\begin{aligned}
4 + 3\log(2 \cdot 5000) &= 4 + 3\log 10{,}000 \\
&= 4 + 3(4) \\
&= 16 \quad ✓
\end{aligned}
$$

Example 10 ■ Solving a Logarithmic Equation Algebraically and Graphically

Solve the equation $\log(x + 2) + \log(x - 1) = 1$ algebraically and graphically.

Solution 1: Algebraic

We first combine the logarithmic terms, using the Laws of Logarithms.

$$
\begin{aligned}
\log[(x + 2)(x - 1)] &= 1 && \text{Law 1} \\
(x + 2)(x - 1) &= 10 && \text{Exponential form (or raise 10 to each side)} \\
x^2 + x - 2 &= 10 && \text{Expand left side} \\
x^2 + x - 12 &= 0 && \text{Subtract 10} \\
(x + 4)(x - 3) &= 0 && \text{Factor} \\
x = -4 \quad &\text{or} \quad x = 3
\end{aligned}
$$

We check these potential solutions in the original equation and find that $x = -4$ is not a solution (because logarithms of negative numbers are undefined), but $x = 3$ is a solution. (See *Check Your Answers* in the margin.)

Check Your Answer

$x = -4$:

$$
\begin{aligned}
&\log(-4 + 2) + \log(-4 - 1) \\
&\quad = \log(-2) + \log(-5) \\
&\qquad \text{undefined} \quad ✗
\end{aligned}
$$

$x = 3$:

$$
\begin{aligned}
&\log(3 + 2) + \log(3 - 1) \\
&\quad = \log 5 + \log 2 = \log(5 \cdot 2) \\
&\quad = \log 10 = 1 \quad ✓
\end{aligned}
$$

Solution 2: Graphical

We first move all terms to one side of the equation:

$$\log(x + 2) + \log(x - 1) - 1 = 0$$

Then we graph

$$y = \log(x + 2) + \log(x - 1) - 1$$

as shown in Figure 2. The solutions are the x-intercepts of the graph. Thus the only solution is $x \approx 3$.

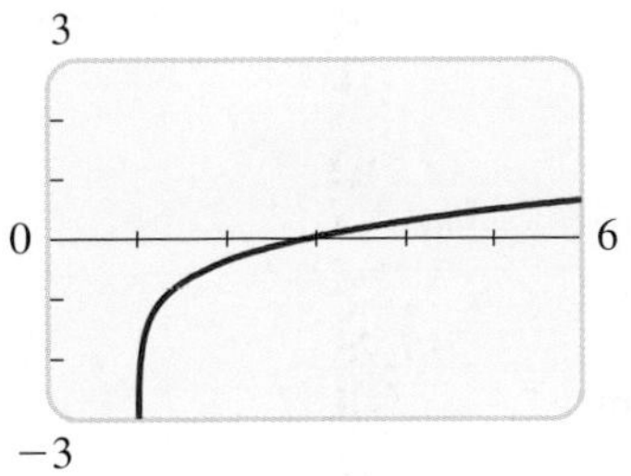

Now Try Exercise 61 ■

Figure 2

Example 11 ■ Solving a Logarithmic Equation Graphically

Solve the equation $x^2 = 2\ln(x + 2)$.

In Example 11 it's not possible to isolate x algebraically, so we must solve the equation graphically.

Solution We first move all terms to one side of the equation.

$$x^2 - 2\ln(x + 2) = 0$$

Then we graph

$$y = x^2 - 2\ln(x + 2)$$

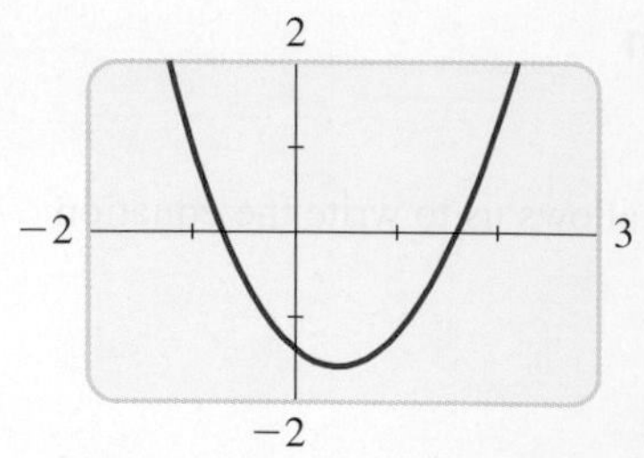

Figure 3

as in Figure 3. The solutions are the x-intercepts of the graph:

$$x \approx -0.71 \quad \text{and} \quad x \approx 1.60$$

Now Try Exercise 67 ■

Logarithmic equations are used in determining the amount of light that reaches various depths in a lake. (This information helps biologists to determine the types of life a lake can support.) As light passes through water (or other transparent materials such as glass or plastic), some of the light is absorbed. It's easy to see that the murkier the water, the more light is absorbed. The exact relationship between light absorption and the distance light travels in a material is described in the next example.

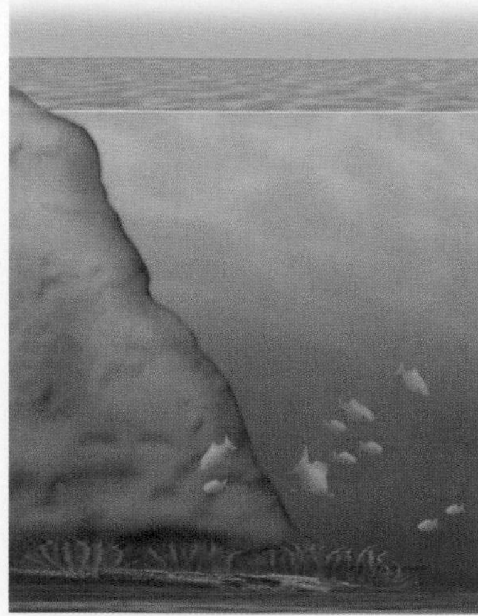

The intensity of light in a lake diminishes with depth. Environmental scientists are interested in the "transparency" of the water in a lake because certain levels of transparency are required to support biodiversity of the submerged macrophyte population.

Example 12 ■ Transparency of a Lake

If I_0 and I denote the intensity of light (in lumens, lm) before and after going through a material and x is the distance (in meters) the light travels in the material, then according to the **Beer-Lambert Law**,

$$-\frac{1}{k}\ln\frac{I}{I_0} = x$$

where k is a constant depending on the type of material.

(a) Solve the equation for I.

(b) For a certain lake $k = 0.025$, and the light intensity is $I_0 = 14$ lumens (lm). Find the light intensity at a depth of 20 m.

Solution

(a) We first isolate the logarithmic term.

$$-\frac{1}{k}\ln\frac{I}{I_0} = x \qquad \text{Given equation}$$

$$\ln\frac{I}{I_0} = -kx \qquad \text{Multiply by } -k$$

$$\frac{I}{I_0} = e^{-kx} \qquad \text{Exponential form}$$

$$I = I_0e^{-kx} \qquad \text{Multiply by } I_0$$

(b) We find I using the formula from part (a).

$$I = I_0e^{-kx} \qquad \text{From part (a)}$$

$$= 14e^{(-0.025)(20)} \qquad I_0 = 14,\ k = 0.025,\ x = 20$$

$$\approx 8.49 \qquad \text{Calculator}$$

The light intensity at a depth of 20 m is about 8.5 lm.

Now Try Exercise 97 ■

■ Compound Interest

Recall the formulas for interest that we found in Section 4.1. If a principal P is invested at an interest rate r for a period of t years, then the amount A of the investment is given by

$$A = P(1 + r) \qquad \text{Simple interest (for one year)}$$

$$A(t) = P\left(1 + \frac{r}{n}\right)^{nt} \qquad \text{Interest compounded } n \text{ times per year}$$

$$A(t) = Pe^{rt} \qquad \text{Interest compounded continuously}$$

Radiocarbon dating is a method that archaeologists use to determine the age of ancient objects. The carbon dioxide in the atmosphere always contains a fixed fraction of radioactive carbon, carbon-14 (^{14}C), with a half-life of about 5730 years. Plants absorb carbon dioxide from the atmosphere, which then makes its way to animals through the food chain. Thus, all living creatures contain the same fixed proportion of ^{14}C to nonradioactive ^{12}C as in the atmosphere.

After an organism dies, it stops assimilating ^{14}C, and the amount of ^{14}C in it begins to decay exponentially. We can determine the time that has elapsed since the death of the organism by measuring the amount of ^{14}C left in it.

For example, if a donkey bone contains 73% as much ^{14}C as a living donkey and it died t years ago, then by the formula for radioactive decay (Section 4.6),

$$0.73 = (1.00)e^{-(t \ln 2)/5730}$$

We solve this exponential equation to find $t \approx 2600$, so the bone is about 2600 years old.

We can use logarithms to determine the time it takes for the principal to increase to a given amount.

Example 13 ■ Finding the Term for an Investment to Double

A sum of \$5000 is invested at an interest rate of 5% per year. Find the time required for the money to double if the interest is compounded according to the following methods.

(a) Semiannual **(b)** Continuous

Solution

(a) We use the formula for compound interest with $P = \$5000$, $A(t) = \$10{,}000$, $r = 0.05$, and $n = 2$, and solve the resulting exponential equation for t.

$$\begin{aligned}
5000\left(1 + \frac{0.05}{2}\right)^{2t} &= 10{,}000 && P\left(1 + \frac{r}{n}\right)^{nt} = A \\
(1.025)^{2t} &= 2 && \text{Divide by 5000} \\
\log(1.025)^{2t} &= \log 2 && \text{Take log of each side} \\
2t \log 1.025 &= \log 2 && \text{Law 3 (bring down the exponent)} \\
t &= \frac{\log 2}{2 \log 1.025} && \text{Divide by 2 log 1.025} \\
t &\approx 14.04 && \text{Calculator}
\end{aligned}$$

The money will double in about 14 years.

(b) We use the formula for continuously compounded interest with $P = \$5000$, $A(t) = \$10{,}000$, and $r = 0.05$, and solve the resulting exponential equation for t.

$$\begin{aligned}
5000e^{0.05t} &= 10{,}000 && Pe^{rt} = A \\
e^{0.05t} &= 2 && \text{Divide by 5000} \\
\ln e^{0.05t} &= \ln 2 && \text{Take ln of each side} \\
0.05t &= \ln 2 && \text{Property of ln} \\
t &= \frac{\ln 2}{0.05} && \text{Divide by 0.05} \\
t &\approx 13.86 && \text{Calculator}
\end{aligned}$$

The money will double in about 13 years 10 months.

Now Try Exercise 87 ■

Example 14 ■ Time Required to Grow an Investment

A sum of \$1000 is invested at an interest rate of 4% per year. Find the time required for the amount to grow to \$5000 if interest is compounded continuously.

Solution We use the formula for continuously compounded interest with $P = \$1000$, $A(t) = \$5000$, and $r = 0.04$, and solve the resulting exponential equation for t.

$$\begin{aligned}
1000e^{0.04t} &= 5000 && Pe^{rt} = A \\
e^{0.04t} &= 5 && \text{Divide by 1000} \\
0.04t &= \ln 5 && \text{Take ln of each side} \\
t &= \frac{\ln 5}{0.04} && \text{Divide by 0.04} \\
t &\approx 40.24 && \text{Calculator}
\end{aligned}$$

The amount will be \$5000 in about 40 years 3 months.

Now Try Exercise 89 ■

4.5 Exercises

Concepts

1. Let's solve the exponential equation $2e^x = 50$.

(a) First, we isolate e^x to get the equivalent equation ________.

(b) Next, we take ln of each side to get the equivalent equation ________.

(c) Now we use a calculator to find $x \approx$ ________.

2. Let's solve the logarithmic equation

$$\log 3 + \log(x - 2) = \log x$$

(a) First, we combine the logarithms on the LHS to get the equivalent equation ________.

(b) Next, we use the fact that log is one-to-one to get the equivalent equation ________.

(c) Now we find $x =$ ________.

Skills

3–10 ■ Exponential Equations Find the solution of the exponential equation, as in Example 1.

3. $5^{x-1} = 625$
4. $e^{x^2} = e^9$

5. $5^{2x-3} = 1$
6. $10^{2x-3} = \frac{1}{10}$

7. $7^{2x-3} = 7^{6+5x}$
8. $e^{1-2x} = e^{3x-5}$

9. $6^{x^2-1} = 6^{1-x^2}$
10. $10^{2x^2-3} = 10^{9-x^2}$

11–36 ■ Exponential Equations **(a)** Find the exact solution of the exponential equation in terms of logarithms. **(b)** Use a calculator to find an approximation to the solution, rounded to six decimal places.

11. $e^x = 16$
12. $e^{-2x} = 5$

13. $10^{-x} = 6$
14. $10^{5x} = 24$

15. $3^{x+5} = 4$
16. $e^{2-3x} = 11$

17. $3 \cdot 6^{1-x} = 15$
18. $5 \cdot 4^{3-2x} = 8$

19. $200(1.02)^{4t} = 1500$
20. $25(1.015)^{12t} = 60$

21. $3e^{5-t} = 12$
22. $5(\frac{1}{2})^{2t-3} = 24$

23. $2^{x/10} = 0.3$
24. $2^{-x/50} = 0.6$

25. $4(1 + 10^{5x}) = 9$
26. $2(5 + 3^{x+1}) = 100$

27. $8 + e^{1-4x} = 20$
28. $1 + e^{4x+1} = 20$

29. $4^x + 2^{1+2x} = 50$
30. $125^x + 5^{3x+1} = 200$

31. $5^t = 10^{3t+2}$
32. $e^t = 3^{1-t}$

33. $5^{x/3} = 3^{x+1}$
34. $3^{2x-1} = 2^{4x+1}$

35. $\dfrac{50}{1 + e^{-x}} = 4$
36. $\dfrac{10}{1 + e^{-x}} = 2$

37–42 ■ Exponential Equations of Quadratic Type Solve the equation.

37. $e^{2x} + 5e^x - 6 = 0$
38. $e^{2x} + 3e^x - 10 = 0$

39. $e^{4x} + 4e^{2x} - 21 = 0$
40. $3^{4x} - 3^{2x} - 6 = 0$

41. $2^x - 10(2^{-x}) + 3 = 0$
42. $e^x + 15e^{-x} - 8 = 0$

43–46 ■ Equations Involving Exponential Functions Solve the equation.

43. $x^2 2^x - 2^x = 0$
44. $x^2 10^x - x10^x = 2(10^x)$

45. $4x^3 e^{-3x} - 3x^4 e^{-3x} = 0$
46. $x^2 e^x + xe^x - e^x = 0$

47–52 ■ Logarithmic Equations Solve the logarithmic equation for x, as in Example 7.

47. $\log(x + 2) + \log(x - 3) = \log(4x)$

48. $\log_5(x + 2) + \log_5(x - 5) = \log_5(6x)$

49. $2 \log x = \log 2 + \log(3x - 4)$

50. $\ln(x - \frac{1}{2}) + \ln 2 = 2 \ln x$

51. $\log_2 3 + \log_2 x = \log_2 5 + \log_2(x - 2)$

52. $\log_4(x + 2) + \log_4 3 = \log_4 5 + \log_4(2x - 3)$

53–66 ■ Logarithmic Equations Solve the logarithmic equation for x.

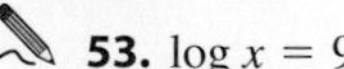

53. $\log x = 9$

54. $\log(x + 3) = 4$

55. $\ln(4 - x) = 1$

56. $\ln(3x + 1) = 0$

57. $\log_3(5 - x) = -1$

58. $\log_{1/2}(3x + 2) = -1$

59. $4 - \log(3 - x) = 3$

60. $\log_2(x^2 - x - 2) = 2$

61. $\log_2 x + \log_2(x - 3) = 2$

62. $\log x + \log(x - 3) = 1$

63. $\log_9(x - 5) + \log_9(x + 3) = 1$

64. $\ln(x - 1) + \ln(x + 1) = 0$

65. $\log_{1/5}(x - 1) - \log_{1/5}(x + 1) = 2$

66. $\log_3(x + 15) - \log_3(x - 1) = 2$

67–74 ■ Solving Equations Graphically Use a graphing device to find all solutions of the equation, rounded to two decimal places.

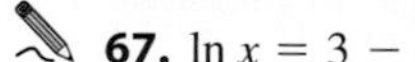

67. $\ln x = 3 - x$
68. $\log x = x^2 - 2$

69. $x^3 - x = \log(x + 1)$
70. $x = \ln(4 - x^2)$

71. $e^x = -x$
72. $2^{-x} = x - 1$

73. $4^{-x} = \sqrt{x}$
74. $e^{x^2} - 2 = x^3 - x$

Skills Plus

75–78 ■ Solving Inequalities Solve the inequality.

75. $\log(x - 2) + \log(9 - x) < 1$

76. $3 \le \log_2 x \le 4$

77. $2 < 10^x < 5$

78. $x^2e^x - 2e^x < 0$

79–82 ■ Inverse Functions Find the inverse function of f.

79. $f(x) = 2^{2x}$

80. $f(x) = 3^{x+1}$

81. $f(x) = \log_2(x - 1)$

82. $f(x) = \log 3x$

83–86 ■ Special Exponential and Logarithmic Equations Find the value(s) of x for which the equation is true.

83. $2^{2/\log_5 x} = \frac{1}{16}$

84. $\log_2(\log_3 x) = 4$

85. $(\log x)^4 + (\log x)^3 = 0$

86. $(\log x)^3 = 3 \log x$

Applications

87. Compound Interest Suppose you invest \$5000 in an account that pays 2.25% interest per year, compounded quarterly.

(a) Find the amount after 5 years.

(b) How long will it take for the investment to double?

88. Compound Interest Suppose you invest \$6500 in an account that pays 4.5% interest per year, compounded continuously.

(a) What is the amount after 4 years?

(b) How long will it take for the amount to be \$8000?

89. Compound Interest Find the time required for an investment of \$5000 to grow to \$8000 at an interest rate of 3.5% per year, compounded quarterly.

90. Compound Interest Find the time required for an investment of \$3000 to grow to \$5000 at an interest rate of 6.5%, compounded semiannually.

91. Doubling an Investment How long will it take for an investment of \$1000 to double in value if the interest rate is 8.5% per year, compounded continuously?

92. Interest Rate A sum of \$1000 was invested for 4 years, and the interest was compounded semiannually. If this sum amounted to \$1435.77 in the given time, what was the interest rate?

93. Radioactive Decay A 15-gram sample of radioactive iodine decays in such a way that the mass remaining after t days is given by

$$m(t) = 15e^{-0.087t}$$

where $m(t)$ is measured in grams. After how many days are there only 5 g remaining?

94. Skydiving The velocity of a skydiver t seconds after jumping is given by

$$v(t) = 24(1 - e^{-0.2t})$$

After how many seconds is the velocity 21 m/s?

95. Fish Population A small lake is stocked with a certain species of fish. The fish population is modeled by the function

$$P = \frac{10}{1 + 4e^{-0.8t}}$$

where P is the number of fish (in thousands) and t is measured in years since the lake was stocked.

(a) Find the fish population after 3 years.

(b) After how many years will the fish population reach 5000 fish?

96. DNA Gel Electrophoresis Gel electrophoresis is a DNA analysis technique in which an electric field is used to propel charged DNA fragments through a porous gel. Shorter strands of DNA move further through the gel than the longer strands. Each DNA sample starts in its own well on one side of the electric field (at the top in the image). The length y of a DNA fragment (the number of base-pairs in the fragment) is inversely proportional to 2^x where x is the distance the fragment travels in the gel. Fragments of different lengths can be distinguished by how far they travel through the gel. (The dark lines in the image are locations where large numbers of identical fragments end up together.) Suppose that fragments of length 200 base-pairs travel 8 cm in the gel.

(a) Find an equation that expresses the inverse relationship of the length y of a DNA fragment and the distance x that the fragment travels in the gel.

(b) What is the length of a fragment that travels 5 cm?

(c) What is the distance traveled by a fragment of length 2000 base-pairs?

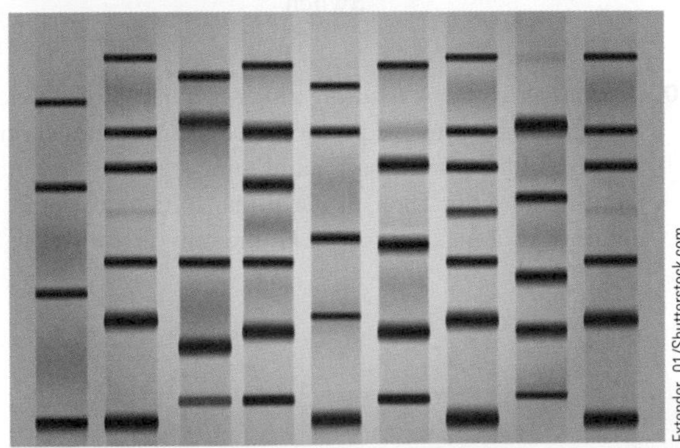

97. Atmospheric Pressure Atmospheric pressure P (in kilopascals, kPa) at altitude h (in kilometers, km) is governed by the formula

$$\ln\frac{P}{P_0} = -\frac{h}{k}$$

where $k = 7$ and $P_0 = 100$ kPa are constants.

(a) Solve the equation for P.

(b) Use part (a) to find the pressure P at an altitude of 4 km.

98. Cooling an Engine Suppose a car is driven on a cold winter day (−6°C outside) and the engine overheats (at about 104°C). When the car is parked, the engine begins to cool down. The temperature T of the engine t minutes after the car is parked satisfies the equation

$$\ln\frac{T-(-6)}{110} = -0.11t$$

(a) Solve the equation for T.

(b) Use part (a) to find the temperature of the engine after 20 min ($t = 20$).

99. Electric Circuits An electric circuit contains a battery that produces a voltage of 60 volts (V), a resistor with a resistance of 13 ohms (Ω), and an inductor with an inductance of 5 henrys (H), as shown in the figure. Using calculus, it can be shown that the current $I = I(t)$ (in amperes, A) t seconds after the switch is closed is

$$I = \tfrac{60}{13}(1 - e^{-13t/5})$$

(a) Use this equation to express the time t as a function of the current I.

(b) After how many seconds is the current 2 A?

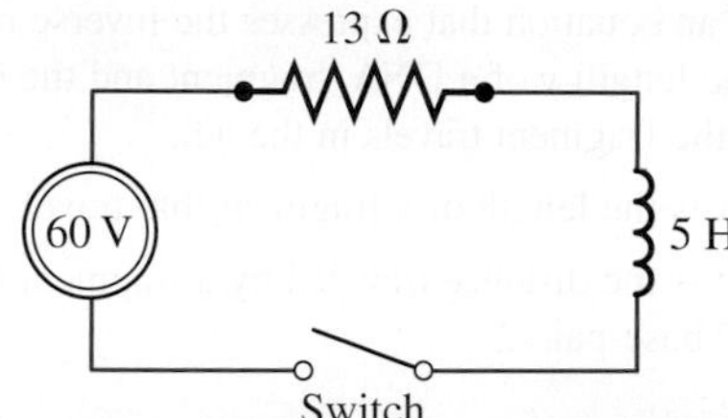

100. Learning Curve A *learning curve* is a graph of a function $P(t)$ that measures the performance of someone learning a skill as a function of the training time t. At first, the rate of learning is rapid. Then, as performance increases and approaches a maximal value M, the rate of learning decreases. It has been found that the function

$$P(t) = M - Ce^{-kt}$$

where k and C are positive constants and $C < M$ is a reasonable model for learning.

(a) Express the learning time t as a function of the performance level P.

(b) For a particular pole-vaulter in training, the learning curve is given by

$$P(t) = 20 - 14e^{-0.024t}$$

where $P(t)$ is the height the pole-vaulter is able to vault after t months. After how many months of training is the pole-vaulter able to vault 3.5 m?

(c) Draw a graph of the learning curve in part (b).

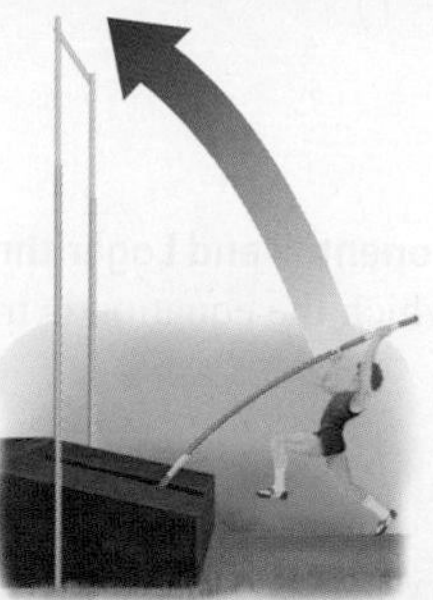

■ Discuss ■ Discover ■ Prove ■ Write

101. Discuss: Estimating a Solution Without actually solving the equation, find two whole numbers between which the solution of $9^x = 20$ must lie. Do the same for $9^x = 100$. Explain how you reached your conclusions.

102. Discuss ■ Discover: A Surprising Equation Take logarithms to show that the equation

$$x^{1/\log x} = 5$$

has no solution. For what values of k does the equation

$$x^{1/\log x} = k$$

have a solution? What does this tell us about the graph of the function $f(x) = x^{1/\log x}$? Confirm your answer using a graphing device.

103. Discuss: Disguised Equations Each of these equations can be transformed into an equation of linear or quadratic type by applying the hint. Solve each equation.

(a) $(x-1)^{\log(x-1)} = 100(x-1)$
[*Hint:* Take log of each side.]

(b) $\log_2 x + \log_4 x + \log_8 x = 11$
[*Hint:* Change all logs to base 2.]

(c) $4^x - 2^{x+1} = 3$
[*Hint:* Write as a quadratic in 2^x.]

4.6 Modeling with Exponential Functions

■ Exponential Growth (Doubling Time) ■ Exponential Growth (Relative Growth Rate)
■ Logistic Growth ■ Radioactive Decay ■ Newton's Law of Cooling

Many processes that occur in nature—such as population growth, radioactive decay, heat diffusion, and numerous others—can be modeled by using exponential functions. In this section we study exponential models.

■ Exponential Growth (Doubling Time)

Suppose we start with a single bacterium, which divides every hour. After one hour we have 2 bacteria, after two hours we have 2^2 or 4 bacteria, after three hours we have 2^3 or 8 bacteria, and so on (see Figure 1). We see that we can model the bacteria population after t hours by $f(t) = 2^t$.

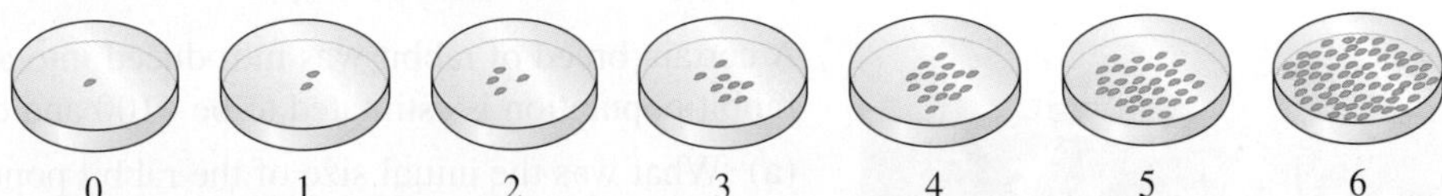

Figure 1 | Bacteria population

If we start with 10 of these bacteria, then the population is modeled by $f(t) = 10 \cdot 2^t$. A slower-growing strain of bacteria doubles every 3 hours; in this case the population is modeled by $f(t) = 10 \cdot 2^{t/3}$. In general, we have the following.

> **Exponential Growth (Doubling Time)**
>
> If the initial size of a population is n_0 and the doubling time is a, then the size of the population at time t is
>
> $$n(t) = n_0 2^{t/a}$$
>
> where a and t are measured in the same time units (minutes, hours, days, years, and so on).

Example 1 ■ Bacteria Population

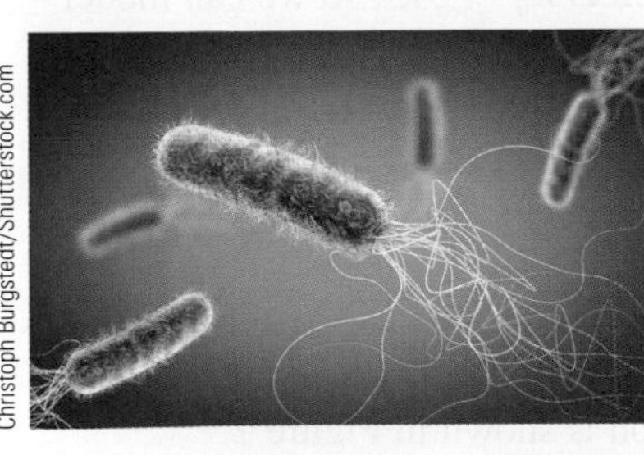

Christoph Burgstedt/Shutterstock.com

Under ideal conditions a certain bacteria population doubles every four hours. Initially, there are 1000 bacteria in a colony.

(a) Find a model for the bacteria population after t hours.

(b) How many bacteria are in the colony after 24 hours?

(c) After how many hours will the bacteria count reach one million?

Solution

(a) The population at time t is modeled by

$$n(t) = 1000 \cdot 2^{t/4} = 1000 \cdot 2^{0.25t}$$

where t is measured in hours.

(b) After 24 hours the number of bacteria is

$$n(24) = 1000 \cdot 2^{0.25(24)} = 64{,}000$$

(c) We set $n(t) = 1{,}000{,}000$ in the model that we found in part (a) and solve the resulting exponential equation for t.

$$1{,}000{,}000 = 1000 \cdot 2^{0.25t} \qquad n(t) = 1000 \cdot 2^{0.25t}$$

$$1000 = 2^{0.25t} \qquad \text{Divide by 1000}$$

$$\log 1000 = \log 2^{0.25t} \qquad \text{Take log of each side}$$

$$3 = (0.25t)\log 2 \qquad \text{Properties of log}$$

$$t = \frac{3}{0.25\log 2} \approx 39.86 \qquad \text{Solve for } t$$

The bacteria level reaches one million in about 40 hours.

 **Now Try Exercise 1** ■

Example 2 ■ Rabbit Population

A certain breed of rabbit was introduced into a grassland habitat. After 8 months the rabbit population is estimated to be 4100 and doubling every 3 months.

(a) What was the initial size of the rabbit population?

(b) Estimate the population 1 year after the rabbits were introduced to the grassland habitat.

(c) Sketch a graph of the rabbit population.

Solution

(a) The doubling time is $a = 3$, so the population after t months is

$$n(t) = n_0 2^{t/3} \qquad \text{Model}$$

where n_0 is the initial population. Since the population is 4100 after 8 months, we have

$$n(8) = n_0 2^{8/3} \qquad \text{From model}$$

$$4100 = n_0 2^{8/3} \qquad \text{Because } n(8) = 4100$$

$$n_0 = \frac{4100}{2^{8/3}} \qquad \text{Divide by } 2^{8/3} \text{ and switch sides}$$

$$n_0 \approx 645 \qquad \text{Calculator}$$

Thus we estimate that 645 rabbits were introduced to the grassland habitat.

(b) From part (a) we know that the initial population is $n_0 = 645$, so we can model the population after t months by

$$n(t) = 645 \cdot 2^{t/3} \qquad \text{Model}$$

After 1 year $t = 12$, so

$$n(12) = 645 \cdot 2^{12/3} = 10{,}320$$

Thus after 1 year there would be about 10,000 rabbits.

20,000

0 20

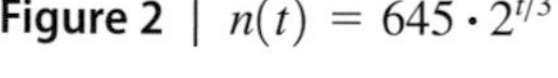

Figure 2 | $n(t) = 645 \cdot 2^{t/3}$

(c) We first note that the domain is $t \geq 0$. The graph is shown in Figure 2.

 **Now Try Exercise 3** ■

■ Exponential Growth (Relative Growth Rate)

We have used an exponential function with base 2 to model population growth (in terms of the doubling time). We could also model the same population with an exponential function with base 3 (in terms of the tripling time). In fact, we can find an exponential

The growth of a population with relative growth rate r is analogous to the growth of an investment with continuously compounded interest rate r.

model with any base. If we use the base e, we get a population model in terms of the **relative growth rate** r: the rate of population growth expressed as a proportion of the population at any time. In this case r is the "instantaneous" growth rate. (In calculus the concept of instantaneous rate is given a precise meaning.) For instance, if $r = 0.02$, then at any time t the growth rate is 2% of the population at time t.

Exponential Growth (Relative Growth Rate)

A population that experiences **exponential growth** increases according to the model

$$n(t) = n_0 e^{rt}$$

where $n(t)$ = population at time t

n_0 = initial size of the population

r = relative rate of growth (expressed as a proportion of the population)

Note The formula for population growth is the same as that for continuously compounded interest. In fact, the same principle is at work in both cases: The growth of a population (or an investment) per time period is proportional to the size of the population (or the amount of the investment). A population of 1,000,000 will increase more in one year than a population of 1000; in exactly the same way, an investment of \$1,000,000 will increase more in one year than an investment of \$1000.

In the following examples we assume that the populations grow exponentially.

Example 3 ■ Predicting the Size of a Population

The initial bacterium count in a culture is 500. A biologist later makes a sample count of bacteria in the culture and finds that the relative rate of growth is 40% per hour.

(a) Find a function that models the number of bacteria after t hours.

(b) What is the estimated count after 10 hours?

(c) After how many hours will the bacteria count reach 80,000?

(d) Sketch a graph of the function $n(t)$.

Solution

(a) We use the exponential growth model with $n_0 = 500$ and $r = 0.4$ to get

$$n(t) = 500e^{0.4t}$$

where t is measured in hours.

(b) Using the function in part (a), we find that the bacterium count after 10 hours is

$$n(10) = 500e^{0.4(10)} = 500e^4 \approx 27{,}300$$

(c) We set $n(t) = 80{,}000$ and solve the resulting exponential equation for t.

$$80{,}000 = 500 \cdot e^{0.4t} \qquad n(t) = 500 \cdot e^{0.4t}$$

$$160 = e^{0.4t} \qquad \text{Divide by 500}$$

$$\ln 160 = 0.4t \qquad \text{Take ln of each side}$$

$$t = \frac{\ln 160}{0.4} \approx 12.68 \qquad \text{Solve for } t$$

The bacteria level reaches 80,000 in about 12.7 hours.

(d) The graph is shown in Figure 3.

Figure 3 | $y = 500e^{0.4t}$

Now Try Exercise 5 ■

The relative growth rate of world population has been declining over the past few decades—from 1.32% in 2000 to 1.04% in 2020.

Standing Room Only

The population of the world was about 7.8 billion in 2020 and was increasing at 1.04% per year. Assuming that each person occupies an average of 0.37 m^2 of the surface of the earth, the exponential model for population growth projects that by the year 3074 there will be standing room only! (The total land surface area of the world is about 167×10^{12} m^2.)

Example 4 ■ Comparing Different Rates of Population Growth

In 2020 the population of the world was 7.8 billion, and the relative rate of growth was 1.04% per year. It is claimed that a rate of 0.50% per year would make a significant difference in the total population in just a few decades. Test this claim by estimating the population of the world in the year 2050 using a relative rate of growth of (a) 1.04% per year and (b) 0.50% per year.

Graph the population functions for the next 100 years for the two relative growth rates in the same viewing rectangle.

Solution

(a) By the exponential growth model we have

$$n(t) = 7.8e^{0.0104t}$$

where $n(t)$ is measured in billions and t is measured in years since 2020. Because the year 2050 is 30 years after 2020, we find

$$n(30) = 7.8e^{0.0104(30)} = 7.8e^{0.312} \approx 10.7$$

The estimated population in the year 2050 is about 10.7 billion.

(b) We use the function

$$n(t) = 7.8e^{0.005t}$$

and find

$$n(30) = 7.8e^{0.005(30)} = 7.8e^{0.15} \approx 9.1$$

The estimated population in the year 2050 is about 9.1 billion.

The graphs in Figure 4 show that a small change in the relative rate of growth will, over time, make a large difference in population size.

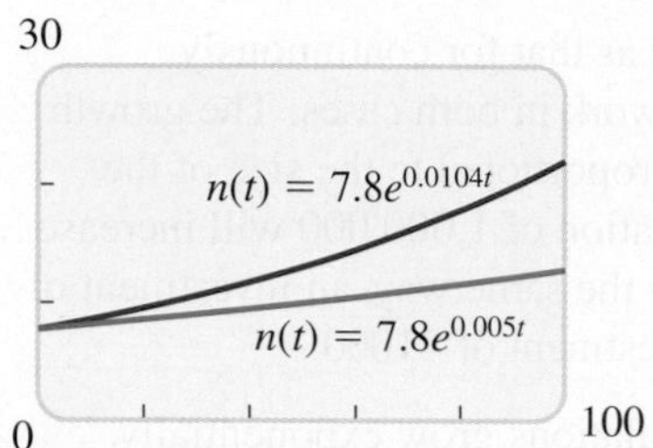

Figure 4

 **Now Try Exercise 7** ■

Example 5 ■ Expressing a Model in Terms of *e*

A population of a certain species of fish starts with 3000 fish and the number doubles every 3.75 years. Find a function of the given form that models the number of fish after t years, and graph the function.

(a) $n(t) = n_0 2^{t/a}$ **(b)** $n(t) = n_0 e^{rt}$

Solution

(a) The initial population is $n_0 = 3000$ and the doubling time is $a = 3.75$ years. The model is

$$n(t) = 3000 \cdot 2^{t/3.75}$$

The model is graphed in Figure 5.

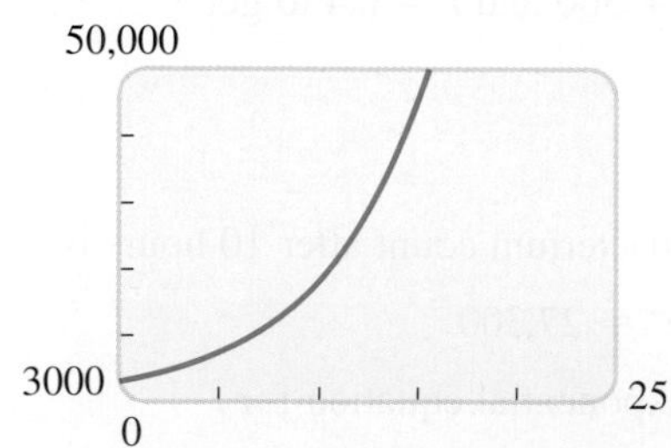

Figure 5 | Graph of $y = 3000 \cdot 2^{t/3.75}$

(b) Since the initial population is $n_0 = 3000$, the model we want has the form $n(t) = 3000e^{rt}$. To find the relative growth rate r we equate the two models and solve for r:

$3000e^{rt} = 3000 \cdot 2^{t/3.75}$	Equate models
$e^{rt} = 2^{t/3.75}$	Divide by 3000
$rt = \ln 2^{t/3.75}$	Take ln of each side
$rt = \dfrac{t \ln 2}{3.75}$	Property of ln
$r = \dfrac{\ln 2}{3.75}$	Divide by t (for $t \neq 0$)
$r \approx 0.1848$	Calculator

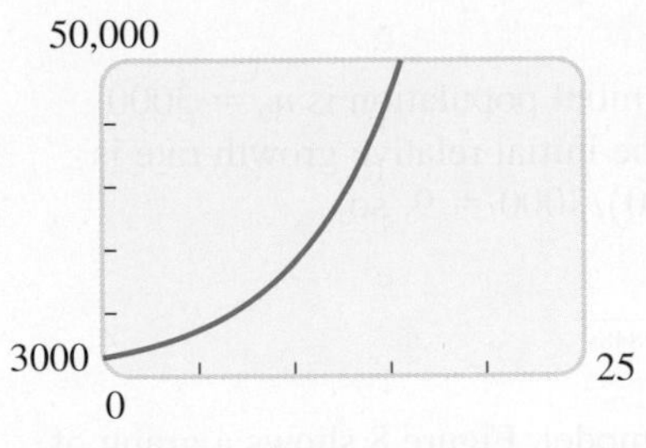

Figure 6 | Graph of $y = 3000 \cdot e^{0.1848t}$

Now that we know the relative growth rate r we can find the model:

$$n(t) = 3000 \cdot e^{0.1848t}$$

The model is graphed in Figure 6.

The graphs of the functions in Figures 5 and 6 are the same because the two functions are just two different ways of expressing the same model. ■

Now Try Exercise 9

Note In general, a model of the form $y = C \cdot b^t$ can be expressed in the form

$$y = C \cdot e^{kt}$$

To find the appropriate value of k, we note that for the models to be the same, we must have $C \cdot b^t = C \cdot e^{kt}$. So

$$b^t = e^{kt} \Leftrightarrow b^t = (e^k)^t \Leftrightarrow b = e^k \Leftrightarrow k = \ln b$$

It follows that the model $y = C \cdot b^t$ is the same as the model

$$y = C \cdot e^{(\ln b)t}$$

■ Logistic Growth

Proportionality is studied in Section 1.12.

The population models that we've studied so far are appropriate only under ideal conditions that allow for unlimited growth. In the real world, a given environment has limited resources (food, water, living space, and so on) and can support a maximum population M, called the *carrying capacity*. Populations tend to first increase exponentially and then level off as they approach M. Under such conditions the rate of growth of the population $P(t)$ at any time t is jointly proportional to $P(t)$ and to $M - P(t)$. It can be shown using calculus (see Stewart *Calculus* 9e, Sections 9.1 and 9.4) that under such conditions, the population is modeled by a *logistic function*, as follows.

Logistic Growth

A population that experiences **logistic growth** increases according to the model

$$n(t) = \frac{M}{1 + Ae^{-rt}}$$

where $n(t) =$ population at time t
$M =$ carrying capacity
$A = (M - n_0)/n_0$, where n_0 is the initial population
$r =$ initial relative growth rate

Example 6 ■ Logistic Population Growth

Suppose that the fish population in Example 5 exists in a small lake that can support a maximum of 30,000 fish. An initial population of 3000 fish is introduced into the lake.

(a) Model the population with a logistic growth model.

(b) Draw a graph of the logistic growth model and compare with the exponential growth model in Example 5.

Solution

(a) The carrying capacity is $M = 30{,}000$ and the initial population is $n_0 = 3000$. From Example 5 we know that for these fish the initial relative growth rate is $r = 0.1848$. We calculate $A = (30{,}000 - 3000)/3000 = 9$, so

$$n(t) = \frac{30{,}000}{1 + 9e^{-0.1848t}}$$

(b) Figure 7 shows a graph of the logistic growth model. Figure 8 shows a graph of both the logistic and exponential growth models for comparison. We see that with the logistic growth model the fish population initially grows exponentially (when plenty of resources are available) and then the rate of growth slows down as the population approaches the carrying capacity of 30,000. With the exponential growth model, the fish population continues to grow exponentially without bound.

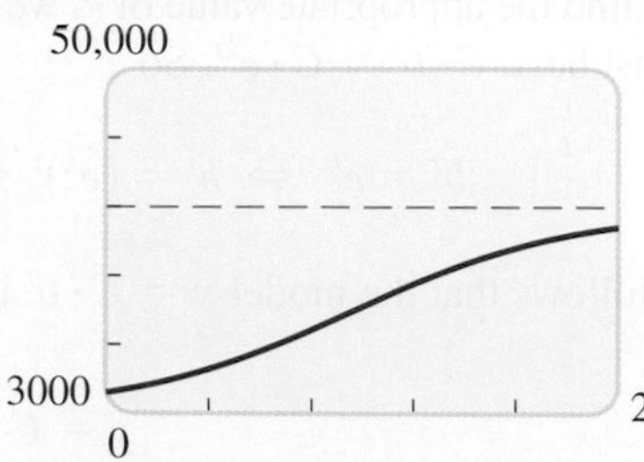

Figure 7 | $n(t) = \dfrac{30{,}000}{1 + 9e^{-0.1848t}}$

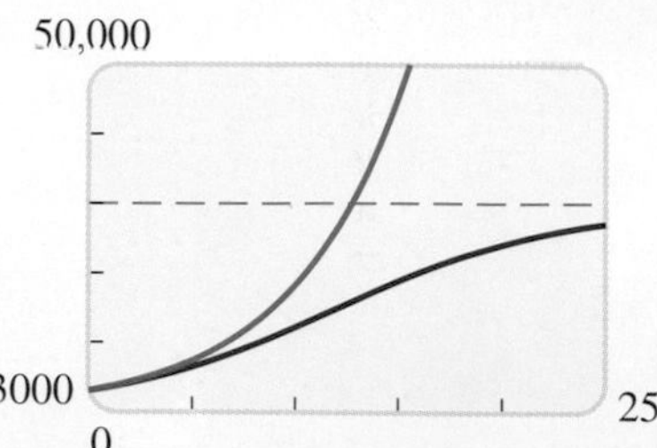

Figure 8 | Comparison of logistic and exponential models

Now Try Exercise 17 ■

■ Radioactive Decay

Radioactive substances decay by spontaneously emitting radiation. The rate of decay is proportional to the mass of the substance. This is analogous to population growth except that the mass *decreases*. Physicists express the rate of decay in terms of **half-life**, the time it takes for a sample of the substance to decay to half its original mass. For example, the half-life of radium-226 is 1600 years, so a 100-gram sample decays to 50 g $\left(\text{or } \frac{1}{2} \times 100 \text{ g}\right)$ in 1600 years, then to 25 g $\left(\text{or } \frac{1}{2} \times \frac{1}{2} \times 100 \text{ g}\right)$ in 3200 years, and so on. In general, for a radioactive substance with mass m_0 and half-life h, the amount remaining at time t is modeled by

$$m(t) = m_0 2^{-t/h}$$

where h and t are measured in the same time units (minutes, hours, days, years, and so on).

Discovery Project ■ **Modeling Radiation with Coins and Dice**

Radioactive elements decay when their atoms spontaneously emit radiation and change into smaller, stable atoms. But if atoms decay randomly, how is it possible to find a function that models their behavior? We'll try to answer this question by experiments with randomly tossed coins and randomly rolled dice. The experiments allow us to experience how a very large number of random events can result in predictable exponential results. You can find the project at **www.stewartmath.com**.

The half-lives of **radioactive elements** vary from very long to very short. Here are some examples.

Element	Half-life
thorium-232	14 billion years
uranium-238	4.5 billion years
uranium-235	704 million years
thorium-230	75,380 years
plutonium-239	24,100 years
carbon-14	5,730 years
radium-226	1,600 years
americium-241	432 years
cesium-137	30 years
strontium-90	28 years
polonium-210	138 days
thorium-234	24 days
iodine-131	8 days
radon-222	3.8 days
lead-211	36 minutes
krypton-91	10 seconds

It is customary to express the exponential model for radioactive decay in the form $m(t) = m_0e^{-rt}$. To do this, we need to find the relative decay rate r. Since h is the half-life, we have

$$m(t) = m_0e^{-rt} \qquad \text{Model}$$

$$\frac{m_0}{2} = m_0e^{-rh} \qquad h \text{ is the half-life}$$

$$\frac{1}{2} = e^{-rh} \qquad \text{Divide by } m_0$$

$$\ln\frac{1}{2} = -rh \qquad \text{Take ln of each side}$$

$$r = \frac{\ln 2}{h} \qquad \text{Solve for } r \ (\ln\left(\tfrac{1}{2}\right) = -\ln 2)$$

This last equation allows us to find the relative decay rate r from the half-life h.

Radioactive Decay Model

If m_0 is the initial mass of a radioactive substance with half-life h, then the mass remaining at time t is modeled by the function

$$m(t) = m_0e^{-rt}$$

where $r = \dfrac{\ln 2}{h}$ is the **relative decay rate**.

Example 7 ■ Radioactive Decay

Polonium-210 (^{210}Po) has a half-life of 138 days. Suppose a sample of this substance has a mass of 300 mg.

(a) Find a function $m(t) = m_02^{-t/h}$ that models the mass remaining after t days.

(b) Find a function $m(t) = m_0e^{-rt}$ that models the mass remaining after t days.

(c) Find the mass remaining after one year.

(d) How long will it take for the sample to decay to a mass of 200 mg?

(e) Draw a graph of the sample mass as a function of time.

Solution

(a) We have $m_0 = 300$ and $h = 138$, so the amount remaining after t days is

$$m(t) = 300 \cdot 2^{-t/138}$$

(b) We have $m_0 = 300$ and $r = \ln 2/138 \approx 0.00502$, so the amount remaining after t days is

$$m(t) = 300 \cdot e^{-0.00502t}$$

In parts (c) and (d) we can also use the model found in part (a). Check that the result is the same using either model.

(c) We use the function we found in part (a) with $t = 365$ (1 year):

$$m(365) = 300e^{-0.00502(365)} \approx 48.014$$

Thus approximately 48 mg of ^{210}Po remains after 1 year.

(d) We use the function that we found in part (b) with $m(t) = 200$ and solve the resulting exponential equation for t:

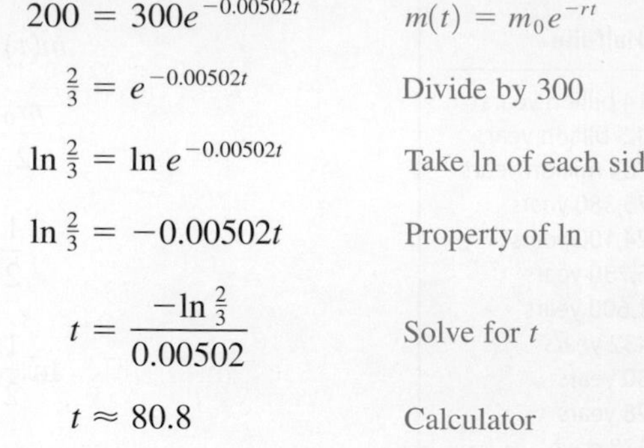

$$
\begin{aligned}
200 &= 300e^{-0.00502t} && m(t) = m_0 e^{-rt} \\
\tfrac{2}{3} &= e^{-0.00502t} && \text{Divide by 300} \\
\ln \tfrac{2}{3} &= \ln e^{-0.00502t} && \text{Take ln of each side} \\
\ln \tfrac{2}{3} &= -0.00502t && \text{Property of ln} \\
t &= \frac{-\ln \frac{2}{3}}{0.00502} && \text{Solve for } t \\
t &\approx 80.8 && \text{Calculator}
\end{aligned}
$$

The time required for the sample to decay to 200 mg is about 81 days.

(e) We can graph the model in part (a) or the one in part (b). The graphs are identical. See Figure 9.

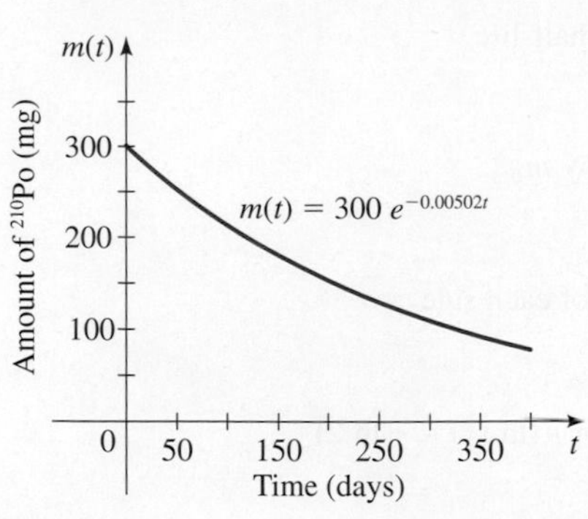

Figure 9

Now Try Exercise 21 ■

■ Newton's Law of Cooling

Newton's Law of Cooling states that the rate at which an object cools is proportional to the temperature difference between the object and its surroundings, provided that the temperature difference is not too large. By using calculus, the following model can be deduced from this law.

> **Newton's Law of Cooling**
>
> If D_0 is the initial temperature difference between an object and its surroundings, and if its surroundings have temperature T_s, then the temperature of the object at time t is modeled by the function
>
> $$T(t) = T_s + D_0 e^{-kt}$$
>
> where k is a positive constant that depends on the type of object.

Example 8 ■ Newton's Law of Cooling

A cup of coffee has a temperature of 93°C and is placed in a room that has a temperature of 21°C. After 10 minutes the temperature of the coffee is 65°C.

(a) Find a function that models the temperature of the coffee after t minutes.

(b) Find the temperature of the coffee after 15 minutes.

(c) After how long will the coffee have cooled to 38°C?

(d) Illustrate by drawing a graph of the temperature function.

Solution

(a) The temperature of the room is $T_s = 21°\text{C}$, and the initial temperature difference is

$$D_0 = 93 - 21 = 72°\text{C}$$

So by Newton's Law of Cooling, the temperature after t minutes is modeled by the function

$$T(t) = 21 + 72e^{-kt}$$

We need to find the constant k associated with this cup of coffee. To do this, we use the fact that when $t = 10$, the temperature is $T(10) = 65$. So

$21 + 72e^{-10k} = 65$	$T_s + D_0e^{-kt} = T(t)$
$72e^{-10k} = 44$	Subtract 21
$e^{-10k} = \frac{44}{72}$	Divide by 72
$-10k = \ln \frac{11}{18}$	Take ln of each side
$k = -\frac{1}{10} \ln \frac{11}{18}$	Solve for k
$k \approx 0.04923$	Calculator

Substituting this value of k into the expression for $T(t)$, we get

$$T(t) = 21 + 72e^{-0.04923t}$$

(b) We use the function that we found in part (a) with $t = 15$:

$$T(15) = 21 + 72e^{-0.04923(15)} \approx 55°\text{C}$$

(c) We use the function that we found in part (a) with $T(t) = 38$ and solve the resulting exponential equation for t.

$21 + 72e^{-0.04923t} = 38$	$T_s + D_0e^{-kt} = T(t)$
$72e^{-0.04923t} = 17$	Subtract 21
$e^{-0.04923t} = \frac{17}{72}$	Divide by 72
$-0.04923t = \ln \frac{17}{72}$	Take ln of each side
$t = \dfrac{\ln \frac{17}{72}}{-0.04855}$	Solve for t
$t \approx 29.3$	Calculator

The coffee will have cooled to 38°C after about half an hour.

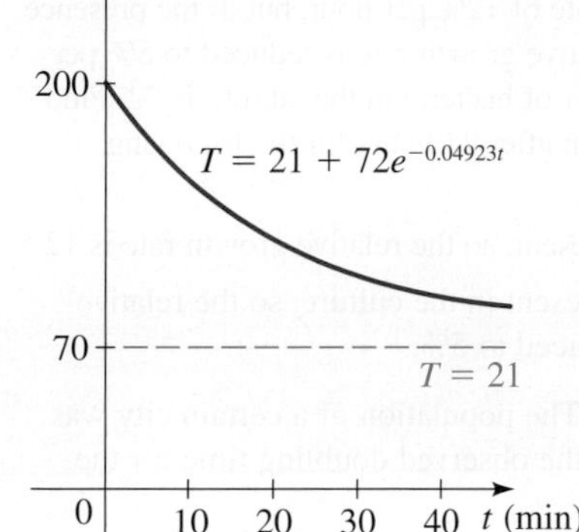

Figure 10 | Temperature of coffee after t minutes

(d) The graph of the temperature function is sketched in Figure 10. Notice that the line $t = 21$ is a horizontal asymptote, so as we would expect, the temperature of the coffee decreases to the temperature of the surroundings.

 **Now Try Exercise 33** ■

4.6 Exercises

Applications

1–16 ■ Exponential Growth These exercises use the exponential growth model.

 **1. Bacteria Culture** A certain culture of the bacterium *Streptococcus A* initially has 10 bacteria and is observed to double every 1.5 hours.

(a) Find an exponential model $n(t) = n_0 2^{t/a}$ for the number of bacteria in the culture after t hours.

(b) Estimate the number of bacteria after 35 hours.

(c) After how many hours will the bacteria count reach 10,000?

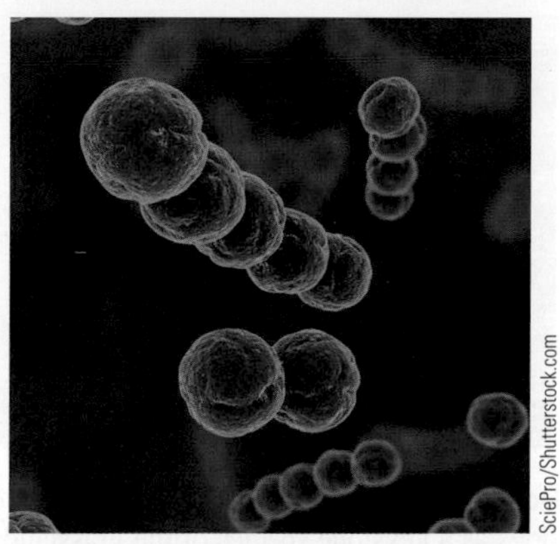

Streptococcus A
(12,000 × magnification)

2. Bacteria Culture A certain culture of the bacterium *Rhodobacter sphaeroides* initially has 25 bacteria and is observed to double every 5 hours.

(a) Find an exponential model

$$n(t) = n_0 2^{t/a}$$

for the number of bacteria in the culture after t hours.

(b) Estimate the number of bacteria after 18 hours.

(c) After how many hours will the bacteria count reach 1 million?

3. Squirrel Population A grey squirrel population was introduced in a certain county of Great Britain 30 years ago. Biologists observe that the population doubles every 6 years, and now the population is 100,000.

(a) What was the initial size of the squirrel population?

(b) Estimate the squirrel population 10 years from now.

(c) Sketch a graph of the squirrel population.

4. Bird Population A species of bird was introduced in a certain county 25 years ago. Biologists observe that the population doubles every 10 years, and now the population is 13,000.

(a) What was the initial size of the bird population?

(b) Estimate the bird population 5 years from now.

(c) Sketch a graph of the bird population.

5. Beaver Population Beavers are sometimes seen as pests, but lately scientists have discovered the importance of this dam-building species to maintaining the viability of freshwater ecosystems. For instance, beaver dams create ponds and wetlands, help store water for farms and ranches, and help filter out water pollution. It is estimated that for a certain northeastern ecosystem a beaver population has a relative growth rate of 12% per year and the population in 2005 was 12,800.

(a) Find a function

$$n(t) = n_0 e^{rt}$$

that models the population t years after 2005.

(b) Use the model from part (a) to estimate the beaver population in 2010.

(c) After how many years will the population reach 50,000?

(d) Sketch a graph of the beaver population function for the years 2005 to 2020.

P Harstela/Shutterstock.com

6. Prairie Dog Population Decline Ecologists have identified prairie dogs as a keystone species of the grasslands ecosystem of the West. Researchers have observed that the population of a certain species of prairie dog is declining at a relative rate of 6% per year in a certain county of South Dakota. (This means the relative growth rate is -0.06.) It is estimated that the population in 2010 was 450,000.

(a) Find a function $n(t) = n_0 e^{rt}$ that models the population t years after 2010.

(b) Use the model from part (a) to estimate the prairie dog population in 2025.

(c) After how many years will the population decrease to 300,000?

(d) Sketch a graph of the prairie dog population function for the years 2010 to 2025.

7. Population of a Country The population of a country has a relative growth rate of 3% per year. The government is trying to reduce the growth rate to 2%. The population in 2011 was approximately 110 million. Find the projected population for the year 2036 for the following conditions.

(a) The relative growth rate remains at 3% per year.

(b) The relative growth rate is reduced to 2% per year.

8. Bacteria Culture It is observed that a certain bacteria culture has a relative growth rate of 12% per hour, but in the presence of an antibiotic the relative growth rate is reduced to 5% per hour. The initial number of bacteria in the culture is 22. Find the projected population after 24 hours for the following conditions.

(a) No antibiotic is present, so the relative growth rate is 12%.

(b) An antibiotic is present in the culture, so the relative growth rate is reduced to 5%.

9. Population of a City The population of a certain city was 112,000 in 2014, and the observed doubling time for the population is 18 years.

(a) Find an exponential model $n(t) = n_0 2^{t/a}$ for the population t years after 2014.

(b) Find an exponential model $n(t) = n_0 e^{rt}$ for the population t years after 2014.

(c) Sketch a graph of the population at time t.

(d) Estimate how long it takes the population to reach 500,000.

10. Bat Population The bat population in a certain Midwestern county was 350,000 in 2012, and the observed doubling time for the population is 25 years.

(a) Find an exponential model $n(t) = n_0 2^{t/a}$ for the population t years after 2012.

(b) Find an exponential model $n(t) = n_0 e^{rt}$ for the population t years after 2012.

(c) Sketch a graph of the population at time t.

(d) Estimate how long it takes the population to reach 2 million.

11. Deer Population The graph shows the deer population in a Pennsylvania county between 2010 and 2014. Assume that the population grows exponentially.

(a) What was the deer population in 2010?

(b) Find a function that models the deer population t years after 2010.

(c) What is the projected deer population in 2018?

(d) Estimate how long it takes the population to reach 100,000.

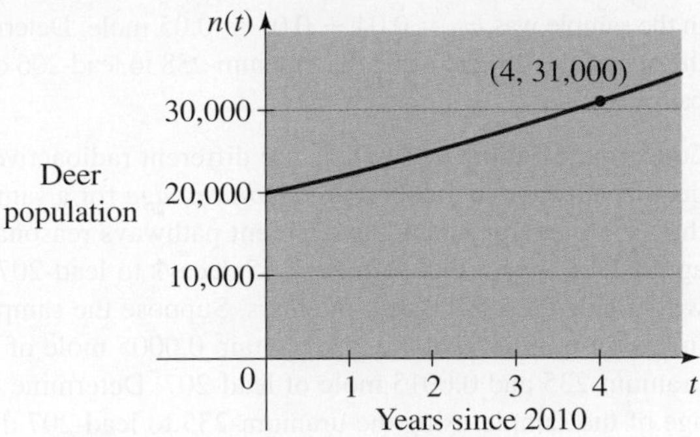

12. Frog Population Some bullfrogs were introduced into a small pond. The graph shows the bullfrog population for the next few years. Assume that the population grows exponentially.

(a) What was the initial bullfrog population?

(b) Find a function that models the bullfrog population t years since the bullfrogs were put into the pond.

(c) What is the projected bullfrog population after 15 years?

(d) Estimate how long it takes the population to reach 75,000.

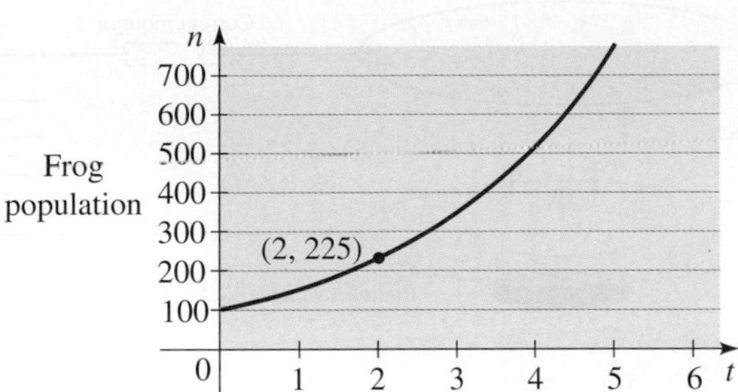

13. Bacteria Culture A culture starts with 8600 bacteria. After 1 hour the count is 10,000.

(a) Find a function that models the number of bacteria $n(t)$ after t hours.

(b) Find the number of bacteria after 2 hours.

(c) After how many hours will the number of bacteria double?

14. Bacteria Culture The count in a culture of bacteria was 400 after 2 hours and 25,600 after 6 hours.

(a) What is the relative rate of growth of the bacteria population? Express your answer as a percentage.

(b) What was the initial size of the culture?

(c) Find a function that models the number of bacteria $n(t)$ after t hours.

(d) Find the number of bacteria after 4.5 hours.

(e) After how many hours will the number of bacteria reach 50,000?

15. Population Decline The population of a certain country was 49 million in 2000 and 44 million in 2019. Assume that the population continues to decline at this rate.

(a) Find a function that models the population (in millions) t years after 2000. Use the model to estimate in what year the population will decline to 35 million.

(b) In how many years will the population be cut in half?

16. World Population The population of the world was 7.8 billion in 2020, and the observed relative growth rate was 1.04% per year. Assume that the world population continues to grow at this rate. Use an exponential model to estimate each of the following.

(a) The year in which year the population will reach 10 billion.

(b) The time required for the population to double.

17–20 ■ Logistic Growth These exercises use the logistic growth model.

17. World Population The relative growth rate of world population has been decreasing steadily in recent years. Based on this, some population models predict that world population will eventually stabilize around 11 billion. In 1950 the world population was 2.5 billion and the observed relative growth rate was 1.89% per year.

(a) Assuming the carrying capacity of the earth is 11 billion, model the population (in billions) with a logistic growth model, where t is the number of years since 1950. Use the model to estimate the year in which the population will reach 10 billion.

(b) Draw a graph of the logistic growth model.

18. Beaver Population Suppose that the beaver population in Exercise 5 exists on an island that can support a maximum of 80,000 beavers. An initial population of 12,800 beavers is introduced onto the island in 2005.

(a) Model the population $n(t)$ with a logistic growth model, where t is the number of years since 2005.

(b) Draw a graph of the logistic growth model and compare with the exponential growth model in Exercise 5 for the years 2005 to 2045.

19. Spread of a Disease An infectious disease begins to spread in a small city of population 10,000. Initially, 8 people in the city are infected with the disease. Researchers have observed that the initial relative growth rate of the disease is 0.57 per day, and the number of people that get infected levels off when 60% of the population have been infected. (So the carrying capacity of the disease is 6000.)

(a) Model the number of infections $n(t)$ with a logistic growth model, where t is the number of days since the initial 8 people were infected.

(b) Draw a graph of the logistic growth model. After how many days does the number of infections reach 5900?

20. Overstocked Pond A pond in a fish farm has a carrying capacity of 10,000 fish of a certain species of fish that has relative growth rate $r = 0.2$ per year. The pond is inadvertently stocked with 18,000 fish.

(a) Find a logistic model $n(t)$ for the fish population t years since the pond was stocked.

(b) Draw a graph of the model you found and use it to describe how the population changes over many years. What does $n(t)$ approach as $t \to \infty$?

21–31 ■ Radioactive Decay These exercises use the radioactive decay model.

21. Radioactive Radium The half-life of radium-226 is 1600 years. Suppose we have a 22-milligram sample.

(a) Find a function $m(t) = m_0 2^{-t/h}$ that models the mass remaining after t years.

(b) Find a function $m(t) = m_0 e^{-rt}$ that models the mass remaining after t years.

(c) How much of the sample will remain after 4000 years?

(d) After how many years will only 18 mg of the sample remain?

22. Radioactive Cesium The half-life of cesium-137 is 30 years. Suppose we have a 10-gram sample.

(a) Find a function $m(t) = m_0 2^{-t/h}$ that models the mass remaining after t years.

(b) Find a function $m(t) = m_0 e^{-rt}$ that models the mass remaining after t years.

(c) How much of the sample will remain after 80 years?

(d) After how many years will only 2 g of the sample remain?

23. Radioactive Strontium The half-life of strontium-90 is 29 years. How long will it take a 50-milligram sample to decay to a mass of 32 mg?

24. Radioactive Radium Radium-221 has a half-life of 30 s. How long will it take for 95% of a sample to decay?

25. Finding Half-Life If 250 mg of a radioactive element decays to 200 mg in 48 hours, find the half-life of the element.

26. Radioactive Radon After 3 days a sample of radon-222 has decayed to 58% of its original amount.

(a) What is the half-life of radon-222?

(b) How long will it take the sample to decay to 20% of its original amount?

27. Carbon-14 Dating A wooden artifact from an ancient tomb contains 65% of the carbon-14 that is present in living trees. How long ago was the artifact made? (The half-life of carbon-14 is 5730 years.)

28. Carbon-14 Dating The burial cloth of an Egyptian mummy is estimated to contain 59% of the carbon-14 it contained originally. How long ago was the mummy buried? (The half-life of carbon-14 is 5730 years.)

29. Uranium-Lead Dating Radioactive uranium-238 atoms decay to lead-206 atoms with a half-life of 4.5 billion years. A sample of zircon crystal contains 0.04 moles of uranium-238 and 0.01 moles of lead-206. Assuming that the lead-206 atoms were the product of the decay of uranium-238 atoms, it follows that the initial number of uranium-238 atoms in the sample was $m_0 = 0.04 + 0.01 = 0.05$ mole. Determine the age of the sample using the uranium-238 to lead-206 decay pathway.

30. Concordant Dating Scientists use different radioactive decay pathways to arrive at a *concordant age* for a sample, that is, an age for which the different pathways reasonably agree. It is known that uranium-235 decays to lead-207 with a half-life of 700 million years. Suppose the sample in Exercise 29 is found to also contain 0.0005 mole of uranium-235 and 0.0015 mole of lead-207. Determine the age of the sample using the uranium-235 to lead-207 decay pathway. Does your answer reasonably agree with the answer to Exercise 29? Would you say that the calculated ages are concordant?

31. Smoke Detectors Smoke alarms use a small amount of radioactive americium-241, which ionizes the air between two electrically charged plates, causing current to flow between the plates. When smoke enters the chamber, it disrupts the current, thus activating the alarm. Suppose that a smoke detector contains 0.29 microgram of americium-241. How long does it take for the radioactive material to decrease to 80% of its initial mass? (The half-life of americium-241 is 432 years.)

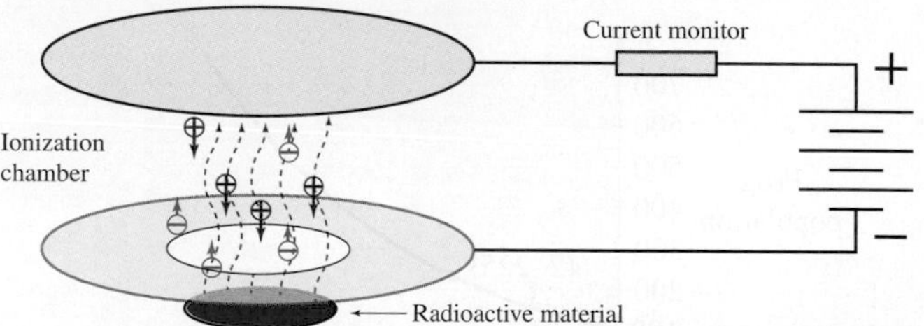

Note: The National Fire Protection Association (NFPA) recommends that smoke detectors be replaced after 10 years and that batteries be replaced every six months. The americium-241 will outlast every other component of the detector.

32–35 ■ Law of Cooling These exercises use Newton's Law of Cooling.

32. Time of Death Newton's Law of Cooling is used in homicide investigations to determine the time of death. The normal body temperature is 37°C. Immediately following death, the body begins to cool. It has been determined experimentally that the constant in Newton's Law of Cooling is approximately $k = 0.1947$, assuming that time is measured in hours. Suppose that the temperature of the surroundings is 15.5°C.

(a) Find a function $T(t)$ that models the temperature t hours after death.

(b) If the temperature of the body is now 22.2°C, how long ago was the time of death?

33. Cooling Soup A hot bowl of soup is served at a dinner party. It starts to cool according to Newton's Law of Cooling, so its temperature at time t is given by

$$T(t) = 18 + 63e^{-0.05t}$$

where t is measured in minutes and T is measured in °C.

(a) What is the initial temperature of the soup?

(b) What is the temperature after 10 min?

(c) After how long will the temperature be 38°C?

34. Cooling Turkey A roasted turkey is taken from an oven when its temperature has reached 85°C and is placed on a table in a room where the temperature is 24°C.

(a) If the temperature of the turkey is 65.5°C after half an hour, what is its temperature after 45 min?

(b) After how many hours has the turkey cooled to 38°C?

35. Boiling Water A kettle full of water is brought to a boil in a room with temperature 20°C. After 15 minutes the temperature of the water has decreased from 100°C to 75°C. Find the temperature after another 10 minutes. Illustrate by graphing the temperature function.

4.7 Logarithmic Scales

■ The pH Scale ■ The Richter Scale ■ The Decibel Scale

Animal	W (kg)	$\log W$
Ant	0.000003	−5.5
Elephant	4000	3.6
Whale	170,000	5.2

When a physical quantity varies over a very large range, it is often convenient to take its logarithm in order to work with more manageable numbers. On a **logarithmic scale**, numbers are represented by their logarithms. For example, the table in the margin gives the weights W of some animals (in kilograms) and their logarithms ($\log W$).

The weights (W) vary enormously, but on a logarithmic scale, the weights are represented by more manageable numbers ($\log W$), as illustrated in Figure 1.

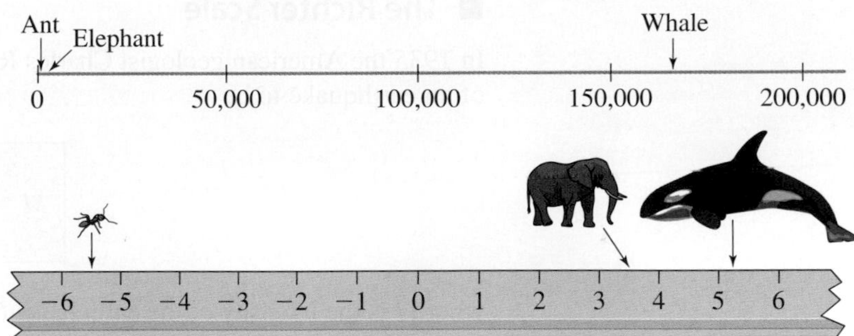

Figure 1 | Weight graphed on the real line (top) and on a logarithmic scale (bottom)

We discuss three commonly used logarithmic scales: the pH scale, which measures acidity; the Richter scale, which measures the intensity of earthquakes; and the decibel scale, which measures the loudness of sounds. Other quantities that are measured on logarithmic scales are light intensity, information capacity, and radiation.

■ The pH Scale

Chemists measured the acidity of a solution by giving its hydrogen ion concentration until Søren Peter Lauritz Sørensen, in 1909, proposed a more convenient measure. He defined

$$\text{pH} = -\log[\text{H}^+]$$

where $[\text{H}^+]$ is the concentration of hydrogen ions measured in moles per liter (M). He did this to avoid very small numbers and negative exponents. For instance,

$$\text{if} \quad [\text{H}^+] = 10^{-4}\text{ M}, \quad \text{then} \quad \text{pH} = -\log_{10}(10^{-4}) = -(-4) = 4$$

Solutions with a pH of 7 are defined as *neutral*, those with pH < 7 are *acidic*, and those with pH > 7 are *basic*. Notice that when the pH increases by one unit, $[\text{H}^+]$ decreases by a factor of 10.

pH for Some Common Substances

Substance	pH
Milk of magnesia	10.5
Seawater	8.0–8.4
Human blood	7.3–7.5
Crackers	7.0–8.5
Hominy	6.9–7.9
Cow's milk	6.4–6.8
Spinach	5.1–5.7
Tomatoes	4.1–4.4
Oranges	3.0–4.0
Apples	2.9–3.3
Limes	1.3–2.0
Battery acid	1.0

Example 1 ■ pH Scale and Hydrogen Ion Concentration

(a) The hydrogen ion concentration of a sample of human blood was measured to be $[\mathrm{H}^+] = 3.16 \times 10^{-8}$ M. Find the pH, and classify the blood as acidic or basic.

(b) The most acidic rainfall ever measured occurred in Scotland in 1974; its pH was 2.4. Find the hydrogen ion concentration.

Solution

(a) A calculator gives

$$\mathrm{pH} = -\log[\mathrm{H}^+] = -\log(3.16 \times 10^{-8}) \approx 7.5$$

Since this is greater than 7, the blood is basic.

(b) To find the hydrogen ion concentration, we need to solve for $[\mathrm{H}^+]$ in the logarithmic equation

$$\log[\mathrm{H}^+] = -\mathrm{pH}$$

so we write it in exponential form:

$$[\mathrm{H}^+] = 10^{-\mathrm{pH}}$$

In this case pH = 2.4, so

$$[\mathrm{H}^+] = 10^{-2.4} \approx 4.0 \times 10^{-3}\ \mathrm{M}$$

 **Now Try Exercises 1 and 3** ■

■ The Richter Scale

In 1935 the American geologist Charles Richter (1900–1985) defined the magnitude M of an earthquake to be

$$M = \log \frac{I}{S}$$

where I is the intensity of the earthquake (measured by the amplitude of a seismograph reading taken 100 km from the epicenter of the earthquake) and S is the intensity of a "standard" earthquake (whose amplitude is 1 micron $= 10^{-4}$ cm). (In practice, seismograph stations may not be exactly 100 km from the epicenter, so appropriate adjustments are made in calculating the magnitude of an earthquake.) The magnitude of a standard earthquake is

$$M = \log \frac{S}{S} = \log 1 = 0$$

Robert Vos/AFP/Getty Images

Discovery Project ■ The Even-Tempered Clavier

Poets, writers, philosophers, and even politicians have extolled the virtues of music—its beauty and its power to communicate emotion. But at the heart of music is a logarithmic scale. The tones that we are familiar with from our everyday listening can all be reproduced by the keys of a piano. The keys of a piano, in turn, are "evenly tempered" using a logarithmic scale. In this project we explore how exponential and logarithmic functions are used in properly tuning a piano. You can find the project at **www.stewartmath.com**.

Largest Earthquakes

Location	Date	Magnitude
Chile	1960	9.5
Alaska	1964	9.2
Japan	2011	9.1
Sumatra	2004	9.1
Kamchatka	1952	9.0
Chile	2010	8.8
Ecuador	1906	8.8
Alaska	1965	8.7
Alaska	1957	8.6
Sumatra	2005	8.6
Sumatra	2012	8.6
Indonesia	1938	8.5
Kamchatka	1923	8.5

Source: US Geological Society

Richter studied many earthquakes that occurred between 1900 and 1950. The largest had magnitude 8.9 on the Richter scale, and the smallest had magnitude 0. This corresponds to a ratio of intensities of 800,000,000, so the Richter scale provides more manageable numbers to work with. For instance, an earthquake of magnitude 6 is ten times stronger than an earthquake of magnitude 5.

Example 2 ■ Magnitude and Intensity

(a) Find the magnitude of an earthquake that has an intensity of 3.75 (that is, the amplitude of the seismograph reading is 3.75 cm).

(b) An earthquake was measured to have a magnitude of 5.1 on the Richter scale. Find the intensity of the earthquake.

Solution

(a) From the definition of magnitude we see that

$$M = \log \frac{I}{S} = \log \frac{3.75}{10^{-4}} = \log 37500 \approx 4.6$$

Thus the magnitude is 4.6 on the Richter scale.

(b) To find the intensity, we need to solve for I in the logarithmic equation

$$M = \log \frac{I}{S}$$

so we write it in exponential form:

$$10^M = \frac{I}{S}$$

In this case $S = 10^{-4}$ and $M = 5.1$, so

$$10^{5.1} = \frac{I}{10^{-4}} \qquad M = 5.1,\ S = 10^{-4}$$

$$(10^{-4})(10^{5.1}) = I \qquad \text{Multiply by } 10^{-4}$$

$$I = 10^{1.1} \approx 12.6 \qquad \text{Add exponents}$$

Thus the intensity of the earthquake is about 12.6, which means that the amplitude of the seismograph reading is about 12.6 cm.

 **Now Try Exercise 9** ■

Example 3 ■ Magnitude of Earthquakes

There are several other logarithmic scales used to calculate the magnitude of earthquakes. For instance, the US Geological Survey uses the *moment magnitude scale.*

The 1906 earthquake in San Francisco had an estimated magnitude of 8.3 on the Richter scale. In the same year a powerful earthquake occurred on the Colombia-Ecuador border that was four times as intense. What was the magnitude of the Colombia-Ecuador earthquake on the Richter scale?

Solution If I is the intensity of the San Francisco earthquake, then from the definition of magnitude we have

$$M = \log \frac{I}{S} = 8.3$$

The intensity of the Colombia-Ecuador earthquake was $4I$, so its magnitude was

$$M = \log \frac{4I}{S} = \log 4 + \log \frac{I}{S} = \log 4 + 8.3 \approx 8.9$$

 **Now Try Exercise 11** ■

Example 4 ■ Intensity of Earthquakes

@Painet Inc./Alamy Stock Photo

The 1989 Loma Prieta earthquake that shook San Francisco had a magnitude of 7.1 on the Richter scale. How many times more intense was the 1906 earthquake (see Example 3) than the 1989 event?

Solution If I_1 and I_2 are the intensities of the 1906 and 1989 earthquakes, then we are required to find I_1/I_2. To relate this to the definition of magnitude, we divide the numerator and denominator by S.

$$\log \frac{I_1}{I_2} = \log \frac{I_1/S}{I_2/S} \qquad \text{Divide numerator and denominator by } S$$

$$= \log \frac{I_1}{S} - \log \frac{I_2}{S} \qquad \text{Law 2 of logarithms}$$

$$= 8.3 - 7.1 = 1.2 \qquad \text{Definition of earthquake magnitude}$$

Therefore

$$\frac{I_1}{I_2} = 10^{\log(I_1/I_2)} = 10^{1.2} \approx 16$$

The 1906 earthquake was about 16 times as intense as the 1989 earthquake.

Now Try Exercise 13 ■

■ The Decibel Scale

The ear is sensitive to an extremely wide range of sound intensities I (measured in W/m^2). We take as a reference intensity $I_0 = 10^{-12}$ W/m^2 (watts per square meter) at a frequency of 1000 hertz, which measures a sound that is just barely audible (the threshold of hearing). The psychological sensation of loudness varies with the logarithm of the intensity (the Weber-Fechner Law), so the **decibel level** B, measured in decibels (dB), is defined as

$$B = 10 \log \frac{I}{I_0}$$

The decibel level of the barely audible reference sound is

$$B = 10 \log \frac{I_0}{I_0} = 10 \log 1 = 0 \text{ dB}$$

Example 5 ■ Decibel Level and Intensity

(a) Find the decibel level of a jet engine at takeoff if the intensity was measured at 100 W/m^2.

(b) Find the intensity level of a motorcycle engine at full throttle if the decibel level was measured at 90 dB.

Solution

(a) From the definition of decibel level we see that

$$B = 10 \log \frac{I}{I_0} = 10 \log \frac{10^2}{10^{-12}} = 10 \log 10^{14} = 140$$

Thus the decibel level is 140 dB.

The **decibel levels of sounds** that we can hear vary from very loud to very soft. Here are some examples of the decibel levels of commonly heard sounds.

Source of Sound	B (dB)
Jet takeoff	140
Jackhammer	130
Rock concert	120
Subway	100
Heavy traffic	80
Ordinary traffic	70
Normal conversation	50
Whisper	30
Rustling leaves	10–20
Threshold of hearing	0

(b) To find the intensity, we need to solve for I in the logarithmic equation

$$B = 10 \log \frac{I}{I_0} \qquad \text{Definition of decibel level}$$

$$\frac{B}{10} = \log I - \log 10^{-12} \qquad \text{Divide by 10, } I_0 = 10^{-12}$$

$$\frac{B}{10} = \log I + 12 \qquad \text{Definition of logarithm}$$

$$\frac{B}{10} - 12 = \log I \qquad \text{Subtract 12}$$

$$\log I = \frac{90}{10} - 12 = -3 \qquad B = 90$$

$$I = 10^{-3} \qquad \text{Exponential form}$$

Thus the intensity is 10^{-3} W/m^2.

 **Now Try Exercises 15 and 17** ■

The table in the margin lists decibel levels for some common sounds ranging from the threshold of human hearing to the jet takeoff of Example 5. The threshold of pain is about 120 dB.

4.7 Exercises

Applications

1. Finding pH The hydrogen ion concentration of a sample of each substance is given. Calculate the pH of the substance.

(a) Lemon juice: $[\text{H}^+] = 5.0 \times 10^{-3}$ M

(b) Tomato juice: $[\text{H}^+] = 3.2 \times 10^{-4}$ M

(c) Seawater: $[\text{H}^+] = 5.0 \times 10^{-9}$ M

2. Finding pH An unknown substance has a hydrogen ion concentration of $[\text{H}^+] = 3.1 \times 10^{-8}$ M. Find the pH and classify the substance as acidic or basic.

3. Ion Concentration The pH reading of a sample of each substance is given. Calculate the hydrogen ion concentration of the substance.

(a) Vinegar: pH = 3.0 **(b)** Milk: pH = 6.5

4. Ion Concentration The pH reading of a glass of liquid is given. Find the hydrogen ion concentration of the liquid.

(a) Beer: pH = 4.6 **(b)** Water: pH = 7.3

5. Finding pH The hydrogen ion concentrations in cheeses range from 4.0×10^{-7} M to 1.6×10^{-5} M. Find the corresponding range of pH readings.

6. Finding pH of Wine The hydrogen ion concentrations for wines vary from 1.58×10^{-4} M to 1.58×10^{-3} M. Find the corresponding range of pH readings.

7. pH of Wine If the pH of a wine is too high, say, 4.0 or above, the wine becomes unstable and has a flat taste.

(a) A certain California red wine has a pH of 3.2, and a certain Italian white wine has a pH of 2.9. Find the corresponding hydrogen ion concentrations of the two wines.

(b) Which wine has the lower hydrogen ion concentration?

8. pH of Saliva The pH of saliva is normally in the range of 6.4 to 7.0. However, when a person is ill, the person's saliva becomes more acidic.

(a) Suppose the pH of the saliva of a patient is 5.5. What is the hydrogen ion concentration of the patient's saliva?

(b) After the patient recovers, the pH of their saliva is 6.5. Was the saliva more acidic or less acidic when the patient was sick?

(c) Will the hydrogen ion concentration of the saliva increase or decrease as the patient recovers?

9. Earthquake Magnitude and Intensity

(a) Find the magnitude of an earthquake that has an intensity of 31.25 (that is, the amplitude of the seismograph reading is 31.25 cm).

(b) An earthquake was measured to have a magnitude of 4.8 on the Richter scale. Find the intensity of the earthquake.

10. Earthquake Magnitude and Intensity

(a) Find the magnitude of an earthquake that has an intensity of 72.1 (that is, the amplitude of the seismograph reading is 72.1 cm).

(b) An earthquake was measured to have a magnitude of 5.8 on the Richter scale. Find the intensity of the earthquake.

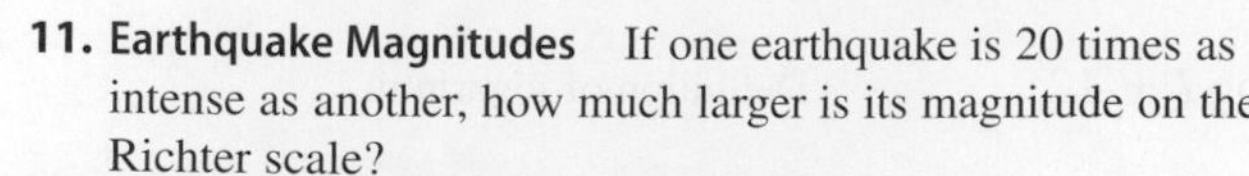

11. Earthquake Magnitudes If one earthquake is 20 times as intense as another, how much larger is its magnitude on the Richter scale?

12. Earthquake Magnitudes The 1906 earthquake in San Francisco had a magnitude of 8.3 on the Richter scale. At the same time in Japan an earthquake with magnitude 4.9 caused only minor damage. How many times more intense was the San Francisco earthquake than the Japan earthquake?

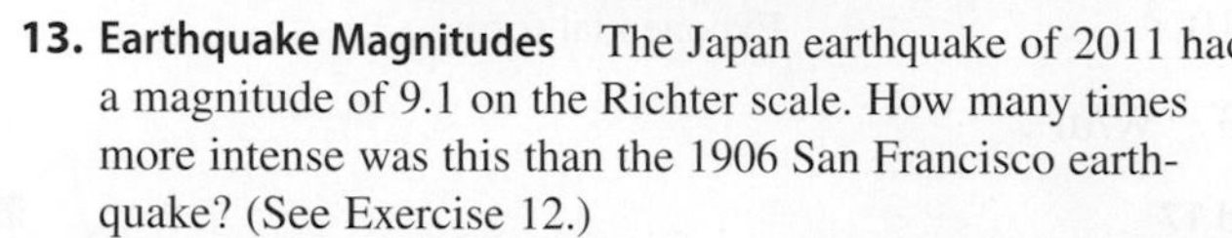

13. Earthquake Magnitudes The Japan earthquake of 2011 had a magnitude of 9.1 on the Richter scale. How many times more intense was this than the 1906 San Francisco earthquake? (See Exercise 12.)

14. Earthquake Magnitudes The Northridge, California, earthquake of 1994 had a magnitude of 6.8 on the Richter scale. A year later, a 7.2-magnitude earthquake struck Kobe, Japan. How many times more intense was the Kobe earthquake than the Northridge earthquake?

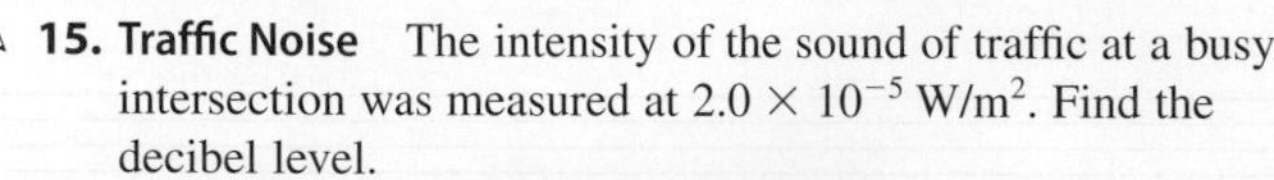

15. Traffic Noise The intensity of the sound of traffic at a busy intersection was measured at 2.0×10^{-5} W/m^2. Find the decibel level.

16. Leaf Blower The intensity of the sound from a certain leaf blower is measured at 3.2×10^{-2} W/m^2. Find the decibel level.

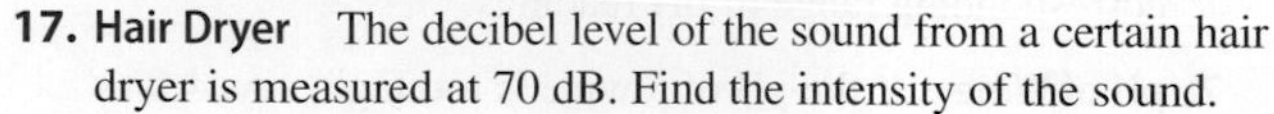

17. Hair Dryer The decibel level of the sound from a certain hair dryer is measured at 70 dB. Find the intensity of the sound.

18. Subway Noise The decibel level of the sound of a subway train was measured at 98 dB. Find the intensity of the sound.

19. Hearing Loss from Headphones Recent research has shown that the use of earbuds can cause permanent hearing loss.

(a) The intensity of the sound from the speakers of a certain audio system is measured at 3.1×10^{-5} W/m^2. Find the decibel level.

(b) If earbuds are used with the audio system in part (a), the decibel level is 90 dB. Find the intensity.

(c) Find the ratio of the intensity of the sound from the audio system with earbuds to that of the sound without earbuds.

20. Comparing Decibel Levels The noise from a power mower was measured at 106 dB. The noise level at a rock concert was measured at 120 dB. Find the ratio of the intensity of the rock music to that of the power mower.

Discuss ■ Discover ■ Prove ■ Write

21. Prove: Inverse Square Law for Sound A law of physics states that the intensity of sound is inversely proportional to the square of the distance d from the source: $I = k/d^2$.

(a) Use this model and the equation

$$B = 10 \log \frac{I}{I_0}$$

(described in this section) to show that the decibel levels B_1 and B_2 at distances d_1 and d_2 from a sound source are related by the equation

$$B_2 = B_1 + 20 \log \frac{d_1}{d_2}$$

(b) The intensity level at a rock concert is 120 dB at a distance 2 m from the speakers. Find the intensity level at a distance of 10 m.

Chapter 4 Review

Properties and Formulas

Exponential Functions | Section 4.1

The **exponential function** f with base a (where $a > 0, a \neq 1$) is defined for all real numbers x by

$$f(x) = a^x$$

The domain of f is $\mathbb{R}$, and the range of f is $(0, \infty)$ The graph of f has one of the following shapes, depending on the value of a:

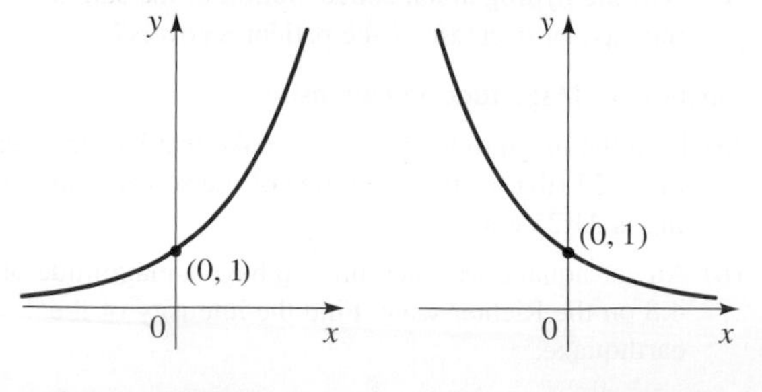

$f(x) = a^x$ for $a > 1$ $\quad$ $f(x) = a^x$ for $0 < a < 1$

The Natural Exponential Function | Section 4.2

The **natural exponential function** is the exponential function with base e:

$$f(x) = e^x$$

The number e is defined to be the number that the expression $(1 + 1/n)^n$ approaches as $n \to \infty$. An approximate value for the irrational number e is

$$e \approx 2.7182818284590\ldots$$

Compound Interest | Sections 4.1 and 4.2

If a principal P is invested in an account paying an annual interest rate r, compounded n times a year, then after t years the **amount** $A(t)$ in the account is

$$A(t) = P\left(1 + \frac{r}{n}\right)^{nt}$$

If the interest is compounded **continuously**, then the amount is

$$A(t) = Pe^{rt}$$

Logarithmic Functions | Section 4.3

The **logarithmic function** $\log_a$ with base a (where $a > 0, a \neq 1$) is defined for $x > 0$ by

$$\log_a x = y \quad \Leftrightarrow \quad a^y = x$$

So $\log_a x$ is the exponent to which the base a must be raised to give y.

The domain of $\log_a$ is $(0, \infty)$, and the range is $\mathbb{R}$. For $a > 1$, the graph of the function $\log_a$ has the following shape:

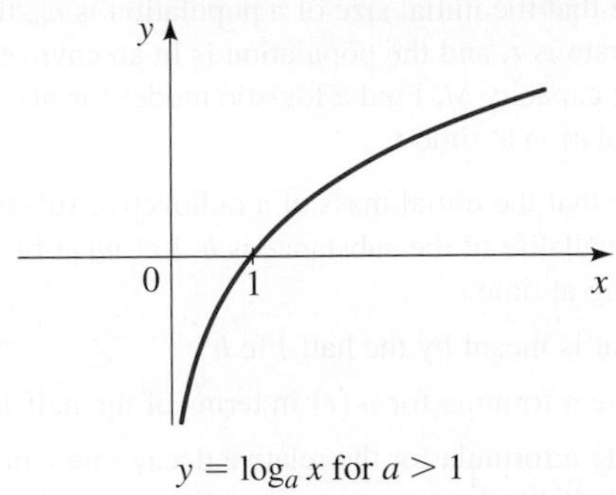

$y = \log_a x$ for $a > 1$

Common and Natural Logarithms | Section 4.3

The logarithm function with base 10 is called the **common logarithm** and is denoted **log**. So

$$\log x = \log_{10} x$$

The logarithm function with base e is called the **natural logarithm** and is denoted **ln**. So

$$\ln x = \log_e x$$

Properties of Logarithms | Section 4.3

1. $\log_a 1 = 0$
2. $\log_a a = 1$
3. $\log_a a^x = x$
4. $a^{\log_a x} = x$

Laws of Logarithms | Section 4.4

Let a be a logarithm base ($a > 0, a \neq 1$), and let A, B, and C be any real numbers or algebraic expressions that represent real numbers, with $A > 0$ and $B > 0$. Then:

1. $\log_a(AB) = \log_a A + \log_a B$
2. $\log_a(A/B) = \log_a A - \log_a B$
3. $\log_a(A^C) = C \log_a A$

Change of Base Formula | Section 4.4

$$\log_b x = \frac{\log_a x}{\log_a b}$$

Guidelines for Solving Exponential Equations | Section 4.5

1. Isolate the exponential term on one side of the equation.
2. Take the logarithm of each side, and use the Laws of Logarithms to "bring down the exponent."
3. Solve for the variable.

Guidelines for Solving Logarithmic Equations | Section 4.5

1. Isolate the logarithmic term(s) on one side of the equation, and use the Laws of Logarithms to combine logarithmic terms if necessary.
2. Rewrite the equation in exponential form.
3. Solve for the variable.

Exponential Growth Model | Section 4.6

A population experiences **exponential growth** if it can be modeled by the exponential function

$$n(t) = n_0 e^{rt}$$

where $n(t)$ is the population at time t, n_0 is the initial population (at time $t = 0$), and r is the relative growth rate (expressed as a proportion of the population).

Logistic Growth Model | Section 4.6

A population experiences **logistic growth** if it can be modeled by a function of the form

$$n(t) = \frac{M}{1 + Ae^{-rt}}$$

where $n(t)$ is the population at time t, r is the initial relative growth rate, M is the **carrying capacity** of the environment, and $A = (M - n_0)/n_0$, where n_0 is the initial population.

Radioactive Decay Model | Section 4.6

If a **radioactive substance** with half-life h has initial mass m_0, then at time t the mass $m(t)$ of the substance that remains is modeled by the exponential function

$$m(t) = m_0 e^{-rt}$$

where $r = \dfrac{\ln 2}{h}$.

Newton's Law of Cooling | Section 4.6

If an object has an initial temperature that is D_0 degrees warmer than the surrounding temperature T_s, then at time t the temperature $T(t)$ of the object is modeled by the function

$$T(t) = T_s + D_0 e^{-kt}$$

where the constant $k > 0$ depends on the size and type of the object.

Logarithmic Scales | Section 4.7

The **pH scale** measures the acidity of a solution:

$$\text{pH} = -\log[\text{H}^+]$$

The **Richter scale** measures the intensity of earthquakes:

$$M = \log \frac{I}{S}$$

The **decibel scale** measures the intensity of sound:

$$B = 10 \log \frac{I}{I_0}$$

Concept Check

1. Let f be the exponential function with base a.

(a) Write an equation that defines f.

(b) Write an equation for the exponential function f with base 3.

2. Let f be the exponential function $f(x) = a^x$, where $a > 0$.

(a) What is the domain of f?

(b) What is the range of f?

(c) Sketch graphs of f for the following cases.

(i) $a > 1$ **(ii)** $0 < a < 1$

3. If x is large, which function grows faster, $f(x) = 2^x$ or $g(x) = x^2$?

4. (a) How is the number e defined?

(b) Give an approximate value of e, correct to four decimal places.

(c) What is the natural exponential function?

5. (a) How is $\log_a x$ defined?

(b) Find $\log_3 9$.

(c) What is the natural logarithm?

(d) What is the common logarithm?

(e) Write the exponential form of the equation $\log_7 49 = 2$.

6. Let f be the logarithmic function $f(x) = \log_a x$.

(a) What is the domain of f?

(b) What is the range of f?

(c) Sketch a graph of the logarithmic function for the case $a > 1$.

7. State the three Laws of Logarithms.

8. (a) State the Change of Base Formula.

(b) Find $\log_7 30$.

9. (a) What is an exponential equation?

(b) How do you solve an exponential equation?

(c) Solve for x: $2^x = 19$

10. (a) What is a logarithmic equation?

(b) How do you solve a logarithmic equation?

(c) Solve for x: $4 \log_3 x = 7$

11. Suppose that an amount P is invested at an interest rate r and $A(t)$ is the amount of the investment after t years. Write a formula for $A(t)$ in the following cases.

(a) Interest is compounded n times per year.

(b) Interest is compounded continuously.

12. Suppose that the initial size of a population is n_0 and the population grows exponentially. Let $n(t)$ be the size of the population at time t.

(a) Write a formula for $n(t)$ in terms of the doubling time a.

(b) Write a formula for $n(t)$ in terms of the relative growth rate r.

13. Suppose that the initial size of a population is n_0, the initial growth rate is r, and the population is in an environment with carrying capacity M. Find a logistic model for $n(t)$, the size of the population at time t.

14. Suppose that the initial mass of a radioactive substance is m_0 and the half-life of the substance is h. Let $m(t)$ be the mass remaining at time t.

(a) What is meant by the half-life h?

(b) Write a formula for $m(t)$ in terms of the half-life h.

(c) Write a formula for the relative decay rate r in terms of the half-life h.

(d) Write a formula for $m(t)$ in terms of the relative decay rate r.

15. Suppose that the initial temperature difference between an object and its surroundings is D_0 and the surroundings have temperature T_s. Let $T(t)$ be the temperature at time t. State Newton's Law of Cooling for $T(t)$.

16. What is a logarithmic scale? If we use a logarithmic scale with base 10, what do the following numbers correspond to on the logarithmic scale?

(i) 100 **(ii)** 100,000 **(iii)** 0.0001

17. (a) What does the pH scale measure?

(b) Define the pH of a substance with hydrogen ion concentration $[H^+]$.

18. (a) What does the Richter scale measure?

(b) Define the magnitude M of an earthquake in terms of the intensity I of the earthquake and the intensity S of a standard earthquake.

19. (a) What does the decibel scale measure?

(b) Define the decibel level B of a sound in terms of the intensity I of the sound and the intensity I_0 of a barely audible sound.

Answers to the Concept Check can be found at the book companion website stewartmath.com.

Exercises

1–4 ■ Evaluating Exponential Functions Use a calculator to find the indicated values of the exponential function, rounded to three decimal places.

1. $f(x) = 5^x$; $f(-1.5), f(\sqrt{2}), f(2.5)$

2. $f(x) = 3 \cdot 2^x$; $f(-2.2), f(\sqrt{7}), f(5.5)$

3. $g(x) = 4e^{x-2}$; $g(-0.7), g(1), g(\pi)$

4. $g(x) = \frac{7}{4}e^{x+1}$; $g(-2), g(\sqrt{3}), g(3.6)$

5–18 ■ Graphing Exponential and Logarithmic Functions Sketch the graph of the function. State the domain, range, and asymptote.

5. $f(x) = 3^x + 1$

6. $f(x) = (\frac{1}{2})^x - 5$

7. $g(x) = 4^{-(x+1)}$

8. $g(x) = -2^{x-1}$

9. $h(x) = -e^{x+2} - 1$

10. $h(x) = 3e^{-x} + 1$

11. $f(x) = \log_3(x - 2)$

12. $f(x) = -\log_4(x + 2)$

13. $f(x) = \log_{1/3} x + 1$

14. $f(x) = 1 - \log_{1/2}(x + 2)$

15. $g(x) = \log_2(-x) - 2$

16. $g(x) = -\log_3(x + 3) + 2$

17. $g(x) = 2 \ln x$

18. $g(x) = \ln(x^2)$

19–22 ■ Domain Find the domain of the function.

19. $f(x) = 10^{x^2} + \log(1 - 2x)$

20. $g(x) = \log(2 + x - x^2)$

21. $h(x) = \ln(x^2 - 4)$

22. $k(x) = \ln|x|$

23–26 ■ Exponential Form Write the equation in exponential form.

23. $\log_2 1024 = 10$

24. $\log_6 37 = x$

25. $\log x = y$

26. $\ln c = 17$

27–30 ■ Logarithmic Form Write the equation in logarithmic form.

27. $2^6 = 64$

28. $49^{-1/2} = \frac{1}{7}$

29. $10^x = 74$

30. $e^k = m$

31–46 ■ Evaluating Logarithmic Expressions Evaluate the expression without using a calculator.

31. $\log_2 128$

32. $\log_8 1$

33. $10^{\log 45}$

34. $\log 0.000001$

35. $\ln e^6$

36. $\log_4 8$

37. $\log_3\left(\frac{1}{27}\right)$

38. $2^{\log_2 13}$

39. $\log_5 \sqrt{5}$

40. $e^{2 \ln 7}$

41. $\log 25 + \log 4$

42. $\log_3 \sqrt{243}$

43. $\log_2 16^{23}$

44. $\log_5 250 - \log_5 2$

45. $\log_8 6 - \log_8 3 + \log_8 2$

46. $\log(\log 10^{100})$

47–52 ■ Expanding Logarithmic Expressions Expand the logarithmic expression.

47. $\log(AB^2C^3)$

48. $\log_2(x\sqrt{x^2 + 1})$

49. $\ln\sqrt{\dfrac{x^2 - 1}{x^2 + 1}}$

50. $\log\left(\dfrac{4x^3}{y^2(x - 1)^5}\right)$

51. $\log_5\left(\dfrac{x^2(1 - 5x)^{3/2}}{\sqrt{x^3 - x}}\right)$

52. $\ln\left(\dfrac{\sqrt[3]{x^4 + 12}}{(x + 16)\sqrt{x - 3}}\right)$

53–58 ■ Combining Logarithmic Expressions Combine into a single logarithm.

53. $\log 6 + 4 \log 2$

54. $\log x + \log(x^2 y) + 3 \log y$

55. $\frac{3}{2} \log_2(x - y) - 2 \log_2(x^2 + y^2)$

56. $\log_5 2 + \log_5(x + 1) - \frac{1}{3} \log_5(3x + 7)$

57. $\log(x - 2) + \log(x + 2) - \frac{1}{2} \log(x^2 + 4)$

58. $\frac{1}{2}[\ln(x - 4) + 5 \ln(x^2 + 4x)]$

59–72 ■ Exponential and Logarithmic Equations Solve the equation. Find the exact solution if possible; otherwise, use a calculator to approximate to two decimal places.

59. $2^{6x-3} = 8$

60. $\left(\frac{1}{3}\right)^{1-x} = \frac{1}{9}$

61. $5^{3x+2} = 2$

62. $10^{4-3x} = 5$

63. $2^{5x+1} = 3^{4-x}$

64. $10^{2x/5} = e^{3x+1}$

65. $x^3 \cdot 3^{4x} - x^2 \cdot 3^{4x} = 6x \cdot 3^{4x}$

66. $e^{2x} - 6e^x + 9 = 0$

67. $\log x + \log(x + 1) = \log 12$

68. $\ln(x - 2) + \ln 3 = \ln(5x - 7)$

69. $\log_2(1 - x) = 4$

70. $\ln(2x - 3) + 1 = 0$

71. $\log_3(x - 8) + \log_3 x = 2$

72. $\log_{1/2}(x - 5) - \log_{1/2}(x + 2) = 1$

73–76 ■ Exponential Equations Use a calculator to find the solution of the equation, rounded to six decimal places.

73. $5^{-2x/3} = 0.63$

74. $2^{3x-5} = 7$

75. $5^{2x+1} = 3^{4x-1}$

76. $e^{-15k} = 10{,}000$

77–80 ■ Local Extrema and Asymptotes Draw a graph of the function and use it to determine the asymptotes and the local maximum and minimum values.

77. $y = e^{x/(x+2)}$

78. $y = 10^x - 5^x$

79. $y = \log(x^3 - x)$

80. $y = 2x^2 - \ln x$

81–82 ■ Solving Equations Find the solution(s) of the equation, rounded to two decimal places.

81. $3 \log x = 6 - 2x$

82. $4 - x^2 = e^{-2x}$

83–84 ■ Solving Inequalities Solve the inequality graphically.

83. $\ln x > x - 2$

84. $e^x < 4x^2$

85. **Increasing and Decreasing** Use a graph of $f(x) = e^x - 3e^{-x} - 4x$ to find, approximately, the intervals on which f is increasing and on which f is decreasing.

86. **Equation of a Line** Find an equation of the line shown in the figure.

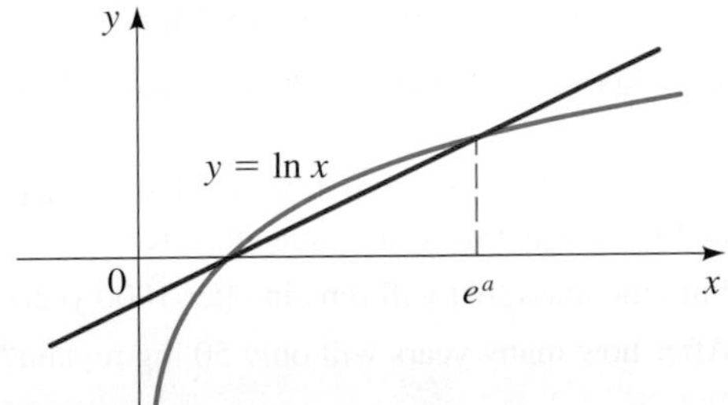

87–90 ■ Change of Base Use the Change of Base Formula to express the given logarithm in terms of common or natural logarithms, and then evaluate, rounded to six decimal places.

87. $\log_4 15$ **88.** $\log_7(\frac{3}{4})$

89. $\log_9 0.28$ **90.** $\log_{100} 250$

91. Comparing Logarithms Which is larger, $\log_4 258$ or $\log_5 620$?

92. Inverse Function Find the inverse of the function $f(x) = 2^{3^x}$, and state its domain and range.

93. Compound Interest If $12,000 is invested at an interest rate of 10% per year, find the amount of the investment at the end of 3 years for each compounding method.

(a) Semiannual (b) Monthly

(c) Daily (d) Continuous

94. Compound Interest A sum of $5000 is invested at an interest rate of $8\frac{1}{2}\%$ per year, compounded semiannually.

(a) Find the amount of the investment after $1\frac{1}{2}$ years.

(b) After what period of time will the investment amount to $7000?

(c) If interest were compounded continously instead of semiannually, how long would it take for the amount to grow to $7000?

95. Compound Interest A money market account pays 5.2% annual interest, compounded daily. If $100,000 is invested in this account, how long will it take for the account to accumulate $10,000 in interest?

96. Compound Interest A retirement savings plan pays 4.5% interest, compounded continuously. How long will it take for an investment in this plan to double?

97–98 ■ APY Determine the annual percentage yield (APY) for the given nominal annual interest rate and compounding frequency.

97. 4.25%; daily

98. 3.2%; monthly

99. Cat Population The stray-cat population in a small town grows exponentially. In 2022 the town had 30 stray cats, and the relative growth rate was 15% per year.

(a) Find a function that models the stray-cat population $n(t)$ after t years.

(b) Find the projected population after 4 years.

(c) Find the number of years required for the stray-cat population to reach 500.

100. Bacterial Growth A culture contains 10,000 bacteria initially. After 1 hour the bacteria count is 25,000.

(a) Find the doubling period.

(b) Find the number of bacteria after 3 hours.

101. Radioactive Decay Thorium-230 has a half-life of 75,380 years.

(a) If a sample has a mass of 150 mg, find a function that models the mass that remains after t years.

(b) Find the mass that will remain after 1000 years.

(c) After how many years will only 50 mg remain?

102. Radioactive Decay A sample of bismuth-210 decayed to 33% of its original mass after 8 days.

(a) Find the half-life of this element.

(b) Find the mass remaining after 12 days.

103. Radioactive Decay The half-life of palladium-100 is 4 days. After 20 days a sample has been reduced to a mass of 0.375 g.

(a) What was the initial mass of the sample?

(b) Find a function that models the mass remaining after t days.

(c) What is the mass after 3 days?

(d) After how many days will only 0.15 g remain?

104. Bird Population The graph shows the population $n(t)$ of a rare species of bird in a nature reserve, where t represents years since 2020.

(a) Find a function that models the bird population at time t.

(b) What is the bird population expected to be in the year 2031?

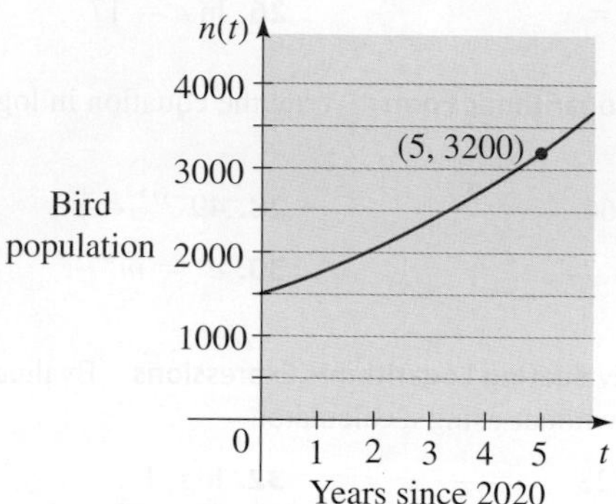

105. Logistic Growth of Yeast The yeast fungus *Saccharomyces cerevisiae* is used in leavening dough for making bread. In a recent research article it was reported that the doubling time of *S. cerevisiae* colonies is about 1.5 hours. About 100 colonies were placed in a Petri dish with a carrying capacity of 1400 colonies.

(a) Find the initial relative growth rate r.

(b) Find a logistic model for the number of yeast colonies $n(t)$ at time t.

(c) Graph the model that you found in part (b) and determine from the graph how long it takes for the yeast colonies to reach half the carrying capacity.

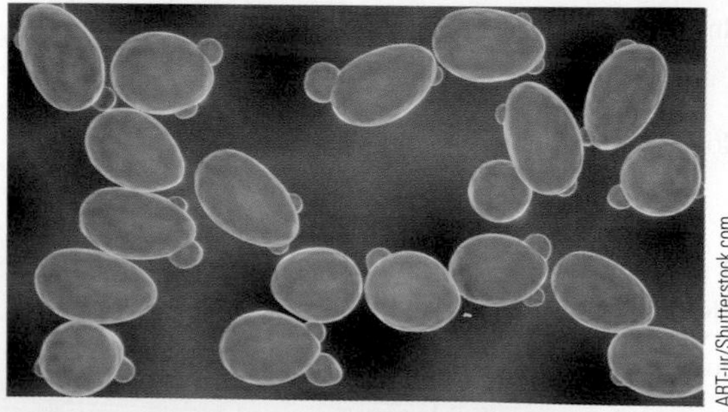

ART-ur/Shutterstock.com

106. Law of Cooling A car engine runs at a temperature of 87°C. When the engine is turned off, it cools according to Newton's Law of Cooling with constant $k = 0.0341$, where the time is measured in minutes. Find the time needed for the engine to cool to 32°C if the surrounding temperature is 15°C.

107. pH Scale The hydrogen ion concentration of fresh egg whites was measured as

$$[H^+] = 1.3 \times 10^{-8}\ M$$

Find the pH, and classify the substance as acidic or basic.

108. pH Scale The pH of lime juice is 1.9. Find the hydrogen ion concentration.

109. Richter Scale If one earthquake has magnitude 6.5 on the Richter scale, what is the magnitude of another quake that is 35 times as intense?

110. Decibel Scale The drilling of a jackhammer was measured at 132 dB. The sound of whispering was measured at 28 dB. Find the ratio of the intensity of the drilling to that of the whispering.

Matching

111. Equations and Their Graphs Match each equation with its graph. Give reasons for your answers. (Don't use a graphing device.)

(a) $y = 2^x$ **(b)** $y = -\ln x$ **(c)** $2x + 3y = 6$ **(d)** $y = 1 - \dfrac{1}{x^3}$

(e) $y = \log_2 x$ **(f)** $y = 4e^{-x^2/4}$ **(g)** $y = 2 - 2x - x^2$ **(h)** $y = 10xe^{-x}$

I

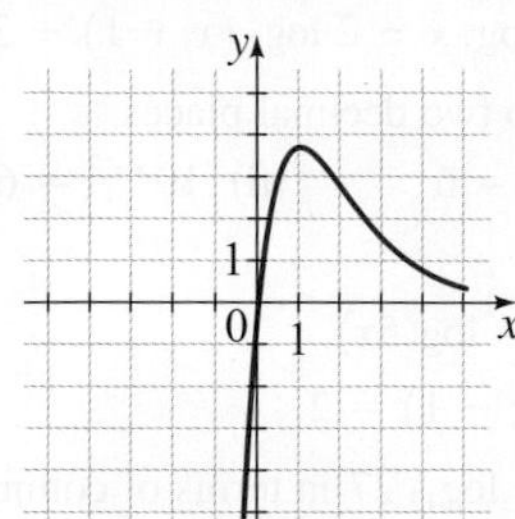

II

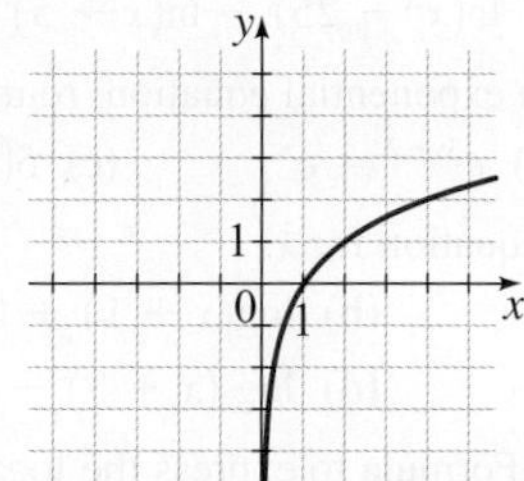

III

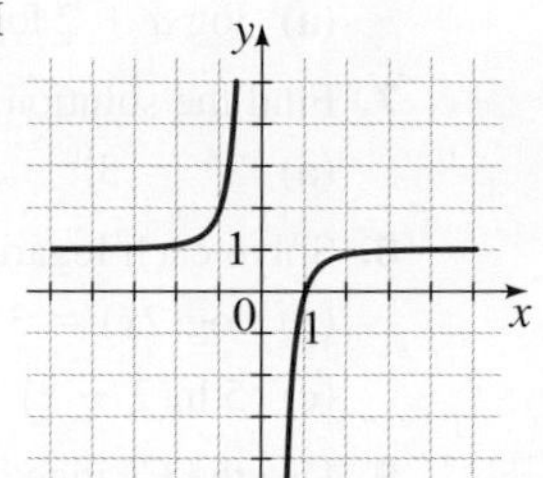

IV

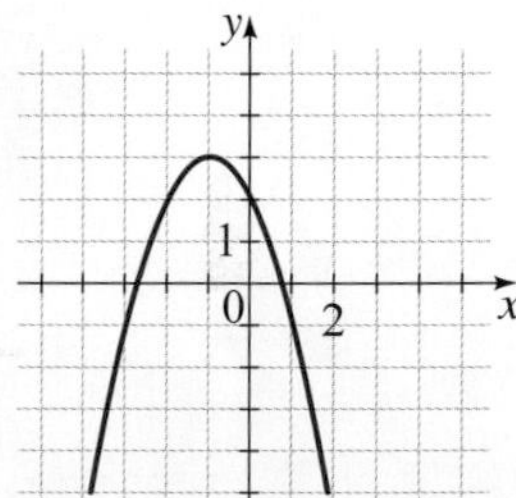

V

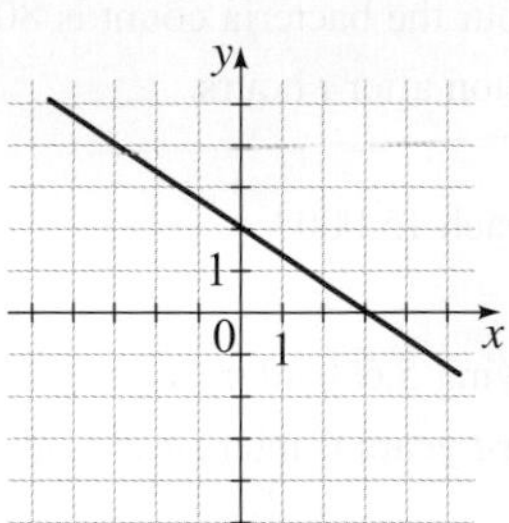

VI

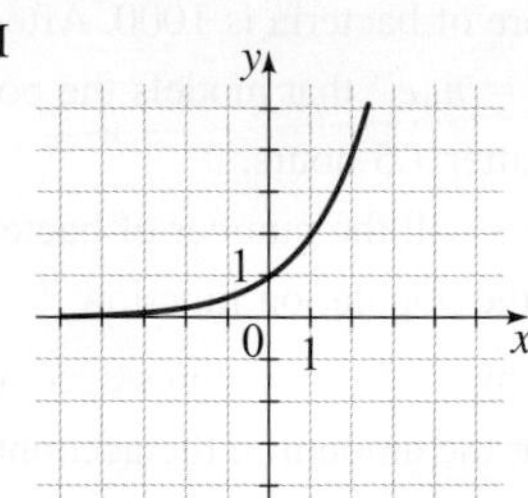

VII

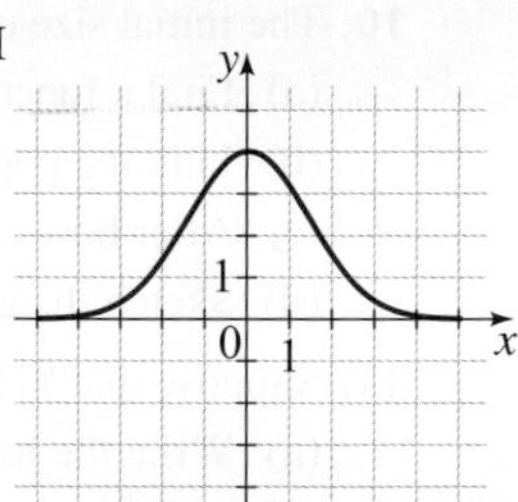

VIII

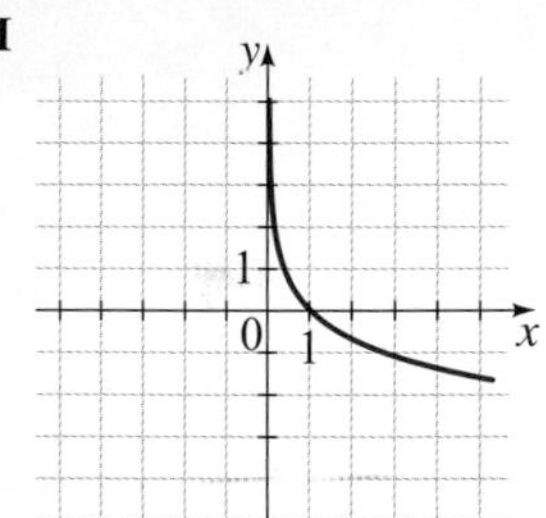

Chapter 4 | Test

1. Sketch the graph of each function, and state its domain, range, and asymptote. Show the x- and y-intercepts on the graph.

(a) $f(x) = 3 - 3^x$ **(b)** $g(x) = \log_3(x + 3)$

2. Find the domain of each function.

(a) $f(t) = \ln(2t - 3)$ **(b)** $g(x) = \log(x^2 - 1)$

3. (a) Write the equation $6^{2x} = 25$ in logarithmic form.

(b) Write the equation $\ln A = 3$ in exponential form.

4. Find the exact value of each expression.

(a) $10^{\log 36}$ **(b)** $\ln e^3$ **(c)** $\log_3 \sqrt{27}$

(d) $\log_2 80 - \log_2 10$ **(e)** $\log_8 4$ **(f)** $\log_6 4 + \log_6 9$

5. Use the Laws of Logarithms to expand each expression.

(a) $\log\left(\dfrac{xy^3}{z^2}\right)$ **(b)** $\ln\sqrt{\dfrac{x}{y}}$ **(c)** $\log\sqrt{\dfrac{x^2 + 1}{x^3(x - 1)}}$

6. Use the Laws of Logarithms to combine each expression into a single logarithm.

(a) $\log a + 2\log b$ **(b)** $\ln(x^2 - 25) - \ln(x + 5)$ **(c)** $\log_3 x - 2\log_3(x + 1) + 3\log_3 y$

7. Find the solution of each exponential equation, rounded to two decimal places.

(a) $3^{4x} = 3^{100}$ **(b)** $e^{3x-2} = e^{x^2}$ **(c)** $5\left(\frac{2}{3}\right)^{3x+1} = 6$ **(d)** $10^{x+3} = 6^{2x}$

8. Solve each logarithmic equation for x.

(a) $\log(2x) = 3$ **(b)** $\log(x + 1) + \log 2 = \log(5x)$

(c) $5\ln(3 - x) = 4$ **(d)** $\log_4(x + 3) - \log_4(x - 1) = 2$

9. Use the Change of Base Formula to express the logarithm $\log_{12} 27$ in terms of common or natural logarithms, and then evaluate your result.

10. The initial size of a culture of bacteria is 1000. After 1 hour the bacteria count is 8000.

(a) Find a function $n(t) = n_0 e^{rt}$ that models the population after t hours.

(b) Find the population after 1.5 hours.

(c) After how many hours will the number of bacteria reach 15,000?

(d) Sketch the graph of the population function.

11. Suppose that \$12,000 is invested in a savings account paying 5.6% interest per year.

(a) Write the formula for the amount in the account after t years if interest is compounded monthly.

(b) Find the amount in the account after 3 years if interest is compounded daily.

(c) How long will it take for the amount in the account to grow to \$20,000 if interest is compounded continuously?

12. The half-life of krypton-91 (^{91}Kr) is 10 s. At time $t = 0$ a heavy canister contains 3 g of this radioactive gas.

(a) Find a function $m(t) = m_0 2^{-t/h}$ that models the amount of ^{91}Kr remaining in the canister after t seconds.

(b) Find a function $m(t) = m_0 e^{-rt}$ that models the amount of ^{91}Kr remaining in the canister after t seconds.

(c) How much ^{91}Kr remains after 1 minute?

(d) After how long will the amount of ^{91}Kr remaining be reduced to 1 μg (1 microgram, or 10^{-6} g)?

13. An earthquake measuring 6.4 on the Richter scale struck Japan in July 2007, causing extensive damage. Earlier that year, a minor earthquake measuring 3.1 on the Richter scale was felt in parts of Pennsylvania. How many times more intense was the Japanese earthquake than the Pennsylvania earthquake?

A Cumulative Review Test for Chapters 2, 3, and 4 can be found at the book companion website www.stewartmath.com.

When we model real-world data, the shape of a scatter plot is one factor that helps us choose an appropriate model. In Figure 1, the first scatter plot suggests a linear model. But what type of function fits the second plot? A quadratic or an exponential? It's not easy to tell by just looking at the scatter plot. In this section we learn how to fit exponential and power curves to data.

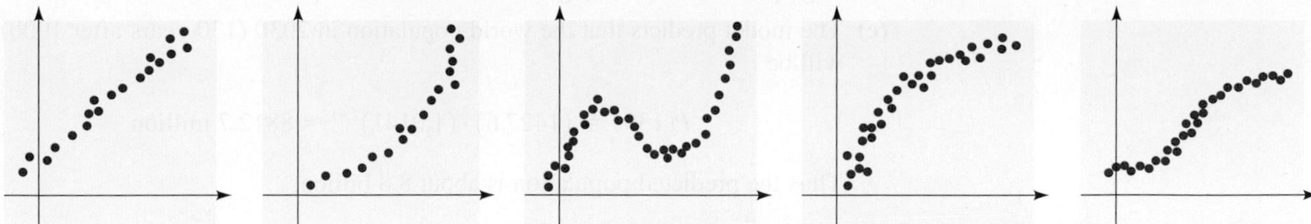

Figure 1

There are other factors to be considered in choosing a type of function to model real-world data. You can explore some of these in the Discovery Project *The Art of Modeling* at **www.stewartmath.com**.

■ Modeling with Exponential Functions

If a scatter plot shows that the data increase rapidly, we might want to model the data using an exponential model, that is, a function of the form

$$f(x) = a \cdot b^x$$

where a and b are constants.

Example 1 ■ An Exponential Model for World Population

Table 1 gives the population of the world since 1900.

(a) Draw a scatter plot of the data. Is a linear model appropriate?

(b) Find an exponential function that models the data and draw a graph of the function that you found, together with the scatter plot.

(c) What does the model predict for world population in the year 2030?

Table 1
World Population

Years Since 1900 (t)	Population in Millions (P)
0	1650
10	1750
20	1860
30	2070
40	2300
50	2520
60	3020
70	3700
80	4450
90	5300
100	6060
110	6920
120	7790

Solution

(a) The scatter plot in Figure 2(a) does not appear to lie along a line, so a linear model is not appropriate.

(b) Using the `ExponentiaL Regression` command on a graphing device [see Figure 2(b)] we get an exponential model for the population in millions.

$$P(t) = (1427.6) \cdot (1.0141)^t$$

A graph of this model together with the scatter plot is shown in Figure 2(c).

(c) The model predicts that the world population in 2030 (130 years after 1900) will be

$$P(130) = (1427.6) \cdot (1.0141)^{130} \approx 8812.7 \text{ million}$$

Thus the predicted population is about 8.8 billion.

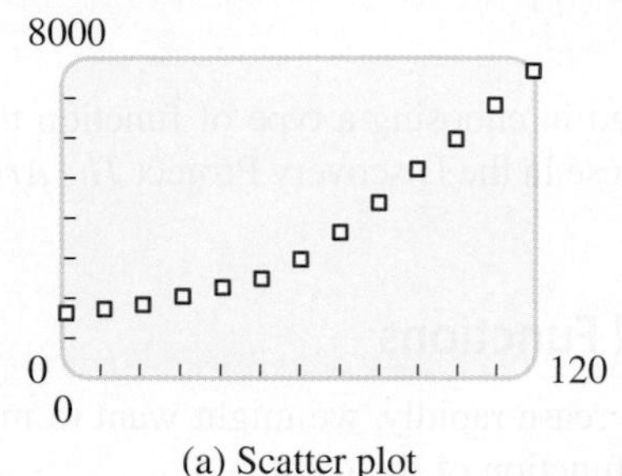

(a) Scatter plot

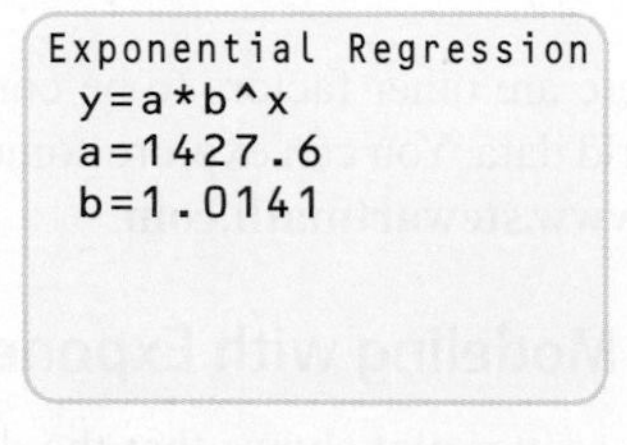

(b) Exponential model

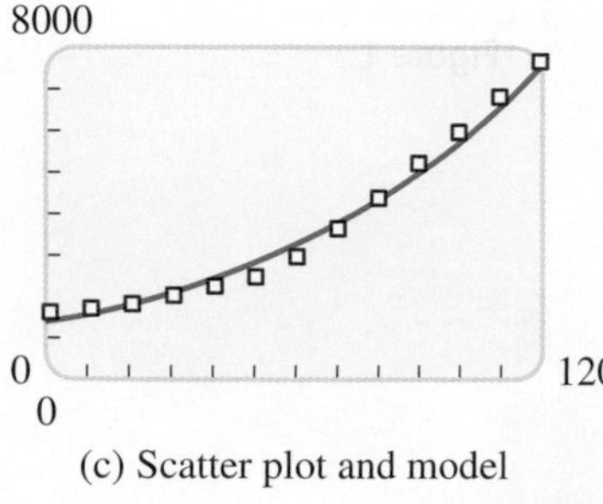

(c) Scatter plot and model

Figure 2

Note The model in Example 1 has the form $y = a \cdot b^t$. If we want to find the relative growth rate, we need to express the model in the form $y = Ce^{kt}$. By the note following Example 4.6.5 we have $k = \ln b$. In this case $k = \ln 1.0141 \approx 0.014$ and the model is

$$P(t) = (1427.6) \cdot e^{0.014t}$$

From this form of the model we see that the relative growth rate is $r = 0.014$, or 1.4%.

■ Modeling with Power Functions

If the scatter plot of the data we are studying resembles the graph of $y = ax^2$, $y = ax^{1.32}$, or some other power function, then we use a *power model*, that is, a function of the form

$$f(x) = a \cdot x^b$$

where a is a positive constant and b is any real number.

In the next example we find a power model that relates the period of a planet to its distance from the sun. Distance in the solar system is measured in astronomical units; an *astronomical unit* (AU) is the mean distance between the earth and the sun. The *period* of a planet is the time it takes the planet to make a complete revolution around the sun (measured in earth years).

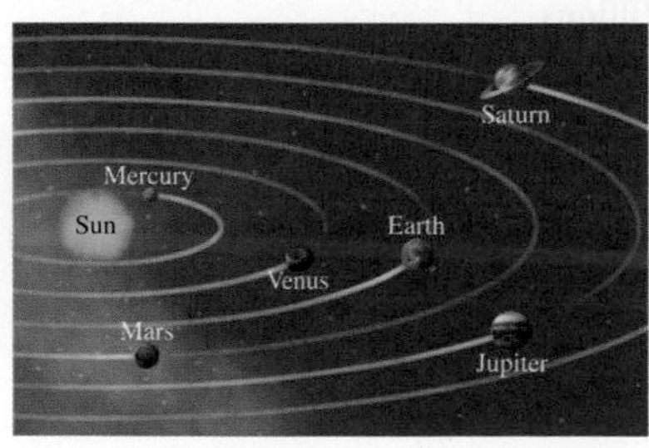

Example 2 ■ A Power Model for Planetary Periods

Table 2 (on the next page) gives the mean distance d of each planet from the sun in astronomical units and its period T in years.

(a) Draw a scatter plot of the data. Is a linear model appropriate?

(b) Find a power function that models the data and draw a graph of the function that you found, together with the scatter plot. How well does the model fit the data?

(c) Use the model to find the period of an asteroid whose mean distance from the sun is 5 AU.

Table 2
Distances and Periods of the Planets

Planet	d	T
Mercury	0.387	0.241
Venus	0.723	0.615
Earth	1.000	1.000
Mars	1.523	1.881
Jupiter	5.203	11.861
Saturn	9.541	29.457
Uranus	19.190	84.008
Neptune	30.086	164.784
Pluto*	39.507	248.350

*Pluto is a "dwarf planet."

Solution

(a) The scatter plot in Figure 3(a) shows that the data do not appear to lie along a line, so a linear model is not appropriate.

(b) Using the `Power Regression` command on a graphing device [see Figure 3(b)] we obtain the power model $T = 1.000396d^{1.49966}$. Correct to two decimal places the model is

$$T = d^{1.5}$$

The graph of the model and scatter plot are shown in Figure 3(c). The model appears to fit the data well.

(c) With $d = 5$ AU the model gives

$$T = 5^{1.5} \approx 11.18$$

The period of the asteroid is about 11.2 years.

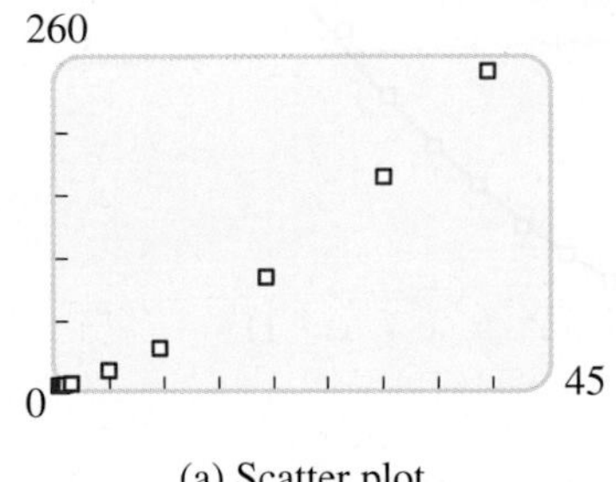

(a) Scatter plot

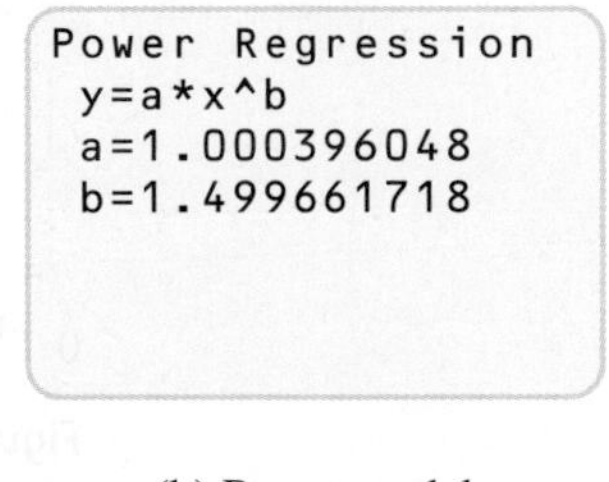

(b) Power model

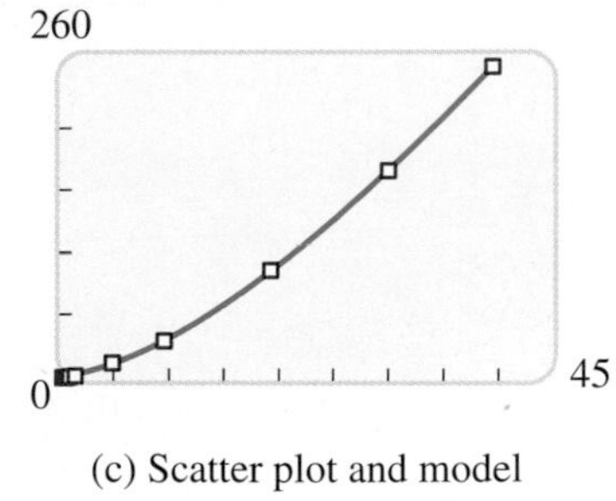

(c) Scatter plot and model

Figure 3

■

Johannes Kepler (see Section 10.4) first discovered the formula in Example 2 by analyzing planetary data. Later Isaac Newton (see Section 12.2) derived the formula from his inverse square law of gravity.

■ An Exponential or Power Model?

It is often difficult to visually determine from a scatter plot whether a power or exponential function best fits the data. To help us decide, we can *linearize the data* by applying a function that "straightens" the scatter plot.

For the data points (x, y), a scatter plot of the points $(x, \ln y)$ is called a **semi-log plot** and a scatter plot of the points $(\ln x, \ln y)$ is called a **log-log plot**. If the data points (x, y) lie on an exponential curve $y = Ce^{kx}$, then a semi-log plot of the data will lie on a line (see Exercise 4.4.81). If the data points (x, y) lie on a power curve $y = a \cdot x^b$, then a log-log plot of the data will lie on a line (see Exercise 4.4.82). In the next example we use semi-log and log-log plots to help us decide whether an exponential or power model fits the data better.

Table 3

x	y	$\ln x$	$\ln y$
1	2	0	0.7
2	6	0.7	1.8
3	14	1.1	2.6
4	22	1.4	3.1
5	34	1.6	3.5
6	46	1.8	3.8
7	64	1.9	4.2
8	80	2.1	4.4
9	102	2.2	4.6
10	130	2.3	4.9

Example 3 ■ An Exponential or Power Model?

Table 3 contains several data points (x, y). The table also has the values of $\ln x$ and $\ln y$ for each data point.

(a) Draw a scatter plot, a semi-log plot, and a log-log-plot of the data.

(b) Is an exponential function or a power function appropriate for modeling these data?

(c) Find and graph an appropriate model for the data.

Solution

(a) We use the values from Table 3 to graph the plots in Figures 4(a–c).

(b) The log-log plot in Figure 4(c) is very nearly linear so a power model is appropriate.

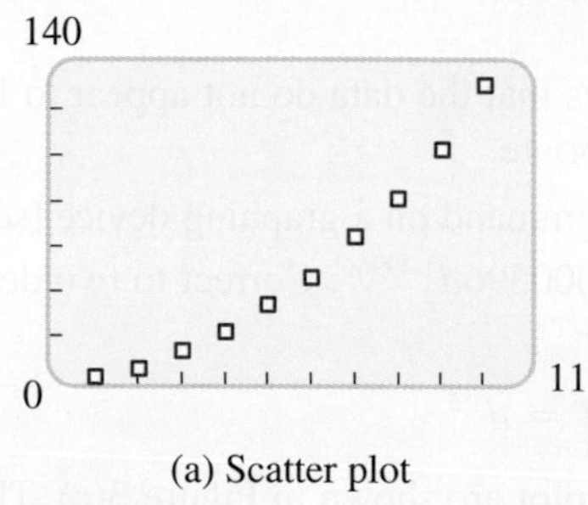

(a) Scatter plot

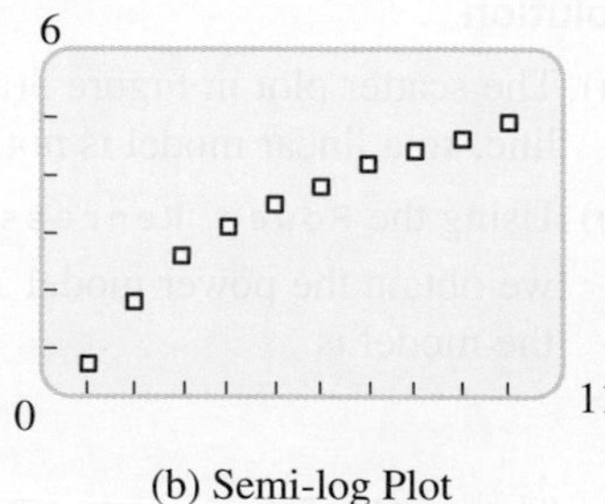

(b) Semi-log Plot

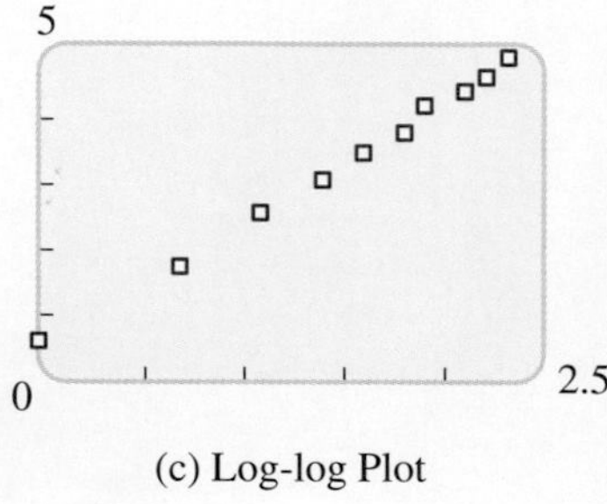

(c) Log-log Plot

Figure 4

(c) Using the `Power Regression` command on a graphing device, we find the power function that best fits the data: $y = 1.85x^{1.82}$. The graph of this function and the scatter plot of the data are shown in Figure 5.

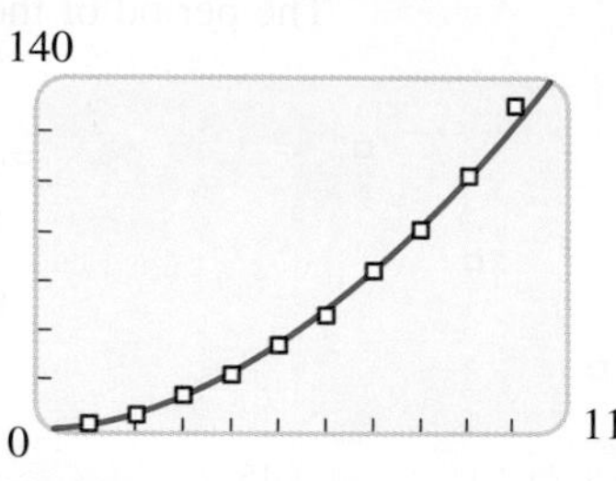

Figure 5

■

You can explore more properties of semi-log and log-log plots in the Discovery Project *Semi-log and Log-Log Plots* at the textbook website: **www.stewartmath.com.**

Problems

1. **Half-Life of Radioactive Iodine** A student is trying to determine the half-life of radioactive iodine-131. The student measures the amount of iodine-131 in a sample solution every 8 hours. The data are shown in the following table.

   (a) Make a scatter plot of the data.

(b) Use a graphing device to find an exponential model.

(c) Use your model to find the half-life of iodine-131.

Time (h)	Amount of ^{131}I (g)
0	4.80
8	4.66
16	4.51
24	4.39
32	4.29
40	4.14
48	4.04

2. A Falling Ball In a physics experiment a lead ball is dropped from a height of 5 m. The students record the distance the ball has fallen every one-tenth of a second. (This can be done by using a camera and a strobe light.) Their data are shown in the table below.

(a) Make a scatter plot of the data.

(b) Use a graphing device to find a power model.

(c) Use your model to predict how far a dropped ball would fall in 3 s.

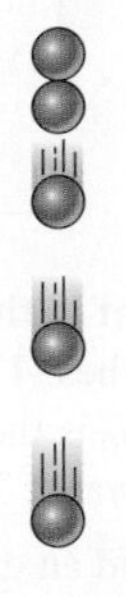

Time (s)	Distance (m)
0.1	0.048
0.2	0.197
0.3	0.441
0.4	0.882
0.5	1.227
0.6	1.765
0.7	2.401
0.8	3.136
0.9	3.969
1.0	4.902

The number of different bat species in a cave is related to the size of the cave by a power function.

3. Modeling the Species-Area Relation The table gives the areas of several caves in central Mexico and the number of bat species that live in each cave.*

(a) Use a graphing device to find a power function that models the data. Draw a graph of the function and a scatter plot of the data on the same screen. Does the model fit the data well?

(b) The cave called El Sapo near Puebla, Mexico, has a surface area of $A = 205 \text{ m}^2$. Use the model to estimate the number of bat species you would expect to find in that cave.

Cave	Area (m^2)	Number of Species
La Escondida	18	1
El Escorpion	19	1
El Tigre	58	1
Misión Imposible	60	2
San Martin	128	5
El Arenal	187	4
La Ciudad	344	6
Virgen	511	7

*A. K. Brunet and R. A. Medallin, "The Species-Area Relationship in Bat Assemblages of Tropical Caves." *Journal of Mammalogy,* 82(4):1114–1122, 2001.

Ostrich and Sandhill crane

4. Bird Flight: Exponential or Power Model? Ornithologists have catalogued the weights and wingspans of many species of birds that fly. The table gives some of the data.

(a) Make semi-log and log-log plots of the data to determine whether an exponential or power function best fits these data, and then find an appropriate model. Graph the model and a scatter plot of the data on the same screen.

(b) Ostriches have a mass of about 136 kg and have a wingspan of about 1.83 meters. How does the model you found explain why ostriches can't fly?

Bird	Weight (kg)	Wingspan (m.)
Turkey vulture	1.99	1.75
Bald eagle	3.09	2.13
Great horned owl	1.40	1.12
Cooper's hawk	0.47	0.71
Sandhill crane	4.09	2.01
Atlantic puffin	0.43	0.61
California condor	8.08	2.77
Common loon	3.09	1.22
Yellow warbler	0.01	0.20
Common grackle	0.09	0.41
Wood stork	2.30	1.60
Mallard	1.10	0.89

Sunlight zone
200 m
Twilight zone
1000 m
Midnight zone
4000 m
The abyss
6000 m
Hadal zone
11000 m

5. Sunlight in the Twilight Zone As sunlight passes through ocean water its intensity diminishes. The intensity I at depth x is modeled by the Beer-Lambert Law: $I = I_0 e^{-kx}$, where I_0 is the intensity at the surface and k is a constant that depends on the "murkiness" of the water. The data in the table were obtained by a marine biologist.

(a) Find an exponential function of the form given by the Beer-Lambert Law to model these data. What is the "murkiness" constant k in the model you found?

(b) Some species of deep-sea fish need light intensity of at least 3×10^{-12} W/m^2 to be able to see.* Can these species thrive in the twilight zone (see the graphic in the margin)?

Depth (m)	0	100	200	300	400	600	1000
Intensity (W/m^2)	300	15.1	0.75	0.04	0.002	4.5×10^{-6}	2.9×10^{-11}

6. Logistic Population Growth The table gives the population of black flies in a closed laboratory container over an 18-day period.

(a) Use the `Logistic` command on a graphing device to find a logistic model for these data. Graph the model and a scatter plot of the data on the same screen. Does the model appear to fit the data well?

(b) Use the model to estimate the carrying capacity of the container.

Time (days)	0	2	4	6	8	10	12	16	18
Number of Flies	10	25	66	144	262	374	446	492	498

*_Seawater: Its Composition, Properties and Behaviour_, The Open University, 1995.

salarko/Shutterstock.com

5 Trigonometric Functions: Unit Circle Approach

In this chapter and the next we introduce two different but equivalent ways of viewing the trigonometric functions: as *functions of real numbers* (Chapter 5) or as *functions of angles* (Chapter 6). The two approaches to trigonometry are independent of each other, so either Chapter 5 or Chapter 6 may be studied first. The applications of trigonometry are numerous, including modeling sound waves, signal processing, digital coding of music and videos, producing CAT scans for medical imaging, finding distances to stars, and many others. These applications are very diverse, and we need to study both approaches to trigonometry because the different approaches are required for different applications.

One of the main applications of trigonometry that we study in this chapter is periodic motion. If you've ever taken a Ferris wheel ride, then you know about periodic motion—that is, motion that repeats over and over. Periodic motion occurs often in nature, as in the daily rising and setting of the sun, the daily variation in tide levels (the photo shows low tide at a harbor), the vibrations of a leaf in the wind, and many more. We will see in this chapter how the trigonometric functions are used to model periodic motion.

5.1 The Unit Circle

■ The Unit Circle ■ Terminal Points on the Unit Circle ■ The Reference Number

In this section we explore some properties of the circle of radius 1 centered at the origin. These properties are used in the next section to define the trigonometric functions.

■ The Unit Circle

The set of points at a distance 1 from the origin is a circle of radius 1 (see Figure 1). In Section 1.9 we learned that the equation of this circle is $x^2 + y^2 = 1$.

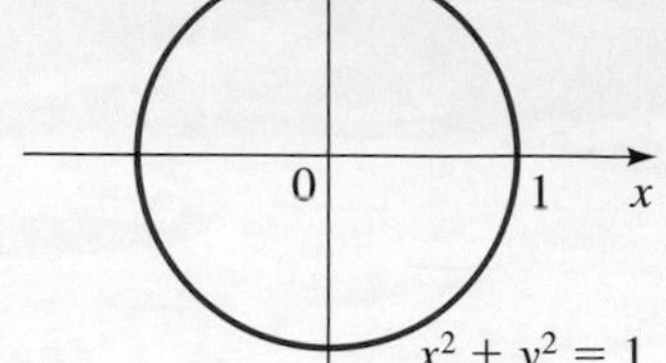

Figure 1 | The unit circle

The Unit Circle

The **unit circle** is the circle of radius 1 centered at the origin in the xy-plane. Its equation is

$$x^2 + y^2 = 1$$

Example 1 ■ A Point on the Unit Circle

Circles are studied in Section 1.9.

Show that the point $P\left(\frac{\sqrt{3}}{3}, \frac{\sqrt{6}}{3}\right)$ is on the unit circle.

Solution We need to show that this point satisfies the equation of the unit circle, that is, $x^2 + y^2 = 1$. Since

$$\left(\frac{\sqrt{3}}{3}\right)^2 + \left(\frac{\sqrt{6}}{3}\right)^2 = \frac{3}{9} + \frac{6}{9} = 1$$

P is on the unit circle.

Now Try Exercise 5 ■

Example 2 ■ Locating a Point on the Unit Circle

The point $P(\sqrt{3}/2, y)$ is on the unit circle in Quadrant IV. Find its y-coordinate.

Solution Since the point is on the unit circle, we have

$$\left(\frac{\sqrt{3}}{2}\right)^2 + y^2 = 1$$

$$y^2 = 1 - \frac{3}{4} = \frac{1}{4}$$

$$y = \pm\frac{1}{2}$$

Because the point is in Quadrant IV, its y-coordinate must be negative, so $y = -\frac{1}{2}$.

Now Try Exercise 11 ■

■ Terminal Points on the Unit Circle

Suppose t is a real number. If $t \geq 0$, let's mark off a distance t along the unit circle, starting at the point $(1, 0)$ and moving in a counterclockwise direction. If $t < 0$, we mark off a distance $|t|$ in a clockwise direction (Figure 2). In this way we arrive at a point $P(x, y)$ on the unit circle. The point $P(x, y)$ obtained in this way is called the **terminal point** determined by the real number t.

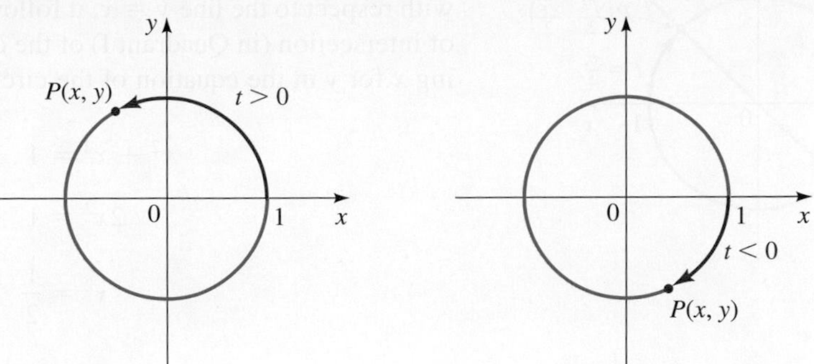

Figure 2

(a) Terminal point $P(x, y)$ determined by $t > 0$

(b) Terminal point $P(x, y)$ determined by $t < 0$

The circumference of the unit circle is $C = 2\pi(1) = 2\pi$. So if a point starts at $(1, 0)$ and moves counterclockwise all the way around the unit circle and returns to $(1, 0)$, it travels a distance of 2π. To move halfway around the circle, it travels a distance of $\frac{1}{2}(2\pi) = \pi$. To move a quarter of the distance around the circle, it travels a distance of $\frac{1}{4}(2\pi) = \pi/2$. Where does the point end up when it travels these distances along the circle? From Figure 3 we see, for example, that when it travels a distance of π starting at $(1, 0)$, its terminal point is $(-1, 0)$.

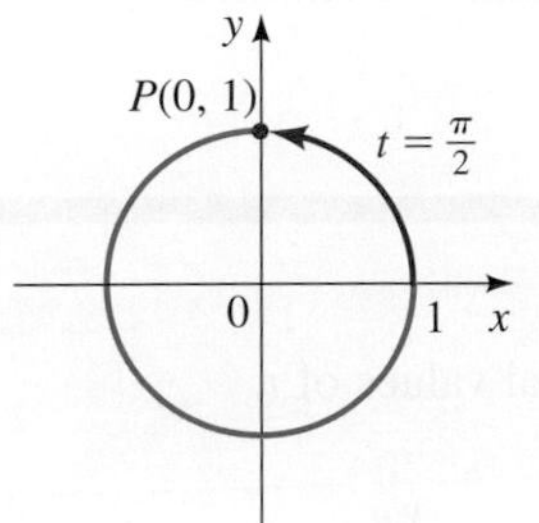

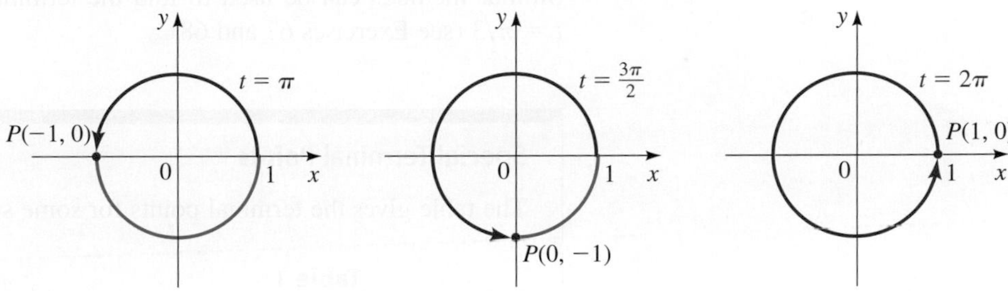

Figure 3 | Terminal points determined by $t = \frac{\pi}{2}, \pi, \frac{3\pi}{2}$, and 2π

Example 3 ■ Finding Terminal Points

Find the terminal point on the unit circle determined by each real number t.

(a) $t = 3\pi$ **(b)** $t = -\pi$ **(c)** $t = -\dfrac{\pi}{2}$

Solution From Figure 4 we get the following:

(a) The terminal point determined by 3π is $(-1, 0)$.

(b) The terminal point determined by $-\pi$ is $(-1, 0)$.

(c) The terminal point determined by $-\pi/2$ is $(0, -1)$.

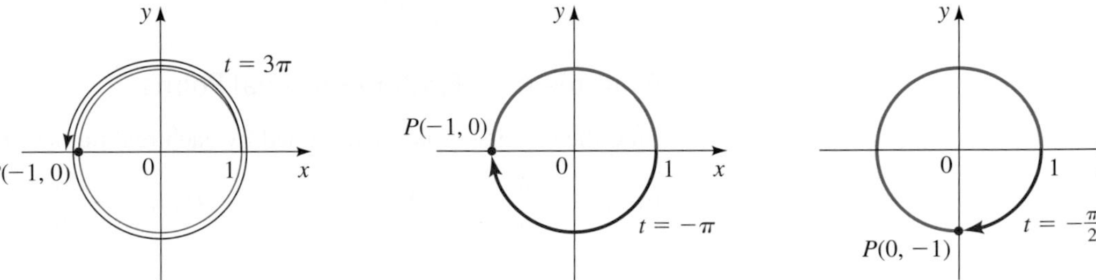

Figure 4

Notice that different values of t can determine the same terminal point.

Now Try Exercise 25 ■

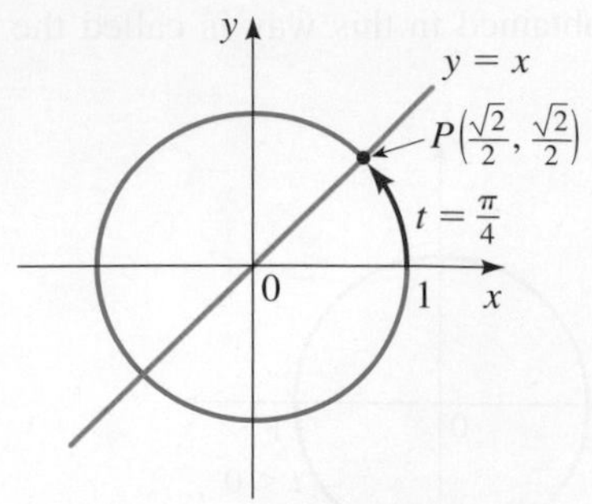

Figure 5

The terminal point $P(x, y)$ determined by $t = \pi/4$ is the same distance from $(1, 0)$ as from $(0, 1)$ along the unit circle (see Figure 5). Since the unit circle is symmetric with respect to the line $y = x$, it follows that P lies on the line $y = x$. So P is the point of intersection (in Quadrant I) of the circle $x^2 + y^2 = 1$ and the line $y = x$. Substituting x for y in the equation of the circle, we get

$$x^2 + x^2 = 1$$

$$2x^2 = 1 \qquad \text{Combine like terms}$$

$$x^2 = \frac{1}{2} \qquad \text{Divide by 2}$$

$$x = \pm\frac{1}{\sqrt{2}} \qquad \text{Take square roots}$$

Since P is in Quadrant I, $x = 1/\sqrt{2}$ and since $y = x$, we have $y = 1/\sqrt{2}$ also. Thus the terminal point determined by $\pi/4$ is

$$P\left(\frac{1}{\sqrt{2}}, \frac{1}{\sqrt{2}}\right) = P\left(\frac{\sqrt{2}}{2}, \frac{\sqrt{2}}{2}\right)$$

Similar methods can be used to find the terminal points determined by $t = \pi/6$ and $t = \pi/3$ (see Exercises 67 and 68).

Special Terminal Points

The table gives the terminal points for some special values of t.

Table 1

t	Terminal Point Determined by t
0	(1,0)
$\frac{\pi}{6}$	$\left(\frac{\sqrt{3}}{2}, \frac{1}{2}\right)$
$\frac{\pi}{4}$	$\left(\frac{\sqrt{2}}{2}, \frac{\sqrt{2}}{2}\right)$
$\frac{\pi}{3}$	$\left(\frac{1}{2}, \frac{\sqrt{3}}{2}\right)$
$\frac{\pi}{2}$	(0,1)
π	(−1,0)
$\frac{3\pi}{2}$	(0,−1)

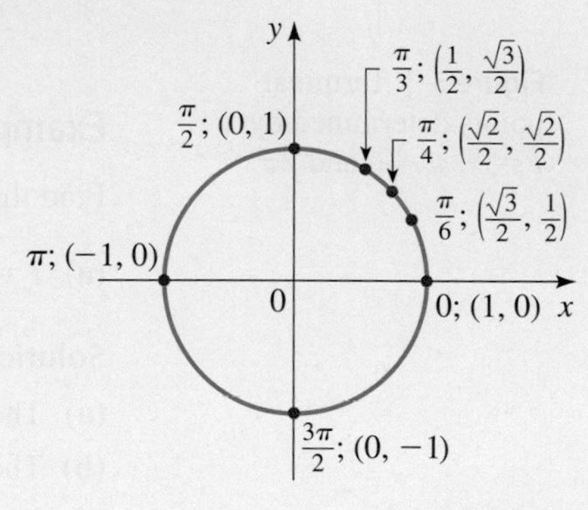

Example 4 ■ Finding Terminal Points

Find the terminal point determined by each real number t.

(a) $t = -\dfrac{\pi}{4}$ **(b)** $t = \dfrac{3\pi}{4}$ **(c)** $t = -\dfrac{5\pi}{6}$

Solution

(a) Let P be the terminal point determined by $-\pi/4$, and let Q be the terminal point determined by $\pi/4$. From Figure 6(a) we see that the point P has the same coordinates as Q except for sign. Since P is in Quadrant IV, its

x-coordinate is positive and its y-coordinate is negative. Thus, the terminal point is $P(\sqrt{2}/2, -\sqrt{2}/2)$.

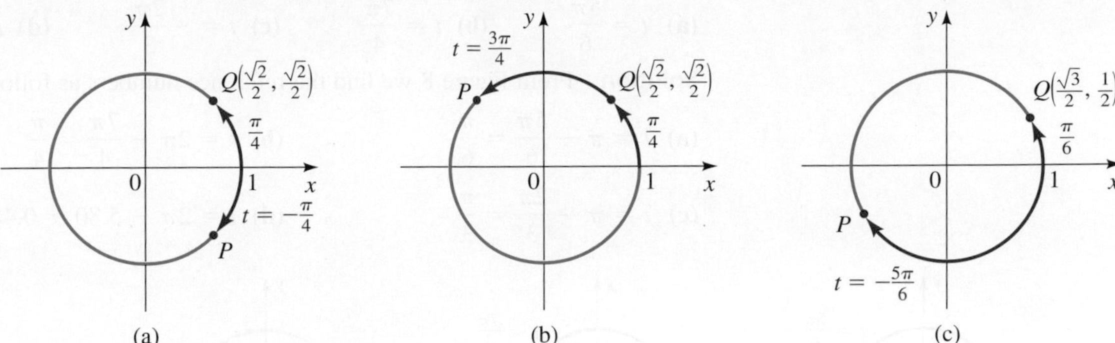

Figure 6

(b) Let P be the terminal point determined by $3\pi/4$, and let Q be the terminal point determined by $\pi/4$. From Figure 6(b) we see that the point P has the same coordinates as Q except for sign. Since P is in Quadrant II, its x-coordinate is negative and its y-coordinate is positive. Thus the terminal point is $P(-\sqrt{2}/2, \sqrt{2}/2)$.

(c) Let P be the terminal point determined by $-5\pi/6$, and let Q be the terminal point determined by $\pi/6$. From Figure 6(c) we see that the point P has the same coordinates as Q except for sign. Since P is in Quadrant III, its coordinates are both negative. Thus the terminal point is $P(-\sqrt{3}/2, -\frac{1}{2})$.

 **Now Try Exercise 33** ■

■ The Reference Number

From Examples 3 and 4 we see that to find a terminal point in any quadrant we need only know the "corresponding" terminal point in the first quadrant. We use the idea of the *reference number* to help us find terminal points.

Reference Number

Let t be a real number. The **reference number** $\bar{t}$ associated with t is the shortest distance along the unit circle between the terminal point determined by t and the x-axis.

Figure 7 shows that to find the reference number $\bar{t}$, it's helpful to know the quadrant in which the terminal point determined by t lies. If the terminal point lies in Quadrant I or IV, where x is positive, we find $\bar{t}$ by moving along the circle to the *positive* x-axis. If it lies in Quadrant II or III, where x is negative, we find $\bar{t}$ by moving along the circle to the *negative* x-axis.

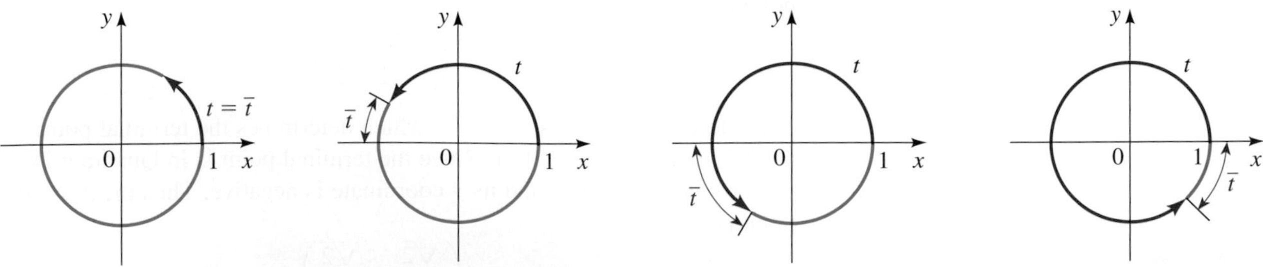

Figure 7 | The reference number $\bar{t}$ for t

Example 5 ■ Finding Reference Numbers

Find the reference number for each value of t.

(a) $t = \dfrac{5\pi}{6}$ **(b)** $t = \dfrac{7\pi}{4}$ **(c)** $t = -\dfrac{2\pi}{3}$ **(d)** $t = 5.80$

Solution From Figure 8 we find the reference numbers as follows.

(a) $\bar{t} = \pi - \dfrac{5\pi}{6} = \dfrac{\pi}{6}$ **(b)** $\bar{t} = 2\pi - \dfrac{7\pi}{4} = \dfrac{\pi}{4}$

(c) $\bar{t} = \pi - \dfrac{2\pi}{3} = \dfrac{\pi}{3}$ **(d)** $\bar{t} = 2\pi - 5.80 \approx 0.48$

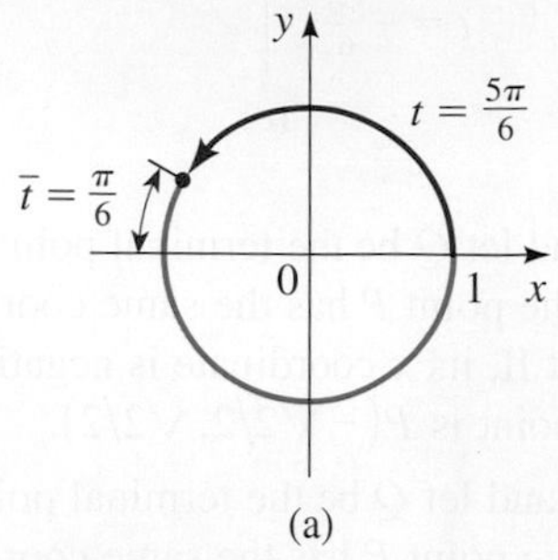

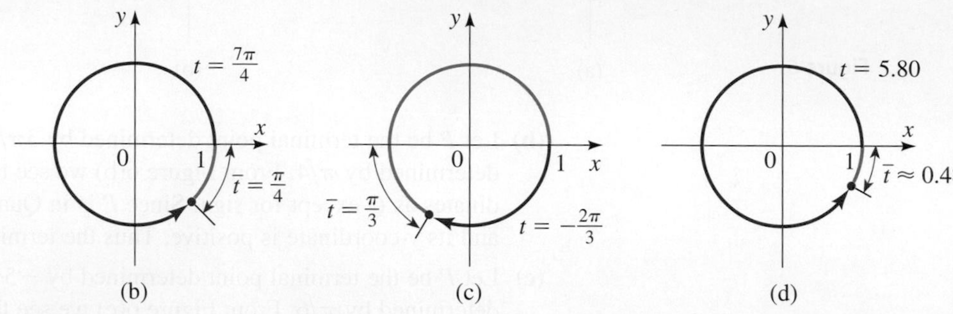

Figure 8

Now Try Exercise 43 ■

Using Reference Numbers to Find Terminal Points

To find the terminal point P determined by any value of t, we use the following steps:

1. Find the reference number $\bar{t}$.
2. Find the terminal point $Q(a, b)$ determined by $\bar{t}$.
3. The terminal point determined by t is $P(\pm a, \pm b)$, where the signs are chosen according to the quadrant in which this terminal point lies.

Example 6 ■ Using Reference Numbers to Find Terminal Points

Find the terminal point determined by each real number t.

(a) $t = \dfrac{5\pi}{6}$ **(b)** $t = \dfrac{7\pi}{4}$ **(c)** $t = -\dfrac{2\pi}{3}$

Solution The reference numbers associated with these values of t were found in Example 5.

(a) The reference number is $\bar{t} = \pi/6$, which determines the terminal point $\left(\sqrt{3}/2, \frac{1}{2}\right)$ from Table 1. Since the terminal point determined by t is in Quadrant II, its x-coordinate is negative and its y-coordinate is positive. Thus the desired terminal point is

$$\left(-\frac{\sqrt{3}}{2}, \frac{1}{2}\right)$$

(b) The reference number is $\bar{t} = \pi/4$, which determines the terminal point $\left(\sqrt{2}/2, \sqrt{2}/2\right)$ from Table 1. Since the terminal point is in Quadrant IV, its x-coordinate is positive and its y-coordinate is negative. Thus the desired terminal point is

$$\left(\frac{\sqrt{2}}{2}, -\frac{\sqrt{2}}{2}\right)$$

(c) The reference number is $\bar{t} = \pi/3$, which determines the terminal point $\left(\frac{1}{2}, \sqrt{3}/2\right)$ from Table 1. Since the terminal point determined by t is in Quadrant III, both of its coordinates are negative. Thus the desired terminal point is

$$\left(-\frac{1}{2}, -\frac{\sqrt{3}}{2}\right)$$

 Now Try Exercise 47 ■

Since the circumference of the unit circle is 2π, the terminal point determined by t is the same as that determined by $t + 2\pi$ or $t - 2\pi$. In general, we can add or subtract 2π any number of times without changing the terminal point determined by t. We use this observation in the next example to find terminal points for large values of t.

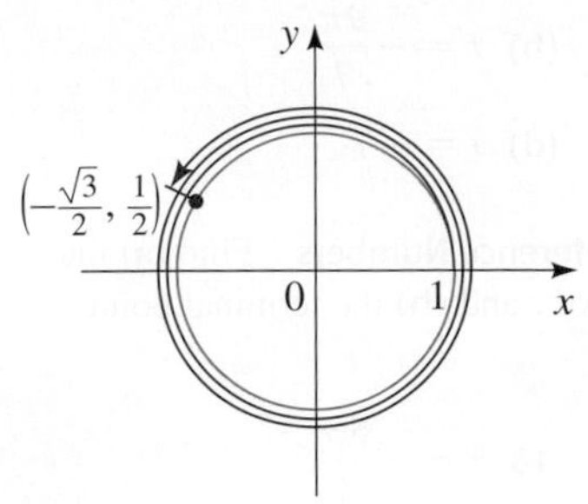

Figure 9

Example 7 ■ Finding the Terminal Point for Large t

Find the terminal point determined by $t = \dfrac{29\pi}{6}$.

Solution Since

$$t = \frac{29\pi}{6} = 4\pi + \frac{5\pi}{6}$$

we see that the terminal point of t is the same as that of $5\pi/6$ (that is, we subtract 4π). So by Example 6(a) the terminal point is $\left(-\sqrt{3}/2, \frac{1}{2}\right)$. (See Figure 9.)

 Now Try Exercise 53 ■

5.1 Exercises

Concepts

1. (a) The unit circle is the circle centered at ______ with radius ______.

(b) The equation of the unit circle is ______.

(c) Suppose the point $P(x, y)$ is on the unit circle. Find the missing coordinate:

(i) $P(1, \square)$ (ii) $P(\square, 1)$

(iii) $P(-1, \square)$ (iv) $P(\square, -1)$

2. (a) If we mark off a distance t along the unit circle, starting at $(1, 0)$ and moving in a counterclockwise direction, we arrive at the ______ point determined by t.

(b) The terminal points determined by $\pi/2$, π, $-\pi/2$, 2π are ______, ______, ______, and ______, respectively.

3. If the terminal point determined by t is $P(a, b)$, then the terminal point determined by $t + 2\pi$ is ______. The terminal point for $t = \pi/3$ is ______ and so the terminal point for $t = 7\pi/3$ is ______.

4. Complete the entries in the following figure. Is it true in general that if the terminal point determined by t is $P(a, b)$, then the terminal point determined by $t + \pi$ is $P(-a, -b)$?

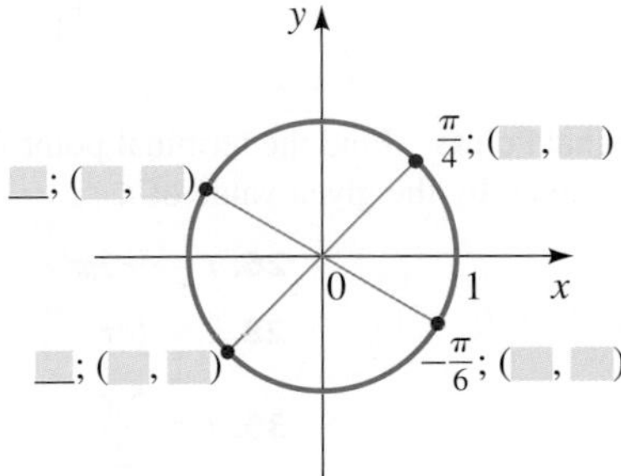

Skills

5–10 ■ Points on the Unit Circle Show that the point is on the unit circle.

 **5.** $\left(\dfrac{3}{5}, -\dfrac{4}{5}\right)$

6. $\left(-\dfrac{24}{25}, -\dfrac{7}{25}\right)$

7. $\left(\dfrac{3}{4}, -\dfrac{\sqrt{7}}{4}\right)$

8. $\left(-\dfrac{5}{7}, -\dfrac{2\sqrt{6}}{7}\right)$

9. $\left(-\dfrac{\sqrt{5}}{3}, \dfrac{2}{3}\right)$

10. $\left(\dfrac{\sqrt{11}}{6}, \dfrac{5}{6}\right)$

11–16 ■ Points on the Unit Circle Find the missing coordinate of P, using the fact that P lies on the unit circle in the given quadrant.

Coordinates	Quadrant
11. $P(-\frac{3}{5}, \quad)$	III
12. $P(\quad , -\frac{7}{25})$	IV
13. $P(\quad , \frac{1}{3})$	II
14. $P(\frac{2}{5}, \quad)$	I
15. $P(\quad , -\frac{2}{7})$	IV
16. $P(-\frac{2}{3}, \quad)$	II

17–22 ■ Points on the Unit Circle The point P is on the unit circle. Find $P(x, y)$ from the given information.

17. The x-coordinate of P is $\frac{5}{13}$, and the y-coordinate is negative.

18. The y-coordinate of P is $-\frac{3}{5}$, and the x-coordinate is positive.

19. The y-coordinate of P is $\frac{2}{3}$, and the x-coordinate is negative.

20. The x-coordinate of P is positive, and the y-coordinate of P is $-\sqrt{5}/5$.

21. The x-coordinate of P is $-\sqrt{2}/3$, and P lies below the x-axis.

22. The x-coordinate of P is $-\frac{2}{5}$, and P lies above the x-axis.

23–24 ■ Terminal Points Find t and the terminal point determined by t for each point in the figure. In Exercise 23, t increases in increments of $\pi/4$; in Exercise 24, t increases in increments of $\pi/6$.

23. **24.**

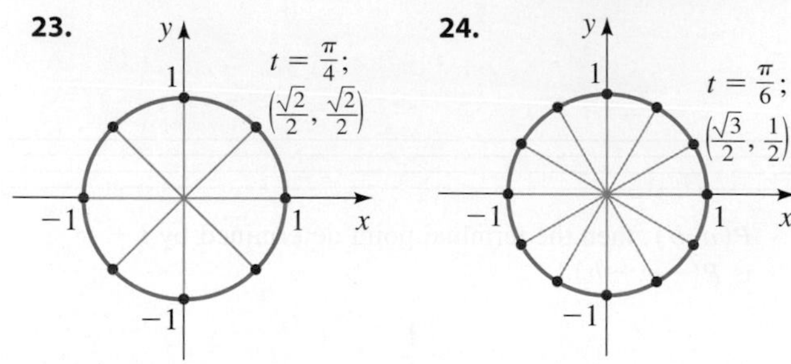

25–42 ■ Terminal Points Find the terminal point $P(x, y)$ on the unit circle determined by the given value of t.

25. $t = 5\pi$ **26.** $t = -3\pi$

27. $t = -4\pi$ **28.** $t = 6\pi$

29. $t = \dfrac{3\pi}{2}$ **30.** $t = \dfrac{5\pi}{2}$

31. $t = -\dfrac{5\pi}{2}$ **32.** $t = -\dfrac{3\pi}{2}$

33. $t = \dfrac{5\pi}{6}$ **34.** $t = -\dfrac{\pi}{3}$

35. $t = -\dfrac{3\pi}{4}$ **36.** $t = \dfrac{5\pi}{4}$

37. $t = -\dfrac{5\pi}{3}$ **38.** $t = -\dfrac{7\pi}{6}$

39. $t = \dfrac{7\pi}{4}$ **40.** $t = \dfrac{2\pi}{3}$

41. $t = \dfrac{7\pi}{6}$ **42.** $t = -\dfrac{7\pi}{4}$

43–46 ■ Reference Numbers Find the reference number for each value of t.

43. (a) $t = \dfrac{4\pi}{3}$ **(b)** $t = \dfrac{5\pi}{3}$

(c) $t = -\dfrac{7\pi}{6}$ **(d)** $t = 3.5$

44. (a) $t = 9\pi$ **(b)** $t = -\dfrac{5\pi}{4}$

(c) $t = \dfrac{25\pi}{6}$ **(d)** $t = 4$

45. (a) $t = \dfrac{5\pi}{7}$ **(b)** $t = -\dfrac{7\pi}{9}$

(c) $t = -3$ **(d)** $t = 5$

46. (a) $t = \dfrac{11\pi}{5}$ **(b)** $t = -\dfrac{9\pi}{7}$

(c) $t = 6$ **(d)** $t = -7$

47–60 ■ Terminal Points and Reference Numbers Find **(a)** the reference number for each value of t and **(b)** the terminal point determined by t.

47. $t = \dfrac{3\pi}{4}$ **48.** $t = -\dfrac{5\pi}{4}$

49. $t = -\dfrac{5\pi}{6}$ **50.** $t = \dfrac{4\pi}{3}$

51. $t = \dfrac{11\pi}{6}$ **52.** $t = -\dfrac{\pi}{6}$

53. $t = \dfrac{13\pi}{4}$ **54.** $t = \dfrac{13\pi}{6}$

55. $t = \dfrac{41\pi}{6}$ **56.** $t = \dfrac{17\pi}{4}$

57. $t = -\dfrac{11\pi}{3}$ **58.** $t = \dfrac{31\pi}{6}$

59. $t = \dfrac{16\pi}{3}$ **60.** $t = -\dfrac{41\pi}{4}$

61–64 ■ Terminal Points The unit circle is graphed in the figure. Use the figure to find the terminal point determined by the real number t, with coordinates rounded to one decimal place.

61. $t = 1$

62. $t = 2.5$

63. $t = -1.1$

64. $t = 4.2$

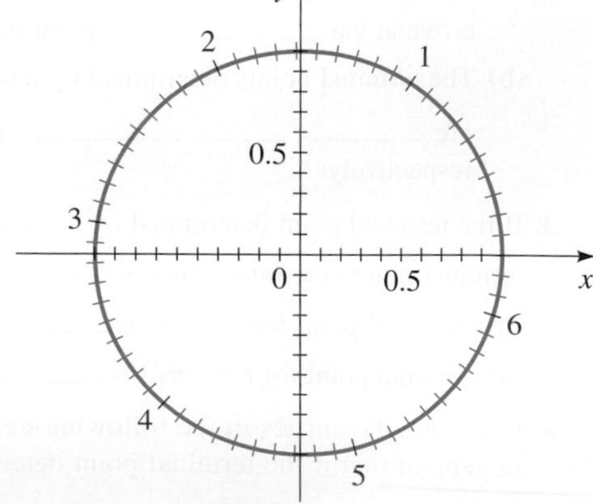

■ Skills Plus

65. Terminal Points Suppose that the terminal point determined by t is the point $\left(\frac{3}{5}, \frac{4}{5}\right)$ on the unit circle. Find the terminal point determined by each of the following.

(a) $\pi - t$ **(b)** $-t$

(c) $\pi + t$ **(d)** $2\pi + t$

66. Terminal Points Suppose that the terminal point determined by t is the point $\left(\frac{3}{4}, \sqrt{7}/4\right)$ on the unit circle. Find the terminal point determined by each of the following.

(a) $-t$ **(b)** $4\pi + t$

(c) $\pi - t$ **(d)** $t - \pi$

■ Discuss ■ Discover ■ Prove ■ Write

67. Discover ■ Prove: Finding the Terminal Point for $\pi/6$ Suppose the terminal point determined by $t = \pi/6$ is $P(x, y)$ and the points Q and R are as shown in the following figure. Why are the distances PQ and PR the same? Use this fact, together with the Distance Formula, to show that the coordinates of P satisfy the equation $2y = \sqrt{x^2 + (y - 1)^2}$. Simplify this equation using the fact that $x^2 + y^2 = 1$. Solve the simplified equation to find $P(x, y)$.

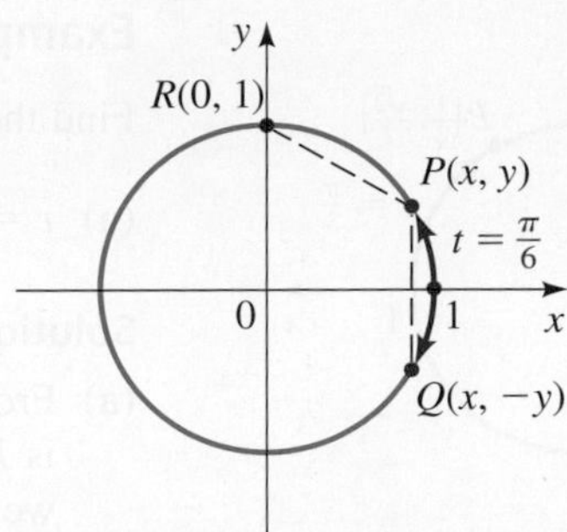

68. Discover ■ Prove: Finding the Terminal Point for $\pi/3$ Now that you know the terminal point determined by $t = \pi/6$, use symmetry to find the terminal point determined by $t = \pi/3$. (See the figure.) Explain your reasoning.

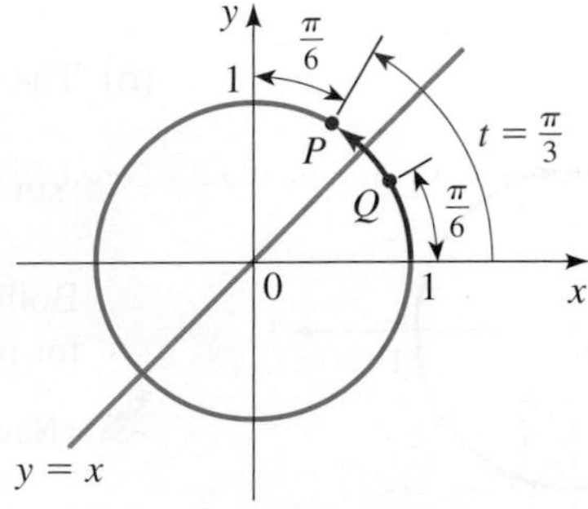

5.2 Trigonometric Functions of Real Numbers

■ The Trigonometric Functions ■ Values of the Trigonometric Functions ■ Fundamental Identities

In this section we use properties of the unit circle from the preceding section to define the trigonometric functions.

■ The Trigonometric Functions

Recall that to find the terminal point $P(x, y)$ for a given real number t, we move a distance $|t|$ along the unit circle, starting at the point $(1, 0)$. We move in a counterclockwise direction if t is positive and in a clockwise direction if t is negative (see Figure 1). We now use the x- and y-coordinates of the point $P(x, y)$ to define several functions. For instance, we define the function called *sine* by assigning to each real number t the y-coordinate of the terminal point $P(x, y)$ determined by t. The trigonometric functions **sine**, **cosine**, **tangent**, **cosecant**, **secant**, and **cotangent** are defined by using the coordinates of $P(x, y)$, as in the following box. The symbols we use for the names of these functions are abbreviations of their full names.

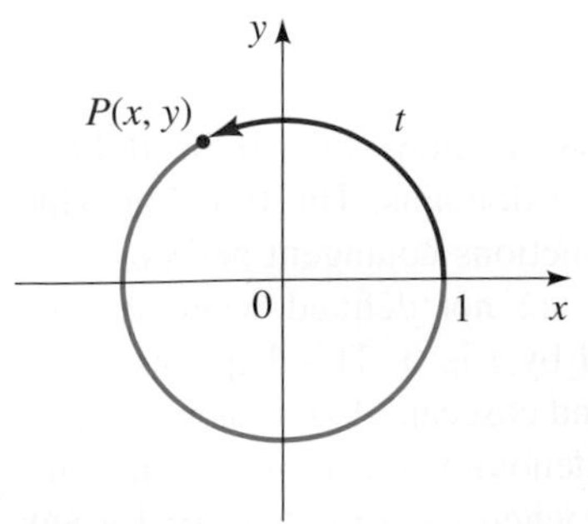

Figure 1

Definition of the Trigonometric Functions

Let t be any real number and let $P(x, y)$ be the terminal point on the unit circle determined by t. We define

$$\sin t = y \qquad \cos t = x \qquad \tan t = \frac{y}{x} \quad (x \neq 0)$$

$$\csc t = \frac{1}{y} \quad (y \neq 0) \qquad \sec t = \frac{1}{x} \quad (x \neq 0) \qquad \cot t = \frac{x}{y} \quad (y \neq 0)$$

Because the trigonometric functions can be defined in terms of the unit circle, they are sometimes called the **circular functions**.

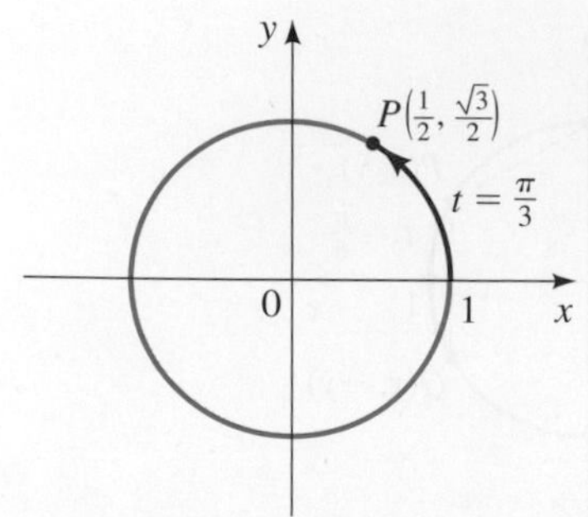

Figure 2

Example 1 ■ Evaluating Trigonometric Functions

Find the six trigonometric functions of each real number t.

(a) $t = \dfrac{\pi}{3}$ **(b)** $t = \dfrac{\pi}{2}$

Solution

(a) From Table 5.1.1 we see that the terminal point determined by $t = \pi/3$ is $P(\frac{1}{2}, \sqrt{3}/2)$. (See Figure 2.) Since the coordinates are $x = \frac{1}{2}$ and $y = \sqrt{3}/2$, we have

$$\sin\frac{\pi}{3} = \frac{\sqrt{3}}{2} \qquad \cos\frac{\pi}{3} = \frac{1}{2} \qquad \tan\frac{\pi}{3} = \frac{\sqrt{3}/2}{1/2} = \sqrt{3}$$

$$\csc\frac{\pi}{3} = \frac{2\sqrt{3}}{3} \qquad \sec\frac{\pi}{3} = 2 \qquad \cot\frac{\pi}{3} = \frac{1/2}{\sqrt{3}/2} = \frac{1}{\sqrt{3}} = \frac{\sqrt{3}}{3}$$

(b) The terminal point determined by $\pi/2$ is $P(0, 1)$. (See Figure 3.) So

$$\sin\frac{\pi}{2} = 1 \qquad \cos\frac{\pi}{2} = 0 \qquad \csc\frac{\pi}{2} = \frac{1}{1} = 1 \qquad \cot\frac{\pi}{2} = \frac{0}{1} = 0$$

Both $\tan \pi/2$ and $\sec \pi/2$ are undefined because $x = 0$ appears in the denominator in each of their definitions.

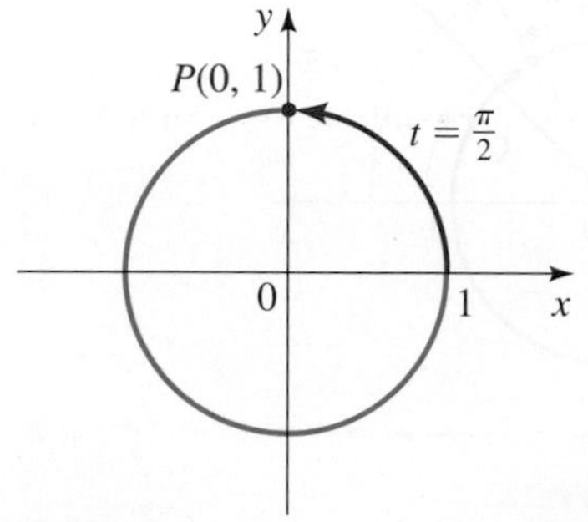

Figure 3

 **Now Try Exercise 5** ■

Some special values of the trigonometric functions are listed in Table 1 in the following box. This table is obtained from Table 5.1.1, together with the definitions of the trigonometric functions.

Special Values of the Trigonometric Functions

The table gives the values of the trigonometric functions for some special values of t.

Table 1

t	$\sin t$	$\cos t$	$\tan t$	$\csc t$	$\sec t$	$\cot t$
0	0	1	0	—	1	—
$\frac{\pi}{6}$	$\frac{1}{2}$	$\frac{\sqrt{3}}{2}$	$\frac{\sqrt{3}}{3}$	2	$\frac{2\sqrt{3}}{3}$	$\sqrt{3}$
$\frac{\pi}{4}$	$\frac{\sqrt{2}}{2}$	$\frac{\sqrt{2}}{2}$	1	$\sqrt{2}$	$\sqrt{2}$	1
$\frac{\pi}{3}$	$\frac{\sqrt{3}}{2}$	$\frac{1}{2}$	$\sqrt{3}$	$\frac{2\sqrt{3}}{3}$	2	$\frac{\sqrt{3}}{3}$
$\frac{\pi}{2}$	1	0	—	1	—	0
π	0	−1	0	—	−1	—
$\frac{3\pi}{2}$	−1	0	—	−1	—	0

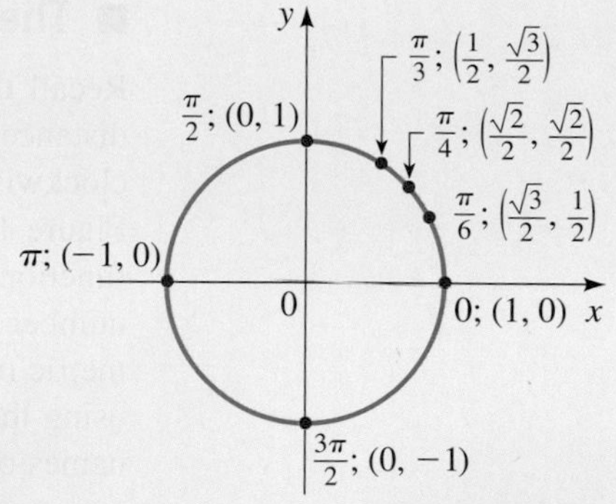

We can remember the special values of the sines and cosines by writing them in the form $\sqrt{■}/2$:

t	$\sin t$	$\cos t$
0	$\sqrt{0}/2$	$\sqrt{4}/2$
$\pi/6$	$\sqrt{1}/2$	$\sqrt{3}/2$
$\pi/4$	$\sqrt{2}/2$	$\sqrt{2}/2$
$\pi/3$	$\sqrt{3}/2$	$\sqrt{1}/2$
$\pi/2$	$\sqrt{4}/2$	$\sqrt{0}/2$

Example 1 shows that some of the trigonometric functions fail to be defined for certain real numbers. So we need to determine their domains. The functions sine and cosine are defined for all values of t. Since the functions cotangent and cosecant have y in the denominator of their definitions, they are not defined whenever the y-coordinate of the terminal point $P(x, y)$ determined by t is 0. This happens when $t = n\pi$ for any integer n, so the domains of cotangent and cosecant do not include these points. The functions tangent and secant have x in the denominator in their definitions, so they are not defined whenever $x = 0$. This happens when $t = (\pi/2) + n\pi$ for any integer n.

Domains of the Trigonometric Functions

Function	Domain
sin, cos	All real numbers
tan, sec	All real numbers other than $\dfrac{\pi}{2} + n\pi$ for any integer n
cot, csc	All real numbers other than $n\pi$ for any integer n

Note The range of the sine and cosine functions is the interval $[-1, 1]$. We can see this from the definition because the values of these functions are coordinates of points on the unit circle. The ranges of the other trigonometric functions will be discussed in Section 5.4.

■ Values of the Trigonometric Functions

To compute values of the trigonometric functions for any real number t, we first determine their signs. The signs of the trigonometric functions depend on the quadrant in which the terminal point of t lies. For example, if the terminal point $P(x, y)$ determined by t lies in Quadrant III, then its coordinates are both negative. So $\sin t$, $\cos t$, $\csc t$, and $\sec t$ are all negative, whereas $\tan t$ and $\cot t$ are positive. You can check the other entries in the following box.

The following mnemonic device will help you remember which trigonometric functions are positive in each quadrant: **A**ll of them, **S**ine, **T**angent, or **C**osine.

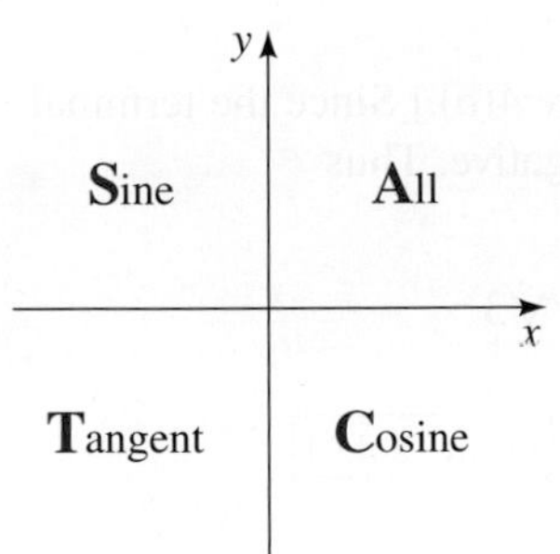

You can remember this as "All Students Take Calculus."

Signs of the Trigonometric Functions

Quadrant	Positive Functions	Negative Functions
I	all	none
II	sin, csc	cos, sec, tan, cot
III	tan, cot	sin, csc, cos, sec
IV	cos, sec	sin, csc, tan, cot

For example $\cos(2\pi/3) < 0$ because the terminal point of $t = 2\pi/3$ is in Quadrant II, whereas $\tan 4 > 0$ because the terminal point of $t = 4$ is in Quadrant III.

In Section 5.1 we used the reference number to find the terminal point determined by a real number t. Since the trigonometric functions are defined in terms of the coordinates of terminal points, we can use the reference number to find values of the trigonometric functions. Suppose that $\bar{t}$ is the reference number for t. Then the terminal point of $\bar{t}$ has the same coordinates, except possibly for sign, as the terminal point of t. So the value of each trigonometric function at t is the same, except possibly for sign, as its value at $\bar{t}$. We illustrate this procedure in the next example.

Evaluating Trigonometric Functions for any Real Number

To find the values of the trigonometric functions for any real number t, we carry out the following steps.

1. **Find the reference number.** Find the reference number $\bar{t}$ associated with t.
2. **Find the sign.** Determine the sign of the trigonometric function of t by noting the quadrant in which the terminal point lies.
3. **Find the value.** The value of the trigonometric function of t is the same, except possibly for sign, as the value of the trigonometric function of $\bar{t}$.

Example 2 ■ Evaluating Trigonometric Functions

Find each value.

(a) $\cos \dfrac{2\pi}{3}$ **(b)** $\tan\left(-\dfrac{\pi}{3}\right)$ **(c)** $\sin \dfrac{19\pi}{4}$

Solution

(a) The reference number for $2\pi/3$ is $\pi/3$. [See Figure 4(a).] Since the terminal point of $2\pi/3$ is in Quadrant II, $\cos(2\pi/3)$ is negative. Thus

$$\cos \frac{2\pi}{3} = -\cos \frac{\pi}{3} = -\frac{1}{2}$$

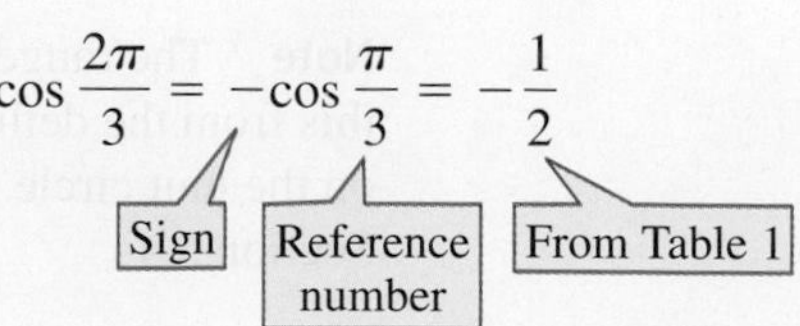

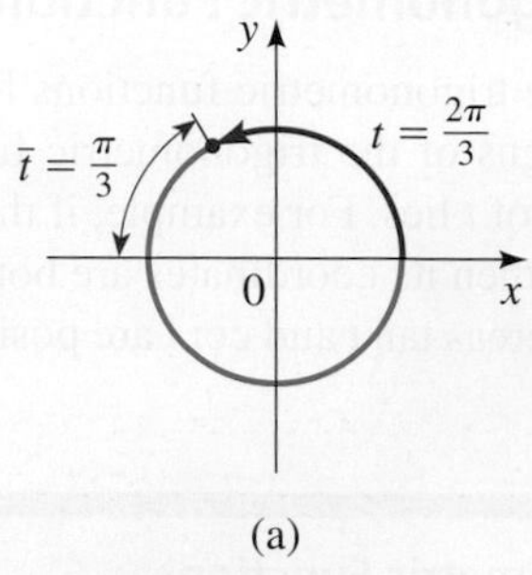

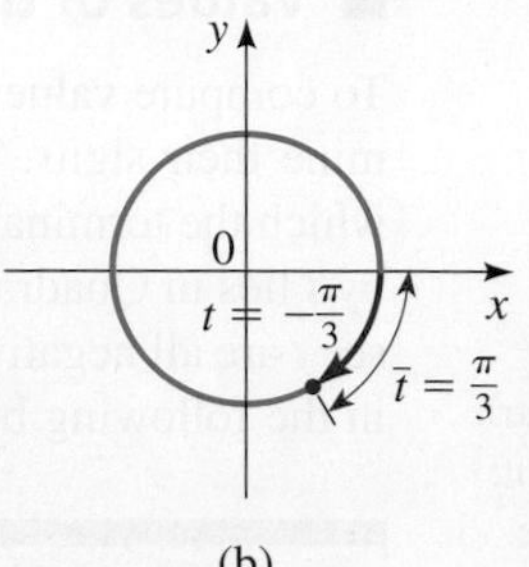

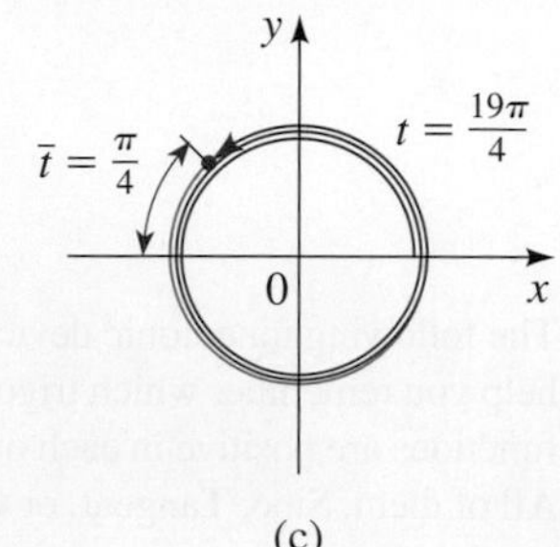

Figure 4 (a) (b) (c)

(b) The reference number for $-\pi/3$ is $\pi/3$. [See Figure 4(b).] Since the terminal point of $-\pi/3$ is in Quadrant IV, $\tan(-\pi/3)$ is negative. Thus

$$\tan\left(-\frac{\pi}{3}\right) = -\tan \frac{\pi}{3} = -\sqrt{3}$$

Sign | Reference number | From Table 1

(c) Since $(19\pi/4) - 4\pi = 3\pi/4$, the terminal points determined by $19\pi/4$ and $3\pi/4$ are the same. The reference number for $3\pi/4$ is $\pi/4$. [See Figure 4(c).] Since the terminal point of $3\pi/4$ is in Quadrant II, $\sin(3\pi/4)$ is positive. Thus

$$\sin \frac{19\pi}{4} = \sin \frac{3\pi}{4} = +\sin \frac{\pi}{4} = \frac{\sqrt{2}}{2}$$

Subtract 4π | Sign | Reference number | From Table 1

 **Now Try Exercise 7** ■

Using a Calculator So far, we were able to compute the values of the trigonometric functions only for certain values of t. In fact, we can compute the values of these functions whenever t is a multiple of $\pi/6$, $\pi/4$, $\pi/3$, and $\pi/2$. How can we compute the trigonometric functions for other values of t? For example, how can we find $\sin 1.5$? One way is to carefully sketch a diagram and read the value (see Exercises 39–46); however, this method is not very accurate. Fortunately, programmed directly into scientific calculators are mathematical procedures (called *numerical methods*) that find the values of *sine*, *cosine*, and *tangent* correct to the number of digits in the display. The calculator must be put in *radian mode* to evaluate these functions.

For an explanation of numerical methods see *Mathematics in the Modern World* in Section 5.4.

(*text continues*)

The Unit Circle Approach and the Right Triangle Approach

If you have already studied the trigonometric functions of *angles* θ defined using ratios of sides of right triangles (Chapter 6), you may be wondering how these are related to the trigonometric functions of *real numbers t* defined using points on the unit circle, as in this chapter. The functions are exactly the same, provided angles are measured in radians.

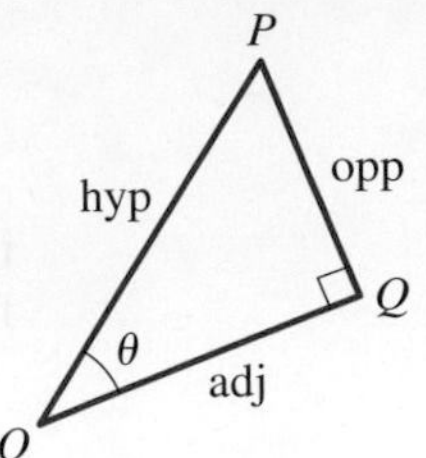

Figure A | Triangle OPQ is a right triangle.

To see how, let's start with a right triangle OPQ as in Figure A. Place the triangle in the coordinate plane with angle θ in standard position and draw a unit circle at the origin, as shown in Figure B. The point $P'(x, y)$ on the unit circle in Figure B is the terminal point determined by the arc of length t. By the definitions in this chapter, we have

$$\sin t = y$$
$$\cos t = x$$

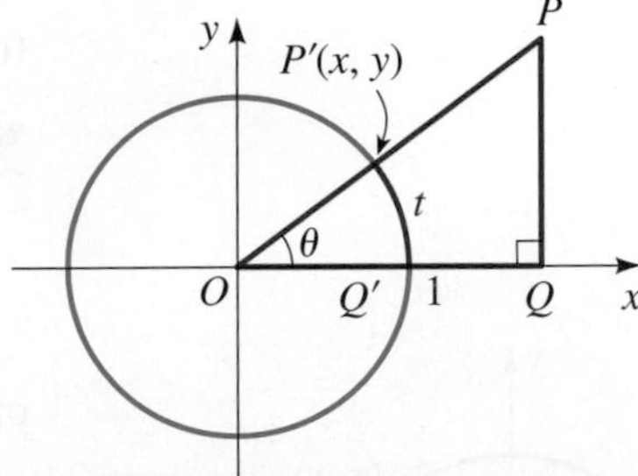

Figure B | The radian measure of the angle θ is t.

Now let's drop a perpendicular from P' to the point Q' on the x-axis, as in Figure C. Observe that triangle OPQ is similar to triangle $OP'Q'$, whose legs have lengths x and y. By the definition of the trigonometric functions of the angle θ (Chapter 6), we have

$$\sin \theta = \frac{\text{opp}}{\text{hyp}} = \frac{PQ}{OP} = \frac{P'Q'}{OP'} = \frac{y}{1} = y = \sin t$$

$$\cos \theta = \frac{\text{adj}}{\text{hyp}} = \frac{OQ}{OP} = \frac{OQ'}{OP'} = \frac{x}{1} = x = \cos t$$

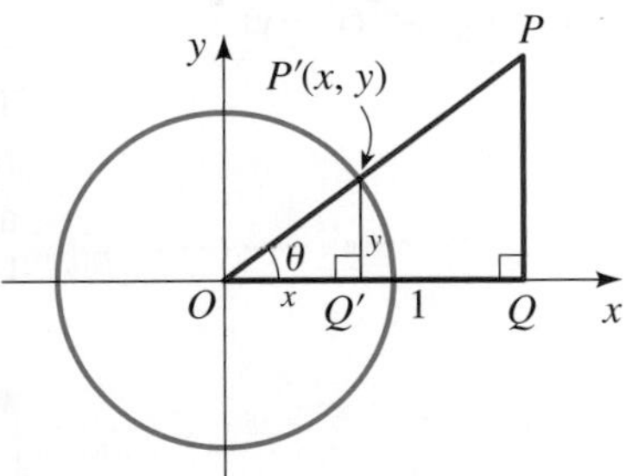

Figure C | Triangle OPQ is similar to triangle $OP'Q'$.

Since the radian measure of θ is t, we see that the trigonometric functions of the angle with radian measure θ are exactly the same as the trigonometric functions defined in terms of the terminal point on the unit circle determined by the real number t. In other words, as functions, they assign identical values to a given real number—the real number is the length t of an arc in one case, or the radian measure of θ in the other.

Why then do we study trigonometry in two different ways? Because different applications require that we view the trigonometric functions differently. (Compare the applications in Section 5.6 with those in Sections 6.5 and 6.6.)

Some calculators only give the values of sine, cosine, and tangent. To find the values of cosecant, secant, and cotangent on such a calculator, we use the following *reciprocal relations*:

$$\csc t = \frac{1}{\sin t} \qquad \sec t = \frac{1}{\cos t} \qquad \cot t = \frac{1}{\tan t}$$

These relations follow from the definitions of the trigonometric functions. For instance, since $\sin t = y$ and $\csc t = 1/y$, we have $\csc t = 1/y = 1/(\sin t)$. The others follow similarly.

Example 3 ■ Using a Calculator to Evaluate Trigonometric Functions

Using a calculator in radian mode, we obtain the following values, rounded to six decimal places. Check to make sure you get these answers on your calculator.

(a) $\sin 2.2 \approx 0.808496$ **(b)** $\cos 1.1 \approx 0.453596$

(c) $\cot 28 = \dfrac{1}{\tan 28} \approx -3.553286$ **(d)** $\csc 0.98 = \dfrac{1}{\sin 0.98} \approx 1.204098$

 **Now Try Exercises 41 and 43** ■

Let's consider the relationship between the trigonometric functions of t and those of $-t$. From Figure 5 we see that

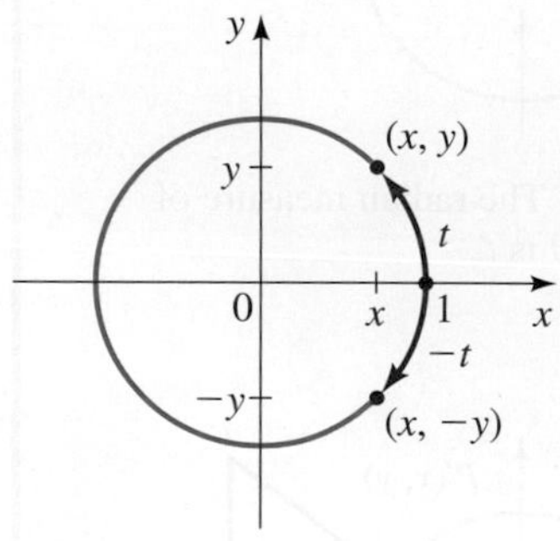

Figure 5

$$\sin(-t) = -y = -\sin t$$
$$\cos(-t) = x = \cos t$$
$$\tan(-t) = \frac{-y}{x} = -\frac{y}{x} = -\tan t$$

These equations show that sine and tangent are odd functions, whereas cosine is an even function. Additionally, the reciprocal of an even function is even and the reciprocal of an odd function is odd. This fact, together with the reciprocal relations, completes our knowledge of the even-odd properties for all the trigonometric functions.

Even and odd functions are defined in Section 2.6.

Even-Odd Properties

Sine, cosecant, tangent, and cotangent are odd functions; cosine and secant are even functions.

$$\sin(-t) = -\sin t \qquad \cos(-t) = \cos t \qquad \tan(-t) = -\tan t$$
$$\csc(-t) = -\csc t \qquad \sec(-t) = \sec t \qquad \cot(-t) = -\cot t$$

Example 4 ■ Even and Odd Trigonometric Functions

Use the even-odd properties of the trigonometric functions to determine each value.

(a) $\sin\left(-\dfrac{\pi}{6}\right)$ **(b)** $\cos\left(-\dfrac{\pi}{4}\right)$

Solution By the even-odd properties and Table 1, we have

(a) $\sin\left(-\dfrac{\pi}{6}\right) = -\sin\dfrac{\pi}{6} = -\dfrac{1}{2}$ Sine is odd

(b) $\cos\left(-\dfrac{\pi}{4}\right) = \cos\dfrac{\pi}{4} = \dfrac{\sqrt{2}}{2}$ Cosine is even

Now Try Exercise 15 ■

■ Fundamental Identities

The trigonometric functions are related to each other through equations called **trigonometric identities**. We give the most important ones in the following box.*

Fundamental Identities

Reciprocal Identities

$$\csc t = \frac{1}{\sin t} \qquad \sec t = \frac{1}{\cos t} \qquad \cot t = \frac{1}{\tan t} \qquad \tan t = \frac{\sin t}{\cos t} \qquad \cot t = \frac{\cos t}{\sin t}$$

Pythagorean Identities

$$\sin^2 t + \cos^2 t = 1 \qquad \tan^2 t + 1 = \sec^2 t \qquad 1 + \cot^2 t = \csc^2 t$$

Proof The reciprocal identities follow immediately from the definitions. We now prove the Pythagorean identities. By definition, $\cos t = x$ and $\sin t = y$, where x and y are the coordinates of a point $P(x, y)$ on the unit circle. Since $P(x, y)$ is on the unit circle, we have $x^2 + y^2 = 1$. Thus

$$\sin^2 t + \cos^2 t = 1$$

Dividing both sides by $\cos^2 t$ (provided that $\cos t \neq 0$), we get

$$\frac{\sin^2 t}{\cos^2 t} + \frac{\cos^2 t}{\cos^2 t} = \frac{1}{\cos^2 t}$$

$$\left(\frac{\sin t}{\cos t}\right)^2 + 1 = \left(\frac{1}{\cos t}\right)^2$$

$$\tan^2 t + 1 = \sec^2 t$$

We have used the reciprocal identities $\sin t/(\cos t) = \tan t$ and $1/(\cos t) = \sec t$. Similarly, dividing both sides of the first Pythagorean identity by $\sin^2 t$ (provided that $\sin t \neq 0$) gives us $1 + \cot^2 t = \csc^2 t$. ■

As their name indicates, the fundamental identities play a central role in trigonometry because we can use them to relate any trigonometric function to any other. So if we know the value of any one of the trigonometric functions at t, then we can find the values of all the others at t.

Example 5 ■ Finding All Trigonometric Functions from the Value of One

If $\cos t = \frac{3}{5}$ and t is in Quadrant IV, find the values of all the trigonometric functions at t.

*We follow the usual convention of writing $\sin^2 t$ for $(\sin t)^2$. In general, we write $\sin^n t$ for $(\sin t)^n$ for all integers n except $n = -1$. The superscript $n = -1$ will be assigned another meaning in Section 5.5. The same convention applies to the other five trigonometric functions.

The Value of π

The number π is the ratio of the circumference of a circle to its diameter. It has been known since ancient times that this ratio is the same for all circles. The first systematic effort to find a numerical approximation for π was made by Archimedes (ca. 240 B.C.), who proved that $\frac{22}{7} < \pi < \frac{223}{71}$ by finding the perimeters of regular polygons inscribed in and circumscribed about a circle.

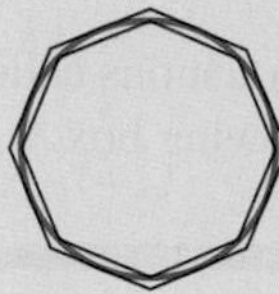

In about A.D. 480, the Chinese scientist and mathematician Zu Chongzhi gave the approximation

$$\pi \approx \frac{355}{113} = 3.141592\ldots$$

which is correct to six decimals. This remained the most accurate estimation of π until the Dutch mathematician Adrianus Romanus (1593) used polygons with more than a billion sides to compute π correct to 15 decimals. In the 17th century, mathematicians began to use infinite series and trigonometric identities in the quest for π. The Englishman William Shanks spent 15 years (1858–1873) using these methods to compute π to 707 decimals, but in 1946 it was found that his figures were wrong beginning with the 528th decimal. Today, with the aid of computers, mathematicians routinely determine π correct to millions of decimal places. In fact, mathematicians have now developed algorithms that can be programmed into computers to calculate π to many trillions of decimal places.

Solution From the Pythagorean identities we have

$$\begin{aligned} \sin^2 t + \cos^2 t &= 1 && \text{Pythagorean identity} \\ \sin^2 t + \left(\tfrac{3}{5}\right)^2 &= 1 && \text{Substitute } \cos t = \tfrac{3}{5} \\ \sin^2 t = 1 - \tfrac{9}{25} &= \tfrac{16}{25} && \text{Solve for } \sin^2 t \\ \sin t &= \pm\tfrac{4}{5} && \text{Take square roots} \end{aligned}$$

Since this point is in Quadrant IV, $\sin t$ is negative, so $\sin t = -\frac{4}{5}$. Now that we know both $\sin t$ and $\cos t$, we can find the values of the other trigonometric functions using the reciprocal identities.

$$\sin t = -\frac{4}{5} \qquad \cos t = \frac{3}{5} \qquad \tan t = \frac{\sin t}{\cos t} = \frac{-\frac{4}{5}}{\frac{3}{5}} = -\frac{4}{3}$$

$$\csc t = \frac{1}{\sin t} = -\frac{5}{4} \qquad \sec t = \frac{1}{\cos t} = \frac{5}{3} \qquad \cot t = \frac{1}{\tan t} = -\frac{3}{4}$$

Now Try Exercise 67 ■

Example 6 ■ Writing One Trigonometric Function in Terms of Another

Write $\tan t$ in terms of $\cos t$, where t is in Quadrant III.

Solution Since $\tan t = \sin t/\cos t$, we need to write $\sin t$ in terms of $\cos t$. By the Pythagorean identities, we have

$$\begin{aligned} \sin^2 t + \cos^2 t &= 1 \\ \sin^2 t &= 1 - \cos^2 t && \text{Solve for } \sin^2 t \\ \sin t &= \pm\sqrt{1 - \cos^2 t} && \text{Take square roots} \end{aligned}$$

Since $\sin t$ is negative in Quadrant III, the negative sign applies here. Thus

$$\tan t = \frac{\sin t}{\cos t} = \frac{-\sqrt{1 - \cos^2 t}}{\cos t}$$

 Now Try Exercise 55 ■

5.2 Exercises

Concepts

1. Let $P(x, y)$ be the terminal point on the unit circle determined by t. Then $\sin t =$ ______, $\cos t =$ ______, and $\tan t =$ ______.

2. If $P(x, y)$ is on the unit circle, then $x^2 + y^2 =$ ______. So for all t we have $\sin^2 t + \cos^2 t =$ ______, and we can write cosine in terms of sine as $\cos t = \pm\sqrt{\blacksquare}$ and sine in terms of cosine as $\sin t = \pm\sqrt{\blacksquare}$.

3–4 Let t_1, t_2, and t_3 be the real numbers whose terminal points are shown on the unit circle in the following figure. Arrange the values of the given trigonometric function in increasing order.

3. $\sin t_1$, $\sin t_2$, $\sin t_3$ **4.** $\cos t_1$, $\cos t_2$, $\cos t_3$

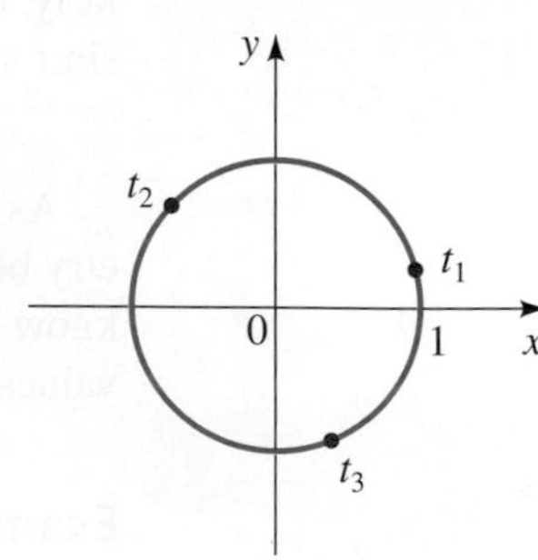

Skills

5–6 ■ Evaluating Trigonometric Functions Find $\cos t$ and $\sin t$ for the values of t whose terminal points are shown on the unit circle in the figure. In Exercise 5, t increases in increments of $\pi/4$;

in Exercise 6, t increases in increments of $\pi/6$. (See Exercises 5.1.23 and 5.1.24.)

5. **6.**

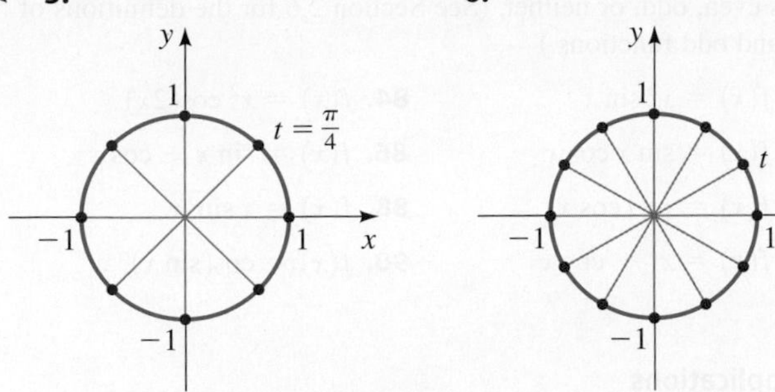

7–24 ■ Evaluating Trigonometric Functions Find the exact value of the trigonometric function at the given real number.

7. (a) $\sin\left(-\dfrac{2\pi}{3}\right)$ (b) $\cos\dfrac{17\pi}{4}$ (c) $\tan\dfrac{17\pi}{6}$

8. (a) $\sin\left(-\dfrac{5\pi}{6}\right)$ (b) $\cos\dfrac{5\pi}{6}$ (c) $\tan\dfrac{14\pi}{3}$

9. (a) $\sin\dfrac{13\pi}{4}$ (b) $\cos\left(-\dfrac{3\pi}{4}\right)$ (c) $\tan\dfrac{7\pi}{6}$

10. (a) $\sin\dfrac{11\pi}{3}$ (b) $\cos\dfrac{11\pi}{6}$ (c) $\tan\dfrac{7\pi}{4}$

11. (a) $\cos\dfrac{3\pi}{4}$ (b) $\cos\dfrac{5\pi}{4}$ (c) $\cos\dfrac{7\pi}{4}$

12. (a) $\sin\dfrac{3\pi}{4}$ (b) $\sin\dfrac{5\pi}{4}$ (c) $\sin\dfrac{7\pi}{4}$

13. (a) $\sin\dfrac{7\pi}{3}$ (b) $\csc\dfrac{7\pi}{3}$ (c) $\cot\dfrac{7\pi}{3}$

14. (a) $\csc\dfrac{5\pi}{4}$ (b) $\sec\dfrac{5\pi}{4}$ (c) $\tan\dfrac{5\pi}{4}$

15. (a) $\cos\left(-\dfrac{\pi}{3}\right)$ (b) $\sec\left(-\dfrac{\pi}{3}\right)$ (c) $\sin\left(-\dfrac{\pi}{3}\right)$

16. (a) $\tan\left(-\dfrac{\pi}{4}\right)$ (b) $\csc\left(-\dfrac{\pi}{4}\right)$ (c) $\cot\left(-\dfrac{\pi}{4}\right)$

17. (a) $\cos\left(-\dfrac{\pi}{6}\right)$ (b) $\csc\left(-\dfrac{\pi}{3}\right)$ (c) $\tan\left(-\dfrac{\pi}{6}\right)$

18. (a) $\sin\left(-\dfrac{\pi}{4}\right)$ (b) $\sec\left(-\dfrac{\pi}{4}\right)$ (c) $\cot\left(-\dfrac{\pi}{6}\right)$

19. (a) $\csc\dfrac{7\pi}{6}$ (b) $\sec\left(-\dfrac{\pi}{6}\right)$ (c) $\cot\left(-\dfrac{5\pi}{6}\right)$

20. (a) $\sec\dfrac{3\pi}{4}$ (b) $\cos\left(-\dfrac{2\pi}{3}\right)$ (c) $\tan\left(-\dfrac{7\pi}{6}\right)$

21. (a) $\sin\dfrac{4\pi}{3}$ (b) $\sec\dfrac{11\pi}{6}$ (c) $\cot\left(-\dfrac{\pi}{3}\right)$

22. (a) $\csc\dfrac{2\pi}{3}$ (b) $\sec\left(-\dfrac{5\pi}{3}\right)$ (c) $\cos\dfrac{10\pi}{3}$

23. (a) $\sin 13\pi$ (b) $\cos 14\pi$ (c) $\tan 15\pi$

24. (a) $\sin\dfrac{25\pi}{2}$ (b) $\cos\dfrac{25\pi}{2}$ (c) $\cot\dfrac{25\pi}{2}$

25–28 ■ Evaluating Trigonometric Functions Find the value of each of the six trigonometric functions (if it is defined) at the given real number t. Use your answers to complete the table.

25. $t = 0$ **26.** $t = \dfrac{\pi}{2}$

27. $t = \pi$ **28.** $t = \dfrac{3\pi}{2}$

t	$\sin t$	$\cos t$	$\tan t$	$\csc t$	$\sec t$	$\cot t$
0	0	1		undefined		
$\frac{\pi}{2}$						
π			0			undefined
$\frac{3\pi}{2}$						

29–38 ■ Evaluating Trigonometric Functions The terminal point $P(x, y)$ determined by a real number t is given. Find $\sin t$, $\cos t$, and $\tan t$.

29. $\left(-\dfrac{4}{5}, \dfrac{3}{5}\right)$ **30.** $\left(\dfrac{3}{5}, -\dfrac{4}{5}\right)$

31. $\left(-\dfrac{\sqrt{3}}{2}, \dfrac{1}{2}\right)$ **32.** $\left(-\dfrac{1}{2}, -\dfrac{\sqrt{3}}{2}\right)$

33. $\left(-\dfrac{6}{7}, \dfrac{\sqrt{13}}{7}\right)$ **34.** $\left(\dfrac{40}{41}, \dfrac{9}{41}\right)$

35. $\left(-\dfrac{5}{13}, -\dfrac{12}{13}\right)$ **36.** $\left(\dfrac{\sqrt{5}}{5}, \dfrac{2\sqrt{5}}{5}\right)$

37. $\left(-\dfrac{20}{29}, \dfrac{21}{29}\right)$ **38.** $\left(\dfrac{24}{25}, -\dfrac{7}{25}\right)$

39–46 ■ Values of Trigonometric Functions Find an approximate value of the given trigonometric function by using **(a)** the figure and **(b)** a calculator. Compare the two values.

39. $\sin 1$

40. $\cos 0.8$

41. $\sin 1.2$

42. $\cos 5$

43. $\tan 0.8$

44. $\tan(-1.3)$

45. $\cos 4.1$

46. $\sin(-5.2)$

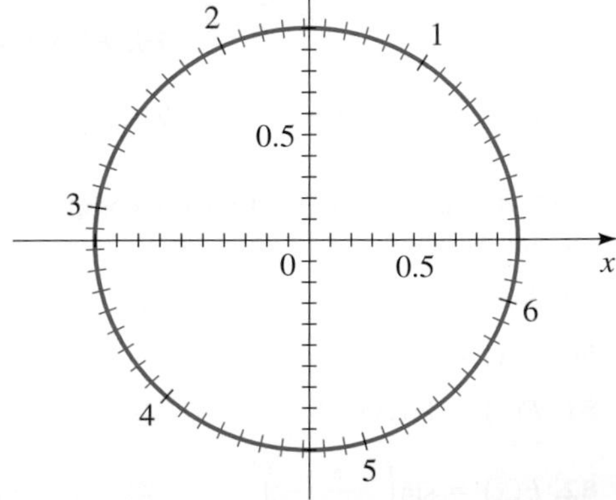

47–50 ■ Sign of a Trigonometric Expression Find the sign of the expression if the terminal point determined by t is in the given quadrant.

47. $\sin t \cos t$, Quadrant IV **48.** $\sin t \tan t$, Quadrant IV

49. $\dfrac{\tan t \sin t}{\cot t}$, Quadrant III **50.** $\cos t \sec t$, any quadrant

51–54 ■ Quadrant of a Terminal Point From the information given, find the quadrant in which the terminal point determined by t lies.

51. $\sin t > 0$ and $\cos t < 0$

52. $\tan t > 0$ and $\sin t < 0$

53. $\csc t > 0$ and $\sec t < 0$

54. $\cos t < 0$ and $\cot t < 0$

55–66 ■ Writing One Trigonometric Expression in Terms of Another Write the first expression in terms of the second if the terminal point determined by t is in the given quadrant.

55. $\cos t$, $\sin t$; Quadrant III **56.** $\sin t$, $\cos t$; Quadrant IV

57. $\sin t$, $\cos t$; Quadrant II **58.** $\tan t$, $\cos t$; Quadrant IV

59. $\tan t$, $\cos t$; Quadrant II **60.** $\tan t$, $\sin t$; Quadrant II

61. $\tan t$, $\sec t$; Quadrant IV **62.** $\sec t$, $\tan t$; Quadrant IV

63. $\csc t$, $\cot t$; Quadrant II **64.** $\sin t$, $\sec t$; Quadrant III

65. $\tan^2 t$, $\sin t$; any quadrant

66. $\sec^2 t \sin^2 t$, $\cos t$; any quadrant

67–74 ■ Using the Pythagorean Identities Find the values of the trigonometric functions of t from the given information.

67. $\sin t = -\frac{4}{5}$, terminal point of t is in Quadrant IV

68. $\cos t = -\frac{7}{25}$, terminal point of t is in Quadrant III

69. $\sec t = 3$, terminal point of t is in Quadrant IV

70. $\tan t = \frac{1}{4}$, terminal point of t is in Quadrant III

71. $\tan t = -\frac{12}{5}$, $\sin t > 0$

72. $\csc t = 5$, $\cos t < 0$

73. $\sin t = -\frac{1}{4}$, $\sec t < 0$

74. $\tan t = -4$, $\csc t > 0$

75–78 ■ Expressing a Function as a Composition Find functions f and g such that $F = f \circ g$.

75. $F(x) = \cos^2 x$ **76.** $F(x) = e^{\sin x}$

77. $F(x) = \sqrt{1 + \tan x}$ **78.** $F(x) = \dfrac{\sin x}{1 - \sin x}$

79–82 ■ Expressing a Function as a Composition Find functions f, g, and h such that $F = f \circ g \circ h$.

79. $F(x) = e^{\sin^2 x}$

80. $F(x) = \sin\sqrt{\ln x}$

81. $F(x) = \ln(\cos^2 x)$

82. $F(x) = \sin\left(\dfrac{e^x}{1 + e^x}\right)$

Skills Plus

83–90 ■ Even and Odd Functions Determine whether the function is even, odd, or neither. (See Section 2.6 for the definitions of even and odd functions.)

83. $f(x) = x^2 \sin x$ **84.** $f(x) = x^2 \cos(2x)$

85. $f(x) = \sin x \cos x$ **86.** $f(x) = \sin x + \cos x$

87. $f(x) = |x| \cos x$ **88.** $f(x) = x \sin^3 x$

89. $f(x) = x^3 + \cos x$ **90.** $f(x) = \cos(\sin x)$

Applications

91. Harmonic Motion The displacement from equilibrium ($y = 0$) of an oscillating mass attached to a spring is given by $y(t) = 4\cos(3\pi t)$ where y is measured in centimeters and t in seconds. Find the displacement at the times indicated in the table.

t	$y(t)$
0	
0.25	
0.50	
0.75	
1.00	
1.25	

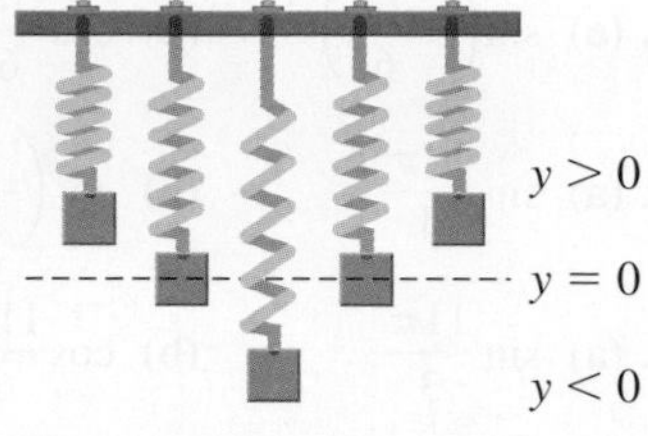

92. Circadian Rhythms Everybody's blood pressure varies over the course of the day. In a certain individual the resting diastolic blood pressure at time t is given by $B(t) = 80 + 7\sin(\pi t/12)$, where t is measured in hours since midnight and $B(t)$ in mmHg (millimeters of mercury). Find this person's resting diastolic blood pressure at each time.

(a) 6:00 A.M. **(b)** 10:30 A.M. **(c)** Noon **(d)** 8:00 P.M.

93. Electric Circuit After the switch is closed in the circuit shown, the current t seconds later is $I(t) = 0.8e^{-3t}\sin(10t)$, where I is measured in amps. Find the current at the times **(a)** $t = 0.1$ s and **(b)** $t = 0.5$ s.

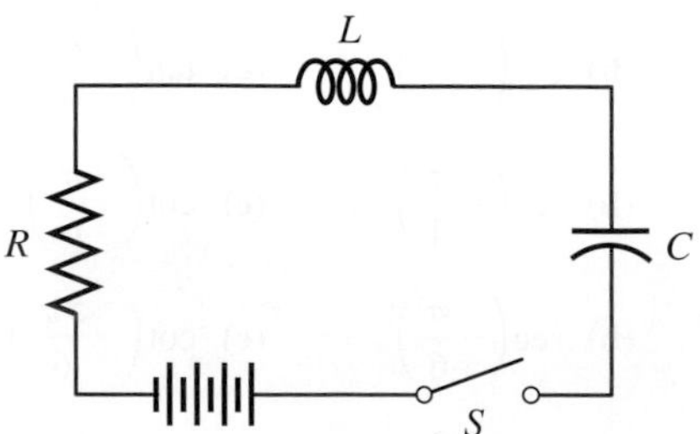

$L = 10^3$ h
$R = 6 \times 10^3\ \Omega$
$C = 9.17\ \mu$F
$E = 4.8 \times 10^3$ V

94. Bungee Jumping A bungee jumper plummets from a high bridge to the river below and then bounces upward, over and over again. At time t seconds after the jump, the jumper's height H (in meters) above the river is given by $H(t) = 100 + 75e^{-t/20}\cos\left(\frac{\pi t}{4}\right)$. Find the height of the jumper above the river at the times indicated in the table.

t	$H(t)$
0	
1	
2	
4	
6	
8	
12	

■ Discuss ■ Discover ■ Prove ■ Write

95. Discuss ■ Discover: A Sum of Sines Find the exact value of

$$\sin\frac{\pi}{100} + \sin\frac{2\pi}{100} + \sin\frac{3\pi}{100} + \cdots + \sin\frac{200\pi}{100}$$

PS *Draw a diagram.* Draw a unit circle to see how the values of the terms are related to each other.

96. Discover ■ Prove: Reduction Formulas A *reduction formula* "reduces" the number of terms in the input for a trigonometric function. Explain how the figure shows that the following reduction formulas are valid:

$$\sin(t + \pi) = -\sin t \qquad \cos(t + \pi) = -\cos t$$

$$\tan(t + \pi) = \tan t$$

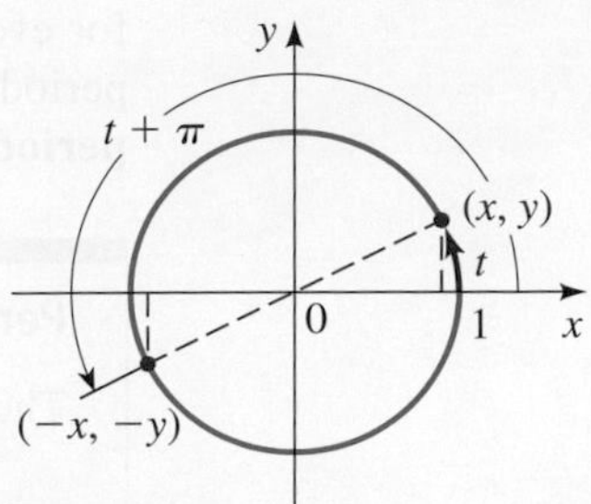

97. Discover ■ Prove: More Reduction Formulas By the Angle-Side-Angle Theorem from geometry (see Appendix A), triangles CDO and AOB in the figure are congruent. Explain how this proves that if B has coordinates (x, y), then D has coordinates $(-y, x)$. Then explain how the figure shows that the following reduction formulas are valid:

$$\sin\left(t + \frac{\pi}{2}\right) = \cos t \qquad \cos\left(t + \frac{\pi}{2}\right) = -\sin t$$

$$\tan\left(t + \frac{\pi}{2}\right) = -\cot t$$

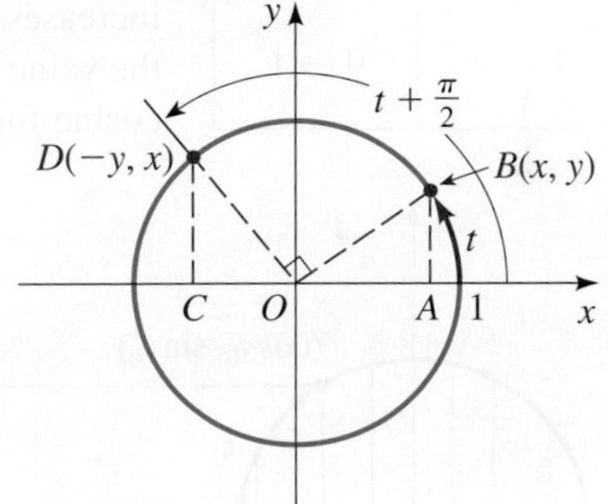

5.3 Trigonometric Graphs

■ **Graphs of Sine and Cosine** ■ **Graphs of Transformations of Sine and Cosine**
■ **Using Graphing Devices to Graph Trigonometric Functions**

The graph of a function gives us a good idea of its behavior. In this section we graph the sine and cosine functions and certain transformations of these functions. The other trigonometric functions are graphed in the next section.

■ Graphs of Sine and Cosine

To help us graph the sine and cosine functions, we first observe that these functions repeat their values in a regular fashion. To see exactly how this happens, recall that the circumference of the unit circle is 2π. It follows that the terminal point $P(x, y)$ determined by the real number t is the same as that determined by $t + 2\pi$. Since the sine and cosine functions are defined in terms of the coordinates of $P(x, y)$, their values are unchanged by the addition of any integer multiple of 2π. In other words,

$$\sin(t + 2n\pi) = \sin t \qquad \text{for any integer } n$$

$$\cos(t + 2n\pi) = \cos t \qquad \text{for any integer } n$$

Thus the sine and cosine functions are *periodic* according to the following definition: A function f is **periodic** if there is a positive number p such that $f(t + p) = f(t)$

for every t. The least such positive number (if it exists) is the **period** of f. If f has period p, then the graph of f on any interval of length p is called **one complete period** of f.

Periodic Properties of Sine and Cosine

The functions sine and cosine have period 2π:

$$\sin(t + 2\pi) = \sin t \qquad \cos(t + 2\pi) = \cos t$$

Table 1

t	$\sin t$	$\cos t$
$0 \to \frac{\pi}{2}$	$0 \to 1$	$1 \to 0$
$\frac{\pi}{2} \to \pi$	$1 \to 0$	$0 \to -1$
$\pi \to \frac{3\pi}{2}$	$0 \to -1$	$-1 \to 0$
$\frac{3\pi}{2} \to 2\pi$	$-1 \to 0$	$0 \to 1$

So the sine and cosine functions repeat their values in any interval of length 2π. To sketch their graphs, we first graph one period. To sketch the graphs on the interval $0 \le t \le 2\pi$, we could try to make a table of values and use those points to draw the graph. Since no such table can be complete, let's look more closely at the definitions of these functions.

Recall that $\sin t$ is the y-coordinate of the terminal point $P(x, y)$ on the unit circle determined by the real number t. How does the y-coordinate of this point vary as t increases? We see that the y-coordinate of $P(x, y)$ increases to 1, then decreases to -1 repeatedly as the point $P(x, y)$ travels around the unit circle (see Figure 1). In fact, as t increases from 0 to $\pi/2$, $y = \sin t$ increases from 0 to 1. As t increases from $\pi/2$ to π, the value of $y = \sin t$ decreases from 1 to 0. Table 1 shows the variation of the sine and cosine functions for t between 0 and 2π.

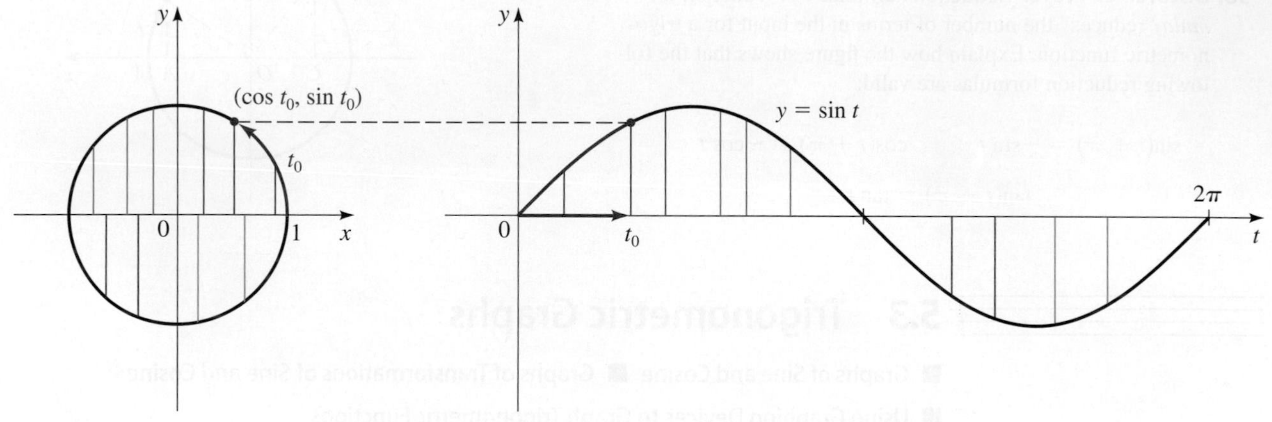

Figure 1

To draw the graphs more accurately, we find a few other values of $\sin t$ and $\cos t$ in Table 2. We could find still other values with the aid of a calculator.

Table 2

t	0	$\frac{\pi}{6}$	$\frac{\pi}{3}$	$\frac{\pi}{2}$	$\frac{2\pi}{3}$	$\frac{5\pi}{6}$	π	$\frac{7\pi}{6}$	$\frac{4\pi}{3}$	$\frac{3\pi}{2}$	$\frac{5\pi}{3}$	$\frac{11\pi}{6}$	2π
$\sin t$	0	$\frac{1}{2}$	$\frac{\sqrt{3}}{2}$	1	$\frac{\sqrt{3}}{2}$	$\frac{1}{2}$	0	$-\frac{1}{2}$	$-\frac{\sqrt{3}}{2}$	-1	$-\frac{\sqrt{3}}{2}$	$-\frac{1}{2}$	0
$\cos t$	1	$\frac{\sqrt{3}}{2}$	$\frac{1}{2}$	0	$-\frac{1}{2}$	$-\frac{\sqrt{3}}{2}$	-1	$-\frac{\sqrt{3}}{2}$	$-\frac{1}{2}$	0	$\frac{1}{2}$	$\frac{\sqrt{3}}{2}$	1

We use the information from Table 1 to graph one period of the functions $\sin t$ and $\cos t$ for t between 0 and 2π in Figures 2(a) and 3(a). The *key points* of the graphs of sine and cosine are the x-intercepts and the maximum and minimum points, as shown on the graphs. Now using the fact that these functions are periodic with period 2π, we get their complete graphs by continuing the same pattern to the left and to the right in every successive interval of length 2π, as shown in Figures 2(b) and 3(b).

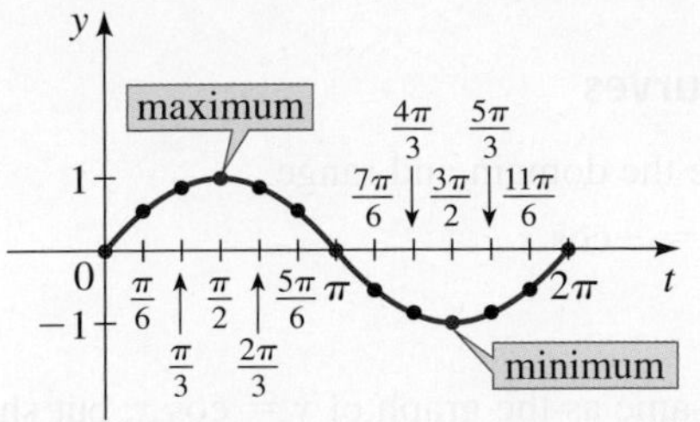

(a) One period of $y = \sin t$, $0 \le t \le 2\pi$

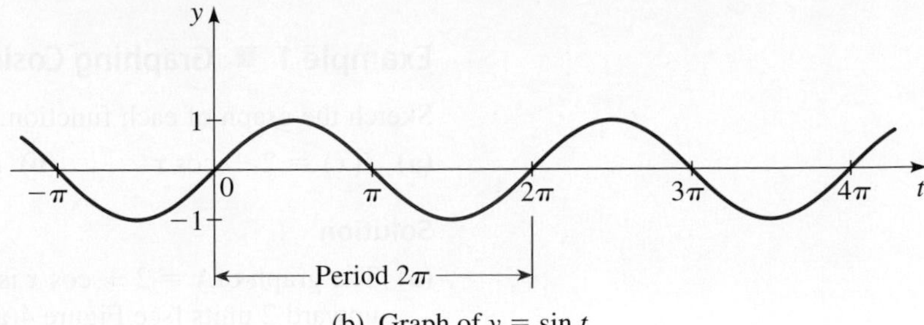

(b) Graph of $y = \sin t$

Figure 2

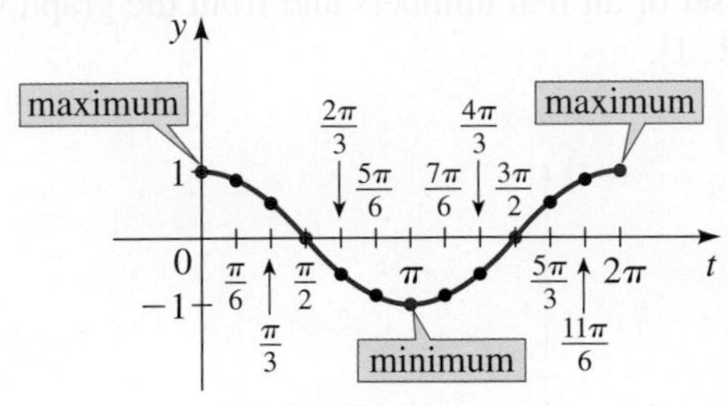

(a) One period of $y = \cos t$, $0 \le t \le 2\pi$

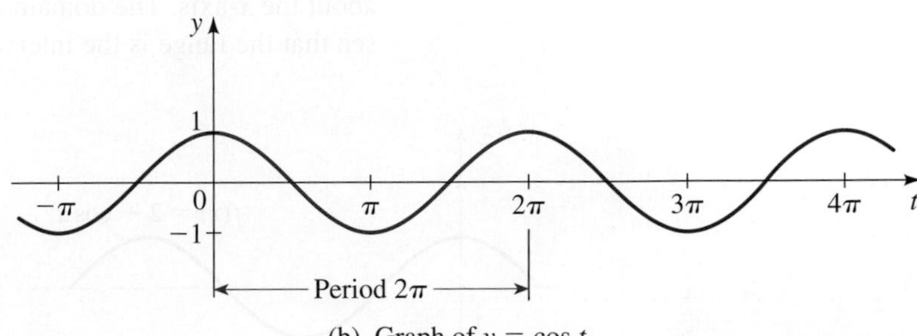

(b) Graph of $y = \cos t$

Figure 3

Note that the x-intercepts of $y = \sin x$ occur at multiples of π, that is, at $x = n\pi$, where n is an integer, and the x-intercepts of $y = \cos x$ occur at $x = (\pi/2) + n\pi$. Each graph has a local maximum or minimum value midway between successive x-intercepts.

Graphs of Sine and Cosine

The trigonometric functions $y = \sin x$ and $y = \cos x$ have domain all real numbers, range $[-1, 1]$, and are periodic with period 2π.

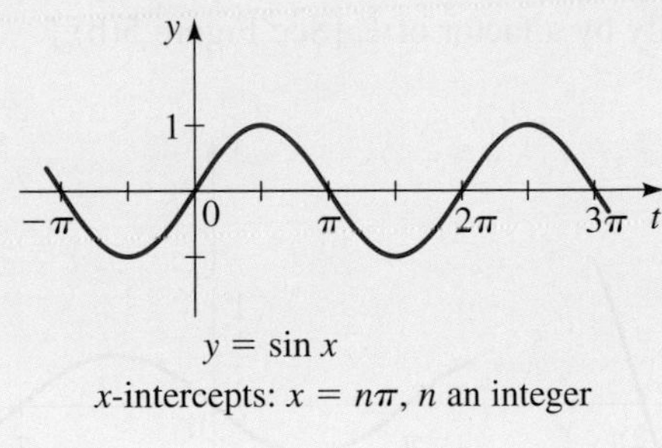

$y = \sin x$

x-intercepts: $x = n\pi$, n an integer

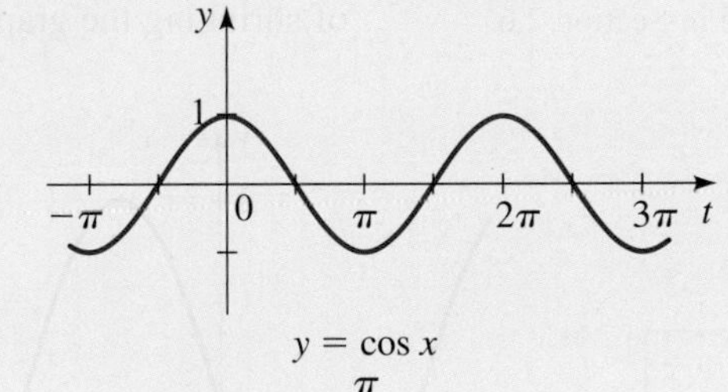

$y = \cos x$

x-intercepts: $x = \frac{\pi}{2} + n\pi$, n an integer

The graph of the sine function is symmetric with respect to the origin. This is expected because sine is an odd function. Since cosine is an even function, its graph is symmetric with respect to the y-axis.

■ Graphs of Transformations of Sine and Cosine

We now consider graphs of functions that are transformations of the sine and cosine functions. Thus the graphing techniques of Section 2.6 are very useful here.

It's traditional to use the letter x to denote the variable in the domain of a function. So from here on we use the letter x and write $y = \sin x$, $y = \cos x$, $y = \tan x$, and so on to denote these functions.

Example 1 ■ Graphing Cosine Curves

Sketch the graph of each function. State the domain and range.

(a) $f(x) = 2 + \cos x$ **(b)** $g(x) = -\cos x$

Solution

(a) The graph of $y = 2 + \cos x$ is the same as the graph of $y = \cos x$, but shifted upward 2 units [see Figure 4(a)]. The domain is the set of all real numbers and from the graph we see that the range is the interval $[1, 3]$.

(b) The graph of $y = -\cos x$ in Figure 4(b) is the reflection of the graph of $y = \cos x$ about the x-axis. The domain is the set of all real numbers and from the graph we see that the range is the interval $[-1, 1]$.

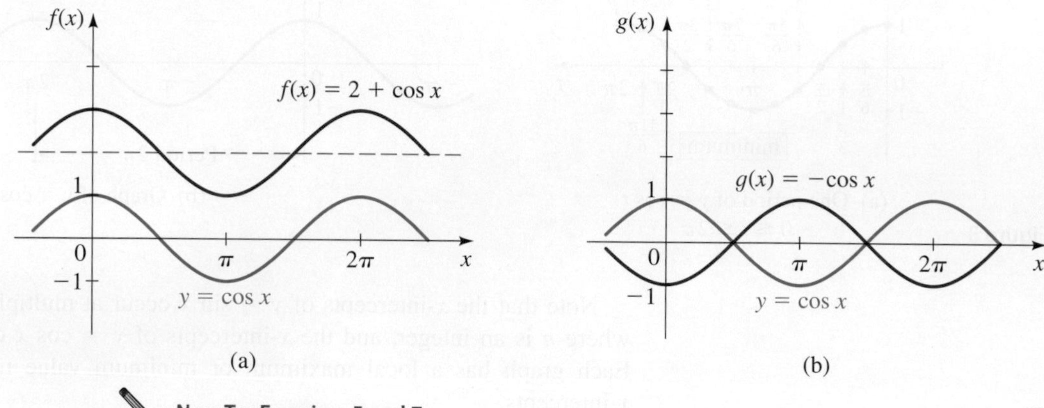

Figure 4

Now Try Exercises 5 and 7 ■

Let's graph $y = 2 \sin x$. We start with the graph of $y = \sin x$ and multiply the y-coordinate of each point on the graph by 2. This has the effect of stretching the graph vertically by a factor of 2. [See Figure 5(a).] To graph $y = \frac{1}{2} \sin x$, we start with the graph of $y = \sin x$ and multiply the y-coordinate of each point by $\frac{1}{2}$. This has the effect of shrinking the graph vertically by a factor of $\frac{1}{2}$. [See Figure 5(b).]

Vertical stretching and shrinking of graphs is discussed in Section 2.6.

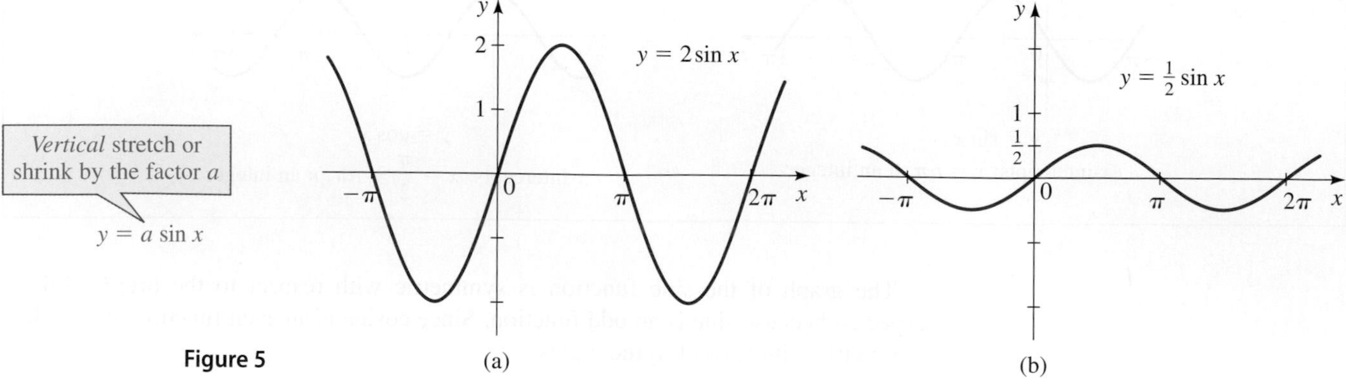

Figure 5

In general, for the functions

$$y = a \sin x \quad \text{and} \quad y = a \cos x$$

the number $|a|$ is called the **amplitude** and is the largest value these functions attain. Graphs of $y = a \sin x$ for several values of a are shown in Figure 6.

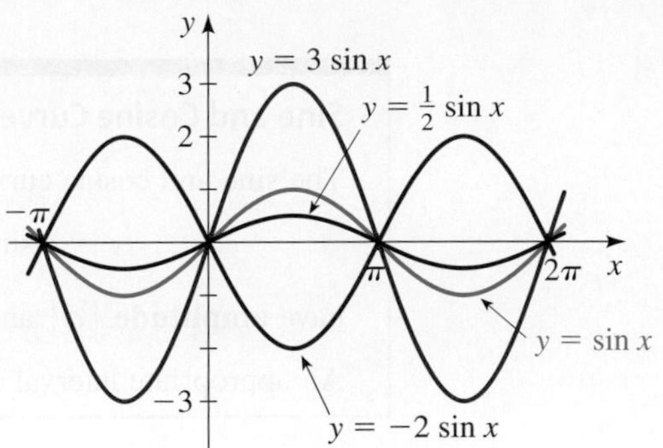

Figure 6

Example 2 ■ A Vertically Stretched Cosine Curve

Find the amplitude of $y = -3 \cos x$, and sketch its graph. State the domain and range.

Solution The amplitude is $|-3| = 3$, so the largest value the graph attains is 3 and the smallest value is -3. To sketch the graph, we begin with the graph of $y = \cos x$, stretch the graph vertically by a factor of 3, and reflect about the x-axis to arrive at the graph in Figure 7. The domain is the set of all real numbers and from the graph we see that the range is the interval $[-3, 3]$.

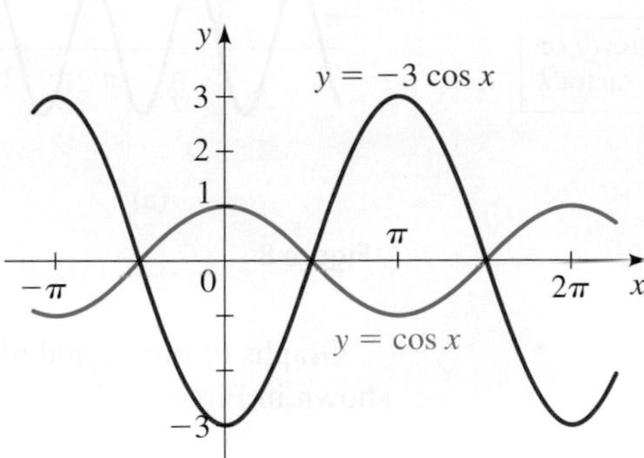

Figure 7

Now Try Exercise 11 ■

Because the sine and cosine functions have period 2π, the functions

$$y = a \sin kx \qquad \text{and} \qquad y = a \cos kx \qquad (k > 0)$$

complete one period as kx varies from 0 to 2π, that is, for $0 \le kx \le 2\pi$ or for $0 \le x \le 2\pi/k$. So these functions complete one period as x varies between 0 and $2\pi/k$

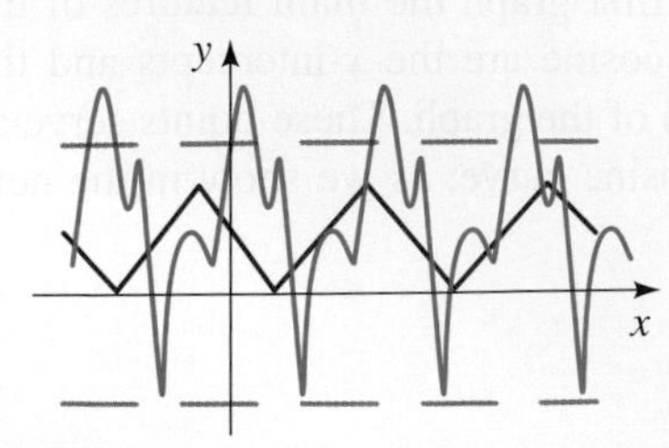

Discovery Project ■ Periodic Functions

We have learned that a function which repeats its values regularly over successive intervals of the same length is called *periodic*. The sine and cosine functions are the fundamental periodic functions, but there are many others. In this project we explore periodic functions graphically, by making up graphs of periodic functions and investigating how transformations of these functions affect their periods. We also investigate conditions under which combinations of periodic functions are again periodic. You can find the project at **www.stewartmath.com**.

and thus have period $2\pi/k$. The graphs of these functions are called **sine curves** and **cosine curves**, respectively. (Sine and cosine curves are often collectively referred to as **sinusoidal** curves.)

> **Sine and Cosine Curves**
>
> The sine and cosine curves
>
> $$y = a \sin kx \qquad \text{and} \qquad y = a \cos kx \qquad (k > 0)$$
>
> have **amplitude** $|a|$ and **period** $2\pi/k$.
>
> An appropriate interval on which to graph one complete period is $[0, 2\pi/k]$.

Horizontal stretching and shrinking of graphs is discussed in Section 2.6.

To see how the value of k affects the graph of $y = \sin kx$, let's graph the sine curve $y = \sin 2x$. Since the period is $2\pi/2 = \pi$, the graph completes one period in the interval $0 \le x \le \pi$. [See Figure 8(a).] For the sine curve $y = \sin \frac{1}{2}x$ the period is $2\pi \div \frac{1}{2} = 4\pi$, so the graph completes one period in the interval $0 \le x \le 4\pi$. [See Figure 8(b).] We see that the effect of k is to *shrink* the graph horizontally if $k > 1$ or to *stretch* the graph horizontally if $k < 1$.

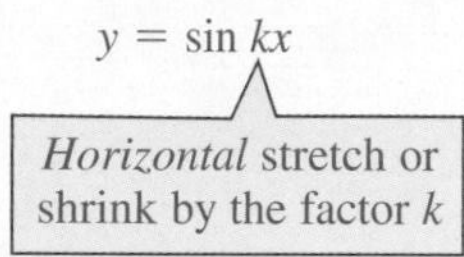

y = sin kx
Horizontal stretch or shrink by the factor k

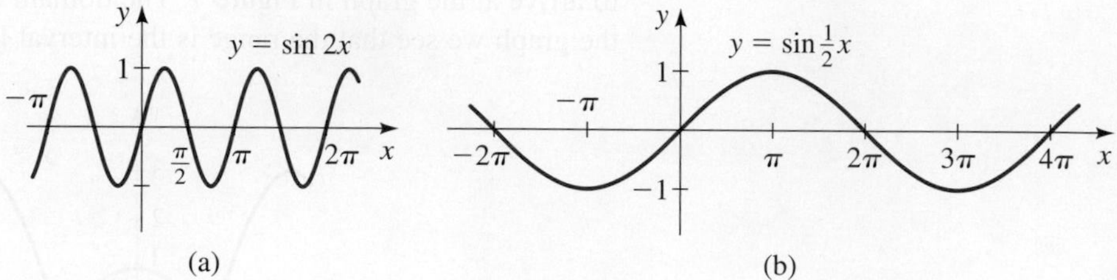

Figure 8

Graphs of one period of the sine curve $y = a \sin kx$ for several values of k are shown in Figure 9.

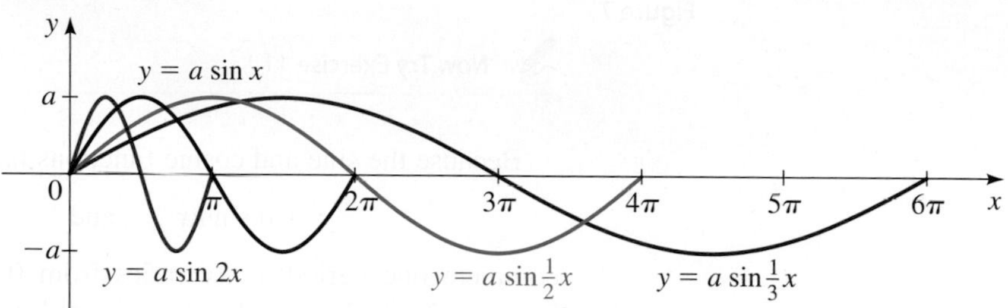

Figure 9

To graph a sine or cosine curve it is helpful to first graph the main features of the curve. The **key points** of the graphs of sine and cosine are the x-intercepts and the maximum and minimum points (peaks and valleys) of the graph. These points serve as guides to graphing an appropriate sine curve or cosine curve, as we show in the next several examples.

Example 3 ■ Amplitude and Period

Find the amplitude and period of each function, and sketch its graph.

(a) $y = 4 \cos 3x$ **(b)** $y = -2 \sin \frac{1}{2}x$

Solution

(a) We get the amplitude and period from the form of the function as follows.

$$\text{amplitude} = |a| = 4$$

$$y = 4\cos 3x$$

$$\text{period} = \frac{2\pi}{k} = \frac{2\pi}{3}$$

The amplitude is 4 and the period is $2\pi/3$.

Sketching the graph. An appropriate interval on which to sketch one complete period is $[0, 2\pi/3]$. To find the key points on this interval we divide the interval into four subintervals, each of length

$$\frac{2\pi}{3} \times \frac{1}{4} = \frac{\pi}{6}$$

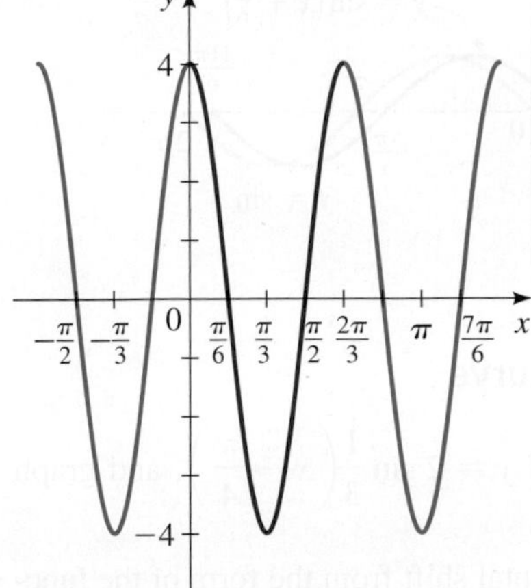

Figure 10 | $y = 4\cos 3x$

Now, we start at the left endpoint of the interval (that is, at $x = 0$), and successively add $\pi/6$ to obtain the x-coordinates of the key points, as shown in the following table of values.

x	0	$\frac{\pi}{6}$	$\frac{\pi}{3}$	$\frac{\pi}{2}$	$\frac{2\pi}{3}$
y	4	0	−4	0	4
	maximum	**intercept**	**minimum**	**intercept**	**maximum**

The graph in Figure 10 is obtained by plotting a cosine curve with amplitude 4 on the interval $[0, 2\pi/3]$, using the points in the table.

(b) For $y = -2\sin\frac{1}{2}x$, we find

$$\text{amplitude} = |a| = |-2| = 2$$

$$\text{period} = \frac{2\pi}{\frac{1}{2}} = 4\pi$$

Sketching the graph. An appropriate interval on which to sketch one complete period is $[0, 4\pi]$. To find the key points we divide the interval into four subintervals, each of length $4\pi/4 = \pi$. Starting at $x = 0$ and successively adding π, we obtain the x-coordinates of the key points and complete the table of values as follows.

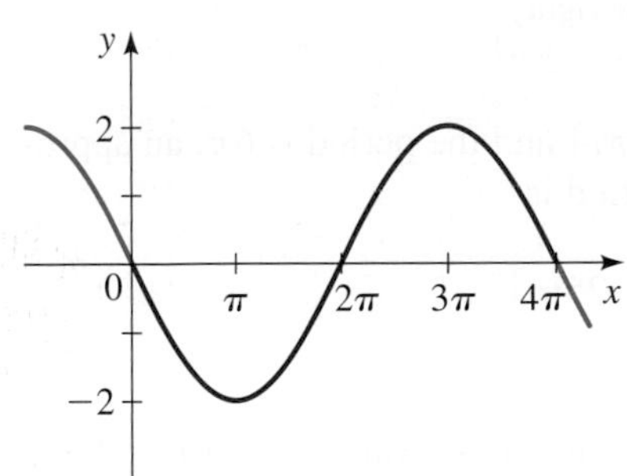

Figure 11 | $y = -2\sin\frac{1}{2}x$

x	0	π	2π	3π	4π
y	0	−2	0	2	0
	intercept	**minimum**	**intercept**	**maximum**	**intercept**

The graph in Figure 11 is obtained by plotting a sine curve with amplitude 2 on the interval $[0, 4\pi]$, using the points in the table.

Now Try Exercises 23 and 25 ■

The graphs of functions of the form $y = a\sin k(x - b)$ and $y = a\cos k(x - b)$ are simply sine and cosine curves shifted horizontally by an amount $|b|$. They are shifted to the right if $b > 0$ or to the left if $b < 0$. We summarize the properties of these functions in the following box.

The related concept of phase shift of a sine curve is discussed in Section 5.6.

Shifted Sine and Cosine Curves

The sine and cosine curves

$$y = a \sin k(x - b) \qquad \text{and} \qquad y = a \cos k(x - b) \qquad (k > 0)$$

have **amplitude** $|a|$, **period** $2\pi/k$, and **horizontal shift** b.

An appropriate interval on which to graph one complete period is $[b, b + (2\pi/k)]$.

The graphs of $y = \sin\left(x - \frac{\pi}{3}\right)$ and $y = \sin\left(x + \frac{\pi}{6}\right)$ are shown in Figure 12.

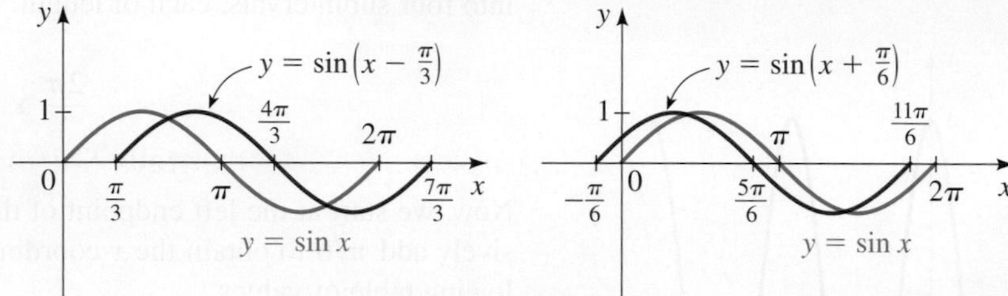

Figure 12 | Horizontal shifts of a sine curve

Example 4 ■ A Horizontally Shifted Sine Curve

Find the amplitude, period, and horizontal shift of $y = 2 \sin \frac{1}{3}\left(x - \frac{\pi}{4}\right)$, and graph one complete period.

Solution We get the amplitude, period, and horizontal shift from the form of the function as follows:

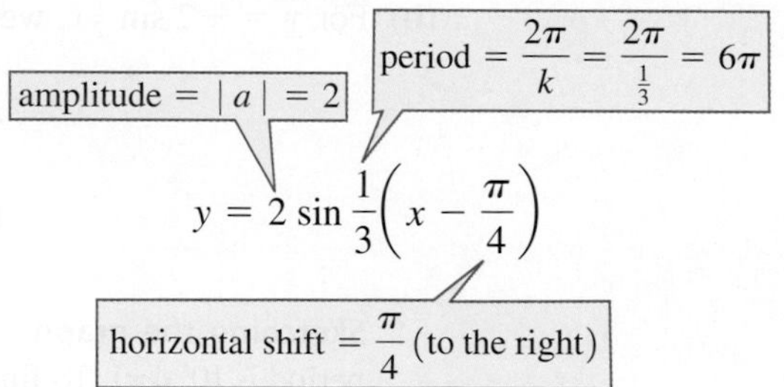

amplitude = $|a| = 2$
period = $\frac{2\pi}{k} = \frac{2\pi}{\frac{1}{3}} = 6\pi$
$y = 2 \sin \frac{1}{3}\left(x - \frac{\pi}{4}\right)$
horizontal shift = $\frac{\pi}{4}$ (to the right)

Sketching the graph. Since the horizontal shift is $\pi/4$ and the period is 6π, an appropriate interval on which to sketch one complete period is

$$\left[\frac{\pi}{4}, \frac{\pi}{4} + 6\pi\right] = \left[\frac{\pi}{4}, \frac{25\pi}{4}\right]$$

Here is another way to find an appropriate interval on which to graph one complete period. Since the period of $y = \sin x$ is 2π, the function $y = 2 \sin \frac{1}{3}(x - \frac{\pi}{4})$ will go through one complete period as $\frac{1}{3}(x - \frac{\pi}{4})$ varies from 0 to 2π.

Start of period:	End of period:
$\frac{1}{3}(x - \frac{\pi}{4}) = 0$	$\frac{1}{3}(x - \frac{\pi}{4}) = 2\pi$
$x - \frac{\pi}{4} = 0$	$x - \frac{\pi}{4} = 6\pi$
$x = \frac{\pi}{4}$	$x = \frac{25\pi}{4}$

So we graph one period on the interval $\left[\frac{\pi}{4}, \frac{25\pi}{4}\right]$.

To find the key points, we divide this interval into four subintervals, each of length $6\pi/4 = 3\pi/2$. Starting at the left endpoint $x = \pi/4$ and successively adding $3\pi/2$, we obtain the x-coordinates of the key points, and complete the table of values as follows.

x	$\frac{\pi}{4}$	$\frac{7\pi}{4}$	$\frac{13\pi}{4}$	$\frac{19\pi}{4}$	$\frac{25\pi}{4}$
y	0	2	0	-2	0
	intercept	**maximum**	**intercept**	**minimum**	**intercept**

The graph in Figure 13 is obtained by plotting a sine curve with amplitude 2 on the interval $[\pi/4, 25\pi/4]$, using the points in the table.

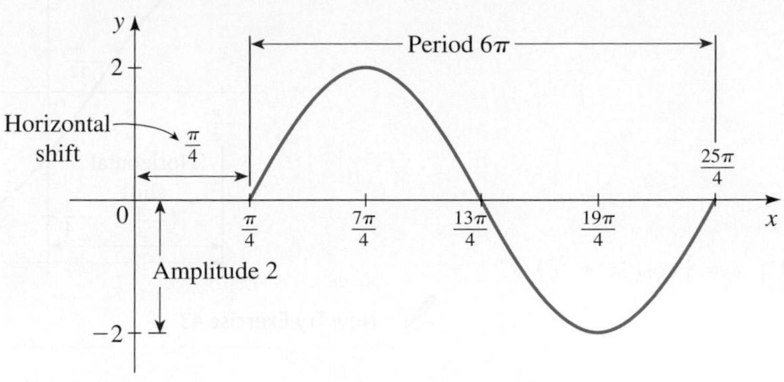

Figure 13 | $y = 2\sin\frac{1}{3}\left(x - \frac{\pi}{4}\right)$

Now Try Exercise 39 ■

Example 5 ■ A Horizontally Shifted Cosine Curve

Find the amplitude, period, and horizontal shift of $y = \frac{3}{4}\cos\left(2x + \frac{2\pi}{3}\right)$, and graph one complete period.

Solution We first write this function in the form $y = a\cos k(x - b)$. To do this, we factor 2 from the expression $2x + \frac{2\pi}{3}$ to get

$$y = \frac{3}{4}\cos 2\left[x - \left(-\frac{\pi}{3}\right)\right]$$

Thus we have

$$\text{amplitude} = |a| = \frac{3}{4}$$

$$\text{period} = \frac{2\pi}{k} = \frac{2\pi}{2} = \pi$$

$$\text{horizontal shift} = b = -\frac{\pi}{3} \qquad \text{Shift } \frac{\pi}{3} \text{ to the } \textit{left}$$

We can also find one complete period as follows:

Start of period:	End of period:
$2x + \frac{2\pi}{3} = 0$	$2x + \frac{2\pi}{3} = 2\pi$
$2x = -\frac{2\pi}{3}$	$2x = \frac{4\pi}{3}$
$x = -\frac{\pi}{3}$	$x = \frac{2\pi}{3}$

So we graph one period on the interval $\left[-\frac{\pi}{3}, \frac{2\pi}{3}\right]$.

Sketching the graph. Because the horizontal shift is $-\pi/3$ and the period is π, an appropriate interval on which to sketch one complete period is

$$\left[-\frac{\pi}{3}, -\frac{\pi}{3} + \pi\right] = \left[-\frac{\pi}{3}, \frac{2\pi}{3}\right]$$

To find the key points, we divide this interval into four subintervals, each of length $\pi/4$. Starting at the left endpoint $x = -\pi/3$ and successively adding $\pi/4$, we obtain the x-coordinates of the key points, and complete the table of values as follows.

x	$-\frac{\pi}{3}$	$-\frac{\pi}{12}$	$\frac{\pi}{6}$	$\frac{5\pi}{12}$	$\frac{2\pi}{3}$
y	$\frac{3}{4}$	0	$-\frac{3}{4}$	0	$\frac{3}{4}$
	maximum	**intercept**	**minimum**	**intercept**	**maximum**

The graph in Figure 14 (on the next page) is obtained by plotting a cosine curve with amplitude $\frac{3}{4}$ on the interval $[-\pi/3, 2\pi/3]$, using the points in the table.

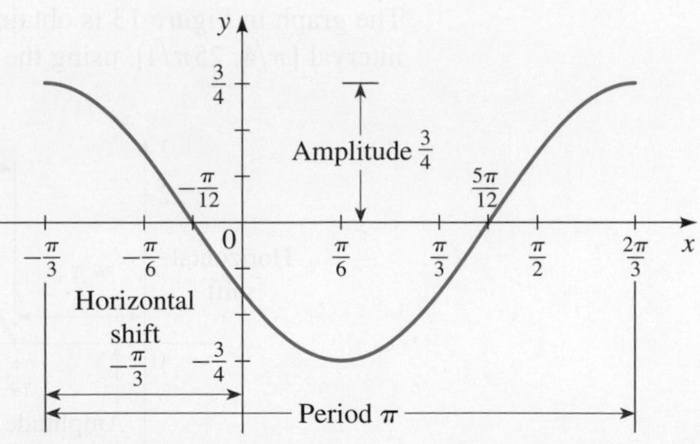

Figure 14 | $y = \frac{3}{4}\cos\left(2x + \frac{2\pi}{3}\right)$

Now Try Exercise 43 ■

■ Using Graphing Devices to Graph Trigonometric Functions

When we use a graphing calculator or a computer to graph a function, it is important to choose the viewing rectangle carefully in order to produce a reasonable graph of the function; this is especially true for trigonometric functions. The next example shows that, if care is not taken, it's easy to produce a misleading graph.

Example 6 ■ Choosing a Viewing Rectangle

Graph the function $f(x) = \sin 50x$ in an appropriate viewing rectangle.

Solution Figure 15 shows the graph of f produced by a graphing calculator in two different viewing rectangles. These calculator outputs are not accurate representations of the graph of f.

The appearance of the graphs in Figure 15 depends on the machine used and on the number of points plotted. The graphs you get with your own graphing device might not look like these figures, but they may also be inaccurate.

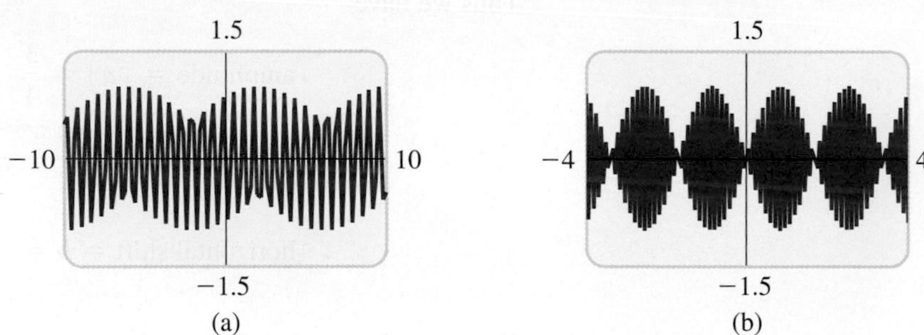

Figure 15 | Graphs of $f(x) = \sin 50x$ in two viewing rectangles

To explain the big differences in appearance of these graphs and to find an appropriate viewing rectangle, we need to find the period of the function $y = \sin 50x$.

$$\text{period} = \frac{2\pi}{50} = \frac{\pi}{25} \approx 0.126$$

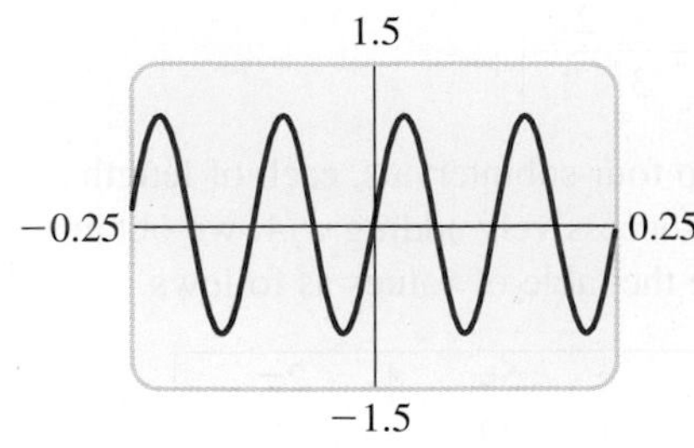

Figure 16 | $f(x) = \sin 50x$

This suggests that we should deal only with small values of x in order to show just a few oscillations of the graph. If we choose the viewing rectangle $[-0.25, 0.25]$ by $[-1.5, 1.5]$, we get the accurate graph shown in Figure 16.

Now we see what went wrong in Figure 15. The oscillations of $y = \sin 50x$ are so rapid that when the calculator plots points and joins them, it misses most of the maximum and minimum points and therefore gives a misleading impression of the graph.

Now Try Exercise 59 ■

Example 7 ■ A Sum of Sine and Cosine Curves

Graph $f(x) = 2 \cos x$, $g(x) = \sin 2x$, and $h(x) = 2 \cos x + \sin 2x$ on a common screen to illustrate the method of graphical addition.

Solution Notice that $h = f + g$, so its graph is obtained by adding the corresponding y-coordinates of the graphs of f and g. The graphs of f, g, and h are shown in Figure 17.

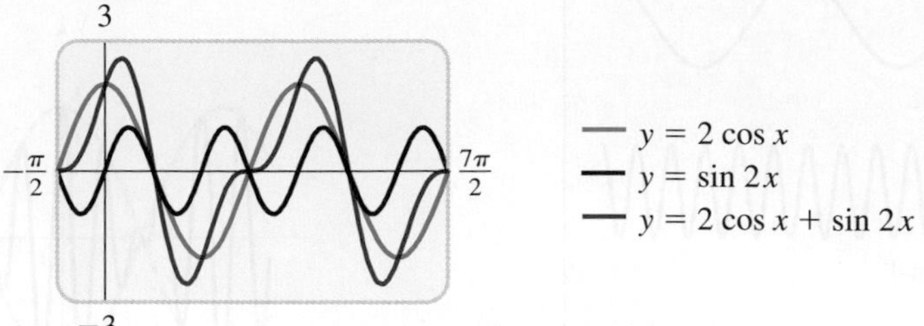

Figure 17

Now Try Exercise 69 ■

Example 8 ■ A Cosine Curve with Variable Amplitude

Graph the functions $y = x^2$, $y = -x^2$, and $y = x^2 \cos 6\pi x$ on a common screen. Comment on and explain the relationship among the graphs.

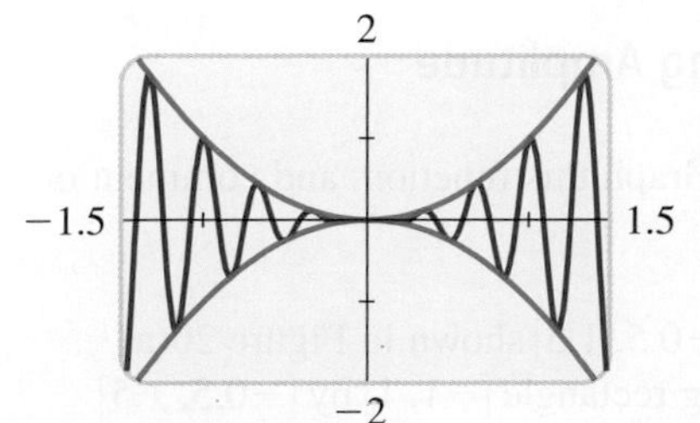

Figure 18 | Graphs of $y = x^2 \cos 6\pi x$ and $y = \pm x^2$

Solution Figure 18 shows all three graphs in the viewing rectangle $[-1.5, 1.5]$ by $[-2, 2]$. It appears that the graph of $y = x^2 \cos 6\pi x$ lies between the graphs of the functions $y = x^2$ and $y = -x^2$.

To understand this, recall that the values of $\cos 6\pi x$ lie between -1 and 1, that is,

$$-1 \le \cos 6\pi x \le 1$$

for all values of x. Multiplying the inequalities by x^2 and noting that $x^2 \ge 0$, we get

$$-x^2 \le x^2 \cos 6\pi x \le x^2$$

This explains why the functions $y = x^2$ and $y = -x^2$ form a boundary for the graph of $y = x^2 \cos 6\pi x$. (Note that the graphs touch when $\cos 6\pi x = \pm 1$.)

Now Try Exercise 73 ■

Example 8 shows that the function $y = x^2$ controls the amplitude of the graph of $y = x^2 \cos 6\pi x$. In general, if $f(x) = a(x) \sin kx$ or $f(x) = a(x) \cos kx$, the function a determines how the amplitude of f varies, and the graph of f lies between the graphs of $y = -a(x)$ and $y = a(x)$. Example 9 illustrates another instance of this behavior.

Jeffrey Lepore/Science Source

Discovery Project ■ Predator-Prey Models

Many animal populations fluctuate regularly in size and so can be modeled by trigonometric functions Predicting population changes allows scientists to detect anomalies and take steps to protect a species. In this project we study the population of a predator species and the population of its prey. If the prey is abundant, the predator population grows, but too many predators tend to deplete the prey. This results in a decrease in the predator population, then the prey population increases, and so on. You can find the project at **www.stewartmath.com**.

Mathematics in the Modern World

Data Transmission with Radio Waves

Radio transmissions consist of sound waves superimposed on a harmonic electromagnetic wave form called the **carrier signal**.

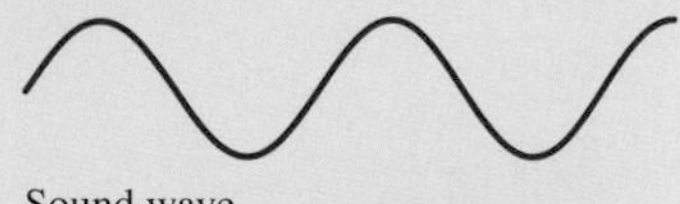

Sound wave

Carrier signal

There are two types of radio transmission, called **amplitude modulation (AM)** and **frequency modulation (FM)**. In AM broadcasting, the sound wave changes, or **modulates**, the amplitude of the carrier, but the frequency remains unchanged.

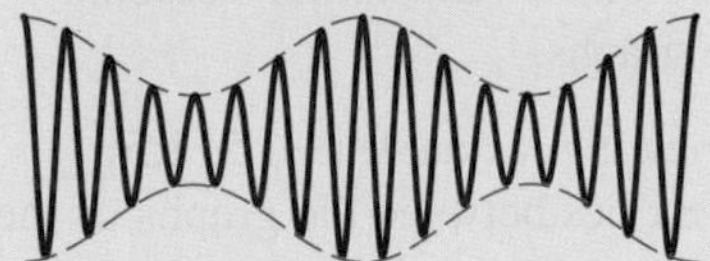

AM signal

In FM broadcasting, the sound wave modulates the frequency, but the amplitude remains the same.

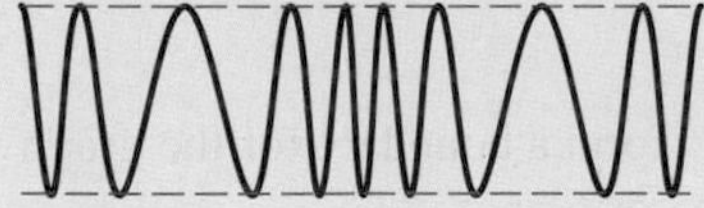

FM signal

Radio waves are also used to transmit digital data in cell phone communication, such as text, voice, and media (see Exercise 5.6.50).

Example 9 ■ A Cosine Curve with Variable Amplitude

Graph the function $f(x) = \cos 2\pi x \cos 16\pi x$.

Solution The graph is shown in Figure 19. Although it was drawn by a computer, we could have drawn it without one, by first sketching the boundary curves $y = \cos 2\pi x$ and $y = -\cos 2\pi x$. The graph of f is a cosine curve that lies between the graphs of these two functions.

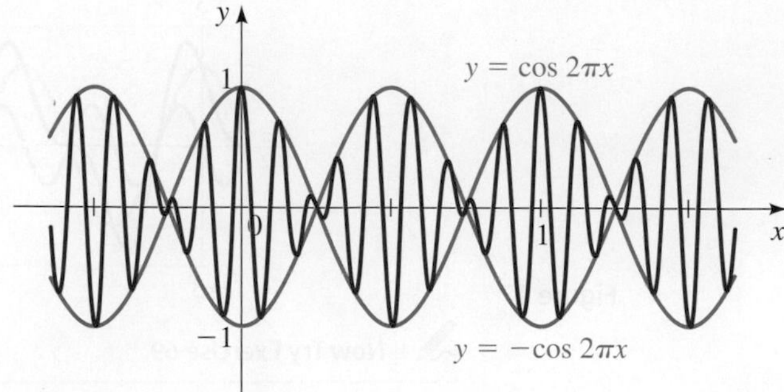

Figure 19 | Graphs of $f(x) = \cos 2\pi x \cos 16\pi x$ and $y = \pm\cos 2\pi x$

Now Try Exercise 75 ■

Example 10 ■ A Sine Curve with Decaying Amplitude

The function $f(x) = \dfrac{\sin x}{x}$ is useful in calculus. Graph this function, and comment on its behavior when x is close to 0.

Solution The viewing rectangle $[-15, 15]$ by $[-0.5, 1.5]$ shown in Figure 20(a) gives a global view of the graph of f. The viewing rectangle $[-1, 1]$ by $[-0.5, 1.5]$ in Figure 20(b) focuses on the behavior of f when $x \approx 0$. Notice that although $f(x)$ is not defined when $x = 0$ (in other words, 0 is not in the domain of f), the values of f seem to approach 1 as x gets close to 0. This fact is used in calculus.

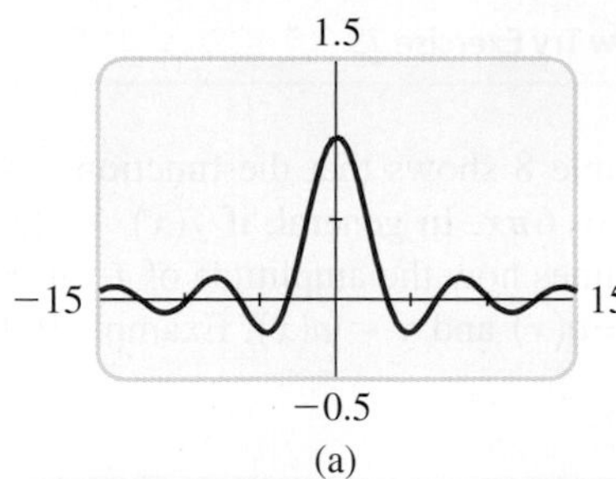

(a)

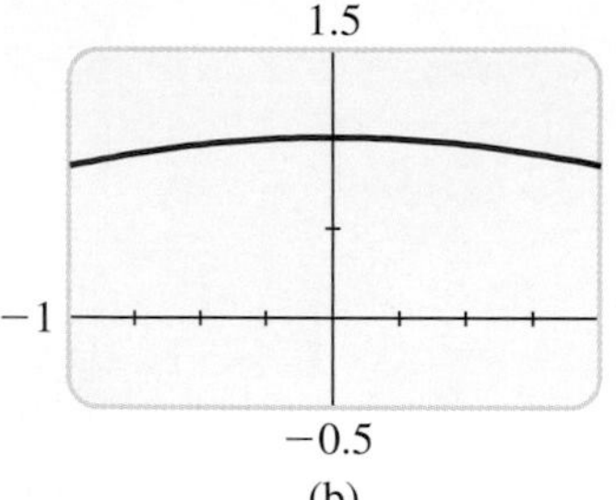

(b)

Figure 20 | Graphs of $f(x) = \dfrac{\sin x}{x}$

Now Try Exercise 85 ■

The function in Example 10 can be written as

$$f(x) = \frac{1}{x}\sin x$$

and may thus be viewed as a sine function whose amplitude is controlled by the function $a(x) = 1/x$. Notice how the amplitude gets smaller, or decays, as $|x|$ increases.

5.3 | Exercises

Concepts

1. If a function f is periodic with period p, then $f(t + p) =$ ________ for every t. The trigonometric functions $y = \sin x$ and $y = \cos x$ are periodic, with period ________ and have amplitude ________. Sketch a graph of each function on the interval $[0, 2\pi]$.

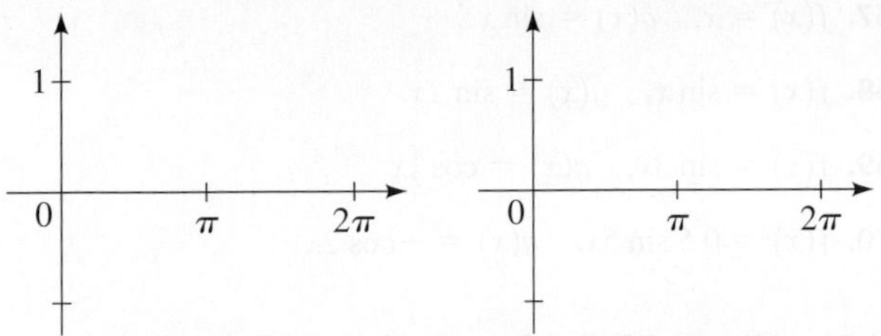

2. To obtain the graph of $y = 5 + \sin x$, we start with the graph of $y = \sin x$, then shift it 5 units ________ (upward/downward). To obtain the graph of $y = -\cos x$, we start with the graph of $y = \cos x$, then reflect it about the ____-axis.

3. (a) The sine and cosine curves $y = a \sin kx$ and $y = a \cos kx$, $k > 0$, have amplitude ________ and period ________. The sine curve $y = 3 \sin 2x$ has amplitude ________ and period ________; an appropriate interval to graph one period is ________.

(b) Graph one period of $y = 3 \sin 2x$ and label the x-coordinates of the key points used for graphing the function.

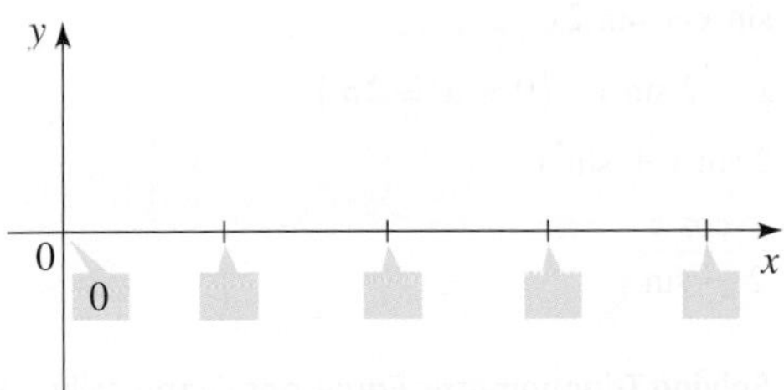

4. (a) The sine curve $y = a \sin k(x - b)$ has amplitude ________, period ________, and horizontal shift ________. The sine curve $y = 4 \sin 3\left(x - \frac{\pi}{6}\right)$ has amplitude ________, period ________, and horizontal shift ________; an appropriate interval on which to graph one period is ________.

(b) Graph one period of $y = 4 \sin 3\left(x - \frac{\pi}{6}\right)$ and label the x-coordinates of the key points used for graphing the function.

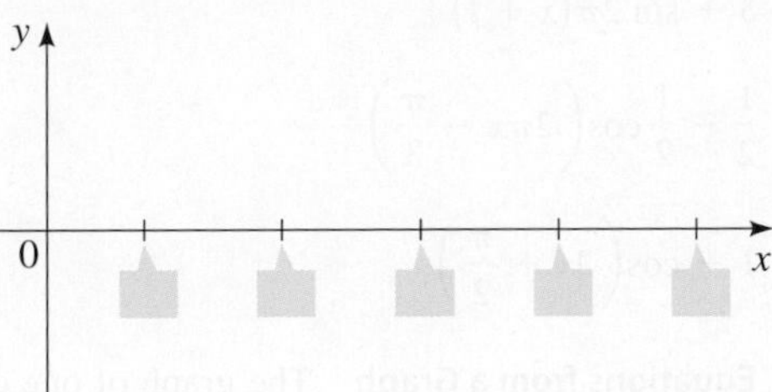

Skills

5–18 ■ Graphing Sine and Cosine Functions Graph the function, and state the domain and range.

5. $f(x) = 2 + \sin x$ **6.** $f(x) = -2 + \cos x$

7. $f(x) = -\sin x$ **8.** $f(x) = 2 - \cos x$

9. $f(x) = -2 + \sin x$ **10.** $f(x) = -1 + \cos x$

11. $g(x) = 3 \cos x$ **12.** $g(x) = 2 \sin x$

13. $g(x) = -\frac{1}{2} \sin x$ **14.** $g(x) = -\frac{2}{3} \cos x$

15. $g(x) = 3 + 3 \cos x$ **16.** $g(x) = 4 - 2 \sin x$

17. $h(x) = |\cos x|$ **18.** $h(x) = |\sin x|$

19–34 ■ Amplitude and Period Find the amplitude and period of the function, and sketch its graph.

19. $y = \cos 2x$ **20.** $y = -\sin 2x$

21. $y = -\cos 4x$ **22.** $y = \sin \pi x$

23. $y = 3 \sin 2\pi x$ **24.** $y = -2 \cos 8x$

25. $y = 10 \sin \frac{1}{2}x$ **26.** $y = 5 \cos \frac{1}{4}x$

27. $y = -\frac{1}{3} \cos \frac{1}{3}x$ **28.** $y = 4 \sin(-2x)$

29. $y = -2 \sin 8\pi x$ **30.** $y = -3 \sin 4\pi x$

31. $y = 2 \sin 3x$ **32.** $y = 4 \cos 6x$

33. $y = 1 + \frac{1}{2} \cos \pi x$ **34.** $y = -2 + \cos 4\pi x$

35–50 ■ Horizontal Shifts Find the amplitude, period, and horizontal shift of the function, and graph one complete period.

35. $y = \cos\left(x - \frac{\pi}{2}\right)$ **36.** $y = 2 \sin\left(x - \frac{\pi}{3}\right)$

37. $y = -2 \sin\left(x - \frac{\pi}{6}\right)$ **38.** $y = 3 \cos\left(x + \frac{\pi}{4}\right)$

39. $y = 4 \sin \pi(x - \frac{1}{2})$ **40.** $y = -2 \cos \pi(x + \frac{1}{4})$

41. $y = 2 \cos 4\left(x - \frac{\pi}{4}\right)$ **42.** $y = -3 \cos 3\left(x + \frac{\pi}{3}\right)$

43. $y = \cos(2x + \pi)$ **44.** $y = \sin\left(3x - \frac{\pi}{2}\right)$

45. $y = 2\sin\left(\frac{2}{3}x - \frac{\pi}{6}\right)$

46. $y = 5\cos\left(3x - \frac{\pi}{4}\right)$

47. $y = 2 - 2\cos 3\left(x + \frac{\pi}{3}\right)$

48. $y = 3 + \sin 2\pi(x + \frac{1}{8})$

49. $y = \frac{1}{2} - \frac{1}{2}\cos\left(2\pi x - \frac{\pi}{3}\right)$

50. $y = 1 + \cos\left(3x + \frac{\pi}{2}\right)$

51–58 ■ Equations from a Graph The graph of one complete period of a sine or cosine curve is given. Find the amplitude and period, and write an equation that represents the curve in the form

$$y = a\sin k(x - b) \quad \text{or} \quad y = a\cos k(x - b)$$

51.

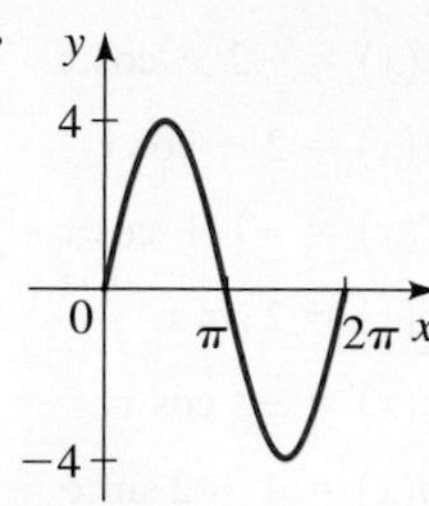

52.

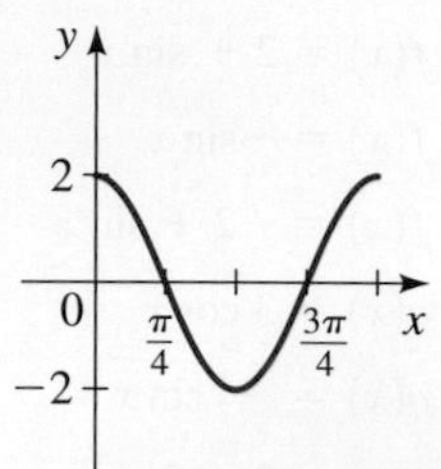

53.

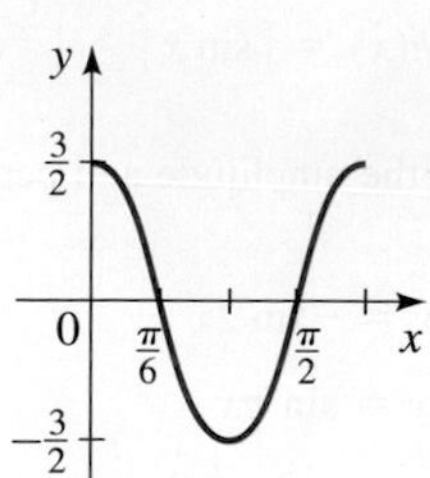

54.

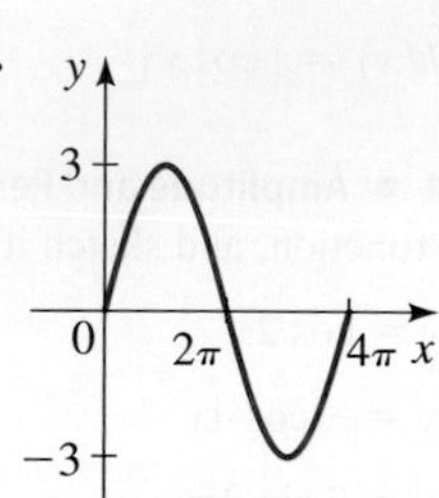

55.

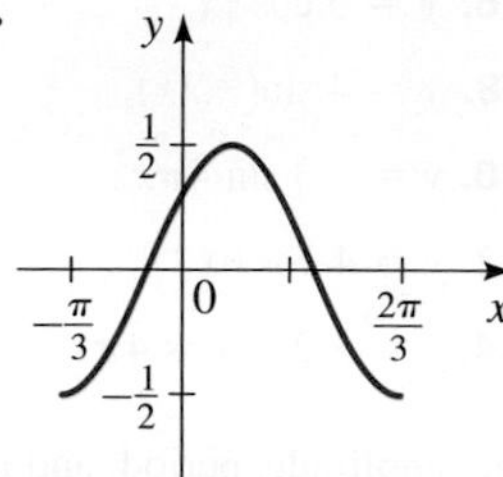

56.

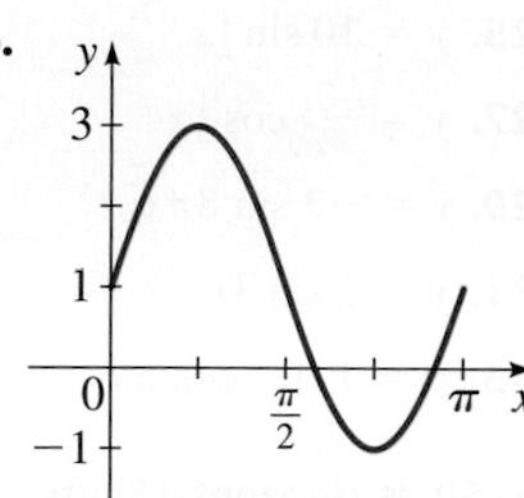

57.

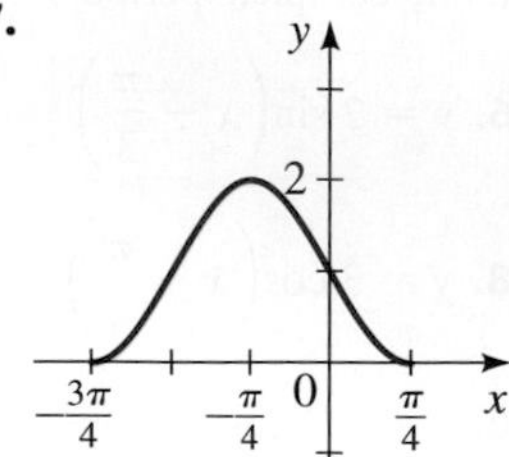

58.

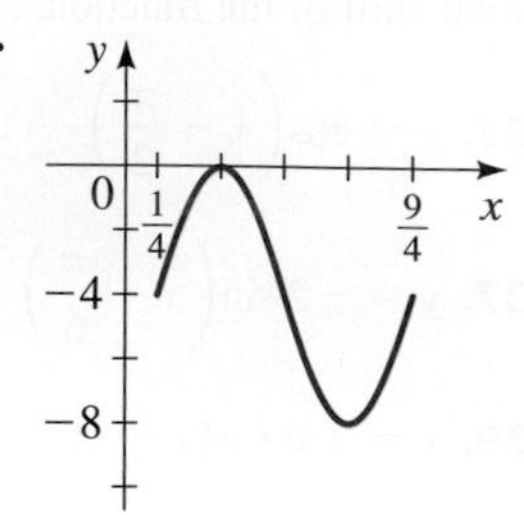

59–66 ■ Graphing Trigonometric Functions Determine an appropriate viewing rectangle for each function, and use it to draw the graph.

59. $f(x) = \cos 100x$

60. $f(x) = 3\sin 120x$

61. $f(x) = \sin(x/40)$

62. $f(x) = \cos(x/80)$

63. $y = \tan 25x$

64. $y = \csc 40x$

65. $y = \sin^2(20x)$

66. $y = \sqrt{\cos 10\pi x}$

67–70 ■ Graphical Addition Graph f, g, and $f + g$ on a common screen to illustrate graphical addition.

67. $f(x) = x, \quad g(x) = \sin x$

68. $f(x) = \sin x, \quad g(x) = \sin 2x$

69. $f(x) = \sin 3x, \quad g(x) = \cos\frac{1}{2}x$

70. $f(x) = 0.5\sin 5x, \quad g(x) = -\cos 2x$

71–76 ■ Sine and Cosine Curves with Variable Amplitude Graph the three functions on a common screen. How are the graphs related?

71. $y = x^2, \quad y = -x^2, \quad y = x^2\sin x$

72. $y = x, \quad y = -x, \quad y = x\cos x$

73. $y = \sqrt{x}, \quad y = -\sqrt{x}, \quad y = \sqrt{x}\sin 5\pi x$

74. $y = \dfrac{1}{1 + x^2}, \quad y = -\dfrac{1}{1 + x^2}, \quad y = \dfrac{\cos 2\pi x}{1 + x^2}$

75. $y = \cos 3\pi x, \quad y = -\cos 3\pi x, \quad y = \cos 3\pi x\cos 21\pi x$

76. $y = \sin 2\pi x, \quad y = -\sin 2\pi x, \quad y = \sin 2\pi x\sin 10\pi x$

Skills Plus

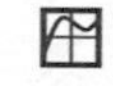

77–80 ■ Maximums and Minimums Find the maximum and minimum values of the function, rounded to two decimal places.

77. $y = \sin x + \sin 2x$

78. $y = x - 2\sin x \quad (0 \le x \le 2\pi)$

79. $y = 2\sin x + \sin^2 x$

80. $y = \dfrac{\cos x}{2 + \sin x}$

81–84 ■ Solving Trigonometric Equations Graphically Find all solutions of the equation that lie in the interval $[0, \pi]$. State each answer rounded to two decimal places. (See Section 1.11.)

81. $\cos x = 0.4$

82. $\tan x = 2$

83. $\csc x = 3$

84. $\cos x = x$

85–86 ■ Limiting Behavior of Trigonometric Functions A function f is given.

(a) Is f even, odd, or neither?

(b) Find the x-intercepts of the graph of f.

(c) Graph f in an appropriate viewing rectangle.

(d) Describe the behavior of the function as $x \to \pm\infty$.

(e) Notice that $f(x)$ is not defined when $x = 0$. What happens as x approaches 0?

85. $f(x) = \dfrac{1 - \cos x}{x}$ **86.** $f(x) = \dfrac{\sin 4x}{2x}$

■ Applications

87. Height of a Wave As a wave passes by an offshore piling, the height of the water is modeled by the function

$$h(t) = 3\cos\left(\frac{\pi}{10}t\right)$$

where $h(t)$ is the height in meters above mean sea level at time t seconds.

(a) Find the period of the wave.

(b) Find the wave height, that is, the vertical distance between the trough and the crest of the wave.

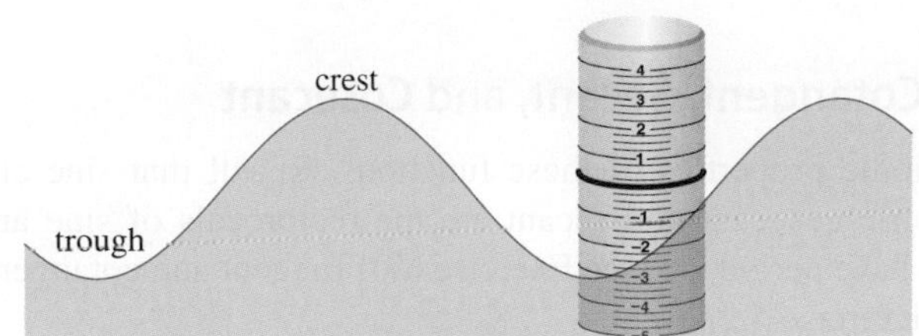

88. Sound Vibrations A tuning fork is struck, producing a pure tone as its tines vibrate. The vibrations can be modeled by the function

$$v(t) = 0.7\sin(880\pi t)$$

where $v(t)$ is the displacement of the tines in millimeters at time t seconds.

(a) Find the period of the vibration.

(b) Find the frequency of the vibration, that is, the number of times the fork vibrates per second.

(c) Graph the function v.

89. Blood Pressure Each time your heart beats, your blood pressure first increases and then decreases as the heart rests between beats. The maximum and minimum blood pressures are called the *systolic* and *diastolic* pressures, respectively. A *blood pressure reading* is written as systolic/diastolic; a reading of 120/80 is considered normal. A certain person's blood pressure is modeled by the function

$$p(t) = 115 + 25\sin(160\pi t)$$

where $p(t)$ is the pressure in mmHg (millimeters of mercury), at time t measured in minutes.

(a) Find the period of p.

(b) Find the number of heartbeats per minute.

(c) Graph the function p.

(d) Find the blood pressure reading. How does this compare to normal blood pressure?

90. Variable Stars Variable stars are ones whose brightness varies periodically. One of the most visible is R Leonis; its brightness can be modeled by the function

$$b(t) = 7.9 - 2.1\cos\left(\frac{\pi}{156}t\right)$$

where t is measured in days.

(a) Find the period of R Leonis.

(b) Find the maximum and minimum brightness.

(c) Graph the function b.

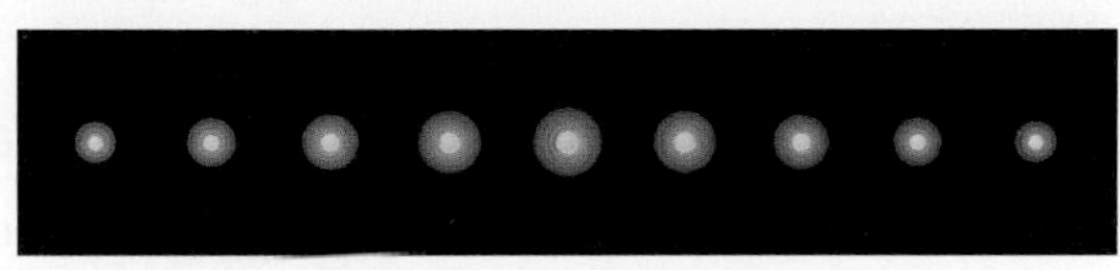

■ Discuss ■ Discover ■ Prove ■ Write

91. Discuss ■ Discover: Number of Solutions Find the number of solutions of the equation

$$\sin x = \frac{x}{100}$$

PS *Draw a diagram.* First try to solve the simpler problem $\sin x = x/10$ by graphing each side of the equation.

92. Discuss: Compositions Involving Trigonometric Functions This exercise explores the effect of the inner function g on a composite function $y = f(g(x))$.

(a) Graph the function $y = \sin\sqrt{x}$ using the viewing rectangle $[0, 400]$ by $[-1.5, 1.5]$. In what ways does this graph differ from the graph of the sine function?

(b) Graph the function $y = \sin(x^2)$ using the viewing rectangle $[-5, 5]$ by $[-1.5, 1.5]$. In what ways does this graph differ from the graph of the sine function?

93. Discuss ■ Discover: Combinations of Trigonometric Functions For each function determine whether it is periodic; if so, state the period and the maximum and minimum values on each period. Is the function even, odd, or neither? Use the information you found to match the function with its graph.

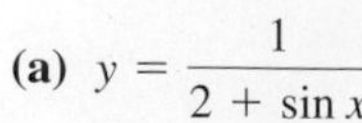

(a) $y = \dfrac{1}{2 + \sin x}$

(b) $y = e^{\sin 2x}$

(c) $y = \dfrac{\sin x}{1 + |x|}$

(d) $y = \cos\left(\dfrac{1}{1 + x^2}\right)$

I

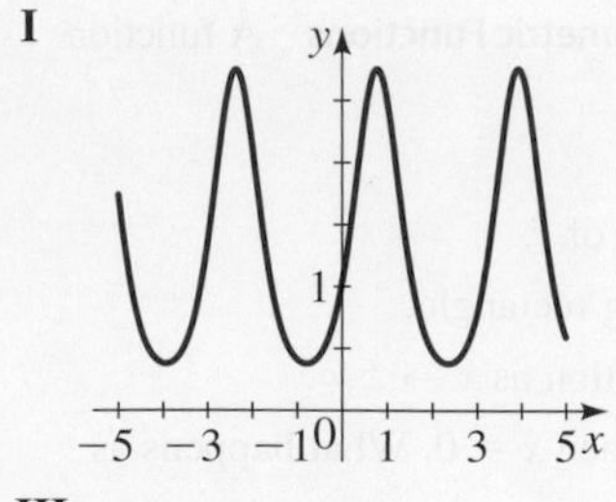

II

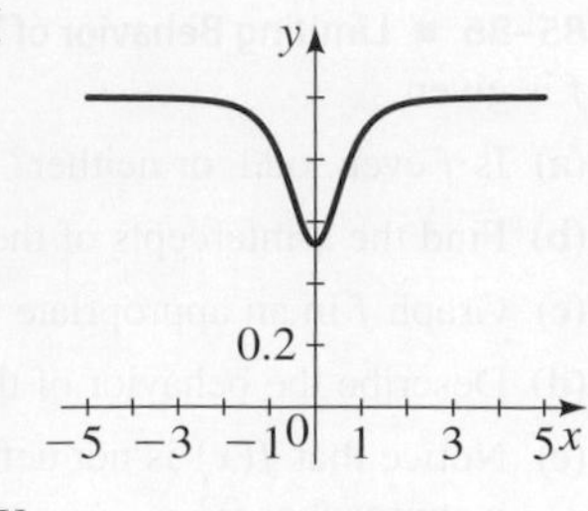

III

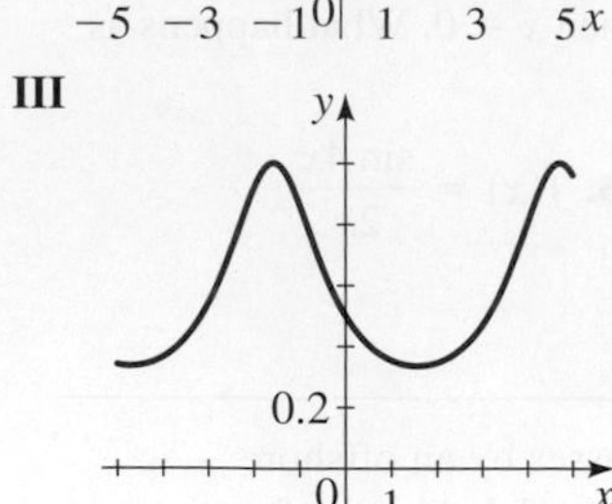

IV

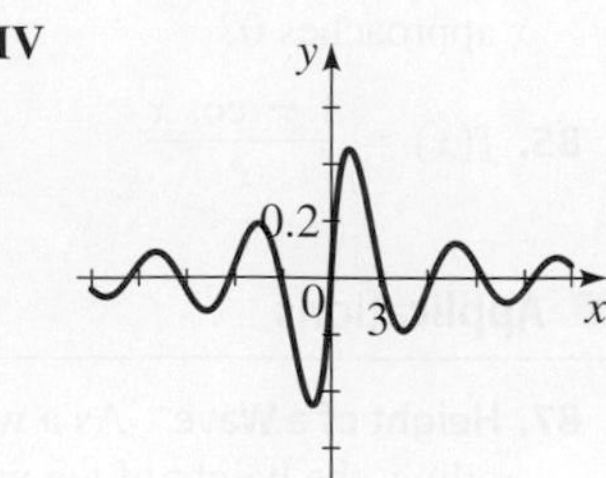

5.4 More Trigonometric Graphs

■ Graphs of Tangent, Cotangent, Secant, and Cosecant ■ Graphs of Transformations of Tangent and Cotangent ■ Graphs of Transformations of Secant and Cosecant

In this section we graph the tangent, cotangent, secant, and cosecant functions as well as transformations of these functions.

■ Graphs of Tangent, Cotangent, Secant, and Cosecant

We begin by stating the periodic properties of these functions. Recall that sine and cosine have period 2π. Because cosecant and secant are the reciprocals of sine and cosine, respectively, they also have period 2π (see Exercise 67). Tangent and cotangent, however, have period π (see Exercise 5.2.96).

Periodic Properties

The functions tangent and cotangent have period π:

$$\tan(x + \pi) = \tan x \qquad \cot(x + \pi) = \cot x$$

The functions secant and cosecant have period 2π:

$$\sec(x + 2\pi) = \sec x \qquad \csc(x + 2\pi) = \csc x$$

We first sketch the graph of tangent. Since it has period π, we need only sketch the graph on *any* interval of length π and then repeat the pattern to the left and to the right. We sketch the graph on the interval $(-\pi/2, \pi/2)$. Since $\tan(\pi/2)$ and $\tan(-\pi/2)$ aren't defined, we need to be careful in sketching the graph at points near $\pi/2$ and $-\pi/2$. As x gets near $\pi/2$ through values less than $\pi/2$, the value of $\tan x$ becomes large. To see this, notice that as x gets close to $\pi/2$, $\cos x$ approaches 0 and $\sin x$ approaches 1 and so $\tan x = \sin x/\cos x$ is large. A table of values of $\tan x$ for x close to $\pi/2$ (≈ 1.570796) is shown in the margin. So as x approaches $\pi/2$ from the left, the value of $\tan x$ increases without bound. We express this by writing

$$\tan x \to \infty \qquad \text{as} \qquad x \to \frac{\pi^-}{2}$$

x	$\tan x$
0	0
$\pi/6$	0.58
$\pi/4$	1.00
$\pi/3$	1.73
1.4	5.80
1.5	14.10
1.55	48.08
1.57	1,255.77
1.5707	10,381.33

In a similar way, as x approaches $-\pi/2$ from the right, the value of $\tan x$ decreases without bound. We write this as

$$\tan x \to -\infty \qquad \text{as} \qquad x \to -\frac{\pi}{2}^{+}$$

Arrow notation and asymptotes are discussed in Section 3.6.

This means that $x = -\pi/2$ and $x = \pi/2$ are vertical asymptotes of the graph of $y = \tan x$.

With this information, we sketch the graph of $y = \tan x$ for $-\pi/2 < x < \pi/2$ in Figure 1. The function $y = \cot x$ is graphed on the interval $(0, \pi)$ by a similar analysis (see Figure 2).

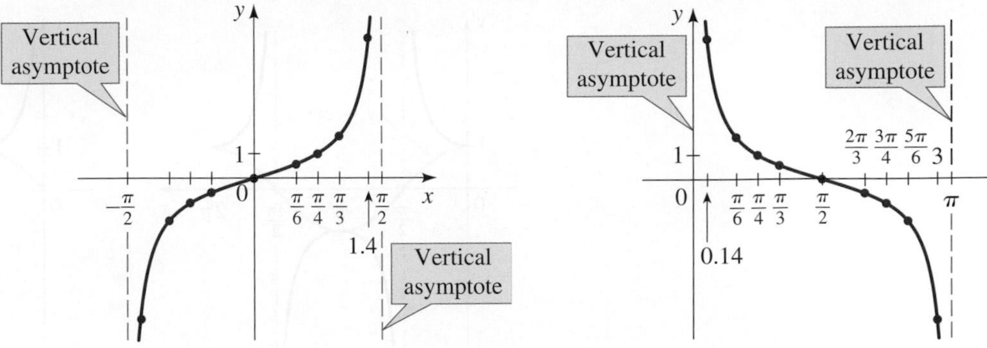

Figure 1 | One period of $y = \tan x$

Figure 2 | One period of $y = \cot x$

The values of x where $\cos x = 0$ (the x-intercepts of $y = \cos x$) are described in Section 5.3.

The complete graphs of tangent and cotangent can now be obtained by using the fact that these functions are periodic with period π; their graphs look the same on successive intervals of length π. Note that because $\tan x = \sin x/\cos x$, the graph has vertical asymptotes at those values of x for which $\cos x = 0$. Similarly, since $\cot x = \cos x/\sin x$, the graph has vertical asymptotes at those values of x for which $\sin x = 0$. We summarize these observations.

Graphs of Tangent and Cotangent

The trigonometric functions $y = \tan x$ and $y = \cot x$ are periodic with period π.

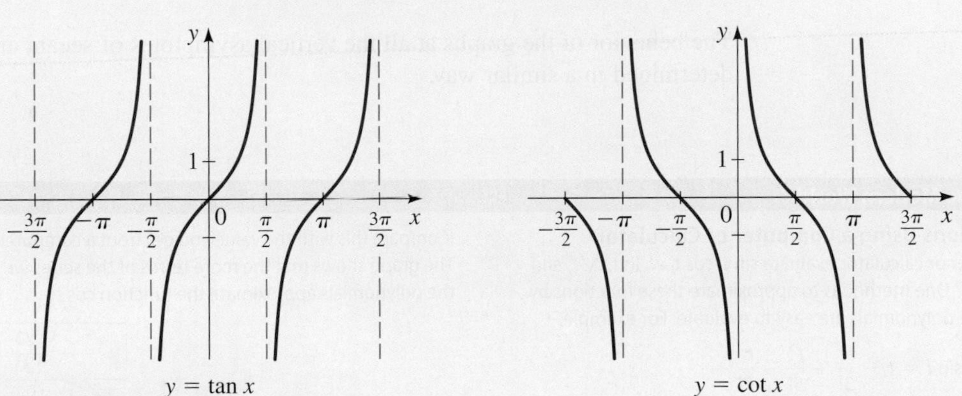

$y = \tan x$ $\qquad$ $y = \cot x$

The vertical asymptotes of $y = \tan x$ are $x = \dfrac{\pi}{2} + n\pi$, where n is an integer.

The vertical asymptotes of $y = \cot x$ are $x = n\pi$, where n is an integer.

From the graphs we see that the range of the tangent and cotangent functions is $(-\infty, \infty)$. Also, both graphs are symmetric about the origin because both are odd functions (see Section 2.6).

Let's graph one period of the secant and cosecant functions on the interval $(0, 2\pi)$. The reciprocal identities

$$\sec x = \frac{1}{\cos x} \qquad \text{and} \qquad \csc x = \frac{1}{\sin x}$$

tell us how we can graph these functions. To graph $y = \sec x$ we take the reciprocals of the y-coordinates of the points of the graph of $y = \cos x$. (See Figure 3.) Similarly, to graph $y = \csc x$ we take the reciprocals of the y-coordinates of the points of the graph of $y = \sin x$. (See Figure 4.)

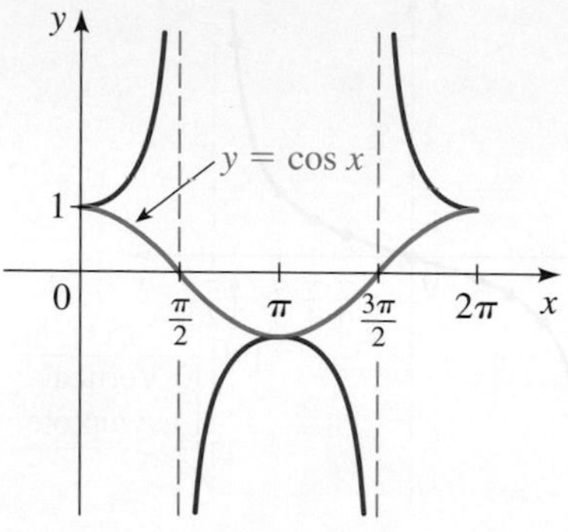

Figure 3 | One period of $y = \sec x$

Figure 4 | One period of $y = \csc x$

Note that the graphs of secant and cosecant have vertical asymptotes at those values of x for which the denominator $\cos x = 0$ or $\sin x = 0$, respectively. The behavior of the graphs near the vertical asymptotes depends on the sign of the denominators. For instance, $y = \sec x$ has vertical asymptotes at $x = \pi/2$ (because $\cos \pi/2 = 0$). Since $\cos x$ is positive to the left of $\pi/2$ and negative to the right, we see from Figure 3 that

$$\sec x \to \infty \quad \text{as} \quad x \to \frac{\pi}{2}^-$$

$$\sec x \to -\infty \quad \text{as} \quad x \to \frac{\pi}{2}^+$$

The behavior of the graphs at all the vertical asymptotes of secant and cosecant can be determined in a similar way.

Mathematics in the Modern World

Evaluating Functions Using a Computer or Calculator

How does a computer or calculator evaluate $\sin t$, $\cos t$, e^t, $\ln t$, $\sqrt{t}$, and other such functions? One method is to approximate these functions by polynomials because polynomials are easy to evaluate. For example,

$$\sin t = t - \frac{t^3}{3!} + \frac{t^5}{5!} - \frac{t^7}{7!} + \cdots$$

$$\cos t = 1 - \frac{t^2}{2!} + \frac{t^4}{4!} - \frac{t^6}{6!} + \cdots$$

where $n! = 1 \cdot 2 \cdot 3 \cdot \cdots \cdot n$. These remarkable formulas were found by the British mathematician Brook Taylor (1685–1731). These formulas are called Taylor series; you will learn to derive them in your calculus course. If we use the first three terms of a Taylor series to find cos 0.4, we get

$$\cos 0.4 \approx 1 - \frac{(0.4)^2}{2!} + \frac{(0.4)^4}{4!} \approx 0.92106667$$

(Compare this with the value you get from a computer or calculator.) The graph shows that the more terms of the series we use, the more closely the polynomials approximate the function $\cos t$.

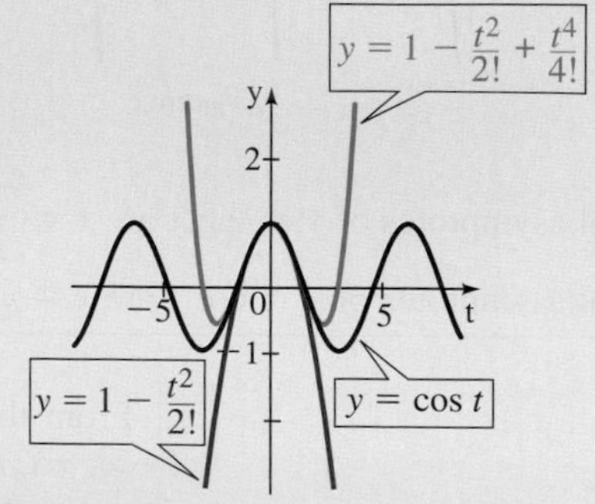

The complete graphs of secant and cosecant can now be obtained from the fact that these functions are periodic with period 2π.

Graphs of Secant and Cosecant

The trigonometric functions $y = \sec x$ and $y = \csc x$ are periodic with period 2π.

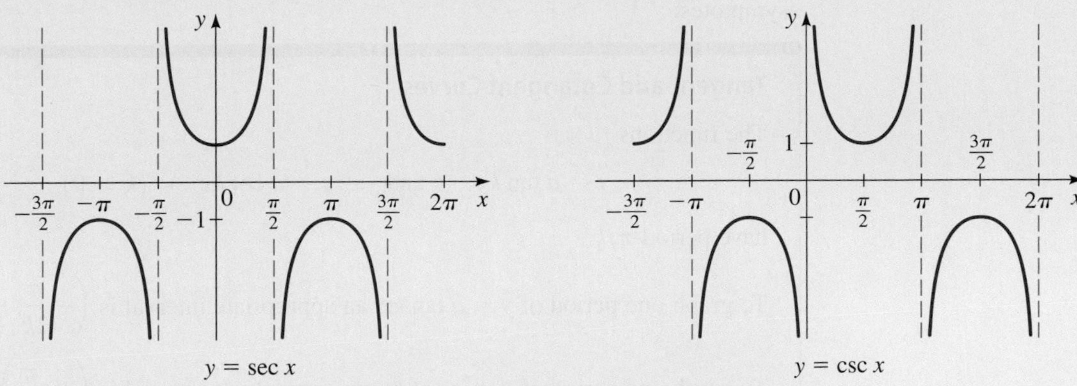

The vertical asymptotes of $y = \sec x$ are $x = \frac{\pi}{2} + n\pi$, where n is an integer.

The vertical asymptotes of $y = \csc x$ are $x = n\pi$, where n is an integer.

From the graphs we see that the range of the secant and cosecant functions is $(-\infty, -1] \cup [1, \infty)$. Also, the graph of secant is symmetric about the y-axis and the graph of cosecant is symmetric about the origin. This is because secant is an even function, whereas cosecant is an odd function (see Section 2.6).

■ Graphs of Transformations of Tangent and Cotangent

We now consider graphs of transformations of the tangent and cotangent functions.

Example 1 ■ Graphing Tangent Curves

Graph each function.

(a) $y = 2 \tan x$ **(b)** $y = -\tan x$

Solution We first graph $y = \tan x$ and then transform it as required.

(a) To graph $y = 2 \tan x$, we multiply the y-coordinate of each point on the graph of $y = \tan x$ by 2. This has the effect of stretching the graph vertically by a factor of 2. The resulting graph is shown in Figure 5(a).

(b) The graph of $y = -\tan x$ in Figure 5(b) is obtained from that of $y = \tan x$ by reflecting about the x-axis.

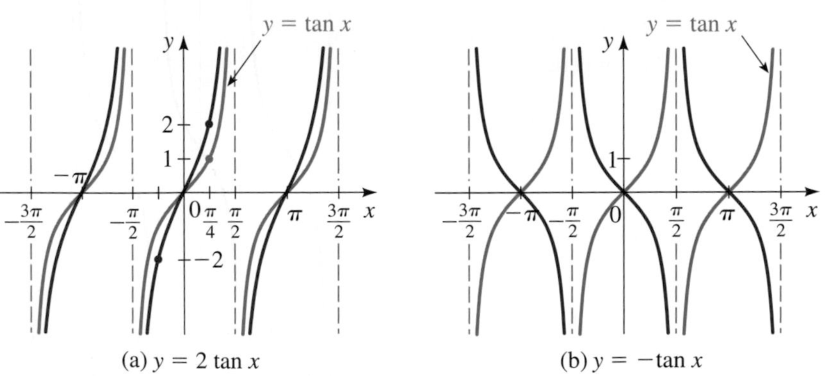

(a) $y = 2 \tan x$ (b) $y = -\tan x$

Figure 5

Now Try Exercises 9 and 11 ■

Since the tangent and cotangent functions have period π, the functions

$$y = a \tan kx \qquad \text{and} \qquad y = a \cot kx \qquad (k > 0)$$

complete one period as kx varies from 0 to π, that is, for $0 \le kx \le \pi$. Solving this inequality, we get $0 \le x \le \pi/k$. So they each have period π/k. To sketch a complete period of these graphs, it's convenient to select an interval between vertical asymptotes.

Tangent and Cotangent Curves

The functions

$$y = a \tan kx \qquad \text{and} \qquad y = a \cot kx \qquad (k > 0)$$

have period π/k.

To graph one period of $y = a \tan kx$, an appropriate interval is $\left(-\dfrac{\pi}{2k}, \dfrac{\pi}{2k}\right)$.

To graph one period of $y = a \cot kx$, an appropriate interval is $\left(0, \dfrac{\pi}{k}\right)$.

Note When we graph one period of a tangent or cotangent curve the asymptotes are at the endpoints of the appropriate interval and the x-intercept is at the midpoint of the interval. The asymptotes and x-intercepts are the **key features** that guide us in graphing tangent and cotangent curves.

Example 2 ■ Graphing Tangent Curves

Graph each function.

(a) $y = \tan 2x$ **(b)** $y = \tan 2\left(x - \dfrac{\pi}{4}\right)$

Solution

(a) The period is $\pi/2$ and an appropriate interval is $(-\pi/4, \pi/4)$. The endpoints $x = -\pi/4$ and $x = \pi/4$ are vertical asymptotes and the x-intercept occurs at $x = 0$, the midpoint of the interval. Thus we graph one complete period of the function on $(-\pi/4, \pi/4)$. The graph has the same shape as that of the tangent function (see Figure 1) but is shrunk horizontally by a factor of $\frac{1}{2}$. We then repeat that portion of the graph to the left and to the right. See Figure 6(a).

(b) The graph is the same as that in part (a), but it is shifted to the right $\pi/4$, as shown in Figure 6(b).

Because $y = \tan x$ completes one period between $x = -\frac{\pi}{2}$ and $x = \frac{\pi}{2}$, the function $y = \tan 2(x - \frac{\pi}{4})$ completes one period as $2(x - \frac{\pi}{4})$ varies from $-\frac{\pi}{2}$ to $\frac{\pi}{2}$.

Start of period:	End of period:
$2(x - \frac{\pi}{4}) = -\frac{\pi}{2}$	$2(x - \frac{\pi}{4}) = \frac{\pi}{2}$
$x - \frac{\pi}{4} = -\frac{\pi}{4}$	$x - \frac{\pi}{4} = \frac{\pi}{4}$
$x = 0$	$x = \frac{\pi}{2}$

So we graph one period on the interval $(0, \frac{\pi}{2})$.

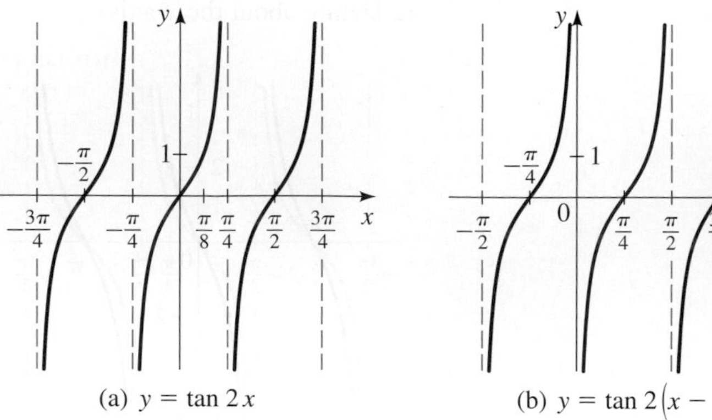

Figure 6

Now Try Exercises 19, 35, and 43 ■

Example 3 ■ A Horizontally Shifted Cotangent Curve

Graph the function $y = 2\cot\left(3x - \dfrac{\pi}{4}\right)$.

Solution We first put the equation in the form $y = a\cot k(x - b)$ by factoring 3 from the expression $3x - \dfrac{\pi}{4}$:

$$y = 2\cot\left(3x - \frac{\pi}{4}\right) = 2\cot 3\left(x - \frac{\pi}{12}\right)$$

Because $y = \cot x$ completes one period between $x = 0$ and $x = \pi$, the function $y = 2\cot(3x - \frac{\pi}{4})$ completes one period as $3x - \frac{\pi}{4}$ varies from 0 to π.

Start of period:	End of period:
$3x - \frac{\pi}{4} = 0$	$3x - \frac{\pi}{4} = \pi$
$3x = \frac{\pi}{4}$	$3x = \frac{5\pi}{4}$
$x = \frac{\pi}{12}$	$x = \frac{5\pi}{12}$

So we graph one period on the interval $\left(\frac{\pi}{12}, \frac{5\pi}{12}\right)$.

Thus the graph is the same as that of $y = 2\cot 3x$ but is shifted to the right $\pi/12$. The period of $y = 2\cot 3x$ is $\pi/3$, and an appropriate interval for graphing one period is $(0, \pi/3)$. To get the corresponding interval for the desired graph, we shift this interval to the right $\pi/12$. So we have

$$\left(0 + \frac{\pi}{12}, \frac{\pi}{3} + \frac{\pi}{12}\right) = \left(\frac{\pi}{12}, \frac{5\pi}{12}\right)$$

Finally, we graph one period in the shape of cotangent (see Figure 2) on the interval $(\pi/12, 5\pi/12)$ and repeat that portion of the graph to the left and to the right (see Figure 7).

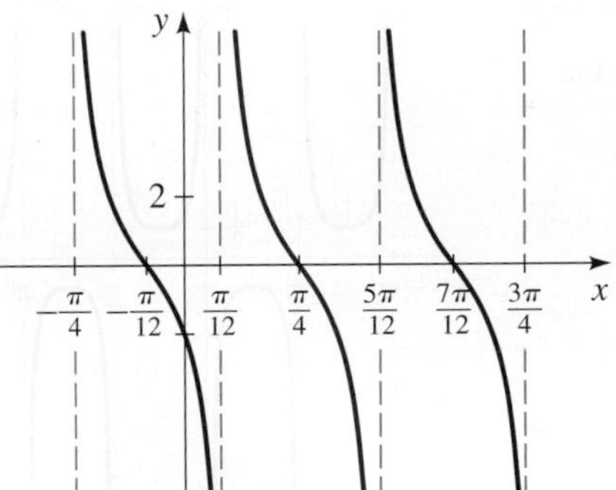

Figure 7 | $y = 2\cot\left(3x - \dfrac{\pi}{4}\right)$

Now Try Exercises 37 and 47 ■

■ Graphs of Transformations of Secant and Cosecant

We have already observed that the secant and cosecant functions are the reciprocals of the cosine and sine functions. Thus the following result is the counterpart of the result for sine and cosine curves in Section 5.3.

> **Secant and Cosecant Curves**
>
> The functions
>
> $$y = a\sec kx \quad \text{and} \quad y = a\csc kx \qquad (k > 0)$$
>
> have period $2\pi/k$.
>
> An appropriate interval on which to graph one complete period is $\left(0, \dfrac{2\pi}{k}\right)$.

To graph $y = a\sec kx$ and $y = a\csc kx$ on the interval $(0, 2\pi/k)$, we sketch on that interval graphs of the same shape as those in Figures 3 and 4, respectively.

Example 4 ■ Graphing Cosecant Curves

Graph each function.

(a) $y = \frac{1}{2} \csc 2x$ **(b)** $y = \frac{1}{2} \csc\left(2x + \frac{\pi}{2}\right)$

Solution

(a) The period is $2\pi/2 = \pi$. An appropriate interval is $(0, \pi)$, and the asymptotes occur whenever $\sin 2x = 0$. So the asymptotes for this period are $x = 0$, $x = \pi/2$, and $x = \pi$. With this information we sketch on the interval $(0, \pi)$ a graph with the same general shape as that of one period of the cosecant function (see Figure 4). The complete graph in Figure 8(a) is obtained by repeating this portion of the graph to the left and to the right.

(b) We first write

$$y = \frac{1}{2} \csc\left(2x + \frac{\pi}{2}\right) = \frac{1}{2} \csc 2\left(x + \frac{\pi}{4}\right)$$

From this we see that the graph is the same as that in part (a) but shifted to the left $\pi/4$. The graph is shown in Figure 8(b).

Because $y = \csc x$ completes one period between $x = 0$ and $x = 2\pi$, the function $y = \frac{1}{2}\csc(2x + \frac{\pi}{2})$ completes one period as $2x + \frac{\pi}{2}$ varies from 0 to 2π.

Start of period:	End of period:
$2x + \frac{\pi}{2} = 0$	$2x + \frac{\pi}{2} = 2\pi$
$2x = -\frac{\pi}{2}$	$2x = \frac{3\pi}{2}$
$x = -\frac{\pi}{4}$	$x = \frac{3\pi}{4}$

So we graph one period on the interval $\left(-\frac{\pi}{4}, \frac{3\pi}{4}\right)$.

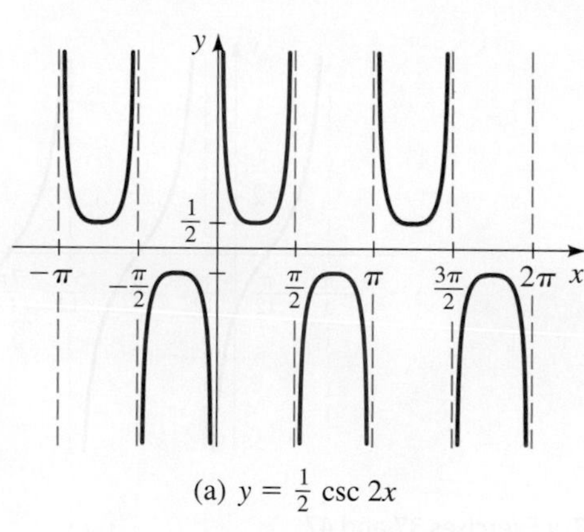

(a) $y = \frac{1}{2} \csc 2x$

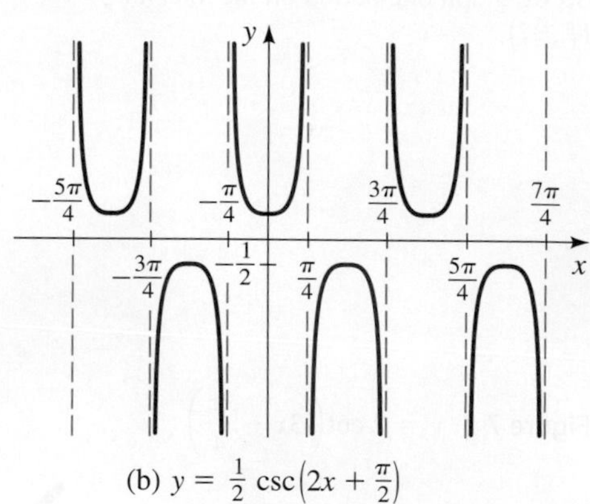

(b) $y = \frac{1}{2} \csc\left(2x + \frac{\pi}{2}\right)$

Figure 8

Now Try Exercises 29 and 49 ■

Example 5 ■ Graphing a Secant Curve

Graph $y = 3 \sec \frac{1}{2}x$.

Solution The period is $2\pi \div \frac{1}{2} = 4\pi$. An appropriate interval is $(0, 4\pi)$, and the asymptotes occur wherever $\cos \frac{1}{2}x = 0$. Thus the asymptotes for this period are $x = \pi$, $x = 3\pi$. With this information we sketch on the interval $(0, 4\pi)$ a graph with the same general shape as that of one period of the secant function (see Figure 3). The complete graph in Figure 9 is obtained by repeating this portion of the graph to the left and to the right.

Figure 9 | $y = 3 \sec \frac{1}{2}x$

Now Try Exercises 31 and 51 ■

5.4 | Exercises

■ Concepts

1. The trigonometric function $y = \tan x$ has period ________ and asymptotes $x =$ ________. Sketch a graph of this function on the interval $(-\pi/2, \pi/2)$.

2. The trigonometric function $y = \csc x$ has period ________ and asymptotes $x =$ ________. Sketch a graph of this function on the interval $(-\pi, \pi)$.

■ Skills

3–8 ■ Graphs of Trigonometric Functions Match the trigonometric function with one of the graphs I–VI.

3. $f(x) = \tan\left(x + \dfrac{\pi}{4}\right)$

4. $f(x) = \sec 2x$

5. $f(x) = \cot 4x$

6. $f(x) = -\tan x$

7. $f(x) = 2\sec x$

8. $f(x) = 1 + \csc x$

I

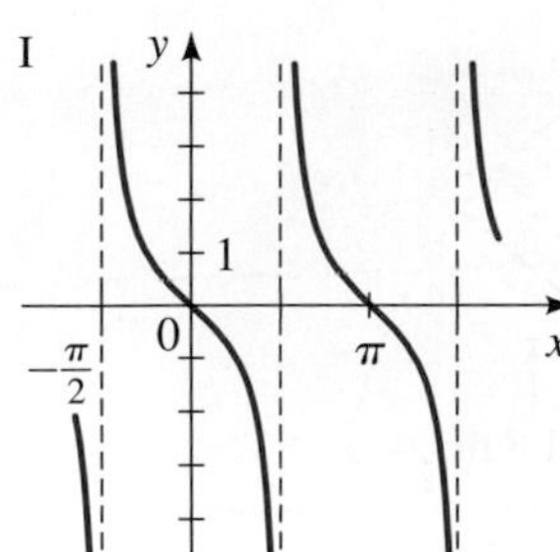

II

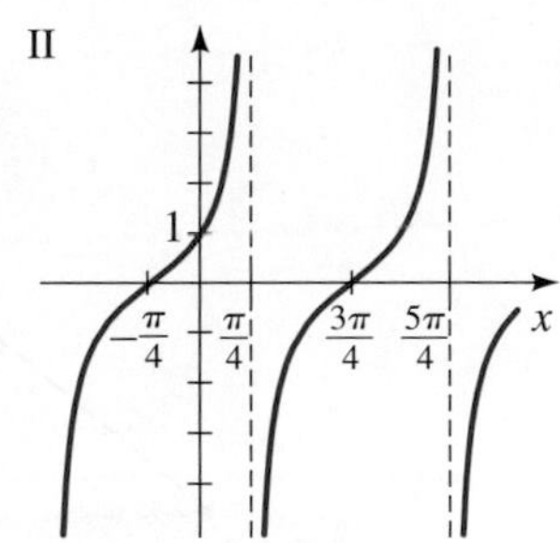

III

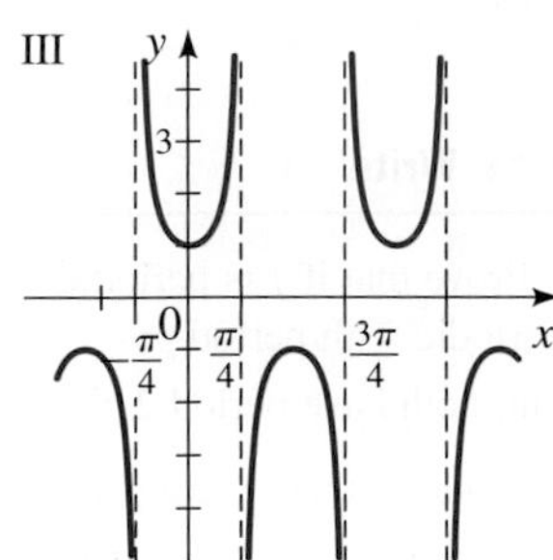

IV

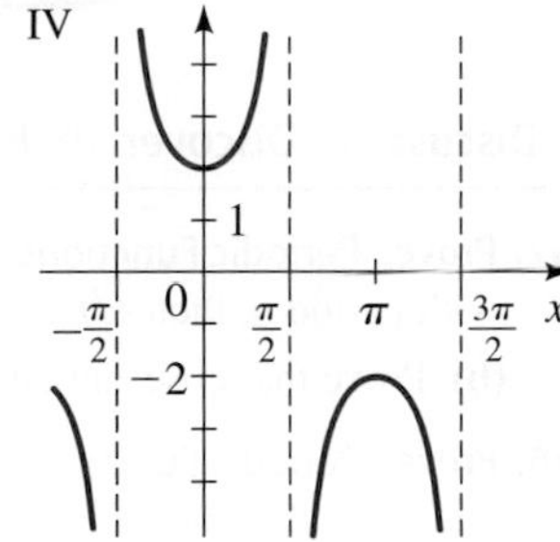

V

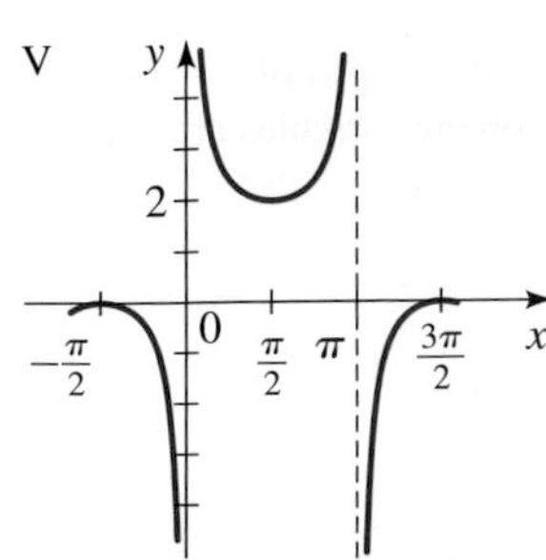

VI

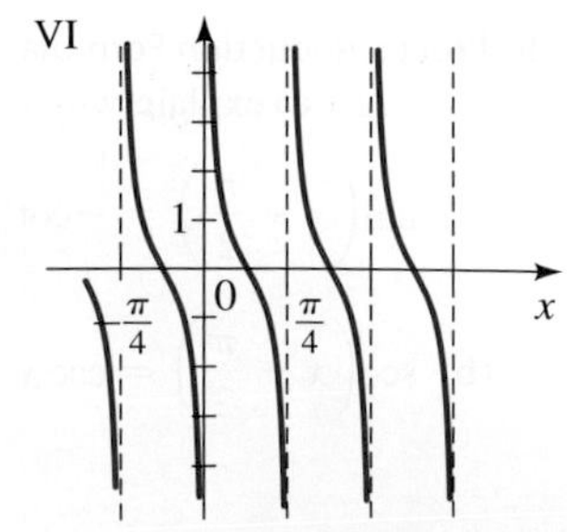

9–18 ■ Graphs of Trigonometric Functions State the period, and graph the function.

9. $y = 3\tan x$

10. $y = -3\tan x$

11. $y = -\frac{3}{2}\tan x$

12. $y = \frac{3}{4}\tan x$

13. $y = -\cot x$

14. $y = 2\cot x$

15. $y = 2\csc x$

16. $y = \frac{1}{2}\csc x$

17. $y = 3\sec x$

18. $y = -3\sec x$

19–34 ■ Graphs of Trigonometric Functions with Different Periods Find the period, and graph the function.

19. $y = \tan 3x$

20. $y = \tan 4x$

21. $y = -5\tan \pi x$

22. $y = -3\tan 4\pi x$

23. $y = 2\cot 3\pi x$

24. $y = 3\cot 2\pi x$

25. $y = \tan \dfrac{\pi}{4}x$

26. $y = \cot \dfrac{\pi}{2}x$

27. $y = 2\tan 3\pi x$

28. $y = 2\tan \dfrac{\pi}{2}x$

29. $y = \csc 4x$

30. $y = 5\csc 3x$

31. $y = \sec 2x$

32. $y = \frac{1}{2}\sec 4\pi x$

33. $y = 5\csc \dfrac{3\pi}{2}x$

34. $y = 5\sec 2\pi x$

35–60 ■ Graphs of Trigonometric Functions with Horizontal Shifts Find the period, and graph the function.

35. $y = \tan\left(x + \dfrac{\pi}{4}\right)$

36. $y = \tan\left(x - \dfrac{\pi}{4}\right)$

37. $y = \cot\left(x + \dfrac{\pi}{4}\right)$

38. $y = 2\cot\left(x - \dfrac{\pi}{3}\right)$

39. $y = \csc\left(x - \dfrac{\pi}{4}\right)$

40. $y = \sec\left(x + \dfrac{\pi}{4}\right)$

41. $y = \dfrac{1}{2}\sec\left(x - \dfrac{\pi}{6}\right)$

42. $y = 3\csc\left(x + \dfrac{\pi}{2}\right)$

43. $y = \tan 2\left(x - \dfrac{\pi}{3}\right)$

44. $y = \cot\left(2x - \dfrac{\pi}{4}\right)$

45. $y = 5\cot\left(3x + \dfrac{\pi}{2}\right)$

46. $y = 4\tan(4x - 2\pi)$

47. $y = \cot\left(2x - \frac{\pi}{2}\right)$

48. $y = \frac{1}{2}\tan(\pi x - \pi)$

49. $y = 2\csc\left(\pi x - \frac{\pi}{3}\right)$

50. $y = 3\sec\left(\frac{1}{4}x - \frac{\pi}{6}\right)$

51. $y = \sec 2\left(x - \frac{\pi}{4}\right)$

52. $y = \csc 2\left(x + \frac{\pi}{2}\right)$

53. $y = 5\sec\left(3x - \frac{\pi}{2}\right)$

54. $y = \frac{1}{2}\sec(2\pi x - \pi)$

55. $y = \tan\left(\frac{2}{3}x - \frac{\pi}{6}\right)$

56. $y = \tan\frac{1}{2}\left(x + \frac{\pi}{4}\right)$

57. $y = 3\sec\pi(x + \frac{1}{2})$

58. $y = \sec\left(3x + \frac{\pi}{2}\right)$

59. $y = -2\tan\left(2x - \frac{\pi}{3}\right)$

60. $y = 2\cot(3\pi x + 3\pi)$

61–64 ■ Graphing Trigonometric Functions Determine an appropriate viewing rectangle for each function, and use it to draw the graph.

61. $y = \tan 30x$

62. $y = \csc 50x$

63. $y = \sqrt{\tan 20\pi x}$

64. $y = \sec^2(10x)$

Applications

65. Lighthouse The beam from a lighthouse completes one rotation every 2 min. At time t, the distance d shown in the figure is

$$d(t) = 3\tan\pi t$$

where t is measured in minutes and d in kilometers.

(a) Find $d(0.15)$, $d(0.25)$, and $d(0.45)$.

(b) Sketch a graph of the function d for $0 \le t < \frac{1}{2}$.

(c) What happens to the distance d as t approaches $\frac{1}{2}$?

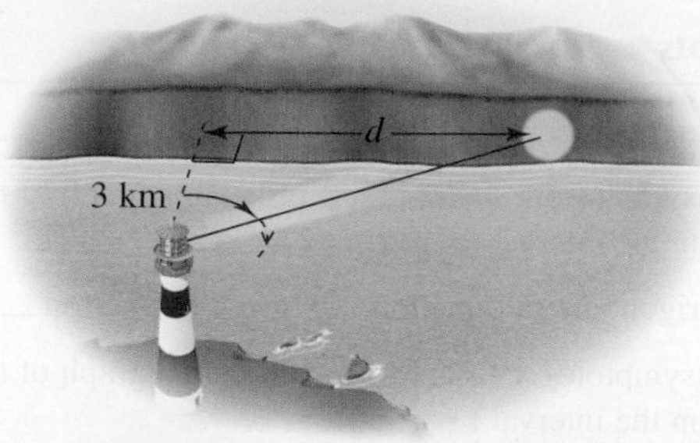

66. Length of a Shadow On a day when the sun passes directly overhead at noon, a 1.8-m-tall person casts a shadow of length

$$S(t) = 1.8\left|\cot\frac{\pi}{12}t\right|$$

where S is measured in meters and t is the number of hours since 6 A.M.

(a) Find the length of the person's shadow at 8:00 A.M., noon, 2:00 P.M., and 5:45 P.M.

(b) Sketch a graph of the function S for $0 < t < 12$.

(c) From the graph, determine the values of t at which the length of the shadow equals the person's height. To what time of day does each of these values correspond?

(d) Explain what happens to the shadow as the time approaches 6 P.M. (that is, as $t \to 12^-$).

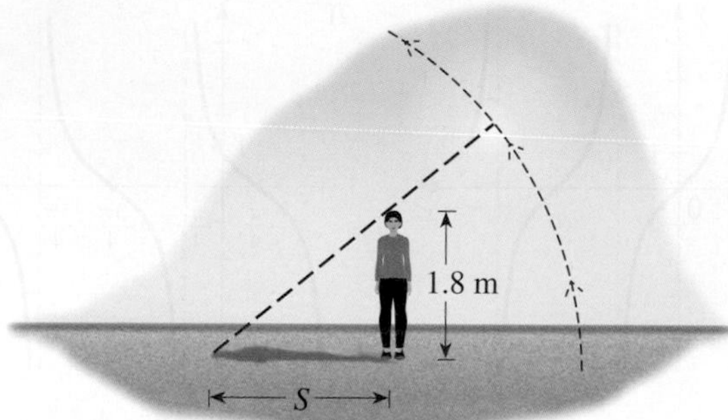

Discuss ■ Discover ■ Prove ■ Write

67. Prove: Periodic Functions **(a)** Prove that if f is periodic with period p, then $1/f$ is also periodic with period p.

(b) Prove that cosecant and secant both have period 2π.

68. Prove: Periodic Functions Prove that if f and g are periodic with period p, then f/g is also periodic but its period could be smaller than p.

69. Prove: Reduction Formulas Use the graphs of $y = \tan x$ and $y = \sec x$ to explain why the following formulas are true.

(a) $\tan\left(x - \frac{\pi}{2}\right) = -\cot x$

(b) $\sec\left(x - \frac{\pi}{2}\right) = \csc x$

5.5 Inverse Trigonometric Functions and Their Graphs

■ The Inverse Sine Function ■ The Inverse Cosine Function ■ The Inverse Tangent Function ■ The Inverse Secant, Cosecant, and Cotangent Functions

We study applications of inverse trigonometric functions to triangles in Sections 6.4–6.6.

Recall from Section 2.8 that the inverse of a function f is a function f^{-1} that reverses the rule of f. For a function to have an inverse, it must be one-to-one. Since the trigonometric functions are not one-to-one, they do not have inverses; however, it is possible to restrict the domains of the trigonometric functions in such a way that the resulting functions are one-to-one.

■ The Inverse Sine Function

Let's consider the sine function. There are many ways to restrict the domain of sine so that the new function is one-to-one. A natural way to do this is to restrict the domain to the interval $[-\pi/2, \pi/2]$. The reason for this choice is that sine is one-to-one on this interval and moreover attains each of the values in its range on this interval. From Figure 1 we see that sine is one-to-one on this restricted domain (by the Horizontal Line Test) and so has an inverse.

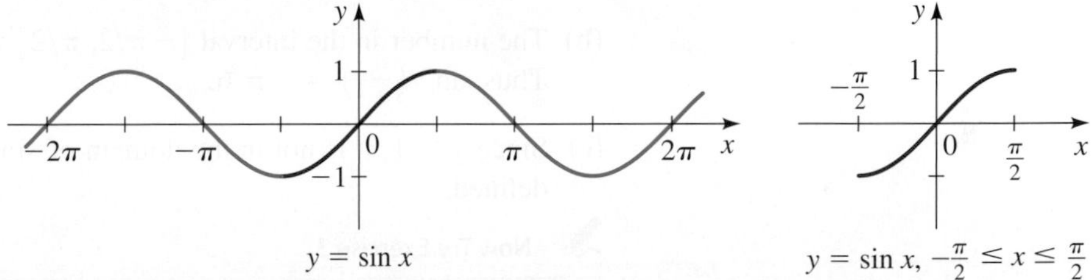

Figure 1 | Graphs of the sine function and the restricted sine function

We can now define an inverse sine function on this restricted domain. The graph of $y = \sin^{-1} x$ is shown in Figure 2; it is obtained by reflecting the graph of $y = \sin x$, $-\pi/2 \le x \le \pi/2$, about the line $y = x$.

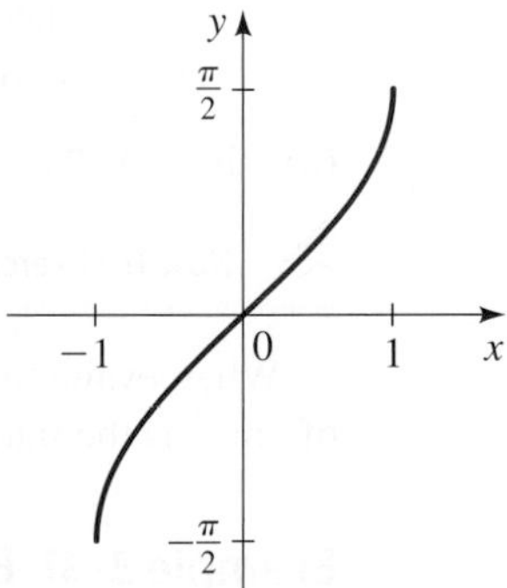

Figure 2 | $y = \sin^{-1} x$

Definition of the Inverse Sine Function

The **inverse sine function** is the function $\sin^{-1}$ with domain $[-1, 1]$ and range $[-\pi/2, \pi/2]$ defined by

$$\sin^{-1} x = y \quad \Leftrightarrow \quad \sin y = x$$

The inverse sine function is also called **arcsine**, denoted by **arcsin**.

Thus $y = \sin^{-1} x$ *is the number in the interval* $[-\pi/2, \pi/2]$ *whose sine is x.* In other words, $\sin(\sin^{-1} x) = x$. In fact, from the general properties of inverse functions studied in Section 2.8, we have the following **cancellation properties**.

$$\sin(\sin^{-1} x) = x \quad \text{for} \quad -1 \le x \le 1$$

$$\sin^{-1}(\sin x) = x \quad \text{for} \quad -\frac{\pi}{2} \le x \le \frac{\pi}{2}$$

Example 1 ■ Evaluating the Inverse Sine Function

Find the exact value.

(a) $\sin^{-1}\left(\frac{1}{2}\right)$ **(b)** $\sin^{-1}\left(-\frac{1}{2}\right)$ **(c)** $\sin^{-1}\left(\frac{3}{2}\right)$

Solution

(a) The number in the interval $[-\pi/2, \pi/2]$ whose sine is $\frac{1}{2}$ is $\pi/6$. Thus $\sin^{-1}(\frac{1}{2}) = \pi/6$.

(b) The number in the interval $[-\pi/2, \pi/2]$ whose sine is $-\frac{1}{2}$ is $-\pi/6$. Thus $\sin^{-1}(-\frac{1}{2}) = -\pi/6$.

(c) Since $\frac{3}{2} > 1$, it is not in the domain of $\sin^{-1} x$, so $\sin^{-1}(\frac{3}{2})$ is not defined.

 Now Try Exercise 3 ■

Example 2 ■ Using a Calculator to Evaluate Inverse Sine

Find approximate values for **(a)** $\sin^{-1}(0.82)$ and **(b)** $\sin^{-1}(\frac{1}{3})$.

Check to make sure you get the same answers as in Example 2 on your calculator.

Solution

We use a calculator to approximate these values. Using the [SIN⁻¹], or [INV] [SIN], or [ARC] [SIN] key(s) on the calculator (with the calculator in radian mode), we get

(a) $\sin^{-1}(0.82) \approx 0.96141$ **(b)** $\sin^{-1}(\frac{1}{3}) \approx 0.33984$

 Now Try Exercises 11 and 21 ■

When evaluating expressions involving $\sin^{-1}$, we need to remember that the range of $\sin^{-1}$ is the interval $[-\pi/2, \pi/2]$.

Example 3 ■ Evaluating Expressions with Inverse Sine

Find the exact value.

(a) $\sin^{-1}\left(\sin\frac{\pi}{3}\right)$ **(b)** $\sin^{-1}\left(\sin\frac{2\pi}{3}\right)$

Solution

(a) Since $\pi/3$ is in the interval $[-\pi/2, \pi/2]$, we can use the cancellation properties of inverse functions:

$$\sin^{-1}\left(\sin\frac{\pi}{3}\right) = \frac{\pi}{3} \qquad \text{Cancellation property: } -\frac{\pi}{2} \le \frac{\pi}{3} \le \frac{\pi}{2}$$

(b) We first evaluate the expression in the parentheses:

$$\sin^{-1}\left(\sin\frac{2\pi}{3}\right) = \sin^{-1}\left(\frac{\sqrt{3}}{2}\right) \qquad \text{Evaluate}$$

$$= \frac{\pi}{3} \qquad \text{Because } \sin\frac{\pi}{3} = \frac{\sqrt{3}}{2}$$

Note: $\sin^{-1}(\sin x) = x$ only if $-\frac{\pi}{2} \le x \le \frac{\pi}{2}$.

Now Try Exercises 31 and 33 ■

■ The Inverse Cosine Function

If the domain of the cosine function is restricted to the interval $[0, \pi]$, then the resulting function is one-to-one and so has an inverse. We choose this interval because on it, cosine attains each of its values exactly once (see Figure 3).

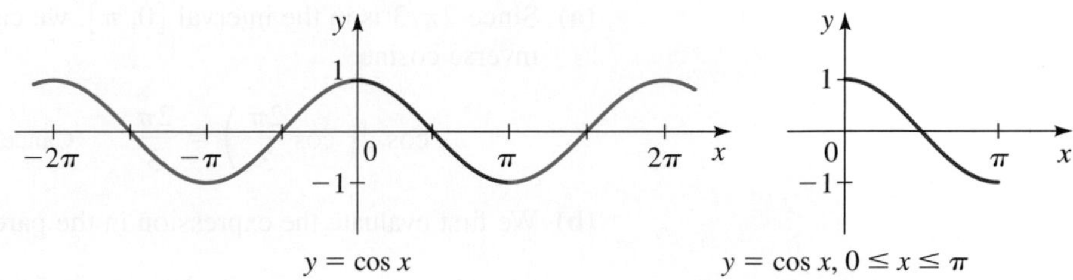

Figure 3 | Graphs of the cosine function and the restricted cosine function

> **Definition of the Inverse Cosine Function**
>
> The **inverse cosine function** is the function $\cos^{-1}$ with domain $[-1, 1]$ and range $[0, \pi]$ defined by
>
> $$\cos^{-1}x = y \quad \Leftrightarrow \quad \cos y = x$$
>
> The inverse cosine function is also called **arccosine**, denoted by **arccos**.

Thus $y = \cos^{-1}x$ *is the number in the interval* $[0, \pi]$ *whose cosine is* x. The following **cancellation properties** follow from the inverse function properties.

$$\cos(\cos^{-1}x) = x \quad \text{for} \quad -1 \le x \le 1$$
$$\cos^{-1}(\cos x) = x \quad \text{for} \quad 0 \le x \le \pi$$

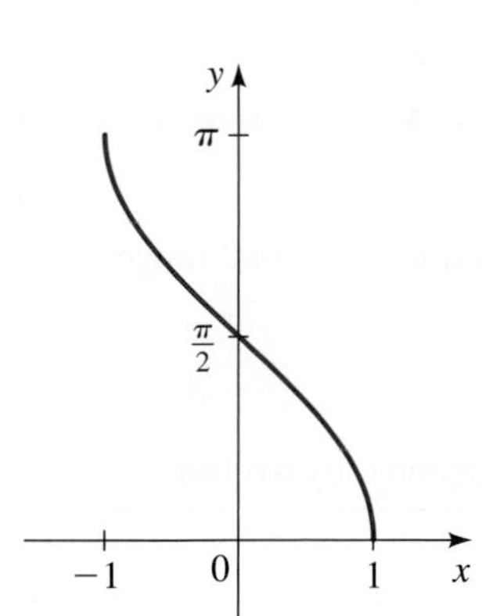

Figure 4 | Graph of $y = \cos^{-1}x$

The graph of $y = \cos^{-1}x$ is shown in Figure 4; it is obtained by reflecting the graph of $y = \cos x$, $0 \le x \le \pi$, about the line $y = x$.

Example 4 ■ Evaluating the Inverse Cosine Function

Find the exact value.

(a) $\cos^{-1}\left(\frac{\sqrt{3}}{2}\right)$ **(b)** $\cos^{-1}0$ **(c)** $\cos^{-1}\left(-\frac{1}{2}\right)$

Solution

(a) The number in the interval $[0, \pi]$ whose cosine is $\sqrt{3}/2$ is $\pi/6$. Thus $\cos^{-1}(\sqrt{3}/2) = \pi/6$.

(b) The number in the interval $[0, \pi]$ whose cosine is 0 is $\pi/2$. Thus $\cos^{-1} 0 = \pi/2$.

(c) The number in the interval $[0, \pi]$ whose cosine is $-\frac{1}{2}$ is $2\pi/3$. Thus $\cos^{-1}(-\frac{1}{2}) = 2\pi/3$. (The graph in Figure 4 shows that if $-1 \leq x < 0$, then $\cos^{-1} x > \pi/2$.)

 **Now Try Exercises 5 and 13** ■

Example 5 ■ Evaluating Expressions Involving Inverse Cosine

Find the exact value.

(a) $\cos^{-1}\left(\cos\dfrac{2\pi}{3}\right)$ **(b)** $\cos^{-1}\left(\cos\dfrac{5\pi}{3}\right)$

Solution

(a) Since $2\pi/3$ is in the interval $[0, \pi]$, we can use the cancellation properties for inverse cosine:

$$\cos^{-1}\left(\cos\frac{2\pi}{3}\right) = \frac{2\pi}{3} \qquad \text{Cancellation property: } 0 \leq \frac{2\pi}{3} \leq \pi$$

(b) We first evaluate the expression in the parentheses:

$$\cos^{-1}\left(\cos\frac{5\pi}{3}\right) = \cos^{-1}\left(\frac{1}{2}\right) \qquad \text{Evaluate}$$

$$= \frac{\pi}{3} \qquad \text{Because } \cos\frac{\pi}{3} = \frac{1}{2}$$

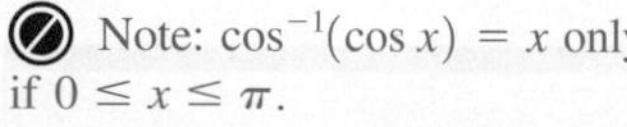
Note: $\cos^{-1}(\cos x) = x$ only if $0 \leq x \leq \pi$.

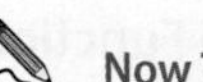 **Now Try Exercises 35 and 37** ■

■ The Inverse Tangent Function

We restrict the domain of the tangent function to the interval $(-\pi/2, \pi/2)$ to obtain a one-to-one function.

Definition of the Inverse Tangent Function

The **inverse tangent function** is the function $\tan^{-1}$ with domain $\mathbb{R}$ and range $(-\pi/2, \pi/2)$ defined by

$$\tan^{-1} x = y \quad \Leftrightarrow \quad \tan y = x$$

The inverse tangent function is also called **arctangent**, denoted by **arctan**.

Thus $y = \tan^{-1} x$ *is the number in the interval* $(-\pi/2, \pi/2)$ *whose tangent is* x. The following **cancellation properties** follow from the inverse function properties.

$$\tan(\tan^{-1} x) = x \quad \text{for} \quad x \in \mathbb{R}$$

$$\tan^{-1}(\tan x) = x \quad \text{for} \quad -\frac{\pi}{2} < x < \frac{\pi}{2}$$

Figure 5 shows the graph of $y = \tan x$ on the interval $(-\pi/2, \pi/2)$ and the graph of its inverse function, $y = \tan^{-1} x$. Note that $\tan^{-1}$ has horizontal asymptotes $y = \dfrac{\pi}{2}$ and $y = -\dfrac{\pi}{2}$.

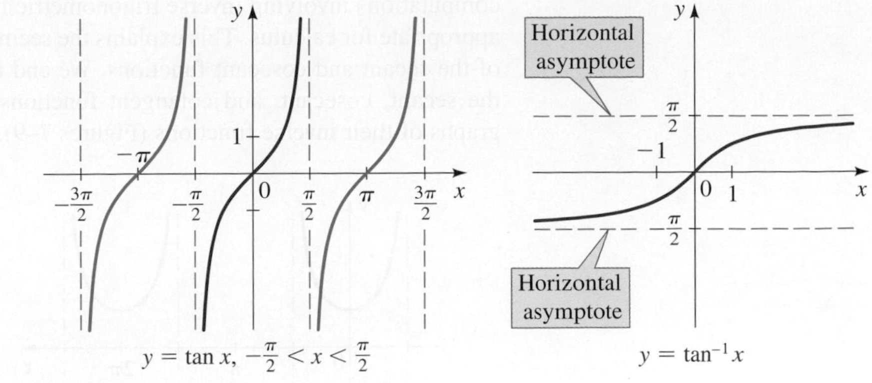

$y = \tan x,\ -\frac{\pi}{2} < x < \frac{\pi}{2}$ $y = \tan^{-1} x$

Horizontal asymptote

Horizontal asymptote

Figure 5 | Graphs of the restricted tangent function and the inverse tangent function

Example 6 ■ Evaluating the Inverse Tangent Function

Find each value.

(a) $\tan^{-1} 1$ **(b)** $\tan^{-1}(-\sqrt{3})$ **(c)** $\tan^{-1} 20$

Solution

(a) The number in the interval $(-\pi/2, \pi/2)$ with tangent 1 is $\pi/4$. Thus $\tan^{-1} 1 = \pi/4$.

(b) The number in the interval $(-\pi/2, \pi/2)$ with tangent $(-\sqrt{3})$ is $-\pi/3$. Thus $\tan^{-1}(-\sqrt{3}) = -\pi/3$.

(c) We use a calculator (in radian mode) to find that $\tan^{-1} 20 \approx 1.52084$.

Now Try Exercises 7 and 17 ■

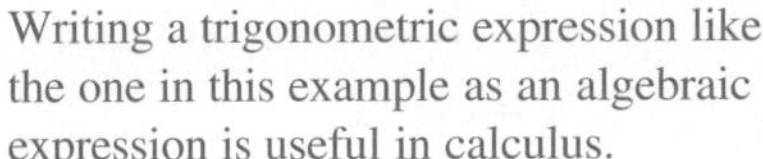
Writing a trigonometric expression like the one in this example as an algebraic expression is useful in calculus.

Example 7 ■ Composing Trigonometric Functions and Their Inverses

Write $\cos(\tan^{-1} x)$ as an algebraic expression in x.

Solution Let $u = \tan^{-1} x$. We need to write the cosine function in terms of the tangent function. Using the reciprocal and Pythagorean identities, we have

$$\cos u = \frac{1}{\sec u} = \frac{1}{\pm\sqrt{1 + \tan^2 u}}$$

Since u is in the interval $[-\pi/2, \pi/2]$ and $\cos u$ is positive on this interval, the $+$ sign is the appropriate choice for the radical. Now, substituting $u = \tan^{-1} x$ and applying the cancellation property $\tan(\tan^{-1} x) = x$, we get

$$\cos(\tan^{-1} x) = \frac{1}{\sqrt{1 + x^2}}$$

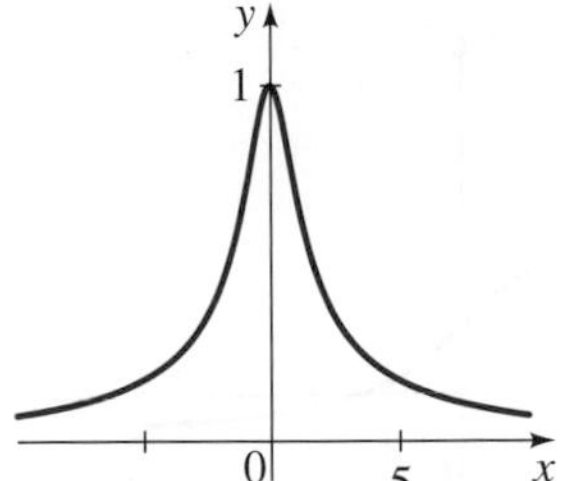

Figure 6 | The graphs of $y = \cos(\tan^{-1} x)$ and $y = 1/\sqrt{1 + x^2}$ are the same.

We can confirm this identity graphically, as shown in Figure 6.

 **Now Try Exercise 51** ■

Note In Example 7 we saw that the $+$ sign was the appropriate choice for the radical in the expression for $\cos(\tan^{-1} x)$. The domains of the inverse trigonometric functions have been chosen in such a way that the positive sign for the radical is appropriate for any expression of the form $S(T^{-1}(x))$, where S and T are any of the six trigonometric functions.

■ The Inverse Secant, Cosecant, and Cotangent Functions

See Exercise 6.4.56 for a method of finding the values of these inverse trigonometric functions on a calculator.

To define the inverse functions of the secant, cosecant, and cotangent functions, we restrict the domain of each function to a set on which it is one-to-one and on which it attains all its values. Although any interval satisfying these criteria is appropriate, we choose to restrict the domains in a way that simplifies the choice of sign in computations involving inverse trigonometric functions. The choices we make are also appropriate for calculus. This explains the seemingly strange restriction for the domains of the secant and cosecant functions. We end this section by displaying the graphs of the secant, cosecant, and cotangent functions with their restricted domains and the graphs of their inverse functions (Figures 7–9).

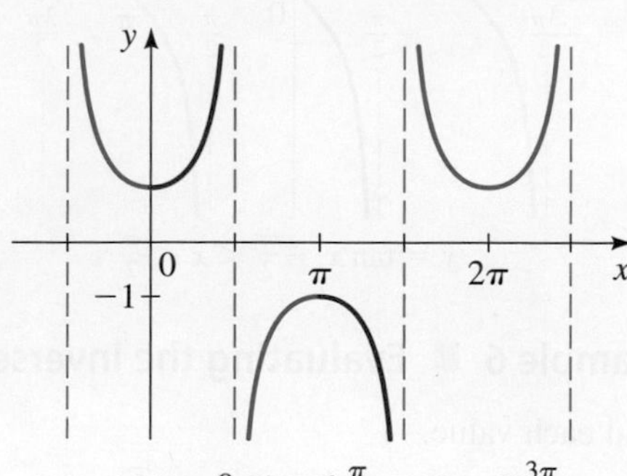

$y = \sec x,\ 0 \le x < \frac{\pi}{2},\ \pi \le x < \frac{3\pi}{2}$

$y = \sec^{-1} x$

Figure 7 | The inverse secant function

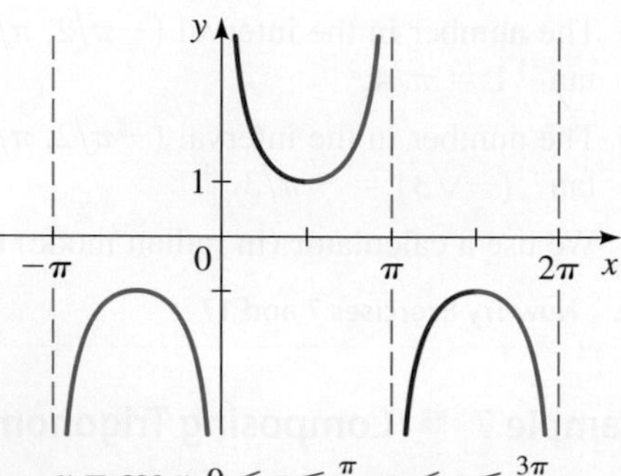

$y = \csc x,\ 0 < x \le \frac{\pi}{2},\ \pi < x \le \frac{3\pi}{2}$

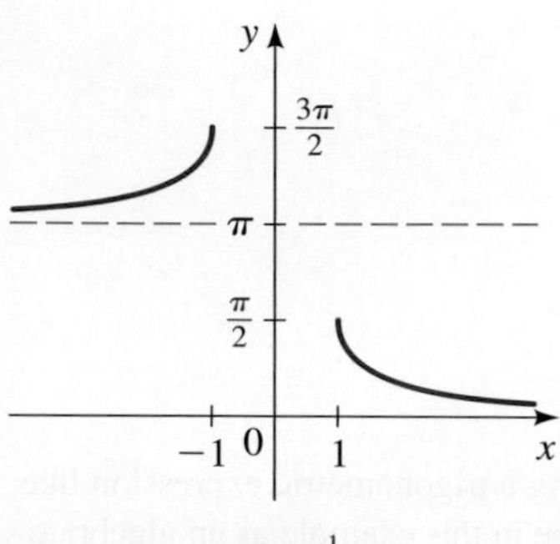

$y = \csc^{-1} x$

Figure 8 | The inverse cosecant function

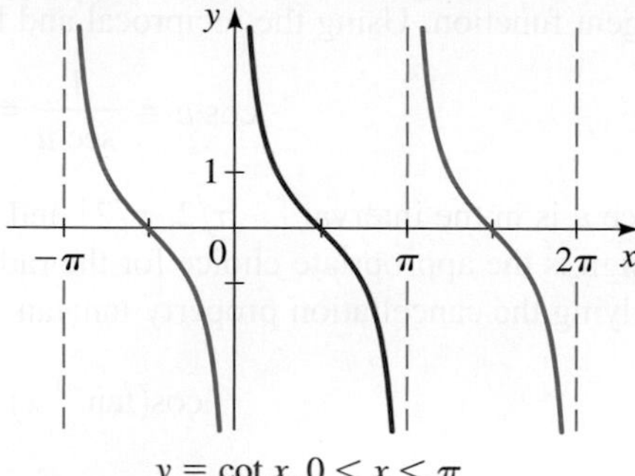

$y = \cot x,\ 0 < x < \pi$

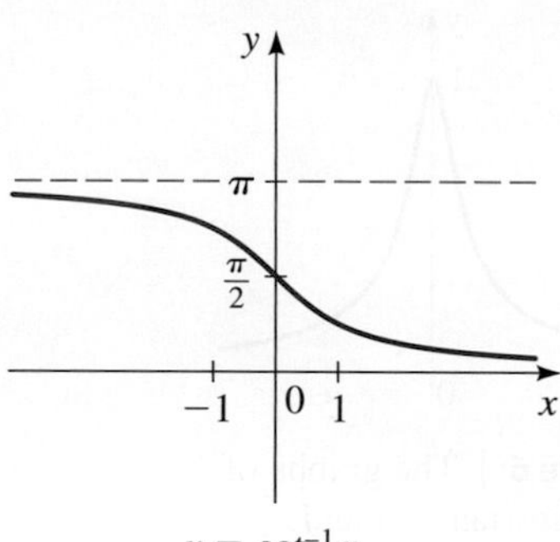

$y = \cot^{-1} x$

Figure 9 | The inverse cotangent function

5.5 | Exercises

■ Concepts

1. (a) To define the inverse sine function, we restrict the domain of sine to the interval ______. On this interval the sine function is one-to-one, and its inverse function $\sin^{-1}$ is defined by $\sin^{-1} x = y \Leftrightarrow \sin$ ____ $=$ ____. For example, $\sin^{-1}\left(\frac{1}{2}\right) =$ ____ because sin ____ $=$ ____.

(b) To define the inverse cosine function, we restrict the domain of cosine to the interval ______. On this interval the cosine function is one-to-one and its inverse function $\cos^{-1}$ is defined by $\cos^{-1} x = y \Leftrightarrow$ cos ____ $=$ ____. For example, $\cos^{-1}\left(\frac{1}{2}\right) =$ ____ because cos ____ $=$ ____.

2. **(a)** The cancellation property $\sin^{-1}(\sin x) = x$ is valid for x in the interval ________. By this property,

$\sin^{-1}\left(\sin\frac{\pi}{4}\right) =$ ________ and

$\sin^{-1}\left(\sin\left(-\frac{\pi}{3}\right)\right) =$ ________.

(b) If x is not in the interval in part (a), then the cancellation property does not apply. For example,

$\sin^{-1}\left(\sin\frac{5\pi}{6}\right) = \sin^{-1}$ (____) = ____.

Skills

3–10 ■ Evaluating Inverse Trigonometric Functions Find the exact value of each expression, if it is defined.

3. **(a)** $\sin^{-1} 1$ **(b)** $\sin^{-1}\left(\frac{\sqrt{3}}{2}\right)$ **(c)** $\sin^{-1} 2$

4. **(a)** $\sin^{-1}(-1)$ **(b)** $\sin^{-1}\left(\frac{\sqrt{2}}{2}\right)$ **(c)** $\sin^{-1}(-2)$

5. **(a)** $\cos^{-1}(-1)$ **(b)** $\cos^{-1}(\frac{1}{2})$ **(c)** $\cos^{-1}\left(-\frac{\sqrt{3}}{2}\right)$

6. **(a)** $\cos^{-1}\left(\frac{\sqrt{2}}{2}\right)$ **(b)** $\cos^{-1} 1$ **(c)** $\cos^{-1}\left(-\frac{\sqrt{2}}{2}\right)$

7. **(a)** $\tan^{-1}(-1)$ **(b)** $\tan^{-1}\sqrt{3}$ **(c)** $\tan^{-1}\left(\frac{\sqrt{3}}{3}\right)$

8. **(a)** $\tan^{-1} 0$ **(b)** $\tan^{-1}(-\sqrt{3})$ **(c)** $\tan^{-1}\left(-\frac{\sqrt{3}}{3}\right)$

9. **(a)** $\cos^{-1}(-\frac{1}{2})$ **(b)** $\sin^{-1}\left(-\frac{\sqrt{2}}{2}\right)$ **(c)** $\tan^{-1} 1$

10. **(a)** $\cos^{-1} 0$ **(b)** $\sin^{-1} 0$ **(c)** $\sin^{-1}(-\frac{1}{2})$

11–22 ■ Inverse Trigonometric Functions with a Calculator Use a calculator to find an approximate value of each expression correct to five decimal places, if it is defined.

11. $\sin^{-1}(\frac{2}{3})$ **12.** $\sin^{-1}(-\frac{8}{9})$

13. $\cos^{-1}(-\frac{3}{7})$ **14.** $\cos^{-1}(\frac{4}{9})$

15. $\cos^{-1}(-0.92761)$ **16.** $\sin^{-1}(0.13844)$

17. $\tan^{-1} 10$ **18.** $\tan^{-1}(-26)$

19. $\tan^{-1}(1.23456)$ **20.** $\cos^{-1}(1.23456)$

21. $\sin^{-1}(-0.25713)$ **22.** $\tan^{-1}(-0.25713)$

23–42 ■ Evaluating Expressions Involving Trigonometric Functions Find the exact value of the expression, if it is defined.

23. $\sin(\sin^{-1}(\frac{1}{4}))$ **24.** $\cos(\cos^{-1}(\frac{2}{3}))$

25. $\tan(\tan^{-1} 5)$ **26.** $\sin(\sin^{-1} 5)$

27. $\sin(\sin^{-1}(\frac{3}{2}))$ **28.** $\tan(\tan^{-1}(\frac{3}{2}))$

29. $\cos(\cos^{-1}(-\frac{1}{5}))$ **30.** $\sin(\sin^{-1}(-\frac{3}{4}))$

31. $\sin^{-1}\left(\sin\frac{\pi}{4}\right)$ **32.** $\cos^{-1}\left(\cos\frac{\pi}{4}\right)$

33. $\sin^{-1}\left(\sin\frac{3\pi}{4}\right)$ **34.** $\cos^{-1}\left(\cos\frac{3\pi}{4}\right)$

35. $\cos^{-1}\left(\cos\frac{5\pi}{6}\right)$ **36.** $\sin^{-1}\left(\sin\frac{5\pi}{6}\right)$

37. $\cos^{-1}\left(\cos\frac{7\pi}{6}\right)$ **38.** $\sin^{-1}\left(\sin\frac{7\pi}{6}\right)$

39. $\tan^{-1}\left(\tan\frac{\pi}{4}\right)$ **40.** $\tan^{-1}\left(\tan\left(-\frac{\pi}{3}\right)\right)$

41. $\tan^{-1}\left(\tan\frac{2\pi}{3}\right)$ **42.** $\sin^{-1}\left(\sin\frac{11\pi}{4}\right)$

43–50 ■ Value of an Expression Find the exact value of the expression.

43. $\sin(\cos^{-1}(\frac{1}{2}))$ **44.** $\tan\left(\sin^{-1}\left(\frac{\sqrt{3}}{2}\right)\right)$

45. $\cos(\sin^{-1} 1)$ **46.** $\tan\left(\cos^{-1}\left(-\frac{\sqrt{2}}{2}\right)\right)$

47. $\sec\left(\tan^{-1}\left(-\frac{\sqrt{3}}{3}\right)\right)$ **48.** $\csc(\tan^{-1}(-\sqrt{3}))$

49. $\csc(\cot^{-1} 1)$ **50.** $\sin(\sec^{-1}(-2))$

51–54 ■ Algebraic Expressions Rewrite the expression as an algebraic expression in x.

51. $\sec(\tan^{-1} x)$

52. $\cos(\sin^{-1} x)$

53. $\tan(\sin^{-1} x)$

54. $\sin(\sec^{-1} x)$

55–58 ■ Expressing a Function as a Composition Find functions f and g such that $F = f \circ g$.

55. $F(x) = e^{\arcsin x}$ **56.** $F(x) = (\tan^{-1} x)^2$

57. $F(x) = \sin^{-1}\left(\frac{1}{x}\right)$ **58.** $F(x) = \frac{1}{1 + \tan^{-1} x}$

59–62 ■ Expressing a Function as a Composition Find functions f, g, and h such that $F = f \circ g \circ h$.

59. $F(x) = e^{\arcsin x^2}$ **60.** $F(x) = \tan^{-1}\sqrt{x^2 + 1}$

61. $F(x) = \tan^{-1}(e^{1-x^2})$ **62.** $F(x) = \ln(\arctan(x^4))$

63–66 ■ Graphing Combinations of Inverse Trigonometric Functions **(a)** Find the domain of the function. **(b)** Use a graphing device to graph the function. Comment on the features of the graph and how they relate to the equation.

63. $f(x) = \tan^{-1}(x^2)$ **64.** $f(x) = \sin^{-1}(x^2)$

65. $f(x) = \sin^{-1}(\cos x)$ **66.** $f(x) = \frac{1}{(\pi/2) + \tan^{-1} x}$

■ **Discuss** ■ **Discover** ■ **Prove** ■ **Write**

67–68 ■ Prove: Identities Involving Inverse Trigonometric Functions **(a)** Graph the function and make a conjecture, and **(b)** prove that your conjecture is true.

67. $y = \sin^{-1} x + \cos^{-1} x$

68. $y = \tan^{-1} x + \tan^{-1}\left(\dfrac{1}{x}\right)$

69. Discuss: Two Different Compositions Let f and g be the functions

$$f(x) = \sin(\sin^{-1} x)$$

and $$g(x) = \sin^{-1}(\sin x)$$

By the cancellation properties, $f(x) = x$ and $g(x) = x$ for suitable values of x. But these functions are not the same for all x. Graph both f and g to show how the functions differ. (Think carefully about the domain and range of $\sin^{-1}$).

5.6 Modeling Harmonic Motion

■ Simple Harmonic Motion ■ Damped Harmonic Motion ■ Phase and Phase Difference

Periodic behavior—behavior that repeats again and again—is common in nature. Perhaps the most familiar example is the daily rising and setting of the sun, which results in the repetitive pattern of day, night, day, night, ... Another example is the daily variation of tide levels at the beach, which results in the repetitive pattern of high tide, low tide, high tide, low tide, ... Certain animal populations increase and decrease in a predictable periodic pattern: A large population exhausts the food supply, which causes the population to dwindle; this in turn results in a more plentiful food supply, which makes it possible for the population to increase; and the pattern then repeats again and again (see *Discovery Project: Predator-Prey Models* referenced in Section 5.3).

Other common examples of periodic behavior involve motion that is caused by vibration or oscillation; a simple example is a mass suspended from a spring that has been compressed and then allowed to oscillate vertically. This back-and-forth motion also occurs in such diverse phenomena as sound waves, light waves, alternating electrical current, and pulsating stars, and many others. In this section we consider the problem of modeling periodic behavior.

■ Simple Harmonic Motion

The trigonometric functions are ideally suited for modeling periodic behavior. A glance at the graphs of the sine and cosine functions, for instance, tells us that each of these functions itself exhibits periodic behavior. Figure 1 shows the graph of $y = \sin t$. If we think of t as time, we see that as time goes on, $y = \sin t$ increases and decreases again and again. Figure 2 shows that the motion of a vibrating mass on a spring is modeled very accurately by $y = \sin t$.

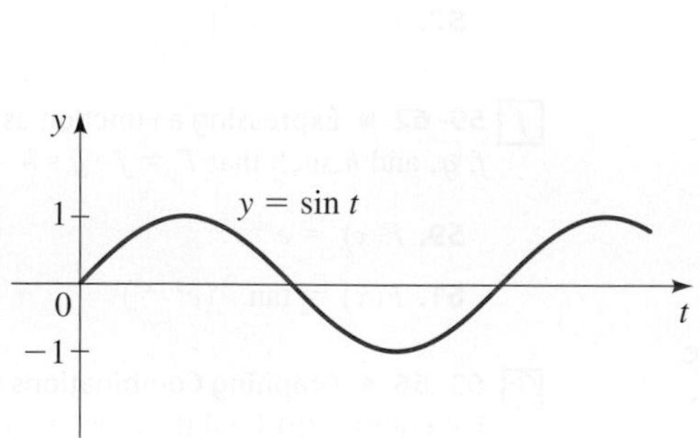

Figure 1 | $y = \sin t$

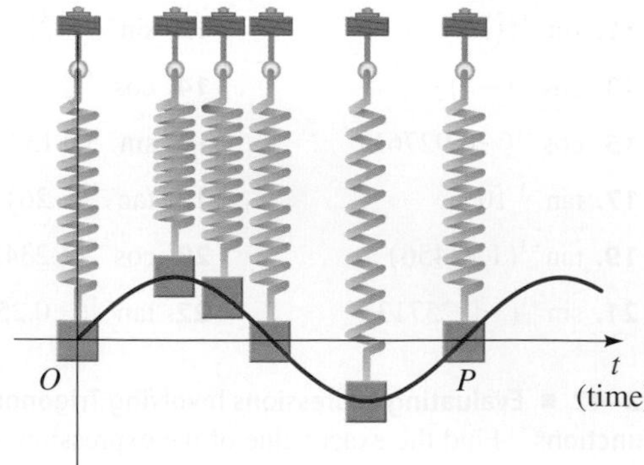

Figure 2 | Motion of a vibrating spring is modeled by $y = \sin t$.

Notice that the mass returns to its original position again and again. A **cycle** is one complete vibration of an object, so the mass in Figure 2 completes one cycle of its motion between O and P. Our observations about how the sine and cosine functions model periodic behavior are summarized in the following box.

The main difference between the two equations describing simple harmonic motion is the starting point. At $t = 0$ we get

$$y = a \sin \omega \cdot 0 = 0$$

$$y = a \cos \omega \cdot 0 = a$$

In the first case the motion "starts" with zero displacement, whereas in the second case the motion "starts" with the displacement at maximum (at the amplitude a).

Simple Harmonic Motion

If the equation describing the displacement y of an object at time t is

$$y = a \sin \omega t \qquad \text{or} \qquad y = a \cos \omega t$$

then the object is in **simple harmonic motion**. In this case,

amplitude $= \lvert a \rvert$	Maximum displacement of the object
period $= \dfrac{2\pi}{\omega}$	Time required to complete one cycle
frequency $= \dfrac{\omega}{2\pi}$	Number of cycles per unit of time

The symbol ω is the lowercase Greek letter "omega," and ν is the letter "nu."

Notice that the functions

$$y = a \sin 2\pi\nu t \qquad \text{and} \qquad y = a \cos 2\pi\nu t$$

have frequency ν because $2\pi\nu/(2\pi) = \nu$. Since we can immediately read the frequency from these equations, we often write equations of simple harmonic motion in this form.

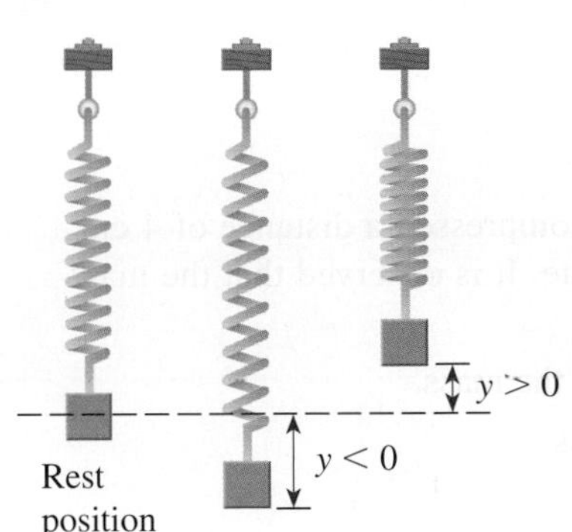

Figure 3

Example 1 ■ A Vibrating Spring

The displacement of a mass suspended by a spring is modeled by the function

$$y = 10 \sin 4\pi t$$

where y is measured in centimeters and t in seconds (see Figure 3).

(a) Find the amplitude, period, and frequency of the motion of the mass.
(b) Sketch a graph of the displacement of the mass.

Solution

(a) From the formulas for amplitude, period, and frequency we get

$$\text{amplitude} = \lvert a \rvert = 10 \text{ cm}$$

$$\text{period} = \frac{2\pi}{\omega} = \frac{2\pi}{4\pi} = \frac{1}{2} \text{ s}$$

$$\text{frequency} = \frac{\omega}{2\pi} = \frac{4\pi}{2\pi} = 2 \text{ cycles per second (Hz)}$$

(b) The graph of the displacement of the mass at time t is shown in Figure 4.

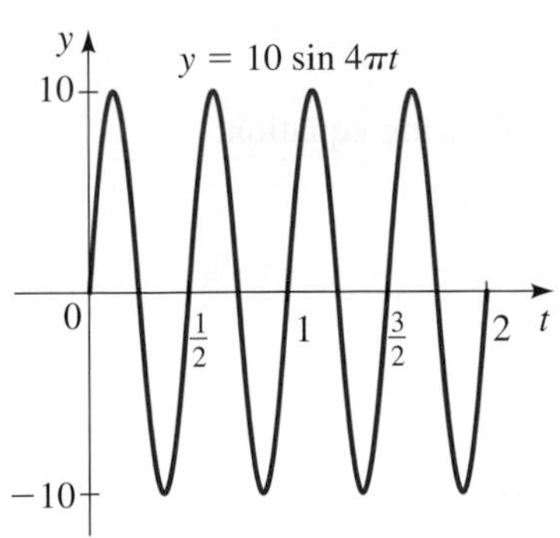

Figure 4

Now Try Exercise 5 ■

Simple harmonic motion occurs in the production of sound. Sound is produced by a regular variation in air pressure from the normal pressure. If the pressure varies in simple harmonic motion, then a pure sound is produced. The tone of the sound depends on the frequency, and the loudness depends on the amplitude.

Example 2 ■ Vibrations of a Musical Note

A sousaphone player plays the note E and sustains the sound for some time. For a pure E the variation in pressure from normal air pressure is given by

$$V(t) = 1.4 \sin 80\pi t$$

where V is measured in kilopascals and t is measured in seconds.

(a) Find the amplitude, period, and frequency of V.

(b) Sketch a graph of V.

(c) If the player increases the loudness of the note, how does the equation for V change?

(d) If the player is playing the note incorrectly and it is a little flat, how does the equation for V change?

Solution

(a) From the formulas for amplitude, period, and frequency we get

$$\text{amplitude} = |1.4| = 1.4$$

$$\text{period} = \frac{2\pi}{80\pi} = \frac{1}{40}$$

$$\text{frequency} = \frac{80\pi}{2\pi} = 40$$

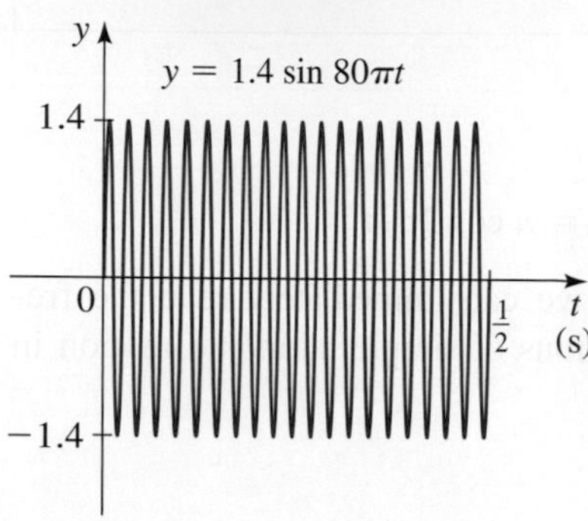

Figure 5

(b) The graph of V is shown in Figure 5.

(c) If the player increases the loudness, then the amplitude increases. So the number 1.4 is replaced by a larger number.

(d) If the note is flat, then the frequency is decreased. Thus the coefficient of t is less than 80π.

 **Now Try Exercise 39** ■

Example 3 ■ Modeling a Vibrating Spring

A mass is suspended from a spring. The spring is compressed a distance of 4 cm and then released, which causes the mass to oscillate. It is observed that the mass returns to the compressed position after $\frac{1}{3}$ s.

(a) Find a function that models the displacement of the mass.

(b) Sketch the graph of the displacement of the mass.

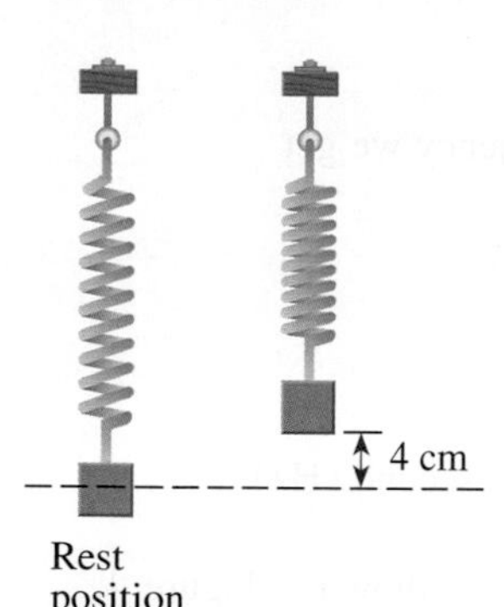

Solution

(a) The motion of the mass is given by one of the equations for simple harmonic motion. The amplitude of the motion is 4 cm. Since this amplitude is reached at time $t = 0$, an appropriate function that models the displacement is of the form

$$y = a \cos \omega t$$

Since the period is $p = \frac{1}{3}$, we can find ω from the following equation:

$$\text{period} = \frac{2\pi}{\omega}$$

$$\frac{1}{3} = \frac{2\pi}{\omega} \qquad \text{Period} = \tfrac{1}{3}$$

$$\omega = 6\pi \qquad \text{Solve for } \omega$$

So the motion of the mass is modeled by the function

$$y = 4 \cos 6\pi t$$

where y is the displacement from the rest position at time t. Notice that when $t = 0$, the displacement is $y = 4$, as we expect.

(b) The graph of the displacement of the mass at time t is shown in Figure 6.

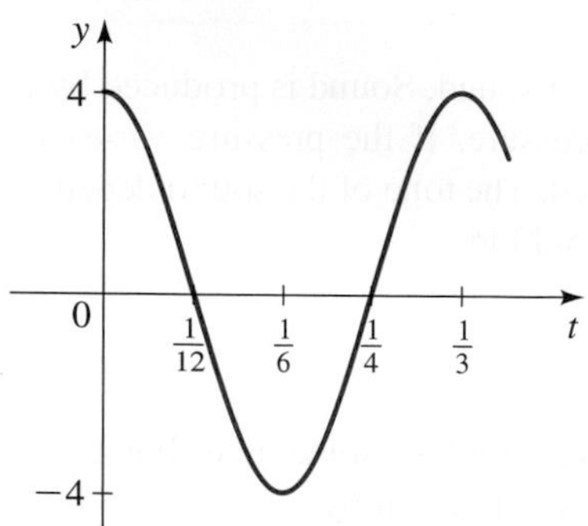

Figure 6 | $y = 4 \cos 6\pi t$

Now Try Exercises 17 and 43 ■

In general, the graphs of the sine and cosine functions representing harmonic motion may be shifted horizontally or vertically. In this case the equations take the form

$$y = a\sin(\omega(t - c)) + b \qquad \text{or} \qquad y = a\cos(\omega(t - c)) + b$$

The vertical shift b indicates that the variation occurs around an average value b. The horizontal shift c indicates the position of the object at $t = c$. (See Figure 7.)

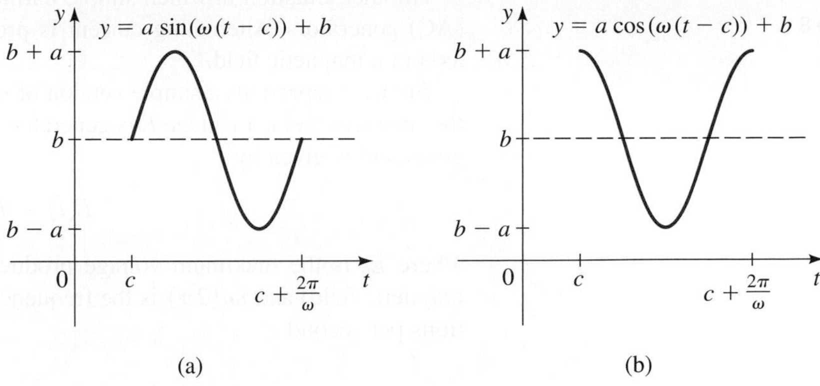

Figure 7

Example 4 ■ Modeling the Brightness of a Variable Star

A variable star is one whose brightness alternately increases and decreases. For the variable star Delta Cephei the time between periods of maximum brightness is 5.4 days. The average brightness (or magnitude) of the star is 4.0, and its brightness varies by ±0.35 magnitude.

(a) Find a function that models the brightness of Delta Cephei as a function of time.

(b) Sketch a graph of the brightness of Delta Cephei as a function of time.

Solution

(a) Let's find a function in the form

$$y = a\cos(\omega(t - c)) + b$$

The amplitude is the maximum variation from average brightness, so the amplitude is $a = 0.35$ magnitudes. We are given that the period is 5.4 days, so

$$\omega = \frac{2\pi}{5.4} \approx 1.16$$

Since the brightness varies from an average value of 4.0 magnitudes, the graph is shifted upward by $b = 4.0$. If we take $t = 0$ to be a time when the star is at

Valentina Rusinova/Shutterstock.com

DAY NIGHT

Discovery Project ■ Hours of Daylight

The number of hours of daylight varies from day to day. In the Northern Hemisphere, the days get longer as summer approaches and shorter in the winter months. At any given latitude, the number of hours of daylight follows a sinusoidal curve. In this project you will find the sinusoidal function that models the hours of daylight for any day of the year at your latitude. You can test the formula yourself by comparing the prediction of the formula with the actual number of hours of daylight hours at your location. You can find the project at **www.stewartmath.com**.

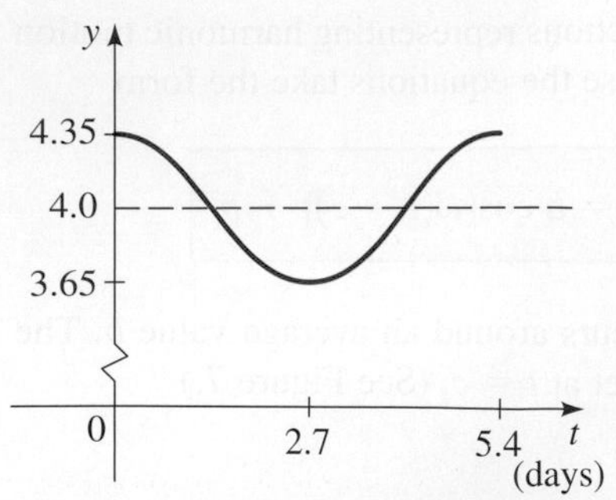

Figure 8

maximum brightness, there is no horizontal shift, so $c = 0$ (because a cosine curve achieves its maximum at $t = 0$). Thus the function is

$$y = 0.35\cos(1.16t) + 4.0$$

where t is the number of days from a time when the star is at maximum brightness.

(b) The graph is sketched in Figure 8.

Now Try Exercise 47 ■

Another situation in which simple harmonic motion occurs is in alternating current (AC) generators. Alternating current is produced when an armature rotates about its axis in a magnetic field.

Figure 9 represents a simple version of such a generator. As the wire passes through the magnetic field, a voltage E is generated in the wire. It can be shown that the voltage generated is given by

$$E(t) = E_0 \cos \omega t$$

where E_0 is the maximum voltage produced (which depends on the strength of the magnetic field) and $\omega/(2\pi)$ is the frequency of the armature, or the number of revolutions per second.

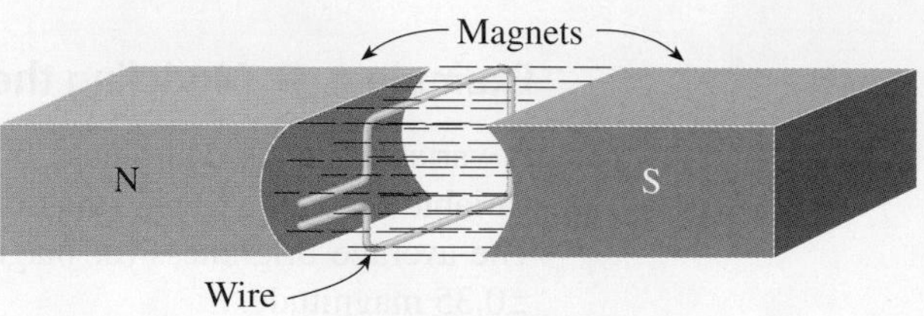

Figure 9

Example 5 ■ Modeling Alternating Current

Ordinary 110-volt household alternating current varies from +155 V to −155 V with a frequency of 60 hertz (cycles per second). Find an equation that describes this variation in voltage.

Solution The variation in voltage follows simple harmonic motion. Since the frequency is 60 cycles per second, we have

$$\frac{\omega}{2\pi} = 60 \qquad \text{or} \qquad \omega = 120\pi$$

Let's take $t = 0$ to be a time when the voltage is +155 V. Then

$$E(t) = a\cos\omega t = 155\cos 120\pi t$$

Now Try Exercise 51 ■

> Why do we say that household current is 110 V when the maximum voltage produced is 155 V? From the symmetry of the cosine function we see that the average voltage produced is zero. This average value would be the same for all AC generators and so gives no information about the voltage generated. To obtain a more informative measure of voltage, engineers use the **root-mean-square** (RMS) method. It can be shown that the RMS voltage is $1/\sqrt{2}$ times the maximum voltage. So for household current the RMS voltage is
>
> $$155 \times \frac{1}{\sqrt{2}} \approx 110 \text{ V}$$

■ Damped Harmonic Motion

The spring in Figure 2 is assumed to be oscillating in a frictionless environment. In this hypothetical case the amplitude of the oscillation will not change. In the presence of friction, however, the motion of the spring eventually "dies down"; that is, the amplitude of the motion decreases with time. Motion of this type is called *damped harmonic motion* and is modeled by a combination of exponential and trigonometric functions.

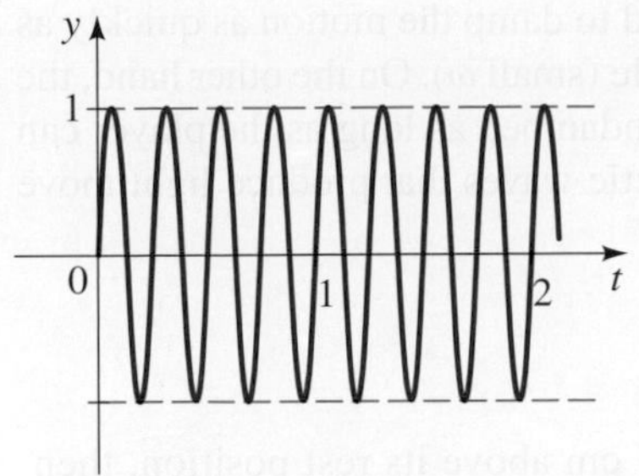

(a) Harmonic motion: $y = \sin 8\pi t$

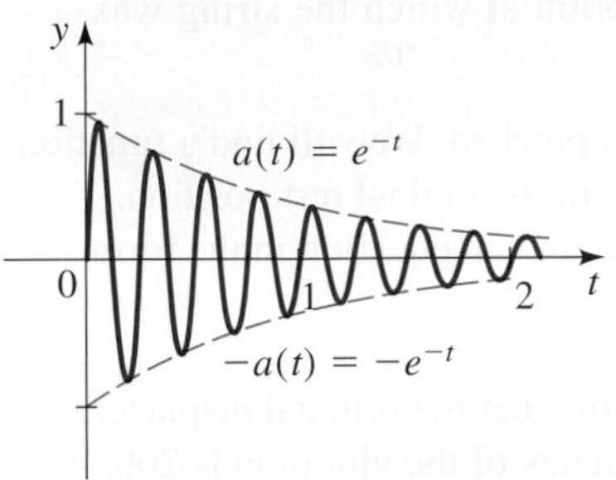

(b) Damped harmonic motion: $y = e^{-t} \sin 8\pi t$

Figure 10

> **Damped Harmonic Motion**
>
> If the equation describing the displacement y of an object at time t is
>
> $$y = ke^{-ct} \sin \omega t \qquad \text{or} \qquad y = ke^{-ct} \cos \omega t \qquad (c > 0)$$
>
> then the object is in **damped harmonic motion**. The constant c is the **damping constant**, k is the initial amplitude, and $2\pi/\omega$ is the period.*

Damped harmonic motion is simply harmonic motion for which the amplitude is governed by the function $a(t) = ke^{-ct}$. Figure 10 shows the difference between harmonic motion and damped harmonic motion.

Example 6 ■ Modeling Damped Harmonic Motion

Two mass-spring systems are experiencing damped harmonic motion, both at 0.5 cycles per second and both with an initial maximum displacement of 10 cm. The first has a damping constant of 0.5, and the second has a damping constant of 0.1.

(a) Find a function of the form $g(t) = ke^{-ct} \cos \omega t$ to model the motion in each case.

(b) Graph the two functions you found in part (a). How do they differ?

Solution

Hz is the abbreviation for hertz. One hertz is one cycle per second.

(a) At time $t = 0$ the displacement is 10 cm. Thus $g(0) = ke^{-c \cdot 0} \cos(\omega \cdot 0) = k$, so $k = 10$. Also, the frequency is $f = 0.5$ Hz, and since $\omega = 2\pi f$, we get $\omega = 2\pi(0.5) = \pi$. Using the given damping constants, we find that the motions of the two springs are given by the functions

$$g_1(t) = 10e^{-0.5t} \cos \pi t \qquad \text{and} \qquad g_2(t) = 10e^{-0.1t} \cos \pi t$$

(b) The functions g_1 and g_2 are graphed in Figure 11. From the graphs we see that in the first case (where the damping constant is larger) the motion dies down quickly, whereas in the second case, perceptible motion continues much longer.

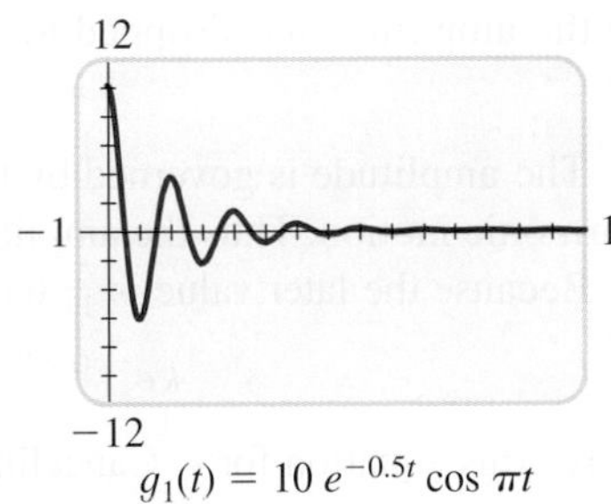

$g_1(t) = 10\,e^{-0.5t} \cos \pi t$

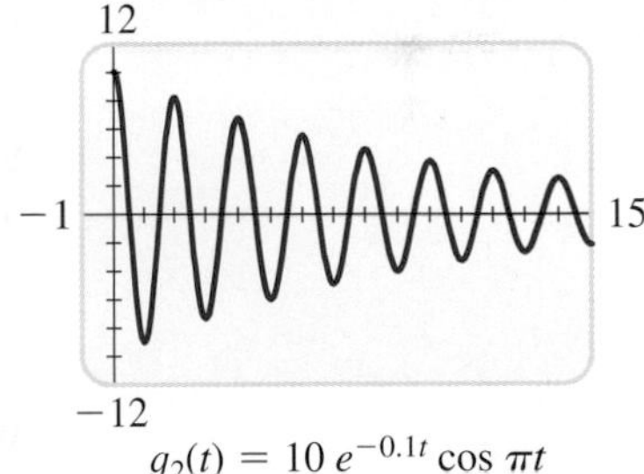

$g_2(t) = 10\,e^{-0.1t} \cos \pi t$

Figure 11

Now Try Exercise 21 ■

As Example 6 indicates, the larger the damping constant c, the more quickly the oscillation dies down. When a guitar string is plucked and then allowed to vibrate freely, a point on that string undergoes damped harmonic motion. We hear the damping of the motion as the sound produced by the vibration of the string fades. How fast the damping of the string occurs (as measured by the size of the constant c) is a property of the size of the string and the material it is made of. Another example of damped harmonic motion is the motion that a car's shock absorber undergoes when the car hits a bump in

*In the case of damped harmonic motion the term *quasi-period* is often used instead of *period* because the motion is not actually periodic—it diminishes with time. However, we will continue to use the term *period* to avoid confusion.

the road. In this case the shock absorber is engineered to damp the motion as quickly as possible (large c) with as small a frequency as possible (small ω). On the other hand, the sound produced by a tuba player playing a note is undamped as long as the player can maintain the loudness of the note. The electromagnetic waves that produce light move in simple harmonic motion that is not damped.

Example 7 ■ A Vibrating Violin String

The G-string on a violin is pulled a distance of 0.5 cm above its rest position, then released and allowed to vibrate. The damping constant c for this string is determined to be 1.4. Suppose that the note produced is a pure G (frequency = 200 Hz). Find an equation that describes the motion of the point at which the string was plucked.

Solution Let P be the point at which the string was plucked. We will find a function $f(t)$ that gives the distance at time t of the point P from its original rest position. Since the maximum displacement occurs at $t = 0$, we find an equation in the form

$$y = ke^{-ct} \cos \omega t$$

From this equation we see that $f(0) = k$. But we know that the original displacement of the string is 0.5 cm. Thus $k = 0.5$. Since the frequency of the vibration is 200, we have $\omega = 2\pi f = 2\pi(200) = 400\pi$. Finally, since we know that the damping constant is 1.4, we get

$$f(t) = 0.5e^{-1.4t} \cos 400\pi t$$

 **Now Try Exercise 53** ■

Example 8 ■ Ripples on a Pond

A stone is dropped in a calm lake, causing waves to form. The up-and-down motion of a point on the surface of the water is modeled by damped harmonic motion. At some time the amplitude of the wave is measured, and 20 s later it is found that the amplitude has dropped to $\frac{1}{10}$ of this value. Find the damping constant c.

Solution The amplitude is governed by the coefficient ke^{-ct} in the equations for damped harmonic motion. Thus the amplitude at time t is ke^{-ct}, and 20 s later, it is $ke^{-c(t+20)}$. Because the later value is $\frac{1}{10}$ the earlier value, we have

$$ke^{-c(t+20)} = \frac{1}{10}ke^{-ct}$$

We now solve this equation for c. Canceling k and using the Laws of Exponents, we get

$$e^{-ct} \cdot e^{-20c} = \frac{1}{10}e^{-ct}$$

$$e^{-20c} = \frac{1}{10} \qquad \text{Cancel } e^{-ct}$$

$$e^{20c} = 10 \qquad \text{Take reciprocals}$$

Taking the natural logarithm of each side gives

$$20c = \ln(10)$$

$$c = \frac{1}{20}\ln(10) \approx \frac{1}{20}(2.30) \approx 0.12$$

Thus the damping constant is $c \approx 0.12$.

 **Now Try Exercise 55** ■

■ Phase and Phase Difference

When two objects are moving in simple harmonic motion with the same frequency, it is often important to determine whether the objects are "moving together" or by how much their motions differ. Let's consider a specific example.

Suppose that an object is rotating counterclockwise along the unit circle and the height y of the object at time t is given by $y = \sin(kt - b)$. When $t = 0$, the height is $y = \sin(-b)$. This means that the motion "starts" at an angle b as shown in Figure 12.

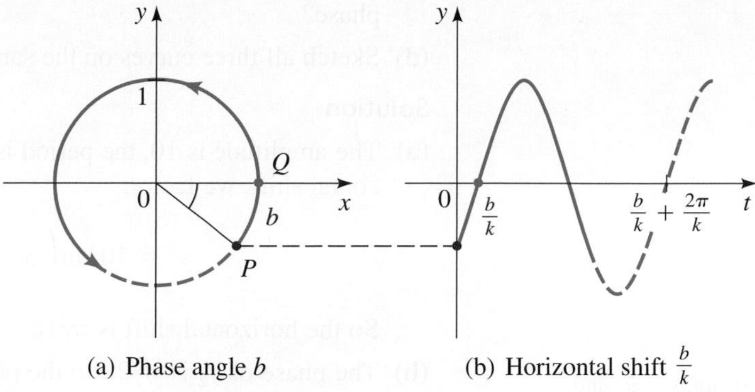

(a) Phase angle b (b) Horizontal shift $\frac{b}{k}$

Figure 12 | Graph of $y = \sin(kt - b)$

The phase angle b depends only on the starting position of the object and not on the frequency. The lag time does depend on the frequency.

We can view the starting point in two ways: as the *angle* between P and Q on the unit circle or as the *time* required for P to "catch up" to Q. The angle b is called the **phase** (or **phase angle**). To find the time required to catch up to Q, we factor out k:

$$y = \sin(kt - b) = \sin k\left(t - \frac{b}{k}\right)$$

We see that P catches up to Q (that is, $y = 0$) when $t = b/k$. This last equation also shows that the graph in Figure 12(b) is **shifted horizontally** b/k (to the right) on the t-axis. The time b/k is called the **lag time** if $b > 0$ (because P is behind, or lags, Q by b/k time units) or the **lead time** if $b < 0$.

Phase

Any sine curve can be expressed in the following equivalent forms:

$y = A\sin(kt - b)$ The **phase** is b.

$y = A\sin k\left(t - \frac{b}{k}\right)$ The **horizontal shift** is $\frac{b}{k}$.

It is often important to know whether two waves with the same period (modeled by sine curves) are *in phase* or *out of phase*. For the curves

$$y_1 = A\sin(kt - b) \quad \text{and} \quad y_2 = A\sin(kt - c)$$

Note that the phase difference depends on the order in which the functions are given.

the **phase difference** between y_1 and y_2 is $b - c$. If the phase difference is a multiple of 2π, the waves are **in phase**; otherwise, the waves are **out of phase**. If two sine curves are in phase, then their graphs coincide.

Example 9 ■ Finding Phase and Phase Difference

Three objects are in harmonic motion modeled by the following curves:

$$y_1 = 10\sin\left(3t - \frac{\pi}{6}\right) \qquad y_2 = 10\sin\left(3t - \frac{\pi}{2}\right) \qquad y_3 = 10\sin\left(3t + \frac{23\pi}{6}\right)$$

(a) Find the amplitude, period, phase, and horizontal shift of the curve y_1.

(b) Find the phase difference between the curves y_1 and y_2. Are the two curves in phase?

(c) Find the phase difference between the curves y_1 and y_3. Are the two curves in phase?

(d) Sketch all three curves on the same axes.

Solution

(a) The amplitude is 10, the period is $2\pi/3$, and the phase is $\pi/6$. To find the horizontal shift, we factor:

$$y_1 = 10\sin\left(3t - \frac{\pi}{6}\right) = 10\sin 3\left(t - \frac{\pi}{18}\right)$$

So the horizontal shift is $\pi/18$.

(b) The phase of y_2 is $\pi/2$. So the phase difference is

$$\frac{\pi}{6} - \frac{\pi}{2} = -\frac{\pi}{3}$$

The phase difference is not a multiple of 2π, so the two curves are out of phase.

(c) The phase of y_3 is $-23\pi/6$. So the phase difference is

$$\frac{\pi}{6} - \left(-\frac{23\pi}{6}\right) = 4\pi = 2(2\pi)$$

The phase difference is a multiple of 2π, so the two curves are in phase.

(d) The graphs are shown in Figure 13. Notice that the curves y_1 and y_3 have the same graph because they are in phase.

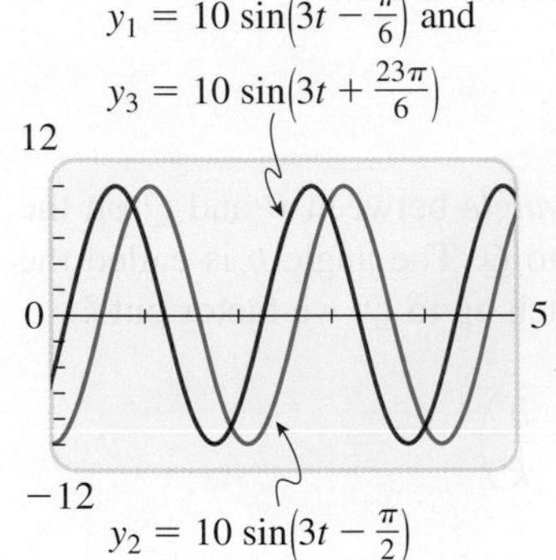

Figure 13

 **Now Try Exercises 29 and 35** ■

Example 10 ■ Using Phase

Aspen, Brady, and Corey are sitting in a stationary Ferris wheel as shown in the figure in the margin. At time $t = 0$ the Ferris wheel starts rotating counterclockwise at the rate of 2 revolutions per minute.

(a) Find sine curves that model the height of each rider above the center line of the Ferris wheel at any time $t > 0$.

(b) Find the phase difference between Aspen and Brady, between Aspen and Corey, and between Brady and Corey.

(c) Find the horizontal shift of Aspen's equation. What is Aspen's lead or lag time (relative to the red seat, R, in the figure)?

Solution

(a) The motion of each rider is modeled by a function of the form $y = A\sin(kt - b)$. From the figure we see that the amplitude is $A = 5$ m. Since the Ferris wheel makes two revolutions per minute, the period is $\frac{1}{2}$ min. So

$$\text{period} = \frac{2\pi}{k} = \frac{1}{2}\text{ min}$$

It follows that $k = 4\pi$. From the figure we see that each rider starts at a different phase. Let's consider Aspen and Brady to be ahead of the red seat, and let's consider Corey to be behind the red seat, R. So the phases of Aspen, Brady, and Corey are $-\pi/2$, $-3\pi/4$, and $\pi/4$, respectively. The equations are as follows.

Aspen	Brady	Corey
$y_A = 5\sin\left(4\pi t + \dfrac{\pi}{2}\right)$	$y_B = 5\sin\left(4\pi t + \dfrac{3\pi}{4}\right)$	$y_C = 5\sin\left(4\pi t - \dfrac{\pi}{4}\right)$

(b) The phase differences are as follows.

Aspen and Brady	Aspen and Corey	Brady and Corey
$-\dfrac{\pi}{2} - \left(-\dfrac{3\pi}{4}\right) = \dfrac{\pi}{4}$	$-\dfrac{\pi}{2} - \dfrac{\pi}{4} = -\dfrac{3\pi}{4}$	$-\dfrac{3\pi}{4} - \dfrac{\pi}{4} = -\pi$

(c) The equation that models Aspen's position above the center line of the Ferris wheel was found in part (a). To find the horizontal shift, we factor Aspen's equation.

$$y_A = 5\sin\left(4\pi t + \frac{\pi}{2}\right) \qquad \text{Aspen's equation}$$

$$y_A = 5\sin 4\pi\left(t + \frac{1}{8}\right) \qquad \text{Factor } 4\pi$$

We see that the horizontal shift is $\frac{1}{8}$ to the left. This means that Aspen's lead time is $\frac{1}{8}$ minutes (so Aspen is $\frac{1}{8}$ of a minute ahead of the red seat). ■

Now Try Exercise 57

5.6 Exercises

Concepts

1. For an object in simple harmonic motion with amplitude a and period $2\pi/\omega$, find an equation that models the displacement y at time t if

(a) $y = 0$ at time $t = 0$: $y =$ ________.

(b) $y = a$ at time $t = 0$: $y =$ ________.

2. For an object in damped harmonic motion with initial amplitude a, period $2\pi/\omega$, and damping constant c, find an equation that models the displacement y at time t if

(a) $y = 0$ at time $t = 0$: $y =$ ________.

(b) $y = a$ at time $t = 0$: $y =$ ________.

3. (a) For an object in harmonic motion modeled by $y = A\sin(kt - b)$ the amplitude is ________, the period is ________, and the phase is ________. To find the horizontal shift, we factor out k to get $y =$ ________. From this form of the equation we see that the horizontal shift is ________.

(b) For an object in harmonic motion modeled by $y = 5\sin(4t - \pi)$ the amplitude is ________, the period is ________, the phase is ________, and the horizontal shift is ________.

4. Objects A and B are in harmonic motion modeled by $y = 3\sin(2t - \pi)$ and $y = 3\sin\left(2t - \dfrac{\pi}{2}\right)$. The phase of A is ________, and the phase of B is ________. The phase difference is ________, so the objects are moving ________ (in phase/out of phase).

Skills

5–12 ■ Simple Harmonic Motion The given function models the displacement of an object moving in simple harmonic motion. **(a)** Find the amplitude, period, and frequency of the motion. **(b)** Sketch a graph of one complete period.

5. $y = 2\sin 3t$

6. $y = 3\cos\frac{1}{2}t$

7. $y = -\cos 0.3t$

8. $y = 2.4\sin 3.6t$

9. $y = -0.25\cos\left(1.5t - \dfrac{\pi}{3}\right)$

10. $y = -\frac{3}{2}\sin(0.2t + 1.4)$

11. $y = 5\cos(\frac{2}{3}t + \frac{3}{4})$ **12.** $y = 1.6\sin(t - 1.8)$

13–16 ■ Simple Harmonic Motion Find a function that models the simple harmonic motion having the given properties. Assume that the displacement is zero at time $t = 0$.

13. amplitude 10 cm, period 3 s

14. amplitude 24 m, period 2 min

15. amplitude 6 cm, frequency $5/\pi$ Hz

16. amplitude 1.2 m, frequency 0.5 Hz

17–20 ■ Simple Harmonic Motion Find a function that models the simple harmonic motion having the given properties. Assume that the displacement is at its maximum at time $t = 0$.

 **17.** amplitude 20 m, period 0.5 min

18. amplitude 35 cm, period 8 s

19. amplitude 2.4 m, frequency 750 Hz

20. amplitude 6.25 cm, frequency 60 Hz

21–28 ■ Damped Harmonic Motion An initial amplitude k, damping constant c, and frequency f or period p are given. (Recall that frequency and period are related by the equation $f = 1/p$.)

(a) Find a function that models the damped harmonic motion. Use a function of the form $y = ke^{-ct}\cos\omega t$ in Exercises 21–24 and of the form $y = ke^{-ct}\sin\omega t$ in Exercises 25–28.

 **(b)** Graph the function.

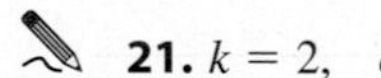 **21.** $k = 2, \quad c = 1.5, \quad f = 3$

22. $k = 15, \quad c = 0.25, \quad f = 0.6$

23. $k = 100, \quad c = 0.05, \quad p = 4$

24. $k = 0.75, \quad c = 3, \quad p = 3\pi$

25. $k = 7, \quad c = 10, \quad p = \pi/6$

26. $k = 1, \quad c = 1, \quad p = 1$

27. $k = 0.3, \quad c = 0.2, \quad f = 20$

28. $k = 12, \quad c = 0.01, \quad f = 8$

29–34 ■ Amplitude, Period, Phase, and Horizontal Shift For each sine curve find the amplitude, period, phase, and horizontal shift.

 **29.** $y = 5\sin\left(2t - \dfrac{\pi}{2}\right)$ **30.** $y = 10\sin\left(t - \dfrac{\pi}{3}\right)$

31. $y = 100\sin(5t + \pi)$ **32.** $y = 50\sin\left(\dfrac{1}{2}t + \dfrac{\pi}{5}\right)$

33. $y = 20\sin 2\left(t - \dfrac{\pi}{4}\right)$ **34.** $y = 8\sin 4\left(t + \dfrac{\pi}{12}\right)$

35–38 ■ Phase and Phase Difference A pair of sine curves with the same period is given. **(a)** Find the phase of each curve. **(b)** Find the phase difference between the first and second curves. **(c)** Determine whether the curves are in phase or out of phase. **(d)** Sketch both curves on the same axes.

35. $y_1 = 10\sin\left(3t - \dfrac{\pi}{2}\right)$; $y_2 = 10\sin\left(3t - \dfrac{5\pi}{2}\right)$

36. $y_1 = 15\sin\left(2t - \dfrac{\pi}{3}\right)$; $y_2 = 15\sin\left(2t - \dfrac{\pi}{6}\right)$

37. $y_1 = 80\sin 5\left(t - \dfrac{\pi}{10}\right)$; $y_2 = 80\sin\left(5t - \dfrac{\pi}{3}\right)$

38. $y_1 = 20\sin 2\left(t - \dfrac{\pi}{2}\right)$; $y_2 = 20\sin 2\left(t - \dfrac{3\pi}{2}\right)$

■ Applications

39. Blood Pressure Each time your heart beats, your blood pressure increases, then decreases as the heart rests between beats. A certain person's blood pressure can be modeled by the function

$$p(t) = 115 + 25\sin(160\pi t)$$

where $p(t)$ is the pressure (in mmHg) at time t, measured in minutes.

(a) Find the amplitude, period, and frequency of p.

(b) Sketch a graph of p.

(c) If a person is exercising, his or her heart beats faster. How does this affect the period and frequency of p?

40. A Bobbing Cork A cork floating in a lake is bobbing in simple harmonic motion. Its displacement above the bottom of the lake can be modeled by

$$y = 0.2\cos 20\pi t + 8$$

where y is measured in meters and t is measured in minutes.

(a) Find the frequency of the motion of the cork.

(b) Sketch a graph of y.

(c) Find the maximum displacement of the cork above the lake bottom.

41. Mass-Spring System A mass attached to a spring is oscillating up and down in simple harmonic motion. The graph gives its displacement $d(t)$ from equilibrium at time t. Express the function d in the form $d(t) = a\sin\omega t$.

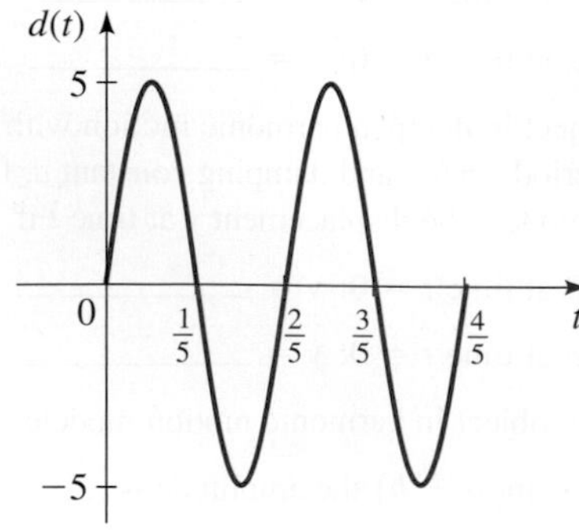

42. Tides The graph shows the variation of the water level relative to mean sea level in Commencement Bay at Tacoma, Washington, for a particular 24-hour period. Assuming that this variation can be modeled by simple harmonic motion, find an equation of the form $y = a \sin \omega t$ that describes the variation in water level as a function of the number of hours after midnight.

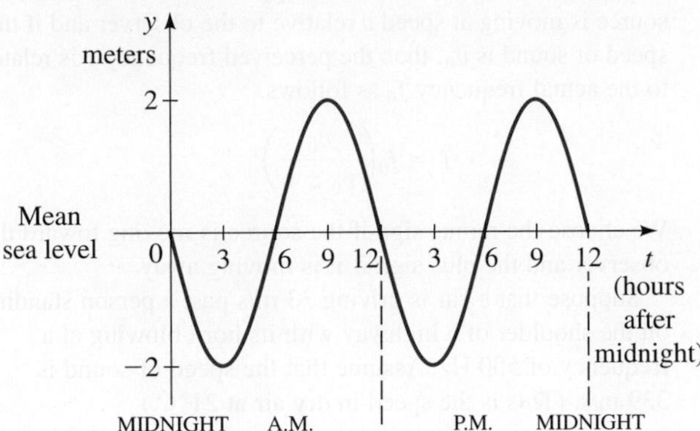

43. Mass-Spring System A mass is suspended on a spring. The spring is compressed so that the mass is located 5 cm above its rest position. The mass is released at time $t = 0$ and allowed to oscillate. It is observed that the mass reaches its lowest point $\frac{1}{2}$ second after it is released. Find an equation that describes the motion of the mass.

44. Mass-Spring System A mass suspended from a spring is pulled down a distance of 1 m from its rest position, as shown in the figure. The mass is released at time $t = 0$ and allowed to oscillate. If the mass returns to the starting position after 1 second, find an equation that describes its motion.

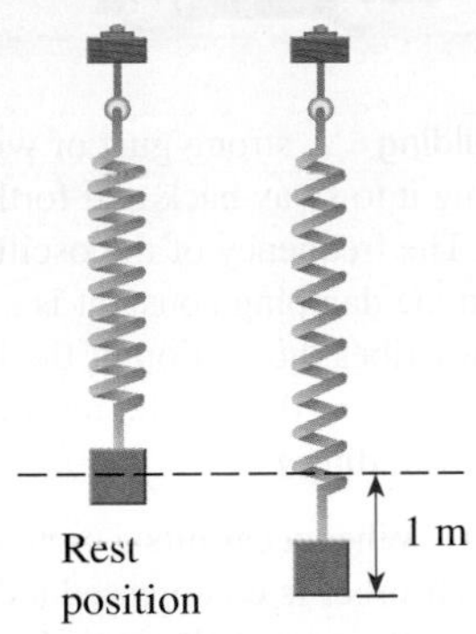

45. Ferris Wheel A Ferris wheel has a radius of 10 m, and the bottom of the wheel passes 1 m above the ground. If the Ferris wheel makes one complete revolution every 20 s, find an equation that gives the height above the ground of a person on the Ferris wheel as a function of time.

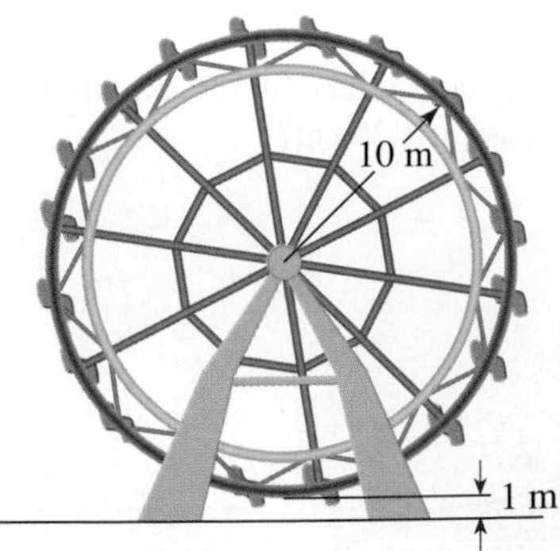

46. Clock Pendulum The pendulum in a grandfather clock makes one complete swing every 2 s. The maximum angle that the pendulum makes with respect to its rest position is 10°. We know from physical principles that the angle θ between the pendulum and its rest position changes in simple harmonic fashion. Find an equation that describes the size of the angle θ as a function of time. (Take $t = 0$ to be a time when the pendulum is vertical.)

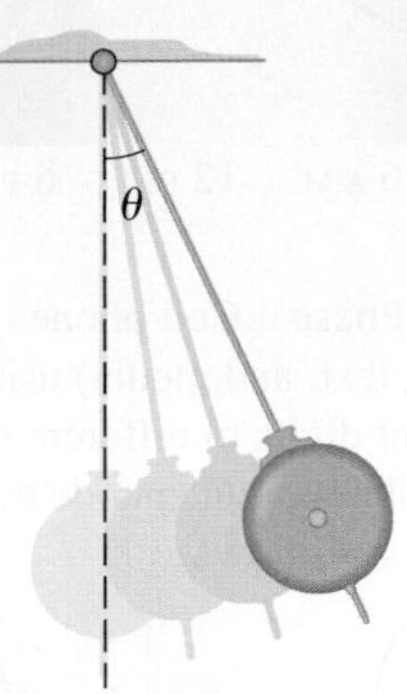

47. Variable Stars The variable star Zeta Gemini has a period of 10 days. The average brightness of the star is 3.8 magnitudes, and the maximum variation from the average is 0.2 magnitudes. Assuming that the variation in brightness follows simple harmonic motion, find an equation of the form

$$y = a \sin \omega t + b$$

that gives the brightness of the star as a function of time.

48. Mass-Spring System The frequency of oscillation of an object suspended on a spring depends on the stiffness k of the spring (called the *spring constant*) and the mass m of the object. If the spring is compressed a distance a and then allowed to oscillate, its displacement is given by

$$f(t) = a \cos \sqrt{k/m}\, t$$

(a) A 10-g mass is suspended from a spring with stiffness $k = 3$. If the spring is compressed a distance 5 cm and then released, find the equation that describes the oscillation of the spring.

(b) Find a general formula for the frequency (in terms of k and m).

(c) How is the frequency affected if the mass is increased? Is the oscillation faster or slower?

(d) How is the frequency affected if a stiffer spring is used (larger k)? Is the oscillation faster or slower?

49. Biological Clocks A *circadian rhythm* is a biological process that oscillates with a period of approximately 24 hours; that is, it is an internal daily biological clock. Blood pressure appears to follow such a rhythm. For a certain individual the average resting blood pressure varies from a maximum of 100 mmHg at 2:00 P.M. to a minimum of 80 mmHg at 2:00 A.M. Find a sine function of the form

$$f(t) = a\sin(\omega(t - c)) + b$$

that models the blood pressure at time t, measured in hours from midnight.

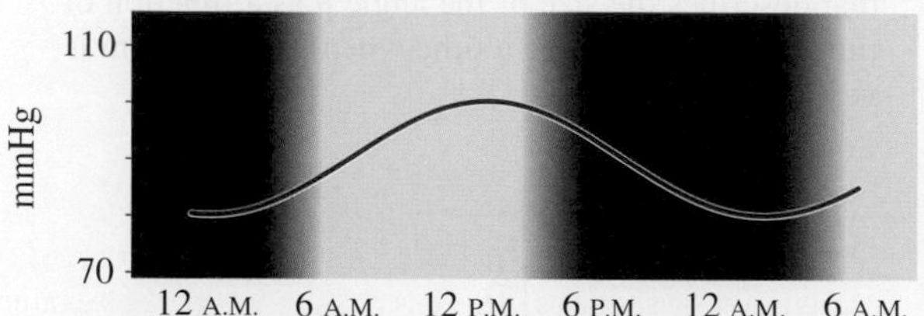

50. Digital Data and Phase Cell phones transmit digital data (including voice, text, and media) using radio waves by encoding different digits to different phases of the carrier wave. For example, we can encode the binary digits 00, 01, 10, 11 by phases as follows.

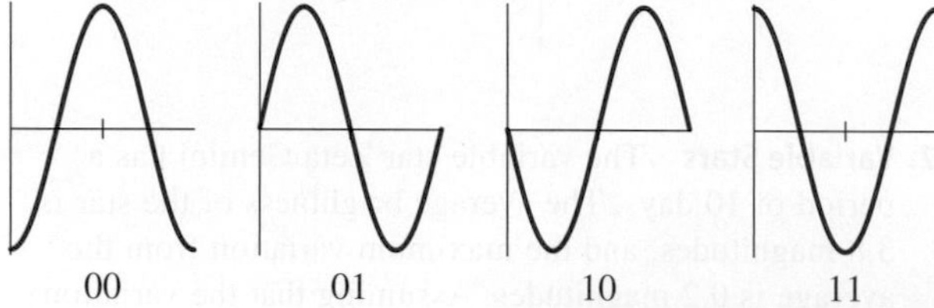

The receiver then uses this code to interpret incoming radio waves. Use the code to find the string of digits that correspond to the following signal.

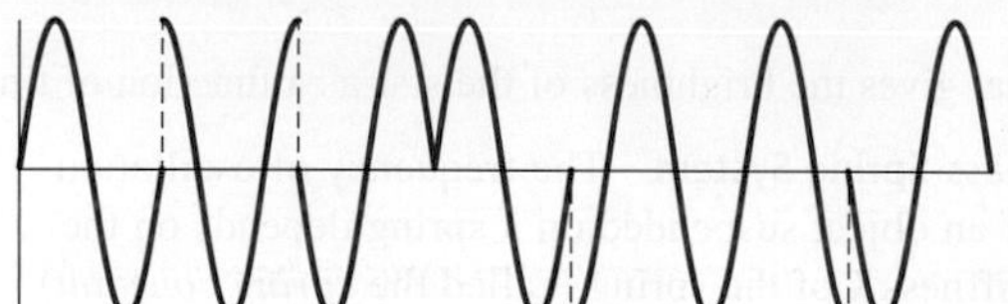

51. Electric Generator The graph shows an oscilloscope reading of the variation in voltage of an AC current produced by a simple generator.

(a) Find the maximum voltage produced.

(b) Find the frequency (cycles per second) of the generator.

(c) How many revolutions per second does the armature in the generator make?

(d) Find a formula that describes the variation in voltage as a function of time.

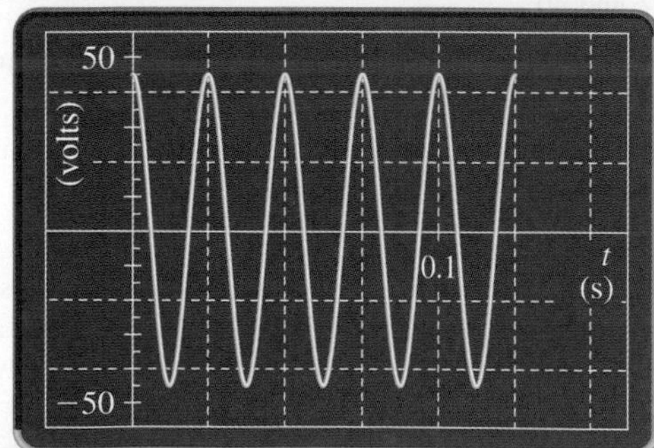

52. Doppler Effect When a car with its horn blowing drives by an observer, the pitch of the horn seems higher as it approaches and lower as it recedes (see the figure below). This phenomenon is called the **Doppler effect**. If the sound source is moving at speed v relative to the observer and if the speed of sound is v_0, then the perceived frequency f is related to the actual frequency f_0 as follows.

$$f = f_0\left(\frac{v_0}{v_0 \pm v}\right)$$

We choose the minus sign if the source is moving toward the observer and the plus sign if it is moving away.

Suppose that a car is driving 33 m/s past a person standing on the shoulder of a highway with its horn blowing at a frequency of 500 Hz. Assume that the speed of sound is 339 m/s. (This is the speed in dry air at 21°C.)

(a) What are the frequencies of the sounds that the person hears as the car approaches and as it moves away?

(b) Let A be the amplitude of the sound. Find functions of the form

$$y = A\sin\omega t$$

that model the perceived sound as the car approaches the person and as it recedes.

53. Motion of a Building A strong gust of wind strikes a tall building, causing it to sway back and forth in damped harmonic motion. The frequency of the oscillation is 0.5 cycles per second, and the damping constant is $c = 0.9$. Find an equation that describes the motion of the building. (Assume that $k = 1$, and take $t = 0$ to be the instant when the gust of wind strikes the building.)

54. Shock Absorber When a car hits a certain bump on the road, its shock absorber is compressed a distance of 15 cm, then released (see the figure). The shock absorber vibrates in damped harmonic motion with a frequency of 2 cycles per second. The damping constant for this particular shock absorber is 2.8.

(a) Find an equation that describes the displacement of the shock absorber from its rest position as a function of time. Take $t = 0$ to be the instant that the shock absorber is released.

(b) How long does it take for the amplitude of the vibration to decrease to 1.25 cm?

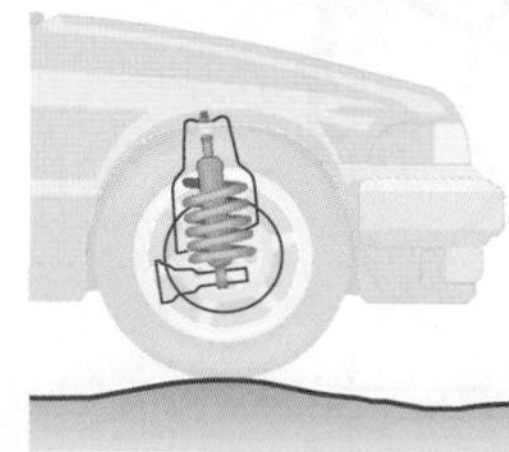

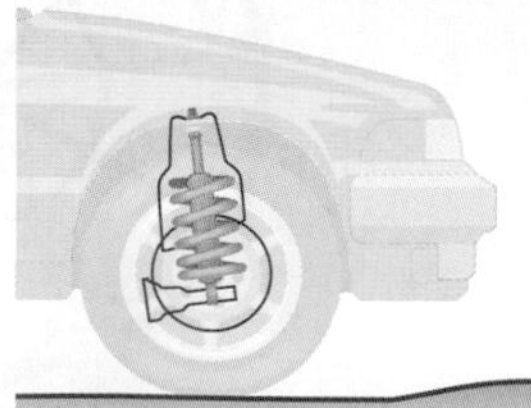

55. Tuning Fork When a tuning fork is struck, it oscillates in damped harmonic motion. The amplitude of the motion is measured, and 3 seconds later it is found that the amplitude has dropped to $\frac{1}{4}$ of this value. Find the damping constant c for this tuning fork.

56. Guitar String A guitar string is pulled at point P a distance of 3 cm above its rest position. It is then released and vibrates in damped harmonic motion with a frequency of 165 cycles per second. After 2 seconds, it is observed that the amplitude of the vibration at point P is 0.6 cm.

(a) Find the damping constant c.

(b) Find an equation that describes the position of point P above its rest position as a function of time. Take $t = 0$ to be the instant that the string is released.

57. Two Fans Electric fans A and B have radius 0.3 m and, when switched on, rotate counterclockwise at the rate of 100 revolutions per minute. Starting with the position shown in the figure, the fans are switched on simultaneously.

(a) For each fan, find an equation that gives the height of the red dot (above the horizontal line shown) t minutes after the fans are switched on.

(b) Are the fans rotating in phase? Through what angle should fan A be rotated counterclockwise in order that the two fans rotate in phase?

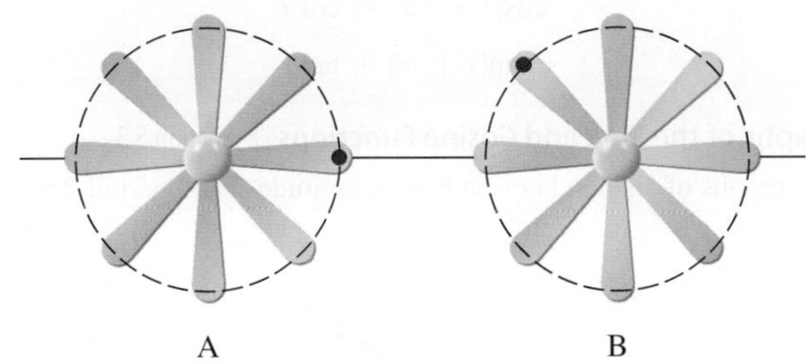

58. Alternating Current Alternating current is produced when an armature rotates about its axis in a magnetic field, as shown in Figure 9. Generators A and B rotate counterclockwise at 60 Hz (cycles per second) and each generator produces a maximum of 50 V. The voltage for each generator is modeled by

$$E_A = 50 \sin 120\pi t \qquad E_B = 50 \sin\left(120\pi t - \frac{5\pi}{4}\right)$$

(a) Find the voltage phase for each generator, and find the phase difference between the first and second curves.

(b) Are the generators producing voltage in phase? Through what angle should the armature in the second generator be rotated counterclockwise in order for the two generators to produce voltage in phase?

■ Discuss ■ Discover ■ Prove ■ Write

59. Discuss: Phases of Sine The phase of a sine curve $y = \sin(kt + b)$ represents a particular location on the graph of the sine function $y = \sin t$. Specifically, when $t = 0$, we have $y = \sin b$, and this corresponds to the point $(b, \sin b)$ on the graph of $y = \sin t$. Observe that each point on the graph of $y = \sin t$ has different characteristics. For example, for $t = \pi/6$, we have $\sin t = \frac{1}{2}$ and the values of sine are increasing, whereas at $t = 5\pi/6$, we also have $\sin t = \frac{1}{2}$ but the values of sine are decreasing. So each point on the graph of sine corresponds to a different "phase" of a sine curve. Complete the descriptions for each label on the graph below.

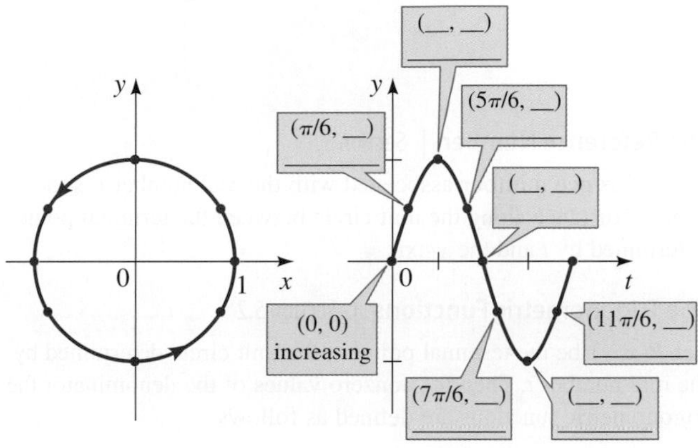

60. Discuss: Phases of the Moon During the course of a lunar cycle (about one month) the moon undergoes the familiar lunar phases. The phases of the moon are completely analogous to the phases of the sine function described in Exercise 59. The figure below shows some phases of the lunar cycle, starting with a "new moon," "waxing crescent moon," "first quarter moon," and so on. The next-to-last phase shown is a "waning crescent moon." Give similar descriptions for the other phases of the moon shown in the figure. What are some events on the earth that follow a monthly cycle and are in phase with the lunar cycle? What are some events that are out of phase with the lunar cycle?

Chapter 5 Review

Properties & Formulas

The Unit Circle | Section 5.1

The **unit circle** is the circle of radius 1 centered at (0, 0). The equation of the unit circle is $x^2 + y^2 = 1$.

Terminal Points on the Unit Circle | Section 5.1

The **terminal point** $P(x, y)$ determined by the real number t is the point obtained by traveling counterclockwise a distance t along the unit circle, starting at $(1, 0)$.

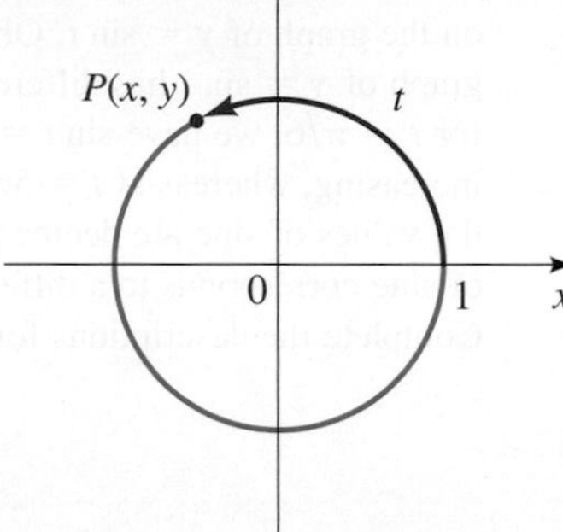

Special terminal points are listed in Table 5.1.1.

The Reference Number | Section 5.1

The **reference number** associated with the real number t is the shortest distance along the unit circle between the terminal point determined by t and the x-axis.

The Trigonometric Functions | Section 5.2

Let $P(x, y)$ be the terminal point on the unit circle determined by the real number t. Then for nonzero values of the denominator the trigonometric functions are defined as follows.

$$\sin t = y \qquad \cos t = x \qquad \tan t = \frac{y}{x}$$

$$\csc t = \frac{1}{y} \qquad \sec t = \frac{1}{x} \qquad \cot t = \frac{x}{y}$$

Special Values of the Trigonometric Functions | Section 5.2

The trigonometric functions have the following values at the special values of t.

t	$\sin t$	$\cos t$	$\tan t$	$\csc t$	$\sec t$	$\cot t$
0	0	1	0	—	1	—
$\frac{\pi}{6}$	$\frac{1}{2}$	$\frac{\sqrt{3}}{2}$	$\frac{\sqrt{3}}{3}$	2	$\frac{2\sqrt{3}}{3}$	$\sqrt{3}$
$\frac{\pi}{4}$	$\frac{\sqrt{2}}{2}$	$\frac{\sqrt{2}}{2}$	1	$\sqrt{2}$	$\sqrt{2}$	1
$\frac{\pi}{3}$	$\frac{\sqrt{3}}{2}$	$\frac{1}{2}$	$\sqrt{3}$	$\frac{2\sqrt{3}}{3}$	2	$\frac{\sqrt{3}}{3}$
$\frac{\pi}{2}$	1	0	—	1	—	0
π	0	−1	0	—	−1	—
$\frac{3\pi}{2}$	−1	0	—	−1	—	0

Basic Trigonometric Identities | Section 5.2

An identity is an equation that is true for all values of the variable. The basic trigonometric identities are as follows.

Reciprocal Identities:

$$\csc t = \frac{1}{\sin t} \qquad \sec t = \frac{1}{\cos t} \qquad \cot t = \frac{1}{\tan t}$$

Pythagorean Identities:

$$\sin^2 t + \cos^2 t = 1$$

$$\tan^2 t + 1 = \sec^2 t$$

$$1 + \cot^2 t = \csc^2 t$$

Even-Odd Properties:

$$\sin(-t) = -\sin t \qquad \cos(-t) = \cos t \qquad \tan(-t) = -\tan t$$

$$\csc(-t) = -\csc t \qquad \sec(-t) = \sec t \qquad \cot(-t) = -\cot t$$

Periodic Properties | Section 5.3

A function f is **periodic** if there is a positive number p such that $f(x + p) = f(x)$ for every x. The least such p is called the **period** of f. The sine and cosine functions have period 2π, and the tangent function has period π.

$$\sin(t + 2\pi) = \sin t$$

$$\cos(t + 2\pi) = \cos t$$

$$\tan(t + \pi) = \tan t$$

Graphs of the Sine and Cosine Functions | Section 5.3

The graphs of sine and cosine have amplitude 1 and period 2π.

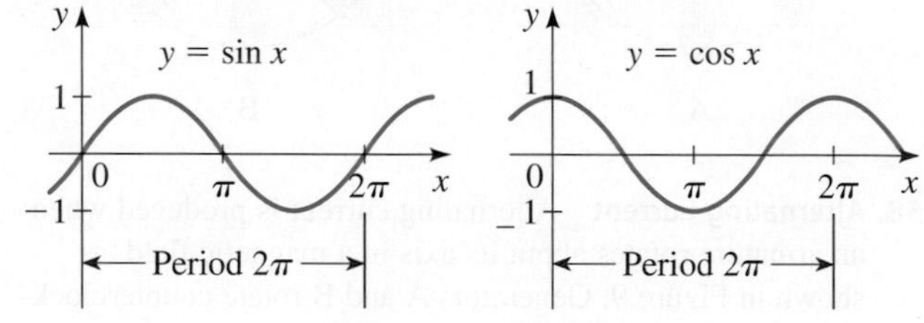

Amplitude 1, Period 2π

Graphs of Transformations of Sine and Cosine | Section 5.3

$y = a \sin k(x - b) \quad (k > 0)$ $\qquad$ $y = a \cos k(x - b) \quad (k > 0)$

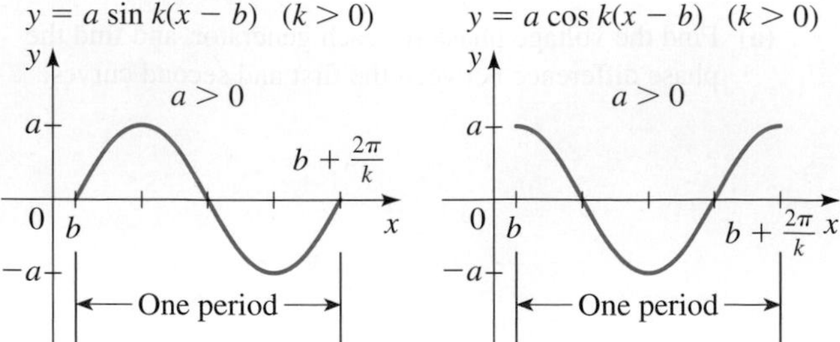

Amplitude a, Period $\frac{2\pi}{k}$, Horizontal shift b

An appropriate interval on which to graph one complete period is $[b, b + (2\pi/k)]$.

Graphs of the Tangent and Cotangent Functions | Section 5.4

These functions have period π.

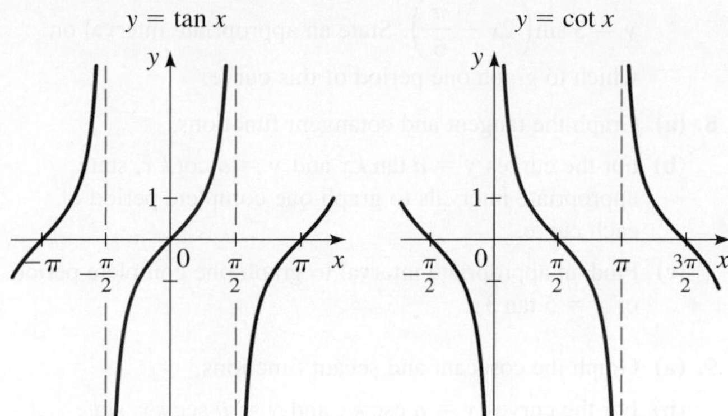

To graph one period of $y = a \tan kx$, an appropriate interval is $\left(-\dfrac{\pi}{2k}, \dfrac{\pi}{2k}\right)$.

To graph one period of $y = a \cot kx$, an appropriate interval is $\left(0, \dfrac{\pi}{k}\right)$.

Graphs of the Secant and Cosecant Functions | Section 5.4

These functions have period 2π.

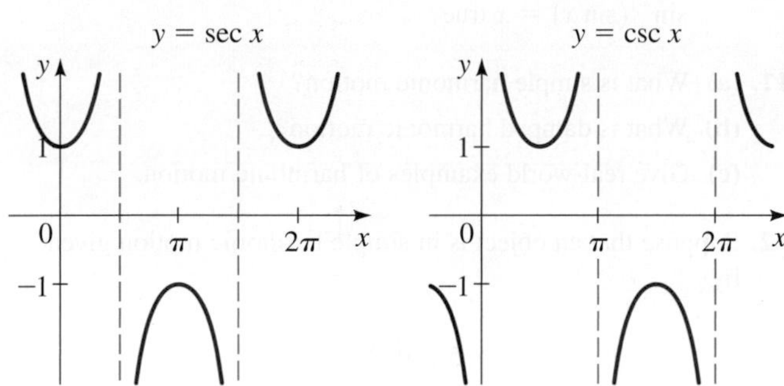

To graph one period of $y = a \sec kx$ or $y = a \csc kx$, an appropriate interval is $\left(0, \dfrac{2\pi}{k}\right)$.

Inverse Trigonometric Functions | Section 5.5

Inverse functions of the trigonometric functions have the following domain and range.

Function	Domain	Range
$\sin^{-1}$	$[-1, 1]$	$\left[-\dfrac{\pi}{2}, \dfrac{\pi}{2}\right]$
$\cos^{-1}$	$[-1, 1]$	$[0, \pi]$
$\tan^{-1}$	$(-\infty, \infty)$	$\left(-\dfrac{\pi}{2}, \dfrac{\pi}{2}\right)$

The inverse trigonometric functions are defined as follows.

$$\sin^{-1} x = y \iff \sin y = x$$

$$\cos^{-1} x = y \iff \cos y = x$$

$$\tan^{-1} x = y \iff \tan y = x$$

Graphs of these inverse functions are shown below.

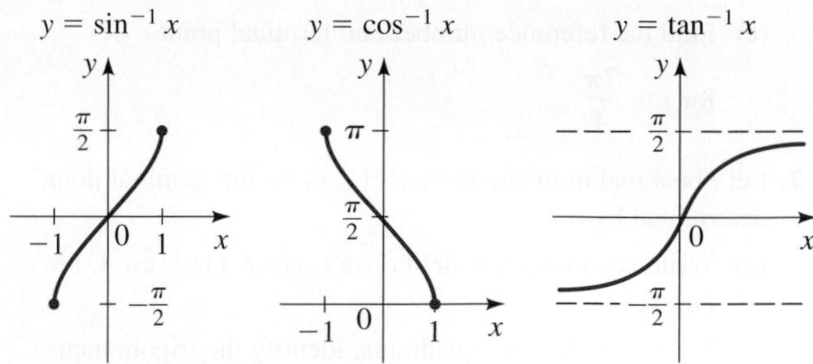

Harmonic Motion | Section 5.6

An object is in **simple harmonic motion** if its displacement y at time t is modeled by $y = a \sin \omega t$ or $y = a \cos \omega t$. In this case the amplitude is $|a|$, the period is $2\pi/\omega$, and the frequency is $\omega/(2\pi)$.

Damped Harmonic Motion | Section 5.6

An object is in **damped harmonic motion** if its displacement y at time t is modeled by $y = ke^{-ct} \sin \omega t$ or $y = ke^{-ct} \cos \omega t$, $c > 0$. In this case c is the damping constant, k is the initial amplitude, and $2\pi/\omega$ is the period.

Phase | Section 5.6

Any sine curve can be expressed in the following equivalent forms:

$y = A \sin(kt - b)$, the **phase** is b

$y = A \sin k\left(t - \dfrac{b}{k}\right)$, the **horizontal shift** is $\dfrac{b}{k}$

The phase (or phase angle) b is the initial angular position of the motion. The number b/k is also called the **lag time** ($b > 0$) or **lead time** ($b < 0$).

Suppose that two objects are in harmonic motion with the same period modeled by

$$y_1 = A \sin(kt - b) \quad \text{and} \quad y_2 = A \sin(kt - c)$$

The **phase difference** between y_1 and y_2 is $b - c$. The motions are "in phase" if the phase difference is a multiple of 2π; otherwise, the motions are "out of phase."

Concept Check

1. (a) What is the unit circle, and what is the equation of the unit circle?

(b) Use a diagram to explain what is meant by the terminal point $P(x, y)$ determined by t.

(c) Find the terminal point for $t = \frac{\pi}{2}$.

(d) What is the reference number associated with t?

(e) Find the reference number and terminal point for $t = \frac{7\pi}{4}$.

2. Let t be a real number, and let $P(x, y)$ be the terminal point determined by t.

(a) Write equations that define $\sin t$, $\cos t$, $\tan t$, $\csc t$, $\sec t$, and $\cot t$.

(b) In each of the four quadrants, identify the trigonometric functions that are positive.

(c) List the special values of sine, cosine, and tangent.

3. (a) Describe the steps we use to find the value of a trigonometric function at a real number t.

(b) Find $\sin \frac{5\pi}{6}$.

4. (a) What is a periodic function?

(b) What are the periods of the six trigonometric functions?

(c) Find $\sin \frac{19\pi}{4}$.

5. (a) What is an even function, and what is an odd function?

(b) Which trigonometric functions are even? Which are odd?

(c) If $\sin t = 0.4$, find $\sin(-t)$.

(d) If $\cos s = 0.7$, find $\cos(-s)$.

6. (a) State the reciprocal identities.

(b) State the Pythagorean identities.

7. (a) Graph the sine and cosine functions.

(b) What are the amplitude, period, and horizontal shift for the sine curve $y = a \sin k(x - b)$ and for the cosine curve $y = a \cos k(x - b)$? Find an appropriate interval to graph one period of these functions.

(c) Find the amplitude, period, and horizontal shift of $y = 3 \sin\left(2x - \frac{\pi}{6}\right)$. State an appropriate interval on which to graph one period of this curve.

8. (a) Graph the tangent and cotangent functions.

(b) For the curves $y = a \tan kx$ and $y = a \cot kx$, state appropriate intervals to graph one complete period of each curve.

(c) Find an appropriate interval to graph one complete period of $y = 5 \tan 3x$.

9. (a) Graph the cosecant and secant functions.

(b) For the curves $y = a \csc kx$ and $y = a \sec kx$, state appropriate intervals to graph one complete period of each curve.

(c) Find an appropriate interval to graph one period of $y = 3 \csc 6x$.

10. (a) Define the inverse sine function, the inverse cosine function, and the inverse tangent function.

(b) Find $\sin^{-1}\left(\frac{1}{2}\right)$, $\cos^{-1}\left(\frac{\sqrt{2}}{2}\right)$, and $\tan^{-1} 1$.

(c) For what values of x is the equation $\sin(\sin^{-1} x) = x$ true? For what values of x is the equation $\sin^{-1}(\sin x) = x$ true?

11. (a) What is simple harmonic motion?

(b) What is damped harmonic motion?

(c) Give real-world examples of harmonic motion.

12. Suppose that an object is in simple harmonic motion given by

$$y = 5 \sin\left(2t - \frac{\pi}{3}\right)$$

(a) Find the amplitude, period, and frequency.

(b) Find the phase and the horizontal shift.

13. Consider the following models of harmonic motion.

$$y_1 = 5 \sin(2t - 1) \qquad y_2 = 5 \sin(2t - 3)$$

Do both motions have the same frequency? What is the phase for each equation? What is the phase difference? Are the objects moving in phase or out of phase?

Answers to the Concept Check can be found at the book companion website stewartmath.com.

Exercises

1–2 ■ Terminal Points A point $P(x, y)$ is given. **(a)** Show that P is on the unit circle. **(b)** Suppose that P is the terminal point determined by t. Find $\sin t$, $\cos t$, and $\tan t$.

1. $P\left(-\frac{\sqrt{3}}{2}, \frac{1}{2}\right)$ **2.** $P\left(\frac{3}{5}, -\frac{4}{5}\right)$

3–6 ■ Reference Number and Terminal Point A real number t is given. **(a)** Find the reference number for t. **(b)** Find the terminal point $P(x, y)$ on the unit circle determined by t. **(c)** Find the six trigonometric functions of t.

3. $t = \frac{2\pi}{3}$ **4.** $t = \frac{5\pi}{3}$

5. $t = -\frac{11\pi}{4}$ **6.** $t = -\frac{7\pi}{6}$

7–16 ■ Values of Trigonometric Functions Find the value of each trigonometric function. If possible, give the exact value; otherwise, use a calculator to find an approximate value rounded to five decimal places.

7. (a) $\sin \frac{3\pi}{4}$ **(b)** $\cos \frac{3\pi}{4}$

8. (a) $\tan \frac{\pi}{3}$ **(b)** $\tan\left(-\frac{\pi}{3}\right)$

9. (a) $\sin 1.1$ **(b)** $\cos 1.1$

10. (a) $\cos \frac{\pi}{5}$ **(b)** $\cos\left(-\frac{\pi}{5}\right)$

11. (a) $\cos \frac{9\pi}{2}$ **(b)** $\sec \frac{9\pi}{2}$

12. (a) $\sin \frac{\pi}{7}$ **(b)** $\csc \frac{\pi}{7}$

13. (a) $\tan \frac{5\pi}{2}$ **(b)** $\cot \frac{5\pi}{2}$

14. (a) $\sin 2\pi$ **(b)** $\csc 2\pi$

15. (a) $\tan \frac{5\pi}{6}$ **(b)** $\cot \frac{5\pi}{6}$

16. (a) $\cos \frac{\pi}{3}$ **(b)** $\sin \frac{\pi}{6}$

17–20 ■ Fundamental Identities Use the fundamental identities to write the first expression in terms of the second.

17. $\frac{\tan t}{\cos t}$, $\sin t$

18. $\tan^2 t \sec t$, $\cos t$

19. $\tan t$, $\sin t$; t in Quadrant IV

20. $\sec t$, $\sin t$; t in Quadrant II

21–24 ■ Values of Trigonometric Functions Find the values of the remaining trigonometric functions at t from the given information.

21. $\sin t = \frac{5}{13}$, $\cos t = -\frac{12}{13}$

22. $\sin t = -\frac{1}{2}$, $\cos t > 0$

23. $\cot t = -\frac{1}{2}$, $\csc t = \sqrt{5}/2$

24. $\cos t = -\frac{3}{5}$, $\tan t < 0$

25–28 ■ Values of Trigonometric Functions Find the values of the trigonometric expression of t from the given information.

25. $\sec t + \cot t$; $\tan t = \frac{1}{4}$, terminal point for t in Quadrant III

26. $\csc t + \sec t$; $\sin t = -\frac{8}{17}$, terminal point for t in Quadrant IV

27. $\tan t + \sec t$; $\cos t = \frac{3}{5}$, terminal point for t in Quadrant I

28. $\sin^2 t + \cos^2 t$; $\sec t = -5$, terminal point for t in Quadrant II

29–36 ■ Horizontal Shifts A trigonometric function is given. **(a)** Find the amplitude, period, and horizontal shift of the function. **(b)** Sketch the graph.

29. $y = 10 \cos \frac{1}{2}x$ **30.** $y = 4 \sin 2\pi x$

31. $y = -\sin \frac{1}{2}x$ **32.** $y = 2 \sin\left(x - \frac{\pi}{4}\right)$

33. $y = 3 \sin(2x - 2)$ **34.** $y = \cos 2\left(x - \frac{\pi}{2}\right)$

35. $y = -\cos\left(\frac{\pi}{2}x + \frac{\pi}{6}\right)$ **36.** $y = 10 \sin\left(2x - \frac{\pi}{2}\right)$

37–40 ■ Functions from a Graph The graph of one period of a function of the form

$$y = a \sin k(x - b) \quad \text{or} \quad y = a \cos k(x - b)$$

is shown. Determine the function.

37.

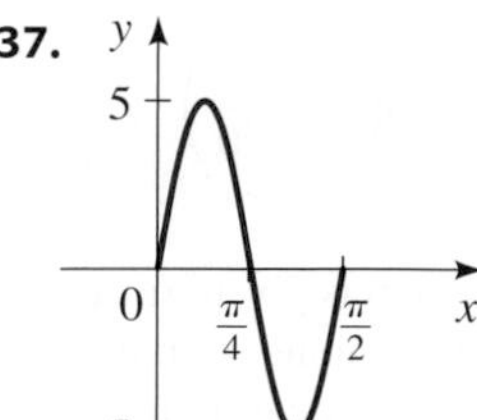

38.

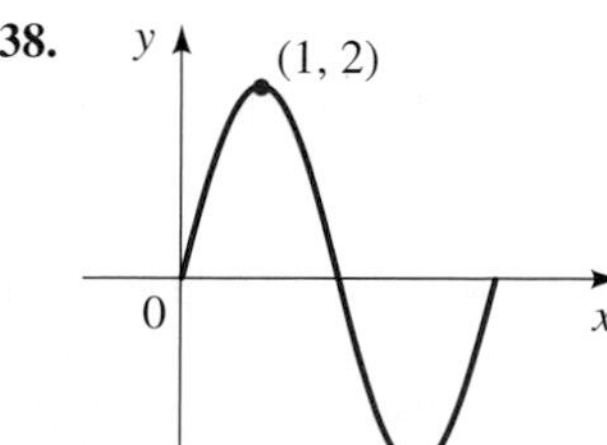

39.

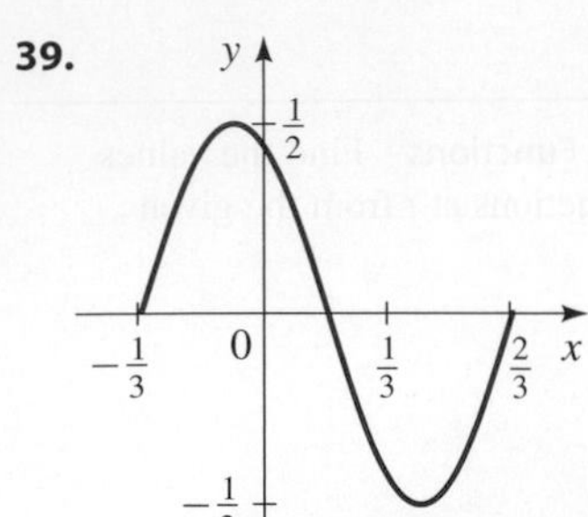

40.

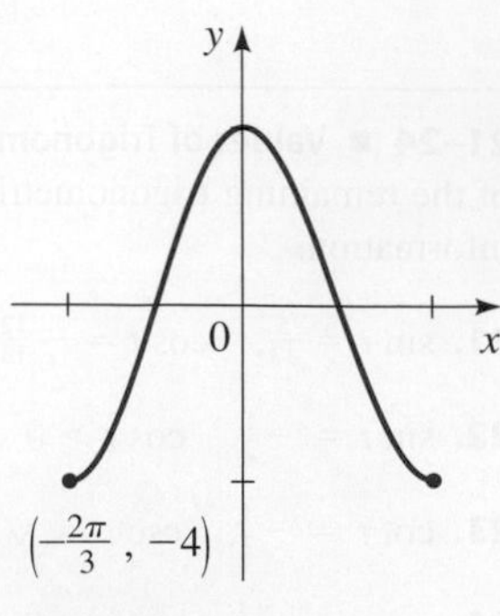

41–48 ■ Graphing Trigonometric Functions Find the period, and sketch the graph.

41. $y = 3 \tan x$

42. $y = \tan \pi x$

43. $y = 2 \cot\left(x - \dfrac{\pi}{2}\right)$

44. $y = \sec\left(\dfrac{1}{2}x - \dfrac{\pi}{2}\right)$

45. $y = 4 \csc(2x + \pi)$

46. $y = \tan\left(x + \dfrac{\pi}{6}\right)$

47. $y = \tan\left(\dfrac{1}{2}x - \dfrac{\pi}{8}\right)$

48. $y = -4 \sec 4\pi x$

49–52 ■ Evaluating Expressions Involving Inverse Trigonometric Functions Find the exact value of the expression, if it is defined.

49. $\sin^{-1} 1$

50. $\cos^{-1}(-\frac{1}{2})$

51. $\sin^{-1}\left(\sin \dfrac{13\pi}{6}\right)$

52. $\tan(\cos^{-1}(\frac{1}{2}))$

53–54 ■ Amplitude, Period, Phase, and Horizontal Shift For each sine curve find the amplitude, period, phase, and horizontal shift.

53. $y = 100 \sin 8\left(t + \dfrac{\pi}{16}\right)$

54. $y = 80 \sin 3\left(t - \dfrac{\pi}{2}\right)$

55–56 ■ Phase and Phase Difference A pair of sine curves with the same period is given. **(a)** Find the phase of each curve. **(b)** Find the phase difference between the first and second curves. **(c)** Determine whether the curves are in phase or out of phase. **(d)** Sketch both curves on the same axes.

55. $y_1 = 25 \sin 3\left(t - \dfrac{\pi}{2}\right)$; $y_2 = 10 \sin\left(3t - \dfrac{5\pi}{2}\right)$

56. $y_1 = 50 \sin\left(10t - \dfrac{\pi}{2}\right)$; $y_2 = 50 \sin 10\left(t - \dfrac{\pi}{20}\right)$

57–62 ■ Even and Odd Functions A function is given. **(a)** Use a graphing device to graph the function. **(b)** Determine from the graph whether the function is periodic and, if so, determine the period. **(c)** Determine from the graph whether the function is odd, even, or neither.

57. $y = |\cos x|$

58. $y = \sin(\cos x)$

59. $y = \cos(2^{0.1x})$

60. $y = 1 + 2^{\cos x}$

61. $y = |x| \cos 3x$

62. $y = \sqrt{x} \sin 3x, \quad x > 0$

63–66 ■ Sine and Cosine Curves with Variable Amplitude Graph the three functions on a common screen. How are the graphs related?

63. $y = x, \quad y = -x, \quad y = x \sin x$

64. $y = 2^{-x}, \quad y = -2^{-x}, \quad y = 2^{-x} \cos 4\pi x$

65. $y = x, \quad y = \sin 4x, \quad y = x + \sin 4x$

66. $y = \sin^2 x, \quad y = \cos^2 x, \quad y = \sin^2 x + \cos^2 x$

67–68 ■ Maxima and Minima Find the maximum and minimum values of the function.

67. $y = \cos x + \sin 2x$

68. $y = \cos x + \sin^2 x$

69–70 ■ Solving Trigonometric Equations Graphically Find all solutions of the equation that lie in the given interval. State each answer rounded to two decimal places.

69. $\sin x = 0.3; \quad [0, 2\pi]$

70. $\cos 3x = x; \quad [0, \pi]$

71. Discover the Period of a Trigonometric Function Let $y_1 = \cos(\sin x)$ and $y_2 = \sin(\cos x)$.

(a) Graph y_1 and y_2 in the same viewing rectangle.

(b) Determine the period of each of these functions from its graph.

(c) Find an inequality between $\sin(\cos x)$ and $\cos(\sin x)$ that is valid for all x.

72. Simple Harmonic Motion A point P moving in simple harmonic motion completes 8 cycles every second. If the amplitude of the motion is 50 cm, find an equation that describes the motion of P as a function of time. Assume that the point P is at its maximum displacement when $t = 0$.

73. Simple Harmonic Motion A mass suspended from a spring oscillates in simple harmonic motion at a frequency of 4 cycles per second. The distance from the highest to the lowest point of the oscillation is 100 cm. Find an equation that describes the distance of the mass from its rest position as a function of time. Assume that the mass is at its lowest point when $t = 0$.

74. Damped Harmonic Motion The top floor of a building undergoes damped harmonic motion after a sudden brief earthquake. At time $t = 0$ the displacement is at a maximum, 16 cm from the normal position. The damping constant is $c = 0.72$, and the building vibrates at 1.4 cycles per second.

(a) Find a function of the form $y = ke^{-ct} \cos \omega t$ to model the motion.

(b) Graph the function you found in part (a).

(c) What is the displacement at time $t = 10$ s?

Matching

75. Equations and Their Graphs Match each equation with its graph. Give reasons for your answers. (Don't use a graphing device.)

(a) $y = \dfrac{x}{4 - x^2}$ **(b)** $y = -4\cos\dfrac{\pi}{2}(x + 1)$ **(c)** $y = \tan\dfrac{\pi x}{4}$ **(d)** $y = x^3$

(e) $y = -x\cos 4x$ **(f)** $y = 1 + 3\sin\dfrac{\pi x}{2}$ **(g)** $y = \sec\dfrac{\pi x}{4}$ **(h)** $y = \dfrac{4}{\pi}\tan^{-1} x$

I

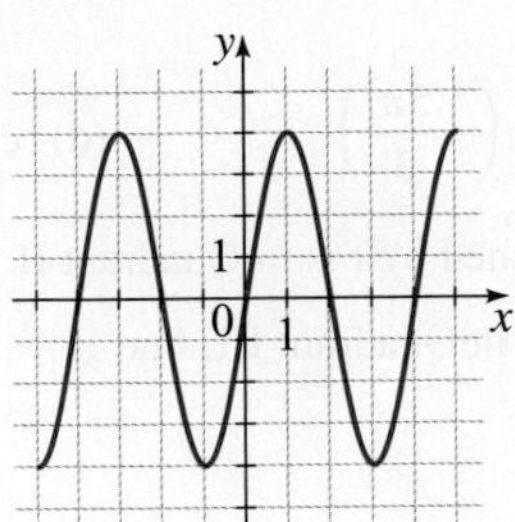

II

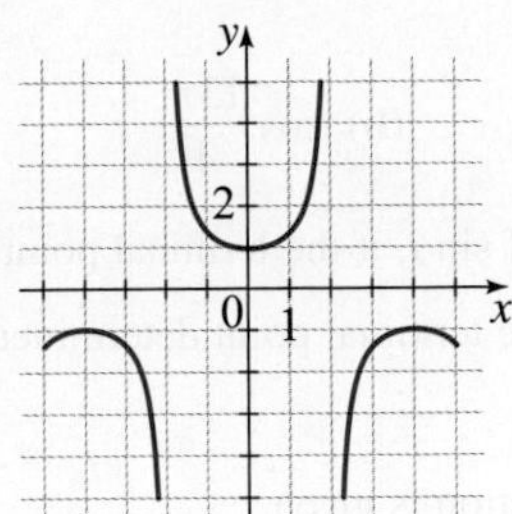

III

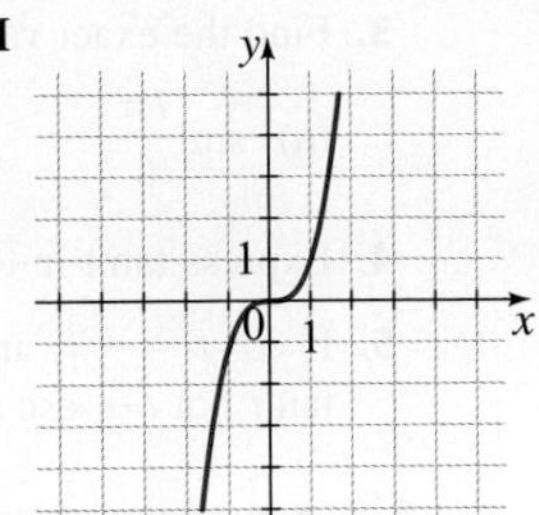

IV

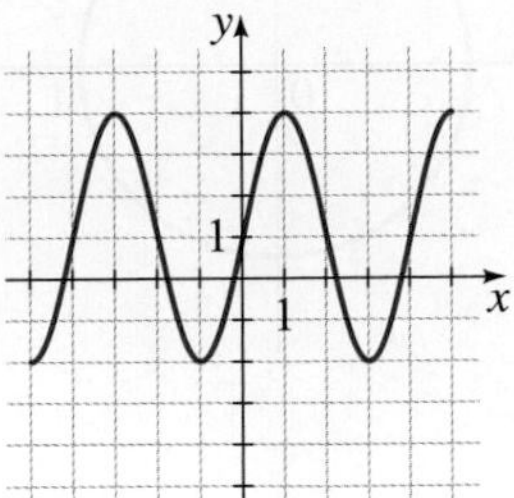

V

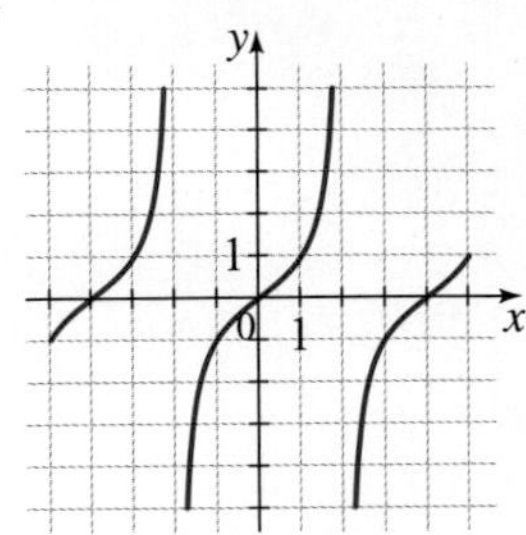

VI

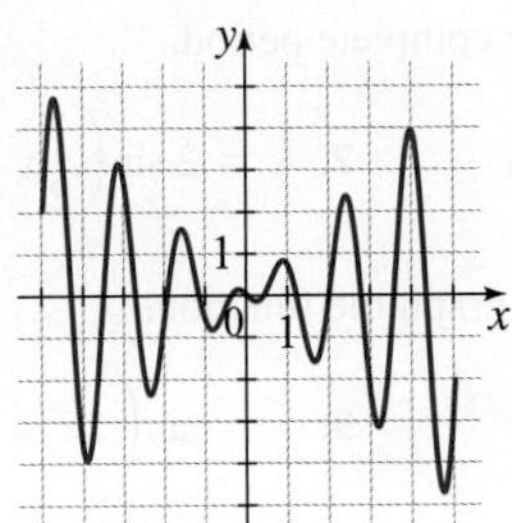

VII

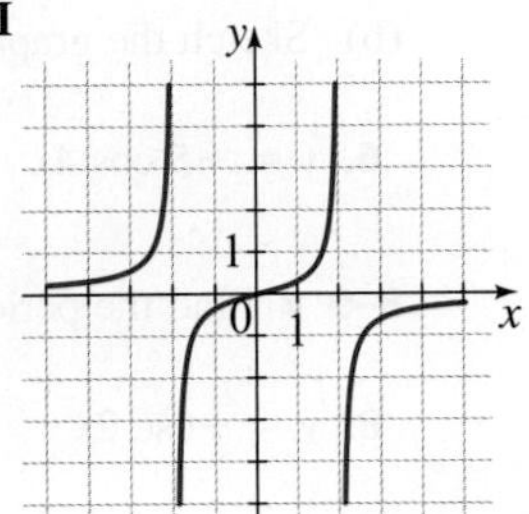

VIII

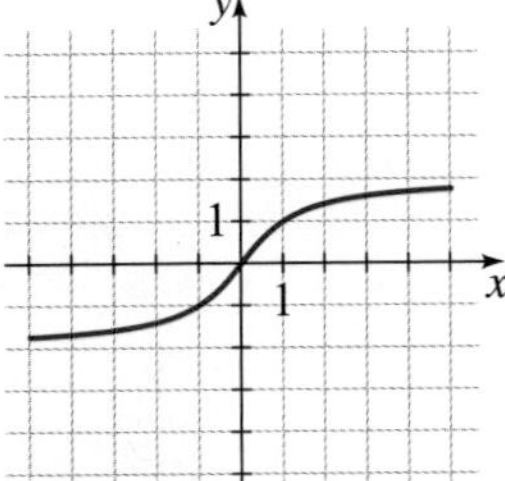

Chapter 5 | Test

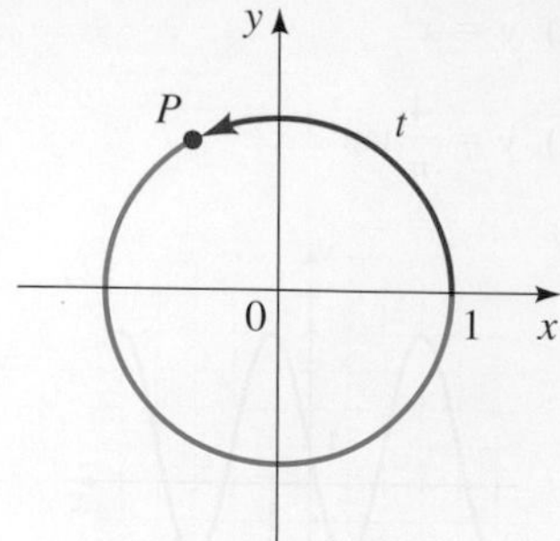

1. The point $P(x, y)$ is on the unit circle in Quadrant IV. If $x = \sqrt{11}/6$, find y.

2. The point P in the figure at the left has y-coordinate $\frac{4}{5}$. Find:

   (a) $\sin t$ **(b)** $\cos t$ **(c)** $\tan t$ **(d)** $\sec t$

3. Find the exact value.

   (a) $\sin \dfrac{7\pi}{6}$ **(b)** $\cos \dfrac{13\pi}{4}$ **(c)** $\tan\left(-\dfrac{5\pi}{3}\right)$ **(d)** $\csc \dfrac{3\pi}{2}$

4. Express $\tan t$ in terms of $\sin t$, if the terminal point determined by t is in Quadrant II.

5. If $\cos t = -\frac{8}{17}$ and if the terminal point determined by t is in Quadrant III, find $\tan t \cot t + \csc t$.

6–7 ■ A trigonometric function is given.

(a) Find the amplitude, period, phase, and horizontal shift of the function.

(b) Sketch the graph of one complete period.

6. $y = -5 \cos 4x$ 7. $y = 2 \sin\left(\dfrac{1}{2}x - \dfrac{\pi}{6}\right)$

8–9 ■ Find the period, and graph the function.

8. $y = -\csc 2x$ 9. $y = \tan\left(2x - \dfrac{\pi}{2}\right)$

10. Find the exact value of each expression, if it is defined.

    (a) $\tan^{-1} 1$ **(b)** $\cos^{-1}\left(-\dfrac{\sqrt{3}}{2}\right)$

    (c) $\tan^{-1}(\tan 3\pi)$ **(d)** $\cos(\tan^{-1}(-\sqrt{3}))$

11. The graph shown at left is one period of a function of the form $y = a \sin k(x - b)$. Determine the function.

12. The sine curves $y_1 = 30 \sin\left(6t - \dfrac{\pi}{2}\right)$ and $y_2 = 30 \sin\left(6t - \dfrac{\pi}{3}\right)$ have the same period.

    (a) Find the phase of each curve.

    (b) Find the phase difference between y_1 and y_2.

    (c) Determine whether the curves are in phase or out of phase.

    (d) Sketch both curves on the same axes.

13. Let $f(x) = \dfrac{\cos x}{1 + x^2}$.

    (a) Use a graphing device to graph f in an appropriate viewing rectangle.

    (b) Determine from the graph whether f is even, odd, or neither.

    (c) Find the minimum and maximum values of f.

14. A mass suspended from a spring oscillates in simple harmonic motion. The mass completes 2 cycles every second, and the distance between the highest point and the lowest point of the oscillation is 10 cm. Find an equation of the form $y = a \sin \omega t$ that gives the distance of the mass from its rest position as a function of time.

15. An object is moving up and down in damped harmonic motion. Its displacement at time $t = 0$ is 16 cm; this is its maximum displacement. The damping constant is $c = 0.1$, and the frequency is 12 Hz.

    (a) Find a function that models this motion.

    (b) Graph the function.

Focus on Modeling | Fitting Sinusoidal Curves to Data

In previous *Focus on Modeling* sections, we learned how to fit linear, polynomial, exponential, and power models to data. Figure 1 shows some scatter plots of data. The scatter plots can help guide us in choosing an appropriate model. (Try to determine what type of function would best model the data in each graph.) If the scatter plot indicates simple harmonic motion, then we might try to model the data with a sine or cosine function. The next example illustrates this process.

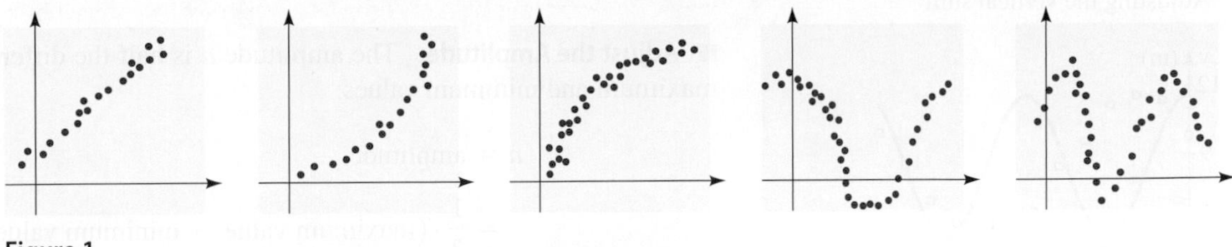

Figure 1

Example 1 ■ Modeling the Height of a Tide

The water depth in a narrow channel varies with the tides. Table 1 shows the water depth over a 12-hour period. A scatter plot of the data is shown in Figure 2.

(a) Find a function that models the water depth with respect to time.

(b) If a boat needs at least 11 m of water depth in order to safely cross the channel, during which times can it do so?

Table 1

Time	Depth (m)
12:00 A.M.	9.8
1:00 A.M.	11.4
2:00 A.M.	11.6
3:00 A.M.	11.2
4:00 A.M.	9.6
5:00 A.M.	8.5
6:00 A.M.	6.5
7:00 A.M.	5.7
8:00 A.M.	5.4
9:00 A.M.	6.0
10:00 A.M.	7.0
11:00 A.M.	8.6
12:00 P.M.	10.0

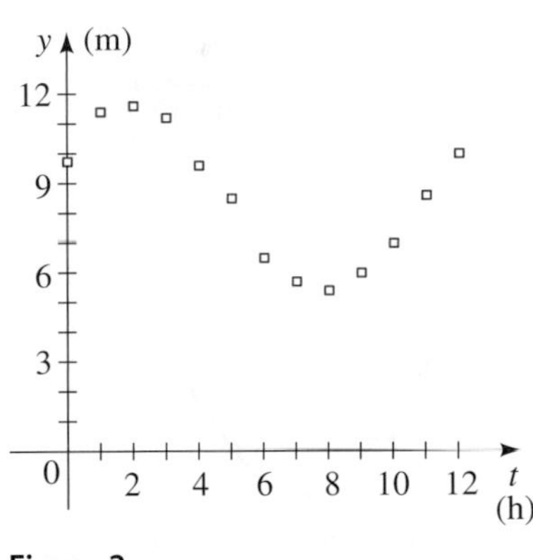

Figure 2

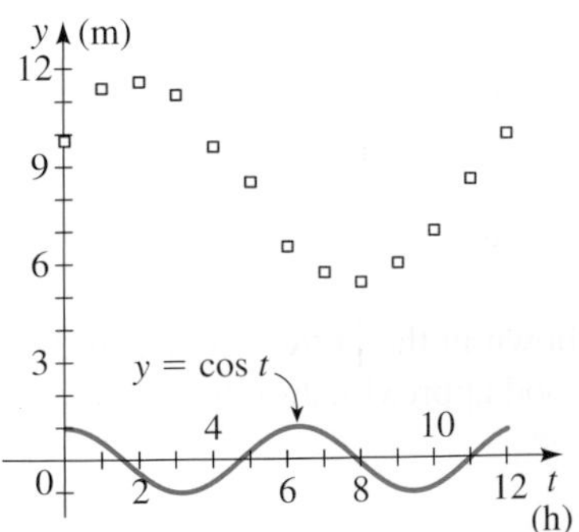

Figure 3

Solution

(a) The data appear to lie on a cosine (or sine) curve. But if we graph $y = \cos t$ on the same graph as the scatter plot, the result in Figure 3 is not even close to the data. To fit the data, we need to adjust the vertical shift, amplitude, period, and horizontal shift of the cosine curve. In other words, we need to find a function of the form

$$y = a\cos(\omega(t - c)) + b$$

We use the following steps, which are illustrated by the graphs in the margin.

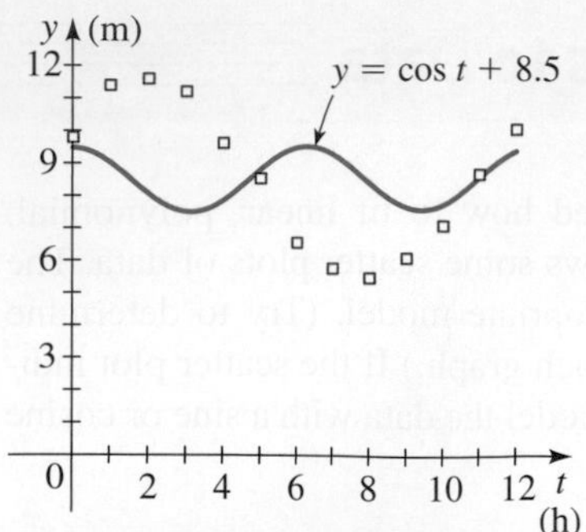

Adjusting the vertical shift

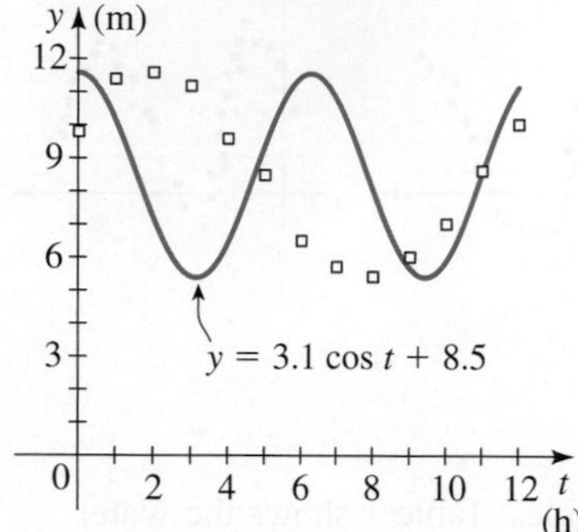

Adjusting the amplitude

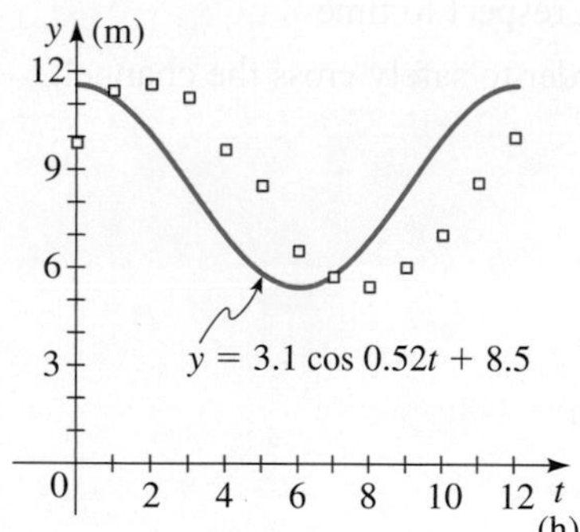

Adjusting the period

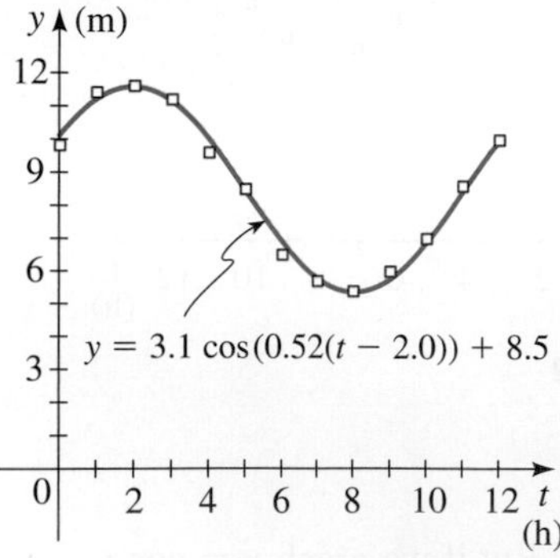

Adjusting the horizontal shift

■ **Adjust the Vertical Shift** The vertical shift b is the average of the maximum and minimum values:

$$\begin{aligned} b &= \text{vertical shift} \\ &= \frac{1}{2} \cdot (\text{maximum value} + \text{minimum value}) \\ &= \frac{1}{2}(11.6 + 5.4) \\ &= 8.5 \end{aligned}$$

■ **Adjust the Amplitude** The amplitude a is half the difference between the maximum and minimum values:

$$\begin{aligned} a &= \text{amplitude} \\ &= \frac{1}{2} \cdot (\text{maximum value} - \text{minimum value}) \\ &= \frac{1}{2}(11.6 - 5.4) \\ &= 3.1 \end{aligned}$$

■ **Adjust the Period** The time between consecutive maximum and minimum values is half of one period. Thus

$$\begin{aligned} \frac{2\pi}{\omega} &= \text{period} \\ &= 2 \cdot (\text{time of maximum value} - \text{time of minimum value}) \\ &= 2(8 - 2) \\ &= 12 \end{aligned}$$

Thus $\omega = 2\pi/12 = 0.52$.

■ **Adjust the Horizontal Shift** Since the maximum value of the data occurs at approximately $t = 2.0$, it represents a cosine curve shifted 2 h to the right. So

$$\begin{aligned} c &= \text{horizontal shift} \\ &= \text{time of maximum value} \\ &= 2.0 \end{aligned}$$

These steps show that a function that models the tides over the given time period is

$$y = 3.1 \cos(0.52(t - 2.0)) + 8.5$$

A graph of the function and the scatter plot are shown in the bottom figure in the margin. It appears that the model we found is a good approximation to the data.

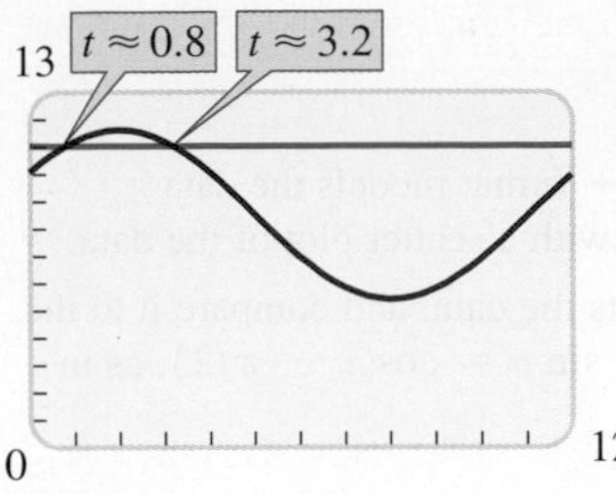

Figure 4

(b) We need to solve the inequality $y \geq 11$. We solve this inequality graphically by graphing $y = 3.1 \cos 0.52(t - 2.0) + 8.5$ and $y = 11$ on the same graph. From the graph in Figure 4 we see that the water depth is higher than 11 m between $t \approx 0.8$ and $t \approx 3.2$. This corresponds to the times 12:48 A.M. to 3:12 A.M. ■

In Example 1 we used the scatter plot to guide us in finding a cosine curve that gives an approximate model of the data. Some graphing devices have a `SinReg` (for sine regression) command that finds the sine curve that best fits the data. The method these devices use is similar to the method of finding a line of best fit, as explained in *Focus on Modeling* following Chapter 1.

Example 2 ■ Fitting a Sine Curve to Data

(a) Use a graphing device to find the sine curve that best fits the depth-of-water data in Table 1. Make a scatter plot of the data and graph the curve you found together with the scatter plot on the same screen.

(b) Compare your result to the model found in Example 1.

Solution

(a) Using the `SinReg` command on a graphing device, we obtain the equation of the sine curve that best fits the data, as shown in Figure 5(b).

$$y = 3.1 \sin(0.53t + 0.55) + 8.42$$

Figure 5(a) shows a scatter plot of the data. From Figure 5(c) we see that the curve appears to fit the data well.

(b) To compare the result found in part (a) with the function we found in Example 1, we change the sine function to a cosine function by using the reduction formula $\sin u = \cos(u - \pi/2)$.

$$\begin{aligned} y &= 3.1 \sin(0.53t + 0.55) + 8.42 \\ &= 3.1 \cos\left(0.53t + 0.55 - \frac{\pi}{2}\right) + 8.42 && \text{Reduction formula} \\ &= 3.1 \cos(0.53t - 1.02) + 8.42 \\ &= 3.1 \cos(0.53(t - 1.92)) + 8.42 && \text{Factor } 0.53 \end{aligned}$$

Comparing this with the function we obtained in Example 1, we see that there are small differences in the coefficients. The rough estimates we made in Example 1 account for these differences.

Figure 5

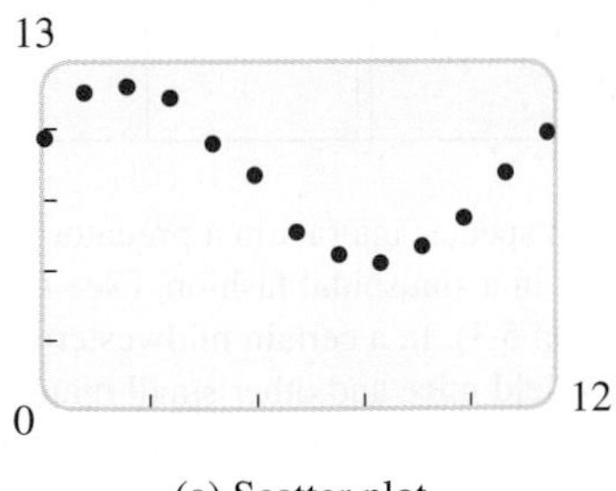

(a) Scatter plot

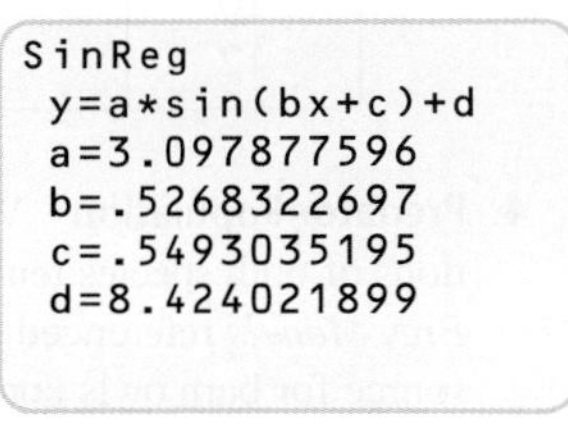

(b) Regression sine curve

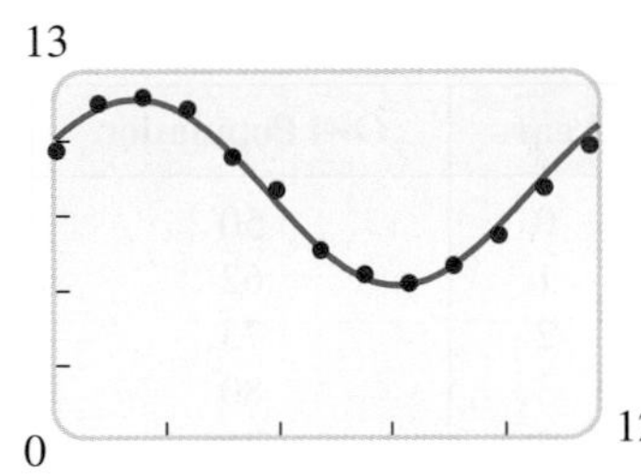

(c) Scatter plot and regression curve

■

Problems

1–2 ■ Modeling Periodic Data A set of data is given.

(a) Find a cosine function of the form $y = a\cos(\omega(t - c)) + b$ that models the data (as in Example 1). Graph the function you found together with a scatter plot of the data.

(b) Use a graphing device to find the sine function that best fits the data, and compare it to the function you found in part (a). [Use the reduction formula $\sin u = \cos(u - \pi/2)$, as in Example 2.]

1.

t	y
0	2.1
2	1.1
4	−0.8
6	−2.1
8	−1.3
10	0.6
12	1.9
14	1.5

2.

t	y
0.0	0.56
0.5	0.45
1.0	0.29
1.5	0.13
2.0	0.05
2.5	−0.10
3.0	0.02
3.5	0.12
4.0	0.26
4.5	0.43
5.0	0.54
5.5	0.63
6.0	0.59

3. Circadian Rhythms Circadian rhythm (from the Latin *circa*—about, and *diem*—day) is the daily biological pattern by which body temperature, blood pressure, and other physiological variables change. The data in the table below show typical changes in human body temperature over a 24-hour period ($t = 0$ corresponds to midnight).

(a) Find a cosine curve that models the data (as in Example 1). Graph the function you found together with a scatter plot of the data.

(b) Use a graphing device to find the sine curve that best fits the data (as in Example 2).

Time	Body Temperature (°C)	Time	Body Temperature (°C)
0	36.8	14	37.3
2	36.7	16	37.4
4	36.6	18	37.3
6	36.7	20	37.2
8	36.8	22	37.0
10	37.0	24	36.8
12	37.2		

4. Predator Population When two species interact in a predator-prey relationship, the populations of both species tend to vary in a sinusoidal fashion. (See *Discovery Project: Predator-Prey Models* referenced in Section 5.3). In a certain midwestern county, the main food source for barn owls consists of field mice and other small mammals. The table gives the population of barn owls in this county every July 1 over a 12-year period.

(a) Find a sine curve that models the data (as in Example 1). Graph the function you found together with a scatter plot of the data.

(b) Use a graphing device to find the sine curve that best fits the data (as in Example 2). Compare to your answer from part (a).

Year	Owl Population
0	50
1	62
2	73
3	80
4	71
5	60
6	51
7	43
8	29
9	20
10	28
11	41
12	49

5. Salmon Survival For reasons that are not yet fully understood, the number of fingerling salmon that survive the trip from their riverbed spawning grounds to the open ocean varies approximately sinusoidally from year to year. The table shows the number of salmon that hatch in a certain British Columbia creek and make their way to the Strait of Georgia. The data are given in thousands of fingerlings, over a period of 16 years.

(a) Find a sine curve that models the data (as in Example 1). Graph the function you found together with a scatter plot of the data.

(b) Use a graphing device to find the sine curve that best fits the data (as in Example 2). Compare to your answer from part (a).

Year	Salmon (× 1000)	Year	Salmon (× 1000)
1985	43	1993	56
1986	36	1994	63
1987	27	1995	57
1988	23	1996	50
1989	26	1997	44
1990	33	1998	38
1991	43	1999	30
1992	50	2000	22

Daniel Collier/Dreamstime.com

6 Trigonometric Functions: Right Triangle Approach

Suppose we want to find the distance from the earth to the sun. Using a tape measure is obviously impractical, so we need something other than simple measurements to tackle this problem. Angles are easier to measure than distances. For example, we can find the angle formed by the sun, earth, and moon by simply pointing to the sun with one arm and to the moon with the other and estimating the angle between them. If we had a way of determining distances from angles, we would be able to find the distance to the sun without having to go there. The trigonometric functions that we study in this chapter provide us with just the tools we need.

The trigonometric functions can be defined in two different but equivalent ways: as functions of real numbers (Chapter 5) or as functions of angles (Chapter 6). The two approaches are independent of each other, so either Chapter 5 or Chapter 6 may be studied first. We study both approaches because the different approaches are required for different applications.

6.1 Angle Measure

■ Angle Measure ■ Angles in Standard Position ■ Length of a Circular Arc
■ Area of a Circular Sector ■ Circular Motion

An **angle** AOB consists of two rays R_1 and R_2 with a common vertex O (see Figure 1). We often interpret an angle as a rotation of the ray R_1 onto R_2. In this case R_1 is called the **initial side**, and R_2 is called the **terminal side** of the angle. If the rotation is counterclockwise, the angle is considered **positive**, and if the rotation is clockwise, the angle is considered **negative**.

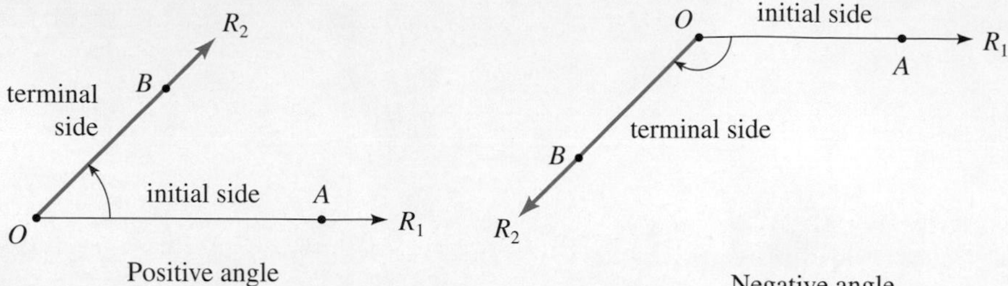

Figure 1

To work with angles we need to give angles names. We usually denote angles by Greek letters, such as α (alpha), β (beta), and θ (theta), or by the uppercase letters A, B, and C.

■ Angle Measure

The **measure** of an angle is the amount of rotation about the vertex required to move R_1 onto R_2. Intuitively, this is how much the angle "opens." One unit of measurement for angles is the **degree**. An angle of measure 1 degree is formed by rotating the initial side $\frac{1}{360}$ of a complete revolution. In calculus and other branches of mathematics a more natural method of measuring angles is used: *radian measure*. The amount an angle opens is measured along the arc of a circle of radius 1 with its center at the vertex of the angle (see Figure 2).

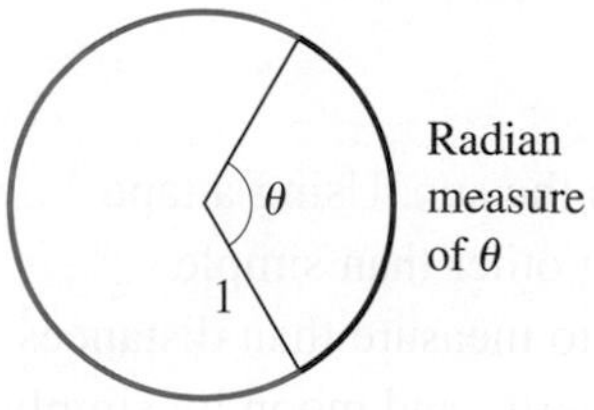

Radian measure of θ

Figure 2

> **Definition of Radian Measure**
>
> If a circle of radius 1 is drawn with the vertex of an angle θ at its center, then the measure of θ in **radians** (abbreviated **rad**) is the length of the arc that subtends the angle.

The circumference of the circle of radius 1 is 2π, so a complete revolution has measure 2π rad, a straight angle has measure π rad, and a right angle has measure $\pi/2$ rad. An angle that is subtended by an arc of length 2 along the unit circle has radian measure 2 (see Figure 3).

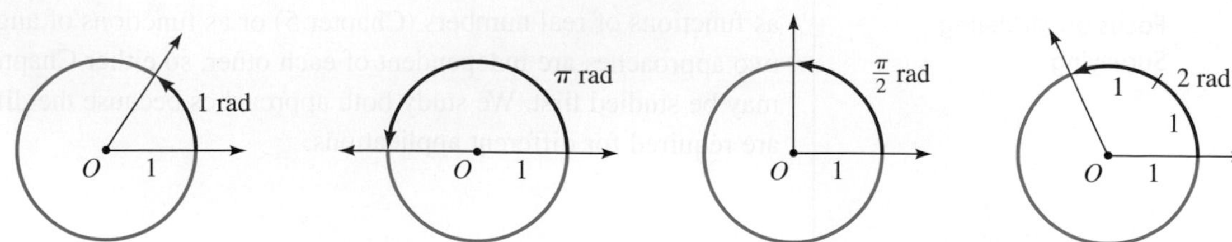

Figure 3 | Radian measure

Because a complete revolution measured in degrees is 360° and measured in radians is 2π rad, we get the following relationship between these two methods of angle measurement.

Relationship Between Degrees and Radians

$$180° = \pi \text{ rad} \qquad 1 \text{ rad} = \left(\frac{180}{\pi}\right)^{\circ} \qquad 1° = \frac{\pi}{180} \text{ rad}$$

1. To convert degrees to radians, multiply by $\dfrac{\pi}{180}$.
2. To convert radians to degrees, multiply by $\dfrac{180}{\pi}$.

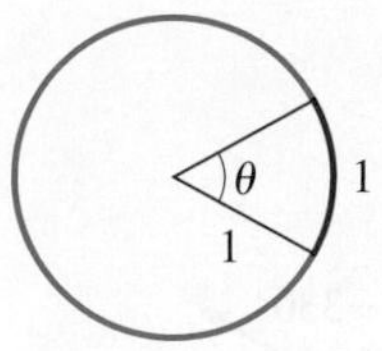

Figure 4 | Measure of $\theta = 1$ rad
Measure of $\theta \approx 57.296°$

To get some idea of the size of a radian, notice that

$$1 \text{ rad} \approx 57.296° \qquad \text{and} \qquad 1° \approx 0.01745 \text{ rad}$$

An angle θ of measure 1 rad is shown in Figure 4.

Example 1 ■ Converting Between Radians and Degrees

(a) Express 60° in radians. **(b)** Express $\dfrac{\pi}{6}$ rad in degrees.

Solution The relationship between degrees and radians gives

(a) $60° = 60\left(\dfrac{\pi}{180}\right) \text{ rad} = \dfrac{\pi}{3} \text{ rad}$ **(b)** $\dfrac{\pi}{6} \text{ rad} = \left(\dfrac{\pi}{6}\right)\left(\dfrac{180}{\pi}\right) = 30°$

Now Try Exercises 5 and 17 ■

Note We often use a phrase such as "a 30° angle" to mean *an angle whose measure is* 30°. Also, for an angle θ we write $\theta = 30°$ or $\theta = \pi/6$ to mean *the measure of* θ *is* 30° *or* $\pi/6$ *rad*. When no unit is given, the angle is assumed to be measured in radians.

■ Angles in Standard Position

An angle is in **standard position** when it is drawn in the xy-plane with its vertex at the origin and its initial side on the positive x-axis. Figure 5 gives examples of angles in standard position.

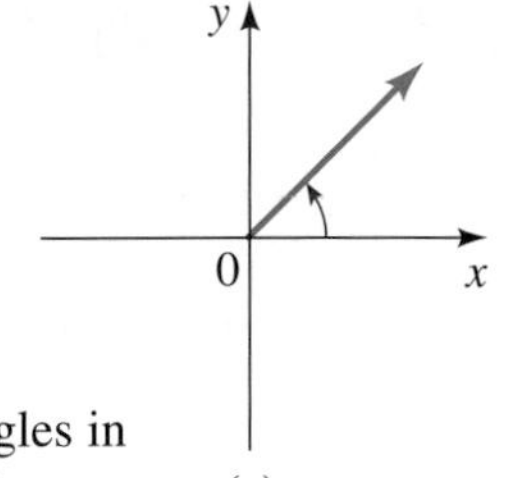
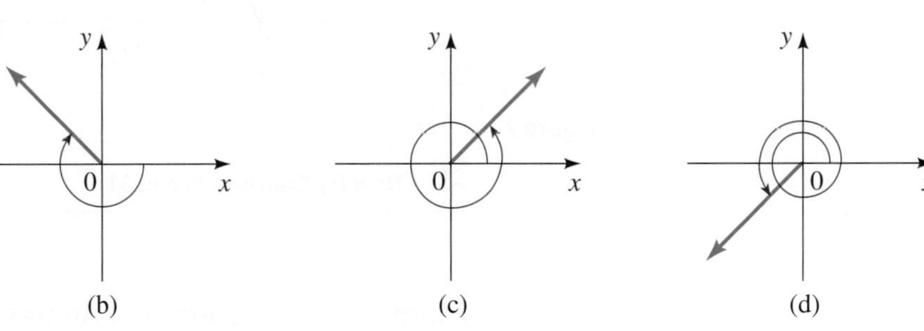

Figure 5 | Angles in standard position

Two angles in standard position are **coterminal** if their terminal sides coincide. In Figure 5 the angles in (a) and (c) are coterminal.

Example 2 ■ Coterminal Angles

(a) Find angles that are coterminal with the angle $\theta = 30°$ in standard position.

(b) Find angles that are coterminal with the angle $\theta = \dfrac{\pi}{3}$ in standard position.

Solution

(a) To find positive angles that are coterminal with θ, we add any multiple of 360°. Thus

$$30° + 360° = 390° \qquad \text{and} \qquad 30° + 720° = 750°$$

are coterminal with $\theta = 30°$. To find negative angles that are coterminal with θ, we subtract any multiple of 360°. Thus

$$30° - 360° = -330° \qquad \text{and} \qquad 30° - 720° = -690°$$

are coterminal with θ. (See Figure 6.)

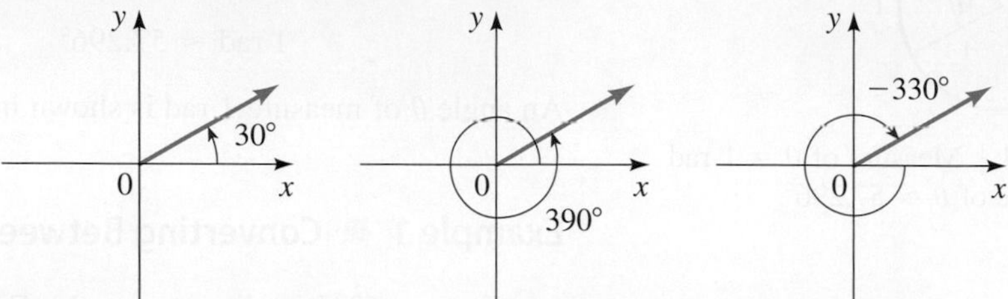

Figure 6

(b) To find positive angles that are coterminal with θ, we add any multiple of 2π. Thus

$$\frac{\pi}{3} + 2\pi = \frac{7\pi}{3} \qquad \text{and} \qquad \frac{\pi}{3} + 4\pi = \frac{13\pi}{3}$$

are coterminal with $\theta = \pi/3$. To find negative angles that are coterminal with θ, we subtract any multiple of 2π. Thus

$$\frac{\pi}{3} - 2\pi = -\frac{5\pi}{3} \qquad \text{and} \qquad \frac{\pi}{3} - 4\pi = -\frac{11\pi}{3}$$

are coterminal with θ. (See Figure 7.)

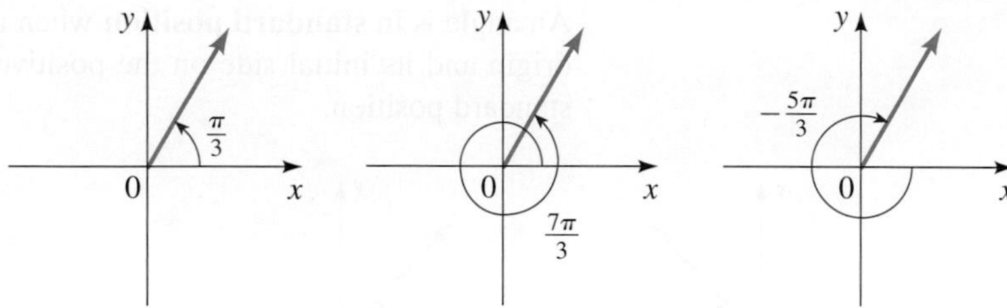

Figure 7

Now Try Exercises 29 and 31 ■

Example 3 ■ Coterminal Angles

Find an angle with measure between 0° and 360° that is coterminal with the angle of measure 1290° in standard position.

Solution We can subtract 360° as many times as we wish from 1290°, and the resulting angle will be coterminal with 1290°. Thus $1290° - 360° = 930°$ is coterminal with 1290°, and so is the angle $1290° - 2(360)° = 570°$.

To find the angle we want between 0° and 360°, we subtract 360° from 1290° as many times as necessary. An efficient way to do this is to determine how many times 360° goes into 1290°, that is, divide 1290 by 360, and the remainder will be the angle we are looking for. We see that 360 goes into 1290 three times with a remainder of 210. Thus 210° is the desired angle (see Figure 8).

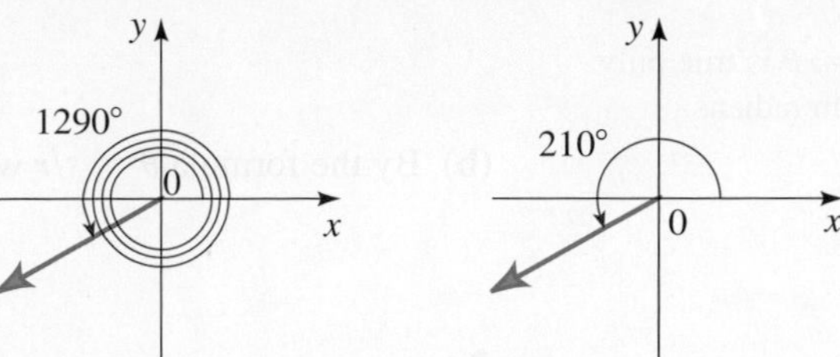

Figure 8

Now Try Exercise 41 ■

■ Length of a Circular Arc

An angle whose radian measure is θ is subtended by an arc that is the fraction $\theta/(2\pi)$ of the circumference of a circle. Thus in a circle of radius r the length s of an arc that subtends the angle θ (see Figure 9) is

$$s = \frac{\theta}{2\pi} \times \text{circumference of circle}$$

$$= \frac{\theta}{2\pi}(2\pi r) = \theta r$$

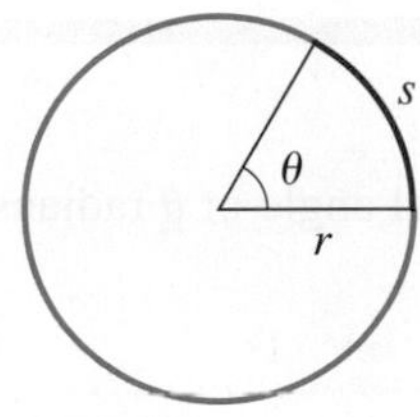

Figure 9 | $s = \theta r$

Length of a Circular Arc

In a circle of radius r the length s of an arc that subtends a central angle of θ radians is

$$s = r\theta$$

Solving for θ, we get the important formula

$$\theta = \frac{s}{r}$$

This formula allows us to define radian measure using a circle of any given radius r: The radian measure of an angle θ is s/r, where s is the length of the circular arc that subtends θ in a circle of radius r. (See Figure 10.)

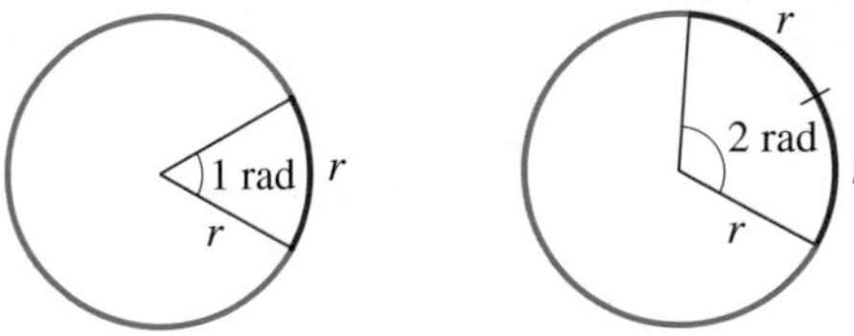

Figure 10 | The radian measure of θ is the number of "radiuses" that can fit in the arc that subtends θ; hence the term *radian*.

Example 4 ■ Arc Length and Angle Measure

(a) Find the length of an arc of a circle with radius 10 m that subtends a central angle of 30°.

(b) A central angle θ in a circle of radius 4 m is subtended by an arc of length 6 m. Find the measure of θ in radians.

Solution

(a) From Example 1(b) we see that $30° = \pi/6$ rad. So the length of the arc is

The formula $s = r\theta$ is true only when θ is measured in radians.

$$s = r\theta = (10)\frac{\pi}{6} = \frac{5\pi}{3} \text{ m}$$

(b) By the formula $\theta = s/r$ we have

$$\theta = \frac{s}{r} = \frac{6}{4} = \frac{3}{2} \text{ rad}$$

Now Try Exercises 57 and 59 ■

■ Area of a Circular Sector

The area of a circle of radius r is $A = \pi r^2$. A sector of this circle with central angle θ has an area that is the fraction $\theta/(2\pi)$ of the area of the entire circle (see Figure 11). So the area of this sector is

$$A = \frac{\theta}{2\pi} \times \text{area of circle}$$

$$= \frac{\theta}{2\pi}(\pi r^2) = \frac{1}{2}r^2\theta$$

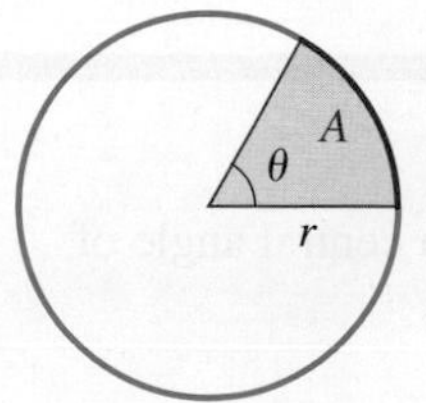

Figure 11 | $A = \frac{1}{2}r^2\theta$

Area of a Circular Sector

In a circle of radius r the area A of a sector with a central angle of θ radians is

$$A = \frac{1}{2}r^2\theta$$

Example 5 ■ Area of a Sector

Find the area of a sector of a circle with central angle 60° if the radius of the circle is 3 m.

Solution To use the formula for the area of a circular sector, we must find the central angle of the sector in radians: $60° = 60(\pi/180)$ rad $= \pi/3$ rad. Thus the area of the sector is

The formula $A = \frac{1}{2}r^2\theta$ is true only when θ is measured in radians.

$$A = \frac{1}{2}r^2\theta = \frac{1}{2}(3)^2\left(\frac{\pi}{3}\right) = \frac{3\pi}{2} \text{ m}^2$$

Now Try Exercise 63 ■

■ Circular Motion

Suppose a point moves along a circle as shown in Figure 12. There are two ways to describe the motion of the point: linear speed and angular speed. **Linear speed** is the rate at which the distance traveled is changing, so linear speed is the distance traveled divided by the time elapsed. **Angular speed** is the rate at which the central angle θ is changing, so angular speed is the number of radians this angle changes divided by the time elapsed.

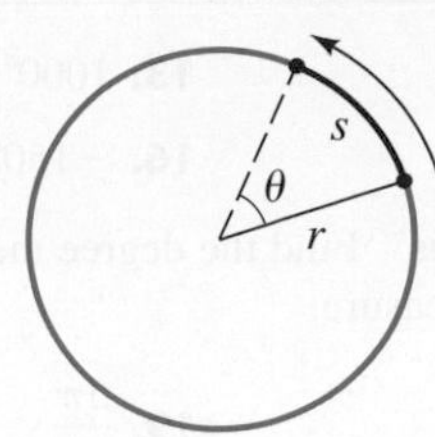

Figure 12

Linear Speed and Angular Speed

Suppose a point moves along a circle of radius r and the ray from the center of the circle to the point traverses θ radians in time t. Let $s = r\theta$ be the distance the point travels in time t. Then the angular and linear speeds of the object are given by

Angular speed $\qquad \omega = \dfrac{\theta}{t}$

Linear speed $\qquad v = \dfrac{s}{t}$

(The symbol ω is the Greek letter "omega.")

Example 6 ■ Finding Linear and Angular Speed

A child rotates a stone in a 1-m-long sling at the rate of 15 revolutions every 10 seconds. Find the angular and linear speeds of the stone.

Solution In 10 s the angle θ changes by $15 \cdot 2\pi = 30\pi$ rad. So the *angular speed* of the stone is

$$\omega = \frac{\theta}{t} = \frac{30\pi \text{ rad}}{10 \text{ s}} = 3\pi \text{ rad/s}$$

The distance traveled by the stone in 10 s is $s = 15 \cdot 2\pi r = 15 \cdot 2\pi \cdot 1 = 30\pi$ m. So the *linear speed* of the stone is

$$v = \frac{s}{t} = \frac{30\pi \text{ m}}{10 \text{ s}} = 3\pi \text{ m/s}$$

 **Now Try Exercise 85** ■

Notice that angular speed does *not* depend on the radius of the circle; it depends only on the angle θ. However, if we know the angular speed ω and the radius r, we can find linear speed as follows: $v = s/t = r\theta/t = r(\theta/t) = r\omega$.

Relationship Between Linear and Angular Speed

If a point moves along a circle of radius r with angular speed ω, then its linear speed v is given by

$$v = r\omega$$

Example 7 ■ Finding Linear Speed from Angular Speed

A cyclist is riding a bicycle with wheels of diameter 65 cm. If the wheels rotate at 125 revolutions per minute (rpm), find the speed (in km/h) at which the cyclist is traveling.

Solution The angular speed of the wheels is $2\pi \cdot 125 = 250\pi$ rad/min. Since the wheels have radius 32.5 cm. (half the diameter), the linear speed is

$$v = r\omega = 32.5 \cdot 250\pi \approx 25{,}525.4 \text{ cm/min}$$

Since there are 100 centimeters per meter, 1000 meters per kilometer, and 60 minutes per hour, the cyclist's speed in kilometers per hour is

In Example 7 we made a conversion to arrive at the final answer in km/h. For an explanation of how to convert units, see the Discovery Project *Real Numbers in the Real World* at **www.stewartmath.com**.

$$\frac{25{,}525.4 \text{ cm/min} \times 60 \text{ min/h}}{100 \text{ cm/m} \times 1000 \text{ m/km}} = \frac{1{,}531{,}526 \text{ cm/h}}{100{,}000 \text{ cm/km}}$$

$$\approx 15.5 \text{ km/h}$$

Now Try Exercise 87 ■

6.1 Exercises

Concepts

1. (a) The radian measure of an angle θ is the length of the ________ that subtends the angle in a circle of radius ________.
   (b) To convert degrees to radians, we multiply by ________.
   (c) To convert radians to degrees, we multiply by ________.
   (d) What does it mean for an angle to be in standard position? Sketch the following angles in standard position: $2\pi/3$, $-\pi/4$, $400°$, $210°$.

2. A central angle θ is drawn in a circle of radius r, as in the figure below.
   (a) The length of the arc subtended by θ is $s =$ ________.
   (b) The area of the sector with central angle θ is $A =$ ________.

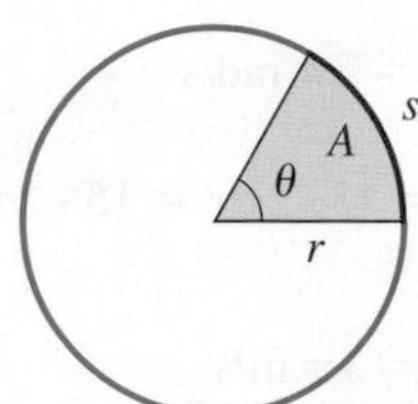

3. Suppose a point moves along a circle with radius r as shown in the figure below. The point travels a distance s along the circle in time t.
   (a) The angular speed of the point is $\omega =$ ________.
   (b) The linear speed of the point is $v =$ ________.
   (c) The linear speed v and the angular speed ω are related by the equation $v =$ ________.

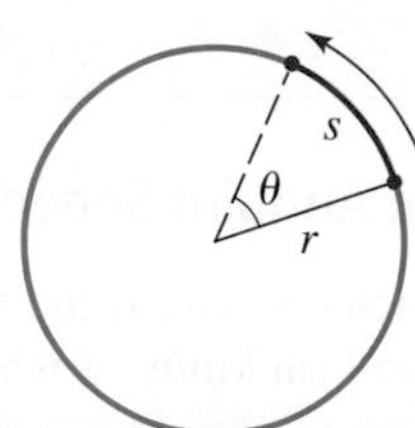

4. Object A is traveling along a circle of radius 2, and Object B is traveling along a circle of radius 5. The objects have the same angular speed. Do the objects have the same linear speed? If not, which object has the greater linear speed?

Skills

5–16 ■ From Degrees to Radians Find the radian measure of the angle with the given degree measure. Round your answer to three decimal places.

5. 20°	**6.** 40°	**7.** 54°
8. 75°	**9.** −45°	**10.** −30°
11. 100°	**12.** 200°	**13.** 1000°
14. 3600°	**15.** −70°	**16.** −150°

17–28 ■ From Radians to Degrees Find the degree measure of the angle with the given radian measure.

17. $\frac{7\pi}{6}$	**18.** $\frac{4\pi}{3}$	**19.** $\frac{5\pi}{6}$
20. $-\frac{3\pi}{2}$	**21.** 3	**22.** −2
23. −3.5	**24.** 1.8	**25.** $\frac{\pi}{10}$
26. $\frac{5\pi}{18}$	**27.** $-\frac{2\pi}{15}$	**28.** $-\frac{13\pi}{12}$

29–34 ■ Coterminal Angles The measure of an angle in standard position is given. Find two positive angles and two negative angles that are coterminal with the given angle.

29. 50°	**30.** 135°	**31.** $\frac{3\pi}{4}$
32. $\frac{11\pi}{6}$	**33.** $-\frac{\pi}{4}$	**34.** −45°

35–40 ■ Coterminal Angles? The measures of two angles in standard position are given. Determine whether the angles are coterminal.

35. 70°, 430°	**36.** −30°, 330°	**37.** $\frac{5\pi}{6}, \frac{17\pi}{6}$
38. $\frac{32\pi}{3}, \frac{11\pi}{3}$	**39.** 155°, 875°	**40.** 50°, 340°

41–46 ■ Finding a Coterminal Angle Find an angle between 0° and 360° that is coterminal with the given angle.

41. 400°	**42.** 375°	**43.** 780°
44. −100°	**45.** −800°	**46.** 1270°

47–52 ■ Finding a Coterminal Angle Find an angle between 0 and 2π that is coterminal with the given angle.

47. $\frac{19\pi}{6}$	**48.** $-\frac{5\pi}{3}$	**49.** 25π
50. 10	**51.** $\frac{17\pi}{4}$	**52.** $\frac{51\pi}{2}$

53–62 ■ Circular Arcs Find the length s of the circular arc, the radius r of the circle, or the central angle θ, as indicated.

53.

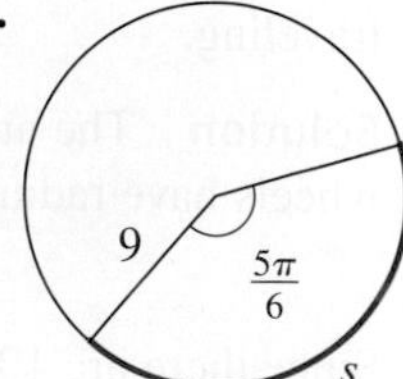

54.

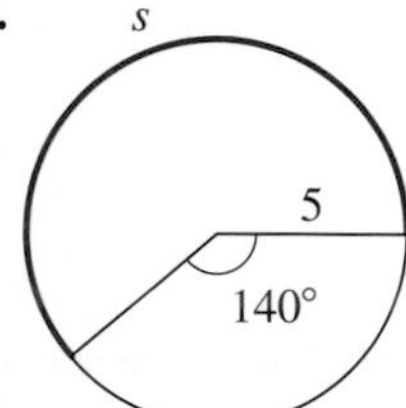

55.

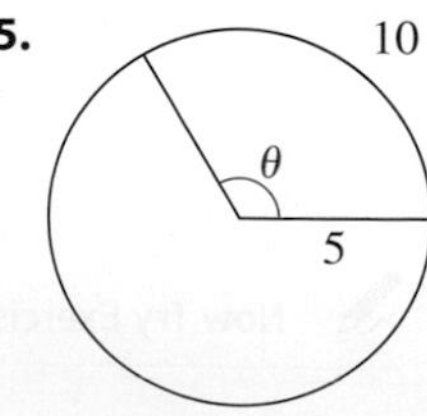

56.

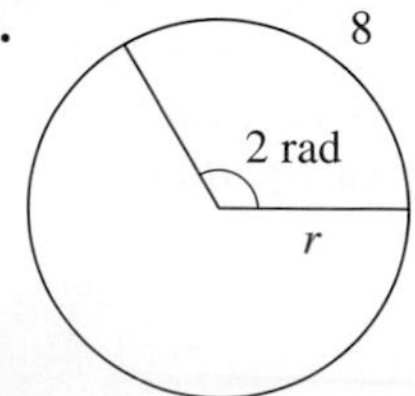

57. Find the length s of the arc that subtends a central angle of measure 2 rad in a circle of radius 4 cm.

58. Find the length s of the arc that subtends a central angle of measure 40° in a circle of radius 12 m.

59. A central angle θ in a circle of radius 9 m is subtended by an arc of length 14 m. Find the measure of θ in degrees and radians.

60. An arc of length 5 m subtends a central angle θ in a circle of radius 3 m. Find the measure of θ in degrees and radians.

61. Find the radius r of the circle if an arc of length 15 m on the circle subtends a central angle of $5\pi/6$.

62. Find the radius r of the circle if an arc of length 20 cm on the circle subtends a central angle of 50°.

63–70 ■ Area of a Circular Sector These exercises involve the formula for the area of a circular sector.

63. Find the area of the sector shown in each figure.

(a)

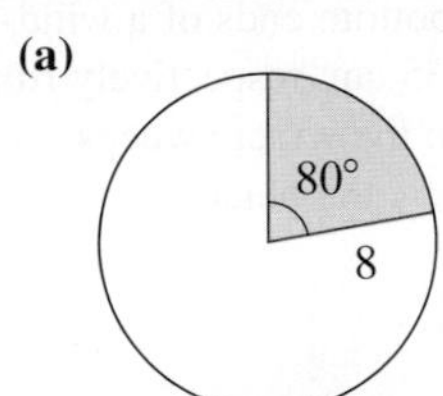

(b)

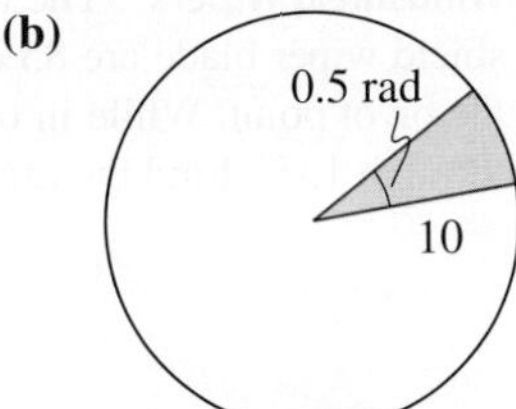

64. Find the radius of each circle if the area of the sector is 12.

(a)

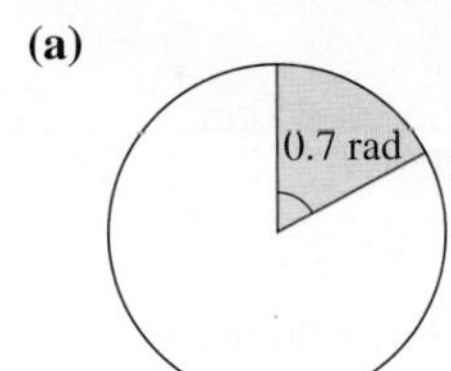

(b)

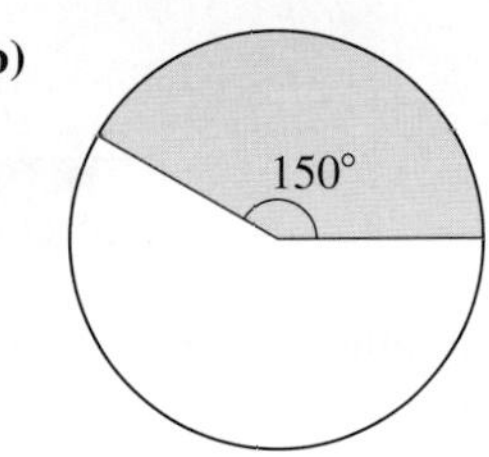

65. Find the area of a sector with central angle $3\pi/4$ rad in a circle of radius 8 m.

66. A sector of a circle has a central angle of 145°. Find the area of the sector if the radius of the circle is 2 m.

67. The area of a sector of a circle with a central angle of 160° is 90 m². Find the radius of the circle.

68. The area of a sector of a circle with a central angle of $5\pi/12$ rad is 20 m². Find the radius of the circle.

69. A sector of a circle of radius 80 km has an area of 1600 km². Find the central angle (in radians) of the sector.

70. The area of a circle is 600 m². Find the area of a sector of this circle that subtends a central angle of 3 rad.

■ Skills Plus

71. Area of a Sector of a Circle Three circles with radii 1, 2, and 3 m are externally tangent to one another, as shown in the figure. Find the area of the sector of the circle of radius 1 that is cut off by the line segments joining the center of that circle to the centers of the other two circles.

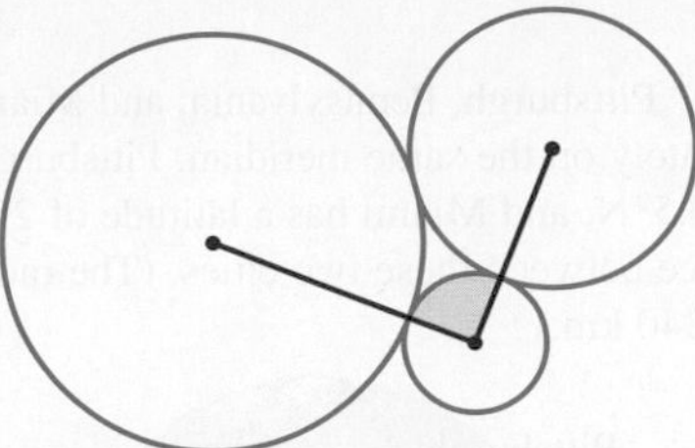

PS *Try to recognize something familiar.* What familiar property does the triangle with vertices at the centers of the circles have?

72. Comparing a Triangle and a Sector of a Circle Two wood sticks and a metal rod, each of length 1, are connected to form a triangle with angle θ_1 at the point P, as shown in the first figure. The rod is then bent to form an arc of a circle with center P, resulting in a smaller angle θ_2 at the point P, as shown in the second figure. Find θ_1, θ_2, and $\theta_1 - \theta_2$.

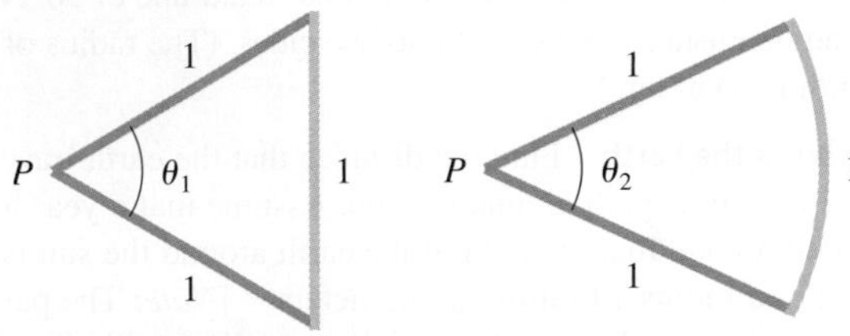

73. Clocks and Angles In one hour the minute hand on a clock moves through a complete circle, and the hour hand moves through $\frac{1}{12}$ of a circle. Through how many radians do the minute hand and the hour hand move between **(a)** 1:00 P.M. and 1:45 P.M. and **(b)** 1:00 P.M. and 6:45 P.M. (on the same day)?

74. Area of a Segment of a Circle The circle shown in the figure has radius 1 and the chord has length 1. Find the area of the shaded *segment of the circle*, that is the region formed by the chord and the circle.

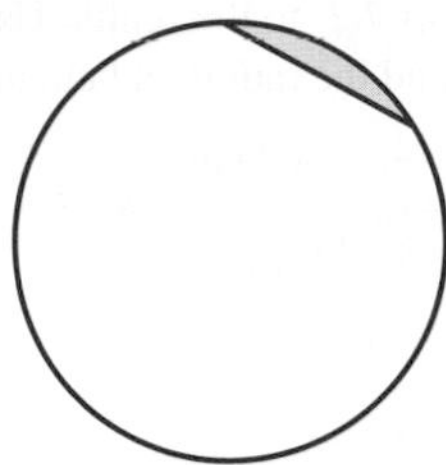

PS *Introduce something extra.* Consider adding radii of the circle to the endpoints of the chord.

■ Applications

75. Travel Distance A car's wheels are 70 cm in diameter. How far (in km) will the car travel if its wheels revolve 10,000 times without slipping?

76. Wheel Revolutions How many revolutions will a car wheel of diameter 75 cm make as the car travels a distance of one kilometer?

77. Latitudes Pittsburgh, Pennsylvania, and Miami, Florida, lie approximately on the same meridian. Pittsburgh has a latitude of 40.5° N, and Miami has a latitude of 25.5° N. Find the distance between these two cities. (The radius of the earth is 6340 km.)

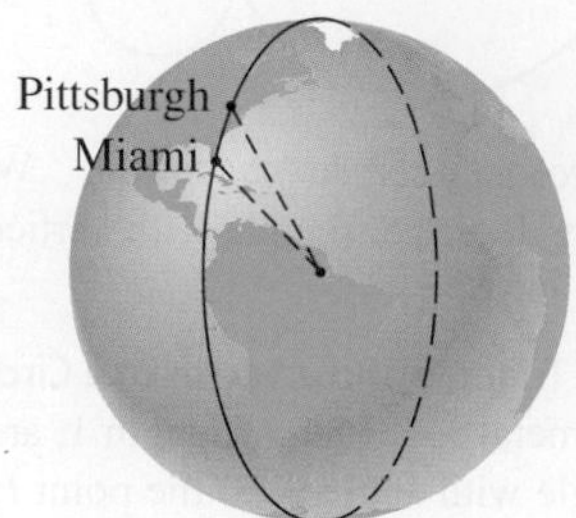

78. Latitudes Memphis, Tennessee, and New Orleans, Louisiana, lie approximately on the same meridian. Memphis has a latitude of 35° N, and New Orleans has a latitude of 30° N. Find the distance between these two cities. (The radius of the earth is 6340 km.)

79. Orbit of the Earth Find the distance that the earth travels in one day in its path around the sun. Assume that a year has 365 days and that the path of the earth around the sun is a circle of radius 149 million kilometers. [*Note:* The path of the earth around the sun is actually an *ellipse* with the sun at one focus (see Section 10.2). This ellipse, however, has very small eccentricity, so it is nearly circular.]

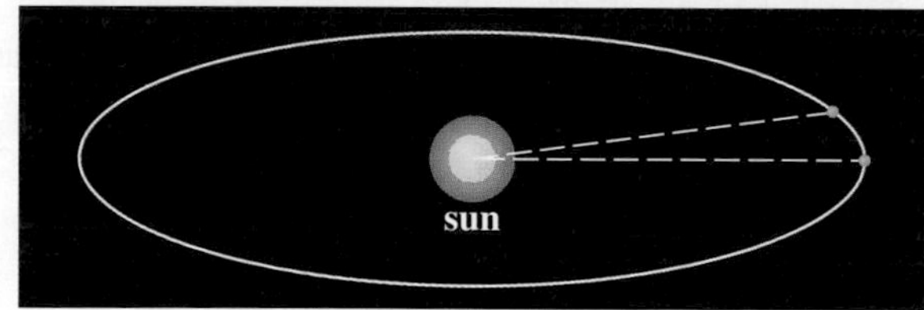

80. Circumference of the Earth The Greek mathematician Eratosthenes (circa 276–195 B.C.) measured the circumference of the earth from the following observations. He noticed that on a certain day the sun shone directly down a deep well in Syene (modern Aswan). At the same time in Alexandria, 800 km to the north (on the same meridian), the rays of the sun shone at an angle of 7.2° to the zenith. Use this information and the figure to find the radius and circumference of the earth.

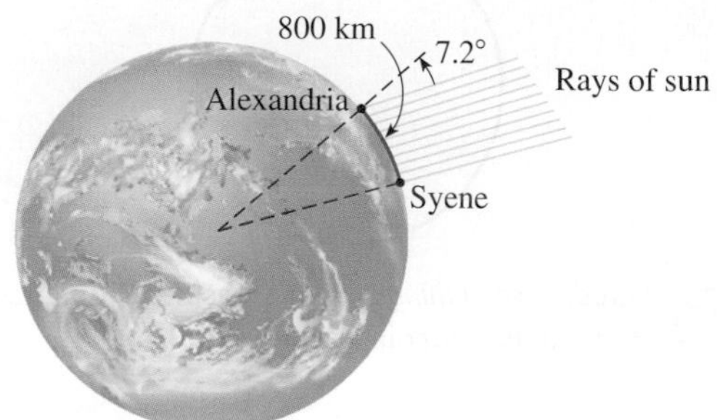

81. Nautical Miles Find the distance along an arc on the surface of the earth that subtends a central angle of 1 minute (1 minute $= \frac{1}{60}$ degree). This distance is called a *nautical mile*. (The radius of the earth is 6340 km.)

82. Irrigation An irrigation system uses a straight sprinkler pipe 90 m long that pivots around a central point at a constant rate, as shown. Because of an obstacle the pipe is allowed to pivot through 280° only. **(a)** Find the area irrigated by this system. **(b)** The water flow on each sprinkler nozzle can be regulated. The nozzle at 30 m from the center is set to spray 150 liters per minute. For uniform irrigation, how many liters per minute should be produced by the nozzle at 90 m from the center?

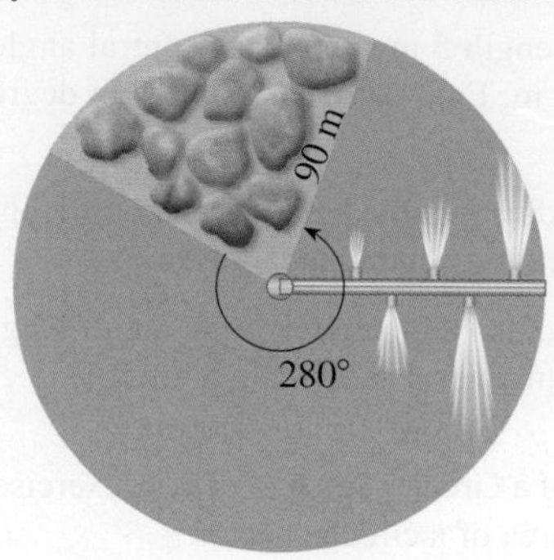

83. Windshield Wipers The top and bottom ends of a windshield wiper blade are 85 cm and 35 cm, respectively from the pivot point. While in operation, the wiper sweeps through 135°. Find the area swept by the blade.

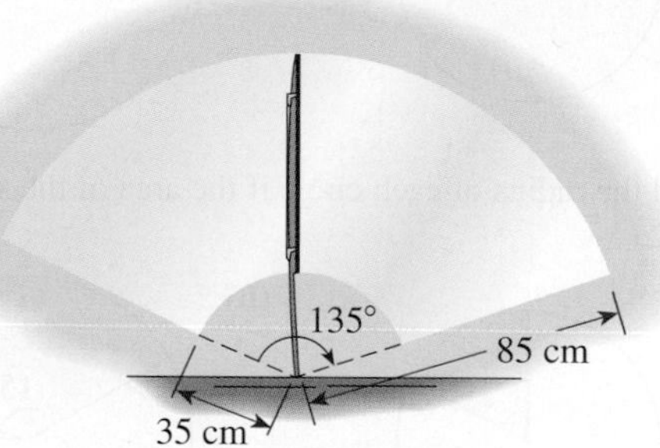

84. A Tethered Cow A cow is tethered by a 30 m rope to the inside corner of an L-shaped building, as shown in the figure. Find the area that the cow can graze.

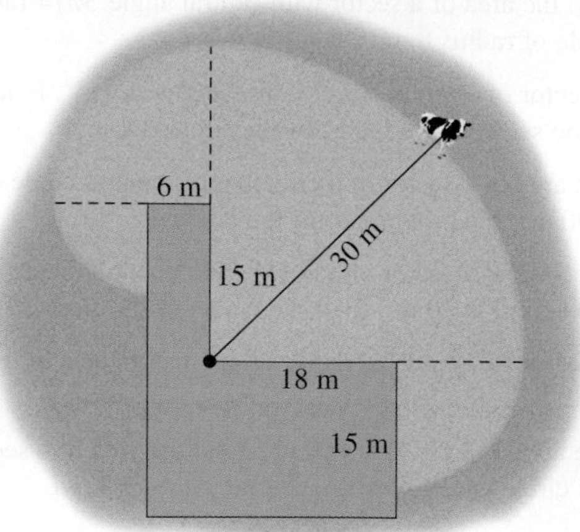

85. Fan A ceiling fan with 40 cm blades rotates at 45 rpm.

(a) Find the angular speed of the fan in rad/min.

(b) Find the linear speed of the tips of the blades in m/min.

86. Miter Saw A miter saw has a blade with a 12 cm radius. Suppose that the blade spins at 6000 rpm.

(a) Find the angular speed of the blade in rad/min.

(b) Find the linear speed of the sawteeth in m/s.

87. Winch A winch of radius 1 m is used to lift heavy loads. If the winch makes 8 revolutions every 15 s, find the speed at which the load is rising.

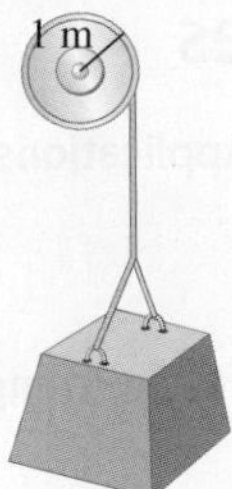

88. Speed of a Car The wheels of a car have radius 28 cm and are rotating at 600 rpm. Find the speed of the car in km/h.

89. Speed at the Equator The earth rotates about its axis once every 23 h 56 min 4 s, and the radius of the earth is 6340 km. Find the linear speed of a point on the equator in km/h.

90. Truck Wheels A truck with 120-cm-diameter wheels is traveling at 80 km/h.

(a) Find the angular speed of the wheels in rad/min.

(b) How many revolutions per minute do the wheels make?

91. Speed of a Current To measure the speed of a current, scientists place a paddle wheel in the stream and observe the rate at which it rotates. If the paddle wheel has radius 0.20 m and rotates at 100 rpm, find the speed of the current in m/s.

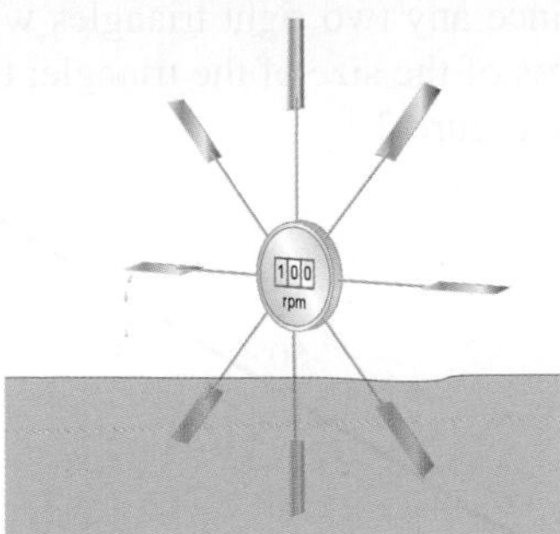

92. Bicycle Wheel The sprockets and chain of a bicycle are shown in the figure. The pedal sprocket has a radius of 10 cm, the wheel sprocket a radius of 5 cm, and the wheel a radius of 32.5 cm. The cyclist pedals at 40 rpm.

(a) Find the angular speed of the wheel sprocket.

(b) Find the linear speed of the bicycle. (Assume that the wheel turns at the same rate as the wheel sprocket.)

93. Conical Cup A conical cup is made from a circular piece of paper with radius 6 cm by cutting out a sector and joining the edges as shown in the figure. Suppose $\theta = 5\pi/3$.

(a) Find the circumference C of the opening of the cup.

(b) Find the radius r of the opening of the cup. [*Hint:* Use $C = 2\pi r$.]

(c) Find the height h of the cup. [*Hint:* Use the Pythagorean Theorem.]

(d) Find the volume of the cup.

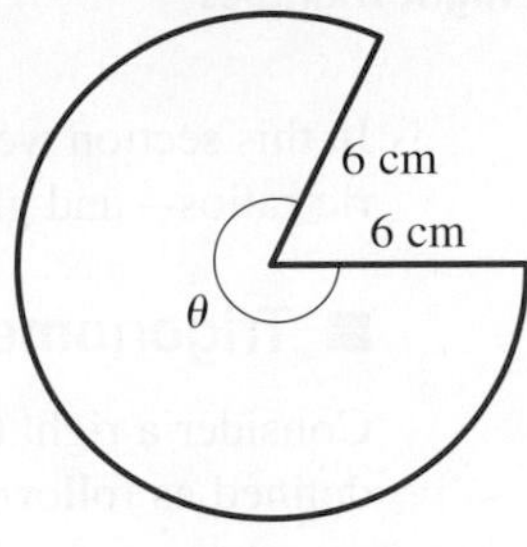

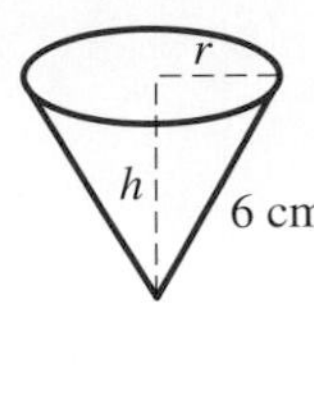

94. Conical Cup In this exercise we find the volume of the conical cup in Exercise 93 for any angle θ.

(a) Follow the steps in Exercise 93 to show that the volume of the cup as a function of θ is

$$V(\theta) = \frac{9}{\pi^2}\theta^2\sqrt{4\pi^2 - \theta^2} \qquad (0 < \theta < 2\pi)$$

(b) Graph the function V.

(c) For what angle θ is the volume of the cup a maximum?

■ Discuss ■ Discover ■ Prove ■ Write

95. Discuss ■ Discover: Regions of Constant Width A region in the plane is said to have *constant width* if any two parallel lines that "enclose" the region (see the figure) have the same distance between them. It may seem at first that the only such region is a circle, but there are others. One example is the Reuleaux "triangle" shown here. Each curved side of the Reuleaux triangle is an arc of a circle centered on one vertex of an equilateral triangle and passing through the other two vertices. Explain why the Reuleaux triangle is a region of constant width. Find the area and perimeter of a Reuleaux triangle drawn on an equilateral triangle of side length 1.

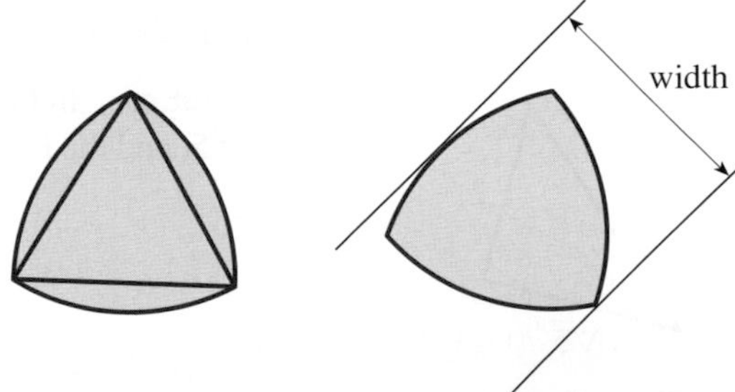

96. Write: Different Ways of Measuring Angles The custom of measuring angles using degrees, with 360° in a circle, dates back to the ancient Babylonians, who used a number system based on groups of 60. Another system of measuring angles divides the circle into 400 units, called *grads*. In this system a right angle is 100 grad, so this fits in with our base 10 number system.

Write a short essay comparing the advantages and disadvantages of these two systems and the radian system of measuring angles. Which system do you prefer? Why?

6.2 Trigonometry of Right Triangles

■ Trigonometric Ratios ■ Special Triangles; Calculators ■ Applications of Trigonometry of Right Triangles

In this section we study certain ratios of the sides of right triangles—called trigonometric ratios—and give several applications.

■ Trigonometric Ratios

Consider a right triangle with θ as one of its acute angles. The trigonometric ratios are defined as follows (see Figure 1).

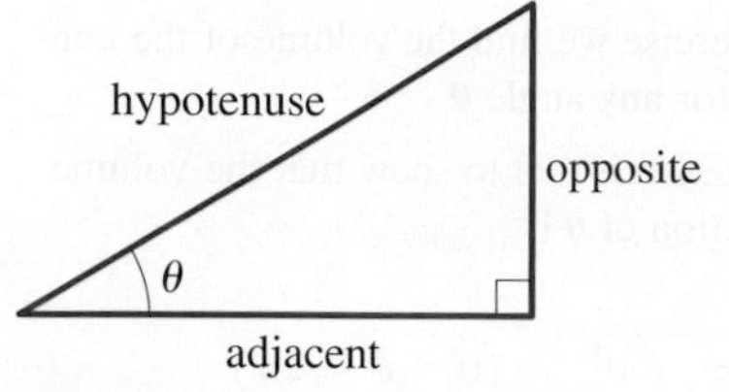

Figure 1

The Trigonometric Ratios

$$\sin\theta = \frac{\text{opposite}}{\text{hypotenuse}} \qquad \cos\theta = \frac{\text{adjacent}}{\text{hypotenuse}} \qquad \tan\theta = \frac{\text{opposite}}{\text{adjacent}}$$

$$\csc\theta = \frac{\text{hypotenuse}}{\text{opposite}} \qquad \sec\theta = \frac{\text{hypotenuse}}{\text{adjacent}} \qquad \cot\theta = \frac{\text{adjacent}}{\text{opposite}}$$

The symbols we use for these ratios are abbreviations for their full names: **sine**, **cosine**, **tangent**, **cosecant**, **secant**, **cotangent**. Since any two right triangles with angle θ are similar, these ratios are the same, regardless of the size of the triangle; the trigonometric ratios depend only on the angle θ. (See Figure 2.)

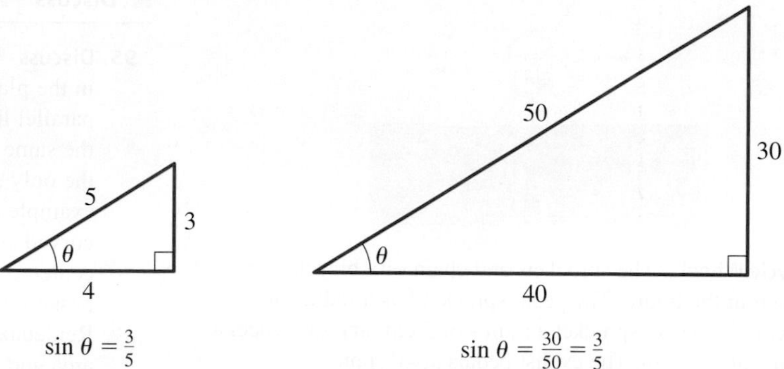

Figure 2

Example 1 ■ Finding Trigonometric Ratios

Find the six trigonometric ratios of the angle θ in Figure 3.

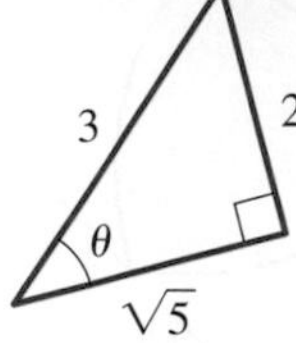

Figure 3

Solution In Figure 3 the side opposite θ has length 2, the side adjacent θ has length $\sqrt{5}$, and the hypotenuse has length 3. So, by the definition of trigonometric ratios, we get

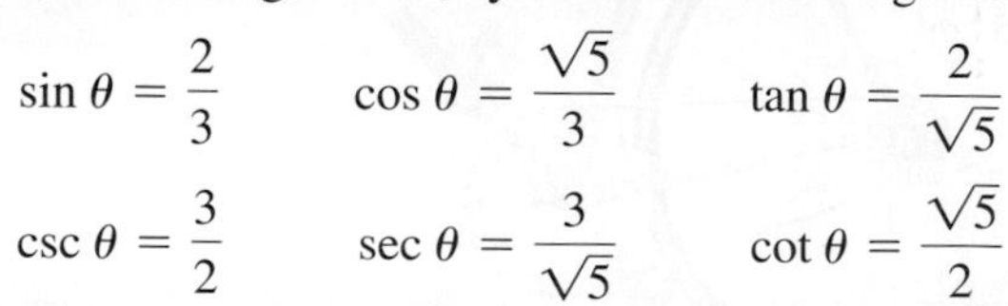

$$\sin\theta = \frac{2}{3} \qquad \cos\theta = \frac{\sqrt{5}}{3} \qquad \tan\theta = \frac{2}{\sqrt{5}}$$

$$\csc\theta = \frac{3}{2} \qquad \sec\theta = \frac{3}{\sqrt{5}} \qquad \cot\theta = \frac{\sqrt{5}}{2}$$

Now Try Exercise 5 ■

Example 2 ■ Finding Trigonometric Ratios

If $\cos\alpha = \frac{3}{4}$, sketch a right triangle with acute angle α, and find the other five trigonometric ratios of α.

Solution Since $\cos\alpha$ is defined as the ratio of the adjacent side to the hypotenuse, we sketch a triangle with hypotenuse of length 4 and a side of length 3 adjacent to α. If the opposite side is x, then by the Pythagorean Theorem, $3^2 + x^2 = 4^2$ or $x^2 = 7$, so $x = \sqrt{7}$. We then use the triangle in Figure 4 to find the ratios.

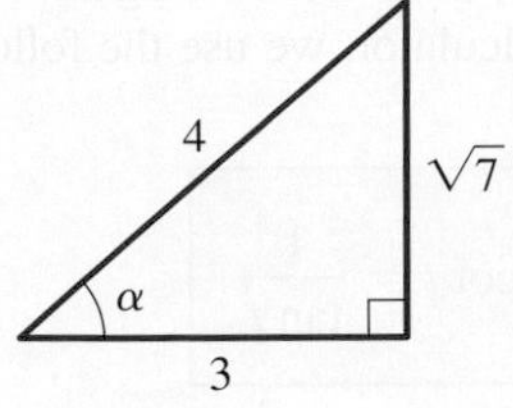

Figure 4

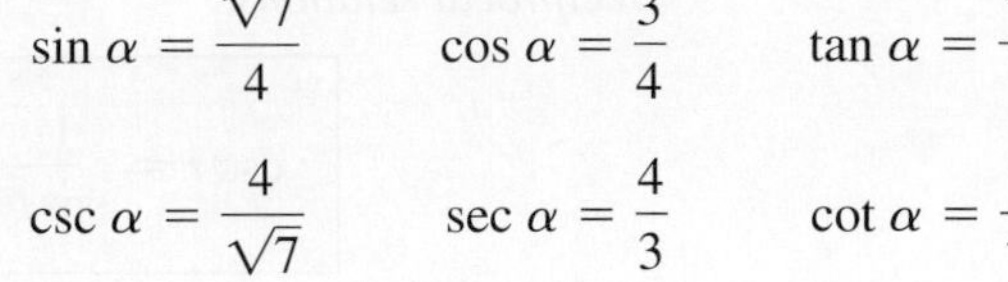

$$\sin\alpha = \frac{\sqrt{7}}{4} \qquad \cos\alpha = \frac{3}{4} \qquad \tan\alpha = \frac{\sqrt{7}}{3}$$

$$\csc\alpha = \frac{4}{\sqrt{7}} \qquad \sec\alpha = \frac{4}{3} \qquad \cot\alpha = \frac{3}{\sqrt{7}}$$

Now Try Exercise 25 ■

■ Special Triangles; Calculators

There are special trigonometric ratios that can be calculated from certain triangles (which we call special triangles). We can also use a calculator to find trigonometric ratios.

Special Ratios Certain right triangles have ratios that can be calculated from the Pythagorean Theorem. Since they are used frequently, we mention them here.

The first triangle is obtained by drawing a diagonal in a square of side 1 (see Figure 5). By the Pythagorean Theorem this diagonal has length $\sqrt{2}$. The resulting triangle has angles 45°, 45°, and 90° (or $\pi/4$, $\pi/4$, and $\pi/2$). To get the second triangle, we start with an equilateral triangle ABC of side 2 and draw the perpendicular bisector DB of the base, as in Figure 6. By the Pythagorean Theorem the length of DB is $\sqrt{3}$. Since DB bisects angle ABC, we obtain a triangle with angles 30°, 60°, and 90° (or $\pi/6$, $\pi/3$, and $\pi/2$).

HIPPARCHUS (circa 190–120 B.C.) is considered the founder of trigonometry. He constructed tables of values for a function closely related to the modern sine function and evaluated for angles at half-degree intervals. These are considered the first trigonometric tables. He used his tables mainly to calculate the paths of the planets through the heavens.

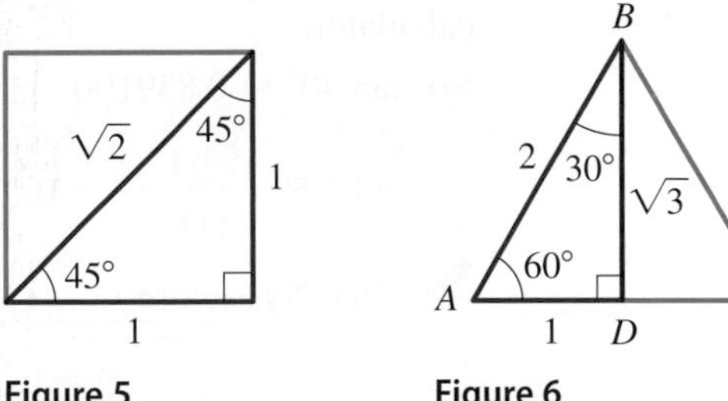

Figure 5 Figure 6

We can now use the special triangles in Figures 5 and 6 to calculate the trigonometric ratios for angles with measures 30°, 45°, and 60° (or $\pi/6$, $\pi/4$, and $\pi/3$). These are listed in the table below.

Special Values of the Trigonometric Ratios

The table gives the values of the trigonometric ratios for the angles of the special triangles.

Table 1

θ in Degrees	θ in Radians	$\sin\theta$	$\cos\theta$	$\tan\theta$	$\csc\theta$	$\sec\theta$	$\cot\theta$
30°	$\frac{\pi}{6}$	$\frac{1}{2}$	$\frac{\sqrt{3}}{2}$	$\frac{\sqrt{3}}{3}$	2	$\frac{2\sqrt{3}}{3}$	$\sqrt{3}$
45°	$\frac{\pi}{4}$	$\frac{\sqrt{2}}{2}$	$\frac{\sqrt{2}}{2}$	1	$\sqrt{2}$	$\sqrt{2}$	1
60°	$\frac{\pi}{3}$	$\frac{\sqrt{3}}{2}$	$\frac{1}{2}$	$\sqrt{3}$	$\frac{2\sqrt{3}}{3}$	2	$\frac{\sqrt{3}}{3}$

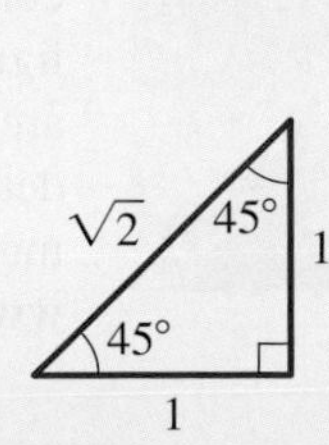

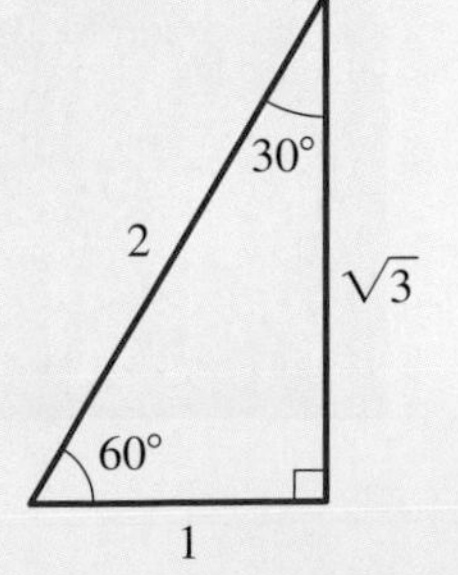

It's useful to remember these special trigonometric ratios because they occur often. Of course, they can be recalled if we remember the triangles from which they are obtained.

Using a Calculator To find the values of the trigonometric ratios for other angles, we use a calculator. Mathematical methods (called *numerical methods*) used in finding the trigonometric ratios are programmed directly into scientific calculators. For instance, when the SIN key is pressed, the calculator computes an approximation to the value of the sine of the given angle. Some calculators give only the values of sine, cosine, and tangent. To find the values of cosecant, secant, and cotangent on such a calculator, we use the following *reciprocal relations*:

For an explanation of numerical methods, see *Mathematics in the Modern World* in Section 5.4.

$$\csc t = \frac{1}{\sin t} \qquad \sec t = \frac{1}{\cos t} \qquad \cot t = \frac{1}{\tan t}$$

You should check that these reciprocal relations follow from the definitions of the trigonometric ratios.

We follow the convention that when we write $\sin t$, *we mean the sine of the angle whose radian measure is* t. For instance, $\sin 1$ means the sine of the angle whose radian measure is 1. When using a calculator to find an approximate value for this number, set your calculator to radian mode; you will find that $\sin 1 \approx 0.841471$. If you want to find the sine of the angle whose measure is 1°, set your calculator to degree mode; you will find that $\sin 1° \approx 0.0174524$.

Example 3 ■ Using a Calculator

Use a calculator in degree mode or radian mode, as appropriate, to find the following. Round to six decimal places.

(a) $\tan 40°$ **(b)** $\cos \frac{2\pi}{5}$ **(c)** $\cot 1.5$ **(d)** $\csc 80°$

Solution We set the calculator in degree mode for parts (a) and (d), and in radian mode for parts (b) and (c). Make sure you get the following answers on your calculator.

(a) $\tan 40° \approx 0.839100$ **(b)** $\cos \frac{2\pi}{5} \approx 0.309017$

(c) $\cot 1.5 = \frac{1}{\tan 1.5} \approx 0.070915$ **(d)** $\csc 80° = \frac{1}{\sin 80°} \approx 1.015427$

 **Now Try Exercise 13** ■

■ Applications of Trigonometry of Right Triangles

A triangle has six parts: three angles and three sides. To **solve a triangle** means to determine all of its parts from the information that is known about the triangle, that is, to determine the lengths of the three sides and the measures of the three angles.

Hulton Archive/Moviepix/Getty Images

Discovery Project ■ Similarity

Similarity of triangles is the basic concept underlying the definition of the trigonometric ratios. The ratios of the sides of a triangle are the same as the corresponding ratios in any similar triangle. But the concept of similarity of figures applies to all shapes, not just triangles. In this project we explore how areas and volumes of similar figures are related. These relationships allow us to determine whether an ape the size of King Kong (that is, an ape similar to, but much larger than, a real ape) can actually exist. You can find the project at **www.stewartmath.com.**

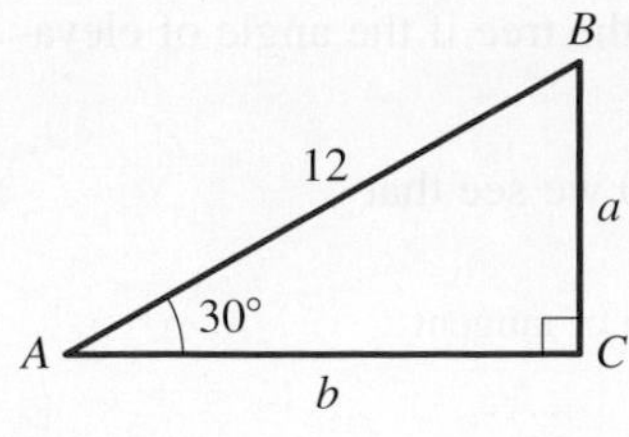

Figure 7

Example 4 ■ Solving a Right Triangle

Solve triangle ABC, shown in Figure 7.

Solution Since the sum of the angles of a triangle is 180 degrees, it follows that $\angle B = 60°$. From Figure 7 we have

$$\sin 30° = \frac{a}{12} \qquad \text{Definition of sine}$$

$$a = 12 \sin 30° \qquad \text{Multiply by 12}$$

$$= 12\left(\tfrac{1}{2}\right) = 6 \qquad \text{Evaluate}$$

Also from Figure 7 we have

$$\cos 30° = \frac{b}{12} \qquad \text{Definition of cosine}$$

$$b = 12 \cos 30° \qquad \text{Multiply by 12}$$

$$= 12\left(\frac{\sqrt{3}}{2}\right) = 6\sqrt{3} \qquad \text{Evaluate}$$

Now Try Exercise 39 ■

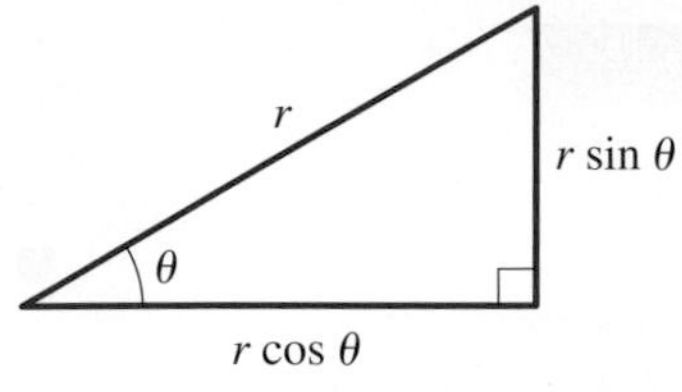

Figure 8

Note Example 4 illustrates the following fact: If a right triangle has hypotenuse r and acute angle θ, then the side opposite θ has length $r \sin \theta$ and the side adjacent to θ has length $r \cos \theta$. This useful fact is illustrated in Figure 8.

The ability to solve right triangles by using the trigonometric ratios is fundamental to many problems in navigation, surveying, astronomy, and the measurement of distances. The applications we consider in this section always involve right triangles, but as we will see in the next three sections, trigonometry is also useful in solving triangles that are not right triangles.

To discuss the next examples, we need some terminology. If an observer is looking at an object, then the line from the eye of the observer to the object is called the **line of sight** (Figure 9). If the object being observed is above the horizontal, then the angle between the line of sight and the horizontal is called the **angle of elevation**. If the object is below the horizontal, then the angle between the line of sight and the horizontal is called the **angle of depression**. In many of the examples and exercises in this chapter, angles of elevation and depression will be given for a hypothetical observer at ground level. If the line of sight follows a physical object, such as an inclined plane or a hillside, we use the term **angle of inclination**.

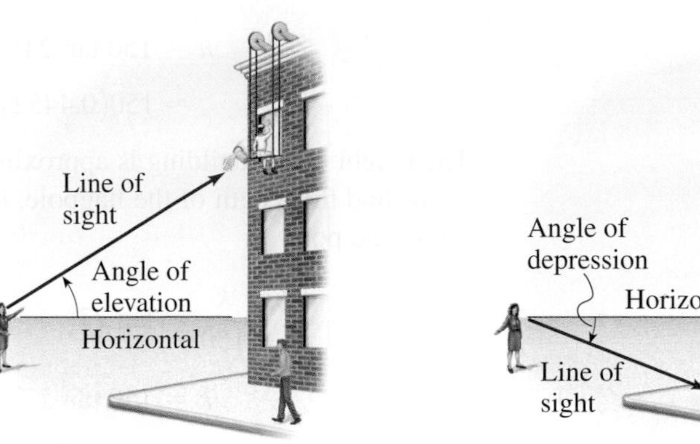

Figure 9

The next example gives an important application of trigonometry to the problem of measurement: We measure the height of a tall tree without having to climb it! Although the example is simple, the result is fundamental to understanding how the trigonometric ratios are applied to such problems.

ARISTARCHUS OF SAMOS (310–230 B.C.) was a famous Greek scientist, musician, astronomer, and geometer. He observed that the angle between the sun and moon can be measured directly (see the figure below). In his book *On the Sizes and Distances of the Sun and the Moon* he estimated the distance to the sun by observing that when the moon is exactly half full, the triangle formed by the sun, the moon, and the earth has a right angle at the moon. His method was similar to the one described in Exercise 69 in this section. Aristarchus was the first to advance the theory that the earth and planets move around the sun, an idea that did not gain full acceptance until after the time of Copernicus, 1800 years later. For this reason Aristarchus is often called "the Copernicus of antiquity."

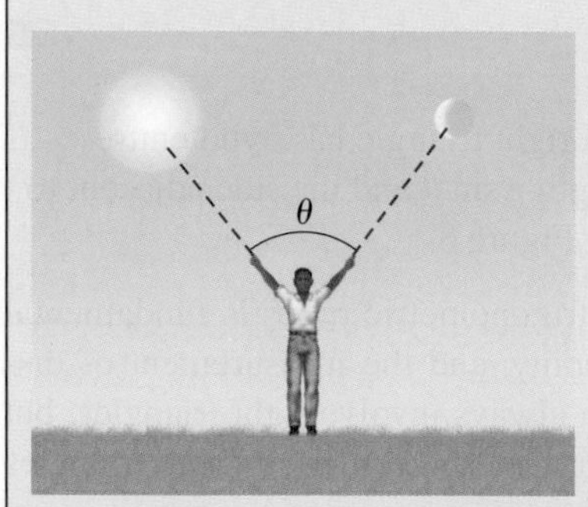

Example 5 ■ Finding the Height of a Tree

A tall tree casts a shadow 160 m long. Find the height of the tree if the angle of elevation of the sun is 25.7°.

Solution Let the height of the tree be h. From Figure 10 we see that

$$\frac{h}{160} = \tan 25.7° \qquad \text{Definition of tangent}$$

$$h = 160 \tan 25.7° \qquad \text{Multiply by 160}$$

$$\approx 160(0.48127) \approx 77 \qquad \text{Use a calculator}$$

Therefore the height of the tree is about 77 m.

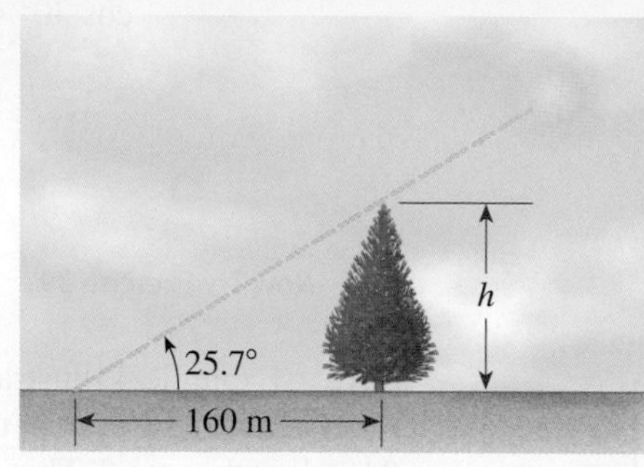

Figure 10

Now Try Exercise 55 ■

Example 6 ■ A Problem Involving Right Triangles

From a point on the ground, 150 m from the base of a building, an observer finds that the angle of elevation to the top of the building is 24° and that the angle of elevation to the top of a flagpole atop the building is 27°. Find the height of the building and the length of the flagpole.

Solution Figure 11 illustrates the situation. The height of the building is found in the same way that we found the height of the tree in Example 5.

$$\frac{h}{150} = \tan 24° \qquad \text{Definition of tangent}$$

$$h = 150 \tan 24° \qquad \text{Multiply by 150}$$

$$\approx 150(0.4452) \approx 67 \qquad \text{Use a calculator}$$

The height of the building is approximately 67 m.

To find the length of the flagpole, let's first find the height from the ground to the top of the pole.

$$\frac{k}{150} = \tan 27° \qquad \text{Definition of tangent}$$

$$k = 150 \tan 27° \qquad \text{Multiply by 150}$$

$$\approx 150(0.5095) \approx 77 \qquad \text{Use a calculator}$$

To find the length of the flagpole, we subtract h from k. So the length of the pole is approximately $77 - 67 = 10$ m.

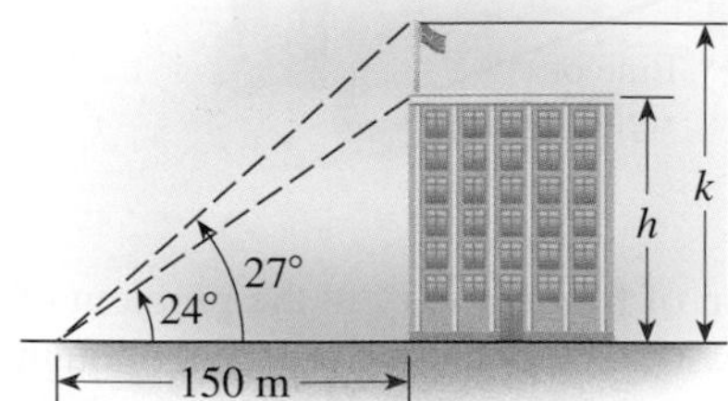

Figure 11

Now Try Exercise 63 ■

6.2 Exercises

Concepts

1. A right triangle with an angle θ is shown in the figure.

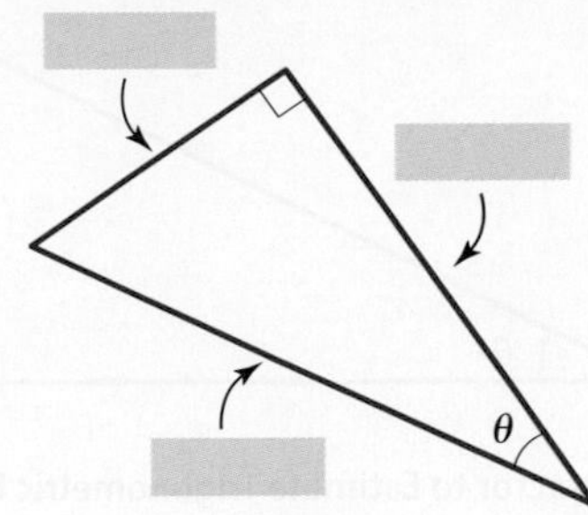

(a) Label the "opposite" and "adjacent" sides of θ and the hypotenuse of the triangle.

(b) The trigonometric ratios of the angle θ are defined as follows:

$\sin\theta = \frac{\square}{\square}$ $\cos\theta = \frac{\square}{\square}$ $\tan\theta = \frac{\square}{\square}$

2. The trigonometric ratios do not depend on the size of the triangle. This is because all right triangles with the same acute angle θ are ________.

3. The reciprocal identities state that

$\csc\theta = \frac{1}{\square}$ $\sec\theta = \frac{1}{\square}$ $\cot\theta = \frac{1}{\square}$

4. Refer to the figure.

(a) Express x and y in terms of r and θ.

(b) If $r = 6$ and $\theta = 30°$, then $x =$ _____ and $y =$ _____.

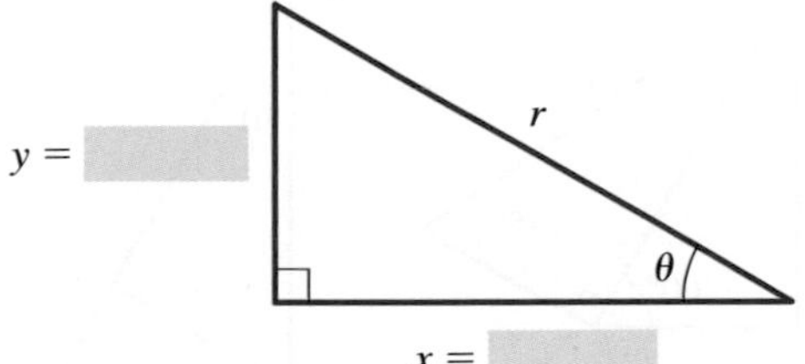

Skills

5–10 ■ Trigonometric Ratios Find the exact values of the six trigonometric ratios of the angle θ in the triangle.

5.

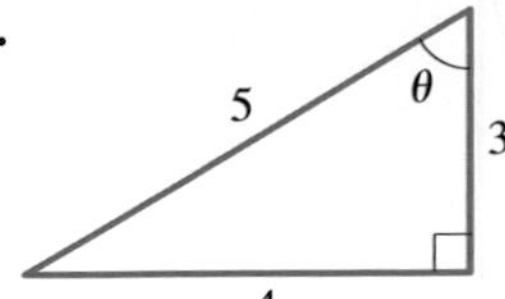

6.

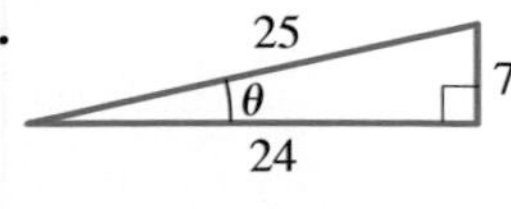

7.

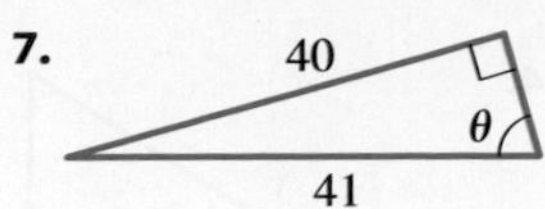

8.

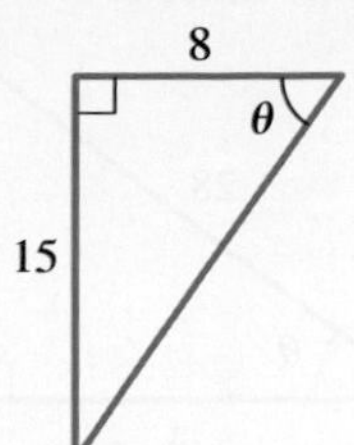

9.

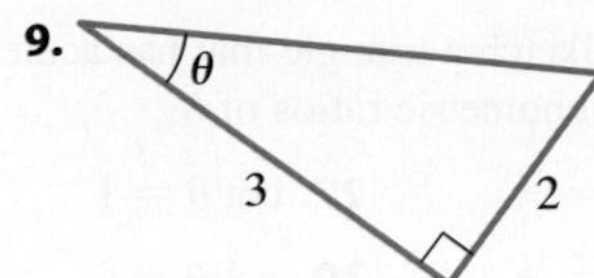

10.

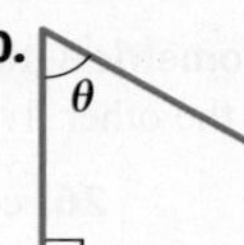

11–12 ■ Trigonometric Ratios Find **(a)** $\sin\alpha$ and $\cos\beta$, **(b)** $\tan\alpha$ and $\cot\beta$, and **(c)** $\sec\alpha$ and $\csc\beta$.

11.

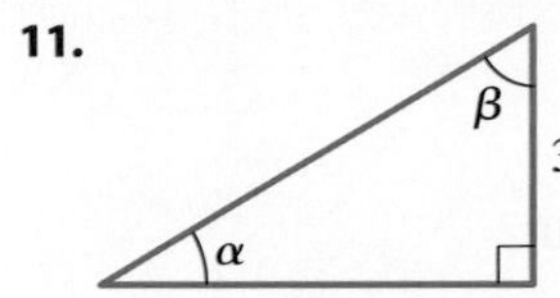

12.

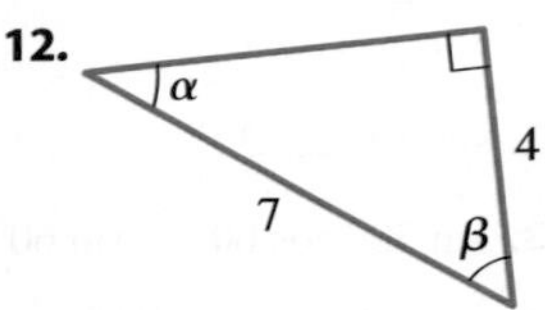

13–16 ■ Using a Calculator Use a calculator in degree or radian mode, as appropriate, to evaluate each expression. Round your answer to five decimal places.

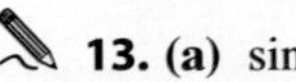

13. (a) $\sin 22°$ **(b)** $\cot\frac{3\pi}{8}$

14. (a) $\cos\frac{4\pi}{5}$ **(b)** $\csc 48°$

15. (a) $\sec 1$ **(b)** $\tan 51°$

16. (a) $\csc 10°$ **(b)** $\sin 3.5$

17–22 ■ Finding an Unknown Side Find the side labeled x. In Exercises 21 and 22 state your answer rounded to five decimal places.

17.

18.

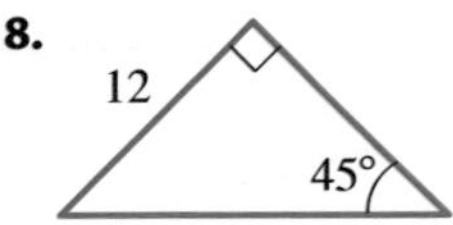

19.

x

60°

13

20.

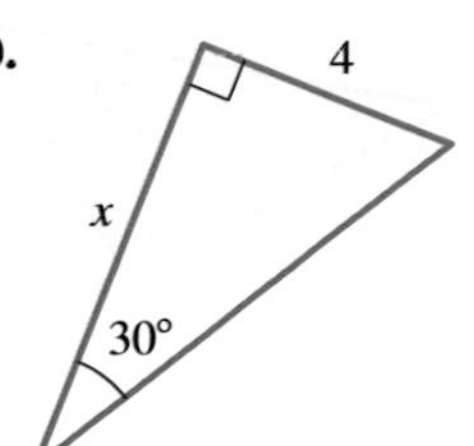

21.

12

36°

x

22.

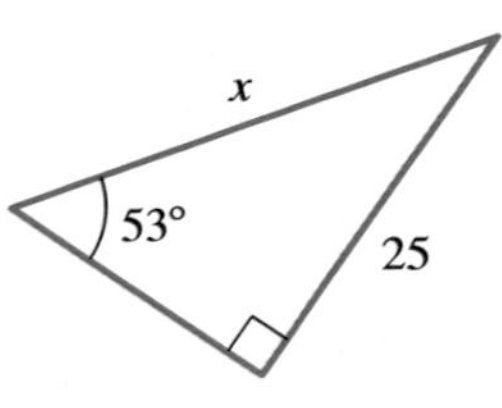

23–24 ■ Trigonometric Ratios Express x and y in terms of trigonometric ratios of θ.

23.

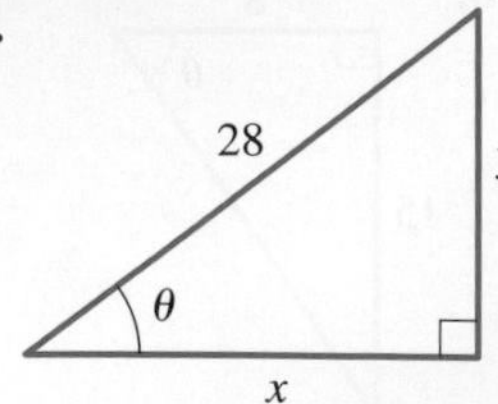

24.

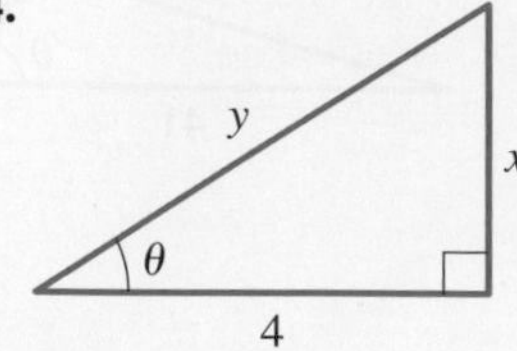

25–30 ■ Trigonometric Ratios Sketch a triangle that has acute angle θ, and find the other five trigonometric ratios of θ.

25. $\tan\theta = \frac{5}{6}$ **26.** $\cos\theta = \frac{12}{13}$ **27.** $\cot\theta = 1$

28. $\tan\theta = \sqrt{3}$ **29.** $\csc\theta = \frac{11}{6}$ **30.** $\cot\theta = \frac{5}{3}$

31–38 ■ Evaluating an Expression Evaluate the expression without using a calculator.

31. $\sin\frac{\pi}{6} + \cos\frac{\pi}{6}$

32. $\sin 30° \csc 30°$

33. $\sin 30° \cos 60° + \sin 60° \cos 30°$

34. $(\sin 60°)^2 + (\cos 60°)^2$

35. $(\cos 30°)^2 - (\sin 30°)^2$

36. $\left(\sin\frac{\pi}{3}\cos\frac{\pi}{4} - \sin\frac{\pi}{4}\cos\frac{\pi}{3}\right)^2$

37. $\left(\cos\frac{\pi}{4} + \sin\frac{\pi}{6}\right)^2$

38. $\left(\sin\frac{\pi}{3}\tan\frac{\pi}{6} + \csc\frac{\pi}{4}\right)^2$

39–46 ■ Solving a Right Triangle Solve the right triangle.

39.

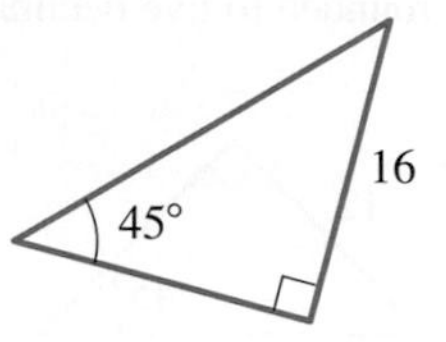

40.

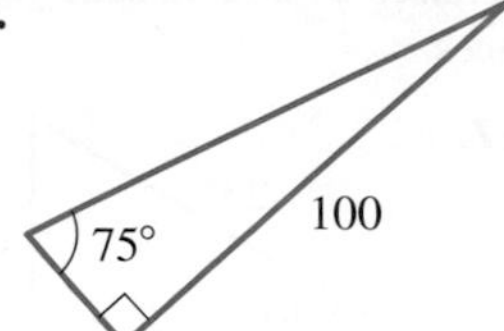

41.

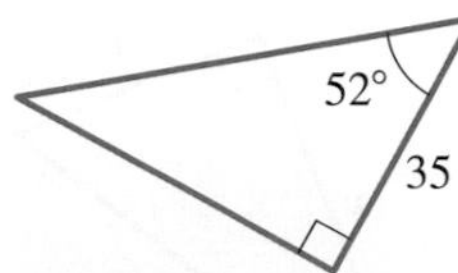

42.

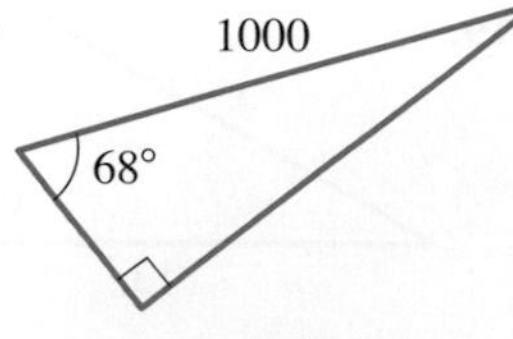

43.

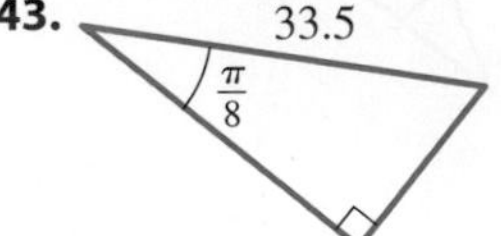

44.

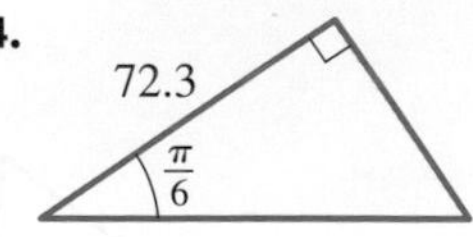

45.

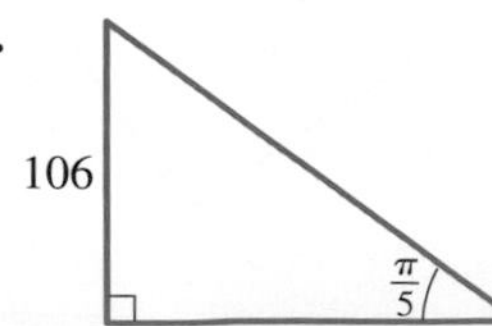

46.

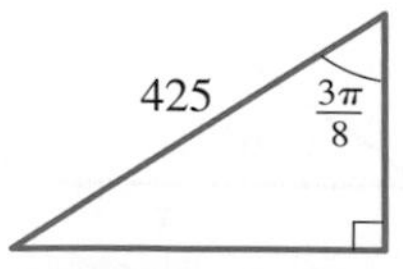

Skills Plus

47. Using a Ruler to Estimate Trigonometric Ratios Use a ruler to carefully measure the sides of the triangle, and then use your measurements to estimate the six trigonometric ratios of θ.

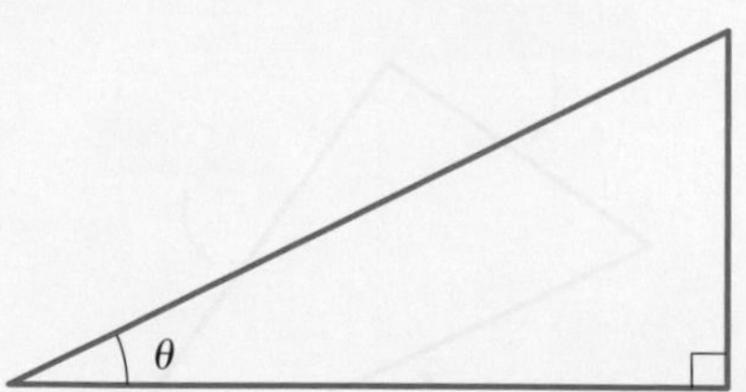

48. Using a Protractor to Estimate Trigonometric Ratios Using a protractor, sketch a right triangle that has the acute angle 40°. Measure the sides carefully, and use your results to estimate the six trigonometric ratios of 40°. Compare your results with those you get from a calculator.

49–52 ■ Finding an Unknown Side Find x rounded to one decimal place.

49.

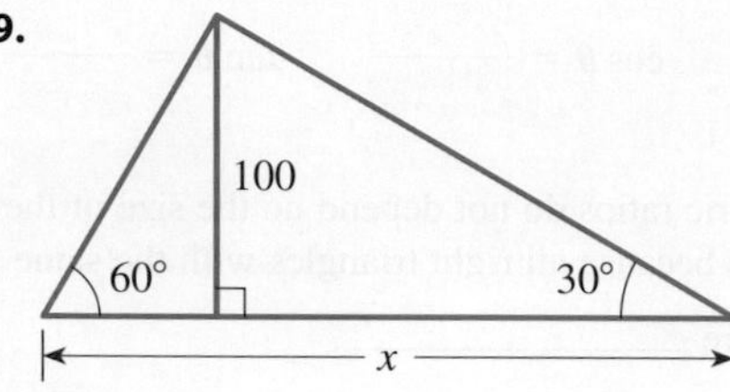

50.

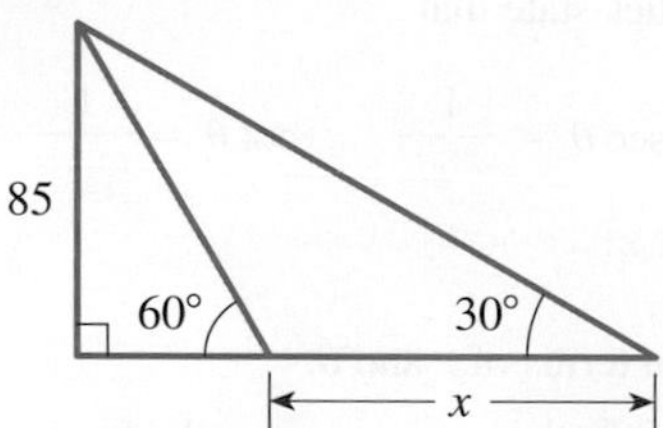

51. **52.**

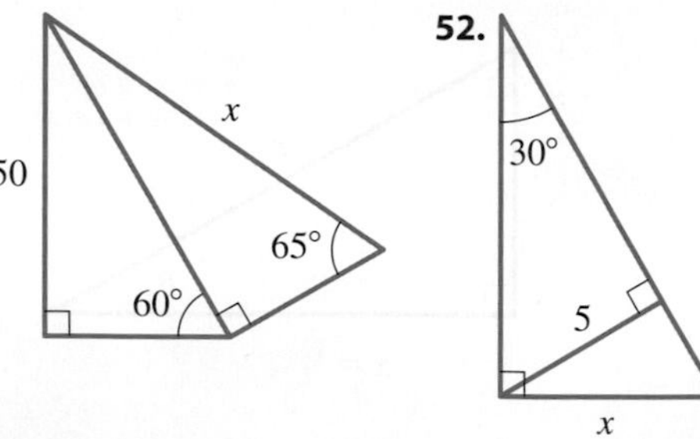

53. Trigonometric Ratios Express the length x in terms of the trigonometric ratios of θ.

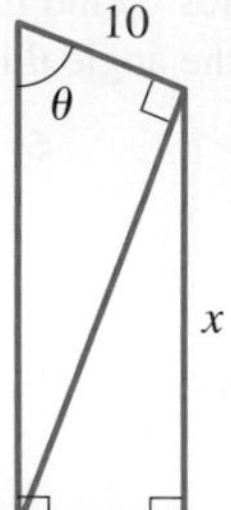

54. Trigonometric Ratios Express the lengths a, b, c, and d in the figure in terms of the trigonometric ratios of θ.

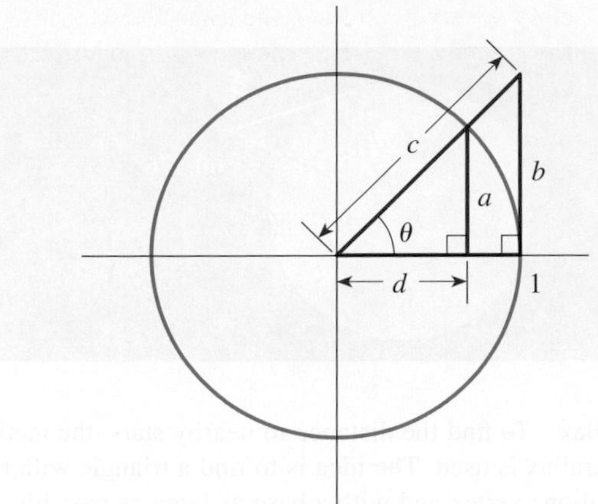

Applications

55. Height of a Building The angle of elevation to the top of the Empire State Building in New York is found to be 11° from the ground at a distance of 1.6 km from the base of the building. Using this information, find the height of the Empire State Building.

56. Landing a Plane A plane is flying within sight of an airport, at an elevation of 10,500 m. The pilot would like to estimate the plane's distance to a point on the runway and finds that the angle of depression to that point is 22°.

(a) What is the distance between the plane and the point on the runway?

(b) What is the distance between a point on the ground directly below the plane and the point on the runway?

57. Misalignment of a Laser Beam A laser beam is to be directed toward the center of the moon, but the beam strays 0.5° from its intended path.

(a) How far has the beam diverged from its assigned target when it reaches the moon? (The distance from the earth to the moon is 384,000 km.)

(b) The radius of the moon is about 1600 km. Will the beam strike the moon?

58. Distance at Sea From the top of a 60-m-tall lighthouse, the angle of depression to a ship in the ocean is 23°. How far is the ship from the base of the lighthouse?

59. Leaning Ladder A 6-m-tall ladder leans against a building so that the angle between the ground and the ladder is 72°. How high does the ladder reach on the building?

60. Height of a Tower A 180-m guy wire is attached to the top of a communications tower. If the wire makes an angle of 65° with the ground, how tall is the communications tower?

61. Elevation of a Kite A tourist is lying on the beach, flying a kite. The tourist holds the end of the kite string at ground level and estimates the angle of elevation of the kite to be 50°. If the string is 135 m long, how high is the kite above the ground?

62. Determining a Distance A hiker standing on a hill sees an 18-m-tall flagpole. The angle of depression to the bottom of the pole is 14°, and the angle of elevation to the top of the pole is 18°. Find the hiker's distance x from the pole.

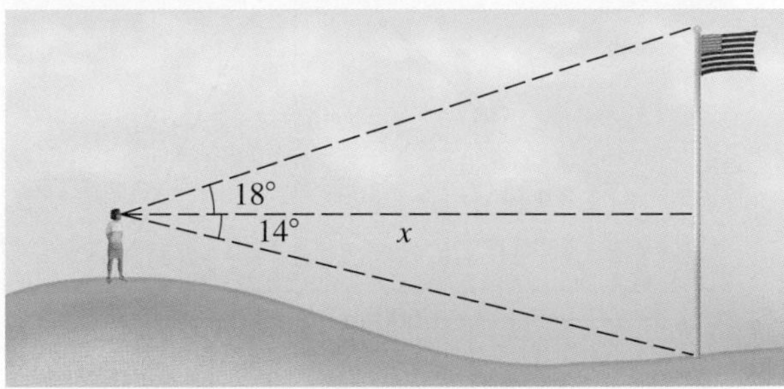

63. Height of a Tower A water tower is located 98 m from a building (see the figure). From a window in the building, an observer notes that the angle of elevation to the top of the tower is 39° and the angle of depression to the bottom of the tower is 25°. How tall is the tower? How high is the window?

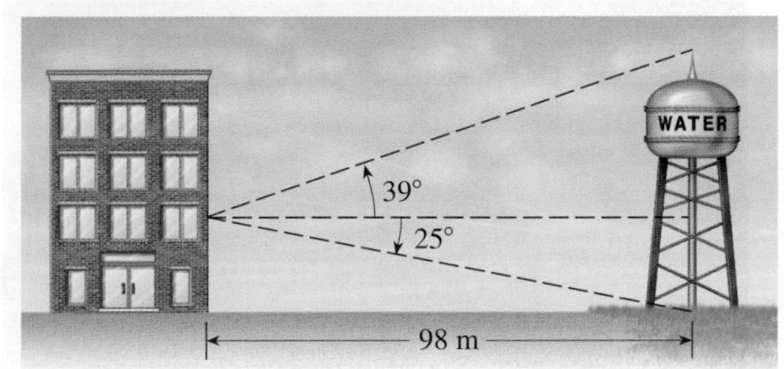

64. Determining a Distance An airplane is flying at an elevation of 1545 m, directly above a straight highway. Two motorists are driving cars on the highway on opposite sides of the plane. The angle of depression to one car is 35°, and that to the other is 52°. How far apart are the cars?

65. Determining a Distance If both cars in Exercise 64 are on one side of the plane and if the angle of depression to one car is 38° and that to the other car is 52°, how far apart are the cars?

66. Height of a Balloon A hot-air balloon is floating above a straight road. To estimate their height above the ground, the balloonists simultaneously measure the angle of depression to two consecutive mileposts on the road on the same side of the balloon. The angles of depression are found to be 20° and 22°. How high is the balloon?

67. Height of a Mountain To estimate the height of a mountain above a level plain, the angle of elevation to the top of the mountain a surveyor measures to be 32°. One thousand meters closer to the mountain along the plain, the angle of elevation is found to be 35°. Estimate the height of the mountain.

68. Height of Cloud Cover To measure the height of the cloud cover at an airport, a worker shines a spotlight upward at an angle 75° from the horizontal. An observer 600 m away measures the angle of elevation to the spot of light to be 45°. Find the height h of the cloud cover.

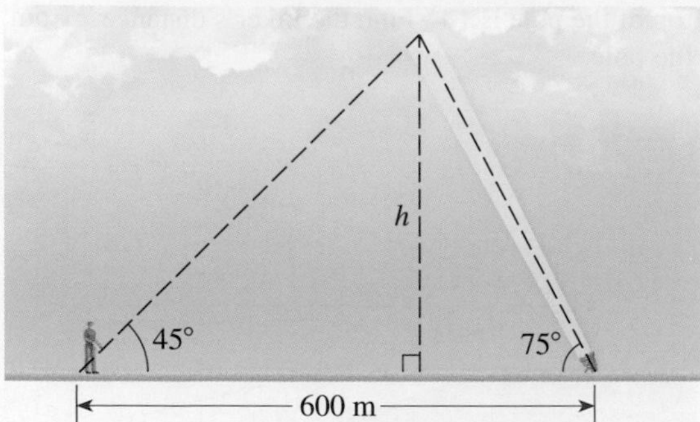

69. Distance to the Sun When the moon is exactly half full, the earth, moon, and sun form a right angle (see the figure). At that time the angle formed by the sun, earth, and moon is measured to be 89.85°. If the distance from the earth to the moon is 384,000 km, estimate the distance from the earth to the sun.

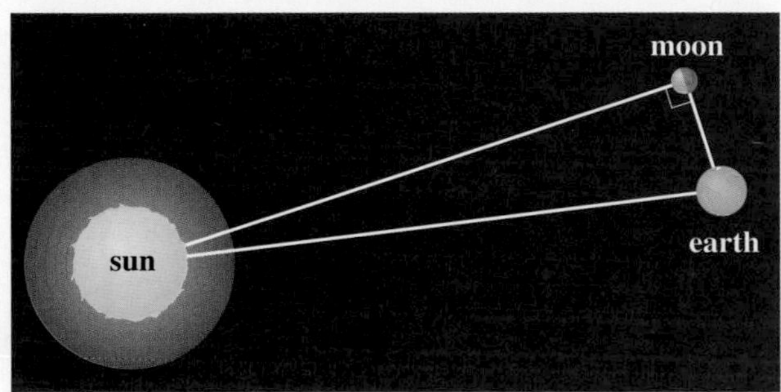

70. Distance to the Moon To find the distance to the sun as in Exercise 69, we needed to know the distance to the moon. Here is a way to estimate that distance: When the moon is seen at its zenith at a point A on the earth, it is observed to be at the horizon from point B (see the figure). Points A and B are 9848 km apart, and the radius of the earth is 6340 km.

(a) Find the angle θ in degrees.

(b) Estimate the distance from point A to the moon.

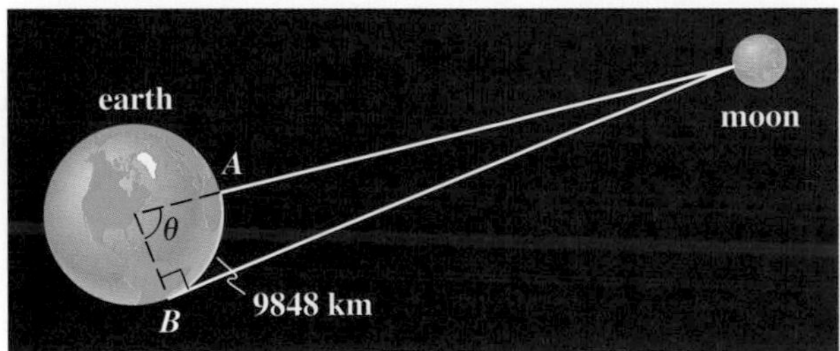

71. Radius of the Earth In Exercise 6.1.80 a method was given for finding the radius of the earth. Here is a more modern method: From a satellite 960 km above the earth it is observed that the angle formed by the vertical and the line of sight to the horizon is 60.276° as shown in the figure. Use this information to find the radius of the earth.

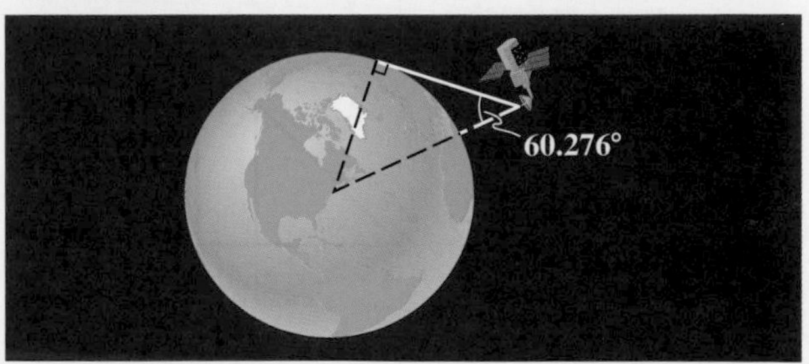

72. Parallax To find the distance to nearby stars, the method of parallax is used. The idea is to find a triangle with the star at one vertex and with a base as large as possible. To do this, the star is observed at two different times exactly 6 months apart, and its apparent change in position is recorded. From these two observations $\angle E_1SE_2$ can be calculated. (The times are chosen so that $\angle E_1SE_2$ is as large as possible, which guarantees that $\angle E_1OS$ is 90°.) The angle E_1SO is called the *parallax* of the star. Alpha Centauri, the star nearest the earth, has a parallax of 0.000211°. (See the figure.) Estimate the distance to this star. (The distance from the earth to the sun is about 14.9×10^7 km.)

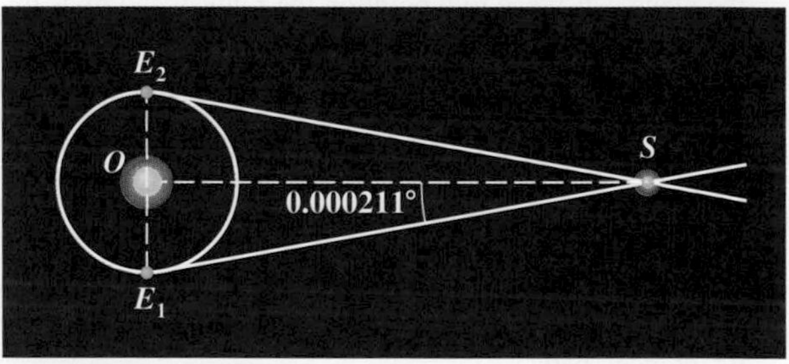

73. Distance from Venus to the Sun The **elongation** α of a planet is the angle formed by the planet, Earth, and sun (see the figure). When Venus achieves its maximum elongation of 46.3°, Earth, Venus, and the sun form a triangle with a right angle at Venus, as shown in the figure. Find the distance between Venus and the sun in astronomical units (AU). (By definition the distance between Earth and the sun is 1 AU.)

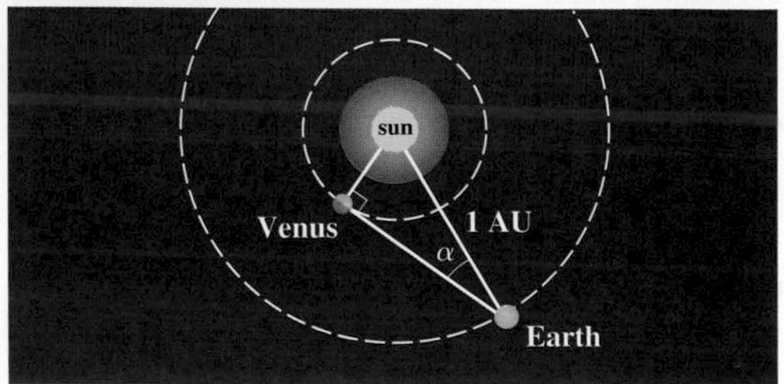

■ **Discuss** ■ **Discover** ■ **Prove** ■ **Write**

74. Discuss: Similar Triangles If two triangles are similar, what properties do they share? Explain how these properties make it possible to define the trigonometric ratios without regard to the size of the triangle.

75. Discuss ■ **Discover:** In the figure, triangle ABC is equilateral and P is the center of a circular arc of radius 1. Find the perimeter of triangle ABC.

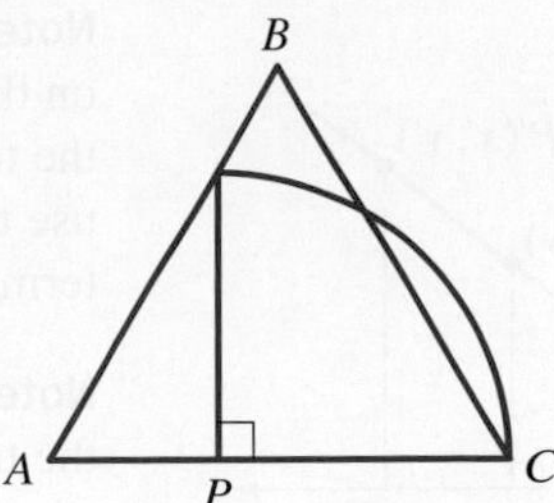

PS *Try to recognize something familiar.* What do you know about the angles in an equilateral triangle?

6.3 Trigonometric Functions of Angles

■ **Trigonometric Functions of Angles** ■ **Evaluating Trigonometric Functions at Any Angle**
■ **Trigonometric Identities** ■ **Areas of Triangles**

In Section 6.2 we defined the *trigonometric ratios* for acute angles. Here we extend the trigonometric ratios to all angles by defining the *trigonometric functions* of angles. With these functions we can solve problems that involve angles that are not acute.

■ Trigonometric Functions of Angles

Let POQ be a right triangle with acute angle θ as shown in Figure 1(a). Place θ in standard position as shown in Figure 1(b).

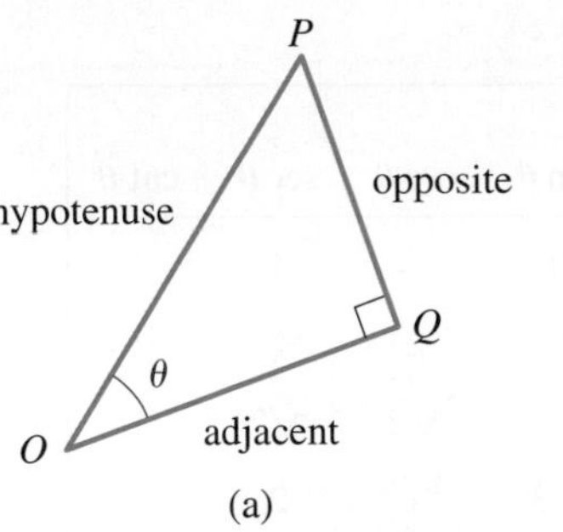

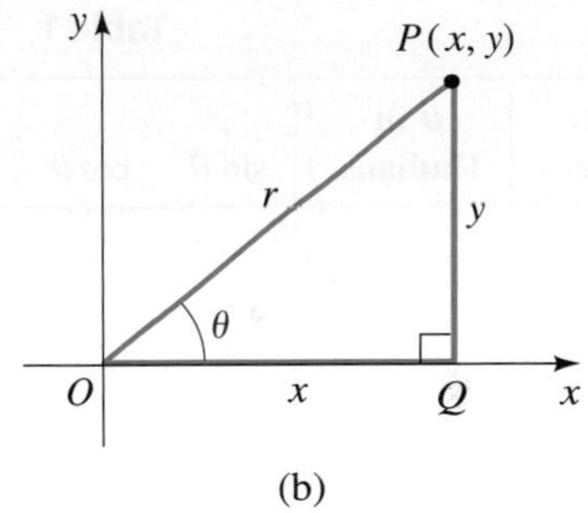

Figure 1 (a) (b)

Then $P = P(x, y)$ is a point on the terminal side of θ. In triangle POQ the opposite side has length y and the adjacent side has length x. Using the Pythagorean Theorem, we see that the hypotenuse has length $r = \sqrt{x^2 + y^2}$. So

$$\sin\theta = \frac{y}{r} \qquad \cos\theta = \frac{x}{r} \qquad \tan\theta = \frac{y}{x}$$

The other trigonometric ratios can be found in the same way.

These observations allow us to extend the trigonometric ratios to any angle. We define the trigonometric functions of angles as follows (see Figure 2).

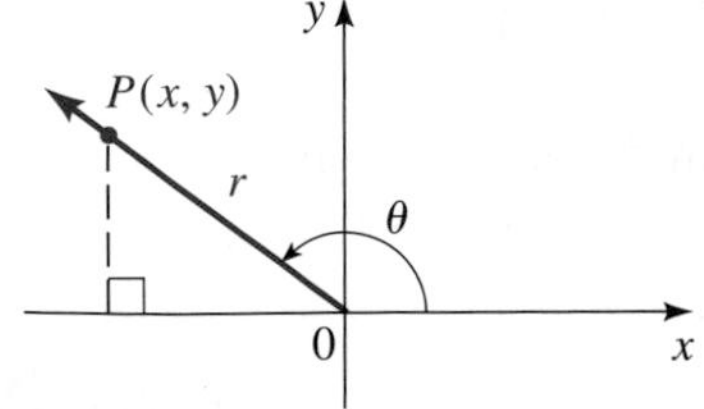

Figure 2

Definition of the Trigonometric Functions

Let θ be an angle in standard position, and let $P(x, y)$ be a point on the terminal side. If $r = \sqrt{x^2 + y^2}$ is the distance from the origin to the point $P(x, y)$, then

$$\sin\theta = \frac{y}{r} \qquad \cos\theta = \frac{x}{r} \qquad \tan\theta = \frac{y}{x} \quad (x \neq 0)$$

$$\csc\theta = \frac{r}{y} \quad (y \neq 0) \qquad \sec\theta = \frac{r}{x} \quad (x \neq 0) \qquad \cot\theta = \frac{x}{y} \quad (y \neq 0)$$

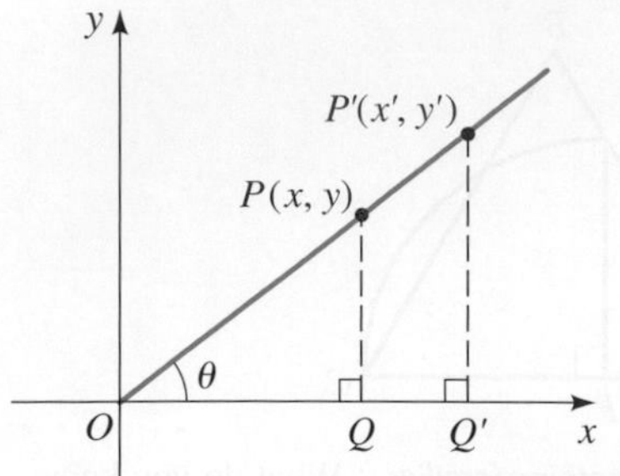

Figure 3

Note 1 It is a crucial fact that the values of the trigonometric functions do *not* depend on the choice of the point $P(x, y)$. This is because if $P'(x', y')$ is any other point on the terminal side, as in Figure 3, then triangles POQ and $P'OQ'$ are similar. We can use this fact to find the trigonometric values of an angle θ if we know *any* point on the terminal side of θ.

Note 2 For an acute angle θ, the trigonometric functions of θ have the same values as the trigonometric ratios for θ defined in Section 6.2. For example, by placing a special triangle in the coordinate plane, as shown in Figure 1, we can see that the trigonometric ratios of the special angles θ have the same values as the trigonometric functions of θ.

Note 3 Since division by 0 is not defined, some trigonometric functions are not defined for certain angles. For example, $\tan 90° = y/x$ is not defined because $x = 0$. The angles for which the trigonometric functions may be undefined are the angles for which either the x- or y-coordinate of a point on the terminal side of the angle is 0. These are **quadrantal angles**—angles that are coterminal with the coordinate axes, The functions tan and sec are not defined for the quadrantal angles $(\pi/2) + n\pi$, where n is any integer, and the functions cot and csc are not defined for the quadrantal angles $n\pi$. The functions sin and cos are defined for all angles.

The following table gives the values of the trigonometric functions at some special angles, including the quadrantal angles.

Special Values of the Trigonometric Functions of Angles

The table gives the values of the trigonometric functions for some special angles.

Table 1

θ in Degrees	θ in Radians	$\sin\theta$	$\cos\theta$	$\tan\theta$	$\csc\theta$	$\sec\theta$	$\cot\theta$
0°	0	0	1	0	—	1	—
30°	$\frac{\pi}{6}$	$\frac{1}{2}$	$\frac{\sqrt{3}}{2}$	$\frac{\sqrt{3}}{3}$	2	$\frac{2\sqrt{3}}{3}$	$\sqrt{3}$
45°	$\frac{\pi}{4}$	$\frac{\sqrt{2}}{2}$	$\frac{\sqrt{2}}{2}$	1	$\sqrt{2}$	$\sqrt{2}$	1
60°	$\frac{\pi}{3}$	$\frac{\sqrt{3}}{2}$	$\frac{1}{2}$	$\sqrt{3}$	$\frac{2\sqrt{3}}{3}$	2	$\frac{\sqrt{3}}{3}$
90°	$\frac{\pi}{2}$	1	0	—	1	—	0
180°	π	0	−1	0	—	−1	—
270°	$\frac{3\pi}{2}$	−1	0	—	−1	—	0

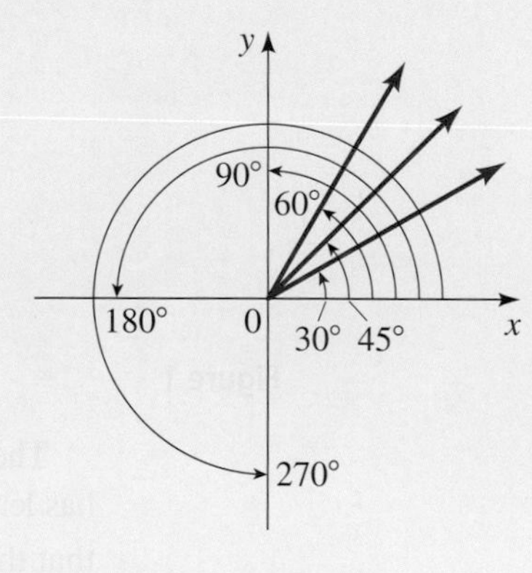

Example 1 ■ Finding Trigonometric Functions of Angles

Find **(a)** $\cos 135°$ and **(b)** $\tan 390°$.

Solution

(a) From Figure 4 (on the next page) we see that $\cos 135° = -x/r$. But $\cos 45° = x/r$, and since $\cos 45° = \sqrt{2}/2$, we have

$$\cos 135° = -\frac{\sqrt{2}}{2}$$

(b) The angles 390° and 30° are coterminal. From Figure 5 (on the next page) it's clear that $\tan 390° = \tan 30°$, and since $\tan 30° = \sqrt{3}/3$, we have

$$\tan 390° = \frac{\sqrt{3}}{3}$$

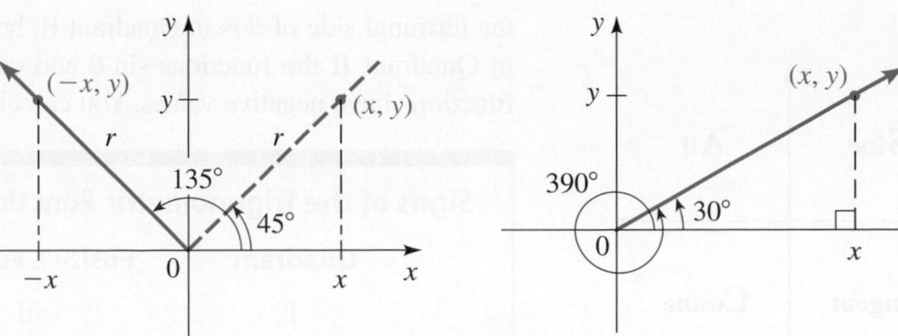

Figure 4

Figure 5

Now Try Exercises 13 and 15 ■

From Example 1 we see that the trigonometric functions of an angle that isn't acute have the same values, except possibly for sign, as the corresponding trigonometric functions of an acute angle. That acute angle will be called the *reference angle*.

Reference Angle

Let θ be an angle in standard position. The **reference angle** $\bar{\theta}$ associated with θ is the acute angle formed by the terminal side of θ and the x-axis.

Figure 6 shows that to find a reference angle $\bar{\theta}$, it's useful to know the quadrant in which the terminal side of the angle θ lies.

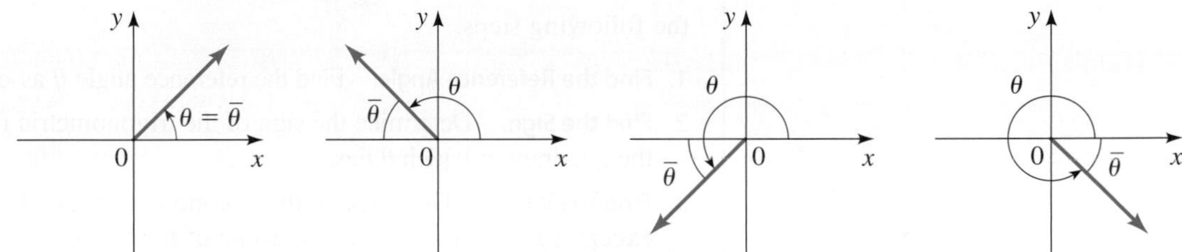

Figure 6 | The reference angle $\bar{\theta}$ for an angle θ

Example 2 ■ Finding Reference Angles

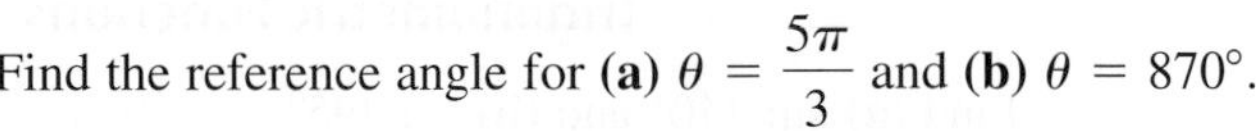

Find the reference angle for **(a)** $\theta = \dfrac{5\pi}{3}$ and **(b)** $\theta = 870°$.

Solution

(a) The reference angle is the acute angle formed by the terminal side of the angle $5\pi/3$ and the x-axis (see Figure 7). Since the terminal side of this angle is in Quadrant IV, the reference angle is

$$\bar{\theta} = 2\pi - \frac{5\pi}{3} = \frac{\pi}{3}$$

Figure 7

(b) The angles 870° and 150° are coterminal [because $870 - 2(360) = 150$]. Thus the terminal side of this angle is in Quadrant II (see Figure 8). So the reference angle is

$$\bar{\theta} = 180° - 150° = 30°$$

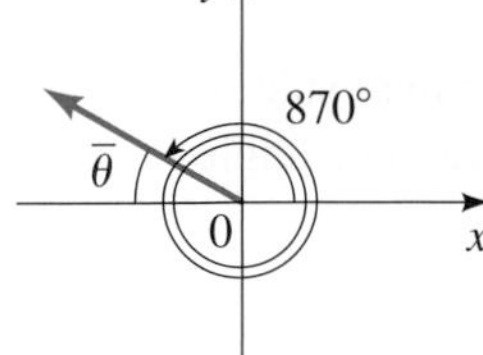

Figure 8

Now Try Exercises 5 and 9 ■

From the definition of the trigonometric functions we see that the values of these functions are all positive if the angle θ has its terminal side in Quadrant I. This is because, in this case the terminal point $P(x, y)$ is in Quadrant I and so both x and y are positive. [Of course, r is always positive, since it is simply the distance from the origin to the point $P(x, y)$.] If

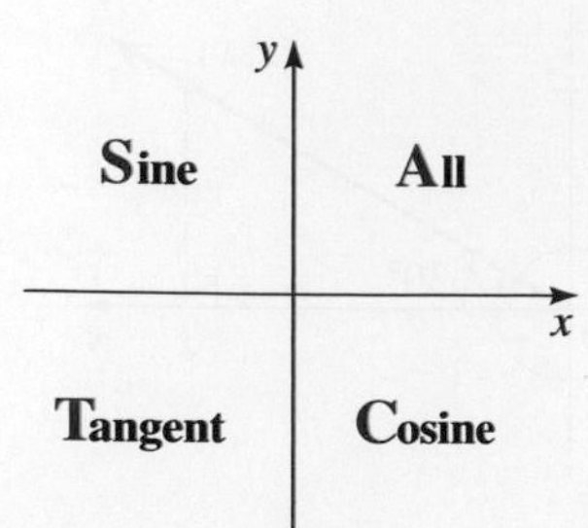

You can remember this as "All Students Take Calculus."

the terminal side of θ is in Quadrant II, however, then x is negative and y is positive. Thus in Quadrant II the functions $\sin\theta$ and $\csc\theta$ are positive and all the other trigonometric functions have negative values. You can check the other entries in the following table.

Signs of the Trigonometric Functions

Quadrant	Positive Functions	Negative Functions
I	all	none
II	sin, csc	cos, sec, tan, cot
III	tan, cot	sin, csc, cos, sec
IV	cos, sec	sin, csc, tan, cot

The diagram in the margin is a mnemonic device for remembering which trigonometric functions are positive in each quadrant: **A**ll of them, **S**in, **T**an, or **C**os. Also, whenever these functions are positive, their respective reciprocals csc, cot, and sec are also positive.

■ Evaluating Trigonometric Functions at Any Angle

A quadrantal angle is coterminal with 0°, 90°, 180°, or 270°, so we can use Table 1 to find values of the trigonometric functions of quadrantal angles.

To evaluate a trigonometric function at an angle θ (excluding quadrantal angles), we first determine the quadrant in which the terminal side of θ lies. Knowing the quadrant allows us to find both the reference angle and the sign of the trigonometric function.

Evaluating Trigonometric Functions for Any Angle

To find the values of the trigonometric functions for any angle θ, we carry out the following steps.

1. **Find the Reference Angle.** Find the reference angle $\bar{\theta}$ associated with the angle θ.
2. **Find the Sign.** Determine the sign of the trigonometric function of θ by noting the quadrant in which θ lies.
3. **Find the Value.** The value of the trigonometric function of θ is the same, except possibly for sign, as the value of the trigonometric function of $\bar{\theta}$.

Example 3 ■ Using the Reference Angle to Evaluate Trigonometric Functions

Find **(a)** $\sin 240°$ and **(b)** $\cot 495°$.

Solution

(a) This angle has its terminal side in Quadrant III, as shown in Figure 9. The reference angle is therefore $240° - 180° = 60°$, and the value of $\sin 240°$ is negative. Thus

$$\sin 240° = -\sin 60° = -\frac{\sqrt{3}}{2}$$

Sign | Reference angle | From Table 1

Figure 9 | $\sin 240°$ is *negative*.

(b) The angle 495° is coterminal with the angle 135°, and the terminal side of this angle is in Quadrant II, as shown in Figure 10. So the reference angle is $180° - 135° = 45°$, and the value of $\cot 495°$ is negative. We have

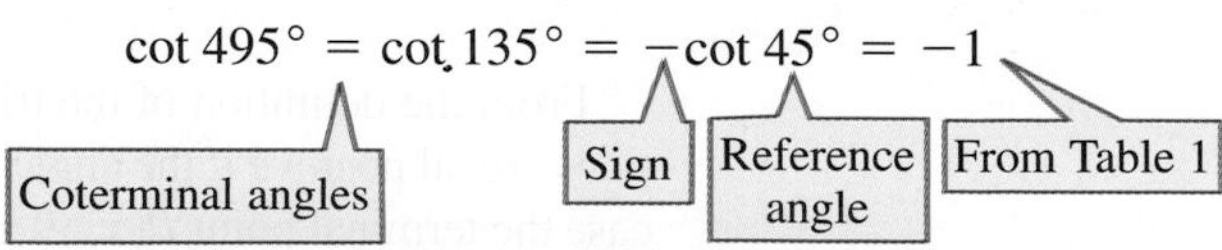

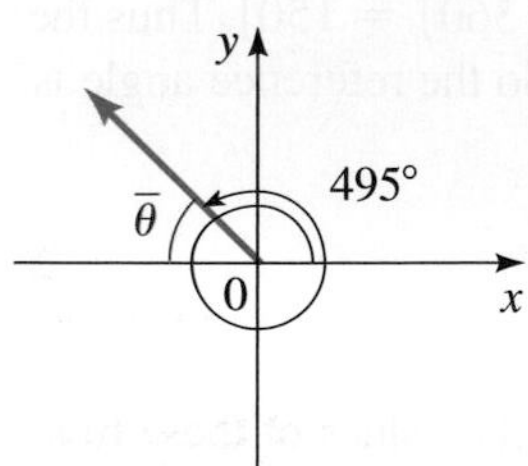

Figure 10 | $\tan 495°$ is *negative*, so $\cot 495°$ is *negative*.

Now Try Exercises 19 and 21 ■

The Right Triangle Approach and The Unit Circle Approach

If you have already studied the trigonometric functions of *real numbers t* defined using points on the unit circle (Chapter 5), you may be wondering how these are related to the trigonometric functions of *angles θ* defined using ratios of sides of right triangles, as in this chapter. The functions are exactly the same, provided angles are measured in radians.

To see how, let's start with a unit circle in the coordinate plane and let $P(x, y)$ be the terminal point determined by an arc of length t on the unit circle, as shown in Figure A. By the definition of the trigonometric functions of real numbers (Chapter 5) we have

$$\sin t = y$$
$$\cos t = x$$

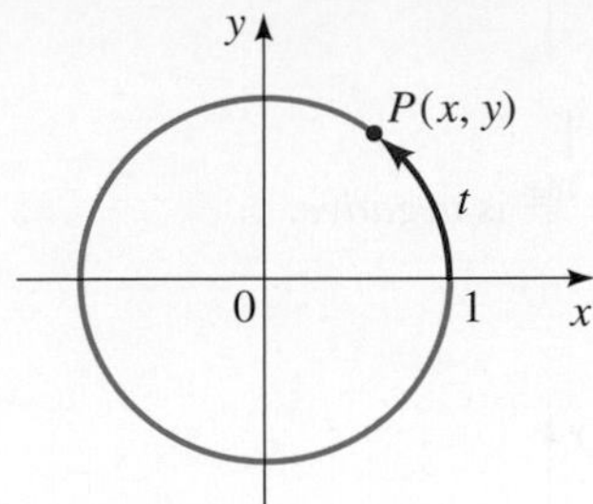

Figure A | $P(x, y)$ is the terminal point determined by t.

Now, the arc of length t subtends an angle θ whose radian measure is t, as in Figure B. Let's drop a perpendicular from P to the point Q on the x-axis and observe that triangle OPQ is a right triangle with legs of length x and y, as in Figure C. So, by the definition of the trigonometric functions of the angle θ, we have

$$\sin \theta = \frac{\text{opp}}{\text{hyp}} = \frac{PQ}{OP} = \frac{y}{1} = y = \sin t$$

$$\cos \theta = \frac{\text{adj}}{\text{hyp}} = \frac{OQ}{OP} = \frac{x}{1} = x = \cos t$$

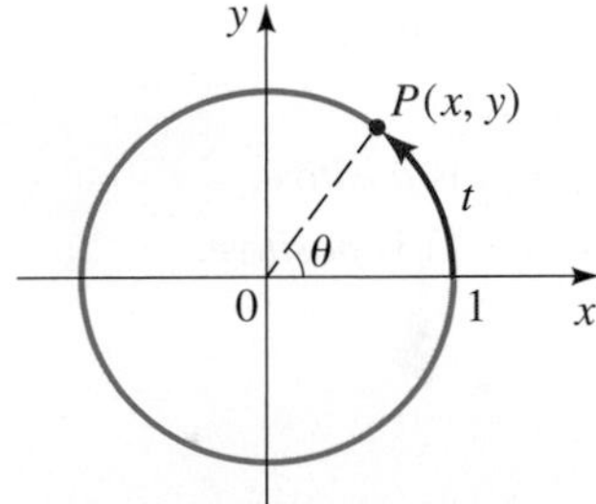

Figure B | The radian measure of the angle θ is t.

Since the radian measure of θ is t, we see that the two different ways of defining the trigonometric function are identical. In other words, as functions, they assign identical values to a given real number—the real number is the radian measure of θ in one case, or the length t of an arc in the other.

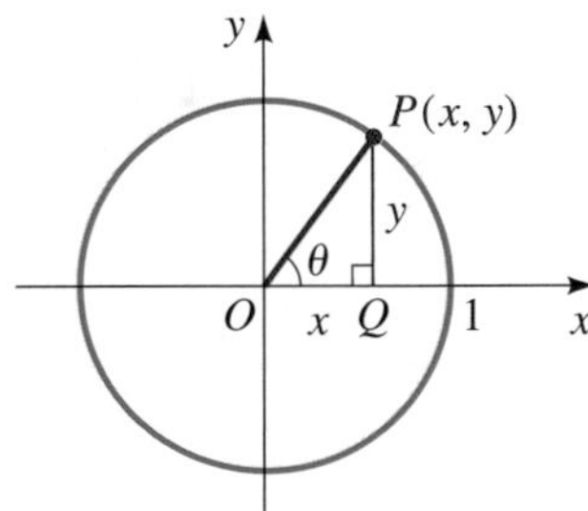

Figure C | Triangle OPQ is a right triangle.

Why then do we study trigonometry in two different ways? Because different applications require that we view the trigonometric functions differently. (Compare the applications in Sections 6.5 and 6.6 with those in Section 5.6.)

Example 4 ■ Using the Reference Angle to Evaluate Trigonometric Functions

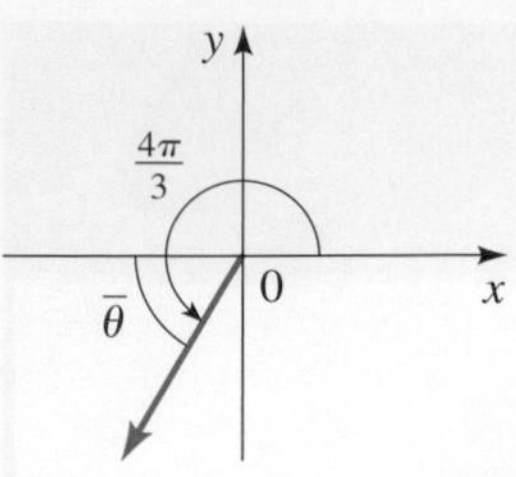

Figure 11 | S A / T C $\sin \frac{16\pi}{3}$ is *negative*.

Find **(a)** $\sin \dfrac{16\pi}{3}$ and **(b)** $\sec\left(-\dfrac{\pi}{4}\right)$.

Solution

(a) The angle $16\pi/3$ is coterminal with $4\pi/3$, and these angles are in Quadrant III (see Figure 11). Thus the reference angle is $(4\pi/3) - \pi = \pi/3$. Since the value of sine is negative in Quadrant III, we have

$$\sin \frac{16\pi}{3} = \sin \frac{4\pi}{3} = -\sin \frac{\pi}{3} = -\frac{\sqrt{3}}{2}$$

Coterminal angles — Sign — Reference angle — From Table 1

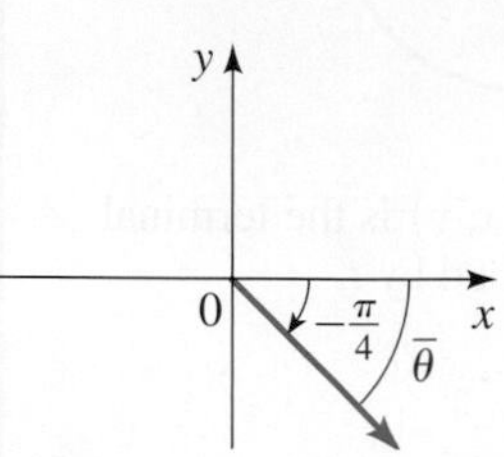

Figure 12 | S A / T C $\cos(-\frac{\pi}{4})$ is *positive*, so $\sec(-\frac{\pi}{4})$ is *positive*.

(b) The angle $-\pi/4$ is in Quadrant IV, and its reference angle is $\pi/4$. (See Figure 12.) Since secant is positive in this quadrant, we have

$$\sec\left(-\frac{\pi}{4}\right) = +\sec \frac{\pi}{4} = \sqrt{2}$$

Sign — Reference angle — From Table 1

Now Try Exercises 25 and 27 ■

■ Trigonometric Identities

The trigonometric functions of angles are related to each other through several important equations called **trigonometric identities**. We've already encountered the reciprocal identities. These identities continue to hold for any angle θ, provided that both sides of the equation are defined. The Pythagorean identities are a consequence of the Pythagorean Theorem.*

Fundamental Identities

Reciprocal Identities

$$\csc \theta = \frac{1}{\sin \theta} \qquad \sec \theta = \frac{1}{\cos \theta} \qquad \cot \theta = \frac{1}{\tan \theta}$$

$$\tan \theta = \frac{\sin \theta}{\cos \theta} \qquad \cot \theta = \frac{\cos \theta}{\sin \theta}$$

Pythagorean Identities

$$\sin^2\theta + \cos^2\theta = 1 \qquad \tan^2\theta + 1 = \sec^2\theta \qquad 1 + \cot^2\theta = \csc^2\theta$$

Proof Let's prove the first Pythagorean identity. Using $x^2 + y^2 = r^2$ (the Pythagorean Theorem) in Figure 13, we have

$$\sin^2\theta + \cos^2\theta = \left(\frac{y}{r}\right)^2 + \left(\frac{x}{r}\right)^2 = \frac{x^2 + y^2}{r^2} = \frac{r^2}{r^2} = 1$$

Thus $\sin^2\theta + \cos^2\theta = 1$. (Although the figure indicates an acute angle, you should check that the proof holds for all angles θ.) ■

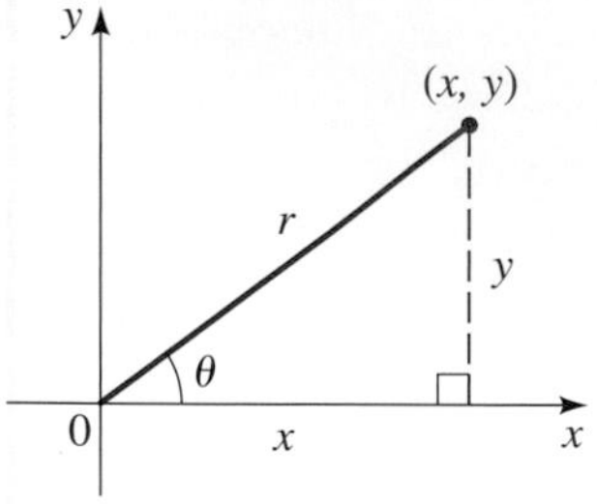

Figure 13

See Exercise 87 for the proofs of the other two Pythagorean identities.

*We follow the usual convention of writing $\sin^2\theta$ for $(\sin \theta)^2$. In general, we write $\sin^n\theta$ for $(\sin \theta)^n$ for all integers n except $n = -1$. The superscript $n = -1$ will be assigned another meaning in Section 6.4. Of course, the same convention applies to the other trigonometric functions.

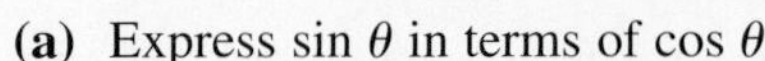

Example 5 ■ Expressing One Trigonometric Function in Terms of Another

(a) Express $\sin\theta$ in terms of $\cos\theta$.

(b) Express $\tan\theta$ in terms of $\sin\theta$, where θ is in Quadrant II.

Solution

(a) From the first Pythagorean identity we get

$$\sin\theta = \pm\sqrt{1-\cos^2\theta}$$

where the sign depends on the quadrant. If θ is in Quadrant I or II, then $\sin\theta$ is positive, so

$$\sin\theta = \sqrt{1-\cos^2\theta}$$

whereas if θ is in Quadrant III or IV, $\sin\theta$ is negative, so

$$\sin\theta = -\sqrt{1-\cos^2\theta}$$

(b) Since $\tan\theta = \sin\theta/(\cos\theta)$, we need to write $\cos\theta$ in terms of $\sin\theta$. By part (a) we have

$$\cos\theta = \pm\sqrt{1-\sin^2\theta}$$

and since $\cos\theta$ is negative in Quadrant II, the negative sign applies here. Thus

$$\tan\theta = \frac{\sin\theta}{\cos\theta} = \frac{\sin\theta}{-\sqrt{1-\sin^2\theta}}$$

 **Now Try Exercise 59** ■

Example 6 ■ Evaluating a Trigonometric Function

If $\tan\theta = \frac{2}{3}$ and θ is in Quadrant III, find $\cos\theta$.

Solution 1 We need to write $\cos\theta$ in terms of $\tan\theta$. From the Pythagorean identity $\tan^2\theta + 1 = \sec^2\theta$ we get $\sec\theta = \pm\sqrt{\tan^2\theta + 1}$. In Quadrant III, $\sec\theta$ is negative, so

$$\sec\theta = -\sqrt{\tan^2\theta + 1}$$

Thus

$$\cos\theta = \frac{1}{\sec\theta} = \frac{1}{-\sqrt{\tan^2\theta+1}}$$

$$= \frac{1}{-\sqrt{\left(\frac{2}{3}\right)^2+1}} = \frac{1}{-\sqrt{\frac{13}{9}}} = -\frac{3}{\sqrt{13}}$$

If you wish to rationalize the denominator, you can express $\cos\theta$ as

$$-\frac{3}{\sqrt{13}}\cdot\frac{\sqrt{13}}{\sqrt{13}} = -\frac{3\sqrt{13}}{13}$$

Solution 2 This problem can be solved more easily by the method of Example 6.2.2. Recall that, except for sign, the values of the trigonometric functions of any angle are the same as those of an acute angle (the reference angle). So, ignoring the sign for the moment, let's sketch a right triangle with an acute angle $\overline{\theta}$ satisfying $\tan\overline{\theta} = \frac{2}{3}$. (See Figure 14.) By the Pythagorean Theorem the hypotenuse of this triangle has length $\sqrt{13}$. From the triangle in Figure 14 we immediately see that $\cos\overline{\theta} = 3/\sqrt{13}$. Since θ is in Quadrant III, $\cos\theta$ is negative, so

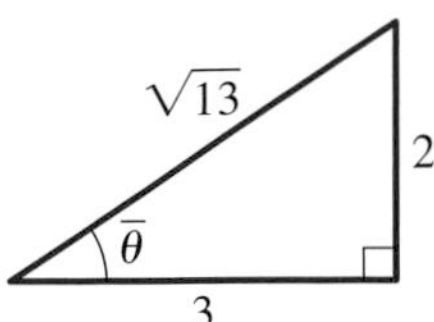

Figure 14

$$\cos\theta = -\frac{3}{\sqrt{13}}$$

Now Try Exercise 47 ■

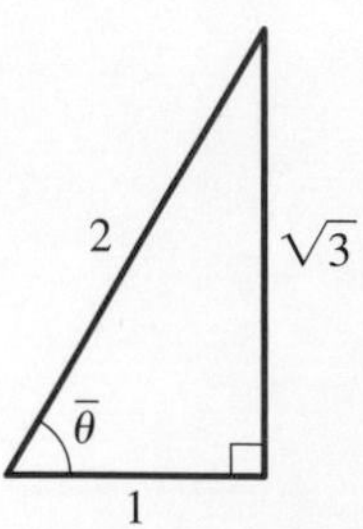

Figure 15

Example 7 ■ Evaluating Trigonometric Functions

If $\sec\theta = 2$ and θ is in Quadrant IV, find the other five trigonometric functions of θ.

Solution We sketch a triangle as in Figure 15 so that $\sec\bar{\theta} = 2$. Taking into account the fact that θ is in Quadrant IV, we get

$$\sin\theta = -\frac{\sqrt{3}}{2} \qquad \cos\theta = \frac{1}{2} \qquad \tan\theta = -\sqrt{3}$$

$$\csc\theta = -\frac{2}{\sqrt{3}} \qquad \sec\theta = 2 \qquad \cot\theta = -\frac{1}{\sqrt{3}}$$

Now Try Exercise 49 ■

■ Areas of Triangles

We conclude this section with an application of the trigonometric functions that involves angles that are not necessarily acute. More extensive applications appear in Sections 6.5 and 6.6.

The area of a triangle is $\mathcal{A} = \frac{1}{2} \times \text{base} \times \text{height}$. If we know the lengths of two sides and the included angle of a triangle, then we can find the height using the trigonometric functions, and from this we can find the area.

If θ is an acute angle, then the height of the triangle in Figure 16(a) is given by $h = b\sin\theta$. Thus the area is

$$\mathcal{A} = \tfrac{1}{2} \times \text{base} \times \text{height} = \tfrac{1}{2}ab\sin\theta$$

(a)

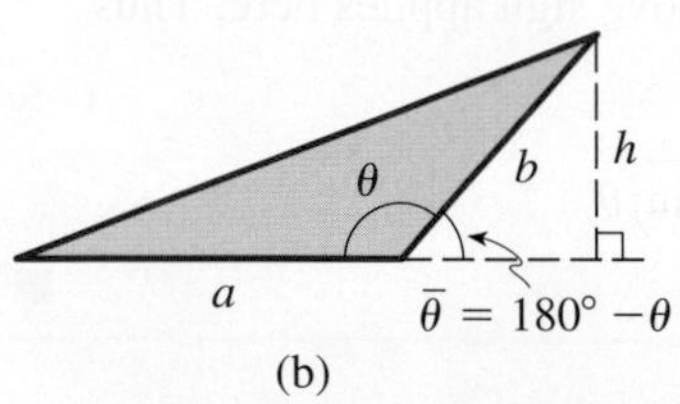

(b)

Figure 16

If the angle θ is not acute, then from Figure 16(b) we see that the height of the triangle is

$$h = b\sin(180^\circ - \theta) = b\sin\theta$$

This is so because the reference angle of θ is the angle $180^\circ - \theta$. Thus in this case also the area of the triangle is

$$\mathcal{A} = \tfrac{1}{2} \times \text{base} \times \text{height} = \tfrac{1}{2}ab\sin\theta$$

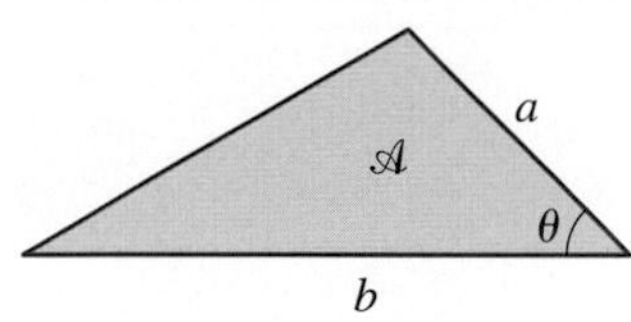

Figure 17

Area of a Triangle

The area $\mathcal{A}$ of a triangle with sides of lengths a and b and with included angle θ (see Figure 17) is

$$\mathcal{A} = \tfrac{1}{2}ab\sin\theta$$

Example 8 ■ Finding the Area of a Triangle

Find the area of triangle ABC shown in Figure 18.

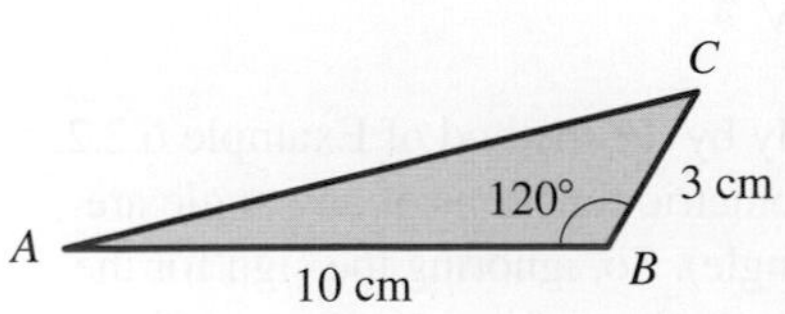

Figure 18

Solution The triangle has sides of length 10 cm and 3 cm, with included angle 120°. Therefore

$$\begin{aligned} \mathcal{A} &= \tfrac{1}{2}ab\sin\theta \\ &= \tfrac{1}{2}(10)(3)\sin 120^\circ \\ &= 15\sin 60^\circ && \text{Reference angle} \\ &= 15\frac{\sqrt{3}}{2} \approx 13\text{ cm}^2 \end{aligned}$$

Now Try Exercise 67 ■

6.3 | Exercises

Concepts

1. If the angle θ is in standard position, as in the figure, $P(x, y)$ is a point on the terminal side of θ, and r is the distance from the origin to P, then

$$\sin\theta = \frac{\square}{\square} \qquad \cos\theta = \frac{\square}{\square} \qquad \tan\theta = \frac{\square}{\square}$$

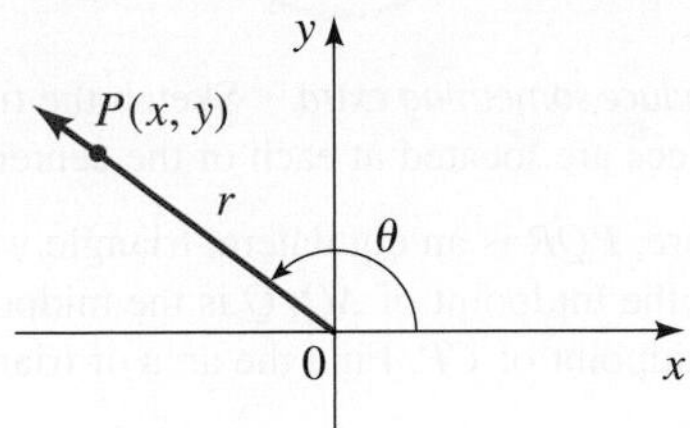

2. The sign of a trigonometric function of θ depends on the ________ in which the terminal side of the angle θ lies.

In Quadrant II, $\sin\theta$ is ________ (positive / negative).

In Quadrant III, $\cos\theta$ is ________ (positive / negative).

In Quadrant IV, $\sin\theta$ is ________ (positive / negative).

3. (a) If θ is in standard position, then the reference angle $\bar{\theta}$ is the acute angle formed by the terminal side of θ and the ________. So the reference angle for $\theta = 100°$ is $\bar{\theta} =$ ________, and that for $\theta = 190°$ is $\bar{\theta} =$ ________.

(b) If θ is any angle, the value of a trigonometric function of θ is the same, except possibly for sign, as the value of the trigonometric function of $\bar{\theta}$. So $\sin 100° = \sin$ ____, and $\sin 190° = -\sin$ ____.

4. The area $\mathscr{A}$ of a triangle with sides of lengths a and b and with included angle θ is given by the formula $\mathscr{A} =$ ________. So the area of the triangle with sides 4 and 7 and included angle $\theta = 30°$ is ________.

Skills

5–12 ■ Reference Angle Find the reference angle for the given angle.

5. (a) $135°$ **(b)** $195°$ **(c)** $300°$

6. (a) $155°$ **(b)** $280°$ **(c)** $390°$

7. (a) $250°$ **(b)** $485°$ **(c)** $-100°$

8. (a) $99°$ **(b)** $-199°$ **(c)** $359°$

9. (a) $\dfrac{7\pi}{10}$ **(b)** $\dfrac{9\pi}{8}$ **(c)** $\dfrac{10\pi}{3}$

10. (a) $\dfrac{5\pi}{6}$ **(b)** $\dfrac{10\pi}{9}$ **(c)** $\dfrac{23\pi}{7}$

11. (a) $\dfrac{5\pi}{7}$ **(b)** -1.4π **(c)** 1.4

12. (a) 2.3π **(b)** 2.3 **(c)** -10π

13–36 ■ Values of Trigonometric Functions Find the exact value of the trigonometric function.

13. $\cos 150°$ **14.** $\sin 210°$ **15.** $\tan 315°$

16. $\cot(-60°)$ **17.** $\csc 300°$ **18.** $\sec 330°$

19. $\sin 225°$ **20.** $\csc 135°$ **21.** $\sec(-420°)$

22. $\tan 120°$ **23.** $\cot 570°$ **24.** $\sin 810°$

25. $\sin\dfrac{3\pi}{2}$ **26.** $\cos\dfrac{4\pi}{3}$ **27.** $\tan\left(-\dfrac{4\pi}{3}\right)$

28. $\cos\left(-\dfrac{11\pi}{6}\right)$ **29.** $\csc\left(-\dfrac{5\pi}{6}\right)$ **30.** $\sec\dfrac{7\pi}{6}$

31. $\sec\dfrac{17\pi}{3}$ **32.** $\csc\dfrac{5\pi}{4}$ **33.** $\cot\left(-\dfrac{\pi}{4}\right)$

34. $\cos\dfrac{7\pi}{4}$ **35.** $\tan\dfrac{5\pi}{2}$ **36.** $\sin\dfrac{11\pi}{6}$

37–40 ■ Quadrant an Angle Lies In From the information given, find the quadrant that θ lies in.

37. $\sin\theta < 0$ and $\cos\theta < 0$

38. $\tan\theta < 0$ and $\sin\theta < 0$

39. $\sec\theta > 0$ and $\tan\theta < 0$

40. $\csc\theta > 0$ and $\cos\theta < 0$

41–46 ■ Values of Trigonometric Functions An angle θ is given in the coordinate plane. Find the exact value of the indicated trigonometric function of θ. (You may find it helpful to use the Pythagorean Theorem.)

41. $\tan\theta$

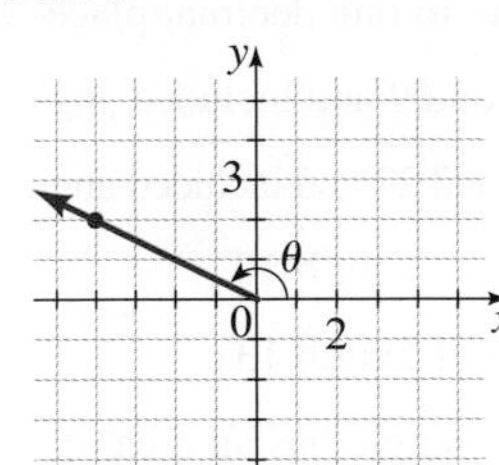

42. $\sin\theta$

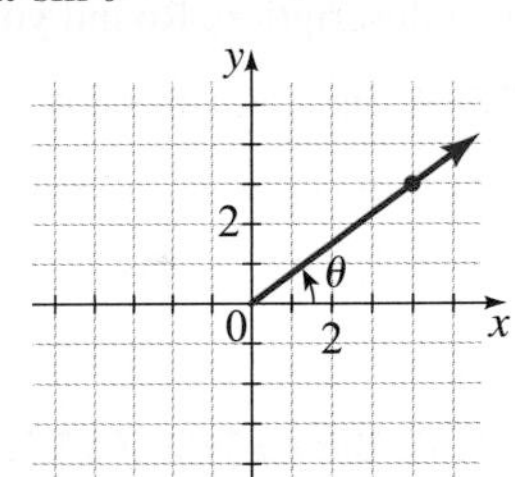

43. $\cos\theta$

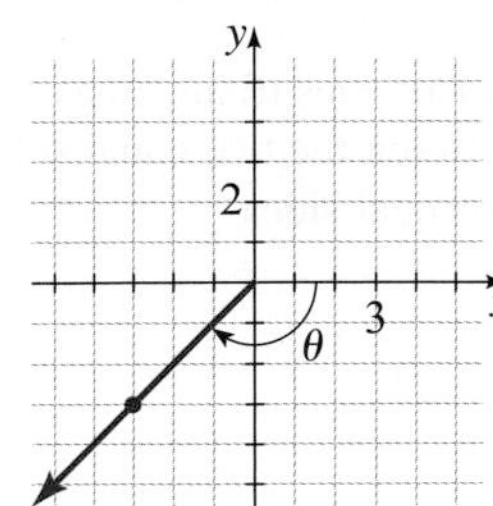

44. $\cot\theta$

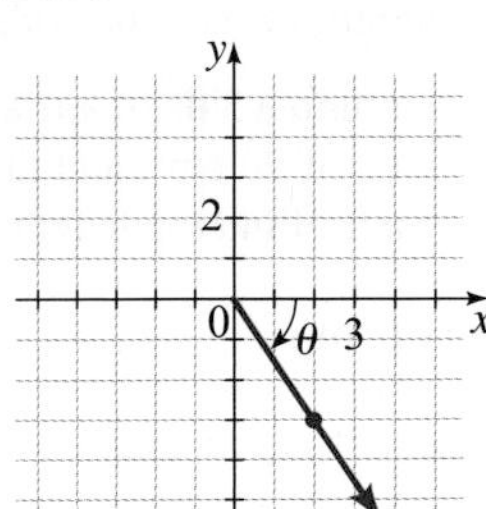

45. $\sec\theta$

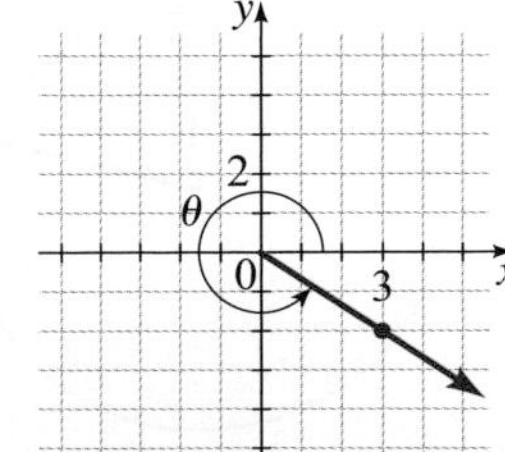

46. $\csc\theta$

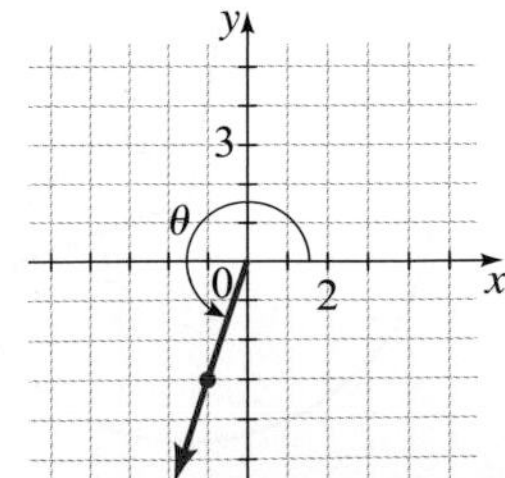

47–58 ■ Values of Trigonometric Functions Find the values of the trigonometric functions of θ from the information given.

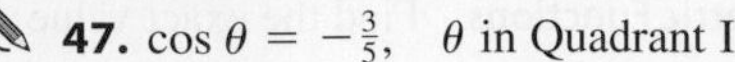

47. $\cos\theta = -\frac{3}{5}$, θ in Quadrant II
48. $\tan\theta = \frac{3}{4}$, θ in Quadrant III
49. $\cot\theta = -2$, θ in Quadrant IV
50. $\csc\theta = 2$, θ in Quadrant I
51. $\sin\theta = -\frac{2}{3}$, θ in Quadrant III
52. $\cot\theta = -\frac{7}{9}$, θ in Quadrant II
53. $\sin\theta = \frac{1}{4}$, $\cos\theta > 0$
54. $\cos\theta = \frac{1}{3}$, $\tan\theta < 0$
55. $\tan\theta = 3$, $\sin\theta < 0$
56. $\csc\theta = -\frac{12}{7}$, $\cos\theta < 0$
57. $\csc\theta = -4$, $\tan\theta < 0$
58. $\sin\theta = \frac{3}{10}$, $\sec\theta > 0$

59–64 ■ Expressing One Trigonometric Function in Terms of Another Write the first trigonometric function in terms of the second for θ in the given quadrant.

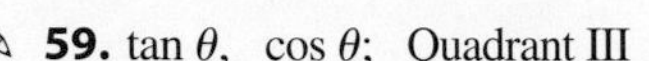

59. $\tan\theta$, $\cos\theta$; Quadrant III
60. $\cot\theta$, $\sin\theta$; Quadrant II
61. $\cos\theta$, $\sin\theta$; Quadrant IV
62. $\sec\theta$, $\sin\theta$; Quadrant I
63. $\sec\theta$, $\tan\theta$; Quadrant II
64. $\csc\theta$, $\cot\theta$; Quadrant III

65–66 ■ Values of an Expression If $\theta = \pi/3$, find the value of each expression.

65. $\sin 2\theta$, $2\sin\theta$
66. $\sin^2\theta$, $\sin(\theta^2)$

67–70 ■ Area of a Triangle Find the area of the triangle with the given description. Round your answers to one decimal place.

67. A triangle with sides of length 7 and 9 and included angle 72°
68. A triangle with sides of length 10 and 22 and included angle 10°
69. An equilateral triangle with side of length 10
70. An equilateral triangle with side of length 13
71. **Finding a Side of a Triangle** A triangle has an area of 16 cm^2. The angle between two sides of the triangle is 36°, where one of these two sides has a length of 5 cm. Find the length of the other side.
72. **Finding a Side of a Triangle** An isosceles triangle has an area of 24 cm^2, and the angle between the two equal sides is $5\pi/6$. Find the length of the two equal sides.

Skills Plus

73–74 ■ Area of a Region Find the area of the shaded region in the figure.

73.

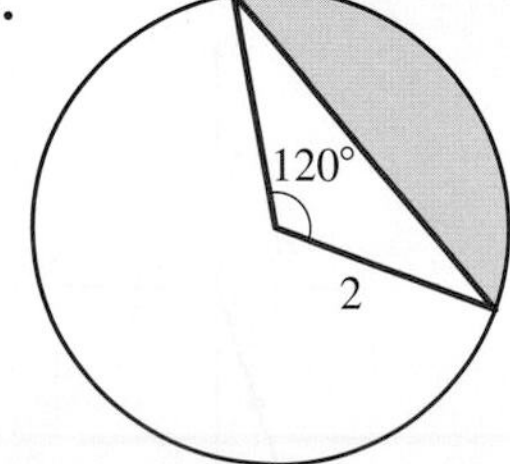

74.

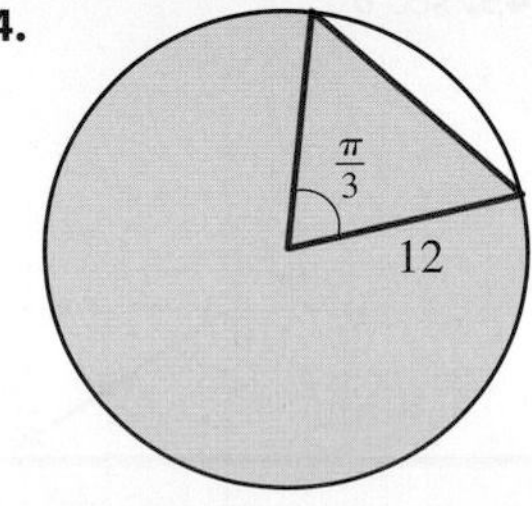

75. The circles in the figure are tangent to each other and each has radius 1. Find the area of the region enclosed by the circles.

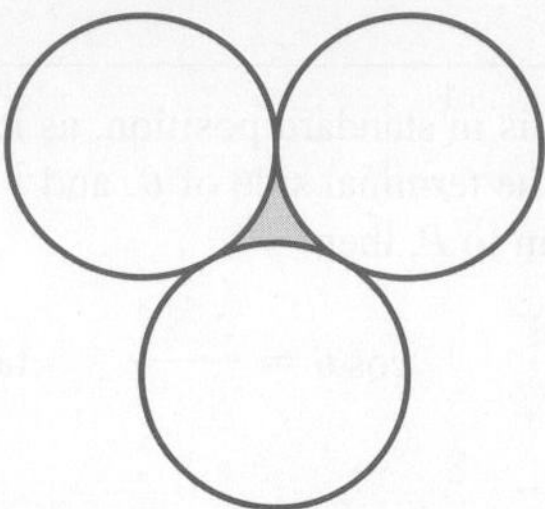

PS *Introduce something extra.* Sketch the triangle whose vertices are located at each of the centers of the circles.

76. In the figure, PQR is an equilateral triangle with area 1. Also, P is the midpoint of AQ, Q is the midpoint of BR, and R is the midpoint of CP. Find the area of triangle ABC.

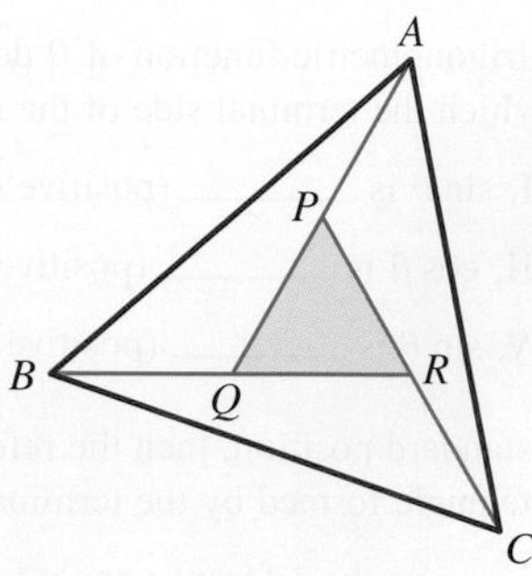

PS *Try to recognize something familiar.* Use the formula for the area of a triangle (which we proved in this section) in different ways.

Applications

77. **Height of a Rocket** A rocket fired straight upward is tracked by an observer on the ground, 1.6 km away.
    (a) Show that when the angle of elevation is θ, the height of the rocket (in m) is $h = 1600\tan\theta$.
    (b) Complete the table to find the height of the rocket at the given angles of elevation.

θ	20°	60°	80°	85°
h				

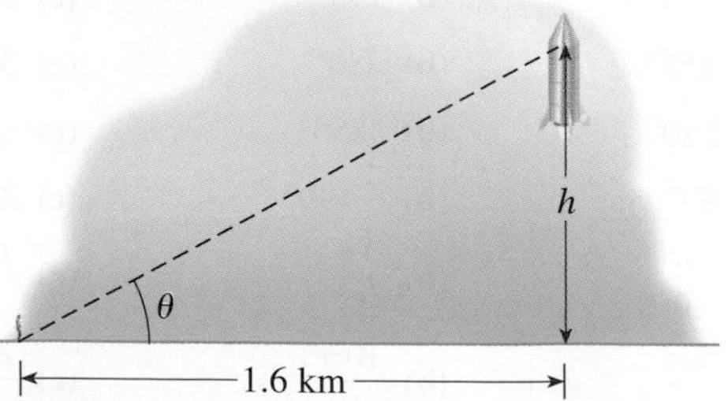

78. **Rain Gutter** A rain gutter is to be constructed from a metal sheet of width 30 cm by bending up one-third of the sheet on each side through an angle θ. (See the following figure.)

(a) Show that the cross-sectional area of the gutter is modeled by the function

$$A(\theta) = 100 \sin \theta + 100 \sin \theta \cos \theta$$

(b) Graph the function A for $0 \le \theta \le \pi/2$.

(c) For what angle θ is the largest cross-sectional area achieved?

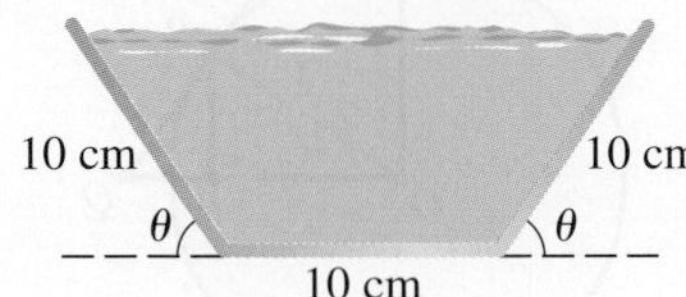

79. Wooden Beam A rectangular beam is to be cut from a cylindrical log of diameter 20 cm. The figures show different ways this can be done.

(a) Express the cross-sectional area of the beam as a function of the angle θ in the figures.

(b) Graph the function you found in part (a).

(c) Find the dimensions of the beam with largest cross-sectional area.

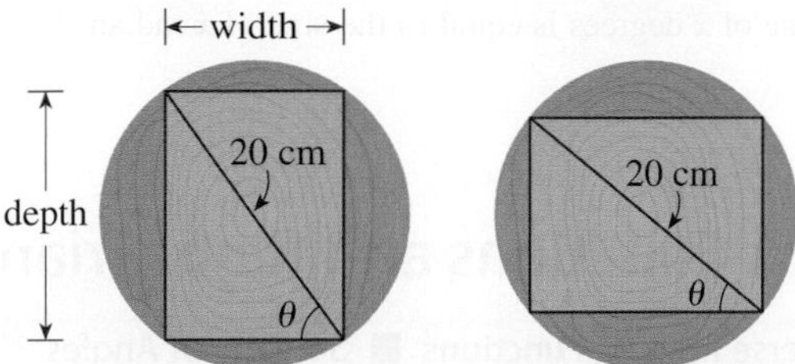

80. Strength of a Beam The strength of a beam is proportional to the width and the square of the depth. A beam is cut from a log as in Exercise 79. Express the strength of the beam as a function of the angle θ in the figures.

81. Throwing a Shot Put The range R and height H of a shot put thrown with an initial velocity of v_0 m/s at an angle θ are given by

$$R = \frac{v_0^2 \sin 2\theta}{g} \qquad \text{and} \qquad H = \frac{v_0^2 \sin^2 \theta}{2g}$$

where g is the acceleration due to gravity. On the earth $g = 9.8 \text{ m/s}^2$ and on the moon $g = 1.6 \text{ m/s}^2$. Find the range and height of a shot put thrown under the given conditions.

(a) On the earth with $v_0 = 3.6$ m/s and $\theta = \pi/6$

(b) On the moon with $v_0 = 3.6$ m/s and $\theta = \pi/6$

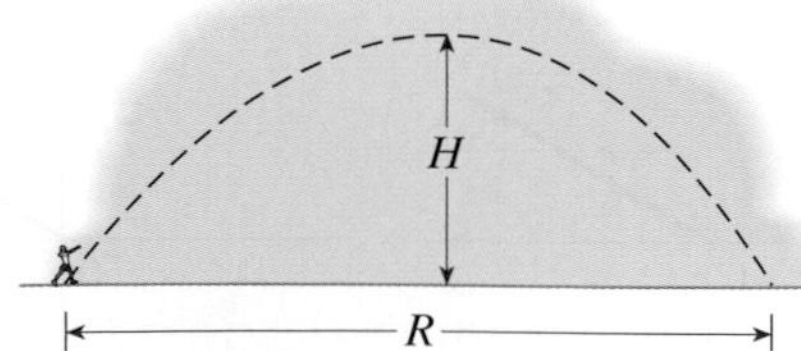

82. Sledding The time (in seconds) that it takes for a sled to slide down a hillside inclined at an angle θ is

$$t = \sqrt{\frac{d}{16 \sin \theta}}$$

where d is the length of the slope (in meters). Find the time it takes to slide down a 600-m slope inclined at 30°.

83. Beehives In a beehive each cell is a regular hexagonal prism, as shown in the figure. The amount of wax W in the cell depends on the apex angle θ and is given by

$$W = 3.02 - 0.38 \cot \theta + 0.65 \csc \theta$$

Bees instinctively choose θ so as to use the least amount of wax possible.

(a) Use a graphing device to graph W as a function of θ for $0 < \theta < \pi$.

(b) For what value of θ does W have its minimum value? [*Note:* Biologists have discovered that bees rarely deviate from this value by more than a degree or two.]

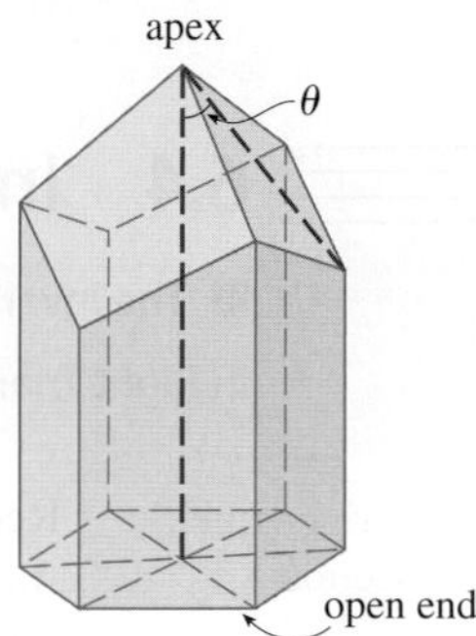

84. Turning a Corner A steel pipe is being carried down a hallway that is 3 m wide. At the end of the hall there is a right-angled turn into a narrower hallway 2 m wide.

(a) Show that the length of the pipe in the figure is modeled by the function

$$L(\theta) = 9 \csc \theta + 6 \sec \theta$$

(b) Graph the function L for $0 < \theta < \pi/2$.

(c) Find the minimum value of the function L.

(d) Explain why the value of L you found in part (c) is the length of the longest pipe that can be carried around the corner.

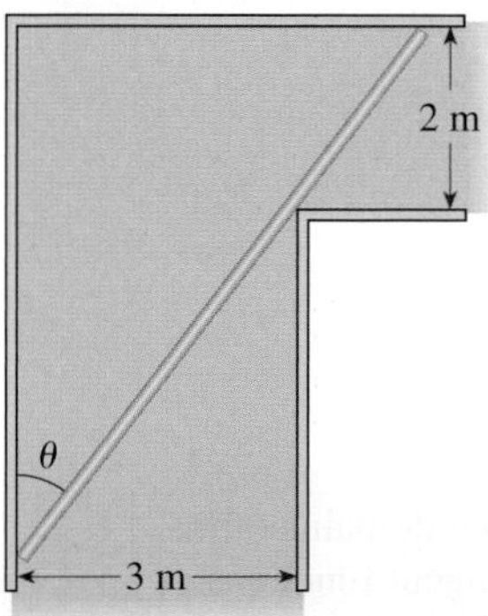

■ Discuss ■ Discover ■ Prove ■ Write

85. Discuss: Using a Calculator To solve a certain problem, you need to find the sine of 4 rad. Your study partners use their calculators and tell you that

$$\sin 4 = 0.0697564737$$

On your calculator you get

$$\sin 4 = -0.7568024953$$

What is wrong? What mistake did your partners make?

86. Discuss ■ Discover: Viète's Trigonometric Diagram In the 16th century the French mathematician François Viète (see Section 1.5) published the following remarkable diagram. Each of the six trigonometric functions of θ is equal to the length of a line segment in the figure. For instance, $\sin\theta = |PR|$, because from triangle OPR we see that

$$\sin\theta = \frac{\text{opp}}{\text{hyp}} = \frac{|PR|}{|OR|} = \frac{|PR|}{1} = |PR|$$

For each of the five other trigonometric functions, find a line segment in the figure whose length equals the value of the function at θ. [*Note:* The radius of the circle is 1, the center is O, segment QS is tangent to the circle at R, and $\angle SOQ$ is a right angle.]

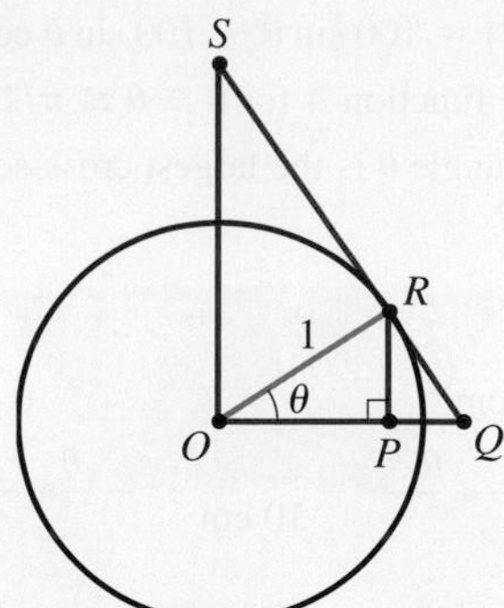

87. Prove: Pythagorean Identities To prove the following Pythagorean identities, start with the first Pythagorean identity, $\sin^2\theta + \cos^2\theta = 1$, which was proved in the text, and then divide both sides by an appropriate trigonometric function of θ.

(a) $\tan^2\theta + 1 = \sec^2\theta$ **(b)** $1 + \cot^2\theta = \csc^2\theta$

88. Discuss ■ Discover: Degrees and Radians What is the smallest positive real number x with the property that the sine of x degrees is equal to the sine of x radians?

6.4 Inverse Trigonometric Functions and Right Triangles

■ The Inverse Sine, Inverse Cosine, and Inverse Tangent Functions ■ Solving for Angles in Right Triangles ■ Evaluating Expressions Involving Inverse Trigonometric Functions

The graphs of the inverse trigonometric functions are studied in Section 5.5.

Recall that for a function to have an inverse, it must be one-to-one. The trigonometric functions are not one-to-one, so we restrict their domains to intervals on which they are one-to-one and on which they attain all their values. The resulting functions have the same range as the original functions but are one-to-one.

■ The Inverse Sine, Inverse Cosine, and Inverse Tangent Functions

Let's first consider the sine function. We restrict the domain of the sine function to angles θ with $-\pi/2 \le \theta \le \pi/2$ (or, equivalently, $-90° \le \theta \le 90°$). From Figure 1 we see that on this domain the sine function attains each of the values in the interval $[-1, 1]$ exactly once and so is one-to-one. Similarly, we restrict the domains of cosine and tangent as shown in Figure 1.

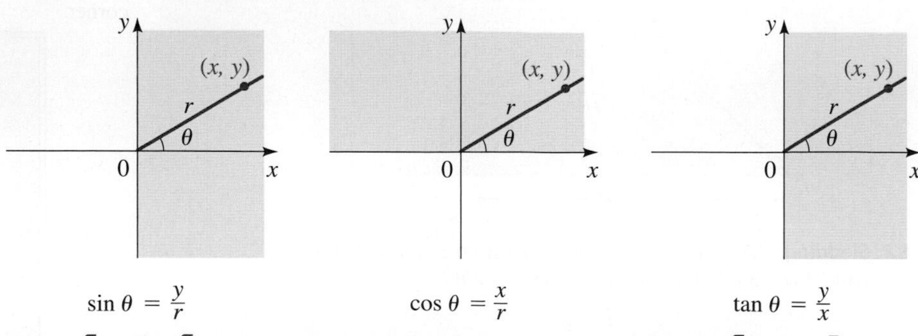

Figure 1 | Restricted domains of the sine, cosine, and tangent functions

On these restricted domains we can define an inverse for each of these functions. By the definition of inverse function we have

$$\sin^{-1} x = y \quad \Leftrightarrow \quad \sin y = x$$
$$\cos^{-1} x = y \quad \Leftrightarrow \quad \cos y = x$$
$$\tan^{-1} x = y \quad \Leftrightarrow \quad \tan y = x$$

We summarize the domains and ranges of the inverse trigonometric functions in the following box.

The Inverse Sine, Inverse Cosine, and Inverse Tangent Functions

The sine, cosine, and tangent functions on the restricted domains $[-\pi/2, \pi/2]$, $[0, \pi]$, and $(-\pi/2, \pi/2)$, respectively, are one-to one and so have inverses. The inverse functions have domain and range as follows.

Function	Domain	Range
$\sin^{-1}$	$[-1, 1]$	$[-\pi/2, \pi/2]$
$\cos^{-1}$	$[-1, 1]$	$[0, \pi]$
$\tan^{-1}$	$\mathbb{R}$	$(-\pi/2, \pi/2)$

The functions $\sin^{-1}$, $\cos^{-1}$, and $\tan^{-1}$ are sometimes called **arcsine, arccosine,** and **arctangent**, respectively.

Since these are inverse functions, they reverse the rule of the original function. For example, since $\sin \pi/6 = \frac{1}{2}$, it follows that $\sin^{-1}(\frac{1}{2}) = \pi/6$. The following example gives further illustrations.

Example 1 ■ Evaluating Inverse Trigonometric Functions

Find the exact value.

(a) $\sin^{-1} \dfrac{\sqrt{3}}{2}$ **(b)** $\cos^{-1}(-\frac{1}{2})$ **(c)** $\tan^{-1} 1$

Solution

(a) The angle in the interval $[-\pi/2, \pi/2]$ whose sine is $\sqrt{3}/2$ is $\pi/3$. Thus $\sin^{-1}(\sqrt{3}/2) = \pi/3$.

(b) The angle in the interval $[0, \pi]$ whose cosine is $-\frac{1}{2}$ is $2\pi/3$. Thus $\cos^{-1}(-\frac{1}{2}) = 2\pi/3$.

(c) The angle in the interval $(-\pi/2, \pi/2)$ whose tangent is 1 is $\pi/4$. Thus $\tan^{-1} 1 = \pi/4$.

Now Try Exercise 5 ■

In many applications of trigonometry—such as surveying and astronomy—measuring angles is often used. In the next example we use a calculator to evaluate inverse trigonometric functions in both degrees and radians.

Example 2 ■ Using a Calculator with Inverse Trigonometric Functions

Find approximate values for the given expression. State your answer in both radians and degrees.

(a) $\sin^{-1}(0.71)$ **(b)** $\tan^{-1} 2$ **(c)** $\cos^{-1} 2$

Solution We use a calculator to approximate these values. Check to make sure you get the same answers on your calculator.

(a) Using the `INV` `SIN`, or `SIN⁻¹`, or `ARC` `SIN` key(s) on the calculator with the calculator in radian mode, we get

$$\sin^{-1}(0.71) \approx 0.790$$

With the calculator in degree mode, we get

$$\sin^{-1}(0.71) \approx 45.235°$$

(b) Using the `INV` `TAN`, or `TAN⁻¹`, or `ARC` `TAN` key(s) on the calculator with the calculator in radian mode, we get

$$\tan^{-1}2 \approx 1.107$$

With the calculator in degree mode, we get

$$\tan^{-1}2 \approx 63.435°$$

(c) Since $2 > 1$, it is not in the domain of $\cos^{-1}$, so $\cos^{-1}2$ is not defined.

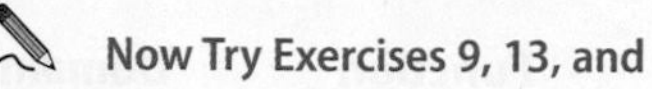

Now Try Exercises 9, 13, and 15 ■

■ Solving for Angles in Right Triangles

In Section 6.2 we solved triangles by using the trigonometric ratios to find the unknown sides. We now use the inverse trigonometric functions to solve for *angles* in a right triangle.

Example 3 ■ Finding an Angle in a Right Triangle

Find the angle θ (in degrees) for the triangle shown in Figure 2.

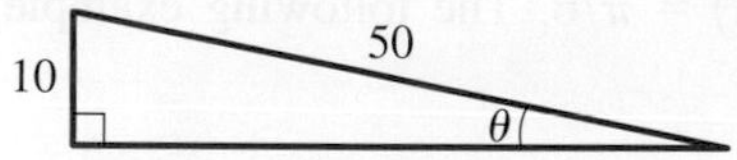

Figure 2

Solution Since θ is the angle opposite the side of length 10 and the hypotenuse has length 50, we have

$$\sin\theta = \frac{10}{50} = \frac{1}{5} \qquad \sin\theta = \frac{\text{opp}}{\text{hyp}}$$

Now we can use $\sin^{-1}$ to find θ.

$$\theta = \sin^{-1}\left(\tfrac{1}{5}\right) \qquad \text{Definition of } \sin^{-1}$$

$$\theta \approx 11.5° \qquad \text{Calculator (in degree mode)}$$

Now Try Exercise 17 ■

Example 4 ■ Solving for an Angle in a Right Triangle

A 12-m ladder leans against a building. If the base of the ladder is 1.8 m from the base of the building, what is the angle formed by the ladder and the building?

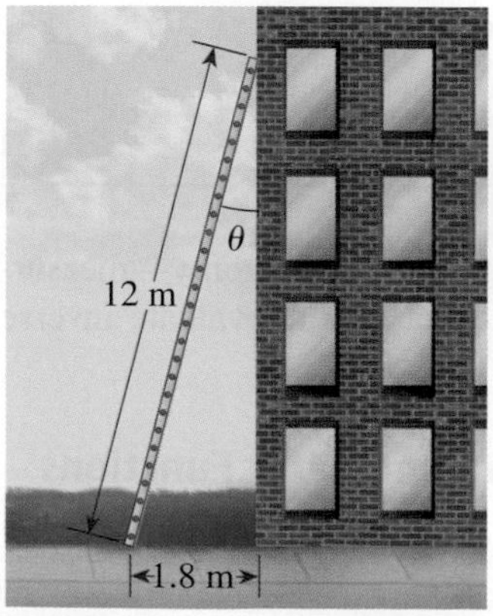

Solution First we sketch a diagram as in Figure 3. If θ is the angle between the ladder and the building, then

$$\sin\theta = \frac{1.8}{12} = 0.15 \qquad \sin\theta = \frac{\text{opp}}{\text{hyp}}$$

Now we use $\sin^{-1}$ to find θ.

$$\theta = \sin^{-1}(0.15) \qquad \text{Definition of } \sin^{-1}$$

$$\theta \approx 8.6° \qquad \text{Calculator (in degree mode)}$$

Figure 3

Now Try Exercise 47 ■

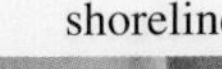

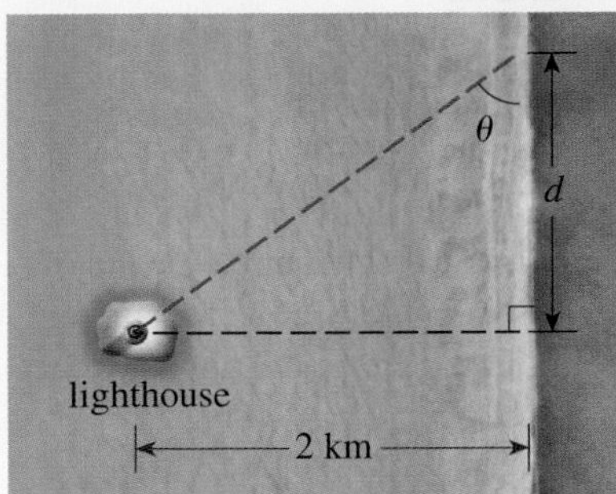

Figure 4

Example 5 ■ The Angle of a Beam of Light

A lighthouse is located on an island that is 2 km off a straight shoreline (see Figure 4). Express the angle formed by the beam of light and the shoreline in terms of the distance d in the figure.

Solution From the figure we see that

$$\tan \theta = \frac{2}{d} \qquad \tan \theta = \frac{\text{opp}}{\text{adj}}$$

Taking the inverse tangent of both sides, we get

$$\tan^{-1}(\tan \theta) = \tan^{-1}\left(\frac{2}{d}\right) \qquad \text{Take } \tan^{-1} \text{ of both sides}$$

$$\theta = \tan^{-1}\left(\frac{2}{d}\right) \qquad \text{Property of inverse functions: } \tan^{-1}(\tan \theta) = \theta$$

 **Now Try Exercise 49** ■

In Sections 6.5 and 6.6 we will learn how to solve any triangle (not necessarily a right triangle). The angles in a triangle are always in the interval $(0, \pi)$ (or between 0° and 180°). We'll see that to solve such triangles, we need to find all angles in the interval $(0, \pi)$ that have a specified sine or cosine. We do this in the next example.

Example 6 ■ Solving a Basic Trigonometric Equation on an Interval

Find all angles θ between 0° and 180° that satisfy the given equation.

(a) $\sin \theta = 0.4$ **(b)** $\cos \theta = 0.4$

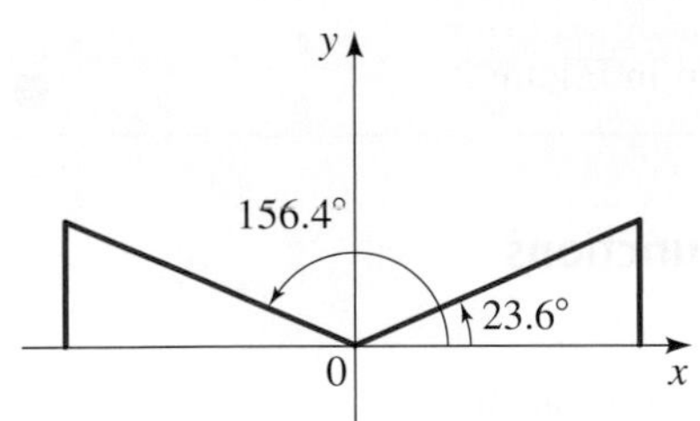

Figure 5

Solution

(a) We use $\sin^{-1}$ to find one solution.

$$\sin \theta = 0.4 \qquad \text{Equation}$$
$$\theta = \sin^{-1}(0.4) \qquad \text{Take } \sin^{-1} \text{ of each side}$$
$$\theta \approx 23.6° \qquad \text{Calculator (in degree mode)}$$

Another solution with θ between 0° and 180° is obtained by taking the supplement of the angle: $180° - 23.6° = 156.4°$. (See Figure 5.) So the solutions of the equation with θ between 0° and 180° are

$$\theta \approx 23.6° \qquad \text{and} \qquad \theta \approx 156.4°$$

The *supplement* of an angle θ (where $0 \le \theta \le 180°$) is the angle $180° - \theta$.

(b) The cosine function is one-to-one on the interval $[0, \pi]$, so there is only one solution of the equation with θ between 0° and 180°. We find that solution by taking $\cos^{-1}$ of each side.

$$\cos \theta = 0.4$$
$$\theta = \cos^{-1}(0.4) \qquad \text{Take } \cos^{-1} \text{ of each side}$$
$$\theta \approx 66.4° \qquad \text{Calculator (in degree mode)}$$

The solution is $\theta \approx 66.4°$

 **Now Try Exercises 27 and 31** ■

■ Evaluating Expressions Involving Inverse Trigonometric Functions

Expressions like $\cos(\sin^{-1} x)$ arise in calculus. We find exact values of such expressions using trigonometric identities or right triangles.

Example 7 ■ Composing Trigonometric Functions and Their Inverses

Find $\cos(\sin^{-1}(\frac{3}{5}))$.

Solution 1 Let $\theta = \sin^{-1}(\frac{3}{5})$. Then θ is the number in the interval $[-\pi/2, \pi/2]$ whose sine is $\frac{3}{5}$. Let's interpret θ as an angle and draw a right triangle with θ as one of its acute angles, with opposite side 3 and hypotenuse 5. (See Figure 6.) The remaining leg of the triangle is found by the Pythagorean Theorem to be 4. From the figure we get

Figure 6 | $\cos\theta = \dfrac{4}{5}$

$$\cos(\sin^{-1}(\tfrac{3}{5})) = \cos\theta \qquad \theta = \sin^{-1}(\tfrac{3}{5})$$

$$= \frac{4}{5} \qquad \cos\theta = \frac{\text{adj}}{\text{hyp}}$$

So $\cos(\sin^{-1}(\frac{3}{5})) = \frac{4}{5}$.

Solution 2 To find $\cos(\sin^{-1}(\frac{3}{5}))$, we first write the cosine function in terms of the sine function so we can use the cancellation properties of inverse functions. Let $u = \sin^{-1}(\frac{3}{5})$. Since $-\pi/2 \le u \le \pi/2$, $\cos u$ is positive, and we can write the following:

$$\cos u = +\sqrt{1 - \sin^2 u} \qquad \cos^2 u + \sin^2 u = 1$$

$$= \sqrt{1 - \sin^2(\sin^{-1}(\tfrac{3}{5}))} \qquad u = \sin^{-1}(\tfrac{3}{5})$$

$$= \sqrt{1 - (\tfrac{3}{5})^2} \qquad \text{Property of inverse functions: } \sin(\sin^{-1}(\tfrac{3}{5})) = \tfrac{3}{5}$$

$$= \sqrt{1 - \tfrac{9}{25}} = \sqrt{\tfrac{16}{25}} = \tfrac{4}{5} \qquad \text{Calculate}$$

So $\cos(\sin^{-1}(\frac{3}{5})) = \frac{4}{5}$.

 **Now Try Exercise 33** ■

Example 8 ■ Composing Trigonometric Functions and Their Inverses

Write $\sin(\cos^{-1}x)$ and $\tan(\cos^{-1}x)$ as algebraic expressions in x.

Solution 1 Let $\theta = \cos^{-1}x$; then $\cos\theta = x$. In Figure 7 we sketch a right triangle with an acute angle θ, adjacent side x, and hypotenuse 1. By the Pythagorean Theorem the remaining leg is $\sqrt{1 - x^2}$. From the figure we have

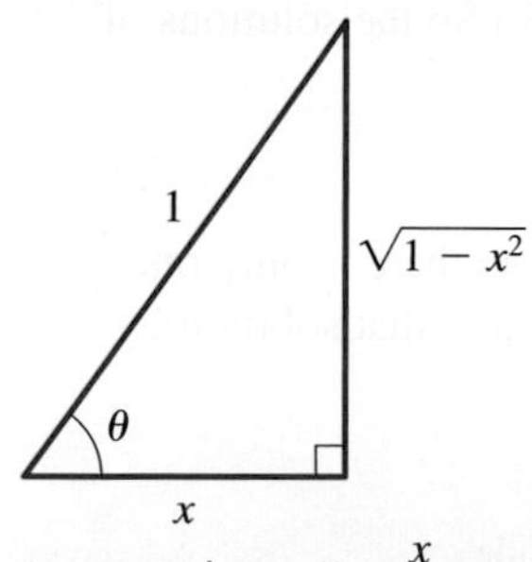

Figure 7 | $\cos\theta = \dfrac{x}{1} = x$

$$\sin(\cos^{-1}x) = \sin\theta = \sqrt{1 - x^2} \qquad \text{and} \qquad \tan(\cos^{-1}x) = \tan\theta = \frac{\sqrt{1 - x^2}}{x}$$

Solution 2 Let $u = \cos^{-1}x$. We need to find $\sin u$ and $\tan u$ in terms of x. As in Example 7 the idea here is to write sine and tangent in terms of cosine. We have

$$\sin u = \pm\sqrt{1 - \cos^2 u} \qquad \text{and} \qquad \tan u = \frac{\sin u}{\cos u} = \frac{\pm\sqrt{1 - \cos^2 u}}{\cos u}$$

To choose the proper signs, note that u lies in the interval $[0, \pi]$ because $u = \cos^{-1}x$. Since $\sin u$ is positive on this interval, the $+$ sign is the correct choice. Substituting $u = \cos^{-1}x$ in the displayed equations and using the cancellation property $\cos(\cos^{-1}x) = x$, we get

$$\sin(\cos^{-1}x) = \sqrt{1 - x^2} \qquad \text{and} \qquad \tan(\cos^{-1}x) = \frac{\sqrt{1 - x^2}}{x}$$

The method of Solution 2 is the same as that used in Example 5.5.7.

Now Try Exercises 41 and 43 ■

Note In Solution 1 of Example 8 it might seem that because we are sketching a triangle, the angle $\theta = \cos^{-1}x$ must be acute. But it turns out that the method of sketching a triangle, which is often used in calculus, works for any x. The domains and ranges of all six inverse trigonometric functions have been chosen in such a way that we can always use a triangle to find $S(T^{-1}x)$, where S and T are any trigonometric functions.

6.4 | Exercises

■ Concepts

1. For a function to have an inverse, it must be ________. To define the inverse sine function, we restrict the ________ of the sine function to the interval ________.

2. (a) The function $\sin^{-1}$ has domain ______ and range ______.

(b) The function $\cos^{-1}$ has domain ______ and range ______.

(c) The function $\tan^{-1}$ has domain ______ and range ______.

3. In the triangle shown we can find the angle θ as follows.

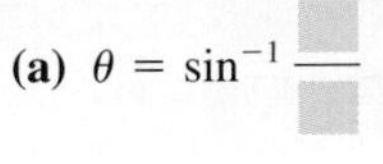

(a) $\theta = \sin^{-1}\frac{\square}{\square}$

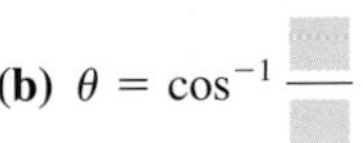

(b) $\theta = \cos^{-1}\frac{\square}{\square}$

(c) $\theta = \tan^{-1}\frac{\square}{\square}$

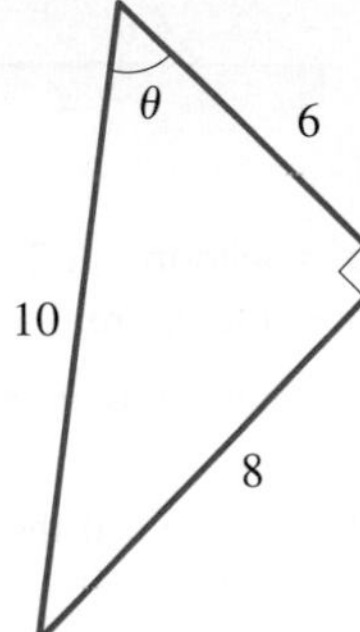

4. To find $\tan\left(\cos^{-1}\left(\frac{5}{13}\right)\right)$, we let $\theta = \cos^{-1}\left(\frac{5}{13}\right)$. So $\cos\theta =$ ______. We then complete the triangle in the figure and use the triangle to find that $\tan\left(\cos^{-1}\left(\frac{5}{13}\right)\right) =$ ______.

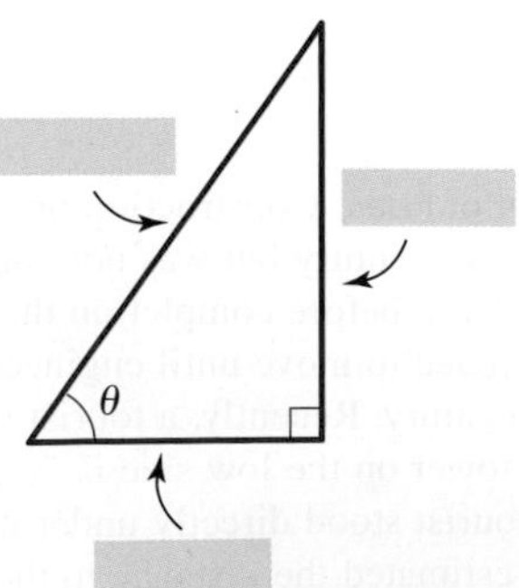

■ Skills

5–8 ■ Evaluating Inverse Trigonometric Functions Find the exact value of each expression, if it is defined. Express your answer in radians.

5. (a) $\sin^{-1}1$ **(b)** $\cos^{-1}0$ **(c)** $\tan^{-1}\sqrt{3}$

6. (a) $\sin^{-1}0$ **(b)** $\cos^{-1}(-1)$ **(c)** $\tan^{-1}0$

7. (a) $\sin^{-1}\left(-\frac{\sqrt{2}}{2}\right)$ **(b)** $\cos^{-1}\left(-\frac{\sqrt{2}}{2}\right)$ **(c)** $\tan^{-1}(-1)$

8. (a) $\sin^{-1}\left(-\frac{\sqrt{3}}{2}\right)$ **(b)** $\cos^{-1}\left(-\frac{1}{2}\right)$ **(c)** $\tan^{-1}(-\sqrt{3})$

9–16 ■ Evaluating Inverse Trigonometric Functions Use a calculator to find an approximate value of each expression rounded to three decimal places, if it is defined. State your answer in radians and degrees.

9. $\sin^{-1}(0.30)$ **10.** $\cos^{-1}(-0.2)$

11. $\cos^{-1}\left(\frac{1}{3}\right)$ **12.** $\sin^{-1}\left(\frac{5}{6}\right)$

13. $\tan^{-1}3$ **14.** $\tan^{-1}(-4)$

15. $\cos^{-1}3$ **16.** $\sin^{-1}(-2)$

17–22 ■ Finding Angles in Right Triangles Find the angle θ in degrees, rounded to one decimal place.

17.

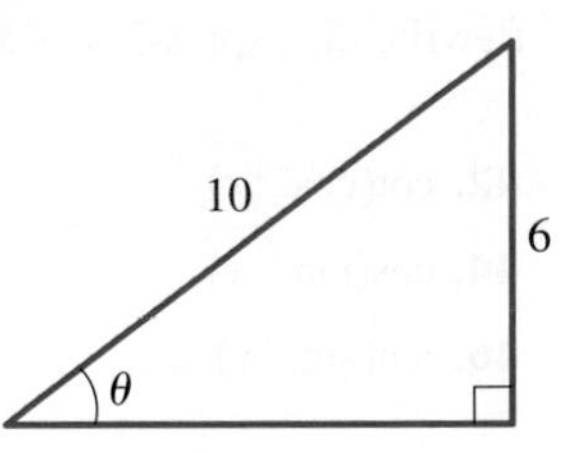

18.

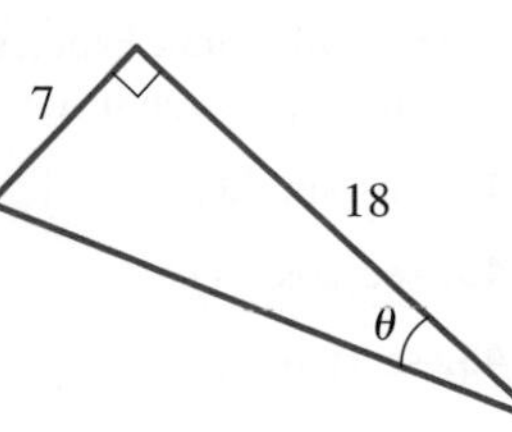

19.

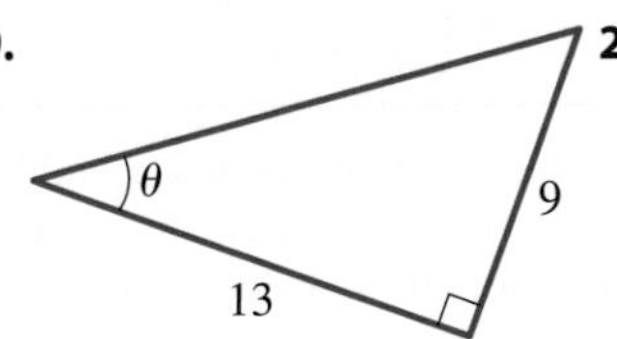

20.

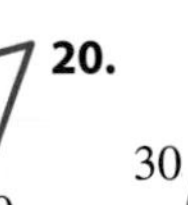

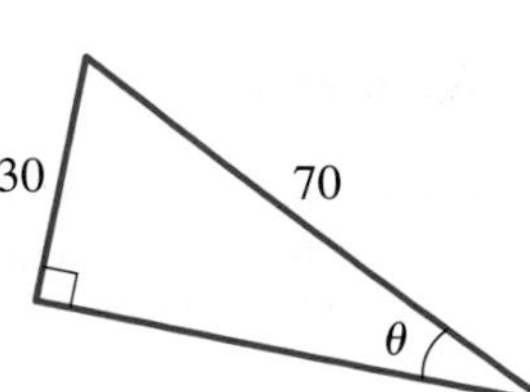

21.

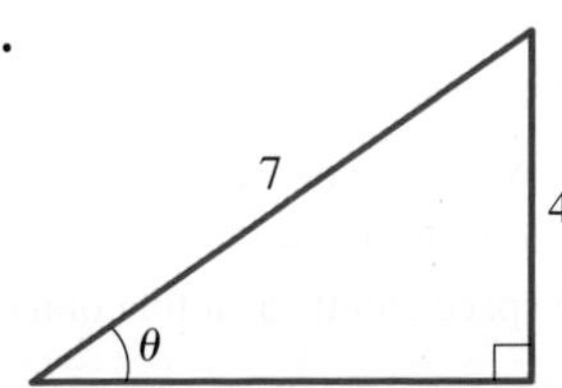

22.

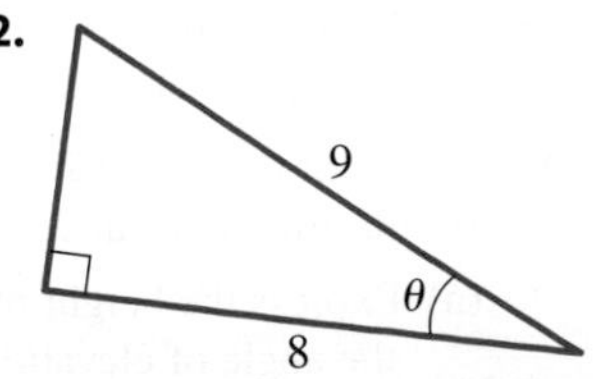

23–26 ■ Finding Angles Find the angle θ in degrees, rounded to one decimal place.

23.

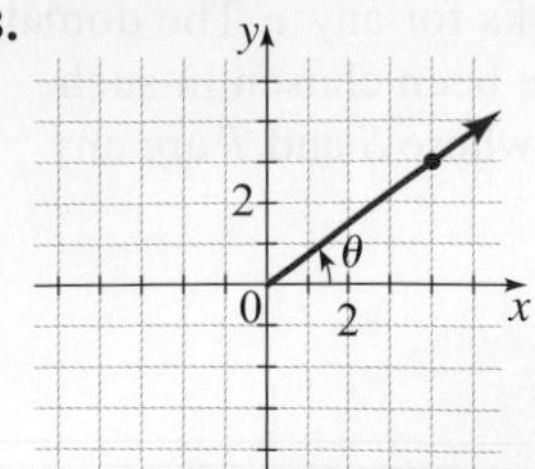

24.

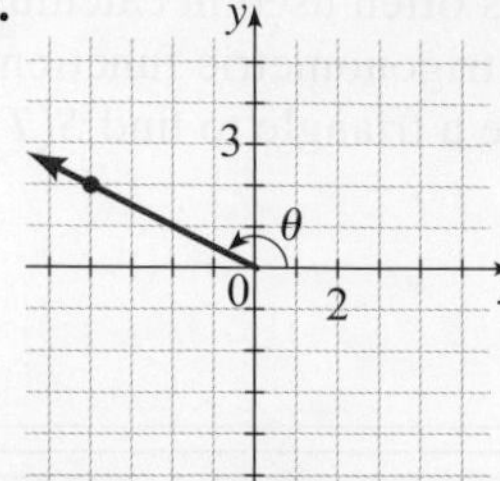

25.

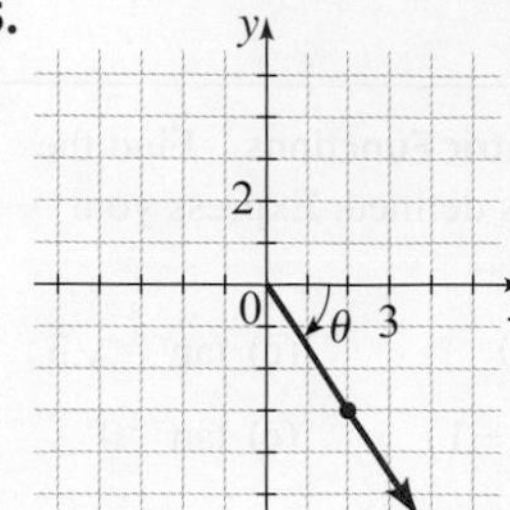

26.

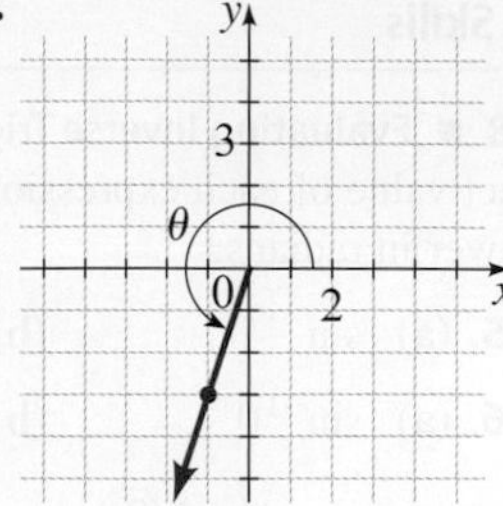

27–32 ■ Basic Trigonometric Equations Find all angles θ between 0° and 180° that satisfy the given equation. Round your answer(s) to one decimal place.

27. $\sin\theta = \frac{2}{3}$ **28.** $\cos\theta = \frac{3}{4}$

29. $\cos\theta = -\frac{2}{5}$ **30.** $\tan\theta = -20$

31. $\tan\theta = 5$ **32.** $\sin\theta = \frac{4}{5}$

33–40 ■ Value of an Expression Find the exact value of the expression.

33. $\cos(\sin^{-1}(\frac{4}{5}))$ **34.** $\cos(\tan^{-1}(\frac{4}{3}))$

35. $\sec(\sin^{-1}(\frac{12}{13}))$ **36.** $\csc(\cos^{-1}(-\frac{7}{25}))$

37. $\sin(\tan^{-1}(-\frac{12}{5}))$ **38.** $\sec(\tan^{-1}(-\frac{2}{3}))$

39. $\sin(\sec^{-1}(-4))$ **40.** $\csc(\cot^{-1}(\frac{3}{4}))$

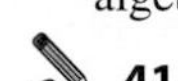

41–46 ■ Algebraic Expressions Rewrite the expression as an algebraic expression in x.

41. $\cos(\sin^{-1}x)$ **42.** $\cot(\cos^{-1}x)$

43. $\sec(\cos^{-1}x)$ **44.** $\cos(\tan^{-1}x)$

45. $\sec(\tan^{-1}x)$ **46.** $\tan(\sin^{-1}x)$

Applications

47. Leaning Ladder A 6-m ladder is leaning against a building. If the base of the ladder is 1.8 m from the base of the building, what is the angle of elevation of the ladder? How high does the ladder reach on the building?

48. Angle of the Sun A 29-m tree casts a shadow that is 36 m long. What is the angle of elevation of the sun?

49. Height of the Space Shuttle An observer views the space shuttle from a distance of 2 km from the launch pad.

(a) Express the height of the space shuttle as a function of the angle of elevation θ.

(b) Express the angle of elevation θ as a function of the height h of the space shuttle.

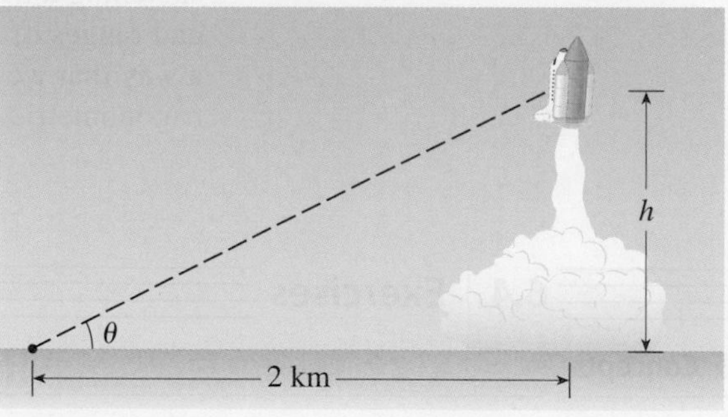

50. Height of a Pole A 15-m pole casts a shadow as shown in the figure.

(a) Express the angle of elevation θ of the sun as a function of the length s of the shadow.

(b) Find the angle θ of elevation of the sun when the shadow is 6 m long.

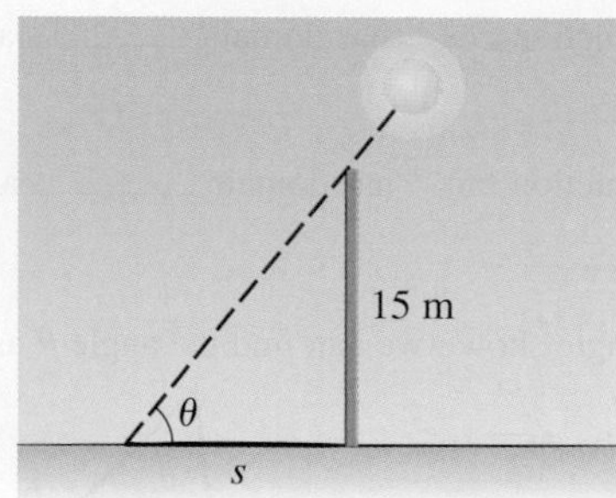

51. Height of a Balloon A 204-m rope anchors a hot-air balloon as shown in the figure.

(a) Express the angle θ as a function of the height h of the balloon.

(b) Find the angle θ if the balloon is 150 m high.

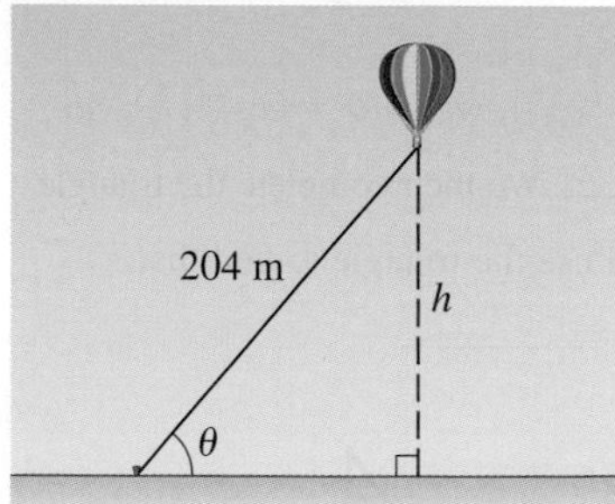

52. Leaning Tower of Pisa Construction on the tower of Pisa began in the 12th century but was not completed until the 14th century. Even before completion the tower began to lean and continued to move until engineers stabilized it in the late 20th century. Recently, a tourist was told that the height of the tower on the low side is 56 m. (See the figure.) The tourist stood directly under the edge of the low side and estimated the distance to the base of the tower to be 3.9 m. Estimate the angle α at which the tower now leans.

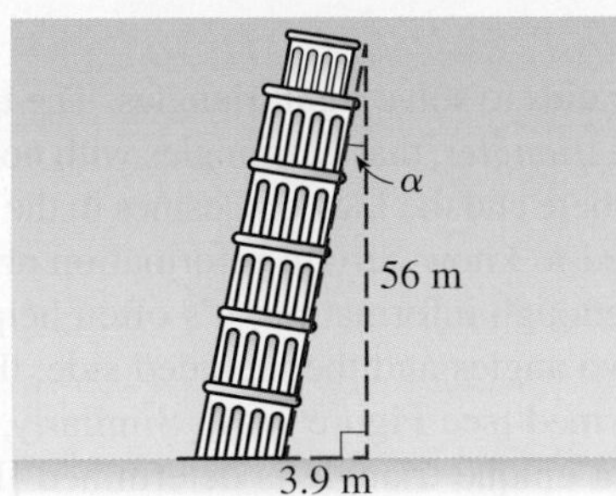

53. View from a Satellite The figures indicate that the higher the orbit of a satellite, the more of the earth the satellite can "see." Let θ, s, and h be as shown in the figure, and assume that the earth is a sphere of radius 6374 km.

(a) Express the angle θ as a function of h.

(b) Express the distance s as a function of θ.

(c) Express the distance s as a function of h. [*Hint:* Find the composition of the functions in parts (a) and (b).]

(d) If the satellite is 160 km above the earth, what is the distance s that it can see?

(e) How high does the satellite have to be to see both Los Angeles and New York, 3920 km apart?

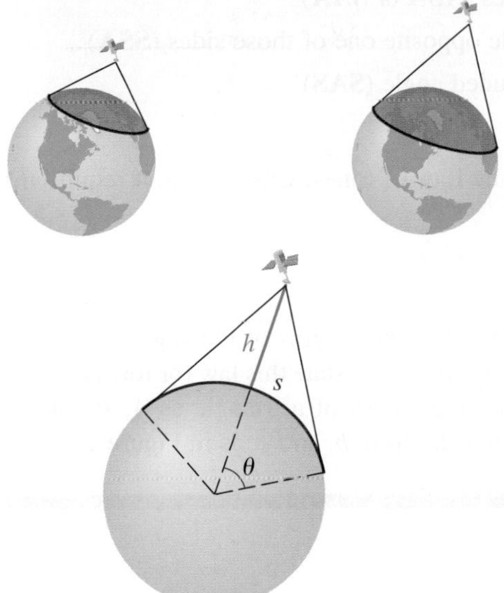

54. Surfing the Perfect Wave For a wave to be surfable, it can't break all at once. Robert Guza and Tony Bowen have shown that a wave has a surfable shoulder if it hits the shoreline at an angle θ given by

$$\theta = \sin^{-1}\left(\frac{1}{(2n + 1)\tan\beta}\right)$$

where β is the angle at which the beach slopes down and where $n = 0, 1, 2, \ldots$.

(a) For $\beta = 10°$, find θ when $n = 3$.

(b) For $\beta = 15°$, find θ when $n = 2$, 3, and 4. Explain why the formula does not give a value for θ when $n = 0$ or 1.

55. Rainbows Rainbows are created when sunlight of different wavelengths (colors) is refracted and reflected in raindrops. The angle of elevation θ of a rainbow is always the same. It can be shown that $\theta = 4\beta - 2\alpha$, where

$$\sin\alpha = k\sin\beta$$

and $\alpha = 59.4°$ and $k = 1.33$ is the index of refraction of water. Use the given information to find the angle of elevation θ of a rainbow. (For a mathematical explanation of rainbows see *Calculus Early Transcendentals,* 9th Edition, by Stewart, Clegg, Watson, page 289–290.)

■ Discuss ■ Discover ■ Prove ■ Write

56. Prove: Inverse Trigonometric Functions on a Calculator Most calculators do not have keys for $\sec^{-1}$, $\csc^{-1}$, or $\cot^{-1}$. Prove the following identities, and then use these identities and a calculator to find $\sec^{-1}2$, $\csc^{-1}3$, and $\cot^{-1}4$.

$$\sec^{-1}x = \cos^{-1}\left(\frac{1}{x}\right) \qquad (x \ge 1)$$

$$\csc^{-1}x = \sin^{-1}\left(\frac{1}{x}\right) \qquad (x \ge 1)$$

$$\cot^{-1}x = \tan^{-1}\left(\frac{1}{x}\right) \qquad (x > 0)$$

6.5 The Law of Sines

■ The Law of Sines ■ The Ambiguous Case

In Section 6.2 we used the trigonometric ratios to solve right triangles. The trigonometric functions can also be used to solve *oblique triangles*, that is, triangles with no right angles. To do this, we first study the Law of Sines here and the Law of Cosines in the next section.

Note that if we know two angles of a triangle, then we know all three angles (because the sum of the angles of any triangle is 180°.)

In general, to solve a triangle, we need to know certain information about its sides and angles. To decide whether we have enough information, it's often helpful to make a sketch. For instance, if we are given two angles and the included side, then it's clear that one and only one triangle can be formed [see Figure 1(a)]. Similarly, if two sides and the included angle are known, then a unique triangle is determined [Figure 1(c)]. But if we know all three angles and no sides, we cannot uniquely determine the triangle because many triangles can have the same three angles. (All these triangles would be similar, of course.) So we won't consider this last case.

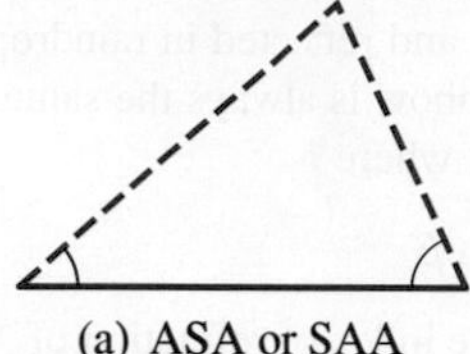
(a) ASA or SAA

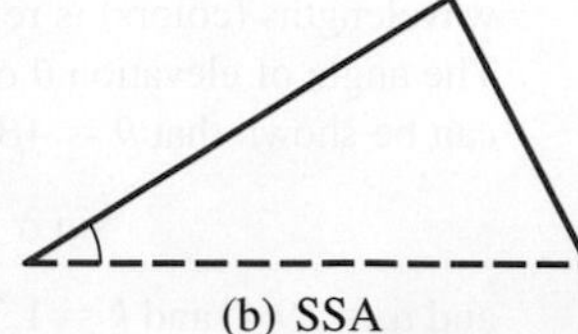
(b) SSA

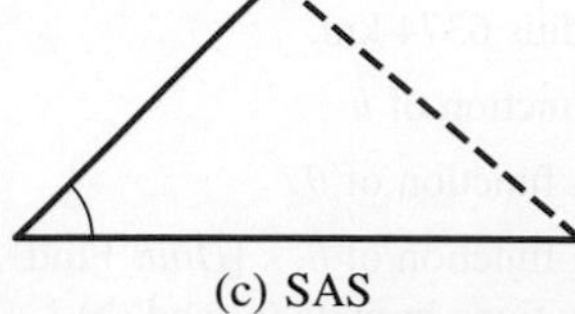
(c) SAS

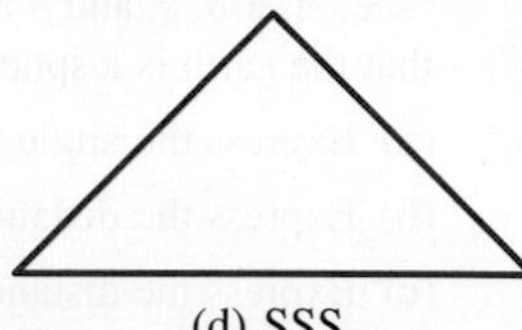
(d) SSS

Figure 1

In general, a triangle is determined by three of its six parts (angles and sides) as long as at least one of these three parts is a side. So the possibilities, illustrated in Figure 1, are as follows.

Case 1 One side and two angles (**ASA** or **SAA**)
Case 2 Two sides and the angle opposite one of those sides (**SSA**)
Case 3 Two sides and the included angle (**SAS**)
Case 4 Three sides (**SSS**)

Cases 1 and 2 are solved by using the Law of Sines; Cases 3 and 4 require the Law of Cosines.

■ The Law of Sines

The **Law of Sines** says that in any triangle the lengths of the sides are proportional to the sines of the corresponding opposite angles. To state this law (or formula) more easily, we follow the convention of labeling the angles of a triangle as A, B, and C and the lengths of the corresponding opposite sides as a, b, and c, as in Figure 2.

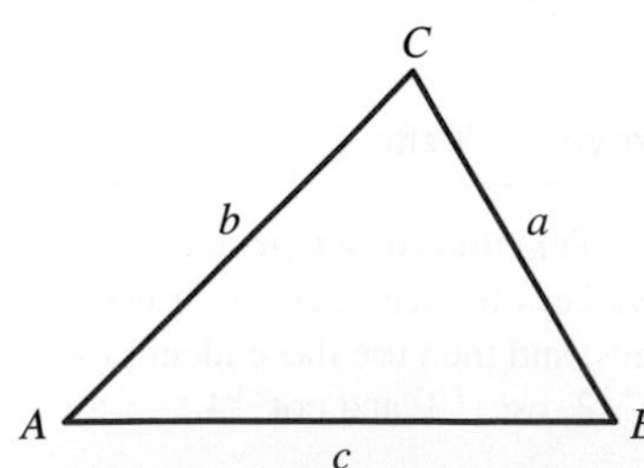

Figure 2

The Law of Sines

In triangle ABC we have

$$\frac{\sin A}{a} = \frac{\sin B}{b} = \frac{\sin C}{c}$$

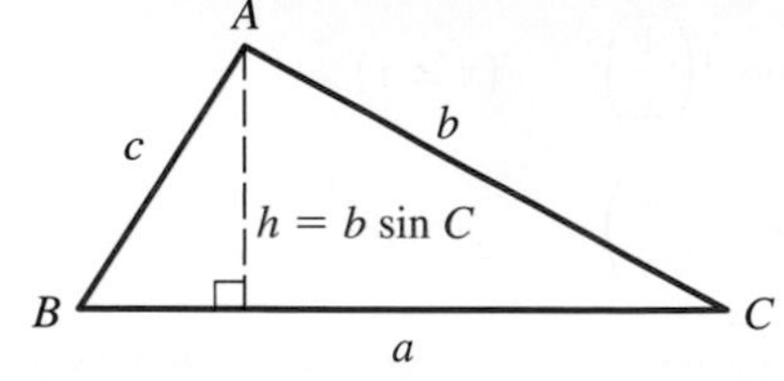

Figure 3

Proof To see why the Law of Sines is true, refer to Figure 3. By the formula given in Section 6.3 the area of triangle ABC is $\frac{1}{2}ab\sin C$. By the same formula the area of this triangle is also $\frac{1}{2}ac\sin B$ and $\frac{1}{2}bc\sin A$. Thus

$$\tfrac{1}{2}bc\sin A = \tfrac{1}{2}ac\sin B = \tfrac{1}{2}ab\sin C$$

Multiplying by $2/(abc)$ gives the Law of Sines. ■

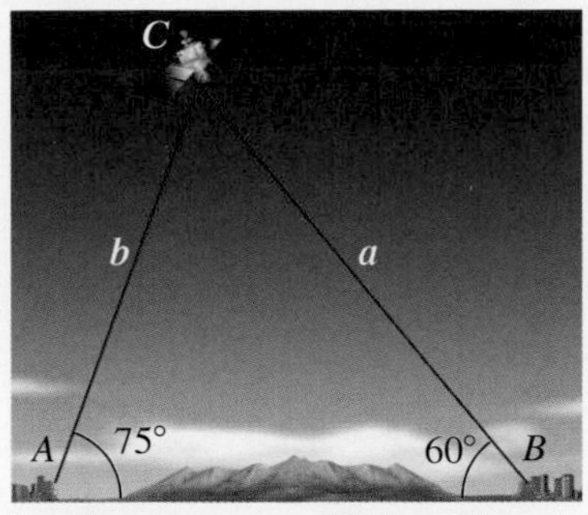

Figure 4

Example 1 ■ Tracking a Satellite (ASA)

A satellite orbiting the earth passes directly overhead at observation stations in Phoenix and Los Angeles, 544 km apart. At an instant when the satellite is between these two stations, its angle of elevation is simultaneously observed to be 60° at Phoenix and 75° at Los Angeles. How far is the satellite from Los Angeles?

Solution We need to find the distance b in Figure 4. Since the sum of the angles in any triangle is 180°, we see that $\angle C = 180° - (75° + 60°) = 45°$. (See Figure 4.)

$$\frac{\sin B}{b} = \frac{\sin C}{c} \qquad \text{Law of Sines}$$

$$\frac{\sin 60°}{b} = \frac{\sin 45°}{544} \qquad \text{Substitute}$$

$$b = \frac{544 \sin 60°}{\sin 45°} \approx 666 \qquad \text{Solve for } b$$

The distance of the satellite from Los Angeles is approximately 666 km.

 **Now Try Exercises 7 and 35** ■

Example 2 ■ Solving a Triangle (SAA)

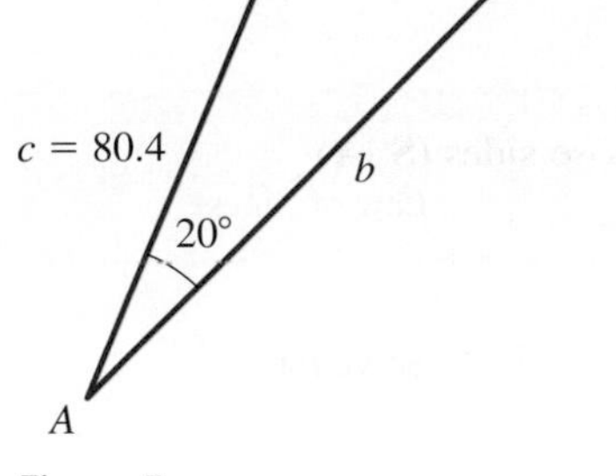

Figure 5

Solve the triangle in Figure 5.

Solution First, $\angle B = 180° - (20° + 25°) = 135°$. Since side c is known, to find side a, we use the relation

$$\frac{\sin A}{a} = \frac{\sin C}{c} \qquad \text{Law of Sines}$$

$$a = \frac{c \sin A}{\sin C} = \frac{80.4 \sin 20°}{\sin 25°} \approx 65.1 \qquad \text{Solve for } a$$

Similarly, to find b, we use

$$\frac{\sin B}{b} = \frac{\sin C}{c} \qquad \text{Law of Sines}$$

$$b = \frac{c \sin B}{\sin C} = \frac{80.4 \sin 135°}{\sin 25°} \approx 134.5 \qquad \text{Solve for } b$$

Now Try Exercise 19 ■

■ The Ambiguous Case

In Case 1 (ASA or SAA) a unique triangle is determined by the given information, as in Examples 1 and 2. But in Case 2 (SSA), the given angle A and the two sides a and b may form two, one, or no triangles, as illustrated in Figure 6.

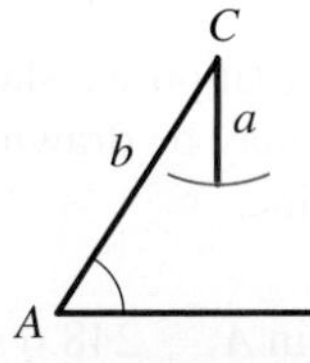

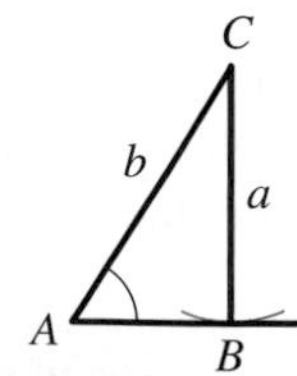

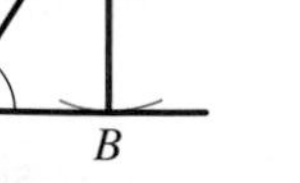

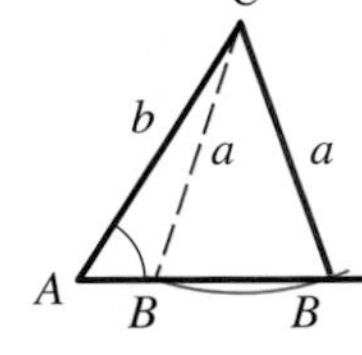

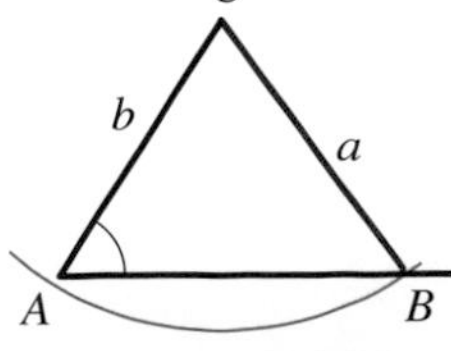

Figure 6 | Case 2: Angle A and sides a, b

Case 2 is called the **ambiguous case** because the given angle and two sides may not determine a unique triangle. Figure 6 shows the possibilities when angle A and sides a and b are given. In part (a) no solution is possible because side a is too short to complete the triangle. In part (b) the solution is a right triangle. In part (c) two solutions are possible and in part (d) there is one triangle with the given properties.

Example 3 ■ SSA, the One-Solution Case

Solve triangle ABC, where $\angle A = 45°$, $a = 7\sqrt{2}$, and $b = 7$.

Solution We first sketch the triangle with the information we have (see Figure 7). Our sketch is necessarily tentative because we don't yet know the other angles. Nevertheless, we can now see the possibilities.

Figure 7

We first find $\angle B$.

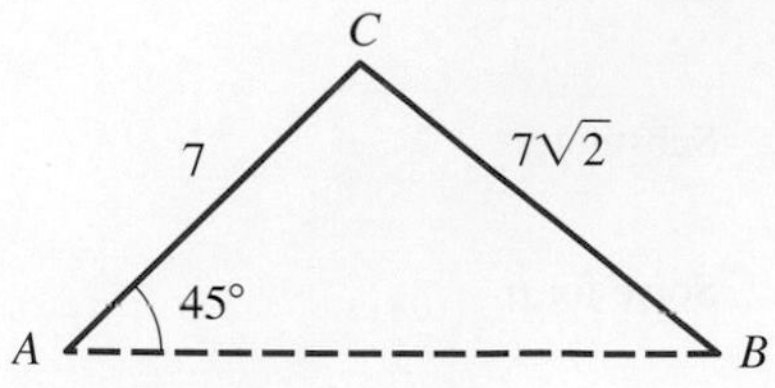

$$\frac{\sin A}{a} = \frac{\sin B}{b} \qquad \text{Law of Sines}$$

$$\sin B = \frac{b \sin A}{a} = \frac{7}{7\sqrt{2}} \sin 45° = \left(\frac{1}{\sqrt{2}}\right)\left(\frac{\sqrt{2}}{2}\right) = \frac{1}{2} \qquad \text{Solve for } \sin B$$

We consider only angles smaller than 180°, since no triangle can contain an angle of 180° or larger.

Which angles B have $\sin B = \frac{1}{2}$? From the preceding section we know that there are two such angles smaller than 180° (they are 30° and 150°). Which of these angles is compatible with what we know about triangle ABC? Since $\angle A = 45°$, we cannot have $\angle B = 150°$, because $45° + 150° > 180°$. So $\angle B = 30°$, and the remaining angle is $\angle C = 180° - (30° + 45°) = 105°$.

Now we can find side c.

$$\frac{\sin B}{b} = \frac{\sin C}{c} \qquad \text{Law of Sines}$$

$$c = \frac{b \sin C}{\sin B} = \frac{7 \sin 105°}{\sin 30°} = \frac{7 \sin 105°}{\frac{1}{2}} \approx 13.5 \qquad \text{Solve for } c$$

Now Try Exercise 21 ■

In Example 3 there were two possibilities for angle B, and one of these was not compatible with the rest of the information. In general, if $\sin A < 1$, we must check the angle and its supplement as possibilities, because any angle smaller than 180° can be in the triangle. To decide whether either possibility works, we check to see whether the resulting sum of the angles exceeds 180°. It can happen, as in Figure 6(c), that both possibilities are compatible with the given information. In that case, two different triangles are solutions to the problem.

The *supplement* of an angle θ (where $0 \le \theta \le 180°$) is the angle $180° - \theta$.

Example 4 ■ SSA, the Two-Solution Case

Solve triangle ABC if $\angle A = 43.1°$, $a = 186.2$, and $b = 248.6$.

Solution From the given information we sketch the triangle shown in Figure 8 (on the next page). Note that side a may be drawn in two possible positions to complete the triangle. From the Law of Sines

$$\sin B = \frac{b \sin A}{a} = \frac{248.6 \sin 43.1°}{186.2} \approx 0.91225$$

Surveying is a method of land measurement used for mapmaking. Surveyors use a process called *triangulation* in which a network of thousands of interlocking triangles is created on the area to be mapped. The process is started by measuring the length of a *baseline* between two surveying stations. Then, with the use of an instrument called a *theodolite*, the angles between these two stations and a third station are measured. The Law of Sines is then used to calculate the two other sides of the triangle formed by the three stations. The calculated sides are used as baselines, and the process is repeated over and over to create a network of triangles. In this method the only distance measured is the initial baseline; all other distances are calculated from the Law of Sines. This method is practical because it is much easier to measure angles than distances.

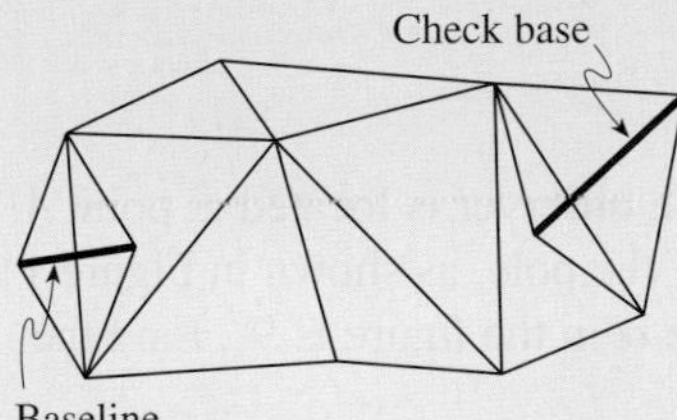

One of the most ambitious mapmaking efforts of all time was the Great Trigonometric Survey of India (see Problem 9 in the *Focus on Modeling* at the end of this chapter). This project required several expeditions and took more than a century to complete. The famous expedition of 1823, led by **Sir George Everest**, lasted 20 years. Ranging over treacherous terrain and encountering the dreaded malaria-carrying mosquito, this expedition reached the foothills of the Himalayas. A later expedition—using triangulation—calculated the height of the highest peak of the Himalayas to be 8840 m. The peak was named in honor of Sir George Everest.

Today, with the use of satellites, the height of Mt. Everest is estimated to be 8848 m. The very close agreement of these two estimates shows the great accuracy of the trigonometric method.

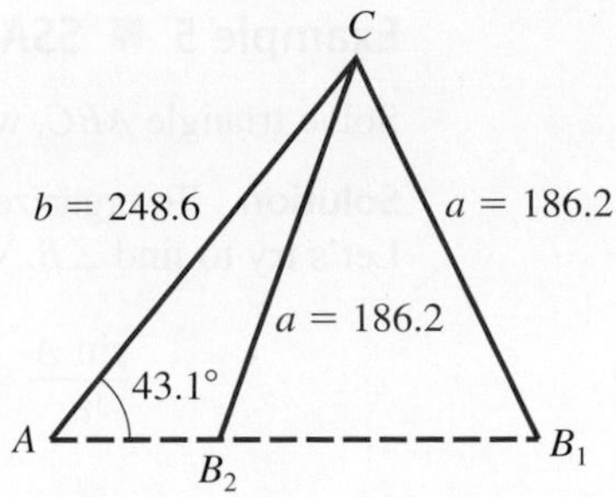

Figure 8

There are two possible angles B between $0°$ and $180°$ such that $\sin B = 0.91225$. Using a calculator (in degree mode), we find that one of the angles is

$$\sin^{-1}(0.91225) \approx 65.8°$$

The other angle is approximately $180° - 65.8° = 114.2°$. We denote these two angles by B_1 and B_2 so that

$$\angle B_1 \approx 65.8° \qquad \text{and} \qquad \angle B_2 \approx 114.2°$$

Thus two triangles satisfy the given conditions: triangle $A_1B_1C_1$ and triangle $A_2B_2C_2$.

Solve triangle $A_1B_1C_1$:

$$\angle C_1 \approx 180° - (43.1° + 65.8°) = 71.1° \qquad \text{Find } \angle C_1$$

Thus $$c_1 = \frac{a_1 \sin C_1}{\sin A_1} \approx \frac{186.2 \sin 71.1°}{\sin 43.1°} \approx 257.8 \qquad \text{Law of Sines}$$

Solve triangle $A_2B_2C_2$:

$$\angle C_2 \approx 180° - (43.1° + 114.2°) = 22.7° \qquad \text{Find } \angle C_2$$

Thus $$c_2 = \frac{a_2 \sin C_2}{\sin A_2} \approx \frac{186.2 \sin 22.7°}{\sin 43.1°} \approx 105.2 \qquad \text{Law of Sines}$$

Triangles $A_1B_1C_1$ and $A_2B_2C_2$ are shown in Figure 9.

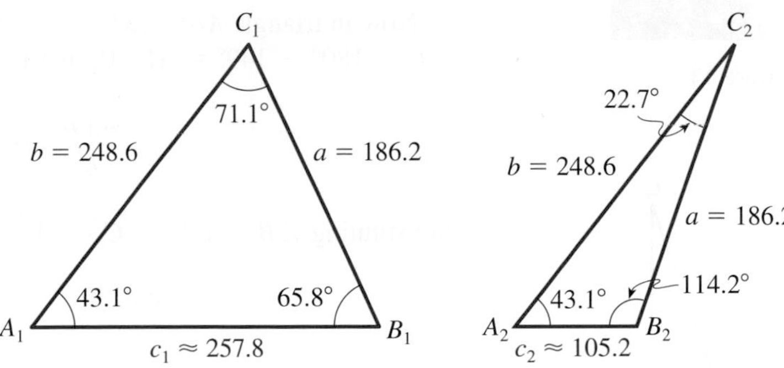

Figure 9 | The two solutions

Now Try Exercise 25 ■

The next example presents a situation for which no triangle is compatible with the given data.

Example 5 ■ SSA, the No-Solution Case

Solve triangle ABC, where $\angle A = 42°$, $a = 70$, and $b = 122$.

Solution To organize the given information, we sketch a tentative diagram in Figure 10. Let's try to find $\angle B$. We have

$$\frac{\sin A}{a} = \frac{\sin B}{b} \qquad \text{Law of Sines}$$

$$\sin B = \frac{b \sin A}{a} = \frac{122 \sin 42°}{70} \approx 1.17 \qquad \text{Solve for } \sin B$$

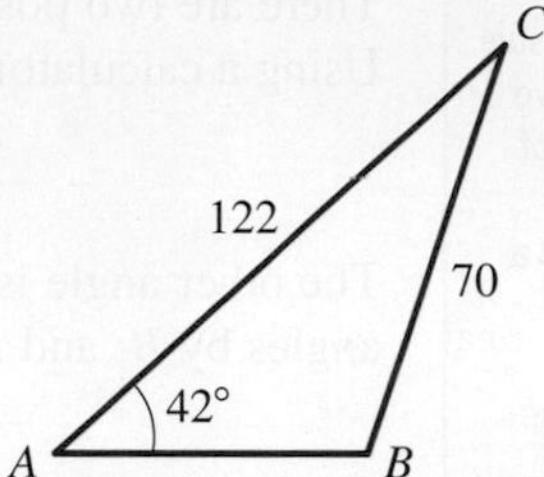

Figure 10

Since the sine of an angle is never greater than 1, we conclude that no triangle satisfies the conditions given in this problem.

 Now Try Exercise 23 ■

Example 6 ■ Calculating a Distance

A bird is perched on top of a pole on a steep hill, and an observer is located at point A on the side of the hill, 110 m downhill from the base of the pole, as shown in Figure 11. The angle of inclination of the hill is 50°, and the angle α in the figure is 9°. Find the distance from the observer to the bird.

Figure 11

Solution We first sketch a diagram as shown in Figure 12. We want to find the distance b in the figure. Triangle ADB is a right triangle, so $\angle DBA = 90° - 50° = 40°$. It follows that $\angle ABC = 180° - 40° = 140°$.

Now in triangle ABC we have $\angle A = 9°$ and $\angle B = 140°$, so $\angle C = 180° - 149° = 31°$. By the Law of Sines we have

$$\frac{\sin B}{b} = \frac{\sin C}{c} \qquad \text{Law of Sines}$$

Substituting $\angle B = 140°$, $\angle C = 31°$, and $c = 110$, we get

$$\frac{\sin 140°}{b} = \frac{\sin 31°}{110}$$

$$b = \frac{110 \sin 140°}{\sin 31°} \qquad \text{Solve for } b$$

$$\approx 137.3 \qquad \text{Calculator}$$

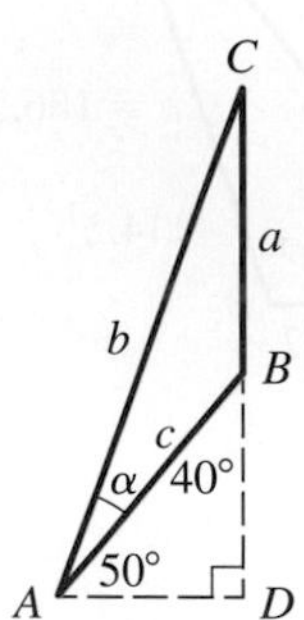

Figure 12

So the distance from the observer to the bird is about 137 m.

Now Try Exercise 41 ■

6.5 | Exercises

■ Concepts

1. In triangle ABC with sides a, b, and c the Law of Sines states that

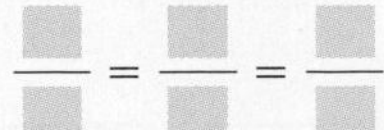

2. The four cases in which we can solve a triangle are

ASA SSA SAS SSS

(a) In which of these cases can we use the Law of Sines to solve the triangle?

(b) Which of the cases listed can lead to more than one solution (the ambiguous case)?

3–4 ■ Two triangles are shown in the figure. Apply the Law of Sines to each triangle to find the indicated side or angle.

3. For triangle ABC we have $\dfrac{\sin 40^\circ}{\square} = \dfrac{\square}{x}$, so $x \approx$ ________.

4. For triangle PQR we have $\dfrac{\sin \theta}{\square} = \dfrac{\square}{\square}$, so $\theta \approx$ ________.

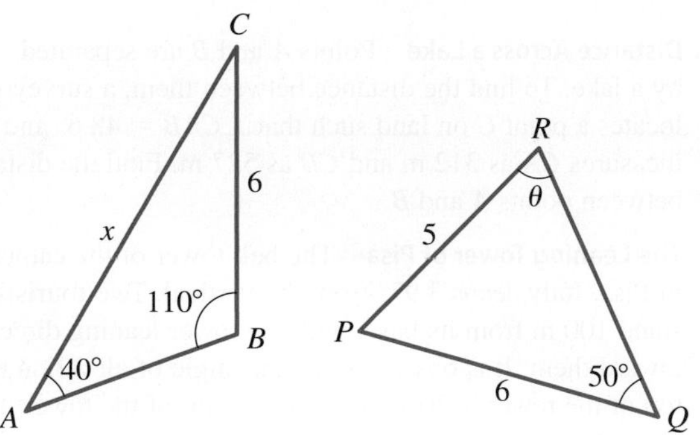

■ Skills

5–10 ■ **Finding an Angle or Side** Use the Law of Sines to find the indicated side x or angle θ.

5.

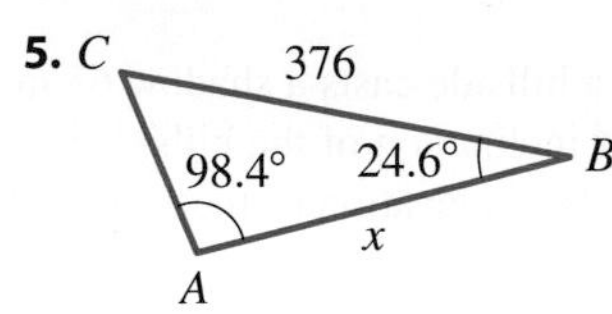

6.

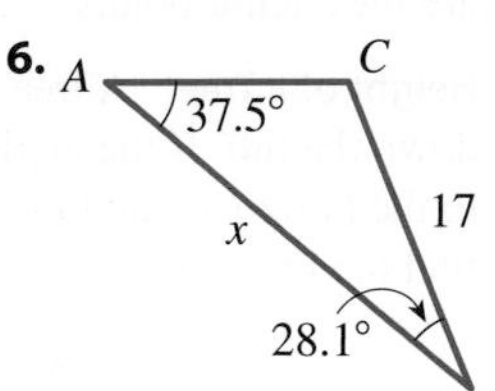

7.

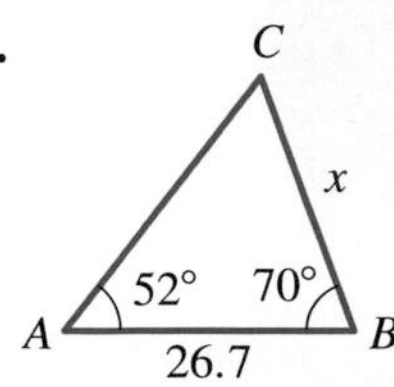

8.

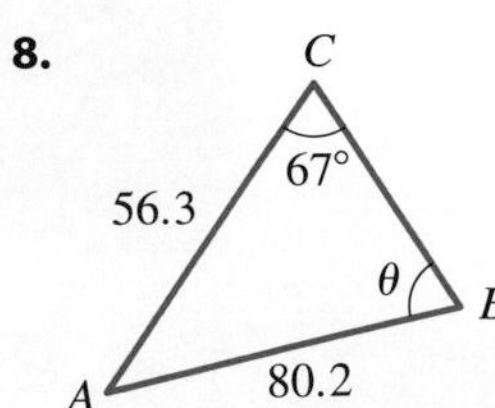

9.

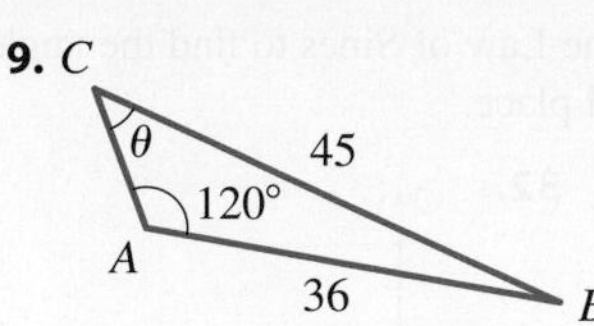

10.

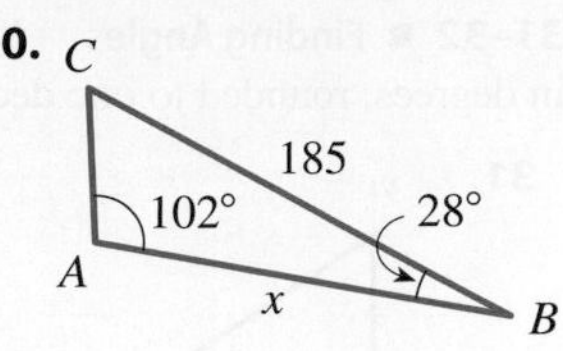

11–14 ■ **Solving a Triangle** Solve the triangle using the Law of Sines.

11. Triangle ABC with $\angle A = 46^\circ$, $\angle B = 20^\circ$, $AB = 65$.

12.

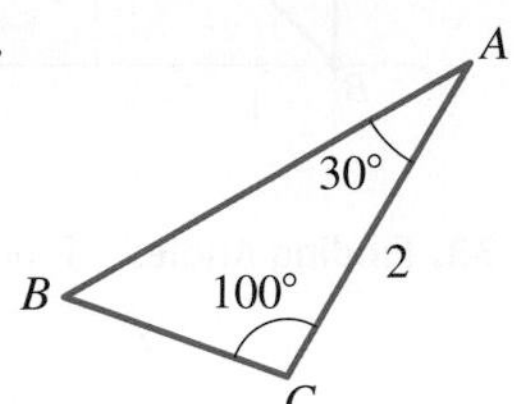

13. Triangle ABC with $AC = 12$, $AB = 12$, $\angle C = 68^\circ$.

14.

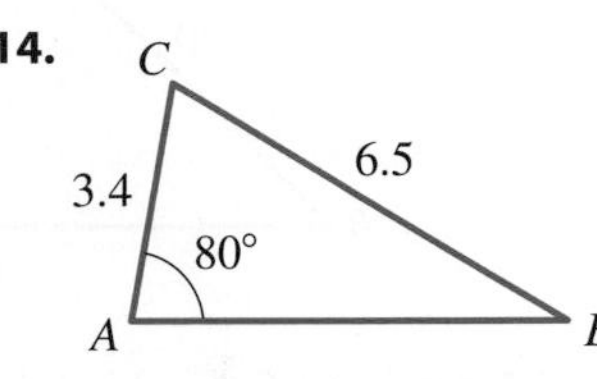

15–20 ■ **Solving a Triangle** Sketch the triangle, and then solve it using the Law of Sines.

15. $\angle A = 50^\circ$, $\angle B = 68^\circ$, $c = 230$

16. $\angle A = 23^\circ$, $\angle B = 110^\circ$, $c = 50$

17. $\angle A = 30^\circ$, $\angle C = 65^\circ$, $b = 10$

18. $\angle A = 22^\circ$, $\angle B = 95^\circ$, $a = 420$

19. $\angle B = 29^\circ$, $\angle C = 51^\circ$, $b = 44$

20. $\angle B = 10^\circ$, $\angle C = 100^\circ$, $c = 115$

21–30 ■ **Solving a Triangle** Use the Law of Sines to solve for all possible triangles that satisfy the given conditions. There may be two triangles, one triangle, or no triangles that satisfy the given conditions.

21. $a = 28$, $b = 15$, $\angle A = 110^\circ$

22. $a = 30$, $c = 40$, $\angle A = 37^\circ$

23. $a = 20$, $c = 45$, $\angle A = 125^\circ$

24. $b = 45$, $c = 42$, $\angle C = 38^\circ$

25. $b = 25$, $c = 30$, $\angle B = 25^\circ$

26. $a = 75$, $b = 100$, $\angle A = 30^\circ$

27. $a = 50$, $b = 100$, $\angle A = 50^\circ$

28. $a = 100$, $b = 80$, $\angle A = 135^\circ$

29. $a = 26$, $c = 15$, $\angle C = 29^\circ$, $\angle A > 90^\circ$

30. $b = 73$, $c = 82$, $\angle B = 58^\circ$, $\angle C < 90^\circ$

Skills Plus

31–32 ■ Finding Angles Use the Law of Sines to find the angle θ in degrees, rounded to one decimal place.

31.

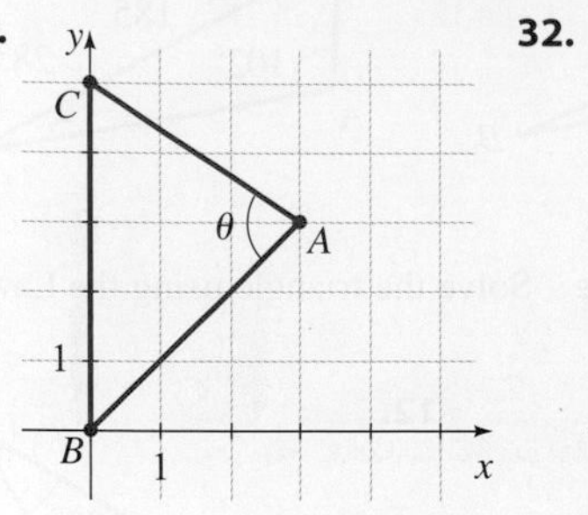

32.

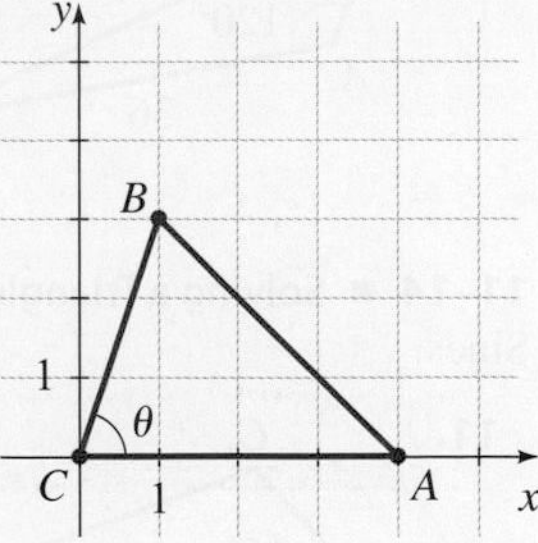

33. Finding Angles Find **(a)** $\angle BCD$ and **(b)** $\angle DCA$.

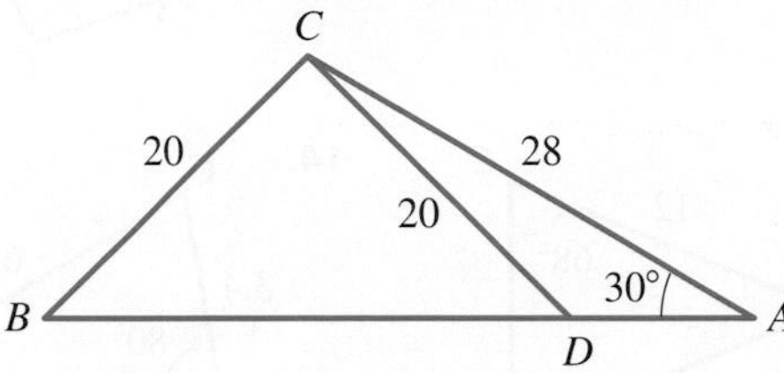

34. Finding a Side Find the length AD.

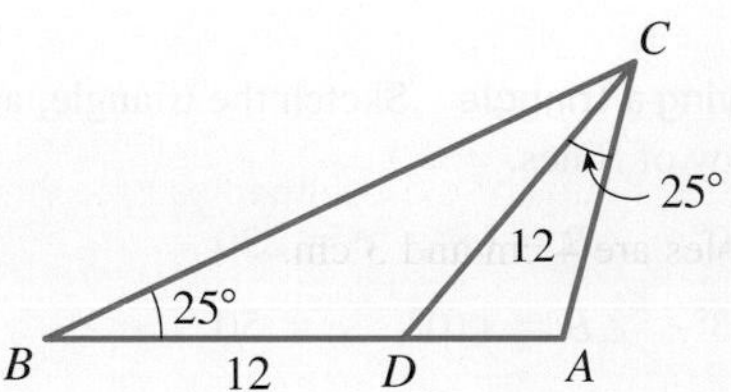

Applications

35. Tracking a Satellite The path of a satellite orbiting the earth passes directly over two tracking stations A and B, 80 km apart. When the satellite is on one side of the two stations, the angles of elevation at A and B are 87.0° and 84.2°, respectively, as shown in the figure.

(a) How far is the satellite from station A?

(b) How high is the satellite above the ground?

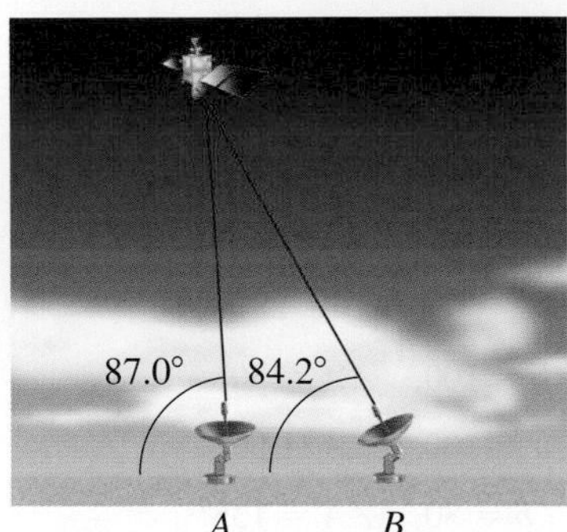

36. Flight of a Plane A pilot who is flying over a straight highway determines the angles of depression to two sign-posts, 8 km apart, to be 32° and 48°, as shown in the figure.

(a) Find the distance of the plane from point A.

(b) Find the elevation of the plane.

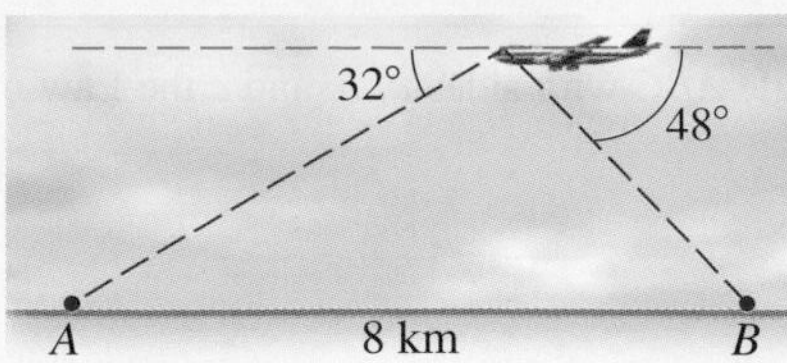

37. Distance Across a River To find the distance across a river, a surveyor chooses points A and B, which are 200 m apart on one side of the river (see the figure). The surveyor then chooses a reference point C on the opposite side of the river and finds that $\angle BAC \approx 82°$ and $\angle ABC \approx 52°$. Approximate the distance from A to C.

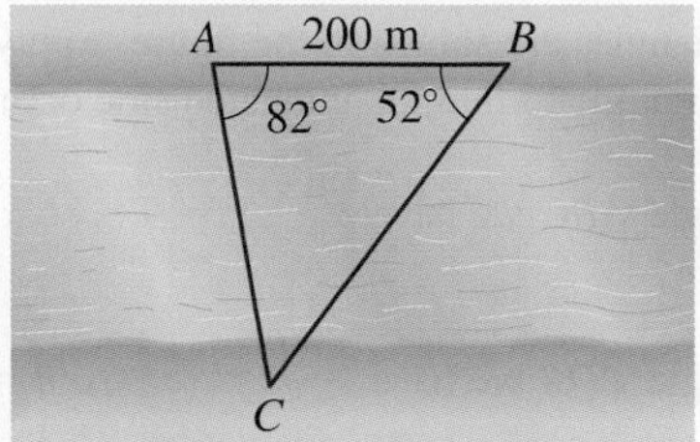

38. Distance Across a Lake Points A and B are separated by a lake. To find the distance between them, a surveyor locates a point C on land such that $\angle CAB = 48.6°$ and then measures CA as 312 m and CB as 527 m. Find the distance between points A and B.

39. The Leaning Tower of Pisa The bell tower of the cathedral in Pisa, Italy, leans 3.97° from the vertical. Two tourists stand 100 m from its base, with the tower leaning directly toward them. It is observed that the angle of elevation to the top of the tower is 30.1°. Find the length of the tower to the nearest meter.

40. Radio Antenna A short-wave radio antenna is supported by two guy wires, 49.5 m and 54 m long. Each wire is attached to the top of the antenna and anchored to the ground at two anchor points on opposite sides of the antenna. The shorter wire makes an angle of 67° with the ground. How far apart are the anchor points?

41. Height of a Tree A tree on a hillside casts a shadow 65 m down the hill. If the angle of inclination of the hillside is 22° to the horizontal and the angle of elevation of the sun is 52°, find the height of the tree.

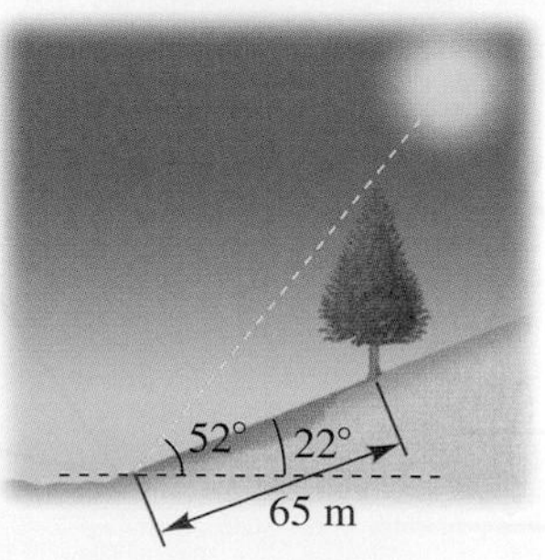

42. Length of a Guy Wire A communications tower is located at the top of a steep hill, as shown in the figure. The angle of inclination of the hill is 58° to the horizontal. A guy wire is to be attached to the top of the tower and to the ground, 100 m downhill from the base of the tower. The angle α in the figure is determined to be 12°. Find the length of cable required for the guy wire.

43. Calculating a Distance Observers at P and Q are located on the side of a hill that is inclined 32° to the horizontal, as shown. The observer at P determines the angle of elevation to a hot-air balloon to be 62°. At the same instant the observer at Q measures the angle of elevation to the balloon to be 71°. If P is 60 m down the hill from Q, find the distance from Q to the balloon.

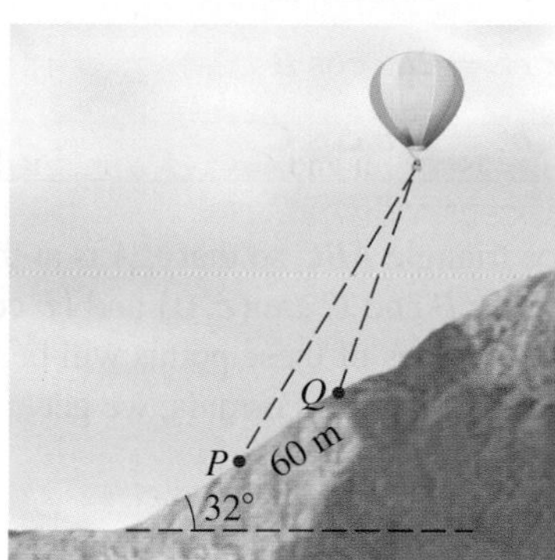

44. Calculating an Angle A water tower 30 m tall is located at the top of a hill. From a distance of 120 m down the hill it is observed that the angle formed between the top and base of the tower is 8°. Find the angle of inclination of the hill.

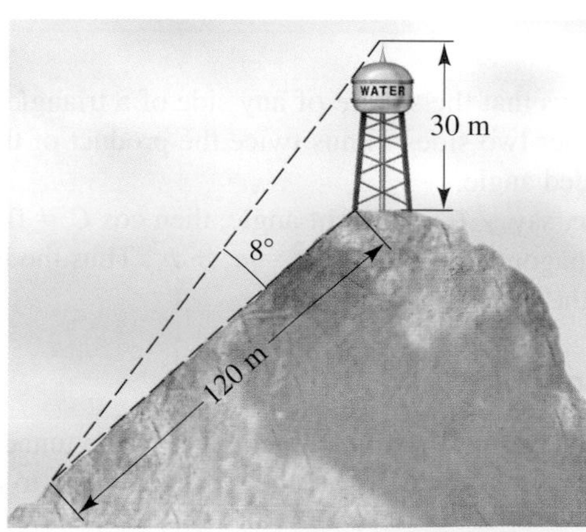

45. Distances to Venus The *elongation* α of a planet is the angle formed by the planet, Earth, and sun (see the figure). It is known that the distance from the sun to Venus is 0.723 AU (see Exercise 6.2.73). At a certain time the elongation of Venus is found to be 39.4°. Find the possible distances from Earth to Venus at that time in astronomical units (AU).

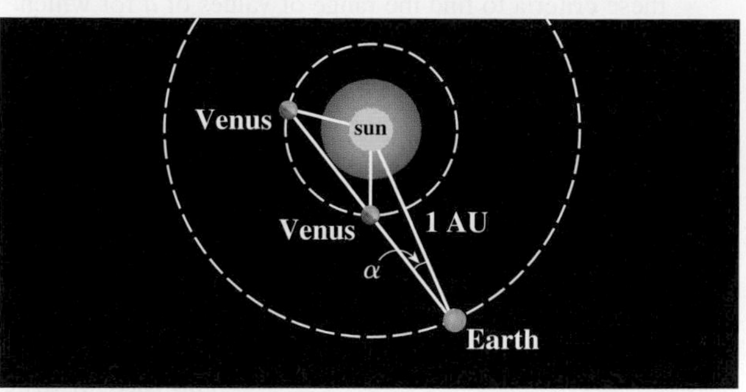

46. Soap Bubbles When two bubbles cling together in midair, their common surface is part of a sphere whose center D lies on the line passing through the centers of the bubbles (see the figure). Also, $\angle ACB$ and $\angle ACD$ each have measure 60°.

(a) Show that the radius r of the common surface is given by

$$r = \frac{ab}{a - b}$$

[*Hint:* Use the Law of Sines together with the fact that an angle θ and its supplement $180° - \theta$ have the same sine.]

(b) Find the radius of the common surface if the radii of the bubbles are 4 cm and 3 cm.

(c) What shape does the common surface take if the two bubbles have equal radii?

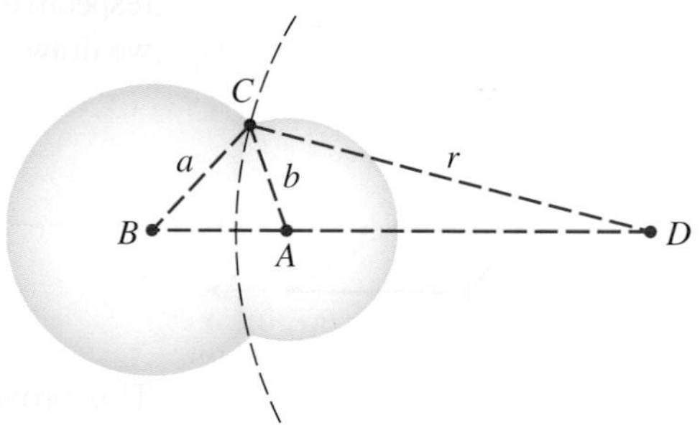

■ Discuss ■ Discover ■ Prove ■ Write

47. Prove: Area of a Triangle Show that, given the three angles A, B, and C of a triangle and one side, say, a, the area of the triangle is

$$\text{area} = \frac{a^2 \sin B \sin C}{2 \sin A}$$

48. Prove: Areas and the Ambiguous Case Suppose we solve a triangle in the ambiguous case. We are given $\angle A$ and sides a and b, and we find the two solutions triangle ABC and triangle $A'B'C'$. Prove that

$$\frac{\text{area of triangle} ABC}{\text{area of triangle } A'B'C'} = \frac{\sin C}{\sin C'}$$

49. Discover: Number of Solutions in the Ambiguous Case We have seen that when the Law of Sines is used to solve a triangle in the SSA case, there may be two solutions, one solution, or no solution. Sketch triangles like those in Figure 6 to verify the criteria in the table for the number of solutions possible if you are given $\angle A$ and sides a and b. If $\angle A = 30°$ and $b = 100$, use these criteria to find the range of values of a for which the triangle ABC has two solutions, one solution, or no solution.

Criterion	Number of Solutions
$a \geq b$	1
$b > a > b \sin A$	2
$a = b \sin A$	1
$a < b \sin A$	0

6.6 The Law of Cosines

■ The Law of Cosines ■ Navigation: Heading and Bearing ■ The Area of a Triangle

■ The Law of Cosines

The Law of Sines cannot be used directly to solve triangles if we know only two sides and the angle between them or if we know only all three sides (these are Cases 3 and 4 of the preceding section). In these two cases the **Law of Cosines** applies.

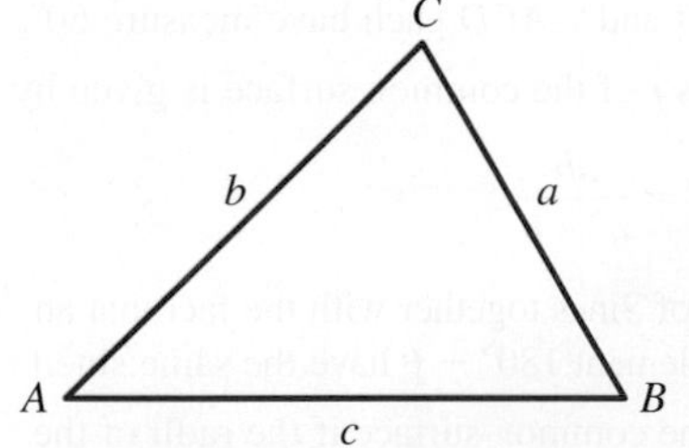

Figure 1

The Law of Cosines

In any triangle ABC (see Figure 1) we have

$$a^2 = b^2 + c^2 - 2bc \cos A$$
$$b^2 = a^2 + c^2 - 2ac \cos B$$
$$c^2 = a^2 + b^2 - 2ab \cos C$$

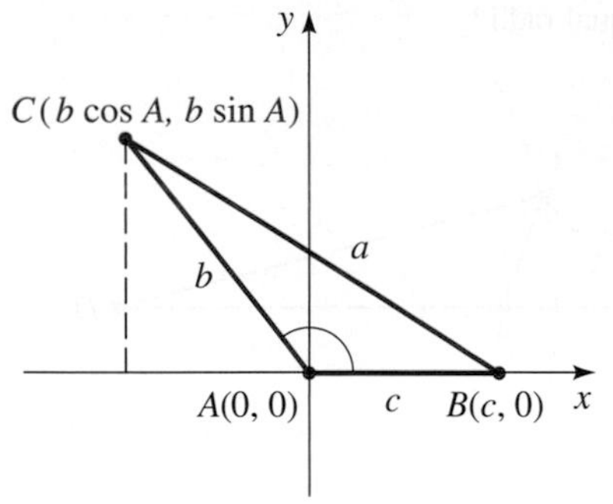

Figure 2

Proof To prove the Law of Cosines, place triangle ABC so that $\angle A$ is at the origin, as shown in Figure 2. The coordinates of vertices B and C are $(c, 0)$ and $(b \cos A, b \sin A)$, respectively. (You should check that the coordinates of these points will be the same if we draw angle A as an acute angle.) Using the Distance Formula, we get

$$\begin{aligned} a^2 &= (b \cos A - c)^2 + (b \sin A - 0)^2 \\ &= b^2 \cos^2 A - 2bc \cos A + c^2 + b^2 \sin^2 A \\ &= b^2(\cos^2 A + \sin^2 A) - 2bc \cos A + c^2 \\ &= b^2 + c^2 - 2bc \cos A \qquad \text{Because } \sin^2 A + \cos^2 A = 1 \end{aligned}$$

This proves the first formula. The other two formulas are obtained in the same way by placing each of the other vertices of the triangle at the origin and repeating the preceding argument. ■

In words, the Law of Cosines says that the square of any side of a triangle is equal to the sum of the squares of the other two sides minus twice the product of those two sides times the cosine of the included angle.

If one of the angles of a triangle, say, $\angle C$, is a right angle, then $\cos C = 0$, and the Law of Cosines reduces to the Pythagorean Theorem, $c^2 = a^2 + b^2$. Thus the Pythagorean Theorem is a special case of the Law of Cosines.

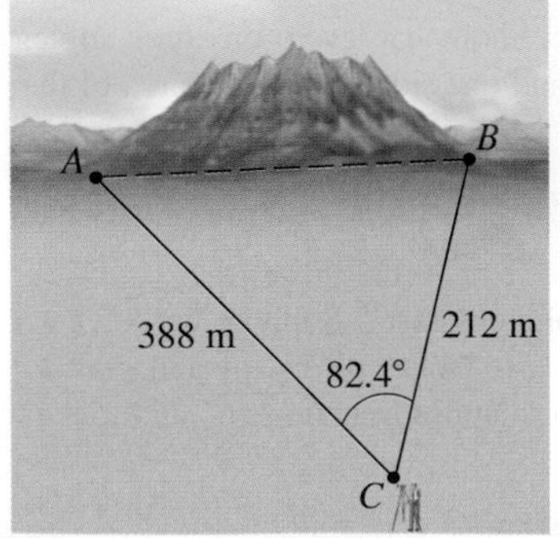

Figure 3

Example 1 ■ Length of a Tunnel

A tunnel is to be built through a mountain. To estimate the length of the tunnel, a surveyor makes the measurements shown in Figure 3. Use the surveyor's data to approximate the length of the tunnel.

THALES OF MILETUS (circa 625–547 B.C.) is the legendary founder of Greek geometry. It is said that he calculated the height of a Greek column by comparing the length of the shadow of his staff with that of the column. Using properties of similar triangles, he argued that the ratio of the height h of the column to the height h' of his staff was equal to the ratio of the length s of the column's shadow to the length s' of the staff's shadow:

$$\frac{h}{h'} = \frac{s}{s'}$$

Since three of these quantities are known, Thales was able to calculate the height of the column.

According to legend, Thales used a similar method to find the height of the Great Pyramid in Egypt, a feat that impressed Egypt's king. Plutarch (circa A.D. 46–119) wrote that "although he [the king of Egypt] admired you [Thales] for other things, yet he particularly liked the manner by which you measured the height of the pyramid without any trouble or instrument." The principle Thales used—the fact that ratios of corresponding sides of similar triangles are equal—is the foundation of the subject of trigonometry.

Solution To approximate the length c of the tunnel, we use the Law of Cosines.

$$\begin{aligned} c^2 &= a^2 + b^2 - 2ab\cos C && \text{Law of Cosines} \\ &= 212^2 + 388^2 - 2(212)(388)\cos 82.4° && \text{Substitute} \\ &\approx 173730.2367 && \text{Use a calculator} \\ c &\approx \sqrt{173730.2367} \approx 416.8 && \text{Take square roots} \end{aligned}$$

Thus the tunnel will be approximately 417 m long.

Now Try Exercises 5 and 41 ■

Example 2 ■ SSS, the Law of Cosines

The sides of a triangle are $a = 5$, $b = 8$, and $c = 12$ (see Figure 4). Find the angles of the triangle.

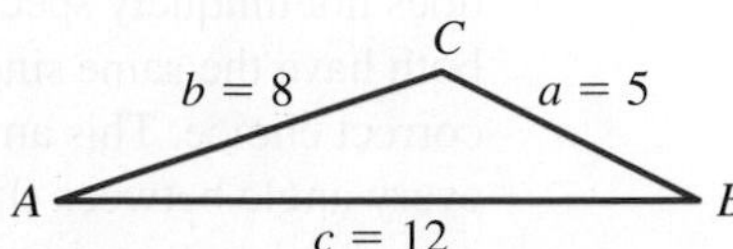

Figure 4

Solution We first find $\angle A$. From the Law of Cosines, $a^2 = b^2 + c^2 - 2bc\cos A$. Solving for $\cos A$, we get

$$\cos A = \frac{b^2 + c^2 - a^2}{2bc} = \frac{8^2 + 12^2 - 5^2}{2(8)(12)} = \frac{183}{192} = 0.953125$$

Using a calculator (in degree mode), we find that $\angle A = \cos^{-1}(0.953125) \approx 18°$. In the same way we get

$$\cos B = \frac{a^2 + c^2 - b^2}{2ac} = \frac{5^2 + 12^2 - 8^2}{2(5)(12)} = 0.875$$

$$\cos C = \frac{a^2 + b^2 - c^2}{2ab} = \frac{5^2 + 8^2 - 12^2}{2(5)(8)} = -0.6875$$

Using a calculator, we find that

$$\angle B = \cos^{-1}(0.875) \approx 29° \quad \text{and} \quad \angle C = \cos^{-1}(-0.6875) \approx 133°$$

Of course, once two angles have been calculated, the third can more easily be found from the fact that the sum of the angles of a triangle is 180°. However, it's a good idea to calculate all three angles using the Law of Cosines and add the three angles as a check on your computations.

Now Try Exercise 9 ■

Example 3 ■ SAS, the Law of Cosines

Solve triangle ABC, where $\angle A = 46.5°$, $b = 10.5$, and $c = 18.0$.

Solution We can find a using the Law of Cosines.

$$\begin{aligned} a^2 &= b^2 + c^2 - 2bc\cos A \\ &= (10.5)^2 + (18.0)^2 - 2(10.5)(18.0)(\cos 46.5°) \approx 174.05 \end{aligned}$$

Thus $a \approx \sqrt{174.05} \approx 13.2$. We also use the Law of Cosines to find $\angle B$ and $\angle C$, by the method we used in Example 2.

$$\cos B = \frac{a^2 + c^2 - b^2}{2ac} = \frac{(13.2)^2 + (18.0)^2 - (10.5)^2}{2(13.2)(18.0)} \approx 0.816477$$

$$\cos C = \frac{a^2 + b^2 - c^2}{2ab} = \frac{(13.2)^2 + (10.5)^2 - (18.0)^2}{2(13.2)(10.5)} \approx -0.142532$$

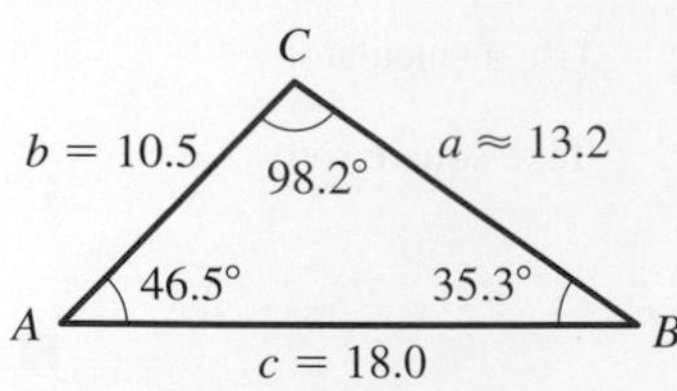

Figure 5

Using a calculator, we find that

$$\angle B = \cos^{-1}(0.816477) \approx 35.3° \quad \text{and} \quad \angle C = \cos^{-1}(-0.142532) \approx 98.2°$$

To summarize: $\angle B \approx 35.3°$, $\angle C \approx 98.2°$, and $a \approx 13.2$. (See Figure 5.)

Now Try Exercise 15 ■

Note We could have used the Law of Sines to find $\angle B$ and $\angle C$ in Example 3 because we knew all three sides and an angle in the triangle. But knowing the sine of an angle does not uniquely specify the angle, since an angle θ and its supplement $180° - \theta$ both have the same sine. Thus we would need to decide which of the two angles is the correct choice. This ambiguity does not arise when we use the Law of Cosines because every angle between 0° and 180° has a unique cosine. So using only the Law of Cosines is preferable in problems like Example 3.

■ Navigation: Heading and Bearing

In navigation a direction is often given as a **bearing**, that is, as an acute angle measured from due north or due south. The bearing N 30° E, for example, indicates a direction that points 30° to the east of due north (see Figure 6).

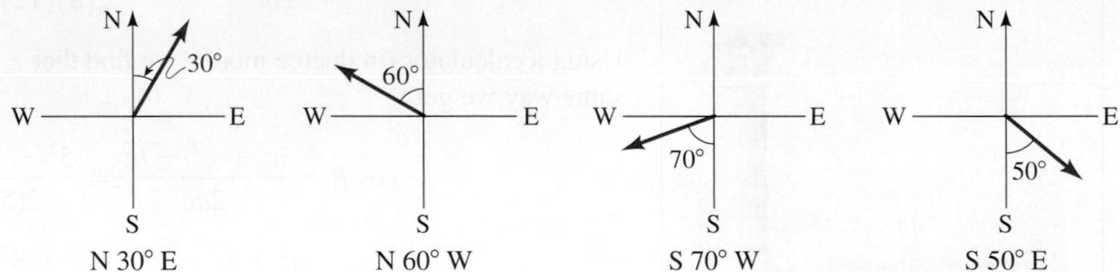

Figure 6

Example 4 ■ Navigation

A pilot takes off from an airport and heads in the direction N 20° E, flying at 320 km/h. After one hour the pilot makes a course correction and heads in the direction N 40° E. Half an hour after that, engine trouble forces the pilot to make an emergency landing.

(a) Find the distance between the airport and the final landing point.

(b) Find the bearing from the airport to the final landing point.

Solution

(a) In one hour the plane travels 320 km, and in half an hour it travels 160 km, so we can plot the pilot's course as in Figure 7. When the pilot makes the course correction, the plane turns 20° to the right, so the angle between the two legs of the trip is $180° - 20° = 160°$. By the Law of Cosines we have

$$\begin{aligned} b^2 &= 320^2 + 160^2 - 2 \cdot 320 \cdot 160 \cos 160° \\ &\approx 224{,}221.19 \end{aligned}$$

Thus $b \approx 473.52$. The pilot lands about 474 km from the takeoff site.

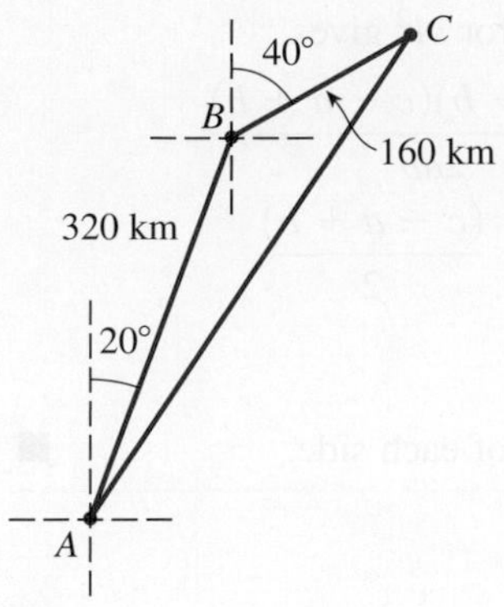

Figure 7

(b) We first use the Law of Sines to find $\angle A$.

$$\frac{\sin A}{160} = \frac{\sin 160°}{473.52}$$

$$\sin A = 160 \cdot \frac{\sin 160°}{473.52}$$

$$\approx 0.11557$$

Using a calculator (in degree mode), we find that $\angle A \approx \sin^{-1}(0.11557) \approx 6.636°$. From Figure 7 we see that the line from the airport to the final landing site points in the direction $20° + 6.636° = 26.636°$ east of due north. Thus the bearing is about N 26.6° E.

Another solution to $\sin A = 0.11557$ is $\angle A = 180° - 6.636° = 173.364°$. But this is too large to be $\angle A$ in triangle *ABC*.

Now Try Exercise 47 ■

■ The Area of a Triangle

An interesting application of the Law of Cosines involves a formula for finding the area of a triangle from the lengths of its three sides (see Figure 8).

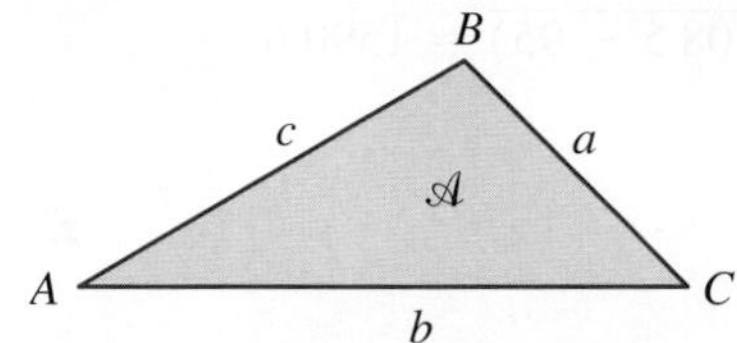

Figure 8

Heron's Formula

The area $\mathcal{A}$ of triangle *ABC* is given by

$$\mathcal{A} = \sqrt{s(s-a)(s-b)(s-c)}$$

where $s = \frac{1}{2}(a + b + c)$ is the **semiperimeter** of the triangle; that is, s is half the perimeter.

Proof We start with the formula $\mathcal{A} = \frac{1}{2}ab \sin C$ from Section 6.3. Thus

$$\mathcal{A}^2 = \tfrac{1}{4}a^2b^2 \sin^2 C$$

$$= \tfrac{1}{4}a^2b^2(1 - \cos^2 C) \qquad \text{Pythagorean identity}$$

$$= \tfrac{1}{4}a^2b^2(1 - \cos C)(1 + \cos C) \qquad \text{Factor}$$

Next, we write the expressions $1 - \cos C$ and $1 + \cos C$ in terms of a, b, and c. By the Law of Cosines we have

$$\cos C = \frac{a^2 + b^2 - c^2}{2ab} \qquad \text{Law of Cosines}$$

$$1 + \cos C = 1 + \frac{a^2 + b^2 - c^2}{2ab} \qquad \text{Add 1}$$

$$= \frac{2ab + a^2 + b^2 - c^2}{2ab} \qquad \text{Common denominator}$$

$$= \frac{(a+b)^2 - c^2}{2ab} \qquad \text{Factor}$$

$$= \frac{(a+b+c)(a+b-c)}{2ab} \qquad \text{Difference of squares}$$

Similarly,

$$1 - \cos C = \frac{(c + a - b)(c - a + b)}{2ab}$$

Substituting these expressions in the formula we obtained for $\mathcal{A}^2$ gives

To see that the factors in the last two products are equal, note for example that

$$\frac{a+b-c}{2} = \frac{a+b+c}{2} - c$$
$$= s - c$$

$$\begin{aligned}\mathcal{A}^2 &= \tfrac{1}{4}a^2b^2\frac{(a+b+c)(a+b-c)}{2ab}\,\frac{(c+a-b)(c-a+b)}{2ab} \\ &= \frac{(a+b+c)}{2}\,\frac{(a+b-c)}{2}\,\frac{(c+a-b)}{2}\,\frac{(c-a+b)}{2} \\ &= s(s-c)(s-b)(s-a)\end{aligned}$$

Heron's Formula now follows from taking the square root of each side. ■

Example 5 ■ Area of a Lot

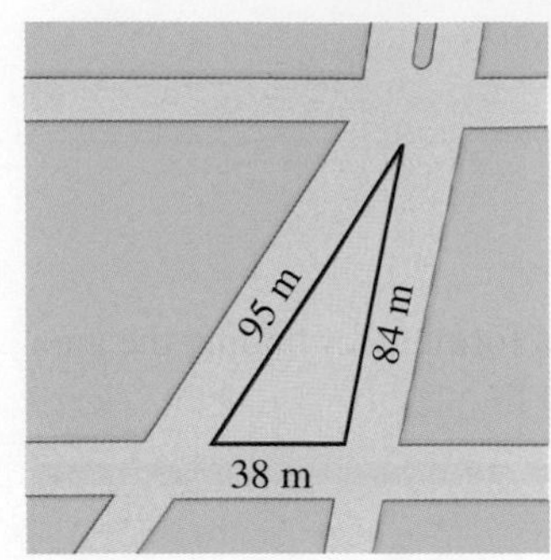

Figure 9

A developer wishes to buy a triangular lot in a busy downtown location (see Figure 9). The lot frontages on the three adjacent streets are 38, 84, and 95 m. Find the area of the lot.

Solution The semiperimeter of the lot is

$$s = \frac{38 + 84 + 95}{2} = 108.5$$

By Heron's Formula the area is

$$\mathcal{A} = \sqrt{108.5(108.5-38)(108.5-84)(108.5-95)} \approx 1590.6$$

Thus the area is approximately 1,591 m^2.

 **Now Try Exercises 31 and 55** ■

6.6 Exercises

■ Concepts

1. For triangle *ABC* with sides *a*, *b*, and *c*, the Law of Cosines states that

$c^2 =$ ______________________.

2. In which of the following cases must the Law of Cosines be used to solve a triangle?

ASA SSS SAS SSA

3–4 ■ Two triangles are shown in the figure. Apply the Law of Cosines to each triangle to find the indicated side or angle.

3. For triangle *ABC* we have $x^2 =$ ___^{2} + ___^{2} − 2 ___ cos ___, so $x \approx$ ________.

4. For triangle *PQR* we have ___^{2} = ___^{2} + ___^{2} − 2 ___ cos θ, so $\cos\theta =$ ________ and $\theta \approx$ ________.

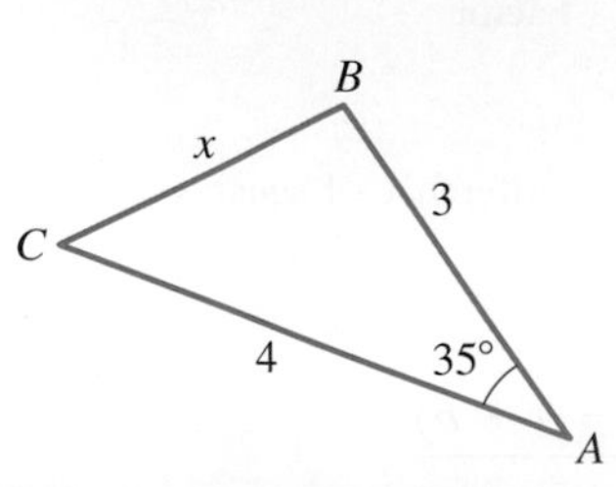

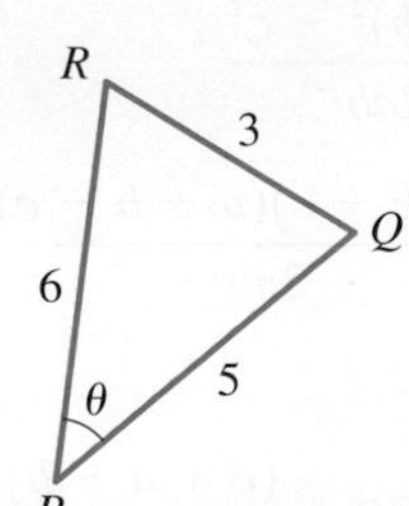

■ Skills

5–12 ■ Finding an Angle or Side Use the Law of Cosines to determine the indicated side x or angle θ.

5.

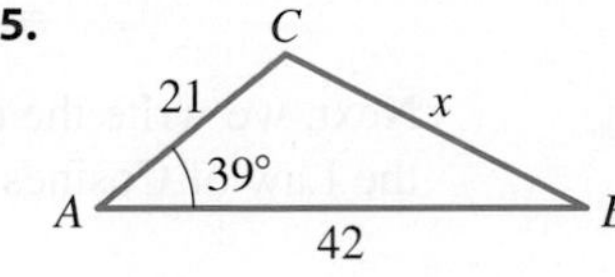

6.

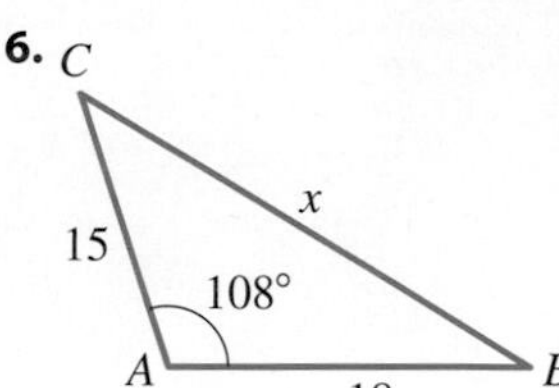

7.

C
25 25
140°
A B
x

8.

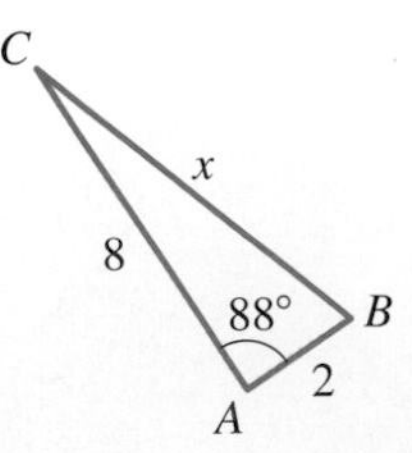

 **9.**

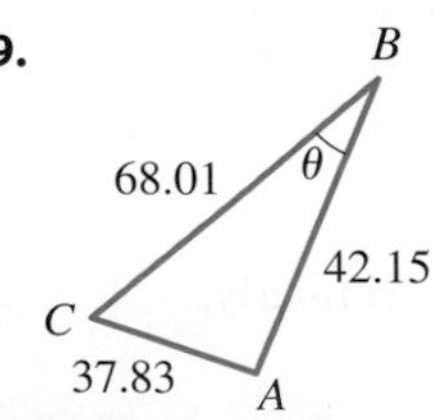

10.

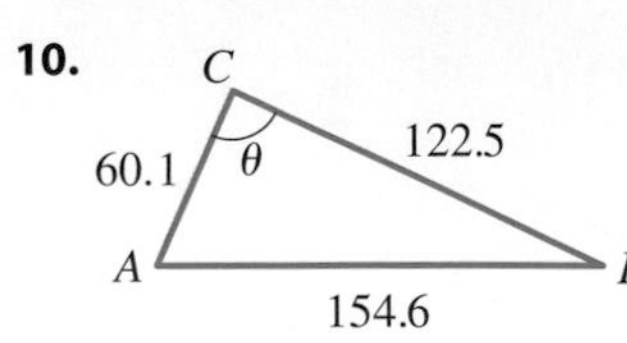

11.

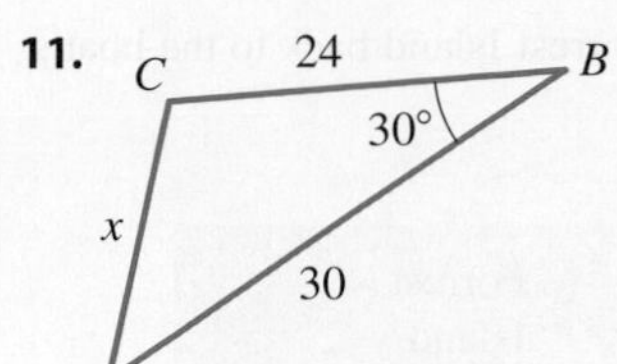

12.

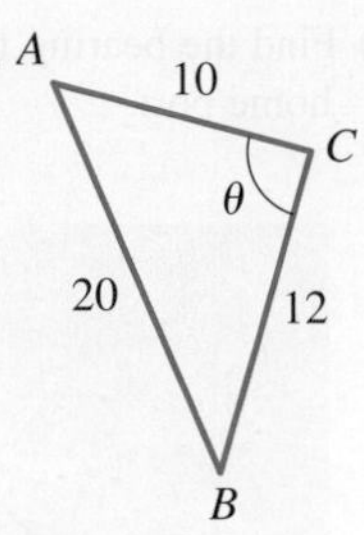

13–22 ■ Solving a Triangle Solve triangle ABC.

13.

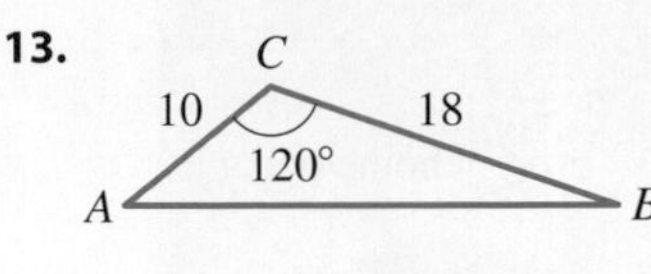

14.

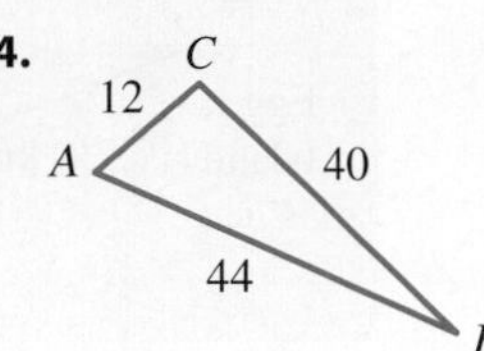

15. $a = 3.0, \quad b = 4.0, \quad \angle C = 53°$

16. $b = 60, \quad c = 30, \quad \angle A = 70°$

17. $a = 20, \quad b = 25, \quad c = 22$

18. $a = 10, \quad b = 12, \quad c = 16$

19. $b = 125, \quad c = 162, \quad \angle B = 40°$

20. $a = 65, \quad c = 50, \quad \angle C = 52°$

21. $a = 50, \quad b = 65, \quad \angle A = 55°$

22. $a = 73.5, \quad \angle B = 61°, \quad \angle C = 83°$

23–30 ■ Law of Sines or Law of Cosines? Find the indicated side x or angle θ. (Use either the Law of Sines or the Law of Cosines, as appropriate.)

23.

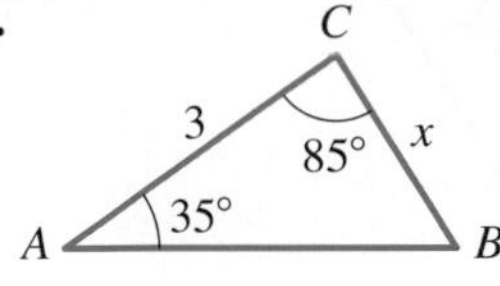

24.

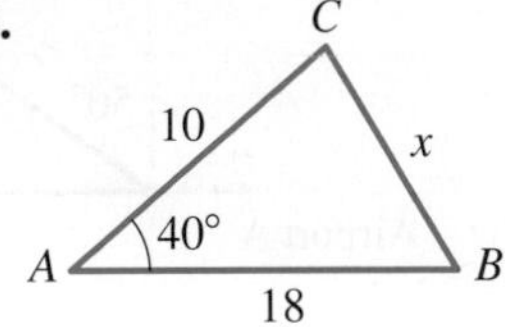

25.

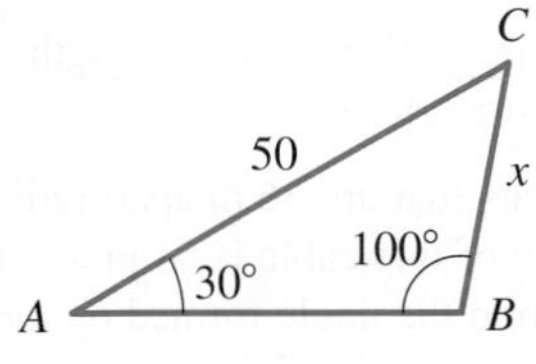

26.

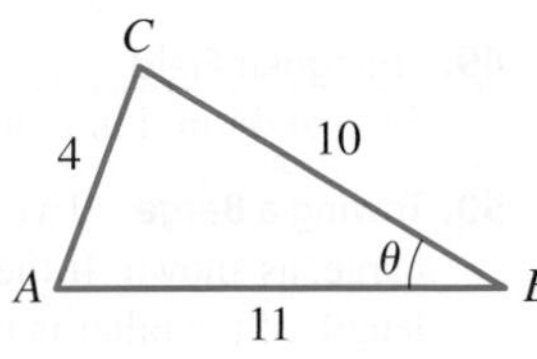

27.

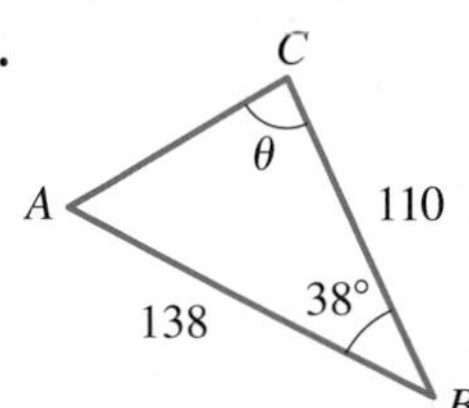

28.

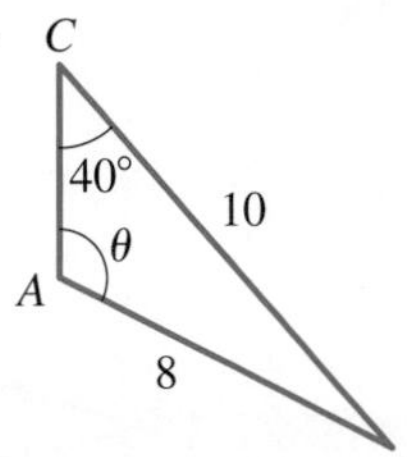

29.

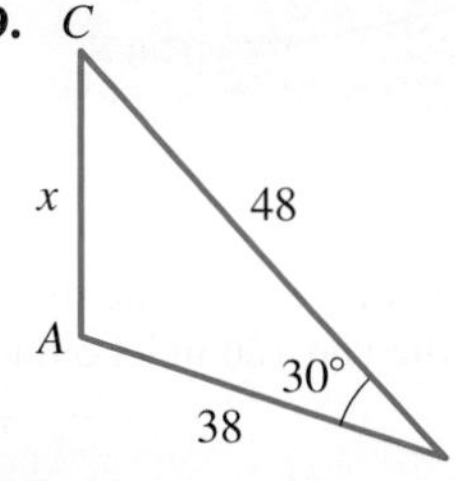

30.

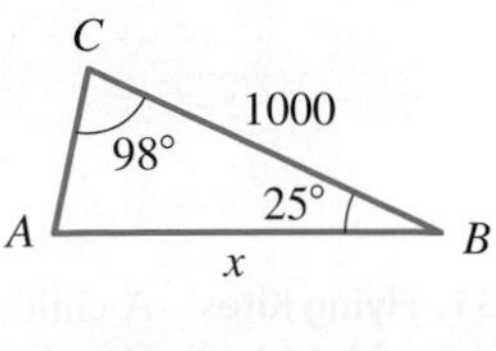

31–34 ■ Heron's Formula Find the area of the triangle whose sides have the given lengths.

31. $a = 9, \quad b = 12, \quad c = 15$

32. $a = 1, \quad b = 2, \quad c = 2$

33. $a = 7, \quad b = 8, \quad c = 9$

34. $a = 11, \quad b = 100, \quad c = 101$

Skills Plus

35–38 ■ Heron's Formula Find the area of the shaded figure, rounded to two decimals.

35.

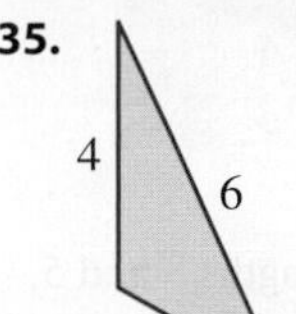

36.

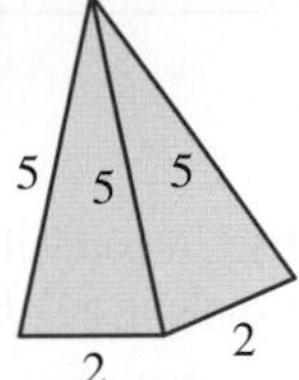

37.

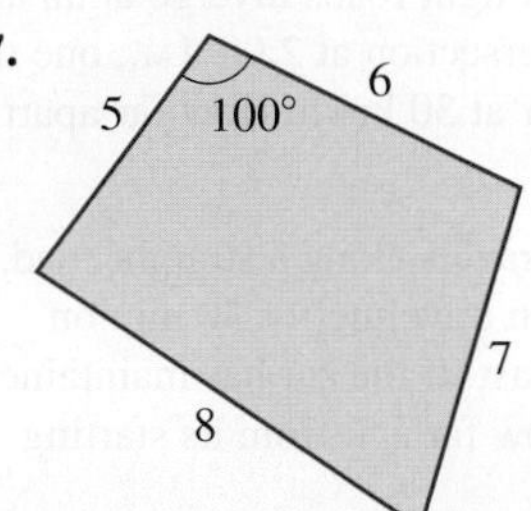

38.

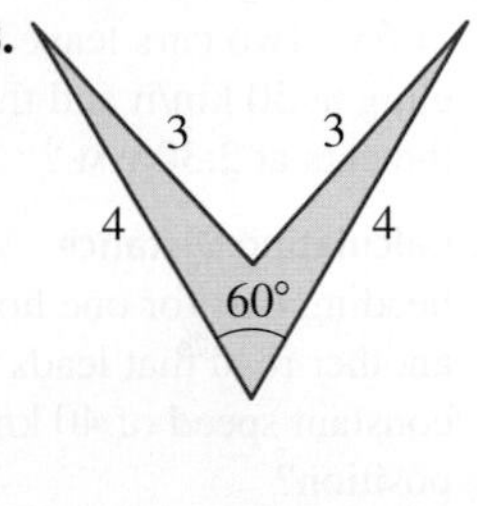

39. Area of a Region Three circles of radii 4, 5, and 6 cm are mutually tangent. Find the shaded area enclosed between the circles, correct to two decimal places.

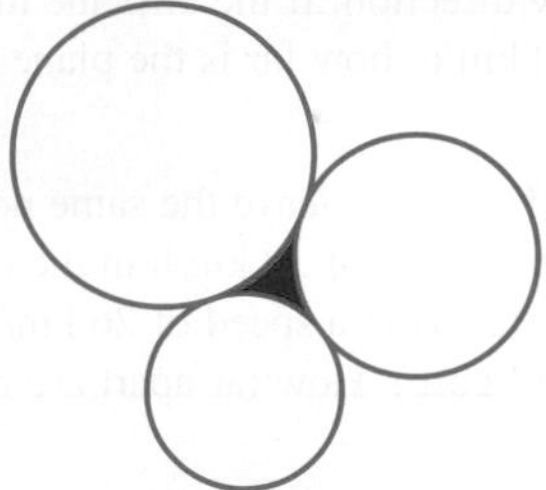

40. Finding a Length In the figure, triangle ABC is a right triangle, $CQ = 6$, and $BQ = 4$. Also, $\angle AQC = 30°$ and $\angle CQB = 45°$. Find the exact length of AQ. [*Hint:* First use the Law of Cosines to find expressions for a^2, b^2, and c^2.]

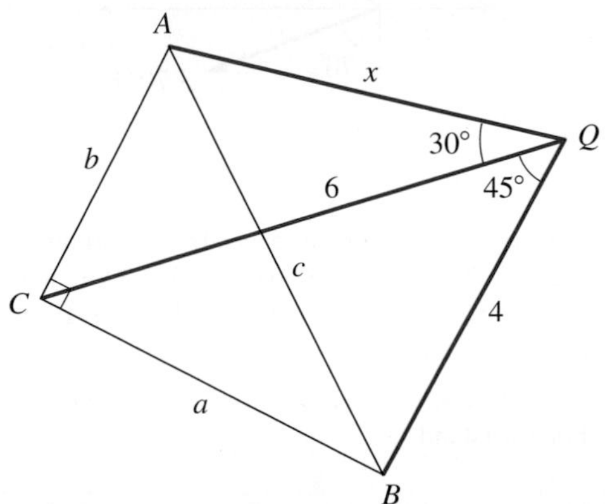

Applications

41. Surveying To find the distance across a small lake, a surveyor has taken the measurements shown. Find the distance across the lake from A to B using this information.

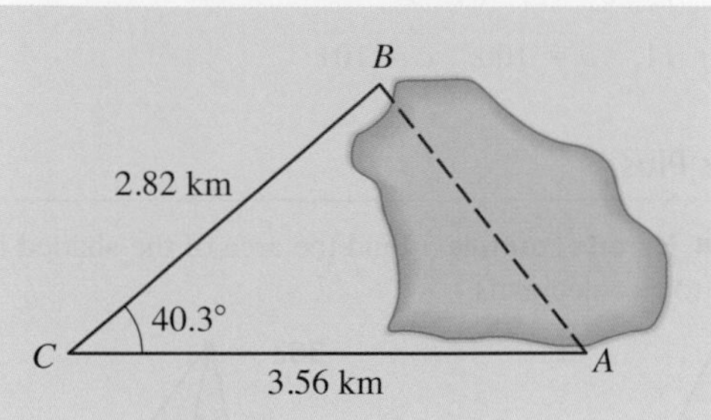

42. Geometry A parallelogram has sides of lengths 3 and 5, and one angle is 50°. Find the lengths of the diagonals.

43. Calculating Distance Two straight roads diverge at an angle of 65°. Two cars leave the intersection at 2:00 P.M., one traveling at 50 km/h and the other at 30 km/h. How far apart are the cars at 2:30 P.M.?

44. Calculating Distance A car travels along a straight road, heading east for one hour, then traveling for 30 min on another road that leads northeast. If the car has maintained a constant speed of 40 km/h, how far is it from its starting position?

45. Dead Reckoning An airplane flies in a straight path for 1 h 30 min. The pilot then makes a course correction, heading 10° to the right of the original course, and flies the plane 2 h in the new direction. If the airplane maintains a constant speed of 1000 km/h, how far is the plane from its starting position?

46. Navigation Two boats leave the same port at the same time. One travels at a speed of 30 km/h in the direction N 50° E, and the other travels at a speed of 26 km/h in a direction S 70° E (see the figure). How far apart are the two boats after one hour?

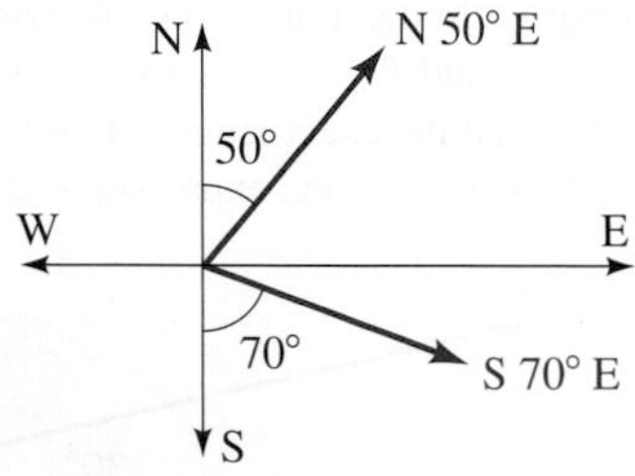

47. Navigation A fishing boat leaves home port and heads in the direction N 70° W. The boat travels 30 km and reaches Egg Island. The next day the boat sails N 10° E for 50 km, reaching Forrest Island.

(a) Find the distance between the boat's home port and Forrest Island.

(b) Find the bearing from Forrest Island back to the boat's home port.

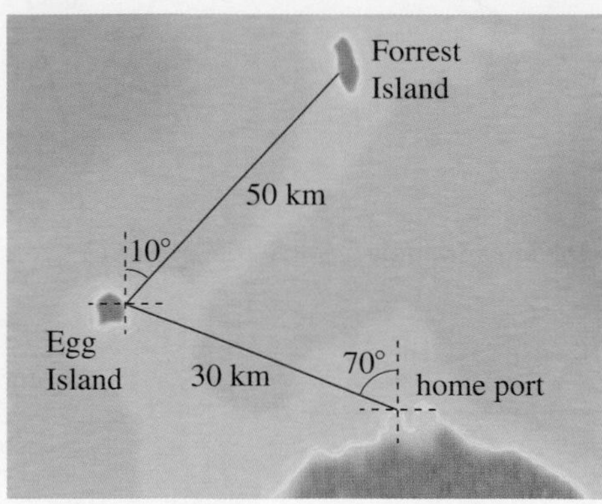

48. Navigation Airport B is 480 km from airport A at a bearing N 50° E (see the figure). A pilot wishing to fly from A to B has mistakenly flown due east at 320 km/h for 30 min when the error is noticed.

(a) How far is the pilot from the desired destination at the time the error is noticed?

(b) At what bearing should the pilot head the plane to arrive at airport B?

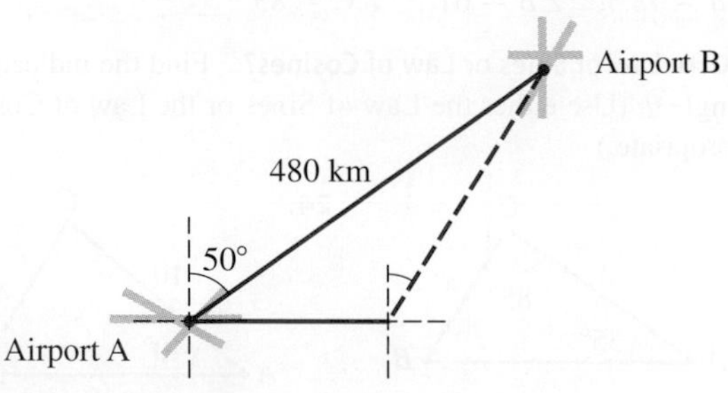

49. Triangular Field A triangular field has sides of lengths 22, 36, and 44 m. Find the largest angle.

50. Towing a Barge Two tugboats that are 36 m apart pull a barge, as shown. If the length of one cable is 64 m and the length of the other is 69 m, find the angle formed by the two cables.

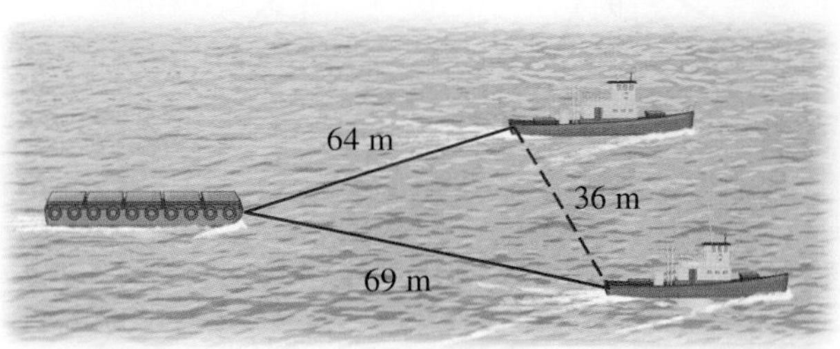

51. Flying Kites A child is flying two kites at the same time, with 114 m of line let out to one kite and 126 m let out to the other. It is estimated that the angle between the two lines is 30°. Approximate the distance between the kites.

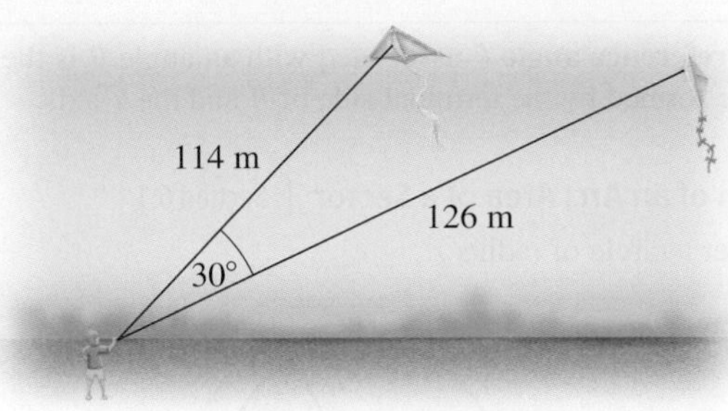

52. Securing a Tower A 38-m tower is located on the side of a mountain that is inclined 32° to the horizontal. A guy wire is to be attached to the top of the tower and anchored at a point 17 m downhill from the base of the tower as shown in the figure. Find the shortest length of wire needed.

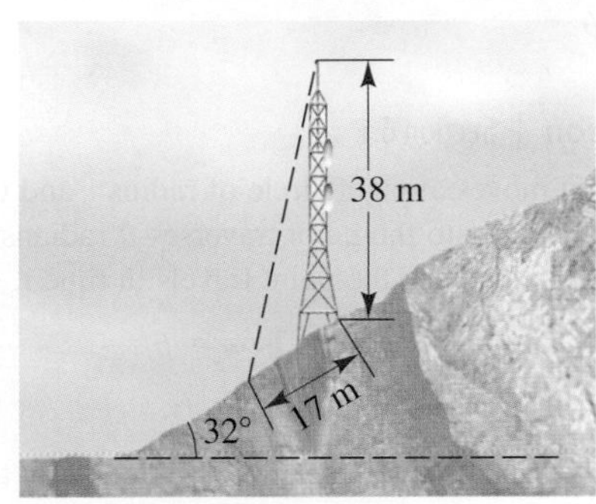

53. Cable Car A steep mountain is inclined 74° to the horizontal and rises 1020 m above the surrounding plain. A cable car is to be installed at a point 240 m from the base and attached to the top of the mountain, as shown in the figure. Find the shortest length of cable needed.

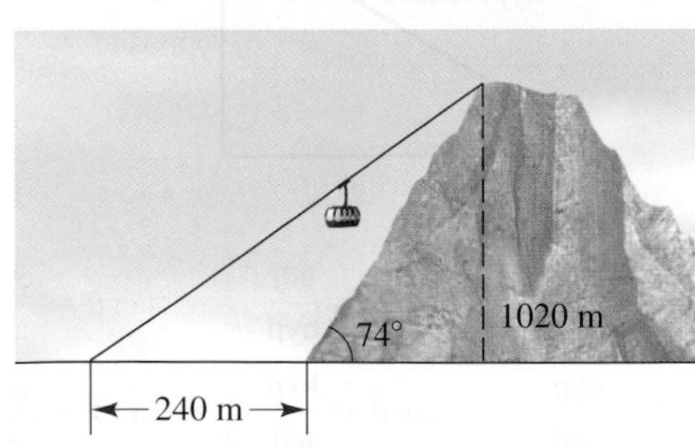

54. CN Tower The CN Tower in Toronto, Canada, is the tallest freestanding structure in North America. A person on the observation deck, 345 m above the ground, wants to determine the distance between two landmarks on the ground below. It is observed that the angle formed by the lines of sight to these two landmarks is 43°. It is also observed that the angle between the vertical and the line of sight to one of the landmarks is 62° and the angle to the other landmark is 54°. Find the distance between the two landmarks.

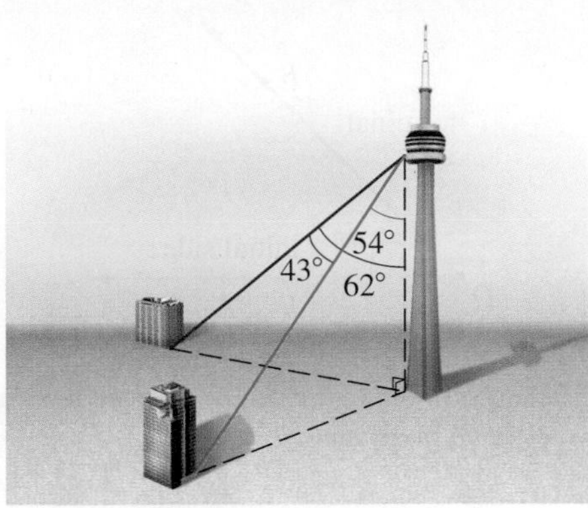

55. Land Value Land in downtown Columbia is valued at \$20 a square meter. What is the value of a triangular lot with sides of lengths 34, 44, and 57 m?

Discuss ■ Discover ■ Prove ■ Write

56. Discuss: Solving for the Angles in a Triangle The note that follows the solution of Example 3 explains an alternative method for finding $\angle B$ and $\angle C$, using the Law of Sines. Use this method to solve the triangle in the example, finding $\angle B$ first and then $\angle C$. Explain how you chose the appropriate value for the measure of $\angle B$. Which method do you prefer for solving an SAS triangle problem: the one explained in Example 3 or the one you used in this exercise?

57. Prove: Projection Laws Prove that in triangle ABC

$$a = b\cos C + c\cos B$$

$$b = c\cos A + a\cos C$$

$$c = a\cos B + b\cos A$$

These are called the *Projection Laws*. [*Hint:* To get the first equation, add the second and third equations in the Law of Cosines and solve for a.]

Chapter 6 Review

Properties and Formulas

Angles | Section 6.1

An **angle** consists of two rays with a common vertex. One of the rays is the **initial side**, and the other the **terminal side**. An angle can be viewed as a rotation of the initial side onto the terminal side. If the rotation is counterclockwise, the angle is **positive**; if the rotation is clockwise, the angle is **negative**.

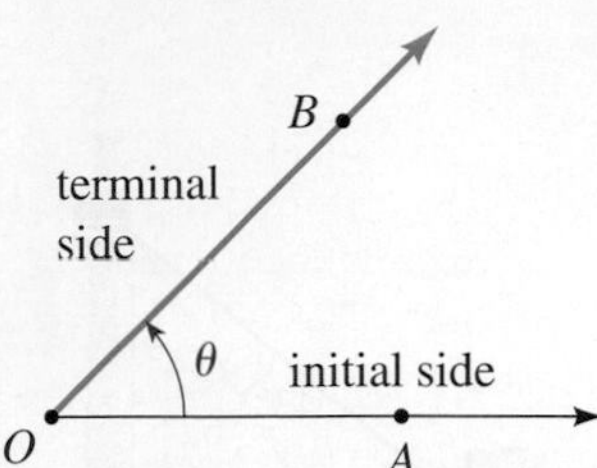

Notation: The angle in the figure can be referred to as angle AOB, or simply as angle O, or as angle θ.

Angle Measure | Section 6.1

The **radian measure** of an angle (abbreviated **rad**) is the length of the arc that the angle subtends in a circle of radius 1, as shown in the figure.

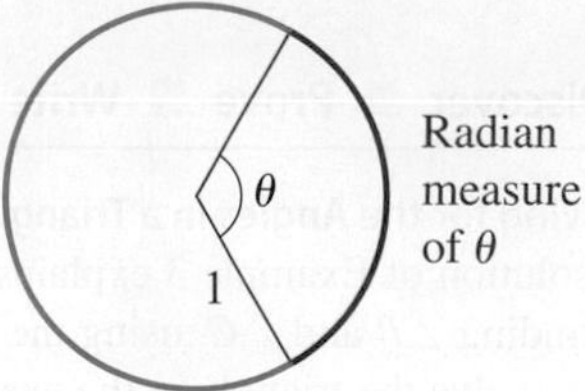

The **degree measure** of an angle is the number of degrees in the angle, where a degree is $\frac{1}{360}$ of a complete circle.

To convert degrees to radians, multiply by $\pi/180$.

To convert radians to degrees, multiply by $180/\pi$.

Angles in Standard Position | Section 6.1 and 6.3

An angle is in **standard position** if it is drawn in the xy-plane with its vertex at the origin and its initial side on the positive x-axis.

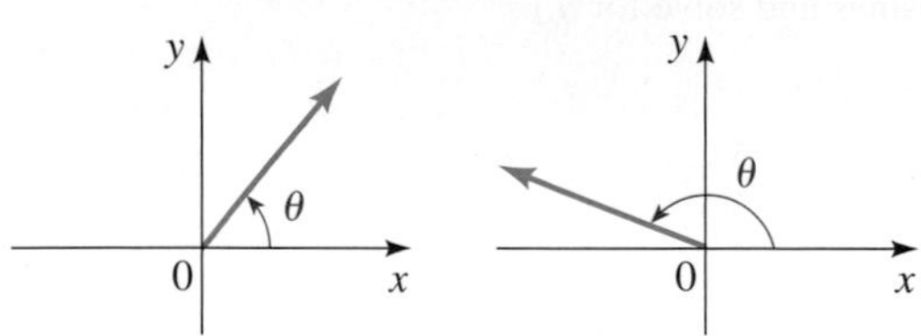

Two angles in standard position are **coterminal** if their terminal sides coincide.

The **reference angle** $\bar{\theta}$ associated with an angle θ is the acute angle formed by the terminal side of θ and the x-axis.

Length of an Arc; Area of a Sector | Section 6.1

Consider a circle of radius r.

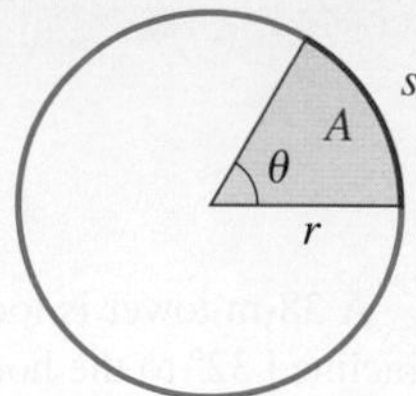

The **length s of an arc** that subtends a central angle of θ radians is $s = r\theta$.

The **area A of a sector** with central angle of θ radians is $A = \frac{1}{2}r^2\theta$.

Circular Motion | Section 6.1

Suppose a point moves along a circle of radius r and the ray from the center of the circle to the point traverses θ radians in time t. Let $s = r\theta$ be the distance the point travels in time t.

The **angular speed** of the point is $\omega = \theta/t$.

The **linear speed** of the point is $v = s/t$.

Linear speed v and angular speed ω are related by the formula $v = r\omega$.

Trigonometric Ratios | Section 6.2

For a right triangle with an acute angle θ the trigonometric ratios are defined as follows.

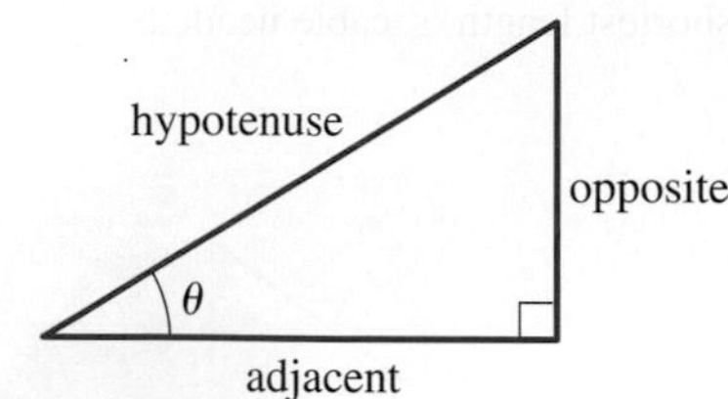

$$\sin\theta = \frac{\text{opp}}{\text{hyp}} \qquad \cos\theta = \frac{\text{adj}}{\text{hyp}} \qquad \tan\theta = \frac{\text{opp}}{\text{adj}}$$

$$\csc\theta = \frac{\text{hyp}}{\text{opp}} \qquad \sec\theta = \frac{\text{hyp}}{\text{adj}} \qquad \cot\theta = \frac{\text{adj}}{\text{opp}}$$

Trigonometric Functions of Angles | Section 6.3

Let θ be an angle in standard position, and let $P(x, y)$ be a point on the terminal side, as shown in the figures. Let $r = \sqrt{x^2 + y^2}$ be the distance from the origin to the point $P(x, y)$.

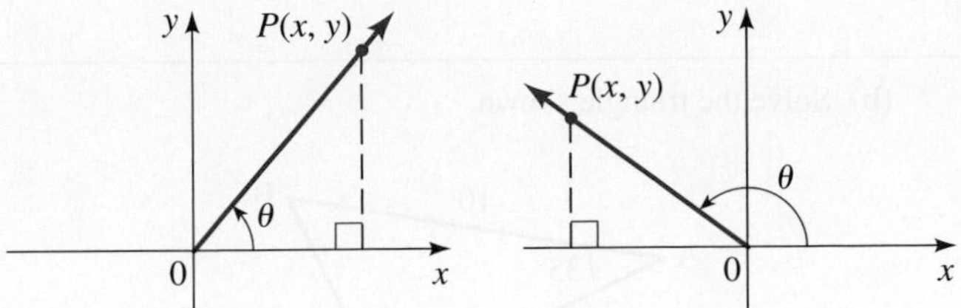

For nonzero values of the denominator the **trigonometric functions** are defined as follows.

$$\sin t = \frac{y}{r} \qquad \cos t = \frac{x}{r} \qquad \tan t = \frac{y}{x}$$

$$\csc t = \frac{r}{y} \qquad \sec t = \frac{r}{x} \qquad \cot t = \frac{x}{y}$$

Special Values of the Trigonometric Functions of Angles | Section 6.3

The following table gives the values of the trigonometric functions at some special angles.

θ	θ	$\sin\theta$	$\cos\theta$	$\tan\theta$	$\csc\theta$	$\sec\theta$	$\cot\theta$
0°	0	0	1	0	—	1	—
30°	$\frac{\pi}{6}$	$\frac{1}{2}$	$\frac{\sqrt{3}}{2}$	$\frac{\sqrt{3}}{3}$	2	$\frac{2\sqrt{3}}{3}$	$\sqrt{3}$
45°	$\frac{\pi}{4}$	$\frac{\sqrt{2}}{2}$	$\frac{\sqrt{2}}{2}$	1	$\sqrt{2}$	$\sqrt{2}$	1
60°	$\frac{\pi}{3}$	$\frac{\sqrt{3}}{2}$	$\frac{1}{2}$	$\sqrt{3}$	$\frac{2\sqrt{3}}{3}$	2	$\frac{\sqrt{3}}{3}$
90°	$\frac{\pi}{2}$	1	0	—	1	—	0
180°	π	0	−1	0	—	−1	—
270°	$\frac{3\pi}{2}$	−1	0	—	−1	—	0

Basic Trigonometric Identities | Section 6.3

An identity is an equation that is true for all values of the variable. The basic trigonometric identities follow.

Reciprocal Identities:

$$\csc\theta = \frac{1}{\sin\theta} \qquad \sec\theta = \frac{1}{\cos\theta} \qquad \cot\theta = \frac{1}{\tan\theta}$$

Pythagorean Identities:

$$\sin^2\theta + \cos^2\theta = 1$$

$$\tan^2\theta + 1 = \sec^2\theta$$

$$1 + \cot^2\theta = \csc^2\theta$$

Area of a Triangle | Section 6.3

The area $\mathscr{A}$ of a triangle with sides of lengths a and b and with included angle θ is

$$\mathscr{A} = \tfrac{1}{2}ab\sin\theta$$

Inverse Trigonometric Functions | Section 6.4

Inverse functions of the trigonometric functions have the domain and range shown in the following table.

Function	Domain	Range
$\sin^{-1}$	$[-1, 1]$	$\left[-\frac{\pi}{2}, \frac{\pi}{2}\right]$
$\cos^{-1}$	$[-1, 1]$	$[0, \pi]$
$\tan^{-1}$	$(-\infty, \infty)$	$\left(-\frac{\pi}{2}, \frac{\pi}{2}\right)$

The inverse trigonometric functions are defined as follows.

$$\sin^{-1}x = y \quad\Leftrightarrow\quad \sin y = x$$

$$\cos^{-1}x = y \quad\Leftrightarrow\quad \cos y = x$$

$$\tan^{-1}x = y \quad\Leftrightarrow\quad \tan y = x$$

The Law of Sines and the Law of Cosines | Sections 6.5 and 6.6

We follow the convention of labeling the angles of a triangle as A, B, C and the lengths of the corresponding opposite sides as a, b, c, as labeled in the figure.

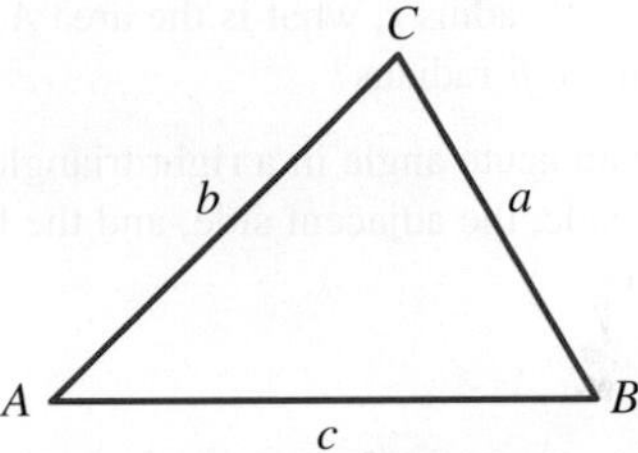

For a triangle ABC we have the following laws.

The **Law of Sines** states that

$$\frac{\sin A}{a} = \frac{\sin B}{b} = \frac{\sin C}{c}$$

The **Law of Cosines** states that

$$a^2 = b^2 + c^2 - 2bc\cos A$$

$$b^2 = a^2 + c^2 - 2ac\cos B$$

$$c^2 = a^2 + b^2 - 2ab\cos C$$

Heron's Formula | Section 6.6

Let ABC be a triangle with sides a, b, and c.

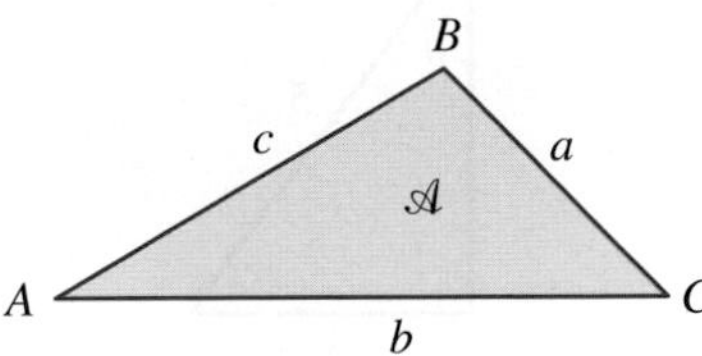

Heron's Formula states that the area $\mathscr{A}$ of triangle ABC is

$$\mathscr{A} = \sqrt{s(s-a)(s-b)(s-c)}$$

where $s = \frac{1}{2}(a + b + c)$ is the semiperimeter of the triangle.

Concept Check

1. (a) How is the degree measure of an angle defined?

(b) How is the radian measure of an angle defined?

(c) How do you convert from degrees to radians? Convert 45° to radians.

(d) How do you convert from radians to degrees? Convert 2 rad to degrees.

2. (a) When is an angle in standard position? Illustrate with a diagram.

(b) When are two angles in standard position coterminal? Illustrate with a diagram.

(c) Are the angles 25° and 745° coterminal?

(d) How is the reference angle for an angle θ defined?

(e) Find the reference angle for 150°.

3. (a) In a circle of radius r, what is the length s of an arc that subtends a central angle of θ radians?

(b) In a circle of radius r, what is the area A of a sector with central angle θ radians?

4. (a) Let θ be an acute angle in a right triangle. Identify the opposite side, the adjacent side, and the hypotenuse in the figure.

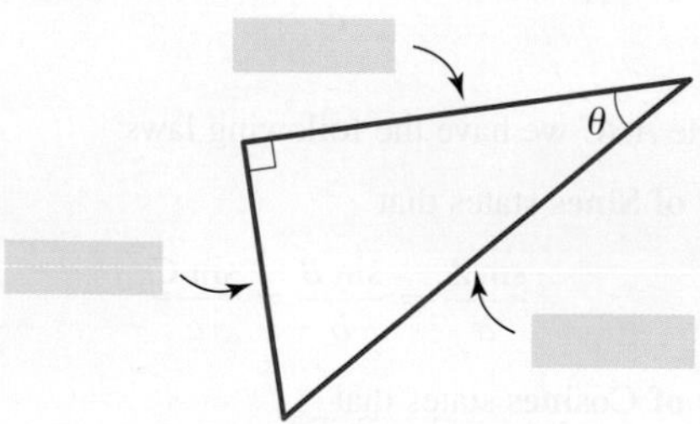

(b) Define the six trigonometric ratios in terms of the adjacent and opposite sides and the hypotenuse.

(c) Find the six trigonometric ratios for the angle θ shown in the figure.

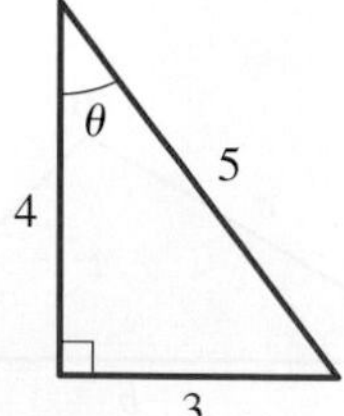

(d) List the special values of sine, cosine, and tangent.

5. (a) What does it mean to solve a triangle?

(b) Solve the triangle shown.

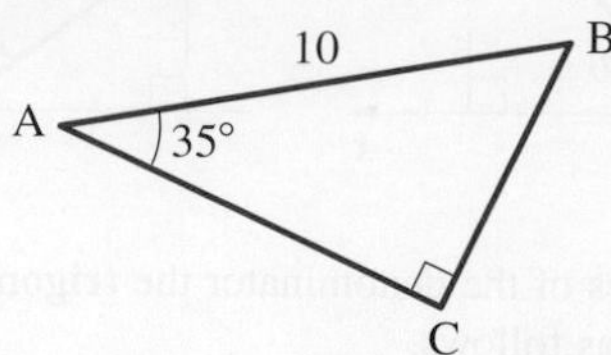

6. (a) Let θ be an angle in standard position, let $P(x, y)$ be a point on the terminal side, and let r be the distance from the origin to P, as shown in the figure. Write expressions for the six trigonometric functions of θ.

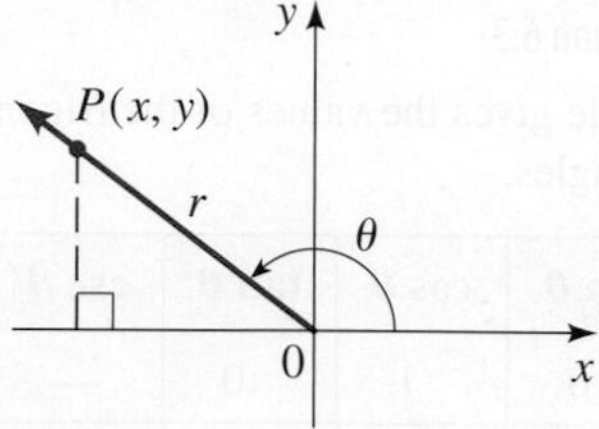

(b) Find the sine, cosine, and tangent for the angle θ shown in the figure.

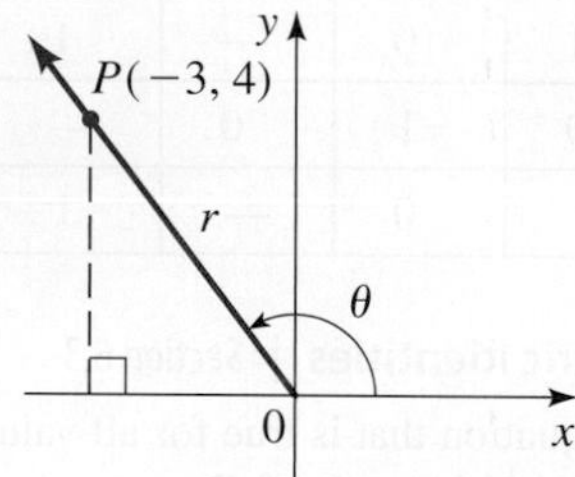

7. In each of the four quadrants, identify the trigonometric functions that are positive.

8. (a) Describe the steps we use to find the value of a trigonometric function of an angle θ.

(b) Find $\sin 5\pi/6$.

9. (a) State the reciprocal identities.

(b) State the Pythagorean identities.

10. (a) What is the area of a triangle with sides of lengths a and b and with included angle θ?

(b) What is the area of a triangle with sides of lengths a, b, and c?

11. (a) Define the inverse sine function, the inverse cosine function, and the inverse tangent function.

(b) Find $\sin^{-1}(\frac{1}{2})$, $\cos^{-1}(\sqrt{2}/2)$, and $\tan^{-1} 1$.

(c) For what values of x is the equation $\sin(\sin^{-1} x) = x$ true? For what values of x is the equation $\sin^{-1}(\sin x) = x$ true?

12. (a) State the Law of Sines.
(b) Find side a in the figure.
(c) Explain the ambiguous case in the Law of Sines.

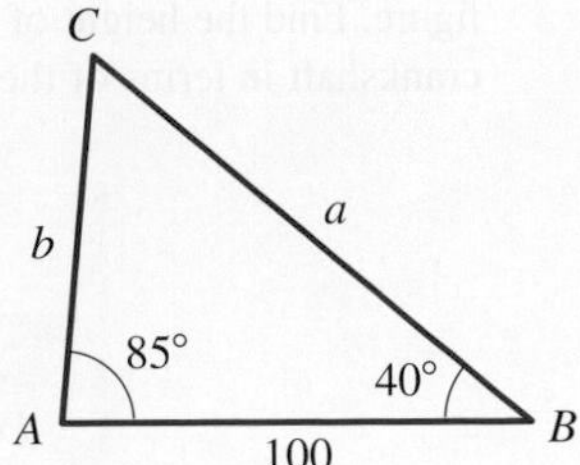

13. (a) State the Law of Cosines.
(b) Find side a in the figure.

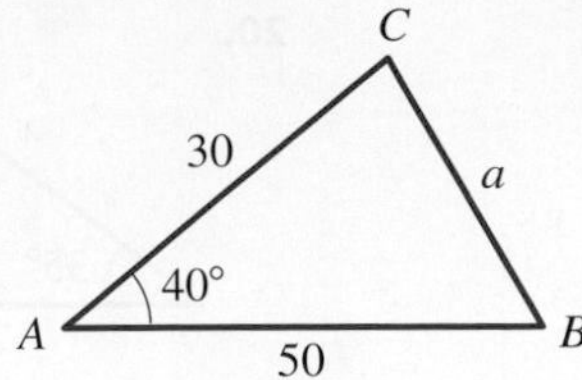

Answers to the Concept Check can be found at the book companion website **stewartmath.com**.

Exercises

1–2 ■ From Degrees to Radians Find the radian measure that corresponds to the given degree measure.

1. (a) $30°$ (b) $150°$ (c) $-20°$ (d) $-225°$

2. (a) $105°$ (b) $72°$ (c) $-405°$ (d) $-315°$

3–4 ■ From Radians to Degrees Find the degree measure that corresponds to the given radian measure.

3. (a) $\frac{5\pi}{6}$ (b) $-\frac{\pi}{9}$ (c) $-\frac{4\pi}{3}$ (d) 4

4. (a) $-\frac{5\pi}{3}$ (b) $\frac{10\pi}{9}$ (c) -5 (d) $\frac{11\pi}{3}$

5–10 ■ Length of a Circular Arc These exercises involve the formula for the length of a circular arc.

5. Find the length of an arc of a circle of radius 10 m if the arc subtends a central angle of $2\pi/5$ rad.

6. A central angle θ in a circle of radius 2.5 cm is subtended by an arc of length 7 cm. Find the measure of θ in degrees and in radians.

7. A circular arc of length 25 m subtends a central angle of $50°$. Find the radius of the circle.

8. A circular arc of length 13π m subtends a central angle of $130°$. Find the radius of the circle.

9. How many revolutions will a car wheel of diameter 70 cm make over a period of half an hour if the car is traveling at 96 km/h?

10. New York and Los Angeles are 3920 km apart. Find the angle that the arc between these two cities subtends at the center of the earth. (The radius of the earth is 6340 km.)

11–14 ■ Area of a Circular Sector These exercises involve the formula for the area of a circular sector.

11. Find the area of a sector with central angle 2 rad in a circle of radius 5 m.

12. Find the area of a sector with central angle $52°$ in a circle of radius 200 m.

13. A sector in a circle of radius 25 m has an area of 125 m^2. Find the central angle of the sector.

14. The area of a sector of a circle with a central angle of $11\pi/6$ radians is 50 m^2. Find the radius of the circle.

15. Angular Speed and Linear Speed A potter's wheel with radius 20 cm spins at 150 rpm. Find the angular and linear speeds of a point on the rim of the wheel.

16. Angular Speed and Linear Speed In an automobile transmission a *gear ratio* g is the ratio

$$g = \frac{\text{angular speed of engine}}{\text{angular speed of wheels}}$$

The angular speed of the engine is shown on the tachometer (in rpm).

A certain sports car has wheels with radius 28 cm. Its gear ratios are shown in the following table. Suppose the car is in fourth gear and the tachometer reads 3500 rpm.

(a) Find the angular speed of the engine.

(b) Find the angular speed of the wheels.

(c) How fast (in km/h) is the car traveling?

Gear	Ratio
1st	4.1
2nd	3.0
3rd	1.6
4th	0.9
5th	0.7

17–18 ■ Trigonometric Ratios Find the values of the six trigonometric ratios of θ.

17. Right triangle with legs 7 and 5, angle θ opposite side 5.

18.

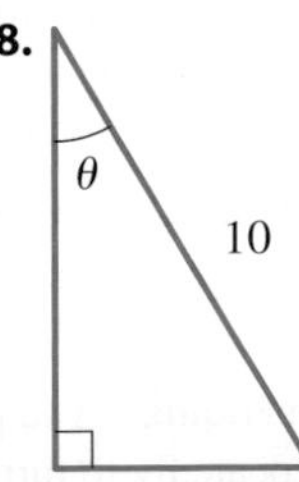

19–22 ■ Finding Sides in Right Triangles Find the sides labeled x and y, rounded to two decimal places.

19.

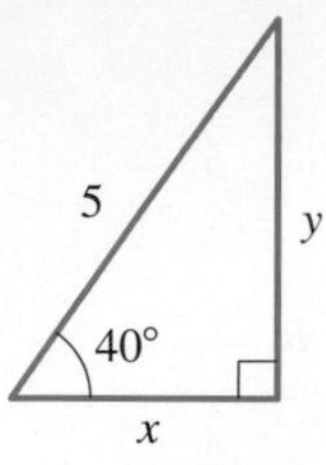

20.

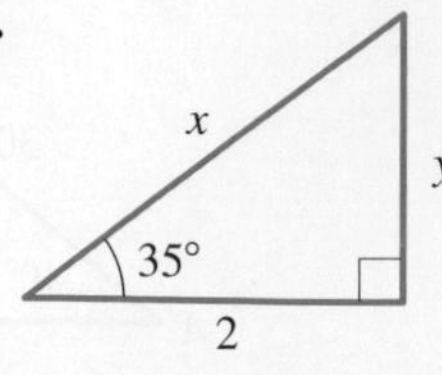

21.

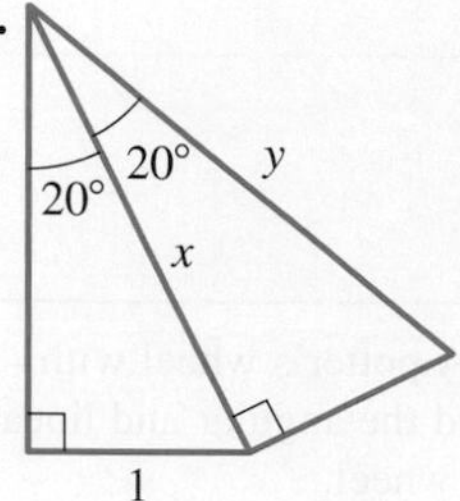

22.

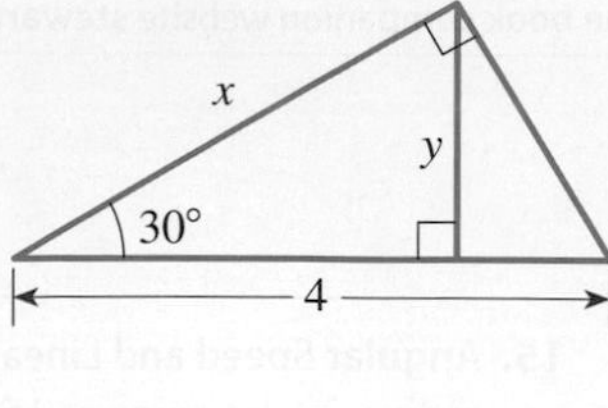

23–26 ■ Solving a Triangle Solve the triangle.

23.

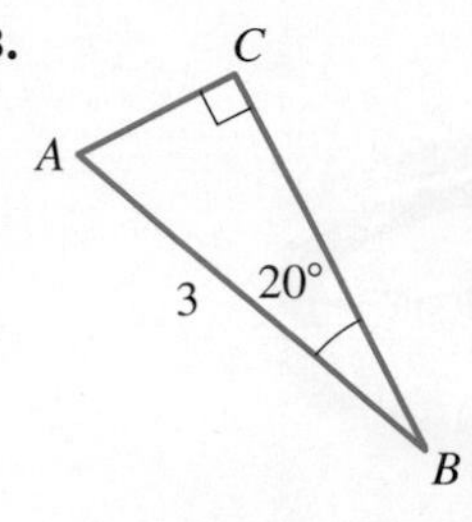

24.

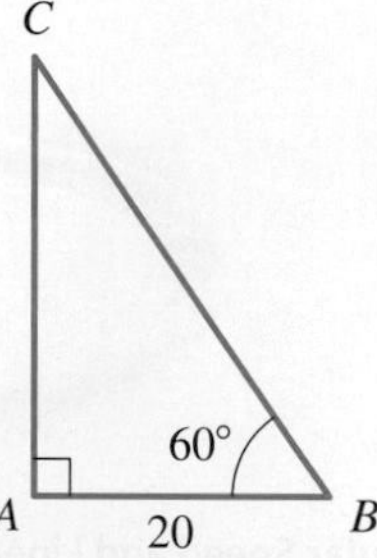

25.

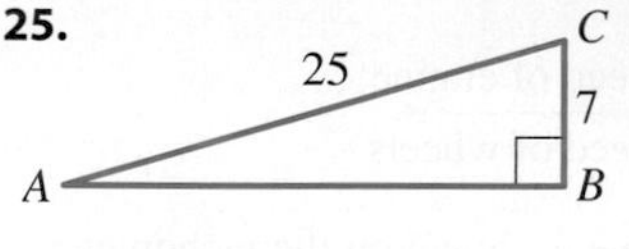

26.

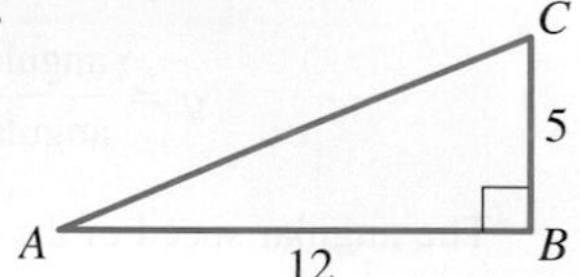

27. Trigonometric Ratios Express the lengths a and b in the figure in terms of the trigonometric ratios of θ.

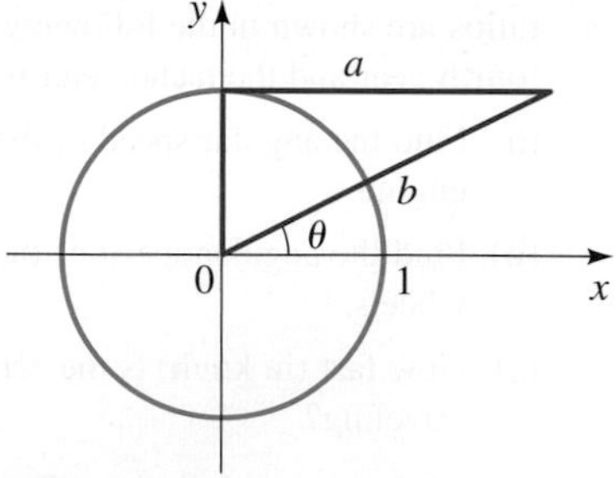

28. CN Tower The highest free-standing tower in North America is the CN Tower in Toronto, Canada. From a distance of 1 km from its base, the angle of elevation to the top of the tower is 28.81°. Find the height of the tower.

29. Perimeter of a Regular Hexagon Find the perimeter of a regular hexagon that is inscribed in a circle of radius 8 m.

30. Pistons of an Engine The pistons in a car engine move up and down repeatedly to turn the crankshaft, as shown in the figure. Find the height of the point P above the center O of the crankshaft in terms of the angle θ.

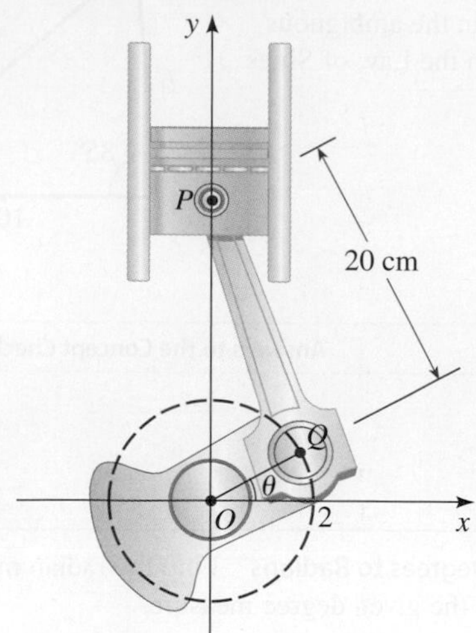

31. Radius of the Moon As viewed from the earth, the angle subtended by the full moon is 0.518°. The distance AB from the earth to the moon is 379,040 km. Use these facts to find the radius of the moon.

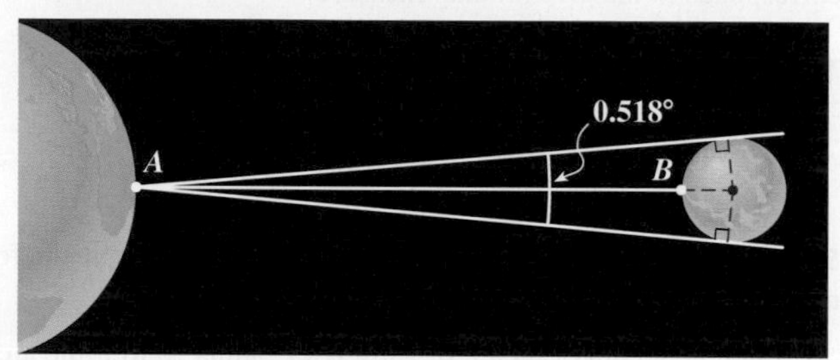

32. Distance Between Two Ships A pilot measures the angles of depression to two ships to be 40° and 52°. (See the figure.) If the pilot is flying at an elevation of 10,500 m, find the distance between the two ships.

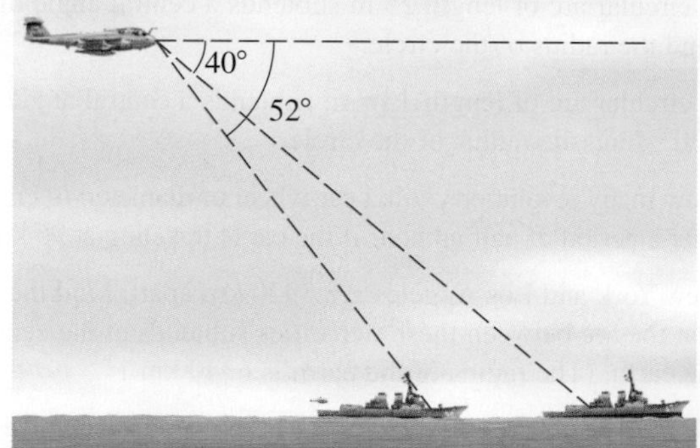

33–44 ■ Values of Trigonometric Functions Find the exact value.

33. $\sin 315°$ **34.** $\csc \dfrac{9\pi}{4}$ **35.** $\tan(-135°)$

36. $\cos \dfrac{5\pi}{6}$ **37.** $\cot\left(-\dfrac{22\pi}{3}\right)$ **38.** $\sin 405°$

39. $\cos 585°$ **40.** $\sec \dfrac{22\pi}{3}$ **41.** $\csc \dfrac{8\pi}{3}$

42. $\sec \dfrac{13\pi}{6}$ **43.** $\cot(-390°)$ **44.** $\tan \dfrac{23\pi}{4}$

45. Values of Trigonometric Functions Find the values of the six trigonometric functions of the angle θ in standard position if the point $(-5, 12)$ is on the terminal side of θ.

46. Values of Trigonometric Functions Find $\sin\theta$ if θ is in standard position and its terminal side intersects the circle of radius 1 centered at the origin at the point $\left(-\sqrt{3}/2, \frac{1}{2}\right)$.

47. Angle Formed by a Line Find the acute angle that is formed by the line $y - \sqrt{3}x + 1 = 0$ and the x-axis.

48. Values of Trigonometric Functions Find the values of the six trigonometric functions of the angle θ in standard position if its terminal side is in Quadrant III and is parallel to the line $4y - 2x - 1 = 0$.

49–52 ■ Expressing One Trigonometric Function in Terms of Another Write the first expression in terms of the second, for θ in the given quadrant.

49. $\tan\theta$, $\cos\theta$; θ in Quadrant II

50. $\sec\theta$, $\sin\theta$; θ in Quadrant III

51. $\tan^2\theta$, $\sin\theta$; θ in any quadrant

52. $\csc^2\theta\cos^2\theta$, $\sin\theta$; θ in any quadrant

53–56 ■ Values of Trigonometric Functions Find the values of the six trigonometric functions of θ from the information given.

53. $\tan\theta = \sqrt{7}/3$, $\sec\theta = \frac{4}{3}$

54. $\sec\theta = \frac{41}{40}$, $\csc\theta = -\frac{41}{9}$

55. $\sin\theta = \frac{3}{5}$, $\cos\theta < 0$

56. $\sec\theta = -\frac{13}{5}$, $\tan\theta > 0$

57–60 ■ Value of an Expression Find the value of the given trigonometric expression.

57. If $\tan\theta = -\frac{1}{2}$ for θ in Quadrant II, find $\sin\theta + \cos\theta$.

58. If $\sin\theta = \frac{1}{2}$ for θ in Quadrant I, find $\tan\theta + \sec\theta$.

59. If $\tan\theta = -1$, find $\sin^2\theta + \cos^2\theta$.

60. If $\cos\theta = -\sqrt{3}/2$ and $\pi/2 < \theta < \pi$, find $\sin 2\theta$.

61–64 ■ Values of Inverse Trigonometric Functions Find the exact value of the expression.

61. $\sin^{-1}(\sqrt{3}/2)$

62. $\tan^{-1}(\sqrt{3}/3)$

63. $\tan\left(\sin^{-1}\left(\frac{2}{5}\right)\right)$

64. $\sin\left(\cos^{-1}\left(\frac{3}{8}\right)\right)$

65–66 ■ Inverse Trigonometric Functions Rewrite the expression as an algebraic expression in x.

65. $\sin(\tan^{-1} x)$

66. $\csc(\sin^{-1} x)$

67–68 ■ Finding an Unknown Side Express θ in terms of x.

67.

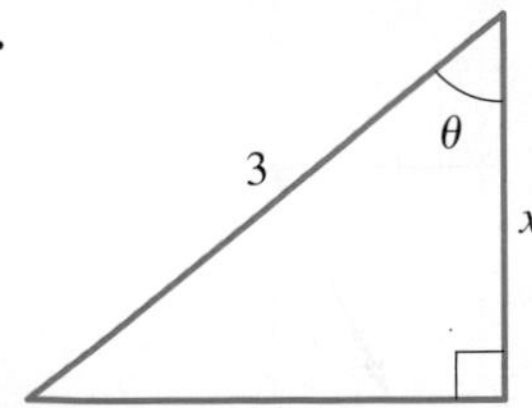

68.

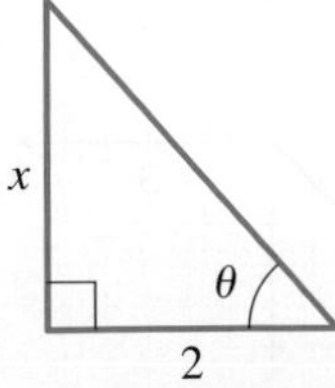

69–78 ■ Law of Sines and Law of Cosines Find the side labeled x or the angle labeled θ.

69. **70.**

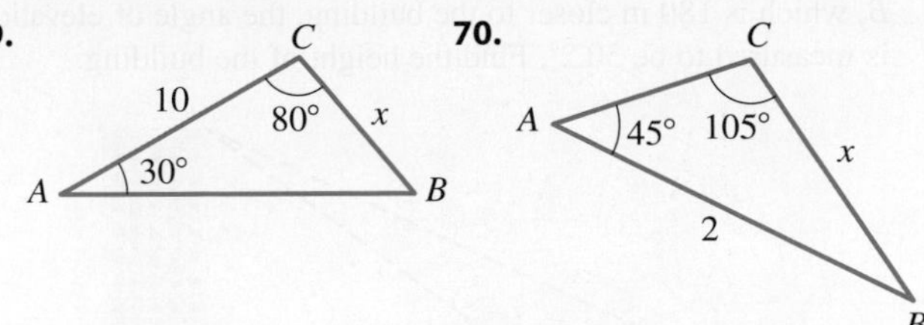

71. **72.**

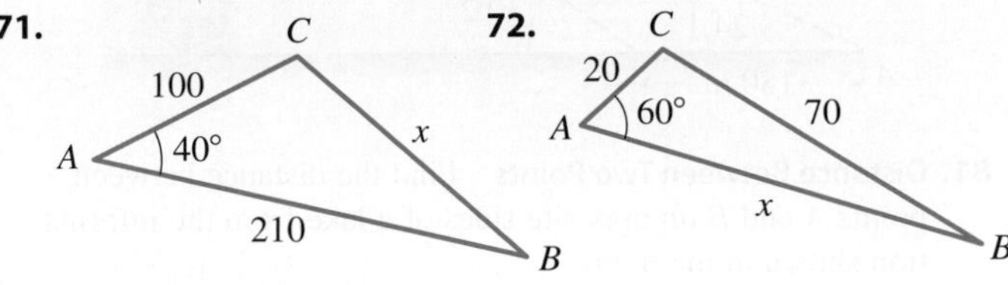

73. **74.**

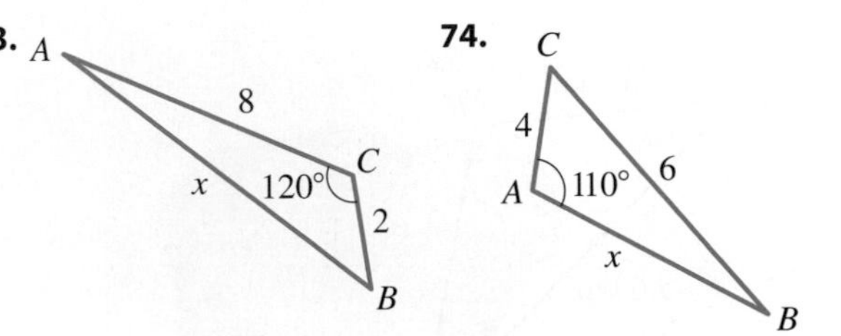

75. **76.**

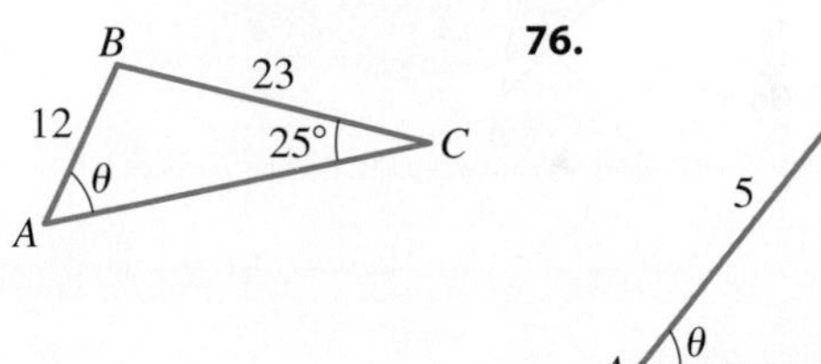

77. **78.**

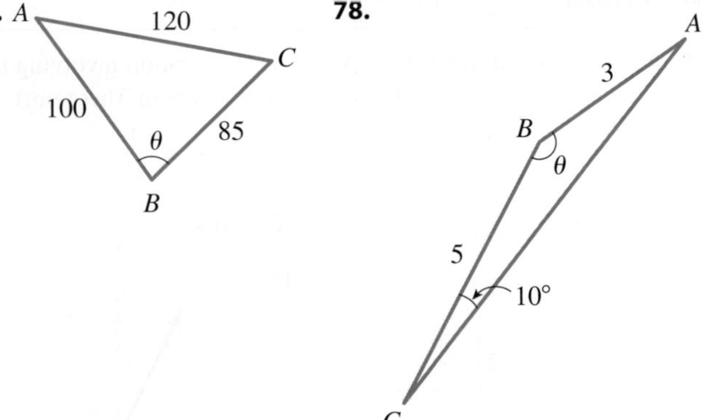

79. Distance Between Two Ships Two ships leave a port at the same time. One travels at 20 km/h in a direction N 32° E, and the other travels at 28 km/h in a direction S 42° E (see the figure). How far apart are the two ships after two hours?

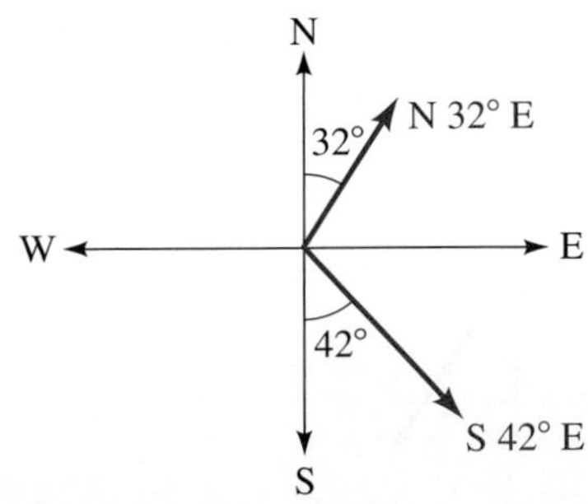

80. Height of a Building From a point A on the ground, the angle of elevation to the top of a tall building is 24.1°. From a point B, which is 180 m closer to the building, the angle of elevation is measured to be 30.2°. Find the height of the building.

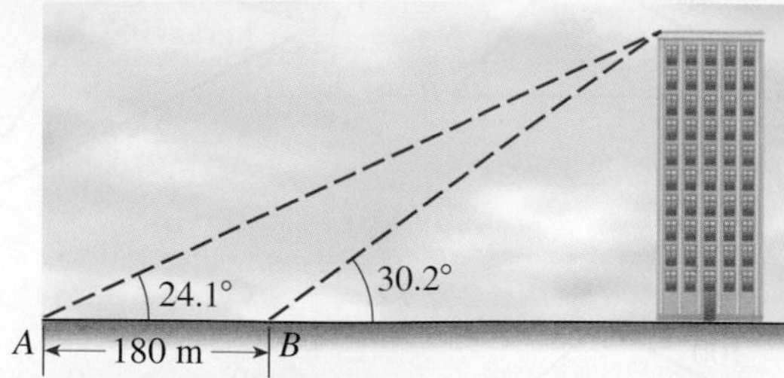

81. Distance Between Two Points Find the distance between points A and B on opposite sides of a lake from the information shown in the figure.

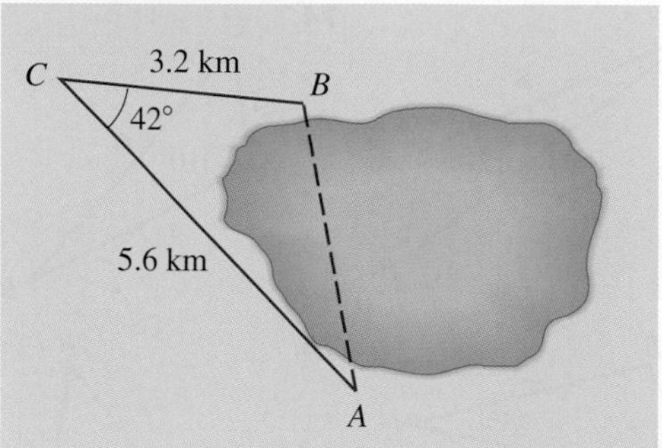

82. Distance Between a Boat and the Shore A boat is cruising the ocean off a straight shoreline. Points A and B are 120 km apart on the shore, as shown in the figure. It is found that $\angle A = 42.3°$ and $\angle B = 68.9°$. Find the shortest distance from the boat to the shore.

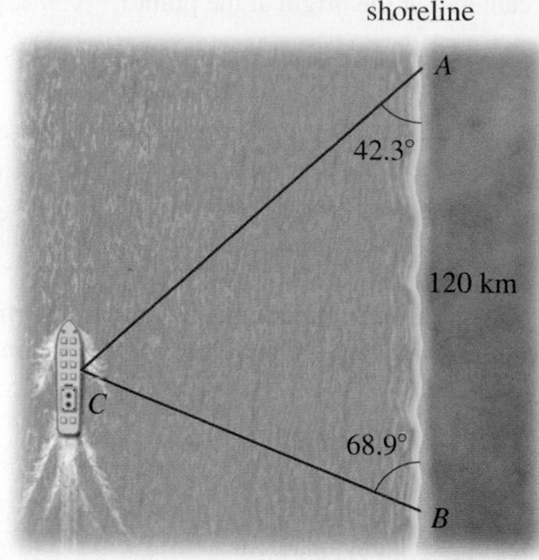

83. Area of a Triangle Find the area of a triangle with sides of length 8 and 14 and included angle 35°.

84. Heron's Formula Find the area of a triangle with sides of length 5, 6, and 8.

Matching

85. Angles and Trigonometry Match each equation involving the angle θ with the appropriate diagram. Give reasons for your answers. (You may find it helpful to use the Pythagorean Theorem.)

(a) $\csc\theta = -\sqrt{2}$ **(b)** $\sin\theta = \dfrac{1}{\sqrt{2}}$ **(c)** $\theta = \sin^{-1}(-\frac{3}{5})$ **(d)** $\tan\theta = \frac{4}{3}$

(e) $\tan\theta = -\frac{3}{2}$ **(f)** $\sin\theta = -\dfrac{2}{\sqrt{5}}$ **(g)** $\theta = \cos^{-1}\left(-\dfrac{1}{\sqrt{5}}\right)$ **(h)** $\sin\theta = \frac{3}{5}$

I

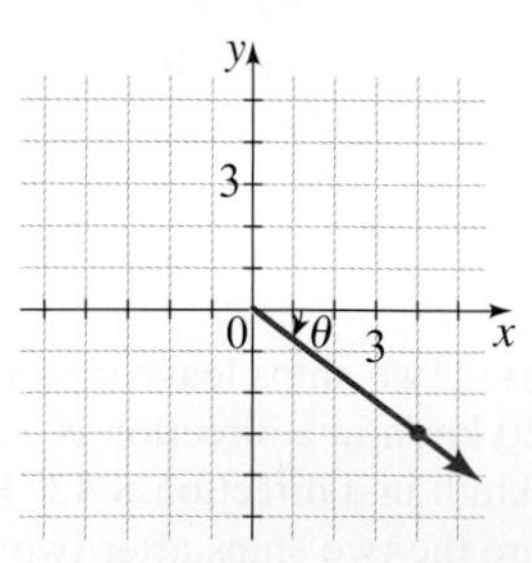

II

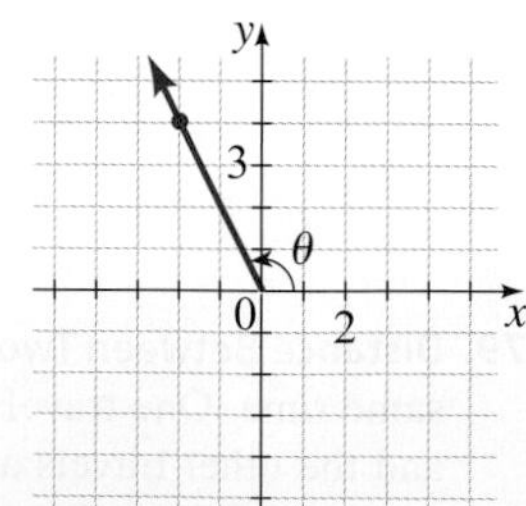

III

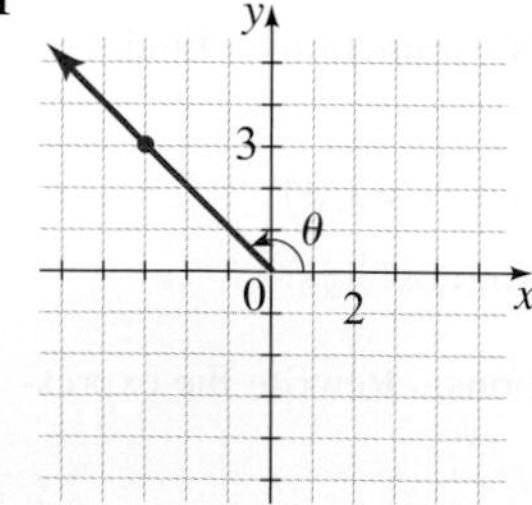

IV

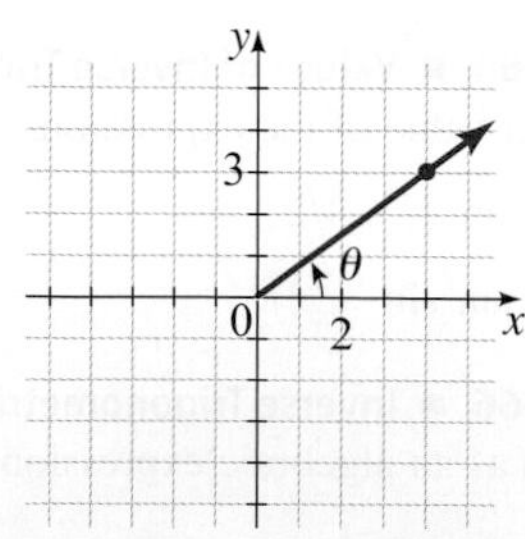

V

VI

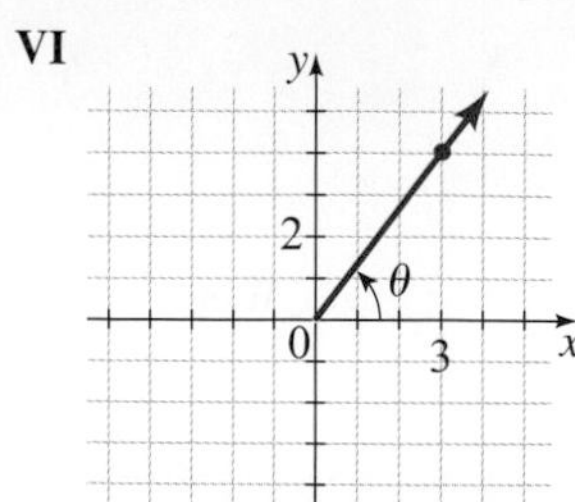

VII

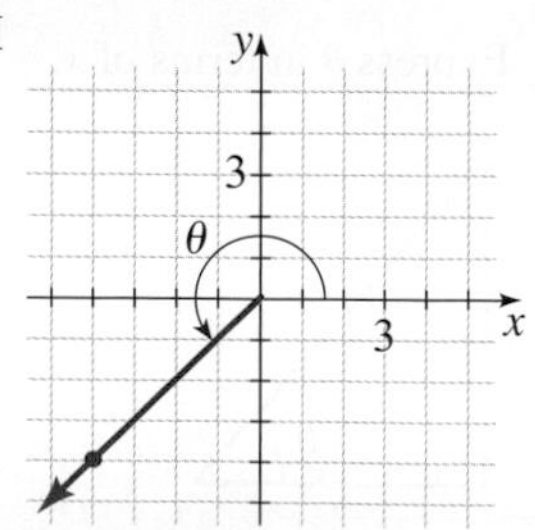

VIII

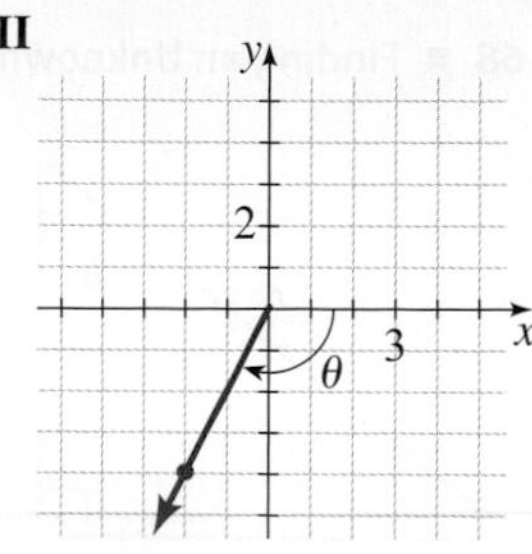

Chapter 6 | Test

1. Find the radian measures that correspond to the degree measures 330° and −135°.

2. Find the degree measures that correspond to the radian measures $4\pi/3$ and -1.3.

3. The rotor blades of a helicopter are 5 m long and are rotating at 120 rpm.

(a) Find the angular speed of the rotor.

(b) Find the linear speed of a point on the tip of a blade.

4. Find the exact value of each trignonometric function.

(a) $\sin 405°$ **(b)** $\tan(-150°)$ **(c)** $\sec \dfrac{5\pi}{3}$ **(d)** $\csc \dfrac{5\pi}{2}$

5. Find $\tan \theta + \sin \theta$ for the angle θ shown in the figure.

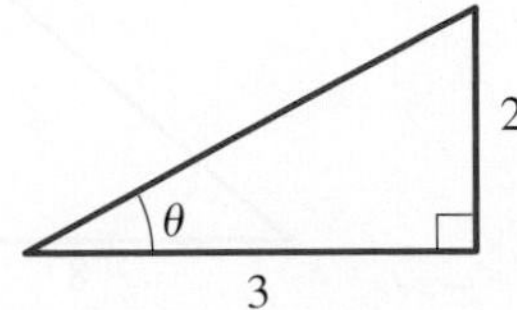

6. Express the lengths a and b shown in the figure in terms of θ.

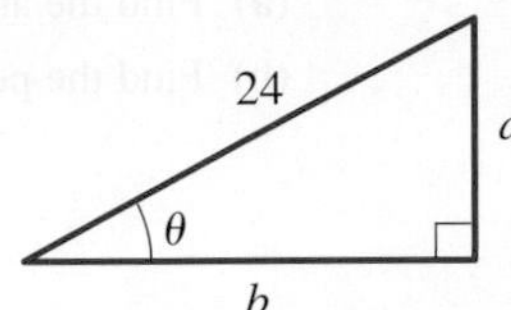

7. If $\cos \theta = -\frac{1}{3}$ and θ is in Quadrant III, find $\tan \theta \cot \theta + \csc \theta$.

8. If $\sin \theta = \frac{5}{13}$ and $\tan \theta = -\frac{5}{12}$, find $\sec \theta$.

9. Express $\tan \theta$ in terms of $\sec \theta$ for θ in Quadrant II.

10. The base of the ladder in the figure is 1.8 m from the building, and the angle formed by the ladder and the ground is 73°. How high up the building does the ladder touch?

11. Express θ in each figure in terms of x.

(a)

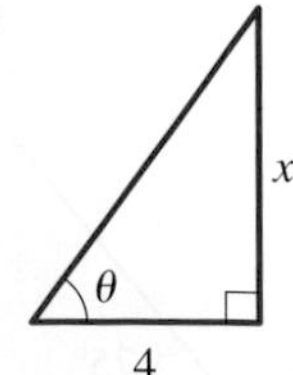

(b)

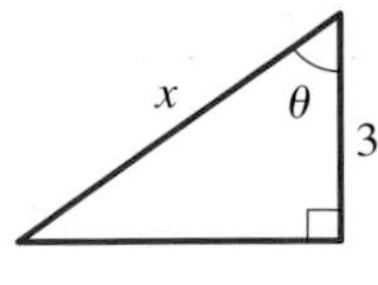

12. Find the exact value of $\cos\left(\tan^{-1}\left(\frac{9}{40}\right)\right)$.

13–18 ■ Find the side labeled x or the angle labeled θ.

13.

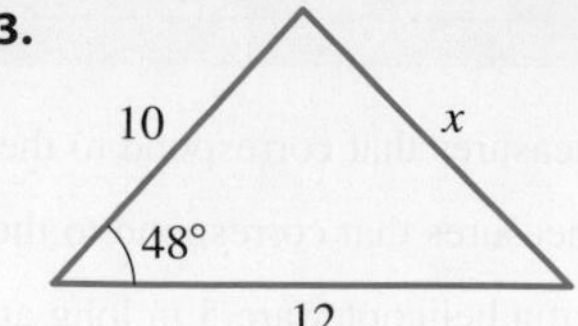

14.

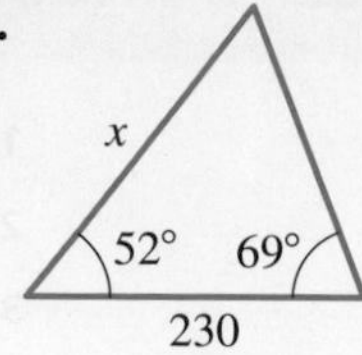

15.

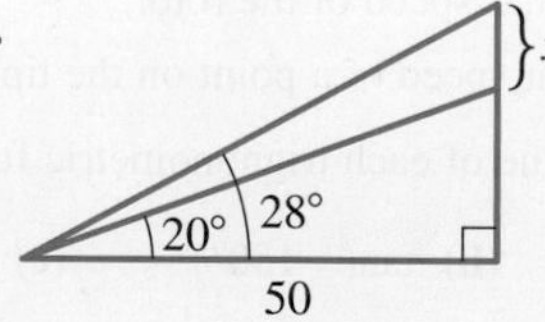

16.

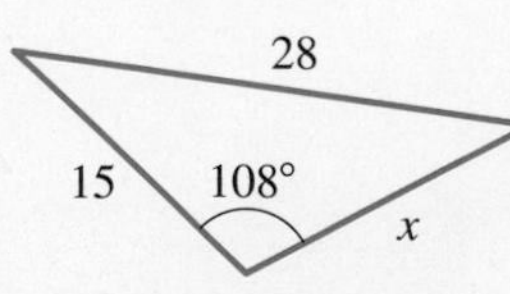

17.

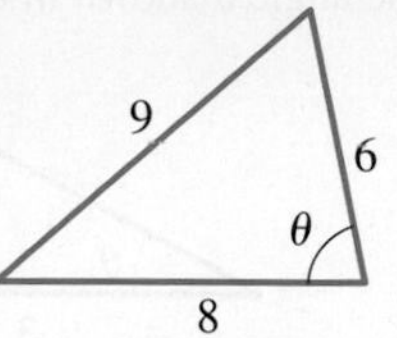

18.

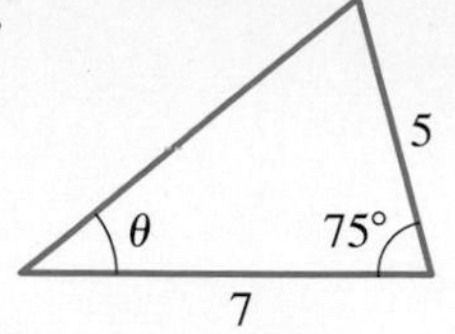

19. Refer to the figure.

(a) Find the area of the shaded region.

(b) Find the perimeter of the shaded region.

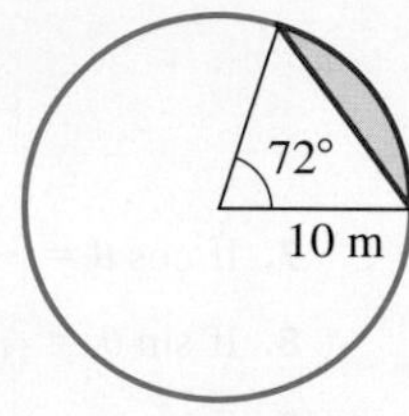

20. Refer to the figure.

(a) Find the angle opposite the longest side.

(b) Find the area of the triangle.

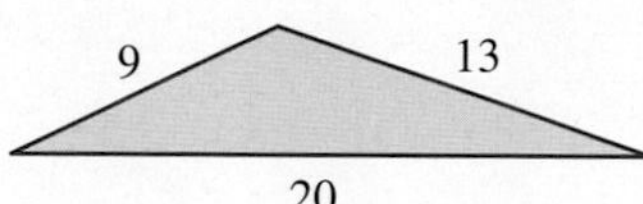

21. Two wires tether a balloon to the ground, as shown in the figure. How high is the balloon above the ground?

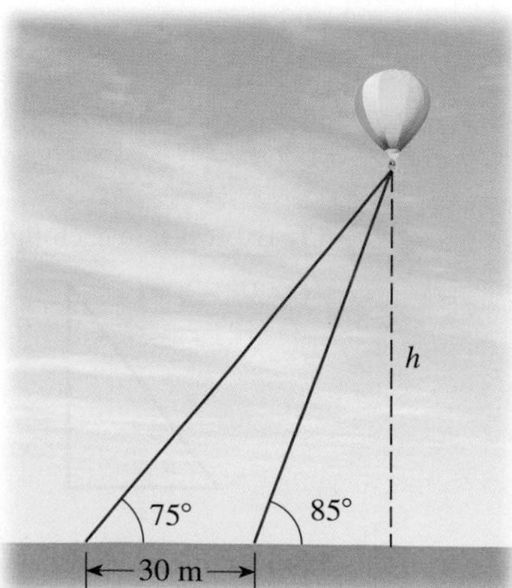

Focus on Modeling | Surveying

How can we measure the height of a mountain or the distance across a lake? Obviously, it may be difficult, inconvenient, or impossible to measure these distances directly (that is, by using a tape measure or a yardstick). On the other hand, it is easy to measure *angles* involving distant objects. That's where trigonometry comes in: Trigonometric ratios relate angles to distances, so they can be used to *calculate* distances from the *measured* angles. In this *Focus* we examine how trigonometry is used to map a town. Mapmaking methods now use satellites and the Global Positioning System (GPS), but mathematics remains at the core of the process.

■ Mapping a Town

A group of students want to draw a map of their hometown. To construct an accurate map (or scale model), they need to find distances between various landmarks in the town. The students make the measurements shown in Figure 1. Note that only one distance is measured: between City Hall and the first bridge. All other measurements are angles.

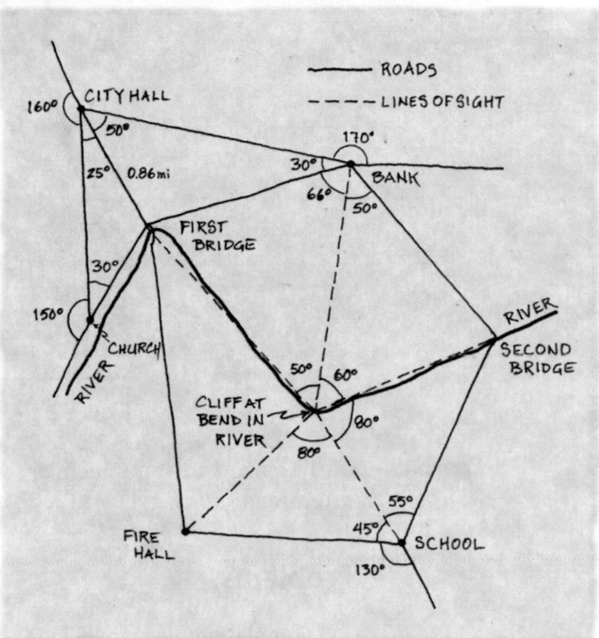

Figure 1

The distances between other landmarks can now be found by using the Law of Sines. For example, the distance x from the bank to the first bridge is calculated by applying the Law of Sines to the triangle with vertices at City Hall, the bank, and the first bridge.

$$\frac{x}{\sin 50^\circ} = \frac{0.86}{\sin 30^\circ} \qquad \text{Law of Sines}$$

$$x = \frac{0.86 \sin 50^\circ}{\sin 30^\circ} \qquad \text{Solve for } x$$

$$\approx 1.32 \text{ km} \qquad \text{Calculator}$$

So the distance between the bank and the first bridge is 1.32 km.

The distance we just found can now be used to find other distances. For instance, we find the distance y between the bank and the cliff as follows:

$$\frac{y}{\sin 64^\circ} = \frac{1.32}{\sin 50^\circ} \qquad \text{Law of Sines}$$

$$y = \frac{1.32 \sin 64^\circ}{\sin 50^\circ} \qquad \text{Solve for } y$$

$$\approx 1.55 \text{ km} \qquad \text{Calculator}$$

Continuing in this fashion, we can calculate all the distances between the landmarks shown in the rough sketch in Figure 1. We can use this information to draw the map shown in Figure 2.

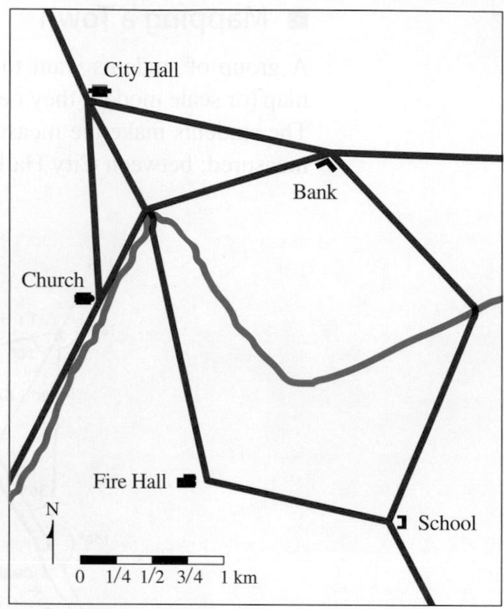

Figure 2

To make a topographic map, we need to measure elevation. This concept is explored in Problems 4–6.

Problems

1. **Completing the Map** Find the distance between the church and City Hall.
2. **Completing the Map** Find the distance between the fire hall and the school. [*Hint:* You will need to find other distances first.]
3. **Determining a Distance** A surveyor standing on the south side of a river wishes to find the distance between points A and B on the north side of the river. The surveyor chooses points C and D on the south side of the river, which are 20 m apart, and measures the angles shown in the figure. Find the distance between A and B.

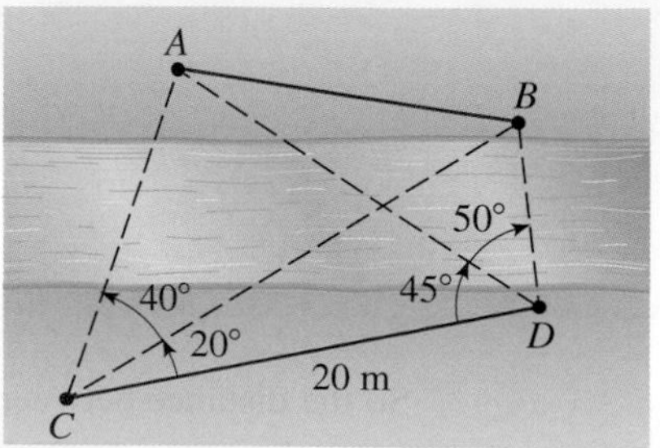

4. Height of a Cliff To measure the height of an inaccessible cliff on the opposite side of a river, a surveyor makes the measurements shown in the figure. Find the height of the cliff.

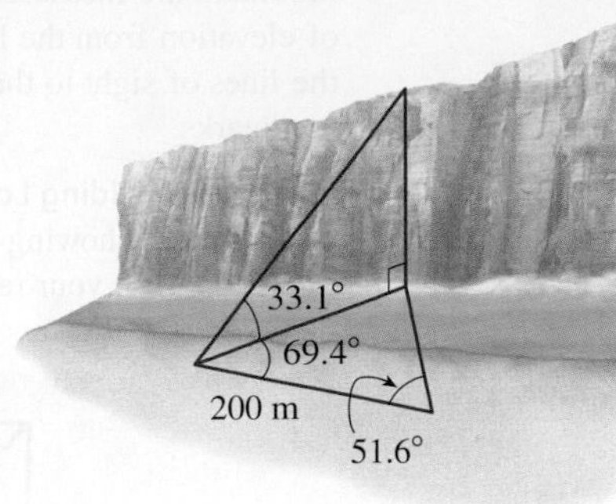

5. Height of a Mountain To calculate the height h of a mountain, angles α and β and distance d are measured, as shown in the figure.

(a) Show that

$$h = \frac{d \tan \alpha \tan \beta}{\tan \beta - \tan \alpha}$$

(b) Use the formula from part (a) to find the height of a mountain if $\alpha = 25°$, $\beta = 29°$, and $d = 800$ m.

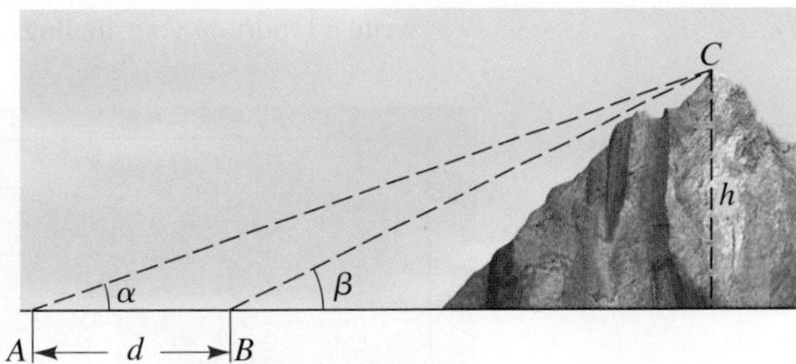

6. Estimating the Radius of the Earth The mathematician al-Biruni (circa 973–1050) devised a method for estimating the radius R of the earth by making just four surveying measurements. He first estimated the height h of a mountain (as described in Problem 5), then climbed the mountain to measure the angle of depression θ to the horizon. Use the diagram to show that

$$R = \frac{h \cos \theta}{1 - \cos \theta}$$

Then estimate the radius of the earth using $h \approx 1.41$ km and $\theta \approx 0.021$ rad.

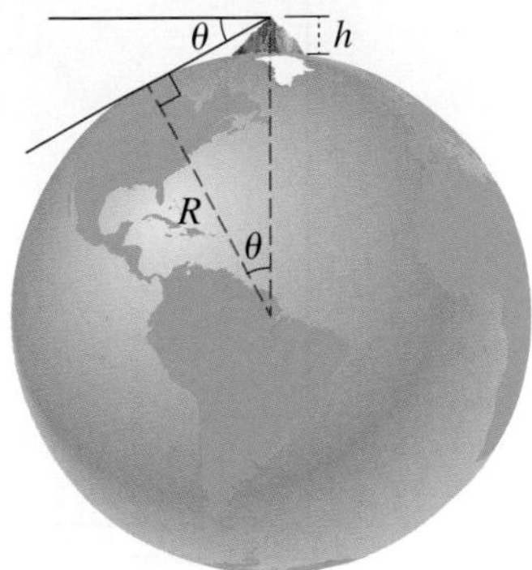

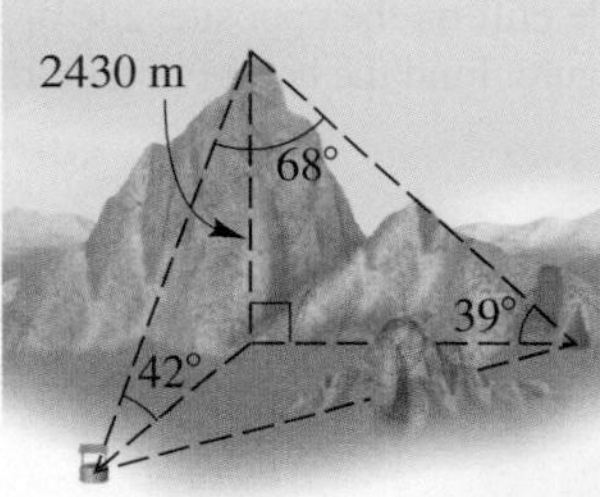

7. Determining a Distance A surveyor has determined that a mountain is 2430 m high. From the top of the mountain the angles of depression to two landmarks at the base of the mountain are measured to be 42° and 39°. (Observe that these are the same as the angles of elevation from the landmarks as shown in the figure at the left.) The angle between the lines of sight to the landmarks is 68°. Calculate the distance between the two landmarks.

8. Surveying Building Lots A surveyor surveys two adjacent lots and makes the following rough sketch showing the measurements obtained. Calculate all the distances shown in the figure, and use your result to draw an accurate map of the two lots.

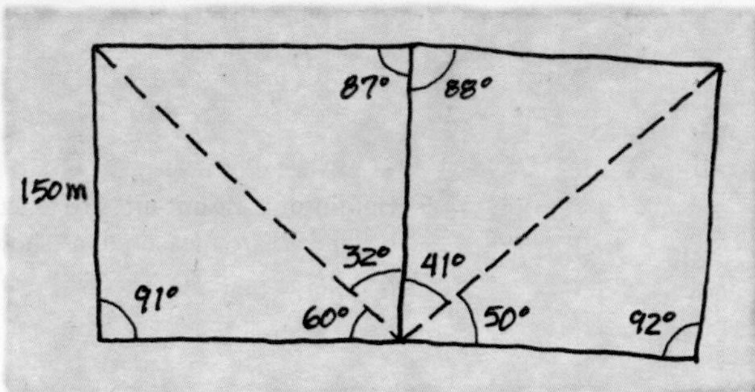

9. Great Survey of India The Great Trigonometric Survey of India was one of the most massive mapping projects ever undertaken (see the margin note on surveying in Section 6.5). Do some research at your library or on the Internet to learn more about the Survey, and write a report on your findings.

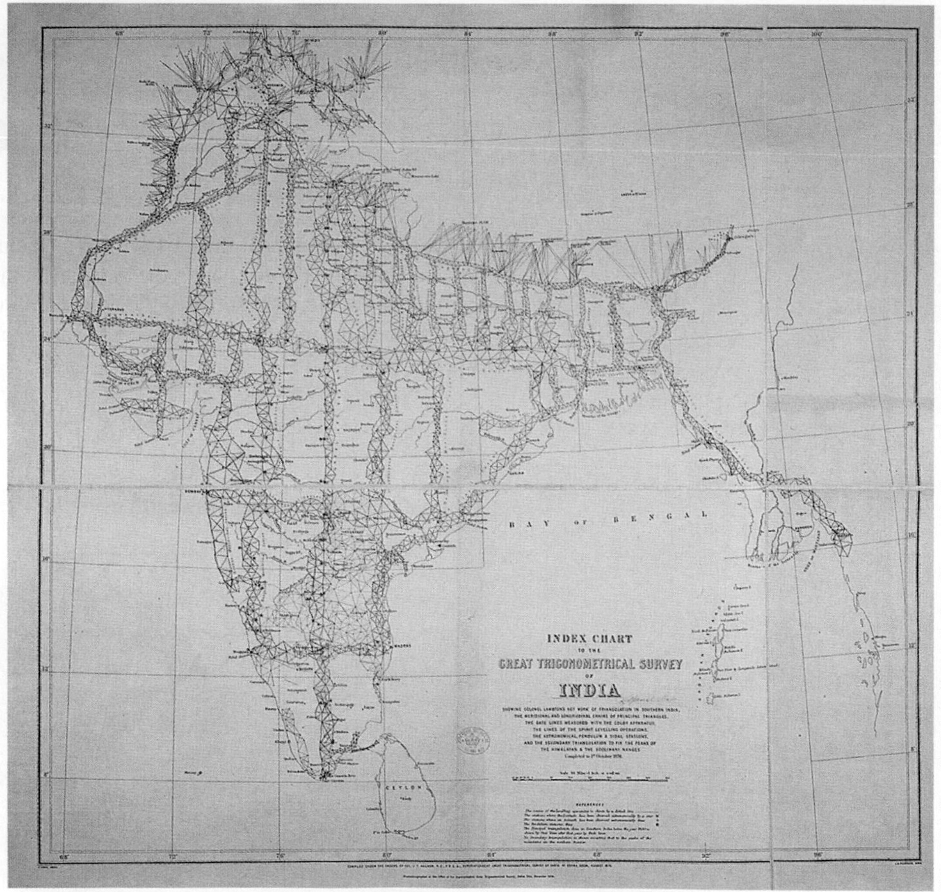

Bilal Seckin/Shutterstock.com

7 Analytic Trigonometry

In Chapters 5 and 6, we studied graphical and geometric properties of the trigonometric functions. In this chapter we study algebraic properties of these functions, that is, simplifying and factoring expressions and solving equations that involve trigonometric functions.

We have used the trigonometric functions to model different real-world phenomena, including periodic motion (such as the sound waves produced by the street musicians in the above photo). To obtain information from a model, we often need to solve equations. If the model involves trigonometric functions, we need to solve trigonometric equations. Solving trigonometric equations often requires using trigonometric identities. We've already encountered some basic trigonometric identities in the preceding chapters. We begin this chapter by finding many new identities.

7.1 Trigonometric Identities

■ Simplifying Trigonometric Expressions ■ Proving Trigonometric Identities

Recall that an **equation** is a statement that two mathematical expressions are equal. For example, the following are equations:

$$x + 2 = 5$$
$$(x + 1)^2 = x^2 + 2x + 1$$
$$\sin^2 t + \cos^2 t = 1$$

An **identity** is an equation that is true for all values of the variable(s) for which both sides of the equation are defined. The last two equations above are identities, but the first one is not, since it is not true for values of x other than 3.

A **trigonometric** identity is an identity involving trigonometric functions. We begin by listing some of the fundamental trigonometric identities. We studied most of these in Chapters 5 and 6; you are asked to prove the cofunction identities in Exercise 116.

Fundamental Trigonometric Identities

Reciprocal Identities

$$\csc x = \frac{1}{\sin x} \qquad \sec x = \frac{1}{\cos x} \qquad \cot x = \frac{1}{\tan x}$$

$$\tan x = \frac{\sin x}{\cos x} \qquad \cot x = \frac{\cos x}{\sin x}$$

Pythagorean Identities

$$\sin^2 x + \cos^2 x = 1 \qquad \tan^2 x + 1 = \sec^2 x \qquad 1 + \cot^2 x = \csc^2 x$$

Even-Odd Identities

$$\sin(-x) = -\sin x \qquad \cos(-x) = \cos x \qquad \tan(-x) = -\tan x$$

Cofunction Identities

$$\sin\left(\frac{\pi}{2} - x\right) = \cos x \qquad \tan\left(\frac{\pi}{2} - x\right) = \cot x \qquad \sec\left(\frac{\pi}{2} - x\right) = \csc x$$

$$\cos\left(\frac{\pi}{2} - x\right) = \sin x \qquad \cot\left(\frac{\pi}{2} - x\right) = \tan x \qquad \csc\left(\frac{\pi}{2} - x\right) = \sec x$$

■ Simplifying Trigonometric Expressions

We can use identities to write the same expression in different ways. It is often possible to rewrite a complicated-looking expression as a much simpler one. To simplify algebraic expressions, we used factoring, common denominators, and the Special Product Formulas. To simplify trigonometric expressions, we use these same techniques together with the fundamental trigonometric identities.

Example 1 ■ Simplifying a Trigonometric Expression

Simplify the expression $\cos t + \tan t \sin t$.

Solution We start by rewriting the expression in terms of sine and cosine.

$$\begin{aligned}
\cos t + \tan t \sin t &= \cos t + \left(\frac{\sin t}{\cos t}\right)\sin t && \text{Reciprocal identity}\\
&= \frac{\cos^2 t + \sin^2 t}{\cos t} && \text{Common denominator}\\
&= \frac{1}{\cos t} && \text{Pythagorean identity}\\
&= \sec t && \text{Reciprocal identity}
\end{aligned}$$

Now Try Exercise 3 ■

Example 2 ■ Simplifying by Combining Fractions

Simplify the expression $\dfrac{\sin\theta}{\cos\theta} + \dfrac{\cos\theta}{1 + \sin\theta}$.

Solution We combine the fractions by using a common denominator.

$$\begin{aligned}
\frac{\sin\theta}{\cos\theta} + \frac{\cos\theta}{1 + \sin\theta} &= \frac{\sin\theta\,(1 + \sin\theta) + \cos^2\theta}{\cos\theta\,(1 + \sin\theta)} && \text{Common denominator}\\
&= \frac{\sin\theta + \sin^2\theta + \cos^2\theta}{\cos\theta\,(1 + \sin\theta)} && \text{Distribute } \sin\theta\\
&= \frac{\sin\theta + 1}{\cos\theta\,(1 + \sin\theta)} && \text{Pythagorean identity}\\
&= \frac{1}{\cos\theta} = \sec\theta && \text{Cancel, and use reciprocal identity}
\end{aligned}$$

Now Try Exercise 25 ■

■ Proving Trigonometric Identities

Many identities follow from the fundamental identities. In the examples that follow, we learn how to prove that a given trigonometric equation is an identity, and in the process we will see how to discover new identities.

First, to decide when a given equation is *not* an identity, all we need to do is show that the equation does not hold for some value of the variable (or variables). Thus the equation

$$\sin x + \cos x = 1$$

is not an identity, because when $x = \pi/4$, we have

$$\sin\frac{\pi}{4} + \cos\frac{\pi}{4} = \frac{\sqrt{2}}{2} + \frac{\sqrt{2}}{2} = \sqrt{2} \neq 1$$

To verify that a trigonometric equation is an identity, we transform one side of the equation into the other side by a series of steps, each of which is itself an identity.

Guidelines for Proving Trigonometric Identities

1. **Start with One Side.** Choose one side of the equation, and write it down. Your goal is to transform it into the other side. It's usually easier to start with the more complicated side.
2. **Use Known Identities.** Use algebra and the identities you know to change the side you started with. Bring fractional expressions to a common denominator, factor, and use the fundamental identities to simplify expressions.
3. **Convert to Sines and Cosines.** If you are stuck, you may find it helpful to first rewrite all functions in terms of sines and cosines.

Warning To prove an identity, we do *not* just perform the same operations on both sides of the equation. For example, if we start with an equation that is not an identity, such as

$$\sin x = -\sin x$$

and square both sides, we get the equation

$$\sin^2 x = \sin^2 x$$

which is clearly an identity. Does this mean that the original equation is an identity? Of course not. The problem here is that the operation of squaring is not **reversible** in the sense that we cannot arrive back at the original equation by taking square roots (reversing the procedure). Only operations that are reversible will necessarily transform an identity into an identity.

Example 3 ■ Proving an Identity by Rewriting in Terms of Sine and Cosine

Consider the equation $\cos\theta\,(\sec\theta - \cos\theta) = \sin^2\theta$.

(a) Verify algebraically that the equation is an identity.

(b) Confirm graphically that the equation is an identity.

Solution

(a) The left-hand side looks more complicated, so we start with it and try to transform it into the right-hand side.

$$\begin{aligned} \text{LHS} &= \cos\theta\,(\sec\theta - \cos\theta) \\ &= \cos\theta\left(\frac{1}{\cos\theta} - \cos\theta\right) && \text{Reciprocal identity} \\ &= 1 - \cos^2\theta && \text{Expand} \\ &= \sin^2\theta = \text{RHS} && \text{Pythagorean identity} \end{aligned}$$

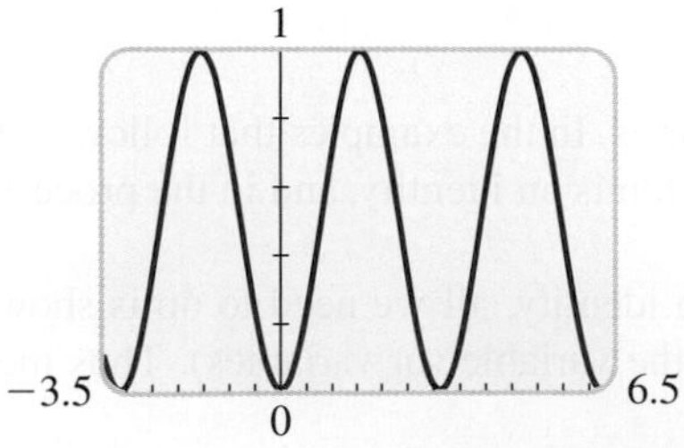

(b) We graph each side of the equation to see whether the graphs coincide. From Figure 1 we see that the graphs of $y = \cos\theta\,(\sec\theta - \cos\theta)$ and $y = \sin^2\theta$ are identical. This confirms that the equation is an identity.

1

−3.5 0 6.5

Figure 1

Now Try Exercise 31 ■

Note In Example 3 it isn't easy to see how to change the right-hand side into the left-hand side, but it's definitely possible: notice that each step is reversible. In other words, if we start with the last expression in the proof and work backward through the steps, the right-hand side is transformed into the left-hand side. You will probably agree, however, that it's more difficult to prove the identity this way. That's why it's often better to change the more complicated side of the identity into the simpler side.

Example 4 ■ Proving an Identity by Combining Fractions

Verify the identity

$$2 \tan x \sec x = \frac{1}{1 - \sin x} - \frac{1}{1 + \sin x}$$

Solution Finding a common denominator and combining the fractions on the right-hand side of this equation, we get

$$\begin{aligned}
\text{RHS} &= \frac{1}{1 - \sin x} - \frac{1}{1 + \sin x} \\
&= \frac{(1 + \sin x) - (1 - \sin x)}{(1 - \sin x)(1 + \sin x)} && \text{Common denominator} \\
&= \frac{2 \sin x}{1 - \sin^2 x} && \text{Simplify} \\
&= \frac{2 \sin x}{\cos^2 x} && \text{Pythagorean identity} \\
&= 2\frac{\sin x}{\cos x}\left(\frac{1}{\cos x}\right) && \text{Factor} \\
&= 2 \tan x \sec x = \text{LHS} && \text{Reciprocal identities}
\end{aligned}$$

Now Try Exercise 67 ■

See the Prologue: *Principles of Problem Solving*

In Example 5 we "introduce something extra" to the problem: We multiply the numerator and the denominator by a trigonometric expression that we have chosen so that we can simplify the result.

Example 5 ■ Proving an Identity by Introducing Something Extra

Verify the identity $\dfrac{\cos u}{1 - \sin u} = \sec u + \tan u$.

Solution We start with the left-hand side and multiply the numerator and denominator by $1 + \sin u$.

We multiply by $1 + \sin u$ because we know by the difference of squares formula that

$$(1 - \sin u)(1 + \sin u) = 1 - \sin^2 u$$

and this is just $\cos^2 u$, a simpler expression.

$$\begin{aligned}
\text{LHS} &= \frac{\cos u}{1 - \sin u} \\
&= \frac{\cos u}{1 - \sin u} \cdot \frac{1 + \sin u}{1 + \sin u} && \text{Multiply numerator and denominator by } 1 + \sin u \\
&= \frac{\cos u\,(1 + \sin u)}{1 - \sin^2 u} && \text{Expand denominator} \\
&= \frac{\cos u\,(1 + \sin u)}{\cos^2 u} && \text{Pythagorean identity} \\
&= \frac{1 + \sin u}{\cos u} && \text{Cancel common factor} \\
&= \frac{1}{\cos u} + \frac{\sin u}{\cos u} && \text{Separate into two fractions} \\
&= \sec u + \tan u = \text{RHS} && \text{Reciprocal identities}
\end{aligned}$$

Now Try Exercise 79 ■

Morphart Creation/Shutterstock.com

EUCLID (circa 300 B.C.) lived in Alexandria, Egypt. His book *Elements* is the most influential scientific book in history. For 2000 years it was the standard introduction to geometry in schools, and for many generations it was considered the best way to develop logical reasoning. Abraham Lincoln, for instance, studied the *Elements* as a way to sharpen his mind. The story is told that King Ptolemy once asked Euclid whether there was a faster way to learn geometry than through the *Elements*. Euclid replied that there is "no royal road to geometry"—meaning by this that mathematics does not respect wealth or social status. Euclid was revered in his own time and was referred to as "The Geometer" or "The Writer of the *Elements*." The greatness of the *Elements* stems from its precise, logical, and systematic treatment of geometry. For dealing with equality, Euclid lists the following rules, which he calls "common notions."

1. Things that are equal to the same thing are equal to each other.
2. If equals are added to equals, the sums are equal.
3. If equals are subtracted from equals, the remainders are equal.
4. Things that coincide with one another are equal.
5. The whole is greater than the part.

Here is another method for proving that an equation is an identity. If we can transform each side of the equation *separately*, by way of identities, to arrive at the same result, then the equation is an identity. Example 6 illustrates this procedure.

Example 6 ■ Proving an Identity by Working with Both Sides Separately

Verify the identity $\dfrac{1+\cos\theta}{\cos\theta} = \dfrac{\tan^2\theta}{\sec\theta - 1}$.

Solution We prove the identity by changing each side separately into the same expression. (You should supply the reasons for each step.)

$$\text{LHS} = \frac{1+\cos\theta}{\cos\theta} = \frac{1}{\cos\theta} + \frac{\cos\theta}{\cos\theta} = \sec\theta + 1$$

$$\text{RHS} = \frac{\tan^2\theta}{\sec\theta - 1} = \frac{\sec^2\theta - 1}{\sec\theta - 1} = \frac{(\sec\theta - 1)(\sec\theta + 1)}{\sec\theta - 1} = \sec\theta + 1$$

It follows that LHS = RHS, so the equation is an identity.

 **Now Try Exercise 85** ■

We conclude this section by describing the technique of *trigonometric substitution*, which we use to convert algebraic expressions to trigonometric ones. This is often useful in calculus, for instance, in finding the area of a circle or an ellipse.

Example 7 ■ Trigonometric Substitution

Substitute $\sin\theta$ for x in the expression $\sqrt{1-x^2}$, and simplify. Assume that $0 \le \theta \le \pi/2$.

Solution Setting $x = \sin\theta$, we have

$\sqrt{1-x^2} = \sqrt{1-\sin^2\theta}$	Substitute $x = \sin\theta$
$= \sqrt{\cos^2\theta}$	Pythagorean identity
$= \cos\theta$	Take square root

The last equality is true because $\cos\theta \ge 0$ for the values of θ in question.

 **Now Try Exercise 91** ■

7.1 | Exercises

■ Concepts

1. An equation is called an identity if it is valid for ______ values of the variable. The equation $2x = x + x$ is an algebraic identity, and the equation $\sin^2 x + \cos^2 x =$ ______ is a trigonometric identity.

2. For any x it is true that $\cos(-x)$ has the same value as $\cos x$. We express this fact as the identity ______.

■ Skills

3–14 ■ Simplifying Trigonometric Expressions Write the trigonometric expression in terms of sine and cosine, and then simplify.

3. $\cos t \tan t$
4. $\cos t \csc t$
5. $\sin\theta \sec\theta$
6. $\tan\theta \csc\theta$
7. $\tan^2 x - \sec^2 x$
8. $\dfrac{\sec x}{\csc x}$

9. $\sin^2\left(\dfrac{\pi}{2} - y\right)\sec y$ **10.** $\tan\left(\dfrac{\pi}{2} - u\right)\sin u$

11. $\sin u + \cot u \cos u$ **12.** $\cos^2\theta\,(1 + \tan^2\theta)$

13. $\dfrac{\sec\theta - \cos\theta}{\sin\theta}$ **14.** $\dfrac{\cot\theta}{\csc\theta - \sin\theta}$

15–30 ■ Simplifying Trigonometric Expressions Simplify the trigonometric expression.

15. $\dfrac{\sin x \sec x}{\tan x}$ **16.** $\dfrac{\cos x \sec x}{\cot x}$

17. $\dfrac{\sin t + \tan t}{\tan t}$ **18.** $\dfrac{1 + \cot A}{\csc A}$

19. $\sin^3\left(\dfrac{\pi}{2} - x\right) + \sin^2 x \cos x$

20. $\sin^4\alpha - \cos^4\alpha + \cos^2\alpha$

21. $\dfrac{\sec^2 x - 1}{\sec^2 x}$ **22.** $\dfrac{\sec x - \cos x}{\tan x}$

23. $\dfrac{1 + \cos y}{1 + \sec y}$ **24.** $\dfrac{1 + \cos(y - \pi/2)}{1 + \csc y}$

25. $\dfrac{1 + \sin u}{\cos u} + \dfrac{\cos u}{1 + \sin u}$ **26.** $\dfrac{\sin t}{1 - \cos t} - \csc t$

27. $\dfrac{\cos(-x)}{\sec(-x) + \tan x}$ **28.** $\dfrac{\cot A - 1}{1 + \tan(-A)}$

29. $\dfrac{1}{1 - \sin\alpha} + \dfrac{1}{1 + \sin\alpha}$ **30.** $\dfrac{2 + \tan^2 x}{\sec^2 x} - 1$

31–32 ■ Proving an Identity Algebraically and Graphically Consider the given equation. **(a)** Verify algebraically that the equation is an identity. **(b)** Confirm graphically that the equation is an identity.

31. $\dfrac{\cos x}{\sec x \sin x} = \csc x - \sin x$ **32.** $\dfrac{\tan y}{\csc y} = \sec y - \cos y$

33–90 ■ Proving Identities Verify the identity using the fundamental trigonometric identities.

33. $\dfrac{\cos\alpha}{\sec\alpha} = \cos^2\alpha$ **34.** $\dfrac{\tan\beta}{\sin\beta} = \sec\beta$

35. $\dfrac{\cos u \sec u}{\tan u} = \cot u$ **36.** $\dfrac{\cot x \sec x}{\csc x} = 1$

37. $\cos^2\left(\dfrac{\pi}{2} - y\right)\csc y = \sin y$

38. $\tan\left(x - \dfrac{\pi}{2}\right)\sin x = \sin\left(x - \dfrac{\pi}{2}\right)$

39. $(\sin x + \cos x)^2 = 1 + 2\sin x \cos x$

40. $\tan\theta + \cot\theta = \sec\theta\csc\theta$

41. $\cos(-x) - \sin(-x) = \cos x + \sin x$

42. $\cot(-\alpha)\cos(-\alpha) + \sin(-\alpha) = -\csc\alpha$

43. $\dfrac{\sec A - 1}{\sec A + 1} = \dfrac{1 - \cos A}{1 + \cos A}$

44. $\dfrac{\cos^2 v}{\sin v} = \csc v - \sin v$

45. $(1 - \cos\beta)(1 + \cos\beta) = \dfrac{1}{\csc^2\beta}$

46. $\dfrac{\cos x}{\sec x} + \dfrac{\sin x}{\csc x} = 1$

47. $\dfrac{1}{1 - \sin^2 y} = 1 + \tan^2 y$

48. $\csc x - \sin x = \cos x \cot x$

49. $(\tan x + \cot x)^2 = \sec^2 x + \csc^2 x$

50. $\sin^2 y + \cos^2 y + \tan^2 y = \sec^2 y$

51. $(1 - \sin^2 t + \cos^2 t)^2 + 4\sin^2 t\cos^2 t = 4\cos^2 t$

52. $\dfrac{2\sin x \cos x}{(\sin x + \cos x)^2 - 1} = 1$

53. $\csc x \cos^2 x + \sin x = \csc x$

54. $\cot^2 t - \cos^2 t = \cot^2 t \cos^2 t$

55. $\dfrac{\sec t - \cos t}{\sec t} = \sin^2 t$

56. $(\cot x - \csc x)(\cos x + 1) = -\sin x$

57. $\cos^2 x - \sin^2 x = 2\cos^2 x - 1$

58. $2\cos^2 x - 1 = 1 - 2\sin^2 x$

59. $\sin^4\theta - \cos^4\theta = \sin^2\theta - \cos^2\theta$

60. $(1 - \cos^2 x)(1 + \cot^2 x) = 1$

61. $\dfrac{(\sin t + \cos t)^2}{\sin t \cos t} = 2 + \sec t \csc t$

62. $\sec t \csc t\,(\tan t + \cot t) = \sec^2 t + \csc^2 t$

63. $\dfrac{1 + \tan^2 u}{1 - \tan^2 u} = \dfrac{1}{\cos^2 u - \sin^2 u}$

64. $\dfrac{1 + \sec^2 x}{1 + \tan^2 x} = 1 + \cos^2 x$

65. $\dfrac{\sec x + \csc x}{\tan x + \cot x} = \sin x + \cos x$

66. $\dfrac{\sin x + \cos x}{\sec x + \csc x} = \sin x \cos x$

67. $\dfrac{1 - \cos x}{\sin x} + \dfrac{\sin x}{1 - \cos x} = 2\csc x$

68. $\dfrac{\csc^2 y - \cot^2 y}{\sec^2 y} = \cos^2 y$

69. $\tan^2 u - \sin^2 u = \tan^2 u \sin^2 u$

70. $\sec^4 x - \tan^4 x = \sec^2 x + \tan^2 x$

71. $\dfrac{1 + \tan x}{1 - \tan x} = \dfrac{\cos x + \sin x}{\cos x - \sin x}$

72. $\dfrac{\cos\theta}{1 - \sin\theta} = \dfrac{\sin\theta - \csc\theta}{\cos\theta - \cot\theta}$

73. $\dfrac{1}{\sec x + \tan x} + \dfrac{1}{\sec x - \tan x} = 2\sec x$

74. $\dfrac{\cos^2 t + \tan^2 t - 1}{\sin^2 t} = \tan^2 t$

75. $\dfrac{1 + \sin x}{1 - \sin x} - \dfrac{1 - \sin x}{1 + \sin x} = 4\tan x \sec x$

76. $\dfrac{\tan x + \tan y}{\cot x + \cot y} = \tan x \tan y$

77. $\dfrac{\sin^3 x + \cos^3 x}{\sin x + \cos x} = 1 - \sin x \cos x$

78. $\dfrac{\sin^3 y - \csc^3 y}{\sin y - \csc y} = \sin^2 y + \csc^2 y + 1$

79. $\dfrac{1 - \cos\alpha}{\sin\alpha} = \dfrac{\sin\alpha}{1 + \cos\alpha}$

80. $\dfrac{\sin x - 1}{\sin x + 1} = \dfrac{-\cos^2 x}{(\sin x + 1)^2}$

81. $\dfrac{\sin w}{\sin w + \cos w} = \dfrac{\tan w}{1 + \tan w}$

82. $\dfrac{\sin A}{1 - \cos A} - \cot A = \csc A$

83. $\dfrac{\sec x}{\sec x - \tan x} = \sec x\,(\sec x + \tan x)$

84. $\sec v - \tan v = \dfrac{1}{\sec v + \tan v}$

85. $\dfrac{(\sin x + \cos x)^2}{\sin^2 x - \cos^2 x} = \dfrac{\sin^2 x - \cos^2 x}{(\sin x - \cos x)^2}$

86. $\dfrac{\tan v \sin v}{\tan v + \sin v} = \dfrac{\tan v - \sin v}{\tan v \sin v}$

87. $\dfrac{1 - \sin x}{1 + \sin x} = (\sec x - \tan x)^2$

88. $\dfrac{1 + \sin x}{1 - \sin x} = (\tan x + \sec x)^2$

89. $\csc x - \cot x = \dfrac{1}{\csc x + \cot x}$

90. $\dfrac{\sec u - 1}{\sec u + 1} = \dfrac{\tan u - \sin u}{\tan u + \sin u}$

91–96 ■ Trigonometric Substitution Make the indicated trigonometric substitution in the given algebraic expression and simplify (see Example 7). Assume that $0 < \theta < \pi/2$.

91. $\dfrac{x}{\sqrt{1 - x^2}}$, $x = \sin\theta$

92. $\sqrt{1 + x^2}$, $x = \tan\theta$

93. $\sqrt{x^2 - 1}$, $x = \sec\theta$

94. $\dfrac{1}{x^2\sqrt{4 + x^2}}$, $x = 2\tan\theta$

95. $\sqrt{9 - x^2}$, $x = 3\sin\theta$

96. $\dfrac{\sqrt{x^2 - 25}}{x}$, $x = 5\sec\theta$

97–100 ■ Determining Identities Graphically Graph f and g in the same viewing rectangle. Do the graphs suggest that the equation $f(x) = g(x)$ is an identity? Prove your answer.

97. $f(x) = \cos^2 x - \sin^2 x$, $g(x) = 1 - 2\sin^2 x$

98. $f(x) = \tan x\,(1 + \sin x)$, $g(x) = \dfrac{\sin x \cos x}{1 + \sin x}$

99. $f(x) = (\sin x + \cos x)^2$, $g(x) = 1$

100. $f(x) = \cos^4 x - \sin^4 x$, $g(x) = 2\cos^2 x - 1$

■ Skills Plus

101–106 ■ Proving More Identities Verify the identity.

101. $(\tan x + \cot x)^2 = \sec^2 x + \csc^2 x$

102. $\dfrac{1 + \cos x + \sin x}{1 + \cos x - \sin x} = \dfrac{1 + \sin x}{\cos x}$

103. $(\sin\alpha - \tan\alpha)(\cos\alpha - \cot\alpha) = (\cos\alpha - 1)(\sin\alpha - 1)$

104. $\sin^6\beta + \cos^6\beta = 1 - 3\sin^2\beta\cos^2\beta$

105. $\tan^6 y = \dfrac{\sin^2 y - \tan^2 y}{\cos^2 y - \cot^2 y}$

106. $\dfrac{1}{(\sin x + \cos x)^2} = \dfrac{\sec^2 x}{(1 + \tan x)^2}$

PS *Try to recognize something familiar.* One method is to expand the denominator of the LHS and then factor $\cos^2 x$ from the denominator.

107–110 ■ Proving Identities Involving Other Functions These identities involve trigonometric functions as well as other functions that we have studied. Verify the identity.

107. $\ln|\tan x \sin x| = 2\ln|\sin x| + \ln|\sec x|$

108. $\ln|\tan x| + \ln|\cot x| = 0$

109. $e^{\sin^2 x}e^{\tan^2 x} = e^{\sec^2 x}e^{-\cos^2 x}$

110. $e^{x + 2\ln|\sin x|} = e^x \sin^2 x$

111. An Identity Involving Three Variables Suppose $x = R\cos\theta\sin\phi$, $y = R\sin\theta\sin\phi$, and $z = R\cos\phi$. Verify the identity $x^2 + y^2 + z^2 = R^2$.

■ Discuss ■ Discover ■ Prove ■ Write

112. Discuss: Is the Equation an Identity? You have encountered many identities in this course. Which of the following equations do you recognize as identities? For those that you think are identities, test several values of the variables to confirm that the equation is true for those values.

(a) $(x + y)^2 = x^2 + 2xy + y^2$ **(b)** $x^2 + y^2 = 1$

(c) $x(y + z) = xy + xz$ **(d)** $e^{\sin^2 x}e^{\cos^2 x} = e$

(e) $xe^{\ln x^2} = x^3$ **(f)** $\sin t + \cos t = 1$

(g) $x^2 - \tan^2 x = 0$

(h) $t^2 - \cos^2 t = (t - \cos t)(t + \cos t)$

113. Discuss: Equations That Are Not Identities How can you tell if an equation is not an identity? Show that the following equations are not identities.

(a) $\sin 2x = 2\sin x$

(b) $\sec^2 x + \csc^2 x = 1$

(c) $\sin(x + y) = \sin x + \sin y$

(d) $\dfrac{1}{\sin x + \cos x} = \csc x + \sec x$

114. Discuss: Graphs and Identities Suppose you graph two functions, f and g, on a graphing device and their graphs appear identical in the viewing rectangle. Does this prove that the equation $f(x) = g(x)$ is an identity? Explain.

115. Discover: Making Up Your Own Identity If you start with a trigonometric expression and rewrite it or simplify it, then setting the original expression equal to the rewritten expression yields a trigonometric identity. For instance, from Example 1 we get the identity

$$\cos t + \tan t \sin t = \sec t$$

Use this technique to make up your own identity, then give it to a classmate to verify.

116. Discuss: Cofunction Identities Use the right triangle shown in the figure to explain why $v = (\pi/2) - u$. Explain how you can obtain all six cofunction identities from this triangle for $0 < u < \pi/2$.

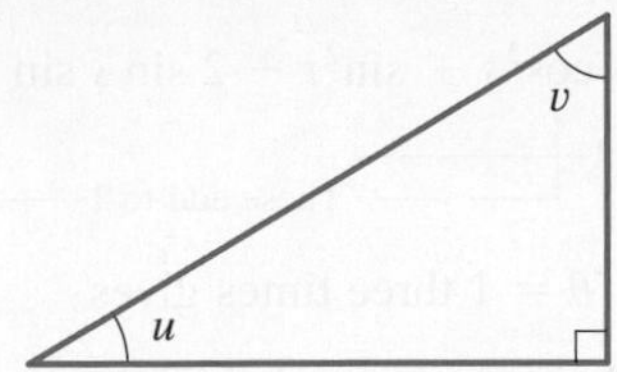

Note that u and v are complementary angles. So the cofunction identities state that "a trigonometric function of an angle u is equal to the corresponding cofunction of the complementary angle v."

117. Discuss ■ Discover: Find the exact value of

$$\sin^{-1}\left(\frac{2}{3}\right) + \cos^{-1}\left(\frac{2}{3}\right) + \tan^{-1}\left(\frac{2}{3}\right) + \cot^{-1}\left(\frac{2}{3}\right)$$

PS *Draw a diagram.* Draw a triangle for each pair of cofunctions.

7.2 Addition and Subtraction Formulas

■ Addition and Subtraction Formulas ■ Expressions Involving Trigonometric Functions of a Sum ■ Expressions of the Form $A \sin x + B \cos x$

■ Addition and Subtraction Formulas

We now derive identities for trigonometric functions of sums and differences.

Addition and Subtraction Formulas

Formulas for Sine:

$$\sin(s + t) = \sin s \cos t + \cos s \sin t$$
$$\sin(s - t) = \sin s \cos t - \cos s \sin t$$

Formulas for Cosine:

$$\cos(s + t) = \cos s \cos t - \sin s \sin t$$
$$\cos(s - t) = \cos s \cos t + \sin s \sin t$$

Formulas for Tangent:

$$\tan(s + t) = \frac{\tan s + \tan t}{1 - \tan s \tan t}$$
$$\tan(s - t) = \frac{\tan s - \tan t}{1 + \tan s \tan t}$$

Proof of Addition Formula for Cosine To prove the formula

$$\cos(s + t) = \cos s \cos t - \sin s \sin t$$

we use Figure 1. In the figure, the distances t, $s + t$, and $-s$ have been marked on the unit circle, starting at $P_0(1, 0)$ and terminating at Q_1, P_1, and Q_0, respectively. The coordinates of these points are as follows:

$$P_0(1, 0) \qquad Q_0(\cos(-s), \sin(-s))$$
$$P_1(\cos(s + t), \sin(s + t)) \qquad Q_1(\cos t, \sin t)$$

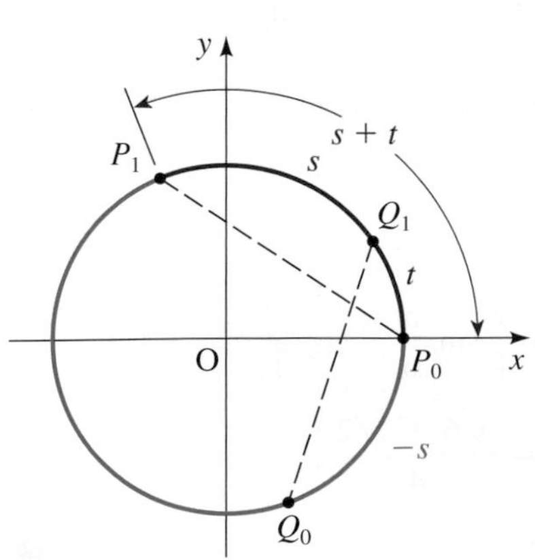

Figure 1

Since $\cos(-s) = \cos s$ and $\sin(-s) = -\sin s$, it follows that the point Q_0 has the coordinates $Q_0(\cos s, -\sin s)$. Notice that the distances between P_0 and P_1 and between Q_0 and Q_1 measured along the arc of the circle are equal. Since equal arcs are subtended by equal chords, it follows that $d(P_0, P_1) = d(Q_0, Q_1)$. Using the Distance Formula, we get

$$\sqrt{[\cos(s + t) - 1]^2 + [\sin(s + t) - 0]^2} = \sqrt{(\cos t - \cos s)^2 + (\sin t + \sin s)^2}$$

Stock Montage/Getty Images

JEAN BAPTISTE JOSEPH FOURIER (1768–1830) is responsible for the most powerful application of the trigonometric functions. He used sums of these functions to describe such physical phenomena as the transmission of sound and the flow of heat.

Orphaned as a young boy, Fourier was educated in a military school, where he became a mathematics teacher at the age of 20. He was later appointed professor at the École Polytechnique but resigned this position to accompany Napoleon on his expedition to Egypt, where Fourier served as governor. After returning to France, he began conducting experiments on heat. The French Academy refused to publish his early papers on this subject because of his lack of rigor. Fourier eventually became Secretary of the Academy and in this capacity had his papers published in their original form. Probably because of his study of heat and his years in the deserts of Egypt, Fourier became obsessed with keeping himself warm—he wore several layers of clothes, even in the summer, and kept his rooms at unbearably high temperatures. Evidently, these habits overburdened his heart and contributed to his death at the age of 62.

Squaring both sides and expanding, we have

$$\underbrace{\cos^2(s+t) - 2\cos(s+t) + 1 + \sin^2(s+t)}_{\cos^2 \text{ and } \sin^2 \text{ terms: These add to 1}}$$

$$= \cos^2 t - 2\cos s\cos t + \cos^2 s + \sin^2 t + 2\sin s\sin t + \sin^2 s$$

(These add to 1: $\cos^2(s+t)$ and $\sin^2(s+t)$; $\cos^2 t$ and $\sin^2 t$; $\cos^2 s$ and $\sin^2 s$.)

Using the Pythagorean identity $\sin^2\theta + \cos^2\theta = 1$ three times gives

$$2 - 2\cos(s+t) = 2 - 2\cos s\cos t + 2\sin s\sin t$$

Finally, subtracting 2 from each side and dividing both sides by -2, we get

$$\cos(s+t) = \cos s\cos t - \sin s\sin t$$

which proves the Addition Formula for Cosine. ■

Proof of Subtraction Formula for Cosine Replacing t with $-t$ in the Addition Formula for Cosine, we obtain

$$\cos(s-t) = \cos(s+(-t))$$

$$= \cos s\cos(-t) - \sin s\sin(-t) \qquad \text{Addition Formula for Cosine}$$

$$= \cos s\cos t + \sin s\sin t \qquad \text{Even-odd identities}$$

This proves the Subtraction Formula for Cosine. ■

See Exercises 77 and 78 for proofs of the other Addition Formulas.

Example 1 ■ Using the Addition and Subtraction Formulas

Find the exact value of each expression.

(a) $\cos 75°$ **(b)** $\cos\dfrac{\pi}{12}$

Solution

(a) Notice that $75° = 45° + 30°$. Since we know the exact values of sine and cosine at 45° and 30°, we use the Addition Formula for Cosine to get

$$\cos 75° = \cos(45° + 30°)$$

$$= \cos 45°\cos 30° - \sin 45°\sin 30°$$

$$= \frac{\sqrt{2}}{2}\frac{\sqrt{3}}{2} - \frac{\sqrt{2}}{2}\frac{1}{2} = \frac{\sqrt{2}\sqrt{3} - \sqrt{2}}{4} = \frac{\sqrt{6} - \sqrt{2}}{4}$$

(b) Since $\dfrac{\pi}{12} = \dfrac{\pi}{4} - \dfrac{\pi}{6}$, the Subtraction Formula for Cosine gives

$$\cos\frac{\pi}{12} = \cos\left(\frac{\pi}{4} - \frac{\pi}{6}\right)$$

$$= \cos\frac{\pi}{4}\cos\frac{\pi}{6} + \sin\frac{\pi}{4}\sin\frac{\pi}{6}$$

$$= \frac{\sqrt{2}}{2}\frac{\sqrt{3}}{2} + \frac{\sqrt{2}}{2}\frac{1}{2} = \frac{\sqrt{6} + \sqrt{2}}{4}$$

Now Try Exercises 3 and 9 ■

Example 2 ■ Using the Addition Formula for Sine

Find the exact value of the expression $\sin 20° \cos 40° + \cos 20° \sin 40°$.

Solution We recognize the expression as the right-hand side of the Addition Formula for Sine with $s = 20°$ and $t = 40°$. So, we have

$$\sin 20° \cos 40° + \cos 20° \sin 40° = \sin(20° + 40°) = \sin 60° = \frac{\sqrt{3}}{2}$$

 **Now Try Exercise 15** ■

Example 3 ■ Proving a Cofunction Identity

Prove the cofunction identity $\cos\left(\frac{\pi}{2} - u\right) = \sin u$.

Solution By the Subtraction Formula for Cosine we have

$$\cos\left(\frac{\pi}{2} - u\right) = \cos\frac{\pi}{2}\cos u + \sin\frac{\pi}{2}\sin u$$

$$= 0 \cdot \cos u + 1 \cdot \sin u = \sin u$$

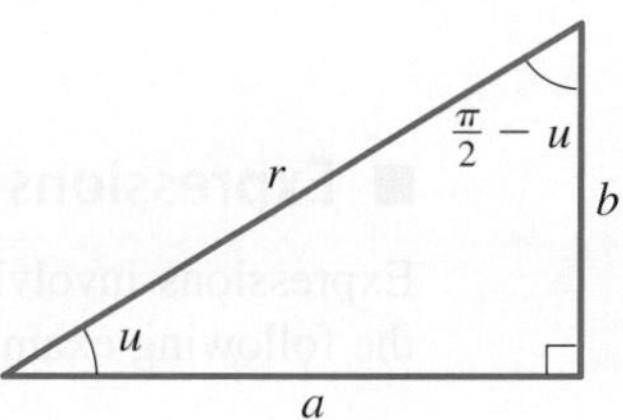

Figure 2 | $\cos(\frac{\pi}{2} - u) = \frac{b}{r} = \sin u$

 **Now Try Exercises 21 and 25** ■

For acute angles, the cofunction identity in Example 3, as well as the other cofunction identities, can also be derived from Figure 2.

Example 4 ■ Proving an Identity

Verify the identity $\dfrac{1 + \tan x}{1 - \tan x} = \tan\left(\dfrac{\pi}{4} + x\right)$.

Solution Starting with the right-hand side and using the Addition Formula for Tangent, we get

$$\text{RHS} = \tan\left(\frac{\pi}{4} + x\right) = \frac{\tan\frac{\pi}{4} + \tan x}{1 - \tan\frac{\pi}{4}\tan x}$$

$$= \frac{1 + \tan x}{1 - \tan x} = \text{LHS}$$

 **Now Try Exercise 33** ■

The next example illustrates a typical use of the Addition and Subtraction Formulas in calculus.

Example 5 ■ An Identity from Calculus

If $f(x) = \sin x$, show that

$$\frac{f(x+h)-f(x)}{h} = \sin x\left(\frac{\cos h - 1}{h}\right) + \cos x\left(\frac{\sin h}{h}\right)$$

Solution

$$\begin{aligned}
\frac{f(x+h)-f(x)}{h} &= \frac{\sin(x+h)-\sin x}{h} && \text{Definition of } f\\
&= \frac{\sin x\cos h + \cos x \sin h - \sin x}{h} && \text{Addition Formula for Sine}\\
&= \frac{\sin x(\cos h - 1) + \cos x \sin h}{h} && \text{Factor}\\
&= \sin x\left(\frac{\cos h - 1}{h}\right) + \cos x\left(\frac{\sin h}{h}\right) && \text{Separate the fraction}
\end{aligned}$$

Now Try Exercise 65 ■

■ Expressions Involving Trigonometric Functions of a Sum

Expressions involving trigonometric functions and their inverses arise in calculus. In the following examples we illustrate how to evaluate such expressions.

Example 6 ■ Simplifying an Expression Involving Inverse Trigonometric Functions

Write $\sin(\cos^{-1}x + \tan^{-1}y)$ as an algebraic expression in x and y, where $-1 \le x \le 1$ and y is any real number.

Solution Let $\theta = \cos^{-1}x$ and $\phi = \tan^{-1}y$. Using the methods of Section 6.4, we sketch triangles with angles θ and ϕ such that $\cos\theta = x$ and $\tan\phi = y$ (see Figure 3). From the triangles we have

$$\sin\theta = \sqrt{1-x^2} \qquad \cos\phi = \frac{1}{\sqrt{1+y^2}} \qquad \sin\phi = \frac{y}{\sqrt{1+y^2}}$$

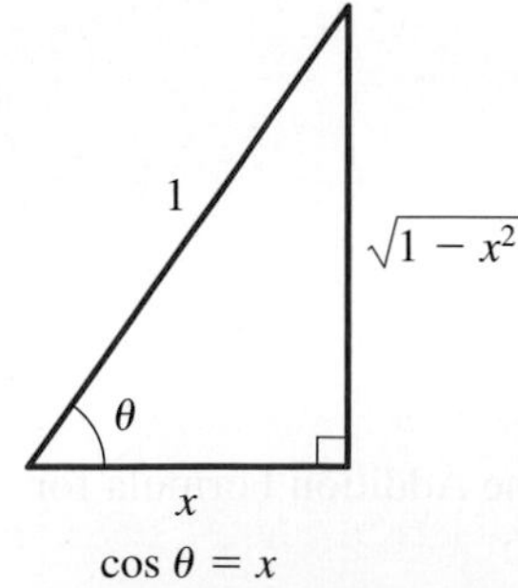

$\cos\theta = x$

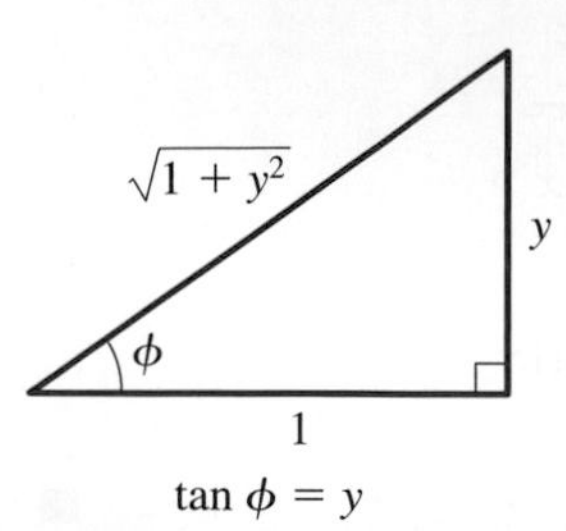

$\tan\phi = y$

Figure 3

From the Addition Formula for Sine we have

$$\begin{aligned}
\sin(\cos^{-1}x + \tan^{-1}y) &= \sin(\theta+\phi)\\
&= \sin\theta\cos\phi + \cos\theta\sin\phi && \text{Addition Formula for Sine}\\
&= \sqrt{1-x^2}\,\frac{1}{\sqrt{1+y^2}} + x\,\frac{y}{\sqrt{1+y^2}} && \text{From triangles}\\
&= \frac{1}{\sqrt{1+y^2}}\left(\sqrt{1-x^2} + xy\right) && \text{Factor } \frac{1}{\sqrt{1+y^2}}
\end{aligned}$$

Now Try Exercises 47 and 51 ■

Example 7 ■ Evaluating an Expression Involving Trigonometric Functions

Evaluate $\sin(\theta + \phi)$, where $\sin\theta = \frac{12}{13}$ with θ in Quadrant II and $\tan\phi = \frac{3}{4}$ with ϕ in Quadrant III.

Solution We first sketch the angles θ and ϕ in standard position with terminal sides in the appropriate quadrants, as shown in Figure 4. Since $\sin\theta = y/r = \frac{12}{13}$, we can label a side and the hypotenuse in the triangle in Figure 4(a). To find the remaining side, we use the Pythagorean Theorem.

$$x^2 + y^2 = r^2 \qquad \text{Pythagorean Theorem}$$
$$x^2 + 12^2 = 13^2 \qquad y = 12, \quad r = 13$$
$$x^2 = 25 \qquad \text{Solve for } x^2$$
$$x = -5 \qquad \text{Because } x < 0 \text{ in Quadrant II}$$

Similarly, since $\tan\phi = y/x = \frac{3}{4}$, we can label two sides of the triangle in Figure 4(b) and then use the Pythagorean Theorem to find the hypotenuse.

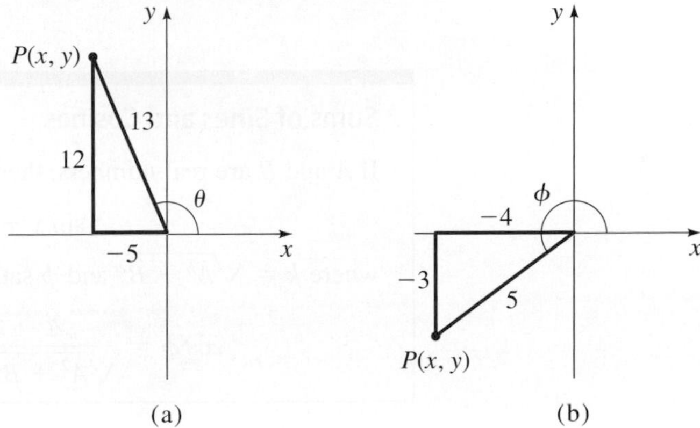

Figure 4

Now, to find $\sin(\theta + \phi)$, we use the Addition Formula for Sine and the triangles in Figure 4.

$$\begin{aligned} \sin(\theta + \phi) &= \sin\theta\cos\phi + \cos\theta\sin\phi && \text{Addition Formula} \\ &= \left(\tfrac{12}{13}\right)\left(-\tfrac{4}{5}\right) + \left(-\tfrac{5}{13}\right)\left(-\tfrac{3}{5}\right) && \text{From triangles} \\ &= -\tfrac{33}{65} && \text{Calculate} \end{aligned}$$

 **Now Try Exercise 55** ■

■ Expressions of the Form $A \sin x + B \cos x$

We can write expressions of the form $A\sin x + B\cos x$ in terms of a single trigonometric function using the Addition Formula for Sine. For example, consider the expression

$$\frac{1}{2}\sin x + \frac{\sqrt{3}}{2}\cos x$$

If we set $\phi = \pi/3$, then $\cos\phi = \frac{1}{2}$ and $\sin\phi = \sqrt{3}/2$, and we can write

$$\begin{aligned} \frac{1}{2}\sin x + \frac{\sqrt{3}}{2}\cos x &= \cos\phi\sin x + \sin\phi\cos x \\ &= \sin(x + \phi) = \sin\left(x + \frac{\pi}{3}\right) \end{aligned}$$

We are able to do this because the coefficients $\frac{1}{2}$ and $\sqrt{3}/2$ are precisely the cosine and sine of a particular number, in this case, $\pi/3$. We can use this same idea in general to write $A \sin x + B \cos x$ in the form $k \sin(x + \phi)$. We start by multiplying the numerator and denominator by $\sqrt{A^2 + B^2}$ to get

$$A \sin x + B \cos x = \sqrt{A^2 + B^2}\left(\frac{A}{\sqrt{A^2 + B^2}} \sin x + \frac{B}{\sqrt{A^2 + B^2}} \cos x\right)$$

We need a number ϕ with the property that

$$\cos \phi = \frac{A}{\sqrt{A^2 + B^2}} \quad \text{and} \quad \sin \phi = \frac{B}{\sqrt{A^2 + B^2}}$$

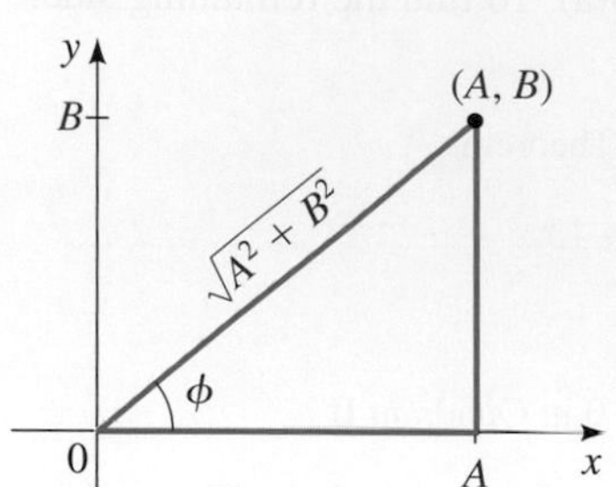

Figure 5

Figure 5 shows that the point (A, B) in the plane determines a number ϕ with precisely these properties. With this ϕ we have

$$\begin{aligned} A \sin x + B \cos x &= \sqrt{A^2 + B^2}(\cos \phi \sin x + \sin \phi \cos x) \\ &= \sqrt{A^2 + B^2} \sin(x + \phi) \end{aligned}$$

We have proved the following theorem.

Sums of Sines and Cosines

If A and B are real numbers, then

$$A \sin x + B \cos x = k \sin(x + \phi)$$

where $k = \sqrt{A^2 + B^2}$ and ϕ satisfies

$$\cos \phi = \frac{A}{\sqrt{A^2 + B^2}} \quad \text{and} \quad \sin \phi = \frac{B}{\sqrt{A^2 + B^2}}$$

Example 8 ■ A Sum of Sine and Cosine Terms

Express $3 \sin x + 4 \cos x$ in the form $k \sin(x + \phi)$.

Solution By the preceding theorem, $k = \sqrt{A^2 + B^2} = \sqrt{3^2 + 4^2} = 5$. The angle ϕ has the property that $\sin \phi = B/k = \frac{4}{5}$ and $\cos \phi = A/k = \frac{3}{5}$, with ϕ in Quadrant I (because $\sin \phi$ and $\cos \phi$ are both positive), so $\phi = \sin^{-1}\left(\frac{4}{5}\right)$. Using a calculator, we get $\phi \approx 53.1°$. Thus

$$3 \sin x + 4 \cos x \approx 5 \sin(x + 53.1°)$$

 **Now Try Exercise 59** ■

Example 9 ■ Graphing a Trigonometric Function

Write the function $f(x) = -\sin 2x + \sqrt{3} \cos 2x$ in the form $k \sin(2x + \phi)$, and use the new form to graph the function.

Solution Since $A = -1$ and $B = \sqrt{3}$, we have $k = \sqrt{A^2 + B^2} = \sqrt{1 + 3} = 2$. The angle ϕ satisfies $\cos \phi = -\frac{1}{2}$ and $\sin \phi = \sqrt{3}/2$. From the signs of these quantities we conclude that ϕ is in Quadrant II. Thus $\phi = 2\pi/3$. By the preceding theorem we can write

$$f(x) = -\sin 2x + \sqrt{3} \cos 2x = 2 \sin\left(2x + \frac{2\pi}{3}\right)$$

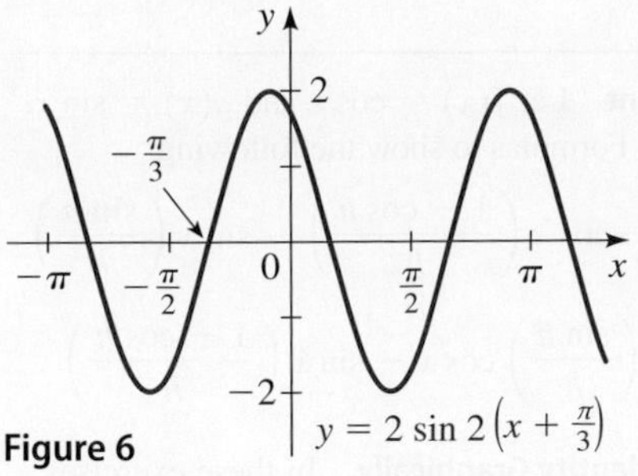

Using the form

$$f(x) = 2\sin 2\left(x + \frac{\pi}{3}\right)$$

we see that the graph is a sine curve with amplitude 2, period $2\pi/2 = \pi$, and horizontal shift $-\pi/3$. The graph is shown in Figure 6.

$y = 2\sin 2\left(x + \frac{\pi}{3}\right)$

Figure 6

Now Try Exercise 63 ■

7.2 Exercises

Concepts

1. If we know the values of the sine and cosine of x and y, we can find the value of $\sin(x + y)$ by using the ________ Formula for Sine. State the formula: $\sin(x + y) =$ ______________.

2. If we know the values of the sine and cosine of x and y, we can find the value of $\cos(x - y)$ by using the ________ Formula for Cosine. State the formula: $\cos(x - y) =$ ______________.

Skills

3–14 ■ Values of Trigonometric Functions Use an Addition or Subtraction Formula to find the exact value of the expression, as demonstrated in Example 1.

3. $\sin 75°$ **4.** $\sin 15°$

5. $\cos 105°$ **6.** $\cos 195°$

7. $\tan 15°$ **8.** $\tan 165°$

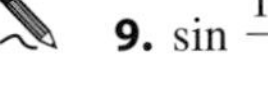

9. $\sin \dfrac{19\pi}{12}$ **10.** $\cos \dfrac{17\pi}{12}$

11. $\tan\left(-\dfrac{\pi}{12}\right)$ **12.** $\sin\left(-\dfrac{5\pi}{12}\right)$

13. $\cos \dfrac{11\pi}{12}$ **14.** $\tan \dfrac{7\pi}{12}$

15–20 ■ Values of Trigonometric Functions Use an Addition or Subtraction Formula to write the expression as a trigonometric function of one number, and then find its exact value.

15. $\cos 23° \cos 67° - \sin 23° \sin 67°$

16. $\sin 35° \cos 25° + \cos 35° \sin 25°$

17. $\sin \dfrac{3\pi}{4} \cos \dfrac{\pi}{4} - \cos \dfrac{3\pi}{4} \sin \dfrac{\pi}{4}$

18. $\dfrac{\tan \dfrac{3\pi}{5} + \tan \dfrac{\pi}{15}}{1 - \tan \dfrac{3\pi}{5} \tan \dfrac{\pi}{15}}$

19. $\dfrac{\tan 55° - \tan 10°}{1 + \tan 55° \tan 10°}$

20. $\cos \dfrac{5\pi}{18} \cos \dfrac{\pi}{9} + \sin \dfrac{5\pi}{18} \sin \dfrac{\pi}{9}$

21–24 ■ Cofunction Identities Prove the cofunction identity using the Addition and Subtraction Formulas for sine and cosine.

21. $\tan\left(\dfrac{\pi}{2} - u\right) = \cot u$ **22.** $\cot\left(\dfrac{\pi}{2} - u\right) = \tan u$

23. $\sec\left(\dfrac{\pi}{2} - u\right) = \csc u$ **24.** $\csc\left(\dfrac{\pi}{2} - u\right) = \sec u$

25–46 ■ Proving Identities Prove the identity.

25. $\sin\left(x - \dfrac{\pi}{2}\right) = -\cos x$

26. $\cos\left(x - \dfrac{\pi}{2}\right) = \sin x$

27. $\sin(x - \pi) = -\sin x$

28. $\cos(x - \pi) = -\cos x$

29. $\tan(x - \pi) = \tan x$

30. $\cot\left(x - \dfrac{\pi}{2}\right) = -\tan x$

31. $\sin\left(\dfrac{\pi}{2} - x\right) = \sin\left(\dfrac{\pi}{2} + x\right)$

32. $\cos\left(x + \dfrac{\pi}{3}\right) + \sin\left(x - \dfrac{\pi}{6}\right) = 0$

33. $\tan\left(x + \dfrac{\pi}{3}\right) = \dfrac{\sqrt{3} + \tan x}{1 - \sqrt{3}\tan x}$

34. $\tan\left(x - \dfrac{\pi}{4}\right) = \dfrac{\tan x - 1}{\tan x + 1}$

35. $\sin(x + y) - \sin(x - y) = 2\cos x \sin y$

36. $\cos(x + y) + \cos(x - y) = 2\cos x \cos y$

37. $\cot(x - y) = \dfrac{\cot x \cot y + 1}{\cot y - \cot x}$

38. $\cot(x + y) = \dfrac{\cot x \cot y - 1}{\cot x + \cot y}$

39. $\tan x - \tan y = \dfrac{\sin(x - y)}{\cos x \cos y}$

40. $1 - \tan x \tan y = \dfrac{\cos(x + y)}{\cos x \cos y}$

41. $\dfrac{\tan x - \tan y}{1 - \tan x \tan y} = \dfrac{\sin(x - y)}{\cos(x + y)}$

42. $\dfrac{\sin(x + y) - \sin(x - y)}{\cos(x + y) + \cos(x - y)} = \tan y$

43. $\cos(x + y)\cos(x - y) = \cos^2 x - \sin^2 y$

44. $\cos(x + y)\cos y + \sin(x + y)\sin y = \cos x$

45. $\sin(x + y + z) = \sin x \cos y \cos z + \cos x \sin y \cos z + \cos x \cos y \sin z - \sin x \sin y \sin z$

46. $\tan(x - y) + \tan(y - z) + \tan(z - x) = \tan(x - y)\tan(y - z)\tan(z - x)$

47–50 ■ Expressions Involving Inverse Trigonometric Functions Write the given expression in terms of x and y only.

 47. $\cos(\sin^{-1}x - \tan^{-1}y)$

48. $\tan(\sin^{-1}x + \cos^{-1}y)$

49. $\sin(\tan^{-1}x - \tan^{-1}y)$

50. $\sin(\sin^{-1}x + \cos^{-1}y)$

51–54 ■ Expressions Involving Inverse Trigonometric Functions Find the exact value of the expression.

 51. $\sin(\cos^{-1}(\frac{1}{2}) + \tan^{-1} 1)$

52. $\cos(\sin^{-1}(\frac{\sqrt{3}}{2}) + \cot^{-1}\sqrt{3})$

53. $\tan(\sin^{-1}(\frac{3}{4}) - \cos^{-1}(\frac{1}{3}))$

54. $\sin(\cos^{-1}(\frac{2}{3}) - \tan^{-1}(\frac{1}{2}))$

55–58 ■ Evaluating Expressions Involving Trigonometric Functions Evaluate each expression under the given conditions.

 55. $\sin(\theta - \phi)$; $\sin\theta = \frac{4}{5}$, θ in Quadrant II, $\tan\phi = \frac{1}{3}$, ϕ in Quadrant III

56. $\cos(\theta + \phi)$; $\tan\theta = \frac{4}{3}$, θ in Quadrant III, $\cos\phi = \frac{2}{3}$, ϕ in Quadrant IV

57. $\sin(\theta + \phi)$; $\sin\theta = \frac{5}{13}$, θ in Quadrant I, $\cos\phi = -2\sqrt{5}/5$, ϕ in Quadrant II

58. $\tan(\theta + \phi)$; $\cos\theta = -\frac{1}{3}$, θ in Quadrant III, $\sin\phi = \frac{1}{4}$, ϕ in Quadrant II

59–62 ■ Expressions in Terms of Sine Write the expression in terms of sine only.

 59. $-\sqrt{3}\sin x + \cos x$

60. $\sin x - \cos x$

61. $5(\sin 2x - \cos 2x)$

62. $3\sin \pi x + 3\sqrt{3}\cos \pi x$

63–64 ■ Graphing a Trigonometric Function **(a)** Express the function in terms of sine only. **(b)** Graph the function.

63. $g(x) = \cos 2x + \sqrt{3}\sin 2x$

64. $f(x) = \sin x + \cos x$

■ Skills Plus

65–66 ■ Difference Quotient Let $f(x) = \cos x$ and $g(x) = \sin x$. Use Addition or Subtraction Formulas to show the following.

65. $\dfrac{f(x + h) - f(x)}{h} = -\cos x\left(\dfrac{1 - \cos h}{h}\right) - \sin x\left(\dfrac{\sin h}{h}\right)$

66. $\dfrac{g(x + h) - g(x)}{h} = \left(\dfrac{\sin h}{h}\right)\cos x - \sin x\left(\dfrac{1 - \cos h}{h}\right)$

67–68 ■ Discovering an Identity Graphically In these exercises we discover an identity graphically and then prove the identity. **(a)** Graph the function and make a conjecture, then **(b)** prove that your conjecture is true.

67. $y = \sin^2\left(x + \dfrac{\pi}{4}\right) + \sin^2\left(x - \dfrac{\pi}{4}\right)$

68. $y = -\frac{1}{2}[\cos(x + \pi) + \cos(x - \pi)]$

69. **Difference of Two Angles** Show that if $\beta - \alpha = \pi/2$, then
$$\sin(x + \alpha) + \cos(x + \beta) = 0$$

70. **Sum of Two Angles** Refer to the figure. Show that $\alpha + \beta = \gamma$, and find $\tan\gamma$.

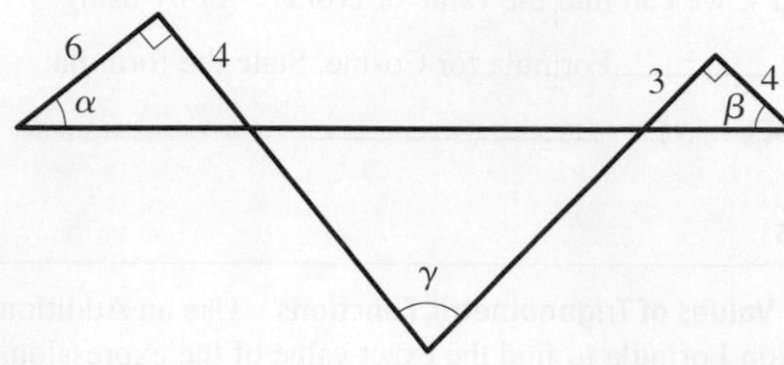

71–72 ■ Identities Involving Inverse Trigonometric Functions Prove the identity.

71. $\tan^{-1}\left(\dfrac{x + y}{1 - xy}\right) = \tan^{-1}x + \tan^{-1}y$

[*Hint:* Let $u = \tan^{-1}x$ and $v = \tan^{-1}y$, so that $x = \tan u$ and $y = \tan v$. Use an Addition Formula to find $\tan(u + v)$.]

72. $\tan^{-1}x + \tan^{-1}\left(\dfrac{1}{x}\right) = \dfrac{\pi}{2}$, $x > 0$ [*Hint:* Let $u = \tan^{-1}x$ and $v = \tan^{-1}\left(\dfrac{1}{x}\right)$, so that $x = \tan u$ and $\dfrac{1}{x} = \tan v$. Use an Addition Formula to find $\cot(u + v)$.]

73. **Angle Between Two Lines** In this exercise we find a formula for the angle formed by two lines in a coordinate plane.

(a) If L is a line in the plane and θ is the angle formed by the line and the x-axis, as shown in the figure, show that the slope m of the line is given by $m = \tan\theta$.

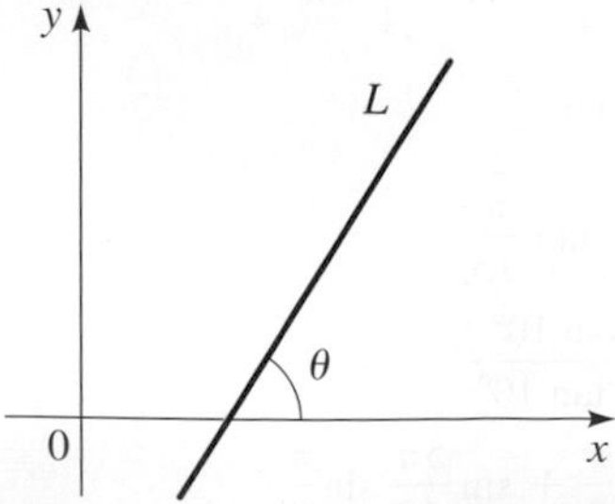

(b) Let L_1 and L_2 be two nonparallel lines in the plane with slopes m_1 and m_2, respectively. Let ψ be the acute angle formed by the two lines (see the figure). Show that

$$\tan \psi = \frac{m_2 - m_1}{1 + m_1 m_2}$$

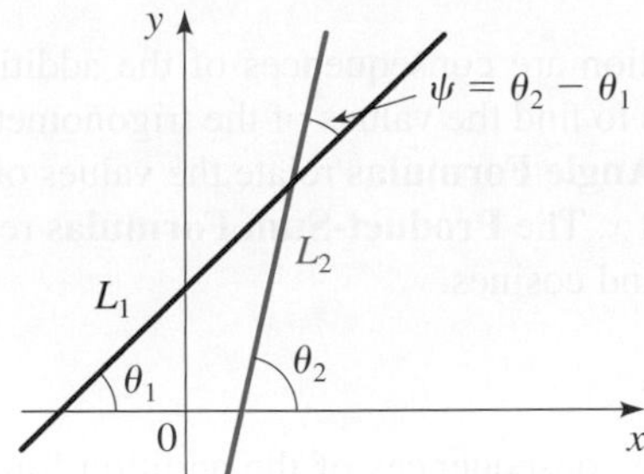

(c) Find the acute angle formed by the two lines

$$y = \tfrac{1}{3}x + 1 \quad \text{and} \quad y = \tfrac{1}{2}x - 3$$

(d) Show that if two lines are perpendicular, then the slope of one is the negative reciprocal of the slope of the other. [*Hint:* First find an expression for $\cot \psi$.]

74. A Sum of Three Angles Find $\angle A + \angle B + \angle C$ in the figure.

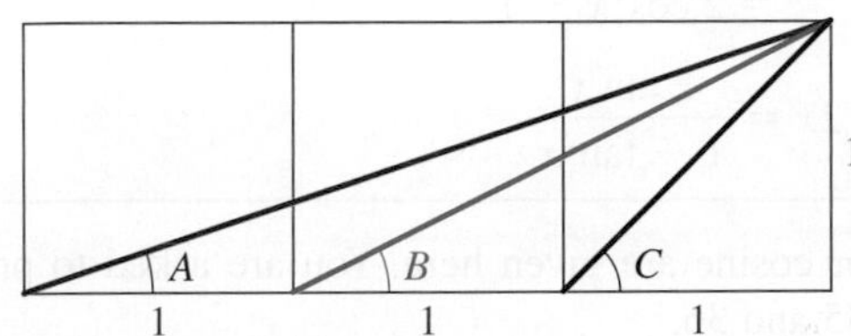

PS *Introduce something extra.* First find $\tan(A + B)$.

Applications

75. Adding an Echo A digital delay device echoes an input signal by repeating it a fixed length of time after it is received. If such a device receives the pure note $f_1(t) = 5 \sin t$ and echoes the pure note $f_2(t) = 5 \cos t$, then the combined sound is $f(t) = f_1(t) + f_2(t)$.

(a) Graph $y = f(t)$, and observe that the graph has the form of a sine curve $y = k \sin(t + \phi)$.

(b) Find k and ϕ.

76. Interference Two identical tuning forks are struck, one a fraction of a second after the other (see the figure). The sounds produced are modeled by $f_1(t) = C \sin \omega t$ and $f_2(t) = C \sin(\omega t + \alpha)$. The two sound waves interfere to produce a single sound modeled by the sum of these functions

$$f(t) = C \sin \omega t + C \sin(\omega t + \alpha)$$

(a) Use the Addition Formula for Sine to show that f can be written in the form $f(t) = A \sin \omega t + B \cos \omega t$, where A and B are constants that depend on α.

(b) Suppose that $C = 10$ and $\alpha = \pi/3$. Find constants k and ϕ so that $f(t) = k \sin(\omega t + \phi)$.

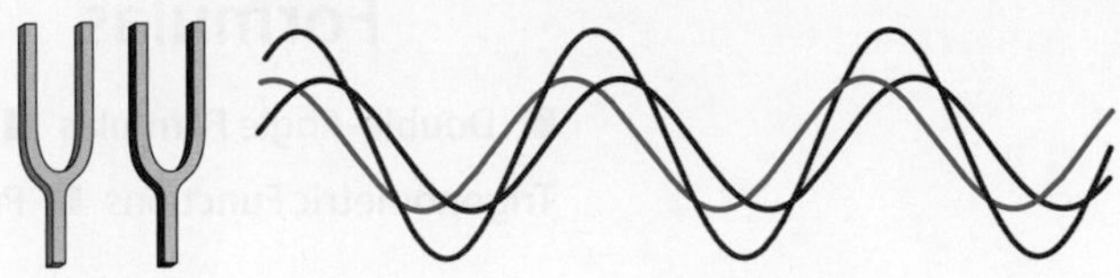

■ Discuss ■ Discover ■ Prove ■ Write

77. Prove: Addition Formula for Sine In the text we proved only the Addition and Subtraction Formulas for Cosine. Use these formulas and the cofunction identities for sine and cosine to prove the Addition Formula for Sine. [*Hint:* To get started, use the cofunction identity for sine to write

$$\sin(s + t) = \cos\left(\frac{\pi}{2} - (s + t)\right)$$

$$= \cos\left(\left(\frac{\pi}{2} - s\right) - t\right)$$

and use the Subtraction Formula for Cosine.]

78. Prove: Addition Formula for Tangent Use the Addition Formulas for Cosine and Sine to prove the Addition Formula for Tangent. [*Hint:* Use

$$\tan(s + t) = \frac{\sin(s + t)}{\cos(s + t)}$$

and divide the numerator and denominator by $\cos s \cos t$.]

79. Prove: An Impossible Triangle The points (m, n) in the coordinate plane, where m and n are integers, are called *lattice points*. Prove that a triangle with vertices at lattice points cannot be equilateral.

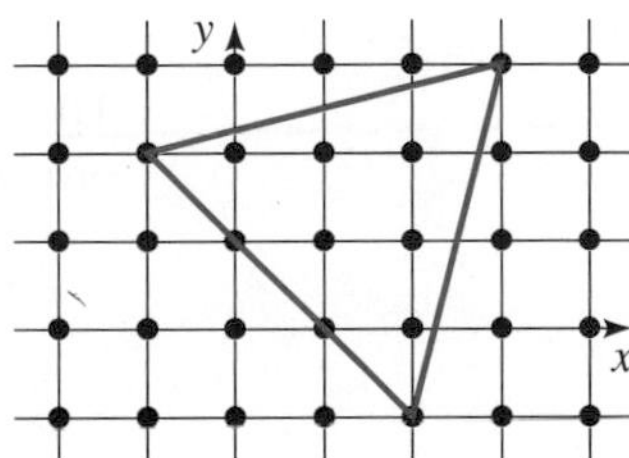

PS *Indirect reasoning.* Assume that such a triangle exists and use the formula in Exercise 73(b) to arrive at a contradiction.

80. Prove: Angles in an Acute Triangle Show that if ABC is an acute triangle, then

$$\tan A + \tan B + \tan C = \tan A \tan B \tan C$$

PS *Try to recognize something familiar.* Use the familiar fact that the sum of the angles of a triangle is 180° and apply the Addition Formula for Tangent.

7.3 Double-Angle, Half-Angle, and Product-Sum Formulas

■ Double-Angle Formulas ■ Half-Angle Formulas ■ Expressions Involving Inverse Trigonometric Functions ■ Product-Sum Formulas

The identities we consider in this section are consequences of the addition formulas. The **Double-Angle Formulas** allow us to find the values of the trigonometric functions at $2x$ from their values at x. The **Half-Angle Formulas** relate the values of the trigonometric functions at $\frac{1}{2}x$ to their values at x. The **Product-Sum Formulas** relate products of sines and cosines to sums of sines and cosines.

■ Double-Angle Formulas

The following formulas are immediate consequences of the addition formulas, which we proved in Section 7.2.

Double-Angle Formulas

Formula for sine: $\sin 2x = 2 \sin x \cos x$

Formulas for cosine:

$$\begin{aligned} \cos 2x &= \cos^2 x - \sin^2 x \\ &= 1 - 2\sin^2 x \\ &= 2\cos^2 x - 1 \end{aligned}$$

Formula for tangent: $\tan 2x = \dfrac{2 \tan x}{1 - \tan^2 x}$

The proofs for the formulas for cosine are given here. You are asked to prove the remaining formulas in Exercises 35 and 36.

Proof of Double-Angle Formulas for Cosine

$$\begin{aligned} \cos 2x &= \cos(x + x) \\ &= \cos x \cos x - \sin x \sin x && \text{Addition Formula} \\ &= \cos^2 x - \sin^2 x \end{aligned}$$

The second and third formulas for $\cos 2x$ are obtained from the formula we just proved and the Pythagorean identity. Substituting $\cos^2 x = 1 - \sin^2 x$ gives

$$\begin{aligned} \cos 2x &= \cos^2 x - \sin^2 x \\ &= (1 - \sin^2 x) - \sin^2 x \\ &= 1 - 2\sin^2 x \end{aligned}$$

The third formula is obtained in the same way, by substituting $\sin^2 x = 1 - \cos^2 x$. ■

Example 1 ■ Using the Double-Angle Formulas

If $\cos x = -\frac{2}{3}$ and x is in Quadrant II, find $\cos 2x$ and $\sin 2x$.

Solution Using one of the Double-Angle Formulas for Cosine, we get

$$\begin{aligned} \cos 2x &= 2\cos^2 x - 1 && \text{Double- Angle Formula} \\ &= 2\left(-\frac{2}{3}\right)^2 - 1 = \frac{8}{9} - 1 = -\frac{1}{9} \end{aligned}$$

To use the formula $\sin 2x = 2 \sin x \cos x$, we need to find $\sin x$ first. We have

$$\sin x = \sqrt{1 - \cos^2 x} = \sqrt{1 - \left(-\tfrac{2}{3}\right)^2} = \frac{\sqrt{5}}{3}$$

where we have used the positive square root because $\sin x$ is positive in Quadrant II. Thus

$$\sin 2x = 2 \sin x \cos x \qquad \text{Double- Angle Formula}$$

$$= 2\left(\frac{\sqrt{5}}{3}\right)\left(-\frac{2}{3}\right) = -\frac{4\sqrt{5}}{9}$$

 **Now Try Exercises 3 and 51** ■

Example 2 ■ A Triple-Angle Formula

Write $\cos 3x$ in terms of $\cos x$.

Solution

$$\begin{aligned}
\cos 3x &= \cos(2x + x) \\
&= \cos 2x \cos x - \sin 2x \sin x && \text{Addition formula} \\
&= (2\cos^2 x - 1)\cos x - (2 \sin x \cos x)\sin x && \text{Double-Angle Formulas} \\
&= 2\cos^3 x - \cos x - 2\sin^2 x \cos x && \text{Expand} \\
&= 2\cos^3 x - \cos x - 2\cos x(1 - \cos^2 x) && \text{Pythagorean identity} \\
&= 2\cos^3 x - \cos x - 2\cos x + 2\cos^3 x && \text{Expand} \\
&= 4\cos^3 x - 3\cos x && \text{Simplify}
\end{aligned}$$

 **Now Try Exercise 109** ■

Example 2 shows that $\cos 3x$ can be written as a polynomial of degree 3 in $\cos x$. The identity $\cos 2x = 2\cos^2 x - 1$ shows that $\cos 2x$ is a polynomial of degree 2 in $\cos x$. In fact, for any natural number n we can write $\cos nx$ as a polynomial in $\cos x$ of degree n (see the note in Exercise 109). The analogous result for $\sin nx$ is not true in general.

Example 3 ■ Proving an Identity

Prove the identity $\dfrac{\sin 3x}{\sin x \cos x} = 4\cos x - \sec x$.

Solution We start with the left-hand side.

$$\begin{aligned}
\text{LHS} &= \frac{\sin 3x}{\sin x \cos x} = \frac{\sin(x + 2x)}{\sin x \cos x} \\
&= \frac{\sin x \cos 2x + \cos x \sin 2x}{\sin x \cos x} && \text{Addition Formula} \\
&= \frac{\sin x(2\cos^2 x - 1) + \cos x(2\sin x \cos x)}{\sin x \cos x} && \text{Double-Angle Formulas} \\
&= \frac{\sin x(2\cos^2 x - 1)}{\sin x \cos x} + \frac{\cos x(2 \sin x \cos x)}{\sin x \cos x} && \text{Separate fraction} \\
&= \frac{2\cos^2 x - 1}{\cos x} + 2\cos x && \text{Cancel} \\
&= 2\cos x - \frac{1}{\cos x} + 2\cos x && \text{Separate fraction} \\
&= 4\cos x - \sec x = \text{RHS} && \text{Reciprocal identity}
\end{aligned}$$

Now Try Exercise 87 ■

■ Half-Angle Formulas

The following formulas allow us to write any trigonometric expression involving even powers of sine and cosine in terms of the first power of cosine only. This technique is important in calculus. The Half-Angle Formulas are immediate consequences of these formulas.

Formulas for Lowering Powers

$$\sin^2 x = \frac{1 - \cos 2x}{2} \qquad \cos^2 x = \frac{1 + \cos 2x}{2}$$

$$\tan^2 x = \frac{1 - \cos 2x}{1 + \cos 2x}$$

Proof The first formula is obtained by solving for $\sin^2 x$ in the Double-Angle Formula $\cos 2x = 1 - 2\sin^2 x$. Similarly, the second formula is obtained by solving for $\cos^2 x$ in the Double-Angle Formula $\cos 2x = 2\cos^2 x - 1$.

The last formula follows from the first two and the reciprocal identities:

$$\tan^2 x = \frac{\sin^2 x}{\cos^2 x} = \frac{\dfrac{1 - \cos 2x}{2}}{\dfrac{1 + \cos 2x}{2}} = \frac{1 - \cos 2x}{1 + \cos 2x}$$

■

Example 4 ■ Lowering Powers in a Trigonometric Expression

Express $\sin^2 x \cos^2 x$ in terms of the first power of cosine.

Solution We use the formulas for lowering powers repeatedly.

$$\begin{aligned}
\sin^2 x \cos^2 x &= \left(\frac{1 - \cos 2x}{2}\right)\left(\frac{1 + \cos 2x}{2}\right) \\
&= \frac{1 - \cos^2 2x}{4} = \frac{1}{4} - \frac{1}{4}\cos^2 2x \\
&= \frac{1}{4} - \frac{1}{4}\left(\frac{1 + \cos(2 \cdot 2x)}{2}\right) = \frac{1}{4} - \frac{1}{8} - \frac{\cos 4x}{8} \\
&= \frac{1}{8} - \frac{1}{8}\cos 4x = \frac{1}{8}(1 - \cos 4x)
\end{aligned}$$

Another way to obtain this identity is to use the Double-Angle Formula for Sine in the form $\sin x \cos x = \frac{1}{2}\sin 2x$. Thus

$$\begin{aligned}
\sin^2 x \cos^2 x &= \frac{1}{4}\sin^2 2x = \frac{1}{4}\left(\frac{1 - \cos 4x}{2}\right) \\
&= \frac{1}{8}(1 - \cos 4x)
\end{aligned}$$

Now Try Exercise 11 ■

Half-Angle Formulas

$$\sin\frac{u}{2} = \pm\sqrt{\frac{1 - \cos u}{2}} \qquad \cos\frac{u}{2} = \pm\sqrt{\frac{1 + \cos u}{2}}$$

$$\tan\frac{u}{2} = \frac{1 - \cos u}{\sin u} = \frac{\sin u}{1 + \cos u}$$

The choice of the $+$ or $-$ sign depends on the quadrant in which $u/2$ lies.

Proof We substitute $x = u/2$ in the formulas for lowering powers and take the square root of each side. This gives the first two Half-Angle Formulas. In the case of the Half-Angle Formula for Tangent we get

$$\tan\frac{u}{2} = \pm\sqrt{\frac{1-\cos u}{1+\cos u}}$$

$$= \pm\sqrt{\left(\frac{1-\cos u}{1+\cos u}\right)\left(\frac{1-\cos u}{1-\cos u}\right)} \qquad \text{Multiply numerator and denominator by } 1-\cos u$$

$$= \pm\sqrt{\frac{(1-\cos u)^2}{1-\cos^2 u}} \qquad \text{Simplify}$$

$$= \pm\frac{|1-\cos u|}{|\sin u|} \qquad \sqrt{A^2} = |A| \text{ and } 1-\cos^2 u = \sin^2 u$$

Now, $1 - \cos u$ is nonnegative for all values of u. It is also true that $\sin u$ and $\tan(u/2)$ always have the same sign. (Verify this.) It follows that

$$\tan\frac{u}{2} = \frac{1-\cos u}{\sin u}$$

The other Half-Angle Formula for Tangent is derived from this by multiplying the numerator and denominator by $1 + \cos u$. ■

Example 5 ■ Using a Half-Angle Formula

Find the exact value of $\sin 22.5°$.

Solution Since 22.5° is half of 45°, we use the Half-Angle Formula for Sine with $u = 45°$. We choose the + sign because 22.5° is in the first quadrant.

$$\sin\frac{45°}{2} = \sqrt{\frac{1-\cos 45°}{2}} \qquad \text{Half-Angle Formula}$$

$$= \sqrt{\frac{1-\sqrt{2}/2}{2}} \qquad \cos 45° = \sqrt{2}/2$$

$$= \sqrt{\frac{2-\sqrt{2}}{4}} \qquad \text{Common denominator}$$

$$= \tfrac{1}{2}\sqrt{2-\sqrt{2}} \qquad \text{Simplify}$$

Now Try Exercise 17 ■

Example 6 ■ Using a Half-Angle Formula

Find $\tan(u/2)$ if $\sin u = \frac{2}{5}$ and u is in Quadrant II.

Solution To use the Half-Angle Formula for Tangent, we first need to find $\cos u$. Since cosine is negative in Quadrant II, we have

$$\cos u = -\sqrt{1-\sin^2 u}$$

$$= -\sqrt{1-\left(\tfrac{2}{5}\right)^2} = -\frac{\sqrt{21}}{5}$$

Thus

$$\tan\frac{u}{2} = \frac{1-\cos u}{\sin u} \qquad \text{Half- Angle Formula}$$

$$= \frac{1+\sqrt{21}/5}{\frac{2}{5}} = \frac{5+\sqrt{21}}{2}$$

Now Try Exercise 37 ■

■ Expressions Involving Inverse Trigonometric Functions

Expressions involving trigonometric functions and their inverses arise in calculus. In the next examples we illustrate how to evaluate such expressions.

Example 7 ■ Evaluating an Expression Involving an Inverse Trigonometric Function

Evaluate $\sin\left(2\cos^{-1}\left(-\frac{2}{5}\right)\right)$.

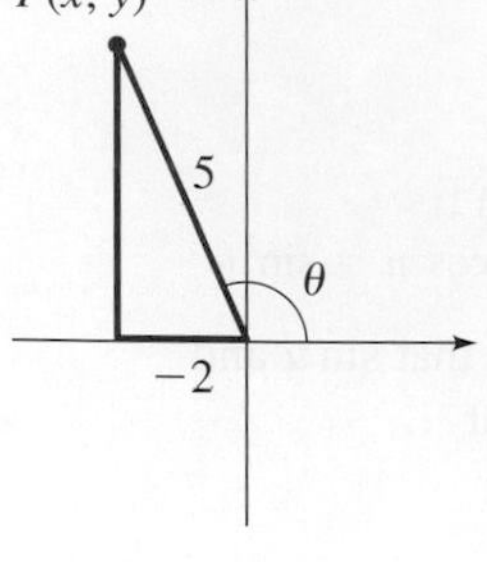

Figure 1

Solution Let $\theta = \cos^{-1}\left(-\frac{2}{5}\right)$. By properties of $\cos^{-1}$, it follows that $\cos\theta = -\frac{2}{5}$ with $0 \le \theta \le \pi$. Thus θ is in Quadrant II. Let's sketch the angle θ in standard position with terminal side in Quadrant II, as shown in Figure 2. Since $\cos\theta = x/r = -\frac{2}{5}$, we can label a side and the hypotenuse of the triangle in Figure 1. To find the remaining side, we use the Pythagorean Theorem.

$$x^2 + y^2 = r^2 \qquad \text{Pythagorean Theorem}$$

$$(-2)^2 + y^2 = 5^2 \qquad x = -2, \quad r = 5$$

$$y = \pm\sqrt{21} \qquad \text{Solve for } y^2$$

$$y = +\sqrt{21} \qquad \text{Because } y > 0$$

We can now use the Double-Angle Formula for Sine.

$$\sin 2\theta = 2\sin\theta\cos\theta \qquad \text{Double-Angle Formula}$$

$$= 2\left(\frac{\sqrt{21}}{5}\right)\left(-\frac{2}{5}\right) = -\frac{4\sqrt{21}}{25} \qquad \text{From the triangle}$$

Now Try Exercise 43 ■

Example 8 ■ Simplifying an Expression Involving an Inverse Trigonometric Function

Write $\sin(2\cos^{-1}x)$ as an algebraic expression in x only, where $-1 \le x \le 1$.

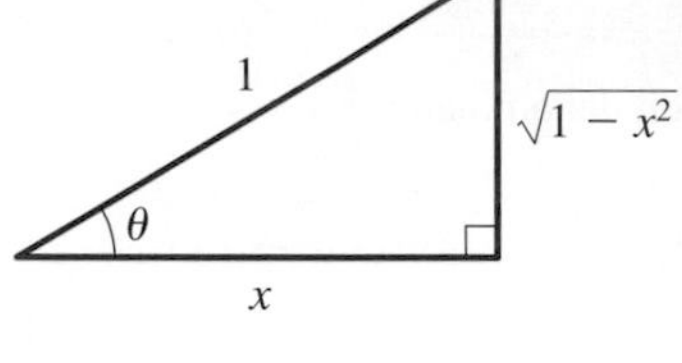

Figure 2

Solution Using the methods of Example 6.4.8, we let $\theta = \cos^{-1}x$, so $\cos\theta = x$, and sketch a triangle as in Figure 2. We need to find $\sin 2\theta$, but from the triangle we can find trigonometric functions of θ only, not 2θ. So we use the Double-Angle Formula for Sine.

$$\sin(2\cos^{-1}x) = \sin 2\theta \qquad \cos^{-1}x = \theta$$

$$= 2\sin\theta\cos\theta \qquad \text{Double-Angle Formula}$$

$$= 2x\sqrt{1 - x^2} \qquad \text{From the triangle}$$

Now Try Exercise 47 ■

Discovery Project ■ Where to Sit at the Movies

To best view a painting or a movie requires that the viewing angle be as large as possible. If the painting or movie screen is at a height above eye level, then being too far away or too close results in a small viewing angle and hence a poor viewing experience. So what is the best distance from which to view a movie or a painting? In this project we use trigonometry to find the best location from which to view a painting or a movie. You can find the project at **www.stewartmath.com**.

■ Product-Sum Formulas

It is possible to write the product $\sin u \cos v$ as a sum of trigonometric functions. To see this, consider the Addition and Subtraction Formulas for Sine:

$$\sin(u + v) = \sin u \cos v + \cos u \sin v$$

$$\sin(u - v) = \sin u \cos v - \cos u \sin v$$

Adding the left- and right-hand sides of these formulas gives

$$\sin(u + v) + \sin(u - v) = 2 \sin u \cos v$$

Dividing by 2 gives the formula

$$\sin u \cos v = \tfrac{1}{2}[\sin(u + v) + \sin(u - v)]$$

The other three **Product-to-Sum Formulas** follow from the Addition Formulas in a similar way.

Product-to-Sum Formulas

$$\sin u \cos v = \tfrac{1}{2}[\sin(u + v) + \sin(u - v)]$$

$$\cos u \sin v = \tfrac{1}{2}[\sin(u + v) - \sin(u - v)]$$

$$\cos u \cos v = \tfrac{1}{2}[\cos(u + v) + \cos(u - v)]$$

$$\sin u \sin v = \tfrac{1}{2}[\cos(u - v) - \cos(u + v)]$$

Example 9 ■ Expressing a Trigonometric Product as a Sum

Express $\sin 3x \sin 5x$ as a sum of trigonometric functions.

Solution Using the fourth Product-to-Sum Formula with $u = 3x$ and $v = 5x$ and the fact that cosine is an even function, we get

$$\begin{aligned}\sin 3x \sin 5x &= \tfrac{1}{2}[\cos(3x - 5x) - \cos(3x + 5x)] \\ &= \tfrac{1}{2}\cos(-2x) - \tfrac{1}{2}\cos 8x \\ &= \tfrac{1}{2}\cos 2x - \tfrac{1}{2}\cos 8x\end{aligned}$$

Now Try Exercise 55 ■

The Product-to-Sum Formulas can also be used as Sum-to-Product Formulas. This is possible because the right-hand side of each Product-to-Sum Formula is a sum and the left side is a product. For example, if we let

$$u = \frac{x + y}{2} \qquad \text{and} \qquad v = \frac{x - y}{2}$$

in the first Product-to-Sum Formula, we get

$$\sin \frac{x + y}{2} \cos \frac{x - y}{2} = \tfrac{1}{2}(\sin x + \sin y)$$

so

$$\sin x + \sin y = 2 \sin \frac{x + y}{2} \cos \frac{x - y}{2}$$

The remaining three of the following **Sum-to-Product Formulas** are obtained in a similar manner.

Sum-to-Product Formulas

$$\sin x + \sin y = 2\sin\frac{x+y}{2}\cos\frac{x-y}{2}$$

$$\sin x - \sin y = 2\cos\frac{x+y}{2}\sin\frac{x-y}{2}$$

$$\cos x + \cos y = 2\cos\frac{x+y}{2}\cos\frac{x-y}{2}$$

$$\cos x - \cos y = -2\sin\frac{x+y}{2}\sin\frac{x-y}{2}$$

Example 10 ■ Expressing a Trigonometric Sum as a Product

Write $\sin 7x + \sin 3x$ as a product.

Solution The first Sum-to-Product Formula gives

$$\sin 7x + \sin 3x = 2\sin\frac{7x+3x}{2}\cos\frac{7x-3x}{2}$$

$$= 2\sin 5x\cos 2x$$

 **Now Try Exercise 61** ■

Example 11 ■ Proving an Identity

Verify the identity $\dfrac{\sin 3x - \sin x}{\cos 3x + \cos x} = \tan x$.

Solution We apply the second Sum-to-Product Formula to the numerator and the third formula to the denominator.

$$\text{LHS} = \frac{\sin 3x - \sin x}{\cos 3x + \cos x} = \frac{2\cos\dfrac{3x+x}{2}\sin\dfrac{3x-x}{2}}{2\cos\dfrac{3x+x}{2}\cos\dfrac{3x-x}{2}} \qquad \text{Sum-to-Product Formulas}$$

$$= \frac{2\cos 2x\sin x}{2\cos 2x\cos x} \qquad \text{Simplify}$$

$$= \frac{\sin x}{\cos x} = \tan x = \text{RHS} \qquad \text{Cancel}$$

 **Now Try Exercise 93** ■

7.3 | Exercises

Concepts

1. If we know the values of $\sin x$ and $\cos x$, we can find the value of $\sin 2x$ by using the ________ Formula for Sine. State the formula: $\sin 2x =$ ________.

2. If we know the value of $\cos x$ and the quadrant in which $x/2$ lies, we can find the value of $\sin(x/2)$ by using the ________ Formula for Sine. State the formula: $\sin(x/2) =$ ________________.

■ Skills

3–10 ■ Double-Angle Formulas Find $\sin 2x$, $\cos 2x$, and $\tan 2x$ from the given information.

3. $\sin x = \frac{5}{13}$, x in Quadrant I

4. $\tan x = -\frac{4}{3}$, x in Quadrant II

5. $\cos x = \frac{4}{5}$, $\csc x < 0$

6. $\csc x = 4$, $\tan x < 0$

7. $\sin x = -\frac{3}{5}$, x in Quadrant III

8. $\sec x = 2$, x in Quadrant IV

9. $\tan x = -\frac{1}{3}$, $\cos x > 0$

10. $\cot x = \frac{2}{3}$, $\sin x > 0$

11–16 ■ Lowering Powers in a Trigonometric Expression Use the formulas for lowering powers to rewrite the expression in terms of the first power of cosine, as in Example 4.

11. $\sin^4 x$ **12.** $\cos^4 x$

13. $\cos^2 x \sin^4 x$ **14.** $\cos^4 x \sin^2 x$

15. $\cos^4 x \sin^4 x$ **16.** $\cos^6 x$

17–28 ■ Half-Angle Formulas Use an appropriate Half-Angle Formula to find the exact value of the expression.

17. $\sin 15°$ **18.** $\tan 15°$

19. $\tan 22.5°$ **20.** $\sin 75°$

21. $\cos 165°$ **22.** $\cos 112.5°$

23. $\tan \dfrac{5\pi}{8}$ **24.** $\cos \dfrac{3\pi}{8}$

25. $\cos \dfrac{\pi}{12}$ **26.** $\tan \dfrac{5\pi}{12}$

27. $\sin \dfrac{9\pi}{8}$ **28.** $\sin \dfrac{11\pi}{12}$

29–34 ■ Double- and Half-Angle Formulas Simplify the expression by using a Double-Angle Formula or a Half-Angle Formula.

29. (a) $2 \sin 16° \cos 16°$ **(b)** $2 \sin 4\theta \cos 4\theta$

30. (a) $\dfrac{2 \tan 5°}{1 - \tan^2 5°}$ **(b)** $\dfrac{2 \tan 5\theta}{1 - \tan^2 5\theta}$

31. (a) $\cos^2 21° - \sin^2 21°$ **(b)** $\cos^2 9\theta - \sin^2 9\theta$

32. (a) $\cos^2\left(\dfrac{\theta}{2}\right) - \sin^2\left(\dfrac{\theta}{2}\right)$ **(b)** $2 \sin\left(\dfrac{\theta}{2}\right) \cos\left(\dfrac{\theta}{2}\right)$

33. (a) $\dfrac{\sin 8°}{1 + \cos 8°}$ **(b)** $\dfrac{1 - \cos 4\theta}{\sin 4\theta}$

34. (a) $\sqrt{\dfrac{1 - \cos 30°}{2}}$ **(b)** $\sqrt{\dfrac{1 - \cos 8\theta}{2}}$

35. Proving a Double-Angle Formula Use the Addition Formula for Sine to prove the Double-Angle Formula for Sine.

36. Proving a Double-Angle Formula Use the Addition Formula for Tangent to prove the Double-Angle Formula for Tangent.

37–42 ■ Using a Half-Angle Formula Find $\sin \dfrac{x}{2}$, $\cos \dfrac{x}{2}$, and $\tan \dfrac{x}{2}$ from the given information.

37. $\sin x = \frac{3}{5}$, $0° < x < 90°$

38. $\cos x = -\frac{4}{5}$, $180° < x < 270°$

39. $\csc x = 3$, $90° < x < 180°$

40. $\tan x = 1$, $0° < x < 90°$

41. $\sec x = \frac{3}{2}$, $270° < x < 360°$

42. $\cot x = 5$, $180° < x < 270°$

43–46 ■ Expressions Involving Inverse Trigonometric Functions Find the exact value of the given expression.

43. $\sin\left(2 \cos^{-1}\left(\frac{7}{25}\right)\right)$ **44.** $\cos\left(2 \tan^{-1}\left(\frac{12}{5}\right)\right)$

45. $\sec\left(2 \sin^{-1}\left(\frac{1}{4}\right)\right)$ **46.** $\tan\left(\frac{1}{2} \cos^{-1}\left(\frac{2}{3}\right)\right)$

47–50 ■ Expressions Involving Inverse Trigonometric Functions Write the given expression as an algebraic expression in x.

47. $\sin(2 \tan^{-1} x)$ **48.** $\tan(2 \cos^{-1} x)$

49. $\sin\left(\frac{1}{2} \cos^{-1} x\right)$ **50.** $\cos(2 \sin^{-1} x)$

51–54 ■ Evaluating an Expression Involving Trigonometric Functions Evaluate each expression under the given conditions.

51. $\cos 2\theta$; $\sin \theta = -\frac{3}{5}$, θ in Quadrant III

52. $\sin(\theta/2)$; $\tan \theta = -\frac{5}{12}$, θ in Quadrant IV

53. $\sin 2\theta$; $\sin \theta = \frac{1}{7}$, θ in Quadrant II

54. $\tan 2\theta$; $\cos \theta = \frac{3}{5}$, θ in Quadrant I

55–60 ■ Product-to-Sum Formulas Write the product as a sum.

55. $\sin 5x \cos 4x$ **56.** $\sin 2x \sin 3x$

57. $\cos x \sin 4x$ **58.** $\cos 5x \cos 3x$

59. $3 \cos 4x \cos 7x$ **60.** $11 \sin \dfrac{x}{2} \cos \dfrac{x}{4}$

61–66 ■ Sum-to-Product Formulas Write the sum as a product.

61. $\sin 7x + \sin 5x$ **62.** $\sin 5x - \sin 4x$

63. $\cos 4x - \cos 6x$ **64.** $\cos 9x + \cos 2x$

65. $\sin 2x - \sin 7x$ **66.** $\sin 3x + \sin 4x$

67–72 ■ Value of a Product or Sum Find the exact value of the product or sum.

67. $2 \sin 52.5° \sin 97.5°$

68. $3 \cos 37.5° \cos 7.5°$

69. $\cos 37.5° \sin 7.5°$

70. $\sin 75° + \sin 15°$

71. $\cos 255° - \cos 195°$

72. $\cos \dfrac{\pi}{12} + \cos \dfrac{5\pi}{12}$

73–96 ■ Proving Identities Prove the identity.

73. $\cos^2 5x - \sin^2 5x = \cos 10x$

74. $\sin 8x = 2 \sin 4x \cos 4x$

75. $(\sin x + \cos x)^2 = 1 + \sin 2x$

76. $\cos^4 x - \sin^4 x = \cos 2x$

77. $\dfrac{2 \tan x}{1 + \tan^2 x} = \sin 2x$

78. $\dfrac{1 - \cos 2x}{\sin 2x} = \tan x$

79. $\tan \dfrac{x}{2} + \cos x \tan \dfrac{x}{2} = \sin x$

80. $\tan \dfrac{x}{2} + \csc x = \dfrac{2 - \cos x}{\sin x}$

81. $\dfrac{\sin 4x}{\sin x} = 4 \cos x \cos 2x$

82. $\dfrac{1 + \sin 2x}{\sin 2x} = 1 + \frac{1}{2} \sec x \csc x$

83. $\dfrac{\cos 2x}{1 + 2 \sin x \cos x} = \dfrac{1 - \tan x}{1 + \tan x}$

84. $\tan x = \dfrac{\sin 2x}{1 + \cos 2x}$

85. $\cot 2x = \dfrac{1 - \tan^2 x}{2 \tan x}$

86. $\sin^4 x + \cos^4 x = \frac{1}{2}(1 + \cos^2 2x)$

87. $\tan 3x = \dfrac{3 \tan x - \tan^3 x}{1 - 3 \tan^2 x}$

88. $\dfrac{2 \sin 2x + \sin 4x}{2 \cos x + 2 \cos 3x} = \tan 2x \cos x$

89. $\dfrac{\sin x + \sin 5x}{\cos x + \cos 5x} = \tan 3x$

90. $\dfrac{\sin 3x + \sin 7x}{\cos 3x - \cos 7x} = \cot 2x$

91. $\dfrac{\sin 10x}{\sin 9x + \sin x} = \dfrac{\cos 5x}{\cos 4x}$

92. $\dfrac{\sin x + \sin 3x + \sin 5x}{\cos x + \cos 3x + \cos 5x} = \tan 3x$

93. $\dfrac{\sin x + \sin y}{\cos x + \cos y} = \tan\left(\dfrac{x + y}{2}\right)$

94. $\tan y = \dfrac{\sin(x + y) - \sin(x - y)}{\cos(x + y) + \cos(x - y)}$

95. $\tan^2\left(\dfrac{x}{2} + \dfrac{\pi}{4}\right) = \dfrac{1 + \sin x}{1 - \sin x}$

96. $(1 - \cos 4x)(2 + \tan^2 x + \cot^2 x) = 8$

97–100 ■ Sum-to-Product Formulas Use a Sum-to-Product Formula to show the following.

97. $\sin 130° - \sin 110° = -\sin 10°$

98. $\cos 100° - \cos 200° = \sin 50°$

99. $\sin 45° + \sin 15° = \sin 75°$

100. $\cos 87° + \cos 33° = \sin 63°$

Skills Plus

101. Proving an Identity Prove the identity

$$\frac{\sin x + \sin 2x + \sin 3x + \sin 4x + \sin 5x}{\cos x + \cos 2x + \cos 3x + \cos 4x + \cos 5x} = \tan 3x$$

102. Proving an Identity Use the identity

$$\sin 2x = 2 \sin x \cos x$$

n times to show that

$$\sin(2^n x) = 2^n \sin x \cos x \cos 2x \cos 4x \cdots \cos 2^{n-1} x$$

103–104 ■ Identities Involving Inverse Trigonometric Functions Prove the identity.

103. $2 \sin^{-1} x = \cos^{-1}(1 - 2x^2), \quad 0 \le x \le 1$
[*Hint:* Let $u = \sin^{-1} x$, so that $x = \sin u$. Use a Double-Angle Formula to show that $1 - 2x^2 = \cos 2u$.]

104. $2 \tan^{-1}\left(\dfrac{1}{x}\right) = \cos^{-1}\left(\dfrac{x^2 - 1}{x^2 + 1}\right)$
[*Hint:* Let $u = \tan^{-1}\left(\dfrac{1}{x}\right)$, so that $x = \dfrac{1}{\tan u} = \cot u$. Use a Double-Angle Formula to show that
$\dfrac{x^2 - 1}{x^2 + 1} = \dfrac{\cot^2 u - 1}{\csc^2 u} = \cos 2u$.]

105–107 ■ Discovering an Identity Graphically In these exercises we discover an identity graphically and then prove the identity.

105. **(a)** Graph $f(x) = \dfrac{\sin 3x}{\sin x} - \dfrac{\cos 3x}{\cos x}$, and make a conjecture.
(b) Prove the conjecture you made in part (a).

106. **(a)** Graph $f(x) = \cos 2x + 2 \sin^2 x$, and make a conjecture.
(b) Prove the conjecture you made in part (a).

107. Let $f(x) = \sin 6x + \sin 7x$.
(a) Graph $y = f(x)$.
(b) Verify that $f(x) = 2 \cos \frac{1}{2}x \sin \frac{13}{2}x$.
(c) Graph $y = 2 \cos \frac{1}{2}x$ and $y = -2 \cos \frac{1}{2}x$, together with the graph in part (a), in the same viewing rectangle. How are these graphs related to the graph of f?

108. A Cubic Equation Let $3x = \pi/3$, and let $y = \cos x$. Use the result of Example 2 to show that y satisfies the equation

$$8y^3 - 6y - 1 = 0$$

[*Note:* This equation has solutions of a certain kind that are used to show that the angle $\pi/3$ cannot be trisected by using a straightedge and compass only.]

109. Tchebycheff Polynomials

(a) Show that there is a polynomial P of degree 4 such that $\cos 4x = P(\cos x)$ (see Example 2).

(b) Show that there is a polynomial Q of degree 5 such that $\cos 5x = Q(\cos x)$.

[*Note:* In general, there is a polynomial P_n of degree n such that $\cos nx = P_n(\cos x)$. These polynomials are called *Tchebycheff polynomials*, after the Russian mathematician P. L. Tchebycheff (1821–1894).]

110. Length of a Bisector In triangle ABC (see the figure) the line segment s bisects angle C. Show that the length of s is given by

$$s = \frac{2ab \cos x}{a + b}$$

[*Hint:* Use the Law of Sines.]

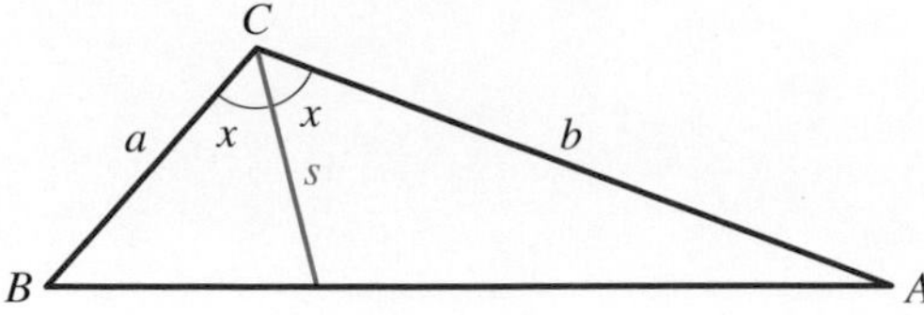

111. Angles of a Triangle If A, B, and C are the angles in a triangle, show that

$$\sin 2A + \sin 2B + \sin 2C = 4 \sin A \sin B \sin C$$

112. Largest Area A rectangle is to be inscribed in a semicircle of radius 5 cm as shown in the figure.

(a) Show that the area of the rectangle is modeled by the function $A(\theta) = 25 \sin 2\theta$.

(b) Find the largest possible area for such an inscribed rectangle. [*Hint:* Use the fact that $\sin u$ achieves its maximum value at $u = \pi/2$.]

(c) Find the dimensions of the inscribed rectangle with the largest possible area.

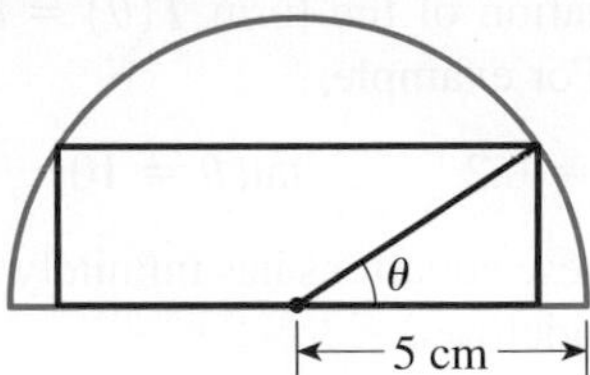

Applications

113. Sawing a Wooden Beam A rectangular beam is to be cut from a cylindrical log of diameter 20 cm.

(a) Show that the cross-sectional area of the beam is modeled by the function $A(\theta) = 200 \sin 2\theta$ where θ is as shown in the figure.

(b) Show that the maximum cross-sectional area of such a beam is 200 cm^2. [*Hint:* Use the fact that $\sin u$ achieves its maximum value at $u = \pi/2$.]

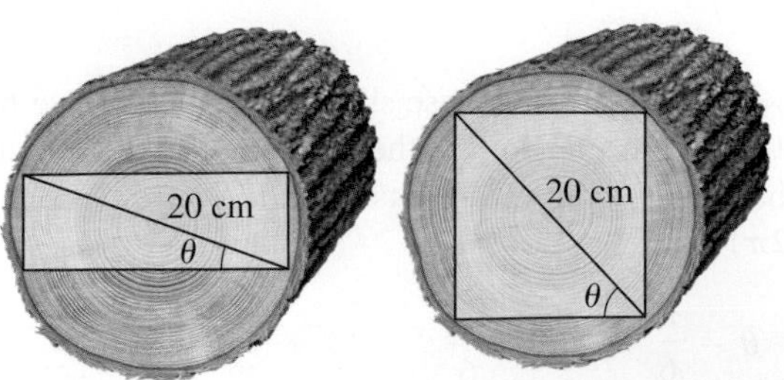

114. Length of a Fold The lower right-hand corner of a long piece of paper 6 cm wide is folded over to the left-hand edge as shown in the figure. The length L of the fold depends on the angle θ. Show that

$$L = \frac{3}{\sin \theta \cos^2 \theta}$$

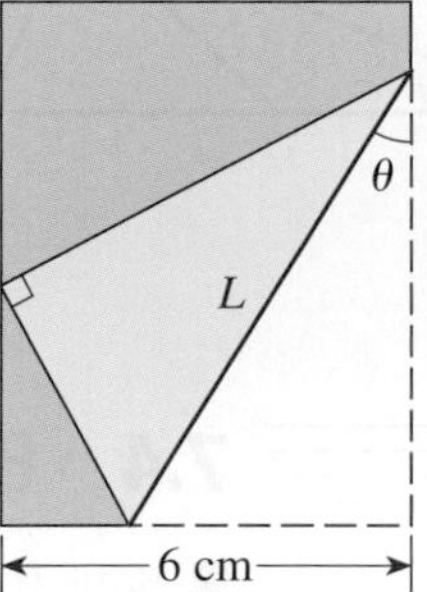

115. Sound Beats When two pure notes that are close in frequency are played together, their sounds interfere to produce *beats*; that is, the loudness (or amplitude) of the sound alternately increases and decreases. If the two notes are given by

$$f_1(t) = \cos 11t \qquad \text{and} \qquad f_2(t) = \cos 13t$$

the resulting sound is $f(t) = f_1(t) + f_2(t)$.

(a) Graph the function $y = f(t)$.

(b) Verify that $f(t) = 2 \cos t \cos 12t$.

(c) Graph $y = 2 \cos t$ and $y = -2 \cos t$, together with the graph in part (a), in the same viewing rectangle. How do these graphs describe the variation in the loudness of the sound?

116. Phone Keypad Tones When a key is pressed on a phone, the keypad generates two pure tones, which combine to produce a sound that uniquely identifies the key. The figure shows the low frequency f_1 and the high frequency f_2 associated with each key. Pressing a key produces the sound wave $y = \sin(2\pi f_1 t) + \sin(2\pi f_2 t)$.

(a) Find the function that models the sound produced when the 4 key is pressed.

(b) Use a Sum-to-Product Formula to express the sound generated by the 4 key as a product of a sine and a cosine function.

(c) Graph the sound wave generated by the 4 key from $t = 0$ to $t = 0.006$ s.

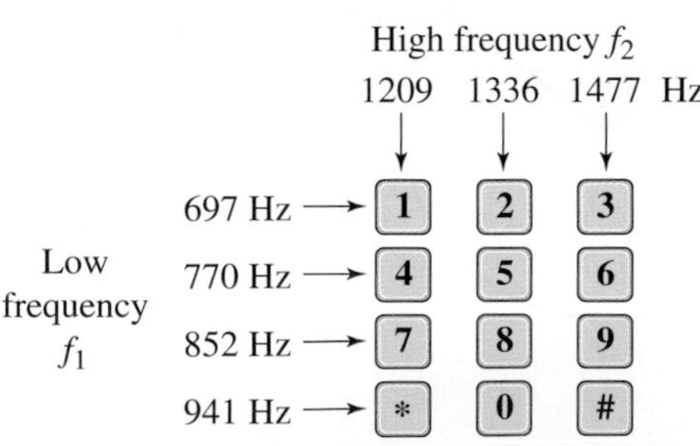

■ Discuss ■ Discover ■ Prove ■ Write

117. Prove: Geometric Proof of a Double-Angle Formula Use the figure to prove that $\sin 2\theta = 2 \sin \theta \cos \theta$.

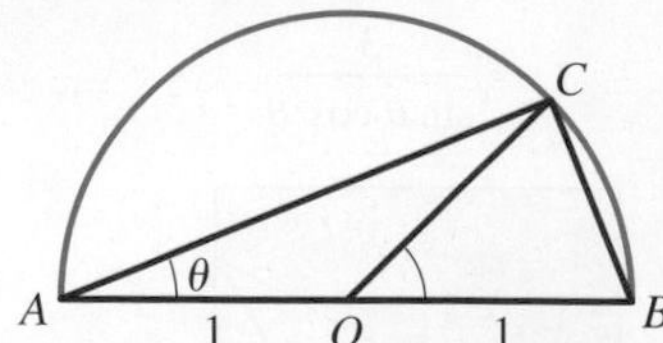

[*Hint:* Find the area of triangle ABC in two different ways. You will need the following facts from geometry. See Appendix A, *Geometry Review*:

An angle inscribed in a semicircle is a right angle, so $\angle ACB$ is a right angle.

The central angle subtended by the chord of a circle is twice the angle subtended by the chord on the circle, so in the figure $\angle BOC$ is 2θ.]

7.4 Basic Trigonometric Equations

■ Basic Trigonometric Equations ■ Solving Trigonometric Equations by Factoring

An equation that contains trigonometric functions is called a **trigonometric equation**. For example, the following are trigonometric equations:

$$\sin^2\theta + \cos^2\theta = 1 \qquad 2\sin\theta - 1 = 0$$

The first equation is an *identity*—that is, it is true for every value of the variable θ. The second equation is true only for certain values of θ. To solve a trigonometric equation, we find all the values of the variable that make the equation true.

■ Basic Trigonometric Equations

A **basic trigonometric equation** is an equation of the form $T(\theta) = c$, where T is a trigonometric function and c is a constant. For example,

$$\sin\theta = 0.5 \qquad \cos\theta = 0.2 \qquad \tan\theta = 10$$

are basic trigonometric equations. Each of these equations has infinitely many solutions. To find all of them, we use the following guidelines.

Solving Basic Trigonometric Equations

1. **Find the Solutions in One Period.** A basic trigonometric equation contains one trigonometric function. Find the solutions of the equation in one period of that trigonometric function.
2. **Find All Solutions.** Find all solutions by adding integer multiples of the period to the solutions you found in Step 1.

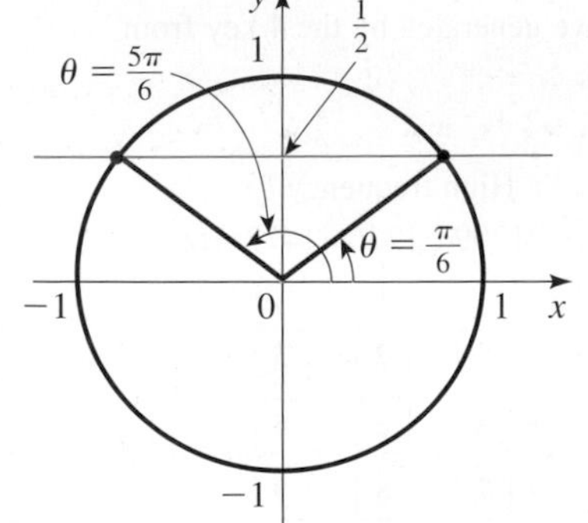

Figure 1

Example 1 ■ Solving a Basic Trigonometric Equation

Solve the equation $\sin\theta = \dfrac{1}{2}$.

Solution **Find the solutions in one period.** Because sine has period 2π, we first find the solutions in any interval of length 2π. To find these solutions, we look at the unit circle in Figure 1. We see that $\sin\theta = \frac{1}{2}$ in Quadrants I and II, so from the figure, the solutions in the interval $[0, 2\pi)$ are

$$\theta = \frac{\pi}{6} \qquad \theta = \frac{5\pi}{6}$$

Find all solutions. Because the sine function repeats its values every 2π units, we get all solutions of the equation by adding integer multiples of 2π to these solutions:

$$\theta = \frac{\pi}{6} + 2k\pi \qquad \theta = \frac{5\pi}{6} + 2k\pi$$

where k is any integer. Figure 2 gives a graphical representation of the solutions.

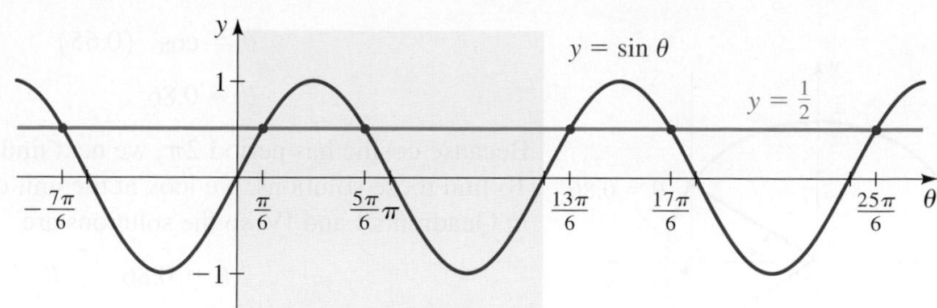

Figure 2

 Now Try Exercise 5 ■

Example 2 ■ Solving a Basic Trigonometric Equation

Solve the equation $\cos\theta = -\dfrac{\sqrt{2}}{2}$, and list six specific solutions.

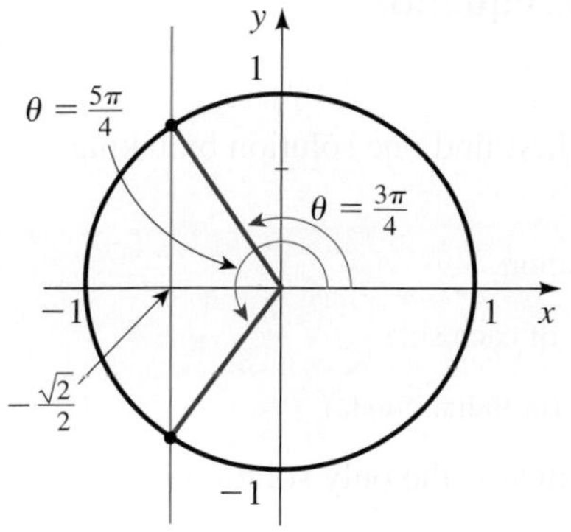

Figure 3

Solution **Find the solutions in one period.** Because cosine has period 2π, we first find the solutions in any interval of length 2π. From the unit circle in Figure 3 we see that $\cos\theta = -\sqrt{2}/2$ in Quadrants II and III, and so the solutions in the interval $[0, 2\pi)$ are

$$\theta = \frac{3\pi}{4} \qquad \theta = \frac{5\pi}{4}$$

Find all solutions. Because the cosine function repeats its values every 2π units, we get all solutions of the equation by adding integer multiples of 2π to these solutions:

$$\theta = \frac{3\pi}{4} + 2k\pi \qquad \theta = \frac{5\pi}{4} + 2k\pi$$

where k is any integer. You can check that for $k = -1, 0, 1$ we get the following specific solutions:

$$\theta = \underbrace{-\frac{5\pi}{4}, -\frac{3\pi}{4}}_{k=-1}, \underbrace{\frac{3\pi}{4}, \frac{5\pi}{4}}_{k=0}, \underbrace{\frac{11\pi}{4}, \frac{13\pi}{4}}_{k=1}$$

Figure 4 gives a graphical representation of the solutions.

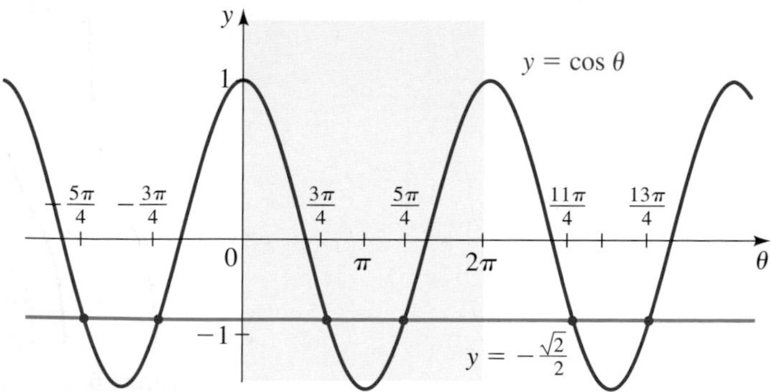

Figure 4

 Now Try Exercise 17 ■

Example 3 ■ Solving a Basic Trigonometric Equation

Solve the equation $\cos\theta = 0.65$.

Solution **Find the solutions in one period.** We first find one solution by taking $\cos^{-1}$ of each side of the equation.

$$\cos\theta = 0.65 \qquad \text{Given equation}$$

$$\theta = \cos^{-1}(0.65) \qquad \text{Take } \cos^{-1} \text{ of each side}$$

$$\theta \approx 0.86 \qquad \text{Calculator (in radian mode)}$$

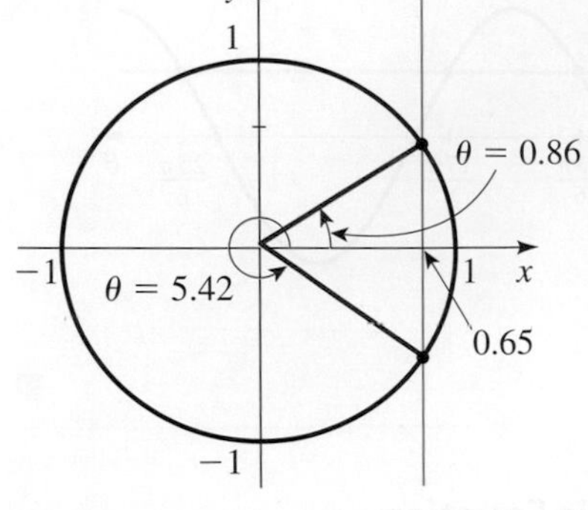

Figure 5

Because cosine has period 2π, we next find the solutions in any interval of length 2π. To find these solutions, we look at the unit circle in Figure 5. We see that $\cos\theta = 0.86$ in Quadrants I and IV, so the solutions are

$$\theta \approx 0.86 \qquad \theta \approx 2\pi - 0.86 \approx 5.42$$

Find all solutions. To gct all solutions of the equation, we add integer multiples of 2π to these solutions:

$$\theta \approx 0.86 + 2k\pi \qquad \theta \approx 5.42 + 2k\pi$$

where k is any integer.

Now Try Exercise 21 ■

Example 4 ■ Solving a Basic Trigonometric Equation

Solve the equation $\tan\theta = 2$.

Solution **Find the solutions in one period.** We first find one solution by taking $\tan^{-1}$ of each side of the equation.

$$\tan\theta = 2 \qquad \text{Given equation}$$

$$\theta = \tan^{-1} 2 \qquad \text{Take } \tan^{-1} \text{ of each side}$$

$$\theta \approx 1.12 \qquad \text{Calculator (in radian mode)}$$

By the definition of $\tan^{-1}$ the solution that we obtained is the only solution in the interval $(-\pi/2, \pi/2)$, which is an interval of length π.

Find all solutions. Since tangent has period π, we get all solutions of the equation by adding integer multiples of π:

$$\theta \approx 1.12 + k\pi$$

where k is any integer. A graphical representation of the solutions is shown in Figure 6. You can check that the solutions shown in the graph correspond to $k = -1, 0, 1, 2, 3$.

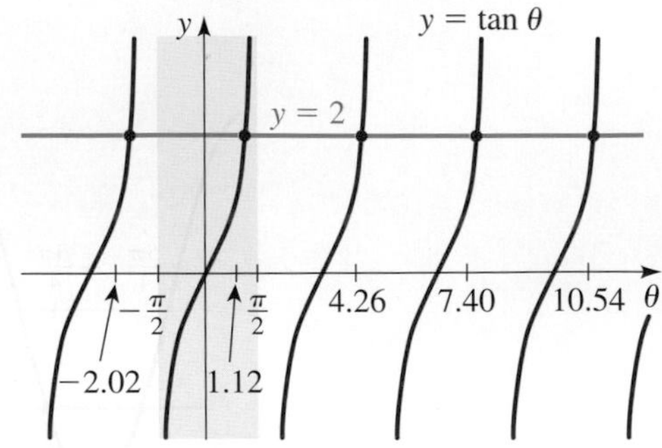

Figure 6

Now Try Exercise 23 ■

In the next example we solve trigonometric equations that are algebraically equivalent to basic trigonometric equations.

Example 5 ■ Solving Trigonometric Equations

Find all solutions of each equation.

(a) $2\sin\theta - 1 = 0$ **(b)** $\tan^2\theta - 3 = 0$

Solution

(a) We start by isolating $\sin\theta$.

$2\sin\theta - 1 = 0$	Given equation
$2\sin\theta = 1$	Add 1
$\sin\theta = \frac{1}{2}$	Divide by 2

This last equation is the equation we solved in Example 1. The solutions are

$$\theta = \frac{\pi}{6} + 2k\pi \qquad \theta = \frac{5\pi}{6} + 2k\pi$$

where k is any integer.

(b) We start by isolating $\tan\theta$.

$\tan^2\theta - 3 = 0$	Given equation
$\tan^2\theta = 3$	Add 3
$\tan\theta = \pm\sqrt{3}$	Take the square root

Because tangent has period π, we first find the solutions in any interval of length π. In the interval $(-\pi/2, \pi/2)$ the solutions are $\theta = \pi/3$ and $\theta = -\pi/3$. To get all solutions, we add integer multiples of π to these solutions:

$$\theta = \frac{\pi}{3} + k\pi \qquad \theta = -\frac{\pi}{3} + k\pi$$

where k is any integer.

 **Now Try Exercises 27 and 33** ■

■ Solving Trigonometric Equations by Factoring

Zero-Product Property

If $AB = 0$, then $A = 0$ or $B = 0$.

Factoring is one of the most useful techniques for solving equations, including trigonometric equations. The idea is to move all terms to one side of the equation, factor, and then use the Zero-Product Property (see Section 1.5), thus reducing the problem to solving basic trigonometric equations.

Example 6 ■ A Trigonometric Equation of Quadratic Type

Solve the equation $2\cos^2\theta - 7\cos\theta + 3 = 0$.

Equation of Quadratic Type

$2C^2 - 7C + 3 = 0$

$(2C - 1)(C - 3) = 0$

Solution We factor the left-hand side of the equation.

		$2\cos^2\theta - 7\cos\theta + 3 = 0$	Given equation
		$(2\cos\theta - 1)(\cos\theta - 3) = 0$	Factor
$2\cos\theta - 1 = 0$	or	$\cos\theta - 3 = 0$	Set each factor equal to 0
$\cos\theta = \frac{1}{2}$	or	$\cos\theta = 3$	Solve for $\cos\theta$

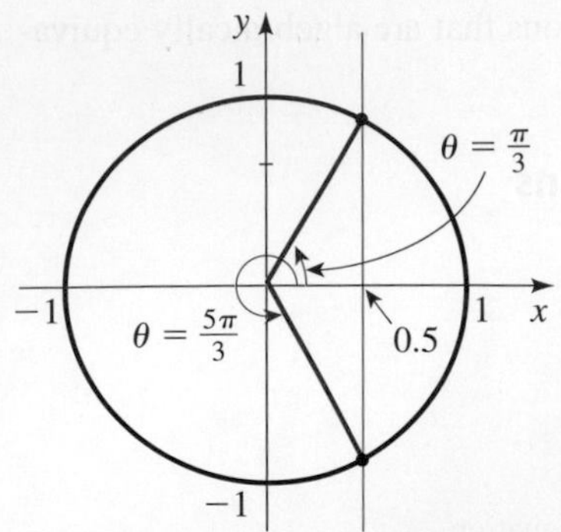

Figure 7

Because cosine has period 2π, we first find the solutions of these two equations in the interval $[0, 2\pi)$. For the first equation the solutions are $\theta = \pi/3$ and $\theta = 5\pi/3$. (See Figure 7.) The second equation has no solution because $\cos\theta$ is never greater than 1. Thus the solutions are

$$\theta = \frac{\pi}{3} + 2k\pi \qquad \theta = \frac{5\pi}{3} + 2k\pi$$

where k is any integer.

Now Try Exercise 41 ■

Example 7 ■ Solving a Trigonometric Equation by Factoring

Solve the equation $5 \sin\theta \cos\theta + 4 \cos\theta = 0$.

Solution We factor the left-hand side of the equation.

$$5 \sin\theta \cos\theta + 4 \cos\theta = 0 \qquad \text{Given equation}$$

$$\cos\theta\,(5 \sin\theta + 4) = 0 \qquad \text{Factor}$$

$$\cos\theta = 0 \quad \text{or} \quad 5\sin\theta + 4 = 0 \qquad \text{Set each factor equal to 0}$$

$$\sin\theta = -0.8 \qquad \text{Solve for } \sin\theta$$

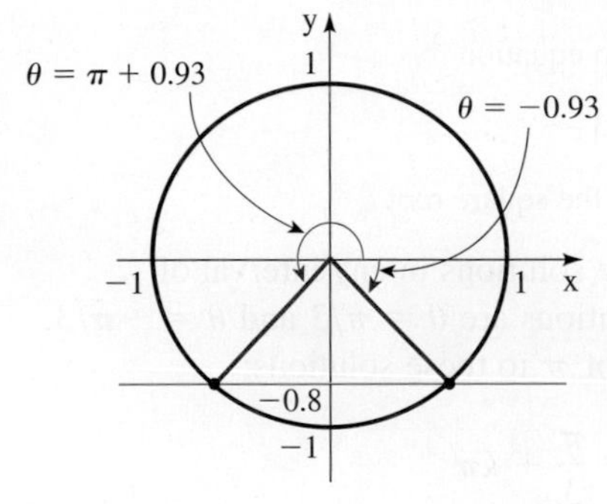

Figure 8

Because sine and cosine have period 2π, we first find the solutions of these two equations in an interval of length 2π. For the first equation the solutions in the interval $[0, 2\pi)$ are $\theta = \pi/2$ and $\theta = 3\pi/2$. To solve the second equation, we take $\sin^{-1}$ of each side.

$$\sin\theta = -0.80 \qquad \text{Second equation}$$

$$\theta = \sin^{-1}(-0.80) \qquad \text{Take } \sin^{-1} \text{ of each side}$$

$$\theta \approx -0.93 \qquad \text{Calculator (in radian mode)}$$

So the solutions in an interval of length 2π are $\theta = -0.93$ and $\theta = \pi + 0.93 \approx 4.07$. (See Figure 8.) We get all the solutions of the equation by adding integer multiples of 2π to these solutions.

$$\theta = \frac{\pi}{2} + 2k\pi \qquad \theta = \frac{3\pi}{2} + 2k\pi \qquad \theta \approx -0.93 + 2k\pi \qquad \theta \approx 4.07 + 2k\pi$$

where k is any integer.

Now Try Exercise 53 ■

7.4 Exercises

Concepts

1. Because the trigonometric functions are periodic, if a basic trigonometric equation has one solution, it has ________ (several/infinitely many) solutions.

2. The basic equation $\sin x = 2$ has ________ (no/one/infinitely many) solution(s), whereas the basic equation $\sin x = 0.3$ has ________ (no/one/infinitely many) solution(s).

3. We can find some of the solutions of $\sin x = 0.3$ graphically by graphing $y = \sin x$ and $y =$ ________. Use the graph below to estimate some of the solutions.

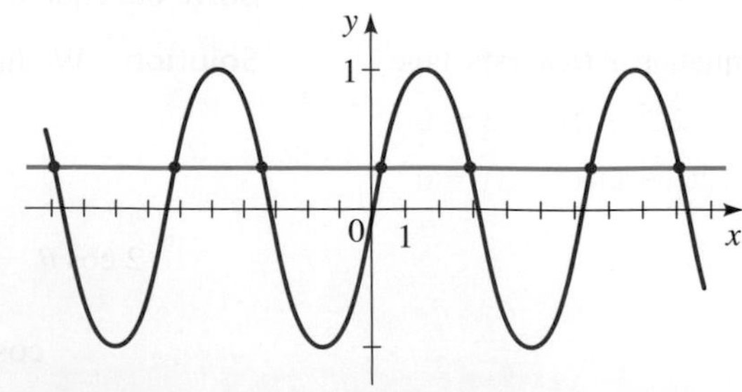

4. We can find the solutions of $\sin x = 0.3$ algebraically.

(a) First we find the solutions in the interval $[0, 2\pi)$. We get one such solution by taking $\sin^{-1}$ to get $x \approx$ ______.
The other solution in this interval is $x \approx$ ______.

(b) We find all solutions by adding multiples of ______ to the solutions in $[0, 2\pi)$. The solutions are $x \approx$ ______ and $x \approx$ ______.

■ Skills

5–16 ■ Solving Basic Trigonometric Equations Solve the given equation. Give exact answers where possible.

5. $\sin\theta = \dfrac{\sqrt{3}}{2}$ **6.** $\sin\theta = -\dfrac{\sqrt{2}}{2}$

7. $\cos\theta = -1$ **8.** $\cos\theta = \dfrac{\sqrt{3}}{2}$

9. $\cos\theta = \frac{1}{4}$ **10.** $\sin\theta = -0.3$

11. $\sin\theta = -0.45$ **12.** $\cos\theta = 0.32$

13. $\tan\theta = -\sqrt{3}$ **14.** $\tan\theta = 1$

15. $\tan\theta = 5$ **16.** $\tan\theta = -\frac{1}{3}$

17–24 ■ Solving Basic Trigonometric Equations Solve the given equation, and list six specific solutions.

17. $\cos\theta = -\dfrac{\sqrt{3}}{2}$ **18.** $\cos\theta = \dfrac{1}{2}$

19. $\sin\theta = \dfrac{\sqrt{2}}{2}$ **20.** $\sin\theta = -\dfrac{\sqrt{3}}{2}$

21. $\cos\theta = 0.28$ **22.** $\tan\theta = 2.5$

23. $\tan\theta = -10$ **24.** $\sin\theta = -0.9$

25–38 ■ Solving Trigonometric Equations Find all solutions of the given equation.

25. $\cos\theta + 1 = 0$ **26.** $\sin\theta + 1 = 0$

27. $2\cos\theta - \sqrt{3} = 0$ **28.** $2\sin\theta + 1 = 0$

29. $3\cos\theta - 1 = 0$ **30.** $10\sin\theta + 3 = 0$

31. $3\tan^2\theta - 1 = 0$ **32.** $\cot\theta + 1 = 0$

33. $2\cos^2\theta - 1 = 0$ **34.** $4\sin^2\theta - 3 = 0$

35. $3\sin^2\theta - 1 = 0$ **36.** $\tan^2\theta - 9 = 0$

37. $\sec^2\theta - 2 = 0$ **38.** $\csc^2\theta - 4 = 0$

39–56 ■ Solving Trigonometric Equations by Factoring Solve the given equation.

39. $(\tan^2\theta - 4)(2\cos\theta + 1) = 0$

40. $(\tan\theta - 2)(16\sin^2\theta - 1) = 0$

41. $4\cos^2\theta - 4\cos\theta + 1 = 0$

42. $2\sin^2\theta - \sin\theta - 1 = 0$

43. $\tan^2\theta - \tan\theta - 6 = 0$

44. $3\cos^4\theta - 5\cos^2\theta + 2 = 0$

45. $2\cos^2\theta - 7\cos\theta + 3 = 0$

46. $\sin^2\theta - \sin\theta - 2 = 0$

47. $\cos^2\theta - \cos\theta - 6 = 0$

48. $2\sin^2\theta + 5\sin\theta - 12 = 0$

49. $\sin^2\theta = 2\sin\theta + 3$

50. $3\tan^3\theta = \tan\theta$

51. $\cos\theta\,(2\sin\theta + 1) = 0$

52. $\sec\theta\,(2\cos\theta - \sqrt{2}) = 0$

53. $\cos\theta\sin\theta - 2\cos\theta = 0$

54. $\tan\theta\sin\theta + \sin\theta = 0$

55. $3\tan\theta\sin\theta - 2\tan\theta = 0$

56. $4\cos\theta\sin\theta + 3\cos\theta = 0$

■ Applications

57. Refraction of Light It has been observed since ancient times that light refracts, or "bends," as it travels from one medium to another (from air to water, for example). If v_1 is the speed of light in one medium and v_2 is its speed in another medium, then according to **Snell's Law**,

$$\frac{\sin\theta_1}{\sin\theta_2} = \frac{v_1}{v_2}$$

where θ_1 is the *angle of incidence* and θ_2 is the *angle of refraction* (see the figure). The number v_1/v_2 is called the *index of refraction*. The index of refraction for several substances is given in the table.

If a ray of light passes through the surface of a lake at an angle of incidence of 70°, what is the angle of refraction?

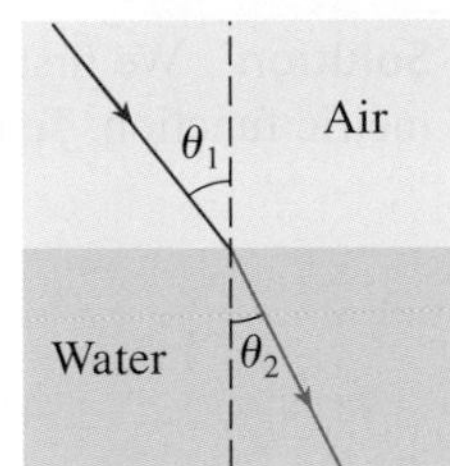

Substance	Refraction from Air to Substance
Water	1.33
Alcohol	1.36
Glass	1.52
Diamond	2.41

58. Total Internal Reflection When light passes from a more-dense to a less-dense medium—from glass to air, for example—the angle of refraction predicted by Snell's Law (see Exercise 57) can be 90° or larger. In this case the light beam is actually reflected back into the denser medium. This phenomenon, called *total internal reflection*, is the principle behind fiber optics. Set $\theta_2 = 90°$ in Snell's Law, and solve for θ_1 to determine the critical angle of incidence at which total internal reflection begins to occur when light passes from glass to air. (*Note:* The index of refraction from glass to air is the reciprocal of the index from air to glass.)

59. Phases of the Moon As the moon revolves around the earth, the side that faces the earth is usually just partially illuminated by the sun. The phases of the moon describe how much of the surface appears to be in sunlight. An astronomical measure of phase is given by the fraction F of the lunar disc that is lit. When the angle between the sun, earth, and moon is θ ($0 \le \theta < 360°$), then

$$F = \tfrac{1}{2}(1 - \cos\theta)$$

Determine the angles θ that correspond to the following phases:

(a) $F = 0$ (new moon)

(b) $F = 0.25$ (a crescent moon)

(c) $F = 0.5$ (first or last quarter)

(d) $F = 1$ (full moon)

■ Discuss ■ Discover ■ Prove ■ Write

60. Discuss ■ Write: Equations and Identities Which of the following statements is true?

A. Every identity is an equation.

B. Every equation is an identity.

Give examples to illustrate your answer. Write a short paragraph to explain the difference between an equation and an identity.

7.5 More Trigonometric Equations

■ Solving Trigonometric Equations by Using Identities ■ Equations with Trigonometric Functions of Multiples of Angles

In this section we solve trigonometric equations by first using identities to simplify the equation. We also solve trigonometric equations in which the terms contain multiples of angles.

■ Solving Trigonometric Equations by Using Identities

In the first two examples we use trigonometric identities to express a trigonometric equation in a form in which it can be factored.

Example 1 ■ Using a Trigonometric Identity

Solve the equation $1 + \sin\theta = 2\cos^2\theta$.

Solution We first need to rewrite this equation so that it contains only one trigonometric function. To do this, we use a trigonometric identity.

$$1 + \sin\theta = 2\cos^2\theta \qquad \text{Given equation}$$

$$1 + \sin\theta = 2(1 - \sin^2\theta) \qquad \text{Pythagorean identity}$$

$$2\sin^2\theta + \sin\theta - 1 = 0 \qquad \text{Put all terms on one side}$$

$$(2\sin\theta - 1)(\sin\theta + 1) = 0 \qquad \text{Factor}$$

$$2\sin\theta - 1 = 0 \quad \text{or} \quad \sin\theta + 1 = 0 \qquad \text{Set each factor equal to 0}$$

$$\sin\theta = \frac{1}{2} \quad \text{or} \quad \sin\theta = -1 \qquad \text{Solve for } \sin\theta$$

$$\theta = \frac{\pi}{6}, \frac{5\pi}{6} \quad \text{or} \quad \theta = \frac{3\pi}{2} \qquad \text{Solve for } \theta \text{ in the interval } [0, 2\pi)$$

Because sine has period 2π, we get all the solutions of the equation by adding integer multiples of 2π to these solutions. Thus the solutions are

$$\theta = \frac{\pi}{6} + 2k\pi \qquad \theta = \frac{5\pi}{6} + 2k\pi \qquad \theta = \frac{3\pi}{2} + 2k\pi$$

where k is any integer.

Now Try Exercises 3 and 11 ■

Example 2 ■ Using a Trigonometric Identity

Solve the equation $\sin 2\theta - \cos\theta = 0$.

Solution The first term is a function of 2θ, and the second is a function of θ, so we begin by using a trigonometric identity to rewrite the first term as a function of θ only.

$$\begin{aligned}
\sin 2\theta - \cos\theta &= 0 && \text{Given equation}\\
2\sin\theta\cos\theta - \cos\theta &= 0 && \text{Double-Angle Formula}\\
\cos\theta\,(2\sin\theta - 1) &= 0 && \text{Factor}\\
\cos\theta = 0 \quad \text{or} \quad 2\sin\theta - 1 &= 0 && \text{Set each factor equal to 0}\\
\sin\theta &= \frac{1}{2} && \text{Solve for } \sin\theta\\
\theta = \frac{\pi}{2}, \frac{3\pi}{2} \quad \text{or} \quad \theta &= \frac{\pi}{6}, \frac{5\pi}{6} && \text{Solve for } \theta \text{ in } [0, 2\pi)
\end{aligned}$$

Both sine and cosine have period 2π, so we get all the solutions of the equation by adding integer multiples of 2π to these solutions. Thus the solutions are

$$\theta = \frac{\pi}{2} + 2k\pi \qquad \theta = \frac{3\pi}{2} + 2k\pi \qquad \theta = \frac{\pi}{6} + 2k\pi \qquad \theta = \frac{5\pi}{6} + 2k\pi$$

where k is any integer.

 **Now Try Exercises 7 and 9** ■

Example 3 ■ Squaring an Equation and Using an Identity

Solve the equation $\cos\theta + 1 = \sin\theta$ in the interval $[0, 2\pi)$.

Solution To get an equation that involves either sine only or cosine only, we square both sides and use a Pythagorean identity.

$$\begin{aligned}
\cos\theta + 1 &= \sin\theta && \text{Given equation}\\
\cos^2\theta + 2\cos\theta + 1 &= \sin^2\theta && \text{Square both sides}\\
\cos^2\theta + 2\cos\theta + 1 &= 1 - \cos^2\theta && \text{Pythagorean identity}\\
2\cos^2\theta + 2\cos\theta &= 0 && \text{Simplify}\\
2\cos\theta\,(\cos\theta + 1) &= 0 && \text{Factor}\\
2\cos\theta = 0 \quad \text{or} \quad \cos\theta + 1 &= 0 && \text{Set each factor equal to 0}\\
\cos\theta = 0 \quad \text{or} \quad \cos\theta &= -1 && \text{Solve for } \cos\theta\\
\theta = \frac{\pi}{2}, \frac{3\pi}{2} \quad \text{or} \quad \theta &= \pi && \text{Solve for } \theta \text{ in } [0, 2\pi)
\end{aligned}$$

Because we squared both sides, we need to check for extraneous solutions. From *Check Your Answers* we see that the solutions of the given equation are $\pi/2$ and π.

Check Your Answers

$\theta = \frac{\pi}{2}$	$\theta = \frac{3\pi}{2}$	$\theta = \pi$
$\cos\frac{\pi}{2} + 1 = \sin\frac{\pi}{2}$	$\cos\frac{3\pi}{2} + 1 = \sin\frac{3\pi}{2}$	$\cos\pi + 1 = \sin\pi$
$0 + 1 = 1$ ✓	$0 + 1 \stackrel{?}{=} -1$ ✗	$-1 + 1 = 0$ ✓

Now Try Exercise 13 ■

Example 4 ■ Finding Intersection Points

Find the values of x for which the graphs of $f(x) = \sin x$ and $g(x) = \cos x$ intersect.

Solution 1: Graphical

The graphs intersect where $f(x) = g(x)$. In Figure 1 we graph $y_1 = \sin x$ and $y_2 = \cos x$ on the same screen, for x between 0 and 2π. Using a graphing device, we see that the two points of intersection in this interval occur where $x \approx 0.785$ and $x \approx 3.927$. Since sine and cosine are periodic with period 2π, the intersection points occur where

$$x \approx 0.785 + 2k\pi \qquad \text{and} \qquad x \approx 3.927 + 2k\pi$$

where k is any integer.

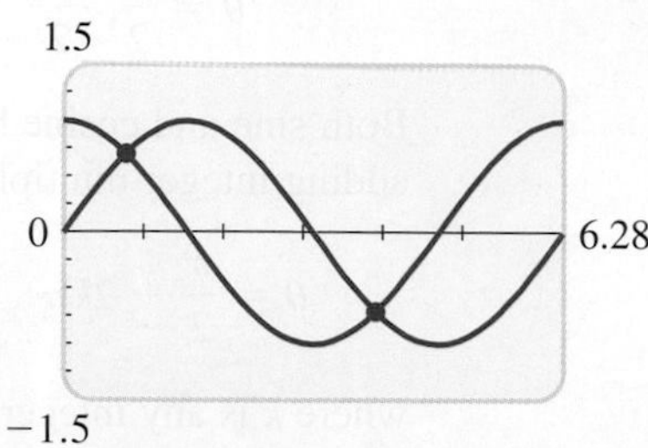

Figure 1

Solution 2: Algebraic

To find the exact solution, we set $f(x) = g(x)$ and solve the resulting equation algebraically:

$$\sin x = \cos x \qquad \text{Equate functions}$$

Since the numbers x for which $\cos x = 0$ are not solutions of the equation, we can divide both sides by $\cos x$:

$$\frac{\sin x}{\cos x} = 1 \qquad \text{Divide by } \cos x$$

$$\tan x = 1 \qquad \text{Reciprocal identity}$$

The only solution of this equation in the interval $(-\pi/2, \pi/2)$ is $x = \pi/4$. Since tangent has period π, we get all solutions of the equation by adding integer multiples of π:

$$x = \frac{\pi}{4} + k\pi$$

where k is any integer. The graphs intersect for these values of x. You should use your calculator to check that, rounded to three decimals, these are the same values that we obtained in Solution 1.

Now Try Exercise 35 ■

■ Equations with Trigonometric Functions of Multiples of Angles

When solving trigonometric equations that involve functions of multiples of angles, we first solve for the multiple of the angle, then divide to solve for the angle.

Example 5 ■ A Trigonometric Equation Involving a Multiple of an Angle

Consider the equation $2 \sin 3\theta - 1 = 0$.

(a) Find all solutions of the equation.

(b) List the solutions in the interval $[0, 2\pi)$.

Solution

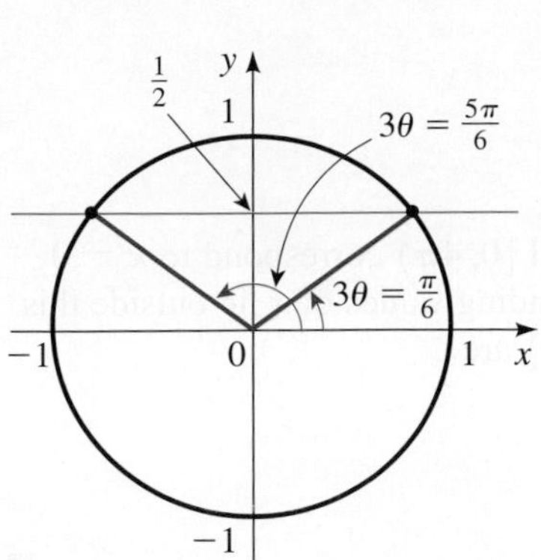

Figure 2

(a) We first isolate $\sin 3\theta$ and then solve for the angle 3θ.

$$2 \sin 3\theta - 1 = 0 \qquad \text{Given equation}$$

$$2 \sin 3\theta = 1 \qquad \text{Add 1}$$

$$\sin 3\theta = \frac{1}{2} \qquad \text{Divide by 2}$$

$$3\theta = \frac{\pi}{6}, \frac{5\pi}{6} \qquad \text{Solve for } 3\theta \text{ in the interval } [0, 2\pi) \text{ (see Figure 2)}$$

To get all solutions, we add integer multiples of 2π to these solutions. So the solutions are of the form

$$3\theta = \frac{\pi}{6} + 2k\pi \qquad 3\theta = \frac{5\pi}{6} + 2k\pi$$

To solve for θ, we divide by 3 to get the solutions

Compare to the solution of Example 7.4.5(a).

$$\theta = \frac{\pi}{18} + \frac{2k\pi}{3} \qquad \theta = \frac{5\pi}{18} + \frac{2k\pi}{3}$$

where k is any integer.

(b) The solutions from part (a) that are in the interval $[0, 2\pi)$ correspond to $k = 0, 1,$ and 2. For all other values of k the corresponding values of θ lie outside this interval. So the solutions in the interval $[0, 2\pi)$ are

$$\theta = \underbrace{\frac{\pi}{18}, \frac{5\pi}{18}}_{k=0}, \underbrace{\frac{13\pi}{18}, \frac{17\pi}{18}}_{k=1}, \underbrace{\frac{25\pi}{18}, \frac{29\pi}{18}}_{k=2}$$

Now Try Exercise 17 ■

Example 6 ■ A Trigonometric Equation Involving a Half-Angle

Consider the equation $\sqrt{3} \tan \frac{\theta}{2} - 1 = 0$.

(a) Find all solutions of the equation.

(b) List the solutions in the interval $[0, 4\pi)$.

Solution

(a) We start by isolating $\tan \frac{\theta}{2}$.

$$\sqrt{3} \tan \frac{\theta}{2} - 1 = 0 \qquad \text{Given equation}$$

$$\sqrt{3} \tan \frac{\theta}{2} = 1 \qquad \text{Add 1}$$

$$\tan \frac{\theta}{2} = \frac{1}{\sqrt{3}} \qquad \text{Divide by } \sqrt{3}$$

$$\frac{\theta}{2} = \frac{\pi}{6} \qquad \text{Solve for } \frac{\theta}{2} \text{ in the interval } \left(-\frac{\pi}{2}, \frac{\pi}{2}\right)$$

Since tangent has period π, to get all solutions we add integer multiples of π to this solution. So the solutions are of the form

$$\frac{\theta}{2} = \frac{\pi}{6} + k\pi$$

Multiplying by 2, we get the solutions

$$\theta = \frac{\pi}{3} + 2k\pi$$

where k is any integer.

(b) The solutions from part (a) that are in the interval $[0, 4\pi)$ correspond to $k = 0$ and $k = 1$. For all other values of k the corresponding values of x lie outside this interval. Thus the solutions in the interval $[0, 4\pi)$ are

$$x = \frac{\pi}{3}, \frac{7\pi}{3}$$

 **Now Try Exercise 23** ■

7.5 Exercises

Concepts

1–2 ■ We can use identities to help us solve trigonometric equations.

1. Using a Pythagorean identity we see that the equation $\sin x + \sin^2 x + \cos^2 x = 1$ is equivalent to the basic equation ________ whose solutions are $x =$ ________.

2. Using a Double-Angle Formula we see that the equation $\sin x + \sin 2x = 0$ is equivalent to the equation ________. Factoring, we see that solving this equation is equivalent to solving the two basic equations ________ and ________.

Skills

3–16 ■ Solving Trigonometric Equations by Using Identities Solve the given equation.

 **3.** $2\cos^2\theta + \sin\theta = 1$

4. $\sin^2\theta = 4 - 2\cos^2\theta$

5. $\tan^2\theta - 2\sec\theta = 2$

6. $\csc^2\theta = \cot\theta + 3$

7. $\sin 2\theta - \sin\theta = 0$

8. $3\sin 2\theta - 2\sin\theta = 0$

9. $3\cos 2\theta - 2\cos^2\theta = 0$

10. $\cos 2\theta = \cos^2\theta - \frac{1}{2}$

11. $2\sin^2\theta - \cos\theta = 1$

12. $\tan\theta - 3\cot\theta = 0$

13. $\sin\theta - 1 = \cos\theta$

14. $\cos\theta - \sin\theta = 1$

15. $\tan\theta + 1 = \sec\theta$

16. $2\tan\theta + \sec^2\theta = 4$

17–30 ■ Solving Trigonometric Equations Involving a Multiple of an Angle An equation is given. **(a)** Find all solutions of the equation. **(b)** List the solutions in the interval $[0, 2\pi)$.

17. $2\cos 3\theta = 1$

18. $2\sin 2\theta = 1$

19. $2\cos 2\theta + 1 = 0$

20. $2\sin 3\theta + 1 = 0$

21. $\sqrt{3}\tan 3\theta + 1 = 0$

22. $\sec 4\theta - 2 = 0$

23. $\cos\frac{\theta}{2} - 1 = 0$

24. $\tan\frac{\theta}{4} + \sqrt{3} = 0$

25. $2\sin\frac{\theta}{3} + \sqrt{3} = 0$

26. $\sec\frac{\theta}{2} = \cos\frac{\theta}{2}$

27. $\sin 2\theta = 3\cos 2\theta$

28. $\csc 3\theta = 5\sin 3\theta$

29. $1 - 2\sin\theta = \cos 2\theta$

30. $\tan 3\theta + 1 = \sec 3\theta$

31–34 ■ Solving Trigonometric Equations by Factoring An equation is given. **(a)** Solve the equation by factoring. **(b)** List the solutions in the interval $[0, 2\pi)$.

31. $3\tan^3\theta - 3\tan^2\theta - \tan\theta + 1 = 0$

32. $4\sin\theta\cos\theta + 2\sin\theta - 2\cos\theta - 1 = 0$

33. $2\sin\theta\tan\theta - \tan\theta = 1 - 2\sin\theta$

34. $\sec\theta\tan\theta - \cos\theta\cot\theta = \sin\theta$

35–38 ■ Finding Intersection Points Graphically **(a)** Graph f and g in the given viewing rectangle and find the intersection points graphically, rounded to two decimal places. **(b)** Find the intersection points of f and g algebraically. Give exact answers.

35. $f(x) = 3\cos x + 1, \quad g(x) = \cos x - 1;$
$[-2\pi, 2\pi]$ by $[-2.5, 4.5]$

36. $f(x) = \sin 2x + 1, \quad g(x) = 2\sin 2x + 1;$
$[-2\pi, 2\pi]$ by $[-1.5, 3.5]$

37. $f(x) = \tan x, \quad g(x) = \sqrt{3};$
$[-\pi/2, \pi/2]$ by $[-10, 10]$

38. $f(x) = \sin x - 1, \quad g(x) = \cos x;$
$[-2\pi, 2\pi]$ by $[-2.5, 1.5]$

39–42 ■ Using Addition or Subtraction Formulas Use an Addition or Subtraction Formula to simplify the equation. Then find all solutions in the interval $[0, 2\pi)$.

39. $\cos\theta\cos 3\theta - \sin\theta\sin 3\theta = 0$

40. $\cos\theta\cos 2\theta + \sin\theta\sin 2\theta = \frac{1}{2}$

41. $\sin 2\theta\cos\theta - \cos 2\theta\sin\theta = \sqrt{3}/2$

42. $\sin 3\theta\cos\theta - \cos 3\theta\sin\theta = 0$

43–52 ■ Using Double- or Half-Angle Formulas Use a Double- or Half-Angle Formula to solve the equation in the interval $[0, 2\pi)$.

43. $\sin 2\theta + \cos\theta = 0$

44. $\tan\dfrac{\theta}{2} - \sin\theta = 0$

45. $\cos 2\theta + \cos\theta = 2$

46. $\tan\theta + \cot\theta = 4\sin 2\theta$

47. $\cos 2\theta - \cos^2\theta = 0$

48. $2\sin^2\theta = 2 + \cos 2\theta$

49. $\cos 2\theta - \cos 4\theta = 0$

50. $\sin 3\theta - \sin 6\theta = 0$

51. $\cos\theta - \sin\theta = \sqrt{2}\sin\dfrac{\theta}{2}$

52. $\sin\theta - \cos\theta = \frac{1}{2}$

53–56 ■ Using Sum-to-Product Formulas Solve the equation by first using a Sum-to-Product Formula.

53. $\sin\theta + \sin 3\theta = 0$

54. $\cos 5\theta - \cos 7\theta = 0$

55. $\cos 4\theta + \cos 2\theta = \cos\theta$

56. $\sin 5\theta - \sin 3\theta = \cos 4\theta$

57–62 ■ Solving Trigonometric Equations Graphically Use a graphing device to find the solutions of the equation, rounded to two decimal places.

57. $\sin 2x = x$

58. $\cos x = \dfrac{x}{3}$

59. $2^{\sin x} = x$

60. $\sin x = x^3$

61. $\dfrac{\cos x}{1 + x^2} = x^2$

62. $\cos x = \frac{1}{2}(e^x + e^{-x})$

Skills Plus

63–64 ■ Equations Involving Inverse Trigonometric Functions Solve the given equation for x.

63. $\tan^{-1}x + \tan^{-1}2x = \dfrac{\pi}{4}$ [*Hint:* Let $u = \tan^{-1}x$ and $v = \tan^{-1}2x$. Solve the equation $u + v = \dfrac{\pi}{4}$ by taking the tangent of each side.]

64. $2\sin^{-1}x + \cos^{-1}x = \pi$ [*Hint:* Take the cosine of each side.]

Applications

65. Range of a Projectile If a projectile is fired with velocity v_0 at an angle θ, then its *range*, the horizontal distance it travels (in m), is modeled by the function

$$R(\theta) = \frac{v_0^2 \sin 2\theta}{9.8}$$

(See Focus on Modeling *The Path of a Projectile* following Chapter 8.) If $v_0 = 660$ m/s, what angle (in degrees) should be chosen for the projectile to hit a target on the ground 1500 m away?

66. Damped Vibrations The displacement of a spring vibrating in damped harmonic motion is given by

$$y = 4e^{-3t}\sin 2\pi t$$

Find the times when the spring is at its equilibrium position $(y = 0)$.

67. Hours of Daylight In Philadelphia the number of hours of daylight on day t (where t is the number of days after January 1) is modeled by the function

$$L(t) = 12 + 2.83\sin\left(\frac{2\pi}{365}(t - 80)\right)$$

(a) Which days of the year have about 10 hours of daylight?

(b) How many days of the year have more than 10 hours of daylight?

68. Belts and Pulleys A thin belt of length L surrounds two pulleys of radii R and r, as shown in the figure.

(a) Show that the angle θ (in rad) where the belt crosses itself satisfies the equation

$$\theta + 2\cot\frac{\theta}{2} = \frac{L}{R + r} - \pi$$

[*Hint:* Express L in terms of R, r, and θ by adding up the lengths of the curved and straight parts of the belt.]

(b) Suppose that $R = 2.42$ m, $r = 1.21$ m, and $L = 27.78$ m. Find θ by solving the equation in part (a) graphically. Express your answer both in radians and in degrees.

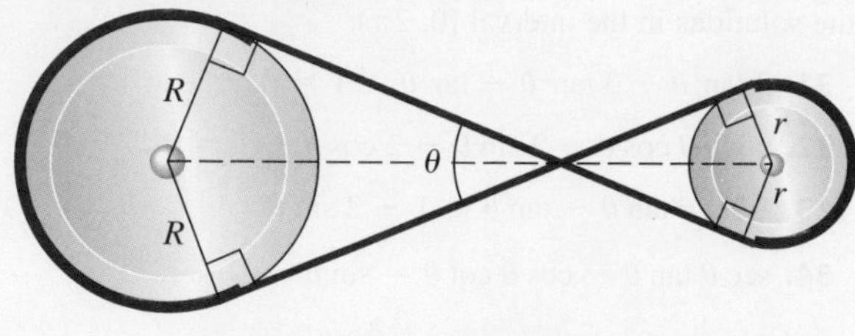

■ Discuss ■ Discover ■ Prove ■ Write

69. Discuss: A Special Trigonometric Equation What makes the equation $\sin(\cos x) = 0$ different from all the other equations we've looked at in this section? Find all solutions of this equation.

Chapter 7 Review

Properties and Formulas

Fundamental Trigonometric Identities | Section 7.1

An **identity** is an equation that is true for all values of the variable(s). A **trigonometric identity** is an identity that involves trigonometric functions. The fundamental trigonometric identities are as follows.

Reciprocal Identities:

$$\csc x = \frac{1}{\sin x} \qquad \sec x = \frac{1}{\cos x} \qquad \cot x = \frac{1}{\tan x}$$

$$\tan x = \frac{\sin x}{\cos x} \qquad \cot x = \frac{\cos x}{\sin x}$$

Pythagorean Identities:

$$\sin^2 x + \cos^2 x = 1$$

$$\tan^2 x + 1 = \sec^2 x$$

$$1 + \cot^2 x = \csc^2 x$$

Even-Odd Identities:

$$\sin(-x) = -\sin x$$

$$\cos(-x) = \cos x$$

$$\tan(-x) = -\tan x$$

Cofunction Identities:

$$\sin\left(\frac{\pi}{2} - x\right) = \cos x \qquad \tan\left(\frac{\pi}{2} - x\right) = \cot x$$

$$\sec\left(\frac{\pi}{2} - x\right) = \csc x$$

$$\cos\left(\frac{\pi}{2} - x\right) = \sin x \qquad \cot\left(\frac{\pi}{2} - x\right) = \tan x$$

$$\csc\left(\frac{\pi}{2} - x\right) = \sec x$$

Proving Trigonometric Identities | Section 7.1

To prove that a trigonometric equation is an identity, we use the following guidelines.

1. **Start with one side.** Select one side of the equation.
2. **Use known identities.** Use algebra and known identities to change the side you started with into the other side.
3. **Convert to sines and cosines.** Sometimes it is helpful to convert all functions in the equation to sines and cosines.

Addition and Subtraction Formulas | Section 7.2

These identities involve the trigonometric functions of a sum or a difference.

Formulas for Sine:

$$\sin(s + t) = \sin s\cos t + \cos s\sin t$$

$$\sin(s - t) = \sin s\cos t - \cos s\sin t$$

Formulas for Cosine:

$$\cos(s + t) = \cos s\cos t - \sin s\sin t$$

$$\cos(s - t) = \cos s\cos t + \sin s\sin t$$

Formulas for Tangent:

$$\tan(s + t) = \frac{\tan s + \tan t}{1 - \tan s\tan t}$$

$$\tan(s - t) = \frac{\tan s - \tan t}{1 + \tan s\tan t}$$

Sums of Sines and Cosines | Section 7.2

If A and B are real numbers, then

$$A\sin x + B\cos x = k\sin(x + \phi)$$

where $k = \sqrt{A^2 + B^2}$ and ϕ satisfies

$$\cos\phi = \frac{A}{\sqrt{A^2 + B^2}} \qquad \sin\phi = \frac{B}{\sqrt{A^2 + B^2}}$$

Double-Angle Formulas | Section 7.3

These identities involve the trigonometric functions of twice the variable.

Formula for Sine:

$$\sin 2x = 2 \sin x \cos x$$

Formulas for Cosine:

$$\begin{aligned}\cos 2x &= \cos^2 x - \sin^2 x \\ &= 1 - 2\sin^2 x \\ &= 2\cos^2 x - 1\end{aligned}$$

Formula for Tangent:

$$\tan 2x = \frac{2 \tan x}{1 - \tan^2 x}$$

Formulas for Lowering Powers | Section 7.3

These formulas allow us to write a trigonometric expression involving even powers of sine and cosine in terms of the first power of cosine only.

$$\sin^2 x = \frac{1 - \cos 2x}{2} \qquad \cos^2 x = \frac{1 + \cos 2x}{2}$$

$$\tan^2 x = \frac{1 - \cos 2x}{1 + \cos 2x}$$

Half-Angle Formulas | Section 7.3

These formulas involve trigonometric functions of half an angle.

$$\sin\frac{u}{2} = \pm\sqrt{\frac{1 - \cos u}{2}} \qquad \cos\frac{u}{2} = \pm\sqrt{\frac{1 + \cos u}{2}}$$

$$\tan\frac{u}{2} = \frac{1 - \cos u}{\sin u} = \frac{\sin u}{1 + \cos u}$$

Product-Sum Formulas | Section 7.3

These formulas involve products and sums of trigonometric functions.

Product-to-Sum Formulas:

$$\begin{aligned}\sin u \cos v &= \tfrac{1}{2}[\sin(u + v) + \sin(u - v)] \\ \cos u \sin v &= \tfrac{1}{2}[\sin(u + v) - \sin(u - v)] \\ \cos u \cos v &= \tfrac{1}{2}[\cos(u + v) + \cos(u - v)] \\ \sin u \sin v &= \tfrac{1}{2}[\cos(u - v) - \cos(u + v)]\end{aligned}$$

Sum-to-Product Formulas:

$$\sin x + \sin y = 2 \sin\frac{x + y}{2}\cos\frac{x - y}{2}$$

$$\sin x - \sin y = 2 \cos\frac{x + y}{2}\sin\frac{x - y}{2}$$

$$\cos x + \cos y = 2 \cos\frac{x + y}{2}\cos\frac{x - y}{2}$$

$$\cos x - \cos y = -2 \sin\frac{x + y}{2}\sin\frac{x - y}{2}$$

Trigonometric Equations | Section 7.4

A **trigonometric equation** is an equation that contains trigonometric functions. A basic trigonometric equation is an equation of the form $T(\theta) = c$, where T is a trigonometric function and c is a constant. For example, $\sin\theta = 0.5$ and $\tan\theta = 2$ are basic trigonometric equations. Solving any trigonometric equation involves solving a basic trigonometric equation.

If a trigonometric equation has a solution, then it has infinitely many solutions.

To find all solutions, we first find the solutions in one period and then add integer multiples of the period.

We can sometimes use trigonometric identities to simplify a trigonometric equation.

Concept Check

1. What is an identity? What is a trigonometric identity?

2. (a) State the Pythagorean identities.

(b) Use a Pythagorean identity to express cosine in terms of sine.

3. (a) State the reciprocal identities for cosecant, secant, and cotangent.

(b) State the even-odd identities for sine and cosine.

(c) State the cofunction identities for sine, tangent, and secant.

(d) Suppose that $\cos(-x) = 0.4$; use the identities in parts (a) and (b) to find $\sec x$.

(e) Suppose that $\sin 10° = a$; use the identities in part (c) to find $\cos 80°$.

4. (a) How do you prove an identity?

(b) Prove the identity $\sin x(\csc x - \sin x) = \cos^2 x$

5. (a) State the Addition and Subtraction Formulas for Sine and Cosine.

(b) Use a formula from part (a) to find $\cos 75°$.

6. (a) State the formula for $A \sin x + B \cos x$.

(b) Express $4 \sin x + 3 \cos x$ as a function of sine only.

7. (a) State the Double-Angle Formula for Sine and the Double-Angle Formulas for Cosine.

(b) Prove the identity $\sec x \sin 2x = 2 \sin x$.

8. (a) State the formulas for lowering powers of sine and cosine.

(b) Prove the identity $4 \sin^2 x \cos^2 x = \sin^2 2x$.

9. (a) State the Half-Angle Formulas for Sine and Cosine.

(b) Find $\cos 15°$.

10. (a) State the Product-to-Sum Formula for the product $\sin u \cos v$.

(b) Express $\sin 5x \cos 3x$ as a sum of trigonometric functions.

11. (a) State the Sum-to-Product Formula for the sum $\sin x + \sin y$.

(b) Express $\sin 5x + \sin 7x$ as a product of trigonometric functions.

12. What is a trigonometric equation? How do we solve a trigonometric equation?

(a) Solve the equation $\cos x = \frac{1}{2}$.

(b) Solve the equation $2 \sin x \cos x = \frac{1}{2}$.

Answers to the Concept Check can be found at the book companion website stewartmath.com.

Exercises

1–22 ■ Proving Identities Verify the identity.

1. $\sin \theta (\cot \theta + \tan \theta) = \sec \theta$

2. $(\sec \theta - 1)(\sec \theta + 1) = \tan^2 \theta$

3. $\cos^2 x \csc x - \csc x = -\sin x$

4. $\dfrac{1}{1 - \sin^2 x} = 1 + \tan^2 x$

5. $\dfrac{\cos^2 x - \tan^2 x}{\sin^2 x} = \cot^2 x - \sec^2 x$

6. $\dfrac{1 + \sec x}{\sec x} = \dfrac{\sin^2 x}{1 - \cos x}$

7. $\dfrac{\cos^2 x}{1 - \sin x} = \dfrac{\cos x}{\sec x - \tan x}$

8. $(1 - \tan x)(1 - \cot x) = 2 - \sec x \csc x$

9. $\sin^2 x \cot^2 x + \cos^2 x \tan^2 x = 1$

10. $(\tan x + \cot x)^2 = \csc^2 x \sec^2 x$

11. $\dfrac{\sin 2x}{1 + \cos 2x} = \tan x$

12. $\dfrac{\cos(x + y)}{\cos x \sin y} = \cot y - \tan x$

13. $\csc x - \tan \dfrac{x}{2} = \cot x$

14. $1 + \tan x \tan \dfrac{x}{2} = \sec x$

15. $\dfrac{\sin 2x}{\sin x} - \dfrac{\cos 2x}{\cos x} = \sec x$

16. $\tan\left(x + \dfrac{\pi}{4}\right) = \dfrac{1 + \tan x}{1 - \tan x}$

17. $\dfrac{\sec x - 1}{\sin x \sec x} = \tan \dfrac{x}{2}$

18. $(\cos x + \cos y)^2 + (\sin x - \sin y)^2 = 2 + 2\cos(x + y)$

19. $\left(\cos \dfrac{x}{2} - \sin \dfrac{x}{2}\right)^2 = 1 - \sin x$

20. $\dfrac{\cos 3x - \cos 7x}{\sin 3x + \sin 7x} = \tan 2x$

21. $\dfrac{\sin(x + y) + \sin(x - y)}{\cos(x + y) + \cos(x - y)} = \tan x$

22. $\sin(x + y)\sin(x - y) = \sin^2 x - \sin^2 y$

23–26 ■ Checking Identities Graphically **(a)** Graph f and g. **(b)** Do the graphs suggest that the equation $f(x) = g(x)$ is an identity? Prove your answer.

23. $f(x) = 1 - \left(\cos \dfrac{x}{2} - \sin \dfrac{x}{2}\right)^2, \quad g(x) = \sin x$

24. $f(x) = \sin x + \cos x, \quad g(x) = \sqrt{\sin^2 x + \cos^2 x}$

25. $f(x) = \tan x \tan \dfrac{x}{2}, \quad g(x) = \dfrac{1}{\cos x}$

26. $f(x) = 1 - 8\sin^2 x + 8\sin^4 x, \quad g(x) = \cos 4x$

27–28 ■ Determining Identities Graphically **(a)** Graph the function(s) and make a conjecture, and **(b)** prove your conjecture.

27. $f(x) = 2\sin^2 3x + \cos 6x$

28. $f(x) = \sin x \cot \dfrac{x}{2}, \quad g(x) = \cos x$

29–46 ■ Solving Trigonometric Equations Solve the equation in the interval $[0, 2\pi)$.

29. $4 \sin \theta - 3 = 0$

30. $5 \cos \theta + 3 = 0$

31. $\cos x \sin x - \sin x = 0$

32. $\sin x - 2\sin^2 x = 0$

33. $2\sin^2 x - 5\sin x + 2 = 0$

34. $\sin x - \cos x - \tan x = -1$

35. $2\cos^2 x - 7\cos x + 3 = 0$

36. $4\sin^2 x + 2\cos^2 x = 3$

37. $\dfrac{1 - \cos x}{1 + \cos x} = 3$

38. $\sin x = \cos 2x$

39. $\tan^3 x + \tan^2 x - 3\tan x - 3 = 0$

40. $\cos 2x \csc^2 x = 2\cos 2x$

41. $\tan \frac{1}{2}x + 2\sin 2x = \csc x$

42. $\cos 3x + \cos 2x + \cos x = 0$

43. $\tan x + \sec x = \sqrt{3}$

44. $2\cos x - 3\tan x = 0$

45. $\cos x = x^2 - 1$

46. $e^{\sin x} = x$

47. Range of a Projectile If a projectile is fired with velocity v_0 at an angle θ, then the maximum height it reaches (in m) is modeled by the function

$$M(\theta) = \frac{v_0^2 \sin^2\theta}{19.2}$$

Suppose $v_0 = 120$ m/s.

(a) At what angle θ should the projectile be fired so that the maximum height it reaches is 600 m?

(b) Is it possible for the projectile to reach a height of 900 m?

(c) Find the angle θ for which the projectile will travel highest.

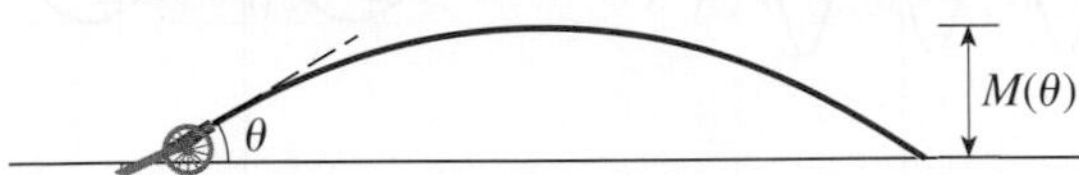

48. Displacement of a Shock Absorber The displacement of an automobile shock absorber is modeled by the function

$$f(t) = 2^{-0.2t} \sin 4\pi t$$

Find the times when the shock absorber is at its equilibrium position [that is, when $f(t) = 0$]. [*Hint:* $2^x > 0$ for all real x.]

49–58 ■ Value of Expressions Find the exact value of the expression.

49. $\cos 15°$

50. $\sin \frac{5\pi}{12}$

51. $\tan \frac{\pi}{8}$

52. $2 \sin \frac{\pi}{12} \cos \frac{\pi}{12}$

53. $\sin 5° \cos 40° + \cos 5° \sin 40°$

54. $\dfrac{\tan 66° - \tan 6°}{1 + \tan 66° \tan 6°}$

55. $\cos^2\left(\frac{\pi}{8}\right) - \sin^2\left(\frac{\pi}{8}\right)$

56. $\frac{1}{2} \cos \frac{\pi}{12} + \frac{\sqrt{3}}{2} \sin \frac{\pi}{12}$

57. $\cos 37.5° \cos 7.5°$

58. $\cos 67.5° + \cos 22.5°$

59–64 ■ Evaluating Expressions Involving Trigonometric Functions Find the exact value of the expression given that $\sec x = \frac{3}{2}$, $\csc y = 3$, and x and y are in Quadrant I.

59. $\sin(x + y)$

60. $\cos(x - y)$

61. $\tan(x + y)$

62. $\sin 2x$

63. $\cos \frac{y}{2}$

64. $\tan \frac{y}{2}$

65–66 ■ Evaluating Expressions Involving Inverse Trigonometric Functions Find the exact value of the expression.

65. $\tan\left(2 \cos^{-1}\left(\frac{3}{7}\right)\right)$

66. $\sin\left(\tan^{-1}\left(\frac{3}{4}\right) + \cos^{-1}\left(\frac{5}{13}\right)\right)$

67–68 ■ Expressions Involving Inverse Trigonometric Functions Write the expression as an algebraic expression in the variable(s).

67. $\tan(2 \tan^{-1} x)$

68. $\cos(\sin^{-1} x + \cos^{-1} y)$

69. Viewing Angle of a Sign A 3-m-wide highway sign is adjacent to a roadway, as shown in the figure. As a driver approaches the sign, the viewing angle θ changes.

(a) Express viewing angle θ as a function of the distance x between the driver and the sign.

(b) The sign is legible when the viewing angle is 2° or greater. At what distance x does the sign first become legible?

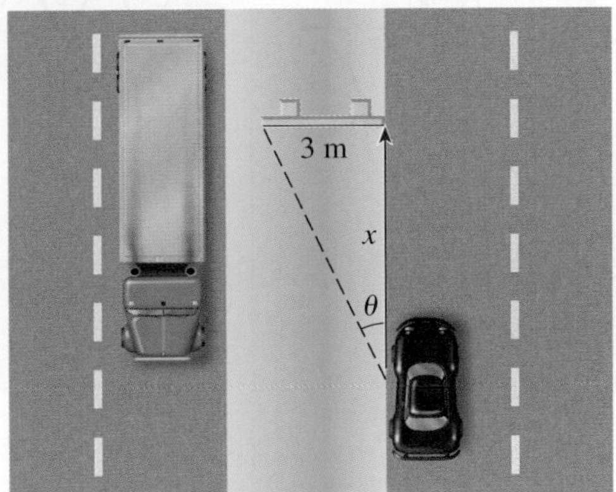

70. Viewing Angle of a Tower A 114-m-tall building supports a 12 m communications tower (see the figure). As a driver approaches the building, the viewing angle θ of the tower changes.

(a) Express the viewing angle θ as a function of the distance x between the driver and the building.

(b) At what distance from the building is the viewing angle θ as large as possible?

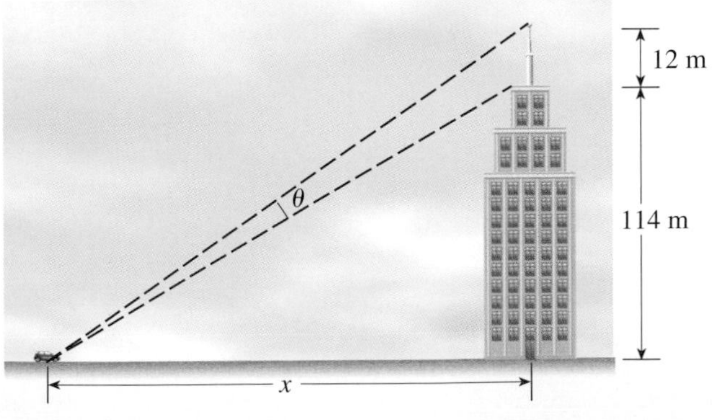

Matching

71. Equations and Their Graphs Match each equation with its graph and give reasons for your answers. Use identities to help recognize the graph. (Don't use a graphing device.)

(a) $y = \sin\left(\dfrac{\pi}{2} - x\right)$ (b) $y = 4 \sin x \cos x$ (c) $y = 1 - \cos 2x$ (d) $y = \dfrac{1 - \cos 2x}{1 + \cos 2x}$

(e) $y = \dfrac{1}{2}\sin x + \dfrac{\sqrt{3}}{2}\cos x$ (f) $y = 1 + \tan^2 x$ (g) $y = \cos^2 x - \sin^2 x$ (h) $y = \dfrac{\sin x}{1 + \cos x}$

I

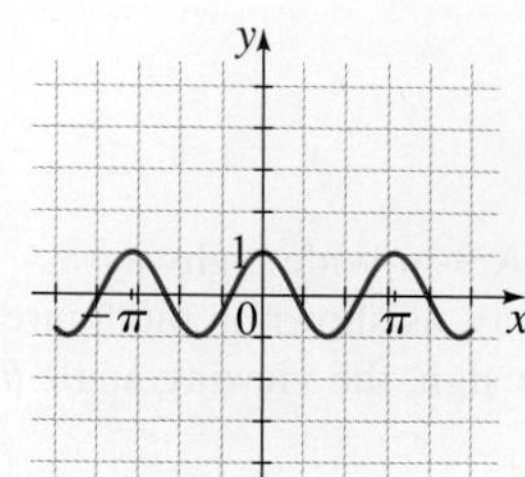

II

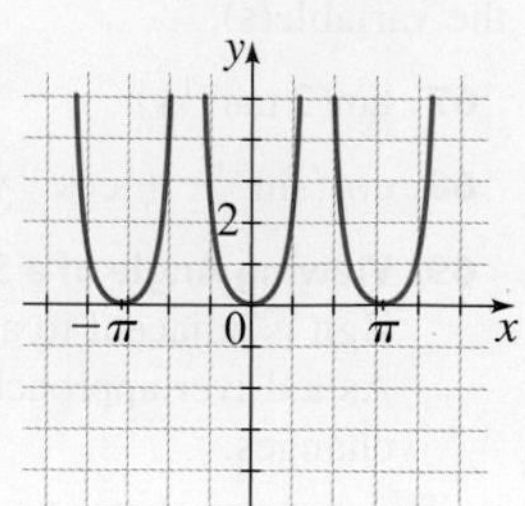

III

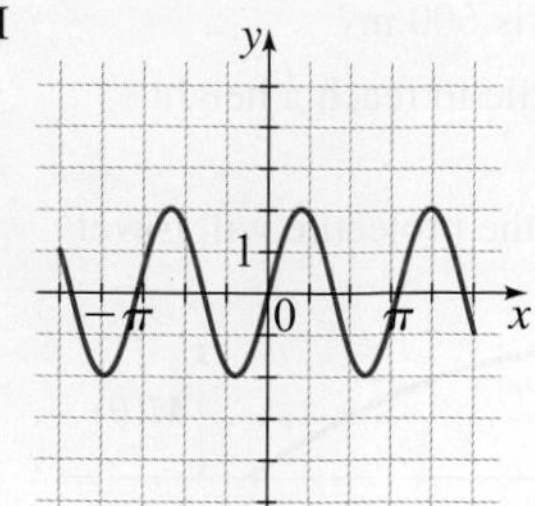

IV

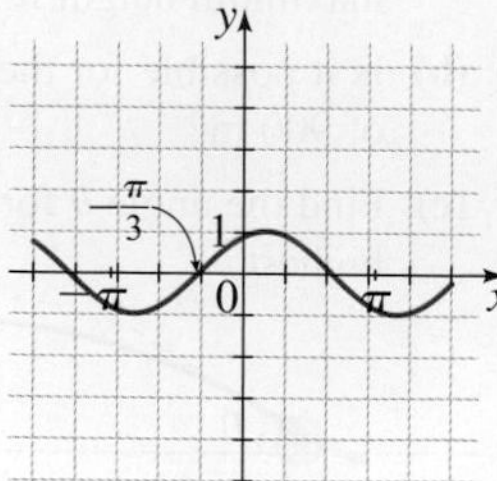

V

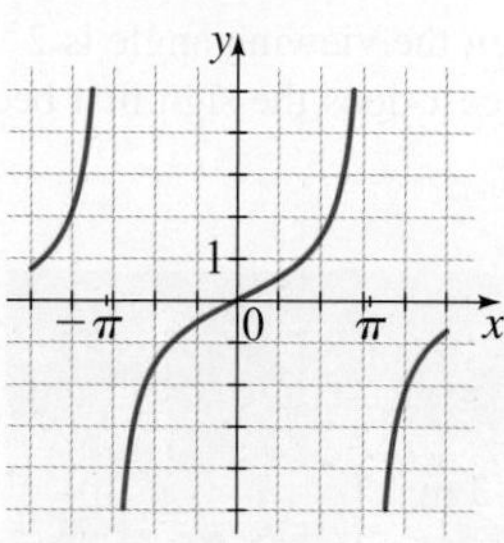

VI

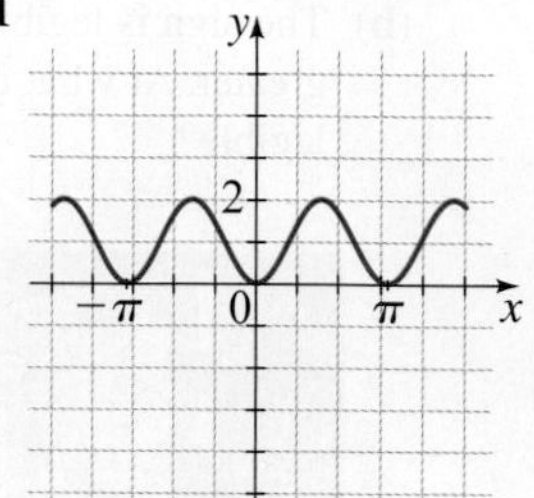

VII

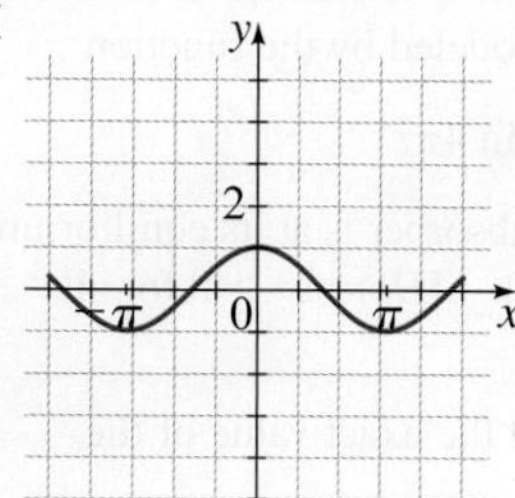

VIII

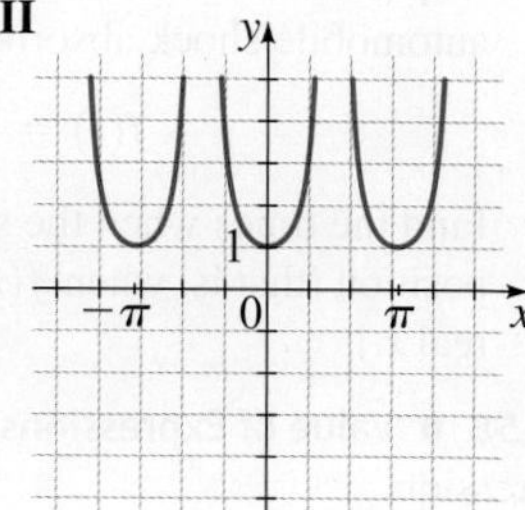

Chapter 7 | Test

1–7 ■ Verify the identity.

1. $\tan\theta\sin\theta + \cos\theta = \sec\theta$

2. $\dfrac{\tan x}{1 - \cos x} = \csc x\,(1 + \sec x)$

3. $\dfrac{2\tan x}{1 + \tan^2 x} = \sin 2x$

4. $\sin x \tan\dfrac{x}{2} = 1 - \cos x$

5. $2\sin^2 3x = 1 - \cos 6x$

6. $\cos 4x = 1 - 8\sin^2 x + 8\sin^4 x$

7. $\left(\sin\dfrac{x}{2} + \cos\dfrac{x}{2}\right)^2 = 1 + \sin x$

8. Let $x = 2\sin\theta$, $-\pi/2 < \theta < \pi/2$. Simplify the expression

$$\frac{x}{\sqrt{4 - x^2}}$$

9. Find the exact value of each expression.

(a) $\sin 8^\circ \cos 22^\circ + \cos 8^\circ \sin 22^\circ$ (b) $\sin 75^\circ$ (c) $\sin\dfrac{\pi}{12}$

10. For the angles α and β in the figures, find $\cos(\alpha + \beta)$.

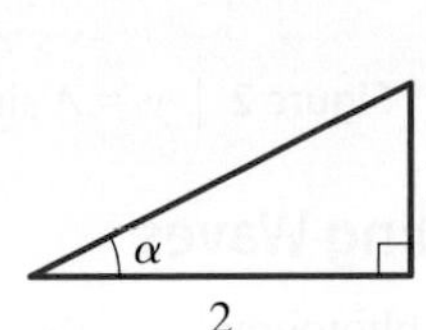

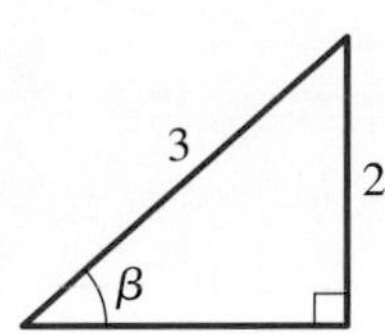

11. Write $\sin 3x \cos 5x$ as a sum of trigonometric functions.

12. Write $\sin 2x - \sin 5x$ as a product of trigonometric functions.

13. If $\sin\theta = -\frac{4}{5}$ and θ is in Quadrant III, find $\tan(\theta/2)$.

14–20 ■ Solve the trigonometric equation in the interval $[0, 2\pi)$. Give the exact value, if possible; otherwise, round your answer to two decimal places.

14. $3\sin\theta - 1 = 0$

15. $(2\cos\theta - 1)(\sin\theta - 1) = 0$

16. $2\cos^2\theta + 5\cos\theta + 2 = 0$

17. $\sin 2\theta - \cos\theta = 0$

18. $5\cos 2\theta = 2$

19. $2\cos^2 x + \cos 2x = 0$

20. $2\tan\dfrac{x}{2} - \csc x = 0$

21. Find the exact value of $\cos\left(2\tan^{-1}\left(\frac{9}{40}\right)\right)$.

22. Rewrite the expression as an algebraic function of x and y: $\sin(\cos^{-1}x - \tan^{-1}y)$.

A Cumulative Review Test for Chapters 5, 6, and 7 can be found at the book companion website www.stewartmath.com.

We've learned that the position of a particle in simple harmonic motion is described by a function of the form $y = A \sin \omega t$ (see Section 5.6). For example, if a string is moved up and down as in Figure 1, then the red dot on the string moves up and down in simple harmonic motion. Of course, the same holds true for each point on the string.

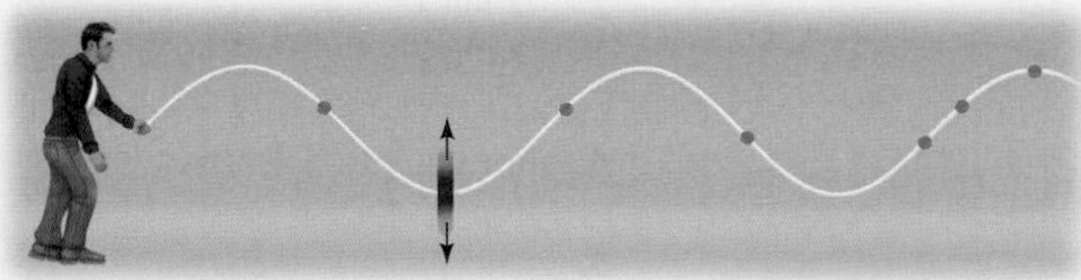

Figure 1

What function describes the shape of the whole string? If we fix an instant in time ($t = 0$) and snap a photograph of the string, we get the shape in Figure 2, which is modeled by

$$y = A \sin kx$$

where y is the height of the string above the x-axis at the point x.

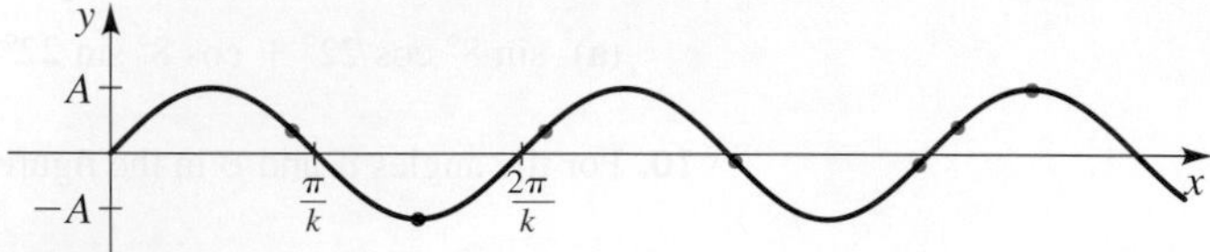

Figure 2 | $y = A \sin kx$

■ Traveling Waves

If we snap photographs of the string at other instants, as in Figure 3, it appears that the waves in the string "travel," or shift to the right.

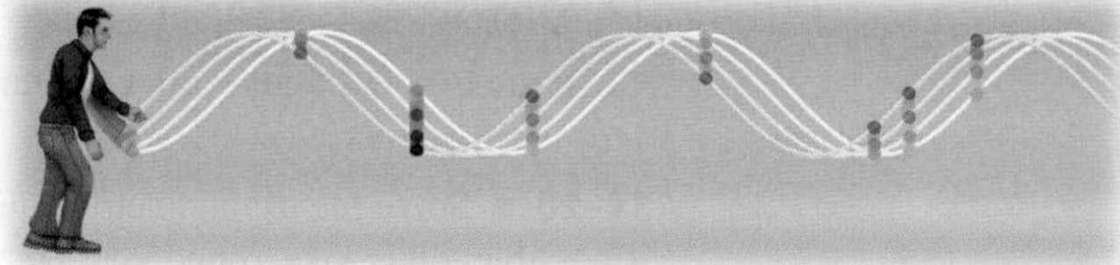

Figure 3

The **velocity** of the wave is the rate at which it moves to the right. If the wave has velocity v, then it moves to the right a distance vt in time t. So the graph of the shifted wave at time t is

$$y(x, t) = A \sin k(x - vt)$$

This function models the position of any point x on the string at any time t. We use the notation $y(x, t)$ to indicate that the function depends on the *two* variables x and t. Here is how this function models the motion of the string.

- **If we fix x**, then $y(x, t)$ is a function of t only, which gives the position of the fixed point x at time t.
- **If we fix t**, then $y(x, t)$ is a function of x only, whose graph is the shape of the string at the fixed time t.

Example 1 ■ A Traveling Wave

A traveling wave is described by the function

$$y(x, t) = 3\sin\left(2x - \frac{\pi}{2}t\right) \qquad (x \geq 0)$$

(a) Find the function that models the position of the point $x = \pi/6$ at any time t. Observe that the point moves in simple harmonic motion.

(b) Sketch the shape of the wave when $t = 0, 0.5, 1.0, 1.5$, and 2.0. Does the wave appear to be traveling to the right?

(c) Find the velocity of the wave.

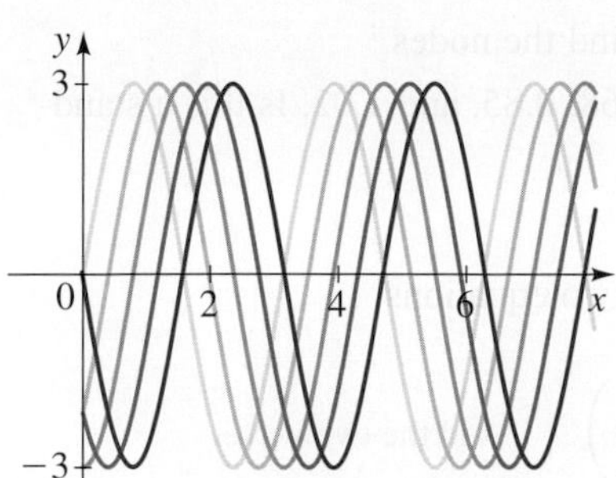

Figure 4 | Traveling wave

Solution

(a) Substituting $x = \pi/6$, we get

$$y\left(\frac{\pi}{6}, t\right) = 3\sin\left(2 \cdot \frac{\pi}{6} - \frac{\pi}{2}t\right) = 3\sin\left(\frac{\pi}{3} - \frac{\pi}{2}t\right)$$

The function $y = 3\sin(\frac{\pi}{3} - \frac{\pi}{2}t)$ describes simple harmonic motion with amplitude 3 and period $2\pi/(\pi/2) = 4$.

(b) The graphs are shown in Figure 4. As t increases, the wave moves to the right.

(c) We express the given function in the standard form $y(x, t) = A\sin k(x - vt)$.

$$y(x, t) = 3\sin\left(2x - \frac{\pi}{2}t\right) \qquad \text{Given}$$

$$= 3\sin 2\left(x - \frac{\pi}{4}t\right) \qquad \text{Factor 2}$$

Comparing this to the standard form, we see that the wave is moving with velocity $v = \pi/4$. ■

■ Standing Waves

If two waves are traveling along the same string, then the movement of the string is determined by the sum of the two waves. For example, if the string is attached to a wall, then the waves bounce back with the same amplitude and speed but in opposite directions. In this case, one wave is described by $y = A\sin k(x - vt)$, and the reflected wave is described by $y = A\sin k(x + vt)$. The resulting wave is

$$y(x, t) = A\sin k(x - vt) + A\sin k(x + vt) \qquad \text{Add the two waves}$$

$$= 2A\sin kx\cos kvt \qquad \text{Sum-to-Product Formula}$$

The points where kx is a multiple of 2π are special because at these points $y = 0$ for any time t. In other words, these points never move. Such points are called **nodes**. Figure 5 shows the graph of the wave for several values of t. We see that the wave does not travel but simply vibrates up and down. Such a wave is called a **standing wave**.

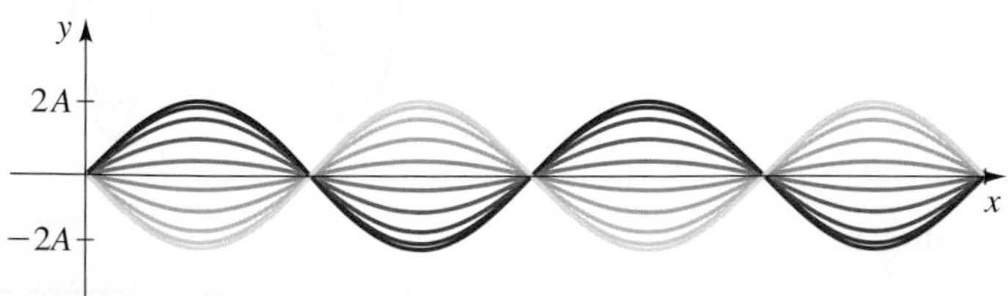

Figure 5 | A standing wave

Example 2 ■ A Standing Wave

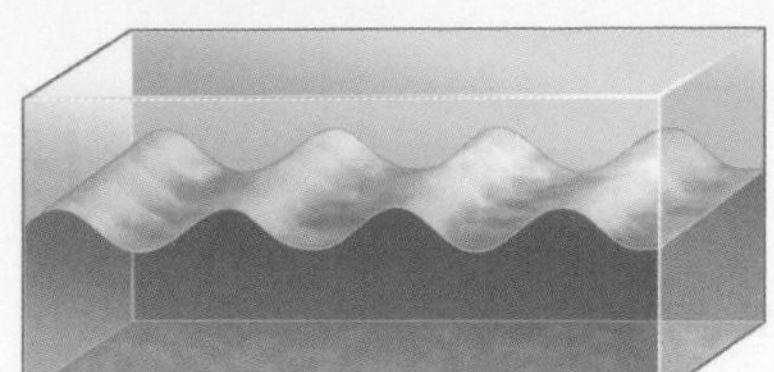

Traveling waves are generated at each end of a wave tank 30 m long, with equations

$$y = 1.5 \sin\left(\frac{\pi}{5}x - 3t\right)$$

and

$$y = 1.5 \sin\left(\frac{\pi}{5}x + 3t\right)$$

(a) Find the equation of the combined wave, and find the nodes.

(b) Sketch the graph for $t = 0, 0.17, 0.34, 0.51, 0.68, 0.85$, and 1.02. Is this a standing wave?

Solution

(a) The combined wave is obtained by adding the two equations.

$$y = 1.5 \sin\left(\frac{\pi}{5}x - 3t\right) + 1.5 \sin\left(\frac{\pi}{5}x + 3t\right) \qquad \text{Add the two waves}$$

$$= 3 \sin\frac{\pi}{5}x \cos 3t \qquad \text{Sum-to-Product Formula}$$

The nodes occur at the values of x for which $\sin\frac{\pi}{5}x = 0$, that is, where $\frac{\pi}{5}x = k\pi$ (k an integer). Solving for x, we get $x = 5k$. So the nodes occur at

$$x = 0, 5, 10, 15, 20, 25, 30$$

(b) The graphs are shown in Figure 6. From the graphs we see that this is a standing wave.

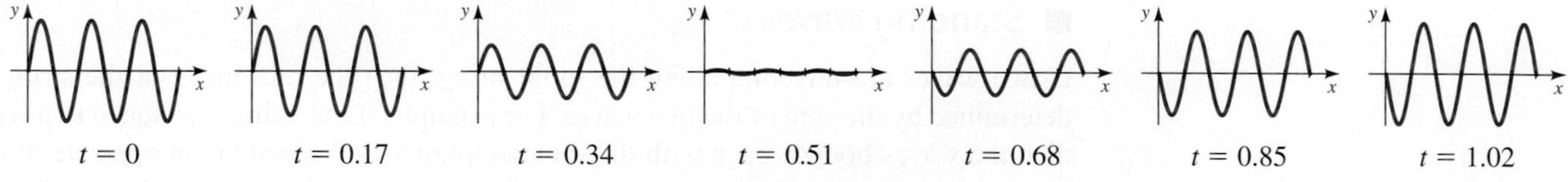

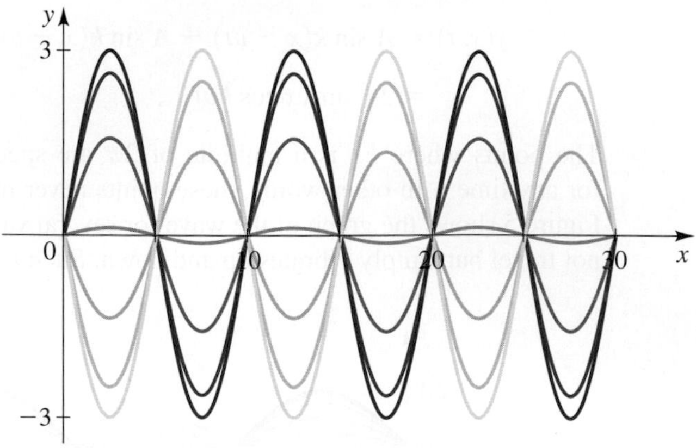

Figure 6 | $y(x, t) = 3 \sin\frac{\pi}{5}x \cos 3t$

■

Problems

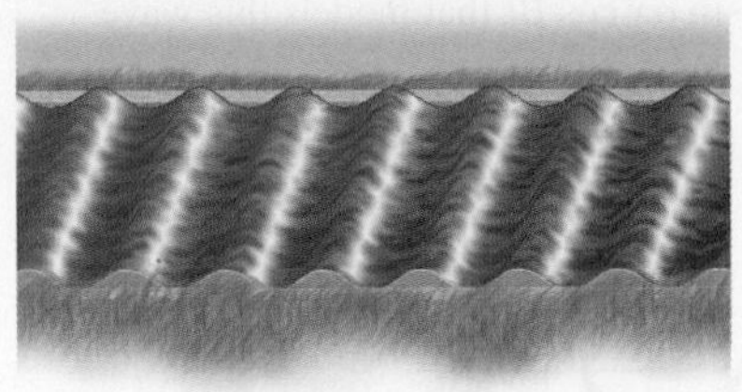

1. Wave on a Canal A wave on the surface of a long canal is described by the function

$$y(x, t) = 5\sin\left(4x - \frac{\pi}{8}t\right) \qquad (x \geq 0)$$

(a) Find the function that models the position of the point $x = 0$ at any time t.

(b) Sketch the shape of the wave when $t = 0$, 1.6, 3.2, 4.8, and 6.4. Is this a traveling wave?

(c) Find the velocity of the wave.

2. Wave in a Rope Traveling waves are generated at each end of a tightly stretched rope 24 m long, with equations

$$y = 0.2\sin(1.047x - 0.524t)$$

$$y = 0.2\sin(1.047x + 0.524t)$$

(a) Find an equation of the combined wave, and find the nodes.

(b) Sketch the graph for $t = 0$, 1, 2, 3, 4, 5, and 6. Is this a standing wave?

3. Traveling Wave A traveling wave is graphed at the instant $t = 0$. If it is moving to the right with velocity 6, find an equation of the form

$$y(x, t) = A\sin(kx - kvt)$$

for this wave.

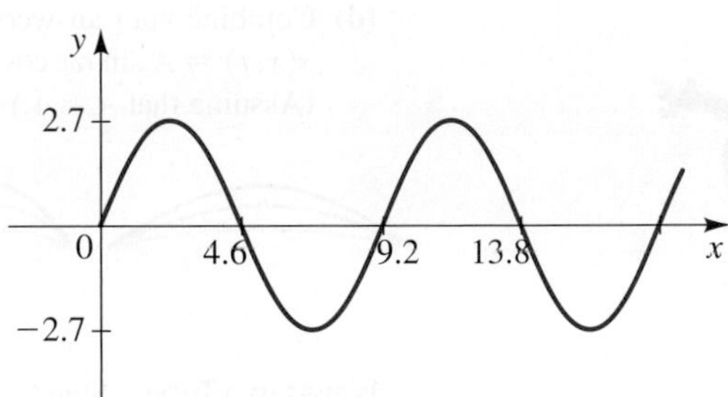

4. Traveling Wave A traveling wave has period $2\pi/3$, amplitude 5, and velocity 0.5.

(a) Find an equation of the wave.

(b) Sketch the graph for $t = 0$, 0.5, 1, 1.5, and 2.

5. Standing Wave A standing wave with amplitude 0.6 is graphed at several times t, as shown in the figure. If the vibration has a frequency of 20 Hz, find an equation of the form

$$y(x, t) = A\sin\alpha x\cos\beta t$$

that models this wave.

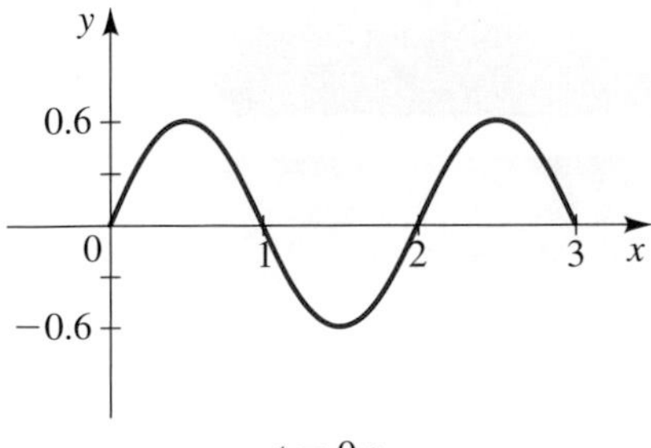

$t = 0$ s

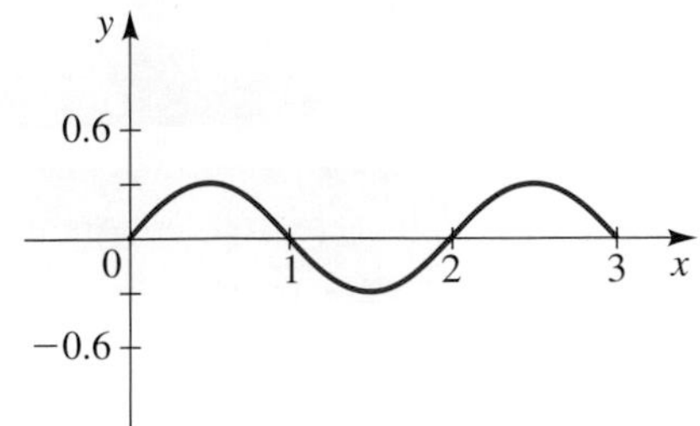

$t = 0.010$ s

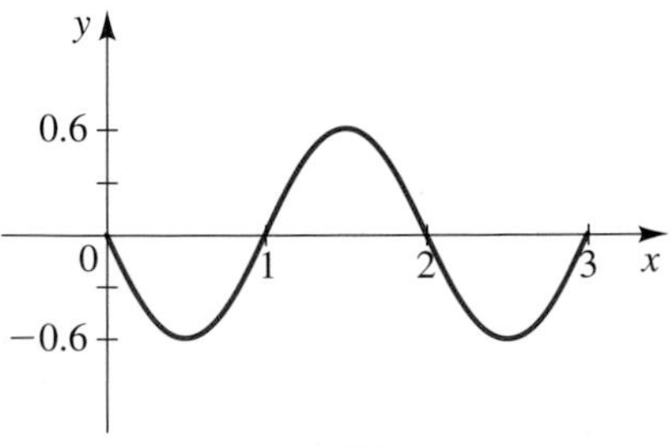

$t = 0.025$ s

6. Standing Wave A standing wave has maximum amplitude 7 and nodes at 0, $\pi/2$, π, $3\pi/2$, 2π, as shown in the figure. Each point that is not a node moves up and down with period 4π. Find a function of the form $y(x, t) = A \sin \alpha x \cos \beta t$ that models this wave.

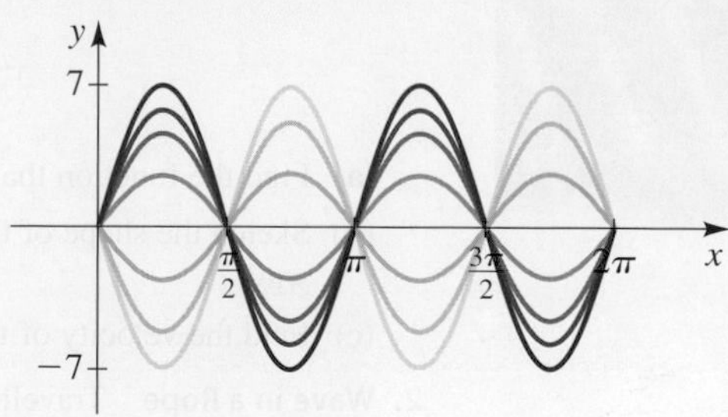

7. Vibrating String When a violin string vibrates, the sound produced results from a combination of standing waves that have evenly placed nodes. The figure illustrates some of the possible standing waves. Let's assume that the string has length π.

(a) For fixed t, the string has the shape of a sine curve $y = A \sin \alpha x$. Find the appropriate value of α for each of the standing waves illustrated.

(b) Do you notice a pattern in the values of α that you found in part (a)? What would the next two values of α be? Sketch rough graphs of the standing waves associated with these new values of α.

(c) Suppose that for fixed t, each point on the string that is not a node vibrates with frequency 440 Hz. Find the value of β for which an equation of the form $y = A \cos \beta t$ would model this motion.

(d) Combine your answers for parts (a) and (c) to find functions of the form $y(x, t) = A \sin \alpha x \cos \beta t$ that model each of the standing waves in the figure. (Assume that $A = 1$.)

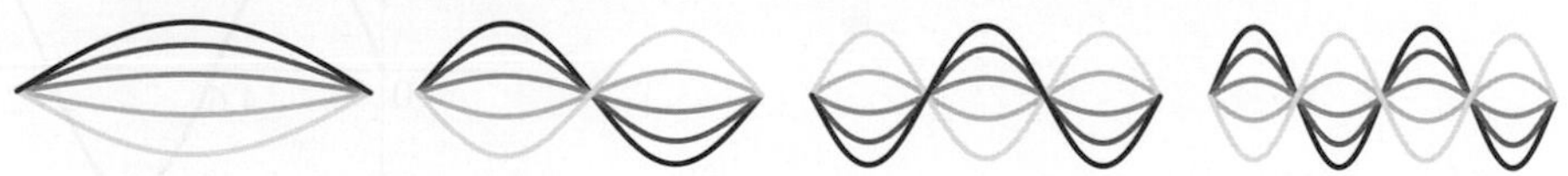

8. Waves in a Tube Standing waves in a violin string must have nodes at the ends of the string because the string is fixed at its endpoints. But this need not be the case with sound waves in a tube (such as a flute or an organ pipe). The figure shows some possible standing waves in a tube.

Suppose that a standing wave in a tube 37.7 m long is modeled by the function

$$y(x, t) = 0.3 \cos \tfrac{1}{2}x \cos 50\pi t$$

Here $y(x, t)$ represents the variation from normal air pressure at the point x meter from the left end of the tube, at time t seconds.

(a) At what points x are the nodes located? Are the endpoints of the tube nodes?

(b) At what frequency does the air vibrate at points that are not nodes?

muratart/Shutterstock.com

8 Polar Coordinates, Parametric Equations, and Vectors

In this chapter we study different ways of describing points and curves in the plane. We are already familiar with *rectangular coordinates* (Section 1.9). Using rectangular coordinates is like describing a location in a city by saying that it's at the corner of 2nd Street and 4th Avenue: such directions would be useful to a person who is driving on the city streets. But we may also describe this same location "as the crow flies;" we can say, for example, that it is 1.5 kilometers northeast of City Hall. These directions would help a drone or a hot-air balloon pilot find the location. That's what we do using *polar coordinates*—we specify the location of a point in the plane by giving its distance and direction from a fixed reference point. We also study *parametric equations*: such equations allow us to track the location of a moving point in the coordinate plane. For example, parametric equations can model the shape of a winding road as well as the location of a car on the road at any time. *Vectors* allow us to describe two quantities simultaneously, such as the magnitude and direction of a force at a given point in the coordinate plane.

8.1 Polar Coordinates

■ Definition of Polar Coordinates ■ Relationship Between Polar and Rectangular Coordinates ■ Polar Equations

In this section we define polar coordinates, and we learn how polar coordinates are related to rectangular coordinates.

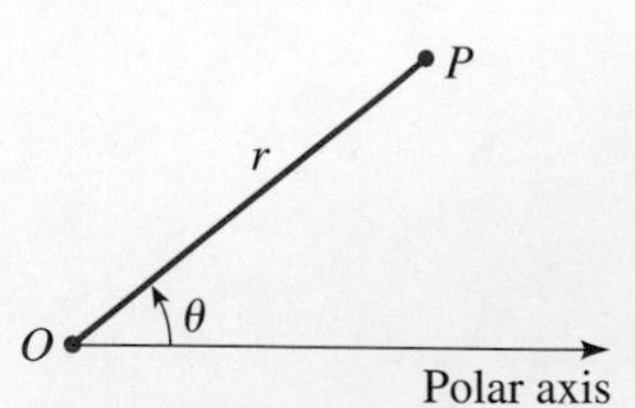

Figure 1

■ Definition of Polar Coordinates

The **polar coordinate system** uses distances and directions to specify the location of a point in the plane. To set up this system, we choose a fixed point O in the plane called the **pole** (or **origin**) and draw from O a ray (half-line) called the **polar axis** as in Figure 1. Then each point P can be assigned polar coordinates $P(r, \theta)$, where

r is the *distance* from O to P

θ is the *angle* between the polar axis and the segment $\overline{OP}$

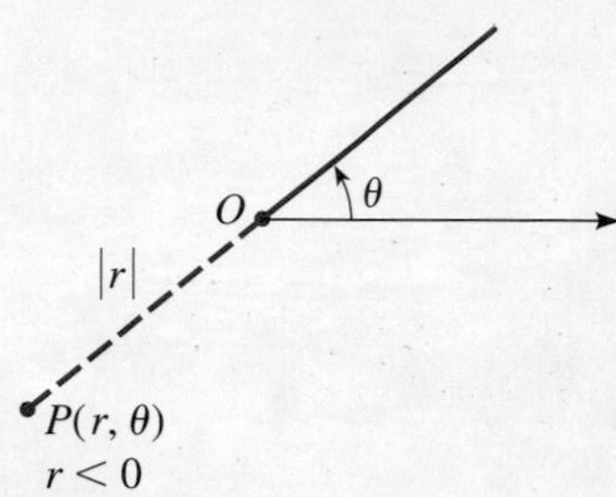

Figure 2

We use the convention that θ is positive if measured in a counterclockwise direction from the polar axis or negative if measured in a clockwise direction. If r is negative, then $P(r, \theta)$ is defined to be the point that lies $|r|$ units from the pole in the direction opposite to that given by θ (see Figure 2).

Example 1 ■ Plotting Points in Polar Coordinates

Plot the points whose polar coordinates are given.

(a) $(1, 3\pi/4)$ **(b)** $(3, -\pi/6)$ **(c)** $(3, 3\pi)$ **(d)** $(-4, \pi/4)$

Solution The points are plotted in Figure 3. The point in part (d) lies 4 units from the origin along the angle $5\pi/4$ because the given value of r is negative.

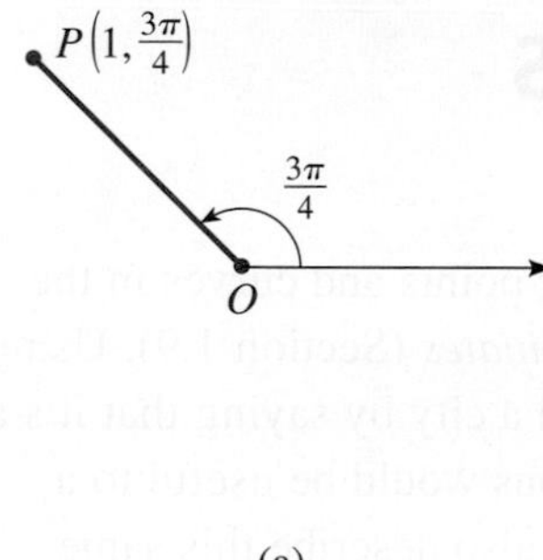

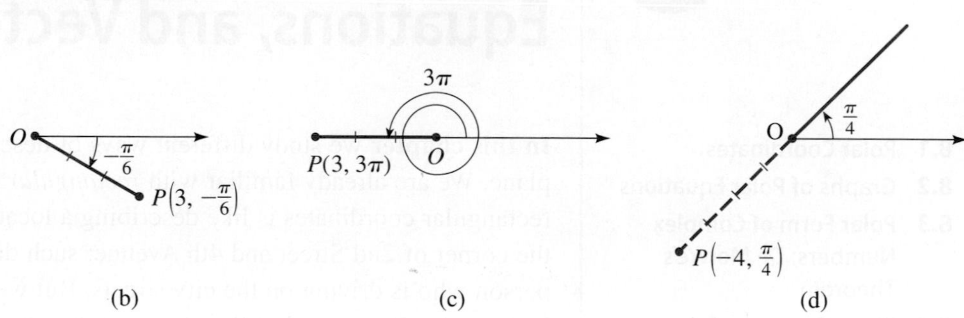

Figure 3

Now Try Exercises 5 and 7 ■

Note that the coordinates (r, θ) and $(-r, \theta + \pi)$ represent the same point, as shown in Figure 4. Moreover, because the angles $\theta + 2n\pi$ (where n is any integer) all have the

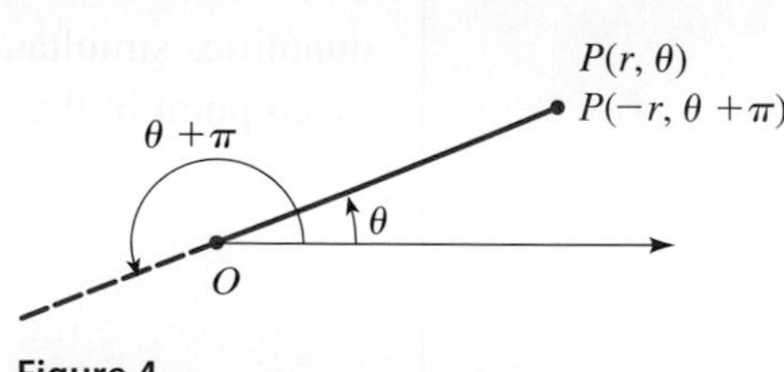

Figure 4

same terminal side as the angle θ, each point in the plane has infinitely many representations in polar coordinates. In fact, any point $P(r, \theta)$ can also be represented by

$$P(r, \theta + 2n\pi) \qquad \text{and} \qquad P(-r, \theta + (2n + 1)\pi)$$

for any integer n.

Example 2 ■ Different Polar Coordinates for the Same Point

(a) Graph the point with polar coordinates $P(2, \pi/3)$.

(b) Find two other polar coordinate representations of P with $r > 0$ and two with $r < 0$.

Solution

(a) The graph is shown in Figure 5(a).

(b) Other representations with $r > 0$ are

$$\left(2, \frac{\pi}{3} + 2\pi\right) = \left(2, \frac{7\pi}{3}\right) \qquad \text{Add } 2\pi \text{ to } \theta$$

$$\left(2, \frac{\pi}{3} - 2\pi\right) = \left(2, -\frac{5\pi}{3}\right) \qquad \text{Add } -2\pi \text{ to } \theta$$

Other representations with $r < 0$ are

$$\left(-2, \frac{\pi}{3} + \pi\right) = \left(-2, \frac{4\pi}{3}\right) \qquad \text{Replace } r \text{ by } -r \text{ and add } \pi \text{ to } \theta$$

$$\left(-2, \frac{\pi}{3} - \pi\right) = \left(-2, -\frac{2\pi}{3}\right) \qquad \text{Replace } r \text{ by } -r \text{ and add } -\pi \text{ to } \theta$$

The graphs in Figure 5 explain why these coordinates represent the same point.

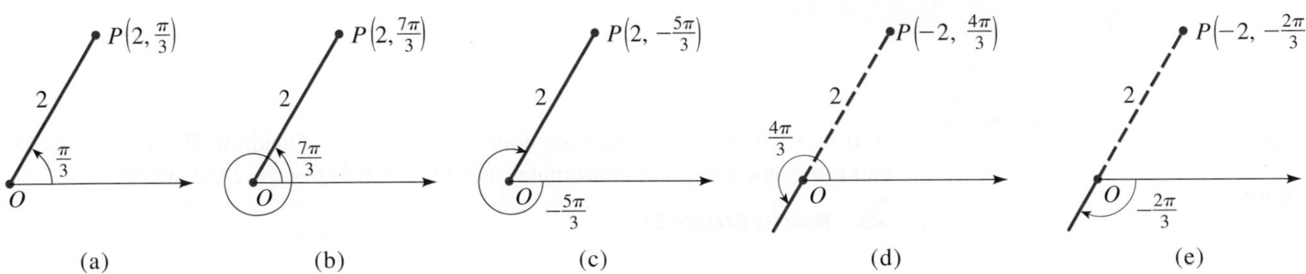

Figure 5

Now Try Exercise 11 ■

■ Relationship Between Polar and Rectangular Coordinates

Situations often arise in which we need to consider polar and rectangular coordinates simultaneously. The connection between the two systems is illustrated in Figure 6 (on the next page), where the polar axis coincides with the positive x-axis. The formulas in the box are obtained from the figure using the definitions of the trigonometric functions and the Pythagorean Theorem. (Although we have pictured the case in which $r > 0$ and θ is acute, the formulas hold for any angle θ and for any value of r.)

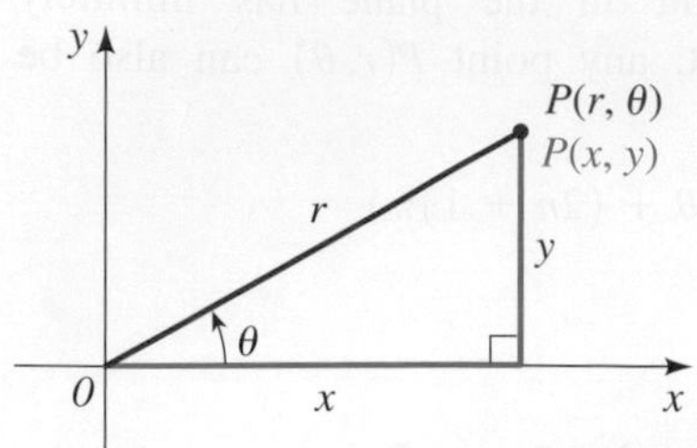

Figure 6

Relationship Between Polar and Rectangular Coordinates

1. To change from polar to rectangular coordinates, use the formulas

$$x = r\cos\theta \qquad \text{and} \qquad y = r\sin\theta$$

2. To change from rectangular to polar coordinates, use the formulas

$$r^2 = x^2 + y^2 \qquad \text{and} \qquad \tan\theta = \frac{y}{x} \quad (x \neq 0)$$

Example 3 ■ Converting Polar Coordinates to Rectangular Coordinates

Find rectangular coordinates for the point that has polar coordinates $(4, 2\pi/3)$.

Solution Since $r = 4$ and $\theta = 2\pi/3$, we have

$$x = r\cos\theta = 4\cos\frac{2\pi}{3} = 4\cdot\left(-\frac{1}{2}\right) = -2$$

$$y = r\sin\theta = 4\sin\frac{2\pi}{3} = 4\cdot\frac{\sqrt{3}}{2} = 2\sqrt{3}$$

Thus the point has rectangular coordinates $(-2, 2\sqrt{3})$.

 **Now Try Exercise 29** ■

Example 4 ■ Converting Rectangular Coordinates to Polar Coordinates

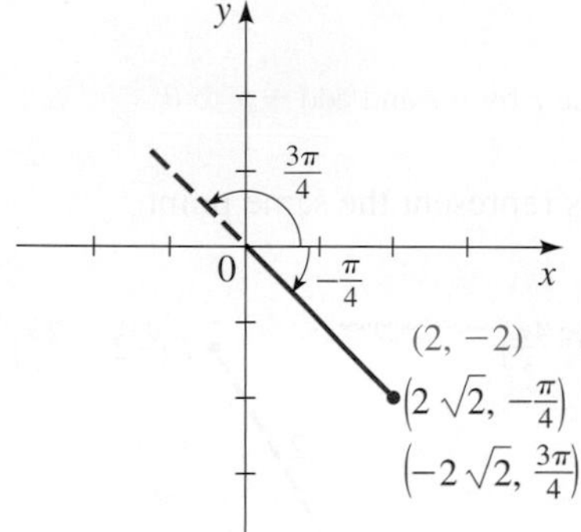

Figure 7

Find polar coordinates for the point that has rectangular coordinates $(2, -2)$.

Solution Using $x = 2$, $y = -2$, we get

$$r^2 = x^2 + y^2 = 2^2 + (-2)^2 = 8$$

so $r = 2\sqrt{2}$ or $-2\sqrt{2}$. Also

$$\tan\theta = \frac{y}{x} = \frac{-2}{2} = -1$$

so $\theta = 3\pi/4$ or $-\pi/4$. Since the point $(2, -2)$ lies in Quadrant IV (see Figure 7), we can represent it in polar coordinates as $(2\sqrt{2}, -\pi/4)$ or $(-2\sqrt{2}, 3\pi/4)$.

 **Now Try Exercise 37** ■

Discovery Project ■ Mapping the World

© maodoltee/Shutterstock.com

In the *Focus on Modeling* following Chapter 6 we learned how surveyors can make a map of a city or town. But mapping the whole world introduces a new difficulty. How is it possible to represent a *spherical* world on a *flat* map? This challenge was faced by Renaissance explorers and their mapmakers, who developed several ingenious solutions. In this project we see how polar coordinates and trigonometry can help us make a map of the whole world on a flat sheet of paper. You can find the project at **www.stewartmath.com.**

RaiDztor/Shutterstock.com

Note that the equations relating polar and rectangular coordinates do not uniquely determine r or θ. When we use these equations to find the polar coordinates of a point, we must be careful that the values we choose for r and θ give us a point in the correct quadrant, as we did in Example 4.

■ Polar Equations

In Examples 3 and 4 we converted points from one coordinate system to the other. Now we consider the same problem for equations. A **polar equation** is an equation in the polar coordinates r and θ; similarly, a **rectangular equation** is an equation in the rectangular coordinates x and y.

Example 5 ■ Converting an Equation from Rectangular Coordinates to Polar Coordinates

Express the equation $x^2 = 4y$ in polar coordinates.

Solution We use the formulas $x = r\cos\theta$ and $y = r\sin\theta$.

$$x^2 = 4y \quad \text{Rectangular equation}$$
$$(r\cos\theta)^2 = 4(r\sin\theta) \quad \text{Substitute } x = r\cos\theta,\ y = r\sin\theta$$
$$r^2\cos^2\theta = 4r\sin\theta \quad \text{Expand}$$
$$r = 4\frac{\sin\theta}{\cos^2\theta} \quad \text{Divide by } r\cos^2\theta$$
$$r = 4\sec\theta\tan\theta \quad \text{Simplify}$$

 Now Try Exercise 47 ■

As Example 5 shows, converting an equation from rectangular coordinates to polar coordinates is straightforward: Just replace x by $r\cos\theta$ and y by $r\sin\theta$, and then simplify. But converting an equation from polar to rectangular form often requires more thought.

Example 6 ■ Converting Equations from Polar Coordinates to Rectangular Coordinates

Express each polar equation in rectangular coordinates. If possible, determine the graph of the equation from its rectangular form.

(a) $r = 5\sec\theta$ **(b)** $r = 2\sin\theta$ **(c)** $r = 2 + 2\cos\theta$

Solution

(a) Since $\sec\theta = 1/\cos\theta$, we multiply both sides by $\cos\theta$.

$$r = 5\sec\theta \quad \text{Polar equation}$$
$$r\cos\theta = 5 \quad \text{Multiply by } \cos\theta$$
$$x = 5 \quad \text{Substitute } x = r\cos\theta$$

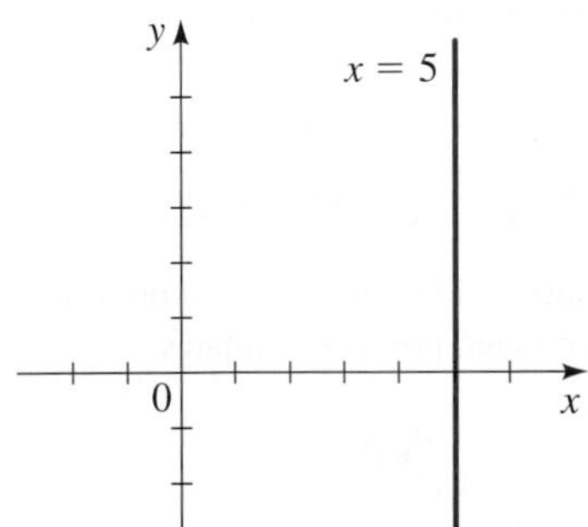

Figure 8

The graph of $x = 5$ is the vertical line in Figure 8.

(b) We multiply both sides of the equation by r, because then we can use the formulas $r^2 = x^2 + y^2$ and $r\sin\theta = y$.

$$r = 2\sin\theta \quad \text{Polar equation}$$
$$r^2 = 2r\sin\theta \quad \text{Multiply by } r$$
$$x^2 + y^2 = 2y \quad r^2 = x^2 + y^2 \text{ and } r\sin\theta = y$$
$$x^2 + y^2 - 2y = 0 \quad \text{Subtract } 2y$$
$$x^2 + (y-1)^2 = 1 \quad \text{Complete the square in } y$$

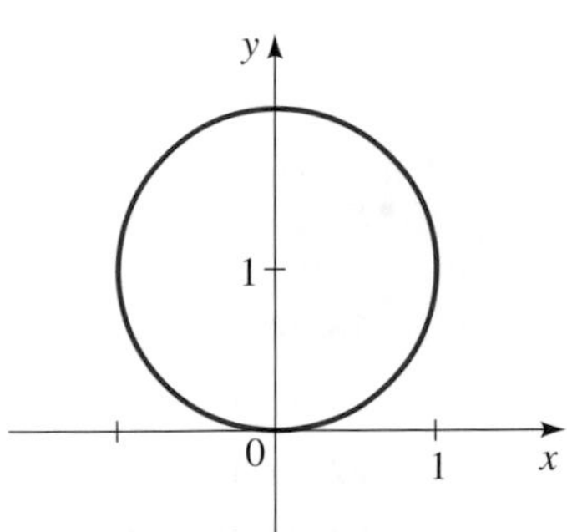

Figure 9

This is the equation of a circle of radius 1 centered at the point $(0, 1)$. It is graphed in Figure 9.

(c) We first multiply both sides of the equation by r:

$$r = 2 + 2\cos\theta$$
$$r^2 = 2r + 2r\cos\theta$$

Using $r^2 = x^2 + y^2$ and $x = r\cos\theta$, we can convert two terms in the equation into rectangular coordinates, but eliminating the remaining r requires more work.

$$x^2 + y^2 = 2r + 2x \qquad r^2 = x^2 + y^2 \text{ and } r\cos\theta = x$$
$$x^2 + y^2 - 2x = 2r \qquad \text{Subtract } 2x$$
$$(x^2 + y^2 - 2x)^2 = 4r^2 \qquad \text{Square both sides}$$
$$(x^2 + y^2 - 2x)^2 = 4(x^2 + y^2) \qquad r^2 = x^2 + y^2$$

In this case the rectangular equation looks more complicated than the polar equation. Although we cannot easily determine the graph of the equation from its rectangular form, we will see in the next section how to graph it using the polar equation.

 **Now Try Exercises 57, 59, and 61** ■

8.1 | Exercises

Concepts

1. We can describe the location of a point in the plane using different ________ systems. The point P shown in the figure has rectangular coordinates (,) and polar coordinates (,).

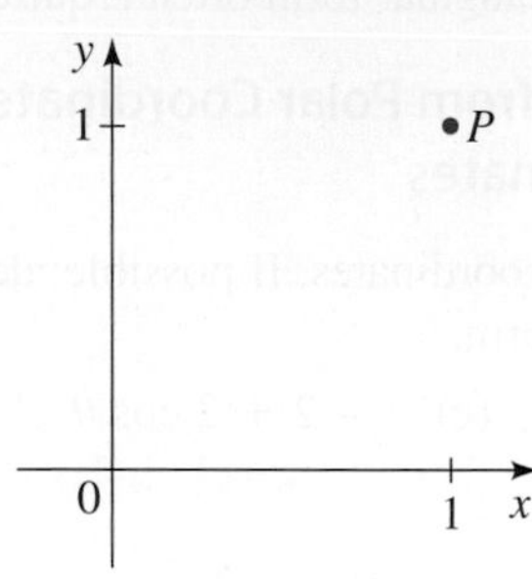

2. Let P be a point in the plane.

(a) If P has polar coordinates (r, θ), then it has rectangular coordinates (x, y) where $x =$ ________ and $y =$ ________.

(b) If P has rectangular coordinates (x, y), then it has polar coordinates (r, θ) where $r^2 =$ ________ and $\tan\theta =$ ________.

3–4 ■ *Yes or No*? If *No*, give a reason.

3. Do the polar coordinates $(2, \pi/6)$ and $(-2, 7\pi/6)$ represent the same point?

4. Do the equations relating polar and rectangular coordinates uniquely determine r and θ?

Skills

5–10 ■ **Plotting Points in Polar Coordinates** Plot the point that has the given polar coordinates.

5. $\left(2, \dfrac{\pi}{2}\right)$ **6.** $(1, 0)$ **7.** $\left(3, -\dfrac{\pi}{4}\right)$

8. $\left(4, -\dfrac{5\pi}{6}\right)$ **9.** $(-2, 4\pi/3)$ **10.** $\left(-3, \dfrac{7\pi}{3}\right)$

11–16 ■ **Different Polar Coordinates for the Same Point** Plot the point that has the given polar coordinates. Then give two other polar coordinate representations of the point, one with $r < 0$ and the other with $r > 0$.

11. $(3, \pi/2)$ **12.** $(2, 3\pi/4)$ **13.** $(-1, 7\pi/6)$

14. $(-2, -\pi/3)$ **15.** $(-5, 0)$ **16.** $(3, 1)$

17–24 ■ **Points in Polar Coordinates** Determine which point in the figure—P, Q, R, or S—has the given polar coordinates.

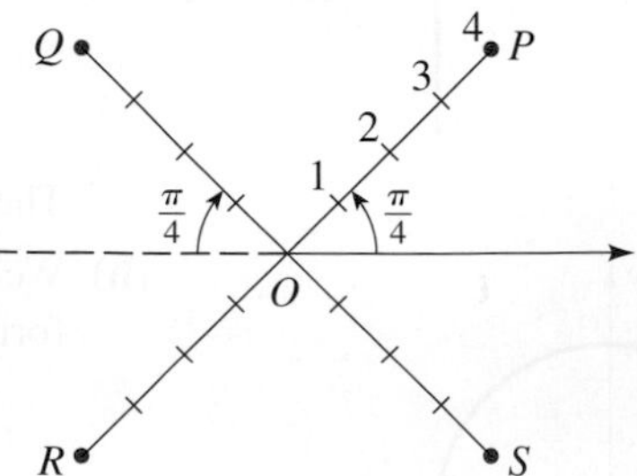

17. $(4, 3\pi/4)$ **18.** $(4, -3\pi/4)$

19. $(-4, -\pi/4)$ **20.** $(-4, 13\pi/4)$

21. $(4, -23\pi/4)$ **22.** $(-4, 23\pi/4)$

23. $(-4, 101\pi/4)$ **24.** $(4, 103\pi/4)$

25–26 ■ Rectangular Coordinates to Polar Coordinates A point is graphed in rectangular form. Find polar coordinates for the point, with $r > 0$ and $0 < \theta < 2\pi$.

25. **26.**

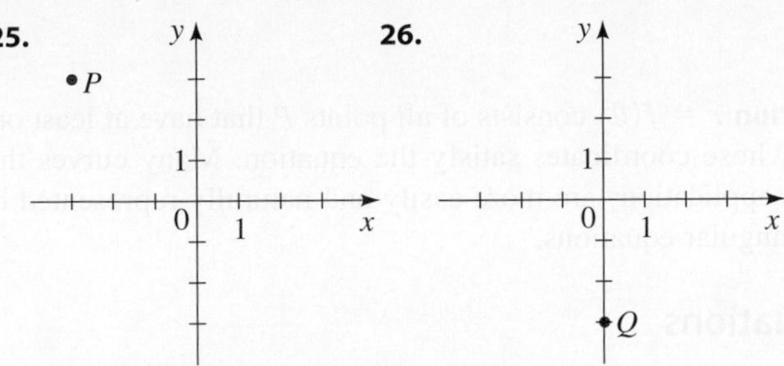

27–28 ■ Polar Coordinates to Rectangular Coordinates A point is graphed in polar form. Find its rectangular coordinates.

27.

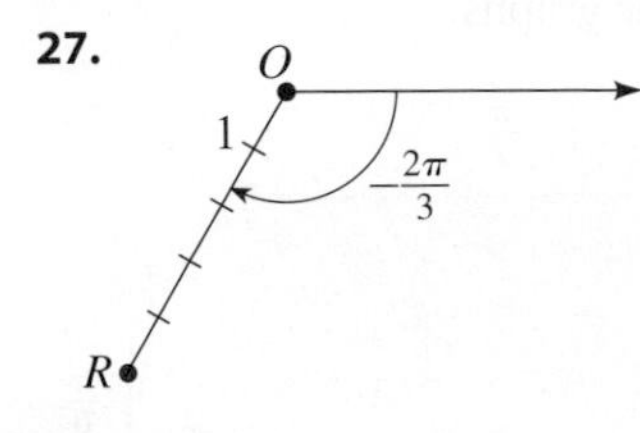

28.

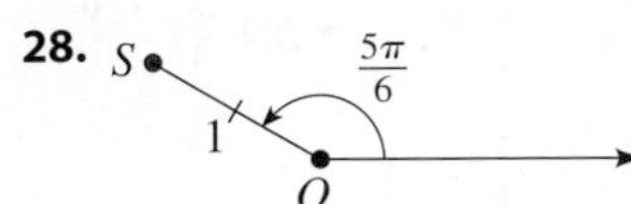

29–36 ■ Polar Coordinates to Rectangular Coordinates Find the rectangular coordinates for the point whose polar coordinates are given.

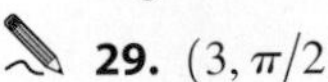

29. $(3, \pi/2)$ **30.** $(6, 2\pi/3)$

31. $(\sqrt{2}, -\pi/4)$ **32.** $(-1, 5\pi/2)$

33. $(5, 5\pi)$ **34.** $(0, 13\pi)$

35. $(\sqrt{3}, -\pi/6)$ **36.** $(2\sqrt{2}, -3\pi/4)$

37–44 ■ Rectangular Coordinates to Polar Coordinates Convert the rectangular coordinates to polar coordinates with $r > 0$ and $0 \le \theta < 2\pi$.

37. $(-1, 1)$ **38.** $(3\sqrt{3}, -3)$

39. $(\sqrt{8}, \sqrt{8})$ **40.** $(-\sqrt{6}, -\sqrt{2})$

41. $(3, 4)$ **42.** $(1, -2)$

43. $(-6, 0)$ **44.** $(0, -\sqrt{3})$

45–52 ■ Rectangular Equations to Polar Equations Convert the equation to polar form.

45. $x = y$ **46.** $x^2 + y^2 = 9$

47. $x = y^2$ **48.** $y = 5$

49. $x = 4$ **50.** $x^2 - y^2 = 1$

51. $x^2 + y^2 = y$ **52.** $(x^2 + y^2)^{3/2} = 6xy$

53–72 ■ Polar Equations to Rectangular Equations Convert the polar equation to rectangular coordinates.

53. $r = 7$ **54.** $r = -3$

55. $\theta = -\dfrac{\pi}{2}$ **56.** $\theta = \pi$

57. $r \cos\theta = 6$ **58.** $r = 2\csc\theta$

59. $r = 4\sin\theta$ **60.** $r = 6\cos\theta$

61. $r = 1 + \cos\theta$ **62.** $r = 3(1 - \sin\theta)$

63. $r = 1 + 2\sin\theta$ **64.** $r = 2 - \cos\theta$

65. $r = \dfrac{1}{\sin\theta - \cos\theta}$ **66.** $r = \dfrac{1}{1 + \sin\theta}$

67. $r = \dfrac{4}{1 + 2\sin\theta}$ **68.** $r = \dfrac{2}{1 - \cos\theta}$

69. $r^2 = \tan\theta$ **70.** $r^2 = \sin 2\theta$

71. $\sec\theta = 2$ **72.** $\cos 2\theta = 1$

■ Discuss ■ Discover ■ Prove ■ Write

73. Discuss ■ Prove: The Distance Formula in Polar Coordinates

(a) Use the Law of Cosines to prove that the distance between the polar points (r_1, θ_1) and (r_2, θ_2) is

$$d = \sqrt{r_1^2 + r_2^2 - 2r_1r_2\cos(\theta_2 - \theta_1)}$$

(b) Find the distance between the points whose polar coordinates are $(3, 3\pi/4)$ and $(-1, 7\pi/6)$, using the formula from part (a).

(c) Convert the points in part (b) to rectangular coordinates. Find the distance between them, using the usual Distance Formula. Do you get the same answer?

74. Discuss: Different Coordinate Systems Certain curves are more naturally described in one coordinate system than in another. In each of the following situations, which coordinate system would be appropriate: rectangular or polar? Give reasons to support your answer.

(a) You need to give driving directions to your home to a friend.

(b) You need to give directions to your home to a homing pigeon.

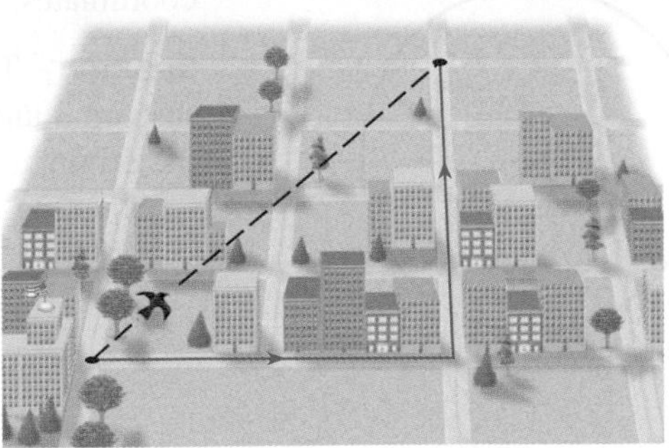

8.2 Graphs of Polar Equations

■ Graphing Polar Equations ■ Symmetry ■ Graphing Polar Equations with Graphing Devices

The **graph of a polar equation** $r = f(\theta)$ consists of all points P that have at least one polar representation (r, θ) whose coordinates satisfy the equation. Many curves that arise in mathematics and its applications are more easily and naturally represented by polar equations than by rectangular equations.

■ Graphing Polar Equations

A rectangular grid is helpful for plotting points in rectangular coordinates (see Figure 1(a)). To plot points in polar coordinates, it is convenient to use a grid consisting of circles centered at the pole and rays emanating from the pole, as shown in Figure 1(b). We will use such grids to help us sketch polar graphs.

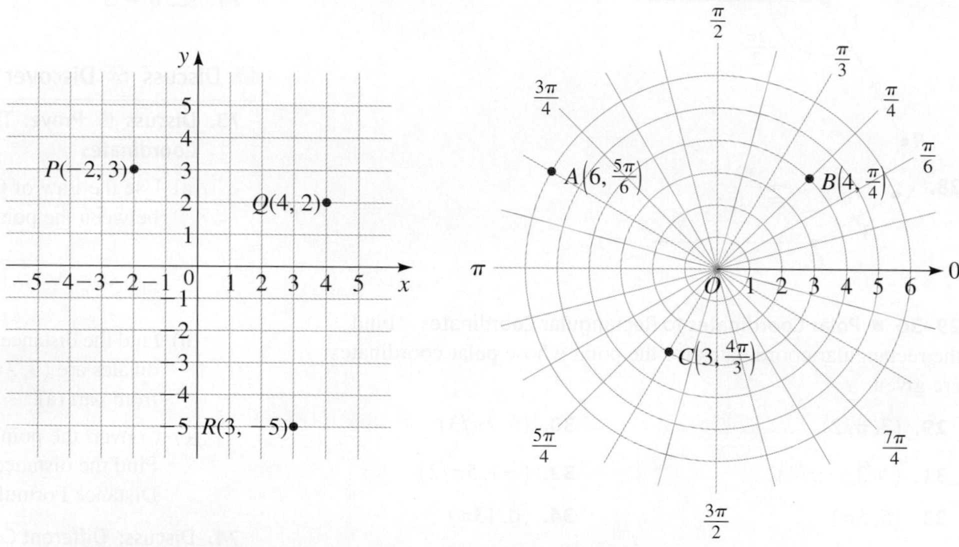

Figure 1 (a) Grid for rectangular coordinates (b) Grid for polar coordinates

In Examples 1 and 2 we see that circles centered at the origin and lines that pass through the origin have particularly simple equations in polar coordinates.

Example 1 ■ Sketching the Graph of a Polar Equation

Sketch a graph of the equation $r = 3$, and express the equation in rectangular coordinates.

Solution The graph consists of all points whose r-coordinate is 3, that is, all points that are 3 units away from the origin. So the graph is a circle of radius 3 centered at the origin, as shown in Figure 2.

Squaring both sides of the equation, we get

$$r^2 = 3^2 \qquad \text{Square both sides}$$

$$x^2 + y^2 = 9 \qquad \text{Substitute } r^2 = x^2 + y^2$$

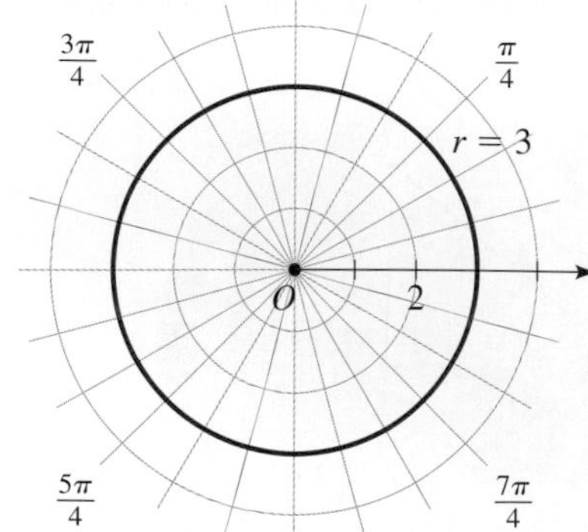

Figure 2

So the equivalent equation in rectangular coordinates is $x^2 + y^2 = 9$.

Now Try Exercise 17 ■

In general, the graph of the equation $r = a$ is a circle of radius $|a|$ centered at the origin. Squaring both sides of this equation, we see that the equivalent equation in rectangular coordinates is $x^2 + y^2 = a^2$.

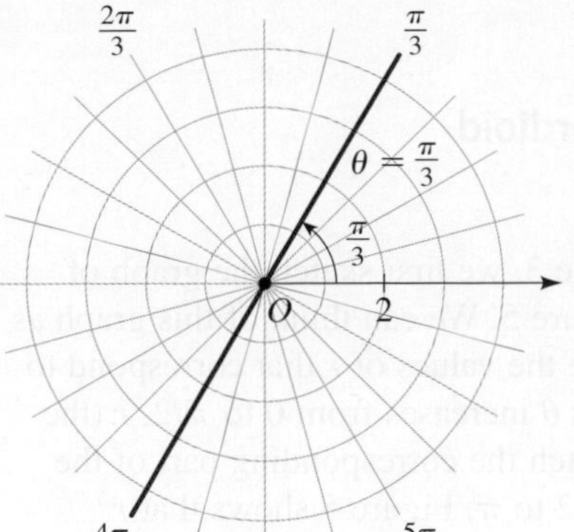

Figure 3

Example 2 ■ Sketching the Graph of a Polar Equation

Sketch a graph of the equation $\theta = \pi/3$, and express the equation in rectangular coordinates.

Solution The graph consists of all points whose θ-coordinate is $\pi/3$. This is the straight line that passes through the origin and makes an angle of $\pi/3$ with the polar axis (see Figure 3). Note that the points $(r, \pi/3)$ on the line with $r > 0$ lie in Quadrant I, whereas those with $r < 0$ lie in Quadrant III. If the point (x, y) lies on this line, then

$$\frac{y}{x} = \tan\theta = \tan\frac{\pi}{3} = \sqrt{3}$$

Thus the rectangular equation of this line is $y = \sqrt{3}x$.

Now Try Exercise 19 ■

To sketch a polar curve whose graph isn't as obvious as the ones in the preceding examples, we plot points calculated for sufficiently many values of θ and then join them in a continuous curve. (This is what we did when we first learned to graph equations in rectangular coordinates.)

Example 3 ■ Sketching the Graph of a Polar Equation

Sketch a graph of the polar equation $r = 2\sin\theta$.

Recall that the functions sine and cosine are periodic with period 2π.

Solution We first use the equation to determine the polar coordinates of several points on the curve. The results are shown in the following table.

θ	0	$\pi/6$	$\pi/4$	$\pi/3$	$\pi/2$	$2\pi/3$	$3\pi/4$	$5\pi/6$	π
$r = 2\sin\theta$	0	1	$\sqrt{2}$	$\sqrt{3}$	2	$\sqrt{3}$	$\sqrt{2}$	1	0

We plot these points in Figure 4 and then join them to sketch the curve. The graph appears to be a circle. We have used values of θ only between 0 and π because the same points (this time expressed with negative r-coordinates) would be obtained if we allowed θ to range from π to 2π.

The polar equation $r = 2\sin\theta$ in rectangular coordinates is

$$x^2 + (y - 1)^2 = 1$$

[see Example 8.1.6(b)]. From the rectangular form of the equation we see that the graph is a circle of radius 1 centered at $(0, 1)$.

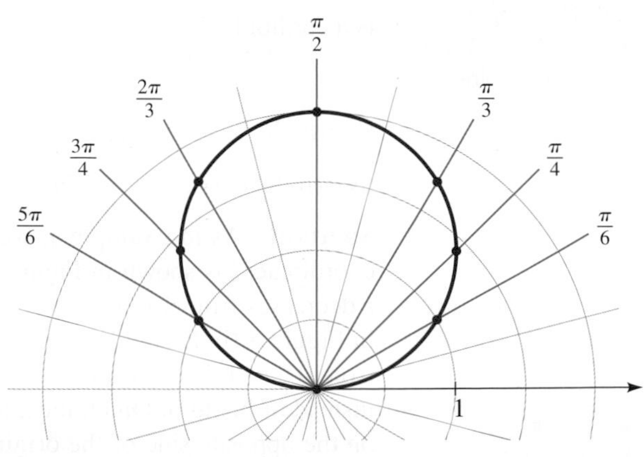

Figure 4 | $r = 2\sin\theta$

Now Try Exercise 21 ■

In general, the graph of an equation of the form

$$r = 2a \sin \theta \qquad \text{or} \qquad r = 2a \cos \theta$$

is a **circle** with radius $|a|$ centered at the points with polar coordinates $(a, \pi/2)$ and $(a, 0)$, respectively.

Example 4 ■ Sketching the Graph of a Cardioid

Sketch a graph of $r = 2 + 2 \cos \theta$.

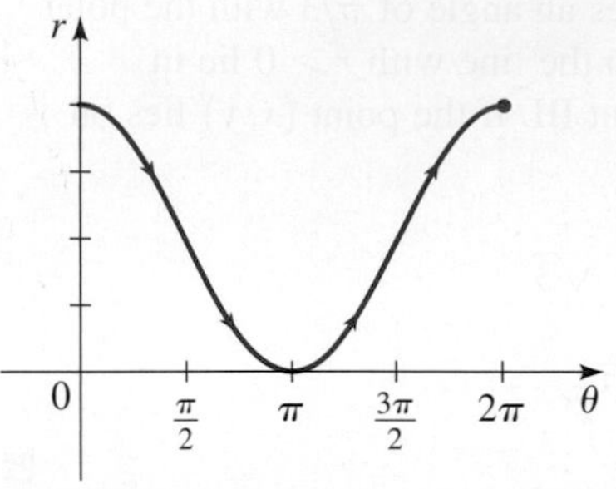

Figure 5 | $r = 2 + 2 \cos \theta$ sketched in rectangular coordinates

Solution Instead of plotting points as in Example 3, we first sketch the graph of $r = 2 + 2 \cos \theta$ in *rectangular* coordinates in Figure 5. We can think of this graph as a table of values that enables us to read at a glance the values of r that correspond to increasing values of θ. For instance, we see that as θ increases from 0 to $\pi/2$, r (the distance from O) decreases from 4 to 2, so we sketch the corresponding part of the polar graph in Figure 6(a). As θ increases from $\pi/2$ to π, Figure 5 shows that r decreases from 2 to 0, so we sketch the next part of the graph as in Figure 6(b). As θ increases from π to $3\pi/2$, r increases from 0 to 2, as shown in part (c). Finally, as θ increases from $3\pi/2$ to 2π, r increases from 2 to 4, as shown in part (d). If we let θ increase beyond 2π or decrease beyond 0, we would simply retrace our path. Combining the portions of the graph from parts (a) through (d) of Figure 6, we sketch the complete graph in part (e).

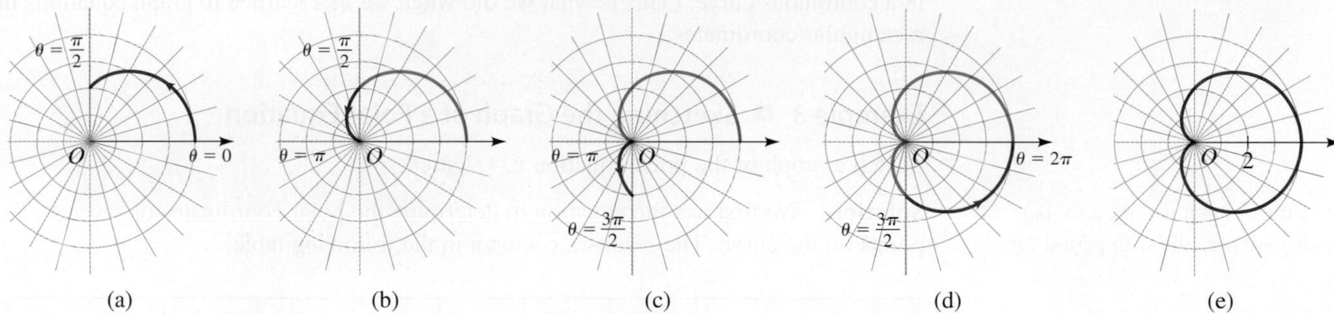

Figure 6 | Steps in sketching $r = 2 + 2 \cos \theta$ in polar coordinates

Now Try Exercise 25 ■

The polar equation $r = 2 + 2 \cos \theta$ in rectangular coordinates is

$$(x^2 + y^2 - 2x)^2 = 4(x^2 + y^2)$$

[see Example 8.1.6(c)]. The simpler form of the polar equation shows that it is more natural to describe cardioids using polar coordinates.

The curve in Figure 6 is called a **cardioid** because it is heart-shaped. In general, the graph of any equation of the form

$$r = a(1 \pm \cos \theta) \qquad \text{or} \qquad r = a(1 \pm \sin \theta)$$

is a cardioid.

Example 5 ■ Sketching the Graph of a Four-Leaved Rose

Sketch the curve $r = \cos 2\theta$.

Solution As in Example 4, we first sketch the graph of $r = \cos 2\theta$ in *rectangular* coordinates, as shown in Figure 7. As θ increases from 0 to $\pi/4$, Figure 7 shows that r decreases from 1 to 0, so we draw the corresponding portion of the polar curve in Figure 8 (indicated by ①). As θ increases from $\pi/4$ to $\pi/2$, the value of r goes from 0 to -1. This means that the distance from the origin increases from 0 to 1, but instead of being in Quadrant I, this portion of the polar curve (indicated by ②) lies on the opposite side of the origin in Quadrant III. The remainder of the curve is drawn in a similar fashion, with the arrows and numbers indicating the order in which the portions are traced out. The resulting curve has four petals and is called a **four-leaved rose.**

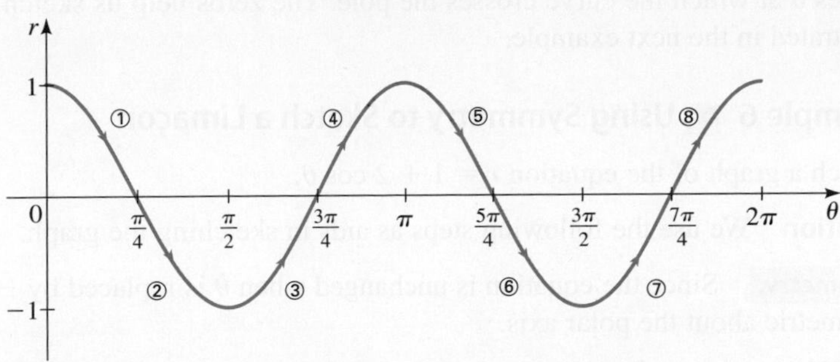

Figure 7 | Graph of $r = \cos 2\theta$ sketched in rectangular coordinates

Figure 8 | Four-leaved rose $r = \cos 2\theta$ sketched in polar coordinates

 **Now Try Exercise 29** ■

In general, the graph of an equation of the form

$$r = a \cos n\theta \qquad \text{or} \qquad r = a \sin n\theta$$

is an ***n*-leaved rose** if n is odd or a $2n$-leaved rose if n is even (as in Example 5).

■ Symmetry

In graphing a polar equation, it's often helpful to take advantage of symmetry. We list three tests for symmetry.

Types of Symmetry

Symmetry	Test	Graph	Property of Graph
With respect to the polar axis	The polar equation is unchanged if we replace θ by $-\theta$.	(r, θ), θ, O, $-\theta$, $(r, -\theta)$	Graph is unchanged when reflected about the polar axis. See Figures 2, 6(e), and 8.
With respect to the pole	The polar equation is unchanged if we replace r by $-r$ or θ by $\theta + \pi$.	(r, θ), O, $(-r, \theta)$	Graph is unchanged when rotated π radians about the pole. See Figure 8.
With respect to the line $\theta = \pi/2$	The polar equation is unchanged if we replace θ by $\pi - \theta$.	$\theta = \frac{\pi}{2}$, $(r, \pi - \theta)$, (r, θ), $\pi - \theta$, θ, O	Graph is unchanged when reflected about the vertical line $\theta = \pi/2$. See Figures 4 and 8.

In rectangular coordinates the zeros of the function $y = f(x)$ correspond to the x-intercepts of the graph. In polar coordinates the zeros of the function $r = f(\theta)$ are the angles θ at which the curve crosses the pole. The zeros help us sketch the graph, as is illustrated in the next example.

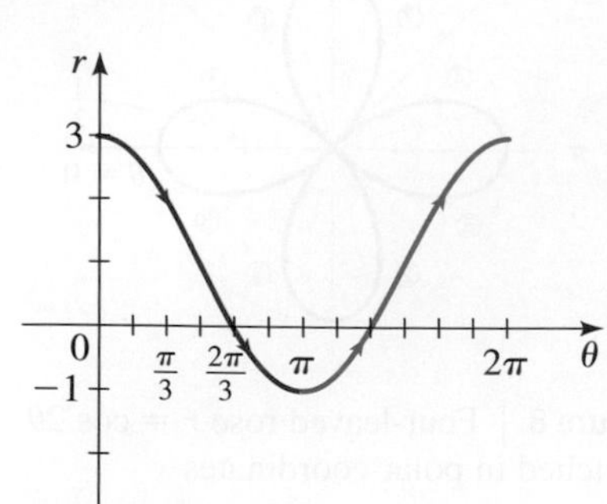

Figure 9

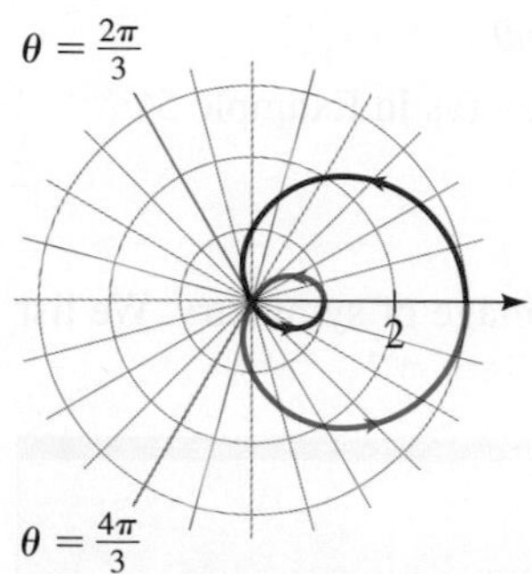

Figure 10 | $r = 1 + 2\cos\theta$

Example 6 ■ Using Symmetry to Sketch a Limaçon

Sketch a graph of the equation $r = 1 + 2\cos\theta$.

Solution We use the following steps as aids in sketching the graph.

Symmetry. Since the equation is unchanged when θ is replaced by $-\theta$, the graph is symmetric about the polar axis.

Zeros. To find the zeros, we solve

$$0 = 1 + 2\cos\theta$$

$$\cos\theta = -\frac{1}{2}$$

$$\theta = \frac{2\pi}{3}, \frac{4\pi}{3}$$

Table of values. As in Example 4, we sketch the graph of $r = 1 + 2\cos\theta$ in *rectangular* coordinates to serve as a table of values (Figure 9).

Now we sketch the polar graph of $r = 1 + 2\cos\theta$ from $\theta = 0$ to $\theta = \pi$ and then use symmetry to complete the graph, as in Figure 10.

 **Now Try Exercise 37** ■

The curve in Figure 10 is called a **limaçon**, after the Middle French word for snail. In general, the graph of an equation of the form

$$r = a \pm b\cos\theta \quad \text{or} \quad r = a \pm b\sin\theta$$

is a limaçon. The shape of the limaçon depends on the relative size of a and b.

■ Graphing Polar Equations with Graphing Devices

Although it's useful to be able to sketch simple polar graphs, we need to use a graphing calculator or computer when the graph is as complicated as the one in Figure 11. Fortunately, most graphing devices are capable of graphing polar equations directly.

Figure 11 | $r = \sin\theta + \sin^3(5\theta/2)$

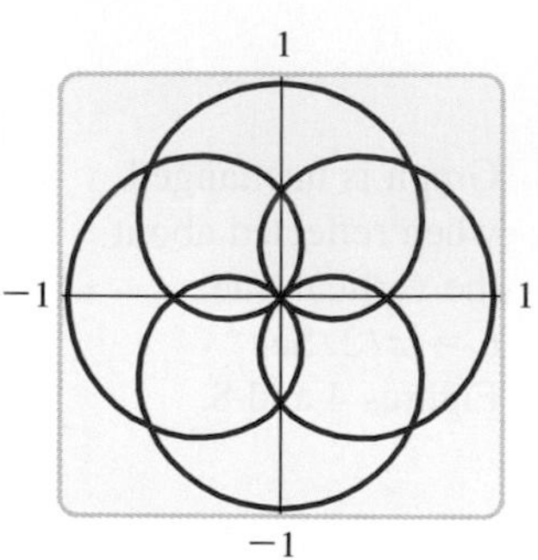

Figure 12 | $r = \cos(2\theta/3)$

Example 7 ■ Drawing the Graph of a Polar Equation

Graph the equation $r = \cos(2\theta/3)$.

Solution We need to determine the domain for θ. So we ask ourselves: How many times must θ go through a complete rotation (2π radians) before the graph starts to repeat itself? The graph repeats itself when the same value of r is obtained at θ and $\theta + 2n\pi$. Thus we need to find an integer n so that

$$\cos\frac{2(\theta + 2n\pi)}{3} = \cos\frac{2\theta}{3}$$

For this equality to hold, $4n\pi/3$ must be a multiple of 2π, and this first happens when $n = 3$. Therefore we obtain the entire graph if we choose values of θ between $\theta = 0$ and $\theta = 0 + 2(3)\pi = 6\pi$. The graph is shown in Figure 12.

 **Now Try Exercise 47** ■

Example 8 ■ A Family of Polar Equations

Graph the family of polar equations $r = 1 + c \sin \theta$ for $c = 3, 2.5, 2, 1.5, 1$. How does the shape of the graph change as c changes?

Solution Figure 13 shows computer-drawn graphs for the given values of c. When $c > 1$, the graph has an inner loop; the loop decreases in size as c decreases. When $c = 1$, the loop disappears, and the graph becomes a cardioid (see Example 4).

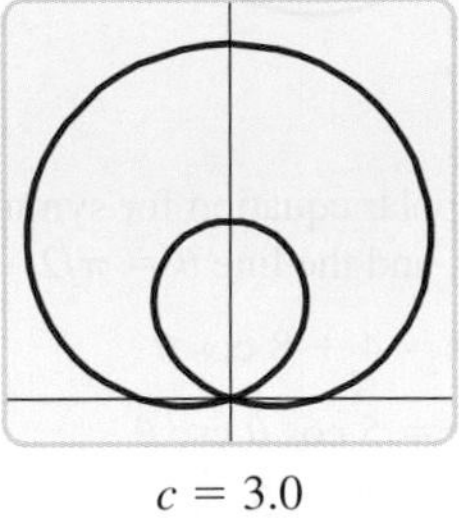
$c = 3.0$

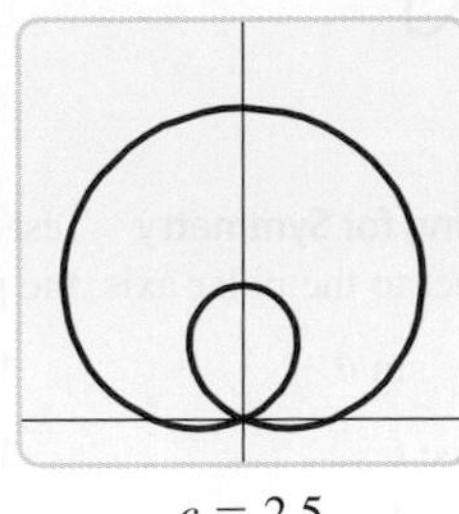
$c = 2.5$

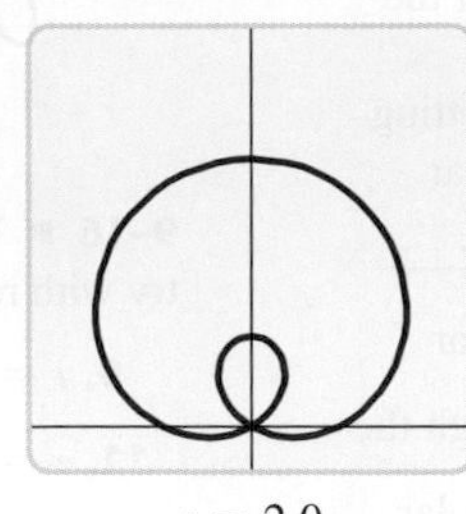
$c = 2.0$

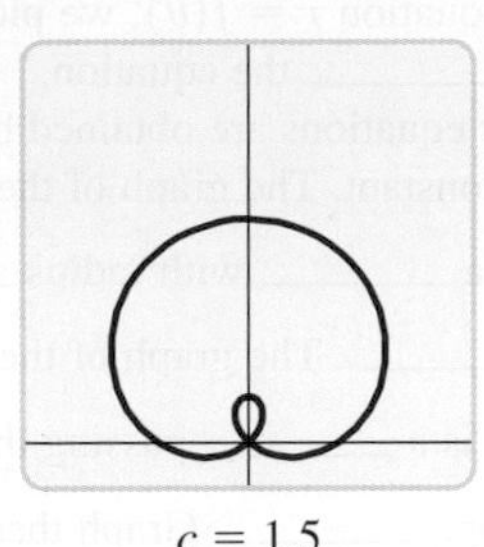
$c = 1.5$

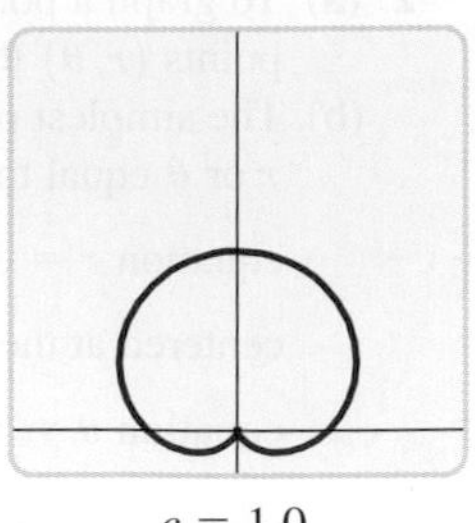
$c = 1.0$

Figure 13 | A family of limaçons $r = 1 + c \sin \theta$ in the viewing rectangle $[-2.5, 2.5]$ by $[-0.5, 4.5]$

Now Try Exercise 51 ■

The box below gives a summary of some of the basic polar graphs used in calculus.

Some Common Polar Curves

Circles and Spiral

$r = a$ circle | $r = a \sin \theta$ circle | $r = a \cos \theta$ circle | $r = a\theta$ spiral

Limaçons

$r = a \pm b \sin \theta$

$r = a \pm b \cos \theta$

$(a > 0, b > 0)$

Orientation depends on the trigonometric function (sine or cosine) and the sign of b.

$a < b$ limaçon with inner loop | $a = b$ cardioid | $b < a < 2b$ dimpled limaçon | $2b \le a$ convex limaçon

Roses

$r = a \sin n\theta$

$r = a \cos n\theta$

n-leaved if n is odd

$2n$-leaved if n is even

$r = a \cos 2\theta$ 4-leaved rose | $r = a \cos 3\theta$ 3-leaved rose | $r = a \cos 4\theta$ 8-leaved rose | $r = a \cos 5\theta$ 5-leaved rose

Lemniscates

Figure-eight-shaped curves

$r^2 = a^2 \sin 2\theta$ lemniscate | $r^2 = a^2 \cos 2\theta$ lemniscate

8.2 | Exercises

Concepts

1. To plot points in polar coordinates, we use a grid consisting of ________ centered at the pole and ________ emanating from the pole.

2. (a) To graph a polar equation $r = f(\theta)$, we plot all the points (r, θ) that ________ the equation.
(b) The simplest polar equations are obtained by setting r or θ equal to a constant. The graph of the polar equation $r = 3$ is a ________ with radius ________ centered at the ________. The graph of the polar equation $\theta = \pi/4$ is a ________ passing through the ________ with slope ________. Graph these polar equations below.

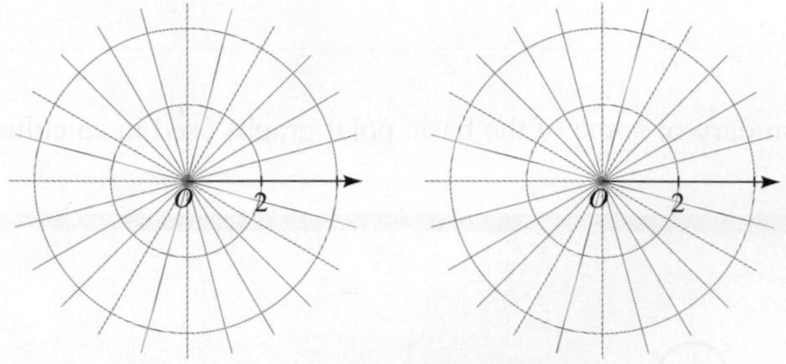

Skills

3–8 ■ Graphs of Polar Equations Match the polar equation with the graphs labeled I–VI. Use the table of common polar curves to help you.

3. $r = 3 \cos \theta$ 4. $r = 3$
5. $r = 2 + 2 \sin \theta$ 6. $r = 1 + 2 \cos \theta$
7. $r = \sin 3\theta$ 8. $r = \sin 4\theta$

I
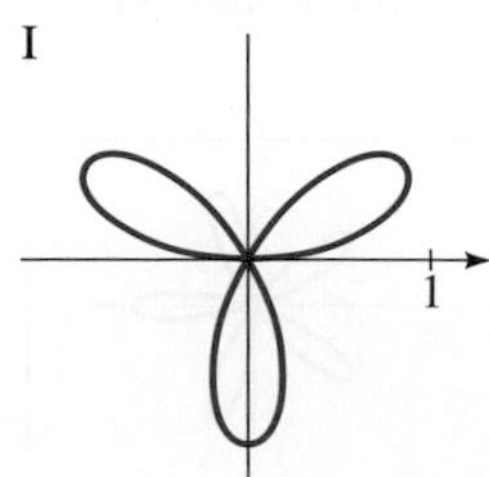

II
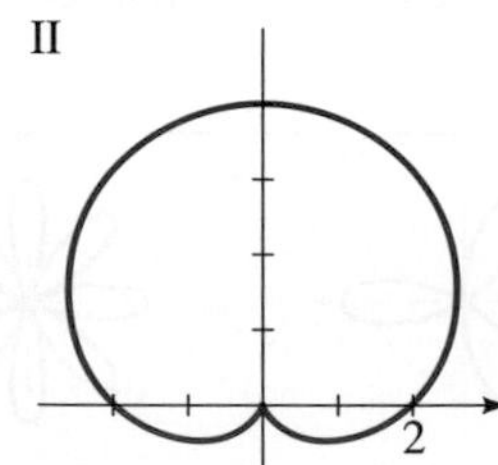

III
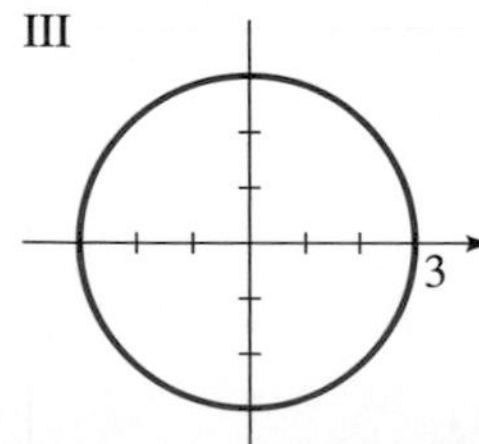

IV
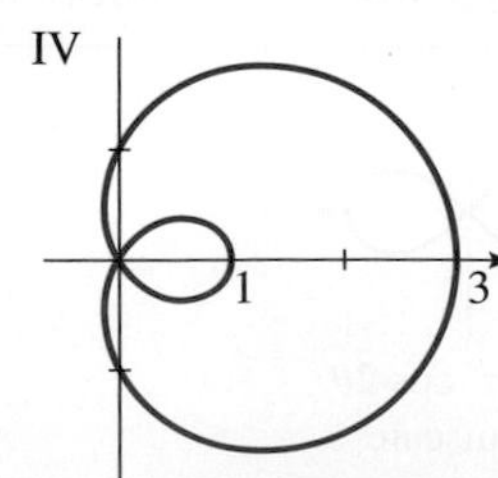

V
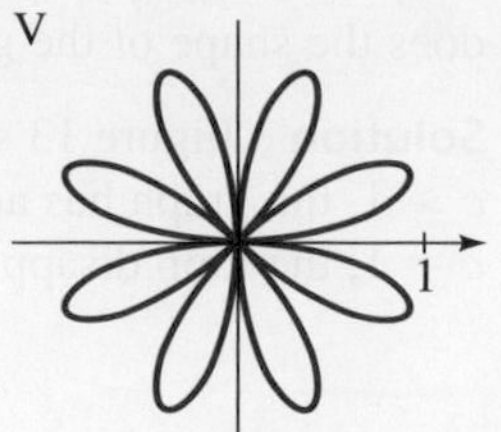

VI
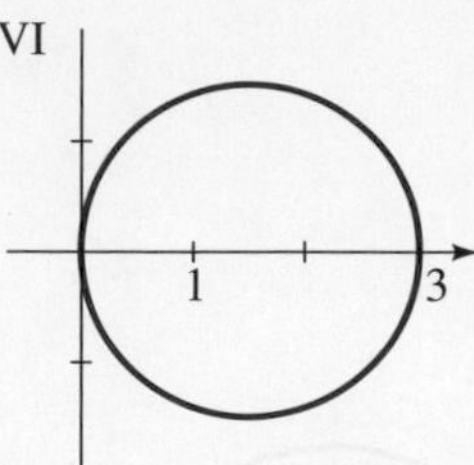

9–16 ■ Testing for Symmetry Test the polar equation for symmetry with respect to the polar axis, the pole, and the line $\theta = \pi/2$.

9. $r = 2 - \sin \theta$ 10. $r = 4 + 8 \cos \theta$
11. $r = 3 \sec \theta$ 12. $r = 5 \cos \theta \csc \theta$
13. $r = \dfrac{4}{3 - 2 \sin \theta}$ 14. $r = \dfrac{5}{1 + 3 \cos \theta}$
15. $r^2 = 4 \cos 2\theta$ 16. $r^2 = 9 \sin \theta$

17–22 ■ Polar to Rectangular Sketch a graph of the polar equation, and express the equation in rectangular coordinates.

17. $r = 2$ 18. $r = -1$
19. $\theta = -\pi/2$ 20. $\theta = 5\pi/6$
21. $r = 6 \sin \theta$ 22. $r = \cos \theta$

23–46 ■ Graphing Polar Equations Sketch a graph of the polar equation.

23. $r = -2 \cos \theta$ 24. $r = 3 \sin \theta$
25. $r = 2 - 2 \cos \theta$ 26. $r = 1 + \sin \theta$
27. $r = -3(1 + \sin \theta)$ 28. $r = \cos \theta - 1$
29. $r = \sin 2\theta$ 30. $r = 2 \cos 3\theta$
31. $r = -\cos 5\theta$ 32. $r = \sin 4\theta$
33. $r = 2 \sin 5\theta$ 34. $r = -3 \cos 4\theta$
35. $r = \sqrt{3} - 2 \sin \theta$ 36. $r = 2 + \sin \theta$
37. $r = \sqrt{3} + \cos \theta$ 38. $r = 1 - 2 \cos \theta$
39. $r = 2 - 2\sqrt{2} \cos \theta$ 40. $r = 3 + 6 \sin \theta$
41. $r^2 = \cos 2\theta$ 42. $r^2 = 4 \sin 2\theta$
43. $r = \theta, \quad \theta \geq 0$ (spiral)
44. $r\theta = 1, \quad \theta > 0$ (reciprocal spiral)
45. $r = 2 + \sec \theta$ (conchoid)
46. $r = \sin \theta \tan \theta$ (cissoid)

47–50 ■ Graphing Polar Equations Use a graphing device to graph the polar equation. Choose the domain of θ to make sure you produce the entire graph.

47. $r = \cos(\theta/2)$ 48. $r = \sin(8\theta/5)$
49. $r = 1 + 2 \sin(\theta/2)$ (nephroid)
50. $r = \sqrt{1 - 0.8 \sin^2\theta}$ (hippopede)

 **51–52 ■ Families of Polar Equations** These exercises involve families of polar equations.

51. Graph the family of polar equations $r = 1 + \sin n\theta$ for $n = 1, 2, 3, 4$, and 5. How is the number of loops related to n?

52. Graph the family of polar equations $r = 1 + c \sin 2\theta$ for $c = 0.3, 0.6, 1, 1.5$, and 2. How does the graph change as c increases?

53–56 ■ Special Polar Equations Match the polar equation with the graphs labeled I–IV. Give reasons for your answers.

53. $r = \sin(\theta/2)$ **54.** $r = 1/\sqrt{\theta}$

55. $r = \theta \sin \theta$ **56.** $r = 1 + 3 \cos(3\theta)$

I

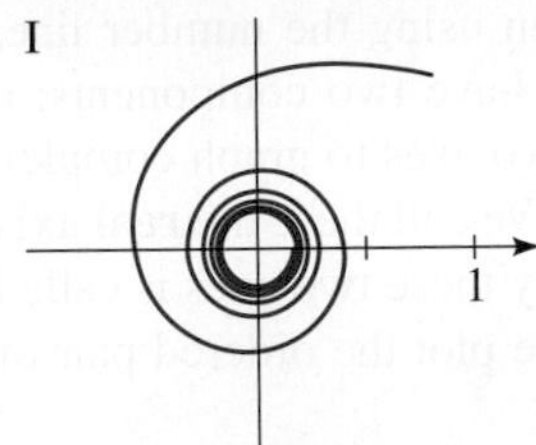

II

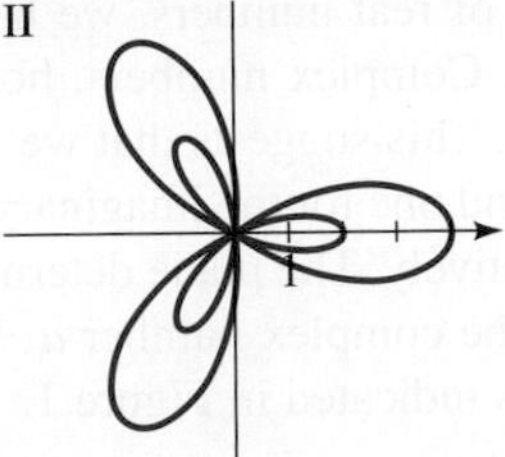

III

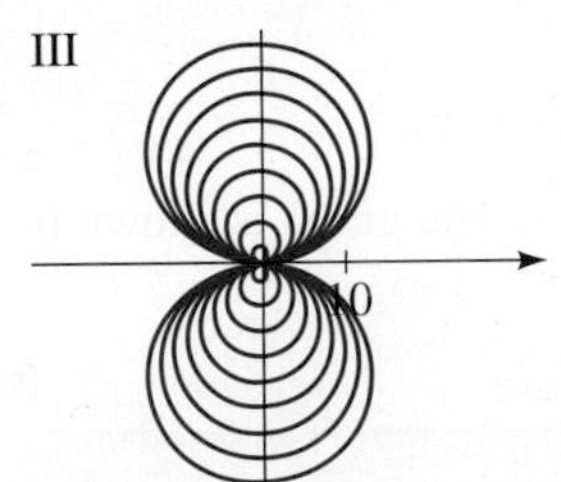

IV

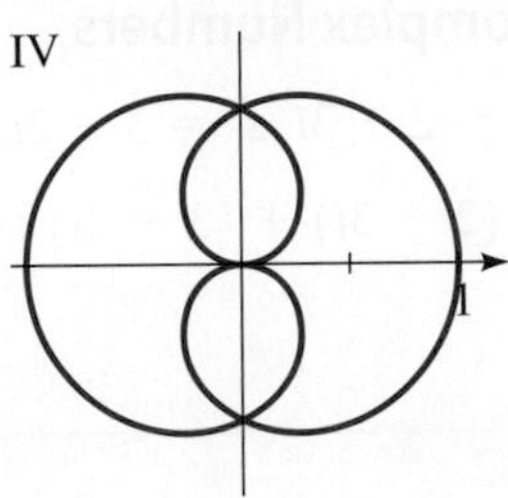

■ Skills Plus

57–60 ■ Rectangular to Polar Sketch a graph of the rectangular equation. [*Hint:* First convert the equation to polar coordinates.]

57. $(x^2 + y^2)^3 = 4x^2y^2$ **58.** $(x^2 + y^2)^3 = (x^2 - y^2)^2$

59. $(x^2 + y^2)^2 = x^2 - y^2$ **60.** $x^2 + y^2 = (x^2 + y^2 - x)^2$

61. A Circle in Polar Coordinates Consider the polar equation $r = a \cos \theta + b \sin \theta$.

(a) Express the equation in rectangular coordinates, and use this to show that the graph of the equation is a circle. What are the center and radius?

(b) Use your answer to part (a) to graph the equation $r = 2 \sin \theta + 2 \cos \theta$.

 **62. A Parabola in Polar Coordinates**

(a) Graph the polar equation $r = \tan \theta \sec \theta$ in the viewing rectangle $[-3, 3]$ by $[-1, 9]$.

(b) Note that your graph in part (a) looks like a parabola (see Section 3.1). Confirm this by converting the equation to rectangular coordinates.

■ Applications

63. Orbit of a Satellite Scientists and engineers often use polar equations to model the motion of satellites in earth orbit. Let's consider a satellite whose orbit is modeled by the equation $r = 36000/(4 - \cos \theta)$, where r is the distance in kilometers between the satellite and the center of the earth and θ is the angle shown in the figure.

(a) On the same viewing screen, graph the circle $r = 6340$ (to represent the earth, which we will assume to be a sphere of radius 6340 km) and the polar equation of the satellite's orbit. Describe the motion of the satellite as θ increases from 0 to 2π.

(b) For what angle θ is the satellite closest to the earth? Find the height of the satellite above the earth's surface for this value of θ.

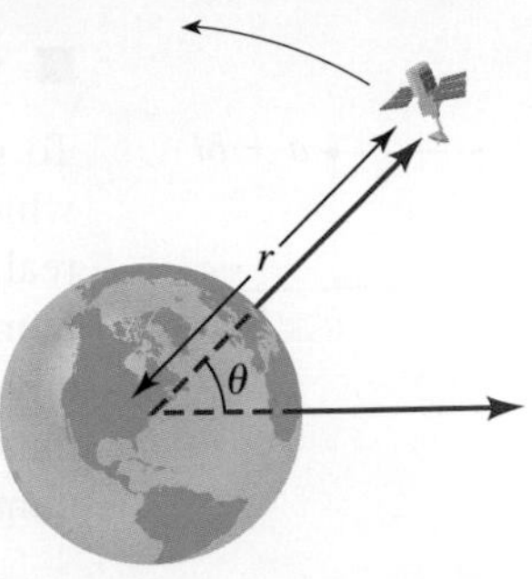

 **64. An Unstable Orbit** The orbit described in Exercise 63 is stable because the satellite traverses the same path over and over as θ increases. Suppose that a meteor strikes the satellite and changes its orbit to

$$r = \frac{36000\left(1 - \dfrac{\theta}{40}\right)}{4 - \cos \theta}$$

(a) On the same viewing screen, graph the circle $r = 6340$ and the new orbit equation, with θ increasing from 0 to 3π. Describe the new motion of the satellite.

(b) Estimate graphically the value of θ at the moment the satellite crashes into the earth.

■ Discuss ■ Discover ■ Prove ■ Write

 **65. Discuss ■ Discover: A Transformation of Polar Graphs** How are the graphs of

$$r = 1 + \sin\left(\theta - \frac{\pi}{6}\right)$$

and
$$r = 1 + \sin\left(\theta - \frac{\pi}{3}\right)$$

related to the graph of $r = 1 + \sin \theta$? In general, how is the graph of $r = f(\theta - \alpha)$ related to the graph of $r = f(\theta)$?

66. Discuss: Choosing a Convenient Coordinate System Compare the polar equation of the circle $r = 2$ with its equation in rectangular coordinates. In which coordinate system is the equation simpler? Do the same for the equation of the four-leaved rose $r = \sin 2\theta$. Which coordinate system would you choose to study these curves?

67. Discuss: Choosing a Convenient Coordinate System Compare the rectangular equation of the line $y = 2$ with its polar equation. In which coordinate system is the equation simpler? Which coordinate system would you choose to study lines?

8.3 Polar Form of Complex Numbers; De Moivre's Theorem

■ Graphing Complex Numbers ■ Polar Form of Complex Numbers
■ De Moivre's Theorem ■ *n*th Roots of Complex Numbers

Complex numbers are discussed in Section 1.6.

In this section we represent complex numbers in polar (or trigonometric) form. This enables us to find the *n*th roots of complex numbers. To describe the polar form of complex numbers, we must first learn to work with complex numbers graphically.

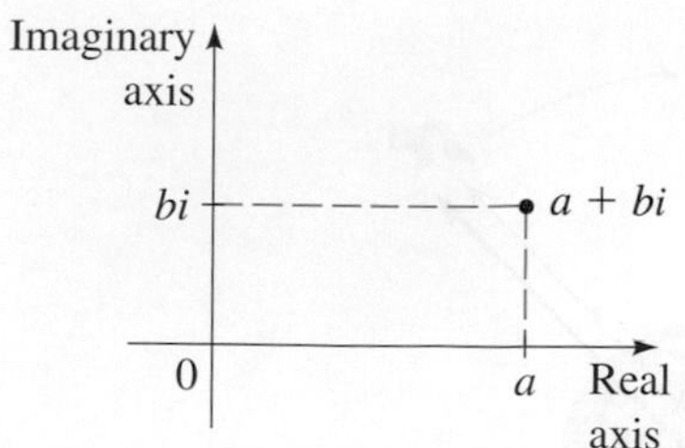

Figure 1

■ Graphing Complex Numbers

To graph real numbers or sets of real numbers, we have been using the number line, which has just one dimension. Complex numbers, however, have two components: a real part and an imaginary part. This suggests that we need two axes to graph complex numbers: one for the real part and one for the imaginary part. We call these the **real axis** and the **imaginary axis**, respectively. The plane determined by these two axes is called the **complex plane**. To graph the complex number $a + bi$, we plot the ordered pair of numbers (a, b) in this plane, as indicated in Figure 1.

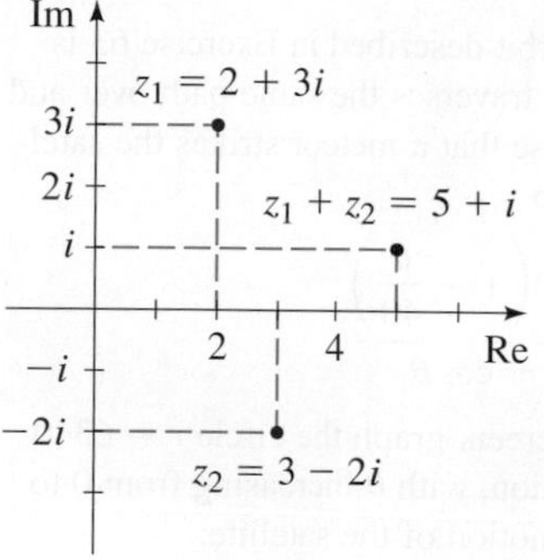

Figure 2

Example 1 ■ Graphing Complex Numbers

Graph the complex numbers $z_1 = 2 + 3i$, $z_2 = 3 - 2i$, and $z_1 + z_2$.

Solution We have $z_1 + z_2 = (2 + 3i) + (3 - 2i) = 5 + i$. The graph is shown in Figure 2.

Now Try Exercise 19 ■

Example 2 ■ Graphing Sets of Complex Numbers

Graph each set of complex numbers.

(a) $S = \{a + bi \mid a \geq 0\}$

(b) $T = \{a + bi \mid a < 1, b \geq 0\}$

Solution

(a) S is the set of complex numbers whose real part is nonnegative. The graph is shown in Figure 3(a).

(b) T is the set of complex numbers for which the real part is less than 1 and the imaginary part is nonnegative. The graph is shown in Figure 3(b).

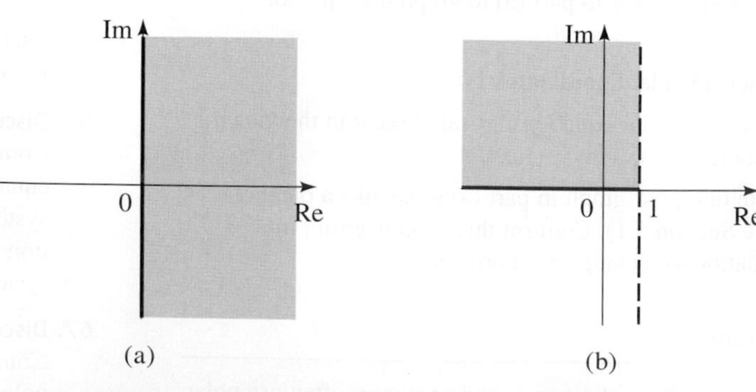

Figure 3

Now Try Exercise 21 ■

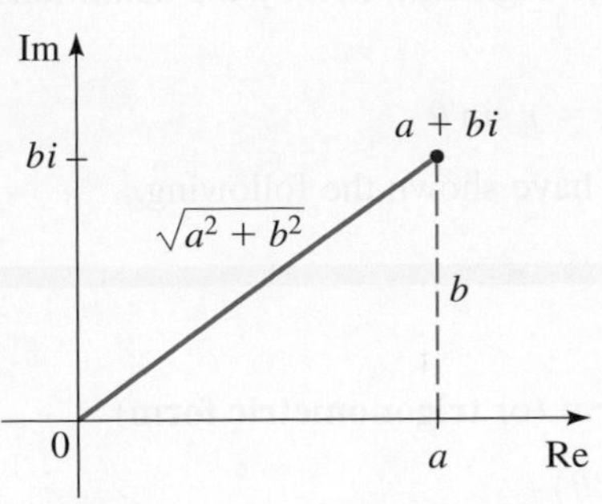

Figure 4

Recall that the absolute value of a real number can be thought of as its distance from the origin on the real number line (see Section 1.1). We define absolute value for complex numbers in a similar fashion. Using the Pythagorean Theorem, we can see from Figure 4 that the distance between $a + bi$ and the origin in the complex plane is $\sqrt{a^2 + b^2}$. This leads to the following definition.

Modulus of a Complex Number

The **modulus** (or **absolute value**) of the complex number $z = a + bi$ is

$$|z| = \sqrt{a^2 + b^2}$$

Example 3 ■ Calculating the Modulus

The plural of *modulus* is *moduli*.

Find the moduli of the complex numbers $3 + 4i$ and $8 - 5i$.

Solution

$$|3 + 4i| = \sqrt{3^2 + 4^2} = \sqrt{25} = 5$$

$$|8 - 5i| = \sqrt{8^2 + (-5)^2} = \sqrt{89}$$

 Now Try Exercise 9 ■

Example 4 ■ Graphing a Set of Complex Numbers

Graph each set of complex numbers.

(a) $C = \{z \mid |z| = 1\}$ **(b)** $D = \{z \mid |z| \le 1\}$

Solution

(a) C is the set of complex numbers whose distance from the origin is 1. Thus C is a circle of radius 1 with center at the origin, as shown in Figure 5.

(b) D is the set of complex numbers whose distance from the origin is less than or equal to 1. Thus D is the disk that consists of all complex numbers on and inside the circle C of part (a), as shown in Figure 6.

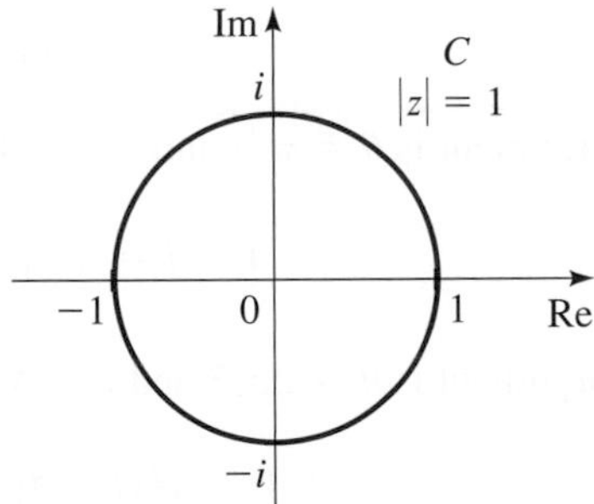

Figure 5

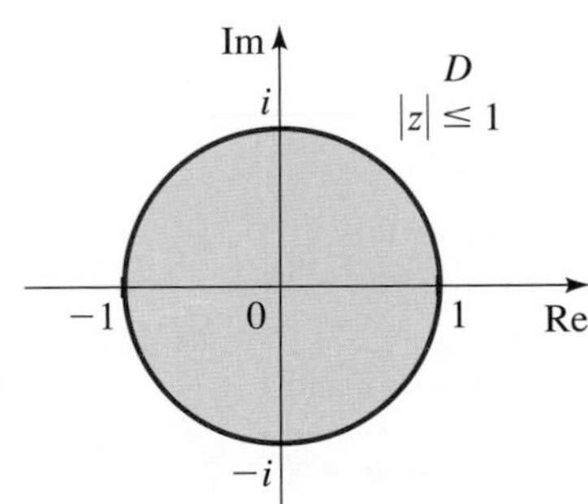

Figure 6

 Now Try Exercises 23 and 25 ■

■ Polar Form of Complex Numbers

Let $z = a + bi$ be a complex number, and in the complex plane let's draw the line segment joining the origin to the point $a + bi$. (See Figure 7 on the next page.) The length of this line segment is $r = |z| = \sqrt{a^2 + b^2}$. If θ is an angle in standard position whose terminal side coincides with this line segment, then by the definitions of sine and cosine (see Section 6.3)

$$a = r\cos\theta \qquad \text{and} \qquad b = r\sin\theta$$

so $z = r\cos\theta + ir\sin\theta = r(\cos\theta + i\sin\theta)$. We have shown the following.

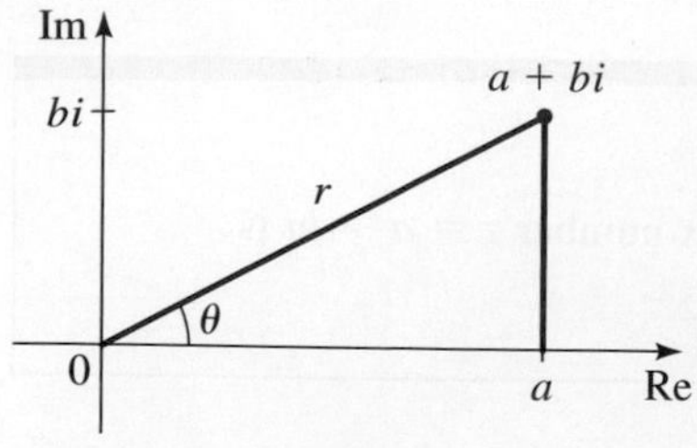

Figure 7

Polar Form of Complex Numbers

A complex number $z = a + bi$ has the **polar form** (or **trigonometric form**)

$$z = r(\cos\theta + i\sin\theta)$$

where $r = |z| = \sqrt{a^2 + b^2}$ and $\tan\theta = b/a$. The number r is the **modulus** of z, and θ is called an **argument** of z.

Note The argument of z is not unique, but any two arguments of z differ by a multiple of 2π. When determining the argument, we must consider the quadrant in which z lies, as we see in the next example.

Example 5 ■ Writing Complex Numbers in Polar Form

Write each complex number in polar form.

(a) $1 + i$ **(b)** $-1 + \sqrt{3}i$ **(c)** $-4\sqrt{3} - 4i$ **(d)** $3 + 4i$

Solution These complex numbers are graphed in Figure 8, in order to help us find their arguments.

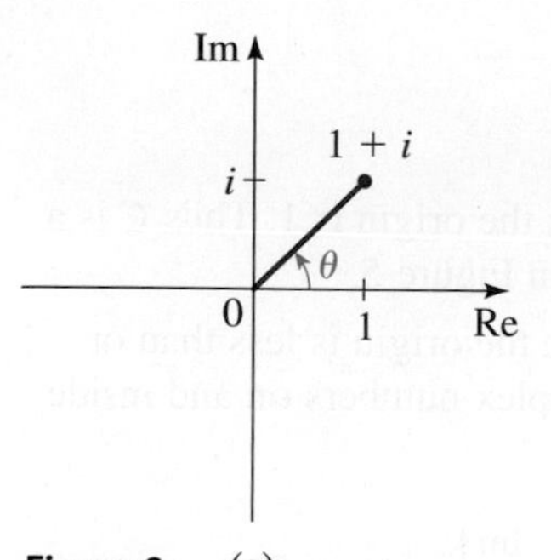

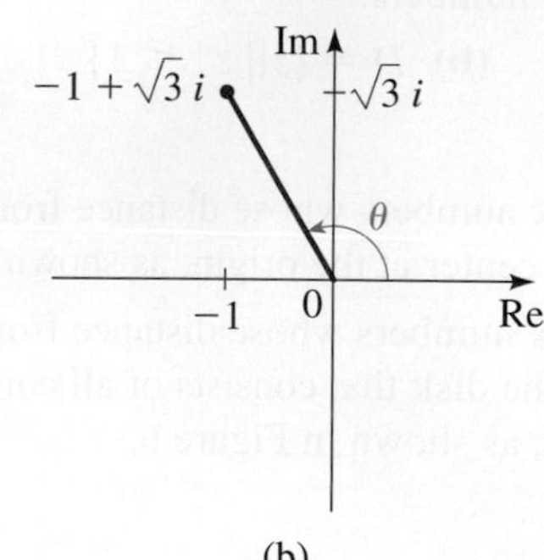

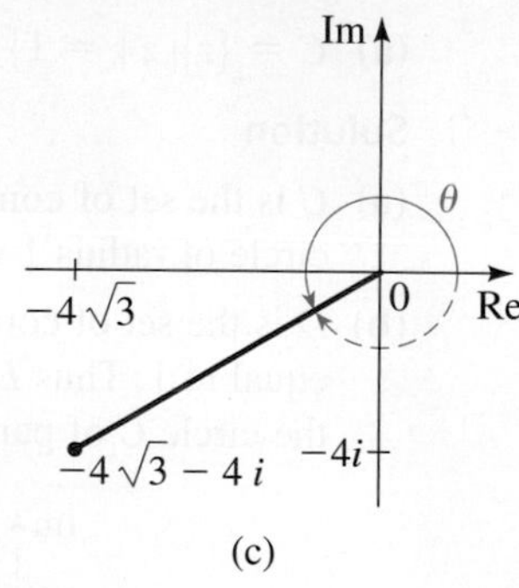

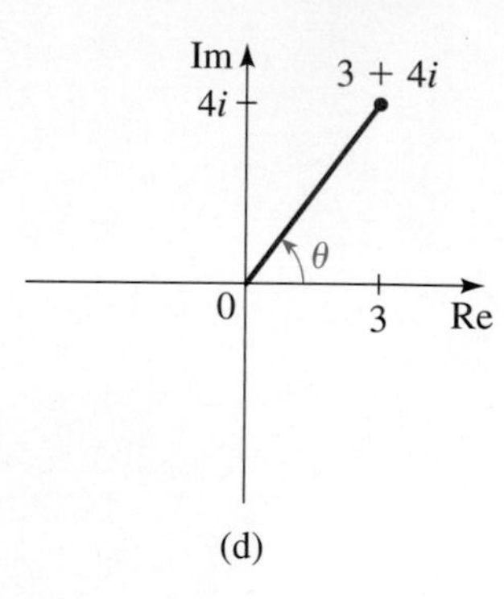

Figure 8 (a) (b) (c) (d)

$\tan\theta = \frac{1}{1} = 1$ and θ in Quadrant I, so $\theta = \frac{\pi}{4}$

(a) An argument is $\theta = \pi/4$ and $r = \sqrt{1 + 1} = \sqrt{2}$. Thus

$$1 + i = \sqrt{2}\left(\cos\frac{\pi}{4} + i\sin\frac{\pi}{4}\right)$$

$\tan\theta = \dfrac{\sqrt{3}}{-1} = -\sqrt{3}$ and θ in Quadrant II, so $\theta = \frac{2\pi}{3}$

(b) An argument is $\theta = 2\pi/3$ and $r = \sqrt{1 + 3} = 2$. Thus

$$-1 + \sqrt{3}i = 2\left(\cos\frac{2\pi}{3} + i\sin\frac{2\pi}{3}\right)$$

$\tan\theta = \dfrac{-4}{-4\sqrt{3}} = \dfrac{1}{\sqrt{3}}$ and θ in Quadrant III, so $\theta = \frac{7\pi}{6}$

(c) An argument is $\theta = 7\pi/6$ (or we could use $\theta = -5\pi/6$), and $r = \sqrt{48 + 16} = 8$. Thus

$$-4\sqrt{3} - 4i = 8\left(\cos\frac{7\pi}{6} + i\sin\frac{7\pi}{6}\right)$$

$\tan\theta = \frac{4}{3}$ and θ in Quadrant I, so $\theta = \tan^{-1}\left(\frac{4}{3}\right)$

(d) An argument is $\theta = \tan^{-1}\left(\frac{4}{3}\right) \approx 0.93$ and $r = \sqrt{3^2 + 4^2} = 5$. So

$$3 + 4i = 5\left[\cos\left(\tan^{-1}\left(\tfrac{4}{3}\right)\right) + i\sin\left(\tan^{-1}\left(\tfrac{4}{3}\right)\right)\right] \approx 5[\cos 0.93 + i\sin 0.93]$$

Now Try Exercises 29, 31, 33, and 43 ■

The Addition Formulas for Sine and Cosine that we discussed in Section 7.2 simplify the multiplication and division of complex numbers in polar form. The following theorem shows how.

Multiplication and Division of Complex Numbers

If the two complex numbers z_1 and z_2 have the polar forms

$$z_1 = r_1(\cos\theta_1 + i\sin\theta_1) \qquad \text{and} \qquad z_2 = r_2(\cos\theta_2 + i\sin\theta_2)$$

then $z_1 z_2$ and z_1/z_2 have the following polar forms:

Multiplication

$$z_1 z_2 = r_1 r_2[\cos(\theta_1 + \theta_2) + i\sin(\theta_1 + \theta_2)]$$

Division

$$\frac{z_1}{z_2} = \frac{r_1}{r_2}[\cos(\theta_1 - \theta_2) + i\sin(\theta_1 - \theta_2)] \qquad (z_2 \neq 0)$$

This theorem says:

To multiply two complex numbers, multiply the moduli and add the arguments.
To divide two complex numbers, divide the moduli and subtract the arguments.

Proof To prove the Multiplication Formula, we multiply the two complex numbers:

$$\begin{aligned} z_1 z_2 &= r_1 r_2(\cos\theta_1 + i\sin\theta_1)(\cos\theta_2 + i\sin\theta_2) \\ &= r_1 r_2[\cos\theta_1\cos\theta_2 - \sin\theta_1\sin\theta_2 + i(\sin\theta_1\cos\theta_2 + \cos\theta_1\sin\theta_2)] \\ &= r_1 r_2[\cos(\theta_1 + \theta_2) + i\sin(\theta_1 + \theta_2)] \end{aligned}$$

In the last step we used the Addition Formulas for Sine and Cosine.

The proof of the Division Formula is left as an exercise. (See Exercise 102.) ■

Example 6 ■ Multiplying and Dividing Complex Numbers

Let

$$z_1 = 2\left(\cos\frac{\pi}{4} + i\sin\frac{\pi}{4}\right) \qquad \text{and} \qquad z_2 = 5\left(\cos\frac{\pi}{3} + i\sin\frac{\pi}{3}\right)$$

Find **(a)** $z_1 z_2$ and **(b)** z_1/z_2.

Solution

(a) By the Multiplication Formula

$$\begin{aligned} z_1 z_2 &= (2)(5)\left[\cos\left(\frac{\pi}{4} + \frac{\pi}{3}\right) + i\sin\left(\frac{\pi}{4} + \frac{\pi}{3}\right)\right] \\ &= 10\left(\cos\frac{7\pi}{12} + i\sin\frac{7\pi}{12}\right) \end{aligned}$$

To approximate the answer, we use a calculator in radian mode and get

$$\begin{aligned} z_1 z_2 &\approx 10(-0.2588 + 0.9659i) \\ &= -2.588 + 9.659i \end{aligned}$$

(b) By the Division Formula

$$\frac{z_1}{z_2} = \frac{2}{5}\left[\cos\left(\frac{\pi}{4} - \frac{\pi}{3}\right) + i\sin\left(\frac{\pi}{4} - \frac{\pi}{3}\right)\right]$$

$$= \frac{2}{5}\left[\cos\left(-\frac{\pi}{12}\right) + i\sin\left(-\frac{\pi}{12}\right)\right]$$

$$= \frac{2}{5}\left(\cos\frac{11\pi}{12} + i\sin\frac{11\pi}{12}\right)$$

Using a calculator in radian mode, we get the approximate answer:

$$\frac{z_1}{z_2} \approx \frac{2}{5}(0.9659 - 0.2588i) = 0.3864 - 0.1035i$$

Now Try Exercise 49 ■

■ De Moivre's Theorem

Repeated use of the Multiplication Formula gives the following useful formula for raising a complex number to a power n for any positive integer n.

De Moivre's Theorem

If $z = r(\cos\theta + i\sin\theta)$, then for any integer n

$$z^n = r^n(\cos n\theta + i\sin n\theta)$$

This theorem says: *To take the nth power of a complex number, take the nth power of the modulus and multiply the argument by n.*

Proof By the Multiplication Formula

$$z^2 = zz = r^2[\cos(\theta + \theta) + i\sin(\theta + \theta)]$$

$$= r^2(\cos 2\theta + i\sin 2\theta)$$

Now we multiply z^2 by z to get

$$z^3 = z^2 z = r^3[\cos(2\theta + \theta) + i\sin(2\theta + \theta)]$$

$$= r^3(\cos 3\theta + i\sin 3\theta)$$

Repeating this argument, we see that for any positive integer n

$$z^n = r^n(\cos n\theta + i\sin n\theta)$$

A similar argument using the Division Formula shows that this also holds for negative integers. ■

Example 7 ■ Finding a Power Using De Moivre's Theorem

Find $\left(\frac{1}{2} + \frac{1}{2}i\right)^{10}$.

Solution Since $\frac{1}{2} + \frac{1}{2}i = \frac{1}{2}(1 + i)$, it follows from Example 5(a) that

$$\frac{1}{2} + \frac{1}{2}i = \frac{\sqrt{2}}{2}\left(\cos\frac{\pi}{4} + i\sin\frac{\pi}{4}\right)$$

So by de Moivre's Theorem

$$\left(\frac{1}{2}+\frac{1}{2}i\right)^{10} = \left(\frac{\sqrt{2}}{2}\right)^{10}\left(\cos\frac{10\pi}{4} + i\sin\frac{10\pi}{4}\right)$$

$$= \frac{2^5}{2^{10}}\left(\cos\frac{5\pi}{2} + i\sin\frac{5\pi}{2}\right) = \frac{1}{32}i$$

 Now Try Exercise 65 ■

■ *n*th Roots of Complex Numbers

An ***n*th root** of a complex number z is any complex number w such that $w^n = z$. De Moivre's Theorem gives us a method for calculating the nth roots of any complex number.

*n*th Roots of Complex Numbers

If $z = r(\cos\theta + i\sin\theta)$ and n is a positive integer, then z has the n distinct nth roots

$$w_k = r^{1/n}\left[\cos\left(\frac{\theta + 2k\pi}{n}\right) + i\sin\left(\frac{\theta + 2k\pi}{n}\right)\right]$$

for $k = 0, 1, 2, \ldots, n - 1$.

Proof To find the nth roots of z, we need to find a complex number w such that

$$w^n = z$$

Let's write z in polar form:

$$z = r(\cos\theta + i\sin\theta)$$

One nth root of z is

$$w = r^{1/n}\left(\cos\frac{\theta}{n} + i\sin\frac{\theta}{n}\right)$$

because by de Moivre's Theorem, $w^n = z$. But the argument θ of z can be replaced by $\theta + 2k\pi$ for any integer k. Since this expression gives a different value of w for $k = 0, 1, 2, \ldots, n - 1$, we have proved the formula in the theorem. ■

The following observations help us use the preceding formula.

Finding the *n*th Roots of $z = r(\cos\theta + i\sin\theta)$

1. **Modulus.** The modulus of each nth root is $r^{1/n}$.
2. **Argument.** The argument of the first root is θ/n.
3. **Roots.** Repeatedly add $2\pi/n$ to get the argument of each successive root.

These observations show that, when graphed, the nth roots of z are spaced equally on the circle of radius $r^{1/n}$.

Example 8 ■ Finding Roots of a Complex Number

Find the six sixth roots of $z = -64$, and graph these roots in the complex plane.

Solution In polar form, $z = 64(\cos \pi + i \sin \pi)$. Applying the formula for nth roots with $n = 6$, we get

$$w_k = 64^{1/6}\left[\cos\left(\frac{\pi + 2k\pi}{6}\right) + i \sin\left(\frac{\pi + 2k\pi}{6}\right)\right]$$

for $k = 0, 1, 2, 3, 4, 5$. Since $64^{1/6} = 2$, we find that the six sixth roots of -64 are

We add $2\pi/6 = \pi/3$ to each argument to get the argument of the next root.

$$w_0 = 2\left(\cos\frac{\pi}{6} + i \sin\frac{\pi}{6}\right) = \sqrt{3} + i \qquad k = 0$$

$$w_1 = 2\left(\cos\frac{\pi}{2} + i \sin\frac{\pi}{2}\right) = 2i \qquad k = 1$$

$$w_2 = 2\left(\cos\frac{5\pi}{6} + i \sin\frac{5\pi}{6}\right) = -\sqrt{3} + i \qquad k = 2$$

$$w_3 = 2\left(\cos\frac{7\pi}{6} + i \sin\frac{7\pi}{6}\right) = -\sqrt{3} - i \qquad k = 3$$

$$w_4 = 2\left(\cos\frac{3\pi}{2} + i \sin\frac{3\pi}{2}\right) = -2i \qquad k = 4$$

$$w_5 = 2\left(\cos\frac{11\pi}{6} + i \sin\frac{11\pi}{6}\right) = \sqrt{3} - i \qquad k = 5$$

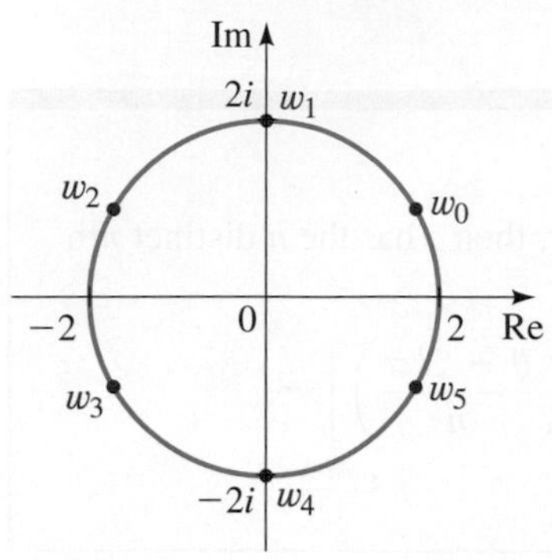

Figure 9 | The six sixth roots of $z = -64$

All these points lie on a circle of radius 2, as shown in Figure 9.

 **Now Try Exercise 81** ■

When finding roots of complex numbers, we sometimes write the argument θ of the complex number in degrees. In this case the nth roots are obtained from the formula

$$w_k = r^{1/n}\left[\cos\left(\frac{\theta + 360°k}{n}\right) + i \sin\left(\frac{\theta + 360°k}{n}\right)\right]$$

for $k = 0, 1, 2, \ldots, n - 1$.

Discovery Project ■ Fractals

Most of the things we model in this book follow regular, predictable patterns. But many real-world phenomena—such as a cloud, a jagged coastline, or a flickering flame—appear to have random or even chaotic shapes. Fractals allow us to model these sorts of shapes. Surprisingly, the extremely complex shapes of fractals and their infinite detail are produced by exceedingly simple rules and endless repetitions that involve iterating simple functions whose inputs and outputs are complex numbers. You can find the project at the book companion website **www.stewartmath.com**.

Example 9 ■ Finding Cube Roots of a Complex Number

Find the three cube roots of $z = 2 + 2i$, and graph these roots in the complex plane.

Solution First we write z in polar form using degrees. We have $r = \sqrt{2^2 + 2^2} = 2\sqrt{2}$ and $\theta = 45°$. Thus

$$z = 2\sqrt{2}(\cos 45° + i \sin 45°)$$

We add $360°/3 = 120°$ to each argument to get the argument of the next root.

Applying the formula for nth roots (in degrees) with $n = 3$, we find that the cube roots of z are of the form

$$w_k = (2\sqrt{2})^{1/3}\left[\cos\left(\frac{45° + 360°k}{3}\right) + i \sin\left(\frac{45° + 360°k}{3}\right)\right]$$

where $k = 0, 1, 2$. Since $(2\sqrt{2})^{1/3} = (2^{3/2})^{1/3} = 2^{1/2} = \sqrt{2}$, the three cube roots are

$$w_0 = \sqrt{2}(\cos 15° + i \sin 15°) \approx 1.366 + 0.366i \qquad k = 0$$

$$w_1 = \sqrt{2}(\cos 135° + i \sin 135°) = -1 + i \qquad k = 1$$

$$w_2 = \sqrt{2}(\cos 255° + i \sin 255°) \approx -0.366 - 1.366i \qquad k = 2$$

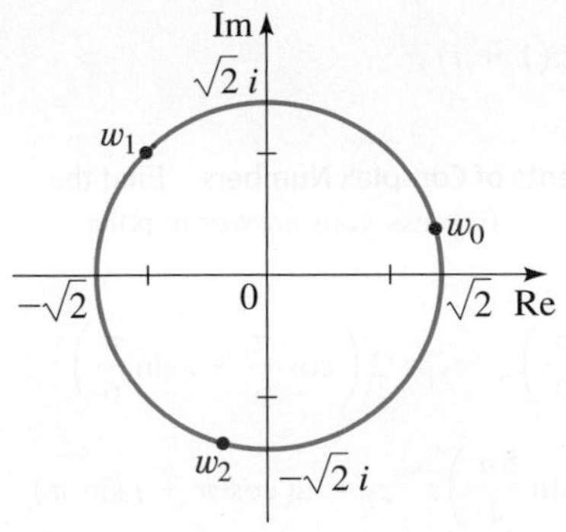

Figure 10 | The three cube roots of $z = 2 + 2i$

The three cube roots of z are graphed in Figure 10. These roots are spaced equally on a circle of radius $\sqrt{2}$.

Now Try Exercise 77 ■

Example 10 ■ Solving an Equation Using the *n*th Roots Formula

Solve the equation $z^6 + 64 = 0$.

Solution This equation can be written as $z^6 = -64$. Thus the solutions are the six sixth roots of -64, which we found in Example 8.

Now Try Exercise 87 ■

8.3 | Exercises

Concepts

1. A complex number $z = a + bi$ has two parts: a is the ________ part, and b is the ________ part. To graph $a + bi$, we graph the ordered pair (____ , ____) in the complex plane.

2. Let $z = a + bi$.
   (a) The modulus of z is $r =$ ________, and an argument of z is an angle θ satisfying $\tan \theta =$ ________.
   (b) We can express z in polar form as $z =$ ________, where r is the modulus of z and θ is the argument of z.

3. (a) The complex number $z = -1 + i$ in polar form is $z =$ ________.
   (b) The complex number $z = 2\left(\cos \frac{\pi}{6} + i \sin \frac{\pi}{6}\right)$ in rectangular form is $z =$ ________.
   (c) The complex number graphed below can be expressed in rectangular form as ________ or in polar form as ________.

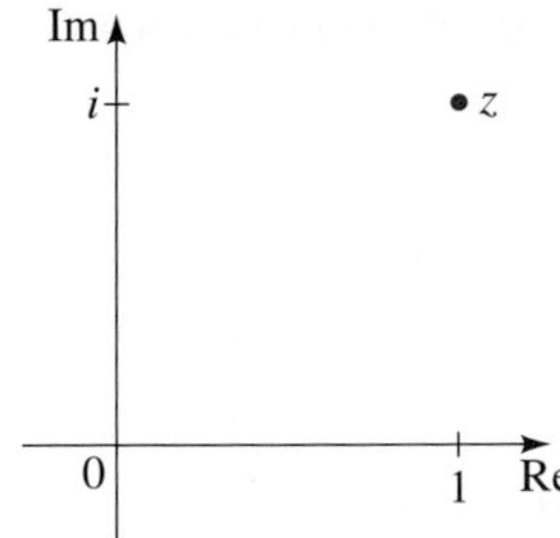

4. How many different nth roots does a nonzero complex number have? ______. The number 16 has ______ fourth roots. These roots are ______, ______, ______, and ______. In the complex plane these roots all lie on a circle of radius ______. Graph the roots on the following graph.

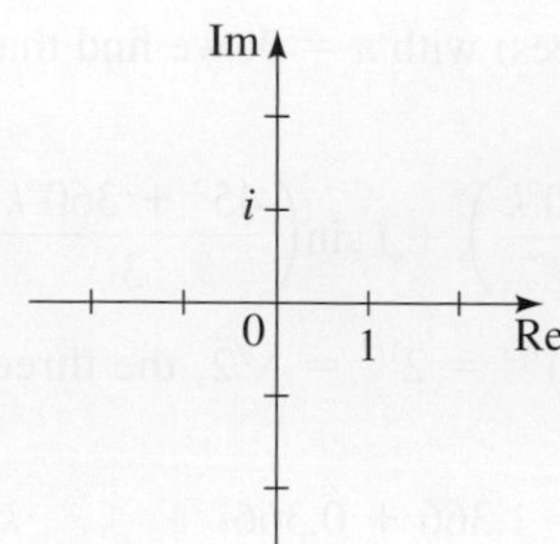

Skills

5–14 ■ A Complex Number and Its Modulus Graph the complex number, and find its modulus.

5. $4i$ **6.** $-3i$

7. -2 **8.** 6

9. $5 + 2i$ **10.** $7 - 3i$

11. $\sqrt{3} + i$ **12.** $-1 - \frac{\sqrt{3}}{3}i$

13. $\frac{3 + 4i}{5}$ **14.** $\frac{-\sqrt{2} + \sqrt{2}i}{2}$

15–16 ■ Graphing Complex Numbers Sketch the complex numbers z, $2z$, $-z$, and $\frac{1}{2}z$ on the same complex plane.

15. $z = 1 + i$ **16.** $z = -1 + \sqrt{3}i$

17–18 ■ Graphing a Complex Number and Its Complex Conjugate Sketch the complex number z and its complex conjugate $\bar{z}$ on the same complex plane.

17. $z = 8 + 2i$ **18.** $z = -5 + 6i$

19–20 ■ Graphing Complex Numbers Sketch z_1, z_2, $z_1 + z_2$, and z_1z_2 on the same complex plane.

19. $z_1 = 2 - i, \quad z_2 = 2 + i$

20. $z_1 = -1 + i, \quad z_2 = 2 - 3i$

21–28 ■ Graphing Sets of Complex Numbers Sketch the set in the complex plane.

21. $\{z = a + bi \mid a \le 0, b \ge 0\}$

22. $\{z = a + bi \mid a > 1, b > 1\}$

23. $\{z \mid |z| = 3\}$ **24.** $\{z \mid |z| \ge 1\}$

25. $\{z \mid |z| < 2\}$ **26.** $\{z \mid 2 \le |z| \le 5\}$

27. $\{z = a + bi \mid a + b < 2\}$

28. $\{z = a + bi \mid a \ge b\}$

29–48 ■ Polar Form of Complex Numbers Write the complex number in polar form with argument θ between 0 and 2π.

29. $1 + i$ **30.** $1 - i$ **31.** $-2 + 2i$

32. $-\sqrt{2} - \sqrt{2}i$ **33.** $-\sqrt{3} - i$ **34.** $-5 + 5\sqrt{3}i$

35. $2\sqrt{3} - 2i$ **36.** $3 + 3\sqrt{3}i$ **37.** $2i$

38. $-5i$ **39.** -3 **40.** $\sqrt{2}$

41. $-\sqrt{6} + \sqrt{2}i$ **42.** $-\sqrt{5} - \sqrt{15}i$ **43.** $4 + 3i$

44. $3 + 2i$ **45.** $4(\sqrt{3} - i)$ **46.** $i(\sqrt{2} - \sqrt{6}i)$

47. $-3(1 - i)$ **48.** $2i(1 + i)$

49–56 ■ Products and Quotients of Complex Numbers Find the product z_1z_2 and the quotient z_1/z_2. Express your answer in polar form.

49. $z_1 = 3\left(\cos\frac{\pi}{3} + i\sin\frac{\pi}{3}\right), \quad z_2 = 2\left(\cos\frac{\pi}{6} + i\sin\frac{\pi}{6}\right)$

50. $z_1 = \sqrt{3}\left(\cos\frac{5\pi}{4} + i\sin\frac{5\pi}{4}\right), \quad z_2 = 2(\cos\pi + i\sin\pi)$

51. $z_1 = \sqrt{2}\left(\cos\frac{5\pi}{3} + i\sin\frac{5\pi}{3}\right),$
$z_2 = 2\sqrt{2}\left(\cos\frac{3\pi}{2} + i\sin\frac{3\pi}{2}\right)$

52. $z_1 = \cos\frac{3\pi}{4} + i\sin\frac{3\pi}{4}, \quad z_2 = \cos\frac{\pi}{3} + i\sin\frac{\pi}{3}$

53. $z_1 = 4(\cos 120° + i\sin 120°),$
$z_2 = 2(\cos 30° + i\sin 30°)$

54. $z_1 = \sqrt{2}(\cos 75° + i\sin 75°),$
$z_2 = 3\sqrt{2}(\cos 60° + i\sin 60°)$

55. $z_1 = 4(\cos 200° + i\sin 200°),$
$z_2 = 25(\cos 150° + i\sin 150°)$

56. $z_1 = \frac{4}{5}(\cos 25° + i\sin 25°),$
$z_2 = \frac{1}{5}(\cos 155° + i\sin 155°)$

57–64 ■ Products and Quotients of Complex Numbers Write z_1 and z_2 in polar form, and then find the product z_1z_2 and the quotients z_1/z_2 and $1/z_1$. Express your answers in polar form.

57. $z_1 = \sqrt{3} + i, \quad z_2 = 1 + \sqrt{3}i$

58. $z_1 = \sqrt{2} - \sqrt{2}i, \quad z_2 = 1 - i$

59. $z_1 = 2\sqrt{3} - 2i, \quad z_2 = -1 + i$

60. $z_1 = -\sqrt{2}i, \quad z_2 = -3 - 3\sqrt{3}i$

61. $z_1 = 5 + 5i, \quad z_2 = 4$

62. $z_1 = 4\sqrt{3} - 4i, \quad z_2 = 8i$

63. $z_1 = -20, \quad z_2 = \sqrt{3} + i$

64. $z_1 = 3 + 4i, \quad z_2 = 2 - 2i$

65–76 ■ Powers Using De Moivre's Theorem Find the indicated power using de Moivre's Theorem.

65. $(-\sqrt{3} + i)^6$

66. $(1 - i)^{10}$

67. $(-\sqrt{2} - \sqrt{2}i)^5$

68. $(1 + i)^7$

69. $\left(\frac{\sqrt{2}}{2} + \frac{\sqrt{2}}{2}i\right)^{12}$

70. $(\sqrt{3} - i)^{-10}$

71. $(2 - 2i)^8$

72. $\left(-\frac{1}{2} - \frac{\sqrt{3}}{2}i\right)^{15}$

73. $(-1 - i)^7$

74. $(3 + \sqrt{3}i)^4$

75. $(2\sqrt{3} + 2i)^{-5}$

76. $(1 - i)^{-8}$

77–86 ■ Roots of Complex Numbers Find the indicated roots, and graph the roots in the complex plane.

77. The square roots of $4\sqrt{3} + 4i$

78. The cube roots of $4\sqrt{3} + 4i$

79. The fourth roots of $-81i$

80. The fifth roots of 32

81. The eighth roots of 1

82. The cube roots of $1 + i$

83. The cube roots of i

84. The fifth roots of i

85. The fourth roots of -1

86. The fifth roots of $-16 - 16\sqrt{3}i$

87–92 ■ Solving Equations Using *n*th Roots Solve the equation.

87. $z^4 + 1 = 0$

88. $z^8 - i = 0$

89. $z^3 - 4\sqrt{3} - 4i = 0$

90. $z^6 - 1 = 0$

91. $z^3 + 1 = -i$

92. $z^3 - 1 = 0$

Skills Plus

93–96 ■ Complex Coefficients and the Quadratic Formula The quadratic formula works whether the coefficients of the equation are real or complex. Solve the following equations using the quadratic formula and, if necessary, de Moivre's Theorem.

93. $z^2 - iz + 1 = 0$

94. $z^2 + iz + 2 = 0$

95. $z^2 - 2iz - 2 = 0$

96. $z^2 + (1 + i)z + i = 0$

97–98 ■ Finding *n*th Roots of a Complex Number Let $w = \cos(2\pi/n) + i\sin(2\pi/n)$, where n is a positive integer.

97. Show that the n distinct roots of 1 are

$$1, w, w^2, w^3, \ldots, w^{n-1}$$

98. If $z \neq 0$ and s is any nth root of z, show that the n distinct roots of z are

$$s, sw, sw^2, sw^3, \ldots, sw^{n-1}$$

99. Properties of the Modulus Verify the property for the complex numbers w and z.

(a) $z\bar{z} = |z|^2$

(b) $|wz| = |w||z|$

(c) $\left|\frac{1}{z}\right| = \frac{1}{|z|}$

(d) $\left|\frac{w}{z}\right| = \frac{|w|}{|z|}$

Discuss ■ Discover ■ Prove ■ Write

100. Discuss: Sums of Roots of Unity Find the exact values of all three cube roots of 1 (see Exercise 97), and then add them. Do the same for the fourth, fifth, sixth, and eighth roots of 1. What do you think is the sum of the nth roots of 1 for any n?

101. Discuss: Products of Roots of Unity Find the product of the three cube roots of 1 (see Exercise 97). Do the same for the fourth, fifth, sixth, and eighth roots of 1. What do you think is the product of the nth roots of 1 for any n?

102. Prove: Division in Polar Form If the two complex numbers z_1 and z_2 have the polar forms

$$z_1 = r_1(\cos\theta_1 + i\sin\theta_1)$$

$$z_2 = r_2(\cos\theta_2 + i\sin\theta_2)$$

show that

$$\frac{z_1}{z_2} = \frac{r_1}{r_2}[\cos(\theta_1 - \theta_2) + i\sin(\theta_1 - \theta_2)]$$

[*Hint:* Multiply numerator and denominator by the complex conjugate of z_2 and simplify.]

103. Prove: Sums of Squares of Integers Use complex numbers to show that the following statement is true: For any integers a and b there exist integers c and d such that

$$c^2 + d^2 = 2(a^2 + b^2)$$

PS *Introduce something extra.* Although the statement involves only integers, one way to prove it is to introduce complex numbers. Let $w = 1 + i$ and $z = a + bi$. Evaluate $|wz|^2$ in two ways using Exercise 99(b).

8.4 Plane Curves and Parametric Equations

■ Plane Curves and Parametric Equations ■ Eliminating the Parameter ■ Finding Parametric Equations for a Curve ■ Using Graphing Devices to Graph Parametric Curves

So far, we have described a curve by giving an equation (in rectangular or polar coordinates) that the coordinates of all the points on the curve must satisfy. But not all curves in the plane can be described in this way. In this section we study parametric equations, which are a general method for describing any curve.

■ Plane Curves and Parametric Equations

We can think of a curve as the path of a point moving in the plane; the x- and y-coordinates of the point are then functions of time. This idea leads to the following definition.

Plane Curves and Parametric Equations

If f and g are functions defined on an interval I, then the set of points $(f(t), g(t))$ is a **plane curve**. The equations

$$x = f(t) \qquad y = g(t)$$

where $t \in I$, are **parametric equations** for the curve, with **parameter** t.

Example 1 ■ Sketching a Plane Curve

Sketch the curve defined by the parametric equations

$$x = t^2 - 3t \qquad y = t - 1$$

Solution For every value of t we get a point on the curve. For example, if $t = 0$, then $x = 0$ and $y = -1$, so the corresponding point is $(0, -1)$. In Figure 1 we plot the points (x, y) determined by the values of t shown in the following table.

The arrows on the curve indicate the direction of the curve for increasing values of t.

t	x	y
−2	10	−3
−1	4	−2
0	0	−1
1	−2	0
2	−2	1
3	0	2
4	4	3
5	10	4

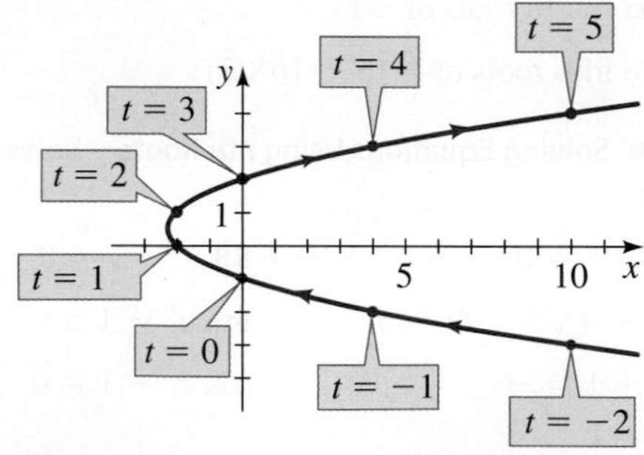

Figure 1

As t increases, a particle whose position is given by the parametric equations moves along the curve in the direction of the arrows.

Now Try Exercise 3 ■

If we replace t by $-t$ in Example 1, we obtain the parametric equations

$$x = t^2 + 3t \qquad y = -t - 1$$

The graph of these parametric equations (see Figure 2) is the same as the curve in Figure 1 but traced out in the opposite direction. On the other hand, if we replace t by

MARIA GAETANA AGNESI (1718–1799) is famous for having written *Instituzioni Analitiche*, one of the first calculus textbooks.

Agnesi was born into a wealthy family in Milan, Italy, the oldest of 21 children. She was a child prodigy, mastering many languages at an early age, including Latin, Greek, and Hebrew. At the age of 20 she published a series of essays on philosophy and natural science. After her mother died, Agnesi took on the task of educating her brothers. In 1748 Agnesi published her famous textbook, which she originally wrote as a text for tutoring her brothers. The book compiled and explained the mathematical knowledge of the day. It contains many carefully chosen examples, one of which is the curve now known as the "witch of Agnesi" (see Exercise 66). One review calls her book an "exposition by examples rather than by theory." The book gained Agnesi immediate recognition. Pope Benedict XIV appointed her to a position at the University of Bologna, writing, "we have had the idea that you should be awarded the well-known chair of mathematics, by which it comes of itself that you should not thank us but we you." This appointment was an extremely high honor for a woman, since very few women then were even allowed to attend university. Just two years later, Agnesi's father died, and she left mathematics completely. She became a nun and devoted the rest of her life and her wealth to caring for sick and dying women, herself dying in poverty at a poorhouse of which she had once been director.

$2t$ in Example 1, we obtain the parametric equations

$$x = 4t^2 - 6t \qquad y = 2t - 1$$

The graph of these parametric equations (see Figure 3) is again the same curve but is traced out "twice as fast." *Thus a parametrization contains more information than just the shape of the curve; it also indicates* how *the curve is being traced out.*

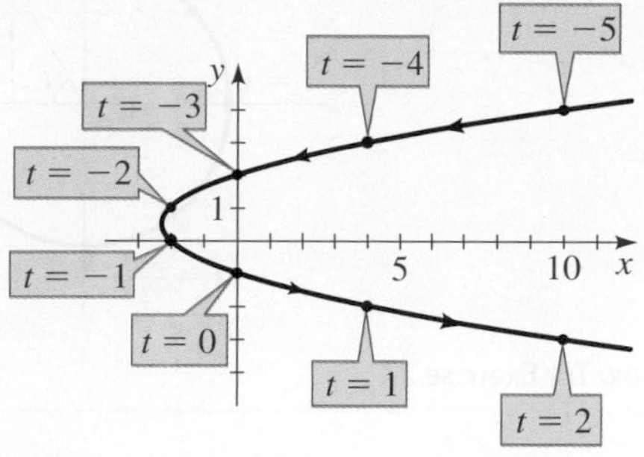

Figure 2 | $x = t^2 + 3t,\ y = -t - 1$

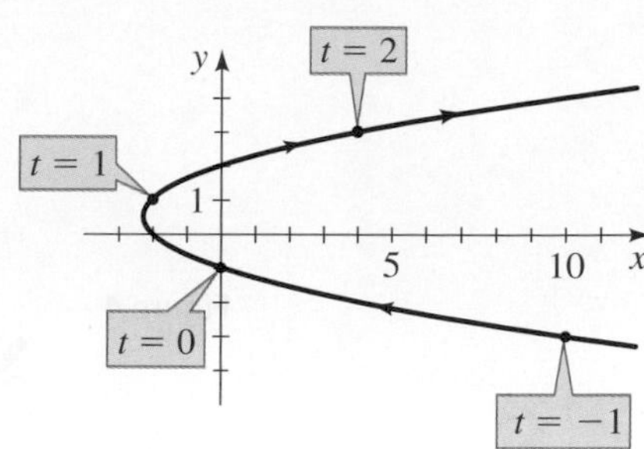

Figure 3 | $x = 4t^2 - 6t,\ y = 2t - 1$

■ Eliminating the Parameter

Often a curve given by parametric equations can also be represented by a single rectangular equation in x and y. The process of finding this equation is called *eliminating the parameter*. One way to do this is to solve for t in one equation, then substitute into the other.

Example 2 ■ Eliminating the Parameter

Eliminate the parameter in the parametric equations of Example 1.

Solution First we solve for t in the simpler equation, then we substitute into the other equation. From the equation $y = t - 1$ we get $t = y + 1$. Substituting into the equation for x, we get

$$x = t^2 - 3t = (y + 1)^2 - 3(y + 1) = y^2 - y - 2$$

Thus the curve in Example 1 has the rectangular equation $x = y^2 - y - 2$, so it is a parabola.

 **Now Try Exercise 5** ■

Eliminating the parameter often helps us identify the shape of a curve, as we see in the next two examples.

Example 3 ■ Modeling Circular Motion

The following parametric equations model the position of a moving object at time t (in seconds):

$$x = \cos t \qquad y = \sin t \qquad t \geq 0$$

Describe and graph the path of the object.

Solution To identify the curve, we eliminate the parameter. Since $\cos^2 t + \sin^2 t = 1$ and since $x = \cos t$ and $y = \sin t$ for every point (x, y) on the curve, we have

$$x^2 + y^2 = (\cos t)^2 + (\sin t)^2 = 1$$

This means that all points on the curve satisfy the equation $x^2 + y^2 = 1$, so the graph is a circle of radius 1 centered at the origin. As t increases from 0 to 2π, the point given by the parametric equations starts at $(1, 0)$ and moves counterclockwise once around the circle, as shown in Figure 4. So the object completes one revolution around the circle in 2π seconds. Notice that the parameter t can be interpreted as the angle shown in the figure.

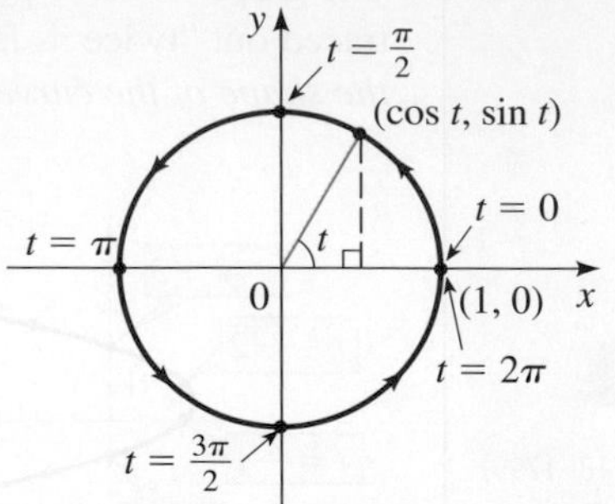

Figure 4

Now Try Exercise 27 ■

Example 4 ■ Sketching a Parametric Curve

Eliminate the parameter, and sketch the graph of the parametric equations

$$x = \sin t \qquad y = 2 - \cos^2 t$$

Solution To eliminate the parameter, we first use the trigonometric identity $\cos^2 t = 1 - \sin^2 t$ to change the second equation:

$$y = 2 - \cos^2 t = 2 - (1 - \sin^2 t) = 1 + \sin^2 t$$

Now we can substitute $\sin t = x$ from the first equation to get

$$y = 1 + x^2$$

so the point (x, y) moves along the parabola $y = 1 + x^2$. However, since $-1 \le \sin t \le 1$, we have $-1 \le x \le 1$, so the parametric equations represent only the part of the parabola between $x = -1$ and $x = 1$. Since $\sin t$ is periodic, the point $(x, y) = (\sin t, 2 - \cos^2 t)$ moves back and forth infinitely often along the parabola between the points $(-1, 2)$ and $(1, 2)$, as shown in Figure 5.

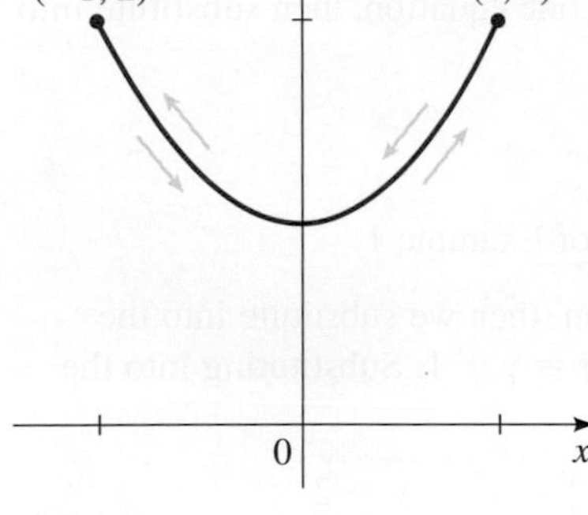

Figure 5

Now Try Exercise 15 ■

■ Finding Parametric Equations for a Curve

It is often possible to find parametric equations for a curve by using some geometric properties that define the curve, as in the next two examples.

Example 5 ■ Finding Parametric Equations for a Graph

Find parametric equations for the line of slope 3 that passes through the point $(2, 6)$.

Solution Let's start at the point $(2, 6)$ and move up and to the right along this line. Because the line has slope 3, for every 1 unit we move to the right, we must move upward 3 units. In other words, if we increase the x-coordinate by t units, we must correspondingly increase the y-coordinate by $3t$ units. This leads to the parametric equations

$$x = 2 + t \qquad y = 6 + 3t$$

To confirm that these equations give the desired line, we eliminate the parameter. We solve for t in the first equation and substitute into the second to get

$$y = 6 + 3(x - 2) = 3x$$

Thus the slope-intercept form of the equation of this line is $y = 3x$, which is a line of slope 3 that does pass through $(2, 6)$ as required. The graph is shown in Figure 6.

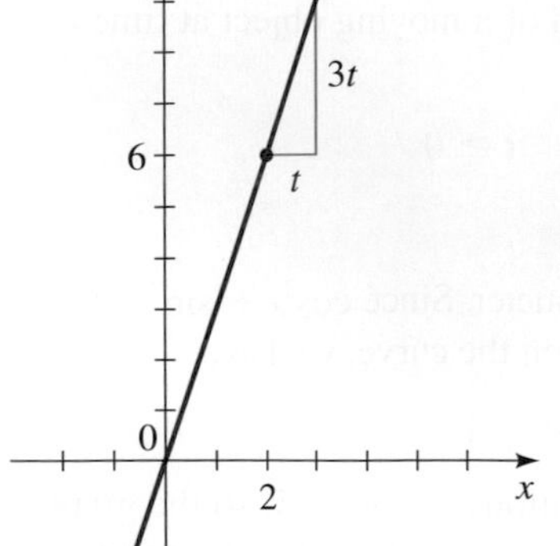

Figure 6

Now Try Exercise 33 ■

Example 6 ■ Parametric Equations for the Cycloid

As a circle rolls along a straight line, the curve traced out by a fixed point P on the circumference of the circle is called a **cycloid** (see Figure 7). If the circle has radius a and rolls along the x-axis, with one position of the point P being at the origin, find parametric equations for the cycloid.

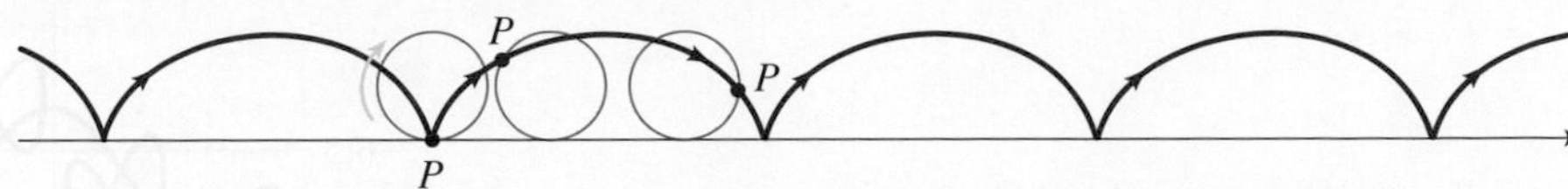

Figure 7

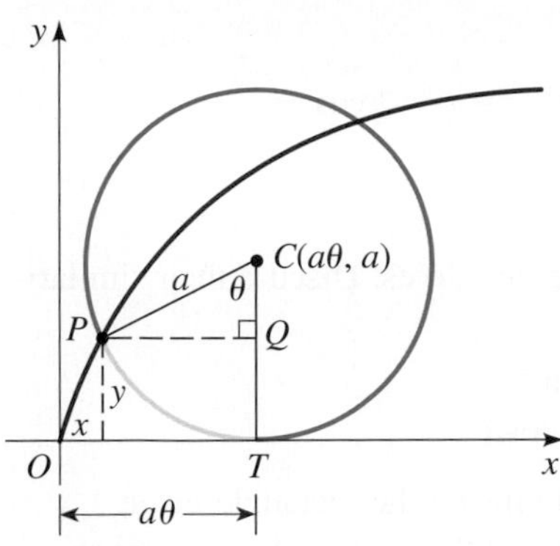

Figure 8

Solution Figure 8 shows the circle and the point P after the circle has rolled through an angle θ (in radians). The distance $d(O, T)$ that the circle has rolled must be the same as the length of the arc PT, which, by the arc length formula, is $a\theta$. (See Section 6.1.) This means that the center of the circle is $C(a\theta, a)$.

Let the coordinates of P be (x, y). Then from Figure 8 (which illustrates the case $0 < \theta < \pi/2$), we see that

$$x = d(O, T) - d(P, Q) = a\theta - a\sin\theta = a(\theta - \sin\theta)$$

$$y = d(T, C) - d(Q, C) = a - a\cos\theta = a(1 - \cos\theta)$$

so parametric equations for the cycloid are

$$x = a(\theta - \sin\theta) \qquad y = a(1 - \cos\theta)$$

Now Try Exercise 53 ■

The cycloid has a number of interesting physical properties. It is the "curve of quickest descent" in the following sense. Let's choose two points P and Q that are not directly above each other and join them with a wire. Suppose we allow a bead to slide down the wire under the influence of gravity (ignoring friction). Of all possible shapes into which the wire can be bent, the bead will slide from P to Q the fastest when the shape is half of an arch of an inverted cycloid (see Figure 9). The cycloid is also the "curve of equal descent" in the sense that no matter where we place a bead B on a cycloid-shaped wire, it takes the same time to slide to the bottom (see Figure 10). These rather surprising properties of the cycloid were proved (using calculus) in the 17th century by several mathematicians and physicists, including Johann Bernoulli, Blaise Pascal, and Christiaan Huygens.

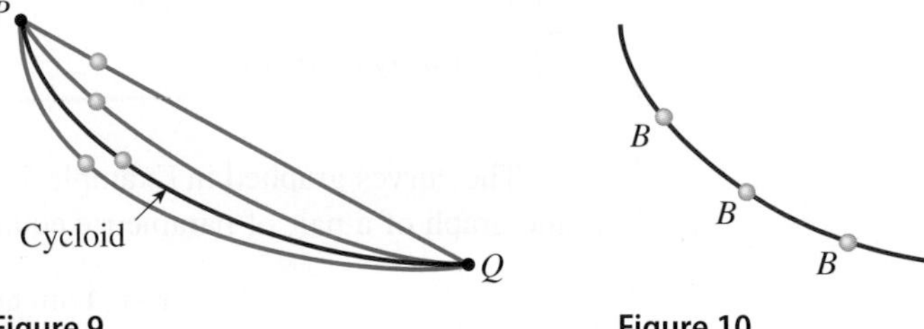

Figure 9

Figure 10

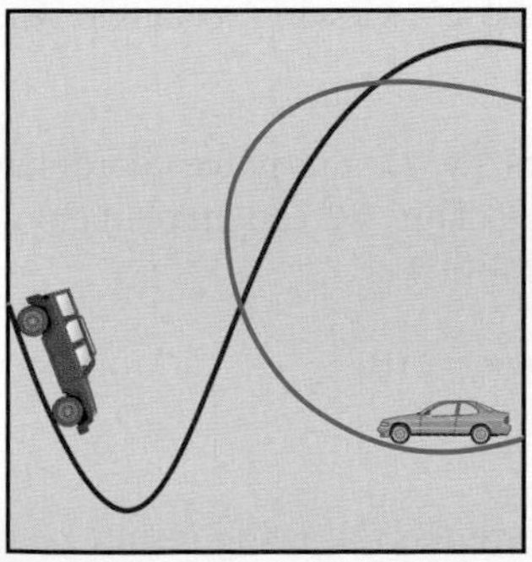

Discovery Project ■ Collision

Two cars are travelling on two different roads. The roads intersect, but will the cars collide? In this project we model the location of each car by a pair of parametric equations, with time t as the parameter. The equations will tell us if the cars collide, that is, if they will be at the same location at the same time. We will also use graphing devices to "animate" the parametric equations, so we can visually confirm whether the cars collide. You can find the project at **www.stewartmath.com**.

■ Using Graphing Devices to Graph Parametric Curves

Most graphing calculators and computer graphing programs can be used to graph parametric equations. Such devices are particularly useful in sketching complicated curves like the one shown in Figure 11.

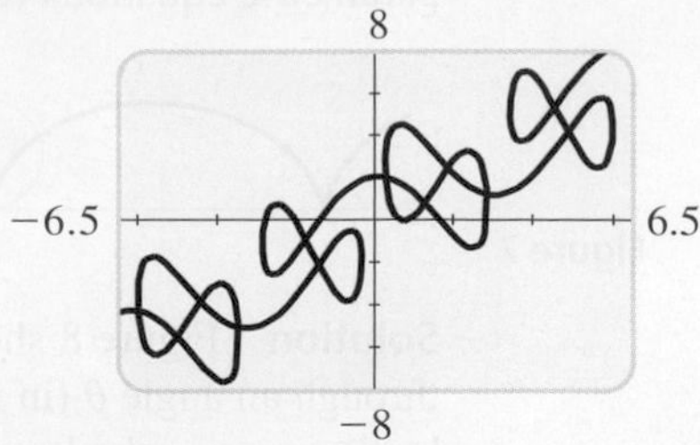

Figure 11 | $x = t + 2 \sin 2t$, $y = t + 2 \cos 5t$

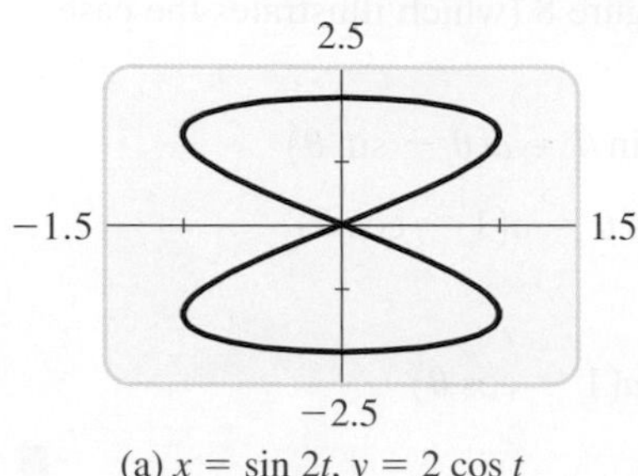

(a) $x = \sin 2t$, $y = 2 \cos t$

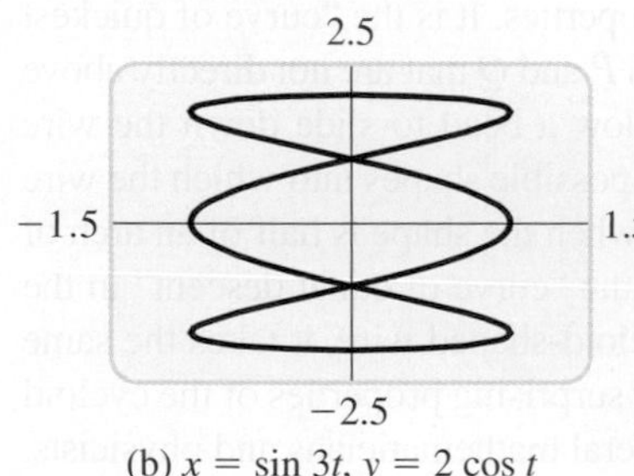

(b) $x = \sin 3t$, $y = 2 \cos t$

Figure 12

Example 7 ■ Graphing Parametric Curves

Use a graphing device to draw the following parametric curves. Discuss their similarities and differences.

(a) $x = \sin 2t$
$y = 2 \cos t$

(b) $x = \sin 3t$
$y = 2 \cos t$

Solution In both parts (a) and (b) the graph will lie inside the rectangle given by $-1 \le x \le 1$, $-2 \le y \le 2$, because both the sine and the cosine of any number will be between -1 and 1. Thus we may use the viewing rectangle $[-1.5, 1.5]$ by $[-2.5, 2.5]$.

(a) Since $2 \cos t$ is periodic with period 2π (see Section 5.3) and since $\sin 2t$ has period π, letting t vary over the interval $0 \le t \le 2\pi$ gives us the complete graph, which is shown in Figure 12(a).

(b) Again, letting t take on values between 0 and 2π gives the complete graph shown in Figure 12(b).

Both graphs are *closed curves*, which means that they form loops with the same starting and ending point; also, both graphs cross over themselves. However, the graph in Figure 12(a) has two loops, like a figure eight, whereas the graph in Figure 12(b) has three loops.

Now Try Exercise 39 ■

The curves graphed in Example 7 are called Lissajous figures. A **Lissajous figure** is the graph of a pair of parametric equations of the form

$$x = A \sin \omega_1 t \qquad y = B \cos \omega_2 t$$

where A, B, ω_1, and ω_2 are positive real constants. Since both $\sin \omega_1 t$ and $\cos \omega_2 t$ are between -1 and 1, a Lissajous figure will lie inside the rectangle determined by $-A \le x \le A$, $-B \le y \le B$. This fact can be used to choose a viewing rectangle when graphing a Lissajous figure, as we did in Example 7.

Recall from Section 8.1 that rectangular coordinates (x, y) and polar coordinates (r, θ) are related by the equations $x = r \cos \theta$, $y = r \sin \theta$. Thus we can graph the polar equation $r = f(\theta)$ by changing it to parametric form as follows.

$$x = r \cos \theta = f(\theta) \cos \theta \qquad \text{Since } r = f(\theta)$$

$$y = r \sin \theta = f(\theta) \sin \theta$$

Replacing θ by the standard parametric variable t, we have the following result.

Polar Equations in Parametric Form

The graph of the polar equation $r = f(\theta)$ is the same as the graph of the parametric equations

$$x = f(t)\cos t \qquad y = f(t)\sin t$$

Example 8 ■ Parametric Form of a Polar Equation

Consider the polar equation $r = \theta$, $1 \le \theta \le 10\pi$.

(a) Express the equation in parametric form.

(b) Draw a graph of the parametric equations from part (a).

Solution Here, $r = f(\theta) = \theta$, so $f(t) = t$.

(a) The given polar equation is equivalent to the parametric equations

$$x = t\cos t \qquad y = t\sin t$$

(b) Since $10\pi \approx 31.42$, we use the viewing rectangle $[-32, 32]$ by $[-32, 32]$, and we let t vary from 1 to 10π. The resulting graph shown in Figure 13 is a *spiral*.

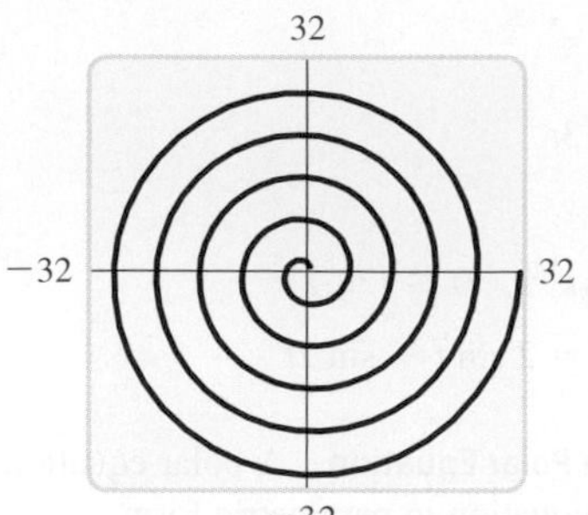

Figure 13 | $x = t\cos t$, $y = t\sin t$

Now Try Exercise 47 ■

8.4 | Exercises

Concepts

1. (a) The parametric equations

$$x = f(t) \qquad y = g(t)$$

give the coordinates of a point $(x, y) = (f(t), g(t))$ for appropriate values of t. The variable t is called a ________.

(b) Suppose that the parametric equations

$$x = t \qquad y = t^2 \qquad t \ge 0$$

model the position of a moving object at time t. When $t = 0$, the object is at (,), and when $t = 1$, the object is at (,).

(c) If we eliminate the parameter in part (b), we get the equation $y =$ ________. We see from this equation that the path of the moving object is a ________.

2. (a) *True or False*? The same curve can be described by parametric equations in many different ways.

(b) The parametric equations

$$x = 2t \qquad y = (2t)^2$$

model the position of a moving object at time t. When $t = 0$, the object is at (,), and when $t = 1$, the object is at (,).

(c) If we eliminate the parameter, we get the equation $y =$ ________, which is the same equation as in Exercise 1(c). So the objects in Exercises 1(b) and 2(b) move along the same ________ but traverse the path differently. Indicate the position of each object when $t = 0$ and when $t = 1$ on the following graph.

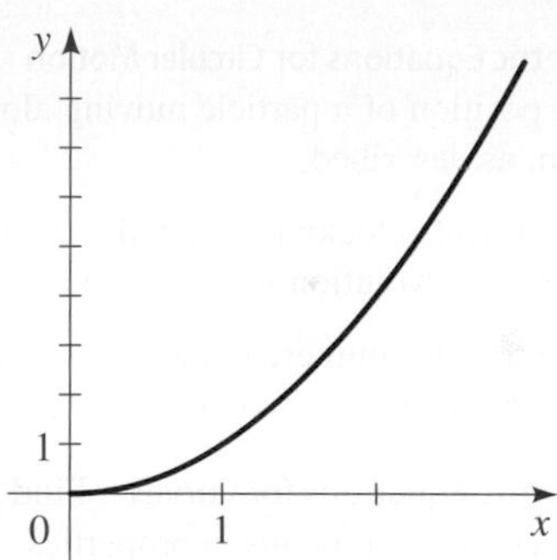

Skills

3–26 ■ Sketching a Curve by Eliminating the Parameter A pair of parametric equations is given. **(a)** Sketch the curve represented by the parametric equations. Use arrows to indicate the direction of the curve as t increases. **(b)** Find an equation in rectangular coordinates for the curve by eliminating the parameter.

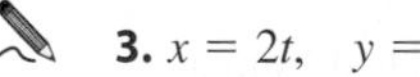

3. $x = 2t, \quad y = t + 6$

4. $x = 6t - 4, \quad y = 3t, \quad t \ge 0$

5. $x = t^2, \quad y = t - 2, \quad 2 \le t \le 4$

6. $x = 2t + 1, \quad y = \left(t + \frac{1}{2}\right)^2$

7. $x = \sqrt{t}, \quad y = 1 - t$

8. $x = t^2, \quad y = t^4 + 1$

9. $x = \dfrac{1}{t}, \quad y = t + 1$

10. $x = t + 1, \quad y = \dfrac{t}{t+1}$

11. $x = 4t^2, \quad y = 8t^3$

12. $x = |t|, \quad y = |1 - |t||$

13. $x = 2 \sin t, \quad y = 2 \cos t, \quad 0 \le t \le \pi$

14. $x = 2 \cos t, \quad y = 3 \sin t, \quad 0 \le t \le 2\pi$

15. $x = \sin^2 t, \quad y = \sin^4 t$

16. $x = \sin^2 t, \quad y = \cos t$

17. $x = \cos t, \quad y = \cos 2t$

18. $x = \cos 2t, \quad y = \sin 2t$

19. $x = \sec t, \quad y = \tan t, \quad 0 \le t < \pi/2$

20. $x = \cot t, \quad y = \csc t, \quad 0 < t < \pi$

21. $x = \tan t, \quad y = \cot t, \quad 0 < t < \pi/2$

22. $x = e^{-t}, \quad y = e^t$

23. $x = e^{2t}, \quad y = e^t$

24. $x = \sec t, \quad y = \tan^2 t, \quad 0 \le t < \pi/2$

25. $x = \cos^2 t, \quad y = \sin^2 t$

26. $x = \cos^3 t, \quad y = \sin^3 t, \quad 0 \le t \le 2\pi$

27–30 ■ Circular Motion The position of an object in circular motion is modeled by the given parametric equations. Describe the path of the object by stating the radius of the circle, the position at time $t = 0$, the orientation of the motion (clockwise or counterclockwise), and the time t that it takes to complete one revolution around the circle.

27. $x = 3 \cos t, \quad y = 3 \sin t$

28. $x = 2 \sin t, \quad y = 2 \cos t$

29. $x = \sin 2t, \quad y = \cos 2t$

30. $x = 4 \cos 3t, \quad y = 4 \sin 3t$

31–32 ■ Parametric Equations for Circular Motion Find parametric equations for the position of a particle moving along a circle centered at the origin, as described.

31. The particle travels clockwise around a circle with radius 5 and completes a revolution in 4π seconds.

32. The particle travels counterclockwise around a circle with radius 1 and completes a revolution in 2 seconds.

33–36 ■ Parametric Equations for Curves Find parametric equations for the curve with the given properties.

33. The line with slope $\frac{1}{2}$, passing through $(4, -1)$

34. The line passing through $(6, 7)$ and $(7, 8)$

35. The circle $x^2 + y^2 = a^2$.

36. The ellipse

$$\frac{x^2}{a^2} + \frac{y^2}{b^2} = 1$$

37. Path of a Projectile If a projectile is launched from a cannon with an initial speed of v_0 m/s at an angle α above the horizontal, then its position after t seconds is given by the parametric equations

$$x = (v_0 \cos \alpha)t \qquad y = (v_0 \sin \alpha)t - 5t^2$$

(where x and y are measured in meters). Show that the path of the projectile is a parabola by eliminating the parameter t.

38. Path of a Projectile Referring to Exercise 37, suppose the projectile is fired into the air with an initial speed of 640 m/s at an angle of 30° to the horizontal.

(a) After how many seconds will the projectile hit the ground?

(b) How far from the cannon will the projectile hit the ground?

(c) What is the maximum height attained by the projectile?

39–44 ■ Graphs of Parametric Equations Use a graphing device to draw the curve represented by the parametric equations.

39. $x = \sin t, \quad y = 2 \cos 3t$

40. $x = 2 \sin t, \quad y = \cos 4t$

41. $x = 3 \sin 5t, \quad y = 5 \cos 3t$

42. $x = \sin 4t, \quad y = \cos 3t$

43. $x = \sin(\cos t), \quad y = \cos t^{3/2}, \quad 0 \le t \le 2\pi$

44. $x = 2 \cos t + \cos 2t, \quad y = 2 \sin t - \sin 2t$

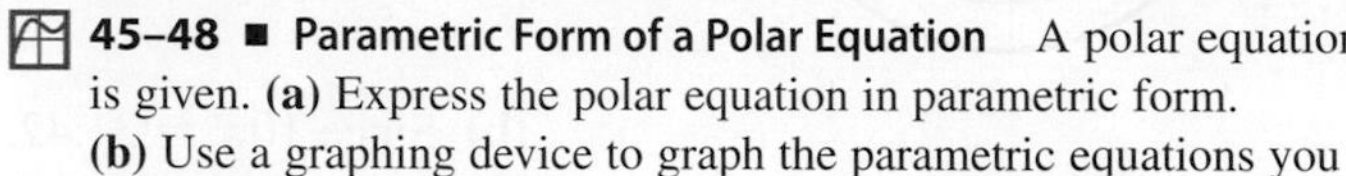

45–48 ■ Parametric Form of a Polar Equation A polar equation is given. **(a)** Express the polar equation in parametric form. **(b)** Use a graphing device to graph the parametric equations you found in part (a).

45. $r = 2^{\theta/12}, \quad 0 \le \theta \le 4\pi$

46. $r = \sin \theta + 2 \cos \theta$

47. $r = \dfrac{4}{2 - \cos \theta}$

48. $r = 2^{\sin \theta}$

49–52 ■ Graphs of Parametric Equations Match the parametric equations with the graphs labeled I–IV. Give reasons for your answers.

49. $x = t^3 - 2t, \quad y = t^2 - t$

50. $x = \sin 3t, \quad y = \sin 4t$

51. $x = t + \sin 2t, \quad y = t + \sin 3t$

52. $x = \sin(t + \sin t), \quad y = \cos(t + \cos t)$

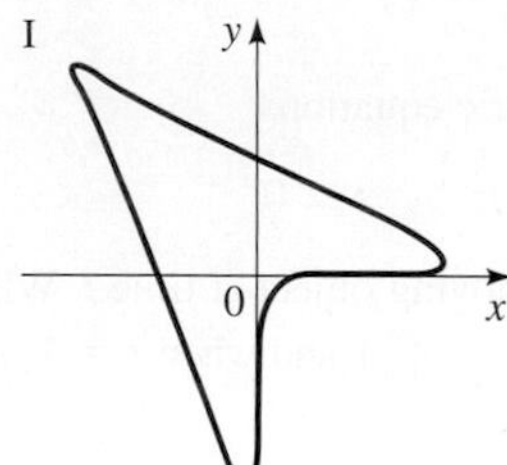

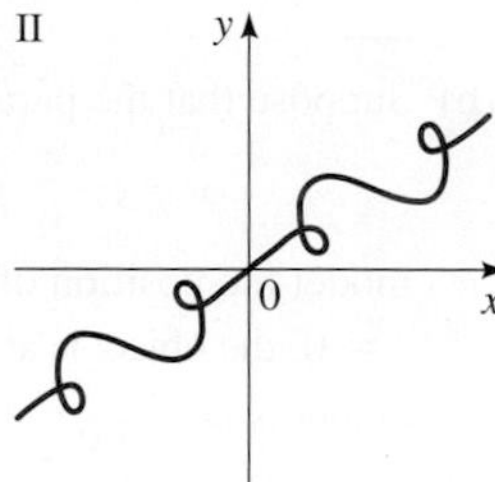

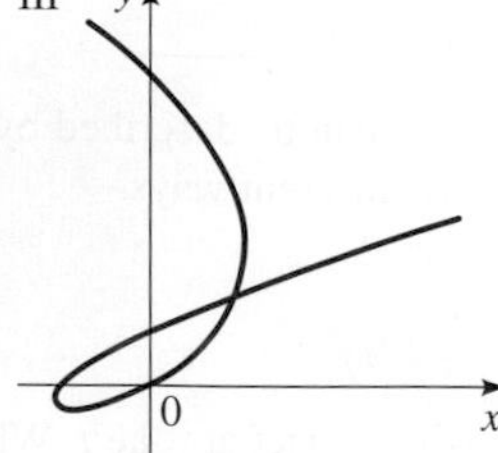

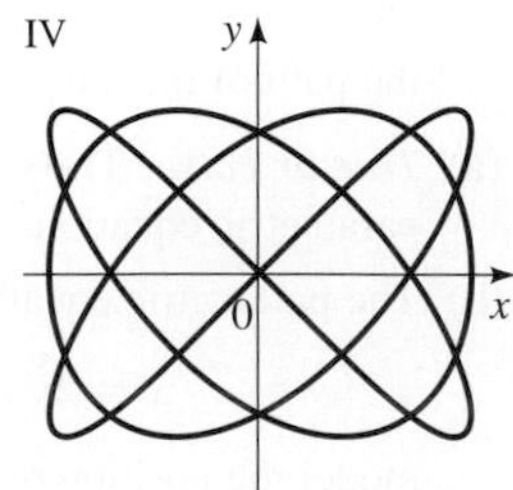

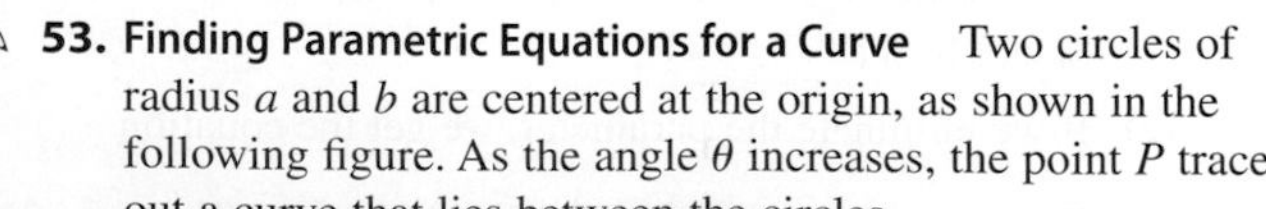

53. Finding Parametric Equations for a Curve Two circles of radius a and b are centered at the origin, as shown in the following figure. As the angle θ increases, the point P traces out a curve that lies between the circles.

(a) Find parametric equations for the curve, using θ as the parameter.

(b) Graph the curve using a graphing device, with $a = 3$ and $b = 2$.

(c) Eliminate the parameter, and identify the curve.

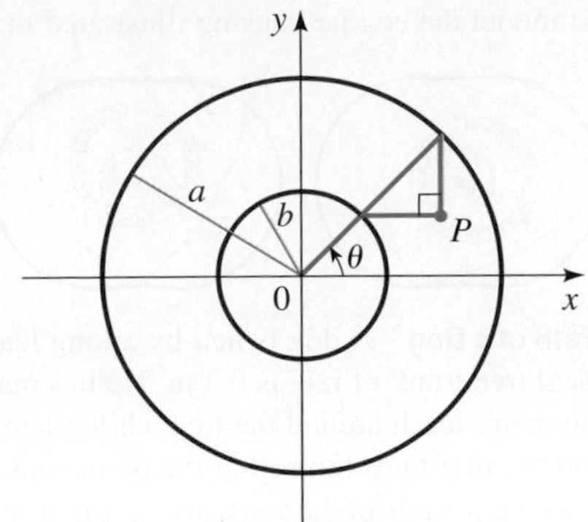

54. Finding Parametric Equations for a Curve Two circles of radius a and b are centered at the origin, as shown in the figure. The line segment AB is tangent to the larger circle so that angle OAB is a right angle. (See Appendix A *Geometry Review*.)

(a) Find parametric equations for the curve traced out by the point P, using the angle θ as the parameter.

(b) Graph the curve using a graphing device, with $a = 3$ and $b = 2$.

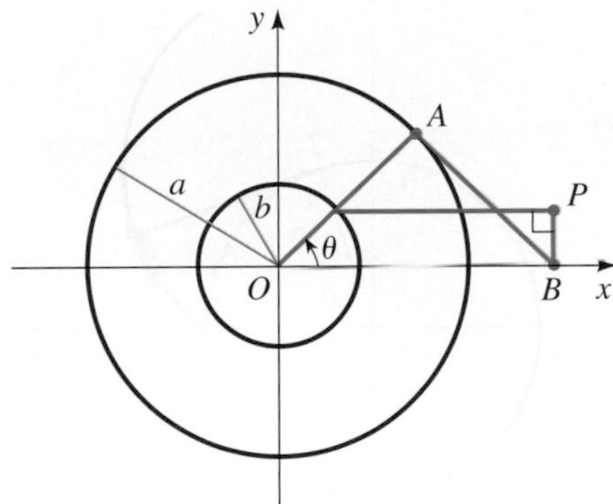

55. Curtate Cycloid

(a) In Example 6, suppose the point P that traces out the curve lies not on the edge of the circle but rather at a fixed point inside the rim, at a distance b from the center (with $b < a$). The curve traced out by P is called a **curtate cycloid** (or **trochoid**). Show that parametric equations for the curtate cycloid are

$$x = a\theta - b\sin\theta \qquad y = a - b\cos\theta$$

(b) Sketch the graph using $a = 3$ and $b = 2$.

56. Prolate Cycloid

(a) In Exercise 55, if the point P lies *outside* the circle at a distance b from the center (with $b > a$), then the curve traced out by P is called a **prolate cycloid**. Show that parametric equations for the prolate cycloid are the same as the equations for the curtate cycloid.

(b) Sketch the graph for the case in which $a = 1$ and $b = 2$.

Skills Plus

57. Parametric Equations of a Hyperbola Eliminate the parameter θ in the following parametric equations. (This curve is called a **hyperbola**; see Section 10.3.)

$$x = a\tan\theta \qquad y = b\sec\theta$$

58. Parametric Equations of a Hyperbola Show that the following parametric equations represent a part of the hyperbola of Exercise 57.

$$x = a\sqrt{t} \qquad y = b\sqrt{t+1}$$

59–62 ■ Graphs of Parametric Equations Sketch the curve given by the parametric equations.

59. $x = t\cos t,\quad y = t\sin t,\quad t \geq 0$

60. $x = \sin t,\quad y = \sin 2t$

61. $x = \dfrac{3t}{1+t^3},\quad y = \dfrac{3t^2}{1+t^3}$

62. $x = \cot t,\quad y = 2\sin^2 t,\quad 0 < t < \pi$

63. Hypocycloid A circle C of radius b rolls on the inside of a larger circle of radius a centered at the origin. Let P be a fixed point on the smaller circle, with initial position at the point $(a, 0)$ as shown in the figure. The curve traced out by P is called a **hypocycloid**.

(a) Show that parametric equations for the hypocycloid are

$$x = (a-b)\cos\theta + b\cos\left(\frac{a-b}{b}\theta\right)$$

$$y = (a-b)\sin\theta - b\sin\left(\frac{a-b}{b}\theta\right)$$

(b) If $a = 4b$, the hypocycloid is called an **astroid**. Show that in this case the parametric equations can be reduced to

$$x = a\cos^3\theta \qquad y = a\sin^3\theta$$

Sketch the curve. Eliminate the parameter to obtain an equation for the astroid in rectangular coordinates.

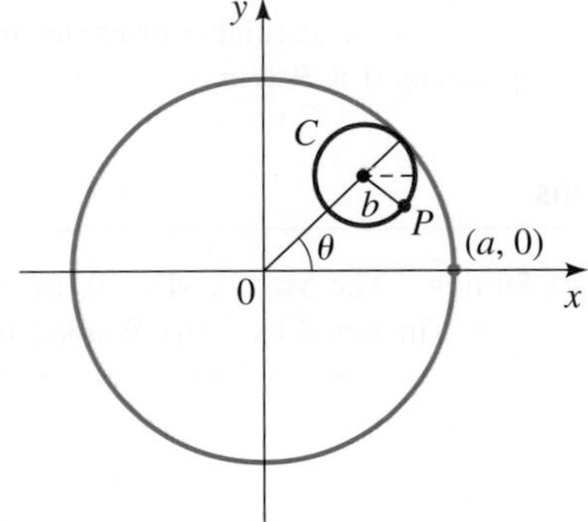

64. Epicycloid If the circle C of Exercise 63 rolls on the *outside* of the larger circle, the curve traced out by P is called an epicycloid. Find parametric equations for the epicycloid.

65. Longbow Curve In the following figure, the circle of radius a is stationary, and for every θ, the point P is the midpoint of the segment QR. The curve traced out by P for $0 < \theta < \pi$

is called the **longbow curve**. Find parametric equations for this curve.

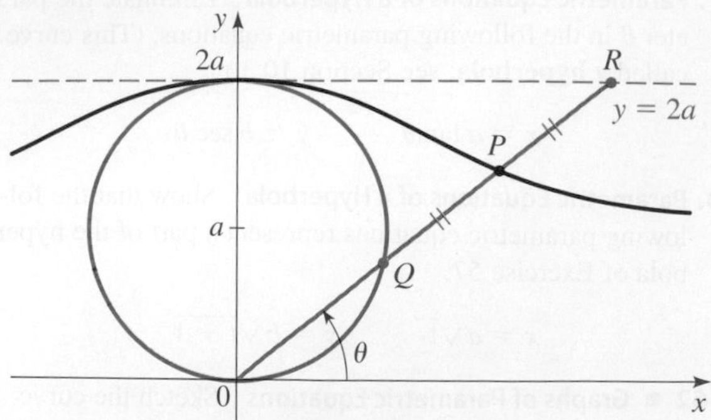

66. The Witch of Agnesi A curve, called a **witch of Agnesi**, consists of all points P determined as shown in the figure.

(a) Show that parametric equations for this curve can be written as

$$x = 2a\cot\theta \qquad y = 2a\sin^2\theta$$

[*Hint:* A triangle inscribed in a semicircle is a right triangle. See Appendix A *Geometry Review*.]

(b) Graph the curve using a graphing device, with $a = 3$.

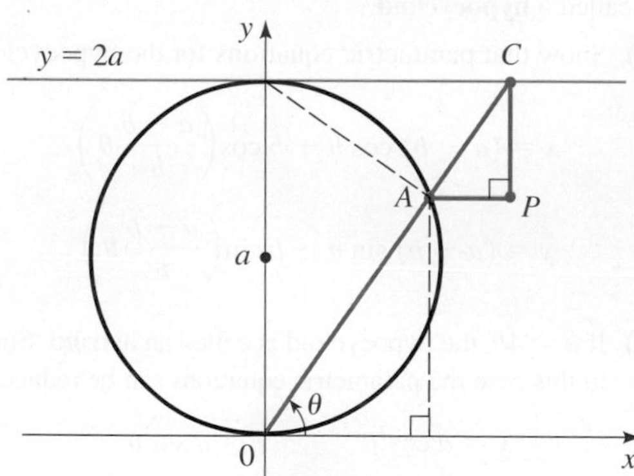

67. Eliminating the Parameter Eliminate the parameter θ in the parametric equations for the cycloid (Example 6) to obtain an equation in rectangular coordinates for the section of the curve given by $0 \le \theta \le \pi$.

Applications

68. The Rotary Engine The Mazda MX-30 uses an unconventional engine (invented by Felix Wankel in 1954) in which the pistons are replaced by a rotor in the shape of a Reuleaux triangle (Exercise 6.1.95). The rotor turns in a special housing as shown in the figure. The vertices of the rotor maintain contact with the housing at all times, while the center of the rotor traces out a circle of radius r, turning the drive shaft. (For an animation go to www.wikipedia.org/wiki/Wankel_engine.) The shape of the housing is given by the following parametric equations (where R is the distance between the vertices and center of the rotor):

$$x = r\cos 3\theta + R\cos\theta \qquad y = r\sin 3\theta + R\sin\theta$$

(a) Suppose that the drive shaft has radius $r = 1$. Graph the curve given by the parametric equations for the following values of R: 0.5, 1, 3, 5.

(b) Which of the four values of R given in part (a) seems to best model the engine housing illustrated in the figure?

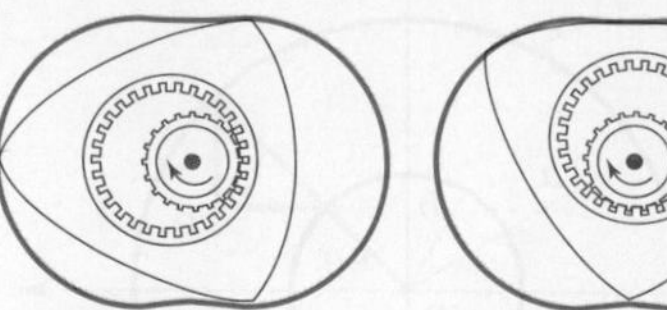

69. Spiral Path of a Dog A dog is tied by a long leash to a cylindrical tree trunk of radius 0.3 m. He has managed to wrap the entire leash around the tree while playing in the yard, and the dog finds himself at the point (0.3, 0) in the figure. Seeing a squirrel, he runs around the tree counterclockwise, unwinding the leash and keeping it taut while chasing the intruder.

(a) Show that parametric equations for the dog's path (called an **involute of a circle**) are

$$x = \cos\theta + \theta\sin\theta \qquad y = \sin\theta - \theta\cos\theta$$

[*Hint:* Note that the leash is always tangent to the tree, so OT is perpendicular to TD.]

(b) Graph the path of the dog for $0 \le \theta \le 4\pi$.

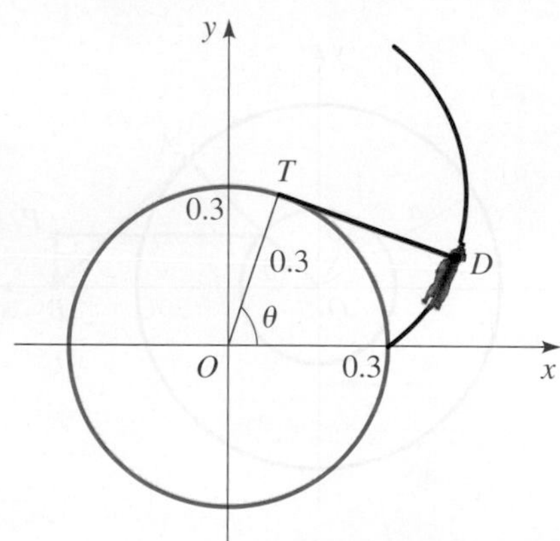

Discuss ■ Discover ■ Prove ■ Write

70. Discover ■ Write: More Information in Parametric Equations In this section we stated that parametric equations contain more information than just the shape of a curve. Write a short paragraph explaining this statement. Use the following example and your answers to parts (a) and (b) below in your explanation.

The position of a particle is given by the parametric equations

$$x = \sin t \qquad y = \cos t$$

where t represents time. We know that the shape of the path of the particle is a circle.

(a) How long does it take the particle to travel once around the circle? Find parametric equations for the case when the particle moves twice as fast around the circle.

(b) Does the particle travel clockwise or counterclockwise around the circle? Find parametric equations for the case when the particle moves in the opposite direction around the circle.

71. Discuss: Different Ways of Tracing Out a Curve The curves C, D, E, and F are defined parametrically as follows, where the parameter t takes on all real values unless otherwise stated:

C: $x = t, \quad y = t^2$

D: $x = \sqrt{t}, \quad y = t, \quad t \geq 0$

E: $x = \sin t, \quad y = \sin^2 t$

F: $x = 3^t, \quad y = 3^{2t}$

(a) Show that the points on all four of these curves satisfy the same equation in rectangular coordinates.

(b) Draw the graph of each curve and explain how the curves differ from one another.

8.5 Vectors

■ **Geometric Description of Vectors** ■ **Vectors in the Coordinate Plane**
■ **Using Vectors to Model Velocity and Force**

In applications of mathematics, certain quantities are determined completely by their magnitude—for example, length, mass, area, temperature, and energy. We speak of a length of 5 m or a mass of 3 kg; only one number is needed to describe each of these quantities. Such a quantity is called a **scalar**.

On the other hand, to describe the displacement of an object, two numbers are required: the *magnitude* and the *direction* of the displacement. To describe the velocity of a moving object, we must specify both the *speed* and the *direction* of travel. Quantities such as displacement, velocity, acceleration, and force that involve magnitude as well as direction are called *directed quantities*. One way to represent such quantities mathematically is through the use of *vectors*.

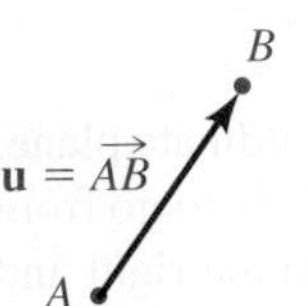

Figure 1

■ Geometric Description of Vectors

A **vector** in the plane is a line segment with an assigned direction. We sketch a vector as shown in Figure 1 with an arrow to specify the direction. We denote this vector by $\overrightarrow{AB}$. Point A is the **initial point**, and B is the **terminal point** of the vector $\overrightarrow{AB}$. The length of the line segment AB is called the **magnitude** or **length** of the vector and is denoted by $|\overrightarrow{AB}|$. We use boldface letters to denote vectors. Thus we write $\mathbf{u} = \overrightarrow{AB}$.

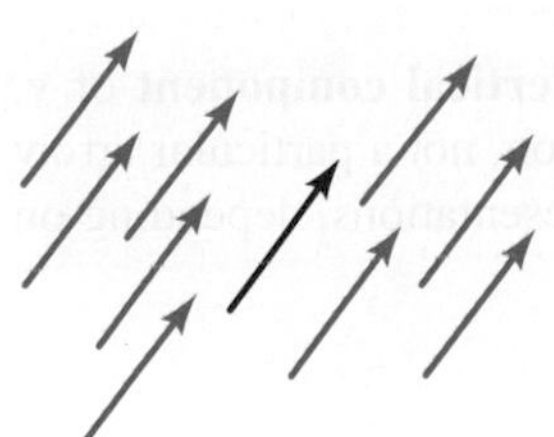
Figure 2

Two vectors are considered **equal** if they have equal magnitude and the same direction. Thus all the vectors in Figure 2 are equal. This definition of equality makes sense if we think of a vector as representing a displacement. Two such displacements are the same if they have equal magnitudes and the same direction. So the vectors in Figure 2 can be thought of as the *same* displacement applied to objects in different locations in the plane.

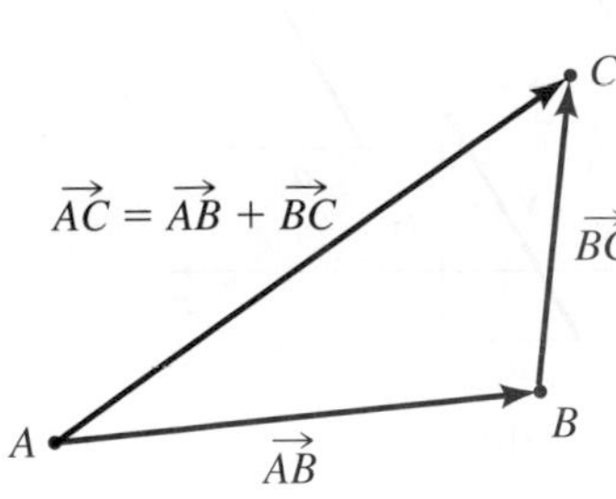

Figure 3

If the displacement $\mathbf{u} = \overrightarrow{AB}$ is followed by the displacement $\mathbf{v} = \overrightarrow{BC}$, then the resulting displacement is $\overrightarrow{AC}$ as shown in Figure 3. In other words, the single displacement represented by the vector $\overrightarrow{AC}$ has the same effect as the other two displacements together. We call the vector $\overrightarrow{AC}$ the **sum** of the vectors $\overrightarrow{AB}$ and $\overrightarrow{BC}$, and we write $\overrightarrow{AC} = \overrightarrow{AB} + \overrightarrow{BC}$. (The **zero vector**, denoted by **0**, represents no displacement.) Thus to find the sum of any two vectors **u** and **v**, we sketch vectors equal to **u** and **v** with the initial point of one at the terminal point of the other [see Figure 4(a)]. If we draw **u** and **v** starting at the same point, then **u** + **v** is the vector that is the diagonal of the parallelogram formed by **u** and **v** shown in Figure 4(b).

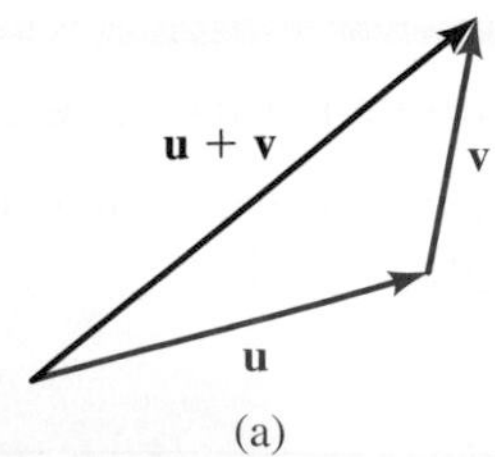

Figure 4 | Addition of vectors

If c is a real number and $\mathbf{v}$ is a vector, we define a new vector $c\mathbf{v}$ as follows: The vector $c\mathbf{v}$ has magnitude $|c||\mathbf{v}|$ and has the same direction as $\mathbf{v}$ if $c > 0$ and the opposite direction if $c < 0$. If $c = 0$, then $c\mathbf{v} = \mathbf{0}$, the zero vector. This process is called **multiplication of a vector by a scalar**. Multiplying a vector by a scalar has the effect of stretching or shrinking the vector. Figure 5 shows graphs of the vector $c\mathbf{v}$ for different values of c. We write the vector $(-1)\mathbf{v}$ as $-\mathbf{v}$. Thus $-\mathbf{v}$ is the vector with the same length as $\mathbf{v}$ but with opposite direction.

The **difference** of two vectors $\mathbf{u}$ and $\mathbf{v}$ is defined by $\mathbf{u} - \mathbf{v} = \mathbf{u} + (-\mathbf{v})$. Figure 6 shows that the vector $\mathbf{u} - \mathbf{v}$ is the other diagonal of the parallelogram formed by $\mathbf{u}$ and $\mathbf{v}$.

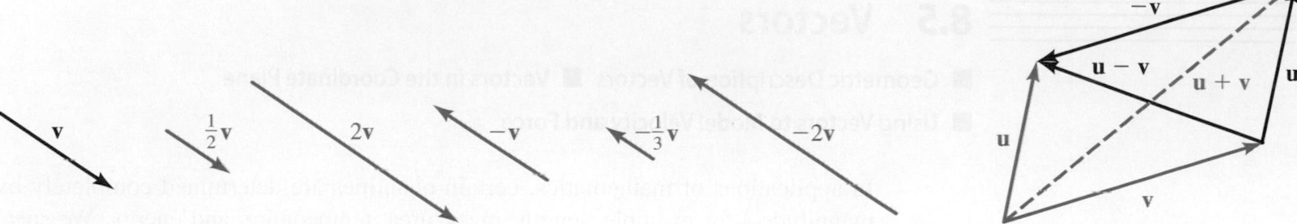

Figure 5 | Multiplication of a vector by a scalar

Figure 6 | Subtraction of vectors

■ Vectors in the Coordinate Plane

So far, we've discussed vectors geometrically. By placing a vector in a coordinate plane, we can describe it analytically (that is, by using components). In Figure 7(a), to go from the initial point of the vector $\mathbf{v}$ to the terminal point, we move a_1 units to the right and a_2 units upward. We represent $\mathbf{v}$ as an ordered pair of real numbers.

Note the distinction between the *vector* $\langle a_1, a_2 \rangle$ and the *point* (a_1, a_2).

$$\mathbf{v} = \langle a_1, a_2 \rangle$$

where a_1 is the **horizontal component** of $\mathbf{v}$ and a_2 is the **vertical component** of $\mathbf{v}$. Remember that a vector represents a magnitude and a direction, not a particular arrow in the plane. Thus the vector $\langle a_1, a_2 \rangle$ has many different representations, depending on its initial point [see Figure 7(b)].

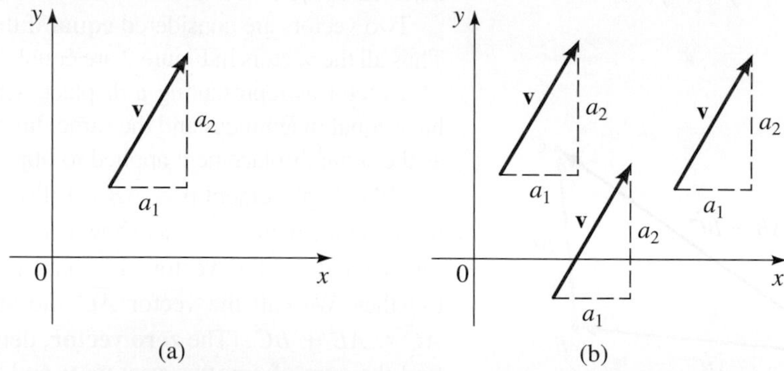

Figure 7

Using Figure 8, we can state the relationship between a geometric representation of a vector and the analytic one as follows.

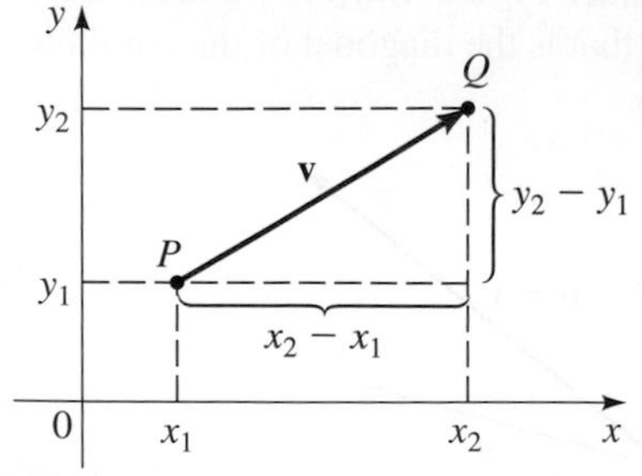

Figure 8

Component Form of a Vector

If a vector $\mathbf{v}$ is represented in the plane with initial point $P(x_1, y_1)$ and terminal point $Q(x_2, y_2)$, then

$$\mathbf{v} = \langle x_2 - x_1, y_2 - y_1 \rangle$$

Example 1 ■ Describing Vectors in Component Form

(a) Find the component form of the vector **u** with initial point $(-2, 5)$ and terminal point $(3, 7)$.

(b) If the vector $\mathbf{v} = \langle 3, 7 \rangle$ is sketched with initial point $(2, 4)$, what is its terminal point?

(c) Sketch representations of the vector $\mathbf{w} = \langle 2, 3 \rangle$ with initial points at $(0, 0)$, $(2, 2)$, $(-2, -1)$, and $(1, 4)$.

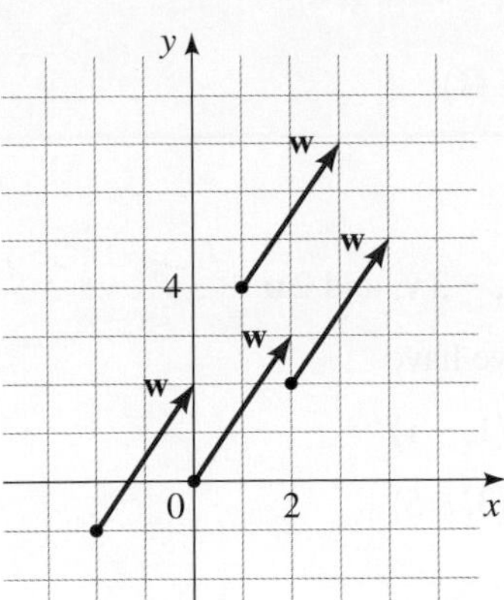

Figure 9

Solution

(a) The desired vector is

$$\mathbf{u} = \langle 3 - (-2), 7 - 5 \rangle = \langle 5, 2 \rangle$$

(b) Let the terminal point of **v** be (x, y). Then

$$\langle x - 2, y - 4 \rangle = \langle 3, 7 \rangle$$

So $x - 2 = 3$ and $y - 4 = 7$, or $x = 5$ and $y = 11$. The terminal point is $(5, 11)$.

(c) Representations of the vector **w** are sketched in Figure 9.

 **Now Try Exercises 11, 19, and 23** ■

We now give analytic definitions of the various operations on vectors that we have described geometrically. Let's start with equality of vectors. We've said that two vectors are equal if they have equal magnitude and the same direction. For the vectors $\mathbf{u} = \langle a_1, a_2 \rangle$ and $\mathbf{v} = \langle b_1, b_2 \rangle$ this means that $a_1 = b_1$ and $a_2 = b_2$. In other words, two vectors are **equal** if and only if their corresponding components are equal. Thus all the arrows in Figure 7(b) represent the same vector, as do all the arrows in Figure 9.

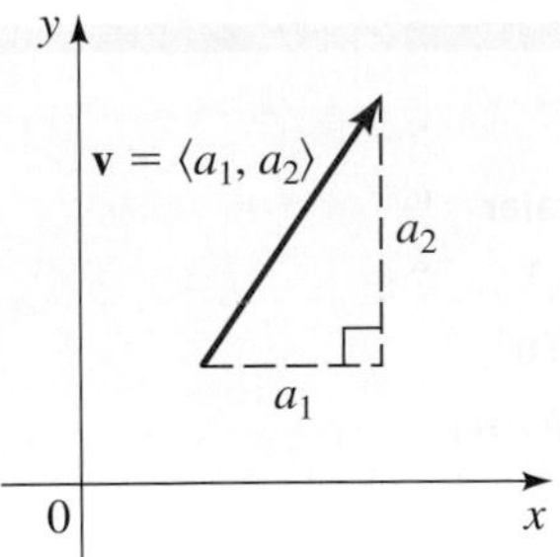

Figure 10

Applying the Pythagorean Theorem to the triangle in Figure 10, we obtain the following formula for the magnitude of a vector.

> **Magnitude of a Vector**
>
> The **magnitude** or **length** of a vector $\mathbf{v} = \langle a_1, a_2 \rangle$ is
>
> $$|\mathbf{v}| = \sqrt{a_1^2 + a_2^2}$$

Example 2 ■ Magnitudes of Vectors

Find the magnitude of each vector.

(a) $\mathbf{u} = \langle 2, -3 \rangle$ **(b)** $\mathbf{v} = \langle 5, 0 \rangle$ **(c)** $\mathbf{w} = \langle \frac{3}{5}, \frac{4}{5} \rangle$

Solution

(a) $|\mathbf{u}| = \sqrt{2^2 + (-3)^2} = \sqrt{13}$

(b) $|\mathbf{v}| = \sqrt{5^2 + 0^2} = \sqrt{25} = 5$

(c) $|\mathbf{w}| = \sqrt{\left(\frac{3}{5}\right)^2 + \left(\frac{4}{5}\right)^2} = \sqrt{\frac{9}{25} + \frac{16}{25}} = 1$

 **Now Try Exercise 37** ■

The following definitions of addition, subtraction, and scalar multiplication of vectors correspond to the geometric descriptions given earlier. Figure 11 (on the next page) shows how the analytic definition of addition corresponds to the geometric one.

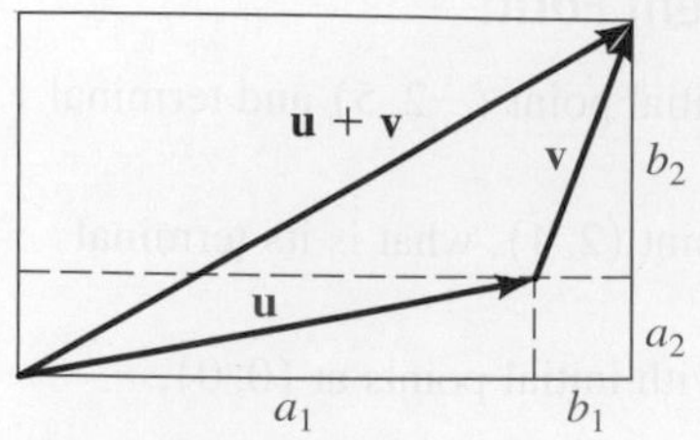

Figure 11

Algebraic Operations on Vectors

If $\mathbf{u} = \langle a_1, a_2\rangle$ and $\mathbf{v} = \langle b_1, b_2\rangle$, then

$$\mathbf{u} + \mathbf{v} = \langle a_1 + b_1, a_2 + b_2\rangle$$
$$\mathbf{u} - \mathbf{v} = \langle a_1 - b_1, a_2 - b_2\rangle$$
$$c\mathbf{u} = \langle ca_1, ca_2\rangle \qquad (c \in \mathbb{R})$$

Example 3 ■ Operations with Vectors

If $\mathbf{u} = \langle 2, -3\rangle$ and $\mathbf{v} = \langle -1, 2\rangle$, find $\mathbf{u} + \mathbf{v}$, $\mathbf{u} - \mathbf{v}$, $2\mathbf{u}$, $-3\mathbf{v}$, and $2\mathbf{u} + 3\mathbf{v}$.

Solution By the definitions of the vector operations we have

$$\mathbf{u} + \mathbf{v} = \langle 2, -3\rangle + \langle -1, 2\rangle = \langle 1, -1\rangle$$
$$\mathbf{u} - \mathbf{v} = \langle 2, -3\rangle - \langle -1, 2\rangle = \langle 3, -5\rangle$$
$$2\mathbf{u} = 2\langle 2, -3\rangle = \langle 4, -6\rangle$$
$$-3\mathbf{v} = -3\langle -1, 2\rangle = \langle 3, -6\rangle$$
$$2\mathbf{u} + 3\mathbf{v} = 2\langle 2, -3\rangle + 3\langle -1, 2\rangle = \langle 4, -6\rangle + \langle -3, 6\rangle = \langle 1, 0\rangle$$

 **Now Try Exercise 31** ■

The following properties for vector operations can be proved from the definitions. The **zero vector** is the vector $\mathbf{0} = \langle 0, 0\rangle$. It plays the same role for addition of vectors as the number 0 does for addition of real numbers.

Properties of Vectors

Vector Addition	**Multiplication by a Scalar**
$\mathbf{u} + \mathbf{v} = \mathbf{v} + \mathbf{u}$	$c(\mathbf{u} + \mathbf{v}) = c\mathbf{u} + c\mathbf{v}$
$\mathbf{u} + (\mathbf{v} + \mathbf{w}) = (\mathbf{u} + \mathbf{v}) + \mathbf{w}$	$(c + d)\mathbf{u} = c\mathbf{u} + d\mathbf{u}$
$\mathbf{u} + \mathbf{0} = \mathbf{u}$	$(cd)\mathbf{u} = c(d\mathbf{u}) = d(c\mathbf{u})$
$\mathbf{u} + (-\mathbf{u}) = \mathbf{0}$	$1\mathbf{u} = \mathbf{u}$
Length of a Vector	$0\mathbf{u} = \mathbf{0}$
$\lvert c\mathbf{u}\rvert = \lvert c\rvert\,\lvert\mathbf{u}\rvert$	$c\mathbf{0} = \mathbf{0}$

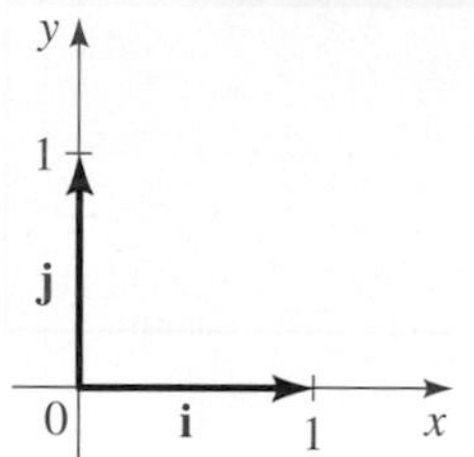

Figure 12

A vector of length 1 is called a **unit vector**. For instance, in Example 2(c) the vector $\mathbf{w} = \langle \frac{3}{5}, \frac{4}{5}\rangle$ is a unit vector. Two useful unit vectors are $\mathbf{i}$ and $\mathbf{j}$, defined by

$$\mathbf{i} = \langle 1, 0\rangle \qquad \mathbf{j} = \langle 0, 1\rangle$$

(See Figure 12.) These vectors are special because any vector can be expressed in terms of them. (See Figure 13.)

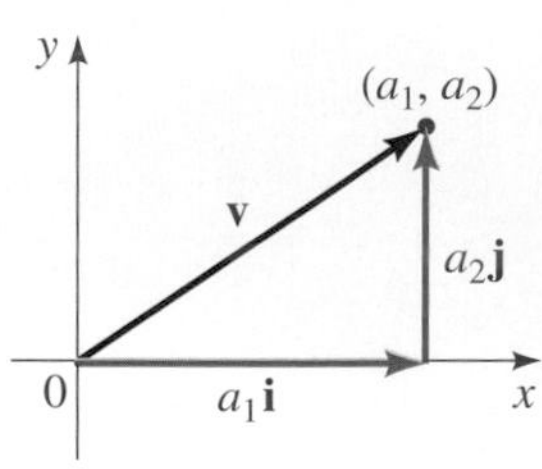

Figure 13

Vectors in Terms of i and j

The vector $\mathbf{v} = \langle a_1, a_2\rangle$ can be expressed in terms of $\mathbf{i}$ and $\mathbf{j}$ by

$$\mathbf{v} = \langle a_1, a_2\rangle = a_1\mathbf{i} + a_2\mathbf{j}$$

Example 4 ■ Vectors in Terms of i and j

(a) Write the vector $\mathbf{u} = \langle 5, -8\rangle$ in terms of $\mathbf{i}$ and $\mathbf{j}$.

(b) If $\mathbf{u} = 3\mathbf{i} + 2\mathbf{j}$ and $\mathbf{v} = -\mathbf{i} + 6\mathbf{j}$, write $2\mathbf{u} + 5\mathbf{v}$ in terms of $\mathbf{i}$ and $\mathbf{j}$.

Solution

(a) $\mathbf{u} = 5\mathbf{i} + (-8)\mathbf{j} = 5\mathbf{i} - 8\mathbf{j}$

(b) The properties of addition and scalar multiplication of vectors show that we can manipulate vectors in the same way as algebraic expressions. Thus

$$\begin{aligned} 2\mathbf{u} + 5\mathbf{v} &= 2(3\mathbf{i} + 2\mathbf{j}) + 5(-\mathbf{i} + 6\mathbf{j}) \\ &= (6\mathbf{i} + 4\mathbf{j}) + (-5\mathbf{i} + 30\mathbf{j}) \\ &= \mathbf{i} + 34\mathbf{j} \end{aligned}$$

Now Try Exercises 27 and 35 ■

Let $\mathbf{v}$ be a vector in the plane with its initial point at the origin. The **direction** of $\mathbf{v}$ is θ, the smallest positive angle in standard position formed by the positive x-axis and $\mathbf{v}$. (See Figure 14.) If we know the magnitude and direction of a vector, then Figure 14 shows that we can find the horizontal and vertical components of the vector.

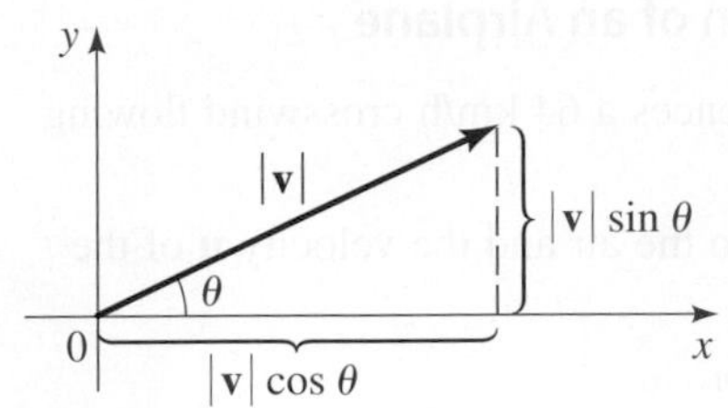

Figure 14

Horizontal and Vertical Components of a Vector

Let $\mathbf{v}$ be a vector with magnitude $|\mathbf{v}|$ and direction θ.

Then $\mathbf{v} = \langle a_1, a_2 \rangle = a_1\mathbf{i} + a_2\mathbf{j}$, where

$$a_1 = |\mathbf{v}| \cos\theta \quad \text{and} \quad a_2 = |\mathbf{v}| \sin\theta$$

Thus we can express $\mathbf{v}$ as

$$\mathbf{v} = |\mathbf{v}| \cos\theta\,\mathbf{i} + |\mathbf{v}| \sin\theta\,\mathbf{j}$$

Example 5 ■ Components and Direction of a Vector

(a) A vector $\mathbf{v}$ has length 8 and direction $\pi/3$. Find the horizontal and vertical components, and write $\mathbf{v}$ in terms of $\mathbf{i}$ and $\mathbf{j}$.

(b) Find the direction of the vector $\mathbf{u} = -\sqrt{3}\,\mathbf{i} + \mathbf{j}$.

Solution

(a) We have $\mathbf{v} = \langle a, b \rangle$, where the components are given by

$$a = 8\cos\frac{\pi}{3} = 4 \quad \text{and} \quad b = 8\sin\frac{\pi}{3} = 4\sqrt{3}$$

Thus $\mathbf{v} = \langle 4, 4\sqrt{3} \rangle = 4\mathbf{i} + 4\sqrt{3}\,\mathbf{j}$.

(b) From Figure 15 we see that the direction θ has the property that

$$\tan\theta = \frac{1}{-\sqrt{3}} = -\frac{\sqrt{3}}{3}$$

Thus the reference angle for θ is $\pi/6$. Since the terminal point of the vector $\mathbf{u}$ is in Quadrant II, it follows that $\theta = 5\pi/6$.

Figure 15

Now Try Exercises 41 and 51 ■

■ Using Vectors to Model Velocity and Force

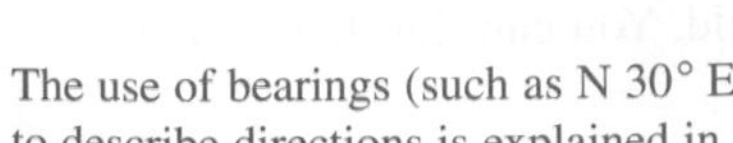
The use of bearings (such as N 30° E) to describe directions is explained in Section 6.6.

The **velocity** of a moving object is modeled by a vector whose direction is the direction of motion and whose magnitude is the speed. Figure 16 on the next page shows some vectors $\mathbf{u}$, representing the velocity of wind flowing in the direction N 30° E, and a vector $\mathbf{v}$, representing the velocity of an airplane flying through this wind at the point P. From our experience we know that wind affects both the speed and the direction of an airplane. The true velocity of the plane (relative to the ground) is given by the vector $\mathbf{w} = \mathbf{u} + \mathbf{v}$, as shown in Figure 17.

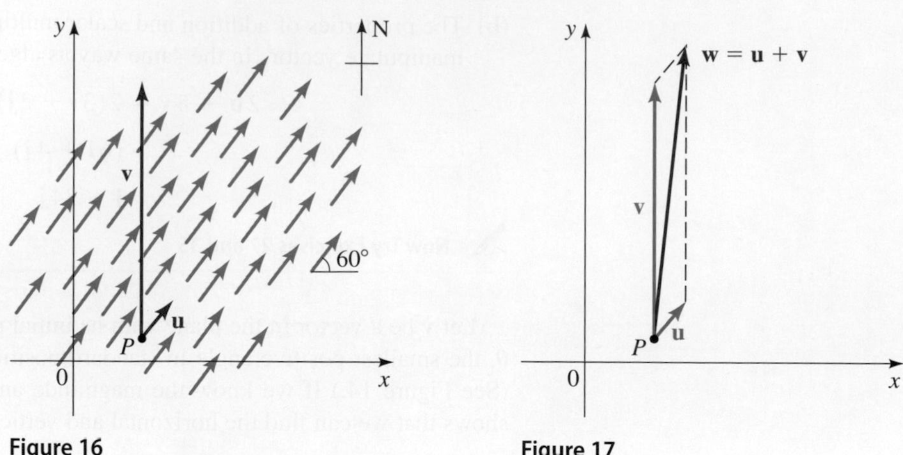

Figure 16 **Figure 17**

Example 6 ■ The True Speed and Direction of an Airplane

An airplane heads due north at 480 km/h. It experiences a 64 km/h crosswind flowing in the direction N 30° E, as shown in Figure 16.

(a) Express the velocity $\mathbf{v}$ of the airplane relative to the air and the velocity $\mathbf{u}$ of the wind, in component form.

(b) Find the true velocity of the airplane as a vector.

(c) Find the true speed and direction of the airplane.

Solution

(a) The velocity of the airplane relative to the air is $\mathbf{v} = 0\mathbf{i} + 300\mathbf{j} = 300\mathbf{j}$. By the formulas for the components of a vector we find that the velocity of the wind is

$$\begin{aligned} \mathbf{u} &= (64 \cos 60°)\mathbf{i} + (64 \sin 60°)\mathbf{j} \\ &= 32\mathbf{i} + 32\sqrt{3}\mathbf{j} \\ &\approx 32\mathbf{i} + 55.43\mathbf{j} \end{aligned}$$

The **true velocity** is the velocity relative to the ground.

(b) The true velocity of the airplane is given by the vector $\mathbf{w} = \mathbf{u} + \mathbf{v}$:

$$\begin{aligned} \mathbf{w} = \mathbf{u} + \mathbf{v} &= (32\mathbf{i} + 32\sqrt{3}\mathbf{j}) + (480\mathbf{j}) \\ &= 32\mathbf{i} + (32\sqrt{3} + 480)\mathbf{j} \\ &\approx 32\mathbf{i} + 535.43\mathbf{j} \end{aligned}$$

(c) The true speed of the airplane is given by the magnitude of $\mathbf{w}$:

$$|\mathbf{w}| \approx \sqrt{(32)^2 + (535.43)^2} \approx 536.4 \text{ km/h}$$

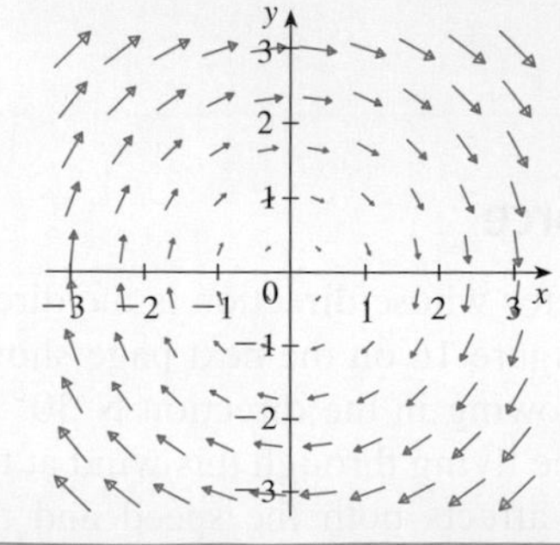

Discovery Project ■ Vector Fields

A *vector field* is a collection of vectors, like the vectors that model wind velocity at each point in some region. In this project we use a graphing device to graph vector fields that are given by a rule. The graph shown here displays at each point (x, y) the vector $y\mathbf{i} - x\mathbf{j}$. You can see that the vectors appear to rotate about the origin. We'll graph different vector fields and then visually determine the path a particle would take when it is put in the field. You can find the project at **www.stewartmath.com**.

The direction of the airplane is the direction θ of the vector **w**. The angle θ has the property that $\tan\theta \approx 535.43/32 = 16.732$, so $\theta \approx 86.6°$. Thus the airplane is heading in the direction N 3.4° E.

 Now Try Exercise 59 ■

Example 7 ■ Calculating a Heading

A boater launches a boat from one shore of a straight river and wants to land at the point directly on the opposite shore. If the speed of the boat (relative to the water) is 16 km/h and the river is flowing east at the rate of 8 km/h, in what direction should the boat be headed in order to arrive at the desired landing point?

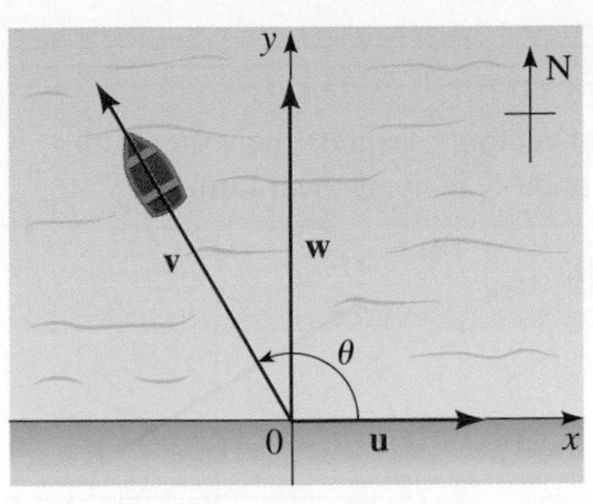

Figure 18

Solution We choose a coordinate system with the origin at the initial position of the boat, as shown in Figure 18. Let **u** and **v** represent the velocities of the river and the boat, respectively. Thus $\mathbf{u} = 5\mathbf{i}$ and, since the speed of the boat is 16 km/h, we have $|\mathbf{v}| = 10$, so

$$\mathbf{v} = (16\cos\theta)\mathbf{i} + (16\sin\theta)\mathbf{j}$$

where the angle θ is as shown in Figure 18. The true course of the boat is given by the vector $\mathbf{w} = \mathbf{u} + \mathbf{v}$. We have

$$\begin{aligned} \mathbf{w} = \mathbf{u} + \mathbf{v} &= 8\mathbf{i} + (16\cos\theta)\mathbf{i} + (16\sin\theta)\mathbf{j} \\ &= (8 + 16\cos\theta)\mathbf{i} + (16\sin\theta)\mathbf{j} \end{aligned}$$

Since the boater wants to land at a point directly across the river, the direction of the boat should have horizontal component 0. In other words, the boat should be pointed at an angle θ in such a way that

$$\begin{aligned} 8 + 16\cos\theta &= 0 \\ \cos\theta &= -\tfrac{1}{2} \\ \theta &= 120° \end{aligned}$$

Thus the boat should be headed in the direction $\theta = 120°$ (or N 30° W).

 Now Try Exercise 61 ■

Force is also represented by a vector. Intuitively, we can think of force as describing a push or a pull on an object, for example, a horizontal push of a book across a table or the downward pull of the earth's gravity on a ball. Force is measured in newtons. If several forces are acting on an object, the **resultant force** experienced by the object is the vector sum of these forces.

Example 8 ■ Resultant Force

Two forces $\mathbf{F}_1$ and $\mathbf{F}_2$ with magnitudes 10 and 20 N, respectively, act on an object at a point P, as shown in Figure 19. Find the resultant force acting at P.

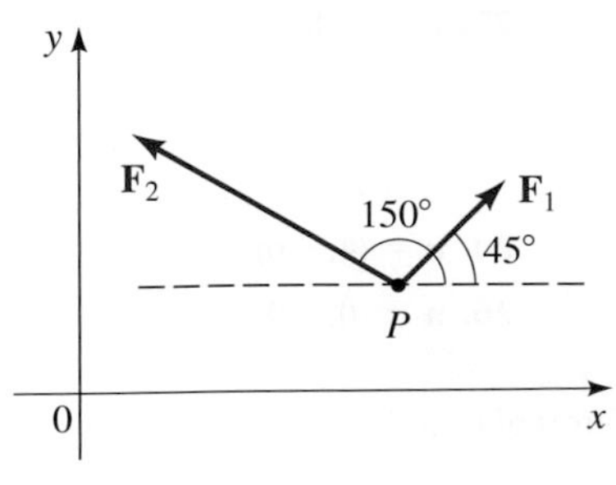

Figure 19

Solution We write $\mathbf{F}_1$ and $\mathbf{F}_2$ in component form:

$$\begin{aligned} \mathbf{F}_1 &= (10\cos 45°)\mathbf{i} + (10\sin 45°)\mathbf{j} = 10\frac{\sqrt{2}}{2}\mathbf{i} + 10\frac{\sqrt{2}}{2}\mathbf{j} \\ &= 5\sqrt{2}\,\mathbf{i} + 5\sqrt{2}\,\mathbf{j} \end{aligned}$$

$$\begin{aligned} \mathbf{F}_2 &= (20\cos 150°)\mathbf{i} + (20\sin 150°)\mathbf{j} = -20\frac{\sqrt{3}}{2}\mathbf{i} + 20\left(\frac{1}{2}\right)\mathbf{j} \\ &= -10\sqrt{3}\,\mathbf{i} + 10\mathbf{j} \end{aligned}$$

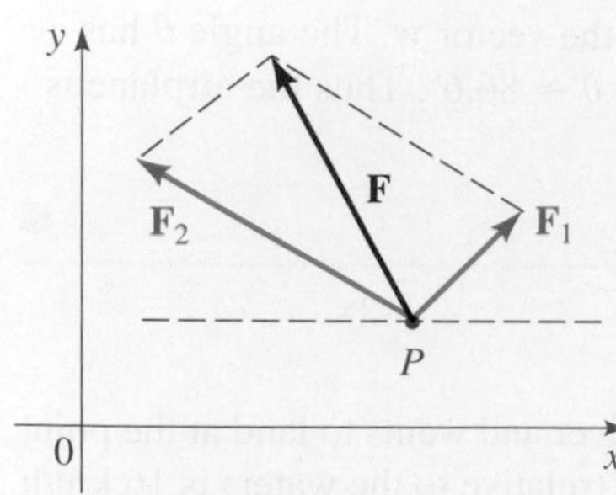

Figure 20

So the resultant force $\mathbf{F}$ is

$$\begin{aligned}\mathbf{F} &= \mathbf{F}_1 + \mathbf{F}_2 \\ &= (5\sqrt{2}\,\mathbf{i} + 5\sqrt{2}\,\mathbf{j}) + (-10\sqrt{3}\,\mathbf{i} + 10\,\mathbf{j}) \\ &= (5\sqrt{2} - 10\sqrt{3})\mathbf{i} + (5\sqrt{2} + 10)\mathbf{j} \\ &\approx -10\,\mathbf{i} + 17\,\mathbf{j}\end{aligned}$$

The resultant force $\mathbf{F}$ is shown in Figure 20.

Now Try Exercise 67 ■

8.5 Exercises

■ Concepts

1. (a) A vector in the plane is a line segment with an assigned direction. In Figure I below, the vector $\mathbf{u}$ has initial point ________ and terminal point ________. Sketch the vectors $2\mathbf{u}$ and $\mathbf{u} + \mathbf{v}$.

   (b) A vector in a coordinate plane is expressed by using components. In Figure II below, the vector $\mathbf{u}$ has initial point(___ , ___) and terminal point(___ , ___). In component form we write $\mathbf{u} = \langle$ ___ , ___ $\rangle$, and $\mathbf{v} = \langle$ ___ , ___ $\rangle$. Then $2\mathbf{u} = \langle$ ___ , ___ $\rangle$ and $\mathbf{u} + \mathbf{v} = \langle$ ___ , ___ $\rangle$.

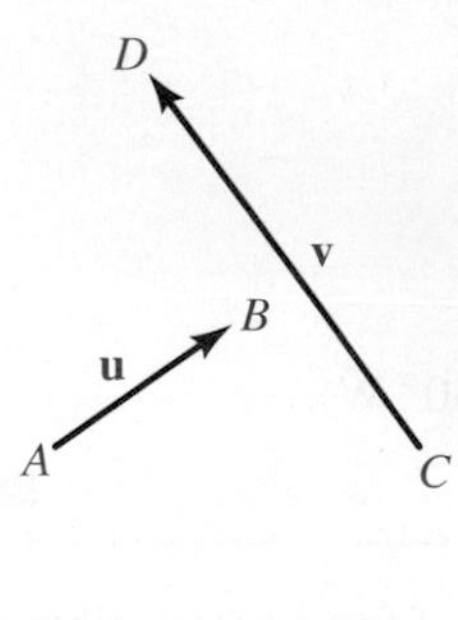

I

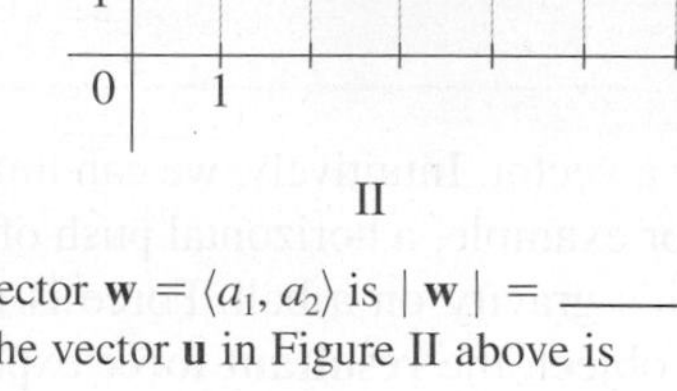

II

2. (a) The length of a vector $\mathbf{w} = \langle a_1, a_2 \rangle$ is $|\mathbf{w}| =$ ________, so the length of the vector $\mathbf{u}$ in Figure II above is $|\mathbf{u}| =$ ________.

   (b) If we know the length $|\mathbf{w}|$ and direction θ of a vector $\mathbf{w}$, then we can express the vector in component form as $\mathbf{w} = \langle$ ________ , ________ $\rangle$.

■ Skills

3–8 ■ Sketching Vectors Sketch the vector indicated. (The vectors $\mathbf{u}$ and $\mathbf{v}$ are shown in the figure.)

3. $2\mathbf{u}$

4. $-\mathbf{v}$
5. $\mathbf{u} + \mathbf{v}$
6. $\mathbf{u} - \mathbf{v}$
7. $\mathbf{v} - 2\mathbf{u}$
8. $2\mathbf{u} + \mathbf{v}$

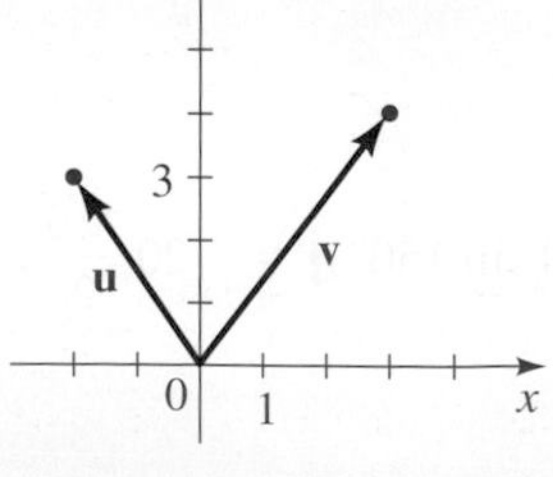

9–18 ■ Component Form of Vectors Express the vector with initial point P and terminal point Q in component form.

9.
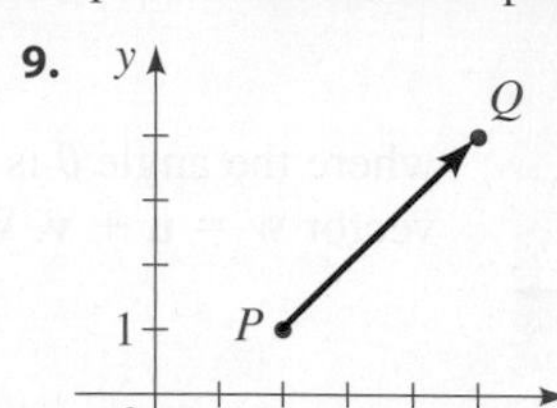

10.
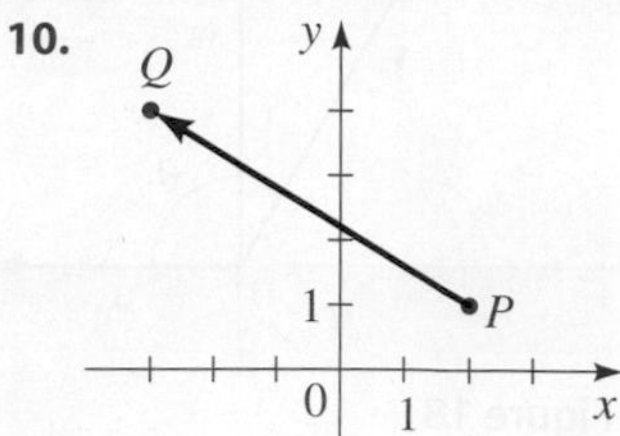

11.
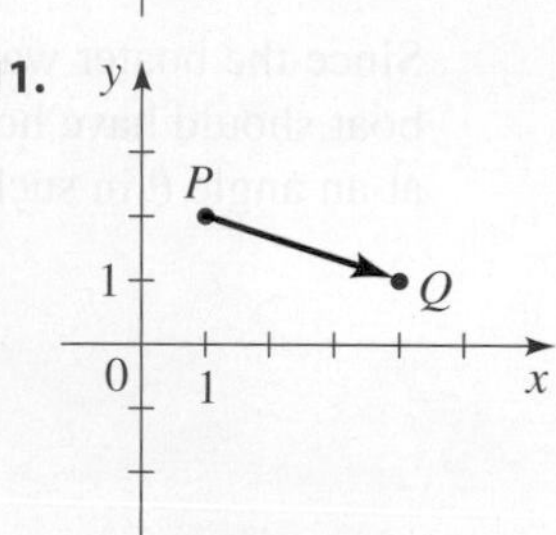

12.
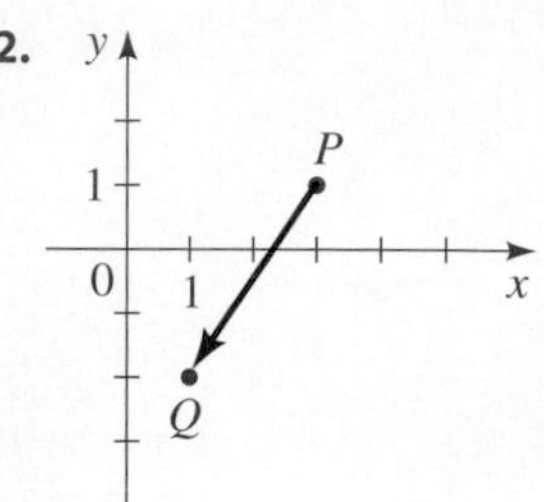

13. $P(1, 3)$, $Q(4, 5)$
14. $P(2, 5)$, $Q(3, 1)$
15. $P(5, 3)$, $Q(1, 0)$
16. $P(-1, 3)$, $Q(-6, -1)$
17. $P(-1, -1)$, $Q(-1, 1)$
18. $P(-8, -6)$, $Q(-1, -1)$

19–22 ■ Sketching Vectors Sketch the given vector with initial point (4, 3), and find the terminal point.

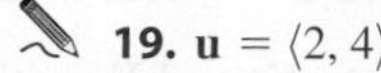
19. $\mathbf{u} = \langle 2, 4 \rangle$
20. $\mathbf{u} = \langle -1, 2 \rangle$
21. $\mathbf{u} = \langle 4, -3 \rangle$
22. $\mathbf{u} = \langle -8, -1 \rangle$

23–26 ■ Sketching Vectors Sketch representations of the given vector with initial points at $(0, 0)$, $(2, 3)$, and $(-3, 5)$.

23. $\mathbf{u} = \langle 3, 5 \rangle$
24. $\mathbf{u} = \langle 4, -6 \rangle$
25. $\mathbf{u} = \langle -7, 2 \rangle$
26. $\mathbf{u} = \langle 0, -9 \rangle$

27–30 ■ Writing Vectors in Terms of i and j Write the given vector in terms of $\mathbf{i}$ and $\mathbf{j}$.

27. $\mathbf{u} = \langle 2, 3 \rangle$
28. $\mathbf{u} = \langle -1, 0 \rangle$
29. $\mathbf{u} = \langle 0, -2 \rangle$
30. $\mathbf{u} = \langle -4, -5 \rangle$

31–36 ■ Operations with Vectors Find $2\mathbf{u}$, $-3\mathbf{v}$, $\mathbf{u}+\mathbf{v}$, and $3\mathbf{u}-4\mathbf{v}$ for the given vectors $\mathbf{u}$ and $\mathbf{v}$.

31. $\mathbf{u} = \langle 1, 4 \rangle, \quad \mathbf{v} = \langle -1, 2 \rangle$ **32.** $\mathbf{u} = \langle -2, 5 \rangle, \quad \mathbf{v} = \langle 2, -8 \rangle$

33. $\mathbf{u} = \langle 0, -1 \rangle, \quad \mathbf{v} = \langle -2, 0 \rangle$ **34.** $\mathbf{u} = \mathbf{i}, \quad \mathbf{v} = -2\mathbf{j}$

35. $\mathbf{u} = 2\mathbf{i} - \mathbf{j}, \quad \mathbf{v} = \mathbf{j}$ **36.** $\mathbf{u} = \mathbf{i} + \mathbf{j}, \quad \mathbf{v} = \mathbf{i} - \mathbf{j}$

37–40 ■ Magnitude of Vectors Find $|\mathbf{u}|$, $|\mathbf{v}|$, $|2\mathbf{u}|$, $|\frac{1}{2}\mathbf{v}|$, $|\mathbf{u}+\mathbf{v}|$, $|\mathbf{u}-\mathbf{v}|$, and $|\mathbf{u}| - |\mathbf{v}|$.

37. $\mathbf{u} = 3\mathbf{i} - \mathbf{j}, \quad \mathbf{v} = 2\mathbf{i} + 3\mathbf{j}$

38. $\mathbf{u} = -2\mathbf{i} + 3\mathbf{j}, \quad \mathbf{v} = \mathbf{i} - 2\mathbf{j}$

39. $\mathbf{u} = \langle 10, -1 \rangle, \quad \mathbf{v} = \langle -2, -2 \rangle$

40. $\mathbf{u} = \langle -6, 6 \rangle, \quad \mathbf{v} = \langle -2, -1 \rangle$

41–46 ■ Components of a Vector Find the horizontal and vertical components of the vector with given length and direction, and write the vector in terms of the vectors $\mathbf{i}$ and $\mathbf{j}$.

41. $|\mathbf{v}| = 10, \quad \theta = 60°$ **42.** $|\mathbf{v}| = 20, \quad \theta = 150°$

43. $|\mathbf{v}| = 1, \quad \theta = 225°$ **44.** $|\mathbf{v}| = 800, \quad \theta = 125°$

45. $|\mathbf{v}| = 4, \quad \theta = 10°$ **46.** $|\mathbf{v}| = \sqrt{3}, \quad \theta = 300°$

47–52 ■ Magnitude and Direction of a Vector Find the magnitude and direction (in degrees) of the vector.

47. $\mathbf{v} = \langle 3, 4 \rangle$ **48.** $\mathbf{v} = \left\langle -\frac{\sqrt{2}}{2}, -\frac{\sqrt{2}}{2} \right\rangle$

49. $\mathbf{v} = \langle -12, 5 \rangle$ **50.** $\mathbf{v} = \langle 40, 9 \rangle$

51. $\mathbf{v} = \mathbf{i} + \sqrt{3}\mathbf{j}$ **52.** $\mathbf{v} = \mathbf{i} + \mathbf{j}$

Applications

53. Components of a Force A landscaper pushes a lawn mower with a force of 136 N exerted at an angle of 30° to the ground. Find the horizontal and vertical components of the force.

54. Components of a Velocity A jet is flying in a direction N 20° E with a speed of 800 km/h. Find the north and east components of the velocity.

55. Velocity A river flows due south at 3 km/h. A swimmer attempting to cross the river heads due east swimming at 2 km/h relative to the water. Find the true velocity of the swimmer as a vector.

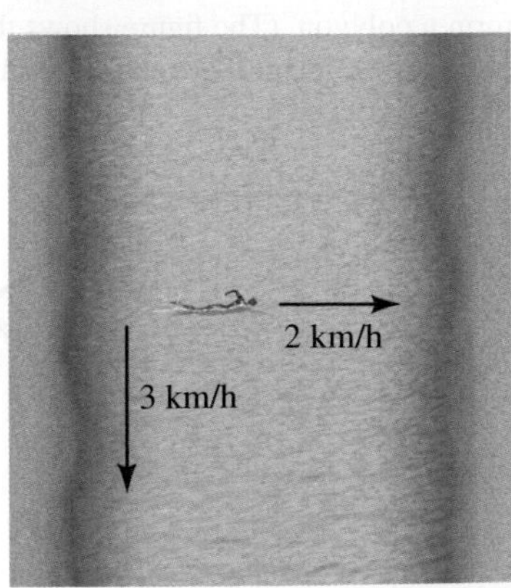

56. Velocity Suppose that in Exercise 55 the current is flowing at 1.2 km/h due south. In what direction should the swimmer head in order to arrive at a landing point due east of the starting point?

57. Velocity A migrating salmon heads in the direction N 45° E, swimming at 8 km/h relative to the water. The prevailing ocean currents flow due east at 5 km/h. Find the true velocity of the fish as a vector.

58. A jet is flying through a wind that is blowing 40 km/h due west. The jet has a speed of 585 km/h relative to the air, and the pilot heads the jet in the direction N 45° W. Find the true speed and direction of the jet.

59. True Velocity of a Jet A pilot heads a jet due east. The jet has a speed of 680 km/h to the air. The wind is blowing due north with a speed of 64 km/h.

(a) Express the velocity of the wind as a vector in component form.

(b) Express the velocity of the jet relative to the air as a vector in component form.

(c) Find the true velocity of the jet as a vector.

(d) Find the true speed and direction of the jet.

60. True Velocity of a Jet A jet is flying through a wind that is blowing with a speed of 88 km/h in the direction N 30° E (see the figure). The jet has a speed of 1225 km/h relative to the air, and the pilot heads the jet in the direction N 45° E.

(a) Express the velocity of the wind as a vector in component form.

(b) Express the velocity of the jet relative to the air as a vector in component form.

(c) Find the true velocity of the jet as a vector.

(d) Find the true speed and direction of the jet.

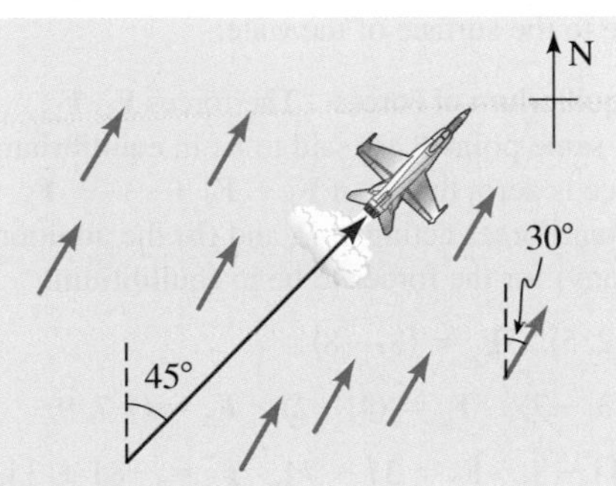

61. True Velocity of a Jet In what direction should the pilot in Exercise 60 head the jet for the true course to be due north?

62. Velocity The speed of an airplane is 480 km/h relative to the air. The wind is blowing due north with a speed of 48 km/h. In what direction should the airplane head in order to arrive at a point due west of its location?

63. Velocity of a Boat A straight river flows east at a speed of 10 km/h. A boater starts at the south shore of the river and heads in a direction 60° from the shore (see the figure). The boat has a speed of 20 km/h relative to the water.

(a) Express the velocity of the river as a vector in component form.

(b) Express the velocity of the boat relative to the water as a vector in component form.

(c) Find the true velocity of the boat.

(d) Find the true speed and direction of the boat.

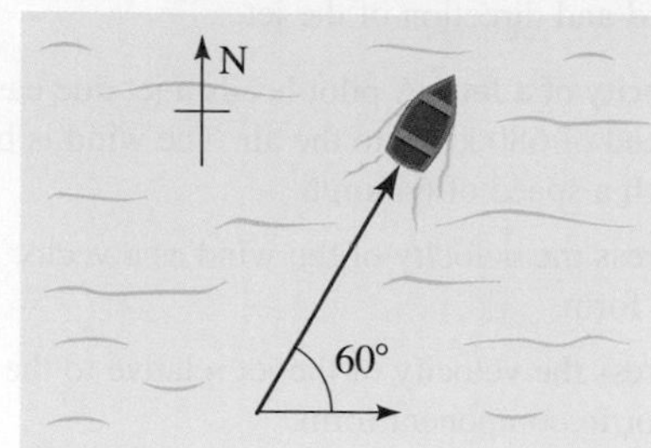

64. Velocity of a Boat The boater in Exercise 63 wants to arrive at a point on the north shore of the river directly opposite the starting point. In what direction should the boat be headed?

65. Velocity of a Boat A boat heads in the direction N 72° E. The speed of the boat relative to the water is 24 km/h. The water is flowing directly south. It is observed that the true direction of the boat is directly east.

(a) Express the velocity of the boat relative to the water as a vector in component form.

(b) Find the speed of the water and the true speed of the boat.

66. Velocity A sailor walks due west on the deck of an ocean liner at 2 km/h. The ocean liner is moving due north at a speed of 25 km/h. Find the speed and direction of the sailor relative to the surface of the water.

67–72 ■ Equilibrium of Forces The forces $\mathbf{F}_1, \mathbf{F}_2, \ldots, \mathbf{F}_n$ acting at the same point P are said to be in equilibrium if the resultant force is zero, that is, if $\mathbf{F}_1 + \mathbf{F}_2 + \cdots + \mathbf{F}_n = \mathbf{0}$. Find **(a)** the resultant forces acting at P, and **(b)** the additional force required (if any) for the forces to be in equilibrium.

67. $\mathbf{F}_1 = \langle 2, 5 \rangle, \quad \mathbf{F}_2 = \langle 3, -8 \rangle$

68. $\mathbf{F}_1 = \langle 3, -7 \rangle, \quad \mathbf{F}_2 = \langle 4, -2 \rangle, \quad \mathbf{F}_3 = \langle -7, 9 \rangle$

69. $\mathbf{F}_1 = 4\mathbf{i} - \mathbf{j}, \quad \mathbf{F}_2 = 3\mathbf{i} - 7\mathbf{j}, \quad \mathbf{F}_3 = -8\mathbf{i} + 3\mathbf{j}, \quad \mathbf{F}_4 = \mathbf{i} + \mathbf{j}$

70. $\mathbf{F}_1 = \mathbf{i} - \mathbf{j}, \quad \mathbf{F}_2 = \mathbf{i} + \mathbf{j}, \quad \mathbf{F}_3 = -2\mathbf{i} + \mathbf{j}$

71.

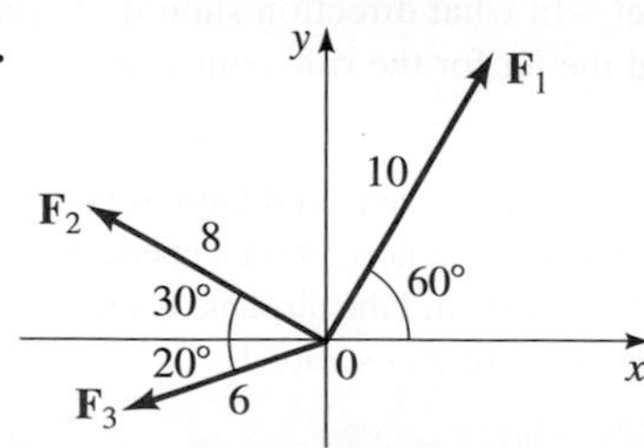

72.

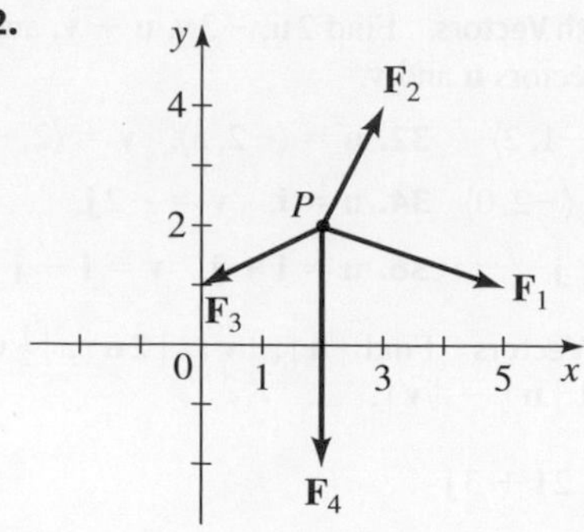

73. Equilibrium of Tensions A 445 N weight hangs from a string as shown in the figure. Find the tensions $\mathbf{T}_1$ and $\mathbf{T}_2$ in the string.

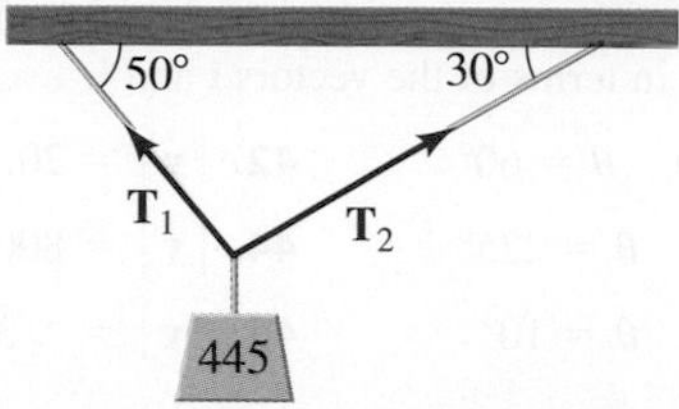

74. Equilibrium of Tensions The cranes in the figure are lifting an object that weighs 81,305 N. Find the tensions $\mathbf{T}_1$ and $\mathbf{T}_2$.

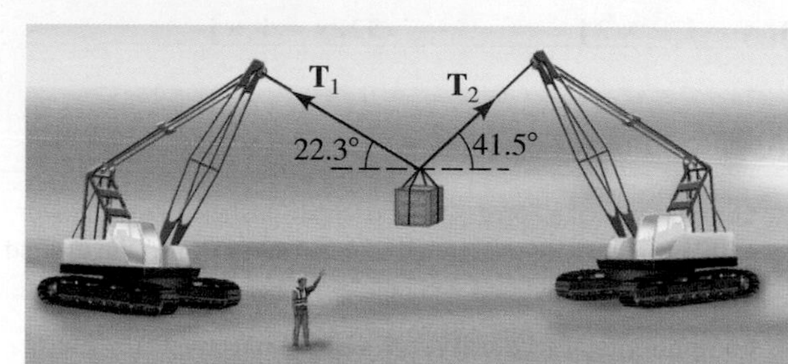

■ Discuss ■ Discover ■ Prove ■ Write

75. Discuss: Vectors That Form a Polygon Suppose that n vectors can be placed head to tail in the plane so that they form a polygon. (The figure shows the case of a hexagon.) Explain why the sum of these vectors is $\mathbf{0}$.

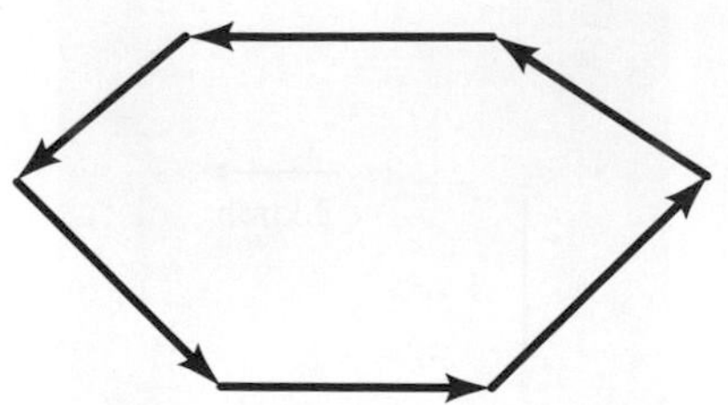

PS *Try to recognize something familiar.* Think about the geometric definition for vector addition.

8.6 The Dot Product

■ The Dot Product of Vectors ■ The Component of u along v
■ The Projection of u onto v ■ Work

In this section we define an operation on vectors called the dot product. This concept is especially useful in calculus and in applications of vectors to physics and engineering.

■ The Dot Product of Vectors

We begin by defining the dot product of two vectors.

Definition of the Dot Product

If $\mathbf{u} = \langle a_1, a_2 \rangle$ and $\mathbf{v} = \langle b_1, b_2 \rangle$ are vectors, then their **dot product**, denoted by $\mathbf{u} \cdot \mathbf{v}$, is defined by

$$\mathbf{u} \cdot \mathbf{v} = a_1 b_1 + a_2 b_2$$

Thus to find the dot product of **u** and **v**, we multiply corresponding components and add. The dot product of vectors is *not* a vector; it is a real number, or scalar.

Example 1 ■ Calculating Dot Products

(a) If $\mathbf{u} = \langle 3, -2 \rangle$ and $\mathbf{v} = \langle 4, 5 \rangle$ then

$$\mathbf{u} \cdot \mathbf{v} = (3)(4) + (-2)(5) = 2$$

(b) If $\mathbf{u} = 2\mathbf{i} + \mathbf{j}$ and $\mathbf{v} = 5\mathbf{i} - 6\mathbf{j}$, then

$$\mathbf{u} \cdot \mathbf{v} = (2)(5) + (1)(-6) = 4$$

Now Try Exercises 5(a) and 11(a) ■

The proof of each of the following properties of the dot product follows from the definition.

Properties of the Dot Product

1. $\mathbf{u} \cdot \mathbf{v} = \mathbf{v} \cdot \mathbf{u}$
2. $(c\mathbf{u}) \cdot \mathbf{v} = c(\mathbf{u} \cdot \mathbf{v}) = \mathbf{u} \cdot (c\mathbf{v})$
3. $(\mathbf{u} + \mathbf{v}) \cdot \mathbf{w} = \mathbf{u} \cdot \mathbf{w} + \mathbf{v} \cdot \mathbf{w}$
4. $|\mathbf{u}|^2 = \mathbf{u} \cdot \mathbf{u}$

Proof We prove only the last property. The proofs of the others are left as exercises. Let $\mathbf{u} = \langle a_1, a_2 \rangle$. Then

$$\mathbf{u} \cdot \mathbf{u} = a_1 a_1 + a_2 a_2 = a_1^2 + a_2^2 = |\mathbf{u}|^2$$ ■

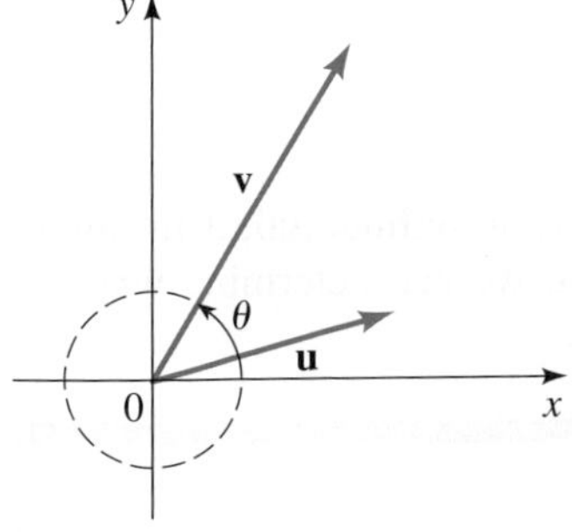

Figure 1

Let **u** and **v** be vectors, and sketch them with initial points at the origin. We define the **angle θ between u and v** to be the smaller of the angles formed by these representations of **u** and **v**. (See Figure 1.) Thus $0 \le \theta \le \pi$. The next theorem relates the angle between two vectors to their dot product.

The Dot Product Theorem

If θ is the angle between two nonzero vectors $\mathbf{u}$ and $\mathbf{v}$, then

$$\mathbf{u}\cdot\mathbf{v} = |\mathbf{u}||\mathbf{v}|\cos\theta$$

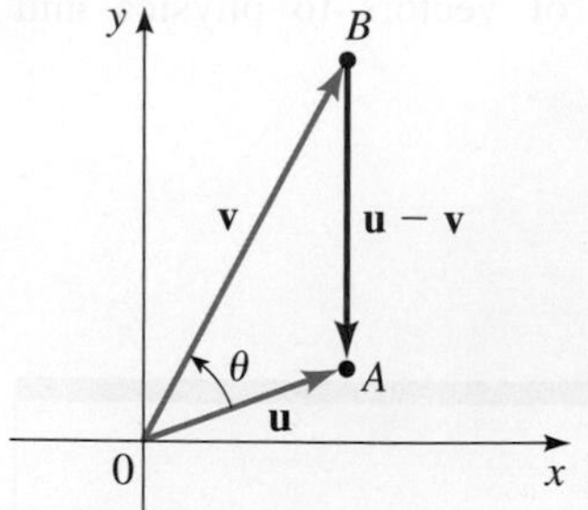

Proof Applying the Law of Cosines to triangle AOB in Figure 2 gives

$$|\mathbf{u}-\mathbf{v}|^2 = |\mathbf{u}|^2 + |\mathbf{v}|^2 - 2|\mathbf{u}||\mathbf{v}|\cos\theta$$

Using the properties of the dot product, we write the left-hand side as follows:

$$\begin{aligned}|\mathbf{u}-\mathbf{v}|^2 &= (\mathbf{u}-\mathbf{v})\cdot(\mathbf{u}-\mathbf{v})\\ &= \mathbf{u}\cdot\mathbf{u} - \mathbf{u}\cdot\mathbf{v} - \mathbf{v}\cdot\mathbf{u} + \mathbf{v}\cdot\mathbf{v}\\ &= |\mathbf{u}|^2 - 2(\mathbf{u}\cdot\mathbf{v}) + |\mathbf{v}|^2\end{aligned}$$

Figure 2

Equating the right-hand sides of the displayed equations, we get

$$\begin{aligned}|\mathbf{u}|^2 - 2(\mathbf{u}\cdot\mathbf{v}) + |\mathbf{v}|^2 &= |\mathbf{u}|^2 + |\mathbf{v}|^2 - 2|\mathbf{u}||\mathbf{v}|\cos\theta\\ -2(\mathbf{u}\cdot\mathbf{v}) &= -2|\mathbf{u}||\mathbf{v}|\cos\theta\\ \mathbf{u}\cdot\mathbf{v} &= |\mathbf{u}||\mathbf{v}|\cos\theta\end{aligned}$$

This proves the theorem. ■

The Dot Product Theorem is useful because it allows us to find the angle between two vectors if we know the components of the vectors. The angle is obtained by solving the equation in the Dot Product Theorem for $\cos\theta$. We state this important result explicitly.

Angle Between two Vectors

If θ is the angle between two nonzero vectors $\mathbf{u}$ and $\mathbf{v}$, then

$$\cos\theta = \frac{\mathbf{u}\cdot\mathbf{v}}{|\mathbf{u}||\mathbf{v}|}$$

Example 2 ■ Finding the Angle between two Vectors

Find the angle between the vectors $\mathbf{u} = \langle 2, 5\rangle$ and $\mathbf{v} = \langle 4, -3\rangle$.

Solution By the formula for the angle between two vectors we have

$$\cos\theta = \frac{\mathbf{u}\cdot\mathbf{v}}{|\mathbf{u}||\mathbf{v}|} = \frac{(2)(4)+(5)(-3)}{\sqrt{4+25}\sqrt{16+9}} = \frac{-7}{5\sqrt{29}}$$

Thus the angle between $\mathbf{u}$ and $\mathbf{v}$ is

$$\theta = \cos^{-1}\left(\frac{-7}{5\sqrt{29}}\right) \approx 105.1°$$

Now Try Exercises 5(b) and 11(b) ■

Two nonzero vectors $\mathbf{u}$ and $\mathbf{v}$ are called **perpendicular**, or **orthogonal**, if the angle between them is $\pi/2$. The following theorem shows that we can determine whether two vectors are perpendicular by finding their dot product.

Orthogonal Vectors

Two nonzero vectors $\mathbf{u}$ and $\mathbf{v}$ are perpendicular if and only if $\mathbf{u}\cdot\mathbf{v} = 0$.

Proof If $\mathbf{u}$ and $\mathbf{v}$ are perpendicular, then the angle between them is $\pi/2$, so

$$\mathbf{u} \cdot \mathbf{v} = |\mathbf{u}||\mathbf{v}| \cos \frac{\pi}{2} = 0$$

Conversely, if $\mathbf{u} \cdot \mathbf{v} = 0$, then

$$|\mathbf{u}||\mathbf{v}| \cos \theta = 0$$

Since $\mathbf{u}$ and $\mathbf{v}$ are nonzero vectors, we conclude that $\cos \theta = 0$, so $\theta = \pi/2$. Thus $\mathbf{u}$ and $\mathbf{v}$ are orthogonal. ■

Example 3 ■ Checking Whether Two Vectors Are Perpendicular

Determine whether the vectors in each pair are perpendicular.

(a) $\mathbf{u} = \langle 3, 5 \rangle$ and $\mathbf{v} = \langle 2, -8 \rangle$ **(b)** $\mathbf{u} = \langle 2, 1 \rangle$ and $\mathbf{v} = \langle -1, 2 \rangle$

Solution

(a) $\mathbf{u} \cdot \mathbf{v} = (3)(2) + (5)(-8) = -34 \neq 0$, so $\mathbf{u}$ and $\mathbf{v}$ are not perpendicular.

(b) $\mathbf{u} \cdot \mathbf{v} = (2)(-1) + (1)(2) = 0$, so $\mathbf{u}$ and $\mathbf{v}$ are perpendicular.

Now Try Exercises 17 and 19 ■

■ The Component of u along v

The **component of u along v** (also called the **component of u in the direction of v** or the **scalar projection of u onto v**) is defined to be

$$|\mathbf{u}| \cos \theta$$

Note that the component of **u** along **v** is a scalar, not a vector.

where θ is the angle between $\mathbf{u}$ and $\mathbf{v}$. Figure 3 gives a geometric interpretation of this concept. Intuitively, the component of $\mathbf{u}$ along $\mathbf{v}$ is the magnitude of the portion of $\mathbf{u}$ that points in the direction of $\mathbf{v}$. Notice that the component of $\mathbf{u}$ along $\mathbf{v}$ is negative if $\pi/2 < \theta \leq \pi$.

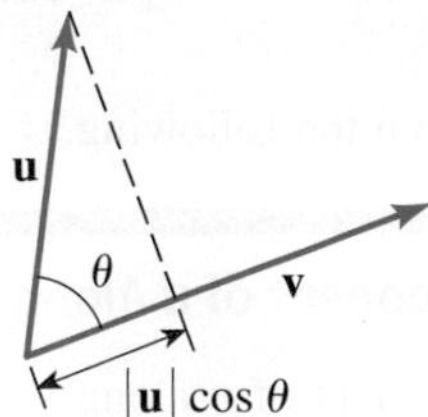

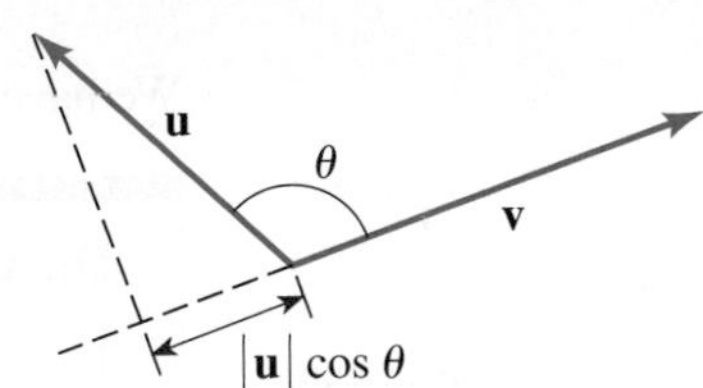

Figure 3

In analyzing forces in physics and engineering, it's often helpful to express a vector as a sum of two vectors lying in perpendicular directions. For example, suppose a car is parked on an inclined driveway as shown in Figure 4. The weight of the car is a vector $\mathbf{w}$ that points directly downward. We can write

$$\mathbf{w} = \mathbf{u} + \mathbf{v}$$

where $\mathbf{u}$ is parallel to the driveway and $\mathbf{v}$ is perpendicular to the driveway. The vector $\mathbf{u}$ is the force that tends to roll the car down the driveway, and $\mathbf{v}$ is the force experienced by the surface of the driveway. The magnitudes of these forces are the components of $\mathbf{w}$ along $\mathbf{u}$ and $\mathbf{v}$, respectively.

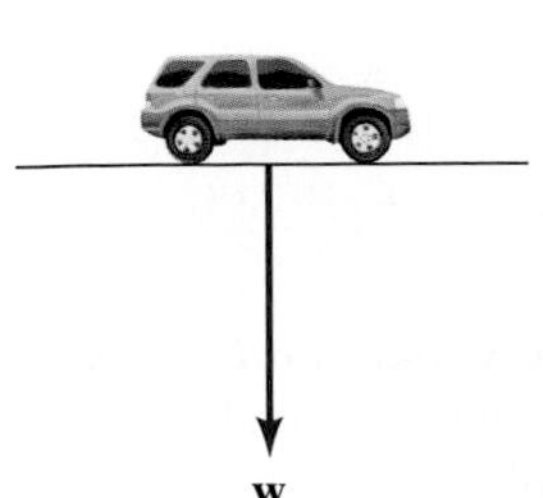

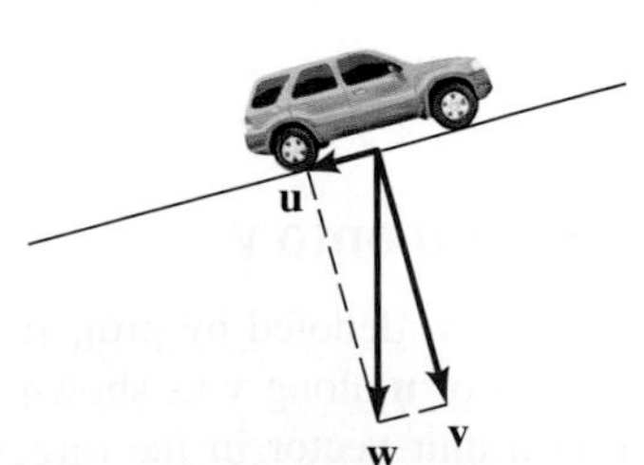

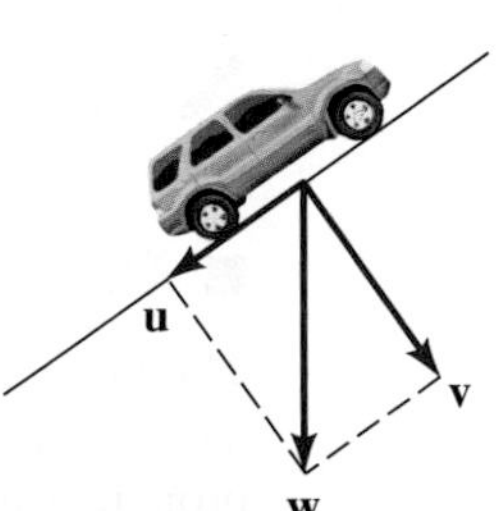

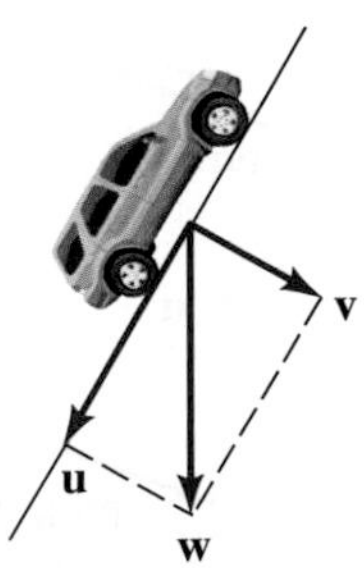

Figure 4

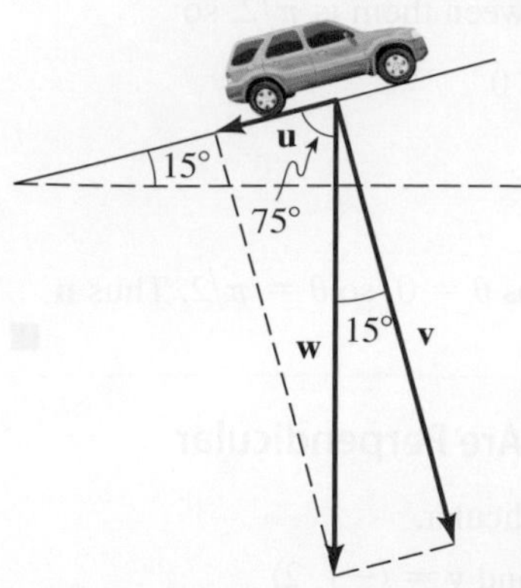

Figure 5

Example 4 ■ Resolving a Force into Components

A car weighing 13000 N is parked on a driveway that is inclined 15° to the horizontal, as shown in Figure 5.

(a) Find the magnitude of the force required to prevent the car from rolling down the driveway.

(b) Find the magnitude of the force experienced by the driveway due to the weight of the car.

Solution The car exerts a force **w** of 13000 N directly downward. We resolve **w** into the sum of two vectors **u** and **v**, one parallel to the surface of the driveway and the other perpendicular to it, as shown in Figure 5.

(a) The magnitude of the part of the force **w** that causes the car to roll down the driveway is

$$|\mathbf{u}| = \text{component of } \mathbf{w} \text{ along } \mathbf{u} = 13000 \cos 75° \approx 3365$$

Thus the force needed to prevent the car from rolling down the driveway is about 3365 N.

(b) The magnitude of the force exerted by the car on the driveway is

$$|\mathbf{v}| = \text{component of } \mathbf{w} \text{ along } \mathbf{v} = 13000 \cos 15° \approx 12560$$

The force experienced by the driveway is about 12,560 N.

 Now Try Exercise 51 ■

The component of **u** along **v** can be computed by using dot products:

$$|\mathbf{u}| \cos \theta = \frac{|\mathbf{v}||\mathbf{u}| \cos \theta}{|\mathbf{v}|} = \frac{\mathbf{u} \cdot \mathbf{v}}{|\mathbf{v}|}$$

We have shown the following.

The component of **u** along **v** is a scalar, not a vector.

The Component of u Along v

The component of **u** along **v** (or the scalar projection of **u** onto **v**) is

$$\text{comp}_{\mathbf{v}}\, \mathbf{u} = \frac{\mathbf{u} \cdot \mathbf{v}}{|\mathbf{v}|}$$

Example 5 ■ Finding Components

Let $\mathbf{u} = \langle 1, 4 \rangle$ and $\mathbf{v} = \langle -2, 1 \rangle$. Find the component of **u** along **v**.

Solution From the formula for the component of **u** along **v** we have

$$\text{comp}_{\mathbf{v}}\, \mathbf{u} = \frac{\mathbf{u} \cdot \mathbf{v}}{|\mathbf{v}|} = \frac{(1)(-2) + (4)(1)}{\sqrt{4 + 1}} = \frac{2}{\sqrt{5}}$$

 Now Try Exercise 27 ■

■ The Projection of u onto v

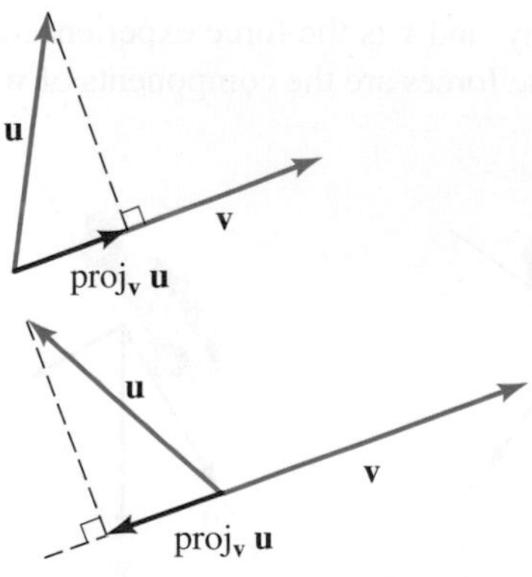

The projection of **u** onto **v**, denoted by $\text{proj}_{\mathbf{v}}\, \mathbf{u}$, is the vector *parallel* to **v** and whose *length* is the component of **u** along **v** as shown in Figure 6. To find an expression for $\text{proj}_{\mathbf{v}}\, \mathbf{u}$, we first find a unit vector in the direction of **v** and then multiply it by the component of **u** along **v**, as follows.

Figure 6

$$\begin{aligned}\text{proj}_{\mathbf{v}}\,\mathbf{u} &= (\text{component of } \mathbf{u} \text{ along } \mathbf{v})(\text{unit vector in direction of } \mathbf{v})\\ &= \left(\frac{\mathbf{u}\cdot\mathbf{v}}{|\mathbf{v}|}\right)\frac{\mathbf{v}}{|\mathbf{v}|} = \left(\frac{\mathbf{u}\cdot\mathbf{v}}{|\mathbf{v}|^2}\right)\mathbf{v}\end{aligned}$$

We often need to **resolve** a vector $\mathbf{u}$ into the sum of two vectors, one parallel to $\mathbf{v}$ and one orthogonal to $\mathbf{v}$. That is, we want to write $\mathbf{u} = \mathbf{u}_1 + \mathbf{u}_2$, where $\mathbf{u}_1$ is parallel to $\mathbf{v}$ and $\mathbf{u}_2$ is orthogonal to $\mathbf{v}$. In this case, $\mathbf{u}_1 = \text{proj}_{\mathbf{v}}\,\mathbf{u}$ and $\mathbf{u}_2 = \mathbf{u} - \text{proj}_{\mathbf{v}}\,\mathbf{u}$. (see Exercise 45.)

The Vector Projection of u Onto v

The **projection of u onto v** is the vector $\text{proj}_{\mathbf{v}}\,\mathbf{u}$ given by

$$\text{proj}_{\mathbf{v}}\,\mathbf{u} = \left(\frac{\mathbf{u}\cdot\mathbf{v}}{|\mathbf{v}|^2}\right)\mathbf{v}$$

If the vector $\mathbf{u}$ is **resolved** into $\mathbf{u}_1$ and $\mathbf{u}_2$, where $\mathbf{u}_1$ is parallel to $\mathbf{v}$ and $\mathbf{u}_2$ is orthogonal to $\mathbf{v}$, then

$$\mathbf{u}_1 = \text{proj}_{\mathbf{v}}\,\mathbf{u} \qquad \text{and} \qquad \mathbf{u}_2 = \mathbf{u} - \text{proj}_{\mathbf{v}}\,\mathbf{u}$$

The projection of **u** onto **v** is a vector, not a scalar.

Example 6 ■ Resolving a Vector into Orthogonal Vectors

Let $\mathbf{u} = \langle -2, 9\rangle$ and $\mathbf{v} = \langle -1, 2\rangle$.

(a) Find $\text{proj}_{\mathbf{v}}\,\mathbf{u}$.

(b) Resolve $\mathbf{u}$ into $\mathbf{u}_1$ and $\mathbf{u}_2$, where $\mathbf{u}_1$ is parallel to $\mathbf{v}$ and $\mathbf{u}_2$ is orthogonal to $\mathbf{v}$.

Solution

(a) By the formula for the projection of one vector onto another we have

$$\begin{aligned}\text{proj}_{\mathbf{v}}\,\mathbf{u} &= \left(\frac{\mathbf{u}\cdot\mathbf{v}}{|\mathbf{v}|^2}\right)\mathbf{v} && \text{Formula for projection}\\ &= \left(\frac{\langle -2, 9\rangle\cdot\langle -1, 2\rangle}{(-1)^2 + 2^2}\right)\langle -1, 2\rangle && \text{Definition of } \mathbf{u} \text{ and } \mathbf{v}\\ &= 4\langle -1, 2\rangle = \langle -4, 8\rangle\end{aligned}$$

(b) By the formula in the preceding box we have $\mathbf{u} = \mathbf{u}_1 + \mathbf{u}_2$, where

$$\begin{aligned}\mathbf{u}_1 &= \text{proj}_{\mathbf{v}}\,\mathbf{u} = \langle -4, 8\rangle && \text{From part (a)}\\ \mathbf{u}_2 &= \mathbf{u} - \text{proj}_{\mathbf{v}}\,\mathbf{u} = \langle -2, 9\rangle - \langle -4, 8\rangle = \langle 2, 1\rangle\end{aligned}$$

Now Try Exercise 31 ■

■ Work

One use of the dot product occurs in calculating work. In everyday use, the term *work* means the total amount of effort required to perform a task. In physics, *work* has a technical meaning that conforms to this intuitive meaning. If a constant force of magnitude F moves an object through a distance d along a straight line, then the **work** done is

$$W = Fd \qquad \text{or} \qquad \text{work} = \text{force} \times \text{distance}$$

If F is measured in newtons and d in meters, then the unit of work is a newton-meter (N-m). For example, how much work is done in lifting a 20-N weight 6 m off the ground? Since a force of 20 N is required to lift this weight and since the weight moves through a distance of 6 m, the amount of work done is

$$W = Fd = (20)(6) = 120 \text{ N-m}$$

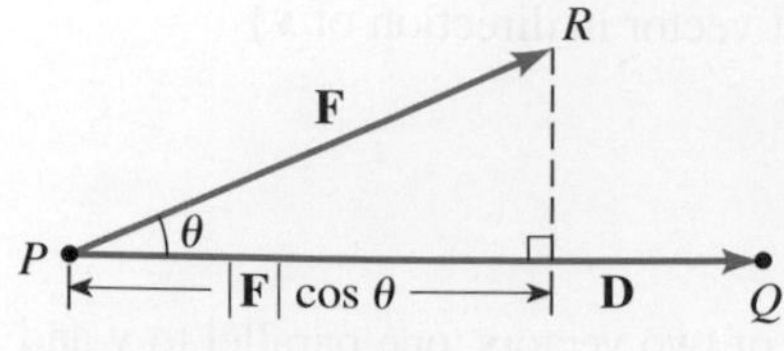

Figure 7

This formula applies only when the force is directed along the direction of motion. In the general case, if the force **F** moves an object from P to Q, as illustrated in Figure 7, then only the component of the force in the direction of $\mathbf{D} = \overrightarrow{PQ}$ affects the object. Thus the effective magnitude of the force on the object is

$$\text{comp}_{\mathbf{D}}\,\mathbf{F} = |\,\mathbf{F}\,|\cos\theta$$

So the work done is

$$W = \text{force} \times \text{distance} = (|\,\mathbf{F}\,|\cos\theta)|\,\mathbf{D}\,| = |\,\mathbf{F}\,||\,\mathbf{D}\,|\cos\theta = \mathbf{F}\cdot\mathbf{D}$$

We have derived the following formula for calculating work.

Work

The **work** W done by a force **F** in moving along a displacement vector **D** is

$$W = \mathbf{F}\cdot\mathbf{D}$$

Example 7 ■ Calculating Work

A force is given by the vector $\mathbf{F} = \langle 2, 3\rangle$ and moves an object from the point $(1, 3)$ to the point $(5, 9)$. Find the work done.

Solution The displacement vector is

$$\mathbf{D} = \langle 5 - 1, 9 - 3\rangle = \langle 4, 6\rangle$$

So the work done is

$$W = \mathbf{F}\cdot\mathbf{D} = \langle 2, 3\rangle\cdot\langle 4, 6\rangle = 26$$

If the unit of force is newtons and the distance is measured in meters, then the work done is 26 N-m.

Now Try Exercise 37 ■

Example 8 ■ Calculating Work

A child pulls a wagon horizontally by exerting a force of 100 N on the handle. If the handle makes an angle of 60° with the horizontal, find the work done in moving the wagon 30 m.

Solution We choose a coordinate system with the origin at the initial position of the wagon (see Figure 8). That is, the wagon moves from the point $P(0, 0)$ to the point $Q(30, 0)$. The vector that represents this displacement is

$$\mathbf{D} = 30\,\mathbf{i}$$

y
60°
Q(30, 0)
P(0, 0)
x

Figure 8

Discovery Project ■ Sailing Against the Wind

Sailors depend on the wind to propel their boats. But what if the wind is blowing in a direction opposite to the direction in which they want to travel? Although it is impossible to sail directly against the wind, it is possible to sail at an angle into the wind so that the sailboat can make headway against the wind. In this project we discover how vectors that model the sail, the keel, and the wind can be combined to find the direction in which the boat will move. You can find the project at **www.stewartmath.com**.

The force on the handle can be written in terms of components (see Section 8.5) as

$$\mathbf{F} = (100\cos 60°)\mathbf{i} + (100\sin 60°)\mathbf{j} = 50\mathbf{i} + 50\sqrt{3}\,\mathbf{j}$$

Thus the work done is

$$W = \mathbf{F}\cdot\mathbf{D} = (50\mathbf{i} + 50\sqrt{3}\,\mathbf{j})\cdot(30\mathbf{i}) = 1500 \text{ N-m}$$

 Now Try Exercise 49 ■

8.6 Exercises

Concepts

1–2 ■ Let $\mathbf{u} = \langle a_1, a_2\rangle$ and $\mathbf{v} = \langle b_1, b_2\rangle$ be nonzero vectors in the plane, and let θ be the angle between them.

1. The dot product of **u** and **v** is defined by

$\mathbf{u}\cdot\mathbf{v} =$ ______________

The dot product of two vectors is a __________, not a vector.

2. The angle θ satisfies

$$\cos\theta = \frac{\square}{\square}$$

So if $\mathbf{u}\cdot\mathbf{v} = 0$, the vectors are __________.

To find the angle θ between the vectors **u** and **v** in the figure, we first find

$$\cos\theta = \frac{\langle \square, \square\rangle\cdot\langle \square, \square\rangle}{|\langle \square, \square\rangle||\langle \square, \square\rangle|} = \text{\_\_\_\_\_\_\_\_\_\_}$$

and so $\theta \approx$ __________, rounded to the nearest degree.

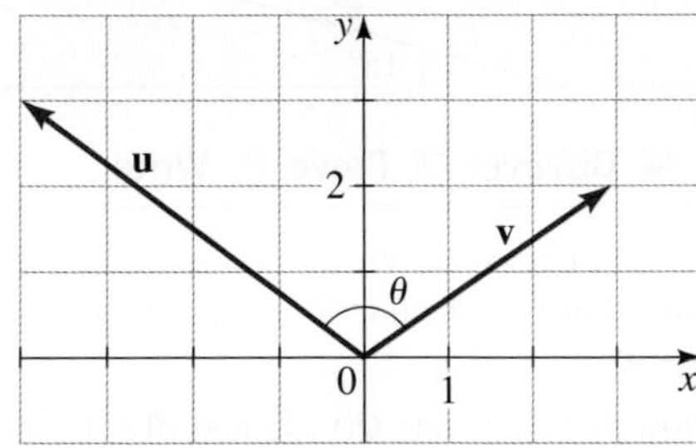

3. (a) The component of **u** along **v** is the scalar $|\mathbf{u}|\cos\theta$ and can be expressed in terms of the dot product as $\text{comp}_{\mathbf{v}}\,\mathbf{u} =$ ______________. Sketch this component in the figure below.

(b) The projection of **u** onto **v** is the vector $\text{proj}_{\mathbf{v}}\,\mathbf{u} =$ ______________. Sketch this projection in the figure below.

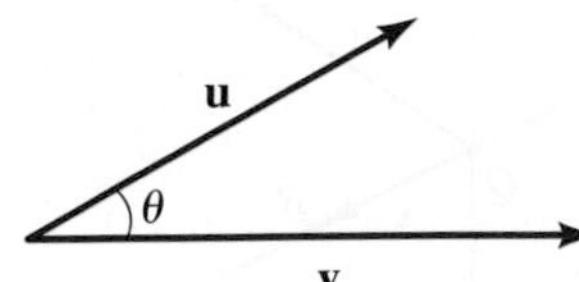

4. The work done by a force **F** in moving an object along a vector **D** is $W =$ ______________.

Skills

5–16 ■ Dot Products and Angles Between Vectors Find **(a)** $\mathbf{u}\cdot\mathbf{v}$ and **(b)** the angle between **u** and **v** to the nearest degree.

5. $\mathbf{u} = \langle 2, 0\rangle, \quad \mathbf{v} = \langle 1, 1\rangle$

6. $\mathbf{u} = \mathbf{i} + \sqrt{3}\,\mathbf{j}, \quad \mathbf{v} = -\sqrt{3}\,\mathbf{i} + \mathbf{j}$

7. $\mathbf{u} = \langle 1, 0\rangle, \quad \mathbf{w} = \langle 1, \sqrt{3}\rangle$

8. $\mathbf{u} = \langle -6, 6\rangle, \quad \mathbf{v} = \langle 1, -1\rangle$

9. $\mathbf{u} = \langle 3, -2\rangle, \quad \mathbf{v} = \langle 1, 2\rangle$

10. $\mathbf{u} = \langle 3, 4\rangle, \quad \mathbf{w} = \langle 4, 3\rangle$

11. $\mathbf{u} = -5\mathbf{j}, \quad \mathbf{v} = -\mathbf{i} - \sqrt{3}\,\mathbf{j}$

12. $\mathbf{u} = \mathbf{i} + \mathbf{j}, \quad \mathbf{v} = \mathbf{i} - \mathbf{j}$

13. $\mathbf{u} = \mathbf{i} + 3\mathbf{j}, \quad \mathbf{v} = 4\mathbf{i} - \mathbf{j}$

14. $\mathbf{u} = 3\mathbf{i} + 4\mathbf{j}, \quad \mathbf{v} = -2\mathbf{i} - \mathbf{j}$

15. $\mathbf{u} = \langle 1, \sqrt{3}\rangle, \quad \mathbf{w} = \langle 1, -\sqrt{3}\rangle$

16. $\mathbf{u} = \langle 6, 8\rangle, \quad \mathbf{w} = \langle 3, 4\rangle$

17–22 ■ Perpendicular Vectors? Determine whether the given vectors are perpendicular.

17. $\mathbf{u} = \langle 6, 4\rangle, \quad \mathbf{v} = \langle -2, 3\rangle$

18. $\mathbf{u} = \langle 0, -5\rangle, \quad \mathbf{v} = \langle 4, 0\rangle$

19. $\mathbf{u} = \langle -2, 6\rangle, \quad \mathbf{v} = \langle 4, 2\rangle$

20. $\mathbf{u} = 2\mathbf{i}, \quad \mathbf{v} = -7\mathbf{j}$

21. $\mathbf{u} = 2\mathbf{i} - 8\mathbf{j}, \quad \mathbf{v} = -12\mathbf{i} - 3\mathbf{j}$

22. $\mathbf{u} = 4\mathbf{i}, \quad \mathbf{v} = -\mathbf{i} + 3\mathbf{j}$

23–26 ■ Dot Products Find the indicated quantity, given that $\mathbf{u} = 2\mathbf{i} + \mathbf{j}$, $\mathbf{v} = \mathbf{i} - 3\mathbf{j}$, and $\mathbf{w} = 3\mathbf{i} + 4\mathbf{j}$.

23. $\mathbf{u}\cdot\mathbf{v} + \mathbf{u}\cdot\mathbf{w}$

24. $\mathbf{u}\cdot(\mathbf{v} + \mathbf{w})$

25. $(\mathbf{u} + \mathbf{v})\cdot(\mathbf{u} - \mathbf{v})$

26. $(\mathbf{u}\cdot\mathbf{v})(\mathbf{u}\cdot\mathbf{w})$

27–30 ■ The Component of u along v Find the component of **u** along **v**.

27. $\mathbf{u} = \langle 4, 6\rangle, \quad \mathbf{v} = \langle 3, -4\rangle$

28. $\mathbf{u} = \langle -3, 5\rangle, \quad \mathbf{v} = \langle 1/\sqrt{2}, 1/\sqrt{2}\rangle$

29. $\mathbf{u} = 7\mathbf{i} - 24\mathbf{j}, \quad \mathbf{v} = \mathbf{j}$

30. $\mathbf{u} = 7\mathbf{i}, \quad \mathbf{v} = 8\mathbf{i} + 6\mathbf{j}$

31–36 ■ Vector Projection of u onto v **(a)** Calculate $\text{proj}_{\mathbf{v}}\,\mathbf{u}$. **(b)** Resolve $\mathbf{u}$ into $\mathbf{u}_1$ and $\mathbf{u}_2$, where $\mathbf{u}_1$ is parallel to $\mathbf{v}$ and $\mathbf{u}_2$ is orthogonal to $\mathbf{v}$.

31. $\mathbf{u} = \langle -2, 4\rangle, \quad \mathbf{v} = \langle 1, 1\rangle$

32. $\mathbf{u} = \langle 7, -4\rangle, \quad \mathbf{v} = \langle 2, 1\rangle$

33. $\mathbf{u} = \langle 1, 2\rangle, \quad \mathbf{v} = \langle 1, -3\rangle$

34. $\mathbf{u} = \langle 11, 3\rangle, \quad \mathbf{v} = \langle -3, -2\rangle$

35. $\mathbf{u} = \langle 2, 9\rangle, \quad \mathbf{v} = \langle -3, 4\rangle$

36. $\mathbf{u} = \langle 1, 1\rangle, \quad \mathbf{v} = \langle 2, -1\rangle$

37–40 ■ Calculating Work Find the work done by the force $\mathbf{F}$ in moving an object from P to Q.

37. $\mathbf{F} = 4\mathbf{i} - 5\mathbf{j}; \quad P(0, 0), Q(3, 8)$

38. $\mathbf{F} = 400\mathbf{i} + 50\mathbf{j}; \quad P(-1, 1), Q(200, 1)$

39. $\mathbf{F} = 10\mathbf{i} + 3\mathbf{j}; \quad P(2, 3), Q(6, -2)$

40. $\mathbf{F} = -4\mathbf{i} + 20\mathbf{j}; \quad P(0, 10), Q(5, 25)$

Skills Plus

41–44 ■ Properties of Vectors Let $\mathbf{u}$, $\mathbf{v}$, and $\mathbf{w}$ be vectors, and let c be a scalar. Prove the given property.

41. $\mathbf{u}\cdot\mathbf{v} = \mathbf{v}\cdot\mathbf{u}$

42. $(c\mathbf{u})\cdot\mathbf{v} = c(\mathbf{u}\cdot\mathbf{v}) = \mathbf{u}\cdot(c\mathbf{v})$

43. $(\mathbf{u} + \mathbf{v})\cdot\mathbf{w} = \mathbf{u}\cdot\mathbf{w} + \mathbf{v}\cdot\mathbf{w}$

44. $(\mathbf{u} - \mathbf{v})\cdot(\mathbf{u} + \mathbf{v}) = |\mathbf{u}|^2 - |\mathbf{v}|^2$

45. Projection Show that $\text{proj}_{\mathbf{v}}\,\mathbf{u}$ and $\mathbf{u} - \text{proj}_{\mathbf{v}}\,\mathbf{u}$ are orthogonal.

46. Projection Show that $\mathbf{v}\cdot\text{proj}_{\mathbf{v}}\,\mathbf{u} = \mathbf{u}\cdot\mathbf{v}$.

Applications

47. Work The force $\mathbf{F} = 4\mathbf{i} - 7\mathbf{j}$ moves an object 4 m along the x-axis in the positive direction. Find the work done if the unit of force is the newton.

48. Work A constant force $\mathbf{F} = \langle 2, 8\rangle$ moves an object along a straight line from the point $(2, 5)$ to the point $(11, 13)$. Find the work done if the distance is measured in meters and the force is measured in newtons.

49. Work A lawn mower is pushed a distance of 60 m along a horizontal path by a constant force of 220 N. The handle of the lawn mower is held at an angle of 30° from the horizontal (see the figure). Find the work done.

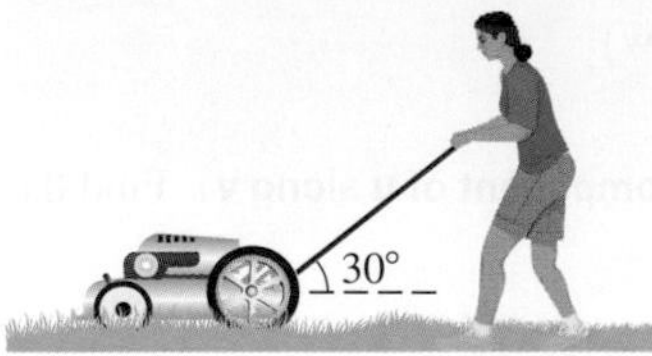

50. Work A car drives 150 m on a road that is inclined 12° to the horizontal, as shown in the following figure. The car weighs 11,000 N. Thus gravity acts straight down on the car with a constant force $\mathbf{F} = -11000\mathbf{j}$. Find the work done by the car in overcoming gravity.

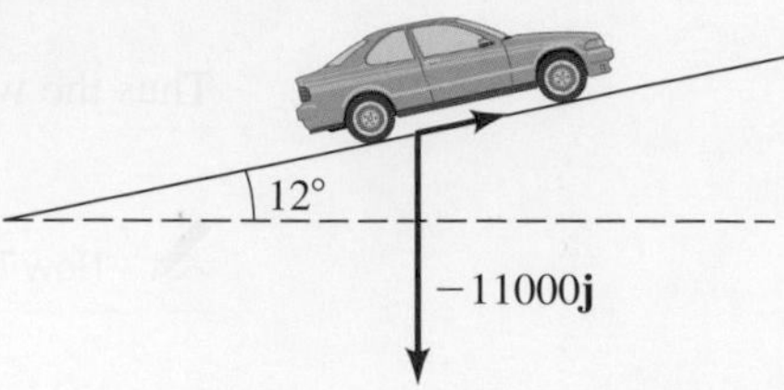

51. Force A car is on a driveway that is inclined 10° to the horizontal. A force of 2180 N is required to keep the car from rolling down the driveway.

(a) Find the weight of the car.

(b) Find the force the car exerts against the driveway.

52. Force A car is on a driveway that is inclined 25° to the horizontal. If the car weighs 12,300 N, find the force required to keep it from rolling down the driveway.

53. Force A package that weighs 200 N is placed on an inclined plane. If a force of 80 N is just sufficient to keep the package from sliding, find the angle of inclination of the plane. (Ignore the effects of friction.)

54. Force A cart weighing 180 N is placed on a ramp inclined at 15° to the horizontal. The cart is held in place by a rope inclined at 60° to the horizontal, as shown in the figure. Find the force that the rope must exert on the cart to keep it from rolling down the ramp.

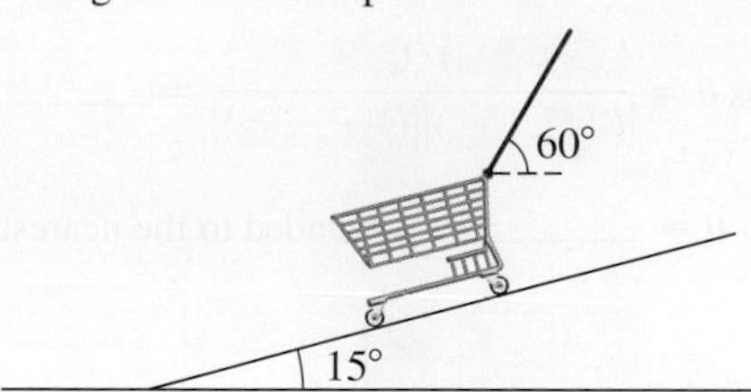

Discuss ■ Discover ■ Prove ■ Write

55. Discuss ■ Discover ■ Write: Distance from a Point to a Line Let L be the line $2x + 4y = 8$, and let P be the point $(3, 4)$.

(a) Show that the points $Q(0, 2)$ and $R(2, 1)$ lie on L.

(b) Let $\mathbf{u} = \overrightarrow{QP}$ and $\mathbf{v} = \overrightarrow{QR}$, as shown in the figure. Find $\mathbf{w} = \text{proj}_{\mathbf{v}}\,\mathbf{u}$.

(c) Sketch a graph that explains why $|\mathbf{u} - \mathbf{w}|$ is the distance from P to L. Find this distance.

(d) Write a short paragraph describing the steps you would take to find the distance from a given point to a given line.

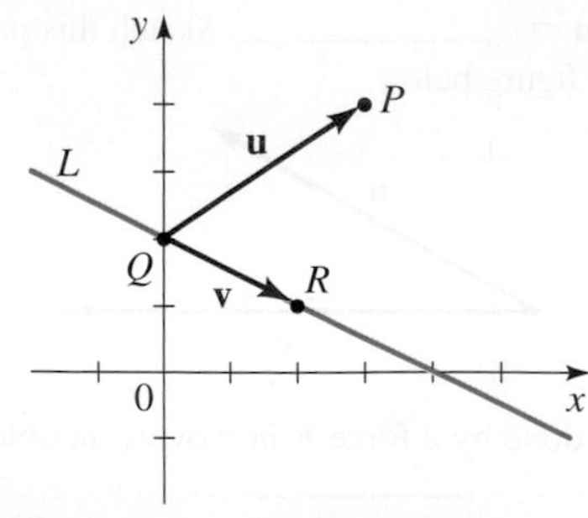

Chapter 8 Review

Properties and Formulas

Polar Coordinates | Section 8.1

In the **polar coordinate** system the location of a point P in the plane is determined by an ordered pair (r, θ), where r is the distance from the pole O to P and θ is the angle formed by the polar axis and the segment $\overrightarrow{OP}$, as shown in the figure.

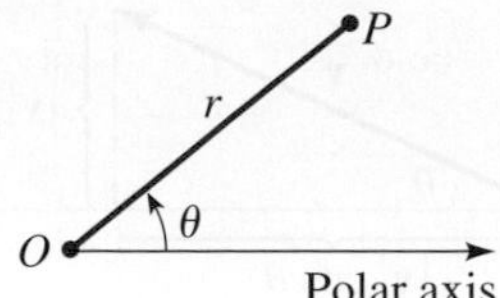

Polar and Rectangular Coordinates | Section 8.1

Any point P in the plane has polar coordinates $P(r, \theta)$ and rectangular coordinates $P(x, y)$, as shown.

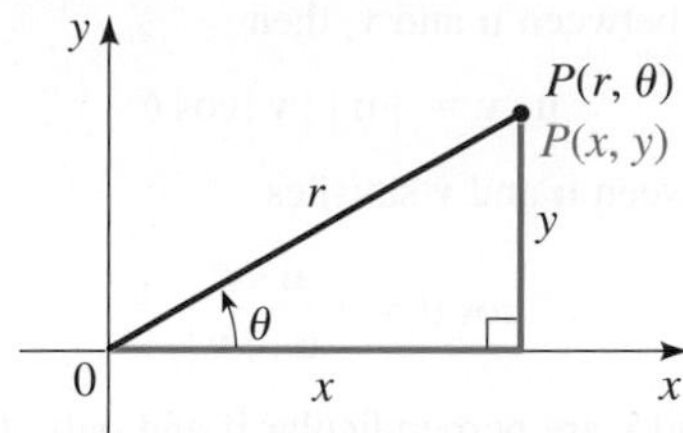

- **To change from polar to rectangular coordinates**, we use the equations

$$x = r\cos\theta \qquad \text{and} \qquad y = r\sin\theta$$

- **To change from rectangular to polar coordinates**, we use the equations

$$r^2 = x^2 + y^2 \qquad \text{and} \qquad \tan\theta = \frac{y}{x}$$

Polar Equations and Graphs | Section 8.2

A **polar equation** is an equation in the variables r and θ. The **graph of a polar equation** $r = f(\theta)$ consists of all points (r, θ) whose coordinates satisfy the equation.

Symmetry in Graphs of Polar Equations | Section 8.2

The graph of a polar equation is

- **symmetric about the polar axis** if the equation is unchanged when we replace θ by $-\theta$;
- **symmetric about the pole** if the equation is unchanged when we replace r by $-r$, or θ by $\theta + \pi$.
- **symmetric about the vertical line** $\theta = \pi/2$ if the equation is unchanged when we replace θ by $\pi - \theta$.

Complex Numbers | Section 8.3

A **complex number** is a number of the form $a + bi$, where $i^2 = -1$ and where a and b are real numbers. For the complex number $z = a + bi$, a is called the **real part** and b is called the **imaginary part**. A complex number $a + bi$ is graphed in the complex plane as shown.

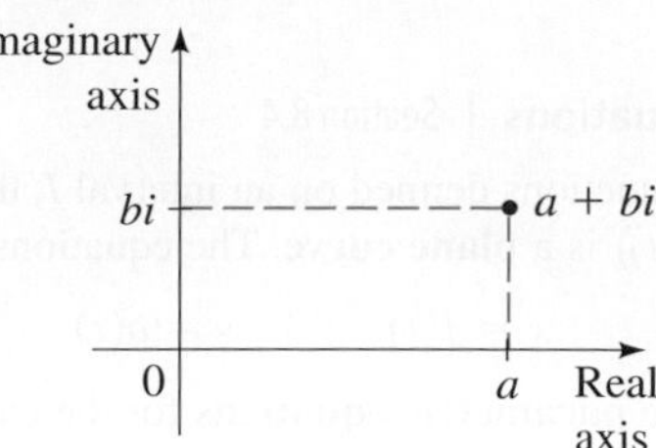

The **modulus** (or **absolute value**) of a complex number $z = a + bi$ is

$$|z| = \sqrt{a^2 + b^2}$$

Polar Form of Complex Numbers | Section 8.3

A complex number $z = a + bi$ has the **polar form** (or **trigonometric form**)

$$z = r(\cos\theta + i\sin\theta)$$

where $r = |z|$ and $\tan\theta = b/a$. The number r is the modulus of z and θ is the argument of z.

Multiplication and Division of Complex Numbers in Polar Form | Section 8.3

Suppose the complex numbers z_1 and z_2 have the following polar form:

$$z_1 = r_1(\cos\theta_1 + i\sin\theta_1)$$

$$z_2 = r_2(\cos\theta_2 + i\sin\theta_2)$$

Then

$$z_1 z_2 = r_1 r_2[\cos(\theta_1 + \theta_2) + i\sin(\theta_1 + \theta_2)]$$

$$\frac{z_1}{z_2} = \frac{r_1}{r_2}[\cos(\theta_1 - \theta_2) + i\sin(\theta_1 - \theta_2)]$$

De Moivre's Theorem | Section 8.3

If $z = r(\cos\theta + i\sin\theta)$ is a complex number in polar form and n is a positive integer, then

$$z^n = r^n(\cos n\theta + i\sin n\theta)$$

*n*th Roots of Complex Numbers | Section 8.3

If $z = r(\cos\theta + i\sin\theta)$ is a complex number in polar form and n is a positive integer, then z has the n distinct nth roots $w_0, w_1, \ldots, w_{n-1}$, where

$$w_k = r^{1/n}\left[\cos\left(\frac{\theta + 2k\pi}{n}\right) + i\sin\left(\frac{\theta + 2k\pi}{n}\right)\right]$$

where $k = 0, 1, 2, \ldots, n - 1$.

Finding the *n*th Roots of *z* | Section 8.3

To find the *n*th roots of $z = r(\cos\theta + i\sin\theta)$, we use the following observations:

1. The modulus of each *n*th root is $r^{1/n}$.
2. The argument of the first root w_0 is θ/n.
3. Repeatedly add $2\pi/n$ to get the argument of each successive root.

Parametric Equations | Section 8.4

If f and g are functions defined on an interval I, then the set of points $(f(t), g(t))$ is a **plane curve**. The equations

$$x = f(t) \qquad y = g(t)$$

where $t \in I$, are **parametric equations** for the curve, with **parameter** t.

Polar Equations in Parametric Form | Section 8.4

The graph of the polar equation $r = f(\theta)$ is the same as the graph of the parametric equations

$$x = f(t)\cos t \qquad y = f(t)\sin t$$

Vectors | Section 8.5

A **vector** is a quantity with both magnitude and direction.

A vector in the coordinate plane is expressed in terms of two coordinates or components

$$\mathbf{v} = \langle a_1, a_2\rangle$$

If a vector $\mathbf{v}$ has its initial point at $P(x_1, y_1)$ and its terminal point at $Q(x_2, y_2)$, then

$$\mathbf{v} = \langle x_2 - x_1, y_2 - y_1\rangle$$

Let $\mathbf{u} = \langle a_1, a_2\rangle$, $\mathbf{v} = \langle b_1, b_2\rangle$, and $c \in \mathbb{R}$. The operations on vectors are defined as follows.

$\mathbf{u} + \mathbf{v} = \langle a_1 + b_1, a_2 + b_2\rangle$	Addition
$\mathbf{u} - \mathbf{v} = \langle a_1 - b_1, a_2 - b_2\rangle$	Subtraction
$c\mathbf{u} = \langle ca_1, ca_2\rangle$	Scalar multiplication

The unit vectors **i** and **j** are defined by

$$\mathbf{i} = \langle 1, 0\rangle \qquad \mathbf{j} = \langle 0, 1\rangle$$

Any vector $\mathbf{v} = \langle a_1, a_2\rangle$ can be expressed as

$$\mathbf{v} = a_1\mathbf{i} + a_2\mathbf{j}$$

Let $\mathbf{v} = \langle a_1, a_2\rangle$. The **magnitude** (or **length**) of **v** is

$$|\mathbf{v}| = \sqrt{a_1^2 + a_2^2}$$

The **direction** of **v** is the smallest positive angle θ in standard position formed by the positive x-axis and **v** (see the figure below).

If $\mathbf{v} = \langle a_1, a_2\rangle$, then the components of **v** satisfy

$$a_1 = |\mathbf{v}|\cos\theta \qquad a_2 = |\mathbf{v}|\sin\theta$$

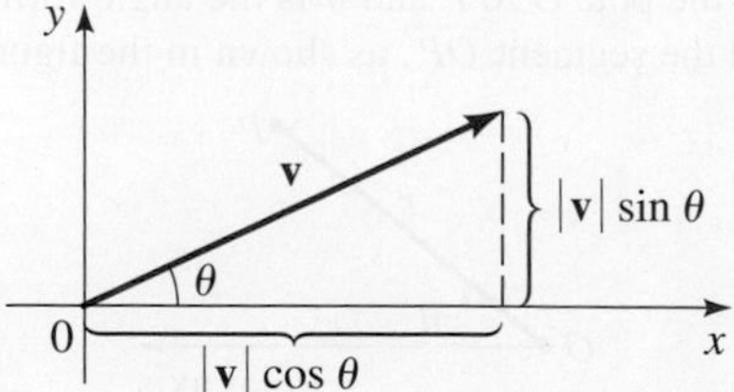

The Dot Product of Vectors | Section 8.6

If $\mathbf{u} = \langle a_1, a_2\rangle$ and $\mathbf{v} = \langle b_1, b_2\rangle$, then their **dot product** is

$$\mathbf{u}\cdot\mathbf{v} = a_1b_1 + a_2b_2$$

If θ is the angle between **u** and **v**, then

$$\mathbf{u}\cdot\mathbf{v} = |\mathbf{u}|\,|\mathbf{v}|\cos\theta$$

The angle θ between **u** and **v** satisfies

$$\cos\theta = \frac{\mathbf{u}\cdot\mathbf{v}}{|\mathbf{u}|\,|\mathbf{v}|}$$

The vectors **u** and **v** are perpendicular if and only if

$$\mathbf{u}\cdot\mathbf{v} = 0$$

The **component of u along v** (a scalar) and the **projection of u onto v** (a vector) are given by

$$\text{comp}_{\mathbf{v}}\,\mathbf{u} = \frac{\mathbf{u}\cdot\mathbf{v}}{|\mathbf{v}|} \qquad \text{proj}_{\mathbf{v}}\,\mathbf{u} = \left(\frac{\mathbf{u}\cdot\mathbf{v}}{|\mathbf{v}|^2}\right)\mathbf{v}$$

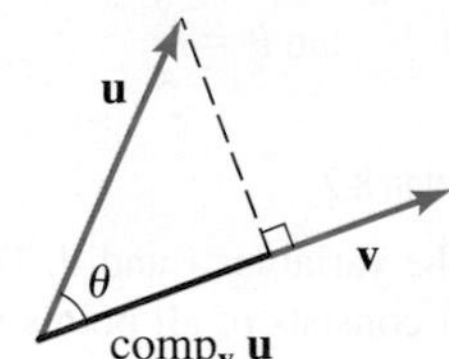

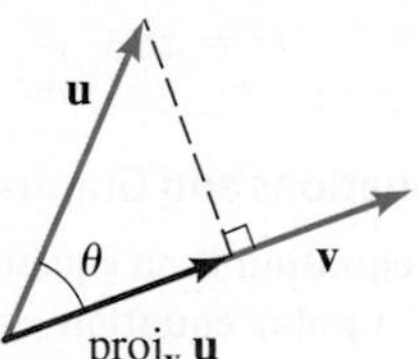

The **work** W done by a force **F** in moving along a vector **D** is

$$W = \mathbf{F}\cdot\mathbf{D}$$

Concept Check

1. (a) Explain the polar coordinate system.
   (b) Graph the points with polar coordinates $(2, \pi/3)$ and $(-1, 3\pi/4)$.
   (c) State the equations that relate the rectangular coordinates of a point to its polar coordinates.
   (d) Find rectangular coordinates for $(2, \pi/3)$.
   (e) Find polar coordinates for $P(-2, 2)$.
2. (a) What is a polar equation?
   (b) Convert the polar equation $r = \sin\theta$ to an equivalent rectangular equation.
3. (a) How do we graph a polar equation?
   (b) Sketch a graph of the polar equation $r = 4 + 4\cos\theta$. What is the graph called?
4. (a) What is the complex plane? How do we graph a complex number $z = a + bi$ in the complex plane?
   (b) What are the modulus and argument of the complex number $z = a + bi$?
   (c) Graph the point $z = \sqrt{3} - i$, and find the modulus and argument of z.

5. (a) How do we express the complex number z in polar form?

(b) Express $z = \sqrt{3} - i$ in polar form.

6. Let $z_1 = 2\left(\cos\dfrac{\pi}{3} + i\sin\dfrac{\pi}{3}\right)$

and $z_2 = 5\left(\cos\dfrac{\pi}{4} + i\sin\dfrac{\pi}{4}\right)$

(a) Find the product $z_1 z_2$.

(b) Find the quotient z_1/z_2.

7. (a) State de Moivre's Theorem.

(b) Use de Moivre's Theorem to find the fifth power of $z = 2\left(\cos\dfrac{\pi}{3} + i\sin\dfrac{\pi}{3}\right)$.

8. (a) State the formula for the nth roots of a complex number $z = r(\cos\theta + i\sin\theta)$.

(b) How do we find the nth roots of a complex number?

(c) Find the three third roots of $z = -8$.

9. (a) What are parametric equations?

(b) Sketch a graph of the following parametric equations, using arrows to indicate the direction of the curve.

$$x = t + 1 \qquad y = t^2 \qquad -2 \le t \le 2$$

(c) Eliminate the parameter to obtain an equation in x and y.

10. (a) What is a vector in the plane? How do we represent a vector in the coordinate plane?

(b) Find the vector with initial point $(2, 3)$ and terminal point $(4, 10)$.

(c) Let $\mathbf{v} = \langle 2, 1\rangle$. If the initial point of $\mathbf{v}$ is placed at $P(1, 1)$, where is its terminal point? Sketch several representations of $\mathbf{v}$.

(d) How is the magnitude of $\mathbf{v} = \langle a_1, a_2\rangle$ defined? Find the magnitude of $\mathbf{w} = \langle 3, 4\rangle$.

(e) What are the vectors $\mathbf{i}$ and $\mathbf{j}$? Express the vector $\mathbf{v} = \langle 5, 9\rangle$ in terms of $\mathbf{i}$ and $\mathbf{j}$.

(f) Let $\mathbf{v} = \langle a_1, a_2\rangle$ be a vector in the coordinate plane. What is meant by the direction θ of $\mathbf{v}$? What are the components of $\mathbf{v}$ in terms of its length and direction? Sketch a figure to illustrate your answer.

(g) Suppose that $\mathbf{v}$ has length $|\mathbf{v}| = 5$ and direction $\theta = \pi/6$. Express $\mathbf{v}$ in terms of its coordinates.

11. (a) Define addition and scalar multiplication for vectors.

(b) If $\mathbf{u} = \langle 2, 3\rangle$ and $\mathbf{v} = \langle 5, 9\rangle$, find $\mathbf{u} + \mathbf{v}$ and $4\mathbf{u}$.

12. (a) Define the dot product of the vectors $\mathbf{u} = \langle a_1, a_2\rangle$ and $\mathbf{v} = \langle b_1, b_2\rangle$, and state the formula for the angle θ between $\mathbf{u}$ and $\mathbf{v}$.

(b) If $\mathbf{u} = \langle 2, 3\rangle$ and $\mathbf{v} = \langle 1, 4\rangle$, find $\mathbf{u}\cdot\mathbf{v}$ and find the angle between $\mathbf{u}$ and $\mathbf{v}$.

13. How much work is done by the force $\mathbf{F}$ in moving an object along a displacement vector $\mathbf{D}$?

Answers to the Concept Check can be found at the book companion website stewartmath.com.

Exercises

1–4 ■ Polar Coordinates to Rectangular Coordinates A point $P(r, \theta)$ is given in polar coordinates. **(a)** Plot the point P. **(b)** Find rectangular coordinates for P.

1. $(12, \pi/6)$ **2.** $(8, -3\pi/4)$

3. $(-3, 7\pi/4)$ **4.** $(-\sqrt{3}, 2\pi/3)$

5–8 ■ Rectangular Coordinates to Polar Coordinates A point $P(x, y)$ is given in rectangular coordinates. **(a)** Plot the point P. **(b)** Find polar coordinates for P with $r \ge 0$. **(c)** Find polar coordinates for P with $r \le 0$.

5. $(8, 8)$ **6.** $(-\sqrt{2}, \sqrt{6})$

7. $(-6\sqrt{2}, -6\sqrt{2})$ **8.** $(4, -4)$

9–12 ■ Rectangular Equations to Polar Equations **(a)** Convert the equation to polar coordinates and simplify. **(b)** Graph the equation. [*Hint:* Use the form of the equation that you find easier to graph.]

9. $x + y = 4$ **10.** $xy = 1$

11. $x^2 + y^2 = 4x + 4y$ **12.** $(x^2 + y^2)^2 = 2xy$

13–20 ■ Polar Equations to Rectangular Equations **(a)** Sketch the graph of the polar equation. **(b)** Express the equation in rectangular coordinates.

13. $r = 3 + 3\cos\theta$ **14.** $r = 3\sin\theta$

15. $r = 2\sin 2\theta$ **16.** $r = 4\cos 3\theta$

17. $r^2 = \sec 2\theta$ **18.** $r^2 = 4\sin 2\theta$

19. $r = \sin\theta + \cos\theta$ **20.** $r = \dfrac{4}{2 + \cos\theta}$

21–24 ■ Graphing Polar Equations Use a graphing device to graph the polar equation. Choose the domain of θ to produce the entire graph.

21. $r = \cos(\theta/3)$ **22.** $r = \sin(9\theta/4)$

23. $r = 1 + 4\cos(\theta/3)$ **24.** $r = \sin(2^\theta) - 2, \quad 0 \le \theta \le 2\pi$

25–30 ■ Complex Numbers A complex number is given. **(a)** Graph the complex number in the complex plane. **(b)** Find the modulus and argument. **(c)** Write the number in polar form.

25. $4 + 4i$ **26.** $-10i$

27. $5 + 3i$ **28.** $1 + \sqrt{3}i$

29. $-1 + i$ **30.** -20

31–34 ■ Powers Using de Moivre's Theorem Use de Moivre's Theorem to find the indicated power.

31. $(1 - \sqrt{3}i)^4$ **32.** $(1 + i)^8$

33. $(\sqrt{3} + i)^{-4}$ **34.** $\left(\dfrac{1}{2} + \dfrac{\sqrt{3}}{2}i\right)^{20}$

35–38 ■ Roots of Complex Numbers Find the indicated roots.

35. The square roots of $-16i$

36. The cube roots of $4 + 4\sqrt{3}i$

37. The sixth roots of 1 **38.** The eighth roots of i

39–42 ■ Parametric Curves A pair of parametric equations is given. **(a)** Sketch the curve represented by the parametric equations. Use arrows to indicate the direction of the curve as t increases. **(b)** Find an equation in rectangular coordinates for the curve by eliminating the parameter.

39. $x = 1 - t^2, \quad y = 1 + t$ **40.** $x = t^2 - 1, \quad y = t^2 + 1$

41. $x = 1 + \cos t, \quad y = 1 - \sin t, \quad 0 \le t \le \pi/2$

42. $x = \dfrac{1}{t} + 2, \quad y = \dfrac{2}{t^2}, \quad 0 < t \le 2$

43–44 ■ Graphs of Parametric Equations Use a graphing device to draw the parametric curve.

43. $x = \cos 2t, \quad y = \sin 3t$

44. $x = \sin(t + \cos 2t), \quad y = \cos(t + \sin 3t)$

45. Finding Parametric Equations for a Curve In the figure, the point P is the midpoint of the segment QR and $0 \le \theta < \pi/2$. Using θ as the parameter, find a parametric representation for the curve traced out by P.

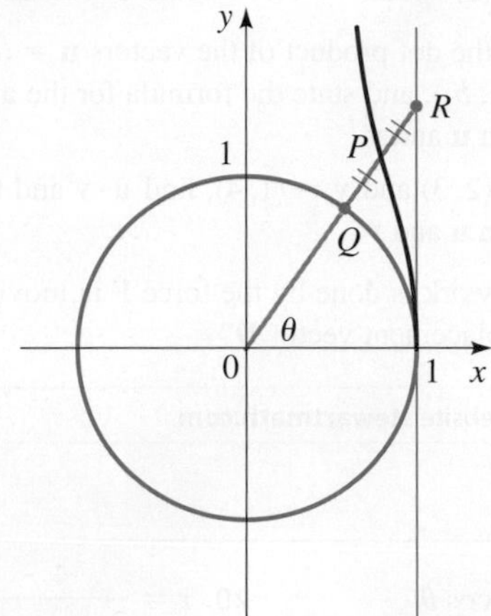

46. Finding Parametric Equations for a Curve Find parametric equations for the curve traced out by the points P shown in the figure, using the angle θ as the parameter. In the figure, $|OR| = |QP|$.

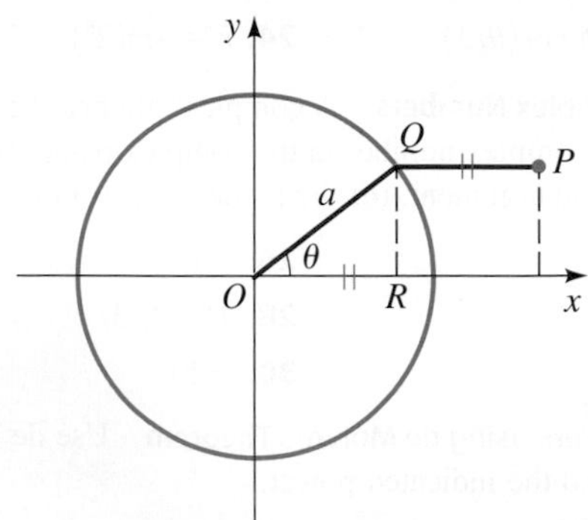

47–48 ■ Operations with Vectors Find $|\mathbf{u}|$, $\mathbf{u} + \mathbf{v}$, $\mathbf{u} - \mathbf{v}$, $2\mathbf{u}$, and $3\mathbf{u} - 2\mathbf{v}$.

47. $\mathbf{u} = \langle -2, 3 \rangle, \quad \mathbf{v} = \langle 8, 1 \rangle$ **48.** $\mathbf{u} = 2\mathbf{i} + \mathbf{j}, \quad \mathbf{v} = \mathbf{i} - 2\mathbf{j}$

49. Component Form of a Vector A vector has initial point $P(0, 3)$ and terminal point $Q(3, -1)$. Express the vector in component form.

50. Terminal Point of a Vector If the vector $5\mathbf{i} - 8\mathbf{j}$ is placed in the plane with its initial point at $P(5, 6)$, find its terminal point.

51–52 ■ Length and Direction of Vectors Find the length and direction of the given vector.

51. $\mathbf{u} = \langle -2, 2\sqrt{3} \rangle$ **52.** $\mathbf{v} = 2\mathbf{i} - 5\mathbf{j}$

53–54 ■ Component Form of a Vector The length $|\mathbf{u}|$ and direction θ of a vector $\mathbf{u}$ are given. Express $\mathbf{u}$ in component form.

53. $|\mathbf{u}| = 20, \quad \theta = 60°$ **54.** $|\mathbf{u}| = 13.5, \quad \theta = 125°$

55. Resultant Force Two tugboats are pulling a barge as shown in the figure. One pulls with a force of 100 kN in the direction N 50° E, and the other pulls with a force of 170 kN in the direction S 75° E.

(a) Find the resultant force on the barge as a vector.

(b) Find the magnitude and direction of the resultant force.

56. True Velocity of a Plane An airplane heads N 60° E at a speed of 960 km/h relative to the air. A wind begins to blow in the direction N 30° W at 80 km/h. (See the figure.)

(a) Find the velocity of the airplane as a vector.

(b) Find the true speed and direction of the airplane.

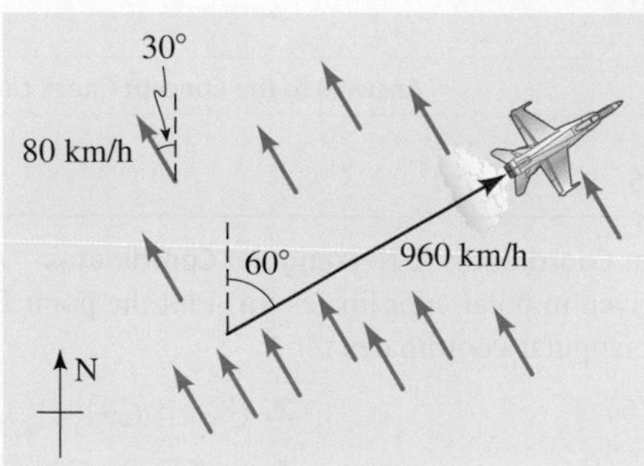

57–60 ■ Dot Products Find the vectors $|\mathbf{u}|$, $\mathbf{u} \cdot \mathbf{u}$, and $\mathbf{u} \cdot \mathbf{v}$.

57. $\mathbf{u} = \langle 4, -3 \rangle, \quad \mathbf{v} = \langle 9, -8 \rangle$ **58.** $\mathbf{u} = \langle 5, 12 \rangle, \quad \mathbf{v} = \langle 10, -4 \rangle$

59. $\mathbf{u} = -2\mathbf{i} + 2\mathbf{j}, \quad \mathbf{v} = \mathbf{i} + \mathbf{j}$ **60.** $\mathbf{u} = 10\mathbf{j}, \quad \mathbf{v} = 5\mathbf{i} - 3\mathbf{j}$

61–64 ■ Orthogonal Vectors Are $\mathbf{u}$ and $\mathbf{v}$ orthogonal? If not, find the angle between them.

61. $\mathbf{u} = \langle -4, 2 \rangle, \quad \mathbf{v} = \langle 3, 6 \rangle$ **62.** $\mathbf{u} = \langle 5, 3 \rangle, \quad \mathbf{v} = \langle -2, 6 \rangle$

63. $\mathbf{u} = 2\mathbf{i} + \mathbf{j}, \quad \mathbf{v} = \mathbf{i} + 3\mathbf{j}$ **64.** $\mathbf{u} = \mathbf{i} - \mathbf{j}, \quad \mathbf{v} = \mathbf{i} + \mathbf{j}$

65–66 ■ Scalar and Vector Projections Two vectors $\mathbf{u}$ and $\mathbf{v}$ are given. **(a)** Find the component of $\mathbf{u}$ along $\mathbf{v}$. **(b)** Find $\text{proj}_{\mathbf{v}} \mathbf{u}$. **(c)** Resolve $\mathbf{u}$ into the vectors $\mathbf{u}_1$ and $\mathbf{u}_2$, where $\mathbf{u}_1$ is parallel to $\mathbf{v}$ and $\mathbf{u}_2$ is perpendicular to $\mathbf{v}$.

65. $\mathbf{u} = \langle 3, 1 \rangle, \quad \mathbf{v} = \langle 6, -1 \rangle$ **66.** $\mathbf{u} = \mathbf{i} + 2\mathbf{j}, \quad \mathbf{v} = 4\mathbf{i} - 9\mathbf{j}$

67. Work A force $\mathbf{F} = 2\mathbf{i} + 9\mathbf{j}$ moves an object from the point $(7, -1)$ to the point $(1, 1)$. Find the work done if the distance is measured in meters and the force is measured in kilonewtons.

68. Work A force $\mathbf{F}$ with magnitude 250 kN moves an object in the direction of a vector $\mathbf{D}$ a distance of 20 m. If the work done is 3800 kN.m, find the angle between $\mathbf{F}$ and $\mathbf{D}$.

Chapter 8 | Test

1. **(a)** Convert the point whose polar coordinates are $(8, 5\pi/4)$ to rectangular coordinates.
   (b) Find two polar coordinate representations for the rectangular coordinate point $(-6, 2\sqrt{3})$, one with $r > 0$ and one with $r < 0$ and both with $0 \le \theta < 2\pi$.
2. **(a)** Graph the polar equation $r = 8\cos\theta$. What type of curve is this?
   (b) Convert the equation to rectangular coordinates.
3. Graph the polar equation $r = 3 + 6\sin\theta$. What type of curve is this?
4. Let $z = 1 + \sqrt{3}i$.
   (a) Graph z in the complex plane.
   (b) Write z in polar form.
   (c) Find the complex number z^9.
5. Let $z_1 = 4\left(\cos\frac{7\pi}{12} + i\sin\frac{7\pi}{12}\right)$ and $z_2 = 2\left(\cos\frac{5\pi}{12} + i\sin\frac{5\pi}{12}\right)$.
   Find z_1z_2 and $\frac{z_1}{z_2}$.
6. Find the cube roots of $27i$, and sketch these roots in the complex plane.
7. **(a)** Sketch the curve represented by the parametric equations below. Use arrows to indicate the direction of the curve as t increases.
   $$x = 3\sin t + 3 \qquad y = 2\cos t \qquad 0 \le t \le \pi$$
   (b) Eliminate the parameter t in part (a) to obtain an equation for this curve in rectangular coordinates.
8. Find parametric equations for the line of slope 2 that passes through the point $(3, 5)$.
9. The position of an object in circular motion is modeled by the parametric equations
   $$x = 3\sin 2t \qquad y = 3\cos 2t$$
   where t is measured in seconds.
   (a) Describe the path of the object by stating the radius of the circle, the position at time $t = 0$, the orientation of motion (clockwise or counterclockwise), and the time t it takes to complete one revolution around the circle.
   (b) Suppose the speed of the object is doubled. Find new parametric equations that model the motion of the object.
   (c) Find an equation in rectangular coordinates for the same curve by eliminating the parameter.
   (d) Find a polar equation for the same curve.
10. Let $\mathbf{u}$ be the vector with initial point $P(3, -1)$ and terminal point $Q(-3, 9)$.
   (a) Graph $\mathbf{u}$ in the coordinate plane.
   (b) Express $\mathbf{u}$ in terms of $\mathbf{i}$ and $\mathbf{j}$.
   (c) Find the length of $\mathbf{u}$.
11. Let $\mathbf{u} = \langle 1, 3\rangle$, and let $\mathbf{v} = \langle -6, 2\rangle$.
   (a) Find $\mathbf{u} - 3\mathbf{v}$.
   (b) Find $|\mathbf{u} + \mathbf{v}|$.
   (c) Find $\mathbf{u} \cdot \mathbf{v}$.
   (d) Are $\mathbf{u}$ and $\mathbf{v}$ perpendicular?

12. Let $\mathbf{u} = \langle -4\sqrt{3}, 4 \rangle$.

(a) Graph $\mathbf{u}$ in the coordinate plane, with initial point $(0, 0)$.

(b) Find the length and direction of $\mathbf{u}$.

13. A river is flowing due east at 8 km/h. A motorboat heads in the direction N 30° E in the river. The speed of the motorboat relative to the water is 12 km/h.

(a) Express the true velocity of the motorboat as a vector.

(b) Find the true speed and direction of the motorboat.

14. Let $\mathbf{u} = 3\mathbf{i} + 2\mathbf{j}$ and $\mathbf{v} = 5\mathbf{i} - \mathbf{j}$.

(a) Find the angle between $\mathbf{u}$ and $\mathbf{v}$.

(b) Find the component of $\mathbf{u}$ along $\mathbf{v}$.

(c) Find $\text{proj}_{\mathbf{v}} \mathbf{u}$.

15. A force $\mathbf{F} = 3\mathbf{i} - 5\mathbf{j}$ moves an object from the point $(2, 2)$ to the point $(7, -13)$. Find the work done if the distance is measured in meters and the force is measured in kilonewtons.

In this section we use parametric equations and vectors to model the motion of a projectile, such as a ball thrown upward, an object launched from a catapult, or a cannonball fired from a cannon.

Parametric Equations for the Path of a Projectile

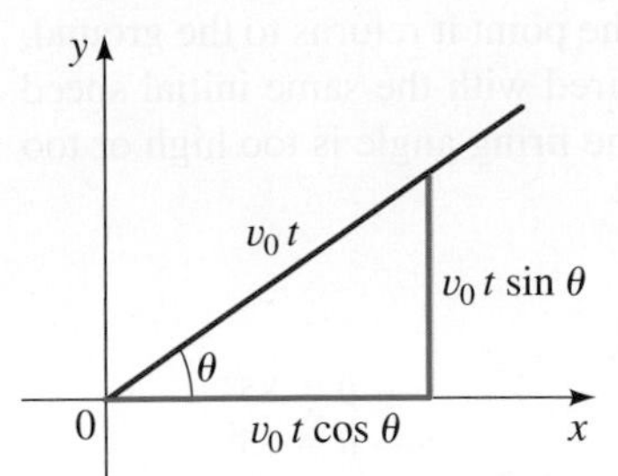

Figure 1 | Path of a projectile with no gravity

Suppose that we fire a projectile into the air from ground level, with an initial speed v_0 and at an angle θ upward from the ground. If there were no gravity (and no air resistance), the projectile would just keep moving indefinitely at the same speed and in the same direction. Since distance = speed × time, at time t the projectile would have traveled a distance $v_0 t$, so its position at time t would be given by the following parametric equations (assuming that the origin of our coordinate system is placed at the initial location of the projectile; see Figure 1):

$$x = (v_0 \cos \theta)t \qquad y = (v_0 \sin \theta)t \qquad \text{No gravity}$$

But, of course, we know that gravity will pull the projectile back to ground level. It can be shown that the effect of gravity after t seconds can be accounted for by subtracting $\frac{1}{2}gt^2$ from the vertical position of the projectile. In this expression, g is the gravitational acceleration: $g \approx 9.8\text{ m/s}^2$. Thus we have the following parametric equations for the path of the projectile:

$$x = (v_0 \cos \theta)t \qquad y = (v_0 \sin \theta)t - \tfrac{1}{2}gt^2 \qquad \text{Position at time } t$$

Example ■ The Path of a Cannonball

Find parametric equations that model the path of a cannonball fired into the air with an initial speed of 150 m/s at a 30° angle of elevation. Sketch the path of the cannonball.

The path of a projectile can be "animated" on a graphing device. See the Discovery Project *Collsion* referenced in Section 8.4.

Solution Substituting the given initial speed and angle into the general parametric equations of the path of a projectile, we get

$$x = (150 \cos 30°)t \qquad y = (150 \sin 30°)t - \tfrac{1}{2}(9.8)t^2 \qquad \text{Substitute } v_0 = 150,\ \theta = 30°,\ g = 9.8$$

$$x \approx 129.9t \qquad y = 75t - 4.9t^2 \qquad \text{Simplify}$$

This path is graphed in Figure 2.

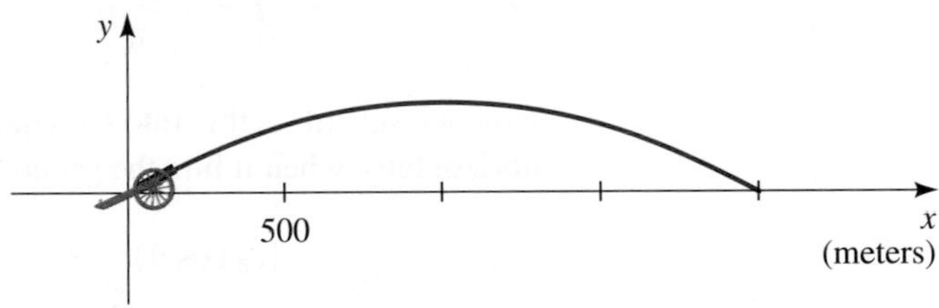

Figure 2 | Path of a cannonball

■

Range of a Projectile

How can we tell where and when the cannonball of the first example hits the ground? Since ground level corresponds to $y = 0$, we substitute this value for y and solve for t.

$$0 = 75t - 4.9t^2 \qquad \text{Set } y = 0$$

$$0 = t(75 - 4.9t) \qquad \text{Factor}$$

$$t = 0 \quad \text{or} \quad t = \frac{75}{4.9} \approx 15.3 \qquad \text{Solve for } t$$

GALILEO GALILEI (1564–1642) was born in Pisa, Italy. He studied medicine but later abandoned this in favor of science and mathematics. At the age of 25, by dropping cannonballs of various sizes from the Leaning Tower of Pisa, he demonstrated that light objects fall at the same rate as heavier ones. This contradicted the then-accepted view of Aristotle that heavier objects fall more quickly. Galileo also showed that the distance an object falls is proportional to the square of the time it has been falling, and from this he was able to prove that the path of a projectile is a parabola.

Galileo constructed the first telescope and, using it, discovered the moons of Jupiter. His advocacy of the Copernican view that the earth revolves around the sun (rather than being stationary) led to his being called before the Inquisition. By then an old man, he was forced to recant his views, but he is said to have muttered under his breath, "Nevertheless, it does move." Galileo revolutionized science by expressing scientific principles in the language of mathematics. He said, "The great book of nature is written in mathematical symbols."

The first solution, $t = 0$, is the time when the cannon was fired; the second solution means that the cannonball hits the ground after 15.3 s of flight. To see *where* this happens, we substitute this value of t into the equation for x, the horizontal location of the cannonball.

$$x \approx 129.9t = 129.9(15.3) \approx 1987.5 \text{ m}$$

The cannonball travels almost 2 km before hitting the ground—that is the *range* is 2 km. In general, for a projectile launched from ground level, the **range of the projectile** is the horizontal distance from the point it is launched to the point it returns to the ground.

Figure 3 shows the paths of several projectiles, all fired with the same initial speed but at different angles. From the graphs we see that if the firing angle is too high or too low, the projectile doesn't travel very far.

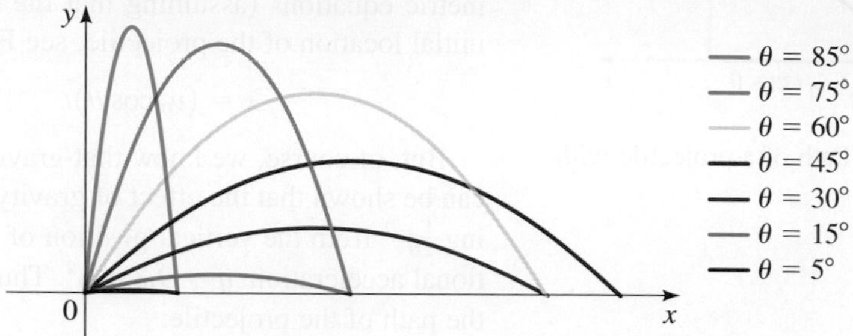

Figure 3 | Paths of projectiles

Let's try to find the optimal firing angle—the angle that shoots the projectile as far as possible. We'll go through the same steps as we did when calculating the range, but we'll use the general parametric equations instead. First, we solve for the time when the projectile hits the ground by substituting $y = 0$.

$$0 = (v_0 \sin\theta)t - \tfrac{1}{2}gt^2 \qquad \text{Substitute } y = 0$$

$$0 = t(v_0 \sin\theta - \tfrac{1}{2}gt) \qquad \text{Factor}$$

$$0 = v_0 \sin\theta - \tfrac{1}{2}gt \qquad \text{Set second factor equal to 0}$$

$$t = \frac{2v_0 \sin\theta}{g} \qquad \text{Solve for } t$$

Now we substitute this into the equation for x to see how far the projectile has traveled horizontally when it hits the ground.

$$x = (v_0 \cos\theta)t \qquad \text{Parametric equation for } x$$

$$= (v_0 \cos\theta)\left(\frac{2v_0 \sin\theta}{g}\right) \qquad \text{Substitute } t = (2v_0 \sin\theta)/g$$

$$= \frac{2v_0^2 \sin\theta \cos\theta}{g} \qquad \text{Simplify}$$

$$= \frac{v_0^2 \sin 2\theta}{g} \qquad \text{Use identity } \sin 2\theta = 2\sin\theta\cos\theta$$

We want to choose θ so that x is as large as possible. The largest value that the sine of an angle can have is 1, the sine of 90°. Thus we want $2\theta = 90°$, or $\theta = 45°$. So to send the projectile as far as possible, it should be launched at an angle of 45°. From the last equation in the preceding display, we see that it will travel a distance $x = v_0^2/g$.

Vector Equation for the Velocity of a Projectile

We found parametric equations for the path of a projectile. Now we find a vector equation for the velocity of the projectile.

A projectile is fired with initial speed v_0 m/s at an angle θ upward from the ground. Figure 4 shows that the initial velocity vector is

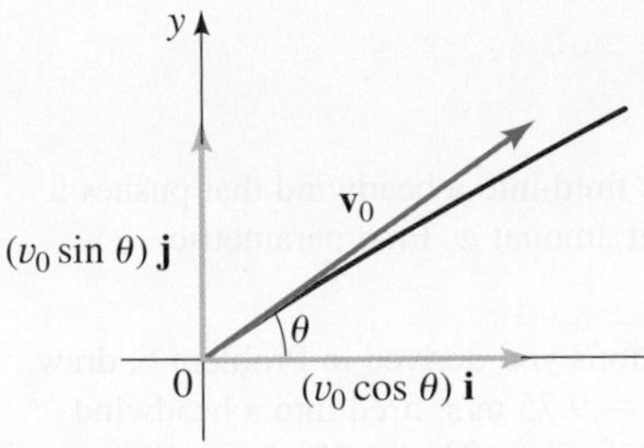

$$\mathbf{v}_0 = (v_0 \cos \theta)\mathbf{i} + (v_0 \sin \theta)\mathbf{j} \qquad \text{Initial velocity}$$

Figure 4 | Initial velocity vector

The horizontal component of the velocity vector does not change with time, but the vertical component decreases as gravity tugs downward on the projectile. Gravity reduces the upward velocity by g m/s every second. Thus, after t seconds the vertical component is reduced by gt m/s. So the velocity $\mathbf{v}_t$ at any time t is given by

$$\mathbf{v}_t = (v_0 \cos \theta)\mathbf{i} + (v_0 \sin \theta - gt)\mathbf{j} \qquad \text{Velocity at time } t$$

For the cannonball in the example, the velocity at time t is

$$\begin{aligned}\mathbf{v}_t &= (150 \cos 30°)\mathbf{i} + (150 \sin 30° - 9.8t)\mathbf{j} \\ &\approx 129.9\mathbf{i} + (75 - 9.8t)\mathbf{j}\end{aligned}$$

Figure 5 shows a graph of the path of the projectile and velocity vectors $\mathbf{v}_t$ (and their component vectors) at four points on the path.

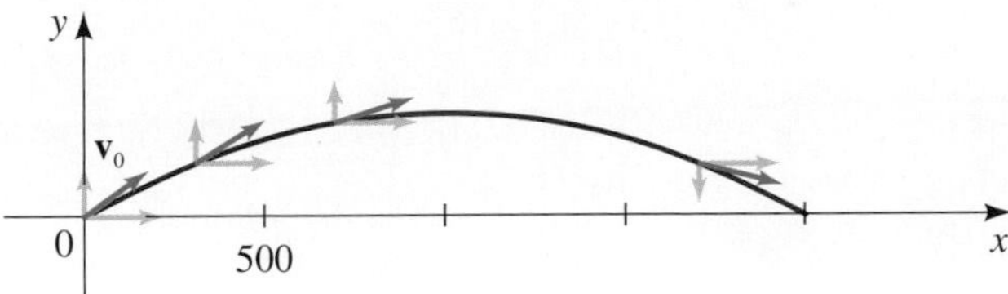

Figure 5 | Path and velocity of a projectile

Problems

1. **Trajectories Are Parabolas** From the graphs in Figure 3 the paths of projectiles appear to be parabolas that open downward. Eliminate the parameter t from the general parametric equations to verify that these are indeed parabolas.

2. **Path of a Baseball** Suppose a baseball is thrown at 10 m/s at a 60° angle to the horizontal from a height of 1 m above the ground.

   (a) Find parametric equations for the path of the baseball, and sketch its graph.

   (b) How far does the baseball travel horizontally, and when does it hit the ground?

3. **Path of a Rocket** Suppose that a rocket is fired at an angle of 5° from the vertical with an initial speed of 300 m/s.

   (a) Find the length of time the rocket is in the air.

   (b) Find the greatest height it reaches.

   (c) Find the horizontal distance it has traveled when it hits the ground.

   (d) Graph the rocket's path.

4. **Firing a Projectile** The initial speed of a projectile is 330 m/s.

   (a) At what angle should the projectile be fired so that it hits a target 10 km away? (You should find that there are two possible angles.) Graph the projectile paths for both angles.

   (b) For which angle is the target hit sooner?

5. Maximum Height Show that the maximum height reached by a projectile as a function of its initial speed v_0 and its firing angle θ is

$$y = \frac{v_0^2 \sin^2\theta}{2g}$$

6. Shooting into the Wind Suppose that a projectile is fired into a headwind that pushes it back so as to reduce its horizontal speed by a constant amount w. Find parametric equations for the path of the projectile.

7. Shooting into the Wind Using the parametric equations you derived in Problem 6, draw graphs of the path of a projectile with initial speed $v_0 = 9.75$ m/s, fired into a headwind of $w = 7.3$ m/s, for the angles $\theta = 5°, 15°, 30°, 40°, 45°, 55°, 60°$, and $75°$. Is it still true that the greatest range is attained when firing at 45°? Draw some more graphs for different angles, and use these graphs to estimate the optimal firing angle.

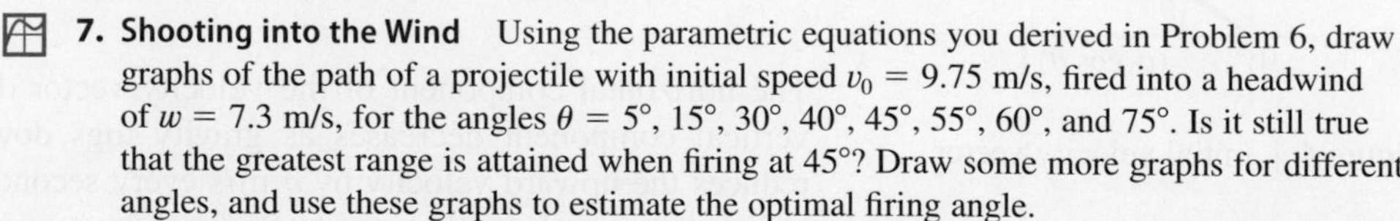

8. Path and Velocity of a Projectile A cannonball is fired from ground level with initial speed 200 m/s at an angle of elevation of 60°.

(a) Find parametric equations for the position of the cannonball t seconds after it is fired.

(b) Find the velocity vectors $\mathbf{v}_t$ of the cannonball at time t. What is the velocity at its highest point? What is the speed when it hits the ground?

(c) Draw a graph of the path of the cannonball and sketch velocity vectors on the graph at the points corresponding to $t = 6, 10, 18, 30$, as shown in Figure 4.

Lukas Davidziuk/Shutterstock.com

9 Systems of Equations and Inequalities

Throughout the preceding chapters we modeled real-world situations by equations. But many real-world situations involve too many variables to be modeled by a single equation. For example, weather depends on the relationships among many variables, including temperature, wind speed, air pressure, and humidity. So to model the weather (and forecast a snowstorm like the one pictured above), scientists use many equations, each having many variables. Such collections of equations, called systems of equations, *work together* to describe the weather. Systems of equations with hundreds of variables are used by airlines to establish consistent flight schedules and by telecommunications companies to find efficient routings for telephone calls. In this chapter we learn how to solve systems of equations that consist of several equations in several variables.

9.1 Systems of Linear Equations in Two Variables

■ Systems of Linear Equations and Their Solutions ■ Substitution Method ■ Elimination Method ■ Graphical Method ■ The Number of Solutions of a Linear System in Two Variables ■ Modeling with Linear Systems

■ Systems of Linear Equations and Their Solutions

A linear equation in two variables is an equation of the form

$$ax + by = c$$

The graph of a linear equation is a line (see Section 1.10).

A **system of equations** is a set of equations that involve the same variables. A **system of linear equations** is a system of equations in which each equation is linear. A **solution** of a system is an assignment of values for the variables that makes *each* equation in the system true. To **solve** a system means to find all solutions of the system.

Here is an example of a system of linear equations in two variables:

$$\begin{cases} 2x - y = 5 & \text{Equation 1} \\ x + 4y = 7 & \text{Equation 2} \end{cases}$$

We can check that $x = 3$ and $y = 1$ is a solution of this system.

Equation 1	**Equation 2**
$2x - y = 5$	$x + 4y = 7$
$2(3) - 1 = 5$ ✓	$3 + 4(1) = 7$ ✓

The solution can also be written as the ordered pair $(3, 1)$.

Note that the graphs of Equations 1 and 2 are lines (as shown in Figure 1). Since the solution $(3, 1)$ satisfies each equation, the point $(3, 1)$ lies on each line. So $(3, 1)$ is the point of intersection of the two lines.

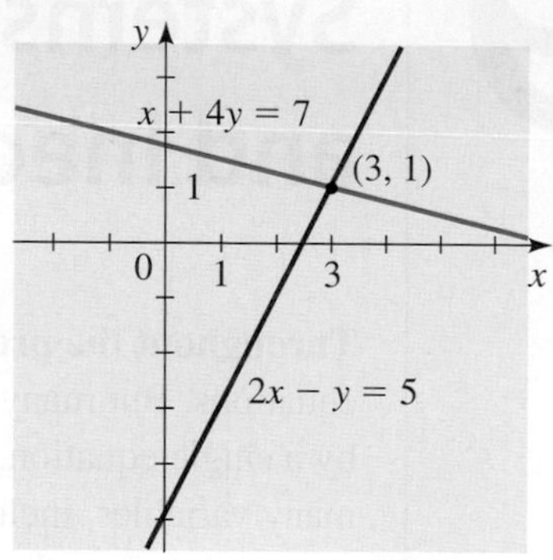

Figure 1

■ Substitution Method

To solve a system using the **substitution method**, we start with one equation in the system and solve for one variable in terms of the other variable.

Substitution Method

1. **Solve for One Variable.** Choose one equation, and solve for one variable in terms of the other variable.
2. **Substitute.** Substitute the expression you found in Step 1 into the other equation to get an equation in one variable, then solve for that variable.
3. **Back-Substitute.** Substitute the value you found in Step 2 back into the expression found in Step 1 to solve for the remaining variable.

Example 1 ■ Substitution Method

Find all solutions of the system.

$$\begin{cases} 2x + y = 1 & \text{Equation 1} \\ 3x + 4y = 14 & \text{Equation 2} \end{cases}$$

Solution **Solve for one variable.** We solve for y in the first equation.

$$y = 1 - 2x \qquad \text{Solve for } y \text{ in Equation 1}$$

Substitute. Now we substitute for y in the second equation and solve for x.

$$\begin{aligned} 3x + 4(1 - 2x) &= 14 && \text{Substitute } y = 1 - 2x \text{ into Equation 2} \\ 3x + 4 - 8x &= 14 && \text{Expand} \\ -5x + 4 &= 14 && \text{Simplify} \\ -5x &= 10 && \text{Subtract 4} \\ x &= -2 && \text{Solve for } x \end{aligned}$$

Back-substitute. Next we back-substitute $x = -2$ into the equation $y = 1 - 2x$.

$$y = 1 - 2(-2) = 5 \qquad \text{Back-substitute}$$

Thus $x = -2$ and $y = 5$, so the solution is the ordered pair $(-2, 5)$. Figure 2 shows that the graphs of the two equations intersect at the point $(-2, 5)$.

Check Your Answer

$x = -2, y = 5$:

$$\begin{cases} 2(-2) + 5 = 1 \\ 3(-2) + 4(5) = 14 \end{cases} \checkmark$$

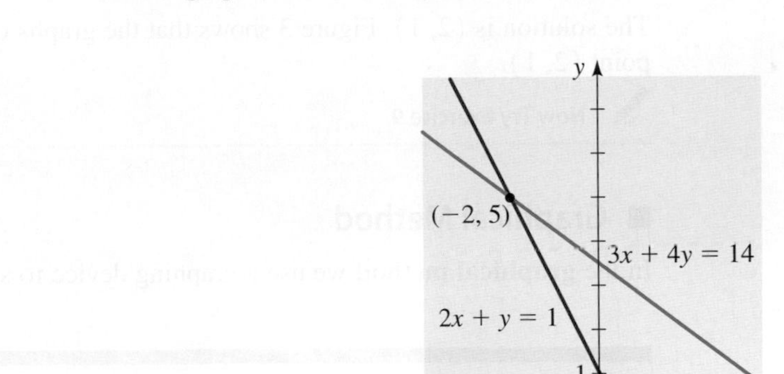

Figure 2

Now Try Exercise 5 ■

■ Elimination Method

To solve a system using the **elimination method**, we try to combine the equations using sums or differences so as to eliminate one of the variables.

Elimination Method

1. **Adjust the Coefficients.** Multiply one or more of the equations by appropriate numbers so that the coefficient of one variable in one equation is the negative of its coefficient in the other equation.
2. **Eliminate a Variable.** Add the two equations to eliminate one variable, then solve for the remaining variable.
3. **Back-Substitute.** Substitute the value that you found in Step 2 back into one of the original equations, and solve for the remaining variable.

Example 2 ■ Elimination Method

Find all solutions of the system.

$$\begin{cases} 3x + 5y = 11 & \text{Equation 1} \\ 2x - y = 3 & \text{Equation 2} \end{cases}$$

Solution **Adjust the coefficients.** We multiply the second equation by 5 to prepare for eliminating y from the equations. The second equation becomes $10x - 5y = 15$.

Eliminate a variable. We add the equations to eliminate y.

$$\begin{array}{rll} 3x + 5y &= 11 & \\ 10x - 5y &= 15 & 5 \times \text{Equation 2} \\ \hline 13x &= 26 & \text{Add} \\ x &= 2 & \text{Solve for } x \end{array}$$

Back-substitute. Now we back-substitute into the first equation and solve for y.

$$\begin{array}{rll} 3(2) + 5y &= 11 & \text{Back-substitute } x = 2 \\ 5y &= 5 & \text{Subtract 6} \\ y &= 1 & \text{Solve for } y \end{array}$$

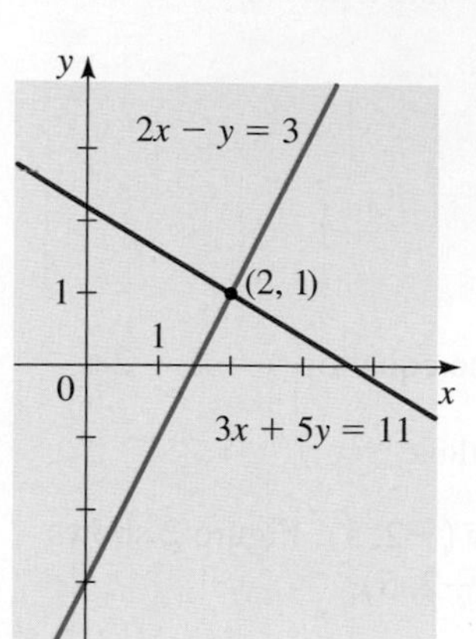

Figure 3

The solution is $(2, 1)$. Figure 3 shows that the graphs of the equations intersect at the point $(2, 1)$.

Now Try Exercise 9 ■

■ Graphical Method

In the **graphical method** we use a graphing device to solve the system of equations.

> **Graphical Method**
>
> 1. **Graph Each Equation.** Use a graphing device to graph the equations on the same screen. To graph the equations using a graphing calculator, you may first need to solve for y as a function of x.
> 2. **Find Intersection Points.** The solutions are the x- and y-coordinates of the point(s) of intersection.

Example 3 ■ Graphical Method

Find all solutions of the system.

$$\begin{cases} 1.35x - 2.13y = -2.36 \\ 2.16x + 0.32y = 1.06 \end{cases}$$

Solution **Graph each equation.** To graph, we solve for y in each equation.

$$\begin{cases} y = 0.63x + 1.11 \\ y = -6.75x + 3.31 \end{cases}$$

where we have rounded the coefficients to two decimals.

Find intersection point. Figure 4 shows that the two lines intersect. From the graph we see that the solution is approximately $(0.30, 1.30)$.

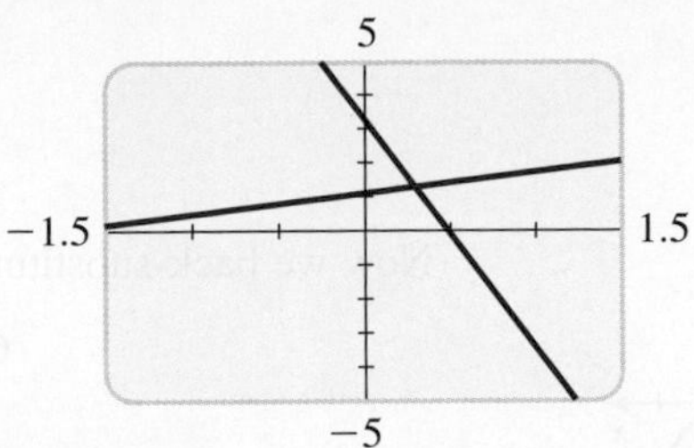

Figure 4

Now Try Exercises 13 and 51 ■

■ The Number of Solutions of a Linear System in Two Variables

The graph of a linear system in two variables is a pair of lines, so to solve the system graphically, we must find the intersection point(s) of the lines. Two lines may intersect in a single point, they may be parallel, or they may coincide, as shown in Figure 5. So there are three possible outcomes in solving such a system.

Number of Solutions of a Linear System in Two Variables

For a system of linear equations in two variables, exactly one of the following is true. (See Figure 5.)

1. The system has exactly one solution.
2. The system has no solution.
3. The system has infinitely many solutions.

A system that has no solution is said to be **inconsistent**. A system with infinitely many solutions is called **dependent**.

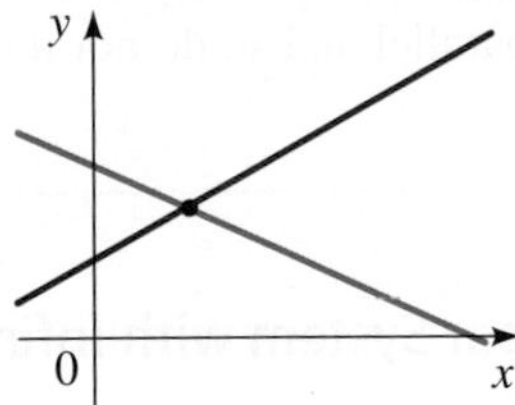

(a) Lines intersect at a single point. The system has one solution.

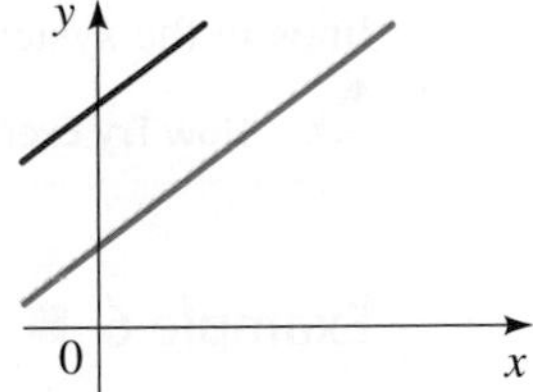

(b) Lines are parallel and do not intersect. The system has no solution.

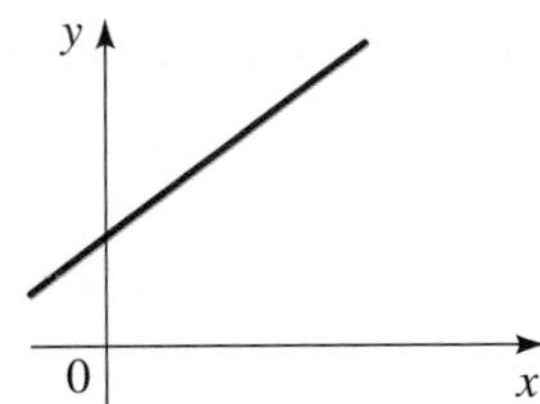

(c) Lines coincide—equations determine the same line. The system has infinitely many solutions.

Figure 5

Example 4 ■ A Linear System with One Solution

Solve the system and graph the lines.

$$\begin{cases} 3x - y = 0 & \text{Equation 1} \\ 5x + 2y = 22 & \text{Equation 2} \end{cases}$$

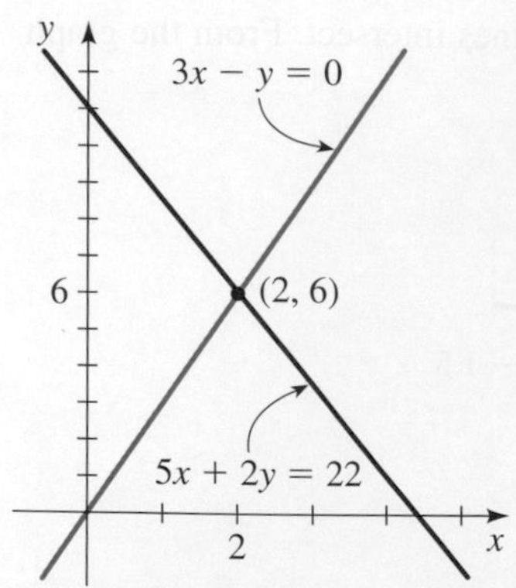

Figure 6 | The lines intersect

Solution We eliminate y from the equations and solve for x.

$$\begin{cases} 6x - 2y = 0 & 2 \times \text{Equation 1} \\ 5x + 2y = 22 & \end{cases}$$

$$11x = 22 \qquad \text{Add}$$

$$x = 2 \qquad \text{Solve for } x$$

Now we back-substitute into the first equation and solve for y:

$$6(2) - 2y = 0 \qquad \text{Back-substitute } x = 2$$

$$-2y = -12 \qquad \text{Subtract 12}$$

$$y = 6 \qquad \text{Solve for } y$$

The solution of the system is the ordered pair (2, 6), that is,

$$x = 2 \qquad y = 6$$

The graph in Figure 6 shows that the lines in the system intersect at the point (2, 6).

Check Your Answer

$x = 2, y = 6$:

$$\begin{cases} 3(2) - (6) = 0 \\ 5(2) + 2(6) = 22 \end{cases} \checkmark$$

Now Try Exercise 23 ■

Example 5 ■ A Linear System with No Solution

Solve the system.

$$\begin{cases} 8x - 2y = 5 & \text{Equation 1} \\ -12x + 3y = 7 & \text{Equation 2} \end{cases}$$

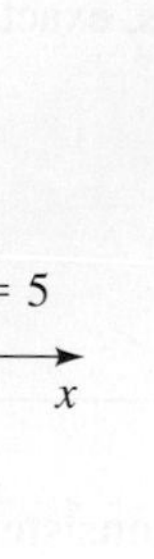

Figure 7 | The lines are parallel

Solution This time we try to find a suitable combination of the two equations to eliminate the variable y. Multiplying the first equation by 3 and the second equation by 2 gives

$$\begin{cases} 24x - 6y = 15 & 3 \times \text{Equation 1} \\ -24x + 6y = 14 & 2 \times \text{Equation 2} \end{cases}$$

$$0 = 29 \qquad \text{Add}$$

Adding the two equations eliminates *both x and y* in this case, and we end up with $0 = 29$, which is obviously false. No matter what values we assign to x and y, we cannot make this statement true, so the system has *no solution*. Figure 7 shows that the lines in the system are parallel and so do not intersect. The system is inconsistent.

Now Try Exercise 37 ■

Example 6 ■ A Linear System with Infinitely Many Solutions

Solve the system.

$$\begin{cases} 3x - 6y = 12 & \text{Equation 1} \\ 4x - 8y = 16 & \text{Equation 2} \end{cases}$$

Solution We multiply the first equation by 4 and the second equation by 3 to prepare for subtracting the equations to eliminate x. The new equations are

$$\begin{cases} 12x - 24y = 48 & 4 \times \text{Equation 1} \\ 12x - 24y = 48 & 3 \times \text{Equation 2} \end{cases}$$

We see that the two equations in the original system are simply different ways of expressing the equation of one single line. The coordinates of any point on this line

give a solution of the system. Writing the equation in slope-intercept form, we have $y = \frac{1}{2}x - 2$. So if we let t represent any real number, we can write the solution as

$$x = t$$
$$y = \tfrac{1}{2}t - 2$$

We can also write the solution in ordered-pair form as

$$\left(t, \tfrac{1}{2}t - 2\right)$$

where t is any real number. The system has infinitely many solutions (see Figure 8).

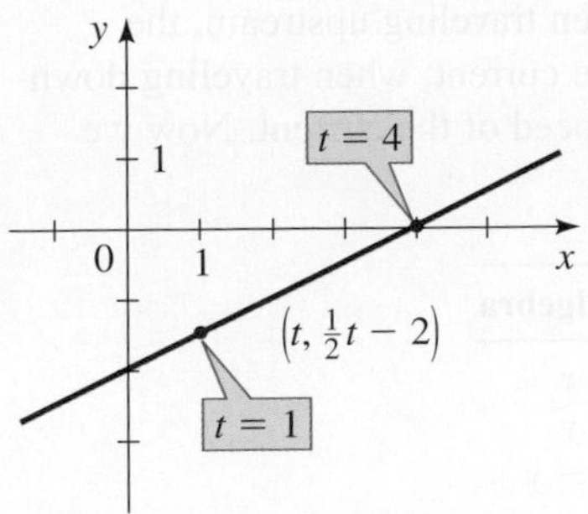

Figure 8 | The lines coincide

Now Try Exercise 39 ■

In Example 3, to get specific solutions we have to assign values to t. For instance, if $t = 1$, we get the solution $\left(1, -\frac{3}{2}\right)$. If $t = 4$, we get the solution $(4, 0)$. For every value of t we get a different solution. (See Figure 8.)

■ Modeling with Linear Systems

When we use equations to solve problems in the sciences or in other areas, we frequently obtain systems like the ones we've been considering. When modeling with systems of equations, we use the following guidelines, which are similar to those given in Section 1.7.

Guidelines for Modeling with Systems of Equations

1. **Identify the Variables.** Identify the quantities that the problem asks you to find. These are usually determined by a careful reading of the question posed at the end of the problem. Introduce notation for the variables (call them x and y, or some other letters).
2. **Express All Unknown Quantities in Terms of the Variables.** Read the problem again, and express all the quantities mentioned in the problem in terms of the variables you defined in Step 1.
3. **Set Up a System of Equations.** Find the crucial facts in the problem that give the relationships between the expressions you found in Step 2. Set up a system of equations (or a model) that expresses these relationships.
4. **Solve the System and Interpret the Results.** Solve the system you found in Step 3, check your solutions, and **state your answer** as a sentence that answers the question posed in the problem.

The next two examples illustrate how to model with systems of equations.

Example 7 ■ A Distance-Speed-Time Problem

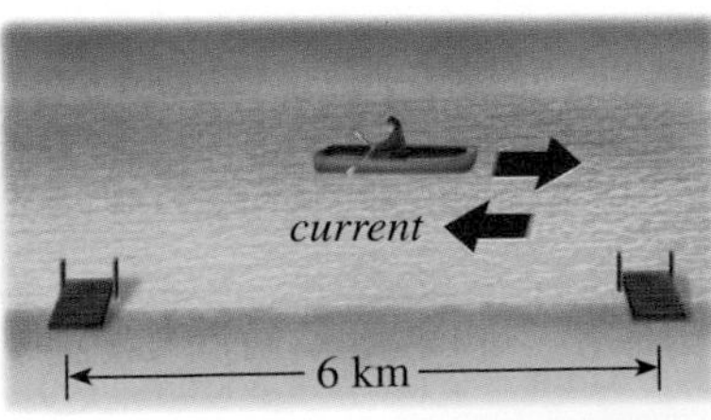

A boater rows a boat upstream from one point on a river to another point 6 km away in $1\frac{1}{2}$ hours. The return trip, traveling with the current, takes only 45 minutes. What is the boater's rowing speed relative to the water, and at what speed is the current flowing?

Solution **Identify the variables.** We are asked to find the rowing speed and the speed of the current, so we let

$$x = \text{rowing speed (km/h)}$$
$$y = \text{current speed (km/h)}$$

Mathematics in the Modern World

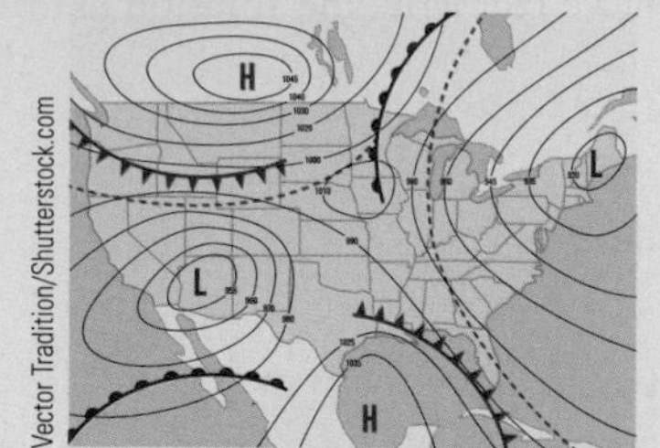

Vector Tradition/Shutterstock.com

Weather Prediction

Today's meteorologists do much more than predict tomorrow's weather. They research long-term weather patterns, depletion of the ozone layer, global warming, and other effects of human activity on the weather. But daily weather prediction is still a major part of meteorology; its value is measured by the innumerable human lives that are saved each year through accurate prediction of hurricanes, blizzards, and other catastrophic weather phenomena. Early in the 20th century, mathematicians proposed to model weather with equations that used the current values of hundreds of atmospheric variables. Although this model worked in principle, it was impossible to predict future weather patterns with it because of the difficulty of measuring all the variables accurately and solving all the equations. Today, new mathematical models combined with high-speed computer simulations and better data have vastly improved weather prediction. As a result, many human as well as economic disasters have been averted. Mathematicians at the National Oceanographic and Atmospheric Administration (NOAA) are continually researching better methods of weather prediction.

Express unknown quantities in terms of the variable. When traveling upstream, the boat's speed is the rowing speed *minus* the speed of the current; when traveling downstream, the boat's speed is the rowing speed *plus* the speed of the current. Now we translate this information into the language of algebra.

In Words	In Algebra
Rowing speed	x
Current speed	y
Speed upstream	$x - y$
Speed downstream	$x + y$

Set up a system of equations. The distance upstream and downstream is 6 km, so using the fact that speed × time = distance for both legs of the trip, we get

$$\boxed{\text{speed upstream}} \times \boxed{\text{time upstream}} = \boxed{\text{distance traveled}}$$

$$\boxed{\text{speed downstream}} \times \boxed{\text{time downstream}} = \boxed{\text{distance traveled}}$$

In algebraic notation this translates into the following equations.

$$(x - y)\tfrac{3}{2} = 6 \quad \text{Equation 1}$$

$$(x + y)\tfrac{3}{4} = 6 \quad \text{Equation 2}$$

(The times have been converted to hours, since we are expressing the speeds in kilometers per *hour*.)

Solve the system. We multiply the equations by 2 and 4, respectively, to clear the denominators.

$$\begin{cases} 3x - 3y = 12 & 2 \times \text{Equation 1} \\ 3x + 3y = 24 & 4 \times \text{Equation 2} \end{cases}$$

$$6x = 36 \quad \text{Add}$$

$$x = 6 \quad \text{Solve for } x$$

Back-substituting this value of x into the first equation (the second works just as well) and solving for y, we get

$$3(6) - 3y = 12 \quad \text{Back-substitute } x = 6$$

$$-3y = 12 - 18 \quad \text{Subtract 18}$$

$$y = 2 \quad \text{Solve for } y$$

The boater's rowing speed relative to the water is 6 km/h, and the current flows at 2 km/h.

Check Your Answer

Speed upstream is

$$\frac{\text{distance}}{\text{time}} = \frac{6\text{ km}}{1\frac{1}{2}\text{ h}} = 4\text{ km/h}$$

and this should equal

$$\text{rowing speed} - \text{current flow}$$
$$= 6\text{ km/h} - 2\text{ km/h} = 4\text{ km/h}$$

Speed downstream is

$$\frac{\text{distance}}{\text{time}} = \frac{6\text{ km}}{\frac{3}{4}\text{ h}} = 8\text{ km/h}$$

and this should equal

$$\text{rowing speed} + \text{current flow}$$
$$= 6\text{ km/h} + 2\text{ km/h} = 8\text{ km/h}$$ ✓

 **Now Try Exercise 65** ■

Example 8 ■ A Mixture Problem

A vintner fortifies wine that contains 10% alcohol by adding a 70% alcohol solution to it. The resulting mixture has an alcoholic strength of 16% and fills 1000 one-liter bottles. How many liters (L) of the wine and of the alcohol solution does the vintner use?

Solution **Identify the variables.** Since we are asked for the amounts of wine and alcohol, we let

$$x = \text{amount of wine used (L)}$$
$$y = \text{amount of alcohol solution used (L)}$$

Express all unknown quantities in terms of the variable. From the fact that the wine contains 10% alcohol and the solution contains 70% alcohol, we get the following.

In Words	In Algebra
Amount of wine used (L)	x
Amount of alcohol solution used (L)	y
Amount of alcohol in wine (L)	$0.10x$
Amount of alcohol in solution (L)	$0.70y$

Set up a system of equations. The volume of the mixture must be the total of the two volumes the vintner is adding together, so

$$x + y = 1000$$

Also, the amount of alcohol in the mixture must be the total of the alcohol contributed by the wine and by the alcohol solution, that is,

$$0.10x + 0.70y - (0.16)1000$$
$$0.10x + 0.70y = 160 \qquad \text{Simplify}$$
$$x + 7y = 1600 \qquad \text{Multiply by 10 to clear decimals}$$

Thus we get the system

$$\begin{cases} x + y = 1000 & \text{Equation 1} \\ x + 7y = 1600 & \text{Equation 2} \end{cases}$$

Solve the system. Subtracting the first equation from the second eliminates the variable x, and we get

$$6y = 600 \qquad \text{Subtract Equation 1 from Equation 2}$$
$$y = 100 \qquad \text{Solve for } y$$

We now back-substitute $y = 100$ into the first equation and solve for x.

$$x + 100 = 1000 \qquad \text{Back-substitute } y = 100$$
$$x = 900 \qquad \text{Solve for } x$$

The vintner uses 900 L of wine and 100 L of the alcohol solution.

Now Try Exercise 67 ■

9.1 Exercises

Concepts

1. The system of equations

$$\begin{cases} 2x + 3y = 7 \\ 5x - y = 9 \end{cases}$$

is a system of two equations in the two variables ________ and ________. To determine whether $(5, -1)$ is a solution of this system, we check whether $x = 5$ and $y = -1$ satisfy each ________ in the system. Which of the following are solutions of this system?

$$(5, -1), \quad (-1, 3), \quad (2, 1)$$

2. A system of equations in two variables can be solved by the ________ method, the ________ method, or the ________ method. Solve the system in Exercise 1 by each of these methods. The solution is (___, ___).

3. A system of two linear equations in two variables can have one solution, ________ solution, or ________ solutions.

4. The following is a system of two linear equations in two variables.

$$\begin{cases} x + y = 1 \\ 2x + 2y = 2 \end{cases}$$

The graph of the first equation is the same as the graph of the second equation, so the system has ________ ________ solutions. We express these solutions by writing

$$x = t \qquad y = \text{\_\_\_\_\_\_\_\_}$$

where t is any real number. Some of the solutions of this system are (1, ___), (−3, ___), and (5, ___).

Skills

5–8 ■ Substitution Method Use the substitution method to find all solutions of the system of equations.

5. $\begin{cases} x + y = 2 \\ 2x - 4y = 16 \end{cases}$ **6.** $\begin{cases} x - y = 8 \\ 5x - 4y = 35 \end{cases}$

7. $\begin{cases} x - 3y = 11 \\ 3x - 5y = 17 \end{cases}$ **8.** $\begin{cases} 2x + y = 7 \\ x + 2y = 2 \end{cases}$

9–12 ■ Elimination Method Use the elimination method to find all solutions of the system of equations.

9. $\begin{cases} 2x - 3y = 7 \\ x - 5y = 0 \end{cases}$ **10.** $\begin{cases} 4x + y = 5 \\ 5x + 2y = 4 \end{cases}$

11. $\begin{cases} 3x - 2y = -13 \\ -6x + 5y = 28 \end{cases}$ **12.** $\begin{cases} 2x - 5y = -18 \\ 3x + 4y = 19 \end{cases}$

13–14 ■ Graphical Method Two equations and their graphs are given. Estimate the intersection point from the graph and check that the point is a solution to the system.

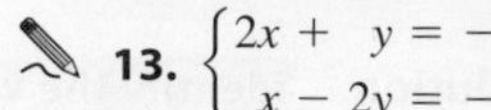

13. $\begin{cases} 2x + y = -1 \\ x - 2y = -8 \end{cases}$

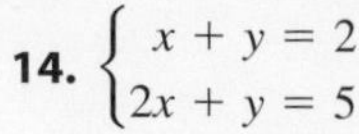

14. $\begin{cases} x + y = 2 \\ 2x + y = 5 \end{cases}$

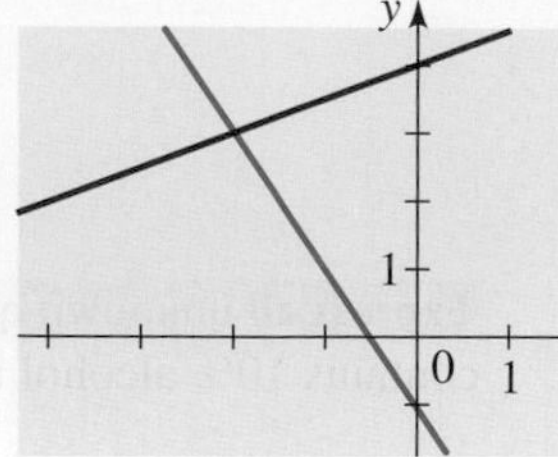

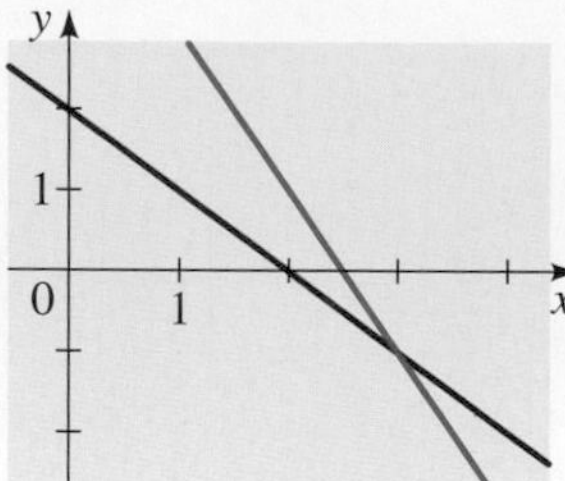

15–20 ■ Number of Solutions Determined Graphically Graph each linear system, either with or without a graphing device. Use the graph to determine whether the system has one solution, no solution, or infinitely many solutions. If there is exactly one solution, use the graph to find it.

15. $\begin{cases} x - y = 4 \\ 2x + y = 2 \end{cases}$ **16.** $\begin{cases} 2x - y = 4 \\ 3x + y = 6 \end{cases}$

17. $\begin{cases} 2x - 3y = 12 \\ -x + \frac{3}{2}y = 4 \end{cases}$ **18.** $\begin{cases} 2x + 6y = 0 \\ -3x - 9y = 18 \end{cases}$

19. $\begin{cases} -x + \frac{1}{2}y = -5 \\ 2x - y = 10 \end{cases}$ **20.** $\begin{cases} 12x + 15y = -18 \\ 2x + \frac{5}{2}y = -3 \end{cases}$

21–50 ■ Solving a System of Equations Solve the system, or show that it has no solution. If the system has infinitely many solutions, express them in the ordered-pair form given in Example 6.

21. $\begin{cases} 5x + 3y = 18 \\ 5x - 3y = 12 \end{cases}$ **22.** $\begin{cases} 2x + y = 10 \\ x - 3y = -16 \end{cases}$

23. $\begin{cases} 2x - 3y = 9 \\ 4x + 3y = 9 \end{cases}$ **24.** $\begin{cases} 3x + 2y = 0 \\ -x - 2y = 8 \end{cases}$

25. $\begin{cases} x + 3y = 5 \\ 2x - y = 3 \end{cases}$ **26.** $\begin{cases} x + y = 7 \\ 2x - 3y = -1 \end{cases}$

27. $\begin{cases} -x + y = 2 \\ 4x - 3y = -3 \end{cases}$ **28.** $\begin{cases} 4x - 3y = 28 \\ 9x - y = -6 \end{cases}$

29. $\begin{cases} x + 2y = 7 \\ 5x - y = 2 \end{cases}$ **30.** $\begin{cases} -4x + 12y = 0 \\ 12x + 4y = 160 \end{cases}$

31. $\begin{cases} -\frac{1}{3}x - \frac{1}{6}y = -1 \\ \frac{2}{3}x + \frac{1}{6}y = 3 \end{cases}$ **32.** $\begin{cases} \frac{3}{4}x + \frac{1}{2}y = 5 \\ -\frac{1}{4}x - \frac{3}{2}y = 1 \end{cases}$

33. $\begin{cases} \frac{1}{2}x + \frac{1}{3}y = 2 \\ \frac{1}{5}x - \frac{2}{3}y = 8 \end{cases}$ **34.** $\begin{cases} 0.2x - 0.2y = -1.8 \\ -0.3x + 0.5y = 3.3 \end{cases}$

35. $\begin{cases} 3x + 2y = 8 \\ x - 2y = 0 \end{cases}$ **36.** $\begin{cases} 4x + 2y = 16 \\ x - 5y = 70 \end{cases}$

37. $\begin{cases} x + 4y = 8 \\ 3x + 12y = 2 \end{cases}$

38. $\begin{cases} -3x + 5y = 2 \\ 9x - 15y = 6 \end{cases}$

39. $\begin{cases} 2x - 6y = 10 \\ -3x + 9y = -15 \end{cases}$

40. $\begin{cases} 2x - 3y = -8 \\ 14x - 21y = 3 \end{cases}$

41. $\begin{cases} 6x + 4y = 12 \\ 9x + 6y = 18 \end{cases}$

42. $\begin{cases} 25x - 75y = 100 \\ -10x + 30y = -40 \end{cases}$

43. $\begin{cases} 8s - 3t = -3 \\ 5s - 2t = -1 \end{cases}$

44. $\begin{cases} u - 30v = -5 \\ -3u + 80v = 5 \end{cases}$

45. $\begin{cases} \frac{1}{2}x + \frac{3}{5}y = 3 \\ \frac{5}{3}x + 2y = 10 \end{cases}$

46. $\begin{cases} \frac{3}{2}x - \frac{1}{3}y = \frac{1}{2} \\ 2x - \frac{1}{2}y = -\frac{1}{2} \end{cases}$

47. $\begin{cases} 0.4x + 1.2y = 14 \\ 12x - 5y = 10 \end{cases}$

48. $\begin{cases} 26x - 10y = -4 \\ -0.6x + 1.2y = 3 \end{cases}$

49. $\begin{cases} \frac{1}{3}x - \frac{1}{4}y = 2 \\ -8x + 6y = 10 \end{cases}$

50. $\begin{cases} -\frac{1}{10}x + \frac{1}{2}y = 4 \\ 2x - 10y = -80 \end{cases}$

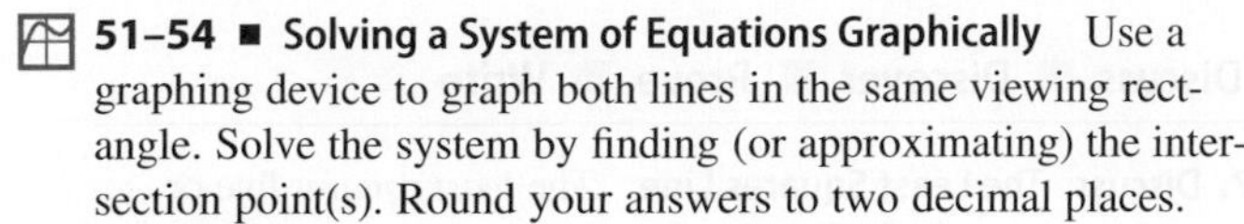

51–54 ■ Solving a System of Equations Graphically Use a graphing device to graph both lines in the same viewing rectangle. Solve the system by finding (or approximating) the intersection point(s). Round your answers to two decimal places.

51. $\begin{cases} 0.21x + 3.17y = 9.51 \\ 2.35x - 1.17y = 5.89 \end{cases}$

52. $\begin{cases} 18.72x - 14.91y = 12.33 \\ 6.21x - 12.92y = 17.82 \end{cases}$

53. $\begin{cases} 2371x - 6552y = 13{,}591 \\ 9815x + 992y = 618{,}555 \end{cases}$

54. $\begin{cases} -435x + 912y = 0 \\ 132x + 455y = 994 \end{cases}$

Skills Plus

55–58 ■ Solving a General System of Equations Find x and y in terms of a and b.

55. $\begin{cases} x + y = 0 \\ x + ay = 1 \end{cases} \quad (a \neq 1)$

56. $\begin{cases} ax + by = 0 \\ x + y = 1 \end{cases} \quad (a \neq b)$

57. $\begin{cases} ax + by = 1 \\ bx + ay = 1 \end{cases} \quad (a^2 - b^2 \neq 0)$

58. $\begin{cases} ax + by = 0 \\ a^2x + b^2y = 1 \end{cases} \quad (a \neq 0, b \neq 0, a \neq b)$

Applications

59. **Number Problem** Find two numbers whose sum is 34 and whose difference is 10.

60. **Number Problem** The sum of two numbers is twice their difference. The larger number is 6 more than twice the smaller. Find the numbers.

61. **Value of Coins** There are 14 coins, all of which are dimes and quarters. If the total value of the coins is \$2.75, how many dimes and how many quarters are there?

62. **Admission Fees** The admission fee at an amusement park is \$1.50 for children and \$4.00 for adults. On a certain day, 2200 people entered the park, and the admission fees that were collected totaled \$5050. How many children and how many adults were admitted?

63. **Gas Station** A gas station sells regular gas for \$2.20 per liter and premium gas for \$3.00 a liter. At the end of a business day 280 liters of gas had been sold, and receipts totaled \$680. How many liters of each type of gas had been sold?

64. **Fruit Stand** A fruit stand sells two varieties of strawberries: standard and deluxe. A box of standard strawberries sells for \$7, and a box of deluxe strawberries sells for \$10. In one day the stand sold 135 boxes of strawberries for a total of \$1110. How many boxes of each type were sold?

65. **Airplane Speed** A pilot flies a small airplane from Fargo to Bismarck, North Dakota—a distance of 288 km. Because the plane is flying into a headwind, the trip takes 2 hours. On the way back, the wind is still blowing at the same speed, so the return trip takes only 1 h 12 min. What is the speed of the airplane in still air, and how fast is the wind blowing?

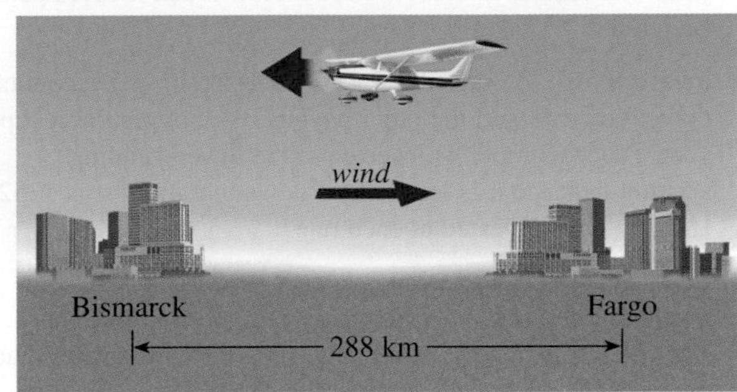

66. **Boat Speed** A boat on a river travels downstream between two points, 20 km apart, in 1 hour. The return trip against the current takes $2\frac{1}{2}$ hours. What is the boat's speed relative to the water, and how fast does the current in the river flow?

67. **Nutrition** A researcher performs an experiment to test a hypothesis that involves the nutrients niacin and retinol. In the experiment, one group of laboratory rats is fed a daily diet of precisely 32 units of niacin and 22,000 units

of retinol, using two types of commercial pellets. Food A contains 0.12 units of niacin and 100 units of retinol per gram. Food B contains 0.20 units of niacin and 50 units of retinol per gram. How many grams of each food is fed to this group of rats each day?

68. Coffee Blends A coffee shop sells two types of coffee: Kenyan, costing \$3.50 a kilogram, and Sri Lankan, costing \$5.60 a kilogram. If 3 kg of a blend of the two types of coffee costs \$11.55, how many kilograms of each kind went into the mixture?

69. Mixture Problem A chemist has two large containers of sulfuric acid solution, with different concentrations of acid in each container. Blending 300 mL of the first solution and 600 mL of the second gives a mixture that is 15% acid, whereas blending 100 mL of the first with 500 mL of the second gives a $12\frac{1}{2}\%$ acid mixture. What are the concentrations of sulfuric acid in the original containers?

70. Mixture Problem A biologist has two brine solutions, one containing 5% salt and another containing 20% salt. How many milliliters of each solution should be mixed together to obtain 1 L of a solution that contains 14% salt?

71. Investments A total of \$20,000 is invested in two accounts, one paying 5% and the other paying 8% simple interest per year. The annual interest is \$1180. How much is invested at each rate?

72. Investments A sum of money is invested in two accounts, one paying 6% and the other paying 10% simple interest per year. Twice as much is invested in the lower-yielding account because it is less risky. The annual interest is \$3520. How much is invested at each rate?

73. Distance, Speed, and Time A truck and an SUV leave a restaurant at the same time, going in opposite directions. The truck travels at 60 km/h and travels 35 km farther than the SUV, which travels at 40 km/h. The SUV's trip takes 15 minutes longer than the truck's trip. For what length of time does each vehicle travel?

74. How Much Gold in the Crown? Archimedes was able to determine the amount of gold in a crown by first finding the crown's volume. (See the vignette *Archimedes* in Section 10.1.) Suppose a crown made of gold and silver weighs 235 grams and has a volume of 14 cubic centimeters (cm^3). Find the weight of the gold and the weight of the silver in the crown. (Use the fact that the density of a substance is its weight divided by its volume: the density of gold is 19.3 g/cm^3 and the density of silver is 10.5 g/cm^3.)

75. Number Problem The sum of the digits of a two-digit number is 7. When the digits are reversed, the number is increased by 27. Find the number.

76. Area of a Triangle Find the area of the triangle that lies in the first quadrant (with its base on the x-axis) and that is bounded by the lines $y = 2x - 4$ and $y = -4x + 20$.

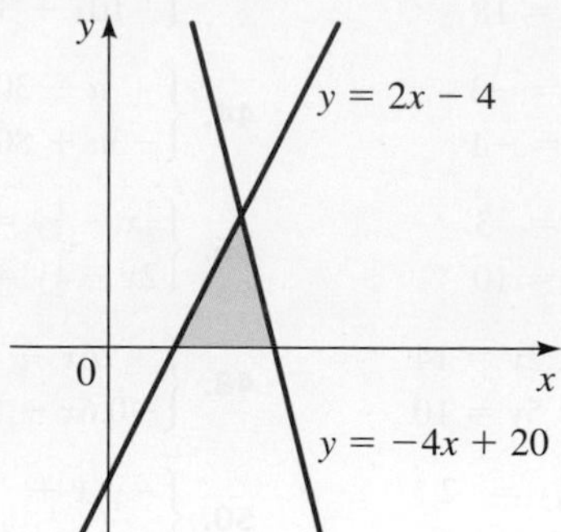

■ Discuss ■ Discover ■ Prove ■ Write

77. Discuss: The Least Squares Line The *least squares* line or *regression* line is the line that best fits a set of points in the plane. We studied this line in the *Focus on Modeling* that follows Chapter 1. By using calculus, it can be shown that the line that best fits the n data points

$$(x_1, y_1), (x_2, y_2), \ldots, (x_n, y_n)$$

is the line $y = ax + b$, where the coefficients a and b satisfy the following pair of linear equations. (The notation $\sum_{k=1}^{n} x_k$ stands for the sum of all the x's. See Section 11.1 for a complete description of sigma (Σ) notation.)

$$\left(\sum_{k=1}^{n} x_k\right) a + nb = \sum_{k=1}^{n} y_k$$

$$\left(\sum_{k=1}^{n} x_k^2\right) a + \left(\sum_{k=1}^{n} x_k\right) b = \sum_{k=1}^{n} x_k y_k$$

Use these equations to find the least squares line for the following data points.

$$(1, 3), \quad (2, 5), \quad (3, 6), \quad (5, 6), \quad (7, 9)$$

Sketch the points and your line to confirm that the line fits these points well. If your calculator computes regression lines, see whether it gives you the same line as the formulas.

9.2 Systems of Linear Equations in Several Variables

■ Solving a Linear System ■ The Number of Solutions of a Linear System ■ Modeling Using Linear Systems

A **linear equation in *n* variables** is an equation that can be put in the form

$$a_1x_1 + a_2x_2 + \cdots + a_nx_n = c$$

where $a_1, a_2, \ldots, a_n$ and c are real numbers, and $x_1, x_2, \ldots, x_n$ are the variables. If we have only three or four variables, we generally use x, y, z, and w instead of x_1, x_2, x_3, and x_4. Such equations are called *linear* because if we have just two variables, the equation is $a_1x + a_2y = c$, which is the equation of a line. Here are some examples of equations in three variables that illustrate the difference between linear and nonlinear equations.

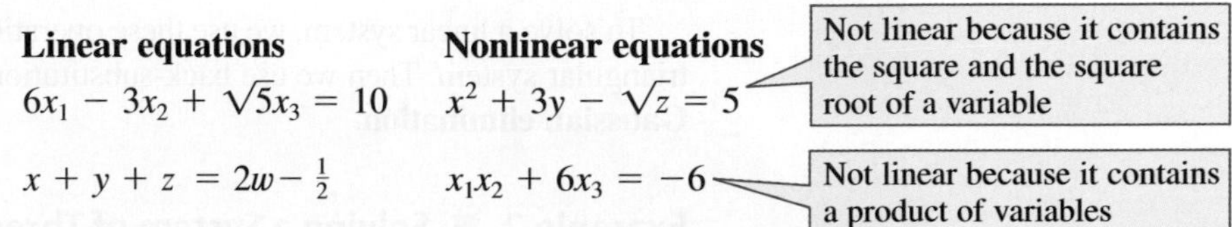

Linear equations
$6x_1 - 3x_2 + \sqrt{5}x_3 = 10$
$x + y + z = 2w - \frac{1}{2}$
Nonlinear equations
$x^2 + 3y - \sqrt{z} = 5$
$x_1x_2 + 6x_3 = -6$
Not linear because it contains the square and the square root of a variable
Not linear because it contains a product of variables

In this section we study systems of linear equations in three or more variables.

■ Solving a Linear System

The following are two examples of systems of linear equations in three variables. The second system is in **triangular form**; that is, the variable x doesn't appear in the second equation, and the variables x and y do not appear in the third equation.

A system of linear equations

$$\begin{cases} x - 2y - z = 1 \\ -x + 3y + 3z = 4 \\ 2x - 3y + z = 10 \end{cases}$$

A system in triangular form

$$\begin{cases} x - 2y - z = 1 \\ y + 2z = 5 \\ z = 3 \end{cases}$$

We can solve a system that is in triangular form by using back-substitution. So our goal in this section is to start with a system of linear equations and change it to a system in triangular form that has the same solutions as the original system. We begin by showing how to use back-substitution to solve a system that is already in triangular form.

Example 1 ■ Solving a Triangular System Using Back-Substitution

Solve the following system using back-substitution:

$$\begin{cases} x - 2y - z = 1 & \text{Equation 1} \\ y + 2z = 5 & \text{Equation 2} \\ z = 3 & \text{Equation 3} \end{cases}$$

Solution From the last equation we know that $z = 3$. We back-substitute this into the second equation and solve for y.

$$y + 2(3) = 5 \qquad \text{Back-substitute } z = 3 \text{ into Equation 2}$$

$$y = -1 \qquad \text{Solve for } y$$

Then we back-substitute $y = -1$ and $z = 3$ into the first equation and solve for x.

$$x - 2(-1) - (3) = 1 \qquad \text{Back-substitute } y = -1 \text{ and } z = 3 \text{ into Equation 1}$$

$$x = 2 \qquad \text{Solve for } x$$

The solution of the system is $x = 2$, $y = -1$, $z = 3$. We can also write the solution as the ordered triple $(2, -1, 3)$.

Now Try Exercise 7 ■

To change a system of linear equations to an **equivalent system** (that is, a system with the same solutions as the original system), we use the elimination method. This means that we can use the following operations.

Operations That Yield an Equivalent System

1. Add a nonzero multiple of one equation to another.
2. Multiply an equation by a nonzero constant.
3. Interchange the positions of two equations.

To solve a linear system, we use these operations to change the system to an equivalent triangular system. Then we use back-substitution as in Example 1. This process is called **Gaussian elimination**.

Example 2 ■ Solving a System of Three Equations in Three Variables

Solve the following system using Gaussian elimination:

$$\begin{cases} x - 2y + 3z = 1 & \text{Equation 1} \\ x + 2y - z = 13 & \text{Equation 2} \\ 3x + 2y - 5z = 3 & \text{Equation 3} \end{cases}$$

Solution We need to change this to a triangular system, so we begin by eliminating the x-term from the second equation.

$$\begin{array}{rl} x + 2y - z = 13 & \text{Equation 2} \\ \underline{x - 2y + 3z = 1} & \text{Equation 1} \\ 4y - 4z = 12 & \text{Equation 2} + (-1) \times \text{Equation 1} = \text{new Equation 2} \end{array}$$

This gives us a new, equivalent system that is one step closer to triangular form.

$$\begin{cases} x - 2y + 3z = 1 & \text{Equation 1} \\ 4y - 4z = 12 & \text{Equation 2} \\ 3x + 2y - 5z = 3 & \text{Equation 3} \end{cases}$$

Now we eliminate the x-term from the third equation.

$$\begin{array}{r} 3x + 2y - 5z = 3 \\ \underline{-3x + 6y - 9z = -3} \\ 8y - 14z = 0 \end{array}$$

$$\begin{cases} x - 2y + 3z = 1 \\ 4y - 4z = 12 \\ 8y - 14z = 0 & \text{Equation 3} + (-3) \times \text{Equation 1} = \text{new Equation 3} \end{cases}$$

Then we eliminate the y-term from the third equation.

$$\begin{array}{r} 8y - 14z = 0 \\ \underline{-8y + 8z = -24} \\ -6z = -24 \end{array}$$

$$\begin{cases} x - 2y + 3z = 1 \\ 4y - 4z = 12 \\ -6z = -24 & \text{Equation 3} + (-2) \times \text{Equation 2} = \text{new Equation 3} \end{cases}$$

The system is now in triangular form, but it will be easier to work with if we divide the second and third equations by the common factors of each term.

$$\begin{cases} x - 2y + 3z = 1 \\ y - z = 3 & \frac{1}{4} \times \text{Equation 2} = \text{new Equation 2} \\ z = 4 & -\frac{1}{6} \times \text{Equation 3} = \text{new Equation 3} \end{cases}$$

Now we use back-substitution to solve the system. From the third equation we get $z = 4$. We back-substitute this into the second equation and solve for y.

$$y - (4) = 3 \quad \text{Back-substitute } z = 4 \text{ into Equation 2}$$

$$y = 7 \quad \text{Solve for } y$$

Now we back-substitute $y = 7$ and $z = 4$ into the first equation and solve for x.

$$x - 2(7) + 3(4) = 1 \quad \text{Back-substitute } y = 7 \text{ and } z = 4 \text{ into Equation 1}$$

$$x = 3 \quad \text{Solve for } x$$

The solution of the system is $x = 3$, $y = 7$, $z = 4$, which we can write as the ordered triple $(3, 7, 4)$.

Check Your Answer

$x = 3, y = 7, z = 4$:

$$(3) - 2(7) + 3(4) = 1$$
$$(3) + 2(7) - (4) = 13$$
$$3(3) + 2(7) - 5(4) = 3$$

Now Try Exercise 17 ■

■ The Number of Solutions of a Linear System

To learn more about three-dimensional space see the additional topic *Three-Dimensional Coordinate Geometry* at the book companion website **www.stewartmath.com.**

The graph of a linear equation in three variables is a plane in three-dimensional space. A system of three equations in three variables represents three planes in space. The solutions of the system are the points where all three planes intersect. Three planes may intersect in a point, in a line, or not at all, or all three planes may coincide. Figure 1 illustrates some of these possibilities. Checking these possibilities we see that there are three possible outcomes when solving such a system.

Number of Solutions of a Linear System

For a system of linear equations, exactly one of the following is true.

1. The system has exactly one solution.
2. The system has no solution.
3. The system has infinitely many solutions.

A system with no solution is said to be **inconsistent**, and a system with infinitely many solutions is said to be **dependent**. As we see in the next example, a linear system has no solution if we end up with a *false equation* after applying Gaussian elimination to the system.

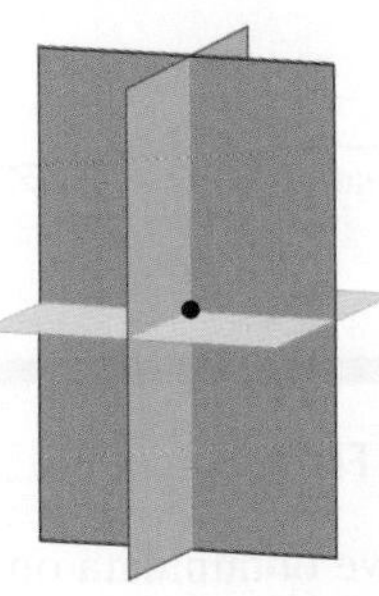

(a) The three planes intersect at a single point. The system has one solution.

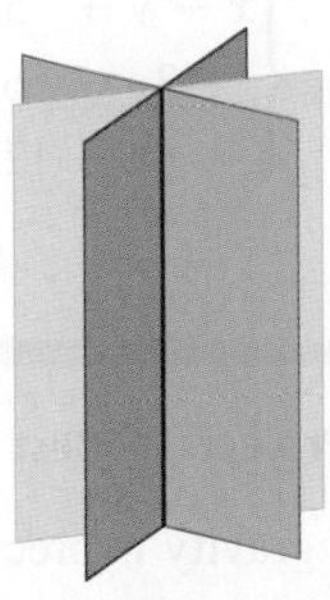

(b) The three planes intersect along a line. The system has infinitely many solutions.

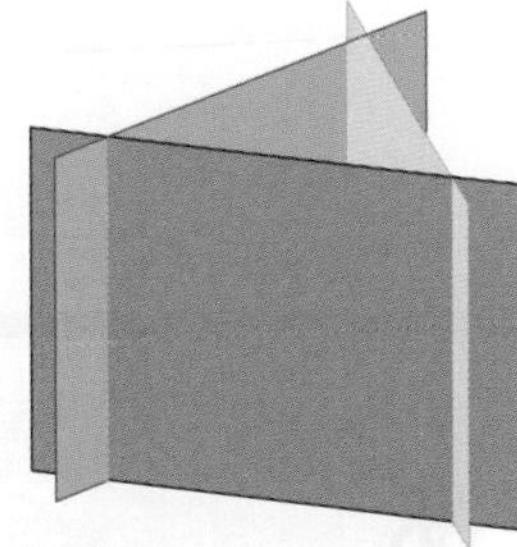

(c) The three planes have no point in common. The system has no solution.

Figure 1

Example 3 ■ A System with No Solution

Solve the following system:

$$\begin{cases} x + 2y - 2z = 1 & \text{Equation 1} \\ 2x + 2y - z = 6 & \text{Equation 2} \\ 3x + 4y - 3z = 5 & \text{Equation 3} \end{cases}$$

Solution To put this in triangular form, we begin by eliminating the x-terms from the second equation and the third equation.

$$\begin{cases} x + 2y - 2z = 1 \\ -2y + 3z = 4 \\ 3x + 4y + 3z = 5 \end{cases} \qquad \text{Equation 2} + (-2) \times \text{Equation 1} = \text{new Equation 2}$$

$$\begin{cases} x + 2y - 2z = 1 \\ -2y + 3z = 4 \\ -2y + 3z = 2 \end{cases} \qquad \text{Equation 3} + (-3) \times \text{Equation 1} = \text{new Equation 3}$$

Now we eliminate the y-term from the third equation.

$$\begin{cases} x + 2y - 2z = 1 \\ -2y + 3z = 4 \\ 0 = -2 \end{cases} \qquad \text{Equation 3} + (-1) \times \text{Equation 2} = \text{new Equation 3}$$

The system is now in triangular form, but the third equation says $0 = -2$, which is false. No matter what values we assign to x, y, and z, the third equation will never be true. This means that the system has *no solution*.

 **Now Try Exercise 29** ■

Example 4 ■ A System with Infinitely Many Solutions

Solve the following system:

$$\begin{cases} x - y + 5z = -2 & \text{Equation 1} \\ 2x + y + 4z = 2 & \text{Equation 2} \\ 2x + 4y - 2z = 8 & \text{Equation 3} \end{cases}$$

Solution To put this in triangular form, we begin by eliminating the x-terms from the second equation and the third equation.

$$\begin{cases} x - y + 5z = -2 \\ 3y - 6z = 6 \\ 2x + 4y - 2z = 8 \end{cases} \qquad \text{Equation 2} + (-2) \times \text{Equation 1} = \text{new Equation 2}$$

$$\begin{cases} x - y + 5z = -2 \\ 3y - 6z = 6 \\ 6y - 12z = 12 \end{cases} \qquad \text{Equation 3} + (-2) \times \text{Equation 1} = \text{new Equation 3}$$

Discovery Project ■ Best Fit Versus Exact Fit

The law of gravity is precise. But when we obtain data on the distance an object falls in a given time, our measurements are not exact. We can, however, find the line (or parabola) that *best* fits our data. Not all of the data points will lie on the line (or parabola). But if we are given just two points, we can find a line of *exact* fit—that is, a line that passes through the two points. Similarly, we can find a parabola through three points. In this project we compare exact data with models of real-world data. You can find the project at the book companion website: **www.stewartmath.com**.

Now we eliminate the y-term from the third equation.

$$\begin{cases} x - y + 5z = -2 \\ 3y - 6z = 6 \\ 0 = 0 \end{cases} \quad \text{Equation 3} + (-2) \times \text{Equation 2} = \text{new Equation 3}$$

The new third equation is true, but it gives us no new information, so we can drop it from the system. Only two equations are left. We can use them to solve for x and y in terms of z, but z can take on any value, so there are infinitely many solutions.

To find the complete solution of the system, we begin by solving for y in terms of z, using the new second equation.

$$3y - 6z = 6 \quad \text{Equation 2}$$

$$y - 2z = 2 \quad \text{Multiply by } \tfrac{1}{3}$$

$$y = 2z + 2 \quad \text{Solve for } y$$

Then we solve for x in terms of z, using the first equation.

$$x - (2z + 2) + 5z = -2 \quad \text{Substitute } y = 2z + 2 \text{ into Equation 1}$$

$$x + 3z - 2 = -2 \quad \text{Simplify}$$

$$x = -3z \quad \text{Solve for } x$$

To describe the complete solution, we let z be any real number t. The solution is

$$x = -3t$$

$$y = 2t + 2$$

$$z = t$$

We can also write this as the ordered triple $(-3t, 2t + 2, t)$.

 **Now Try Exercise 33** ■

In the solution of Example 4 the variable t is called a **parameter**. To get a specific solution, we give a specific value to the parameter t. For instance, if we set $t = 2$, we get

$$x = -3(2) = -6$$

$$y = 2(2) + 2 = 6$$

$$z = 2$$

Thus $(-6, 6, 2)$ is a solution of the system. Here are some other solutions of the system obtained by substituting other values for the parameter t.

Parameter t	Solution $(-3t, 2t + 2, t)$
-1	$(3, 0, -1)$
0	$(0, 2, 0)$
3	$(-9, 8, 3)$
10	$(-30, 22, 10)$

You should check that these points satisfy the original equations. There are infinitely many choices for the parameter t, so the system has infinitely many solutions.

■ Modeling Using Linear Systems

Linear systems are used to model situations that involve several varying quantities. In the next example we consider an application of linear systems to finance.

Example 5 ■ Modeling a Financial Problem Using a Linear System

A student receives an inheritance of \$50,000. A financial advisor suggests that the money be invested in three mutual funds: a money-market fund, a blue-chip stock fund, and a high-tech stock fund. The advisor estimates that the money-market fund will return 5% over the next year, the blue-chip fund 9%, and the high-tech fund 16%. The student wants a total first-year return of \$4000. To avoid excessive risk, three times as much is invested in the money-market fund as in the high-tech stock fund. How much should be invested in each fund?

Solution

Let $x =$ amount invested in the money-market fund

$y =$ amount invested in the blue-chip stock fund

$z =$ amount invested in the high-tech stock fund

We convert each fact given in the problem into an equation.

$$x + y + z = 50{,}000 \qquad \text{Total amount invested is \$50,000}$$

$$0.05x + 0.09y + 0.16z = 4000 \qquad \text{Total investment return is \$4000}$$

$$x = 3z \qquad \text{Money-market amount is } 3 \times \text{high-tech amount}$$

Multiplying the second equation by 100 and rewriting the third, we get the following system, which we solve using Gaussian elimination.

$$\begin{cases} x + y + z = 50{,}000 \\ 5x + 9y + 16z = 400{,}000 & 100 \times \text{Equation 2} \\ x - 3z = 0 & \text{Subtract } 3z \end{cases}$$

$$\begin{cases} x + y + z = 50{,}000 \\ 4y + 11z = 150{,}000 & \text{Equation 2} + (-5) \times \text{Equation 1} = \text{new Equation 2} \\ -y - 4z = -50{,}000 & \text{Equation 3} + (-1) \times \text{Equation 1} = \text{new Equation 3} \end{cases}$$

$$\begin{cases} x + y + z = 50{,}000 \\ -5z = -50{,}000 & \text{Equation 2} + 4 \times \text{Equation 3} = \text{new Equation 2} \\ -y - 4z = -50{,}000 \end{cases}$$

$$\begin{cases} x + y + z = 50{,}000 \\ z = 10{,}000 & (-\tfrac{1}{5}) \times \text{Equation 2} \\ y + 4z = 50{,}000 & (-1) \times \text{Equation 3} \end{cases}$$

$$\begin{cases} x + y + z = 50{,}000 \\ y + 4z = 50{,}000 & \text{Interchange Equations 2 and 3} \\ z = 10{,}000 \end{cases}$$

Now that the system is in triangular form, we use back-substitution to find that $x = 30{,}000$, $y = 10{,}000$, and $z = 10{,}000$. This means that the student should invest

\$30,000 in the money-market fund

\$10,000 in the blue-chip stock fund

\$10,000 in the high-tech stock fund

Now Try Exercise 39 ■

9.2 Exercises

Concepts

1–2 ■ These exercises refer to the following system:

$$\left\{\begin{aligned} x - y + z &= 2 \\ -x + 2y + z &= -3 \\ 3x + y - 2z &= 2 \end{aligned}\right.$$

1. If we add 2 times the first equation to the second equation, the second equation becomes ______ = ____.

2. To eliminate x from the third equation, we add ______ times the first equation to the third equation. The third equation becomes ______ = ____.

Skills

3–6 ■ Is the System of Equations Linear? State whether the equation or system of equations is linear.

3. $6x - \sqrt{3}y + \frac{1}{2}z = 0$

4. $x^2 + y^2 + z^2 = 4$

5. $\left\{\begin{aligned} xy - 3y + z &= 5 \\ x - y^2 + 5z &= 0 \\ 2x + yz &= 3 \end{aligned}\right.$

6. $\left\{\begin{aligned} x - 2y + 3z &= 10 \\ 2x + 5y &= 2 \\ y + 2z &= 4 \end{aligned}\right.$

7–12 ■ Triangular Systems Use back-substitution to solve the triangular system.

7. $\left\{\begin{aligned} x + 2y - z &= -5 \\ y + z &= 2 \\ z &= 4 \end{aligned}\right.$

8. $\left\{\begin{aligned} 3x - 3y + z &= 0 \\ y + 4z &= 10 \\ z &= 3 \end{aligned}\right.$

9. $\left\{\begin{aligned} x + 2y + z &= 7 \\ -y + 3z &= 9 \\ 2z &= 6 \end{aligned}\right.$

10. $\left\{\begin{aligned} x - 2y + 3z &= 10 \\ 2y - z &= 2 \\ 3z &= 12 \end{aligned}\right.$

11. $\left\{\begin{aligned} 2x - y + 6z &= 5 \\ y + 4z &= 0 \\ -2z &= 1 \end{aligned}\right.$

12. $\left\{\begin{aligned} 4x + 3z &= 10 \\ 2y - z &= -6 \\ \tfrac{1}{2}z &= 4 \end{aligned}\right.$

13–16 ■ Eliminating a Variable Perform an operation on the given system that eliminates the indicated variable. Write the new equivalent system.

13. $\left\{\begin{aligned} 3x + y + z &= 4 \\ -x + y + 2z &= 0 \\ x - 2y - z &= -1 \end{aligned}\right.$

Eliminate the x-term from the second equation.

14. $\left\{\begin{aligned} -5x + 2y - 3z &= 3 \\ 10x - 3y + z &= -20 \\ -x + 3y + z &= 8 \end{aligned}\right.$

Eliminate the x-term from the second equation.

15. $\left\{\begin{aligned} 2x + y - 3z &= 5 \\ 2x + 3y + z &= 13 \\ 6x - 5y - z &= 7 \end{aligned}\right.$

Eliminate the x-term from the third equation.

16. $\left\{\begin{aligned} x - 3y + 2z &= -1 \\ y + z &= -1 \\ 2y - z &= 1 \end{aligned}\right.$

Eliminate the y-term from the third equation.

17–38 ■ Solving a System of Equations in Three Variables Find the complete solution of the linear system, or show that the system is inconsistent.

17. $\left\{\begin{aligned} x + 2y + z &= 3 \\ y - z &= -4 \\ -x - 2y + 3z &= 9 \end{aligned}\right.$

18. $\left\{\begin{aligned} x - 5y - 3z &= 4 \\ 3y + 5z &= 11 \\ x - 2y + z &= 11 \end{aligned}\right.$

19. $\left\{\begin{aligned} x - 3y - 2z &= 5 \\ 3x - 2y + z &= 8 \\ y - 3z &= 1 \end{aligned}\right.$

20. $\left\{\begin{aligned} x - 2y + 3z &= -10 \\ 3y + z &= 7 \\ x + y - z &= 7 \end{aligned}\right.$

21. $\left\{\begin{aligned} x + y + z &= 4 \\ x + 3y + 3z &= 10 \\ 2x + y - z &= 3 \end{aligned}\right.$

22. $\left\{\begin{aligned} x + y + z &= 0 \\ -x + 2y + 5z &= 3 \\ 3x - y &= 6 \end{aligned}\right.$

23. $\left\{\begin{aligned} x - 4z &= 1 \\ 2x - y - 6z &= 4 \\ 2x + 3y - 2z &= 8 \end{aligned}\right.$

24. $\left\{\begin{aligned} x - y + 2z &= 2 \\ 3x + y + 5z &= 8 \\ 2x - y - 2z &= -7 \end{aligned}\right.$

25. $\left\{\begin{aligned} 2x + 4y - z &= 2 \\ x + 2y - 3z &= -4 \\ 3x - y + z &= 1 \end{aligned}\right.$

26. $\left\{\begin{aligned} 2x + y - z &= -8 \\ -x + y + z &= 3 \\ -2x + 4z &= 18 \end{aligned}\right.$

27. $\left\{\begin{aligned} 2y + 4z &= -1 \\ -2x + y + 2z &= -1 \\ 4x - 2y &= 0 \end{aligned}\right.$

28. $\left\{\begin{aligned} y - z &= -1 \\ 6x + 2y + z &= 2 \\ -x - y - 3z &= -2 \end{aligned}\right.$

29. $\left\{\begin{aligned} x - 3y + z &= 2 \\ 3x + 4y - 2z &= 1 \\ -2x + 6y - 2z &= 3 \end{aligned}\right.$

30. $\left\{\begin{aligned} -x + 2y + 5z &= 4 \\ x - 2z &= 0 \\ 4x - 2y - 11z &= 2 \end{aligned}\right.$

31. $\left\{\begin{aligned} 2x + 3y - z &= 1 \\ x + 2y &= 3 \\ x + 3y + z &= 4 \end{aligned}\right.$

32. $\left\{\begin{aligned} x - 2y - 3z &= 5 \\ 2x + y - z &= 5 \\ 4x - 3y - 7z &= 5 \end{aligned}\right.$

33. $\left\{\begin{aligned} x + y - z &= 0 \\ x + 2y - 3z &= -3 \\ 2x + 3y - 4z &= -3 \end{aligned}\right.$

34. $\left\{\begin{aligned} x - 2y + z &= 3 \\ 2x - 5y + 6z &= 7 \\ 2x - 3y - 2z &= 5 \end{aligned}\right.$

35. $\left\{\begin{aligned} x + 3y - 2z &= 0 \\ 2x + 4z &= 4 \\ 4x + 6y &= 4 \end{aligned}\right.$

36. $\left\{\begin{aligned} 2x + 4y - z &= 3 \\ x + 2y + 4z &= 6 \\ x + 2y - 2z &= 0 \end{aligned}\right.$

37. $\left\{\begin{aligned} x + z + 2w &= 6 \\ y - 2z &= -3 \\ x + 2y - z &= -2 \\ 2x + y + 3z - 2w &= 0 \end{aligned}\right.$

38. $\left\{\begin{aligned} x + y + z + w &= 0 \\ x + y + 2z + 2w &= 0 \\ 2x + 2y + 3z + 4w &= 1 \\ 2x + 3y + 4z + 5w &= 2 \end{aligned}\right.$

Applications

39. Financial Planning A financial planner invests \$100,000 in three types of bonds: short-term, intermediate-term, and long-term. The short-term bonds pay 4%, the intermediate-term bonds pay 5%, and the long-term bonds pay 6% simple interest per year. The planner wishes to realize a total annual income of 5.1%, with equal amounts invested in short- and intermediate-term bonds. How much should be invested in each type of bond?

40. Financial Planning An amount of \$50,000 is invested in three types of accounts: one paying 3%, one paying $5\frac{1}{2}$%, and one paying 9% simple interest per year. Twice as much is invested in the lowest-yielding, least-risky account as in the highest-yielding account. How much should be invested in each account to achieve a total annual return of \$2540?

41. Agriculture A farmer has 1200 acres of land on which to grow corn, wheat, and soybeans. It costs \$45 per acre to grow corn, \$60 to grow wheat, and \$50 to grow soybeans. Because of market demand, the farmer will grow twice as many acres of wheat as of corn. An amount of \$63,750 is allocated for the cost of growing the three crops. How many acres of each crop should be planted?

42. Gas Station A gas station sells three types of gas: Regular for \$3.00 a liter, Performance Plus for \$3.20 a liter, and Premium for \$3.30 a liter. On a particular day 24,700 liters of gas were sold for a total of \$20,050. Three times as many liters of Regular as Premium gas were sold. How many liters of each type of gas were sold that day?

43. Nutrition A biologist is performing an experiment on the effects of various combinations of vitamins. In the experiment, one group of laboratory rabbits is fed a daily diet of precisely 9 mg of niacin, 14 mg of thiamin, and 32 mg of riboflavin, using three types of commercial rabbit pellets; their vitamin content (per portion) is given in the table. How many portions of each type of food should each rabbit be given daily to satisfy the experiment requirements?

	Type A	Type B	Type C
Niacin (mg/portion)	2	3	1
Thiamin (mg/portion)	3	1	3
Riboflavin (mg/portion)	8	5	7

44. Diet Program A patient started a new diet that requires each meal to have 460 calories, 6 g of fiber, and 11 g of fat. The table shows the fiber, fat, and calorie content of one serving of each of three breakfast foods. How many servings of each food should the patient eat to follow this diet?

Food	Fiber (g)	Fat (g)	Calories
Toast	2	1	100
Cottage cheese	0	5	120
Fruit	2	0	60

45. Juice Blends The Juice Company offers three kinds of smoothies: Midnight Mango, Tropical Torrent, and Pineapple Power. Each smoothie contains the amounts of juices shown in the table.

On a particular day the Juice Company used 24.6 L of mango juice, 20.7 L of pineapple juice, and 13.5 L of orange juice. How many smoothies of each kind were sold that day?

Smoothie	Mango Juice (mL)	Pineapple Juice (mL)	Orange Juice (mL)
Midnight Mango	240	90	90
Tropical Torrent	180	150	90
Pineapple Power	60	240	120

46. Appliance Manufacturing Kitchen Korner produces refrigerators, dishwashers, and stoves at three different factories. The table gives the number of each type of product produced at each factory per day. Kitchen Korner receives an order for 110 refrigerators, 150 dishwashers, and 114 ovens. How many days should each plant be scheduled to fill this order?

Appliance	Factory A	Factory B	Factory C
Refrigerator	8	10	14
Dishwasher	16	12	10
Stove	10	18	6

47. Stock Portfolio An investor owns three stocks: A, B, and C. The closing prices of the stocks on three successive trading days are given in the table.

Despite the volatility in the stock prices, the total value of the investor's stocks remained unchanged at \$74,000 at the end of each of these three days. How many shares of each stock does the investor own?

	Stock A	Stock B	Stock C
Monday	\$10	\$25	\$29
Tuesday	\$12	\$20	\$32
Wednesday	\$16	\$15	\$32

48. Electricity By using Kirchhoff's Laws, it can be shown that the currents I_1, I_2, and I_3 that pass through the three branches of the circuit in the figure satisfy the given linear system. Solve the system to find I_1, I_2, and I_3.

$$\begin{cases} I_1 + I_2 - I_3 = 0 \\ 16I_1 - 8I_2 = 4 \\ 8I_2 + 4I_3 = 5 \end{cases}$$

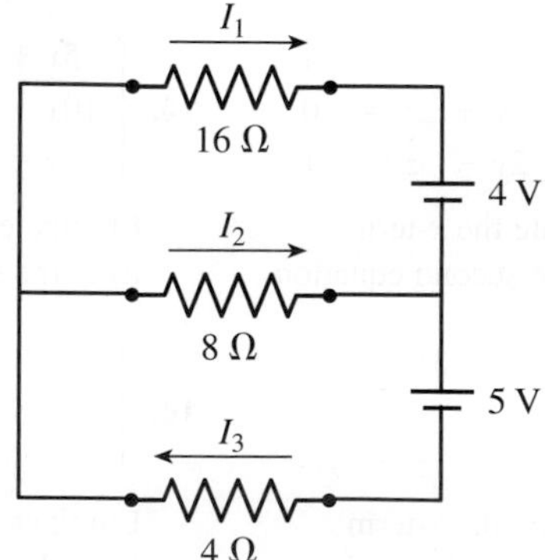

■ Discuss ■ Discover ■ Prove ■ Write

49. Prove: Can a Linear System Have Exactly Two Solutions?

(a) Suppose that (x_0, y_0, z_0) and (x_1, y_1, z_1) are solutions of the system

$$\begin{cases} a_1x + b_1y + c_1z = d_1 \\ a_2x + b_2y + c_2z = d_2 \\ a_3x + b_3y + c_3z = d_3 \end{cases}$$

Show that

$$\left(\frac{x_0 + x_1}{2}, \frac{y_0 + y_1}{2}, \frac{z_0 + z_1}{2}\right)$$

is also a solution.

(b) Use the result of part (a) to prove that if the system has two different solutions, then it has infinitely many solutions.

9.3 Matrices and Systems of Linear Equations

■ Matrices ■ The Augmented Matrix of a Linear System ■ Elementary Row Operations ■ Gaussian Elimination ■ Gauss-Jordan Elimination ■ Inconsistent and Dependent Systems ■ Modeling with Linear Systems

The plural of *matrix* is *matrices*.

A *matrix* is simply a rectangular array of numbers. Matrices are used to organize information into categories that correspond to the rows and columns of the matrix. For example, a scientist might organize information on a population of endangered whales as follows:

$$\begin{array}{r} \\ \textbf{Male} \\ \textbf{Female} \end{array} \begin{array}{c} \begin{array}{ccc} \textbf{Immature} & \textbf{Juvenile} & \textbf{Adult} \end{array} \\ \begin{bmatrix} 12 & 52 & 18 \\ 15 & 42 & 11 \end{bmatrix} \end{array}$$

This is a compact way of saying that there are 12 immature males, 15 immature females, 18 adult males, and so on.

We represent a linear system by a matrix, called the *augmented matrix* of the system.

$$\begin{array}{ccc} \textbf{Linear system} & & \textbf{Augmented matrix} \\ \begin{cases} 2x - y = 5 \\ x + 4y = 7 \end{cases} & \begin{array}{c} \longleftarrow \boxed{\text{Equation 1}} \longrightarrow \\ \longleftarrow \boxed{\text{Equation 2}} \longrightarrow \end{array} & \begin{array}{c} \begin{bmatrix} 2 & -1 & 5 \\ 1 & 4 & 7 \end{bmatrix} \\ \begin{array}{ccc} \uparrow & \uparrow & \\ \boxed{x} & \boxed{y} & \end{array} \end{array} \end{array}$$

The augmented matrix contains the same information as the system but in a simpler form. The operations we learned for solving systems of equations can now be performed on the augmented matrix.

■ Matrices

We begin by defining the various elements that make up a matrix.

Definition of Matrix

An $m \times n$ **matrix** is a rectangular array of numbers with m **rows** and n **columns**.

$$\begin{array}{c} \left.\begin{bmatrix} a_{11} & a_{12} & a_{13} & \cdots & a_{1n} \\ a_{21} & a_{22} & a_{23} & \cdots & a_{2n} \\ a_{31} & a_{32} & a_{33} & \cdots & a_{3n} \\ \vdots & \vdots & \vdots & \ddots & \vdots \\ a_{m1} & a_{m2} & a_{m3} & \cdots & a_{mn} \end{bmatrix} \begin{array}{l} \leftarrow \\ \leftarrow \\ \leftarrow \\ \\ \leftarrow \end{array} \right\} m \text{ rows} \\ \underbrace{\begin{array}{ccccc} \uparrow & \uparrow & \uparrow & & \uparrow \end{array}}_{n \text{ columns}} \end{array}$$

We say that the matrix has **dimension** $m \times n$. The numbers a_{ij} are the **entries** of the matrix. The subscript on the entry a_{ij} indicates that it is in the ith row and the jth column.

Here are some examples of matrices.

Matrix	Dimension	
$\begin{bmatrix} 1 & 3 & 0 \\ 2 & 4 & -1 \end{bmatrix}$	2×3	2 rows by 3 columns
$\begin{bmatrix} 6 & -5 & 0 & 1 \end{bmatrix}$	1×4	1 row by 4 columns

■ The Augmented Matrix of a Linear System

We can write a system of linear equations as a matrix, called the **augmented matrix** of the system, by writing only the coefficients and constants that appear in the equations. Here is an example.

Linear system	Augmented matrix
$\begin{cases} 3x - 2y + z = 5 \\ x + 3y - z = 0 \\ -x \quad\quad + 4z = 11 \end{cases}$	$\begin{bmatrix} 3 & -2 & 1 & 5 \\ 1 & 3 & -1 & 0 \\ -1 & 0 & 4 & 11 \end{bmatrix}$

Notice that a missing variable in an equation corresponds to a 0 entry in the augmented matrix.

Example 1 ■ Finding the Augmented Matrix of a Linear System

Write the augmented matrix of the following system of equations:

$$\begin{cases} 6x - 2y - z = 4 \\ x + 3z = 1 \\ 7y + z = 5 \end{cases}$$

Solution First we write the linear system with the variables lined up in columns.

$$\begin{cases} 6x - 2y - z = 4 \\ x + 3z = 1 \\ 7y + z = 5 \end{cases}$$

The augmented matrix is the matrix whose entries are the coefficients and the constants in this system.

$$\begin{bmatrix} 6 & -2 & -1 & 4 \\ 1 & 0 & 3 & 1 \\ 0 & 7 & 1 & 5 \end{bmatrix}$$

 **Now Try Exercise 11** ■

■ Elementary Row Operations

The operations that we used in Section 9.2 to solve linear systems correspond to operations on the rows of the augmented matrix of the system. For example, adding a multiple of one equation to another corresponds to adding a multiple of one row to another.

> **Elementary Row Operations**
>
> 1. Add a multiple of one row to another.
> 2. Multiply a row by a nonzero constant.
> 3. Interchange two rows.

Performing any of these operations on the augmented matrix of a system does not change its solution. We use the following notation to describe the elementary row operations:

Symbol	Description
$R_i + kR_j \to R_i$	Change the ith row by adding k times row j to it, and then put the result back in row i.
kR_i	Multiply the ith row by k.
$R_i \leftrightarrow R_j$	Interchange the ith and jth rows.

In the next example we compare the two ways of writing systems of linear equations.

Example 2 ■ Using Elementary Row Operations to Solve a Linear System

Solve the following system of linear equations:

$$\begin{cases} x - y + 3z = 4 \\ x + 2y - 2z = 10 \\ 3x - y + 5z = 14 \end{cases}$$

Solution Our goal is to eliminate the x-term from the second equation and the x- and y-terms from the third equation. For comparison we write both the system of equations and its augmented matrix for each step.

System | **Augmented matrix**

$$\begin{cases} x - y + 3z = 4 \\ x + 2y - 2z = 10 \\ 3x - y + 5z = 14 \end{cases} \qquad \begin{bmatrix} 1 & -1 & 3 & 4 \\ 1 & 2 & -2 & 10 \\ 3 & -1 & 5 & 14 \end{bmatrix}$$

Add $(-1) \times$ Equation 1 to Equation 2.
Add $(-3) \times$ Equation 1 to Equation 3.

$$\begin{cases} x - y + 3z = 4 \\ 3y - 5z = 6 \\ 2y - 4z = 2 \end{cases} \qquad \xrightarrow[R_3 - 3R_1 \to R_3]{R_2 - R_1 \to R_2} \begin{bmatrix} 1 & -1 & 3 & 4 \\ 0 & 3 & -5 & 6 \\ 0 & 2 & -4 & 2 \end{bmatrix}$$

Multiply Equation 3 by $\frac{1}{2}$.

$$\begin{cases} x - y + 3z = 4 \\ 3y - 5z = 6 \\ y - 2z = 1 \end{cases} \qquad \xrightarrow{\frac{1}{2}R_3} \begin{bmatrix} 1 & -1 & 3 & 4 \\ 0 & 3 & -5 & 6 \\ 0 & 1 & -2 & 1 \end{bmatrix}$$

Add $(-3) \times$ Equation 3 to Equation 2 (to eliminate y from Equation 2).

$$\begin{cases} x - y + 3z = 4 \\ z = 3 \\ y - 2z = 1 \end{cases} \qquad \xrightarrow{R_2 - 3R_3 \to R_2} \begin{bmatrix} 1 & -1 & 3 & 4 \\ 0 & 0 & 1 & 3 \\ 0 & 1 & -2 & 1 \end{bmatrix}$$

Interchange Equations 2 and 3.

$$\begin{cases} x - y + 3z = 4 \\ y - 2z = 1 \\ z = 3 \end{cases} \qquad \xrightarrow{R_2 \leftrightarrow R_3} \begin{bmatrix} 1 & -1 & 3 & 4 \\ 0 & 1 & -2 & 1 \\ 0 & 0 & 1 & 3 \end{bmatrix}$$

Now we use back-substitution to find that $x = 2$, $y = 7$, and $z = 3$. The solution is $(2, 7, 3)$.

Now Try Exercise 29 ■

■ Gaussian Elimination

In general, to solve a system of linear equations using its augmented matrix, we use elementary row operations to arrive at a matrix in a certain form. This form is described in the following box.

Row-Echelon Form and Reduced Row-Echelon Form of a Matrix

A matrix is in **row-echelon form** if it satisfies the following conditions.

1. The first nonzero number in each row (reading from left to right) is 1. This is called the **leading entry**.
2. The leading entry in each row is to the right of the leading entry in the row immediately above it.
3. All rows consisting entirely of zeros are at the bottom of the matrix.

A matrix is in **reduced row-echelon** form if it is in row-echelon form and also satisfies the following condition.

4. Every number above and below each leading entry is a 0.

In the following matrices the first one is not in row-echelon form. The second one *is* in row-echelon form, and the third one is in reduced row-echelon form. The entries in red are the leading entries.

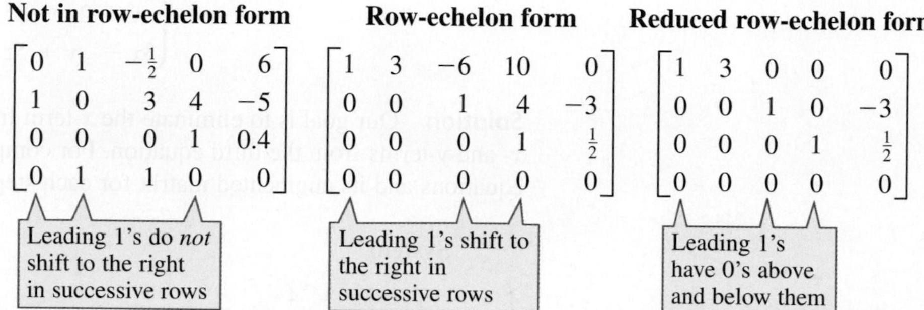

Here is a systematic way to put a matrix into row-echelon form using elementary row operations:

- Start by obtaining 1 in the top left corner. Then obtain zeros below that 1 by adding appropriate multiples of the first row to the rows below it.
- Next, obtain a leading 1 in the next row, and then obtain zeros below that 1.
- At each stage make sure that every leading entry is to the right of the leading entry in the row above it—rearrange the rows if necessary.
- Continue this process until you arrive at a matrix in row-echelon form.

This is how the process might work for a 3×4 matrix:

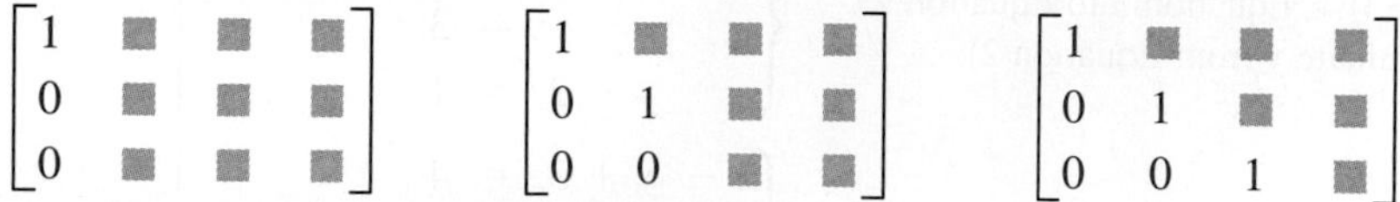

Once an augmented matrix is in row-echelon form, we can solve the corresponding linear system using back-substitution. This technique is called **Gaussian elimination**, in honor of its inventor, the German mathematician C. F. Gauss (see Section 3.5).

Solving a System Using Gaussian Elimination

1. **Augmented Matrix.** Write the augmented matrix of the system.
2. **Row-Echelon Form.** Use elementary row operations to change the augmented matrix to row-echelon form.
3. **Back-Substitution.** Write the new system of equations that corresponds to the row-echelon form of the augmented matrix and solve by back-substitution.

Example 3 ■ Solving a System Using Row-Echelon Form

Solve the following system of linear equations using Gaussian elimination:

$$\begin{cases} 4x + 8y - 4z = 4 \\ 3x + 8y + 5z = -11 \\ -2x + y + 12z = -17 \end{cases}$$

Solution We first write the augmented matrix of the system, and then we use elementary row operations to put it in row-echelon form.

Augmented matrix:

$$\begin{bmatrix} \textcircled{4} & 8 & -4 & 4 \\ 3 & 8 & 5 & -11 \\ -2 & 1 & 12 & -17 \end{bmatrix}$$

Need a 1 here

$$\xrightarrow{\frac{1}{4}R_1} \begin{bmatrix} 1 & 2 & -1 & 1 \\ \textcircled{3} & 8 & 5 & -11 \\ \textcircled{-2} & 1 & 12 & -17 \end{bmatrix}$$

Need 0's here

$$\xrightarrow[R_3 + 2R_1 \to R_3]{R_2 - 3R_1 \to R_2} \begin{bmatrix} 1 & 2 & -1 & 1 \\ 0 & \textcircled{2} & 8 & -14 \\ 0 & 5 & 10 & -15 \end{bmatrix}$$

Need a 1 here

$$\xrightarrow{\frac{1}{2}R_2} \begin{bmatrix} 1 & 2 & -1 & 1 \\ 0 & 1 & 4 & -7 \\ 0 & \textcircled{5} & 10 & -15 \end{bmatrix}$$

Need a 0 here

$$\xrightarrow{R_3 - 5R_2 \to R_3} \begin{bmatrix} 1 & 2 & -1 & 1 \\ 0 & 1 & 4 & -7 \\ 0 & 0 & \textcircled{-10} & 20 \end{bmatrix}$$

Need a 1 here

Row-echelon form:

$$\xrightarrow{-\frac{1}{10}R_3} \begin{bmatrix} 1 & 2 & -1 & 1 \\ 0 & 1 & 4 & -7 \\ 0 & 0 & 1 & -2 \end{bmatrix}$$

We now have an equivalent matrix in row-echelon form, and the corresponding system of equations is

$$\begin{cases} x + 2y - z = 1 \\ y + 4z = -7 \\ z = -2 \end{cases}$$

Back-substitute: We use back-substitution to solve the system.

$$y + 4(-2) = -7 \qquad \text{Back-substitute } z = -2 \text{ into Equation 2}$$

$$y = 1 \qquad \text{Solve for } y$$

$$x + 2(1) - (-2) = 1 \qquad \text{Back-substitute } y = 1 \text{ and } z = -2 \text{ into Equation 1}$$

$$x = -3 \qquad \text{Solve for } x$$

So the solution of the system is $(-3, 1, -2)$.

 **Now Try Exercise 31** ■

Graphing devices are able to work with matrices. Matrices are stored in the memory of the device using names such as [A], [B], [C], ... On graphing devices the `ref` command gives the row-echelon form of a matrix. For the augmented matrix in Example 3 the `ref`

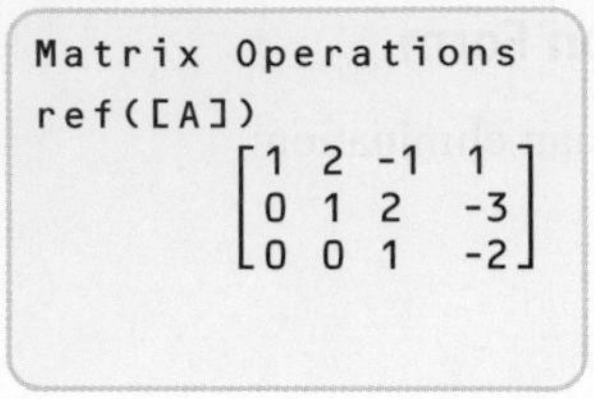

```
Matrix Operations
ref([A])
[1 2 -1  1]
[0 1  2 -3]
[0 0  1 -2]
```

Figure 1

command gives the output shown in Figure 1. Notice that the row-echelon form in Figure 1 differs from the one we got in Example 3. This is because the graphing device used different row operations than we did. You should check that the row-echelon form that your device gives you leads to the same solution as ours.

■ Gauss-Jordan Elimination

If we put the augmented matrix of a linear system in *reduced* row-echelon form, then we don't need to back-substitute to solve the system. To put a matrix in reduced row-echelon form, we use the following steps.

- Use the elementary row operations to put the matrix in row-echelon form.
- Obtain zeros above each leading entry by adding multiples of the row containing that entry to the rows above it. Begin with the last leading entry and work upward.

Here is how the process works for a 3×4 matrix:

$$\begin{bmatrix} 1 & \blacksquare & \blacksquare & \blacksquare \\ 0 & 1 & \blacksquare & \blacksquare \\ 0 & 0 & 1 & \blacksquare \end{bmatrix} \to \begin{bmatrix} 1 & \blacksquare & 0 & \blacksquare \\ 0 & 1 & 0 & \blacksquare \\ 0 & 0 & 1 & \blacksquare \end{bmatrix} \to \begin{bmatrix} 1 & 0 & 0 & \blacksquare \\ 0 & 1 & 0 & \blacksquare \\ 0 & 0 & 1 & \blacksquare \end{bmatrix}$$

Using the reduced row-echelon form to solve a system is called **Gauss-Jordan elimination**. The process is illustrated in the next example.

Example 4 ■ Solving a System Using Reduced Row-Echelon Form

Solve the following system of linear equations using Gauss-Jordan elimination:

$$\begin{cases} 4x + 8y - 4z = 4 \\ 3x + 8y + 5z = -11 \\ -2x + y + 12z = -17 \end{cases}$$

Solution In Example 3 we used Gaussian elimination on the augmented matrix of this system to arrive at an equivalent matrix in row-echelon form. We continue using elementary row operations on the last matrix in Example 3 to arrive at an equivalent matrix in reduced row-echelon form.

$$\begin{bmatrix} 1 & 2 & \textcircled{-1} & 1 \\ 0 & 1 & \textcircled{4} & -7 \\ 0 & 0 & 1 & -2 \end{bmatrix}$$

Need 0's here

$$\xrightarrow[\text{R}_1 + \text{R}_3 \to \text{R}_1]{\text{R}_2 - 4\text{R}_3 \to \text{R}_2} \begin{bmatrix} 1 & \textcircled{2} & 0 & -1 \\ 0 & 1 & 0 & 1 \\ 0 & 0 & 1 & -2 \end{bmatrix}$$

Need a 0 here

$$\xrightarrow{\text{R}_1 - 2\text{R}_2 \to \text{R}_1} \begin{bmatrix} 1 & 0 & 0 & -3 \\ 0 & 1 & 0 & 1 \\ 0 & 0 & 1 & -2 \end{bmatrix}$$

We now have an equivalent matrix in reduced row-echelon form, and the corresponding system of equations is

$$\begin{cases} x = -3 \\ y = 1 \\ z = -2 \end{cases}$$

Since the system is in reduced row-echelon form, back-subsitution is not required to get the solution.

So the solution is $(-3, 1, -2)$.

Now Try Exercise 33 ■

```
Matrix Operations
rref([A])
    [1 0 0 -3]
    [0 1 0 1 ]
    [0 0 1 -2]
```

Figure 2

Graphing devices also have a command that puts a matrix in reduced row-echelon form. (The command is usually `rref`.) For the augmented matrix in Example 4 the `rref` command gives the output shown in Figure 2. The device gives the same reduced row-echelon form as the one we got in Example 4. This is because every matrix has a *unique* reduced row-echelon form.

■ Inconsistent and Dependent Systems

The systems of linear equations that we considered in Examples 1–4 had exactly one solution. But as we know from Section 9.2, a linear system may have one solution, no solution, or infinitely many solutions. Fortunately, the row-echelon form of a system allows us to determine which of these cases applies, as described in the following box.

First we need some terminology: A **leading variable** in a linear system is one that corresponds to a leading entry in the row-echelon form of the augmented matrix of the system.

The Solutions of a Linear System in Row-Echelon Form

Suppose the augmented matrix of a system of linear equations has been transformed by Gaussian elimination into row-echelon form. Then exactly one of the following is true.

1. **No solution.** If the row-echelon form contains a row that represents the equation $0 = c$, where c is not zero, then the system has no solution. A system with no solution is called **inconsistent**.
2. **One solution.** If each variable in the row-echelon form is a leading variable, then the system has exactly one solution, which we find using back-substitution or Gauss-Jordan elimination.
3. **Infinitely many solutions.** If the variables in the row-echelon form are not all leading variables and if the system is not inconsistent, then it has infinitely many solutions. In this case the system is called **dependent**. We solve the system by putting the matrix in reduced row-echelon form and then expressing the leading variables in terms of the nonleading variables. The nonleading variables may take on any real numbers as their values.

The matrices below, all in row-echelon form, illustrate the three cases described in the above box.

No solution

$$\begin{bmatrix} 1 & 2 & 5 & 7 \\ 0 & 1 & 3 & 4 \\ 0 & 0 & 0 & 1 \end{bmatrix}$$

Last equation says $0 = 1$

One solution

$$\begin{bmatrix} 1 & 6 & -1 & 3 \\ 0 & 1 & 2 & -2 \\ 0 & 0 & 1 & 8 \end{bmatrix}$$

Each variable is a leading variable

Infinitely many solutions

$$\begin{bmatrix} 1 & 2 & -3 & 1 \\ 0 & 1 & 5 & -2 \\ 0 & 0 & 0 & 0 \end{bmatrix}$$

z is not a leading variable

Example 5 ■ A System with No Solution

Solve the following system:

$$\begin{cases} x - 3y + 2z = 12 \\ 2x - 5y + 5z = 14 \\ x - 2y + 3z = 20 \end{cases}$$

Solution We transform the system into row-echelon form.

$$\begin{bmatrix} 1 & -3 & 2 & 12 \\ 2 & -5 & 5 & 14 \\ 1 & -2 & 3 & 20 \end{bmatrix} \xrightarrow[R_3 - R_1 \to R_3]{R_2 - 2R_1 \to R_2} \begin{bmatrix} 1 & -3 & 2 & 12 \\ 0 & 1 & 1 & -10 \\ 0 & 1 & 1 & 8 \end{bmatrix}$$

$$\xrightarrow{R_3 - R_2 \to R_3} \begin{bmatrix} 1 & -3 & 2 & 12 \\ 0 & 1 & 1 & -10 \\ 0 & 0 & 0 & 18 \end{bmatrix} \xrightarrow{\frac{1}{18}R_3} \begin{bmatrix} 1 & -3 & 2 & 12 \\ 0 & 1 & 1 & -10 \\ 0 & 0 & 0 & 1 \end{bmatrix}$$

This last matrix is in row-echelon form, so we can stop the Gaussian elimination process. Now if we translate the last row back into equation form, we get $0x + 0y + 0z = 1$, or $0 = 1$, which is false. No matter what values we choose for x, y, and z, the last equation will never be a true statement. This means that the system *has no solution.*

Now Try Exercise 39 ■

```
Matrix Operations
ref([A])
 [1 -2.5  2.5  7 ]
 [0  1    1   -10]
 [0  0    0    1 ]
```

Figure 3

Figure 3 shows the row-echelon form produced by a graphing device for the augmented matrix in Example 5. You should check that the device result agrees with Example 5.

Example 6 ■ A System with Infinitely Many Solutions

Find the complete solution of the following system:

$$\begin{cases} -3x - 5y + 36z = 10 \\ -x + 7z = 5 \\ x + y - 10z = -4 \end{cases}$$

Solution We transform the system into reduced row-echelon form.

$$\begin{bmatrix} -3 & -5 & 36 & 10 \\ -1 & 0 & 7 & 5 \\ 1 & 1 & -10 & -4 \end{bmatrix} \xrightarrow{R_1 \leftrightarrow R_3} \begin{bmatrix} 1 & 1 & -10 & -4 \\ -1 & 0 & 7 & 5 \\ -3 & -5 & 36 & 10 \end{bmatrix}$$

$$\xrightarrow[R_3 + 3R_1 \to R_3]{R_2 + R_1 \to R_2} \begin{bmatrix} 1 & 1 & -10 & -4 \\ 0 & 1 & -3 & 1 \\ 0 & -2 & 6 & -2 \end{bmatrix} \xrightarrow{R_3 + 2R_2 \to R_3} \begin{bmatrix} 1 & 1 & -10 & -4 \\ 0 & 1 & -3 & 1 \\ 0 & 0 & 0 & 0 \end{bmatrix}$$

$$\xrightarrow{R_1 - R_2 \to R_1} \begin{bmatrix} 1 & 0 & -7 & -5 \\ 0 & 1 & -3 & 1 \\ 0 & 0 & 0 & 0 \end{bmatrix}$$

The third row corresponds to the equation $0 = 0$. This equation is always true, no matter what values are used for x, y, and z. Since the equation adds no new information about the variables, we can drop it from the system. So the last matrix corresponds to the system

$$\begin{cases} x - 7z = -5 & \text{Equation 1} \\ y - 3z = 1 & \text{Equation 2} \end{cases}$$

Leading variables

Now we solve for the leading variables x and y in terms of the nonleading variable z.

$x = 7z - 5$ Solve for x in Equation 1

$y = 3z + 1$ Solve for y in Equation 2

To obtain the complete solution, we let z be any real number t, and we express x, y, and z in terms of t.

$$x = 7t - 5$$
$$y = 3t + 1$$
$$z = t$$

We can also write the solution as the ordered triple $(7t - 5, 3t + 1, t)$, where t is any real number.

 **Now Try Exercise 41** ■

In Example 6, to get specific solutions, we give a specific value to t. For example, if $t = 1$, then

$$x = 7(1) - 5 = 2$$
$$y = 3(1) + 1 = 4$$
$$z = 1$$

Here are some other solutions of the system obtained by substituting other values for the parameter t.

Parameter t	Solution $(7t - 5, 3t + 1, t)$
-1	$(-12, -2, -1)$
0	$(-5, 1, 0)$
2	$(9, 7, 2)$
5	$(30, 16, 5)$

Example 7 ■ A System with Infinitely Many Solutions

Find the complete solution of the following system:

$$\begin{cases} x + 2y - 3z - 4w = 10 \\ x + 3y - 3z - 4w = 15 \\ 2x + 2y - 6z - 8w = 10 \end{cases}$$

Solution We transform the system into reduced row-echelon form.

$$\begin{bmatrix} 1 & 2 & -3 & -4 & 10 \\ 1 & 3 & -3 & -4 & 15 \\ 2 & 2 & -6 & -8 & 10 \end{bmatrix} \xrightarrow[\mathrm{R}_3 - 2\mathrm{R}_1 \to \mathrm{R}_3]{\mathrm{R}_2 - \mathrm{R}_1 \to \mathrm{R}_2} \begin{bmatrix} 1 & 2 & -3 & -4 & 10 \\ 0 & 1 & 0 & 0 & 5 \\ 0 & -2 & 0 & 0 & -10 \end{bmatrix}$$

$$\xrightarrow{\mathrm{R}_3 + 2\mathrm{R}_2 \to \mathrm{R}_3} \begin{bmatrix} 1 & 2 & -3 & -4 & 10 \\ 0 & 1 & 0 & 0 & 5 \\ 0 & 0 & 0 & 0 & 0 \end{bmatrix} \xrightarrow{\mathrm{R}_1 - 2\mathrm{R}_2 \to \mathrm{R}_1} \begin{bmatrix} 1 & 0 & -3 & -4 & 0 \\ 0 & 1 & 0 & 0 & 5 \\ 0 & 0 & 0 & 0 & 0 \end{bmatrix}$$

The last matrix is in reduced row-echelon form. Since the last row represents the

Discovery Project ■ Computer Graphics I

Slobodian.Y/Shutterstock.com

One significant use of matrices is in the digital representation of images. A camera or scanner converts an image into a matrix by dividing the image into a rectangular array of elements called pixels (like the pixelated image of a toucan shown here) and then assigning a value to each pixel. The value represents the color or brightness of that pixel. In this project we'll discover how changing the numbers in such a matrix can enhance an image by changing the contrast or brightness. You can find the project at **www.stewartmath.com**.

equation $0 = 0$, we may discard it. So the last matrix corresponds to the system

$$\begin{cases} x \qquad - 3z - 4w = 0 \\ \quad y \qquad\qquad\quad = 5 \end{cases}$$

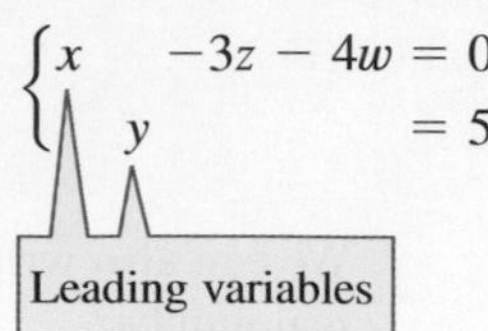

To obtain the complete solution, we solve for the leading variables x and y in terms of the nonleading variables z and w, and we let z and w be any real numbers s and t, respectively. Thus the complete solution is

$$x = 3s + 4t$$
$$y = 5$$
$$z = s$$
$$w = t$$

where s and t are any real numbers.

Now Try Exercise 61 ■

Note that s and t do *not* have to be the *same* real number in the solution for Example 7. We can choose arbitrary values for each if we wish to construct a specific solution to the system. For example, if we let $s = 1$ and $t = 2$, then we get the solution $(11, 5, 1, 2)$. You should check that this does indeed satisfy all three of the original equations in Example 7.

Examples 6 and 7 illustrate this general fact: If a system in row-echelon form has n nonzero equations in m variables $(m > n)$, then the complete solution will have $m - n$ nonleading variables. For instance, in Example 6 we arrived at *two* nonzero equations in the *three* variables x, y, and z, which gave us $3 - 2 = 1$ nonleading variable.

■ Modeling with Linear Systems

Linear equations, often containing hundreds or even thousands of variables, occur frequently in the applications of algebra to the sciences and to other fields. For now, let's consider an example that involves only three variables.

Example 8 ■ Nutritional Analysis Using a System of Linear Equations

A nutritionist is performing an experiment with student volunteers. A subject receives a daily diet that consists of a combination of three commercial diet foods: MiniCal, LiquiFast, and SlimQuick. For the experiment it is important that the subject consume exactly 500 mg of potassium, 75 g of protein, and 1150 units of vitamin D every day. The amounts of these nutrients in 1 serving of each food are given in the table. How many servings of each food should the subject eat every day to satisfy the nutrient requirements exactly?

	MiniCal	LiquiFast	SlimQuick
Potassium (mg)	50	75	10
Protein (g)	5	10	3
Vitamin D (units)	90	100	50

Solution Let x, y, and z represent the number of servings of MiniCal, LiquiFast, and SlimQuick, respectively, that the subject should eat every day. This means that the subject will get $50x$ mg of potassium from MiniCal, $75y$ mg from LiquiFast, and $10z$ mg from SlimQuick, for a total of $50x + 75y + 10z$ mg potassium in all. Since

```
Matrix Operations
rref([A])
       [1 0 0 5 ]
       [0 1 0 2 ]
       [0 0 1 10]
```

Figure 4

the potassium requirement is 500 mg, we get the first equation below. Similar reasoning for the protein and vitamin D requirements leads to the system

$$\begin{cases} 50x + 75y + 10z = 500 & \text{Potassium} \\ 5x + 10y + 3z = 75 & \text{Protein} \\ 90x + 100y + 50z = 1150 & \text{Vitamin D} \end{cases}$$

Dividing the first equation by 5 and the third one by 10 gives the system

$$\begin{cases} 10x + 15y + 2z = 100 \\ 5x + 10y + 3z = 75 \\ 9x + 10y + 5z = 115 \end{cases}$$

Check Your Answer

$x = 5, y = 2, z = 10$:

$$\begin{cases} 10(5) + 15(2) + 2(10) = 100 \\ 5(5) + 10(2) + 3(10) = 75 \\ 9(5) + 10(2) + 5(10) = 115 \end{cases}$$ ✓

We can solve this system using Gaussian elimination, or we can use a graphing device to find the reduced row-echelon form of the augmented matrix of the system. Using the `rref` command on a graphing device, we get the output shown in Figure 4. From the reduced row-echelon form we see that $x = 5$, $y = 2$, $z = 10$. The subject should be served 5 servings of MiniCal, 2 servings of LiquiFast, and 10 servings of SlimQuick every day.

Now Try Exercise 69 ■

A more practical application might involve dozens of foods and nutrients rather than just three. Such problems lead to systems with large numbers of variables and equations. Computers or graphing devices are essential for solving such large systems.

9.3 | Exercises

Concepts

1. If a system of linear equations has infinitely many solutions, then the system is called ______. If a system of linear equations has no solution, then the system is called ______.

2. Write the augmented matrix of the following system of equations.

System

$$\begin{cases} x + y - z = 1 \\ x \qquad + 2z = -3 \\ \qquad 2y - z = 3 \end{cases}$$

Augmented matrix

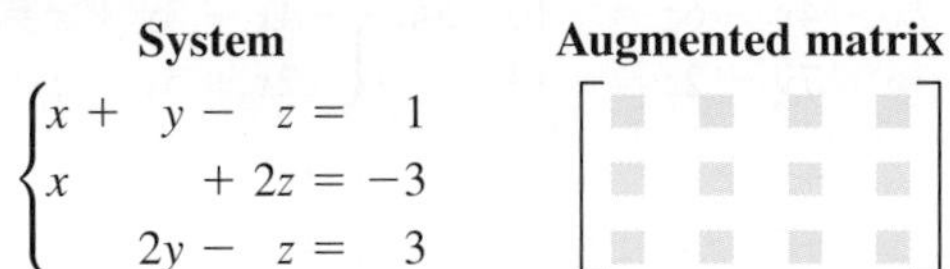

3. The following matrix is the augmented matrix of a system of linear equations in the variables x, y, and z. (It is given in reduced row-echelon form.)

$$\begin{bmatrix} 1 & 0 & -1 & 3 \\ 0 & 1 & 2 & 5 \\ 0 & 0 & 0 & 0 \end{bmatrix}$$

(a) The leading variables are ______.

(b) Is the system inconsistent or dependent? ______

(c) The solution of the system is:

$x =$ ______, $y =$ ______, $z =$ ______

4. The augmented matrix of a system of linear equations is given in reduced row-echelon form. Find the solution of the system.

(a) $\begin{bmatrix} 1 & 0 & 0 & 2 \\ 0 & 1 & 0 & 1 \\ 0 & 0 & 1 & 3 \end{bmatrix}$ **(b)** $\begin{bmatrix} 1 & 0 & 1 & 2 \\ 0 & 1 & 1 & 1 \\ 0 & 0 & 0 & 0 \end{bmatrix}$ **(c)** $\begin{bmatrix} 1 & 0 & 0 & 2 \\ 0 & 1 & 0 & 1 \\ 0 & 0 & 0 & 3 \end{bmatrix}$

$x =$ ______ $x =$ ______ $x =$ ______

$y =$ ______ $y =$ ______ $y =$ ______

$z =$ ______ $z =$ ______ $z =$ ______

Skills

5–10 ■ Dimension of a Matrix State the dimension of the matrix.

5. $\begin{bmatrix} 2 & 7 \\ 0 & -1 \\ 5 & -3 \end{bmatrix}$ **6.** $\begin{bmatrix} -1 & 5 & 4 & 0 \\ 0 & 2 & 11 & 3 \end{bmatrix}$

7. $\begin{bmatrix} 12 \\ 35 \end{bmatrix}$ **8.** $\begin{bmatrix} -3 \\ 0 \\ 1 \end{bmatrix}$

9. $\begin{bmatrix} 1 & 4 & 7 \end{bmatrix}$ **10.** $\begin{bmatrix} 1 & 0 \\ 0 & 1 \end{bmatrix}$

11–12 ■ The Augmented Matrix Write the augmented matrix for the system of linear equations.

11. $\begin{cases} 3x + y - z = 2 \\ 2x - y = 1 \\ x - z = 3 \end{cases}$ **12.** $\begin{cases} -x + z = -1 \\ 3y - 2z = 7 \\ x - y + 3z = 3 \end{cases}$

13–20 ■ Form of a Matrix A matrix is given. **(a)** Determine whether the matrix is in row-echelon form. **(b)** Determine whether the matrix is in reduced row-echelon form. **(c)** Write the system of equations for which the given matrix is the augmented matrix.

13. $\begin{bmatrix} 1 & 0 & -3 \\ 0 & 1 & 5 \end{bmatrix}$ **14.** $\begin{bmatrix} 1 & 3 & -3 \\ 0 & 1 & 5 \end{bmatrix}$

15. $\begin{bmatrix} 1 & 2 & 8 & 0 \\ 0 & 1 & 3 & 2 \\ 0 & 0 & 0 & 0 \end{bmatrix}$ **16.** $\begin{bmatrix} 1 & 0 & -7 & 0 \\ 0 & 1 & 3 & 0 \\ 0 & 0 & 0 & 1 \end{bmatrix}$

17. $\begin{bmatrix} 1 & 0 & 0 & 0 \\ 0 & 0 & 0 & 0 \\ 0 & 1 & 5 & 1 \end{bmatrix}$ **18.** $\begin{bmatrix} 1 & 0 & 0 & 1 \\ 0 & 1 & 0 & 2 \\ 0 & 0 & 1 & 3 \end{bmatrix}$

19. $\begin{bmatrix} 1 & 3 & 0 & -1 & 0 \\ 0 & 0 & 1 & 2 & 0 \\ 0 & 0 & 0 & 0 & 1 \\ 0 & 0 & 0 & 0 & 0 \end{bmatrix}$ **20.** $\begin{bmatrix} 1 & 3 & 0 & 1 & 0 & 0 \\ 0 & 1 & 0 & 4 & 0 & 0 \\ 0 & 0 & 0 & 1 & 1 & 2 \\ 0 & 0 & 0 & 1 & 0 & 0 \end{bmatrix}$

21–24 ■ Elementary Row Operations Perform the indicated elementary row operation.

21. $\begin{bmatrix} -1 & 1 & 2 & 0 \\ 3 & 1 & 1 & 4 \\ 1 & -2 & -1 & -1 \end{bmatrix}$

Add 3 times Row 1 to Row 2.

22. $\left[\begin{array}{ccc|c} -5 & 2 & -3 & 3 \\ 10 & -3 & 1 & -20 \\ -1 & 3 & 1 & 8 \end{array}\right]$

Add 2 times Row 1 to Row 2.

23. $\begin{bmatrix} 2 & 1 & -3 & 5 \\ 2 & 3 & 1 & 13 \\ 6 & -5 & -1 & 7 \end{bmatrix}$

Add -3 times Row 1 to Row 3.

24. $\begin{bmatrix} 1 & -3 & 2 & -1 \\ 0 & 1 & 1 & -1 \\ 0 & 2 & -1 & 1 \end{bmatrix}$

Add -2 times Row 2 to Row 3.

25–28 ■ Back-Substitution A matrix is given in row-echelon form. **(a)** Write the system of equations for which the given matrix is the augmented matrix. **(b)** Use back-substitution to solve the system.

25. $\begin{bmatrix} 1 & -2 & 4 & 3 \\ 0 & 1 & 2 & 7 \\ 0 & 0 & 1 & 2 \end{bmatrix}$ **26.** $\begin{bmatrix} 1 & 1 & -3 & 8 \\ 0 & 1 & -3 & 5 \\ 0 & 0 & 1 & -1 \end{bmatrix}$

27. $\begin{bmatrix} 1 & 2 & 3 & -1 & 7 \\ 0 & 1 & -2 & 0 & 5 \\ 0 & 0 & 1 & 2 & 5 \\ 0 & 0 & 0 & 1 & 3 \end{bmatrix}$

28. $\begin{bmatrix} 1 & 0 & -2 & 2 & 5 \\ 0 & 1 & 3 & 0 & -1 \\ 0 & 0 & 1 & -1 & 0 \\ 0 & 0 & 0 & 1 & -1 \end{bmatrix}$

29–38 ■ Linear Systems with One Solution The system of linear equations has a unique solution. Find the solution using Gaussian elimination or Gauss-Jordan elimination.

29. $\begin{cases} x - 2y + z = 1 \\ y + 2z = 5 \\ x + y + 3z = 8 \end{cases}$ **30.** $\begin{cases} x + y + 6z = 3 \\ x + y + 3z = 3 \\ x + 2y + 4z = 7 \end{cases}$

31. $\begin{cases} x - 2y + z = 1 \\ 3x + y + 2z = 4 \\ -2x - 3y + z = -5 \end{cases}$ **32.** $\begin{cases} x + z = 7 \\ 2x - y - 2z = -5 \\ 5x + 3y - z = 2 \end{cases}$

33. $\begin{cases} x - y - 2z = 0 \\ 2x + 3y + z = -5 \\ 3x + y - 8z = -16 \end{cases}$ **34.** $\begin{cases} x + z = 3 \\ x + 2y + z = 3 \\ 2x - 3y - 2z = 2 \end{cases}$

35. $\begin{cases} x_1 + 2x_2 - x_3 = 9 \\ 2x_1 - x_3 = -2 \\ 3x_1 + 5x_2 + 2x_3 = 22 \end{cases}$ **36.** $\begin{cases} 2x_1 + x_2 = 7 \\ 2x_1 - x_2 + x_3 = 6 \\ 3x_1 - 2x_2 + 4x_3 = 11 \end{cases}$

37. $\begin{cases} 2x - 3y - z = 13 \\ -x + 2y - 5z = 6 \\ 5x - y - z = 49 \end{cases}$ **38.** $\begin{cases} 10x + 10y - 20z = 60 \\ 15x + 20y + 30z = -25 \\ -5x + 30y - 10z = 45 \end{cases}$

39–48 ■ Dependent or Inconsistent Linear Systems Determine whether the system of linear equations is inconsistent or dependent. If it is dependent, find the complete solution.

39. $\begin{cases} x + 2y - z = 3 \\ 3x + 7y + 2z = 5 \\ 2x + 3y - 7z = 4 \end{cases}$ **40.** $\begin{cases} x - 2y + 3z = 4 \\ 3x - z = -3 \\ x + 4y - 7z = 2 \end{cases}$

41. $\begin{cases} 2x - 3y - 9z = -5 \\ x + 3z = 2 \\ -3x + y - 4z = -3 \end{cases}$ **42.** $\begin{cases} x - 2y + 5z = 3 \\ -2x + 6y - 11z = 1 \\ 3x - 16y + 20z = -26 \end{cases}$

43. $\begin{cases} x - y + 3z = 3 \\ 4x - 8y + 32z = 24 \\ 2x - 3y + 11z = 4 \end{cases}$ **44.** $\begin{cases} -2x + 6y - 2z = -12 \\ x - 3y + 2z = 10 \\ -x + 3y + 2z = 6 \end{cases}$

45. $\begin{cases} x + 4y - 2z = -3 \\ 2x - y + 5z = 12 \\ 8x + 5y + 11z = 30 \end{cases}$ **46.** $\begin{cases} 3r + 2s - 3t = 10 \\ r - s - t = -5 \\ r + 4s - t = 20 \end{cases}$

47. $\begin{cases} 2x + y - 2z = 12 \\ -x - \frac{1}{2}y + z = -6 \\ 3x + \frac{3}{2}y - 3z = 18 \end{cases}$ **48.** $\begin{cases} y - 5z = 7 \\ 3x + 2y = 12 \\ 3x + 10z = 80 \end{cases}$

49–64 ■ Solving a Linear System Solve the system of linear equations.

49. $\begin{cases} 4x - 3y + z = -8 \\ -2x + y - 3z = -4 \\ x - y + 2z = 3 \end{cases}$ **50.** $\begin{cases} 2x - 3y + 5z = 14 \\ 4x - y - 2z = -17 \\ -x - y + z = 3 \end{cases}$

51. $\begin{cases} 3x - y + z = 3 \\ x - 2z = 4 \\ 2x + y - 11z = 1 \end{cases}$ **52.** $\begin{cases} x - 3y + 2z = 5 \\ 2x - 3y - 2z = -2 \\ -x + 4z = 7 \end{cases}$

53. $\begin{cases} x + 2y - 3z = -5 \\ -2x - 4y - 6z = 10 \\ 3x + 7y - 2z = -13 \end{cases}$ **54.** $\begin{cases} 3x + y = 2 \\ -4x + 3y + z = 4 \\ 2x + 5y + z = 0 \end{cases}$

55. $\begin{cases} x - y + 6z = 8 \\ x + z = 5 \\ x + 3y - 14z = -4 \end{cases}$ **56.** $\begin{cases} 3x - y + 2z = -1 \\ 4x - 2y + z = -7 \\ -x + 3y - 2z = -1 \end{cases}$

57. $\begin{cases} -x + 2y + z - 3w = 3 \\ 3x - 4y + z + w = 9 \\ -x - y + z + w = 0 \\ 2x + y + 4z - 2w = 3 \end{cases}$

58. $\begin{cases} x + y - z - w = 6 \\ 2x + z - 3w = 8 \\ x - y + 4w = -10 \\ 3x + 5y - z - w = 20 \end{cases}$

59. $\begin{cases} x + y + 2z - w = -2 \\ 3y + z + 2w = 2 \\ x + y + 3w = 2 \\ -3x + z + 2w = 5 \end{cases}$

60. $\begin{cases} x - 3y + 2z + w = -2 \\ x - 2y - 2w = -10 \\ z + 5w = 15 \\ 3x + 2z + w = -3 \end{cases}$

61. $\begin{cases} x - y + w = 0 \\ 3x - z + 2w = 0 \\ x - 4y + z + 2w = 0 \end{cases}$

62. $\begin{cases} 2x - y + 2z + w = 5 \\ -x + y + 4z - w = 3 \\ 3x - 2y - z = 0 \end{cases}$

63. $\begin{cases} x + z + w = 4 \\ y - z = -4 \\ x - 2y + 3z + w = 12 \\ 2x - 2z + 5w = -1 \end{cases}$

64. $\begin{cases} y - z + 2w = 0 \\ 3x + 2y + w = 0 \\ 2x + 4w = 12 \\ -2x - 2z + 5w = 6 \end{cases}$

65–68 ■ Solving a Linear System Using a Graphing Device Solve the system of linear equations by using the `rref` command on a graphing device. State your answer rounded to two decimal places.

65. $\begin{cases} 0.75x - 3.75y + 2.95z = 4.0875 \\ 0.95x - 8.75y = 3.375 \\ 1.25x - 0.15y + 2.75z = 3.6625 \end{cases}$

66. $\begin{cases} 1.31x + 2.72y - 3.71z = -13.9534 \\ -0.21x + 3.73z = 13.4322 \\ 2.34y - 4.56z = -21.3984 \end{cases}$

67. $\begin{cases} 42x - 31y - 42w = -0.4 \\ -6x - 9w = 4.5 \\ 35x - 67z + 32w = 348.8 \\ 31y + 48z - 52w = -76.6 \end{cases}$

68. $\begin{cases} 49x - 27y + 52z = -145.0 \\ 27y + 43w = -118.7 \\ -31y + 42z = -72.1 \\ 73x - 54y = -132.7 \end{cases}$

Applications

69. Nutrition A doctor recommends that a patient take 50 mg each of niacin, riboflavin, and thiamin daily to alleviate a vitamin deficiency. The patient has three brands of vitamin pills. The amounts of the relevant vitamins per pill are given in the table. How many pills of each type should be taken every day to get 50 mg of each vitamin?

	VitaMax	Vitron	VitaPlus
Niacin (mg)	5	10	15
Riboflavin (mg)	15	20	0
Thiamin (mg)	10	10	10

70. Mixtures A chemist has three acid solutions at various concentrations. The first is 10% acid, the second is 20%, and the third is 40%. How many milliliters of each solution should be used to make 100 mL of 18% solution, if four times as much of the 10% solution is used as the 40% solution?

71. Distance, Speed, and Time Athlete A, Athlete B, and Athlete C enter a race in which they have to run, swim, and cycle over a marked course. Their average speeds are given in the table. Athlete C finishes first with a total time of 1 h 45 min. Athlete A comes in second with a time of 2 h 30 min. Athlete B finishes last with a time of 3 h. Find the distance (in km) for each part of the race.

	Average Speed (km/h)		
	Running	Swimming	Cycling
Athlete A	10	4	20
Athlete B	$7\frac{1}{2}$	6	15
Athlete C	15	3	40

72. Classroom Use A small school has 100 students who occupy three classrooms: A, B, and C. After the first period of the school day, half the students in room A move to room B, one-fifth of the students in room B move to room C, and one-third of the students in room C move to room A. Nevertheless, the total number of students in each room is the same for both periods. How many students occupy each room?

73. Manufacturing Furniture A furniture factory makes wooden tables, chairs, and armoires. Each piece of furniture requires three operations: cutting the wood, assembling, and finishing. Each operation requires the number of hours given in the table. The workers in the factory can provide 300 hours of cutting, 400 hours of assembling, and 590 hours of finishing each work week. How many tables, chairs, and armoires should be produced so that all available labor-hours are used? Or is this impossible?

	Table	Chair	Armoire
Cutting (h)	$\frac{1}{2}$	1	1
Assembling (h)	$\frac{1}{2}$	$1\frac{1}{2}$	1
Finishing (h)	1	$1\frac{1}{2}$	2

74. Traffic Flow A section of a city's street network is shown in the figure. The arrows indicate one-way streets, and the numbers show how many cars enter or leave this section of the city via the indicated street in a certain one-hour period. The variables x, y, z, and w represent the number of cars that travel along the portions of First, Second, Avocado, and Birch Streets during this period. Find x, y, z, and w, assuming that none of the cars stop or park on any of the streets shown.

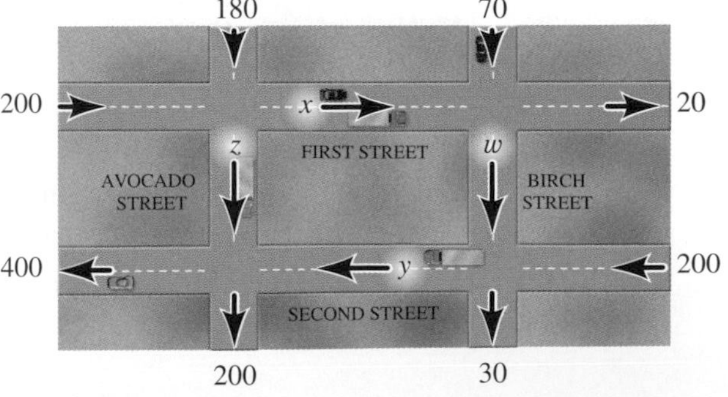

■ Discuss ■ Discover ■ Prove ■ Write

75. Discuss: Polynomials Determined by a Set of Points Two points uniquely determine a line $y = ax + b$ in the coordinate plane. Similarly, three points uniquely determine a quadratic (second-degree) polynomial

$$y = ax^2 + bx + c$$

four points uniquely determine a cubic (third-degree) polynomial

$$y = ax^3 + bx^2 + cx + d$$

and so on. (Some exceptions to this rule occur if the three points actually lie on a line, or the four points lie on a quadratic or line, and so on.) For the following set of five points, find the line that contains the first two points, the quadratic that contains the first three points, the cubic that contains the first four points, and the fourth-degree polynomial that contains all five points.

$$(0, 0), \quad (1, 12), \quad (2, 40), \quad (3, 6), \quad (-1, -14)$$

Graph the points and functions in the same viewing rectangle using a graphing device.

9.4 The Algebra of Matrices

■ Equality of Matrices ■ Addition, Subtraction, and Scalar Multiplication of Matrices ■ Multiplication of Matrices ■ Properties of Matrix Multiplication ■ Applications of Matrix Multiplication

Thus far, we have used matrices simply for notational convenience when solving linear systems. Matrices have many other uses in mathematics and the sciences, and for most of these applications a knowledge of matrix algebra is essential. Like numbers, matrices can be added, subtracted, multiplied, and divided. In this section we learn how to perform these algebraic operations on matrices.

■ Equality of Matrices

Two matrices are equal if they have the same entries in the same positions.

Equal matrices

$$\begin{bmatrix} \sqrt{4} & 2^2 & e^0 \\ 0.5 & 1 & 1-1 \end{bmatrix} = \begin{bmatrix} 2 & 4 & 1 \\ \frac{1}{2} & \frac{2}{2} & 0 \end{bmatrix}$$

Unequal matrices

$$\begin{bmatrix} 1 & 2 \\ 3 & 4 \\ 5 & 6 \end{bmatrix} \neq \begin{bmatrix} 1 & 3 & 5 \\ 2 & 4 & 6 \end{bmatrix}$$

Equality of Matrices

The matrices $A = [a_{ij}]$ and $B = [b_{ij}]$ are **equal** if and only if they have the same dimension $m \times n$, and corresponding entries are equal, that is,

$$a_{ij} = b_{ij}$$

for $i = 1, 2, \ldots, m$ and $j = 1, 2, \ldots, n$.

Example 1 ■ Equal Matrices

Find a, b, c, and d if

$$\begin{bmatrix} a & b \\ c & d \end{bmatrix} = \begin{bmatrix} 1 & 3 \\ 5 & 2 \end{bmatrix}$$

Solution Since the two matrices are equal, corresponding entries must be the same. So we must have $a = 1$, $b = 3$, $c = 5$, and $d = 2$.

Now Try Exercises 5 and 7 ■

Courtesy UC Berkeley Office of Media Relations

JULIA ROBINSON (1919–1985) was born in St. Louis, Missouri, and grew up in Point Loma, California. Because of an illness, Robinson missed two years of school, but later, with the aid of a tutor, she completed fifth, sixth, seventh, and eighth grades, all in one year. Later, at San Diego State University, reading biographies of mathematicians in E. T. Bell's *Men of Mathematics* awakened in her what became a lifelong passion for mathematics. She said, "I cannot overemphasize the importance of such books . . . in the intellectual life of a student." Robinson is famous for her work on Hilbert's tenth problem (see Section 9.6), which asks for a general procedure for determining whether an equation has integer solutions. Her ideas led to a complete answer to the problem: the answer involved certain properties of the Fibonacci numbers (see Section 11.1) discovered by a 22-year-old Russian mathematician named Yuri Matijasevič. As a result of her brilliant work on Hilbert's tenth problem, Robinson was offered a professorship at the University of California, Berkeley, and became the first woman mathematician elected to the National Academy of Sciences. She also served as president of the American Mathematical Society.

■ Addition, Subtraction, and Scalar Multiplication of Matrices

Two matrices can be added or subtracted if they have the same dimension. (Otherwise, their sum or difference is undefined.) We add or subtract the matrices by adding or subtracting corresponding entries. To multiply a matrix by a number, we multiply every element of the matrix by that number. This is called the *scalar product*.

Sum, Difference, and Scalar Product of Matrices

Let $A = [a_{ij}]$ and $B = [b_{ij}]$ be matrices of the same dimension $m \times n$, and let c be any real number.

1. The **sum** $A + B$ is the $m \times n$ matrix obtained by adding corresponding entries of A and B.
$$A + B = [a_{ij} + b_{ij}]$$
2. The **difference** $A - B$ is the $m \times n$ matrix obtained by subtracting corresponding entries of A and B.
$$A - B = [a_{ij} - b_{ij}]$$
3. The **scalar product** cA is the $m \times n$ matrix obtained by multiplying each entry of A by c.
$$cA = [ca_{ij}]$$

Example 2 ■ Performing Algebraic Operations on Matrices

Let
$$A = \begin{bmatrix} 2 & -3 \\ 0 & 5 \\ 7 & -\frac{1}{2} \end{bmatrix} \qquad B = \begin{bmatrix} 1 & 0 \\ -3 & 1 \\ 2 & 2 \end{bmatrix}$$
$$C = \begin{bmatrix} 7 & -3 & 0 \\ 0 & 1 & 5 \end{bmatrix} \qquad D = \begin{bmatrix} 6 & 0 & -6 \\ 8 & 1 & 9 \end{bmatrix}$$

Carry out each indicated operation, or explain why it cannot be performed.

(a) $A + B$ **(b)** $C - D$ **(c)** $C + A$ **(d)** $5A$

Solution

(a) $$A + B = \begin{bmatrix} 2 & -3 \\ 0 & 5 \\ 7 & -\frac{1}{2} \end{bmatrix} + \begin{bmatrix} 1 & 0 \\ -3 & 1 \\ 2 & 2 \end{bmatrix} = \begin{bmatrix} 3 & -3 \\ -3 & 6 \\ 9 & \frac{3}{2} \end{bmatrix}$$

(b) $$C - D = \begin{bmatrix} 7 & -3 & 0 \\ 0 & 1 & 5 \end{bmatrix} - \begin{bmatrix} 6 & 0 & -6 \\ 8 & 1 & 9 \end{bmatrix} = \begin{bmatrix} 1 & -3 & 6 \\ -8 & 0 & -4 \end{bmatrix}$$

(c) $C + A$ is undefined because we can't add matrices of different dimensions.

(d) $$5A = 5\begin{bmatrix} 2 & -3 \\ 0 & 5 \\ 7 & -\frac{1}{2} \end{bmatrix} = \begin{bmatrix} 10 & -15 \\ 0 & 25 \\ 35 & -\frac{5}{2} \end{bmatrix}$$

 **Now Try Exercises 23 and 25** ■

The properties in the box follow from the definitions of matrix addition and scalar multiplication, together with the corresponding properties of real numbers.

Properties of Addition and Scalar Multiplication of Matrices

Let A, B, and C be $m \times n$ matrices and let c and d be scalars.

$A + B = B + A$	Commutative Property of Matrix Addition
$(A + B) + C = A + (B + C)$	Associative Property of Matrix Addition
$c(dA) = cdA$	Associative Property of Scalar Multiplication
$(c + d)A = cA + dA$ $c(A + B) = cA + cB$	Distributive Properties of Scalar Multiplication

Example 3 ■ Solving a Matrix Equation

Solve the matrix equation

$$2X - A = B$$

for the unknown matrix X, where

$$A = \begin{bmatrix} 2 & 3 \\ -5 & 1 \end{bmatrix} \qquad B = \begin{bmatrix} 4 & -1 \\ 1 & 3 \end{bmatrix}$$

Solution We use the properties of matrices to solve for X.

$2X - A = B$	Given equation
$2X = B + A$	Add the matrix A to each side
$X = \frac{1}{2}(B + A)$	Multiply each side by the scalar $\frac{1}{2}$
So $X = \frac{1}{2}\left(\begin{bmatrix} 4 & -1 \\ 1 & 3 \end{bmatrix} + \begin{bmatrix} 2 & 3 \\ -5 & 1 \end{bmatrix}\right)$	Substitute the matrices A and B
$= \frac{1}{2}\begin{bmatrix} 6 & 2 \\ -4 & 4 \end{bmatrix}$	Add matrices
$= \begin{bmatrix} 3 & 1 \\ -2 & 2 \end{bmatrix}$	Multiply by the scalar $\frac{1}{2}$

 **Now Try Exercise 17** ■

■ Multiplication of Matrices

Multiplication of two matrices is more difficult to describe than other matrix operations. In later examples we will see why multiplying matrices involves a rather complex procedure, which we now describe.

First, the product AB (or $A \cdot B$) of two matrices A and B is defined only when the number of columns in A is equal to the number of rows in B. This means that if we write their dimensions side by side, the two inner numbers must match:

Matrices	A	B
Dimensions	$m \times n$	$n \times k$

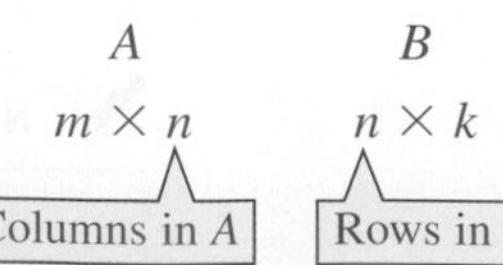

If we think of the row of A and the column of B as vectors, then their inner product is the same as their dot product (see Section 8.6).

If the dimensions of A and B match in this fashion, then the product AB is a matrix of dimension $m \times k$. Before describing how to obtain AB, we first define the **inner product** of a row of A and a column of B to be the number obtained by multiplying corresponding entries and adding the results as follows:

Row of A	**Column of B**	**Inner Product**
$\begin{bmatrix} a_1 & a_2 \cdots a_n \end{bmatrix}$	$\begin{bmatrix} b_1 \\ b_2 \\ \vdots \\ b_n \end{bmatrix}$	$a_1b_1 + a_2b_2 + \cdots + a_nb_n$

For instance, consider the matrices D and B in Example 2. The inner product of the second row in D and the first column in B is

$$8 \cdot 1 + 1 \cdot (-3) + 9 \cdot 2 = 23$$

We now define the **product** AB of two matrices.

Matrix Multiplication

If $A = [a_{ij}]$ is an $m \times n$ matrix and $B = [b_{ij}]$ an $n \times k$ matrix, then their product is the $m \times k$ matrix

$$C = [c_{ij}]$$

where c_{ij} is the inner product of the ith row of A and the jth column of B. We write the product as

$$C = AB$$

This definition of matrix product says that each entry in the matrix AB is obtained from a *row* of A and a *column* of B as follows: The entry c_{ij} in the ith row and jth column of the matrix AB is obtained by multiplying the entries in the ith row of A with the corresponding entries in the jth column of B and adding the results.

jth column of B

Entry in ith row and jth column of AB

ith row of A

$$\begin{bmatrix} \blacksquare & \blacksquare & \blacksquare \end{bmatrix} \cdot \begin{bmatrix} \blacksquare \\ \blacksquare \\ \blacksquare \end{bmatrix} = \begin{bmatrix} c_{ij} \end{bmatrix}$$

Example 4 ■ Multiplying Matrices

Let

$$A = \begin{bmatrix} 1 & 3 \\ -1 & 0 \end{bmatrix} \qquad \text{and} \qquad B = \begin{bmatrix} -1 & 5 & 2 \\ 0 & 4 & 7 \end{bmatrix}$$

Calculate, if possible, the products AB and BA.

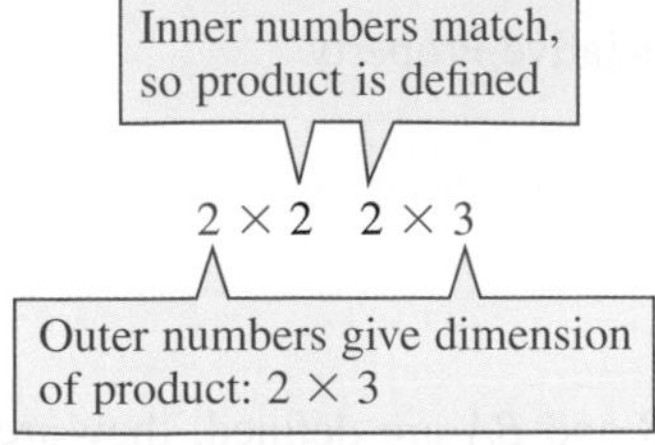

Solution Since A has dimension 2×2 and B has dimension 2×3, the product AB is defined and has dimension 2×3. We can therefore write

$$AB = \begin{bmatrix} 1 & 3 \\ -1 & 0 \end{bmatrix} \begin{bmatrix} -1 & 5 & 2 \\ 0 & 4 & 7 \end{bmatrix} = \begin{bmatrix} ? & ? & ? \\ ? & ? & ? \end{bmatrix}$$

where the question marks must be filled in using the rule defining the product of two matrices. If we define $C = AB = [c_{ij}]$, then the entry c_{11} is the inner product of the first row of A and the first column of B:

$$\begin{bmatrix} 1 & 3 \\ -1 & 0 \end{bmatrix} \begin{bmatrix} -1 & 5 & 2 \\ 0 & 4 & 7 \end{bmatrix} \qquad 1 \cdot (-1) + 3 \cdot 0 = -1$$

Similarly, we calculate the remaining entries of the product as follows.

Entry	**Inner product of:**	**Value**	**Product matrix**
c_{12}	$\begin{bmatrix} 1 & 3 \\ -1 & 0 \end{bmatrix} \begin{bmatrix} -1 & 5 & 2 \\ 0 & 4 & 7 \end{bmatrix}$	$1 \cdot 5 + 3 \cdot 4 = 17$	$\begin{bmatrix} -1 & 17 & \\ & & \end{bmatrix}$
c_{13}	$\begin{bmatrix} 1 & 3 \\ -1 & 0 \end{bmatrix} \begin{bmatrix} -1 & 5 & 2 \\ 0 & 4 & 7 \end{bmatrix}$	$1 \cdot 2 + 3 \cdot 7 = 23$	$\begin{bmatrix} -1 & 17 & 23 \\ & & \end{bmatrix}$
c_{21}	$\begin{bmatrix} 1 & 3 \\ -1 & 0 \end{bmatrix} \begin{bmatrix} -1 & 5 & 2 \\ 0 & 4 & 7 \end{bmatrix}$	$(-1) \cdot (-1) + 0 \cdot 0 = 1$	$\begin{bmatrix} -1 & 17 & 23 \\ 1 & & \end{bmatrix}$
c_{22}	$\begin{bmatrix} 1 & 3 \\ -1 & 0 \end{bmatrix} \begin{bmatrix} -1 & 5 & 2 \\ 0 & 4 & 7 \end{bmatrix}$	$(-1) \cdot 5 + 0 \cdot 4 = -5$	$\begin{bmatrix} -1 & 17 & 23 \\ 1 & -5 & \end{bmatrix}$
c_{23}	$\begin{bmatrix} 1 & 3 \\ -1 & 0 \end{bmatrix} \begin{bmatrix} -1 & 5 & 2 \\ 0 & 4 & 7 \end{bmatrix}$	$(-1) \cdot 2 + 0 \cdot 7 = -2$	$\begin{bmatrix} -1 & 17 & 23 \\ 1 & -5 & -2 \end{bmatrix}$

Not equal, so product is not defined

$2 \times 3 \quad 2 \times 2$

Thus we have
$$AB = \begin{bmatrix} -1 & 17 & 23 \\ 1 & -5 & -2 \end{bmatrix}$$

The product BA is not defined, however, because the dimensions of B and A are

$$2 \times 3 \qquad \text{and} \qquad 2 \times 2$$

The inner two numbers are not the same, so the rows and columns won't match up if we try to calculate the product.

Now Try Exercise 27 ■

Graphing devices are capable of performing matrix algebra. For instance, if we enter the matrices in Example 4 into the matrix variables [A] and [B] on a graphing device, then the device finds their product, as shown in Figure 1.

```
Matrix Operations
[A]          [B]
  [ 1  3]      [-1 5 2]
  [-1  0]      [0  4 7]
[A]*[B]
     [-1  17 23]
     [1   -5 -2]
```

Figure 1

■ Properties of Matrix Multiplication

Although matrix multiplication is not commutative, it does obey the Associative and Distributive properties.

Properties of Matrix Multiplication

Let A, B, and C be matrices for which the following products are defined. Then

$A(BC) = (AB)C$	Associative Property
$A(B + C) = AB + AC$ $(B + C)A = BA + CA$	Distributive Property

The next example shows that even when both AB and BA are defined, they aren't necessarily equal. This proves that matrix multiplication is *not* commutative.

Example 5 ■ Matrix Multiplication Is Not Commutative

Let $A = \begin{bmatrix} 5 & 7 \\ -3 & 0 \end{bmatrix}$ and $B = \begin{bmatrix} 1 & 2 \\ 9 & -1 \end{bmatrix}$

Calculate the products AB and BA.

Solution Since both matrices A and B have dimension 2×2, both products AB and BA are defined, and each product is also a 2×2 matrix.

$$AB = \begin{bmatrix} 5 & 7 \\ -3 & 0 \end{bmatrix}\begin{bmatrix} 1 & 2 \\ 9 & -1 \end{bmatrix} = \begin{bmatrix} 5\cdot 1 + 7\cdot 9 & 5\cdot 2 + 7\cdot(-1) \\ (-3)\cdot 1 + 0\cdot 9 & (-3)\cdot 2 + 0\cdot(-1) \end{bmatrix}$$

$$= \begin{bmatrix} 68 & 3 \\ -3 & -6 \end{bmatrix}$$

$$BA = \begin{bmatrix} 1 & 2 \\ 9 & -1 \end{bmatrix}\begin{bmatrix} 5 & 7 \\ -3 & 0 \end{bmatrix} = \begin{bmatrix} 1\cdot 5 + 2\cdot(-3) & 1\cdot 7 + 2\cdot 0 \\ 9\cdot 5 + (-1)\cdot(-3) & 9\cdot 7 + (-1)\cdot 0 \end{bmatrix}$$

$$= \begin{bmatrix} -1 & 7 \\ 48 & 63 \end{bmatrix}$$

This shows that, in general, $AB \neq BA$. In fact, in this example AB and BA don't even have an entry in common.

 **Now Try Exercise 29** ■

■ Applications of Matrix Multiplication

We now consider some applied examples that give some indication of why mathematicians have chosen to define the matrix product in such an apparently bizarre fashion. Example 6 shows how our definition of matrix product allows us to express a system of linear equations as a single matrix equation.

Example 6 ■ Writing a Linear System as a Matrix Equation

Show that the following matrix equation is equivalent to the system of equations in Example 9.3.2.

Matrix equations like this one are described in more detail in Section 9.5.

$$\begin{bmatrix} 1 & -1 & 3 \\ 1 & 2 & -2 \\ 3 & -1 & 5 \end{bmatrix}\begin{bmatrix} x \\ y \\ z \end{bmatrix} = \begin{bmatrix} 4 \\ 10 \\ 14 \end{bmatrix}$$

Solution If we perform matrix multiplication on the left-hand side of the equation, we get

$$\begin{bmatrix} x - y + 3z \\ x + 2y - 2z \\ 3x - y + 5z \end{bmatrix} = \begin{bmatrix} 4 \\ 10 \\ 14 \end{bmatrix}$$

Because two matrices are equal only if their corresponding entries are equal, we equate entries to get

$$\begin{cases} x - y + 3z = 4 \\ x + 2y - 2z = 10 \\ 3x - y + 5z = 14 \end{cases}$$

This is exactly the system of equations in Example 9.3.2.

Now Try Exercise 47 ■

Courtesy of the Archives, California Institute of Technology

OLGA TAUSSKY-TODD (1906–1995) was instrumental in developing applications of matrix theory. Described as "in love with anything matrices can do," she successfully applied matrices to aerodynamics, a field used in the design of airplanes and rockets. Taussky-Todd was also famous for her work in number theory, which deals with prime numbers and divisibility. Although number theory has often been called the least applicable branch of mathematics, it is now used in significant ways throughout the computer industry.

Taussky-Todd studied mathematics at a time when young women rarely aspired to be mathematicians. She said, "When I entered university I had no idea what it meant to study mathematics." One of the most respected mathematicians of her day, she was for many years a professor of mathematics at Caltech in Pasadena.

Example 7 ■ Representing Demographic Data by Matrices

In a certain city the proportions of voters in each age group who are registered as Democrats, Republicans, or Independents are given by the following matrix.

	Age Group		
	18–30	**31–50**	**Over 50**
Democrat	0.30	0.60	0.50
Republican	0.50	0.35	0.25
Independent	0.20	0.05	0.25

$= A$

The next matrix gives the distribution, by age group and sex, of the voting population of this city.

Age Group		**Male**	**Female**
	18–30	5,000	6,000
	31–50	10,000	12,000
	Over 50	12,000	15,000

$= B$

For this problem, let's make the (highly unrealistic) assumption that within each age group, political preference is not related to gender; that is, the percentage of Democratic males in the 18–30 group, for example, is the same as the percentage of Democratic females in this group.

(a) Calculate the product AB.

(b) How many males are registered as Democrats in this city?

(c) How many females are registered as Republicans?

Solution

(a) $$AB = \begin{bmatrix} 0.30 & 0.60 & 0.50 \\ 0.50 & 0.35 & 0.25 \\ 0.20 & 0.05 & 0.25 \end{bmatrix} \begin{bmatrix} 5{,}000 & 6{,}000 \\ 10{,}000 & 12{,}000 \\ 12{,}000 & 15{,}000 \end{bmatrix} = \begin{bmatrix} 13{,}500 & 16{,}500 \\ 9{,}000 & 10{,}950 \\ 4{,}500 & 5{,}550 \end{bmatrix}$$

(b) When we take the inner product of a row in A with a column in B, we are adding the number of people in each age group who belong to the category in question. For example, the entry c_{21} of AB (the 9000) is obtained by taking the inner product of the Republican row in A with the Male column in B. This number is therefore

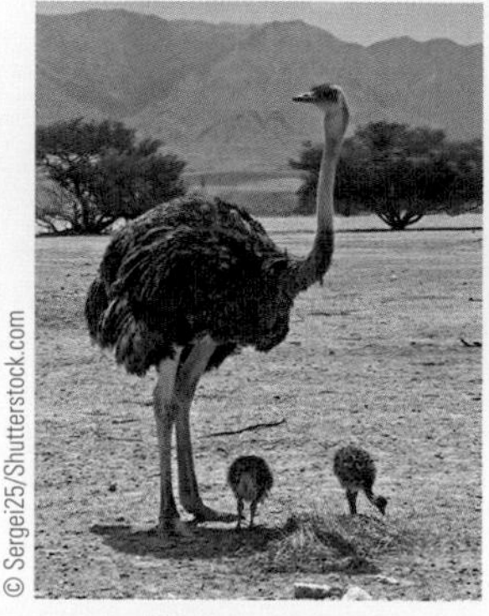

© Sergei25/Shutterstock.com

Discovery Project ■ Will the Species Survive?

To study how a species survives, scientists observe the stages in the life cycle of the species—for example, young, juvenile, adult. The proportion of the population at each stage and the proportion that survives to the next stage in each season are modeled by matrices. In this project we explore how matrix multiplication is used to predict the population proportions for the next season, the season after that, and so on, ultimately predicting the long-term prospects for the survival of the species. You can find the project at the book companion website: **www.stewartmath.com**.

the total number of male Republicans in this city. We can label the rows and columns of AB as follows.

$$\begin{array}{l} \\ \textbf{Democrat} \\ \textbf{Republican} \\ \textbf{Independent} \end{array} \begin{array}{c} \begin{array}{cc} \textbf{Male} & \textbf{Female} \end{array} \\ \begin{bmatrix} 13{,}500 & 16{,}500 \\ 9{,}000 & 10{,}950 \\ 4{,}500 & 5{,}550 \end{bmatrix} \end{array} = AB$$

Thus 13,500 males are registered as Democrats in this city.

(c) There are 10,950 females registered as Republicans.

 **Now Try Exercise 53**

In Example 7 the entries in each column of A add up to 1. (Can you see why this has to be true, given what the matrix describes?) A matrix with this property is called **stochastic**. Stochastic matrices are used extensively in statistics, where they arise frequently in situations like the one described here.

9.4 | Exercises

Concepts

1. We can add (or subtract) two matrices only if they have the same ________.

2. (a) We can multiply two matrices only if the number of ________ in the first matrix is the same as the number of ________ in the second matrix.

(b) If A is a 3×3 matrix and B is a 4×3 matrix, which of the following matrix multiplications are possible?

(i) AB (ii) BA (iii) AA (iv) BB

3. Which of the following operations can we perform for a matrix A of any dimension?

(i) $A + A$ (ii) $2A$ (iii) $A \cdot A$

4. Fill in the missing entries in the product matrix.

$$\begin{bmatrix} 3 & 1 & 2 \\ -1 & 2 & 0 \\ 1 & 3 & -2 \end{bmatrix} \begin{bmatrix} -1 & 3 & -2 \\ 3 & -2 & -1 \\ 2 & 1 & 0 \end{bmatrix} = \begin{bmatrix} 4 & \blacksquare & -7 \\ 7 & -7 & \blacksquare \\ \blacksquare & -5 & -5 \end{bmatrix}$$

Skills

5–6 ■ Equality of Matrices Determine whether the matrices A and B are equal.

 **5.** $A = \begin{bmatrix} 1 & -2 & 0 \\ \frac{1}{2} & 6 & 0 \end{bmatrix} \quad B = \begin{bmatrix} 1 & -2 \\ \frac{1}{2} & 6 \end{bmatrix}$

6. $A = \begin{bmatrix} \frac{1}{4} & \ln 1 \\ 2 & 3 \end{bmatrix} \quad B = \begin{bmatrix} 0.25 & 0 \\ \sqrt{4} & \frac{6}{2} \end{bmatrix}$

7–8 ■ Equality of Matrices Find the values of a and b that make the matrices A and B equal.

 **7.** $A = \begin{bmatrix} 3 & 4 \\ -1 & a \end{bmatrix} \quad B = \begin{bmatrix} b & 4 \\ -1 & -5 \end{bmatrix}$

8. $A = \begin{bmatrix} 3 & 5 & 7 \\ -4 & a & 2 \end{bmatrix} \quad B = \begin{bmatrix} 3 & 5 & b \\ -4 & -5 & 2 \end{bmatrix}$

9–16 ■ Matrix Operations Perform the matrix operation, or if it is impossible, explain why.

9. $\begin{bmatrix} 2 & 6 \\ -5 & 3 \end{bmatrix} + \begin{bmatrix} -1 & -3 \\ 6 & 2 \end{bmatrix}$

10. $\begin{bmatrix} 0 & 1 & 1 \\ 1 & 1 & 0 \end{bmatrix} - \begin{bmatrix} 2 & 1 & -1 \\ 1 & 3 & -2 \end{bmatrix}$

11. $3\begin{bmatrix} 1 & 2 \\ 4 & -1 \\ 1 & 0 \end{bmatrix}$

12. $2\begin{bmatrix} 1 & 1 & 0 \\ 1 & 0 & 1 \\ 0 & 1 & 1 \end{bmatrix} + \begin{bmatrix} 1 & 1 \\ 2 & 1 \\ 3 & 1 \end{bmatrix}$

13. $\begin{bmatrix} 2 & 6 \\ 1 & 3 \\ 2 & 4 \end{bmatrix} \begin{bmatrix} 1 & -2 \\ 3 & 6 \\ -2 & 0 \end{bmatrix}$

14. $\begin{bmatrix} 2 & 1 & 2 \\ 6 & 3 & 4 \end{bmatrix} \begin{bmatrix} 1 & -2 \\ 3 & 6 \\ -2 & 0 \end{bmatrix}$

15. $\begin{bmatrix} 1 & 2 \\ -1 & 4 \end{bmatrix} \begin{bmatrix} 1 & -2 & 3 \\ 2 & 2 & -1 \end{bmatrix}$

16. $\begin{bmatrix} 2 & -3 \\ 0 & 1 \\ 1 & 2 \end{bmatrix} \begin{bmatrix} 5 \\ 1 \end{bmatrix}$

17–22 ■ Matrix Equations Solve the matrix equation for the unknown matrix X, or explain why no solution exists.

$$A = \begin{bmatrix} 4 & 6 \\ 1 & 3 \end{bmatrix} \qquad B = \begin{bmatrix} 2 & 5 \\ 3 & 7 \end{bmatrix}$$

$$C = \begin{bmatrix} 2 & 3 \\ 1 & 0 \\ 0 & 2 \end{bmatrix} \qquad D = \begin{bmatrix} 10 & 20 \\ 30 & 20 \\ 10 & 0 \end{bmatrix}$$

17. $2X + A = B$ **18.** $3X - B = C$

19. $2(B - X) = D$ **20.** $5(X - C) = D$

21. $\frac{1}{5}(X + D) = C$ **22.** $2A = B - 3X$

23–36 ■ Matrix Operations The matrices $A, B, C, D, E, F, G,$ and H are defined as follows.

$$A = \begin{bmatrix} 2 & -5 \\ 0 & 7 \end{bmatrix} \qquad B = \begin{bmatrix} 3 & \frac{1}{2} & 5 \\ 1 & -1 & 3 \end{bmatrix} \qquad C = \begin{bmatrix} 2 & -\frac{5}{2} & 0 \\ 0 & 2 & -3 \end{bmatrix}$$

$$D = \begin{bmatrix} 7 & 3 \end{bmatrix} \qquad E = \begin{bmatrix} 1 \\ 2 \\ 0 \end{bmatrix} \qquad F = \begin{bmatrix} 1 & 0 & 0 \\ 0 & 1 & 0 \\ 0 & 0 & 1 \end{bmatrix}$$

$$G = \begin{bmatrix} 5 & -3 & 10 \\ 6 & 1 & 0 \\ -5 & 2 & 2 \end{bmatrix} \qquad H = \begin{bmatrix} 3 & 1 \\ 2 & -1 \end{bmatrix}$$

Carry out the indicated algebraic operation, or explain why it cannot be performed.

23. (a) $B + C$ **(b)** $B + F$

24. (a) $C - B$ **(b)** $2C - 6B$

25. (a) $5A$ **(b)** $C - 5A$

26. (a) $3B + 2C$ **(b)** $2H + D$

27. (a) AD **(b)** DA

28. (a) DH **(b)** HD

29. (a) AH **(b)** HA

30. (a) BC **(b)** BF

31. (a) GF **(b)** GE

32. (a) B^2 **(b)** F^2

33. (a) A^2 **(b)** A^3

34. (a) $(DA)B$ **(b)** $D(AB)$

35. (a) ABE **(b)** AHE

36. (a) $DB + DC$ **(b)** $BF + FE$

37–42 ■ Matrix Operations The matrices A, B, and C are defined as follows.

$$A = \begin{bmatrix} 0.3 & 1.1 & 2.4 \\ 0.9 & -0.1 & 0.4 \\ -0.7 & 0.3 & -0.5 \end{bmatrix} \qquad B = \begin{bmatrix} 1.2 & -0.1 \\ 0 & -0.5 \\ 0.5 & -2.1 \end{bmatrix}$$

$$C = \begin{bmatrix} -0.2 & 0.2 & 0.1 \\ 1.1 & 2.1 & -2.1 \end{bmatrix}$$

Use a graphing device to carry out the indicated algebraic operation, or explain why it cannot be performed.

37. AB **38.** BA **39.** BC

40. CB **41.** $B + C$ **42.** A^2

43–46 ■ Equality of Matrices Solve for x and y.

43. $\begin{bmatrix} x & 2y \\ 4 & 6 \end{bmatrix} = \begin{bmatrix} 2 & -2 \\ 2x & -6y \end{bmatrix}$

44. $3\begin{bmatrix} x & y \\ y & x \end{bmatrix} = \begin{bmatrix} 6 & -9 \\ -9 & 6 \end{bmatrix}$

45. $2\begin{bmatrix} x & y \\ x + y & x - y \end{bmatrix} = \begin{bmatrix} 2 & -4 \\ -2 & 6 \end{bmatrix}$

46. $\begin{bmatrix} x & y \\ -y & x \end{bmatrix} - \begin{bmatrix} y & x \\ x & -y \end{bmatrix} = \begin{bmatrix} 4 & -4 \\ -6 & 6 \end{bmatrix}$

47–50 ■ Linear Systems as Matrix Equations Write the system of equations as a matrix equation (see Example 6).

47. $\begin{cases} 2x - 5y = 7 \\ 3x + 2y = 4 \end{cases}$

48. $\begin{cases} 6x - y + z = 12 \\ 2x + z = 7 \\ y - 2z = 4 \end{cases}$

49. $\begin{cases} 3x_1 + 2x_2 - x_3 + x_4 = 0 \\ x_1 - x_3 = 5 \\ 3x_2 + x_3 - x_4 = 4 \end{cases}$

50. $\begin{cases} x - y + z = 2 \\ 4x - 2y - z = 2 \\ x + y + 5z = 2 \\ -x - y - z = 2 \end{cases}$

Skills Plus

51. Products of Matrices The matrices A, B, and C are defined as follows.

$$A = \begin{bmatrix} 1 & 0 & 6 & -1 \\ 2 & \frac{1}{2} & 4 & 0 \end{bmatrix}$$

$$B = \begin{bmatrix} 1 & 7 & -9 & 2 \end{bmatrix} \qquad C = \begin{bmatrix} 1 \\ 0 \\ -1 \\ -2 \end{bmatrix}$$

Determine which of the following products are defined, and calculate the ones that are.

ABC ACB BAC

BCA CAB CBA

52. Expanding Matrix Binomials

(a) Prove that if A and B are 2×2 matrices, then

$$(A + B)^2 = A^2 + AB + BA + B^2$$

(b) If A and B are 2×2 matrices, is it necessarily true that

$$(A + B)^2 \stackrel{?}{=} A^2 + 2AB + B^2$$

Applications

53. Education and Income A civic club takes a survey to determine the education and income of its members. Matrix A summarizes the proportions of members in various categories of income levels and years of postsecondary education. Matrix B shows the total number of members in each income category.

(a) Calculate the product matrix AB.

(b) Interpret the entries of the matrix AB.

Income Level

	Less than \$50,000	\$50,000 to \$100,000	\$100,000 or more
None	0.75	0.10	0
1 to 4	0.25	0.70	0.70
More than 4	0	0.20	0.30

$= A$

	Total
Less than \$50,000	4
\$50,000 to \$100,000	20
\$100,000 or more	10

$= B$

54. Exam Scores A large physics class takes a survey of the number of hours the students slept before an exam and their exam scores. Matrix A summarizes the proportions of students in different categories of exam scores and hours of sleep. Matrix B shows the total number of students in three categories of exam scores.

(a) Calculate the product matrix AB.

(b) Interpret the entries of the matrix AB.

Exam Score

	Below 60	60 to 80	Above 80
Less than 4	0.75	0.20	0.05
4 to 7	0.60	0.30	0.10
More than 7	0.40	0.30	0.30

$= A$

	Total
Below 60	80
60 to 80	170
Above 80	40

$= B$

55. Frozen-Food Revenue Some of the frozen foods that Joe's Specialty Foods sells are pesto pizza, spinach ravioli, and macaroni and cheese. The sales distribution for these products is tabulated in matrix A. The retail price (in dollars) for each item is tabulated in matrix B.

(a) Calculate the product matrix AB.

(b) What is the total revenue for Monday?

(c) What is the total revenue from all three days?

Specialty Food

	Pizza	Ravioli	Mac & Cheese
Monday	50	20	15
Tuesday	40	75	20
Wednesday	35	60	100

$= A$

	Price (\$)
Pizza	3.50
Ravioli	5.75
Mac & Cheese	4.25

$= B$

56. Fast-Food Sales A small fast-food chain with restaurants in Santa Monica, Long Beach, and Anaheim sells only hamburgers, hot dogs, and milkshakes. On a certain day, sales were distributed according to the following matrix.

Number of Items Sold

	Santa Monica	Long Beach	Anaheim
Hamburgers	4000	1000	3500
Hot dogs	400	300	200
Milkshakes	700	500	9000

$= A$

The price of each item is given by the following matrix.

Hamburger	Hot Dog	Milkshake
\$0.90	\$0.80	\$1.10

$= B$

(a) Calculate the product BA.

(b) Interpret the entries in the product matrix BA.

57. Car-Manufacturing Profits A specialty-car manufacturer has plants in Auburn, Biloxi, and Chattanooga. Three models are produced, with daily production given in the following matrix.

Cars Produced Each Day

	Model K	Model R	Model W
Auburn	12	10	0
Biloxi	4	4	20
Chattanooga	8	9	12

$= A$

Because of a wage increase, February profits are lower than January profits. The profit per car is tabulated by model in the following matrix.

	January	February
Model K	\$1000	\$500
Model R	\$2000	\$1200
Model W	\$1500	\$1000

$= B$

(a) Calculate the product AB.

(b) Assuming that all cars produced were sold, what was the daily profit in January from the Biloxi plant?

(c) What was the total daily profit (from all three plants) in February?

58. Canning Tomato Products Jaeger Foods produces tomato sauce and tomato paste, canned in small, medium, large, and giant-sized cans. The matrix A gives the size (in kilograms) of each container.

	Small	Medium	Large	Giant
Kilograms	6	10	14	28

$= A$

The matrix B tabulates one day's production of tomato sauce and tomato paste.

	Cans of Sauce	Cans of Paste
Small	2000	2500
Medium	3000	1500
Large	2500	1000
Giant	1000	500

$= B$

(a) Calculate the product AB.

(b) Interpret the entries in the product matrix AB.

59. Produce Sales A farmer's three children, Ashton, Bryn, and Cimeron, run three roadside produce stands during the summer months. One weekend they all sell watermelons, yellow squash, and tomatoes. The matrices A and B tabulate the number of kilograms of each product sold by each sibling on Saturday and Sunday.

Saturday

$$\begin{array}{r} \\ \textbf{Ashton} \\ \textbf{Bryn} \\ \textbf{Cimeron} \end{array}\begin{array}{c} \textbf{Melons} \quad \textbf{Squash} \quad \textbf{Tomatoes} \\ \begin{bmatrix} 120 & 50 & 60 \\ 40 & 25 & 30 \\ 60 & 30 & 20 \end{bmatrix} = A \end{array}$$

Sunday

$$\begin{array}{r} \\ \textbf{Ashton} \\ \textbf{Bryn} \\ \textbf{Cimeron} \end{array}\begin{array}{c} \textbf{Melons} \quad \textbf{Squash} \quad \textbf{Tomatoes} \\ \begin{bmatrix} 100 & 60 & 30 \\ 35 & 20 & 20 \\ 60 & 25 & 30 \end{bmatrix} = B \end{array}$$

The matrix C gives the price per kilogram (in dollars) for each type of produce that they sell.

Price per kilogram

$$\begin{array}{r} \textbf{Melons} \\ \textbf{Squash} \\ \textbf{Tomatoes} \end{array}\begin{bmatrix} 0.10 \\ 0.50 \\ 1.00 \end{bmatrix} = C$$

Perform each of the following matrix operations, and interpret the entries in each result.

(a) AC **(b)** BC **(c)** $A + B$ **(d)** $(A + B)C$

■ Discuss ■ Discover ■ Prove ■ Write

60. Discuss: When Are Both Products Defined? What must be true about the dimensions of the matrices A and B if both products AB and BA are defined?

61. Discover: Powers of a Matrix For the given matrix A, find a formula for A^n, the product of the matrix A with itself n times.

(a) $A = \begin{bmatrix} 1 & 1 \\ 0 & 1 \end{bmatrix}$ **(b)** $A = \begin{bmatrix} 1 & 1 \\ 1 & 1 \end{bmatrix}$

PS *Try to recognize a pattern.* Calculate $A^2, A^3, A^4, \dots$ until you recognize a pattern.

62. Discuss: Square Roots of Matrices A **square root** of a matrix B is a matrix A with the property that $A^2 = B$. (This is the same definition as for a square root of a number.) Find as many square roots as you can of each matrix:

$$\begin{bmatrix} 4 & 0 \\ 0 & 9 \end{bmatrix} \qquad \begin{bmatrix} 1 & 5 \\ 0 & 9 \end{bmatrix}$$

[*Hint:* If $A = \begin{bmatrix} a & b \\ c & d \end{bmatrix}$, write the equations that a, b, c, and d would have to satisfy if A is the square root of the given matrix.]

9.5 Inverses of Matrices and Matrix Equations

■ The Inverse of a Matrix ■ Finding the Inverse of a 2 × 2 Matrix ■ Finding the Inverse of an $n \times n$ Matrix ■ Matrix Equations ■ Modeling with Matrix Equations

In Section 9.4 we saw that when the dimensions are appropriate, matrices can be added, subtracted, and multiplied. In this section we investigate division of matrices. With this operation we can solve equations that involve matrices.

■ The Inverse of a Matrix

First, we define *identity matrices*, which play the same role for matrix multiplication as the number 1 does for ordinary multiplication of numbers; that is, $1 \cdot a = a \cdot 1 = a$ for all numbers a. A **square matrix** is one that has the same number of rows as columns. The **main diagonal** of a square matrix consists of the entries whose row and column numbers are the same. These entries stretch diagonally down the matrix, from top left to bottom right.

Identity Matrix

The **identity matrix** I_n is the $n \times n$ matrix for which each main diagonal entry is a 1 and for which all other entries are 0.

Thus the 2×2, 3×3, and 4×4 identity matrices are

$$I_2 = \begin{bmatrix} 1 & 0 \\ 0 & 1 \end{bmatrix} \qquad I_3 = \begin{bmatrix} 1 & 0 & 0 \\ 0 & 1 & 0 \\ 0 & 0 & 1 \end{bmatrix} \qquad I_4 = \begin{bmatrix} 1 & 0 & 0 & 0 \\ 0 & 1 & 0 & 0 \\ 0 & 0 & 1 & 0 \\ 0 & 0 & 0 & 1 \end{bmatrix}$$

Identity matrices behave like the number 1 in the sense that

$$A \cdot I_n = A \qquad \text{and} \qquad I_n \cdot B = B$$

whenever these products are defined.

Example 1 ■ Identity Matrices

The following matrix products show how multiplying a matrix by an identity matrix of the appropriate dimension leaves the matrix unchanged.

$$\begin{bmatrix} 1 & 0 \\ 0 & 1 \end{bmatrix} \begin{bmatrix} 3 & 5 & 6 \\ -1 & 2 & 7 \end{bmatrix} = \begin{bmatrix} 3 & 5 & 6 \\ -1 & 2 & 7 \end{bmatrix}$$

$$\begin{bmatrix} -1 & 7 & \frac{1}{2} \\ 12 & 1 & 3 \\ -2 & 0 & 7 \end{bmatrix} \begin{bmatrix} 1 & 0 & 0 \\ 0 & 1 & 0 \\ 0 & 0 & 1 \end{bmatrix} = \begin{bmatrix} -1 & 7 & \frac{1}{2} \\ 12 & 1 & 3 \\ -2 & 0 & 7 \end{bmatrix}$$

 Now Try Exercise 1(a), (b) ■

If A and B are $n \times n$ matrices, and if $AB = BA = I_n$, then we say that B is the *inverse* of A, and we write $B = A^{-1}$. The concept of the inverse of a matrix is analogous to that of the reciprocal of a real number.

Inverse of a Matrix

Let A be a square $n \times n$ matrix. If there exists an $n \times n$ matrix A^{-1} with the property that

$$AA^{-1} = A^{-1}A = I_n$$

then we say that A^{-1} is the **inverse** of A. If A has an inverse, then we say that A is **invertible**.

Example 2 ■ Verifying That a Matrix Is an Inverse

Verify that B is the inverse of A, where

$$A = \begin{bmatrix} 2 & 1 \\ 5 & 3 \end{bmatrix} \qquad \text{and} \qquad B = \begin{bmatrix} 3 & -1 \\ -5 & 2 \end{bmatrix}$$

Solution We perform the matrix multiplications to show that $AB = I$ and $BA = I$.

$$\begin{bmatrix} 2 & 1 \\ 5 & 3 \end{bmatrix} \begin{bmatrix} 3 & -1 \\ -5 & 2 \end{bmatrix} = \begin{bmatrix} 2 \cdot 3 + 1(-5) & 2(-1) + 1 \cdot 2 \\ 5 \cdot 3 + 3(-5) & 5(-1) + 3 \cdot 2 \end{bmatrix} = \begin{bmatrix} 1 & 0 \\ 0 & 1 \end{bmatrix}$$

$$\begin{bmatrix} 3 & -1 \\ -5 & 2 \end{bmatrix} \begin{bmatrix} 2 & 1 \\ 5 & 3 \end{bmatrix} = \begin{bmatrix} 3 \cdot 2 + (-1)5 & 3 \cdot 1 + (-1)3 \\ (-5)2 + 2 \cdot 5 & (-5)1 + 2 \cdot 3 \end{bmatrix} = \begin{bmatrix} 1 & 0 \\ 0 & 1 \end{bmatrix}$$

 Now Try Exercise 3 ■

■ Finding the Inverse of a 2 × 2 Matrix

The following rule provides a simple way to find the inverse of a 2×2 matrix, when it exists. For larger matrices there is a more general procedure for finding inverses, which we consider later in this section.

Inverse of a 2 × 2 Matrix

If $A = \begin{bmatrix} a & b \\ c & d \end{bmatrix}$, then

$$A^{-1} = \frac{1}{ad - bc}\begin{bmatrix} d & -b \\ -c & a \end{bmatrix}$$

If $ad - bc = 0$, then A has no inverse.

Example 3 ■ Finding the Inverse of a 2 × 2 Matrix

Let

$$A = \begin{bmatrix} 4 & 5 \\ 2 & 3 \end{bmatrix}$$

Find A^{-1}, and verify that $AA^{-1} = A^{-1}A = I_2$.

Solution Using the rule for the inverse of a 2 × 2 matrix, we get

$$A^{-1} = \frac{1}{4 \cdot 3 - 5 \cdot 2}\begin{bmatrix} 3 & -5 \\ -2 & 4 \end{bmatrix} = \frac{1}{2}\begin{bmatrix} 3 & -5 \\ -2 & 4 \end{bmatrix} = \begin{bmatrix} \frac{3}{2} & -\frac{5}{2} \\ -1 & 2 \end{bmatrix}$$

To verify that this is indeed the inverse of A, we calculate AA^{-1} and $A^{-1}A$:

$$AA^{-1} = \begin{bmatrix} 4 & 5 \\ 2 & 3 \end{bmatrix}\begin{bmatrix} \frac{3}{2} & -\frac{5}{2} \\ -1 & 2 \end{bmatrix} = \begin{bmatrix} 4 \cdot \frac{3}{2} + 5(-1) & 4(-\frac{5}{2}) + 5 \cdot 2 \\ 2 \cdot \frac{3}{2} + 3(-1) & 2(-\frac{5}{2}) + 3 \cdot 2 \end{bmatrix} = \begin{bmatrix} 1 & 0 \\ 0 & 1 \end{bmatrix}$$

$$A^{-1}A = \begin{bmatrix} \frac{3}{2} & -\frac{5}{2} \\ -1 & 2 \end{bmatrix}\begin{bmatrix} 4 & 5 \\ 2 & 3 \end{bmatrix} = \begin{bmatrix} \frac{3}{2} \cdot 4 + (-\frac{5}{2})2 & \frac{3}{2} \cdot 5 + (-\frac{5}{2})3 \\ (-1)4 + 2 \cdot 2 & (-1)5 + 2 \cdot 3 \end{bmatrix} = \begin{bmatrix} 1 & 0 \\ 0 & 1 \end{bmatrix}$$

 Now Try Exercise 7 ■

The quantity $ad - bc$ that appears in the rule for calculating the inverse of a 2 × 2 matrix is called the **determinant** of the matrix. If the determinant is 0, then the matrix does not have an inverse (since we cannot divide by 0).

■ Finding the Inverse of an *n* × *n* Matrix

For 3 × 3 and larger square matrices the following technique provides the most efficient way to calculate the inverse. If A is an $n \times n$ matrix, we first construct the $n \times 2n$ matrix that has the entries of A on the left and of the identity matrix I_n on the right:

$$\left[\begin{array}{cccc|cccc} a_{11} & a_{12} & \cdots & a_{1n} & 1 & 0 & \cdots & 0 \\ a_{21} & a_{22} & \cdots & a_{2n} & 0 & 1 & \cdots & 0 \\ \vdots & \vdots & \ddots & \vdots & \vdots & \vdots & \ddots & \vdots \\ a_{n1} & a_{n2} & \cdots & a_{nn} & 0 & 0 & \cdots & 1 \end{array}\right]$$

We then use the elementary row operations on this new large matrix to change the left side into the identity matrix. (This means that we are changing the large matrix to reduced row-echelon form.) The right side is transformed automatically into A^{-1}. (We omit the proof of this fact.)

Example 4 ■ Finding the Inverse of a 3 × 3 Matrix

Let A be the matrix

$$A = \begin{bmatrix} 1 & -2 & -4 \\ 2 & -3 & -6 \\ -3 & 6 & 15 \end{bmatrix}$$

(a) Find A^{-1}.

(b) Verify that $AA^{-1} = A^{-1}A = I_3$.

ARTHUR CAYLEY (1821–1895) was an English mathematician who was instrumental in developing the theory of matrices. He was the first to use a single symbol such as *A* to represent a matrix, thereby introducing the idea that a matrix is a single entity rather than just a collection of numbers. Cayley practiced law until the age of 42, but his primary interest from adolescence was mathematics, and he published almost 200 articles on the subject in his spare time. In 1863 he accepted a professorship in mathematics at Cambridge, where he taught until his death. Cayley's work on matrices was of purely theoretical interest in his day, but in the 20th century many of his results found application in physics, the social sciences, business, and other fields. One of the most common uses of matrices today is in computers, where matrices are employed for data storage, error correction, image manipulation, and many other purposes.

Solution

(a) We begin with the 3 × 6 matrix whose left half is A and whose right half is the identity matrix I_3.

$$\left[\begin{array}{rrr|rrr} 1 & -2 & -4 & 1 & 0 & 0 \\ 2 & -3 & -6 & 0 & 1 & 0 \\ -3 & 6 & 15 & 0 & 0 & 1 \end{array}\right]$$

We then transform the left half of this new matrix into the identity matrix by performing the following sequence of elementary row operations on the *entire* new matrix.

$$\xrightarrow[\text{R}_3 + 3\text{R}_1 \to \text{R}_3]{\text{R}_2 - 2\text{R}_1 \to \text{R}_2} \left[\begin{array}{rrr|rrr} 1 & -2 & -4 & 1 & 0 & 0 \\ 0 & 1 & 2 & -2 & 1 & 0 \\ 0 & 0 & 3 & 3 & 0 & 1 \end{array}\right]$$

$$\xrightarrow{\frac{1}{3}\text{R}_3} \left[\begin{array}{rrr|rrr} 1 & -2 & -4 & 1 & 0 & 0 \\ 0 & 1 & 2 & -2 & 1 & 0 \\ 0 & 0 & 1 & 1 & 0 & \frac{1}{3} \end{array}\right]$$

$$\xrightarrow{\text{R}_1 + 2\text{R}_2 \to \text{R}_1} \left[\begin{array}{rrr|rrr} 1 & 0 & 0 & -3 & 2 & 0 \\ 0 & 1 & 2 & -2 & 1 & 0 \\ 0 & 0 & 1 & 1 & 0 & \frac{1}{3} \end{array}\right]$$

$$\xrightarrow{\text{R}_2 - 2\text{R}_3 \to \text{R}_2} \left[\begin{array}{rrr|rrr} 1 & 0 & 0 & -3 & 2 & 0 \\ 0 & 1 & 0 & -4 & 1 & -\frac{2}{3} \\ 0 & 0 & 1 & 1 & 0 & \frac{1}{3} \end{array}\right]$$

We have now transformed the left half of this matrix into an identity matrix. (This means that we have put the entire matrix in reduced row-echelon form.) Note that to do this in as systematic a fashion as possible, we first changed the elements below the main diagonal to zeros, just as we would if we were using Gaussian elimination. We then changed each main diagonal element to a 1 by multiplying by the appropriate constant(s). Finally, we completed the process by changing the entries above the main diagonal to zeros.

The right half is now A^{-1}.

$$A^{-1} = \begin{bmatrix} -3 & 2 & 0 \\ -4 & 1 & -\frac{2}{3} \\ 1 & 0 & \frac{1}{3} \end{bmatrix}$$

(b) We calculate AA^{-1} and $A^{-1}A$ and verify that both products give the identity matrix I_3.

$$AA^{-1} = \begin{bmatrix} 1 & -2 & -4 \\ 2 & -3 & -6 \\ -3 & 6 & 15 \end{bmatrix} \begin{bmatrix} -3 & 2 & 0 \\ -4 & 1 & -\frac{2}{3} \\ 1 & 0 & \frac{1}{3} \end{bmatrix} = \begin{bmatrix} 1 & 0 & 0 \\ 0 & 1 & 0 \\ 0 & 0 & 1 \end{bmatrix}$$

$$A^{-1}A = \begin{bmatrix} -3 & 2 & 0 \\ -4 & 1 & -\frac{2}{3} \\ 1 & 0 & \frac{1}{3} \end{bmatrix} \begin{bmatrix} 1 & -2 & -4 \\ 2 & -3 & -6 \\ -3 & 6 & 15 \end{bmatrix} = \begin{bmatrix} 1 & 0 & 0 \\ 0 & 1 & 0 \\ 0 & 0 & 1 \end{bmatrix}$$

 **Now Try Exercises 19** ■

Graphing devices are also able to calculate matrix inverses. On a graphing calculator, to find the inverse of [A], we key in

[A] $\boxed{\text{X}^{-1}}$ $\boxed{\text{ENTER}}$

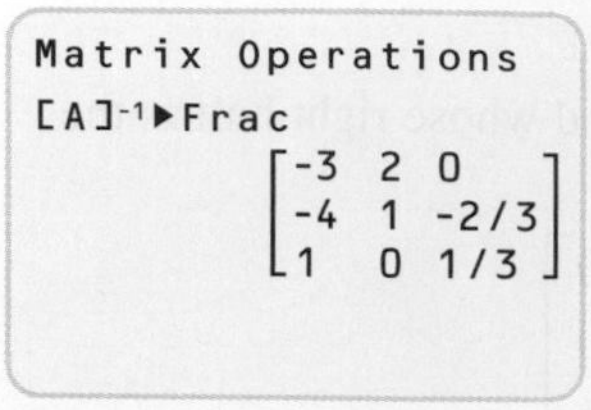

```
Matrix Operations
[A]-1▶Frac
  [-3  2  0   ]
  [-4  1  -2/3]
  [1   0  1/3 ]
```

Figure 1

For the matrix of Example 4 this results in the output shown in Figure 1, where we use the ▶Frac command (or $\frac{\square}{\square}$ command) to display the output in fraction form rather than in decimal form.

The next example shows that not every square matrix has an inverse.

Example 5 ■ A Matrix That Does Not Have an Inverse

Find the inverse of the matrix

$$\begin{bmatrix} 2 & -3 & -7 \\ 1 & 2 & 7 \\ 1 & 1 & 4 \end{bmatrix}$$

Solution We proceed as follows.

$$\left[\begin{array}{ccc|ccc} 2 & -3 & -7 & 1 & 0 & 0 \\ 1 & 2 & 7 & 0 & 1 & 0 \\ 1 & 1 & 4 & 0 & 0 & 1 \end{array}\right] \xrightarrow{R_1 \leftrightarrow R_2} \left[\begin{array}{ccc|ccc} 1 & 2 & 7 & 0 & 1 & 0 \\ 2 & -3 & -7 & 1 & 0 & 0 \\ 1 & 1 & 4 & 0 & 0 & 1 \end{array}\right]$$

$$\xrightarrow[R_3 - R_1 \to R_3]{R_2 - 2R_1 \to R_2} \left[\begin{array}{ccc|ccc} 1 & 2 & 7 & 0 & 1 & 0 \\ 0 & -7 & -21 & 1 & -2 & 0 \\ 0 & -1 & -3 & 0 & -1 & 1 \end{array}\right]$$

$$\xrightarrow{-\frac{1}{7}R_2} \left[\begin{array}{ccc|ccc} 1 & 2 & 7 & 0 & 1 & 0 \\ 0 & 1 & 3 & -\frac{1}{7} & \frac{2}{7} & 0 \\ 0 & -1 & -3 & 0 & -1 & 1 \end{array}\right]$$

$$\xrightarrow[R_1 - 2R_2 \to R_1]{R_3 + R_2 \to R_3} \left[\begin{array}{ccc|ccc} 1 & 0 & 1 & \frac{2}{7} & \frac{3}{7} & 0 \\ 0 & 1 & 3 & -\frac{1}{7} & \frac{2}{7} & 0 \\ 0 & 0 & 0 & -\frac{1}{7} & -\frac{5}{7} & 1 \end{array}\right]$$

At this point we would like to change the 0 in the (3, 3) position of this matrix to a 1 without changing the zeros in the (3, 1) and (3, 2) positions. But there is no way to accomplish this: No matter what multiple of rows 1 and/or 2 we add to row 3, we can't change the third zero in row 3 without changing the first or second zero as well. Thus we cannot change the left half to the identity matrix, so the original matrix doesn't have an inverse.

 **Now Try Exercise 21** ■

If we encounter a row of zeros on the left side when trying to find an inverse, as we did in Example 5, then the original matrix does not have an inverse. (A matrix that has no inverse is called *singular*.) If we try to calculate the inverse of the matrix from Example 5 on a graphing device, we get an error message like the one shown in Figure 2.

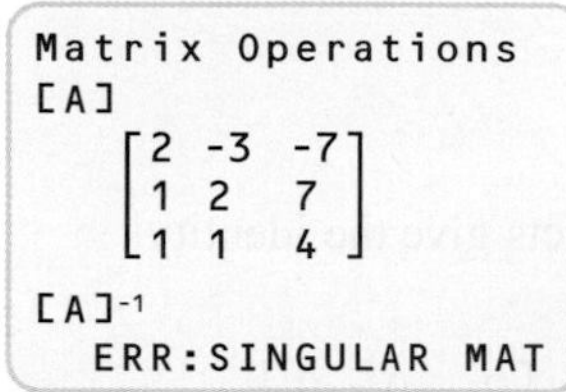

Figure 2

■ Matrix Equations

We saw in Example 9.4.6 that a system of linear equations can be written as a single matrix equation. For example, the system

$$\begin{cases} x - 2y - 4z = 7 \\ 2x - 3y - 6z = 5 \\ -3x + 6y + 15z = 0 \end{cases}$$

is equivalent to the matrix equation

$$\underbrace{\begin{bmatrix} 1 & -2 & -4 \\ 2 & -3 & -6 \\ -3 & 6 & 15 \end{bmatrix}}_{A} \underbrace{\begin{bmatrix} x \\ y \\ z \end{bmatrix}}_{X} = \underbrace{\begin{bmatrix} 7 \\ 5 \\ 0 \end{bmatrix}}_{B}$$

If we let

$$A = \begin{bmatrix} 1 & -2 & -4 \\ 2 & -3 & -6 \\ -3 & 6 & 15 \end{bmatrix} \qquad X = \begin{bmatrix} x \\ y \\ z \end{bmatrix} \qquad B = \begin{bmatrix} 7 \\ 5 \\ 0 \end{bmatrix}$$

then this matrix equation can be written as

$$AX = B$$

The matrix A is called the **coefficient matrix**. We can use matrix operations to solve for the matrix X and we get $X = A^{-1}B$. See the proof below.

In Example 4 we showed that

$$A^{-1} = \begin{bmatrix} -3 & 2 & 0 \\ -4 & 1 & -\frac{2}{3} \\ 1 & 0 & \frac{1}{3} \end{bmatrix}$$

So from $X = A^{-1}B$ we have

$$\underbrace{\begin{bmatrix} x \\ y \\ z \end{bmatrix}}_{X} = \underbrace{\begin{bmatrix} -3 & 2 & 0 \\ -4 & 1 & -\frac{2}{3} \\ 1 & 0 & \frac{1}{3} \end{bmatrix}}_{A^{-1}} \underbrace{\begin{bmatrix} 7 \\ 5 \\ 0 \end{bmatrix}}_{B} = \begin{bmatrix} -11 \\ -23 \\ 7 \end{bmatrix}$$

Thus $x = -11$, $y = -23$, $z = 7$ is the solution of the original system.

Solving a Matrix Equation

If A is a square $n \times n$ matrix that has an inverse A^{-1} and if X is a variable matrix and B is a known matrix, both with n rows, then the solution of the matrix equation

$$AX = B$$

is given by

$$X = A^{-1}B$$

Proof We solve the matrix equation by multiplying each side by the inverse of A.

$AX = B$	Matrix equation
$A^{-1}(AX) = A^{-1}B$	Multiply on left by A^{-1}
$(A^{-1}A)X = A^{-1}B$	Associative Property
$I_3X = A^{-1}B$	Property of inverses
$X = A^{-1}B$	Property of identity matrix

■

Solving the matrix equation $AX = B$ is similar to solving a real-number equation like

$$3x = 12$$

which we do by multiplying each side by the reciprocal (or inverse) of 3.

$$\tfrac{1}{3}(3x) = \tfrac{1}{3}(12)$$

$$x = 4$$

Example 6 ■ Solving a System Using a Matrix Inverse

A system of equations is given.

(a) Write the system of equations as a matrix equation.

(b) Solve the system by solving the matrix equation.

$$\begin{cases} 2x - 5y = 15 \\ 3x - 6y = 36 \end{cases}$$

Mathematics in the Modern World

Mathematical Ecology

In the 1970s humpback whales became a center of controversy. Environmentalists believed that whaling threatened the whales with imminent extinction; whalers saw their livelihood threatened by any attempt to stop whaling. Are whales really threatened to extinction by whaling? What level of whaling is safe to guarantee survival of the whales? These questions motivated mathematicians to study population patterns of whales and other species more closely.

As early as the 1920s Lotka and Volterra had founded the field of mathematical biology by creating predator-prey models. Their models, which draw on a branch of mathematics called differential equations, take into account the rates at which predator eats prey and the rates of growth of each population: As predator eats prey, the prey population decreases. This means less food supply for the predators, so their population begins to decrease. With fewer predators the prey population begins to increase, and so on. Normally, a state of equilibrium develops, and the two populations alternate between a minimum and a maximum value. However, if the predators eat the prey too fast, they will be left without food and will thus ensure their own extinction.

Since Lotka and Volterra's time, more detailed mathematical models of animal populations have been developed. For many species the population is divided into several stages: immature, juvenile, adult, and so on. The proportion of each stage that survives or reproduces in a given time period is entered into a matrix (called a transition matrix); matrix multiplication is then used to predict the population in succeeding time periods. (See *Discovery Project: Will the Species Survive?* at the book companion website: **www.stewartmath.com**.)

The power of mathematics to model and predict is an invaluable tool in the ongoing debate over the environment.

Solution

(a) We write the system as a matrix equation of the form $AX = B$.

$$\underbrace{\begin{bmatrix} 2 & -5 \\ 3 & -6 \end{bmatrix}}_{A}\underbrace{\begin{bmatrix} x \\ y \end{bmatrix}}_{X} = \underbrace{\begin{bmatrix} 15 \\ 36 \end{bmatrix}}_{B}$$

(b) Using the rule for finding the inverse of a 2×2 matrix, we get

$$A^{-1} = \begin{bmatrix} 2 & -5 \\ 3 & -6 \end{bmatrix}^{-1} = \frac{1}{2(-6) - (-5)3}\begin{bmatrix} -6 & -(-5) \\ -3 & 2 \end{bmatrix} = \frac{1}{3}\begin{bmatrix} -6 & 5 \\ -3 & 2 \end{bmatrix}$$

Multiplying each side of the matrix equation by this inverse matrix, we get

$$\underbrace{\begin{bmatrix} x \\ y \end{bmatrix}}_{X} = \underbrace{\frac{1}{3}\begin{bmatrix} -6 & 5 \\ -3 & 2 \end{bmatrix}}_{A^{-1}}\underbrace{\begin{bmatrix} 15 \\ 36 \end{bmatrix}}_{B} = \begin{bmatrix} 30 \\ 9 \end{bmatrix}$$

So $x = 30$ and $y = 9$.

 **Now Try Exercise 39** ■

■ Modeling with Matrix Equations

Suppose we need to solve several systems of equations that have the same coefficient matrix. Then converting the systems to matrix equations provides an efficient way to obtain the solutions, because we need to find the inverse of the coefficient matrix only once. This procedure is particularly convenient if we use a graphing device to perform the matrix operations, as illustrated in the next example.

Example 7 ■ Modeling Nutritional Requirements Using Matrix Equations

A pet-store owner feeds hamsters and gerbils different mixtures of three types of rodent food: KayDee Food, Pet Pellets, and Rodent Chow. The animals should get the correct amount of each brand to satisfy their daily requirements for protein, fat, and carbohydrates. Suppose that hamsters require 340 mg of protein, 280 mg of fat, and 440 mg of carbohydrates, and gerbils need 480 mg of protein, 360 mg of fat, and 680 mg of carbohydrates each day. The amount of each nutrient (in mg) in 1 g of each brand is given in the table. How many grams of each food should the hamsters and gerbils be fed daily to satisfy their daily nutritional requirements?

	KayDee Food	Pet Pellets	Rodent Chow
Protein (mg)	10	0	20
Fat (mg)	10	20	10
Carbohydrates (mg)	5	10	30

Solution We let x_1, x_2, and x_3 be the respective amounts (in grams) of KayDee Food, Pet Pellets, and Rodent Chow that the hamsters should eat, and we let y_1, y_2, and y_3 be the corresponding amounts for the gerbils. Then we want to solve the matrix equations

$$\begin{bmatrix} 10 & 0 & 20 \\ 10 & 20 & 10 \\ 5 & 10 & 30 \end{bmatrix} \begin{bmatrix} x_1 \\ x_2 \\ x_3 \end{bmatrix} = \begin{bmatrix} 340 \\ 280 \\ 440 \end{bmatrix} \qquad \text{Hamster equation}$$

$$\begin{bmatrix} 10 & 0 & 20 \\ 10 & 20 & 10 \\ 5 & 10 & 30 \end{bmatrix} \begin{bmatrix} y_1 \\ y_2 \\ y_3 \end{bmatrix} = \begin{bmatrix} 480 \\ 360 \\ 680 \end{bmatrix} \qquad \text{Gerbil equation}$$

Let

$$A = \begin{bmatrix} 10 & 0 & 20 \\ 10 & 20 & 10 \\ 5 & 10 & 30 \end{bmatrix} \quad B = \begin{bmatrix} 340 \\ 280 \\ 440 \end{bmatrix} \quad C = \begin{bmatrix} 480 \\ 360 \\ 680 \end{bmatrix} \quad X = \begin{bmatrix} x_1 \\ x_2 \\ x_3 \end{bmatrix} \quad Y = \begin{bmatrix} y_1 \\ y_2 \\ y_3 \end{bmatrix}$$

Then we can write these matrix equations as

$$AX = B \qquad \text{Hamster equation}$$

$$AY = C \qquad \text{Gerbil equation}$$

We want to solve for X and Y, so we multiply both sides of each equation by A^{-1}, the inverse of the coefficient matrix. We could find A^{-1} by hand, but it is more convenient to use a graphing device, as shown in Figure 3.

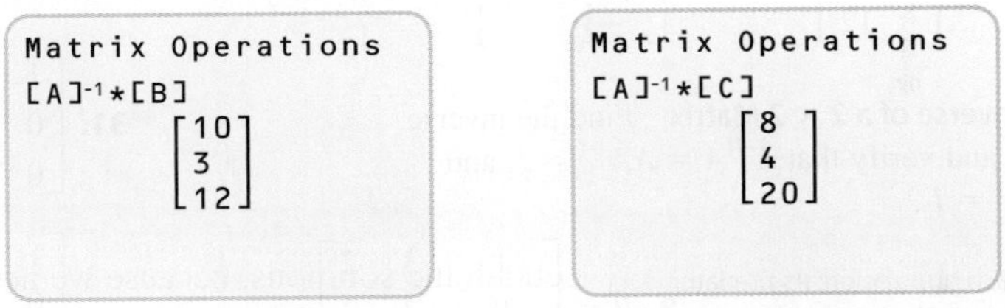

```
Matrix Operations
[A]-1*[B]
        [10]
        [3 ]
        [12]
```

(a)

```
Matrix Operations
[A]-1*[C]
        [8 ]
        [4 ]
        [20]
```

(b)

Figure 3

So

$$X = A^{-1}B = \begin{bmatrix} 10 \\ 3 \\ 12 \end{bmatrix} \qquad Y = A^{-1}C = \begin{bmatrix} 8 \\ 4 \\ 20 \end{bmatrix}$$

Thus each hamster should be fed 10 g of KayDee Food, 3 g of Pet Pellets, and 12 g of Rodent Chow; and each gerbil should be fed 8 g of KayDee Food, 4 g of Pet Pellets, and 20 g of Rodent Chow daily. ■

Now Try Exercise 61

9.5 Exercises

Concepts

1. (a) The matrix $I = \begin{bmatrix} 1 & 0 \\ 0 & 1 \end{bmatrix}$ is called an ________ matrix.

(b) If A is a 2×2 matrix, then $AI =$ ________ and $IA =$ ________.

(c) If A and B are 2×2 matrices with $AB = I$, then B is the ________ of A.

2. (a) Write the following system as a matrix equation $AX = B$.

System	Matrix equation
	$A \cdot X = B$
$5x + 3y = 4$ $3x + 2y = 3$	$\begin{bmatrix} \blacksquare & \blacksquare \\ \blacksquare & \blacksquare \end{bmatrix}\begin{bmatrix} \blacksquare \\ \blacksquare \end{bmatrix} = \begin{bmatrix} \blacksquare \\ \blacksquare \end{bmatrix}$

(b) The inverse of A is $A^{-1} = \begin{bmatrix} \blacksquare & \blacksquare \\ \blacksquare & \blacksquare \end{bmatrix}$.

(c) The solution of the matrix equation is $X = A^{-1}B$.

$$\begin{array}{ccccc} X & = & A^{-1} & B & \\ \begin{bmatrix} x \\ y \end{bmatrix} & = & \begin{bmatrix} \square & \square \\ \square & \square \end{bmatrix} & \begin{bmatrix} \square \\ \square \end{bmatrix} & = \begin{bmatrix} \square \\ \square \end{bmatrix} \end{array}$$

(d) The solution of the system is $x =$ ________, $y =$ ________.

Skills

3–6 ■ Verifying the Inverse of a Matrix Calculate the products AB and BA to verify that B is the inverse of A.

3. $A = \begin{bmatrix} 4 & 1 \\ 7 & 2 \end{bmatrix} \quad B = \begin{bmatrix} 2 & -1 \\ -7 & 4 \end{bmatrix}$

4. $A = \begin{bmatrix} 2 & -3 \\ 4 & -7 \end{bmatrix} \quad B = \begin{bmatrix} \frac{7}{2} & -\frac{3}{2} \\ 2 & -1 \end{bmatrix}$

5. $A = \begin{bmatrix} 1 & 3 & -1 \\ 1 & 4 & 0 \\ -1 & -3 & 2 \end{bmatrix} \quad B = \begin{bmatrix} 8 & -3 & 4 \\ -2 & 1 & -1 \\ 1 & 0 & 1 \end{bmatrix}$

6. $A = \begin{bmatrix} 3 & 2 & 4 \\ 1 & 1 & -6 \\ 2 & 1 & 12 \end{bmatrix} \quad B = \begin{bmatrix} 9 & -10 & -8 \\ -12 & 14 & 11 \\ -\frac{1}{2} & \frac{1}{2} & \frac{1}{2} \end{bmatrix}$

7–8 ■ The Inverse of a 2 × 2 Matrix Find the inverse of the matrix and verify that $A^{-1}A = AA^{-1} = I_2$ and $B^{-1}B = BB^{-1} = I_3$.

7. $A = \begin{bmatrix} 7 & 4 \\ 3 & 2 \end{bmatrix}$

8. $B = \begin{bmatrix} 1 & 3 & 2 \\ 0 & 2 & 2 \\ -2 & -1 & 0 \end{bmatrix}$

9–10 ■ The Inverse of a 2 × 2 Matrix Use a graphing device to find the inverse of the matrix and to verify that $A^{-1}A = AA^{-1} = I_2$ and $B^{-1}B = BB^{-1} = I_3$. Use the appropriate command on your graphing device to obtain the answer in fractions.

9. $A = \begin{bmatrix} 1.2 & 0.3 \\ -1.2 & 0.2 \end{bmatrix}$

10. $B = \begin{bmatrix} 5 & -1 & 3 \\ 6 & -1 & 3 \\ 7 & 1 & -2 \end{bmatrix}$

11–26 ■ Finding the Inverse of a Matrix Find the inverse of the matrix if it exists.

11. $\begin{bmatrix} 3 & 2 \\ 13 & 9 \end{bmatrix}$

12. $\begin{bmatrix} 5 & 7 \\ 3 & 4 \end{bmatrix}$

13. $\begin{bmatrix} 2 & 5 \\ -5 & -13 \end{bmatrix}$

14. $\begin{bmatrix} -7 & 4 \\ 8 & -5 \end{bmatrix}$

15. $\begin{bmatrix} 6 & -3 \\ -8 & 4 \end{bmatrix}$

16. $\begin{bmatrix} \frac{1}{2} & \frac{1}{3} \\ 5 & 4 \end{bmatrix}$

17. $\begin{bmatrix} 0.4 & -1.2 \\ 0.3 & 0.6 \end{bmatrix}$

18. $\begin{bmatrix} 4 & 2 & 3 \\ 3 & 3 & 2 \\ 1 & 0 & 1 \end{bmatrix}$

19. $\begin{bmatrix} 2 & 4 & 1 \\ -1 & 1 & -1 \\ 1 & 4 & 0 \end{bmatrix}$

20. $\begin{bmatrix} 5 & 7 & 4 \\ 3 & -1 & 3 \\ 6 & 7 & 5 \end{bmatrix}$

21. $\begin{bmatrix} 1 & 2 & 3 \\ 4 & 5 & -1 \\ 1 & -1 & -10 \end{bmatrix}$

22. $\begin{bmatrix} 2 & 1 & 0 \\ 1 & 1 & 4 \\ 2 & 1 & 2 \end{bmatrix}$

23. $\begin{bmatrix} 0 & -2 & 2 \\ 3 & 1 & 3 \\ 1 & -2 & 3 \end{bmatrix}$

24. $\begin{bmatrix} 3 & -2 & 0 \\ 5 & 1 & 1 \\ 2 & -2 & 0 \end{bmatrix}$

25. $\begin{bmatrix} 1 & 2 & 0 & 3 \\ 0 & 1 & 1 & 1 \\ 0 & 1 & 0 & 1 \\ 1 & 2 & 0 & 2 \end{bmatrix}$

26. $\begin{bmatrix} 1 & 0 & 1 & 0 \\ 0 & 1 & 0 & 1 \\ 1 & 1 & 1 & 0 \\ 1 & 1 & 1 & 1 \end{bmatrix}$

27–34 ■ Finding the Inverse of a Matrix Use a graphing device to find the inverse of the matrix, if it exists. Use the appropriate command on your graphing device to obtain the answer in fractions.

27. $\begin{bmatrix} -3 & 2 & 3 \\ 0 & -1 & 3 \\ 1 & 0 & -2 \end{bmatrix}$

28. $\begin{bmatrix} -5 & 2 & 1 \\ 5 & 1 & 0 \\ 0 & -1 & -2 \end{bmatrix}$

29. $\begin{bmatrix} -1 & -4 & 0 & 1 \\ 1 & 0 & -1 & 0 \\ 0 & 4 & 1 & -2 \\ 2 & 2 & -2 & 0 \end{bmatrix}$

30. $\begin{bmatrix} -3 & 0 & -1 & 1 \\ 3 & -1 & 1 & -1 \\ 1 & 3 & 0 & 1 \\ -2 & -3 & 1 & 0 \end{bmatrix}$

31. $\begin{bmatrix} 1 & 7 & 3 \\ 0 & 2 & 1 \\ 0 & 0 & 3 \end{bmatrix}$

32. $\begin{bmatrix} 1 & 0 & 0 & 0 \\ 2 & 5 & 0 & 0 \\ 4 & 2 & 3 & 0 \\ 5 & 1 & 2 & 1 \end{bmatrix}$

33. $\begin{bmatrix} 1 & 0 & 0 & 0 \\ 0 & 2 & 0 & 0 \\ 0 & 0 & 4 & 0 \\ 0 & 0 & 0 & 7 \end{bmatrix}$

34. $\begin{bmatrix} -1 & 0 & 0 \\ 0 & 2 & 0 \\ 0 & 0 & -3 \end{bmatrix}$

35–38 ■ Products Involving Matrices and Inverses The matrices A and B are defined as follows.

$$A = \begin{bmatrix} -1 & 0 & 2 \\ 0 & -2 & -1 \\ 4 & 2 & 1 \end{bmatrix} \qquad B = \begin{bmatrix} 2 & -1 & -2 \\ 0 & 3 & 1 \\ -1 & 0 & 2 \end{bmatrix}$$

Use a graphing device to carry out the indicated algebraic operations, or explain why they cannot be performed. Use the appropriate command on your graphing device to obtain the answer in fractions.

35. $A^{-1}B$ **36.** AB^{-1} **37.** BAB^{-1} **38.** $B^{-1}AB$

39–46 ■ Solving a Linear System as a Matrix Equation Solve the system of equations by converting to a matrix equation and using the inverse of the coefficient matrix, as in Example 6. Use the inverses from Exercises 11–14, 19, 20, 23, and 25.

39. $\begin{cases} 3x + 2y = 1 \\ 13x + 9y = 3 \end{cases}$

40. $\begin{cases} 5x + 7y = -9 \\ 3x + 4y = -6 \end{cases}$

41. $\begin{cases} 2x + 5y = 2 \\ -5x - 13y = 20 \end{cases}$

42. $\begin{cases} -7x + 4y = 0 \\ 8x - 5y = 100 \end{cases}$

43. $\begin{cases} 2x + 4y + z = 7 \\ -x + y - z = 0 \\ x + 4y = -2 \end{cases}$

44. $\begin{cases} 5x + 7y + 4z = 1 \\ 3x - y + 3z = 1 \\ 6x + 7y + 5z = 1 \end{cases}$

45. $\begin{cases} -2y + 2z = 12 \\ 3x + y + 3z = -2 \\ x - 2y + 3z = 8 \end{cases}$ **46.** $\begin{cases} x + 2y + 3w = 0 \\ y + z + w = 1 \\ y + w = 2 \\ x + 2y + 2w = 3 \end{cases}$

47–52 ■ Solving a Linear System Solve the system of equations by converting to a matrix equation. Use a graphing device to perform the necessary matrix operations, as in Example 7.

47. $\begin{cases} x + y - 2z = 3 \\ 2x + 5z = 11 \\ 2x + 3y = 12 \end{cases}$ **48.** $\begin{cases} 3x + 4y - z = 2 \\ 2x - 3y + z = -5 \\ 5x - 2y + 2z = -3 \end{cases}$

49. $\begin{cases} 12x + \frac{1}{2}y - 7z = 21 \\ 11x - 2y + 3z = 43 \\ 13x + y - 4z = 29 \end{cases}$ **50.** $\begin{cases} x + \frac{1}{2}y - \frac{1}{3}z = 4 \\ x - \frac{1}{4}y + \frac{1}{6}z = 7 \\ x + y - z = -6 \end{cases}$

51. $\begin{cases} x + y - 3w = 0 \\ x - 2z = 8 \\ 2y - z + w = 5 \\ 2x + 3y - 2w = 13 \end{cases}$

52. $\begin{cases} x + y + z + w = 15 \\ x - y + z - w = 5 \\ x + 2y + 3z + 4w = 26 \\ x - 2y + 3z - 4w = 2 \end{cases}$

Skills Plus

53–54 ■ Solving a Matrix Equation Solve the matrix equation by multiplying each side by the appropriate inverse matrix.

53. $\begin{bmatrix} 3 & -2 \\ -4 & 3 \end{bmatrix} \begin{bmatrix} x & y & z \\ u & v & w \end{bmatrix} = \begin{bmatrix} 1 & 0 & -1 \\ 2 & 1 & 3 \end{bmatrix}$

54. $\begin{bmatrix} 0 & -2 & 2 \\ 3 & 1 & 3 \\ 1 & -2 & 3 \end{bmatrix} \begin{bmatrix} x & u \\ y & v \\ z & w \end{bmatrix} = \begin{bmatrix} 3 & 6 \\ 6 & 12 \\ 0 & 0 \end{bmatrix}$

55–56 ■ Inverses of Special Matrices Find the inverse of the matrix.

55. $\begin{bmatrix} a & -a \\ a & a \end{bmatrix}$ $(a \neq 0)$ **56.** $\begin{bmatrix} a & 0 & 0 & 0 \\ 0 & b & 0 & 0 \\ 0 & 0 & c & 0 \\ 0 & 0 & 0 & d \end{bmatrix}$ $(abcd \neq 0)$

57–60 ■ When Do Matrices Have Inverses? Find the inverse of the matrix. For what value(s) of x, if any, does the matrix have no inverse?

57. $\begin{bmatrix} 2 & x \\ x & x^2 \end{bmatrix}$ **58.** $\begin{bmatrix} e^x & -e^{2x} \\ e^{2x} & e^{3x} \end{bmatrix}$

59. $\begin{bmatrix} 1 & e^x & 0 \\ e^x & -e^{2x} & 0 \\ 0 & 0 & 2 \end{bmatrix}$ **60.** $\begin{bmatrix} x & 1 \\ -x & \frac{1}{x - 1} \end{bmatrix}$

Applications

61. Nutrition A nutritionist is studying the effects of the nutrients folic acid, choline, and inositol. There are three types of food available, and each type contains the following amounts of these nutrients per serving.

	Type A	Type B	Type C
Folic acid (mg)	3	1	3
Choline (mg)	4	2	4
Inositol (mg)	3	2	4

(a) Find the inverse of the matrix

$$\begin{bmatrix} 3 & 1 & 3 \\ 4 & 2 & 4 \\ 3 & 2 & 4 \end{bmatrix}$$

and use it to solve the remaining parts of this problem.

(b) How many servings of each food should the laboratory rats be fed if their daily diet is to contain 10 mg of folic acid, 14 mg of choline, and 13 mg of inositol?

(c) How much of each food is needed to supply 9 mg of folic acid, 12 mg of choline, and 10 mg of inositol?

(d) Will any combination of these foods supply 2 mg of folic acid, 4 mg of choline, and 11 mg of inositol?

62. Nutrition Refer to Exercise 61. Suppose food type C has been improperly labeled, and it actually contains 4 mg of folic acid, 6 mg of choline, and 5 mg of inositol per serving. Would it still be possible to use matrix inversion to solve parts (b), (c), and (d) of Exercise 61? Why or why not?

63. Sales Commissions A salesperson works at a kiosk that offers three different models of cell phones: standard with 64 GB capacity, deluxe with 128 GB capacity, and super-deluxe with 256 GB capacity. For each phone sold the salesperson earns a commission based on the cell phone model. One week 9 standard, 11 deluxe, and 8 super-deluxe are sold and the salesperson makes \$740 in commission. The next week 13 standard, 15 deluxe, and 16 super-deluxe are sold for a \$1204 commission. The third week 8 standard, 7 deluxe, and 14 super-deluxe are sold, earning the salesperson \$828 in commission.

(a) Let x, y, and z represent the commission the saleperson earns on standard, deluxe, and super-deluxe, respectively. Translate the given information into a system of equations in x, y, and z.

(b) Express the system of equations you found in part (a) as a matrix equation of the form $AX = B$.

(c) Find the inverse of the coefficient matrix A and use it to solve the matrix equation in part (b). How much commission does the salesperson earn on each model of cell phone?

Discuss ■ Discover ■ Prove ■ Write

64. Discuss: No Zero-Product Property for Matrices We have used the Zero-Product Property to solve algebraic equations. Matrices do *not* have this property. Let O represent the **2 × 2 zero matrix**

$$O = \begin{bmatrix} 0 & 0 \\ 0 & 0 \end{bmatrix}$$

Find 2×2 matrices $A \neq O$ and $B \neq O$ such that $AB = O$. Can you find a matrix $A \neq O$ such that $A^2 = O$?

65. Prove: The Inverse of a Product of Matrices Let A and B be $n \times n$ invertible matrices. Show that the product matrix AB is invertible, and its inverse is

$$(AB)^{-1} = B^{-1}A^{-1}$$

PS *Try to recognize something familiar.* Apply the definition of the inverse of a matrix to the matrix AB.

9.6 Determinants and Cramer's Rule

■ Determinant of a 2 × 2 Matrix ■ Determinant of an $n \times n$ Matrix ■ Row and Column Transformations ■ Cramer's Rule ■ Areas of Triangles Using Determinants

If a matrix is **square** (that is, it has the same number of rows as columns), then we can assign to it a number called its *determinant*. Determinants can be used to solve systems of linear equations, as we will see later in this section. They are also useful in determining whether a matrix has an inverse.

■ Determinant of a 2 × 2 Matrix

We denote the determinant of a square matrix A by the symbol $\det(A)$ or $|A|$. We first define $\det(A)$ for the simplest cases. If $A = [a]$ is a 1×1 matrix, then $\det(A) = a$. The following box gives the definition of a 2×2 determinant.

We will use both notations, $\det(A)$ and $|A|$, for the determinant of A. Although the symbol $|A|$ looks like the absolute value symbol, it will be clear from the context which meaning is intended.

Determinant of a 2 × 2 Matrix

The **determinant** of the 2×2 matrix $A = \begin{bmatrix} a & b \\ c & d \end{bmatrix}$ is

$$\det(A) = |A| = \begin{vmatrix} a & b \\ c & d \end{vmatrix} = ad - bc$$

Example 1 ■ Determinant of a 2 × 2 Matrix

Evaluate $|A|$ for $A = \begin{bmatrix} 6 & -3 \\ 2 & 3 \end{bmatrix}$.

Solution

To evaluate a 2×2 determinant, we take the product of the diagonal from top left to bottom right and subtract the product from top right to bottom left, as indicated by the arrows.

$$\begin{vmatrix} 6 & -3 \\ 2 & 3 \end{vmatrix} = 6 \cdot 3 - (-3)2 = 18 - (-6) = 24$$

Now Try Exercise 5 ■

■ Determinant of an $n \times n$ Matrix

To define the concept of determinant for an arbitrary $n \times n$ matrix, we need the following terminology.

Minors and Cofactors

Let A be an $n \times n$ matrix.

1. The **minor** M_{ij} of the element a_{ij} is the determinant of the matrix obtained by deleting the ith row and jth column of A.
2. The **cofactor** A_{ij} of the element a_{ij} is

$$A_{ij} = (-1)^{i+j}M_{ij}$$

DAVID HILBERT (1862–1943) was born in Königsberg, Germany, and became a professor at Göttingen University. He is considered by many to be the greatest mathematician of the 20th century. At the International Congress of Mathematicians held in Paris in 1900, Hilbert set the direction of mathematics for the 20th century by posing 23 problems that he believed to be of crucial importance. He said that "these are problems whose solutions we expect from the future." Most of Hilbert's problems have now been solved (see Julia Robinson, Section 9.4, and Alan Turing, Section 2.6), and their solutions have led to important new areas of mathematical research. Yet as of this writing, several of Hilbert's problems remain unsolved. In his work, Hilbert emphasized structure, logic, and the foundations of mathematics. Part of his genius lay in his ability to see the most general possible statement of a problem. For instance, Euler proved that every whole number is the sum of four squares; Hilbert proved a similar statement for all powers of positive integers. Hilbert firmly believed that every mathematical problem had a solution; he famously said "We must know, we will know."

For example, if A is the matrix

$$\begin{bmatrix} 2 & 3 & -1 \\ 0 & 2 & 4 \\ -2 & 5 & 6 \end{bmatrix}$$

then the minor M_{12} is the determinant of the matrix obtained by deleting the first row and second column from A. Thus

$$M_{12} = \begin{vmatrix} 2 & 3 & -1 \\ 0 & 2 & 4 \\ -2 & 5 & 6 \end{vmatrix} = \begin{vmatrix} 0 & 4 \\ -2 & 6 \end{vmatrix} = 0(6) - 4(-2) = 8$$

So the cofactor $A_{12} = (-1)^{1+2}M_{12} = -8$. Similarly,

$$M_{33} = \begin{vmatrix} 2 & 3 & -1 \\ 0 & 2 & 4 \\ -2 & 5 & 6 \end{vmatrix} = \begin{vmatrix} 2 & 3 \\ 0 & 2 \end{vmatrix} = 2 \cdot 2 - 3 \cdot 0 = 4$$

So $A_{33} = (-1)^{3+3}M_{33} = 4$.

Note that the cofactor of a_{ij} is simply the minor of a_{ij} multiplied by either 1 or -1, depending on whether $i + j$ is even or odd. Thus in a 3×3 matrix we obtain the cofactor of any element by prefixing its minor with the sign obtained from the following checkerboard pattern.

$$\begin{bmatrix} + & - & + \\ - & + & - \\ + & - & + \end{bmatrix}$$

We are now ready to define the determinant of any square matrix.

The Determinant of a Square Matrix

If A is an $n \times n$ matrix, then the **determinant** of A is obtained by multiplying each element of the first row by its cofactor and then adding the results. In symbols,

$$\det(A) = |A| = \begin{vmatrix} a_{11} & a_{12} & \cdots & a_{1n} \\ a_{21} & a_{22} & \cdots & a_{2n} \\ \vdots & \vdots & \ddots & \vdots \\ a_{n1} & a_{n2} & \cdots & a_{nn} \end{vmatrix} = a_{11}A_{11} + a_{12}A_{12} + \cdots + a_{1n}A_{1n}$$

Example 2 ■ Determinant of a 3 × 3 Matrix

Evaluate the determinant of the matrix

$$A = \begin{bmatrix} 2 & 3 & -1 \\ 0 & 2 & 4 \\ -2 & 5 & 6 \end{bmatrix}$$

Solution

$$\det(A) = \begin{vmatrix} 2 & 3 & -1 \\ 0 & 2 & 4 \\ -2 & 5 & 6 \end{vmatrix} = 2\begin{vmatrix} 2 & 4 \\ 5 & 6 \end{vmatrix} - 3\begin{vmatrix} 0 & 4 \\ -2 & 6 \end{vmatrix} + (-1)\begin{vmatrix} 0 & 2 \\ -2 & 5 \end{vmatrix}$$

$$= 2(2 \cdot 6 - 4 \cdot 5) - 3[0 \cdot 6 - 4(-2)] - [0 \cdot 5 - 2(-2)]$$

$$= -16 - 24 - 4$$

$$= -44$$

 **Now Try Exercises 21 and 29** ■

In our definition of the determinant we used the cofactors of elements in the first row only. This is called **expanding the determinant by the first row**. In fact, *we can expand the determinant by any row or column in the same way and obtain the same result in each case* (although we won't prove this). The next example illustrates this principle.

Example 3 ■ Expanding a Determinant About a Row and a Column

Let A be the matrix of Example 2. Evaluate the determinant of A by expanding

(a) by the second row.

(b) by the third column.

Verify that each expansion gives the same value.

Solution

(a) Expanding by the second row, we get

$$\det(A) = \begin{vmatrix} 2 & 3 & -1 \\ 0 & 2 & 4 \\ -2 & 5 & 6 \end{vmatrix} = -0\begin{vmatrix} 3 & -1 \\ 5 & 6 \end{vmatrix} + 2\begin{vmatrix} 2 & -1 \\ -2 & 6 \end{vmatrix} - 4\begin{vmatrix} 2 & 3 \\ -2 & 5 \end{vmatrix}$$

$$= 0 + 2[2\cdot 6 - (-1)(-2)] - 4[2\cdot 5 - 3(-2)]$$

$$= 0 + 20 - 64 = -44$$

(b) Expanding by the third column gives

$$\det(A) = \begin{vmatrix} 2 & 3 & -1 \\ 0 & 2 & 4 \\ -2 & 5 & 6 \end{vmatrix}$$

$$= -1\begin{vmatrix} 0 & 2 \\ -2 & 5 \end{vmatrix} - 4\begin{vmatrix} 2 & 3 \\ -2 & 5 \end{vmatrix} + 6\begin{vmatrix} 2 & 3 \\ 0 & 2 \end{vmatrix}$$

$$= -[0\cdot 5 - 2(-2)] - 4[2\cdot 5 - 3(-2)] + 6(2\cdot 2 - 3\cdot 0)$$

$$= -4 - 64 + 24 = -44$$

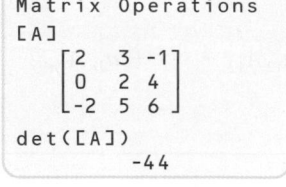

```
Matrix Operations
[A]
 [2   3 -1]
 [0   2  4]
 [-2  5  6]
det([A])
        -44
```

Figure 1

In both cases we obtain the same value for the determinant as when we expanded by the first row in Example 2.

We can also use a graphing device to compute determinants, as shown in Figure 1.

Now Try Exercise 39 ■

The following criterion allows us to determine whether a square matrix has an inverse without actually calculating the inverse. This is one of the most important uses of the determinant in matrix algebra, and it is the reason for the name *determinant*.

Invertibility Criterion

If A is a square matrix, then A has an inverse if and only if $\det(A) \neq 0$.

We will not prove this fact, but from the formula for the inverse of a 2×2 matrix (Section 9.5), you can see why it is true in the 2×2 case.

Pictorial Press Ltd/Alamy Stock Photo

EMMY NOETHER (1882–1935) was one of the foremost mathematicians of the early 20th century. Her groundbreaking work in abstract algebra provided much of the foundation for this field, and her work in invariant theory was essential in the development of Einstein's theory of general relativity. Although women weren't allowed to study at German universities in her time, she audited courses unofficially and went on to receive a doctorate at Erlangen *summa cum laude*, despite the opposition of the academic senate, which declared that women students would "overthrow all academic order." She subsequently taught mathematics at Göttingen, Moscow, and Frankfurt. In 1933 she left Germany to escape Nazi persecution, accepting a position at Bryn Mawr College in suburban Philadelphia. She lectured there and at the Institute for Advanced Study in Princeton, New Jersey, until her untimely death in 1935.

Example 4 ■ Using the Determinant to Show That a Matrix Is Not Invertible

Show that the matrix A has no inverse.

$$A = \begin{bmatrix} 1 & 2 & 0 & 4 \\ 0 & 0 & 0 & 3 \\ 5 & 6 & 2 & 6 \\ 2 & 4 & 0 & 9 \end{bmatrix}$$

Solution We begin by calculating the determinant of A. Since all but one of the elements of the second row is zero, we expand the determinant by the second row. If we do this, we see from the following equation that only the cofactor A_{24} will have to be calculated.

$$\begin{aligned} \det(A) &= \begin{vmatrix} 1 & 2 & 0 & 4 \\ 0 & 0 & 0 & 3 \\ 5 & 6 & 2 & 6 \\ 2 & 4 & 0 & 9 \end{vmatrix} \\ &= -0 \cdot A_{21} + 0 \cdot A_{22} - 0 \cdot A_{23} + 3 \cdot A_{24} = 3A_{24} \\ &= 3\begin{vmatrix} 1 & 2 & 0 \\ 5 & 6 & 2 \\ 2 & 4 & 0 \end{vmatrix} \qquad \text{Expand this by column 3} \\ &= 3(-2)\begin{vmatrix} 1 & 2 \\ 2 & 4 \end{vmatrix} \\ &= 3(-2)(1 \cdot 4 - 2 \cdot 2) = 0 \end{aligned}$$

Since the determinant of A is zero, A cannot have an inverse, by the Invertibility Criterion.

 **Now Try Exercise 25** ■

■ Row and Column Transformations

The preceding example shows that if we expand a determinant about a row or column that contains many zeros, our work is reduced considerably because we don't have to evaluate the cofactors of the elements that are zero. The following principle often simplifies the process of finding a determinant by introducing zeros into the matrix without changing the value of the determinant.

Row and Column Transformations of a Determinant

If A is a square matrix and if the matrix B is obtained from A by adding a multiple of one row to another or a multiple of one column to another, then $\det(A) = \det(B)$.

Example 5 ■ Using Row and Column Transformations to Calculate a Determinant

Find the determinant of the matrix A. Does it have an inverse?

$$A = \begin{bmatrix} 8 & 2 & -1 & -4 \\ 3 & 5 & -3 & 11 \\ 24 & 6 & 1 & -12 \\ 2 & 2 & 7 & -1 \end{bmatrix}$$

Solution If we add −3 times row 1 to row 3, we change all but one element of row 3 to zeros.

$$\begin{bmatrix} 8 & 2 & -1 & -4 \\ 3 & 5 & -3 & 11 \\ 0 & 0 & 4 & 0 \\ 2 & 2 & 7 & -1 \end{bmatrix}$$

This new matrix has the same determinant as A, and if we expand its determinant by the third row, we get

$$\det(A) = 4\begin{vmatrix} 8 & 2 & -4 \\ 3 & 5 & 11 \\ 2 & 2 & -1 \end{vmatrix}$$

Now, adding 2 times column 3 to column 1 in this determinant gives us

$$\det(A) = 4\begin{vmatrix} 0 & 2 & -4 \\ 25 & 5 & 11 \\ 0 & 2 & -1 \end{vmatrix} \qquad \text{Expand this by column 1}$$

$$= 4(-25)\begin{vmatrix} 2 & -4 \\ 2 & -1 \end{vmatrix}$$

$$= 4(-25)[2(-1) - (-4)2] = -600$$

Since the determinant of A is not zero, A does have an inverse.

 Now Try Exercise 35 ■

■ Cramer's Rule

The solutions of linear equations can sometimes be expressed by using determinants. To illustrate, let's solve the following pair of linear equations for the variable x.

$$\begin{cases} ax + by = r \\ cx + dy = s \end{cases}$$

To eliminate the variable y, we multiply the first equation by d and the second by b and subtract.

$$\begin{aligned} adx + bdy &= rd \\ bcx + bdy &= bs \\ \hline adx - bcx &= rd - bs \end{aligned}$$

Slobodian.Y/Shutterstock.com

Discovery Project ■ Computer Graphics II

Matrix algebra is the basic tool used in computer graphics. Properties of each pixel in an image are stored in a large matrix in the computer memory. In this project we discover how matrix multiplication can be used to "move" a point in the plane to a prescribed location. Combining such moves for each pixel in an image enables us to stretch, compress, translate, and otherwise transform an image on a computer screen by using matrix algebra. You can find the project at **www.stewartmath.com**.

Factoring the left-hand side, we get $(ad - bc)x = rd - bs$. Assuming that $ad - bc \neq 0$, we can now solve this equation for x:

$$x = \frac{rd - bs}{ad - bc}$$

Similarly, we find

$$y = \frac{as - cr}{ad - bc}$$

The numerator and denominator of the fractions for x and y are determinants of 2×2 matrices. So we can express the solution of the system using determinants as follows.

Cramer's Rule for Systems in Two Variables

The linear system

$$\begin{cases} ax + by = r \\ cx + dy = s \end{cases}$$

has the solution

$$x = \frac{\begin{vmatrix} r & b \\ s & d \end{vmatrix}}{\begin{vmatrix} a & b \\ c & d \end{vmatrix}} \qquad y = \frac{\begin{vmatrix} a & r \\ c & s \end{vmatrix}}{\begin{vmatrix} a & b \\ c & d \end{vmatrix}}$$

provided that $\begin{vmatrix} a & b \\ c & d \end{vmatrix} \neq 0$.

Using the notation

$$D = \begin{bmatrix} a & b \\ c & d \end{bmatrix} \qquad D_x = \begin{bmatrix} r & b \\ s & d \end{bmatrix} \qquad D_y = \begin{bmatrix} a & r \\ c & s \end{bmatrix}$$

Coefficient matrix

Replace first column of D by r and s

Replace second column of D by r and s

we can write the solution of the system as

$$x = \frac{|D_x|}{|D|} \quad \text{and} \quad y = \frac{|D_y|}{|D|}$$

Example 6 ■ Using Cramer's Rule to Solve a System with Two Variables

Use Cramer's Rule to solve the system.

$$\begin{cases} 2x + 6y = -1 \\ x + 8y = 2 \end{cases}$$

Solution For this system we have

$$|D| = \begin{vmatrix} 2 & 6 \\ 1 & 8 \end{vmatrix} = 2 \cdot 8 - 6 \cdot 1 = 10$$

$$|D_x| = \begin{vmatrix} -1 & 6 \\ 2 & 8 \end{vmatrix} = (-1)8 - 6 \cdot 2 = -20$$

$$|D_y| = \begin{vmatrix} 2 & -1 \\ 1 & 2 \end{vmatrix} = 2 \cdot 2 - (-1)1 = 5$$

The solution is

$$x = \frac{|D_x|}{|D|} = \frac{-20}{10} = -2$$

$$y = \frac{|D_y|}{|D|} = \frac{5}{10} = \frac{1}{2}$$

 **Now Try Exercise 41** ■

Cramer's Rule can be extended to apply to any system of n linear equations in n variables in which the determinant of the coefficient matrix is not zero. As we saw in the preceding section, any such system can be written in matrix form as

$$\begin{bmatrix} a_{11} & a_{12} & \cdots & a_{1n} \\ a_{21} & a_{22} & \cdots & a_{2n} \\ \vdots & \vdots & \ddots & \vdots \\ a_{n1} & a_{n2} & \cdots & a_{nn} \end{bmatrix} \begin{bmatrix} x_1 \\ x_2 \\ \vdots \\ x_n \end{bmatrix} = \begin{bmatrix} b_1 \\ b_2 \\ \vdots \\ b_n \end{bmatrix}$$

By analogy with our derivation of Cramer's Rule in the case of two equations in two unknowns, we let D be the coefficient matrix in this system, and D_{x_i} be the matrix obtained by replacing the ith column of D by the numbers $b_1, b_2, \ldots, b_n$ that appear to the right of the equal sign. The solution of the system is then given by the following rule.

Cramer's Rule

If a system of n linear equations in the n variables $x_1, x_2, \ldots, x_n$ is equivalent to the matrix equation $DX = B$, and if $|D| \neq 0$, then its solutions are

$$x_1 = \frac{|D_{x_1}|}{|D|} \qquad x_2 = \frac{|D_{x_2}|}{|D|} \qquad \cdots \qquad x_n = \frac{|D_{x_n}|}{|D|}$$

where D_{x_i} is the matrix obtained by replacing the ith column of D by the $n \times 1$ matrix B.

Example 7 ■ Using Cramer's Rule to Solve a System with Three Variables

Use Cramer's Rule to solve the system.

$$\begin{cases} 2x - 3y + 4z = 1 \\ x \qquad\quad + 6z = 0 \\ 3x - 2y \qquad\quad = 5 \end{cases}$$

Solution First, we evaluate the determinants that appear in Cramer's Rule. Note that D is the coefficient matrix and that D_x, D_y, and D_z are obtained by replacing the first, second, and third columns of D by the constant terms.

$$|D| = \begin{vmatrix} 2 & -3 & 4 \\ 1 & 0 & 6 \\ 3 & -2 & 0 \end{vmatrix} = -38 \qquad |D_x| = \begin{vmatrix} 1 & -3 & 4 \\ 0 & 0 & 6 \\ 5 & -2 & 0 \end{vmatrix} = -78$$

$$|D_y| = \begin{vmatrix} 2 & 1 & 4 \\ 1 & 0 & 6 \\ 3 & 5 & 0 \end{vmatrix} = -22 \qquad |D_z| = \begin{vmatrix} 2 & -3 & 1 \\ 1 & 0 & 0 \\ 3 & -2 & 5 \end{vmatrix} = 13$$

Now we use Cramer's Rule to get the solution:

$$x = \frac{|D_x|}{|D|} = \frac{-78}{-38} = \frac{39}{19} \qquad y = \frac{|D_y|}{|D|} = \frac{-22}{-38} = \frac{11}{19}$$

$$z = \frac{|D_z|}{|D|} = \frac{13}{-38} = -\frac{13}{38}$$

Now Try Exercise 47 ■

Solving the system in Example 7 using Gaussian elimination would involve matrices whose elements are fractions with fairly large denominators. Thus in cases like Examples 6 and 7, Cramer's Rule gives us an efficient way to solve systems of linear equations. But in systems with more than three equations, evaluating the various determinants that are involved is usually a long and tedious task (unless you are using a graphing device). Moreover, the rule doesn't apply if $|D| = 0$ or if D is not a square matrix. So Cramer's Rule is a useful alternative to Gaussian elimination, but only in some situations.

■ Areas of Triangles Using Determinants

Determinants provide a simple way to calculate the area of a triangle in the coordinate plane.

Area of a Triangle

If a triangle in the coordinate plane has vertices (a_1, b_1), (a_2, b_2), and (a_3, b_3), then its area is

$$\mathcal{A} = \pm\tfrac{1}{2}\begin{vmatrix} a_1 & b_1 & 1 \\ a_2 & b_2 & 1 \\ a_3 & b_3 & 1 \end{vmatrix}$$

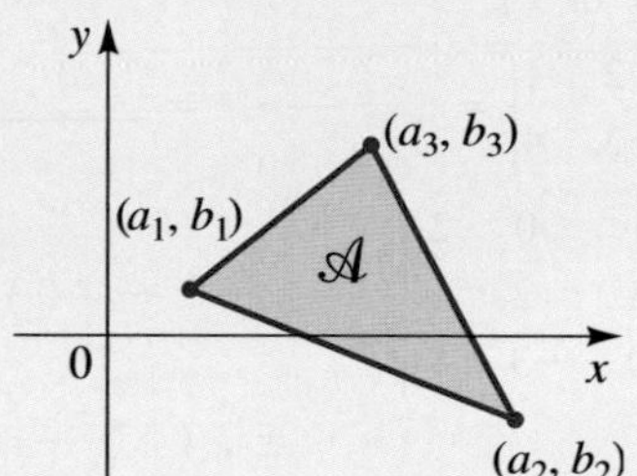

where the sign is chosen to make the area positive.

You are asked to prove this formula in Exercise 74.

Example 8 ■ Area of a Triangle

Find the area of the triangle shown in Figure 2.

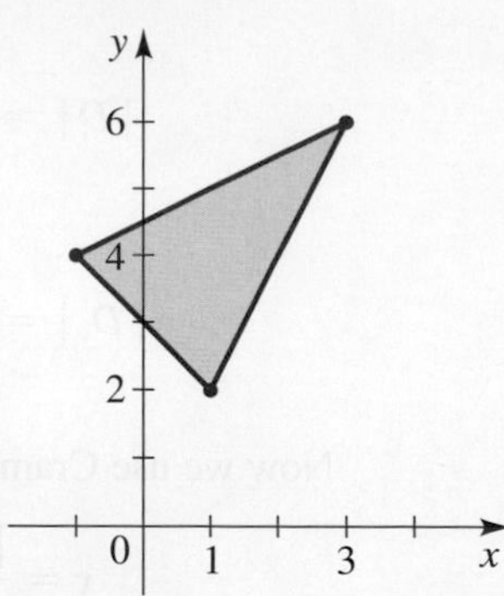

Figure 2

Solution The vertices are $(-1, 4)$, $(3, 6)$, and $(1, 2)$. Using the formula for the area of a triangle, we get

$$\mathscr{A} = \pm\tfrac{1}{2}\begin{vmatrix} -1 & 4 & 1 \\ 3 & 6 & 1 \\ 1 & 2 & 1 \end{vmatrix} = \pm\tfrac{1}{2}(-12)$$

To make the area positive, we choose the negative sign in the formula. Thus the area of the triangle is

$$\mathscr{A} = -\tfrac{1}{2}(-12) = 6$$

Now Try Exercise 57 ■

We can calculate the determinant in Example 8 by using the methods of this section or by using a graphing device.

```
Matrix Operations
[A]
[-1  4  1]
[ 3  6  1]
[ 1  2  1]
det([A])
              -12
```

9.6 Exercises

■ Concepts

1. *True or false?* $\det(A)$ is defined only for a square matrix A.

2. *True or false?* $\det(A)$ is a number, not a matrix.

3. *True or false?* If $\det(A) = 0$, then A is not invertible.

4. Fill in the blanks with appropriate numbers to calculate the determinant. Where there is "±", choose the appropriate sign (+ or −).

(a) $\begin{vmatrix} 2 & 1 \\ -3 & 4 \end{vmatrix} = \blacksquare\blacksquare - \blacksquare\blacksquare = \_\_\_\_\_\_$

(b) $\begin{vmatrix} 1 & 0 & 2 \\ 3 & 2 & 1 \\ 0 & -3 & 4 \end{vmatrix} = \pm\blacksquare(\blacksquare\blacksquare - \blacksquare\blacksquare) \pm \blacksquare(\blacksquare\blacksquare - \blacksquare\blacksquare)$

$\pm\blacksquare(\blacksquare\blacksquare - \blacksquare\blacksquare) = \_\_\_\_\_\_$

■ Skills

5–14 ■ Finding Determinants Find the determinant of the matrix, if it exists.

5. $\begin{bmatrix} 2 & 0 \\ 0 & 3 \end{bmatrix}$

6. $\begin{bmatrix} 0 & -1 \\ 2 & 0 \end{bmatrix}$

7. $\begin{bmatrix} \frac{3}{2} & 1 \\ -1 & -\frac{2}{3} \end{bmatrix}$

8. $\begin{bmatrix} 0.2 & 0.4 \\ -0.4 & -0.8 \end{bmatrix}$

9. $\begin{bmatrix} 4 & 5 \\ 0 & -1 \end{bmatrix}$

10. $\begin{bmatrix} -2 & 1 \\ 3 & -2 \end{bmatrix}$

11. $\begin{bmatrix} 2 & 5 \end{bmatrix}$

12. $\begin{bmatrix} 3 \\ 0 \end{bmatrix}$

13. $\begin{bmatrix} \frac{1}{2} & \frac{1}{8} \\ 1 & \frac{1}{2} \end{bmatrix}$

14. $\begin{bmatrix} 2.2 & -1.4 \\ 0.5 & 1.0 \end{bmatrix}$

15–20 ■ Minors and Cofactors Evaluate the specified minor and cofactor using the matrix A.

$$A = \begin{bmatrix} 1 & 0 & \frac{1}{2} \\ -3 & 5 & 2 \\ 0 & 0 & 4 \end{bmatrix}$$

15. M_{11}, A_{11}

16. M_{33}, A_{33}

17. M_{12}, A_{12}

18. M_{13}, A_{13}

19. M_{23}, A_{23}

20. M_{32}, A_{32}

21–28 ■ Finding Determinants Find the determinant of the matrix. Determine whether the matrix has an inverse, but don't calculate the inverse.

21. $\begin{bmatrix} 2 & 1 & 0 \\ 0 & -2 & 4 \\ 0 & 1 & -3 \end{bmatrix}$ **22.** $\begin{bmatrix} 1 & 2 & 5 \\ -2 & -3 & 2 \\ 3 & 5 & 3 \end{bmatrix}$

23. $\begin{bmatrix} 30 & 0 & 20 \\ 0 & -10 & -20 \\ 40 & 0 & 10 \end{bmatrix}$ **24.** $\begin{bmatrix} -2 & -\frac{3}{2} & \frac{1}{2} \\ 2 & 4 & 0 \\ \frac{1}{2} & 2 & 1 \end{bmatrix}$

25. $\begin{bmatrix} 1 & 3 & 7 \\ 2 & 0 & 8 \\ 0 & 2 & 2 \end{bmatrix}$ **26.** $\begin{bmatrix} 0 & -1 & 0 \\ 2 & 6 & 4 \\ 1 & 0 & 3 \end{bmatrix}$

27. $\begin{bmatrix} 1 & 3 & 3 & 0 \\ 0 & 2 & 0 & 1 \\ -1 & 0 & 0 & 2 \\ 1 & 6 & 4 & 1 \end{bmatrix}$ **28.** $\begin{bmatrix} 1 & 2 & 0 & 2 \\ 3 & -4 & 0 & 4 \\ 0 & 1 & 6 & 0 \\ 1 & 0 & 2 & 0 \end{bmatrix}$

29–34 ■ Finding Determinants Use a graphing device to find the determinant of the matrix. Determine whether the matrix has an inverse, but don't calculate the inverse.

29. $\begin{bmatrix} 1 & 2 & -1 \\ 2 & 2 & 1 \\ 1 & 2 & 2 \end{bmatrix}$

30. $\begin{bmatrix} 10 & -20 & 31 \\ 10 & -11 & 45 \\ -20 & 40 & -50 \end{bmatrix}$

31. $\begin{bmatrix} 1 & 10 & 2 & 7 \\ 2 & 18 & 18 & 13 \\ -3 & -30 & -4 & -24 \\ 1 & 10 & 2 & 10 \end{bmatrix}$

32. $\begin{bmatrix} 1 & 3 & -2 & 5 \\ -3 & -9 & 11 & 5 \\ 2 & 6 & 0 & 31 \\ 5 & 15 & -10 & 39 \end{bmatrix}$

33. $\begin{bmatrix} 4 & 3 & -2 & 10 \\ -8 & -6 & 24 & -1 \\ 20 & 15 & 3 & 27 \\ 12 & 9 & -6 & -1 \end{bmatrix}$

34. $\begin{bmatrix} 2 & 3 & -5 & 10 \\ -2 & -2 & 26 & 3 \\ 6 & 9 & -16 & 45 \\ -8 & -12 & 20 & -36 \end{bmatrix}$

35–38 ■ Determinants Using Row and Column Operations Evaluate the determinant, using row or column operations whenever possible to simplify your work.

35. $\begin{vmatrix} 0 & 0 & 4 & 6 \\ 2 & 1 & 1 & 3 \\ 2 & 1 & 2 & 3 \\ 3 & 0 & 1 & 7 \end{vmatrix}$ **36.** $\begin{vmatrix} -2 & 3 & -1 & 7 \\ 4 & 6 & -2 & 3 \\ 7 & 7 & 0 & 5 \\ 3 & -12 & 4 & 0 \end{vmatrix}$

37. $\begin{vmatrix} 1 & 2 & 3 & 4 & 5 \\ 0 & 2 & 4 & 6 & 8 \\ 0 & 0 & 3 & 6 & 9 \\ 0 & 0 & 0 & 4 & 8 \\ 0 & 0 & 0 & 0 & 5 \end{vmatrix}$ **38.** $\begin{vmatrix} 2 & -1 & 6 & 4 \\ 7 & 2 & -2 & 5 \\ 4 & -2 & 10 & 8 \\ 6 & 1 & 1 & 4 \end{vmatrix}$

39. Calculating a Determinant in Different Ways Consider the matrix

$$B = \begin{bmatrix} 4 & 1 & 0 \\ -2 & -1 & 1 \\ 4 & 0 & 3 \end{bmatrix}$$

(a) Evaluate $\det(B)$ by expanding by the second row.

(b) Evaluate $\det(B)$ by expanding by the third column.

(c) Do your results in parts (a) and (b) agree?

40. Determinant of a Special Matrix Find the determinant of a 10×10 matrix which has a 2 in each main diagonal entry and zeros everywhere else.

41–56 ■ Cramer's Rule Use Cramer's Rule to solve the system.

41. $\begin{cases} 2x - y = -9 \\ x + 2y = 8 \end{cases}$ **42.** $\begin{cases} 6x + 12y = 33 \\ 4x + 7y = 20 \end{cases}$

43. $\begin{cases} x - 6y = 3 \\ 3x + 2y = 1 \end{cases}$ **44.** $\begin{cases} \frac{1}{2}x + \frac{1}{3}y = 1 \\ \frac{1}{4}x - \frac{1}{6}y = -\frac{3}{2} \end{cases}$

45. $\begin{cases} 0.4x + 1.2y = 0.4 \\ 1.2x + 1.6y = 3.2 \end{cases}$ **46.** $\begin{cases} 10x - 17y = 21 \\ 20x - 31y = 39 \end{cases}$

47. $\begin{cases} x - y + 2z = 0 \\ 3x + z = 11 \\ -x + 2y = 0 \end{cases}$ **48.** $\begin{cases} 5x - 3y + z = 6 \\ 4y - 6z = 22 \\ 7x + 10y = -13 \end{cases}$

49. $\begin{cases} 2x_1 + 3x_2 - 5x_3 = 1 \\ x_1 + x_2 - x_3 = 2 \\ 2x_2 + x_3 = 8 \end{cases}$ **50.** $\begin{cases} -2a + c = 2 \\ a + 2b - c = 9 \\ 3a + 5b + 2c = 22 \end{cases}$

51. $\begin{cases} \frac{1}{3}x - \frac{1}{5}y + \frac{1}{2}z = \frac{7}{10} \\ -\frac{2}{3}x + \frac{2}{5}y + \frac{3}{2}z = \frac{11}{10} \\ x - \frac{4}{5}y + z = \frac{9}{5} \end{cases}$ **52.** $\begin{cases} 2x - y = 5 \\ 5x + 3z = 19 \\ 4y + 7z = 17 \end{cases}$

53. $\begin{cases} 3y + 5z = 4 \\ 2x - z = 10 \\ 4x + 7y = 0 \end{cases}$ **54.** $\begin{cases} 2x - 5y = 4 \\ x + y - z = 8 \\ 3x + 5z = 0 \end{cases}$

55. $\begin{cases} x + y + z + w = 0 \\ 2x + w = 0 \\ y - z = 0 \\ x + 2z = 1 \end{cases}$ **56.** $\begin{cases} x + y = 1 \\ y + z = 2 \\ z + w = 3 \\ w - x = 4 \end{cases}$

57–60 ■ Area of a Triangle Sketch the triangle with the given vertices, and use a determinant to find its area.

57. $(0, 0), (6, 2), (3, 8)$

58. $(1, 0), (3, 5), (-2, 2)$

59. $(-1, 3), (2, 9), (5, -6)$

60. $(-2, 5), (7, 2), (3, -4)$

Skills Plus

61–62 ■ Determinants of Special Matrices Evaluate the determinants.

61. $\begin{vmatrix} a & 0 & 0 & 0 & 0 \\ 0 & b & 0 & 0 & 0 \\ 0 & 0 & c & 0 & 0 \\ 0 & 0 & 0 & d & 0 \\ 0 & 0 & 0 & 0 & e \end{vmatrix}$ **62.** $\begin{vmatrix} a & a & a & a & a \\ 0 & a & a & a & a \\ 0 & 0 & a & a & a \\ 0 & 0 & 0 & a & a \\ 0 & 0 & 0 & 0 & a \end{vmatrix}$

63–66 ■ Determinant Equations Solve for x.

63. $\begin{vmatrix} x & 12 & 13 \\ 0 & x-1 & 23 \\ 0 & 0 & x-2 \end{vmatrix} = 0$ **64.** $\begin{vmatrix} x & 1 & 1 \\ 1 & 1 & x \\ x & 1 & x \end{vmatrix} = 0$

65. $\begin{vmatrix} 1 & 0 & x \\ x^2 & 1 & 0 \\ x & 0 & 1 \end{vmatrix} = 0$ **66.** $\begin{vmatrix} a & b & x-a \\ x & x+b & x \\ 0 & 1 & 1 \end{vmatrix} = 0$

67. Using Determinants Show that

$$\begin{vmatrix} 1 & x & x^2 \\ 1 & y & y^2 \\ 1 & z & z^2 \end{vmatrix} = (x-y)(y-z)(z-x)$$

68. Number of Solutions of a Linear System Consider the system

$$\begin{cases} x + 2y + 6z = 5 \\ -3x - 6y + 5z = 8 \\ 2x + 6y + 9z = 7 \end{cases}$$

(a) Verify that $x = -1$, $y = 0$, $z = 1$ is a solution of the system.

(b) Find the determinant of the coefficient matrix.

(c) Without solving the system, determine whether there are any other solutions.

(d) Can Cramer's Rule be used to solve this system? Why or why not?

69. Collinear Points and Determinants

(a) If three points lie on a line, what is the area of the "triangle" that they determine? Use the answer to this question, together with the determinant formula for the area of a triangle, to explain why the points (a_1, b_1), (a_2, b_2), and (a_3, b_3) are collinear if and only if

$$\begin{vmatrix} a_1 & b_1 & 1 \\ a_2 & b_2 & 1 \\ a_3 & b_3 & 1 \end{vmatrix} = 0$$

(b) Use a determinant to check whether each set of points is collinear. Graph them to verify your answer.

(i) $(-6, 4)$, $(2, 10)$, $(6, 13)$

(ii) $(-5, 10)$, $(2, 6)$, $(15, -2)$

70. Determinant Form for the Equation of a Line

(a) Use the result of Exercise 69(a) to show that the equation of the line containing the points (x_1, y_1) and (x_2, y_2) is

$$\begin{vmatrix} x & y & 1 \\ x_1 & y_1 & 1 \\ x_2 & y_2 & 1 \end{vmatrix} = 0$$

(b) Use the result of part (a) to find an equation for the line containing the points $(20, 50)$ and $(-10, 25)$.

Applications

71. Buying Fruit A roadside fruit stand sells apples at 75¢ a kilogram, peaches at 90¢ a kilogram, and pears at 60¢ a kilogram. A customer buys a total of 18 kg of these fruits at a total cost of \$13.80. The peaches and pears together cost \$1.80 more than the apples.

(a) Set up a linear system for the number of kilograms of apples, peaches, and pears the customer purchased.

(b) Solve the system using Cramer's Rule.

72. The Arch of a Bridge The opening of a railway bridge over a roadway is in the shape of a parabola. A surveyor measures the heights of three points on the bridge, as shown in the figure. The shape of the arch can be modeled by an equation of the form

$$y = ax^2 + bx + c$$

(a) Use the surveyed points to set up a system of linear equations for the unknown coefficients a, b, and c.

(b) Solve the system using Cramer's Rule.

73. A Triangular Plot of Land An outdoors club is purchasing land to set up a conservation area. The last remaining piece they need to buy is the triangular plot shown in the figure. Use the determinant formula for the area of a triangle to find the area of the plot.

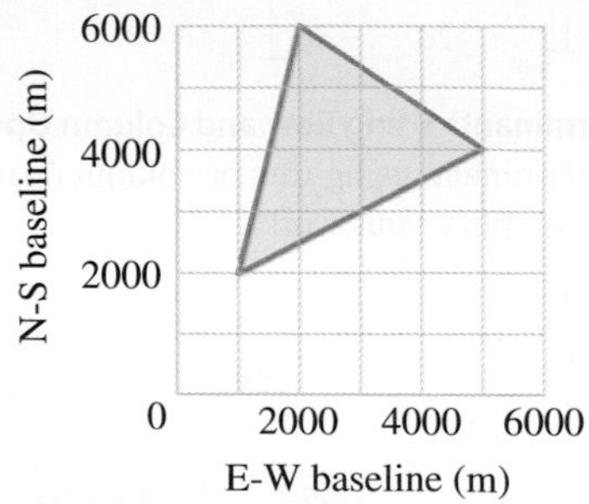

■ Discuss ■ Discover ■ Prove ■ Write

74. Discover ■ Prove: Determinant Formula for the Area of a Triangle The figure shows a triangle in the plane with vertices (a_1, b_1), (a_2, b_2), and (a_3, b_3).

(a) Find the coordinates of the vertices of the surrounding rectangle, and find its area.

(b) Find the area of the red triangle by subtracting the areas of the three blue triangles from the area of the rectangle.

(c) Use your answer to part (b) to show that the area $\mathcal{A}$ of the red triangle is given by

$$\mathcal{A} = \pm\frac{1}{2}\begin{vmatrix} a_1 & b_1 & 1 \\ a_2 & b_2 & 1 \\ a_3 & b_3 & 1 \end{vmatrix}$$

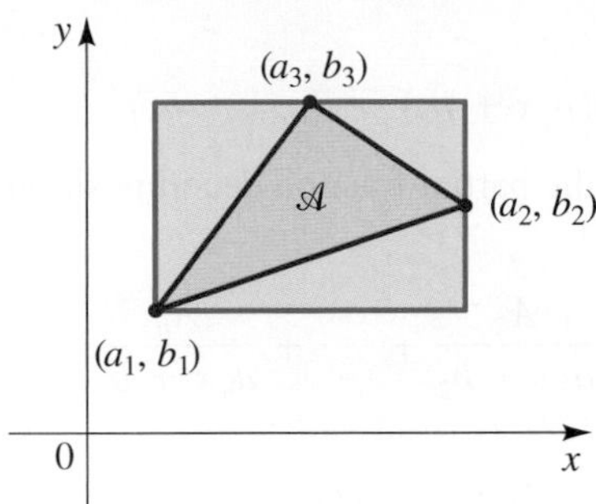

75. Discuss: Matrices with Determinant Zero Use the definition of determinant and the elementary row and column operations to explain why matrices of the following types have determinant 0.

(a) A matrix with a row or column consisting entirely of zeros

(b) A matrix with two rows that are the same or two columns that are the same

(c) A matrix in which one row is a multiple of another row, or one column is a multiple of another column

76. Discuss ■ Write: Solving Linear Systems Suppose you have to solve a linear system with five equations and five variables without the assistance of a calculator or computer. Which method would you prefer: Cramer's Rule or Gaussian elimination? Write a short paragraph explaining the reasons for your answer.

77. Prove: The Determinant of a Product of Matrices Let A and B be $n \times n$ matrices. It is known that

$$\det(AB) = \det(A)\det(B)$$

Use this fact to find a formula for $\det(A^{-1})$ in terms of $\det(A)$.

PS *Try to recognize something familiar.* Use the fact that $AA^{-1} = I_n$ and $\det I_n = 1$.

9.7 Partial Fractions

■ Distinct Linear Factors ■ Repeated Linear Factors ■ Irreducible Quadratic Factors ■ Repeated Irreducible Quadratic Factors

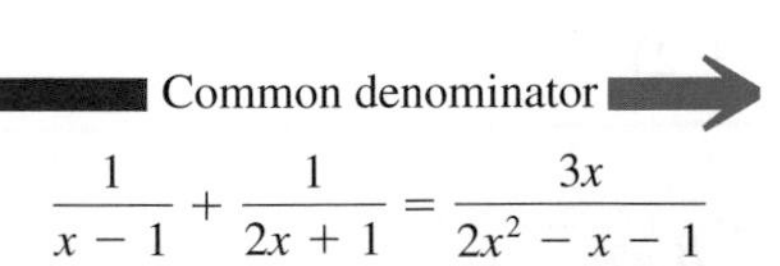

To write a sum or difference of fractional expressions as a single fraction, we bring them to a common denominator. For example,

$$\frac{1}{x-1} + \frac{1}{2x+1} = \frac{(2x+1) + (x-1)}{(x-1)(2x+1)} = \frac{3x}{2x^2 - x - 1}$$

But for some applications of algebra to calculus we must reverse this process—that is, we must express a fraction such as $3x/(2x^2 - x - 1)$ as the sum of the simpler fractions $1/(x-1)$ and $1/(2x+1)$. These simpler fractions are called *partial fractions*; we learn how to find them in this section.

Let r be the rational function

$$r(x) = \frac{P(x)}{Q(x)}$$

where the degree of P is less than the degree of Q. By the Linear and Quadratic Factors Theorem in Section 3.5, every polynomial with real coefficients can be factored completely into linear and irreducible quadratic factors, that is, factors of the form $ax + b$ and $ax^2 + bx + c$, where a, b, and c are real numbers. For instance,

$$x^4 - 1 = (x^2 - 1)(x^2 + 1) = (x - 1)(x + 1)(x^2 + 1)$$

After we have completely factored the denominator Q of r, we can express $r(x)$ as a sum of **partial fractions** of the form

$$\frac{A}{(ax+b)^i} \quad \text{and} \quad \frac{Ax+B}{(ax^2+bx+c)^j}$$

This sum is called the **partial fraction decomposition** of r. Let's examine the details of the four possible cases.

■ Distinct Linear Factors

We first consider the case in which the denominator factors into distinct linear factors.

Case 1: The Denominator Is a Product of Distinct Linear Factors

Suppose that we can factor $Q(x)$ as

$$Q(x) = (a_1x + b_1)(a_2x + b_2)\cdots(a_nx + b_n)$$

with no factor repeated. In this case the partial fraction decomposition of $P(x)/Q(x)$ takes the form

$$\frac{P(x)}{Q(x)} = \frac{A_1}{a_1x + b_1} + \frac{A_2}{a_2x + b_2} + \cdots + \frac{A_n}{a_nx + b_n}$$

The constants $A_1, A_2, \ldots, A_n$ are determined as in the following example.

Example 1 ■ Distinct Linear Factors

Find the partial fraction decomposition of $\dfrac{5x+7}{x^3 + 2x^2 - x - 2}$.

Solution The denominator factors as follows.

$$\begin{aligned} x^3 + 2x^2 - x - 2 &= x^2(x+2) - (x+2) = (x^2 - 1)(x+2) \\ &= (x-1)(x+1)(x+2) \end{aligned}$$

This gives us the partial fraction decomposition

$$\frac{5x+7}{x^3 + 2x^2 - x - 2} = \frac{A}{x-1} + \frac{B}{x+1} + \frac{C}{x+2}$$

Multiplying each side by the common denominator, $(x-1)(x+1)(x+2)$, we get

$$\begin{aligned} 5x + 7 &= A(x+1)(x+2) + B(x-1)(x+2) + C(x-1)(x+1) \\ &= A(x^2 + 3x + 2) + B(x^2 + x - 2) + C(x^2 - 1) && \text{Expand} \\ &= (A + B + C)x^2 + (3A + B)x + (2A - 2B - C) && \text{Combine like terms} \end{aligned}$$

If two polynomials are equal, then their coefficients are equal. Thus because $5x + 7$ has no x^2-term, we have $A + B + C = 0$. Similarly, by comparing the coefficients of x, we see that $3A + B = 5$, and by comparing constant terms, we get $2A - 2B - C = 7$. This leads to the following system of linear equations for A, B, and C.

$$\begin{cases} A + B + C = 0 & \text{Equation 1: Coefficients of } x^2 \\ 3A + B = 5 & \text{Equation 2: Coefficients of } x \\ 2A - 2B - C = 7 & \text{Equation 3: Constant coefficients} \end{cases}$$

THE RHIND PAPYRUS is the oldest known mathematical document. It is an Egyptian scroll written in 1650 B.C. by the scribe Ahmes, who explains that it is an exact copy of a scroll written 200 years earlier. Ahmes claims that his papyrus contains "a thorough study of all things, insight into all that exists, knowledge of all obscure secrets." In fact, the document contains rules for doing arithmetic, including multiplication and division of fractions and several exercises with solutions. The exercise shown below reads: "A heap and its seventh make nineteen; how large is the heap?" In solving problems of this sort, the Egyptians used partial fractions because their number system required all fractions to be written as sums of reciprocals of whole numbers. For example, $\frac{7}{12}$ would be written as $\frac{1}{3} + \frac{1}{4}$.

The papyrus gives a correct formula for the volume of a truncated pyramid, which the ancient Egyptians probably used when building the pyramids at Giza. It also gives the formula $A = \left(\frac{8}{9}d\right)^2$ for the area of a circle with diameter d. How close is this to the actual area?

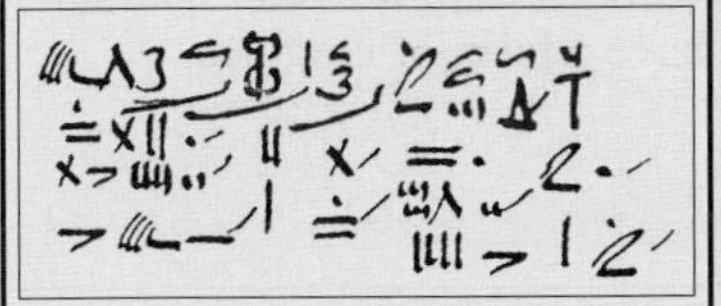

We use Gaussian elimination to solve this system.

$$\begin{cases} A + B + C = 0 \\ \quad -2B - 3C = 5 & \text{Equation 2} + (-3) \times \text{Equation 1} \\ \quad -4B - 3C = 7 & \text{Equation 3} + (-2) \times \text{Equation 1} \end{cases}$$

$$\begin{cases} A + B + C = 0 \\ \quad -2B - 3C = 5 \\ \qquad\quad 3C = -3 & \text{Equation 3} + (-2) \times \text{Equation 2} \end{cases}$$

From the third equation we get $C = -1$. Back-substituting, we find that $B = -1$ and $A = 2$. So the partial fraction decomposition is

$$\frac{5x + 7}{x^3 + 2x^2 - x - 2} = \frac{2}{x - 1} + \frac{-1}{x + 1} + \frac{-1}{x + 2}$$

Now Try Exercises 3 and 13 ■

The same approach works in the remaining cases: Set up the partial fraction decomposition with the unknown constants $A, B, C, \ldots$. Then multiply each side of the resulting equation by the common denominator, combine like terms on the right-hand side of the equation, and equate coefficients. This gives a set of linear equations that will always have a unique solution (provided that the partial fraction decomposition has been set up correctly).

■ Repeated Linear Factors

We now consider the case in which the denominator factors into linear factors, some of which are repeated.

Case 2: The Denominator Is a Product of Linear Factors, Some of Which Are Repeated

Suppose the complete factorization of $Q(x)$ contains the linear factor $ax + b$ repeated k times; that is, $(ax + b)^k$ is a factor of $Q(x)$. Then, corresponding to each such factor, the partial fraction decomposition for $P(x)/Q(x)$ contains

$$\frac{A_1}{ax + b} + \frac{A_2}{(ax + b)^2} + \cdots + \frac{A_k}{(ax + b)^k}$$

Example 2 ■ Repeated Linear Factors

Find the partial fraction decomposition of $\dfrac{x^2 + 1}{x(x - 1)^3}$.

Solution Because the factor $x - 1$ is repeated three times in the denominator, the partial fraction decomposition has the form

$$\frac{x^2 + 1}{x(x - 1)^3} = \frac{A}{x} + \frac{B}{x - 1} + \frac{C}{(x - 1)^2} + \frac{D}{(x - 1)^3}$$

Multiplying each side by the common denominator, $x(x - 1)^3$, gives

$$\begin{aligned} x^2 + 1 &= A(x - 1)^3 + Bx(x - 1)^2 + Cx(x - 1) + Dx \\ &= A(x^3 - 3x^2 + 3x - 1) + B(x^3 - 2x^2 + x) + C(x^2 - x) + Dx && \text{Expand} \\ &= (A + B)x^3 + (-3A - 2B + C)x^2 + (3A + B - C + D)x - A && \text{Combine like terms} \end{aligned}$$

Equating coefficients, we get the following equations.

$$\begin{cases} A + B = 0 & \text{Coefficients of } x^3 \\ -3A - 2B + C = 1 & \text{Coefficients of } x^2 \\ 3A + B - C + D = 0 & \text{Coefficients of } x \\ -A = 1 & \text{Constant coefficients} \end{cases}$$

If we rearrange these equations by putting the last one in the first position, we can see (using substitution) that the solution to the system is $A = -1$, $B = 1$, $C = 0$, $D = 2$, so the partial fraction decomposition is

$$\frac{x^2+1}{x(x-1)^3} = \frac{-1}{x} + \frac{1}{x-1} + \frac{2}{(x-1)^3}$$

 **Now Try Exercises 5 and 29** ■

■ Irreducible Quadratic Factors

We now consider the case in which the denominator has distinct irreducible quadratic factors.

Case 3: The Denominator Has Irreducible Quadratic Factors, None of Which Is Repeated

Suppose the complete factorization of $Q(x)$ contains the quadratic factor $ax^2 + bx + c$ (which can't be factored further). Then, corresponding to this, the partial fraction decomposition of $P(x)/Q(x)$ will have a term of the form

$$\frac{Ax+B}{ax^2+bx+c}$$

Example 3 ■ Distinct Quadratic Factors

Find the partial fraction decomposition of $\dfrac{2x^2 - x + 4}{x^3 + 4x}$.

Solution Since $x^3 + 4x = x(x^2 + 4)$, which can't be factored further, we write

$$\frac{2x^2 - x + 4}{x^3 + 4x} = \frac{A}{x} + \frac{Bx + C}{x^2 + 4}$$

Multiplying by $x(x^2 + 4)$, we get

$$\begin{aligned} 2x^2 - x + 4 &= A(x^2 + 4) + (Bx + C)x \\ &= (A + B)x^2 + Cx + 4A \end{aligned}$$

Equating coefficients gives us the equations

$$\begin{cases} A + B = 2 & \text{Coefficients of } x^2 \\ C = -1 & \text{Coefficients of } x \\ 4A = 4 & \text{Constant coefficients} \end{cases}$$

so $A = 1$, $B = 1$, and $C = -1$. The required partial fraction decomposition is

$$\frac{2x^2 - x + 4}{x^3 + 4x} = \frac{1}{x} + \frac{x - 1}{x^2 + 4}$$

 **Now Try Exercises 7 and 37** ■

■ Repeated Irreducible Quadratic Factors

We now consider the case in which the denominator has irreducible quadratic factors, some of which are repeated.

> **Case 4: The Denominator Has a Repeated Irreducible Quadratic Factor**
>
> Suppose the complete factorization of $Q(x)$ contains the factor $(ax^2 + bx + c)^k$, where $ax^2 + bx + c$ can't be factored further. Then the partial fraction decomposition of $P(x)/Q(x)$ will have the terms
>
> $$\frac{A_1x + B_1}{ax^2 + bx + c} + \frac{A_2x + B_2}{(ax^2 + bx + c)^2} + \cdots + \frac{A_kx + B_k}{(ax^2 + bx + c)^k}$$

Example 4 ■ Repeated Quadratic Factors

Write the form of the partial fraction decomposition of

$$\frac{x^5 - 3x^2 + 12x - 1}{x^3(x^2 + x + 1)(x^2 + 2)^3}$$

Solution The irreducible quadratic factor $x^2 + 2$ is repeated three times.

$$\frac{x^5 - 3x^2 + 12x - 1}{x^3(x^2 + x + 1)(x^2 + 2)^3}$$

$$= \frac{A}{x} + \frac{B}{x^2} + \frac{C}{x^3} + \frac{Dx + E}{x^2 + x + 1} + \frac{Fx + G}{x^2 + 2} + \frac{Hx + I}{(x^2 + 2)^2} + \frac{Jx + K}{(x^2 + 2)^3}$$

Now Try Exercises 11 and 41 ■

To find the values of $A, B, C, D, E, F, G, H, I, J$, and K in Example 4, we would have to solve a system of 11 linear equations. Although possible, this would certainly involve a great deal of work!

The techniques that we have described in this section apply only to rational functions $P(x)/Q(x)$ in which the degree of P is less than the degree of Q. If this isn't the case, we must first use long division to divide Q into P.

Example 5 ■ Using Long Division to Prepare for Partial Fractions

Find the partial fraction decomposition of

$$\frac{2x^4 + 4x^3 - 2x^2 + x + 7}{x^3 + 2x^2 - x - 2}$$

Solution Since the degree of the numerator is larger than the degree of the denominator, we use long division to obtain

$$\begin{array}{r} 2x \\ x^3 + 2x^2 - x - 2 \overline{\smash{)}\,2x^4 + 4x^3 - 2x^2 + x + 7} \\ 2x^4 + 4x^3 - 2x^2 - 4x \\ \hline 5x + 7 \end{array}$$

$$\frac{2x^4 + 4x^3 - 2x^2 + x + 7}{x^3 + 2x^2 - x - 2} = 2x + \frac{5x + 7}{x^3 + 2x^2 - x - 2}$$

The remainder term now satisfies the requirement that the degree of the numerator is less than the degree of the denominator. At this point we proceed as in Example 1 to obtain the decomposition

$$\frac{2x^4 + 4x^3 - 2x^2 + x + 7}{x^3 + 2x^2 - x - 2} = 2x + \frac{2}{x - 1} + \frac{-1}{x + 1} + \frac{-1}{x + 2}$$

Now Try Exercise 43 ■

9.7 Exercises

■ Concepts

1–2 ■ For each rational function r, choose from (i)–(iv) the appropriate form for its partial fraction decomposition.

1. $r(x) = \dfrac{4}{x(x-2)^2}$

(i) $\dfrac{A}{x} + \dfrac{B}{x-2}$ (ii) $\dfrac{A}{x} + \dfrac{B}{(x-2)^2}$

(iii) $\dfrac{A}{x} + \dfrac{B}{x-2} + \dfrac{C}{(x-2)^2}$ (iv) $\dfrac{A}{x} + \dfrac{B}{x-2} + \dfrac{Cx+D}{(x-2)^2}$

2. $r(x) = \dfrac{2x+8}{(x-1)(x^2+4)}$

(i) $\dfrac{A}{x-1} + \dfrac{B}{x^2+4}$

(ii) $\dfrac{A}{x-1} + \dfrac{Bx+C}{x^2+4}$

(iii) $\dfrac{A}{x-1} + \dfrac{B}{x+2} + \dfrac{C}{x^2+4}$

(iv) $\dfrac{Ax+B}{x-1} + \dfrac{Cx+D}{x^2+4}$

■ Skills

3–12 ■ Form of the Partial Fraction Decomposition Write the form of the partial fraction decomposition of the function (as in Example 4). Do not determine the numerical values of the coefficients.

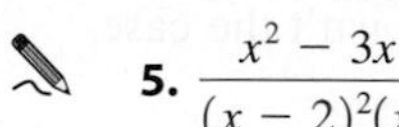

3. $\dfrac{1}{(x-1)(x+2)}$ **4.** $\dfrac{x}{x^2+3x-4}$

5. $\dfrac{x^2-3x+5}{(x-2)^2(x+4)}$ **6.** $\dfrac{1}{x^4-x^3}$

7. $\dfrac{x^2}{(x-3)(x^2+4)}$ **8.** $\dfrac{1}{x^4-1}$

9. $\dfrac{x^3-4x^2+2}{(x^2+1)(x^2+2)}$ **10.** $\dfrac{x^4+x^2+1}{x^2(x^2+4)^2}$

11. $\dfrac{x^3+x+1}{x(2x-5)^3(x^2+2x+5)^2}$ **12.** $\dfrac{1}{(x^3-1)(x^2-1)}$

13–44 ■ Partial Fraction Decomposition Find the partial fraction decomposition of the rational function.

13. $\dfrac{2}{(x-1)(x+1)}$ **14.** $\dfrac{2x}{(x-1)(x+1)}$

15. $\dfrac{5}{(x-1)(x+4)}$ **16.** $\dfrac{x+6}{x(x+3)}$

17. $\dfrac{12}{x^2-9}$ **18.** $\dfrac{x-12}{x^2-4x}$

19. $\dfrac{4}{x^2-4}$ **20.** $\dfrac{2x+1}{x^2+x-2}$

21. $\dfrac{x+14}{x^2-2x-8}$ **22.** $\dfrac{8x-3}{2x^2-x}$

23. $\dfrac{x}{8x^2-10x+3}$ **24.** $\dfrac{7x-3}{x^3+2x^2-3x}$

25. $\dfrac{9x^2-9x+6}{2x^3-x^2-8x+4}$ **26.** $\dfrac{-3x^2-3x+27}{(x+2)(2x^2+3x-9)}$

27. $\dfrac{x^2+1}{x^3+x^2}$ **28.** $\dfrac{3x^2+5x-13}{(3x+2)(x^2-4x+4)}$

29. $\dfrac{2x}{4x^2+12x+9}$ **30.** $\dfrac{x-4}{(2x-5)^2}$

31. $\dfrac{4x^2-x-2}{x^4+2x^3}$ **32.** $\dfrac{x^3-2x^2-4x+3}{x^4}$

33. $\dfrac{-10x^2+27x-14}{(x-1)^3(x+2)}$ **34.** $\dfrac{-2x^2+5x-1}{x^4-2x^3+2x-1}$

35. $\dfrac{3x^3+22x^2+53x+41}{(x+2)^2(x+3)^2}$ **36.** $\dfrac{3x^2+12x-20}{x^4-8x^2+16}$

37. $\dfrac{x-3}{x^3+3x}$ **38.** $\dfrac{3x^2-2x+8}{x^3-x^2+2x-2}$

39. $\dfrac{2x^3+7x+5}{(x^2+x+2)(x^2+1)}$ **40.** $\dfrac{x^2+x+1}{2x^4+3x^2+1}$

41. $\dfrac{x^4+x^3+x^2-x+1}{x(x^2+1)^2}$ **42.** $\dfrac{2x^2-x+8}{(x^2+4)^2}$

43. $\dfrac{x^5-2x^4+x^3+x+5}{x^3-2x^2+x-2}$

44. $\dfrac{x^5-3x^4+3x^3-4x^2+4x+12}{(x-2)^2(x^2+2)}$

■ Skills Plus

45. Partial Fractions Determine A and B in terms of a and b.

$$\frac{ax+b}{x^2-1} = \frac{A}{x-1} + \frac{B}{x+1}$$

46. Partial Fractions Determine A, B, C, and D in terms of a and b.

$$\frac{ax^3+bx^2}{(x^2+1)^2} = \frac{Ax+B}{x^2+1} + \frac{Cx+D}{(x^2+1)^2}$$

■ Discuss ■ Discover ■ Prove ■ Write

47. Discuss: Recognizing Partial Fraction Decompositions For each expression, determine whether it is already a partial fraction decomposition or whether it can be decomposed further.

(a) $\dfrac{x}{x^2+1} + \dfrac{1}{x+1}$ **(b)** $\dfrac{x}{(x+1)^2}$

(c) $\dfrac{1}{x+1} + \dfrac{2}{(x+1)^2}$ **(d)** $\dfrac{x+2}{(x^2+1)^2}$

48. Discuss: Assembling and Disassembling Partial Fractions The following expression is a partial fraction decomposition.

$$\frac{2}{x-1} + \frac{1}{(x-1)^2} + \frac{1}{x+1}$$

Use a common denominator to combine the terms into one fraction. Then use the techniques of this section to find its partial fraction decomposition. Did you get back the original expression?

9.8 Systems of Nonlinear Equations

■ Substitution and Elimination Methods ■ Graphical Method

In this section we solve systems of equations in which the equations are not all linear. The methods we learned in Section 9.1 can also be used to solve nonlinear systems.

■ Substitution and Elimination Methods

To solve a system of nonlinear equations, we can use the substitution or elimination method, as illustrated in the next examples.

Example 1 ■ Substitution Method

Find all solutions of the system.

$$\begin{cases} x^2 + y^2 = 100 & \text{Equation 1} \\ 3x - y = 10 & \text{Equation 2} \end{cases}$$

Solution **Solve for one variable.** We start by solving for y in the second equation.

$$y = 3x - 10 \qquad \text{Solve for } y \text{ in Equation 2}$$

Substitute. Next we substitute for y in the first equation and solve for x.

$$\begin{aligned} x^2 + (3x - 10)^2 &= 100 && \text{Substitute } y = 3x - 10 \text{ into Equation 1} \\ x^2 + (9x^2 - 60x + 100) &= 100 && \text{Expand} \\ 10x^2 - 60x &= 0 && \text{Simplify} \\ 10x(x - 6) &= 0 && \text{Factor} \\ x = 0 \quad \text{or} \quad x &= 6 && \text{Solve for } x \end{aligned}$$

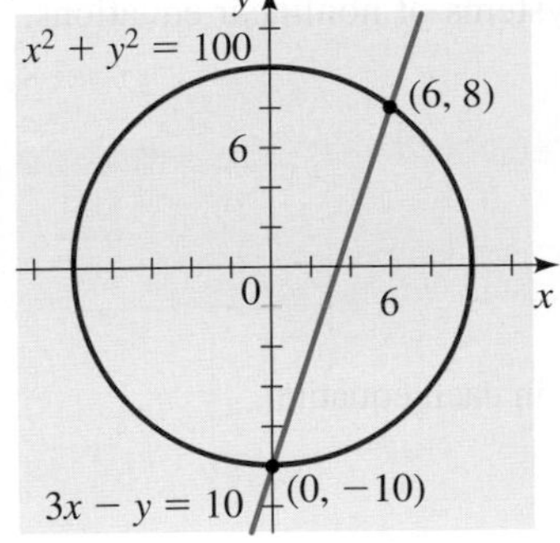

Figure 1

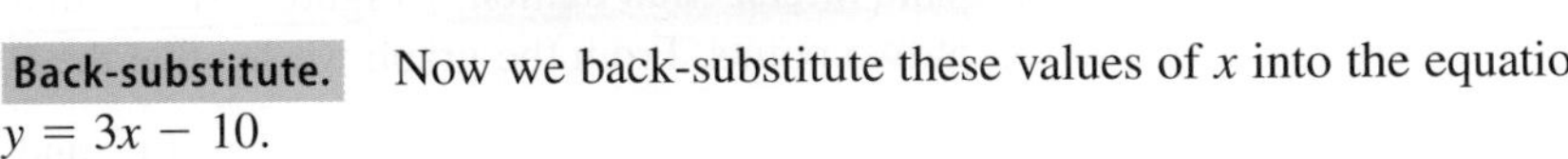

Back-substitute. Now we back-substitute these values of x into the equation $y = 3x - 10$.

$$\begin{aligned} &\text{For } x = 0: && y = 3(0) - 10 = -10 && \text{Back-substitute} \\ &\text{For } x = 6: && y = 3(6) - 10 = 8 && \text{Back-substitute} \end{aligned}$$

So we have two solutions: $(0, -10)$ and $(6, 8)$.

The graph of the first equation is a circle, and the graph of the second equation is a line. Figure 1 shows that the graphs intersect at the two points $(0, -10)$ and $(6, 8)$.

Now Try Exercise 5 ■

Check Your Answers

$x = 0, y = -10$:

$$\begin{cases} (0)^2 + (-10)^2 = 100 \\ 3(0) - (-10) = 10 \end{cases} \checkmark$$

$x = 6, y = 8$:

$$\begin{cases} (6)^2 + (8)^2 = 36 + 64 = 100 \\ 3(6) - (8) = 18 - 8 = 10 \end{cases} \checkmark$$

Example 2 ■ Elimination Method

Find all solutions of the system.

$$\begin{cases} 3x^2 + 2y = 26 & \text{Equation 1} \\ 5x^2 + 7y = 3 & \text{Equation 2} \end{cases}$$

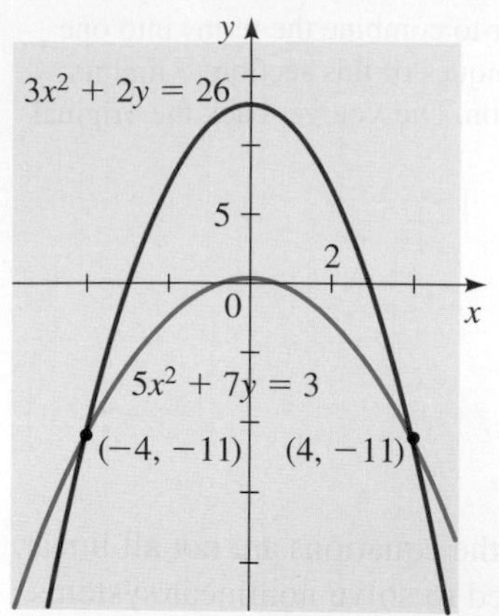

Figure 2

Check Your Answers

$x = -4, y = -11$:

$$\begin{cases} 3(-4)^2 + 2(-11) = 26 \\ 5(-4)^2 + 7(-11) = 3 \end{cases} \checkmark$$

$x = 4, y = -11$:

$$\begin{cases} 3(4)^2 + 2(-11) = 26 \\ 5(4)^2 + 7(-11) = 3 \end{cases} \checkmark$$

Solution **Adjust coefficints.** We choose to eliminate the terms containing x, so we multiply the first equation by 5 and the second equation by -3.

Eliminate a variable. We add the two equations and solve for y.

$$\begin{cases} 15x^2 + 10y = 130 & 5 \times \text{Equation 1} \\ -15x^2 - 21y = -9 & (-3) \times \text{Equation 2} \end{cases}$$

$$-11y = 121 \qquad \text{Add}$$

$$y = -11 \qquad \text{Solve for } y$$

Back-substitute. Now we back-substitute $y = -11$ into one of the original equations, say $3x^2 + 2y = 26$, and solve for x.

$$3x^2 + 2(-11) = 26 \qquad \text{Back-substitute } y = -11 \text{ into Equation 1}$$

$$3x^2 = 48 \qquad \text{Add 22}$$

$$x^2 = 16 \qquad \text{Divide by 3}$$

$$x = -4 \quad \text{or} \quad x = 4 \qquad \text{Solve for } x$$

So we have two solutions: $(-4, -11)$ and $(4, -11)$.

The graphs of both equations are parabolas (see Section 3.1). Figure 2 shows that the graphs intersect at the two points $(-4, -11)$ and $(4, -11)$.

Now Try Exercise 11 ■

■ Graphical Method

The graphical method is particularly useful in solving systems of nonlinear equations.

Example 3 ■ Graphical Method

Find all solutions of the system

$$\begin{cases} x^2 - y = 2 \\ 2x - y = -1 \end{cases}$$

Solution **Graph each equation.** To graph, we solve for y in each equation.

$$\begin{cases} y = x^2 - 2 \\ y = 2x + 1 \end{cases}$$

Find intersection points. Figure 3 shows that the graphs of these equations intersect at two points. From the graph we see that the solutions are

$$(-1, -1) \quad \text{and} \quad (3, 7)$$

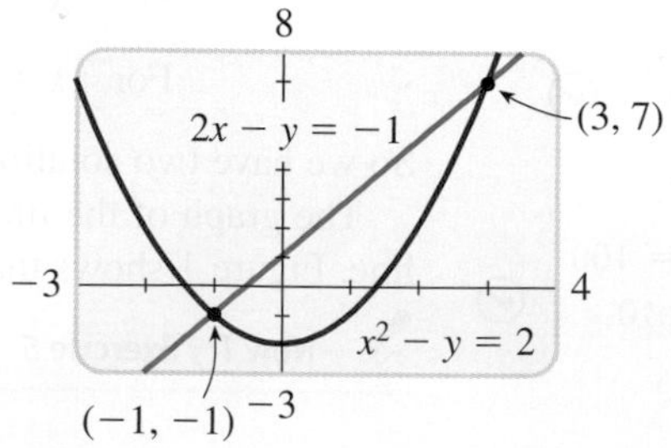

Figure 3

Check Your Answers

$x = -1, y = -1$:

$$\begin{cases} (-1)^2 - (-1) = 2 \\ 2(-1) - (-1) = -1 \end{cases} \checkmark$$

$x = 3, y = 7$:

$$\begin{cases} 3^2 - 7 = 2 \\ 2(3) - 7 = -1 \end{cases} \checkmark$$

Now Try Exercise 33 ■

Example 4 ■ Solving a System of Equations Graphically

Find all solutions of the system, rounded to one decimal place.

$$\begin{cases} x^2 + y^2 = 12 & \text{Equation 1} \\ y = 2x^2 - 5x & \text{Equation 2} \end{cases}$$

Solution **Graph each equation.** The graph of the first equation is a circle, and the graph of the second is a parabola. Using a graphing device, we graph the circle and the parabola on the same screen, as shown in Figure 4.

Find intersection points. The graphs intersect in Quadrants I and II. Zooming in, we see that the intersection points are $(-0.559, 3.419)$ and $(2.847, 1.974)$. There also appears to be an intersection point in Quadrant IV. However, when we zoom in, we see that the curves come close to each other but don't intersect (see Figure 5). Thus the system has two solutions; rounded to the nearest tenth, they are

$$(-0.6, 3.4) \quad \text{and} \quad (2.8, 2.0)$$

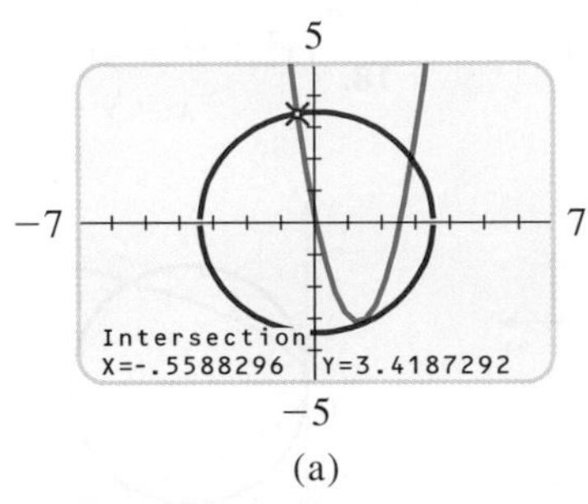

(a)

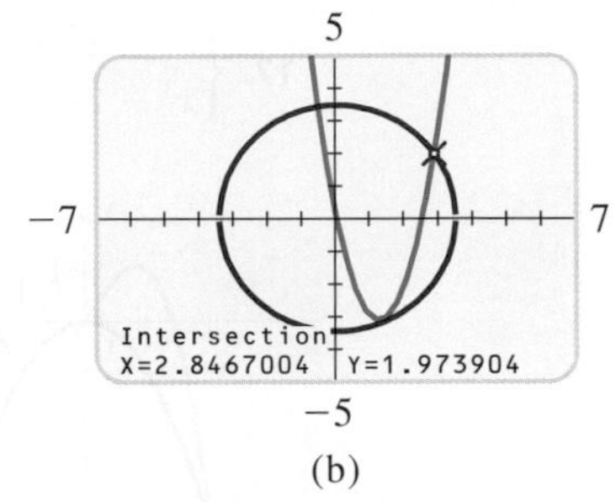

(b)

Figure 4 | $x^2 + y^2 = 12,\ y = 2x^2 - 5x$

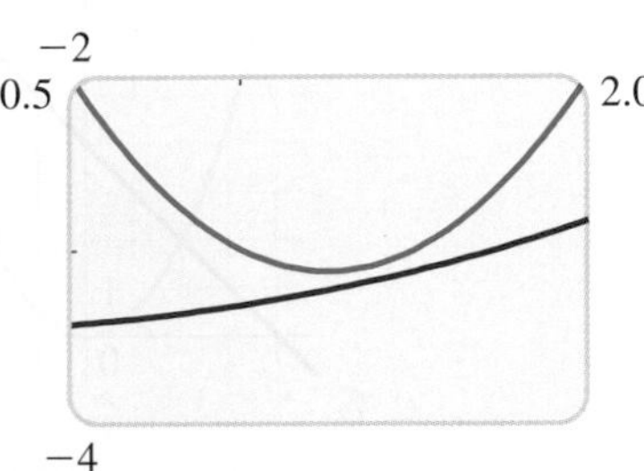

Figure 5 | Zooming in

Now Try Exercise 37 ■

Note Some graphing devices only graph functions, so to graph the circle in Example 4 we must first solve for y in terms of x.

$$x^2 + y^2 = 12$$

$$y^2 = 12 - x^2 \qquad \text{Isolate } y^2 \text{ on LHS}$$

$$y = \pm\sqrt{12 - x^2} \qquad \text{Take square roots}$$

To graph the complete circle, we must graph both functions.

$$y = \sqrt{12 - x^2} \quad \text{and} \quad y = -\sqrt{12 - x^2}$$

Mathematics in the Modern World

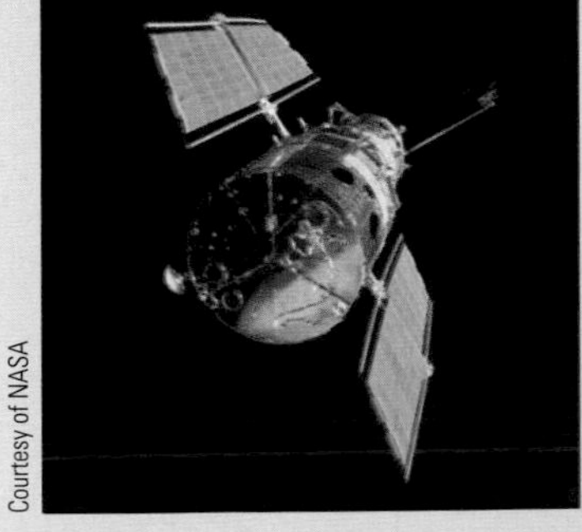

Global Positioning System (GPS)

On a cold, foggy day in 1707 a British naval fleet was sailing home at a fast clip. The fleet's navigators didn't know it, but the fleet was only a few meters from the rocky shores of England. In the ensuing disaster the fleet was totally destroyed. This tragedy could have been avoided had the navigators known their positions. In those days latitude was determined by the position of the North Star (and this could be done only at night in good weather), and longitude was determined by the position of the sun relative to where it would be in England *at that same time.* So navigation required an accurate method of telling time on ships. (The invention of the spring-loaded clock brought about the eventual solution.)

Since then, several different methods have been developed to determine position, and all rely heavily on mathematics (see LORAN, Section 10.3). The latest method, called the Global Positioning System (GPS), uses triangulation. In this system, 24 satellites are strategically located above the surface of the earth. A GPS device measures distance from a satellite, using the travel time of radio signals emitted from the satellite. Knowing the distances to three different satellites tells us that we are at the point of intersection of three different spheres. This uniquely determines our position (see Exercise 51).

9.8 Exercises

Concepts

1–2 ■ The system of equations

$$\begin{cases} 2y - x^2 = 0 \\ y - x = 4 \end{cases}$$

is graphed below.

1. Use the graph to find the solutions of the system.

2. Check that the solutions you found in Exercise 1 satisfy the system.

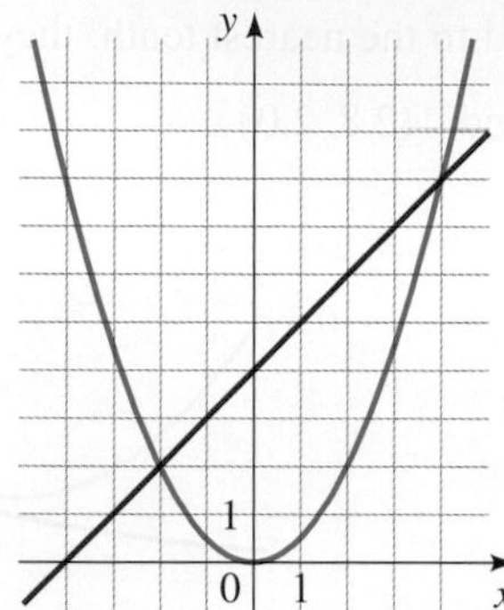

Skills

3–8 ■ Substitution Method Use the substitution method to find all solutions of the system of equations.

3. $\begin{cases} y = x^2 \\ y = x + 12 \end{cases}$

4. $\begin{cases} x^2 + y^2 = 25 \\ y = 2x \end{cases}$

5. $\begin{cases} x^2 + y^2 = 8 \\ x + y = 0 \end{cases}$

6. $\begin{cases} x^2 + y = 9 \\ x - y + 3 = 0 \end{cases}$

7. $\begin{cases} x + y^2 = 0 \\ 2x + 5y^2 = 75 \end{cases}$

8. $\begin{cases} x^2 - y = 1 \\ 2x^2 + 3y = 17 \end{cases}$

9–14 ■ Elimination Method Use the elimination method to find all solutions of the system of equations.

9. $\begin{cases} x^2 - 2y = 1 \\ x^2 + 5y = 29 \end{cases}$

10. $\begin{cases} 3x^2 + 4y = 17 \\ 2x^2 + 5y = 2 \end{cases}$

11. $\begin{cases} 3x^2 - y^2 = 11 \\ x^2 + 4y^2 = 8 \end{cases}$

12. $\begin{cases} 2x^2 + 4y = 13 \\ x^2 - y^2 = \frac{7}{2} \end{cases}$

13. $\begin{cases} x - y^2 + 3 = 0 \\ 2x^2 + y^2 - 4 = 0 \end{cases}$

14. $\begin{cases} x^2 - y^2 = 1 \\ 2x^2 - y^2 = x + 3 \end{cases}$

15–18 ■ Finding Intersection Points Graphically Two equations and their graphs are given. Estimate the intersection point from the graph and check that the point is a solution to the system.

15. $\begin{cases} x^2 + y = 8 \\ x - 2y = -6 \end{cases}$

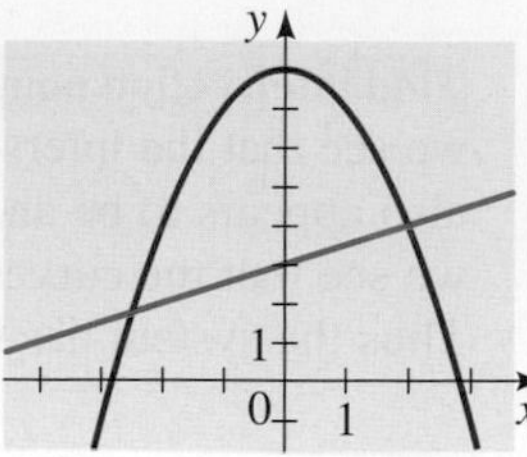

16. $\begin{cases} x - y^2 = -4 \\ x - y = 2 \end{cases}$

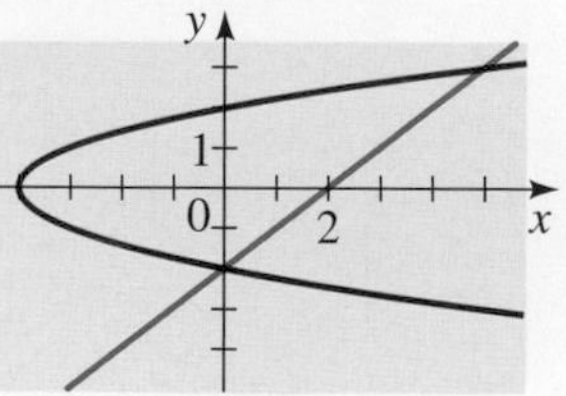

17. $\begin{cases} x^2 + y = 0 \\ x^3 - 2x - y = 0 \end{cases}$

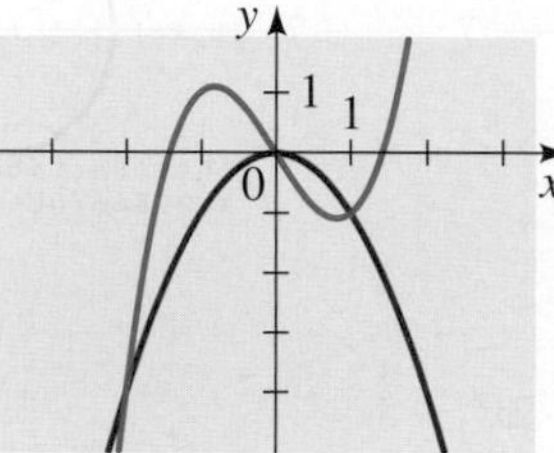

18. $\begin{cases} x^2 + y^2 = 4x \\ x = y^2 \end{cases}$

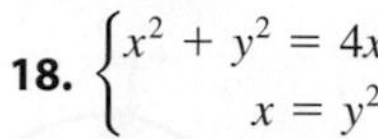

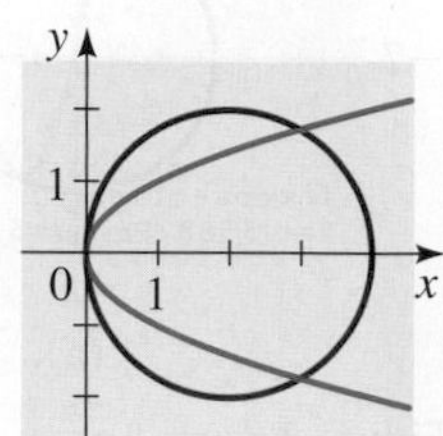

19–32 ■ Solving Nonlinear Systems Find all solutions of the system of equations.

19. $\begin{cases} y + x^2 = 4x \\ y + 4x = 16 \end{cases}$

20. $\begin{cases} x - y^2 = 0 \\ y - x^2 = 0 \end{cases}$

21. $\begin{cases} x - 2y = 2 \\ y^2 - x^2 = 2x + 4 \end{cases}$

22. $\begin{cases} y = 4 - x^2 \\ y = x^2 - 4 \end{cases}$

23. $\begin{cases} x - y = 4 \\ xy = 12 \end{cases}$

24. $\begin{cases} xy = 24 \\ 2x^2 - y^2 + 4 = 0 \end{cases}$

25. $\begin{cases} x^2y = 16 \\ x^2 + 4y + 16 = 0 \end{cases}$

26. $\begin{cases} x + \sqrt{y} = 0 \\ y^2 - 4x^2 = 12 \end{cases}$

27. $\begin{cases} x^2 + y^2 = 9 \\ x^2 - y^2 = 1 \end{cases}$

28. $\begin{cases} x^2 + 2y^2 = 2 \\ 2x^2 - 3y = 15 \end{cases}$

29. $\begin{cases} 2x^2 - 8y^3 = 19 \\ 4x^2 + 16y^3 = 34 \end{cases}$

30. $\begin{cases} x^4 + y^3 = 17 \\ 3x^4 + 5y^3 = 53 \end{cases}$

31. $\begin{cases} \dfrac{2}{x} - \dfrac{3}{y} = 1 \\ -\dfrac{4}{x} + \dfrac{7}{y} = 1 \end{cases}$

32. $\begin{cases} \dfrac{4}{x^2} + \dfrac{6}{y^4} = \dfrac{7}{2} \\ \dfrac{1}{x^2} - \dfrac{2}{y^4} = 0 \end{cases}$

33–40 ■ Graphical Method Use the graphical method to find all solutions of the system of equations, rounded to two decimal places.

33. $\begin{cases} y = x^2 + 8x \\ y = 2x + 16 \end{cases}$ **34.** $\begin{cases} y = x^2 - 4x \\ 2x - y = 2 \end{cases}$

35. $\begin{cases} x^2 + y^2 = 25 \\ x + 3y = 2 \end{cases}$ **36.** $\begin{cases} x^2 + y^2 = 17 \\ x^2 - 2x + y^2 = 13 \end{cases}$

37. $\begin{cases} \dfrac{x^2}{9} + \dfrac{y^2}{18} = 1 \\ y = -x^2 + 6x - 2 \end{cases}$ **38.** $\begin{cases} x^2 - y^2 = 3 \\ y = x^2 - 2x - 8 \end{cases}$

39. $\begin{cases} x^4 + 16y^4 = 32 \\ x^2 + 2x + y = 0 \end{cases}$ **40.** $\begin{cases} y = e^x + e^{-x} \\ y = 5 - x^2 \end{cases}$

Skills Plus

41–44 ■ Some Trickier Systems Follow the hints and solve the systems.

41. $\begin{cases} \log x + \log y = \frac{3}{2} \\ 2 \log x - \log y = 0 \end{cases}$ [*Hint:* Add the equations.]

42. $\begin{cases} 2^x + 2^y = 10 \\ 4^x + 4^y = 68 \end{cases}$ [*Hint:* Note that $4^x = 2^{2x} = (2^x)^2$.]

43. $\begin{cases} x - y = 3 \\ x^3 - y^3 = 387 \end{cases}$ [*Hint:* Factor the left-hand side of the second equation.]

44. $\begin{cases} x^2 + xy = 1 \\ xy + y^2 = 3 \end{cases}$ [*Hint:* Add the equations, and factor the result.]

Applications

45. Dimensions of a Rectangle A rectangle has an area of 180 cm^2 and a perimeter of 54 cm. What are its dimensions?

46. Legs of a Right Triangle A right triangle has an area of 84 m^2 and a hypotenuse 25 m long. What are the lengths of its other two sides?

47. Dimensions of a Rectangle The perimeter of a rectangle is 70, and its diagonal is 25. Find its length and width.

48. Dimensions of a Rectangle A circular piece of sheet metal has a diameter of 20 cm. The edges are to be cut off to form a rectangle of area 160 cm^2. (See the figure.) What are the dimensions of the rectangle?

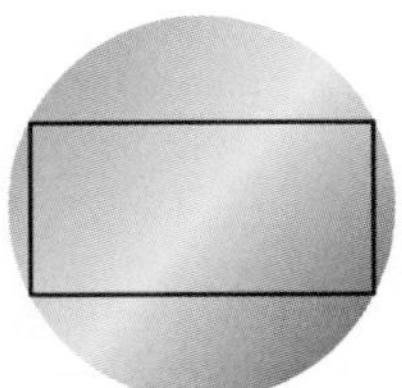

49. Flight of a Rocket A hill is inclined so that its "slope" is $\frac{1}{2}$, as shown in the figure. We introduce a coordinate system with the origin at the base of the hill and with the scales on the axes measured in meters. A rocket is fired from the base of the hill in such a way that its trajectory is the parabola $y = -x^2 + 401x$. At what point does the rocket strike the hillside? How far is this point from the base of the hill (to the nearest centimeter)?

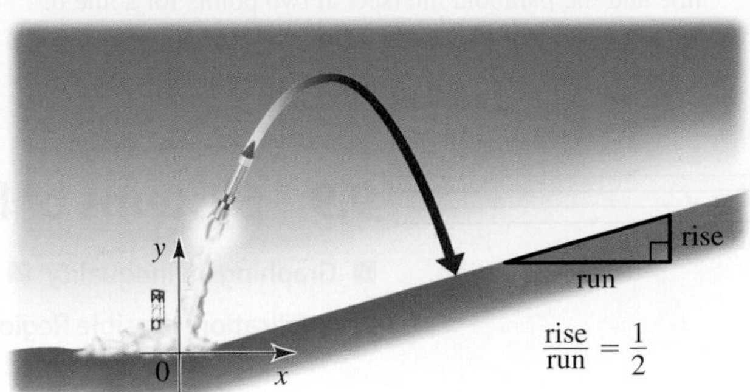

50. Making a Stovepipe A rectangular piece of sheet metal with an area of 1200 in^2 is to be bent into a cylindrical length of stovepipe having a volume of 600 in^3. What are the dimensions of the sheet metal?

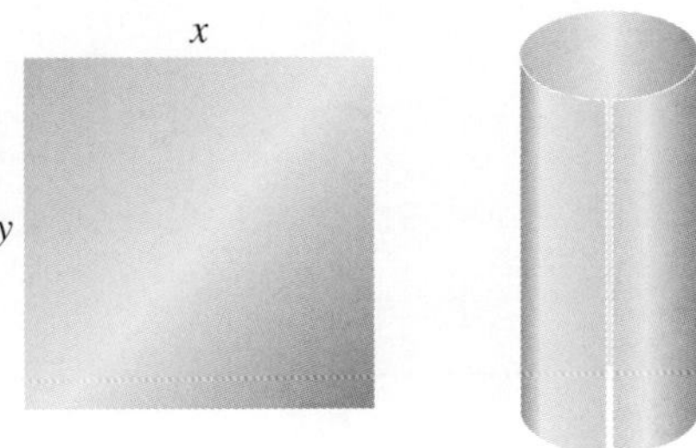

51. Global Positioning System (GPS) The Global Positioning System determines the location of an object from its distances to satellites in orbit around the earth. From the simplified, two-dimensional situation shown in the figure, determine the coordinates of P, using the fact that P is 26 units from satellite A and 20 units from satellite B.

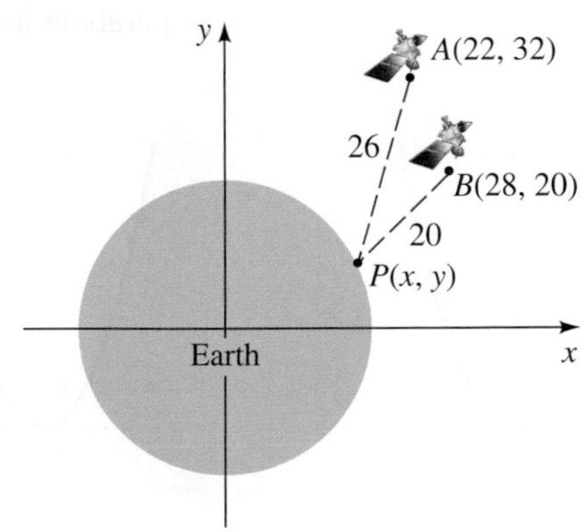

■ **Discuss** ■ **Discover** ■ **Prove** ■ **Write**

52. Discover ■ Prove: Intersection of a Parabola and a Line On a sheet of graph paper or using a graphing device, draw the parabola $y = x^2$. Then draw the graphs of the linear equation $y = x + k$ on the same coordinate plane for various values of k. Try to choose values of k so that the line and the parabola intersect at two points for some of your k's and not for others. For what value of k is there exactly one intersection point? Use the results of your experiment to make a conjecture about the values of k for which the following system has two solutions, one solution, and no solution. Prove your conjecture.

$$\begin{cases} y = x^2 \\ y = x + k \end{cases}$$

9.9 Systems of Inequalities

■ Graphing an Inequality ■ Systems of Inequalities ■ Systems of Linear Inequalities ■ Application: Feasible Regions

In this section we study systems of inequalities in two variables from a graphical point of view.

■ Graphing an Inequality

We begin by considering the graph of a single inequality. We already know that the graph of $y = x^2$, for example, is the *parabola* in Figure 1.

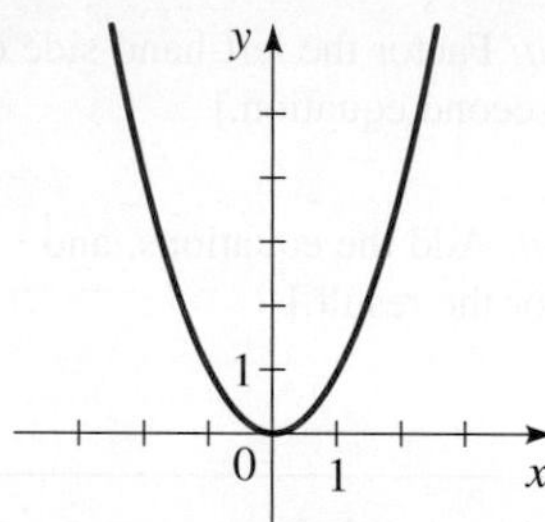

Figure 1 | $y = x^2$

If we replace the equal sign by the symbol $\geq$, we obtain the *inequality* $y \geq x^2$. Its graph consists of not just the parabola in Figure 1, but also every point whose y-coordinate is *larger* than x^2. We indicate the solution in Figure 2(a) by shading the points *above* the parabola.

Similarly, the graph of $y \leq x^2$ in Figure 2(b) consists of all points on and *below* the parabola. However, the graphs of $y > x^2$ and $y < x^2$ do not include the points on the parabola itself, as indicated by the dashed curves in Figures 2(c) and 2(d).

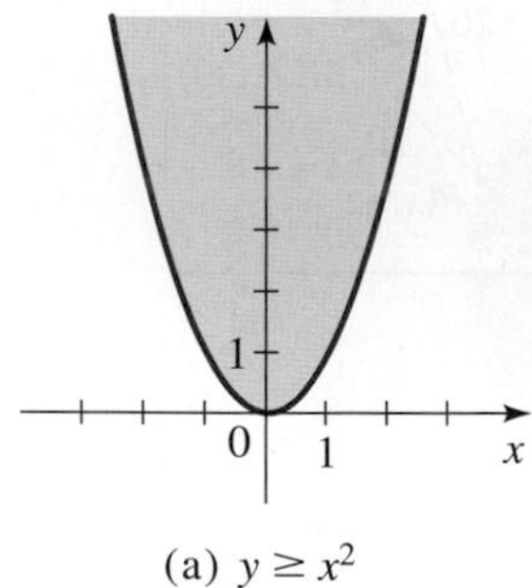

(a) $y \geq x^2$

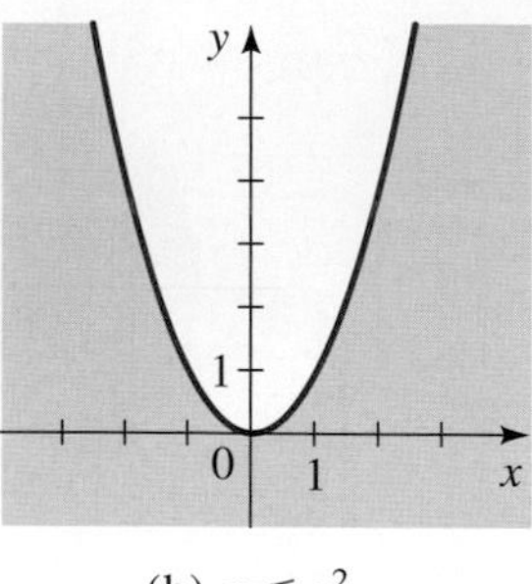

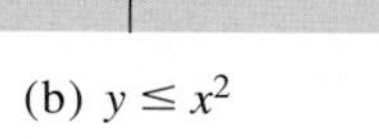

(b) $y \leq x^2$

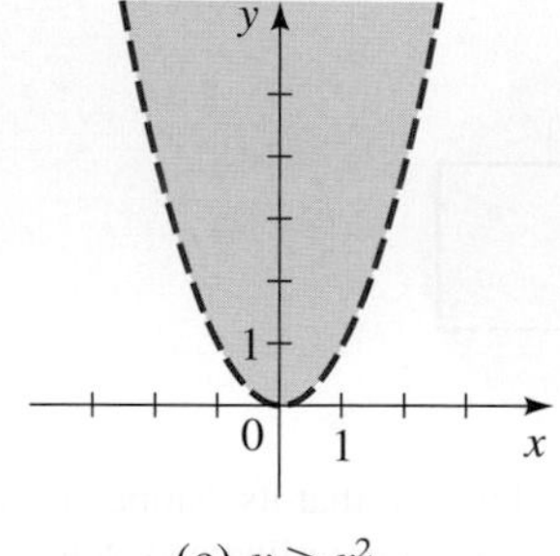

(c) $y > x^2$

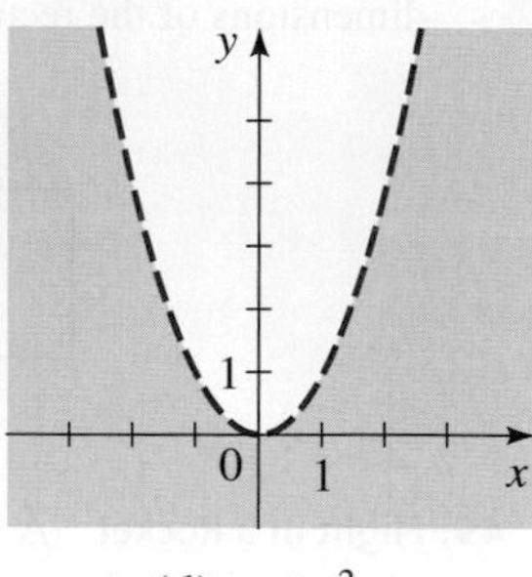

(d) $y < x^2$

Figure 2

The graph of an inequality, in general, consists of a region in the plane whose boundary is the graph of the equation obtained by replacing the inequality sign ($\geq$, $\leq$, $>$, or $<$) with an equal sign. To determine which side of the graph gives the solution set of the inequality, we need only check **test points**.

> **Graphing an Inequality**
>
> To graph an inequality, we carry out the following steps.
>
> 1. **Graph the Equation.** Graph the equation that corresponds to the inequality. Use a dashed curve for $>$ or $<$ and a solid curve for $\leq$ or $\geq$.
> 2. **Graph the Inequality.** The graph of the inequality consists of all the points on one side of the curve that we graphed in Step 1. We use **test points** on either side of the curve to determine whether the points on each side satisfy the inequality. If the point satisfies the inequality, then all the points on that side of the curve satisfy the inequality. In that case, **shade that side of the curve** to indicate that it is part of the graph. If the test point does not satisfy the inequality, then the region isn't part of the graph.

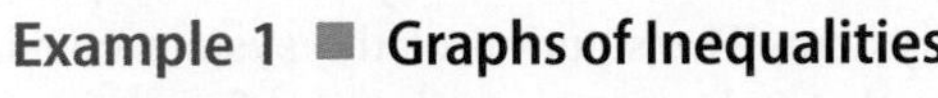

Example 1 ■ Graphs of Inequalities

Graph each inequality.

(a) $x^2 + y^2 < 25$ **(b)** $x + 2y \geq 5$

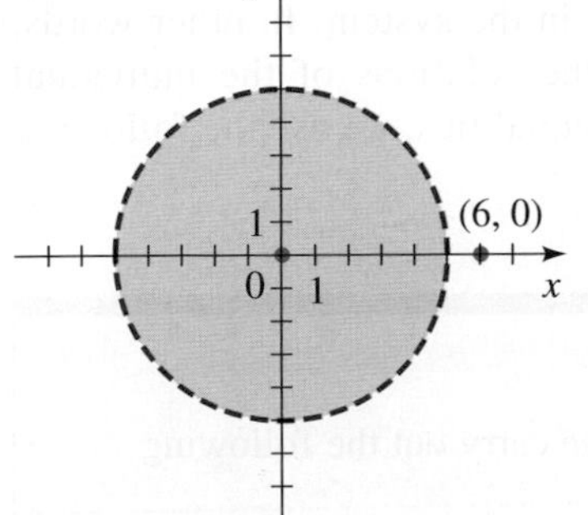

Figure 3 | Graph of $x^2 + y^2 < 25$

Solution We follow the guidelines stated in the preceding box.

(a) **Graph the equation.** The graph of the equation $x^2 + y^2 = 25$ is a circle of radius 5 centered at the origin. The points on the circle itself do not satisfy the inequality because it is of the form $<$, so we graph the circle with a dashed curve, as shown in Figure 3.

Graph the inequality. To determine whether the inside or the outside of the circle satisfies the inequality, we use the test points $(0, 0)$ on the inside and $(6, 0)$ on the outside. To do this, we substitute the coordinates of each point into the inequality and check whether the result satisfies the inequality.

Note that *any* point inside or outside the circle can serve as a test point. We have chosen these points for simplicity.

Test Point	Inequality $x^2 + y^2 < 25$	Conclusion
(0, 0)	$0^2 + 0^2 \overset{?}{<} 25$ ✓	Part of graph
(6, 0)	$6^2 + 0^2 \overset{?}{<} 25$ ✗	Not part of graph

Our check shows that the points *inside* the circle satisfy the inequality. A graph of the inequality is shown in Figure 3.

(b) **Graph the equation.** We first graph the equation $x + 2y = 5$. The graph is the line shown in Figure 4, on the next page.

Graph the inequality. Let's use the test points $(0, 0)$ and $(5, 5)$ on either side of the line.

Test Point	Inequality $x + 2y \geq 5$	Conclusion
(0, 0)	$0 + 2(0) \overset{?}{\geq} 5$ ✗	Not part of graph
(5, 5)	$5 + 2(5) \overset{?}{\geq} 5$ ✓	Part of graph

We can write the inequality in Example 1(b) as

$$y \geq -\tfrac{1}{2}x + \tfrac{5}{2}$$

From this form of the inequality we see that the solution consists of the points with y-values *on* or *above* the line $y = -\frac{1}{2}x + \frac{5}{2}$. So the graph of the inequality is the region *above* the line.

Our check shows that the points *above* the line satisfy the inequality. A graph of the inequality is shown in Figure 4.

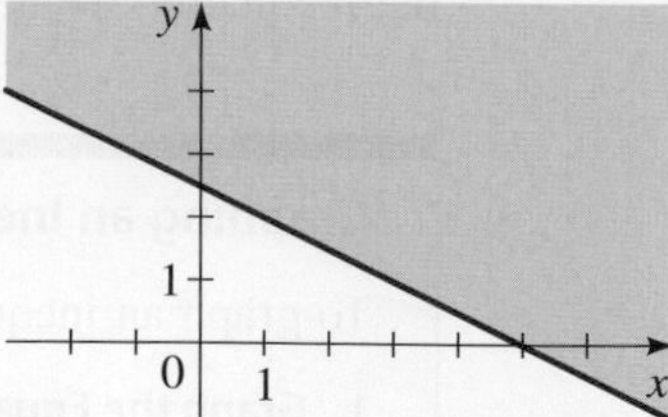

Figure 4 | Graph of $x + 2y \geq 5$

Now Try Exercises 15 and 21 ■

■ Systems of Inequalities

We now consider *systems* of inequalities. The **solution set of a system of inequalities** in two variables is the set of all points in the coordinate plane that satisfy every inequality in the system. The **graph of a system of inequalities** is the graph of the solution set.

To find the solution of a system of inequalities, we first graph each inequality in the system. The solution of the system consists of those points in the coordinate plane that belong to the solution of each inequality in the system. In other words, the solution of the system is the intersection of the solutions of the individual inequalities in the system. So to solve a system of inequalities, we use the following guidelines.

The Solution of a System of Inequalities

To graph the solution of a system of inequalities, we carry out the following steps.

1. **Graph Each Inequality.** Graph each inequality in the system on the same graph.
2. **Graph the Solution of the System.** Shade the region where the graphs of all the inequalities intersect. All the points in this region satisfy each inequality, so they belong to the solution of the system.
3. **Find the Vertices.** Label the vertices of the region that you shaded in Step 2.

Example 2 ■ A System of Two Inequalities

Graph the solution of the system of inequalities, and label its vertices.

$$\begin{cases} x^2 + y^2 < 25 \\ x + 2y \geq 5 \end{cases}$$

Solution These are the two inequalities of Example 1. Here we want to graph only those points that simultaneously satisfy both inequalities.

Graph each inequality. In Figure 5 we graph the solutions of the two inequalities on the same axes (in different colors).

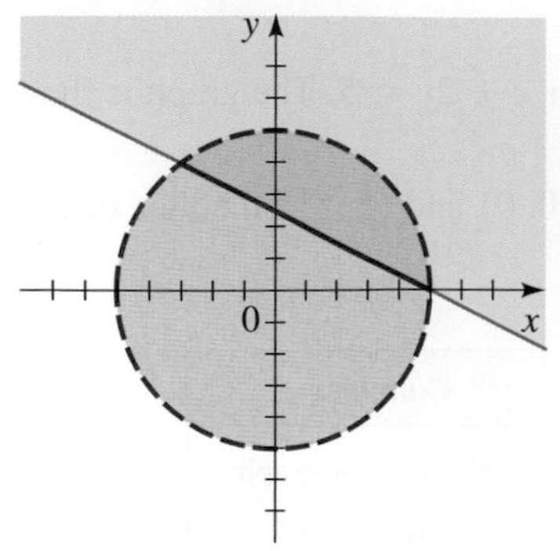

Figure 5

Graph the solution of the system. The solution of the system of inequalities is the intersection of the two graphs. This is the region where the two regions overlap, which is the purple region graphed in Figure 6.

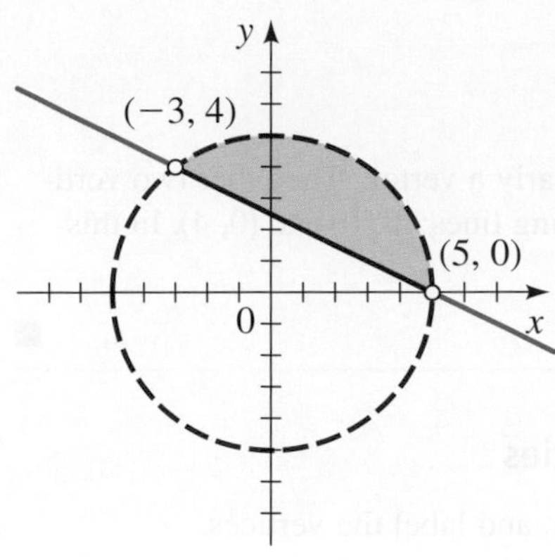

Figure 6 | $\begin{cases} x^2 + y^2 < 25 \\ x + 2y \geq 5 \end{cases}$

Find the Vertices. The points $(-3, 4)$ and $(5, 0)$ in Figure 6 are the **vertices** of the solution set. They are obtained by solving the system of *equations*

$$\begin{cases} x^2 + y^2 = 25 \\ x + 2y = 5 \end{cases}$$

We solve this system of equations by substitution. Solving for x in the second equation gives $x = 5 - 2y$, and substituting this into the first equation gives

$$\begin{aligned} (5 - 2y)^2 + y^2 &= 25 && \text{Substitute } x = 5 - 2y \\ (25 - 20y + 4y^2) + y^2 &= 25 && \text{Expand} \\ -20y + 5y^2 &= 0 && \text{Simplify} \\ -5y(4 - y) &= 0 && \text{Factor} \end{aligned}$$

Thus $y = 0$ or $y = 4$. When $y = 0$, we have $x = 5 - 2(0) = 5$, and when $y = 4$, we have $x = 5 - 2(4) = -3$. So the points of intersection of these curves are $(5, 0)$ and $(-3, 4)$, as shown in Figure 6.

Note that in this case the vertices are not part of the solution set because they don't satisfy the inequality $x^2 + y^2 < 25$ (so they are graphed as open circles in the figure). They simply show where the "corners" of the solution set lie.

Now Try Exercise 43 ■

■ Systems of Linear Inequalities

An inequality is **linear** if it can be put into one of the following forms:

$$ax + by \geq c \qquad ax + by \leq c \qquad ax + by > c \qquad ax + by < c$$

In the next example we graph the solution set of a system of linear inequalities.

Example 3 ■ A System of Four Linear Inequalities

Graph the solution set of the system, and label its vertices.

$$\begin{cases} x + 3y \leq 12 \\ x + y \leq 8 \\ x \geq 0 \\ y \geq 0 \end{cases}$$

Solution **Graph each inequality.** In Figure 6 we first graph the lines given by the equations that correspond to each inequality. To determine the graphs of the first two inequalities, we need to check only one test point. For simplicity let's use the point $(0, 0)$.

Inequality	Test Point (0, 0)	Conclusion
$x + 3y \leq 12$	$0 + 3(0) \overset{?}{\leq} 12$ ✓	Satisfies inequality
$x + y \leq 8$	$0 + 0 \overset{?}{\leq} 8$ ✓	Satisfies inequality

Since $(0, 0)$ is below the line $x + 3y = 12$, our check shows that the region on or below the line must satisfy the inequality. Likewise, since $(0, 0)$ is below the line $x + y = 8$, our check shows that the region on or below this line must satisfy the inequality. The inequalities $x \geq 0$ and $y \geq 0$ say that x and y are nonnegative. These regions are sketched in Figure 7(a).

Graph the solution of the system. The solution of the system of inequalities is the intersection of the graphs. This is the purple region graphed in Figure 7(b).

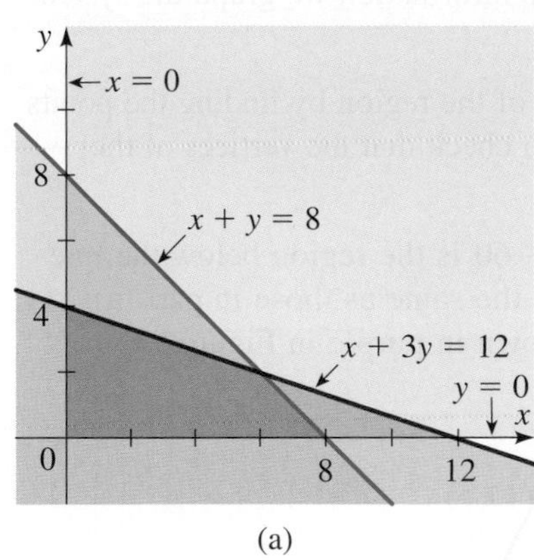

(a)

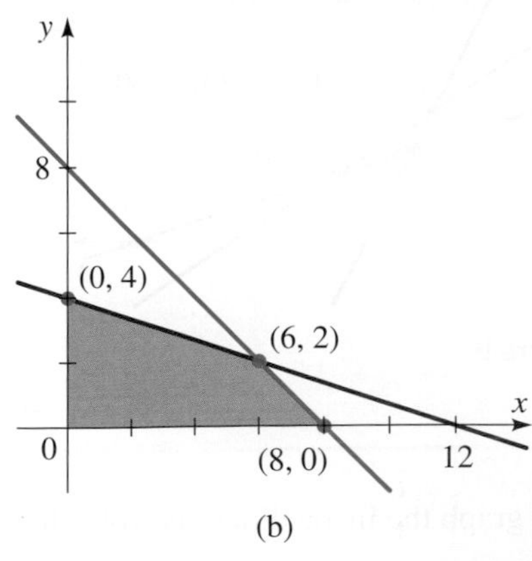

(b)

Figure 7

Find the Vertices. The coordinates of each vertex are obtained by simultaneously solving the equations of the lines that intersect at that vertex. From the system

$$\begin{cases} x + 3y = 12 \\ x + y = 8 \end{cases}$$

we get the vertex (6, 2). The origin (0, 0) is also clearly a vertex. The other two vertices are at the x- and y-intercepts of the corresponding lines: (8, 0) and (0, 4). In this case all the vertices *are* part of the solution set.

 **Now Try Exercise 51** ■

Example 4 ■ A System of Linear Inequalities

Graph the solution set of the system of inequalities, and label the vertices.

(a) $\begin{cases} 10x + 20y \geq 60 \\ 30x + 20y \geq 100 \\ 10x + 40y \geq 80 \\ x \geq 0, \quad y \geq 0 \end{cases}$ **(b)** $\begin{cases} 10x + 20y \leq 60 \\ 30x + 20y \geq 100 \\ 10x + 40y \geq 80 \\ x \geq 0, \quad y \geq 0 \end{cases}$

Solution

(a) **Graph each inequality.** We must graph the lines that correspond to these inequalities and then shade the appropriate regions. The graph of $10x + 20y \geq 60$ is the region above the line $y = 3 - \frac{1}{2}x$. The graph of $30x + 20y \geq 100$ is the region above the line $y = 5 - \frac{3}{2}x$, and the graph of $10x + 40y \geq 80$ is the region above the line $y = 2 - \frac{1}{4}x$.

Graph the solution of the system. The inequalities $x \geq 0$ and $y \geq 0$ indicate that the region is in the first quadrant. With this information we graph the system of inequalities in Figure 8.

Find the vertices. We determine the vertices of the region by finding the points of intersection of the appropriate lines. You can check that the vertices of the region are the ones indicated in Figure 8.

(b) The graph of the first inequality $10x + 20y \leq 60$ is the region below the line $y = 3 - \frac{1}{2}x$, and all the other inequalities are the same as those in part (a), so the solution to the system is the region shown in purple in Figure 9.

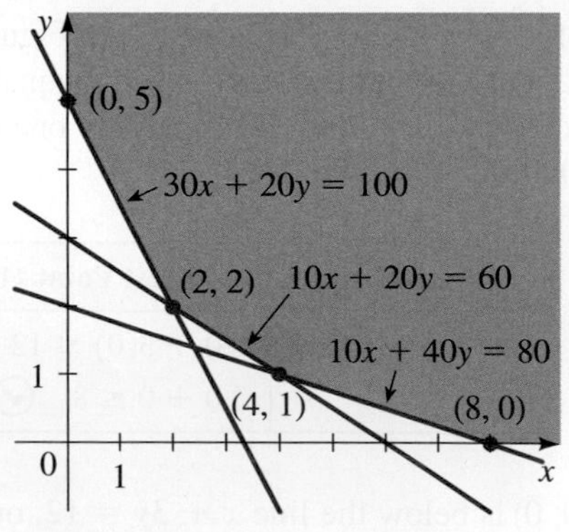

Figure 8

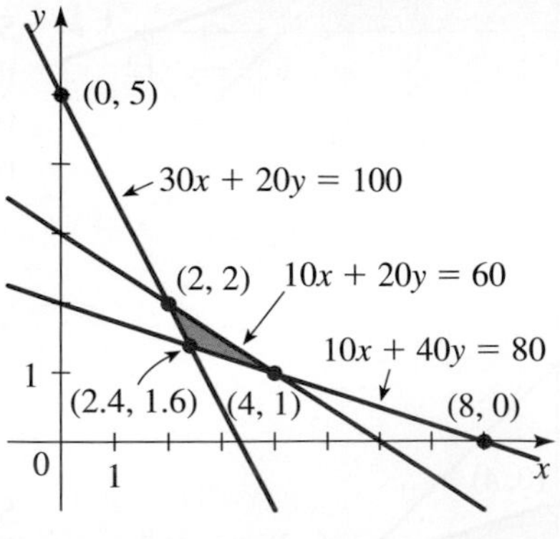

Figure 9

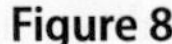 **Now Try Exercises 59 and 63** ■

In the next example we use a graphing device to graph the inequalities and solve the system.

Example 5 ■ A System of Linear Inequalities

Graph the solution set of the system of inequalities, and label the vertices.

$$\begin{cases} x + 2y \geq 8 \\ -x + 2y \leq 4 \\ 3x - 2y \leq 8 \end{cases}$$

Solution **Graph each inequality.** We use a graphing device to obtain the graph in Figure 10(a). The device shades each region in a different pattern (or a different color).

Graph the solution of the system. The solution set is the triangular region that is shaded in all three patterns (or all three colors). The solution set is graphed in Figure 10(b).

Find the vertices. We use the graphing device to find the vertices of the region. The vertices are labeled in Figure 10(b).

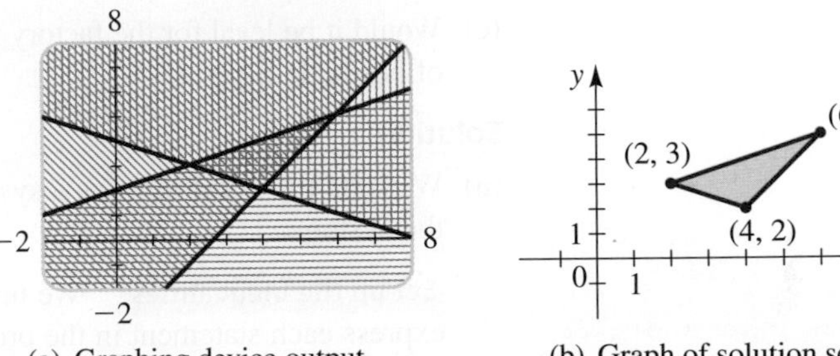

Figure 10 (a) Graphing device output (b) Graph of solution set

Now Try Exercise 65 ■

A region in the plane is called **bounded** if it can be enclosed in a (sufficiently large) circle. A region that is not bounded is called **unbounded**. For example, the regions graphed in Figures 3, 6, 7(b), 9, and 10 are bounded because they can be enclosed in a circle, as illustrated in Figure 11(a). But the regions graphed in Figures 2, 4, and 8 are unbounded because we cannot enclose them in a circle, as illustrated in Figure 11(b).

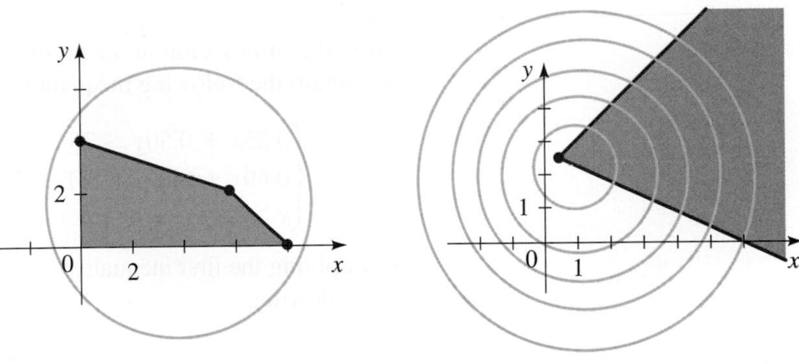

Figure 11 (a) A bounded region can be enclosed in a circle. (b) An unbounded region cannot be enclosed in a circle.

■ Application: Feasible Regions

Many applied problems involve **constraints** on the variables. For instance, a factory manager has only a certain number of workers who can be assigned to perform jobs on the factory floor. A farmer deciding what crops to cultivate has only a certain amount of land that can be seeded. Such constraints or limitations can usually be expressed as systems of inequalities. When dealing with applied inequalities, we usually refer to the solution set of a system as a **feasible region**, because the points in the solution set represent feasible (or possible) values for the quantities being studied.

Example 6 ■ Restricting Pollutant Outputs

A factory produces two agricultural pesticides, A and B. For every barrel of pesticide A, the factory emits 0.25 kg of carbon monoxide (CO) and 0.60 kg of sulfur dioxide (SO_2); and for every barrel of pesticide B, it emits 0.50 kg of CO and 0.20 kg of SO_2. Pollution laws restrict the factory's output of CO to a maximum of 75 kg per day and its output of SO_2 to a maximum of 90 kg per day.

(a) Find a system of inequalities that describes the number of barrels of each pesticide the factory can produce per day and still satisfy the pollution laws. Graph the feasible region.

(b) Would it be legal for the factory to produce 100 barrels of pesticide A and 80 barrels of pesticide B per day?

(c) Would it be legal for the factory to produce 60 barrels of pesticide A and 160 barrels of pesticide B per day?

Solution

(a) We state the constraints as a system of inequalities and then graph the solution of the system.

Set up the inequalities. We first identify and name the variables, and we then express each statement in the problem in terms of the variables. We let the variable x represent the number of barrels of A produced per day and let y be the number of barrels of B produced per day. We can organize the information in the problem as follows.

In Words	In Algebra
Barrels of A produced	x
Barrels of B produced	y
Total CO produced	$0.25x + 0.50y$
Total SO_2 produced	$0.60x + 0.20y$

From the information in the problem and the fact that x and y can't be negative we obtain the following inequalities.

$$\begin{cases} 0.25x + 0.50y \le 75 & \text{At most 75 kg of CO can be produced} \\ 0.60x + 0.20y \le 90 & \text{At most 90 kg of SO}_2\text{ can be produced} \\ x \ge 0, \quad y \ge 0 \end{cases}$$

Multiplying the first inequality by 4 and the second by 5 simplifies the system to the following:

$$\begin{cases} x + 2y \le 300 \\ 3x + y \le 450 \\ x \ge 0, \quad y \ge 0 \end{cases}$$

Graph the solution set. We first graph the equations

$$x + 2y = 300$$
$$3x + y = 450$$

The graphs are the two lines shown in Figure 12. Using the test point $(0, 0)$, we see that the solution set of each of these inequalities is the region below the corresponding line. So the solution to the system is the intersection of these sets, as shown in Figure 12.

(b) Since the point $(100, 80)$ lies inside the feasible region, this production plan is legal (see Figure 12).

(c) Since the point $(60, 160)$ lies outside the feasible region, this production plan is not legal. It violates the CO restriction, although it does not violate the SO_2 restriction (see Figure 12).

Now Try Exercise 69 ■

Figure 12

9.9 Exercises

Concepts

1. If the point $(2, 3)$ is a solution of an inequality in x and y, then the inequality is satisfied when we replace x by ________ and y by ________. Is the point $(2, 3)$ a solution of the inequality $4x - 2y \geq 1$?

2. To graph an inequality, we first graph the corresponding ________. So to graph the inequality $y \leq x + 1$, we first graph the equation ________. To decide which side of the graph of the equation is the graph of the inequality, we use ________ points. Complete the table, and sketch a graph of the inequality by shading the appropriate region of the graph shown below.

Test Point	Inequality $y \leq x + 1$	Conclusion
$(0, 0)$		
$(0, 2)$		

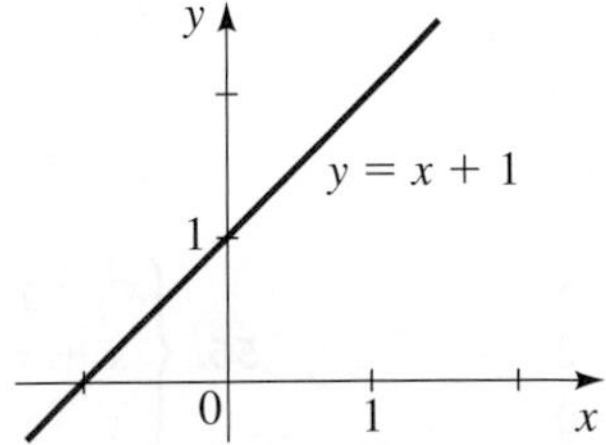

3. If the point $(2, 3)$ is a solution of a *system* of inequalities in x and y, then *each* inequality is satisfied when we replace x by ________ and y by ________. Is the point $(2, 3)$ a solution of the following system?

$$\begin{cases} 2x + 4y \leq 17 \\ 6x + 5y \leq 29 \end{cases}$$

4. Shade the solution of each system of inequalities on the given graph.

(a) $\begin{cases} x - y \geq 0 \\ x + y \geq 2 \end{cases}$ **(b)** 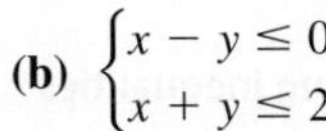

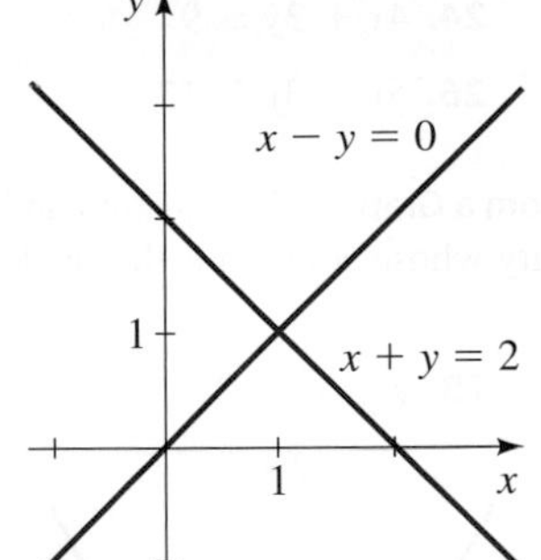

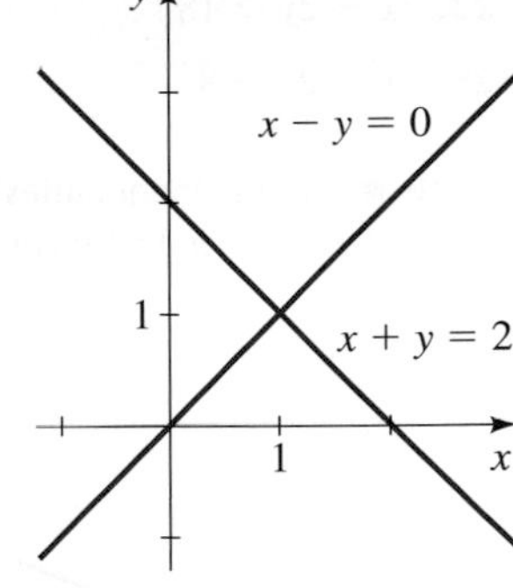

(c) 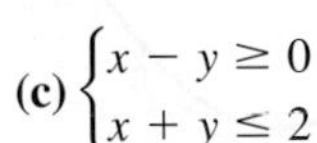

(d) 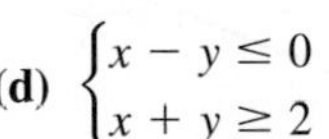

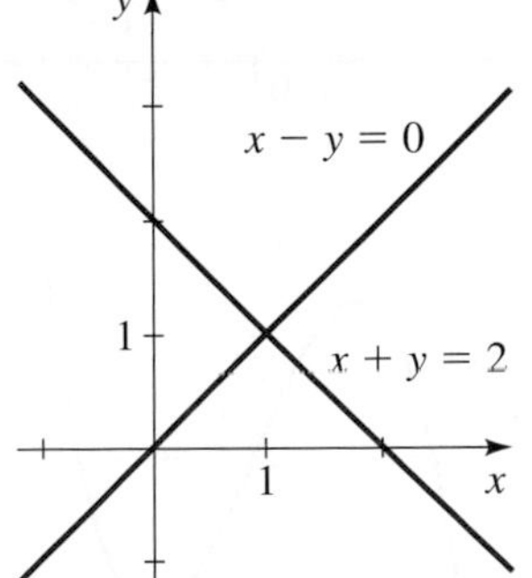

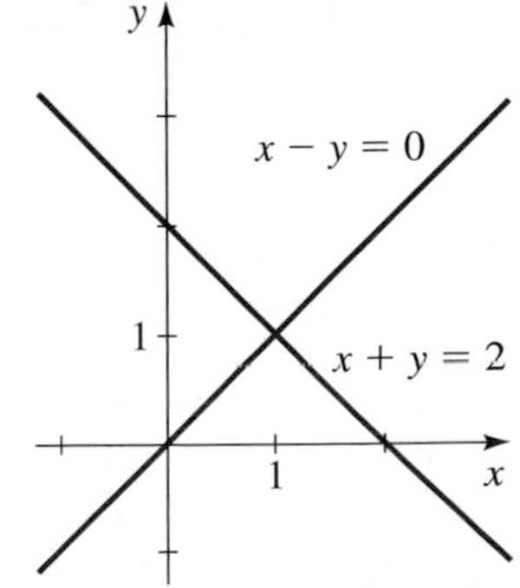

■ Skills

5–6 ■ Solutions of Inequalities An inequality and several points are given. Determine which points are solutions of the inequality.

5. $x - 5y > 3$; $(-1, -2), (1, -2), (1, 2), (8, 1)$

6. $3x + 2y \le 2$; $(-2, 1), (1, 3), (1, -3), (0, 1)$

7–8 ■ Solutions of Systems of Inequalities A system of inequalities and several points are given. Determine which points are solutions of the system.

7. $\begin{cases} 3x - 2y \le 5 \\ 2x + y \ge 3 \end{cases}$; $(0, 0), (1, 2), (1, 1), (3, 1)$

8. $\begin{cases} x + 2y \ge 4 \\ 4x + 3y \ge 11 \end{cases}$; $(0, 0), (1, 3), (3, 0), (1, 2)$

9–22 ■ Graphing Inequalities Graph the inequality.

9. $y < -2x$ **10.** $y \ge 3x$

11. $y \ge 2$ **12.** $x \le -1$

13. $x < 2$ **14.** $y > 1$

15. $y > x - 3$ **16.** $y \le 1 - x$

17. $2x - y \ge -4$ **18.** $3x - y - 9 < 0$

19. $-x^2 + y \ge 5$ **20.** $y > x^2 + 1$

21. $x^2 + y^2 > 9$ **22.** $x^2 + (y - 2)^2 \le 4$

23–26 ■ Graphing Inequalities Use a graphing device to graph the linear inequality.

23. $3x - 2y \ge 18$ **24.** $4x + 3y \le 9$

25. $5x + 2y > 8$ **26.** $5x - 3y \ge 15$

27–30 ■ Finding Inequalities from a Graph An equation and its graph are given. Find an inequality whose solution is the shaded region.

27. $y = \frac{1}{2}x - 1$

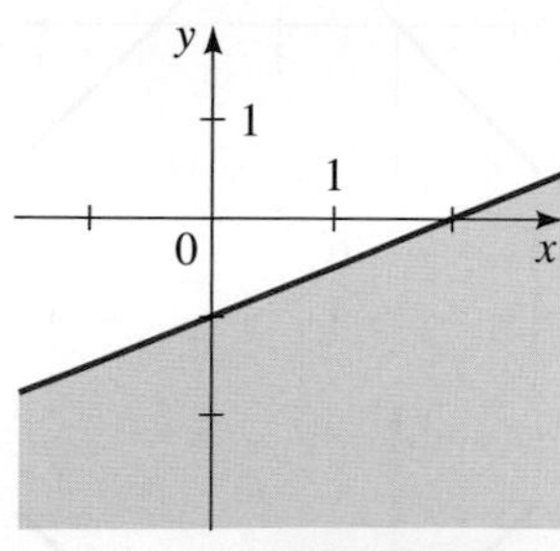

28. $y = x^2 + 2$

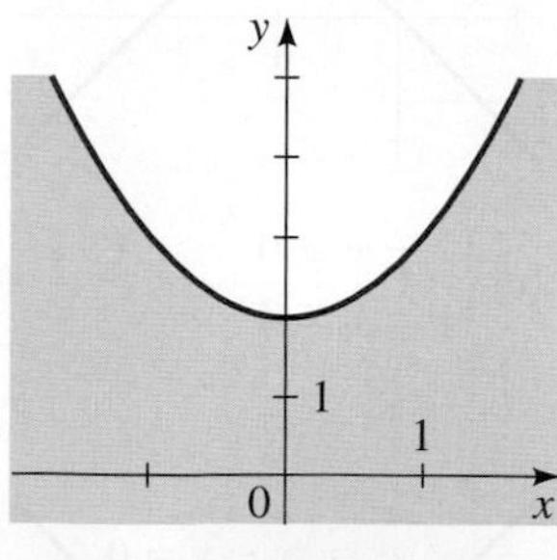

29. $x^2 + y^2 = 4$

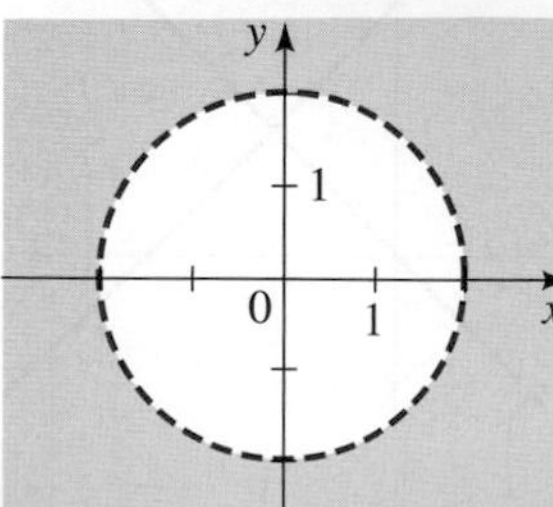

30. $y = x^3 - 4x$

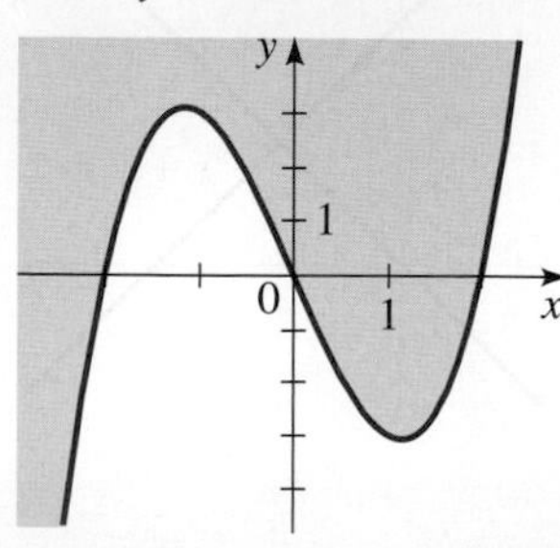

31–58 ■ Systems of Inequalities Graph the solution set of the system of inequalities. Find the coordinates of all vertices, and determine whether the solution set is bounded.

31. $\begin{cases} x + y \le 4 \\ y \ge x \end{cases}$ **32.** $\begin{cases} 2x + 3y > 12 \\ 3x - y < 21 \end{cases}$

33. $\begin{cases} y < \frac{1}{4}x + 2 \\ y \ge 2x - 5 \end{cases}$ **34.** $\begin{cases} x - y > 0 \\ 4 + y \le 2x \end{cases}$

35. $\begin{cases} y \le -2x + 8 \\ y \le -\frac{1}{2}x + 5 \\ x \ge 0, \quad y \ge 0 \end{cases}$ **36.** $\begin{cases} 4x + 3y \le 18 \\ 2x + y \le 8 \\ x \ge 0, \quad y \ge 0 \end{cases}$

37. $\begin{cases} x \ge 0 \\ y \ge 0 \\ 3x + 5y \le 15 \\ 3x + 2y \le 9 \end{cases}$ **38.** $\begin{cases} x > 2 \\ y < 12 \\ 2x - 4y > 8 \end{cases}$

39. $\begin{cases} y \le 9 - x^2 \\ x \ge 0, \quad y \ge 0 \end{cases}$ **40.** $\begin{cases} y \ge x^2 \\ y \le 4 \\ x \ge 0 \end{cases}$

41. $\begin{cases} y < 9 - x^2 \\ y \ge x + 3 \end{cases}$ **42.** $\begin{cases} y \ge x^2 \\ x + y \ge 6 \end{cases}$

43. $\begin{cases} x^2 + y^2 \le 4 \\ x - y > 0 \end{cases}$ **44.** $\begin{cases} x > 0 \\ y > 0 \\ x + y < 10 \\ x^2 + y^2 > 9 \end{cases}$

45. $\begin{cases} x^2 - y \le 0 \\ 2x^2 + y \le 12 \end{cases}$ **46.** $\begin{cases} 2x^2 + y > 4 \\ x^2 - y \le 8 \end{cases}$

47. $\begin{cases} x^2 + y^2 \le 9 \\ 2x + y^2 \ge 1 \end{cases}$ **48.** $\begin{cases} x^2 + y^2 \le 4 \\ x^2 - 2y > 1 \end{cases}$

49. $\begin{cases} x + 2y \le 14 \\ 3x - y \ge 0 \\ x - y \ge 2 \end{cases}$ **50.** $\begin{cases} y < x + 6 \\ 3x + 2y \ge 12 \\ x - 2y \le 2 \end{cases}$

51. $\begin{cases} x \ge 0 \\ y \ge 0 \\ x \le 5 \\ x + y \le 7 \end{cases}$ **52.** $\begin{cases} x \ge 0 \\ y \ge 0 \\ y \le 4 \\ 2x + y \le 8 \end{cases}$

53. $\begin{cases} y > x + 1 \\ x + 2y \le 12 \\ x + 1 > 0 \end{cases}$ **54.** $\begin{cases} x + y > 12 \\ y < \frac{1}{2}x - 6 \\ 3x + y < 6 \end{cases}$

55. $\begin{cases} x^2 + y^2 \le 8 \\ x \ge 2 \\ y \ge 0 \end{cases}$ **56.** $\begin{cases} x^2 - y \ge 0 \\ x + y < 6 \\ x - y < 6 \end{cases}$

57. $\begin{cases} x^2 + y^2 < 9 \\ x + y > 0 \\ x \le 0 \end{cases}$ **58.** $\begin{cases} y \ge x^3 \\ y \le 2x + 4 \\ x + y \ge 0 \end{cases}$

59–64 ■ Systems of Inequalities Graph the system of inequalities, label the vertices, and determine whether the region is bounded or unbounded.

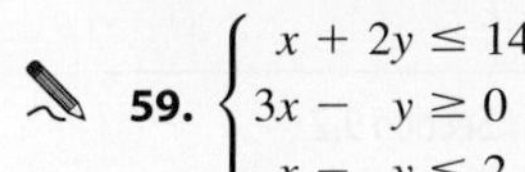

59. $\begin{cases} x + 2y \le 14 \\ 3x - y \ge 0 \\ x - y \le 2 \end{cases}$ **60.** $\begin{cases} x + 2y \le 14 \\ 3x - y \ge 0 \\ x - y \ge 2 \end{cases}$

61. $\begin{cases} x + y \le 12 \\ y \le \frac{1}{2}x - 6 \\ y \le 2x + 6 \end{cases}$ **62.** $\begin{cases} y \ge x + 1 \\ x + 2y \le 12 \\ x + 1 \ge 0 \end{cases}$

63. $\begin{cases} 30x + 10y \ge 50 \\ 10x + 20y \ge 50 \\ 10x + 60y \ge 90 \\ x \ge 0, \quad y \ge 0 \end{cases}$ **64.** $\begin{cases} x + y \ge 6 \\ 4x + 7y \le 39 \\ x + 5y \ge 13 \\ x \ge 0, \quad y \ge 0 \end{cases}$

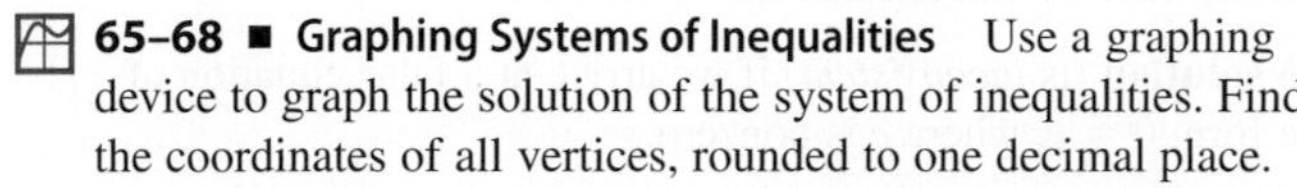

65–68 ■ Graphing Systems of Inequalities Use a graphing device to graph the solution of the system of inequalities. Find the coordinates of all vertices, rounded to one decimal place.

65. $\begin{cases} y \ge x - 3 \\ y \ge -2x + 6 \\ y \le 8 \end{cases}$ **66.** $\begin{cases} x + y \ge 12 \\ 2x + y \le 24 \\ x - y \ge -6 \end{cases}$

67. $\begin{cases} y \le 6x - x^2 \\ x + y \ge 4 \end{cases}$ **68.** $\begin{cases} y \ge x^3 \\ 2x + y \ge 0 \\ y \le 2x + 6 \end{cases}$

Applications

69. Planting Crops A farmer has 500 acres of arable land on which to plant potatoes and corn. The farmer has \$40,000 available for planting and \$30,000 for fertilizer. Planting one acre of potatoes costs \$90, and planting one acre of corn costs \$50. Fertilizer costs \$30 for one acre of potatoes and \$80 for one acre of corn.

(a) Find a system of inequalities that describes the number of acres of each crop that the farmer can plant with the available resources. Graph the feasible region.

(b) Can the farmer plant 300 acres of potatoes and 180 acres of corn?

(c) Can the farmer plant 150 acres of potatoes and 325 acres of corn?

70. Planting Crops A farmer has 300 acres of arable land for planting cauliflower and cabbage. The farmer has \$17,500 available for planting and \$12,000 for fertilizer. Planting one acre of cauliflower costs \$70, and planting one acre of cabbage costs \$35. Fertilizer costs \$25 for one acre of cauliflower and \$55 for one acre of cabbage.

(a) Find a system of inequalities that describes the number of acres of each crop that the farmer can plant with the available resources. Graph the feasible region.

(b) Can the farmer plant 155 acres of cauliflower and 115 acres of cabbage?

(c) Can the farmer plant 115 acres of cauliflower and 175 acres of cabbage?

71. Publishing Books A publishing company publishes a total of no more than 100 books every year. At least 20 of these are nonfiction, but the company always publishes at least as much fiction as nonfiction. Find a system of inequalities that describes the possible numbers of fiction and nonfiction books that the company can produce each year consistent with these policies. Graph the solution set.

72. Furniture Manufacturing A furniture maker manufactures unfinished tables and chairs. Each table requires 3 hours of sawing and 1 hour of assembly. Each chair requires 2 hours of sawing and 2 hours of assembly. The furniture maker can put in up to 12 hours of sawing and 8 hours of assembly work each day. Find a system of inequalities that describes all possible combinations of tables and chairs that the furniture maker can make daily. Graph the solution set.

73. Coffee Blends A coffee merchant sells two different coffee blends. The Standard blend uses 120 g of arabica and 360 g of robusta beans per package; the Deluxe blend uses 300 g of arabica and 180 g of robusta beans per package. The merchant has 35.56 kg of arabica and 40 kg of robusta beans available. Find a system of inequalities that describes the possible number of Standard and Deluxe packages the merchant can make. Graph the solution set.

74. Nutrition A cat-food manufacturer uses fish and beef by-products. The fish contains 12 g of protein and 3 g of fat per ounce. The beef contains 6 g of protein and 9 g of fat per ounce. Each can of cat food must contain at least 60 g of protein and 45 g of fat. Find a system of inequalities that describes the possible number of ounces of fish and beef by-products that can be used in each can to satisfy these minimum requirements. Graph the solution set.

Discuss ■ Discover ■ Prove ■ Write

75. Discuss: Shading Unwanted Regions To graph the solution of a system of inequalities, we have shaded the solution of each inequality in a different color; the solution of the system is the region where all the shaded parts overlap. Here is a different method: For each inequality, shade the region that does *not* satisfy the inequality. Explain why the part of the plane that is left unshaded is the solution of the system. Solve the following system by both methods. Which do you prefer? Why?

$$\begin{cases} x + 2y > 4 \\ -x + y < 1 \\ x + 3y < 9 \\ x < 3 \end{cases}$$

Chapter 9 Review

Properties and Formulas

Systems of Equations | Section 9.1

A **system of equations** is a set of equations that involve the same variables. A **system of linear equations** is a system of equations in which each equation is linear. Systems of linear equations in two variables (x and y) and three variables (x, y, and z) have the following forms:

Linear system 2 variables	Linear system 3 variables
$a_{11}x + a_{12}y = b_1$	$a_{11}x + a_{12}y + a_{13}z = b_1$
$a_{21}x + a_{22}y = b_2$	$a_{21}x + a_{22}y + a_{23}z = b_2$
	$a_{31}x + a_{32}y + a_{33}z = b_3$

A **solution** of a system of equations is an assignment of values for the variables that makes *each* equation in the system true. To **solve** a system means to find all solutions of the system.

Substitution Method | Section 9.1

To solve a pair of equations in two variables by substitution:

1. **Solve for one variable** in terms of the other variable in one equation.
2. **Substitute** into the other equation to get an equation in one variable, and solve for this variable.
3. **Back-substitute** the value(s) of the variable you have found into either original equation, and solve for the remaining variable.

Elimination Method | Section 9.1

To solve a pair of equations in two variables by elimination:

1. **Adjust the coefficients** by multiplying the equations by appropriate constants so that the term(s) involving one of the variables are of opposite signs in the equations.
2. **Add the equations** to eliminate that one variable; this gives an equation in the other variable. Solve for this variable.
3. **Back-substitute** the value(s) of the variable that you have found into either original equation, and solve for the remaining variable.

Graphical Method | Section 9.1

To solve a pair of equations in two variables graphically, first put each equation in function form, $y = f(x)$.

1. **Graph the equations** on a common screen.
2. **Find the points of intersection** of the graphs. The solutions are the x- and y-coordinates of the points of intersection.

Gaussian Elimination | Section 9.2

To use **Gaussian elimination** to solve a system of linear equations, use the following operations to change the system to an **equivalent** simpler system:

1. Add a nonzero multiple of one equation to another.
2. Multiply an equation by a nonzero constant.
3. Interchange the position of two equations in the system.

Number of Solutions of a Linear System | Section 9.2

A system of linear equations can have:

1. A unique solution for each variable.
2. No solution, in which case the system is **inconsistent**.
3. Infinitely many solutions, in which case the system is **dependent**.

How to Determine the Number of Solutions of a Linear System | Section 9.2

When **Gaussian elimination** is used to solve a system of linear equations, then we can tell that the system has:

1. **No solution** (is *inconsistent*) if we arrive at a false equation of the form $0 = c$, where c is nonzero.
2. **Infinitely many solutions** (is *dependent*) if the system is consistent but we end up with fewer equations than variables (after discarding redundant equations of the form $0 = 0$).

Matrices | Section 9.3

A **matrix** A of **dimension** $m \times n$ is a rectangular array of numbers with m **rows** and n **columns**:

$$A = \begin{bmatrix} a_{11} & a_{12} & \cdots & a_{1n} \\ a_{21} & a_{22} & \cdots & a_{2n} \\ \vdots & \vdots & \ddots & \vdots \\ a_{m1} & a_{m2} & \cdots & a_{mn} \end{bmatrix}$$

Augmented Matrix of a System | Section 9.3

The **augmented matrix** of a system of linear equations is the matrix consisting of the coefficients and the constant terms. For example, for the two-variable system

$$a_{11}x + a_{12}x = b_1$$
$$a_{21}x + a_{22}x = b_2$$

the augmented matrix is

$$\begin{bmatrix} a_{11} & a_{12} & b_1 \\ a_{21} & a_{22} & b_2 \end{bmatrix}$$

Elementary Row Operations | Section 9.3

To solve a system of linear equations using the augmented matrix of the system, the following operations can be used to transform the rows of the matrix:

1. Add a nonzero multiple of one row to another.
2. Multiply a row by a nonzero constant.
3. Interchange two rows.

Row-Echelon Form of a Matrix | Section 9.3

A matrix is in **row-echelon form** if its entries satisfy the following conditions:

1. The first nonzero entry in each row (the **leading entry**) is the number 1.

2. The leading entry of each row is to the right of the leading entry in the row above it.

3. All rows consisting entirely of zeros are at the bottom of the matrix.

If the matrix also satisfies the following condition, it is in **reduced row-echelon** form:

4. If a column contains a leading entry, then every other entry in that column is a 0.

Number of Solutions of a Linear System | Section 9.3

If the augmented matrix of a system of linear equations has been reduced to row-echelon form using elementary row operations, then the system has:

1. **No solution** if the row-echelon form contains a row that represents the equation $0 = 1$. In this case the system is **inconsistent**.

2. **One solution** if each variable in the row-echelon form is a leading variable.

3. **Infinitely many solutions** if the system is not inconsistent but not every variable is a leading variable. In this case the system is **dependent**.

Operations on Matrices | Section 9.4

If A and B are $m \times n$ matrices and c is a scalar (real number), then:

1. The **sum** $A + B$ is the $m \times n$ matrix that is obtained by adding corresponding entries of A and B.

2. The **difference** $A - B$ is the $m \times n$ matrix that is obtained by subtracting corresponding entries of A and B.

3. The **scalar product** cA is the $m \times n$ matrix that is obtained by multiplying each entry of A by c.

Multiplication of Matrices | Section 9.4

If A is an $m \times n$ matrix and B is an $n \times k$ matrix (so the number of columns of matrix A is the same as the number of rows of matrix B), then the **matrix product** AB is the $m \times k$ matrix whose ij-entry is the inner product of the ith row of A and the jth column of B.

Properties of Matrix Operations | Section 9.4

If A, B, and C are matrices of compatible dimensions, then the following properties hold:

1. **Commutativity of addition:**

$$A + B = B + A$$

2. **Associativity:**

$$(A + B) + C = A + (B + C)$$
$$(AB)C = A(BC)$$

3. **Distributivity:**

$$A(B + C) = AB + AC$$
$$(B + C)A = BA + CA$$

(Note that matrix *multiplication* is *not* commutative.)

Identity Matrix | Section 9.5

The **identity matrix** I_n is the $n \times n$ matrix whose main diagonal entries are all 1 and whose other entries are all 0:

$$I_n = \begin{bmatrix} 1 & 0 & \cdots & 0 \\ 0 & 1 & \cdots & 0 \\ \vdots & \vdots & \ddots & \vdots \\ 0 & 0 & \cdots & 1 \end{bmatrix}$$

If A is an $m \times n$ matrix, then

$$AI_n = A \quad \text{and} \quad I_m A = A$$

Inverse of a Matrix | Section 9.5

If A is an $n \times n$ matrix, then the inverse of A is the $n \times n$ matrix A^{-1} with the following properties:

$$A^{-1}A = I_n \quad \text{and} \quad AA^{-1} = I_n$$

To find the inverse of a matrix, we use a procedure involving elementary row operations. (Note that *some* square matrices do not have an inverse.)

Inverse of a 2 × 2 Matrix | Section 9.5

For 2×2 matrices the following special rule provides a shortcut for finding the inverse:

$$A = \begin{bmatrix} a & b \\ c & d \end{bmatrix} \quad \Rightarrow \quad A^{-1} = \frac{1}{ad - bc}\begin{bmatrix} d & -b \\ -c & a \end{bmatrix}$$

Writing a Linear System as a Matrix Equation | Section 9.5

A system of n linear equations in n variables can be written as a single matrix equation

$$AX = B$$

where A is the $n \times n$ matrix of coefficients, X is the $n \times 1$ matrix of the variables, and B is the $n \times 1$ matrix of the constants. For example, the linear system of two equations in two variables

$$a_{11}x + a_{12}x = b_1$$
$$a_{21}x + a_{22}x = b_2$$

can be expressed as

$$\begin{bmatrix} a_{11} & a_{12} \\ a_{21} & a_{22} \end{bmatrix}\begin{bmatrix} x \\ y \end{bmatrix} = \begin{bmatrix} b_1 \\ b_2 \end{bmatrix}$$

Solving Matrix Equations | Section 9.5

If A is an invertible $n \times n$ matrix, X is an $n \times 1$ variable matrix, and B is an $n \times 1$ constant matrix, then the matrix equation

$$AX = B$$

has the unique solution

$$X = A^{-1}B$$

Determinant of a 2 × 2 Matrix | Section 9.6

The **determinant** of the matrix

$$A = \begin{bmatrix} a & b \\ c & d \end{bmatrix}$$

is the *number*

$$\det(A) = |A| = ad - bc$$

Minors and Cofactors | Section 9.6

If $A = |a_{ij}|$ is an $n \times n$ matrix, then the **minor** M_{ij} of the entry a_{ij} is the determinant of the matrix obtained by deleting the ith row and the jth column of A.

The **cofactor** A_{ij} of the entry a_{ij} is

$$A_{ij} = (-1)^{i+j} M_{ij}$$

(Thus, the minor and the cofactor of each entry either are the same or are negatives of each other.)

Determinant of an $n \times n$ Matrix | Section 9.6

To find the **determinant** of the $n \times n$ matrix

$$A = \begin{bmatrix} a_{11} & a_{12} & \cdots & a_{1n} \\ a_{21} & a_{22} & \cdots & a_{2n} \\ \vdots & \vdots & \ddots & \vdots \\ a_{n1} & a_{n2} & \cdots & a_{nn} \end{bmatrix}$$

we choose a row or column to **expand**, and then we calculate the number that is obtained by multiplying each element of that row or column by its cofactor and then adding the resulting products. For example, if we choose to expand about the first row, we get

$$\det(A) = |A| = a_{11}A_{11} + a_{12}A_{12} + \cdots + a_{1n}A_{1n}$$

Invertibility Criterion | Section 9.6

A square matrix has an inverse if and only if its determinant is not 0.

Row and Column Transformations | Section 9.6

If we add a nonzero multiple of one row to another row in a square matrix or a nonzero multiple of one column to another column, then the determinant of the matrix is unchanged.

Cramer's Rule | Section 9.6

If a system of n linear equations in the n variables $x_1, x_2, \ldots, x_n$ is equivalent to the matrix equation $DX = B$ and if $|D| \neq 0$, then the solutions of the system are

$$x_1 = \frac{|D_{x_1}|}{|D|} \qquad x_2 = \frac{|D_{x_2}|}{|D|} \qquad \cdots \qquad x_n = \frac{|D_{x_n}|}{|D|}$$

where D_{x_i} is the matrix that is obtained from D by replacing its ith column by the constant matrix B.

Area of a Triangle Using Determinants | Section 9.6

If a triangle in the coordinate plane has vertices (a_1, b_1), (a_2, b_2), and (a_3, b_3), then the area of the triangle is given by

$$\mathcal{A} = \pm\tfrac{1}{2}\begin{vmatrix} a_1 & b_1 & 1 \\ a_2 & b_2 & 1 \\ a_3 & b_3 & 1 \end{vmatrix}$$

where the sign is chosen to make the area positive.

Partial Fractions | Section 9.7

The *partial fraction decomposition* of a rational function

$$r(x) = \frac{P(x)}{Q(x)}$$

(where the degree of P is less than the degree of Q) is a sum of simpler fractional expressions that equal $r(x)$ when brought to a common denominator. The denominator of each simpler fraction is either a linear or quadratic factor of $Q(x)$ or a power of such a linear or quadratic factor. To find the terms of the partial fraction decomposition, we first factor $Q(x)$ into linear and irreducible quadratic factors. The terms then have the following forms, depending on the factors of $Q(x)$.

1. For every **distinct linear factor** $ax + b$ there is a term of the form

$$\frac{A}{ax + b}$$

2. For every **repeated linear factor** $(ax + b)^m$ there are terms of the form

$$\frac{A_1}{ax + b} + \frac{A_2}{(ax + b)^2} + \cdots + \frac{A_m}{(ax + b)^m}$$

3. For every **distinct quadratic factor** $ax^2 + bx + c$ there is a term of the form

$$\frac{Ax + B}{ax^2 + bx + c}$$

4. For every **repeated quadratic factor** $(ax^2 + bx + c)^m$ there are terms of the form

$$\frac{A_1x + B_1}{ax^2 + bx + c} + \frac{A_2x + B_2}{(ax^2 + bx + c)^2} + \cdots + \frac{A_mx + B_m}{(ax^2 + bx + c)^m}$$

Graphing Inequalities | Section 9.9

To graph an inequality:

1. Graph the equation that corresponds to the inequality. This "boundary curve" divides the coordinate plane into separate regions.
2. Use **test points** to determine which region(s) satisfy the inequality.
3. Shade the region(s) that satisfy the inequality, and use a solid line for the boundary curve if it satisfies the inequality ($\leq$ or $\geq$) and a dashed line if it does not ($<$ or $>$).

Graphing Systems of Inequalities | Section 9.9

To graph the solution of a system of inequalities (or **feasible region** determined by the inequalities):

1. Graph all the inequalities on the same coordinate plane.
2. The solution is the intersection of the solutions of all the inequalities, so shade the region that satisfies all the inequalities.
3. Determine the coordinates of the intersection points of all the boundary curves that touch the solution set of the system. These points are the **vertices** of the solution.

Concept Check

1. (a) What are the three methods we use to solve a system of equations?

   (b) Solve the system by the elimination method and by the graphical method.

$$\begin{cases} x + y = 3 \\ 3x - y = 1 \end{cases}$$

2. For a system of two linear equations in two variables:

   (a) How many solutions are possible?

   (b) What is meant by an inconsistent system? a dependent system?

3. What operations can be performed on a linear system so as to arrive at an equivalent system?

4. (a) Explain how Gaussian elimination works.

   (b) Use Gaussian elimination to put the following system in triangular form, and then solve the system.

System **Triangular form**

$$\begin{cases} x + y - 2z = 3 \\ x + 2y + z = 5 \\ 3x - y + 5z = 1 \end{cases}$$

5. What does it mean to say that A is a matrix with dimension $m \times n$?

6. What is the row-echelon form of a matrix? What is a leading entry?

7. (a) What is the augmented matrix of a system? What are leading variables?

   (b) What are the elementary row operations on an augmented matrix?

   (c) How do we solve a system using the augmented matrix?

   (d) Write the augmented matrix of the following system of linear equations.

System **Augmented Matrix**

$$\begin{cases} x + y - 2z = 3 \\ x + 2y + z = 5 \\ 3x - y + 5z = 1 \end{cases}$$

   (e) Solve the system in part (d).

8. Suppose you have used Gaussian elimination to transform the augmented matrix of a linear system into row-echelon form. How can you tell whether the system has exactly one solution? no solution? infinitely many solutions?

9. What is the reduced row-echelon form of a matrix?

10. (a) How do Gaussian elimination and Gauss-Jordan elimination differ?

    (b) Use Gauss-Jordan elimination to solve the linear system in 7(d).

11. If A and B are matrices with the same dimension and k is a real number, how do you find $A + B$ and kA?

12. (a) What must be true of the dimensions of A and B for the product AB to be defined?

    (b) If A has dimension 2×3 and if B has dimension 3×2, is the product AB defined? If so, what is the dimension of AB?

    (c) Find the matrix product.

$$\begin{bmatrix} 2 & 1 \\ 4 & 0 \end{bmatrix}\begin{bmatrix} 3 & 4 & 1 \\ 5 & 1 & 2 \end{bmatrix}$$

13. (a) What is an identity matrix I_n? If A is an $n \times n$ matrix, what are the products AI_n and I_nA?

    (b) If A is an $n \times n$ matrix, what is its inverse matrix?

    (c) Complete the formula for the inverse of a 2×2 matrix

$$A = \begin{bmatrix} a & b \\ c & d \end{bmatrix}$$

    (d) Find the inverse of A.

$$A = \begin{bmatrix} 1 & 1 \\ 3 & -1 \end{bmatrix}$$

14. (a) Express the system in 1(b) as a matrix equation $AX = B$.

    (b) If a linear system is expressed as a matrix equation $AX = B$, how do we solve the system? Solve the system in part (a).

15. (a) Is it true that the determinant $\det(A)$ of a matrix A is defined only if A is a square matrix?

    (b) Find the determinant of the matrix A in 13(d).

    (c) Use Cramer's Rule to solve the system in 1(b).

16. (a) How do we express a rational function r as a partial fraction decomposition?

    (b) Give the form of each partial fraction decomposition.

    (i) $\dfrac{2x}{(x-5)(x-2)^2}$ (ii) $\dfrac{2x}{(x-5)(x^2+1)}$

17. (a) How do we graph an inequality in two variables?

    (b) Graph the solution set of the inequality $x + y \geq 3$.

    (c) Graph the solution set of the system of inequalities: $x + y \geq 3$, $3x - y \geq 1$.

Answers to the Concept Check can be found at the book companion website stewartmath.com.

Exercises

1–6 ■ Systems of Linear Equations in Two Variables Solve the system of equations, and graph the lines.

1. $\begin{cases} 3x - y = 5 \\ 2x + y = 5 \end{cases}$ **2.** $\begin{cases} y = 2x + 6 \\ y = -x + 3 \end{cases}$

3. $\begin{cases} 2x - 7y = 28 \\ y = \frac{2}{7}x - 4 \end{cases}$ **4.** $\begin{cases} 6x - 8y = 15 \\ -\frac{3}{2}x + 2y = -4 \end{cases}$

5. $\begin{cases} 2x - y = 1 \\ x + 3y = 10 \\ 3x + 4y = 15 \end{cases}$ **6.** $\begin{cases} 2x + 5y = 9 \\ -x + 3y = 1 \\ 7x - 2y = 14 \end{cases}$

7–10 ■ Systems of Nonlinear Equations Solve the system of equations.

7. $\begin{cases} y = x^2 + 2x \\ y = 6 + x \end{cases}$ **8.** $\begin{cases} x^2 + y^2 = 8 \\ y = x + 2 \end{cases}$

9. $\begin{cases} 3x + \dfrac{4}{y} = 6 \\ x - \dfrac{8}{y} = 4 \end{cases}$ **10.** $\begin{cases} x^2 + y^2 = 10 \\ x^2 + 2y^2 - 7y = 0 \end{cases}$

11–14 ■ Systems of Nonlinear Equations Use a graphing device to solve the system. Round answers to two decimal places.

11. $\begin{cases} 0.32x + 0.43y = 0 \\ 7x - 12y = 341 \end{cases}$

12. $\begin{cases} \sqrt{12}x - 3\sqrt{2}y = 660 \\ 7137x + 3931y = 20{,}000 \end{cases}$

13. $\begin{cases} x - y^2 = 10 \\ x = \frac{1}{22}y + 12 \end{cases}$ **14.** $\begin{cases} y = 5^x + x \\ y = x^5 + 5 \end{cases}$

15–20 ■ Matrices A matrix is given.

(a) State the dimension of the matrix.
(b) Is the matrix in row-echelon form?
(c) Is the matrix in reduced row-echelon form?
(d) Write the system of equations for which the given matrix is the augmented matrix.

15. $\begin{bmatrix} 1 & 2 & -5 \\ 0 & 1 & 3 \end{bmatrix}$ **16.** $\begin{bmatrix} 1 & 0 & 6 \\ 0 & 1 & 0 \end{bmatrix}$

17. $\begin{bmatrix} 1 & 0 & 8 & 0 \\ 0 & 1 & 5 & -1 \\ 0 & 0 & 0 & 0 \end{bmatrix}$ **18.** $\begin{bmatrix} 1 & 3 & 6 & 2 \\ 2 & 1 & 0 & 5 \\ 0 & 0 & 1 & 0 \end{bmatrix}$

19. $\begin{bmatrix} 0 & 1 & -3 & 4 \\ 1 & 1 & 0 & 7 \\ 1 & 2 & 1 & 2 \end{bmatrix}$ **20.** $\begin{bmatrix} 1 & 8 & 6 & -4 \\ 0 & 1 & -3 & 5 \\ 0 & 0 & 2 & -7 \\ 1 & 1 & 1 & 0 \end{bmatrix}$

21–42 ■ Systems of Linear Equations in Several Variables Find the complete solution of the system, or show that the system has no solution.

21. $\begin{cases} x + y + 2z = 6 \\ 2x + 5z = 12 \\ x + 2y + 3z = 9 \end{cases}$ **22.** $\begin{cases} x - 2y + 3z = 1 \\ x - 3y - z = 0 \\ 2x - 6z = 6 \end{cases}$

23. $\begin{cases} x - 2y + 3z = 1 \\ 2x - y + z = 3 \\ 2x - 7y + 11z = 2 \end{cases}$ **24.** $\begin{cases} x + y + z + w = 2 \\ 2x - 3z = 5 \\ x - 2y + 4w = 9 \\ x + y + 2z + 3w = 5 \end{cases}$

25. $\begin{cases} x + 2y + 2z = 6 \\ x - y = -1 \\ 2x + y + 3z = 7 \end{cases}$ **26.** $\begin{cases} x - y + z = 2 \\ x + y + 3z = 6 \\ 2y + 3z = 5 \end{cases}$

27. $\begin{cases} x - 2y + 3z = -2 \\ 2x - y + z = 2 \\ 2x - 7y + 11z = -9 \end{cases}$ **28.** $\begin{cases} x - y + z = 2 \\ x + y + 3z = 6 \\ 3x - y + 5z = 10 \end{cases}$

29. $\begin{cases} x + y + z + w = 0 \\ x - y - 4z - w = -1 \\ x - 2y + 4w = -7 \\ 2x + 2y + 3z + 4w = -3 \end{cases}$

30. $\begin{cases} x + 3z = -1 \\ y - 4w = 5 \\ 2y + z + w = 0 \\ 2x + y + 5z - 4w = 4 \end{cases}$

31. $\begin{cases} x - 3y + z = 4 \\ 4x - y + 15z = 5 \end{cases}$ **32.** $\begin{cases} 2x - 3y + 4z = 3 \\ 4x - 5y + 9z = 13 \\ 2x + 7z = 0 \end{cases}$

33. $\begin{cases} -x + 4y + z = 8 \\ 2x - 6y + z = -9 \\ x - 6y - 4z = -15 \end{cases}$ **34.** $\begin{cases} x - z + w = 2 \\ 2x + y - 2w = 12 \\ 3y + z + w = 4 \\ x + y - z = 10 \end{cases}$

35. $\begin{cases} x - y + 3z = 2 \\ 2x + y + z = 2 \\ 3x + 4z = 4 \end{cases}$ **36.** $\begin{cases} x - y = 1 \\ x + y + 2z = 3 \\ x - 3y - 2z = -1 \end{cases}$

37. $\begin{cases} x - y + z - w = 0 \\ 3x - y - z - w = 2 \end{cases}$ **38.** $\begin{cases} x - y = 3 \\ 2x + y = 6 \\ x - 2y = 9 \end{cases}$

39. $\begin{cases} x - y + z = 0 \\ 3x + 2y - z = 6 \\ x + 4y - 3z = 3 \end{cases}$ **40.** $\begin{cases} x + 2y + 3z = 2 \\ 2x - y - 5z = 1 \\ 4x + 3y + z = 6 \end{cases}$

41. $\begin{cases} x + y - z - w = 2 \\ x - y + z - w = 0 \\ 2x + 2w = 2 \\ 2x + 4y - 4z - 2w = 6 \end{cases}$

42. $\begin{cases} x - y - 2z + 3w = 0 \\ y - z + w = 1 \\ 3x - 2y - 7z + 10w = 2 \end{cases}$

43. Investments An investor has savings in two accounts, one paying 6% interest per year and the other paying 7%. Twice as much is invested in the 7% account as in the 6% account, and the annual interest income is \$600. How much is invested in each account?

44. Number of Coins A piggy bank contains 50 coins, all of them nickels, dimes, or quarters. The total value of the coins is \$5.60, and the value of the dimes is five times the value of the nickels. How many coins of each type are there?

45. Investments An amount of \$60,000 is invested in money-market accounts at three different banks. Bank A pays 2% interest per year, bank B pays 2.5%, and bank C pays 3%. Twice as much is invested in bank B as in the other two banks combined. After one year, the investment has earned \$1575 in interest. How much is invested in each bank?

46. Number of Fish Caught A commercial fishing boat fishes for haddock, sea bass, and red snapper. The haddock sells for \$1.25/kg, sea bass for \$0.75/kg, and red snapper for \$2.00/kg. Yesterday the fishing boat caught 560 kg of fish worth \$575. The haddock and red snapper together are worth \$320. How many kilograms of each fish did the fishing boat catch?

47–58 ■ Matrix Operations Let

$$A = \begin{bmatrix} 2 & 0 & -1 \end{bmatrix} \qquad B = \begin{bmatrix} 1 & 2 & 4 \\ -2 & 1 & 0 \end{bmatrix}$$

$$C = \begin{bmatrix} \frac{1}{2} & 3 \\ 2 & \frac{3}{2} \\ -2 & 1 \end{bmatrix} \qquad D = \begin{bmatrix} 1 & 4 \\ 0 & -1 \\ 2 & 0 \end{bmatrix}$$

$$E = \begin{bmatrix} 2 & -1 \\ -\frac{1}{2} & 1 \end{bmatrix} \qquad F = \begin{bmatrix} 4 & 0 & 2 \\ -1 & 1 & 0 \\ 7 & 5 & 0 \end{bmatrix} \qquad G = \begin{bmatrix} 5 \end{bmatrix}$$

Carry out the indicated operation, or explain why it cannot be performed.

47. $A + B$

48. $C - D$

49. $2C + 3D$

50. $5B - 2C$

51. GA

52. AG

53. BC

54. CB

55. BF

56. FC

57. $(C + D)E$

58. $F(2C - D)$

59–60 ■ Inverse Matrices Verify that matrices A and B are inverses of each other by calculating the products AB and BA.

59. $A = \begin{bmatrix} 2 & -5 \\ -2 & 6 \end{bmatrix}, \quad B = \begin{bmatrix} 3 & \frac{5}{2} \\ 1 & 1 \end{bmatrix}$

60. $A = \begin{bmatrix} 2 & -1 & 3 \\ 2 & -2 & 1 \\ 0 & 1 & 1 \end{bmatrix}, \quad B = \begin{bmatrix} -\frac{3}{2} & 2 & \frac{5}{2} \\ -1 & 1 & 2 \\ 1 & -1 & -1 \end{bmatrix}$

61–66 ■ Matrix Equations Solve the matrix equation for the unknown matrix X, or show that no solution exists, where

$$A = \begin{bmatrix} 2 & 1 \\ 3 & 2 \end{bmatrix} \qquad B = \begin{bmatrix} 1 & -2 \\ -2 & 4 \end{bmatrix} \qquad C = \begin{bmatrix} 0 & 1 & 3 \\ -2 & 4 & 0 \end{bmatrix}$$

61. $A + 3X = B$

62. $\frac{1}{2}(X - 2B) = A$

63. $2(X - A) = 3B$

64. $2X + C = 5A$

65. $AX = C$

66. $AX = B$

67–74 ■ Determinants and Inverse Matrices Find the determinant and, if possible, the inverse of the matrix.

67. $\begin{bmatrix} 1 & 4 \\ 2 & 9 \end{bmatrix}$

68. $\begin{bmatrix} 2 & 2 \\ 1 & -3 \end{bmatrix}$

69. $\begin{bmatrix} 4 & -12 \\ -2 & 6 \end{bmatrix}$

70. $\begin{bmatrix} 2 & 4 & 0 \\ -1 & 1 & 2 \\ 0 & 3 & 2 \end{bmatrix}$

71. $\begin{bmatrix} 3 & 0 & 1 \\ 2 & -3 & 0 \\ 4 & -2 & 1 \end{bmatrix}$

72. $\begin{bmatrix} 1 & 2 & 3 \\ 2 & 4 & 5 \\ 2 & 5 & 6 \end{bmatrix}$

73. $\begin{bmatrix} 1 & 0 & 0 & 1 \\ 0 & 2 & 0 & 2 \\ 0 & 0 & 3 & 3 \\ 0 & 0 & 0 & 4 \end{bmatrix}$

74. $\begin{bmatrix} 1 & 0 & 1 & 0 \\ 0 & 1 & 0 & 1 \\ 1 & 1 & 1 & 2 \\ 1 & 2 & 1 & 2 \end{bmatrix}$

75–78 ■ Using Inverse Matrices to Solve a System Express the system of linear equations as a matrix equation. Then solve the matrix equation by multiplying each side by the inverse of the coefficient matrix.

75. $\begin{cases} 12x - 5y = 10 \\ 5x - 2y = 17 \end{cases}$

76. $\begin{cases} 6x - 5y = 1 \\ 8x - 7y = -1 \end{cases}$

77. $\begin{cases} 2x + y + 5z = \frac{1}{3} \\ x + 2y + 2z = \frac{1}{4} \\ x + 3z = \frac{1}{6} \end{cases}$

78. $\begin{cases} 2x + 3z = 5 \\ x + y + 6z = 0 \\ 3x - y + z = 5 \end{cases}$

79–82 ■ Using Cramer's Rule to Solve a System Solve the system using Cramer's Rule.

79. $\begin{cases} 2x + 7y = 13 \\ 6x + 16y = 30 \end{cases}$

80. $\begin{cases} 12x - 11y = 140 \\ 7x + 9y = 20 \end{cases}$

81. $\begin{cases} 2x - y + 5z = 0 \\ -x + 7y = 9 \\ 5x + 4y + 3z = -9 \end{cases}$

82. $\begin{cases} 3x + 4y - z = 10 \\ x - 4z = 20 \\ 2x + y + 5z = 30 \end{cases}$

83–84 ■ Area of a Triangle Use the determinant formula for the area of a triangle to find the area of the triangle in the figure.

83.

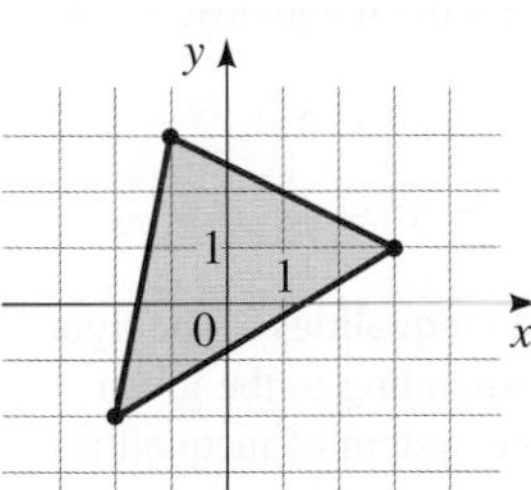

84.

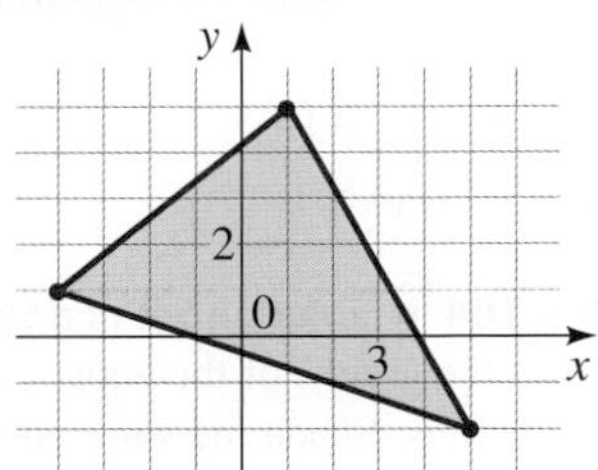

85–90 ■ Partial Fraction Decomposition Find the partial fraction decomposition of the rational expression.

85. $\dfrac{3x + 1}{x^2 - 2x - 15}$

86. $\dfrac{8}{x^3 - 4x}$

87. $\dfrac{2x - 4}{x(x - 1)^2}$

88. $\dfrac{x + 6}{x^3 - 2x^2 + 4x - 8}$

89. $\dfrac{2x - 1}{x^3 + x}$

90. $\dfrac{5x^2 - 3x + 10}{x^4 + x^2 - 2}$

91–94 ■ Intersection Points Two equations and their graphs are given. Estimate the intersection point from the graph and check that the point is a solution to the system.

91. $\begin{cases} 2x + 3y = 7 \\ x - 2y = 0 \end{cases}$

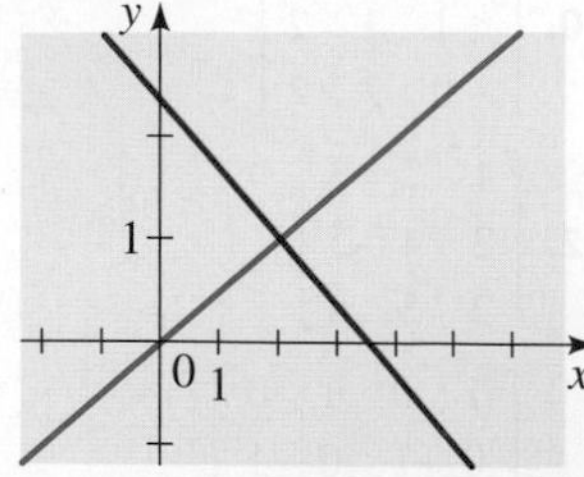

92. $\begin{cases} 3x + y = 8 \\ y = x^2 - 5x \end{cases}$

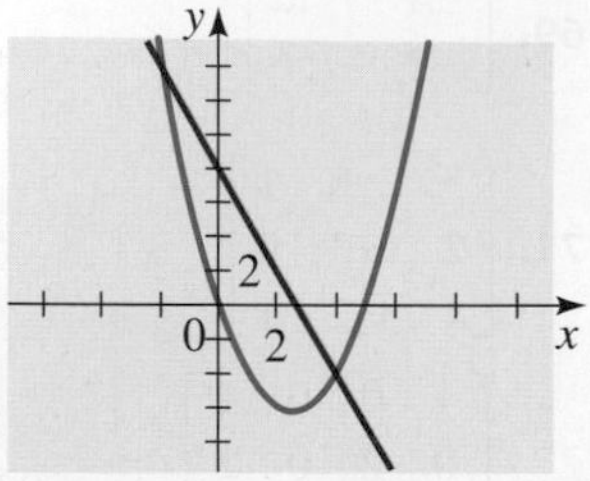

93. $\begin{cases} x^2 + y = 2 \\ x^2 - 3x - y = 0 \end{cases}$

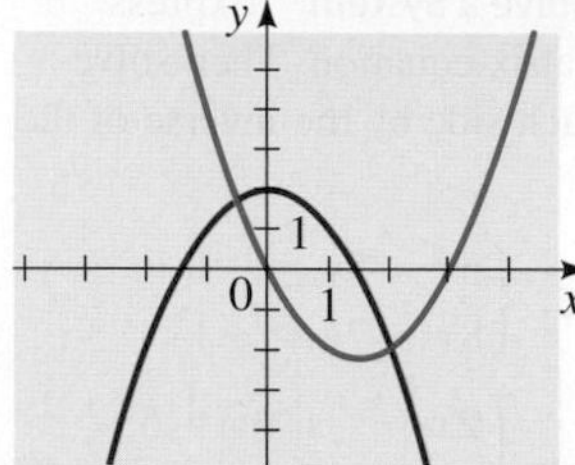

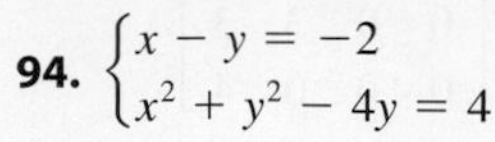

94. $\begin{cases} x - y = -2 \\ x^2 + y^2 - 4y = 4 \end{cases}$

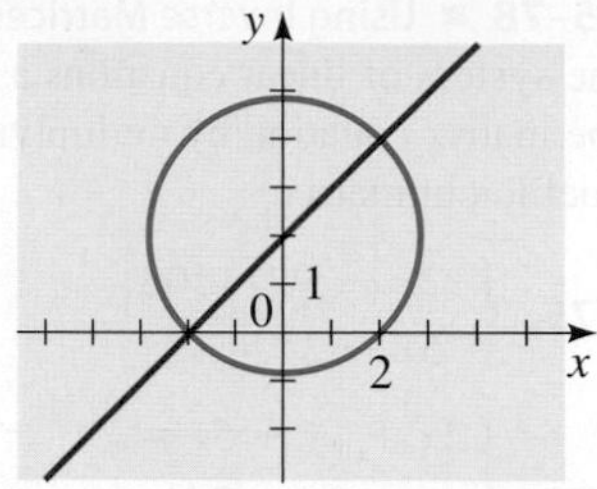

95–96 ■ Finding an Inequality from a Graph An equation and its graph are given. Find an inequality whose solution is the shaded region.

95. $x + y^2 = 4$

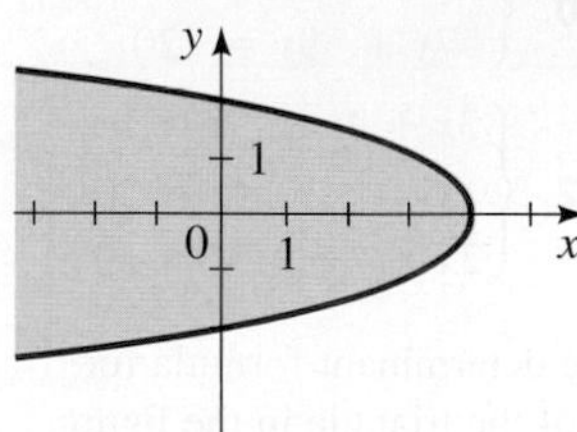

96. $x^2 + y^2 = 8$

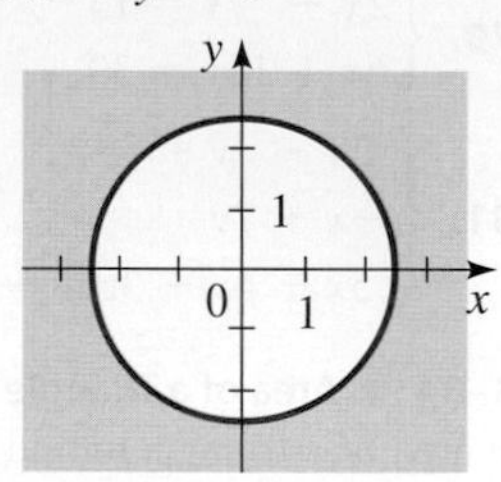

97–100 ■ Graphing Inequalities Graph the inequality.

97. $3x + y \le 6$

98. $y \ge x^2 - 3$

99. $x^2 + y^2 \ge 9$

100. $x - y^2 < 4$

101–104 ■ Solution Set of a System of Inequalities The figure shows the graphs of the equations corresponding to the given inequalities. Shade the solution set of the system of inequalities.

101. $\begin{cases} y \ge x^2 - 3x \\ y \le \frac{1}{3}x - 1 \end{cases}$

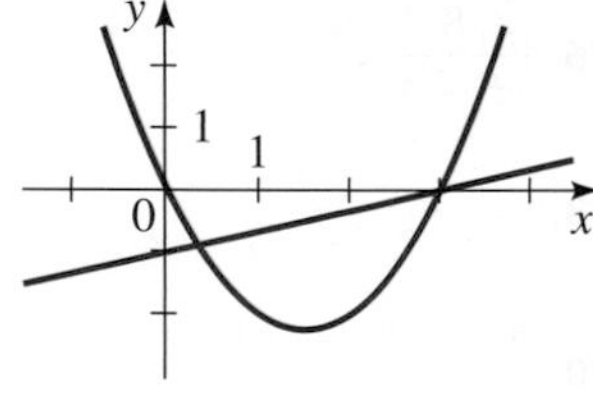

102. $\begin{cases} y \ge x - 1 \\ x^2 + y^2 \le 1 \end{cases}$

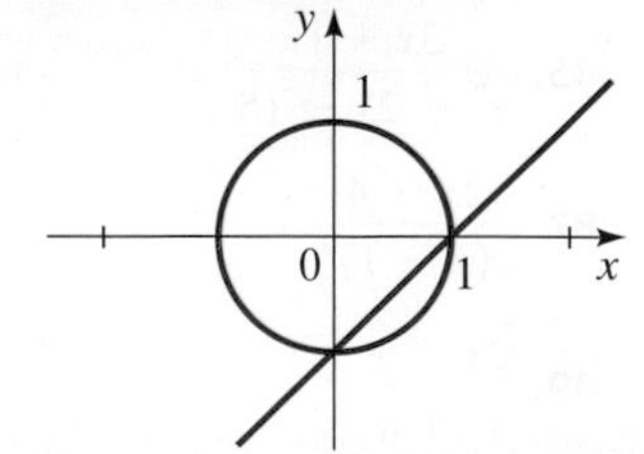

103. $\begin{cases} x + y \ge 2 \\ y - x \le 2 \\ x \le 3 \end{cases}$

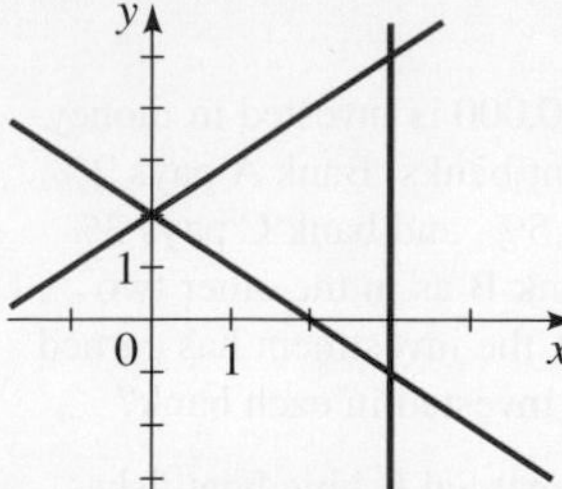

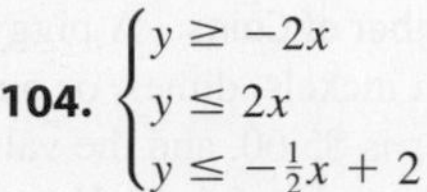

104. $\begin{cases} y \ge -2x \\ y \le 2x \\ y \le -\frac{1}{2}x + 2 \end{cases}$

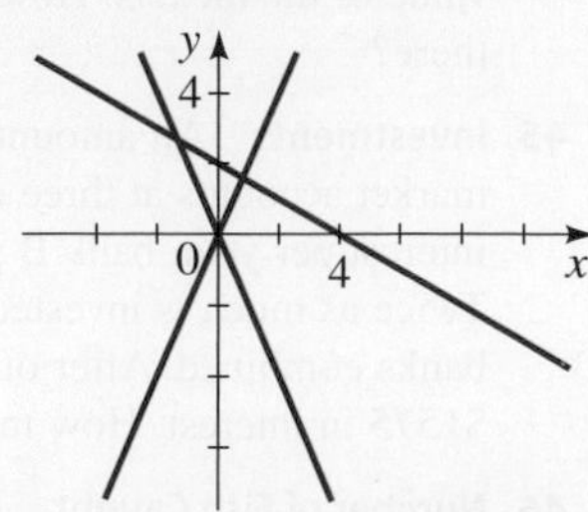

105–108 ■ Systems of Inequalities Graph the solution set of the system of inequalities. Find the coordinates of all vertices, and determine whether the solution set is bounded or unbounded.

105. $\begin{cases} x^2 + y^2 < 9 \\ x + y < 0 \end{cases}$

106. $\begin{cases} y - x^2 \ge 4 \\ y < 20 \end{cases}$

107. $\begin{cases} x \ge 0, \quad y \ge 0 \\ x + 2y \le 12 \\ y \le x + 4 \end{cases}$

108. $\begin{cases} x \ge 4 \\ x + y \ge 24 \\ x \le 2y + 12 \end{cases}$

109–110 ■ General Systems of Equations Solve for x, y, and z in terms of a, b, and c.

109. $\begin{cases} -x + y + z = a \\ x - y + z = b \\ x + y - z = c \end{cases}$

110. $\begin{cases} ax + by + cz = a - b + c \\ bx + by + cz = c \\ cx + cy + cz = c \end{cases}$ $\quad (a \ne b, b \ne c, c \ne 0)$

111. General Systems of Equations For what values of k do the following three lines have a common point of intersection?

$$x + y = 12$$
$$kx - y = 0$$
$$y - x = 2k$$

112. General Systems of Equations For what value of k does the following system have infinitely many solutions?

$$\begin{cases} kx + y + z = 0 \\ x + 2y + kz = 0 \\ -x \qquad + 3z = 0 \end{cases}$$

Chapter 9 | Test

1–2 ■ A system of equations is given. **(a)** Determine whether the system is linear or nonlinear. **(b)** Find all solutions of the system.

1. $\begin{cases} x + 3y = 7 \\ 5x + 2y = -4 \end{cases}$ **2.** $\begin{cases} 6x + y^2 = 10 \\ 3x - y = 5 \end{cases}$

3. Use a graphing device to find all solutions of the system, rounded to two decimal places.

$$\begin{cases} x - 2y = 1 \\ y = x^3 - 2x^2 \end{cases}$$

4. In $2\frac{1}{2}$ hours an airplane travels 600 km against the wind. It takes 50 minutes to travel 300 km with the wind. Find the speed of the wind and the speed of the airplane in still air.

5. Determine whether each matrix is in reduced row-echelon form, row-echelon form, or neither.

(a) $\begin{bmatrix} 1 & 2 & 4 & -6 \\ 0 & 1 & -3 & 0 \end{bmatrix}$ **(b)** $\begin{bmatrix} 1 & 0 & -1 & 0 & 0 \\ 0 & 1 & 3 & 0 & 0 \\ 0 & 0 & 0 & 1 & 0 \\ 0 & 0 & 0 & 0 & 1 \end{bmatrix}$ **(c)** $\begin{bmatrix} 1 & 1 & 0 \\ 0 & 0 & 1 \\ 0 & 1 & 3 \end{bmatrix}$

6. Use Gaussian elimination to find the complete solution of the system, or show that no solution exists.

(a) $\begin{cases} x - y + 2z = 0 \\ 2x - 4y + 5z = -5 \\ 2y - 3z = 5 \end{cases}$ **(b)** $\begin{cases} 2x - 3y + z = 3 \\ x + 2y + 2z = -1 \\ 4x + y + 5z = 4 \end{cases}$

7. Use Gauss-Jordan elimination to find the complete solution of the system.

$$\begin{cases} x + 3y - z = 0 \\ 3x + 4y - 2z = -1 \\ -x + 2y = 1 \end{cases}$$

8. Three friends enter a coffee shop. The first orders two coffees, one juice, and two doughnuts and pays \$6.25. The second orders one coffee and three doughnuts and pays \$3.75. The third orders three coffees, one juice, and four doughnuts and pays \$9.25. Find the price of coffee, juice, and doughnuts at this coffee shop.

9. Let

$$A = \begin{bmatrix} 2 & 3 \\ 2 & 4 \end{bmatrix} \quad B = \begin{bmatrix} 2 & 4 \\ -1 & 1 \\ 3 & 0 \end{bmatrix} \quad C = \begin{bmatrix} 1 & 0 & 4 \\ -1 & 1 & 2 \\ 0 & 1 & 3 \end{bmatrix}$$

Carry out the indicated operation, or explain why it cannot be performed.

(a) $A + B$ **(b)** AB **(c)** $BA - 3B$ **(d)** CBA

(e) A^{-1} **(f)** B^{-1} **(g)** $\det(B)$ **(h)** $\det(C)$

10. (a) Write a matrix equation equivalent to the following system.

$$\begin{cases} 4x - 3y = 10 \\ 3x - 2y = 30 \end{cases}$$

(b) Find the inverse of the coefficient matrix, and use it to solve the system.

11. Only one of the following matrices has an inverse. Find the determinant of each matrix, and use the determinants to identify the one that has an inverse. Then find the inverse.

$$A = \begin{bmatrix} 1 & 4 & 1 \\ 0 & 2 & 0 \\ 1 & 0 & 1 \end{bmatrix} \quad B = \begin{bmatrix} 1 & 4 & 0 \\ 0 & 2 & 0 \\ -3 & 0 & 1 \end{bmatrix}$$

12. Solve using Cramer's Rule:

$$\begin{cases} 2x \qquad\quad - z = 14 \\ 3x - y + 5z = 0 \\ 4x + 2y + 3z = -2 \end{cases}$$

13. Find the partial fraction decomposition of each rational expression.

(a) $\dfrac{4x - 1}{(x - 1)^2(x + 2)}$ **(b)** $\dfrac{2x - 3}{x^3 + 3x}$

14. Graph the solution set of the system of inequalities. Label the vertices with their coordinates.

(a) $\begin{cases} 2x + y \le 8 \\ x - y \ge -2 \\ x + 2y \ge 4 \end{cases}$ **(b)** $\begin{cases} x^2 + y \le 5 \\ y \ge 2x + 5 \end{cases}$

Focus on Modeling | Linear Programming

Linear programming is a modeling technique that is used to determine the optimal allocation of resources in business, the military, and other areas of human endeavor. For example, a manufacturer who makes several different products from the same raw materials can use linear programming to determine how much of each product should be produced to maximize the profit. This modeling technique is probably the most important practical application of systems of linear inequalities. In 1975 Leonid Kantorovich and T. C. Koopmans won the Nobel Prize in economics for their work in the development of this technique.

Linear programming can be applied to complex problems with hundreds or even thousands of variables. Here we consider problems involving two variables (x and y) to which the graphical methods of Section 9.9 can be applied. Each linear programming problem includes restrictions, called **constraints**, that lead to a system of linear inequalities whose solution is called the **feasible region**. The objective of the problem is to maximize or minimize a linear function in the variables x and y, called the **objective function**. This function always attains its largest and smallest values at **vertices** of the feasible region. The following guidelines show the steps used to set up and solve a linear programming problem.

Guidelines for Linear Programming

1. **Choose the Variables.** Decide what variable quantities in the problem should be named x and y.
2. **Find the Objective Function.** Write an expression for the function we want to maximize or minimize.
3. **Graph the Feasible Region.** Express the constraints as a system of inequalities, and graph the solution of this system (the feasible region).
4. **Find the Maximum or Minimum.** Evaluate the objective function at the vertices of the feasible region to determine its maximum or minimum value.

Example 1 ■ Manufacturing for Maximum Profit

A small shoe manufacturer makes two styles of shoes: oxfords and loafers. Two machines are used in the process: a cutting machine and a sewing machine. Each type of shoe requires 15 minutes per pair on the cutting machine. Oxfords require 10 minutes of sewing per pair, and loafers require 20 minutes of sewing per pair. Because the manufacturer can hire only one operator for each machine, each process is available for just 8 hours per day. If the profit is $15 on each pair of oxfords and $20 on each pair of loafers, how many pairs of each type should be produced per day for maximum profit?

Because loafers produce more profit, it would seem best to manufacture only loafers. Surprisingly, this does not turn out to be the most profitable solution.

Solution First we organize the given information into a table. To be consistent, let's convert all times to hours.

	Oxfords	Loafers	Time Available
Time on cutting machine (h)	$\frac{1}{4}$	$\frac{1}{4}$	8
Time on sewing machine (h)	$\frac{1}{6}$	$\frac{1}{3}$	8
Profit	\$15	\$20	

We describe the model and solve the problem in four steps.

Linear Programming helps the telephone industry to determine the most efficient way to route telephone calls. The computerized routing decisions must be made very rapidly so that callers are not kept waiting for connections. Since the database of customers and routes is huge, an extremely fast method for solving linear programming problems is essential. In 1984 the 28-year-old mathematician **Narendra Karmarkar**, working at Bell Labs in Murray Hill, New Jersey, discovered just such a method. His idea is so ingenious and his method so fast that the discovery caused a sensation in the mathematical world. Although mathematical discoveries rarely make the news, this one was reported in *Time*, on December 3, 1984. Today airlines routinely use Karmarkar's technique to minimize costs in scheduling passengers, flight personnel, fuel, baggage, and maintenance workers.

Choose the variables. To make a mathematical model, we first give names to the variable quantities. For this problem we let

$$x = \text{number of pairs of oxfords made daily}$$
$$y = \text{number of pairs of loafers made daily}$$

Find the objective function. Our goal is to determine which values for x and y give maximum profit. Since each pair of oxfords provides \$15 profit and each pair of loafers provides \$20, the total profit is given by

$$P = 15x + 20y$$

This function is the *objective function*.

Graph the feasible region. As x and y increase, so does the profit. But we cannot choose arbitrarily large values for these variables because of the restrictions, or *constraints*, in the problem. Each restriction is an inequality in the variables.

In this problem the total number of cutting hours needed is $\frac{1}{4}x + \frac{1}{4}y$. Since only 8 hours are available on the cutting machine, we have

$$\tfrac{1}{4}x + \tfrac{1}{4}y \le 8$$

Similarly, by considering the amount of time needed and available on the sewing machine, we get

$$\tfrac{1}{6}x + \tfrac{1}{3}y \le 8$$

We cannot produce a negative number of shoes, so we also have

$$x \ge 0 \quad \text{and} \quad y \ge 0$$

Thus x and y must satisfy the constraints

$$\begin{cases} \frac{1}{4}x + \frac{1}{4}y \le 8 \\ \frac{1}{6}x + \frac{1}{3}y \le 8 \\ x \ge 0, \quad y \ge 0 \end{cases}$$

If we multiply the first inequality by 4 and the second by 6, we obtain the simplified system

$$\begin{cases} x + y \le 32 \\ x + 2y \le 48 \\ x \ge 0, \quad y \ge 0 \end{cases}$$

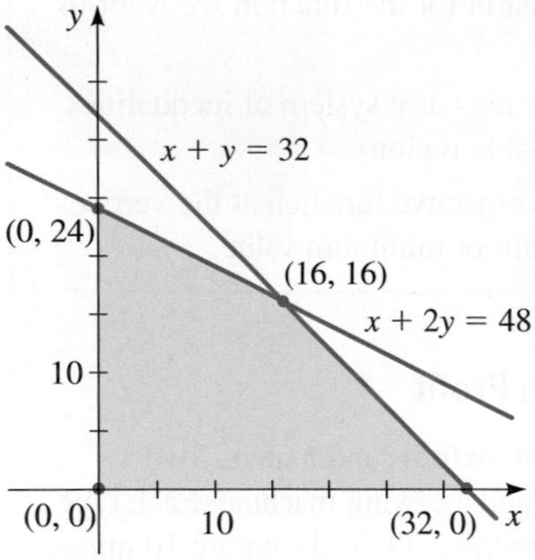

Figure 1

The solution of this system (with vertices labeled) is sketched in Figure 1. The only values that satisfy the restrictions of the problem are the ones that correspond to points of the shaded region in Figure 1. This is the *feasible region* for the problem.

Find the maximum profit. As x or y increases, profit increases as well. Thus it seems reasonable that the maximum profit will occur at a point on one of the outside edges of the feasible region, where it is impossible to increase x or y without going outside the region. In fact, it can be shown that the maximum value occurs at a vertex (see Problem 16). This means that we need to check the profit only at the vertices. From the table we see that the largest value of P occurs at the point (16, 16), where $P =$ \$560. Thus the manufacturer should make 16 pairs of oxfords and 16 pairs of loafers, for a maximum daily profit of \$560.

Vertex	$P = 15x + 20y$
(0, 0)	0
(0, 24)	15(0) + 20(24) = \$480
(16, 16)	15(16) + 20(16) = \$560 ← Maximum profit
(32, 0)	15(32) + 20(0) = \$480

■

Example 2 ■ A Shipping Problem

A car dealer has warehouses in Millville and Trenton and dealerships in Camden and Atlantic City. Every car that is sold at the dealerships must be delivered from one of the warehouses. On a certain day the Camden dealers sell 10 cars, and the Atlantic City dealers sell 12. The Millville warehouse has 15 cars available, and the Trenton warehouse has 10. The cost of shipping one car is \$50 from Millville to Camden, \$40 from Millville to Atlantic City, \$60 from Trenton to Camden, and \$55 from Trenton to Atlantic City. How many cars should be moved from each warehouse to each dealership to fill the orders at minimum cost?

Solution Our first step is to organize the given information. Rather than construct a table, we draw a diagram to show the flow of cars from the warehouses to the dealerships (see Figure 2 below). The diagram shows the number of cars available at each warehouse or required at each dealership and the cost of shipping between these locations.

Choose the variables. The arrows in Figure 2 indicate four possible routes, so the problem seems to involve four variables. We let

x = number of cars to be shipped from Millville to Camden

y = number of cars to be shipped from Millville to Atlantic City

To fill the orders, we must have

$10 - x$ = number of cars shipped from Trenton to Camden

$12 - y$ = number of cars shipped from Trenton to Atlantic City

So the only variables in the problem are x and y.

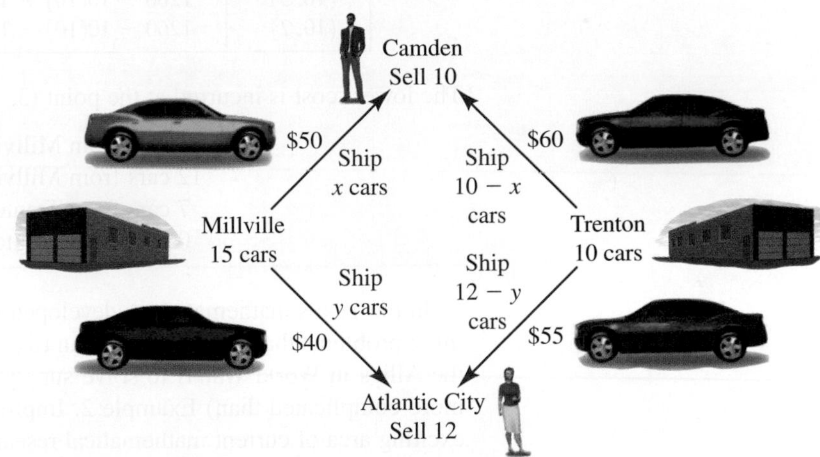

Figure 2

Find the objective function. The objective of this problem is to minimize cost. From Figure 2 we see that the total cost C of shipping the cars is

$$\begin{aligned} C &= 50x + 40y + 60(10 - x) + 55(12 - y) \\ &= 50x + 40y + 600 - 60x + 660 - 55y \\ &= 1260 - 10x - 15y \end{aligned}$$

This is the objective function.

Graph the feasible region. Now we derive the constraint inequalities that define the feasible region. First, the number of cars shipped on each route can't be negative, so we have

$$\begin{aligned} x \geq 0 \qquad & \qquad y \geq 0 \\ 10 - x \geq 0 \qquad & \qquad 12 - y \geq 0 \end{aligned}$$

Second, the total number of cars shipped from each warehouse can't exceed the number of cars available there, so

$$x + y \le 15$$
$$(10 - x) + (12 - y) \le 10$$

Simplifying the latter inequality, we get

$$22 - x - y \le 10$$
$$-x - y \le -12$$
$$x + y \ge 12$$

The inequalities $10 - x \ge 0$ and $12 - y \ge 0$ can be rewritten as $x \le 10$ and $y \le 12$. Thus the feasible region is described by the constraints

$$\begin{cases} x + y \le 15 \\ x + y \ge 12 \\ 0 \le x \le 10 \\ 0 \le y \le 12 \end{cases}$$

The feasible region is graphed in Figure 3.

Figure 3

Find the minimum cost. We check the value of the objective function at each vertex of the feasible region.

Vertex	$C = 1260 - 10x - 15y$
(0, 12)	$1260 - 10(0) - 15(12) = \$1080$
(3, 12)	$1260 - 10(3) - 15(12) = \$1050$ ← Minimum cost
(10, 5)	$1260 - 10(10) - 15(5) = \$1085$
(10, 2)	$1260 - 10(10) - 15(2) = \$1130$

The lowest cost is incurred at the point (3, 12). Thus the dealer should ship

3 cars from Millville to Camden
12 cars from Millville to Atlantic City
7 cars from Trenton to Camden
0 cars from Trenton to Atlantic City ■

In the 1940s mathematicians developed matrix methods for solving linear programming problems that involve more than two variables. These methods were first used by the Allies in World War II to solve supply problems similar to (but, of course, much more complicated than) Example 2. Improving such matrix methods is an active and exciting area of current mathematical research.

Problems

1–4 ■ Find the maximum and minimum values of the given objective function on the indicated feasible region.

1. $M = 200 - x - y$

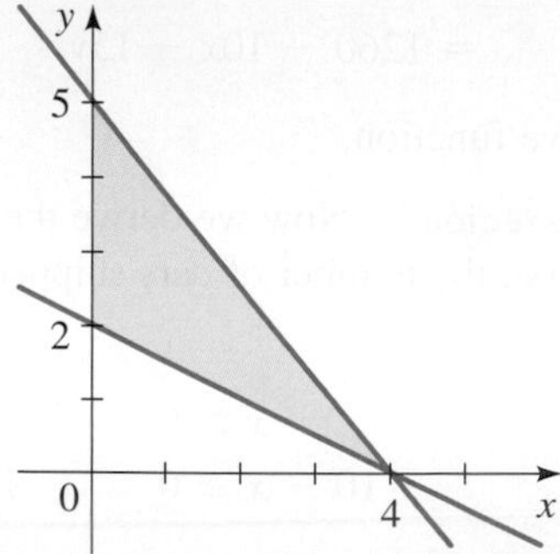

2. $N = \frac{1}{2}x + \frac{1}{4}y + 40$

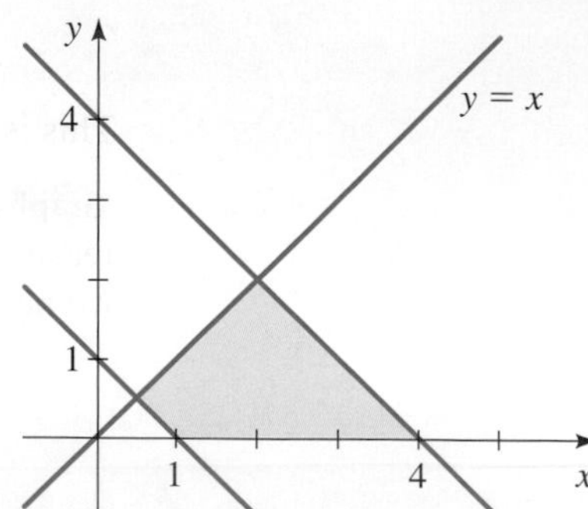

3. $P = 140 - x + 3y$

$$\begin{cases} x \geq 0, y \geq 0 \\ 2x + y \leq 10 \\ 2x + 4y \leq 28 \end{cases}$$

4. $Q = 70x + 82y$

$$\begin{cases} x \geq 0, y \geq 0 \\ x \leq 10, y \leq 20 \\ x + y \geq 5 \\ x + 2y \leq 18 \end{cases}$$

5. Making Furniture A furniture manufacturer makes wooden tables and chairs. The production process involves two basic types of labor: carpentry and finishing. A table requires 2 hours of carpentry and 1 hour of finishing, and a chair requires 3 hours of carpentry and $\frac{1}{2}$ hour of finishing. The profit is \$35 per table and \$20 per chair. The manufacturer's employees can supply a maximum of 108 hours of carpentry work and 20 hours of finishing work per day. How many tables and chairs should be made each day to maximize profit?

6. A Housing Development A housing contractor has subdivided a farm into 100 building lots. There are two types of homes for these lots: colonial and ranch style. A colonial requires \$30,000 of capital and produces a profit of \$4000 when sold. A ranch-style house requires \$40,000 of capital and provides an \$8000 profit. If the contractor has \$3.6 million of capital on hand, how many houses of each type should be built for maximum profit? Will any of the lots be left vacant?

7. Hauling Fruit A trucking company transports citrus fruit from Florida to Montreal. Each crate of oranges is 0.1 m³ in volume and weighs 36 kg. Each crate of grapefruit has a volume of 0.15 m³ and weighs 45 kg. Each company truck has a maximum capacity of 7.5 m³ and can carry no more than 2520 kg. Moreover, each truck is not permitted to carry more crates of grapefruit than crates of oranges. If the profit is \$2.50 on each crate of oranges and \$4 on each crate of grapefruit, how many crates of each fruit should a truck carry for maximum profit?

8. Manufacturing Calculators A manufacturer of calculators produces two models: standard and scientific. The long-term demand for the two models mandates that the company manufacture at least 100 standard and 80 scientific calculators each day. However, because of limitations on production capacity, no more than 200 standard and 170 scientific calculators can be made daily. To satisfy a shipping contract, a total of at least 200 calculators must be shipped every day.

(a) If the production cost is \$5 for a standard calculator and \$7 for a scientific one, how many of each model should be produced daily to minimize this cost?

(b) If each standard calculator results in a \$2 loss but each scientific one produces a \$5 profit, how many of each model should be made daily to maximize profit?

9. Shipping Televisions An electronics discount chain has a sale on a certain brand of 1.5 m high-definition television set. The chain has stores in Santa Monica and El Toro and warehouses in Long Beach and Pasadena. To satisfy rush orders, 15 sets must be shipped from the warehouses to the Santa Monica store, and 19 must be shipped to the El Toro store. The cost of shipping a set is \$5 from Long Beach to Santa Monica, \$6 from Long Beach to El Toro, \$4 from Pasadena to Santa Monica, and \$5.50 from Pasadena to El Toro. If the Long Beach warehouse has 24 sets and the Pasadena warehouse has 18 sets in stock, how many sets should be shipped from each warehouse to each store to fill the orders at a minimum shipping cost?

10. Delivering Plywood A building supply company has two warehouses, one on the east side and one on the west side of a city. Two customers order some 1.25 centimeter plywood. Customer A needs 50 sheets, and customer B needs 70 sheets. The east-side warehouse has 80 sheets, and the west-side warehouse has 45 sheets of this plywood in stock. The east-side warehouse's delivery costs per sheet are \$0.50 to customer A and \$0.60 to customer B. The west-side warehouse's delivery costs per sheet are \$0.40 to customer A and \$0.55 to customer B. How many sheets should be shipped from each warehouse to each customer to minimize delivery costs?

11. Packaging Nuts A confectioner sells two types of nut mixtures. The standard-mixture package contains 100 g of cashews and 200 g of peanuts and sells for \$1.95. The deluxe-mixture package contains 150 g of cashews and 50 g of peanuts and sells for \$2.25. The confectioner has 15 kg of cashews and 20 kg of peanuts available. On the basis of past sales, the confectioner needs to have at least as many standard as deluxe packages available. How many bags of each mixture should be packaged to maximize revenue?

12. Feeding Lab Rabbits A biologist wishes to feed laboratory rabbits a mixture of two types of foods. Type I contains 8 g of fat, 12 g of carbohydrate, and 2 g of protein per serving. Type II contains 12 g of fat, 12 g of carbohydrate, and 1 g of protein per serving. Type I costs \$0.20 per serving and type II costs \$0.30 per serving. Each rabbit receives a daily minimum of 24 g of fat, 36 g of carbohydrate, and 4 g of protein, but get no more than 5 servings of food per day. How many servings of each food type should be fed to each rabbit daily to satisfy the dietary requirements at minimum cost?

13. Investing in Bonds A financial advisor needs to invest \$12,000 in three types of bonds: municipal bonds paying 7% interest per year, bank certificates paying 8%, and high-risk bonds paying 12%. For tax reasons the amount invested in municipal bonds should be at least three times the amount invested in bank certificates. To keep the level of risk manageable, no more than \$2000 should be invested in high-risk bonds. How much should be invested in each type of bond to maximize the annual interest yield? [*Hint:* Let $x =$ amount in municipal bonds and $y =$ amount in bank certificates. Then the amount in high-risk bonds will be $12{,}000 - x - y$.]

14. Annual Interest Yield Refer to Problem 13. Suppose the investor decides to increase the maximum invested in high-risk bonds to \$3000 but leaves the other conditions unchanged. By how much will the maximum possible interest yield increase?

15. Business Strategy A small software company publishes computer games, educational software, and utility software. Their business strategy is to market a total of 36 new programs each year, at least four of these being games. The number of utility programs published is never more than twice the number of educational programs. On average, the company makes an annual profit of \$5000 on each computer game, \$8000 on each educational program, and \$6000 on each utility program. How many of each type of software should the company publish annually for maximum profit?

16. Extreme Values and Vertices This exercise illustrates why the minimum and maximum values of the objective function occur at vertices of the feasible region. The feasible region for the following linear programming problem is graphed in the figure.

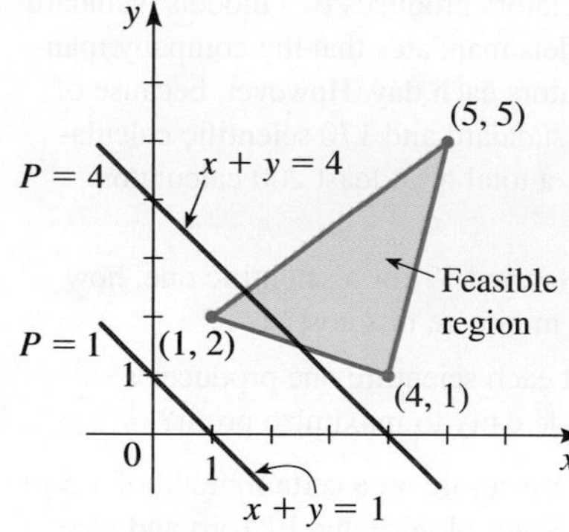

$$\begin{cases} 4y - 3x \le 5 \\ 4x - y \le 15 \\ x + 3y \ge 7 \end{cases}$$

$$P = x + y$$

For each value of P the graph of the objective function is a line; the lines for $P = 1$ and $P = 4$ are shown in the figure. On the given graph, sketch the lines corresponding to increasing values of P, starting at $P = 1$. What are the minimum and maximum values of P for which the line $P = x + y$ intersects the feasible region? Explain why these are the minimum and maximum values of P on the feasible region. At what points of the feasible region do these extreme values occur?

Pol Sole/Shutterstock.com

10 Conic Sections

Conic sections are the curves that are formed when a plane cuts a cone, as shown in the figure. For example, if a cone is cut horizontally, the cross section is a circle. So a circle is a conic section. Other ways of cutting a cone produce ellipses, parabolas, and hyperbolas.

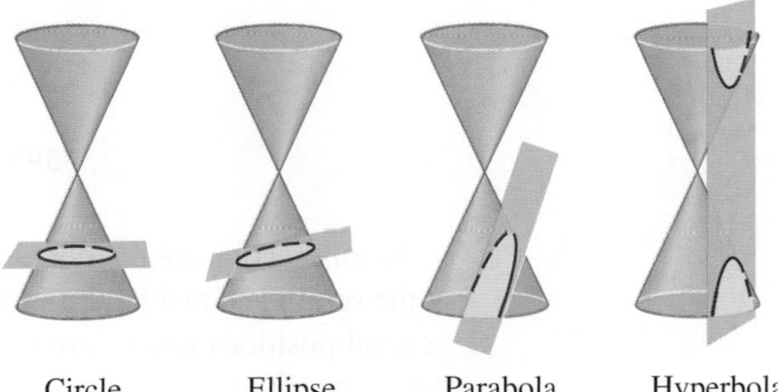

Our goal in this chapter is to find equations whose graphs are conic sections. We will find such equations by analyzing the geometric properties of conic sections. These properties make conic sections useful for many real-world applications. For instance, a reflecting surface with parabolic cross sections concentrates light at a single point. This property of a parabola is used in the construction of reflecting telescopes, as well as in describing the path of planets and comets. In the *Focus on Modeling* at the end of the chapter we explore how these curves are used in architecture.

10.1 Parabolas

■ Geometric Definition of a Parabola ■ Equations and Graphs of Parabolas ■ Applications

■ Geometric Definition of a Parabola

We saw in Section 3.1 that the graph of the equation

$$y = ax^2 + bx + c$$

is a U-shaped curve called a *parabola* that opens either upward or downward, depending on whether the number a is positive or negative.

In this section we study parabolas from a geometric, rather than an algebraic, point of view. We begin with the geometric definition of a parabola and show how this leads to the algebraic formula that we are already familiar with.

> **Geometric Definition of a Parabola**
>
> A **parabola** is the set of all points in the plane that are equidistant from a fixed point F (called the **focus**) and a fixed line l (called the **directrix**).

This definition is illustrated in Figure 1. The **vertex** V of the parabola lies halfway between the focus and the directrix, and the **axis of symmetry** is the line that runs through the focus perpendicular to the directrix.

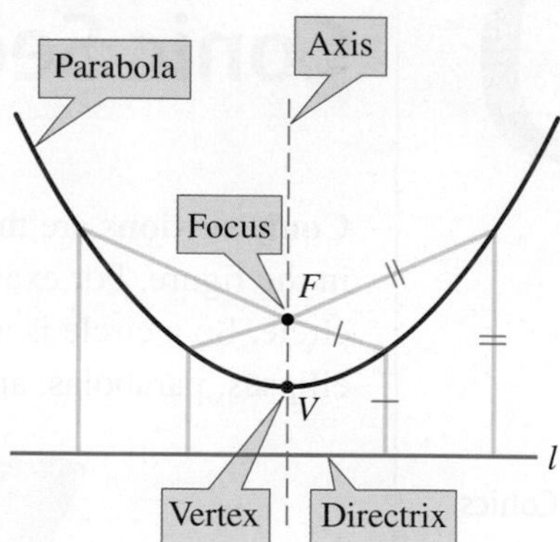

Figure 1

In this section we restrict our attention to parabolas that are situated with the vertex at the origin and that have a vertical or horizontal axis of symmetry. (Parabolas in more general positions will be considered in Section 10.4.) If the focus of such a parabola is the point $F(0, p)$, then the axis of symmetry must be vertical, and the directrix has the equation $y = -p$. Figure 2 illustrates the case $p > 0$.

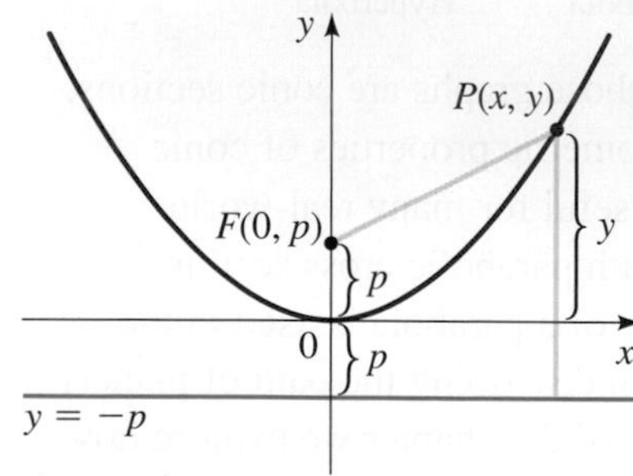

Figure 2

Deriving the Equation of a Parabola If $P(x, y)$ is any point on the parabola, then the distance from P to the focus F (using the Distance Formula) is

$$\sqrt{x^2 + (y - p)^2}$$

The distance from P to the directrix is

$$|y - (-p)| = |y + p|$$

By the definition of a parabola these two distances must be equal.

$$\sqrt{x^2 + (y - p)^2} = |y + p|$$

$$x^2 + (y - p)^2 = |y + p|^2 = (y + p)^2 \qquad \text{Square both sides}$$

$$x^2 + y^2 - 2py + p^2 = y^2 + 2py + p^2 \qquad \text{Expand}$$

$$x^2 - 2py = 2py \qquad \text{Simplify}$$

$$x^2 = 4py$$

This is the **standard equation** of a parabola with vertical axis and vertex at the origin. If $p > 0$, then the parabola opens upward; if $p < 0$, it opens downward. When x is replaced by $-x$, the equation remains unchanged, so the graph is symmetric about the y-axis.

■ Equations and Graphs of Parabolas

The following box summarizes what we have just proved about the equation and features of a parabola with a vertical axis.

Parabola with Vertical Axis

The graph of the equation

$$x^2 = 4py$$

is a parabola with the following properties.

VERTEX	$V(0, 0)$
FOCUS	$F(0, p)$
DIRECTRIX	$y = -p$

The parabola opens upward if $p > 0$ or downward if $p < 0$.

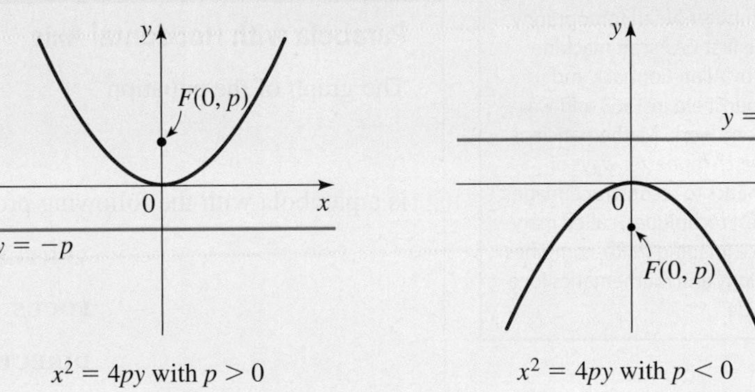

Figure 3

Example 1 ■ Finding the Equation of a Parabola

Find the standard equation for the parabola with vertex $V(0, 0)$ and focus $F(0, 2)$, and sketch its graph.

Solution Since the focus is $F(0, 2)$, we conclude that $p = 2$ (so the directrix is $y = -2$). Thus the standard equation of the parabola is

$$x^2 = 4(2)y \qquad x^2 = 4py \text{ with } p = 2$$

$$x^2 = 8y$$

Since $p = 2 > 0$, the parabola opens upward. See Figure 3.

Now Try Exercises 31 and 49 ■

Mathematics in the Modern World

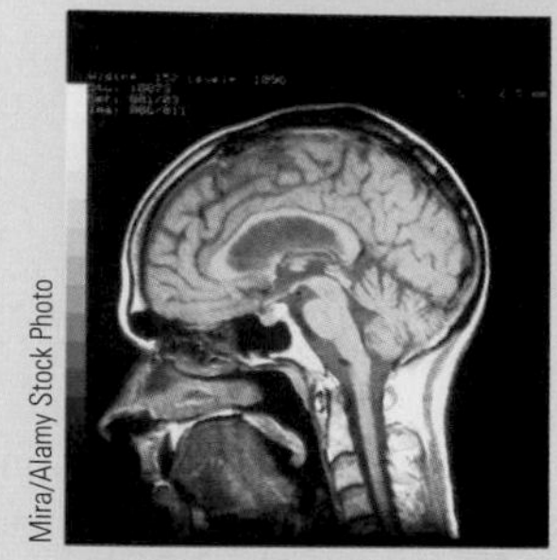
Mira/Alamy Stock Photo

Looking Inside Your Head

Would you like to look inside your head? The idea isn't particularly appealing to most of us, but doctors often need to do just that. If they can look without invasive surgery, all the better. An X-ray doesn't really give a look inside, it simply gives a "graph" of the density of tissue the X-rays must pass through. So an X-ray is a "flattened" view in one direction. Suppose you get an X-ray view from many different directions. Can these "graphs" be used to reconstruct the three-dimensional inside view? This is a purely mathematical problem. The mathematician Johann Radon conceived of and solved this problem only because it was mathematically interesting to him. His solution, published in 1917, had no practical application at that time because reconstructing the inside view requires thousands of tedious computations. Today, high-speed computers make it possible to "look inside" by a process called computer-aided tomography (CAT scan). The first CAT scan machine was invented by Allan Cormack and Sir Godfrey Hounsfield in 1963 and was based on Radon's work. Mathematicians continue to search for better ways of using mathematics to reconstruct images. One of the latest techniques, called magnetic resonance imaging (MRI), combines molecular biology and mathematics for a clear "look inside."

Example 2 ■ Finding the Focus and Directrix of a Parabola from Its Equation

Find the focus and directrix of the parabola $y = -x^2$, and sketch the graph.

Solution To find the focus and directrix, we put the given equation in the standard form $x^2 = -y$. Comparing this to the equation $x^2 = 4py$, we see that $4p = -1$, so $p = -\frac{1}{4}$. Thus the focus is $F\left(0, -\frac{1}{4}\right)$, and the directrix is $y = \frac{1}{4}$. The graph of the parabola, together with the focus and the directrix, is shown in Figure 4(a). We can also draw the graph using a graphing device, as shown in Figure 4(b).

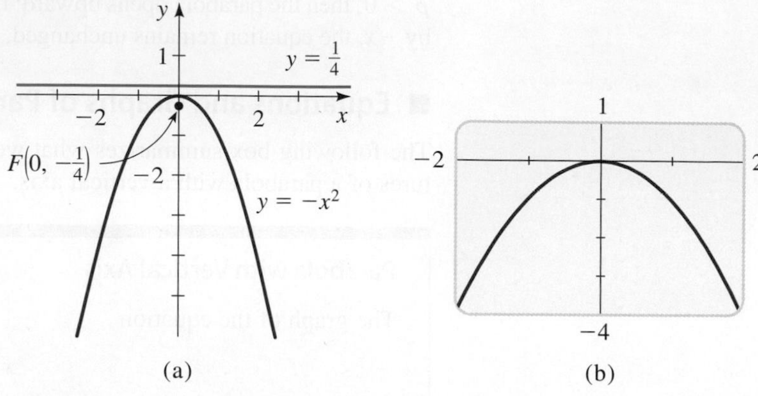

Figure 4

 **Now Try Exercise 11** ■

Reflecting the graph in Figure 2 about the diagonal line $y = x$ has the effect of interchanging the roles of x and y. This results in a parabola with horizontal axis and vertex at the origin, with standard equation $y^2 = 4px$.

Parabola with Horizontal Axis

The graph of the equation

$$y^2 = 4px$$

is a parabola with the following properties.

VERTEX	$V(0, 0)$
FOCUS	$F(p, 0)$
DIRECTRIX	$x = -p$

The parabola opens to the right if $p > 0$ or to the left if $p < 0$.

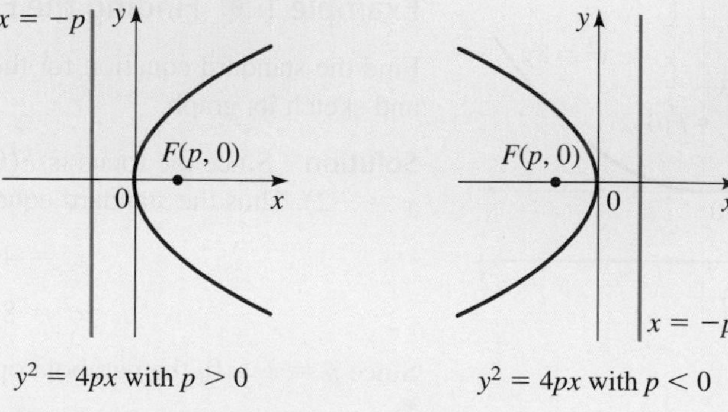

$y^2 = 4px$ with $p > 0$

$y^2 = 4px$ with $p < 0$

Example 3 ■ A Parabola with Horizontal Axis

A parabola has the equation $6x + y^2 = 0$.

(a) Find the focus and directrix of the parabola, and sketch the graph.

(b) Use a graphing device to draw the graph.

Solution

(a) To find the focus and directrix, we put the given equation in the standard form $y^2 = -6x$. Comparing this to the equation $y^2 = 4px$, we see that $4p = -6$, so $p = -\frac{3}{2}$. Thus the focus is $F\left(-\frac{3}{2}, 0\right)$, and the directrix is $x = \frac{3}{2}$. Since $p < 0$, the parabola opens to the left. The graph of the parabola, together with the focus and the directrix, is shown in Figure 5(a).

Equations that define functions are discussed in Section 2.2.

(b) The equation $6x + y^2 = 0$ does not define y as a function of x, but most graphing devices can draw the graph of this equation, as shown in Figure 5(b).

Using a Graphing Calculator Many graphing calculators can only graph equations that define y as a function of x. To use such a graphing calculator we first solve for y.

$$6x + y^2 = 0$$

$$y^2 = -6x \qquad \text{Subtract } 6x$$

$$y = \pm\sqrt{-6x} \qquad \text{Take square roots}$$

The graph of the parabola in Figure 5(b) is obtained by graphing both functions $y = \sqrt{-6x}$ and $y = -\sqrt{-6x}$.

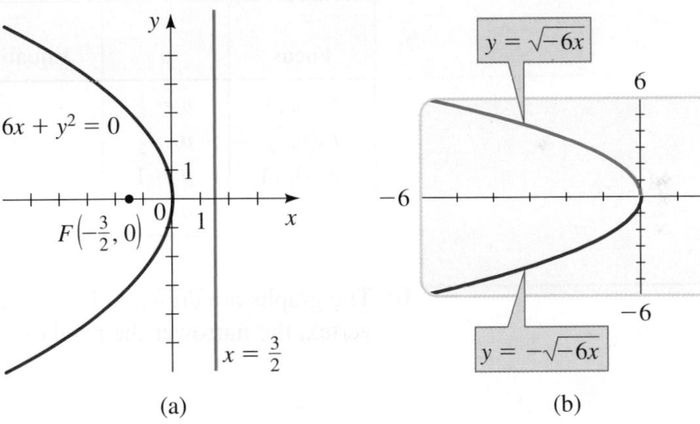

Figure 5

Now Try Exercises 13 and 27 ■

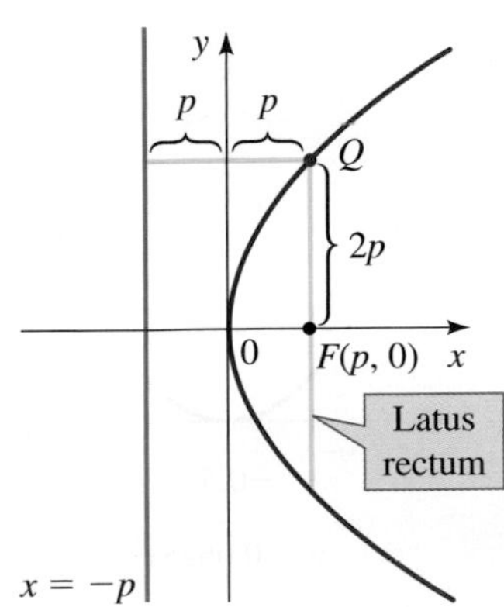

Figure 6

We can use the coordinates of the focus to estimate the "width" of a parabola when sketching its graph. The line segment that runs through the focus perpendicular to the axis, with endpoints on the parabola, is called the **latus rectum**, and its length is the **focal diameter** of the parabola. From Figure 6 we can see that the distance from an endpoint Q of the latus rectum to the directrix is $|2p|$. Thus the distance from Q to the focus must be $|2p|$ as well (by the definition of a parabola), so the focal diameter is $|4p|$. In the next example we use the focal diameter to determine the "width" of a parabola when graphing it.

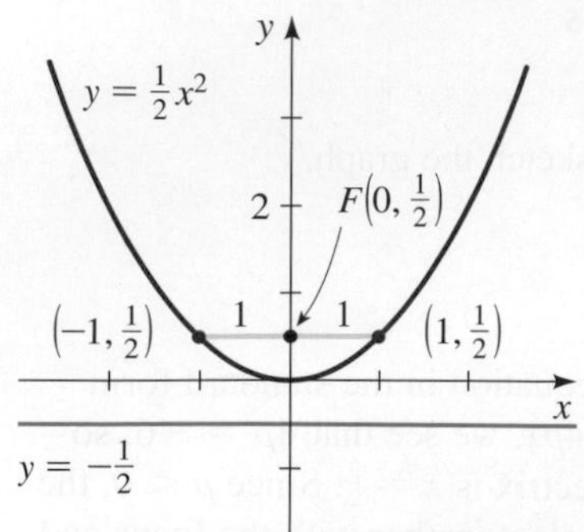

Figure 7

Example 4 ■ The Focal Diameter of a Parabola

Find the focus, directrix, and focal diameter of the parabola $y = \frac{1}{2}x^2$, and sketch its graph.

Solution We first put the equation in the standard form $x^2 = 4py$.

$$y = \tfrac{1}{2}x^2$$

$$x^2 = 2y \qquad \text{Multiply by 2, switch sides}$$

From this equation we see that $4p = 2$, so the focal diameter is 2. Solving for p gives $p = \frac{1}{2}$, so the focus is $(0, \frac{1}{2})$, and the directrix is $y = -\frac{1}{2}$. Since the focal diameter is 2, the latus rectum extends 1 unit to the left and 1 unit to the right of the focus. The graph is sketched in Figure 7.

 **Now Try Exercise 15** ■

In the next example we graph a family of parabolas to show how changing the distance between the focus and the vertex affects the "width" of a parabola.

Example 5 ■ A Family of Parabolas

(a) Find equations for the parabolas with vertex at the origin and foci $F_1(0, \frac{1}{8})$, $F_2(0, \frac{1}{2})$, $F_3(0, 1)$, and $F_4(0, 4)$.

(b) Draw the graphs of the parabolas in part (a). What do you conclude?

Solution

(a) Since the foci are on the positive y-axis, the parabolas open upward and have equations of the form $x^2 = 4py$. This leads to the following equations.

Focus	p	Equation $x^2 = 4py$	Form of the Equation for Graphing Calculator
$F_1(0, \frac{1}{8})$	$p = \frac{1}{8}$	$x^2 = \frac{1}{2}y$	$y = 2x^2$
$F_2(0, \frac{1}{2})$	$p = \frac{1}{2}$	$x^2 = 2y$	$y = 0.5x^2$
$F_3(0, 1)$	$p = 1$	$x^2 = 4y$	$y = 0.25x^2$
$F_4(0, 4)$	$p = 4$	$x^2 = 16y$	$y = 0.0625x^2$

(b) The graphs are drawn in Figure 8. We see that the closer the focus is to the vertex, the narrower the parabola.

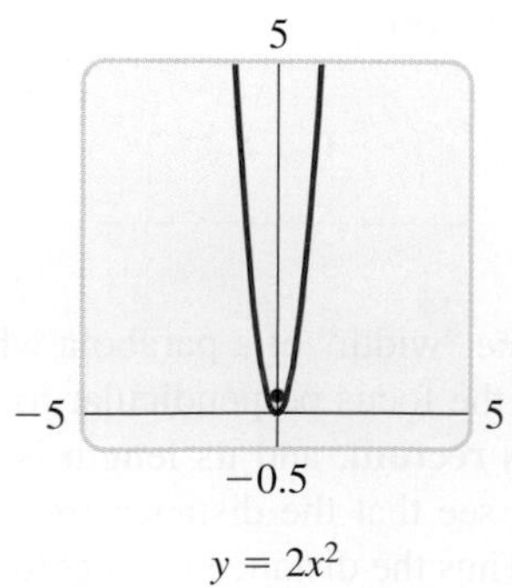

$y = 2x^2$

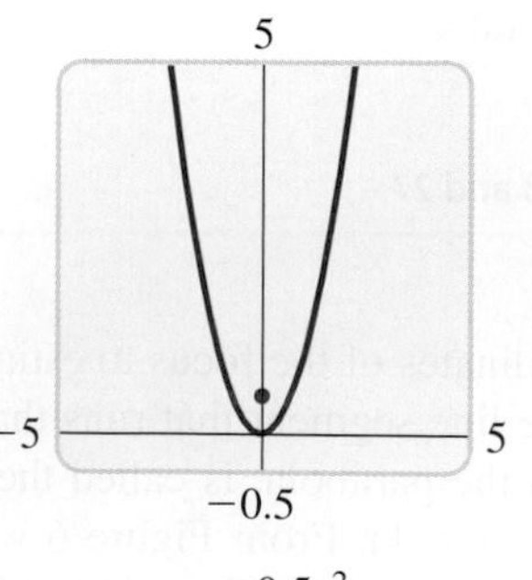

$y = 0.5x^2$

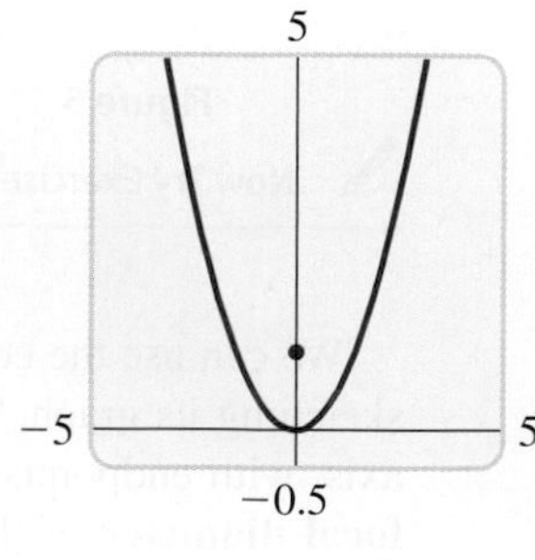

$y = 0.25x^2$

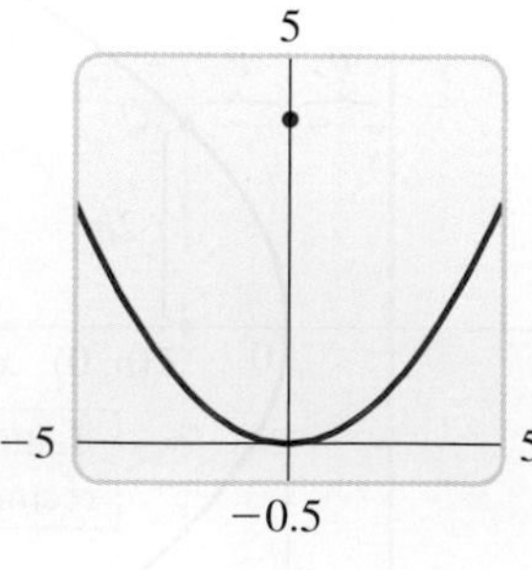

$y = 0.0625x^2$

Figure 8 | A family of parabolas

Now Try Exercise 59 ■

■ Applications

Parabolas have a property that makes them useful as reflectors for lamps and telescopes. Light from a source placed at the focus of a surface with parabolic cross section will be reflected in such a way that it travels parallel to the axis of the parabola (see Figure 9). Thus a parabolic mirror reflects the light into a beam of parallel rays. Conversely, light approaching the reflector in rays parallel to its axis of symmetry is concentrated to the focus. This *reflection property*, which can be proved by using calculus, is used in the construction of reflecting telescopes.

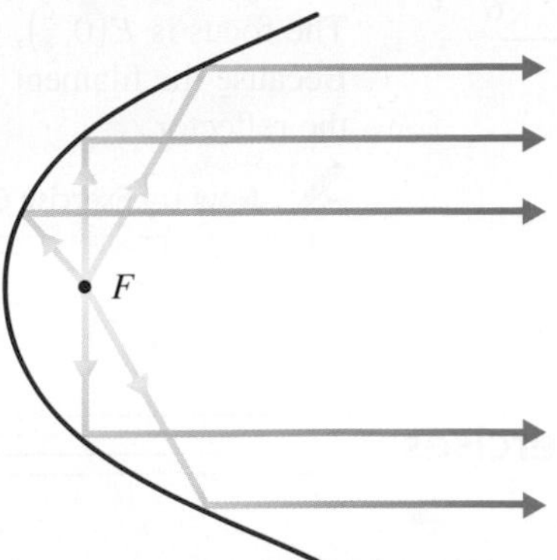

Figure 9 | Parabolic reflector

Example 6 ■ Finding the Focal Point of a Searchlight Reflector

A searchlight has a parabolic reflector that forms a "bowl," which is 30 cm wide from rim to rim and 20 cm deep, as shown in Figure 10. If the filament of the light bulb is located at the focus, how far from the vertex of the reflector is it?

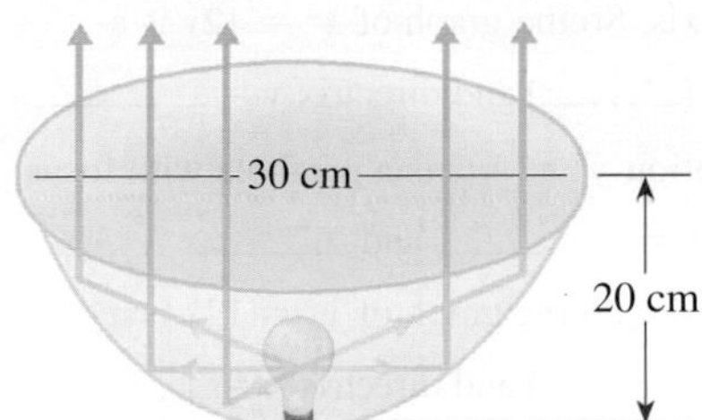

Figure 10 | A parabolic reflector

ARCHIMEDES (287–212 B.C.) was the greatest mathematician of the ancient world. He was born in Syracuse, a Greek colony on Sicily, a generation after Euclid (see Section 7.1). One of his many discoveries is the Law of the Lever (see Exercise 1.7.77). He famously said, "Give me a place to stand and a fulcrum for my lever, and I can lift the earth."

Renowned as a mechanical genius for his many engineering inventions, he designed pulleys for lifting heavy ships and the spiral screw for transporting water to higher levels. He is said to have used parabolic mirrors to concentrate the rays of the sun to set fire to Roman ships attacking Syracuse.

King Hieron II of Syracuse once suspected a goldsmith of keeping part of the gold intended for the king's crown and replacing it with an equal amount of silver. The king asked Archimedes for advice. While in deep thought at a public bath, Archimedes discovered the solution to the king's problem when he noticed that his body's volume was the same as the volume of water it displaced from the tub. Using this insight, he was able to measure the volume of the crown and so determine what the weight of an all-gold crown should be (Exercise 9.1.74). As the story is told, he ran home, forgetting that he was naked, shouting, "Eureka, eureka!" ("I have found it, I have found it!") This incident attests to his enormous powers of concentration.

In spite of his engineering prowess, Archimedes was most proud of his mathematical discoveries. These include the formulas for the volume of a sphere, $\left(V = \frac{4}{3}\pi r^3\right)$ and the surface area of a sphere $\left(S = 4\pi r^2\right)$ and a careful analysis of the properties of parabolas and other conics.

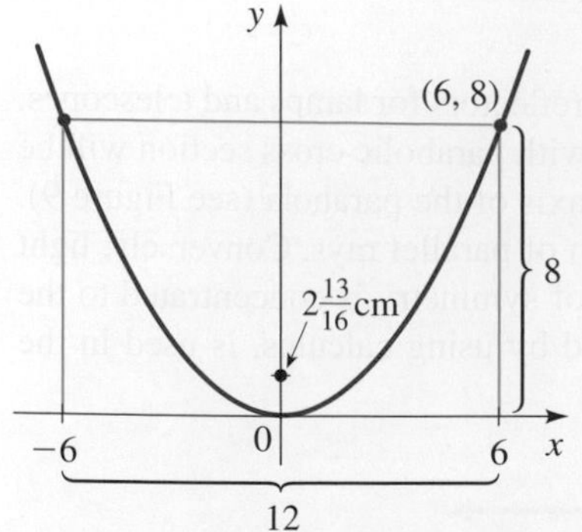

Figure 11

Solution We introduce a coordinate system and place a parabolic cross section of the reflector so that its vertex is at the origin and its axis is vertical (see Figure 11). Then the equation of this parabola has the form $x^2 = 4py$. From Figure 11 we see that the point (6, 8) lies on the parabola. We use this to find p.

$$6^2 = 4p(8) \qquad \text{The point (6, 8) satisfies the equation } x^2 = 4py$$

$$36 = 32p$$

$$p = \tfrac{9}{8}$$

The focus is $F\left(0, \frac{9}{8}\right)$, so the distance between the vertex and the focus is $\frac{45}{16} = 2\frac{13}{16}$ cm. Because the filament is positioned at the focus, it is located $2\frac{13}{16}$ cm from the vertex of the reflector.

Now Try Exercise 61 ■

10.1 Exercises

Concepts

1. A parabola is the set of all points in the plane that are equidistant from a fixed point called the ______ and a fixed line called the ______ of the parabola.

2. The graph of the equation $x^2 = 4py$ is a parabola with focus $F($___, ___$)$, directrix $y =$ ______, and ______ (horizontal/vertical) axis. So the graph of $x^2 = 12y$ is a parabola with focus $F($___, ___$)$ and directrix $y =$ ______.

3. The graph of the equation $y^2 = 4px$ is a parabola with focus $F($___, ___$)$, directrix $x =$ ______, and ______ (horizontal/vertical) axis. So the graph of $y^2 = 12x$ is a parabola with focus $F($___, ___$)$ and directrix $x =$ ______.

4. Label the focus, directrix, and vertex on the graphs given for the parabolas in Exercises 2 and 3.

(a) $x^2 = 12y$ **(b)** $y^2 = 12x$

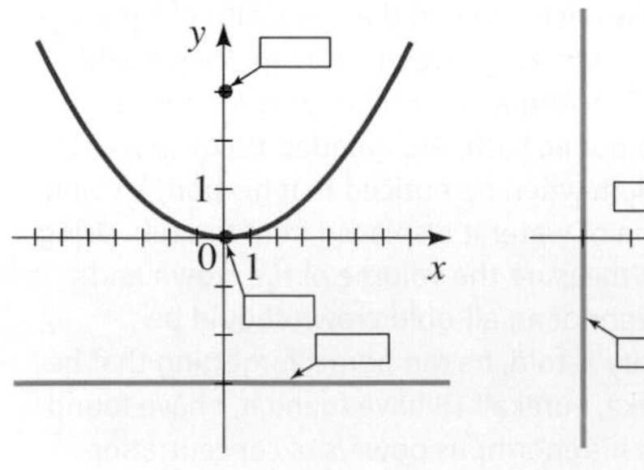

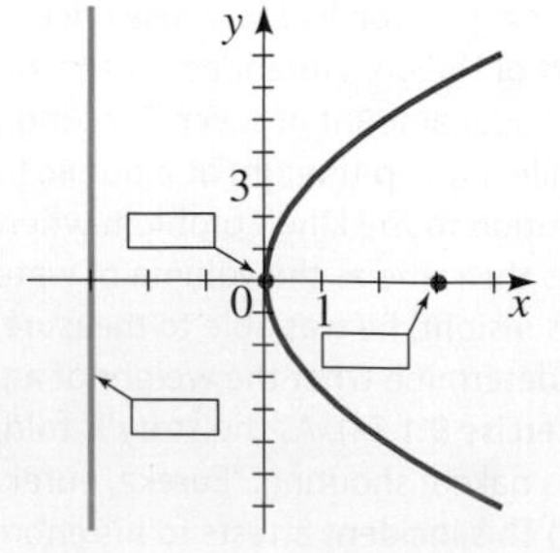

Skills

5–10 ■ Graphs of Parabolas Match the equation with the graphs labeled I–VI. Give reasons for your answers.

5. $y^2 = 2x$ **6.** $y^2 = -\frac{1}{4}x$

7. $x^2 = -6y$ **8.** $2x^2 = y$

9. $y^2 - 8x = 0$ **10.** $12y + x^2 = 0$

I

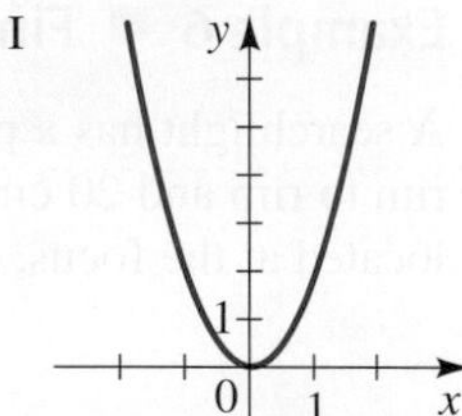

II

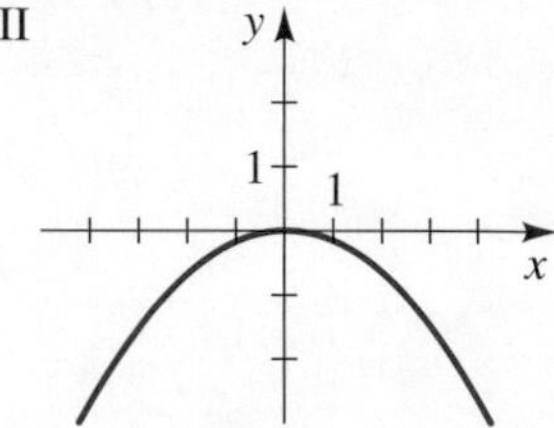

III

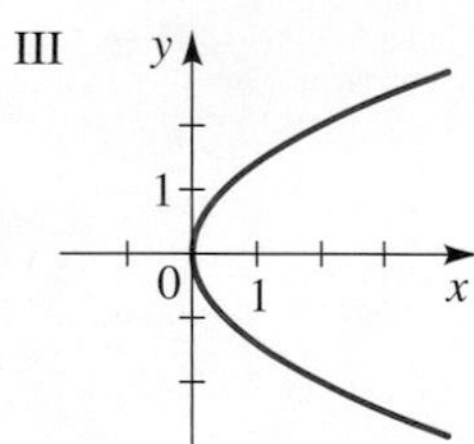

IV

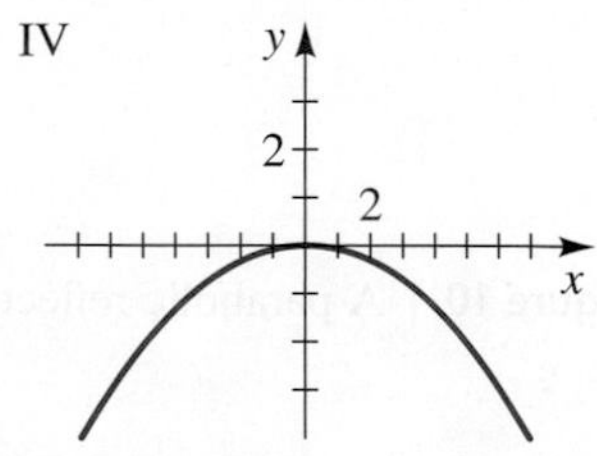

V

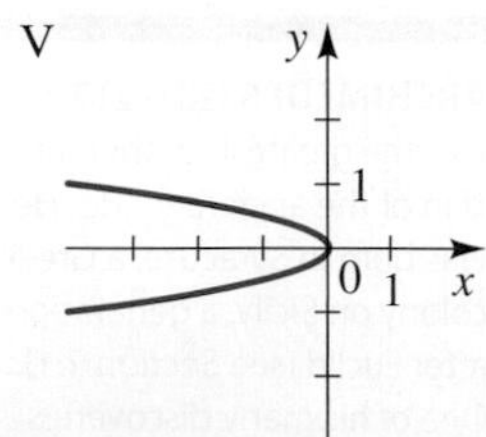

VI

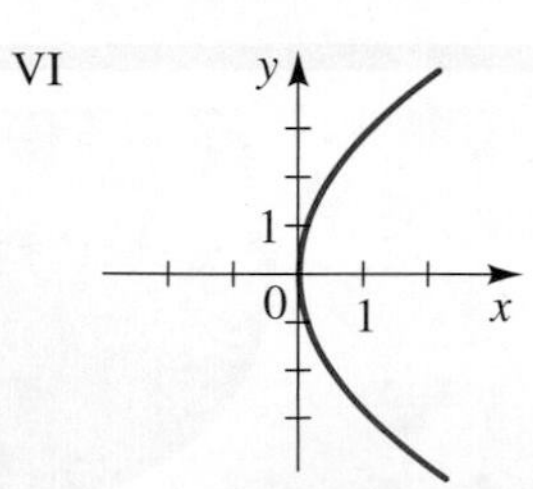

11–24 ■ Graphing Parabolas An equation of a parabola is given. **(a)** Find the focus, directrix, and focal diameter of the parabola. **(b)** Sketch a graph of the parabola and its directrix.

11. $x^2 = 16y$ **12.** $x^2 = -8y$

13. $y^2 = -4x$ **14.** $y^2 = 24x$

15. $x = \frac{1}{16}y^2$ **16.** $y = \frac{1}{2}x^2$

17. $y = -2x^2$ **18.** $x = -\frac{1}{12}y^2$

19. $5y = x^2$ **20.** $9x = y^2$

21. $x^2 + 12y = 0$ **22.** $x + \frac{1}{5}y^2 = 0$

23. $5x + 3y^2 = 0$ **24.** $8x^2 + 12y = 0$

 **25–30 ■ Graphing Parabolas** Use a graphing device to graph the parabola.

25. $x^2 = 20y$ **26.** $x^2 = -8y$

 **27.** $y^2 = -\frac{1}{3}x$ **28.** $8y^2 = x$

29. $4x + y^2 = 0$ **30.** $x - 2y^2 = 0$

31–48 ■ Finding the Equation of a Parabola Find the standard equation for the parabola that has its vertex at the origin and satisfies the given condition(s).

31. Focus: $F(0, 3)$

32. Focus: $F\left(0, -\frac{1}{8}\right)$

33. Focus: $F(-8, 0)$

34. Focus: $F(5, 0)$

35. Focus: $F\left(0, -\frac{3}{4}\right)$

36. Focus: $F\left(-\frac{1}{12}, 0\right)$

37. Directrix: $x = -2$

38. Directrix: $y = \frac{1}{4}$

39. Directrix: $y = \frac{1}{10}$

40. Directrix: $x = -\frac{1}{8}$

41. Directrix: $x = \frac{1}{20}$

42. Directrix: $y = -5$

43. Focus on the positive x-axis, 2 units away from the directrix

44. Focus on the negative y-axis, 6 units away from the directrix

45. Opens downward with focus 10 units away from the vertex

46. Opens upward with focus 5 units away from the vertex

47. Directrix has y-intercept 6

48. Focal diameter 8 and focus on the negative y-axis

49–58 ■ Finding the Equation of a Parabola Find the standard equation of the parabola whose graph is shown.

49.

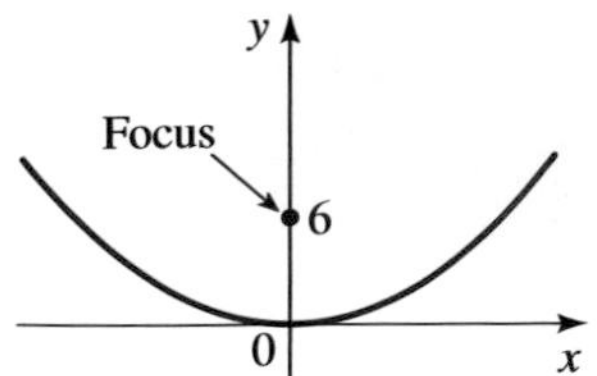

50.

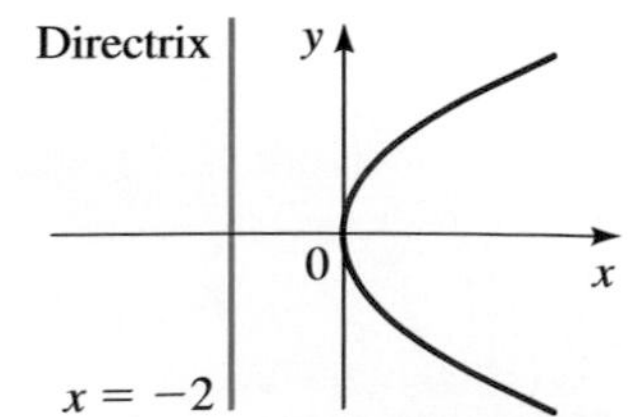

51.

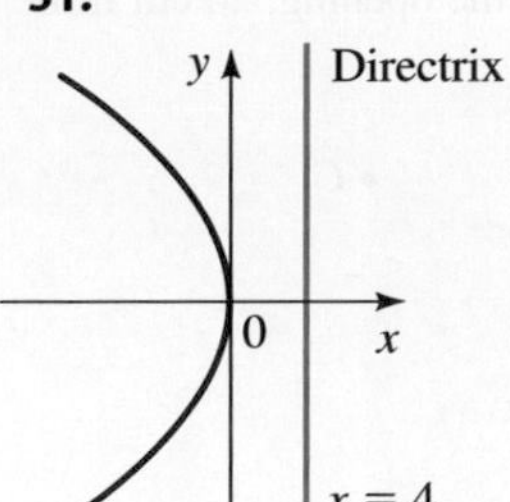

52.

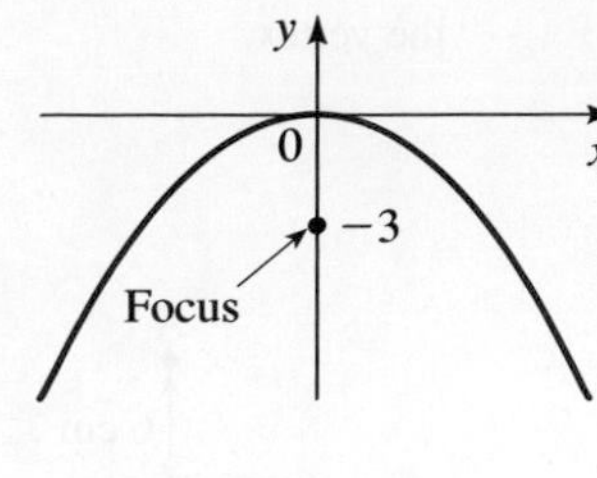

53.

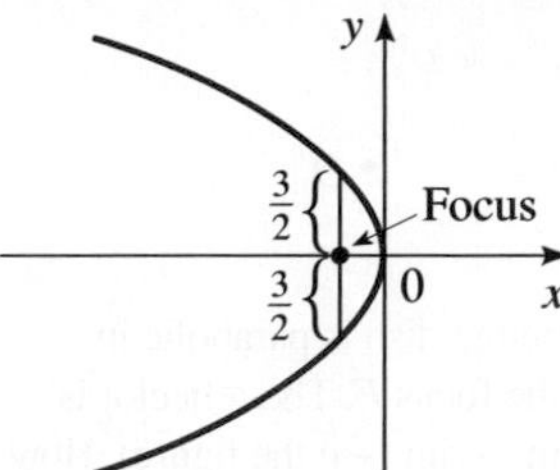

54.

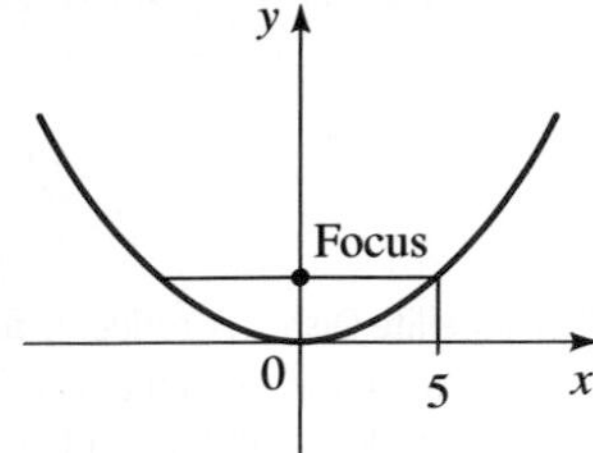

55.

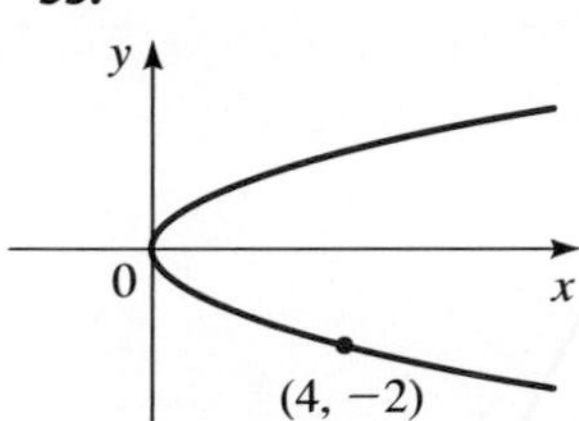

56.

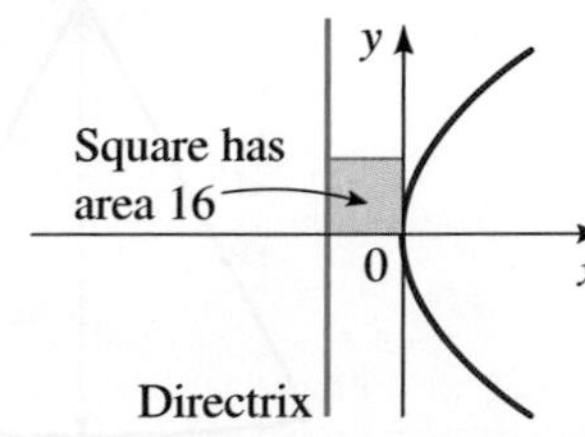

57.

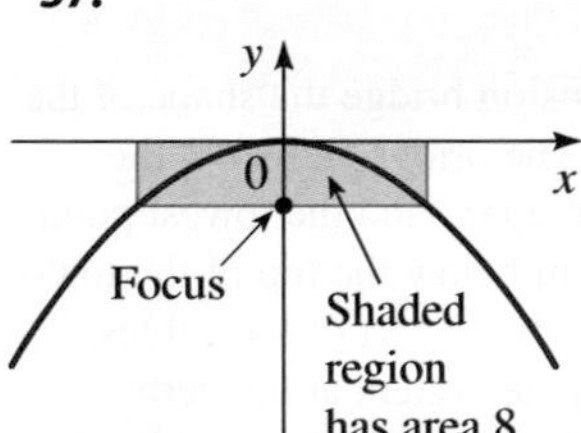

58.

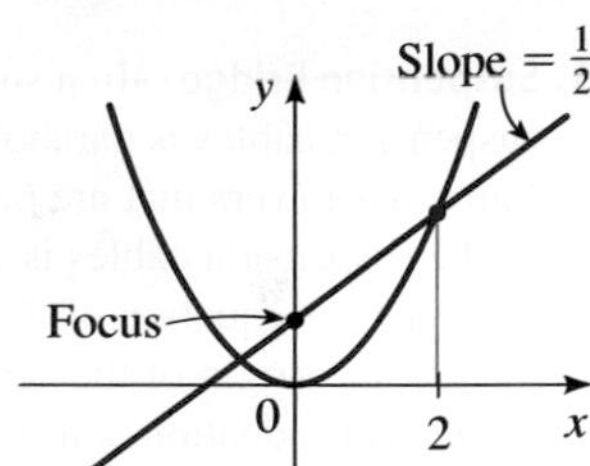

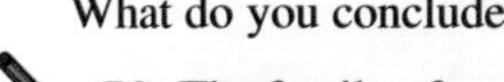 **59–60 ■ Families of Parabolas** **(a)** Find equations for the family of parabolas with the given description. **(b)** Draw the graphs. What do you conclude?

59. The family of parabolas with vertex at the origin and with directrixes $y = \frac{1}{2}$, $y = 1$, $y = 4$, and $y = 8$

60. The family of parabolas with vertex at the origin, focus on the positive y-axis, and with focal diameters 1, 2, 4, and 8

■ Applications

 **61. Parabolic Reflector** A lamp with a parabolic reflector is shown in the figure. The bulb is placed at the focus, and the focal diameter is 12 cm.

(a) Find the standard equation of the parabola.

(b) Find the diameter $d(C, D)$ of the opening, 20 cm from the vertex.

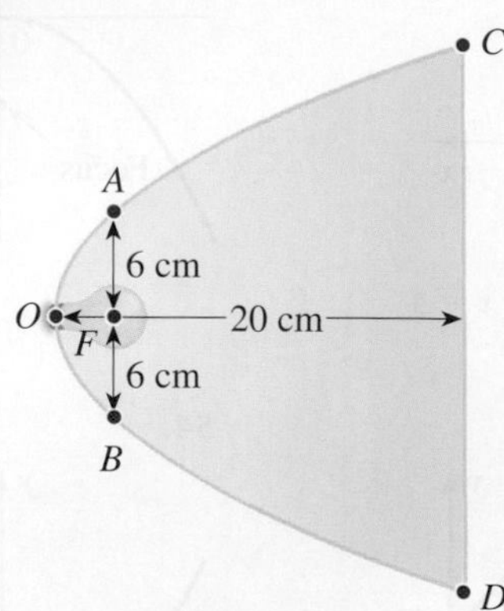

62. Satellite Dish A reflector for a satellite dish is parabolic in cross section, with the receiver at the focus F. The reflector is 30 cm deep and 6 m wide from rim to rim (see the figure). How far is the receiver from the vertex of the parabolic reflector?

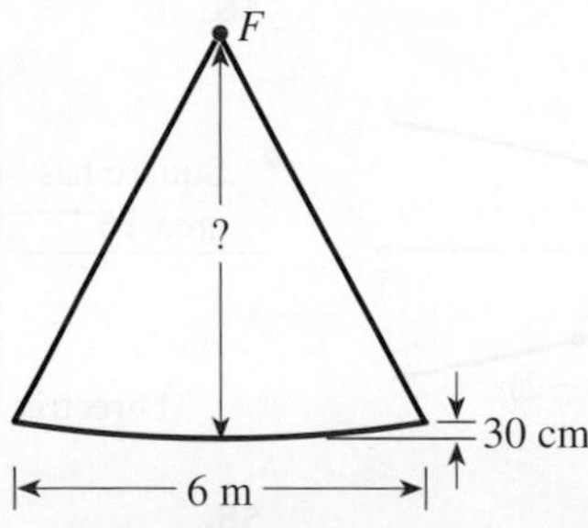

63. Suspension Bridge In a suspension bridge the shape of the suspension cables is parabolic. The bridge shown in the figure has towers that are 600 m apart, and the lowest point of the suspension cables is 150 m below the top of the towers. Find the equation of the parabolic part of the cables, placing the origin of the coordinate system at the vertex. [*Note*: This equation is used to find the length of cable needed in the construction of the bridge.]

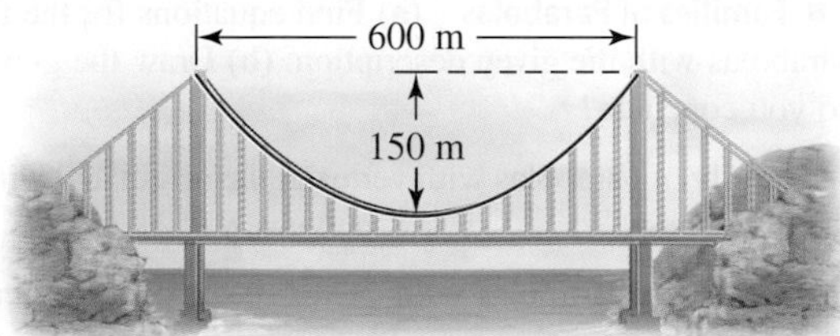

64. Reflecting Telescope The Hale telescope at the Mount Palomar Observatory has a 500 cm mirror, as shown in the figure. The mirror is constructed in a parabolic shape that collects light from the stars and focuses it at the *prime focus*, that is, the focus of the parabola. The mirror is 9.48 cm deep at its center. Find the **focal length** of this parabolic mirror, that is, the distance from the vertex to the focus.

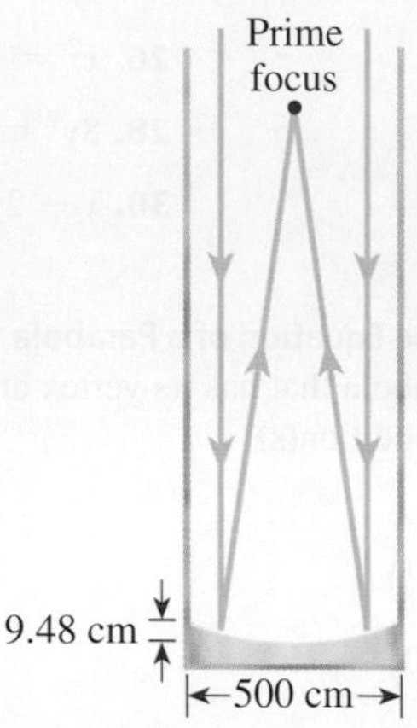

■ Discuss ■ Discover ■ Prove ■ Write

65. Discuss ■ Write: Parabolas in the Real World Several examples of the uses of parabolas are given in the text. Find other situations in which parabolas occur.

66. Discuss: Light Cone from a Flashlight A flashlight is held to form a lighted area on the ground, as shown in the figure. Is it possible to angle the flashlight in such a way that the boundary of the lighted area is a parabola? Explain your answer.

10.2 Ellipses

■ Geometric Definition of an Ellipse ■ Equations and Graphs of Ellipses
■ Eccentricity of an Ellipse

■ Geometric Definition of an Ellipse

An ellipse is an oval curve that looks like an elongated circle. More precisely, we have the following definition.

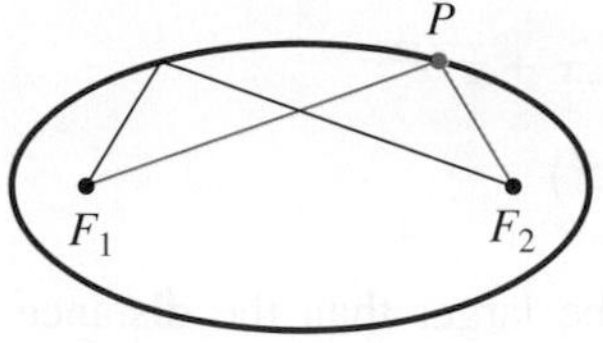

Figure 1

Geometric Definition of an Ellipse

An **ellipse** is the set of all points in the plane the sum of whose distances from two fixed points F_1 and F_2 is a constant. (See Figure 1.) These two fixed points are the **foci** (plural of **focus**) of the ellipse.

The geometric definition suggests a simple method for drawing an ellipse. Place a sheet of paper on a drawing board, and insert thumbtacks at the two points that are to be the foci of the ellipse. Attach the ends of a string to the tacks, as shown in Figure 2(a). With the point of a pencil, hold the string taut. Then carefully move the pencil around the foci, keeping the string taut at all times. The pencil will trace out an ellipse, because the sum of the distances from the point of the pencil to the foci will always equal the length of the string, which is constant.

If the string is only slightly longer than the distance between the foci, then the ellipse that is traced out will be elongated in shape, as in Figure 2(a), but if the foci are close together relative to the length of the string, the ellipse will be almost circular, as shown in Figure 2(b).

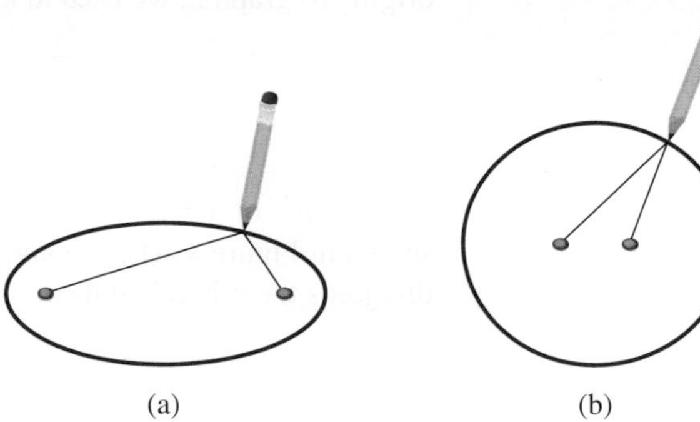

Figure 2

Deriving the Equation of an Ellipse To obtain the simplest equation for an ellipse, we place the foci on the x-axis at $F_1(-c, 0)$ and $F_2(c, 0)$ so that the origin is halfway between them (see Figure 3).

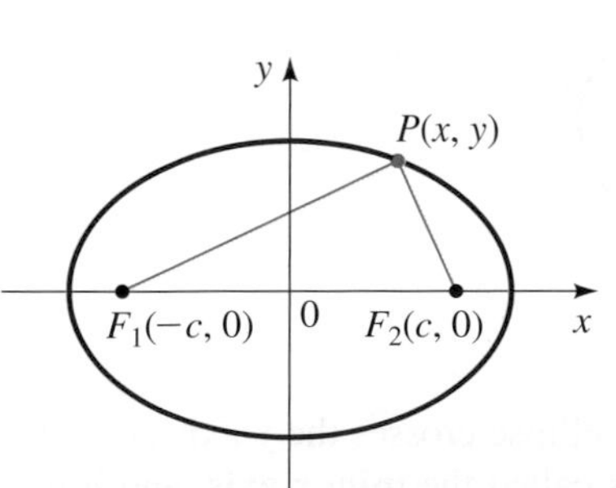

Figure 3

For later convenience we let the sum of the distances from a point on the ellipse to the foci be $2a$. Then if $P(x, y)$ is any point on the ellipse, we have

$$d(P, F_1) + d(P, F_2) = 2a$$

So from the Distance Formula we have

$$\sqrt{(x + c)^2 + y^2} + \sqrt{(x - c)^2 + y^2} = 2a$$

or

$$\sqrt{(x - c)^2 + y^2} = 2a - \sqrt{(x + c)^2 + y^2}$$

The orbits of the planets are ellipses, with the sun at one focus.

Squaring each side and expanding, we get

$$x^2 - 2cx + c^2 + y^2 = 4a^2 - 4a\sqrt{(x+c)^2 + y^2} + (x^2 + 2cx + c^2 + y^2)$$

which simplifies to

$$4a\sqrt{(x+c)^2 + y^2} = 4a^2 + 4cx$$

Dividing each side by 4 and squaring again, we get

$$a^2[(x+c)^2 + y^2] = (a^2 + cx)^2$$

$$a^2x^2 + 2a^2cx + a^2c^2 + a^2y^2 = a^4 + 2a^2cx + c^2x^2$$

$$(a^2 - c^2)x^2 + a^2y^2 = a^2(a^2 - c^2)$$

Since the sum of the distances from P to the foci must be larger than the distance between the foci, we have that $2a > 2c$, or $a > c$. Thus $a^2 - c^2 > 0$; if we divide each side of the preceding equation by $a^2(a^2 - c^2)$, we get

$$\frac{x^2}{a^2} + \frac{y^2}{a^2 - c^2} = 1$$

For convenience, let $b^2 = a^2 - c^2$ (with $b > 0$). Since $b^2 < a^2$, it follows that $b < a$. The preceding equation then becomes

$$\frac{x^2}{a^2} + \frac{y^2}{b^2} = 1 \quad (a > b)$$

This is the **standard equation** of an ellipse with foci on the x-axis and center at the origin. To graph it, we need to know the x- and y-intercepts. Setting $y = 0$, we get

$$\frac{x^2}{a^2} = 1$$

so $x^2 = a^2$, or $x = \pm a$. Thus the ellipse crosses the x-axis at $(a, 0)$ and $(-a, 0)$, as shown in Figure 4. These points are called the **vertices** of the ellipse, and the segment that joins them is called the **major axis**. Its length is $2a$.

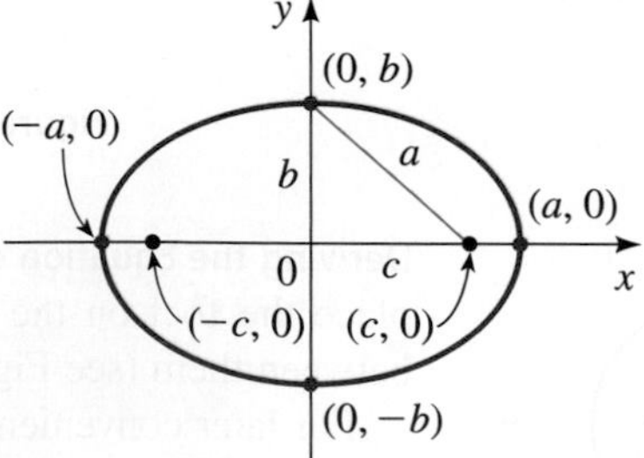

Figure 4 | $\frac{x^2}{a^2} + \frac{y^2}{b^2} = 1$ with $a > b$

Similarly, if we set $x = 0$, we get $y = \pm b$, so the ellipse crosses the y-axis at $(0, b)$ and $(0, -b)$. The segment that joins these points is called the **minor axis**, and it has length $2b$. Note that $2a > 2b$, so the major axis is longer than the minor axis. The origin is the **center** of the ellipse.

If the foci of the ellipse are placed on the y-axis at $(0, \pm c)$ rather than on the x-axis, then the roles of x and y are reversed in the preceding discussion, and we get a vertical ellipse.

■ Equations and Graphs of Ellipses

The following box summarizes what we have just proved about ellipses centered at the origin.

In the standard equation for an ellipse, a^2 is the *larger* denominator, and b^2 is the *smaller*. To find c^2, we subtract: larger denominator minus smaller denominator.

Ellipse with Center at the Origin

The graph of each of the following equations is an ellipse with center at the origin and having the given properties.

EQUATION	$\dfrac{x^2}{a^2} + \dfrac{y^2}{b^2} = 1$ $a > b > 0$	$\dfrac{x^2}{b^2} + \dfrac{y^2}{a^2} = 1$ $a > b > 0$
VERTICES	$(\pm a, 0)$	$(0, \pm a)$
MAJOR AXIS	Horizontal, length $2a$	Vertical, length $2a$
MINOR AXIS	Vertical, length $2b$	Horizontal, length $2b$
FOCI	$(\pm c, 0),\ c^2 = a^2 - b^2$	$(0, \pm c),\ c^2 = a^2 - b^2$
GRAPH	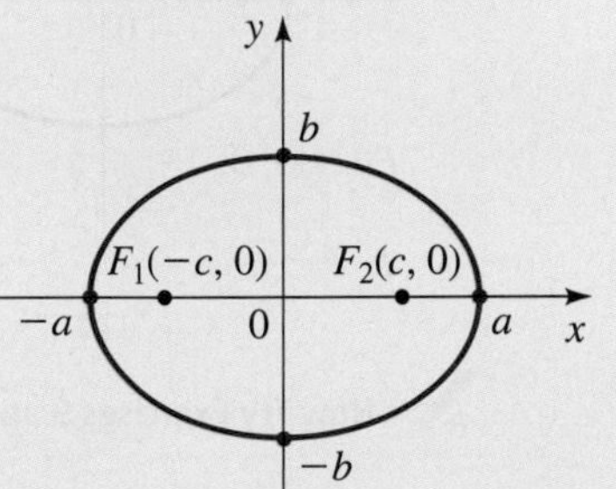	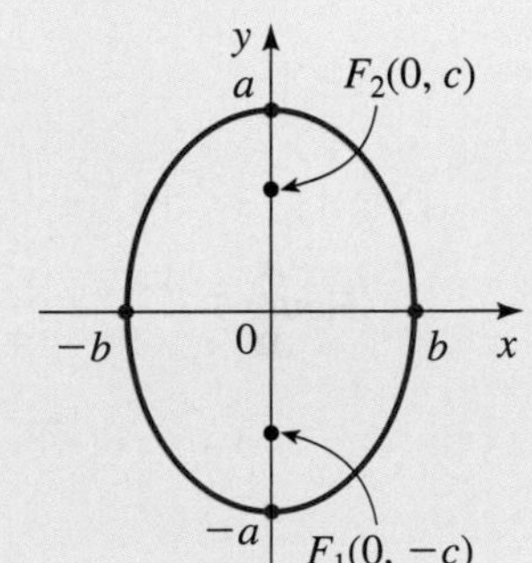

Example 1 ■ Sketching an Ellipse

An ellipse has the standard equation

$$\frac{x^2}{9} + \frac{y^2}{4} = 1$$

(a) Find the foci, the vertices, and the lengths of the major and minor axes, and sketch the graph.

(b) Draw the graph using a graphing device.

Solution

(a) Since the denominator of x^2 is larger, the ellipse has a horizontal major axis. This gives $a^2 = 9$ and $b^2 = 4$, so $c^2 = a^2 - b^2 = 9 - 4 = 5$. Thus $a = 3$, $b = 2$, and $c = \sqrt{5}$.

FOCI	$(\pm\sqrt{5}, 0)$
VERTICES	$(\pm 3, 0)$
LENGTH OF MAJOR AXIS	6
LENGTH OF MINOR AXIS	4

The graph is shown in Figure 5(a) on the next page.

(b) Most graphing devices can draw the graph of this equation, as shown in Figure 5(b).

Using a Graphing Calculator To graph the equation we first solve for y.

Note that the equation of an ellipse does not define y as a function of x (see Section 2.2).

$$\frac{x^2}{9} + \frac{y^2}{4} = 1$$

$$y^2 = 4\left(1 - \frac{x^2}{9}\right) \qquad \text{Subtract } \frac{x^2}{9}\text{, multiply by 4}$$

$$y = \pm 2\sqrt{1 - \frac{x^2}{9}} \qquad \text{Take square roots}$$

The graph of the ellipse in Figure 5(b) is obtained by graphing both functions $y = 2\sqrt{1 - x^2/9}$ and $y = -2\sqrt{1 - x^2/9}$.

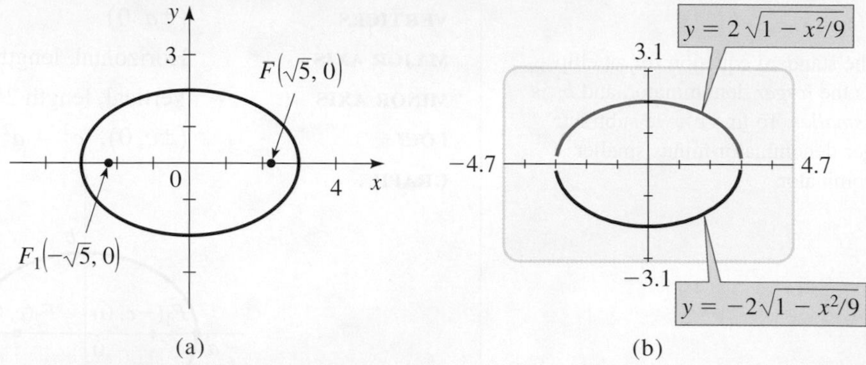

Figure 5 | $\frac{x^2}{9} + \frac{y^2}{4} = 1$

Now Try Exercises 9 and 35 ■

Example 2 ■ Finding the Foci of an Ellipse

Find the foci of the ellipse $16x^2 + 9y^2 = 144$, and sketch its graph.

Solution First we put the equation in standard form. Dividing by 144, we get

$$\frac{x^2}{9} + \frac{y^2}{16} = 1$$

Since $16 > 9$, this is an ellipse with its foci on the y-axis and with $a = 4$ and $b = 3$:

$$c^2 = a^2 - b^2 = 16 - 9 = 7$$

$$c = \sqrt{7}$$

Thus the foci are $(0, \pm\sqrt{7})$. The graph is shown in Figure 6(a). We can also draw the graph using a graphing device as shown in Figure 6(b).

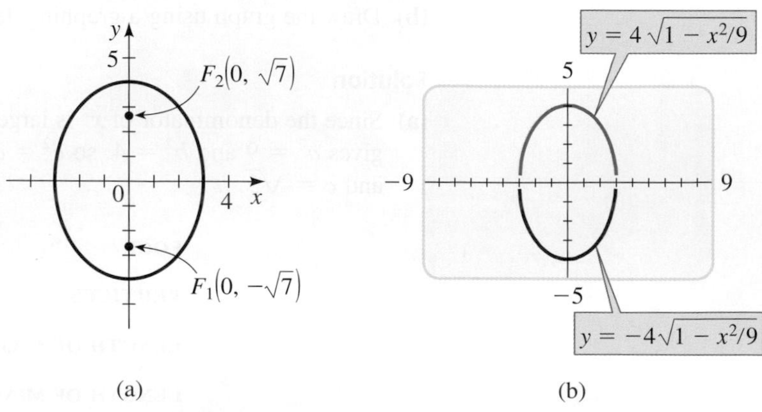

Figure 6 | $16x^2 + 9y^2 = 144$

Now Try Exercises 15 and 37 ■

Example 3 ■ Finding the Equation of an Ellipse

The vertices of an ellipse are $(\pm 4, 0)$, and the foci are $(\pm 2, 0)$. Find the standard equation of the ellipse, and sketch the graph.

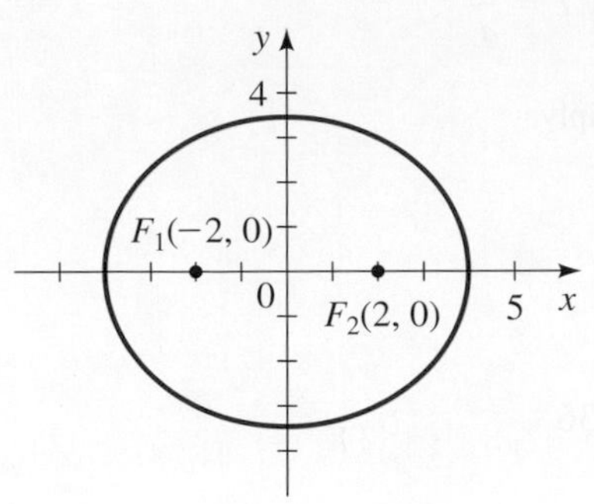

Figure 7 | $\dfrac{x^2}{16} + \dfrac{y^2}{12} = 1$

Solution Since the vertices are $(\pm 4, 0)$, we have $a = 4$, and the major axis is horizontal. The foci are $(\pm 2, 0)$, so $c = 2$. To write the equation, we need to find b. Since $c^2 = a^2 - b^2$, we have

$$2^2 = 4^2 - b^2$$

$$b^2 = 16 - 4 = 12$$

Thus the standard equation of the ellipse is

$$\frac{x^2}{16} + \frac{y^2}{12} = 1$$

The graph is shown in Figure 7.

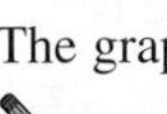 **Now Try Exercises 31 and 39** ■

■ Eccentricity of an Ellipse

We saw earlier in this section (Figure 2) that if $2a$ is only slightly greater than $2c$, the ellipse is long and thin, whereas if $2a$ is much greater than $2c$, the ellipse is almost circular. We measure the deviation of an ellipse from being circular by the ratio of a and c.

If $a = b$ in the equation of an ellipse, then

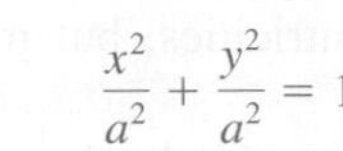

$$\frac{x^2}{a^2} + \frac{y^2}{a^2} = 1$$

so $x^2 + y^2 = a^2$. In this case the "ellipse" is a circle with radius a and eccentricity 0.

Definition of Eccentricity

For the ellipse $\dfrac{x^2}{a^2} + \dfrac{y^2}{b^2} = 1$ or $\dfrac{x^2}{b^2} + \dfrac{y^2}{a^2} = 1$ (with $a > b > 0$), the **eccentricity** e is the number

$$e = \frac{c}{a}$$

where $c = \sqrt{a^2 - b^2}$. The eccentricity of every ellipse satisfies $0 < e < 1$.

Thus if e is close to 1, then c is almost equal to a, and the ellipse is elongated in shape, but if e is close to 0, then the ellipse is close to a circle in shape. The eccentricity is a measure of how "stretched" the ellipse is.

In Figure 8 we show a number of ellipses to demonstrate the effect of varying the eccentricity e.

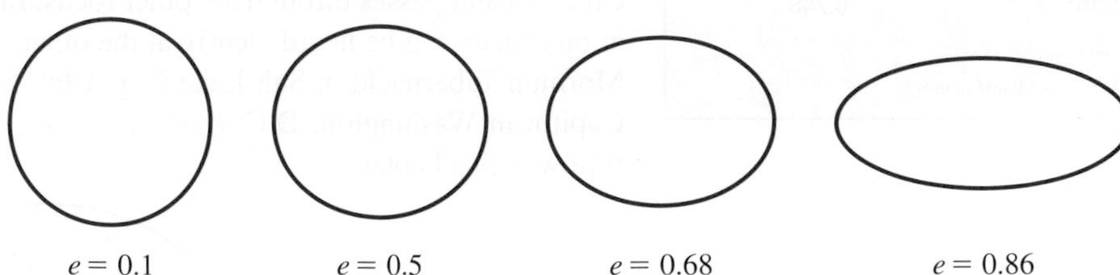

Figure 8 | Ellipses with various eccentricities

Example 4 ■ Finding the Equation of an Ellipse from Its Eccentricity and Foci

Find the standard equation of the ellipse with foci $(0, \pm 8)$ and eccentricity $e = \frac{4}{5}$, and sketch its graph.

Solution We are given $e = \frac{4}{5}$ and $c = 8$. Thus

$$\frac{4}{5} = \frac{8}{a} \qquad \text{Eccentricity } e = \frac{c}{a}$$

$$4a = 40 \qquad \text{Cross-multiply}$$

$$a = 10$$

To find b, we use the fact that $c^2 = a^2 - b^2$.

$$8^2 = 10^2 - b^2$$

$$b^2 = 10^2 - 8^2 = 36$$

$$b = 6$$

Thus the standard equation of the ellipse is

$$\frac{x^2}{36} + \frac{y^2}{100} = 1$$

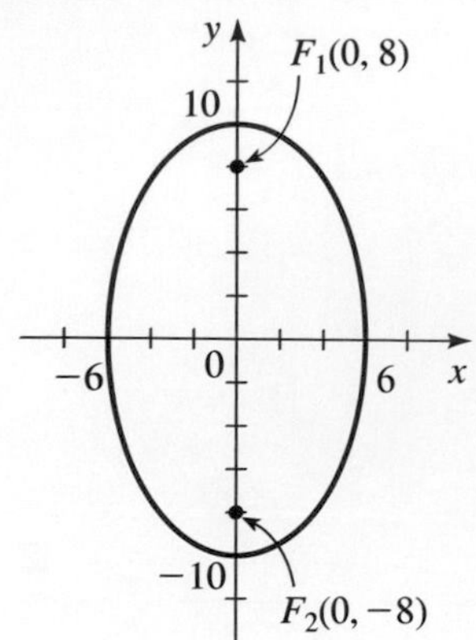

Figure 9 | $\frac{x^2}{36} + \frac{y^2}{100} = 1$

Because the foci are on the y-axis, the ellipse is oriented vertically. To sketch the ellipse, we find the intercepts. The x-intercepts are ± 6, and the y-intercepts are ± 10. The graph is sketched in Figure 9.

 **Now Try Exercise 53** ■

Eccentricities of the Orbits of the Planets

The orbits of the planets are ellipses with the sun at one focus. For most planets these ellipses have very small eccentricity, so they are nearly circular. However, Mercury and Pluto, the innermost and outermost known planets, respectively, have visibly elliptical orbits.

Planet	Eccentricity
Mercury	0.206
Venus	0.007
Earth	0.017
Mars	0.093
Jupiter	0.048
Saturn	0.056
Uranus	0.046
Neptune	0.010
Pluto*	0.248

*Pluto is a "dwarf planet."

Gravitational attraction causes the planets to move in elliptical orbits around the sun with the sun at one focus. This remarkable property was first observed by Johannes Kepler and was later deduced by Isaac Newton from his inverse square Law of Gravity, using calculus. The orbits of the planets have different eccentricities, but most are nearly circular (see the margin).

Ellipses, like parabolas, have an interesting *reflection property* that leads to a number of practical applications. If a light source is placed at one focus of a reflecting surface with elliptical cross sections, then all the light will be reflected off the surface to the other focus, as shown in Figure 10. This principle, which works for sound waves as well as for light, is used in *lithotripsy*, a treatment for kidney stones. The patient is placed in a tub of water with elliptical cross sections in such a way that the kidney stone is accurately located at one focus. High-intensity sound waves generated at the other focus are reflected to the stone and destroy it with minimal damage to surrounding tissue. The patient is spared the trauma of surgery and recovers within days instead of weeks.

The reflection property of ellipses is also used in the construction of *whispering galleries*. Sound coming from one focus bounces off the walls and ceiling of an elliptical room and passes through the other focus. In these rooms even quiet whispers spoken at one focus can be heard clearly at the other. Famous whispering galleries include the Mormon Tabernacle in Salt Lake City, Utah, and the National Statuary Hall of the US Capitol in Washington, D.C. (See the Focus on Modeling *Conics in Architecture* that follows this chapter.)

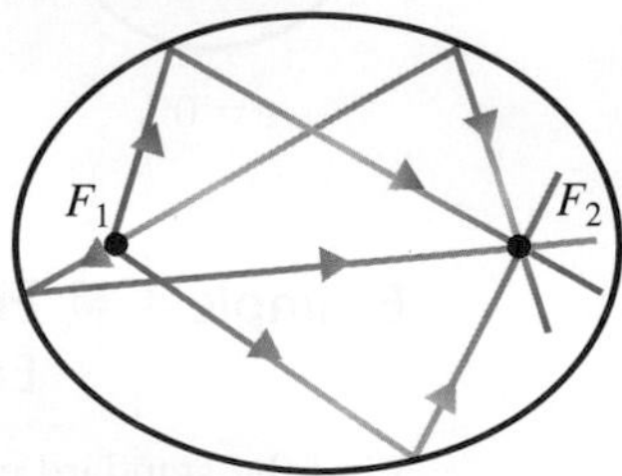

Figure 10

10.2 | Exercises

■ Concepts

1. An ellipse is the set of all points in the plane for which the ________ of the distances from two fixed points F_1 and F_2 is constant. The points F_1 and F_2 are called the ________ of the ellipse.

2. The graph of the equation $\dfrac{x^2}{a^2} + \dfrac{y^2}{b^2} = 1$ with $a > b > 0$ is an ellipse with ________ (horizontal/vertical) major axis, vertices (___, ___) and (___, ___) and foci $(\pm c, 0)$, where $c =$ ________. So the graph of $\dfrac{x^2}{5^2} + \dfrac{y^2}{4^2} = 1$ is an ellipse with vertices (___, ___) and (___, ___) and foci (___, ___) and (___, ___).

3. The graph of the equation $\dfrac{x^2}{b^2} + \dfrac{y^2}{a^2} = 1$ with $a > b > 0$ is an ellipse with ________ (horizontal/vertical) major axis, vertices (___, ___) and (___, ___) and foci $(0, \pm c)$, where $c =$ ________. So the graph of $\dfrac{x^2}{4^2} + \dfrac{y^2}{5^2} = 1$ is an ellipse with vertices (___, ___) and (___, ___) and foci (___, ___) and (___, ___).

4. Label the vertices and foci on the graphs given for the ellipses in Exercises 2 and 3.

(a) $\dfrac{x^2}{5^2} + \dfrac{y^2}{4^2} = 1$ **(b)** $\dfrac{x^2}{4^2} + \dfrac{y^2}{5^2} = 1$

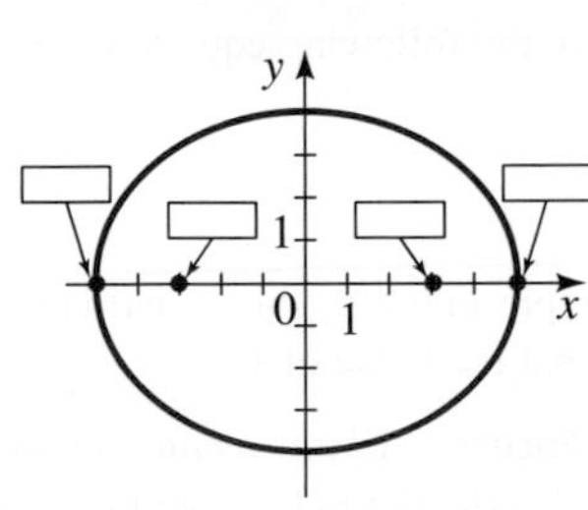

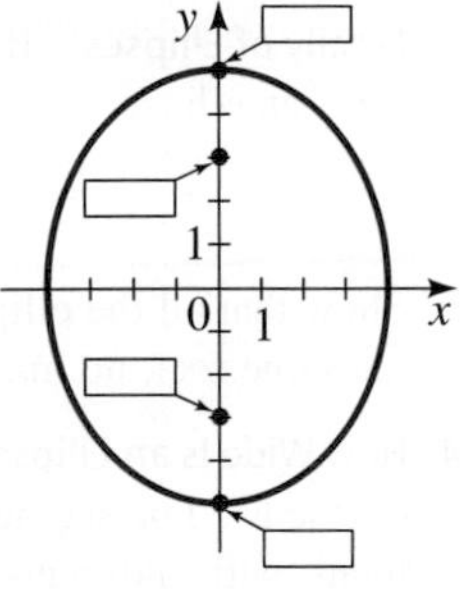

■ Skills

5–8 ■ Graphs of Ellipses Match the equation with the graphs labeled I–IV. Give reasons for your answers.

5. $\dfrac{x^2}{16} + \dfrac{y^2}{4} = 1$

6. $x^2 + \dfrac{y^2}{9} = 1$

7. $4x^2 + y^2 = 4$

8. $16x^2 + 25y^2 = 400$

Graphs for Exercises 5–8:

I

II

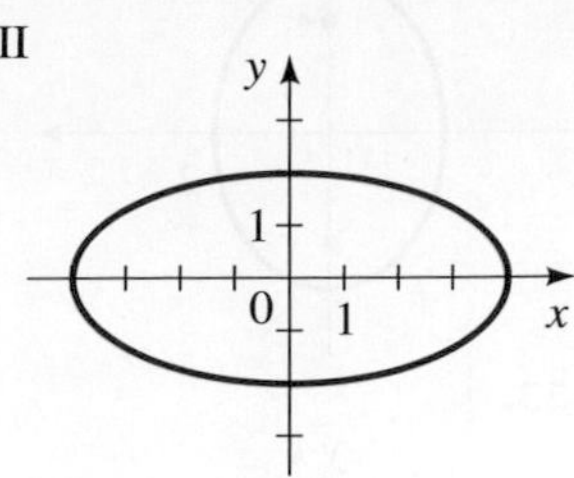

III

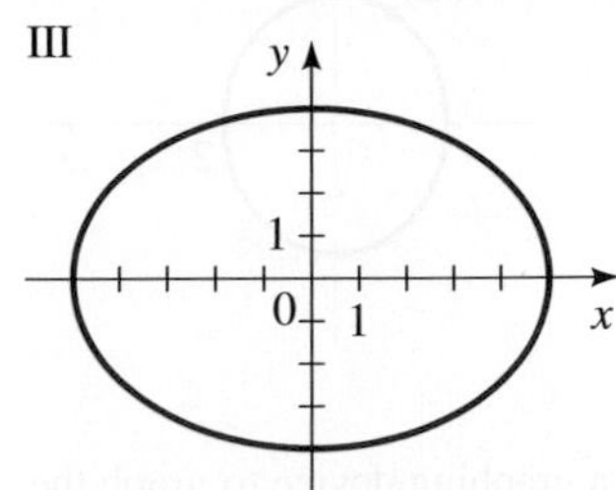

IV

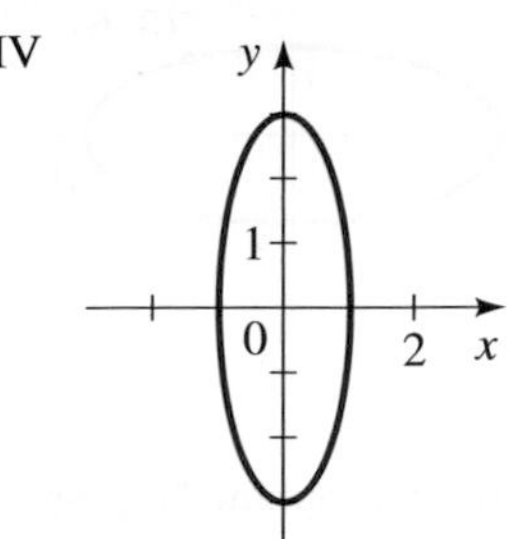

9–28 ■ Graphing Ellipses An equation of an ellipse is given. **(a)** Find the vertices, foci, and eccentricity of the ellipse. **(b)** Determine the lengths of the major and minor axes. **(c)** Sketch a graph of the ellipse.

9. $\dfrac{x^2}{25} + \dfrac{y^2}{9} = 1$ **10.** $\dfrac{x^2}{16} + \dfrac{y^2}{25} = 1$

11. $\dfrac{x^2}{36} + \dfrac{y^2}{81} = 1$ **12.** $\dfrac{x^2}{4} + y^2 = 1$

13. $\dfrac{x^2}{49} + \dfrac{y^2}{25} = 1$ **14.** $\dfrac{x^2}{9} + \dfrac{y^2}{64} = 1$

15. $9x^2 + 4y^2 = 36$ **16.** $4x^2 + 25y^2 = 100$

17. $x^2 + 4y^2 = 16$ **18.** $4x^2 + y^2 = 16$

19. $16x^2 + 25y^2 = 1600$ **20.** $2x^2 + 49y^2 = 98$

21. $3x^2 + y^2 = 9$ **22.** $x^2 + 3y^2 = 9$

23. $2x^2 + y^2 = 4$ **24.** $3x^2 + 4y^2 = 12$

25. $x^2 + 4y^2 = 1$ **26.** $9x^2 + 4y^2 = 1$

27. $x^2 = 4 - 2y^2$ **28.** $y^2 = 1 - 2x^2$

29–34 ■ Finding the Equation of an Ellipse Find the standard equation for the ellipse whose graph is shown.

29.

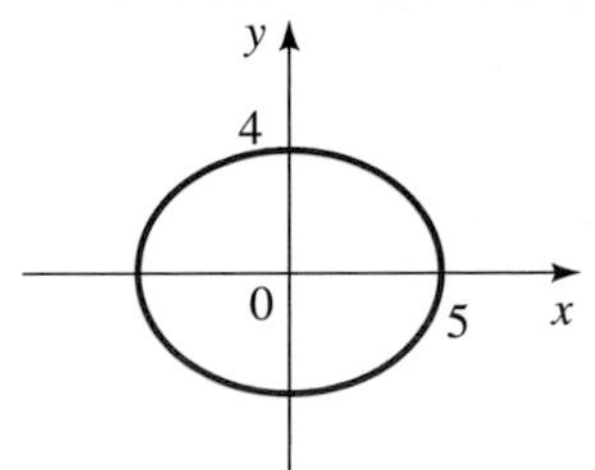

30.

31.

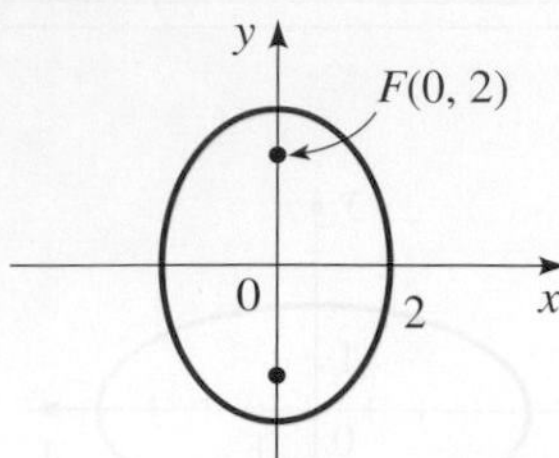

32.

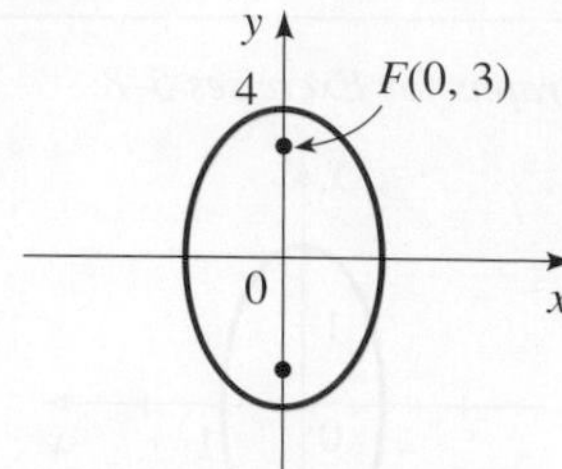

33.

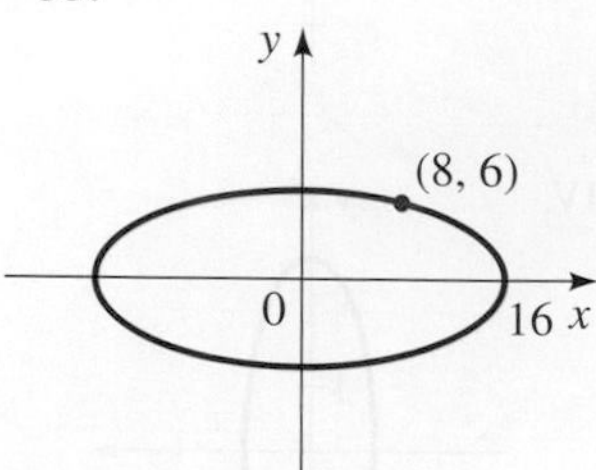

34.

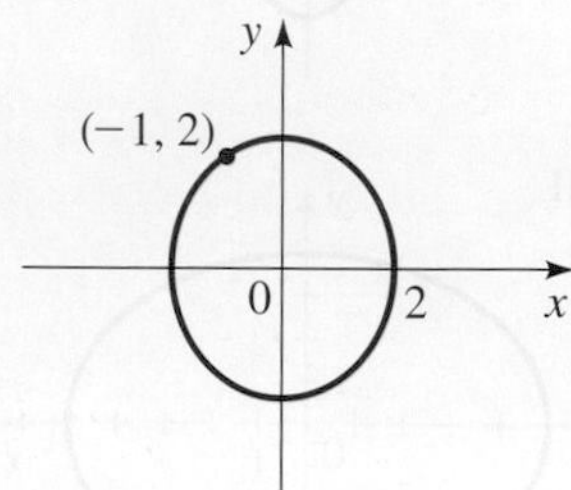

35–38 ■ Graphing Ellipses Use a graphing device to graph the ellipse.

35. $\dfrac{x^2}{25} + \dfrac{y^2}{20} = 1$ **36.** $x^2 + \dfrac{y^2}{12} = 1$

37. $6x^2 + y^2 = 36$ **38.** $x^2 + 2y^2 = 8$

39–56 ■ Finding the Equation of an Ellipse Find the standard equation for the ellipse that satisfies the given conditions.

39. Foci: $(\pm 4, 0)$, vertices: $(\pm 5, 0)$

40. Foci: $(0, \pm 3)$, vertices: $(0, \pm 5)$

41. Foci: $F(\pm 1, 0)$, vertices: $(\pm 2, 0)$

42. Foci: $F(0, \pm 2)$, vertices: $(0, \pm 3)$

43. Foci: $F(0, \pm\sqrt{10})$, vertices: $(0, \pm 7)$

44. Foci: $F(\pm\sqrt{15}, 0)$, vertices: $(\pm 6, 0)$

45. Length of major axis: 4, length of minor axis: 2, foci on y-axis

46. Length of major axis: 6, length of minor axis: 4, foci on x-axis

47. Foci: $(0, \pm 2)$, length of minor axis: 6

48. Foci: $(\pm 5, 0)$, length of major axis: 12

49. Endpoints of major axis: $(\pm 10, 0)$, distance between foci: 6

50. Endpoints of minor axis: $(0, \pm 3)$, distance between foci: 8

51. Length of major axis: 10, foci on x-axis, ellipse passes through the point $(\sqrt{5}, 2)$

52. Length of minor axis: 10, foci on y-axis, ellipse passes through the point $(\sqrt{5}, \sqrt{40})$

53. Eccentricity: $\frac{1}{3}$, foci: $(0, \pm 2)$

54. Eccentricity: 0.75, foci: $(\pm 1.5, 0)$

55. Eccentricity: $\sqrt{3}/2$, foci on y-axis, length of major axis: 4

56. Eccentricity: $\sqrt{5}/3$, foci on x-axis, length of major axis: 12

■ Skills Plus

57–60 ■ Intersecting Ellipses Find the intersection points of the pair of ellipses. Sketch the graphs of each pair of equations on the same coordinate axes, and label the points of intersection.

57. $\begin{cases} 4x^2 + y^2 = 4 \\ 4x^2 + 9y^2 = 36 \end{cases}$ **58.** $\begin{cases} \dfrac{x^2}{16} + \dfrac{y^2}{9} = 1 \\ \dfrac{x^2}{9} + \dfrac{y^2}{16} = 1 \end{cases}$

59. $\begin{cases} 100x^2 + 25y^2 = 100 \\ x^2 + \dfrac{y^2}{9} = 1 \end{cases}$ **60.** $\begin{cases} 25x^2 + 144y^2 = 3600 \\ 144x^2 + 25y^2 = 3600 \end{cases}$

61. Ancillary Circle The *ancillary circle* of an ellipse is the circle with radius equal to half the length of the minor axis and center the same as the ellipse (see the figure). The ancillary circle is thus the largest circle that can fit within an ellipse.

(a) Find an equation for the ancillary circle of the ellipse $x^2 + 4y^2 = 16$.

(b) For the ellipse and ancillary circle of part (a), show that if (s, t) is a point on the ancillary circle, then $(2s, t)$ is a point on the ellipse.

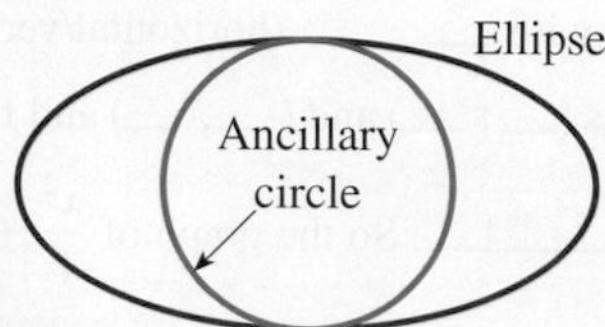

62. Family of Ellipses

(a) Use a graphing device to sketch the top half (the portion in the first and second quadrants) of the family of ellipses $x^2 + ky^2 = 100$ for $k = 4$, 10, 25, and 50.

(b) What do the members of this family of ellipses have in common? How do they differ?

63. Family of Ellipses If $k > 0$, the following equation represents an ellipse:

$$\frac{x^2}{k} + \frac{y^2}{4 + k} = 1$$

Show that all the ellipses represented by this equation have the same foci, no matter what the value of k.

64. How Wide Is an Ellipse at a Focus? A *latus rectum* for an ellipse is a line segment perpendicular to the major axis at a focus, with endpoints on the ellipse, as shown in the figure. Show that the length of a latus rectum is $2b^2/a$ for the ellipse

$$\frac{x^2}{a^2} + \frac{y^2}{b^2} = 1 \qquad (a > b)$$

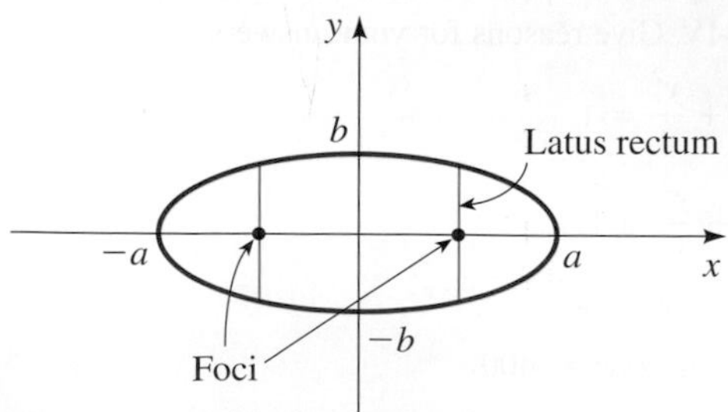

Applications

65. Perihelion and Aphelion The planets move around the sun in elliptical orbits with the sun at one focus. The point in the orbit at which the planet is closest to the sun is called *perihelion*, and the point at which it is farthest is called *aphelion*. These points are the vertices of the orbit. The earth's distance from the sun is 147,000,000 km at perihelion and 153,000,000 km at aphelion. Find an equation for the earth's orbit. (Place the origin at the center of the orbit with the sun on the x-axis.)

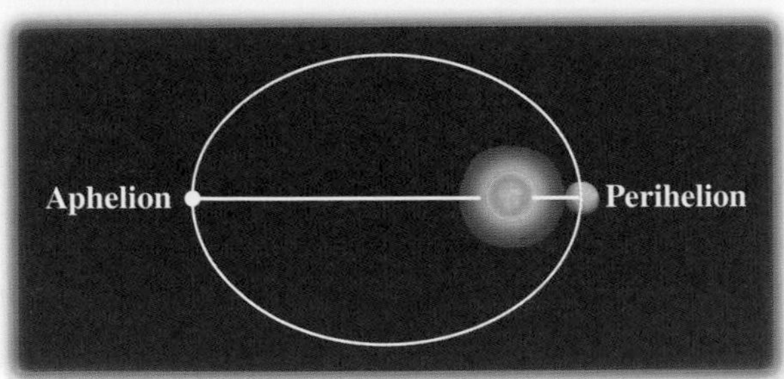

66. The Orbit of Pluto With an eccentricity of 0.25, Pluto's orbit is the most eccentric in the solar system. The length of the minor axis of its orbit is approximately 10,000,000,000 km. Find the distance between Pluto and the sun at perihelion and at aphelion. (See Exercise 65.)

67. Lunar Orbit For an object in an elliptical orbit around the moon, the points in the orbit that are closest to and farthest from the center of the moon are called *perilune* and *apolune*, respectively. These are the vertices of the orbit. The center of the moon is at one focus of the orbit. The *Apollo 11* spacecraft was placed in a lunar orbit with perilune at 109 km and apolune at 312 km above the surface of the moon. Assuming that the moon is a sphere of radius 1720 km, find an equation for the orbit of *Apollo 11*. (Place the coordinate axes so that the origin is at the center of the orbit and the foci are located on the x-axis.)

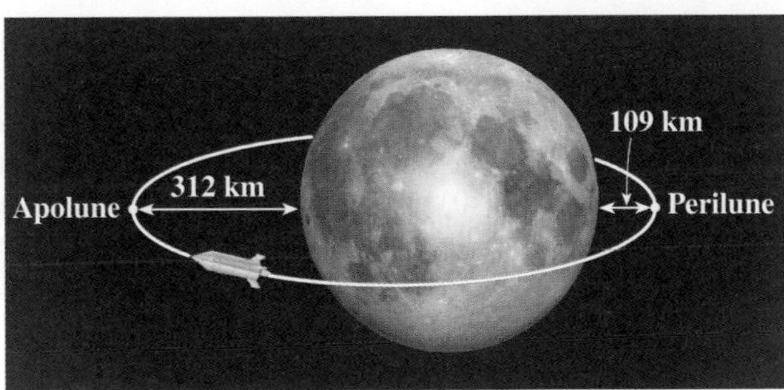

68. Plywood Ellipse A carpenter wishes to construct an elliptical table top from a 1 m by 2 m sheet of plywood, by tracing out the ellipse using the "thumbtack and string" method illustrated in Figures 2 and 3. What length of string should be used, and how far apart should the tacks be located, if the ellipse is to be the largest possible that can be cut out of the plywood sheet?

69. Sunburst Window A "sunburst" window above a doorway is constructed in the shape of the top half of an ellipse, as shown in the figure. The window is 50 cm tall at its highest point and 200 cm wide at the bottom. Find the height of the window 62.5 cm from the center of the base.

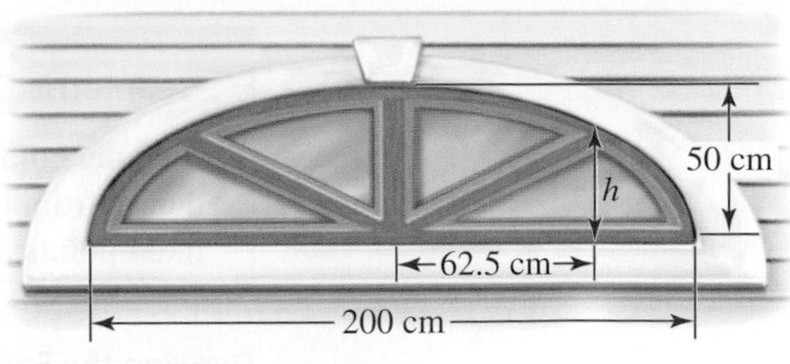

Discuss ■ Discover ■ Prove ■ Write

70. Discuss: Drawing an Ellipse on a Whiteboard Try drawing an ellipse as accurately as possible on a whiteboard. How would a piece of string and two friends help this process?

71. Discuss: Light Cone from a Flashlight A flashlight shines on a wall, as shown in the figure. What is the shape of the boundary of the lighted area? Explain your answer.

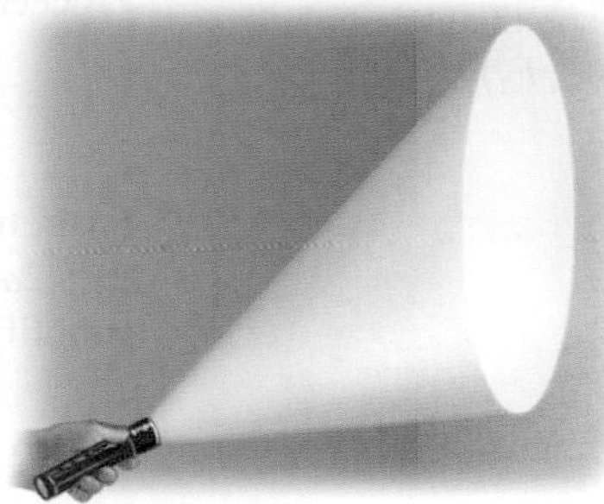

72. Discuss: Is It an Ellipse? A piece of paper is wrapped around a cylindrical bottle, and then a compass is used to draw a circle on the paper, as shown in the figure. When the paper is laid flat, is the shape drawn on the paper an ellipse? (You don't need to prove your answer, but you might want to do the experiment and see what you get.)

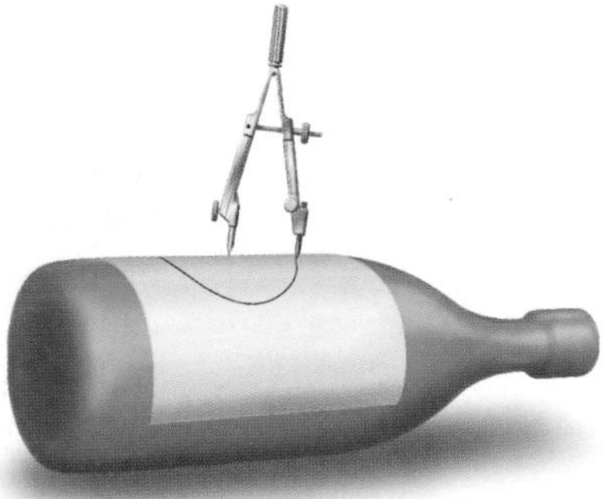

10.3 Hyperbolas

■ Geometric Definition of a Hyperbola ■ Equations and Graphs of Hyperbolas

■ Geometric Definition of a Hyperbola

Although ellipses and hyperbolas have completely different shapes, their definitions and equations are similar. Instead of using the *sum* of distances from two fixed foci, as in the case of an ellipse, we use the *difference* to define a hyperbola.

Geometric Definition of a Hyperbola

A **hyperbola** is the set of all points in the plane, the difference of whose distances from two fixed points F_1 and F_2 is a constant. (See Figure 1.) These two fixed points are the **foci** of the hyperbola.

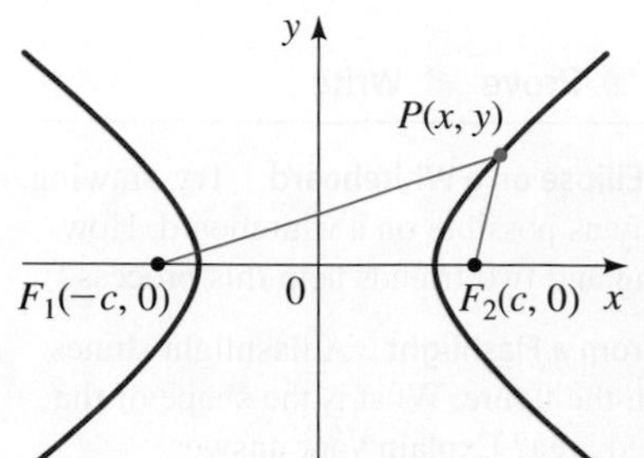

Figure 1 | P is on the hyperbola if $|d(P, F_1) - d(P, F_2)| = 2a$.

Deriving the Equation of a Hyperbola As in the case of the ellipse, we get the simplest equation for the hyperbola by placing the foci on the x-axis at $(\pm c, 0)$, as shown in Figure 1. By definition, if $P(x, y)$ lies on the hyperbola, then either $d(P, F_1) - d(P, F_2)$ or $d(P, F_2) - d(P, F_1)$ must equal some positive constant, which we call $2a$. Thus we have

$$d(P, F_1) - d(P, F_2) = \pm 2a$$

or

$$\sqrt{(x + c)^2 + y^2} - \sqrt{(x - c)^2 + y^2} = \pm 2a$$

Proceeding as we did in the case of the ellipse (Section 10.2), this simplifies to

$$(c^2 - a^2)x^2 - a^2y^2 = a^2(c^2 - a^2)$$

From triangle PF_1F_2 in Figure 1 we see that $|d(P, F_1) - d(P, F_2)| < 2c$. It follows that $2a < 2c$, or $a < c$. Thus $c^2 - a^2 > 0$, so we can set $b^2 = c^2 - a^2$. We then simplify the last displayed equation to get

$$\frac{x^2}{a^2} - \frac{y^2}{b^2} = 1$$

This is the **standard equation** of a hyperbola with foci on the x-axis and center at the origin.

If we replace x by $-x$ or y by $-y$ in this equation, it remains unchanged, so the hyperbola is symmetric about both the x- and y-axes and about the origin. The x-intercepts are $\pm a$, and the points $(a, 0)$ and $(-a, 0)$ are the **vertices** of the hyperbola. There is no y-intercept, because setting $x = 0$ in the equation of the hyperbola leads to $-y^2 = b^2$, which has no real solution. Furthermore, the equation of the hyperbola implies that

$$\frac{x^2}{a^2} = \frac{y^2}{b^2} + 1 \geq 1$$

so $x^2/a^2 \geq 1$; thus $x^2 \geq a^2$, and hence $x \geq a$ or $x \leq -a$. This means that the hyperbola consists of two parts, called its **branches**. The segment joining the two vertices on the separate branches is the **transverse axis** of the hyperbola, and the origin is called its **center**.

If we place the foci of the hyperbola on the y-axis rather than on the x-axis, this has the effect of reversing the roles of x and y in the derivation of the equation of the hyperbola. This leads to a hyperbola with a vertical transverse axis.

■ Equations and Graphs of Hyperbolas

The main properties of hyperbolas are listed in the following box.

Hyperbola with Center at the Origin

The graph of each equation is a hyperbola with center at the origin and having the given properties.

EQUATION	$\dfrac{x^2}{a^2} - \dfrac{y^2}{b^2} = 1 \quad (a > 0, b > 0)$	$\dfrac{y^2}{a^2} - \dfrac{x^2}{b^2} = 1 \quad (a > 0, b > 0)$
VERTICES	$(\pm a, 0)$	$(0, \pm a)$
TRANSVERSE AXIS	Horizontal, length $2a$	Vertical, length $2a$
ASYMPTOTES	$y = \pm \dfrac{b}{a}x$	$y = \pm \dfrac{a}{b}x$
FOCI	$(\pm c, 0), \quad c^2 = a^2 + b^2$	$(0, \pm c), \quad c^2 = a^2 + b^2$
GRAPH	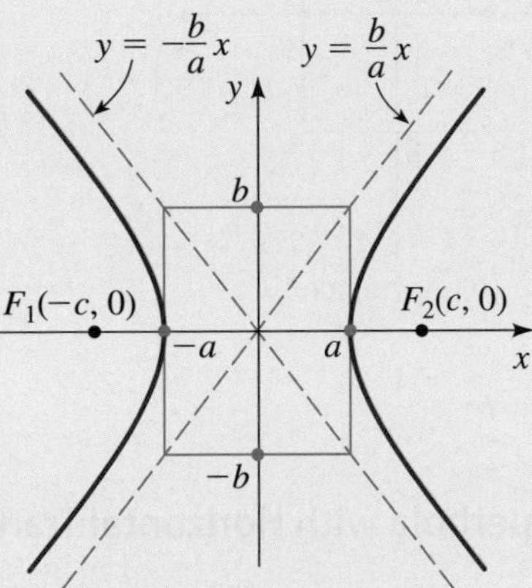	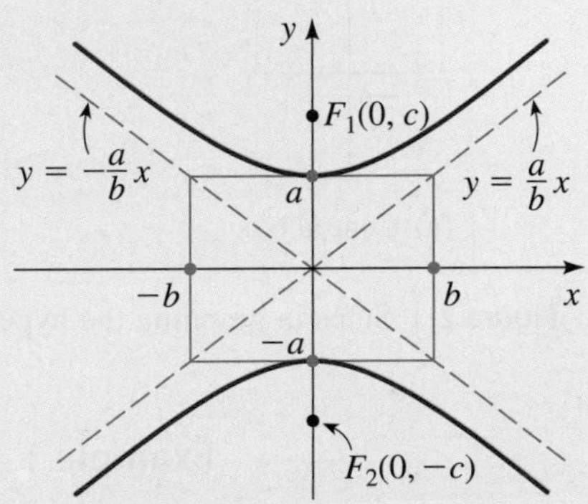

Asymptotes of rational functions are discussed in Section 3.6.

The *asymptotes* mentioned in this box are lines that the hyperbola approaches for large values of x and y. To find the asymptotes in the first case in the box, we solve the equation for y to get

$$
\begin{aligned}
y &= \pm \frac{b}{a}\sqrt{x^2 - a^2} \\
&= \pm \frac{b}{a}x\sqrt{1 - \frac{a^2}{x^2}}
\end{aligned}
$$

As x gets large, a^2/x^2 gets closer to zero. In other words, as $x \to \infty$, we have $a^2/x^2 \to 0$. So for large x the value of y can be approximated as $y = \pm(b/a)x$. This shows that these lines are asymptotes of the hyperbola.

We use the following guidelines to sketch the graph of a hyperbola.

How to Sketch a Hyperbola

1. **Sketch the Central Box.** This is the rectangle centered at the origin, with sides parallel to the axes, that crosses one axis at $\pm a$ and the other at $\pm b$.
2. **Sketch the Asymptotes.** These are the lines obtained by extending the diagonals of the central box.
3. **Plot the Vertices.** These are the two x-intercepts or the two y-intercepts.
4. **Sketch the Hyperbola.** Start at a vertex, and sketch a branch of the hyperbola, approaching the asymptotes. Sketch the other branch in the same way.

Asymptotes are an essential aid for graphing a hyperbola; they help us to determine its shape. A convenient way to find the asymptotes, for a hyperbola with horizontal transverse axis, is to first plot the points $(a, 0)$, $(-a, 0)$, $(0, b)$, and $(0, -b)$. Then sketch horizontal and vertical segments through these points to construct a rectangle, as shown in Figure 2(a). We call this rectangle the **central box** of the hyperbola. The slopes of the diagonals of the central box are $\pm b/a$, so by extending them, we obtain the asymptotes $y = \pm(b/a)x$, as sketched in Figure 2(b). Finally, we plot the vertices and use the asymptotes as a guide in sketching the hyperbola shown in Figure 2(c). (A similar procedure applies to graphing a hyperbola that has a vertical transverse axis.)

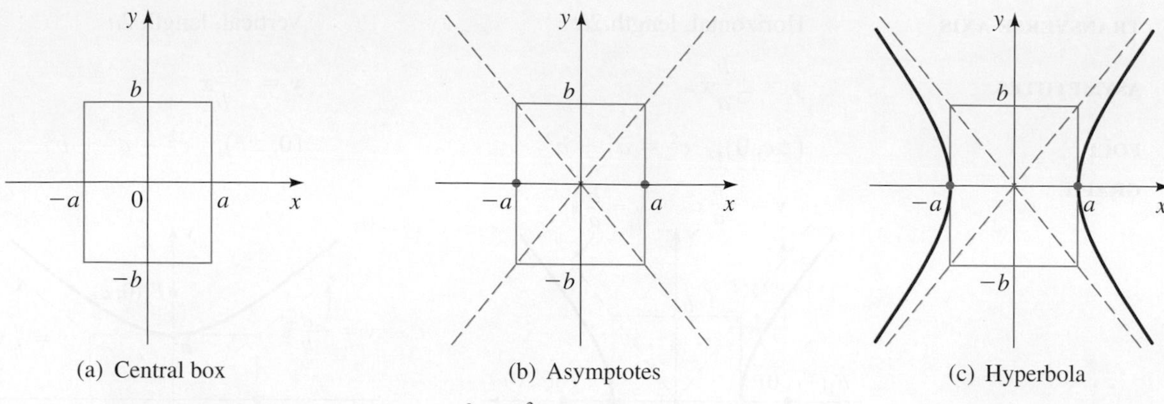

Figure 2 | Steps in graphing the hyperbola $\frac{x^2}{a^2} - \frac{y^2}{b^2} = 1$

Example 1 ■ A Hyperbola with Horizontal Transverse Axis

A hyperbola has the equation

$$9x^2 - 16y^2 = 144$$

(a) Find the vertices, foci, length of the transverse axis, and asymptotes, and sketch the graph.

(b) Draw the graph using a graphing device.

Solution

(a) First we divide both sides of the equation by 144 to put it into standard form:

$$\frac{x^2}{16} - \frac{y^2}{9} = 1$$

Because the x^2-term is positive, the hyperbola has a horizontal transverse axis; its vertices and foci are on the x-axis. Since $a^2 = 16$ and $b^2 = 9$, we get $a = 4$, $b = 3$, and $c = \sqrt{16 + 9} = 5$. Thus we have

VERTICES	$(\pm 4, 0)$
FOCI	$(\pm 5, 0)$
ASYMPTOTES	$y = \pm\frac{3}{4}x$

The length of the transverse axis is $2a = 8$. After sketching the central box and asymptotes, we complete the sketch of the hyperbola as in Figure 3(a).

(b) Most graphing devices can draw the graph of this equation, as in Figure 3(b).

Note that the equation of a hyperbola does not define y as a function of x (see Section 2.2).

Using a Graphing Calculator To graph the equation we first solve for y.

$$9x^2 - 16y^2 = 144$$

$$-16y^2 = -9x^2 + 144 \qquad \text{Subtract } 9x^2$$

$$y^2 = 9\left(\frac{x^2}{16} - 1\right) \qquad \text{Divide by } -16 \text{ and factor } 9$$

$$y = \pm 3\sqrt{\frac{x^2}{16} - 1} \qquad \text{Take square roots}$$

The graph of the hyperbola in Figure 3(b) is obtained by graphing both functions $y = 3\sqrt{(x^2/16) - 1}$ and $y = -3\sqrt{(x^2/16) - 1}$.

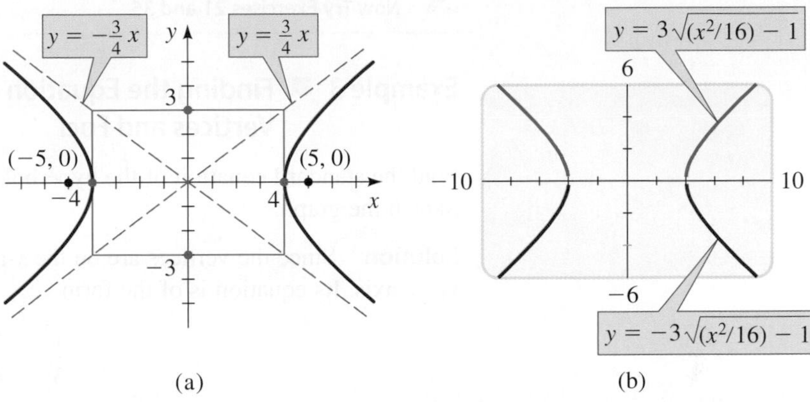

Figure 3 | $9x^2 - 16y^2 = 144$ (a) (b)

Now Try Exercises 9 and 33 ■

Example 2 ■ A Hyperbola with Vertical Transverse Axis

Find the vertices, foci, length of the transverse axis, and asymptotes of the hyperbola, and sketch its graph.

$$x^2 - 9y^2 + 9 = 0$$

Solution We begin by writing the equation in the standard form for a hyperbola:

$$x^2 - 9y^2 = -9$$

$$y^2 - \frac{x^2}{9} = 1 \qquad \text{Divide by } -9$$

Because the y^2-term is positive, the hyperbola has a vertical transverse axis; its foci and vertices are on the y-axis. Since $a^2 = 1$ and $b^2 = 9$, we get $a = 1$, $b = 3$, and $c = \sqrt{1 + 9} = \sqrt{10}$. Thus we have

VERTICES	$(0, \pm 1)$
FOCI	$(0, \pm\sqrt{10})$
ASYMPTOTES	$y = \pm\frac{1}{3}x$

The length of the transverse axis is $2a = 2$. We sketch the central box and asymptotes, then complete the graph, as shown in Figure 4(a) on the next page. We can also draw the graph using a graphing device, as shown in Figure 4(b).

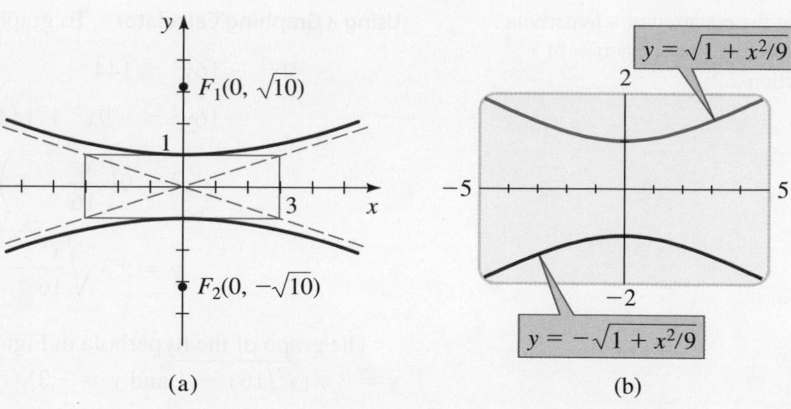

Figure 4 | $x^2 - 9y^2 + 9 = 0$

Now Try Exercises 21 and 35 ■

Example 3 ■ Finding the Equation of a Hyperbola from Its Vertices and Foci

Find the standard equation of the hyperbola with vertices $(\pm 3, 0)$ and foci $(\pm 4, 0)$. Sketch the graph.

Solution Since the vertices are on the x-axis, the hyperbola has a horizontal transverse axis. Its equation is of the form

$$\frac{x^2}{3^2} - \frac{y^2}{b^2} = 1$$

We have $a = 3$ and $c = 4$. To find b, we use the relation $a^2 + b^2 = c^2$.

$$3^2 + b^2 = 4^2$$

$$b^2 = 4^2 - 3^2 = 7$$

$$b = \sqrt{7}$$

Thus the standard equation of the hyperbola is

$$\frac{x^2}{9} - \frac{y^2}{7} = 1$$

The graph is shown in Figure 5.

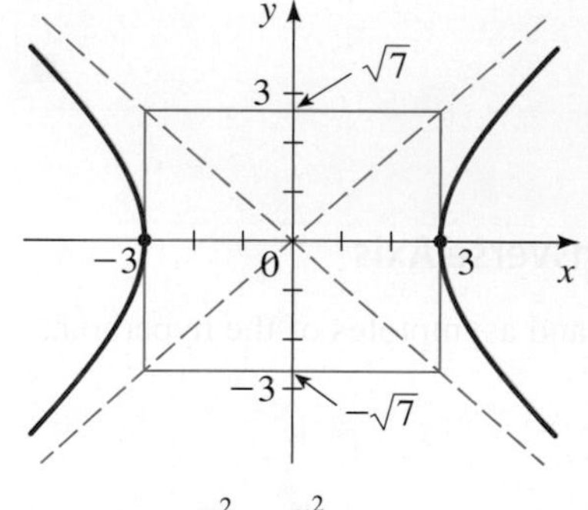

Figure 5 | $\dfrac{x^2}{9} - \dfrac{y^2}{7} = 1$

Now Try Exercises 27 and 37 ■

Example 4 ■ Finding the Equation of a Hyperbola from Its Vertices and Asymptotes

Find the standard equation and the foci of the hyperbola with vertices $(0, \pm 2)$ and asymptotes $y = \pm 2x$. Sketch the graph.

Solution Since the vertices are on the y-axis, the hyperbola has a vertical transverse axis with $a = 2$. From the asymptote equation we see that $a/b = 2$. Since $a = 2$, we get $2/b = 2$, so $b = 1$. Thus the standard equation of the hyperbola is

$$\frac{y^2}{4} - x^2 = 1$$

To find the foci, we calculate $c^2 = a^2 + b^2 = 2^2 + 1^2 = 5$, so $c = \sqrt{5}$. Thus the foci are $(0, \pm\sqrt{5})$. The graph is shown in Figure 6.

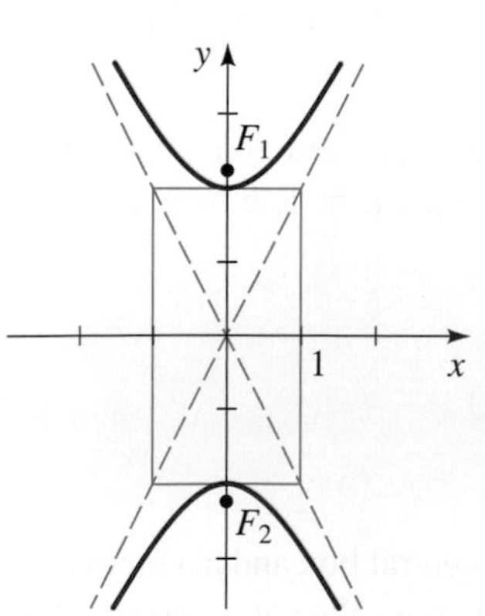

Figure 6 | $\dfrac{y^2}{4} - x^2 = 1$

Now Try Exercises 31 and 41 ■

Like parabolas and ellipses, hyperbolas have an interesting *reflection property*. Light aimed at one focus of a hyperbolic mirror is reflected toward the other focus, as shown in Figure 7. This property is used in the construction of Cassegrain-type telescopes. A hyperbolic mirror is placed in the telescope tube so that light reflected from the primary parabolic reflector is aimed at one focus of the hyperbolic mirror. The light is then refocused at a more accessible point below the primary reflector (Figure 8).

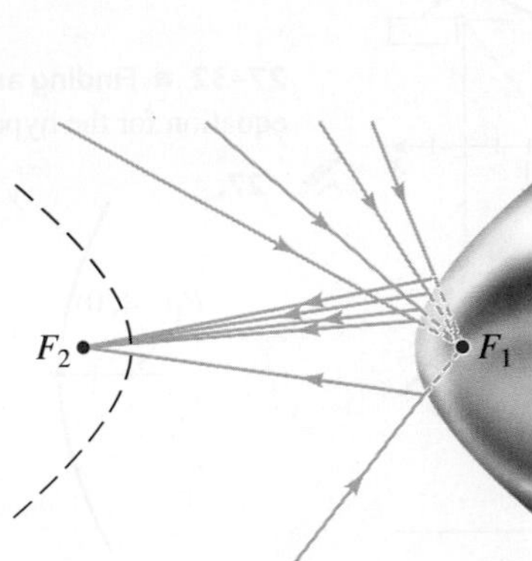

Figure 7 | Reflection property of hyperbolas

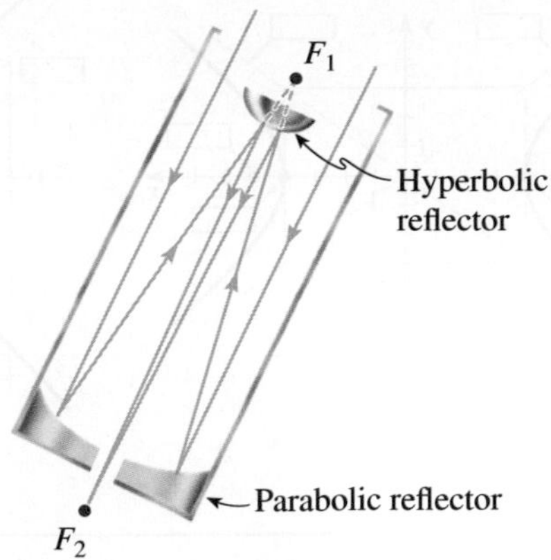

Figure 8 | Cassegrain-type telescope

The LORAN (LOng RAnge Navigation) system was used until the early 1990s; it has now been superseded by the GPS system (see Section 9.8). In the LORAN system, hyperbolas are used onboard a boat to determine its location. In Figure 9 radio stations at A and B transmit signals simultaneously for reception by the boat at P. The onboard computer converts the time difference in reception of these signals into a distance difference $d(P, A) - d(P, B)$. From the definition of a hyperbola this locates the boat on one branch of a hyperbola with foci at A and B (sketched in black in the figure). The same procedure is carried out with two other radio stations at C and D, and this locates the boat on a second hyperbola (shown in red in the figure). (In practice, only three stations are needed because one station can be used as a focus for both hyperbolas.) The coordinates of the intersection point of these two hyperbolas, which can be calculated precisely by the computer, give the location of P.

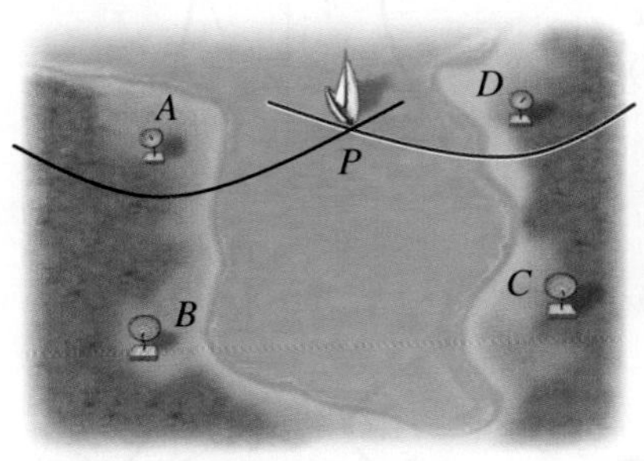

Figure 9 | LORAN system for finding the location of a ship

10.3 | Exercises

Concepts

1. A hyperbola is the set of all points in the plane for which the ________ of the distances from two fixed points F_1 and F_2 is constant. The points F_1 and F_2 are called the ________ of the hyperbola.

2. The graph of the equation $\dfrac{x^2}{a^2} - \dfrac{y^2}{b^2} = 1$ with $a > 0, b > 0$ is a hyperbola with ________ (horizontal/vertical) transverse axis, vertices (__, __) and (__, __) and foci $(\pm c, 0)$, where $c =$ ________. So the graph of $\dfrac{x^2}{4^2} - \dfrac{y^2}{3^2} = 1$ is a hyperbola with vertices (__, __) and (__, __) and foci (__, __) and (__, __).

3. The graph of the equation $\dfrac{y^2}{a^2} - \dfrac{x^2}{b^2} = 1$ with $a > 0, b > 0$ is a hyperbola with ________ (horizontal/vertical) transverse axis, vertices (__, __) and (__, __) and foci $(0, \pm c)$, where $c =$ ________. So the graph of $\dfrac{y^2}{4^2} - \dfrac{x^2}{3^2} = 1$ is a hyperbola with vertices (__, __) and (__, __) and foci (__, __) and (__, __).

4. Label the vertices, foci, and asymptotes on the graphs given for the hyperbolas in Exercises 2 and 3.

(a) $\dfrac{x^2}{4^2} - \dfrac{y^2}{3^2} = 1$ **(b)** $\dfrac{y^2}{4^2} - \dfrac{x^2}{3^2} = 1$

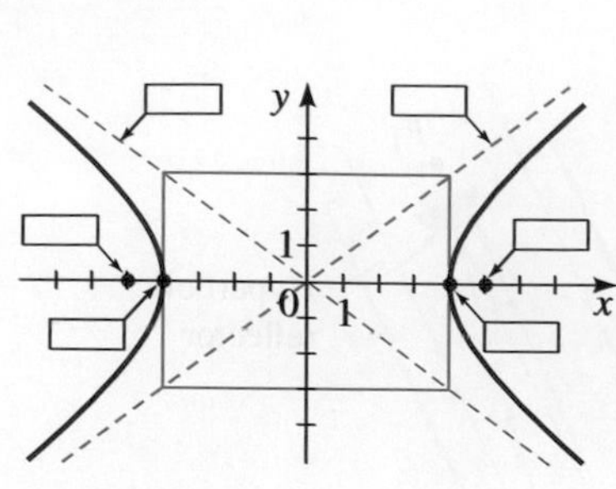

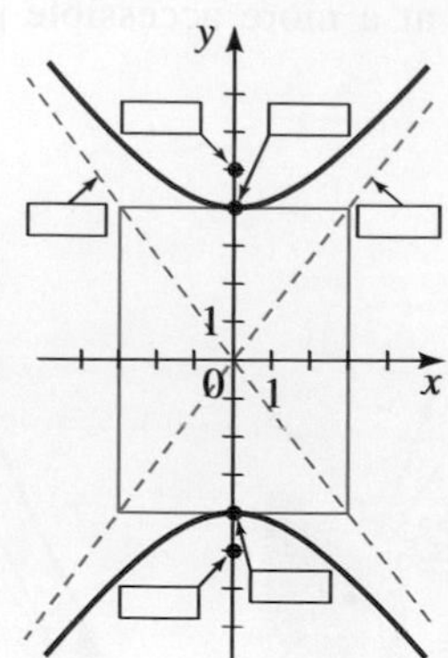

 **Skills**

5–8 ■ Graphs of Hyperbolas Match the equation with the graphs labeled I–IV. Give reasons for your answers.

5. $\dfrac{x^2}{4} - y^2 = 1$

6. $y^2 - \dfrac{x^2}{9} = 1$

7. $16y^2 - x^2 = 144$

8. $9x^2 - 25y^2 = 225$

I
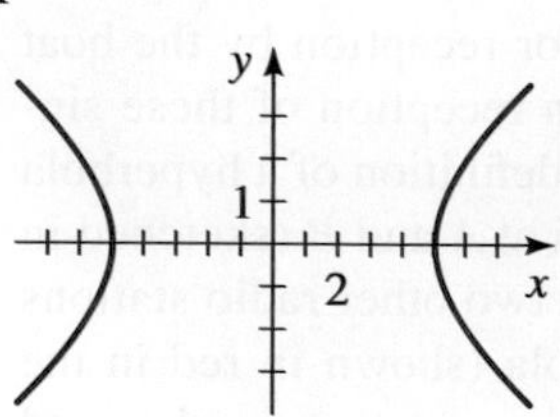

II
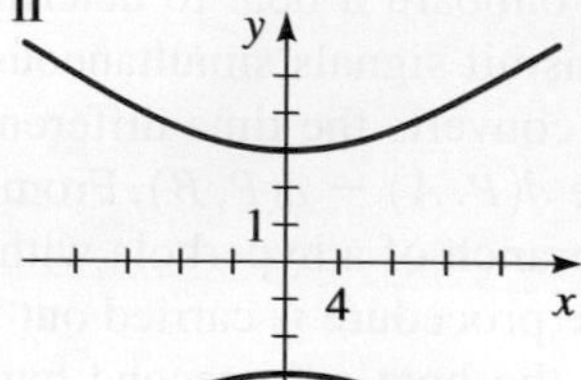

III
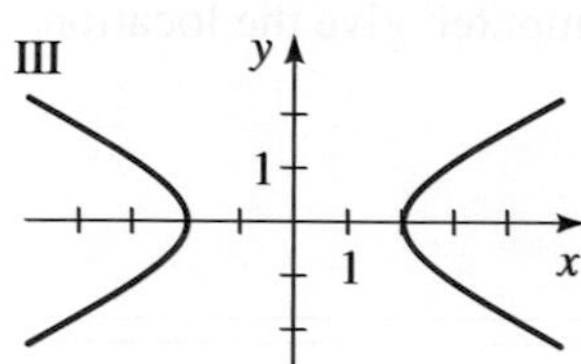

IV
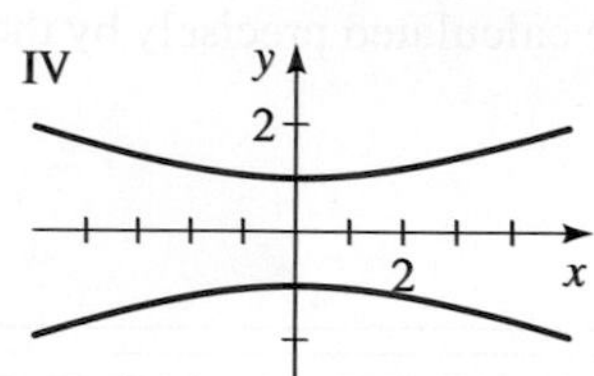

9–26 ■ Graphing Hyperbolas An equation of a hyperbola is given. **(a)** Find the vertices, foci, and asymptotes of the hyperbola. **(b)** Determine the length of the transverse axis. **(c)** Sketch a graph of the hyperbola.

 **9.** $\dfrac{x^2}{4} - \dfrac{y^2}{16} = 1$

10. $\dfrac{y^2}{9} - \dfrac{x^2}{16} = 1$

11. $\dfrac{y^2}{36} - \dfrac{x^2}{4} = 1$

12. $\dfrac{x^2}{9} - \dfrac{y^2}{64} = 1$

13. $y^2 - \dfrac{x^2}{25} = 1$

14. $\dfrac{x^2}{2} - y^2 = 1$

15. $x^2 - y^2 = 1$

16. $\dfrac{x^2}{16} - \dfrac{y^2}{12} = 1$

17. $9x^2 - 4y^2 = 36$

18. $25y^2 - 9x^2 = 225$

19. $4y^2 - 9x^2 = 144$

20. $y^2 - 25x^2 = 100$

 **21.** $x^2 - 4y^2 - 8 = 0$

22. $3y^2 - x^2 - 9 = 0$

23. $x^2 - y^2 + 4 = 0$

24. $x^2 - 3y^2 + 12 = 0$

25. $4y^2 - x^2 = 1$

26. $9x^2 - 16y^2 = 1$

27–32 ■ Finding an Equation of a Hyperbola Find the standard equation for the hyperbola whose graph is shown.

 **27.**

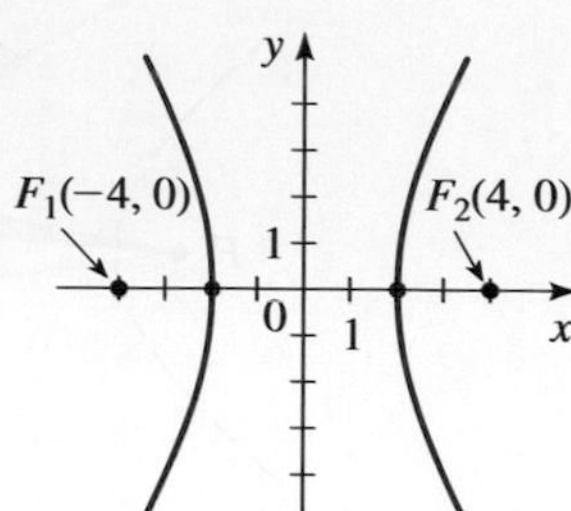

28.

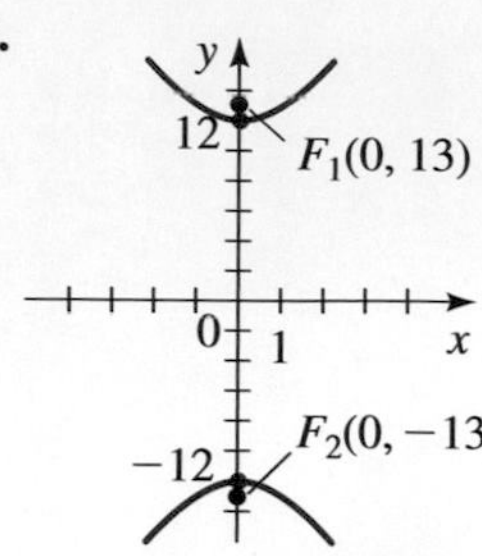

29.

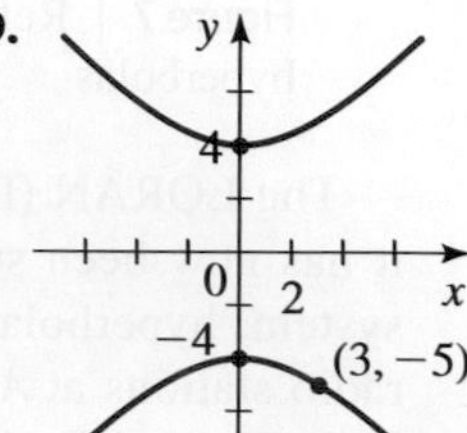

30.

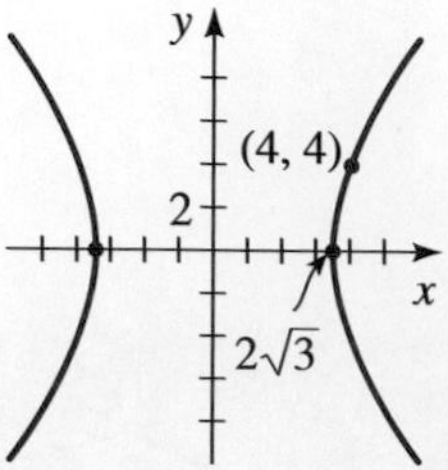

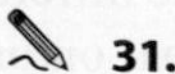 **31.**

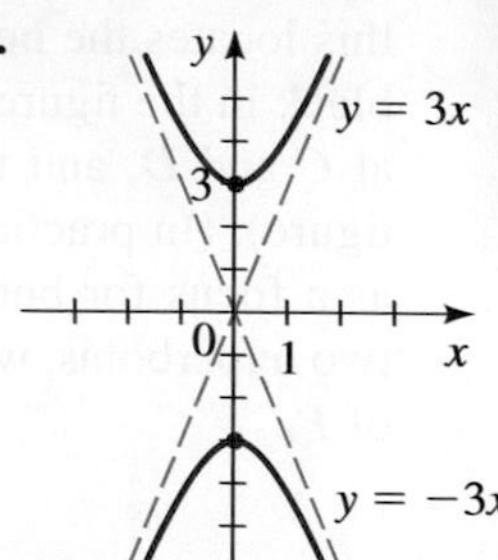

32.

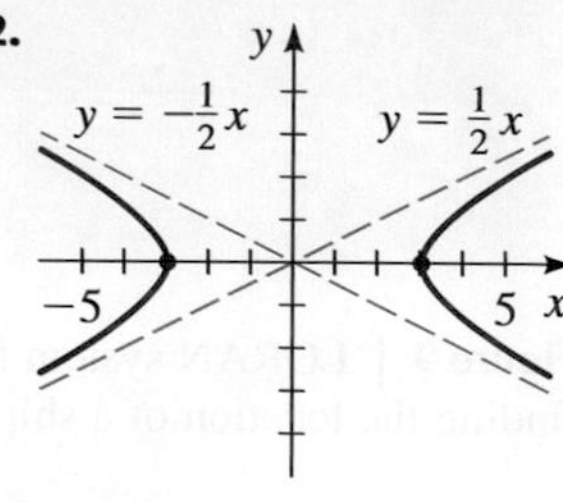

33–36 ■ Graphing Hyperbolas Use a graphing device to graph the hyperbola.

33. $x^2 - 2y^2 = 8$

34. $3y^2 - 4x^2 = 24$

35. $\dfrac{y^2}{2} - \dfrac{x^2}{6} = 1$

36. $\dfrac{x^2}{100} - \dfrac{y^2}{64} = 1$

37–50 ■ Finding the Equation of a Hyperbola Find the standard equation for the hyperbola that satisfies the given conditions.

 **37.** Foci: $(\pm 5, 0)$, vertices: $(\pm 3, 0)$

38. Foci: $(0, \pm 10)$, vertices: $(0, \pm 8)$

39. Foci: $(0, \pm 2)$, vertices: $(0, \pm 1)$

40. Foci: $(\pm 6, 0)$, vertices: $(\pm 2, 0)$

41. Vertices: $(\pm 1, 0)$, asymptotes: $y = \pm 5x$

42. Vertices: $(0, \pm 6)$, asymptotes: $y = \pm\frac{1}{3}x$

43. Vertices: $(0, \pm 6)$, hyperbola passes through $(-5, 9)$

44. Vertices: $(\pm 2, 0)$, hyperbola passes through $(3, \sqrt{30})$

45. Asymptotes: $y = \pm x$, hyperbola passes through $(5, 3)$

46. Asymptotes: $y = \pm x$, hyperbola passes through $(1, 2)$

47. Foci: $(0, \pm 3)$, hyperbola passes through $(1, 4)$

48. Foci: $(\pm\sqrt{10}, 0)$, hyperbola passes through $(4, \sqrt{18})$

49. Foci: $(\pm 5, 0)$, length of transverse axis: 6

50. Foci: $(0, \pm 1)$, length of transverse axis: 1

Skills Plus

51. Perpendicular Asymptotes

(a) Show that the asymptotes of the hyperbola $x^2 - y^2 = 5$ are perpendicular to each other.

(b) Find the standard equation for the hyperbola with foci $(\pm c, 0)$ and with asymptotes perpendicular to each other.

52. Conjugate Hyperbolas The hyperbolas

$$\frac{x^2}{a^2} - \frac{y^2}{b^2} = 1 \quad \text{and} \quad \frac{x^2}{a^2} - \frac{y^2}{b^2} = -1$$

are said to be *conjugate* to each other.

(a) Show that the hyperbolas

$$x^2 - 4y^2 + 16 = 0 \quad \text{and} \quad 4y^2 - x^2 + 16 = 0$$

are conjugate to each other, and sketch their graphs on the same coordinate axes.

(b) What do the hyperbolas of part (a) have in common?

(c) Show that any pair of conjugate hyperbolas have the relationship you discovered in part (b).

53. Equation of a Hyperbola In the derivation of the equation of the hyperbola at the beginning of this section we said that the equation

$$\sqrt{(x + c)^2 + y^2} - \sqrt{(x - c)^2 + y^2} = \pm 2a$$

simplifies to

$$(c^2 - a^2)x^2 - a^2y^2 = a^2(c^2 - a^2)$$

Supply the steps needed to show this.

54. Verifying a Geometric Property of a Hyperbola

(a) For the hyperbola

$$\frac{x^2}{9} - \frac{y^2}{16} = 1$$

determine the values of a, b, and c, and find the coordinates of the foci F_1 and F_2.

(b) Show that the point $P(5, \frac{16}{3})$ lies on this hyperbola.

(c) Find $d(P, F_1)$ and $d(P, F_2)$.

(d) Verify that the difference between $d(P, F_1)$ and $d(P, F_2)$ is $2a$.

55. Confocal Hyperbolas Hyperbolas are called *confocal* if they have the same foci.

(a) Show that the hyperbolas

$$\frac{y^2}{k} - \frac{x^2}{16 - k} = 1 \qquad (0 < k < 16)$$

are confocal.

(b) Use a graphing device to draw the top branches of the family of hyperbolas in part (a) for $k = 1, 4, 8$, and 12. How does the shape of the graph change as k increases?

Applications

56. Navigation In the figure, the LORAN stations at A and B are 800 km apart, and the ship at P receives station A's signal 2640 microseconds (μs) before it receives the signal from station B.

(a) Assuming that radio signals travel at 299 m/μs, find $d(P, A) - d(P, B)$.

(b) Find an equation for the branch of the hyperbola indicated in red in the figure. (Use kilometers as the unit of distance.)

(c) If A is due north of B and if P is due east of A, how far is P from A?

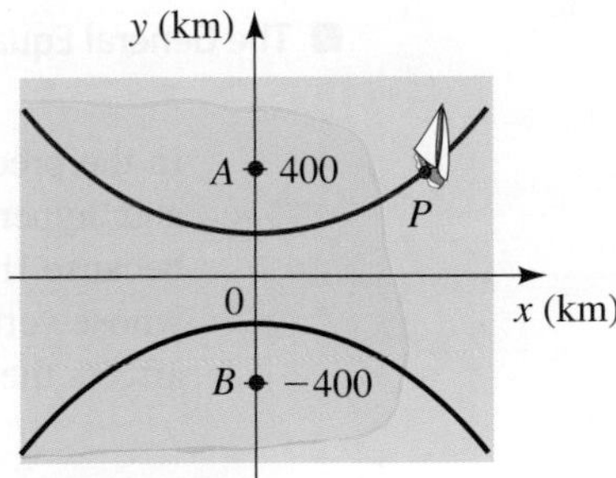

57. Comet Trajectories Some comets, such as Halley's comet, are a permanent part of the solar system, traveling in elliptical orbits around the sun. Other comets pass through the solar system only once, following a hyperbolic path with the sun at a focus. The figure shows the path of such a comet. Find an equation for the path, assuming that the closest the comet comes to the sun is 3.2×10^9 km and that the path the comet was taking before it neared the solar system is at a right angle to the path it continues on after leaving the solar system.

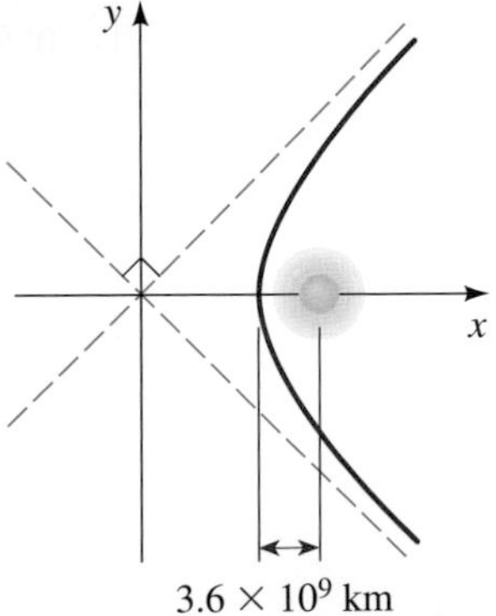

58. Ripples in Pool Two stones are dropped simultaneously into a calm pool of water. The crests of the resulting waves form equally spaced concentric circles, as shown in the figures. The waves interact with each other to create certain interference patterns.

(a) Explain why the red dots lie on an ellipse.

(b) Explain why the blue dots lie on a hyperbola.

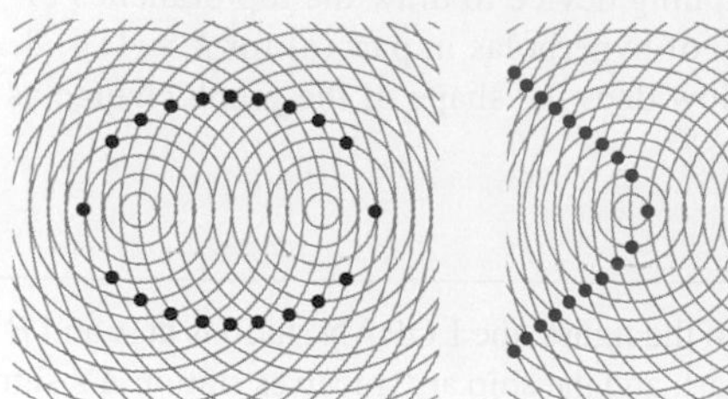

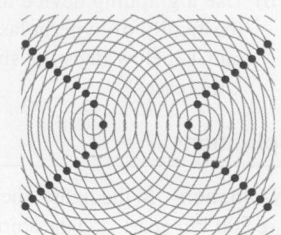

■ **Discuss** ■ **Discover** ■ **Prove** ■ **Write**

59. Discuss ■ Write: Hyperbolas in the Real World Several examples of the uses of hyperbolas are given in the text. Find other situations in which hyperbolas occur.

60. Discuss: Light from a Lamp The light from a lamp forms a lighted area on a wall, as shown in the figure. Why is the boundary of this lighted area a hyperbola? How can one hold a flashlight so that its beam forms a hyperbola on the ground?

10.4 Shifted Conics

■ Shifting Graphs of Equations ■ Shifted Ellipses ■ Shifted Parabolas ■ Shifted Hyperbolas ■ The General Equation of a Shifted Conic

In the preceding sections we studied parabolas with vertices at the origin and ellipses and hyperbolas with centers at the origin. We restricted ourselves to these cases because these equations have the simplest form. In this section we consider conics whose vertices and centers are not necessarily at the origin, and we determine how this affects their equations.

■ Shifting Graphs of Equations

In Section 2.6 we studied transformations of functions that have the effect of shifting their graphs. In general, for any equation in x and y, if we replace x by $x - h$ or by $x + h$, the graph of the new equation is simply the old graph shifted horizontally; if y is replaced by $y - k$ or by $y + k$, the graph is shifted vertically. The following box gives the details.

Shifting Graphs of Equations

If h and k are positive real numbers, then replacing x by $x - h$ or by $x + h$ and replacing y by $y - k$ or by $y + k$ has the following effect(s) on the graph of any equation in x and y.

Replacement	How the graph is shifted
1. x replaced by $x - h$	Right h units
2. x replaced by $x + h$	Left h units
3. y replaced by $y - k$	Upward k units
4. y replaced by $y + k$	Downward k units

JOHANNES KEPLER (1571–1630) was the first to give a correct description of the motion of the planets. The cosmology of his time postulated complicated systems of circles moving on circles to describe these motions. Kepler sought a simpler and more harmonious description. As the official astronomer at the imperial court in Prague, he studied the astronomical observations of the Danish astronomer Tycho Brahe, whose data were the most accurate available at the time. After numerous attempts to find a theory, Kepler made the momentous discovery that the orbits of the planets are elliptical. His three great laws of planetary motion are

1. The orbit of each planet is an ellipse with the sun at one focus.
2. The line segment that joins the sun to a planet sweeps out equal areas in equal time (see the figure).
3. The square of the period of revolution of a planet is proportional to the cube of the length of the major axis of its orbit.

Kepler's formulation of these laws is perhaps the most impressive deduction from empirical data in the history of science.

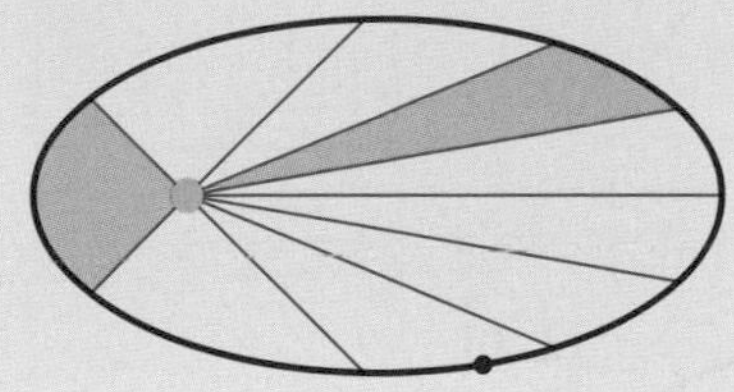

■ Shifted Ellipses

Let's apply horizontal and vertical shifting to the ellipse with the standard equation

$$\frac{x^2}{a^2} + \frac{y^2}{b^2} = 1 \qquad \text{Standard equation of ellipse}$$

whose graph is shown in Figure 1. If we shift it so that its center is at the point (h, k) instead of at the origin, then its standard equation becomes

$$\frac{(x - h)^2}{a^2} + \frac{(y - k)^2}{b^2} = 1 \qquad \text{Standard equation of shifted ellipse}$$

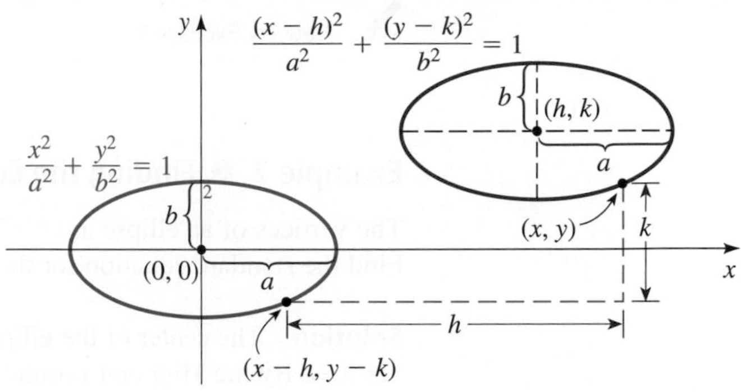

Figure 1 | Shifted ellipse

Example 1 ■ Sketching the Graph of a Shifted Ellipse

Sketch a graph of the ellipse

$$\frac{(x + 1)^2}{4} + \frac{(y - 2)^2}{9} = 1$$

and determine the coordinates of the foci.

Solution The ellipse

$$\frac{(x + 1)^2}{4} + \frac{(y - 2)^2}{9} = 1 \qquad \text{Shifted ellipse}$$

is shifted so that its center is at $(-1, 2)$. It is obtained from the ellipse

$$\frac{x^2}{4} + \frac{y^2}{9} = 1 \qquad \text{Ellipse with center at origin}$$

by shifting it left 1 unit and upward 2 units. The endpoints of the minor and major axes of the ellipse with center at the origin are $(2, 0), (-2, 0), (0, 3), (0, -3)$. We apply the required shifts to these points to obtain the corresponding points on the shifted ellipse.

$$(2, 0) \to (2 - 1, 0 + 2) = (1, 2)$$
$$(-2, 0) \to (-2 - 1, 0 + 2) = (-3, 2)$$
$$(0, 3) \to (0 - 1, 3 + 2) = (-1, 5)$$
$$(0, -3) \to (0 - 1, -3 + 2) = (-1, -1)$$

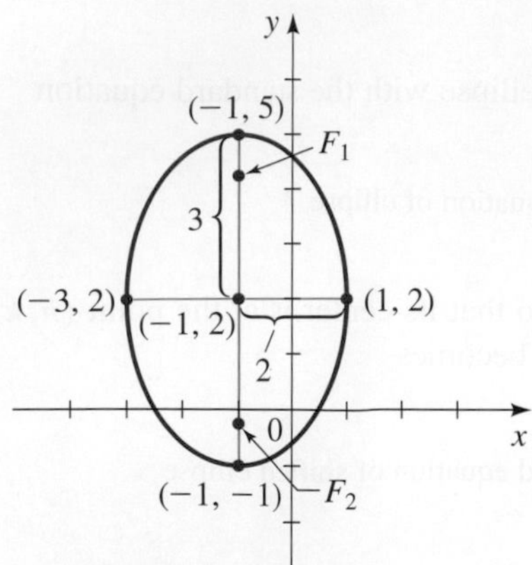

Figure 2 | $\dfrac{(x+1)^2}{4} + \dfrac{(y-2)^2}{9} = 1$

This helps us sketch the graph in Figure 2.

To find the foci of the shifted ellipse, we first find the foci of the ellipse with center at the origin. Since $a^2 = 9$ and $b^2 = 4$, we have $c^2 = 9 - 4 = 5$, so $c = \sqrt{5}$. So the foci are $(0, \pm\sqrt{5})$. Shifting left 1 unit and upward 2 units, we get

$$(0, \sqrt{5}) \to (0 - 1, \sqrt{5} + 2) = (-1, 2 + \sqrt{5})$$

$$(0, -\sqrt{5}) \to (0 - 1, -\sqrt{5} + 2) = (-1, 2 - \sqrt{5})$$

Thus the foci of the shifted ellipse are

$$F_1(-1, 2 + \sqrt{5}) \qquad \text{and} \qquad F_2(-1, 2 - \sqrt{5})$$

Now Try Exercise 7 ■

Example 2 ■ Finding the Equation of a Shifted Ellipse

The vertices of an ellipse are $(-7, 3)$ and $(3, 3)$, and the foci are $(-6, 3)$ and $(2, 3)$. Find the standard equation for the ellipse, and sketch its graph.

Solution The center of the ellipse is the midpoint of the line segment between the vertices. By the Midpoint Formula the center is

The Midpoint Formula is given in Section 1.9.

$$\left(\frac{-7+3}{2}, \frac{3+3}{2}\right) = (-2, 3) \qquad \text{Center}$$

Since the vertices lie on a horizontal line, the major axis is horizontal. The length of the major axis is $3 - (-7) = 10$, so $a = 5$. The distance between the foci is $2 - (-6) = 8$, so $c = 4$. Since $c^2 = a^2 - b^2$, we have

$$4^2 = 5^2 - b^2 \qquad c = 4, a = 5$$

$$b^2 = 25 - 16 = 9 \qquad \text{Solve for } b^2$$

Thus the standard equation of the ellipse is

$$\frac{(x+2)^2}{25} + \frac{(y-3)^2}{9} = 1 \qquad \text{Standard equation of shifted ellipse}$$

The graph is shown in Figure 3.

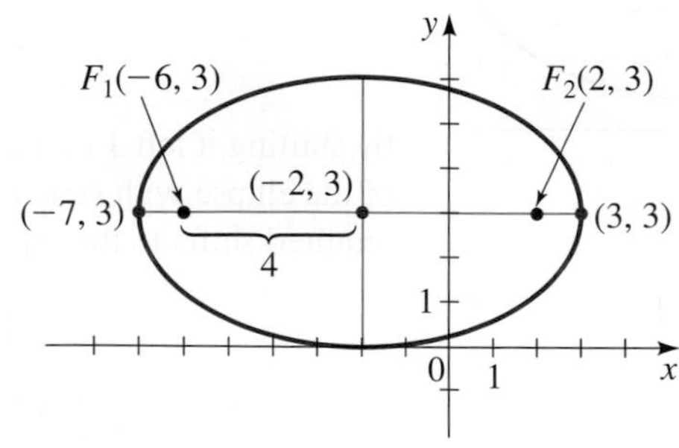

Figure 3 | $\dfrac{(x+2)^2}{25} + \dfrac{(y-3)^2}{9} = 1$

Now Try Exercise 35 ■

■ Shifted Parabolas

Applying shifts to parabolas leads to the standard equations and graphs shown in Figure 4.

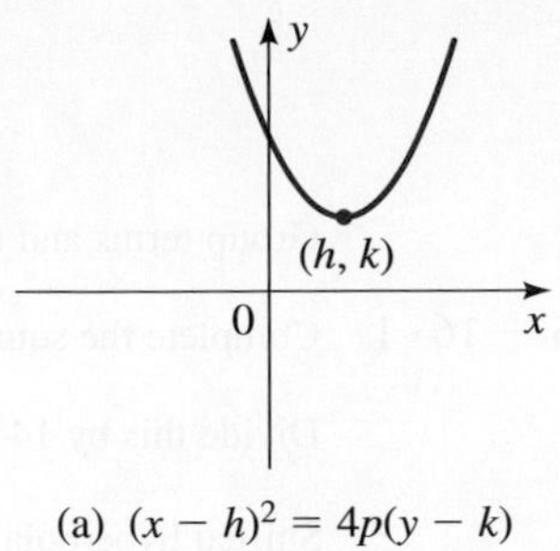

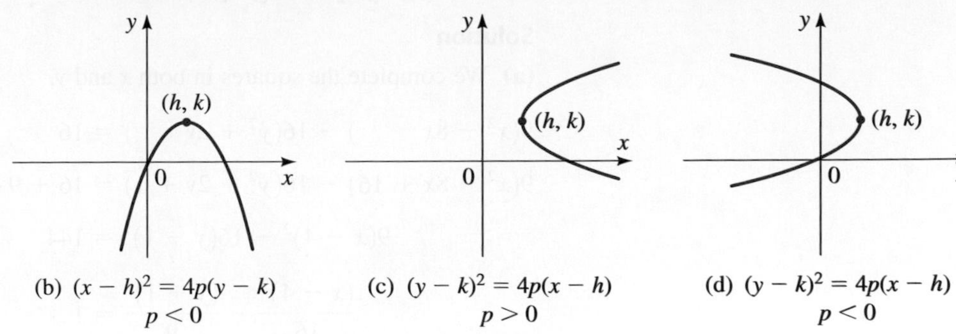

(a) $(x - h)^2 = 4p(y - k)$ $p > 0$

(b) $(x - h)^2 = 4p(y - k)$ $p < 0$

(c) $(y - k)^2 = 4p(x - h)$ $p > 0$

(d) $(y - k)^2 = 4p(x - h)$ $p < 0$

Figure 4 | Shifted parabolas

Example 3 ■ Graphing a Shifted Parabola

Determine the vertex, focus, and directrix, and sketch a graph of the parabola

$$x^2 - 4x = 8y - 28$$

Solution We complete the square in x to put this equation into one of the standard forms given in Figure 4.

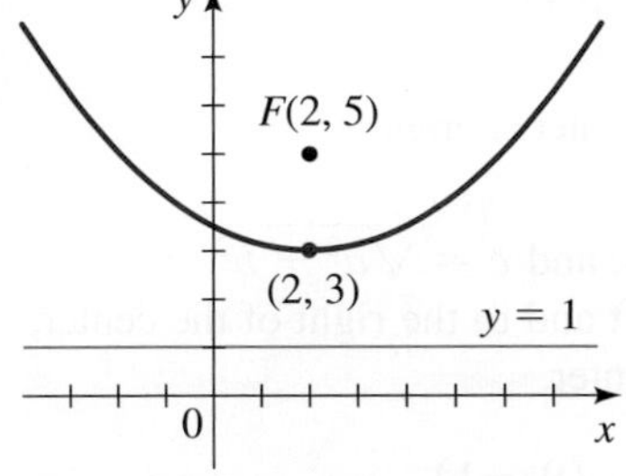

Figure 5 | $x^2 - 4x = 8y - 28$

$$x^2 - 4x + 4 = 8y - 28 + 4 \qquad \text{Add 4 to complete the square}$$

$$(x - 2)^2 = 8y - 24 \qquad \text{Perfect square}$$

$$(x - 2)^2 = 8(y - 3) \qquad \text{Shifted parabola in standard form}$$

This parabola opens upward with vertex $(2, 3)$. It is obtained from the parabola

$$x^2 = 8y \qquad \text{Parabola with vertex at origin}$$

by shifting right 2 units and upward 3 units. Since $4p = 8$, we have $p = 2$, so the focus is 2 units above the vertex and the directrix is 2 units below the vertex. Thus the focus is $(2, 5)$ and the directrix is $y = 1$. The graph is shown in Figure 5.

Now Try Exercises 13 and 19 ■

■ Shifted Hyperbolas

Applying shifts to hyperbolas leads to the equations and graphs shown in Figure 6.

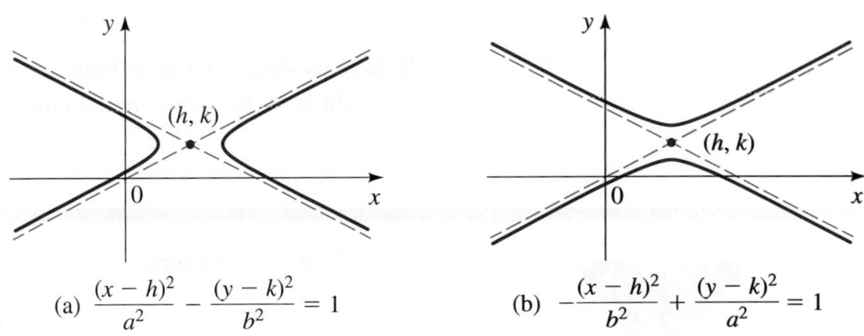

(a) $\dfrac{(x - h)^2}{a^2} - \dfrac{(y - k)^2}{b^2} = 1$

(b) $-\dfrac{(x - h)^2}{b^2} + \dfrac{(y - k)^2}{a^2} = 1$

Figure 6 | Shifted hyperbolas

Example 4 ■ Graphing a Shifted Hyperbola

A shifted conic has the equation $9x^2 - 72x - 16y^2 - 32y = 16$.

(a) Complete the square in x and y to show that the equation represents a hyperbola.

(b) Find the center, vertices, foci, and asymptotes of the hyperbola, and sketch its graph.
(c) Draw the graph using a graphing device.

Solution

(a) We complete the squares in both x and y.

$$9(x^2 - 8x \qquad) - 16(y^2 + 2y \qquad) = 16 \qquad \text{Group terms and factor}$$

$$9(x^2 - 8x + 16) - 16(y^2 + 2y + 1) = 16 + 9 \cdot 16 - 16 \cdot 1 \qquad \text{Complete the squares}$$

$$9(x - 4)^2 - 16(y + 1)^2 = 144 \qquad \text{Divide this by 144}$$

$$\frac{(x - 4)^2}{16} - \frac{(y + 1)^2}{9} = 1 \qquad \text{Shifted hyperbola in standard form}$$

Comparing this to Figure 6(a), we see that this is the equation of a shifted hyperbola.

(b) The shifted hyperbola has center $(4, -1)$ and a horizontal transverse axis.

CENTER $(4, -1)$

Its graph will have the same shape as the unshifted hyperbola

$$\frac{x^2}{16} - \frac{y^2}{9} = 1 \qquad \text{Hyperbola with center at origin}$$

Since $a^2 = 16$ and $b^2 = 9$, we have $a = 4$, $b = 3$, and $c = \sqrt{a^2 + b^2} = \sqrt{16 + 9} = 5$. Thus the foci lie 5 units to the left and to the right of the center, and the vertices lie 4 units to either side of the center.

FOCI $(-1, -1)$ and $(9, -1)$

VERTICES $(0, -1)$ and $(8, -1)$

The asymptotes of the unshifted hyperbola are $y = \pm\frac{3}{4}x$, so the asymptotes of the shifted hyperbola are found as follows.

ASYMPTOTES $y + 1 = \pm\frac{3}{4}(x - 4)$

$$y + 1 = \pm\tfrac{3}{4}x \mp 3$$

$$y = \tfrac{3}{4}x - 4 \qquad \text{and} \qquad y = -\tfrac{3}{4}x + 2$$

To help us sketch the hyperbola, we draw the central box; it extends 4 units left and right from the center and 3 units upward and downward from the center. We

---

Lightspring/Shutterstock.com; Jopelka/Shutterstock.com; Serafima Antipova/Shutterstock.com

Discovery Project ■ Symmetry

We have learned about certain symmetry properties of graphs of equations and how to test an equation for symmetry about the coordinate axes or the origin. There are other types of symmetry. For example, we have the feeling that the starfish in the photo is somehow symmetric, even though it is not symmetric about any axis or about the origin. In this project we investigate the symmetries of figures in the plane, including the shifted and rotated conics. We also investigate symmetry in nature. You can find the project at **www.stewartmath.com**.

then draw the asymptotes and complete the graph of the shifted hyperbola as shown in Figure 7(a).

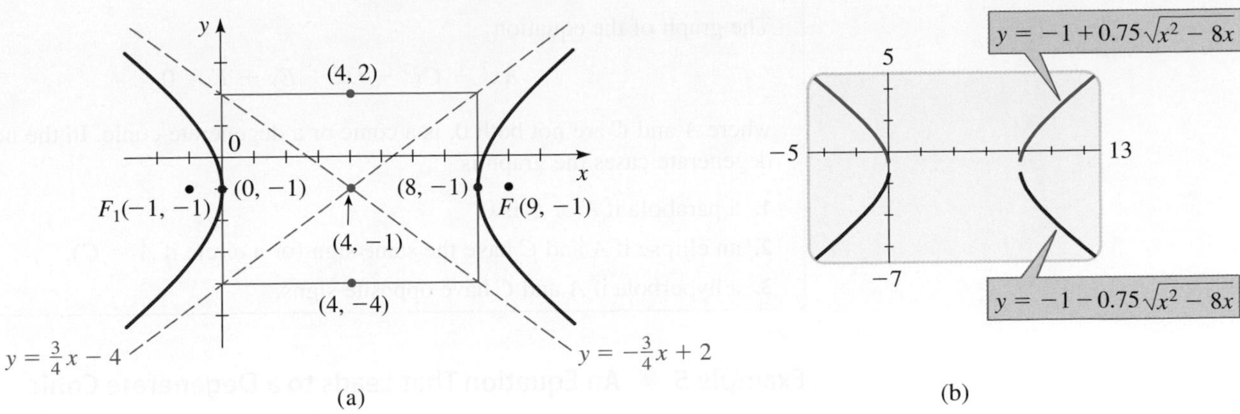

Figure 7 | $9x^2 - 72x - 16y^2 - 32y = 16$

(c) Most graphing devices can draw the graph of this equation, as shown in Figure 7(b).

Note that the equation of a hyperbola does not define y as a function of x (see Section 2.2).

Using a Graphing Calculator To graph the equation we first solve for y. The given equation is a quadratic equation in y, so we use the Quadratic Formula to solve for y. Writing the equation in the form

$$\overbrace{16}^{a}y^2 + \overbrace{32}^{b}y + \overbrace{(-9x^2 + 72x + 16)}^{c} = 0$$

we get

$$y = \frac{-b \pm \sqrt{b^2 - 4ac}}{2a}$$

$$\begin{aligned} y &= \frac{-32 \pm \sqrt{32^2 - 4(16)(-9x^2 + 72x + 16)}}{2(16)} && \text{Quadratic Formula} \\ &= \frac{-32 \pm \sqrt{576x^2 - 4608x}}{32} && \text{Expand} \\ &= \frac{-32 \pm 24\sqrt{x^2 - 8x}}{32} && \text{Factor 576 from under the radical} \\ &= -1 \pm \tfrac{3}{4}\sqrt{x^2 - 8x} && \text{Simplify} \end{aligned}$$

The graph of the hyperbola in Figure 7(b) is obtained by graphing both functions

$$y = -1 + 0.75\sqrt{x^2 - 8x} \qquad \text{and} \qquad y = -1 - 0.75\sqrt{x^2 - 8x}$$

 **Now Try Exercises 21, 27, and 61** ■

■ The General Equation of a Shifted Conic

If we expand and simplify the equations of any of the shifted conics illustrated in Figures 1, 4, and 6, then we will always obtain an equation of the form

$$Ax^2 + Cy^2 + Dx + Ey + F = 0$$

where A and C are not both 0. Conversely, if we begin with an equation of this form, then we can complete the square in x and y to see which type of conic section the equation represents. In some cases the graph of the equation turns out to be just a pair of lines or a single point, or there might be no graph at all. These cases are called **degenerate conics**. If the equation is not degenerate, then we can tell whether it represents a parabola, an ellipse, or a hyperbola by examining the signs of A and C, as described in the following box.

General Equation of a Shifted Conic

The graph of the equation

$$Ax^2 + Cy^2 + Dx + Ey + F = 0$$

where A and C are not both 0, is a conic or a degenerate conic. In the nondegenerate cases the graph is

1. a parabola if A or C is 0,
2. an ellipse if A and C have the same sign (or a circle if $A = C$),
3. a hyperbola if A and C have opposite signs.

Example 5 ■ An Equation That Leads to a Degenerate Conic

Sketch the graph of the equation

$$9x^2 - y^2 + 18x + 6y = 0$$

Solution Because the coefficients of x^2 and y^2 have opposite signs, this equation looks as if it should represent a hyperbola (like the equation of Example 4). To see whether this is in fact the case, we complete the squares.

$$9(x^2 + 2x \quad\quad) - (y^2 - 6y \quad\quad) = 0 \qquad \text{Group terms and factor 9}$$

$$9(x^2 + 2x + 1) - (y^2 - 6y + 9) = 0 + 9 \cdot 1 - 9 \qquad \text{Complete the squares}$$

$$9(x + 1)^2 - (y - 3)^2 = 0 \qquad \text{Factor}$$

$$(x + 1)^2 - \frac{(y - 3)^2}{9} = 0 \qquad \text{Divide by 9}$$

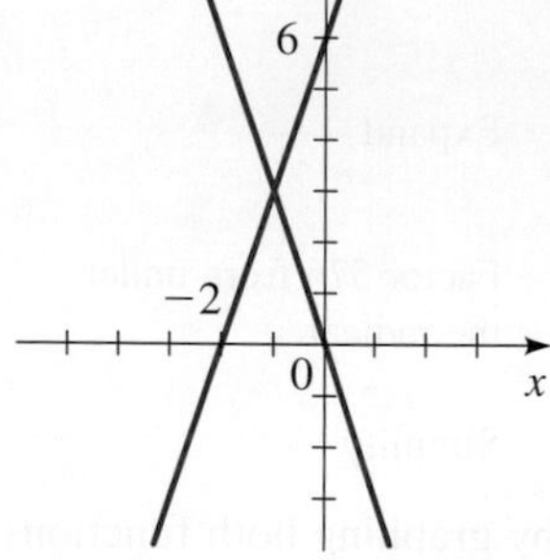

Figure 8 | $9x^2 - y^2 + 18x + 6y = 0$

For this to fit the form of the equation of a hyperbola, we would need a nonzero constant to the right of the equal sign. In fact, further analysis shows that this is the equation of a pair of intersecting lines.

$$(y - 3)^2 = 9(x + 1)^2$$

$$y - 3 = \pm 3(x + 1) \qquad \text{Take square roots}$$

$$y = 3(x + 1) + 3 \quad \text{or} \quad y = -3(x + 1) + 3$$

$$y = 3x + 6 \qquad\qquad y = -3x$$

These lines are graphed in Figure 8.

Now Try Exercise 55 ■

Note Because the equation in Example 5 looked at first glance like the equation of a hyperbola but, in fact, turned out to represent simply a pair of lines, we refer to its graph as a **degenerate hyperbola**. Degenerate ellipses and parabolas can also arise when we complete the square(s) in an equation that seems to represent a conic. For example, the equation

$$4x^2 + y^2 - 8x + 2y + 6 = 0$$

looks as if it should represent an ellipse, because the coefficients of x^2 and y^2 have the same sign. But completing the squares leads to

$$(x - 1)^2 + \frac{(y + 1)^2}{4} = -\frac{1}{4}$$

which has no solution at all (since the sum of two squares cannot be negative). This equation is therefore degenerate.

10.4 Exercises

Concepts

1. Suppose we want to graph an equation in x and y.

(a) If we replace x by $x - 3$, the graph of the equation is shifted to the ________ by 3 units. If we replace x by $x + 3$, the graph of the equation is shifted to the ________ by 3 units.

(b) If we replace y by $y - 1$, the graph of the equation is shifted ________ by 1 unit. If we replace y by $y + 1$, the graph of the equation is shifted ________ by 1 unit.

2. The graphs of $x^2 = 12y$ and $(x - 3)^2 = 12(y - 1)$ are given. Label the focus, directrix, and vertex on each parabola.

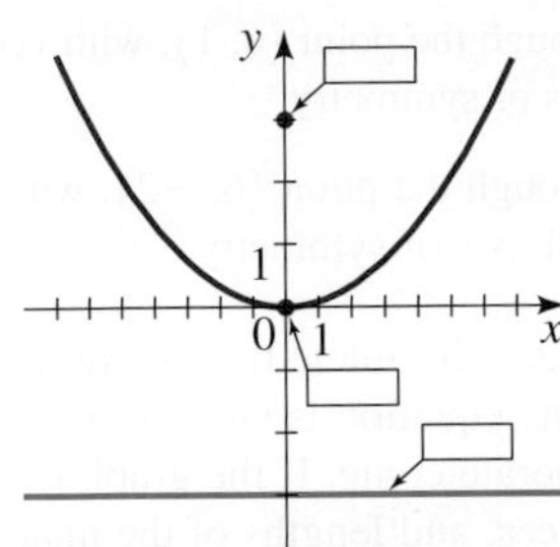

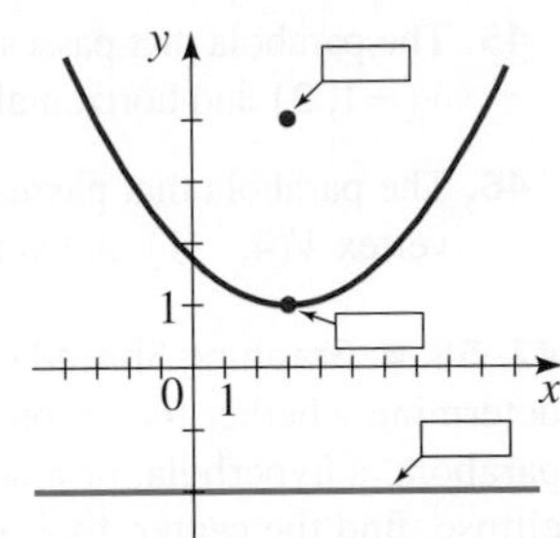

3. The graphs of the ellipses

$$\frac{x^2}{5^2} + \frac{y^2}{4^2} = 1 \quad \text{and} \quad \frac{(x-3)^2}{5^2} + \frac{(y-1)^2}{4^2} = 1$$

are given. Label the vertices and foci on each ellipse.

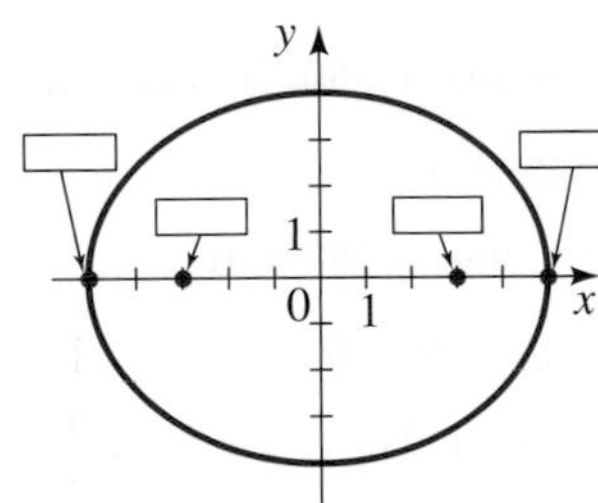

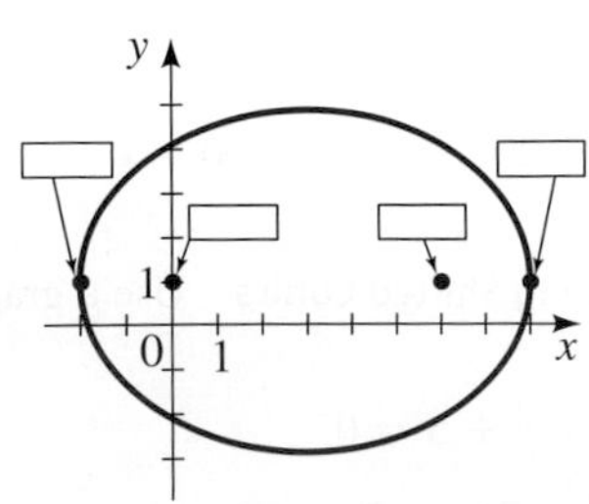

4. The graphs of the hyperbolas

$$\frac{x^2}{4^2} - \frac{y^2}{3^2} = 1 \quad \text{and} \quad \frac{(x-3)^2}{4^2} - \frac{(y-1)^2}{3^2} = 1$$

are given. Label the vertices, foci, and asymptotes on each hyperbola.

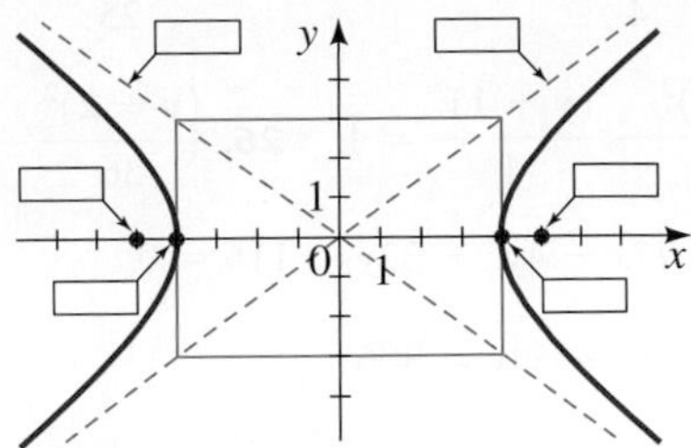

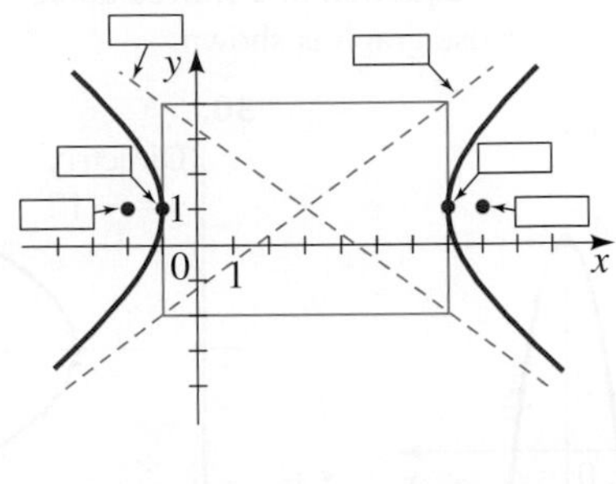

Skills

5–12 ■ Graphing Shifted Ellipses An equation of an ellipse is given. **(a)** Find the center, vertices, and foci of the ellipse. **(b)** Determine the lengths of the major and minor axes. **(c)** Sketch a graph of the ellipse.

5. $\dfrac{(x-2)^2}{9} + \dfrac{(y-1)^2}{4} = 1$

6. $\dfrac{(x-3)^2}{16} + (y+3)^2 = 1$

7. $\dfrac{x^2}{9} + \dfrac{(y+5)^2}{25} = 1$

8. $x^2 + \dfrac{(y+2)^2}{4} = 1$

9. $\dfrac{(x+5)^2}{16} + \dfrac{(y-1)^2}{4} = 1$

10. $\dfrac{(x+1)^2}{36} + \dfrac{(y+1)^2}{64} = 1$

11. $4x^2 + 25y^2 - 50y = 75$

12. $9x^2 - 54x + y^2 + 2y + 46 = 0$

13–20 ■ Graphing Shifted Parabolas An equation of a parabola is given. **(a)** Find the vertex, focus, and directrix of the parabola. **(b)** Sketch a graph showing the parabola and its directrix.

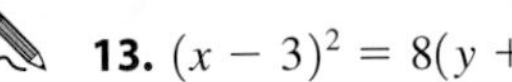

13. $(x-3)^2 = 8(y+1)$

14. $(y+1)^2 = 16(x-3)$

15. $(y+5)^2 = -6x + 12$

16. $y^2 = 16x - 8$

17. $2(x-1)^2 = y$

18. $-4\left(x + \frac{1}{2}\right)^2 = y$

19. $y^2 - 6y - 12x + 33 = 0$

20. $x^2 + 2x - 20y + 41 = 0$

21–28 ■ Graphing Shifted Hyperbolas An equation of a hyperbola is given. **(a)** Find the center, vertices, foci, and asymptotes of the hyperbola. **(b)** Sketch a graph showing the hyperbola and its asymptotes.

21. $\dfrac{(x+1)^2}{9} - \dfrac{(y-3)^2}{16} = 1$ **22.** $(x-8)^2 - (y+6)^2 = 1$

23. $y^2 - \dfrac{(x+1)^2}{4} = 1$ **24.** $\dfrac{(y-1)^2}{25} - (x+3)^2 = 1$

25. $\dfrac{(x+1)^2}{9} - \dfrac{(y+1)^2}{4} = 1$ **26.** $\dfrac{(y+2)^2}{36} - \dfrac{x^2}{64} = 1$

27. $36x^2 + 72x - 4y^2 + 32y + 116 = 0$

28. $25x^2 - 9y^2 - 54y = 306$

29–34 ■ Finding the Equation of a Shifted Conic Find an equation for the conic whose graph is shown.

29.

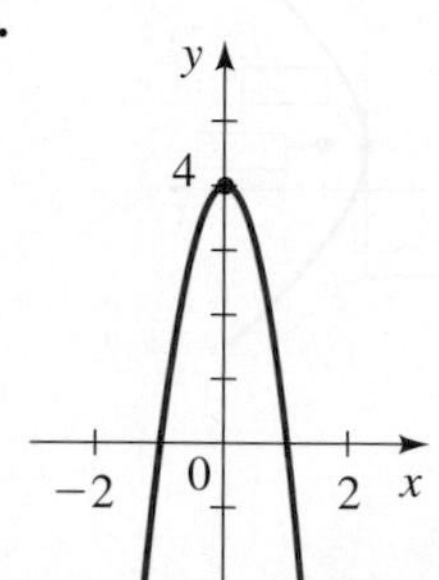

30.

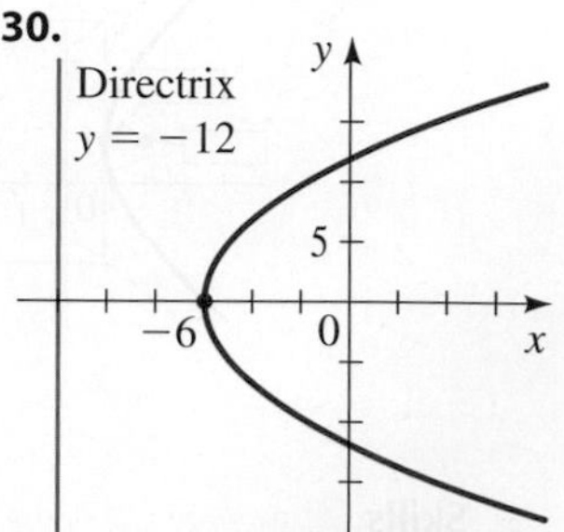

31.

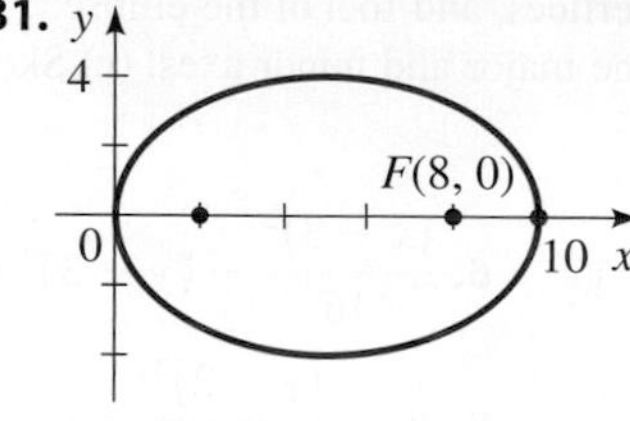

32.

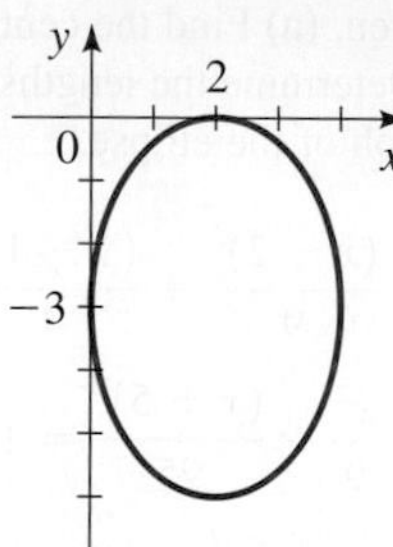

33.

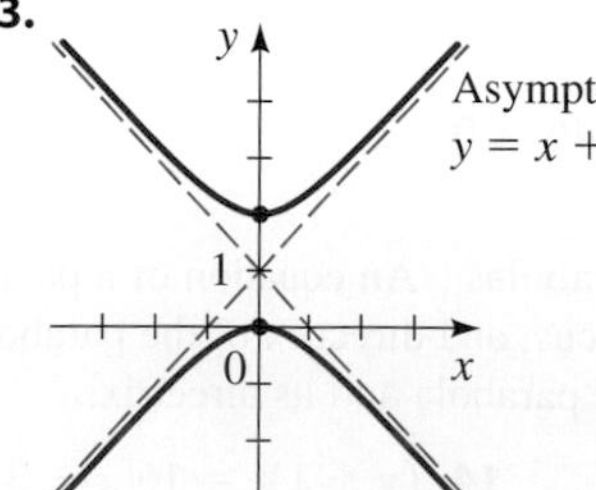

34.

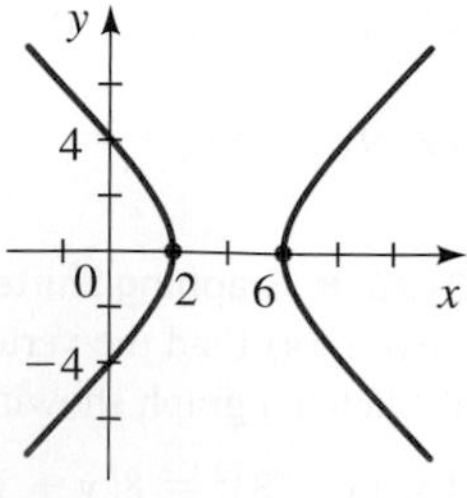

35–46 ■ Finding the Equation of a Shifted Conic Find the standard equation for the conic section with the given properties.

35. The ellipse with center $C(2, -3)$, vertices $V_1(-8, -3)$ and $V_2(12, -3)$, and foci $F_1(-4, -3)$ and $F_2(8, -3)$

36. The ellipse with vertices $V_1(-1, -4)$ and $V_2(-1, 6)$ and foci $F_1(-1, -3)$ and $F_2(-1, 5)$

37. The hyperbola with center $C(-1, 4)$, vertices $V_1(-1, -3)$ and $V_2(-1, 11)$, and foci $F_1(-1, -5)$ and $F_2(-1, 13)$

38. The hyperbola with vertices $V_1(-1, -1)$ and $V_2(5, -1)$ and foci $F_1(-4, -1)$ and $F_2(8, -1)$

39. The parabola with vertex $V(-3, 5)$ and directrix $y = 2$

40. The parabola with focus $F(1, 3)$ and directrix $x = 3$

41. The hyperbola with foci $F_1(1, -5)$ and $F_2(1, 5)$ that passes through the point $(1, 4)$

42. The hyperbola with foci $F_1(-2, 2)$ and $F_2(4, 2)$ that passes through the point $(3, 2)$

43. The ellipse with foci $F_1(1, -4)$ and $F_2(5, -4)$ that passes through the point $(3, 1)$

44. The ellipse with foci $F_1(3, -4)$ and $F_2(3, 4)$, and x-intercepts 0 and 6

45. The parabola that passes through the point $(6, 1)$, with vertex $V(-1, 2)$ and horizontal axis of symmetry

46. The parabola that passes through the point $(6, -2)$, with vertex $V(4, -1)$ and vertical axis of symmetry

47–58 ■ Graphing Shifted Conics Complete the square to determine whether the graph of the equation is an ellipse, a parabola, a hyperbola, or a degenerate conic. If the graph is an ellipse, find the center, foci, vertices, and lengths of the major and minor axes. If it is a parabola, find the vertex, focus, and directrix. If it is a hyperbola, find the center, foci, vertices, and asymptotes. Then sketch the graph of the equation. If the equation has no graph, explain why.

47. $y^2 = 4(x + 2y)$

48. $9x^2 - 36x + 4y^2 = 0$

49. $x^2 - 5y^2 - 2x + 20y = 44$

50. $x^2 + 6x + 12y + 9 = 0$

51. $4x^2 + 25y^2 - 24x + 250y + 561 = 0$

52. $2x^2 + y^2 = 2y + 1$

53. $16x^2 - 9y^2 - 96x + 288 = 0$

54. $4x^2 - 4x - 8y + 9 = 0$

55. $x^2 + 16 = 4(y^2 + 2x)$

56. $x^2 - y^2 = 10(x - y) + 1$

57. $3x^2 + 4y^2 - 6x - 24y + 39 = 0$

58. $x^2 + 4y^2 + 20x - 40y + 300 = 0$

59–62 ■ Graphing Shifted Conics Use a graphing device to graph the conic.

59. $2x^2 - 4x + y + 5 = 0$

60. $4x^2 + 9y^2 - 36y = 0$

61. $9x^2 + 36 = y^2 + 36x + 6y$

62. $x^2 - 4y^2 + 4x + 8y = 0$

Skills Plus

63. **Degenerate Conic** Determine what the value of F must be if the graph of the equation

$$4x^2 + y^2 + 4(x - 2y) + F = 0$$

is **(a)** an ellipse, **(b)** a single point, or **(c)** the empty set.

64. **Common Focus and Vertex** Find an equation for the ellipse that shares a vertex and a focus with the parabola

$$x^2 + y = 100$$

and has its other focus at the origin.

65. **Confocal Parabolas** This exercise deals with *confocal* parabolas, that is, families of parabolas that have the same focus.

(a) Draw graphs of the family of parabolas

$$x^2 = 4p(y + p)$$

for $p = -2, -\frac{3}{2}, -1, -\frac{1}{2}, \frac{1}{2}, 1, \frac{3}{2}, 2$.

(b) Show that each parabola in this family has its focus at the origin.

(c) Describe the effect on the graph of moving the vertex closer to the origin.

Applications

66. **Path of a Cannonball** A cannon fires a cannonball as shown in the figure. The path of the cannonball is a parabola with vertex at the highest point of the path. If the cannonball lands 500 m from the cannon and the highest point it reaches is 1000 m above the ground, find an equation for the path of the cannonball. Place the origin at the location of the cannon.

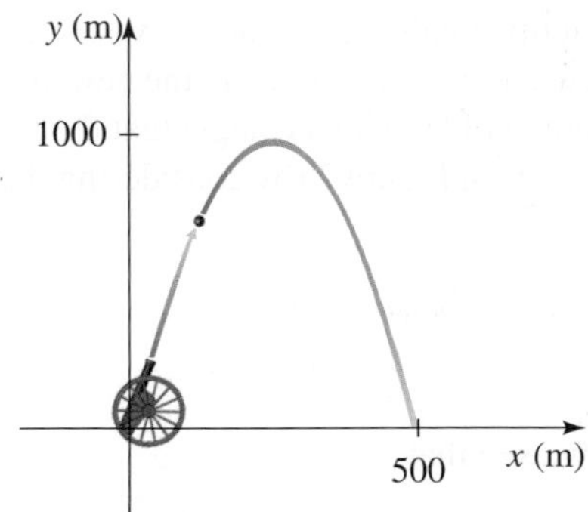

67. **Orbit of a Satellite** A satellite is in an elliptical orbit around the earth with the center of the earth at one focus, as shown in the figure. The height of the satellite above the earth varies between 224 km and 704 km. Assume that the earth is a sphere with radius 6336 km. Find an equation for the path of the satellite with the origin at the center of the earth.

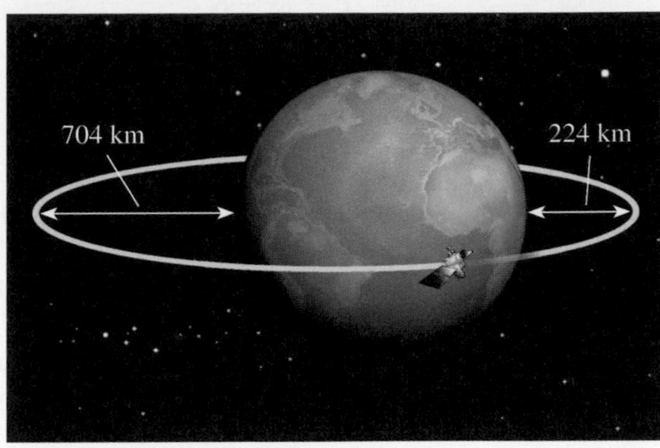

Discuss ■ Discover ■ Prove ■ Write

68. **Discuss: A Family of Confocal Conics** Conics that share a focus are called **confocal**. Consider the family of conics that have a focus at $(0, 1)$ and a vertex at the origin, as shown in the figure.

(a) Find equations of two different ellipses that have these properties.

(b) Find equations of two different hyperbolas that have these properties.

(c) Explain why only one parabola satisfies these properties. Find its equation.

(d) Sketch the conics you found in parts (a), (b), and (c) on the same coordinate axes (for the hyperbolas, sketch the top branches only).

(e) How are the ellipses and hyperbolas related to the parabola?

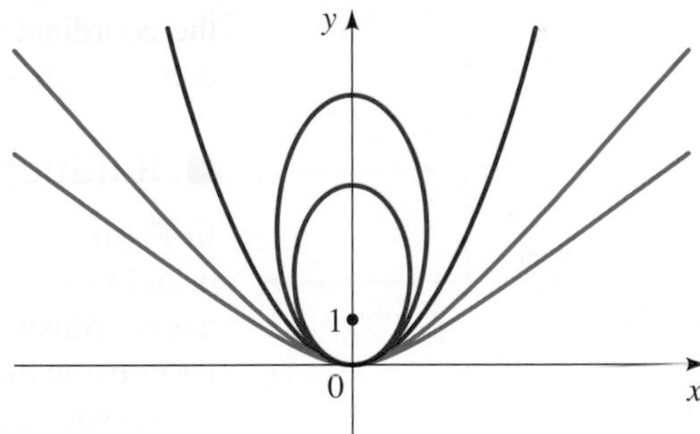

69. **Discuss ■ Discover: Different Forms of a Quadratic Function** We have used several different forms of quadratic functions.

$y = ax^2 + bx + c$	Quadratic function
$y = a(x - r_1)(x - r_2)$	Factored form
$y = a(x - h)^2 + k$	Vertex form
$(x - h)^2 = 4p(y - k)$	Standard form

Any quadratic function can be expressed in each of these equivalent forms. In each case the graph is the same parabola.

(a) Express the quadratic function

$$y = 2x^2 - 8x + 6$$

in each form.

(b) Which form would you use to do each of the following: find the minimum or maximum value of y, find the focus of the parabola, find the real zeros of the function, find the complex zeros of the function.

70. Discuss ■ Prove: Tangent Circles The graph shows two red circles with centers $(-1,0)$ and $(1,0)$ and radii 3 and 5, respectively. Consider the collection of all circles tangent to both of these circles. (Some of these circles are shown in blue.) Show that the centers of all such circles lie on an ellipse with foci $(\pm1,0)$. Find an equation of this ellipse.

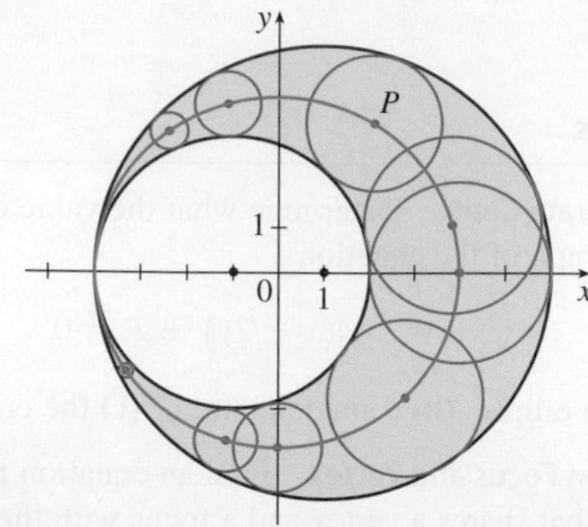

PS *Look for something familiar.* Apply the geometric definition of an ellipse. Observe that for a blue circle with center P and radius r, the distance from one focus to P is $3 + r$ and from the other focus to P is $5 - r$.

10.5 Rotation of Axes

■ Rotation of Axes ■ General Equation of a Conic ■ The Discriminant

In Section 10.4 we studied conics with equations of the form

$$Ax^2 + Cy^2 + Dx + Ey + F = 0$$

We saw that the graph is always an ellipse, parabola, or hyperbola with horizontal or vertical axes (except in the degenerate cases). In this section we study the most general second-degree equation

$$Ax^2 + Bxy + Cy^2 + Dx + Ey + F = 0$$

We will see that the graph of an equation of this form is also a conic. In fact, by rotating the coordinate axes through an appropriate angle, we can eliminate the term Bxy and then use our knowledge of conic sections to analyze the graph.

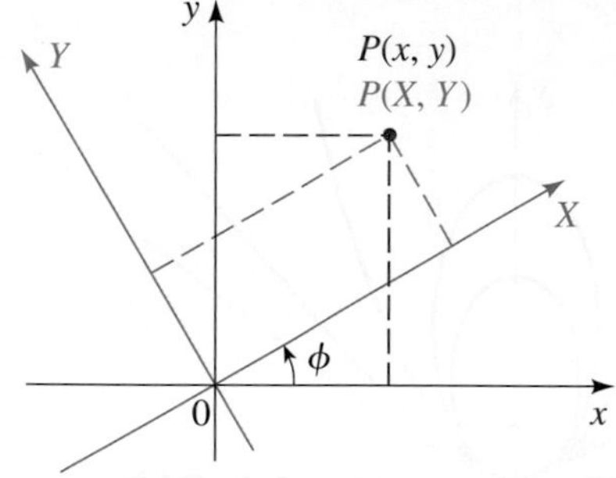

Figure 1

■ Rotation of Axes

In Figure 1 the x- and y-axes have been rotated through an acute angle ϕ about the origin to produce a new pair of axes, which we call the X- and Y-axes. A point P that has coordinates (x, y) in the old system has coordinates (X, Y) in the new system. If we let r denote the distance of P from the origin and let θ be the angle that the segment OP makes with the new X-axis, then we can see from Figure 2 (by considering the two right triangles in the figure) that

$$X = r\cos\theta \qquad Y = r\sin\theta$$

$$x = r\cos(\theta + \phi) \qquad y = r\sin(\theta + \phi)$$

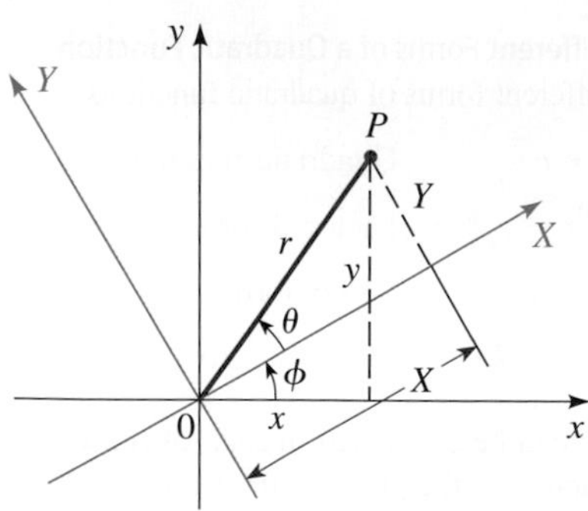

Figure 2

Using the Addition Formula for Cosine, we see that

$$\begin{aligned} x &= r\cos(\theta + \phi) \\ &= r(\cos\theta\cos\phi - \sin\theta\sin\phi) \\ &= (r\cos\theta)\cos\phi - (r\sin\theta)\sin\phi \\ &= X\cos\phi - Y\sin\phi \end{aligned}$$

Similarly, we can apply the Addition Formula for Sine to the expression for y to obtain $y = X \sin \phi + Y \cos \phi$. By treating these equations for x and y as a system of linear equations in the variables X and Y (see Exercise 35), we obtain expressions for X and Y in terms of x and y, as detailed in the following box.

Rotation of Axes Formulas

Suppose the x- and y-axes in a coordinate plane are rotated through the acute angle ϕ to produce the X- and Y-axes, as shown in Figure 1. Then the coordinates (x, y) and (X, Y) of a point in the xy- and the XY-planes are related as follows.

$$x = X \cos \phi - Y \sin \phi \qquad X = x \cos \phi + y \sin \phi$$

$$y = X \sin \phi + Y \cos \phi \qquad Y = -x \sin \phi + y \cos \phi$$

Example 1 ■ Rotation of Axes

If the coordinate axes are rotated through 30°, find the XY-coordinates of the point with xy-coordinates $(2, -4)$.

Solution Using the Rotation of Axes Formulas with $x = 2$, $y = -4$, and $\phi = 30°$, we arrive at

$$X = 2 \cos 30° + (-4) \sin 30° = 2\left(\frac{\sqrt{3}}{2}\right) - 4\left(\frac{1}{2}\right) = \sqrt{3} - 2$$

$$Y = -2 \sin 30° + (-4) \cos 30° = -2\left(\frac{1}{2}\right) - 4\left(\frac{\sqrt{3}}{2}\right) = -1 - 2\sqrt{3}$$

The XY-coordinates are $(-2 + \sqrt{3}, -1 - 2\sqrt{3})$.

 **Now Try Exercise 3** ■

Example 2 ■ Rotating a Hyperbola

Rotate the coordinate axes through 45° to show that the graph of the equation $xy = 2$ is a hyperbola.

Solution We use the Rotation of Axes Formulas with $\phi = 45°$ to obtain

$$x = X \cos 45° - Y \sin 45° = \frac{X}{\sqrt{2}} - \frac{Y}{\sqrt{2}}$$

$$y = X \sin 45° + Y \cos 45° = \frac{X}{\sqrt{2}} + \frac{Y}{\sqrt{2}}$$

Substituting these expressions into the original equation gives

$$\left(\frac{X}{\sqrt{2}} - \frac{Y}{\sqrt{2}}\right)\left(\frac{X}{\sqrt{2}} + \frac{Y}{\sqrt{2}}\right) = 2$$

$$\frac{X^2}{2} - \frac{Y^2}{2} = 2$$

$$\frac{X^2}{4} - \frac{Y^2}{4} = 1$$

We recognize this as a hyperbola with vertices $(\pm 2, 0)$ in the XY-coordinate system. Its asymptotes are $Y = \pm X$, which correspond to the coordinate axes in the xy-system (see Figure 3).

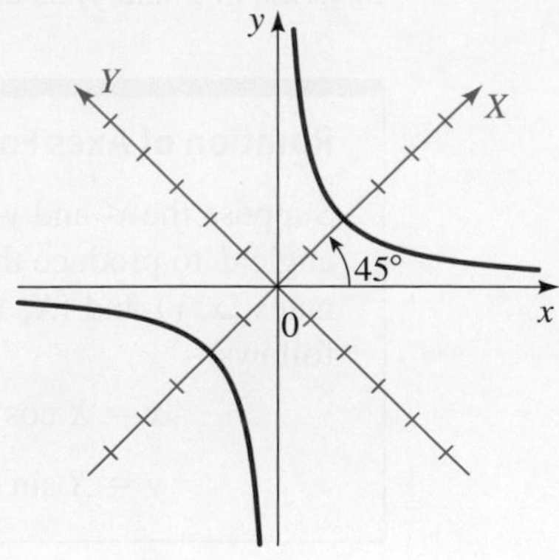

Figure 3 | $xy = 2$

 **Now Try Exercise 11** ■

■ General Equation of a Conic

The method of Example 2 can be used to transform any equation of the form

$$Ax^2 + Bxy + Cy^2 + Dx + Ey + F = 0$$

into an equation in X and Y that doesn't contain an XY-term by choosing an appropriate angle of rotation. To find the angle that works, we rotate the axes through an angle ϕ and substitute for x and y using the Rotation of Axes Formulas.

$$\begin{aligned} A(X\cos\phi - Y\sin\phi)^2 &+ B(X\cos\phi - Y\sin\phi)(X\sin\phi + Y\cos\phi) \\ &+ C(X\sin\phi + Y\cos\phi)^2 + D(X\cos\phi - Y\sin\phi) \\ &+ E(X\sin\phi + Y\cos\phi) + F = 0 \end{aligned}$$

If we expand this and collect like terms, we obtain an equation of the form

$$A'X^2 + B'XY + C'Y^2 + D'X + E'Y + F' = 0$$

where

$$\begin{aligned} A' &= A\cos^2\phi + B\sin\phi\cos\phi + C\sin^2\phi \\ B' &= 2(C - A)\sin\phi\cos\phi + B(\cos^2\phi - \sin^2\phi) \\ C' &= A\sin^2\phi - B\sin\phi\cos\phi + C\cos^2\phi \\ D' &= D\cos\phi + E\sin\phi \\ E' &= -D\sin\phi + E\cos\phi \\ F' &= F \end{aligned}$$

To eliminate the XY-term, we would like to choose ϕ so that $B' = 0$, that is,

Double-Angle Formulas

$\sin 2\phi = 2\sin\phi\cos\phi$

$\cos 2\phi = \cos^2\phi - \sin^2\phi$

$$2(C - A)\sin\phi\cos\phi + B(\cos^2\phi - \sin^2\phi) = 0$$

$$(C - A)\sin 2\phi + B\cos 2\phi = 0 \qquad \text{Double-Angle Formulas for Sine and Cosine}$$

$$B\cos 2\phi = (A - C)\sin 2\phi$$

$$\cot 2\phi = \frac{A - C}{B} \qquad \text{Divide by } B\sin 2\phi$$

The preceding calculation proves the following theorem.

Simplifying the General Conic Equation

To eliminate the xy-term in the general conic equation

$$Ax^2 + Bxy + Cy^2 + Dx + Ey + F = 0$$

rotate the axes through the acute angle ϕ that satisfies

$$\cot 2\phi = \frac{A - C}{B}$$

Example 3 ■ Eliminating the *xy*-Term

Use a rotation of axes to eliminate the xy-term in the equation

$$6\sqrt{3}x^2 + 6xy + 4\sqrt{3}y^2 = 21\sqrt{3}$$

Identify and sketch the curve.

Solution To eliminate the xy-term, we rotate the axes through an angle ϕ that satisfies

$$\cot 2\phi = \frac{A - C}{B} = \frac{6\sqrt{3} - 4\sqrt{3}}{6} = \frac{\sqrt{3}}{3}$$

Thus $2\phi = 60°$, and hence $\phi = 30°$. With this value of ϕ we get

$$x = X\left(\frac{\sqrt{3}}{2}\right) - Y\left(\frac{1}{2}\right) \qquad \text{Rotation of Axes Formulas}$$

$$y = X\left(\frac{1}{2}\right) + Y\left(\frac{\sqrt{3}}{2}\right) \qquad \cos\phi = \frac{\sqrt{3}}{2},\ \sin\phi = \frac{1}{2}$$

Substituting these values for x and y into the given equation leads to

$$6\sqrt{3}\left(\frac{X\sqrt{3}}{2} - \frac{Y}{2}\right)^2 + 6\left(\frac{X\sqrt{3}}{2} - \frac{Y}{2}\right)\left(\frac{X}{2} + \frac{Y\sqrt{3}}{2}\right) + 4\sqrt{3}\left(\frac{X}{2} + \frac{Y\sqrt{3}}{2}\right)^2 = 21\sqrt{3}$$

Expanding and collecting like terms, we get

$$7\sqrt{3}X^2 + 3\sqrt{3}Y^2 = 21\sqrt{3}$$

$$\frac{X^2}{3} + \frac{Y^2}{7} = 1 \qquad \text{Divide by } 21\sqrt{3}$$

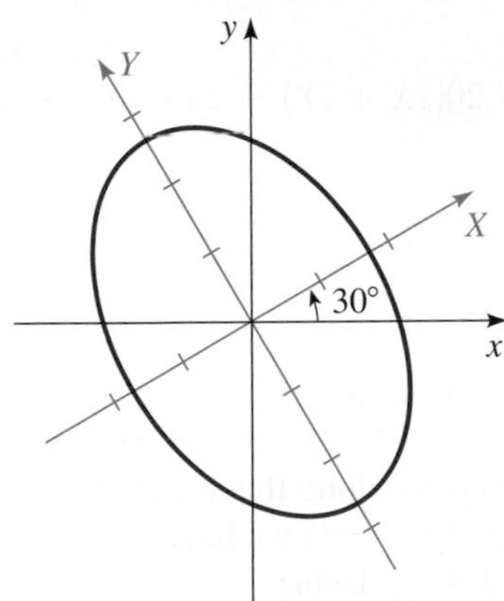

Figure 4 | $6\sqrt{3}x^2 + 6xy + 4\sqrt{3}y^2 = 21\sqrt{3}$

This is the equation of an ellipse in the XY-coordinate system. The foci lie on the Y-axis. Because $a^2 = 7$ and $b^2 = 3$, the length of the major axis is $2\sqrt{7}$, and the length of the minor axis is $2\sqrt{3}$. The ellipse is sketched in Figure 4.

Now Try Exercise 17 ■

Note In the preceding example we were able to determine ϕ without difficulty because we remembered that $\cot 60° = \sqrt{3}/3$. In general, finding ϕ is not quite so easy. The next example illustrates how the following Half-Angle Formulas, which are valid for $0 < \phi < \pi/2$, are useful in determining ϕ. (See Section 7.3.)

$$\cos\phi = \sqrt{\frac{1 + \cos 2\phi}{2}} \qquad \sin\phi = \sqrt{\frac{1 - \cos 2\phi}{2}}$$

Example 4 ■ Graphing a Rotated Conic

A conic has the equation

$$64x^2 + 96xy + 36y^2 - 15x + 20y - 25 = 0$$

(a) Use a rotation of axes to eliminate the xy-term.

(b) Identify and sketch the graph.

(c) Draw the graph using a graphing device.

Solution

(a) To eliminate the xy-term, we rotate the axes through an angle ϕ that satisfies

$$\cot 2\phi = \frac{A - C}{B} = \frac{64 - 36}{96} = \frac{7}{24}$$

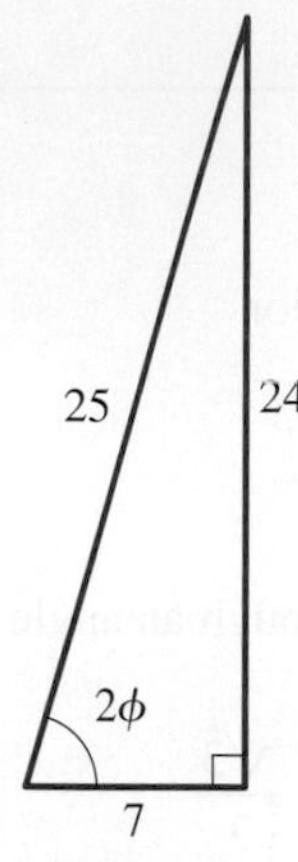

Figure 5

In Figure 5 we sketch a triangle with $\cot 2\phi = \frac{7}{24}$. We see that

$$\cos 2\phi = \tfrac{7}{25}$$

so, using the Half-Angle Formulas, we get

$$\cos \phi = \sqrt{\frac{1 + \frac{7}{25}}{2}} = \sqrt{\frac{16}{25}} = \frac{4}{5}$$

$$\sin \phi = \sqrt{\frac{1 - \frac{7}{25}}{2}} = \sqrt{\frac{9}{25}} = \frac{3}{5}$$

The Rotation of Axes Formulas then give

$$x = \tfrac{4}{5}X - \tfrac{3}{5}Y \qquad \text{and} \qquad y = \tfrac{3}{5}X + \tfrac{4}{5}Y$$

Substituting into the given equation, we have

$$64\left(\tfrac{4}{5}X - \tfrac{3}{5}Y\right)^2 + 96\left(\tfrac{4}{5}X - \tfrac{3}{5}Y\right)\left(\tfrac{3}{5}X + \tfrac{4}{5}Y\right)$$

$$+ 36\left(\tfrac{3}{5}X + \tfrac{4}{5}Y\right)^2 - 15\left(\tfrac{4}{5}X - \tfrac{3}{5}Y\right) + 20\left(\tfrac{3}{5}X + \tfrac{4}{5}Y\right) - 25 = 0$$

Expanding and collecting like terms, we get

$$100X^2 + 25Y - 25 = 0$$

$$4X^2 = -Y + 1 \qquad \text{Simplify}$$

$$X^2 = -\tfrac{1}{4}(Y - 1) \qquad \text{Divide by 4}$$

(b) We recognize this as the equation of a parabola that opens along the negative Y-axis and has vertex $(0, 1)$ in XY-coordinates. Since $4p = -\frac{1}{4}$, we have $p = -\frac{1}{16}$, so the focus is $\left(0, \frac{15}{16}\right)$ and the directrix is $Y = \frac{17}{16}$. Using

$$\phi = \cos^{-1}\left(\tfrac{4}{5}\right) \approx 37°$$

Discovery Project ■ Computer Graphics III

An image on a computer screen is stored in the computer memory as a large matrix. In the *Discovery Projects* Computer Graphics I and II, we used matrix operations to transform an image—adjust contrast, stretch, shrink, reflect, or shear. But rotating an image requires knowledge of the rotation formulas in this section. In this project we express the rotation formulas in matrix form (see Exercise 10.5.37) and experiment with using rotation matrices to rotate an image. You can find the project at **www.stewartmath.com**.

we sketch the graph in Figure 6(a).

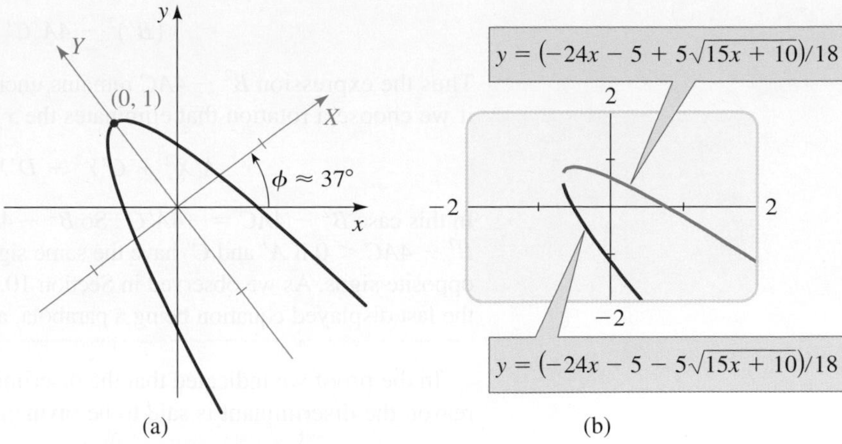

Figure 6 | $64x^2 + 96xy + 36y^2 - 15x + 20y - 25 = 0$

(c) Most graphing devices can draw the graph of this equation, as shown in Figure 6(b).

Using a Graphing Calculator To graph the equation we first solve for y. Writing the equation in the form

$$36y^2 + (96x + 20)y + (64x^2 - 15x - 25) = 0$$

we see that this equation is a quadratic equation in y, so we can use the Quadratic Formula (as in Example 10.4.4) to solve for y. You can check that we get

$$y = (-24x - 5 + 5\sqrt{15x + 10})/18 \qquad \text{and} \qquad y = (-24x - 5 - 5\sqrt{15x + 10})/18$$

The graph of the parabola in Figure 6(b) is obtained by graphing both functions.

 **Now Try Exercise 23** ■

■ The Discriminant

In Examples 3 and 4 we were able to identify the type of conic by rotating the axes. The next theorem gives rules for identifying the type of conic directly from the equation, without rotating axes.

Identifying Conics by the Discriminant

The graph of the equation

$$Ax^2 + Bxy + Cy^2 + Dx + Ey + F = 0$$

is either a conic or a degenerate conic. In the nondegenerate cases the graph is

1. a parabola if $B^2 - 4AC = 0$,
2. an ellipse if $B^2 - 4AC < 0$,
3. a hyperbola if $B^2 - 4AC > 0$.

The quantity $B^2 - 4AC$ is the **discriminant** of the equation.

Proof If we rotate the axes through an angle ϕ, we get an equation of the form

$$A'X^2 + B'XY + C'Y^2 + D'X + E'Y + F' = 0$$

where A', B', C', . . . are given by the formulas in this section. A straightforward calculation shows that

$$(B')^2 - 4A'C' = B^2 - 4AC$$

Thus the expression $B^2 - 4AC$ remains unchanged for any rotation. In particular, if we choose a rotation that eliminates the xy-term ($B' = 0$), we get

$$A'X^2 + C'Y^2 + D'X + E'Y + F' = 0$$

In this case $B^2 - 4AC = -4A'C'$. So $B^2 - 4AC = 0$ if either A' or C' is zero; $B^2 - 4AC < 0$ if A' and C' have the same sign; and $B^2 - 4AC > 0$ if A' and C' have opposite signs. As we observed in Section 10.4, these cases correspond to the graph of the last displayed equation being a parabola, an ellipse, or a hyperbola, respectively. ■

In the proof we indicated that the discriminant is unchanged by any rotation; for this reason the discriminant is said to be **invariant** under rotation.

Example 5 ■ Identifying a Conic by the Discriminant

A conic has the equation

$$3x^2 + 5xy - 2y^2 + x - y + 4 = 0$$

(a) Use the discriminant to identify the conic.

(b) Confirm your answer to part (a) by graphing the conic with a graphing device.

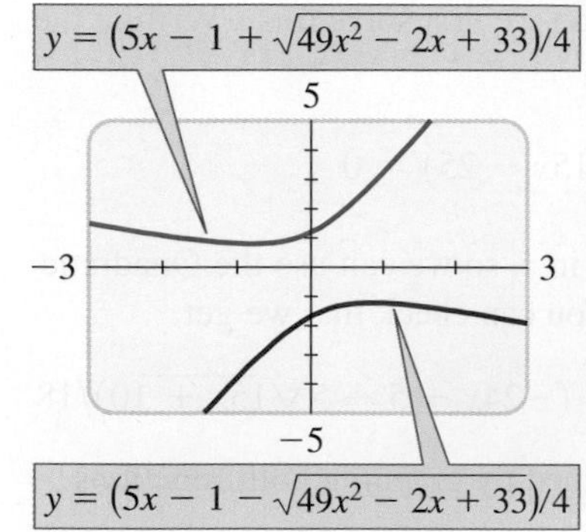

Figure 7

Solution

(a) Since $A = 3$, $B = 5$, and $C = -2$, the discriminant is

$$B^2 - 4AC = 5^2 - 4(3)(-2) = 49 > 0$$

So the conic is a hyperbola.

(b) Using a graphing device, we obtain the graph shown in Figure 7. The graph confirms that the equation represents a hyperbola.

 **Now Try Exercise 29** ■

10.5 | Exercises

■ Concepts

1. Suppose the x- and y-axes are rotated through an acute angle ϕ to produce the new X- and Y-axes. A point P in the plane can be described by its xy-coordinates (x, y) or its XY-coordinates (X, Y). These coordinates are related by the following formulas.

$x =$ ______________ $X =$ ______________

$y =$ ______________ $Y =$ ______________

2. Consider the equation

$$Ax^2 + Bxy + Cy^2 + Dx + Ey + F = 0$$

(a) In general, the graph of this equation is a ____________.

(b) To eliminate the xy-term from this equation, we rotate the axes through an angle ϕ that satisfies

$\cot 2\phi =$ ____________.

(c) The discriminant of this equation is ____________.

If the discriminant is 0, the graph is ____________; if it is negative, the graph is ____________; and if it is positive, the graph is ____________.

■ Skills

3–8 ■ Rotation of Axes Determine the XY-coordinates of the given point if the coordinate axes are rotated through the indicated angle.

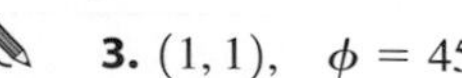

3. $(1, 1)$, $\phi = 45°$

4. $(-2, 1)$, $\phi = 30°$

5. $(3, -\sqrt{3})$, $\phi = 60°$

6. $(2, 0)$, $\phi = 15°$

7. $(0, 2)$, $\phi = 55°$

8. $(\sqrt{2}, 4\sqrt{2})$, $\phi = 45°$

9–14 ■ Finding the Equation for a Rotated Conic Determine the equation of the given conic in *XY*-coordinates when the coordinate axes are rotated through the indicated angle.

9. $x^2 - 3y^2 = 4, \quad \phi = 60°$

10. $y = (x - 1)^2, \quad \phi = 45°$

11. $x^2 - y^2 = 2y, \quad \phi = \cos^{-1}\left(\frac{3}{5}\right)$

12. $x^2 + 2y^2 = 16, \quad \phi = \sin^{-1}\left(\frac{3}{5}\right)$

13. $x^2 + 2\sqrt{3}xy - y^2 = 4, \quad \phi = 30°$

14. $xy = x + y, \quad \phi = \pi/4$

15–28 ■ Graphing a Rotated Conic **(a)** Use the discriminant to determine whether the graph of the equation is a parabola, an ellipse, or a hyperbola. **(b)** Use a rotation of axes to eliminate the *xy*-term. **(c)** Sketch the graph.

15. $xy = 8$

16. $xy + 4 = 0$

17. $x^2 + 2\sqrt{3}xy - y^2 + 2 = 0$

18. $13x^2 + 6\sqrt{3}xy + 7y^2 = 16$

19. $11x^2 - 24xy + 4y^2 + 20 = 0$

20. $21x^2 + 10\sqrt{3}xy + 31y^2 = 144$

21. $\sqrt{3}x^2 + 3xy = 3$

22. $153x^2 + 192xy + 97y^2 = 225$

23. $x^2 + 2xy + y^2 + x - y = 0$

24. $25x^2 - 120xy + 144y^2 - 156x - 65y = 0$

25. $2\sqrt{3}x^2 - 6xy + \sqrt{3}x + 3y = 0$

26. $9x^2 - 24xy + 16y^2 = 100(x - y - 1)$

27. $52x^2 + 72xy + 73y^2 = 40x - 30y + 75$

28. $(7x + 24y)^2 = 600x - 175y + 25$

29–32 ■ Identifying a Conic from Its Discriminant **(a)** Use the discriminant to identify the conic. **(b)** Confirm your answer by graphing the conic using a graphing device.

29. $2x^2 - 4xy + 2y^2 - 5x - 5 = 0$

30. $x^2 - 2xy + 3y^2 = 8$

31. $6x^2 + 10xy + 3y^2 - 6y = 36$

32. $9x^2 - 6xy + y^2 + 6x - 2y = 0$

■ Skills Plus

33. Identifying a Hyperbola Using Rotation of Axes

(a) Use rotation of axes to show that the following equation represents a hyperbola.

$$7x^2 + 48xy - 7y^2 - 200x - 150y + 600 = 0$$

(b) Find the *XY*- and *xy*-coordinates of the center, vertices, and foci.

(c) Find the equations of the asymptotes in *XY*- and *xy*-coordinates.

34. Identifying a Parabola Using Rotation of Axes

(a) Use rotation of axes to show that the following equation represents a parabola.

$$2\sqrt{2}(x + y)^2 = 7x + 9y$$

(b) Find the *XY*- and *xy*-coordinates of the vertex and focus.

(c) Find the equation of the directrix in *XY*- and *xy*-coordinates.

35. Rotation of Axes Formulas Solve the equations

$$x = X\cos\phi - Y\sin\phi$$

$$y = X\sin\phi + Y\cos\phi$$

for X and Y in terms of x and y. [*Hint:* To begin, multiply the first equation by $\cos\phi$ and the second by $\sin\phi$, and then add the two equations to solve for X.]

36. Graphing an Equation Using Rotation of Axes Show that the graph of the equation

$$\sqrt{x} + \sqrt{y} = 1$$

is part of a parabola by rotating the axes through an angle of 45°. [*Hint:* First convert the equation to one that does not involve radicals.]

■ Discuss ■ Discover ■ Prove ■ Write

37. Prove: Matrix Form of Rotation of Axes Formulas Let Z, Z', and R be the matrices

$$Z = \begin{bmatrix} x \\ y \end{bmatrix} \qquad Z' = \begin{bmatrix} X \\ Y \end{bmatrix}$$

$$R = \begin{bmatrix} \cos\phi & -\sin\phi \\ \sin\phi & \cos\phi \end{bmatrix}$$

(a) Show that the Rotation of Axes Formulas can be written as

$$Z = RZ' \qquad \text{and} \qquad Z' = R^{-1}Z$$

(b) Let R_1 and R_2 be matrices that represent rotations through the angles ϕ_1 and ϕ_2, respectively. Show that the product matrix R_1R_2 represents a rotation through an angle $\phi_1 + \phi_2$. [*Hint:* Use the Addition Formulas for Sine and Cosine to simplify the entries of the matrix R_1R_2.]

38. Prove: Algebraic Invariants A quantity is invariant under rotation if it does not change when the axes are rotated. It was stated in the text that for the general equation of a conic the quantity $B^2 - 4AC$ is invariant under rotation.

(a) Use the formulas for A', B', and C' in this section to prove that the quantity $B^2 - 4AC$ is invariant under rotation; that is, show that

$$B^2 - 4AC = B'^2 - 4A'C'$$

(b) Prove that $A + C$ is invariant under rotation.

(c) Is the quantity F invariant under rotation?

39. Discover ■ Prove: Geometric Invariants Do you expect that the distance between two points is invariant under rotation? Prove your answer by comparing the distance $d(P, Q)$ and $d(P', Q')$ where P' and Q' are the images of P and Q under a rotation of axes.

10.6 Polar Equations of Conics

■ A Unified Geometric Description of Conics ■ Polar Equations of Conics

■ A Unified Geometric Description of Conics

Earlier in this chapter, we defined a parabola in terms of a focus and directrix, but we defined the ellipse and hyperbola in terms of two foci. In this section we give a more unified treatment of all three types of conics in terms of a focus and directrix. If we place one focus at the origin, then a conic section has a simple polar equation. Moreover, in polar form, rotation of conics becomes a simple matter. Polar equations of ellipses are crucial in the derivation of Kepler's Laws (see Section 10.4).

Equivalent Description of Conics

Let F be a fixed point (the **focus**), ℓ a fixed line (the **directrix**), and let e be a fixed positive number (the **eccentricity**). The set of all points P such that the ratio of the distance from P to F to the distance from P to ℓ is the constant e is a conic. That is, the set of all points P such that

$$\frac{d(P, F)}{d(P, \ell)} = e$$

is a conic. The conic is a parabola if $e = 1$, an ellipse if $e < 1$, or a hyperbola if $e > 1$.

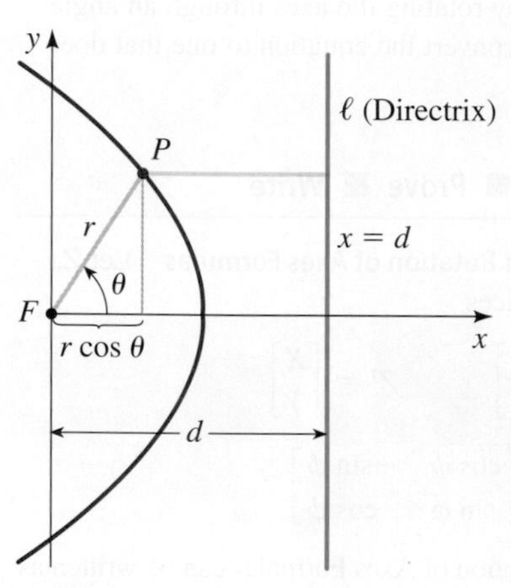

Figure 1

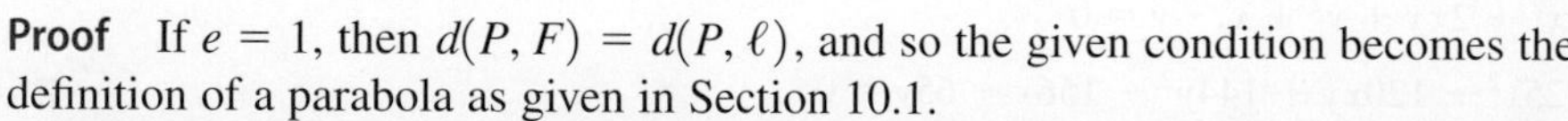

Proof If $e = 1$, then $d(P, F) = d(P, \ell)$, and so the given condition becomes the definition of a parabola as given in Section 10.1.

Now, suppose $e \neq 1$. Let's place the focus F at the origin and the directrix parallel to the y-axis and d units to the right. In this case the directrix has equation $x = d$ and is perpendicular to the polar axis. If the point P has polar coordinates (r, θ), we see from Figure 1 that $d(P, F) = r$ and $d(P, \ell) = d - r\cos\theta$. Thus the condition $d(P, F)/d(P, \ell) = e$, or $d(P, F) = e \cdot d(P, \ell)$, becomes

$$r = e(d - r\cos\theta)$$

If we square both sides of this polar equation and convert to rectangular coordinates, we get

$$x^2 + y^2 = e^2(d - x)^2$$

$$(1 - e^2)x^2 + 2de^2x + y^2 = e^2d^2 \qquad \text{Expand and simplify}$$

$$\left(x + \frac{e^2d}{1 - e^2}\right)^2 + \frac{y^2}{1 - e^2} = \frac{e^2d^2}{(1 - e^2)^2} \qquad \text{Divide by } 1 - e^2 \text{ and complete the square}$$

If $e < 1$, then dividing both sides of this equation by $e^2d^2/(1 - e^2)^2$ gives an equation of the form

$$\frac{(x - h)^2}{a^2} + \frac{y^2}{b^2} = 1$$

where $h = \dfrac{-e^2d}{1 - e^2}$ $\quad a^2 = \dfrac{e^2d^2}{(1 - e^2)^2}$ $\quad b^2 = \dfrac{e^2d^2}{1 - e^2}$

This is the equation of an ellipse with center $(h, 0)$. In Section 10.2 we found that the foci of an ellipse are a distance c from the center, where $c^2 = a^2 - b^2$. In our case

$$c^2 = a^2 - b^2 = \frac{e^4d^2}{(1 - e^2)^2}$$

Thus $c = e^2d/(1 - e^2) = -h$, which confirms that the focus defined in the theorem (namely the origin) is the same as the focus defined in Section 10.2. It also follows that

$$e = \frac{c}{a}$$

If $e > 1$, a similar proof shows that the conic is a hyperbola with $e = c/a$, where $c^2 = a^2 + b^2$. ■

■ Polar Equations of Conics

In the proof we saw that the polar equation of the conic in Figure 1 is $r = e(d - r\cos\theta)$. Solving for r, we get

$$r = \frac{ed}{1 + e\cos\theta}$$

If the directrix is chosen to be to the *left* of the focus ($x = -d$), then we get the equation $r = ed/(1 - e\cos\theta)$. If the directrix is *parallel* to the polar axis ($y = d$ or $y = -d$), then we get $\sin\theta$ instead of $\cos\theta$ in the equation. These observations are summarized in the following box and in Figure 2.

Polar Equations of Conics

A polar equation of the form

$$r = \frac{ed}{1 \pm e\cos\theta} \quad \text{or} \quad r = \frac{ed}{1 \pm e\sin\theta}$$

represents a conic with one focus at the origin and with eccentricity e. The conic is

1. a parabola if $e = 1$,
2. an ellipse if $0 < e < 1$,
3. a hyperbola if $e > 1$.

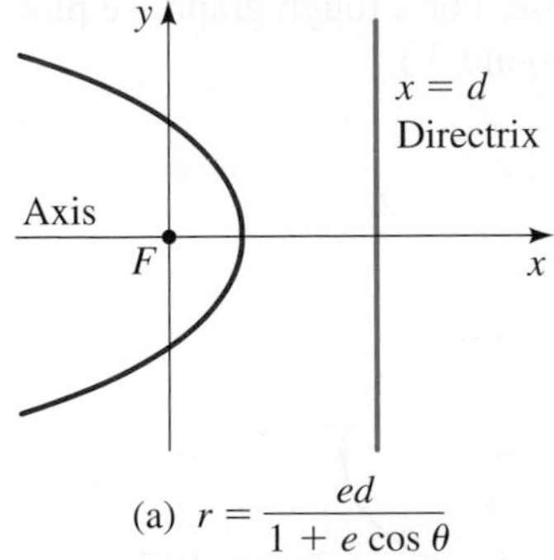

(a) $r = \dfrac{ed}{1 + e\cos\theta}$

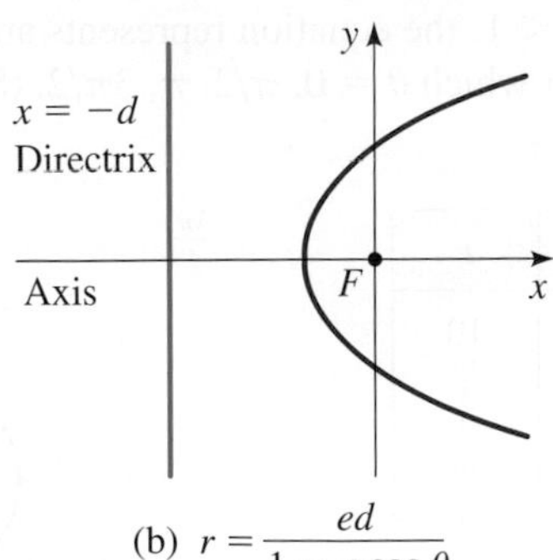

(b) $r = \dfrac{ed}{1 - e\cos\theta}$

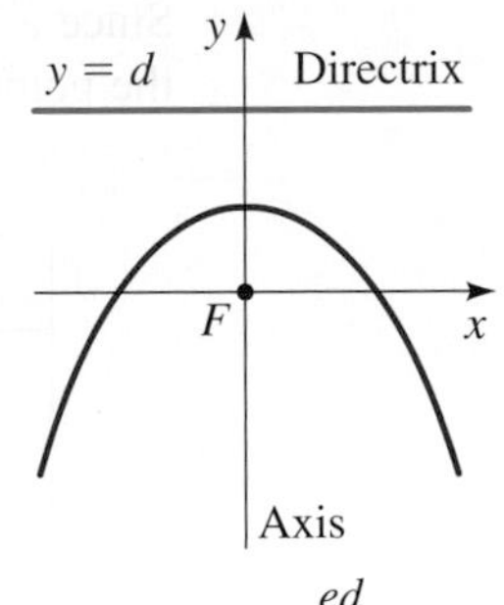

(c) $r = \dfrac{ed}{1 + e\sin\theta}$

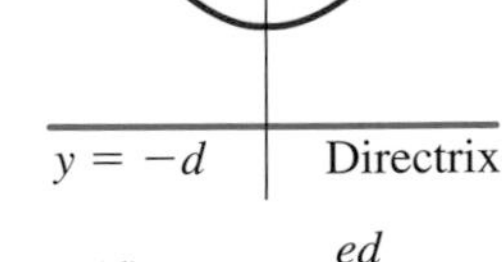

(d) $r = \dfrac{ed}{1 - e\sin\theta}$

Figure 2 | The form of the polar equation of a conic indicates the location of the directrix.

To graph the polar equation of a conic, we first determine the location of the directrix from the form of the equation. The four cases that arise are shown in Figure 2. (The figure shows only the parts of the graphs that are close to the focus at the origin. The shape of the rest of the graph depends on whether the equation represents a parabola, an ellipse, or a hyperbola.) The axis of a conic is perpendicular to the directrix—specifically we have the following:

1. For a parabola the axis of symmetry is perpendicular to the directrix.
2. For an ellipse the major axis is perpendicular to the directrix.
3. For a hyperbola the transverse axis is perpendicular to the directrix.

Example 1 ■ Finding a Polar Equation for a Conic

Find a polar equation for the parabola that has its focus at the origin and whose directrix is the line $y = -6$.

Solution Using $e = 1$ and $d = 6$ and using part (d) of Figure 2, we see that the polar equation of the parabola is

$$r = \frac{6}{1 - \sin\theta}$$

 Now Try Exercise 3 ■

Note To graph a polar conic, it is helpful to plot the points for which $\theta = 0, \pi/2, \pi$, and $3\pi/2$. Using these points and a knowledge of the type of conic (which we obtain from the eccentricity), we get a rough idea of the shape and location of the graph.

Example 2 ■ Identifying and Sketching a Conic

A conic is given by the polar equation

$$r = \frac{10}{3 - 2\cos\theta}$$

(a) Show that the conic is an ellipse, and sketch its graph.

(b) Find the center of the ellipse and the lengths of the major and minor axes.

Solution

(a) To put the polar equation into one of the forms shown in Figure 2, we divide the numerator and denominator by 3:

$$r = \frac{ed}{1 - e\cos\theta}$$

$$r = \frac{\frac{10}{3}}{1 - \frac{2}{3}\cos\theta}$$

Since $e = \frac{2}{3} < 1$, the equation represents an ellipse. For a rough graph we plot the points for which $\theta = 0, \pi/2, \pi, 3\pi/2$. (See Figure 3.)

θ	r
0	10
$\frac{\pi}{2}$	$\frac{10}{3}$
π	2
$\frac{3\pi}{2}$	$\frac{10}{3}$

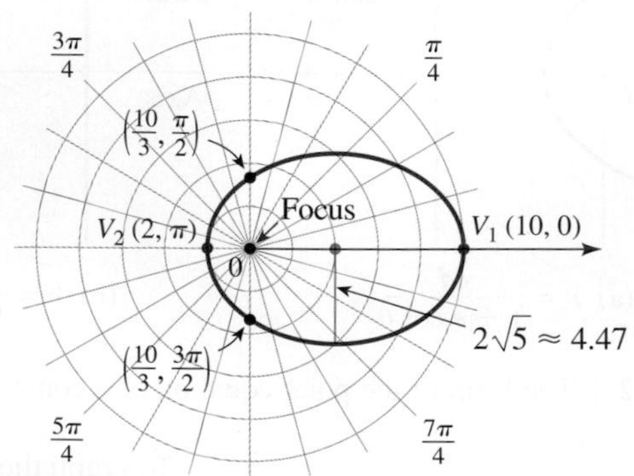

Figure 3 | $r = \dfrac{10}{3 - 2\cos\theta}$

(b) Comparing the equation to the four equations given in Figure 2, we see that the major axis is horizontal. Thus the endpoints of the major axis are $V_1(10, 0)$ and $V_2(2, \pi)$. So the center of the ellipse is at $C(4, 0)$, the midpoint of V_1V_2.

The distance between the vertices V_1 and V_2 is 12; thus the length of the major axis is $2a = 12$, so $a = 6$. To determine the length of the minor axis, we need to find b. From the definition of eccentricity in this section, we have $c = ae = 6\left(\frac{2}{3}\right) = 4$, so

$$b^2 = a^2 - c^2 = 6^2 - 4^2 = 20$$

Thus $b = \sqrt{20} = 2\sqrt{5} \approx 4.47$, and the length of the minor axis is $2b = 4\sqrt{5} \approx 8.94$.

 **Now Try Exercises 17 and 21** ■

Paths of Comets

The path of a comet is an ellipse, a parabola, or a hyperbola with the sun at a focus. This fact can be proved by using calculus and Newton's Laws of Motion.* If the path is a parabola or a hyperbola, the comet will never return. If the path is an ellipse, it can be determined precisely when and where the comet can be seen again. Halley's comet has an elliptical path and returns every 75 years; it was last seen in 1987. The brightest comet of the 20th century was comet Hale-Bopp, seen in 1997. Its orbit is a very eccentric ellipse; it is expected to return to the inner solar system around the year 4385.

*James Stewart, Daniel Clegg, and Saleem Watson, *Calculus: Early Transcendentals*, 9th ed. (Boston, MA: Cengage, 2021), pages 921 and 925.

Example 3 ■ Identifying and Sketching a Conic

A conic is given by the polar equation

$$r = \frac{12}{2 + 4\sin\theta}$$

(a) Show that the conic is a hyperbola, and sketch its graph.

(b) Find the center of the hyperbola, and sketch the asymptotes.

Solution

(a) Dividing the numerator and denominator by 2, we have

$$r = \frac{6}{1 + 2\sin\theta}$$

Since $e = 2 > 1$, the equation represents a hyperbola. For a rough graph we plot the points for which $\theta = 0, \pi/2, \pi, 3\pi/2$. (See Figure 4.)

(b) Comparing the equation to the four equations given in Figure 2, we see that the transverse axis is vertical. Thus the endpoints of the transverse axis (the vertices of the hyperbola) are $V_1(2, \pi/2)$ and $V_2(-6, 3\pi/2) = V_2(6, \pi/2)$. So the center of the hyperbola is $C(4, \pi/2)$, the midpoint of V_1V_2.

To sketch the asymptotes, we need to find a and b. The distance between V_1 and V_2 is 4; thus the length of the transverse axis is $2a = 4$, so $a = 2$. To find b, we first find c. From the definition of eccentricity in this section, we have $c = ae = 2 \cdot 2 = 4$, so

$$b^2 = c^2 - a^2 = 4^2 - 2^2 = 12$$

Thus $b = \sqrt{12} = 2\sqrt{3} \approx 3.46$. Knowing a and b allows us to sketch the central box, from which we obtain the asymptotes shown in Figure 4.

θ	r
0	6
$\frac{\pi}{2}$	2
π	6
$\frac{3\pi}{2}$	-6

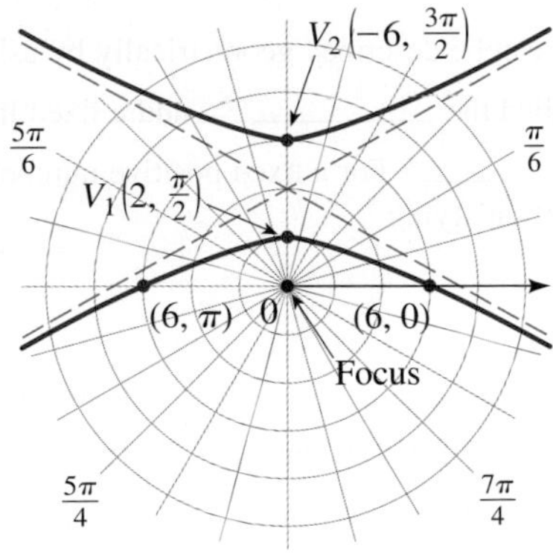

Figure 4 | $r = \dfrac{12}{2 + 4\sin\theta}$

Now Try Exercise 25 ■

To rotate conic sections, it is more convenient to use polar equations than Cartesian equations. We observe that the graph of $r = f(\theta - \alpha)$ is the graph of $r = f(\theta)$ rotated counterclockwise about the origin through an angle α (see Exercise 8.2.65).

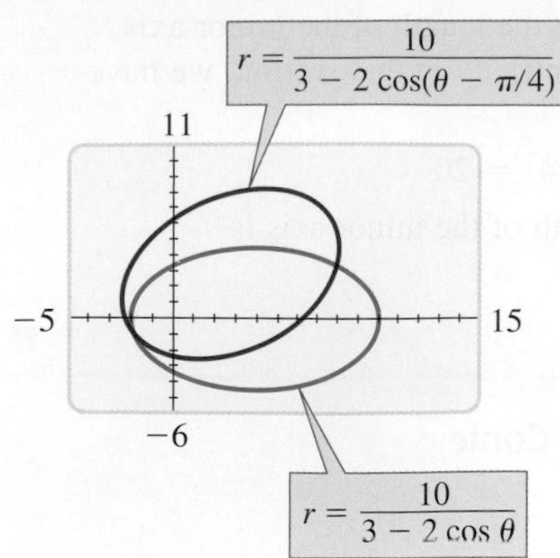

Figure 5

Example 4 ■ Rotating an Ellipse

Suppose the ellipse of Example 2 is rotated through an angle $\pi/4$ about the origin. Find a polar equation for the resulting ellipse, and draw its graph.

Solution We get the equation of the rotated ellipse by replacing θ with $\theta - \pi/4$ in the equation given in Example 2. So the new equation is

$$r = \frac{10}{3 - 2\cos(\theta - \pi/4)}$$

We use this equation to graph the rotated ellipse in Figure 5. Notice that the ellipse has been rotated about the focus at the origin.

Now Try Exercise 37 ■

In Figure 6 we use a computer to sketch a number of conics to demonstrate the effect of varying the eccentricity e. Notice that when e is close to 0, the ellipse is nearly circular, and it becomes more elongated as e increases. When $e = 1$, of course, the conic is a parabola. As e increases beyond 1, the conic is an ever steeper hyperbola.

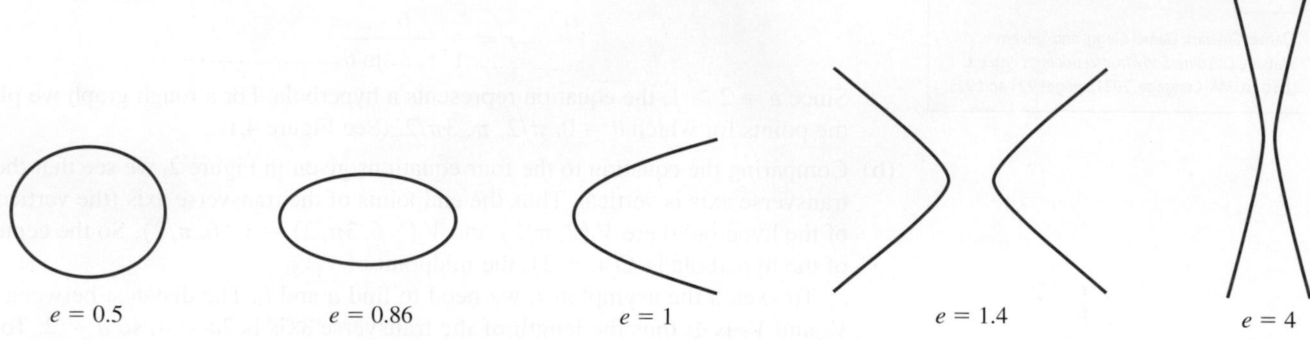

Figure 6

10.6 Exercises

■ Concepts

1. All conics can be described geometrically by using a fixed point F called the ______ and a fixed line ℓ called the ______. For a fixed positive number e the set of all points P satisfying

$$\frac{\boxed{}}{\boxed{}} = e$$

is a ______. If $e = 1$, the conic is a(n) ______; if $e < 1$, the conic is a(n) ______; and if $e > 1$, the conic is a(n) ______. The number e is called the ______ of the conic.

2. The polar equation of a conic with eccentricity e has one of the following forms:

$r =$ ______ or $r =$ ______

■ Skills

3–10 ■ Finding a Polar Equation for a Conic Write a polar equation of a conic that has its focus at the origin and satisfies the given conditions.

3. Ellipse, eccentricity $\frac{2}{3}$, directrix $x = 3$

4. Hyperbola, eccentricity $\frac{4}{3}$, directrix $x = -3$

5. Parabola, directrix $y = 2$

6. Ellipse, eccentricity $\frac{1}{2}$, directrix $y = -4$

7. Hyperbola, eccentricity 4, directrix $r = 5 \sec\theta$

8. Ellipse, eccentricity 0.6, directrix $r = 2 \csc\theta$

9. Parabola, vertex at $(5, \pi/2)$

10. Ellipse, eccentricity 0.4, vertex at $(2, 0)$

11–16 ■ Graphs of Polar Equations of Conics Match the polar equations with the graphs labeled I–VI. Give reasons for your answers.

11. $r = \dfrac{6}{1 + \cos\theta}$ **12.** $r = \dfrac{2}{2 - \cos\theta}$

13. $r = \dfrac{3}{1 - 2\sin\theta}$ **14.** $r = \dfrac{5}{3 - 3\sin\theta}$

15. $r = \dfrac{12}{3 + 2\sin\theta}$ **16.** $r = \dfrac{12}{2 + 3\cos\theta}$

I

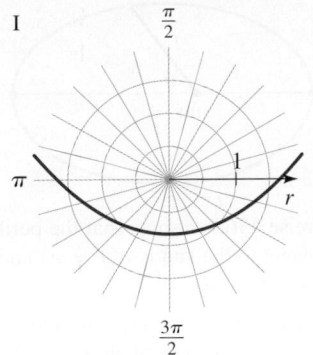

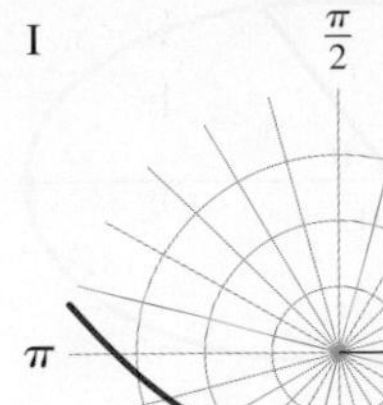

II

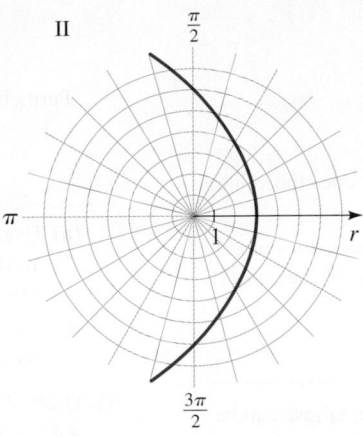

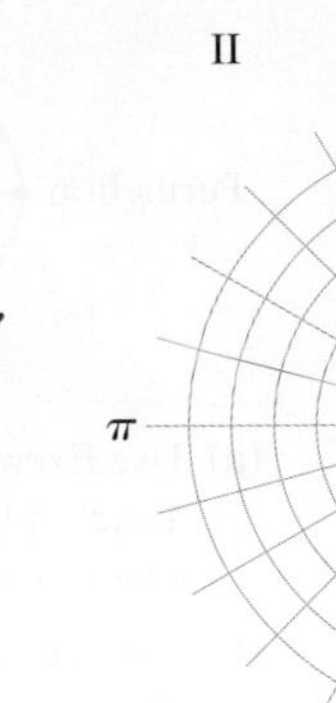

III

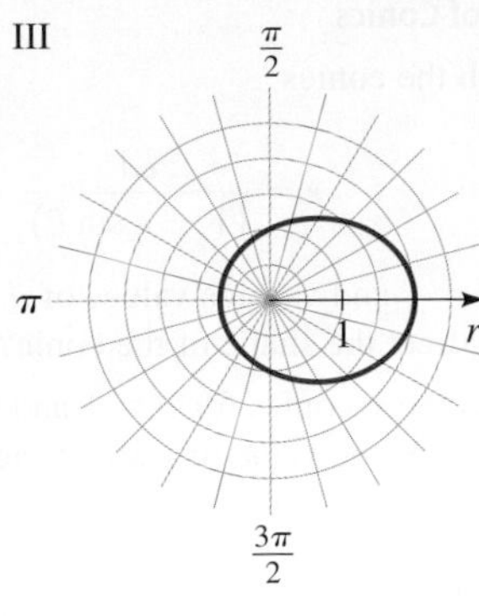

IV

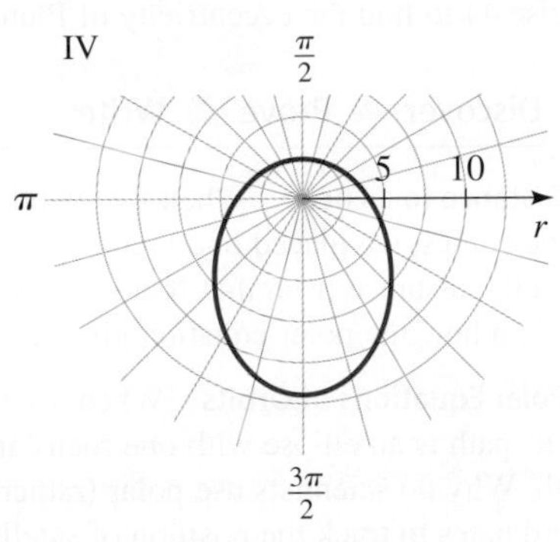

V

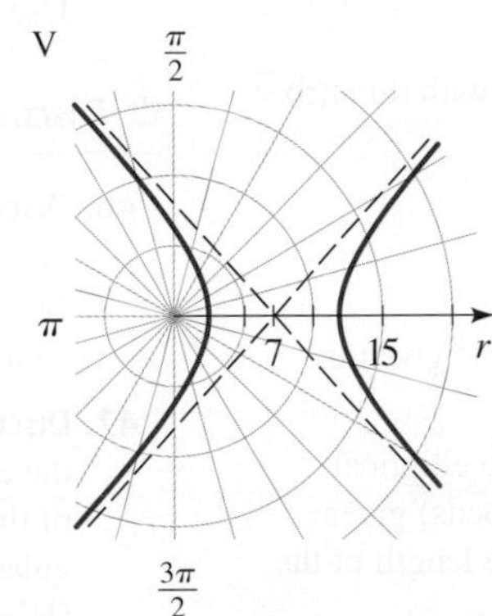

VI

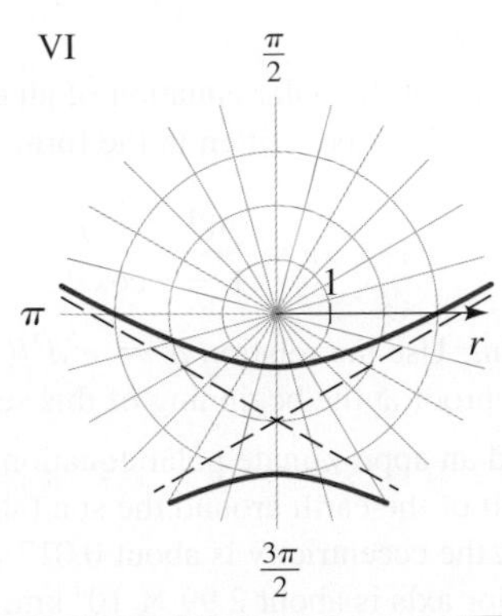

17–20 ■ Polar Equation for a Parabola A polar equation of a conic is given. **(a)** Show that the conic is a parabola, and sketch its graph. **(b)** Find the vertex and directrix, and indicate them on the graph.

17. $r = \dfrac{4}{1 - \sin\theta}$ **18.** $r = \dfrac{3}{2 + 2\sin\theta}$

19. $r = \dfrac{5}{3 + 3\cos\theta}$ **20.** $r = \dfrac{2}{5 - 5\cos\theta}$

21–24 ■ Polar Equation for an Ellipse A polar equation of a conic is given. **(a)** Show that the conic is an ellipse, and sketch its graph. **(b)** Find the vertices and directrix, and indicate them on the graph. **(c)** Find the center of the ellipse and the lengths of the major and minor axes.

21. $r = \dfrac{4}{2 - \cos\theta}$ **22.** $r = \dfrac{6}{3 - 2\sin\theta}$

23. $r = \dfrac{12}{4 + 3\sin\theta}$ **24.** $r = \dfrac{18}{4 + 3\cos\theta}$

25–28 ■ Polar Equation for a Hyperbola A polar equation of a conic is given. **(a)** Show that the conic is a hyperbola, and sketch its graph. **(b)** Find the vertices and directrix, and indicate them on the graph. **(c)** Find the center of the hyperbola, and sketch the asymptotes.

25. $r = \dfrac{8}{1 + 2\cos\theta}$ **26.** $r = \dfrac{10}{1 - 4\sin\theta}$

27. $r = \dfrac{20}{2 - 3\sin\theta}$ **28.** $r = \dfrac{6}{2 + 7\cos\theta}$

29–36 ■ Identifying and Graphing a Conic **(a)** Find the eccentricity, and identify the conic. **(b)** Sketch the conic, and label the vertices.

29. $r = \dfrac{4}{1 + 3\cos\theta}$ **30.** $r = \dfrac{8}{3 + 3\cos\theta}$

31. $r = \dfrac{2}{1 - \cos\theta}$ **32.** $r = \dfrac{10}{3 - 2\sin\theta}$

33. $r = \dfrac{6}{2 + \sin\theta}$ **34.** $r = \dfrac{5}{2 - 3\sin\theta}$

35. $r = \dfrac{7}{2 - 5\sin\theta}$ **36.** $r = \dfrac{8}{3 + \cos\theta}$

37–40 ■ Rotating a Conic A polar equation of a conic is given. **(a)** Find the eccentricity and the directrix of the conic. **(b)** If this conic is rotated about the origin through the given angle θ, write the resulting equation. **(c)** Draw graphs of the original conic and the rotated conic in the same viewing rectangle.

37. $r = \dfrac{1}{4 - 3\cos\theta}$; $\theta = \dfrac{\pi}{3}$

38. $r = \dfrac{2}{5 - 3\sin\theta}$; $\theta = \dfrac{2\pi}{3}$

39. $r = \dfrac{2}{1 + \sin\theta}$; $\theta = -\dfrac{\pi}{4}$

40. $r = \dfrac{9}{2 + 2\cos\theta}$; $\theta = -\dfrac{5\pi}{6}$

Skills Plus

41. Families of Conics Graph the conics $r = e/(1 - e\cos\theta)$ with $e = 0.4, 0.6, 0.8$, and 1.0 in the same viewing rectangle. How does the value of e affect the shape of the curve?

42. Families of Conics

(a) Graph the conics

$$r = \frac{ed}{(1 + e\sin\theta)}$$

for $e = 1$ and various values of d. How does the value of d affect the shape of the conic?

(b) Graph these conics for $d = 1$ and various values of e. How does the value of e affect the shape of the conic?

Applications

43. Orbit of the Earth The polar equation of an ellipse can be expressed in terms of its eccentricity e and the length a of its major axis.

(a) Show that the polar equation of an ellipse with directrix $x = -d$ can be written in the form

$$r = \frac{a(1 - e^2)}{1 - e\cos\theta}$$

[*Hint:* Use the relation $a^2 = e^2d^2/(1 - e^2)^2$ given in the proof at the beginning of this section.]

(b) Find an approximate polar equation for the elliptical orbit of the earth around the sun (at one focus) given that the eccentricity is about 0.017 and the length of the major axis is about 2.99×10^8 km.

44. Perihelion and Aphelion The planets move around the sun in elliptical orbits with the sun at one focus. The positions of a planet that are closest to, and farthest from, the sun are called its **perihelion** and **aphelion**, respectively.

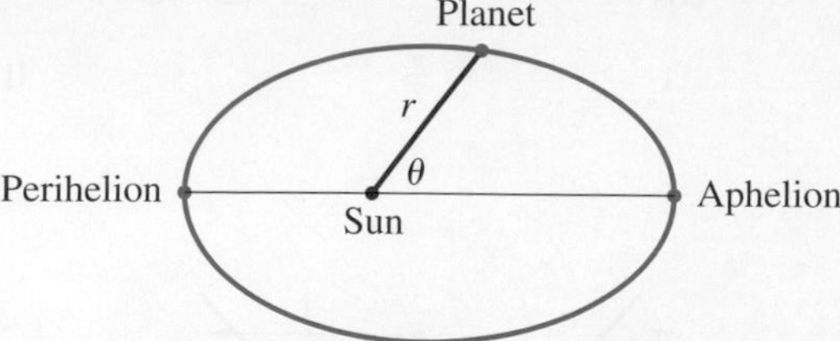

(a) Use Exercise 43(a) to show that the perihelion distance from a planet to the sun is $a(1 - e)$ and the aphelion distance is $a(1 + e)$.

(b) Use the data of Exercise 43(b) to find the distances from the earth to the sun at perihelion and at aphelion.

45. Orbit of Pluto The distance from Pluto to the sun is 4.43×10^9 km at perihelion and 7.37×10^9 km at aphelion. Use Exercise 44 to find the eccentricity of Pluto's orbit.

Discuss ■ Discover ■ Prove ■ Write

46. Discuss: Distance to a Focus When we found polar equations for the conics, we placed one focus at the pole. It's easy to find the distance from that focus to any point on the conic. Explain how the polar equation gives us this distance.

47. Discuss: Polar Equations of Orbits When a satellite orbits the earth, its path is an ellipse with one focus at the center of the earth. Why do scientists use polar (rather than rectangular) coordinates to track the position of satellites? [*Hint:* Your answer to Exercise 46 is relevant here.]

Chapter 10 Review

Properties and Formulas

Geometric Definition of a Parabola | Section 10.1

A **parabola** is the set of points in the plane that are equidistant from a fixed point F (the **focus**) and a fixed line l (the **directrix**).

Graphs of Parabolas with Vertex at the Origin | Section 10.1

A parabola with vertex at the origin has one of the following standard equations.

Vertical Axis

$x^2 = 4py$

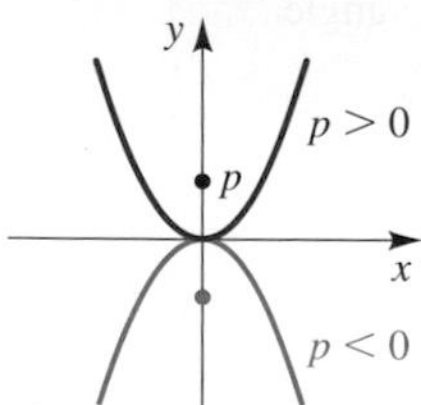

Focus $(0, p)$, directrix $y = -p$

Horizontal Axis

$y^2 = 4px$

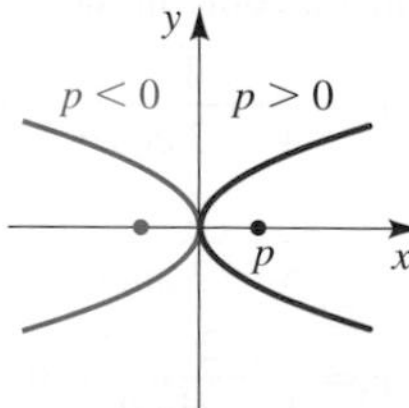

Focus $(p, 0)$, directrix $x = -p$

Geometric Definition of an Ellipse | Section 10.2

An **ellipse** is the set of all points in the plane for which the sum of the distances to each of two given points F_1 and F_2 (the **foci**) is a fixed constant.

Graphs of Ellipses with Center at the Origin | Section 10.2

An ellipse with center at the origin has one of the following standard equations.

Horizontal Axis

$$\frac{x^2}{a^2} + \frac{y^2}{b^2} = 1$$

$(a > b > 0)$

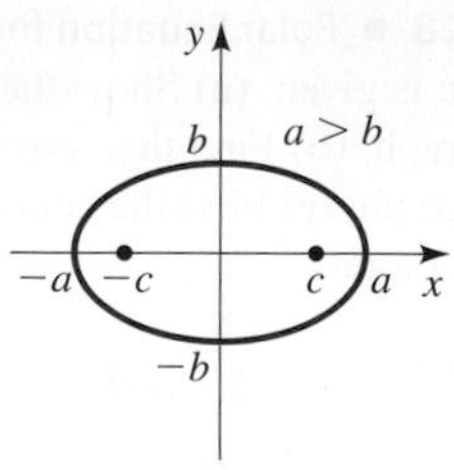

Foci $(\pm c, 0)$, $c^2 = a^2 - b^2$

Vertical Axis

$$\frac{x^2}{b^2} + \frac{y^2}{a^2} = 1$$

$(a > b > 0)$

Foci $(0, \pm c)$, $c^2 = a^2 - b^2$

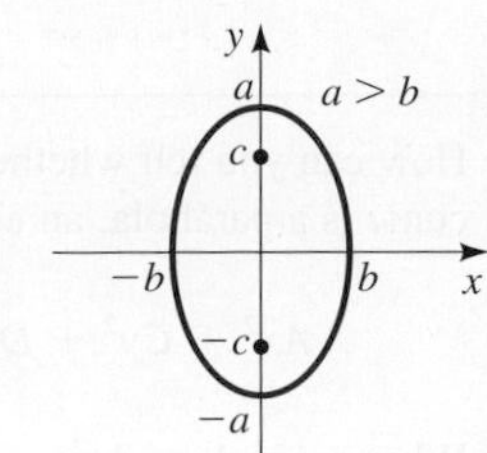

Eccentricity of an Ellipse | Section 10.2

The **eccentricity** of an ellipse with equation $\frac{x^2}{a^2} + \frac{y^2}{b^2} = 1$ or $\frac{x^2}{b^2} + \frac{y^2}{a^2} = 1$ (where $a > b > 0$) is the number

$$e = \frac{c}{a}$$

where $c = \sqrt{a^2 - b^2}$. The eccentricity e of any ellipse is a number between 0 and 1. If e is close to 0, then the ellipse is nearly circular; the closer e gets to 1, the more elongated the ellipse becomes.

Geometric Definition of a Hyperbola | Section 10.3

A **hyperbola** is the set of all points in the plane for which the absolute value of the difference of the distances to each of two given points F_1 and F_2 (the **foci**) is a fixed constant.

Graphs of Hyperbolas with Center at the Origin | Section 10.3

A **hyperbola** with center at the origin has one of the following standard equations.

Horizontal Axis

$$\frac{x^2}{a^2} - \frac{y^2}{b^2} = 1$$

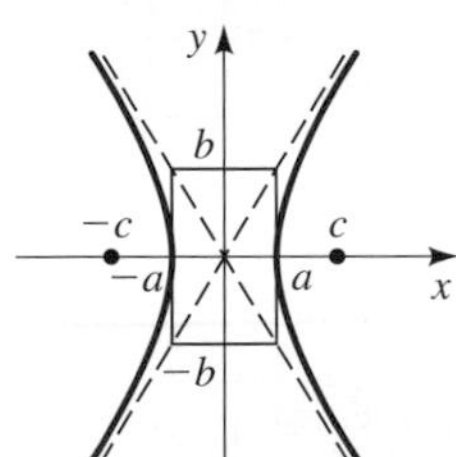

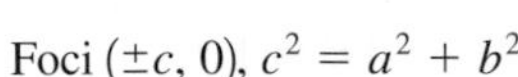

Foci $(\pm c, 0)$, $c^2 = a^2 + b^2$

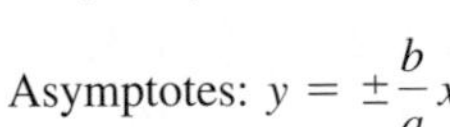

Asymptotes: $y = \pm \frac{b}{a}x$

Vertical Axis

$$-\frac{x^2}{b^2} + \frac{y^2}{a^2} = 1$$

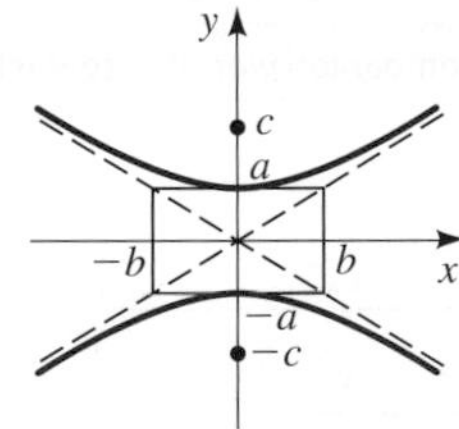

Foci $(0, \pm c)$, $c^2 = a^2 + b^2$

Asymptotes: $y = \pm \frac{a}{b}x$

Shifted Conics | Section 10.4

If the vertex of a parabola or the center of an ellipse or a hyperbola does not lie at the origin but rather at the point (h, k), then we refer to the curve as a **shifted conic**. To find the equation of the shifted conic, we use the "unshifted" form for the appropriate curve and replace x by $x - h$ and y by $y - k$.

General Equation of a Shifted Conic | Section 10.4

The graph of the equation

$$Ax^2 + Cy^2 + Dx + Ey + F = 0$$

(where A and C are not both 0) is either a conic or a degenerate conic. In the nondegenerate cases the graph is

1. a **parabola** if $A = 0$ or $C = 0$,
2. an **ellipse** if A and C have the same sign (or a circle if $A = C$),
3. a **hyperbola** if A and C have opposite signs.

To graph a conic whose equation is given in general form, complete the squares in x and y to put the equation in standard form for a parabola, an ellipse, or a hyperbola.

Rotation of Axes | Section 10.5

Suppose the x- and y-axes in a coordinate plane are rotated through the acute angle ϕ to produce the X- and Y-axes, as shown in the figure. Then the coordinates of a point in the xy- and the XY-planes are related as follows:

$$x = X\cos\phi - Y\sin\phi \qquad X = x\cos\phi + y\sin\phi$$

$$y = X\sin\phi + Y\cos\phi \qquad Y = -x\sin\phi + y\cos\phi$$

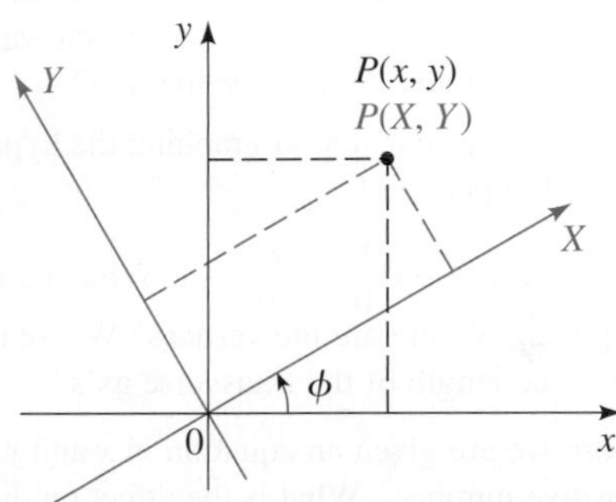

The General Conic Equation | Section 10.5

The general equation of a conic is of the form

$$Ax^2 + Bxy + Cy^2 + Dx + Ey + F = 0$$

The quantity $B^2 - 4AC$ is called the **discriminant** of the equation. The graph is

1. a parabola if $B^2 - 4AC = 0$,
2. an ellipse if $B^2 - 4AC < 0$,
3. a hyperbola if $B^2 - 4AC > 0$.

To eliminate the xy-term in the general equation of a conic, rotate the axes through an angle ϕ that satisfies

$$\cot 2\phi = \frac{A - C}{B}$$

Polar Equations of Conics | Section 10.6

A polar equation of the form

$$r = \frac{ed}{1 \pm e\cos\theta} \quad \text{or} \quad r = \frac{ed}{1 \pm e\sin\theta}$$

represents a conic with one focus at the origin and with eccentricity e. The conic is

1. a parabola if $e = 1$,
2. an ellipse if $0 < e < 1$,
3. a hyperbola if $e > 1$.

Concept Check

1. (a) Give the geometric definition of a parabola.

   (b) Give the standard equation of a parabola with vertex at the origin and with vertical axis. Where is the focus? What is the directrix?

   (c) Graph the equation $x^2 = 8y$. Indicate the focus on the graph.

2. (a) Give the geometric definition of an ellipse.

   (b) Give the standard equation of an ellipse with center at the origin and with major axis along the x-axis. How long is the major axis? How long is the minor axis? Where are the foci? What is the eccentricity of the ellipse?

   (c) Graph the equation $\frac{x^2}{16} + \frac{y^2}{9} = 1$. What are the lengths of the major and minor axes? Where are the foci?

3. (a) Give the geometric definition of a hyperbola.

   (b) Give the standard equation of a hyperbola with center at the origin and with transverse axis along the x-axis. How long is the transverse axis? Where are the vertices? What are the asymptotes? Where are the foci?

   (c) What is a good first step in graphing the hyperbola that is described in part (b)?

   (d) Graph the equation $\frac{x^2}{16} - \frac{y^2}{9} = 1$. What are the asymptotes? Where are the vertices? Where are the foci? What is the length of the transverse axis?

4. (a) Suppose we are given an equation in x and y. Let h and k be positive numbers. What is the effect on the graph of the equation if x is replaced by $x - h$ or $x + h$ and if y is replaced by $y - k$ or $y + k$?

   (b) Sketch a graph of $\frac{(x+2)^2}{16} + \frac{(y-4)^2}{9} = 1$

5. (a) How can you tell whether the following nondegenerate conic is a parabola, an ellipse, or a hyperbola?

$$Ax^2 + Cy^2 + Dx + Ey + F = 0$$

   (b) What conic does $3x^2 - 5y^2 + 4x + 5y - 8 = 0$ represent?

6. (a) Suppose that the x- and y-axes are rotated through an acute angle ϕ to produce the X- and Y-axes. What are the equations that relate the coordinates (x, y) and (X, Y) of a point in the xy-plane and XY-plane, respectively?

   (b) In the equation below, how do you eliminate the xy-term?

$$Ax^2 + Bxy + Cy^2 + Dx + Ey + F = 0$$

   (c) Use a rotation of axes to eliminate the xy-term in the equation

$$25x^2 - 14xy + 25y^2 = 288$$

   Graph the equation.

7. (a) What is the discriminant of the equation in Exercise 6(b)? How can you use the discriminant to determine the type of conic that the equation represents?

   (b) Use the discriminant to identify the equation in Exercise 6(c).

8. (a) Write polar equations that represent a conic with eccentricity e. For what values of e is the conic an ellipse? a hyperbola? a parabola?

   (b) What conic does the polar equation $r = 2/(1 - \cos\theta)$ represent? Graph the conic.

Answers to the Concept Check can be found at the book companion website stewartmath.com.

Exercises

1–12 ■ Graphing Parabolas An equation of a parabola is given. **(a)** Find the vertex, focus, and directrix of the parabola. **(b)** Sketch a graph of the parabola and its directrix.

1. $y^2 = 4x$

2. $x = \frac{1}{12}y^2$

3. $\frac{1}{8}x^2 = y$

4. $x^2 = -8y$

5. $x^2 + 8y = 0$

6. $2x - y^2 = 0$

7. $(y-2)^2 = 4(x+2)$

8. $(x+3)^2 = -20(y+2)$

9. $\frac{1}{2}(y-3)^2 + x = 0$

10. $2(x+1)^2 = y$

11. $\frac{1}{2}x^2 + 2x = 2y + 4$

12. $x^2 = 3(x + y)$

13–24 ■ Graphing Ellipses An equation of an ellipse is given. **(a)** Find the center, vertices, and foci of the ellipse. **(b)** Determine the lengths of the major and minor axes. **(c)** Sketch a graph of the ellipse.

13. $\frac{x^2}{9} + \frac{y^2}{25} = 1$

14. $\frac{x^2}{49} + \frac{y^2}{9} = 1$

15. $\frac{x^2}{49} + \frac{y^2}{4} = 1$

16. $\frac{x^2}{4} + \frac{y^2}{36} = 1$

17. $x^2 + 4y^2 = 16$

18. $9x^2 + 4y^2 = 1$

19. $\frac{(x-3)^2}{9} + \frac{y^2}{16} = 1$

20. $\frac{(x-2)^2}{25} + \frac{(y+3)^2}{16} = 1$

21. $\frac{(x-2)^2}{9} + \frac{(y+3)^2}{36} = 1$

22. $\frac{x^2}{3} + \frac{(y+5)^2}{25} = 1$

23. $4x^2 + 9y^2 = 36y$

24. $2x^2 + y^2 = 2 + 4(x - y)$

25–36 ■ Graphing Hyperbolas An equation of a hyperbola is given. **(a)** Find the center, vertices, foci, and asymptotes of the hyperbola. **(b)** Sketch a graph of the hyperbola.

25. $-\frac{x^2}{9} + \frac{y^2}{16} = 1$

26. $\frac{x^2}{49} - \frac{y^2}{32} = 1$

27. $\frac{x^2}{4} - \frac{y^2}{49} = 1$

28. $\frac{y^2}{25} - \frac{x^2}{4} = 1$

29. $x^2 - 2y^2 = 16$ **30.** $x^2 - 4y^2 + 16 = 0$

31. $\dfrac{(x+4)^2}{16} - \dfrac{y^2}{16} = 1$ **32.** $\dfrac{(x-2)^2}{8} - \dfrac{(y+2)^2}{8} = 1$

33. $\dfrac{(y-3)^2}{4} - \dfrac{(x+1)^2}{36} = 1$ **34.** $\dfrac{(y-3)^2}{3} - \dfrac{x^2}{16} = 1$

35. $9y^2 + 18y = x^2 + 6x + 18$ **36.** $y^2 = x^2 + 6y$

37–42 ■ Finding the Equation of a Conic Find the standard equation for the conic whose graph is shown.

37.

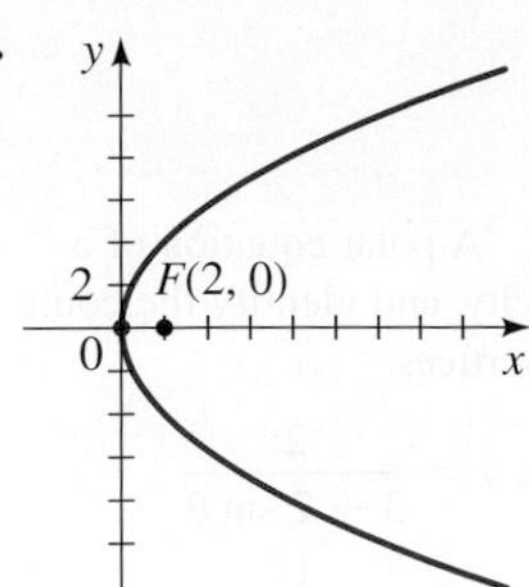

38.

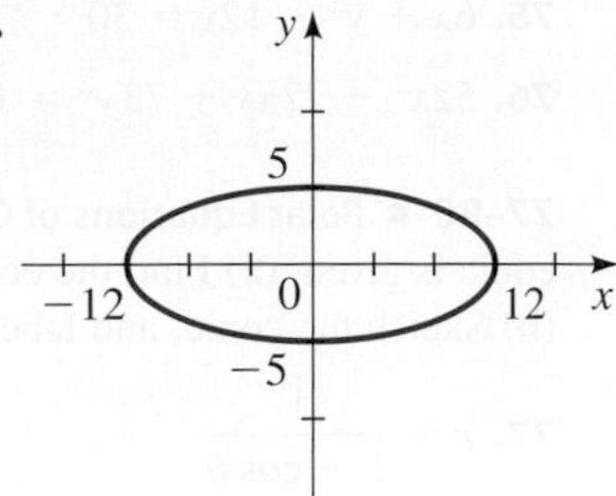

39.

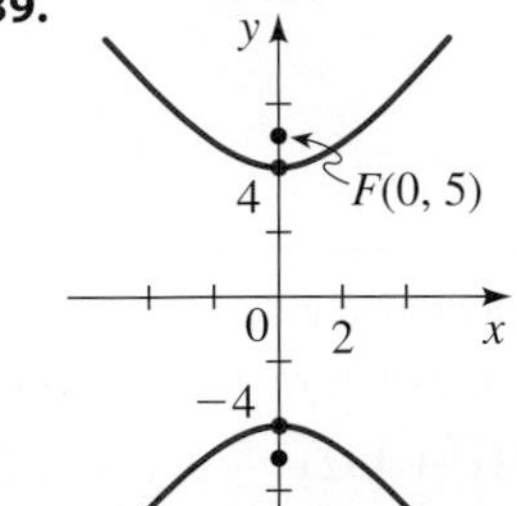

40.

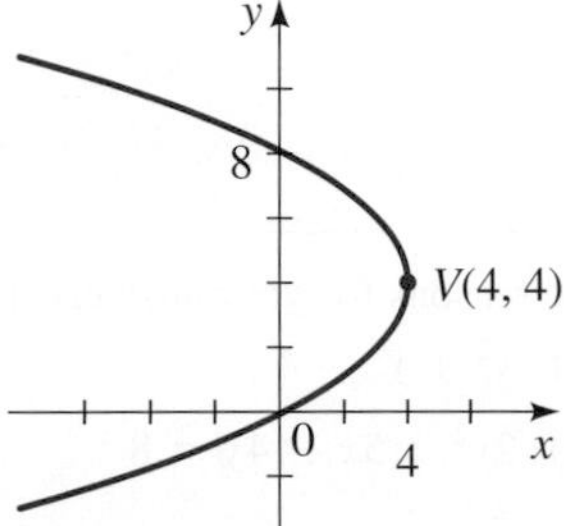

41.

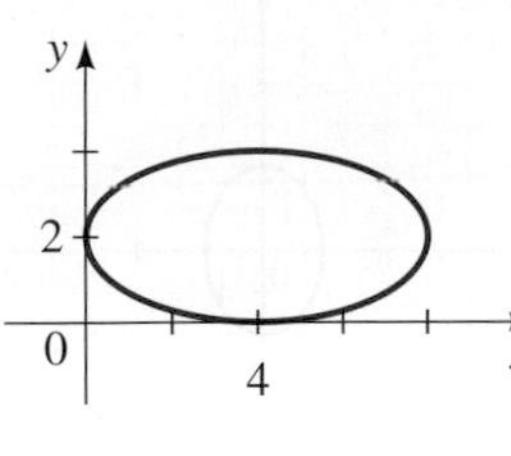

42.

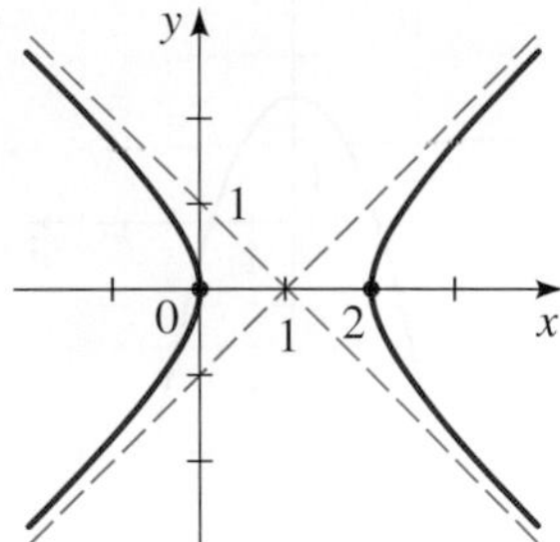

43–54 ■ Identifying and Graphing a Conic Determine whether the equation represents an ellipse, a parabola, a hyperbola, or a degenerate conic. If the graph is an ellipse, find the center, foci, and vertices. If it is a parabola, find the vertex, focus, and directrix. If it is a hyperbola, find the center, foci, vertices, and asymptotes. Then sketch the graph of the equation. If the equation has no graph, explain why.

43. $\dfrac{x^2}{12} + y = 1$ **44.** $\dfrac{x^2}{12} + \dfrac{y^2}{144} = \dfrac{y}{12}$

45. $x^2 - y^2 + 144 = 0$ **46.** $x^2 + 6x = 9y^2$

47. $4x^2 + y^2 = 8(x + y)$ **48.** $3x^2 - 6(x + y) = 10$

49. $x = y^2 - 16y$ **50.** $2x^2 + 4 = 4x + y^2$

51. $2x^2 - 12x + y^2 + 6y + 26 = 0$

52. $36x^2 - 4y^2 - 36x - 8y = 31$

53. $9x^2 + 8y^2 - 15x + 8y + 27 = 0$

54. $x^2 + 4y^2 = 4x + 8$

55–64 ■ Finding the Equation of a Conic Find an equation for the conic section with the given properties.

55. The parabola with focus $F(0, 1)$ and directrix $y = -1$

56. The parabola with vertex at the origin and focus $F(5, 0)$

57. The ellipse with center at the origin and with x-intercepts ± 2 and y-intercepts ± 5

58. The hyperbola with vertices $V(0, \pm 2)$ and asymptotes $y = \pm\frac{1}{2}x$

59. The ellipse with center $C(0, 4)$, foci $F_1(0, 0)$ and $F_2(0, 8)$, and major axis of length 10

60. The hyperbola with center $C(2, 4)$, foci $F_1(2, 1)$ and $F_2(2, 7)$, and vertices $V_1(2, 6)$ and $V_2(2, 2)$

61. The ellipse with foci $F_1(1, 1)$ and $F_2(1, 3)$ and with one vertex on the x-axis

62. The parabola with vertex $V(5, 5)$ and directrix the y-axis

63. The ellipse with vertices $V_1(7, 12)$ and $V_2(7, -8)$ and passing through the point $P(1, 8)$

64. The parabola with vertex $V(-1, 0)$ and horizontal axis of symmetry and crossing the y-axis at $y = 2$

65. Path of the Earth The path of the earth around the sun is an ellipse with the sun at one focus. The ellipse has major axis of length 298,000,000 km and eccentricity 0.017. Find the distance between the earth and the sun when the earth is **(a)** closest to the sun and **(b)** farthest from the sun.

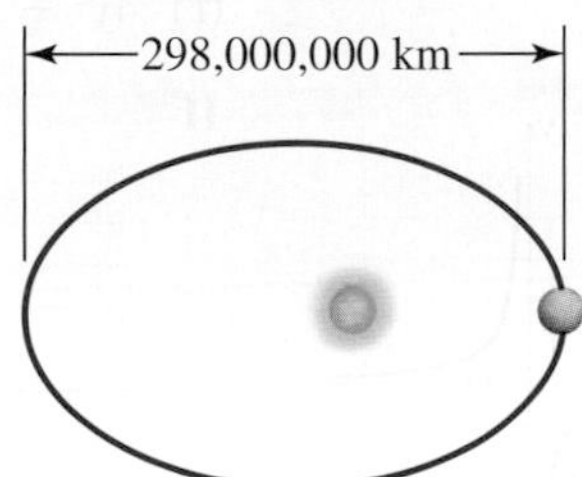

66. LORAN A boat is located 40 km from a straight shoreline. LORAN stations are located at points A and B on the shoreline, 300 km apart. From the LORAN signals, the captain determines that the boat is 80 km closer to A than to B. Find the location of the boat. (Place A and B on the y-axis with the x-axis halfway between them. Find the x- and y-coordinates of the boat.)

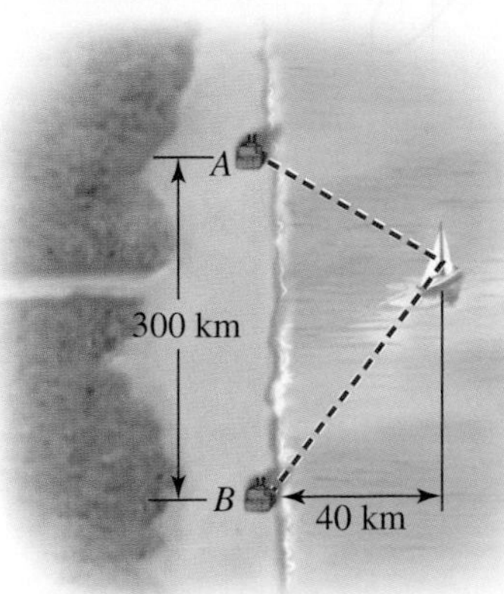

67. Families of Ellipses

(a) Draw graphs of the following family of ellipses for $k = 1, 2, 4,$ and 8.

$$\frac{x^2}{16 + k^2} + \frac{y^2}{k^2} = 1$$

(b) Prove that all the ellipses in part (a) have the same foci.

68. Families of Parabolas

(a) Draw graphs of the following family of parabolas for $k = \frac{1}{2}, 1, 2,$ and 4.

$$y = kx^2$$

(b) Find the foci of the parabolas in part (a).

(c) How does the location of the focus change as k increases?

69–72 ■ Identifying a Conic An equation of a conic is given. **(a)** Use the discriminant to determine whether the graph of the equation is a parabola, an ellipse, or a hyperbola. **(b)** Use a rotation of axes to eliminate the xy-term. **(c)** Sketch the graph.

69. $x^2 + 4xy + y^2 = 1$

70. $5x^2 - 6xy + 5y^2 - 8\sqrt{2}x + 8\sqrt{2}y - 4 = 0$

71. $7x^2 - 6\sqrt{3}xy + 13y^2 - 4\sqrt{3}x - 4y = 0$

72. $9x^2 + 24xy + 16y^2 = 25$

73–76 ■ Identify a Conic from Its Graph Use a graphing device to graph the conic. Identify the type of conic from the graph.

73. $5x^2 + 3y^2 = 60$

74. $9x^2 - 12y^2 + 36 = 0$

75. $6x + y^2 - 12y = 30$

76. $52x^2 - 72xy + 73y^2 = 100$

77–80 ■ Polar Equations of Conics A polar equation of a conic is given. **(a)** Find the eccentricity, and identify the conic. **(b)** Sketch the conic, and label the vertices.

77. $r = \dfrac{1}{1 - \cos\theta}$

78. $r = \dfrac{2}{3 + 2\sin\theta}$

79. $r = \dfrac{4}{1 + 2\sin\theta}$

80. $r = \dfrac{12}{1 - 4\cos\theta}$

Matching

81. Equations and Their Graphs Match the equation with its graph. Give reasons for your answers. (Don't use a graphing device.)

(a) $2x^2 + y^2 = 4$ **(b)** $3x^2 + 4y = 12$ **(c)** $x^2 + y^2 = 4$ **(d)** $x^2 + 2xy + y^2 + x - y = 0$

(e) $xy = 1$ **(f)** $4y^2 - x^2 + 8y = 0$ **(g)** $2y^2 - 5x - 4y = 8$ **(h)** $153x^2 + 192xy + 97y^2 = 225$

I

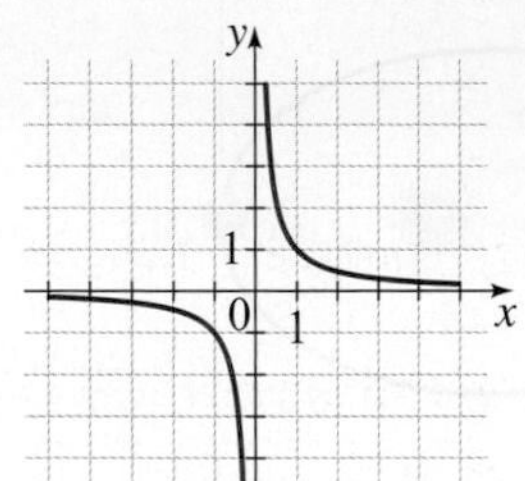

II

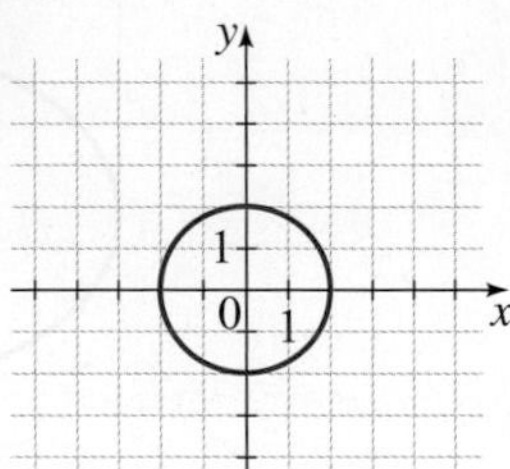

III

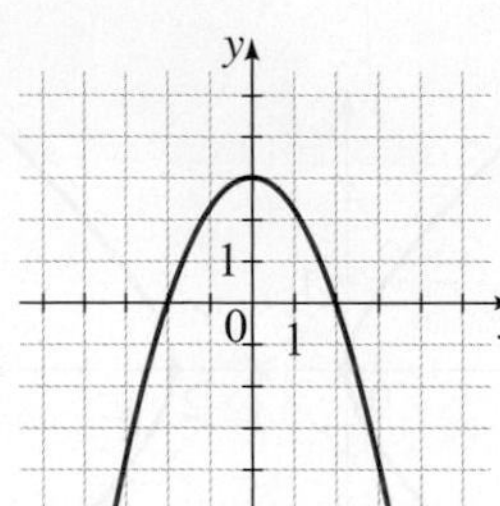

IV

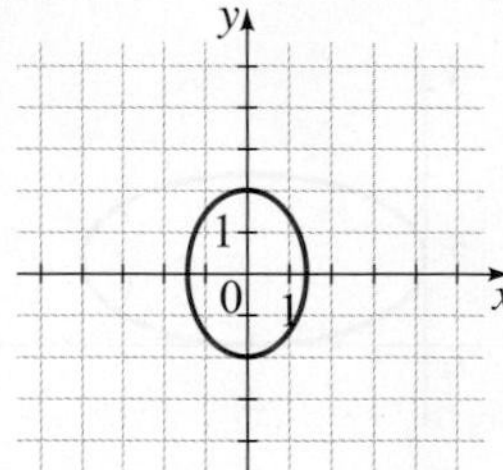

V

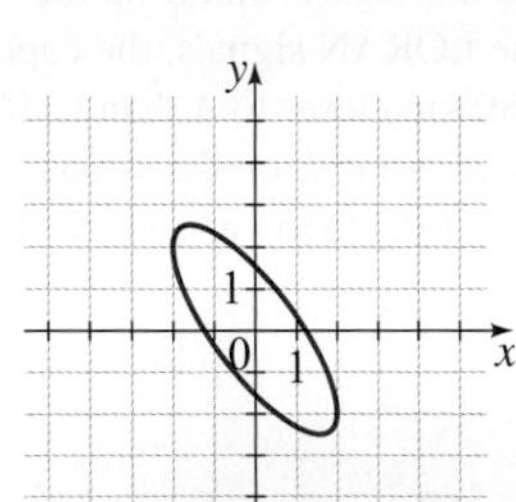

VI

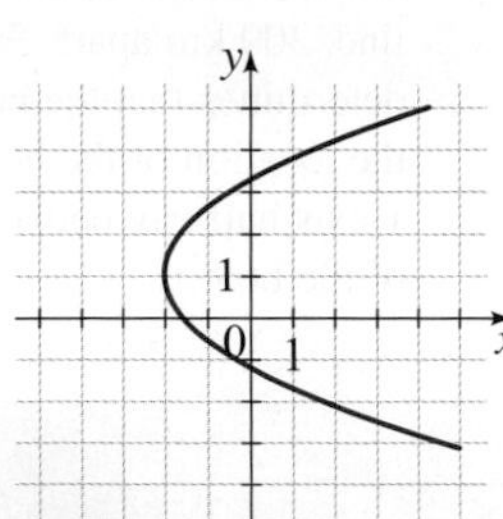

VII

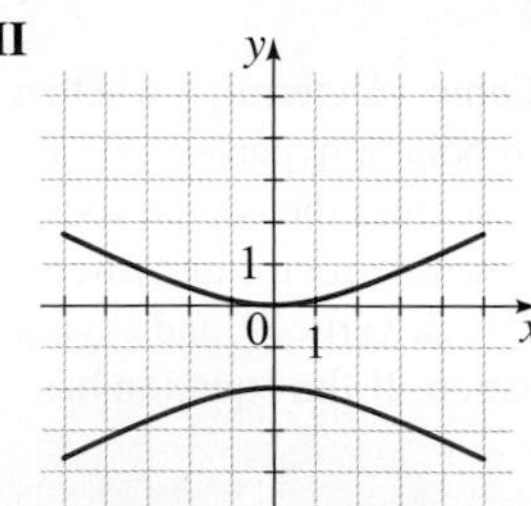

VIII

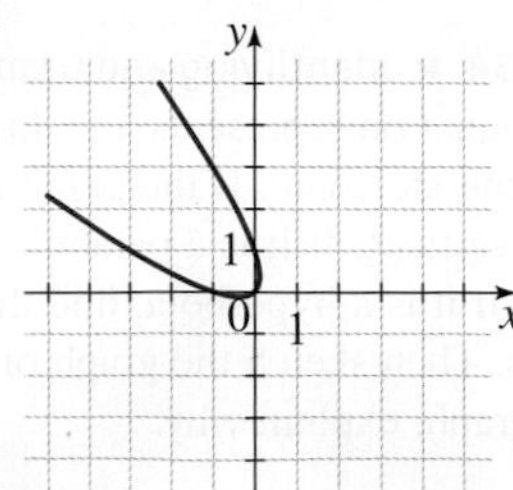

Chapter 10 | Test

1. Find the focus and directrix of the parabola $x^2 = -12y$, and sketch its graph.

2. Find the vertices, foci, and the lengths of the major and minor axes for the ellipse $\dfrac{x^2}{16} + \dfrac{y^2}{4} = 1$. Then sketch its graph.

3. Find the vertices, foci, and asymptotes of the hyperbola $\dfrac{y^2}{9} - \dfrac{x^2}{16} = 1$. Then sketch its graph.

4. Find an equation for the parabola with vertex $(0, 0)$ and focus $(4, 0)$.

5. Find an equation for the ellipse with foci $(\pm 3, 0)$ and vertices $(\pm 4, 0)$.

6. Find an equation for the hyperbola with foci $(0, \pm 5)$ and with asymptotes $y = \pm \frac{3}{4}x$.

7–9 ■ Find the standard equation for the conic whose graph is shown.

7.

y
(−4, 2)
1
−1
0
x

8.

y
(4, 3)
2
0
2
x

9.

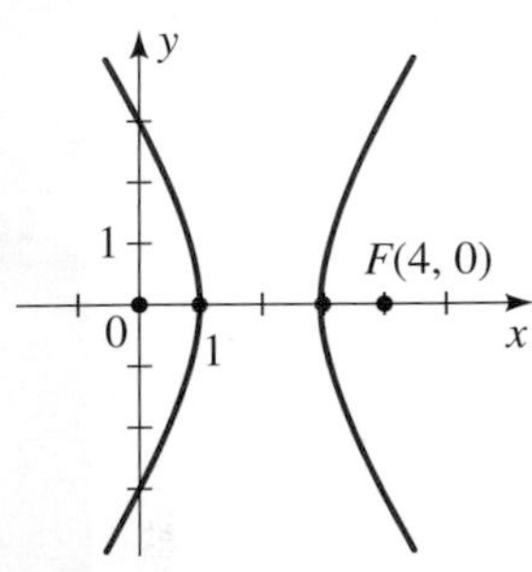

10–12 ■ Determine whether the equation represents an ellipse, a parabola, or a hyperbola. If the graph is an ellipse, find the center, foci, and vertices. If it is a parabola, find the vertex, focus, and directrix. If it is a hyperbola, find the center, foci, vertices, and asymptotes. Then sketch the graph of the equation.

10. $16x^2 + 36y^2 - 96x + 36y + 9 = 0$

11. $9x^2 - 8y^2 + 36x + 64y = 164$

12. $2x + y^2 + 8y + 8 = 0$

13. Find an equation for the ellipse with center $(2, 0)$, foci $(2, \pm 3)$, and major axis of length 8.

14. Find an equation for the parabola with focus $(2, 4)$ and directrix the x-axis.

15. A parabolic reflector for a car headlight forms a bowl shape that is 16 cm wide at its opening and 8 cm deep, as shown in the figure at the left. How far from the vertex should the filament of the bulb be placed if it is to be located at the focus?

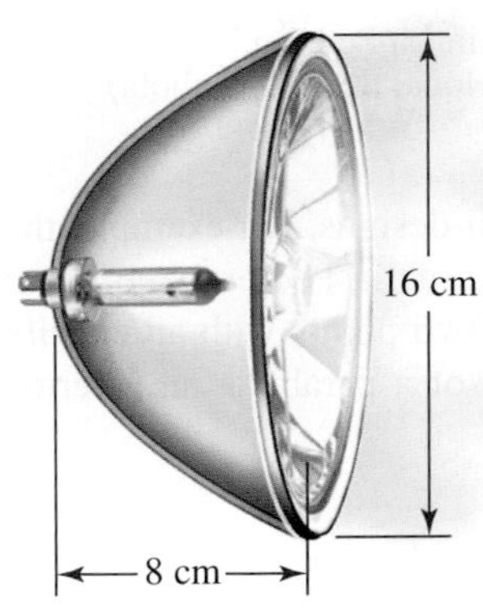

16. (a) Use the discriminant to determine whether the graph of the following equation is a parabola, an ellipse, or a hyperbola:

$$5x^2 + 4xy + 2y^2 = 18$$

(b) Use rotation of axes to eliminate the xy-term in the equation.

(c) Sketch a graph of the equation.

(d) Find the coordinates of the vertices of this conic (in the xy-coordinate system).

17. (a) Find the polar equation of the conic that has a focus at the origin, eccentricity $e = \frac{1}{2}$, and directrix $x = 2$. Sketch a graph of the conic.

(b) What type of conic is represented by the following equation? Sketch its graph.

$$r = \frac{3}{2 - \sin\theta}$$

A Cumulative Review Test for Chapters 8, 9, and 10 can be found at the book companion website www.stewartmath.com.

Focus on Modeling | Conics in Architecture

Many buildings employ conic sections in their design. Architects have various reasons for using these curves, ranging from structural stability to simple beauty. But how can a huge parabola, ellipse, or hyperbola be accurately constructed in concrete and steel? In this *Focus on Modeling,* we will see how the geometric properties of the conics can be used to construct these shapes.

■ Conics in Buildings

In ancient times architecture was part of mathematics, so architects had to be mathematicians. Many of the structures they built—pyramids, temples, amphitheaters, and irrigation projects—still stand. In modern times architects apply even more sophisticated mathematical principles. The photographs below show some structures that employ conic sections in their design.

Roman Amphitheater in Alexandria, Egypt (circle)
Nik Wheeler/Getty Images

Ceiling of Statuary Hall in the US Capitol (ellipse)
Architect of the Capitol

Roof of the Skydome in Toronto, Canada (parabola)
Walter Schmid/The Image Bank/Getty Images

Roof of Washington Dulles Airport (hyperbola and parabola)
Andrew Holt/Getty Images

McDonnell Planetarium, St. Louis, MO (hyperbola)
VisionsofAmerica/Joe Sohm/Getty Images

Attic in La Pedrera, Barcelona, Spain (parabola)
O. Alamany & E. Vicens/Getty Images

Architects have different reasons for using conics in their designs. For example, the Spanish architect Antoni Gaudí used parabolas in the attic of La Pedrera (see photo above). He reasoned that since a rope suspended between two points with an equally distributed load (as in a suspension bridge) has the shape of a parabola, an inverted parabola would provide the best support for a flat roof.

■ Constructing Conics

The equations of the conics are helpful in manufacturing small objects, because a computer-controlled cutting tool can accurately trace a curve given by an equation. But in a building project, how can we construct a portion of a parabola, ellipse, or hyperbola that spans the ceiling or walls of a building? The geometric properties of the conics provide practical ways of constructing them. For example, if you were building a circular tower, you would choose a center point, then make sure that the walls of the tower were

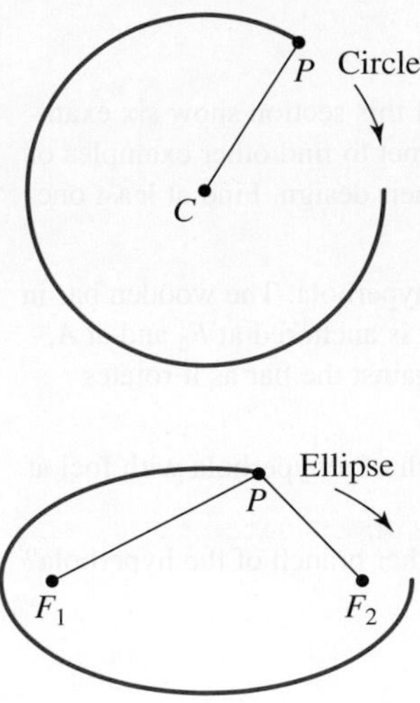

Figure 1 | Constructing a circle and an ellipse

a fixed distance from that point. Elliptical walls can be constructed by using a string anchored at two points, as shown in Figure 1.

To construct a parabola, we can use the apparatus shown in Figure 2. A piece of string of length a is anchored at F and A. The T-square, also of length a, slides along the straight bar L. A pencil at P holds the string taut against the T-square. As the T-square slides to the right, the pencil traces out a curve.

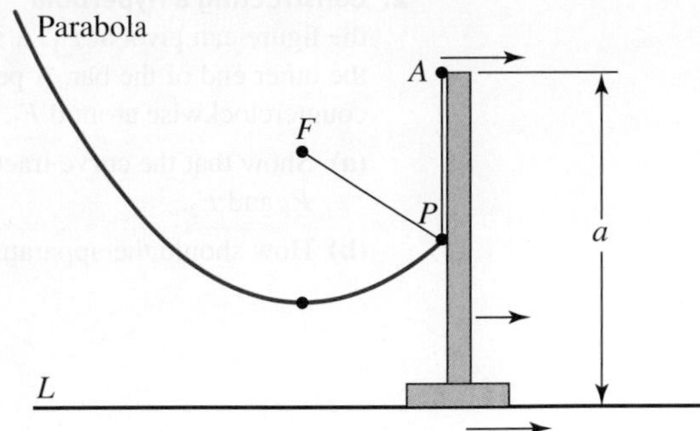

Figure 2 | Constructing a parabola

From the figure we see that

$$d(F, P) + d(P, A) = a \qquad \text{The string is of length } a$$

$$d(L, P) + d(P, A) = a \qquad \text{The T-square is of length } a$$

It follows that $d(F, P) + d(P, A) = d(L, P) + d(P, A)$. Subtracting $d(P, A)$ from each side, we get

$$d(F, P) = d(L, P)$$

The last equation says that the distance from F to P is equal to the distance from P to the line L. Thus the curve is a parabola with focus F and directrix L.

In building projects, it is easier to construct a straight line than a curve. So in some buildings, such as in the Kobe Tower (see Problem 4), a curved surface is produced by using many straight lines. We can also produce a curve using straight lines, such as the parabola shown in Figure 3.

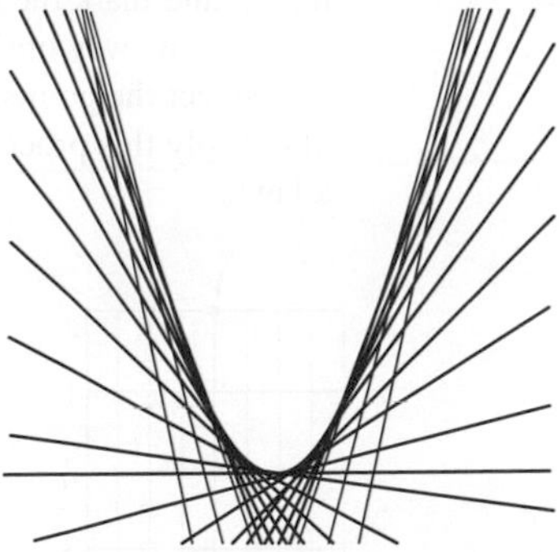

Figure 3 | Tangent lines to a parabola

Each line is **tangent** to the parabola; that is, the line meets the parabola at exactly one point and does not cross the parabola. The line tangent to the parabola $y = x^2$ at the point (a, a^2) is

$$y = 2ax - a^2$$

You are asked to show this in Problem 5. The parabola is called the **envelope** of all such lines.

Problems

1. Conics in Architecture The photographs at the beginning of this section show six examples of buildings that contain conic sections. Search the Internet to find other examples of structures that employ parabolas, ellipses, or hyperbolas in their design. Find at least one example for each type of conic.

2. Constructing a Hyperbola In this problem we construct a hyperbola. The wooden bar in the figure can pivot at F_1. A string that is shorter than the bar is anchored at F_2 and at A, the other end of the bar. A pencil at P holds the string taut against the bar as it rotates counterclockwise around F_1.

(a) Show that the curve traced out by the pencil is one branch of a hyperbola with foci at F_1 and F_2.

(b) How should the apparatus be reconfigured to draw the other branch of the hyperbola?

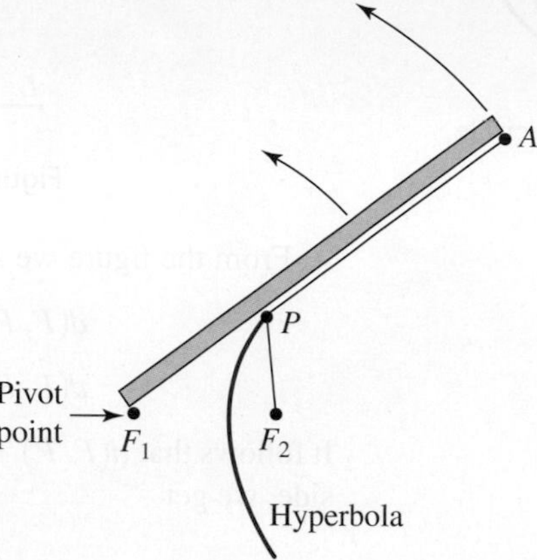

3. A Parabola in a Rectangle The following method can be used to construct a parabola that fits in a given rectangle. The parabola will be approximated by many short line segments.

First, draw a rectangle. Divide the rectangle in half by a vertical line segment, and label the top endpoint V. Next, divide the length and width of each half rectangle into an equal number of parts to form grid lines, as shown in the figure. Draw lines from V to the endpoints of horizontal grid line 1, and mark the points where these lines cross the vertical grid lines labeled 1. Next, draw lines from V to the endpoints of horizontal grid line 2, and mark the points where these lines cross the vertical grid lines labeled 2. Continue in this way until you have used all the horizontal grid lines. Now use line segments to connect the points you have marked to obtain an approximation to the desired parabola. Apply this procedure to draw a parabola that fits into a 6 m by 10 m rectangle on a lawn.

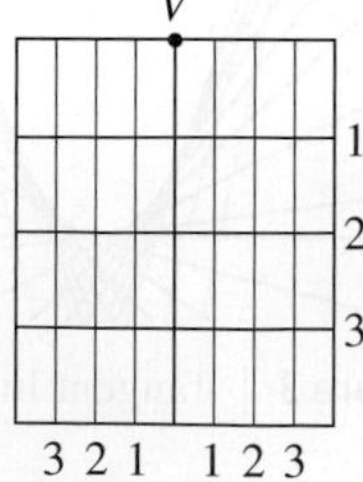

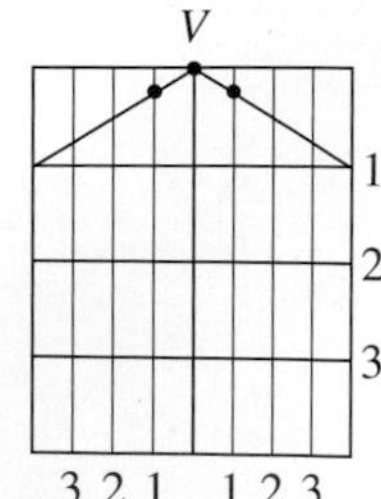

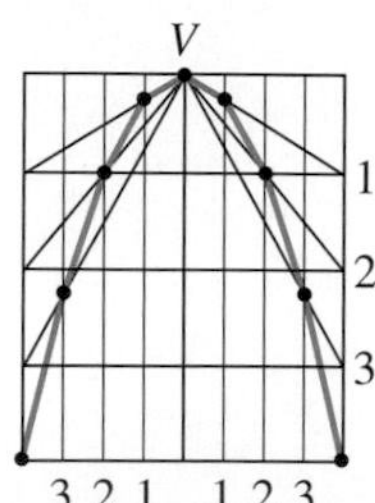

4. Hyperbolas from Straight Lines In this problem we construct hyperbolic shapes using straight lines. Punch equally spaced holes into the edges of two large plastic lids. Connect corresponding holes with strings of equal lengths as shown in the figure on the next page. Holding the strings taut, twist one lid against the other. An imaginary surface passing through the strings has hyperbolic cross sections. (An architectural example of this is the

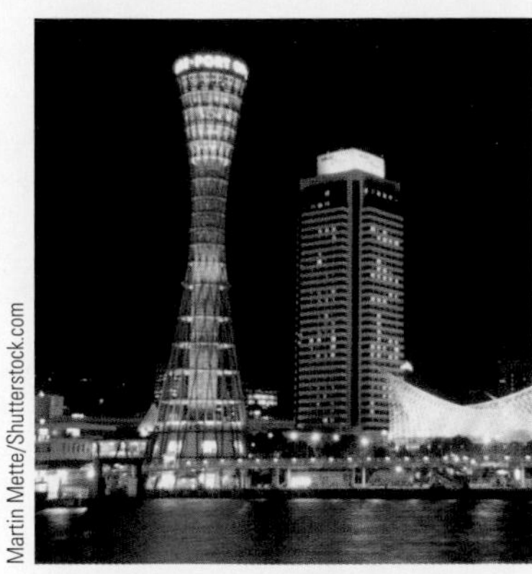

Kobe Tower in Japan, shown in the photograph.) What happens to the vertices of the hyperbolic cross sections as the lids are twisted more?

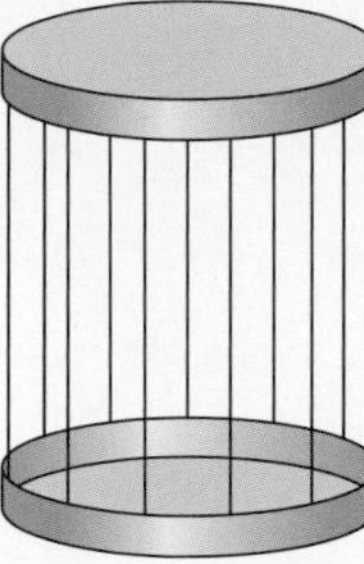

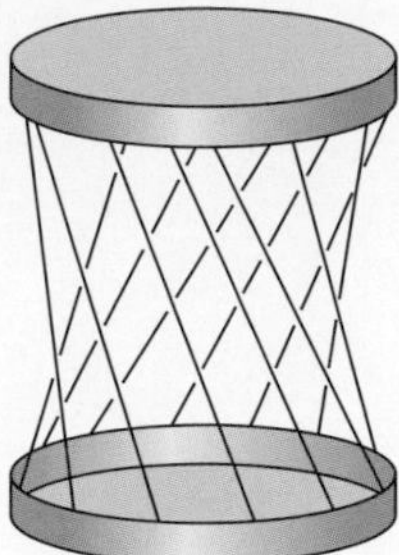

5. Tangent Lines to a Parabola In this problem we show that the line tangent to the parabola $y = x^2$ at the point (a, a^2) has the equation $y = 2ax - a^2$.

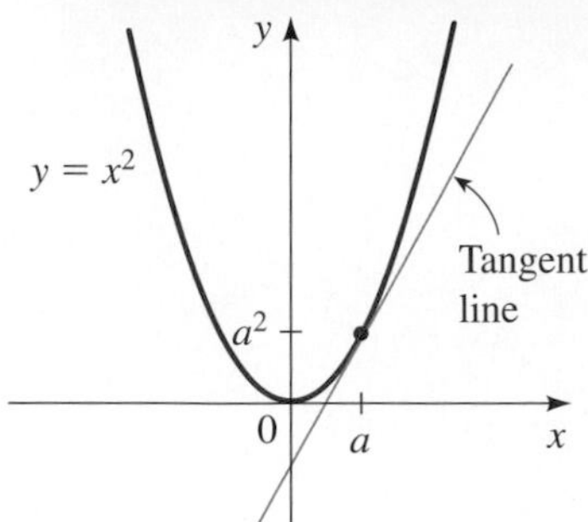

(a) Let m be the slope of the tangent line at (a, a^2). Show that the equation of the tangent line is $y - a^2 = m(x - a)$.

(b) Use the fact that the tangent line intersects the parabola at only one point to show that (a, a^2) is the only solution of the system.

$$\begin{cases} y - a^2 = m(x - a) \\ y = x^2 \end{cases}$$

(c) Eliminate y from the system in part (b) to get a quadratic equation in x. Show that the discriminant of this quadratic is $(m - 2a)^2$. Since the system in part (b) has exactly one solution, the discriminant must equal 0. Find m.

(d) Substitute the value for m you found in part (c) into the equation in part (a), and simplify to get the equation of the tangent line.

6. A Cut Cylinder In this problem we prove that when a cylinder is cut by a plane, an ellipse is formed. An architectural example of this is the Tycho Brahe Planetarium in Copenhagen (see the photograph). In the figure, a cylinder is cut by a plane, resulting in the red curve. Two spheres with the same radius as the cylinder slide inside the cylinder so that they just touch the plane at F_1 and F_2. Choose an arbitrary point P on the curve, and let Q_1 and Q_2 be the two points on the cylinder where a vertical line through P touches the "equator" of each sphere.

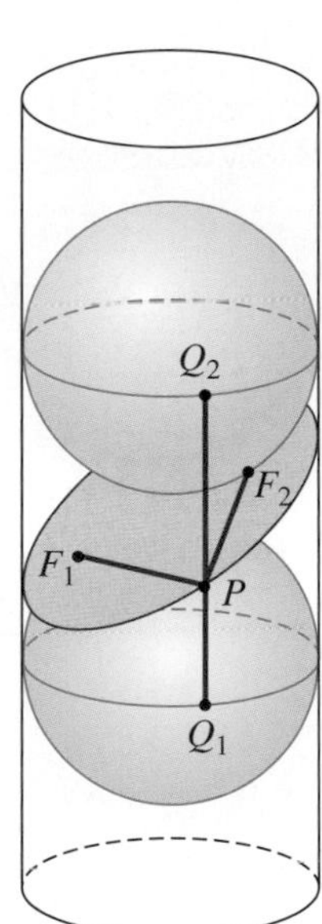

(a) Show that $PF_1 = PQ_1$ and $PF_2 = PQ_2$. [*Hint:* Use the fact that all tangents to a sphere from a given point outside the sphere are of the same length.]

(b) Explain why $PQ_1 + PQ_2$ is the same for all points P on the curve.

(c) Show that $PF_1 + PF_2$ is the same for all points P on the curve.

(d) Conclude that the curve is an ellipse with foci F_1 and F_2.

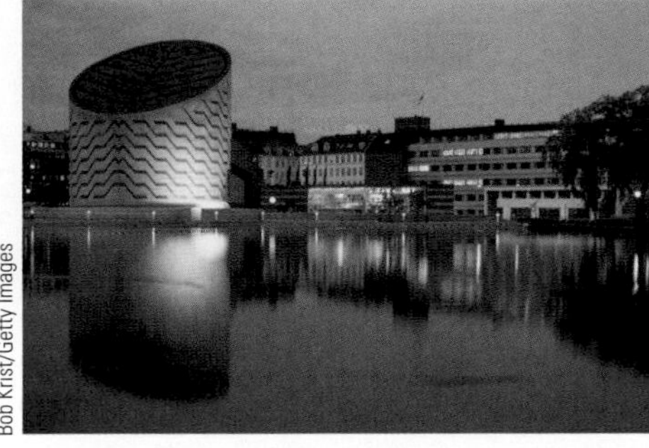

Min C. Chiu/Shutterstock.com

11 Sequences and Series

Throughout this book we have used functions to model real-world situations. The functions we've used have always had real numbers as inputs. But many situations occur in stages: stage 1, 2, 3, ... To model such situations, we need functions whose inputs are the natural numbers 1, 2, 3, . . . (representing the stages). For example, the peaks of a bouncing ball are represented by the natural numbers 1, 2, 3, . . . (representing peak 1, 2, 3, . . .). A function f that models the height of the ball at each peak has natural numbers 1, 2, 3, . . . as inputs and gives the heights as $f(1)$, $f(2)$, $f(3)$, ... In general a function whose inputs are the natural numbers is called a *sequence*. We can think of a sequence as simply a list of numbers written in a specific order.

The applications of sequences are varied. For instance, the amount in a bank account at the end of each month and the number of ancestors for each successive generation are both sequences. Many patterns in nature can also be modeled by sequences. For example, the Fibonacci sequence describes the growth of a rabbit population, the arrangements of leaves on a plant, or the spiral patterns of seeds in a sunflower (pictured above).

11.1 Sequences and Summation Notation

■ Sequences ■ Recursively Defined Sequences ■ The Partial Sums of a Sequence
■ Sigma Notation

Generally speaking, a sequence is an infinite list of numbers. The numbers in the sequence are often written as $a_1, a_2, a_3, \ldots$. The ellipsis (three dots) means that the list continues indefinitely. A simple example is the sequence

$$\begin{array}{ccccc} 5, & 10, & 15, & 20, & 25, \ldots \\ \uparrow & \uparrow & \uparrow & \uparrow & \uparrow \\ a_1 & a_2 & a_3 & a_4 & a_5 \ldots \end{array}$$

We can describe the pattern of the sequence displayed above by the following *formula*:

$$a_n = 5n$$

You may have already thought of a different way to describe the pattern—namely, "you go from one number to the next by adding 5." This natural way of describing the sequence is expressed by the *recursive formula*:

$$a_n = a_{n-1} + 5 \qquad (n = 2, 3, \ldots)$$

starting with $a_1 = 5$. Try substituting $n = 1, 2, 3, \ldots$ in each of these formulas to see how they produce the numbers in the sequence. In this section we see how these different ways are used to describe specific sequences.

■ Sequences

Any ordered list of numbers can be viewed as a function whose input values are 1, 2, 3, . . . and whose output values are the numbers in the list. So we define a sequence as follows.

Definition of a Sequence

A **sequence** is a function a whose domain is the set of natural numbers. The **terms of the sequence** are the function values

$$a(1), a(2), a(3), \ldots, a(n), \ldots$$

We usually write a_n instead of the function notation $a(n)$. So the terms of the sequence are written as

$$a_1, a_2, a_3, \ldots, a_n, \ldots$$

The number a_1 is called the **first term**, a_2 is called the **second term**, and in general, a_n is called the ***n*th term**.

Here is a simple example of a sequence:

$$2, 4, 6, 8, 10, \ldots$$

We can write a sequence in this way when it's clear what the subsequent terms of the sequence are. This sequence consists of even numbers. To be more accurate, however, we need to specify a procedure for finding *all* the terms of the sequence. This can be done by giving a formula for the nth term a_n of the sequence. In this case,

Another way to write this sequence is to use function notation:

$$a(n) = 2n$$

so $a(1) = 2, a(2) = 4, a(3) = 6, \ldots$

$$a_n = 2n$$

and the sequence can be written as

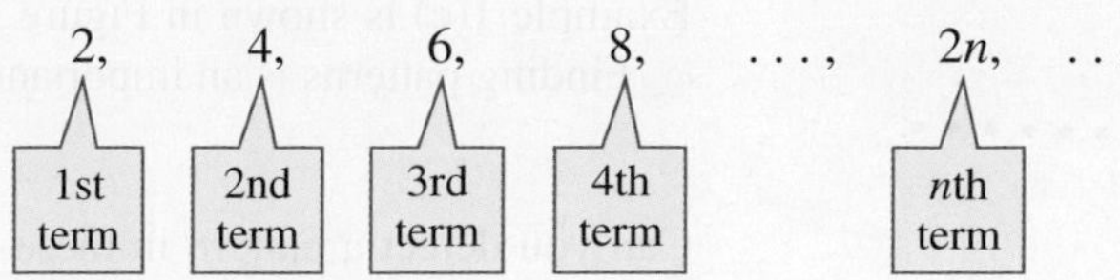

Notice how the formula $a_n = 2n$ gives all the terms of the sequence. For instance, substituting 1, 2, 3, and 4 for n gives the first four terms:

$$a_1 = 2 \cdot 1 = 2 \qquad a_2 = 2 \cdot 2 = 4$$
$$a_3 = 2 \cdot 3 = 6 \qquad a_4 = 2 \cdot 4 = 8$$

To find the 103rd term of this sequence, we use $n = 103$ to get

$$a_{103} = 2 \cdot 103 = 206$$

Example 1 ■ Finding the Terms of a Sequence

Find the first five terms and the 100th term of the sequence defined by each formula.

(a) $a_n = 2n - 1$ **(b)** $c_n = n^2 - 1$

(c) $t_n = \dfrac{n}{n+1}$ **(d)** $r_n = \dfrac{(-1)^n}{2^n}$

Solution To find the first five terms, we substitute $n = 1, 2, 3, 4$, and 5 in the formula for the nth term. To find the 100th term, we substitute $n = 100$. This gives the following.

nth Term	First Five Terms	100th Term
(a) $2n - 1$	1, 3, 5, 7, 9	199
(b) $n^2 - 1$	0, 3, 8, 15, 24	9999
(c) $\dfrac{n}{n+1}$	$\dfrac{1}{2}, \dfrac{2}{3}, \dfrac{3}{4}, \dfrac{4}{5}, \dfrac{5}{6}$	$\dfrac{100}{101}$
(d) $\dfrac{(-1)^n}{2^n}$	$-\dfrac{1}{2}, \dfrac{1}{4}, -\dfrac{1}{8}, \dfrac{1}{16}, -\dfrac{1}{32}$	$\dfrac{1}{2^{100}}$

Now Try Exercises 3, 5, 7, and 9 ■

In Example 1(d) the presence of $(-1)^n$ in the sequence has the effect of making successive terms alternately negative and positive.

It is often useful to picture a sequence by sketching its graph. Since a sequence is a function whose domain is the natural numbers, we can draw its graph in the Cartesian plane. For instance, the graph of the sequence

$$1, \frac{1}{2}, \frac{1}{3}, \frac{1}{4}, \frac{1}{5}, \frac{1}{6}, \ldots, \frac{1}{n}, \ldots$$

is shown in Figure 1.

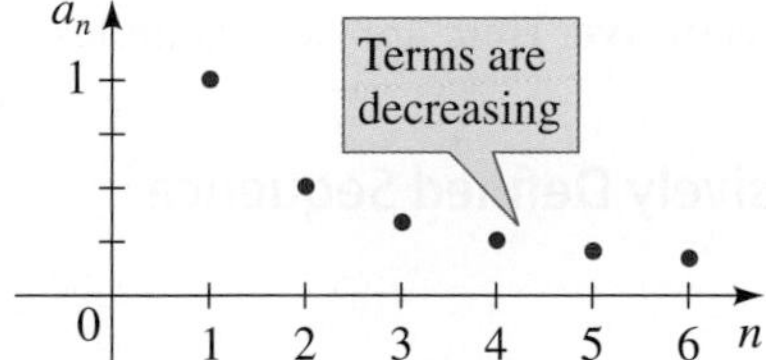

Figure 1

Compare the graph of the sequence shown in Figure 1 to the graph of

$$1, -\frac{1}{2}, \frac{1}{3}, -\frac{1}{4}, \frac{1}{5}, -\frac{1}{6}, \ldots, \frac{(-1)^{n+1}}{n}, \ldots$$

shown in Figure 2. The graph of every sequence consists of isolated points that are *not* connected.

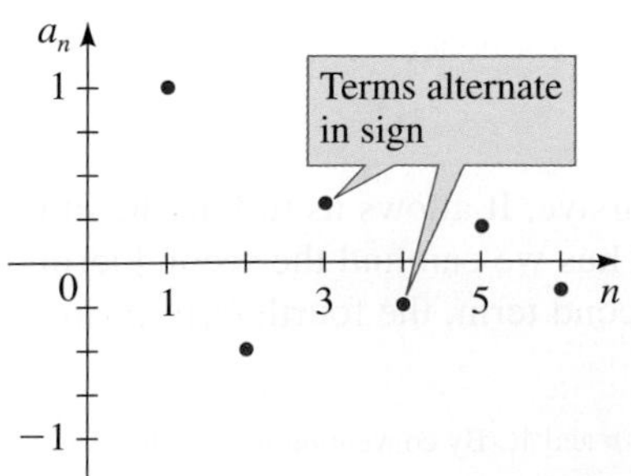

Figure 2

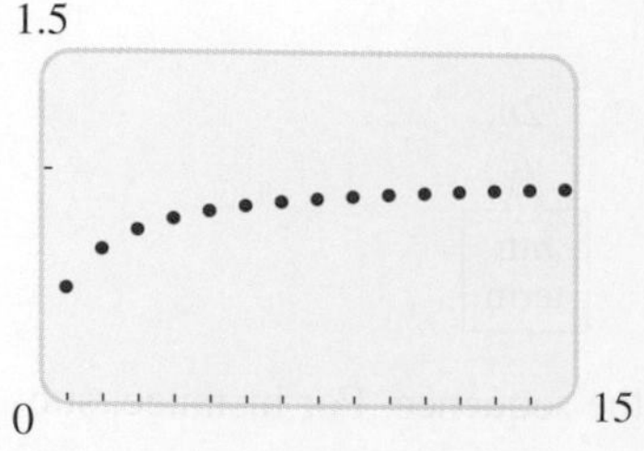

Figure 3 | $t(n) = n/(n + 1)$

Not all sequences can be defined by a formula. For example, there is no known formula for the sequence of prime numbers:*

2, 3, 5, 7, 11, 13, 17, 19, 23, . . .

We can use a graphing device to graph a sequence. The graph of the sequence in Example 1(c) is shown in Figure 3.

Finding patterns is an important part of mathematics. Consider a sequence that begins

$$1, 4, 9, 16, \ldots$$

Can you detect a pattern in these numbers? In other words, can you define a sequence whose first four terms are these numbers? The answer to this question seems straightforward; these numbers are the squares of the numbers 1, 2, 3, 4. Thus the sequence we are looking for is defined by $a_n = n^2$. However, this is not the *only* sequence whose first four terms are 1, 4, 9, 16. In other words, the answer to our problem is not unique (see Exercise 86). In the next example we are interested in finding an *obvious* sequence whose first few terms agree with the given ones.

Example 2 ■ Finding the *n*th Term of a Sequence

Find the nth term of a sequence whose first several terms are given.

(a) $\frac{1}{2}, \frac{3}{4}, \frac{5}{6}, \frac{7}{8}, \ldots$ **(b)** $-2, 4, -8, 16, -32, \ldots$

Solution

(a) We notice that the numerators of these fractions are the odd numbers and the denominators are the even numbers. Even numbers are of the form $2n$, and odd numbers are of the form $2n - 1$ (an odd number differs from an even number by 1). So a sequence that has these numbers for its first four terms is given by

$$a_n = \frac{2n - 1}{2n}$$

(b) These numbers are powers of 2, and they alternate in sign, so a sequence that agrees with these terms is given by

$$a_n = (-1)^n 2^n$$

You should check that these formulas do indeed generate the given terms.

 **Now Try Exercises 29 and 35** ■

ERATOSTHENES (circa 276–195 B.C.) was a renowned Greek geographer, mathematician, and astronomer. He accurately calculated the circumference of the earth by an ingenious method (see Exercise 6.1.80). He is most famous, however, for his method for finding primes, now called the *sieve of Eratosthenes*. The method consists of listing the integers, beginning with 2 (the first prime), and then crossing out all the multiples of 2, which are not prime. The next number remaining on the list is 3 (the second prime), so we again cross out all multiples of it. The next remaining number is 5 (the third prime number), and we cross out all multiples of it, and so on. In this way all numbers that are not prime are crossed out, and the remaining numbers are the primes.

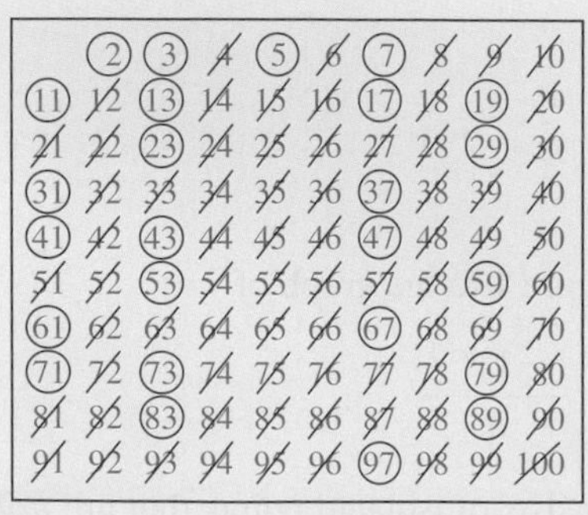

■ Recursively Defined Sequences

Some sequences do not have simple defining formulas like those shown in the preceding example. The nth term of a sequence may depend on some or all of the terms preceding it. A sequence defined in this way is called **recursive**. Here are two examples.

Example 3 ■ Finding the Terms of a Recursively Defined Sequence

A sequence is defined recursively by $a_1 = 1$ and

$$a_n = 3(a_{n-1} + 2)$$

(a) Find the first five terms of the sequence.

(b) Use a graphing device to find the 20th term of the sequence.

Solution

(a) The defining formula for this sequence is recursive. It allows us to find the nth term a_n if we know the preceding term a_{n-1}. Thus we can find the second term from the first term, the third term from the second term, the fourth term from

* A *prime number* is a whole number p whose only divisors are p and 1. (By convention the number 1 is not considered prime.)

Large Prime Numbers
The search for large primes fascinates many people. As of this writing, the largest known prime number is

$$2^{82,589,933} - 1$$

It was discovered by a computer volunteered by Patrick Laroche of Ocala, Florida, to the distributed computing project known as GIMPS (the Great Internet Mersenne Prime Search). In decimal notation this number contains 24,862,048 digits. If it were written in full, it would occupy more than six times as many pages as this book contains. Numbers of the form $2^p - 1$, where p is prime, are called Mersenne numbers and are named for the French monk Marin Mersenne who first studied them in the 1600s. Such numbers are more easily checked for primality than others. That is why the largest known primes are of this form.

the third term, and so on. Since we are given the first term $a_1 = 1$, we can proceed as follows.

$$a_2 = 3(a_1 + 2) = 3(1 + 2) = 9$$
$$a_3 = 3(a_2 + 2) = 3(9 + 2) = 33$$
$$a_4 = 3(a_3 + 2) = 3(33 + 2) = 105$$
$$a_5 = 3(a_4 + 2) = 3(105 + 2) = 321$$

Thus the first five terms of this sequence are

$$1, 9, 33, 105, 321, \ldots$$

(b) Note that to find the 20th term of the recursive sequence, we must first find all 19 preceding terms. This is most easily done by using a graphing device. Figure 4(a) shows how to enter this sequence on the TI-83 calculator. From Figure 4(b) we see that the 20th term of the sequence is

$$a_{20} = 4{,}649{,}045{,}865$$

You can find online calculators for computing recursive sequences.

```
Plot1 Plot2 Plot3
 nMin=1
\u(n)=3(u(n-1)+2)
 u(nMin)={1}
```

(a)

```
u(20)
              4649045865
```

(b)

Figure 4 | $u(n) = 3(u(n - 1) + 2), u(1) = 1$

 **Now Try Exercises 15 and 25** ■

Example 4 ■ The Fibonacci Sequence

Find the first 11 terms of the sequence defined recursively by $F_1 = 1$, $F_2 = 1$, and

$$F_n = F_{n-1} + F_{n-2}$$

Solution To find F_n, we need to find the two preceding terms, F_{n-1} and F_{n-2}. Since we are given F_1 and F_2, we proceed as follows.

$$F_3 = F_2 + F_1 = 1 + 1 = 2$$
$$F_4 = F_3 + F_2 = 2 + 1 = 3$$
$$F_5 = F_4 + F_3 = 3 + 2 = 5$$

It's clear what is happening here. Each term is the sum of the two terms that precede it, so we can write down as many terms as we please. Here are the first 11 terms.

$$1, 1, 2, 3, 5, 8, 13, 21, 34, 55, 89, \ldots$$

Now Try Exercise 19 ■

The sequence in Example 4 is called the **Fibonacci sequence**, named after the 13th century Italian mathematician who used it to solve a problem about the breeding of rabbits (see Exercise 85). The sequence also occurs in numerous other applications in nature. (See Figures 5 and 6.) In fact, so many phenomena behave like the Fibonacci sequence that one mathematical journal, the *Fibonacci Quarterly*, is devoted entirely to its properties.

Figure 5 | The Fibonacci sequence in the branching of a tree

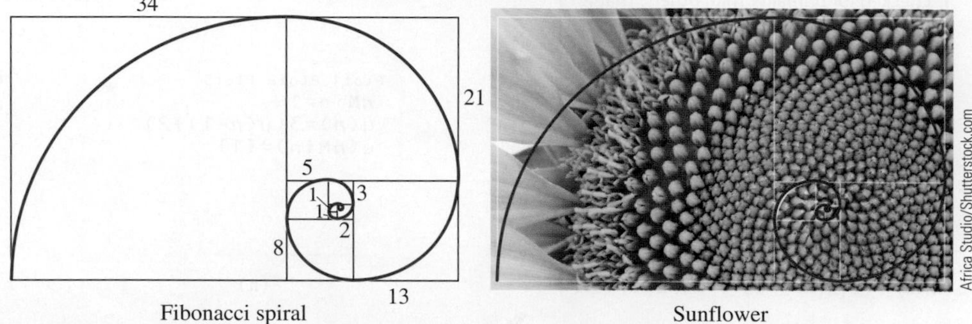

Figure 6 | The Fibonacci sequence in the pattern of seeds in a sunflower

■ The Partial Sums of a Sequence

In calculus we are often interested in adding the terms of a sequence. This leads to the following definition.

The Partial Sums of a Sequence

For the sequence

$$a_1, a_2, a_3, a_4, \ldots, a_n, \ldots$$

the **partial sums** are

$$S_1 = a_1$$
$$S_2 = a_1 + a_2$$
$$S_3 = a_1 + a_2 + a_3$$
$$S_4 = a_1 + a_2 + a_3 + a_4$$
$$\vdots$$
$$S_n = a_1 + a_2 + a_3 + \cdots + a_n$$
$$\vdots$$

S_1 is called the **first partial sum**, S_2 is the **second partial sum**, and so on. S_n is called the ***n*th partial sum**. The sequence $S_1, S_2, S_3, \ldots, S_n, \ldots$ is called the **sequence of partial sums**.

Example 5 ■ Finding the Partial Sums of a Sequence

Find the first four partial sums and the nth partial sum of the sequence given by $a_n = 1/2^n$.

Solution The terms of the sequence are

$$\frac{1}{2}, \frac{1}{4}, \frac{1}{8}, \ldots$$

The first four partial sums are

$$S_1 = \frac{1}{2} \qquad\qquad\qquad\qquad = \frac{1}{2}$$

$$S_2 = \frac{1}{2} + \frac{1}{4} \qquad\qquad\quad = \frac{3}{4}$$

$$S_3 = \frac{1}{2} + \frac{1}{4} + \frac{1}{8} \qquad = \frac{7}{8}$$

$$S_4 = \frac{1}{2} + \frac{1}{4} + \frac{1}{8} + \frac{1}{16} = \frac{15}{16}$$

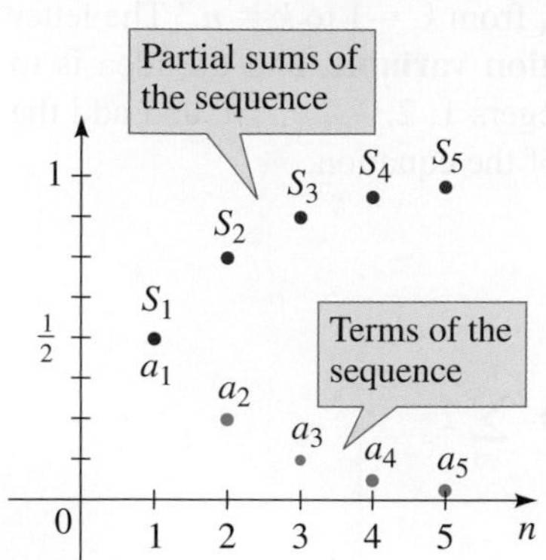

Figure 7 | Graph of the sequence a_n and the sequence of partial sums S_n

Notice that in the value of each partial sum, the denominator is a power of 2 and the numerator is one less than the denominator. In general, the nth partial sum is

$$S_n = \frac{2^n - 1}{2^n} = 1 - \frac{1}{2^n}$$

The first five terms of a_n and S_n are graphed in Figure 7.

Now Try Exercise 43 ■

Example 6 ■ Finding the Partial Sums of a Sequence

Find the first four partial sums and the nth partial sum of the sequence given by

$$a_n = \frac{1}{n} - \frac{1}{n+1}$$

Solution The first four partial sums are

$$S_1 = \left(1 - \frac{1}{2}\right) = 1 - \frac{1}{2}$$

$$S_2 = \left(1 - \frac{1}{2}\right) + \left(\frac{1}{2} - \frac{1}{3}\right) = 1 - \frac{1}{3}$$

$$S_3 = \left(1 - \frac{1}{2}\right) + \left(\frac{1}{2} - \frac{1}{3}\right) + \left(\frac{1}{3} - \frac{1}{4}\right) = 1 - \frac{1}{4}$$

$$S_4 = \left(1 - \frac{1}{2}\right) + \left(\frac{1}{2} - \frac{1}{3}\right) + \left(\frac{1}{3} - \frac{1}{4}\right) + \left(\frac{1}{4} - \frac{1}{5}\right) = 1 - \frac{1}{5}$$

Do you detect a pattern here? Of course. The nth partial sum is

$$S_n = 1 - \frac{1}{n+1}$$

Now Try Exercise 45 ■

■ Sigma Notation

Given a sequence

$$a_1, a_2, a_3, a_4, \ldots$$

we can write the sum of the first n terms using **summation notation**, or **sigma notation**. This notation derives its name from the Greek letter Σ (capital sigma, corresponding to our S for "sum"). Sigma notation is used as follows:

This tells us to end with $k = n$

This tells us to add $\sum_{k=1}^{n} a_k$

This tells us to start with $k = 1$

$$\sum_{k=1}^{n} a_k = a_1 + a_2 + a_3 + a_4 + \cdots + a_n$$

The left side of this expression is read, "The sum of a_k from $k = 1$ to $k = n$." The letter k is called the **index of summation**, or the **summation variable**, and the idea is to replace k in the expression after the sigma by the integers 1, 2, 3, . . . , n, and add the resulting expressions, arriving at the right-hand side of the equation.

Example 7 ■ Sigma Notation

Find each sum.

(a) $\sum_{k=1}^{5} k^2$ **(b)** $\sum_{j=3}^{5} \frac{1}{j}$ **(c)** $\sum_{k=5}^{10} k$ **(d)** $\sum_{i=1}^{6} 2$

Solution

(a) $\sum_{k=1}^{5} k^2 = 1^2 + 2^2 + 3^2 + 4^2 + 5^2 = 55$

(b) $\sum_{j=3}^{5} \frac{1}{j} = \frac{1}{3} + \frac{1}{4} + \frac{1}{5} = \frac{47}{60}$

(c) $\sum_{k=5}^{10} k = 5 + 6 + 7 + 8 + 9 + 10 = 45$

(d) $\sum_{i=1}^{6} 2 = 2 + 2 + 2 + 2 + 2 + 2 = 12$

Now Try Exercises 47 and 49 ■

```
sum(seq(K²,K,1,5,1))
                  55
sum(seq(1/J,J,3,5,
1))▶Frac
               47/60
```

Figure 8

We can use a graphing device to evaluate sums. For instance, Figure 8 shows how the TI-83 can be used to evaluate the sums in parts (a) and (b) of Example 7. You can also find online calculators for computing partial sums of sequences.

The Golden Ratio

The ancient Greeks considered a line segment to be divided into the **golden ratio** if the ratio of the shorter part to the longer part is the same as the ratio of the longer part to the whole segment.

1 x

Thus the segment shown is divided into the golden ratio if

$$\frac{1}{x} = \frac{x}{1 + x}$$

This leads to a quadratic equation whose positive solution is

$$x = \frac{1 + \sqrt{5}}{2} \approx 1.618$$

This ratio occurs naturally in many places. For instance, psychology experiments show that the most pleasing shape of rectangle is one whose sides are in golden ratio. The ancient Greeks agreed with this and built their temples in this ratio.

The golden ratio is related to the Fibonacci sequence (see Exercise 12.4.43). The ratio of two successive Fibonacci numbers

$$\frac{F_{n+1}}{F_n}$$

gets closer to the golden ratio the larger the value of n.

Clark Dunbar/Flame/Corbis

Stefano Bianchetti/Getty Images

FIBONACCI (circa 1170–1250) was born in Pisa, Italy, and was educated in North Africa. He traveled widely in the Mediterranean area and learned the various methods then in use for writing numbers. On returning to Pisa in 1202, Fibonacci advocated the use of the Hindu-Arabic decimal system, the one we use today, over the Roman numeral system that was used in Europe in his time. His most famous book, *Liber Abaci*, expounds on the advantages of the Hindu-Arabic numerals. In fact, multiplication and division were so complicated using Roman numerals that the equivalent of a college degree was necessary to master these skills. Interestingly, in 1299 the city of Florence outlawed the use of the decimal system for merchants and businesses, requiring numbers to be written in Roman numerals or words. One can only speculate about the reasons for this law.

Example 8 ■ Writing Sums in Sigma Notation

Write each sum using sigma notation.

(a) $1^3 + 2^3 + 3^3 + 4^3 + 5^3 + 6^3 + 7^3$ **(b)** $\sqrt{3} + \sqrt{4} + \sqrt{5} + \cdots + \sqrt{77}$

Solution

(a) We can write

$$1^3 + 2^3 + 3^3 + 4^3 + 5^3 + 6^3 + 7^3 = \sum_{k=1}^{7} k^3$$

(b) A natural way to write this sum is

$$\sqrt{3} + \sqrt{4} + \sqrt{5} + \cdots + \sqrt{77} = \sum_{k=3}^{77} \sqrt{k}$$

However, there is no unique way of writing a sum in sigma notation. We could also write this sum as

$$\sqrt{3} + \sqrt{4} + \sqrt{5} + \cdots + \sqrt{77} = \sum_{k=0}^{74} \sqrt{k+3}$$

or

$$\sqrt{3} + \sqrt{4} + \sqrt{5} + \cdots + \sqrt{77} = \sum_{k=1}^{75} \sqrt{k+2}$$

 **Now Try Exercises 67 and 69** ■

The following properties of sums are natural consequences of properties of the real numbers.

Properties of Sums

Let $a_1, a_2, a_3, a_4, \ldots$ and $b_1, b_2, b_3, b_4, \ldots$ be sequences. Then for every positive integer n and any real number c the following properties hold.

1. $\displaystyle\sum_{k=1}^{n} (a_k + b_k) = \sum_{k=1}^{n} a_k + \sum_{k=1}^{n} b_k$ **2.** $\displaystyle\sum_{k=1}^{n} (a_k - b_k) = \sum_{k=1}^{n} a_k - \sum_{k=1}^{n} b_k$

3. $\displaystyle\sum_{k=1}^{n} ca_k = c\left(\sum_{k=1}^{n} a_k\right)$

Proof To prove Property 1, we write out the left-hand side of the equation to get

$$\sum_{k=1}^{n} (a_k + b_k) = (a_1 + b_1) + (a_2 + b_2) + (a_3 + b_3) + \cdots + (a_n + b_n)$$

Because addition is commutative and associative, we can rearrange the terms on the right-hand side to read

$$\sum_{k=1}^{n} (a_k + b_k) = (a_1 + a_2 + a_3 + \cdots + a_n) + (b_1 + b_2 + b_3 + \cdots + b_n)$$

Rewriting the right side using sigma notation gives Property 1. Property 2 is proved in a similar manner. To prove Property 3, we use the Distributive Property:

$$\sum_{k=1}^{n} ca_k = ca_1 + ca_2 + ca_3 + \cdots + ca_n$$

$$= c(a_1 + a_2 + a_3 + \cdots + a_n) = c\left(\sum_{k=1}^{n} a_k\right)$$

■

11.1 Exercises

■ Concepts

1. A sequence is a function whose domain is ________.

2. The nth partial sum of a sequence is the sum of the first ________ terms of the sequence. So for the sequence $a_n = n^2$ the fourth partial sum is $S_4 =$ ___ + ___ + ___ + ___ = ___.

■ Skills

3–14 ■ Terms of a Sequence Find the first four terms and the 100th term of the sequence whose nth term is given.

3. $a_n = n - 3$ **4.** $a_n = 2n - 1$

5. $a_n = \dfrac{1}{3n - 4}$ **6.** $a_n = n^3 + 2$

7. $a_n = 3^n$ **8.** $a_n = \left(\dfrac{-1}{5}\right)^{n-1}$

9. $a_n = \dfrac{(-1)^n}{n^2}$ **10.** $a_n = \dfrac{1}{n^2}$

11. $a_n = 1 + (-1)^n$ **12.** $a_n = (-1)^{n+1}\dfrac{n}{n + 1}$

13. $a_n = n^n$ **14.** $a_n = 3$

15–20 ■ Recursive Sequences A sequence is defined recursively by the given formulas. Find the first five terms of the sequence.

15. $a_n = 2(a_{n-1} + 3)$ and $a_1 = 4$

16. $a_n = \dfrac{a_{n-1}}{6}$ and $a_1 = -24$

17. $a_n = 2a_{n-1} + 1$ and $a_1 = 1$

18. $a_n = \dfrac{1}{1 + a_{n-1}}$ and $a_1 = 1$

19. $a_n = a_{n-1} + a_{n-2}$ and $a_1 = 1, a_2 = 2$

20. $a_n = a_{n-1} + a_{n-2} + a_{n-3}$ and $a_1 = a_2 = a_3 = 1$

21–26 ■ Terms of a Sequence Use a graphing device to do the following. **(a)** Find the first ten terms of the sequence. **(b)** Graph the first ten terms of the sequence.

21. $a_n = 4n + 3$ **22.** $a_n = n^2 + n$

23. $a_n = \dfrac{12}{n}$ **24.** $a_n = 4 - 2(-1)^n$

25. $a_n = \dfrac{1}{a_{n-1}}$ and $a_1 = 2$

26. $a_n = a_{n-1} - a_{n-2}$ and $a_1 = 1, a_2 = 3$

27–38 ■ *n*th term of a Sequence Find the nth term of a sequence whose first several terms are given.

27. $2, 4, 6, 8, \ldots$ **28.** $1, 3, 5, 7, \ldots$

29. $-3, 9, -27, 81, \ldots$ **30.** $-\frac{1}{3}, \frac{1}{9}, -\frac{1}{27}, \frac{1}{81}, \ldots$

31. $4, 9, 14, 19, \ldots$ **32.** $10, 3, -4, -11, \ldots$

33. $5, -25, 125, -625, \ldots$ **34.** $3, 0.3, 0.03, 0.003, \ldots$

35. $1, \frac{3}{4}, \frac{5}{9}, \frac{7}{16}, \frac{9}{25}, \ldots$ **36.** $\frac{3}{4}, \frac{4}{5}, \frac{5}{6}, \frac{6}{7}, \ldots$

37. $0, 2, 0, 2, 0, 2, \ldots$ **38.** $1, \frac{1}{2}, 3, \frac{1}{4}, 5, \frac{1}{6}, \ldots$

39–42 ■ Partial Sums Find the first six partial sums $S_1, S_2, S_3, S_4, S_5, S_6$ of the sequence whose first several terms are given.

39. $2, 4, 6, 8, \ldots$ **40.** $1^2, 2^2, 3^2, 4^2, \ldots$

41. $\dfrac{1}{3}, \dfrac{1}{3^2}, \dfrac{1}{3^3}, \dfrac{1}{3^4}, \ldots$ **42.** $4, -4, 4, -4, \ldots$

43–46 ■ *n*th Partial Sum Find the first four partial sums and the nth partial sum of the sequence a_n.

43. $a_n = \dfrac{2}{3^n}$ **44.** $a_n = \dfrac{1}{n + 1} - \dfrac{1}{n + 2}$

45. $a_n = \sqrt{n} - \sqrt{n + 1}$

46. $a_n = \log\left(\dfrac{n}{n + 1}\right)$ [*Hint:* Use a property of logarithms to write the nth term as a difference.]

47–54 ■ Evaluating a Sum Find the sum.

47. $\displaystyle\sum_{k=1}^{4} k$ **48.** $\displaystyle\sum_{k=1}^{4} k^2$

49. $\displaystyle\sum_{k=1}^{4} \frac{1}{3k}$ **50.** $\displaystyle\sum_{j=1}^{51} (-1)^j$

51. $\displaystyle\sum_{i=1}^{8} [1 + (-1)^i]$ **52.** $\displaystyle\sum_{i=4}^{12} 10$

53. $\displaystyle\sum_{k=1}^{5} 2^{k-1}$ **54.** $\displaystyle\sum_{i=1}^{3} i2^i$

55–60 ■ Evaluating a Sum Use a graphing device to evaluate the sum.

55. $\displaystyle\sum_{k=1}^{10} k^2$ **56.** $\displaystyle\sum_{k=1}^{100} (3k + 4)$

57. $\displaystyle\sum_{j=7}^{20} j^2(1 + j)$ **58.** $\displaystyle\sum_{j=5}^{15} \frac{1}{j^2 + 1}$

59. $\displaystyle\sum_{n=0}^{22} (-1)^n 2n$ **60.** $\displaystyle\sum_{n=1}^{100} \frac{(-1)^n}{n}$

61–66 ■ Sigma Notation Write the sum without using sigma notation.

61. $\displaystyle\sum_{k=1}^{4} k^3$ **62.** $\displaystyle\sum_{j=1}^{4} \sqrt{\frac{j - 1}{j + 1}}$

63. $\displaystyle\sum_{k=0}^{6} \sqrt{k + 4}$ **64.** $\displaystyle\sum_{k=6}^{9} k(k + 3)$

65. $\displaystyle\sum_{k=3}^{100} x^k$ **66.** $\displaystyle\sum_{j=1}^{n} (-1)^{j+1} x^j$

67–74 ■ Sigma Notation Write the sum using sigma notation.

67. $4 + 8 + 12 + 16 + \cdots + 48$

68. $2 + 5 + 8 + \cdots + 29$

69. $1^2 + 2^2 + 3^2 + \cdots + 10^2$

70. $\dfrac{1}{2 \ln 2} - \dfrac{1}{3 \ln 3} + \dfrac{1}{4 \ln 4} - \dfrac{1}{5 \ln 5} + \cdots + \dfrac{1}{100 \ln 100}$

71. $\frac{1}{1 \cdot 2} + \frac{1}{2 \cdot 3} + \frac{1}{3 \cdot 4} + \cdots + \frac{1}{999 \cdot 1000}$

72. $\frac{\sqrt{1}}{1^2} + \frac{\sqrt{2}}{2^2} + \frac{\sqrt{3}}{3^2} + \cdots + \frac{\sqrt{n}}{n^2}$

73. $1 + x + x^2 + x^3 + \cdots + x^{100}$

74. $1 - 2x + 3x^2 - 4x^3 + 5x^4 + \cdots - 100x^{99}$

Skills Plus

75. *n*th Term of a Sequence Find a formula for the nth term of the sequence

$$\sqrt{2},\ \sqrt{2\sqrt{2}},\ \sqrt{2\sqrt{2\sqrt{2}}},\ \sqrt{2\sqrt{2\sqrt{2\sqrt{2}}}}, \ldots$$

[*Hint:* Write each term as a power of 2.]

76. A Formula for the Fibonacci Sequence It is known that the nth term of the Fibonacci sequence is given by the formula

$$F_n = \frac{1}{\sqrt{5}}\left(\frac{(1+\sqrt{5})^n - (1-\sqrt{5})^n}{2^n}\right)$$

Use a calculator to find the first ten terms of the Fibonacci sequence using this formula.

Applications

77. Compound Interest An amount of $2000 is deposited in a savings account that pays 2.4% interest per year compounded monthly. The amount in the account after n months is given by

$$A_n = 2000\left(1 + \frac{0.024}{12}\right)^n$$

(a) Find the first six terms of the sequence.

(b) Find the amount in the account after 3 years.

78. Compound Interest An amount of $100 is deposited at the end of each month into an account that pays 6% interest per year compounded monthly. The amount of interest accumulated after n months is given by

$$I_n = 100\left(\frac{1.005^n - 1}{0.005} - n\right)$$

(a) Find the first six terms of the sequence.

(b) Find the interest accumulated after 5 years.

79. Population of a City A city was incorporated in 2004 with a population of 35,000. It is expected that the population will increase at a rate of 2% per year. The population n years after 2004 is given by

$$P_n = 35{,}000(1.02)^n$$

(a) Find the first five terms of the sequence.

(b) Find the population in 2014.

80. Paying off a Debt A loan of $10,000 is to be repaid in monthly installments of $200. Interest is charged on the balance at a rate of 0.5% per month.

(a) Show that the balance A_n in the nth month is given recursively by $A_0 = 10{,}000$ and

$$A_n = 1.005A_{n-1} - 200$$

(b) Find the balance after 6 months.

81. Fish Farming A fish farm has 5000 catfish in a pond. The number of catfish increases by 8% per month, and 300 catfish are harvested per month.

(a) Show that the catfish population P_n after n months is given recursively by $P_0 = 5000$ and

$$P_n = 1.08P_{n-1} - 300$$

(b) How many fish are there in the pond after 12 months?

82. Price of a House The median price of a house in a certain county increases by about 6% per year. In 2022 the median price was $240,000. Let P_n be the median price n years after 2022.

(a) Find a formula for the sequence P_n.

(b) Find the expected median price in 2030.

83. Salary Increases A management position provides a salary of $45,000 a year with a $2000 raise every year. Let A_n be the salary in the nth year of employment.

(a) Find a recursive definition of A_n.

(b) Find the salary in the fifth year of employment.

84. Concentration of a Solution An expirement is set up to find the optimal salt concentration for the growth of a certain species of mollusk. The experiment begins with a brine solution that has a salt concentration of 4 g/L. The concentration of salt is increased by 10% every day. Let C_0 denote the initial concentration, and let C_n be the concentration after n days.

(a) Find a recursive definition of C_n.

(b) Find the salt concentration after 8 days.

85. Fibonacci's Rabbits Fibonacci posed the following problem: Suppose that rabbits live forever and that every month each pair produces a new pair that becomes productive at age 2 months. If we start with one newborn pair, how many pairs of rabbits will we have in the nth month? Show that the answer is F_n, where F_n is the nth term of the Fibonacci sequence.

Discuss ■ Discover ■ Prove ■ Write

86. Discover ■ Prove: Different Sequences That Start the Same

(a) Show that the first four terms of the sequence defined by $a_n = n^2$ are

$$1, 4, 9, 16, \ldots$$

(b) Show that the first four terms of the sequence defined by $a_n = n^2 + (n-1)(n-2)(n-3)(n-4)$ are also

$$1, 4, 9, 16, \ldots$$

(c) Find a sequence whose first six terms are the same as those of $a_n = n^2$ but whose succeeding terms differ from this sequence.

(d) Find two different sequences that begin

$$2, 4, 8, 16, \ldots$$

87. Discuss: A Recursively Defined Sequence Find the first 40 terms of the sequence defined by

$$a_{n+1} = \begin{cases} \dfrac{a_n}{2} & \text{if } a_n \text{ is an even number} \\ 3a_n + 1 & \text{if } a_n \text{ is an odd number} \end{cases}$$

and $a_1 = 11$. Do the same if $a_1 = 25$. Make a conjecture about this type of sequence. Try several other values for a_1 to test your conjecture.

88. Discuss: A Different Type of Recursion Find the first 10 terms of the sequence defined by

$$a_n = a_{n-a_{n-1}} + a_{n-a_{n-2}}$$

with

$$a_1 = 1 \quad \text{and} \quad a_2 = 1$$

How is this recursive sequence different from the others in this section?

11.2 Arithmetic Sequences

■ Arithmetic Sequences ■ Partial Sums of Arithmetic Sequences

In this section we study a special type of sequence, called an arithmetic sequence.

■ Arithmetic Sequences

Perhaps the simplest way to generate a sequence is to start with a number a and add to it a fixed constant d, over and over again.

Definition of an Arithmetic Sequence

An **arithmetic sequence** is a sequence of the form

$$a, a + d, a + 2d, a + 3d, a + 4d, \ldots$$

The number a is the **first term**, and d is the **common difference** of the sequence. The ***n*th term** of an arithmetic sequence is given by

$$a_n = a + (n - 1)d$$

The number d is called the common difference because any two consecutive terms of an arithmetic sequence differ by d.

Example 1 ■ Arithmetic Sequences

(a) If $a = 2$ and $d = 3$, then we have the arithmetic sequence

$$2, 2 + 3, 2 + 6, 2 + 9, \ldots$$

or

$$2, 5, 8, 11, \ldots$$

Any two consecutive terms of this sequence differ by $d = 3$. The nth term is $a_n = 2 + 3(n - 1)$.

(b) Consider the arithmetic sequence

$$9, 4, -1, -6, -11, \ldots$$

Here the common difference is $d = -5$. The terms of an arithmetic sequence decrease if the common difference is negative. The nth term is $a_n = 9 - 5(n - 1)$.

(c) The graph of the arithmetic sequence $a_n = 1 + 2(n - 1)$ is shown in Figure 1. Notice that the points in the graph lie on the straight line $y = 2x - 1$, which has slope $d = 2$.

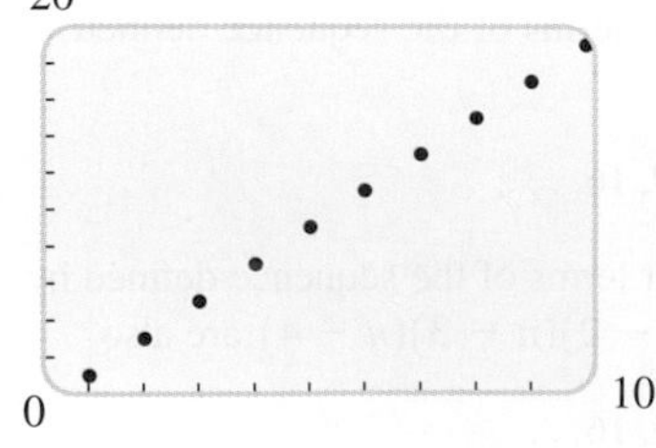

Figure 1

Now Try Exercises 5, 11, and 17 ■

An arithmetic sequence is determined completely by the first term a and the common difference d. Thus if we know the first two terms of an arithmetic sequence, then we can find a formula for the nth term, as the next example shows.

Mathematics in the Modern World

Fair Division of Assets

Dividing an asset fairly among a number of people is of great interest to mathematicians. Problems of this nature include dividing the national budget, disputed land, or assets in divorce cases. In 1994 Brams and Taylor found a mathematical way of dividing things fairly. Their solution has been applied to division problems in political science, legal proceedings, and other areas. To understand the problem, consider the following example. Suppose persons A and B want to divide a property fairly between them. To divide it *fairly* means that both A and B must be satisfied with the outcome of the division. Solution: A gets to divide the property into two pieces, then B gets to choose the piece they want. Since both A and B had a part in the division process, each should be satisfied. The situation becomes much more complicated if three or more people are involved (and that's where mathematics comes in).

Dividing things fairly involves much more than simply cutting things in half; it must take into account the *relative worth* each person attaches to the thing being divided. A story from the Bible illustrates this clearly. Two women appear before King Solomon, each claiming to be the mother of the same newborn baby. To discover which of these two women is the real mother, King Solomon ordered his swordsman to cut the baby in half! The real mother, who attaches far more worth to the baby than anyone else does, immediately gives up her claim to the baby to save the baby's life.

Mathematical solutions to fair-division problems have been applied in an international treaty, the Convention on the Law of the Sea (1982). If a country wants to develop a portion of the sea floor, it is required to divide the portion into two parts, one part to be used by itself and the other by a consortium that will preserve it for later use by a less developed country. The consortium gets first choice.

Example 2 ■ Finding Terms of an Arithmetic Sequence

Find the common difference, the first six terms, the nth term, and the 300th term of the arithmetic sequence

$$13, 7, 1, -5, \ldots$$

Solution Since the first term is 13, we have $a = 13$. The common difference is $d = 7 - 13 = -6$. Thus the nth term of this sequence is

$$a_n = 13 - 6(n - 1)$$

From this we find the first six terms:

$$13, 7, 1, -5, -11, -17, \ldots$$

The 300th term is $a_{300} = 13 - 6(300 - 1) = -1781$.

 **Now Try Exercise 33** ■

The next example shows that an arithmetic sequence is determined completely by *any* two of its terms.

Example 3 ■ Finding Terms of an Arithmetic Sequence

The 11th term of an arithmetic sequence is 52, and the 19th term is 92. Find the 1000th term.

Solution To find the nth term of this sequence, we need to find a and d in the formula

$$a_n = a + (n - 1)d$$

From this formula we get

$$a_{11} = a + (11 - 1)d = a + 10d$$

$$a_{19} = a + (19 - 1)d = a + 18d$$

Since $a_{11} = 52$ and $a_{19} = 92$, we get the following two equations:

$$\begin{cases} 52 = a + 10d \\ 92 = a + 18d \end{cases}$$

Solving this system for a and d, we get $a = 2$ and $d = 5$. (Verify this.) Thus the nth term of this sequence is

$$a_n = 2 + 5(n - 1)$$

The 1000th term is $a_{1000} = 2 + 5(1000 - 1) = 4997$.

 **Now Try Exercise 47** ■

■ Partial Sums of Arithmetic Sequences

Suppose we want to find the sum of the numbers 1, 2, 3, 4, . . . , 100, that is,

$$\sum_{k=1}^{100} k$$

When the now famous mathematician C. F. Gauss (see Section 3.5) was a schoolboy, his teacher posed this problem to the class and expected that it would keep the students busy for a long time. But Gauss answered the question almost immediately. His idea was this: Since we are adding numbers produced according to a fixed pattern, there must also be a pattern (or formula) for finding the sum. He started by writing the numbers

Mathematics in the Modern World

Fair Voting Methods

Mathematics has been applied to voting systems. You may ask, What is the problem with how we vote in elections now? Well, suppose candidates A, B, and C are running for office. The final vote tally is as follows: A got 40%, B got 39%, and C got 21%. So, candidate A wins. But 60% of voters *did not* want A. Moreover, most of the voters who voted for C prefer B over A and would have been willing to change their vote to B so that A would not win. So we have a situation where most of the voters prefer B over A, but A won. Is this fair?

In the 1950s, Kenneth Arrow showed mathematically that no democratic method of voting can be completely fair, and he later won a Nobel Prize for his work. Mathematicians continue to work on finding voting systems that are more fair for the voter. The system most often used in federal, state, and local elections is called *plurality voting* (the candidate with the most votes wins). Other systems include *majority voting* (if no candidate gets a majority, a runoff is held between the top two votegetters), *approval voting* (each voter can vote for as many candidates as they approve of), *ranked-choice voting* (each voter orders the candidates according to his or her preference). In this last system, it is advantageous for candidates to appeal to a wider audience in order to also receive second-choice (or third-choice) votes and thus increase their chances of winning. It is thought that such a system would result in a less polarized electorate.

from 1 to 100 and then below them wrote the same numbers in reverse order. Writing S for the sum and adding corresponding terms give

$$\begin{aligned} S &= 1 + 2 + 3 + \cdots + 98 + 99 + 100 \\ S &= 100 + 99 + 98 + \cdots + 3 + 2 + 1 \\ \hline 2S &= 101 + 101 + 101 + \cdots + 101 + 101 + 101 \end{aligned}$$

It follows that $2S = 100(101) = 10{,}100$, so $S = 5050$.

Of course, the sequence of natural numbers 1, 2, 3, . . . is an arithmetic sequence (with $a = 1$ and $d = 1$), and the method for summing the first 100 terms of this sequence can be used to find a formula for the nth partial sum of any arithmetic sequence. We want to find the sum of the first n terms of the arithmetic sequence whose terms are $a_k = a + (k - 1)d$; that is, we want to find

$$\begin{aligned} S_n &= \sum_{k=1}^{n} [a + (k-1)d] \\ &= a + (a + d) + (a + 2d) + (a + 3d) + \cdots + [a + (n-1)d] \end{aligned}$$

Using Gauss's method, we write

$$\begin{aligned} S_n &= a + (a+d) + \cdots + [a + (n-2)d] + [a + (n-1)d] \\ S_n &= [a + (n-1)d] + [a + (n-2)d] + \cdots + (a+d) + a \\ \hline 2S_n &= [2a + (n-1)d] + [2a + (n-1)d] + \cdots + [2a + (n-1)d] + [2a + (n-1)d] \end{aligned}$$

There are n identical terms on the right side of this equation, so

$$2S_n = n[2a + (n-1)d]$$

$$S_n = \frac{n}{2}[2a + (n-1)d]$$

Notice that $a_n = a + (n - 1)d$ is the nth term of this sequence. So we can write

$$S_n = \frac{n}{2}[a + a + (n-1)d] = n\left(\frac{a + a_n}{2}\right)$$

This last formula says that the sum of the first n terms of an arithmetic sequence is the average of the first and nth terms multiplied by n, the number of terms in the sum. We now summarize this result.

Partial Sums of an Arithmetic Sequence

For the arithmetic sequence given by $a_n = a + (n - 1)d$, the ***n*th partial sum**

$$S_n = a + (a + d) + (a + 2d) + (a + 3d) + \cdots + [a + (n-1)d]$$

is given by either of the following formulas.

1. $S_n = \dfrac{n}{2}[2a + (n-1)d]$ **2.** $S_n = n\left(\dfrac{a + a_n}{2}\right)$

Example 4 ■ Finding a Partial Sum of an Arithmetic Sequence

Find the sum of the first 50 odd numbers.

Solution The odd numbers form an arithmetic sequence with $a = 1$ and $d = 2$. The nth term is $a_n = 1 + 2(n - 1) = 2n - 1$, so the 50th odd number is $a_{50} = 2(50) - 1 = 99$. Substituting in Formula 2 for the partial sum of an arithmetic sequence, we get

$$S_{50} = 50\left(\frac{a + a_{50}}{2}\right) = 50\left(\frac{1 + 99}{2}\right) = 50 \cdot 50 = 2500$$

Now Try Exercise 51 ■

Example 5 ■ Finding a Partial Sum of an Arithmetic Sequence

Find the following partial sum of an arithmetic sequence.

$$3 + 7 + 11 + 15 + \cdots + 159$$

Solution For this sequence $a = 3$ and $d = 4$, so $a_n = 3 + 4(n - 1)$. To find which term of the sequence is the last term 159, we use the formula for the nth term and solve for n.

$$159 = 3 + 4(n - 1) \qquad \text{Set } a_n = 159$$

$$39 = n - 1 \qquad \text{Subtract 3; divide by 4}$$

$$n = 40 \qquad \text{Add 1}$$

To find the partial sum of the first 40 terms, we use Formula 1 for the nth partial sum of an arithmetic sequence:

$$S_{40} = \tfrac{40}{2}[2(3) + 4(40 - 1)] = 3240$$

We can also use Formula 2:

$$S_{40} = 40\left(\frac{3 + 159}{2}\right) = 3240$$

Now Try Exercise 57 ■

Example 6 ■ Finding the Seating Capacity of an Amphitheater

An amphitheater has 50 rows of seats with 30 seats in the first row, 32 in the second, 34 in the third, and so on. Find the total number of seats.

Solution The numbers of seats in the rows form an arithmetic sequence with $a = 30$ and $d = 2$. Since there are 50 rows, the total number of seats is the sum

$$S_{50} = \tfrac{50}{2}[2(30) + 49(2)] \qquad S_n = \frac{n}{2}[2a + (n - 1)d]$$

$$= 3950$$

Thus the amphitheater has 3950 seats.

Now Try Exercise 75 ■

Example 7 ■ Finding the Number of Terms in a Partial Sum

How many terms of the arithmetic sequence 5, 7, 9, . . . must be added to get 572?

Solution We are asked to find n when $S_n = 572$. Substituting $a = 5$, $d = 2$, and $S_n = 572$ in Formula 1 for the partial sum of an arithmetic sequence, we get

$$572 = \frac{n}{2}[2 \cdot 5 + (n - 1)2] \qquad S_n = \frac{n}{2}[2a + (n - 1)d]$$

$$572 = 5n + n(n - 1) \qquad \text{Distributive Property}$$

$$0 = n^2 + 4n - 572 \qquad \text{Expand}$$

$$0 = (n - 22)(n + 26) \qquad \text{Factor}$$

This gives $n = 22$ or $n = -26$. But since n is the *number* of terms in this partial sum, we must have $n = 22$.

Now Try Exercise 65 ■

11.2 Exercises

Concepts

1. An arithmetic sequence is a sequence in which the ________ between successive terms is constant.

2. The sequence given by $a_n = a + (n - 1)d$ is an arithmetic sequence in which a is the first term and d is the ________ ________. So for the arithmetic sequence $a_n = 2 + 5(n - 1)$ the first term is ________, and the common difference is ________.

3–4 ■ *True or False?* If *False*, give a reason.

3. The nth partial sum of an arithmetic sequence is the average of the first and last terms times n.

4. If we know the first and second terms of an arithmetic sequence, then we can find any other term.

Skills

5–10 ■ Terms of an Arithmetic Sequence The nth term of an arithmetic sequence is given. **(a)** Find the first five terms of the sequence. **(b)** What is the common difference d? **(c)** Graph the terms you found in part (a).

5. $a_n = 7 + 3(n - 1)$ **6.** $a_n = -10 + 20(n - 1)$

7. $a_n = -3 - 5(n - 1)$ **8.** $a_n = 7 - 3(n - 1)$

9. $a_n = 1.5 + 0.5(n - 1)$ **10.** $a_n = \frac{1}{2}(n - 1)$

11–16 ■ nth Term of an Arithmetic Sequence Find the nth term of the arithmetic sequence with given first term a and common difference d. What is the 10th term?

11. $a = -10, \quad d = 6$ **12.** $a = 5, \quad d = -2$

13. $a = 0.6, \quad d = -1$ **14.** $a = 1.8, \quad d = -0.2$

15. $a = \frac{5}{2}, \quad d = -\frac{1}{2}$ **16.** $a = \sqrt{3}, \quad d = \sqrt{3}$

17–26 ■ Arithmetic Sequence? The first four terms of a sequence are given. Can these terms be the terms of an arithmetic sequence? If so, find the common difference.

17. 11, 17, 23, 29, . . . **18.** −31, −19, −7, 5, . . .

19. 16, 9, 2, −4, . . . **20.** 100, 68, 36, 4, . . .

21. 2, 4, 8, 16, . . . **22.** 2, 4, 6, 8, . . .

23. $3, \frac{3}{2}, 0, -\frac{3}{2}, \ldots$ **24.** ln 2, ln 4, ln 8, ln 16, . . .

25. 2.6, 4.3, 6.0, 7.7, . . . **26.** $\frac{1}{2}, \frac{1}{3}, \frac{1}{4}, \frac{1}{5}, \ldots$

27–32 ■ Arithmetic Sequence? Find the first five terms of the sequence, and determine whether it is arithmetic. If it is arithmetic, find the common difference, and express the nth term of the sequence in the standard form $a_n = a + (n - 1)d$.

27. $a_n = 4 + 7n$ **28.** $a_n = 4 + 2^n$

29. $a_n = \dfrac{1}{1 + 2n}$ **30.** $a_n = 1 + \dfrac{n}{2}$

31. $a_n = 6n - 10$ **32.** $a_n = 3 + (-1)^n n$

33–44 ■ Terms of an Arithmetic Sequence Determine the common difference, the fifth term, the nth term, and the 100th term of the arithmetic sequence.

33. 6, 8, 10, 12, . . .

34. −5, 0, 5, 10, . . .

35. 29, 11, −7, −25, . . .

36. 64, 49, 34, 19, . . .

37. 4, 9, 14, 19, . . .

38. 11, 8, 5, 2, . . .

39. −12, −8, −4, 0, . . .

40. $\frac{1}{4}, \frac{3}{4}, \frac{5}{4}, \frac{7}{4}, \ldots$

41. 25, 26.5, 28, 29.5, . . .

42. 15, 12.3, 9.6, 6.9, . . .

43. $2, 2 + s, 2 + 2s, 2 + 3s, \ldots$

44. $-t, -t + 3, -t + 6, -t + 9, \ldots$

45–50 ■ Finding Terms of an Arithmetic Sequence Find the indicated term of the arithmetic sequence with the given description.

45. The 50th term is 1000, and the common difference is 6. Find the first and second terms.

46. The 100th term is −750, and the common difference is −20. Find the fifth term.

47. The fourteenth term is $\frac{2}{3}$, and the ninth term is $\frac{1}{4}$. Find the first term and the nth term.

48. The twelfth term is 118, and the eighth term is 146. Find the first term and the nth term.

49. The first term is 25, and the common difference is 18. Which term of the sequence is 601?

50. The first term is 3500, and the common difference is −15. Which term of the sequence is 2795?

51–56 ■ Partial Sums of an Arithmetic Sequence Find the partial sum S_n of the arithmetic sequence that satisfies the given conditions.

51. $a = 3, \quad d = 5, \quad n = 20$

52. $a = 10, \quad d = -8, \quad n = 30$

53. $a = -40, \quad d = 14, \quad n = 15$

54. $a = -2, \quad d = 23, \quad n = 25$

55. $a_1 = 4, \quad a_3 = -2, \quad n = 15$

56. $a_3 = 45, \quad a_7 = 55, \quad n = 49$

57–64 ■ Partial Sums of an Arithmetic Sequence A partial sum of an arithmetic sequence is given. Find the sum.

57. $1 + 5 + 9 + \cdots + 401$

58. $-5 - 2.5 + 0 + 2.5 + \cdots + 60$

59. $250 + 233 + 216 + \cdots + 97$

60. $89 + 85 + 81 + \cdots + 13$

61. $0.7 + 2.7 + 4.7 + \cdots + 56.7$

62. $-10 - 9.9 - 9.8 - \cdots - 0.1$

63. $\sum_{k=0}^{10} (3 + 0.25k)$

64. $\sum_{n=0}^{20} (1 - 2n)$

65–66 ■ Adding Terms of an Arithmetic Sequence Find the number of terms of the arithmetic sequence with the given description that must be added to get a value of 2700.

65. The first term is 5, and the common difference is 2.

66. The first term is 12, and the common difference is 8.

Skills Plus

67. Special Triangle Show that a right triangle whose sides are in arithmetic progression is similar to a 3–4–5 triangle.

68. Product of Numbers Find the product of the numbers

$$10^{1/10}, 10^{2/10}, 10^{3/10}, 10^{4/10}, \ldots, 10^{19/10}$$

69. Harmonic Sequence A sequence is **harmonic** if the reciprocals of the terms of the sequence form an arithmetic sequence. Determine whether the following sequence is harmonic.

$$1, \tfrac{3}{5}, \tfrac{3}{7}, \tfrac{1}{3}, \ldots$$

70. Harmonic Mean The **harmonic mean** of two numbers is the reciprocal of the average of the reciprocals of the two numbers. Find the harmonic mean of 3 and 5.

Applications

71. Depreciation The purchase value of an office computer client-server is \$12,500. Its annual depreciation is \$1875. Find the value of the computer after 6 years.

72. Poles in a Pile Utility poles are being stored in a pile with 25 poles in the first layer, 24 in the second, and so on. If there are 12 layers, how many utility poles does the pile contain?

73. Salary Increases A management position provides a salary of \$45,000 a year with a \$2000 raise every year. Find the total earnings for the first 10 years.

74. Drive-In Theater A drive-in theater has spaces for 20 cars in the first parking row, 22 in the second, 24 in the third, and so on. If there are 21 rows in the theater, find the number of cars that can be parked.

75. Theater Seating An architect designs a theater with 15 seats in the first row, 18 in the second, 21 in the third, and so on. If the theater is to have a seating capacity of 870, how many rows must the architect use in the design?

76. Falling Ball When an object is allowed to fall freely near the surface of the earth, the gravitational pull is such that the object falls 5 m in the first second, 15 m in the next second, 25 m in the next second, and so on.

(a) Find the total distance a ball falls in 6 s.

(b) Find a formula for the total distance a ball falls in n seconds.

77. The Twelve Days of Christmas In the well-known song "The Twelve Days of Christmas," a sweetheart receives k gifts on the kth day for each of the 12 days of Christmas. The sweetheart also receives every previous gift identically on each subsequent day. Thus on the 12th day the sweetheart receives a gift for the first day, 2 gifts for the second, 3 gifts for the third, and so on. Show that the number of gifts received on the 12th day is a partial sum of an arithmetic sequence. Find this sum.

Discuss ■ Discover ■ Prove ■ Write

78. Discuss: Arithmetic Means The **arithmetic mean** (or average) of two numbers a and b is

$$m = \frac{a + b}{2}$$

Note that m is the same distance from a as from b, so a, m, b is an arithmetic sequence. In general, if $m_1, m_2, \ldots, m_k$ are equally spaced between a and b so that

$$a, m_1, m_2, \ldots, m_k, b$$

is an arithmetic sequence, then $m_1, m_2, \ldots, m_k$ are called k arithmetic means between a and b.

(a) Insert two arithmetic means between 10 and 18.

(b) Insert three arithmetic means between 10 and 18.

(c) Suppose a doctor needs to increase a patient's dosage of a certain medicine from 100 mg to 300 mg per day in five equal steps. How many arithmetic means must be inserted between 100 and 300 to give the progression of daily doses, and what are these means?

11.3 Geometric Sequences

■ Geometric Sequences ■ Partial Sums of Geometric Sequences ■ What Is an Infinite Series? ■ Infinite Geometric Series

In this section we study geometric sequences. This type of sequence occurs frequently in applications to finance, population growth, and other fields.

■ Geometric Sequences

Recall that an arithmetic sequence is generated when we repeatedly add a number d to an initial term a. A *geometric* sequence is generated when we start with a number a and repeatedly *multiply* by a fixed nonzero constant r.

Definition of a Geometric Sequence

A **geometric sequence** is a sequence of the form

$$a, ar, ar^2, ar^3, ar^4, \ldots$$

The number a is the **first term**, and r is the **common ratio** of the sequence. The ***n*th term** of a geometric sequence is given by

$$a_n = ar^{n-1}$$

The number r is called the common ratio because the ratio of any two consecutive terms of the sequence is r.

Example 1 ■ Geometric Sequences

(a) If $a = 3$ and $r = 2$, then we have the geometric sequence

$$3, \quad 3 \cdot 2, \quad 3 \cdot 2^2, \quad 3 \cdot 2^3, \quad 3 \cdot 2^4, \quad \ldots$$

or

$$3, 6, 12, 24, 48, \ldots$$

Notice that the ratio of any two consecutive terms is $r = 2$. The nth term is $a_n = 3(2)^{n-1}$.

(b) The sequence

$$2, -10, 50, -250, 1250, \ldots$$

is a geometric sequence with $a = 2$ and $r = -5$. When r is negative, the terms of the sequence alternate in sign. The nth term is $a_n = 2(-5)^{n-1}$.

(c) The sequence

$$1, \frac{1}{3}, \frac{1}{9}, \frac{1}{27}, \frac{1}{81}, \ldots$$

is a geometric sequence with $a = 1$ and $r = \frac{1}{3}$. The nth term is $a_n = 1\left(\frac{1}{3}\right)^{n-1}$.

(d) The graph of the geometric sequence defined by $a_n = \frac{1}{5} \cdot 2^{n-1}$ is shown in Figure 1. Notice that the points in the graph coincide with the graph of the exponential function $y = \frac{1}{5} \cdot 2^{x-1}$.

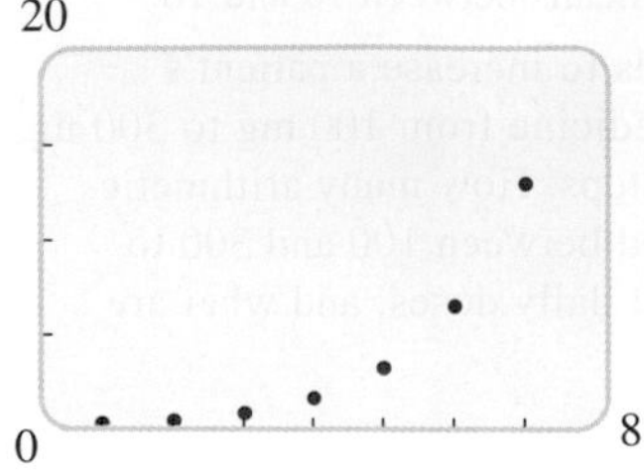

Figure 1

If $0 < r < 1$, then the terms of the geometric sequence ar^{n-1} decrease, whereas if $r > 1$, then the terms increase. (What happens if $r = 1$?)

Now Try Exercises 5, 9, and 13 ■

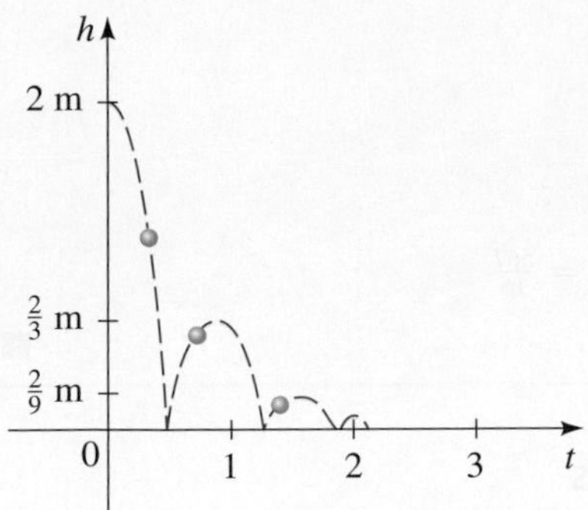

Figure 2

Geometric sequences occur naturally. Here is an example. Suppose a ball has elasticity such that when it is dropped, it bounces up one-third of the distance it has fallen. If this ball is dropped from a height of 2 meters, then it bounces up to a height of $2(\frac{1}{3}) = \frac{2}{3}$ meters. On its second bounce, it returns to a height of $(\frac{2}{3})(\frac{1}{3}) = \frac{2}{9}$ meters, and so on (see Figure 2). Thus the height h_n that the ball reaches on its nth bounce is given by the geometric sequence

$$h_n = \frac{2}{3}\left(\frac{1}{3}\right)^{n-1} = 2\left(\frac{1}{3}\right)^n$$

We can find the nth term of a geometric sequence if we know any two terms, as the following examples show.

Example 2 ■ Finding Terms of a Geometric Sequence

Find the common ratio, the first term, the nth term, and the eighth term of the geometric sequence

$$5, 15, 45, 135, \ldots$$

Solution To find a formula for the nth term of this sequence, we need to find the first term a and the common ratio r. Clearly, $a = 5$. To find r, we find the ratio of any two consecutive terms. For instance, $r = \frac{45}{15} = 3$. Thus

$$a_n = 5(3)^{n-1} \qquad a_n = ar^{n-1}$$

The eighth term is $a_8 = 5(3)^{8-1} = 5(3)^7 = 10{,}935$.

 Now Try Exercise 29 ■

Example 3 ■ Finding Terms of a Geometric Sequence

The third term of a geometric sequence is $\frac{63}{4}$, and the sixth term is $\frac{1701}{32}$. Find the fifth term.

Solution Since this sequence is geometric, its nth term is given by the formula $a_n = ar^{n-1}$. Thus

$$a_3 = ar^{3-1} = ar^2$$

$$a_6 = ar^{6-1} = ar^5$$

From the values we are given for these two terms, we get the following system of equations:

$$\begin{cases} \frac{63}{4} = ar^2 \\ \frac{1701}{32} = ar^5 \end{cases}$$

We solve this system by dividing.

$$\frac{ar^5}{ar^2} = \frac{\frac{1701}{32}}{\frac{63}{4}}$$

$$r^3 = \frac{27}{8} \qquad \text{Simplify}$$

$$r = \frac{3}{2} \qquad \text{Take cube root of each side}$$

Substituting for r in the first equation gives

$$\frac{63}{4} = a\left(\frac{3}{2}\right)^2 \qquad \text{Substitute } r = \tfrac{3}{2} \text{ in } \tfrac{63}{4} = ar^2$$

$$a = 7 \qquad \text{Solve for } a$$

SRINIVASA RAMANUJAN (1887–1920) was born into a poor family in the small town of Kumbakonam in India. Self-taught in mathematics, he worked in virtual isolation from other mathematicians. At the age of 25 he wrote a letter to G. H. Hardy, the leading British mathematician at the time, listing some of his discoveries. His discoveries included the following series for calculating π:

$$\frac{1}{\pi} = \frac{2\sqrt{2}}{9801}\sum_{k=0}^{\infty}\frac{(4k)!(1103 + 26390k)}{(k!)^4 396^{4k}}$$

Hardy immediately recognized Ramanujan's genius, and for the next six years the two worked together in London until Ramanujan fell ill and returned to his hometown in India, where he died a year later. Ramanujan was a genius with a phenomenal ability to see hidden patterns in the properties of numbers. Most of his discoveries were written as complicated infinite series, the importance of which was not recognized until many years after his death. In the last year of his life he wrote 130 pages of mysterious formulas, many of which still defy proof. Hardy tells the story that when he visited Ramanujan in a hospital and arrived in a taxi, he remarked to Ramanujan that the cab's number, 1729, was uninteresting. Ramanujan replied "No, it is a very interesting number. It is the smallest number expressible as the sum of two cubes in two different ways." The 2015 movie *The Man Who Knew Infinity* is a biographical drama about Ramanujan.

It follows that the nth term of this sequence is

$$a_n = 7\left(\tfrac{3}{2}\right)^{n-1}$$

Thus the fifth term is

$$a_5 = 7\left(\tfrac{3}{2}\right)^{5-1} = 7\left(\tfrac{3}{2}\right)^4 = \tfrac{567}{16}$$

 **Now Try Exercise 41** ■

■ Partial Sums of Geometric Sequences

For the geometric sequence $a, ar, ar^2, ar^3, ar^4, \ldots, ar^{n-1}, \ldots$, the nth partial sum is

$$S_n = \sum_{k=1}^{n} ar^{k-1} = a + ar + ar^2 + ar^3 + ar^4 + \cdots + ar^{n-1}$$

To find a formula for S_n, we multiply S_n by r and subtract from S_n.

$$\begin{aligned} S_n &= a + ar + ar^2 + ar^3 + ar^4 + \cdots + ar^{n-1} \\ rS_n &= \phantom{a + {}} ar + ar^2 + ar^3 + ar^4 + \cdots + ar^{n-1} + ar^n \\ \hline S_n - rS_n &= a - ar^n \end{aligned}$$

So

$$S_n(1 - r) = a(1 - r^n)$$

$$S_n = \frac{a(1 - r^n)}{1 - r} \qquad (r \neq 1)$$

We summarize this result.

Partial Sums of a Geometric Sequence

For the geometric sequence defined by $a_n = ar^{n-1}$, the ***n*th partial sum**

$$S_n = a + ar + ar^2 + ar^3 + ar^4 + \cdots + ar^{n-1} \qquad (r \neq 1)$$

is given by

$$S_n = a\frac{1 - r^n}{1 - r}$$

Example 4 ■ Finding a Partial Sum of a Geometric Sequence

Find the following partial sum of a geometric sequence.

$$1 + 4 + 16 + \cdots + 4096$$

Solution For this sequence $a = 1$ and $r = 4$, so $a_n = 4^{n-1}$. Since $4^6 = 4096$, we use the formula for S_n with $n = 7$, and we have

$$S_7 = 1 \cdot \frac{1 - 4^7}{1 - 4} = 5461$$

Thus this partial sum is 5461.

Now Try Exercises 49 and 53 ■

Example 5 ■ Finding a Partial Sum of a Geometric Sequence

Find the sum $\sum_{k=1}^{6} 7\left(-\tfrac{2}{3}\right)^{k-1}$.

Solution The given sum is the sixth partial sum of a geometric sequence with first term $a = 7\left(-\frac{2}{3}\right)^0 = 7$ and $r = -\frac{2}{3}$. Thus by the formula for S_n with $n = 6$ we have

$$S_6 = 7 \cdot \frac{1 - \left(-\frac{2}{3}\right)^6}{1 - \left(-\frac{2}{3}\right)} = 7 \cdot \frac{1 - \frac{64}{729}}{\frac{5}{3}} = \frac{931}{243} \approx 3.83$$

 **Now Try Exercise 59** ■

■ What Is an Infinite Series?

An expression of the form

$$\sum_{k=1}^{\infty} a_k = a_1 + a_2 + a_3 + a_4 + \cdots$$

is called an **infinite series**. The dots mean that we are to continue the addition indefinitely. What meaning can we attach to the sum of infinitely many numbers? It seems at first that it is not possible to add infinitely many numbers and arrive at a finite number. But consider the following problem. You want to eat a cake by first eating half the cake, then eating half of what remains, then again eating half of what remains. This process can continue indefinitely because at each stage, some of the cake remains (see Figure 3).

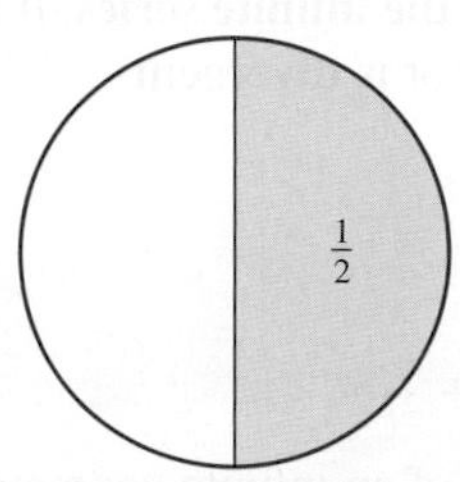

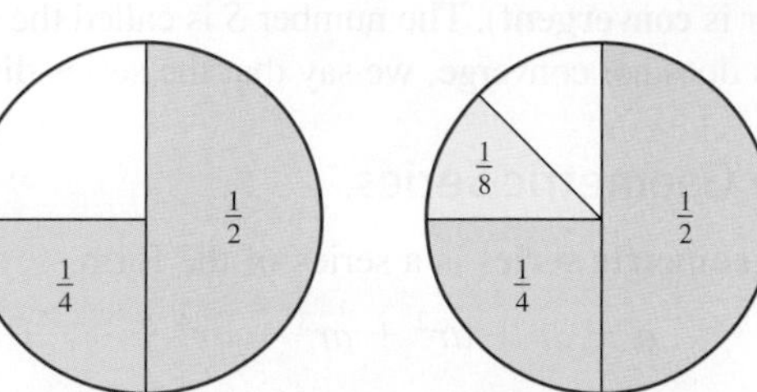

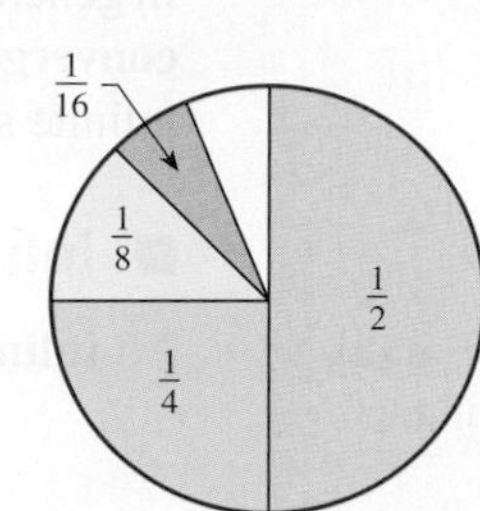

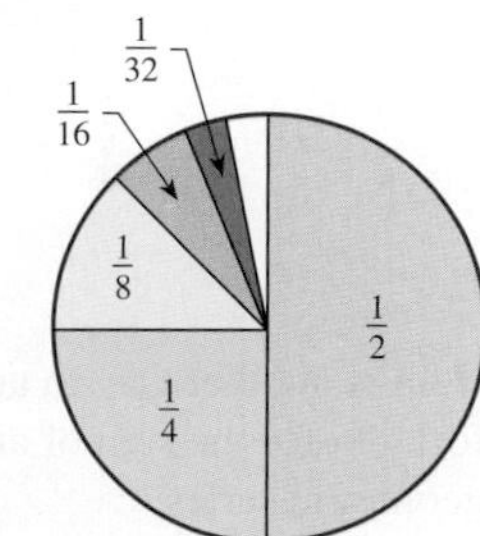

Figure 3

Does this mean that it's impossible to eat all of the cake? Of course not. Let's write down what you have eaten from this cake:

$$\sum_{k=1}^{\infty} \frac{1}{2^k} = \frac{1}{2} + \frac{1}{4} + \frac{1}{8} + \frac{1}{16} + \cdots$$

This is an infinite series, and we note two things about it: First, from Figure 3 it's clear that no matter how many terms of this series we add, the total will never exceed 1. Second, the more terms of this series we add, the closer the sum is to 1. (See Figure 3.) This suggests that the number 1 can be written as the sum of infinitely many smaller numbers:

$$1 = \frac{1}{2} + \frac{1}{4} + \frac{1}{8} + \frac{1}{16} + \cdots + \frac{1}{2^n} + \cdots$$

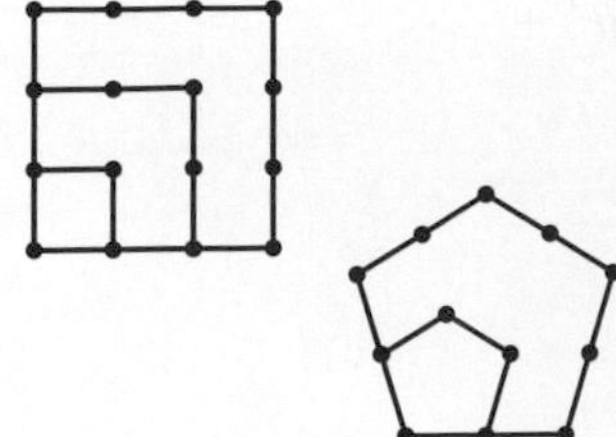

Discovery Project ■ Finding Patterns

Finding patterns in nature is an important part of mathematical modeling. If we can find a pattern (or a formula) that describes the terms of a sequence, then we can use the pattern to predict subsequent terms of the sequence. In this project we investigate difference sequences and how they help us find patterns in triangular, square, pentagonal, and other polygonal numbers. You can find the project at the book companion website **www.stewartmath.com**.

To make this more precise, let's look at the partial sums of this series:

$$S_1 = \frac{1}{2} = \frac{1}{2}$$

$$S_2 = \frac{1}{2} + \frac{1}{4} = \frac{3}{4}$$

$$S_3 = \frac{1}{2} + \frac{1}{4} + \frac{1}{8} = \frac{7}{8}$$

$$S_4 = \frac{1}{2} + \frac{1}{4} + \frac{1}{8} + \frac{1}{16} = \frac{15}{16}$$

and, in general (see Example 11.1.5),

$$S_n = 1 - \frac{1}{2^n}$$

As n gets larger and larger, we are adding more and more terms of this series. Intuitively, as n gets larger, S_n gets closer to the sum of the series. Now notice that as n gets large, $1/2^n$ gets closer and closer to 0. Thus S_n gets close to $1 - 0 = 1$. Using the notation of Section 3.6, we can write

$$S_n \to 1 \quad \text{as} \quad n \to \infty$$

In general, if S_n gets close to a finite number S as n gets large, we say that the infinite series **converges** (or is **convergent**). The number S is called the **sum of the infinite series**. If an infinite series does not converge, we say that the series **diverges** (or is **divergent**).

■ Infinite Geometric Series

An **infinite geometric series** is a series of the form

$$a + ar + ar^2 + ar^3 + ar^4 + \cdots + ar^{n-1} + \cdots$$

We can apply the reasoning that we used earlier to find the sum of an infinite geometric series. The nth partial sum of such a series is given by the formula

$$S_n = a\frac{1 - r^n}{1 - r} \qquad (r \neq 1)$$

It can be shown that if $|r| < 1$, then r^n gets close to 0 as n gets large (you can easily convince yourself of this using a calculator). It follows that S_n gets close to $a/(1 - r)$ as n gets large, or

$$S_n \to \frac{a}{1 - r} \quad \text{as} \quad n \to \infty$$

Thus the sum of this infinite geometric series is $a/(1 - r)$.

Here is another way to arrive at the formula for the sum of an infinite geometric series:

$$\begin{aligned} S &= a + ar + ar^2 + ar^3 + \cdots \\ &= a + r(a + ar + ar^2 + \cdots) \\ &= a + rS \end{aligned}$$

Solve the equation $S = a + rS$ for S to get

$$\begin{aligned} S - rS &= a \\ (1 - r)S &= a \\ S &= \frac{a}{1 - r} \end{aligned}$$

Sum of an Infinite Geometric Series

If $|r| < 1$, then the infinite geometric series

$$\sum_{k=1}^{\infty} ar^{k-1} = a + ar + ar^2 + ar^3 + \cdots$$

converges and has the sum

$$S = \frac{a}{1 - r}$$

If $|r| \geq 1$, the series diverges.

Example 6 ■ Infinite Series

Determine whether the infinite geometric series is convergent or divergent. If it is convergent, find its sum.

(a) $2 + \frac{2}{5} + \frac{2}{25} + \frac{2}{125} + \cdots$ **(b)** $1 + \frac{7}{5} + \left(\frac{7}{5}\right)^2 + \left(\frac{7}{5}\right)^3 + \cdots$

Solution

(a) This is an infinite geometric series with $a = 2$ and $r = \frac{1}{5}$. Since $|r| = \left|\frac{1}{5}\right| < 1$, the series converges. By the formula for the sum of an infinite geometric series, we have

$$S = \frac{2}{1 - \frac{1}{5}} = \frac{5}{2}$$

(b) This is an infinite geometric series with $a = 1$ and $r = \frac{7}{5}$. Since $|r| = \left|\frac{7}{5}\right| > 1$, the series diverges.

 Now Try Exercises 65 and 69 ■

Example 7 ■ Writing a Repeated Decimal as a Fraction

Find the fraction that represents the rational number $2.3\overline{51}$.

Solution This repeating decimal can be written as a series:

$$\frac{23}{10} + \frac{51}{1000} + \frac{51}{100{,}000} + \frac{51}{10{,}000{,}000} + \frac{51}{1{,}000{,}000{,}000} + \cdots$$

After the first term, the terms of this series form an infinite geometric series with

$$a = \frac{51}{1000} \quad \text{and} \quad r = \frac{1}{100}$$

Thus the sum of this part of the series is

$$S = \frac{\frac{51}{1000}}{1 - \frac{1}{100}} = \frac{\frac{51}{1000}}{\frac{99}{100}} = \frac{51}{1000} \cdot \frac{100}{99} = \frac{51}{990}$$

It follows that

$$2.3\overline{51} = \frac{23}{10} + \frac{51}{990} = \frac{2328}{990} = \frac{388}{165}$$

 Now Try Exercise 77 ■

Mathematics in the Modern World

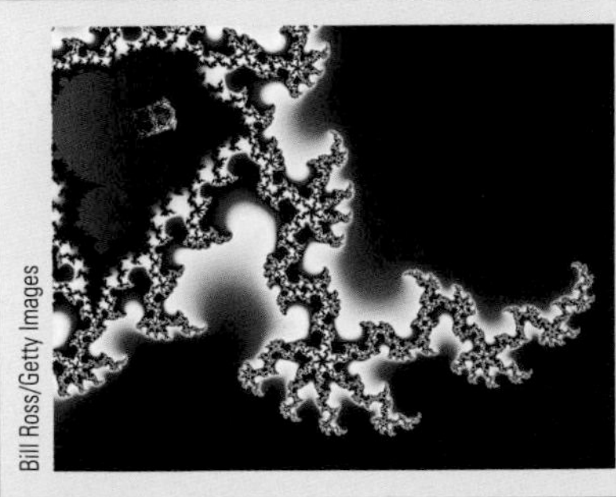
Bill Ross/Getty Images

Fractals

Many of the things we model in this book have regular predictable shapes. But recent advances in mathematics have made it possible to model such seemingly random or even chaotic shapes as those of a cloud, a flickering flame, a mountain, or a jagged coastline. The basic tools in this type of modeling are the fractals invented by the mathematician Benoit Mandelbrot. A *fractal* is a geometric shape built up from a simple basic shape by scaling and repeating the shape indefinitely according to a given rule. Fractals have infinite detail; this means the closer you look, the more you see. They are also *self-similar*; that is, zooming in on a portion of the fractal yields the same detail as the original shape. Because of their beautiful shapes, fractals are used by moviemakers to create fictional landscapes and exotic backgrounds.

Although a fractal is a complex shape, it is produced according to very simple rules. This property of fractals is exploited in a process of storing pictures on a computer called *fractal image compression*.

11.3 | Exercises

Concepts

1. A geometric sequence is a sequence in which the ________ of successive terms is constant.

2. The sequence given by $a_n = ar^{n-1}$ is a geometric sequence in which a is the first term and r is the ________ ________. So for the geometric sequence $a_n = 2(5)^{n-1}$ the first term is ________, and the common ratio is ________.

3. *True or False*? If we know the first and second terms of a geometric sequence, then we can find any other term.

4. (a) The nth partial sum of a geometric sequence $a_n = ar^{n-1}$ is given by $S_n =$ ________.

(b) The series $\sum_{k=1}^{\infty} ar^{k-1} = a + ar + ar^2 + ar^3 + \cdots$ is an infinite ________ series. If $|r| < 1$, then this series ________, and its sum is $S =$ ________. If $|r| \geq 1$, the series ________.

Skills

5–8 ■ *n*th Term of a Geometric Sequence The nth term of a sequence is given. **(a)** Find the first five terms of the sequence. **(b)** What is the common ratio r? **(c)** Graph the terms you found in (a).

5. $a_n = 7(3)^{n-1}$

6. $a_n = 6(-0.5)^{n-1}$

7. $a_n = 8\left(-\frac{1}{4}\right)^{n-1}$

8. $a_n = -\frac{1}{9}(3)^{n-1}$

9–12 ■ *n*th Term of a Geometric Sequence Find the nth term of the geometric sequence with given first term a and common ratio r. What is the fourth term?

9. $a = 7, \quad r = 4$

10. $a = -\frac{3}{2}, \quad r = 3$

11. $a = 5, r = -3$

12. $a = \sqrt{3}, \quad r = \sqrt{3}$

13–22 ■ Geometric Sequence? The first four terms of a sequence are given. Determine whether these terms can be the terms of a geometric sequence. If the sequence is geometric, find the common ratio.

13. $3, 6, 12, 24, \dots$

14. $3, 48, 93, 138, \dots$

15. $\frac{1}{3}, \frac{1}{2}, \frac{2}{3}, \frac{5}{6}, \dots$

16. $432, -144, 48, -16, \dots$

17. $3, \frac{3}{2}, \frac{3}{4}, \frac{3}{8}, \dots$

18. $10, \frac{10}{3}, \frac{10}{9}, \frac{10}{27}, \dots$

19. $-\frac{1}{2}, \frac{1}{2}, -\frac{1}{4}, \frac{1}{4}, \dots$

20. $e^2, e^4, e^6, e^8, \dots$

21. $1.0, 1.1, 1.21, 1.331, \dots$

22. $\frac{1}{2}, \frac{1}{4}, \frac{1}{6}, \frac{1}{8}, \dots$

23–28 ■ Geometric Sequence? Find the first five terms of the sequence, and determine whether it is geometric. If it is geometric, find the common ratio, and express the nth term of the sequence in the standard form $a_n = ar^{n-1}$.

23. $a_n = 2(3)^n$

24. $a_n = 4 + 3^n$

25. $a_n = \dfrac{1}{4^n}$

26. $a_n = (-1)^n 2^n$

27. $a_n = \ln(5^{n-1})$

28. $a_n = n^n$

29–38 ■ Terms of a Geometric Sequence Determine the common ratio, the fifth term, and the nth term of the geometric sequence.

29. $2, 6, 18, 54, \dots$

30. $7, \frac{14}{3}, \frac{28}{9}, \frac{56}{27}, \dots$

31. $0.3, -0.09, 0.027, -0.0081, \dots$

32. $1, \sqrt{2}, 2, 2\sqrt{2}, \dots$

33. $144, -12, 1, -\frac{1}{12}, \dots$

34. $-8, -2, -\frac{1}{2}, -\frac{1}{8}, \dots$

35. $3, 3^{5/3}, 3^{7/3}, 27, \dots$

36. $t, \dfrac{t^2}{2}, \dfrac{t^3}{4}, \dfrac{t^4}{8}, \dots$

37. $1, s^{2/7}, s^{4/7}, s^{6/7}, \dots$

38. $5, 5^{c+1}, 5^{2c+1}, 5^{3c+1}, \dots$

39–46 ■ Finding Terms of a Geometric Sequence Find the indicated term(s) of the geometric sequence with the given description.

39. The first term is 14 and the second term is 4. Find the fourth term.

40. The first term is 8 and the second term is 6. Find the fifth term.

41. The third term is $-\frac{1}{3}$ and the sixth term is 9. Find the first and second terms.

42. The fourth term is 12 and the seventh term is $\frac{32}{9}$. Find the first and nth terms.

43. The third term is -18 and the sixth term is 9216. Find the first and nth terms.

44. The third term is -54 and the sixth term is $\frac{729}{256}$. Find the first and second terms.

45. The common ratio is 0.75 and the fourth term is 729. Find the first three terms.

46. The common ratio is $\frac{1}{6}$ and the third term is 18. Find the first and seventh terms.

47. Which Term? The first term of a geometric sequence is 1536 and the common ratio is $\frac{1}{2}$. Which term of the sequence is 6?

48. Which Term? The second and fifth terms of a geometric sequence are 30 and 3750, respectively. Which term of the sequence is 468,750?

49–52 ■ Partial Sums of a Geometric Sequence Find the partial sum S_n of the geometric sequence that satisfies the given conditions.

49. $a = 5, \ r = 2, \ n = 6$

50. $a = \frac{2}{3}, \quad r = \frac{1}{3}, \quad n = 4$

51. $a_3 = 28, \quad a_6 = 224, \quad n = 6$

52. $a_2 = 0.12, \quad a_5 = 0.00096, \quad n = 4$

53–58 ■ Partial Sums of a Geometric Sequence Find the sum.

53. $1 + 3 + 9 + \cdots + 2187$

54. $1 - \frac{1}{2} + \frac{1}{4} - \frac{1}{8} + \cdots - \frac{1}{512}$

55. $-15 + 30 - 60 + \cdots - 960$

56. $5120 + 2560 + 1280 + \cdots + 20$

57. $1.25 + 12.5 + 125 + \cdots + 12{,}500{,}000$

58. $10800 + 1080 + 108 + \cdots + 0.000108$

59–64 ■ Partial Sums of a Geometric Sequence Find the sum.

59. $\sum_{k=1}^{5} 3(\frac{1}{2})^{k-1}$ **60.** $\sum_{k=1}^{5} 8(-\frac{3}{2})^{k-1}$

61. $\sum_{k=1}^{6} 5(-2)^{k-1}$ **62.** $\sum_{k=1}^{6} 10(5)^{k-1}$

63. $\sum_{k=1}^{5} 3(\frac{2}{3})^{k-1}$ **64.** $\sum_{k=1}^{6} 64(\frac{3}{2})^{k-1}$

65–76 ■ Infinite Geometric Series Determine whether the infinite geometric series is convergent or divergent. If it is convergent, find its sum.

65. $1 + \frac{1}{3} + \frac{1}{9} + \frac{1}{27} + \cdots$ **66.** $1 - \frac{1}{2} + \frac{1}{4} - \frac{1}{8} + \cdots$

67. $1 - \frac{1}{3} + \frac{1}{9} - \frac{1}{27} + \cdots$ **68.** $\frac{2}{5} + \frac{4}{25} + \frac{8}{125} + \cdots$

69. $1 + \frac{3}{2} + \left(\frac{3}{2}\right)^2 + \left(\frac{3}{2}\right)^3 + \cdots$

70. $\frac{1}{3^6} + \frac{1}{3^8} + \frac{1}{3^{10}} + \frac{1}{3^{12}} + \cdots$

71. $3 - \frac{3}{2} + \frac{3}{4} - \frac{3}{8} + \cdots$

72. $1 - 1 + 1 - 1 + \cdots$

73. $3 - 3(1.1) + 3(1.1)^2 - 3(1.1)^3 + \cdots$

74. $-\frac{100}{9} + \frac{10}{3} - 1 + \frac{3}{10} - \cdots$

75. $\frac{1}{\sqrt{2}} + \frac{1}{2} + \frac{1}{2\sqrt{2}} + \frac{1}{4} + \cdots$

76. $1 - \sqrt{2} + 2 - 2\sqrt{2} + 4 - \cdots$

77–82 ■ Repeated Decimal Express the repeating decimal as a fraction.

77. 0.999 . . . **78.** $0.2\overline{53}$ **79.** 0.030303 . . .

80. $2.1\overline{125}$ **81.** $0.\overline{112}$ **82.** 0.123123123 . . .

Skills Plus

83. Geometric Means If the numbers $a_1, a_2, \ldots, a_n$ form a geometric sequence, then $a_2, a_3, \ldots, a_{n-1}$ are **geometric means** between a_1 and a_n. Insert three geometric means between 5 and 80.

84. Partial Sum of a Geometric Sequence Find the sum of the first ten terms of the sequence

$$a + b, a^2 + 2b, a^3 + 3b, a^4 + 4b, \ldots$$

85–86 ■ Arithmetic or Geometric? The first four terms of a sequence are given. Determine whether these terms can be the terms of an arithmetic sequence, a geometric sequence, or neither. If the sequence is arithmetic or geometric, find the next term.

85. (a) $5, -3, 5, -3, \ldots$ (b) $\frac{1}{3}, 1, \frac{5}{3}, \frac{7}{3}, \ldots$

(c) $\sqrt{3}, 3, 3\sqrt{3}, 9, \ldots$ (d) $-3, -\frac{3}{2}, 0, \frac{3}{2}, \ldots$

86. (a) $1, -1, 1, -1, \ldots$ (b) $\sqrt{5}, \sqrt[3]{5}, \sqrt[6]{5}, 1, \ldots$

(c) $2, -1, \frac{1}{2}, 2, \ldots$ (d) $x - 1, x, x + 1, x + 2, \ldots$

Applications

87. Depreciation A construction company purchases a bulldozer for \$160,000. Each year the value of the bulldozer depreciates by 20% of its value in the preceding year. Let V_n be the value of the bulldozer in the nth year. (Let $n = 1$ be the year the bulldozer is purchased.)

(a) Find a formula for V_n.

(b) In what year will the value of the bulldozer be less than \$100,000?

88. Ancestors A person has two parents, four grandparents, eight great-grandparents, and so on. How many ancestors does a person have 15 generations back?

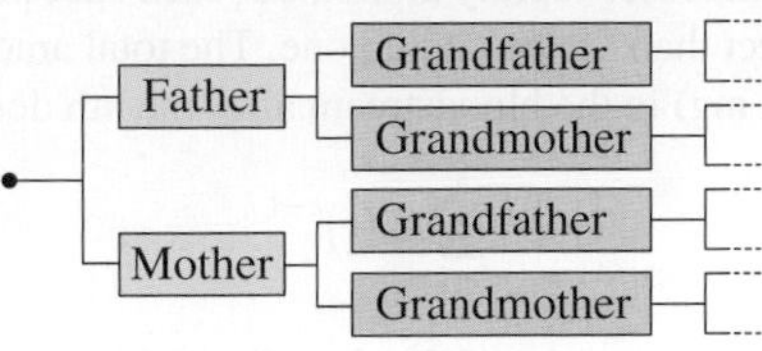

89. Bouncing Ball A ball is dropped from a height of 80 m. The elasticity of this ball is such that it rebounds three-fourths the distance it has fallen. How high does the ball rebound on the fifth bounce? Find a formula for how high the ball rebounds on the nth bounce.

90. Bacteria Culture A culture initially has 5000 bacteria, and its size increases by 8% every hour. How many bacteria are present at the end of 5 hours? Find a formula for the number of bacteria present after n hours.

91. Mixing Coolant A truck radiator holds 5 L and is filled with water. A liter of water is removed from the radiator and replaced with a liter of antifreeze; then a liter of the mixture is removed from the radiator and again replaced by a liter of antifreeze. This process is repeated indefinitely. How much water remains in the tank after this process is repeated 3 times? 5 times? n times?

92. Musical Frequencies The frequencies of musical notes (measured in cycles per second) form a geometric sequence. Middle C has a frequency of 256, and the C that is an octave higher has a frequency of 512. Find the frequency of C two octaves below middle C.

93. Bouncing Ball A ball is dropped from a height of 9 m. The elasticity of the ball is such that it always bounces up one-third the distance it has fallen.

(a) Find the total distance the ball has traveled at the instant it hits the ground the fifth time.

(b) Find a formula for the total distance the ball has traveled at the instant it hits the ground the nth time.

94. Salary Increases A management position provides a salary of \$45,000 in the first year. Each year the salary increases by 5% of the preceding year's salary. Find the total earnings in the first 10 years.

95. St. Ives The following is a well-known children's rhyme:

As I was going to St. Ives,
I met a man with seven wives;
Every wife had seven sacks;
Every sack had seven cats;
Every cat had seven kits;
Kits, cats, sacks, and wives,
How many were going to St. Ives?

Assuming that the entire group is actually going to St. Ives, show that the answer to the question in the rhyme is a partial sum of a geometric sequence, and find the sum.

96. Drug Concentration A certain drug is administered once a day. The concentration of the drug in the patient's bloodstream increases rapidly at first, but each successive dose has less effect than the preceding one. The total amount of the drug (in mg) in the bloodstream after the nth dose is given by

$$\sum_{k=1}^{n} 50\left(\tfrac{1}{2}\right)^{k-1}$$

(a) Find the amount of the drug in the bloodstream after $n = 10$ days.

(b) If the drug is taken on a long-term basis, the amount in the bloodstream is approximated by the infinite series $\sum_{k=1}^{\infty} 50\left(\tfrac{1}{2}\right)^{k-1}$. Find the sum of this series.

97. Bouncing Ball A certain ball rebounds to half the height from which it is dropped. Use an infinite geometric series to approximate the total distance the ball travels after being dropped from 1 meter above the ground until it comes to rest.

98. Bouncing Ball If the ball in Exercise 97 is dropped from a height of 8 m, then 1 second is required for its first complete bounce—from the instant it first touches the ground until it next touches the ground. Each subsequent complete bounce requires $1/\sqrt{2}$ as long as the preceding complete bounce. Use an infinite geometric series to estimate the time interval from the instant the ball first touches the ground until it stops bouncing.

99. Geometry The midpoints of the sides of a square of side 1 are joined to form a new square. This procedure is repeated for each new square (see the figure).

(a) Find the sum of the areas of all the squares.

(b) Find the sum of the perimeters of all the squares.

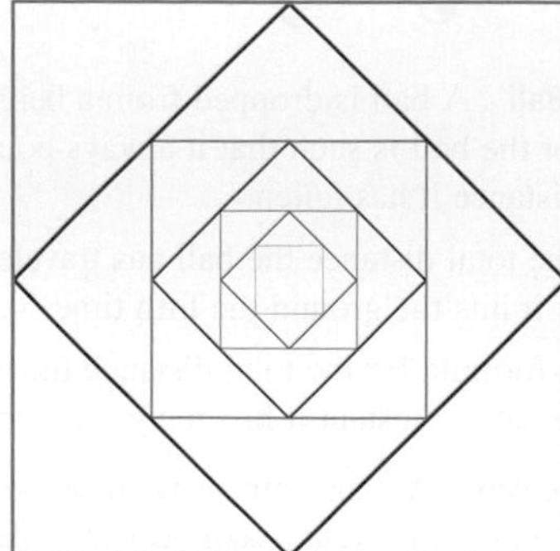

100. Geometry A circular disk of radius R is cut out of paper, as shown in figure (a). Two disks of radius $\frac{1}{2}R$ are cut out of paper and placed on top of the first disk, as in figure (b), and then four disks of radius $\frac{1}{4}R$ are placed on these two disks, as in figure (c). Assuming that this process can be repeated indefinitely, find the total area of all the disks.

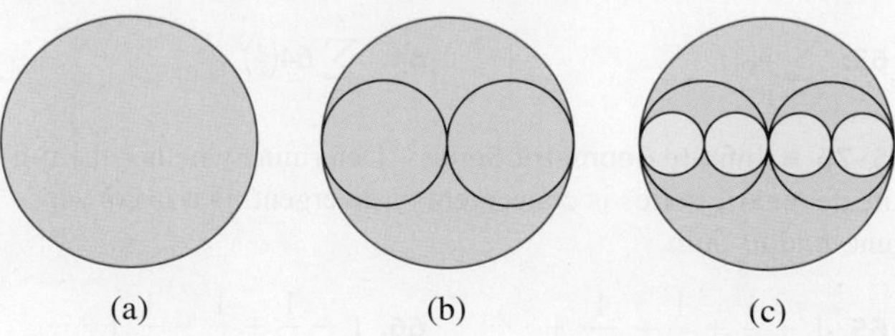

101. Geometry A yellow square of side 1 is divided into nine smaller squares, and the middle square is colored blue as shown in the figure. Each of the smaller yellow squares is in turn divided into nine squares, and each middle square is colored blue. If this process is continued indefinitely, what is the total area that is colored blue?

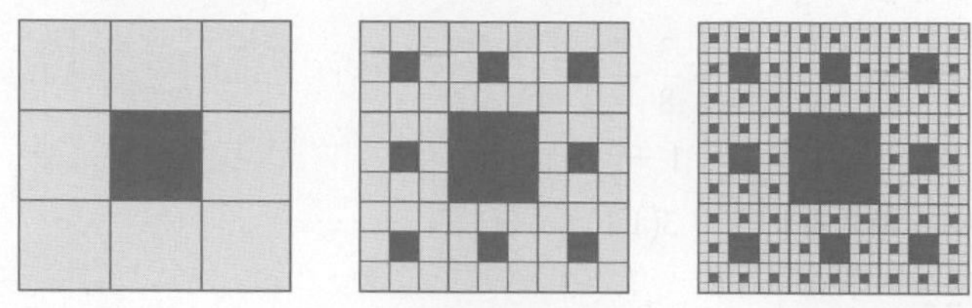

■ Discuss ■ Discover ■ Prove ■ Write

102. Prove: Reciprocals of a Geometric Sequence If $a_1, a_2, a_3, \ldots$ is a geometric sequence with common ratio r, show that the sequence

$$\frac{1}{a_1}, \frac{1}{a_2}, \frac{1}{a_3}, \ldots$$

is also a geometric sequence, and find the common ratio.

103. Prove: Logarithms of a Geometric Sequence If $a_1, a_2, a_3, \ldots$ is a geometric sequence with a common ratio $r > 0$ and $a_1 > 0$, show that the sequence

$$\log a_1, \log a_2, \log a_3, \ldots$$

is an arithmetic sequence, and find the common difference.

104. Prove: Exponentials of an Arithmetic Sequence If $a_1, a_2, a_3, \ldots$ is an arithmetic sequence with common difference d, show that the sequence

$$10^{a_1}, 10^{a_2}, 10^{a_3}, \ldots$$

is a geometric sequence, and find the common ratio.

105. Prove: A Factoring Formula Show that for $r \neq 1$ and k a natural number

$$(1 + r + r^2 + \cdots + r^{2^k-1}) = (1 + r + r^2 + \cdots + r^{2^{k-1}-1})(r^{2^{k-1}} + 1)$$

PS *Look for something familiar.* Use the formula for the sum of a geometric sequence.

11.4 Mathematical Induction

■ Conjecture and Proof ■ Mathematical Induction

There are two aspects to mathematics—discovery and proof—and they are of equal importance. We must discover something before we can attempt to prove it, and we cannot be certain of its truth until it has been proved. In this section we examine the relationship between these two key components of mathematics more closely.

■ Conjecture and Proof

Let's try an experiment. We add more and more of the odd numbers as follows:

$$1 = 1$$
$$1 + 3 = 4$$
$$1 + 3 + 5 = 9$$
$$1 + 3 + 5 + 7 = 16$$
$$1 + 3 + 5 + 7 + 9 = 25$$

What do you notice about the numbers on the right-hand side of these equations? They are, in fact, all perfect squares. These equations say the following:

The sum of the first 1 odd number is 1^2.

The sum of the first 2 odd numbers is 2^2.

The sum of the first 3 odd numbers is 3^2.

The sum of the first 4 odd numbers is 4^2.

The sum of the first 5 odd numbers is 5^2.

This leads naturally to the following question: Is it true that for every natural number n, the sum of the first n odd numbers is n^2? Could this remarkable property be true? We could try a few more numbers and find that the pattern persists for the first 6, 7, 8, 9, and 10 odd numbers. At this point we feel fairly confident that this is always true, so we make a *conjecture*:

The sum of the first n odd numbers is n^2.

Since we know that the nth odd number is $2n - 1$, we can write this statement more precisely as

$$1 + 3 + 5 + \cdots + (2n - 1) = n^2$$

It is important to realize that this is still a conjecture. We cannot conclude by checking a finite number of cases that a property is true for all numbers (there are infinitely many). To see this more clearly, suppose some mathematicians tell us that they have added up the first trillion odd numbers and found that they do *not* add up to 1 trillion squared. What would you tell these mathematicians? It would be silly to say that you're sure it's true because you have already checked the first five cases. You could, however, take out paper and pencil and start checking it yourself, but this task would probably take the rest of your life. The tragedy would be that after completing this task, you would still not be sure of the truth of the conjecture! Do you see why?

Herein lies the power of mathematical proof. A **proof** is a clear argument that demonstrates the truth of a statement beyond doubt.

Consider the polynomial

$$p(n) = n^2 - n + 41$$

Here are some values of $p(n)$:

$p(1) = 41$	$p(2) = 43$
$p(3) = 47$	$p(4) = 53$
$p(5) = 61$	$p(6) = 71$
$p(7) = 83$	$p(8) = 97$

All the values so far are prime numbers. In fact, if you keep going, you will find that $p(n)$ is prime for all natural numbers up to $n = 40$. It might seem reasonable at this point to conjecture that $p(n)$ is prime for *every* natural number n. But that conjecture would be too hasty because it is easily seen that $p(41)$ is *not* prime. This illustrates that we cannot be certain of the truth of a statement no matter how many special cases we check. We need a convincing argument—a *proof*—to determine the truth of a statement.

Christophel Fine Art/Getty Images

BLAISE PASCAL (1623–1662) is considered one of the most versatile minds in modern history. He was a writer and philosopher as well as a gifted mathematician and physicist. Among his contributions that appear in this book are Pascal's triangle and the Principle of Mathematical Induction.

Pascal's father, himself a mathematician, believed that his son should not study mathematics until he was 15 or 16. But at age 12, Blaise insisted on learning geometry and proved most of its elementary theorems himself. At 19 he invented the first mechanical adding machine. In 1647, after writing a major treatise on the conic sections, he abruptly abandoned mathematics because he felt that his intense studies were contributing to his ill health. He devoted himself instead to frivolous recreations such as gambling, but this only served to pique his interest in probability. In 1654 he miraculously survived a carriage accident in which his horses ran off a bridge. Taking this to be a sign from God, Pascal entered a monastery, where he pursued theology and philosophy, writing his famous *Pensées*. He also continued his mathematical research. He valued faith and intuition more than reason as the source of truth, declaring that "the heart has its own reasons, which reason cannot know."

■ Mathematical Induction

Let's consider a special kind of proof called **mathematical induction**. Here is how it works: Suppose we have a statement that says something about all natural numbers n. For example, for any natural number n, let $P(n)$ be the following statement:

$P(n)$: The sum of the first n odd numbers is n^2.

Since this statement is about all natural numbers, it contains infinitely many statements; we will call them $P(1)$, $P(2)$, ...

$P(1)$: The sum of the first 1 odd number is 1^2.

$P(2)$: The sum of the first 2 odd numbers is 2^2.

$P(3)$: The sum of the first 3 odd numbers is 3^2.

⋮ ⋮

How can we prove all of these statements at once? Mathematical induction is a clever way of doing just that.

The crux of the idea is this: Suppose we can prove that whenever one of these statements is true, then the one following it in the list is also true. In other words,

For every k, if $P(k)$ is true, then $P(k + 1)$ is true.

This is called the **induction step** because it leads us from the truth of one statement to the truth of the next. Now suppose that we can also prove that

$P(1)$ is true.

The induction step now leads us through the following chain of statements:

$P(1)$ is true, so $P(2)$ is true.

$P(2)$ is true, so $P(3)$ is true.

$P(3)$ is true, so $P(4)$ is true.

⋮ ⋮

So we see that if both the induction step and $P(1)$ are proved, then statement $P(n)$ is proved for all n. Here is a summary of this important method of proof.

Principle of Mathematical Induction

For each natural number n, let $P(n)$ be a statement depending on n. Suppose that the following two conditions are satisfied.

1. $P(1)$ is true.
2. For every natural number k, if $P(k)$ is true, then $P(k + 1)$ is true.

Then $P(n)$ is true for all natural numbers n.

To apply this principle, there are two steps:

Step 1 Prove that $P(1)$ is true.

Step 2 Assume that $P(k)$ is true, and use this assumption to prove that $P(k + 1)$ is true.

Notice that in Step 2 we do not prove that $P(k)$ is true. We only show that *if* $P(k)$ is true, *then* $P(k+1)$ is also true. The assumption that $P(k)$ is true is called the **induction hypothesis**.

FOOTIES AT THE INDUCTION STEP
IF I WERE ON RUNG NUMBER K THEN IT WOULD BE EASY TO GET TO THE NEXT RUNG.
SURE, AND THEN YOU COULD CLIMB THE WHOLE LADDER.
BUT ONE HAS TO MANAGE TO REACH THE FIRST RUNG TO BEGIN WITH.

We now use mathematical induction to prove that the conjecture that we made at the beginning of this section is true.

Example 1 ■ A Proof by Mathematical Induction

Prove that for all natural numbers n,

$$1 + 3 + 5 + \cdots + (2n - 1) = n^2$$

Solution Let $P(n)$ denote the statement $1 + 3 + 5 + \cdots + (2n - 1) = n^2$.

Step 1 We need to show that $P(1)$ is true. But $P(1)$ is simply the statement that $1 = 1^2$, which is of course true.

Step 2 We assume that $P(k)$ is true. Thus our induction hypothesis is

$$1 + 3 + 5 + \cdots + (2k - 1) = k^2$$

We want to use this to show that $P(k+1)$ is true, that is,

$$1 + 3 + 5 + \cdots + (2k - 1) + [2(k + 1) - 1] = (k + 1)^2$$

[Note that we get $P(k+1)$ by substituting $k+1$ for each n in the statement $P(n)$.] We start with the left-hand side and use the induction hypothesis to obtain the right-hand side of the equation.

$$\begin{aligned}
&1 + 3 + 5 + \cdots + (2k - 1) + [2(k + 1) - 1] \\
&= [1 + 3 + 5 + \cdots + (2k - 1)] + [2(k + 1) - 1] && \text{Group the first } k \text{ terms} \\
&= k^2 + [2(k + 1) - 1] && \text{Induction hypothesis} \\
&= k^2 + [2k + 2 - 1] && \text{Distributive Property} \\
&= k^2 + 2k + 1 && \text{Simplify} \\
&= (k + 1)^2 && \text{Factor}
\end{aligned}$$

This equals k^2 by the induction hypothesis

Thus $P(k+1)$ follows from $P(k)$, and this completes the induction step.

Having proved Steps 1 and 2, we conclude by the Principle of Mathematical Induction that $P(n)$ is true for all natural numbers n.

Now Try Exercise 3 ■

Example 2 ■ A Proof by Mathematical Induction

Prove that for every natural number n,

$$1 + 2 + 3 + \cdots + n = \frac{n(n+1)}{2}$$

Solution Let $P(n)$ be the statement $1 + 2 + 3 + \cdots + n = n(n+1)/2$. We want to show that $P(n)$ is true for all natural numbers n.

Step 1 We need to show that $P(1)$ is true. But $P(1)$ says that

$$1 = \frac{1(1+1)}{2}$$

and this statement is clearly true.

Step 2 Assume that $P(k)$ is true. Thus our induction hypothesis is

$$1 + 2 + 3 + \cdots + k = \frac{k(k+1)}{2}$$

We want to use this to show that $P(k+1)$ is true, that is,

$$1 + 2 + 3 + \cdots + k + (k+1) = \frac{(k+1)[(k+1)+1]}{2}$$

So we start with the left-hand side and use the induction hypothesis to obtain the right-hand side.

$$\begin{aligned} & 1 + 2 + 3 + \cdots + k + (k+1) & \\ &= [1 + 2 + 3 + \cdots + k] + (k+1) & \text{Group the first } k \text{ terms} \\ &= \frac{k(k+1)}{2} + (k+1) & \text{Induction hypothesis} \\ &= (k+1)\left(\frac{k}{2} + 1\right) & \text{Factor } k+1 \\ &= (k+1)\left(\frac{k+2}{2}\right) & \text{Common denominator} \\ &= \frac{(k+1)[(k+1)+1]}{2} & \text{Write } k+2 \text{ as } k+1+1 \end{aligned}$$

This equals $\frac{k(k+1)}{2}$ by the induction hypothesis

Thus $P(k+1)$ follows from $P(k)$, and this completes the induction step.

Having proved Steps 1 and 2, we conclude by the Principle of Mathematical Induction that $P(n)$ is true for all natural numbers n.

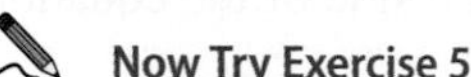

Now Try Exercise 5 ■

The following box gives formulas for the sums of powers of the first n natural numbers. These formulas are important in calculus. Formula 1 is proved in Example 2. The other formulas are also proved by using mathematical induction (see Exercises 6 and 9).

Sums of Powers

0. $\displaystyle\sum_{k=1}^{n} 1 = n$ **1.** $\displaystyle\sum_{k=1}^{n} k = \frac{n(n+1)}{2}$

2. $\displaystyle\sum_{k=1}^{n} k^2 = \frac{n(n+1)(2n+1)}{6}$ **3.** $\displaystyle\sum_{k=1}^{n} k^3 = \frac{n^2(n+1)^2}{4}$

Note that we have numbered each equation with the number that corresponds to the power being summed up.

It might happen that a statement $P(n)$ is false for the first few natural numbers but true from some number on. For example, we might want to prove that $P(n)$ is true for $n \geq 5$. Notice that if we prove that $P(5)$ is true, then this fact, together with the induction step, would imply the truth of $P(5), P(6), P(7), \ldots$. The next example illustrates this point.

Example 3 ■ Proving an Inequality by Mathematical Induction

Prove that $4n < 2^n$ for all $n \geq 5$.

Solution Let $P(n)$ denote the statement $4n < 2^n$.

Step 1 $P(5)$ is the statement that $4 \cdot 5 < 2^5$, or $20 < 32$, which is true.

Step 2 Assume that $P(k)$ is true. Thus our induction hypothesis is

$$4k < 2^k$$

We get $P(k + 1)$ by replacing n by $k + 1$ in the statement $P(n)$.

We want to use this to show that $P(k + 1)$ is true, that is,

$$4(k + 1) < 2^{k+1}$$

So we start with the left-hand side of the inequality and use the induction hypothesis to show that it is less than the right-hand side. For $k \geq 5$ we have

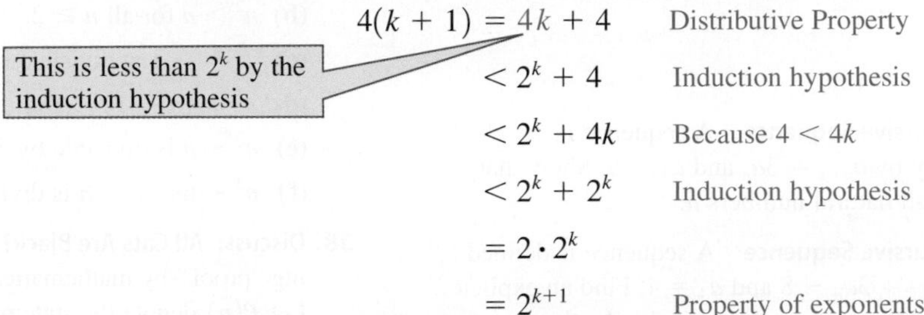

$4(k + 1) = 4k + 4$	Distributive Property
$< 2^k + 4$	Induction hypothesis
$< 2^k + 4k$	Because $4 < 4k$
$< 2^k + 2^k$	Induction hypothesis
$= 2 \cdot 2^k$	
$= 2^{k+1}$	Property of exponents

This is less than 2^k by the induction hypothesis

Thus $P(k + 1)$ follows from $P(k)$, and this completes the induction step.

Having proved Steps 1 and 2, we conclude by the Principle of Mathematical Induction that $P(n)$ is true for all natural numbers $n \geq 5$.

 **Now Try Exercise 21** ■

11.4 Exercises

■ Concepts

1. Mathematical induction is a method of proving that a statement $P(n)$ is true for all ______ numbers n. In Step 1 we prove that ______ is true.

2. Which of the following is true about Step 2 in a proof by mathematical induction?

(i) We prove "$P(k + 1)$ is true."

(ii) We prove "If $P(k)$ is true, then $P(k + 1)$ is true."

■ Skills

3–14 ■ Proving a Formula Use mathematical induction to prove that the formula is true for all natural numbers n.

 **3.** $2 + 4 + 6 + \cdots + 2n = n(n + 1)$

4. $1 + 4 + 7 + \cdots + (3n - 2) = \dfrac{n(3n - 1)}{2}$

5. $5 + 8 + 11 + \cdots + (3n + 2) = \dfrac{n(3n + 7)}{2}$

6. $1^2 + 2^2 + 3^2 + \cdots + n^2 = \dfrac{n(n + 1)(2n + 1)}{6}$

7. $1 \cdot 2 + 2 \cdot 3 + 3 \cdot 4 + \cdots + n(n + 1) = \dfrac{n(n + 1)(n + 2)}{3}$

8. $1 \cdot 3 + 2 \cdot 4 + 3 \cdot 5 + \cdots + n(n + 2) = \dfrac{n(n + 1)(2n + 7)}{6}$

9. $1^3 + 2^3 + 3^3 + \cdots + n^3 = \dfrac{n^2(n + 1)^2}{4}$

10. $1^3 + 3^3 + 5^3 + \cdots + (2n - 1)^3 = n^2(2n^2 - 1)$

11. $2^3 + 4^3 + 6^3 + \cdots + (2n)^3 = 2n^2(n + 1)^2$

12. $\dfrac{1}{1 \cdot 2} + \dfrac{1}{2 \cdot 3} + \dfrac{1}{3 \cdot 4} + \cdots + \dfrac{1}{n(n + 1)} = \dfrac{n}{n + 1}$

13. $1 \cdot 2 + 2 \cdot 2^2 + 3 \cdot 2^3 + 4 \cdot 2^4 + \cdots + n \cdot 2^n = 2[1 + (n - 1)2^n]$

14. $1 + 2 + 2^2 + \cdots + 2^{n-1} = 2^n - 1$

15–24 ■ Proving a Statement Use mathematical induction to show that the given statement is true.

15. $n^2 + n$ is divisible by 2 for all natural numbers n.

16. $5^n - 1$ is divisible by 4 for all natural numbers n.

17. $n^2 - n + 41$ is odd for all natural numbers n.

18. $n^3 - n + 3$ is divisible by 3 for all natural numbers n.

19. $8^n - 3^n$ is divisible by 5 for all natural numbers n.

20. $3^{2n} - 1$ is divisible by 8 for all natural numbers n.

21. $n < 2^n$ for all natural numbers n.

22. $(n + 1)^2 < 2n^2$ for all natural numbers $n \geq 3$.

23. If $x > -1$, then $(1 + x)^n \geq 1 + nx$ for all natural numbers n.

24. $100n \leq n^2$ for all $n \geq 100$.

25. Formula for a Recursive Sequence A sequence is defined recursively by $a_{n+1} = 3a_n$ and $a_1 = 5$. Show that $a_n = 5 \cdot 3^{n-1}$ for all natural numbers n.

26. Formula for a Recursive Sequence A sequence is defined recursively by $a_{n+1} = 3a_n - 8$ and $a_1 = 4$. Find an explicit formula for a_n, and then use mathematical induction to prove that the formula you found is true.

27. Proving a Factorization Show that $x - y$ is a factor of $x^n - y^n$ for all natural numbers n.
[*Hint:* $x^{k+1} - y^{k+1} = x^k(x - y) + (x^k - y^k)y$.]

28. Proving a Factorization Show that $x + y$ is a factor of $x^{2n-1} + y^{2n-1}$ for all natural numbers n.

■ Skills Plus

29–33 ■ Fibonacci Sequence F_n denotes the nth term of the Fibonacci sequence discussed in Section 11.1. Use mathematical induction to prove the statement.

29. F_{3n} is even for all natural numbers n.

30. $F_1 + F_2 + F_3 + \cdots + F_n = F_{n+2} - 1$

31. $F_1^2 + F_2^2 + F_3^2 + \cdots + F_n^2 = F_n F_{n+1}$

32. $F_1 + F_3 + \cdots + F_{2n-1} = F_{2n}$

33. For all $n \geq 2$,

$$\begin{bmatrix} 1 & 1 \\ 1 & 0 \end{bmatrix}^n = \begin{bmatrix} F_{n+1} & F_n \\ F_n & F_{n-1} \end{bmatrix}$$

34. Formula Using Fibonacci Numbers Let a_n be the nth term of the sequence defined recursively by

$$a_{n+1} = \frac{1}{1 + a_n}$$

and let $a_1 = 1$. Find a formula for a_n in terms of the Fibonacci numbers F_n. Prove that the formula you found is valid for all natural numbers n.

35. Discover and Prove an Inequality Let F_n be the nth term of the Fibonacci sequence. Find and prove an inequality relating n and F_n for all natural numbers $n \geq 5$.

36. Discover and Prove an Inequality Find and prove an inequality relating $100n$ and n^3.

■ Discuss ■ Discover ■ Prove ■ Write

37. Discover ■ Prove: True or False? Determine whether each statement is true or false. If you think the statement is true, prove it. If you think it is false, give an example for which it fails.

(a) $p(n) = n^2 - n + 11$ is prime for all n.

(b) $n^2 > n$ for all $n \geq 2$.

(c) $2^{2n+1} + 1$ is divisible by 3 for all $n \geq 1$.

(d) $n^3 \geq (n + 1)^2$ for all $n \geq 2$.

(e) $n^3 - n$ is divisible by 3 for all $n \geq 2$.

(f) $n^3 - 6n^2 + 11n$ is divisible by 6 for all $n \geq 1$.

38. Discuss: All Cats Are Black? What is wrong with the following "proof" by mathematical induction that all cats are black? Let $P(n)$ denote the statement "In any group of n cats, if one cat is black, then they are all black."

Step 1 The statement is clearly true for $n = 1$.

Step 2 Suppose that $P(k)$ is true. We show that $P(k + 1)$ is true.

Suppose we have a group of $k + 1$ cats, one of whom is black; call this cat "Tadpole." Remove some other cat (call it "Sparky") from the group. We are left with k cats, one of whom (Tadpole) is black, so by the induction hypothesis, all k of these are black. Now put Sparky back in the group and take out Tadpole. We again have a group of k cats, all of whom—except possibly Sparky—are black. Then by the induction hypothesis, Sparky must be black too. So all $k + 1$ cats in the original group are black.

Thus by induction $P(n)$ is true for all n. Since everyone has seen at least one black cat, it follows that all cats are black.

11.5 The Binomial Theorem

■ Expanding $(a + b)^n$ ■ The Binomial Coefficients ■ The Binomial Theorem
■ Proof of the Binomial Theorem

An expression of the form $a + b$ is called a **binomial**. Although in principle we can raise $a + b$ to any power, raising it to a very high power would be tedious. In this section we find a formula that gives the expansion of $(a + b)^n$ for any natural number n and then prove the formula using mathematical induction.

■ Expanding $(a + b)^n$

To find a pattern in the expansion of $(a + b)^n$, we first look at some special cases.

$$(a + b)^1 = a + b$$
$$(a + b)^2 = a^2 + 2ab + b^2$$
$$(a + b)^3 = a^3 + 3a^2b + 3ab^2 + b^3$$
$$(a + b)^4 = a^4 + 4a^3b + 6a^2b^2 + 4ab^3 + b^4$$
$$(a + b)^5 = a^5 + 5a^4b + 10a^3b^2 + 10a^2b^3 + 5ab^4 + b^5$$
$$\vdots$$

The following patterns emerge for the expansion of $(a + b)^n$.

1. There are $n + 1$ terms, the first being a^n and the last being b^n.
2. The exponents of a decrease by 1 from term to term, while the exponents of b increase by 1.
3. The sum of the exponents of a and b in each term is n.

For instance, notice how the exponents of a and b behave in the expansion of $(a + b)^5$.

The exponents of *a* decrease:

$$(a + b)^5 = a^{⑤} + 5a^{④}b^1 + 10a^{③}b^2 + 10a^{②}b^3 + 5a^{①}b^4 + b^5$$

The exponents of *b* increase:

$$(a + b)^5 = a^5 + 5a^4b^{①} + 10a^3b^{②} + 10a^2b^{③} + 5a^1b^{④} + b^{⑤}$$

With these observations we can write the form of the expansion of $(a + b)^n$ for any natural number n. For example, writing a question mark for the missing coefficients, we have

$$(a + b)^8 = a^8 + ?a^7b + ?a^6b^2 + ?a^5b^3 + ?a^4b^4 + ?a^3b^5 + ?a^2b^6 + ?ab^7 + b^8$$

To complete the expansion, we need to determine these coefficients. To find a pattern, let's write the coefficients in the expansion of $(a + b)^n$ for the first few values of n in a triangular array as shown in the following array, which is called **Pascal's triangle**.

$(a + b)^0$	1
$(a + b)^1$	1 1
$(a + b)^2$	1 2 1
$(a + b)^3$	① ③ 3 1
$(a + b)^4$	1 ④ ⑥ ④ 1
$(a + b)^5$	1 5 10 ⑩ 5 1

What we now call **Pascal's triangle** appears in this Chinese document by Chu Shih-Chieh, dated 1303. The title reads "The Old Method Chart of the Seven Multiplying Squares." The triangle was rediscovered by Pascal (see Section 11.4).

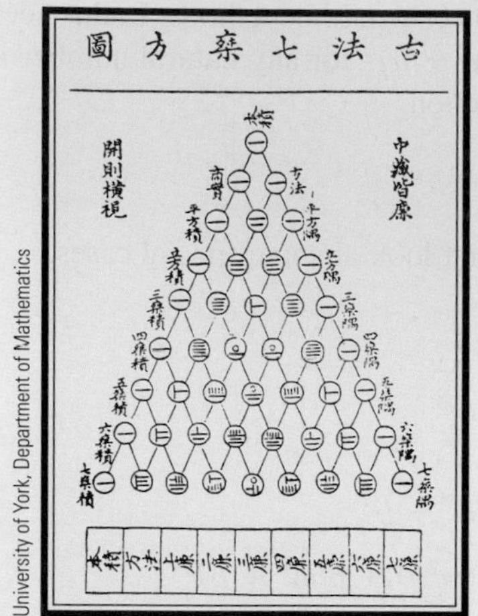

University of York, Department of Mathematics

The row corresponding to $(a + b)^0$ is called the zeroth row and is included to show the symmetry of the array. The key observation about Pascal's triangle is the following property.

Key Property of Pascal's Triangle

Every entry (other than a 1) is the sum of the two entries diagonally above it.

From this property we can find any row of Pascal's triangle from the row above it. For instance, we find the sixth and seventh rows, starting with the fifth row:

$(a+b)^5$			1		5		10		10		5		1		
$(a+b)^6$		1		6		15		20		15		6		1	
$(a+b)^7$	1		7		21		35		35		21		7		1

To see why this property holds, let's consider the following expansions:

$$(a + b)^5 = a^5 + 5a^4b + 10a^3b^2 + \boxed{10a^2b^3} + \boxed{5ab^4} + b^5$$

$$(a + b)^6 = a^6 + 6a^5b + 15a^4b^2 + 20a^3b^3 + \boxed{15a^2b^4} + 6ab^5 + b^6$$

We arrive at the expansion of $(a + b)^6$ by multiplying $(a + b)^5$ by $(a + b)$. Notice, for instance, that the circled term in the expansion of $(a + b)^6$ is obtained via this multiplication from the two circled terms above it. We get this term when the two terms above it are multiplied by b and a, respectively. Thus its coefficient is the sum of the coefficients of these two terms. We will use this observation at the end of this section when we prove the Binomial Theorem.

Having found these patterns, we can now easily obtain the expansion of any binomial, at least to relatively small powers.

Example 1 ■ Expanding a Binomial Using Pascal's Triangle

Find the expansion of $(a + b)^7$ using Pascal's triangle.

Solution The first term in the expansion is a^7, and the last term is b^7. Using the fact that the exponent of a decreases by 1 from term to term and that of b increases by 1 from term to term, we have

$$(a + b)^7 = a^7 + \boxed{?}a^6b + \boxed{?}a^5b^2 + \boxed{?}a^4b^3 + \boxed{?}a^3b^4 + \boxed{?}a^2b^5 + \boxed{?}ab^6 + b^7$$

The appropriate coefficients appear in the seventh row of Pascal's triangle. Thus

$$(a + b)^7 = a^7 + 7a^6b + 21a^5b^2 + 35a^4b^3 + 35a^3b^4 + 21a^2b^5 + 7ab^6 + b^7$$

 **Now Try Exercise 5** ■

Example 2 ■ Expanding a Binomial Using Pascal's Triangle

Use Pascal's triangle to expand $(2 - 3x)^5$.

Solution We find the expansion of $(a + b)^5$ and then substitute 2 for a and $-3x$ for b. Using Pascal's triangle for the coefficients, we get

$$(a + b)^5 = a^5 + 5a^4b + 10a^3b^2 + 10a^2b^3 + 5ab^4 + b^5$$

Substituting $a = 2$ and $b = -3x$ gives

$$\begin{aligned}(2 - 3x)^5 &= (2)^5 + 5(2)^4(-3x) + 10(2)^3(-3x)^2 + 10(2)^2(-3x)^3 + 5(2)(-3x)^4 + (-3x)^5 \\ &= 32 - 240x + 720x^2 - 1080x^3 + 810x^4 - 243x^5\end{aligned}$$

 **Now Try Exercise 13** ■

■ The Binomial Coefficients

Although Pascal's triangle is useful in finding the binomial expansion for reasonably small values of n, it isn't practical for finding $(a + b)^n$ for large values of n. The reason is that the method we use for finding the successive rows of Pascal's triangle is recursive. Thus to find the 100th row of this triangle, we must first find the preceding 99 rows.

We need to examine the pattern in the coefficients more carefully to develop a formula that allows us to calculate directly any coefficient in the binomial expansion. Such a formula exists, and the rest of this section is devoted to finding and proving it. However, to state this formula, we need some notation.

The product of the first n natural numbers is denoted by $\boldsymbol{n!}$ and is called $\boldsymbol{n}$ **factorial**.

$$4! = 1 \cdot 2 \cdot 3 \cdot 4 = 24$$

$$7! = 1 \cdot 2 \cdot 3 \cdot 4 \cdot 5 \cdot 6 \cdot 7 = 5040$$

$$10! = 1 \cdot 2 \cdot 3 \cdot 4 \cdot 5 \cdot 6 \cdot 7 \cdot 8 \cdot 9 \cdot 10 = 3{,}628{,}800$$

$$n! = 1 \cdot 2 \cdot 3 \cdot \cdots \cdot (n-1) \cdot n$$

We also define 0! as follows:

$$0! = 1$$

This definition of 0! makes many formulas involving factorials shorter and easier to write.

The Binomial Coefficient

Let n and r be nonnegative integers with $r \le n$. The **binomial coefficient** is denoted by $\binom{n}{r}$ and is defined by

$$\binom{n}{r} = \frac{n!}{r!(n-r)!}$$

Example 3 ■ Calculating Binomial Coefficients

(a) $$\binom{9}{4} = \frac{9!}{4!(9-4)!} = \frac{9!}{4!\,5!} = \frac{\cancel{1 \cdot 2 \cdot 3 \cdot 4 \cdot 5} \cdot 6 \cdot 7 \cdot 8 \cdot 9}{(1 \cdot 2 \cdot 3 \cdot 4)\cancel{(1 \cdot 2 \cdot 3 \cdot 4 \cdot 5)}} = \frac{6 \cdot 7 \cdot 8 \cdot 9}{1 \cdot 2 \cdot 3 \cdot 4} = 126$$

(b) $$\binom{100}{3} = \frac{100!}{3!(100-3)!} = \frac{\cancel{1 \cdot 2 \cdot 3 \cdot \cdots \cdot 97} \cdot 98 \cdot 99 \cdot 100}{(1 \cdot 2 \cdot 3)\cancel{(1 \cdot 2 \cdot 3 \cdot \cdots \cdot 97)}} = \frac{98 \cdot 99 \cdot 100}{1 \cdot 2 \cdot 3} = 161{,}700$$

(c) $$\binom{100}{97} = \frac{100!}{97!(100-97)!} = \frac{\cancel{1 \cdot 2 \cdot 3 \cdot \cdots \cdot 97} \cdot 98 \cdot 99 \cdot 100}{\cancel{(1 \cdot 2 \cdot 3 \cdot \cdots \cdot 97)}(1 \cdot 2 \cdot 3)} = \frac{98 \cdot 99 \cdot 100}{1 \cdot 2 \cdot 3} = 161{,}700$$

Now Try Exercises 17 and 19 ■

Although the binomial coefficient $\binom{n}{r}$ is defined in terms of a fraction, all the results of Example 3 are natural numbers. In fact, $\binom{n}{r}$ is always a natural number (see Exercise 54).

Notice that the binomial coefficients in parts (b) and (c) of Example 3 are equal. This is a special case of the following relation, which you are asked to prove in Exercise 52.

$$\binom{n}{r} = \binom{n}{n-r}$$

To see the connection between the binomial coefficients and the binomial expansion of $(a + b)^n$, let's calculate the following binomial coefficients:

$$\binom{5}{2} = \frac{5!}{2!(5-2)!} = 10$$

$$\binom{5}{0} = 1 \quad \binom{5}{1} = 5 \quad \binom{5}{2} = 10 \quad \binom{5}{3} = 10 \quad \binom{5}{4} = 5 \quad \binom{5}{5} = 1$$

These are precisely the entries in the fifth row of Pascal's triangle. In fact, we can write Pascal's triangle as follows.

$$\binom{0}{0}$$

$$\binom{1}{0} \quad \binom{1}{1}$$

$$\binom{2}{0} \quad \binom{2}{1} \quad \binom{2}{2}$$

$$\binom{3}{0} \quad \binom{3}{1} \quad \binom{3}{2} \quad \binom{3}{3}$$

$$\binom{4}{0} \quad \binom{4}{1} \quad \binom{4}{2} \quad \binom{4}{3} \quad \binom{4}{4}$$

$$\binom{5}{0} \quad \binom{5}{1} \quad \binom{5}{2} \quad \binom{5}{3} \quad \binom{5}{4} \quad \binom{5}{5}$$

$$\cdot \quad \cdot \quad \cdot \quad \cdot \quad \cdot \quad \cdot \quad \cdot$$

$$\binom{n}{0} \quad \binom{n}{1} \quad \binom{n}{2} \quad \cdot \quad \cdot \quad \cdot \quad \binom{n}{n-1} \quad \binom{n}{n}$$

To demonstrate that this pattern holds, we need to show that any entry in this version of Pascal's triangle is the sum of the two entries diagonally above it. In other words, we must show that each entry satisfies the key property of Pascal's triangle. We now state this property in terms of the binomial coefficients.

Key Property of the Binomial Coefficients

For any nonnegative integers r and k with $r \leq k$,

$$\binom{k}{r-1} + \binom{k}{r} = \binom{k+1}{r}$$

Notice that the two terms on the left-hand side of this equation are adjacent entries in the kth row of Pascal's triangle and the term on the right-hand side is the entry diagonally below them, in the $(k + 1)$st row. Thus this equation is a restatement of the key property of Pascal's triangle in terms of the binomial coefficients. A proof of this formula is outlined in Exercise 53.

■ The Binomial Theorem

We are now ready to state the Binomial Theorem.

The Binomial Theorem

$$(a+b)^n = \binom{n}{0}a^n + \binom{n}{1}a^{n-1}b + \binom{n}{2}a^{n-2}b^2 + \cdots + \binom{n}{n-1}ab^{n-1} + \binom{n}{n}b^n$$

We prove this theorem at the end of this section. First, let's look at some applications of the Binomial Theorem.

Example 4 ■ Expanding a Binomial Using the Binomial Theorem

Use the Binomial Theorem to expand $(x+y)^4$.

Solution By the Binomial Theorem,

$$(x+y)^4 = \binom{4}{0}x^4 + \binom{4}{1}x^3y + \binom{4}{2}x^2y^2 + \binom{4}{3}xy^3 + \binom{4}{4}y^4$$

Verify that

$$\binom{4}{0} = 1 \qquad \binom{4}{1} = 4 \qquad \binom{4}{2} = 6 \qquad \binom{4}{3} = 4 \qquad \binom{4}{4} = 1$$

It follows that

$$(x+y)^4 = x^4 + 4x^3y + 6x^2y^2 + 4xy^3 + y^4$$

 **Now Try Exercise 25** ■

Example 5 ■ Expanding a Binomial Using the Binomial Theorem

Use the Binomial Theorem to expand $(\sqrt{x} - 1)^8$.

Solution We first find the expansion of $(a+b)^8$ and then substitute $\sqrt{x}$ for a and -1 for b. Using the Binomial Theorem, we have

$$(a+b)^8 = \binom{8}{0}a^8 + \binom{8}{1}a^7b + \binom{8}{2}a^6b^2 + \binom{8}{3}a^5b^3 + \binom{8}{4}a^4b^4$$
$$+ \binom{8}{5}a^3b^5 + \binom{8}{6}a^2b^6 + \binom{8}{7}ab^7 + \binom{8}{8}b^8$$

Verify that

$$\binom{8}{0} = 1 \qquad \binom{8}{1} = 8 \qquad \binom{8}{2} = 28 \qquad \binom{8}{3} = 56 \qquad \binom{8}{4} = 70$$

$$\binom{8}{5} = 56 \qquad \binom{8}{6} = 28 \qquad \binom{8}{7} = 8 \qquad \binom{8}{8} = 1$$

So

$$(a+b)^8 = a^8 + 8a^7b + 28a^6b^2 + 56a^5b^3 + 70a^4b^4 + 56a^3b^5$$
$$+ 28a^2b^6 + 8ab^7 + b^8$$

Performing the substitutions $a = x^{1/2}$ and $b = -1$ gives

$$\begin{aligned}(\sqrt{x} - 1)^8 = (x^{1/2})^8 + 8(x^{1/2})^7(-1) + 28(x^{1/2})^6(-1)^2 + 56(x^{1/2})^5(-1)^3 \\ + 70(x^{1/2})^4(-1)^4 + 56(x^{1/2})^3(-1)^5 + 28(x^{1/2})^2(-1)^6 \\ + 8(x^{1/2})(-1)^7 + (-1)^8\end{aligned}$$

This simplifies to

$$(\sqrt{x} - 1)^8 = x^4 - 8x^{7/2} + 28x^3 - 56x^{5/2} + 70x^2 - 56x^{3/2} + 28x - 8x^{1/2} + 1$$

Now Try Exercise 27 ■

The Binomial Theorem can be used to find a particular term of a binomial expansion without having to find the entire expansion.

Recall that

$$\binom{n}{r} = \binom{n}{n-r}$$

General Term of the Binomial Expansion

The term that contains a^r in the expansion of $(a + b)^n$ is

$$\binom{n}{r}a^r b^{n-r}$$

Example 6 ■ Finding a Particular Term in a Binomial Expansion

Find the term that contains x^5 in the expansion of $(2x + y)^{20}$.

Solution The term that contains x^5 is given by the formula for the general term with $a = 2x$, $b = y$, $n = 20$, and $r = 5$. So this term is

$$\binom{20}{5}a^5b^{15} = \frac{20!}{5!(20-5)!}(2x)^5y^{15} = \frac{20!}{5!\,15!}32x^5y^{15} = 496{,}128x^5y^{15}$$

 **Now Try Exercise 39** ■

Example 7 ■ Finding a Particular Term in a Binomial Expansion

Find the coefficient of x^8 in the expansion of $\left(x^2 + \dfrac{1}{x}\right)^{10}$.

Solution Both x^2 and $1/x$ are powers of x, so the power of x in each term of the expansion is determined by both terms of the binomial. To find the required coefficient, we first find the general term in the expansion. By the formula we have $a = x^2$, $b = 1/x$, and $n = 10$, so the general term is

$$\binom{10}{r}(x^2)^r\left(\frac{1}{x}\right)^{10-r} = \binom{10}{r}x^{2r}(x^{-1})^{10-r} = \binom{10}{r}x^{3r-10}$$

Thus the term that contains x^8 is the term in which

$$\begin{aligned}3r - 10 &= 8 \\ r &= 6\end{aligned}$$

So the required coefficient is

$$\binom{10}{6} = 210$$

Now Try Exercise 41 ■

■ Proof of the Binomial Theorem

We now give a proof of the Binomial Theorem using mathematical induction.

Proof Let $P(n)$ denote the statement

$$(a+b)^n = \binom{n}{0}a^n + \binom{n}{1}a^{n-1}b + \binom{n}{2}a^{n-2}b^2 + \cdots + \binom{n}{n-1}ab^{n-1} + \binom{n}{n}b^n$$

Step 1 We show that $P(1)$ is true. But $P(1)$ is just the statement

$$(a+b)^1 = \binom{1}{0}a^1 + \binom{1}{1}b^1 = 1a + 1b = a + b$$

which is certainly true.

Step 2 We assume that $P(k)$ is true. Thus our induction hypothesis is

$$(a+b)^k = \binom{k}{0}a^k + \binom{k}{1}a^{k-1}b + \binom{k}{2}a^{k-2}b^2 + \cdots + \binom{k}{k-1}ab^{k-1} + \binom{k}{k}b^k$$

We use this to show that $P(k+1)$ is true.

$$\begin{aligned}
(a+b)^{k+1} &= (a+b)[(a+b)^k] \\
&= (a+b)\left[\binom{k}{0}a^k + \binom{k}{1}a^{k-1}b + \binom{k}{2}a^{k-2}b^2 + \cdots + \binom{k}{k-1}ab^{k-1} + \binom{k}{k}b^k\right] && \text{Induction hypothesis} \\
&= a\left[\binom{k}{0}a^k + \binom{k}{1}a^{k-1}b + \binom{k}{2}a^{k-2}b^2 + \cdots + \binom{k}{k-1}ab^{k-1} + \binom{k}{k}b^k\right] \\
&\quad + b\left[\binom{k}{0}a^k + \binom{k}{1}a^{k-1}b + \binom{k}{2}a^{k-2}b^2 + \cdots + \binom{k}{k-1}ab^{k-1} + \binom{k}{k}b^k\right] && \text{Distributive Property} \\
&= \binom{k}{0}a^{k+1} + \binom{k}{1}a^kb + \binom{k}{2}a^{k-1}b^2 + \cdots + \binom{k}{k-1}a^2b^{k-1} + \binom{k}{k}ab^k \\
&\quad + \binom{k}{0}a^kb + \binom{k}{1}a^{k-1}b^2 + \binom{k}{2}a^{k-2}b^3 + \cdots + \binom{k}{k-1}ab^k + \binom{k}{k}b^{k+1} && \text{Distributive Property} \\
&= \binom{k}{0}a^{k+1} + \left[\binom{k}{0} + \binom{k}{1}\right]a^kb + \left[\binom{k}{1} + \binom{k}{2}\right]a^{k-1}b^2 \\
&\quad + \cdots + \left[\binom{k}{k-1} + \binom{k}{k}\right]ab^k + \binom{k}{k}b^{k+1} && \text{Group like terms}
\end{aligned}$$

Using the key property of the binomial coefficients, we can write each of the expressions in square brackets as a single binomial coefficient. Also, writing the first and last coefficients as $\binom{k+1}{0}$ and $\binom{k+1}{k+1}$ (these are equal to 1 by Exercise 50) gives

$$(a+b)^{k+1} = \binom{k+1}{0}a^{k+1} + \binom{k+1}{1}a^kb + \binom{k+1}{2}a^{k-1}b^2 + \cdots + \binom{k+1}{k}ab^k + \binom{k+1}{k+1}b^{k+1}$$

But this last equation is precisely $P(k+1)$, and this completes the induction step.

Having proved Steps 1 and 2, we conclude by the Principle of Mathematical Induction that the theorem is true for all natural numbers n. ■

11.5 Exercises

Concepts

1. An algebraic expression of the form $a + b$, which consists of a sum of two terms, is called a ______.

2. We can find the coefficients in the expansion of $(a + b)^n$ from the nth row of ______ triangle. So

$$(a + b)^4 = \square a^4 + \square a^3b + \square a^2b^2 + \square ab^3 + \square b^4$$

3. The binomial coefficients can be calculated directly by using the formula $\binom{n}{k} =$ ______. So $\binom{4}{3} =$ ______.

4. To expand $(a + b)^n$, we can use the ______ Theorem. Using this theorem, we find the expansion $(a + b)^4 =$

$$\binom{\square}{\square}a^4 + \binom{\square}{\square}a^3b + \binom{\square}{\square}a^2b^2 + \binom{\square}{\square}ab^3 + \binom{\square}{\square}b^4$$

Skills

5–16 ■ Pascal's Triangle Use Pascal's triangle to expand the expression.

5. $(x + y)^6$

6. $(2x + 1)^4$

7. $\left(x + \dfrac{1}{x}\right)^4$

8. $(x - y)^5$

9. $(x - 1)^5$

10. $(\sqrt{a} + \sqrt{b})^6$

11. $(x^2y - 1)^5$

12. $(1 + \sqrt{2})^6$

13. $(2x - 3y)^3$

14. $(1 + x^3)^3$

15. $\left(\dfrac{1}{x} - \sqrt{x}\right)^5$

16. $\left(2 + \dfrac{x}{2}\right)^5$

17–24 ■ Calculating Binomial Coefficients Evaluate the expression.

17. $\binom{6}{4}$

18. $\binom{8}{3}$

19. $\binom{100}{98}$

20. $\binom{10}{5}$

21. $\binom{3}{1}\binom{4}{2}$

22. $\binom{5}{2}\binom{5}{3}$

23. $\binom{5}{0} + \binom{5}{1} + \binom{5}{2} + \binom{5}{3} + \binom{5}{4} + \binom{5}{5}$

24. $\binom{5}{0} - \binom{5}{1} + \binom{5}{2} - \binom{5}{3} + \binom{5}{4} - \binom{5}{5}$

25–28 ■ Binomial Theorem Use the Binomial Theorem to expand the expression.

25. $(x + 2y)^4$

26. $(1 - x)^5$

27. $\left(1 + \dfrac{1}{x}\right)^6$

28. $(2A + B^2)^4$

29–42 ■ Terms of a Binomial Expansion Find the indicated terms in the expansion of the given binomial.

29. The first three terms in the expansion of $(x + 2y)^{20}$

30. The first four terms in the expansion of $(x^{1/2} + 1)^{30}$

31. The last two terms in the expansion of $(a^{2/3} + a^{1/3})^{25}$

32. The first three terms in the expansion of

$$\left(x + \frac{1}{x}\right)^{40}$$

33. The middle term in the expansion of $(x^2 + 1)^{18}$

34. The fifth term in the expansion of $(ab - 1)^{20}$

35. The 24th term in the expansion of $(a + b)^{25}$

36. The 28th term in the expansion of $(A - B)^{30}$

37. The 100th term in the expansion of $(1 + y)^{100}$

38. The second term in the expansion of

$$\left(x^2 - \frac{1}{x}\right)^{25}$$

39. The term containing x^4 in the expansion of $(x + 2y)^{10}$

40. The term containing y^3 in the expansion of $(\sqrt{2} + y)^{12}$

41. The term containing b^8 in the expansion of $(a + b^2)^{12}$

42. The term that does not contain x in the expansion of

$$\left(8x + \frac{1}{2x}\right)^8$$

43–46 ■ Factoring Factor using the Binomial Theorem.

43. $x^4 + 4x^3y + 6x^2y^2 + 4xy^3 + y^4$

44. $(x - 1)^5 + 5(x - 1)^4 + 10(x - 1)^3 + 10(x - 1)^2 + 5(x - 1) + 1$

45. $8a^3 + 12a^2b + 6ab^2 + b^3$

46. $x^8 + 4x^6y + 6x^4y^2 + 4x^2y^3 + y^4$

47–48 ■ Simplifying a Difference Quotient Simplify using the Binomial Theorem.

47. $\dfrac{(x + h)^3 - x^3}{h}$

48. $\dfrac{(x + h)^4 - x^4}{h}$

Skills Plus

49–52 ■ Proving a Statement Show that the given statement is true.

49. $(1.01)^{100} > 2$. [*Hint:* Note that $(1.01)^{100} = (1 + 0.01)^{100}$, and use the Binomial Theorem to show that the sum of the first three terms of the expansion is greater than 2.]

50. $\binom{n}{0} = 1$ and $\binom{n}{n} = 1$

51. $\binom{n}{1} = \binom{n}{n-1} = n$

52. $\binom{n}{r} = \binom{n}{n-r}$ for $0 \le r \le n$

53. Proving an Identity In this exercise we prove the identity

$$\binom{n}{r-1} + \binom{n}{r} = \binom{n+1}{r}$$

(a) Write the left-hand side of this equation as the sum of two fractions.

(b) Show that a common denominator of the expression that you found in part (a) is $r!(n-r+1)!$.

(c) Add the two fractions using the common denominator in part (b), simplify the numerator, and note that the resulting expression is equal to the right-hand side of the equation.

54. Proof Using Induction Prove that $\binom{n}{r}$ is an integer for all n and for $0 \le r \le n$. [*Suggestion:* Use induction to show that the statement is true for all n, and use Exercise 53 for the induction step.]

Applications

55. Difference in Volumes of Cubes The volume of a cube of side x centimeters is given by $V(x) = x^3$, so the volume of a cube of side $x + 2$ centimeters is given by $V(x+2) = (x+2)^3$. Use the Binomial Theorem to show that the difference in volume between the larger and smaller cubes is $6x^2 + 12x + 8$ cubic centimeters.

56. Probability of Hitting a Target The probability that an archer hits the target is $p = 0.9$, so the probability that the archer misses the target is $q = 0.1$. It is known that in this situation the probability that the archer hits the target exactly r times in n attempts is given by the term containing p^r in the binomial expansion of $(p+q)^n$. Find the probability that the archer hits the target exactly three times in five attempts.

Discuss ■ Discover ■ Prove ■ Write

57. Discuss: Powers of Factorials Which is larger, $(100!)^{101}$ or $(101!)^{100}$?

PS *Look for something familiar.* Try factoring the expressions. Do they have any common factors?

58. Discover ■ Prove: Sums of Binomial Coefficients Add each of the first five rows of Pascal's triangle, as indicated. Do you see a pattern?

$$1 + 1 = ?$$
$$1 + 2 + 1 = ?$$
$$1 + 3 + 3 + 1 = ?$$
$$1 + 4 + 6 + 4 + 1 = ?$$
$$1 + 5 + 10 + 10 + 5 + 1 = ?$$

On the basis of the pattern you have found, find the sum of the nth row:

$$\binom{n}{0} + \binom{n}{1} + \binom{n}{2} + \cdots + \binom{n}{n}$$

Prove your result by expanding $(1+1)^n$ using the Binomial Theorem.

59. Discover ■ Prove: Alternating Sums of Binomial Coefficients Find the sum

$$\binom{n}{0} - \binom{n}{1} + \binom{n}{2} - \cdots + (-1)^n\binom{n}{n}$$

by finding a pattern as in Exercise 58. Prove your result by expanding $(1-1)^n$ using the Binomial Theorem.

Chapter 11 Review

Properties and Formulas

Sequences | Section 11.1

A **sequence** is a function whose domain is the set of natural numbers. Instead of writing $a(n)$ for the value of the sequence at n, we generally write a_n, and we refer to this value as the ***n*th term** of the sequence. Sequences are often described in list form:

$$a_1, a_2, a_3, \ldots$$

Partial Sums of a Sequence | Section 11.1

For the sequence $a_1, a_2, a_3, \ldots$ the ***n*th partial sum** S_n is the sum of the first n terms of the sequence:

$$S_n = a_1 + a_2 + a_3 + \cdots + a_n$$

The nth partial sum of a sequence can also be expressed by using **sigma notation**:

$$S_n = \sum_{k=1}^{n} a_k$$

Arithmetic Sequences | Section 11.2

An **arithmetic sequence** is a sequence whose terms are obtained by adding the same fixed constant d to each term to get the next term. Thus an arithmetic sequence has the form

$$a, a+d, a+2d, a+3d, \ldots$$

The number a is the **first term** of the sequence, and the number d is the **common difference**. The nth term of the sequence is

$$a_n = a + (n-1)d$$

Partial Sums of an Arithmetic Sequence | Section 11.2

For the arithmetic sequence $a_n = a + (n-1)d$ the nth partial sum $S_n = \sum_{k=1}^{n} [a + (k-1)d]$ is given by either of the following equivalent formulas:

1. $S_n = \frac{n}{2}[2a + (n-1)d]$ **2.** $S_n = n\left(\frac{a + a_n}{2}\right)$

Geometric Sequences | Section 11.3

A **geometric sequence** is a sequence whose terms are obtained by multiplying each term by the same fixed constant r to get the next term. Thus a geometric sequence has the form

$$a, ar, ar^2, ar^3, \ldots$$

The number a is the **first term** of the sequence, and the number r is the **common ratio**. The nth term of the sequence is

$$a_n = ar^{n-1}$$

Partial Sums of a Geometric Sequence | Section 11.3

For the geometric sequence $a_n = ar^{n-1}$ the nth partial sum $S_n = \sum_{k=1}^{n} ar^{k-1}$ (where $r \neq 1$) is given by

$$S_n = a\frac{1 - r^n}{1 - r}$$

Infinite Geometric Series | Section 11.3

An **infinite geometric series** is a series of the form

$$a + ar + ar^2 + ar^3 + \cdots + ar^{n-1} + \cdots$$

An infinite geometric series for which $|r| < 1$ has the sum

$$S = \frac{a}{1 - r}$$

Principle of Mathematical Induction | Section 11.4

For each natural number n, let $P(n)$ be a statement that depends on n. Suppose that each of the following conditions is satisfied.

1. $P(1)$ is true.

2. For every natural number k, if $P(k)$ is true, then $P(k + 1)$ is true.

Then $P(n)$ is true for all natural numbers n.

Sums of Powers | Section 11.4

0. $\sum_{k=1}^{n} 1 = n$

1. $\sum_{k=1}^{n} k = \frac{n(n+1)}{2}$

2. $\sum_{k=1}^{n} k^2 = \frac{n(n+1)(2n+1)}{6}$

3. $\sum_{k=1}^{n} k^3 = \frac{n^2(n+1)^2}{4}$

Binomial Coefficients | Section 11.5

If n and r are positive integers with $n \geq r$, then the **binomial coefficient** $\binom{n}{r}$ is defined by

$$\binom{n}{r} = \frac{n!}{r!(n-r)!}$$

Binomial coefficients satisfy the following properties:

$$\binom{n}{r} = \binom{n}{n-r}$$

$$\binom{k}{r-1} + \binom{k}{r} = \binom{k+1}{r}$$

The Binomial Theorem | Section 11.5

$$(a + b)^n = \binom{n}{0}a^n + \binom{n}{1}a^{n-1}b + \binom{n}{2}a^{n-2}b^2 + \cdots + \binom{n}{n}b^n$$

The term that contains a^r in the expansion of $(a + b)^n$ is $\binom{n}{r}a^r b^{n-r}$.

Concept Check

1. (a) What is a sequence? What notation do we use to denote the terms of a sequence?

(b) Find a formula for the sequence of even numbers and a formula for the sequence of odd numbers.

(c) Find the first three terms and the 10th term of the sequence given by $a_n = n/(n + 1)$.

2. (a) What is a recursively defined sequence?

(b) Find the first four terms of the sequence recursively defined by $a_1 = 3$ and $a_n = n + 2a_{n-1}$.

3. (a) What is meant by the partial sums of a sequence?

(b) Find the first three partial sums of the sequence given by $a_n = 1/n$.

4. (a) What is an arithmetic sequence? Write a formula for the nth term of an arithmetic sequence.

(b) Write a formula for the arithmetic sequence that starts as follows: 3, 8, . . . Write the first five terms of this sequence.

(c) Write two different formulas for the sum of the first n terms of an arithmetic sequence.

(d) Find the sum of the first 20 terms of the sequence in part (b).

5. (a) What is a geometric sequence? Write an expression for the nth term of a geometric sequence that has first term a and common ratio r.

(b) Write an expression for the geometric sequence with first term $a = 3$ and common ratio $r = \frac{1}{2}$. Give the first five terms of this sequence.

(c) Write an expression for the sum of the first n terms of a geometric sequence.

(d) Find the sum of the first five terms of the sequence in part (b).

6. (a) What is an infinite geometric series?

(b) What does it mean for an infinite series to converge? For what values of r does an infinite geometric series converge? If an infinite geometric series converges, then what is its sum?

(c) Write the first four terms of the infinite geometric series with first term $a = 5$ and common ratio $r = 0.4$. Does the series converge? If so, find its sum.

7. (a) Write $1^3 + 2^3 + 3^3 + 4^3 + 5^3$ using sigma notation.

(b) Write $\sum_{k=3}^{5} 2k^2$ without using sigma notation.

8. (a) State the Principle of Mathematical Induction.
   (b) Use mathematical induction to prove that for all natural numbers n, $3^n - 1$ is an even number.

9. (a) Write Pascal's triangle. How are the entries in the triangle related to each other?

   (b) Use Pascal's triangle to expand $(x + c)^3$.

10. (a) What does the symbol $n!$ mean? Find $5!$.
    (b) Define $\binom{n}{r}$, and find $\binom{5}{2}$.

11. (a) State the Binomial Theorem.
    (b) Use the Binomial Theorem to expand $(x + 2)^3$.
    (c) Use the Binomial Theorem to find the term containing x^4 in the expansion of $(x + 2)^{10}$.

Answers to the Concept Check can be found at the book companion website stewartmath.com.

Exercises

1–6 ■ Terms of a Sequence Find the first four terms as well as the tenth term of the sequence with the given nth term.

1. $a_n = \dfrac{n^2}{n+1}$ **2.** $a_n = (-1)^n \dfrac{2^n}{n}$

3. $a_n = \dfrac{(-1)^n + 1}{n^3}$ **4.** $a_n = \dfrac{n(n+1)}{2}$

5. $a_n = \dfrac{(2n)!}{2^n n!}$ **6.** $a_n = \dbinom{n+1}{2}$

7–10 ■ Recursive Sequences A sequence is defined recursively. Find the first seven terms of the sequence.

7. $a_n = a_{n-1} + 2n - 1, \quad a_1 = 1$

8. $a_n = \dfrac{a_{n-1}}{n}, \quad a_1 = 1$

9. $a_n = a_{n-1} + 2a_{n-2}, \quad a_1 = 1, a_2 = 3$

10. $a_n = \sqrt{3a_{n-1}}, \quad a_1 = \sqrt{3}$

11–14 ■ Arithmetic or Geometric Sequence? The nth term of a sequence is given. **(a)** Find the first five terms of the sequence. **(b)** Graph the terms you found in part (a). **(c)** Find the fifth partial sum of the sequence. **(d)** Determine whether the sequence is arithmetic or geometric. Find the common difference or the common ratio.

11. $a_n = 2n + 5$ **12.** $a_n = \dfrac{5}{2^n}$

13. $a_n = \dfrac{3^n}{2^{n+1}}$ **14.** $a_n = 4 - \dfrac{n}{2}$

15–22 ■ Arithmetic or Geometric Sequence? The first four terms of a sequence are given. Determine whether they can be the terms of an arithmetic sequence, a geometric sequence, or neither. If the sequence is arithmetic or geometric, find the fifth term.

15. $5, 5.5, 6, 6.5, \ldots$ **16.** $\sqrt{2}, 2\sqrt{2}, 3\sqrt{2}, 4\sqrt{2}, \ldots$

17. $t - 3, t - 2, t - 1, t, \ldots$ **18.** $\sqrt{2}, 2, 2\sqrt{2}, 4, \ldots$

19. $t^3, t^2, t, 1, \ldots$ **20.** $1, -\frac{3}{2}, 2, -\frac{5}{2}, \ldots$

21. $\frac{3}{4}, \frac{1}{2}, \frac{1}{3}, \frac{2}{9}, \ldots$ **22.** $a, 1, \dfrac{1}{a}, \dfrac{1}{a^2}, \ldots$

23. Proving a Sequence Is Geometric Show that $3, 6i, -12, -24i, \ldots$ is a geometric sequence, and find the common ratio. (Here $i = \sqrt{-1}$.)

24. nth Term of a Geometric Sequence Find the nth term of the geometric sequence $2, 2 + 2i, 4i, -4 + 4i, -8, \ldots$. (Here $i = \sqrt{-1}$.)

25–28 ■ Finding Terms of Arithmetic and Geometric Sequences Find the indicated term of the arithmetic or geometric sequence with the given description.

25. The fourth term of an arithmetic sequence is 11, and the sixth term is 17. Find the second term.

26. The 20th term of an arithmetic sequence is 96, and the common difference is 5. Find the nth term.

27. The third term of a geometric sequence is 9, and the common ratio is $\frac{3}{2}$. Find the fifth term.

28. The second term of a geometric sequence is 10, and the fifth term is $\frac{1250}{27}$. Find the nth term.

29–30 ■ Salary Increases A school advertises two teaching positions.

Position I: Starting salary \$52,000 and each year the salary increases by 4% of the preceding year

Position II: Starting salary \$55,000 and each year the salary increases by \$1600

29. For Position I, find a formula for the salary A_n in the nth year of employment and list the salaries for the first six years of employment.

30. For Position II, find a formula for the salary A_n in the nth year of employment and list the salaries for the first six years of employment. Which teaching position has the larger salary in the sixth year?

31. Bacteria Culture A certain type of bacteria divides every 5 seconds. If three of these bacteria are put into a petri dish, how many bacteria are in the dish at the end of 1 minute?

32. Arithmetic Sequences If $a_1, a_2, a_3, \ldots$ and $b_1, b_2, b_3, \ldots$ are arithmetic sequences, show that $a_1 + b_1, a_2 + b_2, a_3 + b_3, \ldots$ is also an arithmetic sequence.

33. Geometric Sequences If $a_1, a_2, a_3, \ldots$ and $b_1, b_2, b_3, \ldots$ are geometric sequences, show that $a_1b_1, a_2b_2, a_3b_3, \ldots$ is also a geometric sequence.

34. Arithmetic or Geometric Sequence?

(a) If $a_1, a_2, a_3, \ldots$ is an arithmetic sequence, is the sequence $a_1 + 2, a_2 + 2, a_3 + 2, \ldots$ arithmetic?

(b) If $a_1, a_2, a_3, \ldots$ is a geometric sequence, is the sequence $5a_1, 5a_2, 5a_3, \ldots$ geometric?

35. Arithmetic and Geometric Sequences Find the values of x for which the sequence $6, x, 12, \ldots$ is

(a) arithmetic (b) geometric

36. Arithmetic and Geometric Sequences Find the values of x and y for which the sequence $2, x, y, 17, \ldots$ is

(a) arithmetic (b) geometric

37–40 ■ Partial Sums Find the sum.

37. $\sum_{k=3}^{6} (k+1)^2$ **38.** $\sum_{i=1}^{4} \frac{2i}{2i-1}$

39. $\sum_{k=1}^{6} (k+1)2^{k-1}$ **40.** $\sum_{m=1}^{5} 3^{m-2}$

41–44 ■ Sigma Notation Write the sum without using sigma notation. Do not evaluate.

41. $\sum_{k=1}^{10} (k-1)^2$ **42.** $\sum_{j=2}^{100} \frac{1}{j-1}$

43. $\sum_{k=1}^{50} \frac{3^k}{2^{k+1}}$ **44.** $\sum_{n=1}^{10} n^2 2^n$

45–48 ■ Sigma Notation Write the sum using sigma notation. Do not evaluate.

45. $3 + 6 + 9 + 12 + \cdots + 99$

46. $1^2 + 2^2 + 3^2 + \cdots + 100^2$

47. $1 \cdot 2^3 + 2 \cdot 2^4 + 3 \cdot 2^5 + 4 \cdot 2^6 + \cdots + 100 \cdot 2^{102}$

48. $\frac{1}{1 \cdot 2} + \frac{1}{2 \cdot 3} + \frac{1}{3 \cdot 4} + \cdots + \frac{1}{999 \cdot 1000}$

49–54 ■ Sums of Arithmetic and Geometric Sequences Determine whether the expression is a partial sum of an arithmetic or geometric sequence. Then find the sum.

49. $1 + 0.9 + (0.9)^2 + \cdots + (0.9)^5$

50. $3 + 3.7 + 4.4 + \cdots + 10$

51. $\sqrt{5} + 2\sqrt{5} + 3\sqrt{5} + \cdots + 100\sqrt{5}$

52. $\frac{1}{3} + \frac{2}{3} + 1 + \frac{4}{3} + \cdots + 33$

53. $\sum_{n=0}^{6} 3(-4)^n$ **54.** $\sum_{k=0}^{8} 7(5)^{k/2}$

55–60 ■ Infinite Geometric Series Determine whether the infinite geometric series is convergent or divergent. If it is convergent, find its sum.

55. $1 - \frac{2}{5} + \frac{4}{25} - \frac{8}{125} + \cdots$

56. $0.1 + 0.01 + 0.001 + 0.0001 + \cdots$

57. $5 - 5(1.01) + 5(1.01)^2 - 5(1.01)^3 + \cdots$

58. $1 + \frac{1}{3^{1/2}} + \frac{1}{3} + \frac{1}{3^{3/2}} + \cdots$

59. $-1 + \frac{9}{8} - \left(\frac{9}{8}\right)^2 + \left(\frac{9}{8}\right)^3 - \cdots$

60. $a + ab^2 + ab^4 + ab^6 + \cdots, \quad |b| < 1$

61. Terms of an Arithmetic Sequence The first term of an arithmetic sequence is $a = 7$, and the common difference is $d = 3$. How many terms of this sequence must be added to obtain 325?

62. Terms of an Geometric Sequence The sum of the first three terms of a geometric series is 52, and the common ratio is $r = 3$. Find the first term.

63. Ancestors A person has two parents, four grandparents, eight great-grandparents, and so on. What is the total number of a person's ancestors in 15 generations?

64–66 ■ Mathematical Induction Use mathematical induction to prove that the formula is true for all natural numbers n.

64. $1 + 4 + 7 + \cdots + (3n-2) = \frac{n(3n-1)}{2}$

65. $\frac{1}{1 \cdot 3} + \frac{1}{3 \cdot 5} + \frac{1}{5 \cdot 7} + \cdots + \frac{1}{(2n-1)(2n+1)} = \frac{n}{2n+1}$

66. $\left(1 + \frac{1}{1}\right)\left(1 + \frac{1}{2}\right)\left(1 + \frac{1}{3}\right) \cdots \left(1 + \frac{1}{n}\right) = n + 1$

67–69 ■ Proof by Induction Use mathematical induction to show that the given statement is true.

67. $7^n - 1$ is divisible by 6 for all natural numbers n.

68. The Fibonacci number F_{4n} is divisible by 3 for all natural numbers n.

69. Formula for a Recursive Sequence A sequence is defined recursively by $a_{n+1} = 3a_n + 4$ and $a_1 = 4$. Show that $a_n = 2 \cdot 3^n - 2$ for all natural numbers n.

70–73 ■ Binomial Coefficients Evaluate the expression.

70. $\binom{5}{2}\binom{5}{3}$ **71.** $\binom{10}{2} + \binom{10}{6}$

72. $\sum_{k=0}^{5} \binom{5}{k}$ **73.** $\sum_{k=0}^{8} \binom{8}{k}\binom{8}{8-k}$

74–77 ■ Binomial Expansion Expand the expression.

74. $(A - B)^3$ **75.** $(x + 2)^5$

76. $(1 - x^2)^6$ **77.** $(2x + y)^4$

78–80 ■ Terms in a Binomial Expansion Find the indicated terms in the given binomial expansion.

78. Find the 20th term in the expansion of $(a + b)^{22}$.

79. Find the first three terms in the expansion of $(b^{-2/3} + b^{1/3})^{20}$.

80. Find the term containing A^6 in the expansion of $(A + 3B)^{10}$.

Chapter 11 | Test

1. Find the first six terms and the sixth partial sum of the sequence whose nth term is $a_n = 2n^2 - n$.

2. A sequence is defined recursively by $a_{n+1} = 3a_n - n$, $a_1 = 2$. Find the first six terms of the sequence.

3. An arithmetic sequence begins 2, 5, 8, 11, 14, ...

(a) Find the common difference d for this sequence.

(b) Find a formula for the nth term a_n of the sequence.

(c) Find the 35th term of the sequence.

4. A geometric sequence begins 12, 3, $\frac{3}{4}$, $\frac{3}{16}$, $\frac{3}{64}$, ...

(a) Find the common ratio r for this sequence.

(b) Find a formula for the nth term a_n of the sequence.

(c) Find the tenth term of the sequence.

5. The first term of a geometric sequence is 25, and the fourth term is $\frac{1}{5}$.

(a) Find the common ratio r and the fifth term.

(b) Find the sum of the first eight terms.

6. The first term of an arithmetic sequence is 10, and the tenth term is 2.

(a) Find the common difference and the 100th term of the sequence.

(b) Find the sum of the first ten terms.

7. Let $a_1, a_2, a_3, \ldots$ be a geometric sequence with initial term a and common ratio r. Show that $a_1^2, a_2^2, a_3^2, \ldots$ is also a geometric sequence by finding its common ratio.

8. Write the expression without using sigma notation, and then find the sum.

(a) $\sum_{n=1}^{5} (1 - n^2)$ **(b)** $\sum_{n=3}^{6} (-1)^n 2^{n-2}$

9. Find the sum.

(a) $\frac{1}{3} + \frac{2}{3^2} + \frac{2^2}{3^3} + \frac{2^3}{3^4} + \cdots + \frac{2^9}{3^{10}}$

(b) $1 + \frac{1}{2^{1/2}} + \frac{1}{2} + \frac{1}{2^{3/2}} + \cdots$

10. Use mathematical induction to prove that for all natural numbers n,

$$1^2 + 2^2 + 3^2 + \cdots + n^2 = \frac{n(n+1)(2n+1)}{6}$$

11. Expand $(2x + y^2)^5$.

12. Find the term containing x^3 in the binomial expansion of $(3x - 2)^{10}$.

13. A puppy weighs 0.36 kg at birth, and each week he gains 24% in weight. Let a_n be his weight in kilograms at the end of his nth week of life.

(a) Find a formula for a_n.

(b) How much does the puppy weigh when he is 6 weeks old?

(c) Is the sequence $a_1, a_2, a_3, \ldots$ arithmetic, geometric, or neither?

Focus on Modeling | Modeling with Recursive Sequences

Many real-world processes occur in stages. Population growth can be viewed in stages—each new generation represents a new stage. Compound interest is paid in stages—each interest payment creates a new account balance. Many things that change continuously are more easily measured in discrete stages. For example, we can measure the temperature of a continuously cooling object in one-hour intervals. In this *Focus on Modeling* we learn how recursive sequences are used to model such situations. In some cases we can get an explicit formula for a sequence from the recursion relation that defines it by finding a pattern in the terms of the sequence.

■ Recursive Sequences as Models

Suppose you deposit some money in an account that pays 6% interest compounded monthly. The bank has a definite rule for paying interest: At the end of each month the bank adds to your account $\frac{1}{2}\%$ (or 0.005) of the amount in your account at that time. Let's express this rule as follows:

$$\boxed{\text{amount at the end of this month}} = \boxed{\text{amount at the end of last month}} + 0.005 \times \boxed{\text{amount at the end of last month}}$$

Using the Distributive Property, we can write this as

$$\boxed{\text{amount at the end of this month}} = 1.005 \times \boxed{\text{amount at the end of last month}}$$

To model this statement using algebra, let A_0 be the amount of the original deposit, let A_1 be the amount at the end of the first month, let A_2 be the amount at the end of the second month, and so on. So A_n is the amount at the end of the nth month. Thus

$$A_n = 1.005A_{n-1}$$

We recognize this as a recursively defined sequence—it gives us the amount at each stage in terms of the amount at the preceding stage.

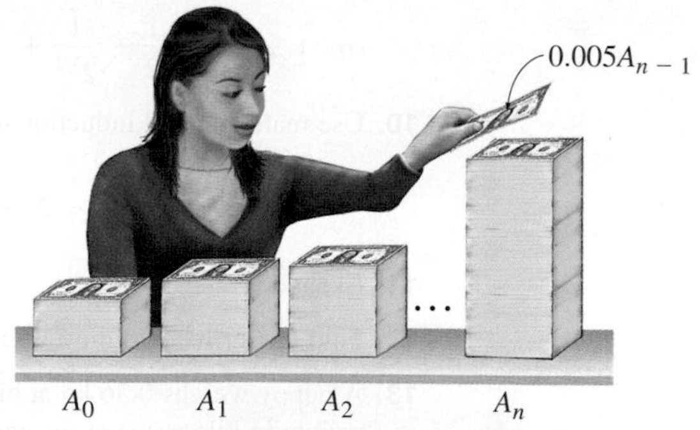

To find a formula for A_n, let's find the first few terms of the sequence and look for a pattern.

$$\begin{aligned}
A_1 &= 1.005A_0 \\
A_2 &= 1.005A_1 = (1.005)^2A_0 \\
A_3 &= 1.005A_2 = (1.005)^3A_0 \\
A_4 &= 1.005A_3 = (1.005)^4A_0
\end{aligned}$$

We can use mathematical induction to prove that the formula we found for A_n is valid for all natural numbers n.

We see that in general, $A_n = (1.005)^nA_0$.

Example 1 ■ Population Growth

A certain animal population grows by 2% each year. The initial population is 5000.

(a) Find a recursive sequence that models the population P_n at the end of the nth year.

(b) Find the first five terms of the sequence P_n.

(c) Find a formula for P_n.

Solution

(a) We can model the population using the following rule:

$$\boxed{\text{population at the end of this year}} = 1.02 \times \boxed{\text{population at the end of last year}}$$

Algebraically, we can write this as the recursion relation

$$P_n = 1.02P_{n-1}$$

(b) Since the initial population is 5000, we have

$$\begin{aligned} P_0 &= 5000 \\ P_1 &= 1.02P_0 = (1.02)5000 \\ P_2 &= 1.02P_1 = (1.02)^2 5000 \\ P_3 &= 1.02P_2 = (1.02)^3 5000 \\ P_4 &= 1.02P_3 = (1.02)^4 5000 \end{aligned}$$

(c) We see from the pattern exhibited in part (b) that $P_n = (1.02)^n 5000$. (Note that P_n is a geometric sequence, with common ratio $r = 1.02$.) ■

Example 2 ■ Daily Drug Dose

A patient is instructed to take a 50-mg pill of a certain drug every morning. It is known that the body eliminates 40% of the drug every 24 hours.

(a) Find a recursive sequence that models the amount A_n of the drug in the patient's body after each pill is taken.

(b) Find the first four terms of the sequence A_n.

(c) Find a formula for A_n.

(d) How much of the drug remains in the patient's body after 5 days? How much will accumulate in the patient's system after prolonged use?

Solution

(a) Each morning, 60% of the drug remains, plus the patient takes an additional 50 mg (the daily dose).

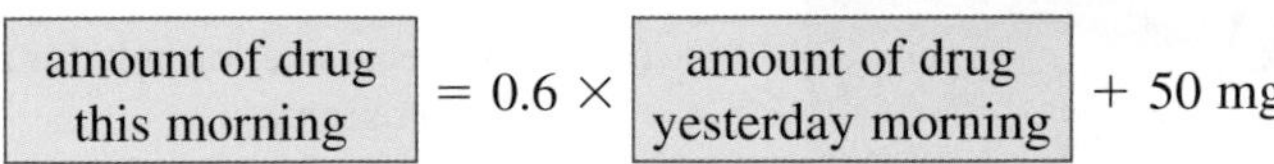

We can express this as a recursion relation

$$A_n = 0.6A_{n-1} + 50$$

(b) Since the initial dose is 50 mg, we have

$$A_0 = 50$$
$$A_1 = 0.6A_0 + 50 = 0.6(50) + 50$$
$$\begin{aligned} A_2 &= 0.6A_1 + 50 = 0.6[0.6(50) + 50] + 50 \\ &= 0.6^2(50) + 0.6(50) + 50 \\ &= 50(0.6^2 + 0.6 + 1) \end{aligned}$$
$$\begin{aligned} A_3 &= 0.6A_2 + 50 = 0.6[0.6^2(50) + 0.6(50) + 50] + 50 \\ &= 0.6^3(50) + 0.6^2(50) + 0.6(50) + 50 \\ &= 50(0.6^3 + 0.6^2 + 0.6 + 1) \end{aligned}$$

(c) From the pattern in part (b) we see that

$$\begin{aligned} A_n &= 50(1 + 0.6 + 0.6^2 + \cdots + 0.6^n) \\ &= 50\left(\frac{1 - 0.6^{n+1}}{1 - 0.6}\right) && \text{Partial sum of a geometric sequence} \\ &= 125(1 - 0.6^{n+1}) && \text{Simplify} \end{aligned}$$

(d) To find the amount remaining after 5 days, we substitute $n = 5$ and get

$$A_5 = 125(1 - 0.6^{5+1}) \approx 119 \text{ mg}$$

To find the amount remaining after prolonged use, we let n become large. As n gets large, 0.6^n approaches 0. That is, $0.6^n \to 0$ as $n \to \infty$ (see Section 4.1). So as $n \to \infty$,

$$A_n = 125(1 - 0.6^{n+1}) \to 125(1 - 0) = 125$$

Thus after prolonged use, the amount of drug in the patient's system approaches 125 mg (see Figure 1, where we have used a graphing device to graph the sequence). ■

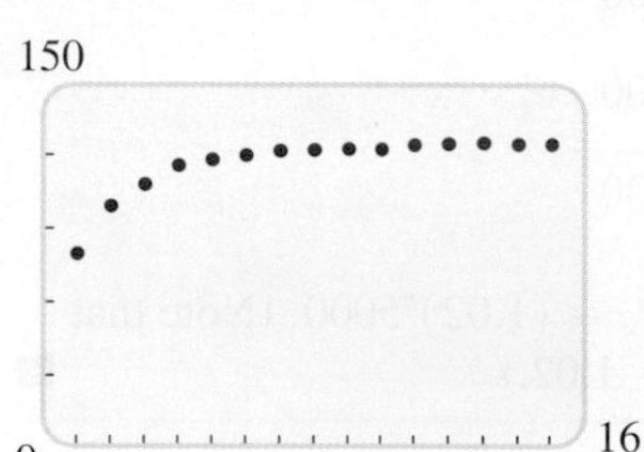

Figure 1

Problems

1. Retirement Accounts Many college professors keep retirement savings with TIAA, the largest annuity program in the world. Interest on these accounts is compounded and credited *daily*. A professor has \$275,000 on deposit with TIAA at the start of 2022 and earns 3.65% interest per year.

(a) Find a recursive sequence that models the amount A_n in the account at the end of the nth day of 2022.

(b) Find the first eight terms of the sequence A_n, rounded to the nearest cent.

(c) Find a formula for A_n.

2. Fitness Program A student decides to embark on a swimming program as the best way to maintain cardiovascular health. The student begins by swimming 5 minutes on the first day, then adds $1\frac{1}{2}$ minutes every day after that.

(a) Find a recursive formula for the number of minutes T_n spent swimming on the nth day of the program.

(b) Find the first six terms of the sequence T_n.

(c) Find a formula for T_n. What kind of sequence is this?

(d) On what day does the student attain the goal of swimming at least 65 minutes a day?

(e) What is the total amount of time the student will have swum after 30 days?

3. Monthly Savings Program A student begins a monthly savings program by depositing \$100 on January 1 in a savings account that pays 3% interest compounded monthly. An amount of \$100 is added to the account at the end of each month, when the interest is credited.

(a) Find a recursive formula for the amount A_n in the account at the end of the nth month. (Include the interest credited for that month and the monthly deposit.)

(b) Find the first five terms of the sequence A_n.

(c) Use the pattern you observed in part (b) to find a formula for A_n. [*Hint:* To find the pattern most easily, it's best *not* to simplify the terms *too* much.]

(d) How much has been saved after five years?

4. Pollution A chemical plant discharges 2400 tons of pollutants every year into an adjacent lake. Through natural runoff, 70% of the pollutants contained in the lake at the beginning of the year are expelled by the end of the year.

(a) Explain why the following sequence models the amount A_n of the pollutant in the lake at the end of the nth year that the plant is operating.

$$A_n = 0.30A_{n-1} + 2400$$

(b) Find the first five terms of the sequence A_n.

(c) Find a formula for A_n.

(d) How much of the pollutant remains in the lake after six years? How much will remain after the plant has been operating for many years?

(e) Verify your answer to part (d) by graphing A_n with a graphing device for $n = 1$ to $n = 20$.

5. Comparing Annual Saving Plans An amount of \$5000 is invested in a one-year Certificate of Deposit (CD) that yields 5% interest per year. At the end of each year, when the CD matures, the total amount (principal and interest) is reinvested, together with an additional amount according to one of the following plans.

Plan I: Add 10% of the total amount at the end of each year.

Plan II: Add \$500n to the total amount at the end of year n.

(a) Explain why the following recursion formulas give the amounts U_n and V_n that is reinvested in the nth year for Plans I and II, respectively.

$$U_n = 1.05U_{n-1} + 0.1(1.05U_{n-1}) \qquad V_n = 1.05V_{n-1} + 500n$$

(b) Calculate several values of U_n and V_n. These are most conveniently calculated using a graphing calculator, as shown in the figure. Observe that Plan II seems to accumulate more savings, but Plan I eventually pulls ahead in this savings race. In what year does this occur?

```
Plot1 Plot2 Plot3
\u(n) ■ 1.05 u(n - 1)
 +0.1(1.05 u(n - 1))
 u(nMin) ■ {5000}
\v(n) ■ 1.05 v(n - 1)
 +500 n
 v(nMin) ■ {5000}
```

Entering the sequences

n	u(n)	v(n)
0	5000	5000
1	5775	5750
2	6670.1	7037.5
3	7704	8889.4
4	8898.1	11334
5	10277	14401
6	11870	18121
n=0		

Table of values of the sequences

Oleksandr Osipov/Shutterstock.com

12 Limits: A Preview of Calculus

In this chapter we study the central idea underlying calculus: the concept of a *limit.* Calculus is used in modeling real-life phenomena, particularly situations that involve change or motion. Limits are used in finding the instantaneous rate of change of a function as well as the area of a region with curved boundary. You will learn in calculus that these two apparently different problems are closely related. In this chapter we see how limits allow us to solve both problems.

In Chapter 2 we learned how to find the average rate of change of a function. For example, to find the average speed (like the speed of a moving soccer ball), we divide the total distance traveled by the total time elasped. But how can we find *instantaneous* speed—that is, thc speed at a given instant? We can't divide the total distance by the total time because in an instant the total distance traveled is zero and the total time spent traveling is also zero! But we can find the average rate of change on smaller and smaller intervals, zooming in on the instant we want. In other words, the instantaneous speed is a *limit* of the average speeds.

In this chapter we also learn how to find areas of regions with curved sides by using the limit process.

12.1 Finding Limits Numerically and Graphically

■ Definition of Limit ■ Estimating Limits Numerically and Graphically ■ Limits That Fail to Exist ■ One-Sided Limits

In this section we use tables of values and graphs of functions to answer the question, What happens to the values $f(x)$ of a function f as the variable x approaches the number a?

■ Definition of Limit

We begin by investigating the behavior of the function f defined by

$$f(x) = x^2 - x + 2$$

for values of x near 2. The following tables give values of $f(x)$ for values of x close to 2 but not equal to 2.

x	$f(x)$
1.0	2.000000
1.5	2.750000
1.8	3.440000
1.9	3.710000
1.95	3.852500
1.99	3.970100
1.995	3.985025
1.999	3.997001

x	$f(x)$
3.0	8.000000
2.5	5.750000
2.2	4.640000
2.1	4.310000
2.05	4.152500
2.01	4.030100
2.005	4.015025
2.001	4.003001

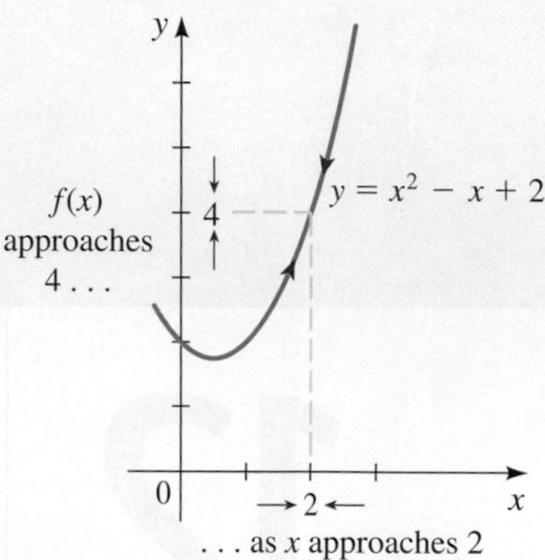

Figure 1

From the table and the graph of f (a parabola) shown in Figure 1 we see that when x is close to 2 (on either side of 2), $f(x)$ is close to 4. In fact, it appears that we can make the values of $f(x)$ as close as we like to 4 by taking x sufficiently close to 2. We express this by saying "the limit of the function $f(x) = x^2 - x + 2$ as x approaches 2 is equal to 4." The notation for this is

$$\lim_{x \to 2} (x^2 - x + 2) = 4$$

In general, we use the following notation.

> **Definition of the Limit of a Function**
>
> We write
>
> $$\lim_{x \to a} f(x) = L$$
>
> and say **"the limit of $f(x)$, as x approaches a,** equals L" if we can make the values of $f(x)$ arbitrarily close to L (as close to L as we like) by taking x to be sufficiently close to a, but not equal to a.

Roughly speaking, this says that the values of $f(x)$ get closer and closer to the number L as x gets closer and closer to the number a (from either side of a) but $x \neq a$.

An alternative notation for $\lim_{x \to a} f(x) = L$ is

$$f(x) \to L \quad \text{as} \quad x \to a$$

which is usually read "$f(x)$ approaches L as x approaches a." This is the notation we used in Section 3.6 when discussing asymptotes of rational functions.

Notice the phrase "but $x \neq a$" in the definition of limit. This means that in finding the limit of $f(x)$ as x approaches a, we never consider $x = a$. In fact, $f(x)$ need not even be defined when $x = a$. The only thing that matters is how f is defined *near a*.

Figure 2 shows the graphs of three functions. Note that in part (c), $f(a)$ is not defined, and in part (b), $f(a) \neq L$. But in each case, regardless of what happens at a, $\lim_{x \to a} f(x) = L$.

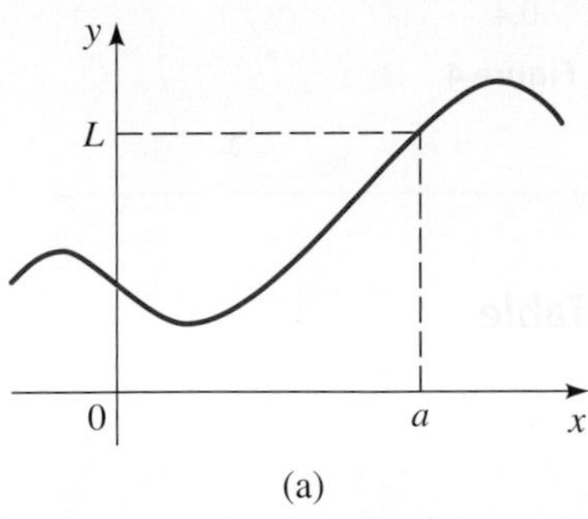

(a)

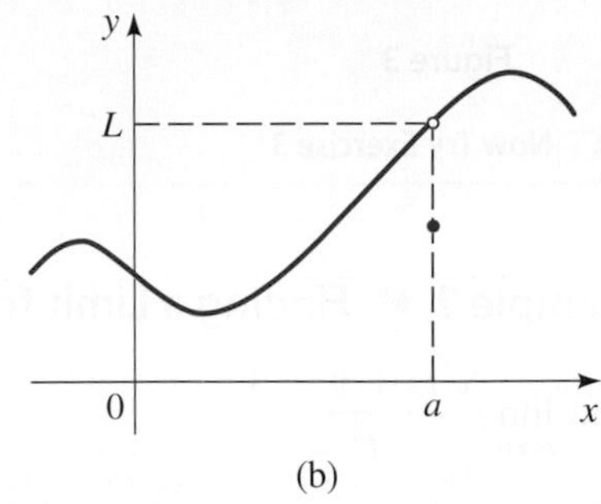

(b)

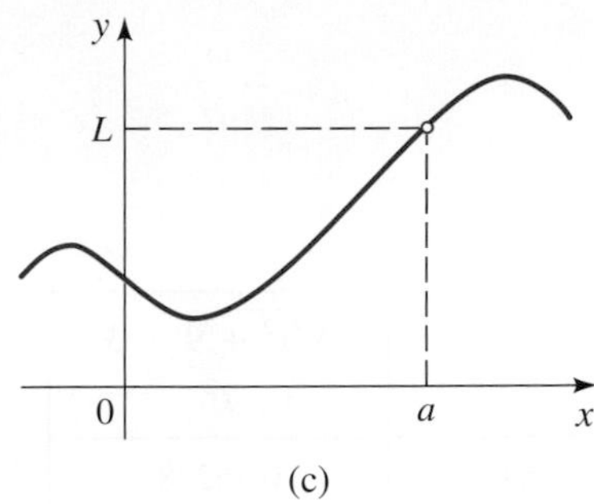

(c)

Figure 2 | $\lim_{x \to a} f(x) = L$ in all three cases

■ Estimating Limits Numerically and Graphically

In Section 12.2 we will develop techniques for finding exact values of limits. For now, we use tables and graphs to estimate limits of functions.

Example 1 ■ Estimating a Limit Numerically and Graphically

Estimate the value of the following limit by making a table of values. Check your work with a graph.

$$\lim_{x \to 1} \frac{x - 1}{x^2 - 1}$$

Solution Notice that the function $f(x) = (x - 1)/(x^2 - 1)$ is not defined when $x = 1$, but this doesn't matter because the definition of $\lim_{x \to a} f(x)$ says that we consider values of x that are close to a but not equal to a. The following tables give values of $f(x)$ (rounded to six decimal places) for values of x that approach 1 (but are not equal to 1).

$x < 1$	$f(x)$
0.5	0.666667
0.9	0.526316
0.99	0.502513
0.999	0.500250
0.9999	0.500025

$x > 1$	$f(x)$
1.5	0.400000
1.1	0.476190
1.01	0.497512
1.001	0.499750
1.0001	0.499975

On the basis of the values in the two tables we make the guess that

$$\lim_{x \to 1} \frac{x - 1}{x^2 - 1} = 0.5$$

As a graphical verification we use a graphing device to produce Figure 3. We see that when x is close to 1, y is close to 0.5. If we zoom in to get a closer look, as shown in Figure 4, we notice that as x gets closer to 1, y becomes closer to 0.5. This reinforces our conclusion.

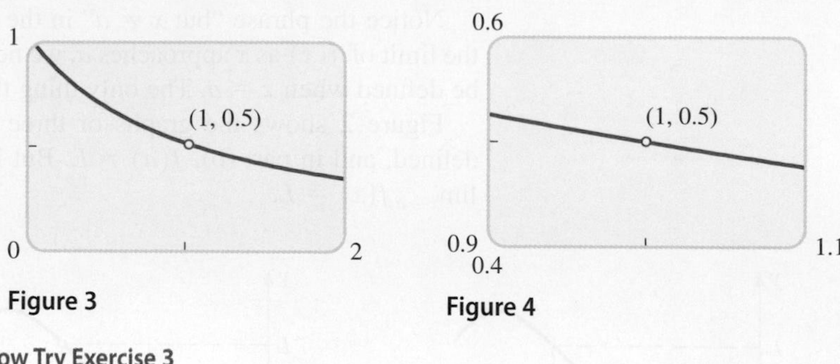

Figure 3

Figure 4

Now Try Exercise 3 ■

Example 2 ■ Finding a Limit from a Table

Find $\lim_{t \to 0} \dfrac{\sqrt{t^2 + 9} - 3}{t^2}$.

t	$\dfrac{\sqrt{t^2 + 9} - 3}{t^2}$
±1.0	0.16228
±0.5	0.16553
±0.1	0.16662
±0.05	0.16666
±0.01	0.16667

Solution The first table in the margin lists values of the function for several values of t near 0. As t approaches 0, the values of the function seem to approach 0.1666666 . . . , so we guess that

$$\lim_{t \to 0} \frac{\sqrt{t^2 + 9} - 3}{t^2} = \frac{1}{6}$$

Now Try Exercise 5 ■

t	$\dfrac{\sqrt{t^2 + 9} - 3}{t^2}$
±0.0005	0.16800
±0.0001	0.20000
±0.00005	0.00000
±0.00001	0.00000

What would have happened in Example 2 if we had taken even smaller values of t? The second table in the margin shows the results from one calculator; you can see that something strange seems to be happening.

If you try these calculations on your own calculator, you might get different values, but eventually, you will get the value 0 if you make t sufficiently small. Does this mean that the answer is really 0 instead of $\frac{1}{6}$? No, the value of the limit is $\frac{1}{6}$, as we will show in the next section. The problem is that the calculator gave false values because $\sqrt{t^2 + 9}$ is very close to 3 when t is small. (In fact, when t is sufficiently small, a calculator's value for $\sqrt{t^2 + 9}$ is 3.000 . . . to as many digits as the calculator is capable of carrying.)

Something similar happens when we try to graph the function of Example 2 on a graphing device. Parts (a) and (b) of Figure 5 show quite accurate graphs of this function, and from these graphs we estimate that the limit is about $\frac{1}{6}$. But if we zoom in too far, as in parts (c) and (d), then we get inaccurate graphs, again because the calculator gave false values.

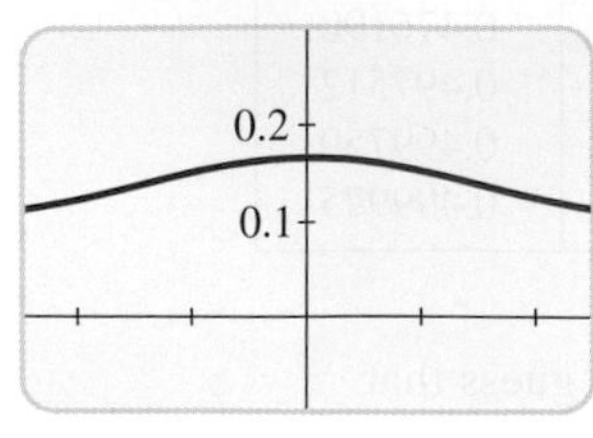

(a) $[-5, 5]$ by $[-0.1, 0.3]$

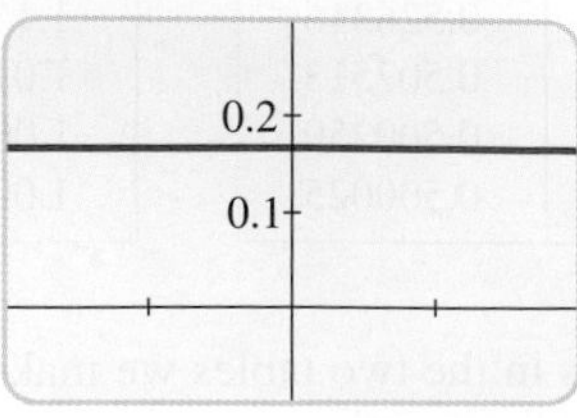

(b) $[-0.1, 0.1]$ by $[-0.1, 0.3]$

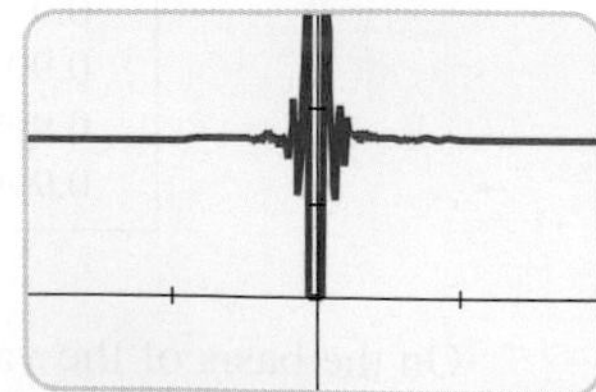

(c) $[-10^{-6}, 10^{-6}]$ by $[-0.1, 0.3]$

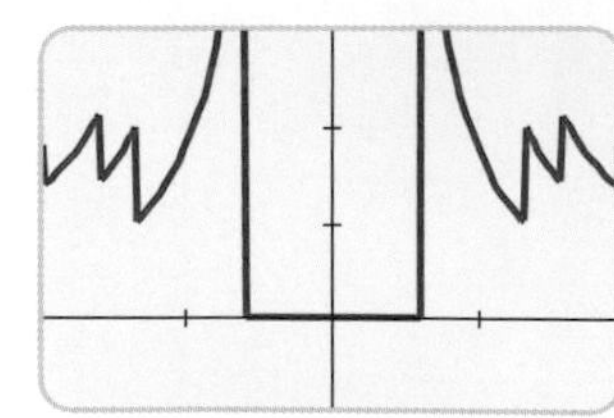

(d) $[-10^{-7}, 10^{-7}]$ by $[-0.1, 0.3]$

Figure 5

■ Limits That Fail to Exist

Functions do not necessarily approach a finite value at every point. In other words, it's possible for a limit not to exist. The next three examples illustrate ways in which this can happen.

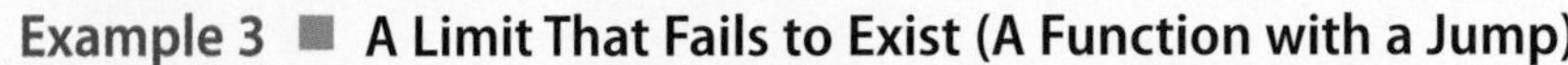

Example 3 ■ A Limit That Fails to Exist (A Function with a Jump)

The Heaviside function H is defined by

$$H(t) = \begin{cases} 0 & \text{if } t < 0 \\ 1 & \text{if } t \geq 0 \end{cases}$$

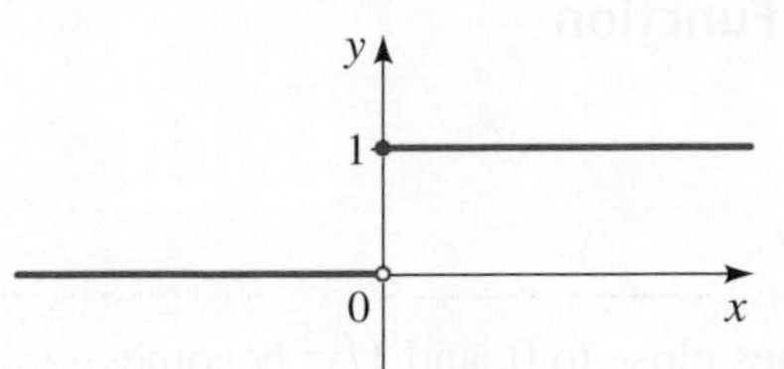

Figure 6

[This function, named after the electrical engineer Oliver Heaviside (1850–1925), can be used to describe an electric current that is switched on at time $t = 0$.] Its graph is shown in Figure 6. Notice the "jump" in the graph at $x = 0$.

As t approaches 0 from the left, $H(t)$ approaches 0. As t approaches 0 from the right, $H(t)$ approaches 1. There is no single number that $H(t)$ approaches as t approaches 0. Therefore $\lim_{t \to 0} H(t)$ does not exist.

Now Try Exercise 27 ■

Example 4 ■ A Limit That Fails to Exist (A Function That Oscillates)

Find $\displaystyle\lim_{x \to 0} \sin \frac{\pi}{x}$.

Solution The function $f(x) = \sin(\pi/x)$ is undefined at 0. Evaluating the function for some small values of x, we get

$$f(1) = \sin \pi = 0 \qquad f(\tfrac{1}{2}) = \sin 2\pi = 0$$

$$f(\tfrac{1}{3}) = \sin 3\pi = 0 \qquad f(\tfrac{1}{4}) = \sin 4\pi = 0$$

$$f(0.1) = \sin 10\pi = 0 \qquad f(0.01) = \sin 100\pi = 0$$

Similarly, $f(0.001) = f(0.0001) = 0$. On the basis of this information we might be tempted to guess that

$$\lim_{x \to 0} \sin \frac{\pi}{x} \stackrel{?}{=} 0$$

but this time our guess is wrong. Note that although $f(1/n) = \sin n\pi = 0$ for any integer n, it is also true that $f(x) = 1$ for infinitely many values of x that approach 0. (See the graph in Figure 7.)

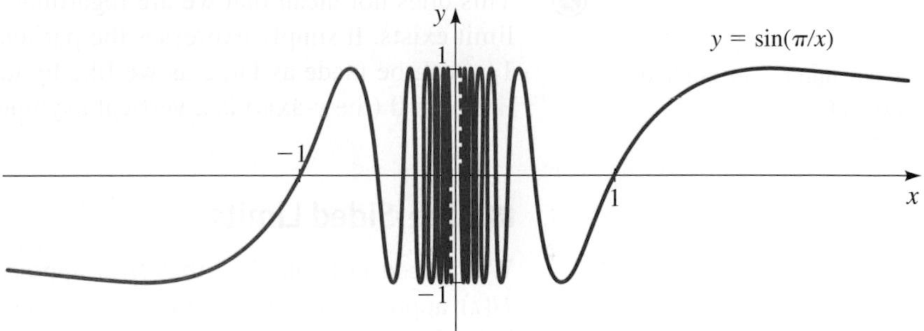

Figure 7

The dashed lines indicate that the values of $\sin(\pi/x)$ oscillate between 1 and -1 infinitely often as x approaches 0. Since the values of $f(x)$ do not approach a fixed number as x approaches 0,

$$\lim_{x \to 0} \sin \frac{\pi}{x} \quad \text{does not exist}$$

Now Try Exercise 25 ■

Example 4 illustrates some of the pitfalls in guessing the value of a limit. It is easy to guess the wrong value if we use inappropriate values of x, but it is difficult to know when to stop calculating values. And as the discussion after Example 2 shows, sometimes calculators and computers give incorrect values. In the next two sections, however, we will develop dependable methods for calculating limits.

Example 5 ■ A Limit That Fails to Exist (A Function with a Vertical Asymptote)

Find $\lim_{x \to 0} \dfrac{1}{x^2}$ if it exists.

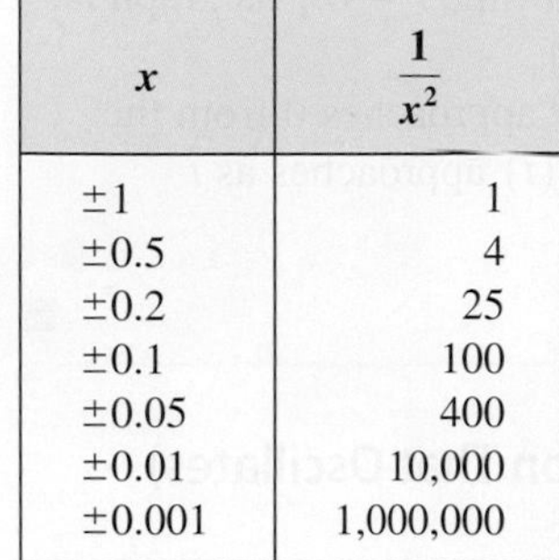

x	$\dfrac{1}{x^2}$
± 1	1
± 0.5	4
± 0.2	25
± 0.1	100
± 0.05	400
± 0.01	10,000
± 0.001	1,000,000

Solution As x becomes close to 0, x^2 also becomes close to 0, and $1/x^2$ becomes very large. (See the table in the margin.) In fact, it appears from the graph of the function $f(x) = 1/x^2$ shown in Figure 8 that the values of $f(x)$ can be made arbitrarily large by taking x close enough to 0. Thus the values of $f(x)$ do not approach a number, so $\lim_{x \to 0}(1/x^2)$ does not exist.

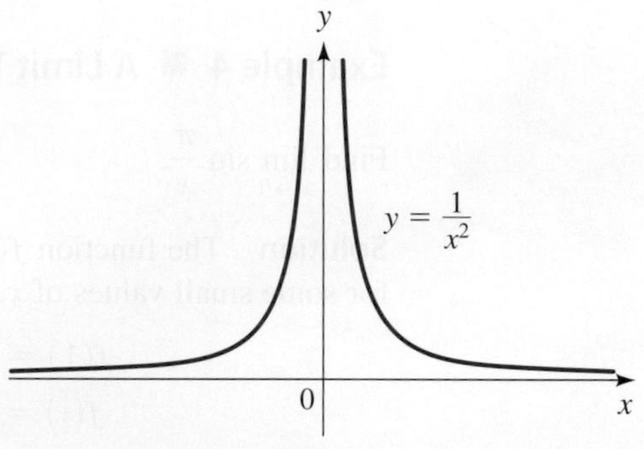

Figure 8

 Now Try Exercise 23 ■

To indicate the kind of behavior exhibited in Example 5, we use the notation

$$\lim_{x \to 0} \frac{1}{x^2} = \infty$$

This does not mean that we are regarding ∞ as a number. Nor does it mean that the limit exists. It simply expresses the particular way in which the limit does not exist: $1/x^2$ can be made as large as we like by taking x close enough to 0. Notice that the line $x = 0$ (the y-axis) is a vertical asymptote.

Vertical asymptotes are studied in Section 3.6.

■ One-Sided Limits

We noticed in Example 3 that $H(t)$ approaches 0 as t approaches 0 from the left and $H(t)$ approaches 1 as t approaches 0 from the right. We indicate this situation symbolically by writing

$$\lim_{t \to 0^-} H(t) = 0 \quad \text{and} \quad \lim_{t \to 0^+} H(t) = 1$$

The symbol "$t \to 0^-$" indicates that we consider only values of t less than 0. Likewise, "$t \to 0^+$" indicates that we consider only values of t greater than 0.

We also say "the limit of $f(x)$, as x approaches a from the left, equals L."

Definition of a One-Sided Limit

We write

$$\lim_{x \to a^-} f(x) = L$$

and say **"the left-hand limit of $f(x)$ as x approaches a**, equals L" if we can make the values of $f(x)$ arbitrarily close to L by taking x to be sufficiently close to a and $x < a$.

Notice that this definition differs from the definition of a two-sided limit only in that we require x to be *less than a.* Similarly, if we require that x be *greater than a,* we write

$$\lim_{x \to a^+} f(x) = L$$

And say, **"the right-hand limit of $f(x)$, as x approaches a**, equals L".

Thus the symbol "$x \to a^+$" means that we consider only $x > a$. These definitions are illustrated in Figure 9.

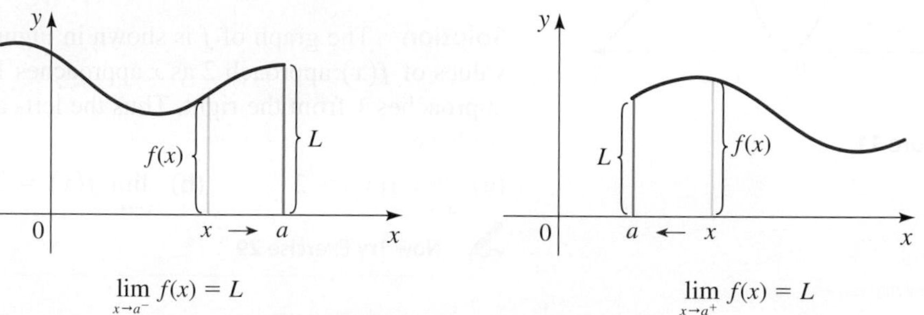

Figure 9

By comparing the definitions of two-sided and one-sided limits, we see that the following is true.

$$\lim_{x \to a} f(x) = L \quad \text{if and only if} \quad \lim_{x \to a^-} f(x) = L \quad \text{and} \quad \lim_{x \to a^+} f(x) = L$$

Thus if the left-hand and right-hand limits are different, the (two-sided) limit does not exist. We use this fact in the next two examples.

Example 6 ■ Limits from a Graph

The graph of a function g is shown in Figure 10. Use it to state the values (if they exist) of the following:

(a) $\lim_{x \to 2^-} g(x)$, $\lim_{x \to 2^+} g(x)$, $\lim_{x \to 2} g(x)$

(b) $\lim_{x \to 5^-} g(x)$, $\lim_{x \to 5^+} g(x)$, $\lim_{x \to 5} g(x)$

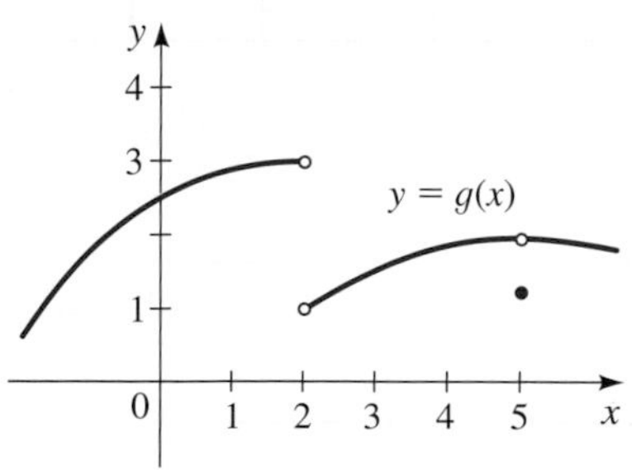

Figure 10

Solution

(a) From the graph we see that the values of $g(x)$ approach 3 as x approaches 2 from the left, but they approach 1 as x approaches 2 from the right. Therefore

$$\lim_{x \to 2^-} g(x) = 3 \qquad \text{and} \qquad \lim_{x \to 2^+} g(x) = 1$$

Since the left- and right-hand limits are different, we conclude that $\lim_{x \to 2} g(x)$ does not exist.

(b) The graph also shows that

$$\lim_{x\to 5^-} g(x) = 2 \qquad \text{and} \qquad \lim_{x\to 5^+} g(x) = 2$$

This time the left- and right-hand limits are the same, so we have

$$\lim_{x\to 5} g(x) = 2$$

Despite this fact, notice that $g(5) \neq 2$.

Now Try Exercise 19 ■

Example 7 ■ A Piecewise-Defined Function

Let f be the function defined by

$$f(x) = \begin{cases} 2x^2 & \text{if } x < 1 \\ 4 - x & \text{if } x \geq 1 \end{cases}$$

Graph f, and use the graph to find the following:

(a) $\lim_{x\to 1^-} f(x)$ **(b)** $\lim_{x\to 1^+} f(x)$ **(c)** $\lim_{x\to 1} f(x)$

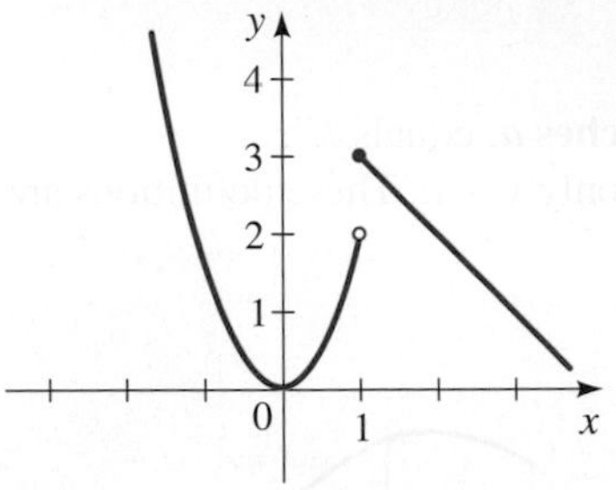

Figure 11

Solution The graph of f is shown in Figure 11. From the graph we see that the values of $f(x)$ approach 2 as x approaches 1 from the left, but they approach 3 as x approaches 1 from the right. Thus the left- and right-hand limits are not equal. So we have

(a) $\lim_{x\to 1^-} f(x) = 2$ **(b)** $\lim_{x\to 1^+} f(x) = 3$ **(c)** $\lim_{x\to 1} f(x)$ does not exist.

Now Try Exercise 29 ■

12.1 Exercises

■ Concepts

1. When we write $\lim_{x\to a} f(x) = L$ then, roughly speaking, the values of $f(x)$ get closer and closer to the number ______ as the values of x get closer and closer to ______. To determine $\lim_{x\to 5} \dfrac{x-5}{x-5}$, we try values for x closer and closer to ______ and find that the limit is ______.

2. We write $\lim_{x\to a^-} f(x) = L$ and say that the ______ of $f(x)$ as x approaches a from the ______ (left/right) is equal to ______. To find the left-hand limit, we try values for x that are ______ (less/greater) than a. A limit exists if and only if both the ______-hand and ______-hand limits exist and are ______.

■ Skills

3–4 ■ Estimating Limits Numerically and Graphically Estimate the value of the limit by making a table of values. Check your work with a graph.

3. $\lim_{x\to 5} \dfrac{x^2-25}{x-5}$ **4.** $\lim_{x\to 3} \dfrac{x^2-x-6}{x-3}$

5–10 ■ Estimating Limits Numerically Complete the table of values (to five decimal places), and use the table to estimate the value of the limit.

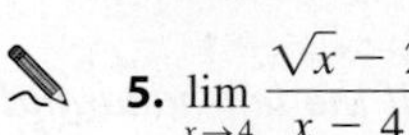

5. $\lim_{x\to 4} \dfrac{\sqrt{x}-2}{x-4}$

x	3.9	3.99	3.999	4.001	4.01	4.1
$f(x)$						

6. $\lim_{x\to 2} \dfrac{x-2}{x^2+x-6}$

x	1.9	1.99	1.999	2.001	2.01	2.1
$f(x)$						

7. $\lim_{x\to 1} \dfrac{x-1}{x^3-1}$

x	0.9	0.99	0.999	1.001	1.01	1.1
$f(x)$						

8. $\displaystyle\lim_{x\to 0} \frac{e^x - 1}{x}$

x	-0.1	-0.01	-0.001	0.001	0.01	0.1
$f(x)$						

9. $\displaystyle\lim_{x\to 0} \frac{\sin x}{x}$

x	± 1	± 0.5	± 0.1	± 0.05	± 0.01
$f(x)$					

10. $\displaystyle\lim_{x\to 0^+} x \ln x$

x	0.1	0.01	0.001	0.0001	0.00001
$f(x)$					

11–16 ■ Estimating Limits Numerically and Graphically Use a table of values to estimate the value of the limit. Then use a graphing device to confirm your result graphically.

11. $\displaystyle\lim_{x\to -4} \frac{x + 4}{x^2 + 7x + 12}$

12. $\displaystyle\lim_{x\to 1} \frac{x^3 - 1}{x^2 - 1}$

13. $\displaystyle\lim_{x\to 0} \frac{5^x - 3^x}{x}$

14. $\displaystyle\lim_{x\to 0} \frac{\sqrt{x + 9} - 3}{x}$

15. $\displaystyle\lim_{x\to 1} \left(\frac{1}{\ln x} - \frac{1}{x - 1}\right)$

16. $\displaystyle\lim_{x\to 0} \frac{\tan 2x}{\tan 3x}$

17–20 ■ Limits from a Graph For the function f whose graph is given, state the value of the given quantity if it exists. If it does not exist, explain why.

17. (a) $\displaystyle\lim_{x\to 1^-} f(x)$ **(b)** $\displaystyle\lim_{x\to 1^+} f(x)$ **(c)** $\displaystyle\lim_{x\to 1} f(x)$

(d) $\displaystyle\lim_{x\to 5} f(x)$ **(e)** $f(5)$

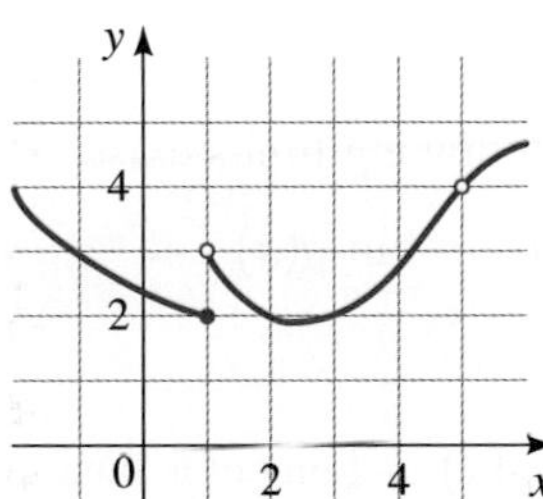

18. (a) $\displaystyle\lim_{x\to 0} f(x)$ **(b)** $\displaystyle\lim_{x\to 3^-} f(x)$ **(c)** $\displaystyle\lim_{x\to 3^+} f(x)$

(d) $\displaystyle\lim_{x\to 3} f(x)$ **(e)** $f(3)$

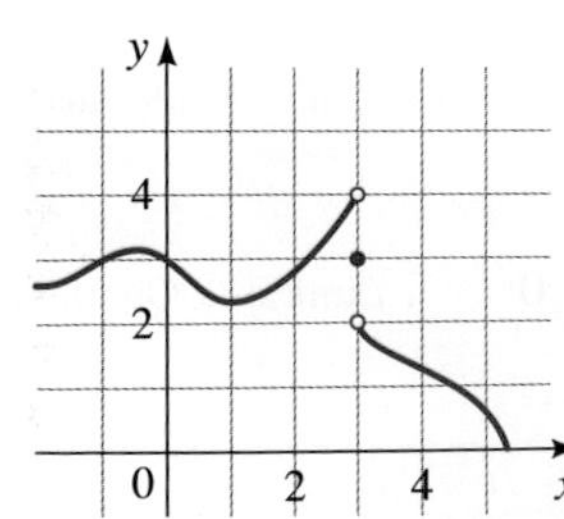

19. (a) $\displaystyle\lim_{t\to 0^-} f(t)$ **(b)** $\displaystyle\lim_{t\to 0^+} f(t)$ **(c)** $\displaystyle\lim_{t\to 0} f(t)$

(d) $\displaystyle\lim_{t\to 2^-} f(t)$ **(e)** $\displaystyle\lim_{t\to 2^+} f(t)$ **(f)** $\displaystyle\lim_{t\to 2} f(t)$

(g) $f(2)$ **(h)** $\displaystyle\lim_{t\to 4} f(t)$

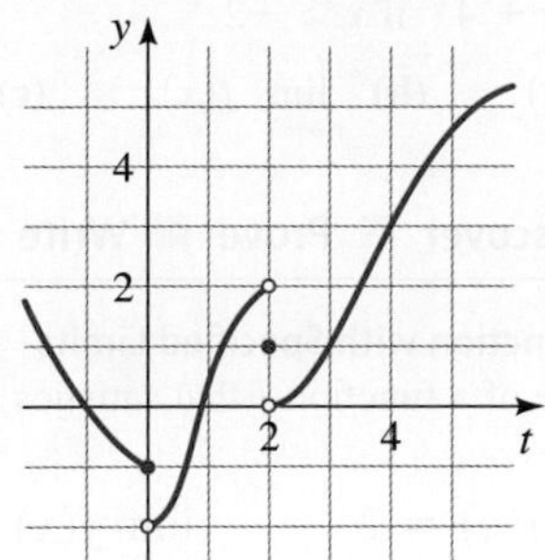

20. (a) $\displaystyle\lim_{x\to 3} f(x)$ **(b)** $\displaystyle\lim_{x\to 1} f(x)$ **(c)** $\displaystyle\lim_{x\to -3} f(x)$

(d) $\displaystyle\lim_{x\to 2^-} f(x)$ **(e)** $\displaystyle\lim_{x\to 2^+} f(x)$ **(f)** $\displaystyle\lim_{x\to 2} f(x)$

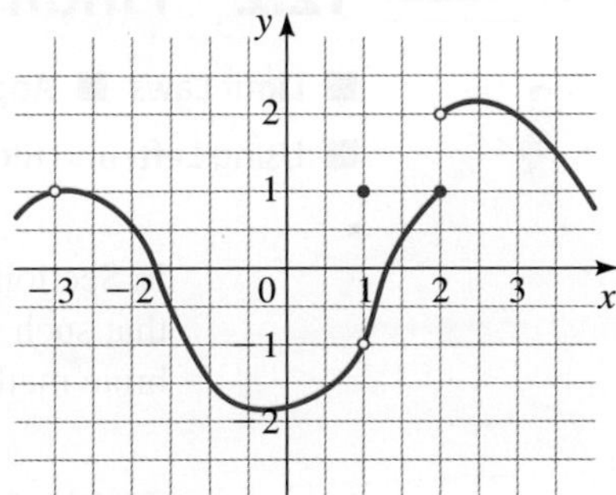

21–28 ■ Estimating Limits Graphically Use a graphing device to determine whether the limit exists. If the limit exists, estimate its value to two decimal places.

21. $\displaystyle\lim_{x\to 1} \frac{x^3 + x^2 + 3x - 5}{2x^2 - 5x + 3}$

22. $\displaystyle\lim_{x\to 0} \frac{x^2}{\cos 5x - \cos 4x}$

23. $\displaystyle\lim_{x\to 0} \ln(\sin^2 x)$

24. $\displaystyle\lim_{x\to 2} \frac{x^3 + 6x^2 - 5x + 1}{x^3 - x^2 - 8x + 12}$

25. $\displaystyle\lim_{x\to 0} \cos\frac{1}{x}$

26. $\displaystyle\lim_{x\to 0} \sin\frac{2}{x}$

27. $\displaystyle\lim_{x\to 3} \frac{|x - 3|}{x - 3}$

28. $\displaystyle\lim_{x\to 0} \frac{1}{1 + e^{1/x}}$

29–32 ■ One-Sided Limits Graph the piecewise-defined function and use your graph to find the values of the limits, if they exist.

29. $f(x) = \begin{cases} x^2 & \text{if } x \le 2 \\ 6 - x & \text{if } x > 2 \end{cases}$

(a) $\displaystyle\lim_{x\to 2^-} f(x)$ **(b)** $\displaystyle\lim_{x\to 2^+} f(x)$ **(c)** $\displaystyle\lim_{x\to 2} f(x)$

30. $f(x) = \begin{cases} 2 & \text{if } x < 0 \\ x + 1 & \text{if } x \ge 0 \end{cases}$

(a) $\displaystyle\lim_{x\to 0^-} f(x)$ **(b)** $\displaystyle\lim_{x\to 0^+} f(x)$ **(c)** $\displaystyle\lim_{x\to 0} f(x)$

31. $f(x) = \begin{cases} -x + 3 & \text{if } x < -1 \\ 3 & \text{if } x \geq -1 \end{cases}$

(a) $\lim_{x \to -1^-} f(x)$ **(b)** $\lim_{x \to -1^+} f(x)$ **(c)** $\lim_{x \to -1} f(x)$

32. $f(x) = \begin{cases} 2x + 10 & \text{if } x \leq -2 \\ -x + 4 & \text{if } x > -2 \end{cases}$

(a) $\lim_{x \to -2^-} f(x)$ **(b)** $\lim_{x \to -2^+} f(x)$ **(c)** $\lim_{x \to -2} f(x)$

■ Discuss ■ Discover ■ Prove ■ Write

33. Discuss: A Function with Specified Limits Sketch the graph of an example of a function f that satisfies all of the following conditions.

$$\lim_{x \to 0^-} f(x) = 2 \qquad \lim_{x \to 0^+} f(x) = 0$$

$$\lim_{x \to 2} f(x) = 1 \qquad f(0) = 2 \qquad f(2) = 3$$

How many such functions are there?

34. Discuss: Graphing Device Pitfalls

(a) Evaluate

$$h(x) = \frac{\tan x - x}{x^3}$$

for $x = 1, 0.5, 0.1, 0.05, 0.01$, and 0.005.

(b) Guess the value of $\lim_{x \to 0} \dfrac{\tan x - x}{x^3}$.

(c) Evaluate $h(x)$ for successively smaller values of x until you finally get a value of 0 for $h(x)$. Are you still confident that your guess in part (b) is correct? Explain why you eventually got a value of 0 for $h(x)$.

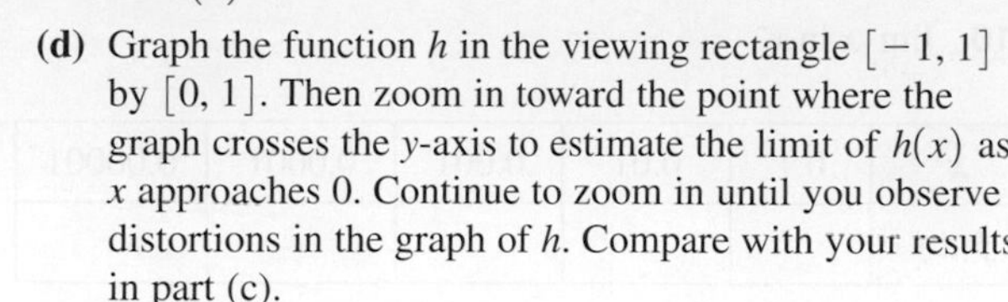

(d) Graph the function h in the viewing rectangle $[-1, 1]$ by $[0, 1]$. Then zoom in toward the point where the graph crosses the y-axis to estimate the limit of $h(x)$ as x approaches 0. Continue to zoom in until you observe distortions in the graph of h. Compare with your results in part (c).

12.2 Finding Limits Algebraically

■ Limit Laws ■ Applying the Limit Laws ■ Finding Limits Using Algebra and the Limit Laws ■ Using Left- and Right-Hand Limits

In Section 12.1 we used calculators and graphs to guess the values of limits, but we saw that such methods don't always lead to the correct answer. In this section we use algebraic methods to find limits exactly.

■ Limit Laws

We use the following properties of limits, called the *Limit Laws*, to calculate limits.

Limit Laws

Suppose that c is a constant and that the following limits exist:

$$\lim_{x \to a} f(x) \quad \text{and} \quad \lim_{x \to a} g(x)$$

Then

1. $\lim_{x \to a} [f(x) + g(x)] = \lim_{x \to a} f(x) + \lim_{x \to a} g(x)$ Limit of a Sum

2. $\lim_{x \to a} [f(x) - g(x)] = \lim_{x \to a} f(x) - \lim_{x \to a} g(x)$ Limit of a Difference

3. $\lim_{x \to a} [cf(x)] = c \lim_{x \to a} f(x)$ Limit of a Constant Multiple

4. $\lim_{x \to a} [f(x)g(x)] = \lim_{x \to a} f(x) \cdot \lim_{x \to a} g(x)$ Limit of a Product

5. $\lim_{x \to a} \dfrac{f(x)}{g(x)} = \dfrac{\lim_{x \to a} f(x)}{\lim_{x \to a} g(x)}$ if $\lim_{x \to a} g(x) \neq 0$ Limit of a Quotient

These five laws can be stated verbally as follows:

Limit of a Sum **1.** The limit of a sum is the sum of the limits.

Limit of a Difference **2.** The limit of a difference is the difference of the limits.

Limit of a Constant Multiple **3.** The limit of a constant times a function is the constant times the limit of the function.

Limit of a Product **4.** The limit of a product is the product of the limits.

Limit of a Quotient **5.** The limit of a quotient is the quotient of the limits (provided that the limit of the denominator is not 0).

It's easy to believe that these properties are true. For instance, if $f(x)$ is close to L and $g(x)$ is close to M, it is reasonable to conclude that $f(x) + g(x)$ is close to $L + M$. This gives us an intuitive basis for believing that Law 1 is true.

If we use Law 4 (Limit of a Product) repeatedly with $g(x) = f(x)$, we obtain the following Law 6 for the limit of a power. A similar law holds for roots.

Limit Laws

6. $\lim_{x\to a} [f(x)]^n = [\lim_{x\to a} f(x)]^n$ where n is a positive integer — Limit of a Power

7. $\lim_{x\to a} \sqrt[n]{f(x)} = \sqrt[n]{\lim_{x\to a} f(x)}$ where n is a positive integer — Limit of a Root

[If n is even, we assume that $\lim_{x\to a} f(x) > 0$.]

In words, these laws say the following:

Limit of a Power **6.** The limit of a power is the power of the limit.

Limit of a Root **7.** The limit of a root is the root of the limit.

Example 1 ■ Using the Limit Laws

Use the Limit Laws and the graphs of f and g in Figure 1 to evaluate the following limits if they exist.

(a) $\lim_{x\to -2} [f(x) + 5g(x)]$ **(b)** $\lim_{x\to 1} [f(x)g(x)]$

(c) $\lim_{x\to 2} \dfrac{f(x)}{g(x)}$ **(d)** $\lim_{x\to 1} [f(x)]^3$

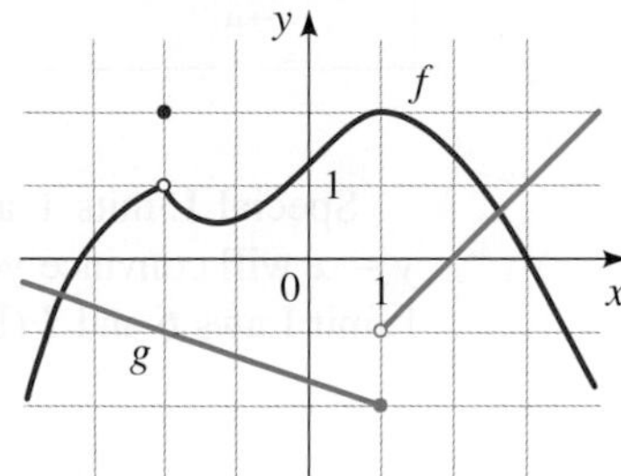

Figure 1

Solution

(a) From the graphs of f and g we see that

$$\lim_{x\to -2} f(x) = 1 \quad \text{and} \quad \lim_{x\to -2} g(x) = -1$$

Therefore we have

$$\begin{aligned}\lim_{x\to-2}[f(x)+5g(x)] &= \lim_{x\to-2} f(x) + \lim_{x\to-2}[5g(x)] && \text{Limit of a Sum}\\ &= \lim_{x\to-2} f(x) + 5\lim_{x\to-2} g(x) && \text{Limit of a Constant Multiple}\\ &= 1 + 5(-1) = -4\end{aligned}$$

(b) We see that $\lim_{x\to1} f(x) = 2$. But $\lim_{x\to1} g(x)$ does not exist because the left- and right-hand limits are different:

$$\lim_{x\to1^-} g(x) = -2 \qquad \lim_{x\to1^+} g(x) = -1$$

So we can't use Law 4 (Limit of a Product). The given limit does not exist because the left-hand limit is not equal to the right-hand limit.

(c) The graphs show that

$$\lim_{x\to2} f(x) \approx 1.4 \qquad \text{and} \qquad \lim_{x\to2} g(x) = 0$$

Because the limit of the denominator is 0, we can't use Law 5 (Limit of a Quotient). The given limit does not exist because the denominator approaches 0 whereas the numerator approaches a nonzero number.

(d) Since $\lim_{x\to1} f(x) = 2$, we use Law 6 to get

$$\begin{aligned}\lim_{x\to1}[f(x)]^3 &= [\lim_{x\to1} f(x)]^3 && \text{Limit of a Power}\\ &= 2^3 = 8\end{aligned}$$

 **Now Try Exercise 3** ■

■ Applying the Limit Laws

In applying the Limit Laws, we need to use four special limits.

Some Special Limits

1. $\lim_{x\to a} c = c$
2. $\lim_{x\to a} x = a$
3. $\lim_{x\to a} x^n = a^n$ where n is a positive integer
4. $\lim_{x\to a} \sqrt[n]{x} = \sqrt[n]{a}$ where n is a positive integer and $a > 0$

Special Limits 1 and 2 are intuitively clear—looking at the graphs of $y = c$ and $y = x$ will convince you of their validity. Special Limits 3 and 4 are particular cases of Limit Laws 6 and 7 (Limits of a Power and of a Root).

Example 2 ■ Using the Limit Laws

Evaluate the following limits, and justify each step.

(a) $\lim_{x\to5}(2x^2 - 3x + 4)$ **(b)** $\lim_{x\to-2}\dfrac{x^3 + 2x^2 - 1}{5 - 3x}$

Solution

(a) $\lim_{x\to 5}(2x^2 - 3x + 4) = \lim_{x\to 5}(2x^2) - \lim_{x\to 5}(3x) + \lim_{x\to 5} 4$ — Limits of a Difference and Sum

$= 2\lim_{x\to 5} x^2 - 3\lim_{x\to 5} x + \lim_{x\to 5} 4$ — Limit of a Constant Multiple

$= 2(5^2) - 3(5) + 4$ — Special Limits 3, 2, and 1

$= 39$

(b) We start by using Law 5, but its use is fully justified only at the final stage when we see that the limits of the numerator and denominator exist and the limit of the denominator is not 0.

$$\lim_{x\to -2}\frac{x^3 + 2x^2 - 1}{5 - 3x} = \frac{\lim_{x\to -2}(x^3 + 2x^2 - 1)}{\lim_{x\to -2}(5 - 3x)}$$ Limit of a Quotient

$$= \frac{\lim_{x\to -2} x^3 + 2\lim_{x\to -2} x^2 - \lim_{x\to -2} 1}{\lim_{x\to -2} 5 - 3\lim_{x\to -2} x}$$ Limits of Sums, Differences, and Constant Multiples

$$= \frac{(-2)^3 + 2(-2)^2 - 1}{5 - 3(-2)}$$ Special Limits 3, 2, and 1

$$= -\frac{1}{11}$$

Now Try Exercises 9 and 11 ■

If we let $f(x) = 2x^2 - 3x + 4$, then $f(5) = 39$. In Example 2(a) we found that $\lim_{x\to 5} f(x) = 39$. In other words, we would have gotten the correct answer by substituting 5 for x. Similarly, direct substitution provides the correct answer in part (b). The functions in Example 2 are a polynomial and a rational function, respectively, and similar use of the Limit Laws proves that direct substitution always works for such functions. We state this fact as follows.

Limits by Direct Substitution

If f is a polynomial or a rational function and a is in the domain of f, then

$$\lim_{x\to a} f(x) = f(a)$$

Functions with this direct substitution property are called **continuous at** a. You will learn more about continuous functions when you study calculus.

Example 3 ■ Finding Limits by Direct Substitution

Evaluate the following limits.

(a) $\lim_{x\to 3}(2x^3 - 10x - 8)$ **(b)** $\lim_{x\to -1}\frac{x^2 + 5x}{x^4 + 2}$

Solution

(a) The function $f(x) = 2x^3 - 10x - 8$ is a polynomial, so we can find the limit by direct substitution.

$$\lim_{x \to 3} (2x^3 - 10x - 8) = 2(3)^3 - 10(3) - 8 = 16$$

(b) The function $f(x) = (x^2 + 5x)/(x^4 + 2)$ is a rational function, and $x = -1$ is in its domain (because the denominator is not zero for $x = -1$). Thus we can find the limit by direct substitution.

$$\lim_{x \to -1} \frac{x^2 + 5x}{x^4 + 2} = \frac{(-1)^2 + 5(-1)}{(-1)^4 + 2} = -\frac{4}{3}$$

 Now Try Exercise 13 ■

■ Finding Limits Using Algebra and the Limit Laws

As we saw in Example 3, evaluating limits by direct substitution is straightforward. But not all limits can be evaluated this way. In fact, most of the situations in which limits are useful require us to work harder to evaluate the limit. The next three examples illustrate how we can use algebra to find limits.

Example 4 ■ Finding a Limit by Canceling a Common Factor

Find $\lim_{x \to 1} \dfrac{x - 1}{x^2 - 1}$.

Solution Let $f(x) = (x - 1)/(x^2 - 1)$. We can't find the limit by substituting $x = 1$ because $f(1)$ isn't defined. Nor can we apply Law 5 (Limit of a Quotient) because the limit of the denominator is 0. Instead, we need to do some preliminary algebra. We factor the denominator as a difference of squares:

$$\frac{x - 1}{x^2 - 1} = \frac{x - 1}{(x - 1)(x + 1)}$$

The numerator and denominator have a common factor of $x - 1$. When we take the limit as x approaches 1, we have $x \neq 1$, and so $x - 1 \neq 0$. Therefore we can cancel the common factor and compute the limit as follows.

$$\begin{aligned} \lim_{x \to 1} \frac{x - 1}{x^2 - 1} &= \lim_{x \to 1} \frac{x - 1}{(x - 1)(x + 1)} && \text{Factor} \\ &= \lim_{x \to 1} \frac{1}{x + 1} && \text{Cancel} \\ &= \frac{1}{1 + 1} = \frac{1}{2} && \text{Let } x \to 1 \end{aligned}$$

This calculation confirms algebraically the answer we got numerically and graphically in Example 12.1.1.

 Now Try Exercise 19 ■

Example 5 ■ Finding a Limit by Simplifying

Evaluate $\lim_{h \to 0} \dfrac{(3 + h)^2 - 9}{h}$.

B. Sanerson/Science Source

SIR ISAAC NEWTON (1642–1727) is universally regarded as one of the giants of physics and mathematics. He is well known for discovering the laws of motion and gravity and for inventing calculus, but he also proved the Binomial Theorem and the laws of optics, and he developed methods for solving polynomial equations to any desired accuracy. He was born a few months after the death of his father. After an unhappy childhood, he entered Cambridge University, where he learned mathematics by studying the writings of Euclid and Descartes.

During the plague years of 1665 and 1666, when the university was closed, Newton thought and wrote about ideas that, once published, instantly revolutionized the sciences. Imbued with a pathological fear of criticism, he published these writings only after many years of encouragement from Edmund Halley (who discovered the famous comet) and other colleagues.

Newton's works brought him enormous fame and prestige. Even poets were moved to praise; Alexander Pope wrote:

Nature and Nature's Laws
lay hid in Night.
God said, "Let Newton be"
and all was Light.

Newton was far more modest about his accomplishments. He said, "I do not know what I may appear to the world, but to myself I seem to have been only like a boy playing on the seashore . . . while the great ocean of truth lay all undiscovered before me." Newton was knighted by Queen Anne in 1705 and was buried with great honor in Westminster Abbey.

Solution We can't use direct substitution to evaluate this limit because the limit of the denominator is 0. So we first simplify the limit algebraically.

$$\begin{aligned}\lim_{h\to 0}\frac{(3+h)^2-9}{h} &= \lim_{h\to 0}\frac{(9+6h+h^2)-9}{h} && \text{Expand}\\ &= \lim_{h\to 0}\frac{6h+h^2}{h} && \text{Simplify}\\ &= \lim_{h\to 0}(6+h) && \text{Cancel } h\\ &= 6 && \text{Let } h\to 0\end{aligned}$$

 Now Try Exercise 25 ■

Example 6 ■ Finding a Limit by Rationalizing

Find $\displaystyle\lim_{t\to 0}\frac{\sqrt{t^2+9}-3}{t^2}$.

Solution We can't apply Law 5 (Limit of a Quotient) immediately, since the limit of the denominator is 0. Here the preliminary algebra consists of rationalizing the numerator.

$$\begin{aligned}\lim_{t\to 0}\frac{\sqrt{t^2+9}-3}{t^2} &= \lim_{t\to 0}\frac{\sqrt{t^2+9}-3}{t^2}\cdot\frac{\sqrt{t^2+9}+3}{\sqrt{t^2+9}+3} && \text{Rationalize numerator}\\ &= \lim_{t\to 0}\frac{(t^2+9)-9}{t^2\left(\sqrt{t^2+9}+3\right)} = \lim_{t\to 0}\frac{t^2}{t^2\left(\sqrt{t^2+9}+3\right)}\\ &= \lim_{t\to 0}\frac{1}{\sqrt{t^2+9}+3} = \frac{1}{\sqrt{\lim_{t\to 0}(t^2+9)}+3} = \frac{1}{3+3} = \frac{1}{6}\end{aligned}$$

This calculation confirms the guess that we made in Example 12.1.2.

 Now Try Exercise 27 ■

■ Using Left- and Right-Hand Limits

Some limits are best calculated by first finding the left- and right-hand limits. The following theorem is a reminder of what we discovered in Section 12.1. It says that *a two-sided limit exists if and only if both of the one-sided limits exist and are equal.*

$$\lim_{x\to a} f(x) = L \quad \text{if and only if} \quad \lim_{x\to a^-} f(x) = L = \lim_{x\to a^+} f(x)$$

When computing one-sided limits, we use the fact that the Limit Laws also hold for one-sided limits.

Example 7 ■ Comparing Right and Left Limits

Show that $\displaystyle\lim_{x\to 0}|x| = 0$.

The result of Example 7 looks plausible from Figure 2.

Solution Recall that

$$|x| = \begin{cases} x & \text{if } x \geq 0 \\ -x & \text{if } x < 0 \end{cases}$$

Since $|x| = x$ for $x > 0$, we have

$$\lim_{x \to 0^+} |x| = \lim_{x \to 0^+} x = 0$$

For $x < 0$ we have $|x| = -x$, so

$$\lim_{x \to 0^-} |x| = \lim_{x \to 0^-} (-x) = 0$$

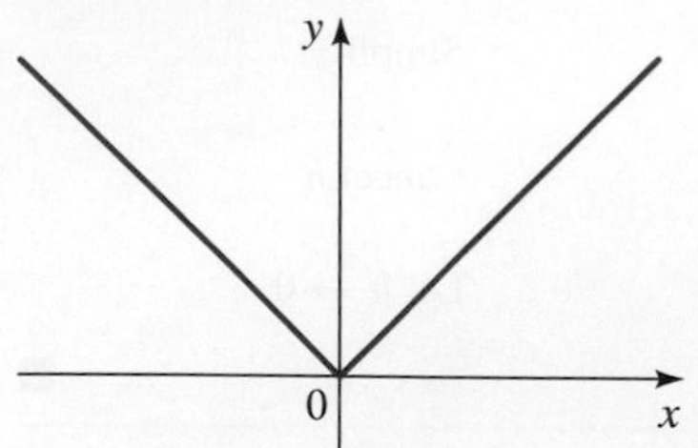

Figure 2 | $y = |x|$

Because the left- and right-hand limits exist and are equal, we have

$$\lim_{x \to 0} |x| = 0$$

Now Try Exercise 37 ■

Example 8 ■ Comparing Right and Left Limits

Prove that $\lim_{x \to 0} \dfrac{|x|}{x}$ does not exist.

Solution Since $|x| = x$ for $x > 0$ and $|x| = -x$ for $x < 0$, we have

$$\lim_{x \to 0^+} \frac{|x|}{x} = \lim_{x \to 0^+} \frac{x}{x} = \lim_{x \to 0^+} 1 = 1$$

$$\lim_{x \to 0^-} \frac{|x|}{x} = \lim_{x \to 0^-} \frac{-x}{x} = \lim_{x \to 0^-} (-1) = -1$$

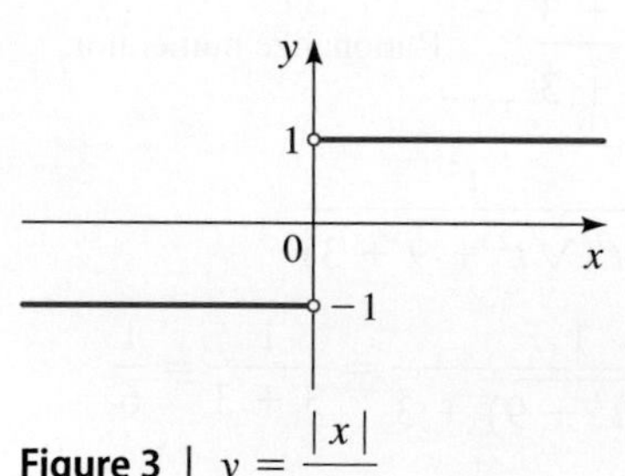

Figure 3 | $y = \dfrac{|x|}{x}$

Since the left-hand and right-hand limits exist and are different, it follows that $\lim_{x \to 0} |x|/x$ does not exist. The graph of the function $f(x) = |x|/x$ shown in Figure 3 confirms the limits that we found.

Now Try Exercise 39 ■

Example 9 ■ The Limit of a Piecewise-Defined Function

Let

$$f(x) = \begin{cases} \sqrt{x - 4} & \text{if } x > 4 \\ 8 - 2x & \text{if } x < 4 \end{cases}$$

Determine whether $\lim_{x \to 4} f(x)$ exists.

Solution Since $f(x) = \sqrt{x - 4}$ for $x > 4$, we have

$$\lim_{x \to 4^+} f(x) = \lim_{x \to 4^+} \sqrt{x - 4} = \sqrt{4 - 4} = 0$$

Since $f(x) = 8 - 2x$ for $x < 4$, we have

$$\lim_{x \to 4^-} f(x) = \lim_{x \to 4^-} (8 - 2x) = 8 - 2 \cdot 4 = 0$$

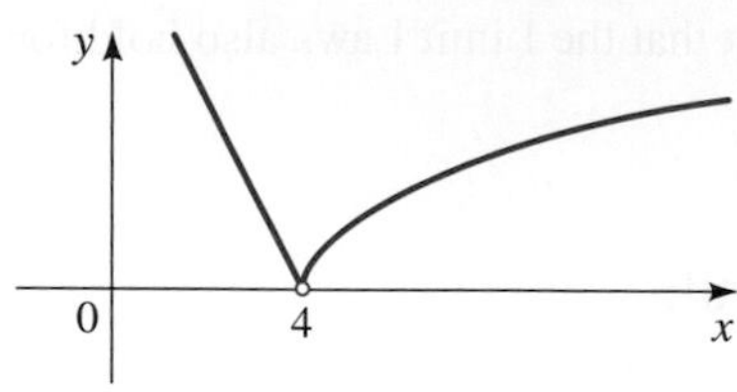

Figure 4

The left- and right-hand limits are equal. Thus the limit exists, and

$$\lim_{x \to 4} f(x) = 0$$

The graph of f is shown in Figure 4.

Now Try Exercise 43 ■

12.2 | Exercises

Concepts

1. Suppose the following limits exist:

$$\lim_{x\to a} f(x) \quad \text{and} \quad \lim_{x\to a} g(x)$$

Then $\lim_{x\to a}[f(x)+g(x)] =$ ________ + ________, and

$\lim_{x\to a}[f(x)g(x)] =$ ________ · ________.

These formulas can be stated verbally as follows: The limit of a sum is the ________ of the limits, and the limit of a product is the ________ of the limits.

2. If f is a polynomial or a rational function and a is in the domain of f, then $\lim_{x\to a} f(x) =$ ________.

Skills

3. Limits from a Graph The graphs of f and g are given. Use them to evaluate each limit if it exists. If the limit does not exist, explain why.

(a) $\lim_{x\to 2}[f(x)+g(x)]$ **(b)** $\lim_{x\to 1}[f(x)+g(x)]$

(c) $\lim_{x\to 0}[f(x)g(x)]$ **(d)** $\lim_{x\to -1}\dfrac{f(x)}{g(x)}$

(e) $\lim_{x\to 2} x^3 f(x)$ **(f)** $\lim_{x\to 1}\sqrt{3+f(x)}$

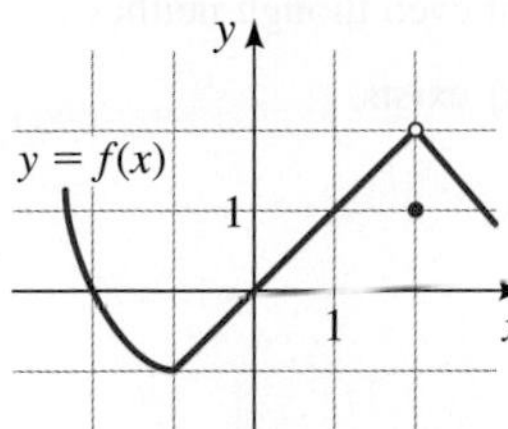

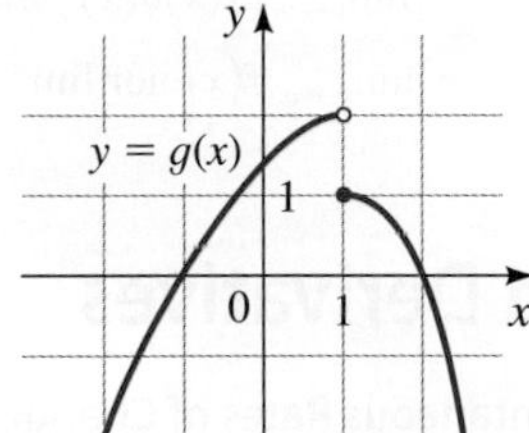

4. Using Limit Laws Suppose that

$$\lim_{x\to a} f(x) = -3 \qquad \lim_{x\to a} g(x) = 0 \qquad \lim_{x\to a} h(x) = 8$$

Find the value of the given limit. If the limit does not exist, explain why.

(a) $\lim_{x\to a}[f(x)+h(x)]$ **(b)** $\lim_{x\to a}[f(x)]^2$

(c) $\lim_{x\to a}\sqrt[3]{h(x)}$ **(d)** $\lim_{x\to a}\dfrac{1}{f(x)}$

(e) $\lim_{x\to a}\dfrac{f(x)}{h(x)}$ **(f)** $\lim_{x\to a}\dfrac{g(x)}{f(x)}$

(g) $\lim_{x\to a}\dfrac{f(x)}{g(x)}$ **(h)** $\lim_{x\to a}\dfrac{2f(x)}{h(x)-f(x)}$

5–18 ■ Using Limit Laws Evaluate the limit and justify each step by indicating the appropriate Limit Law(s).

5. $\lim_{x\to 5} x$ **6.** $\lim_{x\to 0} 3$

7. $\lim_{t\to 3} 4t$ **8.** $\lim_{t\to 2}(1-3t)$

9. $\lim_{x\to 4}(5x^2-2x+3)$ **10.** $\lim_{x\to 0}(3x^3-2x^2+5)$

11. $\lim_{x\to -1}\dfrac{x-2}{x^2+4x-3}$ **12.** $\lim_{x\to 2}\dfrac{2-x}{x^2+1}$

13. $\lim_{x\to 3}(x^3+2)(x^2-5x)$ **14.** $\lim_{t\to -2}(t+1)^9(t^2-1)$

15. $\lim_{x\to 1}\left(\dfrac{x^4+x^2-6}{x^4+2x+3}\right)^2$

16. $\lim_{x\to 0}\left(\dfrac{-5x^{20}-2x^2+3000}{x^2-1}\right)^{1/3}$

17. $\lim_{x\to 12}\left(\sqrt{x^2+25}-\sqrt{3x}\right)$ **18.** $\lim_{u\to -2}\sqrt{u^4+3u+6}$

19–32 ■ Finding Limits Evaluate the limit, if it exists.

19. $\lim_{x\to 2}\dfrac{x^2+x-6}{x-2}$ **20.** $\lim_{x\to -4}\dfrac{x^2+5x+4}{x^2+3x-4}$

21. $\lim_{x\to -2}\dfrac{x^2-x+6}{x+2}$ **22.** $\lim_{x\to 1}\dfrac{x^3-1}{x^2-1}$

23. $\lim_{t\to -3}\dfrac{t^2-9}{2t^2+7t+3}$ **24.** $\lim_{x\to 2}\dfrac{x^4-16}{x-2}$

25. $\lim_{h\to 0}\dfrac{(2+h)^2-4}{h}$ **26.** $\lim_{h\to 0}\dfrac{(2+h)^3-8}{h}$

27. $\lim_{x\to 7}\dfrac{\sqrt{x+2}-3}{x-7}$ **28.** $\lim_{h\to 0}\dfrac{\sqrt{1+h}-1}{h}$

29. $\lim_{x\to -4}\dfrac{\frac{1}{4}+\frac{1}{x}}{4+x}$ **30.** $\lim_{t\to 0}\left(\dfrac{1}{t}-\dfrac{1}{t^2+t}\right)$

31. $\lim_{h\to 0}\dfrac{(3+h)^{-1}-3^{-1}}{h}$ **32.** $\lim_{t\to 4}\dfrac{\frac{1}{\sqrt{t}}-\frac{1}{2}}{t-4}$

33–36 ■ Finding Limits Find the limit, and use a graphing device to confirm your result graphically.

33. $\lim_{x\to 1}\dfrac{x^2-1}{\sqrt{x}-1}$ **34.** $\lim_{x\to 0}\dfrac{(4+x)^3-64}{x}$

35. $\lim_{x\to -1}\dfrac{x^2-x-2}{x^3-x}$ **36.** $\lim_{x\to 1}\dfrac{x^8-1}{x^5-1}$

37–42 ■ Does the Limit Exist? Find the limit, if it exists. If the limit does not exist, explain why.

37. $\lim_{x\to -4}|x+4|$ **38.** $\lim_{x\to -4^-}\dfrac{|x+4|}{x+4}$

39. $\lim_{x\to 2}\dfrac{|x-2|}{x-2}$ **40.** $\lim_{x\to 1.5}\dfrac{2x^2-3x}{|2x-3|}$

41. $\lim_{x\to 0^-}\left(\dfrac{1}{x}-\dfrac{1}{|x|}\right)$ **42.** $\lim_{x\to 0^+}\left(\dfrac{1}{x}-\dfrac{1}{|x|}\right)$

43. Does the Limit Exist? Let

$$f(x) = \begin{cases} x-1 & \text{if } x < 2 \\ x^2-4x+6 & \text{if } x \ge 2 \end{cases}$$

(a) Find $\lim_{x\to 2^-} f(x)$ and $\lim_{x\to 2^+} f(x)$.

(b) Does $\lim_{x\to 2} f(x)$ exist?

(c) Sketch the graph of f.

44. Does the Limit Exist? Let

$$h(x) = \begin{cases} x & \text{if } x < 0 \\ x^2 & \text{if } 0 < x \le 2 \\ 8 - x & \text{if } x > 2 \end{cases}$$

(a) Evaluate each limit if it exists.

(i) $\lim_{x \to 0^+} h(x)$ (ii) $\lim_{x \to 0} h(x)$

(iii) $\lim_{x \to 1} h(x)$ (iv) $\lim_{x \to 2^-} h(x)$

(v) $\lim_{x \to 2^+} h(x)$ (vi) $\lim_{x \to 2} h(x)$

(b) Sketch the graph of h.

■ Skills Plus

45. Finding Limits Numerically and Graphically

(a) Estimate the value of

$$\lim_{x \to 0} \frac{x}{\sqrt{1 + 3x} - 1}$$

by graphing the function $f(x) = x/(\sqrt{1 + 3x} - 1)$.

(b) Make a table of values of $f(x)$ for x close to 0, and guess the value of the limit.

(c) Use the Limit Laws to prove that your guess is correct.

46. Finding Limits Numerically and Graphically

(a) Use a graph of

$$f(x) = \frac{\sqrt{3 + x} - \sqrt{3}}{x}$$

to estimate the value of $\lim_{x \to 0} f(x)$ to two decimal places.

(b) Use a table of values of $f(x)$ to estimate the limit to four decimal places.

(c) Use the Limit Laws to find the exact value of the limit.

■ Discuss ■ Discover ■ Prove ■ Write

47. Discuss: Cancellation and Limits

(a) What is wrong with the following equation?

$$\frac{x^2 + x - 6}{x - 2} = x + 3$$

(b) In view of part (a), explain why the equation

$$\lim_{x \to 2} \frac{x^2 + x - 6}{x - 2} = \lim_{x \to 2} (x + 3)$$

is correct.

48. Discuss: The Lorentz Contraction In the theory of relativity the Lorentz contraction formula

$$L = L_0 \sqrt{1 - v^2/c^2}$$

expresses the length L of an object as a function of its velocity v with respect to an observer, where L_0 is the length of the object at rest and c is the speed of light. Find $\lim_{v \to c^-} L$, and interpret the result. Why is a left-hand limit necessary?

49. Discuss ■ Prove: Limits of Sums and Products

(a) Show by means of an example that $\lim_{x \to a} [f(x) + g(x)]$ may exist even though neither $\lim_{x \to a} f(x)$ nor $\lim_{x \to a} g(x)$ exists.

(b) Show by means of an example that $\lim_{x \to a} [f(x)g(x)]$ may exist even though neither $\lim_{x \to a} f(x)$ nor $\lim_{x \to a} g(x)$ exists.

12.3 Tangent Lines and Derivatives

■ Tangent Lines ■ Derivatives ■ Instantaneous Rates of Change

In this section we see how limits arise when we attempt to find the tangent line to a curve or the instantaneous rate of change of a function.

■ Tangent Lines

A *tangent line* is a line that *just* touches a curve. For instance, Figure 1 shows the parabola $y = x^2$ and the tangent line t that touches the parabola at the point $P(1, 1)$. We will be able to find an equation of the tangent line t as soon as we know its slope m. The difficulty is that we know only one point, P, on t, whereas we need two points to compute the slope. But observe that we can compute an approximation to m by choosing a nearby point $Q(x, x^2)$ on the parabola (as in Figure 2) and computing the slope m_{PQ} of the secant line PQ.

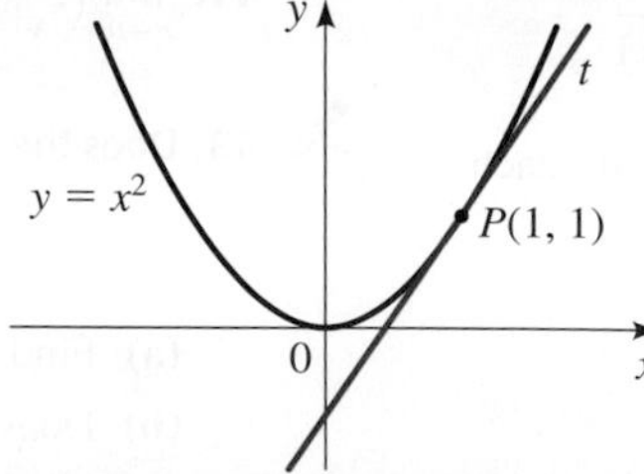

Figure 1

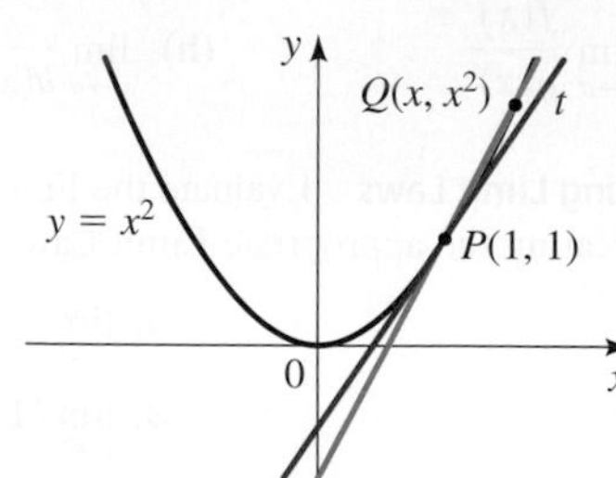

Figure 2

We choose $x \neq 1$ so that $Q \neq P$. Then

$$m_{PQ} = \frac{x^2 - 1}{x - 1}$$

Now we let x approach 1, so Q approaches P along the parabola. Figure 3 shows how the corresponding secant lines rotate about P and approach the tangent line t.

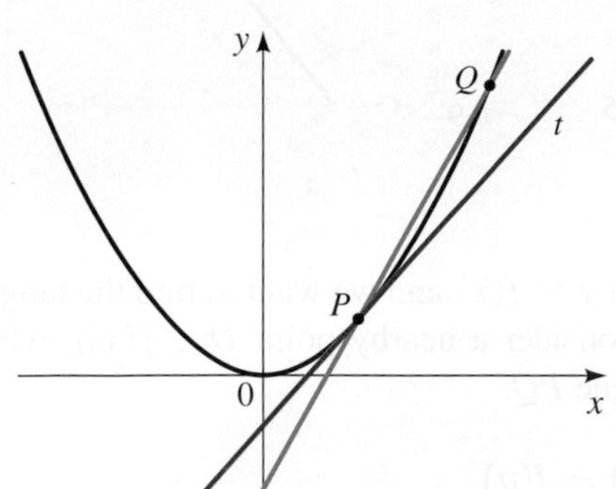

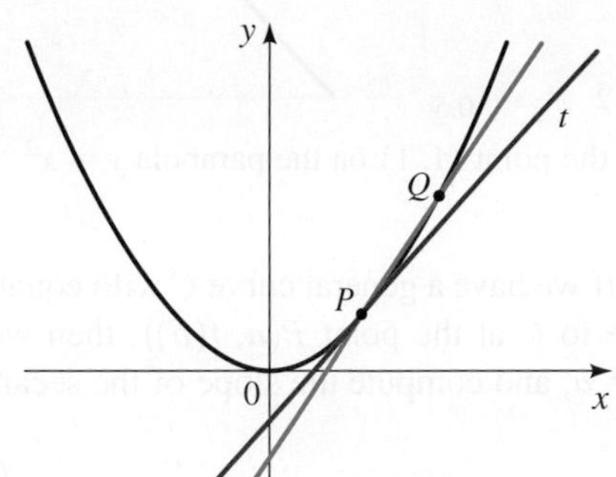

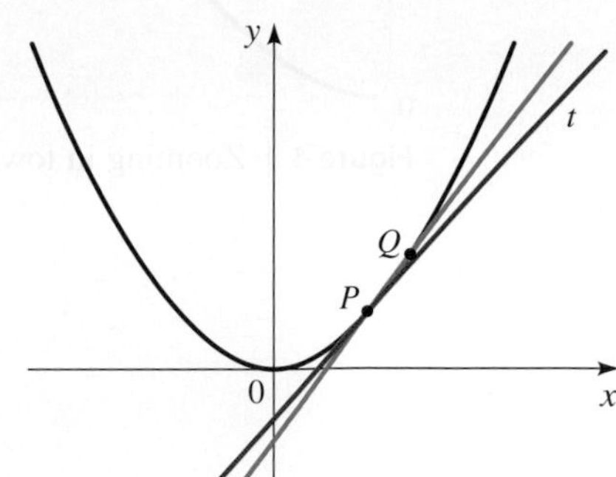

Q approaches *P* from the right

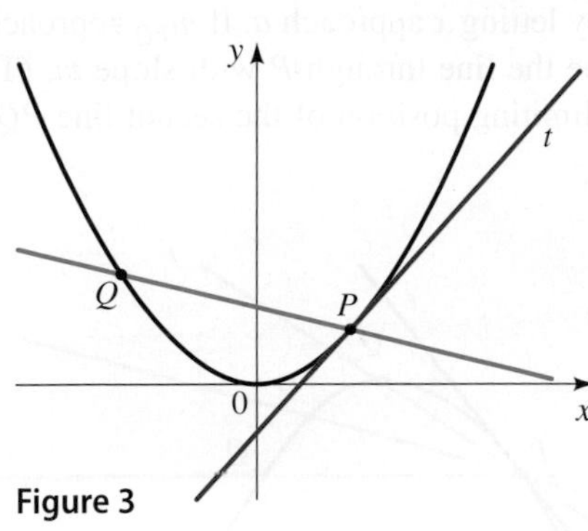

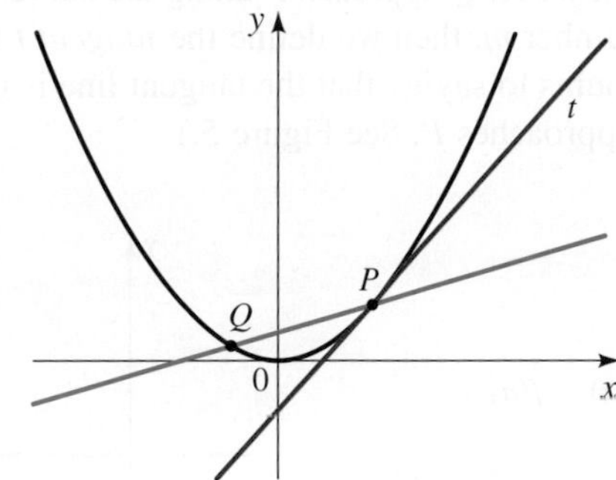

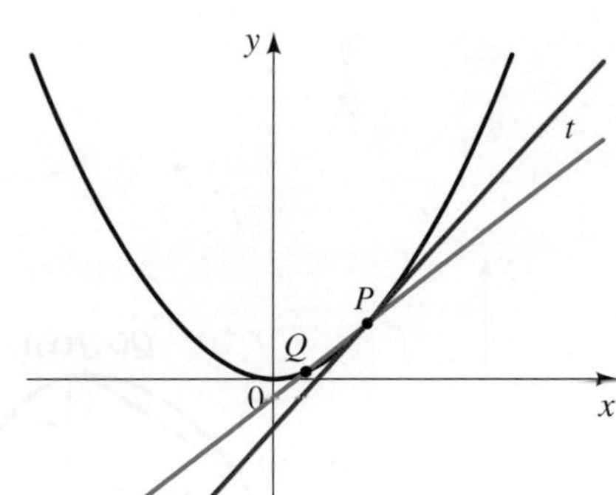

Figure 3

Q approaches *P* from the left

The slope of the tangent line is the limit of the slopes of the secant lines:

$$m = \lim_{Q \to P} m_{PQ}$$

So using the method of Section 12.2, we have

$$\begin{aligned} m &= \lim_{x \to 1} \frac{x^2 - 1}{x - 1} = \lim_{x \to 1} \frac{(x - 1)(x + 1)}{x - 1} \\ &= \lim_{x \to 1} (x + 1) = 1 + 1 = 2 \end{aligned}$$

The point-slope form for the equation of a line through the point (x_1, y_1) with slope m is

$$y - y_1 = m(x - x_1)$$

(See Section 1.10.)

Now that we know the slope of the tangent line is $m = 2$, we can use the point-slope form of the equation of a line to find its equation.

$$y - 1 = 2(x - 1) \qquad \text{or} \qquad y = 2x - 1$$

We sometimes refer to the slope of the tangent line to a curve at a point as the **slope of the curve** at the point. The idea is that if we zoom in far enough toward the point, the curve looks almost like a straight line. Figure 4 illustrates this procedure for the curve $y = x^2$. The more we zoom in, the more the parabola looks like a line. In other words, the curve becomes almost indistinguishable from its tangent line.

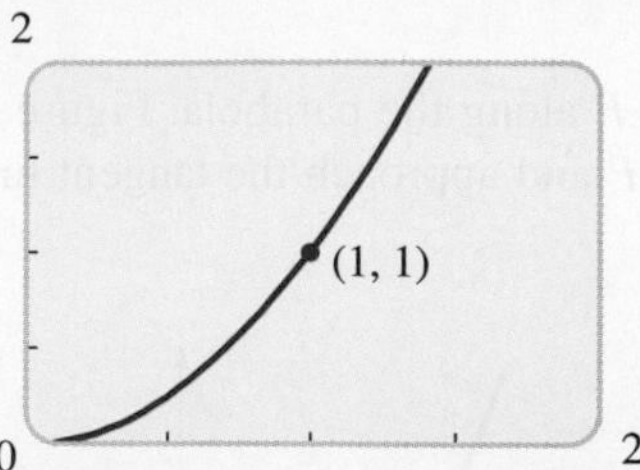

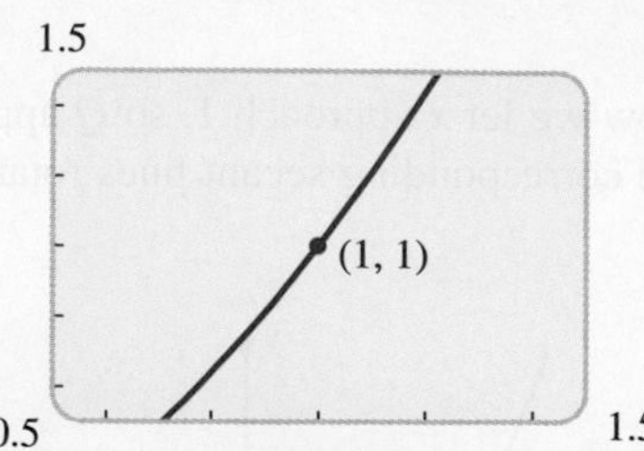

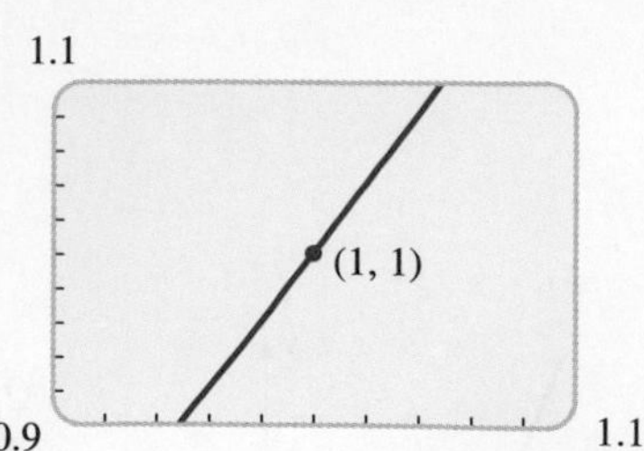

Figure 4 | Zooming in toward the point (1, 1) on the parabola $y = x^2$

If we have a general curve C with equation $y = f(x)$ and we want to find the tangent line to C at the point $P(a, f(a))$, then we consider a nearby point $Q(x, f(x))$, where $x \neq a$, and compute the slope of the secant line PQ.

$$m_{PQ} = \frac{f(x) - f(a)}{x - a}$$

Then we let Q approach P along the curve C by letting x approach a. If m_{PQ} approaches a number m, then we define the *tangent* t to be the line through P with slope m. (This amounts to saying that the tangent line is the limiting position of the secant line PQ as Q approaches P. See Figure 5.)

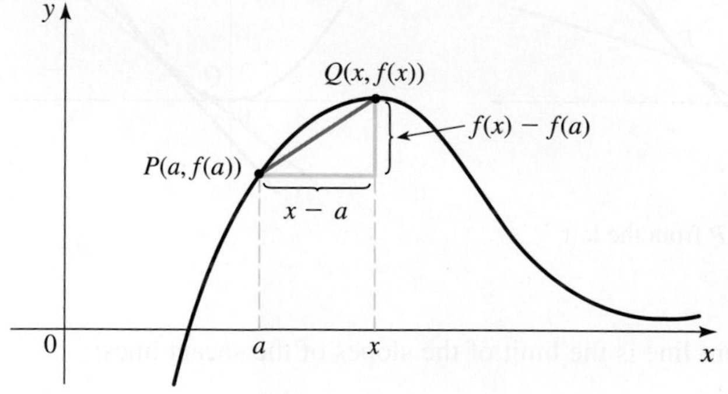

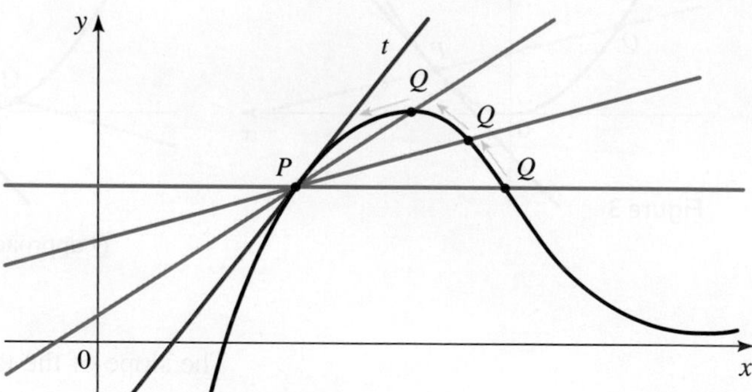

Figure 5

Definition of a Tangent Line

The **tangent line** to the curve $y = f(x)$ at the point $P(a, f(a))$ is the line through P with slope

$$m = \lim_{x \to a} \frac{f(x) - f(a)}{x - a}$$

provided that this limit exists.

Example 1 ■ Finding a Tangent Line to a Hyperbola

Find an equation of the tangent line to the hyperbola $y = 3/x$ at the point $(3, 1)$.

Solution Let $f(x) = 3/x$. Then the slope of the tangent line at $(3, 1)$ is

$$
\begin{aligned}
m &= \lim_{x\to 3} \frac{f(x) - f(3)}{x - 3} && \text{Definition of } m \\
&= \lim_{x\to 3} \frac{\dfrac{3}{x} - 1}{x - 3} && f(x) = \frac{3}{x} \\
&= \lim_{x\to 3} \frac{3 - x}{x(x - 3)} && \text{Multiply numerator and denominator by } x \\
&= \lim_{x\to 3}\left(-\frac{1}{x}\right) && \text{Cancel } x - 3 \\
&= -\frac{1}{3} && \text{Let } x \to 3
\end{aligned}
$$

Therefore an equation of the tangent line at the point $(3, 1)$ is

$$y - 1 = -\tfrac{1}{3}(x - 3)$$

which simplifies to

$$x + 3y - 6 = 0$$

The hyperbola and its tangent at the point $(3, 1)$ are shown in Figure 6.

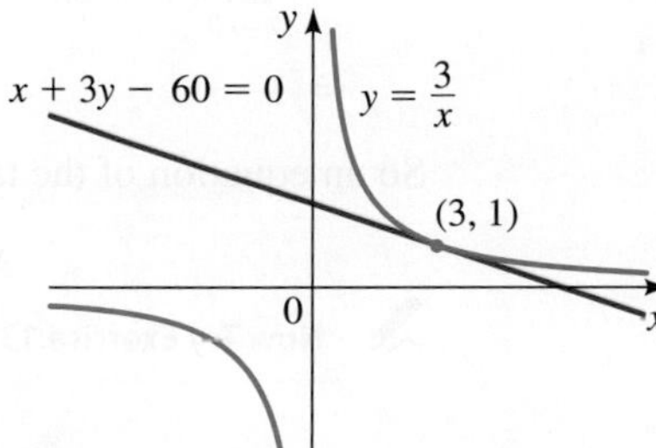

Figure 6

Now Try Exercises 3 and 11 ■

There is another expression for the slope of a tangent line that is sometimes easier to use. Let $h = x - a$. Then $x = a + h$, so the slope of the secant line PQ is

$$m_{PQ} = \frac{f(a + h) - f(a)}{h}$$

See Figure 7, in which the case $h > 0$ is illustrated and Q is to the right of P. If it happened that $h < 0$, however, Q would be to the left of P.

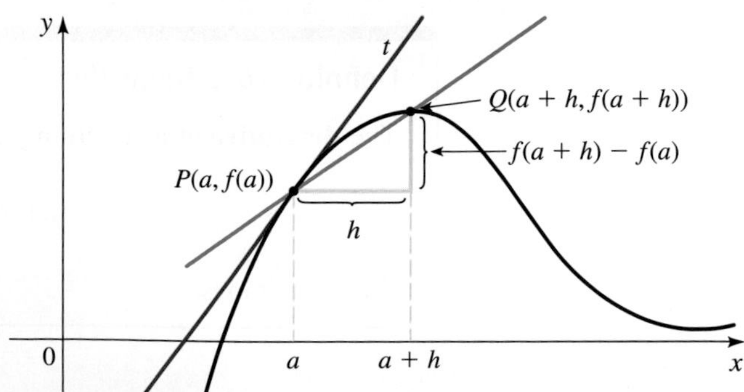

Figure 7

Notice that as x approaches a, h approaches 0 (because $h = x - a$), so the expression for the slope of the tangent line is the limit of difference quotients:

$$m = \lim_{h \to 0} \frac{f(a + h) - f(a)}{h}$$

Example 2 ■ Finding a Tangent Line

Find an equation of the tangent line to the curve $y = x^3 - 2x + 3$ at the point $(1, 2)$.

Solution If $f(x) = x^3 - 2x + 3$, then the slope of the tangent line where $a = 1$ is

$$\begin{aligned} m &= \lim_{h \to 0} \frac{f(1 + h) - f(1)}{h} && \text{Definition of } m \\ &= \lim_{h \to 0} \frac{[(1 + h)^3 - 2(1 + h) + 3] - [1^3 - 2(1) + 3]}{h} && f(x) = x^3 - 2x + 3 \\ &= \lim_{h \to 0} \frac{1 + 3h + 3h^2 + h^3 - 2 - 2h + 3 - 2}{h} && \text{Expand numerator} \\ &= \lim_{h \to 0} \frac{h + 3h^2 + h^3}{h} && \text{Simplify} \\ &= \lim_{h \to 0} (1 + 3h + h^2) && \text{Cancel } h \\ &= 1 && \text{Let } h \to 0 \end{aligned}$$

So an equation of the tangent line at $(1, 2)$ is

$$y - 2 = 1(x - 1) \qquad \text{or} \qquad y = x + 1$$

 **Now Try Exercise 13** ■

■ Derivatives

Recall from Section 2.4 that the expression

$$\frac{f(a + h) - f(a)}{h}$$

is called a difference quotient and represents the average rate of change of f between $x = a$ and $x = a + h$.

We have seen that the slope of the tangent line to the curve $y = f(x)$ at the point $(a, f(a))$ can be written as

$$\lim_{h \to 0} \frac{f(a + h) - f(a)}{h}$$

It turns out that this expression arises in many other contexts as well, such as finding velocities and other rates of change. Because this type of limit occurs so widely, it is given a special name and notation.

Definition of a Derivative

The **derivative of a function f at a number a**, denoted by $f'(a)$, is

$$f'(a) = \lim_{h \to 0} \frac{f(a + h) - f(a)}{h}$$

if this limit exists.

Example 3 ■ Finding a Derivative at a Point

Find the derivative of the function $f(x) = 5x^2 + 3x - 1$ at the number 2.

Solution According to the definition of a derivative, with $a = 2$, we have

$$f'(2) = \lim_{h \to 0} \frac{f(2 + h) - f(2)}{h} \qquad \text{Definition of } f'(2)$$

$$= \lim_{h \to 0} \frac{[5(2 + h)^2 + 3(2 + h) - 1] - [5(2)^2 + 3(2) - 1]}{h} \qquad f(x) = 5x^2 + 3x - 1$$

$$= \lim_{h \to 0} \frac{20 + 20h + 5h^2 + 6 + 3h - 1 - 25}{h} \qquad \text{Expand}$$

$$= \lim_{h \to 0} \frac{23h + 5h^2}{h} \qquad \text{Simplify}$$

$$= \lim_{h \to 0} (23 + 5h) \qquad \text{Cancel } h$$

$$= 23 \qquad \text{Let } h \to 0$$

Now Try Exercise 19 ■

Note We see from the definition of a derivative that the number $f'(a)$ is the slope of the tangent line to the curve $y = f(x)$ at the point $(a, f(a))$. So the result of Example 3 shows that the slope of the tangent line to the parabola $y = 5x^2 + 3x - 1$ at the point $(2, 25)$ is $f'(2) = 23$.

Example 4 ■ Finding a Derivative

Let $f(x) = \sqrt{x}$.

(a) Find $f'(a)$. **(b)** Find $f'(1)$, $f'(4)$, and $f'(9)$.

Solution

(a) We use the definition of the derivative at a.

$$f'(a) = \lim_{h \to 0} \frac{f(a + h) - f(a)}{h} \qquad \text{Definition of derivative}$$

$$= \lim_{h \to 0} \frac{\sqrt{a + h} - \sqrt{a}}{h} \qquad f(x) = \sqrt{x}$$

$$= \lim_{h \to 0} \frac{\sqrt{a + h} - \sqrt{a}}{h} \cdot \frac{\sqrt{a + h} + \sqrt{a}}{\sqrt{a + h} + \sqrt{a}} \qquad \text{Rationalize numerator}$$

$$= \lim_{h \to 0} \frac{(a + h) - a}{h\left(\sqrt{a + h} + \sqrt{a}\right)} \qquad \text{Difference of squares}$$

$$= \lim_{h \to 0} \frac{h}{h\left(\sqrt{a + h} + \sqrt{a}\right)} \qquad \text{Simplify numerator}$$

$$= \lim_{h \to 0} \frac{1}{\sqrt{a + h} + \sqrt{a}} \qquad \text{Cancel } h$$

$$= \frac{1}{\sqrt{a} + \sqrt{a}} = \frac{1}{2\sqrt{a}} \qquad \text{Let } h \to 0$$

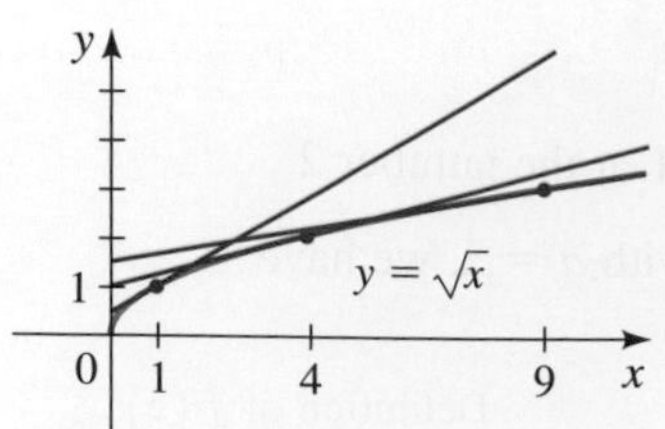

Figure 8

(b) Substituting $a = 1$, $a = 4$, and $a = 9$ into the result of part (a), we get

$$f'(1) = \frac{1}{2\sqrt{1}} = \frac{1}{2} \qquad f'(4) = \frac{1}{2\sqrt{4}} = \frac{1}{4} \qquad f'(9) = \frac{1}{2\sqrt{9}} = \frac{1}{6}$$

These values of the derivative are the slopes of the tangent lines shown in Figure 8.

Now Try Exercises 25 and 27 ■

■ Instantaneous Rates of Change

Why are we interested in *instantaneous* speed? Because, for example, if we drop a ball from a high cliff, the ball falls faster and faster, its speed increasing at each *instant*. In order to model the motion of the ball we need to know its speed at each *instant*. Knowing this information allows us to completely describe all quantities related to the motion of the ball, including its speed, acceleration, and the distance it has traveled at any given time. You will learn more about these concepts in your Calculus course.

In Section 2.4 we defined the average rate of change of a function f between the numbers a and x as

$$\text{average rate of change} = \frac{\text{change in } y}{\text{change in } x} = \frac{f(x) - f(a)}{x - a}$$

Suppose we consider the average rate of change over smaller and smaller intervals by letting x approach a. The limit of these average rates of change is called the instantaneous rate of change.

Instantaneous Rate of Change

If $y = f(x)$, the **instantaneous rate of change of y with respect to x** at $x = a$ is the limit of the average rates of change as x approaches a:

$$\text{instantaneous rate of change} = \lim_{x \to a} \frac{f(x) - f(a)}{x - a} = f'(a)$$

Notice that we now have two ways of interpreting the derivative:

- $f'(a)$ is the slope of the tangent line to $y = f(x)$ at $x = a$
- $f'(a)$ is the instantaneous rate of change of y with respect to x at $x = a$

In the special case in which $x = t = \text{time}$ and $s = f(t) = \text{displacement}$ (directed distance) at time t of an object traveling in a straight line, the instantaneous rate of change is called the **instantaneous velocity**.

Example 5 ■ Instantaneous Velocity of a Falling Object

If an object is dropped from a height of 1000 m, its distance above the ground (in meters) after t seconds is given by $h(t) = 1000 - 5t^2$. Find the object's instantaneous velocity after 4 seconds.

Discovery Project ■ Designing a Roller Coaster

To ensure an exhilarating ride, a roller coaster ought to consist of steep rises and drops joined by thrilling curves. For a safe ride, these curves must fit together "smoothly." In designing a roller coaster, you can choose where to locate the ascents and drops. We'll explore how the derivative can help us join these ascents and drops smoothly. You can find the project at **www.stewartmath.com**.

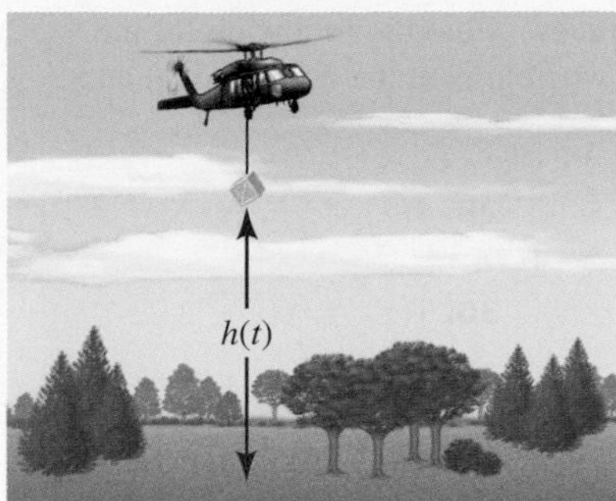

Solution After 4 seconds have elapsed, the height is $h(4) = 920$ m. The instantaneous velocity is

$$h'(4) = \lim_{t \to 4} \frac{h(t) - h(4)}{t - 4} \qquad \text{Definition of } h'(4)$$

$$= \lim_{t \to 4} \frac{1000 - 5t^2 - 920}{t - 4} \qquad h(t) = 1000 - 5t^2$$

$$= \lim_{t \to 4} \frac{80 - 5t^2}{t - 4} \qquad \text{Simplify}$$

$$= \lim_{t \to 4} \frac{5(4 - t)(4 + t)}{t - 4} \qquad \text{Factor numerator}$$

$$= \lim_{t \to 4} -5(4 + t) \qquad \text{Cancel } t - 4$$

$$= -5(4 + 4) = -40 \text{ m/s} \qquad \text{Let } t \to 4$$

The negative sign indicates that the height is *decreasing* at a rate of 40 m/s.

Now Try Exercise 37 ■

US Population

t	$P(t)$ (millions)
2012	313.9
2014	318.4
2016	323.1
2018	326.8
2020	329.5

Source: US Census Bureau

t	$\dfrac{P(t) - P(2016)}{t - 2016}$
2012	2.30
2014	2.35
2018	1.85
2020	1.60

Here, we have estimated the derivative by averaging the slopes of two secant lines. Another method is to plot the population function and estimate the slope of the tangent line when $t = 2016$.

Example 6 ■ Estimating an Instantaneous Rate of Change

Let $P(t)$ be the population of the United States at time t. The first table in the margin gives approximate values of this function by providing midyear population estimates from 2012 to 2020. Interpret and estimate the value of $P'(2016)$.

Solution The derivative $P'(2016)$ means the rate of change of P with respect to t when $t = 2016$, that is, the rate of increase of the population in 2016.

According to the definition of a derivative, we have

$$P'(2016) = \lim_{t \to 2016} \frac{P(t) - P(2016)}{t - 2016}$$

So we compute and tabulate values of the difference quotient (the average rates of change) as shown in the second table in the margin. We see that $P'(2016)$ lies somewhere between 2.35 and 1.85 million. (Here we are making the reasonable assumption that the population didn't fluctuate wildly between 2012 and 2020.) We estimate that the rate of increase of the US population in 2016 was the average of these two numbers, namely,

$$P'(2016) \approx 2.10 \text{ million people/year}$$

Now Try Exercise 43 ■

12.3 | Exercises

■ Concepts

1. The derivative of a function f at a number a is

$$f'(a) = \lim_{h \to 0} \frac{\square - \square}{\square}$$

if the limit exists. The derivative $f'(a)$ is the ________ of the tangent line to the curve $y = f(x)$ at the point ($\square$, $\square$).

2. If $y = f(x)$, the average rate of change of f between the numbers x and a is $\dfrac{\square - \square}{\square - \square}$. The limit of the average rates of change as x approaches a is the ________ rate of change of y with respect to x at $x = a$; this is also the derivative $f'(\square)$.

Skills

3–10 ■ Slope of a Tangent Line Find the slope of the tangent line to the graph of f at the given point.

3. $f(x) = 3x + 4$, at $(1, 7)$

4. $f(x) = 5 - 2x$, at $(-3, 11)$

5. $f(x) = 4x^2 - 3x$, at $(-1, 7)$

6. $f(x) = 1 + 2x - 3x^2$, at $(1, 0)$

7. $f(x) = 2x^3$, at $(2, 16)$

8. $f(x) = x^3 + 1$, at $(2, 9)$

9. $f(x) = \dfrac{5}{x + 2}$, at $(3, 1)$

10. $f(x) = \dfrac{6}{x + 1}$, at $(2, 2)$

11–18 ■ Equation of a Tangent Line Find an equation of the tangent line to the curve at the given point. Graph the curve and the tangent line.

11. $f(x) = -2x^2 + 1$, at $(2, -7)$

12. $f(x) = 4x^2 - 3$, at $(-1, 1)$

13. $y = x + x^2$, at $(-1, 0)$

14. $y = 2x - x^3$, at $(1, 1)$

15. $y = \dfrac{x}{x - 1}$, at $(2, 2)$

16. $y = \dfrac{1}{x^2}$, at $(-1, 1)$

17. $y = \sqrt{x + 3}$, at $(1, 2)$

18. $y = \sqrt{1 + 2x}$, at $(4, 3)$

19–26 ■ The Derivative at a Number Find the derivative of the function at the given number.

19. $f(x) = 1 - 3x^2$, at 2

20. $f(x) = 2 - 3x + x^2$, at -1

21. $f(x) = x - 3x^2$, at -1

22. $f(x) = x + x^3$, at 1

23. $f(x) = \dfrac{1}{x + 1}$, at 2

24. $f(x) = \dfrac{x}{2 - x}$, at -3

25. $F(x) = \dfrac{1}{\sqrt{x}}$, at 4

26. $G(x) = 1 + 2\sqrt{x}$, at 4

27–30 ■ Evaluating Derivatives Find the following for the given function f: **(a)** $f'(a)$, where a is in the domain of f, and **(b)** $f'(3)$ and $f'(4)$.

27. $f(x) = x^2 + 2x$

28. $f(x) = -\dfrac{1}{x^2}$

29. $f(x) = \dfrac{x}{x + 1}$

30. $f(x) = \sqrt{x - 2}$

Skills Plus

31. Tangent Lines

(a) If $f(x) = x^3 - 2x + 4$, find $f'(a)$.

(b) Find equations of the tangent lines to the graph of f at the points whose x-coordinates are 0, 1, and 2.

(c) Graph f and the three tangent lines.

32. Tangent Lines

(a) If $g(x) = 1/(2x - 1)$, find $g'(a)$.

(b) Find equations of the tangent lines to the graph of g at the points whose x-coordinates are -1, 0, and 1.

(c) Graph g and the three tangent lines.

33–36 ■ Which Derivative Does the Limit Represent? The given limit represents the derivative of a function f at a number a. Find f and a.

33. $\displaystyle\lim_{h \to 0} \frac{(1 + h)^{10} - 1}{h}$

34. $\displaystyle\lim_{x \to 5} \frac{2^x - 32}{x - 5}$

35. $\displaystyle\lim_{t \to 1} \frac{\sqrt{t + 1} - \sqrt{2}}{t - 1}$

36. $\displaystyle\lim_{h \to 0} \frac{\cos(\pi + h) + 1}{h}$

Applications

37. Velocity of a Ball If a ball is thrown straight up with a velocity of 12 m/s, its height (in m) after t seconds is given by $y = 12t - 4.8t^2$. Find the instantaneous velocity when $t = 2$.

38. Velocity on the Moon If an arrow is shot upward on the moon with a velocity of 58 m/s, its height (in meters) after t seconds is given by $H = 58t - 0.83t^2$.

(a) Find the instantaneous velocity of the arrow after 1 second.

(b) Find the instantaneous velocity of the arrow when $t = a$.

(c) At what time t will the arrow hit the moon?

(d) With what velocity will the arrow hit the moon?

39. Velocity of a Particle The displacement s (in meters) of a particle moving in a straight line is given by the equation of motion $s = 4t^3 + 6t + 2$, where t is measured in seconds. Find the instantaneous velocity of the particle s at times $t = a$, $t = 1$, $t = 2$, $t = 3$.

40. Inflating a Balloon A spherical balloon is being inflated. Find the rate of change of the surface area $(S = 4\pi r^2)$ with respect to the radius r when $r = 2$ m.

41. Temperature Change A roast turkey is taken from an oven when its temperature has reached 185°F and is placed on a table in a room where the temperature is 75°F. The graph shows how the temperature of the turkey decreases and eventually approaches room temperature. By measuring the slope of the tangent, estimate the rate of change of the temperature after 60 minutes.

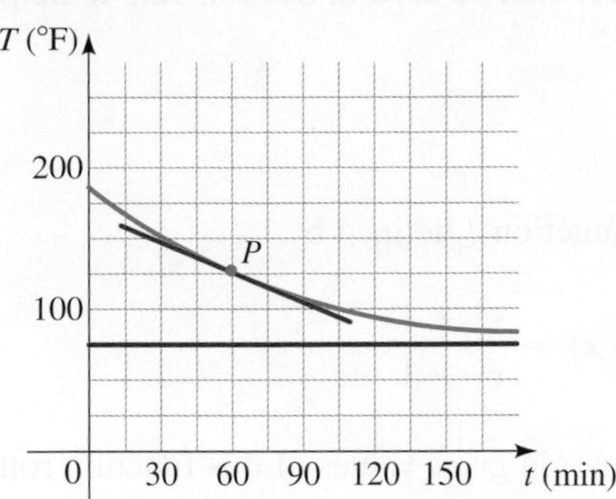

42. Heart Rate A cardiac monitor is used to measure the heart rate of a patient after surgery. It compiles the number of heartbeats after t min. When the data in the table are graphed, the slope of the tangent line represents the heart rate in beats per minute.

t (min)	Heartbeats
36	2530
38	2661
40	2806
42	2948
44	3080

(a) Find the average heart rates (slopes of the secant lines) over the time intervals $[40, 42]$ and $[42, 44]$.

(b) Estimate the patient's heart rate after 42 min by averaging the slopes of these two secant lines.

43. Water Flow A tank holds 1000 L of water, which drains from the bottom of the tank in half an hour. The values in the table show the volume V of water remaining in the tank (in L) after t minutes.

t (min)	V (L)
5	694
10	444
15	250
20	111
25	28
30	0

(a) Find the average rates at which water flows from the tank (slopes of secant lines) for the time intervals $[10, 15]$ and $[15, 20]$.

(b) The slope of the tangent line at the point $(15, 250)$ represents the rate at which water is flowing from the tank after 15 min. Estimate this rate by averaging the slopes of the secant lines in part (a).

44. World Population Growth The table gives approximate values for the world population by providing midyear population estimates for the years 1900–2020. Estimate the rate of population growth in 1920 and in 2010 by averaging the slopes of two secant lines.

Year	Population (millions)	Year	Population (millions)
1900	1650	1970	3710
1910	1750	1980	4450
1920	1860	1990	5290
1930	2070	2000	6090
1940	2300	2010	6870
1950	2560	2020	7757
1960	3040		

Source: US Census Bureau

■ Discuss ■ Discover ■ Prove ■ Write

45. Discuss: Estimating Derivatives from a Graph For the function g whose graph is given, arrange the following numbers in increasing order, and explain your reasoning.

$$0 \qquad g'(-2) \qquad g'(0) \qquad g'(2) \qquad g'(4)$$

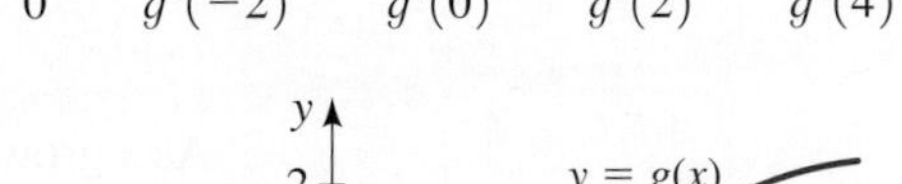
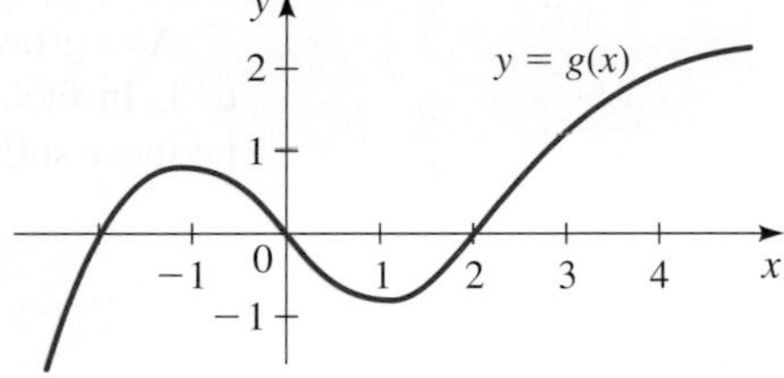

46. Discuss: Estimating Velocities from a Graph The graph shows the position function of a car. Use the shape of the graph to explain your answers to the following questions.

(a) What was the initial velocity of the car?

(b) Was the car going faster at B or at C?

(c) Was the car slowing down or speeding up at A, B, and C?

(d) What happened between D and E?

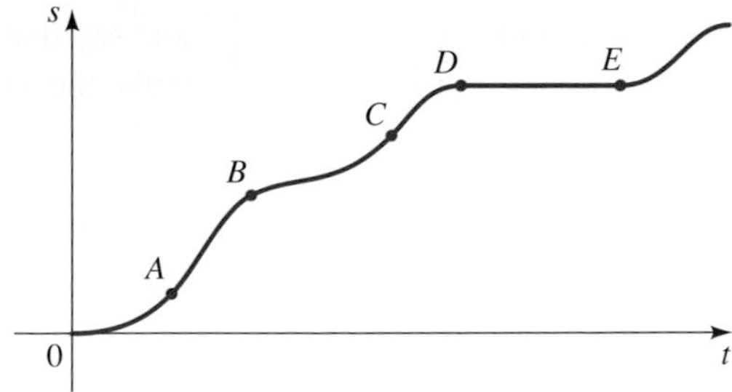

12.4 Limits at Infinity; Limits of Sequences

■ Limits at Infinity ■ Limits of Sequences

In this section we study a special kind of limit called a *limit at infinity*. We examine the limit of a function $f(x)$ as x becomes large. We also examine the limit of a sequence a_n as n becomes large. Limits of sequences will be used in Section 12.5 to help us find the area under the graph of a function.

■ Limits at Infinity

Let's investigate the behavior of the function f defined by

$$f(x) = \frac{x^2 - 1}{x^2 + 1}$$

x	$f(x)$
0	−1.000000
±1	0.000000
±2	0.600000
±3	0.800000
±4	0.882353
±5	0.923077
±10	0.980198
±50	0.999200
±100	0.999800
±1000	0.999998

as x becomes large. The table in the margin gives values of this function rounded to six decimal places, and the graph of f has been drawn by a computer in Figure 1.

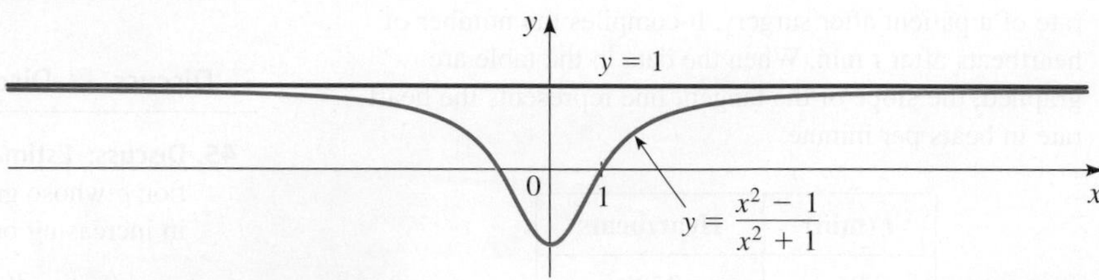

Figure 1

As x grows larger and larger, you can see that the values of $f(x)$ get closer and closer to 1. In fact, it seems that we can make the values of $f(x)$ as close as we like to 1 by taking x sufficiently large. This situation is expressed symbolically by writing

$$\lim_{x \to \infty} \frac{x^2 - 1}{x^2 + 1} = 1$$

In general, we use the notation $\lim_{x \to \infty} f(x) = L$ to indicate that the values of $f(x)$ become closer and closer to L as x becomes larger and larger.

Definition of a Limit at Infinity

We write

$$\lim_{x \to \infty} f(x) = L$$

and say that **"the limit of $f(x)$, as x approaches infinity**, equals L" if we can make the values of $f(x)$ arbitrarily close to L by taking x to be sufficiently large.

We also say "the limit of $f(x)$, as x increases without bound, equals L."

Another notation for $\lim_{x \to \infty} f(x) = L$ is

$$f(x) \to L \quad \text{as} \quad x \to \infty$$

Limits at infinity are also discussed in Section 3.6.

Graphical illustrations of limits at infinity are shown in Figure 2. Notice that there are many ways for the graph of f to approach the line $y = L$ as we look to the far right of each graph.

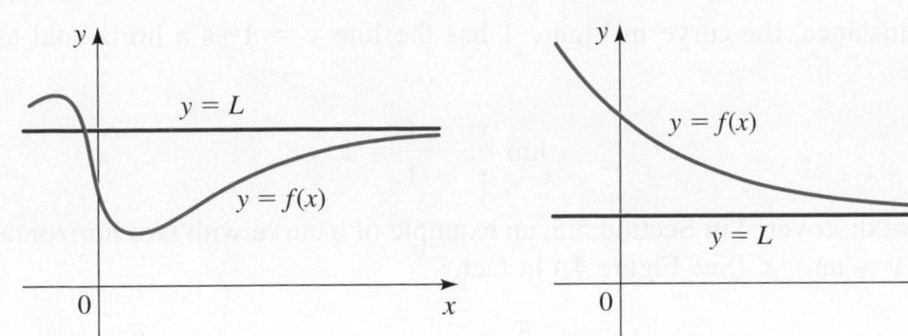

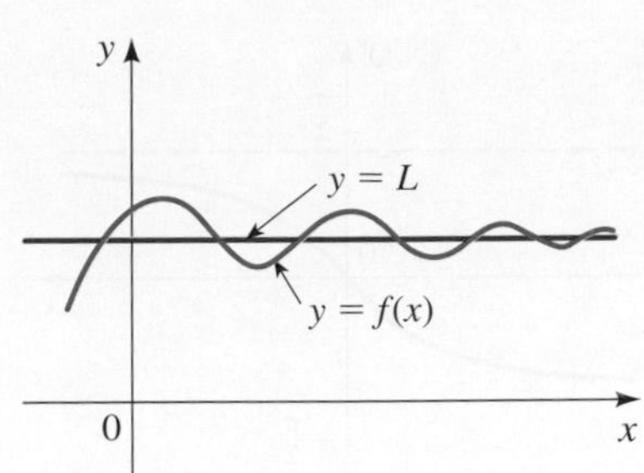

Figure 2 | Examples illustrating $\lim_{x\to\infty} f(x) = L$

Referring back to Figure 1, we see that for numerically large negative values of x, the values of $f(x)$ are close to 1. This is expressed by writing

$$\lim_{x\to-\infty} \frac{x^2 - 1}{x^2 + 1} = 1$$

We also say "the limit of $f(x)$, as x decreases (through negative values) without bound, equals L."

Definition of a Limit at Negative Infinity

We write

$$\lim_{x\to-\infty} f(x) = L$$

and say that **"the limit of $f(x)$, as x approaches negative infinity**, equals L" if we can make the values of $f(x)$ arbitrarily close to L by taking x to be sufficiently large negative.

This definition is illustrated in Figure 3. Notice how each graph approaches the line $y = L$ as we look to the far left.

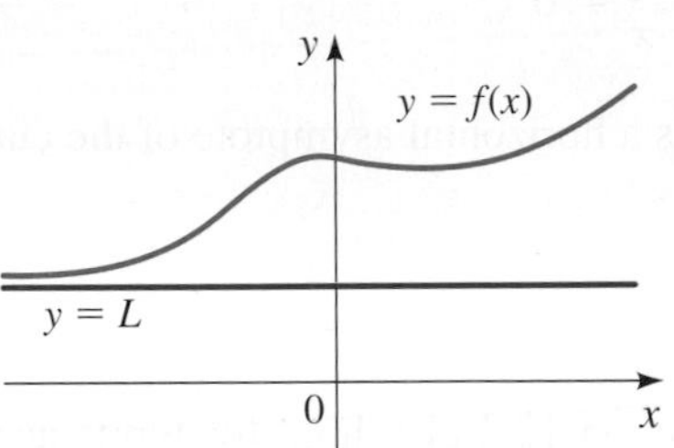

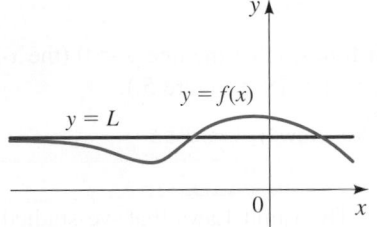

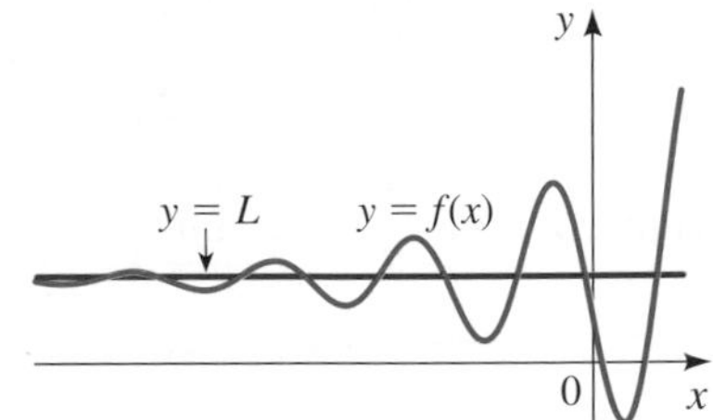

Figure 3 | Examples illustrating $\lim_{x\to-\infty} f(x) = L$

Note The symbols ∞ and $-\infty$ do not represent numbers. When we write "$x \to \infty$" we mean that x increases without bound. Graphically, this means that x is allowed to move to the right on the real line indefinitely. Similarly, when we write "$x \to -\infty$" we mean that x decreases (through negative numbers) without bound. Graphically, this means that x can move to the left indefinitely.

In Section 3.6 we studied horizontal asymptotes of rational function. We now define the concept of a horizontal asymptote for any function by using limits.

Horizontal Asymptote

The line $y = L$ is called a **horizontal asymptote** of the curve $y = f(x)$ if either

$$\lim_{x\to\infty} f(x) = L \qquad \text{or} \qquad \lim_{x\to-\infty} f(x) = L$$

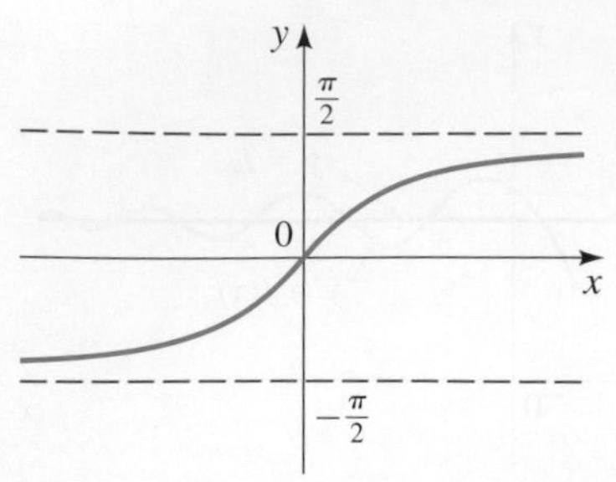

Figure 4 | $y = \tan^{-1}x$

For instance, the curve in Figure 1 has the line $y = 1$ as a horizontal asymptote because

$$\lim_{x\to\infty} \frac{x^2 - 1}{x^2 + 1} = 1$$

As we discovered in Section 5.5, an example of a curve with two horizontal asymptotes is $y = \tan^{-1}x$. (See Figure 4.) In fact,

$$\lim_{x\to-\infty} \tan^{-1}x = -\frac{\pi}{2} \qquad \text{and} \qquad \lim_{x\to\infty} \tan^{-1}x = \frac{\pi}{2}$$

so both of the lines $y = -\pi/2$ and $y = \pi/2$ are horizontal asymptotes.

Example 1 ■ Limits at Infinity

Find $\lim_{x\to\infty} \frac{1}{x}$ and $\lim_{x\to-\infty} \frac{1}{x}$.

We first investigated horizontal asymptotes and limits at infinity for rational functions in Section 3.6.

Solution Observe that when x is large, $1/x$ is small. For instance,

$$\frac{1}{100} = 0.01 \qquad \frac{1}{10{,}000} = 0.0001 \qquad \frac{1}{1{,}000{,}000} = 0.000001$$

In fact, by taking x large enough, we can make $1/x$ as close to 0 as we please. Therefore

$$\lim_{x\to\infty} \frac{1}{x} = 0$$

Similar reasoning shows that when x is large negative, $1/x$ is small negative, so we also have

$$\lim_{x\to-\infty} \frac{1}{x} = 0$$

It follows that the line $y = 0$ (the x-axis) is a horizontal asymptote of the curve $y = 1/x$. (See Figure 5.)

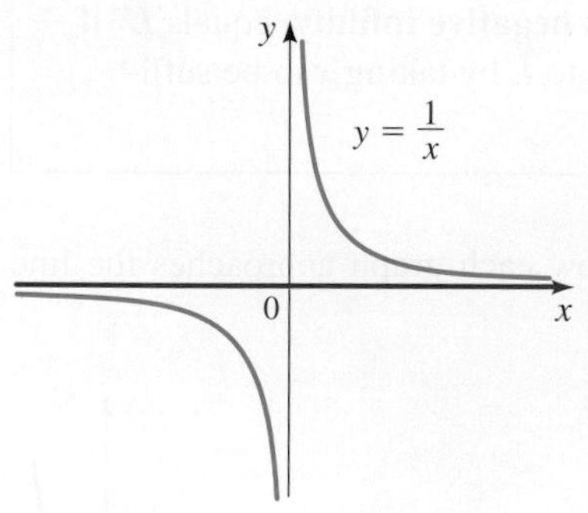

Figure 5 | $\lim_{x\to\infty} \frac{1}{x} = 0, \ \lim_{x\to-\infty} \frac{1}{x} = 0$

 Now Try Exercise 5 ■

The Limit Laws that we studied in Section 12.2 also hold for limits at infinity. In particular, if we combine Law 6 (Limit of a Power) with the results of Example 1, we obtain the following important rule for calculating limits.

> If k is any positive integer, then
>
> $$\lim_{x\to\infty} \frac{1}{x^k} = 0 \qquad \text{and} \qquad \lim_{x\to-\infty} \frac{1}{x^k} = 0$$

Example 2 ■ Finding a Limit at Infinity

Evaluate $\lim_{x\to\infty} \frac{3x^2 - x - 2}{5x^2 + 4x + 1}$.

Solution To evaluate the limit at infinity of a rational function, we first divide both the numerator and denominator by the highest power of x that occurs in the denominator. (We may assume that $x \neq 0$, since we are interested only in large values of x.) In this case the highest power of x in the denominator is x^2, so we have

$$\lim_{x \to \infty} \frac{3x^2 - x - 2}{5x^2 + 4x + 1} = \lim_{x \to \infty} \frac{3 - \dfrac{1}{x} - \dfrac{2}{x^2}}{5 + \dfrac{4}{x} + \dfrac{1}{x^2}}$$ Divide numerator and denominator by x^2

$$= \frac{\lim\limits_{x \to \infty} \left(3 - \dfrac{1}{x} - \dfrac{2}{x^2}\right)}{\lim\limits_{x \to \infty} \left(5 + \dfrac{4}{x} + \dfrac{1}{x^2}\right)}$$ Limit of a Quotient

$$= \frac{\lim\limits_{x \to \infty} 3 - \lim\limits_{x \to \infty} \dfrac{1}{x} - 2 \lim\limits_{x \to \infty} \dfrac{1}{x^2}}{\lim\limits_{x \to \infty} 5 + 4 \lim\limits_{x \to \infty} \dfrac{1}{x} + \lim\limits_{x \to \infty} \dfrac{1}{x^2}}$$ Limits of Sums, Differences, and Constant Multiples

$$= \frac{3 - 0 - 0}{5 + 0 + 0} = \frac{3}{5}$$ Let $x \to \infty$

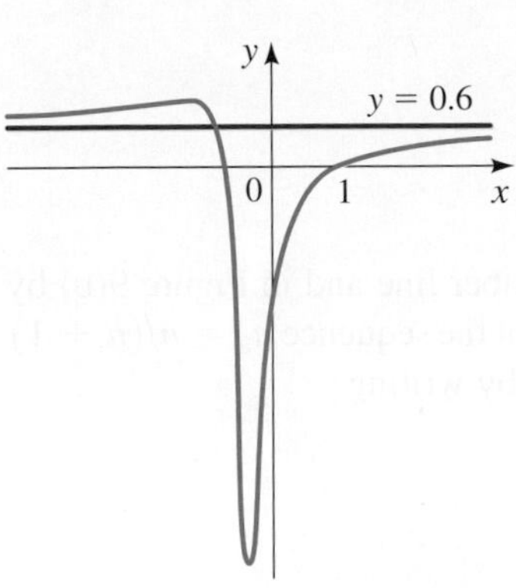

Figure 6

Similarly, the limit as $x \to -\infty$ is also $\frac{3}{5}$. Figure 6 confirms that the graph of the given rational function approaches the horizontal asymptote $y = \frac{3}{5}$.

Now Try Exercise 9 ■

Example 3 ■ A Limit at Negative Infinity

Use numerical and graphical methods to find $\lim\limits_{x \to -\infty} e^x$.

Solution From the graph of the natural exponential function $y = e^x$ in Figure 7 and the corresponding table of values we see that

$$\lim_{x \to -\infty} e^x = 0$$

It follows that the line $y = 0$ (the x-axis) is a horizontal asymptote.

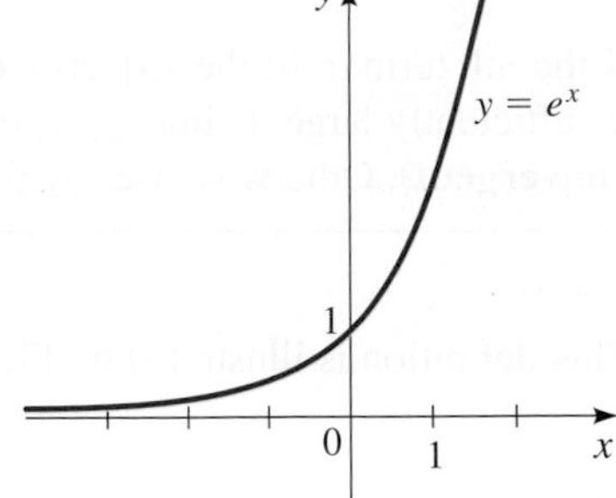

x	e^x
0	1.00000
−1	0.36788
−2	0.13534
−3	0.04979
−5	0.00674
−8	0.00034
−10	0.00005

Figure 7

Now Try Exercise 19 ■

Example 4 ■ A Function with No Limit at Infinity

Evaluate $\lim\limits_{x \to \infty} \sin x$.

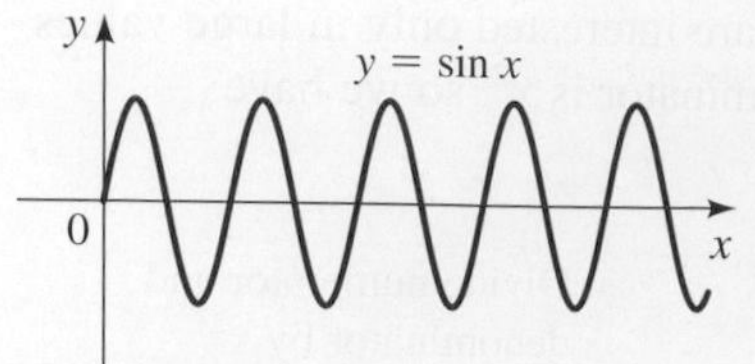

Figure 8

Solution From the graph in Figure 8 and the periodic nature of the sine function we see that as x increases, the values of $\sin x$ oscillate between 1 and -1 infinitely often, so they don't approach any definite number. Therefore $\lim_{x \to \infty} \sin x$ does not exist.

Now Try Exercise 17 ■

■ Limits of Sequences

In Section 11.1 we introduced the idea of a sequence of numbers $a_1, a_2, a_3, \ldots$. Here we are interested in their behavior as n becomes large. For instance, the sequence defined by

$$a_n = \frac{n}{n+1}$$

is pictured in Figure 9(a) by plotting its terms on a number line and in Figure 9(b) by plotting its graph. From Figure 9 it appears that the terms of the sequence $a_n = n/(n+1)$ are approaching 1 as n becomes large. We indicate this by writing

$$\lim_{n \to \infty} \frac{n}{n+1} = 1$$

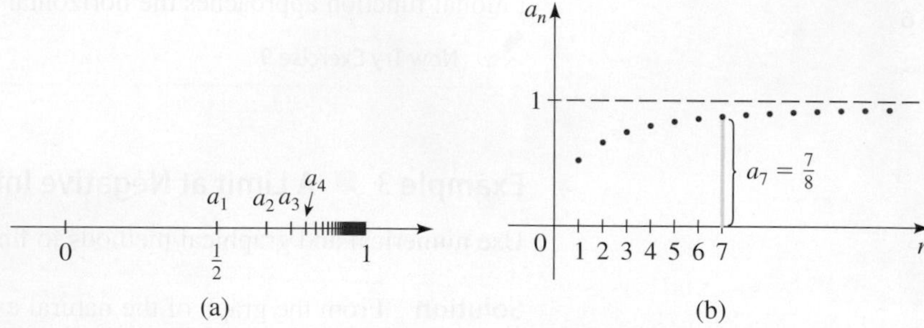

Figure 9

Definition of the Limit of a Sequence

A sequence $a_1, a_2, a_3, \ldots$ has the **limit** L and we write

$$\lim_{n \to \infty} a_n = L \qquad \text{or} \qquad a_n \to L \quad \text{as} \quad n \to \infty$$

if the nth term a_n of the sequence can be made arbitrarily close to L by taking n sufficiently large. If $\lim_{n \to \infty} a_n$ exists, we say the sequence **converges** (or is **convergent**). Otherwise, we say the sequence **diverges** (or is **divergent**).

This definition is illustrated by Figure 10.

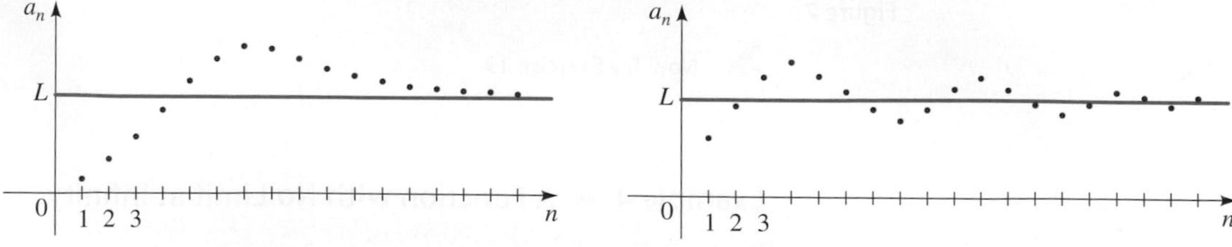

Figure 10 | Graphs of two sequences with $\lim_{n \to \infty} a_n = L$

If we compare the definitions of $\lim_{n\to\infty} a_n = L$ and $\lim_{x\to\infty} f(x) = L$, we see that the only difference is that n is required to be an integer. Thus we get the following result which is illustrated in Figure 11.

> If $\lim\limits_{x\to\infty} f(x) = L$ and $f(n) = a_n$ when n is an integer, then $\lim\limits_{n\to\infty} a_n = L$.

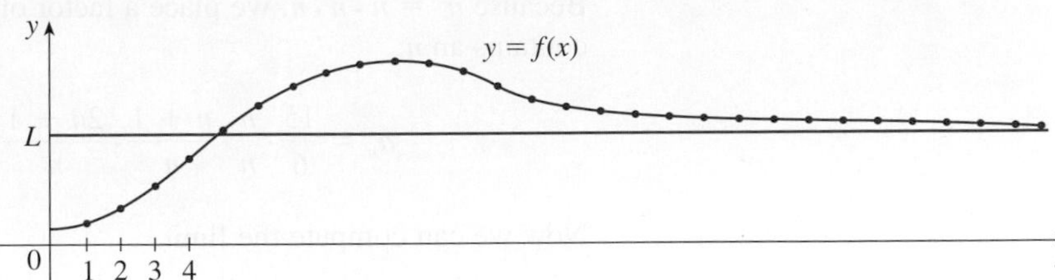

Figure 11 | $\lim\limits_{x\to\infty} f(x) = L$ and $\lim\limits_{x\to\infty} f(n) = L$

In particular, since we know that $\lim_{x\to\infty}(1/x^k) = 0$ when k is a positive integer, we have

$$\lim_{n\to\infty} \frac{1}{n^k} = 0 \qquad \text{if } k \text{ is a positive integer}$$

Note that the Limit Laws given in Section 12.2 also hold for limits of sequences.

Example 5 ■ Finding the Limit of a Sequence

Find $\lim\limits_{n\to\infty} \dfrac{n}{n+1}$.

Solution The method is similar to the one we used in Example 2: Divide the numerator and denominator by the highest power of n, and then use the Limit Laws.

$$\lim_{n\to\infty} \frac{n}{n+1} = \lim_{n\to\infty} \frac{1}{1 + \dfrac{1}{n}} \qquad \text{Divide numerator and denominator by } n$$

$$= \frac{\lim\limits_{n\to\infty} 1}{\lim\limits_{n\to\infty} 1 + \lim\limits_{n\to\infty} \dfrac{1}{n}} \qquad \text{Limits of a Quotient and a Sum}$$

$$= \frac{1}{1+0} = 1 \qquad \text{Let } n \to \infty$$

This result shows that the guesses we made earlier from Figure 9(a) and Figure 9(b) were correct.

Therefore the sequence $a_n = n/(n+1)$ is convergent.

Now Try Exercise 23 ■

Example 6 ■ A Sequence That Diverges

Determine whether the sequence $a_n = (-1)^n$ is convergent or divergent.

Solution If we write out the terms of the sequence, we obtain

$$-1, 1, -1, 1, -1, 1, -1, \ldots$$

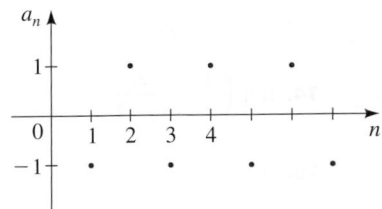

The graph of this sequence is shown in Figure 12. Since the terms oscillate between 1 and -1 infinitely often, a_n does not approach any number. Thus $\lim_{n\to\infty}(-1)^n$ does not exist; that is, the sequence $a_n = (-1)^n$ is divergent.

Figure 12

Now Try Exercise 29 ■

Example 7 ■ Finding the Limit of a Sequence

Find the limit of the sequence given by

$$a_n = \frac{15}{n^3}\left[\frac{n(n+1)(2n+1)}{6}\right]$$

Solution Before calculating the limit, let's first simplify the expression for a_n. Because $n^3 = n \cdot n \cdot n$, we place a factor of n beneath each factor in the numerator that contains an n:

$$a_n = \frac{15}{6} \cdot \frac{n}{n} \cdot \frac{n+1}{n} \cdot \frac{2n+1}{n} = \frac{5}{2} \cdot 1 \cdot \left(1 + \frac{1}{n}\right)\left(2 + \frac{1}{n}\right)$$

Now we can compute the limit.

$$\lim_{n\to\infty} a_n = \lim_{n\to\infty} \frac{5}{2}\left(1 + \frac{1}{n}\right)\left(2 + \frac{1}{n}\right) \qquad \text{Definition of } a_n$$

$$= \frac{5}{2} \lim_{n\to\infty}\left(1 + \frac{1}{n}\right) \lim_{n\to\infty}\left(2 + \frac{1}{n}\right) \qquad \text{Limit of a Product}$$

$$= \frac{5}{2}(1)(2) = 5 \qquad \text{Let } n \to \infty$$

Now Try Exercise 31 ■

12.4 Exercises

Concepts

1. Let f be a function defined on some interval (a, ∞). Then

$$\lim_{x\to\infty} f(x) = L$$

means that the values of $f(x)$ can be made arbitrarily close to ________ by taking ________ sufficiently large. In this case the line $y = L$ is called a ________ ________ of the function $y = f(x)$. For example, $\lim_{x\to\infty} \frac{1}{x} =$ ________, and the line $y =$ ________ is a horizontal asymptote.

2. A sequence $a_1, a_2, a_3, \ldots$ has the limit L if the nth term a_n of the sequence can be made arbitrarily close to ________ by taking n to be sufficiently ________. If the limit exists, we say that the sequence ________; otherwise, the sequence ________.

Skills

3–4 ■ Limits from a Graph
(a) Use the graph of f to find the following limits.
(i) $\lim_{x\to\infty} f(x)$ (ii) $\lim_{x\to-\infty} f(x)$
(b) State the equations of the horizontal asymptotes.

3.

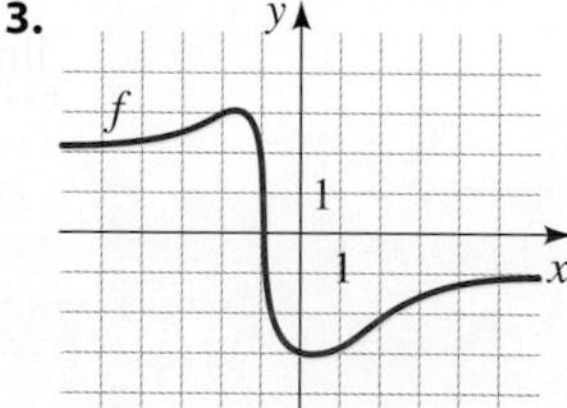

4.

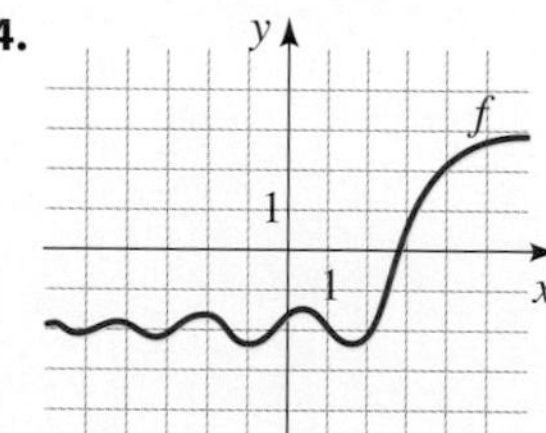

5–18 ■ Limits at Infinity Find the limit.

5. $\lim_{x\to\infty} \frac{6}{x}$

6. $\lim_{x\to\infty} \frac{3}{x^4}$

7. $\lim_{x\to\infty} \frac{2x+1}{5x-1}$

8. $\lim_{x\to\infty} \frac{2-3x}{4x+5}$

9. $\lim_{x\to-\infty} \frac{4x^2+1}{2+3x^2}$

10. $\lim_{x\to-\infty} \frac{x^2+2}{x^3+x+1}$

11. $\lim_{t\to\infty} \frac{8t^3+t}{(2t-1)(2t^2+1)}$

12. $\lim_{r\to\infty} \frac{4r^3-r^2}{(r+1)^3}$

13. $\lim_{x\to\infty} \frac{x^4}{1-x^2+x^3}$

14. $\lim_{t\to\infty} \left(\frac{1}{t} - \frac{2t}{t-1}\right)$

15. $\lim_{x\to-\infty} \left(\frac{x-1}{x+1} + 6\right)$

16. $\lim_{x\to-\infty} \left(\frac{3-x}{3+x} - 2\right)$

17. $\lim_{x\to\infty} \cos x$

18. $\lim_{x\to\infty} \sin^2 x$

19–22 ■ Estimating Limits Numerically and Graphically Use a table of values to estimate the limit. Then use a graphing device to confirm your result graphically.

19. $\lim_{x\to-\infty} \dfrac{\sqrt{x^2+4x}}{4x+1}$ **20.** $\lim_{x\to\infty} \left(\sqrt{9x^2+x}-3x\right)$

21. $\lim_{x\to\infty} \dfrac{x^5}{e^x}$ **22.** $\lim_{x\to\infty} \left(1+\dfrac{2}{x}\right)^{3x}$

23–34 ■ Limits of Sequences If the sequence with the given nth term is convergent, find its limit. If it is divergent, explain why.

23. $a_n = \dfrac{1+n}{n+n^2}$

24. $a_n = \dfrac{5n}{n+5}$

25. $a_n = \dfrac{n^2}{n+1}$

26. $a_n = \dfrac{n-1}{n^3+1}$

27. $a_n = \dfrac{1}{3^n}$

28. $a_n = \dfrac{(-1)^n}{n}$

29. $a_n = \sin(n\pi/2)$

30. $a_n = \cos n\pi$

31. $a_n = \dfrac{3}{n^2}\left[\dfrac{n(n+1)}{2}\right]$

32. $a_n = \dfrac{12}{n^4}\left[\dfrac{n(n+1)}{2}\right]^2$

33. $a_n = \dfrac{24}{n^3}\left[\dfrac{n(n+1)(2n+1)}{6}\right]$

34. $a_n = \dfrac{5}{n}\left(n+\dfrac{4}{n}\left[\dfrac{n(n+1)}{2}\right]\right)$

Skills Plus

35–36 ■ A Function from a Description Find a formula for a function f that satisfies the following conditions.

35. Vertical asymptotes $x = 1$ and $x = 3$ and horizontal asymptote $y = 1$

36. $\lim_{x\to\infty} f(x) = 0, \quad \lim_{x\to0} f(x) = -\infty, \quad f(2) = 0,$

$\lim_{x\to3^-} f(x) = \infty, \quad \lim_{x\to3^+} f(x) = -\infty$

37. Asymptote Behavior How close to -3 do we have to take x so that

$$\frac{1}{(x+3)^2} > 10{,}000$$

38. Equivalent Limits Show that

$$\lim_{x\to\infty} f(x) = \lim_{t\to0^+} f\left(\frac{1}{t}\right)$$

and

$$\lim_{x\to-\infty} f(x) = \lim_{t\to0^-} f\left(\frac{1}{t}\right)$$

if these limits exist.

Applications

39. Salt Concentration

(a) A tank contains 5000 liters of pure water. Brine that contains 30 grams of salt per liter of water is pumped into the tank at a rate of 25 L/min. Show that the concentration of salt after t minutes (in g/L) is

$$C(t) = \frac{30t}{200+t}$$

(b) What happens to the concentration as $t \to \infty$?

40. Velocity of a Raindrop The downward velocity (in m/s) of a falling raindrop at time t is modeled by the function

$$v(t) = 9.1(1 - e^{-1.2t})$$

(a) Find the terminal velocity of the raindrop by evaluating $\lim_{t\to\infty} v(t)$.

(b) Graph $v(t)$, and use the graph to estimate how long it takes for the velocity of the raindrop to reach 99% of its terminal velocity.

Discuss ■ Discover ■ Prove ■ Write

41. Discuss ■ Discover: The Tail of a Sequence Let a_n be a sequence. Explain why

$$\lim_{n\to\infty} a_n = \lim_{n\to\infty} a_{n+1}$$

That is, removing the first term of a sequence does not affect the limit. Does removing the first 100 terms of a sequence affect its limit? Conclude that convergence and divergence are properties of the "tail" of a sequence.

42. Discuss ■ Discover: Limit of a Recursive Sequence A sequence is defined recursively by $a_1 = 0$ and $a_{n+1} = \sqrt{2 + a_n}$. Calculate several terms of the sequence; do you think the sequence converges? Assume that the sequence does converge and that $\lim \lim_{n\to\infty} a_n = L$. Use Exercise 41 to show that L satisfies the equation $L = \sqrt{2 + L}$. Solve this equation to find the limit L.

43. Discover ■ Prove: The Fibonacci Sequence and the Golden Ratio Let F_n denote the nth term of the Fibonacci sequence (Section 11.1). Assume that the sequence F_{n+1}/F_n converges. Show that limit of this sequence is the Golden Ratio:

$$\lim_{n\to\infty} \frac{F_{n+1}}{F_n} = \frac{1+\sqrt{5}}{2}$$

PS *Introduce something extra.* Assume that the limit is L. Use the fact that $F_{n+1} = F_{n-1} + F_n$ and the property in Exercise 41 to find an equation that L must satisfy.

12.5 Areas

■ The Area Problem ■ Definition of Area

We have seen that limits are needed to compute the slope of a tangent line or an instantaneous rate of change. Here we will see that they are also needed to find the area of a region with a curved boundary. The problem of finding such areas has consequences far beyond simply finding area. (See the *Focus on Modeling* at the end of the chapter.)

■ The Area Problem

One of the central problems in calculus is the *area problem*: Find the area of the region S that lies under the curve $y = f(x)$ from a to b. This means that S, illustrated in Figure 1, is bounded by the graph of a function f (where $f(x) \geq 0$), the vertical lines $x = a$ and $x = b$, and the x-axis.

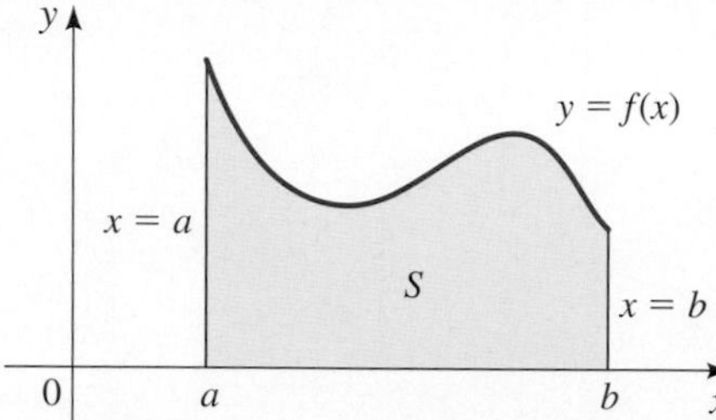

Figure 1

In trying to solve the area problem, we have to ask ourselves: What is the meaning of the word *area*? Let's start by answering this question for regions with straight sides. For a rectangle, the area is defined as the product of the length and the width. The area of a triangle is half the base times the height. The area of a polygon is found by dividing it into triangles (as in Figure 2) and adding the areas of the triangles.

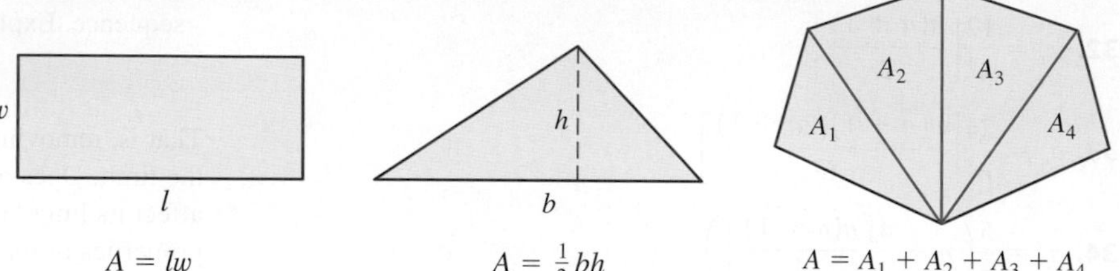

Figure 2 $A = lw$ $A = \frac{1}{2}bh$ $A = A_1 + A_2 + A_3 + A_4$

However, it is not so easy to find the area of a region with curved sides. We all have an intuitive idea of what the area of a region is. But part of the area problem is to make this intuitive idea precise by giving an exact definition of area.

Recall that in defining a tangent, we first approximated the slope of the tangent line by slopes of secant lines, and then we took the limit of these approximations. We pursue a similar idea for areas. We first approximate the region S by rectangles, and then we take the limit of the areas of these rectangles as we increase the number of rectangles. The following example illustrates the procedure.

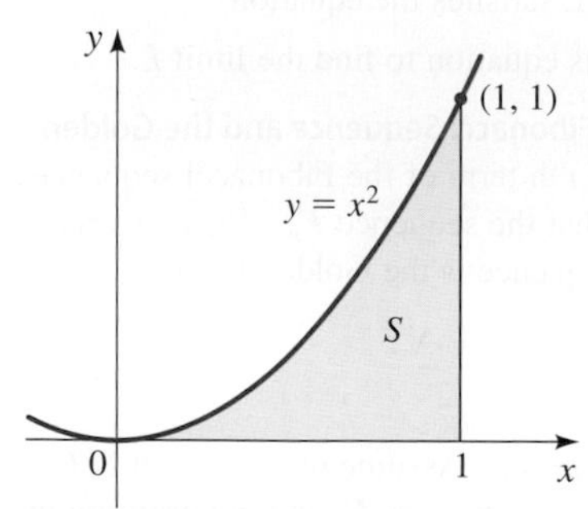

Figure 3

Example 1 ■ Estimating an Area Using Rectangles

Use rectangles to estimate the area under the parabola $y = x^2$ from 0 to 1 (the parabolic region S illustrated in Figure 3).

Solution We first notice that the area of S must be somewhere between 0 and 1 because S is contained in a square with side length 1, but we can certainly do better than that. Suppose we divide S into four strips $S1$, $S2$, $S3$, and $S4$ by drawing the vertical lines $x = \frac{1}{4}$, $x = \frac{1}{2}$, and $x = \frac{3}{4}$, as in Figure 4(a). We can approximate each strip by a rectangle whose base is the same as the strip and whose height is the same as the right edge of the strip [see Figure 4(b)]. In other words, the heights of these rectangles are the values of the function $f(x) = x^2$ at the right endpoints of the following subintervals: $\left[0, \frac{1}{4}\right]$, $\left[\frac{1}{4}, \frac{1}{2}\right]$, $\left[\frac{1}{2}, \frac{3}{4}\right]$, and $\left[\frac{3}{4}, 1\right]$.

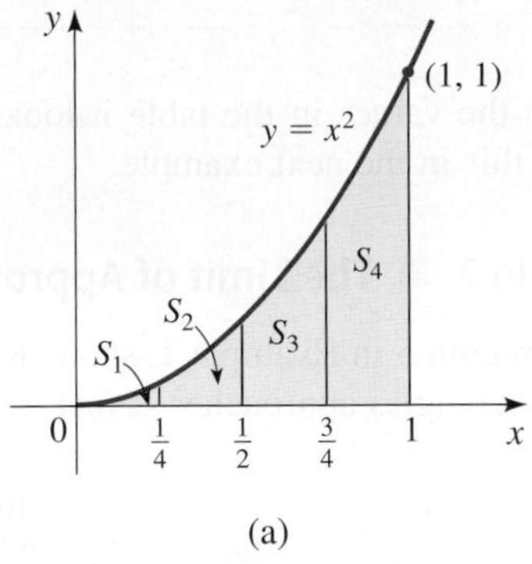

(a)

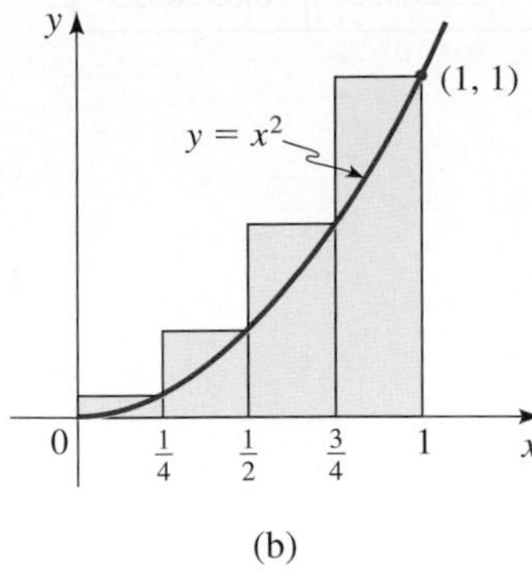

(b)

Figure 4

Each rectangle has width $\frac{1}{4}$, and the heights are $\left(\frac{1}{4}\right)^2$, $\left(\frac{1}{2}\right)^2$, $\left(\frac{3}{4}\right)^2$, and 1^2. If we let R_4 be the sum of the areas of these approximating rectangles, we get

$$R_4 = \tfrac{1}{4}\cdot\left(\tfrac{1}{4}\right)^2 + \tfrac{1}{4}\cdot\left(\tfrac{1}{2}\right)^2 + \tfrac{1}{4}\cdot\left(\tfrac{3}{4}\right)^2 + \tfrac{1}{4}\cdot 1^2 = \tfrac{15}{32} = 0.46875$$

From Figure 4(b) we see that the area A of S is less than R_4, so

$$A < 0.46875$$

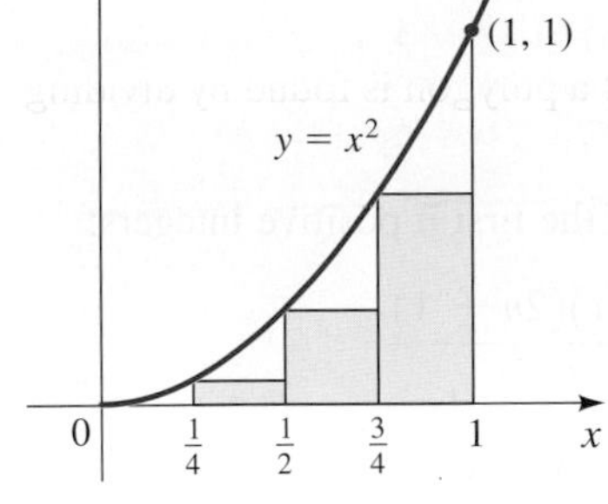

Figure 5

Instead of using the rectangles in Figure 4(b), we could use the smaller rectangles in Figure 5 whose heights are the values of f at the left endpoints of the subintervals. (The leftmost rectangle has collapsed because its height is 0.) The sum of the areas of these approximating rectangles is

$$L_4 = \tfrac{1}{4}\cdot 0^2 + \tfrac{1}{4}\cdot\left(\tfrac{1}{4}\right)^2 + \tfrac{1}{4}\cdot\left(\tfrac{1}{2}\right)^2 + \tfrac{1}{4}\cdot\left(\tfrac{3}{4}\right)^2 = \tfrac{7}{32} = 0.21875$$

We see that the area of S is larger than L_4, so we have lower and upper estimates for A:

$$0.21875 < A < 0.46875$$

We can repeat this procedure with a larger number of strips. Figure 6 shows what happens when we divide the region S into eight strips of equal width. By computing the sum of the areas of the smaller rectangles (L_8) and the sum of the areas of the larger rectangles (R_8), we obtain better lower and upper estimates for A:

$$0.2734375 < A < 0.3984375$$

So one possible answer to the question is to say that the true area of S lies somewhere between 0.2734375 and 0.3984375.

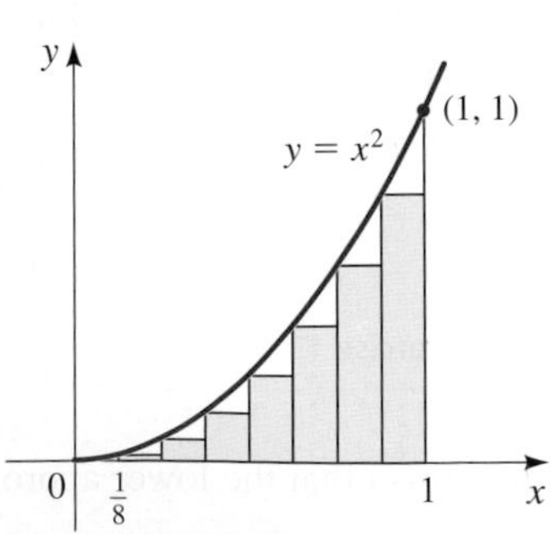

(a) Using left endpoints

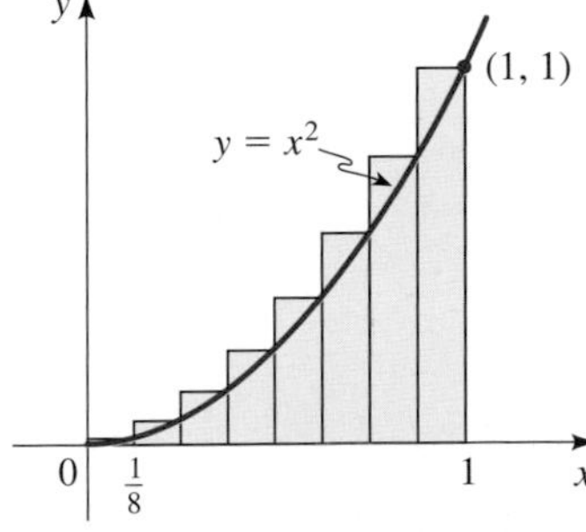

(b) Using right endpoints

Figure 6 | Approximating S with eight rectangles

n	L_n	R_n
10	0.2850000	0.3850000
20	0.3087500	0.3587500
30	0.3168519	0.3501852
50	0.3234000	0.3434000
100	0.3283500	0.3383500
1000	0.3328335	0.3338335

We could obtain better estimates by increasing the number of strips. The table in the margin shows the results of similar calculations (with a computer) using n rectangles whose heights are found with left endpoints (L_n) or right endpoints (R_n). In particular, we see by using 50 strips that the area lies between 0.3234 and 0.3434. With 1000 strips we narrow it down even more: A lies between 0.3328335 and 0.3338335. A good estimate is obtained by averaging these numbers: $A \approx 0.3333335$.

Now Try Exercise 3 ■

From the values in the table it looks as if R_n is approaching $\frac{1}{3}$ as n increases. We confirm this in the next example.

Example 2 ■ The Limit of Approximating Sums

For the region S in Example 1, show that the sum of the areas of the upper approximating rectangles approaches $\frac{1}{3}$, that is,

$$\lim_{n\to\infty} R_n = \tfrac{1}{3}$$

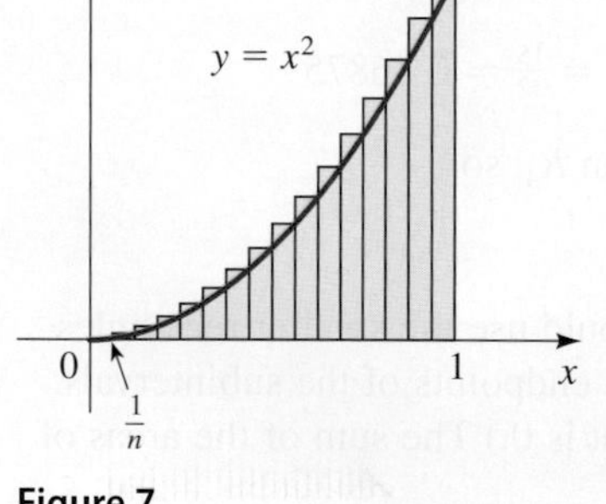

Figure 7

Solution Let R_n be the sum of the areas of the n rectangles shown in Figure 7. Each rectangle has width $1/n$, and the heights are the values of the function $f(x) = x^2$ at the points $1/n, 2/n, 3/n, \ldots, n/n$. That is, the heights are $(1/n)^2, (2/n)^2, (3/n)^2, \ldots, (n/n)^2$. Thus

$$R_n = \frac{1}{n}\left(\frac{1}{n}\right)^2 + \frac{1}{n}\left(\frac{2}{n}\right)^2 + \frac{1}{n}\left(\frac{3}{n}\right)^2 + \cdots + \frac{1}{n}\left(\frac{n}{n}\right)^2$$

$$= \frac{1}{n}\cdot\frac{1}{n^2}(1^2 + 2^2 + 3^2 + \cdots + n^2)$$

$$= \frac{1}{n^3}(1^2 + 2^2 + 3^2 + \cdots + n^2)$$

Here we need the formula for the sum of the squares of the first n positive integers:

This formula was discussed in Section 11.4.

$$1^2 + 2^2 + 3^2 + \cdots + n^2 = \frac{n(n+1)(2n+1)}{6}$$

Putting the preceding formula into our expression for R_n, we get

$$R_n = \frac{1}{n^3}\cdot\frac{n(n+1)(2n+1)}{6} = \frac{(n+1)(2n+1)}{6n^2}$$

Thus we have

$$\lim_{n\to\infty} R_n = \lim_{n\to\infty}\frac{(n+1)(2n+1)}{6n^2}$$

$$= \lim_{n\to\infty}\frac{1}{6}\left(\frac{n+1}{n}\right)\left(\frac{2n+1}{n}\right)$$

$$= \lim_{n\to\infty}\frac{1}{6}\left(1+\frac{1}{n}\right)\left(2+\frac{1}{n}\right)$$

$$= \tfrac{1}{6}\cdot 1\cdot 2 = \tfrac{1}{3}$$

Now Try Exercise 13 ■

It can be shown that the lower approximating sums also approach $\frac{1}{3}$, that is,

$$\lim_{n\to\infty} L_n = \tfrac{1}{3}$$

From Figures 8 and 9 it appears that as n increases, both R_n and L_n become better and better approximations to the area of S. Therefore we *define* the area A to be the limit of the sums of the areas of the approximating rectangles, that is,

$$A = \lim_{n\to\infty} R_n = \lim_{n\to\infty} L_n = \tfrac{1}{3}$$

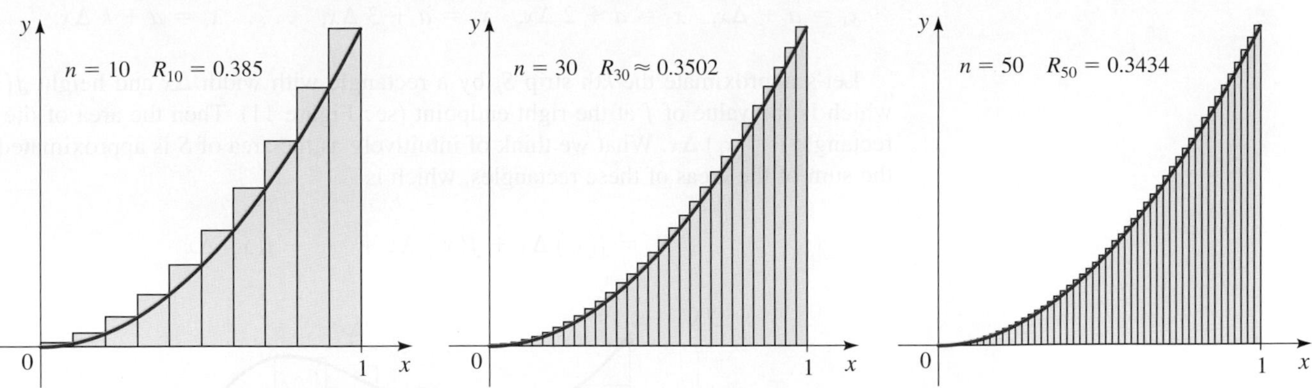

Figure 8

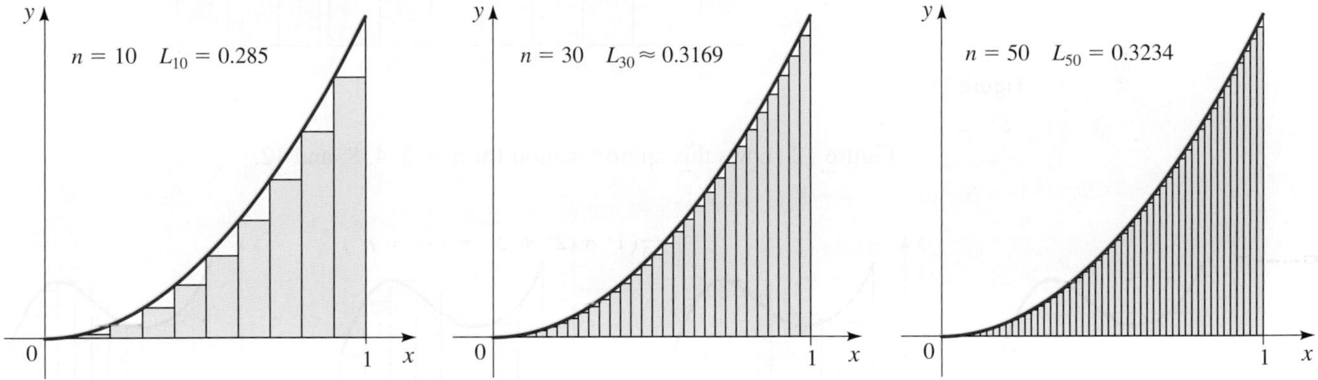

Figure 9

■ Definition of Area

Let's apply the idea of Examples 1 and 2 to the more general region S of Figure 1. We start by subdividing S into n strips $S_1, S_2, \ldots, S_n$ of equal width, as shown in Figure 10.

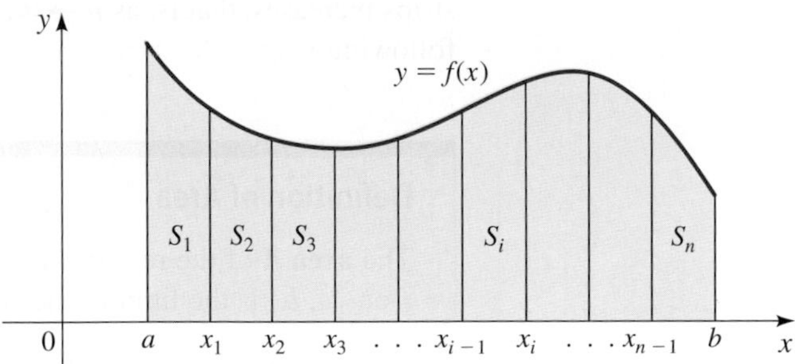

Figure 10

The width of the interval $[a, b]$ is $b - a$, so the width of each of the n strips is

$$\Delta x = \frac{b - a}{n}$$

These strips divide the interval $[a, b]$ into n subintervals

$$[x_0, x_1], \quad [x_1, x_2], \quad [x_2, x_3], \quad \ldots, \quad [x_{n-1}, x_n]$$

where $x_0 = a$ and $x_n = b$. The right endpoints of the subintervals are

$$x_1 = a + \Delta x, \quad x_2 = a + 2\,\Delta x, \quad x_3 = a + 3\,\Delta x, \quad \ldots, \quad x_k = a + k\,\Delta x, \quad \ldots$$

Let's approximate the kth strip S_k by a rectangle with width Δx and height $f(x_k)$, which is the value of f at the right endpoint (see Figure 11). Then the area of the kth rectangle is $f(x_k)\,\Delta x$. What we think of intuitively as the area of S is approximated by the sum of the areas of these rectangles, which is

$$R_n = f(x_1)\,\Delta x + f(x_2)\,\Delta x + \cdots + f(x_n)\,\Delta x$$

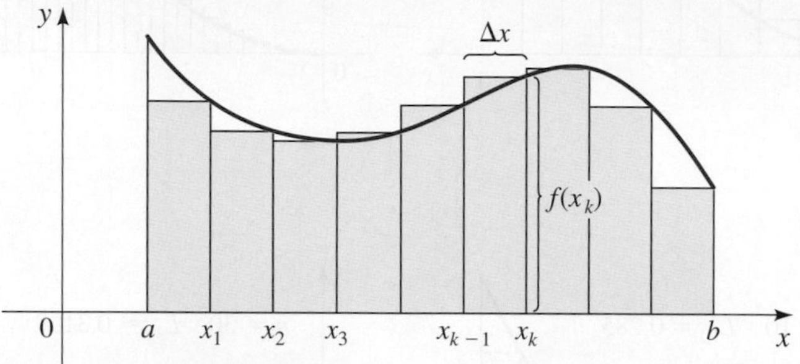

Figure 11

Figure 12 shows this approximation for $n = 2$, 4, 8, and 12.

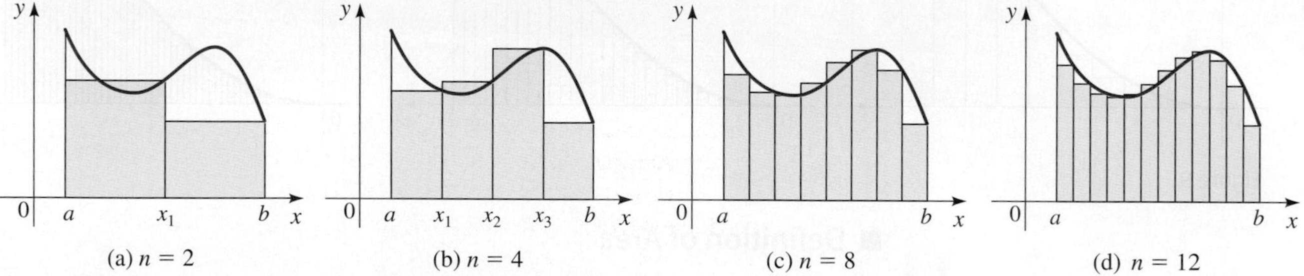

Figure 12

Notice that this approximation appears to become better and better as the number of strips increases, that is, as $n \to \infty$. Therefore we define the area A of the region S in the following way.

Definition of Area

The **area** A of the region S that lies under the graph of the continuous function f on $[a, b]$ is the limit of the sum of the areas of approximating rectangles:

$$A = \lim_{n \to \infty} R_n = \lim_{n \to \infty} [f(x_1)\,\Delta x + f(x_2)\,\Delta x + \cdots + f(x_n)\,\Delta x]$$

Using sigma notation, we write this as follows:

$$A = \lim_{n \to \infty} \sum_{k=1}^{n} f(x_k)\,\Delta x$$

In using this formula for area, remember that Δx is the width of an approximating rectangle, x_k is the right endpoint of the kth rectangle, and $f(x_k)$ is its height. So

$$\text{Width:} \qquad \Delta x = \frac{b-a}{n}$$

$$\text{Right endpoint:} \qquad x_k = a + k\,\Delta x$$

$$\text{Height:} \qquad f(x_k) = f(a + k\,\Delta x)$$

When working with sums, we will need the following properties from Section 11.1.

$$\sum_{k=1}^{n}(a_k \pm b_k) = \sum_{k=1}^{n} a_k \pm \sum_{k=1}^{n} b_k \qquad \sum_{k=1}^{n} ca_k = c\sum_{k=1}^{n} a_k$$

We will also need the following formulas for the sums of the powers of the first n natural numbers from Section 11.4.

$$\sum_{k=1}^{n} c = nc \qquad \sum_{k=1}^{n} k = \frac{n(n+1)}{2}$$

$$\sum_{k=1}^{n} k^2 = \frac{n(n+1)(2n+1)}{6} \qquad \sum_{k=1}^{n} k^3 = \frac{n^2(n+1)^2}{4}$$

Example 3 ■ Finding the Area Under a Curve

Find the area of the region that lies under the parabola $y = x^2$, where $0 \le x \le 5$.

Solution The region is graphed in Figure 13. To find the area, we first find the dimensions of the approximating rectangles at the nth stage.

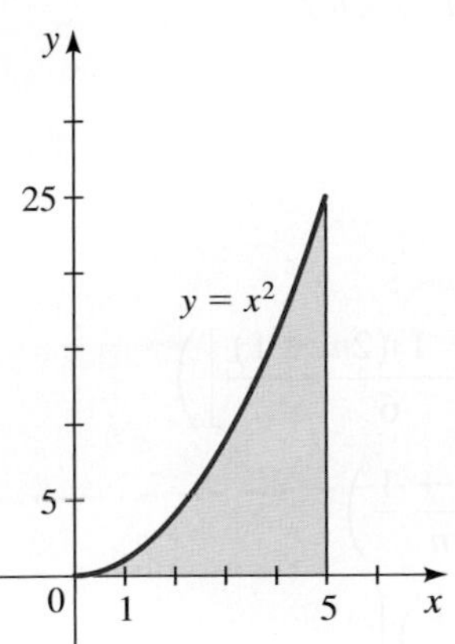

Figure 13

$$\text{Width:} \qquad \Delta x = \frac{b-a}{n} = \frac{5-0}{n} = \frac{5}{n}$$

$$\text{Right endpoint:} \qquad x_k = a + k\,\Delta x = 0 + k\left(\frac{5}{n}\right) = \frac{5k}{n}$$

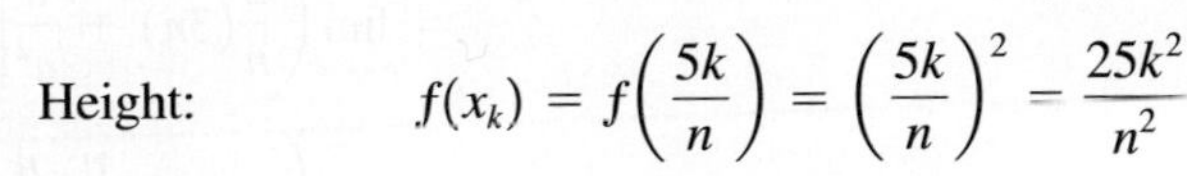

$$\text{Height:} \qquad f(x_k) = f\left(\frac{5k}{n}\right) = \left(\frac{5k}{n}\right)^2 = \frac{25k^2}{n^2}$$

Now we substitute these values into the definition of area.

$$A = \lim_{n\to\infty}\sum_{k=1}^{n} f(x_k)\,\Delta x \qquad \text{Definition of area}$$

$$= \lim_{n\to\infty}\sum_{k=1}^{n}\frac{25k^2}{n^2}\cdot\frac{5}{n} \qquad f(x_k) = \frac{25k^2}{n^2},\ \Delta x = \frac{5}{n}$$

$$= \lim_{n\to\infty}\sum_{k=1}^{n}\frac{125k^2}{n^3} \qquad \text{Simplify}$$

$$= \lim_{n\to\infty}\frac{125}{n^3}\sum_{k=1}^{n} k^2 \qquad \text{Factor } \frac{125}{n^3}$$

$$= \lim_{n\to\infty}\frac{125}{n^3}\cdot\frac{n(n+1)(2n+1)}{6} \qquad \text{Sum of Squares Formula}$$

$$= \lim_{n\to\infty}\frac{125(2n^2+3n+1)}{6n^2} \qquad \text{Cancel } n \text{, and expand the numerator}$$

$$= \lim_{n\to\infty}\frac{125}{6}\left(2 + \frac{3}{n} + \frac{1}{n^2}\right) \qquad \text{Divide the numerator and denominator by } n^2$$

$$= \frac{125}{6}(2+0+0) = \frac{125}{3} \qquad \text{Let } n \to \infty$$

We can also calculate the limit by writing

$$\frac{125}{n^3}\cdot\frac{n(n+1)(2n+1)}{6} = \frac{125}{6}\left(\frac{n}{n}\right)\left(\frac{n+1}{n}\right)\left(\frac{2n+1}{n}\right)$$

as in Example 2.

Thus the area of the region is $\frac{125}{3} \approx 41.7$.

Now Try Exercise 15 ■

Example 4 ■ Finding the Area Under a Curve

The figure below shows the region whose area is computed in Example 4.

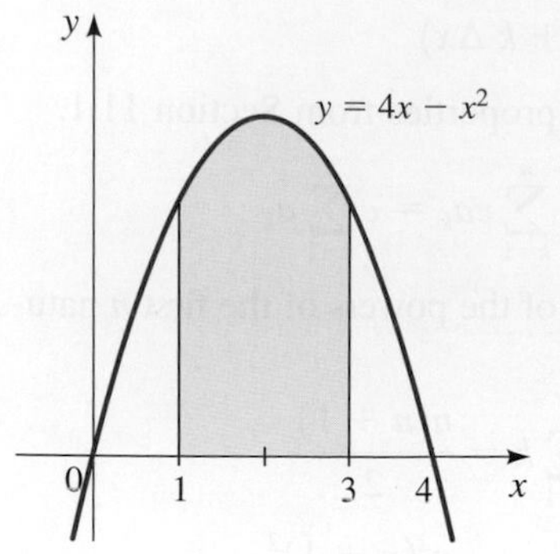

Find the area of the region that lies under the parabola $y = 4x - x^2$, where $1 \le x \le 3$.

Solution We start by finding the dimensions of the approximating rectangles at the nth stage.

$$\text{Width:} \qquad \Delta x = \frac{b-a}{n} = \frac{3-1}{n} = \frac{2}{n}$$

$$\text{Right endpoint:} \qquad x_k = a + k\,\Delta x = 1 + k\left(\frac{2}{n}\right) = 1 + \frac{2k}{n}$$

$$\text{Height:} \qquad f(x_k) = f\left(1 + \frac{2k}{n}\right) = 4\left(1 + \frac{2k}{n}\right) - \left(1 + \frac{2k}{n}\right)^2$$

$$= 4 + \frac{8k}{n} - 1 - \frac{4k}{n} - \frac{4k^2}{n^2}$$

$$= 3 + \frac{4k}{n} - \frac{4k^2}{n^2}$$

Thus according to the definition of area, we get

$$A = \lim_{n\to\infty} \sum_{k=1}^{n} f(x_k)\,\Delta x = \lim_{n\to\infty} \sum_{k=1}^{n} \left(3 + \frac{4k}{n} - \frac{4k^2}{n^2}\right)\left(\frac{2}{n}\right)$$

$$= \lim_{n\to\infty} \left(\sum_{k=1}^{n} 3 + \frac{4}{n}\sum_{k=1}^{n} k - \frac{4}{n^2}\sum_{k=1}^{n} k^2\right)\left(\frac{2}{n}\right)$$

$$= \lim_{n\to\infty} \left(\frac{2}{n}\sum_{k=1}^{n} 3 + \frac{8}{n^2}\sum_{k=1}^{n} k - \frac{8}{n^3}\sum_{k=1}^{n} k^2\right)$$

$$= \lim_{n\to\infty} \left(\frac{2}{n}(3n) + \frac{8}{n^2}\left[\frac{n(n+1)}{2}\right] - \frac{8}{n^3}\left[\frac{n(n+1)(2n+1)}{6}\right]\right)$$

$$= \lim_{n\to\infty} \left(6 + 4\cdot\frac{n}{n}\cdot\frac{n+1}{n} - \frac{4}{3}\cdot\frac{n}{n}\cdot\frac{n+1}{n}\cdot\frac{2n+1}{n}\right)$$

$$= \lim_{n\to\infty} \left[6 + 4\left(1 + \frac{1}{n}\right) - \frac{4}{3}\left(1 + \frac{1}{n}\right)\left(2 + \frac{1}{n}\right)\right]$$

$$= 6 + 4\cdot 1 - \frac{4}{3}\cdot 1\cdot 2 = \frac{22}{3}$$

Now Try Exercise 17 ■

12.5 Exercises

Concepts

1–2 ■ The graph of a function f is shown below.

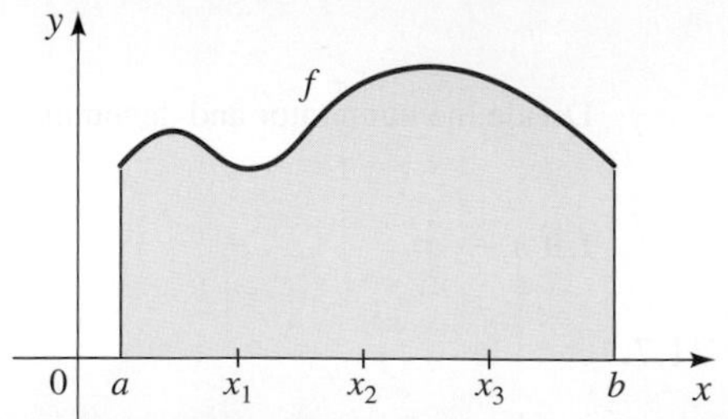

1. To find the area under the graph of f, we first approximate the area by ________. Approximate the area by drawing four rectangles. The area R_4 of this approximation is

$$R_4 = \underline{\qquad} + \underline{\qquad} + \underline{\qquad} + \underline{\qquad}$$

2. Let R_n be the approximation obtained by using n rectangles of equal width. The exact area under the graph of f is

$$A = \lim_{n\to\infty} \underline{\qquad}$$

■ Skills

3. Estimating an Area Using Rectangles

(a) By reading values from the given graph of f, use five rectangles to find a lower estimate and an upper estimate for the area under the given graph of f from $x = 0$ to $x = 10$. In each case, sketch the rectangles that you use.

(b) Find new estimates using 10 rectangles in each case.

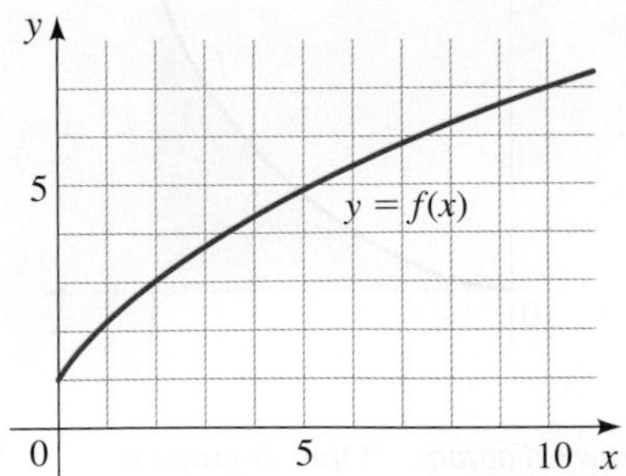

4. Estimating an Area Using Rectangles

(a) Use six rectangles to find estimates of each type for the area under the given graph of f from $x = 0$ to $x = 12$.
(i) L_6 (using left endpoints)
(ii) R_6 (using right endpoints)

(b) Is L_6 an underestimate or an overestimate of the true area?

(c) Is R_6 an underestimate or an overestimate of the true area?

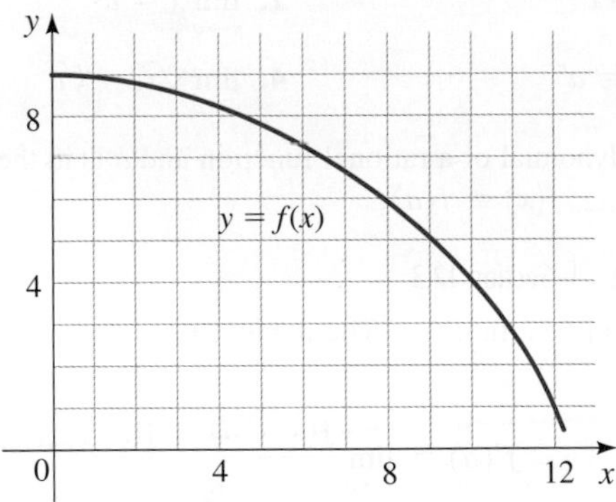

5–8 ■ Estimating Areas Using Rectangles Approximate the area of the shaded region under the graph of the given function by using the indicated rectangles. (The rectangles have equal width.)

5. $f(x) = \frac{1}{2}x + 2$

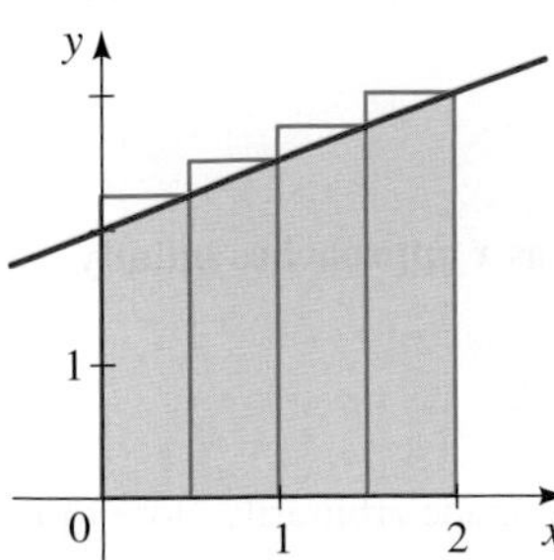

6. $f(x) = 4 - x^2$

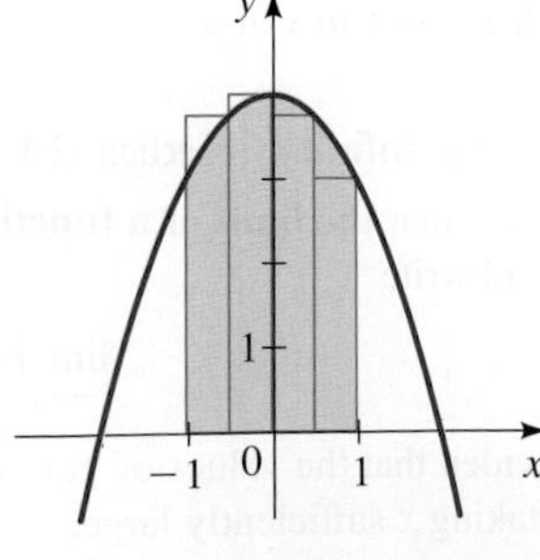

7. $f(x) = \dfrac{4}{x}$

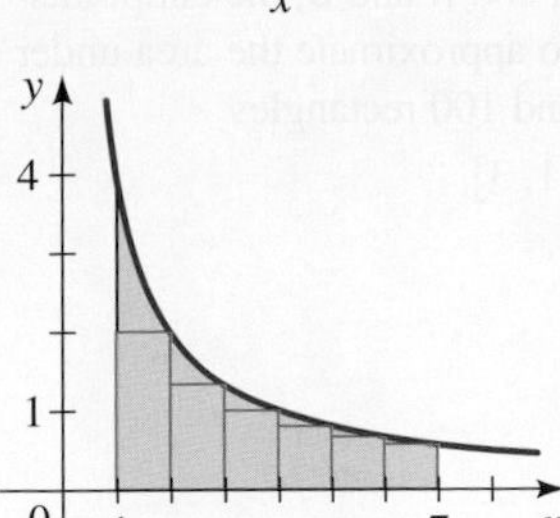

8. $f(x) = 9x - x^3$

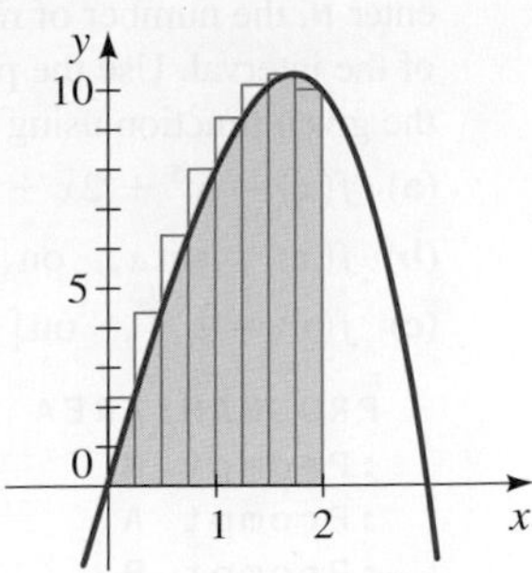

9–12 ■ Estimating Areas Using Rectangles In these exercises we estimate the area under the graph of a function by using rectangles.

9. (a) Estimate the area under the graph of $f(x) = 1/x$ from $x = 1$ to $x = 5$ using four approximating rectangles and right endpoints. Sketch the graph and the rectangles. Is your estimate an underestimate or an overestimate?

(b) Repeat part (a), using left endpoints.

10. (a) Estimate the area under the graph of $f(x) = 25 - x^2$ from $x = 0$ to $x = 5$ using five approximating rectangles and right endpoints. Sketch the graph and the rectangles. Is your estimate an underestimate or an overestimate?

(b) Repeat part (a) using left endpoints.

11. (a) Estimate the area under the graph of $f(x) = 1 + x^2$ from $x = -1$ to $x = 2$ using three rectangles and right endpoints. Then improve your estimate by using six rectangles. Sketch the curve and the approximating rectangles.

(b) Repeat part (a) using left endpoints.

12. (a) Estimate the area under the graph of $f(x) = e^{-x}$, $0 \le x \le 4$, using four approximating rectangles and
(i) right endpoints
(ii) left endpoints
In each case, sketch the curve and the rectangles.

(b) Improve your estimates in part (a) by using eight rectangles.

13–14 ■ Finding the Area Under A Curve Use the definition of area as a limit to find the area of the region that lies under the curve. Check your answer by sketching the region and using geometry.

13. $y = 3x, \quad 0 \le x \le 5$

14. $y = 2x + 1, \quad 1 \le x \le 3$

15–20 ■ Finding the Area Under a Curve Find the area of the region that lies under the graph of f over the given interval.

15. $f(x) = 3x^2, \quad 0 \le x \le 2$

16. $f(x) = x + x^2, \quad 0 \le x \le 1$

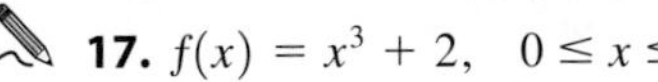

17. $f(x) = x^3 + 2, \quad 0 \le x \le 5$

18. $f(x) = 4x^3, \quad 2 \le x \le 5$

19. $f(x) = x + 6x^2, \quad 1 \le x \le 4$

20. $f(x) = 20 - 2x^2, \quad 2 \le x \le 3$

■ Discuss ■ Discover ■ Prove ■ Write

21. Discuss: Approximating Area with a Calculator The following TI-84 program finds the approximate area under the graph of f on the interval $[a, b]$ using n rectangles. To use the program, first store the function f in Y_1. The program prompts you to enter N, the number of rectangles, and A and B, the endpoints of the interval. Use the program to approximate the area under the given function using 10, 20, and 100 rectangles.

(a) $f(x) = x^5 + 2x + 3$, on $[1, 3]$

(b) $f(x) = \sin x$, on $[0, \pi]$

(c) $f(x) = e^{-x^2}$, on $[-1, 1]$

```
PROGRAM:AREA
:Prompt N
:Prompt A
:Prompt B
:(B-A)/N→D
:0→S
:A→X
:For (K,1,N)
:X+D→X
:S+D*Y₁→S
:End
:Disp "AREA IS"
:Disp S
```

22. Discuss ■ Prove: Area Under the Graph of a Function Let f be a continuous one-to-one function defined on $[0, 1]$ such that $f(0) = 0$ and $f(1) = 1$, as shown in the figure. Show that

$$(\text{Area under } f) + (\text{Area under } f^{-1}) = 1$$

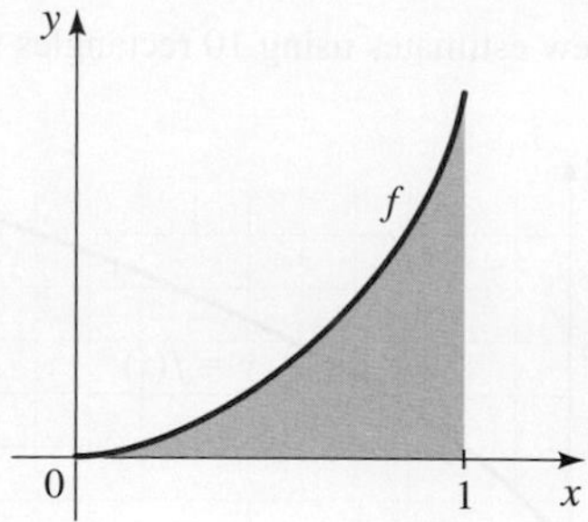

PS *Draw a diagram.* Draw a graph of f^{-1}. Argue from the graphs of f and f^{-1} that the given equation is true.

Chapter 12 Review

Properties and Formulas

Limits | Section 12.1

We say that the **limit of a function** f, as x approaches a, equals L, and we write

$$\lim_{x \to a} f(x) = L$$

provided that the values of $f(x)$ can be made arbitrarily close to L by taking x to be sufficiently close to a.

The **left-hand** and **right-hand** limits of f, as x approaches a, are defined similarly:

$$\lim_{x \to a^-} f(x) = L \qquad \lim_{x \to a^+} f(x) = L$$

The limit of f, as x approaches a, exists if and only if both left- and right-hand limits exist: $\lim_{x \to a} f(x) = L$ if and only if $\lim_{x \to a^-} f(x) = L$ and $\lim_{x \to a^+} f(x) = L$.

Algebraic Properties of Limits | Section 12.2

The following **Limit Laws** hold.

1. $\lim_{x \to a} [f(x) + g(x)] = \lim_{x \to a} f(x) + \lim_{x \to a} g(x)$

2. $\lim_{x \to a} [f(x) - g(x)] = \lim_{x \to a} f(x) - \lim_{x \to a} g(x)$

3. $\lim_{x \to a} cf(x) = c \lim_{x \to a} f(x)$

4. $\lim_{x \to a} [f(x)g(x)] = \lim_{x \to a} f(x) \cdot \lim_{x \to a} g(x)$

5. $\lim_{x \to a} \dfrac{f(x)}{g(x)} = \dfrac{\lim_{x \to a} f(x)}{\lim_{x \to a} g(x)}$, if $\lim_{x \to a} g(x) \neq 0$

6. $\lim_{x \to a} [f(x)]^n = \left[\lim_{x \to a} f(x)\right]^n$ **7.** $\lim_{x \to a} \sqrt[n]{f(x)} = \sqrt[n]{\lim_{x \to a} f(x)}$

The following **special limits** hold.

1. $\lim_{x \to a} c = c$ **2.** $\lim_{x \to a} x = a$

3. $\lim_{x \to a} x^n = a^n$ **4.** $\lim_{x \to a} \sqrt[n]{x} = \sqrt[n]{a}$

If f is a polynomial or a rational function and a is in the domain of f, then $\lim_{x \to a} f(x) = f(a)$.

Derivatives | Section 12.3

Let $y = f(x)$ be a function. The **derivative of f at a**, denoted by $f'(a)$, is

$$f'(a) = \lim_{h \to 0} \frac{f(x + h) - f(x)}{h}$$

Equivalently, the derivative $f'(a)$ is

$$f'(a) = \lim_{x \to a} \frac{f(x) - f(a)}{x - a}$$

The derivative of f at a is the **slope of the tangent line** to the curve $y = f(x)$ at the point $P(a, f(a))$.

The derivative of f at a is the **instantaneous rate of change of y with respect to x** at $x = a$.

Limits at Infinity | Section 12.4

We say that **the limit of a function** f, as x **approaches infinity**, is L, and write

$$\lim_{x \to \infty} f(x) = L$$

provided that the values of $f(x)$ can be made arbitrarily close to L by taking x sufficiently large.

We say that **the limit of a function f, as x approaches negative infinity**, is L, and we write

$$\lim_{x\to-\infty} f(x) = L$$

provided that the values of $f(x)$ can be made arbitrarily close to L by taking x sufficiently large negative.

The line $y = L$ is a horizontal asymptote of the curve $y = f(x)$ if either

$$\lim_{x\to\infty} f(x) = L \quad \text{or} \quad \lim_{x\to-\infty} f(x) = L$$

The following special limits hold, where $k > 0$:

$$\lim_{x\to\infty} \frac{1}{x^k} = 0 \quad \text{and} \quad \lim_{x\to-\infty} \frac{1}{x^k} = 0$$

Limits of Sequences | Section 12.4

We say that a sequence $a_1, a_2, a_3, \ldots$ has the limit L, and we write

$$\lim_{n\to\infty} a_n = L$$

provided that the nth term a_n of the sequence can be made arbitrarily close to L by taking n sufficiently large.

If $\lim_{x\to\infty} f(x) = L$ and if $f(n) = a_n$ when n is an integer, then $\lim_{n\to\infty} a_n = L$.

Area | Section 12.5

Let f be a continuous function defined on the interval $[a, b]$. The area A of the region that lies under the graph of f is the limit of the sum of the areas of approximating rectangles:

$$A = \lim_{n\to\infty} [f(x_1)\,\Delta x + f(x_2)\,\Delta x + \cdots + f(x_n)\,\Delta x]$$

$$= \lim_{n\to\infty} \sum_{k=1}^{n} f(x_k)\,\Delta x$$

where

$$\Delta x = \frac{b-a}{n} \quad \text{and} \quad x_k = a + k\,\Delta x$$

Summation Formulas | Section 12.5

The following summation formulas are useful for calculating areas.

$$\sum_{k=1}^{n} c = nc \qquad \sum_{k=1}^{n} k = \frac{n(n+1)}{2}$$

$$\sum_{k=1}^{n} k^2 = \frac{n(n+1)(2n+1)}{6} \qquad \sum_{k=1}^{n} k^3 = \frac{n^2(n+1)^2}{4}$$

Concept Check

1. (a) Explain what is meant by $\lim_{x\to a} f(x) = L$.

(b) If $\lim_{x\to2} f(x) = 5$, is it possible that $f(2) = 3$?

(c) Find $\lim_{x\to2} x^2$.

2. To evaluate the limit of a function, we often need to first rewrite the function using the rules of algebra. What is the logical first step in evaluating each of the following limits?

(a) $\displaystyle\lim_{x\to2} \frac{x^2-4}{x-2}$ **(b)** $\displaystyle\lim_{h\to0} \frac{(5+h)^2-25}{h}$

(c) $\displaystyle\lim_{x\to3} \frac{\sqrt{x+1}-2}{x-3}$ **(d)** $\displaystyle\lim_{x\to7} \frac{\left(\frac{1}{7}-\frac{1}{x}\right)}{x-7}$

3. (a) Explain what it means to say:

$$\lim_{x\to3^-} f(x) = 5 \qquad \lim_{x\to3^+} f(x) = 10$$

(b) If the two equations in part (a) are true, is it possible that $\lim_{x\to3} f(x) = 5$?

(c) Find $\lim_{x\to2^-} f(x)$ and $\lim_{x\to2^+} f(x)$, where f is defined as follows.

$$f(x) = \begin{cases} 1 & \text{if } x \le 2 \\ x & \text{if } x > 2 \end{cases}$$

(d) For f as given in part (c), does $\lim_{x\to2} f(x)$ exist?

4. (a) Define the derivative $f'(a)$ of a function f at $x = a$.

(b) State an equivalent formulation for $f'(a)$.

(c) Find the derivative of $f(x) = x^2$ at $x = 3$.

5. (a) Give two different interpretations of the derivative of the function $y = f(x)$ at $x = a$.

(b) For the function $f(x) = x^2$, find the slope of the tangent line to the graph of f at the point $(3, 9)$ on the graph.

(c) For the function $y = x^2$, find the instantaneous rate of change of y with respect to x when $x = 3$.

(d) Write expressions for the average rate of change of y with respect to x between a and x and for the instantaneous rate of change of y with respect to x at $x = a$.

6. (a) Explain what is meant by $\lim_{x\to\infty} f(x) = L$. Draw sketches to illustrate different ways in which this can happen.

(b) Find $\displaystyle\lim_{x\to\infty} \frac{3x^2+x}{x^2+1}$.

(c) Explain why $\lim_{x\to\infty} \sin x$ does not exist.

7. (a) If $a_1, a_2, a_3, \ldots$ is a sequence, what is meant by $\lim_{n\to\infty} a_n = L$? What is a convergent sequence?

(b) Find $\lim_{n\to\infty} (-1)^n/n$.

8. (a) Suppose S is the region under the graph of the function $y = f(x)$ and above the x-axis, where $a \le x \le b$. Explain how this area is approximated by rectangles, and write an expression for the area of S as a limit of sums.

(b) Find the area under the graph of $f(x) = x^2$ and above the x-axis, between $x = 0$ and $x = 3$.

Answers to the Concept Check can be found at the book companion website stewartmath.com.

Exercises

1–6 ■ Estimating Limits Numerically and Graphically Use a table of values to estimate the value of the limit. Then use a graphing device to confirm your result graphically.

1. $\lim_{x \to 2} \dfrac{x-2}{x^2 - 3x + 2}$

2. $\lim_{t \to -1} \dfrac{t+1}{t^3 - t}$

3. $\lim_{x \to 0} \dfrac{2^x - 1}{x}$

4. $\lim_{x \to 0} \dfrac{\sin 2x}{x}$

5. $\lim_{x \to 1^+} \ln \sqrt{x - 1}$

6. $\lim_{x \to 0^-} \dfrac{\tan x}{|x|}$

7. **Limits from a Graph** The graph of f is shown in the figure. Find each limit, or explain why it does not exist.

(a) $\lim_{x \to 2^+} f(x)$
(b) $\lim_{x \to -3^+} f(x)$
(c) $\lim_{x \to -3^-} f(x)$
(d) $\lim_{x \to -3} f(x)$
(e) $\lim_{x \to 4} f(x)$
(f) $\lim_{x \to \infty} f(x)$
(g) $\lim_{x \to -\infty} f(x)$
(h) $\lim_{x \to 0} f(x)$

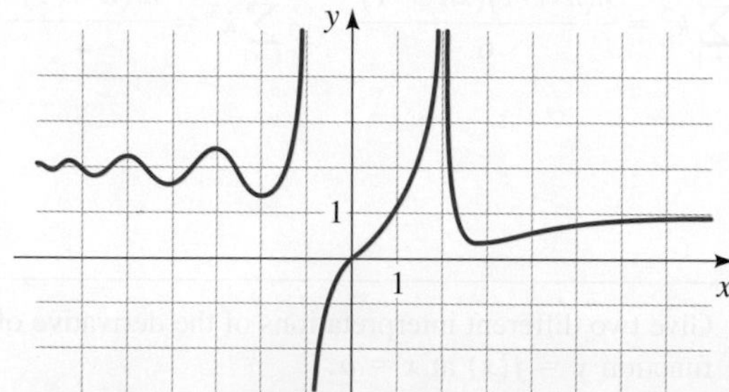

8. **One-Sided Limits** Let

$$f(x) = \begin{cases} 2 & \text{if } x < -1 \\ x^2 & \text{if } -1 \le x \le 2 \\ x + 2 & \text{if } x > 2 \end{cases}$$

Find each limit, or explain why it does not exist.

(a) $\lim_{x \to -1^-} f(x)$
(b) $\lim_{x \to -1^+} f(x)$
(c) $\lim_{x \to -1} f(x)$
(d) $\lim_{x \to 2^-} f(x)$
(e) $\lim_{x \to 2^+} f(x)$
(f) $\lim_{x \to 2} f(x)$
(g) $\lim_{x \to 0} f(x)$
(h) $\lim_{x \to 3} (f(x))^2$

9–20 ■ Finding Limits Evaluate the limit, if it exists. Use the Limit Laws when possible.

9. $\lim_{x \to 2} \dfrac{x+1}{x-3}$

10. $\lim_{t \to 1} (t^3 - 3t + 6)$

11. $\lim_{x \to 3} \dfrac{x^2 + x - 12}{x - 3}$

12. $\lim_{x \to -2} \dfrac{x^2 - 4}{x^2 + x - 2}$

13. $\lim_{u \to 0} \dfrac{(u+1)^2 - 1}{u}$

14. $\lim_{z \to 9} \dfrac{\sqrt{z} - 3}{z - 9}$

15. $\lim_{x \to 3^-} \dfrac{x-3}{|x-3|}$

16. $\lim_{x \to 0} \left(\dfrac{1}{x} + \dfrac{2}{x^2 - 2x} \right)$

17. $\lim_{x \to \infty} \dfrac{2x}{x - 4}$

18. $\lim_{x \to \infty} \dfrac{x^2 + 1}{x^4 - 3x + 6}$

19. $\lim_{x \to \infty} \cos^2 x$

20. $\lim_{t \to -\infty} \dfrac{t^4}{t^3 - 1}$

21–24 ■ Derivative of a Function Find the derivative of the function at the given number.

21. $f(x) = 3x - 5$, at 4

22. $g(x) = 2x^2 - 1$, at -1

23. $f(x) = \sqrt{x}$, at 16

24. $f(x) = \dfrac{x}{x+1}$, at 1

25–28 ■ Evaluating Derivatives (a) Find $f'(a)$. (b) Find $f'(2)$ and $f'(-2)$.

25. $f(x) = 6 - 2x$

26. $f(x) = x^2 - 3x$

27. $f(x) = \sqrt{x + 6}$

28. $f(x) = \dfrac{4}{x}$

29–30 ■ Equation of a Tangent Line Find an equation of the tangent line shown in the figure.

29.

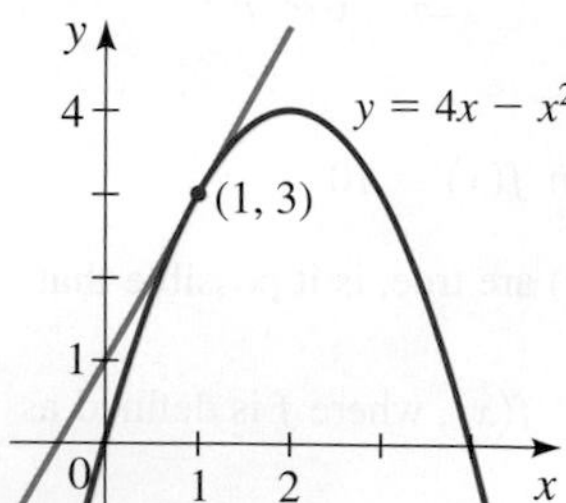

30.

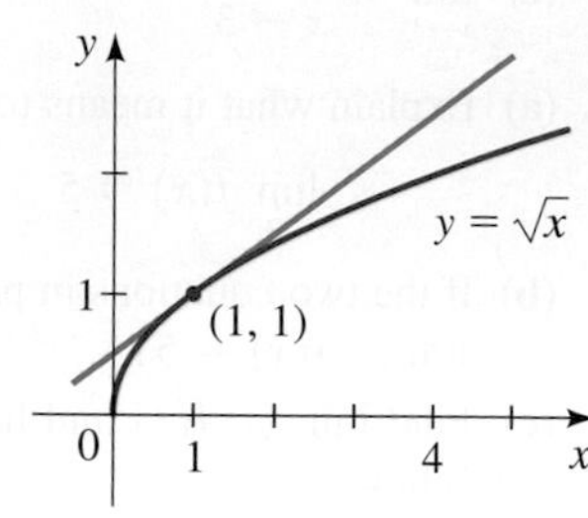

31–34 ■ Equation of a Tangent Line Find an equation of the line tangent to the graph of f at the given point.

31. $f(x) = 2x$, at $(3, 6)$

32. $f(x) = x^2 - 3$, at $(2, 1)$

33. $f(x) = \dfrac{1}{x}$, at $\left(2, \dfrac{1}{2}\right)$

34. $f(x) = \sqrt{x + 1}$, at $(3, 2)$

35. Velocity of a Dropped Stone A stone is dropped from the roof of a building 200 m above the ground. The height of the stone (in m) after t seconds is given by $h(t) = 200 - 5t^2$.

(a) Find the velocity of the stone when $t = 2$.

(b) Find the velocity of the stone when $t = a$.

(c) At what time t will the stone hit the ground?

(d) With what velocity will the stone hit the ground?

36. Instantaneous Rate of Change If a gas is confined in a fixed volume, then according to Boyle's Law the product of the pressure P and the temperature T is a constant. For a certain gas, $PT = 700$, where P is measured in kPa and T is measured in kelvins (K).

(a) Express P as a function of T.

(b) Find the instantaneous rate of change of P with respect to T when $T = 300$ K.

37–42 ■ Limit of a Sequence If the sequence is convergent, find its limit. If it is divergent, explain why.

37. $a_n = \dfrac{n}{5n + 1}$

38. $a_n = \dfrac{n^3}{n^3 + 1}$

39. $a_n = \dfrac{n(n + 1)}{2n^2}$

40. $a_n = \dfrac{n^3}{2n + 6}$

41. $a_n = \cos\left(\dfrac{n\pi}{2}\right)$

42. $a_n = \dfrac{10}{3^n}$

43–44 ■ Estimating Areas Using Rectangles Approximate the area of the shaded region under the graph of the given function by using the indicated rectangles. (The rectangles have equal width.)

43. $f(x) = \sqrt{x}$

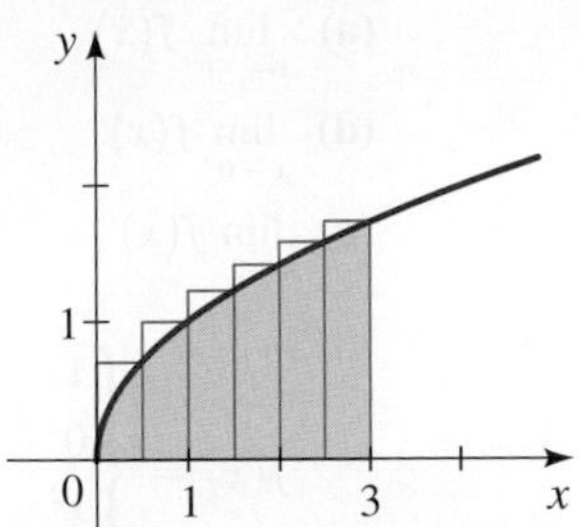

44. $f(x) = 4x - x^2$

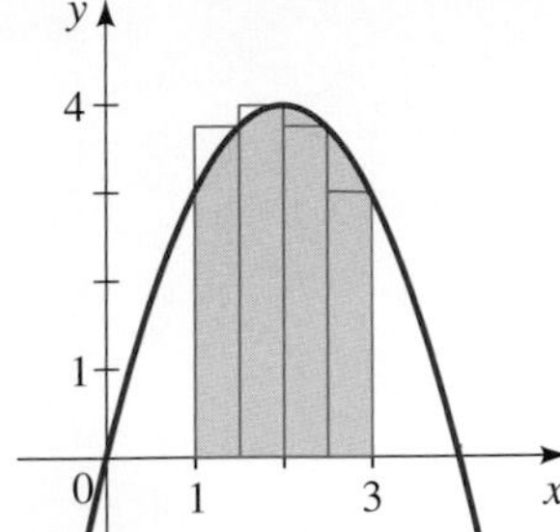

45–48 ■ Area Under a Curve Use the limit definition of area to find the area of the region that lies under the graph of f over the given interval.

45. $f(x) = 2x + 3, \quad 0 \le x \le 2$

46. $f(x) = x^2 + 1, \quad 0 \le x \le 3$

47. $f(x) = x^2 - x, \quad 1 \le x \le 2$

48. $f(x) = x^3, \quad 1 \le x \le 2$

Chapter 12 | Test

1. (a) Use a table of values to estimate the limit

$$\lim_{x \to 0} \frac{x}{\sin 2x}$$

(b) Use a graphing device to confirm your answer graphically.

2. For the piecewise-defined function f whose graph is shown, find:

(a) $\lim_{x \to -1^-} f(x)$ **(b)** $\lim_{x \to -1^+} f(x)$ **(c)** $\lim_{x \to -1} f(x)$

(d) $\lim_{x \to 0^-} f(x)$ **(e)** $\lim_{x \to 0^+} f(x)$ **(f)** $\lim_{x \to 0} f(x)$

(g) $\lim_{x \to 2^-} f(x)$ **(h)** $\lim_{x \to 2^+} f(x)$ **(i)** $\lim_{x \to 2} f(x)$

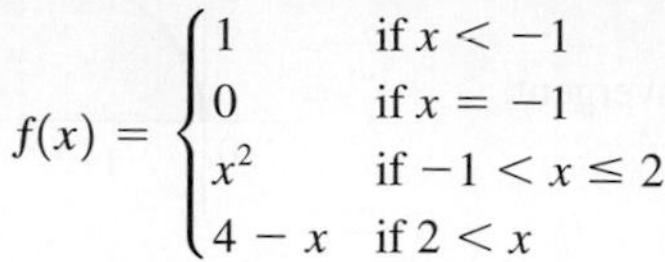

$$f(x) = \begin{cases} 1 & \text{if } x < -1 \\ 0 & \text{if } x = -1 \\ x^2 & \text{if } -1 < x \le 2 \\ 4 - x & \text{if } 2 < x \end{cases}$$

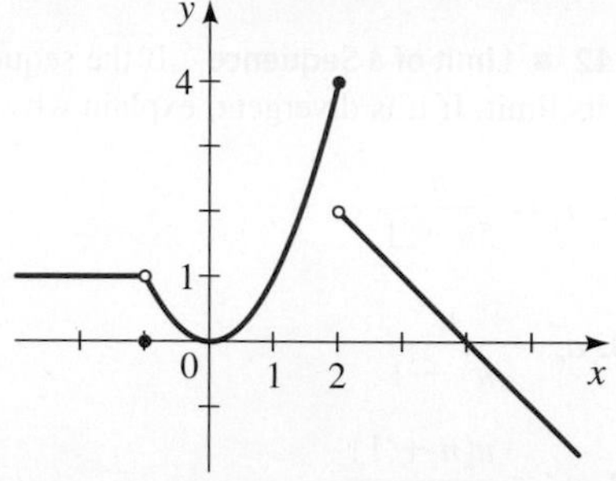

3. Evaluate the limit, if it exists.

(a) $\lim_{x \to 2} \frac{x^2 + 2x - 8}{x - 2}$ **(b)** $\lim_{x \to 2} \frac{x^2 - 2x - 8}{x + 2}$ **(c)** $\lim_{x \to 2} \frac{1}{x - 2}$

(d) $\lim_{x \to 2} \frac{x - 2}{|x - 2|}$ **(e)** $\lim_{x \to 4} \frac{\sqrt{x} - 2}{x - 4}$ **(f)** $\lim_{x \to \infty} \frac{2x^2 - 4}{x^2 + x}$

4. Let $f(x) = x^2 - 2x$. Find:

(a) $f'(a)$ **(b)** $f'(-1), f'(1), f'(2)$

5. Find the equation of the line tangent to the graph of $f(x) = \sqrt{x}$ at the point where $x = 9$.

6. Find the limit of the sequence.

(a) $a_n = \frac{n}{n^2 + 4}$ **(b)** $a_n = \sec n\pi$

7. The region sketched in the figure in the margin lies under the graph of $f(x) = 4 - x^2$, above the interval $0 \le x \le 1$.

(a) Approximate the area of the region with five rectangles, equally spaced along the x-axis, using right endpoints to determine the heights of the rectangles.

(b) Use the limit definition of area to find the exact value of the area of the region.

y
4
y = 4 − x²
1
0
1
x

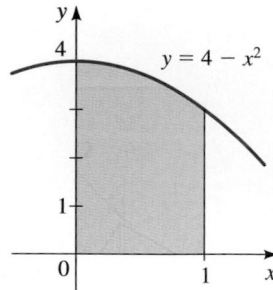

A Cumulative Review Test for Chapters 11 and 12 can be found at the book companion website stewartmath.com.

Focus on Modeling | Interpretations of Area

The area under the graph of a function is used to model many quantities in physics, economics, engineering, and other fields. That is why the area problem is so important. Here, we will show how the concept of work (see Section 8.6) is modeled by area. Several other applications are explored in the problems.

Recall that the work W done in moving an object is the product of the force F applied to the object and the distance d that the object moves:

$$W = Fd \qquad \text{work} = \text{force} \times \text{distance}$$

This formula is used if the force is *constant*. For example, suppose you are pushing a crate across a floor, moving along the positive x-axis from $x = a$ to $x = b$, and you apply a constant force $F = k$. The graph of F as a function of the distance x is shown in Figure 1(a). Notice that the work done is $W = Fd = k(b - a)$, which is the area under the graph of F. [See Figure 1(b).]

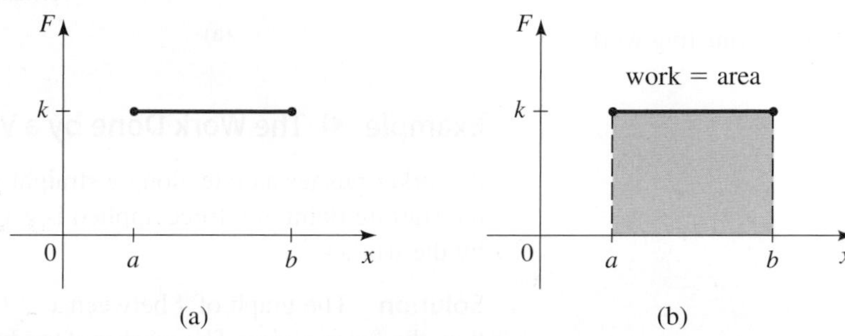

Figure 1 | A constant force F

But what if the force is *not* constant? For example, suppose the force you apply to the crate varies with distance (you push harder at certain places than you do at others). More precisely, suppose that you push the crate along the x-axis in the positive direction, from $x = a$ to $x = b$, and at each point x between a and b you apply a force $f(x)$ to the crate. Figure 2 shows a graph of the force f as a function of the distance x.

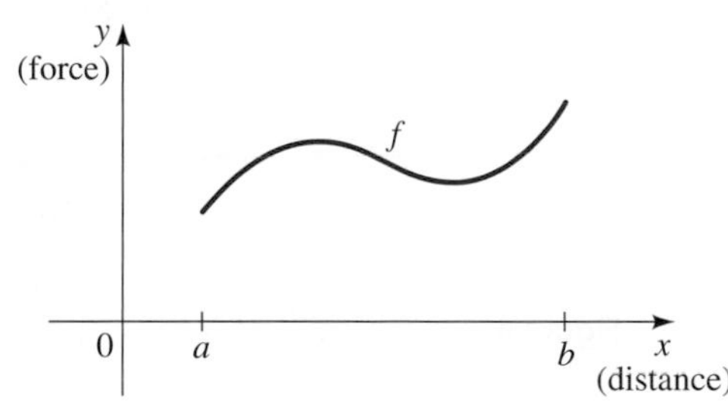

Figure 2 | A variable force

How much work was done? We can't apply the formula for work directly because the force is not constant. So let's divide the interval $[a, b]$ into n subintervals with endpoints $x_0, x_1, \ldots, x_n$ and equal width Δx, as shown in Figure 3(a) on the next page. The force at the right endpoint of the interval $[x_{k-1}, x_k]$ is $f(x_k)$. If n is large, then Δx is small, so the values of f don't change very much over the interval $[x_{k-1}, x_k]$. In other words f is almost constant on the interval, so the work W_k that is done in moving the crate from x_{k-1} to x_k is approximately

$$W_k \approx f(x_k)\,\Delta x$$

Thus we can approximate the work done in moving the crate from $x = a$ to $x = b$ by

$$W \approx \sum_{k=1}^{n} f(x_k)\,\Delta x$$

It seems that this approximation becomes better as we make n larger (and so make the interval $[x_{k-1}, x_k]$ smaller). Therefore we define the work done in moving an object from a to b as the limit of this quantity as $n \to \infty$:

$$W = \lim_{n\to\infty} \sum_{k=1}^{n} f(x_k)\,\Delta x$$

Notice that this is precisely the area under the graph of f between $x = a$ and $x = b$ as defined in Section 12.5. [See Figure 3(b).]

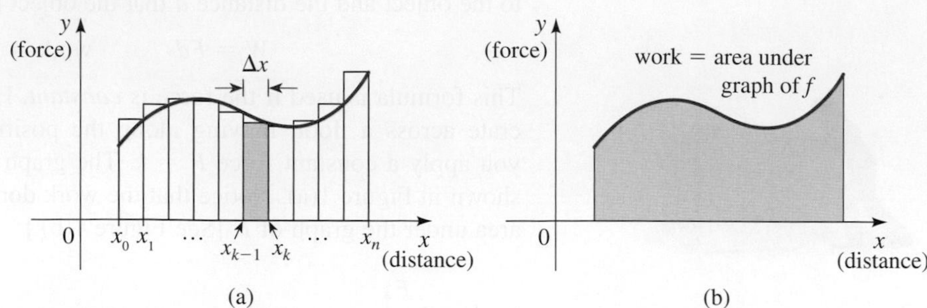

Figure 3 | Approximating work

Example ■ The Work Done by a Variable Force

A worker pushes a crate along a straight path a distance of 18 m. At a distance x from the starting point, the force applied is given by $f(x) = 340 - x^2$. Find the work done by the worker.

Solution The graph of f between $x = 0$ and $x = 18$ is shown in Figure 4. Notice how the force varies: The worker starts by pushing with a force of 340 N but steadily applies less force.

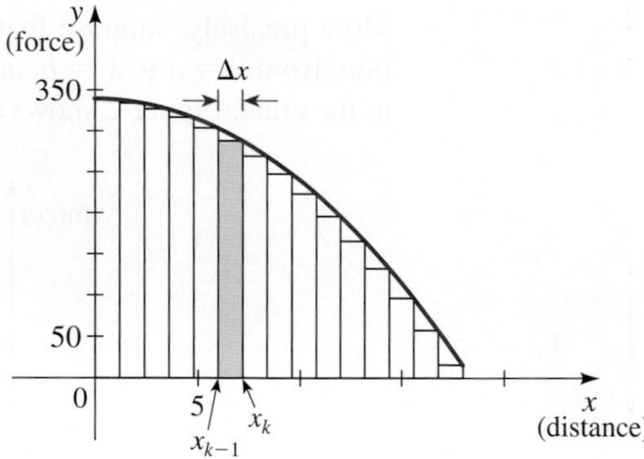

Figure 4

The work done is the area under the graph of f on the interval $[0, 18]$. To find this area, we start by finding the dimensions of the approximating rectangles at the nth stage.

Width: $\Delta x = \dfrac{b-a}{n} = \dfrac{18-0}{n} = \dfrac{18}{n}$

Right endpoint: $x_k = a + k\,\Delta x = 0 + k\left(\dfrac{18}{n}\right) = \dfrac{18k}{n}$

Height: $f(x_k) = f\left(\dfrac{18k}{n}\right) = 340 - \left(\dfrac{18k}{n}\right)^2$

$$= 340 - \frac{324k^2}{n^2}$$

Thus according to the definition of work, we get

$$W = \lim_{n\to\infty} \sum_{k=1}^{n} f(x_k)\,\Delta x = \lim_{n\to\infty} \sum_{k=1}^{n} \left(340 - \frac{324k^2}{n^2}\right)\left(\frac{18}{n}\right)$$

$$= \lim_{n\to\infty} \left(\frac{18}{n} \sum_{k=1}^{n} 340 - \frac{(18)(324)}{n^3} \sum_{k=1}^{n} k^2\right)$$

$$= \lim_{n\to\infty} \left(\frac{18}{n} 340n - \frac{5832}{n^3}\left[\frac{n(n+1)(2n+1)}{6}\right]\right)$$

$$= \lim_{n\to\infty} \left(6120 - 972 \cdot \frac{n}{n} \cdot \frac{n+1}{n} \cdot \frac{2n+1}{n}\right)$$

$$= 6120 - 972 \cdot 1 \cdot 1 \cdot 2 = 4176$$

So the work done by the worker in moving the crate is 4176 N-m. ■

Problems

1. **Work Done by a Winch** A motorized winch is being used to pull a felled tree to a logging truck. The motor exerts a force of $f(x) = 1500 + 10x - \frac{1}{2}x^2$ N on the tree at the instant when the tree has moved x m. The tree must be moved a distance of 40 m, from $x = 0$ to $x = 40$. How much work is done by the winch in moving the tree?

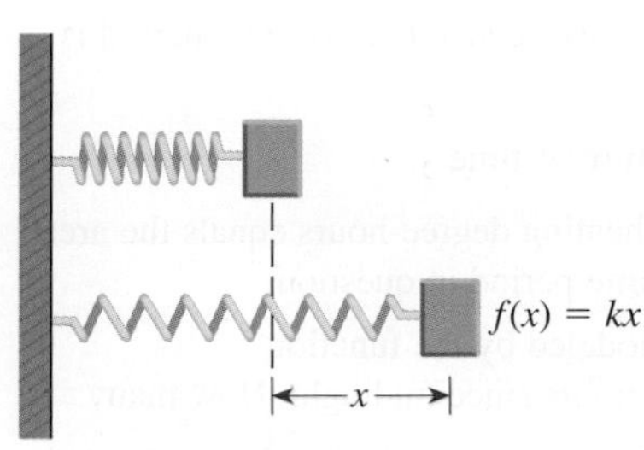

2. **Work Done by a Spring** Hooke's law states that when a spring is stretched, it pulls back with a force proportional to the amount of the stretch. The constant of proportionality is a characteristic of the spring known as the **spring constant**. Thus a spring with spring constant k exerts a force $f(x) = kx$ when it is stretched a distance x.

   A certain spring has spring constant $k = 20$ N/m. Find the work done when the spring is pulled so that the amount by which it is stretched increases from $x = 0$ to $x = 2$ m.

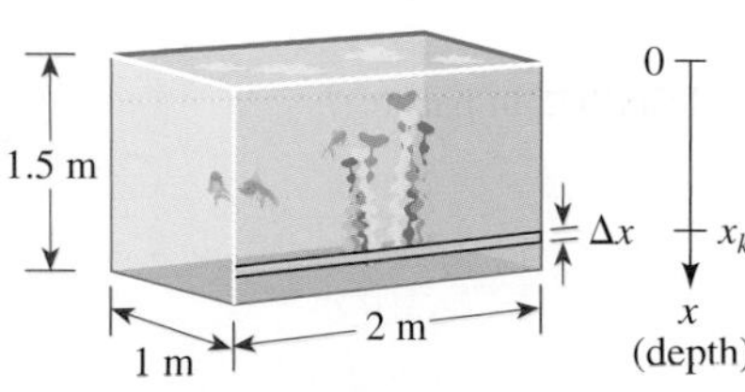

3. **Force of Water** As any diver knows, an object submerged in water experiences pressure, and as depth increases, so does the water pressure. At a depth of x m, the water pressure is $p(x) = 2992.5x$ Pa. To find the force exerted by the water on a surface, we multiply the pressure by the area of the surface:

$$\text{force} = \text{pressure} \times \text{area}$$

   Suppose an aquarium that is 1 m wide, 2 m long, and 1.5 m high is full of water. The bottom of the aquarium has area $1 \times 2 = 2$ m^2, and it experiences water pressure of $p(1.5) = 2992.5 \times 1.5 = 4489$ Pa. Thus the total force exerted by the water on the bottom is $4489 \times 2 = 8978$ N.

   The water also exerts a force on the sides of the aquarium, but this is not as easy to calculate because the pressure increases from top to bottom. To calculate the force on one of the 1.5 m by 2 m sides, we divide its area into n thin horizontal strips of width Δx, as shown in the figure. The area of each strip is

$$\text{length} \times \text{width} = 2\,\Delta x$$

   If the bottom of the kth strip is at the depth x_k, then it experiences water pressure of approximately $p(x_k) = 2992.5$ Pa—the thinner the strip, the more accurate the approximation. Thus on each strip, the water exerts a force of

$$\text{pressure} \times \text{area} = 2992.5x_k \times 2\,\Delta x = 5985x_k\,\Delta x \text{ N}$$

   (a) Explain why the total force exerted by the water on the 1.5 m by 2 m sides of the aquarium is

$$\lim_{n\to\infty} \sum_{k=1}^{n} 5985x_k\,\Delta x$$

   where $\Delta x = 1.5/n$ and $x_k = 1.5k/n$.

(b) What area does the limit in part (a) represent?

(c) Evaluate the limit in part (a) to find the force exerted by the water on one of the 1.5 m by 2 m sides of the aquarium.

(d) Use the same technique to find the force exerted by the water on one of the 1.5 m by 1 m sides of the aquarium.

[*Note:* Engineers use the technique outlined in this problem to find the total force exerted on a dam by the water in the reservoir behind the dam.]

4. Distance Traveled by a Car Since distance = speed × time, a car moving, say, at a constant speed of 70 km/h for 5 hours will travel a distance of 350 km. But what if the speed varies, as it usually does in practice?

(a) Suppose the speed of a moving object at time t is $v(t)$. Explain why the distance traveled by the object between times $t = a$ and $t = b$ is the area under the graph of v between $t = a$ and $t = b$.

(b) The speed of a car t seconds after it starts moving is given by the function

$$v(t) = 6t + 0.1t^3 \text{ m/s}$$

Find the distance traveled by the car from $t = 0$ to $t = 5$ seconds.

5. Heating Capacity If the outdoor temperature reaches a maximum of 32°C one day and only 27°C the next, then we would probably say that the first day was hotter than the second. Suppose, however, that on the first day the temperature was below 15°C for most of the day, reaching the high only briefly, whereas on the second day the temperature stayed above 25°C all the time. Now which day is the hotter one? To better measure how hot a particular day is, scientists use the concept of **heating degree-hour**. If the temperature is a constant D degrees for t hours, then the "heating capacity" generated over this period is Dt heating degree-hours.

$$\text{heating degree-hours} = \text{temperature} \times \text{time}$$

If the temperature is not constant, then the number of heating degree-hours equals the area under the graph of the temperature function over the time period in question.

(a) On a particular day the temperature (in °C) was modeled by the function $D(t) = 30 + \frac{3}{5}t - \frac{1}{50}t^2$, where t was measured in hours since midnight. How many heating degree-hours were experienced on this day, from $t = 0$ to $t = 24$?

(b) What was the maximum temperature on the day described in part (a)?

(c) On another day the temperature (in °C) was modeled by the function $E(t) = 25 + 3t - \frac{1}{8}t^2$. How many heating degree-hours were experienced on this day?

(d) What was the maximum temperature on the day described in part (c)?

(e) Which day was "hotter"?

Appendix A | Geometry Review

■ Congruent Triangles ■ Similar Triangles ■ The Pythagorean Theorem ■ Parallel Lines ■ Circles

In this appendix we review some concepts from geometry. Several of these involve triangles, so let's recall the names of some special types of triangles.

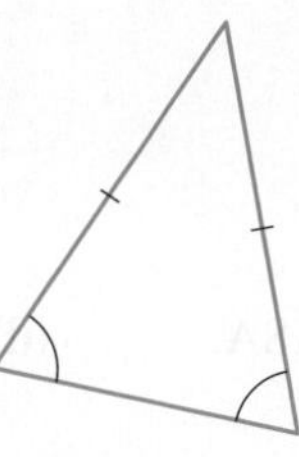

Isosceles triangle
Two sides equal

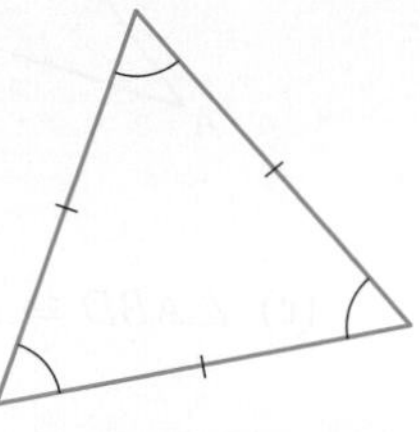

Equilateral triangle
All three sides equal

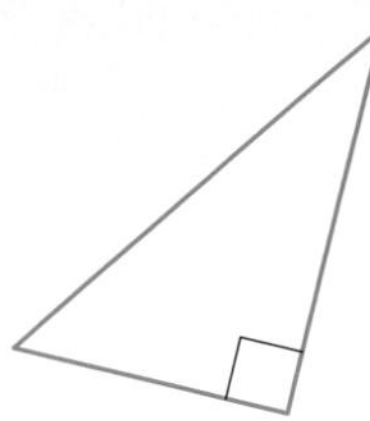

Right triangle
Has one right angle

An angle is **acute** if its measure is between 0 and 90 degrees.

In an isosceles triangle the angles opposite the equal sides are equal. In an equilateral triangle all three angles are equal to each other. In a right triangle one angle is a right angle and the other two angles are acute.

■ Congruent Triangles

Two geometric figures are congruent if they have the same shape and size. In particular, two line segments are congruent if they have the same length, and two angles are congruent if they have the same measure. For triangles we have the following definition.

Congruent Triangles

Two triangles are **congruent** if their vertices can be matched up so that corresponding sides and angles are congruent.

We write $\triangle ABC \cong \triangle PQR$ to mean that triangle ABC is congruent to triangle PQR and that the sides and angles correspond as follows.

$AB = PQ \qquad \angle A = \angle P$

$BC = QR \qquad \angle B = \angle Q$

$AC = PR \qquad \angle C = \angle R$

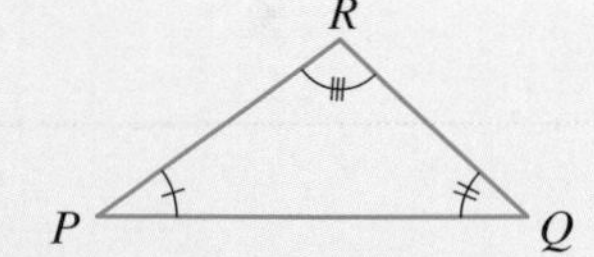

To prove that two triangles are congruent, we don't need to show that all six corresponding parts (side and angles) are congruent. For instance, if all three sides are congruent, then all three angles must also be congruent. You can verify that the following properties lead to congruent triangles.

- **Side-Side-Side (SSS).** If each side of one triangle is congruent to the corresponding side of another triangle, then the two triangles are congruent. See Figure 1(a).
- **Side-Angle-Side (SAS).** If two sides and the included angle in one triangle are congruent to the corresponding sides and angle in another triangle, then the two triangles are congruent. See Figure 1(b).
- **Angle-Side-Angle (ASA).** If two angles and the included side in one triangle are congruent to the corresponding angles and side in another triangle, then the triangles are congruent. See Figure 1(c).

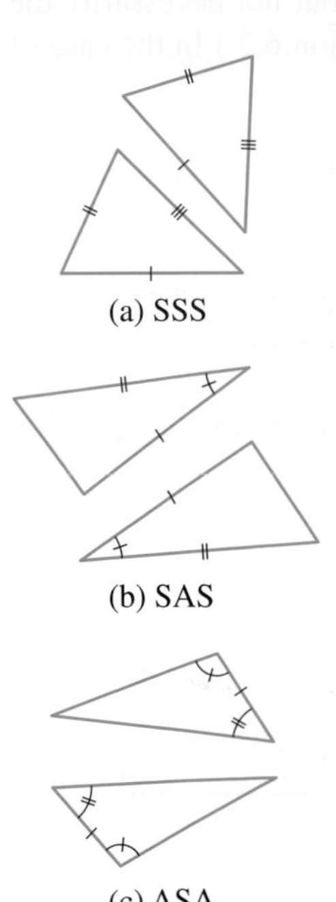

Figure 1

Example 1 ■ Congruent Triangles

(a) $\triangle ADB \cong \triangle CBD$ by SSS.

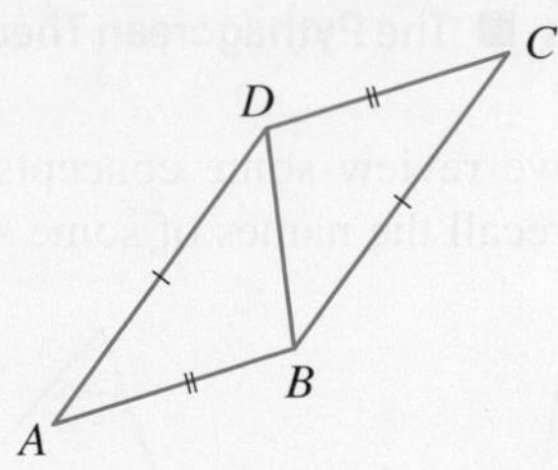

(b) $\triangle ABE \cong \triangle CBD$ by SAS.

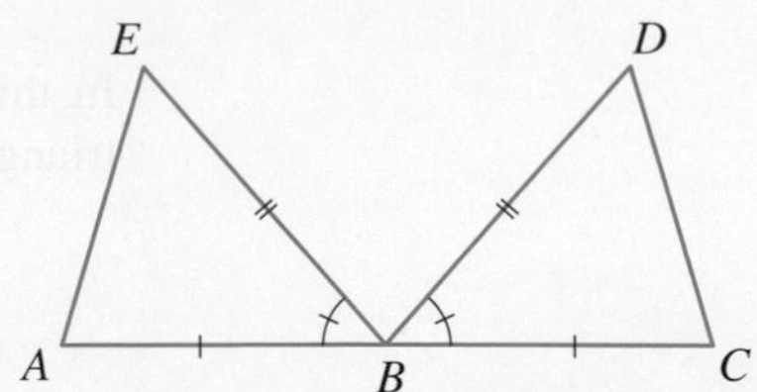

(c) $\triangle ABD \cong \triangle CBD$ by ASA.

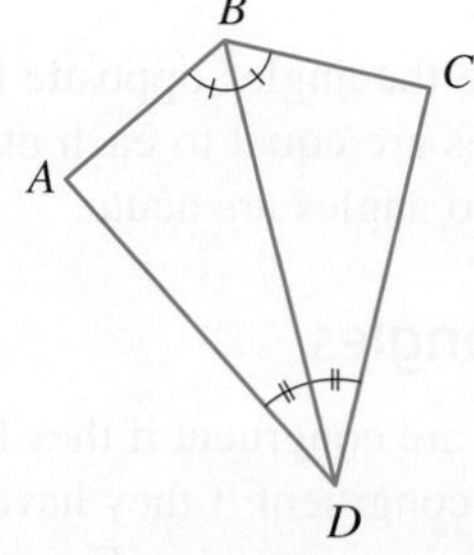

(d) These triangles are not necessarily congruent. "Side-side-angle" does *not* determine congruence.

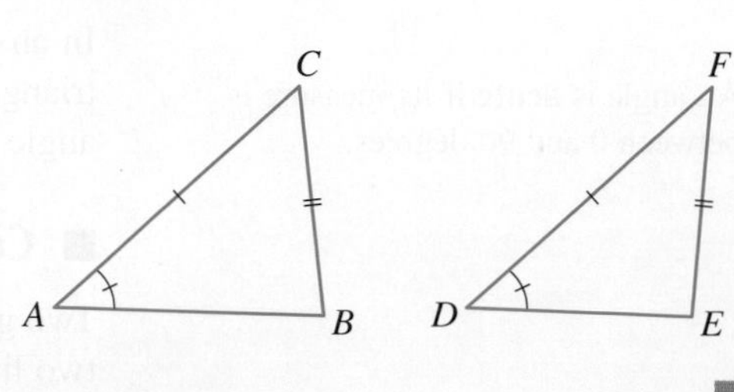

■

■ Similar Triangles

Two geometric figures are similar if they have the same shape, but not necessarily the same size. (See *Discovery Project: Similarity* referenced in Section 6.2.) In the case of triangles we can define similarity as follows.

Similar Triangles

Two triangles are **similar** if their vertices can be matched up so that corresponding angles are congruent. In this case corresponding sides are proportional.

We write $\triangle ABC \sim \triangle PQR$ to mean that triangle ABC is similar to triangle PQR and that the following conditions hold.

The angles correspond as follows:

$$\angle A = \angle P, \quad \angle B = \angle Q, \quad \angle C = \angle R$$

The sides are proportional as follows:

$$\frac{AB}{PQ} = \frac{BC}{QR} = \frac{AC}{PR}$$

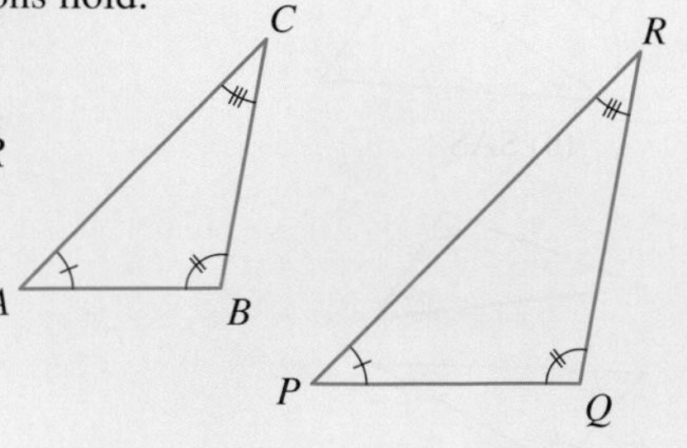

The sum of the angles in any triangle is 180°. [See Example 7(a).] So if we know two angles in a triangle, the third is determined. Thus to prove that two triangles are similar, we need only show that two angles in one triangle are congruent to two angles in the other.

Example 2 ■ Similar Triangles

Find all pairs of similar triangles in the figures.

(a)

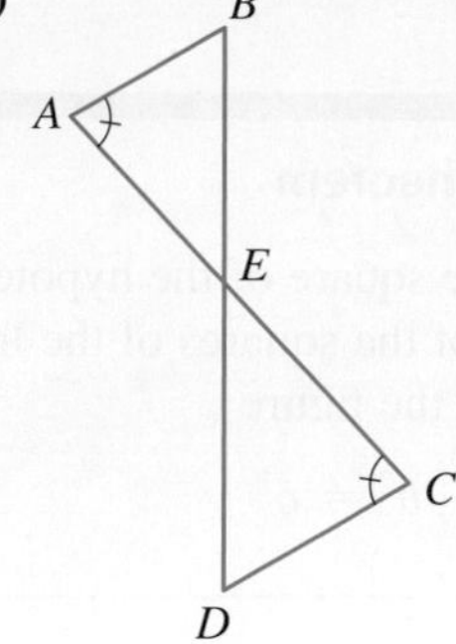

(b)

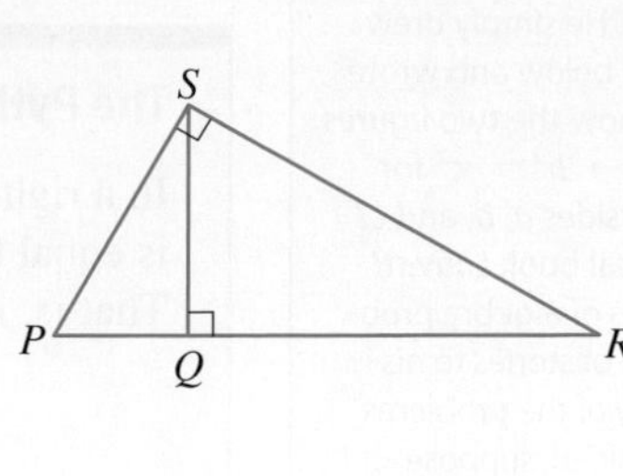

Solution

(a) Since $\angle AEB$ and $\angle CED$ are opposite angles, they are equal. Thus

$$\triangle AEB \sim \triangle CED$$

(b) Since all three triangles in the figure are right triangles, we have

$$\angle QSR + \angle QRS = 90°$$
$$\angle QSR + \angle QSP = 90°$$

Subtracting these equations, we find that $\angle QSP = \angle QRS$. Thus

$$\triangle PQS \sim \triangle SQR \sim \triangle PSR$$

■

Example 3 ■ Proportional Sides in Similar Triangles

Given that the triangles in the figure are similar, find the lengths x and y.

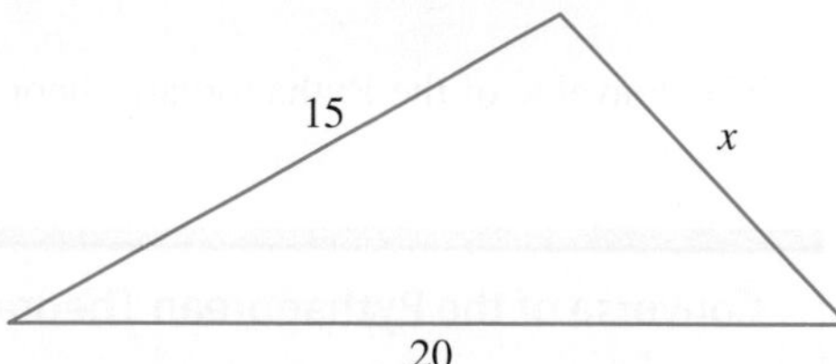

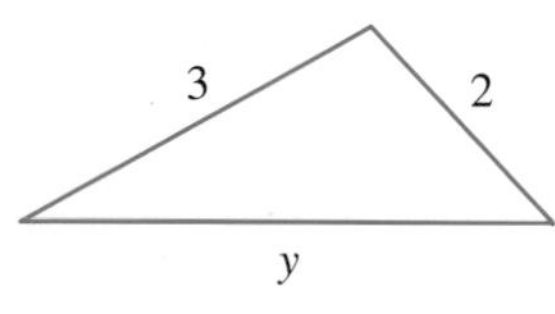

Solution By similarity, we know that the lengths of corresponding sides in the triangles are proportional. First we find x.

$$\frac{x}{2} = \frac{15}{3} \qquad \text{Corresponding sides are proportional}$$

$$x = \frac{2 \cdot 15}{3} = 10 \qquad \text{Solve for } x$$

Now we find y.

$$\frac{15}{3} = \frac{20}{y} \qquad \text{Corresponding sides are proportional}$$

$$y = \frac{20 \cdot 3}{15} = 4 \qquad \text{Solve for } y$$

■

BHASKARA (1114–1185 A.D.) was an Indian mathematician, astronomer, and astrologer. Among his many accomplishments was an ingenious proof of the Pythagorean Theorem. He simply drew the two figures shown below and wrote "Behold!" Do you see how the two figures together show that $a^2 + b^2 = c^2$ for the right triangle with sides a, b, and c? Bhaskara's mathematical book *Līlāvatī* (The Beautiful) consists of algebra problems posed in the form of stories to his daughter Lilavati. Many of the problems begin "Oh beautiful maiden, suppose . . .". The story is told that, using astrology, Bhaskara had determined that great misfortune would befall his daughter if she married at any time other than at a certain hour on a certain day. On her wedding day, as she was anxiously watching the water clock—unbeknownst to her—a pearl fell from her headdress, stopping the flow of water in the clock and causing her to miss the opportune moment for marriage. Bhaskara's *Līlāvatī* was written to console her.

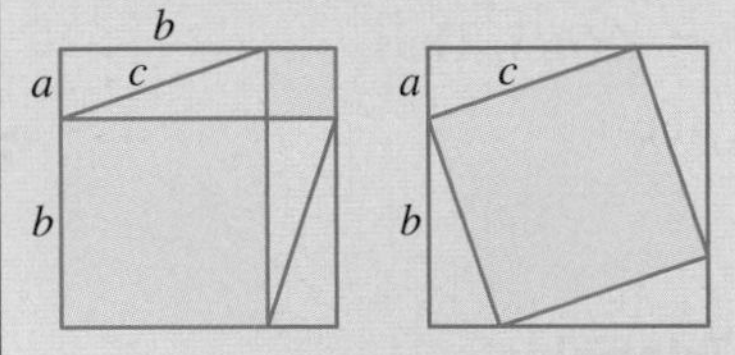

■ The Pythagorean Theorem

In a right triangle the side opposite the right angle is called the **hypotenuse**, and the other two sides are called the **legs**.

The Pythagorean Theorem

In a right triangle the square of the hypotenuse is equal to the sum of the squares of the legs. That is, in $\triangle ABC$ in the figure

$$a^2 + b^2 = c^2$$

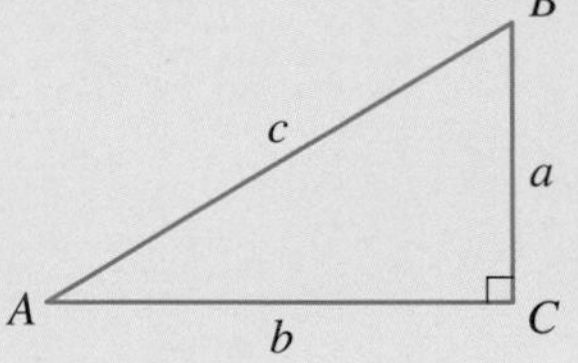

Example 4 ■ Using the Pythagorean Theorem

Find the lengths x and y in the right triangles shown.

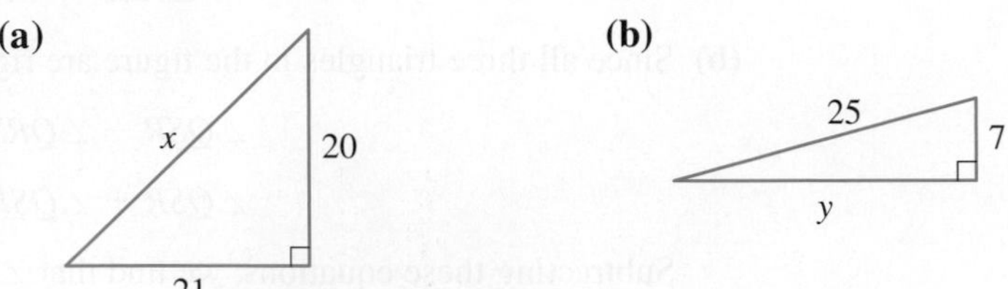

Solution

(a) We use the Pythagorean Theorem with $a = 20$, $b = 21$, and $c = x$. Then $x^2 = 20^2 + 21^2 = 841$. So $x = \sqrt{841} = 29$.

(b) We use the Pythagorean Theorem with $c = 25$, $a = 7$, and $b = y$. Then $25^2 = 7^2 + y^2$, so $y^2 = 25^2 - 7^2 = 576$. Thus $y = \sqrt{576} = 24$. ■

The converse of the Pythagorean Theorem is also true.

Converse of the Pythagorean Theorem

If the square of one side of a triangle is equal to the sum of the squares of the other two sides, then the triangle is a right triangle.

Example 5 ■ Proving That a Triangle Is a Right Triangle

Prove that the triangle with sides of length 8, 15, and 17 is a right triangle.

Solution You can check that $8^2 + 15^2 = 17^2$. So the triangle must be a right triangle by the converse of the Pythagorean Theorem. ■

■ Parallel Lines

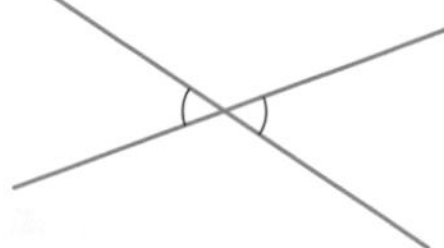

Figure 2 | Opposite angles are equal

Recall that if two lines intersect, then **opposite angles** (or **vertex angles**) formed by the lines are equal (see Figure 2). Two lines that never intersect are called *parallel*. To determine whether two lines are parallel we first draw a **transversal**—that is, a line that intersects both lines. We identify pairs of angles formed by the lines and the transversal, as shown in Figure 3.

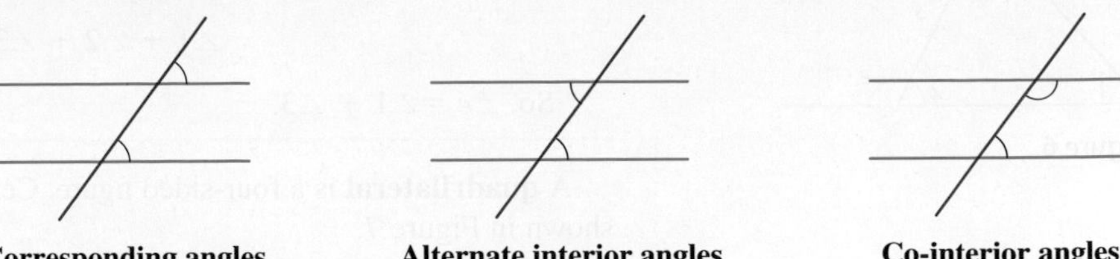

Figure 3 Corresponding angles Alternate interior angles Co-interior angles

Parallel Lines

Two lines in the plane are called **parallel** if they do not intersect. We can determine if two lines are parallel by using the following theorem:

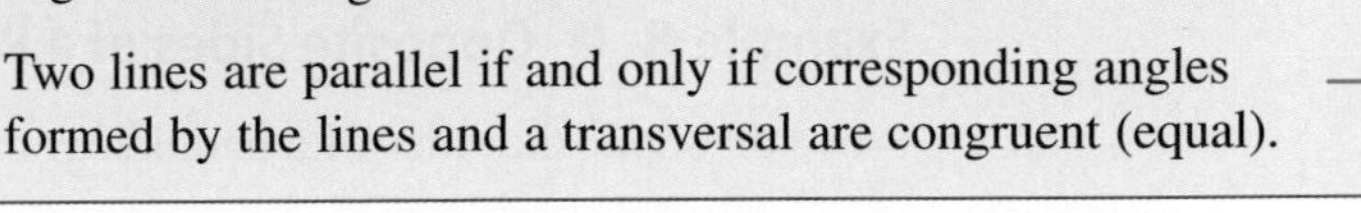

Two lines are parallel if and only if corresponding angles formed by the lines and a transversal are congruent (equal).

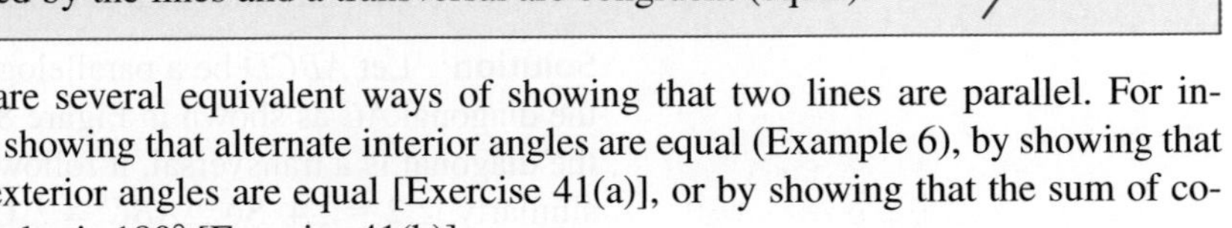

There are several equivalent ways of showing that two lines are parallel. For instance, by showing that alternate interior angles are equal (Example 6), by showing that alternate exterior angles are equal [Exercise 41(a)], or by showing that the sum of co-interior angles is 180° [Exercise 41(b)].

Example 6 ■ Parallel Lines

A transversal intersects two lines. Prove that if alternate interior angles are equal, then the lines are parallel. Also prove the converse: if the lines are parallel, then alternate interior angles are equal.

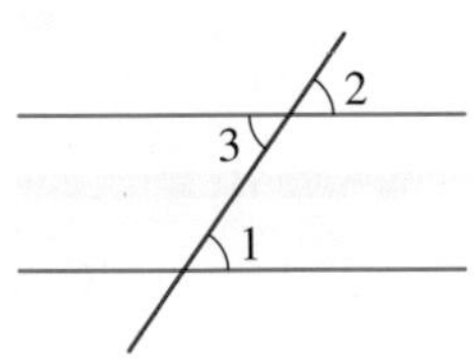

Figure 4

Solution We use the theorem about parallel lines stated in the preceding box. In Figure 4, $\angle 1$ and $\angle 2$ are corresponding angles and $\angle 1$ and $\angle 3$ are alternate interior angles.

($\Rightarrow$) If alternate interior angles are equal, then $\angle 1 = \angle 3$. But $\angle 2 = \angle 3$ because they are opposite angles. It follows that $\angle 1 = \angle 2$. Thus corresponding angles are equal and so the lines are parallel by the preceding theorem.

($\Leftarrow$) Conversely, if the lines are parallel, then by the theorem, corresponding angles are equal; that is, $\angle 1 = \angle 2$. But since $\angle 2 = \angle 3$ (opposite angles), it follows that $\angle 1 = \angle 3$. That is, alternate interior angles are equal. ■

An **exterior angle** of a triangle is an angle between a side of the triangle and an outward extended adjacent side. Part (b) in the next example is called the **Exterior Angle Theorem**.

Example 7 ■ Exterior Angle Theorem

Prove each of the following.

(a) The sum of the angles of a triangle is 180°.

(b) An exterior angle of a triangle is equal to the sum of the two opposite interior angles.

Solution

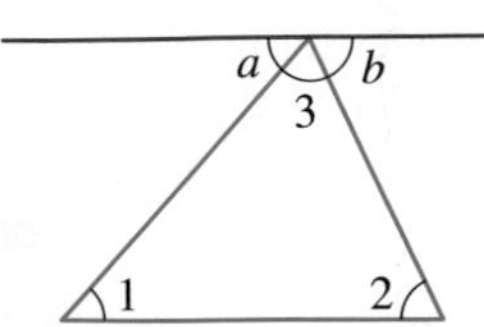

Figure 5

(a) Let's introduce a line parallel to one side of the triangle and passing through the opposite vertex, as shown in Figure 5. By Example 6, $\angle 1 = \angle a$ (alternate interior angles). Similarly, $\angle 2 = \angle b$. So, the sum of the angles of the triangle is

$$\angle 1 + \angle 2 + \angle 3 = \angle a + \angle b + \angle 3 = 180°$$

because $\angle a$, $\angle 3$, and $\angle b$ together form a straight angle (180° angle).

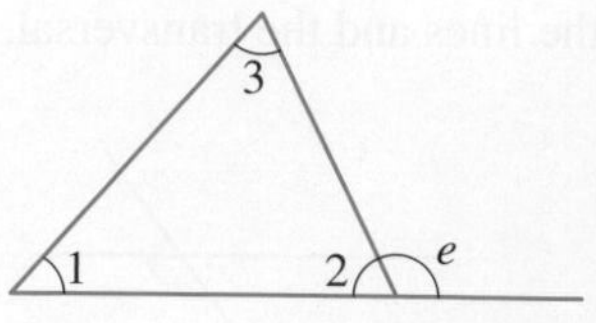

Figure 6

(b) In Figure 6, $\angle e$ is an exterior angle. By part (a), $\angle 1 + \angle 2 + \angle 3 = 180°$. Also, $\angle 2 + \angle e = 180°$ (because they form a straight angle). It follows that

$$\angle 1 + \angle 2 + \angle 3 = \angle e + \angle 2$$

So, $\angle e = \angle 1 + \angle 3$. ■

A **quadrilateral** is a four-sided figure. Certain quadrilaterals have special names, as shown in Figure 7.

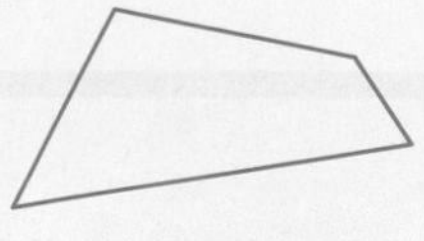
Quadrilateral
Four-sided figure

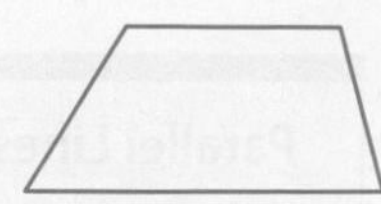
Trapezoid
One pair of parallel sides

Parallelogram
Two pairs of parallel sides

Figure 7

Example 8 ■ Opposite Sides of a Parallelogram Are Congruent

Prove that opposite sides of a parallelogram are congruent.

Solution Let $ABCD$ be a parallelogram, as sketched in Figure 8(a). Let's introduce the diagonal AC as shown in Figure 8(b). Now, since opposite sides are parallel and the diagonal is a transversal, it follows that $\angle 1 = \angle 3$ (alternate interior angles) and similarly, $\angle 2 = \angle 4$. So $\triangle ABC \cong \triangle CDA$ by ASA because the side AC is common to both triangles. Thus $AB = CD$ and $DA = BC$.

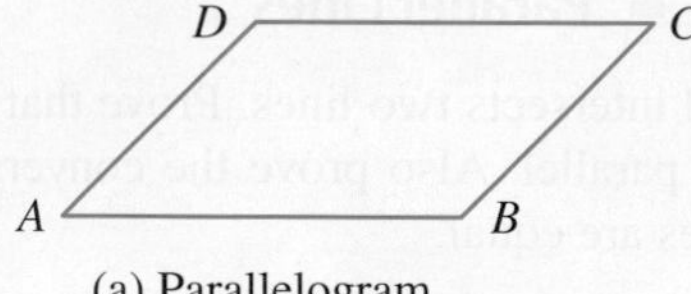

(a) Parallelogram

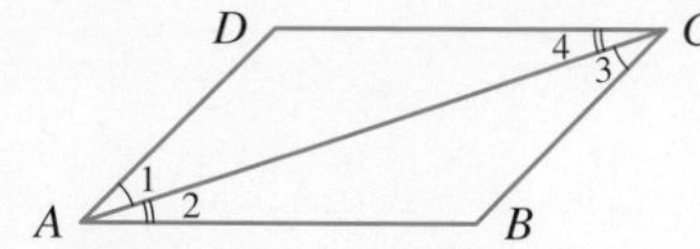

(b) Parallelogram with diagonal

Figure 8

■

Diagonals of a Parallelogram

We can use the following theorem to test whether a quadrilateral is a parallelogram.

A quadrilateral is a parallelogram if and only if the diagonals bisect each other.

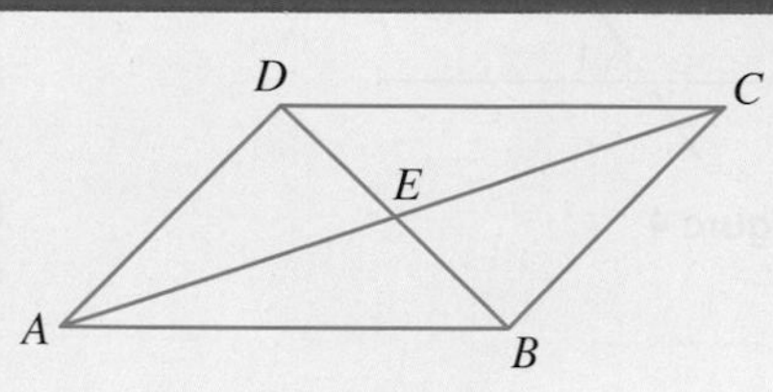

Proof In the figure, the diagonals of quadrilateral $ABCD$ intersect at E.

($\Rightarrow$) If $ABCD$ is a parallelogram, then AB is parallel to DC and the diagonals are transversals. We see that $\triangle AEB \cong \triangle CED$ by ASA because $\angle ABE = \angle CDE$ and $\angle BAE = \angle DCE$ (alternate interior angles), and $AB = CD$ by Example 8. So $AE = CE$ and $BE = DE$ (corresponding sides of congruent triangles). Thus the diagonals bisect each other.

($\Leftarrow$) Conversely, suppose that the diagonals bisect each other. Then $AE = CE$ and $BE = DE$. Also, $\angle AEB = \angle CED$ because they are opposite angles. So $\triangle AEB \cong \triangle CED$ by SAS. But then $\angle ABE = \angle CDE$ and these are alternate interior angles formed by the lines AB and CD and the transversal AC. It follows that AB is parallel to CD. By an analogous argument using triangles AED and CEB we can show that AD is parallel to CB. Thus $ABCD$ is a parallelogram. ■

■ Circles

A **chord** of a circle is a line segment whose endpoints lie on the circle, as shown in Figure 9(a). The angles shown in Figure 9(b) and 9(c) are said to be **subtended by the chord** (shown in red). The angle in Figure 9(b) is said to be subtended by the chord *at the center of the circle*. The angles in Figure 9(c), are subtended by the chord *on the circle* (or *on the circumference of the circle*).

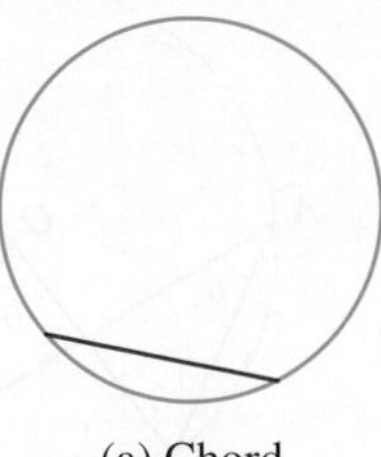

(a) Chord

(b) Angle subtended at center of circle

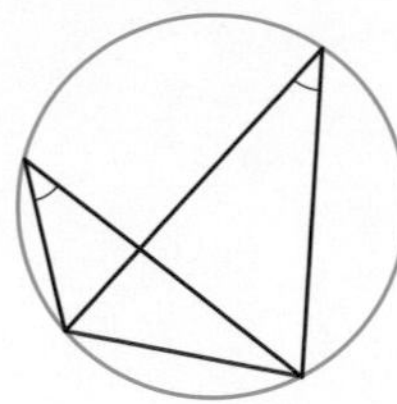

(c) Angles subtended on the circle

Figure 9

Circles and Chords

The angle at the center of a circle subtended by a chord is twice any angle on the circle subtended by the same chord.

Chords of equal length subtend equal angles.

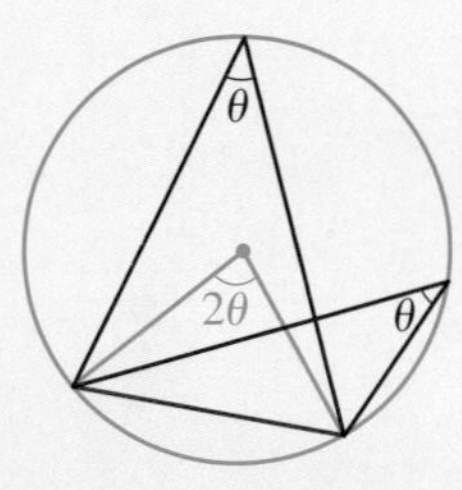

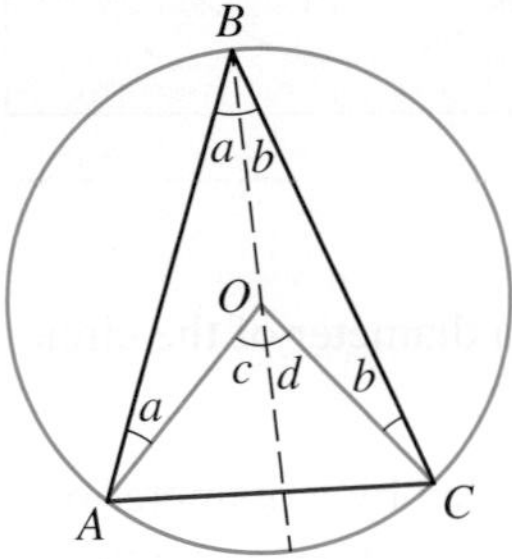

Figure 10

Proof We prove the case illustrated in Figure 10, where the center of the circle is inside $\angle ABC$. The other cases are proved similarly. Let's introduce the diameter BD to the figure. Then triangle AOB is isosceles because two sides are radii of the circle, so the base angles are equal; each is labeled $\angle a$ in the figure. By the Exterior Angle Theorem [Example 7(b)], it follows that $\angle c = \angle a + \angle a = 2\angle a$. Similarly, $\angle d = 2\angle b$. From the figure we see that

$$\begin{aligned}\angle AOC &= \angle c + \angle d \\ &= 2\angle a + 2\angle b \qquad \text{Exterior Angle Thorem} \\ &= 2\,(\angle a + \angle b) \\ &= 2\angle ABC\end{aligned}$$

Thus $\angle AOC = 2\angle ABC$ and this completes the proof. ■

When a chord is a diameter of a circle, we get the following important special case of the preceding result. A proof is outlined in Exercise 48.

Triangle Inscribed in a Semicircle

An angle subtended by a diameter of a circle is a right angle. So a triangle inscribed in a semicircle is a right triangle.

In the figure, AC is a diameter of the circle, so angles B and B' are right angles and $\triangle ABC$ and $\triangle AB'C$ are right triangles.

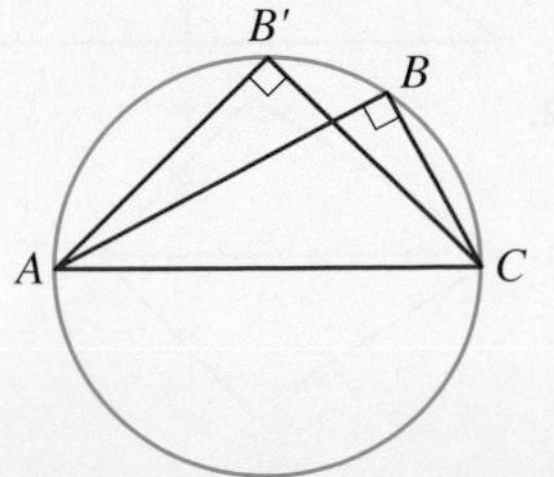

To illustrate the preceding theorems, consider the circle and chords in Figure 11. In the figure, $\angle a$ is subtended by the chord BD on the circle, whereas $\angle b$ is subtended by the same chord BD at the center of the circle. So by the theorem on Circles and Chords, $\angle b = 2\angle a$. Also, both $\angle a$ and $\angle c$ are subtended by the chord BC on the circle, so $\angle a = \angle c$. By the theorem for a Triangle Inscribed in a Semicircle, $\angle BDC$ is a right angle because it is subtended by a diameter of the circle.

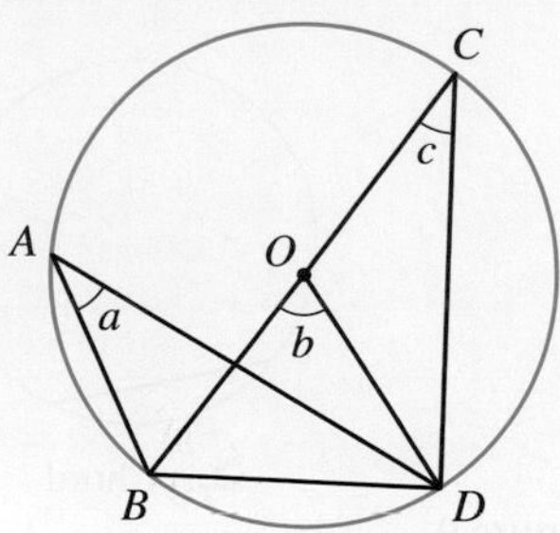

Figure 11

Intuitively, a tangent line to a curve at a given point is a line that just touches (but does not cross) the curve at that point.

Tangent Lines to Circles

A line that intersects a circle at exactly one point is said to be **tangent** to the circle at that point. The following theorem gives a key property of tangents.

A line is tangent to a circle at a point P if and only if the line is perpendicular to the radius of the circle drawn to the point P.

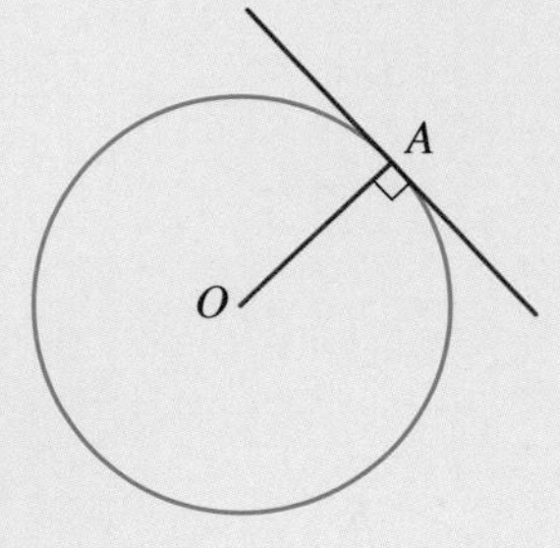

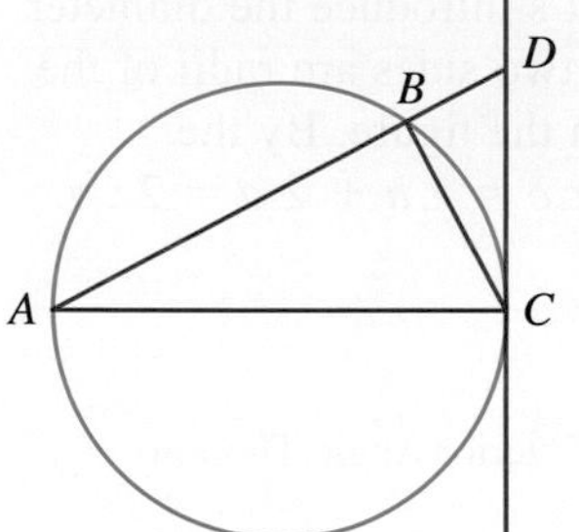

Figure 12

Example 10 ■ Tangent to a Circle

In Figure 12 the line CD is tangent to the circle at C and AC is a diameter of the circle. Show that $\triangle ACD$ is similar to $\triangle ABC$.

Solution Note that $\angle ABC$ is a right angle because it is subtended by a diameter of the circle. Also, $\angle ACD$ is a right angle because the line CD is tangent to the circle. Both triangles share $\angle BAC$. So two angles in $\triangle ACD$ are congruent to two angles in $\triangle ABC$. It follows that the triangles are similar. ■

Appendix A Exercises

1–4 ■ Congruent Triangles? Determine whether the pair of triangles is congruent. If so, state the congruence principle you are using.

1.

2.

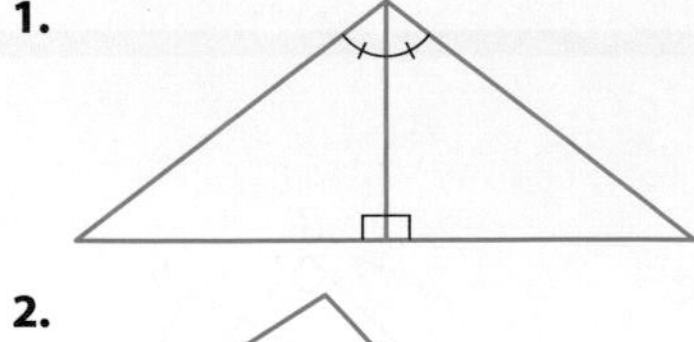

3.

4.

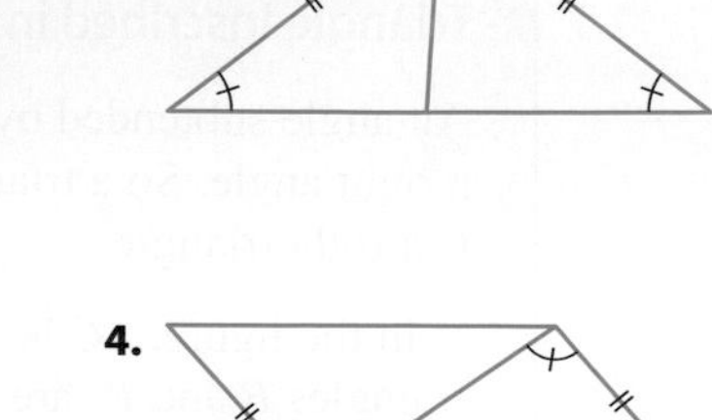

5–8 ■ Similar Triangles? Determine whether the pair of triangles is similar.

5.

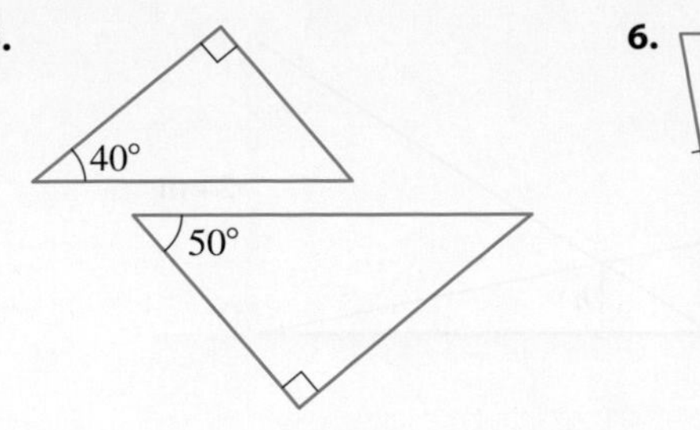

6.

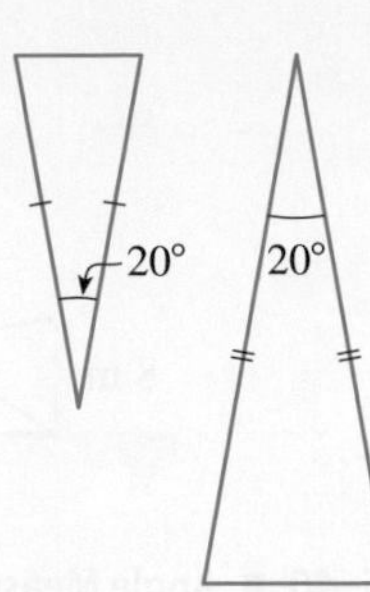

7.

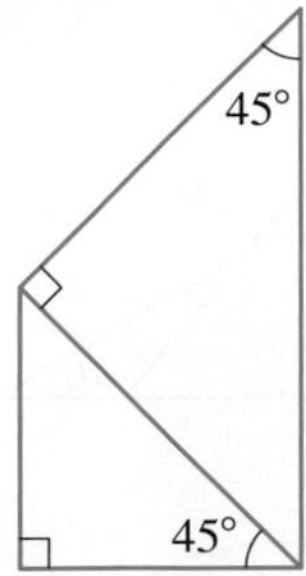

8.

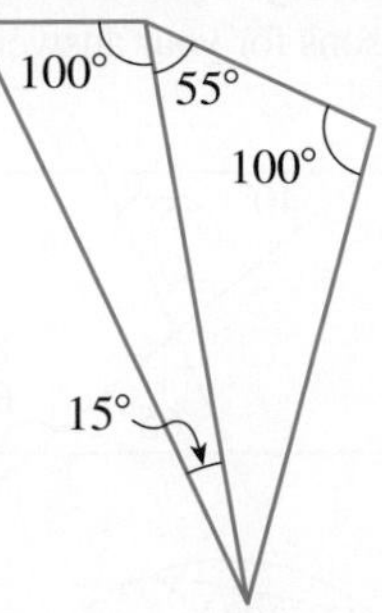

9–12 ■ Similar Triangles Given that the pair of triangles is similar, find the length(s) x and/or y.

9.

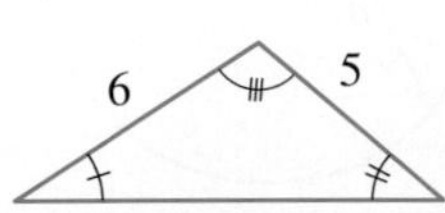

10.

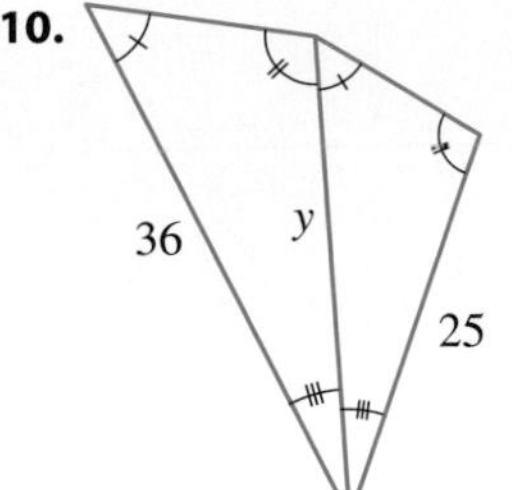

11.

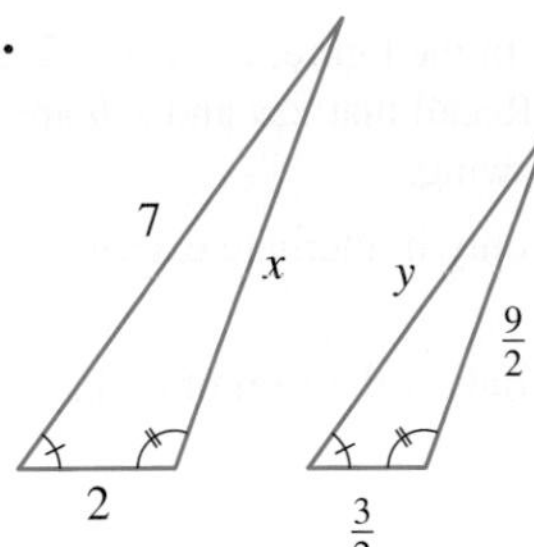

12.

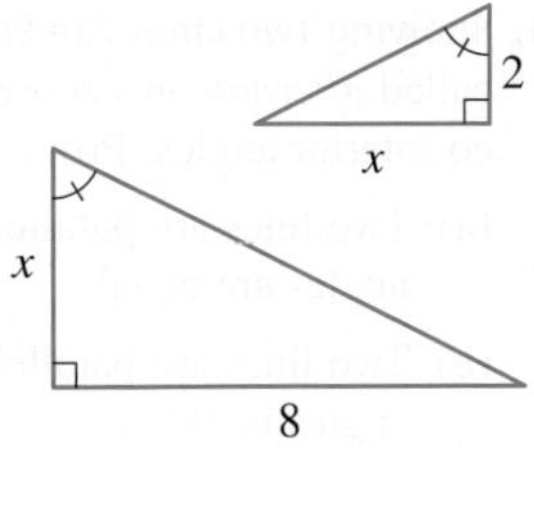

13–14 ■ Using Similarity Express x in terms of a, b, and c.

13.

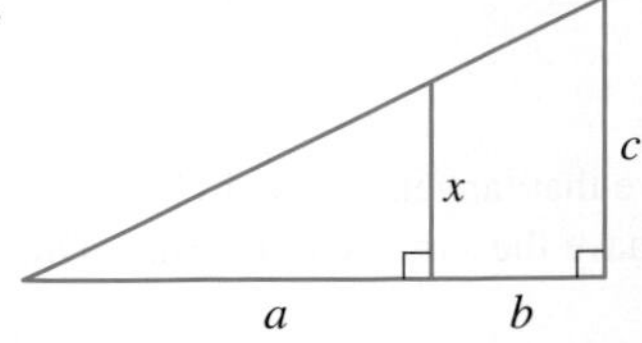

14.

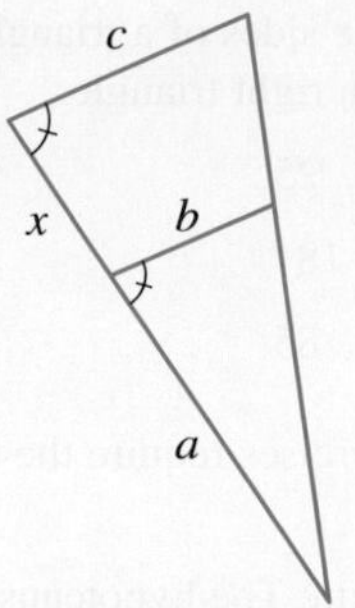

15. Proving Similarity In the figure $CDEF$ is a rectangle. Prove that $\triangle ABC \sim \triangle AED \sim \triangle EBF$.

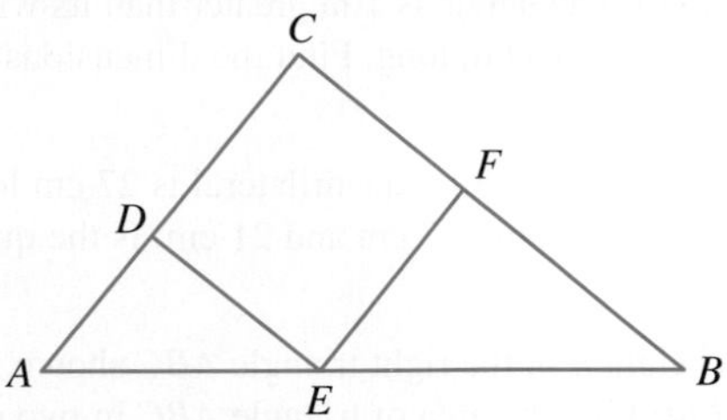

16. Proving Similarity In the figure $DEFG$ is a square. Prove the following:

(a) $\triangle ADG \sim \triangle GCF$

(b) $\triangle ADG \sim \triangle FEB$

(c) $AD \cdot EB = DG \cdot FE$

(d) $DE = \sqrt{AD \cdot EB}$

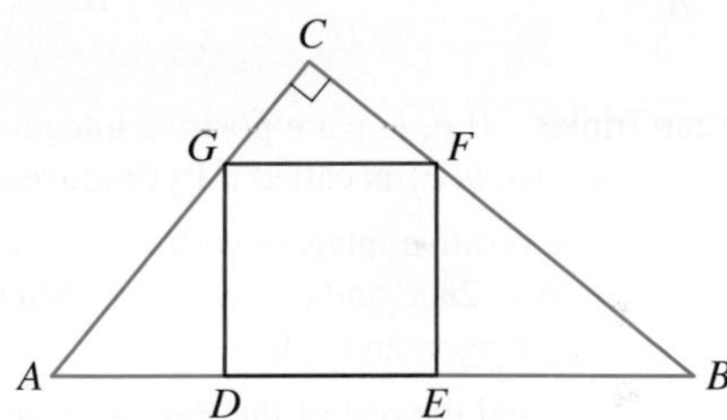

17–22 ■ Pythagorean Theorem In the given right triangle, find the side labeled x.

17.

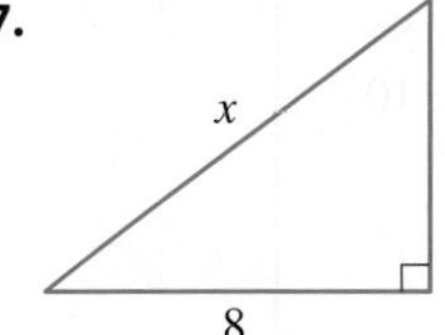

18.

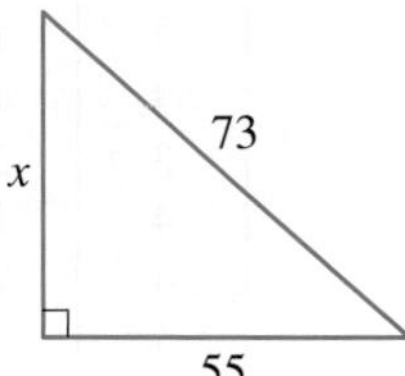

19.

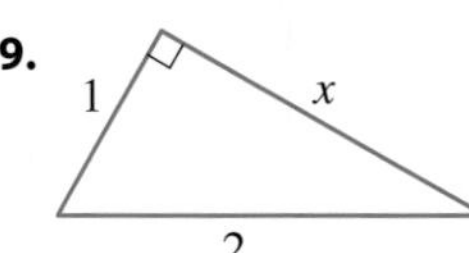

20.

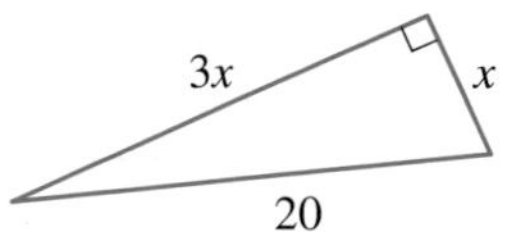

21.

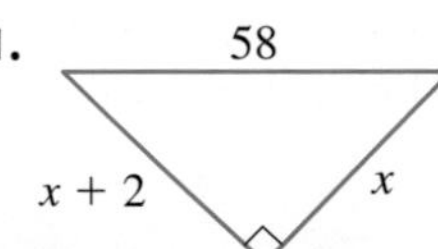

22.

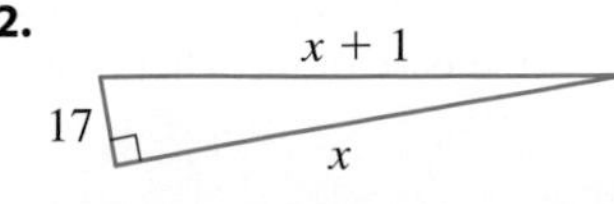

23–28 ■ Right Triangle? The lengths of the sides of a triangle are given. Determine whether the triangle is a right triangle.

23. 5, 12, 13

24. 15, 20, 25

25. 8, 10, 12

26. 6, 17, 18

27. 48, 55, 73

28. 13, 84, 85

29–32 ■ Pythagorean Theorem These exercises require the use of the Pythagorean Theorem.

29. One leg of a right triangle measures 11 cm. The hypotenuse is 1 cm longer than the other leg. Find the length of the hypotenuse.

30. The length of a rectangle is 1 m greater than its width. Each diagonal is 169 m long. Find the dimensions of the rectangle.

31. Each of the diagonals of a quadrilateral is 27 cm long. Two adjacent sides measure 17 cm and 21 cm. Is the quadrilateral a rectangle?

32. Find the height h of the right triangle ABC shown in the figure. [*Hint:* Find the area of triangle ABC in two different ways.]

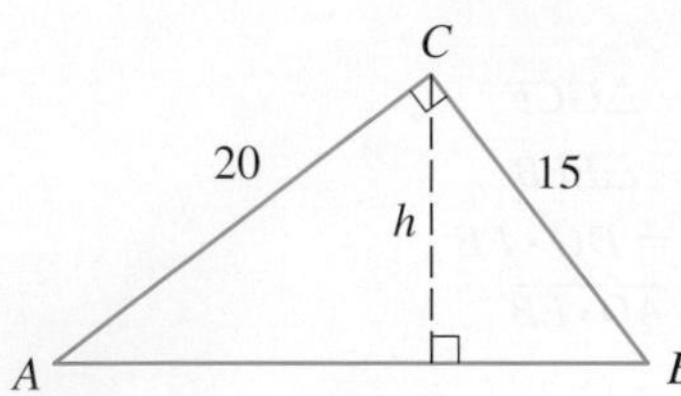

33. Pythagorean Triples If a, b, c are positive integers such that $a^2 + b^2 = c^2$, then (a, b, c) is called a **Pythagorean triple**.

(a) Let m and n be positive integers with $m > n$. Let $a = m^2 - n^2$, $b = 2mn$, and $c = m^2 + n^2$. Show that (a, b, c) is a Pythagorean triple.

(b) Use part (a) to find the rest of the Pythagorean triples given in the table.

m	n	(a, b, c)
2	1	(3, 4, 5)
3	1	(8, 6, 10)
3	2	
4	1	
4	2	
4	3	
5	1	
5	2	
5	3	
5	4	

34. Finding a Length Two vertical poles, one 8 m tall and the other 24 m tall, have ropes stretched from the top of each to the base of the other (see the figure). How high above the ground is the point where the ropes cross? [*Hint:* Use similarity.]

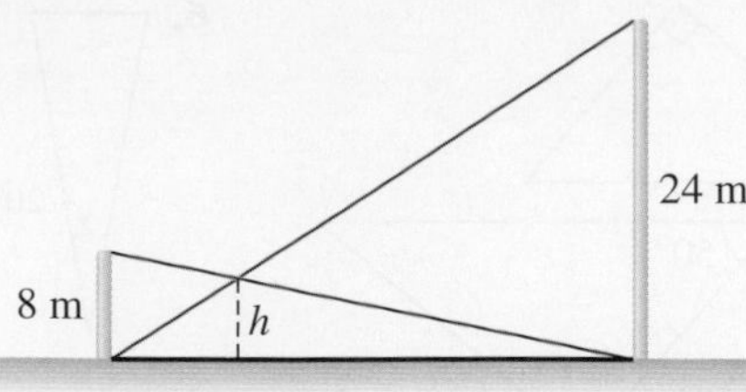

35–40 ■ Angle Measure Find the measure of the angle labeled θ. Give reasons for your answer.

35.

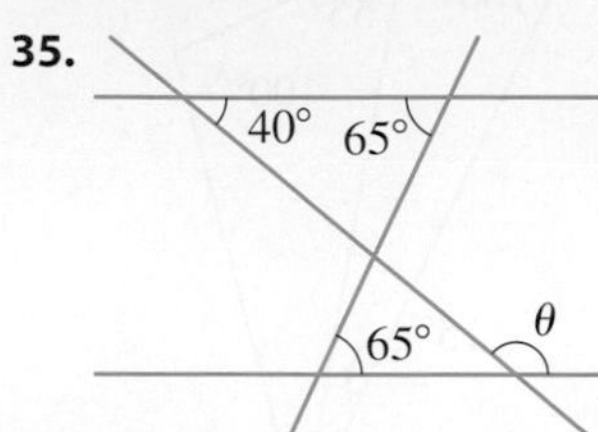

36.

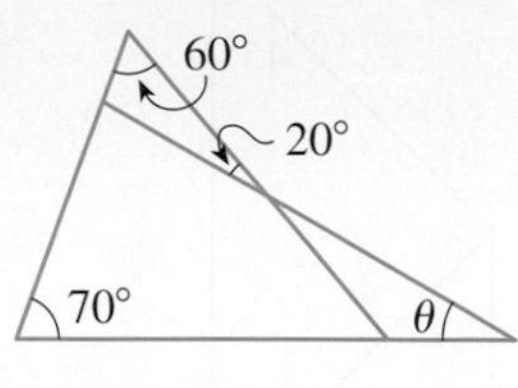

37.

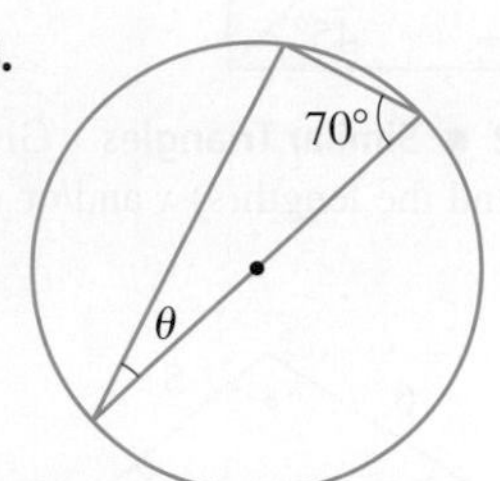

38.

39.

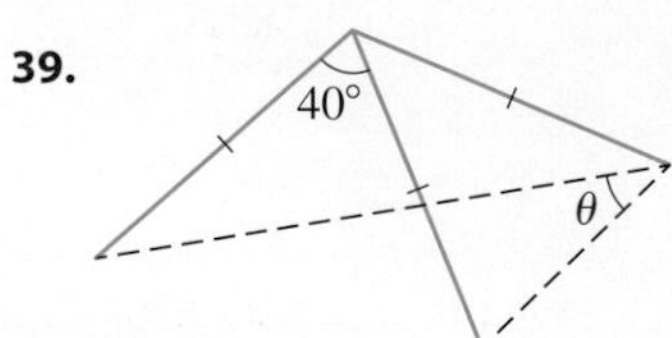

40.

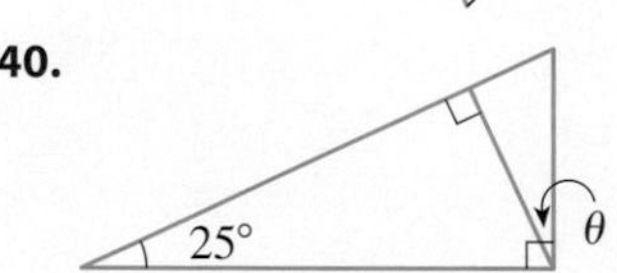

41. Showing Two Lines Are Parallel In the figure, $\angle 1$ and $\angle 2$ are called *alternate exterior angles*. Recall that $\angle a$ and $\angle b$ are co-interior angles. Prove the following.

(a) Two lines are parallel if and only if alternate exterior angles are equal.

(b) Two lines are parallel if and only if the sum of co-interior angles is 180°.

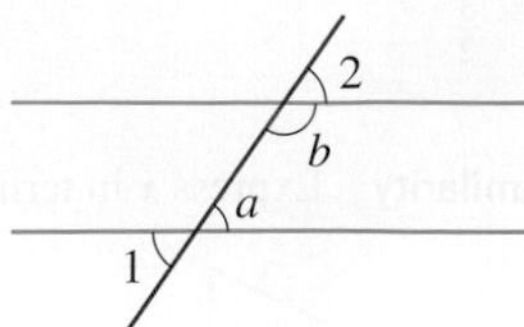

42. Tangents to a Circle Prove that tangents to a circle from a point P outside the circle have the same length. [*Hint:* Draw radii of the circle to the points of tangency and draw the line segment OP.]

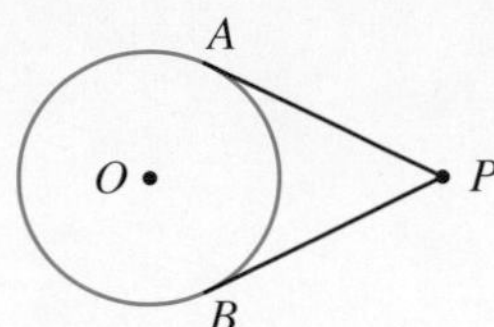

43. Diameter and Chord Prove that if a diameter of a circle bisects a chord, then it is perpendicular to the chord.

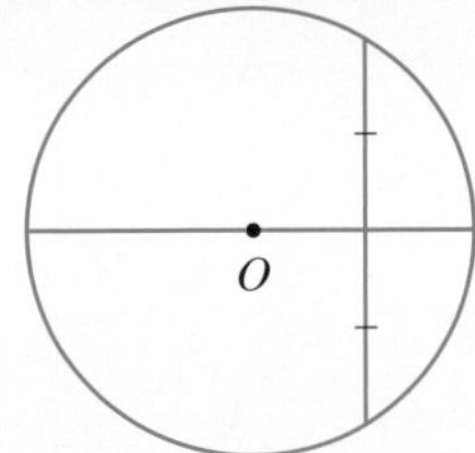

44. Altitude In the figure, triangle ABC is inscribed in a semicircle. Show that the altitude is $h = \sqrt{ab}$.

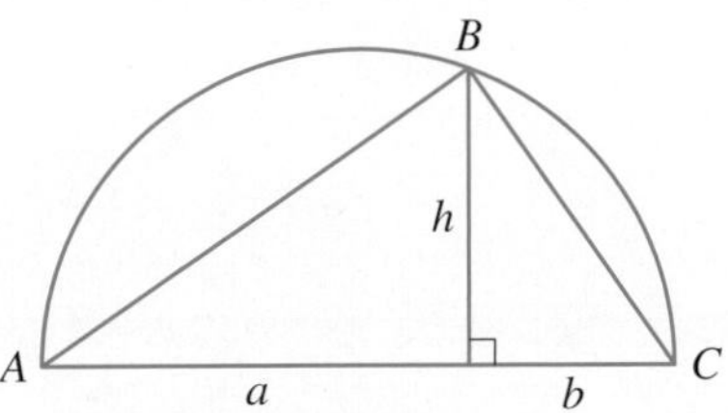

45. Cyclic Quadrilateral A quadrilateral is called *cyclic* if it can be inscribed in a circle (see the figure). Prove that if a quadrilateral is cyclic, then the sum of each pair of opposite angles is 180°. [*Hint*: Draw the diagonals and use the fact that the sum of the angles of a quadrilateral is 360°.]

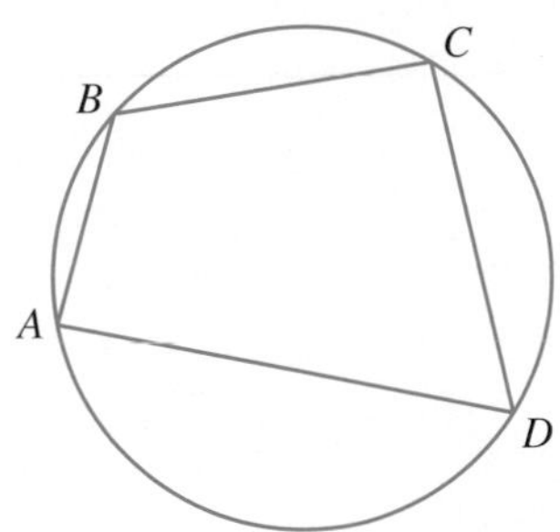

46. Chords Theorem The figure shows a circle and two intersecting chords. Show that $ab = cd$. [*Hint*: First show that $\triangle ABE \sim \triangle CDE$.]

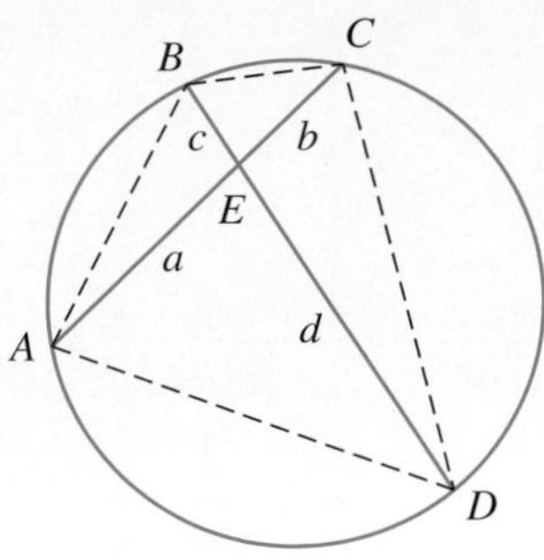

47. Tangent and Chord In the figure, the line is tangent to the circle at A. Prove that $\angle\alpha = \angle\beta$. [*Hint*: Add radii of the circle to points A and C.]

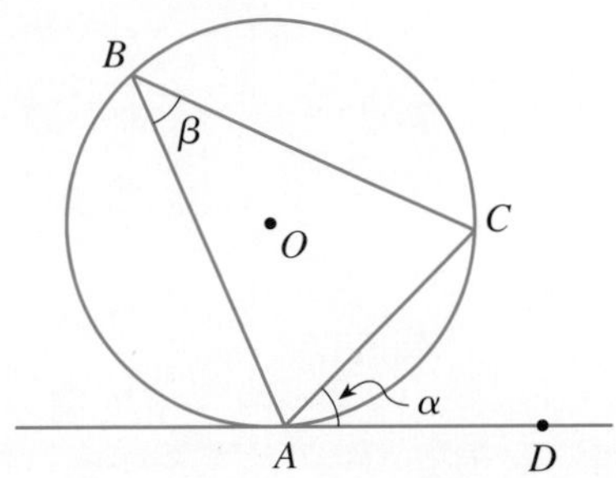

48. Triangle Inscribed in a Semicircle In the figure, $\triangle ABC$ is inscribed in a semicircle. Prove that $\angle ABC$ is a right angle. [*Hint:* Apply the Exterior Angle Theorem to the isosceles triangles formed by introducing the radius OB.]

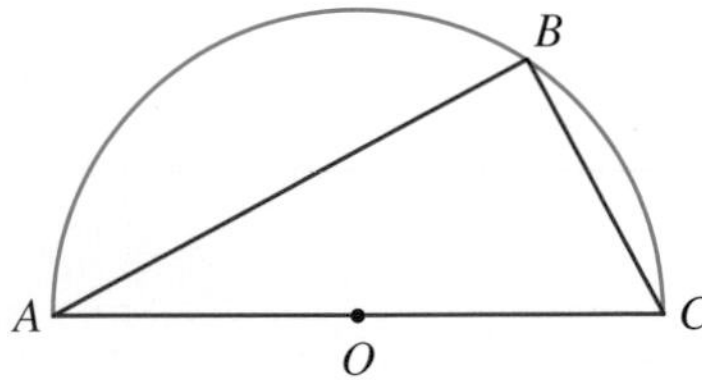

The following appendixes can be found at **www.stewartmath.com.**

APPENDIX B: *Calculations and Significant Figures*

APPENDIX C: *Graphing with a Graphing Calculator*

APPENDIX D: *Using the TI-83/84 Graphing Calculator*

APPENDIX E: *Three-Dimensional Coordinate Geometry*

APPENDIX F: *Mathematics of Finance*

APPENDIX G: *Probability and Statistics*

Answers to Selected Exercises and Chapter Tests

Prologue ■ Page P4

1. It can't go fast enough. **2.** 40% discount **3.** 427, $3n + 1$ **4.** 57 min **5.** No, not necessarily **6.** The same amount **7.** 2π **8.** The North Pole is one such point; there are infinitely many others near the South Pole.

Chapter 1

Section 1.1 ■ Page 10

1. Answers may vary. Examples: **(a)** 2 **(b)** -3 **(c)** $\frac{3}{2}$ **(d)** $\sqrt{2}$ **2. (a)** ba; Commutative **(b)** $(a + b) + c$; Associative **(c)** $ab + ac$; Distributive **3. (a)** $\{x \mid -3 < x < 5\}$ **(b)** $(-3, 5)$ **(c)** −3 5 **4.** absolute-value; positive **5.** $|b - a|$; 7 **6. (a)** Yes **(b)** No **7. (a)** No **(b)** No **8. (a)** Yes **(b)** Yes **9. (a)** 100 **(b)** 0, 100, -8 **(c)** $-1.5, 0, \frac{5}{2}, 2.71, 3.1\overline{4}, 100, -8$ **(d)** $\sqrt{7}, -\pi$ **11.** Commutative Property of Addition **13.** Associative Property of Addition **15.** Distributive Property **17.** Commutative Property of Multiplication **19.** $3 + x$ **21.** $4A + 4B$ **23.** $-2x - 2y$ **25.** $10xy$ **27.** $-5x + 10y$ **29. (a)** $\frac{29}{21}$ **(b)** $\frac{1}{24}$ **31. (a)** 3 **(b)** $\frac{13}{20}$ **33. (a)** $<$ **(b)** $>$ **(c)** $=$ **35. (a)** False **(b)** True **37. (a)** True **(b)** False **39. (a)** $x > 0$ **(b)** $t < 4$ **(c)** $a \geq \pi$ **(d)** $-5 < x < \frac{1}{3}$ **(e)** $|3 - p| \leq 5$ **41. (a)** $\{1, 2, 3, 4, 5, 6, 7, 8\}$ **(b)** $\{2, 4, 6\}$ **43. (a)** $\{1, 2, 3, 4, 5, 6, 7, 8, 9, 10\}$ **(b)** $\{7\}$ **45. (a)** $\{x \mid x \leq 5\}$ **(b)** $\{x \mid -1 < x < 4\}$

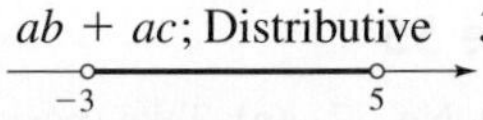

47. $-3 < x < 0$ (−3 0) **49.** $2 \leq x < 8$ (2 8)

51. $x \geq 2$ (2) **53.** $(-\infty, 1]$ (1)

55. $(-2, 1]$ (−2 1)

57. $(-1, \infty)$ (−1)

59. (a) $[-3, 5]$ **(b)** $(-3, 5]$ **(c)** $(-3, \infty)$

61.  (−2 1) **63.**  (0 6)

65.  (−4 4)

67. (a) 50 **(b)** 13 **69. (a)** 2 **(b)** -1 **71. (a)** 12 **(b)** 5 **73.** 5 **75. (a)** 15 **(b)** 24 **(c)** $\frac{67}{40}$ **77. (a)** $\frac{7}{9}$ **(b)** $\frac{13}{45}$ **(c)** $\frac{19}{33}$ **79.** $\pi - 3$ **81.** $b - a$ **83. (a)** $-$ **(b)** $+$ **(c)** $+$ **(d)** $-$ **85.** Distributive Property

Section 1.2 ■ Page 21

1. (a) 5^6 **(b)** base, exponent **2. (a)** add, 3^9 **(b)** subtract, 3^3 **3.** exponent; $\dfrac{1}{a^2}, b^2, \dfrac{1}{a^3b^2}, 6a^2b^3$ **4. (a)** $5^{1/3}$ **(b)** $\sqrt{5}$ **(c)** No **5.** $(4^{1/2})^3 = 8, (4^3)^{1/2} = 8$ **6.** $\dfrac{1}{\sqrt{3}} = \dfrac{1}{\sqrt{3}} \cdot \dfrac{\sqrt{3}}{\sqrt{3}} = \dfrac{\sqrt{3}}{3}$ **7.** $\frac{2}{3}$ **8. (a)** Yes **(b)** No **(c)** No **(d)** No **9. (a)** -64 **(b)** 64 **(c)** $-\frac{27}{25}$ **11. (a)** $\frac{1}{2}$ **(b)** $\frac{1}{8}$ **(c)** $\frac{9}{4}$ **13. (a)** 625 **(b)** 25 **(c)** 64 **15. (a)** $6\sqrt[3]{2}$ **(b)** $\dfrac{\sqrt{2}}{3}$ **(c)** $\dfrac{3\sqrt{3}}{2}$ **17. (a)** $3\sqrt{5}$ **(b)** 4 **(c)** $6\sqrt[3]{2}$ **19. (a)** t^7 **(b)** $16z^6$ **(c)** x^2 **21. (a)** $\dfrac{1}{x^2}$ **(b)** $\dfrac{1}{w}$ **(c)** x^6 **23. (a)** a^6 **(b)** a^{18} **(c)** $\dfrac{5x^9}{8}$ **25. (a)** $6x^3y^5$ **(b)** $\dfrac{25w^4}{z}$ **27. (a)** $\dfrac{x^7}{y}$ **(b)** $\dfrac{a^9}{8b^6}$ **29. (a)** $\dfrac{a^{19}b}{c^9}$ **(b)** $\dfrac{v^{10}}{u^{11}}$ **31. (a)** $\dfrac{4a^8}{b^9}$ **(b)** $\dfrac{125}{x^6y^3}$ **33. (a)** $|x|$ **(b)** $2x^2$ **35. (a)** $2x^3y$ **(b)** $2x^2|y|$ **37. (a)** $7\sqrt{2}$ **(b)** $9\sqrt{3}$ **39. (a)** $(3a + 1)\sqrt{a}$ **(b)** $(4 + x^2)\sqrt{x}$ **41. (a)** $6x\sqrt{1 + x^2}$ **(b)** $9\sqrt{x^2 + y^2}$ **43.** $10^{1/2}$ **45.** $\sqrt[5]{7^3}$ **47.** $5^{-1/2}$ **49.** $\dfrac{1}{\sqrt{y^3}}$ **51. (a)** 2 **(b)** -2 **(c)** $\frac{1}{3}$ **53. (a)** 4 **(b)** $\frac{3}{2}$ **(c)** $\frac{8}{27}$ **55. (a)** 5 **(b)** $\sqrt[5]{3}$ **(c)** 4 **57. (a)** x^2 **(b)** y^2 **59. (a)** $w^{5/3}$ **(b)** $729x^3y^2$ **61. (a)** $4a^4b$ **(b)** $\dfrac{8b^9}{a^6}$ **63. (a)** $\dfrac{9}{y^4}$ **(b)** $\dfrac{4w^2}{81z^5}$ **65. (a)** $x^{3/2}$ **(b)** $x^{6/5}$ **67. (a)** $y^{3/2}$ **(b)** $10x^{7/12}$ **69. (a)** $2st^{11/6}$ **(b)** x **71. (a)** $y^{1/2}$ **(b)** $\dfrac{3u}{v}$ **73. (a)** $\dfrac{\sqrt{6}}{6}$ **(b)** $\dfrac{\sqrt{6}}{2}$ **(c)** $\dfrac{9\sqrt[4]{8}}{2}$ **75. (a)** $\dfrac{\sqrt{5x}}{5x}$ **(b)** $\dfrac{\sqrt{5x}}{5}$ **(c)** $\dfrac{\sqrt[5]{x^2}}{x}$ **77. (a)** $\dfrac{1}{4}$ **(b)** $\dfrac{\sqrt{2}}{4}$ **79. (a)** y **(b)** $-9wz$ **81. (a)** 6.93×10^7 **(b)** 7.2×10^{12} **(c)** 2.8536×10^{-5} **(d)** 1.213×10^{-4} **83. (a)** 319,000 **(b)** 272,100,000 **(c)** 0.000 000 026 70 **(d)** 0.000 000 009 999 **85. (a)** 9.5×10^{12} km **(b)** 4×10^{-13} cm **(c)** 3.3×10^{19} molecules **87.** 1.3×10^{-20} **89.** 1.429×10^{19} **91.** 7.4×10^{-14} **93. (a)** Negative **(b)** Positive **(c)** Negative **(d)** Negative **(e)** Positive **(f)** Negative **95.** 4.1×10^{13} km **97.** 1.3×10^{21} L **99.** 3.18×10^{80} atoms **101. (a)** $\approx$28 km/h **(b)** $\approx$167 m

Section 1.3 ■ Page 33

1. $6x$; $6x(3x^2 + 5)$ **2. (a)** 3; $2x^3, 3x^2, 10x$ **(b)** x; $x(2x^2 + 3x + 10)$ **3.** 12, 8; 2, 6; $(x + 2)(x + 6)$ **4.** $A^2 + 2AB + B^2$; $4x^2 + 12x + 9$ **5.** $A^2 - B^2$; $36 - x^2$ **6.** $(A + B)(A - B)$; $(7x + 3)(7x - 3)$ **7.** $(A + B)^2$; $(x + 5)^2$ **8. (a)** No **(b)** Yes **(c)** Yes **(d)** No **9.** Binomial; $5x^3$, 6; 3 **11.** Monomial; -8; 0 **13.** Four terms; $-x^4, x^3, -x^2, x$; 4 **15.** $7x + 5$ **17.** $x^2 + 2x - 3$ **19.** $5x^3 + 3x^2 - 10x - 2$

21. $9x + 103$ **23.** $-x^5 + 3x^4 + 6x^3$ **25.** $2x^3$
27. $21t^2 - 26t + 8$ **29.** $6x^2 + 7x - 5$ **31.** $2x^2 + 5xy - 3y^2$
33. $16x^2 + 24x + 9$ **35.** $y^2 - 6xy + 9x^2$
37. $4x^2 + 12xy + 9y^2$ **39.** $w^2 - 49$ **41.** $9x^2 - 16$
43. $x - 4$ **45.** $y^3 + 6y^2 + 12y + 8$ **47.** $x^3 + 4x^2 + 7x + 6$
49. $2x^3 - 7x^2 + 7x - 5$ **51.** $x\sqrt{x} - x$ **53.** $y^2 + y$
55. $x - 2\sqrt{xy} + y$ **57.** $x^4 - a^4$ **59.** $a - b^2$
61. $-x^4 + x^2 - 2x + 1$ **63.** $4x^2 + 4xy + y^2 - 9$
65. $x^2(2x - 1)(2x + 1)$ **67.** $(y - 6)(y + 9)$
69. $2xy^2(2x^2 - 3y + 4xy^2)$ **71.** $(x + 3)^4(x + 2)^2$
73. $(z - 2)(z - 9)$ **75.** $(2x - 3)(5x - 2)$
77. $(3x - 1)(x - 5)$ **79.** $(3x + 4)(3x + 8)$
81. $(6a - 7)(6a + 7)$ **83.** $(3x + y)(9x^2 - 3xy + y^2)$
85. $(2s - 5t)(4s^2 + 10st + 25t^2)$ **87.** $(x + 6)^2$
89. $(x + 4)(x^2 + 1)$ **91.** $(x^2 + 1)(5x + 1)$
93. $(x + 1)(x^2 + 1)$ **95.** $x^{2/3}(1 + 3x)$
97. $x^{-3/2}(x^2 - x + 1)$ **99.** $(x^2 + 1)^{-1/2}(x^2 + 3)$
101. $2x(1 + 6x^2)$ **103.** $(x - 4)(x + 2)$
105. $(2x + 3)(x + 1)$ **107.** $9(x - 5)(x + 1)$
109. $(7 - 2y)(7 + 2y)$ **111.** $(t - 3)^2$ **113.** $(y - 5z)^2$
115. $4ab$ **117.** $(x - 1)(x + 1)(x - 3)(x + 3)$
119. $(2x - 5)(4x^2 + 10x + 25)$ **121.** $x(x + 1)^2$
123. $x^2y^3(x + y)(x - y)$ **125.** $(x - 2)(x + 2)(3x - 1)$
127. $x^{-3/2}(1 + x)^2$ **129.** $3(x - 1)(x + 2)$
131. $(a - 1)(a + 1)(a - 2)(a + 2)$
133. $(x - 1)^3(x + 2)^2(x^2 - 4x + 2)$
135. $2(x^2 + 4)^4(x - 2)^3(7x^2 - 10x + 8)$
137. $(x^2 + 3)^{-4/3}(\frac{1}{3}x^2 + 3)$
141. $(a + b + c)(a + b - c)(a - b + c)(-a + b + c)$

Section 1.4 ■ Page 42

1. (a), (c) **2.** numerator; denominator; $\dfrac{x + 1}{x + 3}$
3. numerators; denominators; $\dfrac{2x}{x^2 + 4x + 3}$
4. (a) 3 **(b)** $x(x + 1)^2$ **(c)** $\dfrac{-2x^2 + 1}{x(x + 1)^2}$
5. (a) Yes **(b)** No **6. (a)** Yes **(b)** No
7. $\mathbb{R}$ **9.** $\{x \mid x \neq 3\}$ **11.** $\{x \mid x \geq -3\}$
13. $\{x \mid x \neq -1, 2\}$ **15.** $\{x \mid x \geq 2\}$ **17.** $\frac{1}{2}(x + 5)$
19. $\dfrac{1}{x + 2}$ **21.** $\dfrac{x + 1}{x - 2}$ **23.** $\dfrac{y}{y - 1}$ **25.** $\dfrac{x(2x + 3)}{2x - 3}$
27. $\dfrac{1}{4(x - 2)}$ **29.** $\dfrac{x - 3}{x + 2}$ **31.** $\dfrac{1}{t^2 + 9}$ **33.** $\dfrac{x - 4}{x + 4}$
35. $\dfrac{x + 5}{(2x + 3)(x + 4)}$ **37.** $x^2(x + 1)$ **39.** $\dfrac{x}{yz}$
41. $\dfrac{x + 4}{x + 3}$ **43.** $\dfrac{3x + 7}{(x - 3)(x + 5)}$ **45.** $\dfrac{2x + 5}{(x + 1)(x + 2)}$
47. $\dfrac{2(5x - 9)}{(2x - 3)^2}$ **49.** $\dfrac{u^2 + 3u + 1}{u + 1}$ **51.** $\dfrac{2x + 1}{x^2(x + 1)}$
53. $\dfrac{2x + 7}{(x + 3)(x + 4)}$ **55.** $\dfrac{x - 2}{(x + 3)(x - 3)}$
57. $\dfrac{5x - 6}{x(x - 1)}$ **59.** $\dfrac{1}{x^3}$ **61.** $\dfrac{x + 1}{1 - 2x}$ **63.** $\dfrac{x + 3}{x + 1}$
65. $\dfrac{2}{(x - 1)(x + 3)}$ **67.** $\dfrac{x^2(y - 1)}{y^2(x - 1)}$
69. $-xy$ **71.** $\dfrac{y - x}{xy}$ **73.** $\dfrac{1}{1 - x}$ **75.** $\dfrac{-1}{(1 + x)(1 + x + h)}$
77. $-\dfrac{2x + h}{x^2(x + h)^2}$ **79.** $\dfrac{1}{\sqrt{1 - x^2}}$ **81.** $\dfrac{(x - 3)(19 - x)}{(x + 5)^4}$
83. $\dfrac{x + 2}{(x + 1)^{3/2}}$ **85.** $\dfrac{2x + 3}{(x + 1)^{4/3}}$ **87.** $\sqrt{10} - 3$
89. $\sqrt{5} + \sqrt{3}$ **91.** $\dfrac{y\sqrt{3} - y\sqrt{y}}{3 - y}$ **93.** $\dfrac{-1}{5(2 + \sqrt{5})}$
95. $\dfrac{r - 2}{5(\sqrt{r} - \sqrt{2})}$ **97.** $\dfrac{1}{\sqrt{x^2 + 1} + x}$
99. (a) $\dfrac{R_1R_2}{R_1 + R_2}$ **(b)** $\frac{20}{3} \approx 6.7$ ohms

Section 1.5 ■ Page 55

1. (a) Yes **(b)** Yes **(c)** No **2. (a)** Take (positive and negative) square roots of both sides. **(b)** Subtract 5 from both sides. **(c)** Subtract 2 from both sides.
3. (a) Factor the left side to $(x + 2)(x - 8)$, and use the Zero-Product Property. **(b)** Add 16 to each side, then complete the square by adding 9 to both sides. **(c)** Insert coefficients into the Quadratic Formula. **4. (a)** 0 **(b)** 0, 4 **(c)** factor
5. (a) $\sqrt{2x} = -x$ **(b)** $2x = x^2$ **(c)** 0, 2 **(d)** 0
6. quadratic; $x + 1$; $W^2 - 5W + 6 = 0$
7. $x(x + 2)$; $3(x + 2) + 5x = 2x(x + 2)$
8. square; $(2x + 1)^2 = x + 1$ **9. (a)** No **(b)** Yes
11. (a) Yes **(b)** No **13.** -1 **15.** 18 **17.** $\frac{3}{5}$
19. $-\frac{27}{4}$ **21.** $-\frac{3}{4}$ **23.** 30 **25.** $\frac{13}{6}$ **27.** $-\frac{1}{3}$
29. $m = \dfrac{2E}{v^2}$ **31.** $w = \dfrac{P - 2l}{2}$ **33.** $x = \dfrac{2d - b}{a - 2c}$
35. $x = \dfrac{1 - a}{a^2 - a - 1}$ **37.** $r = \pm\sqrt{\dfrac{3V}{\pi h}}$
39. $b = \pm\sqrt{c^2 - a^2}$ **41.** $-4, 3$ **43.** $-15, 2$
45. $-\frac{3}{2}, \frac{5}{2}$ **47.** ± 2 **49.** $-2, 7$ **51.** $-5 \pm \sqrt{2}$
53. $-5 \pm \sqrt{23}$ **55.** $3 \pm 2\sqrt{5}$ **57.** $-1 \pm \dfrac{\sqrt{30}}{5}$
59. $0, \frac{1}{4}$ **61.** $-3, 5$ **63.** $-\frac{3}{2}, 1$ **65.** $-1 \pm \frac{2\sqrt{6}}{3}$
67. $-\frac{2}{3}$ **69.** $-\frac{9}{2}, \frac{1}{2}$ **71.** No real solution **73.** 2
75. 1 **77.** No real solution **79.** $-50, 100$ **81.** 0, 3
83. $-\frac{7}{5}, 2$ **85.** 7 **87.** 4 **89.** 4 **91.** $\pm 2\sqrt{2}, \pm\sqrt{5}$
93. $-4, -3, -1, 0$ **95.** $\pm 3\sqrt{3}, \pm 2\sqrt{2}$ **97.** 2
99. $-2, -\frac{4}{3}$ **101.** 3.99, 4.01 **103.** -2 **105.** $-1, \frac{4}{3}$
107. $\frac{21}{11}$ **109.** $\dfrac{-3 + 3\sqrt{5}}{2}$ **111.** 256 **113.** $-\frac{19}{3}, 13$
115. 4, 6 **117.** ± 2 **119.** $-\frac{1}{2}$ **121.** 20 **123.** $-3, \dfrac{1 \pm \sqrt{13}}{2}$
125. $\pm\sqrt{a}, \pm 2\sqrt{a}$ **127.** $\sqrt{a^2 + 36}$ **129.** ≈ 4.24 s
131. (a) After 1 s and $1\frac{1}{2}$ s **(b)** Never **(c)** 7.5 m **(d)** After $1\frac{1}{4}$ s **(e)** After $2\frac{1}{2}$ s **133. (a)** 0.00055; ≈ 12.018 m **(b)** 234.375 kg/m^3 **135. (a)** 211,810 **(b)** 160,760 Pa; 194,000 Pa **137.** 215,000 km

Section 1.6 ■ Page 63

1. -1 **2.** 3, 4 **3. (a)** $3 - 4i$ **(b)** $9 + 16 = 25$ **4.** $3 - 4i$
5. Yes **6.** Yes **7.** Real part 3, imaginary part -8
9. Real part $-\frac{2}{3}$, imaginary part $-\frac{5}{3}$ **11.** Real part 3, imaginary part 0 **13.** Real part 0, imaginary part $-\frac{2}{3}$ **15.** Real part $\sqrt{3}$, imaginary part 2 **17.** $3 + 7i$ **19.** $1 - 10i$ **21.** $3 + 5i$
23. $2 - 2i$ **25.** $-19 + 4i$ **27.** $-4 + 8i$ **29.** $26 + 7i$
31. $27 - 8i$ **33.** 13 **35.** $5 - 12i$ **37.** $-i$ **39.** $-1 - i$
41. $-4 + 2i$ **43.** $2 - \frac{4}{3}i$ **45.** $-i$ **47.** $-i$ **49.** $243i$
51. 1 **53.** $5i$ **55.** -6 **57.** $(6 + \sqrt{3}) + (3 - 2\sqrt{3})i$ **59.** 2
61. $\pm 5i$ **63.** $3 \pm 2i$ **65.** $\frac{1}{2} \pm \frac{3}{2}i$ **67.** $-\frac{1}{2} \pm \frac{\sqrt{3}}{2}i$
69. $\frac{2}{9} \pm \frac{4\sqrt{2}}{9}i$ **71.** $-1 \pm \frac{\sqrt{6}}{6}i$ **73.** $8 + 2i$ **75.** 25

Section 1.7 ■ Page 75

2. principal; interest rate; time in years
3. (a) x^2 **(b)** lw **(c)** πr^2 **4.** 1.6 **5.** $\frac{1}{x}$
6. $r = \frac{d}{t}, t = \frac{d}{r}$ **7.** $3n + 3$ **9.** $3n + 6$ **11.** $\frac{160 + s}{3}$
13. $0.025x$ **15.** $4w^2$ **17.** $\frac{d}{55}$ **19.** $\frac{750}{3 + x}$ **21.** 400 km
23. 86 **25.** \$8400 at $2\frac{1}{2}$% and \$3600 at 3% **27.** 7.5%
29. \$14,400 **31.** 6 h **33.** 40 years old **35.** 7 nickels, 7 dimes, 7 quarters **37.** 45 m **39.** 120 m by 120 m
41. 25 m by 35 m **43.** 60 m by 40 m **45.** 120 m
47. (a) 9 cm **(b)** 5 cm **49.** 10 cm **51.** 5.54 m **53.** 5 m
55. 200 mL **57.** 18 g **59.** 0.6 L **61.** 35% **63.** 14 min 35 s
65. 3.5 h **67.** you 3 h, roommate $4\frac{1}{2}$ h **69.** 4 h
71. 800 km/h **73.** 80 km/h (or 384 km/h) **75.** 6 km/h
77. 1.58 m from the fulcrum **79.** 2 m by 6 m by 15 m
81. 13 cm by 13 cm **83.** 0.88 m **85.** 16 km; no **87.** 2.28 m
89. 5.7 m **91.** 4.55 m

Section 1.8 ■ Page 88

1. (a) $<$ **(b)** $\le$ **(c)** $\le$ **(d)** $>$
2. $-1, 2$

Interval	$(-\infty, -1)$	$(-1, 2)$	$(2, \infty)$
Sign of $x + 1$ Sign of $x - 2$	− −	+ −	+ +
Sign of $(x + 1)/(x - 2)$	+	−	+

yes, 2; $[-1, 2)$

3. (a) $[-3, 3]$ **(b)** $(-\infty, -3], [3, \infty)$

4. (a) < 3 **(b)** > 3 **5. (a)** No **(b)** No **6. (a)** Divide by 3 **(b)** Add 2 **(c)** Rewrite as $-8 \le 3x + 2 \le 8$

7. $\{\frac{5}{6}, 1, \sqrt{5}, 3, 5\}$ **9.** $\{3, 5\}$ **11.** $\{-5, -1, \sqrt{5}, 3, 5\}$

13. $(-\infty, \frac{7}{2}]$ **15.** $(4, \infty)$

17. $(-\infty, 2]$ **19.** $(-\infty, -\frac{1}{2})$

21. $(-3, \infty)$ **23.** $(3, \infty)$

25. $(-\infty, -18)$ **27.** $(-\infty, -1]$

29. $[-3, -1)$ **31.** $(2, 6)$

33. $[\frac{9}{2}, 5)$ **35.** $(\frac{15}{2}, \frac{21}{2}]$

37. $(-2, 3)$ **39.** $(-\infty, -\frac{7}{2}] \cup [0, \infty)$

41. $[-3, 6]$ **43.** $(-\infty, -2] \cup [\frac{1}{3}, \infty)$

45. $(-1, 4)$ **47.** $(-\infty, -3) \cup (6, \infty)$

49. $(-2, 2)$ **51.** $(-\infty, -2] \cup [1, 3]$

53. $(-\infty, -2) \cup (-2, 4)$ **55.** $(-\infty, -5] \cup \{-3\} \cup [2, \infty)$

57. $(-2, 0) \cup (2, \infty)$ **59.** $(-\infty, -1) \cup (1, \infty)$

61. $(-\infty, -1) \cup [3, \infty)$ **63.** $(2, \frac{5}{2})$

65. $(-\infty, 5) \cup [16, \infty)$ **67.** $(-2, 0) \cup (2, \infty)$

69. $[-2, -1) \cup (0, 1]$ **71.** $(-3, -\frac{1}{2}) \cup (2, \infty)$

73. $(-4, 4)$ **75.** $(-\infty, -\frac{7}{2}) \cup (\frac{7}{2}, \infty)$

77. $[-7, 13]$ **79.** $(-2, \frac{2}{3})$

81. $(-\infty, -1] \cup [\frac{7}{3}, \infty)$ **83.** $(-4, 8)$

85. $(-6.001, -5.999)$ **87.** $[-\frac{1}{2}, \frac{3}{2}]$

89. $(-\infty, \frac{5}{2})$ **91.** $[-4, 0] \cup [4, \infty)$

93. $(-\infty, -3) \cup (-2, \infty)$ **95.** $(-1, \infty)$

97. $[1, 4]$

$\bullet$———$\bullet$——→ (1, 4)

99. $|x| < 3$ **101.** $|x - 7| \geq 5$

103. $|x| \leq 2$ **105.** $|x| > 3$

107. $|x - 1| \leq 3$ **109.** $x \leq -3$ or $x \geq 3$

111. $x < -2$ or $x > 5$ **113.** $x \geq \dfrac{(a+b)c}{ab}$

115. $x \leq \dfrac{ac - 4a + d}{ab}$ or $x \geq \dfrac{ac + 4a - d}{ab}$

117. $68 \leq F \leq 86$ **119.** More than 200 km
121. Between 12,000 km and 14,000 km
123. **(a)** $-\frac{1}{3}P + \frac{560}{3}$ **(b)** From \$215 to \$290
125. Distances between 20,000 km and 100,000 km
127. **(a)** Acceleration greater than 26.4 m/s^2 **(b)** ≈ 9.1 s
129. Between 0 and 35 km/h
131. Between 20 and 40 m
133. Between 156 and 185 cm

Section 1.9 ■ Page 101

1. **(a)** $(3, -5)$ **(b)** y-axis **2.** $\sqrt{(c-a)^2 + (d-b)^2}$; 10

3. $\left(\dfrac{a+c}{2}, \dfrac{b+d}{2}\right)$; $(4, 6)$

4. 2; 3; No

x	y	(x, y)
-2	$-\frac{1}{2}$	$(-2, -\frac{1}{2})$
-1	0	$(-1, 0)$
0	$\frac{1}{2}$	$(0, \frac{1}{2})$
1	1	$(1, 1)$
2	$\frac{3}{2}$	$(2, \frac{3}{2})$

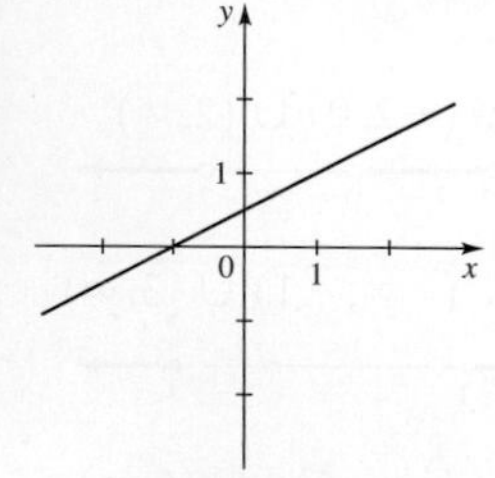

5. **(a)** y; x; -1 **(b)** x; y; $\frac{1}{2}$ **6.** **(a)** $(1, 2)$; 3 **(b)** $(x-3)^2 + (y-4)^2 = 9$ **7.** **(a)** $(a, -b)$ **(b)** $(-a, b)$ **(c)** $(-a, -b)$ **8.** **(a)** $-5, 3$; ± 2 **(b)** x-axis **9.** Yes
10. No **11.** $A(5, 1)$, $B(1, 2)$, $C(-2, 6)$, $D(-6, 2)$, $E(-4, -1)$, $F(-2, 0)$, $G(-1, -3)$, $H(2, -2)$

13.

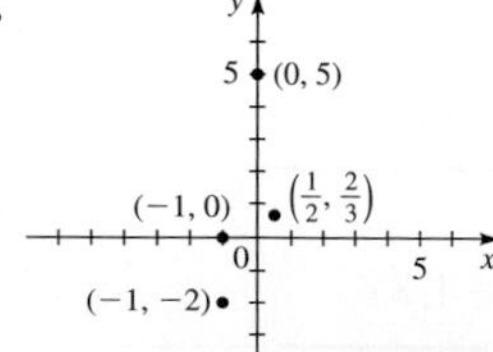

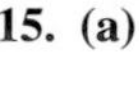
15. **(a)**

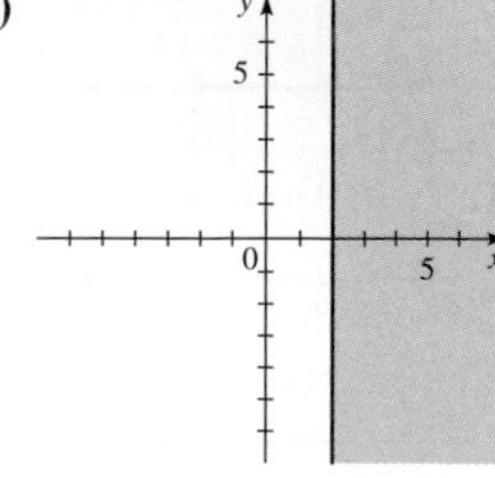

(b)

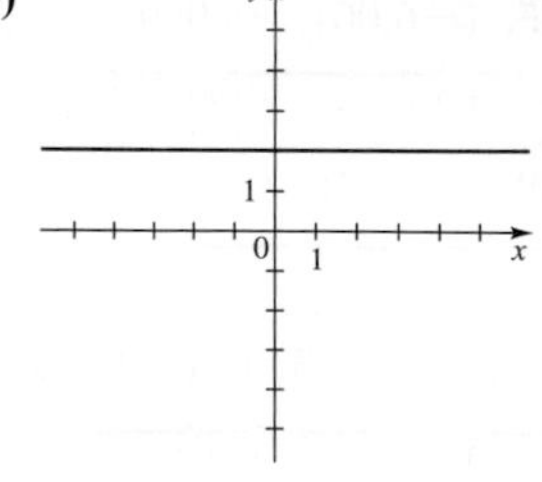

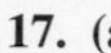
17. **(a)**

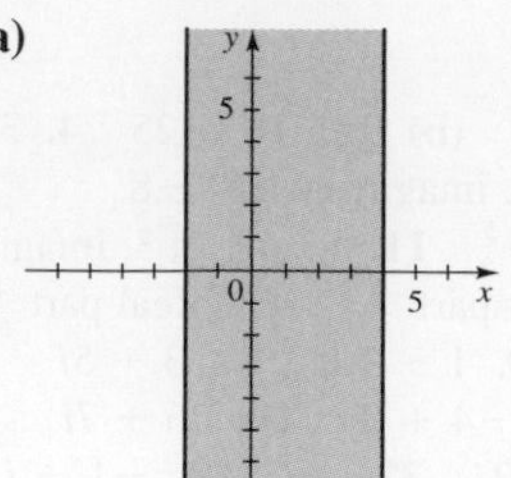

(b)

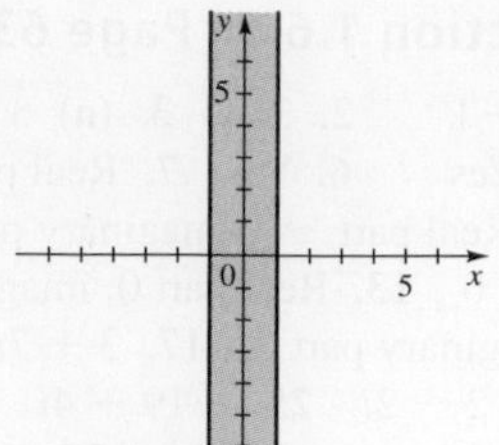

19. **(a)**

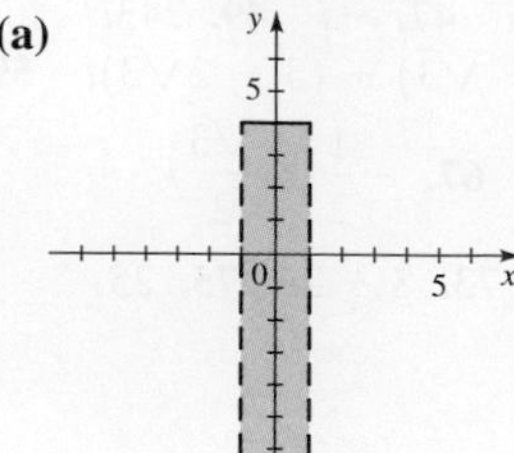

(b)

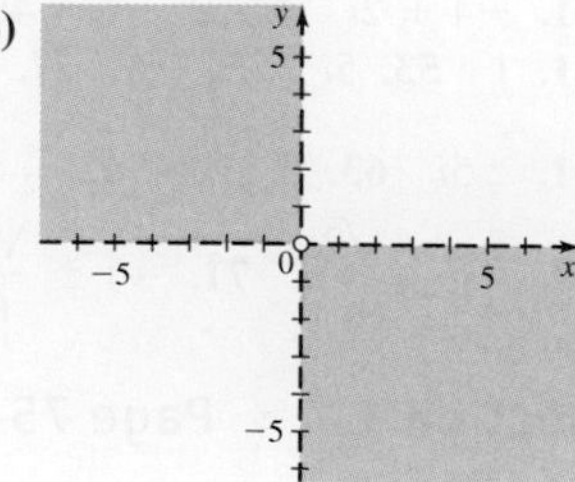

21. **(a)** $\sqrt{13}$ **(b)** $(\frac{3}{2}, 1)$ **23.** **(a)** 10 **(b)** $(1, 0)$

25. **(a)**

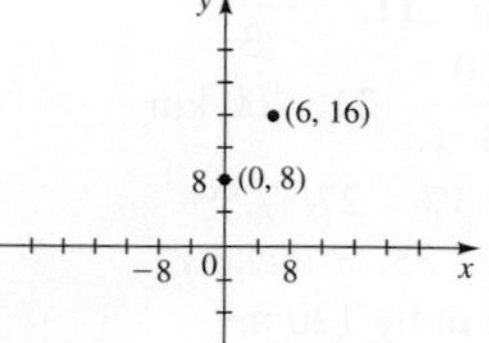

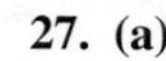
27. **(a)**

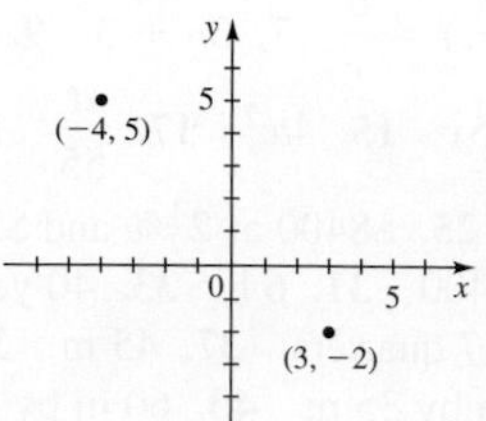

(b) 10 **(c)** $(3, 12)$

(b) $7\sqrt{2}$ **(c)** $(-\frac{1}{2}, \frac{3}{2})$

29. **(a)**

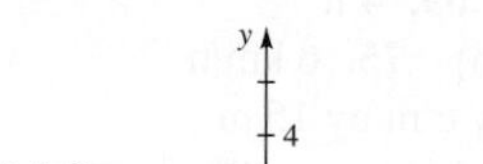

(b) $4\sqrt{10}$ **(c)** $(0, 0)$

31. 24

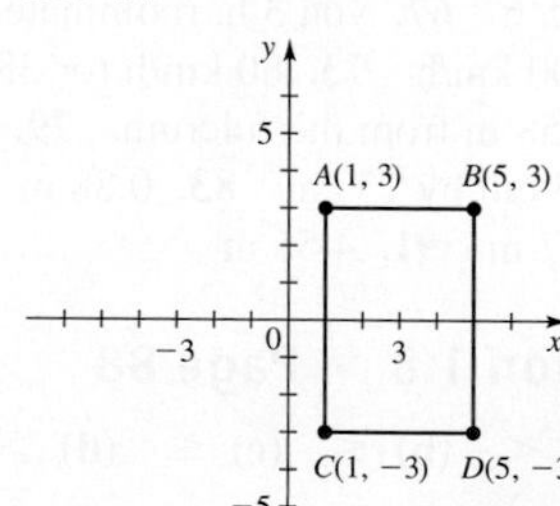

33. Trapezoid, 9

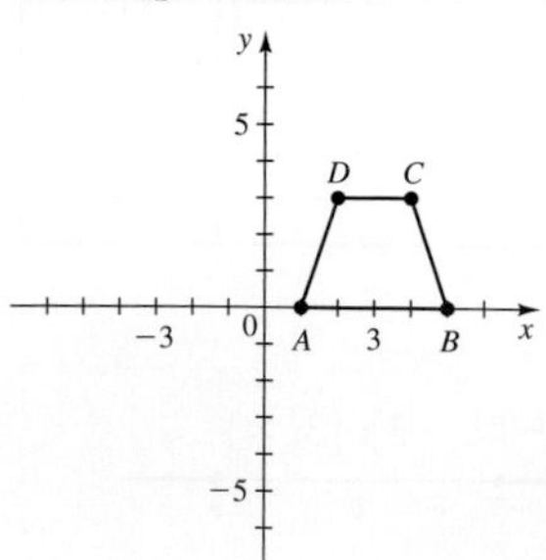

35. $A(6, 7)$ **37.** $Q(-1, 3)$ **41.** **(b)** 10 **45.** $(0, -4)$

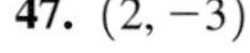
47. $(2, -3)$

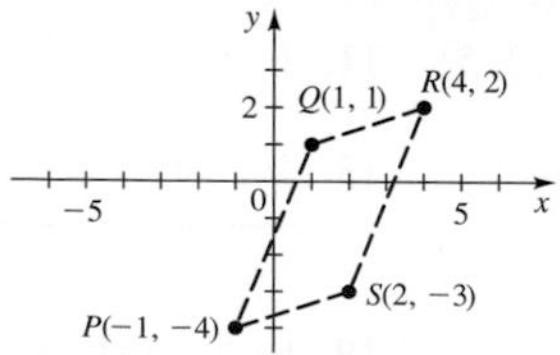

49. (a)

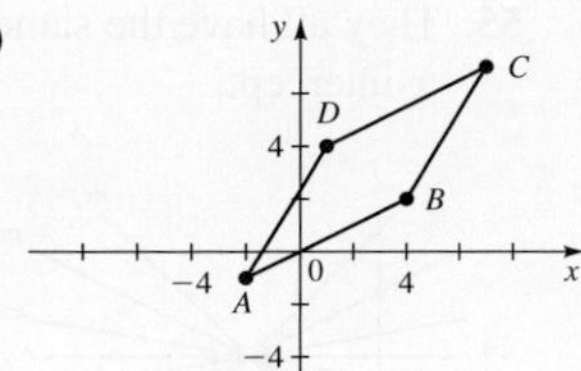

(b) $\left(\frac{5}{2}, 3\right), \left(\frac{5}{2}, 3\right)$

51. Yes, no, yes **53.** Yes, no, yes

55.

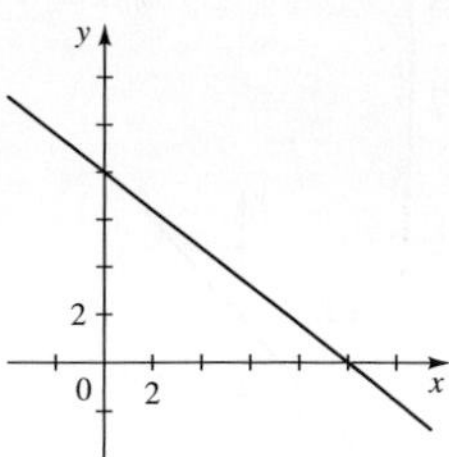

57.

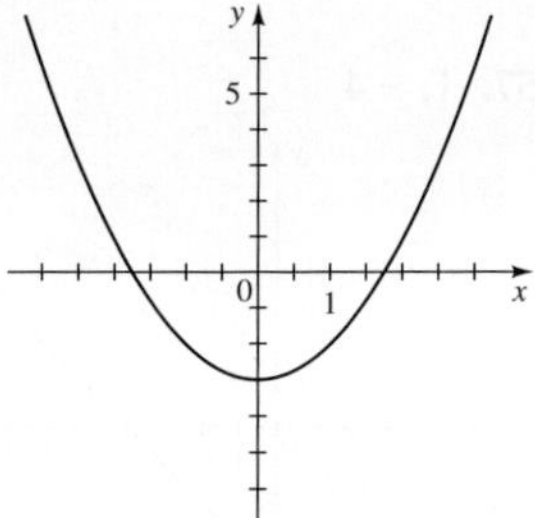

59.

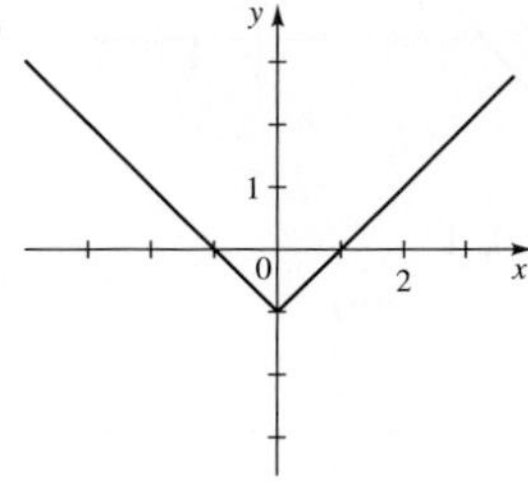

61. (a) x-intercept 3, y-intercept -6, no symmetry

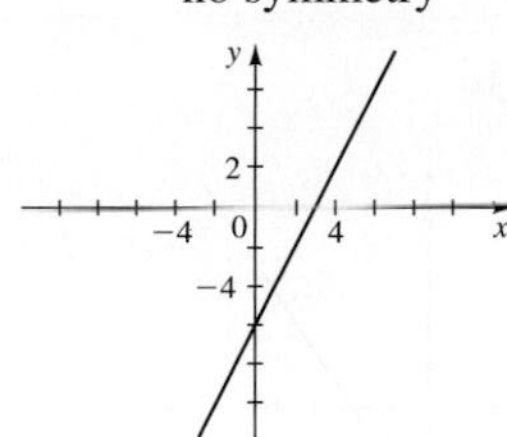

(b) x-intercept 1, y-intercept 2, no symmetry

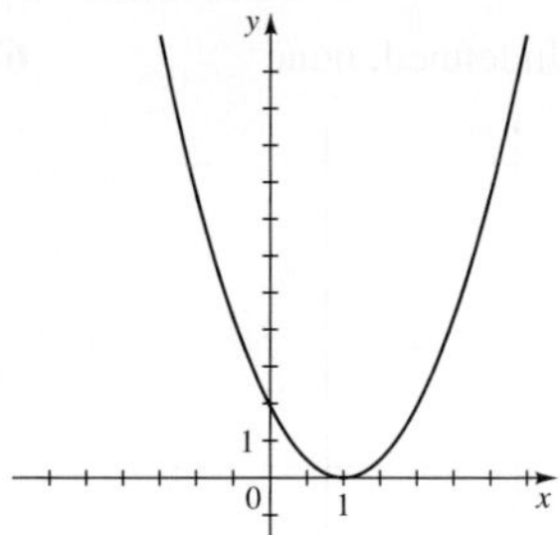

63. (a) No x-intercept, y-intercept 2, no symmetry

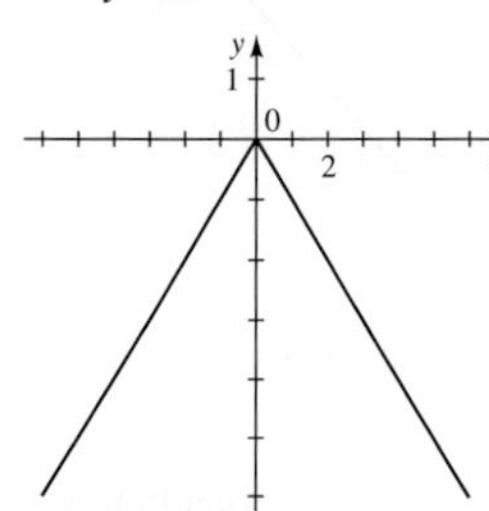

(b) x-intercept 0, y-intercept 0, symmetry with respect to y-axis

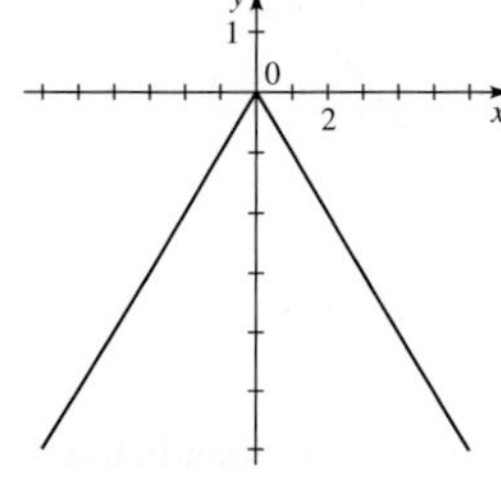

65. (a) x-intercepts ± 2, y-intercept 2, symmetry with respect to y-axis

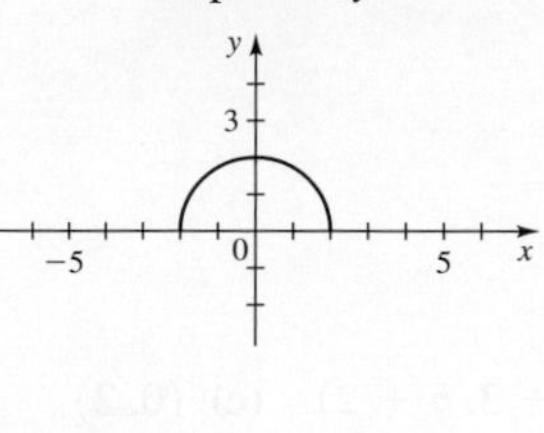

(b) x-intercepts 0 and ± 2, y-intercept 0, symmetry with respect to origin

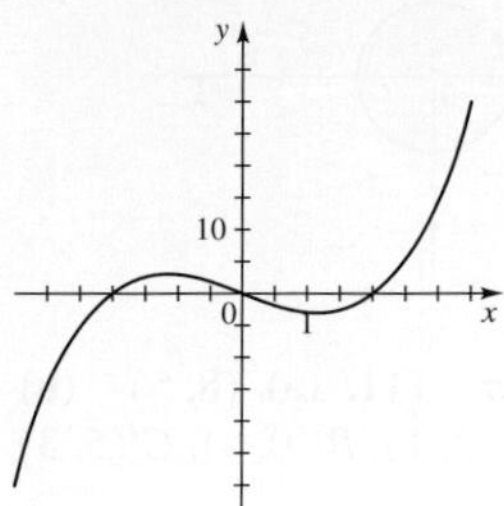

67. (a) x-intercept -6; y-intercept 6
(b) x-intercepts $\pm\sqrt{5}$; y-intercept -5
69. (a) x-intercepts ± 2; no y-intercept
(b) x-intercept $\frac{1}{4}$; y-intercept 1
71. x-intercepts 0, 4; y-intercept 0
73. x-intercepts -2, 2; y-intercepts -4, 4
75. $(0, 0)$, 3

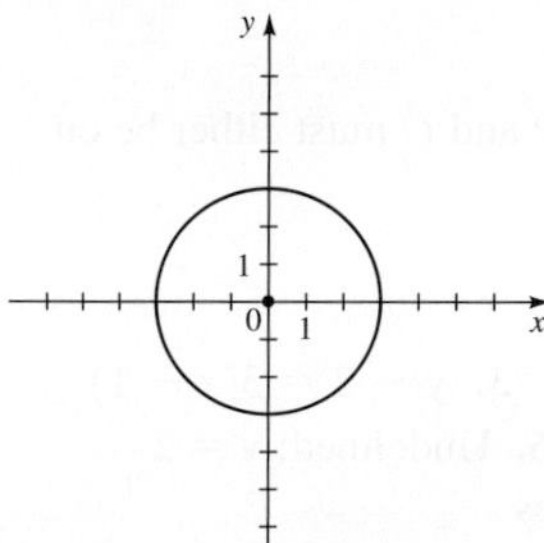

77. $(2, 0)$, 3

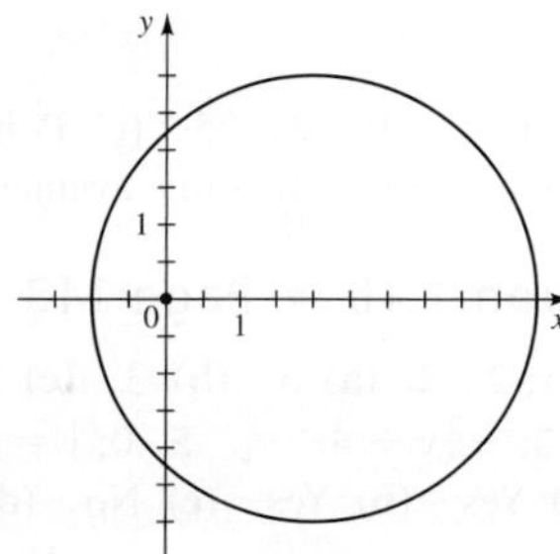

79. $(-3, 4)$, 5

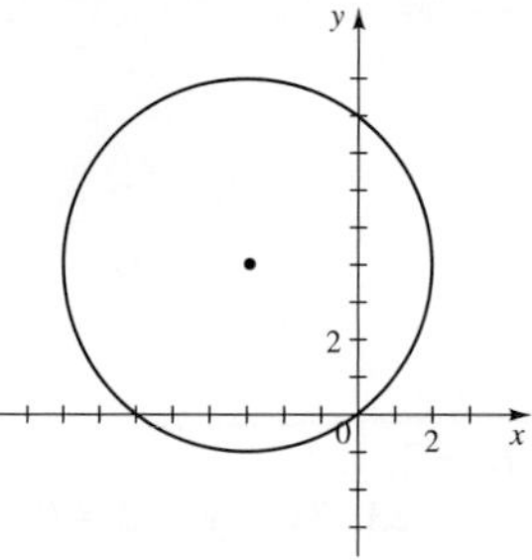

81. $(x + 3)^2 + (y - 1)^2 = 4$ **83.** $x^2 + y^2 = 65$
85. $(x - 2)^2 + (y - 5)^2 = 25$ **87.** $(x - 7)^2 + (y + 3)^2 = 9$
89. $(x + 2)^2 + (y - 2)^2 = 4$ **91.** $(-2, 3), 1$ **93.** $\left(\frac{1}{4}, -\frac{1}{4}\right), \frac{1}{2}$
95. $\left(\frac{3}{4}, 0\right), \frac{3}{4}$ **97.** Symmetry about y-axis **99.** Symmetry with respect to x-axis, y-axis, and origin
101. Symmetry with respect to origin
103.

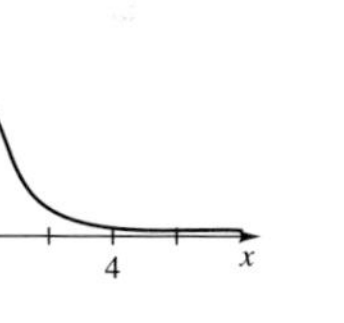

105.

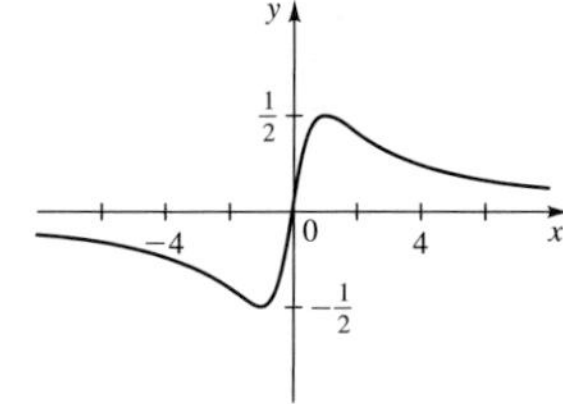

107.

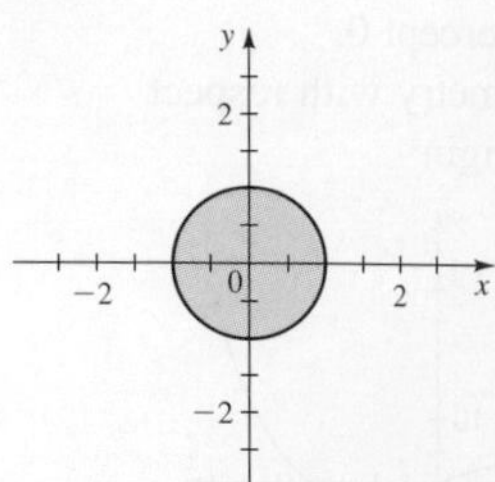

109. 12π **111. (a)** $(8, 5)$ **(b)** $(a + 3, b + 2)$ **(c)** $(0, 2)$ **(d)** $A'(-2, 1), B'(0, 4), C'(5, 3)$

113. (a)

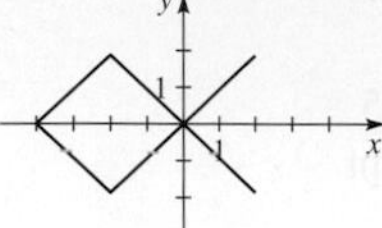

(b)

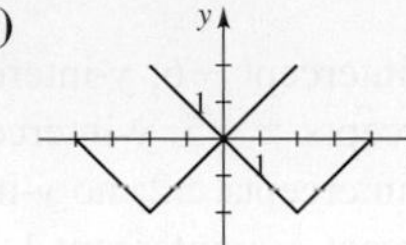

(c)

115. (a) 5 **(b)** 31; 25 **(c)** Points P and Q must either be on the same street or the same avenue.

Section 1.10 ■ Page 113

1. y; x; 2 **2. (a)** 3 **(b)** 3 **(c)** $-\frac{1}{3}$ **3.** $y - 2 = 3(x - 1)$ **4.** 6, 4; $-\frac{2}{3}x + 4$; $-\frac{2}{3}$ **5.** 0; $y = 3$ **6.** Undefined; $x = 2$ **7. (a)** Yes **(b)** Yes **(c)** No **(d)** Yes

8. Yes

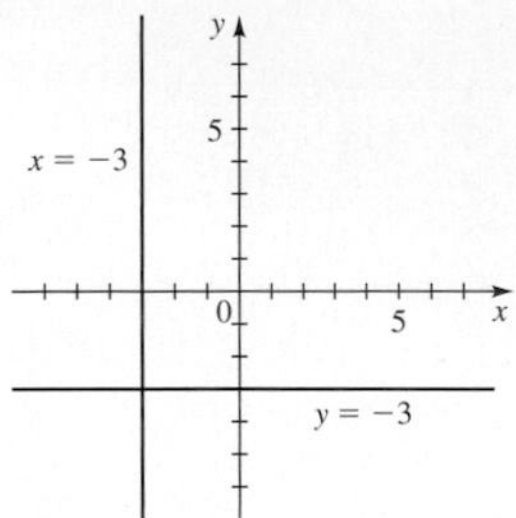

9. -2 **11.** $-\frac{5}{6}$ **13.** 0 **15.** -1 **17.** $-2, \frac{1}{2}, 3, -\frac{1}{4}$
19. $x + y - 4 = 0$ **21.** $3x - 2y - 6 = 0$ **23.** $3x - y - 2 = 0$
25. $3x - y - 11 = 0$ **27.** $2x - 3y + 19 = 0$
29. $5x + y - 11 = 0$ **31.** $2x - y - 9 = 0$
33. $3x - y - 3 = 0$ **35.** $y = 3$ **37.** $x = 2$
39. $2x - y + 6 = 0$ **41.** $y = 5$ **43.** $3x + 2y + 17 = 0$
45. $x = -1$ **47.** $4x - 3y + 11 = 0$ **49.** $x - y + 6 = 0$
51. (a)

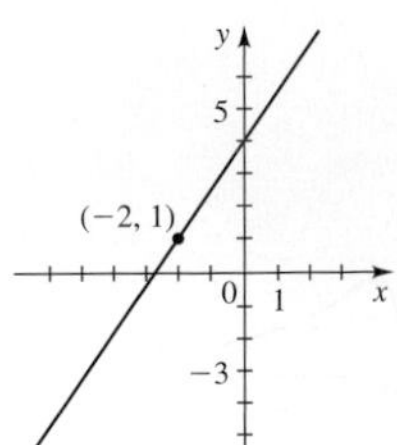

(b) $3x - 2y + 8 = 0$

53. They all have the same slope.

55. They all have the same x-intercept.

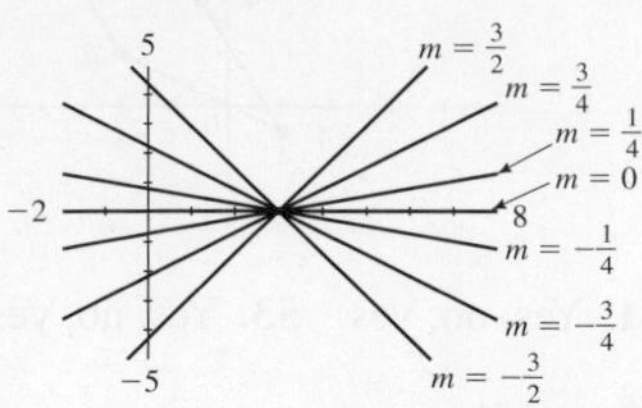

57. 1, -4

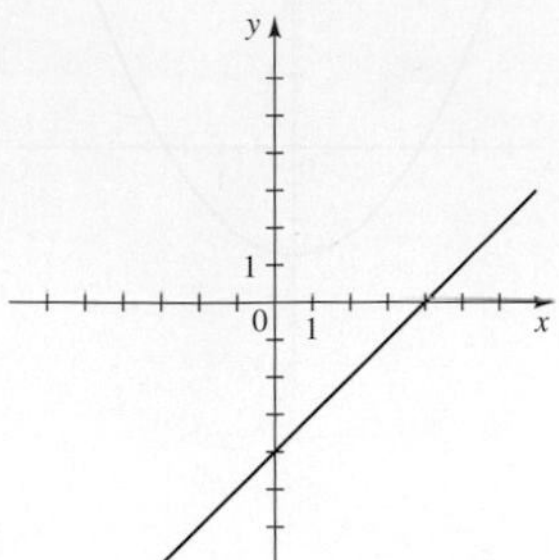

59. 2, 7

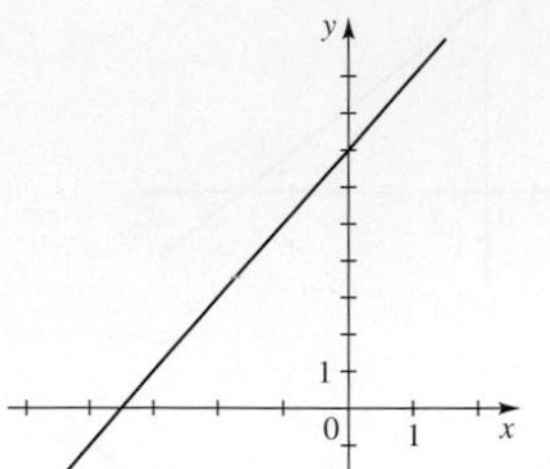

61. $-\frac{4}{5}$, 2

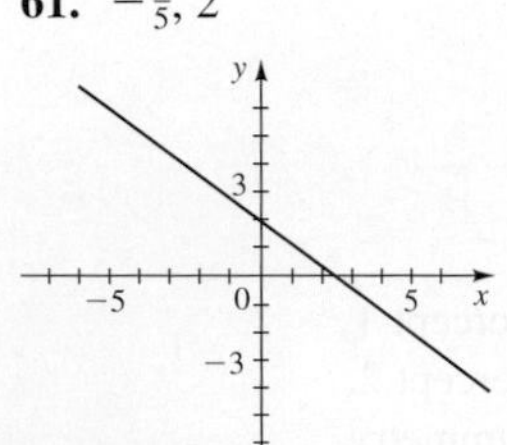

63. 0, 4

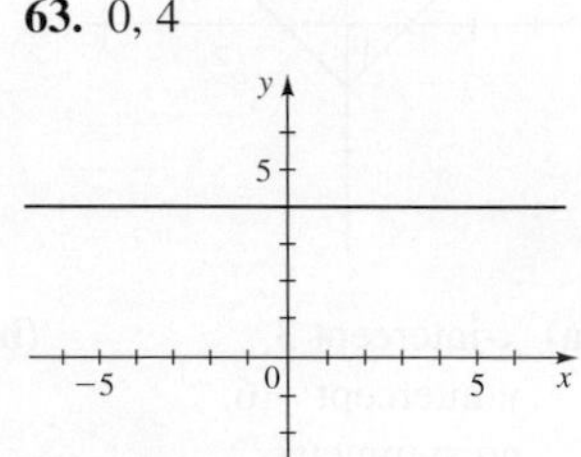

65. Undefined, none

67. 2, -3

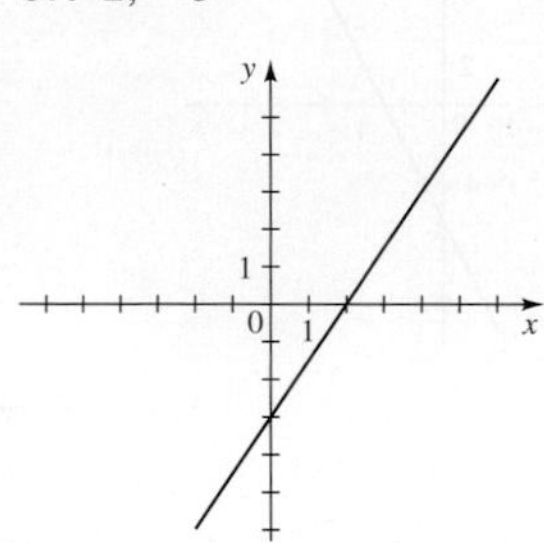

69. -2, 3

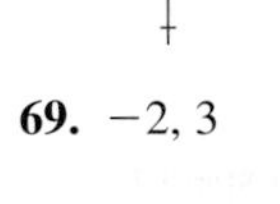

71. $-\frac{2}{3}$, 4

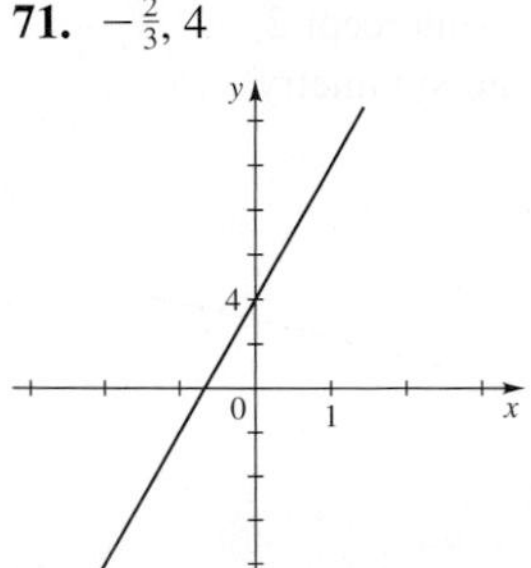

73. Parallel **75.** Perpendicular **77.** Neither
83. $x - y - 3 = 0$ **85. (b)** $4x - 3y - 24 = 0$
89. (a) 8.34; the slope represents an increase of 8.34 mg in dosage for each year of increase in age.
(b) 8.34 mg

91. (a)

(b) The slope represents a cost of \$6 for each toaster oven produced, and the y-intercept represents monthly fixed costs of \$3000.
93. (a) $t = \frac{5}{24}n + 45$ **(b)** 76°F
95. (a) $P = 0.434d + 15$, where P is pressure in kPa and d is depth in meters
(b)

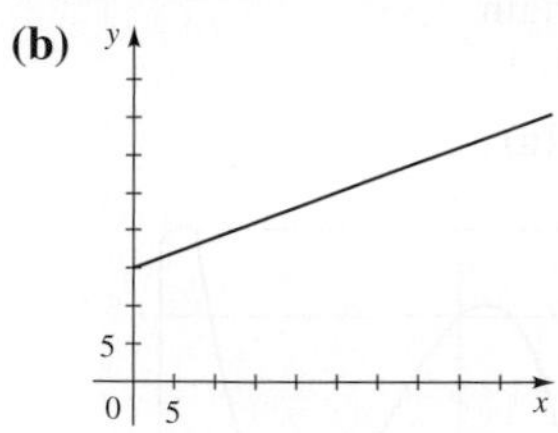

(c) The slope represents an increase of 0.434 kPa in pressure for each one meter increase in depth, and the d-intercept is the air pressure at the surface. **(d)** ≈ 196 m

Section 1.11 ■ Page 122

1. x **2.** above **3. (a)** $x = -1, 0, 1, 3$ **(b)** $[-1, 0] \cup [1, 3]$
4. (a) $x = 1, 4$ **(b)** $(1, 4)$
5. (a)

(b) x-intercepts 0, 1
y-intercept 0
(c) No symmetry

7. (a)

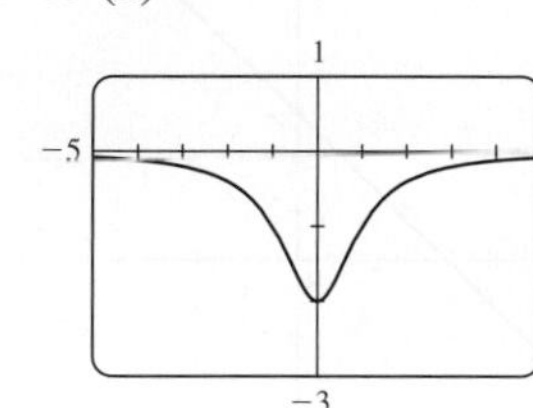

(b) No x-intercept
y-intercept -2
(c) Symmetry with respect to y-axis

9. No **11.** Yes; 2 **13.** 3 **15.** $\frac{5}{14} \approx 0.36$ **17.** $\pm\sqrt{2} \approx \pm 1.41$
19. No solution **21.** $\pm\frac{4}{3} \approx 1.33$
23. $5 + 2\sqrt[4]{5} \approx 7.99$, $5 - 2\sqrt[4]{5} \approx 2.01$ **25.** 5, 6
27. 1.00, 2.00, 3.00 **29.** 1.62 **31.** -1.00, 0.00, 1.00
33. 4 **35.** 4 **37.** 2.55 **39.** -2.05, 0, 1.05
41. $[-2.00, 5.00]$ **43.** $(-\infty, 1.00] \cup [2.00, 3.00]$
45. $(-1.00, 0) \cup (1.00, \infty)$ **47.** $(-\infty, 0)$ **49.** $(-1, 4)$
51. $(-\infty, -5] \cup \{-3\} \cup [2, \infty)$ **53.** 2.27
55. (a)

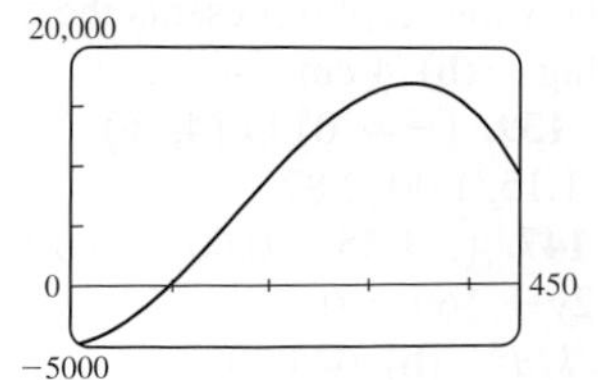

(b) 101 cooktops
(c) $279 < x < 400$

Section 1.12 ■ Page 128

1. directly proportional; proportionality **2.** inversely proportional; proportionality **3.** directly proportional; inversely proportional **4.** $\frac{1}{2}xy$
5. (a) Directly proportional **(b)** Not proportional
6. (a) Not proportional **(b)** Inversely proportional
7. $T = kx$ **9.** $v = k/z$ **11.** $y = ks/t$ **13.** $z = k\sqrt{y}$
15. $V = klwh$ **17.** $R = \dfrac{kP^2t^2}{b^3}$ **19.** $y = 4x$ **21.** $A = \dfrac{75}{r}$
23. $A = \dfrac{18x}{t}$ **25.** $W = 216/r^2$ **27.** $C = 16lwh$
29. $R = \dfrac{27.5}{\sqrt{x}}$ **31. (a)** $z = k\dfrac{x^3}{y^2}$ **(b)** $\frac{27}{4}$
33. (a) $z = kx^3y^5$ **(b)** 864 **35. (a)** $F = kx$ **(b)** 7.5 N/cm **(c)** 45 N **37. (a)** $P = ks^3$ **(b)** 0.012 W/(km/h)^3 **(c)** 324 W
39. ≈ 46 km/h **41.** ≈ 5.3 km/h **43. (a)** $P = kT/V$ **(b)** 8.3 **(c)** ≈ 51.9 kPa **45. (a)** $L = k/d^2$ **(b)** 7000 **(c)** $\frac{1}{4}$ **(d)** 4
47. (a) $R = kL/d^2$ **(b)** $0.00291\overline{6}$ **(c)** $R \approx 137$ ohms **(d)** $\frac{3}{4}$
49. (a) 160,000 **(b)** 1,930,670,340
51. (a) $f = k/L$ **(b)** Halves it **53.** 296 km; 781 km
55. 3.47×10^{-14} W/m^2

Chapter 1 Review ■ Page 135

1. Commutative Property of Addition
3. Distributive Property
5. $-2 \le x < 6$
7. $[5, \infty)$
9. 3 **11.** 4 **13.** $\frac{1}{6}$ **15.** 11 **17. (a)** b^{14} **(b)** $12xy^8$
19. (a) x^2y^2 **(b)** $w^4|z|^5$ **21.** 7.825×10^{10}
23. 1.65×10^{-32} **25.** $(x + 7)(x - 2)$
27. $(x - 1)^2(x + 1)^2$ **29.** $-4(t - 2)(t + 2)$
31. $(x - 1)(x^2 + x + 1)(x + 1)(x^2 - x + 1)$
33. $x^{-1/2}(5x - 3)(x + 1)$ **35.** $(x + 3)(5x^2 - 1)$
37. $(a + b - 5)(a + b + 2)$ **39.** $4y^2 - 49$
41. $2x^3 - 6x^2 + 4x$ **43.** $\dfrac{x + 6}{x + 5}$ **45.** $\dfrac{3x^2 - 7x + 8}{x(x - 2)^2}$
47. $-\dfrac{1}{2x}$ **49.** $\dfrac{x + 4 - 4\sqrt{x}}{x - 4}$ **51.** $\dfrac{\sqrt{11}}{11}$ **53.** $5\sqrt{2} - 5$
55. 5 **57.** No solution **59.** 2, 7 **61.** $-1, \frac{1}{2}$ **63.** $0, \pm\frac{5}{2}$
65. $\dfrac{-2 \pm \sqrt{7}}{3}$ **67.** -5 **69.** 2, 7 **71.** 3, 11 **73. (a)** $3 + i$ **(b)** $8 - i$ **75. (a)** $\frac{6}{5} + \frac{8}{5}i$ **(b)** 2 **77.** $\pm 4i$ **79.** $-3 \pm i$
81. $\pm 4, \pm 4i$ **83.** 20 kg raisins, 30 kg nuts
85. $\frac{1}{4}(\sqrt{329} - 3) \approx 3.78$ km/h **87.** 1 h 50 min
89. $(-3, \infty)$
91. $(-\infty, -1) \cup (8, \infty)$

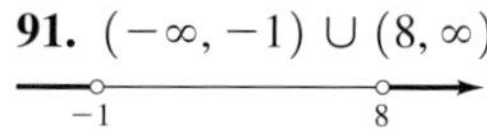

93. $(-\infty, -2) \cup (2, 4]$
95. $[2, 8]$

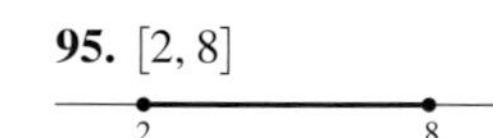

97. (a)

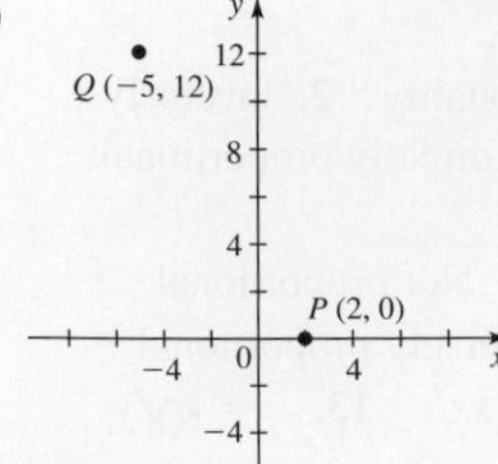

(b) $\sqrt{193}$ **(c)** $\left(-\frac{3}{2}, 6\right)$

(d) $y = -\frac{12}{7}x + \frac{24}{7}$ **(e)** $(x - 2)^2 + y^2 = 193$

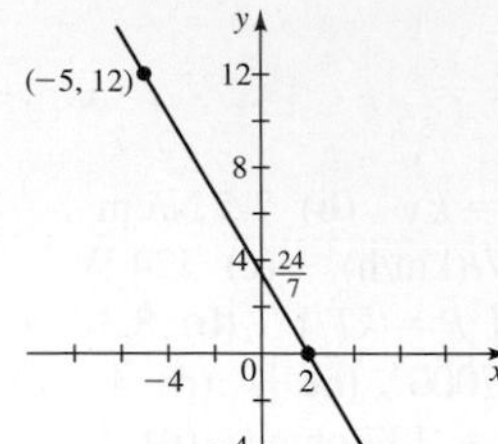

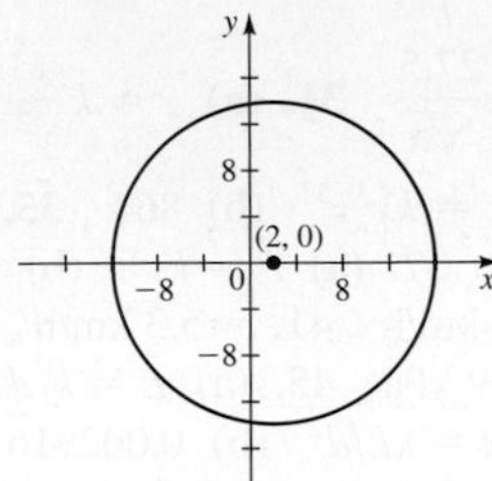

99.

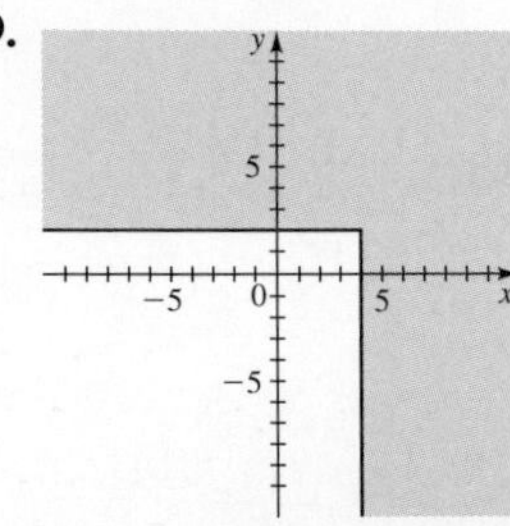

101. *B* **103.** $(x + 5)^2 + (y + 1)^2 = 26$

105. (a) Circle

(b) Center $(4, -1)$, radius 2

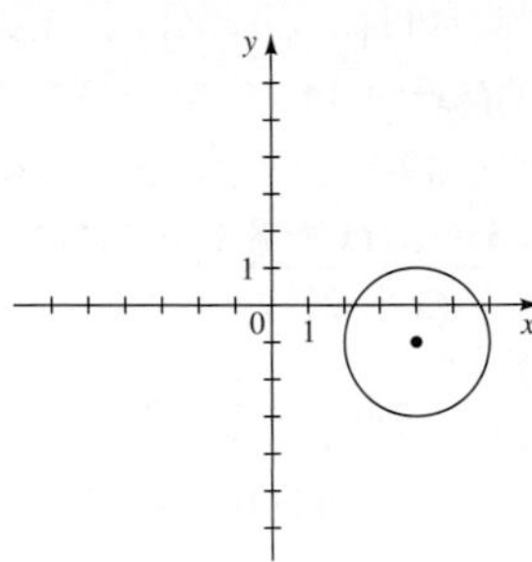

107. (a) No graph

109.

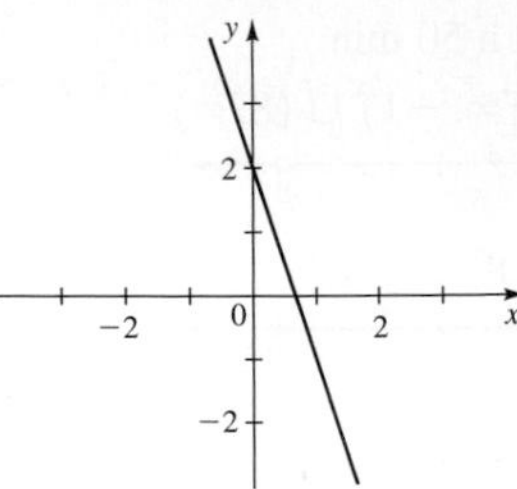

111.

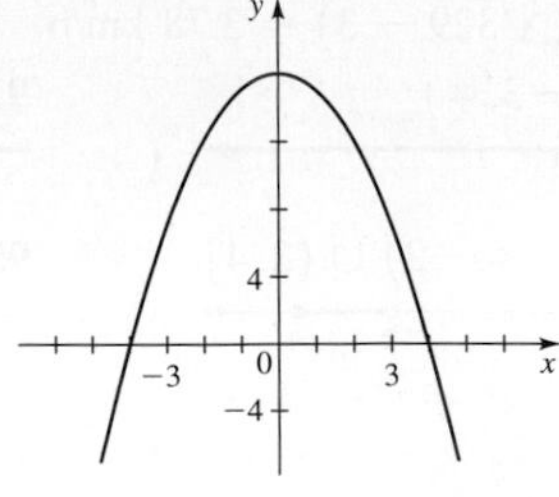

113.

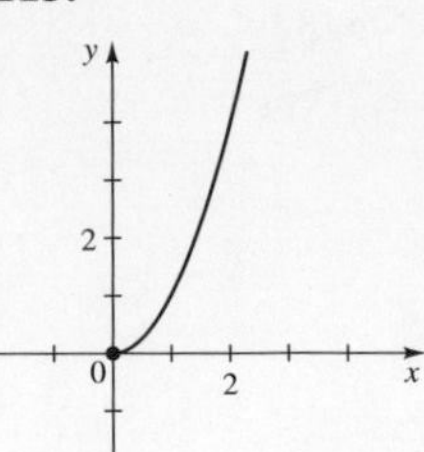

115. (a) Symmetric with respect to *x*-axis

(b) *x*-intercept 16; *y*-intercepts −4, 4

117. (a) Symmetric with respect to *y*-axis

(b) *x*-intercepts −3, 3; *y*-intercept −1

119. (a) Symmetric with respect to origin

(b) *x*-intercepts −1, 1; *y*-intercepts −1, 1

121. (a)

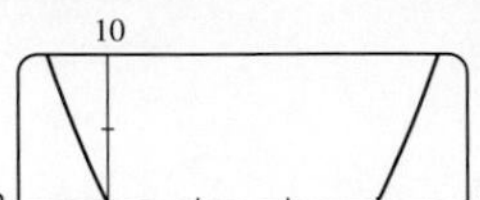

(b) *x*-intercepts 0, 6; *y*-intercept 0

123. (a)

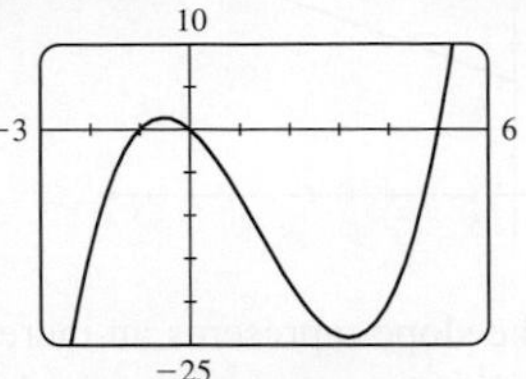

(b) *x*-intercepts −1, 0, 5; *y*-intercept 0

125. (a) $y = 2x + 6$

(b) $2x - y + 6 = 0$

(c)

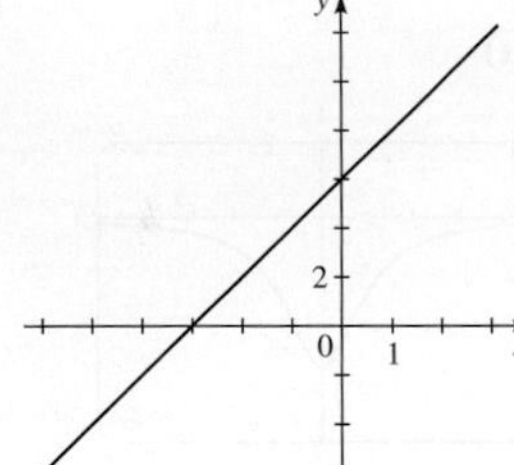

127. (a) $y = \frac{3}{2}x + \frac{5}{2}$

(b) $3x - 2y + 5 = 0$

(c)

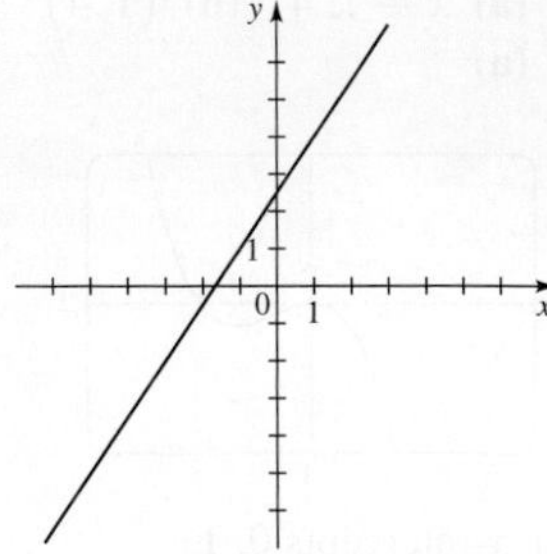

129. (a) $x = 3$

(b) $x - 3 = 0$

(c)

131. (a) $y = -4x$

(b) $4x + y = 0$

(c)

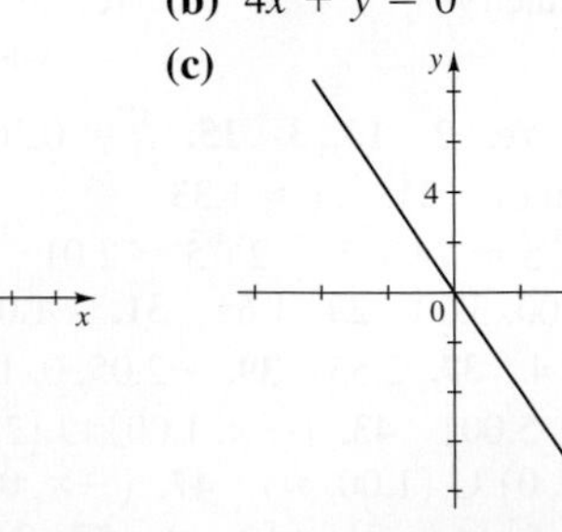

133. (a) The slope represents a stretch of 0.3 cm for each one-kg increase in weight. The *s*-intercept represents the unstretched length of the spring. **(b)** 4 cm

135. −1, 6 **137.** $[-1, 6]$ **139.** $(-\infty, 0] \cup [4, \infty)$

141. −1, 7 **143.** −2.72, −1.15, 1.00, 2.87

145. $(-\infty, -6) \cup (2, \infty)$ **147.** $(-1.85, -0.60) \cup (0.45, 2.00)$

149. $x^2 + y^2 = 169$, $5x - 12y + 169 = 0$

151. $M = 8z$ **153. (a)** $I = k/d^2$ **(b)** 64,000

(c) 160 candles **155.** 11.0 km/h **157.** 1460 m/s

159. 2.4×10^5 km/s; ≈11,538 Mly

160. **(a)** III **(b)** V **(c)** II **(d)** IV **(e)** I **(f)** VII **(g)** VIII **(h)** VI

Chapter 1 Test ■ Page 139

1. **(a)**

(b) $(-\infty, 3], [-1, 4)$ **(c)** 16 **2.** **(a)** 81 **(b)** -81 **(c)** $\frac{1}{81}$ **(d)** 27 **(e)** $\frac{9}{4}$ **(f)** $\frac{1}{8}$ **3.** **(a)** 1.86×10^{11} **(b)** 3.965×10^{-7}

4. **(a)** $6\sqrt{2}$ **(b)** $48a^5b^7$ **(c)** $\dfrac{y^2}{2x^4}$

5. **(a)** $3z$ **(b)** $4x^2 + 7x - 15$ **(c)** $a - b$ **(d)** $4x^2 + 12x + 9$ **(e)** $x^3 + 6x^2 + 12x + 8$

6. **(a)** $(2x - 5)(2x + 5)$ **(b)** $(2x - 3)(x + 4)$ **(c)** $(x - 3)(x - 2)(x + 2)$ **(d)** $x(x + 3)(x^2 - 3x + 9)$ **(e)** $2x^{-1/2}(x + 5)(x - 1)$ **(f)** $x^2y^2(x - 3)(x + 3)$

7. **(a)** $\dfrac{w + 3}{w - 3}$ **(b)** $\dfrac{1}{x - 2}$ **(c)** $-(x + y)$ **8.** $\dfrac{1 + 2\sqrt{2}}{7}$

9. **(a)** 6 **(b)** 1 **(c)** $-3, 4$ **(d)** $-1 \pm \dfrac{\sqrt{2}}{2}$ **(e)** No real solution **(f)** $\pm 1, \pm\sqrt{2}$ **(g)** $\frac{2}{3}, \frac{22}{3}$ **10.** **(a)** $7 + i$ **(b)** $-1 - 5i$ **(c)** $18 + i$ **(d)** $\frac{6}{25} - \frac{17}{25}i$ **(e)** 1 **(f)** $6 - 2i$

11. $-1 \pm \dfrac{\sqrt{2}}{2}i$ **12.** 120 km **13.** 50 m by 120 m

14. **(a)** $[-4, 3)$

(b) $(-2, 0) \cup (1, \infty)$

(c) $(1, 7)$

(d) $(1, 2) \cup (3, \infty)$

15. Between 41°F and 50°F **16.** $0 \le x \le 6$

17. **(a)** $S(3, 6)$ **(b)** 18

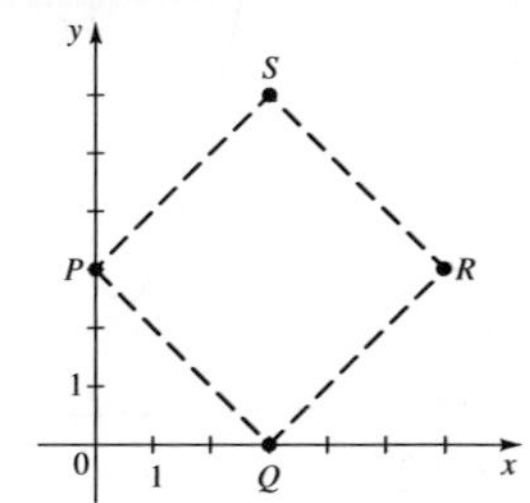

18. **(a)**

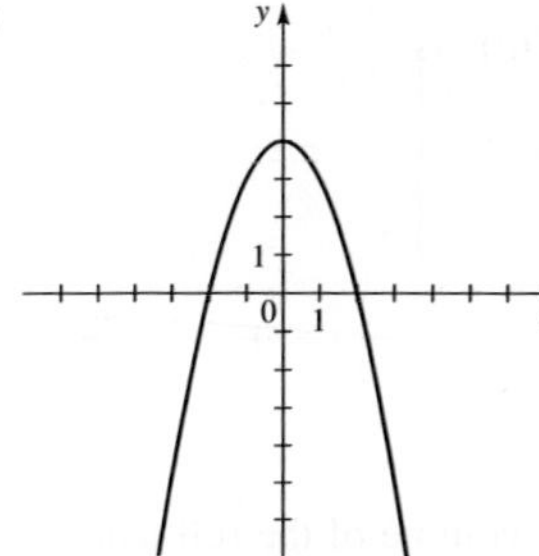

(b) x-intercepts $-2, 2$
y-intercept 4
(c) Symmetric with respect to y-axis

19. **(a)**

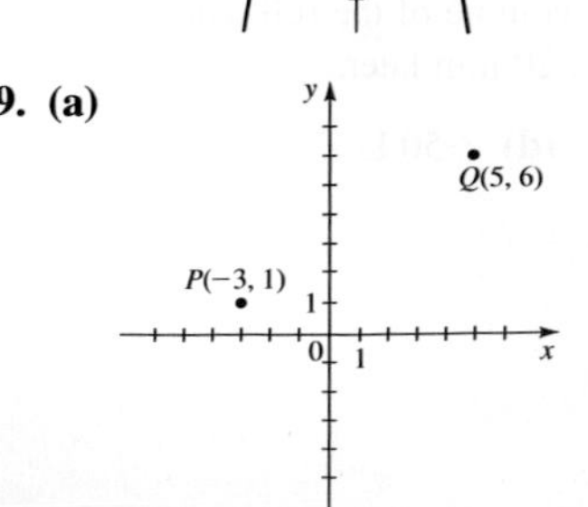

(b) $\sqrt{89}$ **(c)** $\left(1, \frac{7}{2}\right)$ **(d)** $(x - 1)^2 + \left(y - \frac{7}{2}\right)^2 = \frac{89}{4}$

20. **(a)** $C(0, 0)$; $r = \sqrt{5}$ **(b)** $C(-1, 3)$; $r = 2$

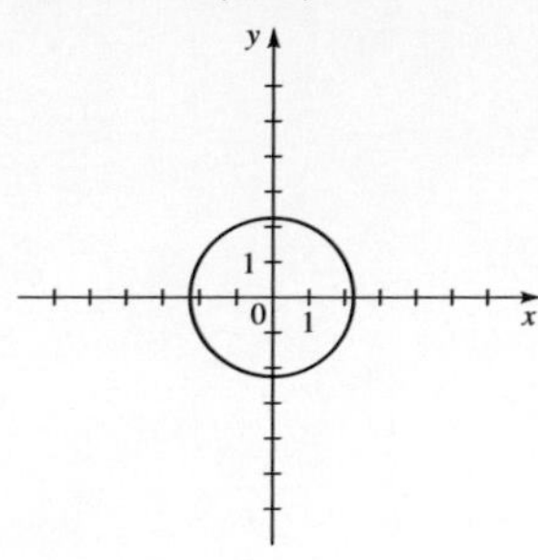

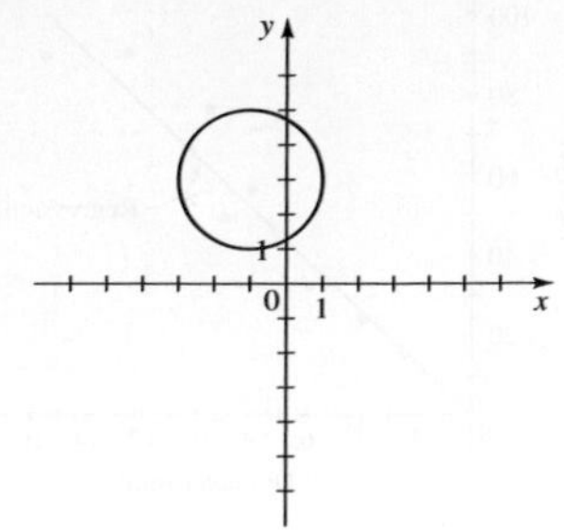

(c) $C(5, 0)$; $r = 3$

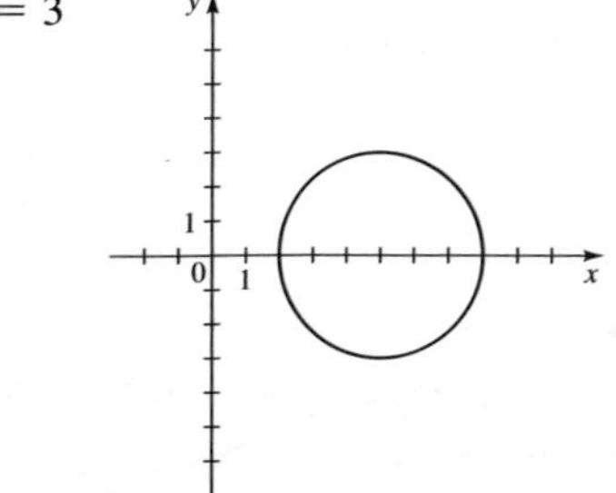

21. $y = \frac{2}{3}x - 5$

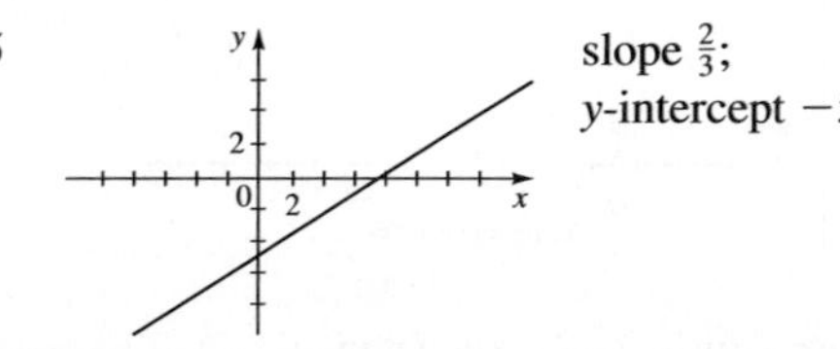

slope $\frac{2}{3}$; y-intercept -5

22. **(a)** $2x - y - 5 = 0$ **(b)** $3x + y - 3 = 0$ **(c)** $2x + 3y - 12 = 0$

23. **(a)** 4°C **(b)**

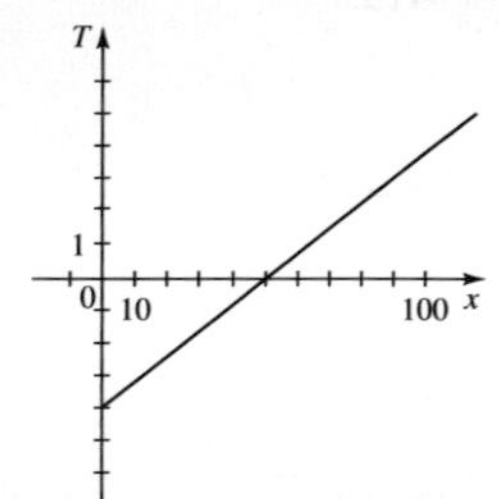

(c) The slope represents an increase of 0.08°C for each one-centimeter increase in depth, the x-intercept is the depth at which the temperature is 0°C, and the T-intercept is the temperature at ground level.

24. **(a)** $-2.94, -0.11, 3.05$ **(b)** $[-1, 2]$

25. **(a)** $M = kwh^2/L$ **(b)** 400 **(c)** 12,000 N

Focus on Modeling ■ Page 144

1. **(a)** $y = 1.8807x + 82.65$

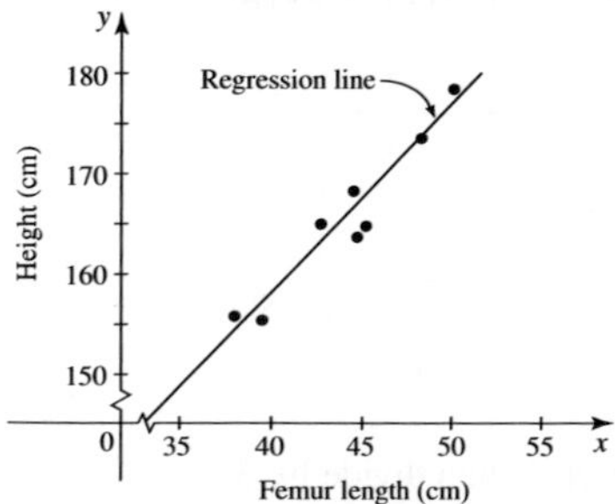

(b) 191.7 cm

3. (a) $y = 6.451x - 0.1523$

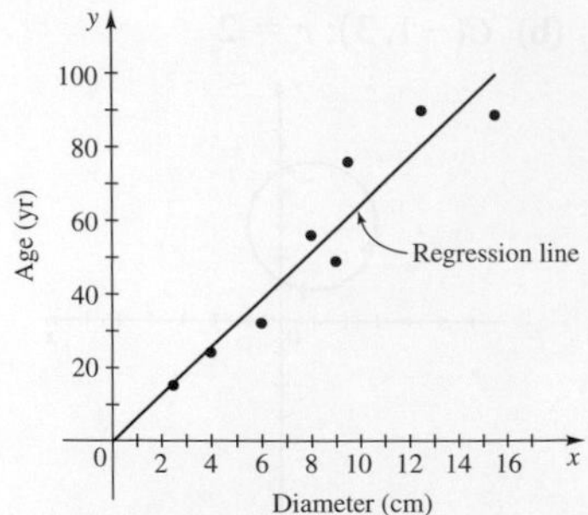

(b) 116 years

5. (a) $y = -0.13198x + 7.2514$

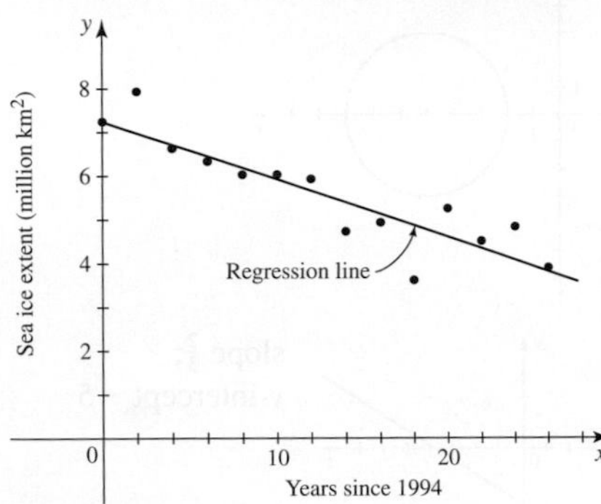

(b) 3.95 million km^2 **(c)** Unlikely to be accurate far into the future

7. (a) $y = -3.9018x + 419.7$

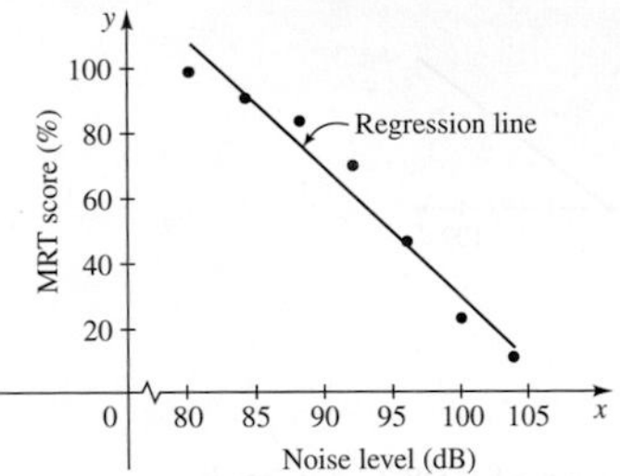

(b) The correlation coefficient is -0.98, so a linear model is appropriate. **(c)** 53%

Chapter 2

Section 2.1 ■ Page 155

1. (a) $f(-1) = 0$ **(b)** $f(2) = 9$ **(c)** $f(2) - f(-1) = 9$
2. domain, range **3. (a)** f and g **(b)** $f(5) = 10, g(5) = 0$
4. (a) square, add 3

(b)

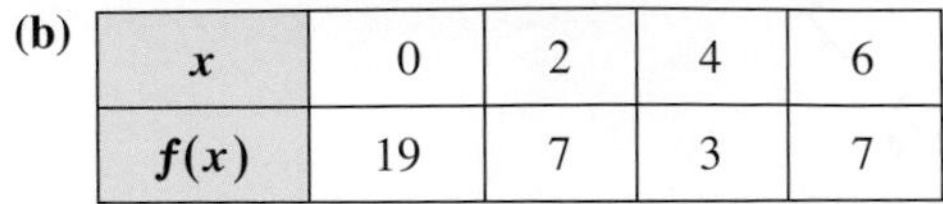

x	0	2	4	6
$f(x)$	19	7	3	7

5. one; (b) **6. (a)** 4, 4 **(b)** Yes **7.** Yes **8.** No
9. $f(x) = 3x - 5$ **11.** $f(x) = \sqrt{x^2 + 1}$
13. Multiply by 5, then add 1
15. Take the square root, subtract 4, then divide by 3

17.

(input)		(output)
1	subtract 1, take square root	0
2	subtract 1, take square root	1
5	subtract 1, take square root	2

19.

x	$f(x)$
−1	8
0	2
1	0
2	2
3	8

21. $13, 13, 1, \frac{4}{3}, 16$ **23.** $\frac{3}{5}, \frac{1}{5}, -\frac{1}{5}, \frac{1+a}{5}, \frac{1-x^2}{5}, \frac{3-a}{5}$

25. $0, 15, 3, a^2 + 2a, x^2 - 2x, \frac{1}{a^2} + \frac{2}{a}$

27. $-\frac{1}{3}$, undefined, $\frac{1}{3}, \frac{1-a}{1+a}, \frac{2-a}{a}, \frac{2-x^2}{x^2}$

29. $5, 1, 11, 16 - \sqrt{5}, 3a^2 - 7a + 5, 3x^4 - x^2 + 1$
31. $6, 2, 1, 2, 2|x|, 2(x^2 + 1)$ **33.** $-14, 1, 2, 24, 35$
35. $8, -\frac{3}{4}, -1, 0, -1$ **37.** $x^2 + 4x + 5, x^2 + 6$
39. $x^2 + 4, x^2 + 8x + 16$ **41.** 12 **43.** -21
45. $3 - a, 3 - a - h, -1$ **47.** 5, 5, 0

49. $\frac{a}{a+1}, \frac{a+h}{a+h+1}, \frac{1}{(a+h+1)(a+1)}$

51. $3 - 5a + 4a^2, 3 - 5a - 5h + 4a^2 + 8ah + 4h^2$, $-5 + 8a + 4h$ **53.** $(-\infty, \infty), (-\infty, \infty)$ **55.** $(-\infty, \infty), [3, \infty)$
57. $[-2, 6], [-6, 18]$ **59.** $\{x \mid x \neq -3\}$ **61.** $\{x \mid x \neq \pm 1\}$
63. $(-\infty, 2]$ **65.** $(-\infty, \infty)$ **67.** $(-\infty, -5] \cup [5, \infty)$
69. $[-2, 3) \cup (3, \infty)$ **71.** $(-\infty, 0] \cup [6, \infty)$
73. $(-\infty, 2)$ **75.** $(\frac{1}{2}, \infty)$
77. (a) $f(x) = x^2 + 1$

(b)

x	$f(x)$
−2	5
−1	2
0	1
1	2
2	5

(c)

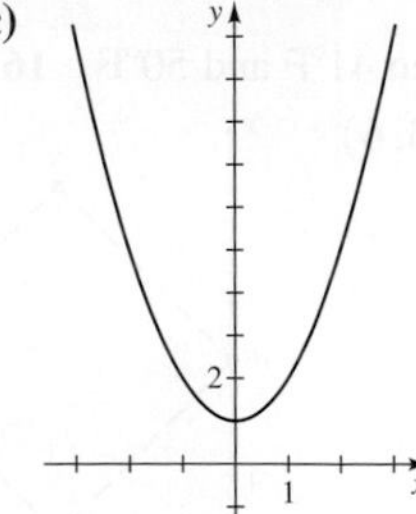

79. (a) $T(x) = 0.08x$

(b)

x	$T(x)$
2	0.16
4	0.32
6	0.48
8	0.64

(c)

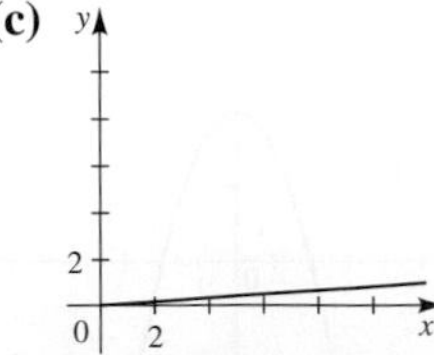

81. $(-\infty, \infty), \{1, 5\}$
83. (a) 50, 0 **(b)** $V(0)$ is the volume of the full tank, and $V(20)$ is the volume of the empty tank, 20 min later.

(c)

x	$V(x)$
0	50
5	28.125
10	12.5
15	3.125
20	0

(d) -50 L

85. **(a)** 8.66 m, 6.61 m, 4.36 m **(b)** The object will appear to get shorter.
87. **(a)** 2 mm, 1.66 mm, 1.48 mm

(b)

x	$R(x)$
1	2
10	1.66
100	1.48
200	1.44
500	1.41
1000	1.39

(c) −0.18 mm

89. **(a)** 35.7 km, 50.5 km **(b)** 66.4 km **(c)** 374.6 km **(d)** 194.3 km **91.** **(a)** 0, 160, 1550 **(b)** The amount of tax paid on incomes of 5000, 12,000, and 25,000 dollars

93. **(a)** $T(x) = \begin{cases} 114x & \text{if } 0 \le x \le 2 \\ 228 + 99(x - 2) & \text{if } x > 2 \end{cases}$

(b) \$228, \$327, \$525 **(c)** Total cost of staying at the hotel

95.

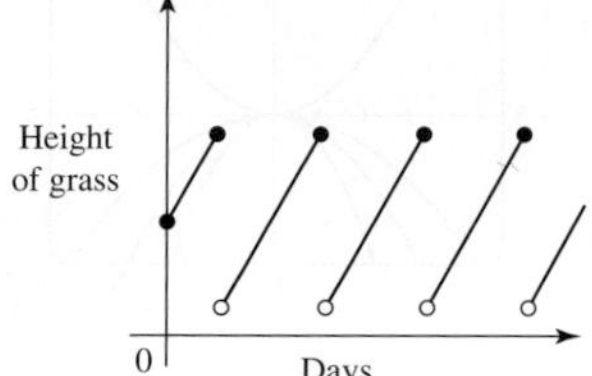

97.

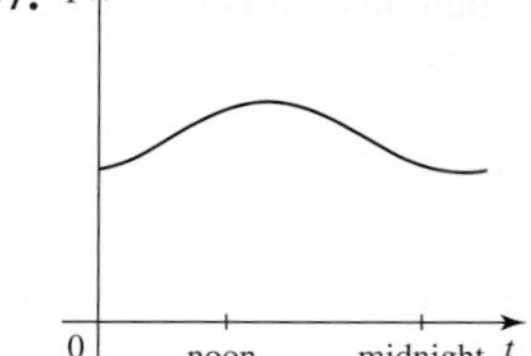

Section 2.2 ■ Page 168

1. $f(x), x^2 - 2, 7, 7$

x	$y = f(x)$	(x, y)
−2	2	(−2, 2)
−1	−1	(−1, −1)
0	−2	(0, −2)
1	−1	(1, −1)
2	2	(2, 2)

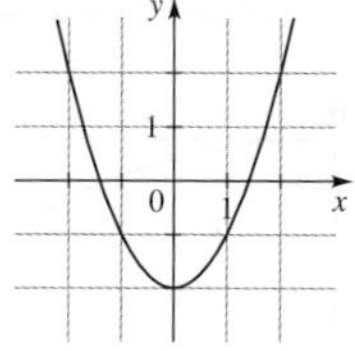

2. 10 **3.** 7 **4.** **(a)** IV **(b)** II **(c)** I **(d)** III
5. The input 1 has two different outputs, 1 and 2.
6. There are 2 different y-values that correspond to an x-value. For instance, $x = 1$ corresponds to $y = \frac{1}{2}$ and to $y = -\frac{1}{2}$.
7. The input 10 is mapped to two different outputs, 10 and 15.
8. The curve does not pass the Vertical Line Test.

9.

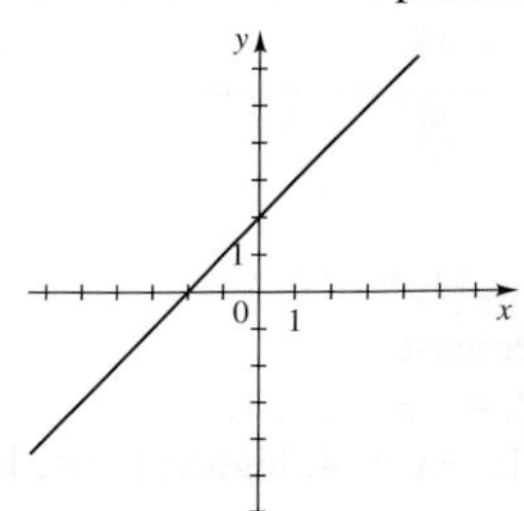

11.

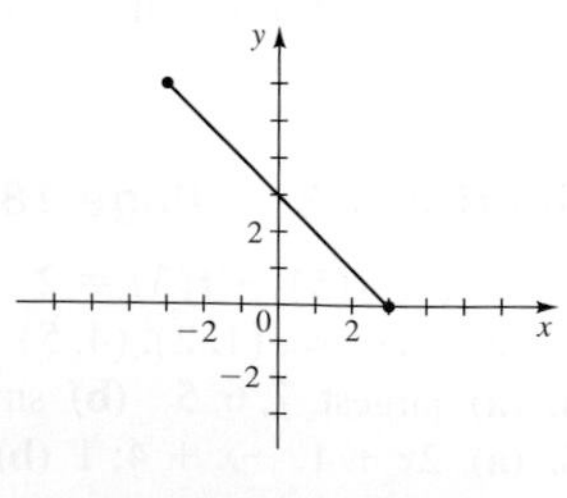

13.

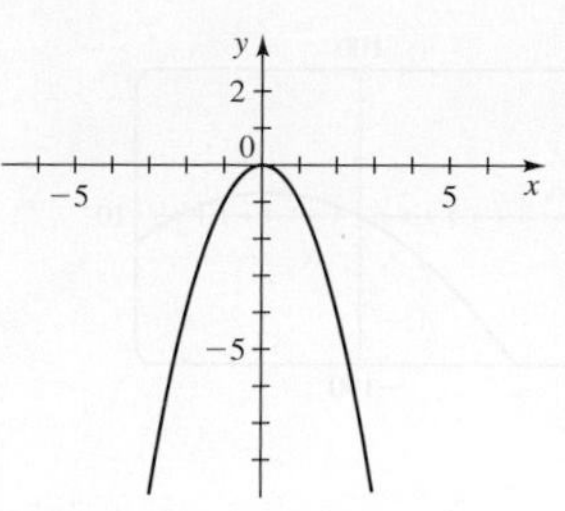

15.

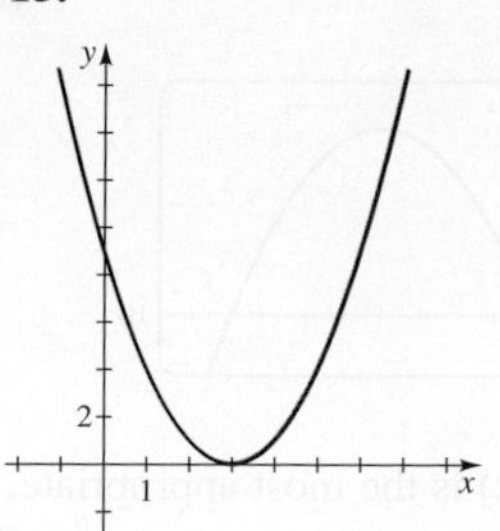

17.

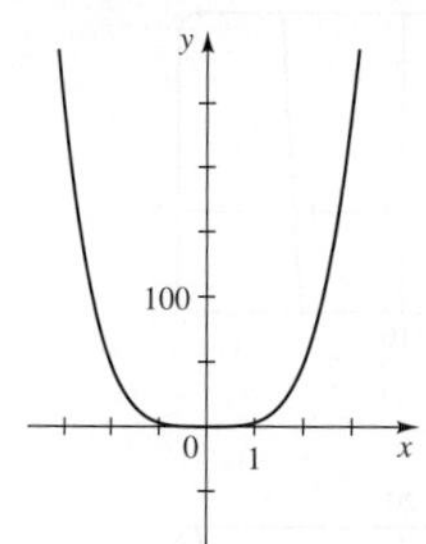

19.

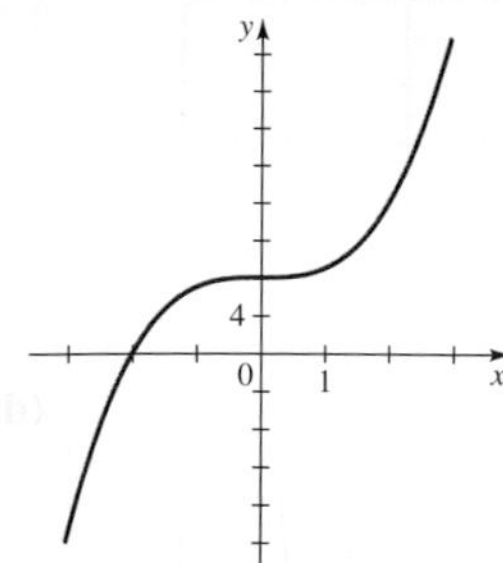

21.

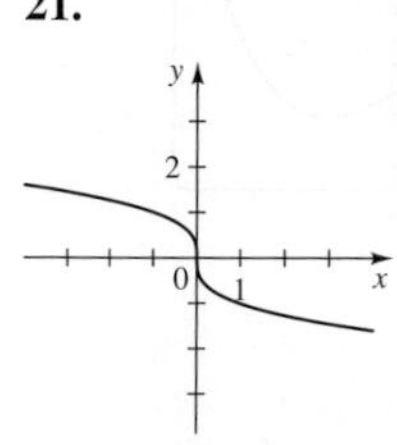

23.

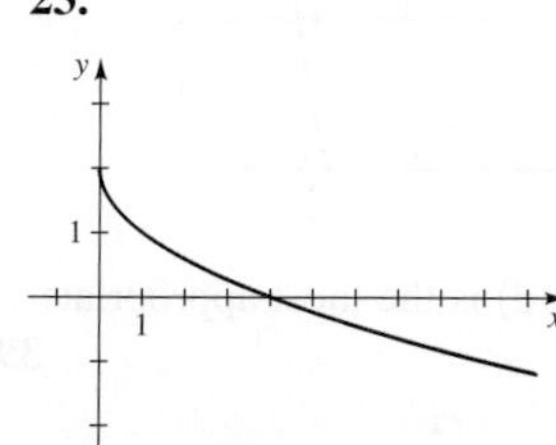

25.

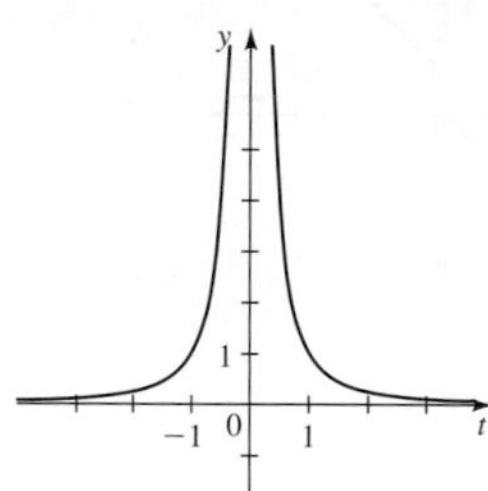

27.

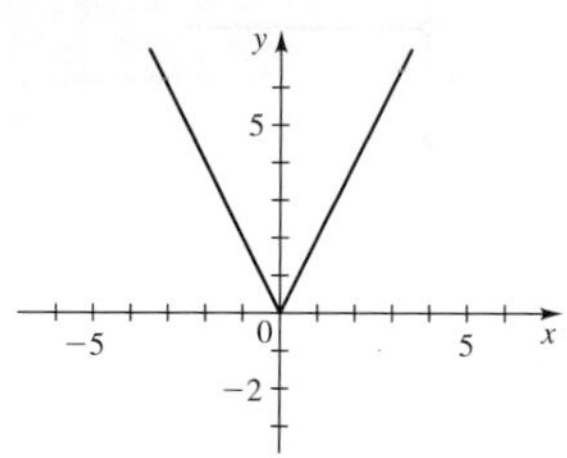

29.

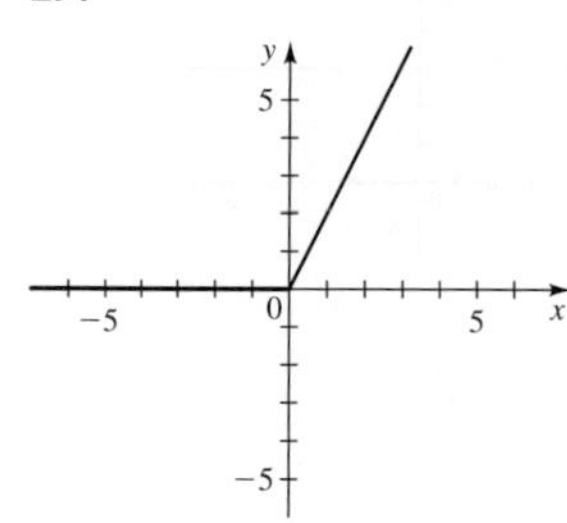

31.

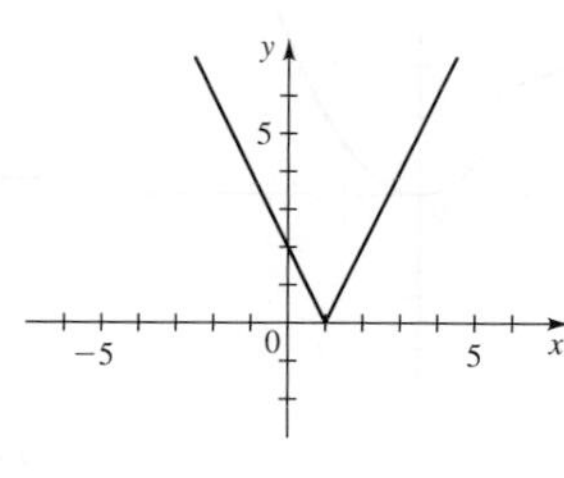

33. **(a)**

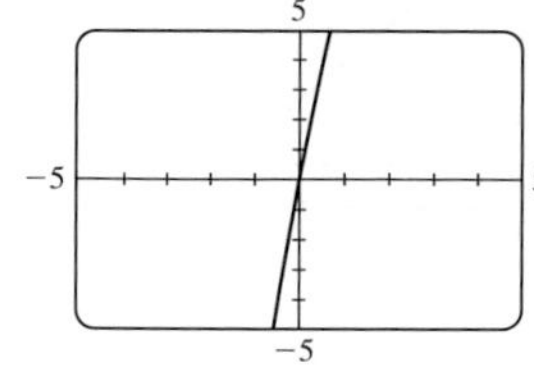

(b)

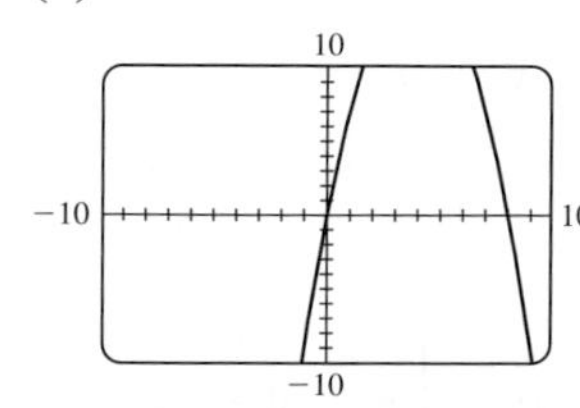

(c)

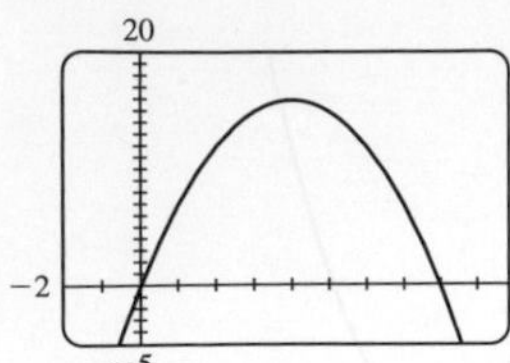

(d)

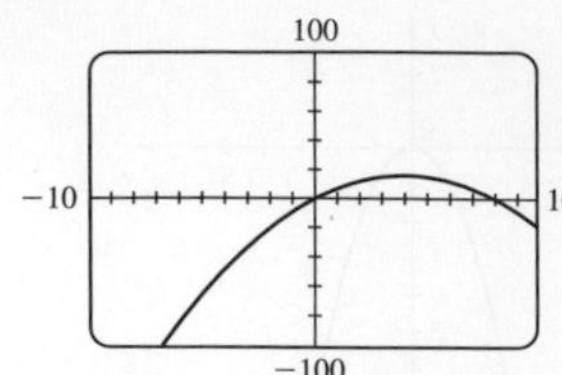

Graph (c) is the most appropriate.

35. (a)

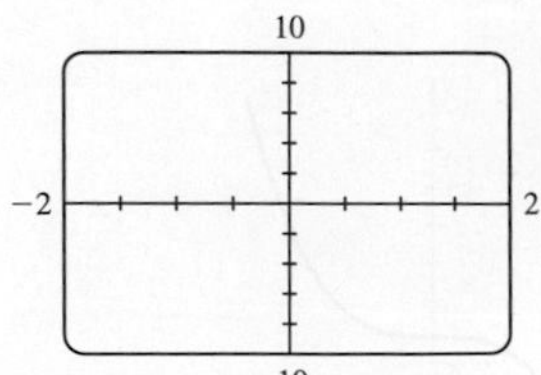

(b)

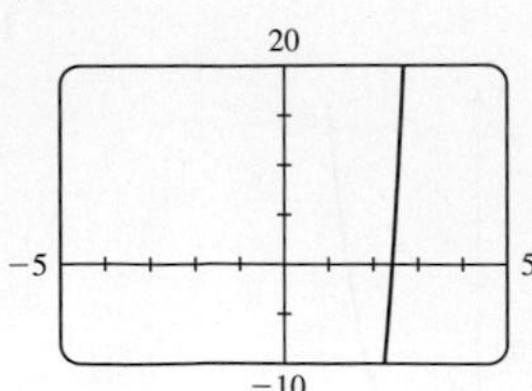

(c)

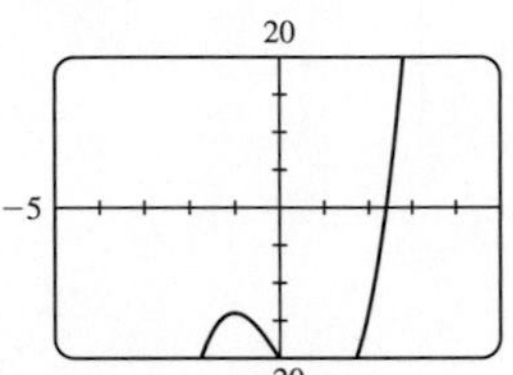

(d)

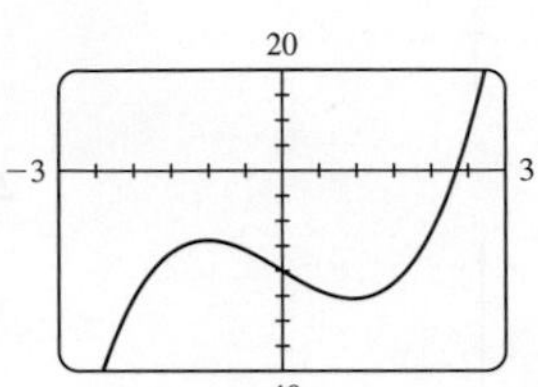

Graph (d) is the most appropriate.

37. **39.**

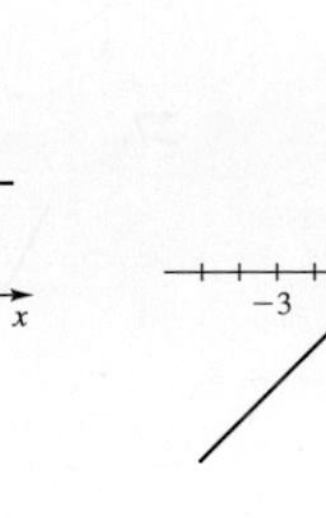

41.

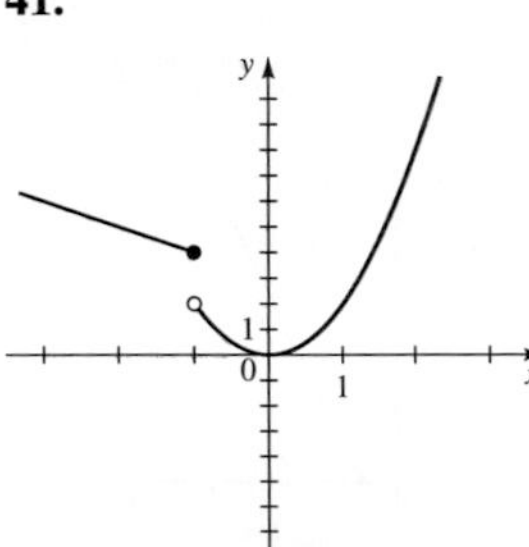

43.

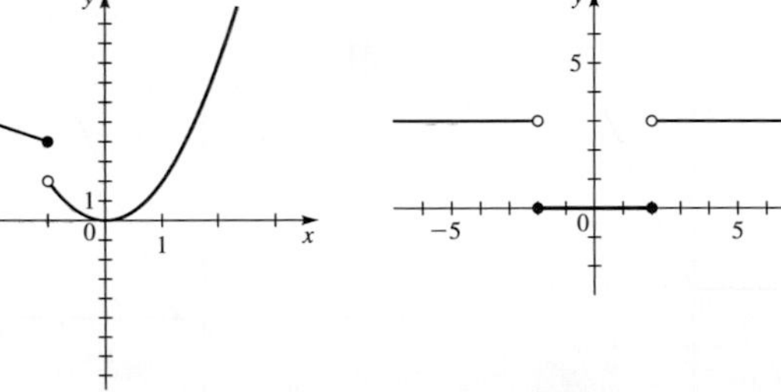

45.

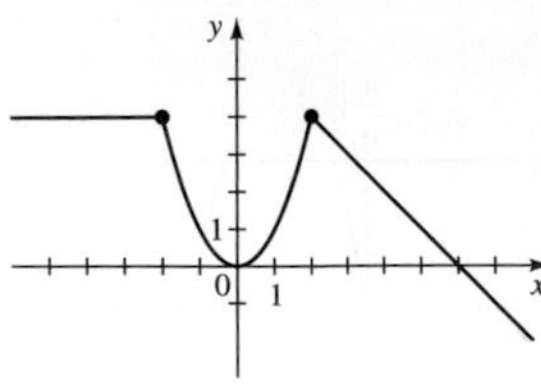

47.

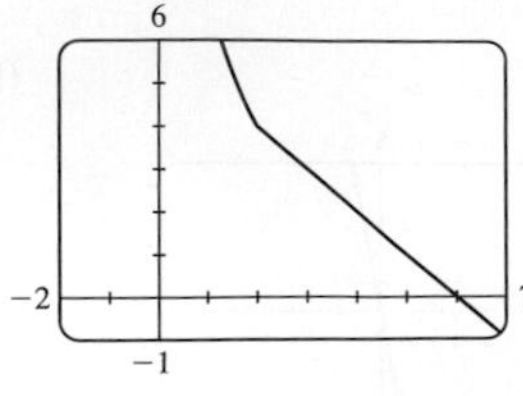

49. $f(x) = \begin{cases} -2 & \text{if } x < -2 \\ x & \text{if } -2 \le x \le 2 \\ 2 & \text{if } x > 2 \end{cases}$

51. (a) Yes **(b)** No **(c)** Yes **(d)** No **53.** Yes
55. Yes **57.** No **59.** No **61.** No **63.** Yes **65.** Yes
67. Not a function; domain $\{0, 1, 4, 5, 6\}$, range $\{1, 2, 3\}$
69. Function

71. (a)

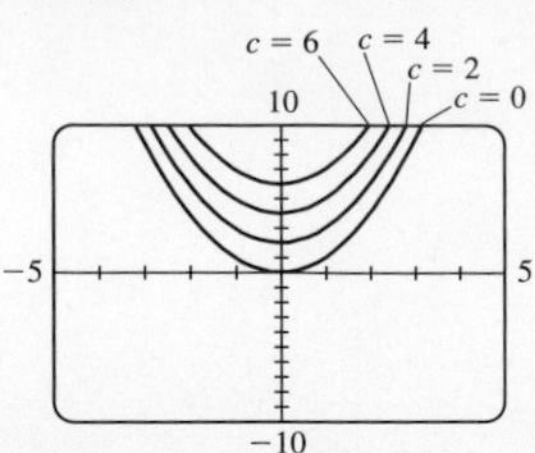

(b)

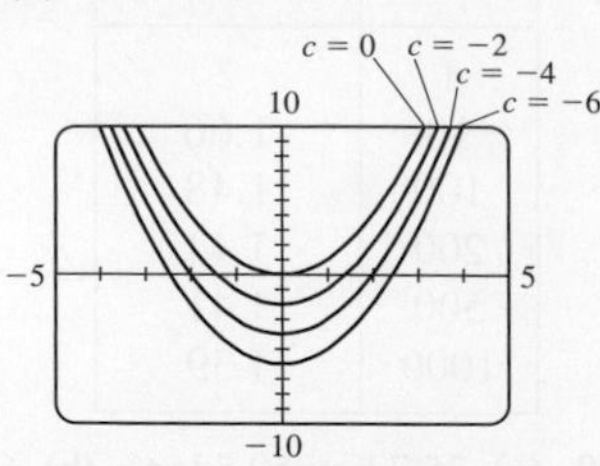

(c) If $c > 0$, then the graph of $f(x) = x^2 + c$ is the same as the graph of $y = x^2$ shifted upward c units. If $c < 0$, then the graph of $f(x) = x^2 + c$ is the same as the graph of $y = x^2$ shifted downward c units.

73. (a)

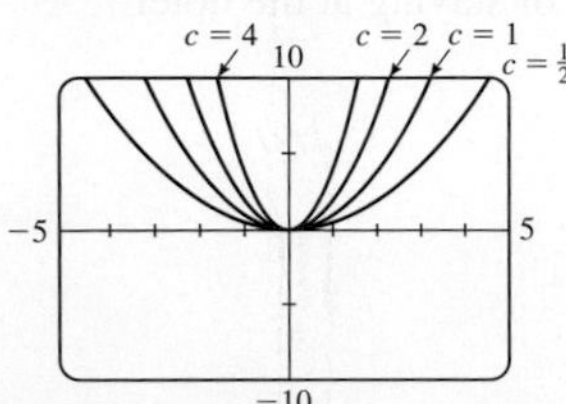

(b)

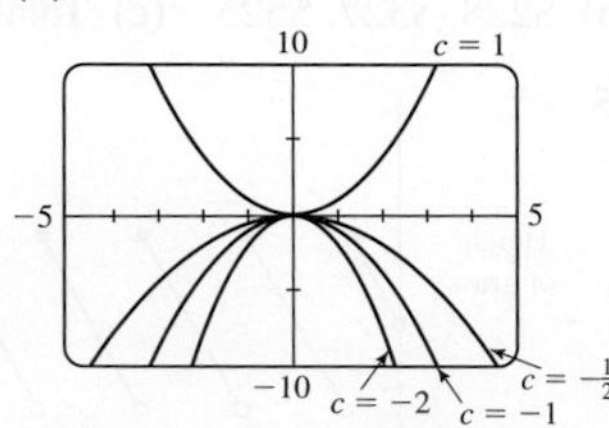

(c) As $|c|$, increases, the graph of $f(x) = cx^2$ is stretched vertically. As $|c|$ decreases, the graph of f is flattened. When $c < 0$, the graph is reflected about the x-axis.

75. $f(x) = -\frac{7}{6}x - \frac{4}{3}$, $-2 \le x \le 4$

77. $f(x) = \sqrt{9 - x^2}$, $-3 \le x \le 3$

79.

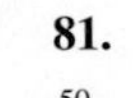

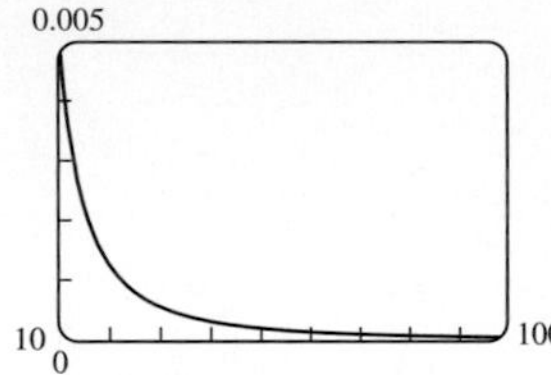

81.

83.

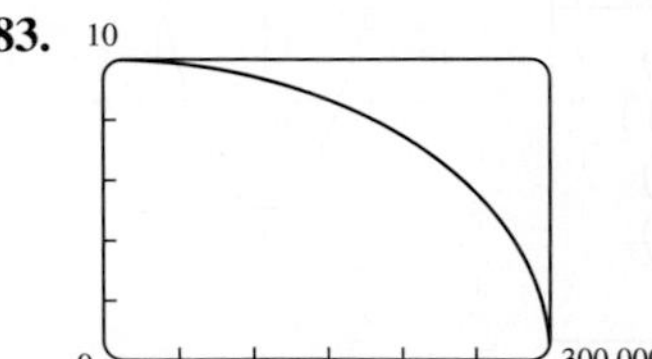

85. $P(x) = \begin{cases} 0.60 & \text{if } 0 < x \le 1 \\ 0.84 & \text{if } 1 < x \le 2 \\ 1.08 & \text{if } 2 < x \le 3 \\ 1.32 & \text{if } 3 < x \le 3.5 \end{cases}$

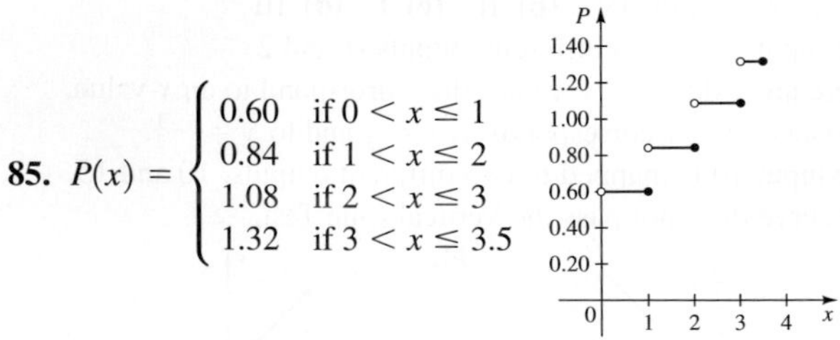

Section 2.3 ▪ Page 180

1. a, 4, 6, $f(5) - f(3) = 2$ **2.** x, y, $[1, 7]$, $[0, 7]$
3. (a) increase, $(1, 2)$, $(4, 5)$ **(b)** decrease, $(2, 4)$, $(5, 7)$
4. (a) largest, 7, 6, 5 **(b)** smallest, 2, 4 **5.** x; x; 1, 7, $[1, 7]$
6. (a) $2x + 1$, $-x + 4$; 1 **(b)** $2x + 1$, $-x + 4$, higher; $(-\infty, 1)$

7. (a) $1, -1, 3, 4$ **(b)** Domain $[-3, 4]$, range $[-1, 4]$ **(c)** $-3, 2, 4$ **(d)** $-3 \le x \le 2$ and $x = 4$ **(e)** 1 **9. (a)** $f(0)$ **(b)** $f(-1)$ **(c)** $-2, 2$ **(d)** $\{x \mid -4 \le x \le -2 \text{ or } 2 \le x \le 4\}$ **(e)** $\{x \mid -2 < x < 2\}$ **11.** Domain $(-3, 3]$, range $[-2, 3]$ **13.** Domain $[-3, 3]$, range $\{-3, -2, 3\}$

15. (a)

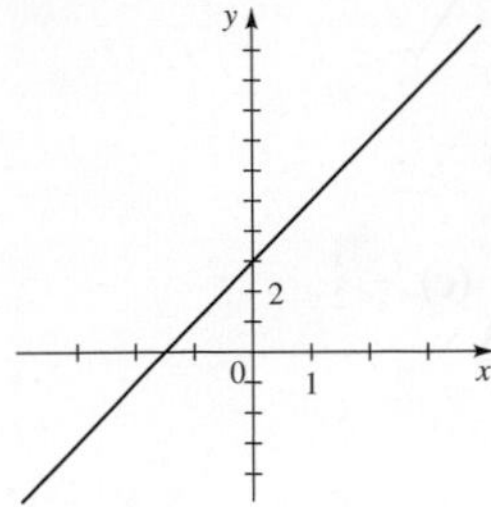

(b) $(-\infty, \infty), (-\infty, \infty)$

17. (a)

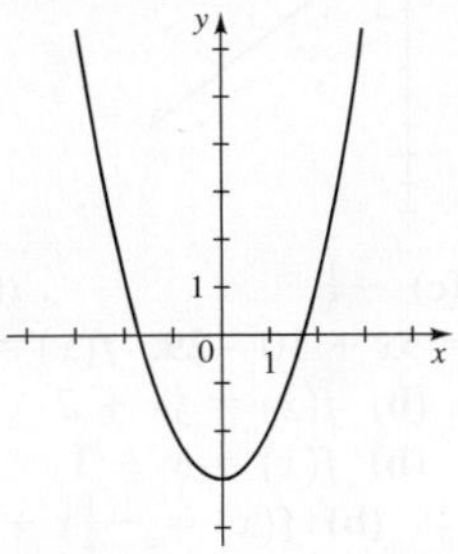

(b) $(-\infty, \infty), [-3, \infty)$

19. (a)

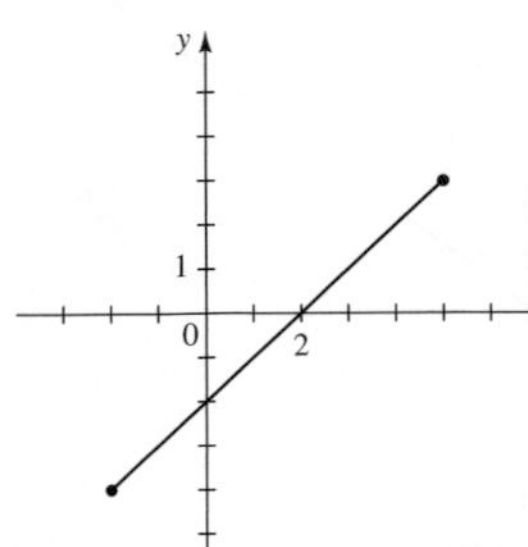

(b) $[-2, 5], [-4, 3]$

21. (a)

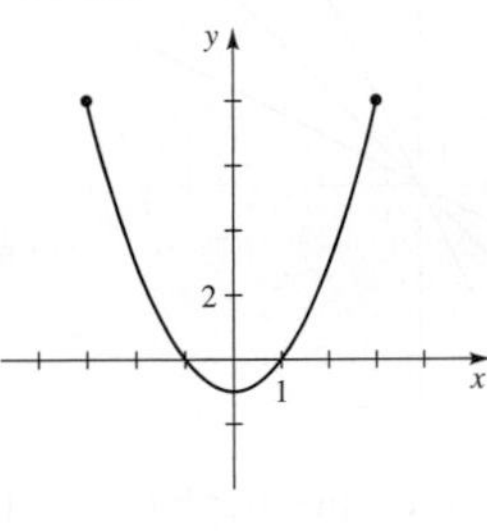

(b) $[-3, 3], [-1, 8]$

23. (a)

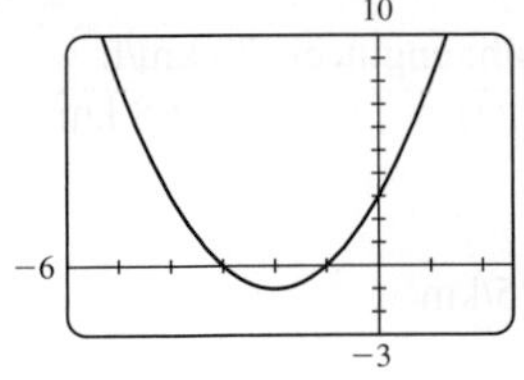

(b) $(-\infty, \infty), [-1, \infty)$

25. (a)

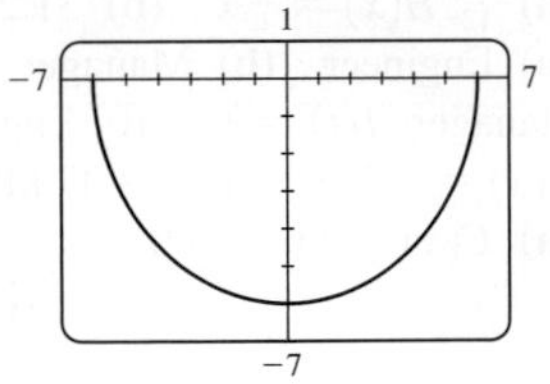

(b) $[-6, 6], [-6, 0]$

27. (a)

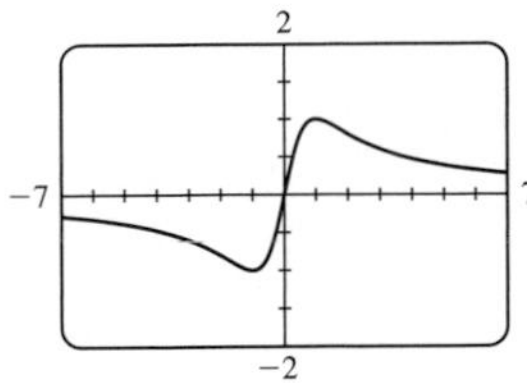

(b) $(-\infty, \infty), [-1, 1]$

29. (a) $x = 2$ **(b)** $x < 2$

31. (a) $x = -2, 1$ **(b)** $-2 \le x \le 1$

33. (a) $x \approx -4.32, -1.12, 1.44$ **(b)** $-4.32 \le x \le -1.12$ or $x \ge 1.44$

35. (a) $x = -1, -0.25, 0.25$ **(b)** $-1 \le x \le -0.25$ or $x \ge 0.25$

37. (a) Domain $[-1, 4]$, range $[-1, 3]$ **(b)** Increasing on $(-1, 1)$ and $(2, 4)$, decreasing on $(1, 2)$

39. (a) Domain $[-3, 3]$, range $[-2, 2]$ **(b)** Increasing on $(-2, -1)$ and $(1, 2)$, decreasing on $(-3, -2)$, $(-1, 1)$, and $(2, 3)$

41. (a)

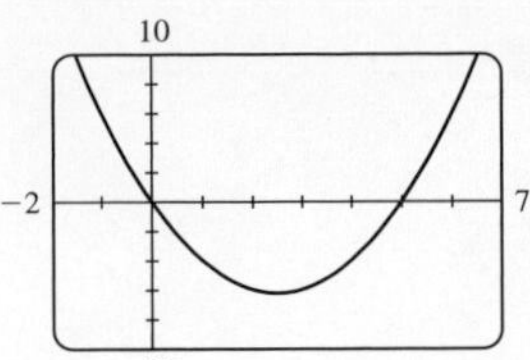

(b) Domain $(-\infty, \infty)$, range $[-6.25, \infty)$ **(c)** Increasing on $(2.5, \infty)$; decreasing on $(-\infty, 2.5)$

43. (a)

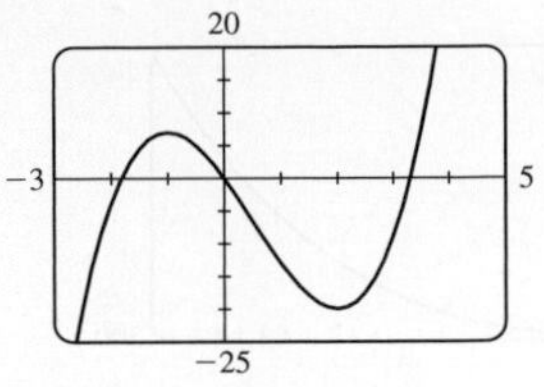

(b) Domain $(-\infty, \infty)$, range $(-\infty, \infty)$ **(c)** Increasing on $(-\infty, -1)$, $(2, \infty)$; decreasing on $(-1, 2)$

45. (a)

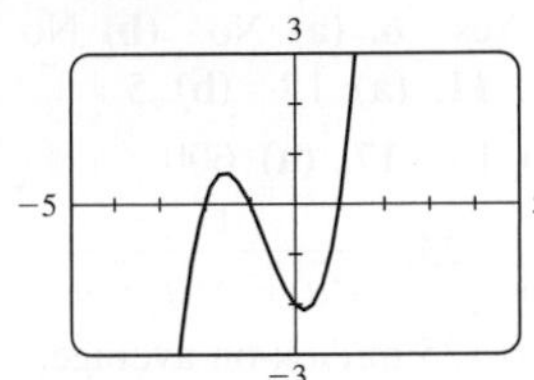

(b) Domain $(-\infty, \infty)$, range $(-\infty, \infty)$ **(c)** Increasing on $(-\infty, -1.55)$, $(0.22, \infty)$; decreasing on $(-1.55, 0.22)$

47. (a)

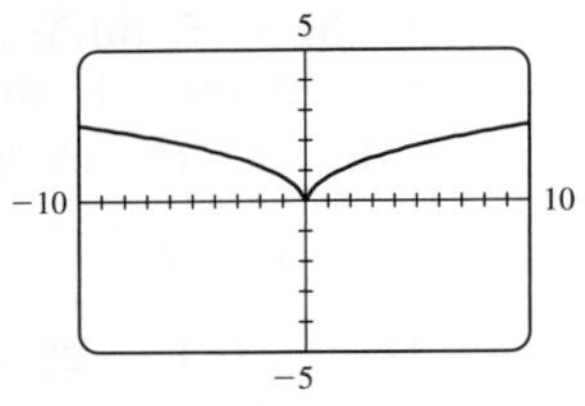

(b) Domain $(-\infty, \infty)$, range $[0, \infty)$ **(c)** Increasing on $(0, \infty)$; decreasing on $(-\infty, 0)$

49. (a)

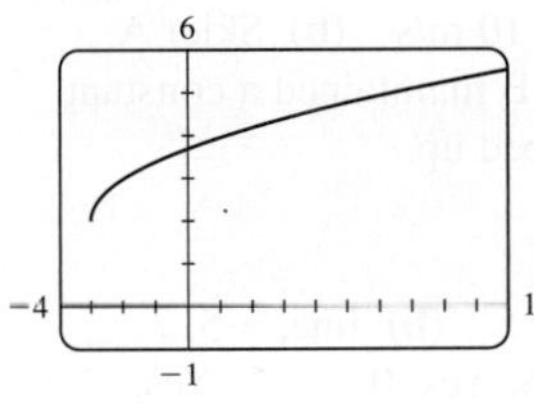

(b) Domain $[-3, \infty)$, range $[2, \infty)$ **(c)** Increasing on $(-3, \infty)$

51. (a) Local maximum 3 when $x = -1$, local maximum 4 when $x = 3$; local minimum -3 when $x = 1$ **(b)** Increasing on $(-\infty, -1), (1, 3)$; decreasing on $(-1, 1), (3, \infty)$

53. (a) Local maximum 3 when $x = 0$; local minimum -1 when $x = -2$, local minimum 1 when $x = 1$ **(b)** Increasing on $(-2, 0), (1, \infty)$; decreasing on $(-\infty, -2), (0, 1)$

55. (a) Local maximum ≈ 0.38 when $x \approx -0.58$; local minimum ≈ -0.38 when $x \approx 0.58$ **(b)** Increasing on $(-\infty, -0.58), (0.58, \infty)$; decreasing on $(-0.58, 0.58)$

57. (a) Local maximum ≈ 0 when $x = 0$; local minimum ≈ -13.61 when $x \approx -1.71$, local minimum ≈ -73.32 when $x \approx 3.21$ **(b)** Increasing on $(-1.71, 0), (3.21, \infty)$; decreasing on $(-\infty, -1.71), (0, 3.21)$ **59. (a)** Local maximum ≈ 5.66 when $x \approx 4.00$ **(b)** Increasing on $(-\infty, 4.00)$; decreasing on $(4.00, 6.00)$ **61. (a)** Local maximum ≈ 0.38 when $x \approx -1.73$; local minimum ≈ -0.38 when $x \approx 1.73$ **(b)** Increasing on $(-\infty, -1.73), (1.73, \infty)$; decreasing on $(-1.73, 0), (0, 1.73)$ **63. (a)** ≈ 11 gigawatts, ≈ 14 gigawatts **(b)** $\approx$3:00 to 4:30 A.M.; $\approx$7:00 P.M. **(c)** ≈ 3 gigawatts

65. (a) Increasing on $\approx(0, 30)$, $\approx(32, 68)$; decreasing on $\approx(30, 32)$ **(b)** The person went on a crash diet and lost weight, only to regain it again later. **(c)** ≈ 40 kg **67. (a)** Increasing on $\approx(0, 150)$, $\approx(300, 365)$; decreasing on $\approx(150, 300)$ **(b)** Local maximum when $x = 150$; local minimum when $x = 300$ **(c)** -16 m **69.** Runner A won the race. All runners finished. Runner B fell but got up again to finish second.

71. (a)

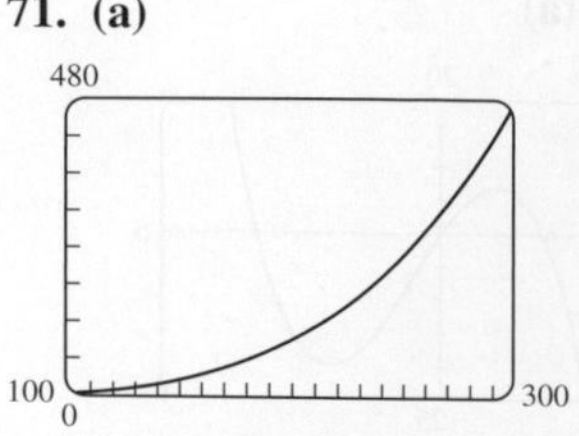

(b) Increases **73.** ≈ 7.5 km/h

Section 2.4 ■ Page 190

1. $\dfrac{100 \text{ km}}{2 \text{ h}} = 50$ km/h **2.** $\dfrac{f(b) - f(a)}{b - a}$ **3.** $\dfrac{25 - 1}{5 - 1} = 6$
4. (a) secant **(b)** 3 **5. (a)** Yes **(b)** Yes **6. (a)** No **(b)** No
7. (a) 2 **(b)** $\frac{2}{3}$ **9. (a)** -4 **(b)** $-\frac{4}{5}$ **11. (a)** 15 **(b)** 5
13. (a) -8 **(b)** $-\frac{1}{2}$ **15. (a)** 26 **(b)** 13 **17. (a)** 600
(b) 60 **19.** $8a + 4h$ **21.** $-\dfrac{1}{a(a + h)}$ **23.** $\dfrac{1}{\sqrt{a} + \sqrt{a + h}}$
25. (a) $\frac{1}{2}$ **27.** f; g; 0, 1.5 **29. (a)** -0.125 m/day; on average, the water level was decreasing at a rate of 0.125 m/day between $x = 100$ and $x = 200$. **(b)** Answers may vary. For instance, on the interval [200, 350] the average rate of change is 0.
31. (a) 376.5 persons/year **(b)** -341.75 persons/year
(c) 2002–2012 **(d)** 2012–2020 **33. (a)** 248.3 cakes/year
(b) -404 cakes/year **(c)** 507 cakes/year **(d)** 2016–2017, 2015–2016 **35.** First 20 minutes: -4.05°F/min, next 20 minutes: -1.5°F/min; first interval **37. (a)** All 10 m/s **(b)** Skier A started quickly and slowed down, skier B maintained a constant speed, and skier C started slowly and sped up.

Section 2.5 ■ Page 198

1. (a) linear, a, b **(b)** line **2. (a)** -5 **(b)** line, -5, 7
3. 15 **4.** 15 L/min **5.** Upward **6.** Yes, 0, 0 **7.** Yes, $f(x) = 2x + \sqrt{5}$ **9.** Yes, $f(x) = -\frac{1}{5}x + 4$ **11.** No **13.** No
15. -2

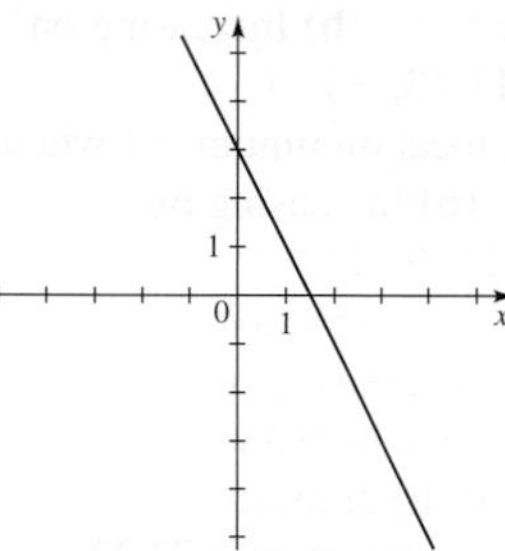

17. $-\frac{2}{3}$

19. (a)

(b) 2 **(c)** 2

21. (a)

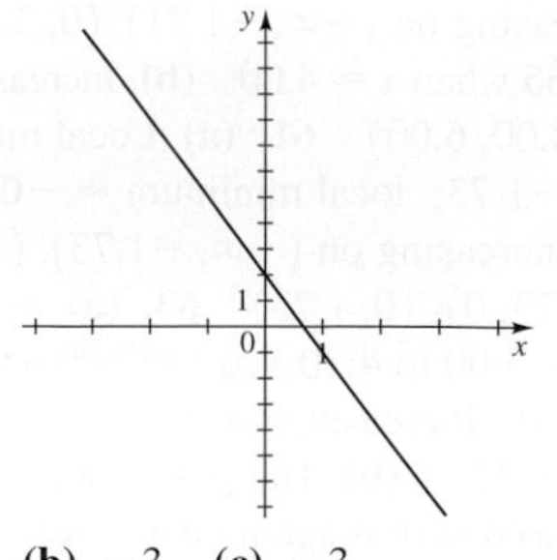

(b) -3 **(c)** -3

23. (a)

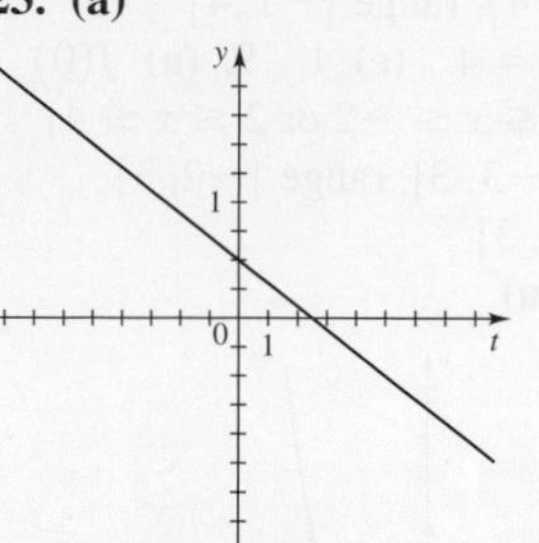

(b) $-\frac{1}{5}$ **(c)** $-\frac{1}{5}$

25. (a)

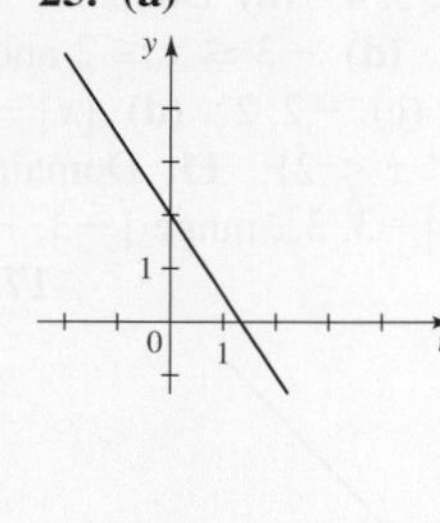

(b) $-\frac{3}{2}$ **(c)** $-\frac{3}{2}$
27. $f(x) = 5x + 10$ **29.** $f(x) = \frac{1}{2}x + 3$
31. (a) $\frac{3}{2}$ **(b)** $f(x) = \frac{3}{2}x + 7$
33. (a) 1 **(b)** $f(x) = x + 3$
35. (a) $-\frac{1}{2}$ **(b)** $f(x) = -\frac{1}{2}x + 2$
37.

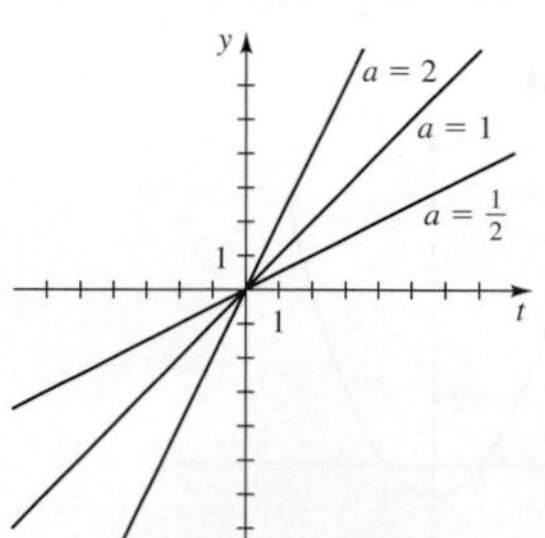

As a increases, the graph of f becomes steeper and the rate of change increases.

39. (a)

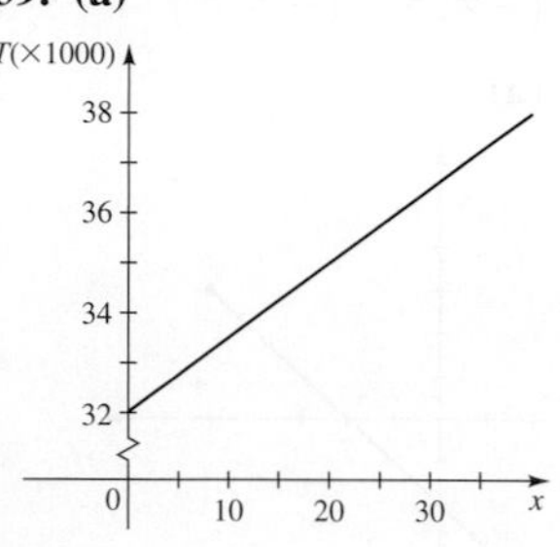

(b) 150
(c) 150,000 tons/year

41. (a) $V(t) = 0.5t + 2$ **(b)** 26 s
43. (a) $\frac{1}{12}$, $H(x) = \frac{1}{12}x$ **(b)** 31.25 cm
45. (a) Engineer **(b)** Manager: 60 km/h; engineer: 70 km/h
(c) Manager: $f(t) = t + 10$; Engineer $g(t) = \frac{7}{6}t$ **47.** 5.08 km
49. $f(x) = -12x + 100$; ≈ 47 kPa
51. (a) $C(x) = \frac{1}{4}x + 260$
(b) $\frac{1}{4}$ **(c)** \$0.25/km

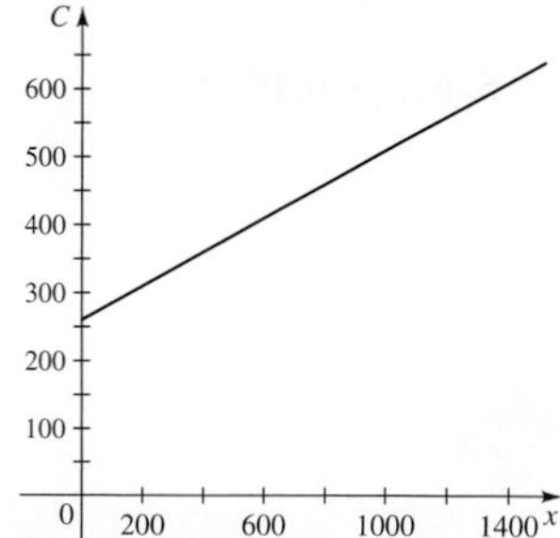

Section 2.6 ■ Page 209

1. (a) upward **(b)** left **2. (a)** downward **(b)** right
3. (a) x-axis **(b)** y-axis **4. (a)** II **(b)** I **(c)** III **(d)** IV
5. Symmetry with respect to the y-axis **6.** Symmetric with respect to the origin **7. (a)** Shift upward 11 units **(b)** Shift 8 units to the left **9. (a)** Reflect about the y-axis, then shrink vertically by a factor of $\frac{1}{4}$ **(b)** Reflect about the x-axis, then stretch vertically by a factor of 5 **11. (a)** Shift 1 unit to the right, then downward 5 units **(b)** Shift 2 units to the left, then downward 4 units **13. (a)** Reflect about the y-axis, then shift upward 5 units **(b)** Shift 2 units to the left, shrink vertically by a factor of $\frac{1}{2}$, reflect about the x-axis, then shift upward 3 units

15. (a) Shrink horizontally by a factor of $\frac{1}{5}$, reflect about the x-axis, then shift upward 2 units **(b)** Stretch horizontally by a factor of 2, shift 1 unit to the left, then shift upward 1 unit
17. (a) Shift to the left 2 units **(b)** Shift upward 2 units
19. (a) Shift to the left 2 units, then shift downward 2 units **(b)** Shift to the right 2 units, then shift upward 2 units
21. (a)

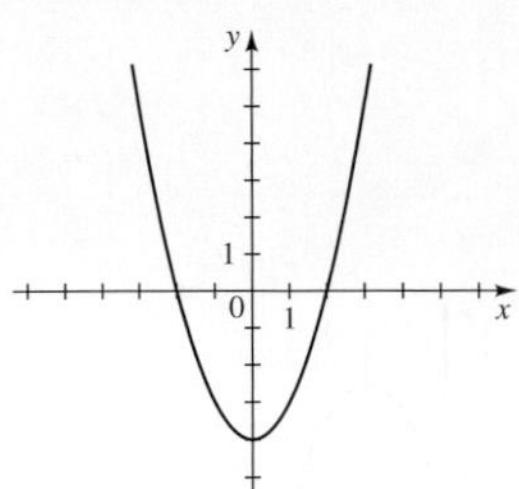

(b)

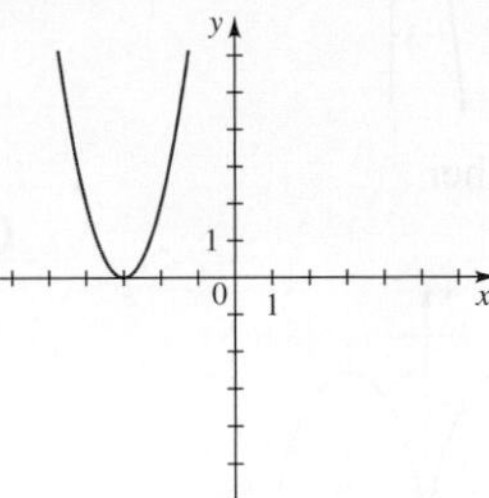

(c)

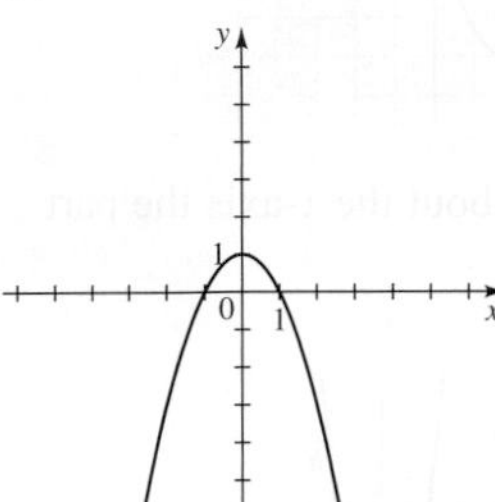

(d)

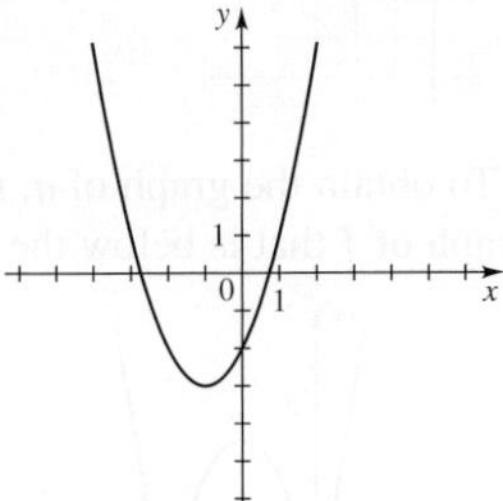

23. II; range $[0, \infty)$ **25.** I; range $[-1, \infty)$
27.

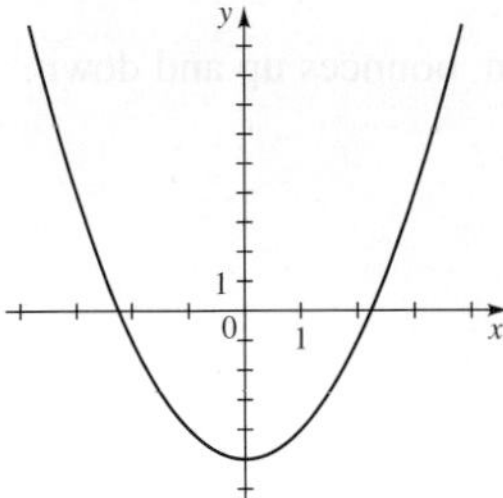

29.

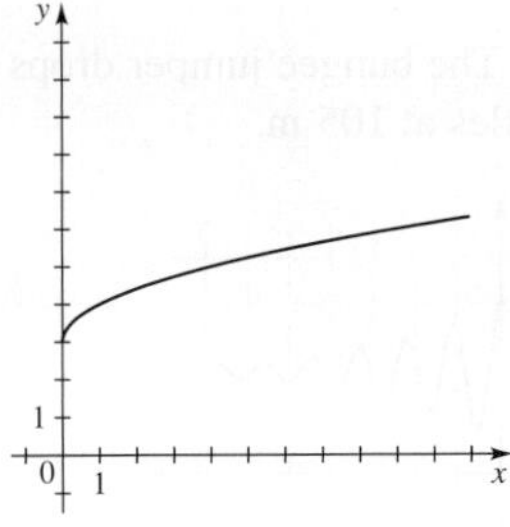

31.

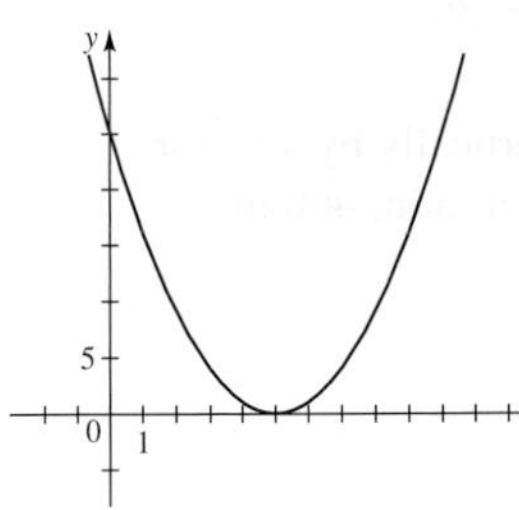

33.

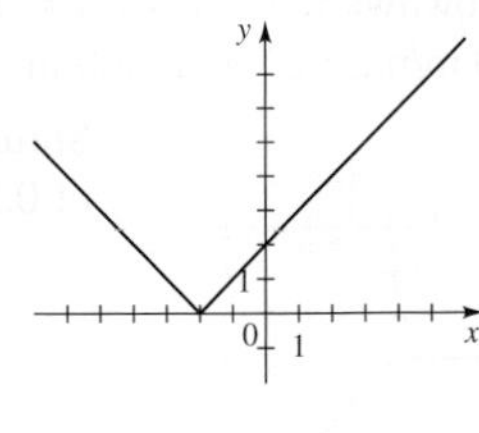

35.

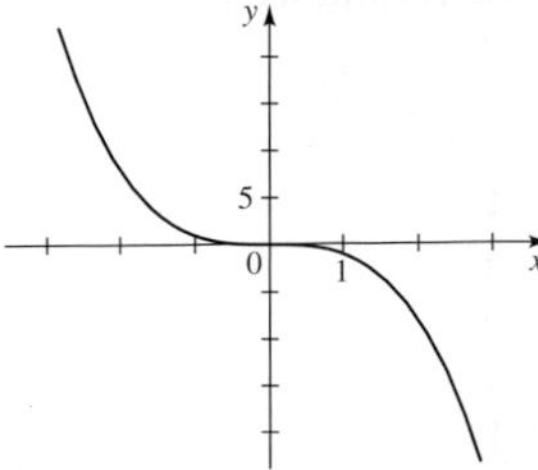

37.

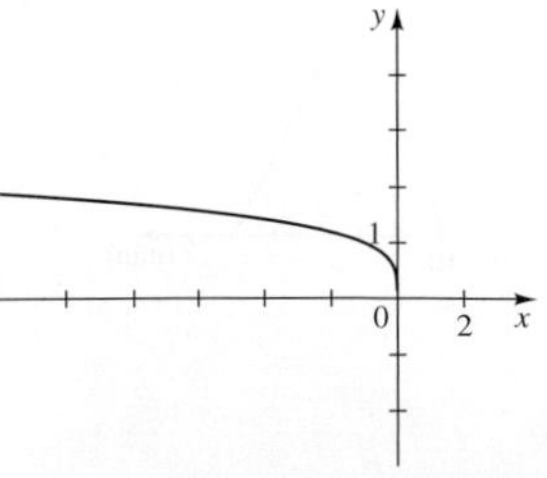

39.

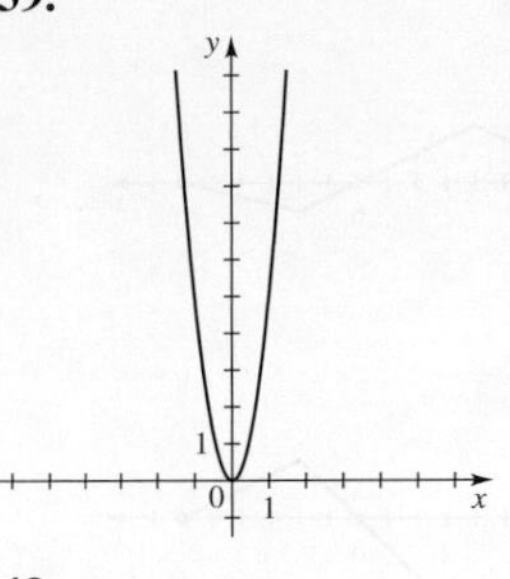

41.

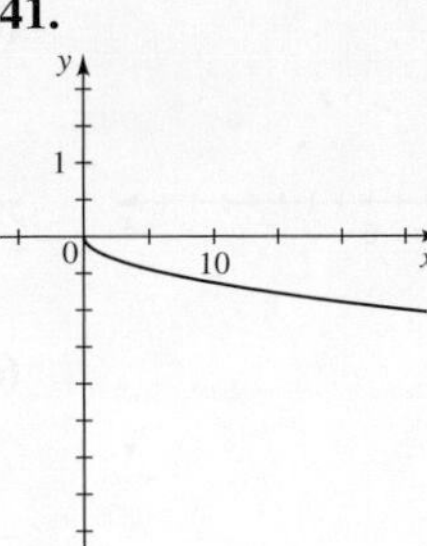

43.

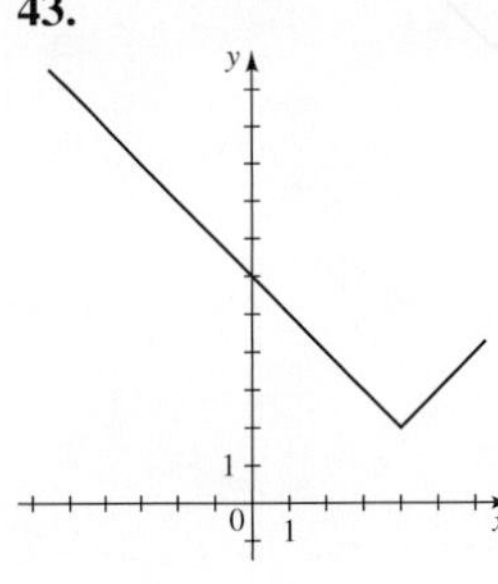

45.

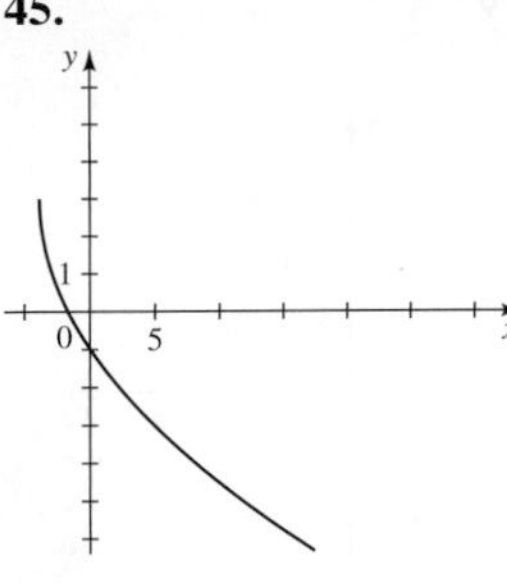

47.

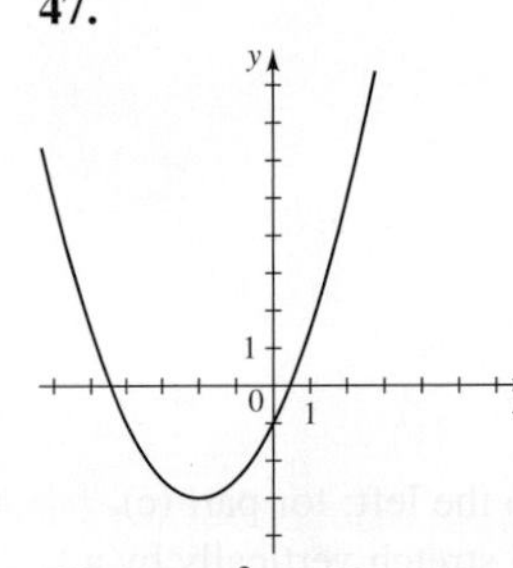

49.

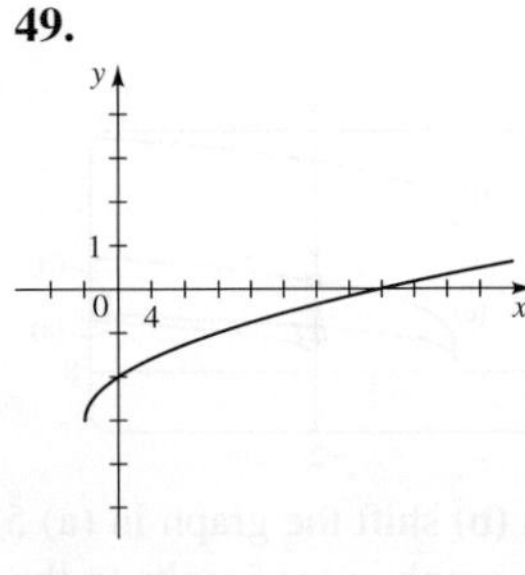

51. $y = x^2 + 10$ **53.** $y = (x - 3)^4$ **55.** $y = |x + 2| - 5$
57. $y = \sqrt[4]{-x} + 1$ **59.** $y = 2(x - 3)^2 - 2$
61. $g(x) = (x - 2)^2$ **63.** $g(x) = |x + 1| + 2$
65. $g(x) = -\sqrt{x + 2}$ **67. (a)** 3 **(b)** 1 **(c)** 2 **(d)** 4
69. (a)

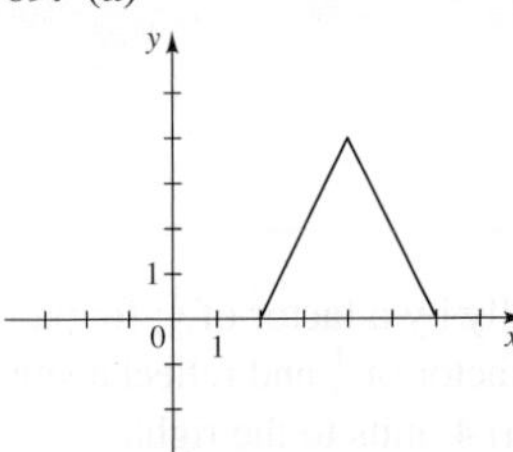

(b)

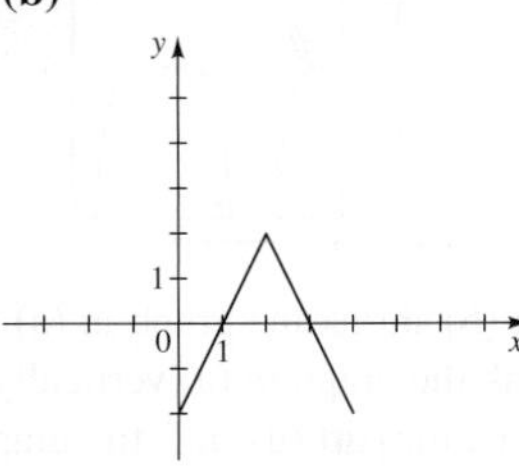

(c)

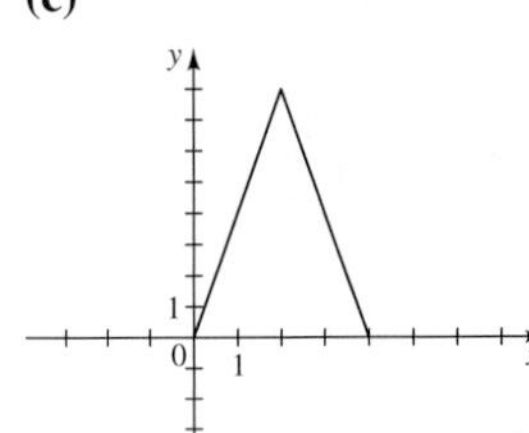

(d)

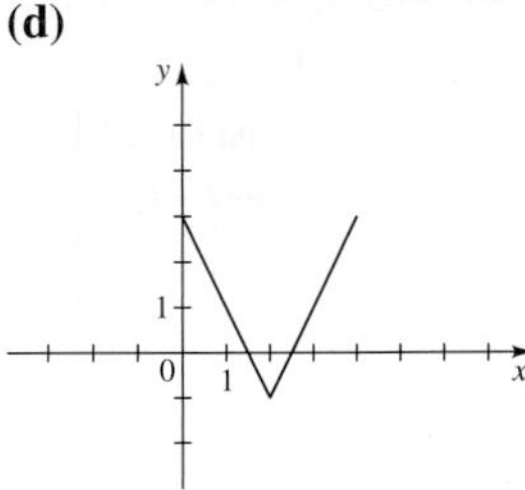

(e)

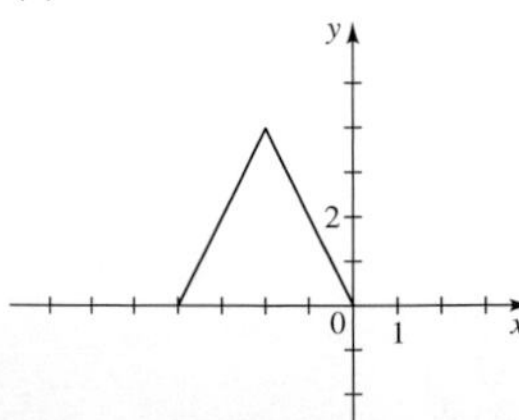

(f)

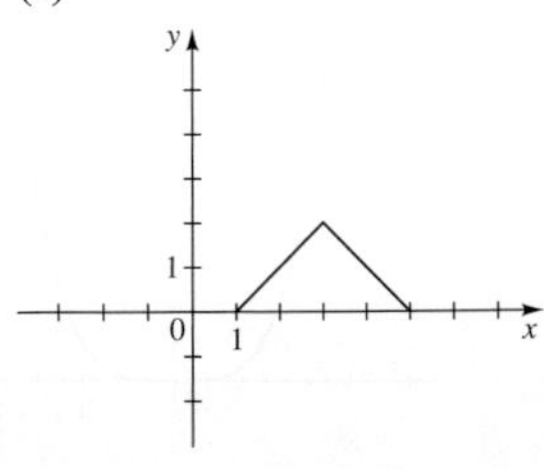

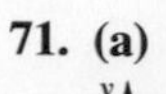

71. (a)

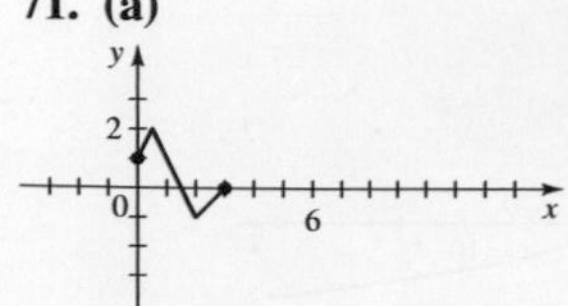

(b)

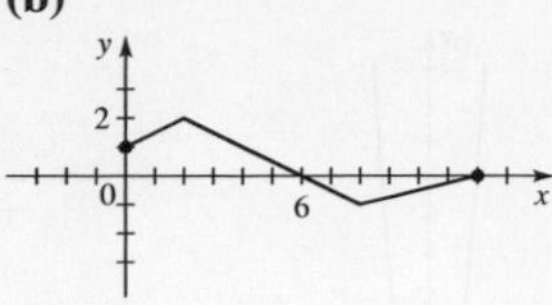

(c)

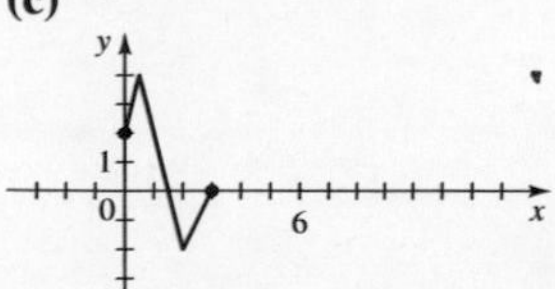

(d)

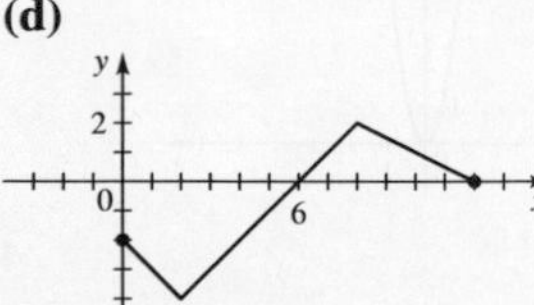

73.

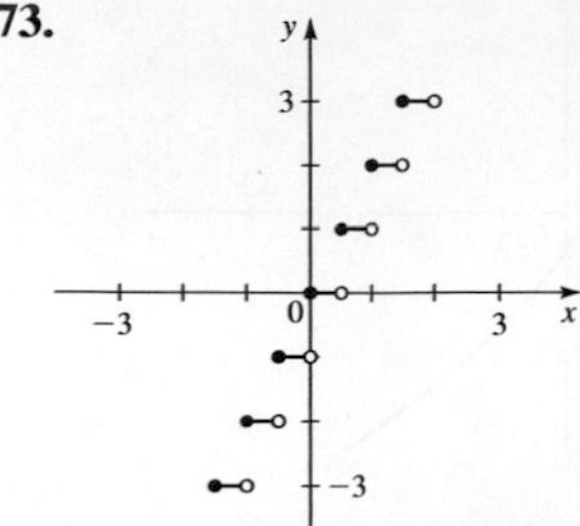

75.

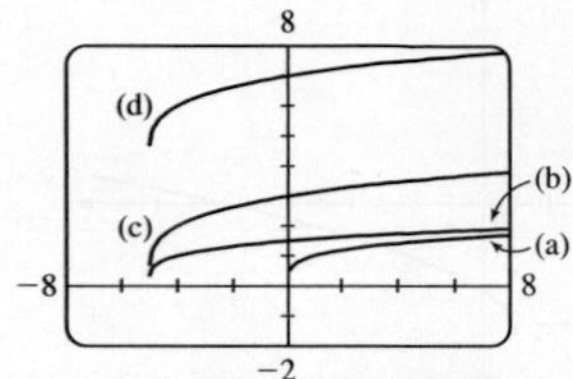

For part (b) shift the graph in (a) 5 units to the left; for part (c) shift the graph in (a) 5 units to the left and stretch vertically by a factor of 2; for part (d) shift the graph in (a) 5 units to the left, stretch vertically by a factor of 2, and then shift upward 4 units.

77.

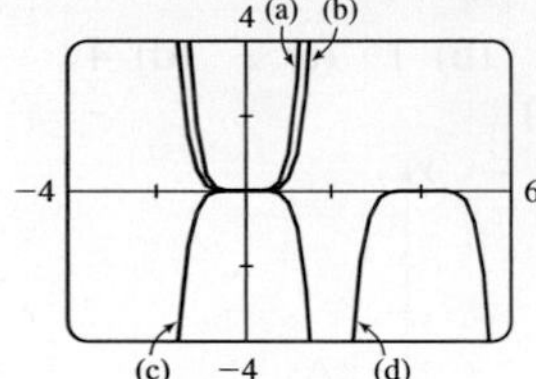

For part (b) shrink the graph in (a) vertically by a factor of $\frac{1}{3}$; for part (c) shrink the graph in (a) vertically by a factor of $\frac{1}{3}$ and reflect about the x-axis; for part (d) shift the graph in (a) 4 units to the right, shrink vertically by a factor of $\frac{1}{3}$, and then reflect about the x-axis.

79.

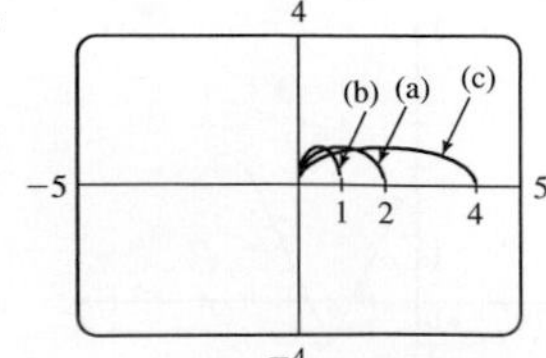

The graph in part (b) is shrunk horizontally by a factor of $\frac{1}{2}$ and the graph in part (c) is stretched horizontally by a factor of 2.

81. Even

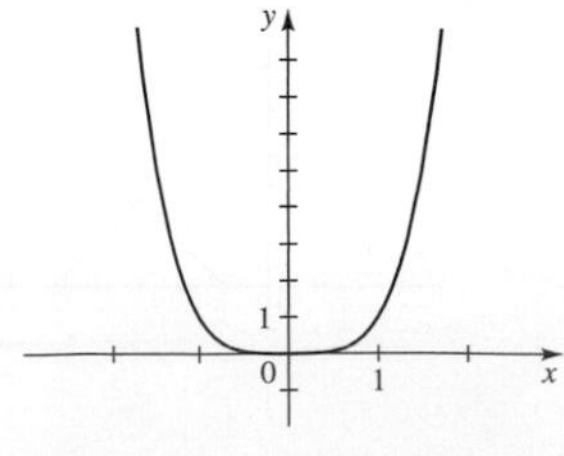

83. Neither

85. Odd

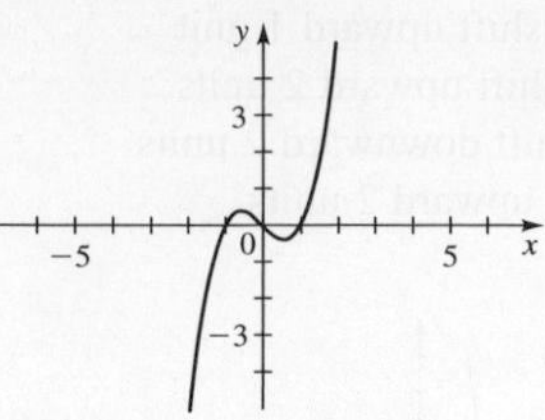

87. Neither

89. (a)

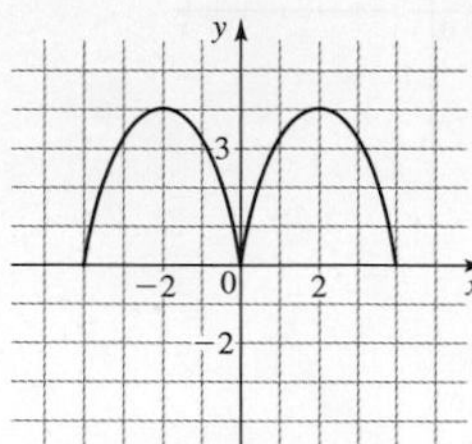

(b)

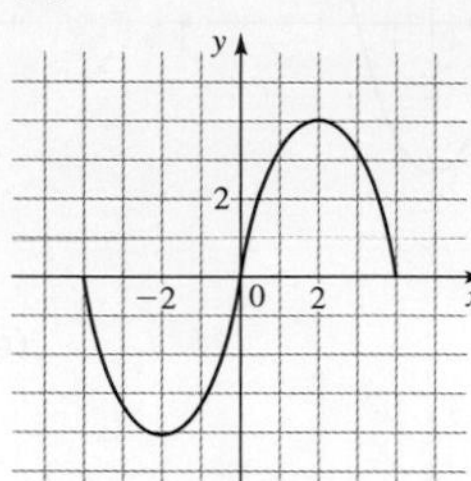

91. (a) To obtain the graph of g, reflect about the x-axis the part of the graph of f that is below the x-axis.

(b)

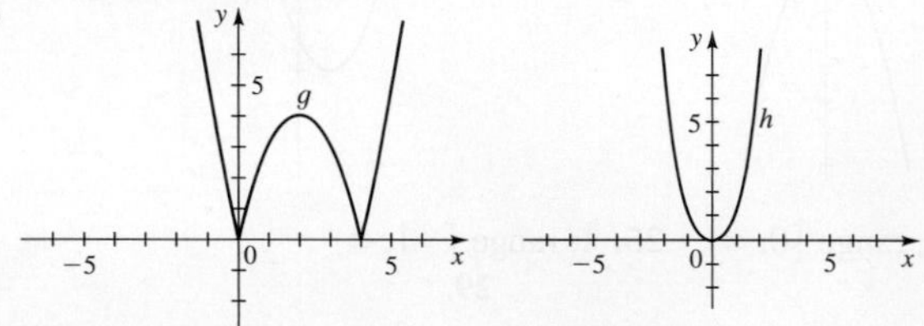

93. (a) The bungee jumper drops to 60 m, bounces up and down, then settles at 105 m.

(b)

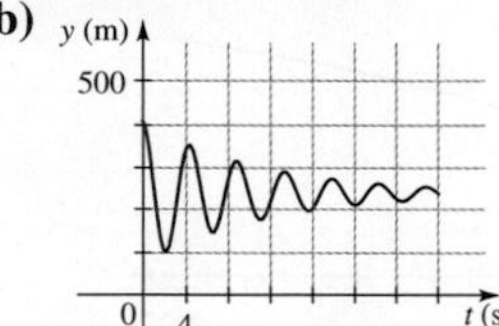

(c) Shift downward 30 m; $H(t) = h(t) - 30$

95. (a) 80 m/min; 20 min; 800 m

(b)

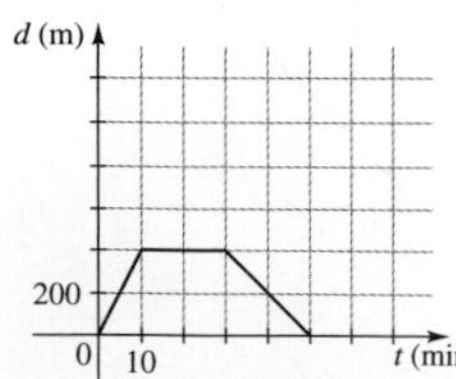

Shrunk vertically by a factor of 0.5; 40 m/min; 400 m

(c)

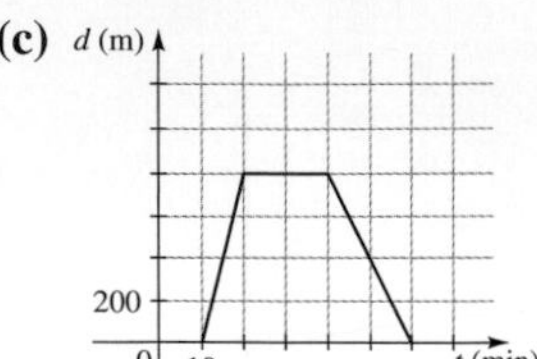

Shifted 10 min to the right; the class left 10 min later

Section 2.7 ■ Page 220

1. $(f+g)(2) = 8, (f-g)(2) = -2, (fg)(2) = 15, \left(\frac{f}{g}\right)(2) = \frac{3}{5}$

2. $f(g(x))$, 12 **3.** Multiply by 2, then add 1; Add 1, then multiply by 2 **4.** $f(x) = x + 1, g(x) = 2x, (f \circ g)(x) = 2x + 1,$ $(g \circ f)(x) = 2(x + 1)$ **5. (a)** f, g **(b)** f, g **(c)** $f, g, 0$
6. g, f **7.** $(f+g)(x) = 2x + 1, (-\infty, \infty);$
$(f-g)(x) = 4x - 1, (-\infty, \infty); (fg)(x) = 3x - 3x^2, (-\infty, \infty);$
$\left(\frac{f}{g}\right)(x) = \frac{3x}{1-x}, (-\infty, 1) \cup (1, \infty)$
9. $(f+g)(x) = x^3 + 2x^2, (-\infty, \infty);$
$(f-g)(x) = x^3, (-\infty, \infty); (fg)(x) = x^5 + x^4, (-\infty, \infty);$
$\left(\frac{f}{g}\right)(x) = x + 1, (-\infty, 0) \cup (0, \infty)$
11. $(f+g)(x) = x^2 - 4x + 5, (-\infty, \infty);$
$(f-g)(x) = -x^2 + 2x + 5, (-\infty, \infty);$
$(fg)(x) = -x^3 + 8x^2 - 15x, (-\infty, \infty);$
$\left(\frac{f}{g}\right)(x) = \frac{5-x}{x^2 - 3x}, (-\infty, 0) \cup (0, 3) \cup (3, \infty)$
13. $(f+g)(x) = \sqrt{25 - x^2} + \sqrt{x+3}, [-3, 5];$
$(f-g)(x) = \sqrt{25 - x^2} - \sqrt{x+3}, [-3, 5];$
$(fg)(x) = \sqrt{(25 - x^2)(x+3)}, [-3, 5];$
$\left(\frac{f}{g}\right)(x) = \sqrt{\frac{25 - x^2}{x+3}}, (-3, 5]$
15. $(f+g)(x) = \frac{4x+1}{x^2 - x - 2}, x \neq -1, x \neq 2;$
$(f-g)(x) = \frac{-2x-5}{x^2 - x - 2}, x \neq -1, x \neq 2;$
$(fg)(x) = \frac{3}{x^2 - x - 2}, x \neq -1, x \neq 2;$
$\left(\frac{f}{g}\right)(x) = \frac{x-2}{3x+3}, x \neq -1, x \neq 2$
17. $[0, 3]$ **19.** $(3, \infty)$
21.

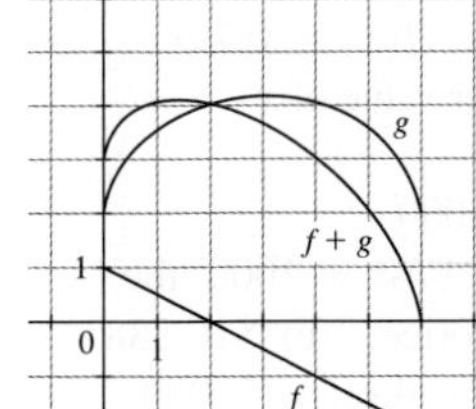

23.

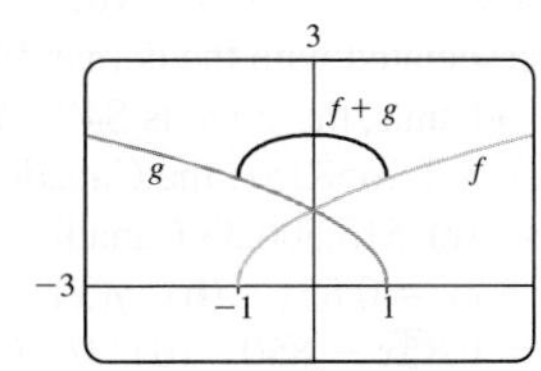

25.

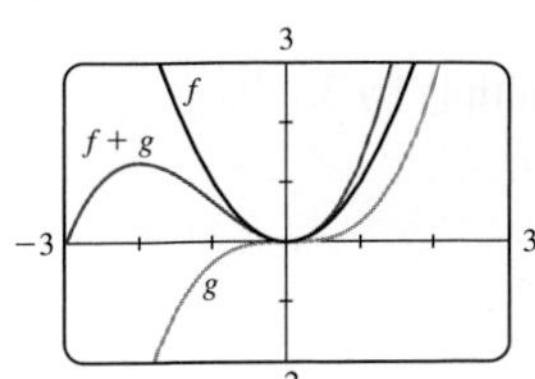

27. (a) 17 **(b)** 83 **29. (a)** 29 **(b)** 3
31. (a) $4x^2 + 13$ **(b)** $16x^2 + 40x + 27$ **33.** 4

35. 5 **37.** 4 **39.** 6 **41.** 3 **43.** 1 **45.** 3
47. $(f \circ g)(x) = 8x + 1, (-\infty, \infty)$;
$(g \circ f)(x) = 8x + 11, (-\infty, \infty); (f \circ f)(x) = 4x + 9, (-\infty, \infty);$
$(g \circ g)(x) = 16x - 5, (-\infty, \infty)$
49. $(f \circ g)(x) = (x+1)^2, (-\infty, \infty);$
$(g \circ f)(x) = x^2 + 1, (-\infty, \infty); (f \circ f)(x) = x^4, (-\infty, \infty);$
$(g \circ g)(x) = x + 2, (-\infty, \infty)$
51. $(f \circ g)(x) = \frac{1}{x} + 1, (0, \infty);$
$(g \circ f)(x) = \frac{1}{\sqrt{x^2 + 1}}, (-\infty, \infty);$
$(f \circ f)(x) = x^4 + 2x^2 + 2, (-\infty, \infty); (g \circ g)(x) = \sqrt[4]{x}, (0, \infty)$
53. $(f \circ g)(x) = \frac{2x - 1}{2x}, x \neq 0;$
$(g \circ f)(x) = \frac{2x}{x+1} - 1, x \neq -1;$
$(f \circ f)(x) = \frac{x}{2x+1}, x \neq -1, x \neq -\frac{1}{2};$
$(g \circ g)(x) = 4x - 3, (-\infty, \infty)$
55. $(f \circ g)(x) = \frac{2x + 4}{x}, x \neq -2, x \neq 0;$
$(g \circ f)(x) = \frac{1}{1 + x}, \quad x \neq -1, x \neq 0;$
$(f \circ f)(x) = x, x \neq 0;$
$(g \circ g)(x) = \frac{x}{3x + 4}, x \neq -2, x \neq -\frac{4}{3}$
57. $(f \circ g)(x) = \frac{1}{\sqrt{x^2 - 4x}}, (-\infty, 0) \cup (4, \infty);$
$(g \circ f)(x) = \frac{1}{x} - \frac{4}{\sqrt{x}}, (0, \infty); (f \circ f)(x) = \sqrt[4]{x}, (0, \infty);$
$(g \circ g)(x) = x^4 - 8x^3 + 12x^2 + 16x, (-\infty, \infty)$
59. $(f \circ g)(x) = 1 - \sqrt[6]{x}, [0, \infty);$
$(g \circ f)(x) = \sqrt[3]{1 - \sqrt{x}}, [0, \infty);$
$(f \circ f)(x) = 1 - \sqrt{1 - \sqrt{x}}, [0, 1]; (g \circ g)(x) = \sqrt[9]{x}, (-\infty, \infty)$
61. $(f \circ g \circ h)(x) = \sqrt{x - 1} - 1$
63. $(f \circ g \circ h)(x) = (\sqrt{x} - 5)^4 + 1$
For Exercises 65–78, there are many possible answers,
65. $g(x) = x - 9, f(x) = x^5$ **67.** $g(x) = x^2, f(x) = x/(x+4)$
69. $g(x) = 1 - x^3, f(x) = |x|$
71. $g(x) = x^3 + 1, f(x) = 1 - \sqrt{x}$
73. $h(x) = x^2, g(x) = x + 1, f(x) = 1/x$
75. $h(x) = \sqrt[3]{x}, g(x) = 4 + x, f(x) = x^9$
77. $h(x) = \sqrt{x}, g(x) = \frac{x}{x-1}, f(x) = x^3$
79. Yes; $m_1 m_2$ **81.** $R(x) = 0.15x - 0.000002x^2$
83. (a) $g(t) = 60t$ **(b)** $f(r) = \pi r^2$ **(c)** $(f \circ g)(t) = 3600\pi t^2$; area as a function of time **85.** $A(t) = 16\pi t^2$
87. (a) $f(x) = 0.80x$ **(b)** $g(x) = x - 50$
(c) $(f \circ g)(x) = 0.80x - 40; (g \circ f)(x) = 0.80x - 50$; applying the 20% discount, then \$50 coupon $(g \circ f)$ gives the lower price
89. (a) $s = \sqrt{1 + d^2}$ **(b)** $d = 350t$
(c) $s(t) = \sqrt{1 + 122{,}500t^2}$

Section 2.8 ▪ Page 231

1. different, Horizontal Line **2. (a)** one-to-one, $g(x) = x^3$ **(b)** $g^{-1}(x) = x^{1/3}$ **3. (a)** Take the cube root, subtract 5, then divide the result by 3. **(b)** $f(x) = (3x + 5)^3$, $f^{-1}(x) = \dfrac{\sqrt[3]{x} - 5}{3}$
4. Yes, 4, 5 **5.** (4, 3) **6. (a)** False **(b)** True **7.** No
9. Yes **11.** No **13.** Yes **15.** Yes **17.** No **19.** No **21.** Yes
23. No **25. (a)** 5 **(b)** 10 **27.** 1 **29. (a)** 6 **(b)** 2
(c) 0 **31.** 4 **33.** 1 **35.** 2 **49.** $f^{-1}(x) = \frac{1}{3}x - 5$
51. $f^{-1}(x) = \frac{4}{3}x + 16$ **53.** $f^{-1}(x) = \sqrt[3]{\frac{1}{4}(5 - x)}$
55. $f^{-1}(x) = \dfrac{1}{x} - 2$ **57.** $f^{-1}(x) = \dfrac{2x}{x + 1}$
59. $f^{-1}(x) = \dfrac{7x + 5}{x - 2}$ **61.** $f^{-1}(x) = \dfrac{x - 3}{5x + 2}$
63. $f^{-1}(x) = \sqrt[3]{3x - 1}$ **65.** $f^{-1}(x) = (x - 2)^3$
67. $f^{-1}(x) = (x - 1)^{2/3}$
69. (a), (b)

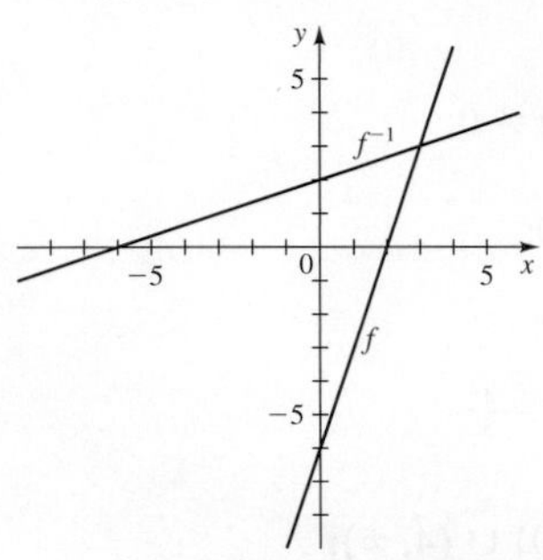

(c) $f^{-1}(x) = \frac{1}{3}(x + 6)$
71. (a), (b)

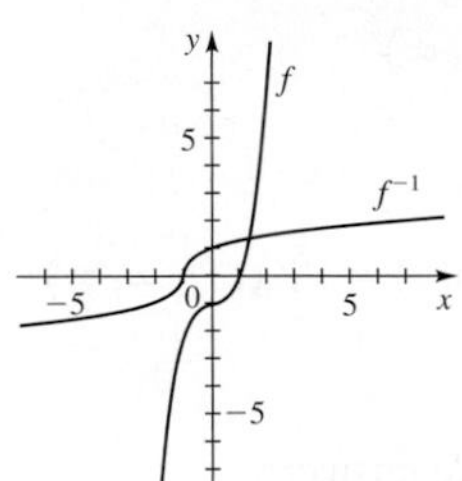

(c) $f^{-1}(x) = \sqrt[3]{x + 1}$

73. (a), (b)

(c) $f^{-1}(x) = (x - 3)^2 + 1, x \geq 3$

75. Not one-to-one

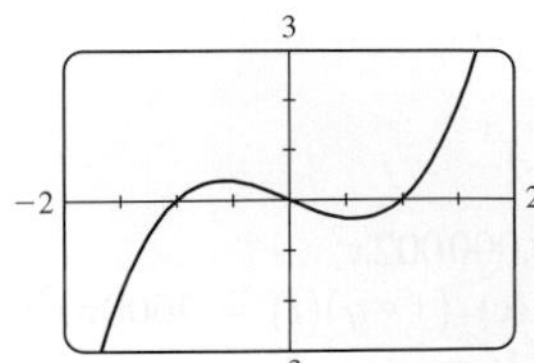

77. One-to-one

79. Not one-to-one

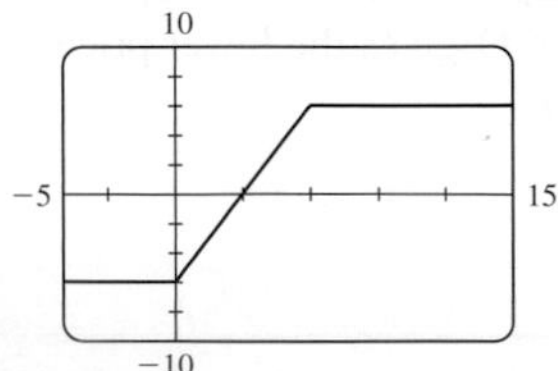

81. (a) $f^{-1}(x) = x - 2$
(b)

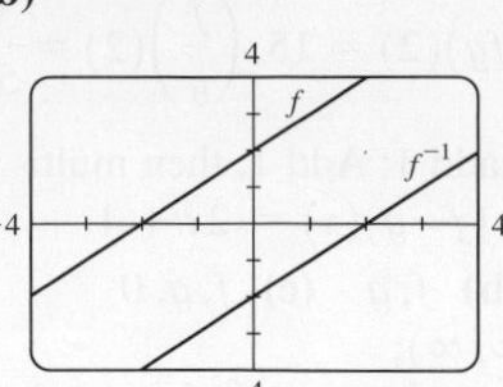

83. (a) $g^{-1}(x) = x^2 - 3, x \geq 0$
(b)

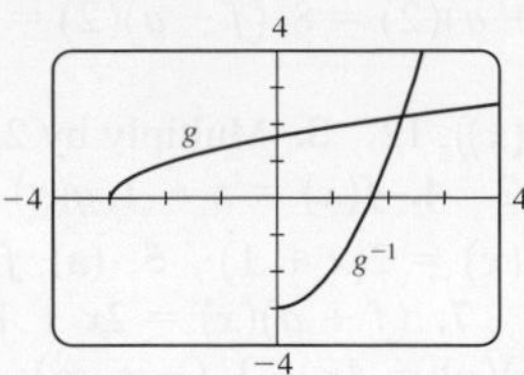

85. $x \geq 0$, $f^{-1}(x) = \sqrt{4 - x}$ **87.** $x \geq -2$, $h^{-1}(x) = \sqrt{x} - 2$
89.

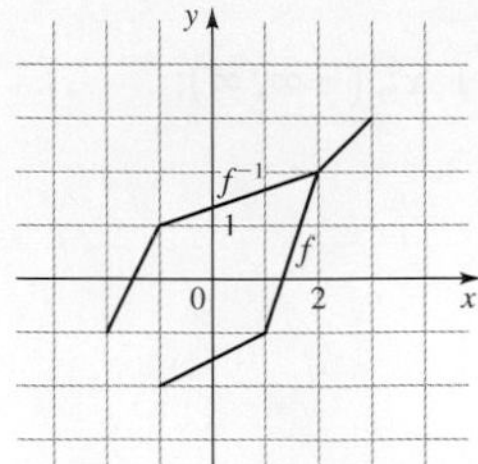

91. (a) $f^{-1}(x) = \sqrt{x + 9}, x \geq -9$
93. (a) $f^{-1}(x) = \dfrac{1}{\sqrt[4]{x}}, x > 0$ **95. (a)** $f^{-1}(x) = x^2, 0 \leq x \leq 3$
97. (a)

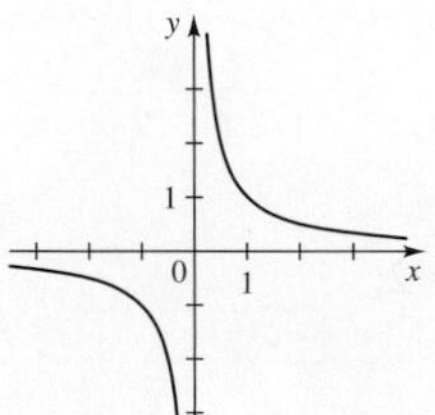

(b) Yes **(c)** $f^{-1}(x) = \dfrac{1}{x}$
99. (a) $f(n) = 16 + 1.5n$ **(b)** $f^{-1}(x) = \frac{2}{3}(x - 16)$; the number of toppings on a pizza that costs x dollars **(c)** 6
101. (a) $f^{-1}(V) = 40 - 4\sqrt{V}$, time elapsed when V liters of water remain **(b)** ≈24.5 min; in 24.5 min the tank has 15 liters of water remaining **103. (a)** $f^{-1}(D) = 50 - \frac{1}{3}D$; the price associated with the demand D **(b)** \$40; when the demand is 30 units, the price is \$40 **105. (a)** $f(x) = 0.79x$ **(b)** $f^{-1}(x) = 1.265823x$; the Canadian dollar value of x US dollars **(c)** \$15,506.33 Canadian
107. (a) $f(x) = 0.85x$ **(b)** $g(x) = x - 1000$ **(c)** $H(x) = 0.85x - 850$ **(d)** $H^{-1}(x) = 1.176x + 1000$, the original sticker price for a given discounted price **(e)** \$16,288, the original price of the car when the discounted price (\$1000 rebate, then 15% off) is \$13,000

Chapter 2 Review ▪ Page 236

1. $f(x) = x^2 - 5$ **3.** Add 10, then multiply by 3.

5.

x	$g(x)$
−1	5
0	0
1	−3
2	−4
3	−3

7. (a) $C(1000) = \$34{,}000$, $C(10{,}000) = \$205{,}000$ **(b)** The costs of printing 1000 and 10,000 copies of the book **(c)** $C(0) = \$5000$; fixed costs **(d)** \$171,000; \$19/copy
9. $6, 2, 18, a^2 - 4a + 6, a^2 + 4a + 6, x^2 - 2x + 3, 4x^2 - 8x + 6$
11. $a^2 + 8, a^2 + 2ah + h^2 + 8, 2a + h$
13. (a) Not a function **(b)** Function **(c)** Function, one-to-one **(d)** Not a function **15.** Domain $[5, \infty)$, range $[0, \infty)$
17. $(-\infty, \infty)$ **19.** $(-\infty, \infty)$ **21.** $\{x \mid x \neq -2, -1, 0\}$
23. $(-\infty, -1] \cup [1, 4]$

25.

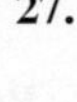

27.

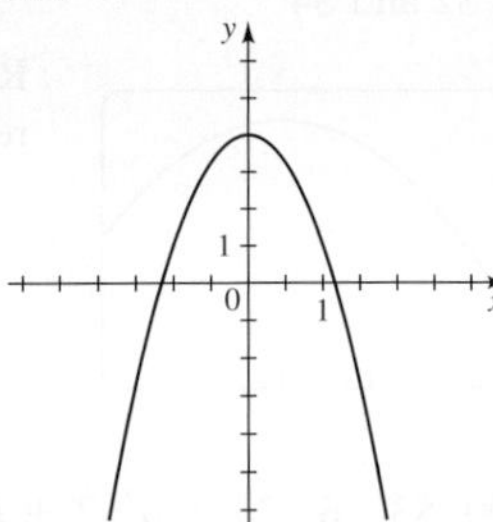

29.

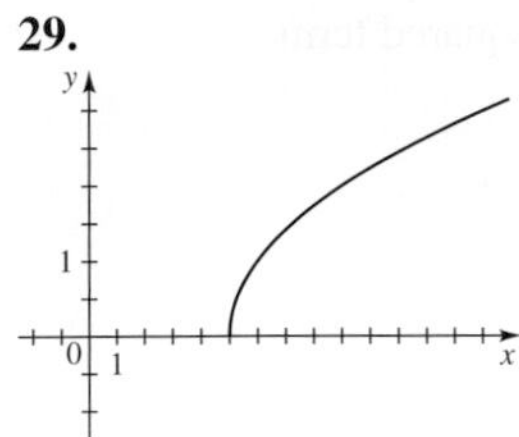

31.

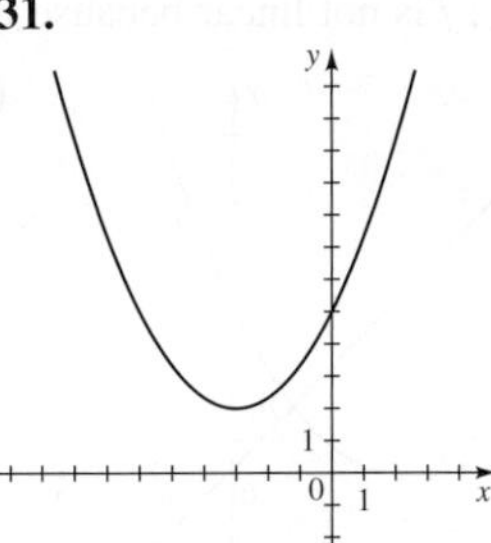

33.

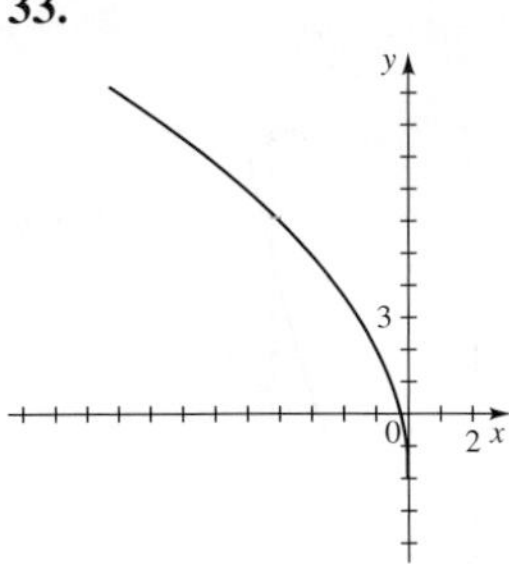

35.

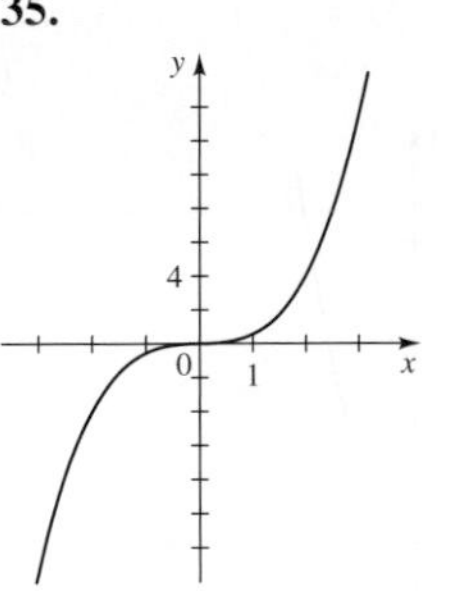

37.

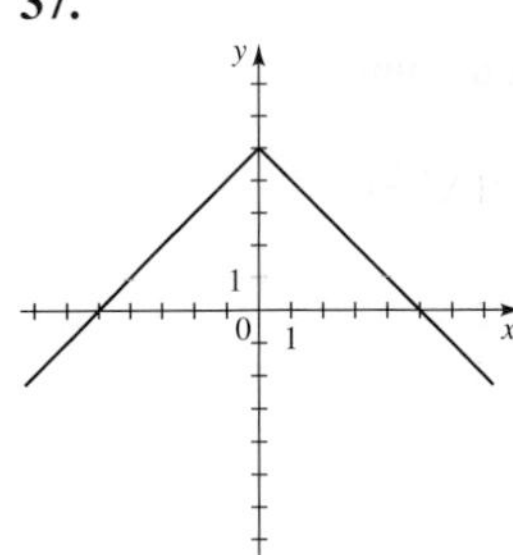

39.

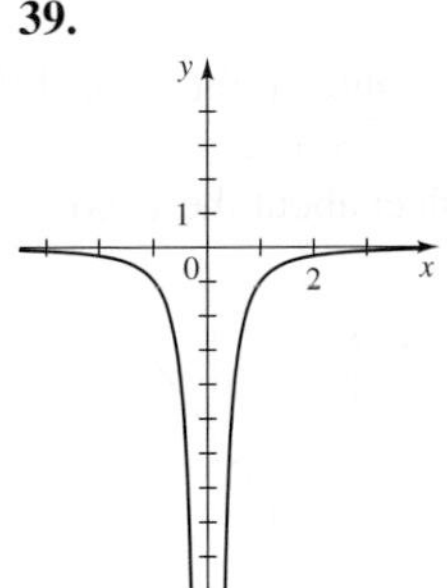

41.

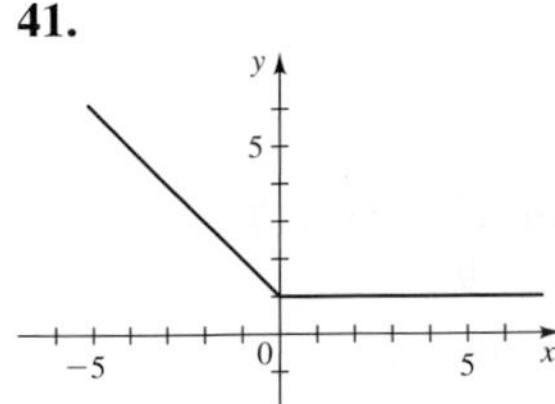

43. No **45.** Yes

47. (a) $(-3, -3), (-2, 0), (0, -1), (2, 3), (3, -3)$

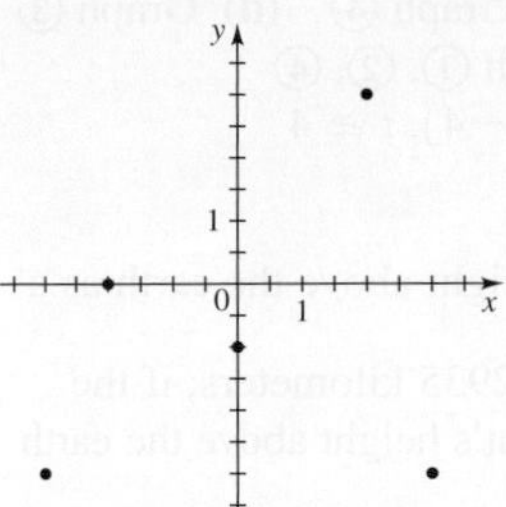

Yes, y is a function of x; domain $\{-3, -2, 0, 2, 3\}$, range $\{-3, -1, 0, 3\}$
(b) $(-3, 3), (2, 1), (0, -2), (2, 5), (3, 3)$

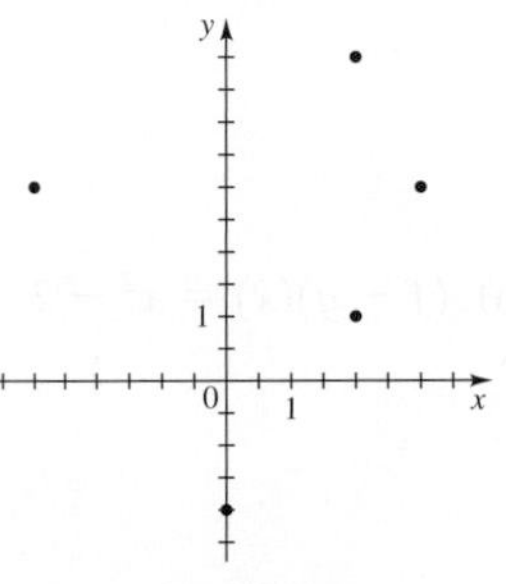

No, y is not a function of x; domain $\{-3, 0, 2, 3\}$, range $\{-2, 1, 3, 5\}$
49. (a) Domain $[-3, 3]$, range $[0, 3]$ **(b)** $x = \pm 3$ **(c)** $(-2.83, 2.83)$
51. (a) Domain $[-2.11, 0.25] \cup [1.86, \infty)$, range $[0, \infty)$ **(b)** $x \approx -2.11, 0.25, 1.86$ **(c)** $(-2, 0) \cup (2, \infty)$
53. (a) Local minimum $= 3$ when $x = 1$ **(b)** Increasing on $(1, \infty)$, decreasing on $(-\infty, 1)$
55. (a) Local maximum ≈ 2.81 when $x \approx -0.46$, local minimum ≈ 3.79 when $x \approx 0.46$ **(b)** Increasing on $(-0.46, 0.46)$, decreasing on $(-\infty, -0.46)$, $(0.46, \infty)$
57. (a) Local maximum ≈ 3.175 when $x \approx 4.00$, local minimum $= 0$ when $x \approx 0$ **(b)** Increasing on $(0, 4)$, decreasing on $(-\infty, 0)$, $(4, \infty)$ **59.** $-4, -1$ **61.** $4, \frac{4}{3}$ **63.** $9, 3$ **65.** No
67. (a)

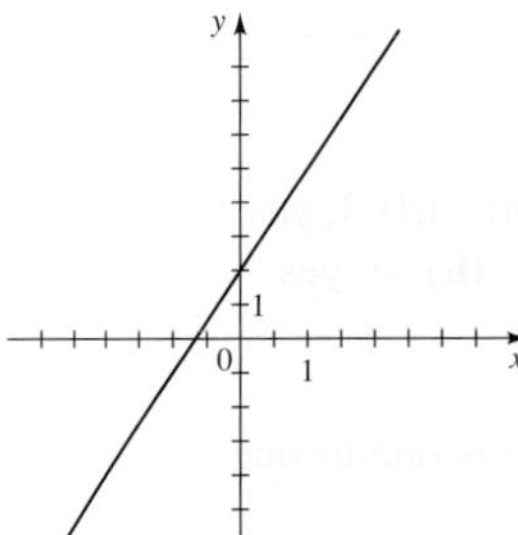

(b) 3 **(c)** 3
69. $f(x) = -2x + 3$ **71.** $f(x) = 2x + 3$
73. $f(x) = -\frac{1}{2}x + 4$ **75. (a)** $\frac{1}{2}, \frac{1}{2}$ **(b)** Yes **(c)** Yes, $\frac{1}{2}$
77. (a) 0, 63 **(b)** No **(c)** No
79. (a) (i) Shift upward 8 units **(ii)** $y = x^3 + 8$ **(b) (i)** Shift left 8 units **(ii)** $y = (x + 8)^3$ **(c) (i)** Stretch vertically by a factor of 2, then shift upward 1 unit **(ii)** $y = 1 + 2x^3$ **(d) (i)** Shift right 2 units, then shift downward 2 units **(ii)** $y = (x - 2)^3 - 2$ **(e) (i)** Reflect about the y-axis **(ii)** $y = -x^3$ **(f) (i)** Reflect about the y-axis, then about the x-axis **(ii)** $y = x^3$ **(g) (i)** Reflect about the x-axis **(ii)** $y = -x^3$ **(h) (i)** Reflect about the line $y = x$ **(ii)** $y = \sqrt[3]{x}$

81. **(a)** Neither **(b)** Odd **(c)** Even **(d)** Neither
83. **(a)** Graph ② **(b)** Graph ⑤ **(c)** Graph ④ **(d)** Graph ③ **(e)** Graph ① **(f)** Graph ③ **(g)** Graph ①, ②, ④
85. **(a)** $g(t) = 1.6t$ **(b)** $g(t) = 0.8(t - 4), t \ge 4$ **(c)** $g(t) = 5 + 0.8t$
87. **(a)** $w^{-1}(x) = 6374\left(\frac{25}{\sqrt{x}} - 1\right)$; height above the earth as a function of the astronaut's weight **(b)** 2935 kilometers; if the astronaut weighs 120 N, then the astronaut's height above the earth is 2935 km
89.

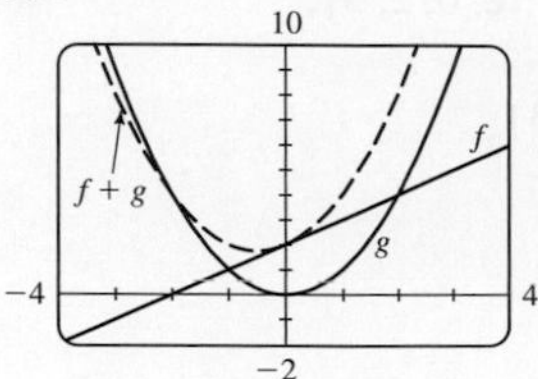

91. **(a)** $(f + g)(x) = x^2 - 6x + 6$ **(b)** $(f - g)(x) = x^2 - 2$
(c) $(fg)(x) = -3x^3 + 13x^2 - 18x + 8$
(d) $(f/g)(x) = (x^2 - 3x + 2)/(4 - 3x)$
(e) $(f \circ g)(x) = 9x^2 - 15x + 6$
(f) $(g \circ f)(x) = -3x^2 + 9x - 2$
93. $(f \circ g)(x) = \sqrt{x - x^2} + 1, [0, 1]$;
$(g \circ f)(x) = -(\sqrt{x} + x), [0, \infty)$;
$(f \circ f)(x) = \sqrt{\sqrt{x} + 1} + 1, [0, \infty)$;
$(g \circ g)(x) = -x^4 + 2x^3 - 2x^2 + x, (-\infty, \infty)$
95. $(f \circ g \circ h)(x) = 1 + \sqrt{x}$ **97.** Yes **99.** No **101.** No
103. $f^{-1}(x) = \dfrac{x + 2}{3}$ **105.** $f^{-1}(x) = \sqrt[3]{x} - 1$ **107.** Yes, 1, 3
109. **(a), (b)**

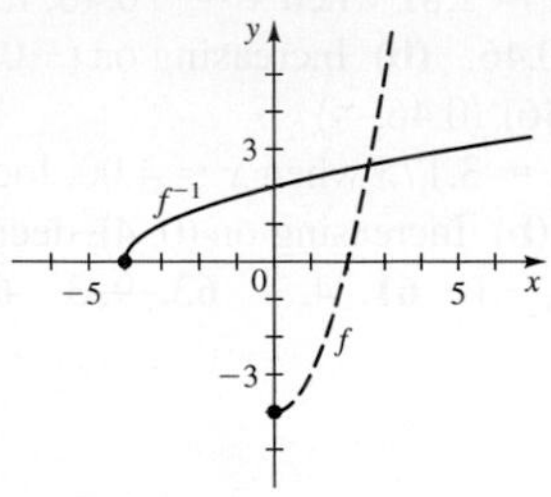

(c) $f^{-1}(x) = \sqrt{x + 4}$
111. **(a)** VI, yes **(b)** IV, yes **(c)** V, no **(d)** I, yes **(e)** VIII, no **(f)** III, yes **(g)** VII, yes **(h)** II, yes

Chapter 2 Test ■ Page 241

1. (a) and (b) are graphs of functions, (a) is one-to-one
2. **(a)** $0, \dfrac{\sqrt{2}}{3}, \dfrac{\sqrt{a + 2}}{a + 3}$ **(b)** $[0, \infty)$
(c) $\dfrac{3\sqrt{10} - 11\sqrt{2}}{264} \approx -0.023$
3. **(a)** $f(x) = (x - 2)^3$
(b)

x	$f(x)$
−1	−27
0	−8
1	−1
2	0
3	1
4	8

(c)

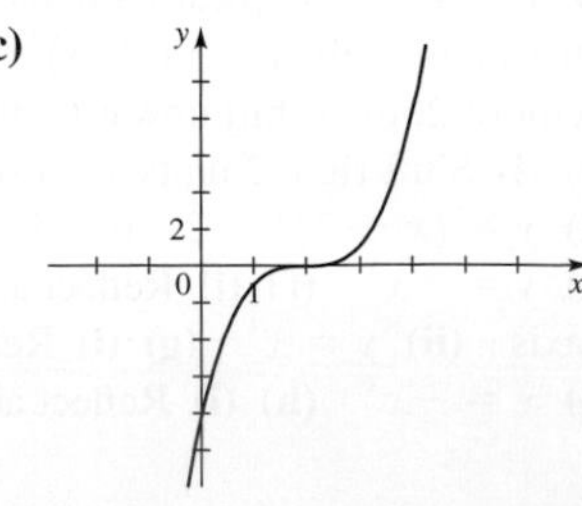

(d) By the Horizontal Line Test; take the cube root, then add 2
(e) $f^{-1}(x) = \sqrt[3]{x} + 2$
4. **(a)** −2, 3 **(b)** 5, 1 **(c)** Domain $[-5, 5]$, range $[-4, 4]$ **(d)** Increasing on $(-5, -4)$, $(-1, 3)$, decreasing on $(-4, -1)$, $(3, 5)$ **(e)** Local maximum = −1 when $x = -4$, local maximum = 4 when $x = 3$; local minimum = −4 when $x = -1$ **(f)** No. The function f does not pass the Horizontal Line Test.
5. **(a)** $R(2) = \$4000$, $R(4) = \$4000$; total sales revenue with prices of \$2 and \$4
(b)

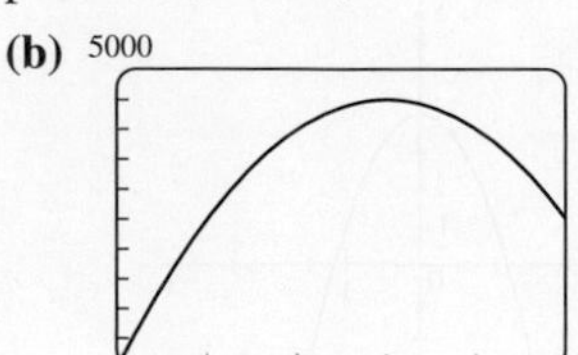

Revenue increases until price reaches \$3, then decreases

(c) \$4500; \$3 **6.** $2h + h^2, 2 + h$
7. **(a)** g; f is not linear because it has a squared term
(b)

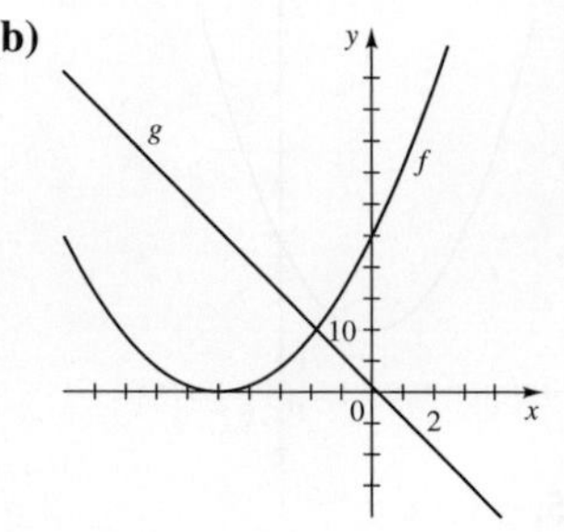

(c) −5

8. **(a)** **(b)**

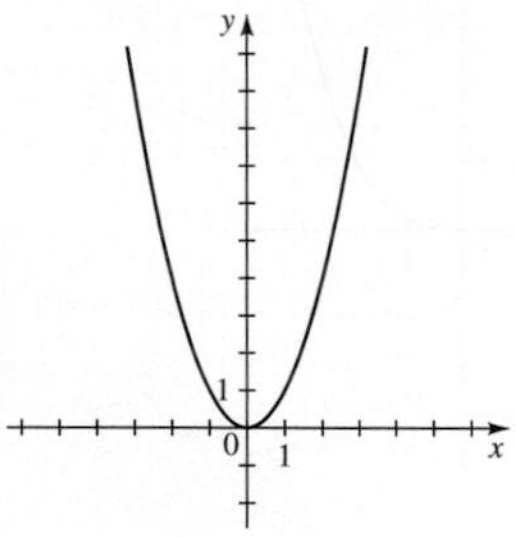

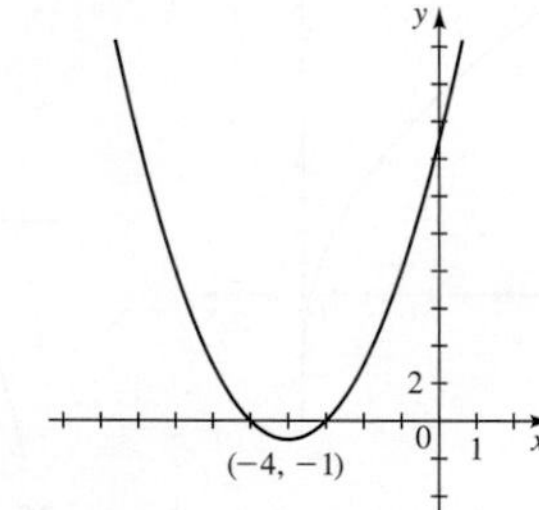

9. **(a)** Shift 3 units right, then shift upward 2 units
(b) $y = \sqrt{x - 3} + 2$
10. **(a)** Reflect about the y-axis **(b)** $y = \sqrt{-x}$
11. **(a)** 3, 0
(b)

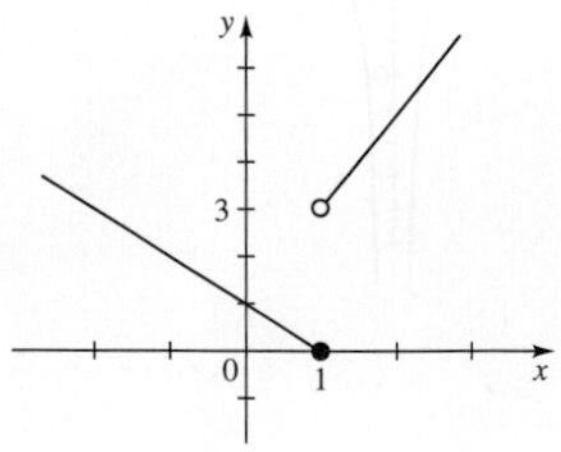

12. **(a)** $x^2 + 2x - 2$ **(b)** $x^2 + 4$ **(c)** $x^2 - 5x + 7$ **(d)** $x^2 + x - 2$ **(e)** 1 **(f)** 4 **(g)** $x - 9$
13. **(a)** Yes **(b)** No **15.** $f^{-1}(x) = -\dfrac{5x + 3}{2x - 1}$

16. (a) $f^{-1}(x) = 3 - x^2, x \geq 0$
(b)

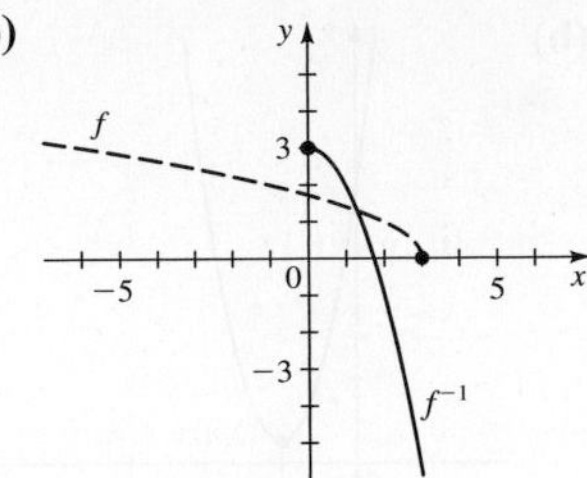

17. Domain $[0, 6]$, range $[1, 7]$ **18.** 1, 3
19.

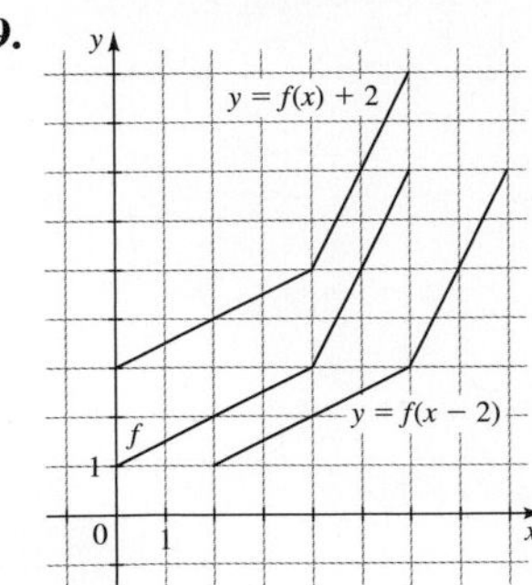

20. $5, \frac{5}{4}$ **21.** 0, 4
22.

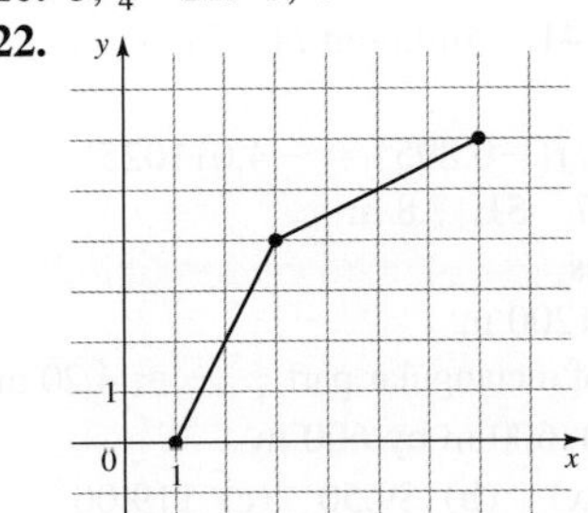

23. (a) **(b)** No

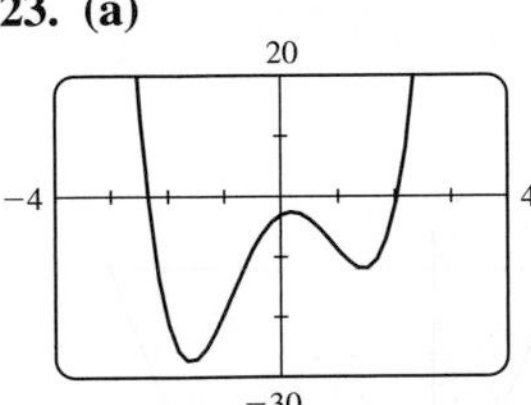

(c) local maximum ≈ -2.55 when $x \approx 0.18$; Local minimum ≈ -27.18 when $x \approx -1.61$, local minimum ≈ -11.93 when $x \approx 1.43$; **(d)** $[-27.18, \infty)$ **(e)** Increasing on $(-1.61, 0.18), (1.43, \infty)$; decreasing on $(-\infty, -1.61), (0.18, 1.43)$

Focus on Modeling ■ Page 246

1. $A(w) = 3w^2, w > 0$ **3.** $V(w) = \frac{1}{2}w^3, w > 0$
5. $A(x) = 10x - x^2, 0 < x < 10$ **7.** $A(x) = (\sqrt{3}/4)x^2, x > 0$
9. $r(A) = \sqrt{A/\pi}, A > 0$ **11.** $S(x) = 2x^2 + \dfrac{240}{x}, x > 0$
13. $D(t) = 25t, t \geq 0$ **15.** $A(b) = b\sqrt{4 - b}, 0 < b < 4$
17. $A(h) = 2h\sqrt{100 - h^2}, 0 < h < 10$
19. (b) $p(x) = x(19 - x)$ **(c)** 9.5, 9.5
21. (b) $A(x) = x(2400 - 2x)$ **(c)** 600 m by 1200 m
23. (a) $f(x) = 8x + (7200/x)$ **(b)** length along road is 30 m, width is 40 m **(c)** 15 m to 60 m
25. (a) $A(x) = 15x - \left(\dfrac{\pi + 4}{8}\right)x^2$
(b) Width ≈ 2.8 m, height of rectangular part ≈ 1.4 m
27. (a) $A(x) = x^2 + \dfrac{48}{x}$
(b) Height ≈ 1.44 m, width ≈ 2.88 m
29. (a) $L(x) = 2x + \dfrac{200}{x}$ **(b)** 10 m by 10 m
31. (a) $T(x) = \frac{1}{2}\sqrt{x^2 - 14x + 53} + \frac{1}{5}x$ **(b)** ≈ 6.13 km from B
33. (b) horizontal is ≈ 9.23, vertical is ≈ 13.00

Chapter 3

Section 3.1 ■ Page 257

1. square **2. (a)** (h, k) **(b)** upward, minimum **(c)** downward, maximum **3.** upward, $(2, -6)$, -6, minimum
4. downward, $(2, -6)$, -6, maximum
5. (a) $(3, 4)$; x-intercepts 1, 5; y-intercept -5
(b) Maximum $f(3) = 4$ **(c)** $\mathbb{R}, (-\infty, 4]$
7. (a) $(1, -3)$; x-intercepts $\dfrac{2 \pm \sqrt{6}}{2}$; y-intercept -1
(b) Minimum $f(-1) = -3$ **(c)** $\mathbb{R}, [-3, \infty)$

9. (a) $f(x) = (x - 2)^2 + 5$
(b) Vertex $(2, 5)$
no x-intercept
y-intercept 9
(c)

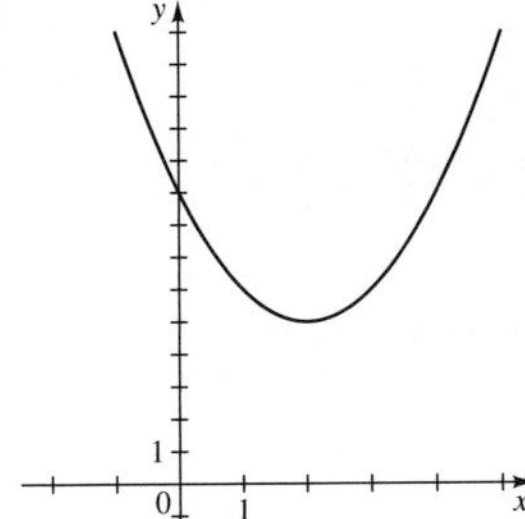

(d) $\mathbb{R}, [5, \infty)$

11. (a) $f(x) = (x - 3)^2 - 9$
(b) Vertex $(3, -9)$
x-intercepts 0, 6
y-intercept 0
(c)

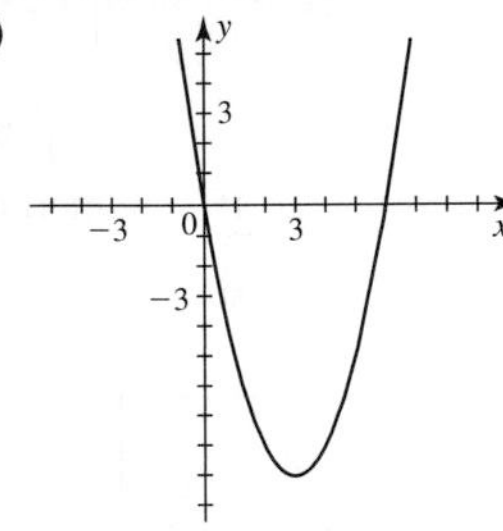

(d) $\mathbb{R}, [-9, \infty)$

13. (a) $f(x) = 3(x + 1)^2 - 3$
(b) Vertex $(-1, -3)$
x-intercepts $-2, 0$
y-intercept 0
(c)

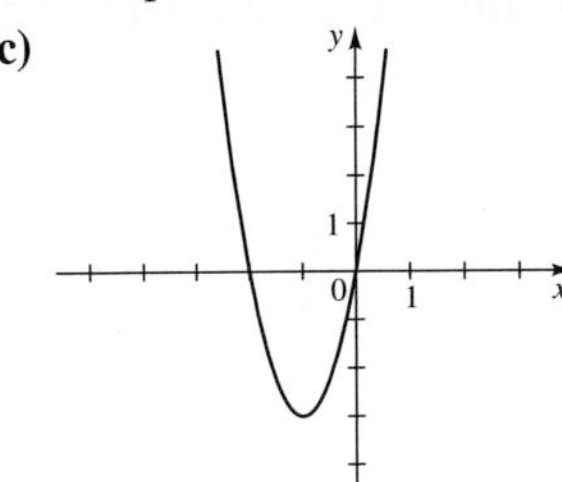

(d) $\mathbb{R}, [-3, \infty)$

15. (a) $f(x) = (x + 2)^2 - 1$
(b) Vertex $(-2, -1)$
x-intercepts $-1, -3$
y-intercept 3
(c)

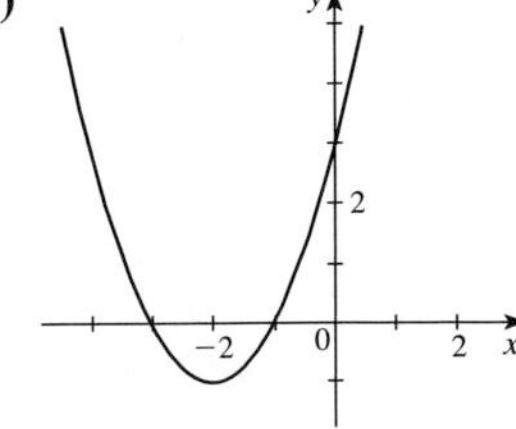

(d) $\mathbb{R}, [-1, \infty)$

17. (a) $f(x) = -(x + 5)^2 + 10$
(b) Vertex $(-5, 10)$; x-intercepts $-5 \pm \sqrt{10}$; y-intercept -15
(c)

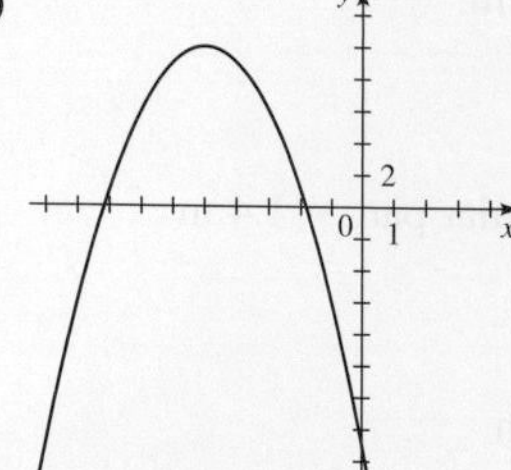

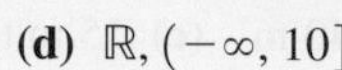
(d) $\mathbb{R}, (-\infty, 10]$

19. (a) $f(x) = 3(x - 1)^2 + 4$
(b) Vertex $(1, 4)$; no x-intercept; y-intercept 7
(c)

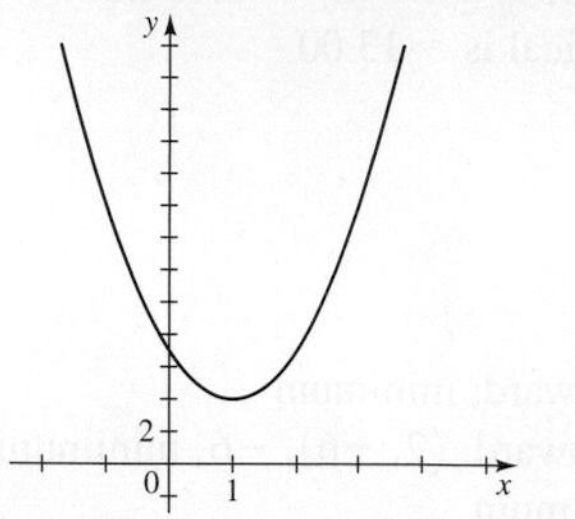

(d) $\mathbb{R}, [4, \infty)$

21. (a) $f(x) = 0.5(x + 6)^2 - 2$
(b) Vertex $(-6, -2)$; x-intercepts $-8, -4$; y-intercept 16
(c)

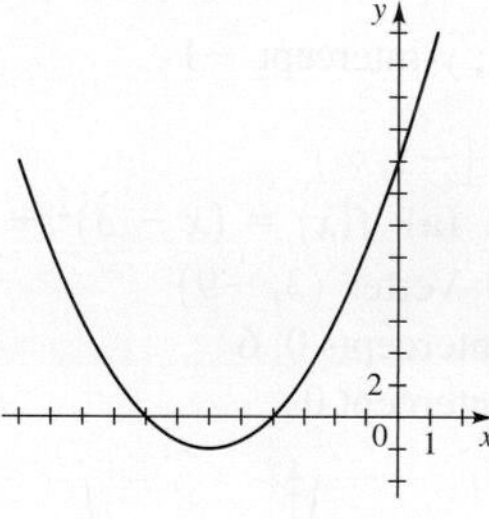

(d) $\mathbb{R}, [-2, \infty)$

23. (a) $f(x) = -4\left(x + \frac{3}{2}\right)^2 + 10$
(b) Vertex $\left(-\frac{3}{2}, 10\right)$; x-intercepts $-\frac{3}{2} - \frac{\sqrt{10}}{2}, -\frac{3}{2} + \frac{\sqrt{10}}{2}$;
y-intercept 1
(c)

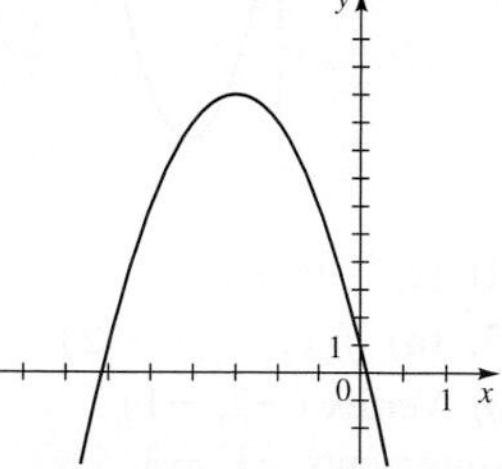

(d) $\mathbb{R}, (-\infty, 10]$

25. (a) $f(x) = (x + 1)^2 - 2$
(b)

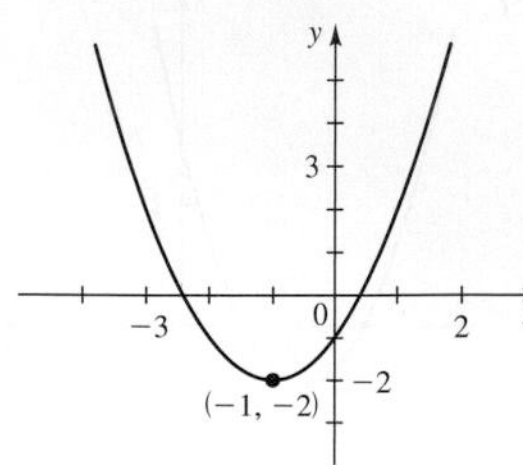

(c) Minimum $f(-1) = -2$

27. (a) $f(x) = 4(x - 1)^2 - 5$
(b)

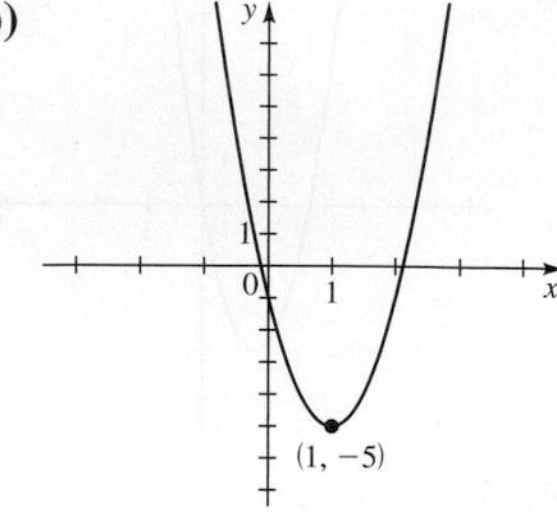

(c) Minimum $f(1) = -5$

29. (a) $f(x) = -\left(x + \frac{3}{2}\right)^2 + \frac{21}{4}$
(b)

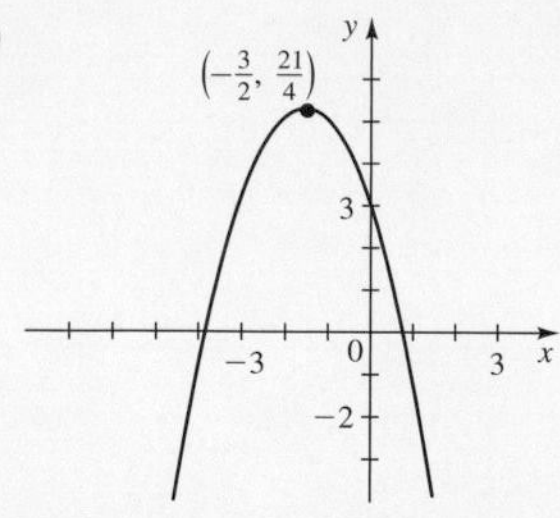

(c) Maximum $f\left(-\frac{3}{2}\right) = \frac{21}{4}$

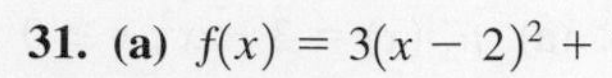
31. (a) $f(x) = 3(x - 2)^2 + 1$
(b)

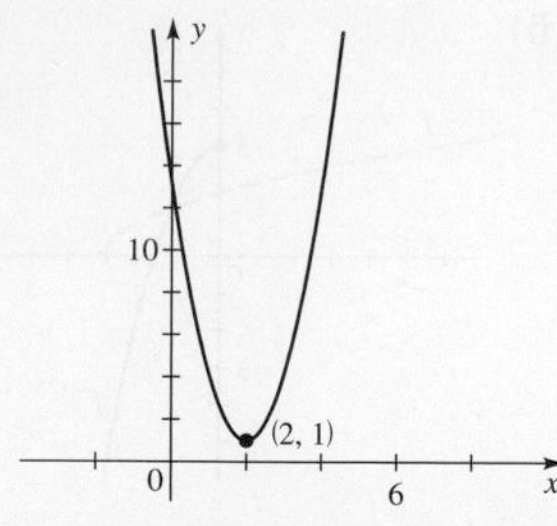

(c) Minimum $f(2) = 1$

33. (a) $f(x) = -\left(x + \frac{1}{2}\right)^2 + \frac{5}{4}$
(b)

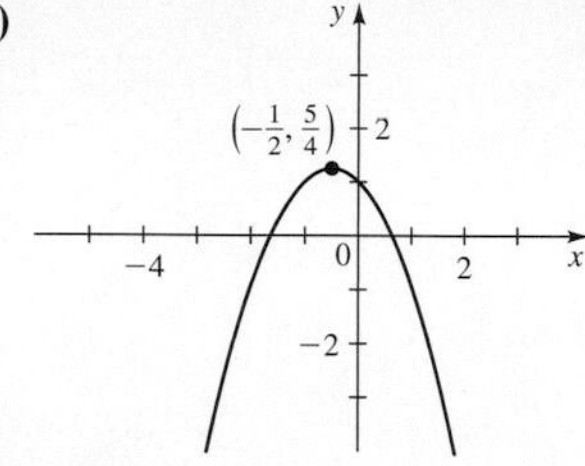

(c) Maximum $f\left(-\frac{1}{2}\right) = \frac{5}{4}$
35. Maximum $f(1) = 2$ **37.** Minimum $f(5) = 10$
39. Minimum $f(0.6) = 15.64$ **41.** Minimum $h(-2) = -8$
43. Maximum $f(-1) = \frac{7}{2}$
45. (a) $f(-0.90) \approx -4.01$ **(b)** $f(-0.895) = -4.011025$
47. $f(x) = 4(x - 2)^2 - 3$ **49.** 7 **51.** 7.8 m
53. \$4000, 100 units **55.** 30 times
57. 50 trees/acre **59.** 600 m by 1200 m
61. Width $\frac{60}{4 + \pi} \approx 8.40$ m, height of rectangular part $\frac{30}{4 + \pi} \approx 4.20$ m
63. (a) $f(x) = x(1200 - x)$ **(b)** 600 m by 600 m
65. (a) $R(x) = x(57{,}000 - 3000x)$ **(b)** \$9.50 **(c)** \$19.00

Section 3.2 ■ Page 271

1. II **2. (a)** $-\infty, \infty$ **(b)** $-\infty, -\infty$
3. (a) 0 **(b)** factor **(c)** x **4.** (a)
5. (a)

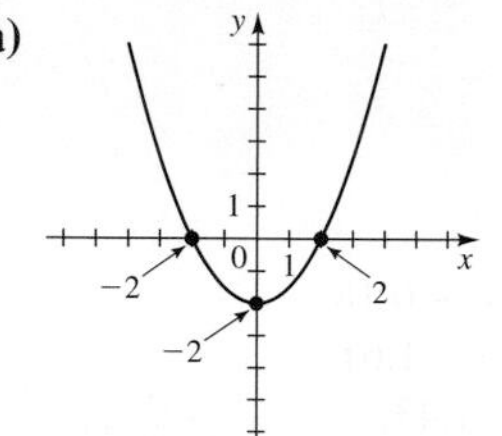

Domain $(-\infty, \infty)$,
range $[-2, \infty)$

(b)

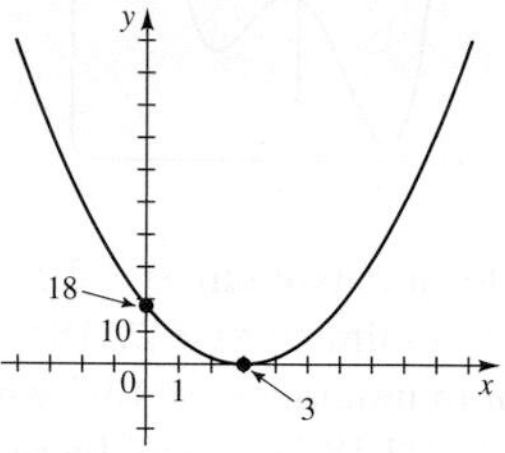

Domain $(-\infty, \infty)$,
range $[0, \infty)$

(c)

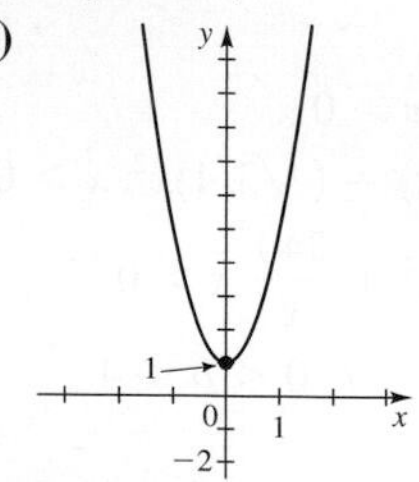

Domain $(-\infty, \infty)$,
range $[1, \infty)$

(d)

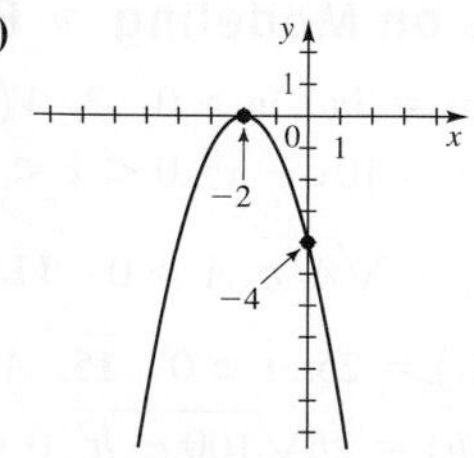

Domain $(-\infty, \infty)$,
range $(-\infty, 0]$

7. (a)

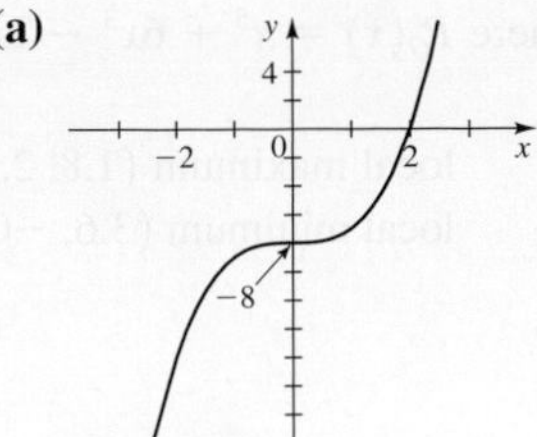

Domain $(-\infty, \infty)$, range $(-\infty, \infty)$

(b)

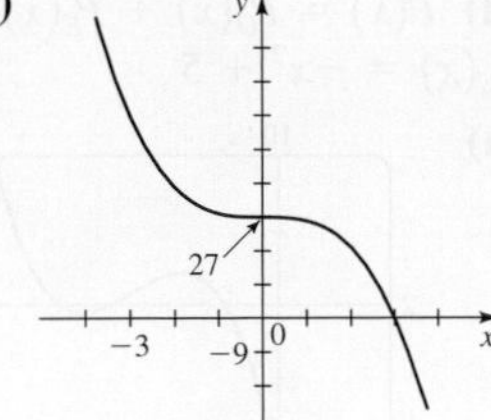

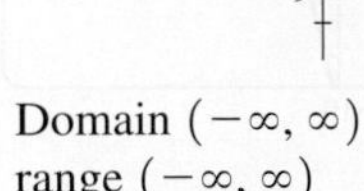

Domain $(-\infty, \infty)$, range $(-\infty, \infty)$

(c)

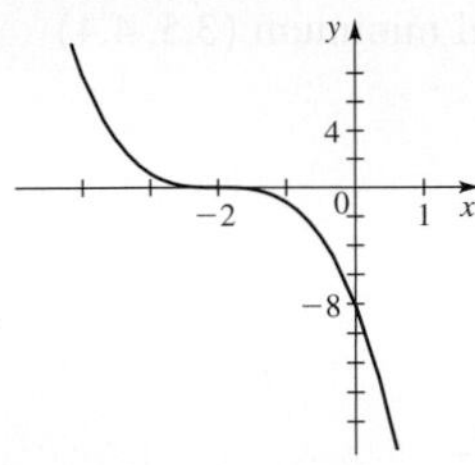

Domain $(-\infty, \infty)$, range $(-\infty, \infty)$

(d)

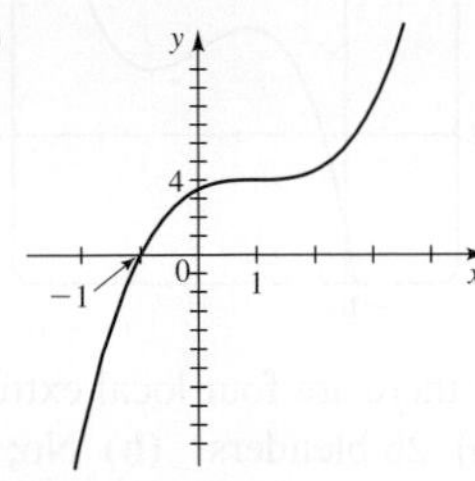

Domain $(-\infty, \infty)$, range $(-\infty, \infty)$

9. (a) $y \to \infty$ as $x \to \infty, y \to -\infty$ as $x \to -\infty$ **(b)** III

11. (a) $y \to -\infty$ as $x \to \infty, y \to \infty$ as $x \to -\infty$ **(b)** V

13. (a) $y \to \infty$ as $x \to \infty, y \to \infty$ as $x \to -\infty$ **(b)** VI

15.

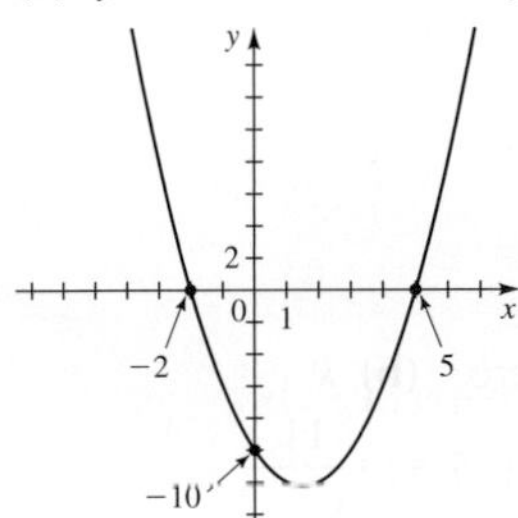

17.

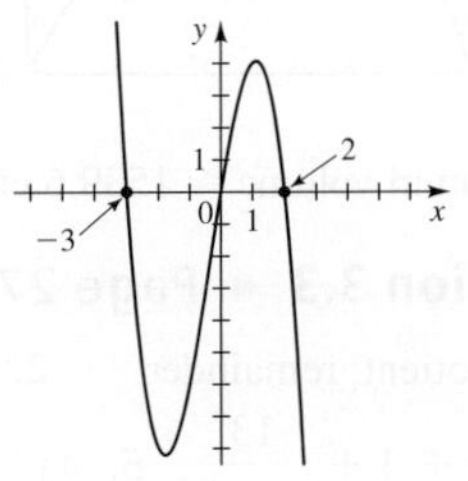

19.

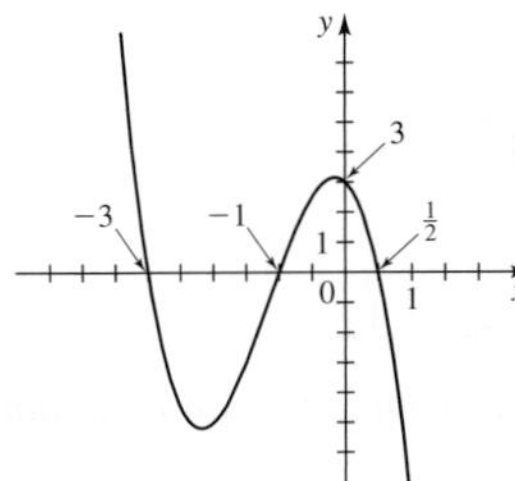

21.

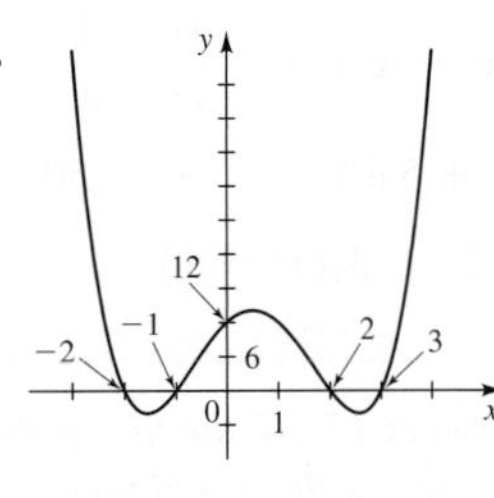

23.

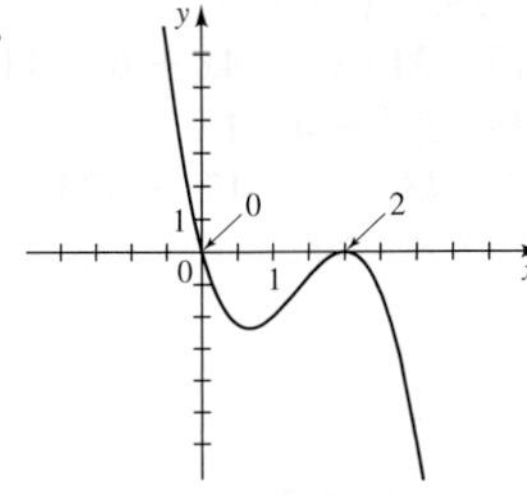

25.

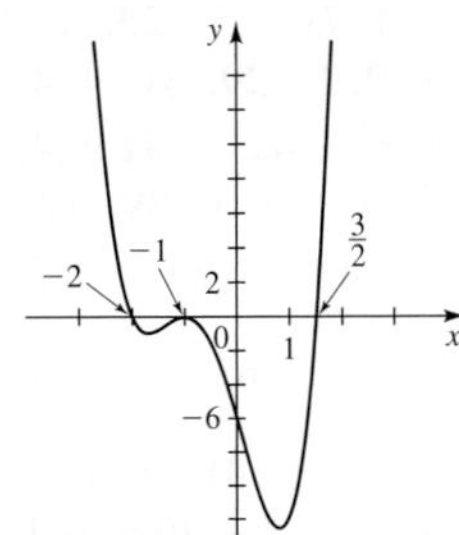

27.

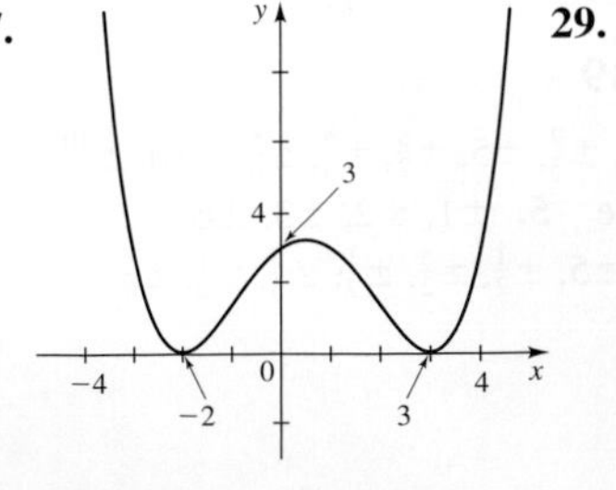

29.

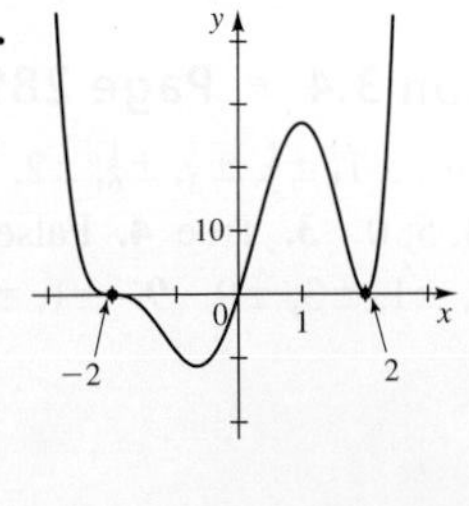

31. $P(x) = x(x + 2)(x - 3)$

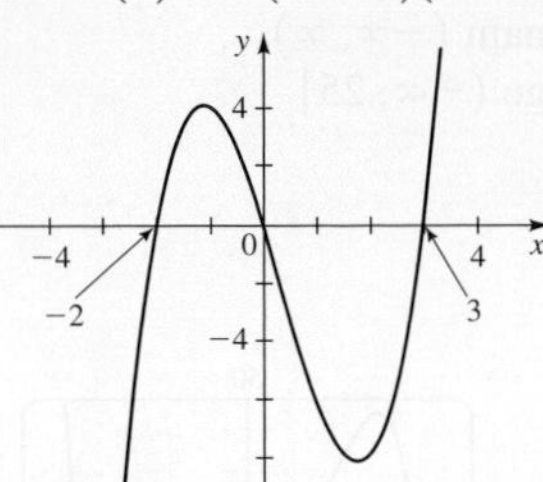

33. $P(x) = -x(x + 3)(x - 4)$

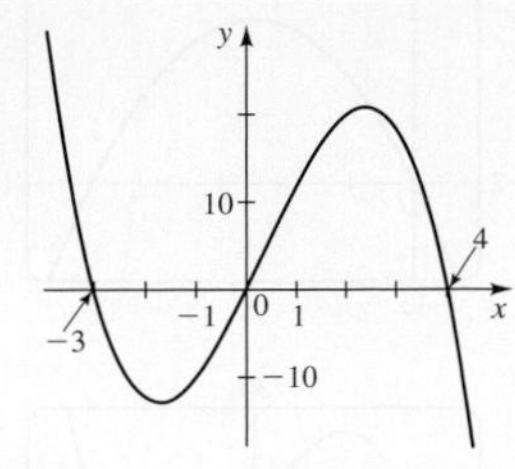

35. $P(x) = x^2(x - 1)(x - 2)$

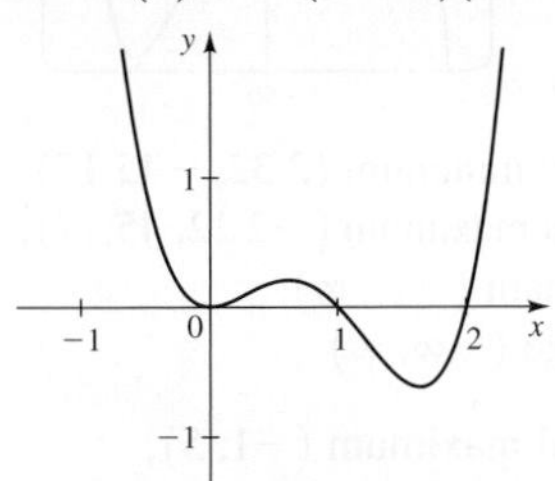

37. $P(x) = (x + 1)^2(x - 1)$

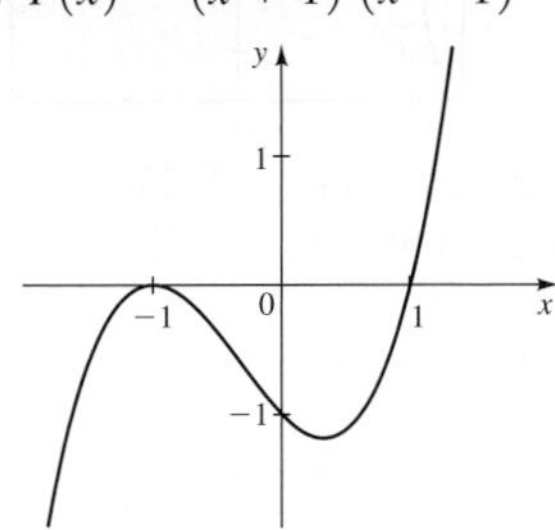

39. $P(x) = (2x - 1)(x + 3)(x - 3)$

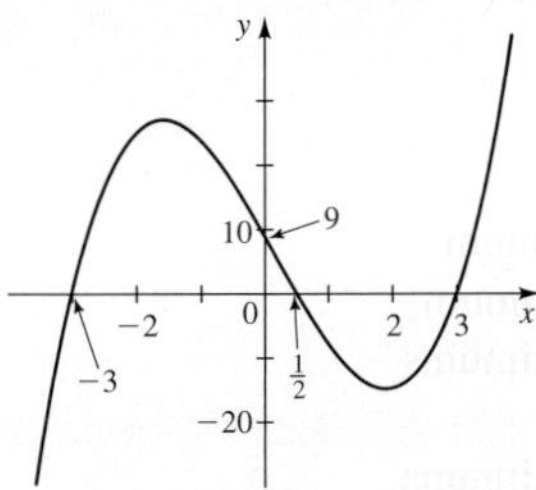

41. $P(x) = (x - 2)^2(x^2 + 2x + 4)$

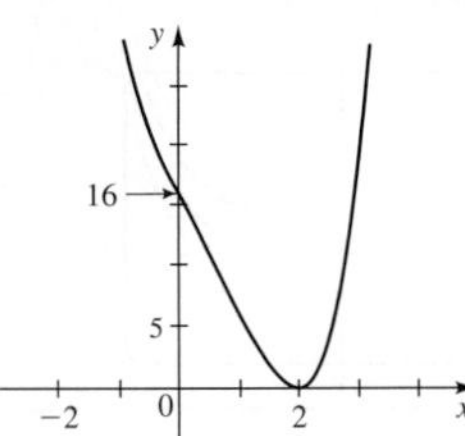

43. $P(x) = (x^2 + 1)(x + 2)(x - 2)$

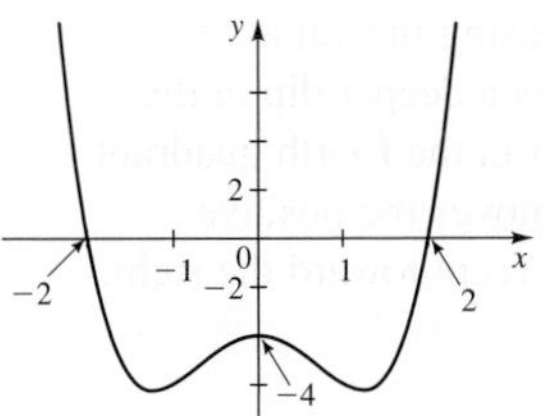

45. $y \to \infty$ as $x \to \infty, y \to -\infty$ as $x \to -\infty$

47. $y \to \infty$ as $x \to \pm\infty$

49. $y \to \infty$ as $x \to \infty, y \to -\infty$ as $x \to -\infty$

51. (a) x-intercepts 0, 4; y-intercept 0 **(b)** Local maximum $(2, 4)$ **(c)** $(-\infty, \infty), (-\infty, 4]$

53. (a) x-intercepts -2, 1; y-intercept -1 **(b)** Local minimum $(-1, -2)$, local maximum $(1, 0)$ **(c)** $(-\infty, \infty), (-\infty, \infty)$

55. 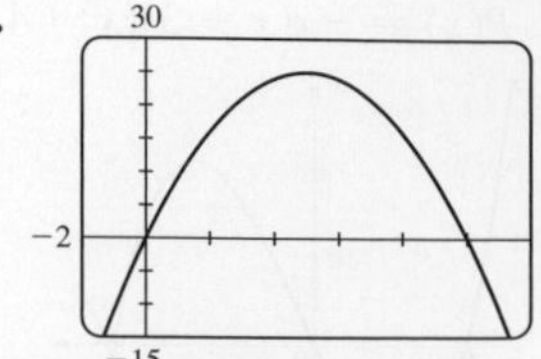

local maximum (5, 25), domain $(-\infty, \infty)$, range $(-\infty, 25]$

57. 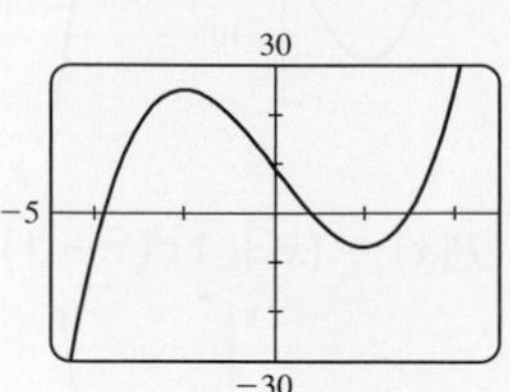

local maximum (−2, 25), local minimum (2, −7), domain $(-\infty, \infty)$, range $(-\infty, \infty)$

59. local minimum (2.32, −45.17), local maximum (−2.32, 45.17), domain $(-\infty, \infty)$, range $(-\infty, \infty)$

61. 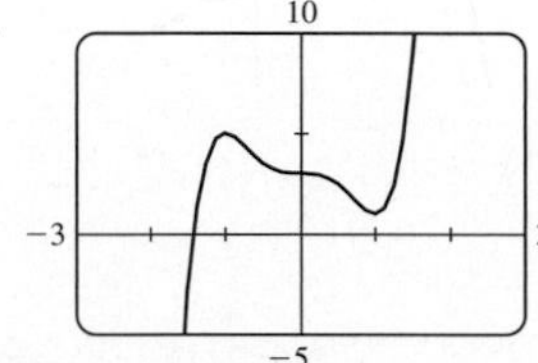

local maximum (−1, 5), local minimum (1, 1), domain $(-\infty, \infty)$, range $(-\infty, \infty)$

63. One local maximum, no local minimum
65. One local maximum, one local minimum
67. One local maximum, two local minimums
69. No local extrema
71. One local maximum, two local minimums

73. 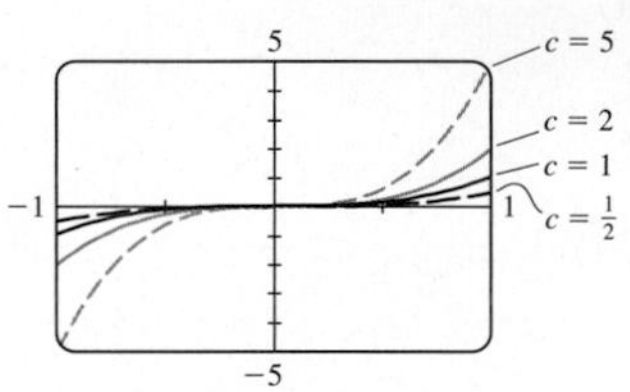

Increasing the value of c stretches the graph vertically.

75. Increasing the value of c shifts the graph upward.

77. 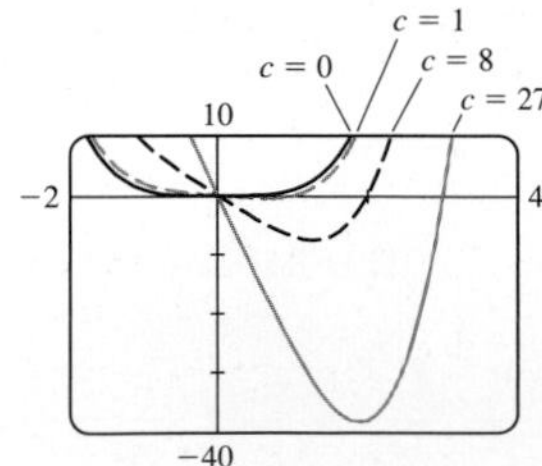

Increasing the value of c causes a deeper dip in the graph in the fourth quadrant and moves the positive x-intercept toward the right.

79. (a) 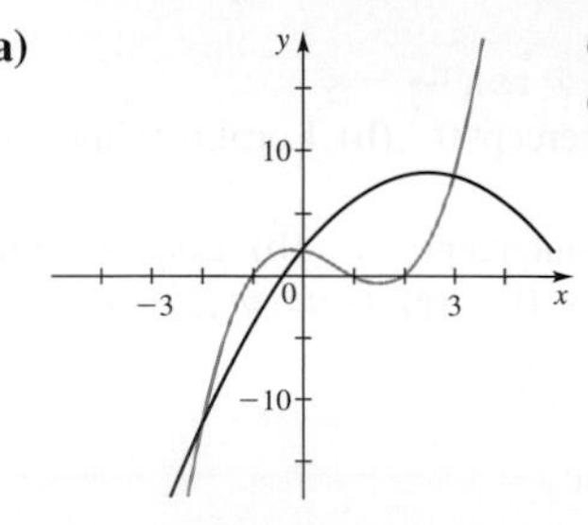

(b) Three **(c)** (0, 2), (3, 8), (−2, −12)

81. (d) $P(x) = P_O(x) + P_E(x)$, where $P_O(x) = x^5 + 6x^3 - 2x$ and $P_E(x) = -x^2 + 5$

83. (a) 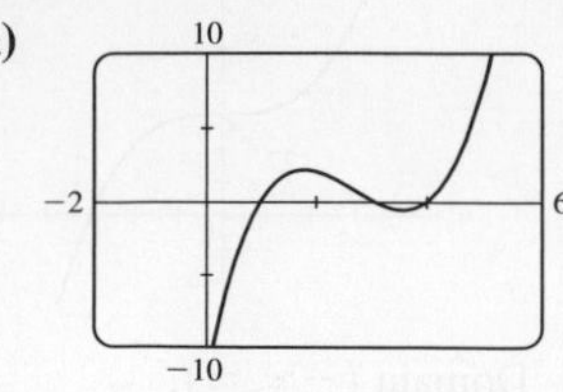

local maximum (1.8, 2.1), local minimum (3.6, −0.6)

(b) local maximum (1.8, 7.1), local minimum (3.5, 4.4)

85. 5; there are four local extrema
87. (a) 26 blenders **(b)** No; $3276.22
89. (a) $V(x) = 4x^3 - 120x^2 + 800x$ **(b)** $0 < x < 10$
(c) 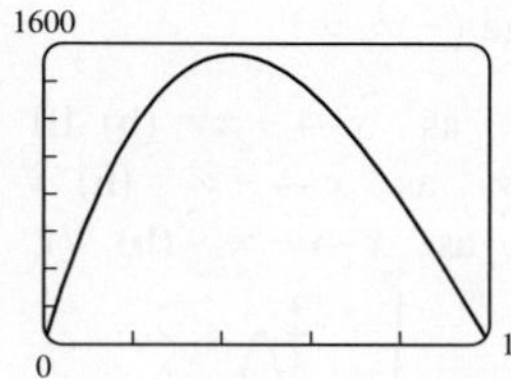

maximum volume ≈ 1539.6 cm^3

Section 3.3 ■ Page 279

1. quotient, remainder **2. (a)** zero **(b)** k
3. $3x + 3 + \dfrac{13}{x - 3}$ **5.** $4x^2 - 8x + 5 - \dfrac{11}{3x + 2}$
7. $2x^2 - x + 1 + \dfrac{4x - 4}{x^2 + 4}$
9. $(x + 5)(3x^2 - 10x + 50) - 245$
11. $(2x - 3)(x^2 - 1) - 3$
13. $(2x^2 + 1)(4x^2 + 2x + 1) + (-2x - 1)$

In answers 15–37 the first polynomial given is the quotient, and the second is the remainder.
15. $x - 1, 5$ **17.** $3x^2 - x, 1$ **19.** $4x - 2, 6x - 5$
21. $3x + 1, 7x - 5$ **23.** $x^4 + 1, 0$ **25.** $2x + 1, 6$
27. $3x - 2, 2$ **29.** $3x^2 + 4x + 9, 13$ **31.** $x^2 - 4x + 6, -11$
33. $x^4 + x^3 + 4x^2 + 4x + 4, -2$ **35.** $2x^2 + 4x, 1$
37. $x^2 + 3x + 9, 0$ **39.** 17 **41.** 12 **43.** −7 **45.** −483
47. 2159 **49.** $\frac{7}{3}$ **51.** −8.279 **57.** 2, 5 **59.** $-1 \pm \sqrt{6}$
61. $\dfrac{5 \pm \sqrt{37}}{6}$ **63.** $x^3 - 3x^2 - x + 3$
65. $x^4 - 8x^3 + 14x^2 + 8x - 15$
67. $-2x^4 + 4x^3 + 10x^2 - 12x$ **69.** $3x^4 - 9x^2 + 6$
71. $(x + 1)(x - 1)(x - 2)$ **73.** $(x + 2)^2(x - 1)^2$

Section 3.4 ■ Page 289

1. $a_0, a_n, \pm 1, \pm\frac{1}{2}, \pm\frac{1}{3}, \pm\frac{1}{6}, \pm 2, \pm\frac{2}{3}, \pm 5, \pm\frac{5}{2}, \pm\frac{5}{3}, \pm\frac{5}{6}, \pm 10, \pm\frac{10}{3}$
2. 1, 3, 5; 0 **3.** True **4.** False **5.** $\pm 1, \pm 2, \pm 3, \pm 6$
7. $\pm\frac{1}{3}, \pm 1, \pm 3, \pm 9$ **9.** $\pm 1, \pm 5, \pm\frac{1}{2}, \pm\frac{5}{2}, \pm\frac{1}{3}, \pm\frac{5}{3}, \pm\frac{1}{6}, \pm\frac{5}{6}$

11. **(a)** $\pm 1, \pm\frac{1}{5}$ **(b)** $-1, 1, \frac{1}{5}$ **13.** **(a)** $\pm 1, \pm 3, \pm\frac{1}{2}, \pm\frac{3}{2}$
(b) $-\frac{1}{2}, 1, 3$ **15.** $-2, 1, 6; P(x) = (x + 2)(x - 1)(x - 6)$
17. $-1, 3; P(x) = (x + 1)(x - 3)^2$
19. $2; P(x) = (x - 2)^3$
21. $-6, 3; P(x) = (x + 6)(x - 3)^2$
23. $-3, -1, 1; P(x) = (x + 3)(x + 1)(x - 1)$
25. $\pm 1, \pm 2; P(x) = (x - 2)(x + 2)(x - 1)(x + 1)$
27. $-4, -2, -1, 1; P(x) = (x + 4)(x + 2)(x - 1)(x + 1)$
29. $\pm 3, \pm\frac{1}{3}; P(x) = (3x + 1)(3x - 1)(x + 3)(x - 3)$
31. $\pm 1, -\frac{3}{2}, \frac{1}{3}; P(x) = (2x + 3)(3x - 1)(x + 1)(x - 1)$
33. $-1, \pm\frac{1}{2}; P(x) = (x + 1)(2x - 1)(2x + 1)$
35. $-\frac{3}{2}, \frac{1}{2}, 1; P(x) = (x - 1)(2x + 3)(2x - 1)$
37. $-\frac{2}{3}, -\frac{1}{2}, \frac{3}{4}; P(x) = (3x + 2)(2x + 1)(4x - 3)$
39. $-3, -\frac{3}{2}, \frac{1}{3}, 2; P(x) = (2x + 3)(x + 3)(3x - 1)(x - 2)$
41. $-3, -2, 1, 3; P(x) = (x + 3)(x + 2)^2(x - 1)(x - 3)$
43. $-1, -\frac{1}{3}, 2, 5; P(x) = (x + 1)^2(x - 2)(x - 5)(3x + 1)$

45. $-1, \dfrac{-1 \pm \sqrt{13}}{3}$ **47.** $-1, 4, \dfrac{3 \pm \sqrt{13}}{2}$

49. $3, \dfrac{1 \pm \sqrt{5}}{2}$ **51.** $\frac{1}{2}, \dfrac{1 \pm \sqrt{3}}{2}$ **53.** $-1, -\frac{1}{2}, -3 \pm \sqrt{10}$

55. **(a)** $-3, \pm 1$ **(b)**

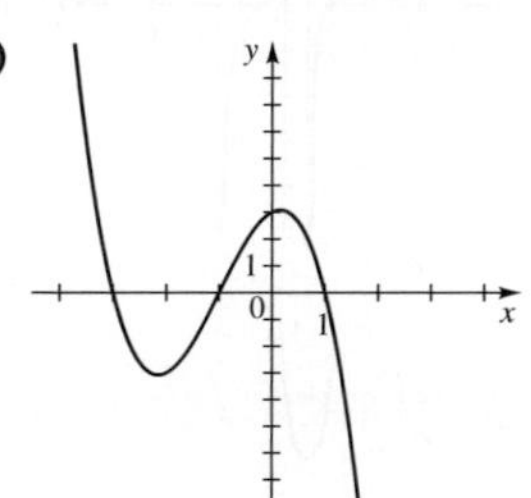

57. **(a)** $-\frac{1}{2}, 2$ **(b)**

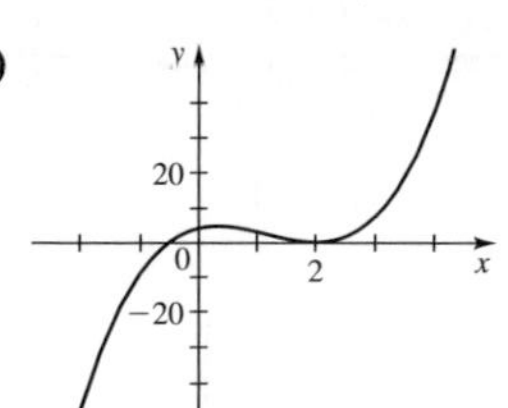

59. **(a)** $-1, 2$ **(b)**

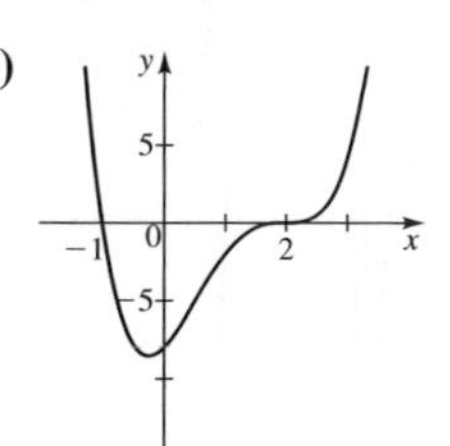

61. **(a)** $-1, 2$ **(b)**

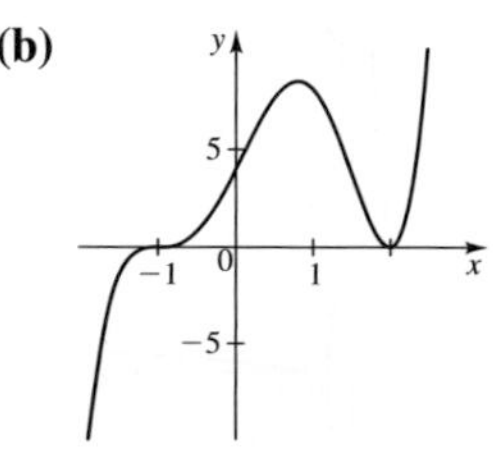

63. 1 positive, 2 or 0 negative; 3 or 1 real **65.** 1 positive, 1 negative; 2 real **67.** 2 or 0 positive, 0 negative; 3 or 1 real (since 0 is a zero but is neither positive nor negative) **77.** $3, -2$ **79.** $3, -1$ **81.** $-2, \frac{1}{2}, \pm 1$ **83.** $\pm\frac{1}{2}, \pm\sqrt{5}$ **85.** $-2, 1, 3, 4$ **91.** $-2, 2, 3$ **93.** $-\frac{3}{2}, -1, 1, 4$ **95.** $-1.28, 1.53$ **97.** -1.50 **99.** 3.35 m **101.** 2.76 m **103.** 88 cm (or 3.20 cm)

Section 3.5 ■ Page 299

1. $6; -7; 2, 3$ **2.** **(a)** $x - a$ **(b)** $(x - a)^m$ **3.** n **4.** $a - bi$; $3 - i$ **5.** **(a)** True **(b)** True **(c)** False, $x^4 + 1 > 0$ for all real x **6.** **(a)** False, $x^2 + 1$ has no real zeros **(b)** True **(c)** False, $x^2 + 1$ factors into linear factors with complex coefficients **7.** **(a)** $0, \pm 2i$ **(b)** $x^2(x - 2i)(x + 2i)$
9. **(a)** $0, 1 \pm i$ **(b)** $x(x - 1 - i)(x - 1 + i)$
11. **(a)** $\pm i$ **(b)** $(x - i)^2(x + i)^2$
13. **(a)** $\pm 2, \pm 2i$ **(b)** $(x - 2)(x + 2)(x - 2i)(x + 2i)$
15. **(a)** $-2, 1 \pm \sqrt{3}i$
(b) $(x + 2)(x - 1 - \sqrt{3}i)(x - 1 + \sqrt{3}i)$
17. **(a)** $\pm 1, -\frac{1}{2} \pm \frac{\sqrt{3}}{2}i, \frac{1}{2} \pm \frac{\sqrt{3}}{2}i$
(b) $(x - 1)(x + 1)\left(x - \frac{1}{2} - \frac{\sqrt{3}}{2}i\right)\left(x - \frac{1}{2} + \frac{\sqrt{3}}{2}i\right) \times \left(x + \frac{1}{2} - \frac{\sqrt{3}}{2}i\right)\left(x + \frac{1}{2} + \frac{\sqrt{3}}{2}i\right)$

In answers 19–35 the factored form is given first, then the zeros are listed with the multiplicity of each in parentheses.
19. $x^2(x + 4i)(x - 4i); 0\,(2), \pm 4i\,(1)$
21. $x^4(x + 1 + i)(x + 1 - i); 0\,(4), -1 \pm i\,(1)$
23. $x(x - 2i)(x + 2i); 0\,(1), 2i\,(1), -2i\,(1)$
25. $(x - 1)(x + 1)(x - i)(x + i); 1\,(1), -1\,(1), i\,(1), -i\,(1)$
27. $16\left(x - \frac{3}{2}\right)\left(x + \frac{3}{2}\right)\left(x - \frac{3}{2}i\right)\left(x + \frac{3}{2}i\right); \frac{3}{2}\,(1), -\frac{3}{2}\,(1), \frac{3}{2}i\,(1), -\frac{3}{2}i\,(1)$ **29.** $(x + 1)(x - 3i)(x + 3i); -1\,(1), 3i\,(1), -3i\,(1)$
31. $x^2(x + \sqrt{5}i)^2(x - \sqrt{5}i)^2; 0\,(2), \pm\sqrt{5}i\,(2)$
33. $(x - 1)(x + 1)(x - 2i)(x + 2i); 1\,(1), -1\,(1), 2i(1), -2i(1)$
35. $x(x - \sqrt{3}i)^2(x + \sqrt{3}i)^2; 0\,(1), \sqrt{3}i\,(2), -\sqrt{3}i\,(2)$
37. $P(x) = x^2 - 2x + 2$ **39.** $Q(x) = x^3 - 3x^2 + 4x - 12$
41. $P(x) = x^3 - 2x^2 + x - 2$
43. $R(x) = x^4 - 4x^3 + 10x^2 - 12x + 5$
45. $T(x) = 6x^4 - 12x^3 + 18x^2 - 12x + 12$ **47.** $2, -1 \pm i$

49. $1, \dfrac{1 \pm \sqrt{3}i}{2}$ **51.** $2, \dfrac{1 \pm \sqrt{3}i}{2}$ **53.** $-\frac{3}{2}, -1 \pm \sqrt{2}i$

55. $-2, 1, \pm 3i$ **57.** $1, \pm 2i, \pm\sqrt{3}i$ **59.** 3 (multiplicity 2), $\pm 2i$
61. $-\frac{1}{2}$ (multiplicity 2), $\pm i$ **63.** 1 (multiplicity 3), $\pm 3i$
65. **(a)** $(x - 5)(x^2 + 4)$ **(b)** $(x - 5)(x - 2i)(x + 2i)$
67. **(a)** $(x - 1)(x + 1)(x^2 + 9)$
(b) $(x - 1)(x + 1)(x - 3i)(x + 3i)$
69. **(a)** $(x - 2)(x + 2)(x^2 - 2x + 4)(x^2 + 2x + 4)$
(b) $(x - 2)(x + 2)[x - (1 + \sqrt{3}i)][x - (1 - \sqrt{3}i)] \times [x + (1 + \sqrt{3}i)][x + (1 - \sqrt{3}i)]$
71. **(a)** 4 real **(b)** 2 real, 2 non-real **(c)** 4 non-real

Section 3.6 ■ Page 314

1. $-\infty, \infty$ **2.** 2 **3.** $-1, 2$ **4.** $\frac{1}{3}$ **5.** $-2, 3$ **6.** 1
7. Vertical asymptote $x = 3$; horizontal asymptote $y = 4$
8. Vertical asymptote $x = 2$; horizontal asymptote $y = -1$
9. **(a)** True **(b)** False **(c)** False **(d)** True **10.** True
11. **(a)** $-3, -19, -199, -1999$; $5, 21, 201, 2001$; $1.2500, 1.0417, 1.0204, 1.0020$; $0.8333, 0.9615, 0.9804, 0.9980$
(b) $r(x) \to -\infty$ as $x \to 2^-$; $r(x) \to \infty$ as $x \to 2^+$
(c) Horizontal asymptote $y = 1$
13. **(a)** $-22, -430, -40{,}300, -4{,}003{,}000$; $-10, -370, -39{,}700, -3{,}997{,}000$; $0.3125, 0.0608, 0.0302, 0.0030$; $-0.2778, -0.0592, -0.0298, -0.0030$
(b) $r(x) \to -\infty$ as $x \to 2^-$; $r(x) \to -\infty$ as $x \to 2^+$
(c) Horizontal asymptote $y = 0$

15.

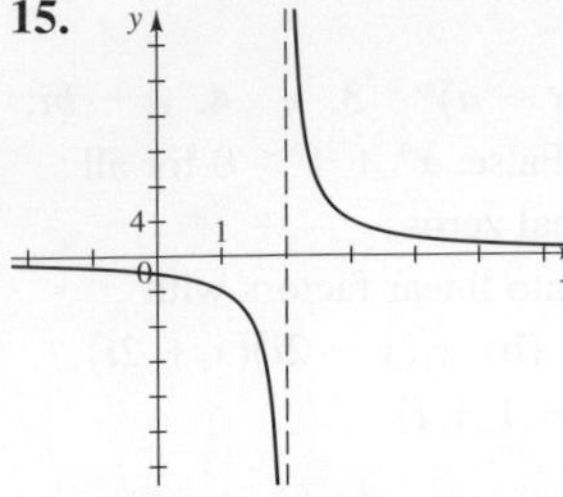

domain $\{x \mid x \neq 2\}$
range $\{y \mid y \neq 0\}$

17.

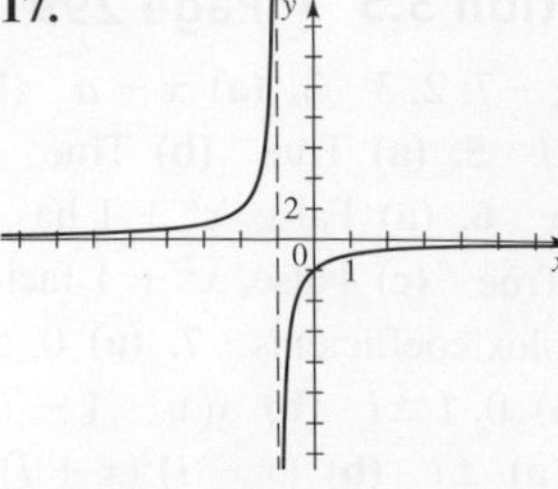

domain $\{x \mid x \neq -1\}$
range $\{y \mid y \neq 0\}$

19.

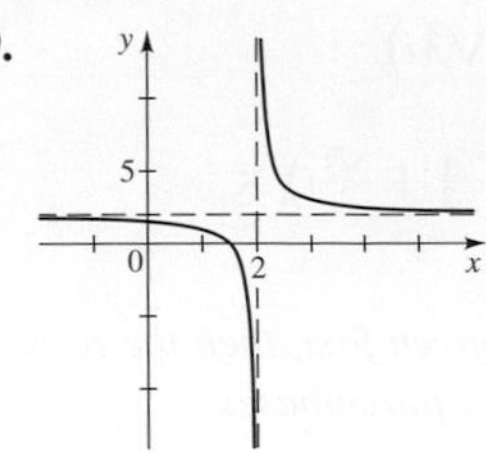

domain $\{x \mid x \neq 2\}$
range $\{y \mid y \neq 2\}$

21.

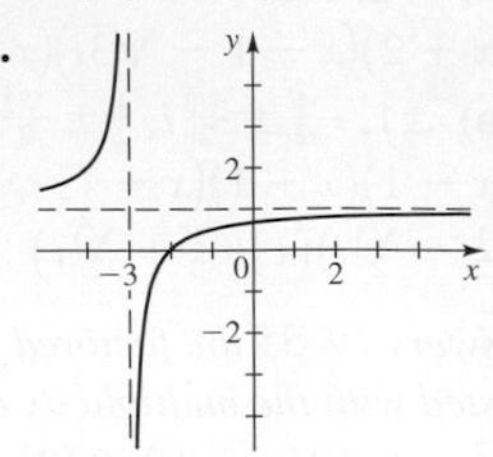

domain $\{x \mid x \neq -3\}$
range $\{y \mid y \neq 1\}$

23. x-intercept 1, y-intercept $-\frac{1}{4}$ **25.** x-intercepts $-1, 2$; y-intercept $\frac{1}{3}$ **27.** x-intercepts $-3, 3$; no y-intercept **29.** x-intercept 3, y-intercept 3, vertical $x = 2$; horizontal $y = 2$ **31.** x-intercepts $-1, 1$; y-intercept $\frac{1}{4}$; vertical $x = -2, x = 2$; horizontal $y = 1$ **33.** Vertical $x = 2$; horizontal $y = 0$ **35.** Horizontal $y = 0$ **37.** Vertical $x = 0, x = -1, x = 1$; horizontal $y = 10$ **39.** Vertical $x = -\frac{7}{4}, x = 2$; horizontal $y = \frac{1}{2}$ **41.** Vertical $x = 0$; horizontal $y = 3$ **43.** Vertical $x = 1$

45.

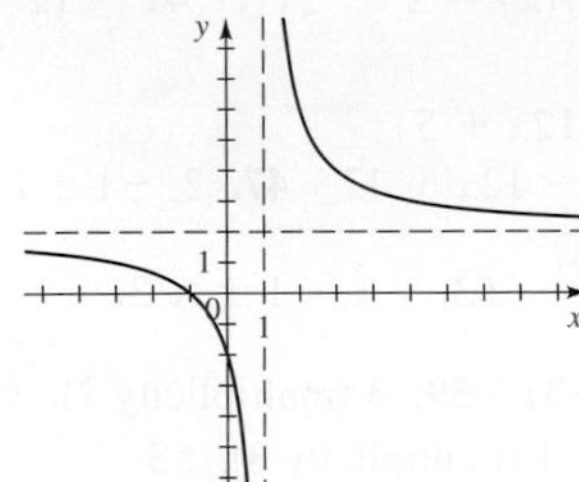

x-intercept -1
y-intercept -2
vertical $x = 1$
horizontal $y = 2$
domain $\{x \mid x \neq 1\}$
range $\{y \mid y \neq 2\}$

47.

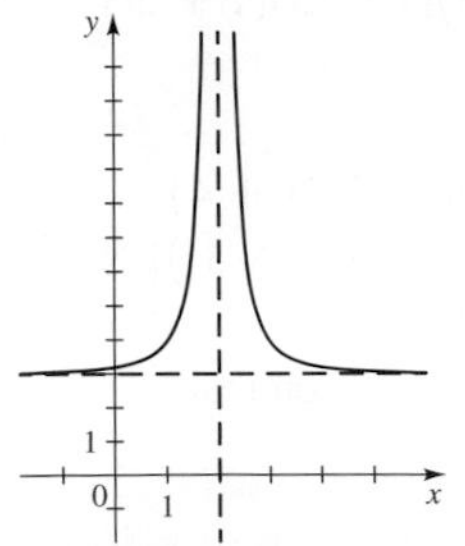

No x-intercept
y-intercept $\frac{13}{4}$
vertical $x = 2$
horizontal $y = 3$
domain $\{x \mid x \neq 2\}$
range $\{y \mid y > 3\}$

49.

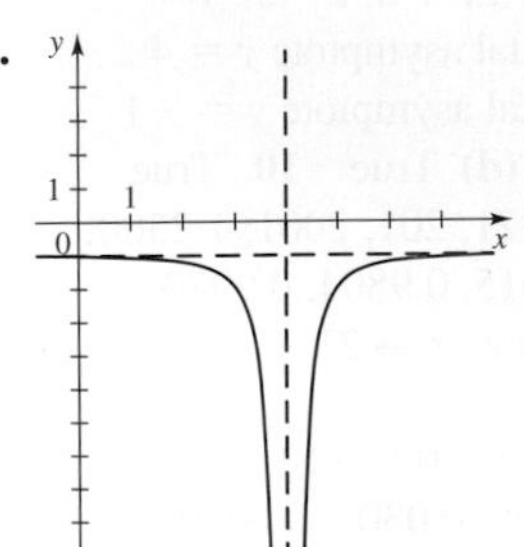

No x-intercept
y-intercept $-\frac{9}{8}$
vertical $x = 4$
horizontal $y = -1$
domain $\{x \mid x \neq 4\}$
range $\{y \mid y < -1\}$

51.

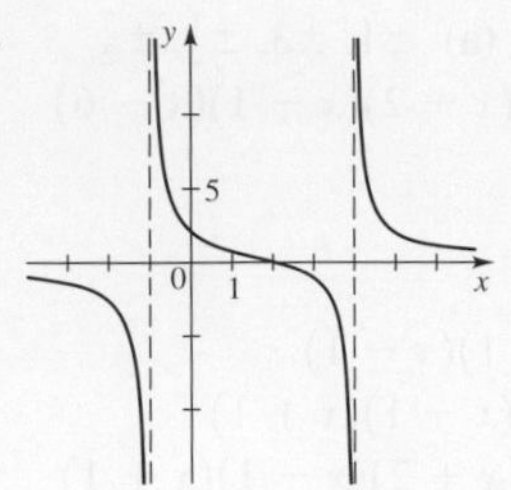

x-intercept 2
y-intercept 2
vertical $x = -1, x = 4$
horizontal $y = 0$
domain $\{x \mid x \neq -1, 4\}$
range $\mathbb{R}$

53.

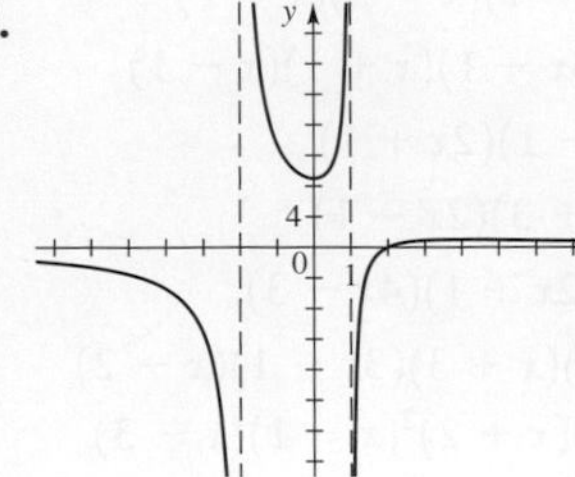

x-intercept 2
y-intercept 9
vertical $x = -2, x = 1$
horizontal $y = 0$
domain $\{x \mid x \neq -2, 1\}$
range $(-\infty, 1] \cup [9, \infty)$

55.

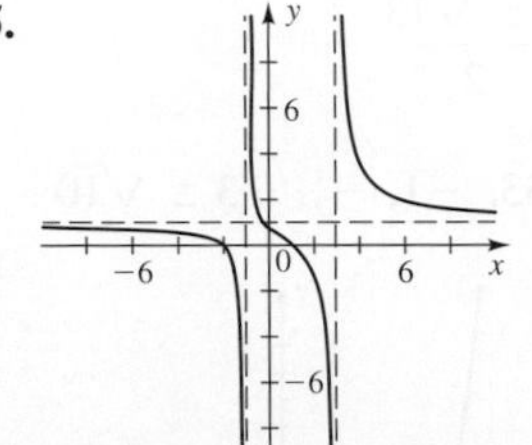

x-intercepts $-2, 1$
y-intercept $\frac{2}{3}$
vertical $x = -1, x = 3$
horizontal $y = 1$
domain $\{x \mid x \neq -1, 3\}$
range $\mathbb{R}$

57.

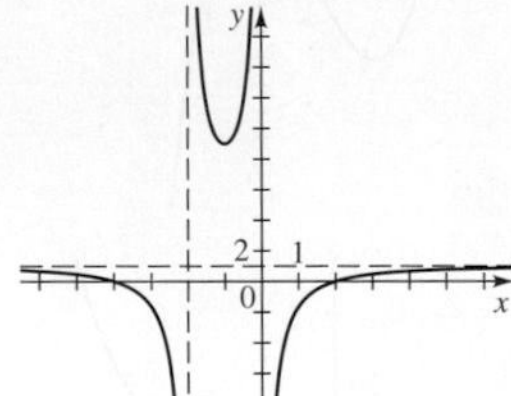

x-intercepts $-4, 2$
y-intercept none
vertical $x = -2, x = 0$
horizontal $y = 1$
domain $\{x \mid x \neq -2, 0\}$
range $\{y \mid y < 1 \text{ or } y \geq 9\}$

59.

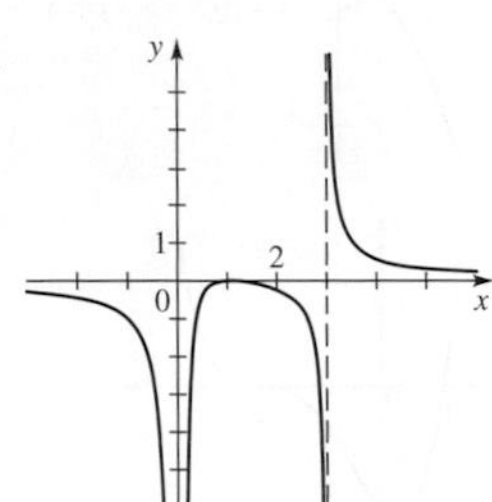

x-intercept 1
y-intercept none
vertical $x = 0, x = 3$
horizontal $y = 0$
domain $\{x \mid x \neq 0, 3\}$
range $\mathbb{R}$

61.

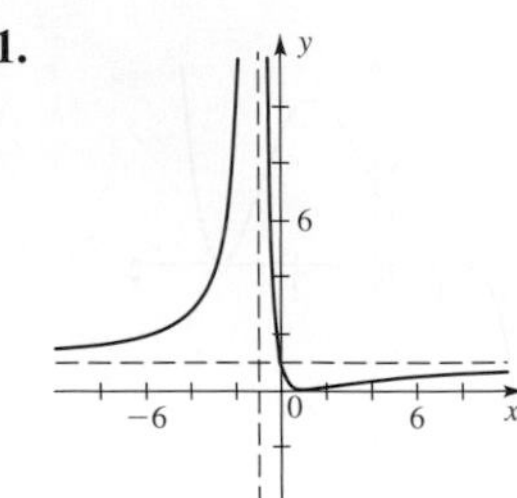

x-intercept 1
y-intercept 1
vertical $x = -1$
horizontal $y = 1$
domain $\{x \mid x \neq -1\}$
range $\{y \mid y \geq 0\}$

63. 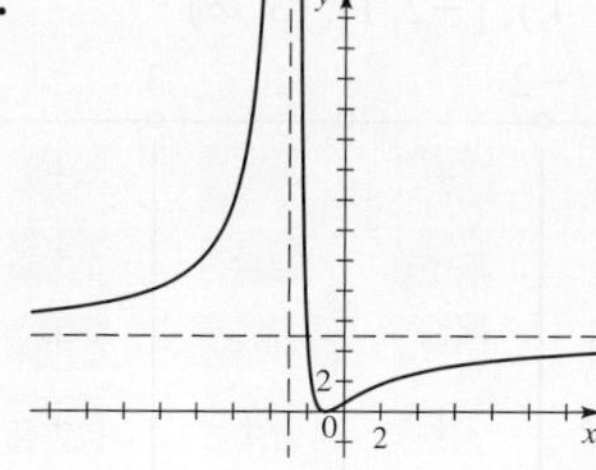
x-intercept -1
y-intercept $\frac{5}{9}$
vertical $x = -3$
horizontal $y = 5$
domain $\{x \mid x \neq -3\}$
range $\{y \mid y \geq 0\}$

65. 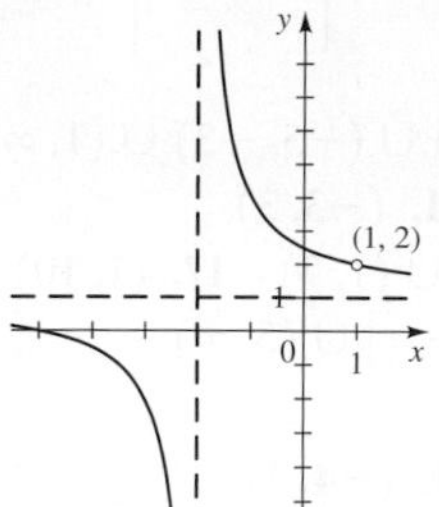

x-intercept -5
y-intercept $\frac{5}{2}$
vertical $x = -2$
horizontal $y = 1$
domain $\{x \mid x \neq -2, 1\}$
range $\{y \mid y \neq 1, 2\}$

67. 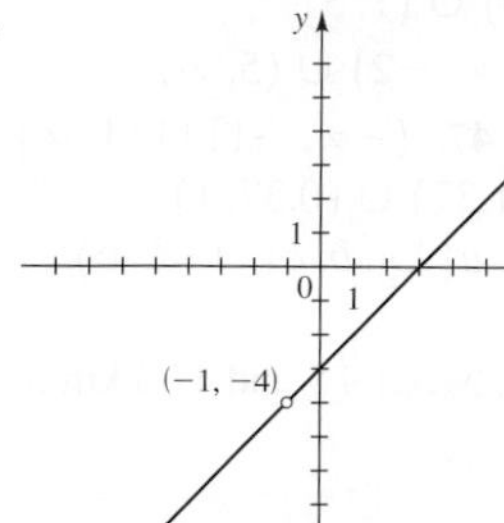

x-intercept 3
y-intercept -3
no asymptote
domain $\{x \mid x \neq -1\}$
range $\{y \mid y \neq -4\}$

69. 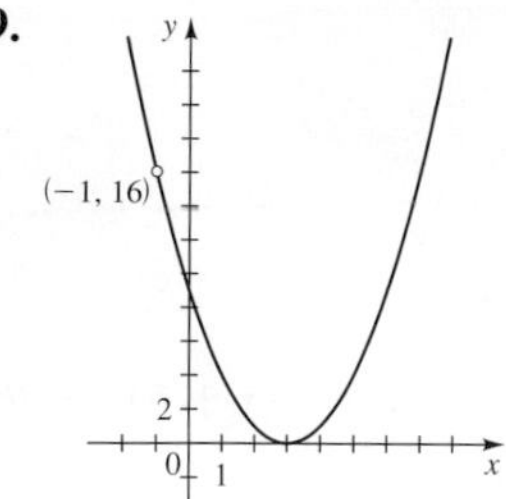

x-intercept 3
y-intercept 9
no asymptote
domain $\{x \mid x \neq -1\}$
range $\{y \mid y \geq 0\}$

71. 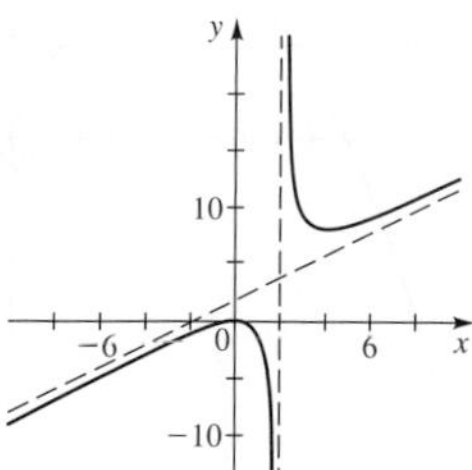
slant $y = x + 2$
vertical $x = 2$

73. 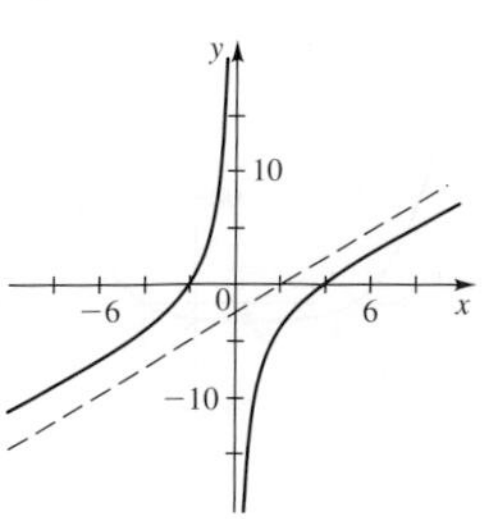
slant $y = x - 2$
vertical $x = 0$

75. 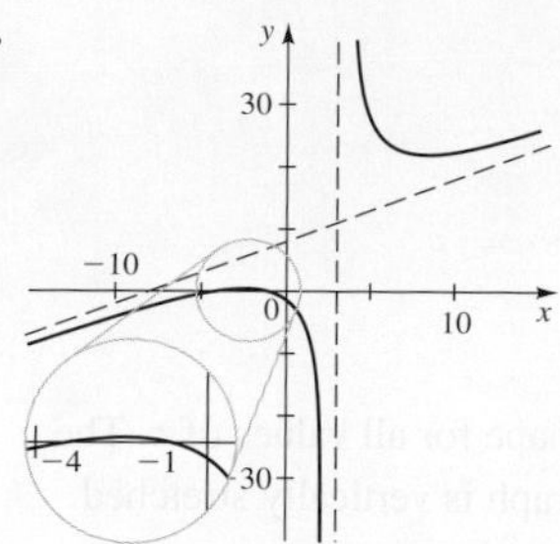
slant $y = x + 8$
vertical $x = 3$

77. 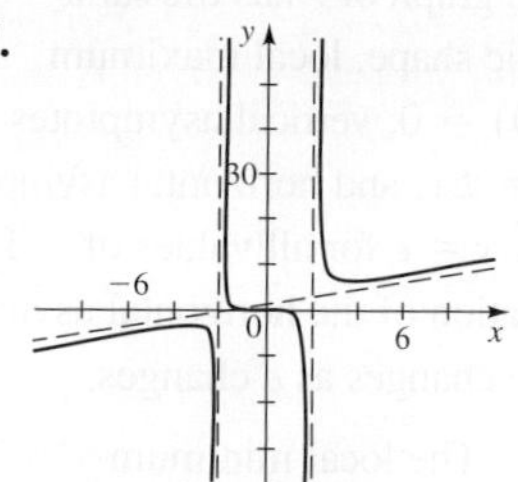
slant $y = x + 1$
vertical $x = 2, x = -2$

79. 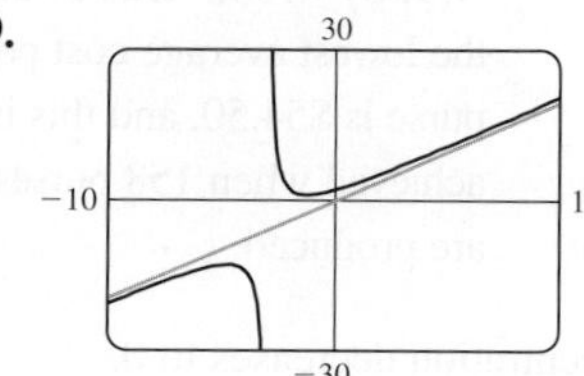
vertical $x = -3$

81. 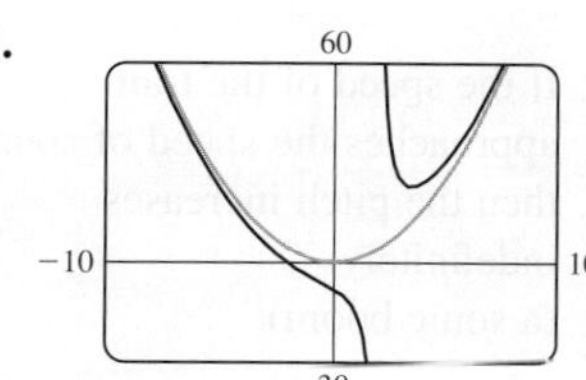
vertical $x = 2$

83. 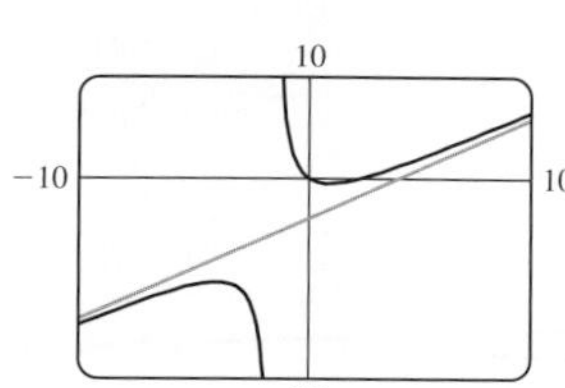
vertical $x = -1.5$
x-intercepts 0, 2.5
y-intercept 0, local maximum $(-3.9, -10.4)$
local minimum $(0.9, -0.6)$
end behavior $y = x - 4$

85. 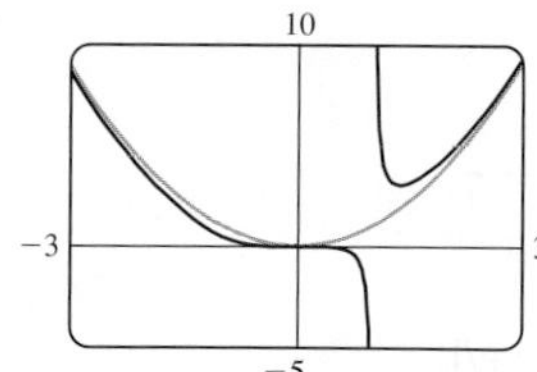
vertical $x = 1$
x-intercept 0
y-intercept 0
local minimum $(1.4, 3.1)$
end behavior $y = x^2$

87. 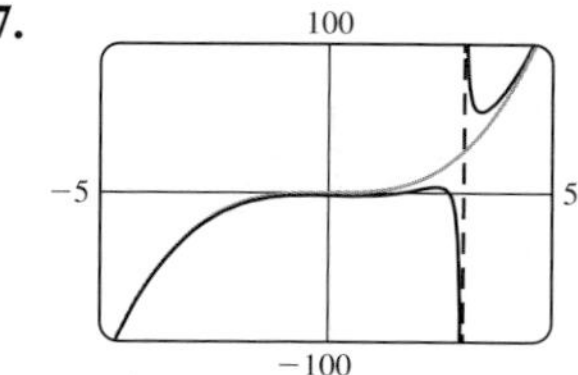
vertical $x = 3$
x-intercepts 1.6, 2.7
y-intercept -2
local maximums $(-0.4, -1.8)$, $(2.4, 3.8)$,
local minimums $(0.6, -2.3)$, $(3.4, 54.3)$
end behavior $y = x^3$

89.

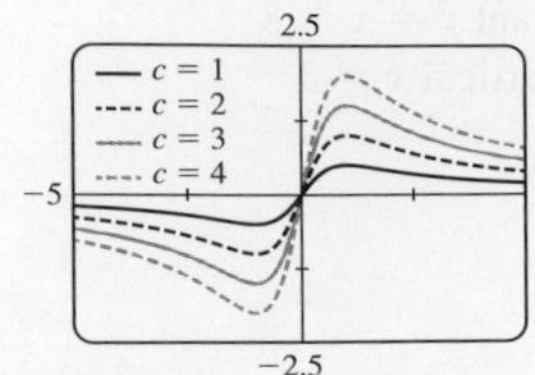

The graph of r has the same basic shape for all values of c. The larger the value of c, the more the graph is vertically stretched.

91.

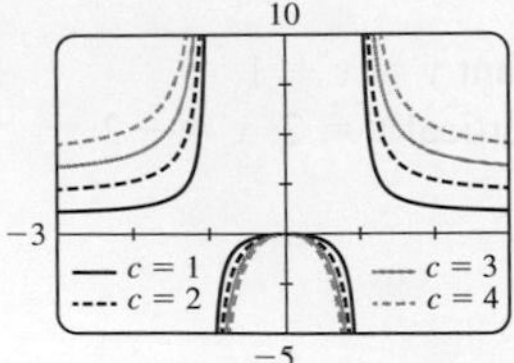

The graph of r has the same basic shape, local maximum $r(0) = 0$, vertical asymptotes $x = \pm 1$, and horizontal asymptote $y = c$ for all values of c. The location of the horizontal asymptote changes as c changes.

93. (b)

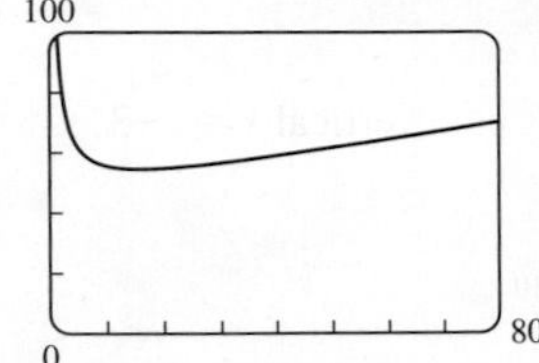

The local minimum $A(158) \approx 54.5$ tells us that the lowest average cost per purse is \$54.50, and this is achieved when 158 purses are produced.

95. (a) 2.50 mg/L **(b)** The concentration decreases to 0. **(c)** ≈ 16.61 h

97.

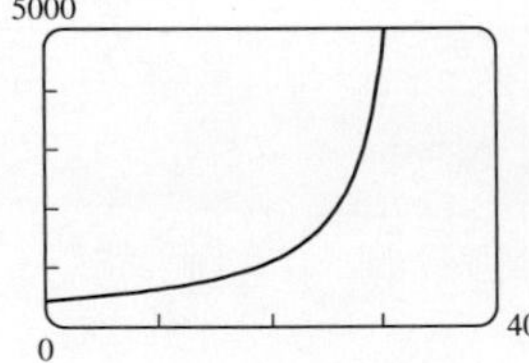

If the speed of the train approaches the speed of sound, then the pitch increases indefinitely (a sonic boom).

99. (a) $C(t) = \dfrac{4 + 5t}{100 + 50t}$

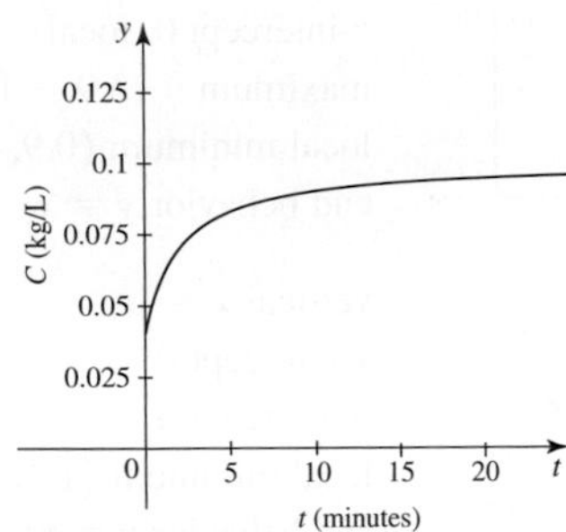

(b) 0.09 kg/L; 0.094 kg/L **(c)** 0.1 kg/L

Section 3.7 ■ Page 323

1. zeros; zeros; $[-2, 0]$, $[1, \infty)$

Sign of	(−∞, −2)	(−2, 0)	(0, 1)	(1, ∞)
x	−	−	+	+
$x + 2$	−	+	+	+
$x - 1$	−	−	−	+
$x(x + 2)(x - 1)$	−	+	−	+

2. zeros; zeros; cut points; $(-\infty, -4,)$, $[-2, 1]$, $(3, \infty)$

Sign of	(−∞, −4)	(−4, −2)	(−2, 1)	(1, 3)	(3, ∞)
$x + 2$	−	−	+	+	+
$x - 1$	−	−	−	+	+
$x - 3$	−	−	−	−	+
$x + 4$	−	+	+	+	+
$\dfrac{(x + 2)(x - 1)}{(x - 3)(x + 4)}$	+	−	+	−	+

3. $(-\infty, -5) \cup (-\frac{5}{2}, 3)$ **5.** $(-\infty, -5) \cup (-5, -3) \cup (1, \infty)$ **7.** $[-4, -2] \cup [2, \infty)$ **9.** $(-\infty, \frac{1}{2})$ **11.** $(-3, 3)$ **13.** $[-5, 1] \cup [3, \infty)$ **15.** $(-\infty, -1) \cup (1, 7)$ **17.** $(1, 10)$ **19.** $[-8, -\frac{5}{2})$ **21.** $(-\frac{5}{2}, 3]$ **23.** $(-7, -\frac{5}{2}] \cup (5, \infty)$ **25.** $(-5, 2)$ **27.** $(-1, 1) \cup (1, \infty)$ **29.** $(-\infty, -3) \cup (-\frac{2}{3}, 1) \cup (3, \infty)$ **31.** $(-4, 3]$ **33.** $(-\infty, -2) \cup (-1, 1) \cup (1, \infty)$ **35.** $[-2, -1) \cup [9, \infty)$ **37.** $[-2, 0) \cup (1, 3]$ **39.** $(-\infty, -2) \cup (-2, -1)$ **41.** $(-\infty, -2) \cup (5, \infty)$ **43.** $(-\frac{1}{2}, 0) \cup (\frac{1}{2}, \infty)$ **45.** $[-2, 3]$ **47.** $(-\infty, -1] \cup [1, \infty)$ **49.** $[-2, 1] \cup [3, \infty)$ **51.** $(-\infty, -1.37) \cup (0.37, 1)$ **53.** $(0, 1.60)$ **55.** $(0, 1]$ **57.** $(-\infty, a] \cup [b, c] \cup [d, \infty)$ **59.** More than 2.66 m

61.

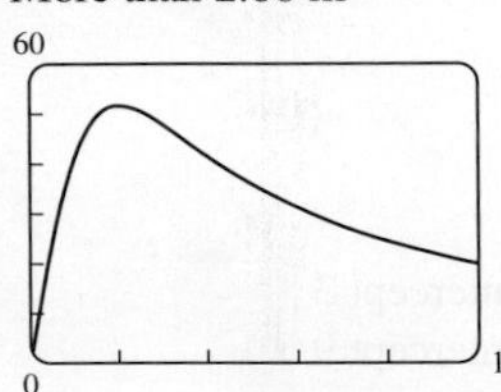

Between 9.5 and 42.3 km/h

Chapter 3 Review ■ Page 327

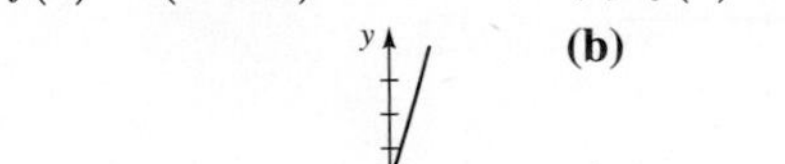

1. (a) $f(x) = (x + 3)^2 - 7$ **(b)**

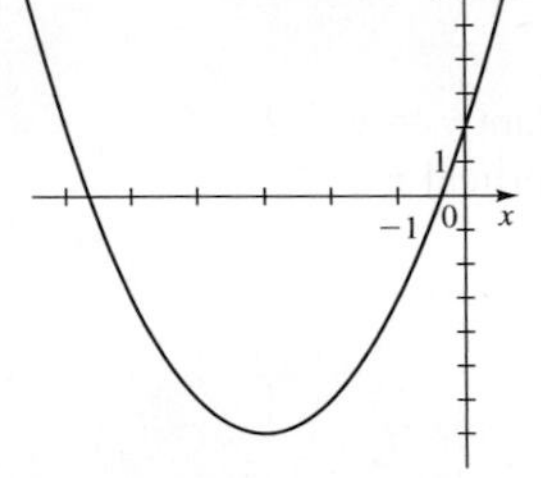

3. (a) $f(x) = -(x + 5)^2 + 26$ **(b)**

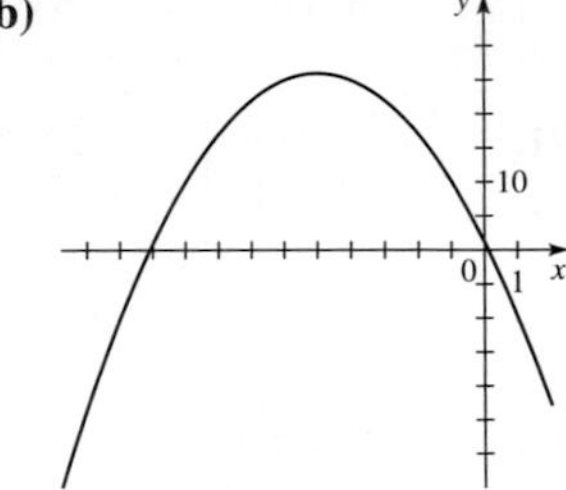

5. Maximum $f(\frac{3}{2}) = \frac{5}{4}$ **7.** 21.25 m

9.

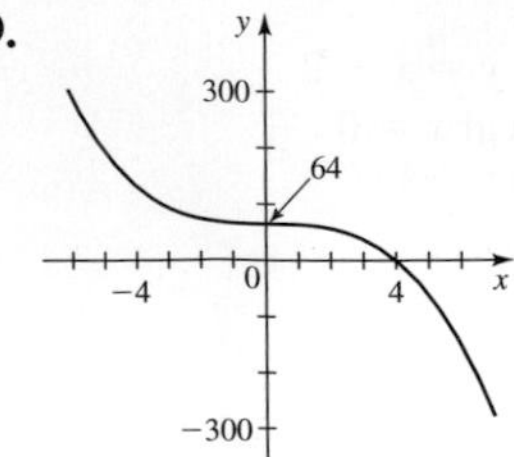

domain $(-\infty, \infty)$, range $(-\infty, \infty)$

11.

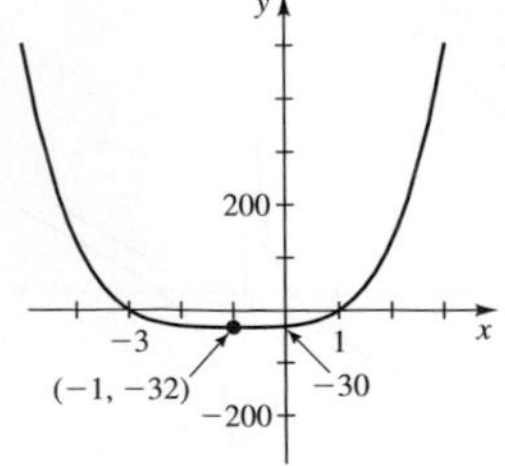

domain $(-\infty, \infty)$, range $[-32, \infty)$

13.

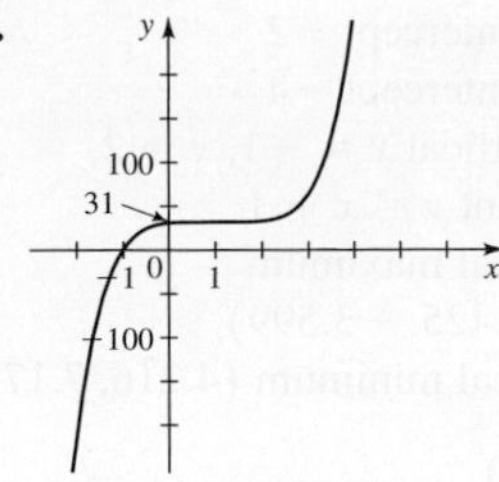

domain $(-\infty, \infty)$, range $(-\infty, \infty)$

15. (a) $y \to \infty$ as $x \to \infty$, $y \to -\infty$ as $x \to -\infty$

17. (a) $y \to -\infty$ as $x \to \infty$, $y \to \infty$ as $x \to -\infty$

15. (b)

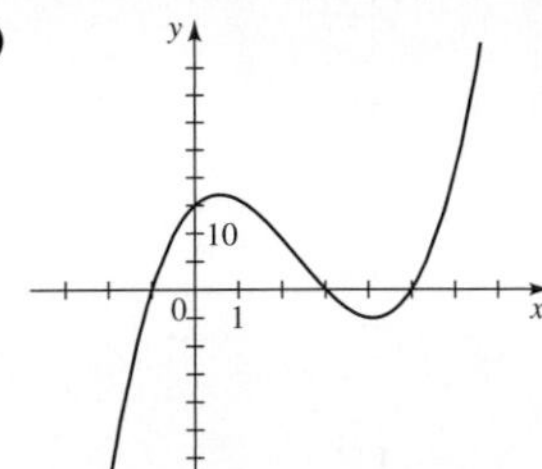

17. (b)

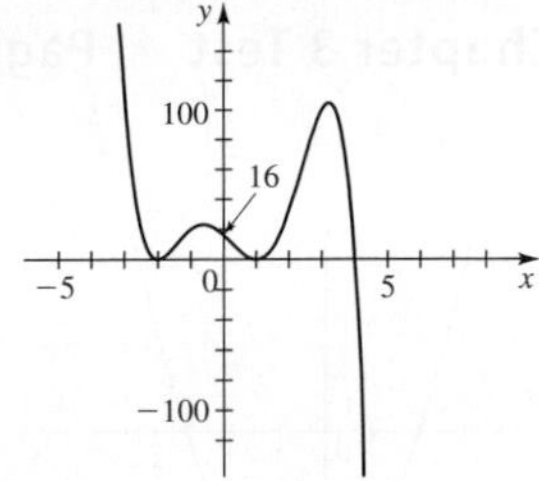

19. (a) 0 (multiplicity 3), 2 (multiplicity 2)

(b)

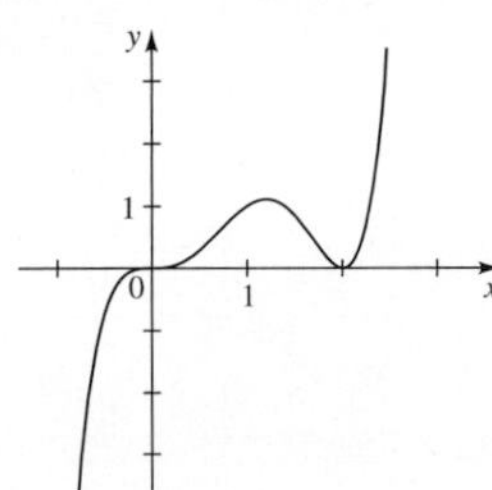

21.

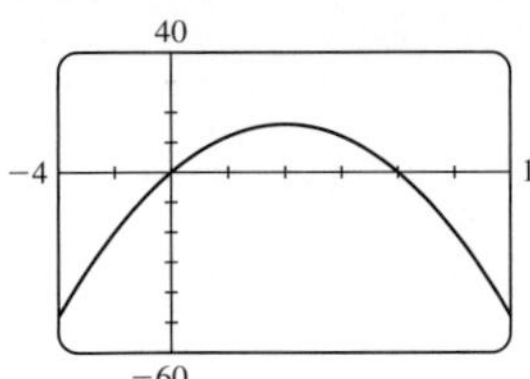

x-intercepts 0, 8
y-intercept 0
local maximum (4, 16)
end behavior $y \to -\infty$ as $x \to \infty$, $y \to -\infty$ as $x \to -\infty$

23.

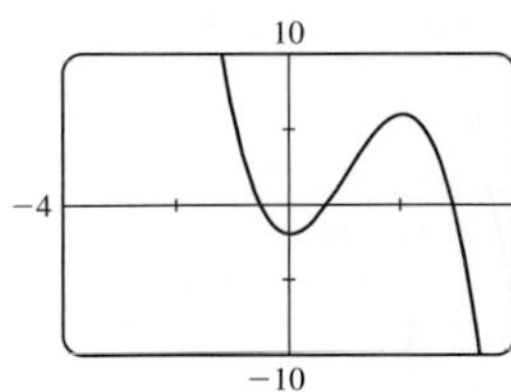

x-intercepts -0.5, 0.7, 2.9
y-intercept -2
local maximum (2, 6)
local minimum $(0, -2)$
end behavior $y \to -\infty$ as $x \to \infty$, $y \to \infty$ as $x \to -\infty$

25.

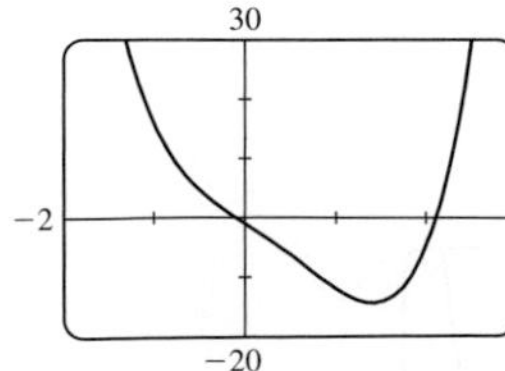

x-intercepts -0.1, 2.1
y-intercept -1
local minimum $(1.4, -14.5)$
end behavior $y \to \infty$ as $x \to \infty$, $y \to \infty$ as $x \to -\infty$

27. (a) $S = 13.8x(100 - x^2)$ **(b)** $0 \le x \le 10$

(c)

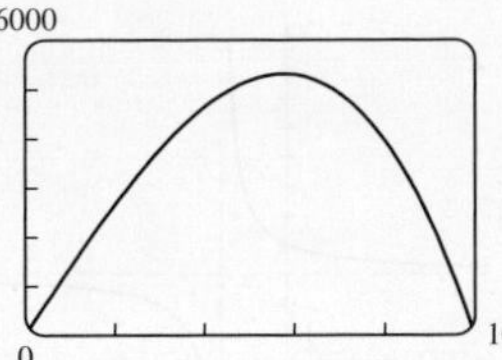

(d) $\approx$5.8 cm

In answers 29–36 the first polynomial is the quotient, and the second is the remainder.

29. $x - 2, -4$ **31.** $2x^2 - 11x + 58, -294$
33. $x^3 - 5x^2 + 17x - 83, 422$ **35.** $2x - 3, 12$ **37.** 3 **39.** 8
43. (a) $\pm1, \pm2, \pm3, \pm6, \pm9, \pm18$ **(b)** 2 or 0 positive; 3 or 1 negative **45. (a)** $\pm1, \pm2, \pm4, \pm8, \pm\frac{1}{3}, \pm\frac{2}{3}, \pm\frac{4}{3}, \pm\frac{8}{3}$
(b) 2 or 0 positive; 3 or 1 negative
47. (a) $-4, 0, 4$ **(b)**

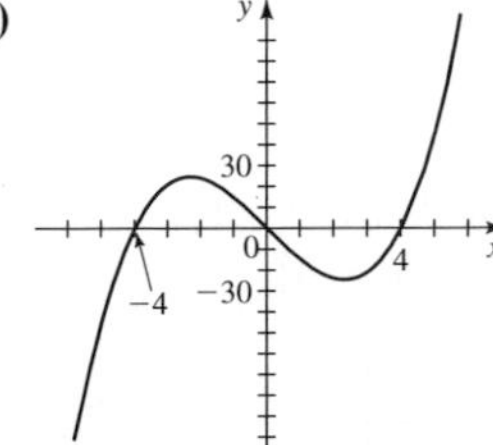

49. (a) $-2, 0$ (multiplicity 2), 1 **(b)**

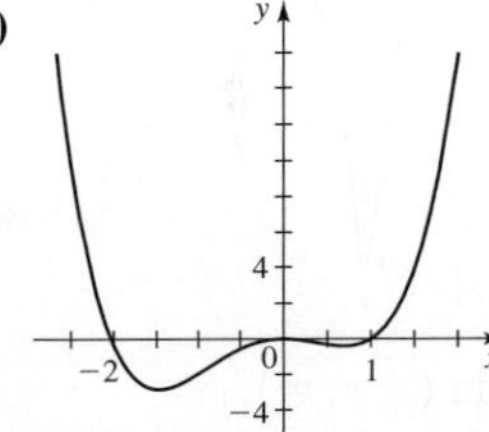

51. (a) $-2, -1, 2, 3$
(b)

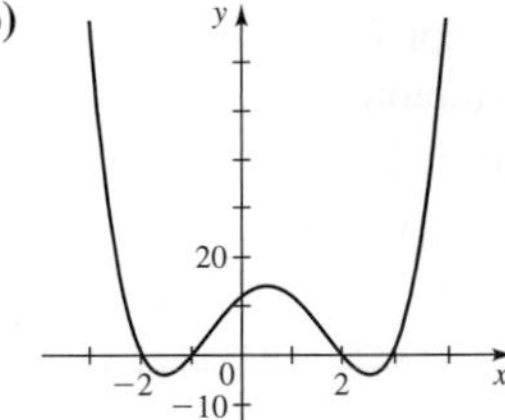

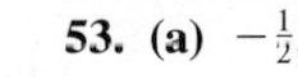
53. (a) $-\frac{1}{2}, 1$
(b)

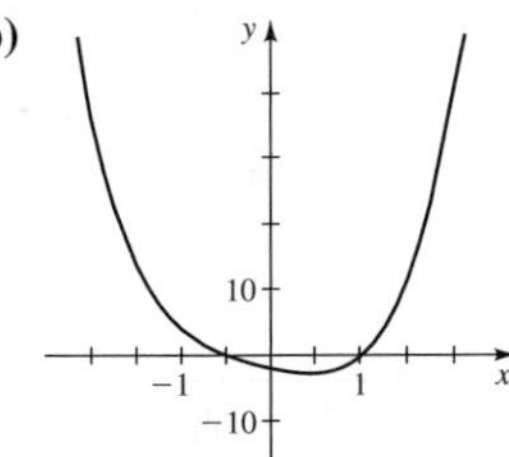

55. $P(x) = 4x^3 - 18x^2 + 14x + 12$
57. No; since the complex conjugates of imaginary zeros will also be zeros, the polynomial would have 8 zeros, contradicting the requirement that it have degree 4.
59. $1, \pm i$ **61.** $-3, 1, 5$ **63.** $-1 \pm 2i, -2$ (multiplicity 2)
65. ±2, 1 (multiplicity 3) **67.** $\pm2, 1 \pm \sqrt{3}i, -1 \pm \sqrt{3}i$
69. $1, 3,$  $\dfrac{-1 \pm \sqrt{7}i}{2}$ **71.** $x = -0.5, 3$ **73.** $x \approx -0.24, 4.24$
75. $2, P(x) = (x - 2)(x^2 + 2x + 2)$
77. (a) Vertical asymptote $x = -4$, horizontal asymptote $y = 0$, no x-intercept, y-intercept $\frac{3}{4}$, domain $\{x \mid x \neq -4\}$ range $\{y \mid y \neq 0\}$ **(b)**

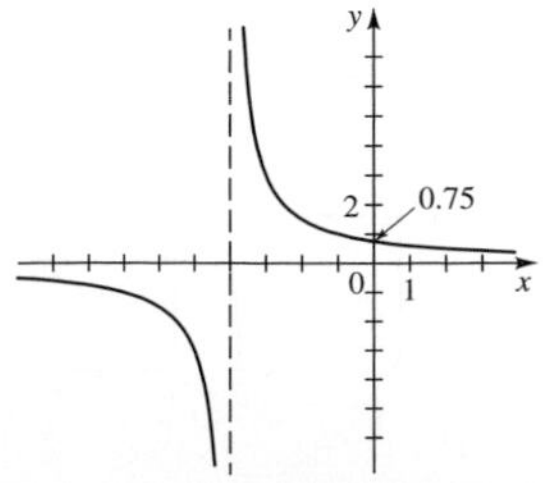

79. **(a)** Vertical asymptote $x = 1$, horizontal asymptote $y = 3$, x-intercept $\frac{4}{3}$, y-intercept 4, domain $\{x \mid x \neq 1\}$ range $\{y \mid y \neq 3\}$ **(b)**

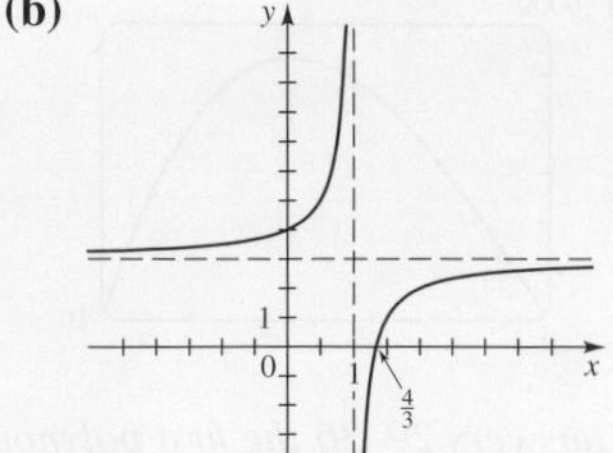

81.

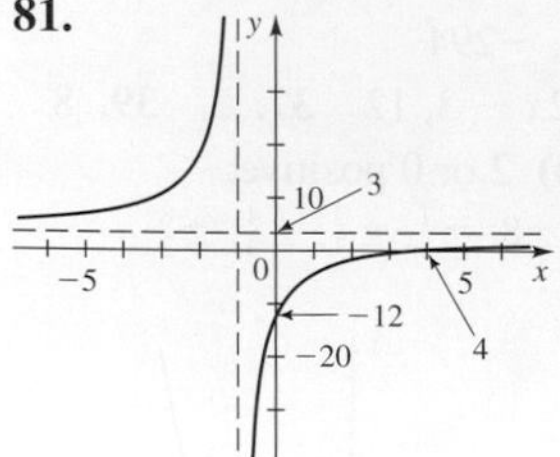

Domain $\{x \mid x \neq -1\}$, range $\{y \mid y \neq 3\}$

83.

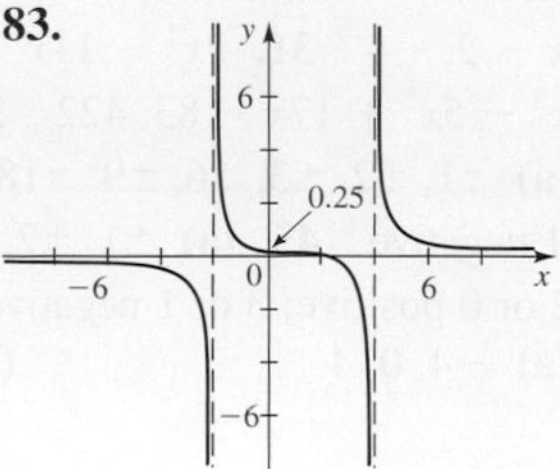

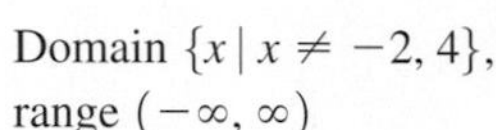

Domain $\{x \mid x \neq -2, 4\}$, range $(-\infty, \infty)$

85.

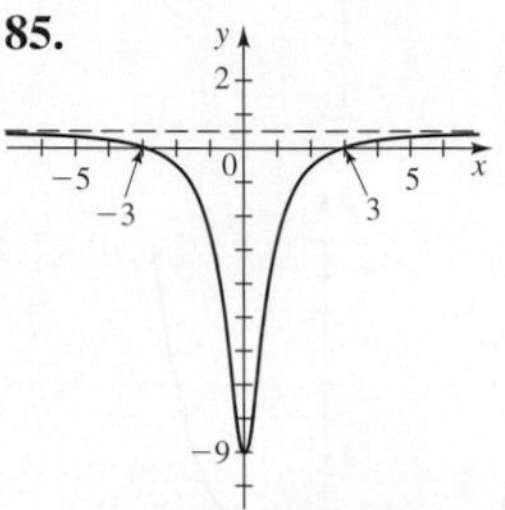

Domain $(-\infty, \infty)$, range $\{y \mid -9 \leq y < \frac{1}{2}\}$

87.

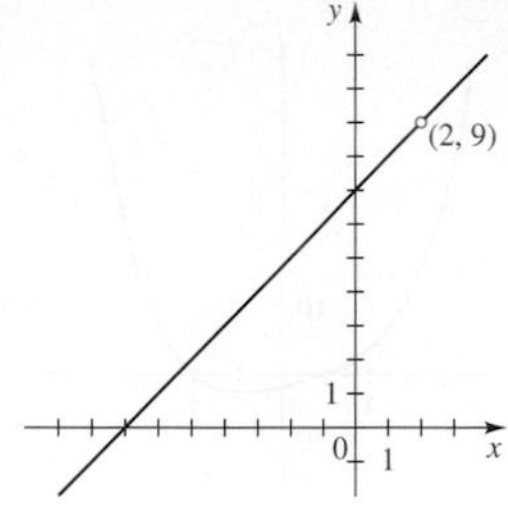

x-intercept -7
y-intercept 7
no asymptote
domain $\{x \mid x \neq 2\}$
range $\{y \mid y \neq 9\}$

89.

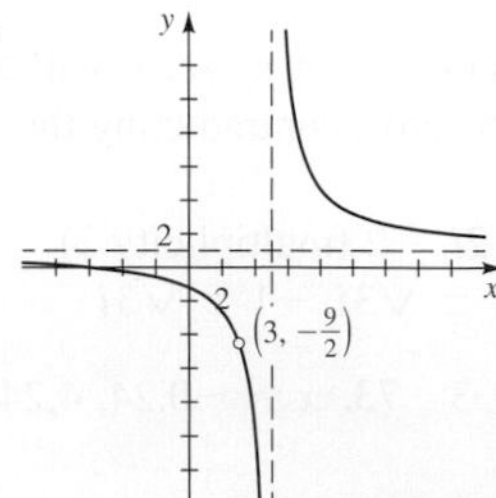

x-intercept -6
y-intercept $-\frac{6}{5}$
vertical $x = 5$
horizontal $y = 1$
domain $\{x \mid x \neq 3, 5\}$
range $\{y \mid y \neq 1, -\frac{9}{2}\}$

91.

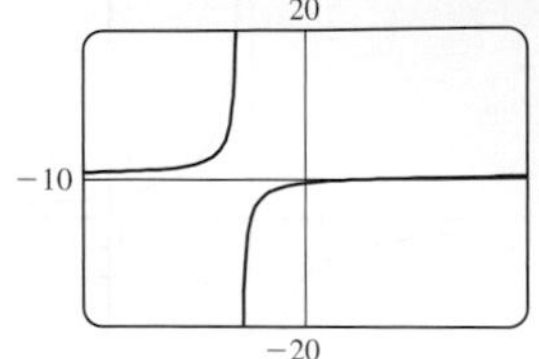

x-intercept 3
y-intercept -0.5
vertical $x = -3$
horizontal $y = 0.5$
no local extrema

93.

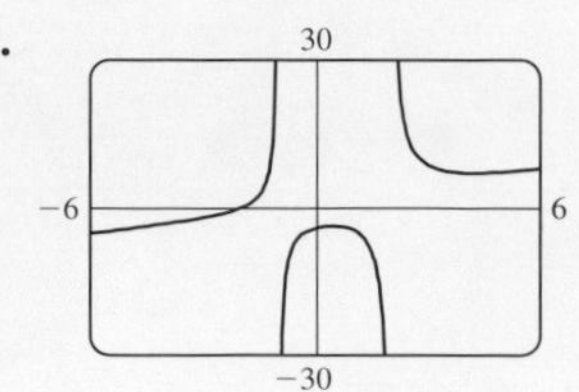

x-intercept -2
y-intercept -4
vertical $x = -1, x = 2$
slant $y = x + 1$
local maximum $(0.425, -3.599)$
local minimum $(4.216, 7.175)$

95. $(-\infty, -1] \cup [\frac{3}{2}, \infty)$ **97.** $(-3, 3)$
99. $(-\infty, -2) \cup (1, 2)$ **101.** $(-3, 0) \cup (2, \frac{9}{2}]$
103. $[-3, \frac{8}{3}]$ **105.** $[0.74, 1.95]$ **109.** **(a)** VII **(b)** V **(c)** III **(d)** I **(e)** IV **(f)** VIII **(g)** VI **(h)** II

Chapter 3 Test ■ Page 330

1. $f(x) = (x - \frac{1}{2})^2 - \frac{25}{4}$

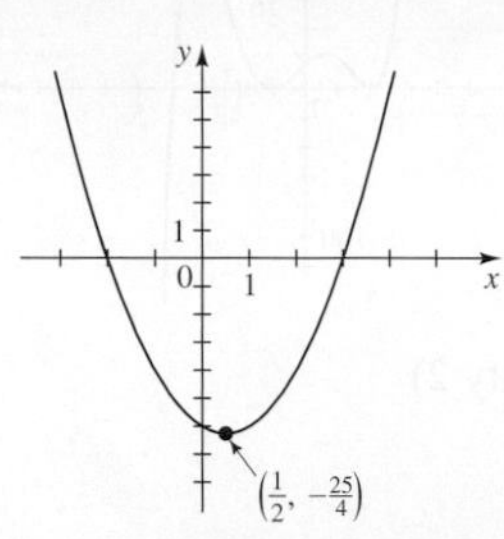

domain $(-\infty, \infty)$, range $[-\frac{25}{4}, \infty)$

2. Minimum $g(-\frac{3}{2}) = -\frac{3}{2}$

3. **(a)** 2500 m **(b)** 1000 m

4.

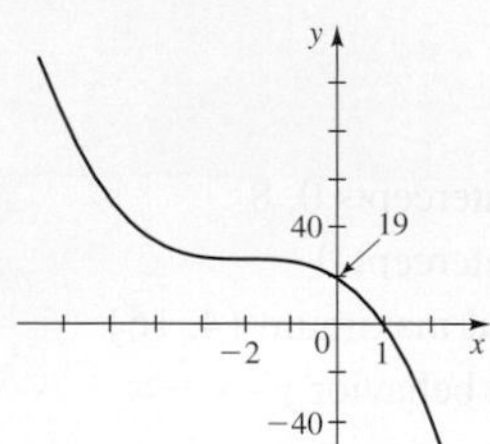

5. **(a)** $x^3 + 2x^2 + 2, 9$ **(b)** $x^3 + 2x^2 + \frac{1}{2}, \frac{15}{2}$

6. **(a)** $\pm1, \pm3, \pm\frac{1}{2}, \pm\frac{3}{2}$ **(b)** $2(x - 3)(x - \frac{1}{2})(x + 1)$ **(c)** $-1, \frac{1}{2}, 3$ **(d)**

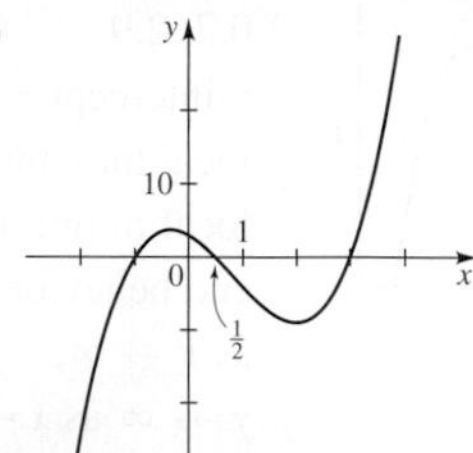

7. $3, -1 \pm i$ **8.** $P(x) = (x - 1)^2(x - 2i)(x + 2i)$

9. $P(x) = x^4 + 2x^3 + 10x^2 + 18x + 9$

10. **(a)** 4, 2, or 0 positive; 0 negative **(c)** 0.17, 3.93

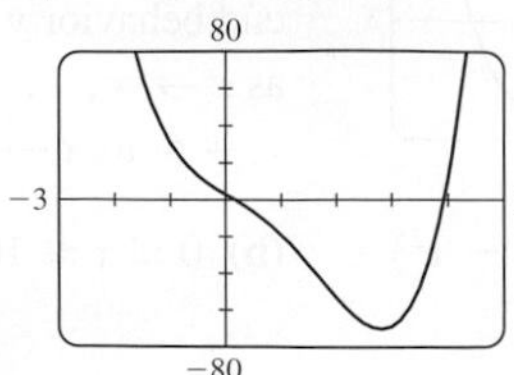

(d) Local minimum $(2.82, -70.31)$

11. **(a)** R **(b)** P **(c)** Q **(d)** T **(e)** S

12. (a) r, u **(b)** s **(c)** s, w **(d)** w

(e) Vertical $x = -1, x = 2$; horizontal $y = 0$

(f)

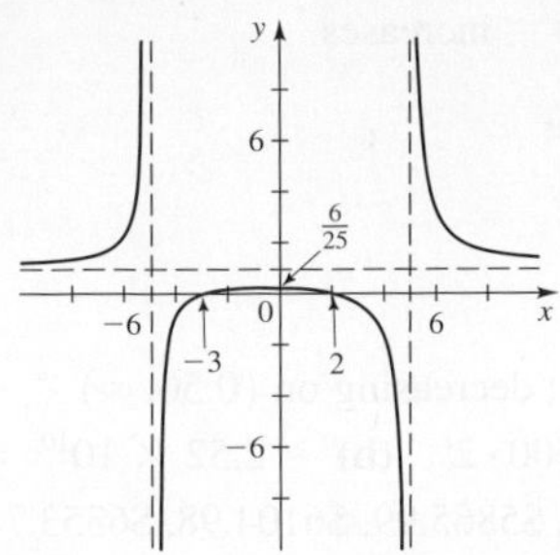

(g) $P(x) = x^2 - 2x - 5$

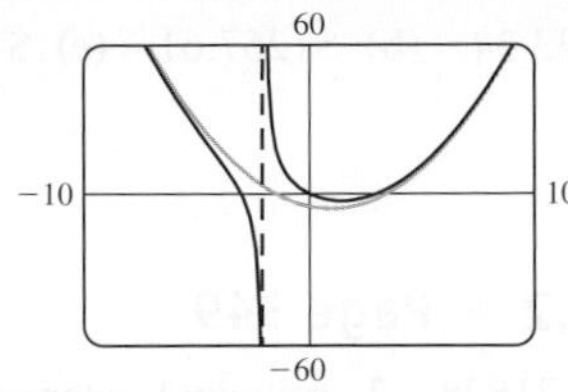

13. $\{x \mid x \le -1 \text{ or } \frac{5}{2} < x \le 3\}$

14. $\{x \mid -1 - \sqrt{5} < x < -1 + \sqrt{5}\}$

15. (a)

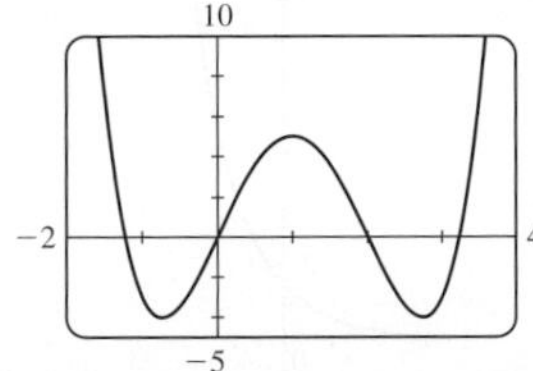

x-intercepts $-1.24, 0, 2, 3.24$; local maximum $P(1) = 5$; local minimums $P(-0.73) = P(2.73) = -4$

(b) $(-\infty, -1.24] \cup [0, 2] \cup [3.24, \infty)$

Focus on Modeling ■ Page 334

1. (a) $y = -0.005864x^2 + 2.900675x - 278.1956$, (where kilometers are measured in thousands)

(b)

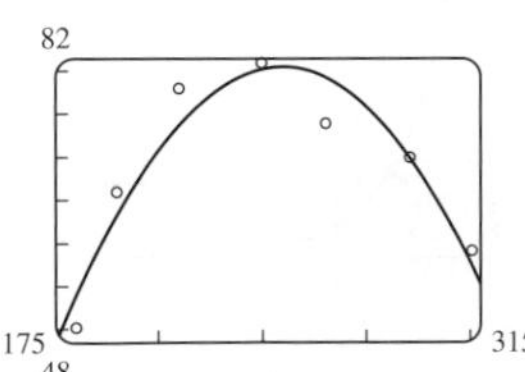

(c) 248 kPa

3. (a) $y = 0.00203709x^3 - 0.104522x^2 + 1.966206x + 1.45576$

(b)

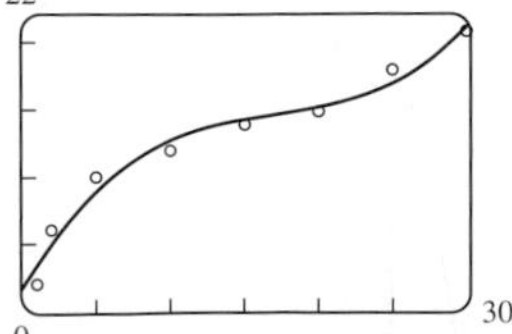

(c) 43 vegetables **(d)** 2.0 s

5. (a) $y = 0.0120536x^2 - 0.490357x + 4.96571$

(b)

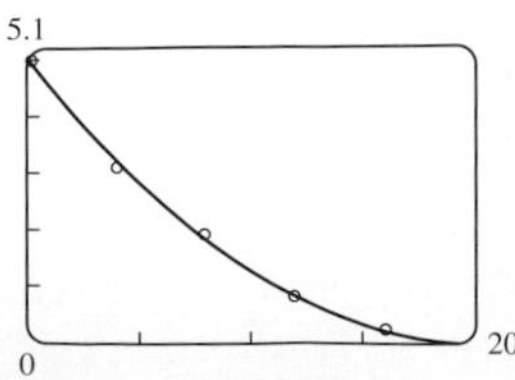

(c) 19.0 min

Chapter 4

Section 4.1 ■ Page 344

1. 5; $\frac{1}{25}$; 1; 25; 15,625 **2. (a)** III **(b)** I **(c)** II **(d)** IV

3. (a) downward **(b)** right **4.** principal, interest rate per year, number of times interest is compounded per year, number of years, amount accumulated after t years; \$112.65

5. horizontal, 0; 0 **6.** horizontal, 3; 3

7. 2.000, 22.195, 0.063, 1.516 **9.** 0.192, 0.070, 15.588, 1.552

11.

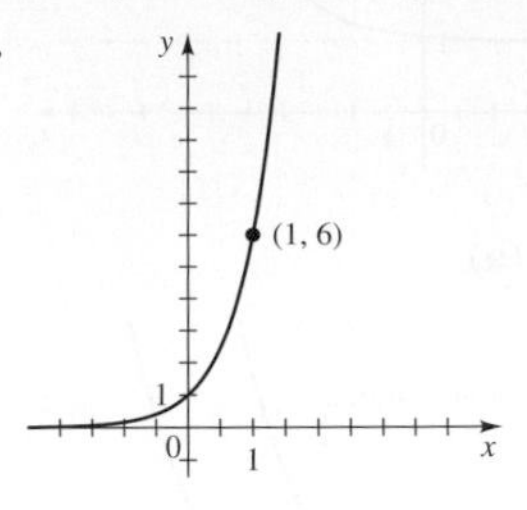

13.

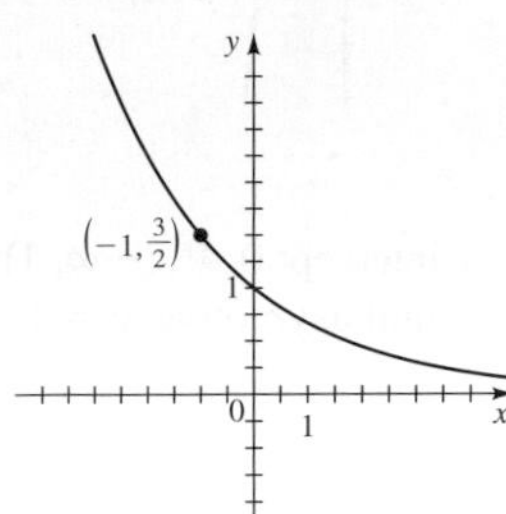

15.

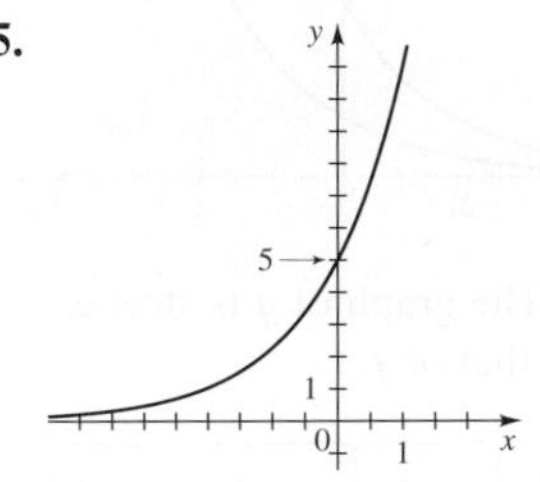

17.

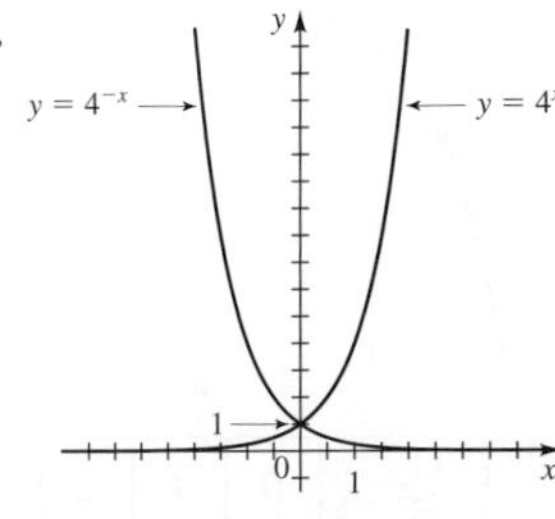

19.

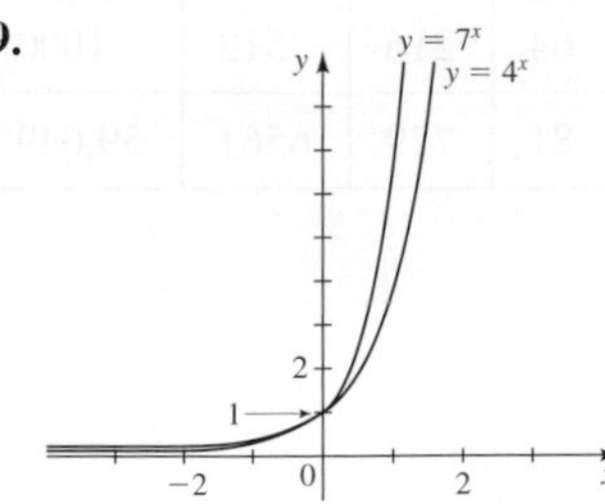

21. $f(x) = 3^x$ **23.** $f(x) = (\frac{1}{4})^x$ **25.** II

27. y-intercept 2, $\mathbb{R}$, $(1, \infty)$, horizontal asymptote $y = 1$

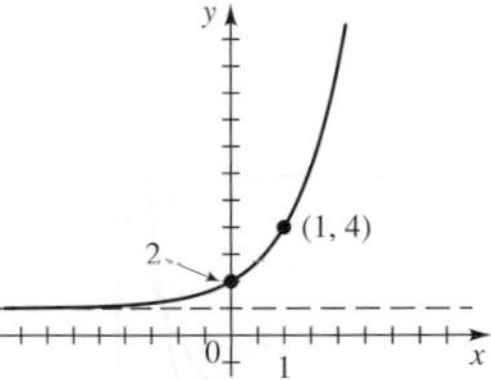

29. y-intercept -1, $\mathbb{R}$, $(-\infty, 0)$, horizontal asymptote $y = 0$

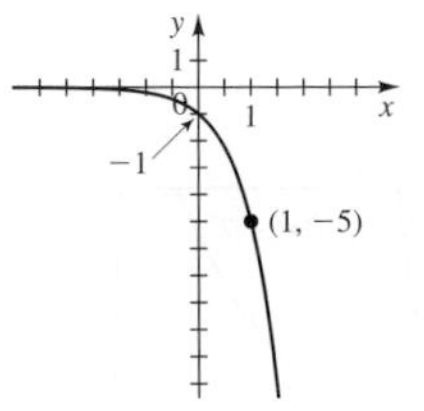

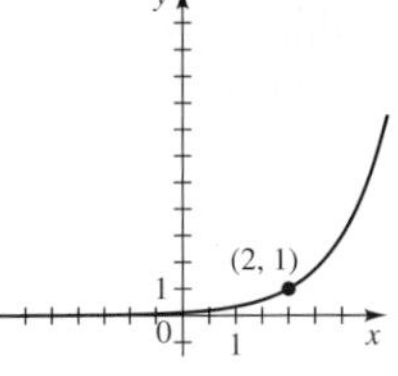

31. y-intercept $\frac{1}{9}$, $\mathbb{R}$, $(0, \infty)$, horizontal asymptote $y = 0$

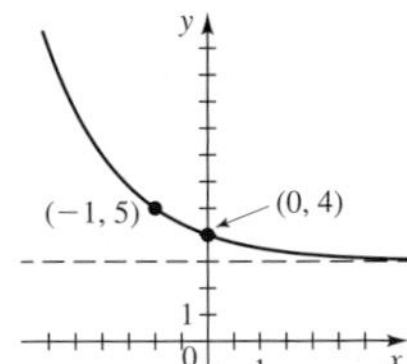

33. y-intercept 4, $\mathbb{R}$, $(3, \infty)$, horizontal asymptote $y = 3$

35. y-intercept -2, $\mathbb{R}$, $(-\infty, -1)$, horizontal asymptote $y = -1$

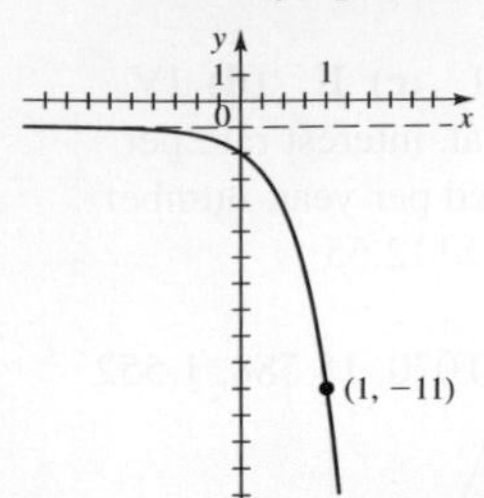

37. y-intercept $\frac{17}{16}$, $\mathbb{R}$, $(1, \infty)$, horizontal asymptote $y = 1$

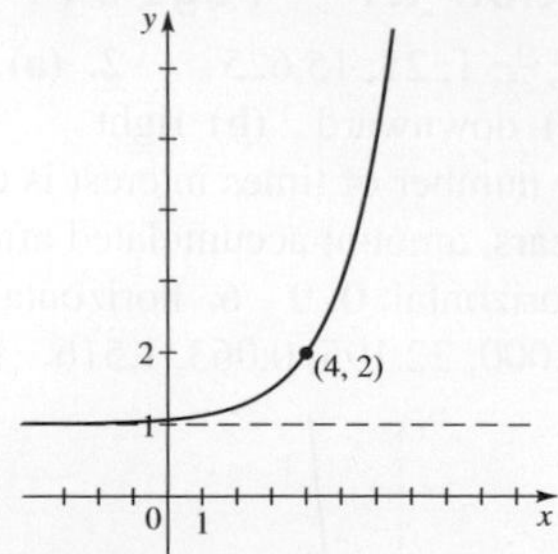

39. y-intercept 0, $\mathbb{R}$, $(-\infty, 1)$, horizontal asymptote $y = 1$

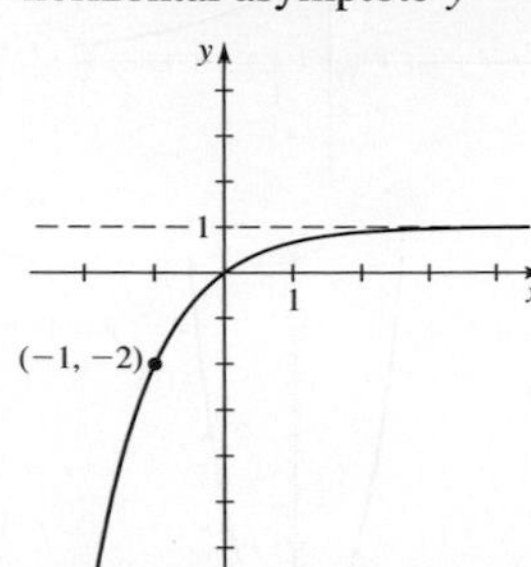

41. (a)

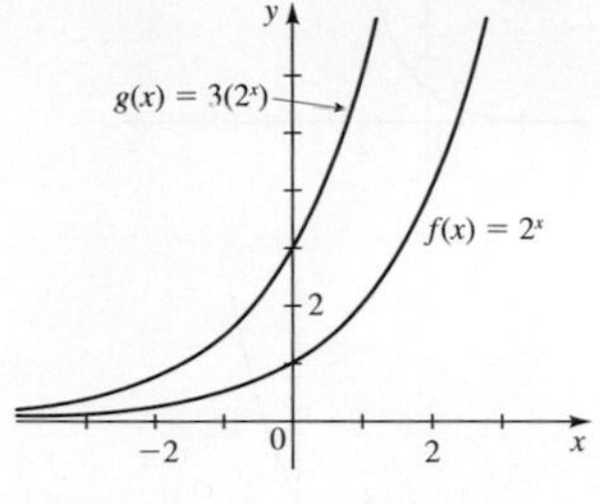

(b) The graph of g is steeper than that of f.

43.

x	0	1	2	3	4	6	8	10
$f(x)$	0	1	8	27	64	216	512	1000
$g(x)$	1	3	9	27	81	729	6561	59,049

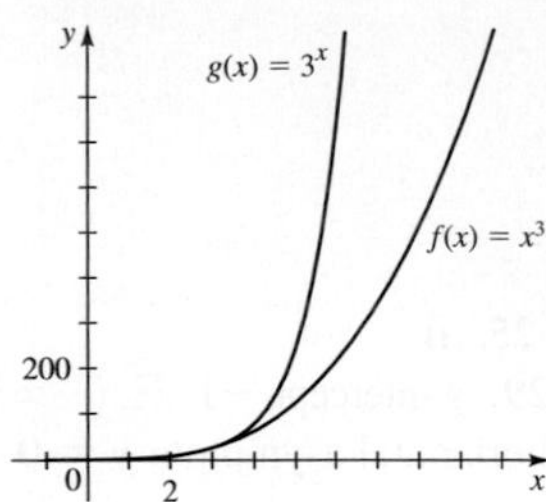

45. (a)

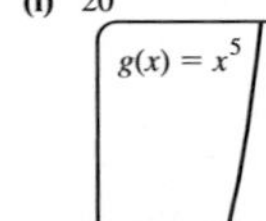

(i)

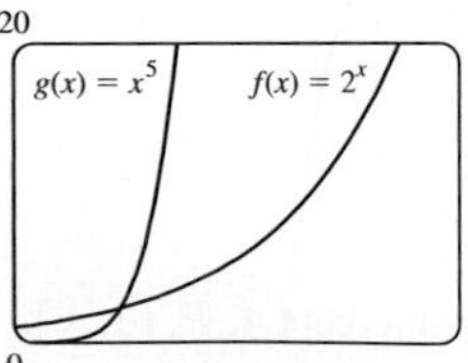

(ii)

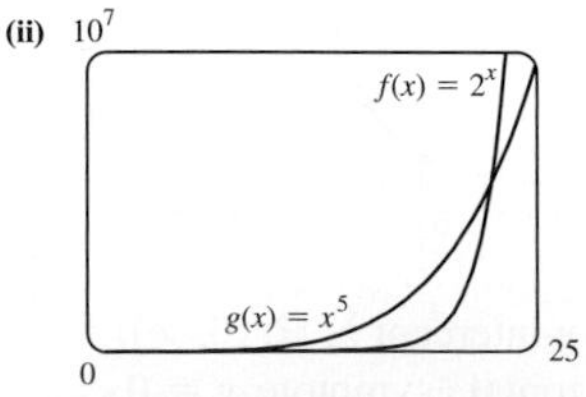

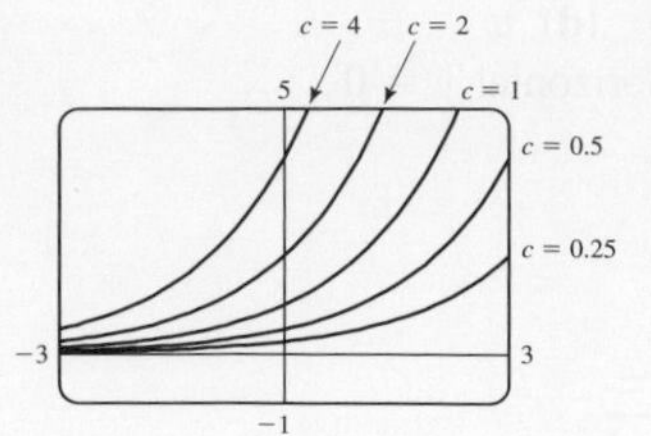

(iii)

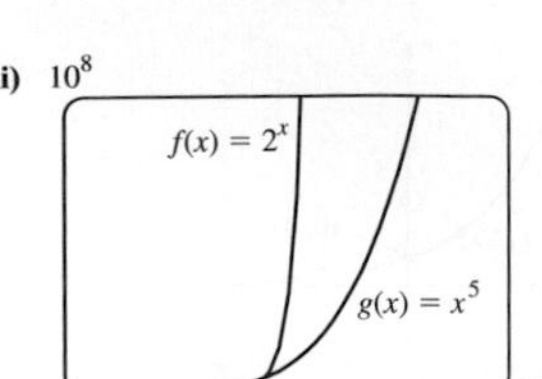

The graph of f ultimately increases much more quickly than that of g.

(b) 1.2, 22.4

47. The larger the value of c, the more rapidly the graph increases.

49. (a) Increasing on $(-\infty, 0.50)$; decreasing on $(0.50, \infty)$ **(b)** $(0, 1.78]$ **53. (a)** $N(t) = 1500 \cdot 2^t$ **(b)** $\approx 2.52 \times 10^{10}$
55. \$5203.71, \$5415.71, \$5636.36, \$5865.99, \$6104.98, \$6353.71
57. (a) \$10,882.52 **(b)** \$14,803.66 **(c)** \$20,137.65
59. (a) \$1233.34 **(b)** \$1267.61 **(c)** \$1578.35 **61.** \$7678.96
63. 8.30%

Section 4.2 ▪ Page 349

1. natural; 2.71828 **2.** principal, interest rate per year, number of years, amount accumulated after t years; \$112.75
3. 2.718, 23.141, 0.050, 4.113

5.

x	$y = f(x)$
−2	0.20
−1	0.55
−0.5	0.91
0	1.5
0.5	2.47
1	4.08
2	11.08

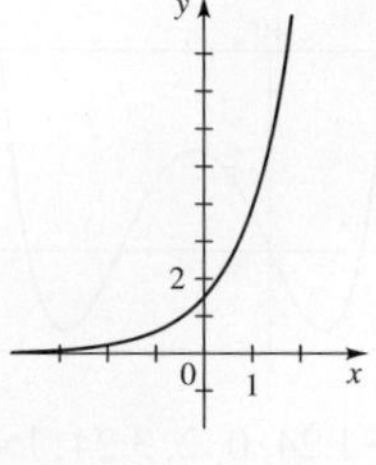

7. y-intercept 4, $\mathbb{R}$, $(3, \infty)$, horizontal asymptote $y = 3$

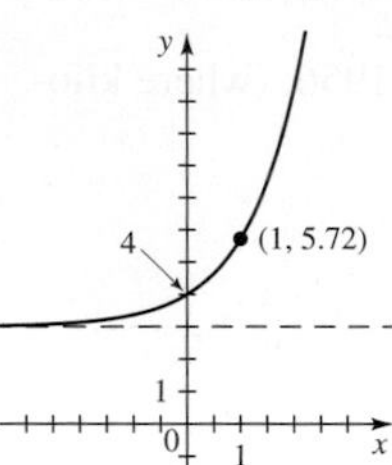

9. y-intercept -2, $\mathbb{R}$, $(-2, \infty)$, horizontal asymptote $y = -3$

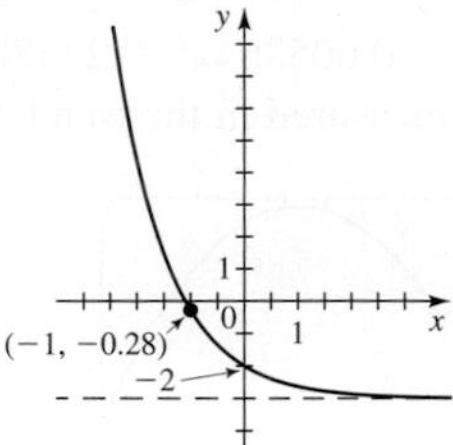

11. y-intercept 2, $\mathbb{R}$, $(-\infty, 3)$, horizontal asymptote $y = 3$

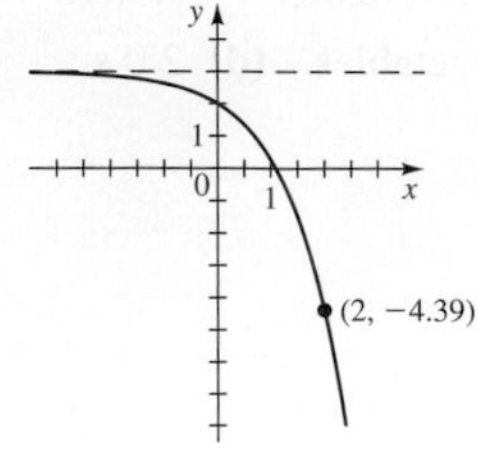

13. y-intercept 1, $\mathbb{R}$, $(-\infty, 4)$, horizontal asymptote $y = 4$

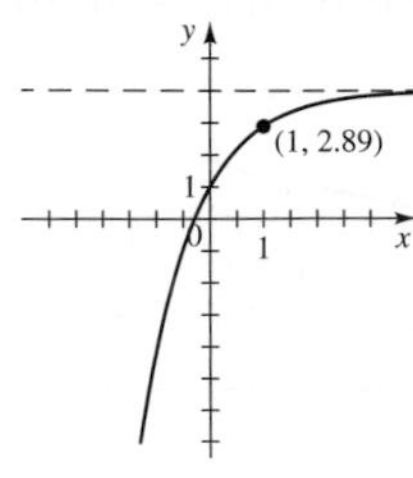

15. y-intercept $e^{-2} \approx 0.14$, $\mathbb{R}$, $(0, \infty)$, horizontal asymptote $y = 0$

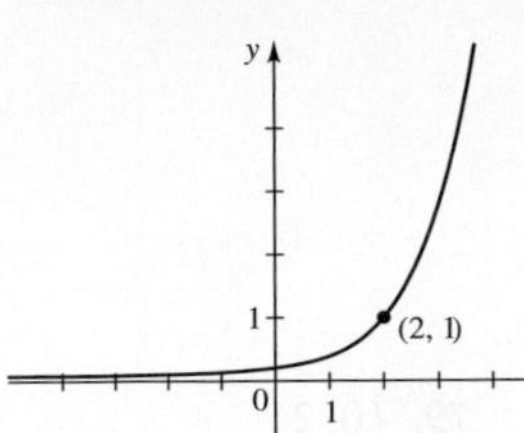

17. y-intercept $e - 3 \approx -0.28$, $\mathbb{R}$, $(-3, \infty)$, horizontal asymptote $y = -3$

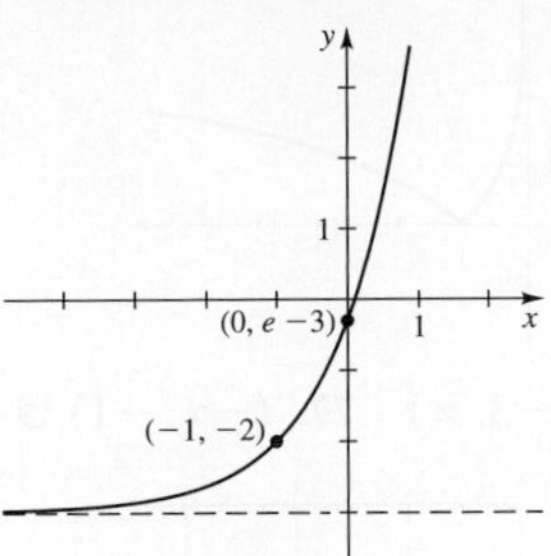

19.

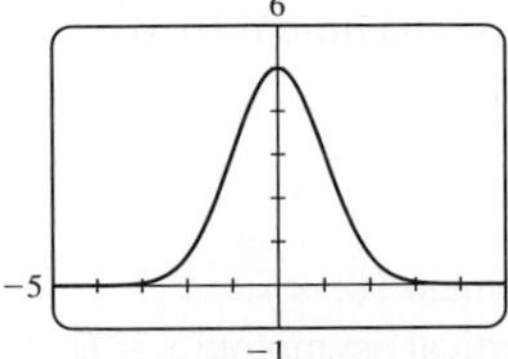

y-intercept 5; horizontal asymptote $y = 0$, local maximum $(0, 5)$

21.

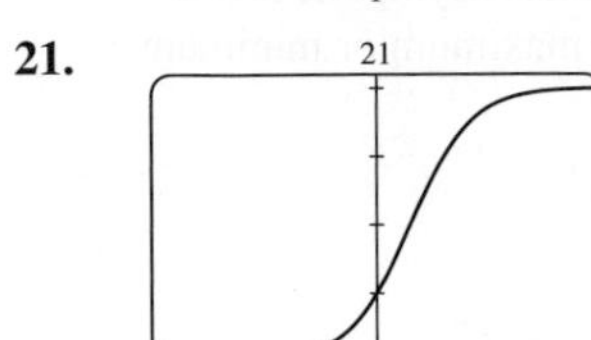

y-intercept 5; horizontal asymptotes $y = 0$ and $y = 20$, no local extrema

23. y-intercept 0; horizontal asymptote $y = 0$, local maximum $(0.67, 0.098)$

Answers to Exercises 25–27 will vary.

25. $g(x) = (x - 10)^2$, $f(x) = 2e^x$

27. $g(x) = 1 + e^x$, $f(x) = \sqrt{x}$

29. (a)

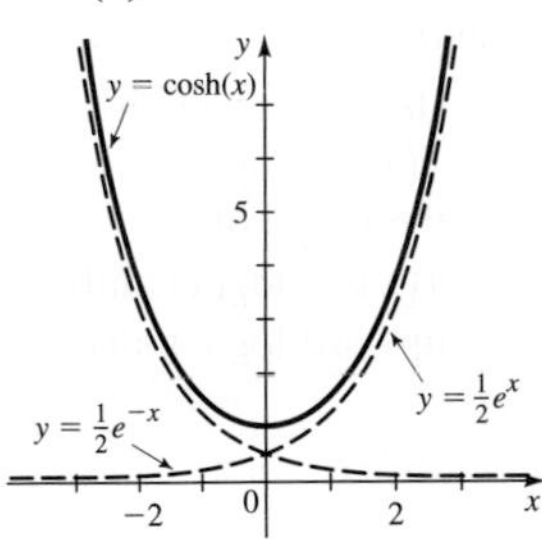

31. (a)

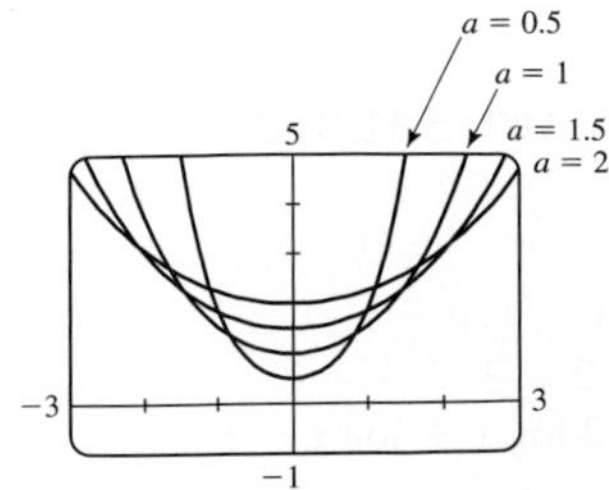

(b) As the value of a increases, the graph flattens out and the y-intercept increases.

33. Local minimum $(0.37, 0.69)$ **35.** 27.4 mg

37. (a)

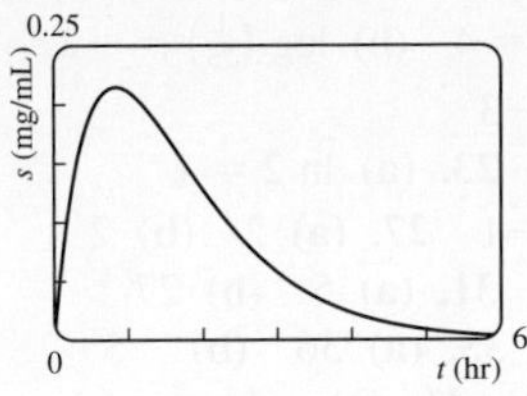

(b) ≈ 50 min **(c)** ≈ 4.86 h

39. (a) 0 **(b)** 34.13 m/s, 46.69 m/s

(c)

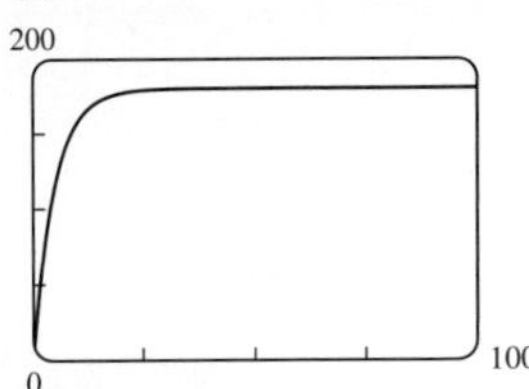

(d) 54 m/s

41. (a) 125 **(b)**

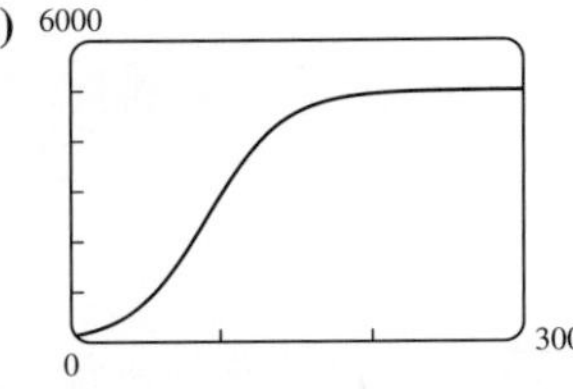

(c) 5000

43. \$7213.18, \$7432.86, \$7659.22, \$7892.48, \$8132.84, \$8380.52

45. (a) \$2145.02 **(b)** \$2300.55 **(c)** \$3043.92

47. (a) \$768.05 **(b)** \$769.22 **(c)** \$769.82 **(d)** \$770.42

49. (a) is best.

51. (a) $A(t) = 5000e^{0.09t}$ **(b)**

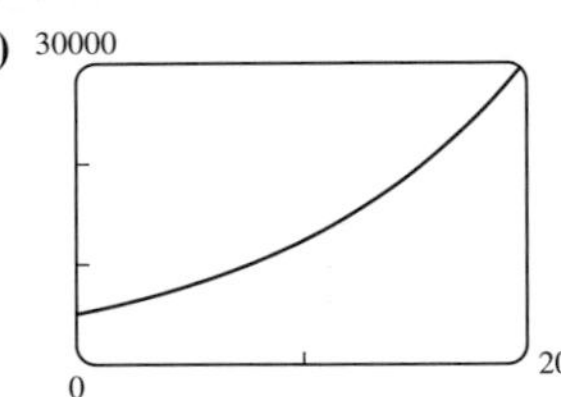

(c) After 17.88 years

Section 4.3 ■ Page 359

1. x

x	10^3	10^2	10^1	10^0	10^{-1}	10^{-2}	10^{-3}	$10^{1/2}$
$\log x$	3	2	1	0	-1	-2	-3	$\frac{1}{2}$

2. 9; 1, 0, -1, 2, $\frac{1}{2}$ **3. (a)** $\log_5 125 = 3$ **(b)** $5^2 = 25$

4. (a) III **(b)** II **(c)** I **(d)** IV **5.** vertical, 0 **6.** vertical, 1

7.

Logarithmic Form	Exponential Form
$\log_8 8 = 1$	$8^1 = 8$
$\log_8 64 = 2$	$8^2 = 64$
$\log_8 4 = \frac{2}{3}$	$8^{2/3} = 4$
$\log_8 512 = 3$	$8^3 = 512$
$\log_8(\frac{1}{8}) = -1$	$8^{-1} = \frac{1}{8}$
$\log_8(\frac{1}{64}) = -2$	$8^{-2} = \frac{1}{64}$

9. **(a)** $3^4 = 81$ **(b)** $\left(\frac{1}{3}\right)^0 = 1$ **11.** **(a)** $8^{1/3} = 2$ **(b)** $10^{-2} = 0.01$
13. **(a)** $3^x = 5$ **(b)** $\left(\frac{1}{6}\right)^3 = 2y$ **15.** **(a)** $e^{2y} = 10$
(b) $e^{-2} = 3x + 1$ **17.** **(a)** $\log_{10} 10{,}000 = 4$ **(b)** $\log_5\left(\frac{1}{25}\right) = -2$
19. **(a)** $\log_8\left(\frac{1}{8}\right) = -1$ **(b)** $\log_2\left(\frac{1}{8}\right) = -3$
21. **(a)** $\log_4 70 = x$ **(b)** $\log_{1/2} w = 3$ **23.** **(a)** $\ln 2 = x$
(b) $\ln y = 3$ **25.** **(a)** 1 **(b)** 0 **(c)** -1 **27.** **(a)** 2 **(b)** 2
(c) 10 **29.** **(a)** -3 **(b)** -3 **(c)** $\frac{1}{2}$ **31.** **(a)** 5 **(b)** 27
(c) 10 **33.** **(a)** $-\frac{2}{3}$ **(b)** 4 **(c)** -1 **35.** **(a)** 36 **(b)** -3
37. **(a)** e^3 **(b)** 2 **39.** **(a)** -3 **(b)** $\frac{1}{8}$ **41.** **(a)** -1
(b) $\frac{1}{1000}$ **43.** **(a)** 2 **(b)** 4 **45.** **(a)** 0.3010 **(b)** 1.5465
(c) -0.1761 **47.** **(a)** 1.6094 **(b)** 3.2308 **(c)** 1.0051

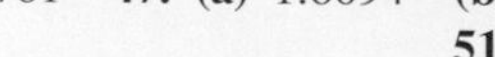

49.

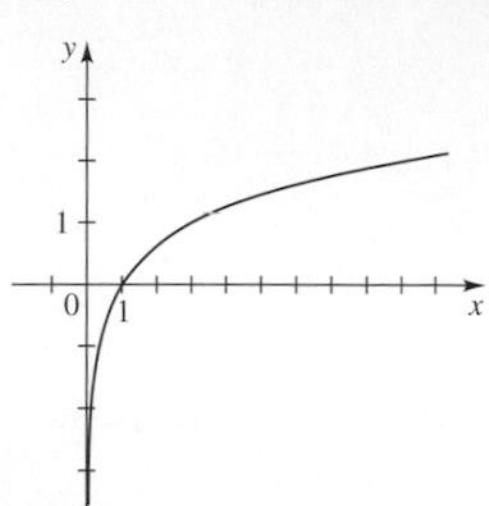

51.

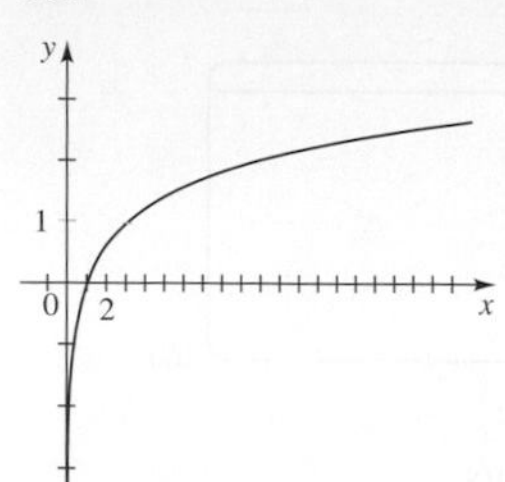

53.

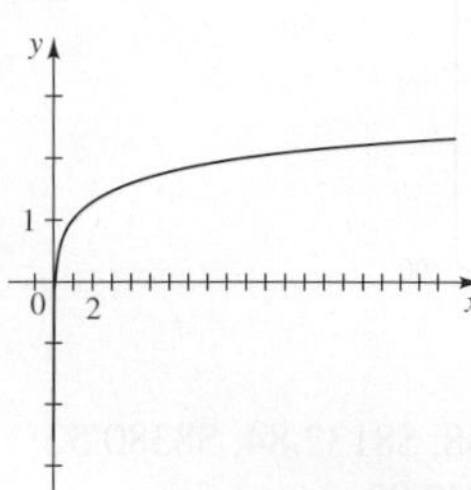

55. $y = \log_5 x$ **57.** $y = \log_9 x$ **59.** I

61.

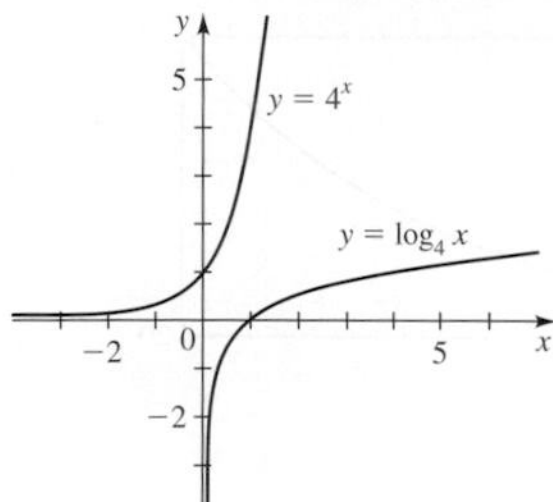

63. $(-\infty, 0), \mathbb{R}, x = 0$

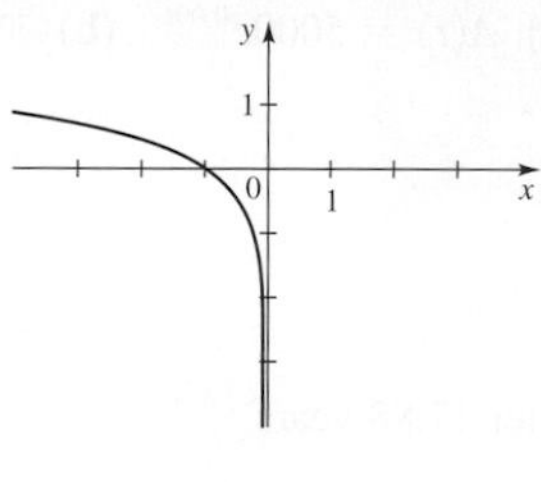

65. $(4, \infty), \mathbb{R}, x = 4$

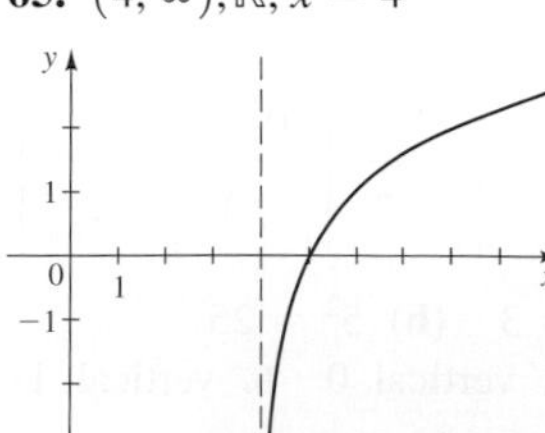

67. $(-5, \infty), \mathbb{R}, x = -5$

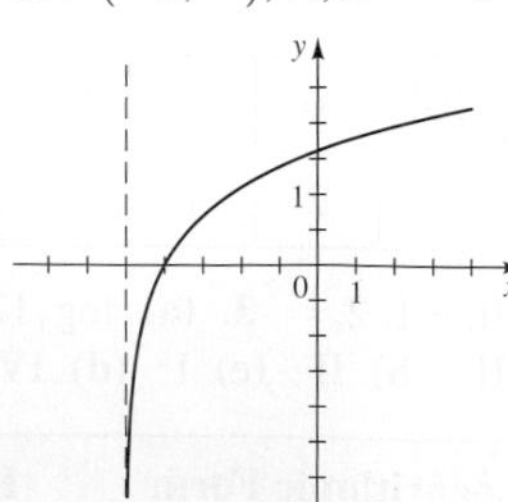

69. $(0, \infty), \mathbb{R}, x = 0$

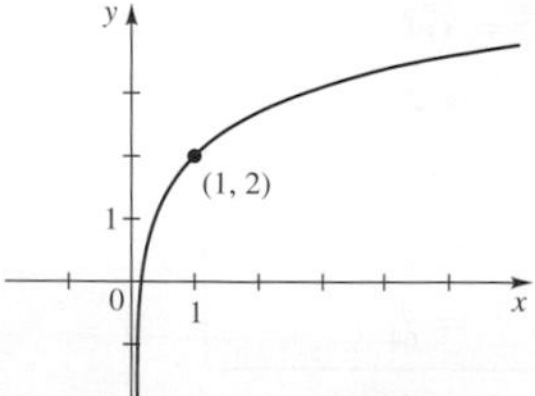

71. $(1, \infty), \mathbb{R}, x = 1$

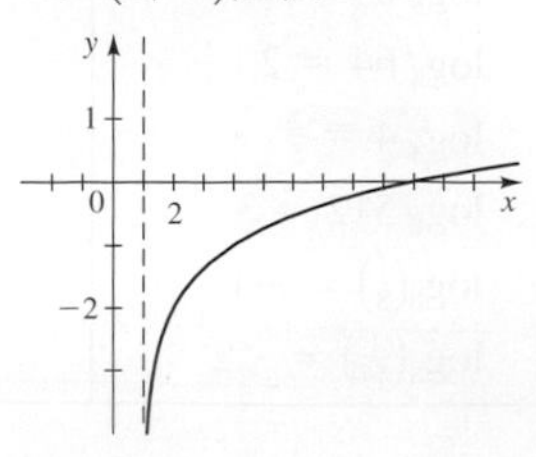

73. $(0, \infty), [0, \infty), x = 0$

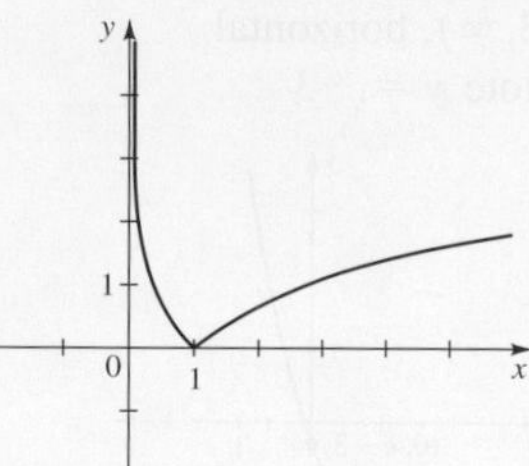

75. $(-3, \infty)$ **77.** $(-\infty, -1) \cup (1, \infty)$ **79.** $(0, 2)$

81.

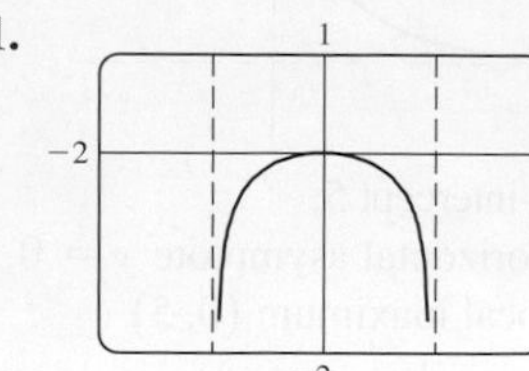

domain $(-1, 1)$
vertical asymptotes $x = 1$, $x = -1$
local maximum $(0, 0)$

83.

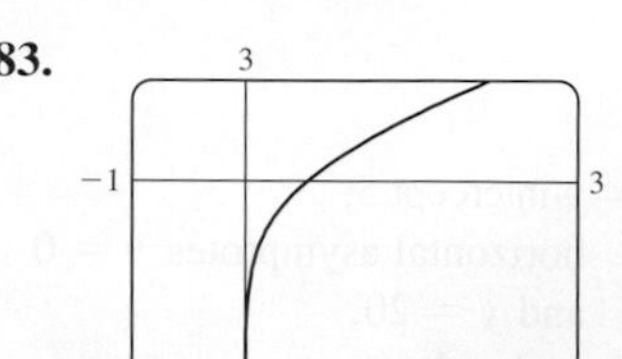

domain $(0, \infty)$
vertical asymptote $x = 0$
no maximum or minimum

85.

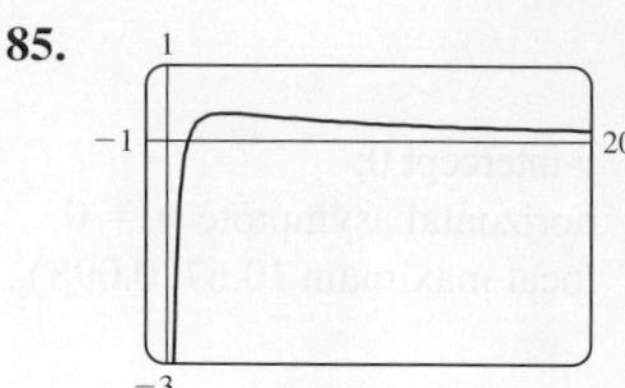

domain $(0, \infty)$
vertical asymptote $x = 0$
horizontal asymptote $y = 0$
local maximum
$\approx (2.72, 0.37)$

Answers to Exercises 87–89 will vary.
87. $g(x) = x^2 + 1, f(x) = \ln x$
89. $g(x) = 1 + |\ln x|, f(x) = \sqrt{x}$
91. $(f \circ g)(x) = 2^{x+1}, (-\infty, \infty)$; $(g \circ f)(x) = 2^x + 1, (-\infty, \infty)$
93. $(f \circ g)(x) = \log_2(x - 2), (2, \infty)$;
$(g \circ f)(x) = (\log_2 x) - 2, (0, \infty)$
95. The graph of f grows more slowly than g.
97. **(a)**

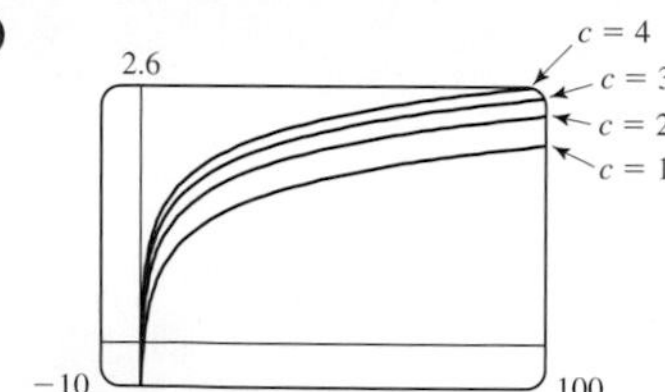

(b) The graph of $f(x) = \log(cx)$ is the graph of $f(x) = \log(x)$ shifted upward $\log c$ units.

99. **(a)** $(1, \infty)$ **(b)** $f^{-1}(x) = 10^{2^x}$
101. **(a)** $f^{-1}(x) = \log_2\left(\dfrac{x}{1 - x}\right)$ **(b)** $(0, 1)$ **103.** 2602 years
105. 11.6 years, 9.9 years, 8.7 years **107.** 5.32, 4.32

Section 4.4 ■ Page 366

1. sum; $\log_5 25 + \log_5 125 = 2 + 3$
2. difference; $\log_5 25 - \log_5 125 = 2 - 3$
3. power; $10 \cdot \log_5 25 = 10 \cdot 2$ **4.** $2 \log x + \log y - \log z$

5. $\log\dfrac{x^2y}{z}$ **6. (a)** $\log_7 12 = \dfrac{\log 12}{\log 7} \approx 1.277$ **(b)** Yes
7. (a) False **(b)** True **8. (a)** True **(b)** False
9. 4 **11.** 2 **13.** 1 **15.** $\frac{1}{2}$ **17.** 3 **19.** 200
21. 4 **23.** $\log_3 8 + \log_3 x$ **25.** $\log_3 2 + \log_3 x + \log_3 y$
27. $3 \ln a$ **29.** $\frac{1}{2}(\log_3 x + \log_3 y + \log_3 z)$ **31.** $3 \ln a + 2 \ln b$
33. $2 + \log_2 a - \log_2 b$ **35.** $3 \log_8 a + 2 \log_8 b - \log_8 c$
37. $\frac{1}{2} + \frac{5}{2}\log_3 x - \log_3 y$ **39.** $3 \log x + 4 \log y - 6 \log z$
41. $\frac{1}{2}\ln(x^4 + 2)$ **43.** $\frac{1}{2}[\log(x + z) - \log y]$
45. $\frac{1}{3}[\ln(x^2 + y^2) - \ln(x + y)]$
47. $\frac{1}{2}[\log(x^2 + 4) - \log(x^2 + 1) - 2\log(x^3 - 7)]$
49. $\log_4 294$ **51.** $\log\dfrac{x^2}{(x+1)^3}$ **53.** $\log\dfrac{x^2-1}{x^3}$
55. $\log_5\sqrt{\dfrac{x+2}{x^3+4x}}$ **57.** $\log\dfrac{x^2}{x-3}$ **59.** 1.430677
61. 0.630930 **63.** 0.493008 **65.** 3.482892
67.

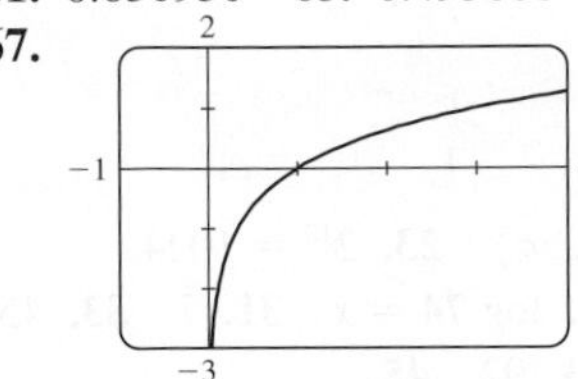

73. (a) $P = c/W^k$ **(b)** 1866, 64
75. (a) $M = -2.5 \log B + 2.5 \log B_0$

Section 4.5 ■ Page 376

1. (a) $e^x = 25$ **(b)** $x = \ln 25$ **(c)** 3.219
2. (a) $\log 3(x - 2) = \log x$ **(b)** $3(x - 2) = x$ **(c)** 3 **3.** 5
5. $\frac{3}{2}$ **7.** −3 **9.** −1, 1 **11. (a)** 4 ln 2 **(b)** 2.772589
13. (a) −log 6 **(b)** −0.778151 **15. (a)** $\dfrac{\ln 4}{\ln 3} - 5$
(b) −3.738140 **17. (a)** $1 - \dfrac{\ln 5}{\ln 6}$ **(b)** 0.101756
19. (a) $\dfrac{\ln 7.5}{4 \ln 1.02}$ **(b)** 25.437319 **21. (a)** 5 − ln 4
(b) 3.613706 **23. (a)** $\dfrac{10 \ln 0.3}{\ln 2}$ **(b)** −17.369656
25. (a) $\frac{1}{5}\log(\frac{5}{4})$ **(b)** 0.019382 **27. (a)** $\dfrac{1 - \ln 12}{4}$
(b) −0.371227 **29. (a)** $\dfrac{\ln(50/3)}{2 \ln 2}$ **(b)** 2.029447
31. (a) $\dfrac{2}{\log 5 - 3}$ **(b)** −0.869176 **33. (a)** $\dfrac{3 \ln 3}{\ln 5 - 3 \ln 3}$
(b) −1.954364 **35. (a)** −ln 11.5 **(b)** −2.442347 **37.** 0
39. $\frac{1}{2}\ln 3 \approx 0.5493$ **41.** 1 **43.** ±1 **45.** 0, $\frac{4}{3}$ **47.** 6
49. 2, 4 **51.** 5 **53.** 10^9 **55.** $4 - e \approx 1.2817$ **57.** $\frac{14}{3}$
59. −7 **61.** 4 **63.** 6 **65.** $\frac{13}{12}$ **67.** 2.21 **69.** 0.00, 1.14
71. −0.57 **73.** 0.36 **75.** $2 < x < 4$ or $7 < x < 9$
77. $\log 2 < x < \log 5$ **79.** $f^{-1}(x) = \dfrac{\ln x}{2 \ln 2}$
81. $f^{-1}(x) = 2^x + 1$ **83.** $1/\sqrt{5} \approx 0.4472$ **85.** 0, $\frac{1}{10}$
87. (a) \$5593.60 **(b)** about 30 years and 10 months
89. about 13 years and 6 months **91.** 8.15 years **93.** 13 days
95. (a) 7337 **(b)** 1.73 years **97. (a)** $P = P_0e^{-h/k}$
(b) 56.47 kPa **99. (a)** $t = -\frac{5}{13}\ln(1 - \frac{13}{60}I)$ **(b)** 0.218 s

Section 4.6 ■ Page 387

1. (a) $n(t) = 10 \cdot 2^{2t/3}$ **(b)** 1.06×10^8 **(c)** 14.9
3. (a) 3125 **(b)** 317,480
(c)

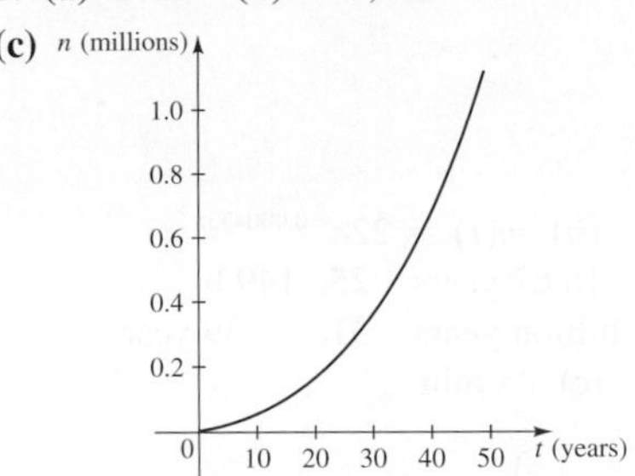

5. (a) $n(t) = 12{,}800e^{0.12t}$ **(b)** 23,300 beavers **(c)** 11.35 years
(d)

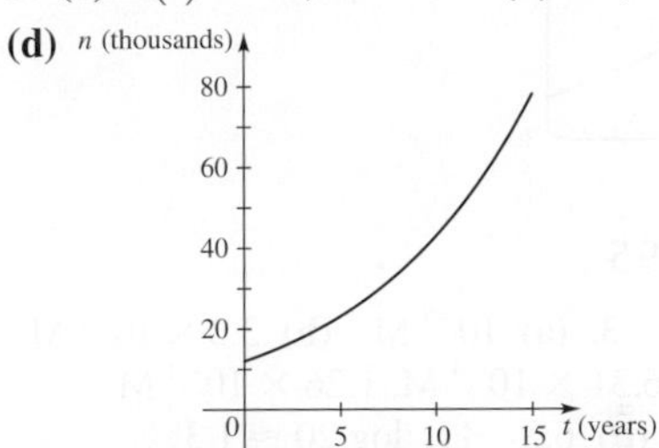

7. (a) 233 million **(b)** 181 million
9. (a) $n(t) = 112{,}000 \cdot 2^{t/18}$ **(b)** $n(t) = 112{,}000e^{0.0385t}$
(c) **(d)** 38.9 years

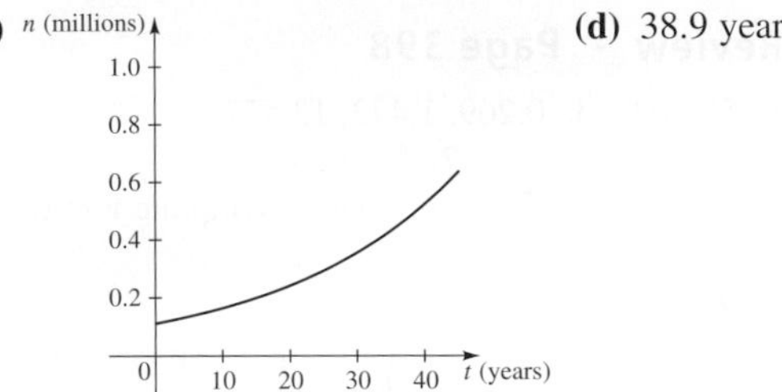

11. (a) 20,000 **(b)** $n(t) = 20{,}000e^{0.1096t}$ **(c)** About 48,000
(d) 14.7 years
13. (a) $n(t) = 8600e^{0.1508t}$ **(b)** About 11,600 **(c)** 4.6 h
15. (a) $n(t) = 49e^{-0.00566t}$ million; 2059 **(b)** 122.5 years
17. (a) $n(t) = \dfrac{11}{1 + 3.4e^{-0.0189t}}$ billion; 2136
(b)

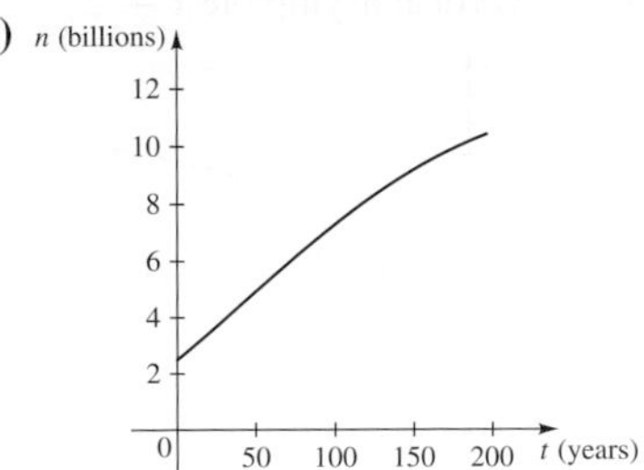

19. (a) $n(t) = \dfrac{6000}{1 + 749e^{-0.57t}}$

(b) About 19 days

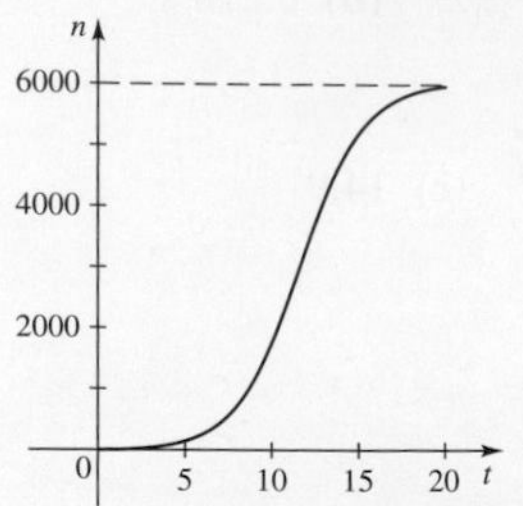

21. (a) $m(t) = 22 \cdot 2^{-t/1600}$ **(b)** $m(t) = 22e^{-0.000433t}$
(c) 3.9 mg **(d)** 463.4 **23.** 18.67 years **25.** 149 h
27. 3560 years **29.** ≈1.45 billion years **31.** ≈139 years
33. (a) 210°F **(b)** 153°F **(c)** 28 min
35. 63°C

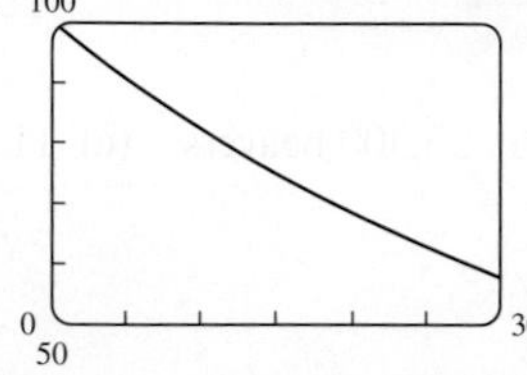

Section 4.7 ■ Page 395

1. (a) 2.3 **(b)** 3.5 **(c)** 8.3 **3. (a)** 10^{-3} M **(b)** 3.2×10^{-7} M
5. $4.8 \le \text{pH} \le 6.4$ **7. (a)** 6.31×10^{-4} M, 1.26×10^{-3} M
(b) red wine **9. (a)** 5.49 **(b)** 6.3 **11.** $\log 20 \approx 1.3$
13. Six times as intense **15.** 73 dB **17.** 10^{-5} W/m^2
19. (a) 75 dB **(b)** 10^{-3} W/m^2 **(c)** 32.3

Chapter 4 Review ■ Page 398

1. 0.089, 9.739, 55.902 **3.** 0.269, 1.472, 12.527

5. $\mathbb{R}$, $(1, \infty)$, horizontal asymptote $y = 1$

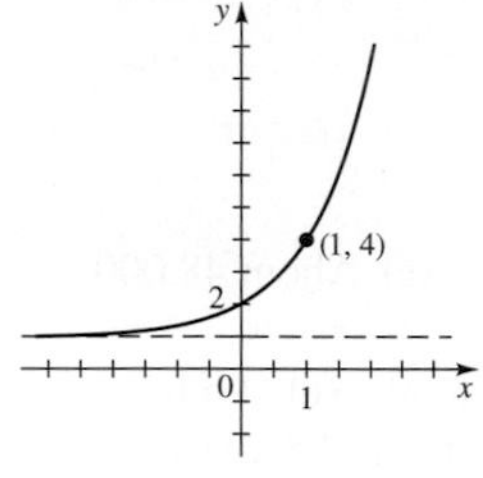

7. $\mathbb{R}$, $(0, \infty)$, horizontal asymptote $y = 0$

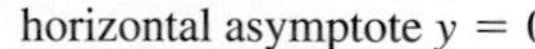

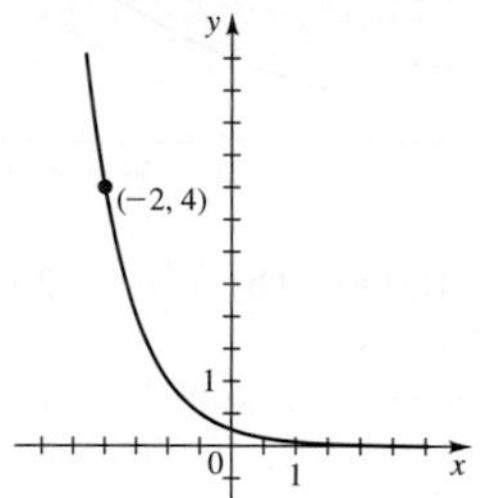

9. $\mathbb{R}$, $(-\infty, -1)$, horizontal asymptote $y = -1$

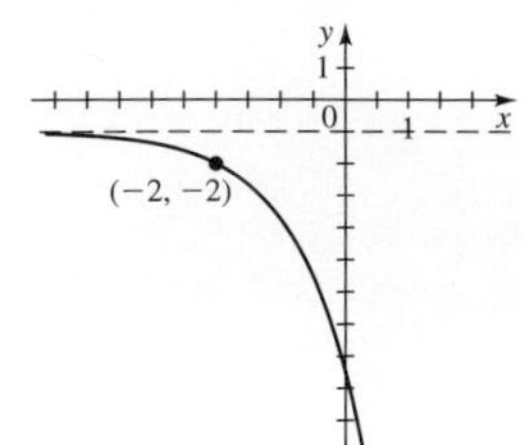

11. $(2, \infty)$, $\mathbb{R}$, vertical asymptote $x = 2$

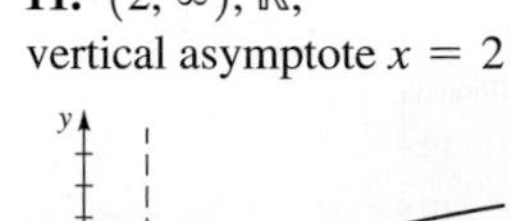

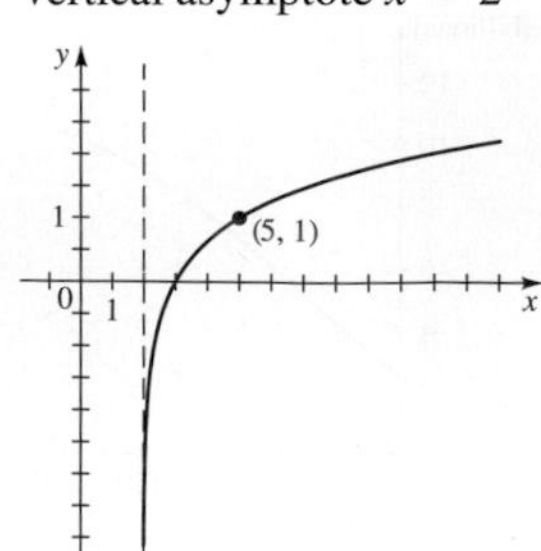

13. $(0, \infty)$, $\mathbb{R}$, vertical asymptote $x = 0$

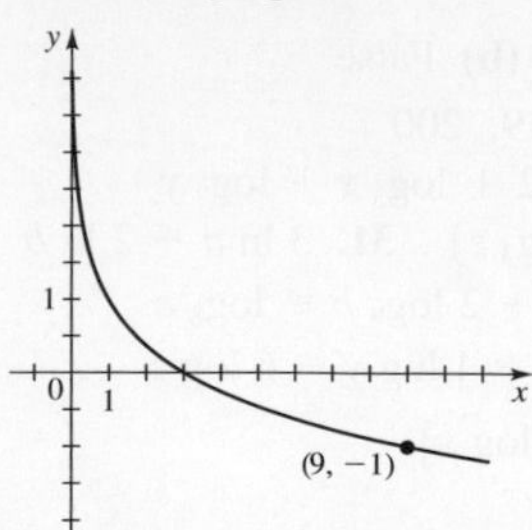

15. $(-\infty, 0)$, $\mathbb{R}$, vertical asymptote $x = 0$

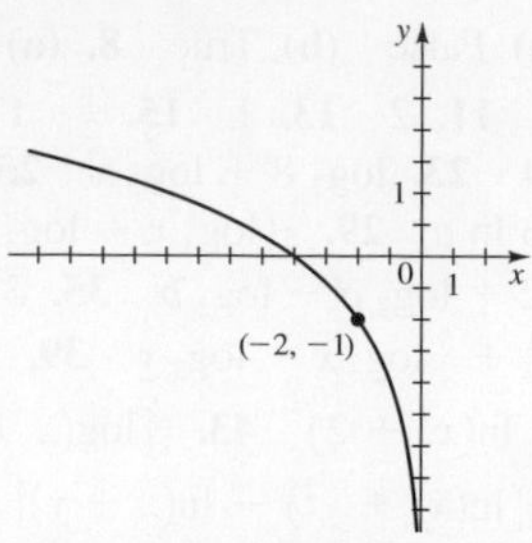

17. $(0, \infty)$, $\mathbb{R}$, vertical asymptote $x = 0$

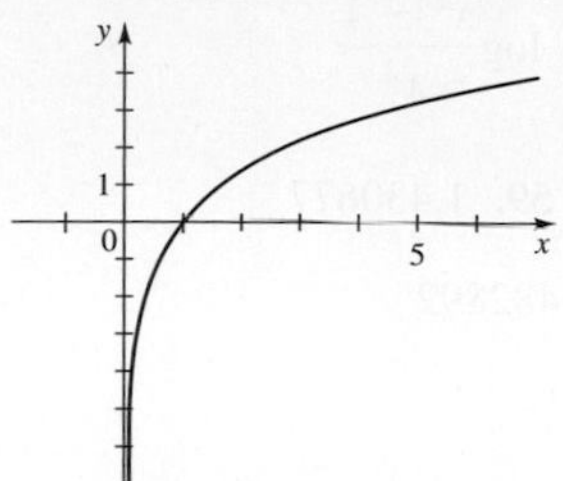

19. $\left(-\infty, \frac{1}{2}\right)$ **21.** $(-\infty, -2) \cup (2, \infty)$ **23.** $2^{10} = 1024$
25. $10^y = x$ **27.** $\log_2 64 = 6$ **29.** $\log 74 = x$ **31.** 7 **33.** 45
35. 6 **37.** −3 **39.** $\frac{1}{2}$ **41.** 2 **43.** 92 **45.** $\frac{2}{3}$
47. $\log A + 2 \log B + 3 \log C$
49. $\frac{1}{2}[\ln(x - 1) + \ln(x + 1) - \ln(x^2 + 1)]$
51. $2 \log_5 x + \frac{3}{2} \log_5(1 - 5x) - \frac{1}{2}[\log_5 x + \log_5(x - 1) + \log_5(x + 1)]$
53. $\log 96$ **55.** $\log_2 \dfrac{(x - y)^{3/2}}{(x^2 + y^2)^2}$ **57.** $\log \dfrac{x^2 - 4}{\sqrt{x^2 + 4}}$
59. 1 **61.** $\dfrac{1}{3}\left(\dfrac{\ln 2}{\ln 5} - 2\right) \approx -0.52$ **63.** $\dfrac{\ln(81/2)}{5 \ln 2 + \ln 3} \approx 0.81$
65. −2, 0, 3 **67.** 3 **69.** −15 **71.** 9 **73.** 0.430618
75. 2.303600
77.

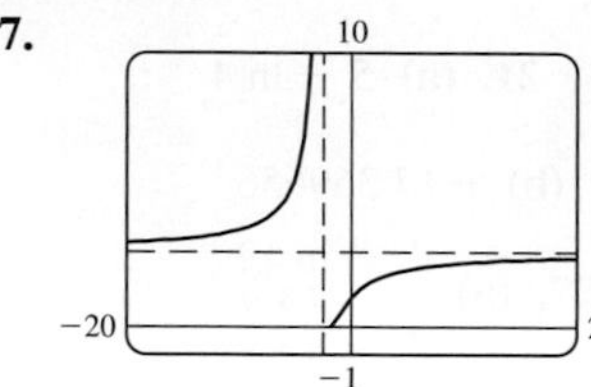

vertical asymptote $x = -2$
horizontal asymptote $y \approx 2.72$
no maximum or minimum

79. vertical asymptotes $x = -1, x = 0, x = 1$
local maximum $\approx(-0.58, -0.41)$

81. 2.42 **83.** $0.16 < x < 3.15$
85. Increasing on $(-\infty, 0)$ and $(1.10, \infty)$, decreasing on $(0, 1.10)$
87. 1.953445 **89.** −0.579352 **91.** $\log_4 258$
93. (a) \$16,081.15 **(b)** \$16,178.18 **(c)** \$16,197.64
(d) \$16,198.31 **95.** 1.83 years **97.** 4.341%
99. (a) $n(t) = 30e^{0.15t}$ **(b)** 55 **(c)** 19 years

101. **(a)** $n(t) = 150 \cdot 2^{-t/75{,}380}$ **(b)** 148.63 mg **(c)** 119,474 years **103.** **(a)** 12 g **(b)** $m(t) = 12e^{-0.173t} = 12 \cdot 2^{-t/4}$ **(c)** 7.1 g **(d)** 25 days

105. **(a)** 0.462 **(b)** $n(t) = \dfrac{1400}{1 + 13e^{-0.462t}}$ **(c)** 5.55 h

107. 7.9, basic **109.** 8.0 **111.** **(a)** VI **(b)** VIII **(c)** V **(d)** III **(e)** II **(f)** VII **(g)** IV **(h)** I

Chapter 4 Test ▪ Page 402

1. **(a)** $\mathbb{R}, (-\infty, 3)$, horizontal asymptote $y = 3$ **(b)** $(-3, \infty), \mathbb{R}$, vertical asymptote $x = -3$

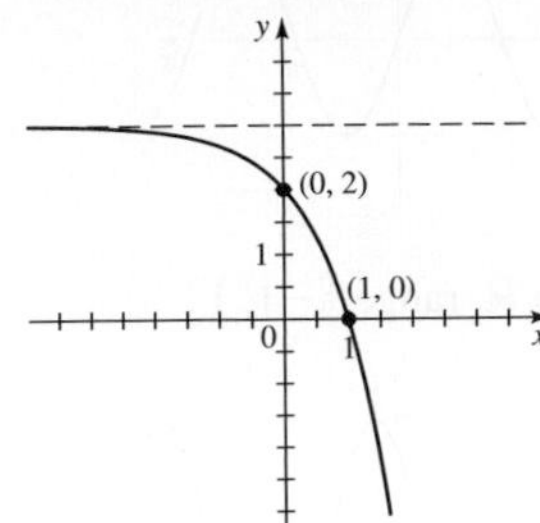

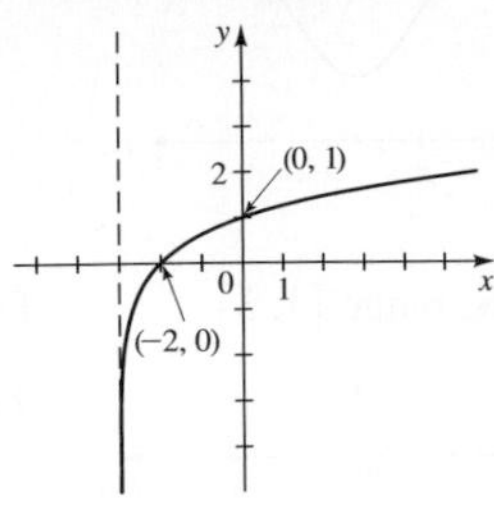

2. **(a)** $(\frac{3}{2}, \infty)$ **(b)** $(-\infty, -1) \cup (1, \infty)$

3. **(a)** $\log_6 25 = 2x$ **(b)** $e^3 = A$

4. **(a)** 36 **(b)** 3 **(c)** $\frac{3}{2}$ **(d)** 3 **(e)** $\frac{2}{3}$ **(f)** 2

5. **(a)** $\log x + 3\log y - 2\log z$ **(b)** $\frac{1}{2}\ln x - \frac{1}{2}\ln y$ **(c)** $\frac{1}{3}[\log(x^2+1) - 3\log x - \log(x-1)]$

6. **(a)** $\log(ab^2)$ **(b)** $\ln(x-5)$ **(c)** $\log_3 \dfrac{xy^3}{(x+1)^2}$

7. **(a)** 25 **(b)** 1, 2 **(c)** $\dfrac{\ln(9/5)}{3\ln(2/3)} \approx -0.48$ **(d)** 5.39

8. **(a)** 500 **(b)** $\frac{2}{3}$ **(c)** $3 - e^{4/5} \approx 0.774$ **(d)** $\frac{19}{15}$

9. $\dfrac{\log 27}{\log 12}$ or $\dfrac{\ln 27}{\ln 12} \approx 1.326$

10. **(a)** $n(t) = 1000e^{2.07944t}$ **(b)** 22,600 **(c)** 1.3

(d)

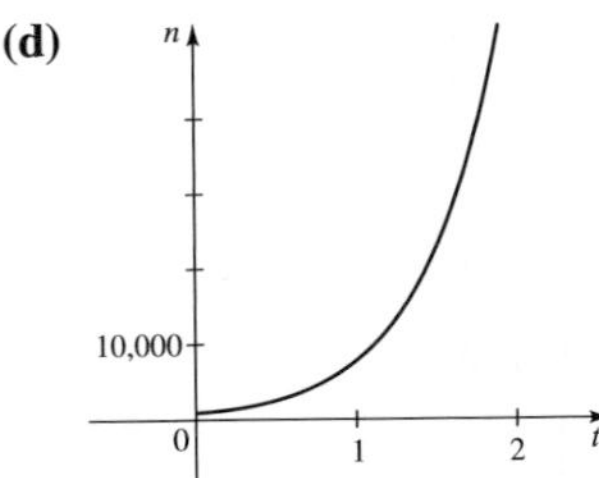

11. **(a)** $A(t) = 12{,}000\left(1 + \dfrac{0.056}{12}\right)^{12t}$ **(b)** \$14,195.06

(c) 9.12 years **12.** **(a)** $m(t) = 3 \cdot 2^{-t/10}$ **(b)** $m(t) = 3e^{-0.0693t}$ **(c)** 0.047 g **(d)** After 3.6 min **13.** 1995 times more intense

Focus on Modeling ▪ Page 406

1. **(a)**

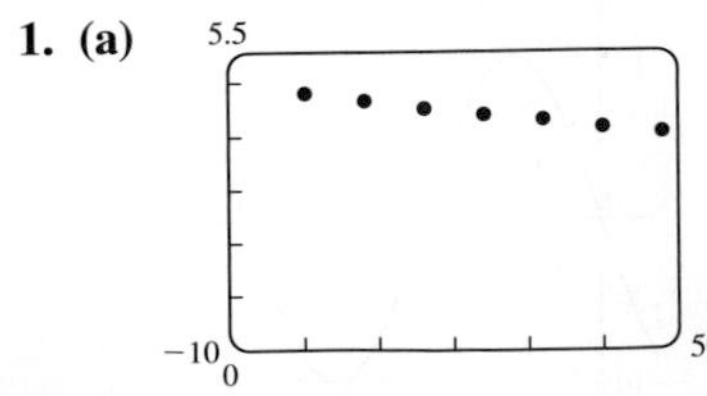

(b) $y = ab^t$, where $a = 4.79246$ and $b = 0.99642$ **(c)** 192.8 h

3. **(a)** $S = 0.14A^{0.64}$

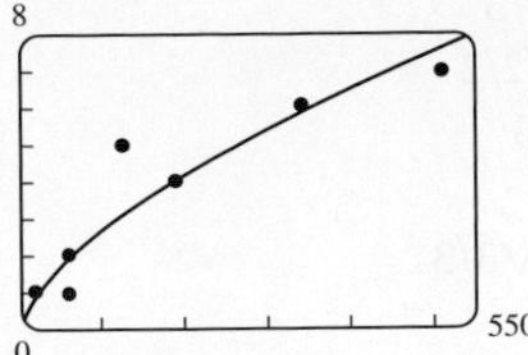

(b) 4 species

5. **(a)** $I = 306.9687 \cdot e^{-0.02999x}$, $k = 0.02999$

(b) Yes; according to the model, the light intensity in the twilight zone is at least 2.911×10^{-11} W/m^2.

Chapter 5

Section 5.1 ▪ Page 415

1. **(a)** $(0, 0)$, 1 **(b)** $x^2 + y^2 = 1$ **(c)** (i) 0 (ii) 0 (iii) 0 (iv) 0 **2.** **(a)** terminal **(b)** $(0, 1), (-1, 0), (0, -1), (1, 0)$

3. P; $(\frac{1}{2}, \frac{\sqrt{3}}{2})$, $(\frac{1}{2}, \frac{\sqrt{3}}{2})$

4.

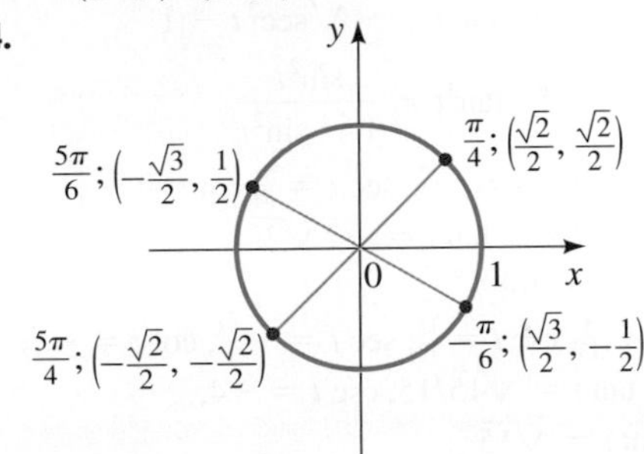

Yes, the statement is true.

11. $-\frac{4}{5}$ **13.** $-2\sqrt{2}/3$ **15.** $3\sqrt{5}/7$ **17.** $P(\frac{5}{13}, -\frac{12}{13})$

19. $P(-\sqrt{5}/3, \frac{2}{3})$ **21.** $P(-\sqrt{2}/3, -\sqrt{7}/3)$

23. $t = \pi/4, (\sqrt{2}/2, \sqrt{2}/2)$; $t = \pi/2, (0, 1)$; $t = 3\pi/4, (-\sqrt{2}/2, \sqrt{2}/2)$; $t = \pi, (-1, 0)$; $t = 5\pi/4, (-\sqrt{2}/2, -\sqrt{2}/2)$; $t = 3\pi/2, (0, -1)$; $t = 7\pi/4, (\sqrt{2}/2, -\sqrt{2}/2)$; $t = 2\pi, (1, 0)$

25. $(-1, 0)$ **27.** $(1, 0)$ **29.** $(0, -1)$ **31.** $(0, -1)$

33. $(-\sqrt{3}/2, \frac{1}{2})$ **35.** $(-\sqrt{2}/2, -\sqrt{2}/2)$ **37.** $(\frac{1}{2}, \sqrt{3}/2)$

39. $(\sqrt{2}/2, -\sqrt{2}/2)$ **41.** $(-\sqrt{3}/2, -\frac{1}{2})$

43. **(a)** $\pi/3$ **(b)** $\pi/3$ **(c)** $\pi/6$ **(d)** $3.5 - \pi \approx 0.36$

45. **(a)** $2\pi/7$ **(b)** $2\pi/9$ **(c)** $\pi - 3 \approx 0.14$ **(d)** $2\pi - 5 \approx 1.28$ **47.** **(a)** $\pi/4$ **(b)** $(-\sqrt{2}/2, \sqrt{2}/2)$

49. **(a)** $\pi/6$ **(b)** $(-\sqrt{3}/2, -\frac{1}{2})$ **51.** **(a)** $\pi/6$ **(b)** $(\sqrt{3}/2, -\frac{1}{2})$

53. **(a)** $\pi/4$ **(b)** $(-\sqrt{2}/2, -\sqrt{2}/2)$

55. **(a)** $\pi/6$ **(b)** $(-\sqrt{3}/2, \frac{1}{2})$ **57.** **(a)** $\pi/3$ **(b)** $(\frac{1}{2}, \sqrt{3}/2)$

59. **(a)** $\pi/3$ **(b)** $(-\frac{1}{2}, -\sqrt{3}/2)$ **61.** $(0.5, 0.8)$ **63.** $(0.5, -0.9)$

65. **(a)** $(-\frac{3}{5}, \frac{4}{5})$ **(b)** $(\frac{3}{5}, -\frac{4}{5})$ **(c)** $(-\frac{3}{5}, -\frac{4}{5})$ **(d)** $(\frac{3}{5}, \frac{4}{5})$

Section 5.2 ▪ Page 424

1. $y, x, y/x$ **2.** 1; 1; $\cos t = \pm\sqrt{1 - \sin^2 t}$, $\sin t = \pm\sqrt{1 - \cos^2 t}$

3. $\sin t_3, \sin t_1, \sin t_2$ **4.** $\cos t_2, \cos t_3, \cos t_1$

5. $\pi/4$, $\cos t = \sqrt{2}/2$, $\sin t = \sqrt{2}/2$; $\pi/2$, $\cos t = 0$, $\sin t = 1$; $3\pi/4$, $\cos t = -\sqrt{2}/2$, $\sin t = -\sqrt{2}/2$; π, $\cos t = -1$, $\sin t = 0$; $5\pi/4$, $\cos t = -\sqrt{2}/2$, $\sin t = -\sqrt{2}/2$; $3\pi/2$, $\cos t = 0$, $\sin t = -1$; $7\pi/4$, $\cos t = \sqrt{2}/2$, $\sin t = -\sqrt{2}/2$; 2π, $\cos t = 1$, $\sin t = 0$

7. **(a)** $-\sqrt{3}/2$ **(b)** $\sqrt{2}/2$ **(c)** $-\sqrt{3}/3$
9. **(a)** $-\sqrt{2}/2$ **(b)** $-\sqrt{2}/2$ **(c)** $\sqrt{3}/3$
11. **(a)** $-\sqrt{2}/2$ **(b)** $-\sqrt{2}/2$ **(c)** $\sqrt{2}/2$
13. **(a)** $\sqrt{3}/2$ **(b)** $2\sqrt{3}/3$ **(c)** $\sqrt{3}/3$
15. **(a)** $\frac{1}{2}$ **(b)** 2 **(c)** $-\sqrt{3}/2$
17. **(a)** $\sqrt{3}/2$ **(b)** $-2\sqrt{3}/3$ **(c)** $-\sqrt{3}/3$
19. **(a)** -2 **(b)** $2\sqrt{3}/3$ **(c)** $\sqrt{3}$
21. **(a)** $-\sqrt{3}/2$ **(b)** $2\sqrt{3}/3$ **(c)** $-\sqrt{3}/3$
23. **(a)** 0 **(b)** 1 **(c)** 0
25. $\sin 0 = 0$, $\cos 0 = 1$, $\tan 0 = 0$, $\sec 0 = 1$, others undefined
27. $\sin \pi = 0$, $\cos \pi = -1$, $\tan \pi = 0$, $\sec \pi = -1$, others undefined
29. $\frac{3}{5}, -\frac{4}{5}, -\frac{3}{4}$ **31.** $\frac{1}{2}, -\sqrt{3}/2, -\sqrt{3}/3$
33. $\sqrt{13}/7, -\frac{6}{7}, -\sqrt{13}/6$ **35.** $-\frac{12}{13}, -\frac{5}{13}, \frac{12}{5}$ **37.** $\frac{21}{29}, -\frac{20}{29}, -\frac{21}{20}$
39. **(a)** 0.8 **(b)** 0.84147 **41.** **(a)** 0.9 **(b)** 0.93204
43. **(a)** 1 **(b)** 1.02964 **45.** **(a)** -0.6 **(b)** -0.57482
47. Negative **49.** Negative **51.** II **53.** II
55. $\cos t = -\sqrt{1 - \sin^2 t}$ **57.** $\sin t = \sqrt{1 - \cos^2 t}$
59. $\tan t = \dfrac{\sqrt{1 - \cos^2 t}}{\cos t}$ **61.** $\tan t = -\sqrt{\sec^2 t - 1}$
63. $\csc t = \sqrt{1 + \cot^2 t}$ **65.** $\tan^2 t = \dfrac{\sin^2 t}{1 - \sin^2 t}$
67. $\cos t = \frac{3}{5}$, $\tan t = -\frac{4}{3}$, $\csc t = -\frac{5}{4}$, $\sec t = \frac{5}{3}$, $\cot t = -\frac{3}{4}$
69. $\sin t = -2\sqrt{2}/3$, $\cos t = \frac{1}{3}$, $\tan t = -2\sqrt{2}$, $\csc t = -\frac{3}{4}\sqrt{2}$, $\cot t = -\sqrt{2}/4$
71. $\sin t = \frac{12}{13}$, $\cos t = -\frac{5}{13}$, $\csc t = \frac{13}{12}$, $\sec t = -\frac{13}{5}$, $\cot t = -\frac{5}{12}$
73. $\cos t = -\sqrt{15}/4$, $\tan t = \sqrt{15}/15$, $\csc t = -4$, $\sec t = -4\sqrt{15}/15$, $\cot t = \sqrt{15}$

For Exercises 75–81, there are many possible answers.
75. $g(x) = \cos x$, $f(x) = x^2$ **77.** $g(x) = 1 + \tan x$, $f(x) = \sqrt{x}$
79. $h(x) = \sin x$, $g(x) = x^2$, $f(x) = e^x$
81. $h(x) = \cos x$, $g(x) = x^2$, $f(x) = \ln x$ **83.** Odd
85. Odd **87.** Even **89.** Neither
91. $y(0) = 4$, $y(0.25) = -2.828$, $y(0.50) = 0$, $y(0.75) = 2.828$, $y(1.00) = -4$, $y(1.25) = 2.828$
93. **(a)** 0.499 amp **(b)** -0.171 amp

Section 5.3 ■ Page 439

1. $f(t)$; 2π, 1

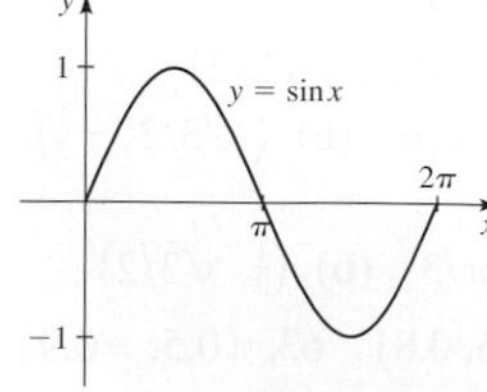

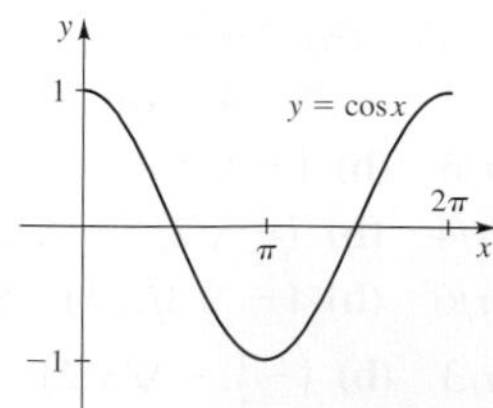

2. upward; x **3.** **(a)** $|a|, 2\pi/k$; 3, π; $[0, \pi]$
(b)

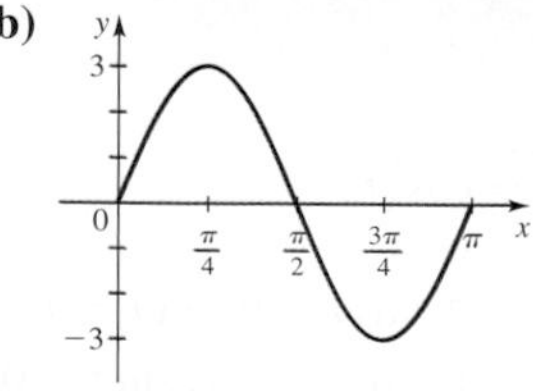

4. **(a)** $|a|, 2\pi/k, b$; 4, $2\pi/3$, $\pi/6$; $[\pi/6, 5\pi/6]$
(b)

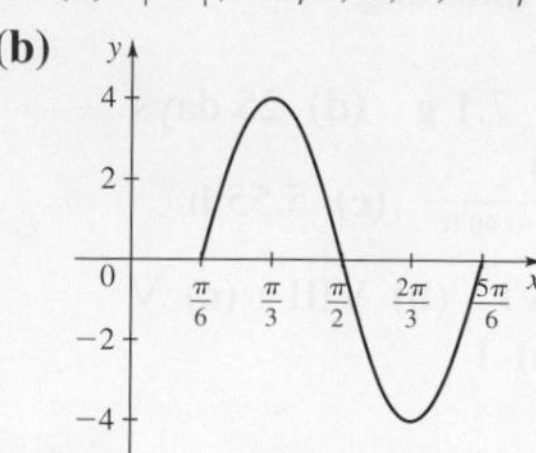

5.

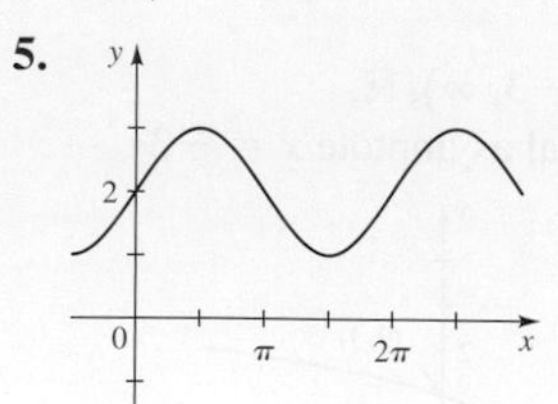

Domain $\mathbb{R}$, range $[1, 3]$

7.

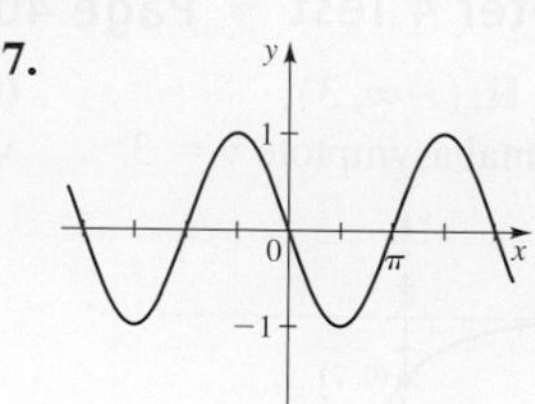

Domain $\mathbb{R}$, range $[-1, 1]$

9.

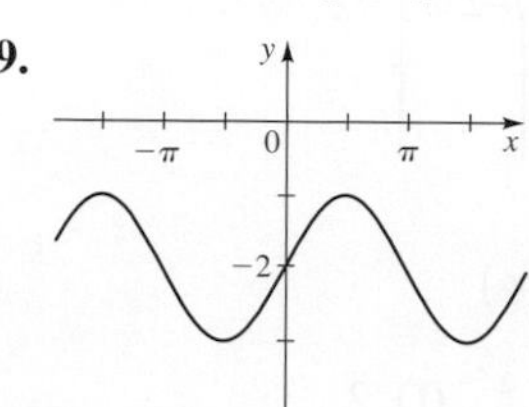

Domain $\mathbb{R}$, range $[-3, -1]$

11.

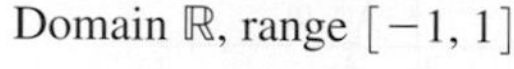

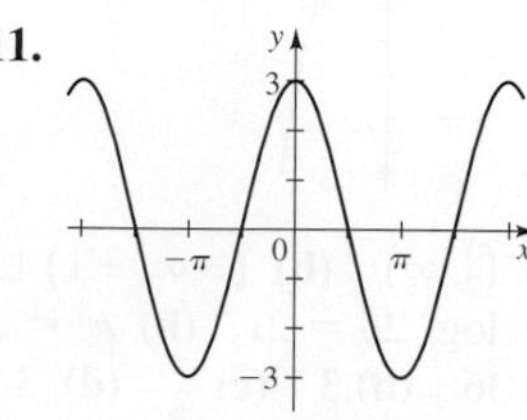

Domain $\mathbb{R}$, range $[-3, 3]$

13.

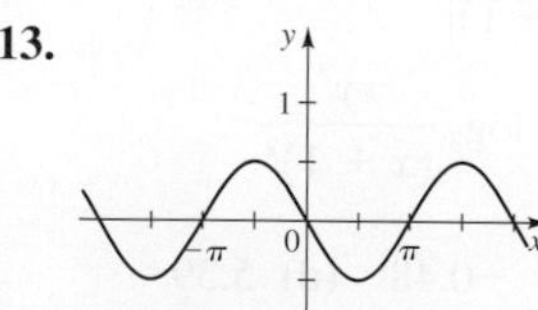

Domain $\mathbb{R}$, range $[-\frac{1}{2}, \frac{1}{2}]$

15.

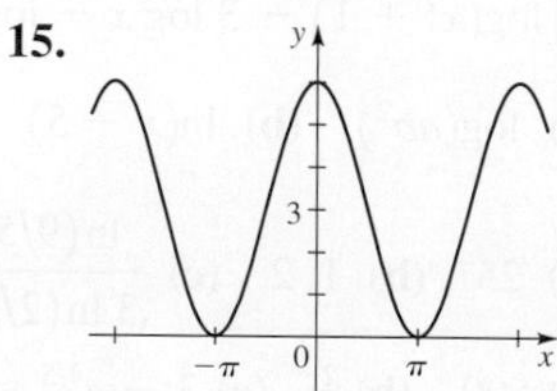

Domain $\mathbb{R}$, range $[0, 6]$

17.

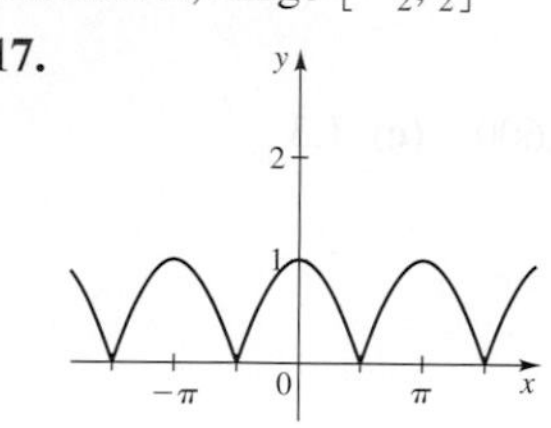

Domain $\mathbb{R}$, range $[0, 1]$

19. 1, π

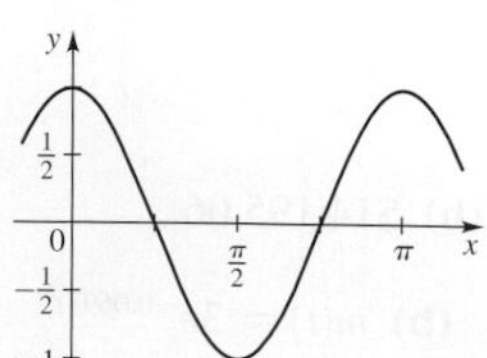

21. 1, $\pi/2$

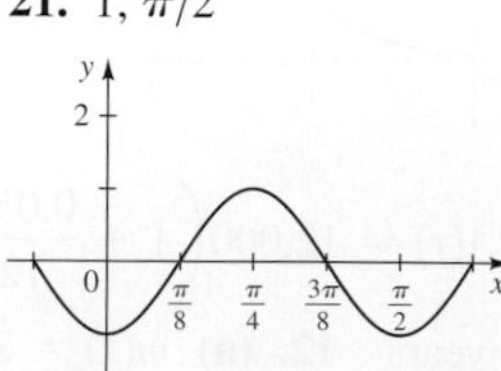

23. 3, 1

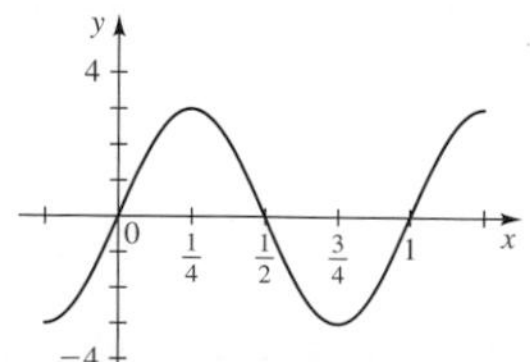

25. 10, 4π

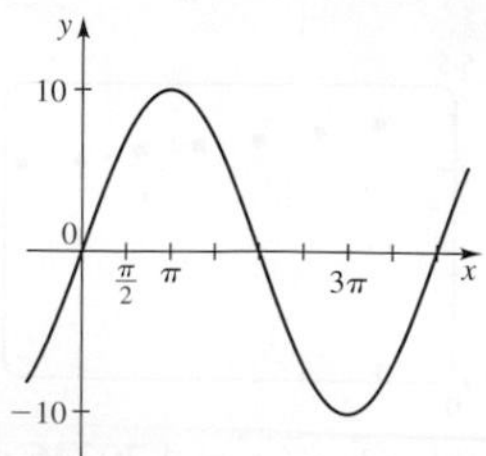

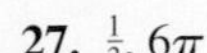

27. $\frac{1}{3}$, 6π

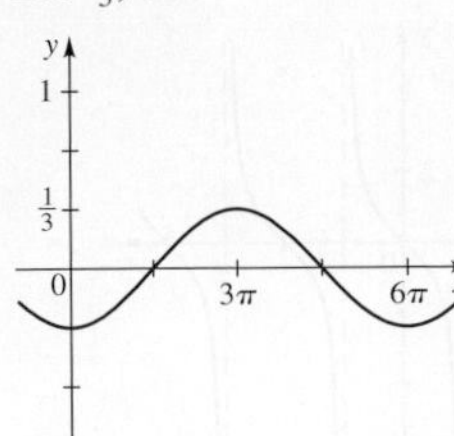

29. 2, $\frac{1}{4}$

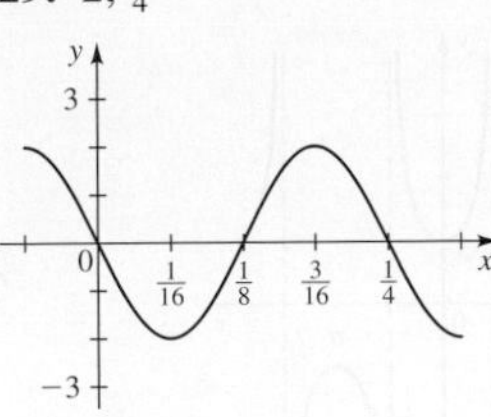

31. 2, $2\pi/3$

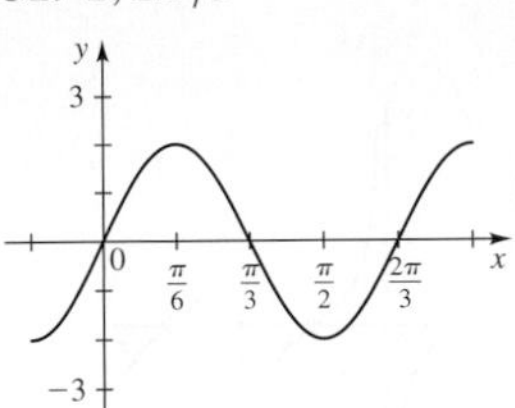

33. $\frac{1}{2}$, 2

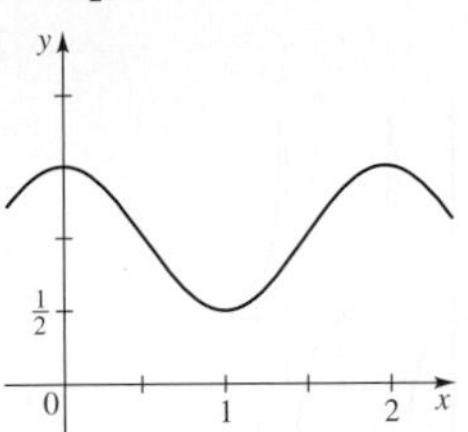

35. 1, 2π, $\pi/2$

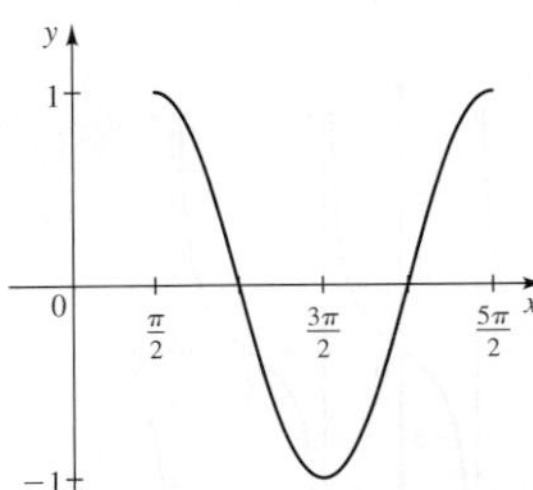

37. 2, 2π, $\pi/6$

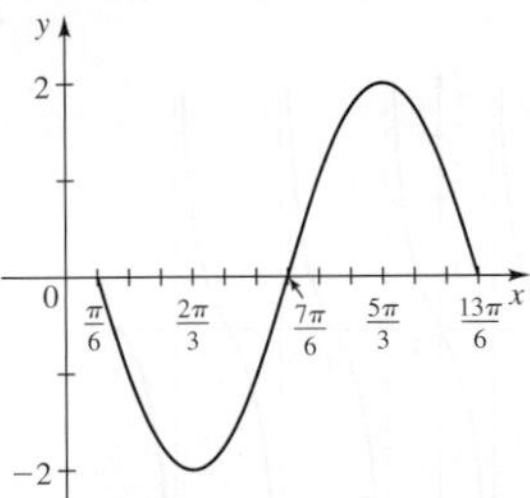

39. 4, 2, $\frac{1}{2}$

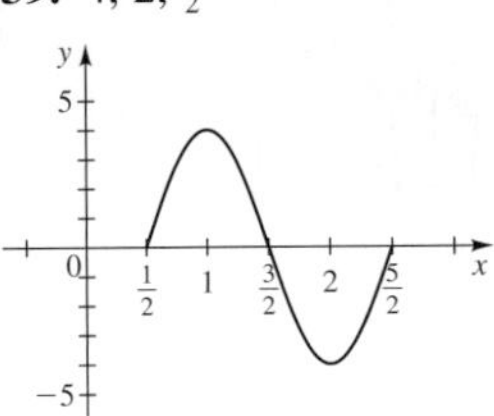

41. 2, $\pi/2$, $\pi/4$

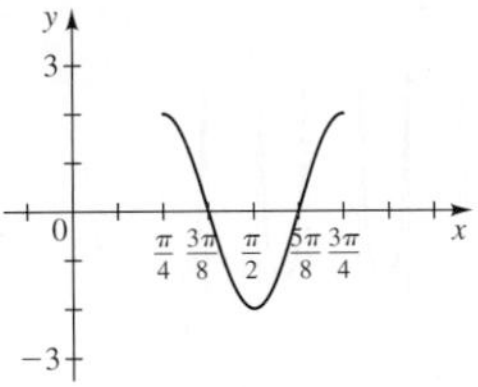

43. 1, π, $-\pi/2$

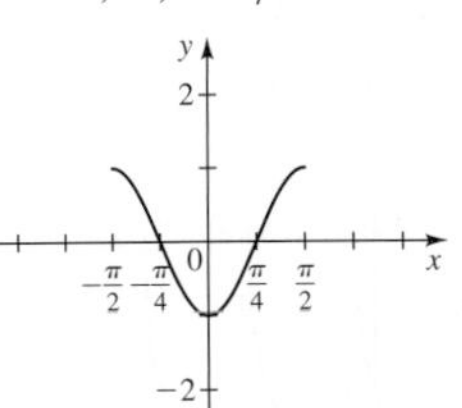

45. 2, 3π, $\pi/4$

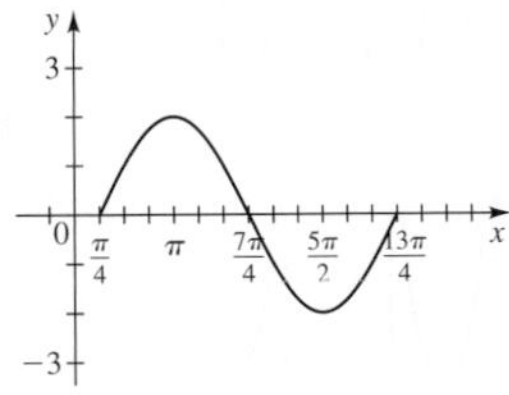

47. 2, $2\pi/3$, $-\pi/3$

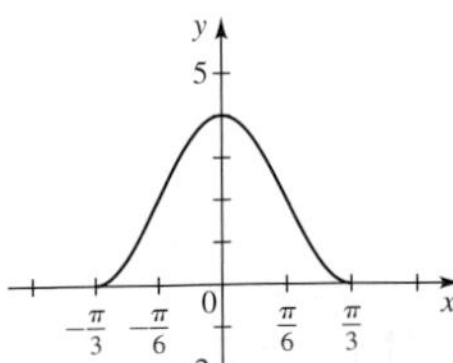

49. $\frac{1}{2}$, 1, $\frac{1}{6}$

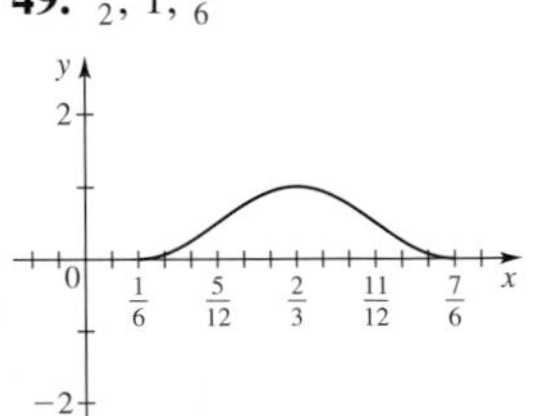

For Exercises 51–57, there are many possible answers.

51. 4, 2π; $y = 4\sin x$, $y = 4\cos\left(x - \frac{\pi}{2}\right)$

53. $\frac{3}{2}$, $2\pi/3$; $y = \frac{3}{2}\sin 3\left(x + \frac{\pi}{6}\right)$, $y = \frac{3}{2}\cos 3x$

55. $\frac{1}{2}$, π; $y = -\frac{1}{2}\sin 2\left(x + \frac{7\pi}{12}\right)$, $y = -\frac{1}{2}\cos 2\left(x + \frac{\pi}{3}\right)$

57. 1, π; $y = 1 + \sin 2\left(x - \frac{\pi}{2}\right)$, $y = 1 + \cos 2\left(x - \frac{3\pi}{4}\right)$

59.

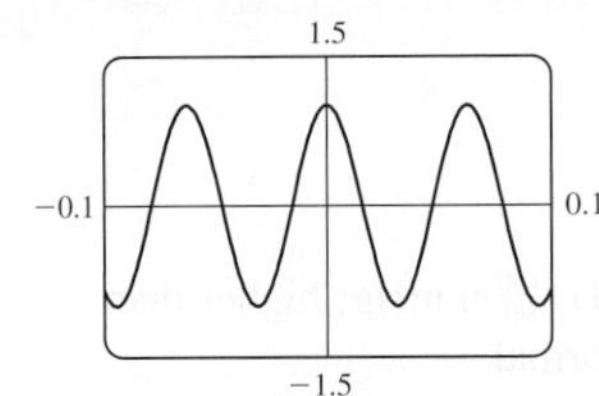

61.

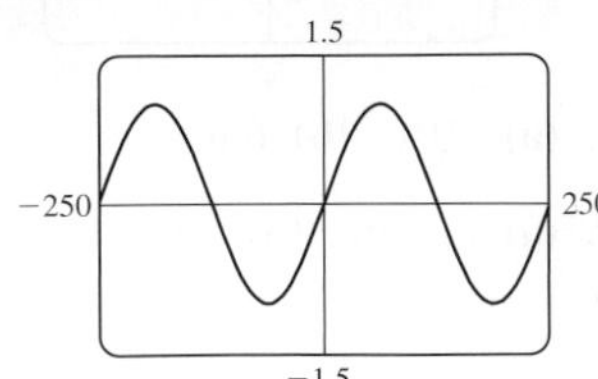

63.

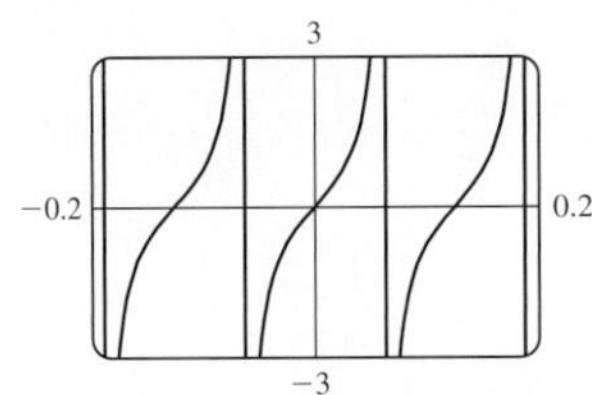

65.

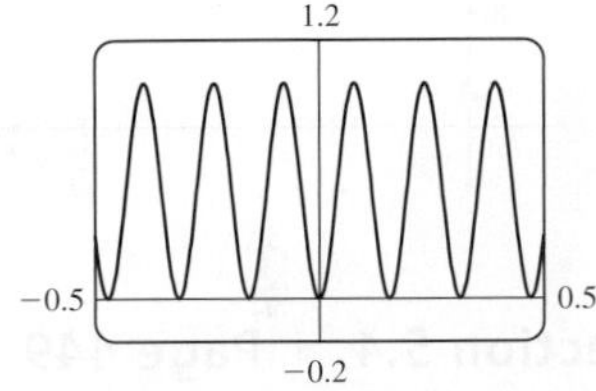

67.

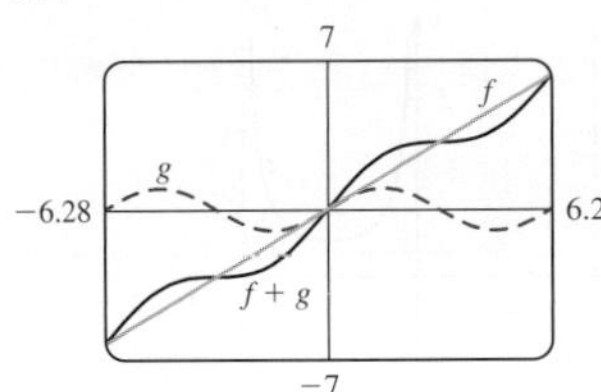

69.

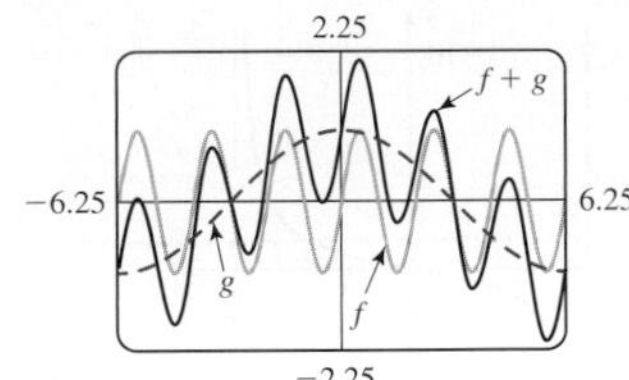

71.

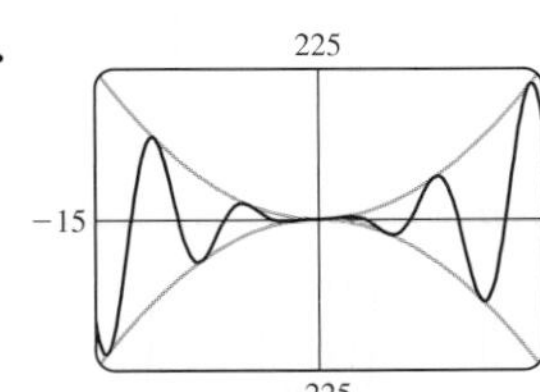

$y = x^2 \sin x$ is a sine curve that lies between the graphs of $y = x^2$ and $y = -x^2$

73.

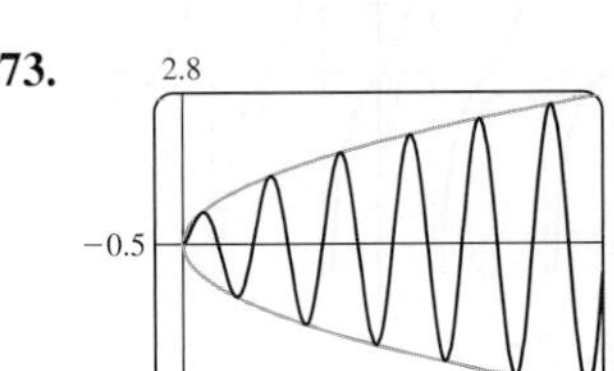

$y = \sqrt{x}\sin 5\pi x$ is a sine curve that lies between the graphs of $y = \sqrt{x}$ and $y = -\sqrt{x}$

75.

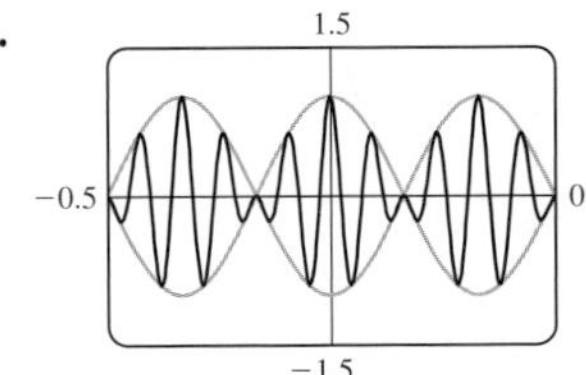

$y = \cos 3\pi x \cos 21\pi x$ is a cosine curve that lies between the graphs of $y = \cos 3\pi x$ and $y = -\cos 3\pi x$

77. Maximum value 1.76 when $x \approx 0.94 + 2n\pi$, minimum value -1.76 when $x \approx -0.94 + 2n\pi$, n any integer.
79. Maximum value 3.00 when $x \approx 1.57 + 2n\pi$, minimum value -1.00 when $x \approx -1.57 + 2n\pi$, n any integer.
81. 1.16 **83.** 0.34, 2.80
85. **(a)** Odd **(b)** $\pm 2\pi, \pm 4\pi, \pm 6\pi, \ldots$

(c)

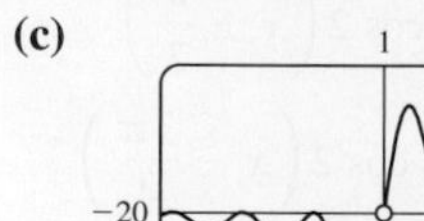
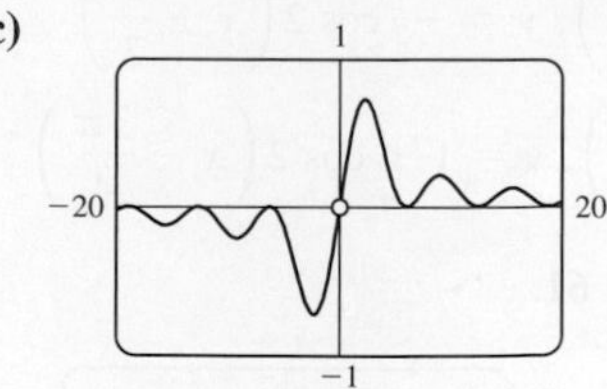

(d) $f(x)$ approaches 0
(e) $f(x)$ approaches 0

87. **(a)** 20 s **(b)** 6 m

89. **(a)** $\frac{1}{80}$ min **(b)** 80

(c)

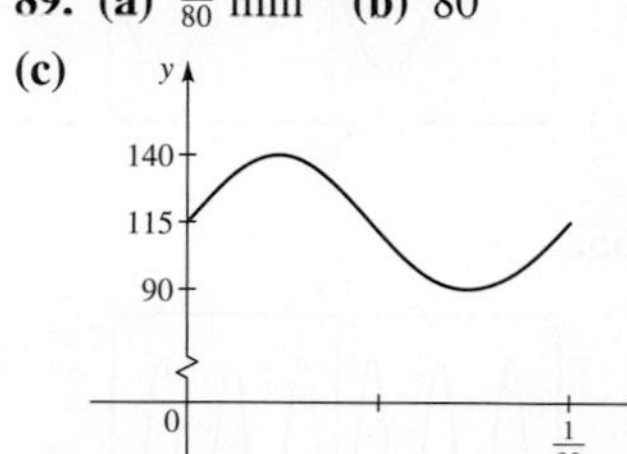

(d) $\frac{140}{90}$ mmHg; higher than normal

Section 5.4 ■ Page 449

1. π; $\frac{\pi}{2} + n\pi$, n an integer

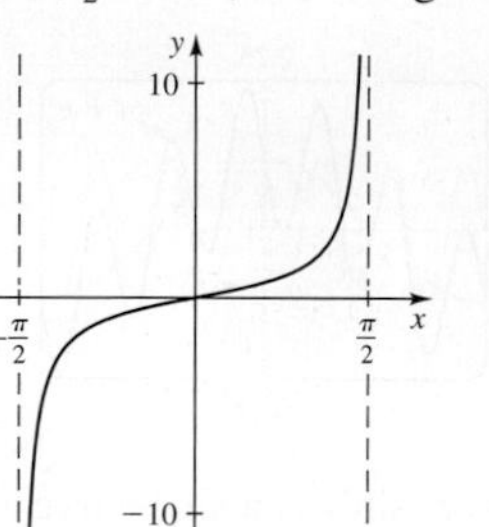

2. 2π; $n\pi$, n an integer

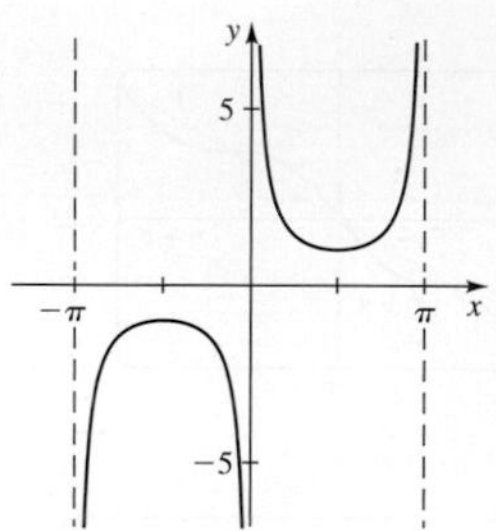

3. II **5.** VI **7.** IV

9. π

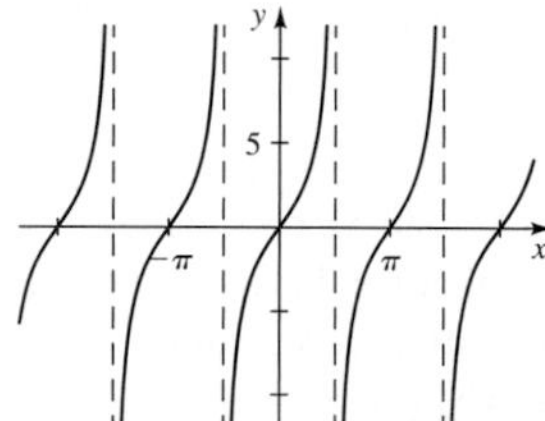

11. π

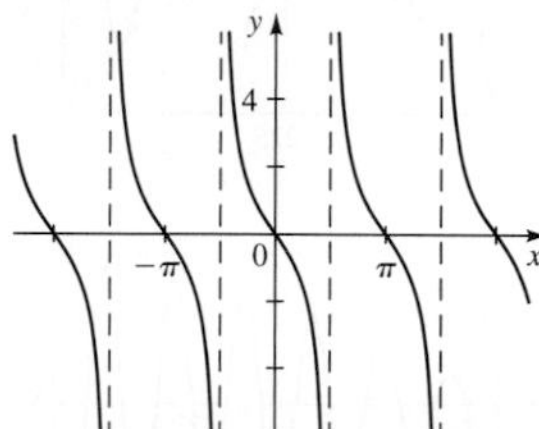

13. π

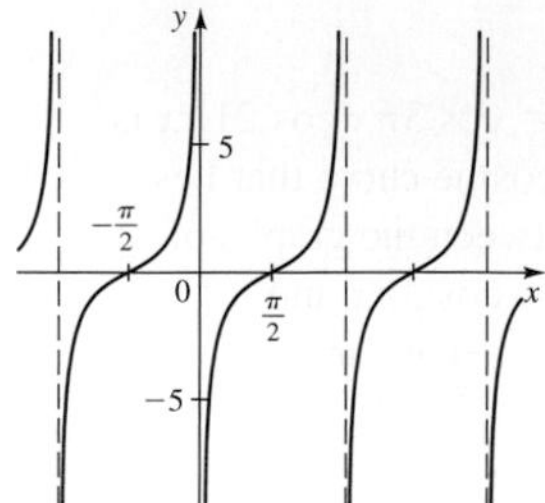

15. 2π

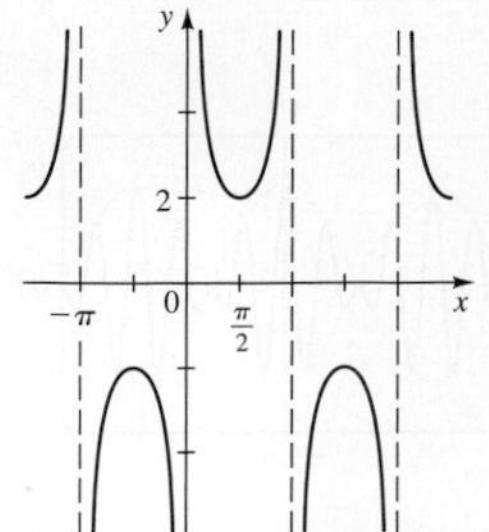

17. 2π

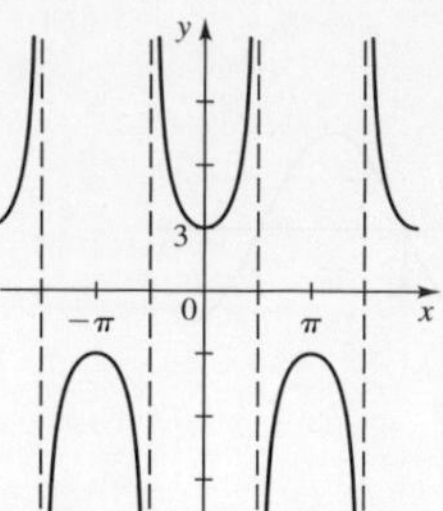

19. $\pi/3$

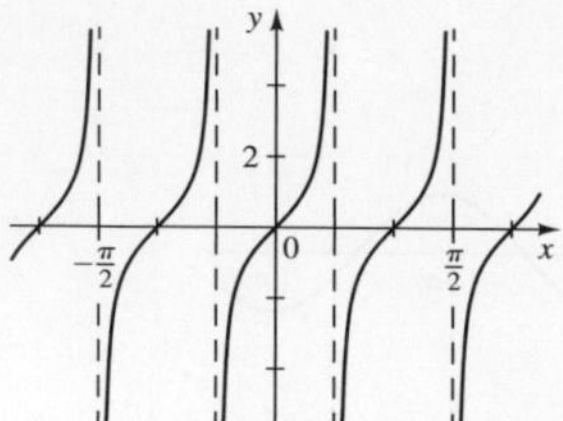

21. 1

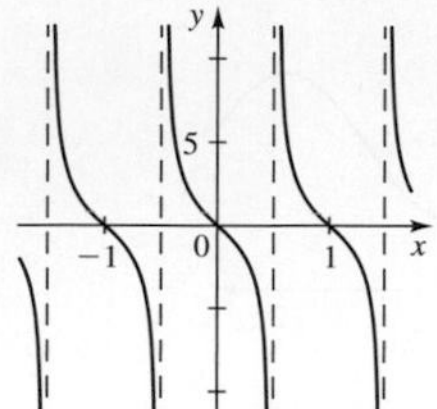

23. $\frac{1}{3}$

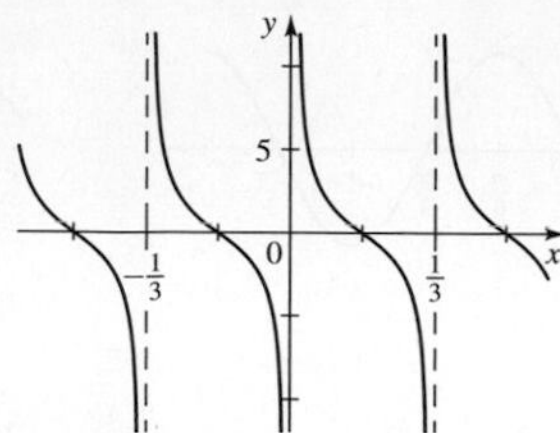

25. 4

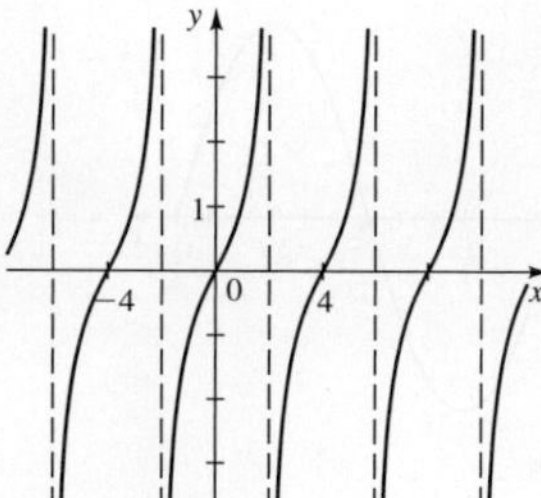

27. $\frac{1}{3}$

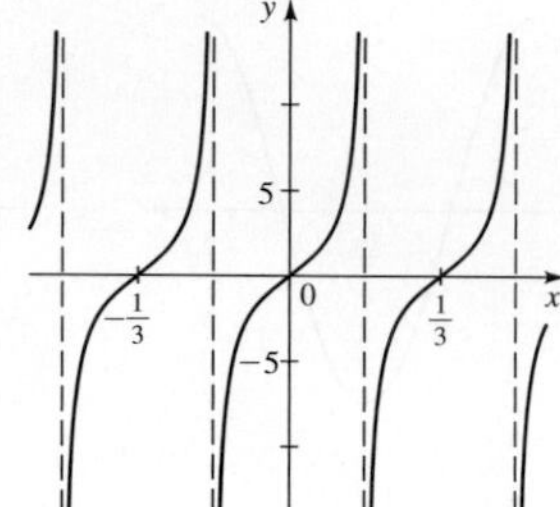

29. $\pi/2$

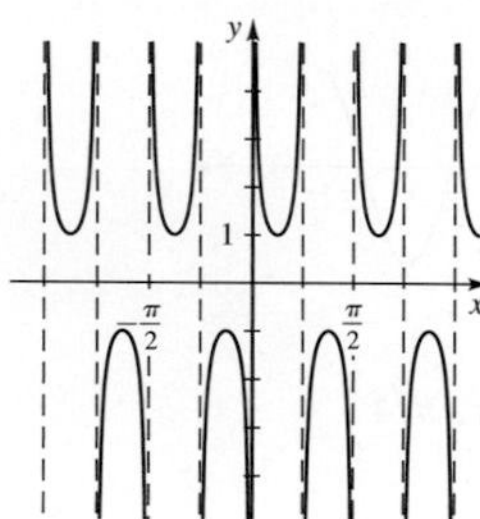

31. π

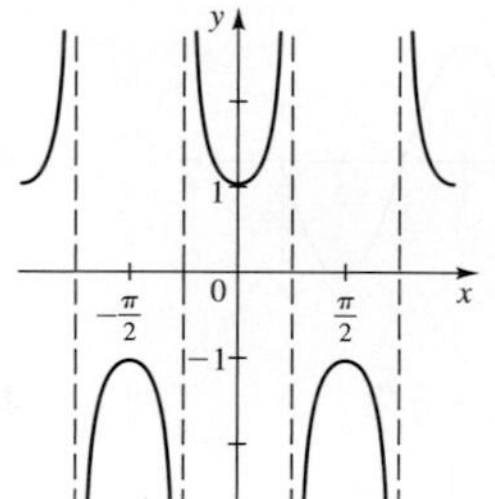

33. $\frac{4}{3}$

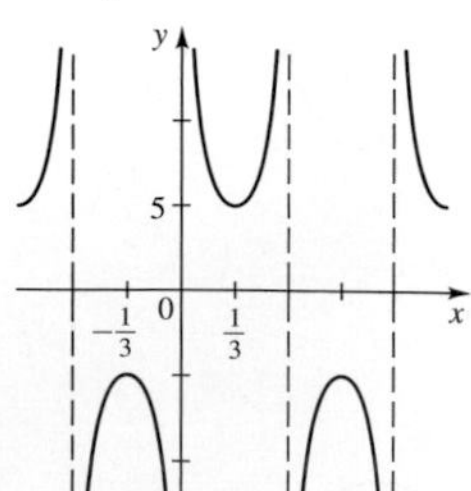

35. π

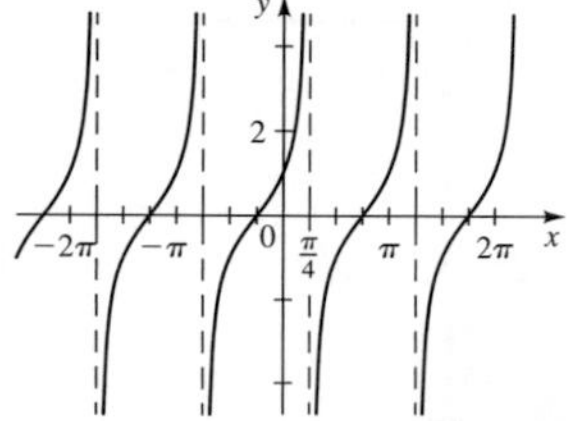

37. π

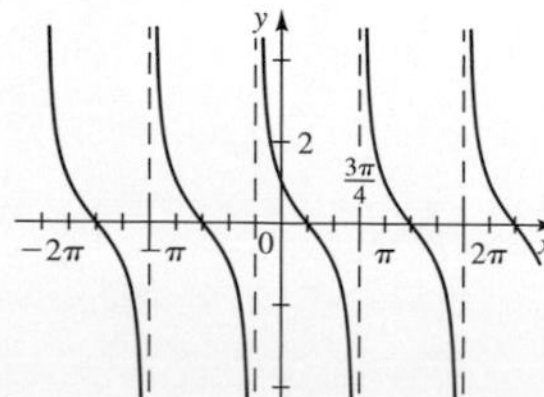

39. 2π

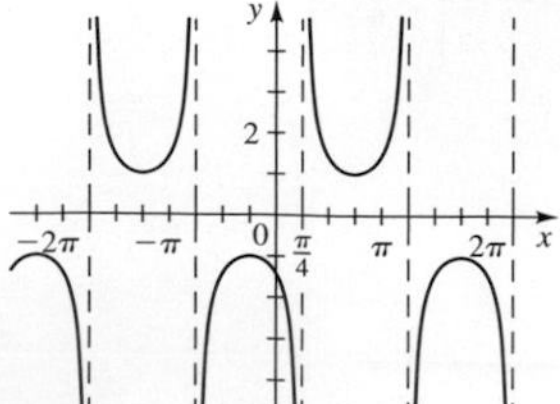

41. 2π

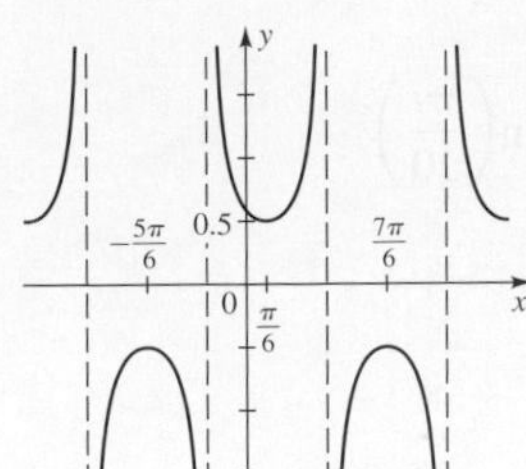

43. $\pi/2$

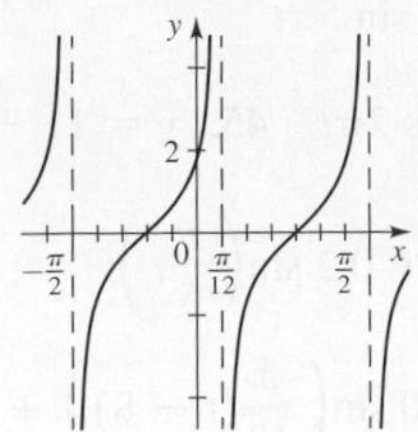

45. $\pi/3$

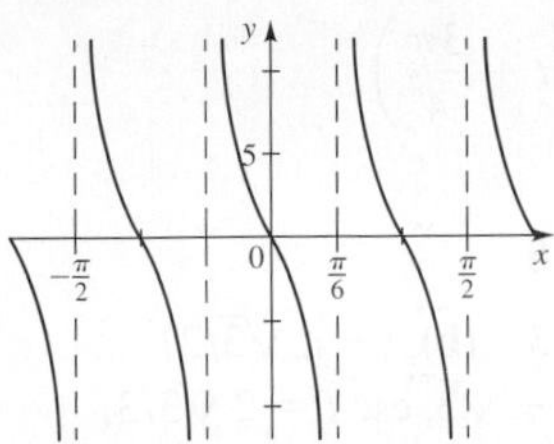

47. $\pi/2$

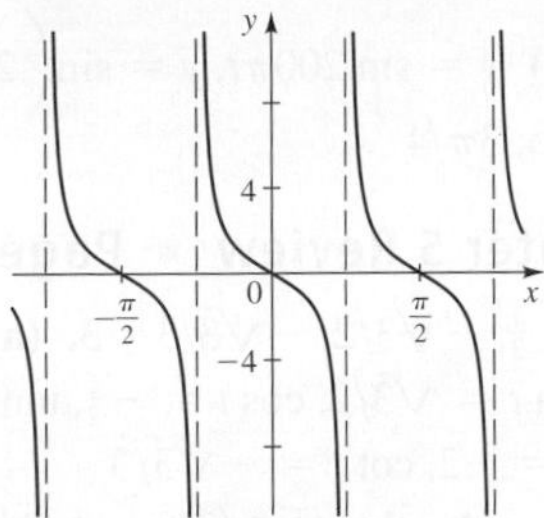

49. 2

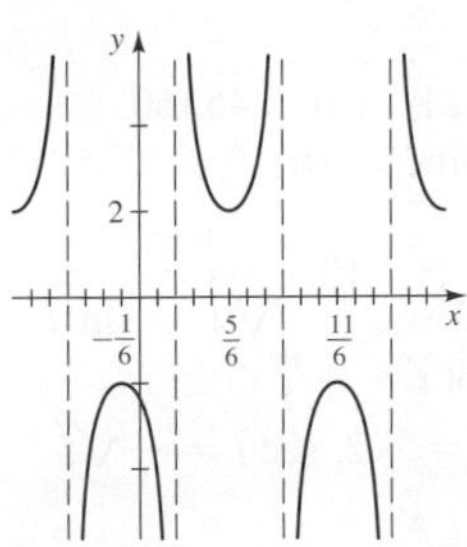

51. π

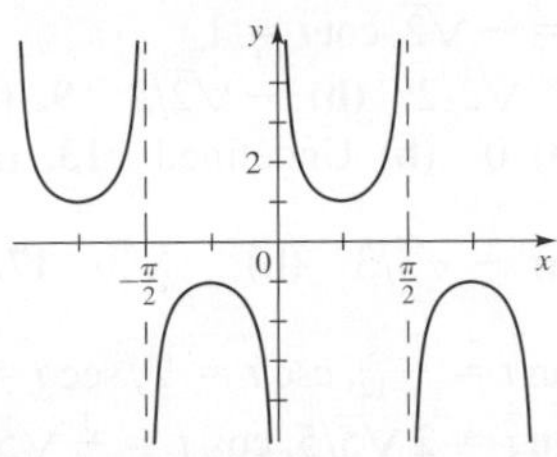

53. $2\pi/3$

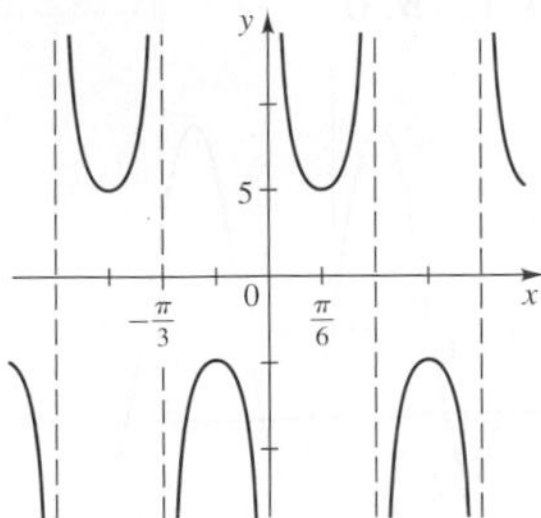

55. $3\pi/2$

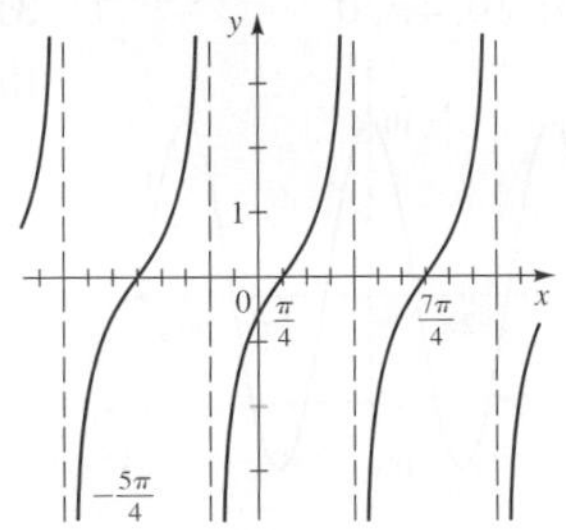

57. 2

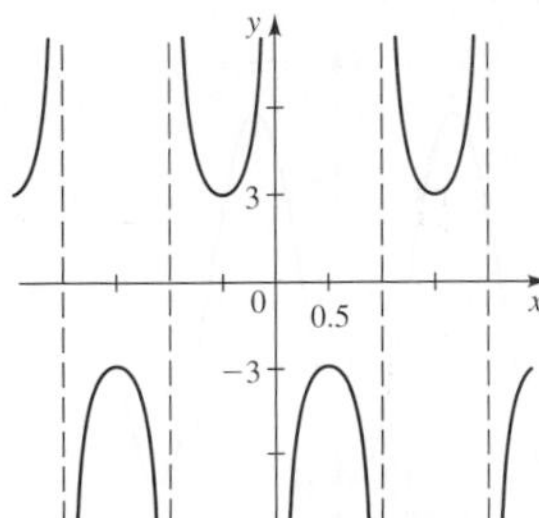

59. $\pi/2$

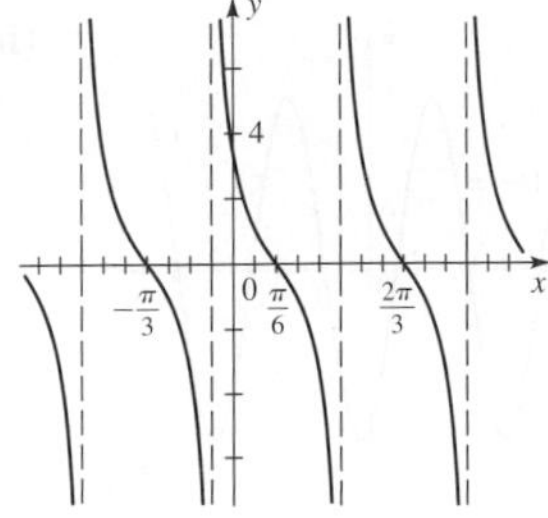

61.

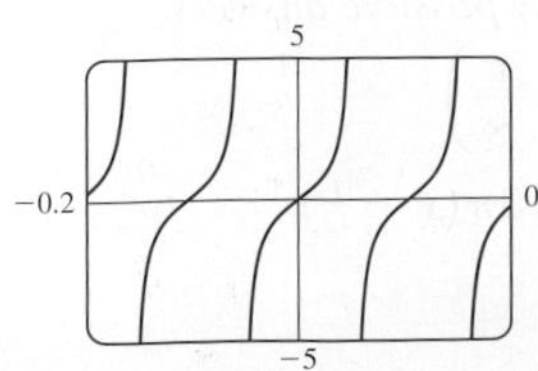

63.

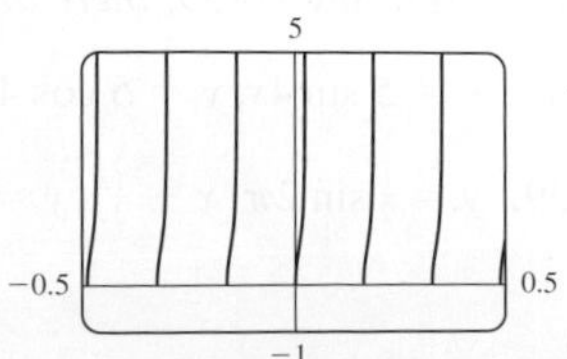

65. (a) 1.53 km, 3.00 km, 31.57 km
(b)

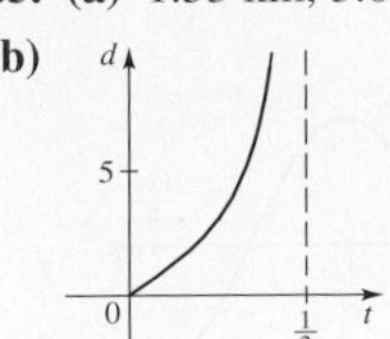

(c) $d(t)$ approaches ∞

Section 5.5 ▪ Page 456

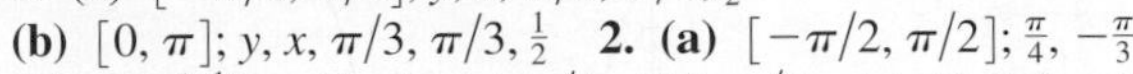

1. (a) $[-\pi/2, \pi/2]$, y, x, $\pi/6$, $\pi/6$, $\frac{1}{2}$
(b) $[0, \pi]$; y, x, $\pi/3$, $\pi/3$, $\frac{1}{2}$ **2. (a)** $[-\pi/2, \pi/2]$; $\frac{\pi}{4}$, $-\frac{\pi}{3}$
(b) $\sin^{-1}(\frac{1}{2}) = \frac{\pi}{6}$ **3. (a)** $\pi/2$ **(b)** $\pi/3$ **(c)** Undefined
5. (a) π **(b)** $\pi/3$ **(c)** $5\pi/6$ **7. (a)** $-\pi/4$ **(b)** $\pi/3$
(c) $\pi/6$ **9. (a)** $2\pi/3$ **(b)** $-\pi/4$ **(c)** $\pi/4$ **11.** 0.72973
13. 2.01371 **15.** 2.75876 **17.** 1.47113 **19.** 0.88998
21. −0.26005 **23.** $\frac{1}{4}$ **25.** 5 **27.** Undefined **29.** $-\frac{1}{5}$
31. $\pi/4$ **33.** $\pi/4$ **35.** $5\pi/6$ **37.** $5\pi/6$ **39.** $\pi/4$
41. $-\pi/3$ **43.** $\sqrt{3}/2$ **45.** 0 **47.** $2\sqrt{3}/3$ **49.** $\sqrt{2}$
51. $\sqrt{1 + x^2}$ **53.** $\dfrac{x}{\sqrt{1 - x^2}}$

For Exercises 55–61, there are many possible answers.
55. $g(x) = \arcsin x$, $f(x) = e^x$ **57.** $g(x) = 1/x$, $f(x) = \sin^{-1}x$
59. $h(x) = x^2$, $g(x) = \arcsin x$, $f(x) = e^x$
61. $h(x) = 1 - x^2$, $g(x) = e^x$, $f(x) = \tan^{-1}x$
63. (a) $\mathbb{R}$ **(b)**

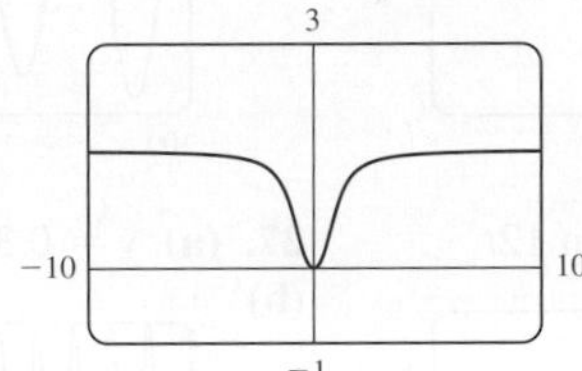

65. (a) $\mathbb{R}$ **(b)**

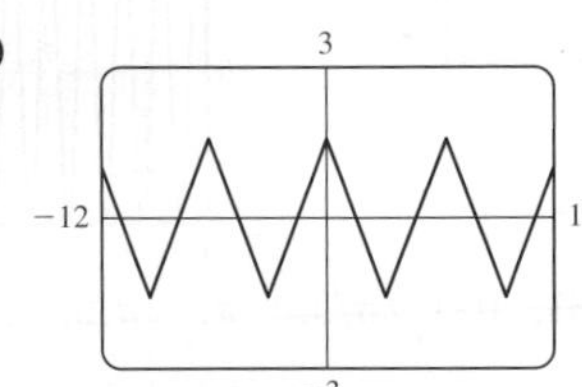

Section 5.6 ▪ Page 467

1. (a) $a \sin \omega t$ **(b)** $a \cos \omega t$
2. (a) $ae^{-ct} \sin \omega t$ **(b)** $ae^{-ct} \cos \omega t$
3. (a) $|A|$, $2\pi/k$, b; $A \sin k\left(t - \frac{b}{k}\right)$; b/k **(b)** 5, $\pi/2$, π, $\pi/4$
4. π, $\pi/2$; $\pi/2$, out of phase
5. (a) 2, $2\pi/3$, $3/(2\pi)$
(b)

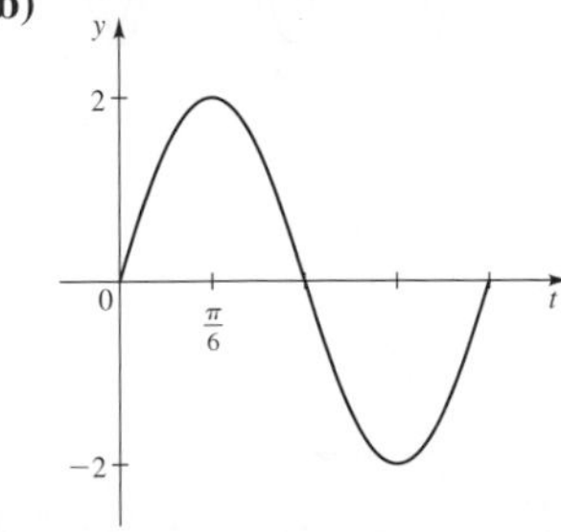

7. (a) 1, $20\pi/3$, $3/(20\pi)$
(b)

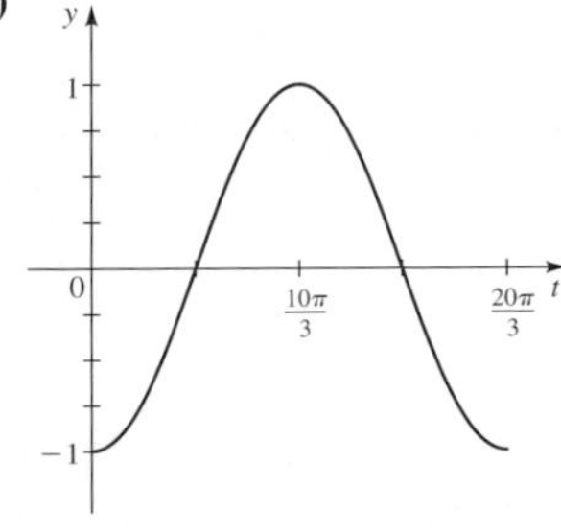

9. (a) $\frac{1}{4}, 4\pi/3, 3/(4\pi)$ **(b)**

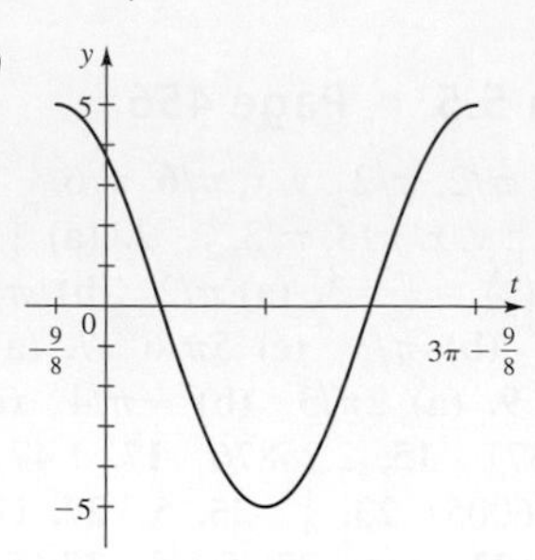

11. (a) $5, 3\pi, 1/(3\pi)$ **(b)**

13. $y = 10 \sin\left(\frac{2\pi}{3}t\right)$ **15.** $y = 6 \sin(10t)$

17. $y = 60 \cos(4\pi t)$ **19.** $y = 2.4 \cos(1500\pi t)$

21. (a) $y = 2e^{-1.5t} \cos 6\pi t$
(b)

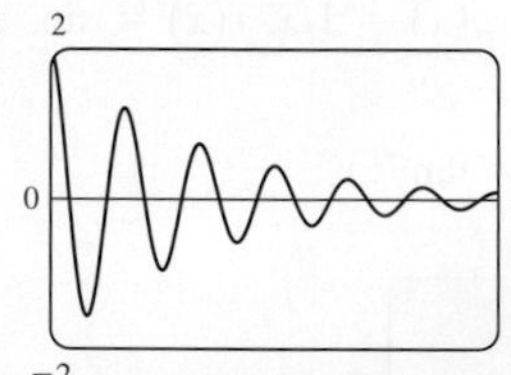

23. (a) $y = 100e^{-0.05t} \cos \frac{\pi}{2}t$
(b)

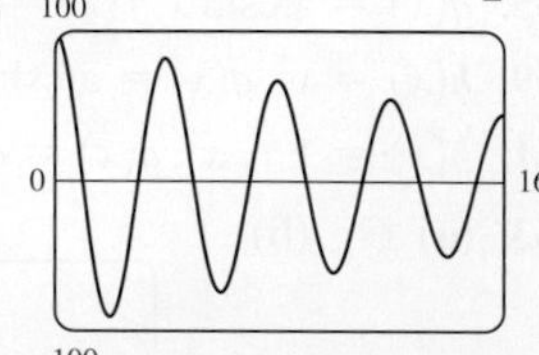

25. (a) $y = 7e^{-10t} \sin 12t$
(b)

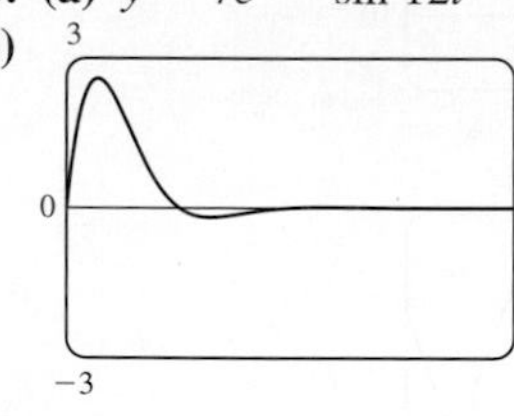

27. (a) $y = 0.3e^{-0.2t} \sin(40\pi t)$
(b)

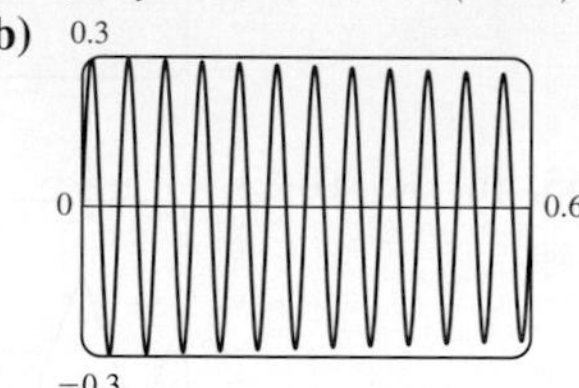

29. $5, \pi, \pi/2, \pi/4$ **31.** $100, 2\pi/5, -\pi, -\pi/5$

33. $20, \pi, \pi/2, \pi/4$

35. (a) $\pi/2, 5\pi/2$
(b) -2π
(c) In phase
(d)

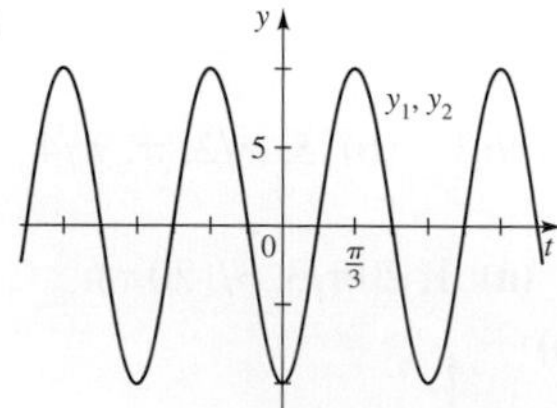

37. (a) $\pi/2, \pi/3$
(b) $\pi/6$
(c) Out of phase
(d)

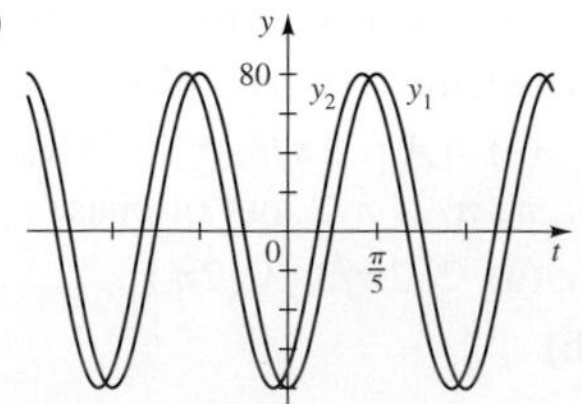

39. (a) $25, \frac{1}{80}, 80$ **(b)**

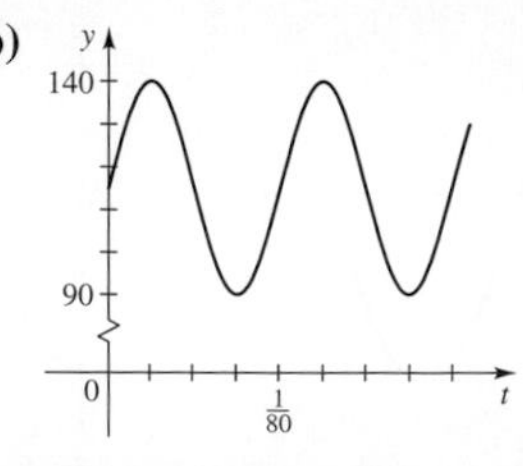

(c) The period decreases and the frequency increases.

41. $d(t) = 5 \sin 5\pi t$

43. $y = 5 \cos 2\pi t$ **45.** $y = 11 + 10 \sin\left(\frac{\pi t}{10}\right)$

47. $y = 3.8 + 0.2 \sin\left(\frac{\pi}{5}t\right)$

49. $f(t) = 10 \sin\left(\frac{\pi}{12}(t - 8)\right) + 90$

51. (a) 45 V **(b)** 40 **(c)** 40 **(d)** $E(t) = 45 \cos 80\pi t$

53. $f(t) = e^{-0.9t} \sin \pi t$ **55.** $c = \frac{1}{3} \ln 4 \approx 0.46$

57. (a) $y = \sin 200\pi t, y = \sin\left(200\pi t + \frac{3\pi}{4}\right)$
(b) No; $3\pi/4$

Chapter 5 Review ■ Page 475

1. (b) $\frac{1}{2}, -\sqrt{3}/2, -\sqrt{3}/3$ **3. (a)** $\pi/3$ **(b)** $\left(-\frac{1}{2}, \sqrt{3}/2\right)$
(c) $\sin t = \sqrt{3}/2, \cos t = -\frac{1}{2}, \tan t = -\sqrt{3}, \csc t = 2\sqrt{3}/3,$ $\sec t = -2, \cot t = -\sqrt{3}/3$

5. (a) $\pi/4$ **(b)** $(-\sqrt{2}/2, -\sqrt{2}/2)$
(c) $\sin t = -\sqrt{2}/2, \cos t = -\sqrt{2}/2, \tan t = 1, \csc t = -\sqrt{2},$ $\sec t = -\sqrt{2}, \cot t = 1$

7. (a) $\sqrt{2}/2$ **(b)** $-\sqrt{2}/2$ **9. (a)** 0.89121 **(b)** 0.45360

11. (a) 0 **(b)** Undefined **13. (a)** Undefined **(b)** 0

15. (a) $-\sqrt{3}/3$ **(b)** $-\sqrt{3}$ **17.** $\dfrac{\sin t}{1 - \sin^2 t}$ **19.** $\dfrac{\sin t}{\sqrt{1 - \sin^2 t}}$

21. $\tan t = -\frac{5}{12}, \csc t = \frac{13}{5}, \sec t = -\frac{13}{12}, \cot t = -\frac{12}{5}$

23. $\sin t = 2\sqrt{5}/5, \cos t = -\sqrt{5}/5, \tan t = -2, \sec t = -\sqrt{5}$

25. $-\dfrac{\sqrt{17}}{4} + 4$ **27.** 3

29. (a) $10, 4\pi, 0$
(b)

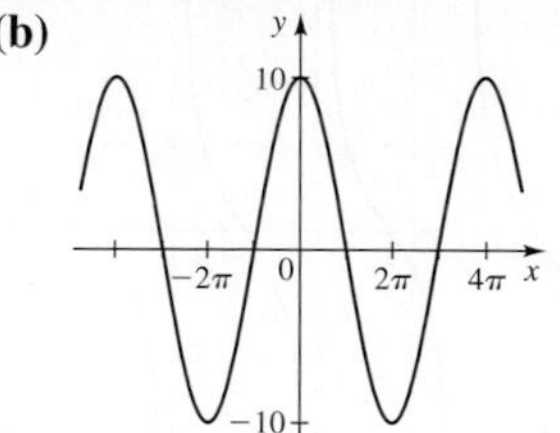

31. (a) $1, 4\pi, 0$
(b)

33. (a) $3, \pi, 1$
(b)

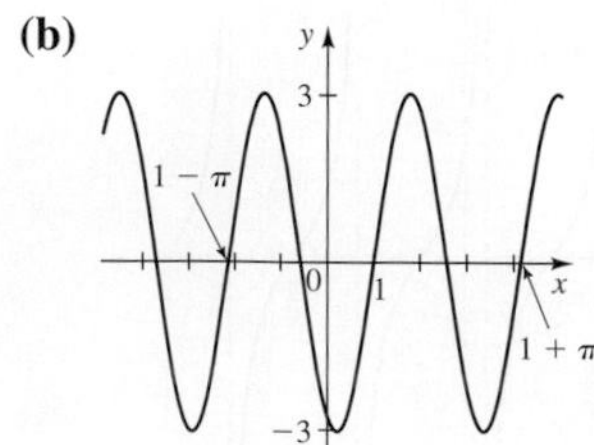

35. (a) $1, 4, -\frac{1}{3}$
(b)

For Exercises 37–39, there are many possible answers.

37. $y = 5 \sin 4x, y = 5 \cos 4\left(x - \frac{\pi}{8}\right)$

39. $y = \frac{1}{2} \sin 2\pi(x + \frac{1}{3}), y = \frac{1}{2} \cos 2\pi(x + \frac{1}{12})$

41. π

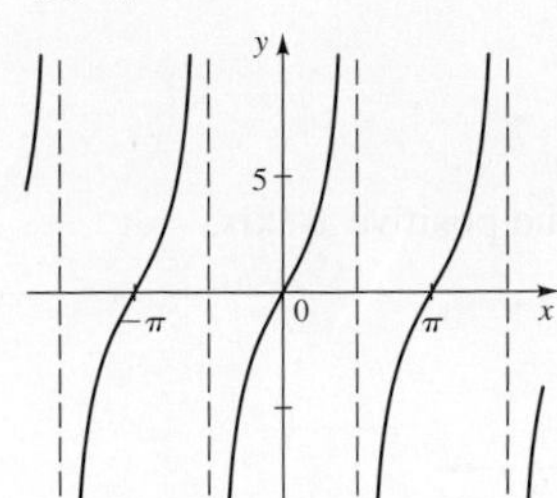

43. π

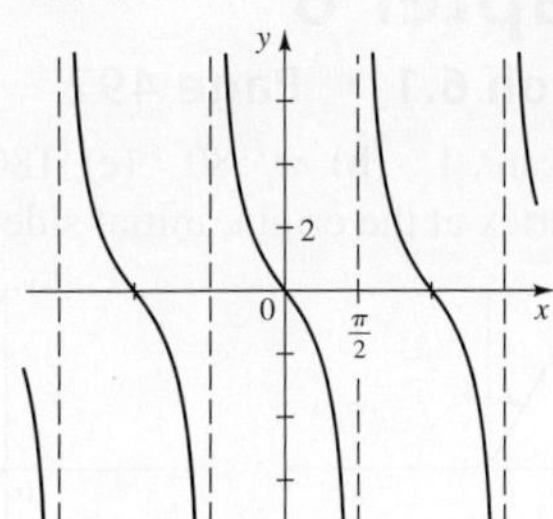

45. π

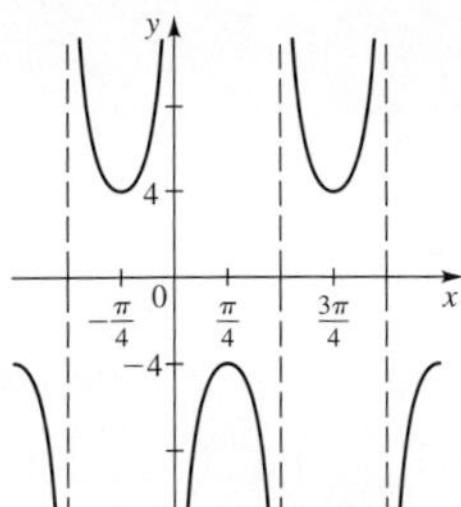

47. 2π

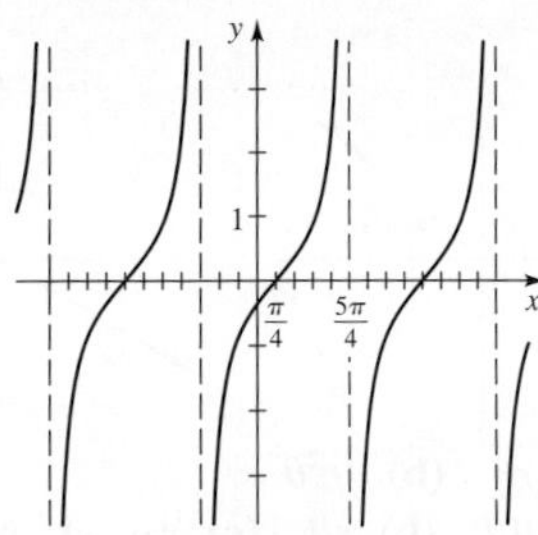

49. $\pi/2$ **51.** $\pi/6$ **53.** $100, \pi/4, -\pi/2, -\pi/16$

55. **(a)** $3\pi/2, 5\pi/2$ **(b)** $-\pi$ **(c)** Out of phase

(d)

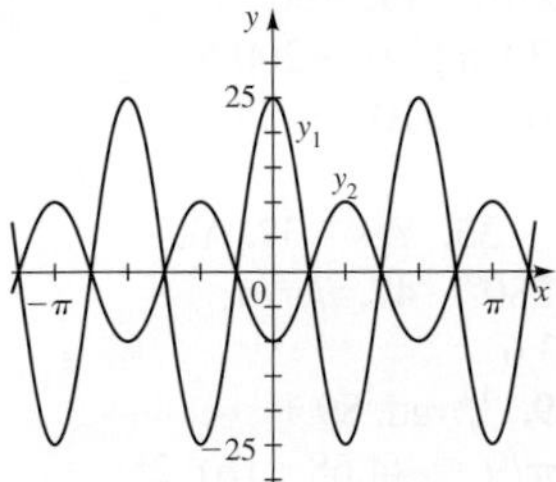

57. **(a)**

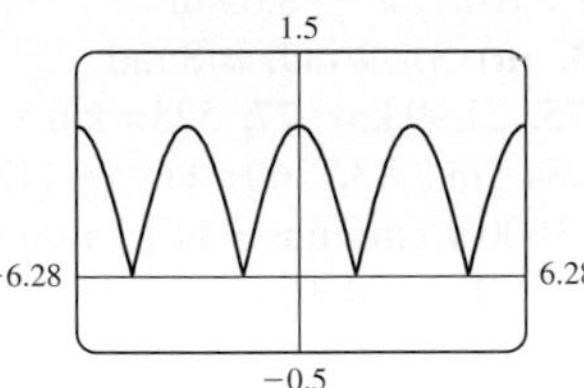

(b) Period π
(c) Even

59. **(a)**

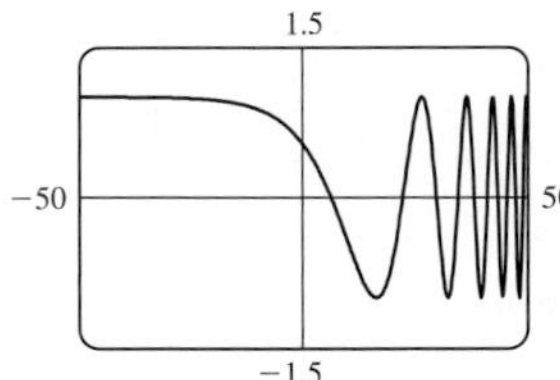

(b) Not periodic
(c) Neither

61. **(a)**

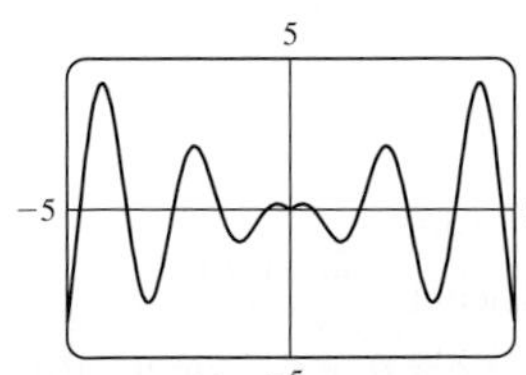

(b) Not periodic
(c) Even

63.

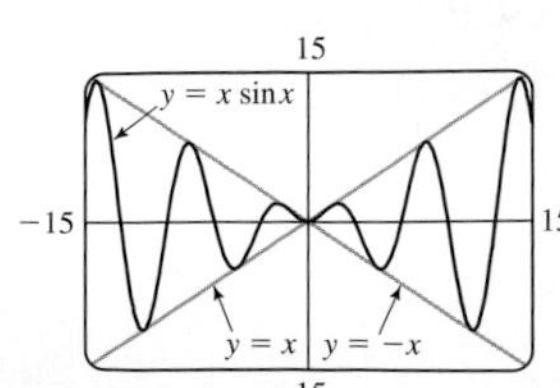

$y = x \sin x$ is a sine function whose graph lies between those of $y = x$ and $y = -x$

65.

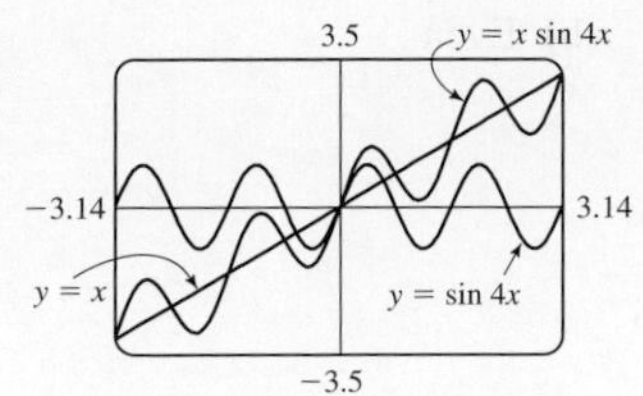

The graphs are related by graphical addition.

67. $1.76, -1.76$ **69.** $0.30, 2.84$

71. **(a)**

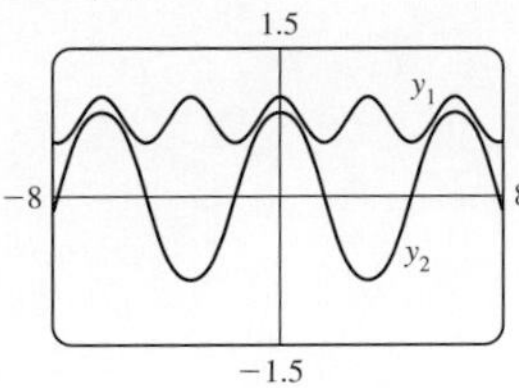

(b) y_1 has period π, y_2 has period 2π
(c) $\sin(\cos x) < \cos(\sin x)$, for all x

73. $y = -50 \cos 8\pi t$ **75.** **(a)** VII **(b)** I **(c)** V **(d)** III **(e)** VI **(f)** IV **(g)** II **(h)** VIII

Chapter 5 Test ▪ Page 478

1. $y = -\frac{5}{6}$ **2.** **(a)** $\frac{4}{5}$ **(b)** $-\frac{3}{5}$ **(c)** $-\frac{4}{3}$ **(d)** $-\frac{5}{3}$

3. **(a)** $-\frac{1}{2}$ **(b)** $-\sqrt{2}/2$ **(c)** $\sqrt{3}$ **(d)** -1

4. $\tan t = -\dfrac{\sin t}{\sqrt{1 - \sin^2 t}}$ **5.** $-\frac{2}{15}$

6. **(a)** $5, \pi/2, 0, 0$

(b)

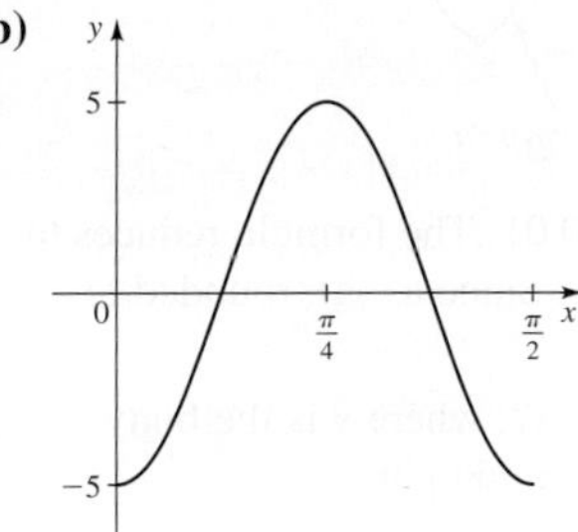

7. **(a)** $2, 4\pi, \pi/6, \pi/3$

(b)

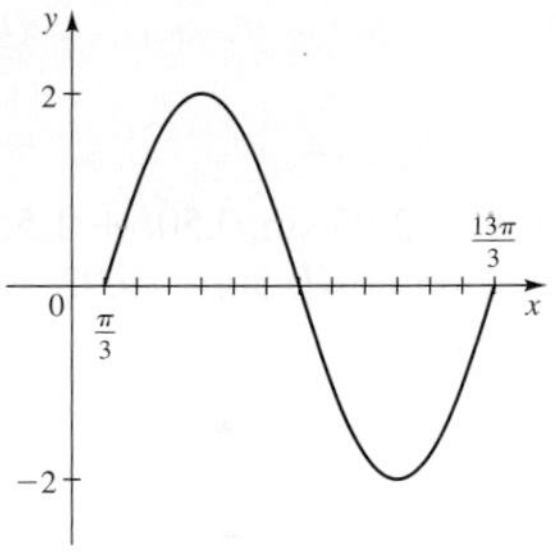

8. π

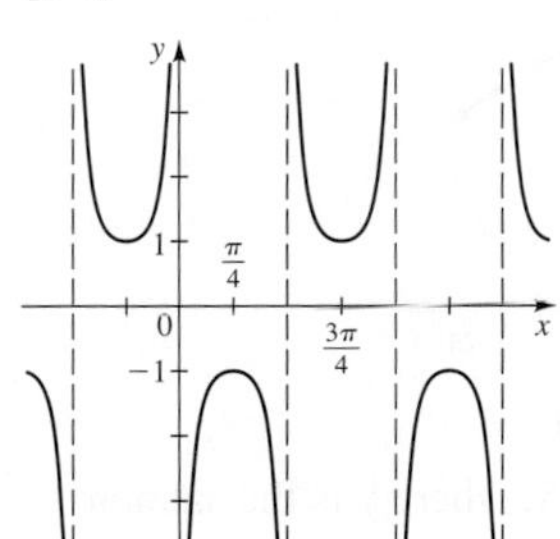

9. $\pi/2$

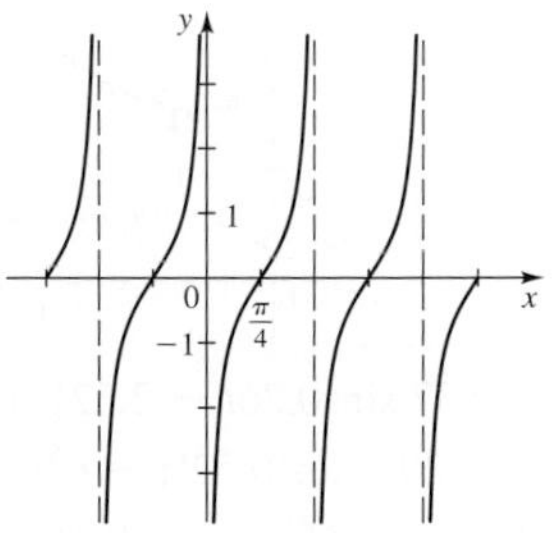

10. **(a)** $\pi/4$ **(b)** $5\pi/6$ **(c)** 0 **(d)** $\frac{1}{2}$

11. $y = 2 \sin 2(x + \pi/3)$

12. **(a)** $\pi/2, \pi/3$ **(b)** $\pi/6$ **(c)** Out of phase

(d)

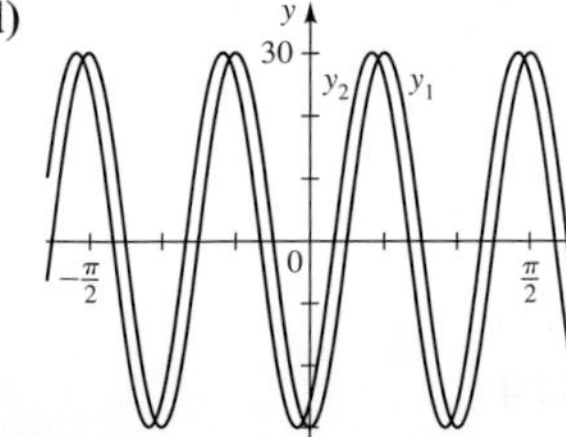

13. (a) **(b)** Even

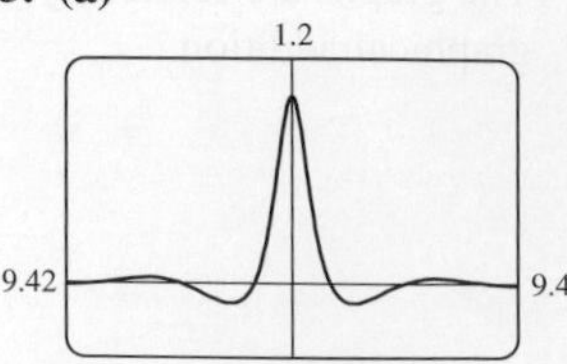

(c) Minimum value −0.11 when $x \approx \pm 2.54$, maximum value 1 when $x = 0$
14. $y = 5 \sin 4\pi t$
15. (a) $y = 16e^{-0.1t} \cos 24\pi t$
(b)

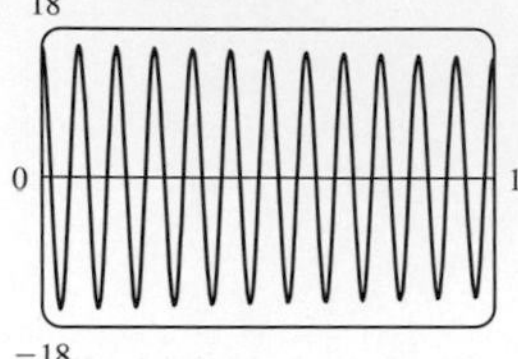

Focus on Modeling ▪ Page 482

1. (a) $y = 2.1 \cos 0.52t$

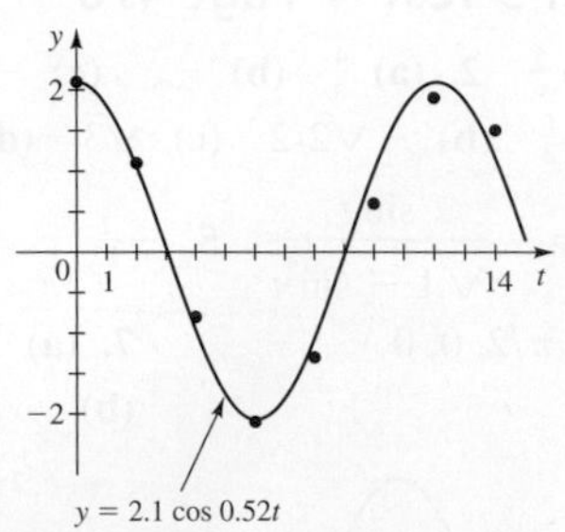

(b) $y = 2.05 \sin(0.50t + 1.55) - 0.01$. The formula reduces to $y = 2.05 \cos(0.50t - 0.02) - 0.01$. Same as (a), rounded to one decimal.
3. (a) $y = 0.4 \cos(0.26(t - 16)) + 37$, where y is the body temperature (°C) and t is hours since midnight

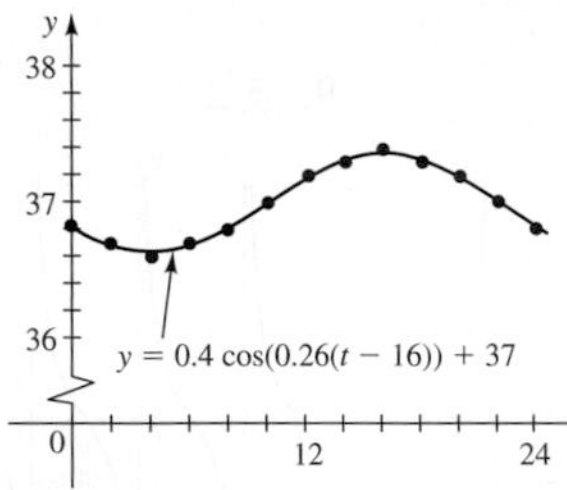

(b) $y = 0.37 \sin(0.26t - 2.62) + 37.0$
5. (a) $y = 20.5 \sin(0.52(t - 6)) + 42.5$, where y is the salmon population (× 1000), and t is years since 1985

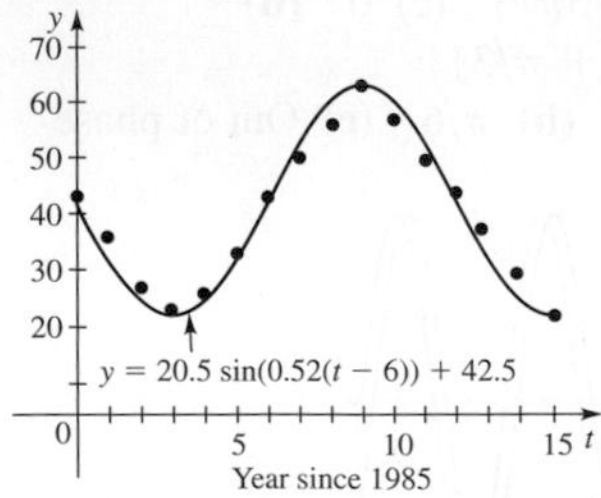

(b) $y = 17.8 \sin(0.52t + 3.11) + 42.4$

Chapter 6

Section 6.1 ▪ Page 492

1. (a) arc, 1 **(b)** $\pi/180$ **(c)** $180/\pi$
(d) Vertex at the origin, initial side on the positive x-axis.

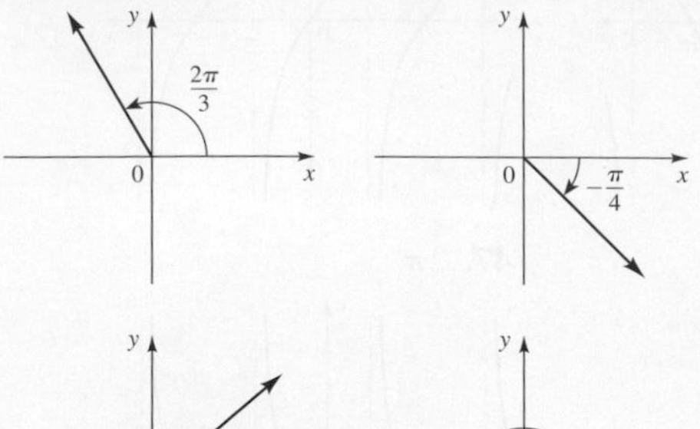

2. (a) $r\theta$ **(b)** $\frac{1}{2}r^2\theta$
3. (a) θ/t **(b)** s/t **(c)** $r\omega$ **4.** No, B **5.** $\pi/9 \approx 0.349$ rad
7. $3\pi/10 \approx 0.942$ rad **9.** $-\pi/4 \approx -0.785$ rad
11. $5\pi/9 \approx 1.745$ rad **13.** $50\pi/9 \approx 17.453$ rad
15. $-7\pi/18 \approx -1.222$ rad **17.** 210° **19.** 150°
21. $(540/\pi)° \approx 171.9°$ **23.** $(-630/\pi)° \approx -200.5°$
25. 18° **27.** −24° **29.** 410°, 770°, −310°, −670°
31. $11\pi/4, 19\pi/4, -5\pi/4, -13\pi/4$
33. $7\pi/4, 15\pi/4, -9\pi/4, -17\pi/4$ **35.** Yes **37.** Yes
39. Yes **41.** 40° **43.** 60° **45.** 280° **47.** $7\pi/6$
49. π **51.** $\pi/4$ **53.** $15\pi/2 \approx 23.6$
55. 2 rad ≈ 114.6° **57.** 8 cm **59.** $\frac{14}{9}$ rad, 89.1°
61. $18/\pi \approx 5.73$ m **63. (a)** $128\pi/9 \approx 44.68$ **(b)** 25
65. $24\pi \approx 75.4\text{ m}^2$ **67.** $9\sqrt{10\pi}/2\pi \approx 8.03$ m
69. $\frac{1}{2}$ rad **71.** $\pi/4\text{ m}^2$ **73. (a)** $3\pi/2$ rad, $\pi/8$ rad
(b) $23\pi/2$ rad, $23\pi/24$ rad **75.** 21.99 km **77.** 528π km ≈ 1660 km
79. 2.56 million km **81.** 1.84 km **83.** $360\pi\text{ cm}^2 \approx 1130.97\text{ cm}^2$
85. (a) 90π rad/min **(b)** 3600π cm/min ≈ 113.1 m/min
87. $16\pi/15$ m/s ≈ 3.35 m/s **89.** 1664.36 km/h **91.** 2.1 m/s
93. (a) 10π cm ≈ 31.4 cm **(b)** 5 cm **(c)** 3.32 cm
(d) 86.8 cm^3

Section 6.2 ▪ Page 501

1. (a)

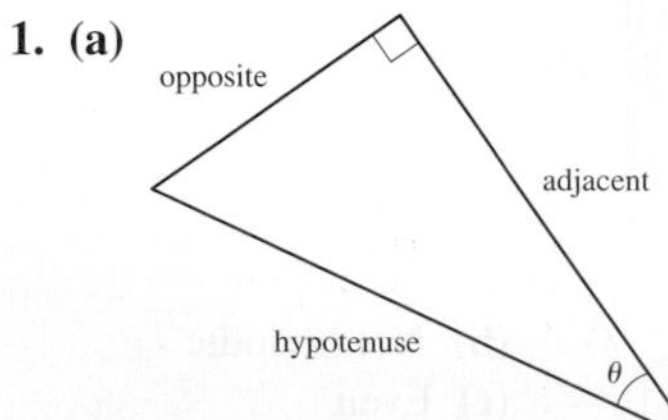

(b) $\dfrac{\text{opposite}}{\text{hypotenuse}}, \dfrac{\text{adjacent}}{\text{hypotenuse}}, \dfrac{\text{opposite}}{\text{adjacent}}$ **2.** similar
3. $\sin\theta, \cos\theta, \tan\theta$ **4. (a)** $x = r\cos\theta, y = r\sin\theta$ **(b)** $3\sqrt{3}, 3$
5. $\sin\theta = \frac{4}{5}, \cos\theta = \frac{3}{5}, \tan\theta = \frac{4}{3}, \csc\theta = \frac{5}{4}, \sec\theta = \frac{5}{3}, \cot\theta = \frac{3}{4}$
7. $\sin\theta = \frac{40}{41}, \cos\theta = \frac{9}{41}, \tan\theta = \frac{40}{9}, \csc\theta = \frac{41}{40}, \sec\theta = \frac{41}{9}$, $\cot\theta = \frac{9}{40}$

9. $\sin\theta = 2\sqrt{13}/13$, $\cos\theta = 3\sqrt{13}/13$, $\tan\theta = \frac{2}{3}$, $\csc\theta = \sqrt{13}/2$, $\sec\theta = \sqrt{13}/3$, $\cot\theta = \frac{3}{2}$
11. **(a)** $3\sqrt{34}/34$, $3\sqrt{34}/34$ **(b)** $\frac{3}{5}, \frac{3}{5}$ **(c)** $\sqrt{34}/5$, $\sqrt{34}/5$
13. **(a)** 0.37461 **(b)** 0.41421 **15.** **(a)** 1.85082 **(b)** 1.23490
17. $\frac{25}{2}$ **19.** $13\sqrt{3}/2$ **21.** 16.51658
23. $x = 28\cos\theta$, $y = 28\sin\theta$
25. $\sin\theta = 5\sqrt{61}/61$, $\cos\theta = 6\sqrt{61}/61$, $\csc\theta = \sqrt{61}/5$, $\sec\theta = \sqrt{61}/6$, $\cot\theta = \frac{6}{5}$

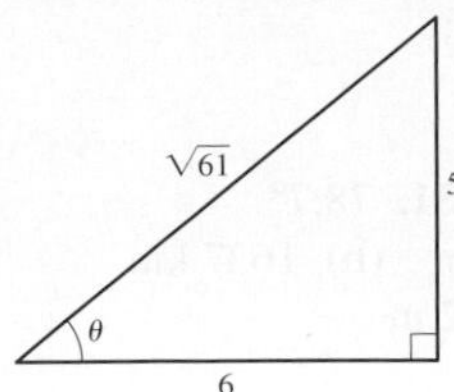

27. $\sin\theta = \sqrt{2}/2$, $\cos\theta = \sqrt{2}/2$, $\tan\theta = 1$, $\csc\theta = \sqrt{2}$, $\sec\theta = \sqrt{2}$

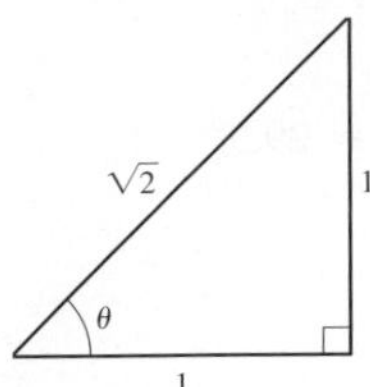

29. $\sin\theta = \frac{6}{11}$, $\cos\theta = \sqrt{85}/11$, $\tan\theta = 6\sqrt{85}/85$, $\sec\theta = 11\sqrt{85}/85$, $\cot\theta = \sqrt{85}/6$

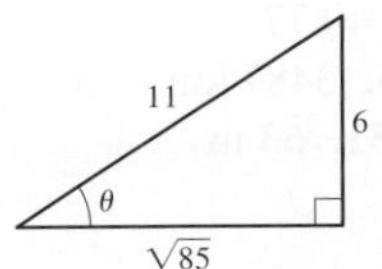

31. $(1 + \sqrt{3})/2$ **33.** 1 **35.** $\frac{1}{2}$ **37.** $\frac{3}{4} + (\sqrt{2}/2)$

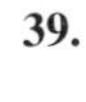

39.

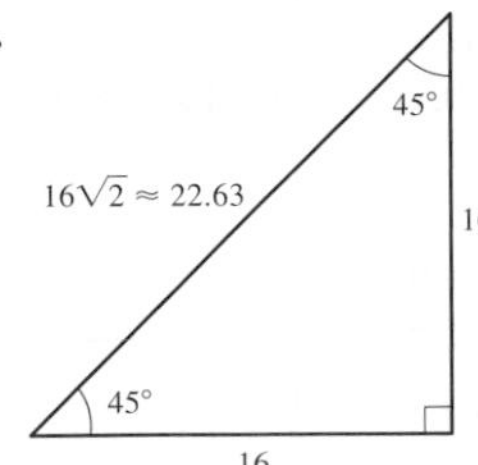

41.

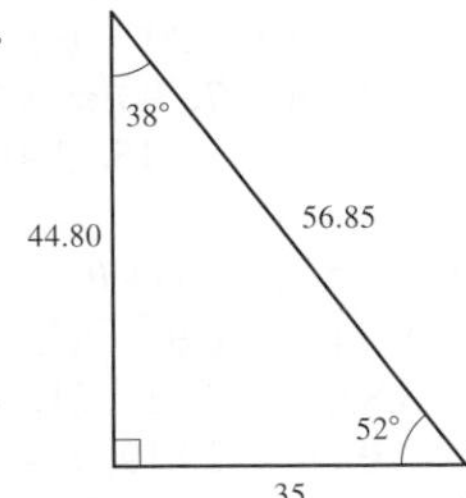

43.

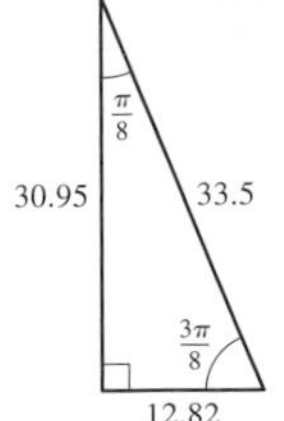

45.

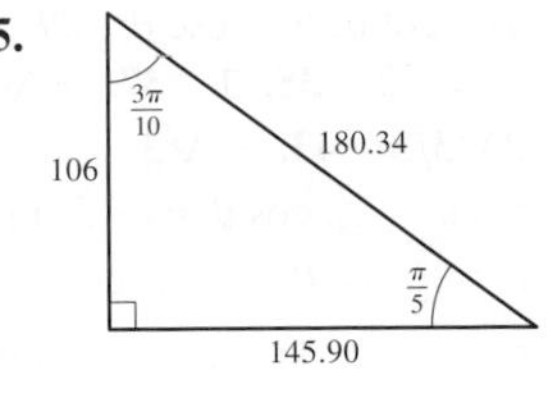

47. $\sin\theta \approx 0.44$, $\cos\theta \approx 0.89$, $\tan\theta = 0.50$, $\csc\theta \approx 2.25$, $\sec\theta \approx 1.125$, $\cot\theta = 2.00$ **49.** 230.9 **51.** 63.7
53. $x = 10\tan\theta\sin\theta$ **55.** 311 m
57. **(a)** 2100 mi **(b)** No **59.** 5.7 m **61.** 103 m
63. 125 m, 46 m **65.** 770 m **67.** 5808 m
69. 146.7 million km **71.** 6340 km **73.** 0.723 AU

Section 6.3 ■ Page 513

1. y/r, x/r, y/x **2.** quadrant; positive; negative; negative
3. **(a)** x-axis; 80°, 10° **(b)** 80°; 10° **4.** $\frac{1}{2}ab\sin\theta$; 7
5. **(a)** 45° **(b)** 15° **(c)** 60° **7.** **(a)** 70° **(b)** 55° **(c)** 80° **9.** **(a)** $3\pi/10$ **(b)** $\pi/8$ **(c)** $\pi/3$
11. **(a)** $2\pi/7$ **(b)** 0.4π **(c)** 1.4 **13.** $-\sqrt{3}/2$ **15.** -1
17. $-2\sqrt{3}/3$ **19.** $-\sqrt{2}/2$ **21.** 2 **23.** $\sqrt{3}$ **25.** -1
27. $-\sqrt{3}$ **29.** -2 **31.** 2 **33.** -1 **35.** Undefined
37. III **39.** IV **41.** $-\frac{1}{2}$ **43.** $-\sqrt{2}/2$ **45.** $\sqrt{13}/3$
47. $\sin\theta = \frac{4}{5}$, $\tan\theta = -\frac{4}{3}$, $\csc\theta = \frac{5}{4}$, $\sec\theta = -\frac{5}{3}$, $\cot\theta = -\frac{3}{4}$
49. $\sin\theta = -\sqrt{5}/5$, $\cos\theta = 2\sqrt{5}/5$, $\tan\theta = -\frac{1}{2}$, $\csc\theta = -\sqrt{5}$, $\sec\theta = \sqrt{5}/2$
51. $\cos\theta = -\sqrt{5}/3$, $\tan\theta = 2\sqrt{5}/5$, $\csc\theta = -\frac{3}{2}$, $\sec\theta = -3\sqrt{5}/5$, $\cot\theta = \sqrt{5}/2$
53. $\cos\theta = \sqrt{15}/4$, $\tan\theta = \sqrt{15}/15$, $\csc\theta = 4$, $\sec\theta = 4\sqrt{15}/15$, $\cot\theta = \sqrt{15}$
55. $\sin\theta = -3\sqrt{10}/10$, $\cos\theta = -\sqrt{10}/10$, $\csc\theta = -\sqrt{10}/3$, $\sec\theta = -\sqrt{10}$, $\cot\theta = \frac{1}{3}$
57. $\sin\theta = -\frac{1}{4}$, $\cos\theta = \sqrt{15}/4$, $\tan\theta = -\sqrt{15}/15$, $\sec\theta = 4\sqrt{15}/15$, $\cot\theta = -\sqrt{15}$
59. $\tan\theta = -\dfrac{\sqrt{1 - \cos^2\theta}}{\cos\theta}$
61. $\cos\theta = \sqrt{1 - \sin^2\theta}$ **63.** $\sec\theta = -\sqrt{1 + \tan^2\theta}$
65. $\sqrt{3}/2$, $\sqrt{3}$ **67.** 30.0 **69.** $25\sqrt{3} \approx 43.3$
71. 10.9 cm **73.** $(4\pi/3) - \sqrt{3} \approx 2.46$ **75.** $\sqrt{3} - \pi/2 \approx 0.16$
77. **(b)**

θ	20°	60°	80°	85°
h	1922	9145	29,944	60,351

79. **(a)** $A(\theta) = 400\sin\theta\cos\theta$
(b)

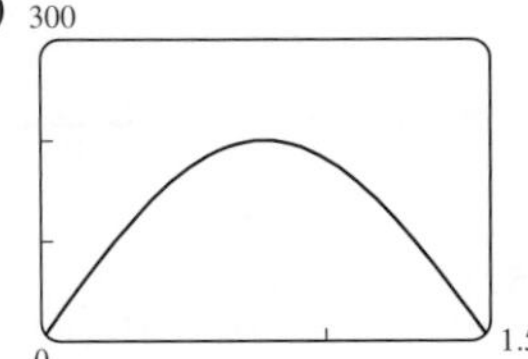

(c) Width = depth ≈ 14.14 cm
81. **(a)** 1.145 m, 0.1653 m
(b) 7.015 m, 1.013 m
83. **(a)**

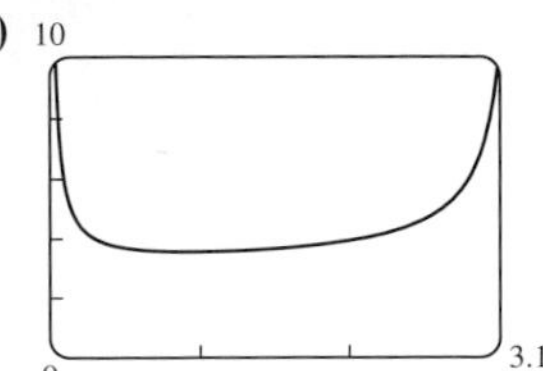

(b) 0.946 rad or 54°

Section 6.4 ▪ Page 521

1. one-to-one; domain, $[-\pi/2, \pi/2]$
2. (a) $[-1, 1], [-\pi/2, \pi/2]$ **(b)** $[-1, 1], [0, \pi]$ **(c)** $\mathbb{R}, (-\pi/2, \pi/2)$ **3. (a)** $\frac{8}{10}$ **(b)** $\frac{6}{10}$ **(c)** $\frac{8}{6}$
4. $\frac{5}{13}, \frac{12}{5}$

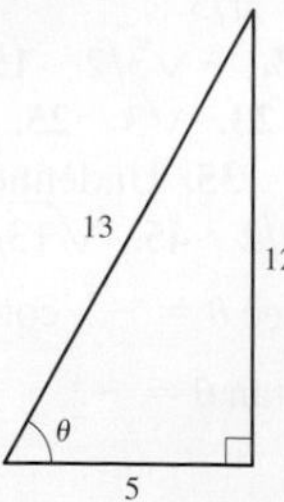

5. (a) $\pi/2$ **(b)** $\pi/2$ **(c)** $\pi/3$
7. (a) $-\pi/4$ **(b)** $3\pi/4$ **(c)** $-\pi/4$ **9.** 0.305, 17.458°
11. 1.231, 70.529° **13.** 1.249, 71.565° **15.** Undefined
17. 36.9° **19.** 34.7° **21.** 34.8° **23.** 36.9° **25.** −56.3°
27. 41.8°, 138.2° **29.** 113.6° **31.** 78.7° **33.** $\frac{3}{5}$ **35.** $\frac{13}{5}$
37. $-\frac{12}{13}$ **39.** $\sqrt{15}/4$ **41.** $\sqrt{1 - x^2}$ **43.** $\dfrac{1}{x}$ **45.** $\sqrt{x^2 + 1}$
47. 72.5°, 5.7 m **49. (a)** $h = 2 \tan \theta$ **(b)** $\theta = \tan^{-1}(h/2)$
51. (a) $\theta = \sin^{-1}(h/680)$ **(b)** $\theta = 47.3°$
53. (a) $\theta = \cos^{-1}\left(\dfrac{6374}{h + 6374}\right)$ **(b)** $s = 12{,}748\theta$
(c) $s = 12{,}748 \cos^{-1}\left(\dfrac{6374}{h + 6374}\right)$ **(d)** 2827 km **(e)** 313.7 km
55. 42°

Section 6.5 ▪ Page 529

1. $\dfrac{\sin A}{a} = \dfrac{\sin B}{b} = \dfrac{\sin C}{c}$ **2. (a)** ASA, SSA **(b)** SSA
3. $\dfrac{\sin 40°}{6} = \dfrac{\sin 110°}{x}, x \approx 8.8$
4. $\dfrac{\sin \theta}{6} = \dfrac{\sin 50°}{5}, \theta \approx 66.8°$
5. 318.8 **7.** 24.8 **9.** 43.9°
11. $\angle C = 114°, a \approx 51.2, b \approx 24.3$
13. $\angle A = 44°, \angle B = 68°, a \approx 8.99$
15. $\angle C = 62°, a \approx 199.5, b \approx 241.5$

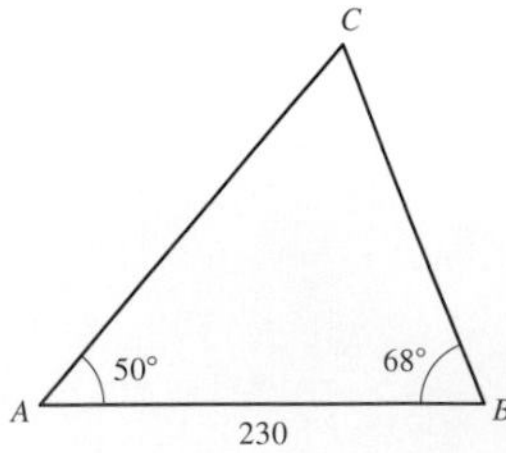

17. $\angle B = 85°, a \approx 5.0, c \approx 9.1$

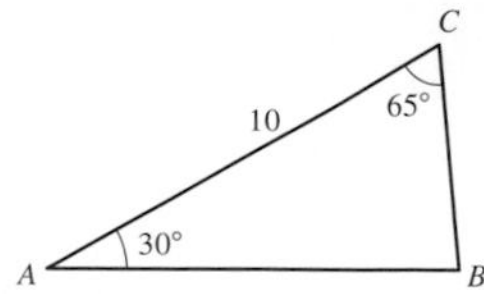

19. $\angle A = 100°, a \approx 89.4, c \approx 70.5$

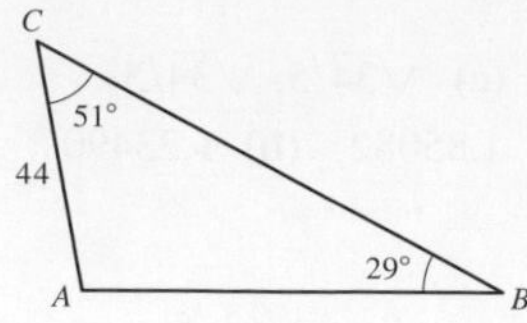

21. $\angle B \approx 30.2°, \angle C \approx 39.8°, c \approx 19.1$ **23.** No solution
25. $\angle A_1 \approx 124.5°, \angle C_1 \approx 30.5°, a_1 \approx 48.7;$
$\angle A_2 \approx 5.5°, \angle C_2 \approx 149.5°, a_2 \approx 5.6$
27. No solution
29. $\angle A \approx 122.8°, \angle B \approx 28.2°, b \approx 14.6$ **31.** 78.7°
33. (a) 91.1° **(b)** 14.4° **35. (a)** 1629 km **(b)** 1627 km
37. 219 m **39.** 56 m **41.** 53 m **43.** 192 m
45. 0.427 AU, 1.119 AU

Section 6.6 ▪ Page 536

1. $a^2 + b^2 - 2ab \cos C$ **2.** SSS, SAS
3. $x^2 = 3^2 + 4^2 - 2 \cdot 3 \cdot 4 \cdot \cos 35°, x \approx 2.3$
4. $3^2 = 6^2 + 5^2 - 2 \cdot 6 \cdot 5 \cdot \cos \theta, \cos \theta = \frac{13}{15}, \theta \approx 29.9°$
5. 28.9 **7.** 47 **9.** 29.89° **11.** 15.1
13. $\angle A \approx 39.4°, \angle B \approx 20.6°, c \approx 24.6$
15. $\angle A \approx 47.5°, \angle B \approx 79.5°, c \approx 3.2$
17. $\angle A \approx 49.9°, \angle B \approx 72.9°, \angle C \approx 57.2°$
19. $\angle A_1 \approx 83.6°, \angle C_1 \approx 56.4°, a_1 \approx 193.2;$
$\angle A_2 \approx 16.4°, \angle C_2 \approx 123.6, a_2 \approx 55.0$
21. No such triangle **23.** 2.0 **25.** 25.4 **27.** 89.2°
29. 24.3 **31.** 54 **33.** 26.83 **35.** 5.33 **37.** 40.77
39. 3.85 cm^2 **41.** 2.30 km **43.** 23.1 km **45.** 3487 km
47. (a) 62.6 km **(b)** S 18.2° E **49.** 95.7° **51.** 63 m
53. 1151 m **55.** $165,554

Chapter 6 Review ▪ Page 543

1. (a) $\pi/6$ **(b)** $5\pi/6$ **(c)** $-\pi/9$ **(d)** $-5\pi/4$
3. (a) 150° **(b)** −20° **(c)** −240° **(d)** 229.2°
5. $4\pi \approx 12.6$ m **7.** $90/\pi \approx 28.6$ m **9.** 21,609 **11.** 25 m^2
13. 0.4 rad ≈ 22.9° **15.** 300π rad/min ≈ 942.5 rad/min, 188.5 rad/min
17. $\sin \theta = 5/\sqrt{74}, \cos \theta = 7/\sqrt{74}, \tan \theta = \frac{5}{7}, \csc \theta = \sqrt{74}/5, \sec \theta = \sqrt{74}/7, \cot \theta = \frac{7}{5}$
19. $x \approx 3.83, y \approx 3.21$ **21.** $x \approx 2.92, y \approx 3.11$
23. $A = 70°, a \approx 2.819, b \approx 1.026$
25. $A \approx 16.3°, C \approx 73.7°, c = 24$
27. $a = \cot \theta, b = \csc \theta$ **29.** 48 m **31.** 1722 km
33. $-\sqrt{2}/2$ **35.** 1 **37.** $-\sqrt{3}/3$ **39.** $-\sqrt{2}/2$
41. $2\sqrt{3}/3$ **43.** $-\sqrt{3}$
45. $\sin \theta = \frac{12}{13}, \cos \theta = -\frac{5}{13}, \tan \theta = -\frac{12}{5}, \csc \theta = \frac{13}{12}, \sec \theta = -\frac{13}{5}, \cot \theta = -\frac{5}{12}$ **47.** 60°
49. $\tan \theta = \sqrt{1 - \cos^2\theta}/\cos \theta$
51. $\tan^2\theta = \sin^2\theta/(1 - \sin^2\theta)$
53. $\sin \theta = \sqrt{7}/4, \cos \theta = \frac{3}{4}, \csc \theta = 4\sqrt{7}/7, \cot \theta = 3\sqrt{7}/7$
55. $\cos \theta = -\frac{4}{5}, \tan \theta = -\frac{3}{4}, \csc \theta = \frac{5}{3}, \sec \theta = -\frac{5}{4}, \cot \theta = -\frac{4}{3}$ **57.** $-\sqrt{5}/5$ **59.** 1 **61.** $\pi/3$ **63.** $2/\sqrt{21}$
65. $x/\sqrt{1 + x^2}$ **67.** $\theta = \cos^{-1}(x/3)$ **69.** 5.32 **71.** 148.07
73. 9.17 **75.** 54.1° **77.** 80.4° **79.** 77.3 km **81.** 3.9 km
83. 32.12 **85. (a)** VII **(b)** III **(c)** I **(d)** VI **(e)** V **(f)** VIII **(g)** II **(h)** IV

Chapter 6 Test ■ Page 547

1. $11\pi/6, -3\pi/4$ **2.** 240°, −74.5° **3. (a)** 240π rad/min ≈ 753.98 rad/min **(b)** 3769.9 m/min = 226 km/h **4. (a)** $\sqrt{2}/2$ **(b)** $\sqrt{3}/3$ **(c)** 2 **(d)** 1 **5.** $(26 + 6\sqrt{13})/39$ **6.** $a = 24 \sin\theta, b = 24\cos\theta$ **7.** $(4 - 3\sqrt{2})/4$ **8.** $-\frac{13}{12}$ **9.** $\tan\theta = -\sqrt{\sec^2\theta - 1}$ **10.** 5.9 m **11. (a)** $\theta = \tan^{-1}(x/4)$ **(b)** $\theta = \cos^{-1}(3/x)$ **12.** $\frac{40}{41}$ **13.** 9.1 **14.** 250.5 **15.** 8.4 **16.** 19.5 **17.** 78.6° **18.** 40.2° **19. (a)** 15.3 m^2 **(b)** 24.3 m **20. (a)** 129.9° **(b)** 44.9 **21.** 166 m

Focus on Modeling ■ Page 550

1. 1.41 km **3.** 14.3 m **5. (b)** 2350 m **7.** 4194 m

Chapter 7

Section 7.1 ■ Page 558

1. all; 1 **2.** $\cos(-x) = \cos x$ **3.** $\sin t$ **5.** $\tan\theta$ **7.** −1 **9.** $\cos y$ **11.** $\csc u$ **13.** $\tan\theta$ **15.** 1 **17.** $\cos t + 1$ **19.** $\cos x$ **21.** $\sin^2 x$ **23.** $\cos y$ **25.** $2\sec u$ **27.** $1 - \sin x$ **29.** $2\sec^2\alpha$

31. (a) $\text{LHS} = \dfrac{1 - \sin^2 x}{\sin x} = \text{RHS}$

33. $\text{LHS} = \cos\alpha \dfrac{1}{\dfrac{1}{\cos\alpha}} = \cos\alpha\cos\alpha = \text{RHS}$

35. $\text{LHS} = \cos u \cdot \dfrac{1}{\cos u} \cdot \cot u = \text{RHS}$

37. $\text{LHS} = \cos^2\left(\dfrac{\pi}{2} - y\right)\dfrac{1}{\sin y} = \sin^2 y \cdot \dfrac{1}{\sin y} = \text{RHS}$

39. $\text{LHS} = \sin^2 x + 2\sin x\cos x + \cos^2 x = \text{RHS}$

41. $\text{LHS} = \cos x - (-\sin x) = \text{RHS}$

43. $\text{LHS} = \dfrac{\sec A - 1}{\sec A + 1} \cdot \dfrac{\cos A}{\cos A} = \dfrac{1 - \cos A}{1 + \cos A} = \text{RHS}$

45. $\text{LHS} = 1 - \cos^2\beta = \sin^2\beta = \text{RHS}$

47. $\text{LHS} = \dfrac{1}{\cos^2 y} = \sec^2 y = \text{RHS}$

49. $\text{LHS} = \tan^2 x + 2\tan x\cot x + \cot^2 x = \tan^2 x + 2 + \cot^2 x$
$= (\tan^2 x + 1) + (\cot^2 x + 1) = \text{RHS}$

51. $\text{LHS} = (2\cos^2 t)^2 + 4\sin^2 t\cos^2 t$
$= 4\cos^2 t \cdot (\cos^2 t + \sin^2 t) = \text{RHS}$

53. $\text{LHS} = \dfrac{\cos^2 x}{\sin x} + \dfrac{\sin^2 x}{\sin x} = \dfrac{1}{\sin x} = \text{RHS}$

55. $\text{LHS} = \dfrac{1/(\cos t) - \cos t}{1/(\cos t)} \cdot \dfrac{\cos t}{\cos t} = \dfrac{1 - \cos^2 t}{1} = \text{RHS}$

57. $\text{LHS} = \cos^2 x - (1 - \cos^2 x) = 2\cos^2 x - 1 = \text{RHS}$

59. $\text{LHS} = (\sin^2\theta)^2 - (\cos^2\theta)^2$
$= (\sin^2\theta - \cos^2\theta)(\sin^2\theta + \cos^2\theta) = \text{RHS}$

61. $\text{LHS} = \dfrac{\sin^2 t + 2\sin t\cos t + \cos^2 t}{\sin t\cos t}$
$= \dfrac{\sin^2 t + \cos^2 t}{\sin t\cos t} + \dfrac{2\sin t\cos t}{\sin t\cos t} = \dfrac{1}{\sin t\cos t} + 2$
$= \text{RHS}$

63. $\text{LHS} = \dfrac{1 + \dfrac{\sin^2 u}{\cos^2 u}}{1 - \dfrac{\sin^2 u}{\cos^2 u}} \cdot \dfrac{\cos^2 u}{\cos^2 u} = \dfrac{\cos^2 u + \sin^2 u}{\cos^2 u - \sin^2 u} = \text{RHS}$

65. $\text{LHS} = \dfrac{\frac{1}{\cos x} + \frac{1}{\sin x}}{\frac{\sin x}{\cos x} + \frac{\cos x}{\sin x}} \cdot \dfrac{\sin x\cos x}{\sin x\cos x} = \dfrac{\sin x + \cos x}{\sin^2 x + \cos^2 x} = \text{RHS}$

67. $\text{LHS} = \dfrac{1 - \cos x}{\sin x} \cdot \dfrac{1 - \cos x}{1 - \cos x} + \dfrac{\sin x}{1 - \cos x} \cdot \dfrac{\sin x}{\sin x}$
$= \dfrac{1 - 2\cos x + \cos^2 x + \sin^2 x}{\sin x\,(1 - \cos x)} = \dfrac{2 - 2\cos x}{\sin x\,(1 - \cos x)}$
$= \dfrac{2(1 - \cos x)}{\sin x\,(1 - \cos x)} = \text{RHS}$

69. $\text{LHS} = \dfrac{\sin^2 u}{\cos^2 u} - \dfrac{\sin^2 u\cos^2 u}{\cos^2 u} = \dfrac{\sin^2 u}{\cos^2 u} \cdot (1 - \cos^2 u) = \text{RHS}$

71. $\text{LHS} = \dfrac{1 + \frac{\sin x}{\cos x}}{1 - \frac{\sin x}{\cos x}} \cdot \dfrac{\cos x}{\cos x} = \dfrac{\cos x + \sin x}{\cos x - \sin x} = \text{RHS}$

73. $\text{LHS} = \dfrac{\sec x - \tan x + \sec x + \tan x}{(\sec x + \tan x)(\sec x - \tan x)}$
$= \dfrac{2\sec x}{\sec^2 x - \tan^2 x} = \text{RHS}$

75. $\text{LHS} = \dfrac{(1 + \sin x)^2 - (1 - \sin x)^2}{(1 - \sin x)(1 + \sin x)}$
$= \dfrac{1 + 2\sin x + \sin^2 x - 1 + 2\sin x - \sin^2 x}{1 - \sin^2 x}$
$= \dfrac{4\sin x}{\cos^2 x} = 4\dfrac{\sin x}{\cos x} \cdot \dfrac{1}{\cos x} = \text{RHS}$

77. $\text{LHS} = \dfrac{(\sin x + \cos x)(\sin^2 x - \sin x\cos x + \cos^2 x)}{\sin x + \cos x}$
$= \sin^2 x - \sin x\cos x + \cos^2 x = \text{RHS}$

79. $\text{LHS} = \dfrac{1 - \cos\alpha}{\sin\alpha} \cdot \dfrac{1 + \cos\alpha}{1 + \cos\alpha}$
$= \dfrac{1 - \cos^2\alpha}{\sin\alpha\,(1 + \cos\alpha)} = \dfrac{\sin^2\alpha}{\sin\alpha\,(1 + \cos\alpha)} = \text{RHS}$

81. $\text{LHS} = \dfrac{\sin w}{\sin w + \cos w} \cdot \dfrac{\frac{1}{\cos w}}{\frac{1}{\cos w}} = \dfrac{\frac{\sin w}{\cos w}}{\frac{\sin w}{\cos w} + \frac{\cos w}{\cos w}} = \text{RHS}$

83. $\text{LHS} = \dfrac{\sec x}{\sec x - \tan x} \cdot \dfrac{\sec x + \tan x}{\sec x + \tan x}$
$= \dfrac{\sec x\,(\sec x + \tan x)}{\sec^2 x - \tan^2 x} = \text{RHS}$

85. $\text{LHS} = \dfrac{(\sin x + \cos x)^2}{(\sin x + \cos x)(\sin x - \cos x)} = \dfrac{\sin x + \cos x}{\sin x - \cos x}$
$= \dfrac{(\sin x + \cos x)(\sin x - \cos x)}{(\sin x - \cos x)(\sin x - \cos x)} = \text{RHS}$

87. $\text{LHS} = \dfrac{1 - \sin x}{1 + \sin x} \cdot \dfrac{1 - \sin x}{1 - \sin x} = \dfrac{1 - 2\sin x + \sin^2 x}{1 - \sin^2 x}$
$= \dfrac{1}{\cos^2 x} - \dfrac{2\sin x}{\cos^2 x} + \dfrac{\sin^2 x}{\cos^2 x}$
$= \sec^2 x - 2\sec x\tan x + \tan^2 x$
$= (\sec x - \tan x)^2 = \text{RHS}$

89. $\text{LHS} = \dfrac{1}{\sin x} - \dfrac{\cos x}{\sin x} = \dfrac{(1-\cos x)(1+\cos x)}{\sin x\,(1+\cos x)}$

$= \dfrac{\sin^2 x}{\sin x\,(1+\cos x)} = \dfrac{1}{\frac{1}{\sin x} + \frac{\cos x}{\sin x}} = \text{RHS}$

91. $\tan\theta$ **93.** $\tan\theta$ **95.** $3\cos\theta$

97. Yes

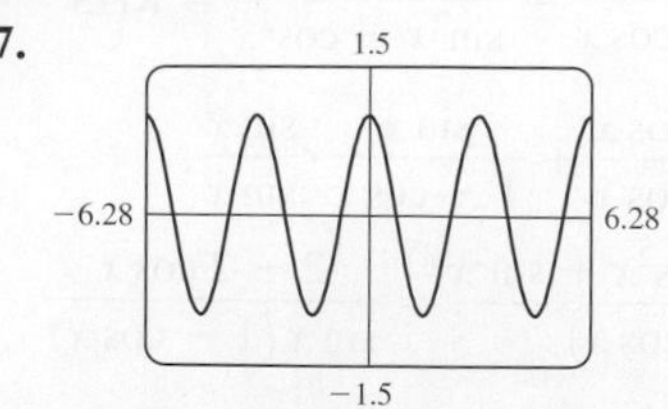

99. No

101. $\text{LHS} = \tan^2 x + 2\tan x\cot x + \cot^2 x$
$= \sec^2 x - 1 + 2 + \csc^2 x - 1 = \text{RHS}$

103. $\text{LHS} = \left(\sin\alpha - \dfrac{\sin\alpha}{\cos\alpha}\right)\left(\cos\alpha - \dfrac{\cos\alpha}{\sin\alpha}\right)$

$= \sin\alpha\left(1 - \dfrac{1}{\cos\alpha}\right)\cdot\cos\alpha\left(1 - \dfrac{1}{\sin\alpha}\right)$

$= \cos\alpha\left(1 - \dfrac{1}{\cos\alpha}\right)\cdot\sin\alpha\left(1 - \dfrac{1}{\sin\alpha}\right)$

$= (\cos\alpha - 1)(\sin\alpha - 1) = \text{RHS}$

105. $\text{RHS} = \dfrac{\sin^2 y - \tan^2 y}{\cos^2 y - \cot^2 y}\cdot\dfrac{\tan^2 y}{\tan^2 y}$

$= \dfrac{\sin^2 y\tan^2 y - \tan^4 y}{\sin^2 y - 1}$

$= \dfrac{\sin^2 y\tan^2 y - \tan^4 y}{-\cos^2 y}$

$= \dfrac{-\sin^2 y\tan^2 y}{\cos^2 y} + \dfrac{\tan^4 y}{\cos^2 y}$

$= -\tan^4 y + \dfrac{\tan^4 y}{\cos^2 y}$

$= \tan^4 y\,(-1 + \sec^2 y) = \text{LHS}$

107. $\text{LHS} = \ln|\tan x| + \ln|\sin x| = \ln\left|\dfrac{\sin x}{\cos x}\right| + \ln|\sin x|$

$= \ln|\sin x| + \ln\left|\dfrac{1}{\cos x}\right| + \ln|\sin x| = \text{RHS}$

109. $\text{LHS} = e^{1-\cos^2 x}e^{\sec^2 x - 1} = e^{1-\cos^2 x + \sec^2 x - 1} = \text{RHS}$

Section 7.2 ■ Page 567

1. Addition; $\sin x\cos y + \cos x\sin y$
2. Subtraction; $\cos x\cos y + \sin x\sin y$

3. $\dfrac{\sqrt{6}+\sqrt{2}}{4}$ **5.** $\dfrac{\sqrt{2}-\sqrt{6}}{4}$ **7.** $2-\sqrt{3}$ **9.** $-\dfrac{\sqrt{6}+\sqrt{2}}{4}$

11. $\sqrt{3}-2$ **13.** $-\dfrac{\sqrt{6}+\sqrt{2}}{4}$ **15.** $\cos 90^\circ = 0$

17. $\sin\frac{\pi}{2} = 1$ **19.** $\tan 45^\circ = 1$

21. $\text{LHS} = \dfrac{\sin(\frac{\pi}{2} - u)}{\cos(\frac{\pi}{2} - u)} = \dfrac{\sin\frac{\pi}{2}\cos u - \cos\frac{\pi}{2}\sin u}{\cos\frac{\pi}{2}\cos u + \sin\frac{\pi}{2}\sin u}$

$= \dfrac{\cos u}{\sin u} = \text{RHS}$

23. $\text{LHS} = \dfrac{1}{\cos(\frac{\pi}{2} - u)} = \dfrac{1}{\cos\frac{\pi}{2}\cos u + \sin\frac{\pi}{2}\sin u}$

$= \dfrac{1}{\sin u} = \text{RHS}$

25. $\text{LHS} = \sin x\cos\frac{\pi}{2} - \cos x\sin\frac{\pi}{2} = \text{RHS}$

27. $\text{LHS} = \sin x\cos\pi - \cos x\sin\pi = \text{RHS}$

29. $\text{LHS} = \dfrac{\tan x - \tan\pi}{1 + \tan x\tan\pi} = \text{RHS}$

31. $\text{LHS} = \sin\left(\dfrac{\pi}{2} - x\right) = \sin\dfrac{\pi}{2}\cos x - \cos\dfrac{\pi}{2}\sin x = \cos x$

$\text{RHS} = \sin\left(\dfrac{\pi}{2} + x\right) = \sin\dfrac{\pi}{2}\cos x + \cos\dfrac{\pi}{2}\sin x = \cos x$

33. $\text{LHS} = \dfrac{\tan x + \tan\frac{\pi}{3}}{1 - \tan x\tan\frac{\pi}{3}} = \text{RHS}$

35. $\text{LHS} = \sin x\cos y + \cos x\sin y$
$- (\sin x\cos y - \cos x\sin y) = \text{RHS}$

37. $\text{LHS} = \dfrac{1}{\tan(x-y)} = \dfrac{1 + \tan x\tan y}{\tan x - \tan y}$

$= \dfrac{1 + \frac{1}{\cot x}\frac{1}{\cot y}}{\frac{1}{\cot x} - \frac{1}{\cot y}}\cdot\dfrac{\cot x\cot y}{\cot x\cot y} = \text{RHS}$

39. $\text{LHS} = \dfrac{\sin x}{\cos x} - \dfrac{\sin y}{\cos y} = \dfrac{\sin x\cos y - \cos x\sin y}{\cos x\cos y} = \text{RHS}$

41. $\text{LHS} = \dfrac{(\tan x - \tan y)(\cos x\cos y)}{(1 - \tan x\tan y)(\cos x\cos y)}$

$= \dfrac{\sin x\cos y - \cos x\sin y}{\cos x\cos y - \sin x\sin y} = \text{RHS}$

43. $\text{LHS} = (\cos x\cos y - \sin x\sin y)(\cos x\cos y + \sin x\sin y)$
$= \cos^2 x\cos^2 y - \sin^2 x\sin^2 y$
$= \cos^2 x(1 - \sin^2 y) - (1 - \cos^2 x)\sin^2 y$
$= \cos^2 x - \sin^2 y\cos^2 x + \sin^2 y\cos^2 x - \sin^2 y = \text{RHS}$

45. $\text{LHS} = \sin((x+y)+z)$
$= \sin(x+y)\cos z + \cos(x+y)\sin z$
$= \cos z[\sin x\cos y + \cos x\sin y]$
$+ \sin z[\cos x\cos y - \sin x\sin y] = \text{RHS}$

47. $\dfrac{\sqrt{1-x^2} + xy}{\sqrt{1+y^2}}$ **49.** $\dfrac{x-y}{\sqrt{1+x^2}\sqrt{1+y^2}}$

51. $\frac{1}{4}(\sqrt{6}+\sqrt{2})$ **53.** $\dfrac{3-2\sqrt{14}}{\sqrt{7}+6\sqrt{2}}$ **55.** $-3\sqrt{10}/10$

57. $2\sqrt{5}/65$ **59.** $2\sin\left(x + \dfrac{5\pi}{6}\right)$ **61.** $5\sqrt{2}\sin 2\left(x + \dfrac{7\pi}{8}\right)$

63. (a) $g(x) = 2 \sin 2\left(x + \frac{\pi}{12}\right)$

(b)

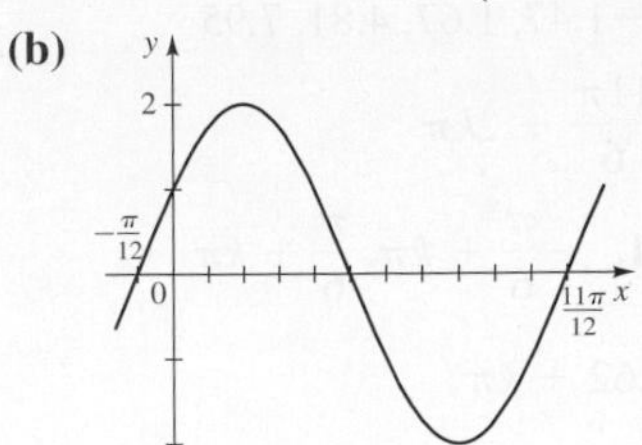

67. (a)

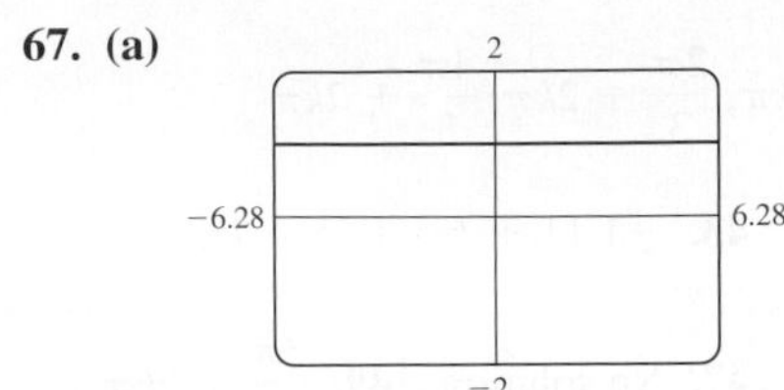

$\sin^2\left(x + \frac{\pi}{4}\right) + \sin^2\left(x - \frac{\pi}{4}\right) = 1$

71. $\text{LHS} = \tan^{-1}\left(\frac{\tan u + \tan v}{1 - \tan u \tan v}\right) = \tan^{-1}(\tan(u + v))$

$= u + v = \text{RHS}$

73. (c) 8.1°

75. (a)

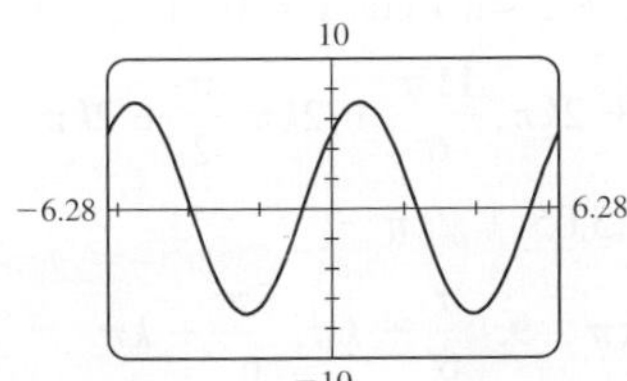

(b) $k = 5\sqrt{2}, \phi = \pi/4$

Section 7.3 ■ Page 576

1. Double-Angle; $2 \sin x \cos x$

2. Half-Angle; $\pm\sqrt{(1 - \cos x)/2}$

3. $\frac{120}{169}, \frac{119}{169}, \frac{120}{119}$ **5.** $-\frac{24}{25}, \frac{7}{25}, -\frac{24}{7}$ **7.** $\frac{24}{25}, \frac{7}{25}, \frac{24}{7}$

9. $-\frac{3}{5}, \frac{4}{5}, -\frac{3}{4}$ **11.** $\frac{1}{2}\left(\frac{3}{4} - \cos 2x + \frac{1}{4}\cos 4x\right)$

13. $\frac{1}{16}(1 - \cos 2x - \cos 4x + \cos 2x \cos 4x)$

15. $\frac{1}{32}\left(\frac{3}{4} - \cos 4x + \frac{1}{4}\cos 8x\right)$

17. $\frac{1}{2}\sqrt{2 - \sqrt{3}}$ **19.** $\sqrt{2} - 1$ **21.** $-\frac{1}{2}\sqrt{2 + \sqrt{3}}$

23. $-\sqrt{2} - 1$ **25.** $\frac{1}{2}\sqrt{2 + \sqrt{3}}$ **27.** $-\frac{1}{2}\sqrt{2 - \sqrt{2}}$

29. (a) sin 32° **(b)** $\sin 8\theta$ **31. (a)** cos 42° **(b)** $\cos 18\theta$

33. (a) tan 4° **(b)** $\tan 2\theta$ **37.** $\sqrt{10}/10, 3\sqrt{10}/10, \frac{1}{3}$

39. $\sqrt{(3 + 2\sqrt{2})/6}, \sqrt{(3 - 2\sqrt{2})/6}, 3 + 2\sqrt{2}$

41. $\sqrt{6}/6, -\sqrt{30}/6, -\sqrt{5}/5$ **43.** $\frac{336}{625}$ **45.** $\frac{8}{7}$ **47.** $\frac{2x}{1 + x^2}$

49. $\sqrt{\frac{1 - x}{2}}$ **51.** $\frac{7}{25}$ **53.** $-8\sqrt{3}/49$ **55.** $\frac{1}{2}(\sin 9x + \sin x)$

57. $\frac{1}{2}(\sin 5x + \sin 3x)$ **59.** $\frac{3}{2}(\cos 11x + \cos 3x)$

61. $2 \sin 6x \cos x$ **63.** $2 \sin 5x \sin x$ **65.** $-2 \cos \frac{9}{2}x \sin \frac{5}{2}x$

67. $(\sqrt{2} + \sqrt{3})/2$ **69.** $\frac{1}{4}(\sqrt{2} - 1)$ **71.** $\sqrt{2}/2$

73. $\text{LHS} = \cos(2 \cdot 5x) = \text{RHS}$

75. $\text{LHS} = \sin^2 x + 2 \sin x \cos x + \cos^2 x$
$= 1 + 2 \sin x \cos x = \text{RHS}$

77. $\text{LHS} = \frac{2 \tan x}{\sec^2 x} = 2 \cdot \frac{\sin x}{\cos x} \cos^2 x = 2 \sin x \cos x = \text{RHS}$

79. $\text{LHS} = \frac{1 - \cos x}{\sin x} + \cos x\left(\frac{1 - \cos x}{\sin x}\right)$
$= \frac{1 - \cos x + \cos x - \cos^2 x}{\sin x} = \frac{\sin^2 x}{\sin x} = \text{RHS}$

81. $\text{LHS} = \frac{2 \sin 2x \cos 2x}{\sin x} = \frac{2(2 \sin x \cos x)(\cos 2x)}{\sin x} = \text{RHS}$

83. $\text{LHS} = \frac{\cos^2 x - \sin^2 x}{(\sin^2 x + \cos^2 x) + 2 \sin x \cos x}$
$= \frac{(\cos x - \sin x)(\cos x + \sin x)}{(\cos x + \sin x)^2}$
$= \frac{(\cos x - \sin x)}{(\cos x + \sin x)} \cdot \frac{\frac{1}{\cos x}}{\frac{1}{\cos x}} = \text{RHS}$

85. $\text{LHS} = \frac{1}{\tan 2x} = \frac{1}{\frac{2 \tan x}{1 - \tan^2 x}} = \text{RHS}$

87. $\text{LHS} = \tan(2x + x) = \frac{\tan 2x + \tan x}{1 - \tan 2x \tan x}$
$= \frac{\frac{2 \tan x}{1 - \tan^2 x} + \tan x}{1 - \frac{2 \tan x}{1 - \tan^2 x} \tan x}$
$= \frac{2 \tan x + \tan x (1 - \tan^2 x)}{1 - \tan^2 x - 2 \tan x \tan x} = \text{RHS}$

89. $\text{LHS} = \frac{2 \sin 3x \cos 2x}{2 \cos 3x \cos 2x} = \frac{\sin 3x}{\cos 3x} = \text{RHS}$

91. $\text{LHS} = \frac{2 \sin 5x \cos 5x}{2 \sin 5x \cos 4x} = \text{RHS}$

93. $\text{LHS} = \frac{2 \sin\left(\frac{x + y}{2}\right) \cos\left(\frac{x - y}{2}\right)}{2 \cos\left(\frac{x + y}{2}\right) \cos\left(\frac{x - y}{2}\right)}$
$= \frac{\sin\left(\frac{x + y}{2}\right)}{\cos\left(\frac{x + y}{2}\right)} = \text{RHS}$

95. $\text{LHS} = \frac{1 - \cos 2\left(\frac{x}{2} + \frac{\pi}{4}\right)}{1 + \cos 2\left(\frac{x}{2} + \frac{\pi}{4}\right)} = \frac{1 - \cos\left(x + \frac{\pi}{2}\right)}{1 + \cos\left(x + \frac{\pi}{2}\right)}$
$= \frac{1 - (-\sin x)}{1 + (-\sin x)} = \text{RHS}$

101. $\text{LHS} = \dfrac{(\sin x + \sin 5x) + (\sin 2x + \sin 4x) + \sin 3x}{(\cos x + \cos 5x) + (\cos 2x + \cos 4x) + \cos 3x}$
$= \dfrac{2 \sin 3x \cos 2x + 2 \sin 3x \cos x + \sin 3x}{2 \cos 3x \cos 2x + 2 \cos 3x \cos x + \cos 3x}$
$= \dfrac{\sin 3x\,(2 \cos 2x + 2 \cos x + 1)}{\cos 3x\,(2 \cos 2x + 2 \cos x + 1)} = \text{RHS}$

103. $\text{RHS} = \cos^{-1}(1 - 2\sin^2 u) = \cos^{-1}(\cos 2u) = 2u = \text{LHS}$

105. **(a)** $\dfrac{\sin 3x}{\sin x} - \dfrac{\cos 3x}{\cos x} = 2$

107. **(a)**

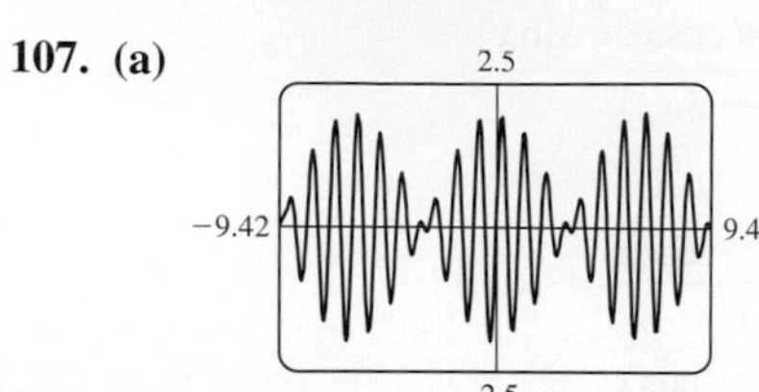

(c)

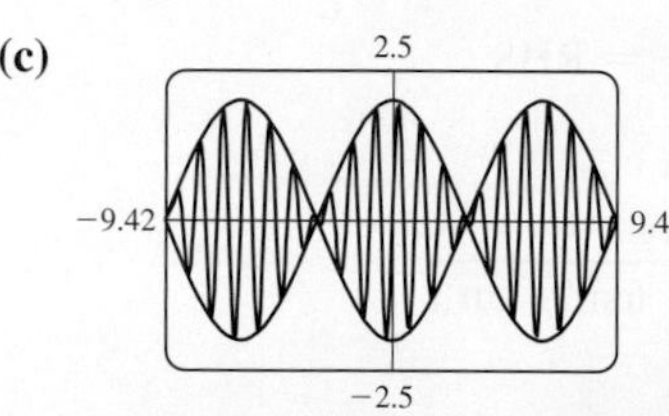

The graph of $y = f(x)$ lies between the two other graphs.

109. **(a)** $P(t) = 8t^4 - 8t^2 + 1$ **(b)** $Q(t) = 16t^5 - 20t^3 + 5t$

115. **(a)** and **(c)**

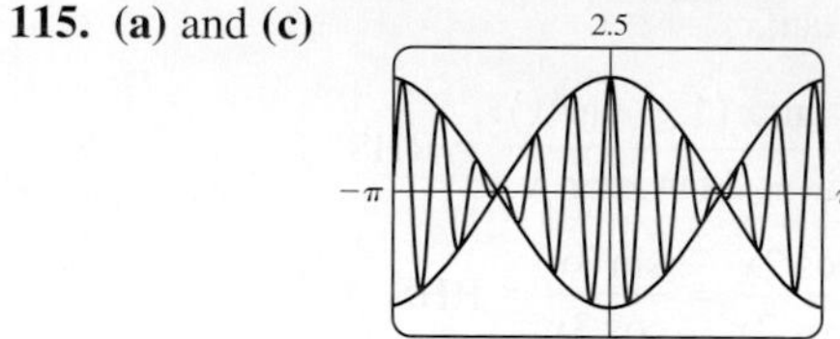

The graph of f lies between the graphs of $y = 2 \cos t$ and $y = -2 \cos t$. Thus, the loudness of the sound varies between $y = \pm 2 \cos t$.

Section 7.4 ■ Page 584

1. infinitely many **2.** no, infinitely many
3. 0.3; $x \approx -9.7, -6.0, -3.4, 0.3, 2.8, 6.6, 9.1$
4. **(a)** 0.30, 2.84 **(b)** 2π; $0.30 + 2k\pi, 2.84 + 2k\pi$
5. $\dfrac{\pi}{3} + 2k\pi, \dfrac{2\pi}{3} + 2k\pi$
7. $(2k + 1)\pi$ **9.** $1.32 + 2k\pi, 4.97 + 2k\pi$
11. $3.61 + 2k\pi, 5.82 + 2k\pi$ **13.** $-\dfrac{\pi}{3} + k\pi$
15. $1.37 + k\pi$ **17.** $\dfrac{5\pi}{6} + 2k\pi, \dfrac{7\pi}{6} + 2k\pi$; $-7\pi/6, -5\pi/6, 5\pi/6, 7\pi/6, 17\pi/6, 19\pi/6$
19. $\dfrac{\pi}{4} + 2k\pi, \dfrac{3\pi}{4} + 2k\pi$; $-7\pi/4, -5\pi/4, \pi/4, 3\pi/4, 9\pi/4, 11\pi/4$
21. $1.29 + 2k\pi, 5.00 + 2k\pi$; $-5.00, -1.29, 1.29, 5.00, 7.57, 11.28$
23. $-1.47 + k\pi$; $-7.75, -4.61, -1.47, 1.67, 4.81, 7.95$
25. $(2k + 1)\pi$ **27.** $\dfrac{\pi}{6} + 2k\pi, \dfrac{11\pi}{6} + 2k\pi$
29. $1.23 + 2k\pi, 5.05 + 2k\pi$ **31.** $-\dfrac{\pi}{6} + k\pi, \dfrac{\pi}{6} + k\pi$
33. $\dfrac{\pi}{4} + k\pi, \dfrac{3\pi}{4} + k\pi$ **35.** $\pm 0.62 + k\pi$
37. $\dfrac{\pi}{4} + k\pi, \dfrac{3\pi}{4} + k\pi$
39. $-1.11 + k\pi, 1.11 + k\pi, \dfrac{2\pi}{3} + 2k\pi, \dfrac{4\pi}{3} + 2k\pi$
41. $\dfrac{\pi}{3} + 2k\pi, \dfrac{5\pi}{3} + 2k\pi$ **43.** $-1.11 + k\pi, 1.25 + k\pi$
45. $\dfrac{\pi}{3} + 2k\pi, \dfrac{5\pi}{3} + 2k\pi$ **47.** No solution **49.** $\dfrac{3\pi}{2} + 2k\pi$
51. $\dfrac{\pi}{2} + k\pi, \dfrac{7\pi}{6} + 2k\pi, \dfrac{11\pi}{6} + 2k\pi$ **53.** $\dfrac{\pi}{2} + k\pi$
55. $k\pi, 0.73 + 2k\pi, 2.41 + 2k\pi$ **57.** 44.95°
59. **(a)** 0° **(b)** 60°, 300° **(c)** 90°, 270° **(d)** 180°

Section 7.5 ■ Page 590

1. $\sin x = 0, k\pi$ **2.** $\sin x + 2 \sin x \cos x = 0, \sin x = 0, 1 + 2 \cos x = 0$ **3.** $\dfrac{7\pi}{6} + 2k\pi, \dfrac{11\pi}{6} + 2k\pi, \dfrac{\pi}{2} + 2k\pi$
5. $\pi + 2k\pi, 1.23 + 2k\pi, 5.05 + 2k\pi$
7. $k\pi, \dfrac{\pi}{3} + 2k\pi, \dfrac{5\pi}{3} + 2k\pi$ **9.** $\dfrac{\pi}{6} + k\pi, \dfrac{5\pi}{6} + k\pi$
11. $\dfrac{\pi}{3} + 2k\pi, \dfrac{5\pi}{3} + 2k\pi, (2k + 1)\pi$
13. $(2k + 1)\pi, \dfrac{\pi}{2} + 2k\pi$ **15.** $2k\pi$
17. **(a)** $\dfrac{\pi}{9} + \dfrac{2k\pi}{3}, \dfrac{5\pi}{9} + \dfrac{2k\pi}{3}$ **(b)** $\pi/9, 5\pi/9, 7\pi/9, 11\pi/9, 13\pi/9, 17\pi/9$
19. **(a)** $\dfrac{\pi}{3} + k\pi, \dfrac{2\pi}{3} + k\pi$ **(b)** $\pi/3, 2\pi/3, 4\pi/3, 5\pi/3$
21. **(a)** $\dfrac{5\pi}{18} + \dfrac{k\pi}{3}$ **(b)** $5\pi/18, 11\pi/18, 17\pi/18, 23\pi/18, 29\pi/18, 35\pi/18$ **23.** **(a)** $4k\pi$ **(b)** 0
25. **(a)** $4\pi + 6k\pi, 5\pi + 6k\pi$ **(b)** None
27. **(a)** $0.62 + \dfrac{k\pi}{2}$ **(b)** 0.62, 2.19, 3.76, 5.33
29. **(a)** $k\pi, \dfrac{\pi}{2} + 2k\pi$ **(b)** $0, \pi/2, \pi$
31. **(a)** $\dfrac{\pi}{6} + k\pi, \dfrac{\pi}{4} + k\pi, \dfrac{5\pi}{6} + k\pi$
(b) $\pi/6, \pi/4, 5\pi/6, 7\pi/6, 5\pi/4, 11\pi/6$
33. **(a)** $\dfrac{\pi}{6} + 2k\pi, \dfrac{5\pi}{6} + 2k\pi, \dfrac{3\pi}{4} + k\pi$
(b) $\pi/6, 3\pi/4, 5\pi/6, 7\pi/4$

35. (a)

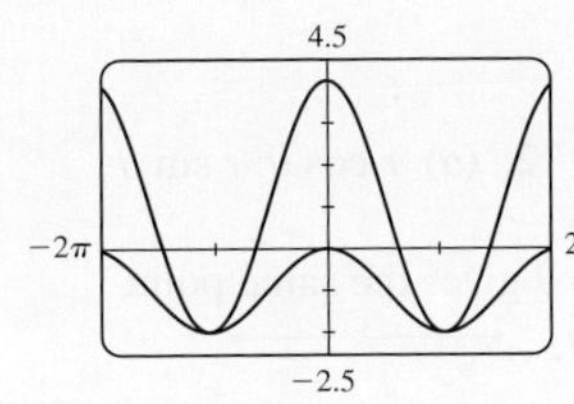

$(\pm 3.14, -2)$

(b) $((2k+1)\pi, -2)$

37. (a)

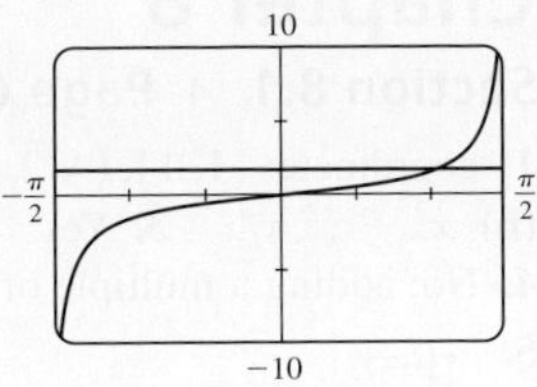

$(1.05, 1.73)$

(b) $\left(\frac{\pi}{3} + k\pi, \sqrt{3}\right)$

39. $\pi/8, 3\pi/8, 5\pi/8, 7\pi/8, 9\pi/8, 11\pi/8, 13\pi/8, 15\pi/8$
41. $\pi/3, 2\pi/3$ **43.** $\pi/2, 7\pi/6, 3\pi/2, 11\pi/6$ **45.** 0
47. $0, \pi$ **49.** $0, \pi/3, 2\pi/3, \pi, 4\pi/3, 5\pi/3$ **51.** $\pi/6, 3\pi/2$

53. $k\pi/2$ **55.** $\frac{\pi}{2} + k\pi, \frac{\pi}{9} + \frac{2k\pi}{3}, \frac{5\pi}{9} + \frac{2k\pi}{3}$

57. $0, \pm 0.95$ **59.** 1.92 **61.** ± 0.71

63. $\frac{\sqrt{17}-3}{4}$ **65.** $0.95°$ or $89.1°$

67. (a) 34th day (February 3), 308th day (November 4)
(b) 275 days

Chapter 7 Review ■ Page 594

1. $\text{LHS} = \sin\theta\left(\frac{\cos\theta}{\sin\theta} + \frac{\sin\theta}{\cos\theta}\right) = \cos\theta + \frac{\sin^2\theta}{\cos\theta}$

$= \frac{\cos^2\theta + \sin^2\theta}{\cos\theta} = \text{RHS}$

3. $\text{LHS} = (1 - \sin^2 x)\csc x - \csc x$

$= \csc x - \sin^2 x \csc x - \csc x$

$= -\sin^2 x \cdot \frac{1}{\sin x} = \text{RHS}$

5. $\text{LHS} = \frac{\cos^2 x}{\sin^2 x} - \frac{\tan^2 x}{\sin^2 x} = \cot^2 x - \frac{1}{\cos^2 x} = \text{RHS}$

7. $\text{LHS} = \frac{\cos x}{\frac{1}{\cos x}(1 - \sin x)} = \frac{\cos x}{\frac{1}{\cos x} - \frac{\sin x}{\cos x}} = \text{RHS}$

9. $\text{LHS} = \sin^2 x \cdot \frac{\cos^2 x}{\sin^2 x} + \cos^2 x \cdot \frac{\sin^2 x}{\cos^2 x} = \cos^2 x + \sin^2 x = \text{RHS}$

11. $\text{LHS} = \frac{2\sin x\cos x}{1 + 2\cos^2 x - 1} = \frac{2\sin x\cos x}{2\cos^2 x} = \frac{\sin x}{\cos x} = \text{RHS}$

13. $\text{LHS} = \csc x - \frac{1 - \cos x}{\sin x}$

$= \csc x - (\csc x - \cot x) = \text{RHS}$

15. $\text{LHS} = \frac{2\sin x\cos x}{\sin x} - \frac{2\cos^2 x - 1}{\cos x}$

$= 2\cos x - 2\cos x + \frac{1}{\cos x} = \text{RHS}$

17. $\text{LHS} = \frac{\frac{1}{\cos x} - 1}{\sin x \cdot \frac{1}{\cos x}} = \left(\frac{1}{\cos x} - 1\right)\frac{\cos x}{\sin x}$

$= \frac{1 - \cos x}{\sin x} = \text{RHS}$

19. $\text{LHS} = \cos^2\frac{x}{2} - 2\sin\frac{x}{2}\cos\frac{x}{2} + \sin^2\frac{x}{2}$

$= 1 - \sin\left(2\cdot\frac{x}{2}\right) = \text{RHS}$

21. $\text{LHS} = \frac{2\sin\left(\frac{(x+y)+(x-y)}{2}\right)\cos\left(\frac{(x+y)-(x-y)}{2}\right)}{2\cos\left(\frac{(x+y)+(x-y)}{2}\right)\cos\left(\frac{(x+y)-(x-y)}{2}\right)}$

$= \frac{2\sin x\cos y}{2\cos x\cos y} = \text{RHS}$

23. (a)

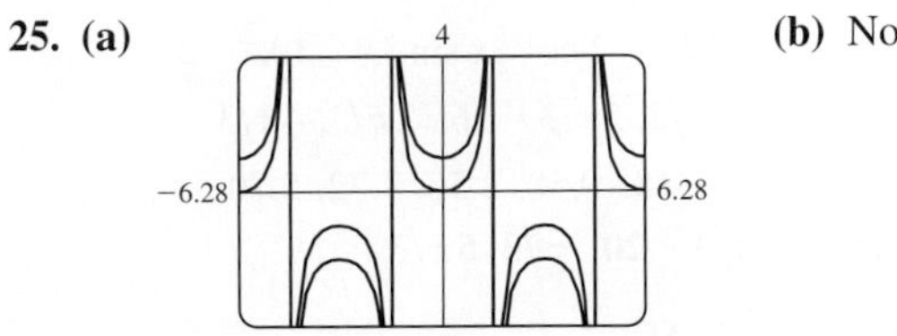

(b) Yes

25. (a)

(b) No

27. (a)

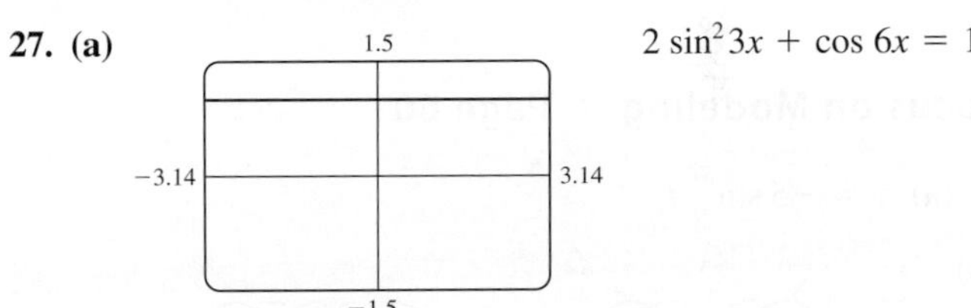

$2\sin^2 3x + \cos 6x = 1$

29. $0.85, 2.29$ **31.** $0, \pi$ **33.** $\pi/6, 5\pi/6$ **35.** $\pi/3, 5\pi/3$
37. $2\pi/3, 4\pi/3$ **39.** $\pi/3, 2\pi/3, 3\pi/4, 4\pi/3, 5\pi/3, 7\pi/4$
41. $\pi/6, \pi/2, 5\pi/6, 7\pi/6, 3\pi/2, 11\pi/6$ **43.** $\pi/6$
45. 1.18 **47. (a)** $63.4°$ **(b)** No **(c)** $90°$

49. $\frac{\sqrt{2}+\sqrt{6}}{4}$ or $\frac{1}{2}\sqrt{2+\sqrt{3}}$ **51.** $\sqrt{2} - 1$ **53.** $\sqrt{2}/2$

55. $\sqrt{2}/2$ **57.** $\frac{\sqrt{2}+\sqrt{3}}{4}$ **59.** $\frac{2}{9}(\sqrt{10} + 1)$

61. $\frac{2}{3}(\sqrt{2} + \sqrt{5})$ **63.** $\sqrt{(3 + 2\sqrt{2})/6}$ **65.** $-\frac{12\sqrt{10}}{31}$

67. $\frac{2x}{1 - x^2}$ **69. (a)** $\theta = \tan^{-1}\left(\frac{3}{x}\right)$ **(b)** 85.9 m

71. (a) VII **(b)** III **(c)** VI **(d)** II **(e)** IV **(f)** VIII
(g) I **(h)** V

Chapter 7 Test ■ Page 597

1. $\text{LHS} = \frac{\sin\theta}{\cos\theta}\cdot\sin\theta + \cos\theta = \frac{\sin^2\theta + \cos^2\theta}{\cos\theta} = \text{RHS}$

2. $\text{LHS} = \frac{\tan x}{1 - \cos x}\cdot\frac{1 + \cos x}{1 + \cos x} = \frac{\tan x\,(1 + \cos x)}{1 - \cos^2 x}$

$= \frac{\frac{\sin x}{\cos x}(1 + \cos x)}{\sin^2 x} = \frac{1}{\sin x}\cdot\frac{1 + \cos x}{\cos x} = \text{RHS}$

3. $\text{LHS} = \frac{2\tan x}{\sec^2 x} = \frac{2\sin x}{\cos x}\cdot\cos^2 x = 2\sin x\cos x = \text{RHS}$

4. $\text{LHS} = \sin x \tan\dfrac{x}{2} = \sin x \cdot \dfrac{1 - \cos x}{\sin x} = \text{RHS}$

5. $\text{LHS} = 2\left(\dfrac{1 - \cos 6x}{2}\right) = \text{RHS}$

6. $\text{LHS} = 1 - 2\sin^2 2x = 1 - 2(2\sin x \cos x)^2$
$= 1 - 8\sin^2 x\,(1 - \sin^2 x) = \text{RHS}$

7. $\text{LHS} = \sin^2\left(\dfrac{x}{2}\right) + 2\sin\dfrac{x}{2}\cos\dfrac{x}{2} + \cos^2\left(\dfrac{x}{2}\right)$
$= 1 + \sin\left[2\cdot\left(\dfrac{x}{2}\right)\right] = \text{RHS}$

8. $\tan\theta$ **9. (a)** $\frac{1}{2}$ **(b)** $\dfrac{\sqrt{2}+\sqrt{6}}{4}$ or $\frac{1}{2}\sqrt{2+\sqrt{3}}$
(c) $\dfrac{\sqrt{6}-\sqrt{2}}{4}$ or $\frac{1}{2}\sqrt{2-\sqrt{3}}$

10. $(10 - 2\sqrt{5})/15$
11. $\frac{1}{2}(\sin 8x - \sin 2x)$ **12.** $-2\cos\frac{7}{2}x\sin\frac{3}{2}x$ **13.** -2
14. 0.34, 2.80 **15.** $\pi/3, \pi/2, 5\pi/3$ **16.** $2\pi/3, 4\pi/3$
17. $\pi/6, \pi/2, 5\pi/6, 3\pi/2$ **18.** 0.58, 2.56, 3.72, 5.70
19. $\pi/3, 2\pi/3, 4\pi/3, 5\pi/3$ **20.** $\pi/3, 5\pi/3$

21. $\frac{1519}{1681}$ **22.** $\dfrac{\sqrt{1-x^2} - xy}{\sqrt{1+y^2}}$

Focus on Modeling ■ Page 601

1. (a) $y = -5\sin\dfrac{\pi}{8}t$
(b)

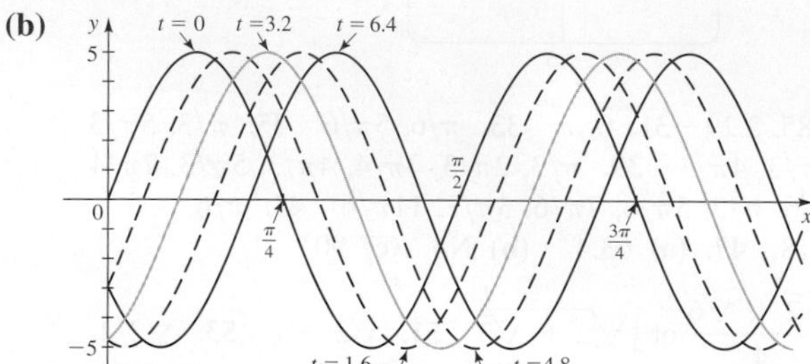

Yes, it is a traveling wave.

(c) $v = \pi/32$
3. $y(x, t) = 2.7\sin(0.68x - 4.10t)$
5. $y(x, t) = 0.6\sin\pi x\cos 40\pi t$
7. (a) 1, 2, 3, 4
(b) 5:

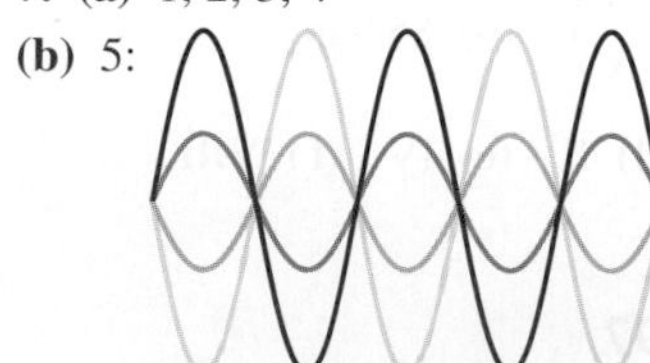

6:

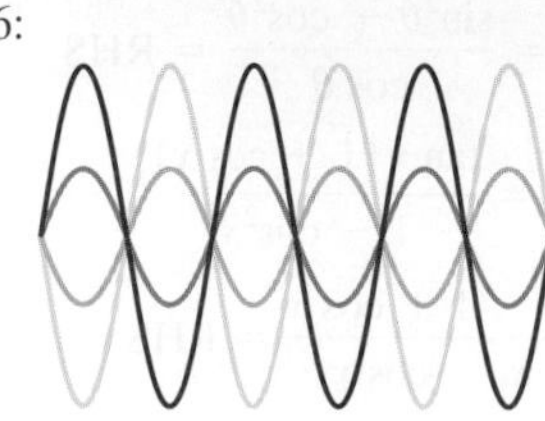

(c) 880π **(d)** $y(x, t) = \sin x\cos 880\pi t$;
$y(x, t) = \sin 2x\cos 880\pi t$; $y(x, t) = \sin 3x\cos 880\pi t$;
$y(x, t) = \sin 4x\cos 880\pi t$

Chapter 8

Section 8.1 ■ Page 608

1. coordinate; $(1, 1), (\sqrt{2}, \pi/4)$ **2. (a)** $r\cos\theta, r\sin\theta$
(b) $x^2 + y^2, y/x$ **3.** Yes
4. No; adding a multiple of 2π to θ gives the same point

5.

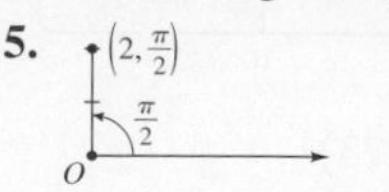

7.

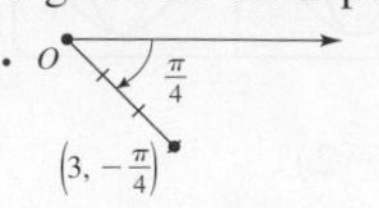

9.

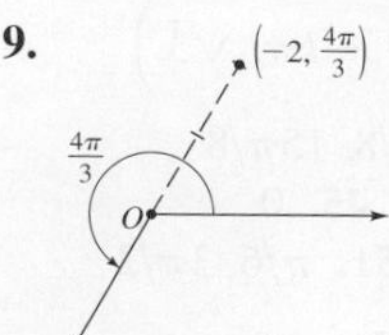

11.

13.

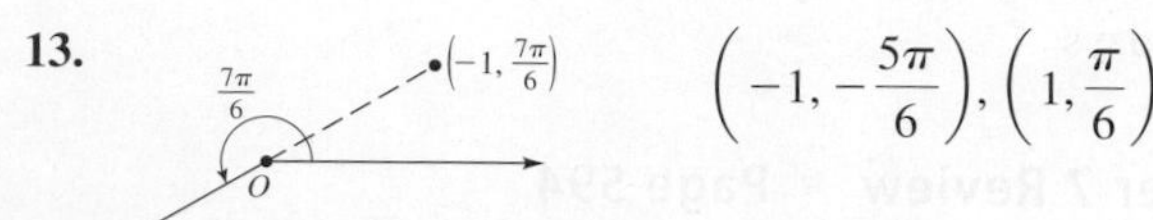

15. (−5, 0) O $(-5, 2\pi), (5, \pi)$

17. Q **19.** Q **21.** P **23.** P **25.** $(3\sqrt{2}, 3\pi/4)$
27. $\left(-\dfrac{5}{2}, -\dfrac{5\sqrt{3}}{2}\right)$ **29.** $(0, 3)$ **31.** $(1, -1)$ **33.** $(-5, 0)$
35. $(3/2, -\sqrt{3}/2)$ **37.** $(\sqrt{2}, 3\pi/4)$ **39.** $(4, \pi/4)$
41. $\left(5, \tan^{-1}\left(\frac{4}{3}\right)\right)$ **43.** $(6, \pi)$ **45.** $\theta = \pi/4$
47. $r = \cot\theta\csc\theta$ **49.** $r = 4\sec\theta$ **51.** $r = \sin\theta$
53. $x^2 + y^2 = 49$ **55.** $x = 0$ **57.** $x = 6$ **59.** $x^2 + y^2 = 4y$
61. $x^2 + y^2 = (x^2 + y^2 - x)^2$ **63.** $(x^2 + y^2 - 2y)^2 = x^2 + y^2$
65. $y - x = 1$ **67.** $x^2 - 3y^2 + 16y - 16 = 0$
69. $x^2 + y^2 = \dfrac{y}{x}$ **71.** $y^2 - 3x^2 = 0$

Section 8.2 ■ Page 616

1. circles, rays **2. (a)** satisfy **(b)** circle, 3, pole; line, pole, 1
3. VI **5.** II **7.** I **9.** Symmetric about $\theta = \pi/2$
11. Symmetric about the polar axis
13. Symmetric about $\theta = \pi/2$
15. All three types of symmetry

17. **19.**

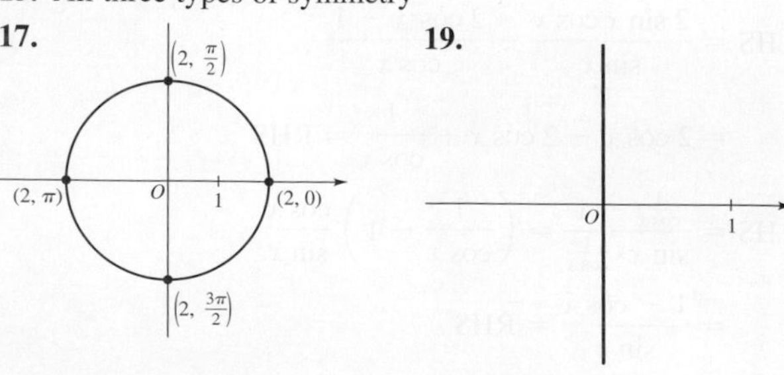

$x^2 + y^2 = 4$ $x = 0$

21.

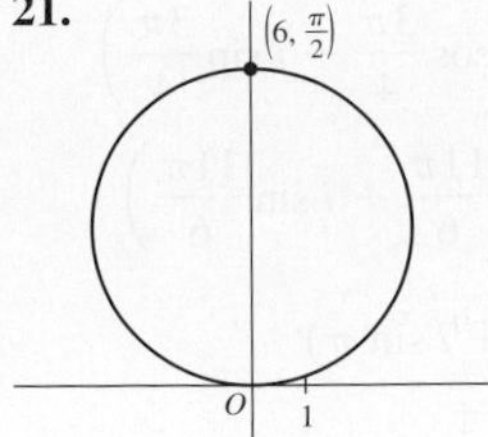

$x^2 + (y - 3)^2 = 9$

23.

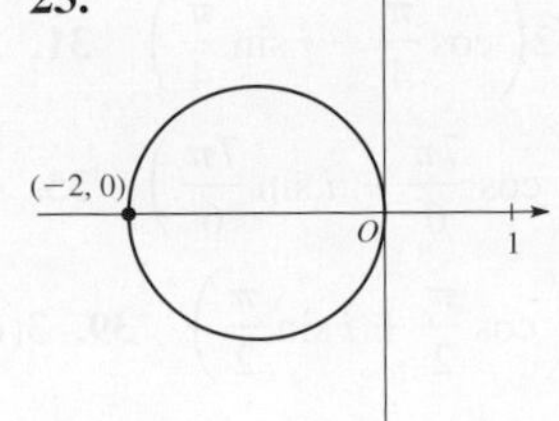

25.

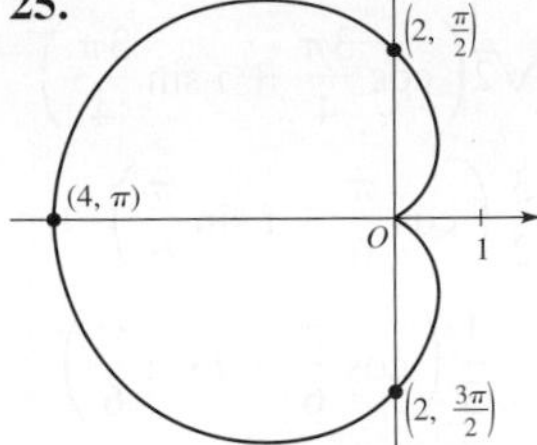

27.

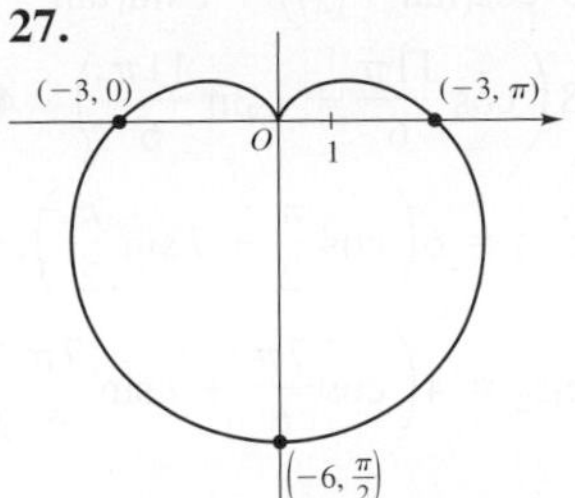

29.

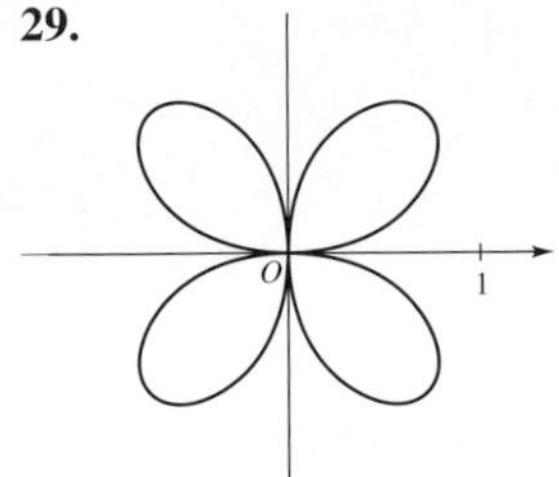

31.

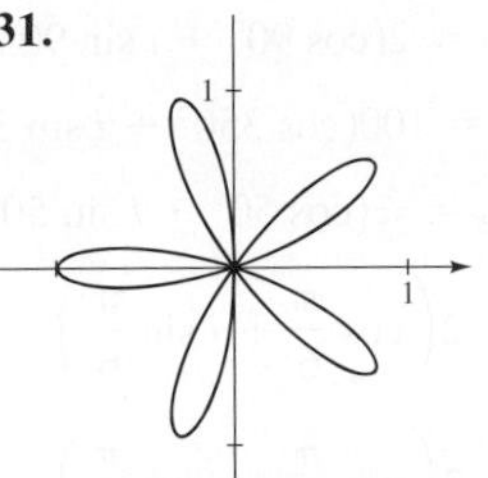

33.

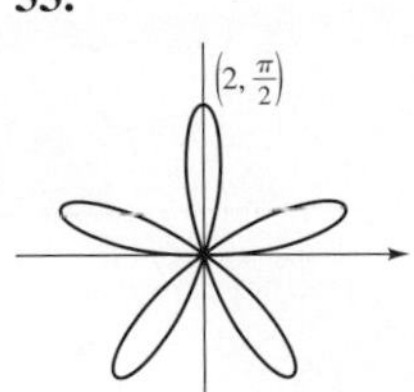

35.

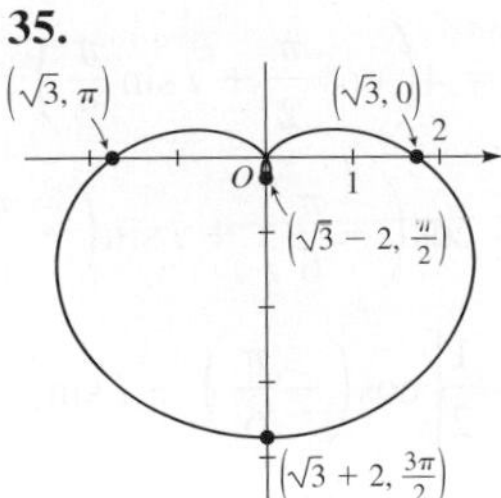

37.

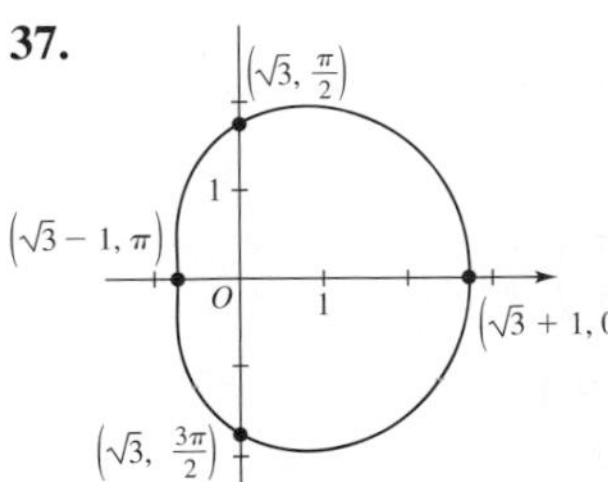

39.

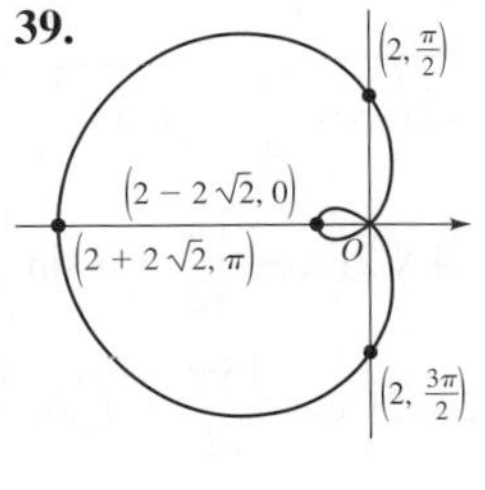

41.

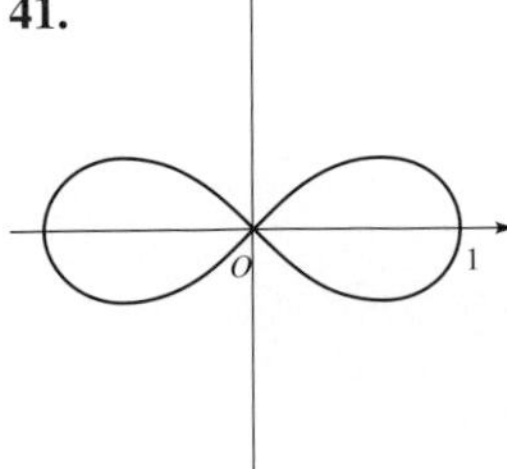

43.

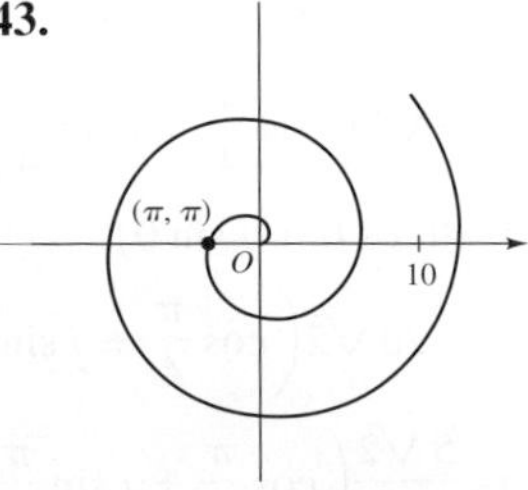

45.

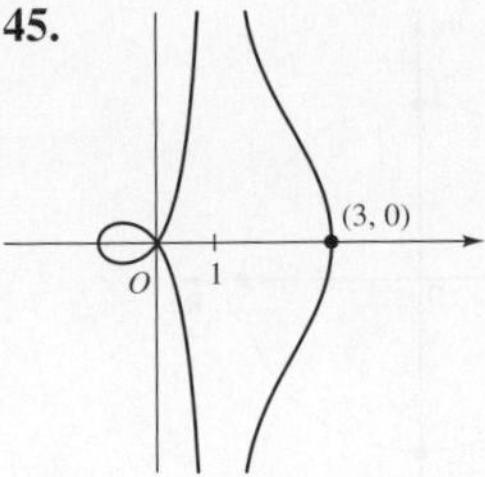

47. $0 \le \theta \le 4\pi$

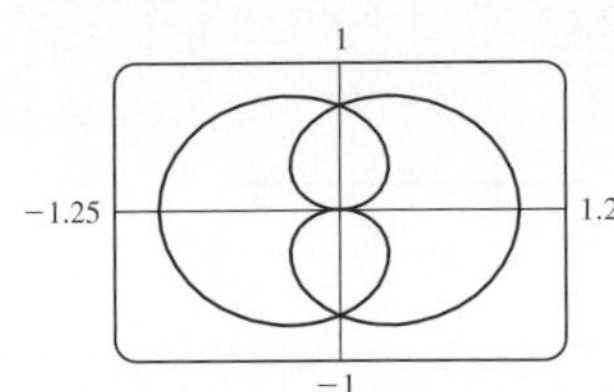

49. $0 \le \theta \le 4\pi$

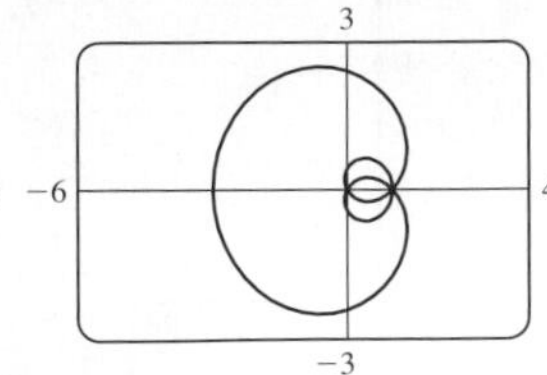

51. The graph of $r = 1 + \sin n\theta$ has n loops.

53. IV **55.** III

57.

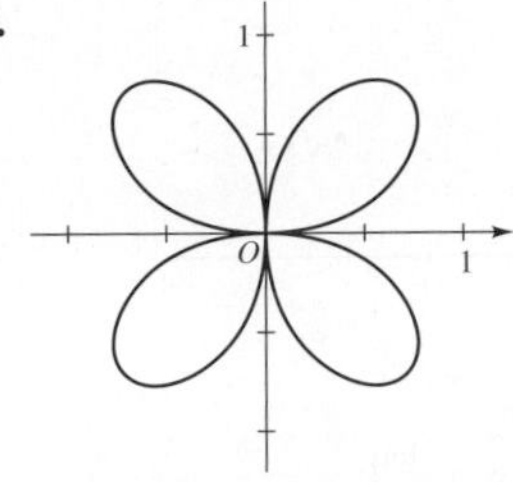

59.

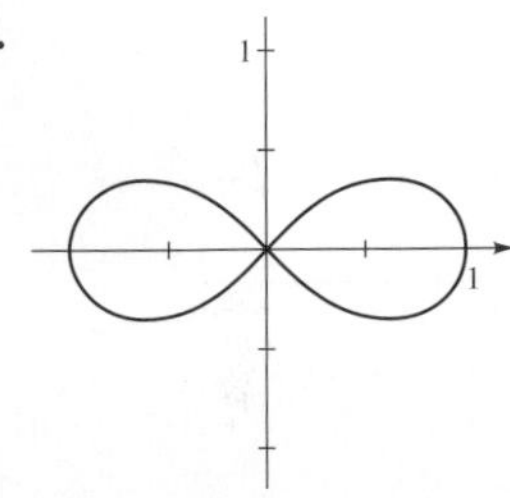

61. **(a)** $\left(x - \dfrac{a}{2}\right)^2 + \left(y - \dfrac{b}{2}\right)^2 = \dfrac{a^2 + b^2}{4}$

$\left(\dfrac{a}{2}, \dfrac{b}{2}\right), \dfrac{1}{2}\sqrt{a^2 + b^2}$

(b)

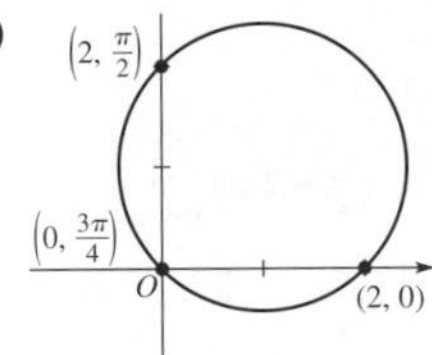

63. **(a)** Elliptical

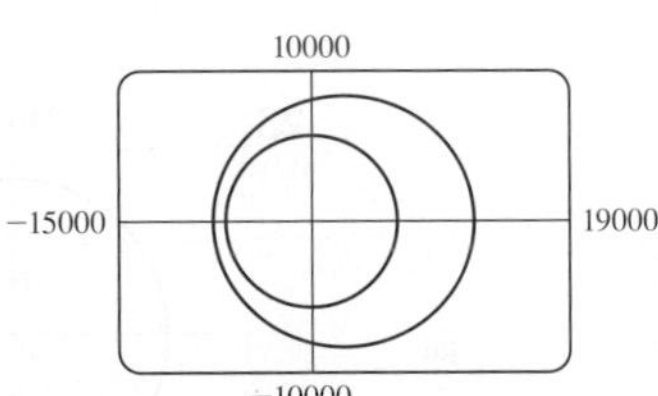

(b) π; 860 km

Section 8.3 ■ Page 625

1. real, imaginary, (a, b) **2.** **(a)** $\sqrt{a^2 + b^2}, b/a$
(b) $r(\cos\theta + i\sin\theta)$

3. **(a)** $\sqrt{2}\left(\cos\dfrac{3\pi}{4} + i\sin\dfrac{3\pi}{4}\right)$ **(b)** $\sqrt{3} + i$

(c) $1 + i, \sqrt{2}\left(\cos\dfrac{\pi}{4} + i\sin\dfrac{\pi}{4}\right)$

4. n; four; $2, 2i, -2, -2i$; 2

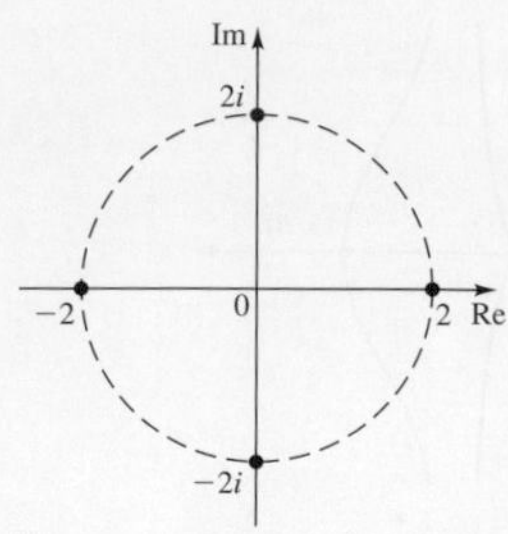

5. 4

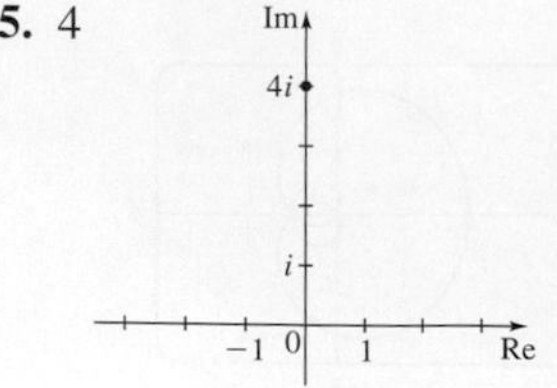

7. 2

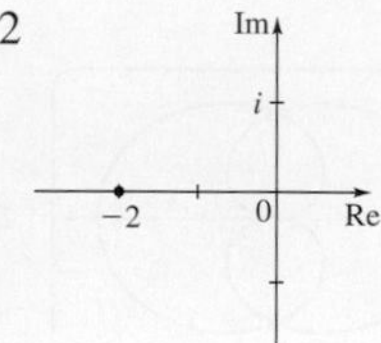

9. $\sqrt{29}$

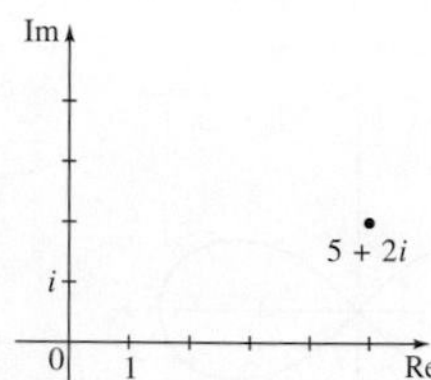

11. 2

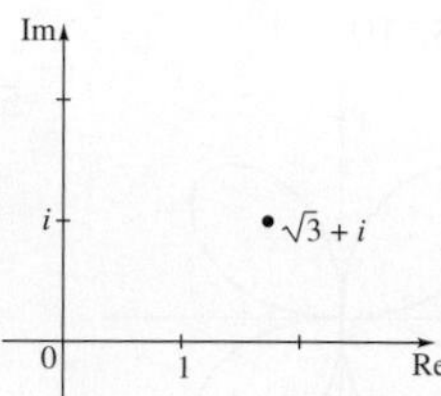

13. 1

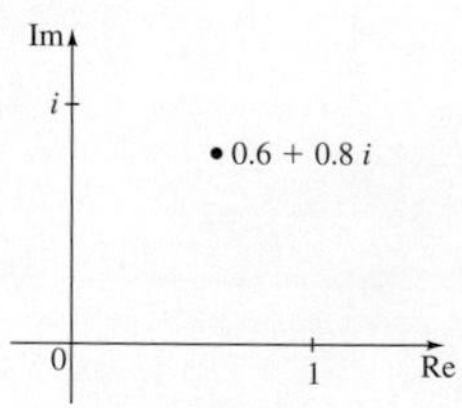

15.

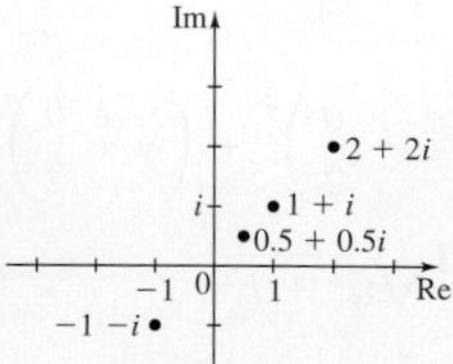

17.

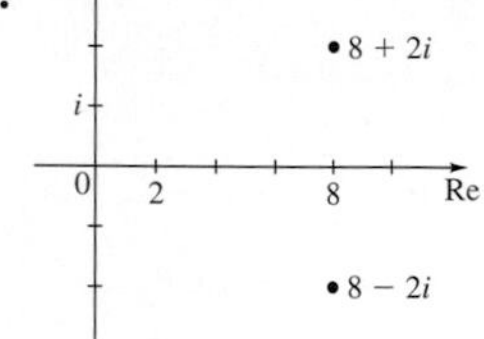

19.

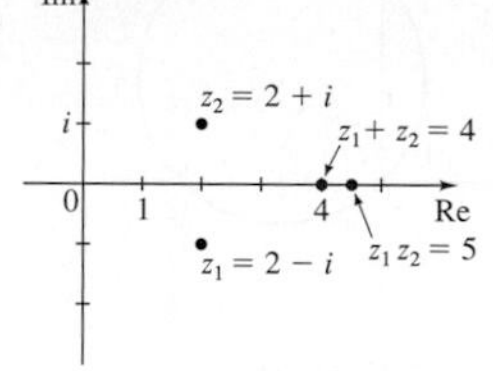

21.

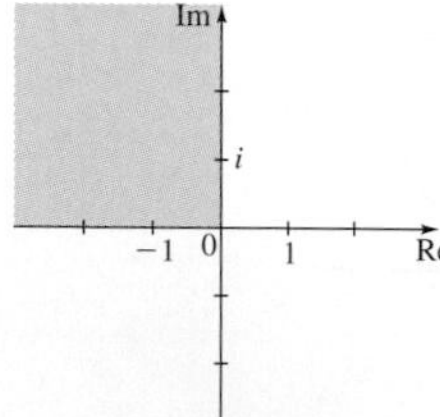

23.

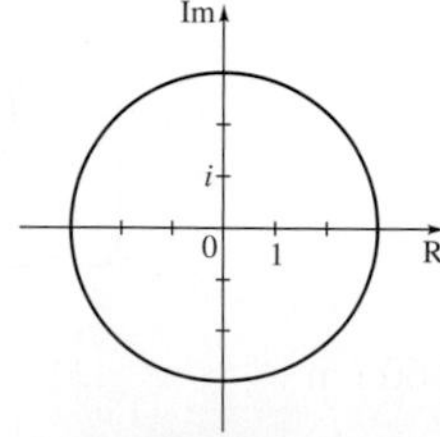

25.

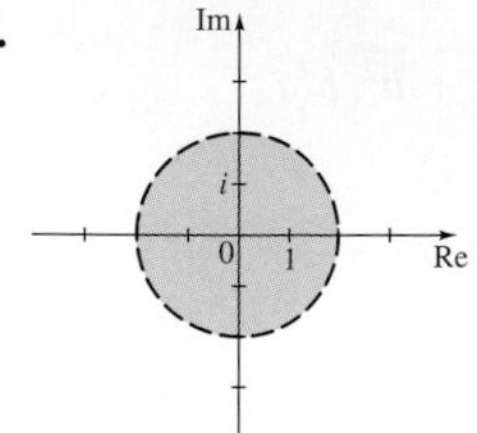

27.

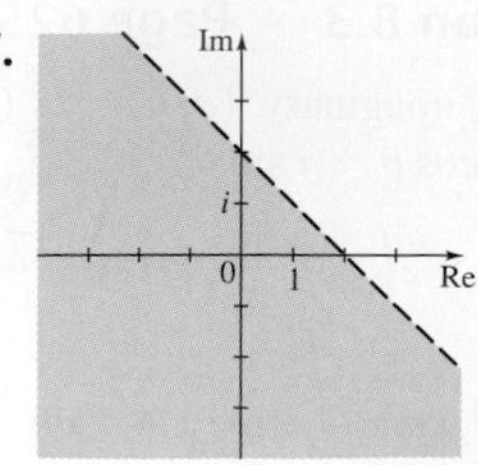

29. $\sqrt{2}\left(\cos\frac{\pi}{4} + i\sin\frac{\pi}{4}\right)$ 31. $2\sqrt{2}\left(\cos\frac{3\pi}{4} + i\sin\frac{3\pi}{4}\right)$

33. $2\left(\cos\frac{7\pi}{6} + i\sin\frac{7\pi}{6}\right)$ 35. $4\left(\cos\frac{11\pi}{6} + i\sin\frac{11\pi}{6}\right)$

37. $2\left(\cos\frac{\pi}{2} + i\sin\frac{\pi}{2}\right)$ 39. $3(\cos\pi + i\sin\pi)$

41. $2\sqrt{2}\left(\cos\frac{5\pi}{6} + i\sin\frac{5\pi}{6}\right)$

43. $5\left[\cos\left(\tan^{-1}\left(\frac{3}{4}\right)\right) + i\sin\left(\tan^{-1}\left(\frac{3}{4}\right)\right)\right]$

45. $8\left(\cos\frac{11\pi}{6} + i\sin\frac{11\pi}{6}\right)$ 47. $3\sqrt{2}\left(\cos\frac{3\pi}{4} + i\sin\frac{3\pi}{4}\right)$

49. $z_1z_2 = 6\left(\cos\frac{\pi}{2} + i\sin\frac{\pi}{2}\right), \frac{z_1}{z_2} = \frac{3}{2}\left(\cos\frac{\pi}{6} + i\sin\frac{\pi}{6}\right)$

51. $z_1z_2 = 4\left(\cos\frac{7\pi}{6} + i\sin\frac{7\pi}{6}\right), \frac{z_1}{z_2} = \frac{1}{2}\left(\cos\frac{\pi}{6} + i\sin\frac{\pi}{6}\right)$

53. $z_1z_2 = 8(\cos 150° + i\sin 150°)$

$z_1/z_2 = 2(\cos 90° + i\sin 90°)$

55. $z_1z_2 = 100(\cos 350° + i\sin 350°)$

$z_1/z_2 = \frac{4}{25}(\cos 50° + i\sin 50°)$

57. $z_1 = 2\left(\cos\frac{\pi}{6} + i\sin\frac{\pi}{6}\right)$

$z_2 = 2\left(\cos\frac{\pi}{3} + i\sin\frac{\pi}{3}\right)$

$z_1z_2 = 4\left(\cos\frac{\pi}{2} + i\sin\frac{\pi}{2}\right)$

$\frac{z_1}{z_2} = \cos\left(-\frac{\pi}{6}\right) + i\sin\left(-\frac{\pi}{6}\right)$

$\frac{1}{z_1} = \frac{1}{2}\left[\cos\left(-\frac{\pi}{6}\right) + i\sin\left(-\frac{\pi}{6}\right)\right]$

59. $z_1 = 4\left(\cos\frac{11\pi}{6} + i\sin\frac{11\pi}{6}\right)$

$z_2 = \sqrt{2}\left(\cos\frac{3\pi}{4} + i\sin\frac{3\pi}{4}\right)$

$z_1z_2 = 4\sqrt{2}\left(\cos\frac{7\pi}{12} + i\sin\frac{7\pi}{12}\right)$

$\frac{z_1}{z_2} = 2\sqrt{2}\left(\cos\frac{13\pi}{12} + i\sin\frac{13\pi}{12}\right)$

$\frac{1}{z_1} = \frac{1}{4}\left(\cos\frac{\pi}{6} + i\sin\frac{\pi}{6}\right)$

61. $z_1 = 5\sqrt{2}\left(\cos\frac{\pi}{4} + i\sin\frac{\pi}{4}\right)$

$z_2 = 4(\cos 0 + i\sin 0)$

$z_1z_2 = 20\sqrt{2}\left(\cos\frac{\pi}{4} + i\sin\frac{\pi}{4}\right)$

$\frac{z_1}{z_2} = \frac{5\sqrt{2}}{4}\left(\cos\frac{\pi}{4} + i\sin\frac{\pi}{4}\right)$

$\frac{1}{z_1} = \frac{\sqrt{2}}{10}\left[\cos\left(-\frac{\pi}{4}\right) + i\sin\left(-\frac{\pi}{4}\right)\right]$

63. $z_1 = 20(\cos \pi + i \sin \pi)$

$z_2 = 2\left(\cos \frac{\pi}{6} + i \sin \frac{\pi}{6}\right)$

$z_1 z_2 = 40\left(\cos \frac{7\pi}{6} + i \sin \frac{7\pi}{6}\right)$

$\frac{z_1}{z_2} = 10\left(\cos \frac{5\pi}{6} + i \sin \frac{5\pi}{6}\right)$

$\frac{1}{z_1} = \frac{1}{20}(\cos \pi + i \sin \pi)$

65. -64 **67.** $16\sqrt{2} + 16\sqrt{2}i$ **69.** -1 **71.** 4096

73. $8(-1 + i)$ **75.** $\frac{1}{2048}(-\sqrt{3} - i)$

77. $2\sqrt{2}\left(\cos \frac{\pi}{12} + i \sin \frac{\pi}{12}\right)$,

$2\sqrt{2}\left(\cos \frac{13\pi}{12} + i \sin \frac{13\pi}{12}\right)$

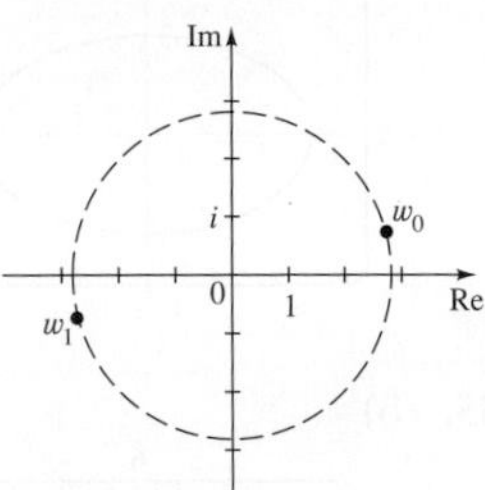

79. $3\left(\cos \frac{3\pi}{8} + i \sin \frac{3\pi}{8}\right)$,

$3\left(\cos \frac{7\pi}{8} + i \sin \frac{7\pi}{8}\right)$,

$3\left(\cos \frac{11\pi}{8} + i \sin \frac{11\pi}{8}\right)$,

$3\left(\cos \frac{15\pi}{8} + i \sin \frac{15\pi}{8}\right)$

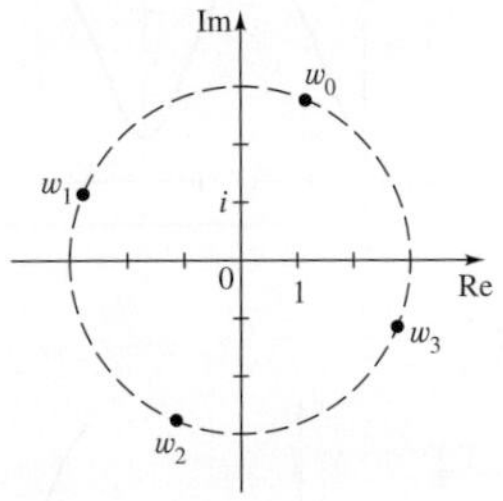

81. $\pm 1, \pm i, \frac{\sqrt{2}}{2} \pm \frac{\sqrt{2}}{2}i$,

$-\frac{\sqrt{2}}{2} \pm \frac{\sqrt{2}}{2}i$

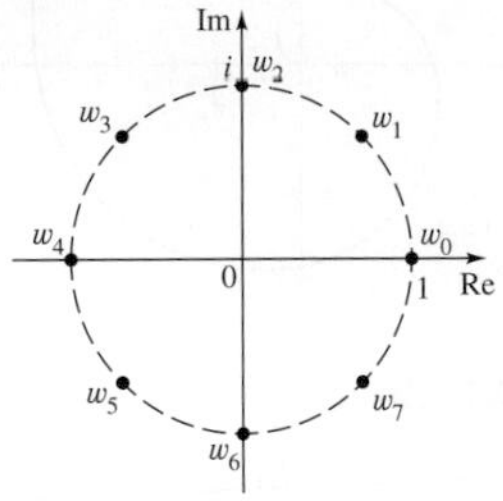

83. $\frac{\sqrt{3}}{2} + \frac{1}{2}i, -\frac{\sqrt{3}}{2} + \frac{1}{2}i, -i$

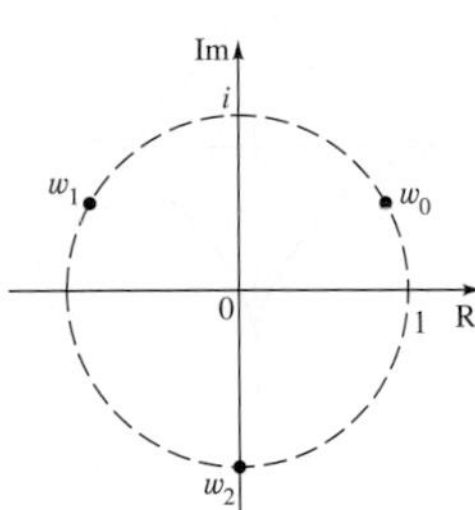

85. $\frac{\sqrt{2}}{2} \pm \frac{\sqrt{2}}{2}i$,

$-\frac{\sqrt{2}}{2} \pm \frac{\sqrt{2}}{2}i$

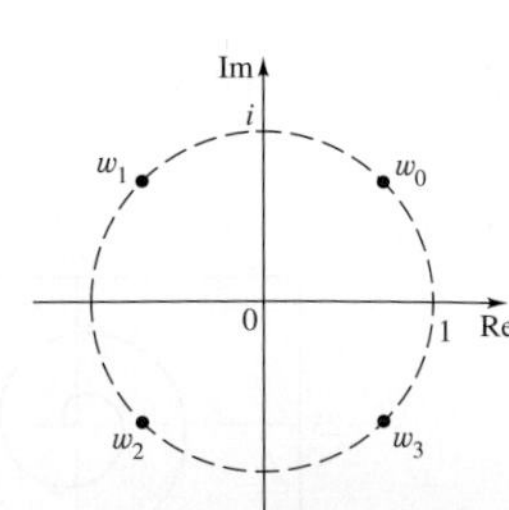

87. $\frac{\sqrt{2}}{2} \pm \frac{\sqrt{2}}{2}i, -\frac{\sqrt{2}}{2} \pm \frac{\sqrt{2}}{2}i$

89. $2\left(\cos \frac{\pi}{18} + i \sin \frac{\pi}{18}\right), 2\left(\cos \frac{13\pi}{18} + i \sin \frac{13\pi}{18}\right)$,

$2\left(\cos \frac{25\pi}{18} + i \sin \frac{25\pi}{18}\right)$

91. $2^{1/6}\left(\cos \frac{5\pi}{12} + i \sin \frac{5\pi}{12}\right), 2^{1/6}\left(\cos \frac{13\pi}{12} + i \sin \frac{13\pi}{12}\right)$,

$2^{1/6}\left(\cos \frac{21\pi}{12} + i \sin \frac{21\pi}{12}\right)$

93. $\frac{1 \pm \sqrt{5}}{2}i$ **95.** $1 + i, -1 + i$

Section 8.4 ▪ Page 633

1. (a) parameter **(b)** $(0, 0), (1, 1)$ **(c)** x^2; parabola

2. (a) True **(b)** $(0, 0), (2, 4)$

(c) x^2; path

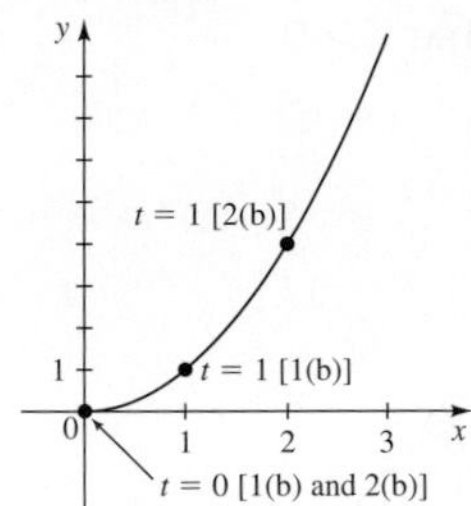

3. (a)

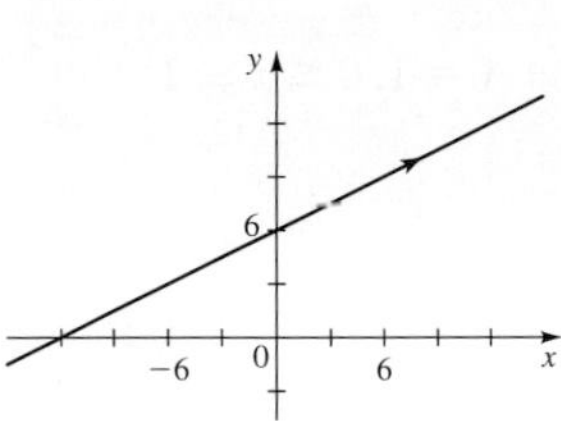

(b) $x - 2y + 12 = 0$

5. (a)

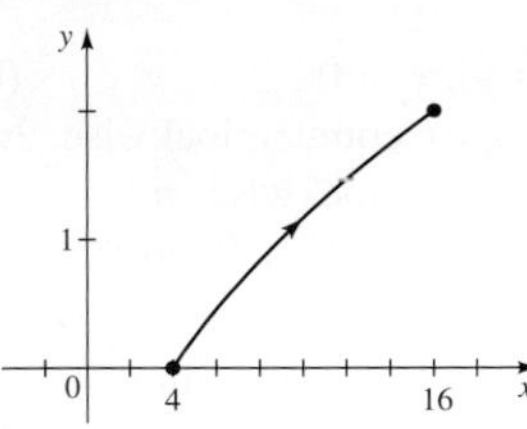

(b) $x = (y + 2)^2$

7. (a)

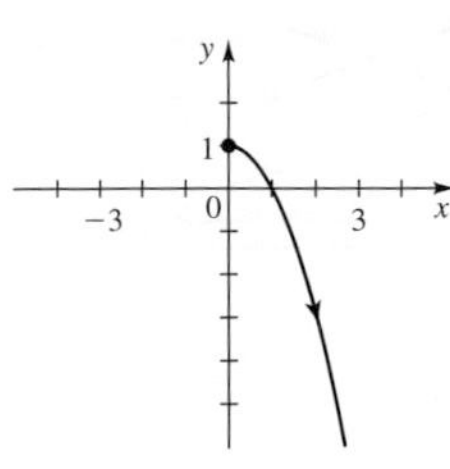

(b) $x = \sqrt{1 - y}$

9. (a)

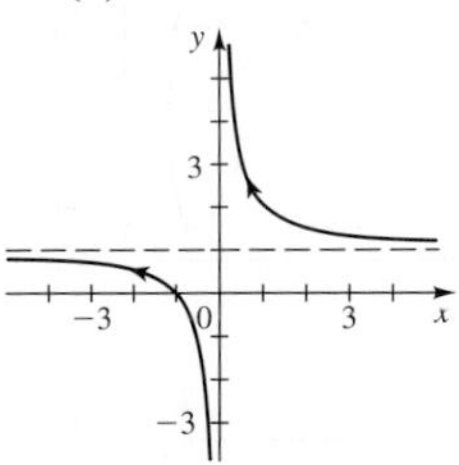

(b) $y = \frac{1}{x} + 1$

11. (a)

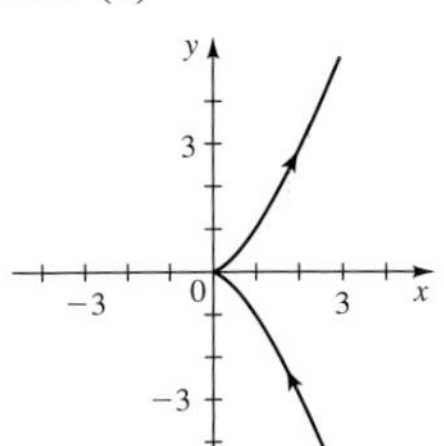

(b) $x^3 = y^2$

13. (a)

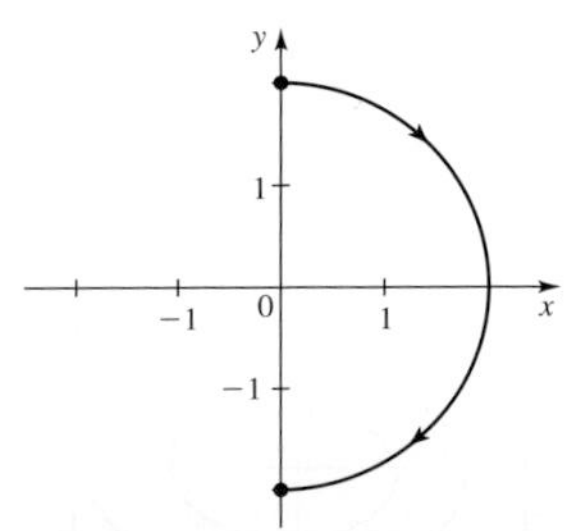

(b) $x^2 + y^2 = 4, x \geq 0$

15. (a)

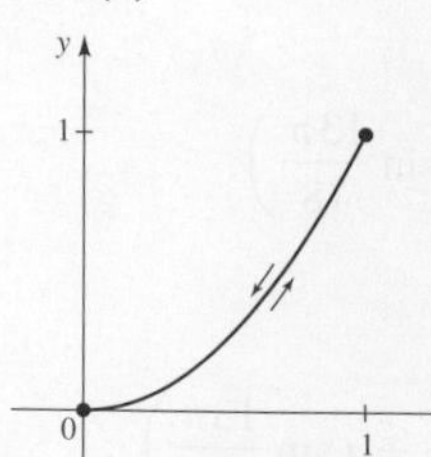

(b) $y = x^2, 0 \le x \le 1$

17. (a)

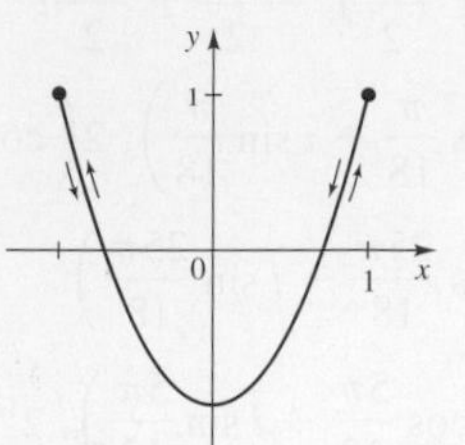

(b) $y = 2x^2 - 1, -1 \le x \le 1$

19. (a)

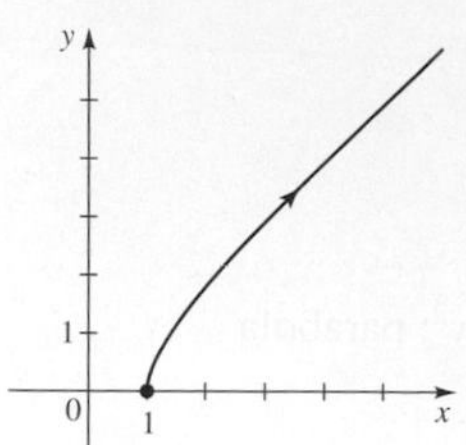

(b) $x^2 - y^2 = 1, x \ge 1, y \ge 0$

21. (a)

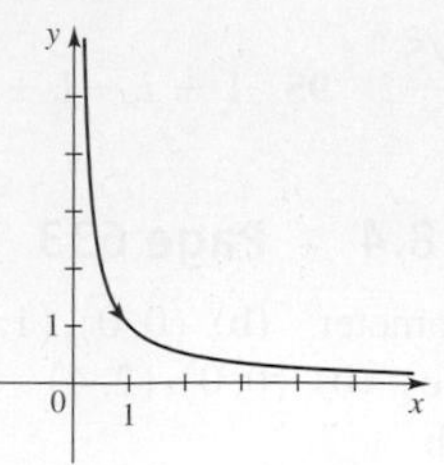

(b) $y = 1/x, x > 0$

23. (a)

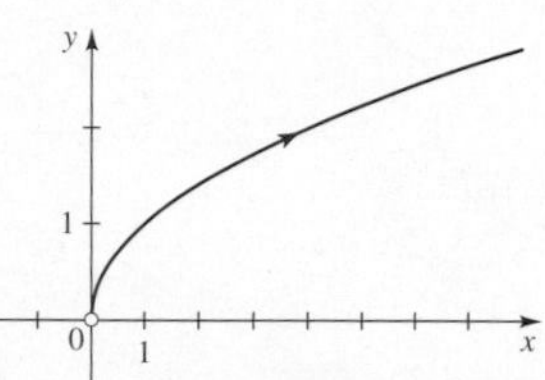

(b) $x = y^2, y > 0$

25. (a)

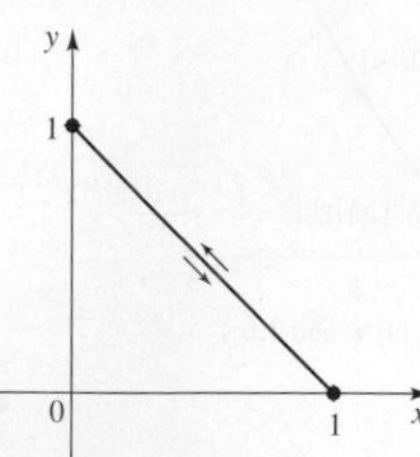

(b) $x + y = 1, 0 \le x \le 1$

27. 3, (3, 0), counterclockwise, 2π

29. 1, (0, 1), clockwise, π

Answers to #31-35 will vary.

31. $x = 5 \sin \frac{1}{2}t, y = 5 \cos \frac{1}{2}t$ **33.** $x = 4 + t, y = -1 + \frac{1}{2}t$

35. $x = a \cos t, y = a \sin t$

39.

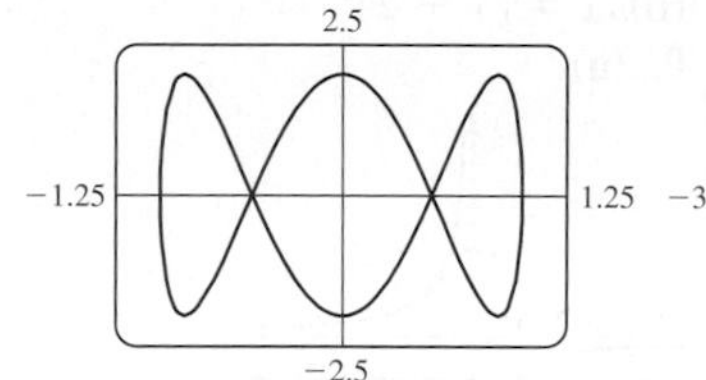

41.

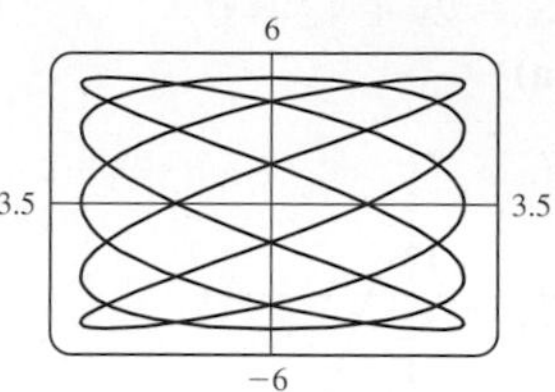

43.

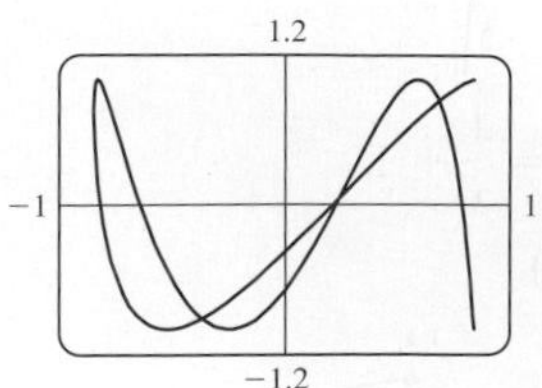

45. (a) $x = 2^{t/12} \cos t, y = 2^{t/12} \sin t$

(b)

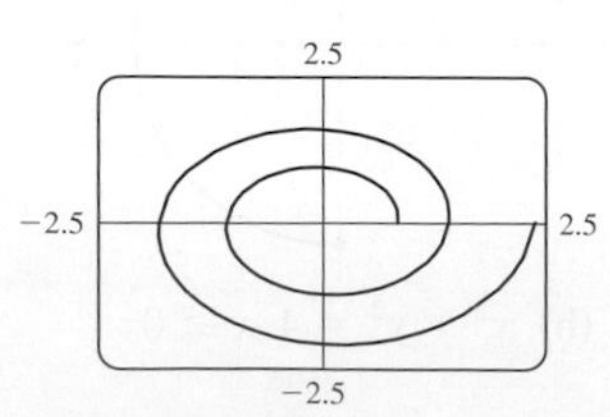

47. (a) $x = \dfrac{4 \cos t}{2 - \cos t}, y = \dfrac{4 \sin t}{2 - \cos t}$

(b)

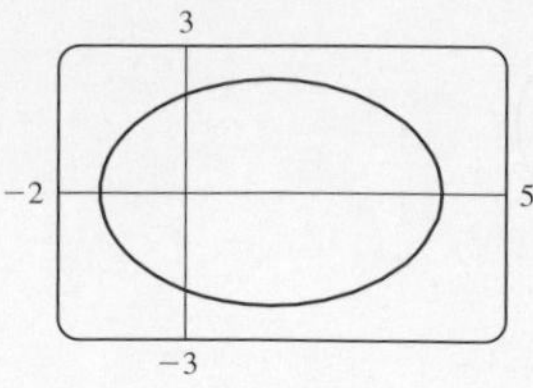

49. III **51.** II

53. (a) $x = a \cos \theta, y = b \sin \theta$

(b)

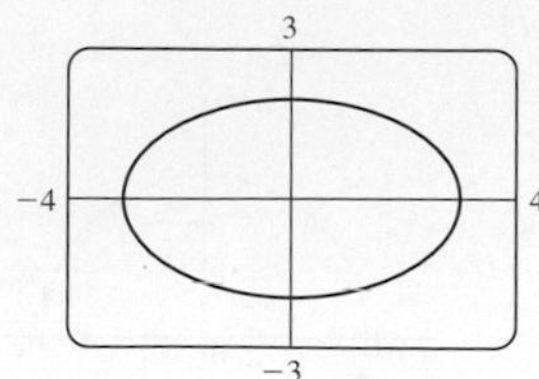

(c) $\dfrac{x^2}{a^2} + \dfrac{y^2}{b^2} = 1$

55. (b)

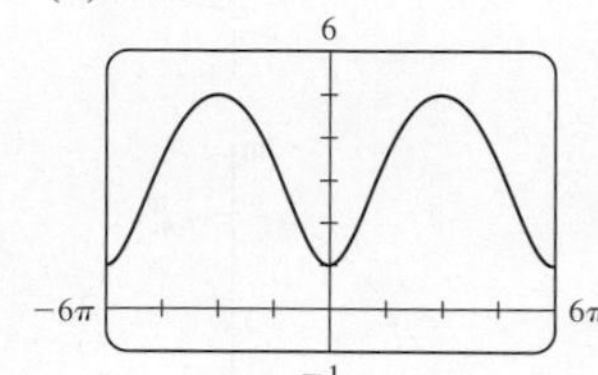

57. $\dfrac{y^2}{b^2} - \dfrac{x^2}{a^2} = 1$

59.

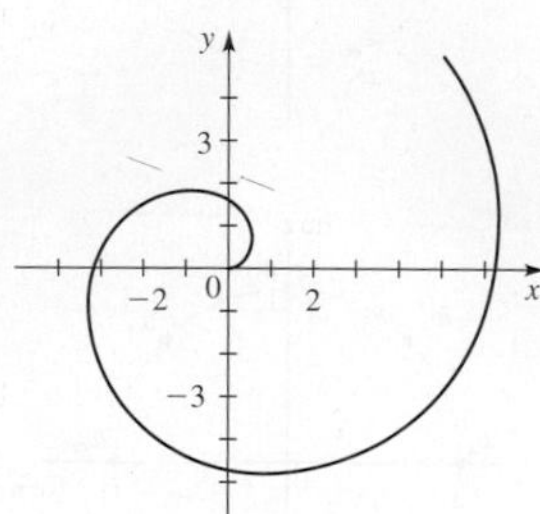

61.

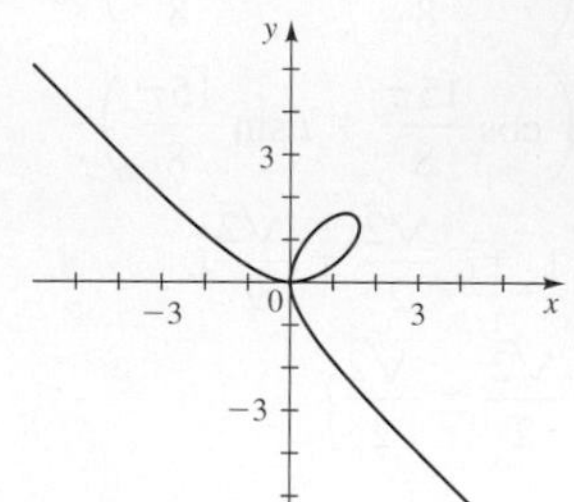

63. (b) $x^{2/3} + y^{2/3} = a^{2/3}$

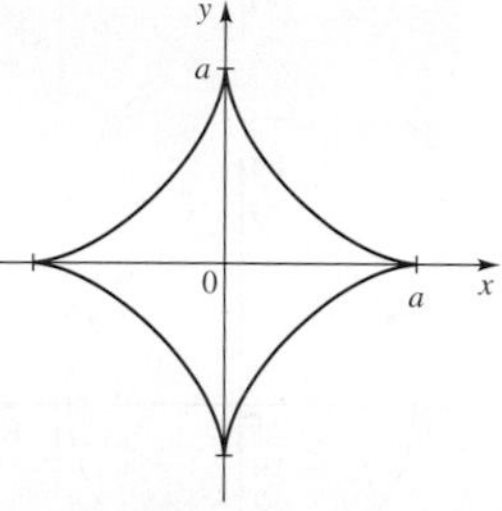

65. $x = a(\sin \theta \cos \theta + \cot \theta), y = a(1 + \sin^2 \theta)$

67. $y = a - a \cos\left(\dfrac{x + \sqrt{2ay - y^2}}{a}\right)$

69. (b)

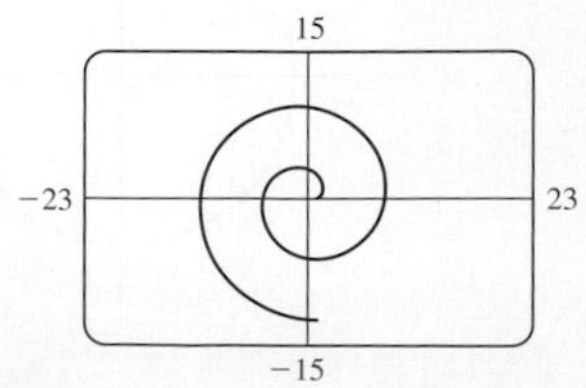

Section 8.5 ■ Page 644

1. (a) A, B

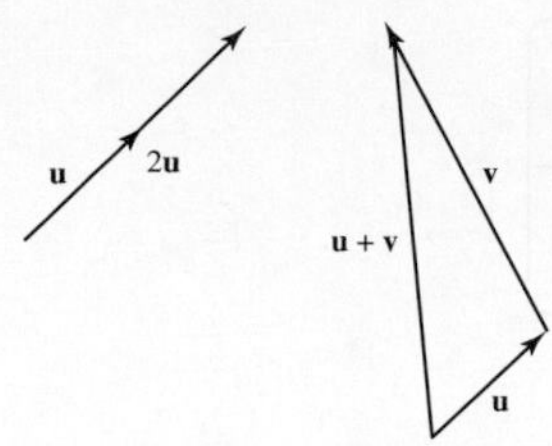

(b) $(2, 1), (4, 3), \langle 2, 2\rangle, \langle -3, 6\rangle, \langle 4, 4\rangle, \langle -1, 8\rangle$

2. (a) $\sqrt{a_1^2 + a_2^2}, 2\sqrt{2}$ **(b)** $\langle |\mathbf{w}| \cos\theta, |\mathbf{w}| \sin\theta\rangle$

3.

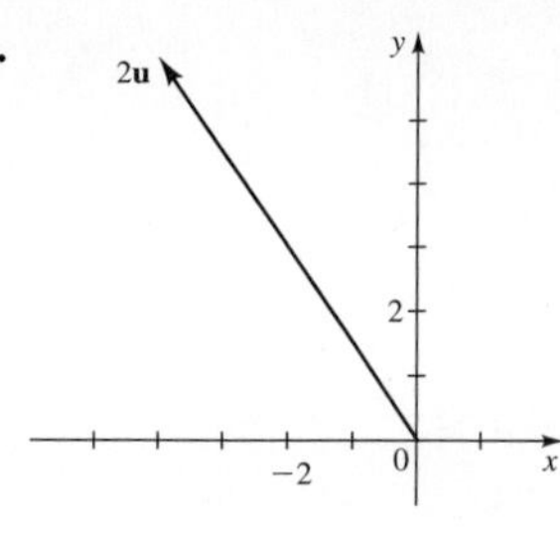

5.

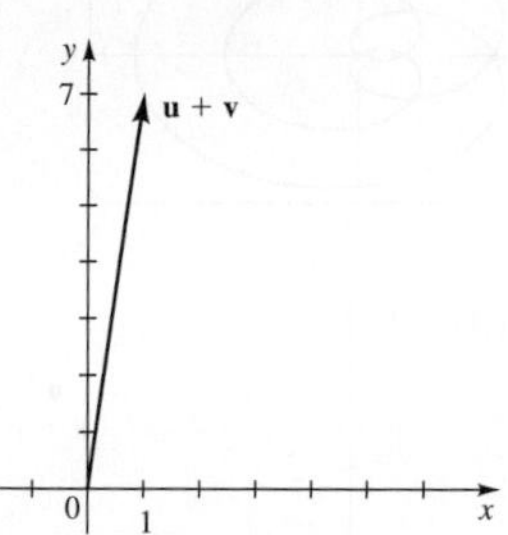

7.

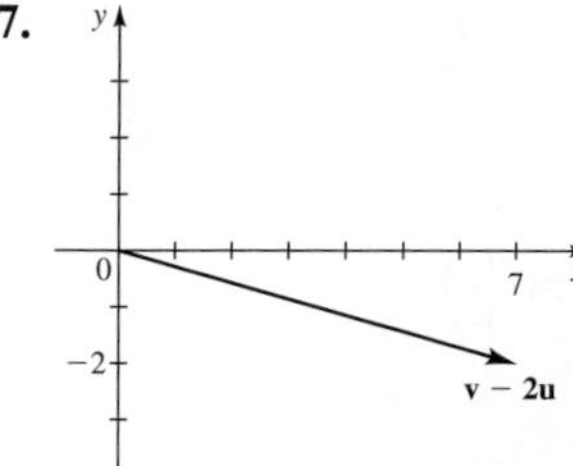

9. $\langle 3, 3\rangle$ **11.** $\langle 3, -1\rangle$ **13.** $\langle 3, 2\rangle$ **15.** $\langle -4, -3\rangle$ **17.** $\langle 0, 2\rangle$

19. **21.**

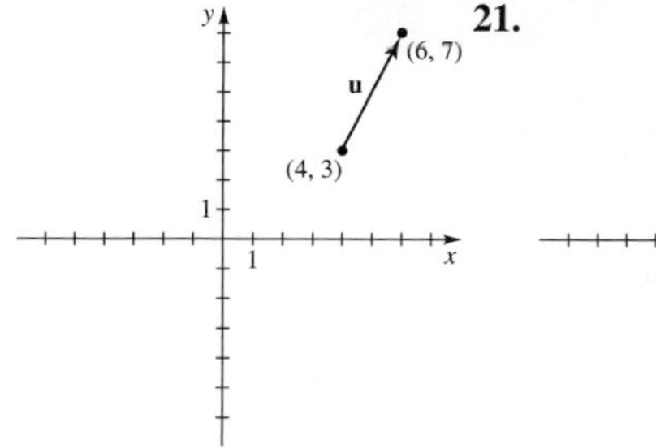

23.

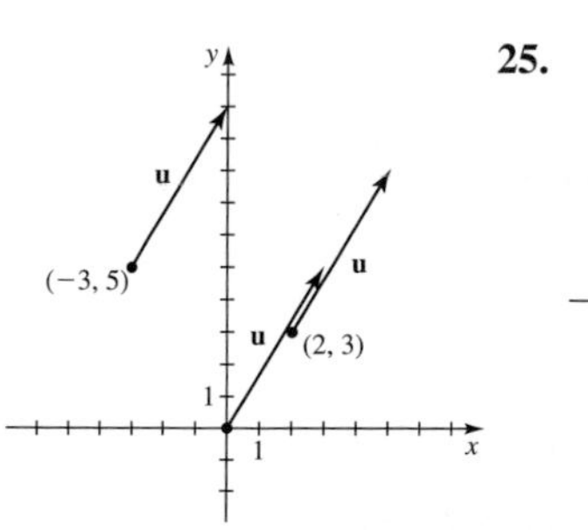

25.

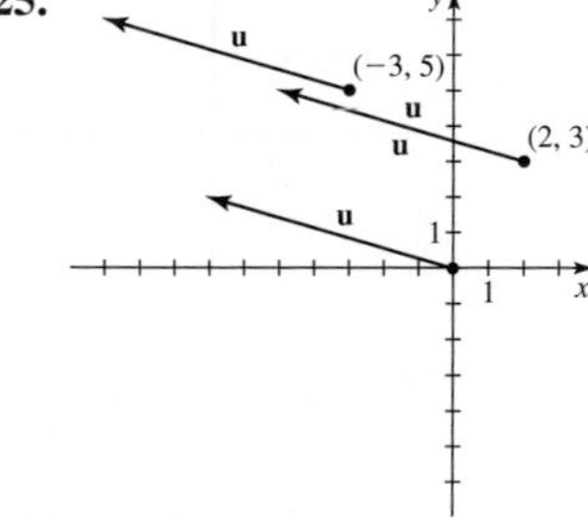

27. $2\mathbf{i} + 3\mathbf{j}$ **29.** $-2\mathbf{j}$ **31.** $\langle 2, 8\rangle, \langle 3, -6\rangle, \langle 0, 6\rangle, \langle 7, 4\rangle$

33. $\langle 0, -2\rangle, \langle 6, 0\rangle, \langle -2, -1\rangle, \langle 8, -3\rangle$

35. $4\mathbf{i} - 2\mathbf{j}, -3\mathbf{j}, 2\mathbf{i}, 6\mathbf{i} - 7\mathbf{j}$

37. $\sqrt{10}, \sqrt{13}, 2\sqrt{10}, \sqrt{13}/2, \sqrt{29}, \sqrt{17}, \sqrt{10} - \sqrt{13}$

39. $\sqrt{101}, 2\sqrt{2}, 2\sqrt{101}, \sqrt{2}, \sqrt{73}, \sqrt{145}, \sqrt{101} - 2\sqrt{2}$

41. $5\mathbf{i} + 5\sqrt{3}\mathbf{j}$ **43.** $-\dfrac{\sqrt{2}}{2}\mathbf{i} - \dfrac{\sqrt{2}}{2}\mathbf{j}$

45. $4 \cos 10°\mathbf{i} + 4 \sin 10°\mathbf{j} \approx 3.94\mathbf{i} + 0.69\mathbf{j}$

47. 5, 53.13° **49.** 13, 157.38° **51.** 2, 60° **53.** $68\sqrt{3}, -68$

55. $2\mathbf{i} - 3\mathbf{j}$ **57.** $\left(\dfrac{5\sqrt{2}}{2} + 3\right)\mathbf{i} + \left(\dfrac{5\sqrt{2}}{2}\right)\mathbf{j}$ **59. (a)** $64\mathbf{j}$

(b) $680\mathbf{i}$ **(c)** $680\mathbf{i} + 64\mathbf{j}$ **(d)** 683 km/h, N 84.6° E

61. 794 km/h, N 26.6° W **63. (a)** $10\mathbf{i}$ **(b)** $10\mathbf{i} + 10\sqrt{3}\mathbf{j}$

(c) $20\mathbf{i} + 10\sqrt{3}\mathbf{j}$ **(d)** 26.5 km/h, N 49.1° E

65. (a) $22.8\mathbf{i} + 7.4\mathbf{j}$ **(b)** 7.4 km/h, 22.8 km/h

67. (a) $\langle 5, -3\rangle$ **(b)** $\langle -5, 3\rangle$ **69. (a)** $-4\mathbf{j}$ **(b)** $4\mathbf{j}$

71. (a) $\langle -7.57, 10.61\rangle$ **(b)** $\langle 7.57, -10.61\rangle$

73. $\mathbf{T}_1 \approx -56.5\mathbf{i} + 67.4\mathbf{j}$, $\mathbf{T}_2 \approx 56.5\mathbf{i} + 32.6\mathbf{j}$

Section 8.6 ■ Page 653

1. $a_1b_1 + a_2b_2$; real number or scalar

2. $\dfrac{\mathbf{u}\cdot\mathbf{v}}{|\mathbf{u}||\mathbf{v}|}$; perpendicular; $\dfrac{\langle -4, 3\rangle\cdot\langle 3, 2\rangle}{|\langle -4, 3\rangle||\langle 3, 2\rangle|} = \dfrac{-6}{5\sqrt{13}}$; 109°

3. (a) $\dfrac{\mathbf{u}\cdot\mathbf{v}}{|\mathbf{v}|}$ **(b)** $\left(\dfrac{\mathbf{u}\cdot\mathbf{v}}{|\mathbf{v}|^2}\right)\mathbf{v}$

4. $\mathbf{F}\cdot\mathbf{D}$ **5. (a)** 2 **(b)** 45° **7. (a)** 1 **(b)** 60°

9. (a) −1 **(b)** 97° **11. (a)** $5\sqrt{3}$ **(b)** 30°

13. (a) 1 **(b)** 86° **15. (a)** −2 **(b)** 120° **17.** Yes

19. No **21.** Yes **23.** 9 **25.** −5 **27.** $-\frac{12}{5}$ **29.** −24

31. (a) $\langle 1, 1\rangle$ **(b)** $\mathbf{u}_1 = \langle 1, 1\rangle, \mathbf{u}_2 = \langle -3, 3\rangle$

33. (a) $\langle -\frac{1}{2}, \frac{3}{2}\rangle$ **(b)** $\mathbf{u}_1 = \langle -\frac{1}{2}, \frac{3}{2}\rangle, \mathbf{u}_2 = \langle \frac{3}{2}, \frac{1}{2}\rangle$

35. (a) $\langle -\frac{18}{5}, \frac{24}{5}\rangle$ **(b)** $\mathbf{u}_1 = \langle -\frac{18}{5}, \frac{24}{5}\rangle, \mathbf{u}_2 = \langle \frac{28}{5}, \frac{21}{5}\rangle$

37. −28 **39.** 25 **47.** 16 kN.m or N-m **49.** 11,431 N

51. (a) 12,554 N **(b)** 12,364 N **53.** 23.6°

Chapter 8 Review ■ Page 657

1. (a)

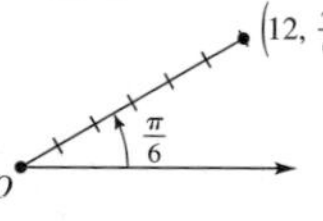

(b) $(6\sqrt{3}, 6)$

3. (a)

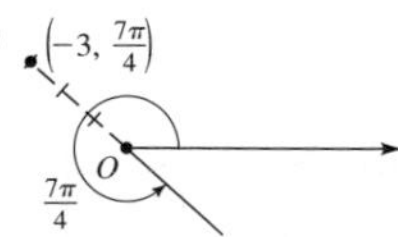

(b) $\left(\dfrac{-3\sqrt{2}}{2}, \dfrac{3\sqrt{2}}{2}\right)$

5. (a)

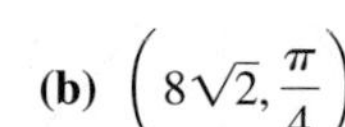

(b) $\left(8\sqrt{2}, \dfrac{\pi}{4}\right)$

(c) $\left(-8\sqrt{2}, \dfrac{5\pi}{4}\right)$

7. (a)

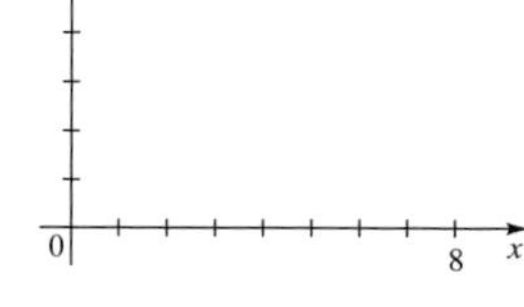

(b) $\left(12, \dfrac{5\pi}{4}\right)$

(c) $\left(-12, \dfrac{\pi}{4}\right)$

9. (a) $r = \dfrac{4}{\cos\theta + \sin\theta}$ **(b)**

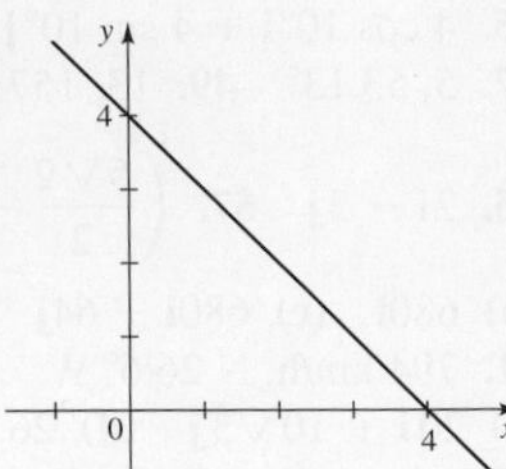

11. (a) $r = 4(\cos\theta + \sin\theta)$ **(b)**

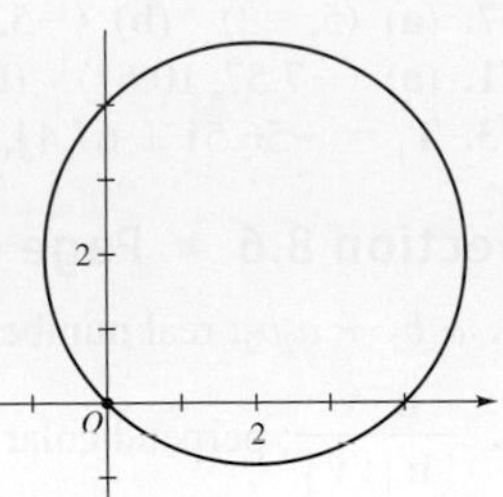

13. (a)

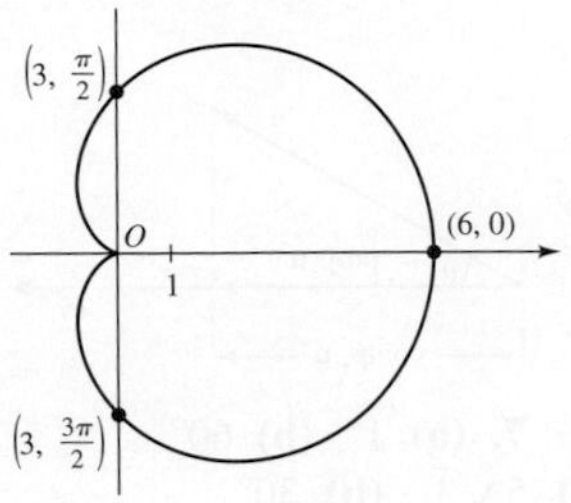

(b) $(x^2 + y^2 - 3x)^2 = 9(x^2 + y^2)$

15. (a)

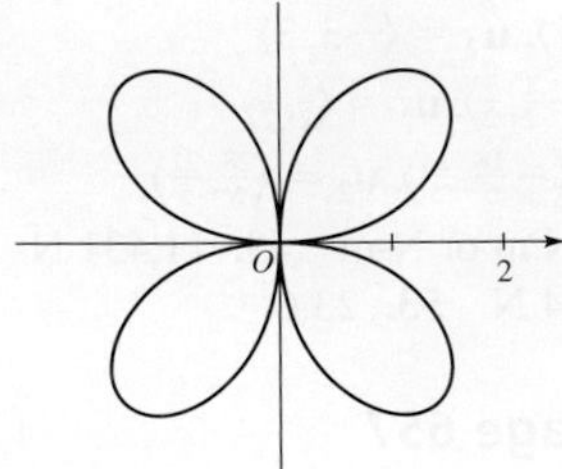

(b) $(x^2 + y^2)^3 = 16x^2y^2$

17. (a)

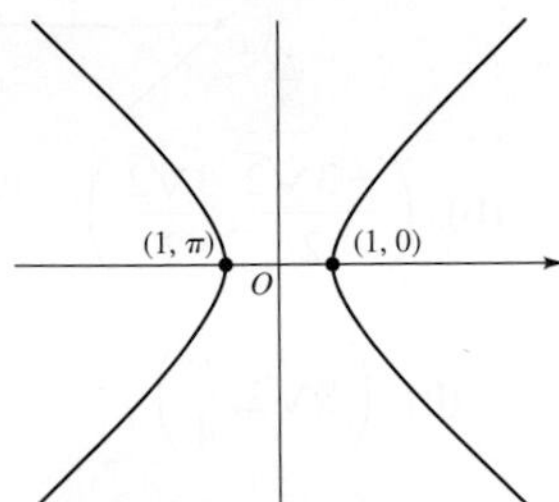

(b) $x^2 - y^2 = 1$

19. (a)

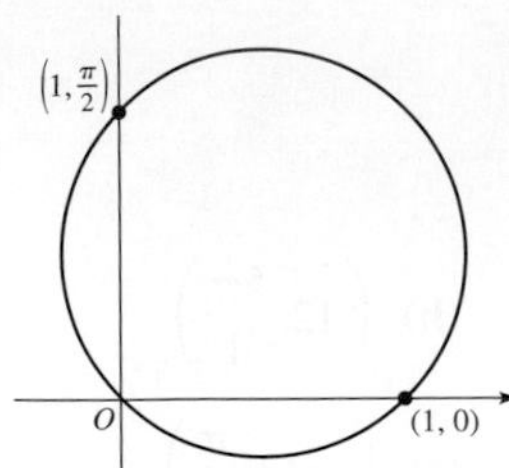

(b) $x^2 + y^2 = x + y$

21. $0 \le \theta \le 3\pi$

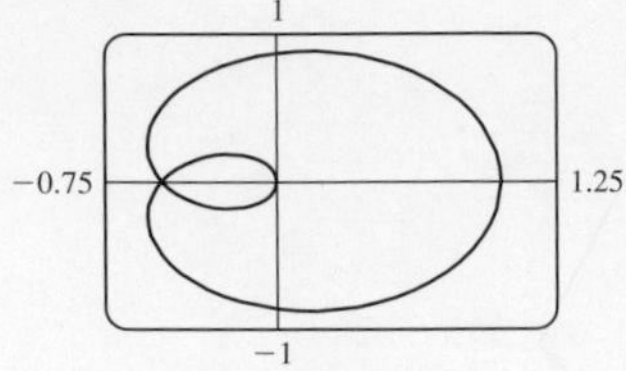

23. $0 \le \theta \le 6\pi$

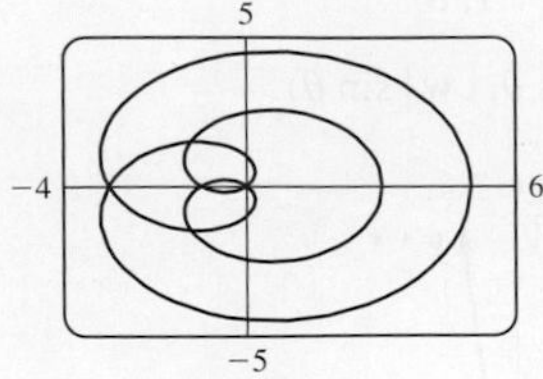

25. (a)

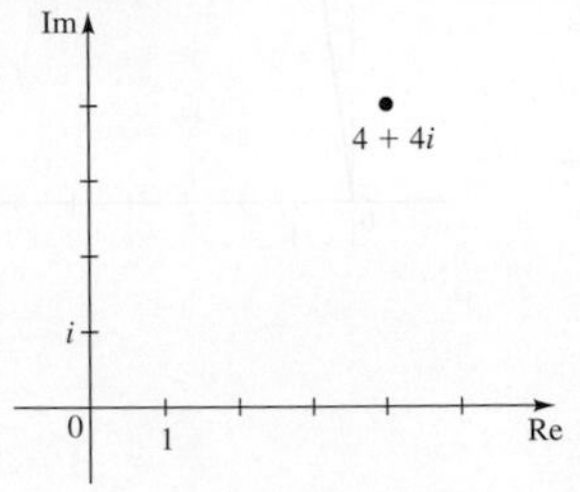

(b) $4\sqrt{2}, \dfrac{\pi}{4}$ **(c)** $4\sqrt{2}\left(\cos\dfrac{\pi}{4} + i\sin\dfrac{\pi}{4}\right)$

27. (a)

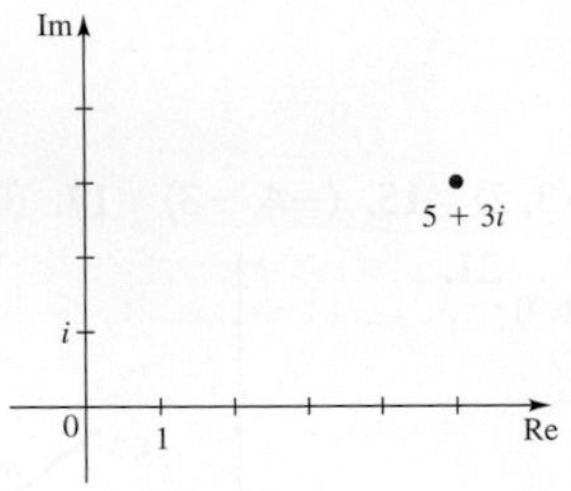

(b) $\sqrt{34}, \tan^{-1}\left(\frac{3}{5}\right)$ **(c)** $\sqrt{34}\left[\cos\left(\tan^{-1}\left(\frac{3}{5}\right)\right) + i\sin\left(\tan^{-1}\left(\frac{3}{5}\right)\right)\right]$

29. (a)

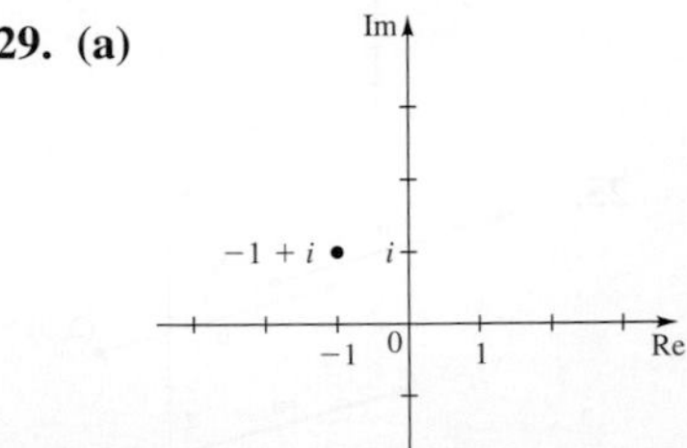

(b) $\sqrt{2}, \dfrac{3\pi}{4}$ **(c)** $\sqrt{2}\left(\cos\dfrac{3\pi}{4} + i\sin\dfrac{3\pi}{4}\right)$

31. $8(-1 + i\sqrt{3})$ **33.** $-\frac{1}{32}(1 + i\sqrt{3})$

35. $2\sqrt{2}(-1 + i), 2\sqrt{2}(1 - i)$

37. $\pm 1, \dfrac{1}{2} \pm \dfrac{\sqrt{3}}{2}i, -\dfrac{1}{2} \pm \dfrac{\sqrt{3}}{2}i$

39. (a)

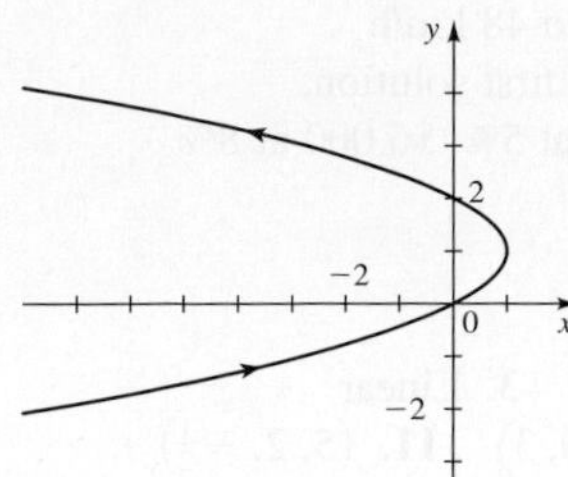

(b) $x = 2y - y^2$

41. (a)

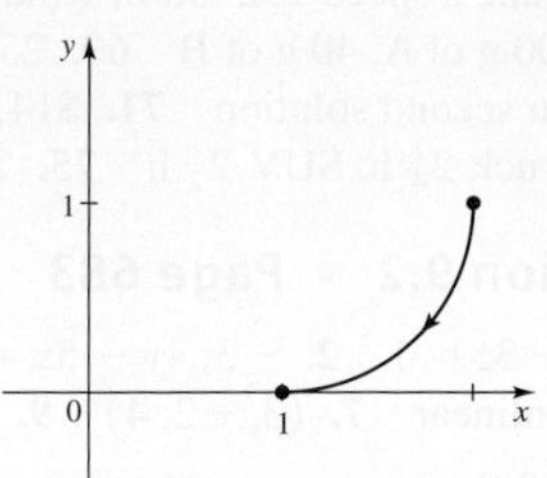

(b) $(x - 1)^2 + (y - 1)^2 = 1,$
$(1 \le x \le 2, 0 \le y \le 1)$

43.

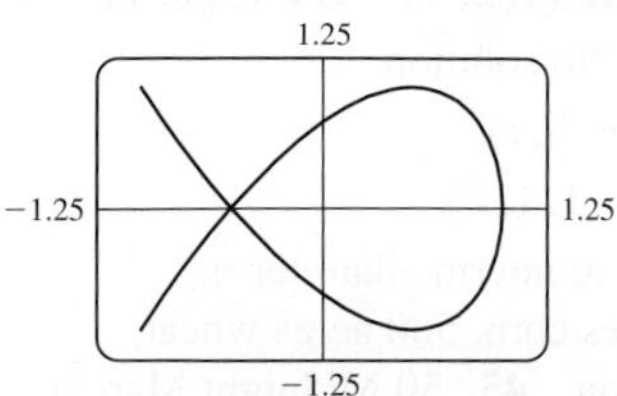

45. $x = \frac{1}{2}(1 + \cos\theta), y = \frac{1}{2}(\sin\theta + \tan\theta)$

47. $\sqrt{13}, \langle 6, 4\rangle, \langle -10, 2\rangle, \langle -4, 6\rangle, \langle -22, 7\rangle$

49. $\langle 3, -4\rangle$ **51.** 4, 120° **53.** $\langle 10, 10\sqrt{3}\rangle$

55. (a) $(240\mathbf{i} + 2\mathbf{j})$ **(b)** 240 kN, N 85.2° E

57. 5, 25, 60 **59.** $2\sqrt{2}$, 8, 0 **61.** Yes **63.** No, 45°

65. (a) $\frac{17\sqrt{37}}{37}$ **(b)** $\langle \frac{102}{37}, -\frac{17}{37}\rangle$ **(c)** $\mathbf{u}_1 = \langle \frac{102}{37}, -\frac{17}{37}\rangle, \mathbf{u}_2 = \langle \frac{9}{37}, \frac{54}{37}\rangle$

67. 6 kN.m

Chapter 8 Test ■ Page 659

1. (a) $(-4\sqrt{2}, -4\sqrt{2})$ **(b)** $(4\sqrt{3}, 5\pi/6), (-4\sqrt{3}, 11\pi/6)$

2. (a) Circle

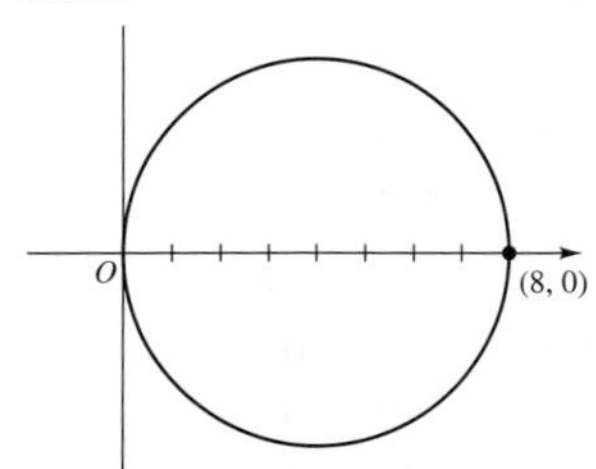

(b) $(x - 4)^2 + y^2 = 16$

3. Limaçon

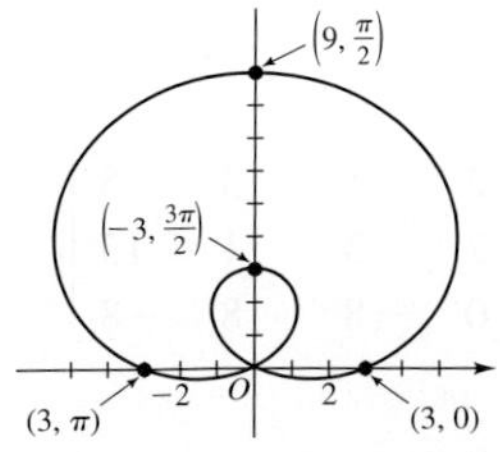

4. (a)

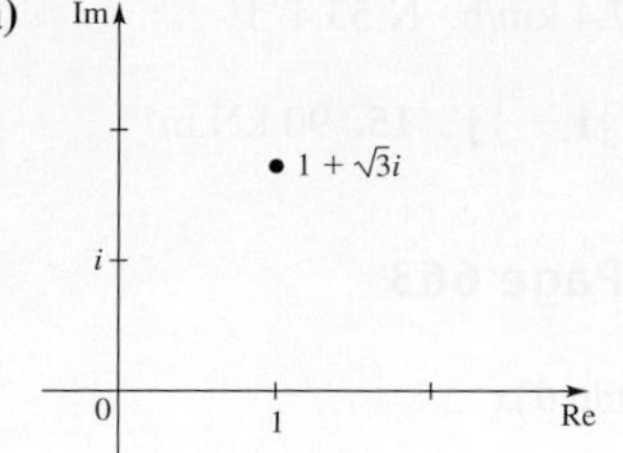

(b) $2\left(\cos\frac{\pi}{3} + i\sin\frac{\pi}{3}\right)$

(c) -512 **5.** $-8, \sqrt{3} + i$

6. $-3i, 3\left(\pm\frac{\sqrt{3}}{2} + \frac{1}{2}i\right)$

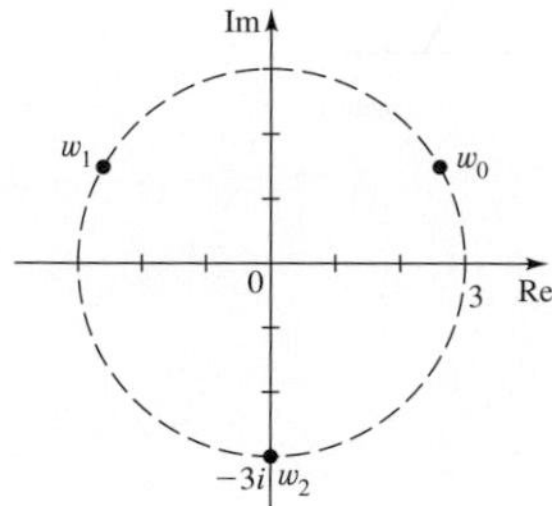

7. (a)

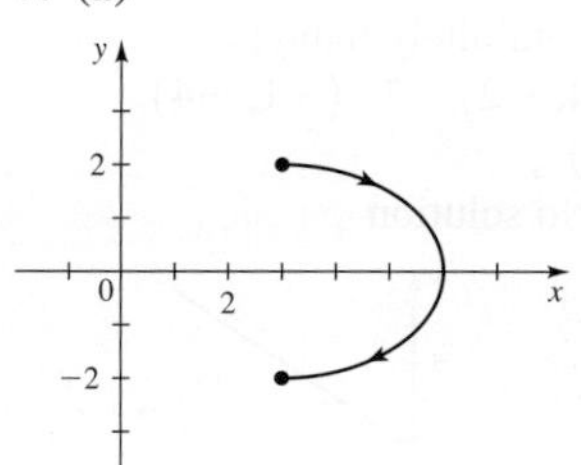

(b) $\frac{(x - 3)^2}{9} + \frac{y^2}{4} = 1 \; (x \ge 3)$

8. $x = 3 + t, y = 5 + 2t$

9. (a) 3, (0, 3), clockwise, π **(b)** $x = 3\sin 4t, y = 3\cos 4t$
(c) $x^2 + y^2 = 9$ **(d)** $r = 3$

10. (a)

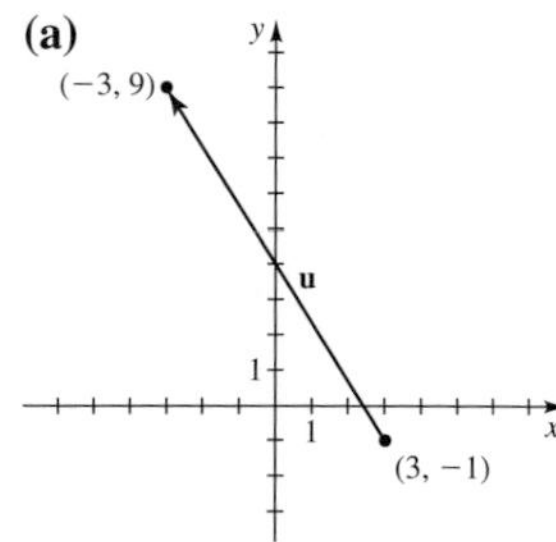

(b) $-6\mathbf{i} + 10\mathbf{j}$ **(c)** $2\sqrt{34}$

11. (a) $\langle 19, -3\rangle$ **(b)** $5\sqrt{2}$ **(c)** 0 **(d)** Yes

12. (a)

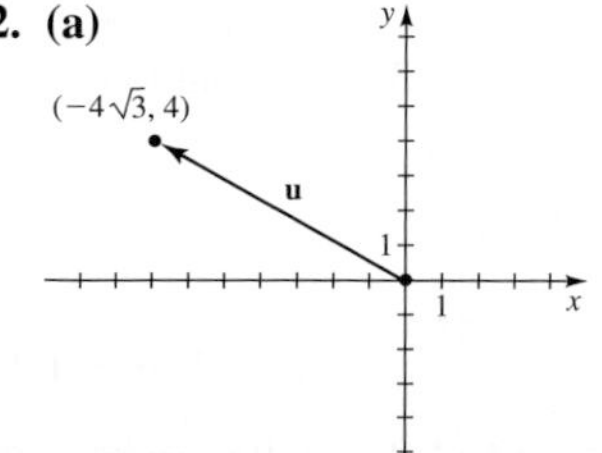

(b) 8, 150°

13. **(a)** $14\mathbf{i} + 6\sqrt{3}\mathbf{j}$ **(b)** 17.4 km/h, N 53.4° E

14. **(a)** 45° **(b)** $\dfrac{\sqrt{26}}{2}$ **(c)** $\frac{5}{2}\mathbf{i} - \frac{1}{2}\mathbf{j}$ **15.** 90 kN.m

Focus on Modeling ■ Page 663

1. $y = -\left(\dfrac{g}{2v_0^2\cos^2\theta}\right)x^2 + (\tan\theta)x$

3. **(a)** 62.26 s **(b)** 4561 m **(c)** 1628 m

(d)

7. No, $\theta \approx 23°$

Chapter 9

Section 9.1 ■ Page 674

1. x, y; equation; (2, 1) **2.** substitution, elimination, graphical **3.** no, infinitely many **4.** infinitely many; $1 - t$; (1, 0), (−3, 4), (5, −4) **5.** (4, −2) **7.** (−1, −4) **9.** (5, 1) **11.** (−3, 2) **13.** (−2, 3)

15. (2, −2)

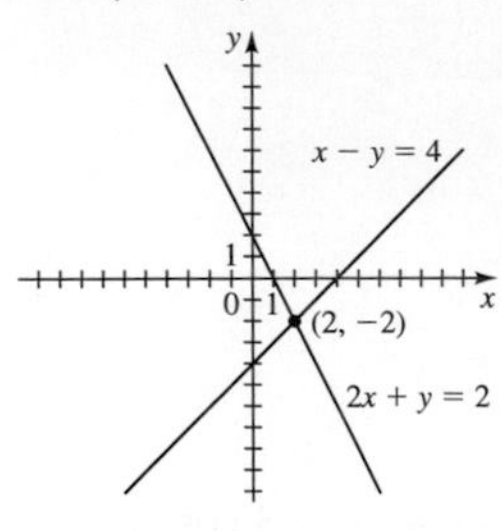

17. No solution

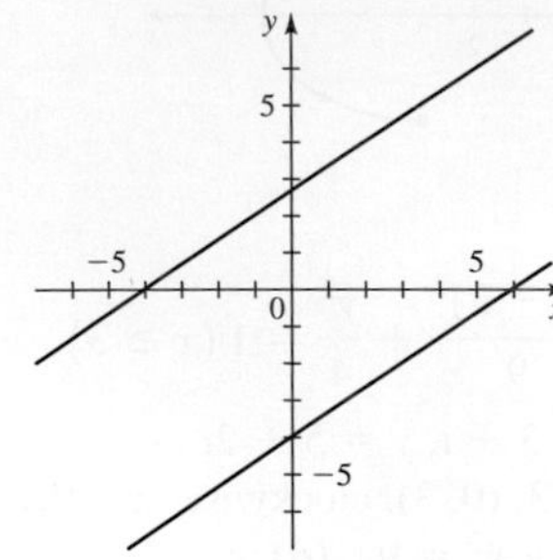

19. Infinitely many solutions

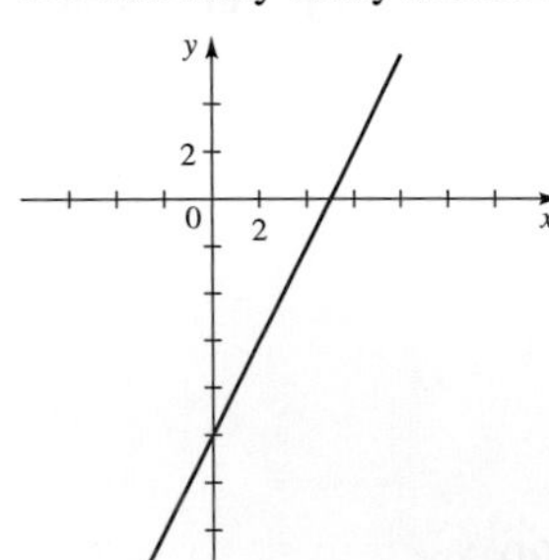

21. (3, 1) **23.** (3, −1) **25.** (2, 1) **27.** (3, 5) **29.** (1, 3) **31.** (6, −6) **33.** (10, −9) **35.** (2, 1) **37.** No solution **39.** $\left(t, \frac{1}{3}t - \frac{5}{3}\right)$ **41.** $\left(t, 3 - \frac{3}{2}t\right)$ **43.** (−3, −7) **45.** $\left(t, 5 - \frac{5}{6}t\right)$ **47.** (5, 10) **49.** No solution **51.** (3.87, 2.74) **53.** (61.00, 20.00) **55.** $\left(-\dfrac{1}{a-1}, \dfrac{1}{a-1}\right)$

57. $\left(\dfrac{1}{a+b}, \dfrac{1}{a+b}\right)$ **59.** 22, 12 **61.** 5 dimes, 9 quarters

63. 200 liters of regular gas, 80 liters of premium gas **65.** Plane's speed 192 km/h, wind speed 48 km/h **67.** 200 g of A, 40 g of B **69.** 25% in first solution, 10% in second solution **71.** \$14,000 at 5%, \$6,000 at 8% **73.** Truck $2\frac{1}{4}$ h, SUV $2\frac{1}{2}$ h **75.** 25

Section 9.2 ■ Page 683

1. $x + 3z = 1$ **2.** −3; $4y - 5z = -4$ **3.** Linear **5.** Nonlinear **7.** (3, −2, 4) **9.** (4, 0, 3) **11.** $\left(5, 2, -\frac{1}{2}\right)$

13. $\begin{cases} 3x + y + z = 4 \\ -y + z = -1 \\ x - 2y - z = -1 \end{cases}$ **15.** $\begin{cases} 2x + y - 3z = 5 \\ 2x + 3y + z = 13 \\ -8y + 8z = -8 \end{cases}$

17. (2, −1, 3) **19.** (2, −1, 0) **21.** (1, 2, 1) **23.** (5, 0, 1) **25.** (0, 1, 2) **27.** $\left(\frac{1}{4}, \frac{1}{2}, -\frac{1}{2}\right)$ **29.** No solution **31.** No solution **33.** $(3 - t, -3 + 2t, t)$ **35.** $\left(2 - 2t, -\frac{2}{3} + \frac{4}{3}t, t\right)$ **37.** (1, −1, 1, 2) **39.** \$30,000 in short-term, \$30,000 in intermediate-term, \$40,000 in long-term **41.** 250 acres corn, 500 acres wheat, 450 acres soybeans **43.** No solution **45.** 50 Midnight Mango, 60 Tropical Torrent, 30 Pineapple Power **47.** 1500 shares of A, 1200 shares of B, 1000 shares of C

Section 9.3 ■ Page 695

1. dependent, inconsistent

2. $\begin{bmatrix} 1 & 1 & -1 & 1 \\ 1 & 0 & 2 & -3 \\ 0 & 2 & -1 & 3 \end{bmatrix}$

3. **(a)** x and y **(b)** dependent **(c)** $x = 3 + t, y = 5 - 2t, z = t$ **4.** **(a)** $x = 2, y = 1, z = 3$ **(b)** $x = 2 - t, y = 1 - t, z = t$ **(c)** No solution **5.** 3 × 2 **7.** 2 × 1 **9.** 1 × 3

11. $\begin{bmatrix} 3 & 1 & -1 & 2 \\ 2 & -1 & 0 & 1 \\ 1 & 0 & -1 & 3 \end{bmatrix}$

13. **(a)** Yes **(b)** Yes **(c)** $\begin{cases} x = -3 \\ y = 5 \end{cases}$

15. **(a)** Yes **(b)** No **(c)** $\begin{cases} x + 2y + 8z = 0 \\ y + 3z = 2 \\ 0 = 0 \end{cases}$

17. **(a)** No **(b)** No **(c)** $\begin{cases} x = 0 \\ 0 = 0 \\ y + 5z = 1 \end{cases}$

19. **(a)** Yes **(b)** Yes **(c)** $\begin{cases} x + 3y - w = 0 \\ z + 2w = 0 \\ 0 = 1 \\ 0 = 0 \end{cases}$

21. $\begin{bmatrix} -1 & 1 & 2 & 0 \\ 0 & 4 & 7 & 4 \\ 1 & -2 & -1 & -1 \end{bmatrix}$ **23.** $\begin{bmatrix} 2 & 1 & -3 & 5 \\ 2 & 3 & 1 & 13 \\ 0 & -8 & 8 & -8 \end{bmatrix}$

25. **(a)** $\begin{cases} x - 2y + 4z = 3 \\ y + 2z = 7 \\ z = 2 \end{cases}$ **(b)** (1, 3, 2)

27. (a) $\begin{cases} x + 2y + 3z - w = 7 \\ y - 2z = 5 \\ z + 2w = 5 \\ w = 3 \end{cases}$ **(b)** $(7, 3, -1, 3)$

29. $(1, 1, 2)$ **31.** $(2, 0, -1)$ **33.** $(1, -3, 2)$ **35.** $(-1, 5, 0)$
37. $(10, 3, -2)$ **39.** No solution **41.** $(2 - 3t, 3 - 5t, t)$
43. No solution **45.** $(-2t + 5, t - 2, t)$
47. $(-\frac{1}{2}s + t + 6, s, t)$ **49.** $(-2, 1, 3)$ **51.** No solution
53. $(-9, 2, 0)$ **55.** $(5 - t, -3 + 5t, t)$ **57.** $(0, -3, 0, -3)$
59. $(-1, 0, 0, 1)$ **61.** $(\frac{1}{3}s - \frac{2}{3}t, \frac{1}{3}s + \frac{1}{3}t, s, t)$
63. $(\frac{7}{4} - \frac{7}{4}t, -\frac{7}{4} + \frac{3}{4}t, \frac{9}{4} + \frac{3}{4}t, t)$
65. $x = 1.25, y = -0.25, z = 0.75$
67. $x = 1.2, y = 3.4, z = -5.2, w = -1.3$
69. 2 VitaMax, 1 Vitron, 2 VitaPlus **71.** 5-km run, 2-km swim, 30-km cycle **73.** Impossible

Section 9.4 ■ Page 705

1. dimension **2. (a)** columns, rows **(b)** (ii), (iii) **3.** (i), (ii)

4. $\begin{bmatrix} 4 & 9 & -7 \\ 7 & -7 & 0 \\ 4 & -5 & -5 \end{bmatrix}$ **5.** No **7.** $a = -5, b = 3$

9. $\begin{bmatrix} 1 & 3 \\ 1 & 5 \end{bmatrix}$ **11.** $\begin{bmatrix} 3 & 6 \\ 12 & -3 \\ 3 & 0 \end{bmatrix}$ **13.** Impossible

15. $\begin{bmatrix} 5 & 2 & 1 \\ 7 & 10 & -7 \end{bmatrix}$ **17.** $\begin{bmatrix} -1 & -\frac{1}{2} \\ 1 & 2 \end{bmatrix}$ **19.** Impossible

21. $\begin{bmatrix} 0 & -5 \\ -25 & -20 \\ -10 & 10 \end{bmatrix}$ **23. (a)** $\begin{bmatrix} 5 & -2 & 5 \\ 1 & 1 & 0 \end{bmatrix}$ **(b)** Impossible

25. (a) $\begin{bmatrix} 10 & -25 \\ 0 & 35 \end{bmatrix}$ **(b)** Impossible

27. (a) Impossible **(b)** $\begin{bmatrix} 14 & -14 \end{bmatrix}$

29. (a) $\begin{bmatrix} -4 & 7 \\ 14 & -7 \end{bmatrix}$ **(b)** $\begin{bmatrix} 6 & -8 \\ 4 & -17 \end{bmatrix}$

31. (a) $\begin{bmatrix} 5 & -3 & 10 \\ 6 & 1 & 0 \\ -5 & 2 & 2 \end{bmatrix}$ **(b)** $\begin{bmatrix} -1 \\ 8 \\ -1 \end{bmatrix}$

33. (a) $\begin{bmatrix} 4 & -45 \\ 0 & 49 \end{bmatrix}$ **(b)** $\begin{bmatrix} 8 & -335 \\ 0 & 343 \end{bmatrix}$

35. (a) $\begin{bmatrix} 13 \\ -7 \end{bmatrix}$ **(b)** Impossible **37.** $\begin{bmatrix} 1.56 & -5.62 \\ 1.28 & -0.88 \\ -1.09 & 0.97 \end{bmatrix}$

39. $\begin{bmatrix} -0.35 & 0.03 & 0.33 \\ -0.55 & -1.05 & 1.05 \\ -2.41 & -4.31 & 4.46 \end{bmatrix}$ **41.** Impossible

43. $x = 2, y = -1$ **45.** $x = 1, y = -2$

47. $\begin{bmatrix} 2 & -5 \\ 3 & 2 \end{bmatrix} \begin{bmatrix} x \\ y \end{bmatrix} = \begin{bmatrix} 7 \\ 4 \end{bmatrix}$

49. $\begin{bmatrix} 3 & 2 & -1 & 1 \\ 1 & 0 & -1 & 0 \\ 0 & 3 & 1 & -1 \end{bmatrix} \begin{bmatrix} x_1 \\ x_2 \\ x_3 \\ x_4 \end{bmatrix} = \begin{bmatrix} 0 \\ 5 \\ 4 \end{bmatrix}$

51. Only ACB is defined. $ACB = \begin{bmatrix} -3 & -21 & 27 & -6 \\ -2 & -14 & 18 & -4 \end{bmatrix}$

53. (a) $\begin{bmatrix} 5 \\ 22 \\ 7 \end{bmatrix}$

(b) Five members have no postsecondary education, 22 have 1 to 4 years, and seven have more than 4 years.

55. (a) $\begin{bmatrix} 353.75 \\ 656.25 \\ 892.50 \end{bmatrix}$ **(b)** \$353.75 **(c)** \$1902.50

57. (a) $\begin{bmatrix} \$32{,}000 & \$18{,}000 \\ \$42{,}000 & \$26{,}800 \\ \$44{,}000 & \$26{,}800 \end{bmatrix}$ **(b)** \$42,000 **(c)** \$71,600

59. (a) $\begin{bmatrix} 97.00 \\ 46.50 \\ 41.00 \end{bmatrix}$ Ashton's stand sold \$97 of produce on Saturday. Bryn's stand sold \$46.50. Cimeron's stand sold \$41.

(b) $\begin{bmatrix} 70.00 \\ 33.50 \\ 48.50 \end{bmatrix}$ Ashton's stand sold \$70 of produce on Sunday. Bryn's stand sold \$33.50. Cimeron's stand sold \$48.50.

(c) $\begin{bmatrix} 220 & 110 & 90 \\ 75 & 45 & 50 \\ 120 & 55 & 50 \end{bmatrix}$ This represents the total numbers of melons, squash, and tomatoes sold during the weekend.

(d) $\begin{bmatrix} 167.00 \\ 80.00 \\ 89.50 \end{bmatrix}$ During the weekend Ashton's stand sold \$167, Bryn's stand sold \$80, and Cimeron's stand sold \$89.50 of produce.

Section 9.5 ■ Page 715

1. (a) identity **(b)** A, A **(c)** inverse

2. (a) $\underset{A}{\begin{bmatrix} 5 & 3 \\ 3 & 2 \end{bmatrix}} \underset{X}{\begin{bmatrix} x \\ y \end{bmatrix}} = \underset{B}{\begin{bmatrix} 4 \\ 3 \end{bmatrix}}$ **(b)** $\begin{bmatrix} 2 & -3 \\ -3 & 5 \end{bmatrix}$

(c) $\underset{A^{-1}}{\begin{bmatrix} 2 & -3 \\ -3 & 5 \end{bmatrix}} \underset{B}{\begin{bmatrix} 4 \\ 3 \end{bmatrix}} = \begin{bmatrix} -1 \\ 3 \end{bmatrix}$ **(d)** $x = -1, y = 3$

7. $\begin{bmatrix} 1 & -2 \\ -\frac{3}{2} & \frac{7}{2} \end{bmatrix}$ **9.** $\begin{bmatrix} \frac{1}{3} & -\frac{1}{2} \\ 2 & 2 \end{bmatrix}$ **11.** $\begin{bmatrix} 9 & -2 \\ -13 & 3 \end{bmatrix}$

13. $\begin{bmatrix} 13 & 5 \\ -5 & -2 \end{bmatrix}$ **15.** No inverse **17.** $\begin{bmatrix} 1 & 2 \\ -\frac{1}{2} & \frac{2}{3} \end{bmatrix}$

19. $\begin{bmatrix} -4 & -4 & 5 \\ 1 & 1 & -1 \\ 5 & 4 & -6 \end{bmatrix}$ **21.** No inverse

23. $\begin{bmatrix} -\frac{9}{2} & -1 & 4 \\ 3 & 1 & -3 \\ \frac{7}{2} & 1 & -3 \end{bmatrix}$ **25.** $\begin{bmatrix} 0 & 0 & -2 & 1 \\ -1 & 0 & 1 & 1 \\ 0 & 1 & -1 & 0 \\ 1 & 0 & 0 & -1 \end{bmatrix}$

27. $\begin{bmatrix} \frac{2}{3} & \frac{4}{3} & 3 \\ 1 & 1 & 3 \\ \frac{1}{3} & \frac{2}{3} & 1 \end{bmatrix}$ **29.** $\begin{bmatrix} -2 & 3 & -1 & -2 \\ 0 & -1 & 0 & \frac{1}{2} \\ -2 & 2 & -1 & -2 \\ -1 & -1 & -1 & 0 \end{bmatrix}$

31. $\begin{bmatrix} 1 & -\frac{7}{2} & \frac{1}{6} \\ 0 & \frac{1}{2} & -\frac{1}{6} \\ 0 & 0 & \frac{1}{3} \end{bmatrix}$ **33.** $\begin{bmatrix} 1 & 0 & 0 & 0 \\ 0 & \frac{1}{2} & 0 & 0 \\ 0 & 0 & \frac{1}{4} & 0 \\ 0 & 0 & 0 & \frac{1}{7} \end{bmatrix}$

35. $\begin{bmatrix} -\frac{1}{4} & \frac{3}{4} & \frac{3}{4} \\ -\frac{7}{16} & -\frac{23}{16} & -\frac{3}{16} \\ \frac{7}{8} & -\frac{1}{8} & -\frac{5}{8} \end{bmatrix}$ **37.** $\begin{bmatrix} -7 & -3 & -4 \\ \frac{22}{7} & -\frac{2}{7} & \frac{16}{7} \\ \frac{50}{7} & \frac{26}{7} & \frac{37}{7} \end{bmatrix}$

39. $x = 3, y = -4$ **41.** $x = 126, y = -50$
43. $x = -38, y = 9, z = 47$ **45.** $x = -20, y = 10, z = 16$
47. $x = 3, y = 2, z = 1$ **49.** $x = 3, y = -2, z = 2$
51. $x = 8, y = 1, z = 0, w = 3$

53. $\begin{bmatrix} 7 & 2 & 3 \\ 10 & 3 & 5 \end{bmatrix}$ **55.** $\frac{1}{2a}\begin{bmatrix} 1 & 1 \\ -1 & 1 \end{bmatrix}$

57. $\begin{bmatrix} 1 & -\frac{1}{x} \\ -\frac{1}{x} & \frac{2}{x^2} \end{bmatrix}$; inverse does not exist for $x = 0$

59. $\frac{1}{2}\begin{bmatrix} 1 & e^{-x} & 0 \\ e^{-x} & -e^{-2x} & 0 \\ 0 & 0 & 1 \end{bmatrix}$; inverse exists for all x

61. (a) $\begin{bmatrix} 0 & 1 & -1 \\ -2 & \frac{3}{2} & 0 \\ 1 & -\frac{3}{2} & 1 \end{bmatrix}$ **(b)** 1 serving type A, 1 serving type B, 2 servings type C
(c) 2 servings type A, 0 servings type B, 1 serving type C **(d)** No

63. (a) $\begin{cases} 9x + 11y + 8z = 740 \\ 13x + 15y + 16z = 1204 \\ 8x + 7y + 14z = 828 \end{cases}$

(b) $\begin{bmatrix} 9 & 11 & 8 \\ 13 & 15 & 16 \\ 8 & 7 & 14 \end{bmatrix}\begin{bmatrix} x \\ y \\ z \end{bmatrix} = \begin{bmatrix} 740 \\ 1204 \\ 828 \end{bmatrix}$

(c) $A^{-1} = \begin{bmatrix} \frac{7}{4} & -\frac{7}{4} & 1 \\ -\frac{27}{28} & \frac{31}{28} & -\frac{5}{7} \\ -\frac{29}{56} & \frac{25}{56} & -\frac{1}{7} \end{bmatrix}$

The commission is \$16 on a standard model, \$28 on a deluxe model, and \$36 on a super-deluxe model.

Section 9.6 ■ Page 726

1. True **2.** True **3.** True **4. (a)** $2\cdot 4 - 1\cdot(-3) = 11$
(b) $+1(2\cdot 4 - 1\cdot(-3)) - 0(3\cdot 4 - 1\cdot 0) + 2(3\cdot(-3) - 2\cdot 0) = -7$
5. 6 **7.** 0 **9.** -4 **11.** Does not exist **13.** $\frac{1}{8}$ **15.** 20, 20
17. -12, 12 **19.** 0, 0 **21.** 4, has an inverse
23. 5000, has an inverse **25.** 0, does not have an inverse
27. -4, has an inverse **29.** -6, has an inverse
31. -12, has an inverse **33.** 0, does not have an inverse
35. -18 **37.** 120 **39. (a)** -2 **(b)** -2 **(c)** Yes
41. $(-2, 5)$ **43.** $(0.6, -0.4)$ **45.** $(4, -1)$ **47.** $(4, 2, -1)$
49. $(1, 3, 2)$ **51.** $(0, -1, 1)$ **53.** $\left(\frac{189}{29}, -\frac{108}{29}, \frac{88}{29}\right)$
55. $\left(\frac{1}{2}, \frac{1}{4}, \frac{1}{4}, -1\right)$ **57.** 21 **59.** $\frac{63}{2}$ **61.** $abcde$ **63.** 0, 1, 2
65. 1, -1 **69. (a)** 0 **(b)** (i) Yes, (ii) No

71. (a) $\begin{cases} x + y + z = 18 \\ 75x + 90y + 60z = 1380 \\ -75x + 90y + 60z = 180 \end{cases}$

(b) 8 apples, 6 peaches, 4 pears
73. 7 million m^2

Section 9.7 ■ Page 734

1. (iii) **2.** (ii) **3.** $\dfrac{A}{x-1} + \dfrac{B}{x+2}$

5. $\dfrac{A}{x-2} + \dfrac{B}{(x-2)^2} + \dfrac{C}{x+4}$

7. $\dfrac{A}{x-3} + \dfrac{Bx+C}{x^2+4}$ **9.** $\dfrac{Ax+B}{x^2+1} + \dfrac{Cx+D}{x^2+2}$

11. $\dfrac{A}{x} + \dfrac{B}{2x-5} + \dfrac{C}{(2x-5)^2} + \dfrac{D}{(2x-5)^3} + \dfrac{Ex+F}{x^2+2x+5} + \dfrac{Gx+H}{(x^2+2x+5)^2}$

13. $\dfrac{1}{x-1} - \dfrac{1}{x+1}$ **15.** $\dfrac{1}{x-1} - \dfrac{1}{x+4}$

17. $\dfrac{2}{x-3} - \dfrac{2}{x+3}$ **19.** $\dfrac{1}{x-2} - \dfrac{1}{x+2}$

21. $\dfrac{3}{x-4} - \dfrac{2}{x+2}$ **23.** $\dfrac{-\frac{1}{2}}{2x-1} + \dfrac{\frac{3}{2}}{4x-3}$

25. $\dfrac{2}{x-2} + \dfrac{3}{x+2} - \dfrac{1}{2x-1}$ **27.** $\dfrac{2}{x+1} - \dfrac{1}{x} + \dfrac{1}{x^2}$

29. $\dfrac{1}{2x+3} - \dfrac{3}{(2x+3)^2}$ **31.** $\dfrac{2}{x} - \dfrac{1}{x^3} - \dfrac{2}{x+2}$

33. $\dfrac{4}{x+2} - \dfrac{4}{x-1} + \dfrac{2}{(x-1)^2} + \dfrac{1}{(x-1)^3}$

35. $\dfrac{3}{x+2} - \dfrac{1}{(x+2)^2} - \dfrac{1}{(x+3)^2}$ **37.** $\dfrac{x+1}{x^2+3} - \dfrac{1}{x}$

39. $\dfrac{2x-5}{x^2+x+2} + \dfrac{5}{x^2+1}$ **41.** $\dfrac{1}{x^2+1} - \dfrac{x+2}{(x^2+1)^2} + \dfrac{1}{x}$

43. $x^2 + \dfrac{3}{x-2} - \dfrac{x+1}{x^2+1}$ **45.** $A = \dfrac{a+b}{2}, B = \dfrac{a-b}{2}$

Section 9.8 ■ Page 738

1. $(4, 8), (-2, 2)$ **3.** $(4, 16), (-3, 9)$ **5.** $(2, -2), (-2, 2)$
7. $(-25, 5), (-25, -5)$ **9.** $(-3, 4)$ $(3, 4)$
11. $(-2, -1), (-2, 1), (2, -1), (2, 1)$
13. $(-1, \sqrt{2}), (-1, -\sqrt{2}), \left(\frac{1}{2}, \sqrt{\frac{7}{2}}\right), \left(\frac{1}{2}, -\sqrt{\frac{7}{2}}\right)$
15. $(2, 4), \left(-\frac{5}{2}, \frac{7}{4}\right)$ **17.** $(0, 0), (1, -1), (-2, -4)$
19. $(4, 0)$ **21.** $(-2, -2)$ **23.** $(6, 2), (-2, -6)$
25. No solution
27. $(\sqrt{5}, 2), (\sqrt{5}, -2), (-\sqrt{5}, 2), (-\sqrt{5}, -2)$
29. $\left(3, -\frac{1}{2}\right), \left(-3, -\frac{1}{2}\right)$ **31.** $\left(\frac{1}{5}, \frac{1}{3}\right)$
33. $(2.00, 20.00), (-8.00, 0)$
35. $(-4.51, 2.17), (4.91, -0.97)$

37. $(1.23, 3.87), (-0.35, -4.21)$
39. $(-2.30, -0.70), (0.48, -1.19)$ **41.** $(\sqrt{10}, 10)$
43. $(-5, -8), (8, 5)$ **45.** 12 cm by 15 cm
47. Length 15, width 20 **49.** (400.50, 200.25), 447.77 m
51. (12, 8)

Section 9.9 ■ Page 747

1. 2, 3; yes
2. equation; $y = x + 1$; test

Test Point	Inequality $y \le x + 1$	Conclusion
(0, 0)	$0 \overset{?}{\le} 0 + 1$ ✓	Part of graph
(0, 2)	$2 \overset{?}{\le} 0 + 1$ ✗	Not part of graph

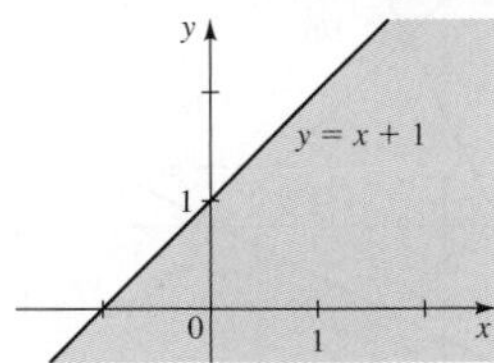

3. 2, 3; yes
4. **(a)**

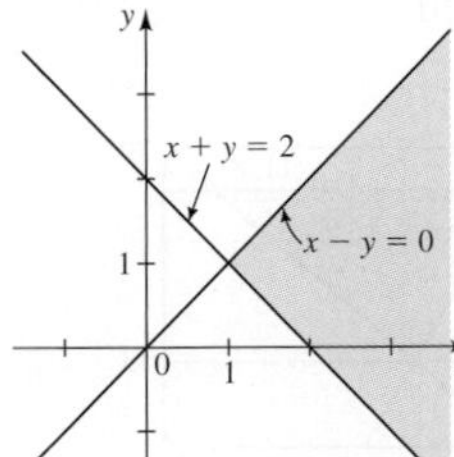

(b)

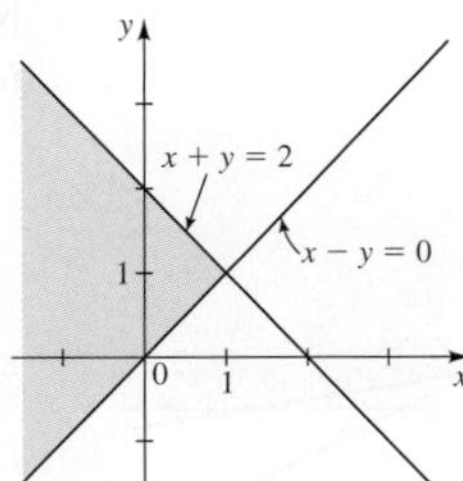

(c)

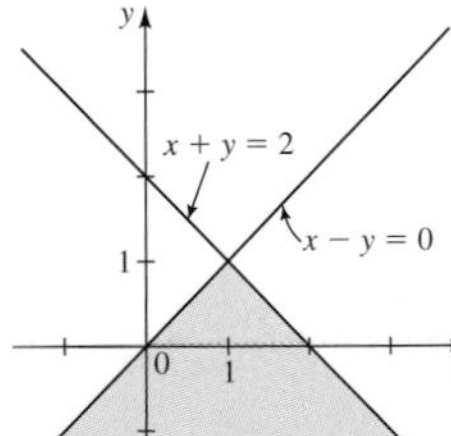

(d)

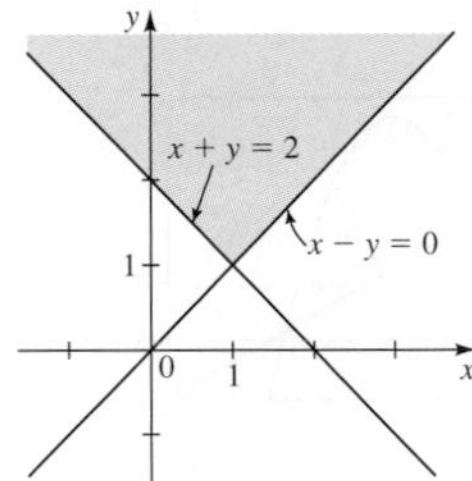

5. $(-1, -2), (1, -2)$ **7.** $(1, 2), (1, 1)$
9.

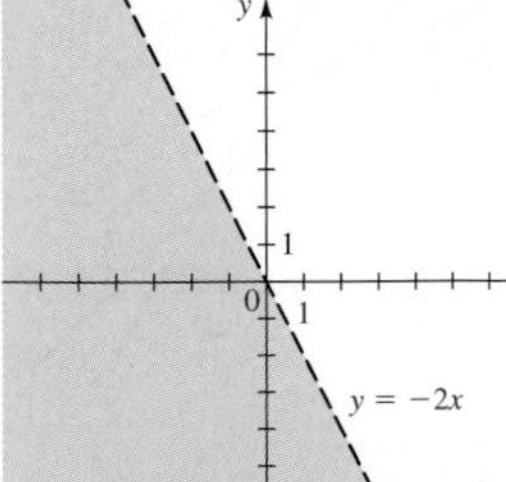

11.

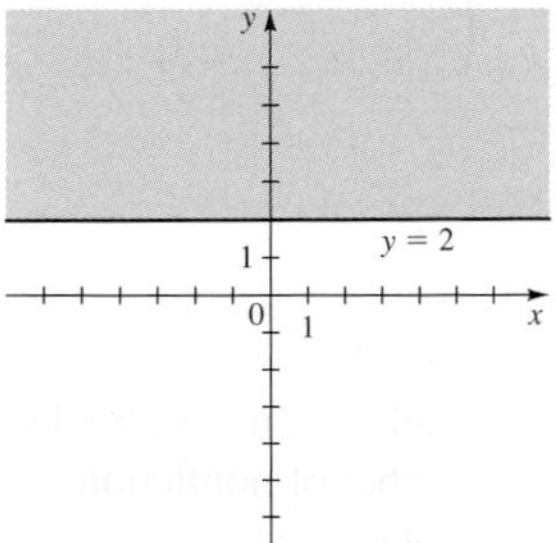

13.

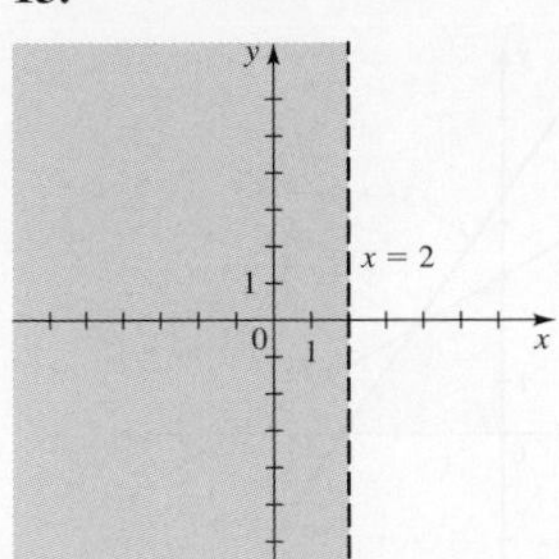

15.

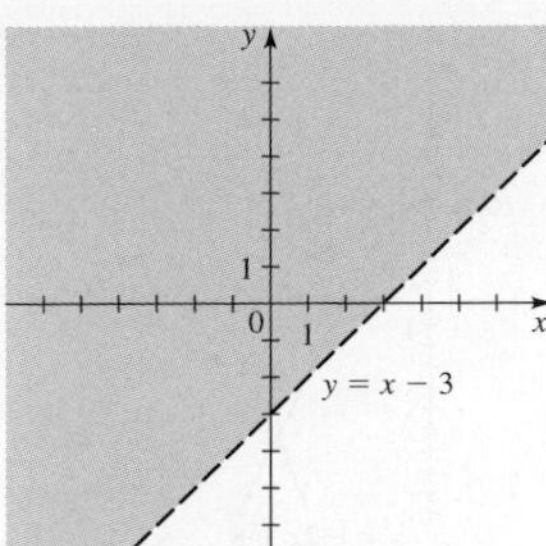

17.

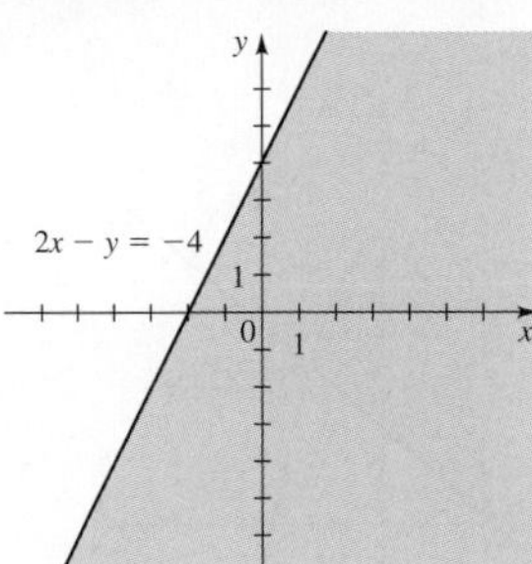

19.

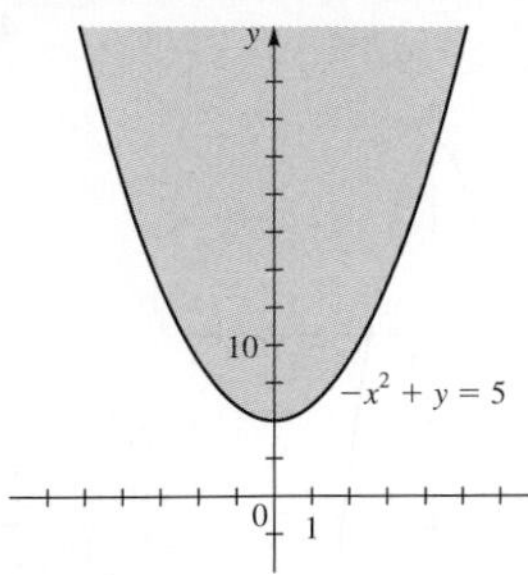

21.

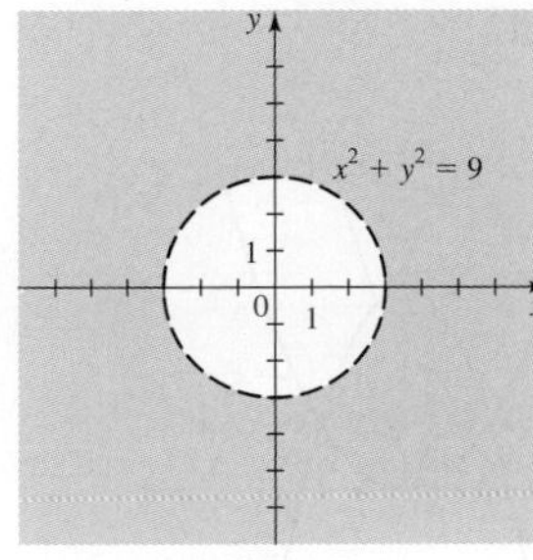

23.

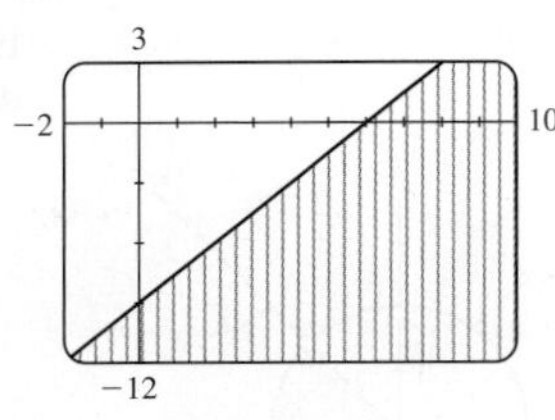

25.

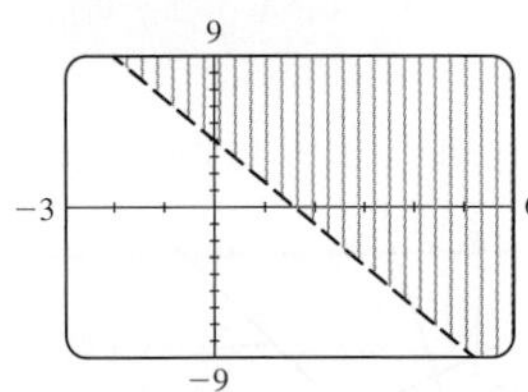

27. $y \le \frac{1}{2}x - 1$ **29.** $x^2 + y^2 > 4$
31.

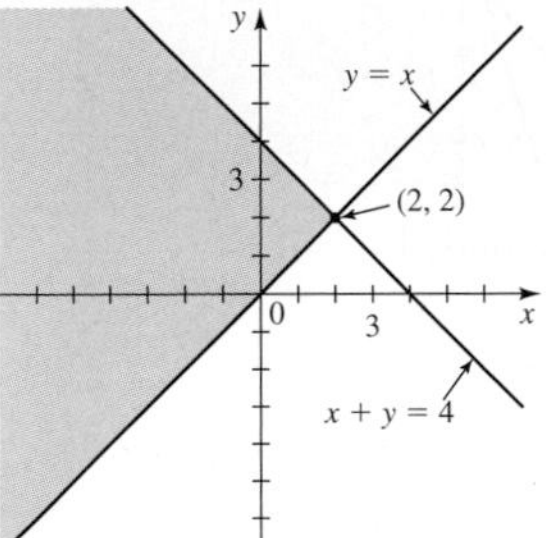

Not bounded

33.

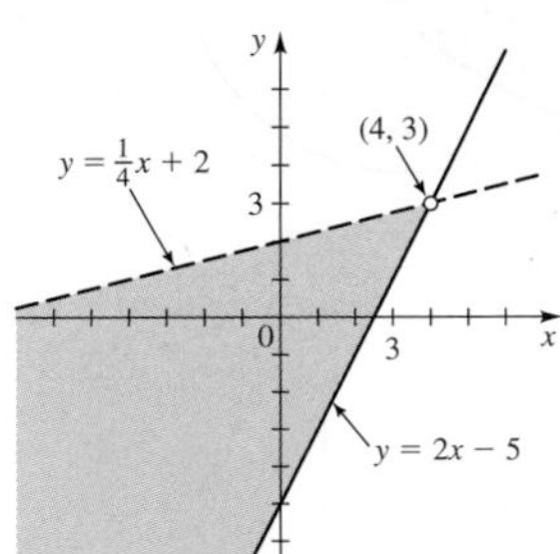

Not bounded

35.

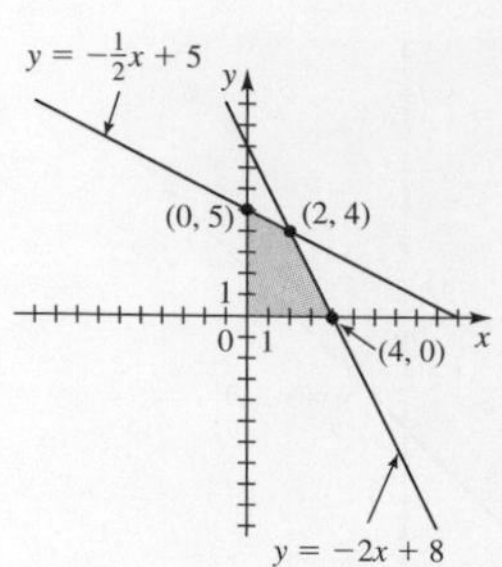

Bounded

37.

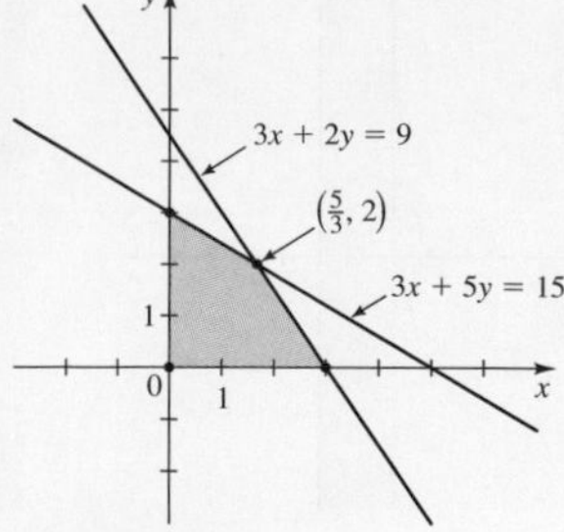

Bounded

39.

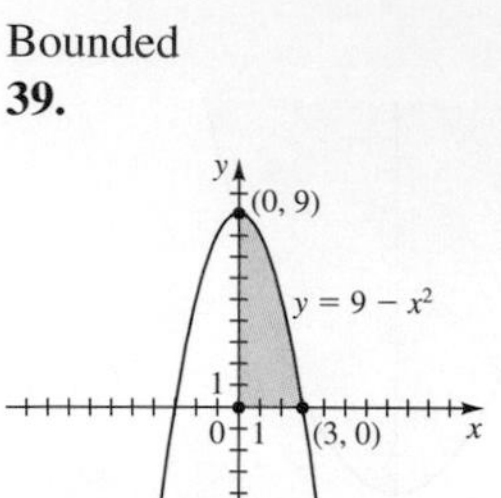

Bounded

41.

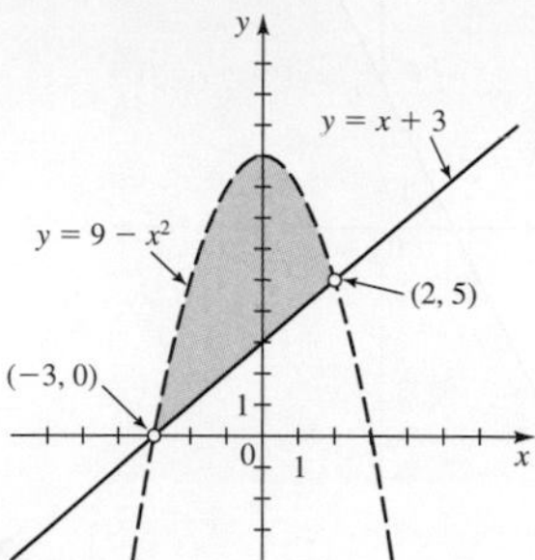

Bounded

43.

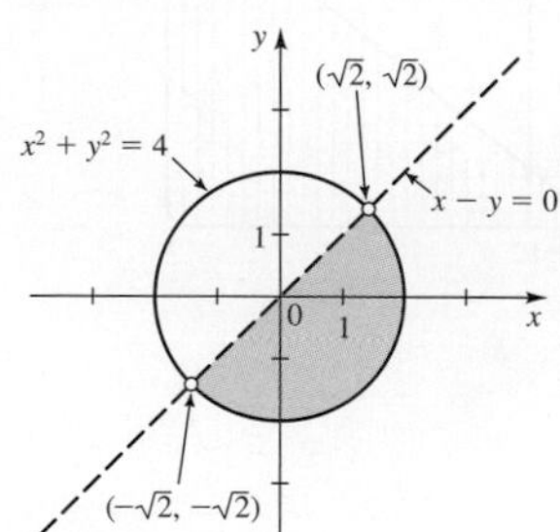

Bounded

45.

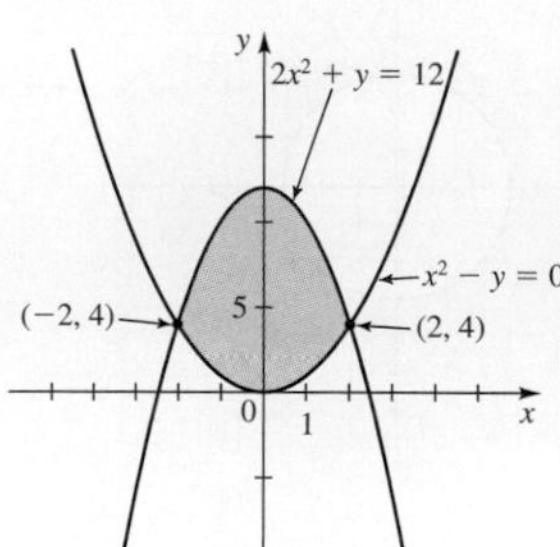

Bounded

47.

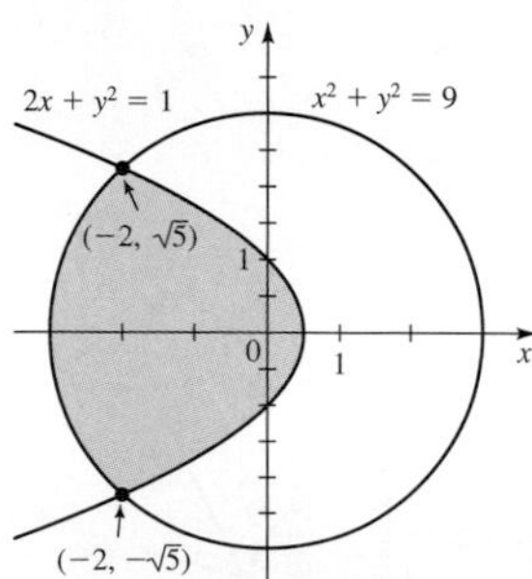

Bounded

49.

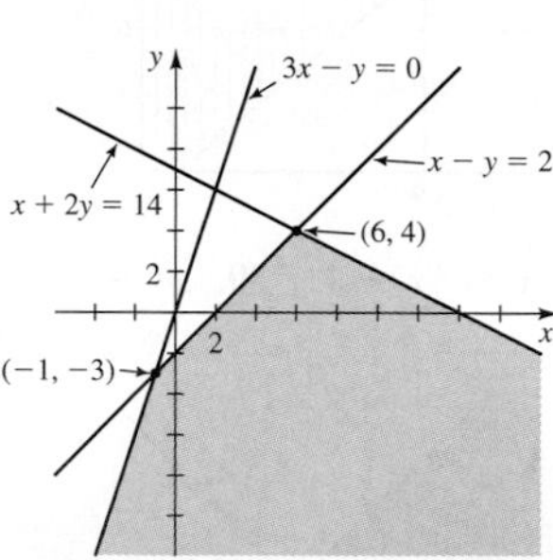

Not bounded

51.

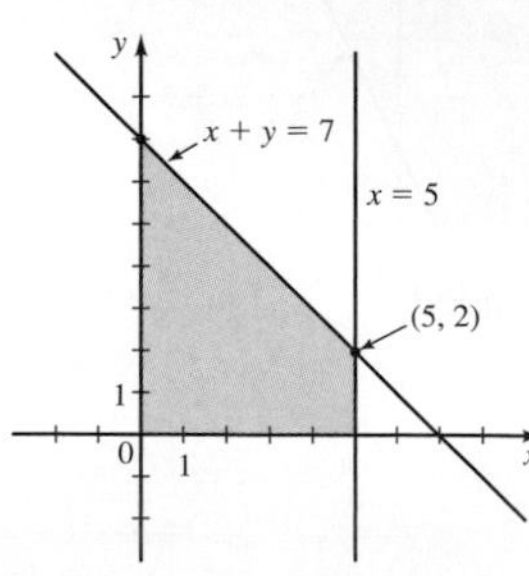

Bounded

53.

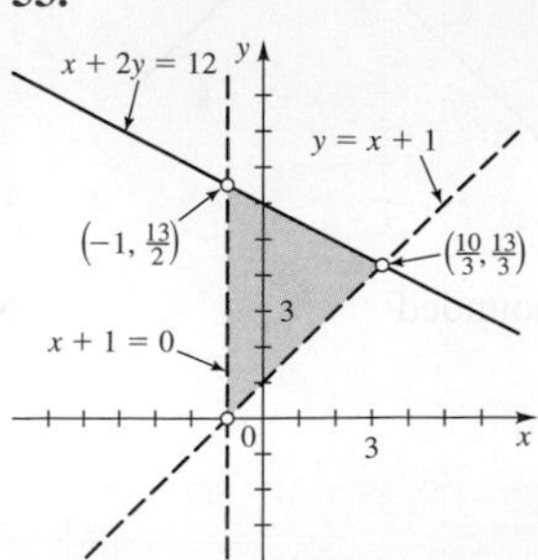

Bounded

55.

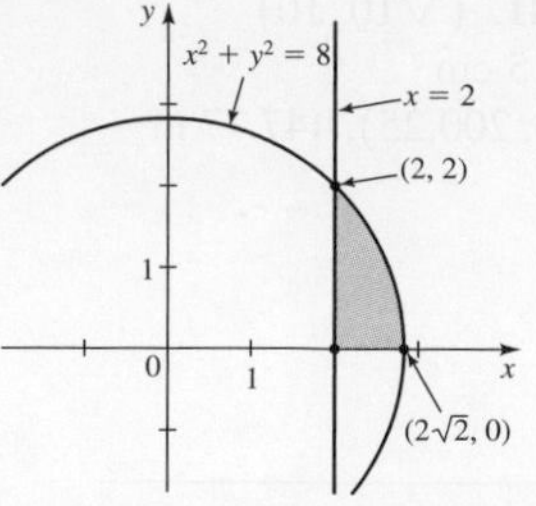

Bounded

57.

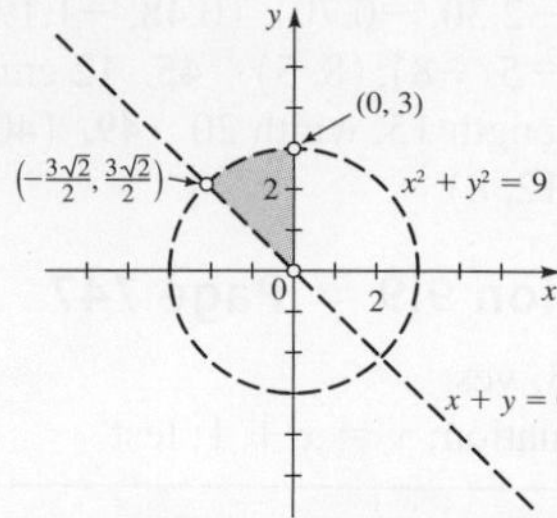

Bounded

59.

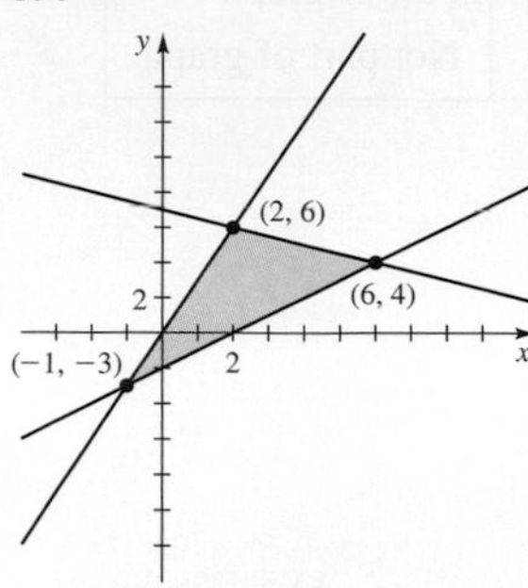

Bounded

61.

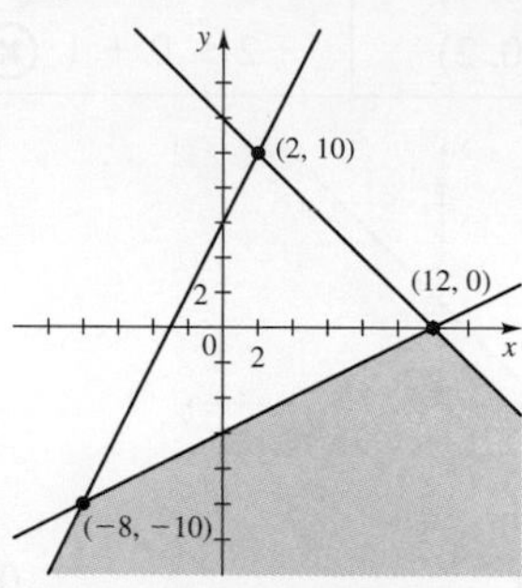

Not bounded

63.

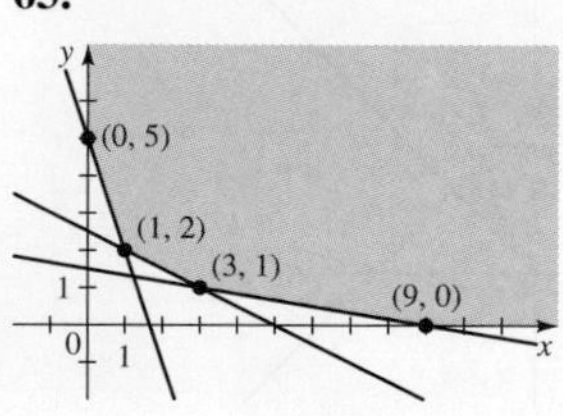

Not bounded

65.

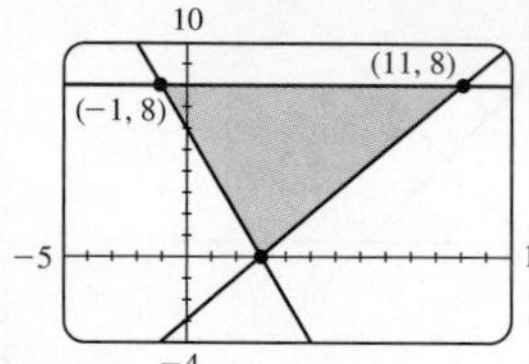

67.

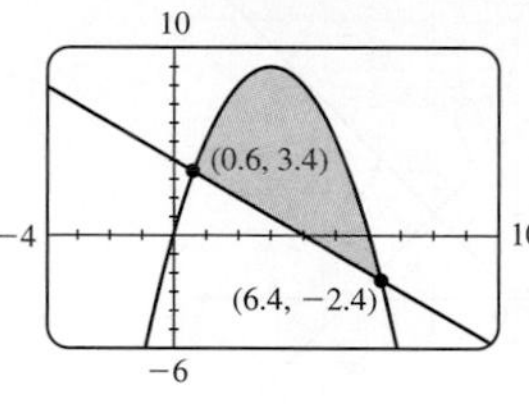

69. (a) $\begin{cases} x + \quad y \le 500 \\ 90x + 50y \le 40{,}000 \\ 30x + 80y \le 30{,}000 \\ x \ge 0, \quad y \ge 0 \end{cases}$

(0, 375) (200, 300) (375, 125) $\left(\frac{4000}{9}, 0\right)$

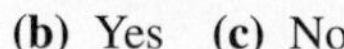

(b) Yes **(c)** No

71. x = number of fiction books
y = number of nonfiction books

$\begin{cases} x + y \le 100 \\ 20 \le y, \quad x \ge y \\ x \ge 0, \quad y \ge 0 \end{cases}$

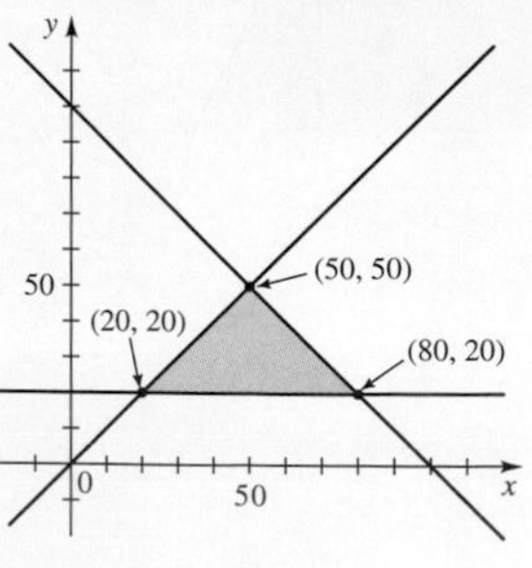

73. x = number of Standard packages
y = number of Deluxe packages

$$\begin{cases} \frac{1}{4}x + \frac{5}{8}y \le 80 \\ \frac{3}{4}x + \frac{3}{8}y \le 90 \\ x \ge 0, \quad y \ge 0 \end{cases}$$

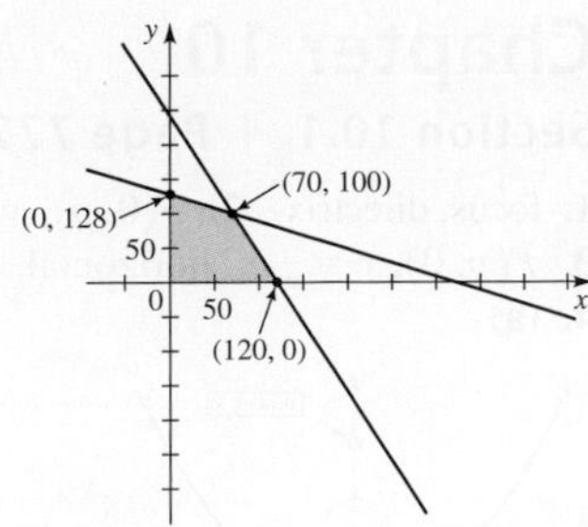

Chapter 9 Review ■ Page 754

1. $(2, 1)$

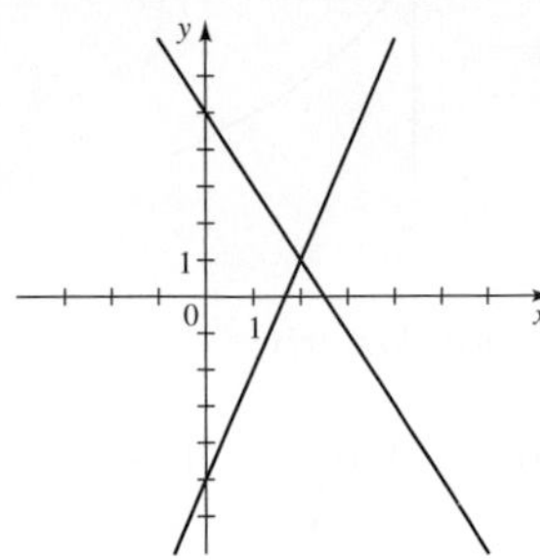

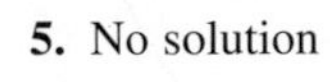

3. x = any number t
$y = \frac{2}{7}t - 4$

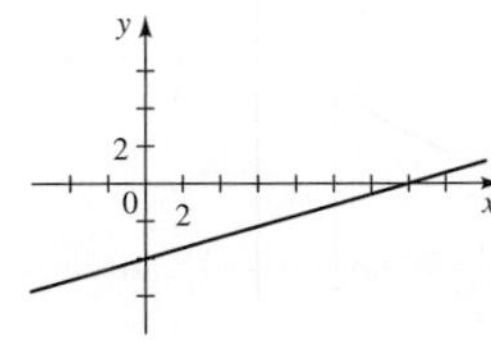

5. No solution

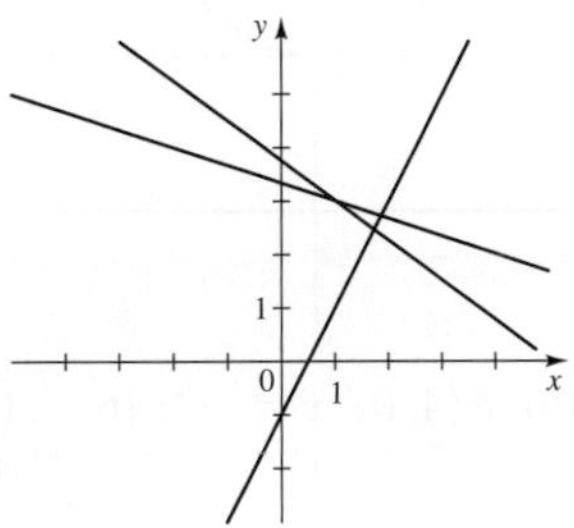

7. $(-3, 3), (2, 8)$ **9.** $\left(\frac{16}{7}, -\frac{14}{3}\right)$ **11.** $(21.41, -15.93)$
13. $(11.94, -1.39), (12.07, 1.44)$
15. **(a)** 2×3 **(b)** Yes **(c)** No

(d) $\begin{cases} x + 2y = -5 \\ \quad\;\; y = \;\;\, 3 \end{cases}$

17. **(a)** 3×4 **(b)** Yes **(c)** Yes

(d) $\begin{cases} x \quad\; + 8z = \;\; 0 \\ \quad y + 5z = -1 \\ \qquad\quad 0 = \;\; 0 \end{cases}$

19. **(a)** 3×4 **(b)** No **(c)** No

(d) $\begin{cases} \quad\;\; y - 3z = 4 \\ x + \;\, y \qquad = 7 \\ x + 2y + \;\, z = 2 \end{cases}$

21. $(1, 1, 2)$ **23.** No solution **25.** $(0, 1, 2)$ **27.** No solution
29. $(1, 0, 1, -2)$ **31.** $(-4t + 1, -t - 1, t)$
33. $(6 - 5t, \frac{1}{2}(7 - 3t), t)$ **35.** $\left(-\frac{4}{3}t + \frac{4}{3}, \frac{5}{3}t - \frac{2}{3}, t\right)$
37. $(s + 1, 2s - t + 1, s, t)$ **39.** No solution
41. $(1, t + 1, t, 0)$ **43.** \$3000 at 6%, \$6000 at 7%
45. \$2500 in bank A, \$40,000 in bank B, \$17,500 in bank C
47. Impossible

49. $\begin{bmatrix} 4 & 18 \\ 4 & 0 \\ 2 & 2 \end{bmatrix}$ **51.** $[10 \quad 0 \quad -5]$ **53.** $\begin{bmatrix} -\frac{7}{2} & 10 \\ 1 & -\frac{9}{2} \end{bmatrix}$

55. $\begin{bmatrix} 30 & 22 & 2 \\ -9 & 1 & -4 \end{bmatrix}$ **57.** $\begin{bmatrix} -\frac{1}{2} & \frac{11}{2} \\ \frac{15}{4} & -\frac{3}{2} \\ -\frac{1}{2} & 1 \end{bmatrix}$ **61.** $\frac{1}{3}\begin{bmatrix} -1 & -3 \\ -5 & 2 \end{bmatrix}$

63. $\begin{bmatrix} \frac{7}{2} & -2 \\ 0 & 8 \end{bmatrix}$ **65.** $\begin{bmatrix} 2 & -2 & 6 \\ -4 & 5 & -9 \end{bmatrix}$ **67.** $1, \begin{bmatrix} 9 & -4 \\ -2 & 1 \end{bmatrix}$

69. 0, no inverse **71.** $-1, \begin{bmatrix} 3 & 2 & -3 \\ 2 & 1 & -2 \\ -8 & -6 & 9 \end{bmatrix}$

73. $24, \begin{bmatrix} 1 & 0 & 0 & -\frac{1}{4} \\ 0 & \frac{1}{2} & 0 & -\frac{1}{4} \\ 0 & 0 & \frac{1}{3} & -\frac{1}{4} \\ 0 & 0 & 0 & \frac{1}{4} \end{bmatrix}$ **75.** $(65, 154)$ **77.** $\left(-\frac{1}{12}, \frac{1}{12}, \frac{1}{12}\right)$

79. $\left(\frac{1}{5}, \frac{9}{5}\right)$ **81.** $\left(-\frac{87}{26}, \frac{21}{26}, \frac{3}{2}\right)$ **83.** 11 **85.** $\dfrac{2}{x - 5} + \dfrac{1}{x + 3}$

87. $\dfrac{-4}{x} + \dfrac{4}{x - 1} + \dfrac{-2}{(x - 1)^2}$ **89.** $\dfrac{-1}{x} + \dfrac{x + 2}{x^2 + 1}$

91. $(2, 1)$ **93.** $\left(-\frac{1}{2}, \frac{7}{4}\right), (2, -2)$ **95.** $x + y^2 \le 4$

97.

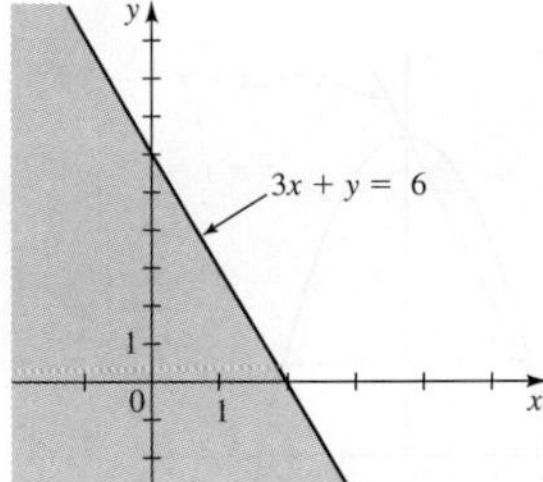

99.

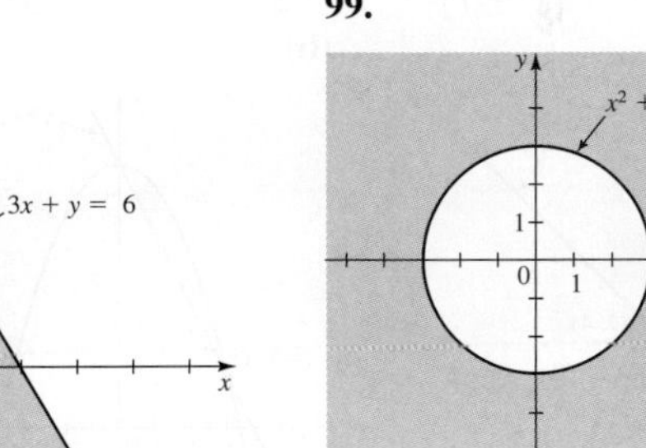

101.

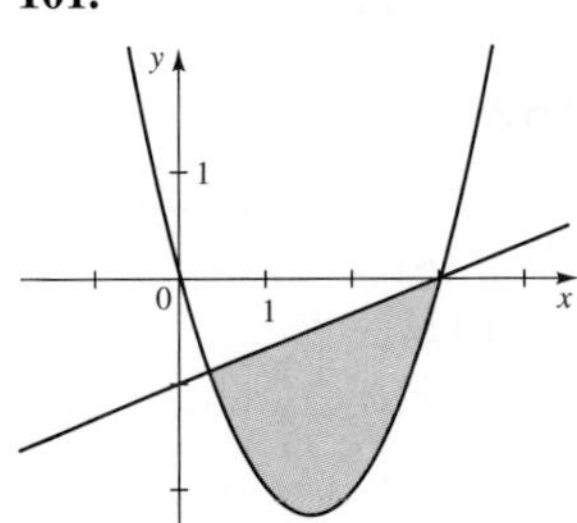

103.

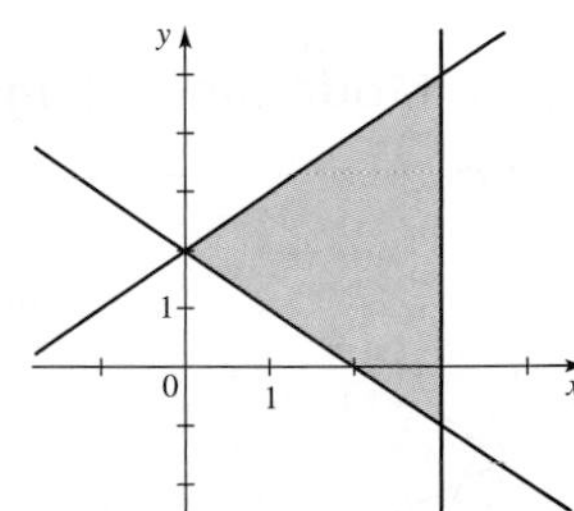

105.

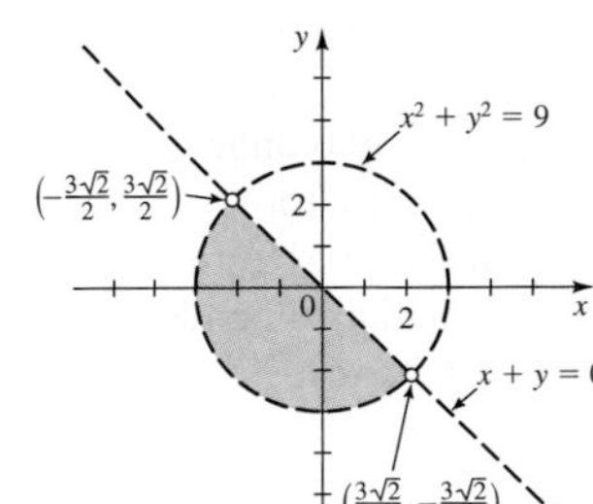

Bounded

107.

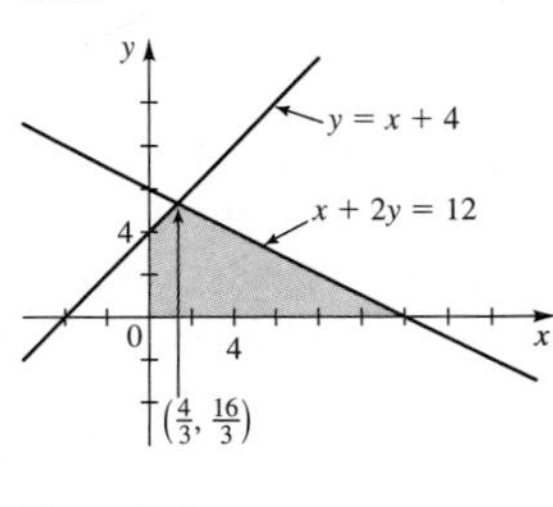

Bounded

109. $x = \dfrac{b + c}{2}, y = \dfrac{a + c}{2}, z = \dfrac{a + b}{2}$ **111.** 2, 3

Chapter 9 Test ▪ Page 757

1. (a) Linear **(b)** $(-2, 3)$ **2. (a)** Nonlinear **(b)** $(1, -2), (\frac{5}{3}, 0)$
3. $(-0.55, -0.78), (0.43, -0.29), (2.12, 0.56)$
4. Wind 60 km/h, airplane 300 km/h
5. (a) Row-echelon form **(b)** Reduced row-echelon form **(c)** Neither **6. (a)** $(\frac{5}{2}, \frac{5}{2}, 0)$ **(b)** No solution
7. $(-\frac{3}{5} + \frac{2}{5}t, \frac{1}{5} + \frac{1}{5}t, t)$
8. Coffee \$1.50, juice \$1.75, donut \$0.75
9. (a) Incompatible dimensions
(b) Incompatible dimensions
(c) $\begin{bmatrix} 6 & 10 \\ 3 & -2 \\ -3 & 9 \end{bmatrix}$ **(d)** $\begin{bmatrix} 36 & 58 \\ 0 & -3 \\ 18 & 28 \end{bmatrix}$ **(e)** $\begin{bmatrix} 2 & -\frac{3}{2} \\ -1 & 1 \end{bmatrix}$
(f) B is not square **(g)** B is not square **(h)** -3
10. (a) $\begin{bmatrix} 4 & -3 \\ 3 & -2 \end{bmatrix}\begin{bmatrix} x \\ y \end{bmatrix} = \begin{bmatrix} 10 \\ 30 \end{bmatrix}$ **(b)** $(70, 90)$
11. $|A| = 0, |B| = 2, B^{-1} = \begin{bmatrix} 1 & -2 & 0 \\ 0 & \frac{1}{2} & 0 \\ 3 & -6 & 1 \end{bmatrix}$
12. $(5, -5, -4)$
13. (a) $\dfrac{1}{x-1} + \dfrac{1}{(x-1)^2} - \dfrac{1}{x+2}$ **(b)** $-\dfrac{1}{x} + \dfrac{x+2}{x^2+3}$
14. (a)

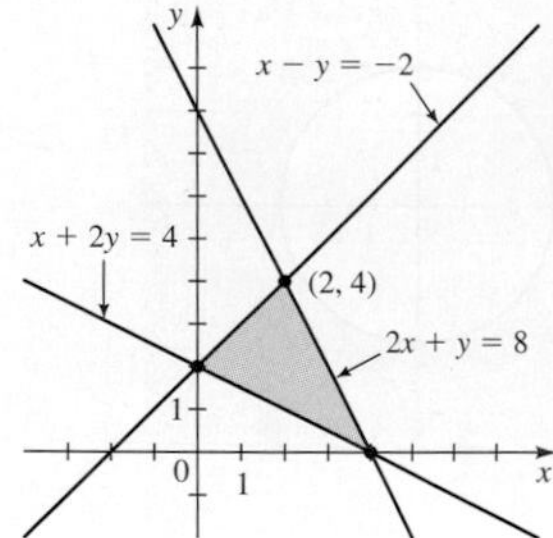

(b)

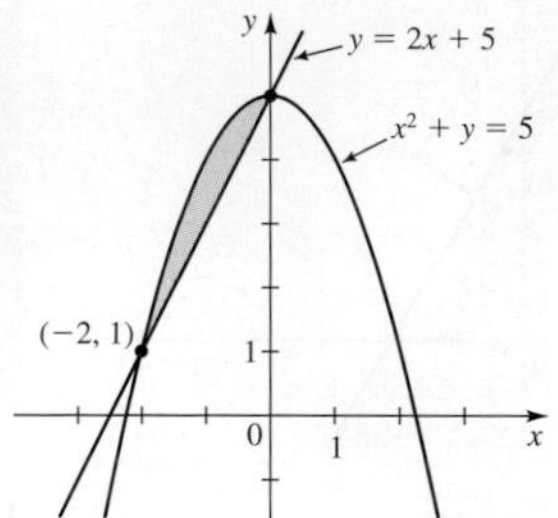

Focus on Modeling ▪ Page 762

1. 198, 195
3.

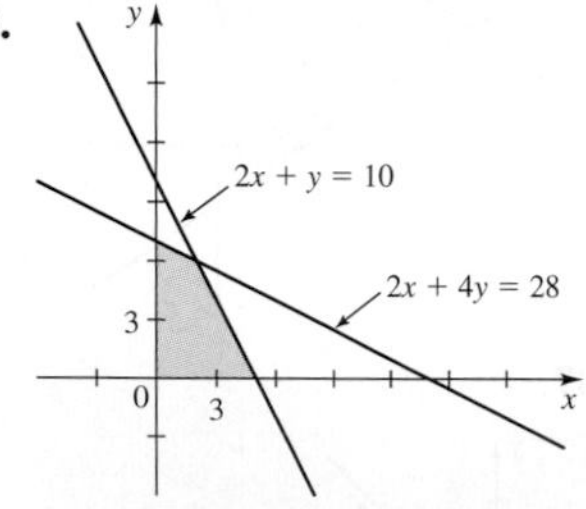

maximum 161
minimum 135

5. 3 tables, 34 chairs **7.** 30 grapefruit crates, 30 orange crates
9. 15 Pasadena to Santa Monica, 3 Pasadena to El Toro, 0 Long Beach to Santa Monica, 16 Long Beach to El Toro
11. 90 standard, 40 deluxe **13.** \$7500 in municipal bonds, \$2500 in bank certificates, \$2000 in high-risk bonds
15. 4 games, 32 educational, 0 utility

Chapter 10

Section 10.1 ▪ Page 772

1. focus, directrix **2.** $F(0, p)$, $y = -p$, vertical, $F(0, 3)$, $y = -3$
3. $F(p, 0)$, $x = -p$, horizontal, $F(3, 0)$, $x = -3$
4. (a)

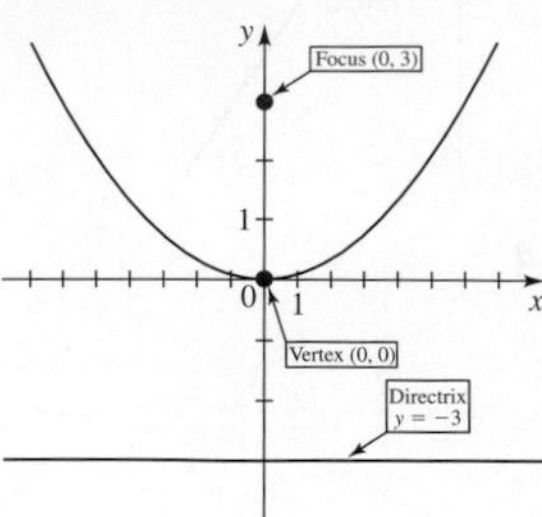

(b)

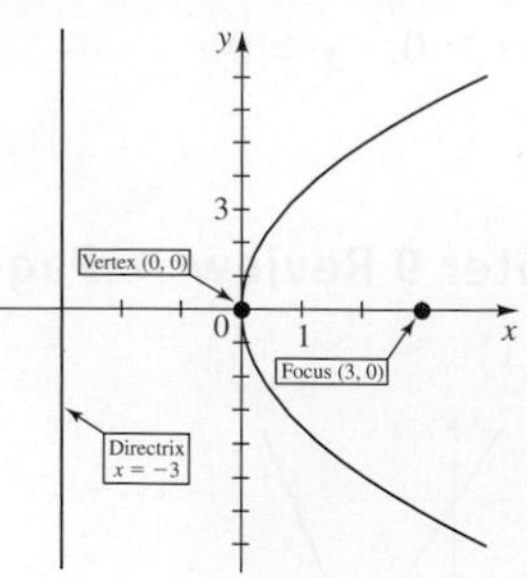

5. III **7.** II **9.** VI

Order of answers for 11–23, part (a): focus; directrix; focal diameter

11. (a) $F(0, 4)$; $y = -4$; 16
(b)

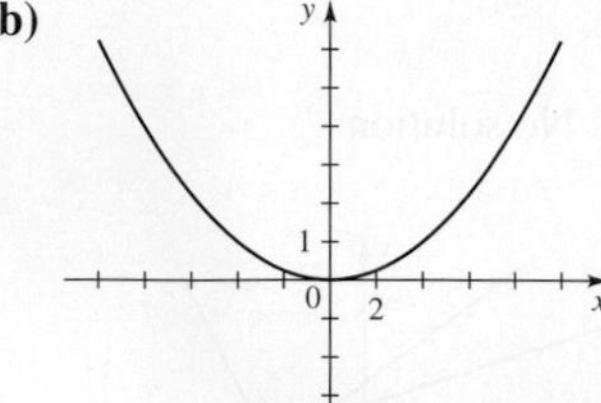

13. (a) $F(-1, 0)$; $x = 1$; 4
(b)

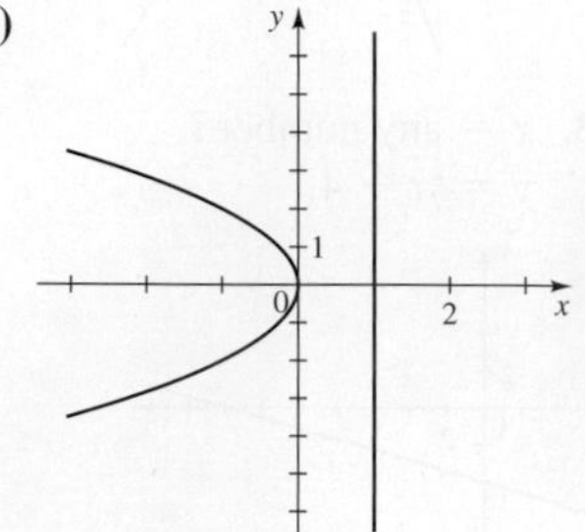

15. (a) $F(4, 0)$; $x = -4$; 16
(b)

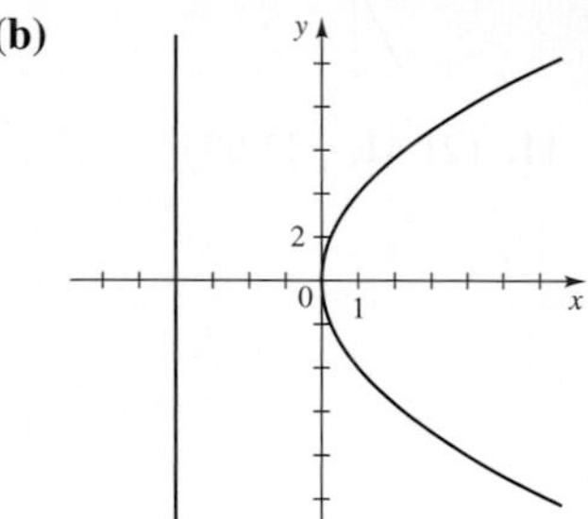

17. (a) $F(0, -\frac{1}{8})$; $y = \frac{1}{8}$; $\frac{1}{2}$
(b)

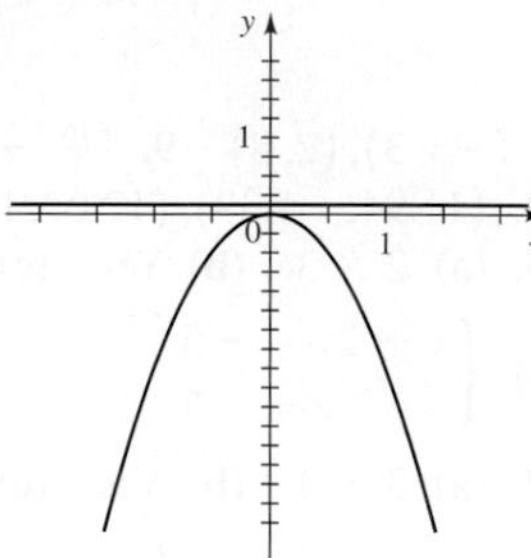

19. (a) $F(0, \frac{5}{4})$; $y = -\frac{5}{4}$; 5
(b)

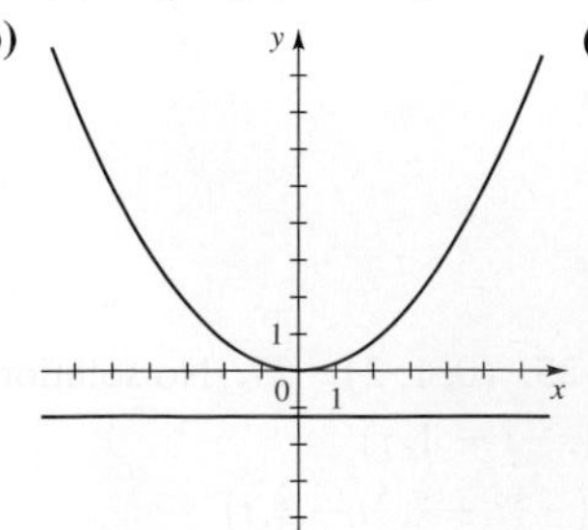

21. (a) $F(0, -3)$; $y = 3$; 12
(b)

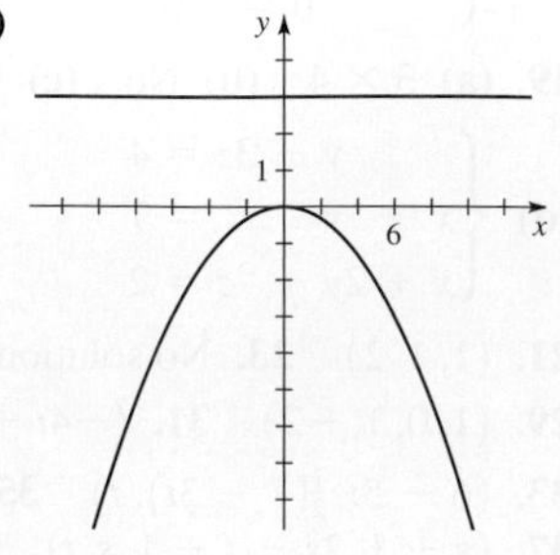

23. (a) $F\left(-\frac{5}{12}, 0\right)$; $x = \frac{5}{12}$; $\frac{5}{3}$
(b)

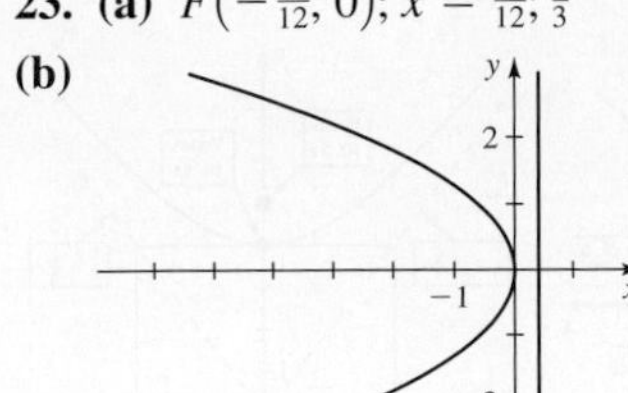

25.

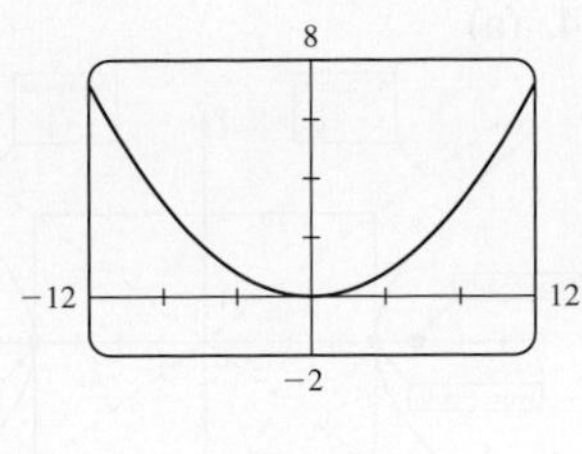

27.

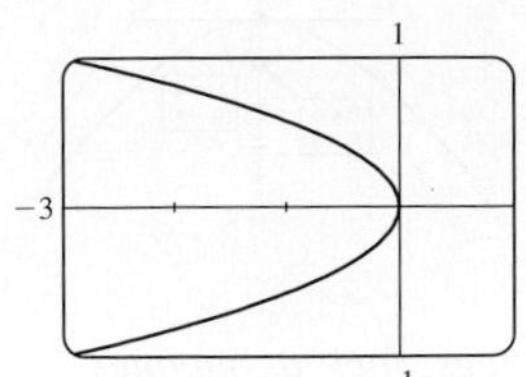

29.

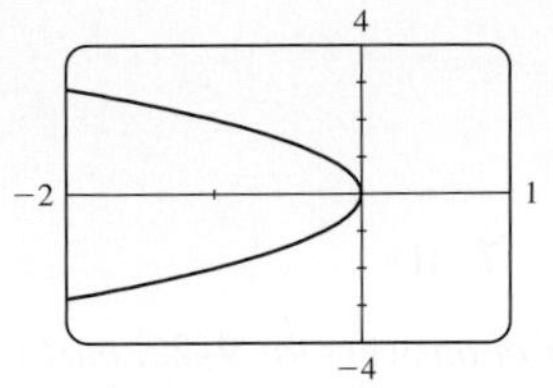

31. $x^2 = 12y$ **33.** $y^2 = -32x$ **35.** $x^2 = -3y$ **37.** $y^2 = 8x$
39. $x^2 = -\frac{2}{5}y$ **41.** $y^2 = -\frac{1}{5}x$ **43.** $y^2 = 4x$ **45.** $x^2 = -40y$
47. $x^2 = -24y$ **49.** $x^2 = 24y$ **51.** $y^2 = -16x$
53. $y^2 = -3x$ **55.** $x = y^2$ **57.** $x^2 = -4\sqrt{2}y$
59. (a) $x^2 = -4py$, $p = \frac{1}{2}$, 1, 4, and 8
(b) The closer the directrix to the vertex, the steeper the parabola.

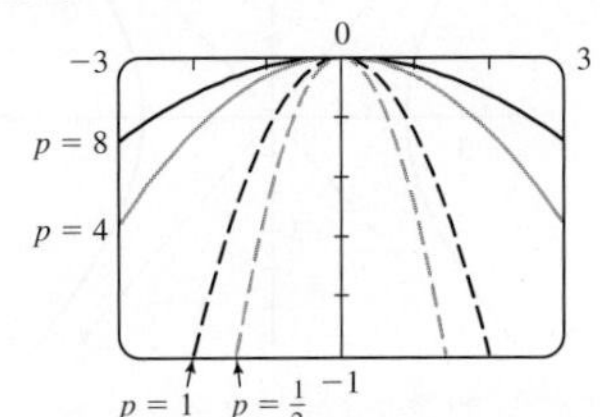

61. (a) $y^2 = 12x$ **(b)** $8\sqrt{15} \approx 31$ cm **63.** $x^2 = 600y$

Section 10.2 ■ Page 781

1. sum; foci
2. horizontal, $(a, 0)$, $(-a, 0)$; $c = \sqrt{a^2 - b^2}$; $(5, 0)$, $(-5, 0)$, $(3, 0)$, $(-3, 0)$
3. vertical, $(0, a)$, $(0, -a)$; $c = \sqrt{a^2 - b^2}$; $(0, 5)$, $(0, -5)$, $(0, 3)$, $(0, -3)$
4. (a)

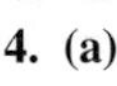

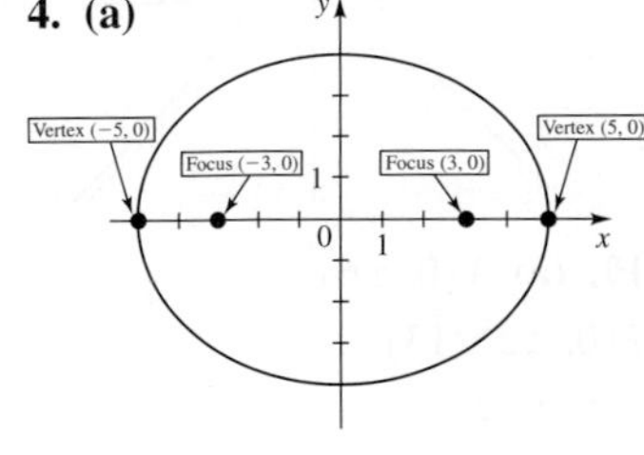

(b)

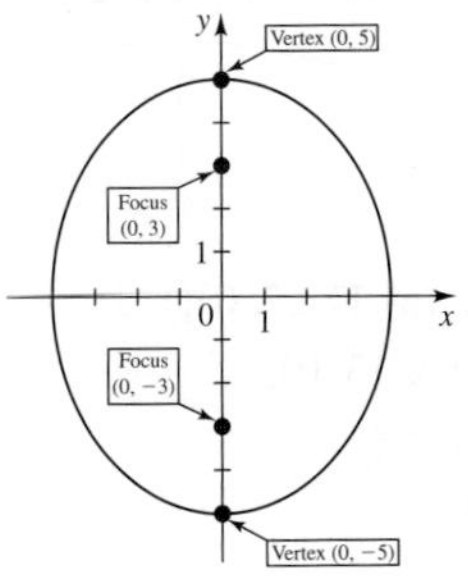

5. II **7.** I

Order of answers for 9–27 part (a): vertices; foci; eccentricity

9. (a) $V(\pm 5, 0)$; $F(\pm 4, 0)$; $\frac{4}{5}$
(b) 10, 6
(c)

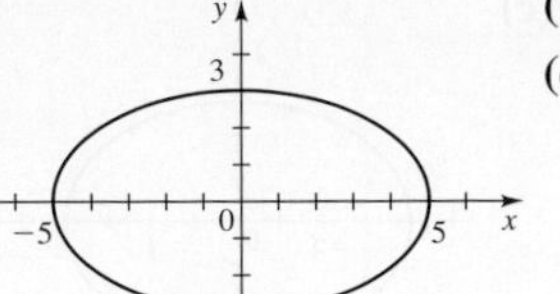

11. (a) $V(0, \pm 9)$; $F(0, \pm 3\sqrt{5})$; $\sqrt{5}/3$
(b) 18, 12
(c)

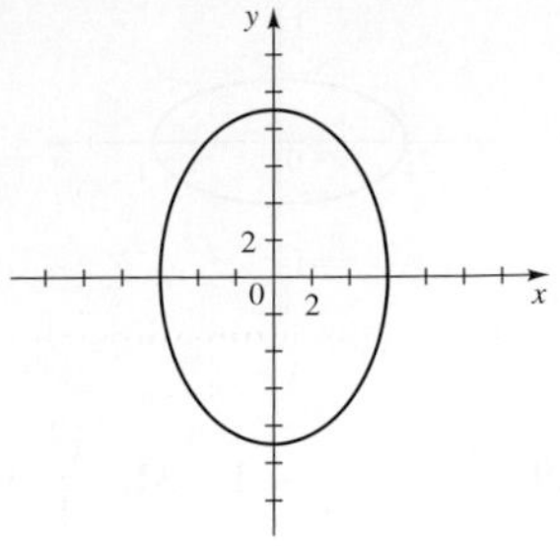

13. (a) $V(\pm 7, 0)$; $F(\pm 2\sqrt{6}, 0)$; $2\sqrt{6}/7$
(b) 14, 10
(c)

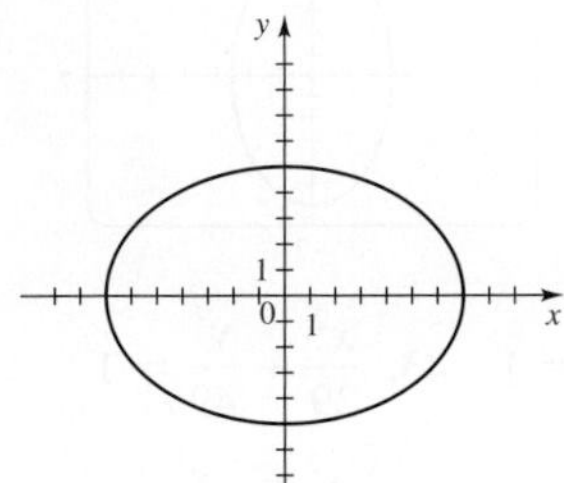

15. (a) $V(0, \pm 3)$; $F(0, \pm\sqrt{5})$; $\sqrt{5}/3$
(b) 6, 4
(c)

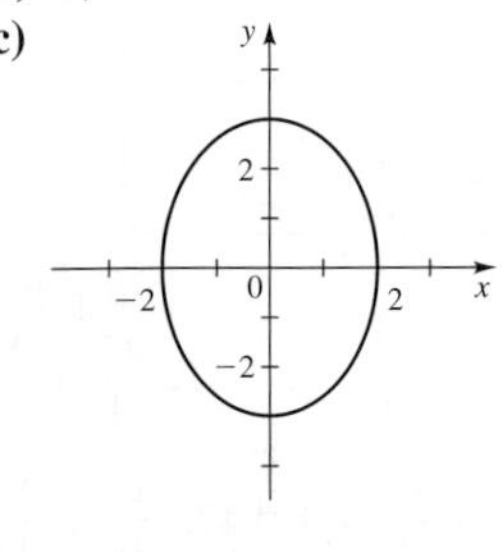

17. (a) $V(\pm 4, 0)$; $F(\pm 2\sqrt{3}, 0)$; $\sqrt{3}/2$
(b) 8, 4
(c)

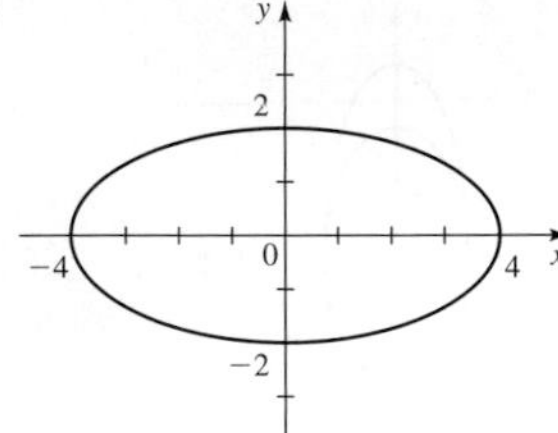

19. (a) $V(\pm 10, 0)$; $F(\pm 6, 0)$; $\frac{3}{5}$
(b) 20, 16
(c)

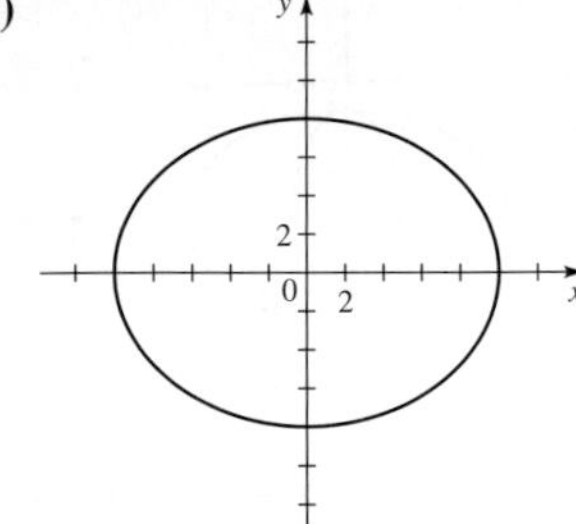

21. (a) $V(0, \pm 3)$; $F(0, \pm\sqrt{6})$; $\sqrt{6}/3$
(b) 6, $2\sqrt{3}$
(c)

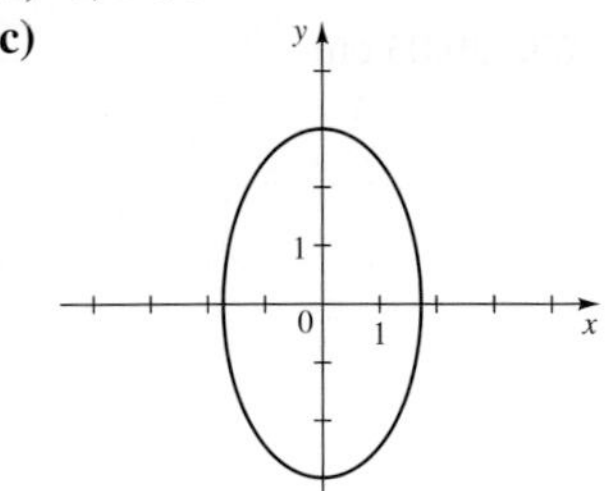

23. (a) $V(0, \pm 2)$; $F(0, \pm\sqrt{2})$; $\sqrt{2}/2$
(b) 4, $2\sqrt{2}$
(c)

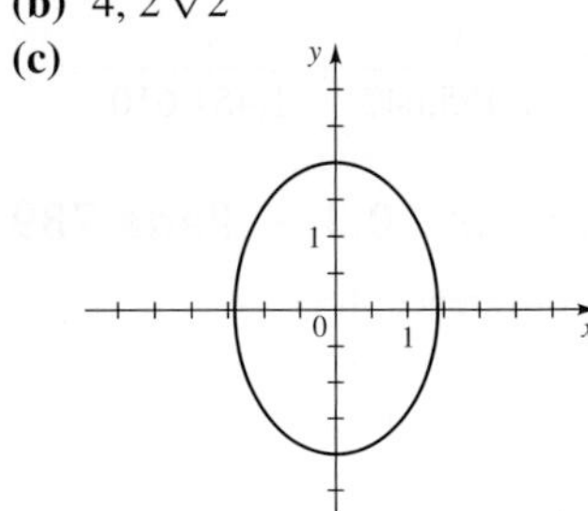

25. **(a)** $V(\pm 1, 0)$; $F(\pm\sqrt{3}/2, 0)$; $\sqrt{3}/2$
(b) 2, 1
(c)

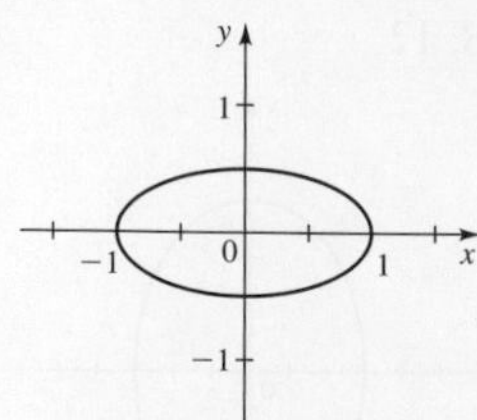

27. **(a)** $V(\pm 2, 0)$; $F(\pm\sqrt{2}, 0)$; $\sqrt{2}/2$
(b) $4, 2\sqrt{2}$
(c)

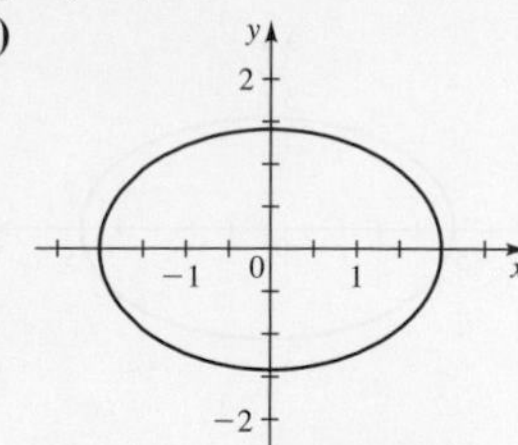

29. $\frac{x^2}{25} + \frac{y^2}{16} = 1$ **31.** $\frac{x^2}{4} + \frac{y^2}{8} = 1$ **33.** $\frac{x^2}{256} + \frac{y^2}{48} = 1$

35.

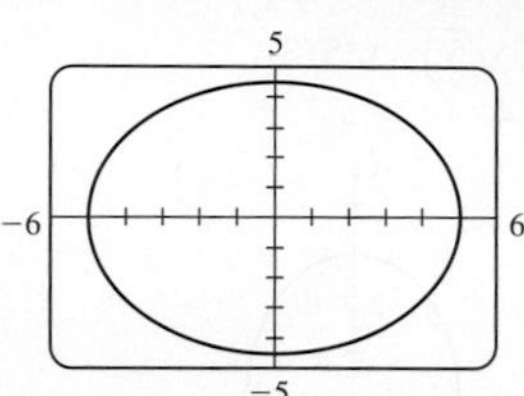

37.

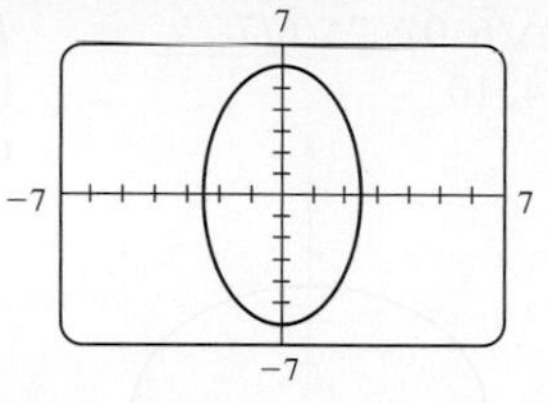

39. $\frac{x^2}{25} + \frac{y^2}{9} = 1$ **41.** $\frac{x^2}{4} + \frac{y^2}{3} = 1$ **43.** $\frac{x^2}{39} + \frac{y^2}{49} = 1$

45. $x^2 + \frac{y^2}{4} = 1$ **47.** $\frac{x^2}{9} + \frac{y^2}{13} = 1$ **49.** $\frac{x^2}{100} + \frac{y^2}{91} = 1$

51. $\frac{x^2}{25} + \frac{y^2}{5} = 1$ **53.** $\frac{x^2}{32} + \frac{y^2}{36} = 1$ **55.** $x^2 + \frac{y^2}{4} = 1$

57. $(0, \pm 2)$

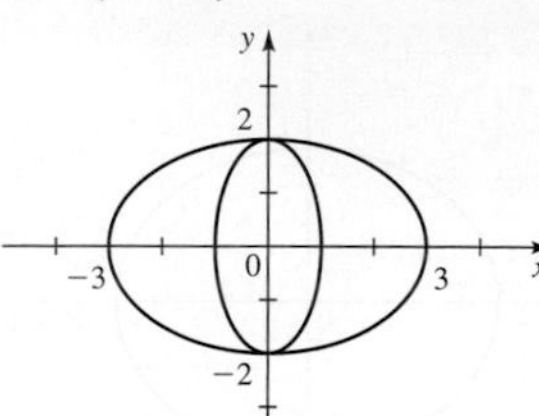

59. $(\pm 1, 0)$

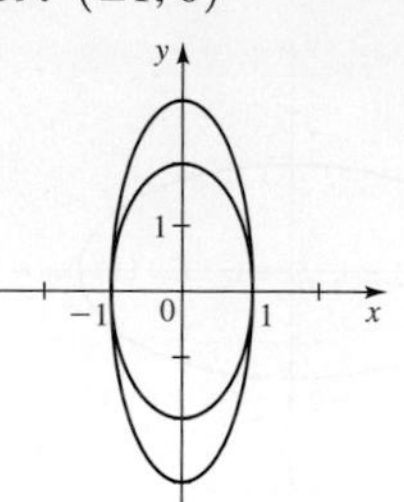

61. **(a)** $x^2 + y^2 = 4$

65. $\frac{x^2}{2.2500 \times 10^{16}} + \frac{y^2}{2.2491 \times 10^{16}} = 1$

67. $\frac{x^2}{1{,}455{,}642} + \frac{y^2}{1{,}451{,}610} = 1$ **69.** 39.03 cm

Section 10.3 ■ Page 789

1. difference; foci
2. horizontal; $(-a, 0), (a, 0)$; $\sqrt{a^2 + b^2}$; $(-4, 0), (4, 0), (-5, 0), (5, 0)$
3. vertical; $(0, -a), (0, a)$; $\sqrt{a^2 + b^2}$; $(0, -4), (0, 4), (0, -5), (0, 5)$

4. **(a)**

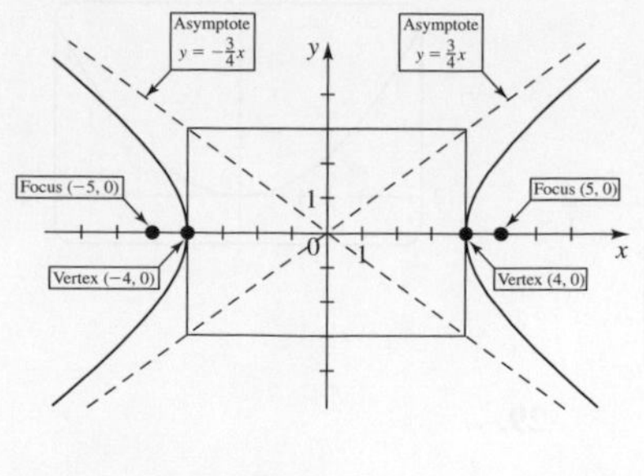

(b)

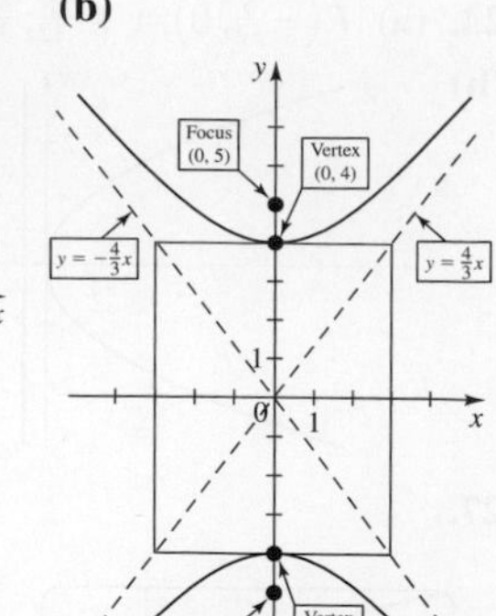

5. III **7.** II

Order of answers for 9–25, part (a): vertices; foci; asymptotes

9. **(a)** $V(\pm 2, 0)$; $F(\pm 2\sqrt{5}, 0)$; $y = \pm 2x$
(b) 4
(c)

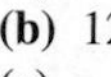

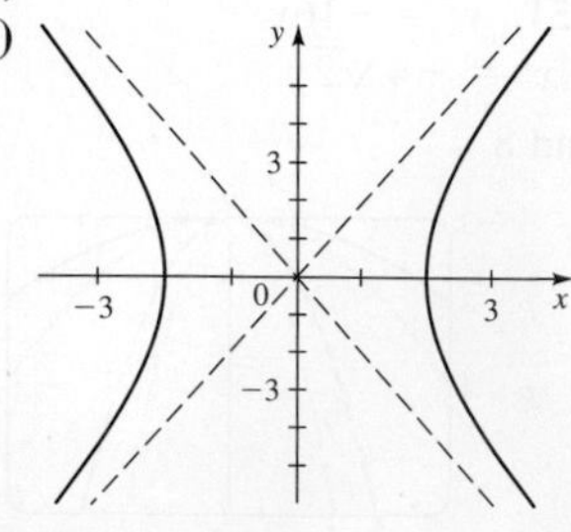

11. **(a)** $V(0, \pm 6)$; $F(0, \pm 2\sqrt{10})$; $y = \pm 3x$
(b) 12
(c)

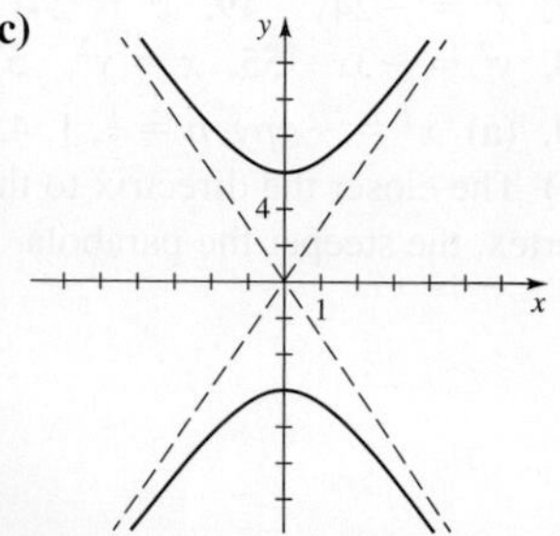

13. **(a)** $V(0, \pm 1)$; $F(0, \pm\sqrt{26})$; $y = \pm\frac{1}{5}x$
(b) 2
(c)

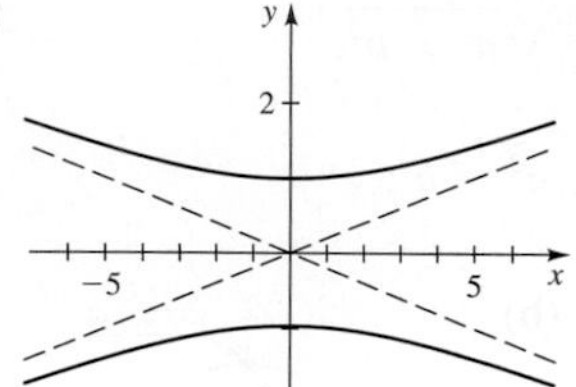

15. **(a)** $V(\pm 1, 0)$; $F(\pm\sqrt{2}, 0)$; $y = \pm x$
(b) 2
(c)

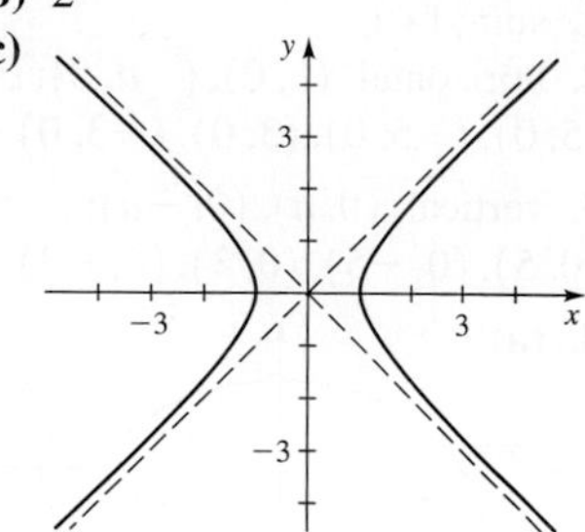

17. **(a)** $V(\pm 2, 0)$; $F(\pm\sqrt{13}, 0)$; $y = \pm\frac{3}{2}x$
(b) 4
(c)

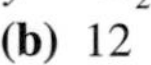

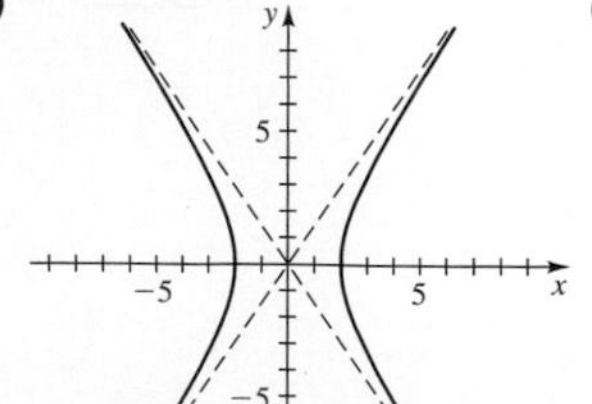

19. **(a)** $V(0, \pm 6)$; $F(0, \pm 2\sqrt{13})$; $y = \pm\frac{3}{2}x$
(b) 12
(c)

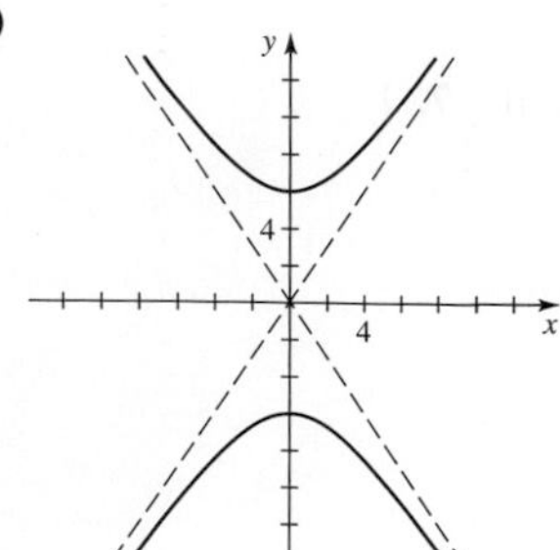

21. (a) $V(\pm 2\sqrt{2}, 0)$; $F(\pm\sqrt{10}, 0)$; $y = \pm\frac{1}{2}x$
(b) $4\sqrt{2}$
(c)

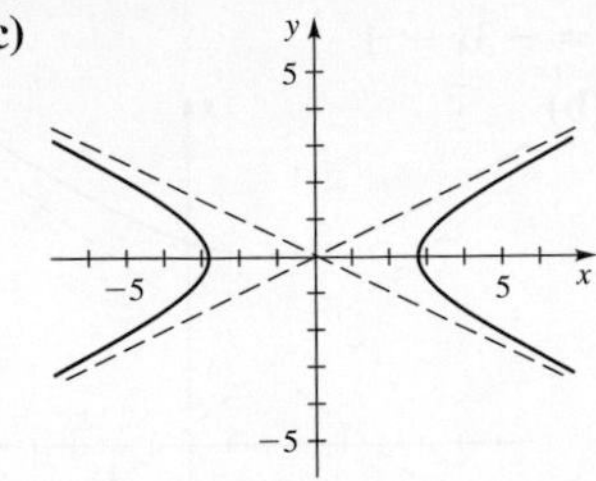

23. (a) $V(0, \pm 2)$; $F(0, \pm 2\sqrt{2})$; $y = \pm x$
(b) 4
(c)

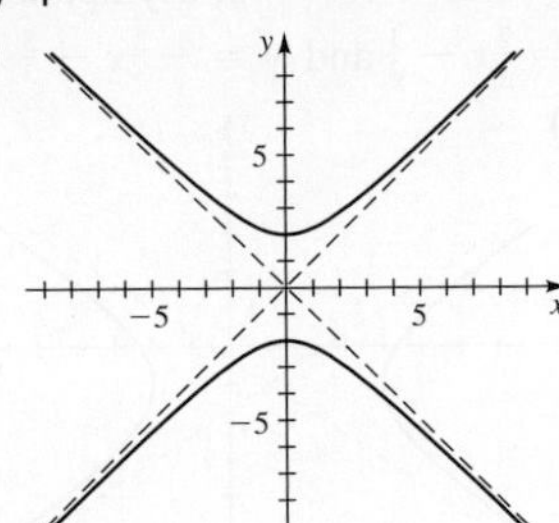

25. (a) $V(0, \pm\frac{1}{2})$; $F(0, \pm\sqrt{5}/2)$; $y = \pm\frac{1}{2}x$
(b) 1
(c)

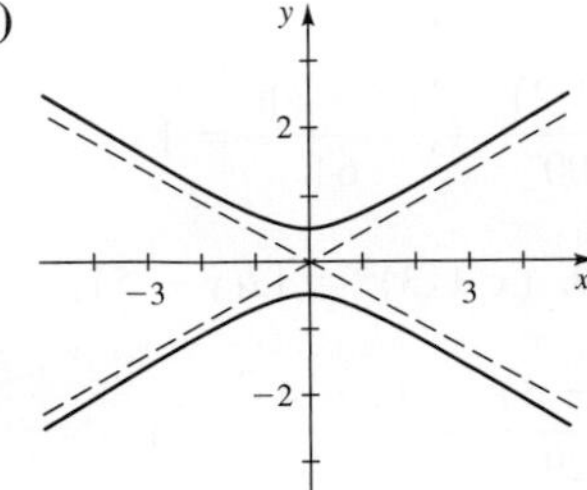

27. $\dfrac{x^2}{4} - \dfrac{y^2}{12} = 1$ **29.** $\dfrac{y^2}{16} - \dfrac{x^2}{16} = 1$ **31.** $\dfrac{y^2}{9} - x^2 = 1$

33.

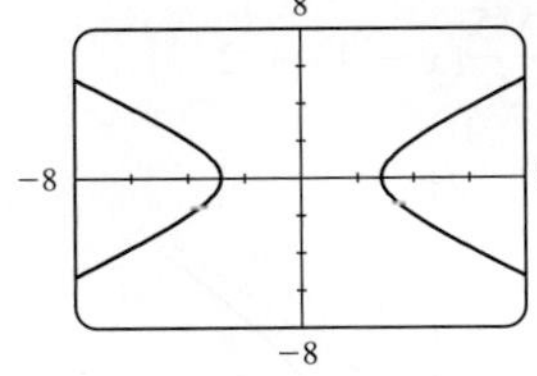

35.

37. $\dfrac{x^2}{9} - \dfrac{y^2}{16} = 1$ **39.** $y^2 - \dfrac{x^2}{3} = 1$ **41.** $x^2 - \dfrac{y^2}{25} = 1$

43. $\dfrac{y^2}{36} - \dfrac{x^2}{20} = 1$ **45.** $\dfrac{x^2}{16} - \dfrac{y^2}{16} = 1$ **47.** $\dfrac{y^2}{8} - x^2 = 1$

49. $\dfrac{x^2}{9} - \dfrac{y^2}{16} = 1$ **51. (b)** $x^2 - y^2 = c^2/2$

55. (b)

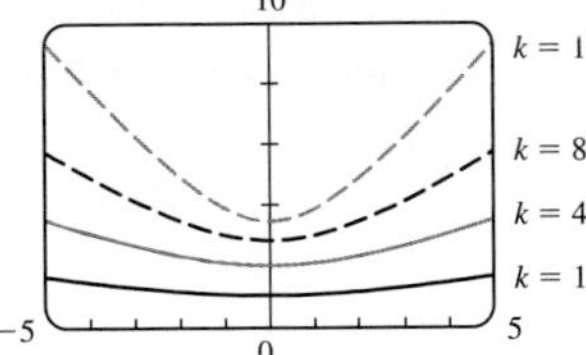

As k increases, the asymptotes get steeper.

57. $x^2 - y^2 = 2.3 \times 10^{19}$

Section 10.4 ■ Page 799

1. (a) right; left **(b)** upward; downward

2.

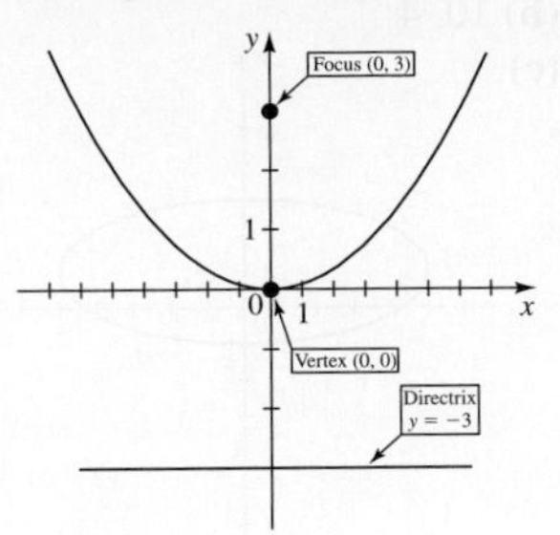

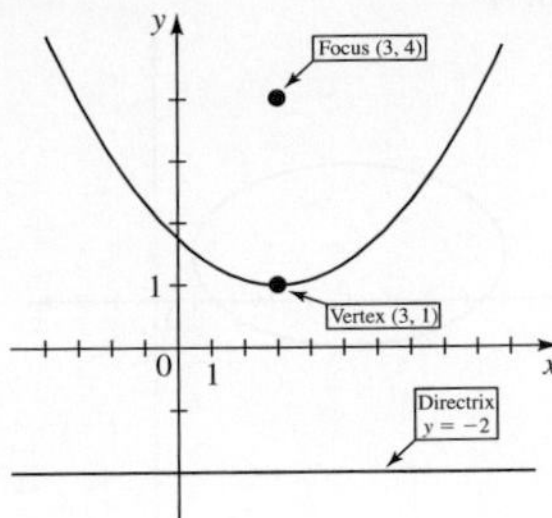

3.

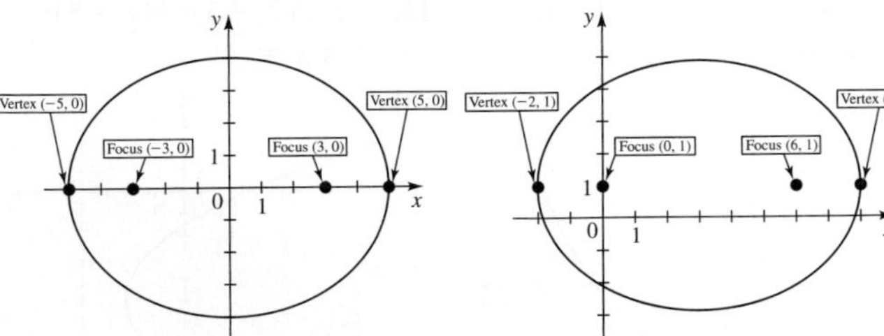

4.

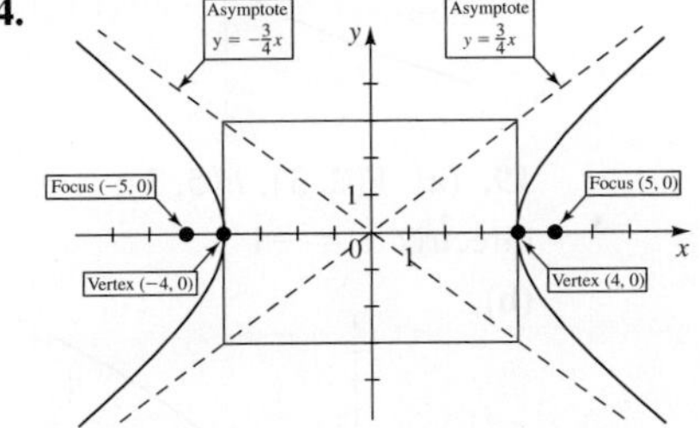

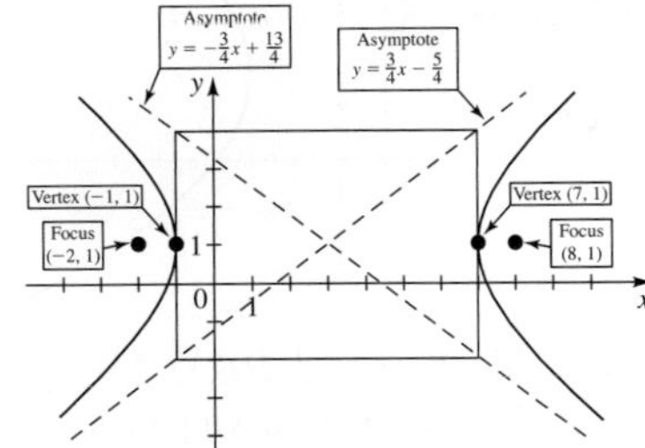

5. (a) $C(2, 1)$; $V_1(-1, 1)$, $V_2(5, 1)$; $F(2 \pm \sqrt{5}, 1)$
(b) 6, 4
(c)

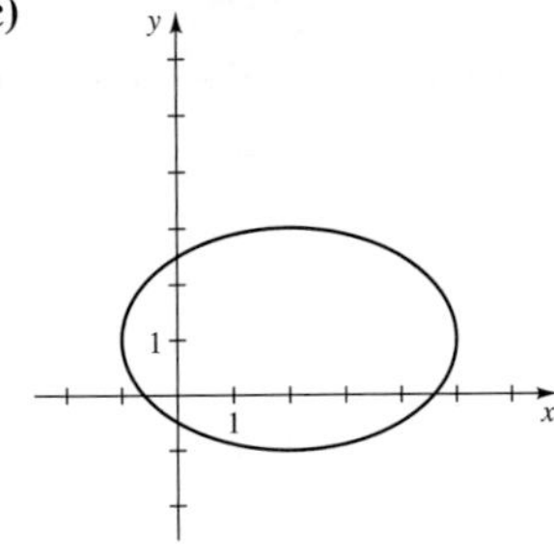

7. (a) $C(0, -5)$; $V_1(0, -10)$, $V_2(0, 0)$; $F_1(0, -9)$, $F_2(0, -1)$
(b) 10, 6
(c)

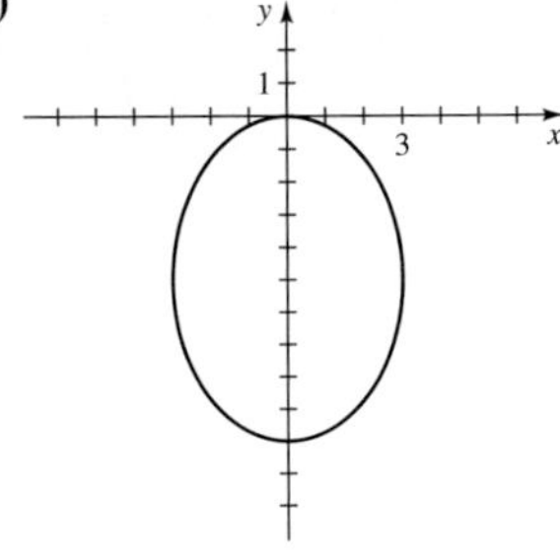

9. (a) $C(-5, 1)$; $V_1(-9, 1)$, $V_2(-1, 1)$; $F(-5 \pm 2\sqrt{3}, 1)$
(b) 8, 4
(c)

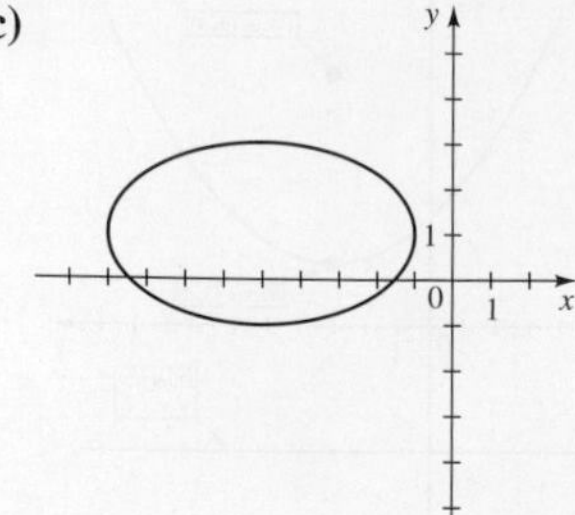

11. (a) $C(0, 1)$; $V(\pm 5, 1)$; $F(\pm\sqrt{21}, 1)$
(b) 10, 4
(c)

13. (a) $V(3, -1)$; $F(3, 1)$; directrix $y = -3$
(b)

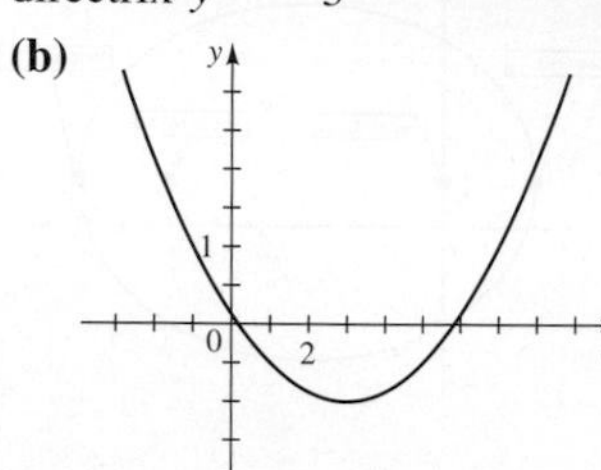

15. (a) $V(2, -5)$; $F(\frac{1}{2}, -5)$; directrix $x = \frac{7}{2}$
(b)

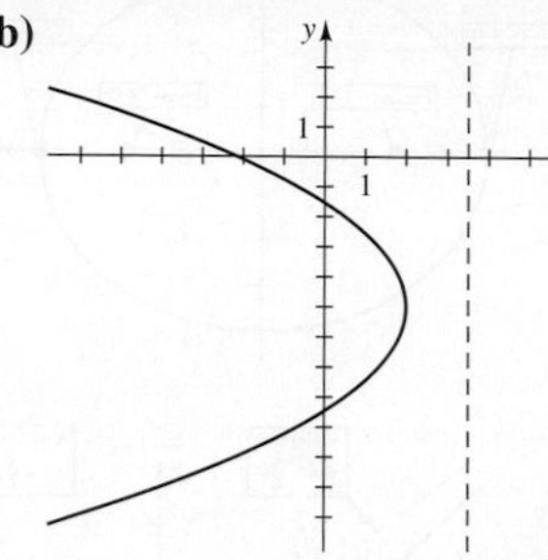

17. (a) $V(1, 0)$; $F\left(1, \frac{1}{8}\right)$; directrix $y = -\frac{1}{8}$
(b)

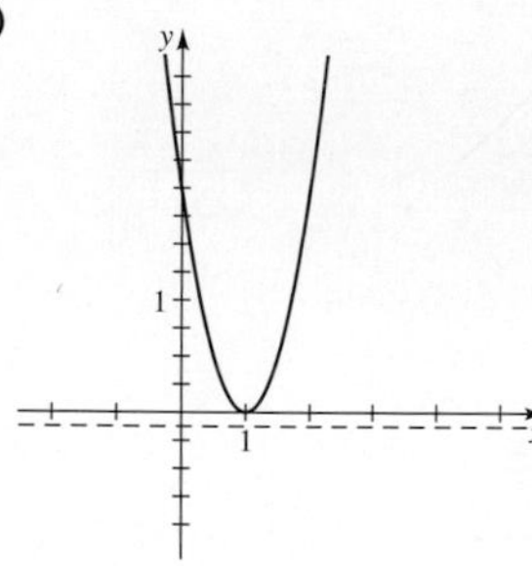

19. (a) $V(2, 3)$; $F(5, 3)$; directrix $x = -1$
(b)

21. (a) $C(-1, 3)$; $V_1(-4, 3)$, $V_2(2, 3)$; $F_1(-6, 3)$, $F_2(4, 3)$; asymptotes $y = \frac{4}{3}x + \frac{13}{3}$ and $y = -\frac{4}{3}x + \frac{5}{3}$
(b)

23. (a) $C(-1, 0)$; $V(-1, \pm 1)$; $F(-1, \pm\sqrt{5})$; asymptotes $y = \frac{1}{2}x + \frac{1}{2}$ and $y = -\frac{1}{2}x - \frac{1}{2}$
(b)

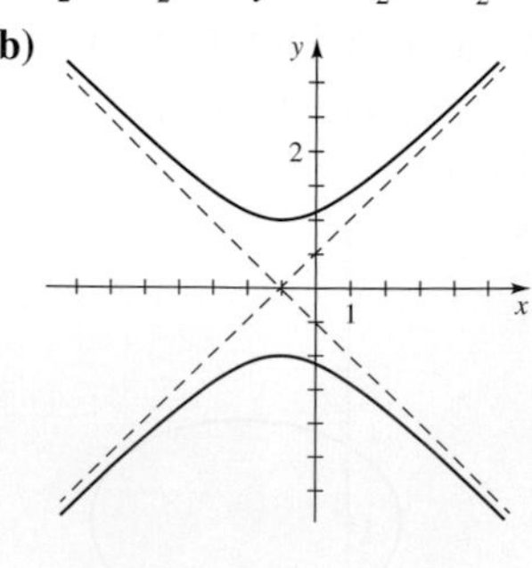

25. (a) $C(-1, -1)$; $V_1(-4, -1)$, $V_2(2, -1)$; $F(-1 \pm \sqrt{13}, -1)$; asymptotes $y = \frac{2}{3}x - \frac{1}{3}$ and $y = -\frac{2}{3}x - \frac{5}{3}$
(b)

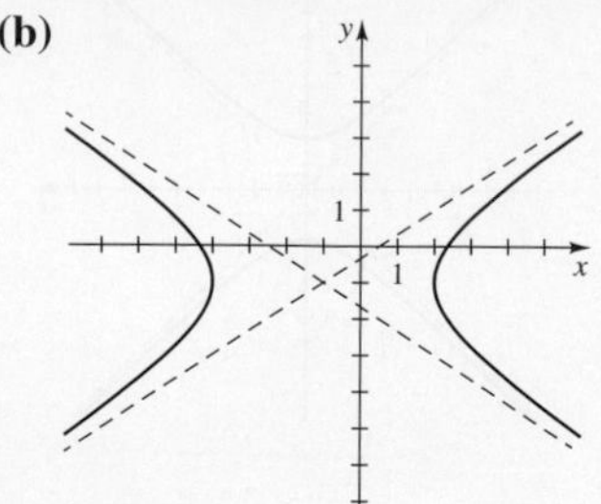

27. (a) $C(-1, 4)$; $V_1(-1, -2)$, $V_2(-1, 10)$; $F(-1, 4 \pm 2\sqrt{10})$; asymptotes $y = 3x + 7$ and $y = -3x + 1$
(b)

29. $x^2 = -\frac{1}{4}(y - 4)$ **31.** $\dfrac{(x-5)^2}{25} + \dfrac{y^2}{16} = 1$

33. $(y - 1)^2 - x^2 = 1$ **35.** $\dfrac{(x-2)^2}{100} + \dfrac{(y+3)^2}{64} = 1$

37. $\dfrac{(y-4)^2}{49} - \dfrac{(x+1)^2}{32} = 1$ **39.** $(x + 3)^2 = 12(y - 5)$

41. $\dfrac{y^2}{16} - \dfrac{(x-1)^2}{9} = 1$ **43.** $\dfrac{(x-3)^2}{29} + \dfrac{(y+4)^2}{25} = 1$

45. $(y - 2)^2 = \frac{1}{7}(x + 1)$

47. Parabola; $V(-4, 4)$; $F(-3, 4)$; directrix $x = -5$

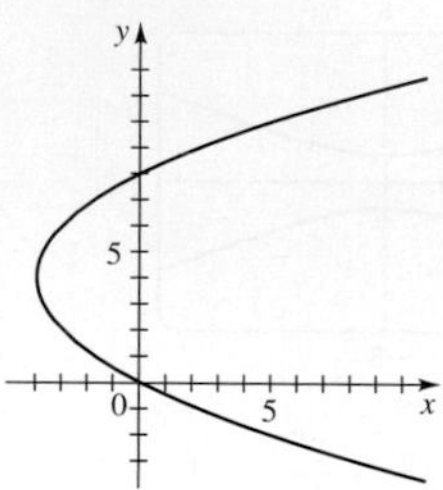

49. Hyperbola; $C(1, 2)$; $F(1 \pm \sqrt{30}, 2)$; $V_1(-4, 2)$, $V_2(6, 2)$; asymptotes

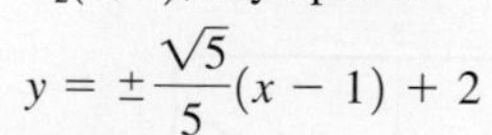

$$y = \pm\frac{\sqrt{5}}{5}(x - 1) + 2$$

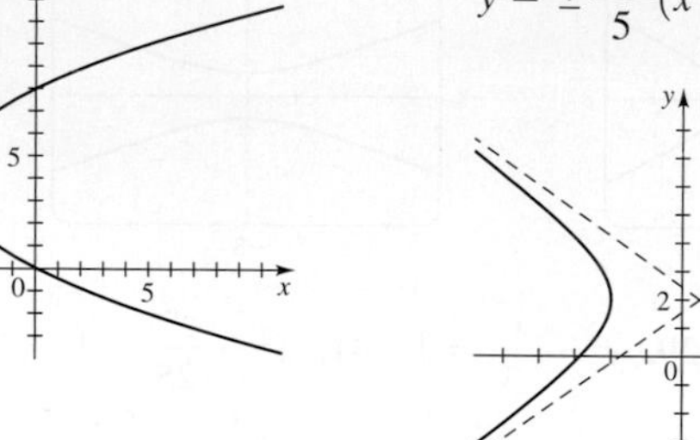

51. Ellipse; $C(3, -5)$; $F\left(3 \pm \sqrt{21}, -5\right)$; $V_1(-2, -5)$, $V_2(8, -5)$; major axis 10, minor axis 4

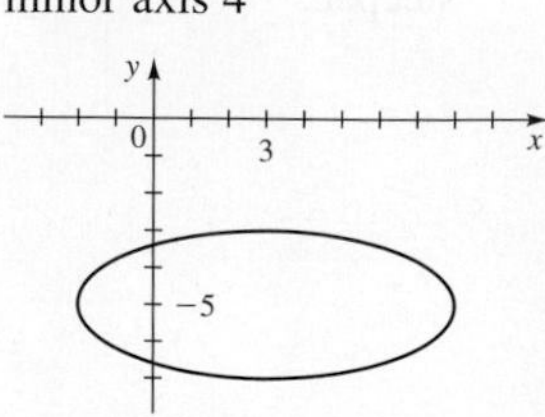

53. Hyperbola; $C(3, 0)$; $F(3, \pm 5)$; $V(3, \pm 4)$; asymptotes $y = \pm\frac{4}{3}(x - 3)$

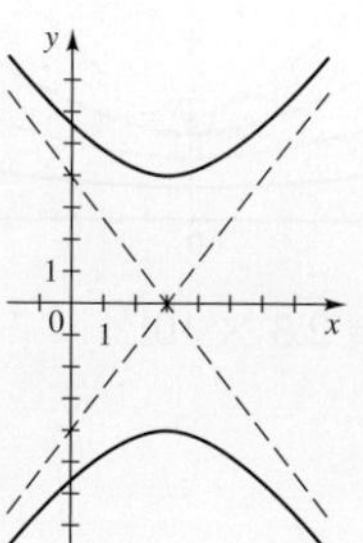

55. Degenerate conic (pair of lines),

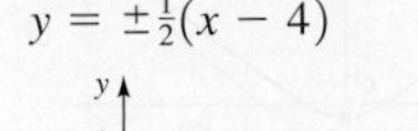

$y = \pm\frac{1}{2}(x - 4)$

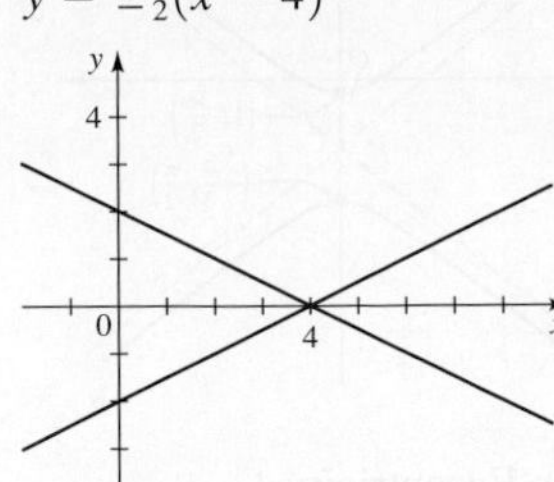

57. Point $(1, 3)$

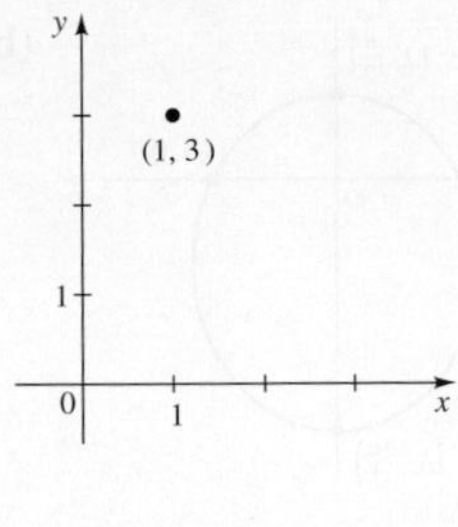

59.

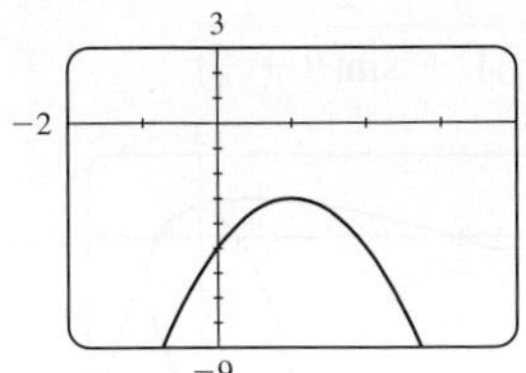

61.

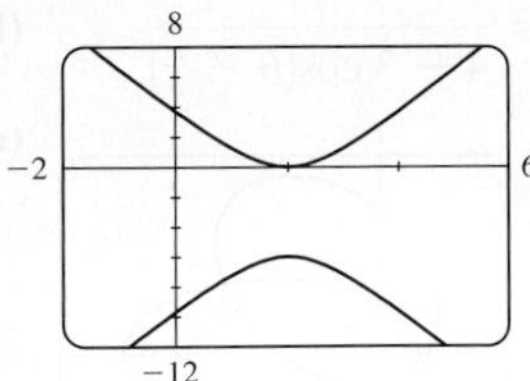

63. **(a)** $F < 17$ **(b)** $F = 17$ **(c)** $F > 17$

65. **(a)**

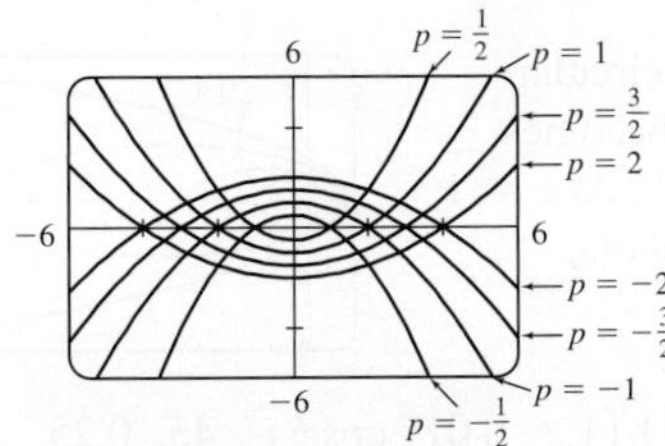

(c) The parabolas become narrower.

67. $\dfrac{(x + 150)^2}{18{,}062{,}500} + \dfrac{y^2}{18{,}040{,}000} = 1$

Section 10.5 ■ Page 808

1. $x = X\cos\phi - Y\sin\phi$, $y = X\sin\phi + Y\cos\phi$, $X = x\cos\phi + y\sin\phi$, $Y = -x\sin\phi + y\cos\phi$
2. **(a)** conic section **(b)** $(A - C)/B$ **(c)** $B^2 - 4AC$, a parabola, an ellipse, a hyperbola **3.** $(\sqrt{2}, 0)$ **5.** $(0, -2\sqrt{3})$
7. $(1.6383, 1.1472)$ **9.** $X^2 + \sqrt{3}XY + 2 = 0$
11. $7Y^2 - 48XY - 7X^2 - 40X - 30Y = 0$ **13.** $X^2 - Y^2 = 2$
15. **(a)** Hyberbola
(b) $X^2 - Y^2 = 16$
(c) $\phi = 45°$

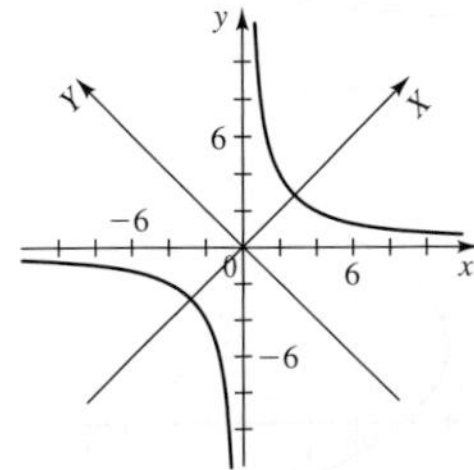

17. **(a)** Hyberbola
(b) $Y^2 - X^2 = 1$
(c) $\phi = 30°$

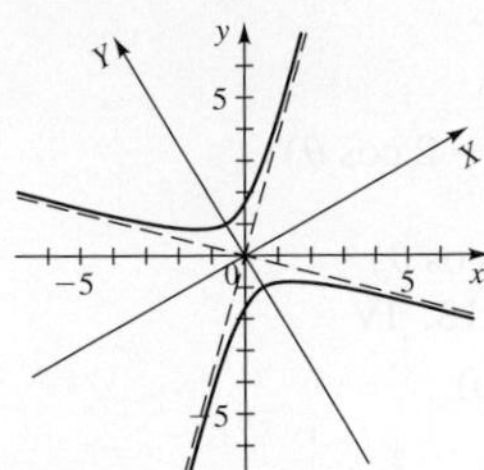

19. **(a)** Hyberbola
(b) $\dfrac{X^2}{4} - Y^2 = 1$
(c) $\phi \approx 53°$

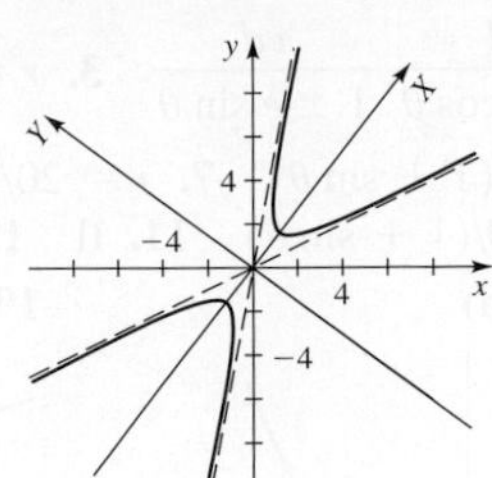

21. **(a)** Hyberbola
(b) $3X^2 - Y^2 = 2\sqrt{3}$
(c) $\phi = 30°$

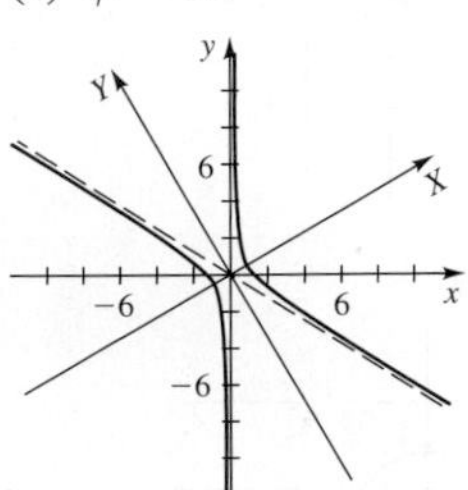

23. **(a)** Parabola
(b) $Y = \sqrt{2}X^2$
(c) $\phi = 45°$

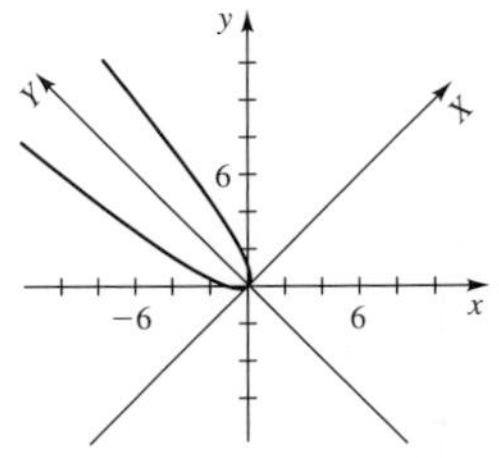

25. **(a)** Hyberbola
(b) $(X - 1)^2 - 3Y^2 = 1$
(c) $\phi = 60°$

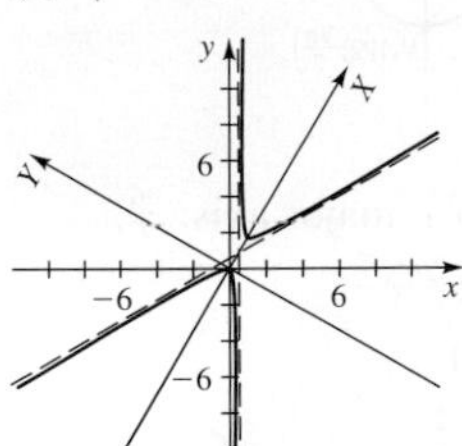

27. **(a)** Ellipse
(b) $X^2 + \dfrac{(Y + 1)^2}{4} = 1$
(c) $\phi \approx 53°$

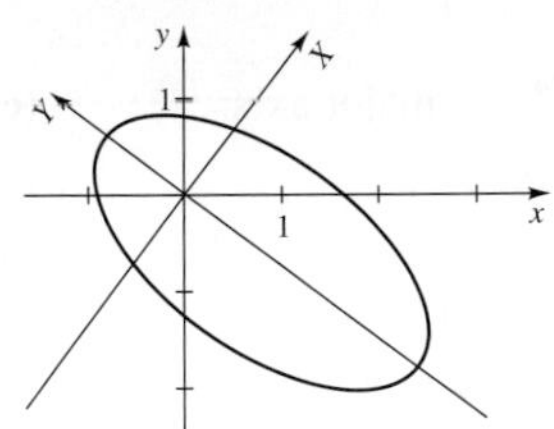

29. **(a)** Parabola
(b)

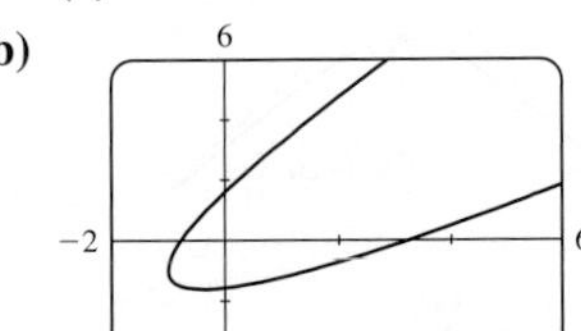

31. **(a)** Hyperbola
(b)

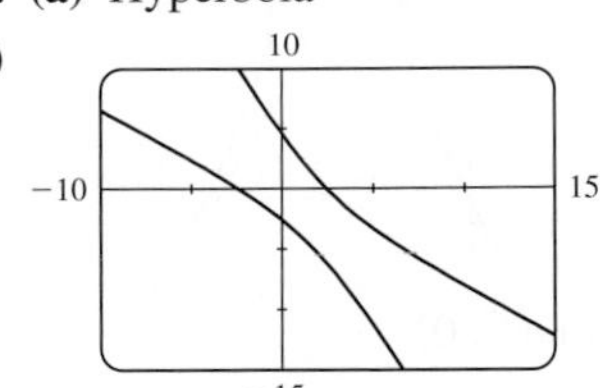

33. **(a)** $(X - 5)^2 - Y^2 = 1$
(b) XY-coordinates: $C(5, 0)$; $V_1(6, 0)$, $V_2(4, 0)$; $F(5 \pm \sqrt{2}, 0)$; xy-coordinates:
$C(4, 3)$; $V_1(\frac{24}{5}, \frac{18}{5})$, $V_2(\frac{16}{5}, \frac{12}{5})$; $F_1(4 + \frac{4}{5}\sqrt{2}, 3 + \frac{3}{5}\sqrt{2})$,
$F_2(4 - \frac{4}{5}\sqrt{2}, 3 - \frac{3}{5}\sqrt{2})$
(c) $Y = \pm(X - 5)$; $7x - y - 25 = 0$, $x + 7y - 25 = 0$
35. $X = x\cos\phi + y\sin\phi$; $Y = -x\sin\phi + y\cos\phi$

Section 10.6 ■ Page 814

1. focus, directrix; $\dfrac{\text{distance from } P \text{ to } F}{\text{distance from } P \text{ to } \ell}$, conic section; parabola, ellipse, hyperbola, eccentricity

2. $\dfrac{ed}{1 \pm e\cos\theta}, \dfrac{ed}{1 \pm e\sin\theta}$ **3.** $r = 6/(3 + 2\cos\theta)$

5. $r = 2/(1 + \sin\theta)$ **7.** $r = 20/(1 + 4\cos\theta)$

9. $r = 10/(1 + \sin\theta)$ **11.** II **13.** VI **15.** IV

17. (a), (b)

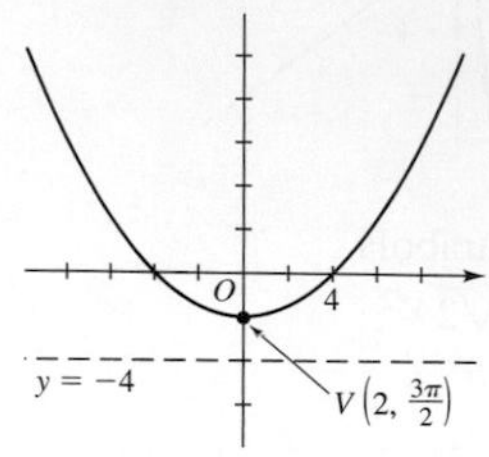

19. (a), (b)

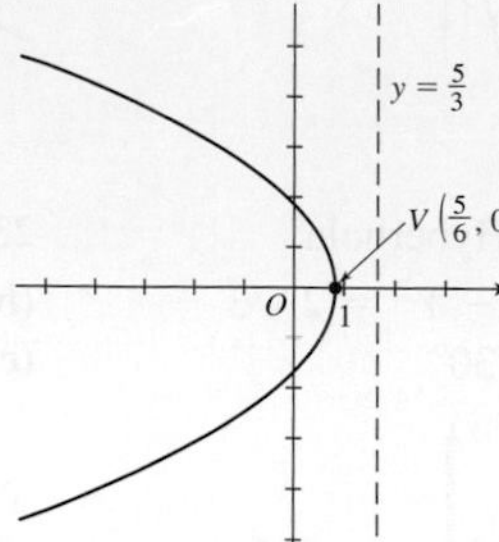

21. (a), (b)

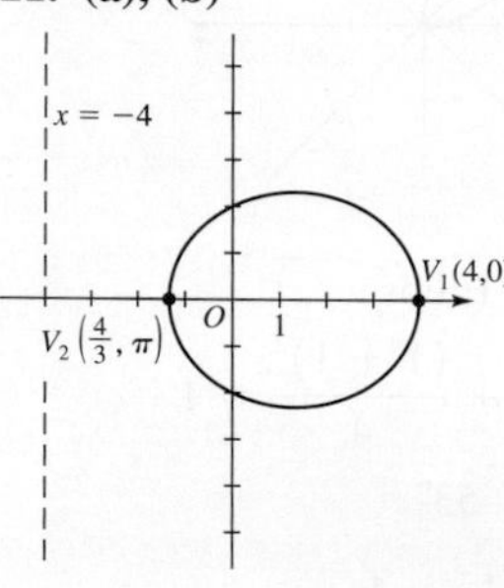

23. (a), (b)

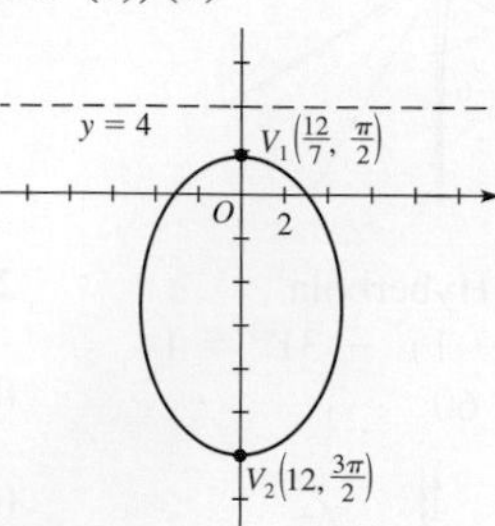

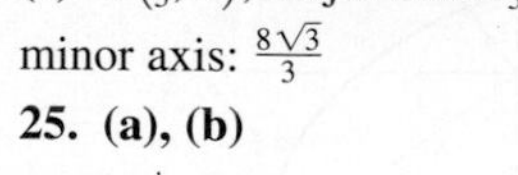

(c) $C(\frac{4}{3}, 0)$, major axis: $\frac{16}{3}$, minor axis: $\frac{8\sqrt{3}}{3}$

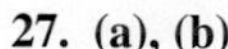

(c) $C(\frac{36}{7}, \frac{3\pi}{2})$, major axis: $\frac{96}{7}$, minor axis: $\frac{24\sqrt{7}}{7}$

25. (a), (b)

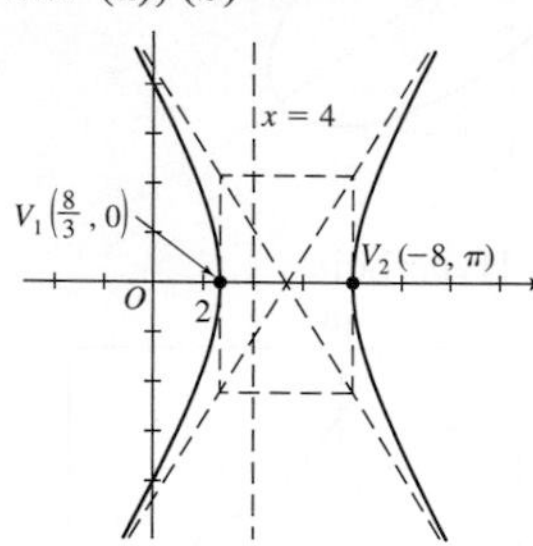

27. (a), (b)

(c) $(\frac{16}{3}, 0)$

(c) $(12, \frac{3\pi}{2})$

29. (a) 3, hyperbola

(b)

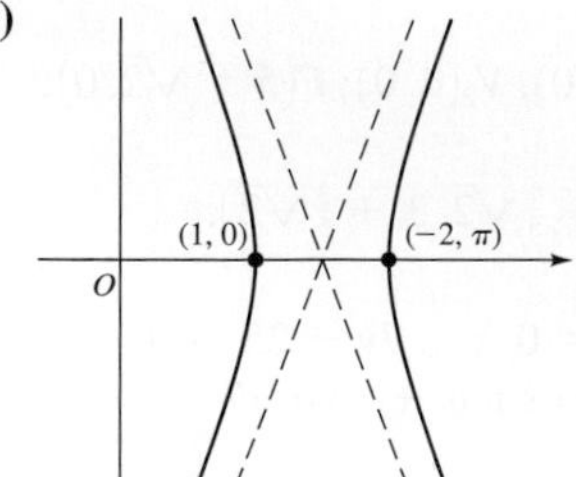

31. (a) 1, parabola

(b)

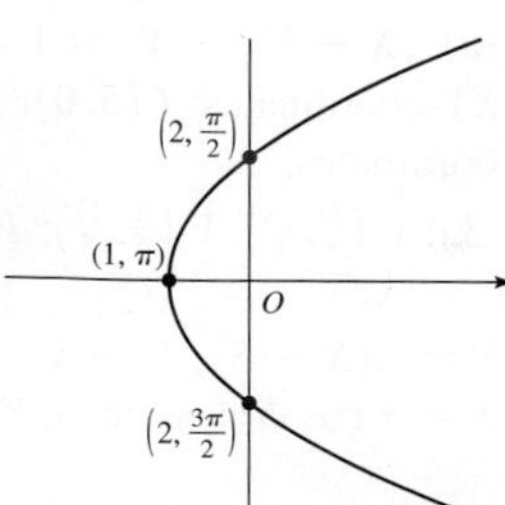

33. (a) $\frac{1}{2}$, ellipse

(b)

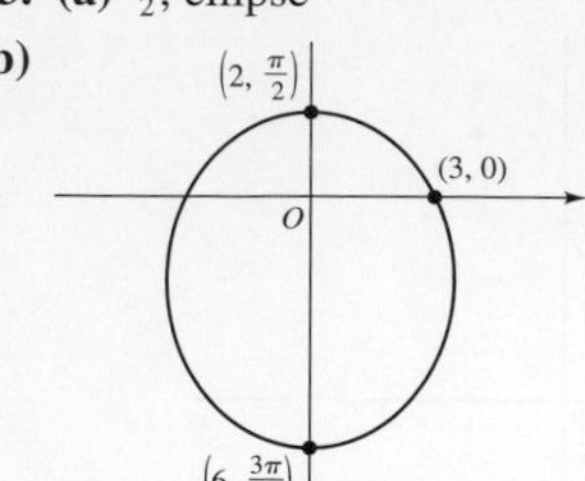

35. (a) $\frac{5}{2}$, hyperbola

(b)

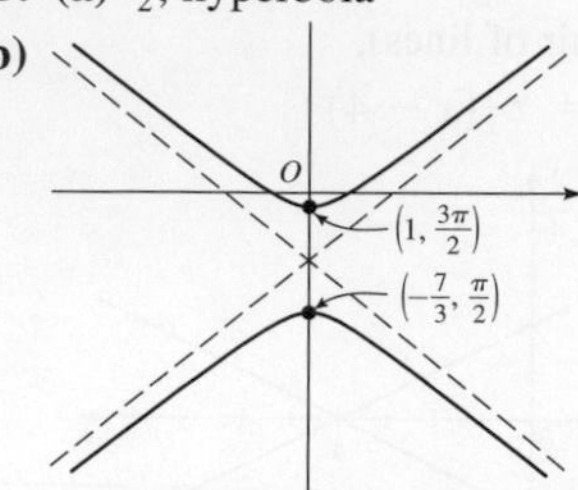

37. (a) Eccentricity $\frac{3}{4}$, directrix $x = -\frac{1}{3}$

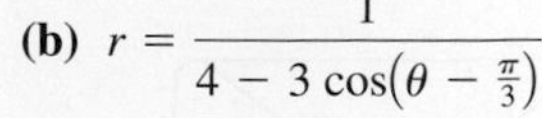

(b) $r = \dfrac{1}{4 - 3\cos(\theta - \frac{\pi}{3})}$

(c)

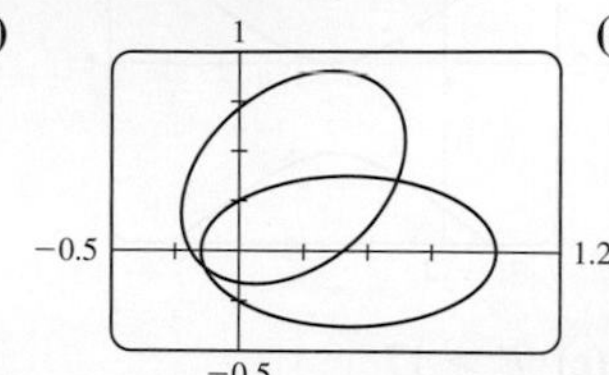

39. (a) Eccentricity 1, directrix $y = 2$

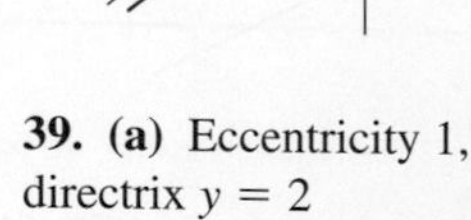
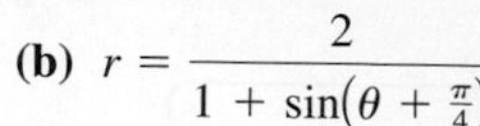

(b) $r = \dfrac{2}{1 + \sin(\theta + \frac{\pi}{4})}$

(c)

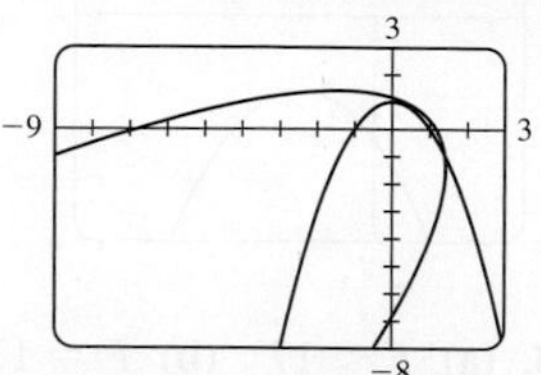

41. The ellipse is nearly circular when e is close to 0 and becomes more elongated as $e \to 1^-$. At $e = 1$ the curve becomes a parabola.

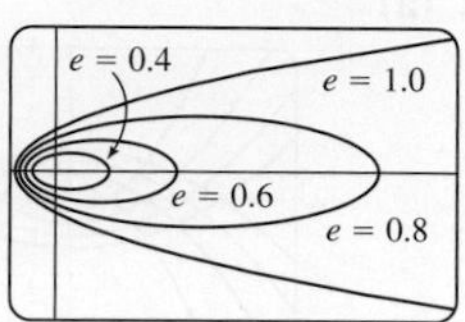

43. (b) $r = (1.49 \times 10^8)/(1 - 0.017\cos\theta)$ **45.** 0.25

Chapter 10 Review ■ Page 818

1. (a) $V(0, 0)$; $F(1, 0)$; directrix $x = -1$

(b)

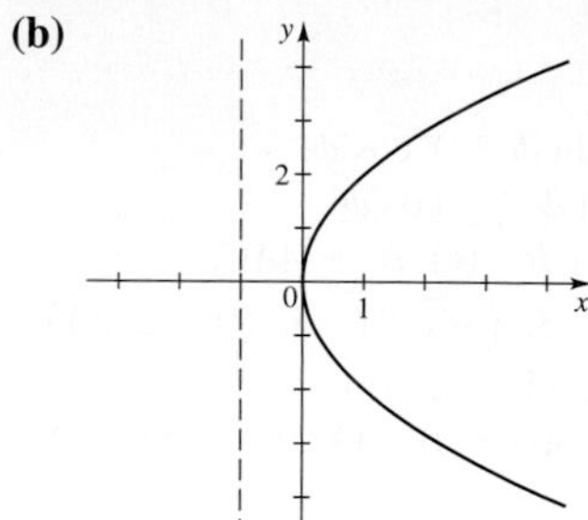

3. (a) $V(0, 0)$; $F(0, 2)$; directrix $y = -2$

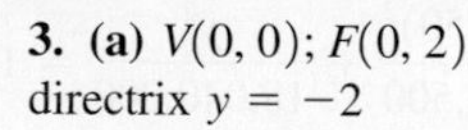

(b)

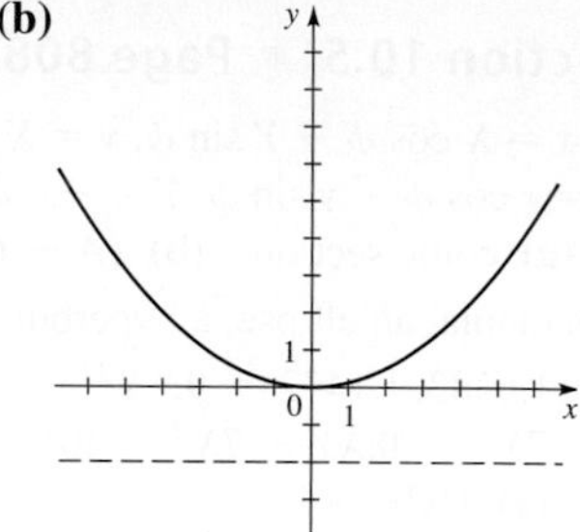

5. (a) $V(0, 0)$; $F(0, -2)$; directrix $y = 2$

(b)

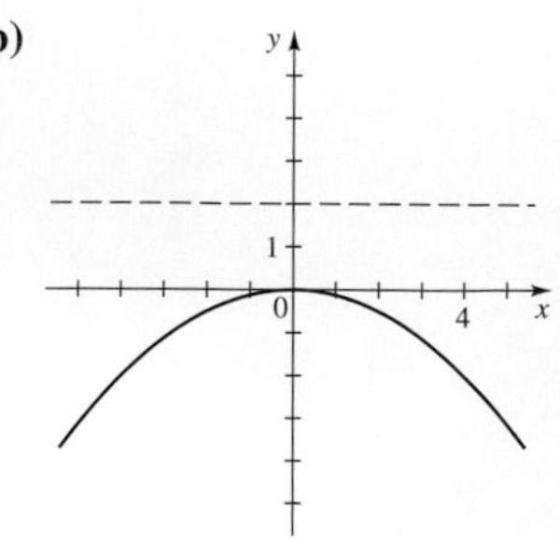

7. (a) $V(-2, 2)$; $F(-1, 2)$; directrix $x = -3$

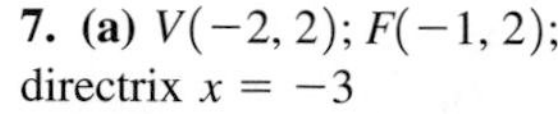

(b)

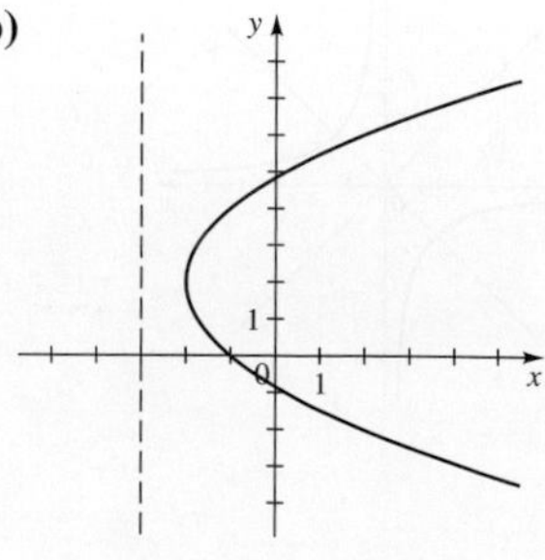

9. (a) $V(0, 3)$; $F\left(-\frac{1}{2}, 3\right)$; directrix $x = \frac{1}{2}$

(b)

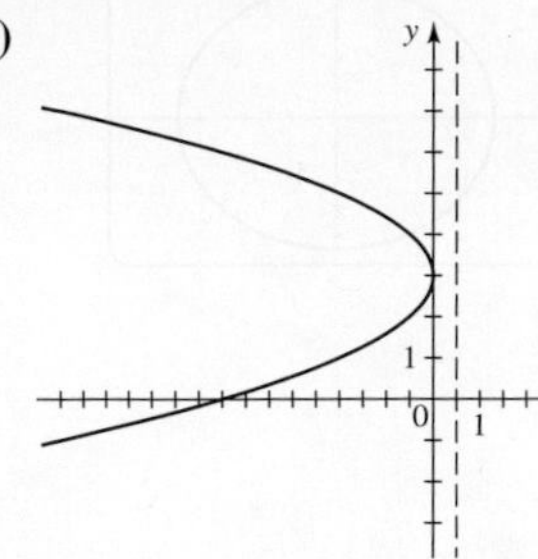

11. (a) $V(-2, -3)$; $F(-2, -2)$; directrix $y = -4$

(b)

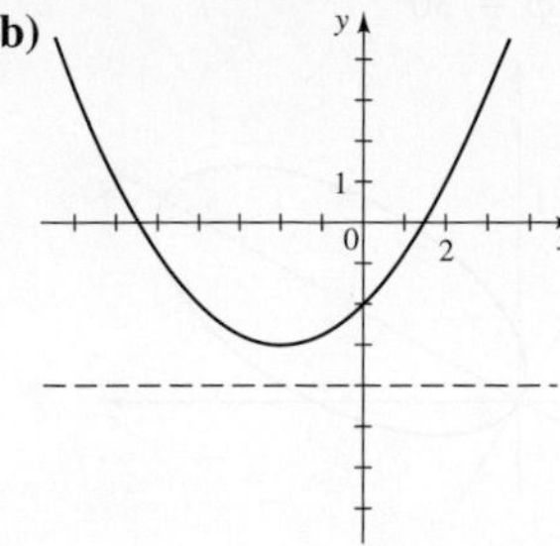

13. (a) $C(0, 0)$; $V(0, \pm 5)$; $F(0, \pm 4)$

(b) 10, 6

(c)

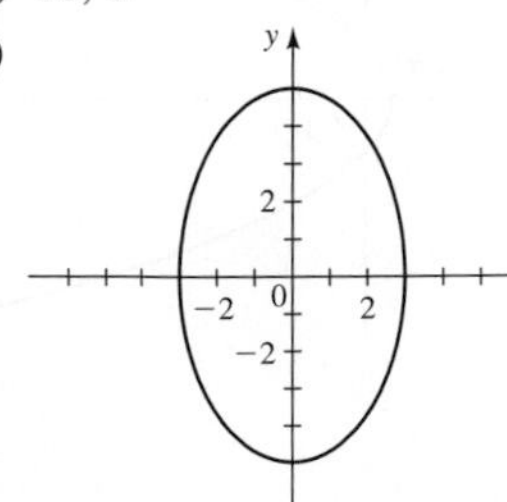

15. (a) $C(0, 0)$; $V(\pm 7, 0)$; $F(\pm 3\sqrt{5}, 0)$

(b) 14, 4

(c)

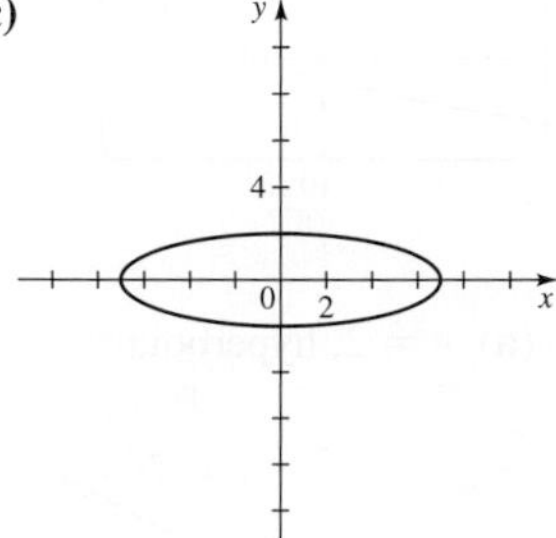

17. (a) $C(0, 0)$; $V(\pm 4, 0)$; $F(\pm 2\sqrt{3}, 0)$

(b) 8, 4

(c)

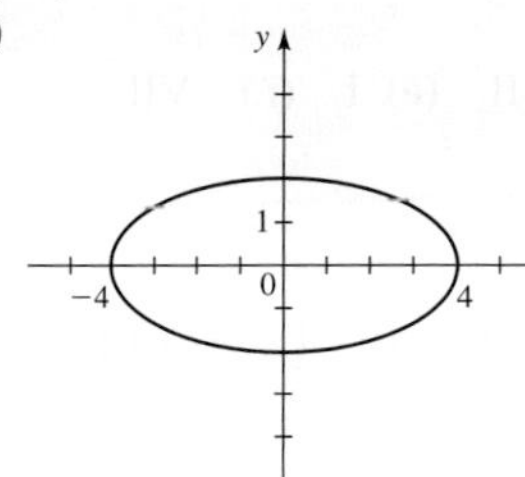

19. (a) $C(3, 0)$; $V(3, \pm 4)$; $F(3, \pm\sqrt{7})$

(b) 8, 6

(c)

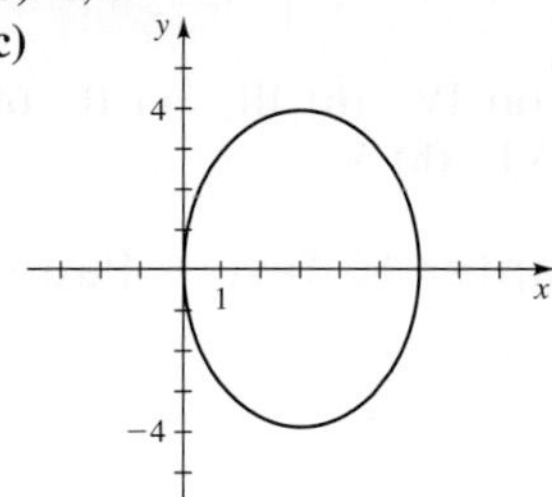

21. (a) $C(2, -3)$; $V_1(2, -9)$, $V_2(2, 3)$; $F(2, -3 \pm 3\sqrt{3})$

(b) 12, 6

(c)

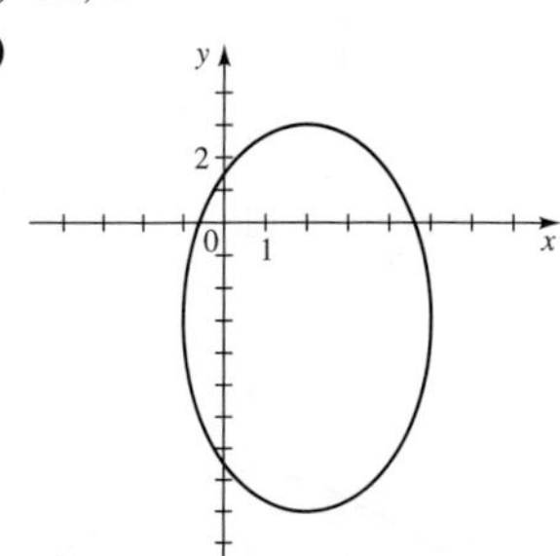

23. (a) $C(0, 2)$; $V(\pm 3, 2)$; $F(\pm\sqrt{5}, 2)$

(b) 6, 4

(c)

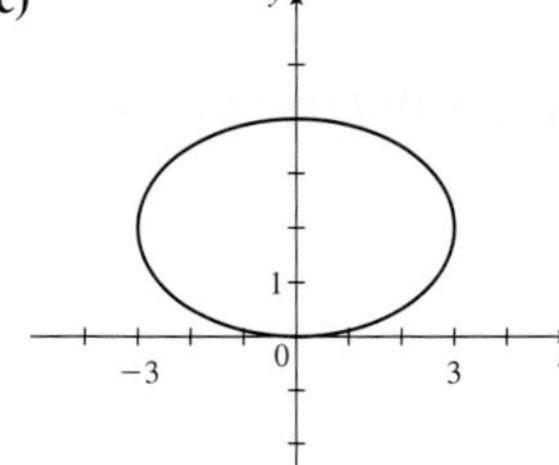

25. (a) $C(0, 0)$; $V(0, \pm 4)$; $F(0, \pm 5)$; asymptotes $y = \pm\frac{4}{3}x$

(b)

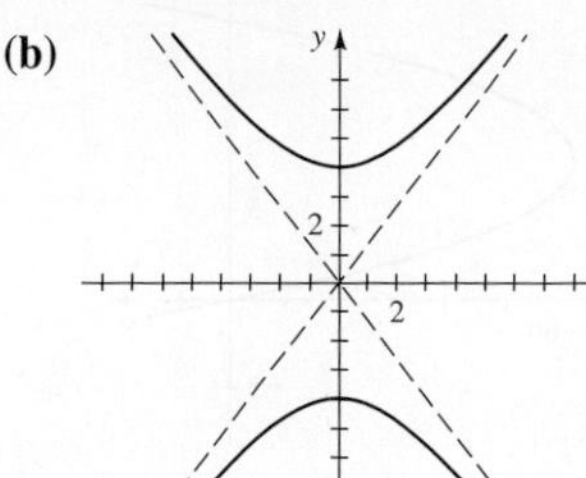

27. (a) $C(0, 0)$; $V(\pm 2, 0)$; $F(\pm\sqrt{53}, 0)$; asymptotes $y = \pm\frac{7}{2}x$

(b)

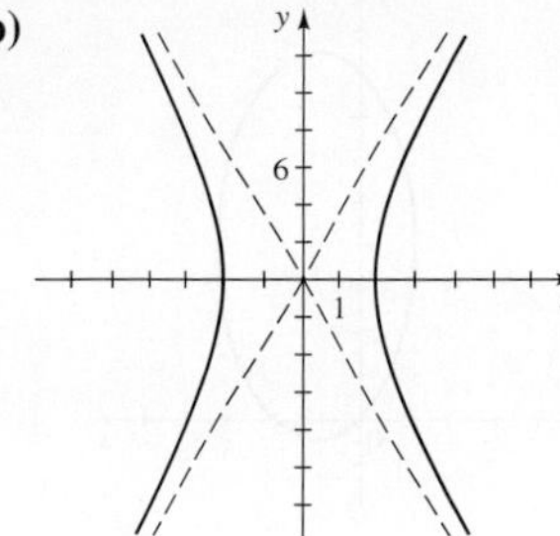

29. (a) $C(0, 0)$; $V(\pm 4, 0)$; $F(\pm 2\sqrt{6}, 0)$; asymptotes $y = \pm\dfrac{1}{\sqrt{2}}x$

(b)

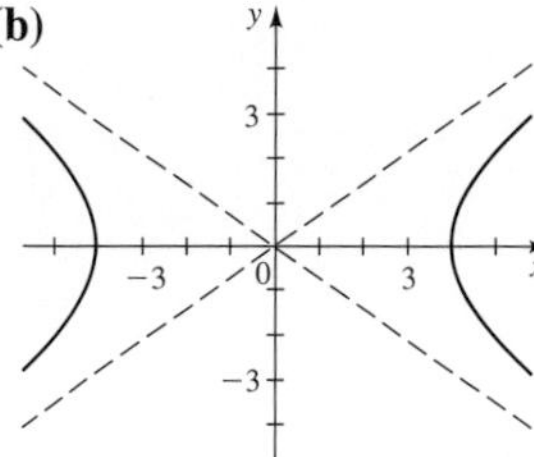
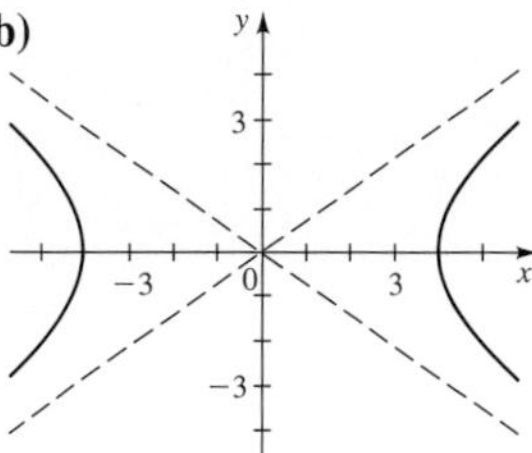

31. (a) $C(-4, 0)$; $V_1(-8, 0)$, $V_2(0, 0)$; $F(-4 \pm 4\sqrt{2}, 0)$; asymptotes $y = \pm(x + 4)$

(b)

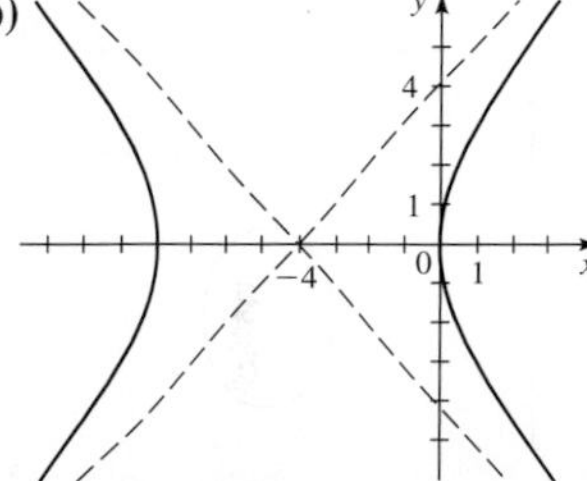

33. (a) $C(-1, 3)$; $V_1(-1, 1)$, $V_2(-1, 5)$; $F(-1, 3 \pm 2\sqrt{10})$; asymptotes $y = \frac{1}{3}x + \frac{10}{3}$ and $y = -\frac{1}{3}x + \frac{8}{3}$

(b)

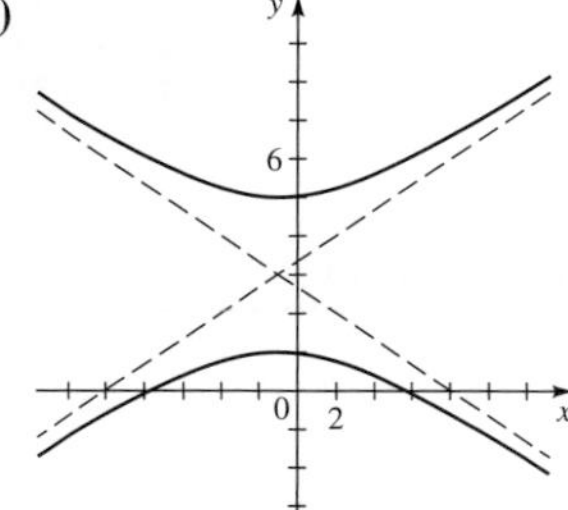

35. (a) $C(-3, -1)$; $V(-3, -1 \pm \sqrt{2})$; $F(-3, -1 \pm 2\sqrt{5})$; asymptotes $y = \frac{1}{3}x$, $y = -\frac{1}{3}x - 2$

(b)

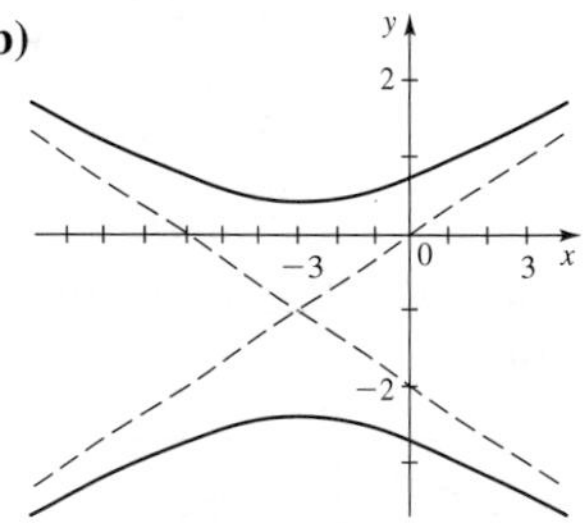

37. $y^2 = 8x$ **39.** $\dfrac{y^2}{16} - \dfrac{x^2}{9} = 1$ **41.** $\dfrac{(x-4)^2}{16} + \dfrac{(y-2)^2}{4} = 1$

43. Parabola; $V(0, 1)$; $F(0, -2)$; directrix $y = 4$

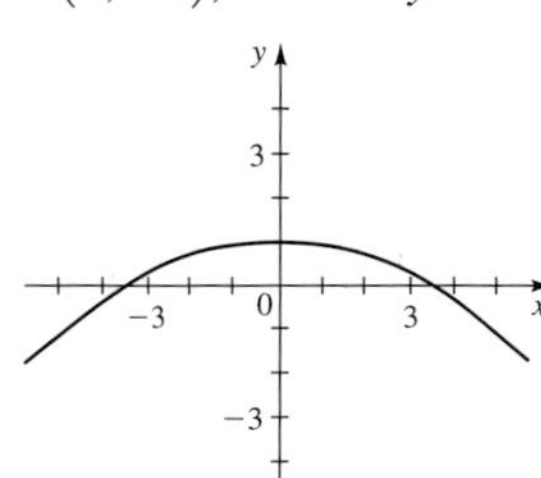

45. Hyperbola; $C(0, 0)$; $F(0, \pm 12\sqrt{2})$; $V(0, \pm 12)$; asymptotes $y = \pm x$

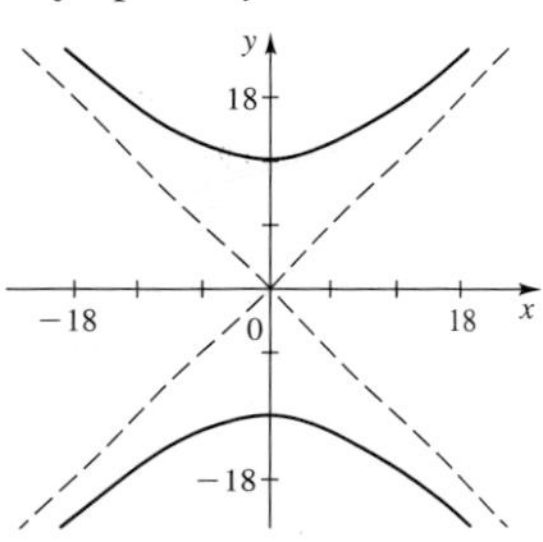

47. Ellipse; $C(1, 4)$; $F(1, 4 \pm \sqrt{15})$; $V(1, 4 \pm 2\sqrt{5})$

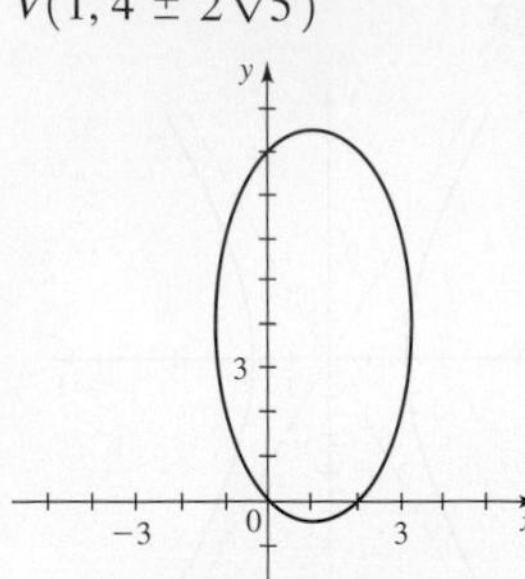

49. Parabola; $V(-64, 8)$; $F(-\frac{255}{4}, 8)$; directrix $x = -\frac{257}{4}$

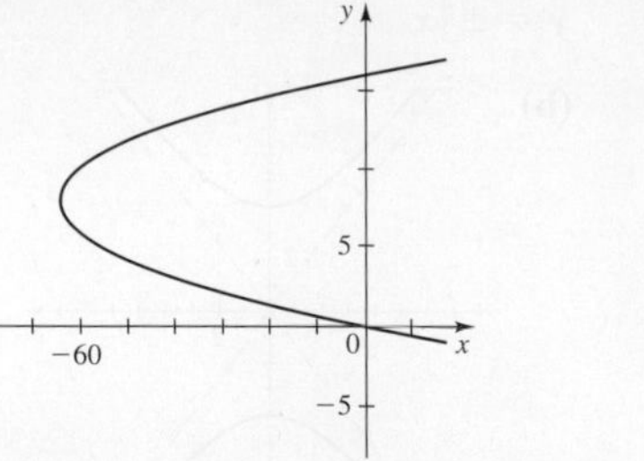

51. Ellipse; $C(3, -3)$; $F\left(3, -3 \pm \dfrac{\sqrt{2}}{2}\right)$; $V_1(3, -4)$, $V_2(3, -2)$

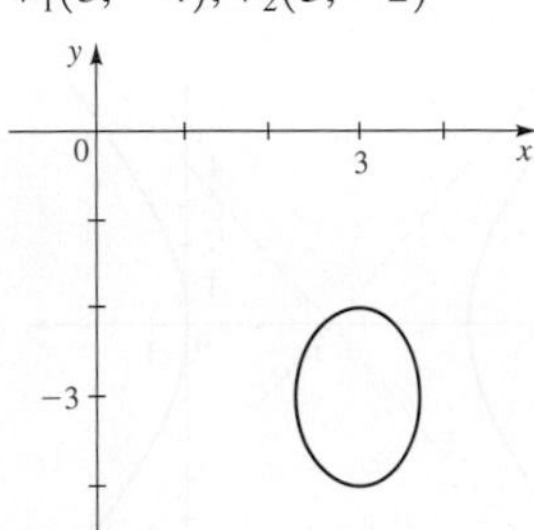

53. Has no graph

55. $x^2 = 4y$ **57.** $\dfrac{x^2}{4} + \dfrac{y^2}{25} = 1$

59. $\dfrac{x^2}{9} + \dfrac{(y-4)^2}{25} = 1$

61. $\dfrac{(x-1)^2}{3} + \dfrac{(y-2)^2}{4} = 1$

63. $\dfrac{4(x-7)^2}{225} + \dfrac{(y-2)^2}{100} = 1$

65. (a) 146,470,400 km **(b)** 151,529,600 km

67. (a)

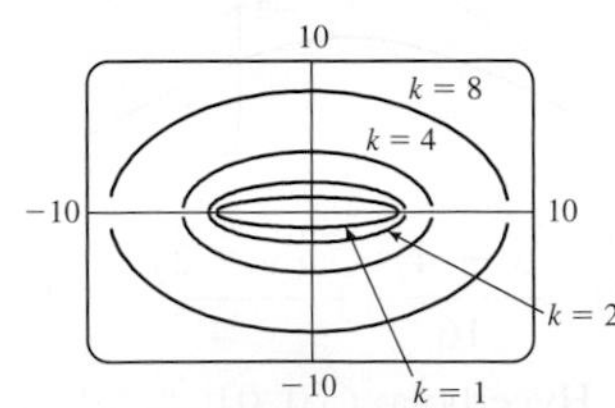

69. (a) Hyperbola **(b)** $3X^2 - Y^2 = 1$

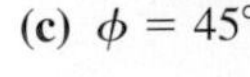
(c) $\phi = 45°$

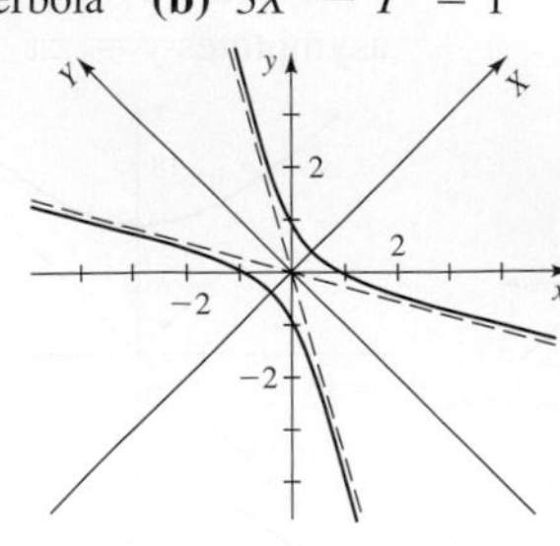

71. (a) Ellipse
(b) $(X-1)^2 + 4Y^2 = 1$
(c) $\phi = 30°$

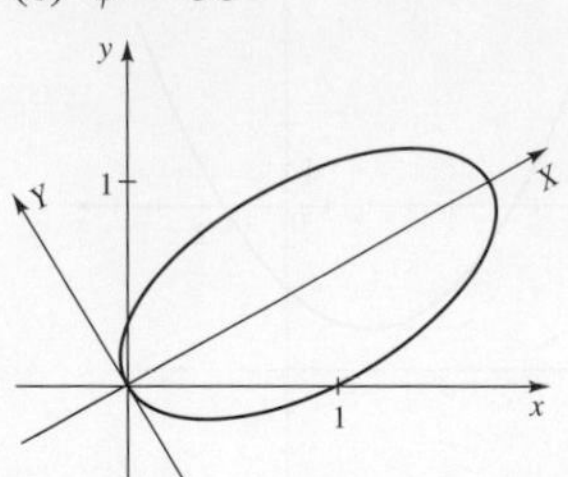

73. Ellipse

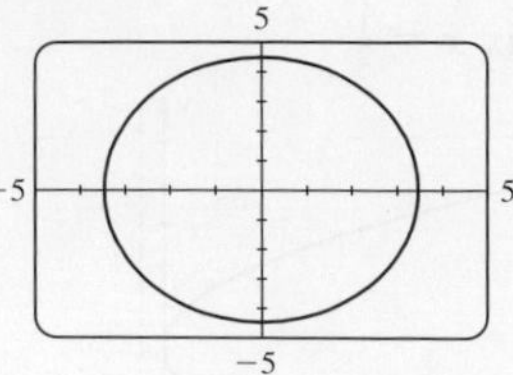

75. Parabola

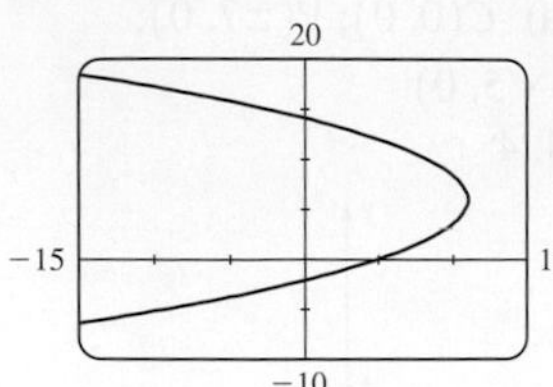

77. (a) $e = 1$, parabola
(b)

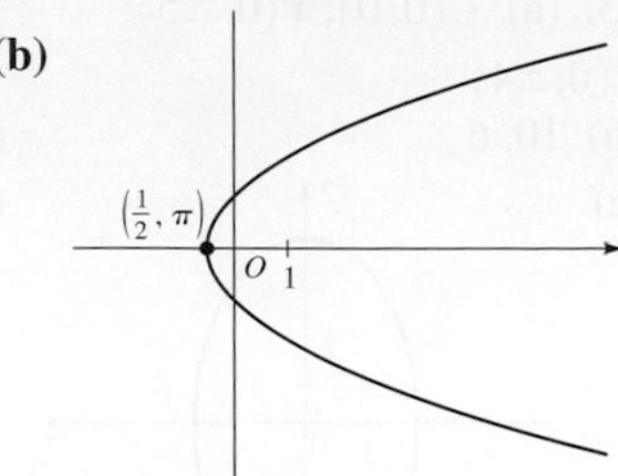

79. (a) $e = 2$, hyperbola

(b)

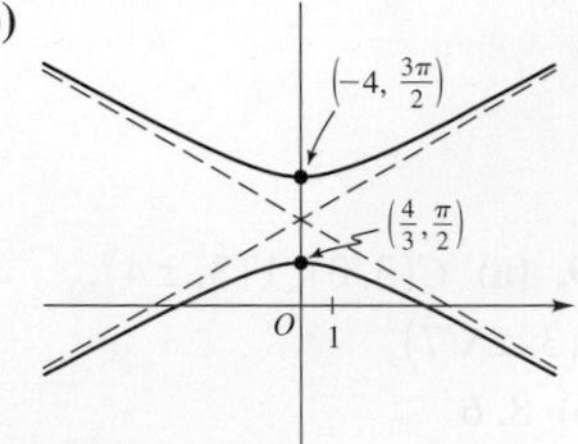

81. (a) IV **(b)** III **(c)** II **(d)** VIII **(e)** I **(f)** VII **(g)** VI **(h)** V

Chapter 10 Test ■ Page 821

1. $F(0, -3)$, $y = 3$

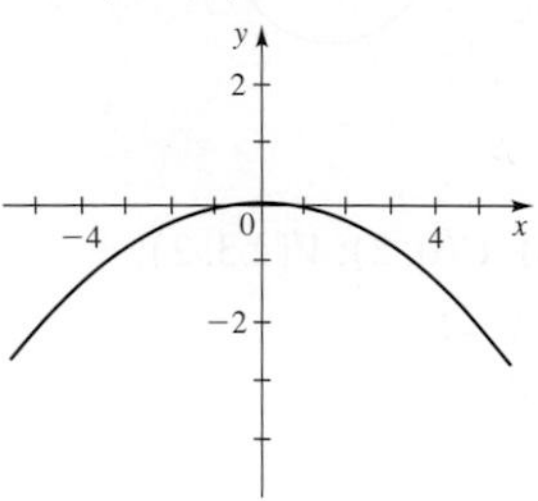

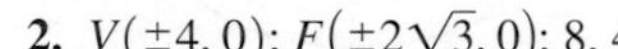
2. $V(\pm 4, 0)$; $F(\pm 2\sqrt{3}, 0)$; 8, 4

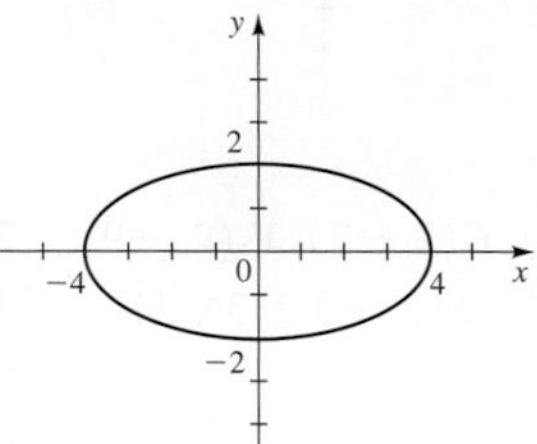

3. $V(0, \pm 3)$; $F(0, \pm 5)$; $y = \pm\frac{3}{4}x$

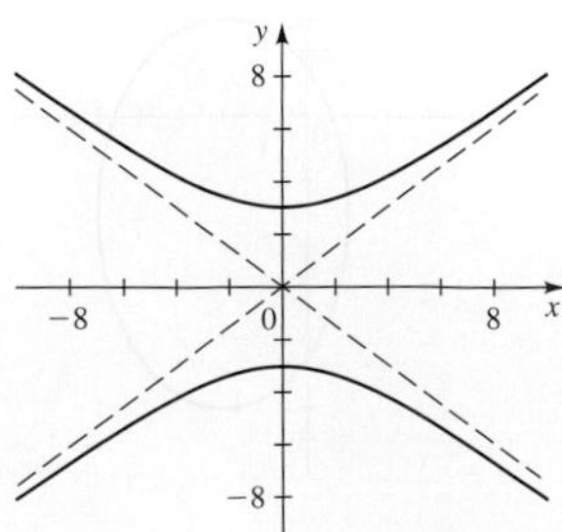

4. $y^2 = 16x$ **5.** $\dfrac{x^2}{16} + \dfrac{y^2}{7} = 1$ **6.** $\dfrac{y^2}{9} - \dfrac{x^2}{16} = 1$

7. $y^2 = -x$ **8.** $\dfrac{x^2}{16} + \dfrac{(y-3)^2}{9} = 1$ **9.** $(x-2)^2 - \dfrac{y^2}{3} = 1$

10. Ellipse; $C(3, -\frac{1}{2})$; $F(3 \pm \sqrt{5}, -\frac{1}{2})$; $V_1(0, -\frac{1}{2})$, $V_2(6, -\frac{1}{2})$

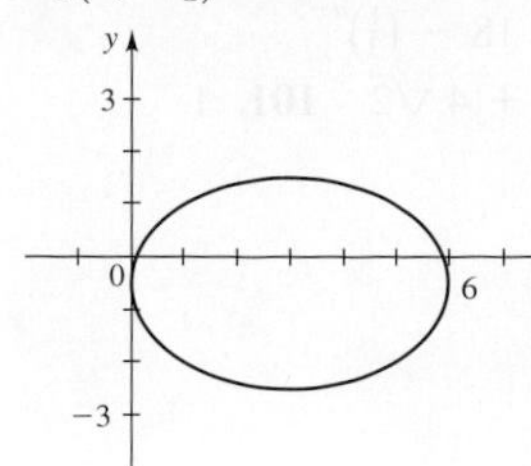

11. Hyperbola; $C(-2, 4)$, $F(-2 \pm \sqrt{17}, 4)$, $V(-2 \pm 2\sqrt{2}, 4)$, asymptotes $y - 4 = \pm\dfrac{3\sqrt{2}}{4}(x + 2)$

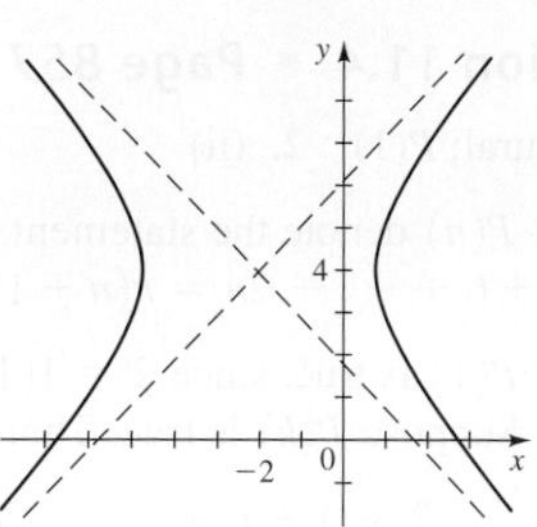

12. Parabola; $V(4, -4)$; $F(\frac{7}{2}, -4)$; directrix $x = \frac{9}{2}$

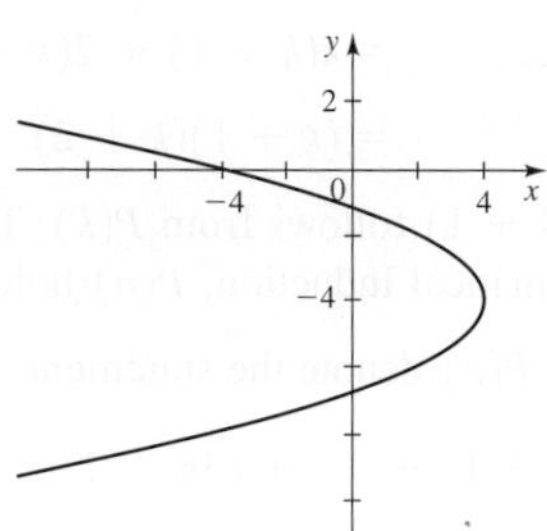

13. $\dfrac{(x-2)^2}{7} + \dfrac{y^2}{16} = 1$ **14.** $(x-2)^2 = 8(y-2)$ **15.** 2 cm

16. **(a)** Ellipse **(b)** $\dfrac{X^2}{3} + \dfrac{Y^2}{18} = 1$

(c) $\phi \approx 27°$

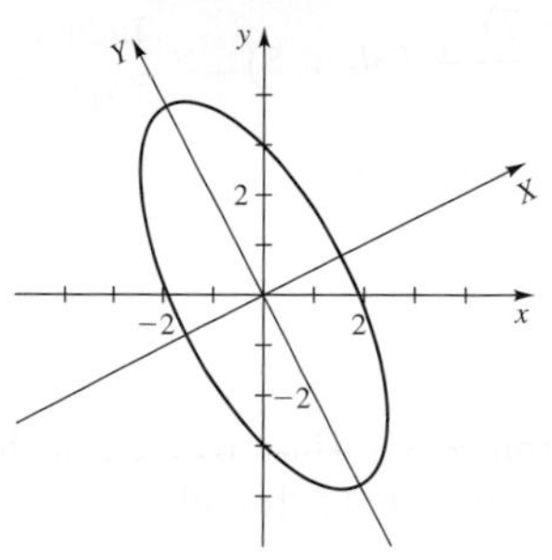

(d) $(-3\sqrt{2/5}, 6\sqrt{2/5}), (3\sqrt{2/5}, -6\sqrt{2/5})$

17. **(a)** $r = \dfrac{1}{1 + 0.5\cos\theta}$ **(b)** Ellipse

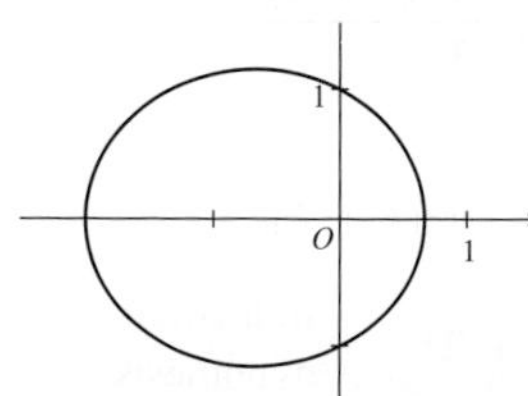

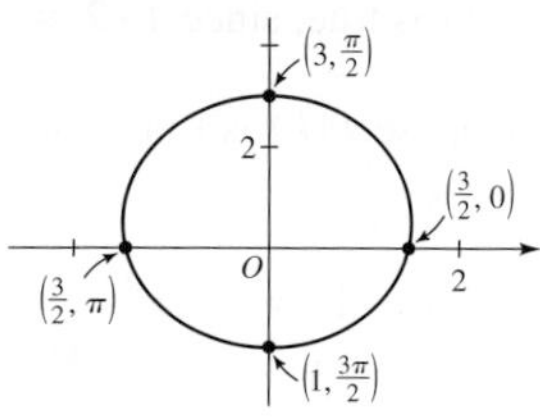

Focus on Modeling ■ Page 824

5. **(c)** $x^2 - mx + (ma - a^2) = 0$, discriminant $m^2 - 4ma + 4a^2 = (m - 2a)^2$, $m = 2a$

Chapter 11

Section 11.1 ■ Page 836

1. the natural numbers **2.** n; $1^2 + 2^2 + 3^2 + 4^2 = 30$
3. $-2, -1, 0, 1; 97$ **5.** $-1, \frac{1}{2}, \frac{1}{5}, \frac{1}{8}; \frac{1}{296}$ **7.** $3, 9, 27, 81; 3^{100}$
9. $-1, \frac{1}{4}, -\frac{1}{9}, \frac{1}{16}; \frac{1}{10{,}000}$ **11.** $0, 2, 0, 2; 2$
13. $1, 4, 27, 256; 100^{100}$ **15.** $4, 14, 34, 74, 154$
17. $1, 3, 7, 15, 31$ **19.** $1, 2, 3, 5, 8$
21. **(a)** 7, 11, 15, 19, 23, 27, 31, 35, 39, 43
(b)

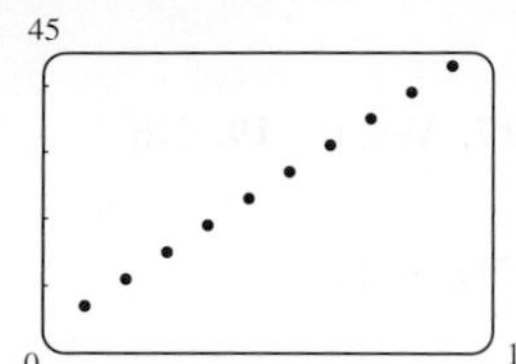

23. **(a)** $12, 6, 4, 3, \frac{12}{5}, 2, \frac{12}{7}, \frac{3}{2}, \frac{4}{3}, \frac{6}{5}$
(b)

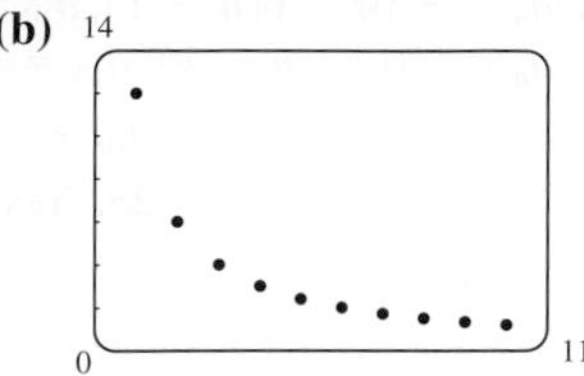

25. **(a)** $2, \frac{1}{2}, 2, \frac{1}{2}, 2, \frac{1}{2}, 2, \frac{1}{2}, 2, \frac{1}{2}$
(b)

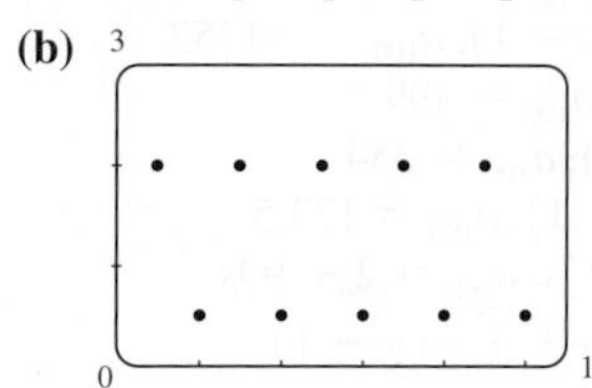

27. $a_n = 2n$ **29.** $a_n = (-3)^n$ **31.** $a_n = 5n - 1$
33. $a_n = (-1)^{n+1}5^n$ **35.** $a_n = (2n-1)/n^2$
37. $a_n = 1 + (-1)^n$ **39.** 2, 6, 12, 20, 30, 42
41. $\frac{1}{3}, \frac{4}{9}, \frac{13}{27}, \frac{40}{81}, \frac{121}{243}, \frac{364}{729}$ **43.** $\frac{2}{3}, \frac{8}{9}, \frac{26}{27}, \frac{80}{81}; S_n = 1 - \dfrac{1}{3^n}$
45. $1 - \sqrt{2}, 1 - \sqrt{3}, -1, 1 - \sqrt{5}; S_n = 1 - \sqrt{n+1}$
47. 10 **49.** $\frac{25}{36}$ **51.** 8 **53.** 31 **55.** 385 **57.** 46,438
59. 22 **61.** $1^3 + 2^3 + 3^3 + 4^3$
63. $\sqrt{4} + \sqrt{5} + \sqrt{6} + \sqrt{7} + \sqrt{8} + \sqrt{9} + \sqrt{10}$
65. $x^3 + x^4 + \cdots + x^{100}$ **67.** $\displaystyle\sum_{k=1}^{12} 4k$ **69.** $\displaystyle\sum_{k=1}^{10} k^2$
71. $\displaystyle\sum_{k=1}^{999} \frac{1}{k(k+1)}$ **73.** $\displaystyle\sum_{k=0}^{100} x^k$ **75.** $2^{(2^n-1)/2^n}$
77. **(a)** 2004.00, 2008.01, 2012.02, 2016.05, 2020.08, 2024.12 **(b)** \$2149.16 **79.** **(a)** 35,700; 36,414; 37,142; 37,885; 38,643 **(b)** 42,665 **81.** **(b)** 6898 **83.** **(a)** $A_n = A_{n-1} + 2000$, $A_1 = 45{,}000$ **(b)** \$53,000

Section 11.2 ■ Page 842

1. difference **2.** common difference; 2, 5 **3.** True **4.** True
5. **(a)** 7, 10, 13, 16, 19
(b) 3
(c)

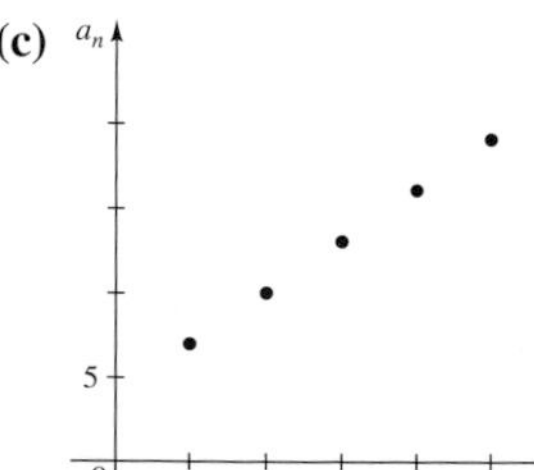

7. **(a)** −3, −8, −13, −18, −23
(b) −5
(c)

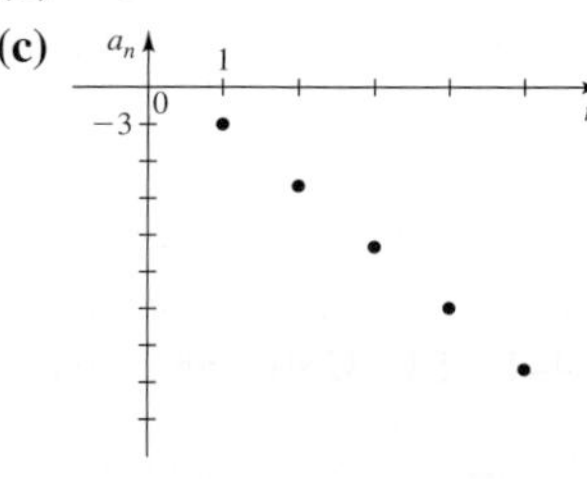

9. (a) 1.5, 2, 2.5, 3, 3.5
(b) 0.5
(c)

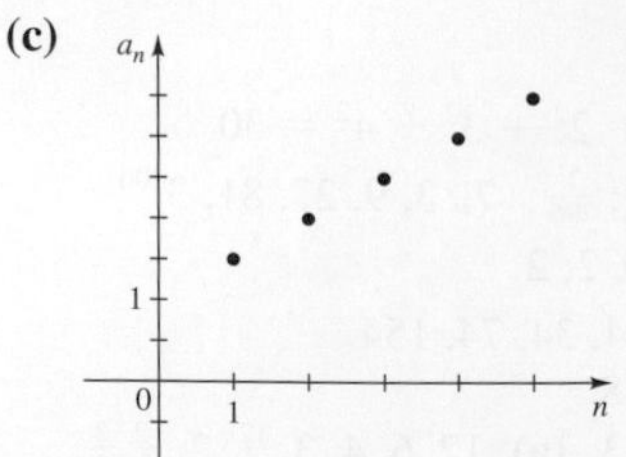

11. $a_n = -10 + 6(n-1), a_{10} = 44$
13. $a_n = 0.6 - (n-1), a_{10} = -8.4$
15. $a_n = \frac{5}{2} - \frac{1}{2}(n-1), a_{10} = -2$ **17.** Yes, 6 **19.** No
21. No **23.** Yes, $-\frac{3}{2}$ **25.** Yes, 1.7
27. 11, 18, 25, 32, 39; 7; $a_n = 11 + 7(n-1)$
29. $\frac{1}{3}, \frac{1}{5}, \frac{1}{7}, \frac{1}{9}, \frac{1}{11}$; not arithmetic
31. $-4, 2, 8, 14, 20; 6; a_n = -4 + 6(n-1)$
33. $2, a_5 = 14, a_n = 6 + 2(n-1), a_{100} = 204$
35. $-18, a_5 = -43, a_n = 29 - 18(n-1), a_{100} = -1753$
37. $5, a_5 = 24, a_n = 4 + 5(n-1), a_{100} = 499$
39. $4, a_5 = 4, a_n = -12 + 4(n-1), a_{100} = 384$
41. $1.5, a_5 = 31, a_n = 25 + 1.5(n-1), a_{100} = 173.5$
43. $s, a_5 = 2 + 4s, a_n = 2 + (n-1)s, a_{100} = 2 + 99s$
45. 706, 712 **47.** $a_1 = -\frac{5}{12}, a_n = -\frac{5}{12} + \frac{1}{12}(n-1)$
49. 33rd **51.** 1010 **53.** 870 **55.** −255 **57.** 20,301
59. 1735 **61.** 832.3 **63.** 46.75 **65.** 50 **69.** Yes
71. \$1250 **73.** \$540,000 **75.** 20 **77.** 78

Section 11.3 ■ Page 850

1. ratio **2.** common ratio; 2, 5 **3.** True **4. (a)** $a\left(\frac{1-r^n}{1-r}\right)$
(b) geometric; converges, $a/(1-r)$; diverges
5. (a) 7, 21, 63, 189, 567
(b) 3
(c)

7. (a) $8, -2, \frac{1}{2}, -\frac{1}{8}, \frac{1}{32}$
(b) $-\frac{1}{4}$
(c)

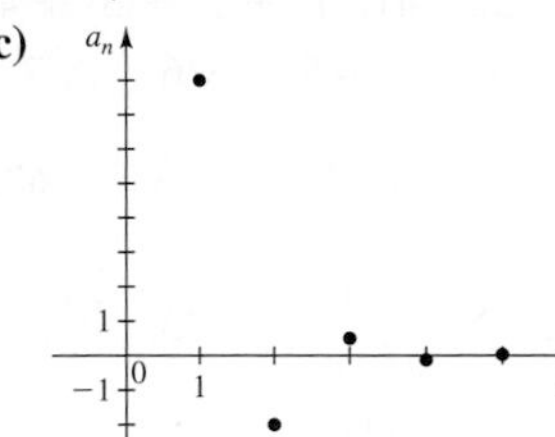

9. $a_n = 7(4)^{n-1}, a_4 = 448$ **11.** $a_n = 5(-3)^{n-1}, a_4 = -135$
13. Yes, 2 **15.** No **17.** Yes, $\frac{1}{2}$ **19.** No **21.** Yes, 1.1
23. 6, 18, 54, 162, 486; geometric, common ratio 3; $a_n = 6 \cdot 3^{n-1}$
25. $\frac{1}{4}, \frac{1}{16}, \frac{1}{64}, \frac{1}{256}, \frac{1}{1024}$; geometric, common ratio $\frac{1}{4}$; $a_n = \frac{1}{4}\left(\frac{1}{4}\right)^{n-1}$
27. 0, ln 5, 2 ln 5, 3 ln 5, 4 ln 5; not geometric
29. $3, a_5 = 162, a_n = 2 \cdot 3^{n-1}$
31. $-0.3, a_5 = 0.00243, a_n = (0.3)(-0.3)^{n-1}$
33. $-\frac{1}{12}, a_5 = \frac{1}{144}, a_n = 144\left(-\frac{1}{12}\right)^{n-1}$
35. $3^{2/3}, a_5 = 3^{11/3}, a_n = 3^{(2n+1)/3}$
37. $s^{2/7}, a_5 = s^{8/7}, a_n = s^{2(n-1)/7}$ **39.** $a_4 = \frac{16}{49}$
41. $a_1 = -\frac{1}{27}, a_2 = \frac{1}{9}$ **43.** $a_1 = -\frac{9}{32}, a_n = -\frac{9}{32}(-8)^{n-1}$
45. $a_1 = 1728, a_2 = 1296, a_3 = 972$ **47.** Ninth **49.** 315
51. 441 **53.** 3280 **55.** −645 **57.** 13,888,888.75
59. $\frac{93}{16}$ **61.** 105 **63.** $\frac{211}{27}$ **65.** $\frac{3}{2}$ **67.** $\frac{3}{4}$
69. Divergent **71.** 2 **73.** Divergent **75.** $\sqrt{2} + 1$
77. 1 **79.** $\frac{1}{33}$ **81.** $\frac{112}{999}$ **83.** 10, 20, 40 **85. (a)** Neither
(b) Arithmetic, 3 **(c)** Geometric, $9\sqrt{3}$ **(d)** Arithmetic, 3
87. (a) $V_n = 160{,}000(0.80)^{n-1}$ **(b)** 4th year **89.** 19 m, $80\left(\frac{3}{4}\right)^n$
91. $\frac{64}{25}, \frac{1024}{625}, 5\left(\frac{4}{5}\right)^n$ **93. (a)** $17\frac{8}{9}$ m **(b)** $18 - \left(\frac{1}{3}\right)^{n-3}$
95. 2801 **97.** 3 m **99. (a)** 2 **(b)** $8 + 4\sqrt{2}$ **101.** 1

Section 11.4 ■ Page 857

1. natural; $P(1)$ **2.** (ii)

3. Let $P(n)$ denote the statement
$2 + 4 + 6 + \cdots + 2n = n(n+1)$.

Step 1 $P(1)$ is true, since $2 = 1(1+1)$.
Step 2 Suppose $P(k)$ is true. Then

$$\begin{aligned} 2 + 4 + 6 + \cdots + 2k + 2(k+1) &\\ = k(k+1) + 2(k+1) & \quad \text{Induction hypothesis} \\ = (k+1)(k+2) & \end{aligned}$$

So $P(k+1)$ follows from $P(k)$. Thus, by the Principle of Mathematical Induction, $P(n)$ holds for all n.

5. Let $P(n)$ denote the statement
$$5 + 8 + 11 + \cdots + (3n+2) = \frac{n(3n+7)}{2}.$$

Step 1 $P(1)$ is true, since $5 = \frac{1(3 \cdot 1 + 7)}{2}$

Step 2 Suppose $P(k)$ is true. Then

$$\begin{aligned} 5 + 8 + 11 + \cdots + (3k+2) + [3(k+1) + 2] &\\ = \frac{k(3k+7)}{2} + (3k+5) & \quad \text{Induction hypothesis} \\ = \frac{3k^2 + 13k + 10}{2} & \\ = \frac{(k+1)[3(k+1) + 7]}{2} & \end{aligned}$$

So $P(k+1)$ follows from $P(k)$. Thus, by the Principle of Mathematical Induction, $P(n)$ holds for all n.

7. Let $P(n)$ denote the statement
$$1 \cdot 2 + 2 \cdot 3 + \cdots + n(n+1) = \frac{n(n+1)(n+2)}{3}.$$

Step 1 $P(1)$ is true, since $1 \cdot 2 = \frac{1 \cdot (1+1) \cdot (1+2)}{3}$.

Step 2 Suppose $P(k)$ is true. Then

$$\begin{aligned} 1 \cdot 2 + 2 \cdot 3 + \cdots + k(k+1) + (k+1)(k+2) &\\ = \frac{k(k+1)(k+2)}{3} + (k+1)(k+2) & \quad \text{Induction hypothesis} \\ = \frac{(k+1)(k+2)(k+3)}{3} & \end{aligned}$$

So $P(k+1)$ follows from $P(k)$. Thus, by the Principle of Mathematical Induction, $P(n)$ holds for all n.

9. Let $P(n)$ denote the statement
$1^3 + 2^3 + 3^3 + \cdots + n^3 = \dfrac{n^2(n+1)^2}{4}$.

Step 1 $P(1)$ is true, since $1^3 = \dfrac{1^2 \cdot (1+1)^2}{4}$.

Step 2 Suppose $P(k)$ is true. Then

$$\begin{aligned} &1^3 + 2^3 + 3^3 + \cdots + k^3 + (k+1)^3 \\ &= \frac{k^2(k+1)^2}{4} + (k+1)^3 \quad \text{Induction hypothesis} \\ &= \frac{(k+1)^2[k^2 + 4(k+1)]}{4} \\ &= \frac{(k+1)^2(k+2)^2}{4} \end{aligned}$$

So $P(k+1)$ follows from $P(k)$. Thus, by the Principle of Mathematical Induction, $P(n)$ holds for all n.

11. Let $P(n)$ denote the statement
$2^3 + 4^3 + 6^3 + \cdots + (2n)^3 = 2n^2(n+1)^2$.

Step 1 $P(1)$ is true, since $2^3 = 2 \cdot 1^2(1+1)^2$.
Step 2 Suppose $P(k)$ is true. Then

$$\begin{aligned} &2^3 + 4^3 + 6^3 + \cdots + (2k)^3 + [2(k+1)]^3 \\ &= 2k^2(k+1)^2 + [2(k+1)]^3 \quad \text{Induction hypothesis} \\ &= (k+1)^2(2k^2 + 8k + 8) \\ &= 2(k+1)^2(k+2)^2 \end{aligned}$$

So $P(k+1)$ follows from $P(k)$. Thus, by the Principle of Mathematical Induction, $P(n)$ holds for all n.

13. Let $P(n)$ denote the statement
$1 \cdot 2 + 2 \cdot 2^2 + \cdots + n \cdot 2^n = 2[1 + (n-1)2^n]$.

Step 1 $P(1)$ is true, since $1 \cdot 2 = 2[1 + 0]$.
Step 2 Suppose $P(k)$ is true. Then

$$\begin{aligned} &1 \cdot 2 + 2 \cdot 2^2 + 3 \cdot 2^3 + \cdots + k \cdot 2^k + (k+1) \cdot 2^{k+1} \\ &= 2[1 + (k-1)2^k] + (k+1) \cdot 2^{k+1} \quad \text{Induction hypothesis} \\ &= 2 + (k-1)2^{k+1} + (k+1) \cdot 2^{k+1} \\ &= 2 + 2k2^{k+1} = 2(1 + k2^{k+1}) \end{aligned}$$

So $P(k+1)$ follows from $P(k)$. Thus, by the Principle of Mathematical Induction, $P(n)$ holds for all n.

15. Let $P(n)$ denote the statement $n^2 + n$ is divisible by 2.

Step 1 $P(1)$ is true, since $1^2 + 1$ is divisible by 2.
Step 2 Suppose $P(k)$ is true. Now

$$\begin{aligned} (k+1)^2 + (k+1) &= k^2 + 2k + 1 + k + 1 \\ &= (k^2 + k) + 2(k+1) \end{aligned}$$

But $k^2 + k$ is divisible by 2 (by the induction hypothesis), and $2(k+1)$ is clearly divisible by 2, so $(k+1)^2 + (k+1)$ is divisible by 2. So $P(k+1)$ follows from $P(k)$. Thus, by the Principle of Mathematical Induction, $P(n)$ holds for all n.

17. Let $P(n)$ denote the statement $n^2 - n + 41$ is odd.

Step 1 $P(1)$ is true, since $1^2 - 1 + 41$ is odd.
Step 2 Suppose $P(k)$ is true. Now

$$(k+1)^2 - (k+1) + 41 = (k^2 - k + 41) + 2k$$

But $k^2 - k + 41$ is odd (by the induction hypothesis), and $2k$ is clearly even, so their sum is odd. So $P(k+1)$ follows from $P(k)$. Thus, by the Principle of Mathematical Induction, $P(n)$ holds for all n.

19. Let $P(n)$ denote the statement $8^n - 3^n$ is divisible by 5.

Step 1 $P(1)$ is true, since $8^1 - 3^1$ is divisible by 5.
Step 2 Suppose $P(k)$ is true. Now

$$\begin{aligned} 8^{k+1} - 3^{k+1} &= 8 \cdot 8^k - 3 \cdot 3^k \\ &= 8 \cdot 8^k - (8 - 5) \cdot 3^k = 8 \cdot (8^k - 3^k) + 5 \cdot 3^k \end{aligned}$$

which is divisible by 5 because $8^k - 3^k$ is divisible by 5 (by the induction hypothesis) and $5 \cdot 3^k$ is clearly divisible by 5. So $P(k+1)$ follows from $P(k)$. Thus, by the Principle of Mathematical Induction, $P(n)$ holds for all n.

21. Let $P(n)$ denote the statement $n < 2^n$.

Step 1 $P(1)$ is true, since $1 < 2^1$.
Step 2 Suppose $P(k)$ is true. Then

$$\begin{aligned} k + 1 &< 2^k + 1 \quad &&\text{Induction hypothesis} \\ &< 2^k + 2^k &&\text{Because } 1 < 2^k \\ &= 2 \cdot 2^k = 2^{k+1} \end{aligned}$$

So $P(k+1)$ follows from $P(k)$. Thus, by the Principle of Mathematical Induction, $P(n)$ holds for all n.

23. Let $P(n)$ denote the statement $(1+x)^n \ge 1 + nx$ for $x > -1$.

Step 1 $P(1)$ is true, since $(1+x)^1 \ge 1 + 1 \cdot x$.
Step 2 Suppose $P(k)$ is true. Then

$$\begin{aligned} (1+x)^{k+1} &= (1+x)(1+x)^k \\ &\ge (1+x)(1+kx) \quad \text{Induction hypothesis} \\ &= 1 + (k+1)x + kx^2 \\ &\ge 1 + (k+1)x \end{aligned}$$

So $P(k+1)$ follows from $P(k)$. Thus, by the Principle of Mathematical Induction, $P(n)$ holds for all n.

25. Let $P(n)$ denote the statement $a_n = 5 \cdot 3^{n-1}$.

Step 1 $P(1)$ is true, since $a_1 = 5 \cdot 3^0 = 5$.
Step 2 Suppose $P(k)$ is true. Then

$$\begin{aligned} a_{k+1} &= 3 \cdot a_k \quad &&\text{Definition of } a_{k+1} \\ &= 3 \cdot 5 \cdot 3^{k-1} &&\text{Induction hypothesis} \\ &= 5 \cdot 3^k \end{aligned}$$

So $P(k+1)$ follows from $P(k)$. Thus, by the Principle of Mathematical Induction, $P(n)$ holds for all n.

27. Let $P(n)$ denote the statement $x - y$ is a factor of $x^n - y^n$.

Step 1 $P(1)$ is true, since $x - y$ is a factor of $x^1 - y^1$.
Step 2 Suppose $P(k)$ is true. Now

$$\begin{aligned} x^{k+1} - y^{k+1} &= x^{k+1} - x^k y + x^k y - y^{k+1} \\ &= x^k(x - y) + (x^k - y^k)y \end{aligned}$$

But $x^k(x - y)$ is clearly divisible by $x - y$, and $(x^k - y^k)y$ is divisible by $x - y$ (by the induction hypothesis), so their sum is divisible by $x - y$. So $P(k + 1)$ follows from $P(k)$. Thus, by the Principle of Mathematical Induction, $P(n)$ holds for all n.

29. Let $P(n)$ denote the statement F_{3n} is even.

Step 1 $P(1)$ is true, since $F_{3\cdot 1} = 2$, which is even.
Step 2 Suppose $P(k)$ is true. Now, by the definition of the Fibonacci sequence,

$$\begin{aligned} F_{3(k+1)} = F_{3k+3} &= F_{3k+2} + F_{3k+1} \\ &= F_{3k+1} + F_{3k} + F_{3k+1} \\ &= F_{3k} + 2 \cdot F_{3k+1} \end{aligned}$$

But F_{3k} is even (by the induction hypothesis), and $2 \cdot F_{3k+1}$ is clearly even, so $F_{3(k+1)}$ is even. So $P(k + 1)$ follows from $P(k)$. Thus, by the Principle of Mathematical Induction, $P(n)$ holds for all n.

31. Let $P(n)$ denote the statement $F_1^2 + F_2^2 + F_3^2 + \cdots + F_n^2 = F_n \cdot F_{n+1}$.

Step 1 $P(1)$ is true, since $F_1^2 = F_1 \cdot F_2$ (because $F_1 = F_2 = 1$).
Step 2 Suppose $P(k)$ is true. Then

$$\begin{aligned} &F_1^2 + F_2^2 + F_3^2 + \cdots + F_k^2 + F_{k+1}^2 \\ &= F_k \cdot F_{k+1} + F_{k+1}^2 && \text{Induction hypothesis} \\ &= F_{k+1}(F_k + F_{k+1}) && \text{Definition of the Fibonacci sequence} \\ &= F_{k+1} \cdot F_{k+2} \end{aligned}$$

So $P(k + 1)$ follows from $P(k)$. Thus, by the Principle of Mathematical Induction, $P(n)$ holds for all n.

33. Let $P(n)$ denote the statement

$$\begin{bmatrix} 1 & 1 \\ 1 & 0 \end{bmatrix}^n = \begin{bmatrix} F_{n+1} & F_n \\ F_n & F_{n-1} \end{bmatrix}.$$

Step 1 $P(2)$ is true, since

$$\begin{bmatrix} 1 & 1 \\ 1 & 0 \end{bmatrix}^2 = \begin{bmatrix} 2 & 1 \\ 1 & 1 \end{bmatrix} = \begin{bmatrix} F_3 & F_2 \\ F_2 & F_1 \end{bmatrix}.$$

Step 2 Suppose $P(k)$ is true. Then

$$\begin{aligned} \begin{bmatrix} 1 & 1 \\ 1 & 0 \end{bmatrix}^{k+1} &= \begin{bmatrix} 1 & 1 \\ 1 & 0 \end{bmatrix}^k \begin{bmatrix} 1 & 1 \\ 1 & 0 \end{bmatrix} \\ &= \begin{bmatrix} F_{k+1} & F_k \\ F_k & F_{k-1} \end{bmatrix} \begin{bmatrix} 1 & 1 \\ 1 & 0 \end{bmatrix} && \text{Induction hypothesis} \\ &= \begin{bmatrix} F_{k+1} + F_k & F_{k+1} \\ F_k + F_{k-1} & F_k \end{bmatrix} \\ &= \begin{bmatrix} F_{k+2} & F_{k+1} \\ F_{k+1} & F_k \end{bmatrix} && \text{Definition of the Fibonacci sequence} \end{aligned}$$

So $P(k + 1)$ follows from $P(k)$. Thus, by the Principle of Mathematical Induction, $P(n)$ holds for all $n \geq 2$.

35. Let $P(n)$ denote the statement $F_n \geq n$.

Step 1 $P(5)$ is true, since $F_5 \geq 5$ (because $F_5 = 5$).
Step 2 Suppose $P(k)$ is true. Now

$$\begin{aligned} F_{k+1} &= F_k + F_{k-1} && \text{Definition of the Fibonacci sequence} \\ &\geq k + F_{k-1} && \text{Induction hypothesis} \\ &\geq k + 1 && \text{Because } F_{k-1} \geq 1 \end{aligned}$$

So $P(k + 1)$ follows from $P(k)$. Thus, by the Principle of Mathematical Induction, $P(n)$ holds for all $n \geq 5$.

Section 11.5 ■ Page 866

1. binomial **2.** Pascal's; 1, 4, 6, 4, 1

3. $\dfrac{n!}{k!(n-k)!}$; $\dfrac{4!}{3!(4-3)!} = 4$

4. Binomial; $\binom{4}{0}, \binom{4}{1}, \binom{4}{2}, \binom{4}{3}, \binom{4}{4}$

5. $x^6 + 6x^5y + 15x^4y^2 + 20x^3y^3 + 15x^2y^4 + 6xy^5 + y^6$

7. $x^4 + 4x^2 + 6 + \dfrac{4}{x^2} + \dfrac{1}{x^4}$

9. $x^5 - 5x^4 + 10x^3 - 10x^2 + 5x - 1$
11. $x^{10}y^5 - 5x^8y^4 + 10x^6y^3 - 10x^4y^2 + 5x^2y - 1$
13. $8x^3 - 36x^2y + 54xy^2 - 27y^3$

15. $\dfrac{1}{x^5} - \dfrac{5}{x^{7/2}} + \dfrac{10}{x^2} - \dfrac{10}{x^{1/2}} + 5x - x^{5/2}$

17. 15 **19.** 4950 **21.** 18 **23.** 32
25. $x^4 + 8x^3y + 24x^2y^2 + 32xy^3 + 16y^4$

27. $1 + \dfrac{6}{x} + \dfrac{15}{x^2} + \dfrac{20}{x^3} + \dfrac{15}{x^4} + \dfrac{6}{x^5} + \dfrac{1}{x^6}$

29. $x^{20}, 40x^{19}y, 760x^{18}y^2$ **31.** $25a^{26/3}, a^{25/3}$
33. $48{,}620x^{18}$ **35.** $300a^2b^{23}$ **37.** $100y^{99}$ **39.** $13{,}440x^4y^6$
41. $495a^8b^8$ **43.** $(x + y)^4$ **45.** $(2a + b)^3$
47. $3x^2 + 3xh + h^2$

Chapter 11 Review ■ Page 869

1. $\frac{1}{2}, \frac{4}{3}, \frac{9}{4}, \frac{16}{5}; \frac{100}{11}$ **3.** $0, \frac{1}{4}, 0, \frac{1}{32}; \frac{1}{500}$
5. 1, 3, 15, 105; 654,729,075
7. 1, 4, 9, 16, 25, 36, 49
9. 1, 3, 5, 11, 21, 43, 85

11. (a) 7, 9, 11, 13, 15 **(b)**

13. (a) $\frac{3}{4}, \frac{9}{8}, \frac{27}{16}, \frac{81}{32}, \frac{243}{64}$ **(b)**

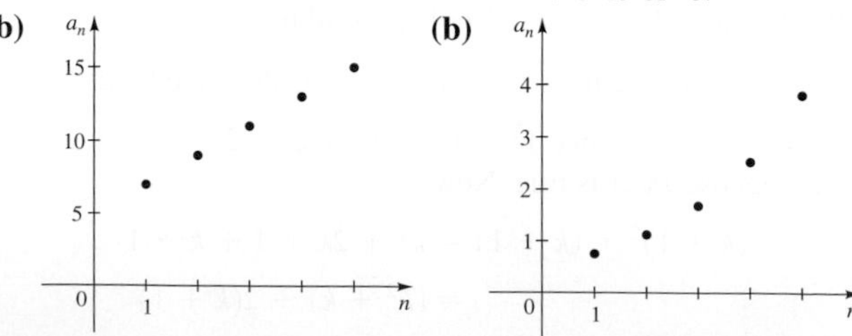

11. (c) 55 **(d)** Arithmetic, common difference 2

13. (c) $\frac{633}{64}$ **(d)** Geometric, common ratio $\frac{3}{2}$

15. Arithmetic, 7 **17.** Arithmetic, $t + 1$ **19.** Geometric, $\frac{1}{t}$

21. Geometric, $\frac{4}{27}$ **23.** $2i$ **25.** $a_2 = 5$ **27.** $a_5 = \frac{81}{4}$

29. $A_n = 52{,}000(1.04)^{n-1}$; Salary: \$52,000; \$54,080; \$56,243.20; \$58,492.93; \$60,832.65; \$63,265.95

31. 12,288 **35. (a)** 9 **(b)** $\pm 6\sqrt{2}$ **37.** 126

39. 384 **41.** $0^2 + 1^2 + 2^2 + \cdots + 9^2$

43. $\frac{3}{2^2} + \frac{3^2}{2^3} + \frac{3^3}{2^4} + \cdots + \frac{3^{50}}{2^{51}}$ **45.** $\sum_{k=1}^{33} 3k$ **47.** $\sum_{k=1}^{100} k2^{k+2}$

49. Geometric; 4.68559 **51.** Arithmetic, $5050\sqrt{5}$

53. Geometric, 9831 **55.** $\frac{5}{7}$ **57.** Divergent

59. Divergent **61.** 13 **63.** 65,534

65. Let $P(n)$ denote the statement

$$\frac{1}{1\cdot 3} + \frac{1}{3\cdot 5} + \cdots + \frac{1}{(2n-1)(2n+1)} = \frac{n}{2n+1}.$$

Step 1 $P(1)$ is true, since $\frac{1}{1\cdot 3} = \frac{1}{2\cdot 1 + 1}$.

Step 2 Suppose $P(k)$ is true. Then

$$\frac{1}{1\cdot 3} + \frac{1}{3\cdot 5} + \cdots + \frac{1}{(2k-1)(2k+1)} + \frac{1}{(2k+1)(2k+3)}$$

$$= \frac{k}{2k+1} + \frac{1}{(2k+1)(2k+3)} \quad \text{Induction hypothesis}$$

$$= \frac{2k^2 + 3k + 1}{(2k+1)(2k+3)} = \frac{k+1}{2k+3}$$

So $P(k+1)$ follows from $P(k)$. Thus, by the Principle of Mathematical Induction, $P(n)$ holds for all n.

67. Let $P(n)$ denote the statement that $7^n - 1$ is divisible by 6.

Step 1 $P(1)$ is true, since $7^1 - 1 = 6$.

Step 2 Suppose $P(k)$ is true. We have

$$7^{k+1} - 1 = 7(7^k - 1) + 6$$

Now $7^k - 1$ is divisible by 6 (induction hypothesis), and so is 6, and hence $7(7^k - 1) + 6$ is also divisible by 6. So $P(k+1)$ follows from $P(k)$. Thus, by the Principle of Mathematical Induction, $P(n)$ holds for all n.

69. Let $P(n)$ denote the statement $a_n = 2 \cdot 3^n - 2$.

Step 1 $P(1)$ is true, since $a_1 = 2 \cdot 3^1 - 2 = 4$.

Step 2 Suppose $P(k)$ is true. Then

$$a_{k+1} = 3a_k + 4$$

$$= 3(2\cdot 3^k - 2) + 4 \quad \text{Induction hypothesis}$$

$$= 2\cdot 3^{k+1} - 2$$

So $P(k+1)$ follows from $P(k)$. Thus, by the Principle of Mathematical Induction, $P(n)$ holds for all n.

71. 255 **73.** 12,870

75. $x^5 + 10x^4 + 40x^3 + 80x^2 + 80x + 32$

77. $16x^4 + 32x^3y + 24x^2y^2 + 8xy^3 + y^4$

79. $b^{-40/3}, 20b^{-37/3}, 190b^{-34/3}$

Chapter 11 Test ■ Page 871

1. 1, 6, 15, 28, 45, 66; 161 **2.** 2, 5, 13, 36, 104, 307

3. (a) 3 **(b)** $a_n = 2 + (n-1)3$ **(c)** 104

4. (a) $\frac{1}{4}$ **(b)** $a_n = 12(\frac{1}{4})^{n-1}$ **(c)** $3/4^8$

5. (a) $r = \frac{1}{5}, a_5 = \frac{1}{25}$ **(b)** $\frac{5^8 - 1}{12{,}500}$

6. (a) $d = -\frac{8}{9}, a_{100} = -78$ **(b)** 60

8. (a) $(1 - 1^2) + (1 - 2^2) + (1 - 3^2) + (1 - 4^2) + (1 - 5^2) = -50$

(b) $(-1)^3 2^1 + (-1)^4 2^2 + (-1)^5 2^3 + (-1)^6 2^4 = 10$

9. (a) $\frac{58{,}025}{59{,}049}$ **(b)** $2 + \sqrt{2}$

10. Let $P(n)$ denote the statement

$$1^2 + 2^2 + 3^2 + \cdots + n^2 = \frac{n(n+1)(2n+1)}{6}.$$

Step 1 $P(1)$ is true, since $1^2 = \frac{1(1+1)(2\cdot 1+1)}{6}$.

Step 2 Suppose $P(k)$ is true. Then

$$1^2 + 2^2 + 3^2 + \cdots + k^2 + (k+1)^2$$

$$= \frac{k(k+1)(2k+1)}{6} + (k+1)^2 \quad \text{Induction hypothesis}$$

$$= \frac{k(k+1)(2k+1) + 6(k+1)^2}{6}$$

$$= \frac{(k+1)[k(2k+1) + 6(k+1)]}{6}$$

$$= \frac{(k+1)(2k^2 + 7k + 6)}{6}$$

$$= \frac{(k+1)[(k+1)+1][2(k+1)+1]}{6}$$

So $P(k+1)$ follows from $P(k)$. Thus, by the Principle of Mathematical Induction, $P(n)$ holds for all n.

11. $32x^5 + 80x^4y^2 + 80x^3y^4 + 40x^2y^6 + 10xy^8 + y^{10}$

12. $\binom{10}{3}(3x)^3(-2)^7 = -414{,}720x^3$

13. (a) $a_n = (0.85)(1.24)^n$ **(b)** 1.31 kg **(c)** Geometric

Focus on Modeling ■ Page 874

1. (a) $A_n = 1.0001A_{n-1}, A_0 = 275{,}000$ **(b)** $A_0 = 275{,}000$, $A_1 = 275{,}027.50$, $A_2 = 275{,}055.00$, $A_3 = 275{,}082.51$, $A_4 = 275{,}110.02$, $A_5 = 275{,}137.53$, $A_6 = 275{,}165.04$, $A_7 = 275{,}192.56$ **(c)** $A_n = 1.0001^n(275{,}000)$

3. (a) $A_n = 1.0025A_{n-1} + 100, A_0 = 100$ **(b)** $A_0 = 100$, $A_1 = 200.25$, $A_2 = 300.75$, $A_3 = 401.50$, $A_4 = 502.51$ **(c)** $A_n = 100[(1.0025^{n+1} - 1)/0.0025]$ **(d)** \$6580.83

5. (b) In the 32nd year

Chapter 12

Section 12.1 ■ Page 884

1. L, a; 5, 1 **2.** limit, left, L; less; left, right, equal
3. 10 **5.** $\frac{1}{4}$ **7.** $\frac{1}{3}$ **9.** 1 **11.** -1 **13.** 0.51 **15.** $\frac{1}{2}$
17. (a) 2 **(b)** 3 **(c)** Does not exist **(d)** 4 **(e)** Not defined
19. (a) -1 **(b)** -2 **(c)** Does not exist **(d)** 2 **(e)** 0
(f) Does not exist **(g)** 1 **(h)** 3 **21.** -8
23. Does not exist **25.** Does not exist **27.** Does not exist
29. (a) 4 **(b)** 4 **(c)** 4

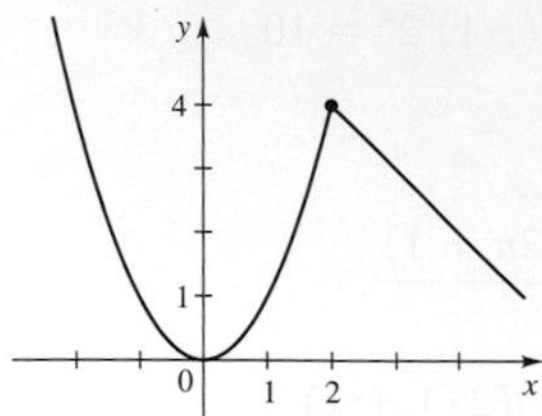

31. (a) 4 **(b)** 3 **(c)** Does not exist

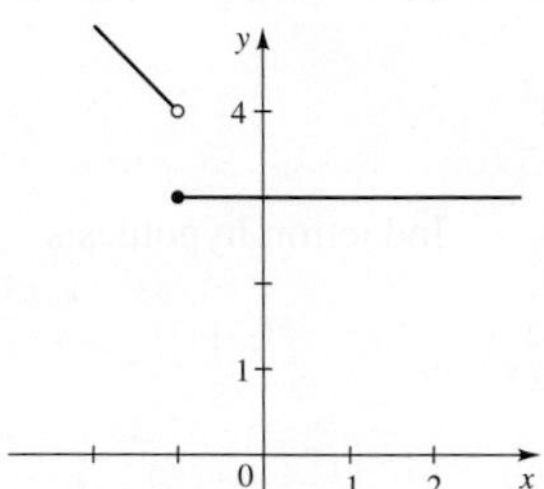

Section 12.2 ■ Page 893

1. $\lim_{x\to a} f(x) + \lim_{x\to a} g(x)$, $\lim_{x\to a} f(x) \cdot \lim_{x\to a} g(x)$; sum, product
2. $f(a)$ **3. (a)** 2 **(b)** Does not exist **(c)** 0
(d) Does not exist **(e)** 16 **(f)** 2
5. 5 **7.** 12 **9.** 75 **11.** $\frac{1}{2}$ **13.** -174 **15.** $\frac{4}{9}$ **17.** 7 **19.** 5
21. Does not exist **23.** $\frac{6}{5}$ **25.** 4 **27.** $\frac{1}{6}$ **29.** $-\frac{1}{16}$ **31.** $-\frac{1}{9}$
33. 4 **35.** $-\frac{3}{2}$

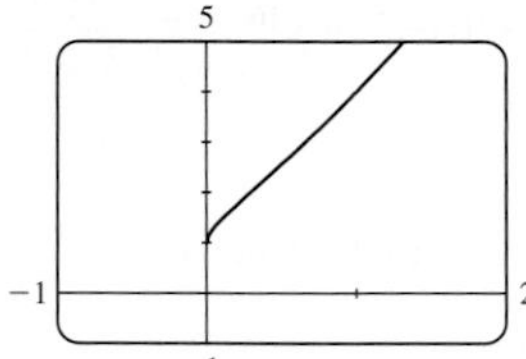

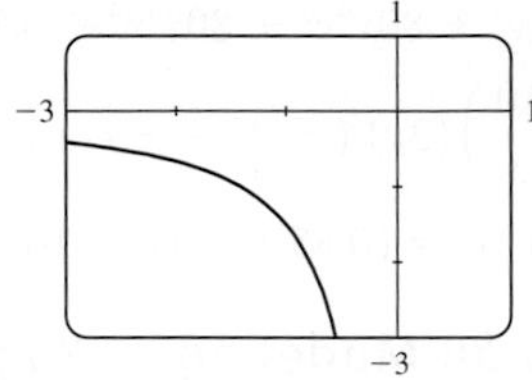

37. 0 **39.** Does not exist **41.** Does not exist
43. (a) 1, 2 **(b)** Does not exist
(c)

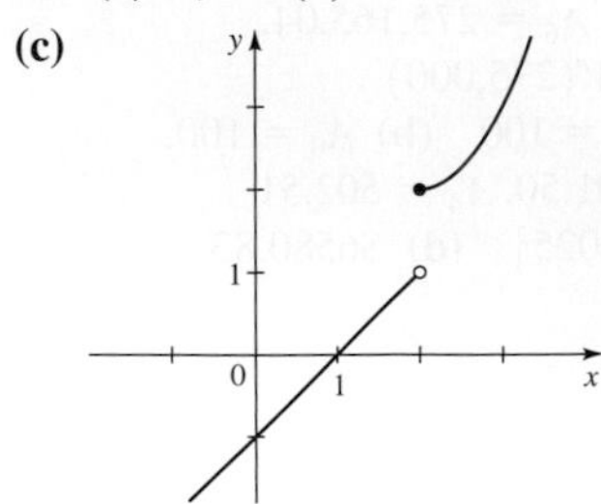

45. (a) 0.667

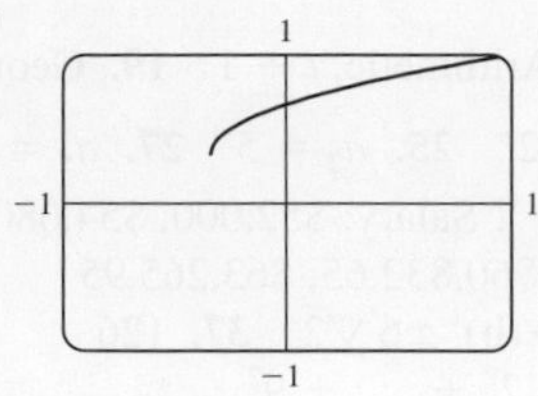

(b) 0.667

x	$f(x)$
0.1	0.71339
0.01	0.67163
0.001	0.66717
0.0001	0.66672

x	$f(x)$
-0.1	0.61222
-0.01	0.66163
-0.001	0.66617
-0.0001	0.66662

(c) $\frac{2}{3}$

Section 12.3 ■ Page 901

1. $\dfrac{f(a+h) - f(a)}{h}$; slope, $(a, f(a))$
2. $\dfrac{f(x) - f(a)}{x - a}$, instantaneous, a **3.** 3 **5.** -11 **7.** 24 **9.** $-\frac{1}{5}$
11. $y = -8x + 9$ **13.** $y = -x - 1$

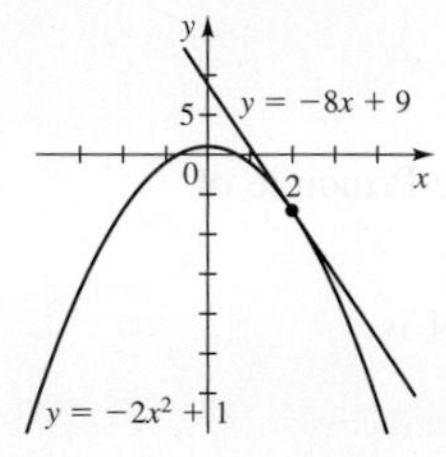

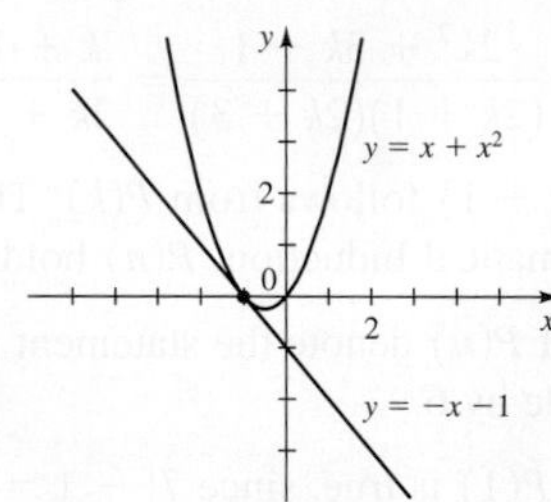

15. $y = -x + 4$ **17.** $y = \frac{1}{4}x + \frac{7}{4}$

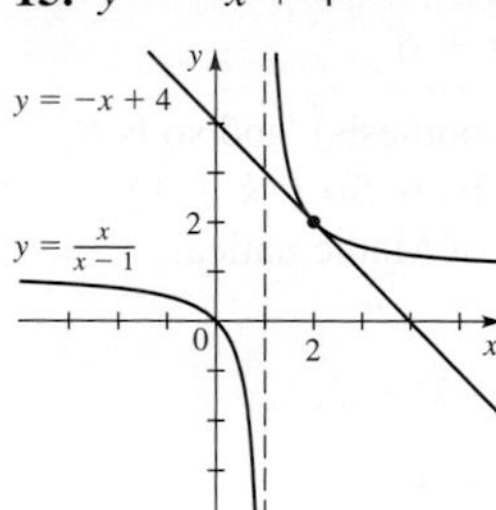

19. $f'(2) = -12$ **21.** $f'(-1) = 7$ **23.** $f'(2) = -\frac{1}{9}$
25. $F'(4) = -\frac{1}{16}$ **27. (a)** $2a + 2$ **(b)** 8, 10
29. (a) $\dfrac{1}{(a+1)^2}$ **(b)** $\frac{1}{16}, \frac{1}{25}$
31. (a) $f'(a) = 3a^2 - 2$
(b) $y = -2x + 4$, $y = x + 2$, $y = 10x - 12$
(c)

33. $f(x) = x^{10}$, $a = 1$ **35.** $f(t) = \sqrt{t+1}$, $a = 1$ **37.** -7.2 m/s

39. $12a^2 + 6$ m/s, 18 m/s, 54 m/s, 114 m/s
41. $-0.8°$F/min **43. (a)** -38.8 L/min, -27.8 L/min
(b) -33.3 L/min

Section 12.4 ■ Page 910

1. L, x; horizontal asymptote; 0, 0
2. L, large; converges, diverges
3. (a) (i) -1 (ii) 2 **(b)** $y = -1$, $y = 2$ **5.** 0
7. $\frac{2}{5}$ **9.** $\frac{4}{3}$ **11.** 2 **13.** Does not exist **15.** 7
17. Does not exist **19.** $-\frac{1}{4}$ **21.** 0 **23.** 0
25. Divergent **27.** 0 **29.** Divergent **31.** $\frac{3}{2}$ **33.** 8
35. $f(x) = \dfrac{x^2}{(x-1)(x-3)}$ [Other answers are possible.]
37. Within 0.01 **39. (b)** approaches 30 g/L

Section 12.5 ■ Page 918

1. rectangles;
$f(x_1)(x_1 - a) + f(x_2)(x_2 - x_1) + f(x_3)(x_3 - x_2) + f(b)(b - x_3)$
2. $\sum_{k=1}^{n} f(x_k)\,\Delta x$
3. (a) 40, 52

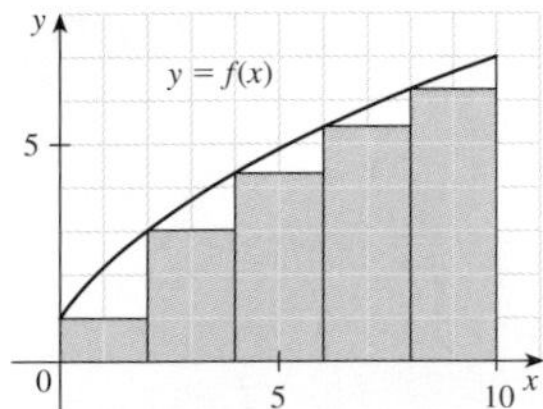

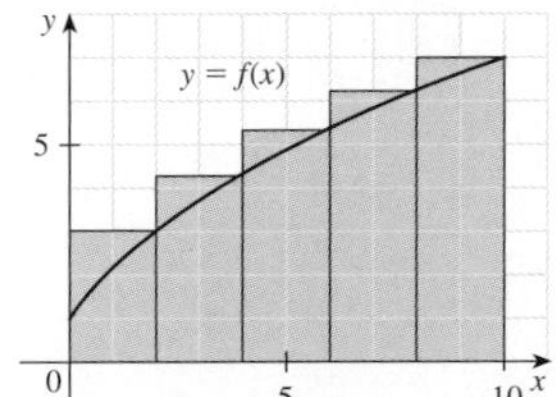

(b) 43, 49
5. 5.25 **7.** $\frac{223}{35}$
9. (a) $\frac{77}{60}$, underestimate **(b)** $\frac{25}{12}$, overestimate

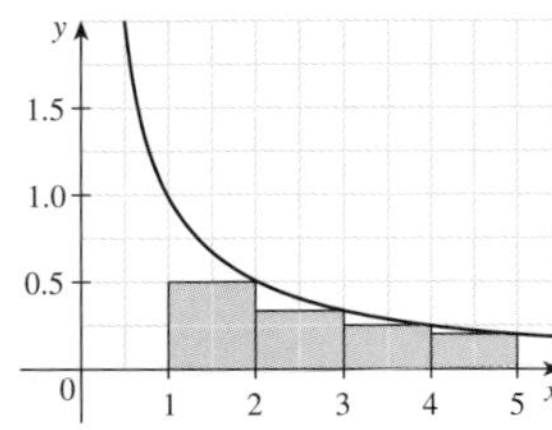

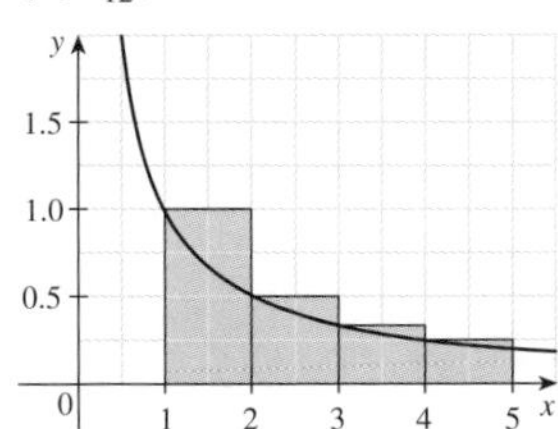

11. (a) 8, 6.875 **(b)** 5, 5.375

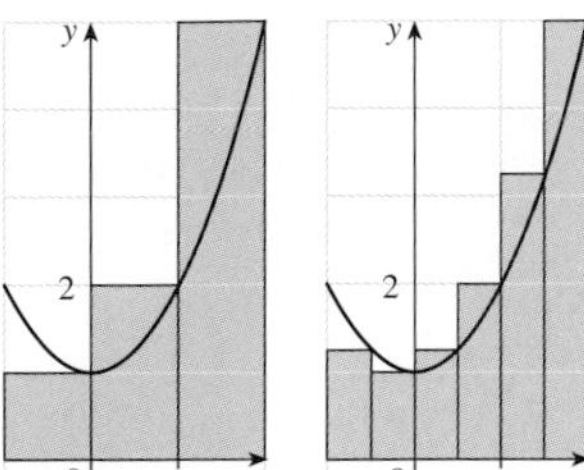

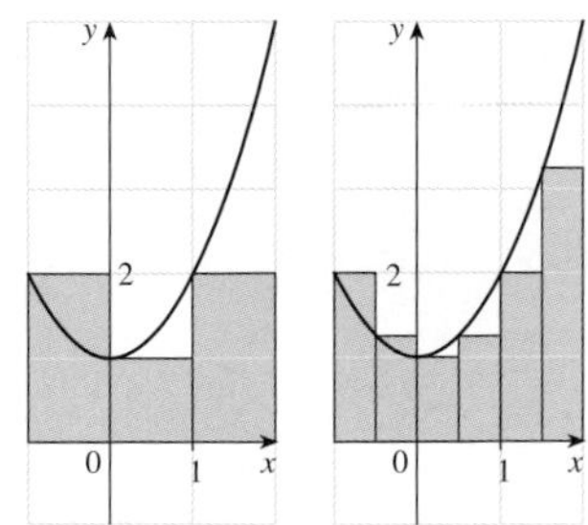

13. 37.5 **15.** 8 **17.** 166.25 **19.** 133.5

Chapter 12 Review ■ Page 922

1. 1 **3.** 0.69 **5.** Does not exist
7. (a) Does not exist **(b)** 2.4 **(c)** 2.4 **(d)** 2.4 **(e)** 0.5
(f) 1 **(g)** 2 **(h)** 0 **9.** -3 **11.** 7 **13.** 2 **15.** -1 **17.** 2
19. Does not exist **21.** $f'(4) = 3$ **23.** $f'(16) = \frac{1}{8}$
25. (a) $f'(a) = -2$ **(b)** $-2, -2$
27. (a) $f'(a) = 1/(2\sqrt{a+6})$ **(b)** $1/(4\sqrt{2})$, $1/4$
29. $y = 2x + 1$ **31.** $y = 2x$ **33.** $y = -\frac{1}{4}x + 1$
35. (a) -20 m/s **(b)** $-10a$ m/s **(c)** $\sqrt{40} \approx 6.32$ s
(d) -63.2 m/s **37.** $\frac{1}{5}$ **39.** $\frac{1}{2}$ **41.** Divergent **43.** 3.83
45. 10 **47.** $\frac{5}{6}$

Chapter 12 Test ■ Page 924

1. (a) $\frac{1}{2}$ **(b)**

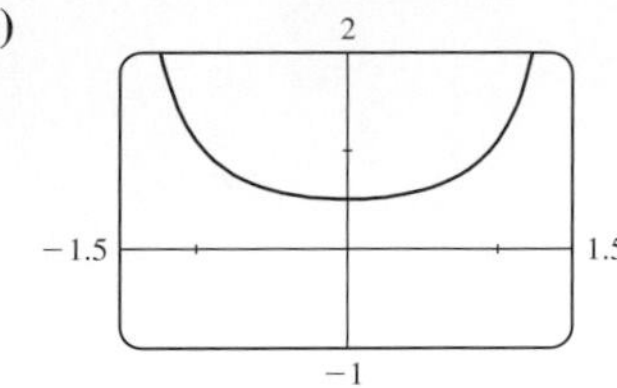

2. (a) 1 **(b)** 1 **(c)** 1 **(d)** 0 **(e)** 0 **(f)** 0 **(g)** 4 **(h)** 2
(i) Does not exist
3. (a) 6 **(b)** -2 **(c)** Does not exist
(d) Does not exist **(e)** $\frac{1}{4}$ **(f)** 2
4. (a) $f'(a) = 2a - 2$ **(b)** $-4, 0, 2$
5. $y = \frac{1}{6}x + \frac{3}{2}$ **6. (a)** 0 **(b)** Does not exist
7. (a) 3.56 **(b)** $\frac{11}{3}$

Focus on Modeling ■ Page 927

1. $57{,}333\frac{1}{3}$ N-m **3. (b)** Area under the graph of $p(x) = 375x$ between $x = 0$ and $x = 4$ **(c)** 6733 N **(d)** 3365 N
5. (a) 1625.28 heating degree-hours **(b)** 70°F
(c) 1488 heating degree-hours **(d)** 75°F
(e) The day in part (a)

Appendix A ■ Page 936

1. Congruent, ASA **2.** Congruent, SSS
3. Not necessarily congruent **4.** Congruent, SAS
5. Similar **6.** Similar **7.** Similar **8.** Not similar
9. $x = 125$ **10.** $y = 30$ **11.** $x = 6$, $y = \dfrac{21}{4}$
12. $x = 4$ **13.** $x = \dfrac{ac}{a+b}$ **14.** $x = \dfrac{ac}{b} - a$
17. $x = 10$ **18.** $x = 48$ **19.** $x = \sqrt{3}$
20. $x = 2\sqrt{10}$ **21.** $x = 40$ **22.** $x = 144$ **23.** Yes
24. Yes **25.** No **26.** No **27.** Yes **28.** Yes **29.** 61 cm
30. 119 m by 120 m **31.** No **32.** 12

33. (b)

m	n	(a, b, c)
2	1	(3, 4, 5)
3	1	(8, 6, 10)
3	2	(5, 12, 13)
4	1	(15, 8, 17)
4	2	(12, 16, 20)
4	3	(7, 24, 25)
5	1	(24, 10, 26)
5	2	(21, 20, 29)
5	3	(16, 30, 34)
5	4	(9, 40, 41)

34. $h = 6$ m

35. 140° **36.** 30° **37.** 30° **38.** 20° **39.** 20° **40.** 25°

Index

Sequences and Series

Arithmetic

$$a, a + d, a + 2d, a + 3d, \ldots \qquad \text{or} \qquad a_n = a + (n - 1)d$$

$$S_n = \sum_{k=1}^{n} a_k = \frac{n}{2}[2a + (n - 1)d] = n\left(\frac{a + a_n}{2}\right)$$

Geometric

$$a, ar, ar^2, ar^3, \ldots \qquad \text{or} \qquad a_n = ar^{n-1}$$

$$S_n = \sum_{k=1}^{n} a_k = a\frac{1 - r^n}{1 - r}$$

Infinite geometric series

$$\sum_{k=1}^{\infty} ar^{k-1} = a + ar + ar^2 + ar^3 + \cdots$$

If $|r| < 1$ the series converges and its sum is

$$S = \frac{a}{1 - r}$$

If $|r| \geq 1$ the series diverges.

The Binomial Theorem

Binomial Theorem

$$(a + b)^n = \binom{n}{0}a^n + \binom{n}{1}a^{n-1}b + \cdots + \binom{n}{n-1}ab^{n-1} + \binom{n}{n}b^n$$

Binomial coefficients

$$\binom{n}{r} = \frac{n!}{r!(n - r)!}\ (r \leq n), \text{ where } n! = 1 \cdot 2 \cdot 3 \cdots (n - 1)n$$

Finance

Compound interest

$$A = P\left(1 + \frac{r}{n}\right)^{nt}$$

where A is the amount after t years, P is the principal, r is the interest rate, and the interest is compounded n times per year.

Continuously compounded interest

$$A = Pe^{rt}$$

where A is the amount after t years, P is the principal, r is the interest rate, and the interest is compounded continuously.

Conic Sections

Circles

$$(x - h)^2 + (y - k)^2 = r^2$$

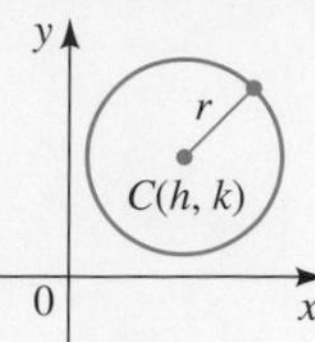

Parabolas

$$x^2 = 4py$$

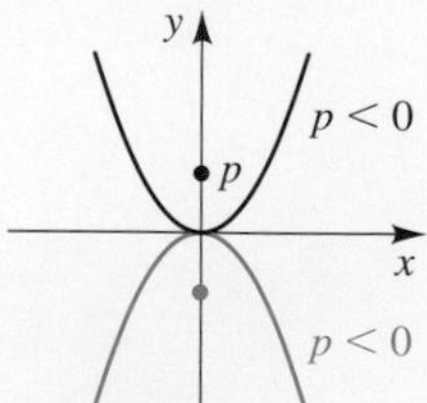

Focus $(0, p)$, directrix $y = -p$

$$y^2 = 4px$$

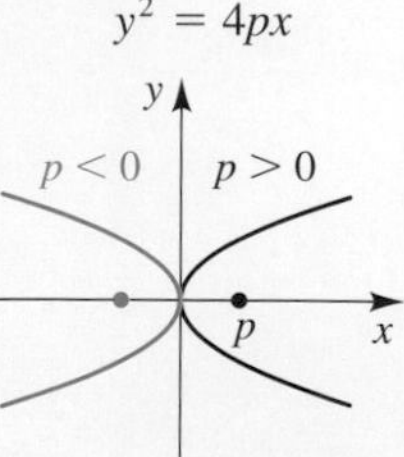

Focus $(p, 0)$, directrix $x = -p$

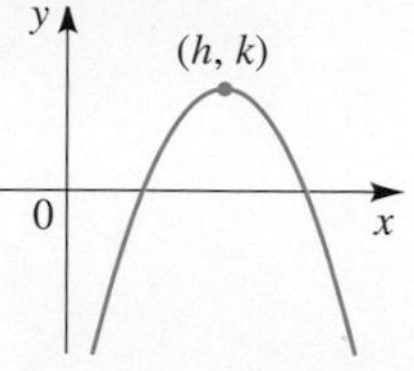

$y = a(x - h)^2 + k,$
$a < 0, \quad h > 0, \quad k > 0$

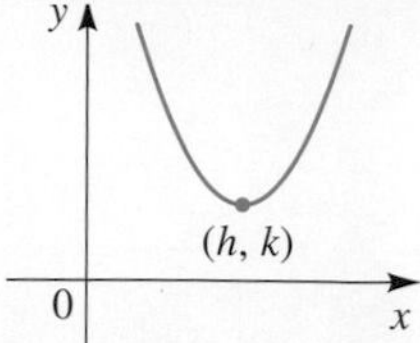

$y = a(x - h)^2 + k,$
$a > 0, \quad h > 0, \quad k > 0$

Ellipses

$$\frac{x^2}{a^2} + \frac{y^2}{b^2} = 1$$

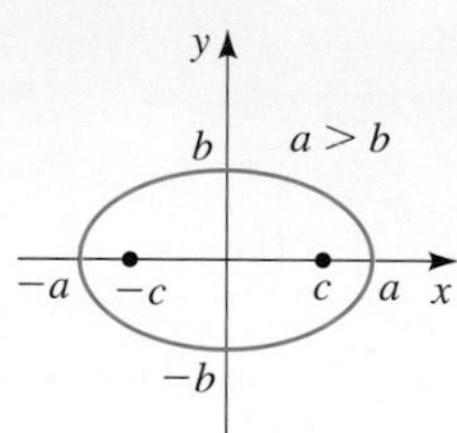

Foci $(\pm c, 0)$, $c^2 = a^2 - b^2$

$$\frac{x^2}{b^2} + \frac{y^2}{a^2} = 1$$

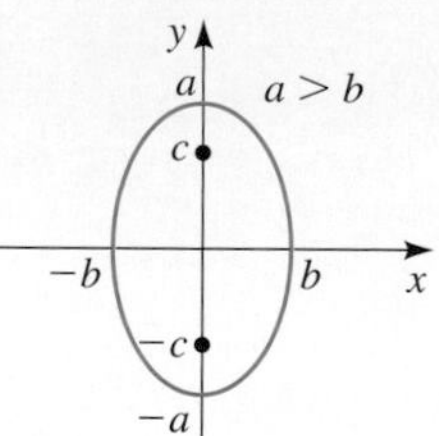

Foci $(0, \pm c)$, $c^2 = a^2 - b^2$

Hyperbolas

$$\frac{x^2}{a^2} - \frac{y^2}{b^2} = 1$$

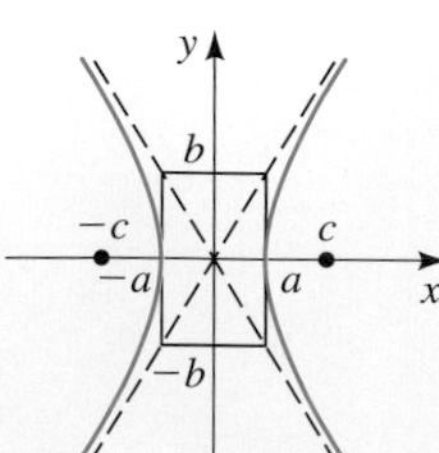

Foci $(\pm c, 0)$, $c^2 = a^2 + b^2$

Asymptotes: $y = \pm\frac{b}{a}x$

$$-\frac{x^2}{b^2} + \frac{y^2}{a^2} = 1$$

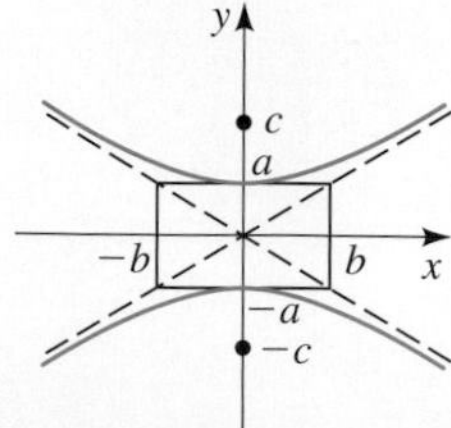

Foci $(0, \pm c)$, $c^2 = a^2 + b^2$

Asymptotes: $y = \pm\frac{a}{b}x$

Angle Measurement

π radians $= 180°$

$1° = \dfrac{\pi}{180}$ rad $\qquad 1 \text{ rad} = \dfrac{180°}{\pi}$

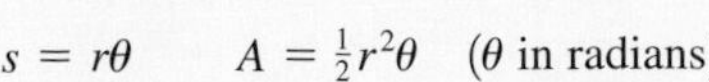

$s = r\theta \qquad A = \frac{1}{2}r^2\theta \quad (\theta \text{ in radians})$

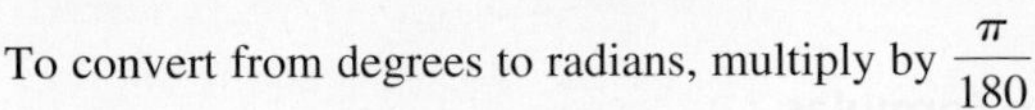

To convert from degrees to radians, multiply by $\dfrac{\pi}{180}$.

To convert from radians to degrees, multiply by $\dfrac{180}{\pi}$.

Trigonometric Functions of Real Numbers

$\sin t = y \qquad \csc t = \dfrac{1}{y}$

$\cos t = x \qquad \sec t = \dfrac{1}{x}$

$\tan t = \dfrac{y}{x} \qquad \cot t = \dfrac{x}{y}$

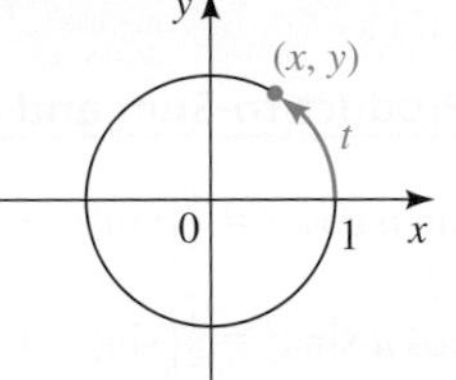

Trigonometric Functions of Angles

$\sin\theta = \dfrac{y}{r} \qquad \csc\theta = \dfrac{r}{y}$

$\cos\theta = \dfrac{x}{r} \qquad \sec\theta = \dfrac{r}{x}$

$\tan\theta = \dfrac{y}{x} \qquad \cot\theta = \dfrac{x}{y}$

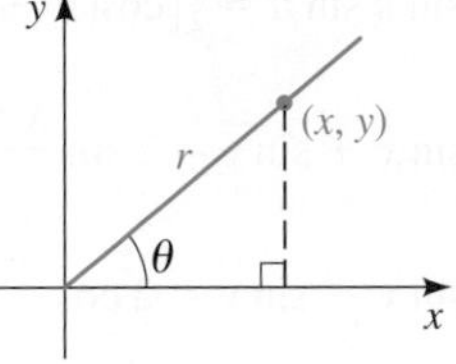

Right Angle Trigonometry

$\sin\theta = \dfrac{\text{opp}}{\text{hyp}} \qquad \csc\theta = \dfrac{\text{hyp}}{\text{opp}}$

$\cos\theta = \dfrac{\text{adj}}{\text{hyp}} \qquad \sec\theta = \dfrac{\text{hyp}}{\text{adj}}$

$\tan\theta = \dfrac{\text{opp}}{\text{adj}} \qquad \cot\theta = \dfrac{\text{adj}}{\text{opp}}$

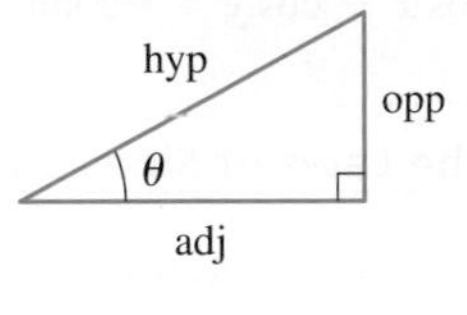

Special Values of the Trigonometric Functions

θ	radians	$\sin\theta$	$\cos\theta$	$\tan\theta$
0°	0	0	1	0
30°	$\pi/6$	$1/2$	$\sqrt{3}/2$	$\sqrt{3}/3$
45°	$\pi/4$	$\sqrt{2}/2$	$\sqrt{2}/2$	1
60°	$\pi/3$	$\sqrt{3}/2$	$1/2$	$\sqrt{3}$
90°	$\pi/2$	1	0	—
180°	π	0	-1	0
270°	$3\pi/2$	-1	0	—

Special Triangles

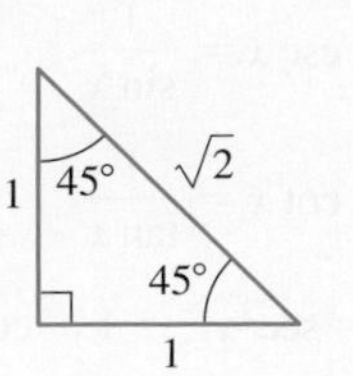

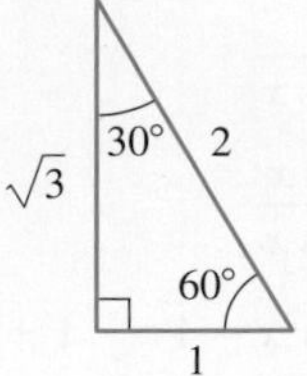

Graphs of the Trigonometric Functions

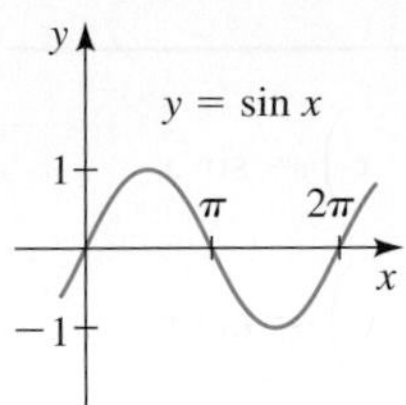

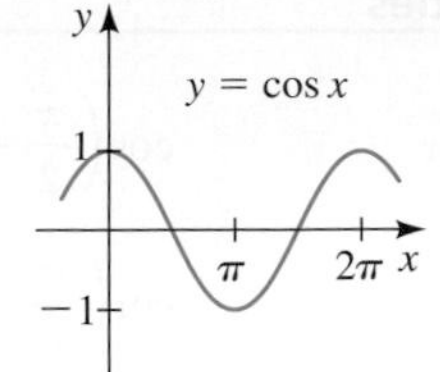

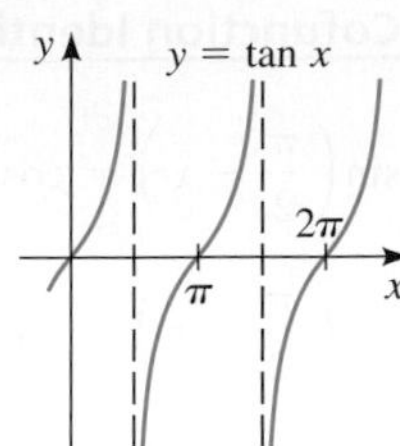

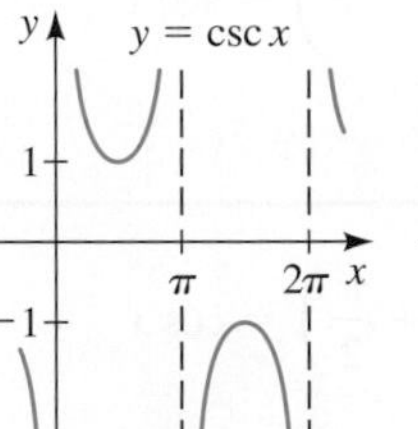

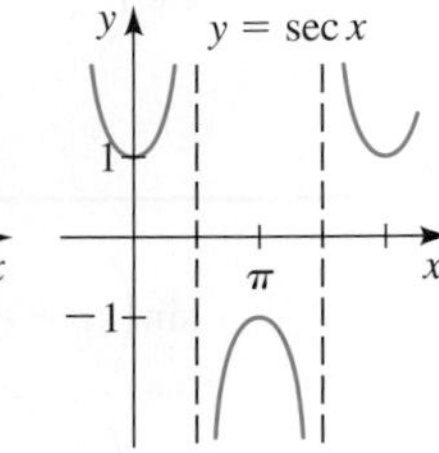

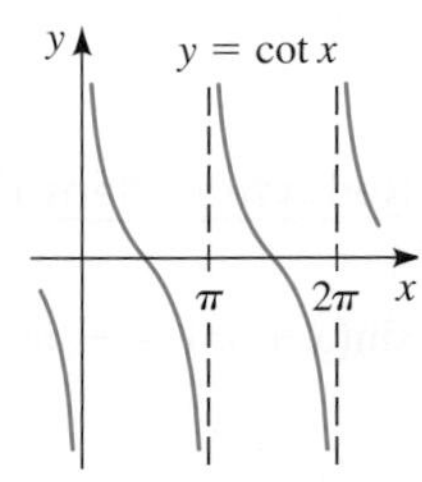

Sine and Cosine Curves

$y = a\sin k(x - b) \quad (k > 0) \qquad y = a\cos k(x - b) \quad (k > 0)$

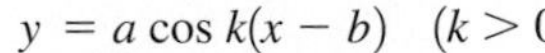

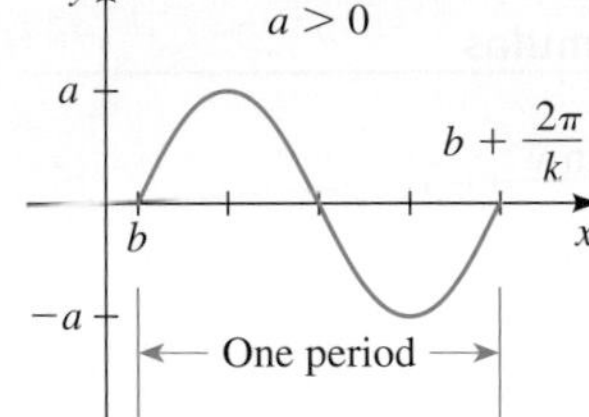

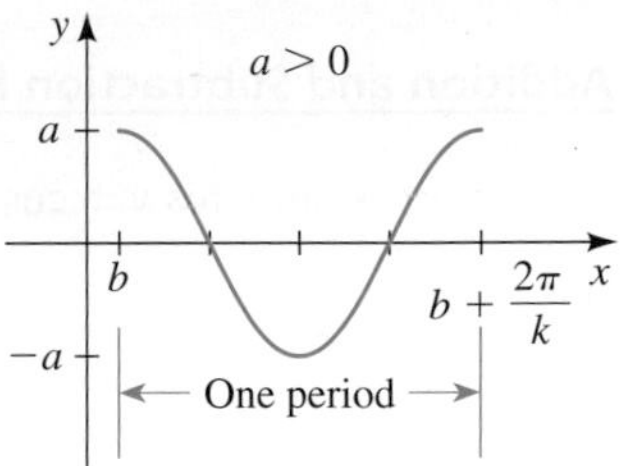

amplitude: $|a|$ period: $2\pi/k$ horizontal shift: b

Graphs of the Inverse Trigonometric Functions

$y = \sin^{-1}x \qquad y = \cos^{-1}x \qquad y = \tan^{-1}x$

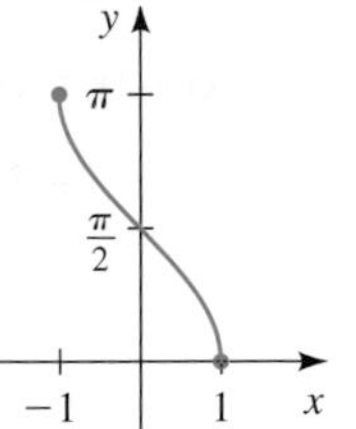

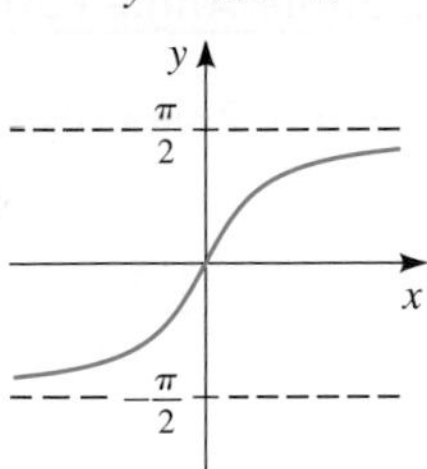

Fundamental Identities

$$\sec x = \frac{1}{\cos x} \qquad \csc x = \frac{1}{\sin x}$$

$$\tan x = \frac{\sin x}{\cos x} \qquad \cot x = \frac{1}{\tan x}$$

$$\sin^2 x + \cos^2 x = 1 \qquad 1 + \tan^2 x = \sec^2 x \qquad 1 + \cot^2 x = \csc^2 x$$

$$\sin(-x) = -\sin x \qquad \cos(-x) = \cos x \qquad \tan(-x) = -\tan x$$

Cofunction Identities

$$\sin\left(\frac{\pi}{2} - x\right) = \cos x \qquad \cos\left(\frac{\pi}{2} - x\right) = \sin x$$

$$\tan\left(\frac{\pi}{2} - x\right) = \cot x \qquad \cot\left(\frac{\pi}{2} - x\right) = \tan x$$

$$\sec\left(\frac{\pi}{2} - x\right) = \csc x \qquad \csc\left(\frac{\pi}{2} - x\right) = \sec x$$

Reduction Identities

$$\sin(x + \pi) = -\sin x \qquad \sin\left(x + \frac{\pi}{2}\right) = \cos x$$

$$\cos(x + \pi) = -\cos x \qquad \cos\left(x + \frac{\pi}{2}\right) = -\sin x$$

$$\tan(x + \pi) = \tan x \qquad \tan\left(x + \frac{\pi}{2}\right) = -\cot x$$

Addition and Subtraction Formulas

$$\sin(x + y) = \sin x \cos y + \cos x \sin y$$

$$\sin(x - y) = \sin x \cos y - \cos x \sin y$$

$$\cos(x + y) = \cos x \cos y - \sin x \sin y$$

$$\cos(x - y) = \cos x \cos y + \sin x \sin y$$

$$\tan(x + y) = \frac{\tan x + \tan y}{1 - \tan x \tan y} \qquad \tan(x - y) = \frac{\tan x - \tan y}{1 + \tan x \tan y}$$

Double-Angle Formulas

$$\sin 2x = 2 \sin x \cos x \qquad \cos 2x = \cos^2 x - \sin^2 x$$

$$= 2\cos^2 x - 1$$

$$\tan 2x = \frac{2 \tan x}{1 - \tan^2 x} \qquad = 1 - 2\sin^2 x$$

Formulas for Reducing Powers

$$\sin^2 x = \frac{1 - \cos 2x}{2} \qquad \cos^2 x = \frac{1 + \cos 2x}{2}$$

$$\tan^2 x = \frac{1 - \cos 2x}{1 + \cos 2x}$$

Half-Angle Formulas

$$\sin\frac{u}{2} = \pm\sqrt{\frac{1 - \cos u}{2}} \qquad \cos\frac{u}{2} = \pm\sqrt{\frac{1 + \cos u}{2}}$$

$$\tan\frac{u}{2} = \frac{1 - \cos u}{\sin u} = \frac{\sin u}{1 + \cos u}$$

Product-to-Sum and Sum-to-Product Identities

$$\sin u \cos v = \tfrac{1}{2}[\sin(u + v) + \sin(u - v)]$$

$$\cos u \sin v = \tfrac{1}{2}[\sin(u + v) - \sin(u - v)]$$

$$\cos u \cos v = \tfrac{1}{2}[\cos(u + v) + \cos(u - v)]$$

$$\sin u \sin v = \tfrac{1}{2}[\cos(u - v) - \cos(u + v)]$$

$$\sin x + \sin y = 2 \sin\frac{x + y}{2} \cos\frac{x - y}{2}$$

$$\sin x - \sin y = 2 \cos\frac{x + y}{2} \sin\frac{x - y}{2}$$

$$\cos x + \cos y = 2 \cos\frac{x + y}{2} \cos\frac{x - y}{2}$$

$$\cos x - \cos y = -2 \sin\frac{x + y}{2} \sin\frac{x - y}{2}$$

The Laws of Sines and Cosines

The Law of Sines

$$\frac{\sin A}{a} = \frac{\sin B}{b} = \frac{\sin C}{c}$$

The Law of Cosines

$$a^2 = b^2 + c^2 - 2bc \cos A$$

$$b^2 = a^2 + c^2 - 2ac \cos B$$

$$c^2 = a^2 + b^2 - 2ab \cos C$$

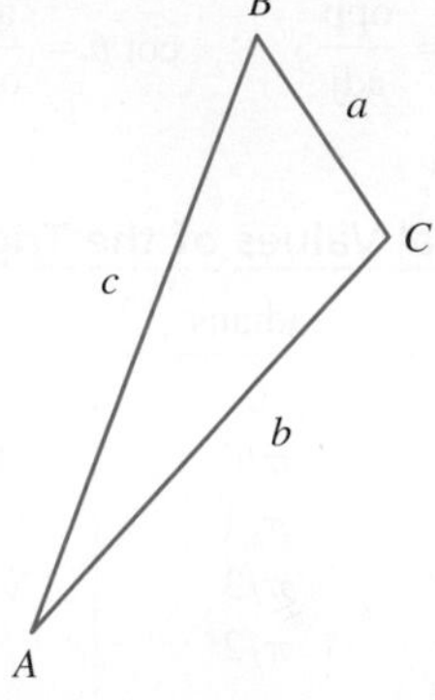